Fighting Ships
2005-2006

Edited by Commodore Stephen Saunders RN

One Hundred and Eighth Edition

Founded in 1897 by Fred T Jane

Total number of entries	4,354	New and updated entries	2,372
Total number of images	4,891	New images	1,630

Visit jfs.janes.com and view the list of latest updates that have been added to the online version of *Jane's Fighting Ships* subsequent to this print edition.

Bookmark jfs.janes.com today!
Jane's Fighting Ships online site gives you details of the additional information
that is unique to online subscribers and the many benefits of upgrading to an online subscription.
Don't delay, visit jfs.janes.com today and view the list of latest updates to this online service.

ISBN 0 7106 2692 4
"Jane's" is a registered trade mark

Copyright © 2005 by Jane's Information Group Limited, Sentinel House, 163 Brighton Road, Coulsdon, Surrey CR5 2YH, UK

In the US and its dependencies
Jane's Information Group Inc, 110 N. Royal Street, Suite 200, Alexandria, Virginia 22314, US

Our solutions open the oceans

Thales is the main international partner of more than 50 naval forces throughout the world and provides interoperable solutions to meet their every requirement. If the sea is your element, Thales will open the oceans with you.

THALES

Prime Contracting

Support & Services

Systems Integration

www.thalesgroup.com

Contents

Jane's Fighting Ships website: jfs.janes.com

ADMINISTRATION

Director: Ian Kay, e-mail: Ian.Kay@janes.com

New Media Publishing Director: Sean Howe, e-mail: Sean.Howe@janes.com

Content Services Director: Anita Slade, e-mail: Anita.Slade@janes.com

Content Systems Manager: Jo Agius, e-mail: Jo.Agius@janes.com

Pre-Press Manager: Christopher Morris, e-mail: Christopher.Morris@janes.com

Team Leader: Melanie Rovery, e-mail: Melanie.Rovery@janes.com

Content Editor: Emma Donald, e-mail: Emma.Donald@janes.com

Production Controller: Laura-Jane Walker, e-mail: Laura-JaneWalker@janes.com

Content Update: Jacqui Beard, Information Collection Team Leader
Tel: +44 (0) 20 8700 3808   Fax: +44 (0) 20 8700 3959
e-mail: yearbook@janes.com

Jane's Information Group Limited, Sentinel House, 163 Brighton Road, Coulsdon,
Surrey CR5 2YH, UK
Tel: +44 (0) 20 8700 3700   Fax: +44 (0) 20 8700 3900

SALES OFFICE

Send Europe and Africa enquiries to: *Mike Gwynn – Head of Information Sales*
Jane's Information Group Limited, Sentinel House, 163 Brighton Road, Coulsdon,
Surrey CR5 2YH, UK
Tel: +44 (0) 20 8700 3750   Fax: +44 (0) 20 8700 3751
e-mail: customerservices.uk@janes.com

Send US/Latin America/Canada enquiries to: *Greg Wallis – VP Information Sales*
Jane's Information Group Inc, 110 N Royal Street, Suite 200, Alexandria, Virginia
22314, US
Tel: (+1 703) 683 37 00   Fax: (+1 703) 836 02 97   Telex: 6819193
Tel: (+1 800) 824 07 68   Fax: (+1 800) 836 02 97
e-mail: customerservices.us@janes.com

Send Asia enquiries to: *David Fisher – Group Business Manager*
Jane's Information Group Asia, 78 Shenton Way, #10-02, Singapore 079120,
Singapore
Tel: (+65) 63 25 08 66   Fax: (+65) 62 26 11 85
e-mail: asiapacific@janes.com

Send Australia/New Zealand enquiries to: *Russel Smith – Business Manager*
Jane's Information Group, PO Box 3502, Rozelle Delivery Centre, New South Wales
2039, Australia
Tel: (+61 2) 85 87 79 00   Fax: (+61 2) 85 87 79 01
e-mail: oceania@janes.com

Send Middle East enquiries to: *Ali Abdellatif Siali – Regional Sales Manager*
Jane's Information Group, PO Box 502138, Dubai, United Arab Emirates
Tel: (+971 4) 390 23 36   Fax: (+971 4) 390 88 48
e-mail: mideast@janes.com

Send Japan enquiries to: *Norihisa Fukuyama – Information Consultant*
Jane's Information Group, Palaceside Building, 5F, 1-1-1, Hitotsubashi, Chiyoda-ku,
Tokyo 100-0003, Japan
Tel: (+81 3) 52 18 76 82   Fax: (+81 3) 52 22 12 80
e-mail: japan@janes.com

Send India enquiries to: *T C Martin – Information Consultant*
Jane's Information Group, PO Box 3806, New Delhi 110049, India
Tel/Fax: (+91 11) 26 51 61 05
e-mail: india@janes.com

ADVERTISEMENT SALES OFFICES

(Head Office)
Jane's Information Group
Sentinel House, 163 Brighton Road,
Coulsdon, Surrey CR5 2YH, UK
Tel: (+44 20) 87 00 37 00
Fax: (+44 20) 87 00 38 59/37 44
e-mail: defadsales@janes.com

Richard West, Senior Key Accounts Manager
Tel: (+44 1892) 72 55 80 Fax: (+44 1892) 72 55 81
e-mail: richard.west@janes.com

Nicky Eakins, Advertising Sales Executive
Tel: (+44 20) 87 00 38 53 Fax: (+44 20) 87 00 38 59/37 44
e-mail: nicky.eakins@janes.com

(US/Canada office)
Jane's Information Group
110 N Royal Street, Suite 200,
Alexandria, Virginia 22314, USA
Tel: (+1 703) 683 37 00
Fax: (+1 703) 836 55 37
e-mail: defadsales@janes.com

USA and Canada
Katie Taplett, US Advertising Sales Director
Tel: (+1 703) 683 37 00 Fax: (+1 703) 836 55 37
e-mail: katie.taplett@janes.com

Northern US and Eastern Canada
Linda Hewish, Northeast Region Advertising Sales Manager
Tel: (+1 703) 683 37 00 Fax: (+1 703) 836 55 37
e-mail: linda.hewish@janes.com

Southeastern US
Kristin D Schulze, Advertising Sales Manager
PO Box 270190, Tampa, Florida 33688-0190, USA
Tel: (+1 813) 961 81 32 Fax: (+1 813) 961 96 42
e-mail: Kristin.Schulze@janes.com

Western US and Western Canada
Richard L Ayer
127 Avenida del Mar, Suite 2A, San Clemente, California 92672, USA
Tel: (+1 949) 366 84 55 Fax: (+1 949) 366 92 89
e-mail: ayercomm@earthlink.com

Australia: *Richard West* (see UK Head Office)

Benelux: *Nicky Eakins* (see UK Head Office)

Brazil: *Katie Taplett* (see US address)

Eastern Europe (excl. Poland): MCW Media & Consulting Wehrstedt
Dr Uwe H Wehrstedt
Hagenbreite 9, D-06463 Ermsleben, Germany
Tel: (+49) 07 00/WEHRSTEDT / (+49 03) 47 43/620 90
Fax: (+49 03) 47 43/620 91
e-mail: info@Wehrstedt.org

France: Patrice Février
BP 418, 35 avenue MacMahon,
F-75824 Paris Cedex 17, France
Tel: (+33 1) 45 72 33 11 Fax: (+33 1) 45 72 17 95
e-mail: patrice.fevrier@wanadoo.fr

Germany and Austria: *MCW Media & Consulting Wehrstedt* (see Eastern Europe)

Greece: *Nicky Eakins* (see UK Head Office)

Hong Kong: *Nicky Eakins* (see UK Head Office)

India: *Nicky Eakins* (see UK Head Office)

Iran: Eideh Info Company
Ali Jahangard
19 4th Street, Ghaem Magham Avenue, Tehran, Iran
Tel: (+98 21) 873 59 23
e-mail: eideh@mavara.com

Israel: Oreet – International Media
15 Kinneret Street, IL-51201 Bene Berak, Israel
Tel: (+972 3) 570 65 27 Fax: (+972 3) 570 65 27
e-mail: admin@oreet-marcom.com
Defence: Liat Shaham
e-mail: liat_s@oreet-marcom.com

Italy and Switzerland: Ediconsult Internazionale Srl
Piazza Fontane Marose 3, I-16123 Genoa, Italy
Tel: (+39 010) 58 36 84 Fax: (+39 010) 56 65 78
e-mail: genova@ediconsult.com

Middle East: *Nicky Eakins* (see UK Head Office)

Pakistan: *Nicky Eakins* (see UK Head Office)

Poland: *Nicky Eakins* (see UK Head Office)

Russia: Vladimir N Usov, PO Box 98, 622018 Nizhniy Tagil,
Sverdlovsk Region, Russian Federation
Tel/Fax: (+7 3435) 23 02 68
e-mail: uvn125@uraltelecom.ru

Scandinavia: The Falsten Partnership
PO Box 27, Portslade, East Sussex BN41 2XA, UK
Tel: (+44 1273) 77 10 20 Fax: (+ 44 1273) 77 00 70
e-mail: sales@falsten.com

Singapore: *Richard West/Nicky Eakins* (see UK Head Office)

South Africa: *Richard West* (see UK Head Office)

South Korea: Infonet Group Inc
Sanbu Rennaissance Tower 902, 456 Gongdukdong, Mapogu, Seoul, South Korea
Contact: Mr Jongseog Lee
Tel: (+82 2) 716 99 22
Fax: (+82 2) 716 95 31
e-mail: jslee@infonetgroup.co.kr

Spain: Via Exclusivas SL
e-mail: viaexclusivas@viaexclusivas.com

Turkey: *Richard West* (see UK Head Office)

ADVERTISING COPY
Linda Letori (Jane's UK Head Office)
Tel: (+44 20) 87 00 37 42 Fax: (+44 20) 87 00 38 59/37 44
e-mail: linda.letori@janes.com

For North America, South America and Caribbean only:
Lia Johns (Jane's US/Canada Office)
Tel: (+1 703) 683 37 00 Fax: (+1 703) 836 55 37
e-mail: lia.johns@janes.com

Quality Policy

Jane's Information Group is the world's leading unclassified information integrator for military, government and commercial organisations worldwide. To maintain this position, the Company will strive to meet and exceed customers' expectations in the design, production and fulfilment of goods and services.

Information published by Jane's is renowned for its accuracy, authority and impartiality, and the Company is committed to seeking ongoing improvement in both products and processes.

Jane's will at all times endeavour to respond directly to market demands and will also ensure that customer satisfaction is measured and employees are encouraged to question and suggest improvements to working practices.

Jane's will continue to invest in its people through training and development to meet the Investor in People standards and changing customer requirements.

www.janes.com

Jane's
Intelligence and Insight You Can Trust

FREE ENTRY/CONTENT IN THIS PUBLICATION

Having your products and services represented in our titles means that they are being seen by the professionals who matter – both by those involved in procurement and by those working for the companies that are likely to affect your business. We therefore feel that it is very much in the interests of your organisation, as well as Jane's, to ensure your data is current and accurate.

- **Don't forget** – You may be missing out on business if your entry in a Jane's product is incorrect because you have not supplied the latest information to us.

- **Ask yourself** – Can you afford not to be represented in Jane's printed and electronic products? And if you are listed, can you afford for your information to be out of date?

- **And most importantly** – The best part of all is that your entries in Jane's products are TOTALLY FREE OF CHARGE.

Please provide (using a photocopy of this form) the information on the following categories where appropriate:

1. Organisation name: _____

2. Division name: _____

3. Location address: _____

4. Mailing address if different: _____

5. Telephone (please include switchboard and main department contact numbers, for example Public Relations, Sales, and so on): _____

6. Facsimile: _____

7. e-mail: _____

8. Web sites: _____

9. Contact name and job title: _____

10. A brief description of your organisation's activities, products and services: _____

11. Jane's publications in which you would like to be included: _____

Please send this information to:
Jacqui Beard, Information Collection, Jane's Information Group,
Sentinel House, 163 Brighton Road, Coulsdon, Surrey, CR5 2YH, UK
Tel: (+44 20) 87 00 38 08
Fax: (+44 20) 87 00 39 59
e-mail: yearbook@janes.com

Copyright enquiries:
Contact: Keith Faulkner
Tel/Fax: (+44 1342) 30 50 32
e-mail: copyright@janes.com

Please tick this box if you do not wish your organisation's staff to be included in Jane's mailing lists ☐

JFS

Glossary

Type abbreviations are listed at head of Pennant List

AAW	Anti-Air Warfare
ACDS	Advanced Combat Direction System
ADCAP	ADvanced CAPabilities
AEW	Airborne Early Warning
AIP	Air Independent Propulsion
ALSC	Afloat Logistics and Sealift Capability
ARCI	Acoustic Rapid COTS Insertion
ARM	Anti-Radiation Missile
ASDS	Advanced Swimmer Delivery System
A/S, ASW	Anti-Submarine (Warfare)
ASM	Air-to-Surface Missile
ASROC	Rocket assisted torpedo, part of whose trajectory is in the air
ASV	Air-to-Surface Vessel
AUV	Autonomous Underwater Vehicle
BPDMS	Base Point Defence Missile System
Cal	Calibre – the diameter of a gun barrel; also used for measuring length of the barrel eg a 6 in gun 50 calibres long (6 in/50) would be 25 ft long
CEC	Co-operative Engagement Capability
CIWS	Close-In Weapon System
CODAG, CODOG, CODLAG, COGAG, COGOG, COSAG, COGAL	Descriptions of mixed propulsion systems: combined diesel and gas turbine, diesel-electric and gas turbine, diesel or gas turbine, gas turbine and gas turbine, gas turbine or gas turbine, steam and gas turbine, gas turbine and electric
CONAS	Combined nuclear and steam
COTS	Commercial Off-The-Shelf
cp	controllable pitch (propellers)
DDS	Dry Dock Shelter
DP	Dual Purpose (gun) for surface or AA use
Displacement	Basically the weight of water displaced by a ship's hull when floating: (a) Light: without fuel, water or ammunition (b) Normal: used for Japanese MSA ships. Similar to 'standard' (c) Standard: as defined by Washington Naval Conference 1922 – fully manned and stored but without fuel or reserve feed-water (d) Full load: fully laden with all stores, ammunition, fuel and water
dwt	deadweight tonnage (see tonnage)
EARS	Electromagnetic Aircraft Recovery System
ECM	Electronic countermeasures, for example, jamming
ECCM	Electronic counter-countermeasures
EEZ	Exclusive Economic Zone
EHF	Extreme High Frequency
ELF	Extreme Low Frequency radio

ELINT	Electronic intelligence, for example, recording radar, W/T and so on
EMALS	Electromagnetic Aircraft Launching System
ERGM	Extended-Range Guided Munitions
ESM	Electronic Support Measures for example, intercept
ESSM	Evolved Sea Sparrow Missile
EW	Electronic Warfare
FLIR	Forward-Looking Infra-Red
FRAM	Fleet Rehabilitation And Modernisation programme
GCCS	Global Command and Control System
GFCS	Gun Fire-Control System
GPS	Global Positioning System
grt	gross registered tonnage (see tonnage)
HDTI	High Definition Thermal Imager
HIFR	Helicopter In-Flight Refuelling
HF	High Frequency
Horsepower (hp) or (hp(m))	Power developed or applied: (a) bhp: brake horsepower = power available at the crankshaft (b) shp: shaft horsepower = power delivered to the propeller shaft (c) ihp: indicated horsepower = power produced by expansion of gases in the cylinders of reciprocating steam engines (d) 1 kW = 1.341 hp = 1.360 metric hp 1 hp = 0.746 kW = 1.014 metric hp 1 metric hp = 0.735 kW = 0.968 hp (e) Sustained horsepower may be different for similar engines in different conditions
IFF	Identification Friend/Foe
IRST	Infra-Red Search and Track
JMCIS	Joint Maritime Command Information System
JTIDS	Joint Tactical Information Distribution System
kT	kiloton
kW	kilowatt
LAMPS	Light Airborne Multipurpose System
LAMS	Local Area Missile System
Length	Expressed in various ways: (a) oa: overall = length between extremities (b) pp: between perpendiculars = between fore side of the stem and after side of the rudderpost (c) wl: waterline = between extremities on the water-line
LF	Low Frequency
MAD	Magnetic Anomaly Detector
MDF	Maritime Defence Force
Measurement	See Tonnage
MF	Medium Frequency
MFCS	Missile Fire-Control System

SAZANAMI

2/2005*, Hachiro Nakai / 1121128

[7]

MG	Machine Gun	**SSM**	Surface-to-Surface Missile
MIDAS	Mine and Ice Detection Avoidance System	**SSTDS**	Surface Ship Torpedo Defence System
MIRV	Multiple, Independently targetable Re-entry Vehicle	**STIR**	Surveillance Target Indicator Radar
MPA	Maritime Patrol Aircraft	**STOBAR**	Short Take Off and Barrier Arrested Recovery
MSA	Japan Maritime Safety Agency	**STOVL**	Short Take Off and Vertical Landing
MSC	US Military Sealift Command	**SUM**	Surface-to-Underwater Missile
MW	Megawatt	**SURTASS**	Surface Towed Array Surveillance System
NBC	Nuclear, Biological and Chemical (warfare)	**SWATH**	Small Waterplane Area Twin Hull
net	net registered tonnage (see tonnage)	**TACAN**	TACtical Air Navigation beacon
n mile	nautical mile (mean value 1.8532 km)	**TACTASS**	TACtical Towed Acoustic Sensor System
NMRS	Near-term Mine Reconnaissance System	**TAINS**	Tercom Aided Inertial Navigation System
NTDS	Naval Tactical Direction System	**TAS**	Target Acquisition System
oa	overall length	**TASM**	Tomahawk Anti-Ship Missile
OTC	Officer in Tactical Command	**TASS**	Towed Array Surveillance System
OTHT	Over The Horizon Targeting	**TBMD**	Theatre Ballistic Missile Defence
PAAMS	Principal Anti-Air Missile System	**Tercom**	Terrain Contour Matching
PAP	Poisson Auto Propulse	**TLAM**	Tomahawk Land Attack Missile
PDMS	Point Defence Missile System	**Tonnage**	Measurement tons, computed on capacity of a ship's hull rather than its 'displacement' (see above):
PWR	Pressurised Water Reactor		(a) Gross: the internal volume of all spaces within the hull and all permanently enclosed spaces above decks
QRCC	Quick Reaction Combat Capability		that are available for cargo, stores and accommodation.
RAIDS	Rapid Anti-ship missile Integrated Defence System		The result in cubic feet divided by 100 = gross tonnage
RAM	Radar Absorbent Material		(b) Net: gross minus all those spaces used for machinery,
RAM	Rolling Airframe Missile		accommodation etc ('non-earning' spaces)
RAS	Replenishment At Sea		(c) Deadweight (dwt): the amount of cargo, bunkers,
RAST	Recovery, Assist, Secure and Traverse system		stores etc that a ship can carry at her load draught
RBU	Anti-submarine rocket launcher	**Tonne**	1,000 kilos = 2,204.6 lb
RIB	Rigid Inflatable Boat		Imperial (long) ton = 1.016 tonne or 2,240 lb
Ro-Ro	Roll-on/Roll-off		US (short) ton = 0.9072 tonne or 2,000 lb
ROV	Remote Operated Vehicle	**UAV**	Unmanned Aerial Vehicle
rpm	revolutions per minute	**UCAV**	Unmanned Combat Aerial Vehicle
SAM	Surface-to-Air Missile	**UHF**	Ultra-High Frequency
SAR	Search And Rescue	**USM**	Underwater-to-surface missile
SATCOM	SATellite COMmunications	**UUV**	Unmanned Undersea Vehicle
SAWCS	Submarine Acoustic Warfare Countermeasures System	**VDS**	Variable Depth Sonar, can be lowered to best listening depth. In helicopters called 'dunking sonar'.
SES	Surface Effect Ship		
SHF	Super High Frequency	**Vertrep**	Vertical replenishment
SINS	Ship's Inertial Navigation System	**VLF**	Very Low Frequency radio
SLBM	Submarine-Launched Ballistic Missile	**VLS**	Vertical Launch System
SLCM	Ship-Launched Cruise Missile	**VSTOL**	Vertical or Short Take-Off/Landing
SLEP	Service Life Extension Programme	**VSV**	Very Slender Vessel
SMCS	Submarine Command System	**VTOL**	Vertical Take-Off/Landing
SRBOC	Super Rapid Blooming Offboard Chaff	**wl**	waterline length
SSDE/SSE	Submerged Signal and Decoy Ejector		
SSDS	Ship Self-Defence System		

IRON DUKE

5/2004*, H M Steele / 1121129

HHI — Building a better future
Global Leader

Helping you navigate the 21st century

We know that advanced navies in the world
require more to reinforce their defense system
in the 21st Century.

Hyundai will always take the initiative
in meeting those requirements,
but with the top quality and efficiency
we are known so far.

HYUNDAI
HEAVY INDUSTRIES CO.,LTD.

SPECIAL & NAVAL SHIPBUILDING

www.hhi.co.kr
Tel : 82-52-230-0101
Fax : 82-52-230-0100
e-mail : e120hhi@hhi.co.kr

How to use *Jane's Fighting Ships*

(see also Glossary and Type abbreviations)

1) Details of major warships are grouped under six separate non-printable headings. These are:

(a) **Number and Class name**. Totals of vessels per class are listed as 'in service + building (proposed)' or 'in service + transfer (proposed)'.

(b) **Building programme**. This includes builders' names and key dates. In general the 'laid down' column reflects keel laying but modern shipbuilding techniques make it difficult to be specific about the start date of actual construction. Launching and christening can be similarly confusing, now that many ships are lowered into the water and formally christened some time later. Some nations commission their ships on completion of building, others after the ships have completed trials. In this hardcopy edition any date after April 2005 is projected or estimated and therefore liable to change.

(c) **Hull**. This section tends to have only specification and performance parameters and contains little free text. Hull related details such as **Military lift** and **Cargo capacity** may be included when appropriate. **Displacement** and **Measurement** tonnages, **Dimensions**, **Horsepower** and so on, are defined in the Glossary. Throughout the life of a ship its displacement tends to creep upwards as additional equipment is added and redundant fixtures and fittings are left in place. For the same reasons, ships of the same class, active in different navies, frequently have different displacements and other dissimilar characteristics. Unless otherwise stated the lengths and widths given are overall and the draught is at full load. Sustained maximum horsepower is given where the information is available and may not be the same for similar engines operating in different hulls under different conditions. **Speed** is the maximum obtainable under trials conditions.

(d) **Weapon systems**. This section contains operational details and some free text on weapons and sensors which are laid out in a consistent order using the same subheadings throughout the book. The titles are: **Missiles** (subdivided into SLBM, SSM, SAM, A/S); **Guns** (numbers of barrels are given and the rate of fire is 'per barrel' unless stated otherwise); **Torpedoes**; **A/S mortars**; **Depth charges**; **Mines**; **Countermeasures**; **Combat data systems**; **Weapons control**; **Radars**; **Sonars**. The Weapons control heading is used for weapons' direction equipment. In most cases the performance specifications are those of the manufacturer and may therefore be considered to be at the top end of the spectrum of effective performance. So-called 'operational effectiveness' is difficult to define, depends upon many variables and in the context of range may be considerably less than the theoretical maximum. Numbers inserted in the text refer to similar numbers included on line drawings.

(e) **Aircraft**. Only the types and numbers are included here. Where appropriate each country has a separate section listing overall numbers and operational parameters of front-line shipborne and land-based maritime aircraft, normally included after the Frigate section if there is one.

(f) **General comments**. A maximum of six sub-headings are used to sweep up the variety of additional information which is available but has no logical place in the other sections. These headings are: **Programmes**; **Modernisation**; **Structure**; **Operational**; **Sales** and **Opinion**. The last of these allows space for informed comment. Some ships remain theoretically in the order of battle in some navies even though they never go to sea and could be more accurately described as in reserve. Where this is known comment is made under **Operational**.

2) Minor or less important ship entries follow the same format except that there is often much less detail in the first four headings and all additional remarks are put together under the single heading of **Comment**. The distinction between major and minor depends upon editorial judgement and is primarily a function of firepower. The age of the ship or class and its relative importance within the Navy concerned is also taken into account.

3) The space devoted to front-line maritime aircraft reflects the importance of air power as an addition to the naval weapon systems armoury, but the format used is necessarily brief and covers only numbers, roles and operational characteristics. Greater detail can be found in *Jane's All the World's Aircraft* and the appropriate volume of the *Jane's Weapon Systems* series.

4) Other than for coastal navies, tables are included at the front of each country section with such things as strength of the fleet, senior appointments, personnel numbers, bases and so on. There is also a list of pennant numbers and a deletions column covering the previous three years. If you cannot find your favourite ship, always look in the **Deletions** list first.

5) No addenda is included because modern typesetting technology allows changes to the main text to be made up to a few weeks before publication.

6) Shipbuilding companies and weapons manufacturers frequently change their names by merger or takeover. As far as possible the published name shows the title when the ship was built or weapon system installed. It is therefore historically accurate.

7) Like many descriptive terms in international naval nomenclature, differences between Coast Guards, Maritime Police, Customs and other paramilitary maritime forces are often indistinct and particular to an individual nation. Such vessels are usually included if they have a paramilitary function and are armed.

8) When selecting photographs for inclusion, priority is given to those that have been taken most recently. A glossy picture five years old may look nice but often does not show the ship as it is now.

9) The Navies by country section is geared to the professional user who needs to be able to make an assessment of the fighting characteristics of a Navy or class of ship without having to cross refer to other Navies and sections of the book. Much effort has also been made to prevent entries spilling across from one page to another.

10) Regular updates can be found on Jane's Online.

11) To help users of this title evaluate the published data, entries have been divided into three categories:

(a) *VERIFIED* The editor has made a detailed examination of the entry's content checking its relevancy and accuracy for publication to the new edition to the best of his knowledge.

(b) *UPDATED* During the verification process, changes to content or photographs have been made to reflect the latest position known to Jane's at time of publication.

(c) *NEW ENTRY* A ship class appearing for the first time in the title.

(12) Photographs are dated and where * appears a new or re-scanned photograph has been substituted or added. Many are followed by a seven digit number to ease identification.

| Total number of entries | 4,354 | New and updated entries | 2,372 |
| Total number of images | 4,891 | New images | 1,630 |

Any update to the content of this product will appear online as it occurs (see jfs.janes.com) for the additional benefits of an online subscription to Jane's Fighting Ships and details of our free online trial) and will be incorporated annually into future print editions.

Copyright enquiries
Contact: Keith Faulkner, Tel/Fax: +44 (0) 1342 305032, e-mail: copyright@janes.com

British Library Cataloguing-in-Publication Data.
A catalogue record for this book is available from the British Library.

Printed and bound in Great Britain by Thanet Press

Jane's Libraries

To assist your information gathering and to save you money, Jane's has grouped some related subject matter together to form 'ready-made' libraries, which you can access in whichever way suits you best – online, on CD-ROM, via Jane's EIS or through Jane's Data Service.

The entire contents of each library can be cross-searched, to ensure you find every reference to the subjects you are looking for. All Jane's libraries are updated according to the delivery service you choose and can stand alone or be networked throughout your organisation.

Jane's Defence Equipment Library

Aero-Engines
Air-Launched Weapons
Aircraft Upgrades
All the World's Aircraft
Ammunition Handbook
Armour and Artillery
Armour and Artillery Upgrades
Avionics
C4I Systems
Electro-Optic Systems
Explosive Ordnance Disposal
Fighting Ships
Infantry Weapons
Land-Based Air Defence
Military Communications
Military Vehicles and Logistics
Mines and Mine Clearance
Naval Weapon Systems
Nuclear, Biological and Chemical Defence
Radar and Electronic Warfare Systems
Strategic Weapon Systems
Underwater Warfare Systems
Unmanned Aerial Vehicles and Targets

Jane's Defence Magazines Library

Defence Industry
Defence Weekly
Foreign Report
Intelligence Digest
Intelligence Review
International Defence Review
Islamic Affairs Analyst
Missiles and Rockets
Navy International
Terrorism and Security Monitor

Jane's Market Intelligence Library

Aircraft Component Manufacturers
All the World's Aircraft
Defence Industry
Defence Weekly
Electronic Mission Aircraft
Fighting Ships
Helicopter Markets and Systems
International ABC Aerospace Directory
International Defence Directory
Marine Propulsion
Naval Construction and Retrofit Markets
Police and Security Equipment
Simulation and Training Systems
Space Directory
Underwater Technology
World Armies
World Defence Industry

Jane's Security Library

Amphibious and Special Forces
Chemical-Biological Defense Guidebook
Facility Security
Fighting Ships
Intelligence Digest
Intelligence Review
Intelligence Watch Report
Islamic Affairs Analyst
Police and Security Equipment
Police Review
Terrorism & Security Monitor
Terrorism Watch Report
World Air Forces
World Armies
World Insurgency and Terrorism

Jane's Sentinel Library

Central Africa
Central America and the Caribbean
Central Europe and the Baltic States
China and Northeast Asia
Eastern Mediterranean
North Africa
North America
Oceania
Russia and the CIS
South America
South Asia
Southeast Asia
Southern Africa
The Balkans
The Gulf States
West Africa
Western Europe

Jane's Transport Library

Aero-Engines
Air Traffic Control
Aircraft Component Manufacturers
Aircraft Upgrades
Airport Review
Airports and Handling Agents –
    Central and Latin America (inc. the Caribbean)
    Europe
    Far East, Asia and Australasia
    Middle East and Africa
    United States and Canada
Airports, Equipment and Services
All the World's Aircraft
Avionics
High-Speed Marine Transportation
Marine Propulsion
Merchant Ships
Naval Construction and Retrofit Markets
Simulation and Training Systems
Transport Finance
Urban Transport Systems
World Airlines
World Railways

Jane's
Intelligence and Insight You Can Trust

104100405

Invisible but Invincible, DSME

You may never see us but, We will always be there

Beyond your imagination, there is always DSME
Whatever you are looking for, it can be realized.
As your trustworthy companion, DSME will keep your confidence.

The happiness of you and your country is our greatest concern.
The presence of your robust warships will make you feel safe and
secure. Make a reservation for security under the name of DSME.
DSME is your true partner in maritime security.

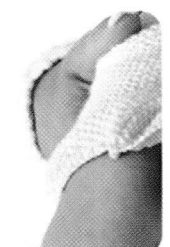

TRUST
PASSION

DSME, your partner with trust and passion

www.dsme.co.kr

DSME

*DAEWOO SHIPBUILDING &
MARINE ENGINEERING CO.,LTD.*

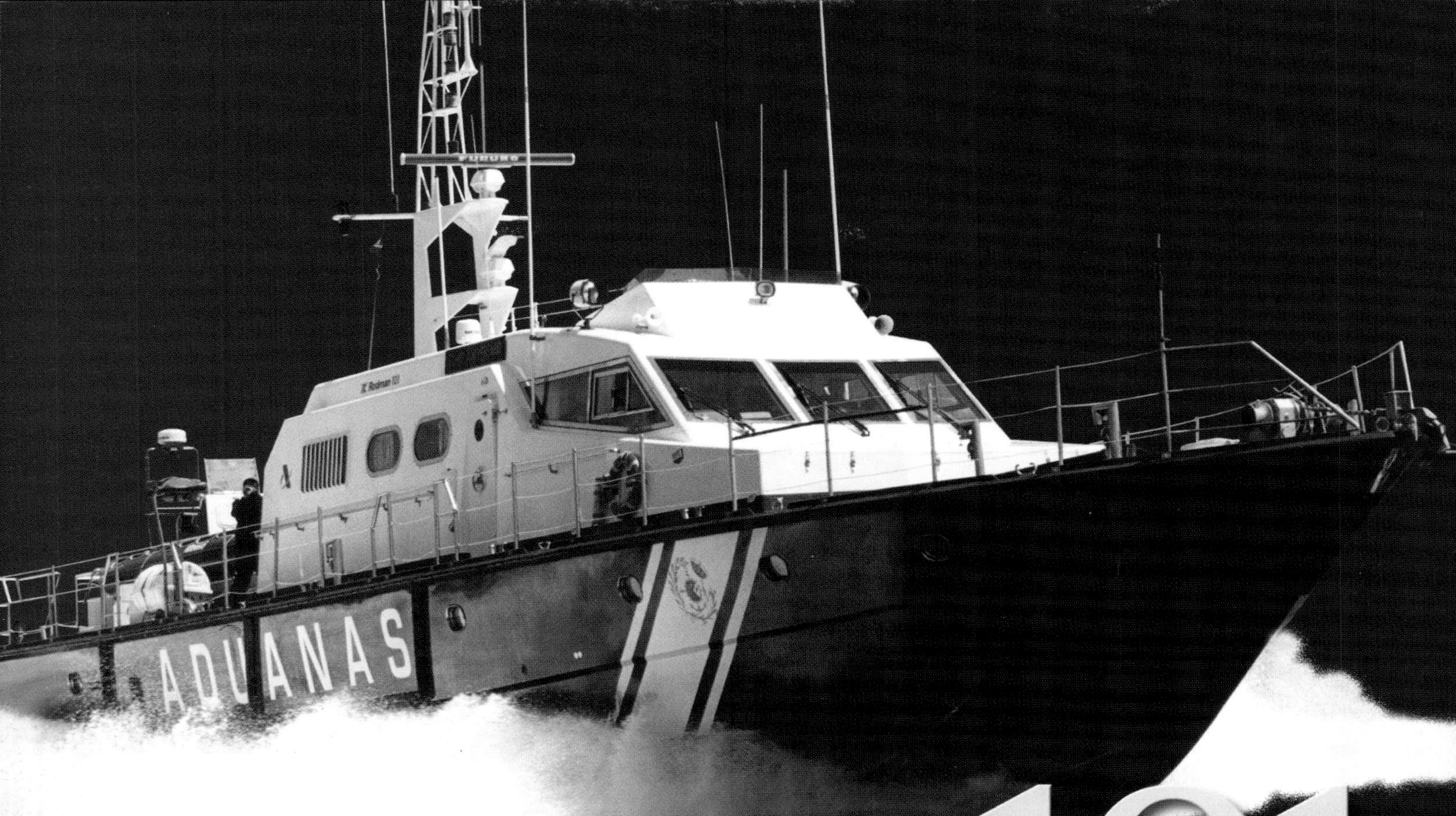

Rodman 101

Rodman 101, a patrol boat with high performance capabilities.

For surveillance, monitoring, security tasks, etc. Navies, Coast Guards, fishery protection services and marine police forces all over the world rely on Rodman Polyships to design and build their units in order to carry out their operational requirements meeting the highest standards with minimum servicing and crew requirements. Rodman, a wide range of units built with composite materials from 9 to 36 metres.

TECHNICAL FEATURES	
Overall length	30 m.
Mould breadth	6 m.
Draught	1,10 m.
Operating range	600 / 1.000 miles
Crew	9/13 men
Power	1.400 / 2.366 HP
Max. speed	35 knots
Cruising speed	28 knots

 **Rodman**

Rios - Teis, s/nº • P. O. BOX 501 • 36200 VIGO - Spain • Tel: +34 986 81 18 11 • Fax: +34 986 81 18 21 • e-mail: vpresidencia@rodman.es

Alphabetical list of advertisers

ALPHABETICAL LIST OF ADVERTISERS

L

Lürssen Werft
Zum Alten Speicher 11, D-28759 Bremen, Germany 269

R

Rodman Polyships
PO Box 501, E-36200 Vigo, Spain .... [15]

S

S.E.M.T Pielstick
Paris Nord II, 22 avenue des Nations, BP 50049 Villepinte,
F-95946 Roissy CDG Cedex, France .... [16]

T

Thales Nederland BV
ta v. de Crediteuradministratie/Corporate Communications,
Postbus 42, GD Hengelo, Netherlands .... [2]

Thyssenkrupp Marine Systems
Werfstrasse 112-114, D-24143 Kiel,
Germany .... *Facing inside front cover*

DISCLAIMER

Jane's Information Group gives no warranties, conditions, guarantees or representations, express or implied, as to the content of any advertisements, including but not limited to compliance with description and quality or fitness for purpose of the product or service. Jane's Information Group will not be liable for any damages, including without limitation, direct, indirect or consequential damages arising from any use of products or services or any actions or omissions taken in direct reliance on information contained in advertisements.

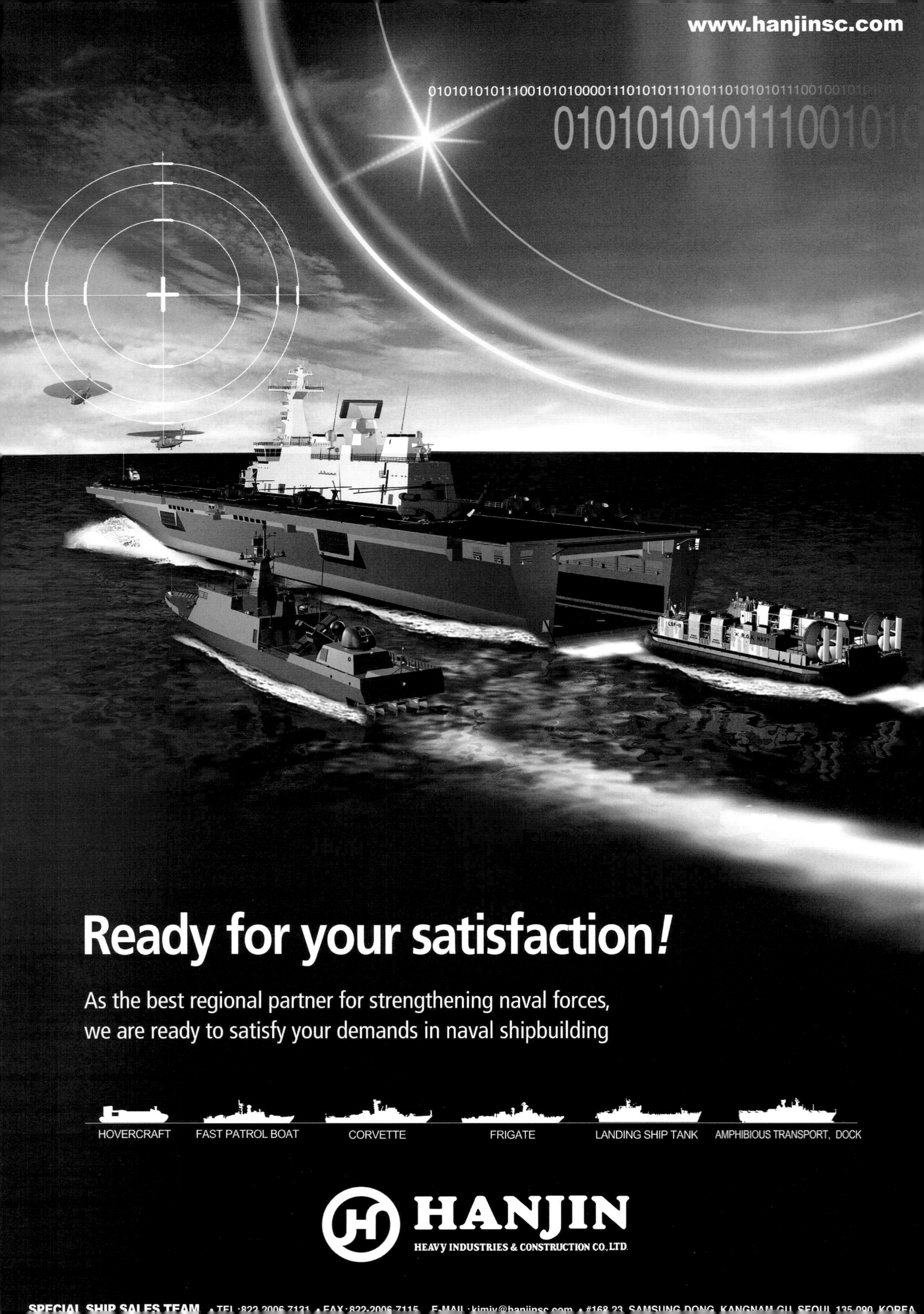

www.hanjinsc.com
Ready for your satisfaction!
As the best regional partner for strengthening naval forces,
we are ready to satisfy your demands in naval shipbuilding
HOVERCRAFT
FAST PATROL BOAT
CORVETTE
FRIGATE
LANDING SHIP TANK
AMPHIBIOUS TRANSPORT, DOCK
HANJIN
HEAVY INDUSTRIES & CONSTRUCTION CO., LTD.
SPECIAL SHIP SALES TEAM • TEL:822 2006 7131 • FAX:822-2006 7115 E-MAIL:kimjy@hanjinsc.com • #168 23, SAMSUNG DONG, KANGNAM GU, SEOUL 135 090, KOREA

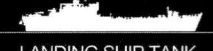

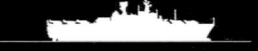

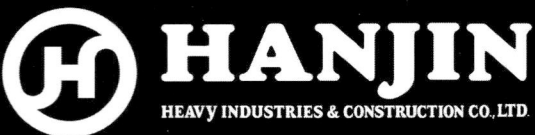

Ensigns and flags of the world's navies

In cases where countries do not have ensigns their warships normally fly the national flag.

Albania
Ensign

Algeria
Ensign

Angola
National Flag and Ensign

Anguilla
National Flag

Antigua and Barbuda
Ensign

Argentina
National Flag and Ensign

Australia
Ensign

Austria
Ensign

Azerbaijan
Ensign

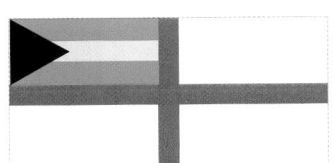

Bahamas
Ensign

Bahrain
National Flag and Ensign

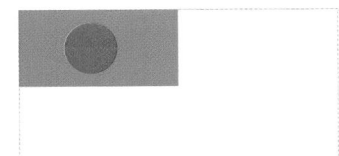

Bangladesh
Ensign

Barbados
Ensign

Belgium
Ensign

Belize
National Flag and Ensign

Benin
National Flag and Ensign

Bermuda
Ensign

Bolivia
Ensign

Brazil
National Flag and Ensign

Brunei
Ensign

Bulgaria
Ensign

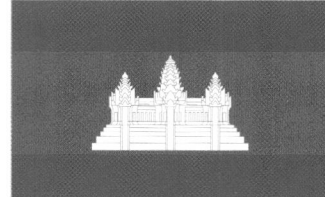

Cambodia
National Flag and Ensign

Cameroon
National Flag and Ensign

Canada
National Flag and Ensign

Cape Verde
National Flag and Ensign

Cayman Islands
National Flag

Chile
National Flag and Ensign

China
Ensign

Colombia
Ensign

Comoros
National Flag and Ensign

Congo-Brazzaville
National Flag and Ensign

Democratic Republic of Congo
National Flag and Ensign

Cook Islands
National Flag

Costa Rica
Ensign

Côte d'Ivoire
National Flag and Ensign

Croatia
Ensign

Cuba
National Flag and Ensign

Cyprus
National Flag and Ensign

Cyprus, Turkish Republic of Northern
(Not recognised by United Nations)
National Flag and Ensign

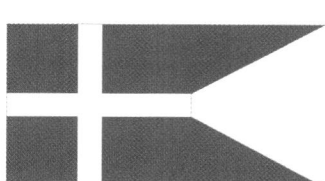

Denmark
Ensign

Djibouti
National Flag and Ensign

Dominica
National Flag and Ensign

Dominican Republic
Ensign

East Timor
National Flag and Ensign

Ecuador
Ensign

Egypt
Ensign

El Salvador
National Flag and Ensign

Equatorial Guinea
National Flag and Ensign

Eritrea
National Flag and Ensign

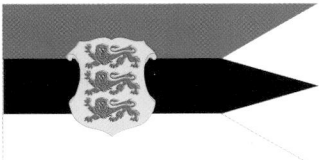

Estonia
Ensign

European Union
Flag of the European Union

Falkland Islands
Falkland Islands Flag

Faroe Islands
Territory Flag

Fiji
Ensign

Finland
Ensign

France
National Flag and Ensign

Gabon
National Flag and Ensign

Gambia
National Flag and Ensign

Georgia
Ensign

Germany
Ensign

Ghana
Ensign

Greece
National Flag and Ensign

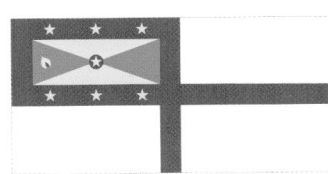

Grenada
Ensign

Guatemala
National Flag and Ensign

Guinea
National Flag and Ensign

Guinea-Bissau
National Flag and Ensign

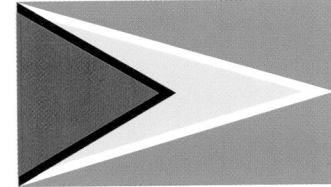

Guyana
National Flag and Ensign

Honduras
Ensign

ENSIGNS AND FLAGS OF THE WORLD'S NAVIES

Hong Kong
Regional Flag and Ensign

Hungary
National Flag

Iceland
Ensign

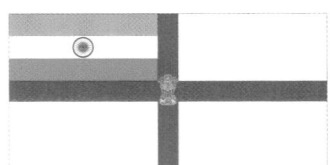

India
Ensign

Indonesia
National Flag and Ensign

Iran
National Flag and Ensign

Iraq
National Flag and Ensign

Ireland
National Flag and Ensign

Israel
Ensign

Italy
Ensign

Jamaica
Ensign

Japan
Japan (Navy) Ensign

Japan
Japan (MSA) Ensign

Jordan
Ensign

Kazakhstan
Ensign

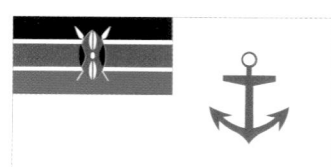

Kenya
Ensign

Kiribati
National Flag and Ensign

Korea, North
National Flag and Ensign

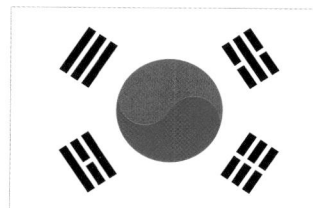

Korea, South
National Flag and Ensign

Kuwait
National Flag and Ensign

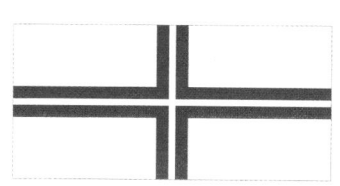

Latvia
Ensign

Lebanon
National Flag and Ensign

Liberia
National Flag and Ensign

Libya
Ensign

Flag Images courtesy of The Flag Institute, © 2004 Graham Bartram. All rights reserved.

Jane's Fighting Ships 2005-2006

[22]

jfs.janes.com

Lithuania
Ensign

Macedonia, Former Yugoslav Republic of
National Flag

Madagascar
National Flag and Ensign

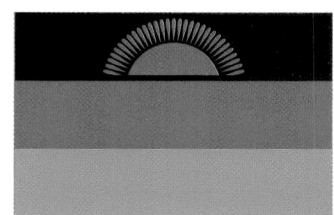

Malawi
National Flag

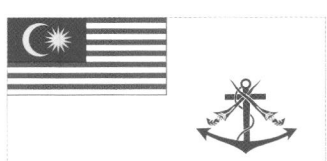

Malaysia
Ensign

Maldives
National Flag and Ensign

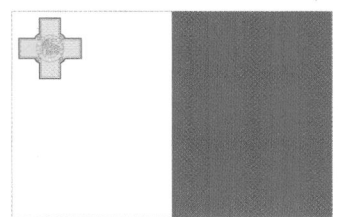

Malta
National Flag and Ensign

Marshall Islands
National Flag and Ensign

Mauritania
National Flag and Ensign

Mauritius
Ensign

Mexico
National Flag and Ensign

Federated States of Micronesia
Flag of the Federation

Morocco
Ensign

Mozambique
National Flag and Ensign

Myanmar
Ensign

Namibia
National Flag and Ensign

NATO
Flag of the North Atlantic Treaty Organisation

Netherlands
National Flag and Ensign

New Zealand
Ensign

Nicaragua
National Flag and Ensign

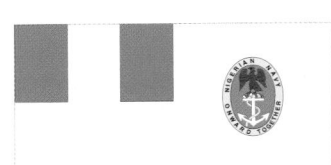

Nigeria
Ensign

Norway
Ensign

Oman
Ensign

Pakistan
Ensign

ENSIGNS AND FLAGS OF THE WORLD'S NAVIES

Palau
National Flag and Ensign

Panama
National Flag and Ensign

Papua New Guinea
Ensign

Paraguay
National Flag and Ensign

Paraguay
*National Flag and Ensign
(reverse)*

Peru
National Flag and Ensign

Philippines
National Flag and Ensign

Poland
Ensign

Portugal
National Flag and Ensign

Qatar
National Flag and Ensign

Romania
National Flag and Ensign

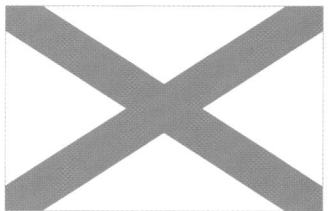

Russian Federation
Ensign

Russian Federation
Border Guard Ensign

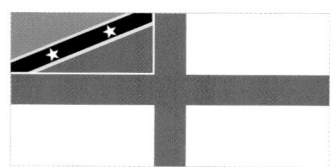

St Kitts and Nevis
Ensign

St Lucia
Ensign

**St Vincent and the
Grenadines**
National Flag and Ensign

Samoa
National Flag and Ensign

Saudi Arabia
National Flag and Ensign

Senegal
National Flag and Ensign

Serbia and Montenegro
Ensign

Seychelles
National Flag

Sierra Leone
Ensign

Singapore
Ensign

Slovenia
National Flag and Ensign

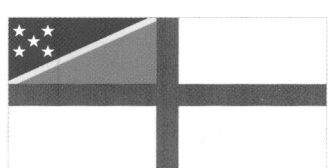

Solomon Islands
Ensign

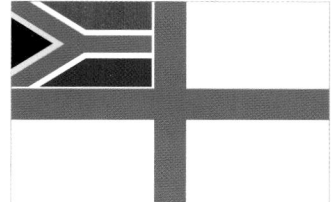

South Africa
Ensign

Spain
National Flag and Ensign

Sri Lanka
Ensign

Sudan
National Flag and Ensign

Suriname
National Flag and Ensign

Sweden
Ensign

Switzerland
National Flag

Syria
National Flag and Ensign

Taiwan
National Flag and Ensign

Tanzania
National Flag and Ensign

Thailand
Ensign

Togo
National Flag and Ensign

Tonga
Ensign

Trinidad and Tobago
Ensign

Tunisia
National Flag and Ensign

Turkey
National Flag and Ensign

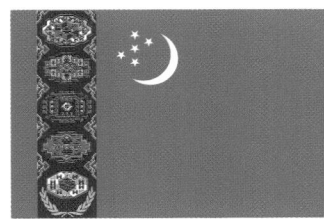

Turkmenistan
National Flag

Tuvalu
National Flag and Ensign

Ukraine
Ensign

United Arab Emirates
National Flag and Ensign

United Kingdom
Ensign

United Nations
*Flag of the United Nations
Organisation*

United States
National Flag and Ensign

Uruguay
National Flag and Ensign

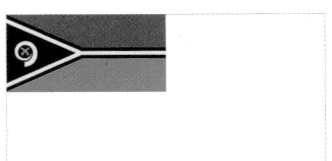

Vanuatu
Ensign

Venezuela
Ensign

Vietnam
National Flag and Ensign

Virgin Islands (UK)
National Flag

Yemen
National Flag and Ensign

Zimbabwe
National Flag

PAPANIKOLIS

2/2005*, Michael Nitz / 1121127

Executive Overview

Introduction

If the sweeping tide of 'transformation' did not come to an abrupt halt in 2005, it was certainly interrupted by a 'reality check' as budgetary measures in the United States and amongst its allies forced unwelcome cuts in naval programmes. Whether this is to mean the end, or merely a pause, in the so-called Revolution in Military Affairs (RMA) remains to be seen but, if some of the wilder claims made on its behalf have now been laid to rest, then a more severe fiscal environment has served a useful purpose. At the heart of 'transformation' is the introduction of Network-Centric Warfare or Network Enabled Capability, as it is known in the United Kingdom. Generally speaking, naval people have been more sceptical about the revolutionary potential of this process than their colleagues in the other services. At sea, the idea of networking is not new. Ships have communicated and shared information for centuries, by visual means (flags and lights), radio and more recently by datalink. All these methods have sought to improve collective effectiveness by superior co-ordination of force sensors and weapons. Thus, while it is undeniable that the application of information technology is enabling huge improvements to be made to situational awareness and operational effectiveness, this should be regarded as a continuation of an evolutionary process that builds on the command-centres and systems that have been around for several decades.

It is also worth observing that 'revolution' is a term that has been applied sparingly in naval circles; it has been reserved for those changes that have immediately rendered comparable platforms or systems obsolete. For example, since the beginning of the 20th century, it might have been applied to the Dreadnought battleships, the submarine, the aircraft carrier, radar, missile technology and nuclear power. Today, it could be argued that the only genuine innovation at the heart of contemporary developments is the Global Positioning System that, for the first time, enabled precise navigation of platforms and weapons. However, it is certainly not yet clear that the net effect of modern advances will prove to be revolutionary; neither will they necessarily deliver the 'force-multiplier' effects that have been claimed.

However, it would be wrong to suggest that navies are less than enthusiastic about technological change. If anything, they have been at the forefront of this process for many years although, as a result, tend to make a more sober assessment of the potential benefits. There is also a sense that the principal focus of the current 'transformation' initiative has been on land warfare where priority has been given to the development of smaller, more agile forces delivered and supported from the sea. The requirements of expeditionary forces and of sea-basing are having a profound effect on naval force structures while the institution of a joint battlespace, to replace the previously distinct land and sea domains, is also likely to have far-reaching effects. However, the fact that change is being implemented to cope with an unpredictable future security environment gives rise to considerable uncertainty. It is possible to imagine a world that contains several super-powers but a pattern of small, frequent military operations involving counter-terrorism, rogue states and non-state actors seems much more likely. The requirement to plan for all of these scenarios against a background of scarcer resources is concentrating the minds of planners.

United States and Canada

The United States Navy found itself in troubled waters in early 2005 as FY06 budget realities enforced some difficult decisions. The most high profile of these was the proposal to retire the carrier *John F Kennedy* that had been planned to undergo a second complex overhaul. Had she remained in service until 2018, she would have been 50 years old. The overall carrier force level is thus to be reduced from 12 to 11, the first such cut since 1994. With the next generation aircraft carrier, CVN 21, also now delayed by a year, there are fears that the total could drop still further to ten. Other programmes to be affected are the future destroyer DD(X), the numbers of which have been reduced to five, the San Antonio class LPDs which have been capped at nine and the Virginia class SSNs whose building rate is not to exceed one per year. The rate of production of the Littoral Combat Ships has also been slowed. However all is not gloom and doom. The future big-deck amphibious ship, LHA(R), is to be based on the LHD 8 design but with improved aviation facilities. The first is expected to enter service in 2012 and follow-on ships are likely to reflect further developments. Meanwhile, the first of up to 18 new Maritime Prepositioning Force (Future) (MPF(F)), could begin construction in FY08. The requirement for these vessels has promoted considerable debate, given the potential need to offload containerised loads and other equipment, to act as hospitals and command platforms and to be able to conduct sustained flight operations. These vessels are, in turn, to be served by high-speed intra-theatre and seashore connectors that will be used to transport people and material to and from the seabase. Somewhere into this matrix is also to be accommodated the Army's plans for three Regional Flotillas (ARFs), to be based around Guam and Saipan in the Pacific Ocean, Diego Garcia in the Indian Ocean and in the Mediterranean.

While there has been disappointment in the US Navy, particularly over the loss of a carrier, a deceleration in a somewhat frenetic defence agenda could prove to be a blessing in disguise. The 'transformation' process has been open to misinterpretation – a cult for some, evolutionary progress for others. Overall, there has been a tendency to regard advanced technology as an end in itself, rather than the means to achieve a desirable operational benefit. It is not surprising, therefore, that development of the future joint force structure has at times been difficult to comprehend. Fortunately, there is a chance to put this right. The Quadrennial Defence Review (QDR) was initiated on 4 November 2004 and is to be completed by February 2006 in order to help shape the defence budget for FY07 and beyond. This is the first such

GEORGE WASHINGTON

4/2004*, US Navy / 1121126

reassessment to be made since the events of 11 September 2001 and, if it does its job properly, should bring some much needed fresh thinking to bear on the problems posed by the complex pattern of threats of the modern world. Current US forces are optimised to conduct major combat operations against peer competitors but, while this remains an enduring requirement, they need also to be capable of contributing to operations in support of the Global War On Terror (GWOT). Moreover, they may also be needed to support Homeland Defence and Humanitarian Assistance efforts such as those following the Asian tsunami of 26 December 2004.

Perhaps the best service that the QDR could perform is to publish some practical criteria against which force structures can be planned. The most obvious conclusions to be drawn by potential opponents from both the 1991 and 2003 Gulf Wars is that it would be unwise to give the US military-industrial machine sufficient time to build up forces and establish a supply chain. The US is unlikely to be afforded the same luxury in the future. It follows that principal parameters of US forces must include the capability to react to contingencies quickly, accurately and decisively. This has already been reflected in the so-called 10-30-30 strategy (the ability to respond to a crisis within 10 days and resolve it within 30 days) and calls for a flexible, versatile force that is not hampered by political constraints. 'Sea Power 21' meets these requirements very well but already it has become clear that 'Sea Basing', initially a subset of the doctrine, has emerged to become the key thread which links requirements for US forces to cope with a wide range of scenarios. The concept, whose essence is flexibility, involves the staging of manoeuvre forces, supported by logistics and combat fire support, in offshore bases in or near the theatre of operations. While every situation will be different, core components are likely to comprise an Expeditionary Strike Group (ESG), a Carrier Strike Group (CSG) and a Maritime Prepositioning Group (MPG) supported by a Combat Logistics Force and from intermediate bases such as Guam or Diego Garcia. The precise size and shape of these forces is to be fully worked out but it is self-evident that the yardstick against which programmes should be assessed is their potential to contribute to sea base operations. If this benchmark were to be applied, it would become obvious that the Littoral Combat Ship is far too narrowly focused and that its roles should be expanded to include defence of the sea base and its high-speed connectors as well as theatre access operations. The reconfigurable, modular approach adopted by LCS should make this relatively easy to facilitate. DD(X), the latest iteration of the Arsenal Ship, seems an extraordinarily expensive way to deploy 80 cruise missiles, particularly in light of the SSGN conversion programme and of the option to distribute these missiles around other surface platforms. It would be better to spend the US$2.5 billion that each of these ships is expected to cost on further development and refinement of DD(X) cutting-edge technologies. These should then be taken forward in the CG(X) programme.

The submarine USS *San Francisco* collided with an underwater seamount in the Pacific Ocean on 7 January 2005 and sustained substantial damage. Reportedly, the boat was travelling at over 30 kt at the time and the force of the impact destroyed the bow-dome and ripped open the port side ballast tanks forward of the fin. Despite these devastating effects, the nuclear propulsion plant was not damaged and the submarine managed to surface. While, sadly, one crewmember was killed in the incident, the survival of the submarine was a testament to its exceptional design and build qualities.

The increase to the Canadian Defence budget, announced on 23 February 2005, "represents the most significant investment in the military in the last 20 years" according to the Department of National Defence. It provides nearly C$13 billion dollars in new funding over 5 years although cynics have observed that most of the money will not become available until 2008-10 and that there is a danger that the measure will never be fully implemented. About a quarter of the money is to be spent on new recruitment while a further quarter is to be spent on improving sustainability and readiness. The remainder is to be reserved for equipment in the longer term. There were no specific pledges to build new ships and it is assumed that such investments are pending the outcome of the defence review. However, whether this review will be the all-encompassing policy-led process, with public input, that is so badly needed remains to be seen. Reports that it is to be an internal departmental document are not encouraging. At the very least, there is a need to put Canadian Navy plans in the context of, say, the need to develop a rapid deployment task group. Such a concept would include already announced plans to procure three multirole Joint Support Ships, replacement of Sea King helicopters by the Sikorsky H-92 Cyclone and the upgrade programme for the Halifax class frigates. It would also re-endorse the requirement for diesel submarines, which are particularly suited to littoral operations, despite hysterical reaction to the *Chicoutimi* fire on 5 October 2004. It might also pave the way for one or two amphibious assault ships, which would have wide utility in future operations, and the replacement of the air-defence capability of the Iroquois class destroyers. But above all, the review must end the hand-to-mouth existence that has bedevilled the Canadian Armed Forces for far too long.

China

The rise of China has been forecast for many decades and recent economic performance has given firm indications as to the nation's potential. But 2004 seems to have been the year in which it became obvious that China's strength was no longer just a matter for academic debate; the country had finally emerged as a major economic and political power. It is difficult to pinpoint a particular reason for this but the rise in the oil price due to increasing Chinese demand, calls for revaluation of the Chinese currency and even the forthcoming Olympic Games have all contributed to perceptions that a once remote and enigmatic state had moved to centre-stage. Using purchasing-power parities for comparison, the economy is already the second biggest in the world.

The extent to which China's industrial revolution is to be matched by modernisation of the military remains to be seen and so, when China's latest paper on defence, *National Defence in 2004*, was published on 27 December 2004, it was read with particular interest. The fifth of its kind to be issued since 1995, the document was again disappointingly light on meaningful details about defence funding (the overall budget for 2005 has been raised by 12.6 per cent to 247.7 billion Yuan (US$29.9 Bn)) and force development, reflecting the secretive nature of the Chinese government. However, the paper does highlight primary Chinese defence concerns which include US military presence in the region, a more demonstrative Japanese military posture, including the development of a ballistic missile defence system, and North Korean nuclear issues. Nevertheless, it is clear that the principal focus of defence policy continues to be Taiwan, an issue on which the paper adopts a notably more belligerent stance than the measured tones of the 2002 document. Not only does it state that "it is the sacred responsibility of the Chinese armed forces to stop the 'Taiwan independence' forces from splitting the country", but also it goes on to say that "should the Taiwan authorities go so far as to make a reckless attempt that constitutes a major incident of 'Taiwan independence', the Chinese people and armed forces will resolutely and thoroughly crush it at any cost". This was further reinforced by the passing of an anti-secession law by the Chinese People's Congress on 7 March 2005. Given this unequivocal policy and legal guidance, it is not surprising that "while continuing to attach importance to the building of the Army, the PLA gives priority to the building of the Navy, Air Force and Second Artillery Force (responsible for ballistic missiles) to seek balanced

SONG CLASS

7/2004*, Ian Edwards / 1121125

development of the combat force structure''. In March 2004, the central role that the navy now plays in Chinese Defence thinking was given concrete expression by the announcement that the head of the PLA(N) had been elevated to take a permanent seat in the Central Military Commission, China's highest national security decision-making body.

While the paper contains few specific references to force structure, it reiterates the now established 'active-defence' doctrine while noting that the PLA Navy is "giving prominence to the building of maritime combat forces, especially amphibious combat forces. It also speeds up the process of updating its weaponry and equipment with priority given to the development of new combat ships as well as various kinds of special-purpose aircraft and relevant equipment. At the same time, the weaponry is increasingly informationalised and long-range precision strike capability raised''.

Naval events of 2004 have included a few surprises and perhaps their most interesting feature has been that, while construction of more capable power-projection and sea-denial forces has continued apace, the scope of modernisation has now been broadened to include hitherto neglected areas such as coastal defence and mine-countermeasures. The biggest surprise of the year was the launch of the first of a new class of conventional submarine at Wuhan on 31 May 2004. It is fair to say that the intelligence community was caught completely unawares by the emergence of the Yuan class and no doubt there are lessons to be learned about the dangers of clinging too firmly to western assumptions. The accepted hypothesis had been that conventional submarine procurement was being taken forward by two parallel programmes: the domestic Song class and the Russian Kilo class. Both are at a mature stage with production of the Song class apparently accelerating and delivery of the first of eight Type 636 variants of the Kilo class having taken place in late 2004. It is unclear how the Yuan class will affect future plans but one possibility is that the new submarine represents a merging of all available technology and signals the end of foreign procurement. Whether the foreign technology is taken from the Kilo class or from the more modern Amur 1650 class is uncertain. The two nuclear submarine programmes have also made progress. The first two Type 093 SSNs have been launched and a third is believed to be under construction. The first of class is expected to enter service in 2006 and is likely to be a marked improvement on the Han class that they will replace. The first Type 094 SSBN was launched at Huludao Shipyard in July 2004. A stretched version of the Type 093, it is now believed to contain 12 rather than 16 missile tubes. The 8,000 km range of the JL-2 missile suggests that the new SSBNs are likely to adopt a 'bastion' patrol philosophy.

Surface-ship construction has also been busy. Both the new Type 052B and Type 052C air-defence destroyers will have entered service by 2006 while the first of yet another new class of destroyer, the Type 051C class, was launched at Dalian Shipyard on 28 December 2004. A second of class will probably follow. Surprisingly, the design looks less stealthy than the Type 052B/C ships, suggesting that the design may predate them. This may be due to delays in procuring the 120 km RIF missile system with which it is reported to be equipped. The entry into service of these six ships, together with the procurement of four Sovremenny class destroyers from Russia and recent upgrades to the two Luhu class represent a significant upgrade in capability over the last five years. Attention is also turning to the replacement of the obsolete Jianghu class frigates. The first two Type 054 frigates have entered service and a modified Type 054A design is reported to be at an advanced stage. Construction of amphibious ships has also continued with the Yuting II class LST, the Yunshu class LSM and Yubai class LCU all in series production in at least three yards. Persistent rumours of plans to build a larger amphibious ship, such as a Landing Platform Dock, have continued to abound and this would be a logical next step in surface-fleet development.

The launch of a new coastal patrol craft with a wave-piercing catamaran hull form, probably based on Australian technology, was another surprising development of 2004. The class already appears to be in production at several yards. If so, this represents a bold step by the Chinese Navy. Speed and economy are among the many attributes of the catamaran design but damage-control, particularly in an aluminium hull, could prove to be a shortcoming. Finally, it would appear that replacement of an obsolete mine-countermeasures force has begun with the launch of a new class of MCMV in April 2004.

United Kingdom

As the Royal Navy celebrates the 200th anniversary of the Battle of Trafalgar, it finds itself in a tough fight for resources. The demands of current land operations, and the necessity to honour the full order for the Royal Air Force's Typhoon interceptor, are both demanding the lion's share of available funding. The Defence Command Paper 'Delivering Security in a Changing World – Future Capabilities', published on 22 July 2004, introduced the most severe platform cuts since the post Cold War Defence Review of 1993. The number of nuclear submarines is to be cut to eight, the escort flotilla is to be reduced from 31 to 25 and mine-countermeasures vessels from 22 to 16. The number of Type 45 destroyers the Navy can expect has been cut to eight. It is also likely that the next order for Astute Class submarines will be limited to a single boat, in addition to the three already being assembled, even though the final anticipated order remains for seven boats. Furthermore, development of the next-generation frigate, the Future Surface Combatant, has been delayed. These sacrifices have been essential more to satisfy short-term measures necessary to balance Defence accounts than, as is commonly thought, to safeguard funding for two new aircraft carriers that are to be the centrepiece of future expeditionary operations.

ILLUSTRIOUS and INVINCIBLE

11/2004*, Royal Navy / 1121124

However, despite political assurances, the aircraft carrier programme (CVF) could also yet fall victim to post UK general election decisions by a new government and are not even guaranteed to survive UK MOD infighting. The Navy has not been helped by the prolonged industrial squabbles that have plagued the project and, not for the first time, there is exasperation at all levels of the Service that the operational requirement, identified in the 1998 Defence Review, seems to be the last thing on official minds. To make matters worse, commissioning of the first of the new class of Astute submarines and of the first Type 45 destroyer, HMS *Daring*, have been delayed until 2009. There has also been a reprofiling of the programme for the new afloat support vessels and slippage in the provision of the new Casualty Treatment Ship. Meanwhile, construction of a new class of landing ships is taking much longer than originally expected. Given this uncertain state of affairs, it is not surprising that the Chief of Naval Staff described himself as "uncomfortable" with the situation while observing that "there is a risk with these reductions and my concern overall is that we are taking risk on risk". In marked contrast to painfully optimistic messages of the past, it was refreshing to hear such a frank assessment from the naval hierarchy. In the relentlessly 'on-message' world in which we live, it amounted to no less than serious criticism of government decisions.

The most serious operational risk is in air-defence, also identified as a major concern by the House of Commons Defence Committee in its report of 17 March 2005. The Sea Harrier fighter is to have been withdrawn from service by 2006 and it is not until 2013, by when the first six Type 45 destroyers (equipped with the PAAMS missile system) will have been commissioned and when the Joint Strike Fighter enters service, that the capability gap will be bridged. The Chief of Naval Staff has again been candid in his assessment that "we will be taking risks in certain types of operations and I would not be too happy being in a very high air threat". The report added that the decision to procure only eight new destroyers could also create attrition problems. There is certainly a suspicion that operational analysis has given insufficient weight to military judgement in reaching its conclusions. To suggest that only 12 destroyers and frigates need be earmarked for large scale operations flies in the face of experience gained in the Falklands War in which 23 such ships were used; four of these were sunk and eight were damaged. Lack of versatility in land-attack capabilities is also troubling. Military support of land-operations will in most cases be discharged by carrier-borne aircraft; the weight of offensive air power offered by the Harrier GR.9 is the principal argument in support of the Sea Harrier decision. However, while Tomahawk was originally procured by the Royal Navy to serve merely as a means of deterrence and coercion, its extensive use in Kosovo, Afghanistan and Iraq has demonstrated its much broader utility. The judgement that a quick, precise reaction could prove to be the key determinant of many situations is widely shared and it is not surprising that procurement of land-attack cruise missiles is now under active consideration by several other navies including those of India, France, Italy, Spain, and Denmark. It is disappointing, therefore, that there is no plan to equip Type 45 destroyers with Tactical Tomahawk given that its virtues are so obvious and cost relatively cheap. It would certainly go some way towards improving operational flexibility that has suffered as a result of force reductions. A force level of only 25 frigates and destroyers means that, at any one time, only 21 are immediately available for operations. Similarly, of an overall force level of eight SSNs, only four are held at full operational readiness. Given that future contingencies are unlikely to afford as much preparation time as recent operations, it follows that every deployed vessel must be as capable and versatile as possible. Any ideas that this might be achieved by a high/low mix of ships, as has been mooted, should be dismissed.

Not all news over the last year has been bad and the Royal Navy can point to some success stories, albeit against an unsettled background. The second new assault ship, HMS *Bulwark*, was commissioned on 28 April 2005 and, once the Bay class landing ships have all entered service in 2006, the United Kingdom will at last possess a modern and effective amphibious force which will be, with the new carriers, the core units of an intervention force. The entry into service of the Apache attack-helicopter will be a powerful addition to these units. Procured to be part of the army's 16 Air Assault Brigade, eight aircraft have been earmarked to replace the Lynx AH Mk 7 helicopters as part of the Commando Helicopter Force. The UK is the first nation to test the Apache at sea and, following successful safety trials in March 2004, it is planned that four aircraft will deploy in HMS *Ocean* for Exercise Argonaut 05 in September 2005. In other helicopter developments, the AgustaWestland Future Lynx was selected in April 2005 as the preferred option to fulfil the navy's maritime surface attack helicopter requirements as well as the army's scout/utility role. The aircraft is likely to have much in common with the Super Lynx helicopters being delivered to Oman. Meanwhile, the contract for a new Falkland Islands Patrol Vessel was awarded to VT Shipbuilding on 25 February 2005. The ship, to be called HMS *Clyde*, is to replace the two ageing Castle class vessels that have fulfilled the role for two decades. The first warship to be built at Portsmouth since the 1960s, the ship will be similar to the three River class and is to enter

service in 2006. Anti-Submarine Warfare (ASW) capabilities were significantly enhanced by installation of the first Low Frequency Active Sonar (Type 2087) in HMS *Westminster*. This system is to be fitted to eight Type 23 frigates which will also embark the Merlin ASW helicopter. Finally, the newly refitted and reconfigured HMS *Illustrious* is to spearhead the transition of the carrier force from the anti-submarine warfare role, for which it was originally designed, to the maritime strike role. To be joined by the similarly modified HMS *Ark Royal* in 2007, the two ships will pave the way to a new era of carrier aviation that includes the Queen Elizabeth class and the Joint Strike Fighter. It is an exciting future albeit one that is still clouded by lot of risk.

The mood of the Royal Navy in the bicentenary year of the Battle of Trafalgar is, on the whole, upbeat. It usually is. But there is no disguising the sense of unease about the future that lies not far below the surface and it would be a great mistake for politicians of any political party to take its goodwill for granted. The men and women of the Royal Navy are not fools and their mood was well summarised by the former Chief of Naval Staff and Chief of the Defence Staff, Lord Boyce, during the House of Lords debate on 13 January 2005. "The Armed Forces know that they are under-funded and under-resourced to do what is expected of them. They know there is no beef for them behind the high-sounding phrases. They know that significant cuts are now being taken in the front line, under the disingenuous reason that we need less because of Network Enabled Capability ..I exhort the Government to give real and honest attention to the hollowing out that is going on, to think about what the Armed Forces do for this country – upon which the Government ride high – and to stop cheeseparing and trading on the good will and professionalism of the men and women of the Navy, Army and Air Force".

Europe and the Mediterranean

As an old era ended, a new one began. The Standing Naval Force Atlantic (STANAVFORLANT) ceased to exist at the end of 2004 as it became the Standing NATO Response Force Maritime Group 1 (SNMG1) on 1st January 2005. At the same time, the Standing Naval Force Mediterranean evolved into SNMG2 while the MCM forces became SNMCMG1 and SNMCMG2. The origins of STANAVFORLANT go back to 1960 when it was first proposed that a NATO Anti-Submarine Warfare Task Group should be formed. This was followed in 1965 by the initiation of the *Matchmaker* series of exercises that led to the formation of a permanent squadron on 13 January 1968. The continuation in service of all four squadrons is a testament to the success of the multinational concept. Not only have they served as visible demonstrations of NATO solidarity but also, at a practical level, they have fostered the harmonisation of procedures and habit of co-operation that are the cornerstones of today's NATO and coalition naval operations. In addition to being standing elements of the NRF, both SNMG 1 and SNMG2 have been involved in Operation 'Active Endeavour', whose mission is to conduct naval operations in support of the international campaign against terrorism. These have included compliant boardings and the escort of merchant ships through the Strait of Gibraltar. In the future, nations of the Euro-Atlantic Partnership Council and Mediterranean Dialogue may also be asked to contribute to these activities.

Meanwhile, the NRF is on track to achieve full worldwide operational capability by October 2006 by when the first schedule of rotating operational commanders will have been completed. In common with the Joint Force and Air commands, the two-star naval command changes once a year and so, after the first full year of command, the UK Maritime Force Commander is to hand over to his Italian counterpart in July 2005. The Spanish Maritime Battle Staff will resume command in July 2006. One of the principal responsibilities of the maritime commander is the training and validation of all the forces assigned. These might include a carrier battle group, amphibious forces and the associated surface and subsurface force components although not all of these forces would necessarily be required for every operation. Force composition will always be tailored to the mission.

The requirement of European forces, whether under NATO or EU auspices, to be capable of power projection operations continues to be the principal influence on naval procurement. Of the leading expeditionary nations, France has one of the most vibrant building programmes and a number of projects had reached significant milestones by early 2005. *Le Vigilant*, a new ballistic missile submarine, was commissioned on 26 November 2004 and is to enter operational service in 2005 when *L'Indomptable* is to be withdrawn. Much of the technology of the strategic submarines is to be reflected in the next generation attack submarine, the Barracuda class. A contract for initial three boats is expected in early 2006 and construction of the first of class is to start at Cherbourg during the same year to meet an in-service date of 2013. The first of a new generation of amphibious ships, *Mistral*, was launched on 6 October 2004 and is to enter service in 2005. She is to be followed a year later by her sister ship, *Tonnerre*. Meanwhile, the final design stage for a second aircraft carrier (PA2) is to begin in mid-2005 when preparatory work has been completed. This second ship is planned to enter service in 2014 before *Charles de Gaulle* starts her

MISTRAL   *10/2004*, J Y Robert /* 1121123

refuel/refit in 2015. However, there are to be few similarities between the two ships. The new carrier is to be based on the so-called Romeo design that, at 60,000 tons, is considerably larger than *Charles de Gaulle*. The ship is also to have an all-electric (rather than nuclear) power and propulsion architecture. Both the new destroyer and frigate programmes are to be the products of Franco-Italian co-operation. *Forbin*, the first of the French Horizon class anti-air warfare destroyers, was launched at Lorient on 10 March 2005 and is to begin sea-trials in 2006. Her sister ship *Chevalier Paul* is to follow a year later. Equipped with the PAAMS anti-air missile system, the ships are to replace the Suffren class. Meanwhile, launch of the first Italian destroyer, *Andrea Doria*, is expected in mid-2005 with her sister ship *Caio Duilio*, to follow in 2009. These ships are to replace the Audace class. The European Multi-Mission Frigate (FREMM) programme was given the go-ahead on 25 October 2004 and orders for the first batch of eight (of 17) ships for France and for the first batch of six (of ten) ships for Italy are expected in 2005. Of other Italian programmes, the centre and stern sections of the new carrier, *Cavour*, were launched at Riva Trigoso on 24 July 2004 and, after being joined to the bow section at Muggiano, sea trials are expected to start in 2006. Other activities at Muggiano included launch on 18 December 2004 of *Sciré*, the second Type 212 submarine with air-independent propulsion (AIP). A contract for a new class of four AIP submarines for the Spanish Navy was awarded to IZAR on 25 March 2004. The boats are to be derived from the Scorpene class of which variants have been sold to Chile, Malaysia and (to be confirmed) India. Construction of the first of class started in early 2005 for delivery in 2011. Following a re-appraisal of other programmes, the new shape of the Spanish Navy is beginning to emerge. A high-capability force is to comprise two carriers, two landing platform docks ships and up to six air-defence destroyers while constabulary tasks are to be undertaken by a new class of eight multirole offshore patrol vessels. Construction of a second fleet replenishment ship to support these forces is to start in 2005.

Perhaps the biggest changes are being effected in Scandinavia where transformation of cold-war era self-defence forces into blue-water navies capable of undertaking modern missions is well underway. The most drastic reorganisation has occurred in Denmark where, following withdrawal from the Viking submarine programme, the submarine service was dis-established at the end of 2004. Instead, the government has opted to consolidate its fleet around surface ships capable of contributing to international crisis management and peace support operations. Principal units are to include two new 6,000 ton combat support ships, *Absalon* and *Esbern Snare*. The largest ships ever to serve in the Royal Danish Navy, the ships are to be capable of command, logistic support and limited sealift operations. A Ro-Ro cargo deck can embark a wide range of payloads including two high-speed insertion craft, variants of the Swedish Combatboat 90E. Three new 5,000 ton frigates are to replace the 1,350 ton Niels Juel class from 2010. These are to be derived from the combat support ship and are likely to have a general-purpose capability including land-attack. *Fridtjof Nansen*, the first of a new class of 5,000 ton destroyers for the Norwegian Navy was launched at

Ferrol on 5 May 2004 and is expected to be commissioned in October 2005. This class will replace the venerable Oslo class which have given sterling service since the mid-1960s. In Sweden, the submarine service is to be comprised entirely of AIP submarines following the conversions of *Södermanland* and *Östergötland*. It is of particular interest that the United States Navy is to be lent assistance in improving its anti-submarine warfare capability. The submarine *Götland* is to deploy to the US for most of 2005 and is to exercise with USN units on both coasts. Sweden has co-operated with NATO as a Partnership for Peace participant for some years and this event does not signify an abandonment of its neutrality policy. It is, nevertheless, further evidence of Sweden's willingness to look beyond its own borders.

The accession of Romania and Bulgaria to NATO and their probable admission to the European Union in 2007 has focused attention on the unstable region of the Black Sea. To the east, much of the Caucasus is at war while, to the north, the election of a pro-western government in Ukraine could lead to increased friction with Russia, particularly over the use of the naval bases in Crimea. Meanwhile, the US is seeking to forward-base troops, probably at Romanian and Bulgarian bases. Both these latter countries have made significant steps to improve their naval inventories over the last year. *Regele Ferdinand*, an ex-UK frigate, arrived in Constanta on 10 December 2004 and is to be followed by her sistership *Regina Maria* in mid-2005. Ownership of these two ships will play a vital role in making NATO membership a reality. In neighbouring Bulgaria, it is likely that the ex-Belgian frigate *Wandelaar* will be acquired in 2005 and this may be followed by up to three tripartite minehunters.

Ten years after the institution of the Mediterranean Dialogue, designed to improve confidence building and co-operation between NATO and its seven Mediterranean partners: Algeria, Egypt, Israel, Jordan. Mauritania, Morocco and Tunisia, the pact was elevated to a genuine partnership at the Istanbul summit in June 2004. This was followed by a visit by the NATO Secretary-General to Algeria in November 2004 and by a meeting at ministerial level between NATO and Dialogue countries in Brussels in December 2004. From a naval perspective, this is likely to improve intelligence sharing and maritime co-operation, particularly as they relate to Operation Active Endeavour. Naval equipment sales may also get a boost. To date, Germany has been one of the most active export countries. Following the sale of Type 148 fast patrol craft to Egypt in 2003, six decommissioned Type 143 class craft are likely to be transferred to Tunisia in 2005. Egypt may also acquire Type 206A submarines to replace its obsolete Romeo class while Israel has reportedly been in negotiations to procure further Dolphin class boats although funding problems may prove to be difficult to overcome.

Overall, there is the appearance of considerable activity in the European naval scene. However, whether there is enough work to sustain the large number of shipbuilders remains to be seen. General consolidation in the defence industry has yet to extend to shipbuilding companies where national pride continues to play a strong role. Franco-Italian co-operation on Horizon

TAMBOV

7/2004* / 1121122

and FREMM and the merger on 5 January 2005 of the German companies of TNSW, HDW and Blohm + Voss, Kockums (Sweden) and Hellenic Shipyards (Greece) under the industrial management of ThyssenKrupp Marine Systems might point to a reversal in trend but these only represent a streamlining of activity rather than a reduction in capacity. In contrast, naval construction in the US is already confined to a few yards and there may be further reductions if construction of the DD(X) destroyer is put out to competition. The problem of over-capacity in the European naval shipbuilding industry will be a formidable problem for the newly formed European Defence Agency.

Russia

Judging by the number of out of area exercises, the Russian Navy had a busy year in 2004. In the Pacific, the destroyer *Marshal Shaposhnikov* and the Ivan Rogov class landing ship *Mitrofan Moskalenko* visited Kure in Japan while the Baltic fleet based landing ship *Minsk,* carrying an embarked force of marine infantry, visited Devonport, UK where both anti-terrorism and disaster-relief exercises were carried out. A Black Sea flotilla of ships, under the command of fleet commander Admiral Masorin, included the cruiser *Moskva*, the destroyer *Smetlivy*, the frigate *Pytlivy*, the landing ship *Azov* and the Boris Chilikin class tanker *Ivan Bubnov*. These visited the Mediterranean in September to conduct exercises with the Italian and US navies while visits were also made to Turkey, Greece and Malta. By far the most ambitious deployment was from the Northern Fleet. The group comprised the aircraft carrier *Admiral Kuznetsov*, the cruiser *Pyotr Velikiy*, the destroyers *Admiral Levchenko*, *Severomorsk* and *Admiral Ushakov*, the frigate *Admiral Chabanenko*, and numerous vessels as well as Oscar and Akula class submarines. During a deployment in the North Atlantic, exercises were conducted by some units with the French and US navies and the Akula class submarine *Vepr* paid a historic visit to Brest, the first call by a Russian nuclear submarine to a foreign port.

Following a low point in Russian naval activity in 2002, 2003 was clearly characterised by a 'back to sea' policy. However, if the 2004 programme was an attempt to demonstrate that it was 'back in business' it was a failure. If anything, the series of exercises served only to show that the Russian Navy has neither regained its former glory nor come to terms with the challenges of the 21st century. The first deployment by Russia's only aircraft carrier since 1996 should have been a showpiece event. Instead it was a tentative affair which barely left the Norwegian Sea and which was marred by poor reliability and a Su-33 accident. Moreover, the setting for the national exercise in which the *Kuznetsov* group simulated an attack by an enemy carrier battle group might have been lifted straight out of the cold war. It is extraordinary that the Russian navy should contemplate such a scenario while at the same time conducting some eleven exercises with NATO countries during the year. It is not surprising that, with this mindset, Admiral Masorin is quoted as saying that his fleet was still learning how to co-operate with foreign navies, given its relatively limited experience of bilateral operations. However, it would be unfair to over-criticise the seagoing navy that is still trying to overcome a decade of neglect during the 1990s and which seems to be caught between two differing visions of what it should be doing. On the one hand it is trying to follow a policy agenda in which the Russian Navy co-operates with other countries and contributes to the general good; on the other it is bound by a doctrine which is based on the defence of its ballistic

missile submarines and approaches to the homeland. The responsibility for deconflicting these guidelines is entirely that of the naval staff and its commander-in chief. Unless it can clearly articulate what the Russian Navy is for, it is difficult to see how a coherent shipbuilding plan can be developed to replace what is already an ageing fleet.

Indian Ocean, Gulf and Caspian

The main surprise about *Indian Maritime Doctrine*, which was adopted in May 2004, is that it had not been published before. India clearly abandoned its earlier defensive doctrine some years ago and its ambitions to become a regional power, capable of projecting force throughout the Indian Ocean, have been reflected in recent procurement decisions. There is a danger, therefore, that the document might be dismissed by critics as a mere 'tidying-up' exercise. Certainly, its main weakness is its naval authorship; it remains something of a 'wish-list' without official endorsement by the United Progressive Alliance government. Nevertheless, it does draw together all the current strands of Indian naval procurement against the contemporary backdrop of littoral warfare and counter-terrorism in a coherent manner. It also draws on the lessons of recent conflicts by highlighting the need to strengthen its command, control, communications and intelligence systems.

One of the core themes of the paper is to build a submarine-based "non-provocative strategic capability". The so-called Advanced Technology Vessel (ATV) project was initiated in the 1980s and is likely now to become a multipurpose submarine capable of adopting a strategic role in addition to traditional attack submarine functions. However, there is doubt as to whether the strategic missile system is to be a land-attack cruise or a short-range ballistic missile. The Sagarica cruise missile has reportedly been under development since 1991 but it is unclear whether the project has since been terminated or delayed in order to incorporate latest technology. Meanwhile Prithvi III, a navalised version of the short-range ballistic missile, was successfully launched at Balasore on 27 October 2004. Such missiles might be mounted in vertical launchers aft of the fin. While it is not surprising that such a complex project is taking time to reach fruition, there have been few indications of progress and sea trials of the first of class are not expected before 2009. Meanwhile the long-expected lease of one or two Russian Akula class submarines has also failed to materialise. A contract for these would be a good indication as to the health of the ATV project given their role in providing engineering and operational training as well as acting as a stop-gap measure. Progress in improving the inventory of conventional submarines has also been slow. While *Sindhudhvaj* is to be the sixth Kilo class to be equipped with ship/land-attack missiles when she undergoes refit in Russia from 2005, plans to acquire/build French Scorpene and/or Russian Amur class submarines were not advanced during 2004. The 30-year programme to construct 24 conventional submarines appears to be marking time.

The future Indian surface fleet is to be centred round at least two carrier battle groups and a modern force of destroyers and frigates. Refit work on the ex-*Admiral Gorshkov* is reported to be progressing well and the ship appears to be on track to enter operational service in 2008. Meanwhile, after several false dawns, work on the indigenous carrier (Air Defence Ship (ADS)) is finally set to start at Cochin in 2005 following two contracts signed in mid-2004 with Fincantieri to finalise the ADS design and its ancillary propulsion systems and main power plants. Fincantieri is also likely to

TABAR   *5/2004*, *Michael Nitz* / 1121121

provide further assistance during the vessel's construction, tests and sea trials. *Tabar*, the third Russian-built Talwar class frigate entered service in 2004. These ships are likely to be fitted with the 150 n mile range supersonic BrahMoS anti-ship cruise missile and an order for a further three ships is expected. After ten years under construction at Garden Reach, *Betwa*, the second Brahmaputra class frigate was commissioned on 7 July 2004 shortly after *Satpura*, the second Shivalik class frigate was launched at Mazagon Dock where the first of three project 15A modified Delhi class destroyers is also being built.

Overall, the capability of the Indian Navy has grown steadily in recent years but, while it contains some impressive ships and weapon systems, procurement decisions have emerged slowly and operations are likely to be constrained for some time by weaknesses in amphibious shipping, afloat support, reconnaissance aircraft and C4I infrastructure. It is clear that, at the moment, *Indian Maritime Doctrine* is not underpinned by the necessary funding. However, all this could change if the focus is shifted away from its traditional rival, Pakistan, to the growth of another ambitious navy, that of the People's Republic of China.

The growth of the Pakistan Navy has also been hampered by lack of funding. The smallest of the three armed services, it receives about 25 per cent of the country's US$3.4 billion annual defence budget. In contrast to India's expansionist vision, Pakistan's naval doctrine is based on the requirement to defend its coast, particularly Karachi and associated sea lines of communication. It is noteworthy that efforts to decrease vulnerability to surprise attack are well advanced. Reliance on the fleet base at Karachi is being reduced by the basing of submarines at Ormara (Jinnah Naval Base), which has been open since 2000. Meanwhile, facilities for surface ships are being created at Gwadar, 250 n miles west of Karachi near the Iranian border, where a deep-water port is under construction. As this is a joint venture with China, it is more than likely that the PLA(N) has been guaranteed rights to use the base, thereby providing China with a strategic foothold in the Arabian Sea. The profile of the port could increase still further if the port were to be connected directly to Xinjiang province by overland route.

Building on close ties with China, a contract to procure new Jiangwei II class frigates was finally concluded on 4 April 2005 although these ships could well be fitted with western as well as Chinese systems. The first ship is to be built in China, work on the second is to be shared while the last two are to be built at Karachi. Initially, these frigates are expected to be in addition to the Type 21 class and so overall frigate numbers could increase to ten over the next few years. The principal offensive arm of the Pakistan navy, its submarine service, should receive a boost in late 2005 with the launch of *Hamza*, its first submarine to be equipped with air-independent propulsion. It should become operational in 2006. Given Pakistani concerns about the vulnerability of Karachi and its sea links, acquisition of additional maritime reconnaissance assets should also be a high priority.

Despite the depressing internal situation in Iraq, there have been some flickers of hope as the rebuilding of the Iraqi armed services achieved some important milestones. The Iraqi Border Riverine Service, established in June 2003, evolved into The Iraqi Coastal Defence Force (ICDF), part of the New Iraqi Army, when it took formal possession of a fleet of five Chinese built 27 m patrol craft at a ceremony in June 2004. This coincided with the commissioning and handover of the newly constructed naval base at Umm Qasr. Training, coordinated by the Royal Navy with assistance from Australian, Dutch and US naval personnel, has progressed well and overall strength is expected to reach over 400 personnel. The force is to be further augmented by the activation of the two Assad class corvettes that have been in the hands of a caretaker team at La Spezia since 1986. Following an announcement by the Italian government in December 2004, both ships are to be refitted for delivery to Iraq in 2006. The principal tasks of the ICDF are to protect the country's coastline, territorial waters (including offshore oil terminals) and major internal waters.

Elsewhere in the Gulf, the impact of higher oil prices is yet to feed through to military budgets and there have been relatively few developments over the last year. The United Arab Emirates operates one of the busier navies in the area while, in parallel, efforts are being made to develop a domestic naval shipbuilding and repair capability with a view to attracting regional customers. Project Baynunah is a joint venture between CMN and Abu Dhabi Shipbuilding (ADSB) to build four corvettes for delivery by 2012. Three are to be built in the UAE. Further contracts for ADSB in the year included the construction of four fast supply vessels, similar to the 12 Ghannatha amphibious craft already in service, construction of a second 42 m landing craft and refit of the two Mubarraz class fast attack craft. The Royal Navy of Oman is another active navy where it is hoped that Project Khareef for three new 80 m offshore patrol vessels will be finalised in 2005. There is also a requirement for a 40 m high-speed sealift vessel that would be utilised along Oman's 1,100 n mile coastline. Progress in Iran has been patchy. On the one hand there have been no developments in the refit of Kilo class submarines and there must be doubts about their operational effectiveness. Neither has there been any sign of their long-awaited move to Char Bahar although a jetty has been extended to facilitate operations from the port. There have also been no reports of building progress of the Mowj corvettes, believed to be derived from the Vosper Mk 5 corvettes. On the other hand, the Qadir class mini-submarines, whose most likely area of operations is the Strait of Hormuz, have been shown on Iranian television and the launch of *Peykan*, the first of probably three new Combattante class fast attack craft built for operations in the Caspian Sea, was well publicised.

It has been a good year for South Africa. The rebuilding of the navy is in full swing with the launch of the first 209 class submarine at Kiel in June 2004 and a second is expected to follow in 2005. The fourth and last Valour class frigate *Mendi* arrived in Simon's Town on Friday 17 September 2004, an occasion marked by a fleet review in False Bay during which all four corvettes were seen at sea simultaneously for the first time. The Department of Environmental affairs also took delivery of *Sarah Baartman*, an 83 m offshore patrol vessel.

BALLARAT

6/2004*, RAN / 1121120

East Asia and Australasia

First steps by the US to build a ballistic missile defence network against possible attacks from countries such as North Korea were taken in October 2004 when an Aegis destroyer, USS *Curtis Wilbur,* began missile surveillance patrols in the Sea of Japan. Such ships are capable of providing instantaneous queuing and target data to command and control and ground-based elements of what is to be a layered defence system; its deployment coincided with the activation of interceptor missiles at Fort Greely, Alaska. The patrols, which are to be conducted on a continuous basis, are to move to the next stage in 2005 when an Aegis cruiser, armed with the Standard SM-3 missile, will field the capability to intercept both short- and medium-range ballistic missiles. A third stage, the stationing of up to ten SM-3 armed destroyers worldwide, is to begin in 2006. The North Korean missile currently comprises the No Dong 1 and 2 and Taepo Dong 1 and 2 missiles. These, with ranges up to 2,300 n miles, pose considerable dangers to the region. However, the longer (up to 4,350 n miles) range Taepo Dong 3 version, reported to be under development, could also threaten Hawaii, Alaska and parts of the US north-west coast. There has also been speculation about the existence of a sea-based ballistic-missile programme, based on the Russian R-27 (SS-N-6) missile. However, even if such technology transfer, denied by the Russian Defence Minister, had taken place, there would be formidable technical and operational problems to overcome. Development of a suitable platform (submarine or surface ship) would be difficult enough but the assumption of a strategic role by an ageing, coastal navy would also be a huge step.

Both North Korea and China were cited as threats to Japanese security by the new National Defence Programme Outline, which provides policy underpinning for defence procurement over the next ten years. The paper, only the third such document to be published since 1976, was agreed by the Japanese cabinet on 10 December 2004 and marked a firm departure from the restricted self-defence policy that the country had maintained since the end of the Second World War. Fittingly, the 50th anniversary of the Japanese Self-Defense Forces had been celebrated just a few months before. There had already been tentative steps over the last few years to take part in international operations but their terms of reference had always been cautiously prescribed. This situation is now set to end with the adoption of a new defence role, "improvement of the international security environment", which should pave the way for more active participation in international peacekeeping and efforts to combat terrorism. In the words of Yoshinori Ono, the Japanese Defence Minister, "from a deterrent to a responsive force, that is the future direction of our defence posture". For the Japanese navy, an era in which it was 'all dressed up and nowhere to go' is now to give way to a period of transformation to its new role. The size and shape of the force is set

to change as overall numbers of destroyers drop to 47 while new more capable ships enter service. The first new helicopter carrier is to be commissioned in 2009 with a second likely to follow two years later; the sea-based segment of ballistic-missile defence capability is to be met by the installation of Standard SM-3 missiles in the 'Kongou' class destroyers; submarine capabilities are to be significantly upgraded by the acquisition of the 'Improved Oyashio' class with air-independent propulsion. However, enthusiasm for Japan's broader military role is likely to be tempered by the reality that there is unlikely to be a surge in out-of-area deployments. Setting aside the political sensitivities, fleet operations will be constrained for some years by a relative lack of afloat support and strategic sea-lift. Small discrete operations, such as the disaster relief deployment to Aceh in January 2005, are likely to be the norm. Neither can the home-defence role be ignored. Japan has a long coastline and numerous offshore islands, including the disputed Senkaku islands, to protect. The incursion of a Chinese nuclear submarine into Japanese territorial waters off Okinawa on 10 November 2004 will also have served as a timely reminder of the need for adequate numbers of maritime patrol aircraft.

More than a quarter of the world's trade, including almost all of Japan and China's oil imports, pass through the Malacca Strait. Therefore, it was disappointing to learn from the annual piracy report of the International Maritime Bureau, published on 7 February 2005, that these waters together with those of Indonesia, remained the world's worst piracy hot-spot. This is despite a worldwide reduction in reported attacks on merchant shipping in 2004 from 445 in 2003 to 325. In recognition of this problem and of increased fears of terrorism, there have been concerted efforts over the last year to improve security. The three littoral states, Malaysia, Singapore and Indonesia, agreed on 29 June 2004 to commit forces to a pool of about 15 vessels which would operate year-round co-ordinated naval patrols. These would be allowed to cross maritime boundaries. While US forces are likely to avoid any direct involvement, without invitation, they are expected to play a crucial role in the provision of intelligence, technical assistance and training. In another initiative, the nations of the Five Power Defence Agreement (Australia, Malaysia, New Zealand, Singapore and UK) agreed to expand the scope of the treaty to include counter-terrorism issues. This should result in an increase in intelligence exchanges and the conduct of training exercises.

Perhaps the most interesting maritime security developments in the region, if not in the world, came from Australia. The announcement on 15 December 2004 that Australia would establish from March 2005 a 1,000 n mile Maritime Identification Zone caused a certain amount of consternation amongst its neighbours. Two-thirds of Indonesian territorial waters fall within the new zone and concerns were voiced that the proposal would not only violate sovereignty but also contravene international law. On closer

SAO PAULO and RONALD REAGAN   *6/2004*, Ships of the World /* 1121119

inspection, however, the initiative is less alarming. The overall aim is to provide a much higher level of maritime awareness in that, within the 1,000 n mile zone, ''vessels destined for Australian ports will be required to provide detailed information on the ship's identity, crew, cargo manifest, location, course, speed and intended route, port of arrival and schedule''. The implication is that uncooperative vessels will be denied access to Australian ports. Reassurance came from Robert Hill, Australian Minister of Defence, when he stated that the zone was ''not an extension of jurisdiction'' but rather ''an extension of geography within which we would like to know the nature of ships that intend to either transit Australian waters or intend to land in Australian ports''. The Australian Maritime Identification Zone is to be managed by another newly created organisation, the Joint Offshore Protection Command (JOPC) which is to be ''responsible for the implementation, co-ordination and management of offshore maritime security''. Australian Defence Force assets are to be linked with those of Customs, particularly the civil surveillance agency Coastwatch. The Director General JOPC is to be responsible to the Chief of the Defence Force and to the chief executive of Customs for military and civilian functions respectively.

New Zealand is some 5,500 n miles from North Korea but is far from immune from today's security threats including international terrorism. If the bomb attack in Bali on 12 October 2002 was a wake-up call, a trend of instability and relative economic decline in the South Pacific, its principal area of interest, has also been a persistent worry. Therefore, the reshaping of New Zealand's navy has been founded as much on the potential requirement to provide assistance to south Pacific states as the need to defend its own extensive waters. The centrepiece of 'Project Protector' is a new multirole vessel with the capability to provide limited tactical sealift for a range of operations ranging from military intervention to disaster relief. The ship is to enter service in 2006. The following year, two new offshore patrol vessels will join the fleet. Their prime duty is to be able to patrol New Zealand's EEZ and the southern ocean but they will also have the capability to assist south Pacific States if required. Four inshore patrol vessels are to patrol New Zealand's inshore zone from 2007. Not that the capability to contribute to operations at the high end of the conflict spectrum is to be abandoned. Two relatively new ANZAC frigates are to remain in service and a mid-life modernisation programme is likely to be implemented from about 2008.

Latin America

Economic performance continued to improve in the principal Latin American countries over the last year. However, given domestic priorities and the need to keep a firm grip on government spending, it is unlikely that military projects will benefit from increased funding in the short term.

In Brazil, where GDP growth was 4.9 per cent in 2004, the overall defence budget has been maintained at the same level and aspirations to become South America's leading naval power continue to be frustrated by financial realities. In fact, the frigate force has had to be reduced by two (to ten) and modernisation of an ageing inventory of surface ships has been restricted to the Niterói class. Despite these difficulties, the country made a substantial contribution to the United Nations Stabilisation Mission in Haiti (MINUSTAH). The employment of the amphibious ships *Mattoso Maia* and *Ceará* to transport equipment, humanitarian aid and 230 marines will have underlined both the utility of these types of ships and of the pressing need to

replace two 50-year old vessels. Priority continues to be given to the submarine programme and it is likely that *Tikuna,* which has been under construction for ten years, will be launched during 2005. It is unclear whether there is to be a second boat of this class in view of the follow-on S-MB-10 programme for five boats. Brazil has traditionally relied on the West for the supply of military technology and so it was interesting to note that President Putin, during his visit in November 2004, announced Russian readiness to ''develop long-term cooperation in high technology and science-intensive fields. This includes sectors such as aircraft building, energy, space and military-technical co-operation''. It would be particularly significant if this were to include Brazil's stalled nuclear submarine programme.

In Argentina, where growth was an impressive 8.3 per cent in 2004, the overall picture of restraint in the defence sector is much the same. Nevertheless, efforts to modernise the fleet have continued with the introduction into service of the sixth Meko 140 Espora class frigate, *Gomez Roca*, while local upgrade of the Meko 360 destroyers is reported to be under consideration. There have also been persistent reports that the navy is to acquire an amphibious capability through the procurement of *Ouragan* and *Orage* when they leave French service in 2005 and 2006. The submarine flotilla received a boost when the Domecq García shipyards, closed since the 1990s, were officially reopened in September 2004. This will enable future maintenance and modernisation to be conducted in-country rather than in Brazil where *Santa Cruz* completed a mid-life refit in 2002. Work to upgrade *Salta* with new batteries began in 2004 while the mid-life refit of *San Juan* is scheduled to start in 2005. Early in 2004 the government confirmed the PAM (Patrulleros de Alta Mar) programme for up to ten 80 m offshore patrol vessels. These are likely to be built in a domestic shipyard but collaboration with Chile, which had been mooted, is thought to be unlikely. The most significant deficiency in the Argentine Navy is the lack of a mine-countermeasures capability following the retirement of the two Ton class minesweepers in 2002.

Chile, which also achieved solid growth of 6.8 per cent in 2004, has by far the most busy naval programme in South America, as befits a country with an Exclusive Economic Zone that is almost twice the size of its land area. *O'Higgins*, the first new Scorpene class submarine is expected to be formally commissioned in 2005 while the second boat, *Carrera*, is planned to start sea trials late in the same year. Although original plans called for the acquisition of two further new submarines by 2014, this appears to have been superseded by a major programme to upgrade the two Thomson class boats which is to start in 2005. *Simpson* is to start modernisation in 2005 and *Thomson* is to follow some 18 months later. Following this work, Chile will own the most modern submarine force on the continent. After several false starts, it has become clear that renewal of the surface fleet is to be achieved by the procurement of second-hand vessels. Two ex-Netherlands frigates are to enter service in 2005 with a further two vessels to follow in 2006 and 2007. There is also a strong possibility that three ex-UK Type 23 frigates will be acquired in 2005 as they are decommissioned from the Royal Navy. Perhaps the biggest constraint on operations is a lack of afloat support and procurement of at least one replenishment tanker is now likely to be a high priority.

Elsewhere, the Peruvian Navy acquired two ex-Italian Lupo class frigates in 2004, bringing its total of ships of this type to six. Transfer of two further ships in the next few years is likely and will probably be matched by the retirement of both of its 1950s vintage ships, the cruiser *Almirante Grau* and destroyer *Ferré*. There are no plans at present to upgrade its ageing submarine force. In Ecuador, the operational state of the navy is poor with both its surface ship and submarine arms plagued by lack of funding. In contrast, the Uruguayan navy is generally serviceable although maintenance of obsolete ships and equipment is becoming increasingly difficult. Although acquisition of Descubierta class frigates from Spain is unlikely to go ahead, there is clearly an interest in replacing its Commandant Rivière class frigates. Finally, it was a mixed year in Mexico where the loss of the Sierra class corvette *Benito Juarez* in a fire was offset by the acquisition of two former Aliya class patrol craft from Israel.

In Conclusion

In reviewing naval developments around the world, it is evident that change is underway in many countries. But, while the United States is setting the agenda, it is clear that not all countries have reached the same conclusions about the future, are faced by the same threats or can afford to keep pace with every technological development. This is not surprising in view of the lack of a common military or scientific challenge. It is also fair to say that while the US is probably not completely wrong about the course it has set for itself, it may not be completely right. The establishment of the NATO Allied Command Transformation is one forum in which the views of allies can be heard but it is to be hoped that the United States will take time to consult with potential coalition partners, both traditional and non-traditional, during the QDR process. The military capability of the US is unchallenged in a conventional sense but the unconventional nature, breadth and scope of the GWOT is going to require a concerted global effort. This will include the sharing of ideas and intelligence, the challenging of assumptions and wider co-operation.

It is certainly too difficult to reach any firm conclusions about the Revolution in Military Affairs. The consensus view is probably that this is an ongoing process that has many years to run. There may also be further factors that have not been fully taken into account. At sea, these include the integration of unmanned air vehicles and, in due course, directed-energy weapons. The potential threat posed by information warfare may also temper the reliance that can be placed on C4I systems. In the meantime, governments would do well to tackle the more mundane challenge posed by institutional culture. In far too many countries, not least the US, the battle for defence resources still manifests itself as a turf battle between the individual armed services. If the current vision of military synchronisation is to be achieved, the navies, armies, airforces and marine corps of the world need to think and procure equipment in a genuinely joint way. That would be 'transformation' indeed.

Stephen Saunders   **May 2005**

Acknowledgements

Jane's Fighting Ships continues to be available in four versions: hardback; CD-ROM, data files and Online. It is a pleasure to report that all four remain in demand, reflecting the different needs of customers.

The business of collecting information and recording change has always been a continuous process, but up to a few years ago its presentation had been cyclical. *Jane's Fighting Ships* hard-copy book remains annual, but for those users more impatient for change as it happens, the Online product, which is updated monthly, is ideal. The *Jane's Fighting Ships* microsite (http://jfs.janes.com) offers a dedicated portal into the electronic environment. Amongst the many offerings on the microsite is the NewsEdge service providing a regular feed of naval related news from hundreds of sources around the world. Feedback on the microsite is always useful and amongst refinements made over the last year, ship silhouettes are in the process of being re-introduced for Online customers.

To the many anonymous people in government and industry who make data collection such a pleasure, my warmest thanks. We are not interested in secrets, but only in ensuring that open discussion on defence is based on reliable facts.

The change to full colour photography was made five years ago. Only a few black and white images now remain and any assistance in replacing them would be much appreciated. Either prints or digital photos can be accepted. In the case of the latter, these should be, ideally, at 300-dpi resolution although, exceptionally, lower quality images of rarely photographed ships will be considered for printing.

The importance of the US Navy in maritime affairs merits a special contributor in Tom Philpott who is the editor of *Military Update* in Washington D.C. Ian Sturton's excellent scale line drawings have long been a major feature of the publication while the Indexes have been compiled for the first time this year by Duncan and Diana Burns. Navies continue to adjust their Ranks and Insignia, and changes in that section are provided by W Maitland Thornton, who is an international expert, with artwork by Derrick Ballington. Similarly, updates to Ensigns and Flags are required each year and these have been provided by Graham Bartram, General Secretary of the Flag Institute, one of the world's main research and documentation centres for flags and vexicology.

Other individual contributors who are at the heart of the updating process, and who wish to be acknowledged include:
Captain M Annati, Mr D Baker, Mr G de Bakker, Mr D Boey, Mr C Borgenstam, Herr S Breyer, Mr J Brodie, Señor C Busquets i Vilanova, Señor A Campanera i Rovira, Herr H Carstens, Mr M Chapman, Mr R Cheung, Dr C Chung, Mr J Cislak, Mr W Clements, Mr G Davies, Mr M Declerck, Mr D Dervissis, Mr I Edwards, Herr H Ehlers, Mr S Emre, Mr R Fildes, Herr F Findler, Mr D Fox, Commander A Fraccaroli, Dr Z Freivogel, Señor A E Galarce, Signor G Ghiglione, Colonel W Globke, Commodore J Goldrick, Lieutenant Commander J Green, Dr E Grove, Mr V Jeffery, Mr M Kadota, Mr G Koop, Mr P Körnefeldt, Mr A A de Kruijf, Colonel J Kürsener, Mr E Laursen, Mr M Laursen, Lieutenant Commander Sue Lloyd, Mr C D Maginley, Monsieur P Marsan, Mr M Mazumdar, Mr S Millar, Mr M Mokrus, Mr R Montchai, Mr J Montes, Captain J E Moore, Mr S Morison, Mr J Mortimer, Mr H Nakai, Mr L-G Nilsson, Herr M Nitz, Mr T Okano, Señor A Ortigueira Gil, Mr R Pabst, Mr F Philips, Mr I J Plokker, Mr M Prendergast, Captain B Prézelin, Señor D Quevedo Carmona, Mr A J R Risseeuw, Monsieur J Y Robert, Mr F Sadek, Mr S San, Mr W Sartori, Mr A Sarup, Mr C Sattler, Mr M Schiele, Mr E Searle, Dr A Sharma, Captain R Sharpe, Monsieur A Sheldon-Duplaix, Mr H M Steele, Mr B Sullivan, Commodore P Swan, Captain T Tamura, Mr G Toremans, Mr N Trudeau, Mr J Webber, Prof A Wessels, Herr M Winter, Mr J Wise, Ms M Wright, Mr C D Yaylali.

Jane's staff at Coulsdon ease the production process, and no praise can be high enough for Emma Donald (content editor), Melanie Rovery (team leader), Jack Brenchley (senior compositor), Andrew Ruffle and Lynette Murphy (compositors), Kevan Box, Wayne Sudbury and Frank Baker (scanning team), Mike Johnson and Andreas Schindler (image archivists), and the production controller, Laura-Jane Walker. Closer to home, my wife Ann is an indispensable member of the year-round editorial and administrative effort.

Cross referencing to other Jane's publications is made easy by Jane's Online service which includes, inter alia: *Jane's All the World's Aircraft, Jane's Amphibious and Special Forces, Jane's Air-Launched Weapons, Jane's C4I Systems, Jane's Electro-Optic Systems, Jane's International Defence Directory, Jane's Marine Propulsion, Jane's Military Biographies, Jane's Naval Weapon Systems, Jane's Naval Construction and Retrofit Markets, Jane's Radar and Electronic Warfare Systems, Jane's Strategic Weapon Systems, Jane's Underwater Warfare Systems, Jane's Underwater Technology, Jane's Unmanned Aerial Vehicles and Targets* and *Jane's World Air Forces. Jane's Sentinel Security Assessments* are an excellent source of politico-military information while Jane's magazines provide up to the minute reports on defence issues. These include *Jane's Defence Weekly, Jane's International Defense Review, Jane's Foreign Report, Jane's Intelligence Review, Jane's Defence Industry, Jane's Missiles and Rockets* and *Jane's Navy International.* Amongst many other publications the Japanese magazine *Ships of the World* is also a source of useful data.

The focus of *Jane's Fighting Ships* remains people at sea, whether on the bridge or in the operations room. The aim is to provide the operational capabilities of a ship or navy in a consistent and concise format. Individual entries are composed so that there is no need to turn a page or cross-refer to other sections. It is always a pleasure to get feedback from those at sea.

All updating material should be sent to:

Commodore Stephen Saunders
Jane's Information Group
Sentinel House
163 Brighton Road
Coulsdon
Surrey CR5 2YH

Fax number: (+44 20) 87 00 39 59
e-mail: yearbook@janes.com

Note: No illustration from this book may be reproduced without the publisher's permission, but the press may reproduce information and governmental photographs, provided that *Jane's Fighting Ships* is acknowledged as the source. Photographs credited to other than official organisations must not be reproduced without permission from the originator.

STEPHEN SAUNDERS

During a 32-year career in the Royal Navy, Stephen Saunders travelled extensively and worked with many different navies. A surface ship officer and anti-submarine warfare specialist, he served in most classes of warship from Mine Countermeasures vessels to Aircraft Carriers. He commanded the frigate HMS *Sirius* and, as Captain 1st Frigate Squadron, HMS *Coventry*; in the latter role he also commanded the Royal Navy's Armilla patrol when deployed to the Gulf. His broad staff experience included attachment to the NATO staff of Commander US 6th Fleet and several tours in the Ministry of Defence, London. Appointments in Naval Operational Requirements and Defence Concepts led to his final job as Director Force Development within the Defence Policy Division. He graduated from the National Defence College, Latimer, in 1982 and the Royal College of Defence Studies in 1994. Since leaving the Royal Navy in 1998, he has worked in the shipbuilding industry and as a defence consultant.

Ranks and insignia of the world's navies

The aim of this section is to show the ranks and insignia that are likely to be worn by officers on formal occasions. In most cases, sleeve ranks are depicted but shoulder boards are shown where appropriate. Where possible the rank titles are described in the language of the relevant country followed by the equivalent rank in English.

The shoulder boards of Flag Officers (which in some cases include Commodores or Rear Admirals Lower Half) are in two principal styles. The use of stars to depict rank (Commodore — 1 star to Admiral of the Fleet — 5 stars) has been widely adopted but these are in many cases embellished with national emblems, crowns, anchors, crossed batons or swords and/or other insignia. Alternatively, shoulder boards bearing the same lace as sleeves may be worn.

Sleeves are drawn to one scale and shoulder insignia to another allowing respective comparisons of size to be made. The exception to this is Croatia where badges worn on the right breast of the uniforms are depicted.

Albania

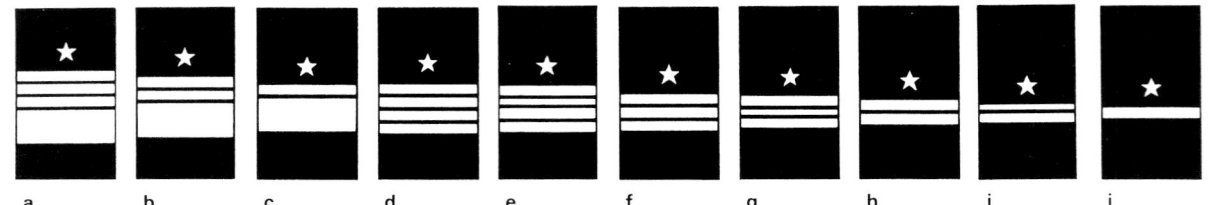

a: *Admiral,* Admiral   **b:** *Nenadmiral,* Vice Admiral   **c:** *Kunderadmiral,* Rear Admiral   **d:** *Kapiten I Rangut Te Pare,* Captain   **e:** *Kapiten I Rangut Te Dyte,* Commander
f: *Kapiten I Rangut Te Trete,* Lieutenant Commander   **g:** *Kapiten Leitnant I Pare,* Senior Lieutenant   **h:** *Kapiten Leitnant,* Lieutenant   **i:** *Leitnant,* Sub Lieutenant
j: *Nen Leitnant,* Acting Sub Lieutanant
Gold on navy blue.

Algeria (Marine de la République Algérienne)

a: *'Liwa',* Rear Admiral   **b:** *'Amid,* Commodore   **c:** *'Aqid,* Captain   **d:** *Muqaddam,* Commander   **e:** *Ra'id,* Lieutenant Commander   **f:** *Naqid,* Lieutenant
g: *Mulazim Awwal,* Sub Lieutenant   **h:** *Mulazim,* Acting Sub Lieutenant
Gold on navy blue.

Angola (Marinha da Guerra)

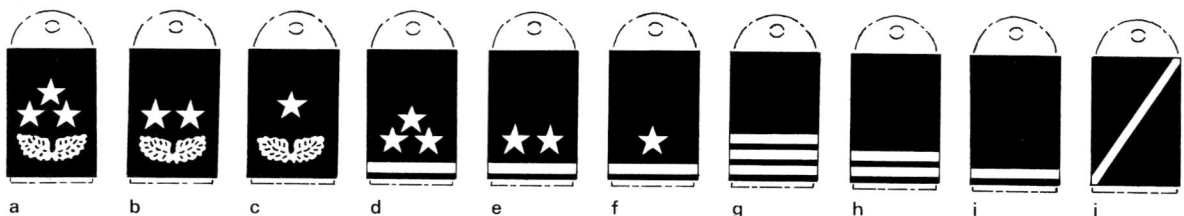

a: *Almirante,* Admiral   **b:** *Vice-Almirante,* Vice Admiral   **c:** *Contra-Almirante,* Rear Admiral   **d:** *Capitão-de-Mar-e-Guerra,* Captain   **e:** *Capitão-de-Fragata,* Commander
f: *Capitão-de-Corveta,* Lieutenant Commander   **g:** *Tenente-de-Navio,* Lieutenant   **h:** *Tenente-de-Fragata,* Sub Lieutenant   **i:** *Tenente-de-Corveta,* Acting Sub Lieutenant
j: *Aspirante,* Cadet
Admiral to Lieutenant Commander, gold on navy blue. Lieutenant to Sub Lieutenant, silver on navy blue. Midshipman and Cadet, light blue on navy blue.

Argentina (Armada Argentina)

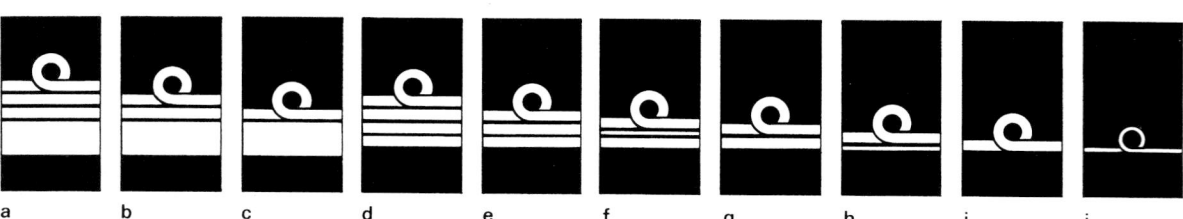

a: *Almirante,* Admiral   **b:** *Vicealmirante,* Vice Admiral   **c:** *Contraalmirante,* Rear Admiral   **d:** *Capitán de Navío,* Captain   **e:** *Capitán de Fragata,* Commander
f: *Capitán de Corbeta,* Lieutenant Commander   **g:** *Teniente de Navío,* Lieutenant   **h:** *Teniente de Fragata,* Sub Lieutenant   **i:** *Teniente de Corbeta,* Acting Lieutenant
j: *Guardiamarina,* Midshipman
Gold on navy blue.

Argentina (Coast Guard) (Prefectura Naval Argentina)

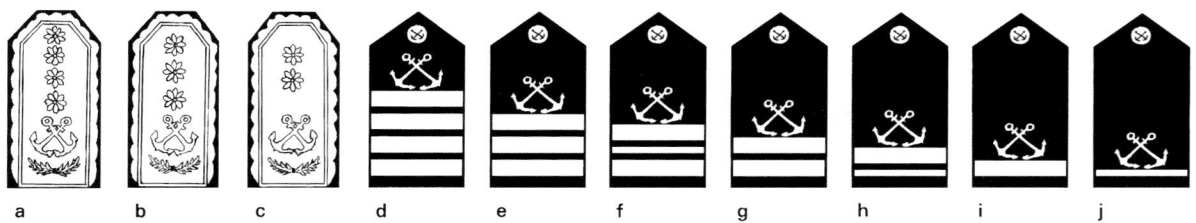

a: *Prefecto Nacional,* Admiral   b: *Subprefecto Nacional,* Vice Admiral   c: *Prefecto General,* Rear Admiral   d: *Prefecto Major,* Captain   e: *Prefecto Principal,* Commander
f: *Prefecto,* Lieutenant Commander   g: *Subprefecto,* Lieutenant   h: *Oficial Principal,* Sub Lieutenant   i: *Oficial Auxilias,* Acting Sub Lieutenant
j: *Oficial Ayudante,* Midshipman
Gold on navy blue. Flag ranks gold shoulder boards, navy blue edging and silver devices

Australia

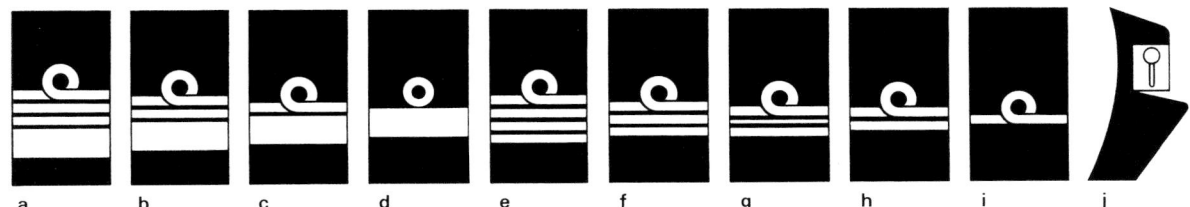

a: Admiral   b: Vice Admiral   c: Rear Admiral   d: Commodore   e: Captain   f: Commander   g: Lieutenant Commander   h: Lieutenant   i: Sub Lieutenant
j: Midshipman
Gold on navy blue.

Bahamas (Royal Bahamas Defence Force (Naval Division))

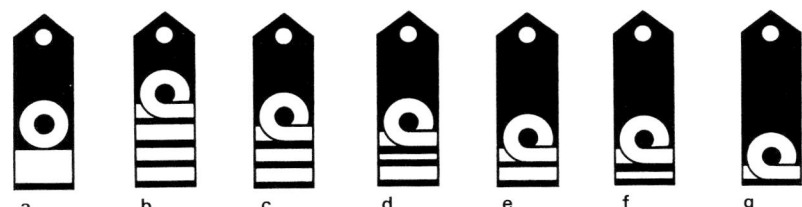

a: Commodore   b: Captain   c: Commander   d: Lieutenant Commander   e: Lieutenant   f: Junior Lieutenant   g: Sub Lieutenant
Gold on black.

Bahrain Coast Guard

a: *'Aqid,* Colonel   b: *Muqaddam,* Lieutenant Colonel   c: *Ra'id,* Major   d: *Naqib,* Captain   e: *Mulazim Awwal,* Lieutenant   f: *Mulazim Thani,* Second Lieutenant
Gold and red on navy blue.

Bangladesh

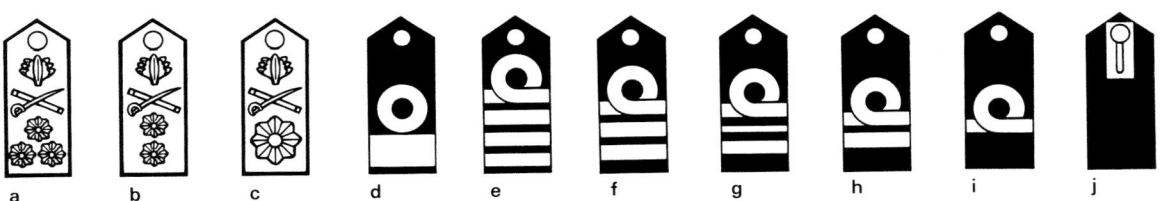

a: Admiral   b: Vice Admiral   c: Rear Admiral   d: Commodore   e: Captain   f: Commander   g: Lieutenant Commander   h: Lieutenant   i: Sub Lieutenant
j: Midshipman
Gold on navy blue. Flag ranks, gold edged blue. silver devices. White patch on midshipman's shoulder strap.

Barbados (Coast Guard)

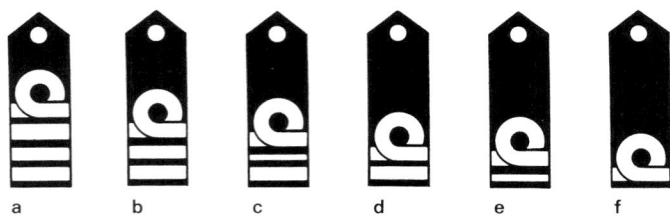

a: Captain **b:** Commander **c:** Lieutenant Commander **d:** Lieutenant **e:** Junior Lieutenant **f:** Sub Lieutenant
Gold on black.

Belgium (Zeemacht/La Force Navale)

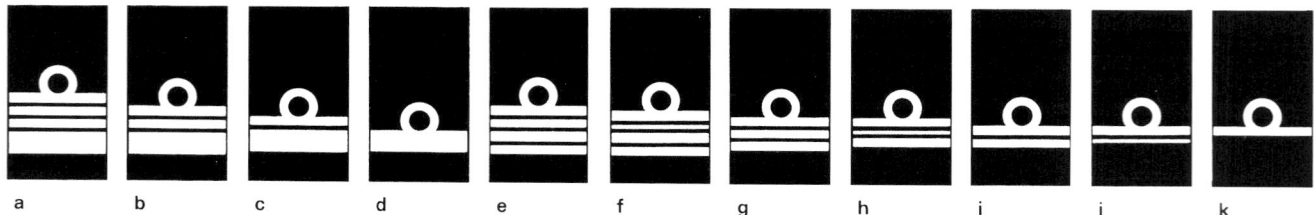

a: *Admiraal/Amiral*, Admiral **b:** *Vice-Admiraal/Vice Amiral*, Vice Admiral **c:** *Divisie-Admiraal/Amiral de Division*, Rear Admiral **d:** *Commodore/Commodore*, Commodore **e:** *Kapitein-ter-Zee/Capitaine de Vaisseau*, Captain **f:** *Fregatkapitein/Capitaine de Frégate*, Commander **g:** *Corvetkapitein/Capitaine de Corvette*, Lieutenant Commander **h:** *Luitenant-ter-Zee 1ste Klasse/Lieutenant de Vaisseau 1re Classe*, Senior Lieutenant **i:** *Luitenant-ter-Zee/Lieutenant de Vaisseau*, Lieutenant **j:** *Vaandrig-ter-Zee/Enseigne de Vaisseau*, Sub Lieutenant **k:** *Vaandrig-ter-Zee 2e Klasse/Enseigne de Vaisseau 2e Classe*, Acting Sub Lieutenant
Ranks given in Flemish, French and English. Rank of Admiral held by the monarch only.
Gold on navy blue.

Benin (Marine du Bénin)

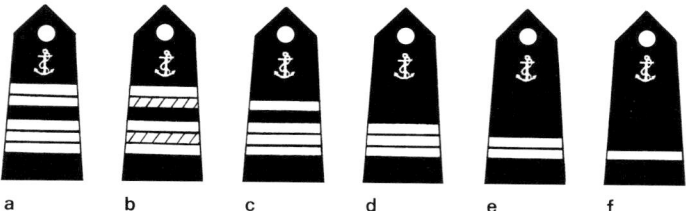

a: *Capitaine de Vaisseau*, Captain **b:** *Capitaine de Frégate*, Commander **c:** *Capitaine de Corvette*, Lieutenant Commander **d:** *Lieutenant de Vaisseau*, Lieutenant **e:** *Enseigne de Vaisseau 1re Classe*, Sub Lieutenant **f:** *Enseigne de Vaisseau 2e Classe*, Acting Sub Lieutenant
Gold on black. Commander, three gold, two silver stripes.

Bolivia

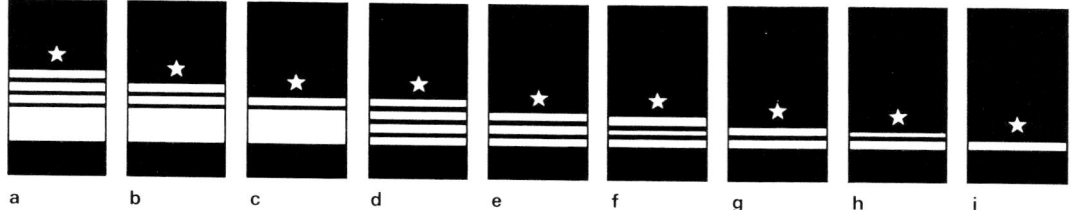

a: *Almirante*, Admiral **b:** *Vicealmirante*, Vice Admiral **c:** *Contraalmirante*, Rear Admiral **d:** *Capitán de Navío*, Captain **e:** *Capitán de Fragata*, Commander **f:** *Capitán de Corbeta*, Lieutenant Commander **g:** *Teniente de Navío*, Lieutenant **h:** *Teniente de Fragata*, Sub Lieutenant **i:** *Alférez*, Acting Sub Lieutenant
Gold on navy blue.

Brazil (Marinha do Brasil)

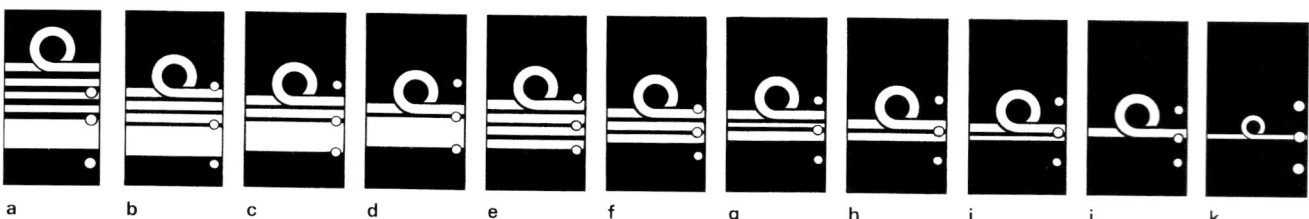

a: *Almirante*, Admiral of the fleet **b:** *Almirante de Esquadra*, Admiral **c:** *Vice-Almirante*, Vice Admiral **d:** *Contra-Almirante*, Rear Admiral **e:** *Capitão-de-Mar-e-Guerra*, Captain **f:** *Capitão-de-Fragata*, Commander **g:** *Capitão-de-Corveta*, Lieutenant Commander **h:** *Capitão-Tenente*, Lieutenant **i:** *Primero Tenente*, Sub Lieutenant **j:** *Segundo Tenente*, Acting Sub Lieutenant **k:** *Guarda-Marinha*, Midshipman
Gold on dark blue.

Brunei (Angkatan Tentera Laut Diraja Brunei)

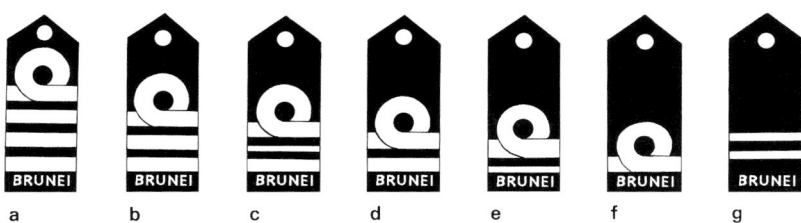

a: *Colonel,* Captain   **b:** *Lieutenant Colonel,* Commander   **c:** *Major,* Lieutenant Commander   **d:** *Captain,* Lieutenant
e: *Lieutenant,* Sub Lieutenant   **f:** *2nd Lieutenant,* Acting Sub Lieutenant   **g:** *Officer Cadet,* Midshipman
Gold on navy blue.

Bulgaria (Voennomorski Sili)

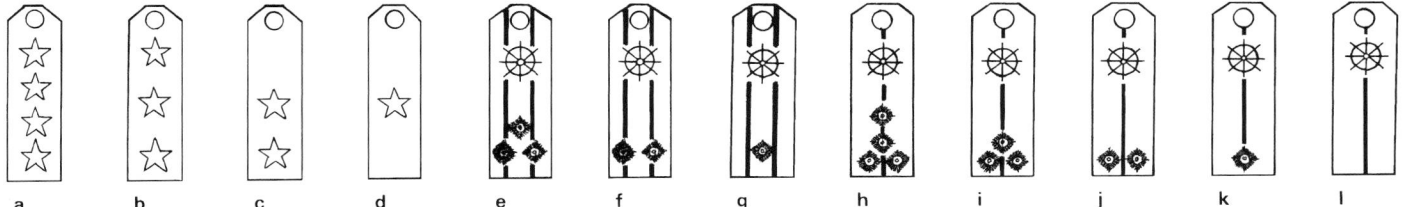

a: *Admiral,* Admiral   **b:** *Vitseadmiral,* Vice Admiral   **c:** *Kontraadmiral,* Rear Admiral   **d:** *Brigadier Admiral,* Commodore   **e:** *Kapitan I Rang,* Captain
f: *Kapitan II Rang,* Commander   **g:** *Kapitan III Rang,* Lieutenant Commander   **h:** *Kapitan Leytenant,* Lieutenant   **i:** *Starshi Leytenant,* First Lieutenant
j: *Leytenant,* Sub Lieutenant   **k:** *Mladshi Leytenant,* Lieutenant Junior Grade   **l:** Warrant Officer
Gold brocade shoulder straps. Insignia are gold (Flag Officers) and brass (all other ranks). Two black stripes (Lieutenant Commander to Captain) and one black stripe (junior officers).

Cambodia

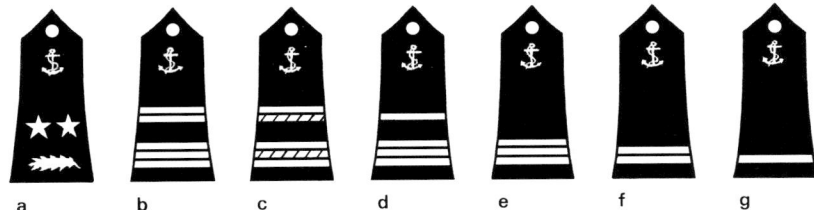

a: *Contre-Amiral,* Rear Admiral   **b:** *Capitaine de Vaisseau,* Captain   **c:** *Capitaine de Frégate,* Commander   **d:** *Capitaine de Corvette,* Lieutenant Commander
e: *Lieutenant de Vaisseau,* Lieutenant   **f:** *Enseigne de Vaisseau 1re Classe,* Sub Lieutenant   **g:** *Enseigne de Vaisseau 2e Classe,* Acting Sub Lieutenant
Silver stars, gold anchor and palm on navy blue.

Cameroon (Marine Nationale République du Cameroun)

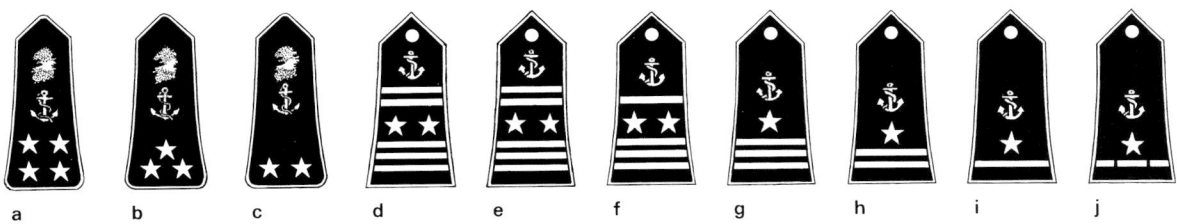

a: *Vice-Amiral d'Escadre,* Admiral   **b:** *Vice-Amiral,* Vice Admiral   **c:** *Contre-Amiral,* Rear Admiral   **d:** *Capitaine de Vaisseau,* Captain   **e:** *Capitaine de Frégate,* Commander
f: *Capitaine de Corvette,* Lieutenant Commander   **g:** *Lieutenant de Vaisseau,* Lieutenant   **h:** *Enseigne de Vaisseau 1re Classe,* Sub Lieutenant   **i:** *Enseigne de Vaisseau 2e
Class,* Acting Sub Lieutenant   **j:** *Aspirant,* Midshipman
Navy blue shoulder straps. Edging: gold (Flag Officers); green (other officers). Insignia: white lion and silver stars (Flag Officers); gold (other officers); top two stripes for
Commander are silver.

Canada (Maritime Command)

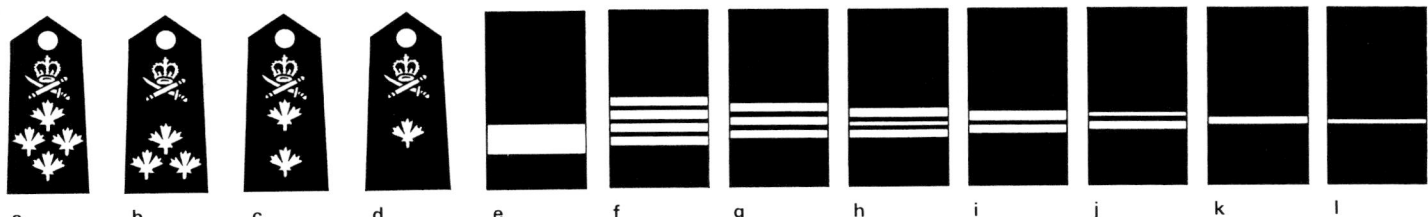

a: Admiral   **b:** Vice Admiral   **c:** Rear Admiral   **d:** Commodore   **e:** All Flag Ranks   **f:** Captain   **g:** Commander   **h:** Lieutenant Commander   **i:** Lieutenant   **j:** Sub Lieutenant
k: Acting Sub Lieutenant   **l:** Officer Cadet
Gold on black.

Canada (Coast Guard)

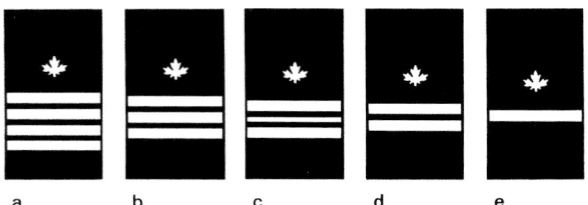

a: Commanding Officer   b: Chief Officer   c: First Officer   d: Second Officer   e: Third Officer
Rank structure for large ships only.
Gold on navy blue.

Chile (Armada de Chile)

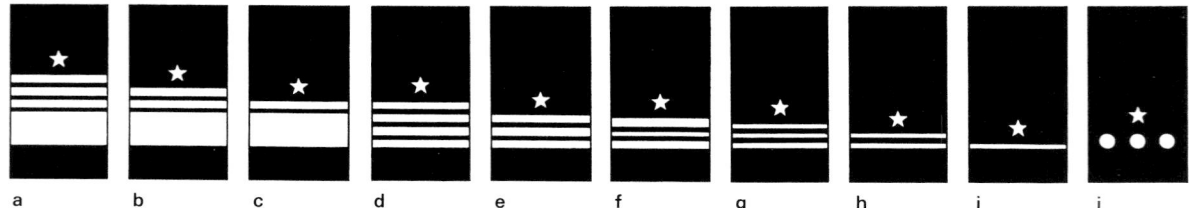

a: *Almirante*, Admiral   b: *Vicealmirante*, Vice Admiral   c: *Contraalmirante*, Rear Admiral   d: *Capitán de Navío*, Captain   e: *Capitán de Fragata*, Commander
f: *Capitán de Corbeta*, Lieutenant Commander   g: *Teniente Primero*, Lieutenant   h: *Teniente Segundo*, Junior Lieutenant   i: *Sub Teniente*, Sub Lieutenant
j: *Guardia Marina*, Midshipman
Gold on black.

China (People's Liberation Army Navy)

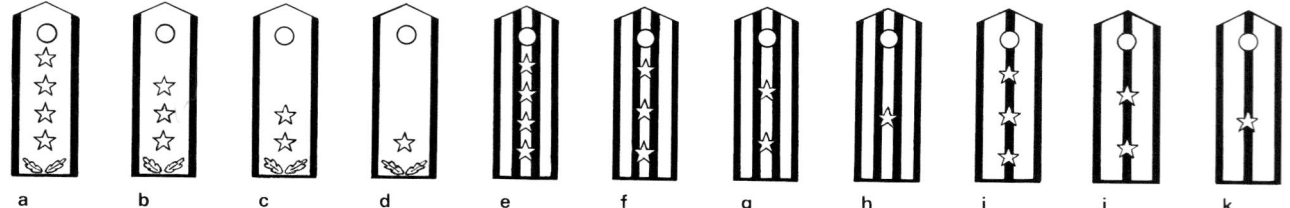

a: *Senior General*, Admiral of the Fleet   b: *General*, Admiral   c: *Lieutenant General*, Vice Admiral   d: *Major General*, Rear Admiral   e: *Senior Colonel*, Commodore
f: *Colonel*, Captain   g: *Lieutenant Colonel*, Commander   h: *Major Colonel*, Lieutenant Commander   i: *Captain*, Lieutenant   j: *Lieutenant*, Sub Lieutenant   k: *Second*
Lieutenant, Acting Sub Lieutenant
Generals, gold edged dark blue. Gold button. Silver stars. Senior Colonel to Major Colonel two dark blue stripes. Captain to Second Lieutenant one dark blue stripe.

Colombia (Armada de la República)

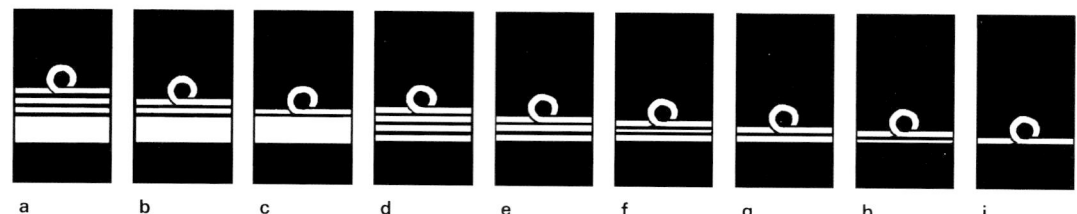

a: *Almirante*, Admiral   b: *Vicealmirante*, Vice Admiral   c: *Contraalmirante*, Rear Admiral   d: *Capitán de Navío*, Captain   e: *Capitán de Fragata*, Commander
f: *Capitán de Corbeta*, Lieutenant Commander   g: *Teniente de Navío*, Lieutenant   h: *Teniente de Fragata*, Sub Lieutenant   i: *Teniente de Corbeta*, Acting Sub Lieutenant
Gold on black.

Democratic Republic of Congo

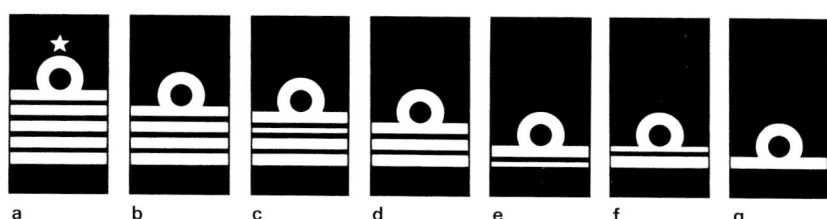

a: *Contre-Amiral*, Rear Admiral   b: *Capitaine de Vaisseau*, Captain   c: *Capitaine de Frégate*,
Commander   d: *Capitaine de Corvette*, Lieutenant Commander   e: *Lieutenant de Vaisseau*, Lieutenant
f: *Enseigne de Vaisseau 1re Classe*, Sub Lieutenant   g: *Enseigne de Vaisseau 2e Classe*, Acting Sub
Lieutenant
Gold on black.

Costa Rica (Guardia Civil Sección Maritime)

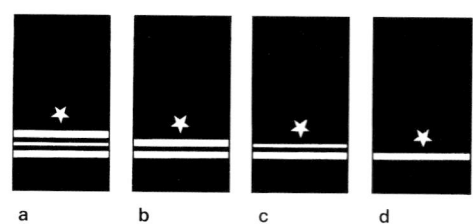

a: *Major*, Lieutenant Commander   b: *Capitan*, Lieutenant
c: *Teniente*, Sub Lieutenant   d: *Sub Teniente*, Acting Sub
Lieutenant
Gold on navy blue.

Côte d'Ivoire (Marine Côte D'Ivoire)

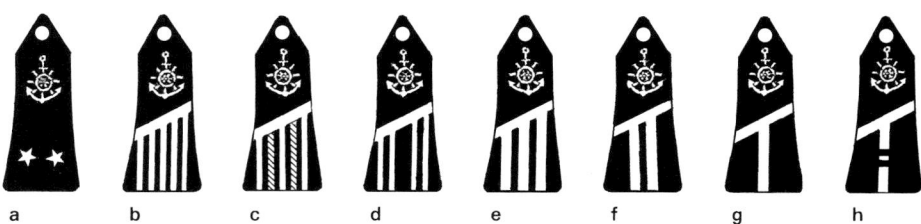

a: *Contre-Amiral*, Rear Admiral   b: *Capitaine de Vaisseau*, Captain   c: *Capitaine de Frégate*, Commander   d: *Capitaine de Corvette*, Lieutenant Commander   e: *Lieutenant de Vaisseau*, Lieutenant   f: *Enseigne de Vaisseau de 1re Classe*, Sub Lieutenant
g: *Enseigne de Vaisseau de 2e Classe*, Acting Sub Lieutenant   h: *Aspirant*, Midshipman
Gold on navy blue. Commander, gold and silver. Stars, silver.

Croatia (Hrvatska Ratna Mornarica)

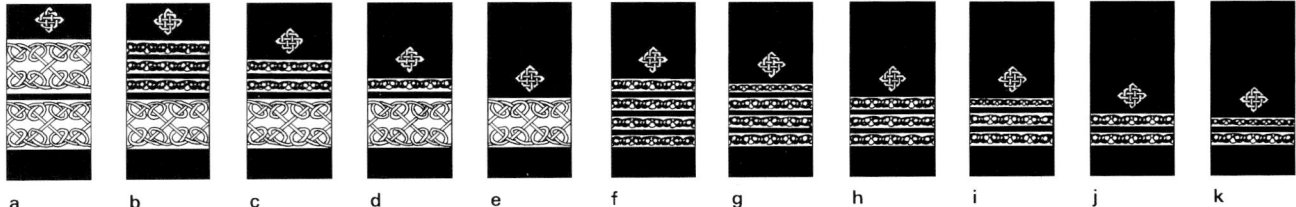

a: *Admiral Flote*, Admiral of the Fleet   b: *Admiral*, Admiral   c: *Viceadmiral*, Vice Admiral   d: *Kontraadmiral*, Rear Admiral   e: *Komodor*, Commodore   f: *Kapetan Bojnog Broda*, Captain   g: *Kapetan Fregate*, Commander   h: *Kapetan Korvete*, Lieutenant Commander   i: Poručnik Bojnog Broda, Lieutenant   j: *Poručnik Fregate*, Sub Lieutenant   k: *Poručnik Korvete*, Acting Sub Lieutenant
Gold on navy blue.

Cuba (Marina de Guerra Revolucionaria)

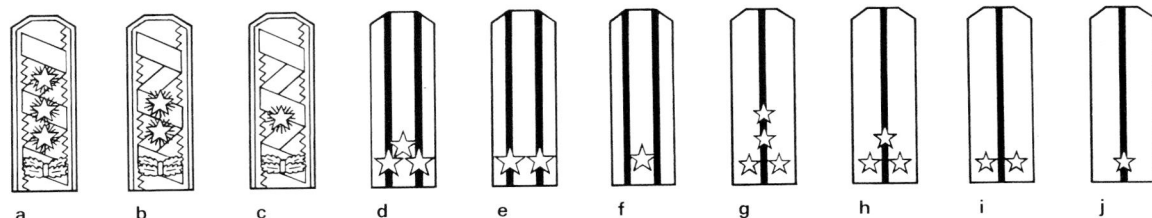

a: *Almirante*, Admiral   b: *Vicealmirante*, Vice Admiral   c: *Contraalmirante*, Rear Admiral   d: *Capitán de Navío*, Captain   e: *Capitán de Frégata*, Commander
f: *Capitán de Corbeta*, Lieutenant Commander   g: *Teniente de Navío*, Senior Lieutenant   h: *Teniente de Frégata*, Lieutenant   i: *Teniente de Corbeta*, Sub Lieutenant
j: *Alférez*, Acting Sub Lieutenant
Black stripes. Admirals, gold stars on blue design.

Cyprus

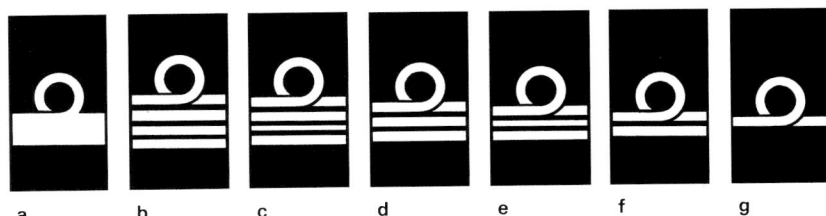

a: *Archipliarchos*, Commodore   b: *Pilarchos*, Captain   c: *Antipliarchos*, Commander   d: *Plotarchis*, Lieutenant Commander
e: *Ypopliarchos*, Lieutenant   f: *Anthypopliarchos*, Sub Lieutenant   g: *Simaioforos*, Acting Sub Lieutenant
Gold on navy blue.

Denmark (Den Kongelige Danske Marine)

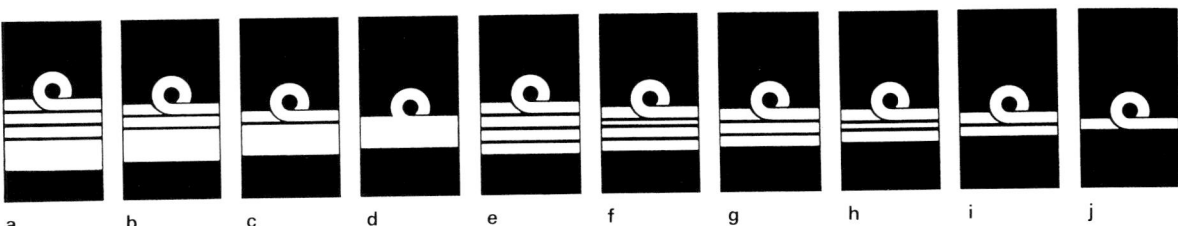

a: *Admiral*, Admiral   b: *Viceadmiral*, Vice Admiral   c: *Kontreadmiral*, Rear Admiral   d: *Flotilleadmiral*, Commodore   e: *Kommandør*, Captain   f: *Kommandørkaptajn*, Senior
Commander   g: *Orlogskaptajn*, Commander   h: *Kaptajnløjtnant*, Lieutenant Commander   i: *Premierløjtnant*, Lieutenant   j: *Løjtnant*, Junior Grade Lieutenant
Gold on black. The Danish Home Guard use the same insignia up to the rank of Captain.

Djibouti

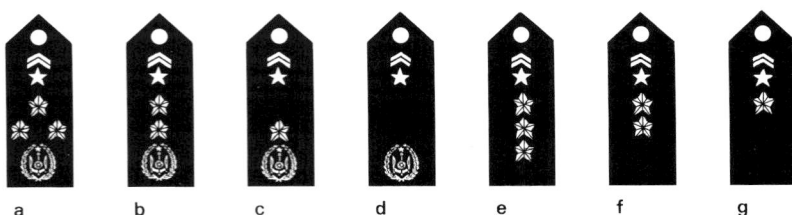

a b c d e f g

a: *Brigadier,* Commodore   **b:** *Colonel,* Captain   **c:** *Lieutenant Colonel,* Commander
d: *Major,* Lieutenant Commander   **e:** *Captain,* Lieutenant   **f:** *Lieutenant,* Sub Lieutenant   **g:** *2nd Lieutenant,* Acting Sub Lieutenant
Gold crest, gold stripes, red star. Rank device silver.

Dominican Republic (Marina de Guerra)

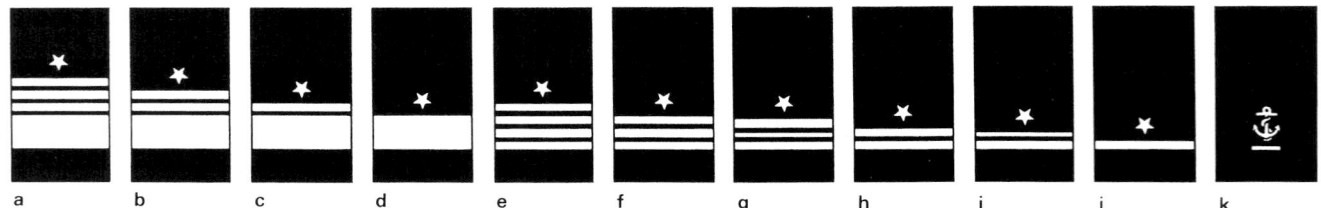

a b c d e f g h i j k

a: *Almirante,* Admiral   **b:** *Vicealmirante,* Vice Admiral   **c:** *Contraalmirante,* Rear Admiral   **d:** *Comodoro,* Commodore   **e:** *Capitán de Navío,* Captain
f: *Capitán de Fragata,* Commander   **g:** *Capitán de Corbeta,* Lieutenant Commander   **h:** *Teniente de Navío,* Lieutenant   **i:** *Alférez de Navío,* Sub Lieutenant
j: *Alférez de Fragata,* Acting Sub Lieutenant   **k:** *Guardiamarina,* Midshipman
Gold on black.

Ecuador (Armada de Guerra)

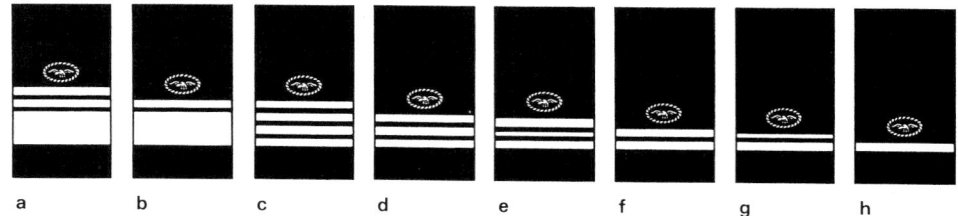

a b c d e f g h

a: *Vicealmirante,* Vice Admiral   **b:** *Contraalmirante,* Rear Admiral   **c:** *Capitán de Navío,* Captain   **d:** *Capitán de Fragata,* Commander
e: *Capitán de Corbeta,* Lieutenant Commander   **f:** *Teniente de Fragata,* Lieutenant   **g:** *Alférez de Navío,* Sub Lieutenant
h: *Alférez de Fragata,* Acting Sub Lieutenant
Gold on black.

Egypt

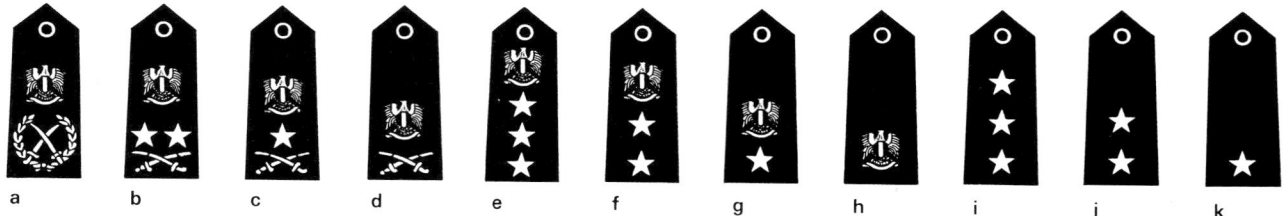

a b c d e f g h i j k

a: *Mushir,* Admiral of the Fleet   **b:** *Fariq Awwal,* Admiral   **c:** *Fariq,* Vice Admiral   **d:** *Liwa',* Rear Admiral   **e:** *'Amid,* Commodore   **f:** *'Aqid,* Captain
g: *Muqaddam,* Commander   **h:** *Ra'id,* Lieutenant Commander   **i:** *Naqib,* Lieutenant   **j:** *Mulazim Awwal,* Sub Lieutenant   **k:** *Mulazim,* Acting Sub Lieutenant
Gold on black. Shield on eagle's breast black, white and red.
A blue uniform with traditional gold lace sleeve ranks is sometimes worn on formal occasions and/or by officers serving abroad in cool climates. Ranks and lace are the same
as Royal Navy with the exception of Sub Lieutenants (Mulazim Awwal) who have a half width lace below the single stripe.

El Salvador (Fuerza Naval de El Salvador)

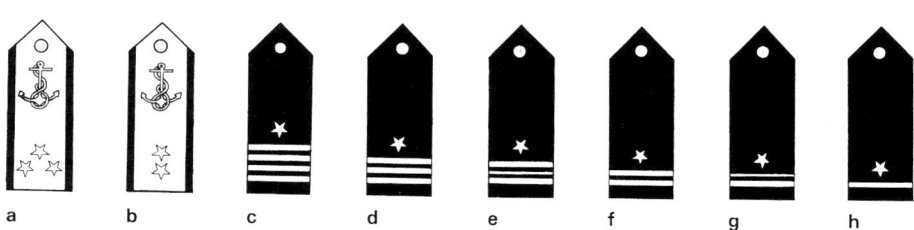

a b c d e f g h

a: *Vice Almirante,* Vice Admiral   **b:** *Contra Almirante,* Rear Admiral   **c:** *Capitan de Navio,* Captain   **d:** *Capitan de Fragata,* Commander
e: *Capitan de Corbeta,* Lieutenant Commander   **f:** *Teniente de Navio,* Lieutenant   **g:** *Teniente de Fragata,* Sub Lieutenant
h: *Teniente de Corbeta,* Acting Sub Lieutenant
Gold on navy blue. Admirals shoulder straps are gold with navy blue edges. Stars and anchors are silver.

Estonia (Eesti Merevägi)

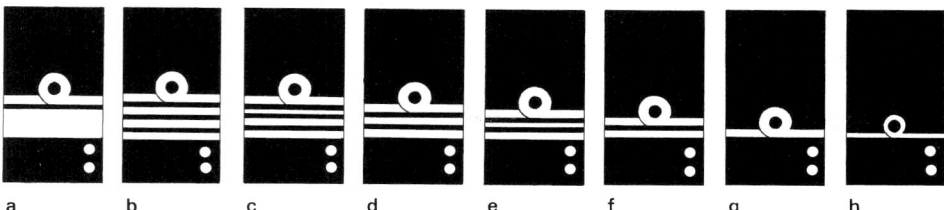

a: *Kontradmiral*, Rear Admiral   b: *Merevāe-kapten*, Captain   c: *Kapten-leitnant*, Commander   d: *Kapten-major*, Lieutenant Commander
e: *Vanem-leitnant*, Senior Lieutenant   f: *Leitnant*, Lieutenant   g: *Noorem-leitnant*, Sub Lieutenant   h: *Lipnik*, Acting Sub Lieutenant

Fiji (Republic of Fiji Military Forces Navy Element)

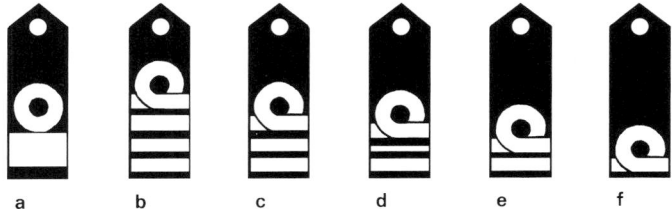

a: Commodore   b: Captain   c: Commander   d: Lieutenant Commander   e: Lieutenant   f: Sub Lieutenant

Finland (Suomen Merivoimat)

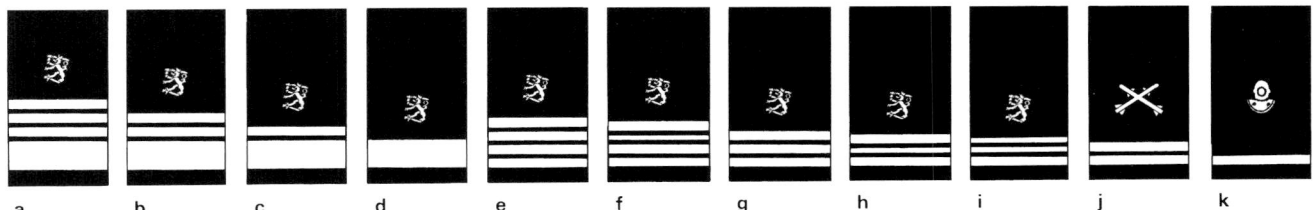

a: *Amiraali*, Admiral   b: *Vara-amiraali*, Vice Admiral   c: *Kontra-amiraali*, Rear Admiral   d: *Lippue-amiraali*, Commodore   e: *Kommodori*, Captain
f: *Komentaja*, Commander   g: *Komentajakapteeni*, Lieutenant Commander   h: *Kapteeniluutnantti*, Senior Lieutenant   i: *Yliluutnantti*, Lieutenant
j: *Luutnantti*, Junior Lieutenant   k: *Aliluutnantti*, Sub Lieutenant
Gold on black. The Coast Guard has the same uniforms and ranks with the exception of Commodore.

France (Marine Nationale)

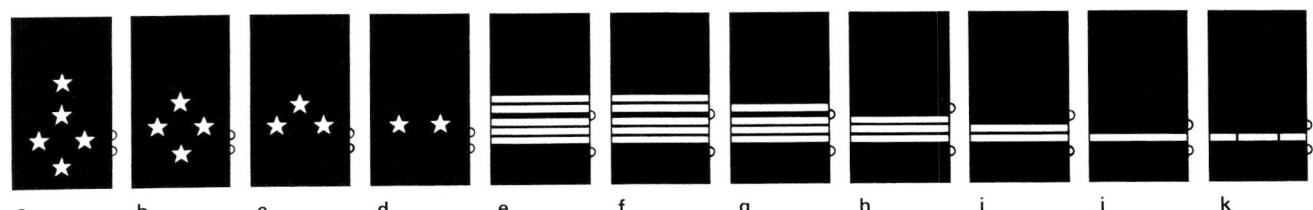

a: *Amiral*, Admiral of the Fleet   b: *Vice-Amiral d'Escadre*, Admiral   c: *Vice-Amiral*, Vice Admiral   d: *Contre-Amiral*, Rear Admiral   e: *Capitaine de Vaisseau*, Captain
f: *Capitaine de Frégate*, Commander   g: *Capitaine de Corvette*, Lieutenant Commander   h: *Lieutenant de Vaisseau*, Lieutenant   i: *Enseigne de Vaisseau de 1re Classe*,
Sub Lieutenant   j: *Enseigne de Vaisseau de 2e Classe*, Acting Sub Lieutenant   k: *Aspirant*, Midshipman
Flag ranks, silver stars. Captain, gold. Commander, three gold two silver. Lieutenant Commander to Midshipman, gold. Vertical stripes on Midshipman's lace, mid blue.
All on dark blue.

Gabon (Marine Gabonaise)

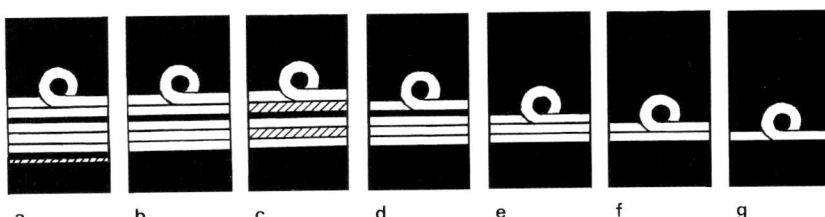

a: *Capitaine de Vaisseau Major*, Commodore   b: *Capitaine de Vaisseau*, Captain   c: *Capitaine de Frégate*, Commander
d: *Capitaine de Corvette*, Lieutenant Commander   e: *Lieutenant de Vaisseau*, Lieutenant   f: *Ensigne de Vaisseau*, Sub Lieutenant
g: *Ensign de Vaisseau 2e classe*, Acting Sub Lieutenant
Gold on navy blue. Commodore and Commander have silver rope lace.

Georgia (Navy)

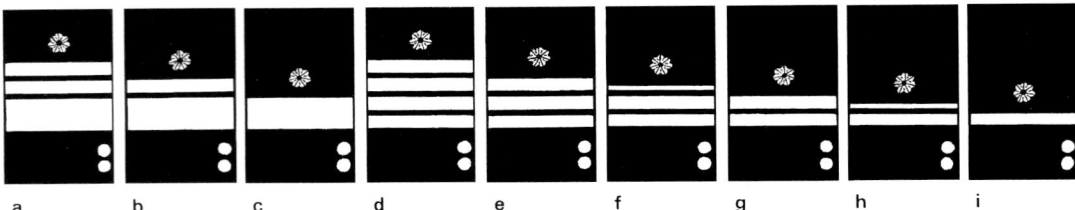

a: *Vitse-Admiral,* Vice Admiral   **b:** *Kontr-Admiral,* Rear Admiral   **c:** *Kapitan Pervogo Ranga,* Captain   **d:** *Kapitan Vtorogo Ranga,* Commander
e: *Kapitan Tretyego Ranga,* Lieutenant Commander   **f:** *Kapitan-Leytenant,* Lieutenant   **g:** *Starshiy Leytenant,* Junior Lieutenant   **h:** *Leytenant,* Sub Lieutenant
i: *Mladshiy Leytenant,* Acting Sub Lieutenant
Gold on navy blue.

Germany (Deutsche Marine)

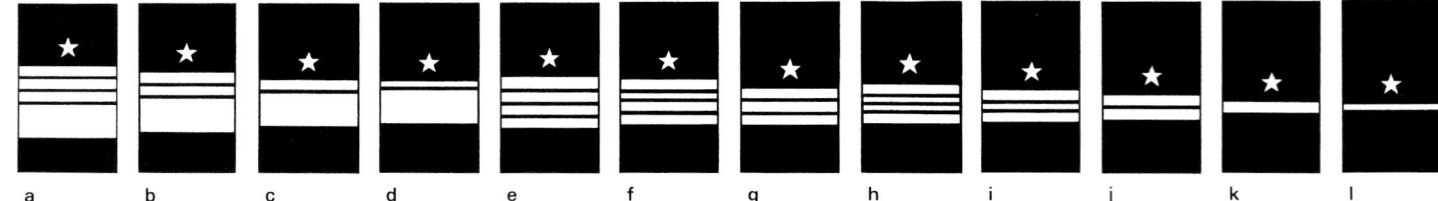

a: *Admiral,* Admiral   **b:** *Viceadmiral,* Vice Admiral   **c:** *Konteradmiral,* Rear Admiral   **d:** *Flottillenadmiral,* Commodore   **e:** *Kapitan zur See,* Captain   **f:** *Fregattenkapitan,*
Commander   **g:** *Korvettenkapitan,* Lieutenant Commander   **h:** *Stabskapitanleutnant,* Senior Lieutenant   **i:** *Kapitanleutnant,* Lieutenant   **j:** *Oberleutnant zur See,* Sub Lieutenant
k: *Leutnant zur See,* Acting Sub Lieutenant   **l:** *Oberfahnrich zur See,* Midshipman
Gold on navy blue.

Germany Coast Guard (Bundesgrenzschutz—See)

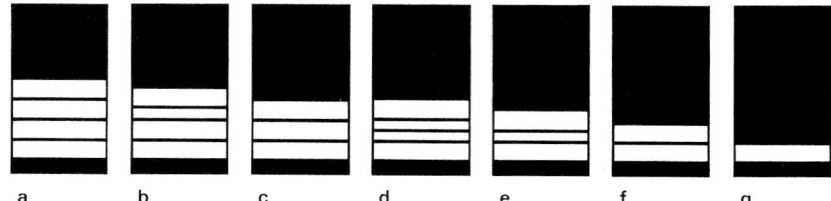

a: *Polizeidirektor im BGS,* Captain   **b:** *Polizeioberrat im BGS,* Commander   **c:** *Polizeirat im BGS,* Lieutenant Commander
d: *Erster Polizeihauptkommissar im BGS,* Senior Lieutenant   **e:** *Polizeihauptkommissar im BGS,* Lieutenant
f: *Polizeioberkommissar im BGS,* Sub Lieutenant   **g:** *Polizeikommissar im BGS,* Acting Sub Lieutenant
Gold on navy blue.

Ghana

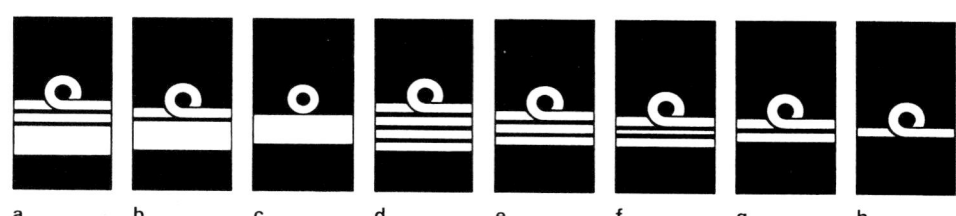

a: Vice Admiral   **b:** Rear Admiral   **c:** Commodore   **d:** Captain   **e:** Commander   **f:** Lieutenant Commander   **g:** Lieutenant   **h:** Sub Lieutenant
Gold on navy blue.

Greece (Hellenic Navy)

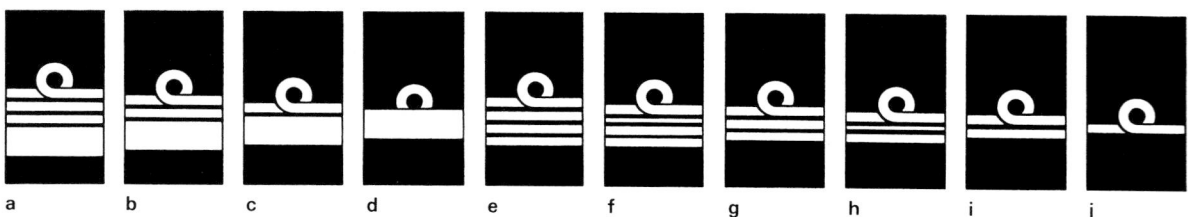

a: *Navarchos,* Admiral   **b:** *Antinavarchos,* Vice Admiral   **c:** *Iponavarchos,* Rear Admiral   **d:** *Archiploiarchos,* Commodore   **e:** *Ploiarchos,* Captain
f: *Antiploiarchos,* Commander   **g:** *Plotarchis,* Lieutenant Commander   **h:** *Ipoploiarchos,* Lieutenant   **i:** *Antipoploiarchos,* Sub Lieutenant
j: *Simeoforos, Acting* Sub Lieutenant
Gold on navy blue.

Greece (Coast Guard)

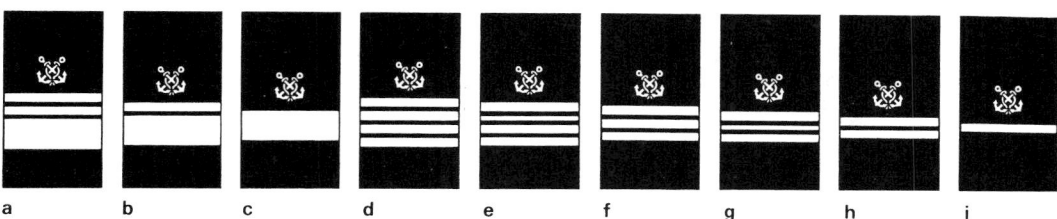

a: *Antinavarchos*, Vice Admiral   b: *Iponavarchos*, Rear Admiral   c: *Archiploiarchos*, Commodore   d: *Ploiarchos*, Captain   e: *Antiploiarchos*, Commander
f: *Plotarchis*, Lieutenant Commander   g: *Ipoploiarchos*, Lieutenant   h: *Antipoploiarchos*, Sub Lieutenant   i: *Simeoforos*, *Acting* Sub Lieutenant
Gold on navy blue.

Guatemala (Marina de Guatemala)

a: *Vicealmirante*, Admiral   b: *Contraalmirante*, Vice Admiral   c: *Capitán de Navío*, Captain   d: *Capitán de Fragata*, Commander
e: *Capitán de Corbeta*, Lieutenant Commander   f: *Teniente de Navío*, Lieutenant   g: *Teniente de Fragata*, Sub Lieutenant
h: *Alférez de Navío*, Sub Lieutenant (JG)   i: *Alférez de Fragata*, Acting Sub Lieutenant
Gold on navy blue. Fouled anchor and quetzal bird gold for officers – silver for admirals.

Guinea (Marine du Guinea)

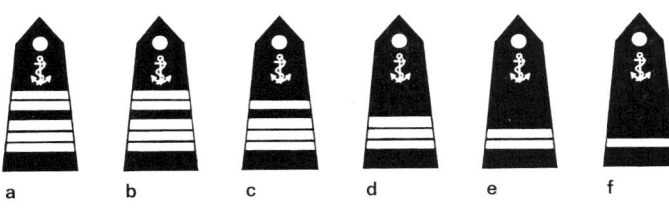

a: *Capitaine de Vaisseau*, Captain   b: *Capitaine de Frégate*, Commander   c: *Capitaine de Corvette*,
Lieutenant Commander   d: *Lieutenant de Vaisseau*, Lieutenant   e: *Enseigne de Vaisseau*
1re Classe, Sub Lieutenant   f: *Enseigne de Vaisseau 2e Classe*, Acting Sub Lieutenant
Gold on black. Commander, three gold and two silver stripes.

Guinea-Bissau

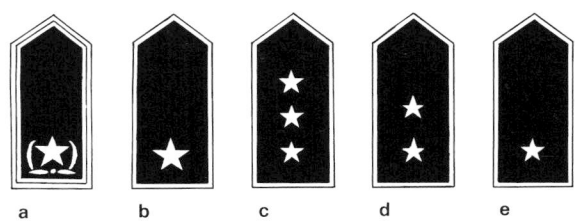

a: *Comandante*, Commander   b: *Major*, Lieutenant Commander
c: *Capitao*, Lieutenant   d: *Tenente*, Sub Lieutenant
e: *Alférez*, Acting Sub Lieutenant
Stars are silver. Piping and wreath in gold. Backing dark blue.

Honduras (Fuerza Naval Republica de Honduras)

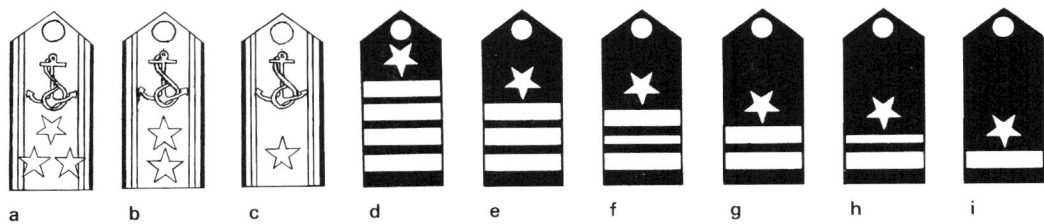

a: *Almirante*, Admiral   b: *Vicealmirante*, Vice Admiral   c: *Contralmirante*, Rear Admiral   d: *Capitán de Navío*, Captain   e: *Capitán de Fragata*, Commander
f: *Capitán de Corbeta*, Lieutenant Commander   g: *Teniente de Navío*, Lieutenant   h: *Teniente de Fragata*, Sub Lieutenant   i: *Alferez de Fragata*, Acting Sub Lieutenant
Gold on navy blue. Flag ranks: gold shoulder boards with silver devices and gold buttons.

Hong Kong (Police Marine Region)

a: *Assistant Commissioner of Police*, Rear Admiral   b: *Chief Superintendent*, Commodore   c: *Senior Superintendent*, Captain
d: *Superintendent*, Commander   e: *Chief Inspector*, Lieutenant Commander   f: *Senior Inspector*, Lieutenant   g: *Inspector*, Lieutenant Junior Grade
h: *Probationary Inspector*, Sub Lieutenant
Silver on dark blue.

Iceland (Landhelgisgaeslan)

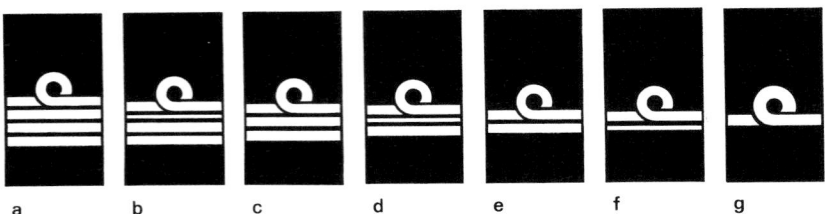

a b c d e f g

a: *Director General,* Captain   **b:** *Chief of Operations,* Senior Commander   **c:** *Captain and Chief Engineer,* Commander
d: *Chief Mate and First Engineer,* Lieutenant Commander   **e:** *First Mate and Second Engineer,* Lieutenant
f: *Second Mate and officers with less than six years' service,* Sub Lieutenant   **g:** *Officers with less than two years' service,* Acting Sub Lieutenant
Gold on navy blue.

India

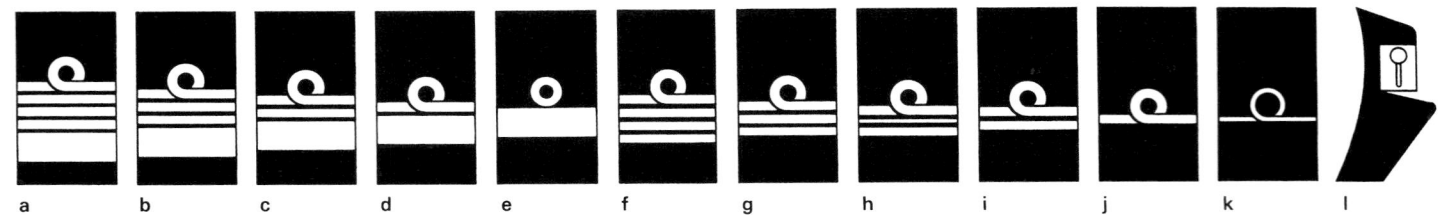

a b c d e f g h i j k l

a: Admiral of the Fleet   **b:** Admiral   **c:** Vice Admiral   **d:** Rear Admiral   **e:** Commodore   **f:** Captain   **g:** Commander   **h:** Lieutenant Commander   **i:** Lieutenant
j: Sub Lieutenant   **k:** Commissioned Officer   **l:** Midshipman (Lapel)
Gold on navy blue.

India (Coast Guard)

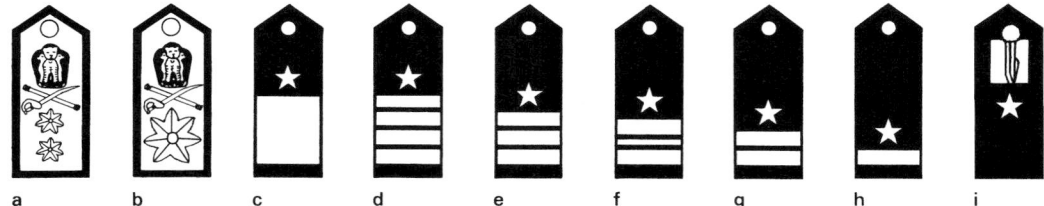

a b c d e f g h i

a: *Director General,* Vice Admiral   **b:** *Inspector General,* Rear Admiral   **c:** *Deputy Inspector General with 3 years' seniority,* Commodore   **d:** *Deputy Inspector General,*
Captain   **e:** *Commandant,* Commander   **f:** *Deputy Commandant,* Lieutenant Commander   **g:** *Assistant Commandant,* Lieutenant   **h:** *Assistant Commandant under training
after completion of Phase III afloat training and during sub courses,* Acting Lieutenant   **i:** *Assistant Commandant under training after completion of Phase II afloat training,*
Midshipman
Gold on navy blue. Silver sword crossed with silver baton, silver embroidered star with thin blue edgings.

Indonesia (Tentara Nasional)

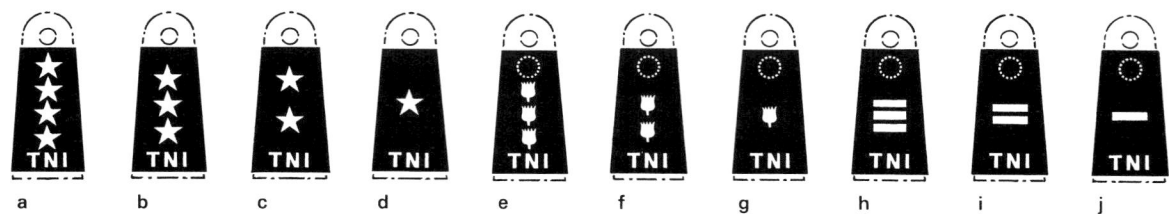

a b c d e f g h i j

a: *Laksamana,* Admiral   **b:** *Laksdya,* Vice Admiral   **c:** *Laksda,* Rear Admiral   **d:** *Laksma,* Commodore   **e:** *Kolonel,* Captain   **f:** *Letnan Kolonel,* Commander
g: *Mayor,* Lieutenant Commander   **h:** *Kapten,* Senior Lieutenant   **i:** *Letnan Satu,* Lieutenant   **j:** *Letnan Dua,* Sub Lieutenant
Gold on medium blue.

Iran

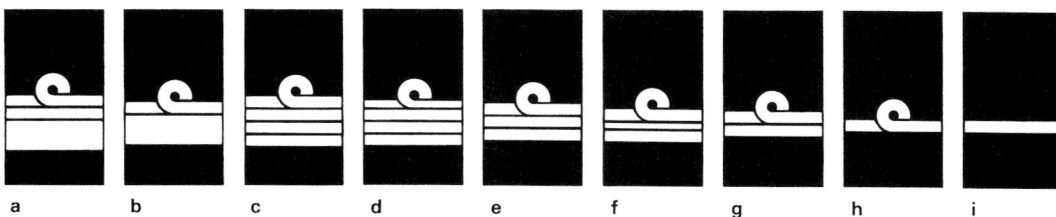

a b c d e f g h i

a: *Daryaban,* Vice Admiral   **b:** *Daryadar,* Rear Admiral   **c:** *Nakhoda Yekom,* Captain   **d:** *Nakhoda Dovom,* Commander
e: *Nakhoda Sevom,* Lieutenant Commander   **f:** *Navsarvan,* Lieutenant   **g:** *Navban Yekom,* Junior Lieutenant
h: *Navban Dovom,* Sub Lieutenant   **i:** *Navban Sevom,* Midshipman.
Gold on navy blue.

Ireland (An Seirbhis Chabhlaigh)

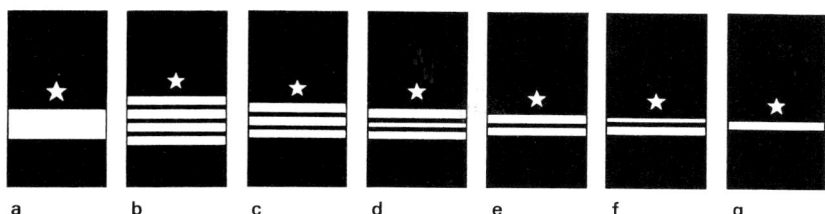

a b c d e f g

a: Commodore   b: Captain   c: Commander   d: Lieutenant Commander   e: Lieutenant   f: Sub Lieutenant   g: Ensign
Gold on navy blue.

Israel (Heyl Hayam)

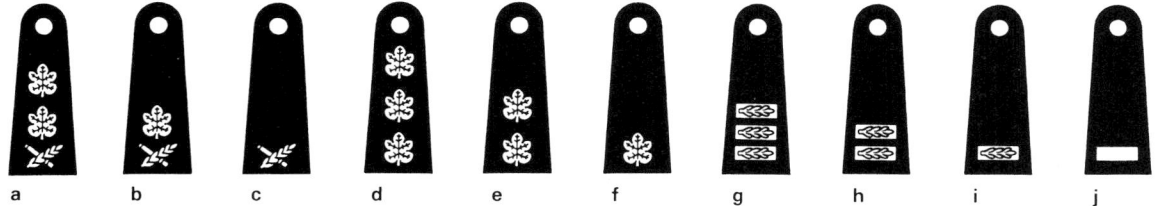

a b c d e f g h i j

a: *General (Rav-Aluf),* Admiral   b: *Major General (Aluf),* Vice Admiral   c: *Brigadier (Tat-Aluf),* Rear Admiral   d: *Colonel (Alut-Mishneh),* Captain
e: *Lieutenant Colonel (Sgan-Aluf),* Commander   f: *Major, (Rav-Seren),* Lieutenant Commander   g: *Captain (Seren),* Lieutenant   h: *First Lieutenant (Segen),* Sub Lieutenant
i: *Second Lieutenant (Segen-Mishneh),* Acting Sub Lieutenant   j: *Officer Aspirant (Mamak),* Officer Candidate
Bright brass or gold generally on dark blue or black. Officer Candidate, white bar.

Italy (Marina Militare)

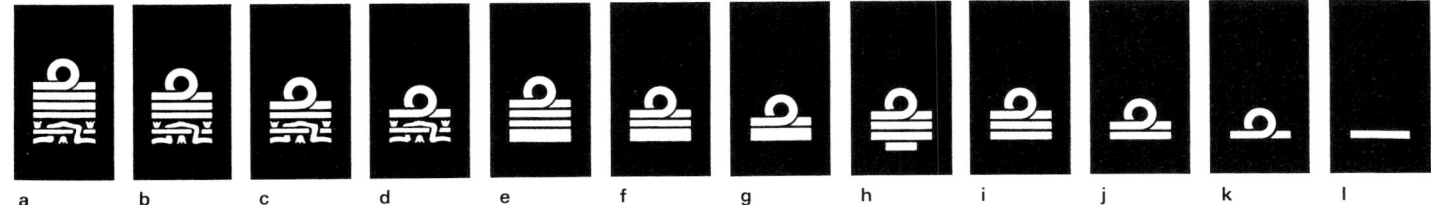

a b c d e f g h i j k l

a: *Ammiraglio di Squadra con Incarichi Speciali,* Admiral Commanding Navy   b: *Ammiraglio di Squadra e Ammiraglio Ispettore Capo,* Admiral and Senior Inspector General
(Navy)   c: *Ammiraglio di Divisione e Ammiraglio Ispettore,* Vice Admiral and Inspector General (Navy)   d: *Contrammiraglio,* Rear Admiral   e: *Capitano di Vascello,* Captain
f: *Capitano di Fregata,* Commander   g: *Capitano di Corvetta,* Lieutenant Commander   h: *1° Tenente di Vascello,* First Lieutenant   i: *Tenente di Vascello,* Lieutenant
j: *Sottotenente di Vascello,* Sub Lieutenant   k: *Guardiamarina,* Acting Sub Lieutenant   l: *Aspirante Guardiamarina,* Midshipman
Gold on dark blue.

Jamaica (Defence Force Coast Guard)

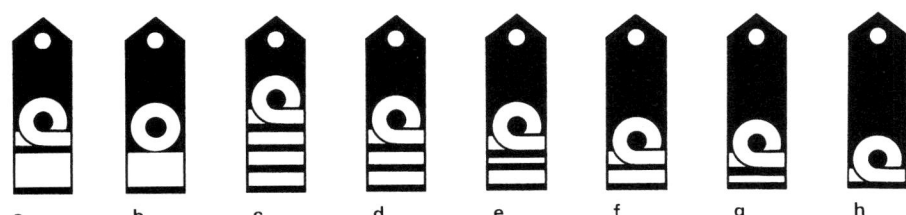

a b c d e f g h

a: Rear Admiral   b: Commodore   c: Captain   d: Commander   e: Lieutenant Commander   f: Lieutenant   g: Junior Lieutenant   h: Ensign
Gold on black.

Japan (Maritime Self Defence Force)

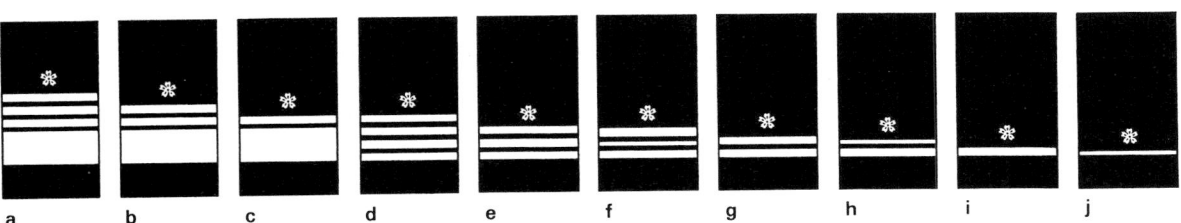

a b c d e f g h i j

a: Admiral   b: Vice Admiral   c: Rear Admiral   d: Captain   e: Commander   f: Lieutenant Commander   g: Lieutenant   h: Sub Lieutenant
i: Acting Sub Lieutenant   j: Warrant Officer
Gold on navy blue.

Japan (Coast Guard)

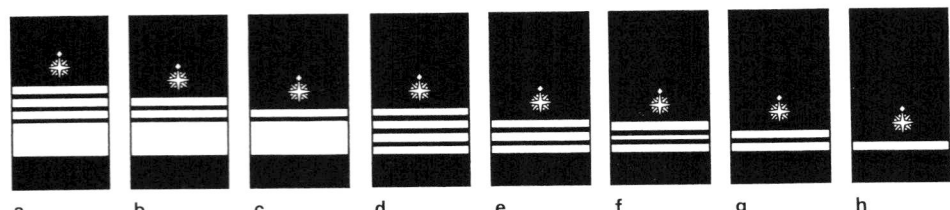

a: Commandant   b: Vice Commandant   c: Superintendent First Grade   d: Superintendent Second Grade   e: Superintendent Third Grade
f: Officer First Grade   g: Officer Second Grade   h: Officer Third Grade
Gold on navy blue.

Jordan

a: *'Amid,* Commodore   b: *'Aqid,* Captain   c: *Muqaddam,* Commander   d: *Ra'id,* Lieutenant Commander
e: *Naqib,* Lieutenant   f: *Mulazim Awwal,* Sub Lieutenant   g: *Mulazim,* Acting Sub Lieutenant
Khaki shoulder straps.

Kenya

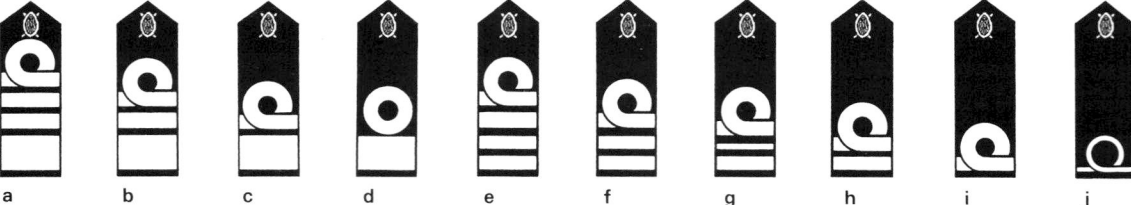

a: *General,* Admiral   b: *Lieutenant General,* Vice Admiral   c: *Major General,* Rear Admiral   d: *Brigadier,* Commodore   e: *Colonel,* Captain
f: *Lieutenant Colonel,* Commander   g: *Major,* Lieutenant Commander   h: *Captain,* Lieutenant   i: *Lieutenant,* Sub Lieutenant   j: *Second Lieutenant,* Acting Sub Lieutenant
Gold embroidery on navy blue.

Korea, North (People's Democratic Republic)

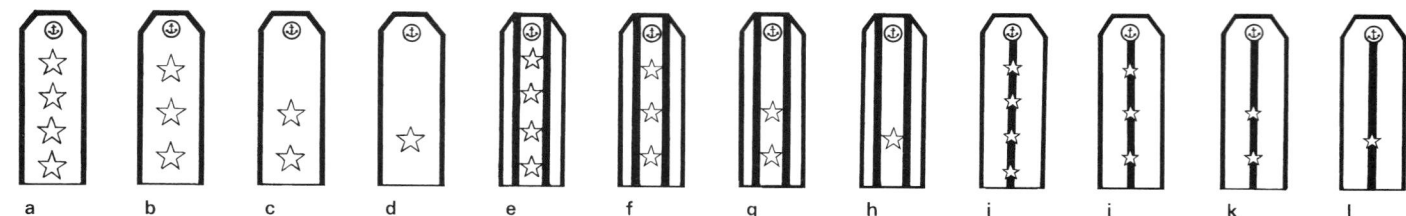

a: Admiral of the Fleet   b: Admiral   c: Vice Admiral   d: Rear Admiral   e: Commodore   f: Captain   g: Commander   h: Lieutenant Commander   i: Senior Lieutenant
j: Lieutenant   k: Sub Lieutenant   l: Acting Sub Lieutenant
Black stripes, silver stars on gold.

Korea, South Republic

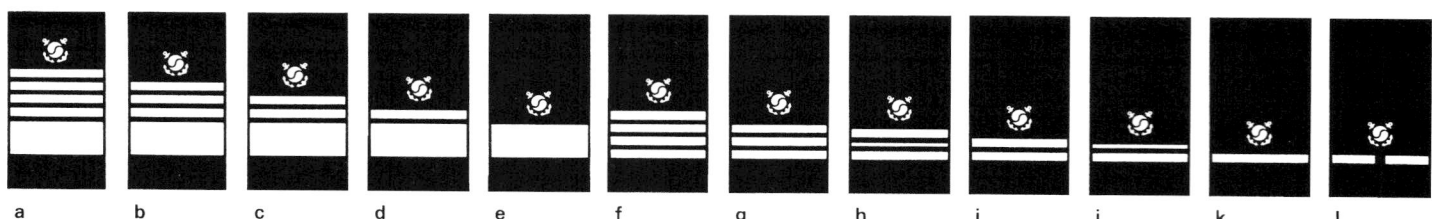

a: Admiral of the Fleet (Fleet Admiral)   b: Admiral   c: Vice Admiral   d: Rear Admiral   e: Commodore   f: Captain   g: Commander   h: Lieutenant Commander   i: Lieutenant
j: Sub Lieutenant   k: Acting Sub Lieutenant   l: Warrant Officer
Gold on navy blue.

Kuwait

a: *Fariq*, Vice Admiral b: *Liwa'*, Rear Admiral c: *'Amid*, Commodore d: *'Aqid*, Captain e: *Muqaddam*, Commander
f: *Ra'id*, Lieutenant Commander g: *Naqib*, Lieutenant h: *Mulazim Awwal*, Sub Lieutenant i: *Mulazim*, Acting Sub Lieutenant
Usually gold on tan. Can be gold on dark green or dark blue.

Latvia (Latvijas Juras Speki)

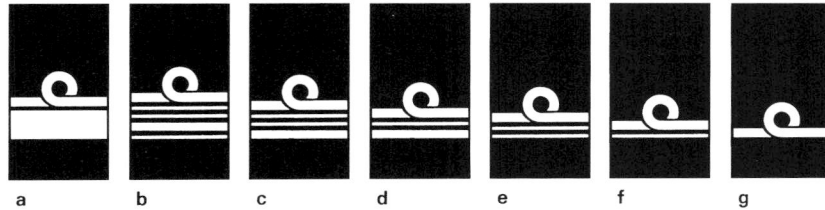

a: *Admirãlis*, Rear Admiral b: Jūraskapteinis, Captain c: *Komandkapteinis*, Commander Senior Grade
d: *Kapteinis*, Commander Junior Grade e: *Kapteinleitnants*, Lieutenant Commander f: *Virsleitnants*, Lieutenant
g: *Leitnants*, Lieutenant Junior Grade
Gold on navy blue

Lebanon

a: *'Imad*, Vice Admiral b: *Liwa'*, Rear Admiral c: *'Amid*, Commodore d: *'Aqid*, Captain e: *Muqaddam*, Commander
f: *Ra'id*, Lieutenant Commander g: *Ra'is*, Lieutenant h: *Mulazim Awwal*, Sub Lieutenant i: *Mulazim*, Acting Sub Lieutenant
Gold on black.

Libya

a: *'Aqid*, Captain b: *Muqaddam*, Commander c: *Ra'id*, Lieutenant Commander
d: *Naqib*, Lieutenant e: *Mulazim Awwal*, Sub Lieutenant f: *Mulazim*, Acting Sub Lieutenant
Gold on navy blue.

Lithuania (Karines Juru Pajegos)

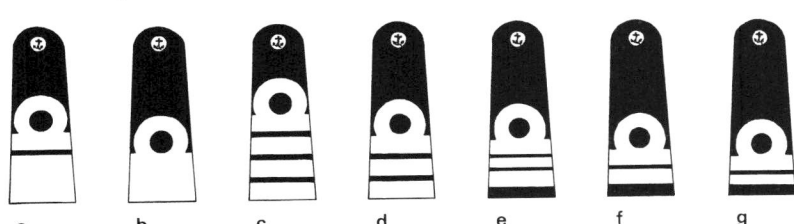

a: *Kontradmirolas*, Rear Admiral b: *Komandoras*, Commodore c: *Komandoras-Leitenantas*, Captain
d: *Jūru Kapitonas*, Commander e: *Kapitonas-Leitenantas*, Lieutenant Commander
f: *Jūru Vyresnysis Leitenantas*, Lieutenant g: *Jūru Leitenantas*, Sub Lieutenant
Gold on black.

Madagascar (Malagasy Republic Marine)

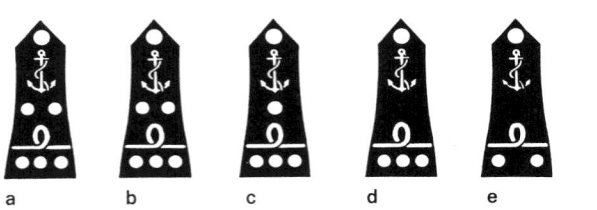

a: *Capitaine de Vaisseau*, Captain   b: *Capitaine de Frégate*, Commander
c: *Capitaine de Corvette*, Lieutenant Commander   d: *Lieutenant de Vaisseau*, Lieutenant
e: *Enseigne de Vaisseau 1re Classe*, Sub Lieutenant
f: *Enseigne de Vaisseau 2e Classe*, Acting Sub Lieutenant
Gold on black. Commander, top two discs silver.

Malawi

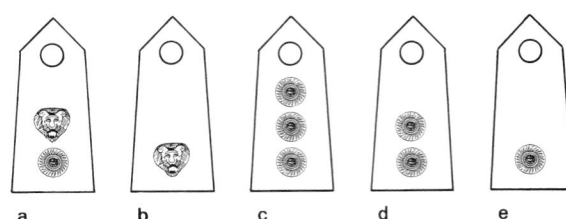

a: *Lieutenant Colonel*, Commander   b: *Major*, Lieutenant Commander
c: *Captain*, Lieutenant   d: *Lieutenant*, Sub Lieutenant
e: *2nd Lieutenant*, Acting Sub Lieutenant
Black on khaki.

Malaysia (Tentera Laut Dira Ja)

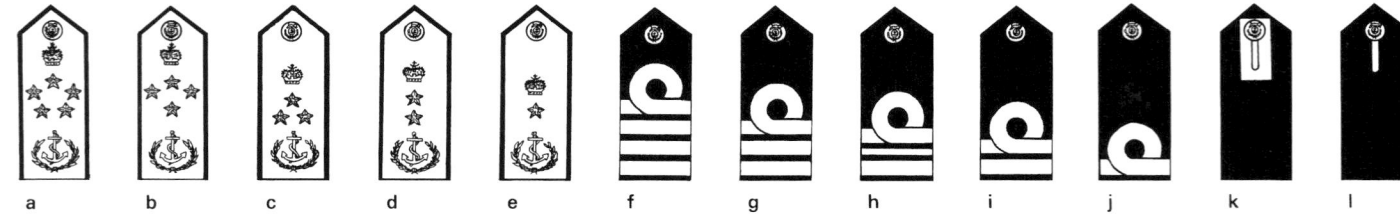

a: *Laksamana Armada*, Admiral of the Fleet   b: *Laksamana*, Admiral   c: *Laksamana Madya*, Vice Admiral   d: *Laksamana Muda*, Rear Admiral   e: *Laksamana Pertama*,
Commodore   f: *Kapeten*, Captain   g: *Komander*, Commander   h: *Leftenan Komander*, Lieutenant Commander   i: *Leftenan*, Lieutenant
j: *Leftenan Madya and Leftenan Muda*, Sub Lieutenant and Acting Sub Lieutenant   k: *Kadet Kanan*, Midshipman   l: *Kadet*, Cadet
Vice Admiral to Commodore, silver on gold. Remainder gold on navy blue, plus midshipman's white patch.

Malta (Maritime Squadron, AFM)

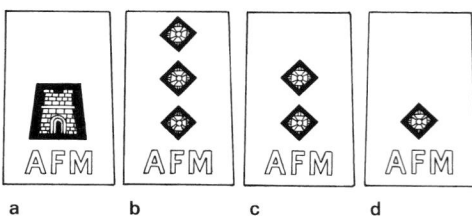

a: *Major*, Lieutenant Commander   b: *Captain*, Lieutenant
c: *Lieutenant*, Sub Lieutenant   d: *2nd Lieutenant*, Acting Sub Lieutenant
White on dark blue.

Mauritania (Marine Mauritanienne)

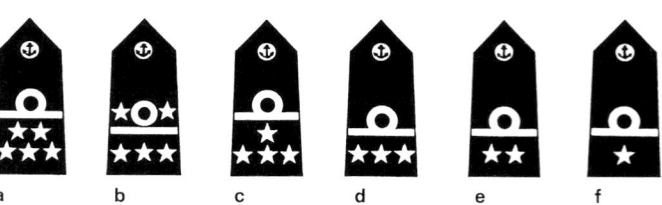

a: *Colonel*, Captain   b: *Lieutenant Colonel*, Commander   c: *Major*, Lieutenant Commander
d: *Captain*, Lieutenant   e: *Lieutenant*, Sub Lieutenant   f: *2nd Lieutenant*, Acting Sub Lieutenant
Gold on blue or green. Exception is two silver stars above lace for commander.

Mexico (Marina Nacional)

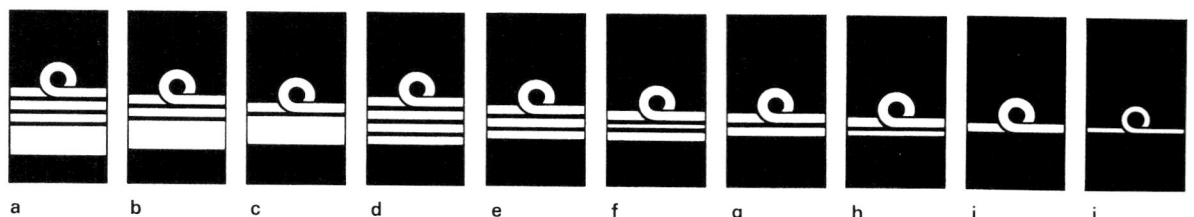

a: *Almirante*, Admiral   b: *Vicealmirante*, Vice Admiral   c: *Contraalmirante*, Rear Admiral   d: *Capitán de Navío*, Captain   e: *Capitán de Fragata*, Commander
f: *Capitán de Corbeta*, Lieutenant Commander   g: *Teniente de Navío*, Lieutenant   h: *Teniente de Fragata*, Sub Lieutenant   i: *Teniente de Corbeta*, Acting Sub Lieutenant
j: *Guardiamarina*, Midshipman
Gold on navy blue.

Morocco (Marine Royale Marocaine)

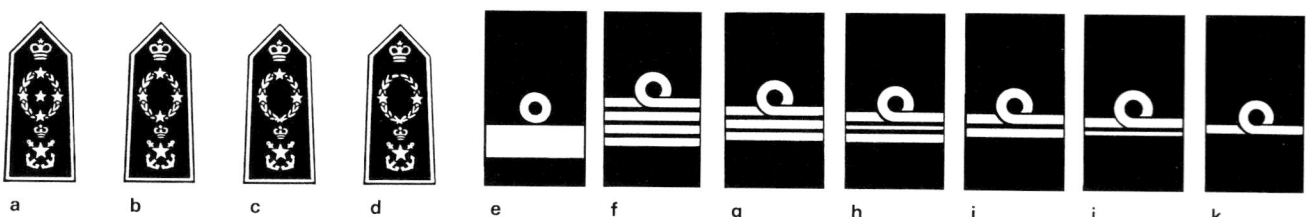

a: *Amiral*, Admiral of the Fleet   b: *Amiral d'Escadre*, Admiral   c: *Vice Amiral*, Vice Admiral   d: *Contre Amiral*, Rear Admiral   e: *Capitaine de Vaisseau Major*, Commodore
f: *Capitaine de Vaisseau*, Captain   g: *Capitaine de Frégate*, Commander   h: *Capitaine de Corvette*, Lieutenant Commander   i: *Lieutenant de Vaisseau*, Lieutenant
j: *Enseigne de Vaisseau 1re Classe*, Sub Lieutenant   k: *Enseigne de Vaisseau 2e Classe*, Acting Sub Lieutenant
Gold on black. Flag ranks silver stars.

Mozambique (Marina Moçambique)

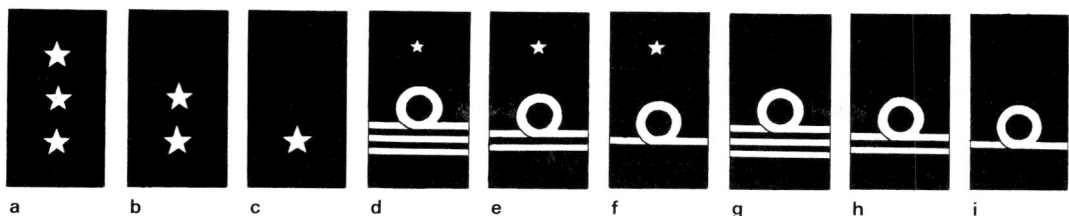

a: *Almirante*, Admiral   b: *Vice-Almirante*, Vice Admiral   c: *Contra-Almirante*, Rear Admiral   d: *Capitão-de-Mar-e-Guerra*, Captain   e: *Capitão-de-Fragate*, Commander
f: *Capitão-Tenente*, *Lieutenant Commander*   g: *Primerio-Tenente*, Lieutenant   h: *Segundo-Tenente*, Sub Lieutenant   i: *Guarda-Marinha*, Midshipman
Gold insignia on dark blue slip-ons.

Myanmar (Tatmadaw Yay)

a: Admiral   b: Vice Admiral   c: Rear Admiral   d: Commodore   e: Captain   f: Commander   g: Lieutenant Commander   h: Lieutenant   i: Sub Lieutenant
j: Acting Sub Lieutenant
Gold on dark blue. All three services have same rank insignia based on the army.

Netherlands

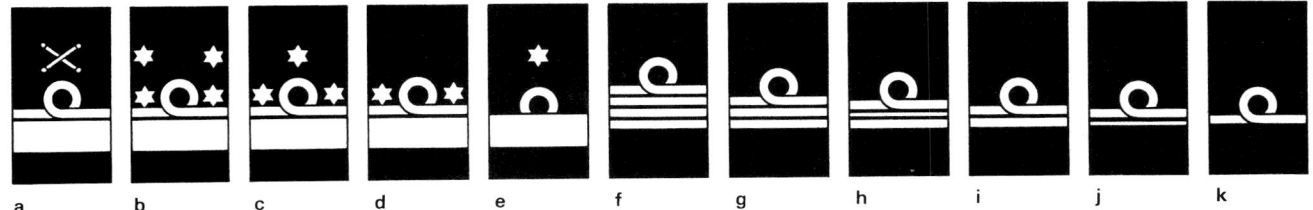

a: *Admiraal*, Admiral of the Fleet   b: *Luitenant-Admiraal*, Admiral   c: *Vice-Admiraal*, Vice Admiral   d: *Schout-bij-nacht*, Rear Admiral   e: *Commandeur*, Commodore
f: *Kapitein ter zee*, Captain   g: *Kapitein-luitenant ter zee*, Commander   h: *Luitenant ter zee der eerste klasse*, Lieutenant Commander   i: *Luitenant ter zee der tweede klasse
oudste categorie*, Lieutenant   j: *Luitenant ter zee der tweede klasse*, Sub Lieutenant   k: *Luitenant ter zee der derde klasse*, Acting Sub Lieutenant
Gold on navy blue. Stars and crossed batons, silver.

New Zealand

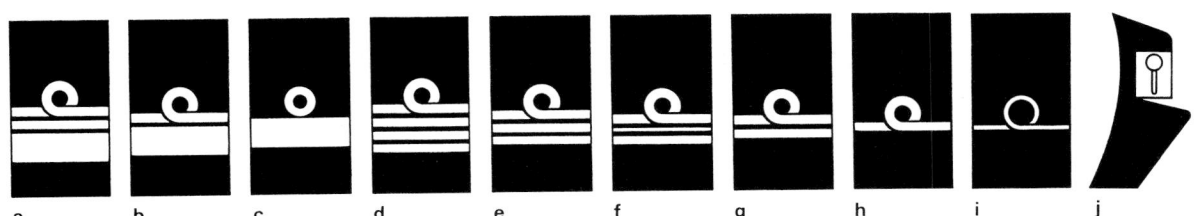

a: Vice Admiral   b: Rear Admiral   c: Commodore   d: Captain   e: Commander   f: Lieutenant Commander   g: Lieutenant   h: Sub Lieutenant
i: Ensign   j: Midshipman
Gold on navy blue.

Nicaragua (la Fuerza Naval)

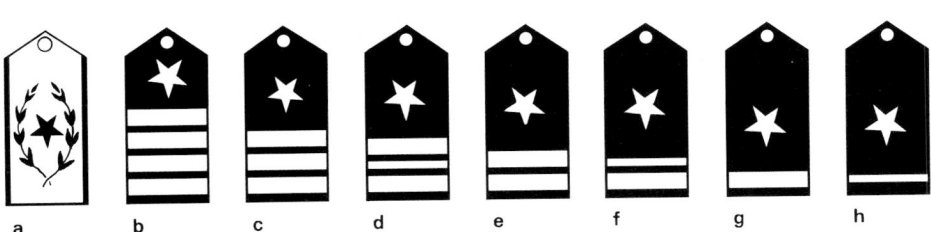

a: *Contraalmirante*, Rear Admiral   b: *Capitán de Navío*, Captain   c: *Capitán de Fragata*, Commander   d: *Capitán de Corbeta*, Lieutenant Commander
e: *Teniente de Navío*, Lieutenant   f: *Teniente de Fragata*, Sub Lieutenant   g: *Teniente de Corbeta*, Acting Sub Lieutenant   h: *Alférez*, Midshipman
Gold button and star. Green wreath on gold brocade, navy blue edges. Gold lace and stars on navy blue.

Nigeria

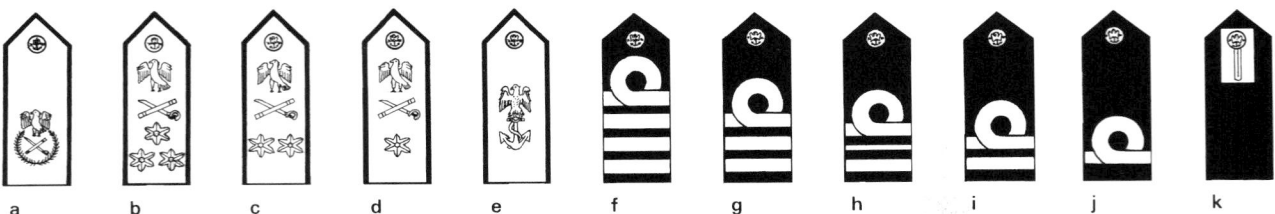

a: Admiral of the Fleet   b: Admiral   c: Vice Admiral   d: Rear Admiral   e: *Brigadier,* Commodore   f: *Colonel,* Captain   g: *Lieutenant Colonel,* Commander
h: *Major,* Lieutenant Commander   i: *Captain,* Lieutenant   j: *Lieutenant,* Sub Lieutenant   k: *Second Lieutenant,* Midshipman
Gold on navy blue. Eagles, red. Stars and crossed battons, silver.

Norway

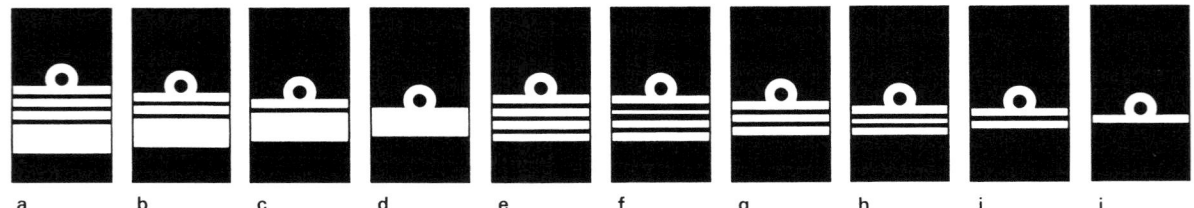

a: *Admiral,* Admiral   b: *Viseadmiral,* Vice Admiral   c: *Kontreadmiral,* Rear Admiral   d: *Flaggkommandør,* Commodore   e: *Kommandør,* Captain
f: *Kommandørkaptein,* Commander Senior Grade   g: *Orlogskaptein,* Commander   h: *Kapteinløytnant,* Lieutenant Commander   i: *Løytnant,* Lieutenant
j: *Fenrik,* Sub Lieutenant
Gold on navy blue. The Coast Guard is manned by naval personnel on secondment. They wear a distinguishing shoulder insignia on the upper left sleeve.

Oman

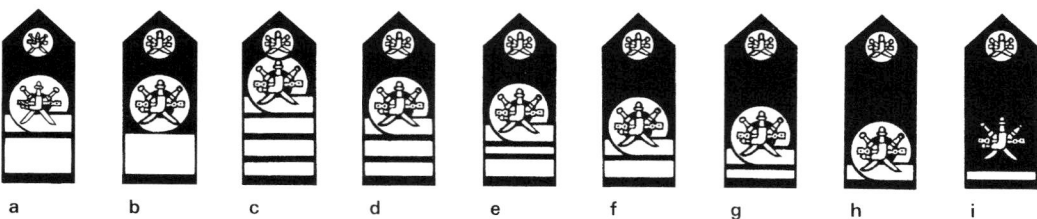

a: *Liwaa Bahry,* Rear Admiral   b: *'Amid Bahry,* Commodore   c: *'Aqid Bahry,* Captain   d: *Muqaddam Bahry,* Commander   e: *Ra'id Bahry,* Lieutenant Commander
f: *Naqib Bahry,* Lieutenant   g: *Mulazim Awwal Bahry,* Sub Lieutenant   h: *Mulazim Tanin Bahry,* Acting Sub Lieutenant   i: *Dabit Murashshah,* Midshipman
Gold on navy blue. White stripe, midshipman.

Pakistan

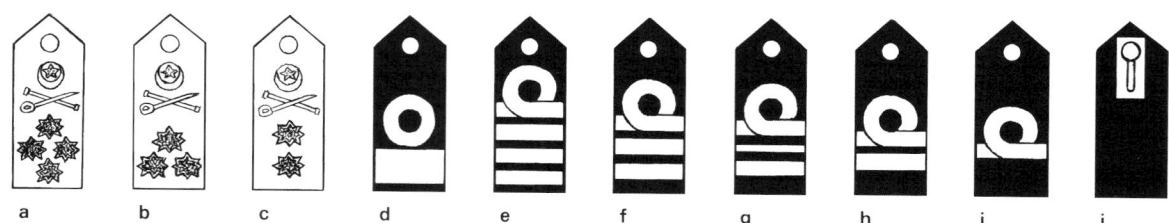

a: Admiral   b: Vice Admiral   c: Rear Admiral   d: Commodore   e: Captain   f: Commander   g: Lieutenant Commander   h: Lieutenant   i: Sub Lieutenant   j: Midshipman
Gold on dark blue shoulder boards. Flag ranks: gold shoulder boards edged blue, silver devices.

Panama (Servicio Maritime Nacional)

a: *Director General,* Rear Admiral   b: *Capitán de Navío,* Captain   c: *Capitán de Fragata,* Commander   d: *Capitán de Corbeta,* Lieutenant Commander
e: *Teniente de Navío,* Lieutenant   f: *Teniente de Fragata,* Lieutenant (JG)   g: *Alférez de Navío,* Sub Lieutenant
Gold on navy blue.

Paraguay (Armada Nacional)

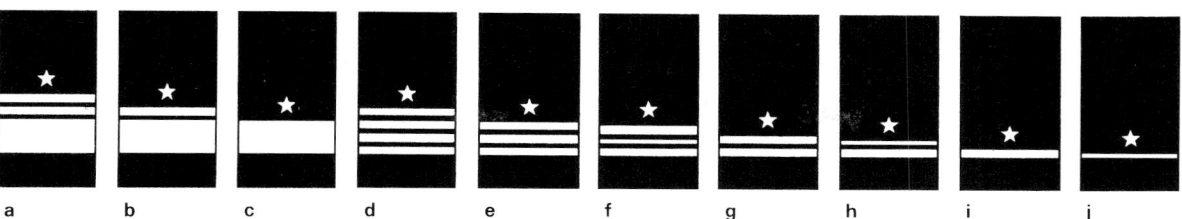

a   b   c   d   e   f   g   h   i   j

a: *Vicealmirante,* Vice Admiral   **b:** *Contralmirante,* Rear Admiral   **c:** *Contraalmirante Medio Inferior,* Rear Admiral Lower Half (Commodore)   **d:** *Capitán de Navío,* Captain
e: *Capitán de Fragata,* Commander   **f:** *Capitán de Corbeta,* Lieutenant Commander   **g:** *Teniente de Navío,* Lieutenant   **h:** *Teniente de Fragata,* Sub Lieutenant
i: *Teniente de Corbeta,* Acting Sub Lieutenant   **j:** *Guardiamarinha,* Midshipman
Gold on navy blue.

Peru (Armada Perúana)

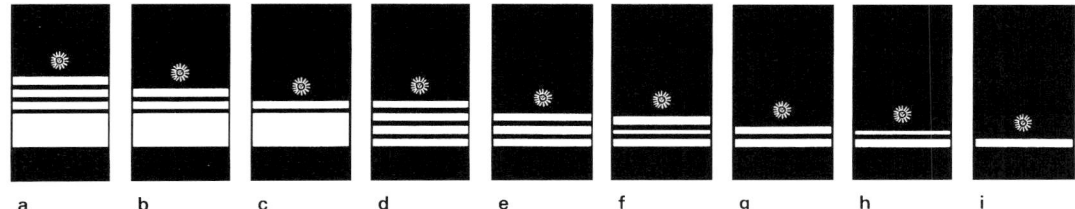

a   b   c   d   e   f   g   h   i

a: *Almirante,* Admiral   **b:** *Vicealmirante,* Vice Admiral   **c:** *Contraalmirante,* Rear Admiral   **d:** *Capitán de Navío,* Captain   **e:** *Capitán de Fragata,* Commander
f: *Capitán de Corbeta,* Lieutenant Commander   **g:** *Teniente Primero,* Lieutenant   **h:** *Teniente Segundo,* Sub Lieutenant   **i:** *Alférez de Fragata,* Acting Sub Lieutenant
Gold on navy blue.

Philippines

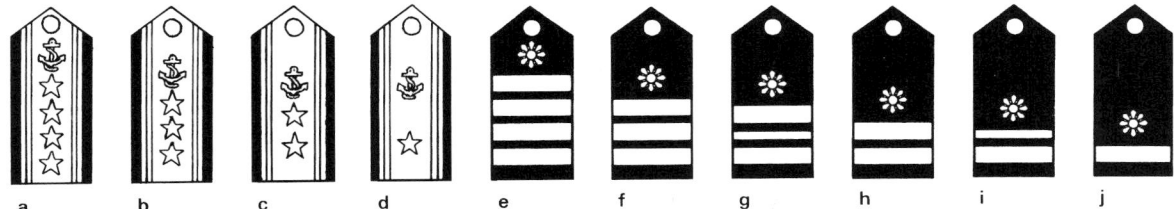

a   b   c   d   e   f   g   h   i   j

a: Admiral   **b:** Vice Admiral   **c:** Rear Admiral   **d:** Commodore   **e:** Captain   **f:** Commander   **g:** Lieutenant Commander   **h:** Lieutenant
i: Lieutenant Junior Grade   **j:** Ensign
Gold on black. Commodore, dark blue edged, silver devices on gold.

Philippines (Coast Guard)

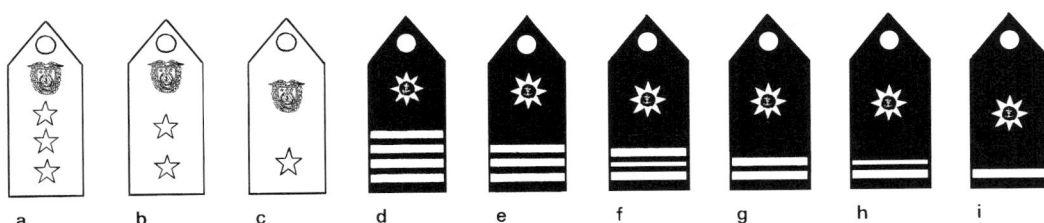

a   b   c   d   e   f   g   h   i

a: Vice Admiral   **b:** Rear Admiral   **c:** Commodore   **d:** Captain   **e:** Commander   **f:** Lieutenant Commander   **g:** Lieutenant Senior Grade
h: Lieutenant Junior Grade   **i:** Ensign

Poland (Marynarka Wojenna)

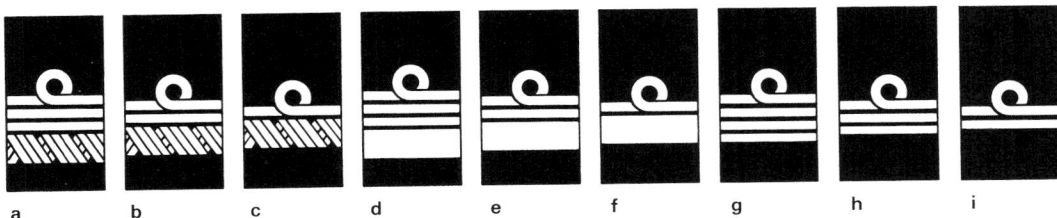

a   b   c   d   e   f   g   h   i

a: *Admiral,* Admiral   **b:** *Vice-Admiral,* Vice Admiral   **c:** *Kontradmiral,* Rear Admiral   **d:** *Komandor,* Captain   **e:** *Komandor Porucznik,* Commander
f: *Komandor Podporucznik,* Lieutenant Commander   **g:** *Kapitan Marynarki,* Lieutenant   **h:** *Porucznik Marynarki,* Sub Lieutenant
i: *Podporucznik Marynarki,* Acting Sub Lieutenant
Gold on dark blue. The Sea Department of the Border Guard (MOSG) use the same insignia up to the rank of Rear Admiral.

Portugal (Marinha Portuguesa)

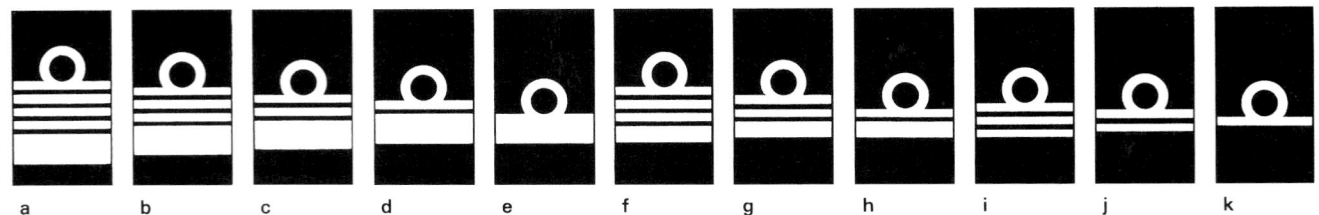

a: *Almirante da Armada*, Admiral of the Fleet   b: *Almirante*, Admiral   c: *Vice-Almirante*, Vice Admiral   d: *Contra-Almirante*, Rear Admiral   e: *Comodoro*, Commodore
f: *Capitão-de-Mar-e-Guerra*, Captain   g: *Capitão-de-Fragata*, Commander   h: *Capitão-Tenente*, Lieutenant Commander   i: *Primeiro-Tenente*, Lieutenant
j: *Segundo-Tenente*, Sub Lieutenant   k: *Guarda-Marinha-ou-Subtenente*, Midshipman or Acting Sub Lieutenant
Gold on navy blue.

Qatar

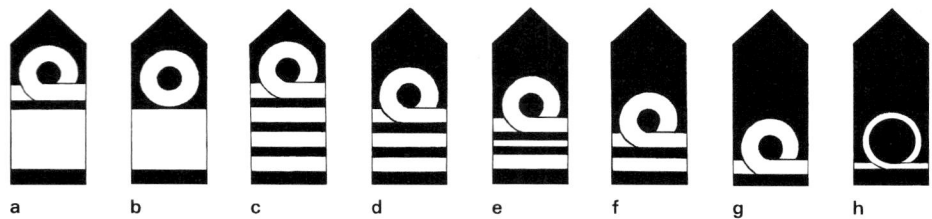

a: Rear Admiral   b: Commodore   c: Captain   d: Commander   e: Lieutenant Commander   f: Lieutenant   g: Sub Lieutenant
h: Acting Sub Lieutenant
Gold bullion lace on navy blue.

Romania (Marină Română)

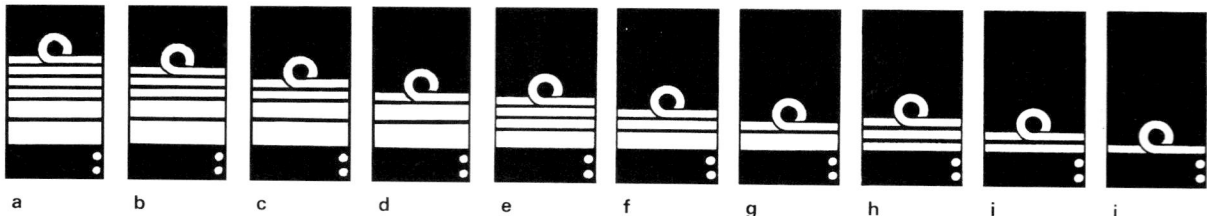

a: *Amiral*, Admiral   b: *Viceamiral*, Vice Admiral   c: *Contraamiral*, Rear Admiral   d: *Contraamiral de Flotilă*, Commodore   e: *Comandor*, Captain
f: *Căpitan Commander*, Commander   g: *Locotenent Comandor*, Lieutenant Commander   h: *Căpitan*, Lieutenant   i: *Locotenent*, Sub Lieutenant
j: *Aspirant*, Midshipman
Gold on dark blue.

Russian Federation (Rosiyskiy Voennomorsky Flot) (Seaman and Marine Engineer Officers)

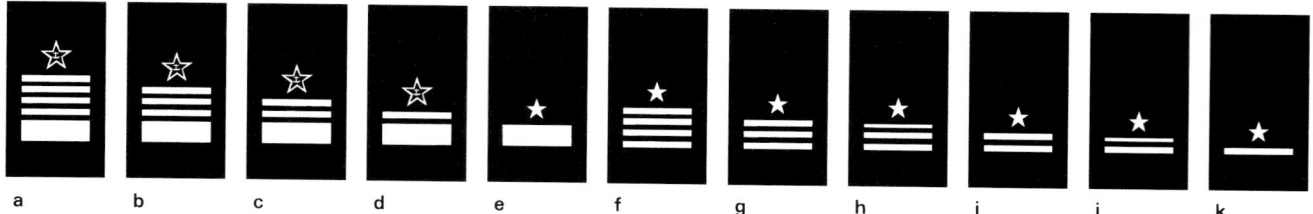

a: *Admiral Flota*, Admiral of the Fleet   b: *Admiral*, Admiral   c: *Vitse-Admiral*, Vice Admiral   d: *Kontr-Admiral*, Rear Admiral   e: *Kapitan Pervogo Ranga*, Captain
f: *Kapitan Vtorogo Ranga*, Commander   g: *Kapitan Tretyego Ranga*, Lieutenant Commander   h: *Kapitan-Leytenant*, Lieutenant   i: *Starshiy Leytenant*, Junior Lieutenant
j: *Leytenant*, Sub Lieutenant   k: *Mladshiy Leytenant*, Acting Sub Lieutenant
Gold on black. Not worn by aviation and specialist officers.

Russian Federation (Aviation and specialist officers, shoulder insignia)

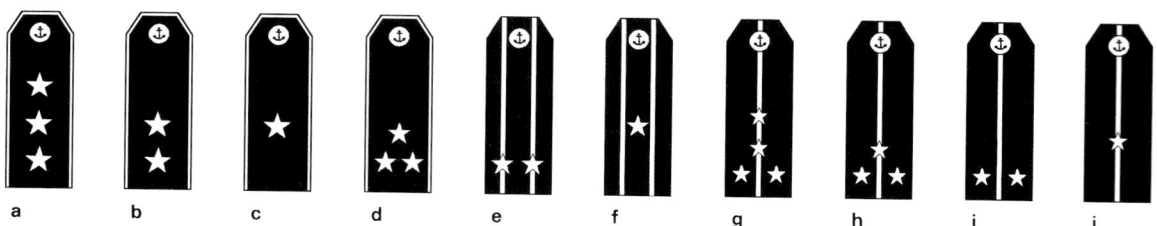

a: *Colonel General*, Admiral   b: *Lieutenant General*, Vice Admiral   c: *Major General*, Rear Admiral   d: *Colonel*, Captain   e: *Lieutenant Colonel*, Commander
f: *Major*, Lieutenant Commander   g: *Captain*, Lieutenant   h: *Senior Lieutenant*, Junior Lieutenant   i: *Lieutenant*, Sub Lieutenant   j: *Junior Lieutenant*, Acting Sub Lieutenant

Saudi Arabia (Royal Saudi Naval Forces)

a | b | c | d | e | f | g | h | i

a: *Lieutenant General (Navy)*, Vice Admiral   b: *Major General (Navy)*, Rear Admiral   c: *Brigadier General (Navy)*, Commodore   d: *Colonel (Navy)*, Captain
e: *Lieutenant Colonel (Navy)*, Commander   f: *Major (Navy)*, Lieutenant Commander   g: *Captain (Navy)*, Lieutenant   h: *Lieutenant (Navy)*, Sub Lieutenant
i: *Second Lieutenant (Navy)*, Acting Sub Lieutenant
Gold buttons, sabres and Arabic titles, light green stars and crowns on black.

Senegal (Marine Sénégalaise)

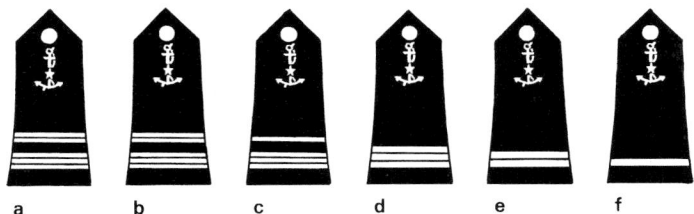

a | b | c | d | e | f

a: *Contre-Amiral*, Rear Admiral   b: *Capitaine de Vaisseau*, Captain   c: *Capitaine de Frégate*, Commander
d: *Capitaine de Corvette*, Lieutenant Commander   e: *Lieutenant de Vaisseau*, Lieutenant
f: *Enseigne de Vaisseau*, Sub Lieutenant
Gold on black. Captain, three gold and two silver stripes.

Serbia and Montenegro

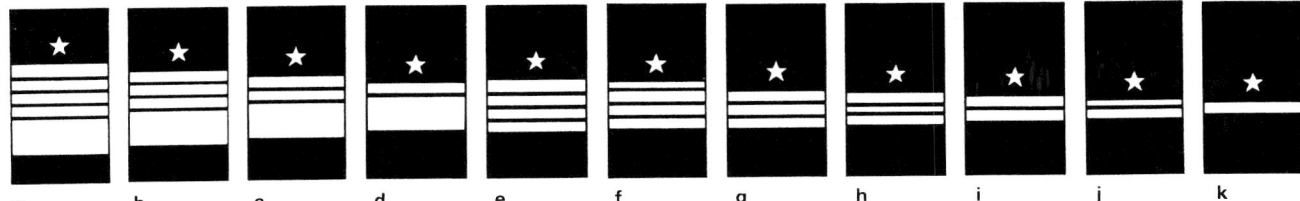

a | b | c | d | e | f | g | h | i | j | k

a: *Admiral Flote*, Admiral of the Fleet   b: *Admiral*, Admiral   c: *Viceadmiral*, Vice Admiral   d: *Kontraadmiral*, Rear Admiral   e: *Kapetan Bojnog Broda*, Captain
f: *Kapetan Fregate*, Commander   g: *Kapetan Korvete*, Lieutenant Commander   h: *Poručnik Bojnog Broda*, Lieutenant (Senior)   i: *Poručnik Fregate*, Lieutenant
j: *Poručnik Korvete*, Sub Lieutenant   k: *Potporučnik*, Acting Sub Lieutenant
Gold on dark blue.

Seychelles

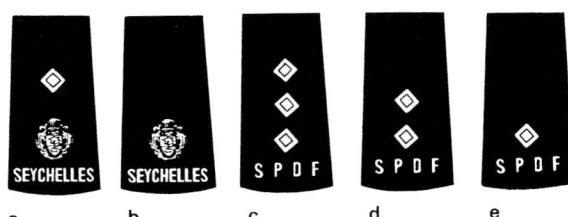

a | b | c | d | e

a: *Lieutenant Colonel*, Commander   b: *Major*, Lieutenant Commander   c: *Captain*, Lieutenant
d: *Lieutenant*, Sub Lieutenant   e: *Second Lieutenant*, Acting Sub Lieutenant
Gold embroidered on dark blue.

Singapore (Republic of Singapore Navy)

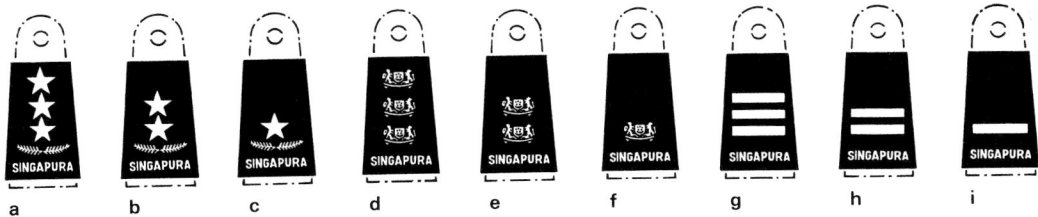

a | b | c | d | e | f | g | h | i

a: Vice Admiral   b: Rear Admiral   c: Commodore   d: Colonel   e: Lieutenant Colonel   f: Major   g: Captain   h: Lieutenant   i: Second Lieutenant
Gold on navy blue. Senior officers only have naval titles.

Slovenia (Slovenska Mornarical)

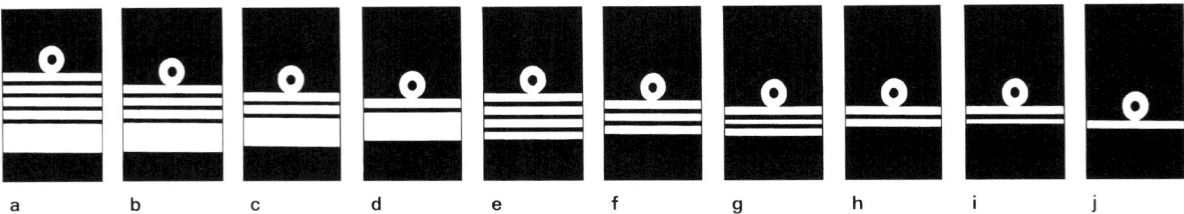

a   b   c   d   e   f   g   h   i   j

a: *Chief Admiral*, Admiral of the Fleet   b: *Admiral*, Admiral   c: *Viceadmiral*, Vice Admiral   d: *Kapitan*, Commodore   e: *Kapitan Bojne Ladje*, Captain
f: *Kapitan Fregate*, Commander   g: *Kapitan Korvete*, Lieutenant Commander   h: *Poročnik Fregate*, Lieutenant   i: *Poročnik Korvete*, Sub Lieutenant
j: *Podporočnik*, Acting Sub Lieutenant
Gold on dark blue.

South Africa

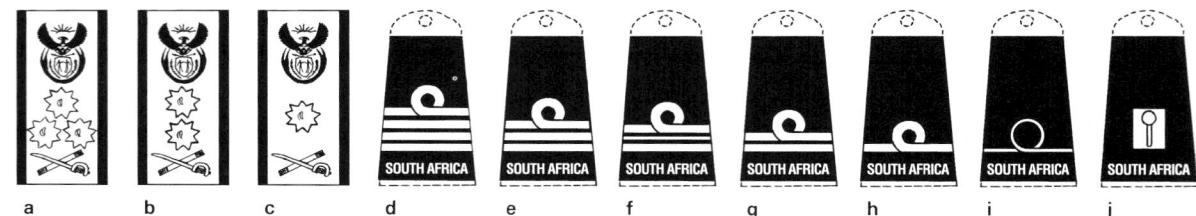

a   b   c   d   e   f   g   h   i   j

a: Vice Admiral   b: Rear Admiral   c: Rear Admiral (JG)   d: Captain   e: Commander   f: Lieutenant Commander   g: Lieutenant   h: Sub Lieutenant
i: Ensign   j: Midshipman
Gold on navy blue. Flag Officers gold brocade with blue piping and silver devices.

Spain (Armada Española)

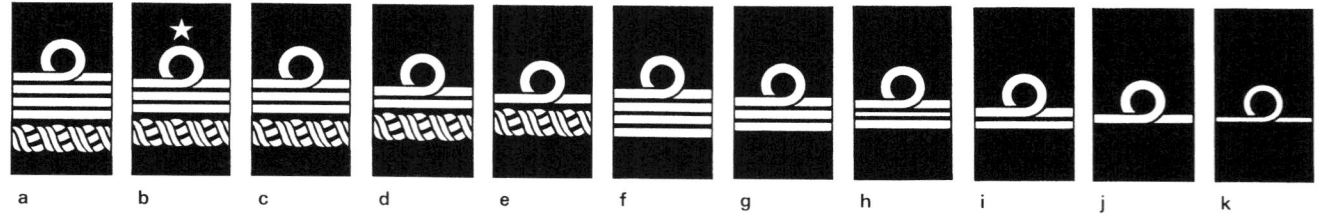

a   b   c   d   e   f   g   h   i   j   k

a: *Capitán General a la Armada*, Captain General   b: *Almirante General*, Admiral   c: *Almirante*, Admiral   d: *Vicealmirante*, Vice Admiral   e: *Contraalmirante*, Rear
Admiral   f: *Capitán de Navío*, Captain   g: *Capitán de Fragata*, Commander   h: *Capitán de Corbeta*, Lieutenant Commander   i: *Teniente de Navío*, Lieutenant
j: *Alférez de Navío*, Sub Lieutenant   k: *Alférez de Fragata*, Acting Sub Lieutenant
Gold on navy blue. Rank of Capitán General held by the monarch only. Rank of Almirante General held by Chief of Naval Staff and Chief of Defence Staff when post held by a
naval officer.

Sri Lanka

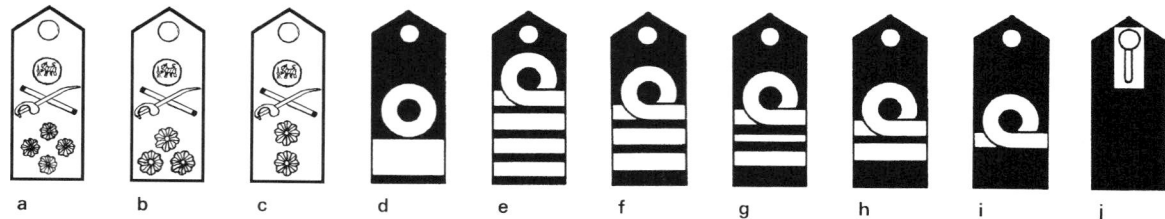

a   b   c   d   e   f   g   h   i   j

a: Admiral   b: Vice Admiral   c: Rear Admiral   d: Commodore   e: Captain   f: Commander   g: Lieutenant Commander   h: Lieutenant   i: Sub Lieutenant
j: Midshipman
Gold on navy blue.

Sudan

a   b   c   d   e   f   g

a: *'Amid*, Commodore   b: *'Aqid*, Captain   c: *Muqaddam*, Commander   d: *Ra'id Lieutenant*, Commander
e: *Naqib*, Lieutenant   f: *Mulazim Awwal*, Sub Lieutenant   g: *Mulazim Thani*, Acting Sub Lieutenant
Gold on black.

Suriname

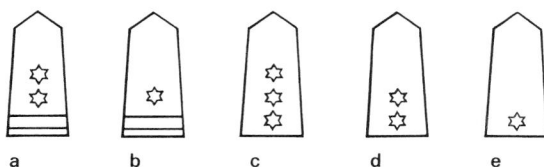

a   b   c   d   e

a: *Kapitein Ter Zee,* Commander   **b:** *Kapitein-Luitenant Ter Zee,* Lieutenant Commander
c: *Luitenant Ter Zee Der 1e Klasse,* Lieutenant   **d:** *Luitenant Ter Zee Der 2e Klasse Oudste Categorie,* Sub Lieutenant
e: *Luitenant Ter Zee Der 3e Klasse,* Acting Sub Lieutenant
Gold on white.

Sweden (Svenska Marinen)

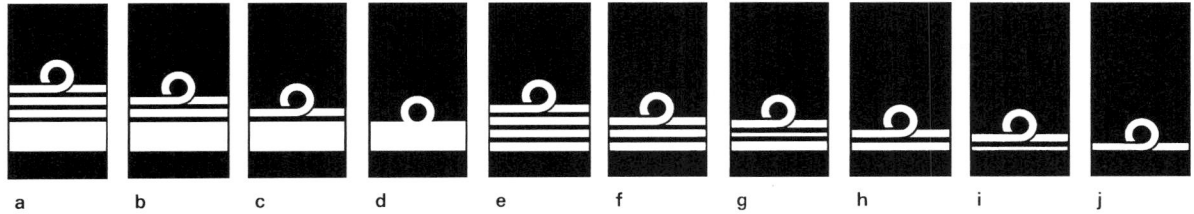

a   b   c   d   e   f   g   h   i   j

a: *Amiral,* Admiral   **b:** *Viceamiral,* Vice Admiral   **c:** *Konteramiral,* Rear Admiral   **d:** *Kommendör av 1. gr,* Commodore   **e:** *Kommendör,* Captain
f: *Kommendörkapten,* Commander   **g:** *Örlogskapten,* Lieutenant Commander   **h:** *Kapten, Lieutenant*   **i:** *Löjnant,* Sub Lieutenant   **j:** *Fänrik,* Acting Sub Lieutenant
Gold on dark blue.

Sweden Coast Guard (Kustbevakning)

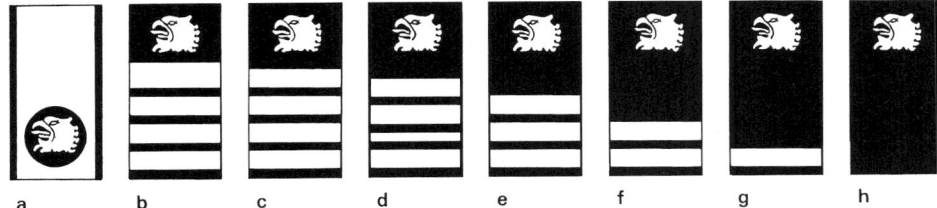

a   b   c   d   e   f   g   h

a: *Generaldirektör,* Rear Admiral   **b:** *Kustbevakningsdirektör,* Commodore   **c:** *Kustbevakningsöverinspectör Överingenjör,* Captain and Senior Engineer Officer
d: *Förste Kustbevakningsinspektör,* Commander   **e:** *Kustbevakningsinspektör,* Lieutenant Commander   **f:** *Kustbevakningassistent,* Lieutenant
g: *Kustuppsyningsman,* Sub Lieutenant   **h:** *Kustbevakningsaspirant,* Midshipman
Gold on navy blue.

Syria

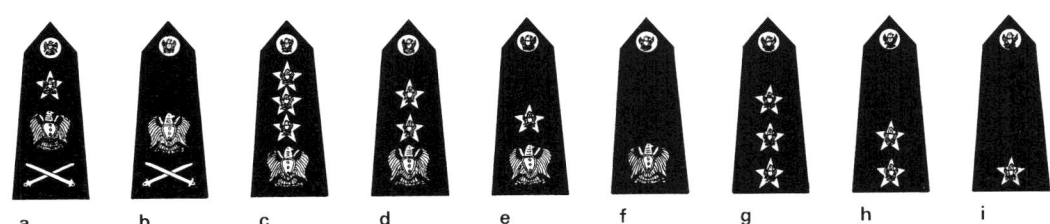

a   b   c   d   e   f   g   h   i

a: *Fariq,* Vice-Admiral   **b:** *Liwa,* Rear Admiral   **c:** *Amid,* Commodor   **d:** *'Aqid,* Captain   **e:** *Muqaddam,* Commander   **f:** *Ra'id,* Lieutenant Commander
g: *Naqib,* Lieutenant   **h:** *Mulazim Awwal,* Sub Lieutenant   **i:** *Mulazim,* Acting Sub Lieutenant
Gold on black.

Taiwan (Republic of China)

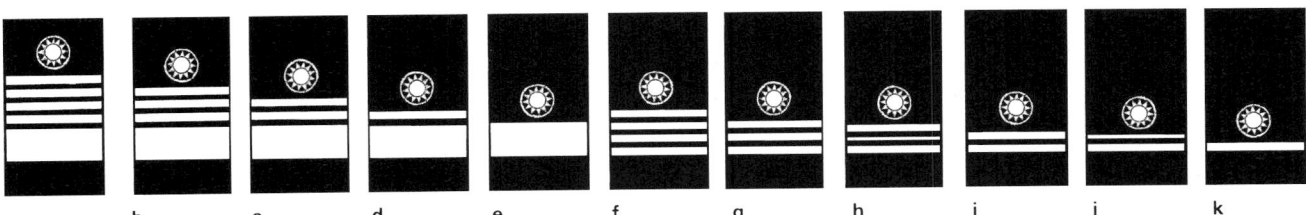

a   b   c   d   e   f   g   h   i   j   k

a: Admiral of the Fleet   **b:** Admiral   **c:** Vice Admiral   **d:** Rear Admiral   **e:** Commodore   **f:** Captain   **g:** Commander   **h:** Lieutenant Commander   **i:** Lieutenant
j: Lieutenant JG (II)   **k:** Sub Lieutenant
Gold on navy blue.

Tanzania

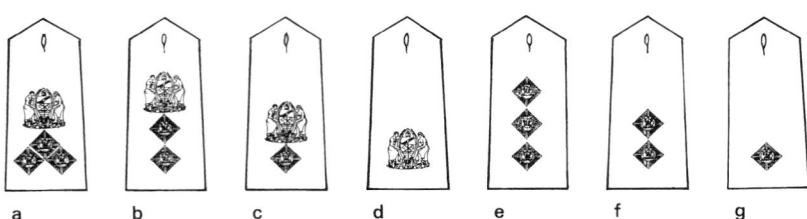

a: *Brigadier Head of the Navy,* Commodore   b: *Colonel,* Captain   c: *Lieutenant Colonel,* Commander
d: *Major,* Lieutenant Commander   e: *Captain,* Lieutenant   f: *Lieutenant,* Sub Lieutenant   g: *Second Lieutenant,* Acting Sub Lieutenant
Gold coloured devices on light tan shoulder strapes.

Thailand (Royal Thai Navy)

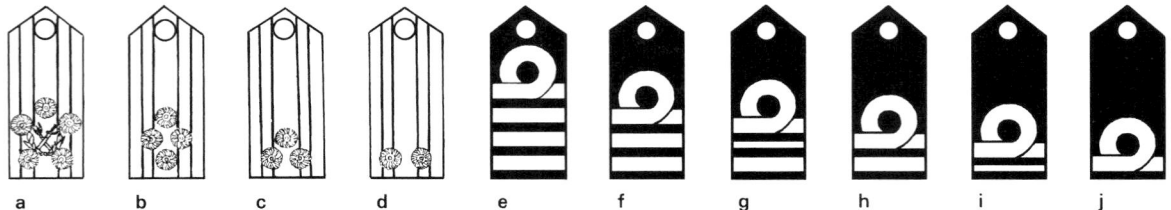

a: Admiral of the Fleet   b: Admiral   c: Vice Admiral   d: Rear Admiral   e: Captain   f: Commander   g: Lieutenant Commander   h: Lieutenant   i: Sub Lieutenant
j: Acting Sub Lieutenant
Admirals wear gold brocaded shoulder straps with silver insignia. Gold lace and buttons on black.

Togo (Marine du Togo)

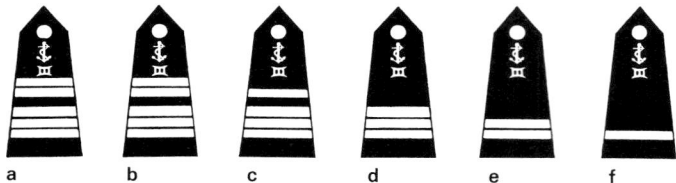

a: *Capitaine de Vaisseau,* Captain   b: *Capitaine de Frégate,* Commander   c: *Capitaine de Corvette,*
Lieutenant Commander   d: *Lieutenant de Vaisseau,* Lieutenant   e: *Enseigne de Vaisseau 1re Classe,* Sub Lieutenant
f: *Enseigne de Vaisseau 2e Classe,* Acting Sub Lieutenant
Gold on black. Commander, three gold two silver stripes.

Tonga

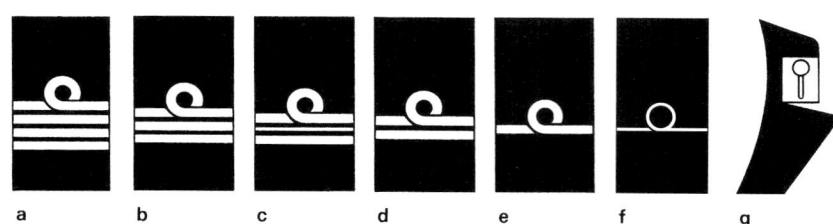

a: Captain   b: Commander   c: Lieutenant Commander   d: Lieutenant   e: Sub Lieutenant   f: Ensign   g: Midshipman
Gold on navy blue.

Trinidad and Tobago Coast Guard

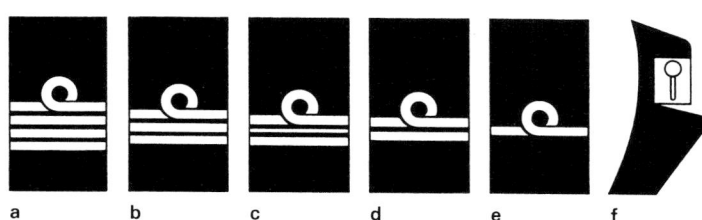

a: Captain   b: Commander   c: Lieutenant Commander   d: Lieutenant   e: Sub Lieutenant   f: Midshipman
Gold on navy blue.

Tunisia

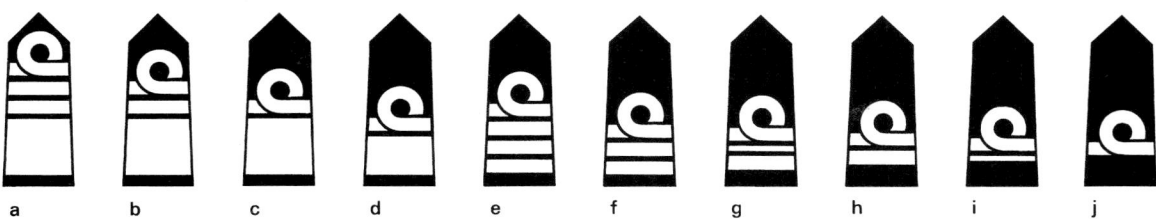

a: *Vice-Amiral d'Escadre*, Admiral   b: *Vice-Amiral*, Vice Admiral   c: *Contre-Amiral*, Rear Admiral   d: *Capitaine de Vaisseau Major*, Commodore   e: *Capitaine de Vaisseau*, Captain   f: *Capitaine de Frégate*, Commander   g: *Capitaine de Corvette*, Lieutenant Commander   h: *Lieutenant de Vaisseau*, Lieutenant   i: *Enseigne de Vaisseau 1ere Classe*, Sub Lieutenant   j: *Enseigne de Vaisseau 2eme Classe*, Acting Sub Lieutenant
Gold on navy blue. Silver on navy blue (Lieutenants and below).

Turkey (Türk Deniz Kuvvetleri)

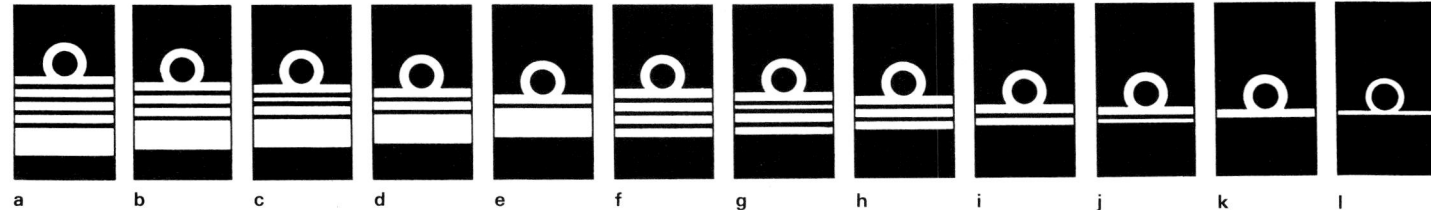

a: *Büyükamiral*, Admiral of the Fleet   b: *Oramiral*, Admiral   c: *Koramiral*, Vice Admiral   d: *Tümamiral*, Rear Admiral   e: *Tugamiral*, Commodore   f: *Albay*, Captain
g: *Yarbay*, Commander   h: *Binbasi*, Lieutenant Commander   i: *Yüzbasi*, Lieutenant   j: *Üstegmen*, Sub Lieutenant   k: *Tegmen*, Acting Sub Lieutenant
l: *Astegmen*, Warrant Officer
Gold on black. The Coast Guard is manned by naval personnel on secondment. They wear a distinguishing shoulder title 'Sahil Guvenlik Kiligi' at the top of each sleeve.

Ukraine

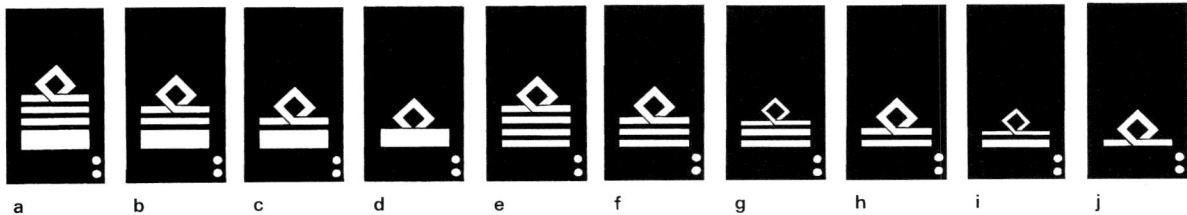

a: *Admiral*, Admiral   b: *Vitse-Admiral*, Vice Admiral   c: *Kontr-Admiral*, Rear Admiral   d: *Kapitan Pervogo Ranga*, Captain   e: *Kapitan Vtorogo Ranga*, Commander
f: *Kapitan Tretyego Ranga*, Lieutenant Commander   g: *Kapitan-Leytenant*, Lieutenant   h: *Starshiy-Leytenant*, Junior Lieutenant   i: *Leytenant*, Sub Lieutenant
j: *Mladshiy-Leytenant*, Acting Sub Lieutenant
Gold on black.

United Arab Emirates

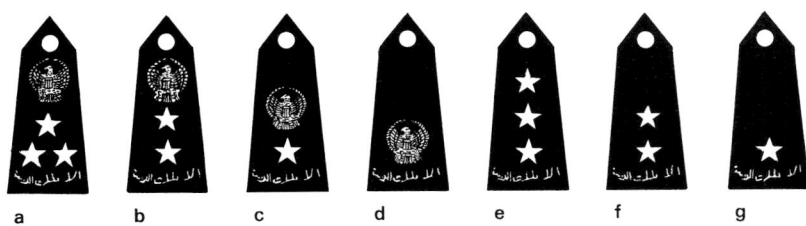

a: *'Amid*, Commodore   b: *'Aqid*, Captain   c: *Muqaddam*, Commander   d: *Ra'id*, Lieutenant Commander
e: *Naqib*, Lieutenant   f: *Mulazim Awwal*, Sub Lieutenant   g: *Mulazim Thani*, Acting Sub Lieutenant
Gold on black.

United Kingdom

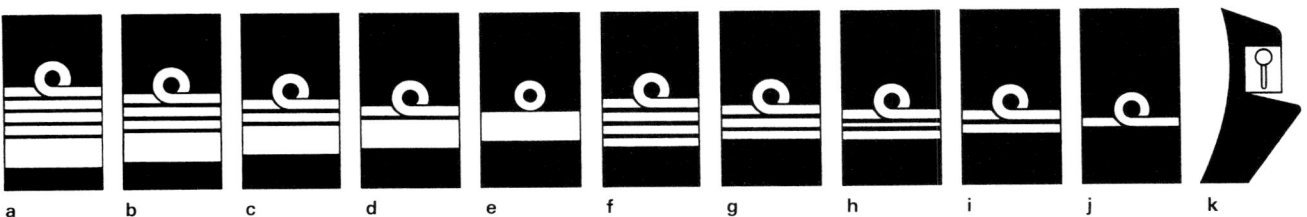

a: Admiral of the Fleet   b: Admiral   c: Vice Admiral   d: Rear Admiral   e: Commodore   f: Captain   g: Commander   h: Lieutenant Commander   i: Lieutenant
j: Sub Lieutenant   k: Midshipman
Gold on navy blue. Rank a: This rank is now in abeyance in peacetime.

United Kingdom (Royal Fleet Auxiliary)

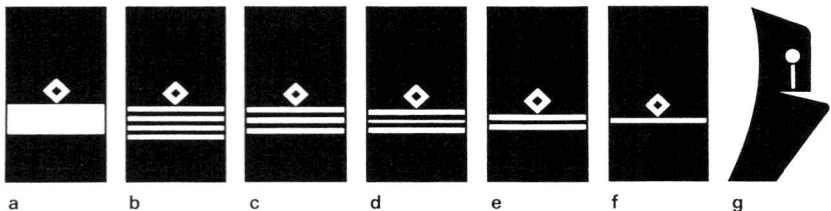

a: Commodore   b: Captain   c: Chief Officer   d: First Officer   e: 2nd Officer   f: 3rd Officer   g: Deck Cadet
Gold on navy blue.

United States

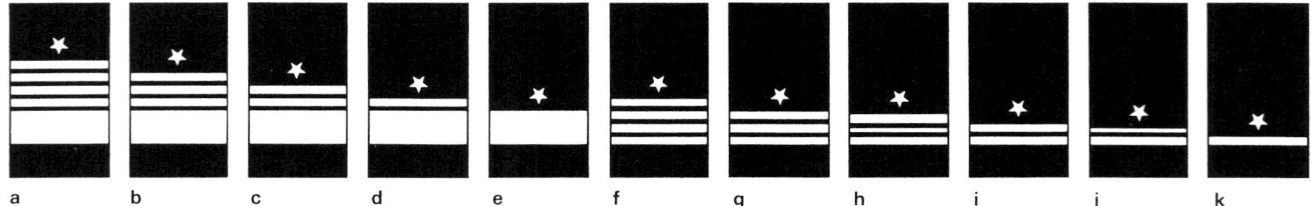

a: *Fleet Admiral,* Admiral of the Fleet   b: *Admiral,* Admiral   c: *Vice Admiral,* Vice Admiral   d: *Rear Admiral (Upper Half),* Rear Admiral   e: *Rear Admiral (Lower Half),*
Commodore   f: *Captain,* Captain   g: *Commander,* Commander   h: *Lieutenant Commander,* Lieutenant Commander   i: *Lieutenant,* Lieutenant   j: *Lieutenant Junior Grade,*
Sub Lieutenant   k: *Ensign,* Acting Sub Lieutenant
Gold on navy blue.

United States Coast Guard

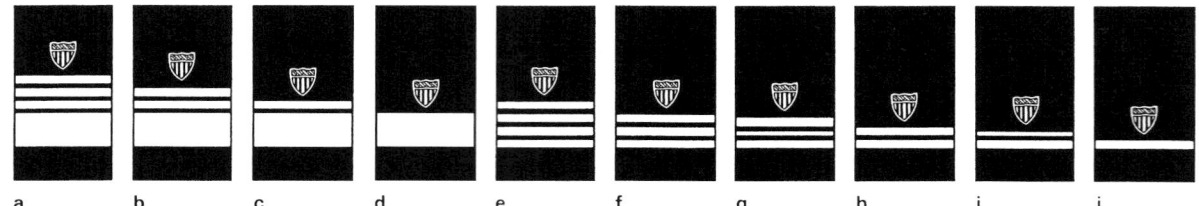

a: *Admiral,* Admiral   b: *Vice Admiral,* Vice Admiral   c: *Rear Admiral,* Rear Admiral   d: *Rear Admiral Lower Half,* Commodore   e: *Captain,* Captain   f: *Commander,*
Commander   g: *Lieutenant Commander,* Lieutenant Commander   h: *Lieutenant,* Lieutenant   i: *Lieutenant Junior Grade,* Sub Lieutenant   j: *Ensign,* Acting Sub Lieutenant
Gold on navy blue.

Uruguay (Armada Nacional)

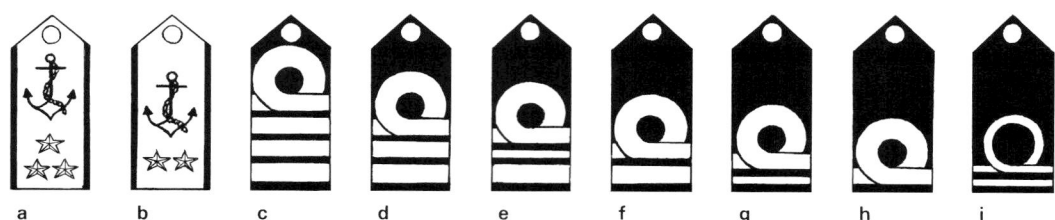

a: *Vicealmirante,* Vice Admiral   b: *Contraalmirante,* Rear Admiral   c: *Capitán de Navío,* Captain   d: *Capitán de Fragata,* Commander   e: *Capitán de Corbeta,* Lieutenant
Commander   f: *Teniente de Navío,* Lieutenant   g: *Alférez de Navío,* Sub Lieutenant   h: *Alférez de Fragata,* Acting Sub Lieutenant   i: *Guardiamarina,* Midshipman
Gold on navy blue. Flag Officer's insignia in silver on gold brocade

Venezuela (Armada de Venezuela)

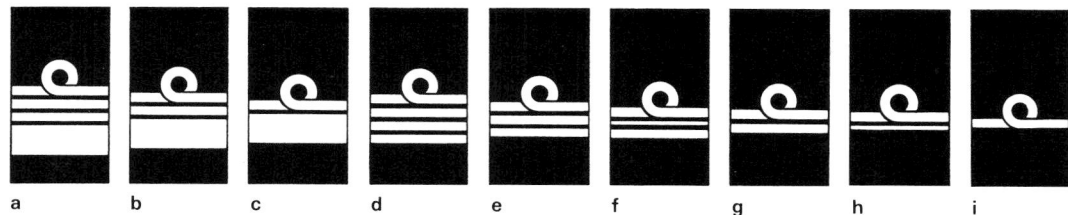

a: *Almirante,* Admiral   b: *Vicealmirante,* Vice Admiral   c: *Contraalmirante,* Rear Admiral   d: *Capitán de Navío,* Captain   e: *Capitán de Fragata,* Commander
f: *Capitán de Corbeta,* Lieutenant Commander   g: *Teniente de Navío,* Lieutenant   h: *Teniente de Fragata,* Sub Lieutenant   i: *Alférez de Navío,* Acting Sub Lieutenant
Gold on navy blue.

Vietnam

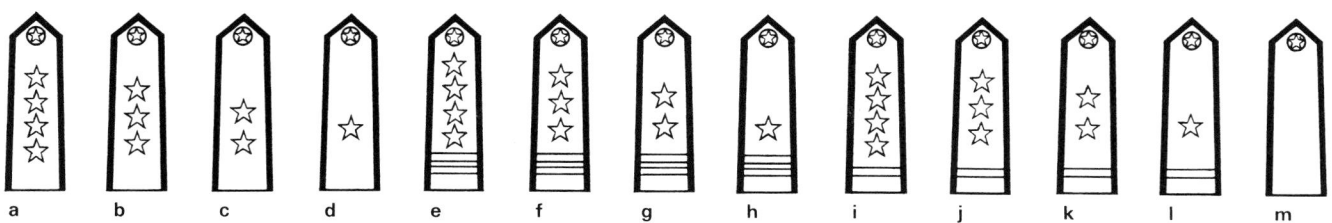

a: *Senior General,* Admiral of the Fleet   b: *Colonel General,* Admiral   c: *Lieutenant General,* Vice Admiral   d: *Major General,* Rear Admiral   e: *Senior Colonel,* Commodore   f: *Colonel,* Captain   g: *Lieutenant Colonel,* Commander   h: *Major,* Lieutenant Commander   i: *Senior Captain,* Senior Lieutenant   j: *Captain,* Lieutenant
k: *Senior Lieutenant,* Sub Lieutenant   l: *2nd Lieutenant,* Acting Sub Lieutenant   m: *Student Officer,* Midshipman
Gold shoulder straps. Generals, edged red gold stars. Remainder, silver stars and lace.

Virgin Islands (UK) (Police – Marine Branch)

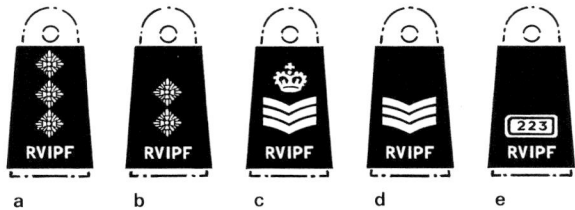

a: Chief Inspector   b: Inspector   c: Station Sergeant   d: Sergeant   e: Constable
Silver on black.

Yemen

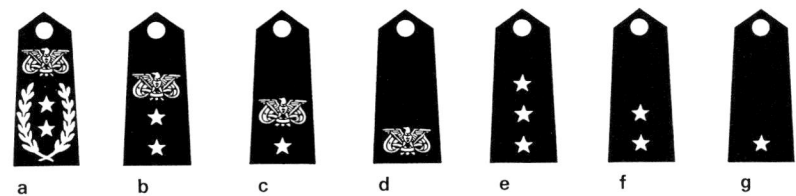

a: *'Amid,* Commodore   b: *'Aqid,* Captain   c: *Muqaddam,* Commander   d: *Ra'id,* Lieutenant Commander
e: *Naqib,* Lieutenant   f: *Mulazim Awwal,* Sub Lieutenant   g: *Mulazim Thani,* Acting Sub Lieutenant
Gold on black.

Pennant list of major surface ships

Type abbreviations

Notes: Designations specific to one nationality are followed by Country abbreviations.
The prefix W denotes a vessel of the Coastguard Service. Suffixes to type indicators are as follows:
F denotes a vessel capable of speeds in excess of 35 kt.
G denotes a vessel with a force guided missile system, including SAM, USM and SUM, usually with a range exceeding 20 miles.
H denotes a vessel equipped with a helicopter, or with a platform for operating one.
J denotes an air cushion or surface effect design.
K denotes a vessel equipped with hydrofoils.
M denotes a Combatant vessel with a close-range guided missile system.

Submarines

AGSS	Submarine, auxiliary, nuclear-powered (USA)
DSRV	Deep submergence rescue vehicle
DSV	Deep submergence vehicle
SDV	Swimmer delivery vehicle
SNA	Submarine, attack, nuclear-powered (Fra)
SNLE	Ballistic missile nuclear-powered submarine (Fra)
SS	Submarine, general
SSA(N)	Submarine, auxiliary, nuclear-powered
SSA	Submarine with ASW capability (Jpn)
SSB	Ballistic missile submarine (CPR)
SSBN	Ballistic missile nuclear-powered submarine
SSC	Submarine, coastal
SSGN	Submarine, surface-to-surface missile, nuclear-powered
SSK	Patrol submarine with ASW capability
SSW	Submarine, midget
SSN	Submarine, attack, nuclear-powered

Aircraft Carriers

CV (M)	Aircraft carrier (guided missile system)
CVH (G)	Helicopter carrier (guided missile system)
CVN (M)	Aircraft carrier, nuclear-powered guided missile system
PAN	Aircraft carrier, nuclear-powered (Fra)

Cruisers

CG	Guided missile cruiser
CGH	Guided missile cruiser with helicopter
CGN	Guided missile cruiser, nuclear-powered
CLM	Guided missile cruiser (Per)

Destroyers

DD	Destroyer
DDG (M)	Guided missile destroyer
DDGH (M)	Guided missile destroyer with helicopter, or helicopter platform
DDK	Destroyer (Jpn)

Frigates

DE	Destroyer escort (Jpn)
FF (L) (H)	Frigate (Light) (Helicopter)
FFG (M)	Guided missile frigate
FFGH (M)	Guided missile frigate with helicopter, or helicopter platform
FS (G) (H) (M)	Corvette (guided missile) (helicopter) (missile)

Patrol Forces

CF	River gunboat (Per)
CM	Corvette (guided missile) (Per)
HSIC	High Speed Interception Craft with speeds in excess of 55 kt
PB	Coastal patrol vessel under 45 m without heavy armament
PB (F) (I) (R)	Patrol boat (fast) (inshore) (river)
PBO (H)	Offshore patrol vessel between 45 and 60 m (helicopter)
PC	Vessel 35-55 m primarily for ASW role
PCK	As for PC but fitted with hydrofoils
PG	Vessel 45-85 m equipped with at least 76 mm (3 in) gun
PGG	As for PG but with force guided missile system
PGGJ	As for PGG but air cushion or ground effect design
PGGK	As for PGG but fitted with hydrofoils
PSO (H)	Offshore patrol vessel over 60 m (helicopter)
PTK	Attack boat torpedo fitted with hydrofoils

Patrol Forces

PTGK	Attack boat guided missile fitted with hydrofoils
SOC	Special operations craft (US)

Landing Ships

AAAV	Advanced Amphibious Assault Vehicle
ACV	Landing craft air cushion (Rus)
AGC	Amphibious command ship (RoC)
ASDS	Advanced Swimmer-Seal Delivery System
EDCG	Landing craft, utility (Brz)
LCA	Landing craft, assault
LCAC	Landing craft air cushion
LCC	Amphibious command ship
LCH	Landing craft, heavy (Aust)
LCM	Landing craft, mechanised
LCP (L)	Landing craft, personnel (large)
LCT	Landing craft, tank
LCU	Landing craft, utility
LCVP	Landing craft, vehicle/personnel with bow ramp
LHA	Amphibious assault ship general purpose with flooded well
LDW	Swimmer delivery vehicle
LHD (M)	Amphibious assault ship (multipurpose), can operate VSTOL aircraft and helicopters
LKA	Amphibious cargo ship with own landing craft
LLP	Assault ship, personnel
LPD	Amphibious transport, dock with own LCMs and helicopter deck
LPH	Amphibious assault ship, helicopter
LSD (H)	Landing ship dock with own landing craft, helicopter
LSL (H)	Landing ship logistic (Aust, UK, Sin), helicopter
LSM (H)	Landing ship medium with bow doors and/or landing ramp, helicopter
LST (H)	Landing ship tank with bow doors and/or landing ramp, helicopter
LSV	Landing ship vehicle with bow doors and/or landing ramp
RCL	Ramped craft, logistic (UK)
TCD	Landing ship, dock (Fra)
UCAC	Utility craft air cushion

Mine Warfare Ships

MCAC	Mine clearance air cushion
MCD	Mine countermeasures vessel, diving support
MCDV	Maritime coast defence vessel (Can)
MCMV	Mine countermeasures vessel
MCS	Mine countermeasures support ship
MH (I) (C) (O)	Minehunter (inshore) (coastal) (ocean)
MHCD	Minehunter coastal with drone
MHSC	Minehunter/sweeper coastal
ML (I) (C) (A)	Minelayer (inshore) (coastal) (auxiliary)
MS (I) (C) (R)	Minesweeper (inshore) (coastal) (river)
MSA (T)	Minesweeper, auxiliary (tug)
MSB	Minesweeper, boat
MSCD	Coastal minesweeper capable of controlling drones
MSD	Minesweeper, drone
MSO	Minesweeper, ocean
SRMH	Single role minehunter (UK)

Auxiliaries

ABU (H)	Buoy tender (helicopter)
AD	Destroyer tender
ADG	Degaussing/deperming ship
AE (L)	Ammunition ship capable of underway replenishment (small)
AEM	Missile support ship
AET (L)	Ammunition transport (small)
AF (L)	Stores ship (small)
AFS	Combat stores ship, capable of underway replenishment

Auxiliaries

AG (H)	Auxiliary miscellaneous (helicopter)
AGB	Icebreaker
AGDS	Deep submergence support ship
AGE (H)	Research ship (helicopter)
AGF (H)	Auxiliary Flag or command ship (helicopter)
AGI (H)	Intelligence collection ship (helicopter)
AGM (H)	Missile range instrumentation ship (helicopter)
AGOB	Polar research ship
AGOR (H)	Oceanographic research ship (helicopter)
AGOS (H)	Ocean surveillance ship (helicopter)
AGP	Patrol craft tender
AGS (C) (H)	Surveying ship (coastal) (helicopter)
AH	Hospital ship
AK (L) (R) (H)	Cargo ship (light) (Ro-Ro) (helicopter)
AKE	Armament stores carrier
AKR	Roll on/roll off sealift ship
AKS (L) (H)	Stores ship (light) (helicopter)
ANL	Boom defence/cable/netlayer
AO	Replenishment oiler (US)
AOE	Fast combat support ship, primarily for POL replenishment
AOR (L) (H)	Replenishment oiler (small) (helicopter)
AOT (L)	Transport oiler (small)
AP (H)	Personnel transport (helicopter)
APB	Barracks ship
APCR	Primary casualty receiving ship
AR (L)	Repair ship (small)
ARC	Submarine cable repair ship
ARS (D) (H)	Salvage ship (heavy lift) (helicopter)
AS (L)	Submarine tender (small)
ASE	Research ship (Jpn)
ASR	Submarine rescue ship
ATA	Auxiliary ocean tug
ATF	Fleet ocean tug and supply ship
ATR	Fleet ocean tug (firefighting and rescue)
ATS	Salvage and rescue ship
AVB	Aviation support ship
AVM	Aviation and missile support
AWT (L)	Water tanker (small)
AX (L) (H)	Training ship (small) (helicopter)
AXS	Sail training ship
AXT	Training tender
HSS	Helicopter support ship
HSV	High speed logistic support vessel (catamaran)
TV	Training ship (Jpn)

Service Craft

ASY	Auxiliary yacht (Jpn)
SAR	Search and rescue vessel
WFL	Water/fuel lighter (Aust)
YAC	Royal yacht
YAG	Service craft, miscellaneous
YAGK	Surface effect craft, experimental
YDG	Degaussing vessel
YDT	Diving tender
YE	Ammunition lighter
YF	Covered personnel transport under 40 m
YFB (H)	Ferry (helicopter)
YFL	Launch
YFRT	Range safety vesel
YFU	Former LCU used for cargo
YGS	Survey launch
YH	Ambulance boat
YM	Dredging craft
YO (G)	Fuel barge (gasolene)
YP	Harbour patrol craft
YPB	Floating barracks
YPC	Oil pollution control vessel
YPT	Torpedo recovery vessel
YT (B) (M) (L)	Harbour tug (large) (medium) (small)
YTR	Harbour fire/rescue craft with several monitors
YTT	Torpedo trials craft
YW	Water barge

Pennant numbers of major surface ships in numerical order

Number	Ship's name	Type	Country	Page	Number	Ship's name	Type	Country	Page
001	Guria	LCU	Georgia	260	L 05	President El Hadj Omar Bongo	LSTH	Gabon	258
001	President H I Remeliik	PB	Palau	547	P 05	Samana	PB	Bahamas	39
001	San Juan	WPBO	Philippines	567	P 05	Itaipú	PBR	Paraguay	551
002	Atia	LCU	Georgia	260	Q 05	Huwar	PGGFM	Qatar	591
002	Edsa II	WPBO	Philippines	567	R 05	Invincible	CV	UK	818
003	Pampanga	WPBO	Philippines	567	06	Newcastle	FFGHM	Australia	25
004	Batangas	WPBO	Philippines	567	06	Almirante Condell	FFGHM	Chile	108
01	Adelaide	FFGHM	Australia	25	BI 06	Rio Ondo	AGS	Mexico	489
01	Pohjanmaa	ML	Finland	217	P 06	Capitan Ortiz	PBF	Paraguay	552
01	Tarangau	PB	Papua New Guinea	550	Q 06	Al Udeid	PGGFM	Qatar	591
A 01	Paluma	AGSC	Australia	32	R 06	Illustrious	CV	UK	818
A 01	Contramaestre Casado	APH	Spain	701	07	Almirante Lynch	FFGHM	Chile	108
AMP 01	Huasteco	APH/AK/AH	Mexico	490	P 07	Général d'Armée Ba Oumar	PBO	Gabon	257
ARE 01	Otomi	ATF	Mexico	491	P 07	Teniente Robles	PBF	Paraguay	552
ATQ 01	Aguascalientes	YOG/YO	Mexico	490	Q 07	Al Deebel	PGGFM	Qatar	591
ATR 01	Maya	AKS	Mexico	490	R 07	Ark Royal	CV	UK	818
BE 01	Cuauhtemoc	AXS	Mexico	491	08	Ministro Zenteno	FFGHM	Chile	108
BL 01	Manuel José Arce	AGP	El Salvador	207	P 08	Colonel Djoue Dabany	PBO	Gabon	257
BI 01	Alejandro de Humbolt	AGOR	Mexico	489	A 09	Manawanui	YDT	New Zealand	519
F 01	Abu Dhabi	FFGHM	UAE	804	HQ 09	Petya class	FFL	Vietnam	928
FM 01	Presidente Eloy Alfaro	FFGHM	Ecuador	188	011	Lieutenant General Dimo Hamaambo	PB	Namibia	503
FSM 01	Palikir	PB	Micronesia	492	011	Varyag	CGHM	Russian Federation	617
HQ 01	Pham Ngu Lao	FF	Vietnam	929	011	Otvazhniy	PGG	Russian Federation	650
M 01	Viesturs	MSC	Latvia	451	012	Olenegorskiy Gorniak	LSTM	Russian Federation	632
P 01	Trident	PB	Barbados	53	013	Korsakov	PGG	Russian Federation	650
P 01	Salamis	PBM	Cyprus	169	014	Podolsk	PGG	Russian Federation	650
P 01	Zibens	PB	Latvia	451	016	Georgiy Pobedonosets	LSTM	Russian Federation	632
P 01	Capitan Cabral	PBR	Paraguay	552	016	Ural	PBO	Russian Federation	648
P 01	Alphonse Reynolds	PB	St Lucia	654	018	Murmansk	PGH	Russian Federation	648
PB 01	Tyrrel Bay	PB	Grenada	298	020	Mitrofan Moskalenko	LPDHM	Russian Federation	632
MRF 01	Sökaren	MSD	Sweden	723	021	Tolyatti	PCM	Russian Federation	649
Q 01	Damsah	PGGF	Qatar	590	022	Tver	PBO	Russian Federation	648
S 01	Gladan	AXS	Sweden	728	023	Nakhodka	PCM	Russian Federation	649
SVG 01	Captain Mulzac	PB	St Vincent	655	023	Nevelsk	PGG	Russian Federation	650
02	Canberra	FFGHM	Australia	25	024	Kaliningrad	PCM	Russian Federation	649
02	Hämeenmaa	ML	Finland	217	026	Yuzhno-Sakhalinsk	PGG	Russian Federation	650
02	Dreger	PB	Papua New Guinea	550	027	Kondopoga	LSTM	Russian Federation	632
02	Tukoro	PB	Vanuatu	921	028	Sochi	PGG	Russian Federation	650
A 02	Mermaid	AGSC	Australia	32	030	Yenisey	PBO	Russian Federation	648
AMP 02	Zapoteco	APH/AK/AH	Mexico	490	031	Alexander Otrakovskiy	LSTM	Russian Federation	632
ARE 02	Yaqui	ATF	Mexico	491	031	Yaroslavl	PCM	Russian Federation	649
ATQ 02	Tlaxcala	YOG/YO	Mexico	490	037	Yastreb	PCM	Russian Federation	649
BE 02	Aldebaran	AKS	Mexico	491	038	Zapolarye	PBO	Russian Federation	648
BI 02	Onjuku	AGS	Mexico	488	LP 039	Miguel Ela Edjodjomo	PB	Equatorial Guinea	208
CAMR 02	Suboficial Oliveira	AGSC	Brazil	74	040	Sarych	PCM	Russian Federation	649
F 02	Al Emirat	FFGHM	UAE	804	041	Grif	PCM	Russian Federation	649
FM 02	Moran Valverde	FFGHM	Ecuador	188	LP 041	Hipolito Micha	PB	Equatorial Guinea	208
FSM 02	Micronesia	PB	Micronesia	492	042	Orlan	PCM	Russian Federation	649
M 02	Imanta	MSC	Latvia	451	042	Siktivkar	PGG	Russian Federation	650
P 02	Waspada	PTG	Brunei	79	043	Amur	PBO	Russian Federation	648
P 02	Lode	PB	Latvia	451	LP 043	Gaspar Obiang Esono	PBR	Equatorial Guinea	208
P 02	Nanawa	PBR	Paraguay	551	LP 043	Novorossiysk	PCM	Russian Federation	648
Q 02	Al Ghariyah	PGGF	Qatar	590	LP 045	Fernando Nuara Engonda	PBR	Equatorial Guinea	208
S 02	Falken	AXS	Sweden	728	052	Cheboksary	PCM	Russian Federation	649
03	Sydney	FFGHM	Australia	25	052	Vorovsky	FFHM	Russian Federation	647
03	Lomor	PB	Marshall Islands	477	054	Eisk MPK 217	FFLM	Russian Federation	625
03	Seeadler	PB	Papua New Guinea	550	055	Marshal Ustinov	CGHM	Russian Federation	617
03	Lata	PB	Solomon Islands	679	055	Kasimov MPD 199	FFLM	Russian Federation	625
A 03	Shepparton	AGSC	Australia	32	055	BDK 98	LSTM	Russian Federation	632
ARE 03	Seri	ATF	Mexico	491	058	Ladoga	PBO	Russian Federation	648
ATR 03	Tarasco	AK	Mexico	490	059	MPK 49	FFLM	Russian Federation	625
BI 03	Altair	AGOR	Mexico	489	060	Anadyr	FFHM	Russian Federation	647
FSM 03	Paluwlap	PB	Micronesia	492	060	Vladimirets	PGK	Russian Federation	630
KA 03	Cometa	WPB	Latvia	452	063	Sokol	PCM	Russian Federation	649
M 03	Nemejs	MHC	Latvia	452	064	Admiral Kutzetsov	CVGM	Russian Federation	615
P 03	Pejuang	PTG	Brunei	79	064	Muromets MPD 134	FFLM	Russian Federation	625
P 03	Yellow Elder	PB	Bahamas	39	065	Briz	PGG	Russian Federation	650
P 03	Linga	PB	Latvia	451	065	Minsk	PCM	Russian Federation	649
Q 03	Rbigah	PGGF	Qatar	590	066	Mukhtar Avezov	LST	Russian Federation	632
04	Darwin	FFGHM	Australia	25	067	Chukotka	PBO	Russian Federation	648
04	Basilisk	PB	Papua New Guinea	550	070	Bobruysk	LSTM	Russian Federation	632
04	Auki	PB	Solomon Islands	679	071	Suzdalets MPK 118	FFLM	Russian Federation	625
A 04	Benalla	AGSC	Australia	32	073	Zabaykalye	PBO	Russian Federation	648
A 04	Kahu	AXL	New Zealand	519	077	Nikolay Korsakov	LSTM	Russian Federation	632
A 04	Martín Posadillo	AKLH	Spain	701	077	Nikolay Kaplunov	PCM	Russian Federation	649
ARE 04	Cora	ATF	Mexico	491	078	MPK 127	FFLM	Russian Federation	625
BI 04	Antares	AGOR	Mexico	489	SSV 080	Pribaltika	AGIM	Russian Federation	634
FSM 04	Constitution	PB	Micronesia	492	081	Nikolay Vilkov	LSTM	Russian Federation	633
M 04	Carlskrona	AXH/MLH	Sweden	722	C 087	Ranger I	PBF	St Kitts and Nevis	654
P 04	Port Nelson	PB	Bahamas	39	C 088	Ranger II	PBF	St Kitts and Nevis	654
P 04	Seteria	PTG	Brunei	79	088	Cholmsk	PGG	Russian Federation	650
P 04	Bulta	PB	Latvia	451	090	Nikolay Sipyagin	AK	Russian Federation	652
P 04	Teneinte Farina	PBR	Paraguay	551	097	Dzerzhinsky	FFHM	Russian Federation	647
Q 04	Barzan	PGGFM	Qatar	591	099	Pyotr Velikiy	CGHMN	Russian Federation	616
05	Melbourne	FFGHM	Australia	25	1	Uruguay	FF	Uruguay	917
05	Uusimaa	ML	Finland	217					
A 05	El Camino Español	AKR	Spain	702					
BI 05	Rio Suchiate	AKS	Mexico	489					
FSM 05	Independence	PB	Micronesia	492					

PENNANT LIST

Number	Ship's name	Type	Country	Page	Number	Ship's name	Type	Country	Page
1/508	Petya III class	FFL	Syria	732	CG 8	Crown Point	PB	Trinidad and Tobago	769
A 1	Comandante General Irigoyen	PSO	Argentina	16	FFG 8	McInerney	FFGHM	USA	874
B 1	Patagonia	AORH	Argentina	18	LHD 8	Makin Island	LHDM	USA	882
BA 1	Tortuguero	ABU	Dominican Republic	185	LPD 8	Dubuque	LPD	USA	885
C 1	Paraguay	PGR	Paraguay	551	LSV 8	Major General Robert Smails	LSV	USA	887
HSV-X1	Joint Venture	HSVH	USA	894	MCM 8	Scout	MCM/MHSO	USA	888
LCC 1	Kao Hsiung	AGF	Taiwan	742	P 8	Paul Bogle	PB	Jamaica	392
LHA 1	Tarawa	LHAM	USA	884	T-AFS 8	Sirius	AFSH	USA	898
LHD 1	Wasp	LHDM	USA	882	T-AOE 8	Arctic	AOEH	USA	897
LSV 1	General Frank S Besson Jr	LSV	USA	887	Z 8	Meduza	AOT	Poland	578
MCM 1	Avenger	MCM/MHSO	USA	888	A 9	Alferez Sobral	PBO	Argentina	16
PCL 1	Ning Hai	PCF	Taiwan	741	A 9	Al Doghas	LCM	Oman	534
S 1	Shabab Oman	AXS	Oman	535	B 9	Astra Valentina	AKS	Argentina	18
SB 1	Ho Chie	LCU	Taiwan	743	CG 9	Galera Point	PB	Trinidad and Tobago	769
T 1	Al Sultana	AKS	Oman	535					
T-AKE 1	Lewis and Clark	AKEH	USA	897	LPD 9	Denver	LPD	USA	885
TSV-1X	Spearhead	HSV	USA	894	MCM 9	Pioneer	MCM/MHSO	USA	888
Z 1	Al Bushra	PBO	Oman	533	PC 9	Chinook	PBFM	USA	889
Z 1	Baltyk	AORL	Poland	578	T-AFS 9	Spica	AFSH	USA	898
2	General Artigas	FF	Uruguay	917	10	Al Riffa	PB	Bahrain	42
2/508	Al Hirasa	FFL	Syria	732	10	Colonia	PB	Uruguay	918
A 2	Teniente Olivieri	PBO	Argentina	16	A 10	Al Temsah	LCM	Oman	534
A 2	Nasr Al Bahr	LSTH	Oman	534	CG 10	Barcolet Point	PB	Trinidad and Tobago	769
BA 2	Capotillo	ABU	Dominican Republic	185					
AOE 2	Camden	AOEHM	USA	891	D 10	Almirante Brown	DDGHM	Argentina	12
LHA 2	Saipan	LHAM	USA	884	L 10	Guarapari	EDCG/LCU	Brazil	71
LHD 2	Essex	LHDM	USA	882	LPD 10	Juneau	LPD	USA	885
LSV 2	CW3 Harold C Clinger	LSV	USA	887	MCM 10	Warrior	MCM/MHSO	USA	888
MCM 2	Defender	MCM/MHSO	USA	888	P 10	Piratini	PB	Brazil	70
PCL 2	An Hai	PCF	Taiwan	741	P 10	Général Nazaire Boulingui	PTM	Gabon	257
Q 2	Libertad	AXS	Argentina	18	PC 10	Firebolt	PBFM	USA	889
SB 2	Ho Ten	LCU	Taiwan	743	T-AFS 10	Saturn	AFSH	USA	898
T-AKE 2	Sacagawea	AKEH	USA	897	T-AOE 10	Bridge	AOEH	USA	897
Z 2	Al Mansoor	PBO	Oman	533	U 10	Aspirante Nascimento	AX	Brazil	75
3	Montevideo	FF	Uruguay	917	Z 10	Dhofar	PGGF	Oman	534
A 3	Francisco de Gurruchaga	PSO	Argentina	16	11	Hawar	PB	Bahrain	42
T-AFS 3	Niagara Falls	AFSH	USA	897	11	Smeli	FFLM	Bulgaria	81
B 3	Canal Beagle	AKS	Argentina	19	11	Capitán Prat	DDGHM	Chile	106
LHA 3	Belleau Wood	LHAM	USA	884	11	Kralj Petar Kresimir IV	FSG	Croatia	164
LHD 3	Kearsage	LHD/MCSM	USA	882	11	Mahamiru	MHC	Malaysia	470
LSV 3	General Brehon B Somervell	LSV	USA	887	11	Rio Negro	PB	Uruguay	918
MCM 3	Sentry	MCM/MHSO	USA	888	A 11	Endeavour	AORH	New Zealand	519
PC 3	Hurricane	PBFM	USA	889	A 11	Marqués de la Ensenada	AORLH	Spain	702
Z 3	Al Najah	PBO	Oman	533	AGF 11	Coronado	AGFH	USA	880
B 4	Bahia San Blas	AKS	Argentina	19	AGS 11	Sunjin	AGE	Korea, South	445
LHA 4	Nassau	LHAM	USA	884	BE 11	Simon Bolivar	AXS	Venezuela	925
LHD 4	Boxer	LHDM	USA	882	BO 11	Punta Brava	AGOR	Venezuela	924
LPD 4	Austin	LPD	USA	885	CF 11	Amazonas	CF/PGR	Peru	558
LSV 4	Lt General William B Bunker	LSV	USA	887	CM 11	Esmeraldas	FSGHM	Ecuador	189
					D 11	La Argentina	DDGHM	Argentina	12
MCM 4	Champion	MCM/MHSO	USA	888	F 11	Zemaitis	FFLM	Lithuania	458
5	15 de Noviembre	PBO	Uruguay	918	GC 11	Almirante Clemente	WFS	Venezuela	926
B 5	Cabo de Hornos	AKS	Argentina	19	HQ 11	Petya class	FFL	Vietnam	928
CG 5	Barracuda	PB	Trinidad and Tobago	769	K 11	Felinto Perry	ASRH	Brazil	75
					K 11	Stockholm	FSG	Sweden	719
LHA 5	Peleliu	LHAM	USA	884	L 11	Tambaú	EDCG/LCU	Brazil	71
LHD 5	Bataan	LHDM	USA	882	M 11	Diana	MCS	Spain	695
LPD 5	Ogden	LPD	USA	885	M 11	Styrsö	MHSDI	Sweden	723
LSV 5	Major General Charles P Gross	LSV	USA	887	MCM 11	Gladiator	MCM/MHSO	USA	888
					P 11	Pirajá	PB	Brazil	70
MCM 5	Guardian	MCM/MHSO	USA	888	P 11	Barceló	PB	Spain	696
PC 5	Typhoon	PBFM	USA	889	PC 11	Whirlwind	PBFM	USA	889
Q 5	Almirante Irizar	AGB/AGOBH	Argentina	19	PC 11	Constitutión	PG	Venezuela	923
T-AFS 5	Concord	AFSH	USA	897	PF 11	Rajah Humabon	FF	Philippines	562
6	25 de Agosto	PBO	Uruguay	918	Q 11	Comodoro Rivadavia	AGOR	Argentina	18
A 6	Suboficial Castillo	PSO	Argentina	16	R 11	Gniewko	ATS	Poland	579
CG 6	Cascadura	PB	Trinidad and Tobago	769	R 11	Principe de Asturias	CV	Spain	688
					SD 11	Wrona	YDG	Poland	579
LHD 6	Bonhomme Richard	LHDM	USA	882	U 11	Guarda Marinha Jensen	AXL	Brazil	75
LPD 6	Duluth	LPD	USA	885	Z 11	Al Sharqiyah	PGGF	Oman	534
LSV 6	Specialist 4 James A Loux	LSV	USA	887	12	Almirante Cochrane	DDGHM	Chile	106
					12	Kralj Dmitar Zvonimir	FSG	Croatia	164
MCM 6	Devastator	MCM/MHSO	USA	888	12	Jerai	MHC	Malaysia	470
PC 6	Sirocco	PBFM	USA	889	12	Paysandu	PB	Uruguay	918
T-AOE 6	Supply	AOEH	USA	897	A 12	São Paulo	CVM	Brazil	62
7	Comodoro Coé	PBO	Uruguay	918	CF 12	Loreto	CF/PGR	Peru	558
CG 7	Corozal Point	PB	Trinidad and Tobago	769	CM 12	Manabi	FSGHM	Ecuador	189
					D 12	Heroina	DDGHM	Argentina	12
LHD 7	Iwo Jima	LHDM	USA	882	F 12	Aukstaitis	FFLM	Lithuania	458
LPD 7	Cleveland	LPD	USA	885	K12	Malmö	FSG	Sweden	719
LSV 7	SSGT Robert T Kuroda	LSV	USA	887	L 12	Camboriú	EDCG/LCU	Brazil	71
MCM 7	Patriot	MCM/MHSO	USA	888	L 12	Ocean	LPH	UK	830
P 7	Fort Charles	PB	Jamaica	391	LPD 12	Shreveport	LPD	USA	885
PC 7	Squall	PBFM	USA	889	GC 12	General José Trinidad Moran	WFS	Venezuela	926
T-AFS 7	San Jose	AFSH	USA	897					
T-AOE 7	Rainier	AOEH	USA	897	M 12	Spårö	MHSDI	Sweden	723
T-ARC 7	Zeus	ARC	USA	899	MCM 12	Ardent	MCM/MHSO	USA	888
A 8	Saba Al Bahr	LCM	Oman	534	MUL 12	Arkösund	MLC	Sweden	722
B 8	Astra Federico	AKS	Argentina	18	P 12	Pampeiro	PB	Brazil	70

Number	Ship's name	Type	Country	Page
P 12	Laya	PB	Spain	696
PC 12	Thunderbolt	PBFM	USA	889
PC 12	Federación	PBG	Venezuela	923
SD 12	Rys	YDG	Poland	579
U 12	Guarda Marinha Brito	AX	Brazil	75
Z 12	Al Bat'nah	PGGF	Oman	534
13	Reshitelni	FSM	Bulgaria	82
13	Ledang	MHC	Malaysia	470
A 13	Tunas Samudera	AXS	Malaysia	472
B 13	Ingeniero Julio Krause	AOTL	Argentina	18
CF 13	Marañon	CF/PGR	Peru	557
CM 13	Los Rios	FSGHM	Ecuador	189
D 13	Sarandi	DDGHM	Argentina	12
HQ 13	Petya class	FFL	Vietnam	928
LPD 13	Nashville	LPD	USA	885
M 13	Skaftö	MHSDI	Sweden	723
MCM 13	Dextrous	MCM/MHSO	USA	888
P 13	Parati	PB	Brazil	70
P 13	Javier Quiroga	PB	Spain	696
PC 13	Independencia	PG	Venezuela	923
14	Bodri	FSM	Bulgaria	82
14	Kinabalu	MHC	Malaysia	470
A 14	Resolution	AGS	New Zealand	519
A 14	Patiño	AORH	Spain	701
AGOR 14	Melville	AGOR	USA	895
CF 14	Ucayali	CF/PGR	Peru	557
CM 14	El Oro	FSGHM	Ecuador	189
J 14	Nirupak	AGSH	India	326
L 14	Albion	LPD	UK	831
LPD 14	Trenton	LPD	USA	885
M 14	Sturkö	MHSDI	Sweden	723
MCM 14	Chief	MCM/MHSO	USA	888
P 14	Penedo	PB	Brazil	70
P 14	Bolong Kauta	PB	Gambia	258
P 14	Ordóñez	PB	Spain	696
PC 14	Libertad	PBG	Venezuela	923
R 14	Zbyszko	ARS	Poland	579
Z 14	Mussandam	PGGF	Oman	534
A 15	Nireekshak	ASR	India	328
AGOR 15	Knorr	AGOR	USA	895
CM 15	Los Galapagos	FSGHM	Ecuador	189
DF 15	Recalada	WAGH/AH	Argentina	22
F 15	Abu Bakr	FFT	Bangladesh	47
G 15	Paraguassú	AP	Brazil	75
HQ 15	Petya class	FFL	Vietnam	928
J 15	Investigator	AGSH	India	326
L 15	Bulwark	LPD	UK	831
LPD 15	Ponce	LPD	USA	885
M 15	Aratú	MSC	Brazil	72
MUL 15	Grundsund	MLC	Sweden	722
P 15	Poti	PB	Brazil	70
P 15	Acevedo	PB	Spain	696
PC 15	Patria	PG	Venezuela	923
Q 15	Cormoran	AGSC	Argentina	18
R 15	Macko	ARS	Poland	579
V 15	Imperial Marinheiro	PG/ATR	Brazil	69
CM 16	Loja	FSGHM	Ecuador	189
F 16	Umar Farooq	FF	Bangladesh	45
J 16	Jamuna	AGSH	India	326
L 16	Absalon	AGF/AKR/AH	Denmark	179
M 16	Anhatomirim	MSC	Brazil	72
P 16	Cándido Pérez	PB	Spain	696
PC 16	Victoria	PBG	Venezuela	923
U 16	Doutor Montenegro	AH	Brazil	76
F 17	Ali Haider	FF	Bangladesh	47
G 17	Potengi	AG	Brazil	77
HQ 17	Petya class	FFL	Vietnam	928
J 17	Sutlej	AGSH	India	326
L 17	Esbern Snare	AGF/AKR/AH	Denmark	179
L 17	Sharabh	LSM	India	325
LPD 17	San Antonio	LPDM	USA	881
M 17	Atalaia	MSC	Brazil	72
U 17	Parnaiba	PGRH	Brazil	70
A 18	Perkons	ATA	Latvia	452
F 18	Osman	FFG	Bangladesh	46
H 18	Comandante Varella	ABU	Brazil	73
J 18	Sandhayak	AGSH	India	326
L 18	Cheetah	LSM/LSMH	India	325
LPD 18	New Orleans	LPDM	USA	881
M 18	Araçatuba	MSC	Brazil	72
MUL 18	Fårösund	MLC	Sweden	722
P 18	Armatolos	PGG	Greece	290
U 18	Oswaldo Cruz	AHH	Brazil	76
19	Almirante Williams	FFHM	Chile	107
H 19	Tenente Castelo	ABU	Brazil	73
J 19	Nirdeshak	AGSH	India	326
L 19	Mahish	LSM/LSMH	India	325
LCC 19	Blue Ridge	LCCH/AGFH	USA	880
LPD 19	Mesa Verde	LPDM	USA	881
M 19	Abrolhas	MSC	Brazil	72
P 19	Navmachos	PGG	Greece	290
PS 19	Miguel Malvar	FS	Philippines	563
T-AGOS 19	Victorious	AGOS	USA	899
T-AH 19	Mercy	AHH	USA	898
20	Ahmad El Fateh	PGGF	Bahrain	42
20	Capitán Miranda	AXS	Uruguay	920
A 20	Moawin	AORH	Pakistan	545
A 20	Neptuno	AGDS	Spain	702
CG 20	Nelson	PBO	Trinidad and Tobago	769
F 20	Godavari	FFGHM	India	315
H 20	Comandante Manhães	ABU	Brazil	73
L 20	Magar	LSTH	India	325
LCC 20	Mount Whitney	LCCH/AGFH	USA	880
LPD 20	Green Bay	LPDM	USA	881
M 20	Albardão	MSC	Brazil	72
MUL 20	Furusund	MLC	Sweden	721
P 20	Murature	AX	Argentina	16
P 20	Pedro Texiera	PBR	Brazil	70
P 20	Anthypoploiarchos Laskos	PGGF/PPG	Greece	289
PS 20	Magat Salamat	FS	Philippines	563
Q 20	Puerto Deseado	AGOB	Argentina	17
T-AH 20	Comfort	AHH	USA	898
U 20	Cisne Branco	AXS	Brazil	75
21	Al Jabiri	PGGF	Bahrain	42
21	Haixun	PBOH	China	150
21	Sibenik	PTFG	Croatia	165
21	Hejaz	LST	Iran	356
21	Cheong Hae Jin	ARS	Korea, South	444
21	Sour	LST	Lebanon	454
21	Vigilant	PSOH	Mauritius	478
21	Sirius	ABU	Uruguay	920
A 21	Kalmat	AOTL	Pakistan	545
CM 21	Velarde	CM/PGGFM	Peru	557
F 21	Gomati	FFGHM	India	315
F 21	Mariscal Sucre	FFGHM	Venezuela	922
G 21	Ary Parreiras	AKSH	Brazil	76
H 21	Sirius	AGSH	Brazil	73
J 21	Darshak	AGSH	India	326
K 21	Göteborg	FSG	Sweden	718
L 21	Guldar	LSM/LSMH	India	325
LM 21	Quito	PGGF	Ecuador	190
LPD 21	New York	LPDM	USA	881
P 21	King	AX	Argentina	16
P 21	Raposo Tavares	PBR	Brazil	70
P 21	Giorgi Toreli	WPBF	Georgia	261
P 21	Plotarchis Blessas	PGGF/PGG	Greece	289
P 21	Emer	PSO	Ireland	360
P 21	Anaga	PB	Spain	696
R 21	Tritão	ATA	Brazil	77
RP 21	Fernando Gomez	AKSL	Venezuela	927
T-AGOS 21	Effective	AGOS	USA	899
Y 21	Oilpress	AOTL	UK	840
22	Abdul Rahman Al Fadel	PGGF	Bahrain	42
22	Karabala	LST	Iran	356
22	Damour	LST	Lebanon	454
22	Oyarvide	AGS	Uruguay	919
CM 22	Santillana	CM/PGGFM	Peru	557
F 22	Ganga	FFGHM	India	315
F 22	Almirante Brión	FFGHM	Venezuela	922
J 22	Sarvekshak	AGSH	India	326
K 22	Gävle	FSG	Sweden	718
L 22	Kumbhir	LSM/LSMH	India	325
LPD 22	San Diego	LPDM	USA	881
P 22	Ayety	WPBO	Georgia	261
P 22	Ypoploiarchos Mikonios	PGGF	Greece	289
P 22	Aoife	PSO	Ireland	360
P 22	Tagomago	PB	Spain	696
PS 22	Sultan Kudarat	FS	Philippines	563
R 22	Tridente	ATA	Brazil	77
R 22	Viraat	CVM	India	311
RM 22	Enriquillo	PG/ATA	Dominican Republic	184
T-AGOS 22	Loyal	AGOS	USA	899
23	Al Taweelah	PGGF	Bahrain	42
23	Amir	LST	Iran	356
23	Maldonaldo	AG	Uruguay	920
A 23	Antares	AGS	Spain	700
AGOR 23	Thomas G Thompson	AGOR	USA	895
CM 23	De los Heros	CM/PGGFM	Peru	557
F 23	General Urdaneta	FFGHM	Venezuela	922
G 23	Almirante Gastão Motta	AOR	Brazil	76
K 23	Kalmar	FSG	Sweden	718
L 23	Gharial	LSTH	India	325
LM 23	Guayaquil	PGGF	Ecuador	190
LPD 23	Anchorage	LPDM	USA	881
P 23	Ypoploiarchos Troupakis	PGGF/PGG	Greece	289
P 23	Aisling	PSO	Ireland	360
P 23	Marola	PB	Spain	696
PS 23	Datu Marikudo	FS	Philippines	563
R 23	Triunfo	ATA	Brazil	77
T-AGM 23	Observation Island	AGM	USA	899
T-AGOS 23	Impeccable	AGOS	USA	900
24	Farsi	LST	Iran	355
24	Lieutenant Remus Lepri	MSC	Romania	597

Number	Ship's name	Type	Country	Page
A 24	Rigel	AGS	Spain	700
AGOR 24	Roger Revelle	AGOR	USA	895
CM 24	Herrera	CM/PGGFM	Peru	557
F 24	General Soublette	FFGHM	Venezuela	922
K 24	Sundsvall	FSG	Sweden	718
LM 24	Cuenca	PGGF	Ecuador	190
LPD 24	Arlington	LPDM	USA	881
P 24	Simeoforos Kavaloudis	PGGF/PGG	Greece	289
P 24	Mouro	PB	Spain	696
R 24	Almirante Guilhem	ATF	Brazil	77
25	Sardasht	LST	Iran	355
25	Kasturi	FSGH	Malaysia	465
25	Lieutenant Lupu Dunescu	MSC	Romania	597
AGOR 25	Atlantis	AGOR	USA	895
AT 25	Ang Pangulo	AP	Philippines	566
CM 25	Larrea	CM/PGGFM	Peru	557
F 25	DW 200H	FFG	Bangladesh	46
F 25	General Salom	FFGHM	Venezuela	922
H 25	Tenente Boanerges	ABU	Brazil	73
LPD 25	Somerset	LPDM	USA	881
P 25	Grosa	PB	Spain	696
R 25	Almirante Guillobel	ATF	Brazil	77
26	Sab Sahel	LST	Iran	355
26	Lekir	FSGH	Malaysia	465
26	Vanguardia	ARS	Uruguay	920
AGOR 26	Kilo Moana	AGOR	USA	895
CM 26	Sanchez Carrión	CM/PGGFM	Peru	557
F 26	Almirante Garcia	FFGHM	Venezuela	922
H 26	Faroleiro Mário Seixas	ABU	Brazil	74
P 26	Dzata	PBO	Ghana	284
P 26	Ypoploiarchos Degiannis	PGGF/PGG	Greece	289
P 26	Medas	PB	Spain	696
R 26	Trinidade	ATA	Brazil	77
T-AE 26	Kilauea	AEH	USA	897
27	Pyong Taek	ATS	Korea, South	444
27	Banco Ortiz	YTB	Uruguay	921
D 27	Pará	FFHM	Brazil	66
G 27	Marajo	AOR	Brazil	77
P 27	Sebo	PBO	Ghana	284
P 27	Simeoforos Xenos	PGGF/PGG	Greece	289
P 27	Izaro	PB	Spain	696
PN 27	Sipa	AOTL	Serbia and Montenegro	669
U 27	Brasil	AXH	Brazil	74
28	Nakhoda Ragam	FSGH	Brunei	78
28	Kwang Yang	ATS	Korea, South	444
FFG 28	Boone	FFGHM	USA	874
G 28	Mattoso Maia	LSTH	Brazil	71
P 28	Achimota	PG	Ghana	283
P 28	Simeoforos Simitzopoulos	PGGF/PGG	Greece	289
P 28	Tabarca	PB	Spain	696
PS 28	Cebu	FS	Philippines	563
T-AE 28	Santa Barbara	AEH	USA	897
29	Bendahara Sakam	FSGH	Brunei	78
29	Jebat	FFGHM	Malaysia	464
29	Lieutenant Dmitrie Nicolescu	MSC	Romania	597
FFG 29	Stephen W Groves	FFGHM	USA	874
M 29	Brecon	MHSC/PP	UK	828
P 29	Yogaga	PG	Ghana	283
P 29	Simeoforos Starakis	PGGF/PGG	Greece	289
PS 29	Negros Occidental	FS	Philippines	563
30	Al Jarim	PB	Bahrain	42
30	Jerambak	FSGH	Brunei	78
30	Lekiu	FFGHM	Malaysia	464
30	Sub Lieutenant Alexandru Axente	MSC	Romania	597
G 30	Ceará	LSDH	Brazil	71
LM 30	Casma	PGG	Chile	110
M 30	Ledbury	MHSC/PP	UK	828
P 30	Roirama	PBR	Brazil	70
P 30	Anzone	PBO	Ghana	283
P 30	Bergantín	PB	Spain	696
Q 30	Al Mabrukah	FS/AXL/AGS	Oman	533
V 30	Inhaúma	FSGH	Brazil	67
31	Drummond	FFG	Argentina	14
31	Al Jasrah	PB	Bahrain	42
31	Iskar	MSC	Bulgaria	83
31	Salamaua	LCM	Papua New Guinea	550
31	Temerario	MSC	Uruguay	919
A 31	Malaspina	AGS	Spain	700
BB 31	Gorgona	AGSC	Colombia	158
CN 31	Rosca Fina	AXL	Brazil	75
F 31	Brahmaputra	FFGHM	India	317
G 31	Rio de Janeiro	LSDH	Brazil	71
H 31	Argus	AGS	Brazil	73
K 31	Visby	FSGH	Sweden	717
LG 31	25 de Julio	WPB	Ecuador	193
LM 31	Chipana	PGG	Chile	110
M 31	Segura	MHC	Spain	699
M 31	Cattistock	MHSC/PP	UK	828
P 31	Rondõnia	PBR	Brazil	70
P 31	Ureca	PB	Equatorial Guinea	208
P 31	Bonsu	PBO	Ghana	283
P 31	Eithne	PSOH	Ireland	359
P 31	Dzükas	PB	Lithuania	459
P 31	Conejera	PB	Spain	696
PS 31	Pangasinan	FS	Philippines	563
Q 31	Qahir Al Amwaj	FSGMH	Oman	532
V 31	Jaceguay	FSGH	Brazil	67
32	Guerrico	FFG	Argentina	14
32	Zibar	MSC	Bulgaria	83
32	Buna	LSM	Papua New Guinea	550
A 32	Tofiño	AGS	Spain	700
CN 32	Voga Picada	AXL	Brazil	75
D 32	Daring	DDGHM	UK	822
F 32	Betwa	FFGHM	India	317
FFG 32	John L Hall	FFGHM	USA	874
K 32	Helsingborg	FSGH	Sweden	717
M 32	Sella	MHC	Spain	699
M 32	Cottesmore	MHSC/PP	UK	828
P 32	Amapá	PBR	Brazil	70
P 32	Selis	PB	Lithuania	459
P 32	Dragonera	PB	Spain	696
PS 32	Iloilo	FS	Philippines	563
Q 32	Al Mua'zzar	FSGMH	Oman	532
T-AE 32	Flint	AEH	USA	897
V 32	Julio de Noronha	FSGH	Brazil	67
Y 32	Moorhen	ARS	UK	840
33	Granville	FFG	Argentina	14
33	Dobrotich	MSC	Bulgaria	83
33	Fortuna	MSC	Uruguay	919
33	Kotor	FFGM	Serbia and Montenegro	666
A 33	Hespérides	AGOBH	Spain	700
AW 33	Lake Bulusan	AWT	Philippines	566
CN 33	Leva Arriba	AXL	Brazil	75
D 33	Dauntless	DDGHM	UK	822
F 33	Beas	FFGHM	India	317
F 33	Infanta Elena	FSGM	Spain	695
FFG 33	Jarrett	FFGHM	USA	874
J 3	Meen	AGS	India	326
K 33	Härnösand	FSGH	Sweden	717
L 33	Teraban	LCU	Brunei	79
M 33	Tambre	MHC	Spain	699
M 33	Viksten	MSI	Sweden	723
M 33	Brocklesby	MHSC/PP	UK	828
P 33	Abhay	FSM	India	320
P 33	Skalvis	PB	Lithuania	459
P 33	Espalmador	PB	Spain	696
RA 33	Miguel Rodriguez	AGS	Venezuela	924
T-AE 33	Shasta	AEH	USA	897
V 33	Frontin	FSGH	Brazil	67
Y 33	Moorfowl	ARS	UK	840
34	Evstati Vinarov	MSC	Bulgaria	83
34	Kris	PB	Malaysia	470
34	Audaz	MSC	Uruguay	919
34	Novi Sad	FFGM	Serbia and Montenegro	666
AW 34	Lake Paoay	AWT	Philippines	566
D 34	Diamond	DDGHM	UK	822
F 34	Himgiri	FFH	India	318
H 34	Almirante Graça Aranha	ABUH	Brazil	73
K 34	Nyköping	FSGH	Sweden	717
J 34	Mithun	AGS	India	326
L 34	Serasa	LCU	Brunei	79
L 34	Vasco Da Gama	LSM	India	325
LM 34	Angamos	PGG	Chile	110
M 34	Turia	MHC	Spain	699
M 34	Middleton	MHSC/PP	UK	828
P 34	Ajay	FSM	India	320
P 34	Alcanada	PB	Spain	696
T-AE 34	Mount Baker	AEH	USA	897
V 34	Barroso	FSGH	Brazil	66
D 35	Defemder	DDGHM	UK	822
F 35	Udaygiri	FFH	India	318
H 35	Amorim Do Valle	AGS	Brazil	73
K 35	Karlstad	FSGH	Sweden	717
L 35	Mk 2/3	LSM	India	325
LG 35	5 de Agosto	WPB	Ecuador	192
M 35	Duero	MHC	Spain	699
M 35	Dulverton	MHSC/PP	UK	828
P 35	Akshay	FSM	India	320
PS 35	Emilio Jacinto	FS	Philippines	562
T-AE 35	Kiska	AEH	USA	897
36	Sundang	PB	Malaysia	470
D 36	Dragon	DDGHM	UK	822
F 36	Dunagiri	FFH	India	318
FFG 36	Underwood	FFGHM	USA	874
H 36	Taurus	AGS	Brazil	73
L 36	Mk 2/3	LSM	India	325
LG 36	27 de Febrero	WPB	Ecuador	192

Number	Ship's name	Type	Country	Page	Number	Ship's name	Type	Country	Page
LM 36	Riquelme	PGG	Chile	110	H 44	Ary Rongel	AGOBH	Brazil	72
M 36	Tajo	MHC	Spain	699	K 44	Nirghat	FSGM	India	323
P 36	Agray	FSM	India	320	LSD 44	Gunston Hall	LSDHM	USA	886
PS 36	Apolinario Mabini	FS	Philippines	562	P 44	Guajará	PBO	Brazil	69
37	Badek	PB	Malaysia	470	P 44	Kirpan	FSGHM	India	322
D 37	Duncan	DDGHM	UK	822	PO 44	Valle Del Cauca	PSOH	Colombia	154
F 37	Beas	FFGHM	India	317	45	Robinson	FFGH	Argentina	13
FFG 37	Crommelin	FFGHM	USA	874	45	Jaradah	LCU	Bahrain	43
H 37	Garnier Sampaio	ABU	Brazil	73	45	Kelewang	PB	Malaysia	470
L 37	Mk 2/3	LSM	India	325	45	Mikhail Kogalniceanu	PGR	Romania	597
LG 37	9 De Octubre	WPBF	Ecuador	192	A 45	Gama	YTB	Pakistan	545
LM 37	Orella	PGG	Chile	110	F 45	União	FFGHM	Brazil	65
M 37	Chiddingfold	MHSC/PP	UK	828	FFG 45	De Wert	FFGHM	USA	874
PS 37	Artemio Ricarte	FS	Philippines	562	K 45	Vibhuti	FSGM	India	323
38	Renchong	PB	Malaysia	470	LSD 45	Comstock	LSDHM	USA	886
FFG 38	Curts	FFGHM	USA	874	P 45	Guaporé	PBO	Brazil	69
L 38	Midhur	LSM	India	325	T-AG 45	Waters	AGM	USA	899
L 38	Galana	LCM	Kenya	428	46	Gomez Roca	FFGH	Argentina	13
LG 38	27 De Octubre	WPBF	Ecuador	192	46	Rentaka	PB	Malaysia	470
LM 38	Serrano	PGG	Chile	110	46	I C Bratianu	PGR	Romania	597
M 38	Atherstone	MHSC/PP	UK	828	AE 46	Cape Bojeador	ABU	Philippines	567
PS 38	General Mariano Alvares	PB	Philippines	563	AP 46	Contre-Almirante Oscar Viel Toro	AGS/AGOBH	Chile	111
39	Tombak	PB	Malaysia	470	F 46	Greenhalgh	FFGHM	Brazil	64
AS 39	Emory S Land	ASH	USA	891	F 46	Krishna	AXH	India	327
FFG 39	Doyle	FFGHM	USA	874	FFG 46	Rentz	FFGHM	USA	874
GS 39	Syzran	AGI	Russian Federation	635	K 46	Vipul	FSGM	India	323
L 39	Mangala	LSM	India	325	LSD 46	Tortuga	LSDHM	USA	886
L 39	Tana	LCM	Kenya	428	P 46	Kuthar	FSGHM	India	322
LM 39	Uribe	PGG	Chile	110	P 46	Gurupá	PBO	Brazil	69
M 39	Hurworth	MHSC/PP	UK	828	47	Sri Perlis	AXL	Malaysia	470
40	Al Zubara	LCU	Bahrain	43	47	Lascar Catargiu	PGR	Romania	597
40	Lembing	AXL	Malaysia	470	A 47	Nasr	AORH	Pakistan	545
A 40	Attock	AOT	Pakistan	545	FFG 47	Nicholas	FFGHM	USA	874
AS 40	Frank Cable	ASH	USA	891	K 47	Vinash	FSGM	India	323
F 40	Niteroi	FFGHM	Brazil	65	LSD 47	Rushmore	LSDHM	USA	886
F 40	Talwar	FFGHM	India	316	P 47	Khanjar	FSGHM	India	322
FFG 40	Halyburton	FFGHM	USA	874	P 47	Gurupi	PBO	Brazil	69
G 40	Atlântico Sud	AKS	Brazil	75	F 48	Bosisio	FFGHM	Brazil	64
H 40	Antares	AGS	Brazil	72	FFG 48	Vandegrift	FFGHM	USA	874
K 40	Veer	FSGM	India	323	K 48	Vidyut	FSGM	India	323
P 40	Grajaú	PBO	Brazil	69	LSD 48	Ashland	LSDHM	USA	886
41	Espora	FFGH	Argentina	13	P 48	Guanabara	PBO	Brazil	69
41	Ajeera	YFU	Bahrain	42	SV 48	Behr Paima	AGS/AGOR	Pakistan	544
41	Letyashti	FS	Bulgaria	81	49	Sri Johor	AXL	Malaysia	470
41	Serampang	PB	Malaysia	470	A 49	Gwadar	AOTL	Pakistan	545
A 41	Vetra	AGOR/AX	Lithuania	459	CG 49	Vincennes	CGHM	USA	865
AP 41	Aquiles	APH	Chile	112	F 49	Rademaker	FFGHM	Brazil	64
F 41	Defensora	FFGHM	Brazil	65	FFG 49	Robert G Bradley	FFGHM	USA	874
F 41	Taragiri	FFH	India	318	LSD 49	Harpers Ferry	LSDHM	USA	886
FFG 41	McClusky	FFGHM	USA	874	P 49	Guaruja	PBO	Brazil	69
K 41	Nirbhik	FSGM	India	323	P 49	Khukri	FSGHM	India	322
L 41	Hernán Cortés	LSTH	Spain	697	50	Al Manama	FSGH	Bahrain	41
LSD 41	Whidbey Island	LSDHM	USA	886	50	Kiisla	PB	Finland	216
M 41	Quorn	MHSC	UK	828	A 50	Alster	AGI	Germany	278
P 41	Guaiba	PBO	Brazil	69	ARS 50	Safeguard	ARS	USA	891
P 41	Orla	PSO	Ireland	360	CG 50	Valley Forge	CGHM	USA	865
PO 41	Espartana	PBO	Colombia	154	FFG 50	Taylor	FFGHM	USA	874
42	Rosales	FFGH	Argentina	13	L 50	Tobruk	LSLH	Australia	30
42	Mashtan	LCU	Bahrain	43	LSD 50	Carter Hall	LSDHM	USA	886
42	Bditelni	FS	Bulgaria	81	N 50	Tyr	ML	Norway	529
42	Merino	AGP/ASH	Chile	112	P 50	Guaratuba	PBO	Brazil	69
42	Panah	PB	Malaysia	470	P 50	Hesperos	PTF	Greece	288
F 42	Constituição	FFGHM	Brazil	65	P 50	Sukanya	PSOH	India	324
F 42	Vindhyagiri	FFH	India	318	51	Al Muharraq	FSGH	Bahrain	41
FFG 42	Klakring	FFGHM	USA	874	51	Kurki	PB	Finland	216
K 42	Nipat	FSGM	India	323	A 51	Mahón	ATA	Spain	703
L 42	Pizarro	LSTH	Spain	697	ARS 51	Grasp	ARS	USA	891
LSD 42	Germantown	LSDHM	USA	886	CG 51	Thomas S Gates	CGHM	USA	865
P 42	Graúna	PBO	Brazil	69	D 51	Rajput	DDGHM	India	314
P 42	Ciara	PSO	Ireland	360	DDG 51	Arleigh Burke	DDGHM	USA	868
PO 42	Capitán Pablo José De Porto	PBO	Colombia	154	FFG 51	Gary	FFGHM	USA	874
43	Spiro	FFGH	Argentina	13	FL 51	Almirante Padilla	FLGHM	Colombia	152
43	Rubodh	LCU	Bahrain	43	FM 51	Carvajal	FFGHM	Peru	556
43	Bezstrashni	FS	Bulgaria	81	L 51	Kanimbla	LCCH/LLP	Australia	29
43	Esmeralda	AXS	Chile	112	L 51	Galicia	LPD	Spain	698
43	Kerambit	AXL	Malaysia	470	L 51	Al Feyi	LCU	UAE	807
F 43	Liberal	FFGHM	Brazil	65	LSD 51	Oak Hill	LSDHM	USA	886
F 43	Trishul	FFGHM	India	316	M 51	Kuršis	MHC	Lithuania	459
FFG 43	Thach	FFGHM	USA	874	MHC 51	Osprey	MHC	USA	889
K 43	Nishank	FSGM	India	323	P 51	Gravataí	PBO	Brazil	69
LSD 43	Fort McHenry	LSDHM	USA	886	P 51	Subhadra	PSOH	India	324
P 43	Goiana	PBO	Brazil	69	P 51	Roisin	PSO	Ireland	359
PO 43	Capitán Jorge Enrique Marquez Duran	PBO	Colombia	154	P 51	Marine Protector class	PB	Malta	475
44	Parker	FFGH	Argentina	13	PL 51	Protector class	WPB	Hong Kong	304
44	Suwad	LCU	Bahrain	43	T-AGS 51	John McDonnell	AGS	USA	900
44	Khrabri	FS	Bulgaria	81	052	Vorovsky	FFHM	Russian Federation	64
44	Beladau	AXL	Malaysia	470	A 52	Oste	AGI	Germany	278
A 44	Bholu	YTB	Pakistan	545	A 52	Las Palmas	AGOB	Spain	700
F 44	Independencia	FFGHM	Brazil	65	ARS 52	Salvor	ARS	USA	891
F 44	Tabar	FFGHM	India	316	B 52	Hercules	DDGHM	Argentina	11
					CG 52	Bunker Hill	CGHM	USA	865
					D 52	Rana	DDGHM	India	314

Number	Ship's name	Type	Country	Page	Number	Ship's name	Type	Country	Page
DDG 52	Barry	DDGHM	USA	868	T-AGS 60	Pathfinder	AGS	USA	900
FFG 52	Carr	FFGHM	USA	874	61	Briz	MSC	Bulgaria	83
FL 52	Caldas	FLGHM	Colombia	152	61	Novigrad	PCM	Croatia	165
FM 52	Villavisencio	FFGHM	Peru	556	61	Turku	PTGM	Finland	216
L 52	Manoora	LCCH/LLP	Australia	29	61	Ashdod	LCT	Israel	366
L 52	Castilla	LPD	Spain	698	CG 61	Monterey	CGHM	USA	865
L 52	Dayyinah	LCU	UAE	807	D 61	Delhi	DDGHM	India	312
LSD 52	Pearl Harbor	LSDHM	USA	886	DDG 61	Ramage	DDGHM	USA	868
M 52	Sūduvis	MHC	Lithuania	459	FFG 61	Ingraham	FFGHM	USA	874
MHC 52	Heron	MHC	USA	889	M 61	Evniki	MHC/MSC	Greece	292
P 52	Marine Protector class	PB	Malta	475	M 61	Pondicherry	MSO	India	325
P 52	Suvarna	PSOH	India	324	MHC 61	Raven	MHC	USA	889
P 52	Niamh	PSO	Ireland	359	P 61	Baradero	PB	Argentina	16
PL 52	Protector class	WPB	Hong Kong	304	P 61	Diciotti class	PBO	Malta	476
A 53	Oker	AGI	Germany	278	P 61	Nassau	PBO	Bahamas	39
A 53	Matanga	ATA/ATR	India	329	P 61	Benevente	PBO	Brazil	70
A 53	La Graña	ATA	Spain	73	P 61	Polemistis	PGG	Greece	290
A 53	Virsaitis	MCCS/AG	Latvia	452	P 61	Kora	FSGHM	India	321
AO 53	Araucano	AOR	Chile	112	P 61	Chilreu	PSO	Spain	696
ARS 53	Grapple	ARS	USA	891	PG 61	Agusan	PB	Philippines	568
CG 53	Mobile Bay	CGHM	USA	865	Q 61	Ciudad De Zarate	ABU	Argentina	18
D 53	Ranjit	DDGHM	India	314	T 61	Trinkat	PBO	India	324
DDG 53	John Paul Jones	DDGHM	USA	868	T 61	Capana	LSTH	Venezuela	924
FFG 53	Hawes	FFGHM	USA	874	T-AGS 61	Sumner	AGS	USA	900
FL 53	Antioquia	FLGHM	Colombia	152	TR 61	Hualcopo	LST	Ecuador	190
FM 53	Montero	FFGHM	Peru	556	62	Shkval	MSC	Bulgaria	83
L 53	Jananah	LCU	UAE	807	62	Solta	PCM	Croatia	165
MHC 53	Pelican	MHC	USA	889	62	Oulu	PTGM	Finland	216
P 53	Kyklon	PTF	Greece	288	CG 62	Chancellorsville	CGHM	USA	865
P 53	Savitri	PSOH	India	324	D 62	Mumbai	DDGHM	India	312
PL 53	Protector class	WPB	Hong Kong	304	DDG 62	Fitzgerald	DDGHM	USA	868
A 54	Amba	ASH	India	327	M 62	Evropi	MHSC	Greece	292
CG 54	Antietam	CGHM	USA	865	M 62	Porbandar	MSO	India	325
D 54	Ranvir	DDGHM	India	314	MHC 62	Shrike	MHC	USA	889
DDG 54	Curtis Wilbur	DDGHM	USA	868	P 62	Barranqueras	PB	Argentina	16
FFG 54	Ford	FFGHM	USA	874	P 62	Bocaina	PBO	Brazil	70
FL 54	Independiente	FLGHM	Colombia	152	P 62	Niki	FS	Greece	288
FM 54	Mariategui	FFGHM	Peru	556	P 62	Kirch	FSGHM	India	321
MHC 54	Robin	MHC	USA	889	P 62	Alboran	PSOH	Spain	696
P 54	Lelaps	PTF	Greece	288	PG 62	Catanduanes	PB	Philippines	568
PL 54	Protector class	WPB	Hong Kong	304	Q 62	Ciudad De Rosario	ABU	Argentina	18
CG 55	Leyte Gulf	CGHM	USA	865	T 62	Tillan Chang	PBO	India	324
D 55	Ranvijay	DDGHM	India	314	T 62	Esequibo	LSTH	Venezuela	924
DDG 55	Stout	DDGHM	USA	868	T-AGS 62	Bowditch	AGS	USA	900
FFG 55	Elrod	FFGHM	USA	874	TR 62	Calicuchima	AETL	Ecuador	191
FM 55	Aguirre	FFGHM	Peru	556	63	Priboy	MSC	Bulgaria	83
FV 55	Indaw	PBO	Myanmar	499	63	Cavtat	PCM	Croatia	165
MHC 55	Oriole	MHC	USA	889	63	Kotka	PTGM	Finland	216
P 55	Sharada	PSOH	India	324	063	Admiral Kuznetsov	CVGM	Russian Federation	615
PL 55	Protector class	WPB	Hong Kong	304	BRS 63	George Slight Marshall	ABU	Chile	111
56	Kajava	AX	Finland	219	CG 63	Cowpens	CGHM	USA	865
CG 56	San Jacinto	CGHM	USA	865	CV 63	Kitty Hawk	CVM	USA	862
DDG 56	John S McCain	DDGHM	USA	868	DDG 63	Stethem	DDGHM	USA	868
FFG 56	Simpson	FFGHM	USA	874	M 63	Kallisto	MHSC	Greece	292
FM 56	Palacios	FFGHM	Peru	556	M 63	Bedi	MSO	India	325
MHC 56	Kingfisher	MHC	USA	889	P 63	Clorinda	PB	Argentina	16
P 56	Tyfon	PTF	Greece	288	P 63	Babitonga	PBO	Brazil	70
P 56	Sujata	PSOH	India	324	P 63	Doxa	FS	Greece	288
PL 56	Protector class	WPB	Hong Kong	304	P 63	Kulish	FSGHM	India	321
57	Lokki	AX	Finland	219	P 63	Arnomendi	PSOH	Spain	696
57	Chun Jee	AORH	Korea, South	445	PG 63	Romblon	PB	Philippines	568
A 57	Shakti	AORH	India	328	Q63	Punta Alta	ABU	Argentina	18
CG 57	Lake Champlain	CGHM	USA	865	T 63	Tarasa	PBO	India	324
DDG 57	Mitscher	DDGHM	USA	868	T 63	Goajira	LSTH	Venezuela	924
FFG 57	Reuben James	FFGHM	USA	874	T-AGS 63	Henson	AGS	USA	900
FV 57	Inya	PBO	Myanmar	499	TR 63	Atahualpa	AWT	Ecuador	191
MHC 57	Cormorant	MHC	USA	889	64	Shtorm	MSC	Bulgaria	83
P 57	Pyrpolitis	PGG	Greece	290	64	Hrvatska Kostajnica	PCM	Croatia	165
58	Dae Chung	AORH	Korea, South	445	CG 64	Gettysburg	CGHM	USA	865
A 58	Jyoti	AORH	India	327	DDG 64	Carney	DDGHM	USA	868
CG 58	Philippine Sea	CGHM	USA	865	M 64	Bhavnagar	MSO	India	325
DDG 58	Laboon	DDGHM	USA	868	P 64	Concepción del Uruguay	PB	Argentina	16
FFG 58	Samuel B Roberts	FFGHM	USA	874					
MHC 58	Black Hawk	MHC	USA	889	P 64	Eleftheria	FS	Greece	288
59	Hwa Chun	AORH	Korea, South	445	P 64	Karmukh	FSGHM	India	321
A 59	Aditya	AORH/AS	India	328	P 64	Tarifa	PSOH	Spain	696
CG 59	Princeton	CGHM	USA	865	PG 64	Palawan	PB	Philippines	568
DDG 59	Russell	DDGHM	USA	868	T 64	Tarmugli	PBO	India	324
FFG 59	Kauffman	FFGHM	USA	874	T 64	Los Llanos	LSTH	Venezuela	924
MHC 59	Falcon	MHC	USA	889	T-AGS 64	Bruce C Heezen	AGS	USA	900
60	Vidal Gormaz	AGOR	Chile	111	TR 64	Quisquis	AWT	Ecuador	192
60	Helsinki	PTGM	Finland	216	A 65	Marinero Jarano	AWT	Spain	702
060	Anadyr	FFHM	Russian Federation	647	CG 65	Chosin	CGHM	USA	865
060	Vladimirets	PGK	Russian Federation	630	CVN 65	Enterprise	CVNM	USA	864
A 60	Gorch Fock	AXS	Germany	279	DDG 65	Benfold	DDGHM	USA	868
CG 60	Normandy	CGHM	USA	865	M 65	Alleppey	MSO	India	325
D 60	Mysore	DDGHM	India	312	P 65	Punta Magotes	PB	Argentina	17
DDG 60	Paul Hamilton	DDGHM	USA	868	T 65	Bangaram	PBO	India	324
FFG 60	Rodney M Davis	FFGHM	USA	874	T-AGS 65	Mary Spears	AGS	USA	900
M 60	Erato	MHC/MSC	Greece	292	TR 65	Taurus	AOTL	Ecuador	191
MHC 60	Cardinal	MHC	USA	889	A 66	Condestable Zaragoza	AWT	Spain	702
P 60	Bahamas	PBO	Bahamas	39	ATF 66	Galvarino	ATF	Chile	113
P 60	Bracui	PBO	Brazil	70	CG 66	Hue City	CGHM	USA	865

Number	Ship's name	Type	Country	Page	Number	Ship's name	Type	Country	Page
DDG 66	Gonzalez	DDGHM	USA	868	F 75	Extremadura	FFGM	Spain	692
M 66	Ratnagiri	MSO	India	325	M 75	Vinga	MHSCDM	Sweden	722
P 66	Rio Santiago	PB	Argentina	17	P 75	Plotarchis Maridakis	PGGF	Greece	290
P 66	Agon	FS	Greece	288	P 75	Descubierta	PSOH	Spain	695
T 66	Bitra	PBO	India	324	76	Hang Tuah	FFH/AX	Malaysia	472
ATF 67	Lautaro	ATF	Chile	113	A 76	Giralda	AXS	Spain	703
CG 67	Shiloh	CGHM	USA	865	CVN 76	Ronald Reagan	CVNM	USA	859
CV 67	John F Kennedy	CVM	USA	862	DDG 76	Higgins	DDGHM	USA	868
DDG 67	Cole	DDGHM	USA	868	M 76	Ven	MHSCDM	Sweden	722
M 67	Karwar	MSO	India	325	P 76	Ypoploiarchos Tournas	PGGF	Greece	290
P 67	Roussen	PGGM	Greece	289	P 76	Sea Wolf	PTGFM	Singapore	674
P 67	Mzizi	PB	Tanzania	748	77	Cabrales	PB	Chile	110
68	Formidable	FFGHM	Singapore	673	A 77	Sálvora	AXS	Spain	703
ATF 68	Leucoton	AFL/ATF	Chile	113	CVN 77	George H W Bush	CVNM	USA	859
CG 68	Anzio	CGHM	USA	865	DDG 77	O'Kane	DDGHM	USA	868
CVN 68	Nimitz	CVNM	USA	859	F 77	Te Kaha	FFHM	New Zealand	517
DDG 68	The Sullivans	DDGHM	USA	868	M 77	Ulvön	MHSCDM	Sweden	722
M 68	Cannanore	MSO	India	325	P 77	Plotarchis Sakipis	PGGF	Greece	290
P 68	Daniolos	PGGM	Greece	289	P 77	Sea Lion	PTGFM	Singapore	674
P 68	Mzia	PB	Tanzania	748	P 77	Infanta Cristina	PSOH	Spain	695
69	Intrepid	FFGHM	Singapore	673	78	Sibbald	PB	Chile	110
CG 69	Vicksburg	CGHM	USA	865	AF 78	Lake Buhi	YO	Philippines	566
CVN 69	Dwight D Eisenhower	CVNM	USA	859	CVN 78	Future Carrier	CVN	USA	858
DDG 69	Milius	DDGHM	USA	868	DDG 78	Porter	DDGHM	USA	868
M 69	Cuddalore	MSO	India	325	F 78	Kent	FFGHM	UK	824
P 69	Kristallidis	PGGM	Greece	289	P 78	Sea Dragon	PTGFM	Singapore	674
70	Rauma	PTGM	Finland	215	P 78	Cazadora	PSOH	Spain	695
70	Steadfast	FFGHM	Singapore	673	AE 79	Limasawa	ABU	Philippines	567
CG 70	Lake Erie	CGHM	USA	865	CVN 79	Future Carrier	CVN	USA	858
CVN 70	Carl Vinson	CVNM	USA	859	DDG 79	Oscar Austin	DDGHM	USA	870
DDG 70	Hopper	DDGHM	USA	868	F 79	Portland	FFGHM	UK	824
M 70	Kakinada	MSO	India	325	P 79	Sea Tiger	PTGFM	Singapore	674
PS 70	Quezon	FS	Philippines	563	P 79	Vencedora	PSOH	Spain	695
RA 70	Chimborazo	ATF	Ecuador	192	80	Hamina	PTGM	Finland	216
71	Micalvi	AEM	Chile	110	DDG 80	Roosevelt	DDGHM	USA	870
71	Raahe	PTGM	Finland	215	F 80	Grafton	FFGHM	UK	824
71	Alvand	FFG	Iran	350	P 80	Sea Hawk	PTGFM	Singapore	674
71	Tenacious	FFGHM	Singapore	673	081	Nikolay Vilkov	LSTM	Russian Federation	633
A 71	Juan Sebastian de Elcano	AXS	Spain	703	81	Zhenghe	AXH	China	141
					81	Cetina	LCT/ML	Croatia	166
AT 71	Mangyan	ABU	Philippines	567	81	Tornio	PTGM	Finland	216
CG 71	Cape St George	CGHM	USA	865	81	Bayandor	FS	Iran	349
CVN 71	Theodore Roosevelt	CVNM	USA	859	A 81	Brambleleaf	AOR	UK	836
DDG 71	Ross	DDGHM	USA	868	CLM 81	Almirante Grau	CG/CLM	Peru	555
M 71	Kozhikode	MSO	India	325	DDG 81	Winston Churchill	DDGHM	USA	870
M 71	Landsort	MHSCDM	Sweden	722	F 81	Santa María	FFGHM	Spain	691
P 71	Serviola	PSOH	Spain	694	F 81	Sutherland	FFGHM	UK	824
T 71	Margarita	LCU	Venezuela	924	P 81	Sea Scorpion	PTGFM	Singapore	674
72	Ortiz	AEM	Chile	110	P 81	Toralla	PB	Spain	697
72	Andrija Mohorovicic	AX	Croatia	166	T 81	Ciudad Bolivar	AORH	Venezuela	924
72	Porvoo	PTGM	Finland	215	82	Huon	MHC	Australia	31
72	Alborz	FFG	Iran	350	82	Shichang	HSS/AHH	China	141
72	Stalwart	FFGHM	Singapore	673	82	Krka	LCT/ML	Croatia	166
A 72	Arosa	AXS	Spain	703	82	Naghdi	FS	Iran	349
AF 72	Lake Taal	YO	Philippines	566	82	Resilience	PGM	Singapore	674
CG 72	Vella Gulf	CGHM	USA	865	DDG 82	Lassen	DDGHM	USA	870
CVN 72	Abraham Lincoln	CVNM	USA	859	F 82	Victoria	FFGHM	Spain	691
DDG 72	Mahan	DDGHM	USA	868	F 82	Somerset	FFGHM	UK	824
F 72	Andalucia	FFGM	Spain	692	P 82	Formentor	PB	Spain	697
M 72	Konkan	MSO	India	325	83	Armidale	PB	Australia	31
M 72	Arholma	MHSCDM	Sweden	722	83	Hawkesbury	MHC	Australia	31
P 72	Ypoploiarchos Votsis	PGGF	Greece	290	83	Unity	PGM	Singapore	674
P 72	Centinela	PSOH	Spain	694	DDG 83	Howard	DDGHM	USA	870
T 72	La Orchila	LCU	Venezuela	924	F 83	Numancia	FFGHM	Spain	691
73	Isaza	PB	Chile	110	F 83	St Albans	FFGHM	UK	824
73	Faust Vrancic	ASR	Croatia	167	K 83	Nashak	FSGM	India	323
73	Naantali	PTGM	Finland	215	M 83	Mahé	MSI	India	326
73	Sabalan	FFG	Iran	350	84	Norman	MHC	Australia	31
73	Supreme	FFGHM	Singapore	673	84	Sovereignty	PGM	Singapore	674
CG 73	Port Royal	CGHM	USA	865	DDG 84	Bulkeley	DDGHM	USA	870
CVN 73	George Washington	CVNM	USA	859	F 84	Enymiri	FSM	Nigeria	521
DDG 73	Decatur	DDGHM	USA	868	F 84	Reina Sofía	FFGHM	Spain	691
M 73	Koster	MHSCDM	Sweden	722	85	Gascoyne	MHC	Australia	31
P 73	Anthypoploiarchos Pezopoulos	PGGF	Greece	290	85	Justice	PGM	Singapore	674
					DDG 85	McCampbell	DDGHM	USA	870
P 73	Vigía	PSOH	Spain	694	F 85	Navarra	FFGHM	Spain	691
74	Morel	PB	Chile	110	F 85	Cumberland	FFGHM	UK	823
A 74	Sagardhwani	AGORH	India	326	P 85	Intrepida	PGGF	Argentina	17
A 74	La Graciosa	AXS	Spain	703	86	Diamantina	MHC	Australia	31
CVN 74	John C Stennis	CVNM	USA	859	86	Freedom	PGM	Singapore	674
DDG 74	McFaul	DDGHM	USA	868	A 86	Tir	AXH	India	326
DM 74	Ferré	DDGH	Peru	555	DDG 86	Shoup	DDGHM	USA	870
F 74	Asturias	FFGM	Spain	692	F 86	Canarias	FFGHM	Spain	691
M 74	Kullen	MHSCDM	Sweden	722	F 86	Campbeltown	FFGHM	UK	823
P 74	Plotarchis Vlahavas	PGGF	Greece	290	LT 86	Zamboanga del Sur	LST	Philippines	565
P 74	Atalaya	PSOH	Spain	694	M 86	Malpe	MSI	India	326
PS 74	Rizal	FS	Philippines	563	P 86	Indomita	PGGF	Argentina	17
75	Grigore Antipa	AGOR/AGI	Romania	598	87	Yarra	MHC	Australia	31
A 75	Tarangini	AXS	India	327	87	Independence	PGM	Singapore	674
A 75	Sisargas	AXS	Spain	703	DDG 87	Mason	DDGHM	USA	870
AU 75	Bessang Pass	PB	Philippines	567	F 87	Chatham	FFGHM	UK	823
CVN 75	Harry S Truman	CVNM	USA	859	H 87	Echo	AGS	UK	834
DDG 75	Donald Cook	DDGHM	USA	868	LT 87	South Cotabato	LST	Philippines	565

Number	Ship's name	Type	Country	Page
DDG 88	Preble	DDGHM	USA	870
H 88	Enterprise	AGS	UK	834
P 88	Victory	FSGM	Singapore	672
AG 89	Kalinga	AKLH	Philippines	567
D 89	Exeter	DDGH	UK	820
DDG 89	Mustin	DDGHM	USA	870
F 89	Aradu	FFGHM	Nigeria	521
P 89	Valour	FSGM	Singapore	672
90	Sabha	FFGHM	Bahrain	40
90	Elicura	LSM	Chile	112
A 90	Varonis	AKS/AXL	Latvia	452
AC 90	Mactan	AK	Philippines	566
D 90	Southampton	DDGH	UK	820
DDG 90	Chaffee	DDGHM	USA	870
P 90	Vigilance	FSGM	Singapore	672
ASY 91	Hashidate	ASY/YAC	Japan	412
BE 91	Guayas	AXS	Ecuador	191
BI 91	Orion	YGS	Ecuador	191
D 91	Nottingham	DDGH	UK	820
DDG 91	Pinckney	DDGHM	USA	870
K 91	Pralaya	FSGM	India	323
M 91	Sagar	MSO	Bangladesh	50
P 91	Valiant	FSGM	Singapore	672
PO 91	Lubin	AKR	Serbia and Montenegro	669
R 91	Charles de Gaulle	CVNM/PAN	France	230
92	Rancagua	LSTH	Chile	112
92	Putsaari	ANL	Finland	221
D 92	Liverpool	DDGH	UK	820
DDG 92	Momsen	DDGHM	USA	870
K 92	Prabal	FSGM	India	323
P 92	Vigour	FSGM	Singapore	672
93	Valdivia	LSTH	Chile	111
DDG 93	Chung-Hoon	DDGHM	USA	87
P 93	Vengeance	FSGM	Singapore	672
94	Orompello	LSM	Chile	112
94	Fearless	PCM	Singapore	674
DDG 94	Nitze	DDGHM	USA	870
95	Chacabuco	LSTH	Chile	112
95	Brave	PCM	Singapore	674
95-1	Astronauta Franklin Chang	PB	Costa Rica	161
D 95	Manchester	DDGH	UK	821
DDG 95	James E Williams	DDGHM	USA	87
M 95	Shapla	MSO/PBO	Bangladesh	50
D 96	Gloucester	DDGH	UK	821
DDG 96	Bainbridge	DDGHM	USA	870
M 96	Saikat	MSO/PBO	Bangladesh	50
097	Dzerzhinsky	FFHM	Russian Federation	647
97	Gallant	PCM	Singapore	674
D 97	Edinburgh	DDGH	UK	821
DDG 97	Halsey	DDGHM	USA	870
M 97	Surovi	MSO/PBO	Bangladesh	50
R 97	Jeanne d'Arc	CVHG	France	229
98	Mursu	AKSL	Finland	219
98	Daring	PCM	Singapore	674
D 98	York	DDGH	UK	821
DDG 98	Forrest Sherman	DDGHM	USA	870
K 98	Prahar	FSGM	India	323
M 98	Shaibal	MHSC/AGS	Bangladesh	50
099	Pyotr Velikiy	CGHMN	Russian Federation	616
99	Kustaanmiekka	AGF/AGI	Finland	219
99	Dauntless	PCM	Singapore	674
DDG 99	Farragut	DDGHM	USA	870
F 99	Cornwall	FFGHM	UK	823
100	Stavropol	PGG	Russian Federation	250
AU 100	Tirad Pass	PB	Philippines	567
DDG 100	Kidd	DDGHM	USA	870
101	Mulniya	FSGM	Bulgaria	81
101	Levuka	PB	Fiji	215
101	Fouque	LSL	Iran	355
101	Al Hussein	PB	Jordan	425
A 101	Mar Caribe	ATF	Spain	702
DD 101	Murasame	DDGHM	Japan	401
DG 101	Gridley	DDGHM	USA	870
F 101	Alvaro de Bazán	FFGHM	Spain	690
FNH 101	Guaymuras	PB	Honduras	301
P 101	Oecussi	PB	East Timor	186
P 101	Kagitingan	PB	Philippines	564
102	Uragon	PTFG	Bulgaria	82
102	Lautoka	PB	Fiji	215
102	Akhmeta	PB	Georgia	260
102	Al Hassan	PB	Jordan	425
102	Kaliningrad	LSTM	Russian Federation	632
D 102	Netzahualcoyotl	DDH	Mexico	481
DD 102	Harusame	DDGHM	Japan	401
DDG 102	Sampson	DDGHM	USA	870
F 102	Almirante Don Juan De Borbón	FFGHM	Spain	690
FNH 102	Honduras	PB	Honduras	301
GC 102	Betelgeuse	PB	Dominican Republic	184
P 102	Atauro	PB	East Timor	186
PG 102	Bagong Lakas	PB	Philippines	564
PM 102	Rafael del Castillo y Rada	PB	Colombia	155
PO 102	Juan De La Barrera	PG/PGH	Mexico	486
103	Burya	PTFG	Bulgaria	82
103	Al Maks	ATA	Egypt	203
103	King Abdullah	PB	Jordan	425
103	Don	PBO	Russian Federation	648
103	Karelia	PBO	Russian Federation	648
103	Kedrov	FFHM	Russian Federation	647
DD 103	Yuudachi	DDGHM	Japan	401
DDG 103	Truxtun	DDGHM	USA	870
F 103	Blas de Lezo	FFGHM	Spain	690
FNH 103	Hibueras	PB	Honduras	301
H 103	Guama	ABU	Cuba	169
P 103	L'Audacieux	PBO	Cameroon	86
PB 103	Alimamy Rassin	PB	Sierra Leone	671
PM 103	José Maria Palas	PB	Colombia	155
PVL 103	Pikker	PB	Estonia	212
PO 103	Mariano Escobedo	PGH/PG	Mexico	486
104	Grum	PTFG	Bulgaria	82
104	Pskov	FFHM	Russian Federation	647
DD 104	Kirisame	DDGHM	Japan	401
DDG 104	Sterett	DDGHM	USA	870
F 104	Mendez Nuñez	FFGHM	Spain	690
FNH 104	Tegucigalpa	PB	Honduras	302
M 104	Walney	MHC/SRMH	UK	829
P 104	Bakassi	PBO	Cameroon	86
PG 104	Bagong Silang	PB	Philippines	564
PM 104	Medardo Monzon Coronado	PB	Colombia	155
PO 104	Manuel Doblado	PG/PGH	Mexico	486
R 104	Ronald H Brown	AGOR	USA	895
105	Jinan	DDGHM	China	127
105	Al Agami	ATA	Egypt	203
105	Baykal	PBO	Russian Federation	648
105	Ivan Yevteyev	AK	Russian Federation	652
105-1	Isla Del Coco	PB	Costa Rica	161
DD 105	Inazuma	DDGHM	Japan	401
DDG 105	Dewey	DDGHM	USA	870
L 105	Arromanches	RCL	UK	843
M 105	Bedok	MHC	Singapore	675
PM 105	Jaime Gómez Castro	PB	Colombia	155
PVL 105	Torm	PB	Estonia	212
106	Xian	DDGM	China	127
106	Brest	PBO	Russian Federation	648
DD 106	Samidare	DDGHM	Japan	401
M 106	Kallang	MHC	Singapore	675
M 106	Penzance	MHC/SRMH	UK	829
PM 106	Juan Nepomuceno Peña	PB	Colombia	155
PO 106	Santos Degollado	PG/PGH	Mexico	486
PVL 106	Maru	PB	Estonia	211
107	Yinchuan	DDGM	China	127
107	Al Antar	ATA	Egypt	203
DD 107	Ikazuchi	DDGHM	Japan	401
L 107	Andalsnes	RCL	UK	843
M 107	Katong	MHC	Singapore	675
M 107	Pembroke	MHC/SRMH	UK	829
PVL 107	Kou	PBO	Estonia	211
108	Xining	DDGM	China	127
DD 108	Akebono	DDGHM	Japan	401
M 108	Punggol	MHC	Singapore	675
M 108	Grimsby	MHC/SRMH	UK	829
PO 108	Juan N Alvares	PG/PGH	Mexico	486
109	Kaifeng	DDG	China	122
109	Al Dekheila	ATA	Egypt	203
A 109	Valvas	AGF	Estonia	211
A 109	Bayleaf	AOR	UK	836
DD 109	Ariake	DDGHM	Japan	401
L 109	Akyab	RCL	UK	843
M 109	Bangor	MHC/SRMH	UK	829
PO 109	Manuel Gutierrez Zamora	PG/PGH	Mexico	486
PVL 109	Valvas	AGF	Estonia	211
110	Dalian	DDG	China	122
110	Alexander Shabalin	LSTM	Russian Federation	632
A 110	Orangeleaf	AOR	UK	836
DD 110	Takanami	DDGHM	Japan	399
L 110	Aachen	RCL	UK	843
M 110	Ramsey	MHC/SRMH	UK	829
PG 110	Tomas Batilo	PBF	Philippines	564
PO 110	Valentin Gomez Farias	PG/PGH	Mexico	486
111	Svetkavitsa	PTFG	Bulgaria	82
111	Al Iskandarani	ATA	Egypt	203
111	Al Tiyar	MSO	Libya	457
111	Marasesti	FFGH	Romania	594
A 111	Alerta	AGI/AGOR	Spain	700
A 111	Oakleaf	AOR	UK	836
D 111	Comodoro Manuel Azueta	FF/AX	Mexico	481
DD 111	Oonami	DDGHM	Japan	399
F 111	Te Mana	FFHM	New Zealand	517
L 111	Arezzo	RCL	UK	843

Number	Ship's name	Type	Country	Page	Number	Ship's name	Type	Country	Page
M 111	T 43 class	MSO	Albania	3	131	Ibn Marwan	LCT	Libya	457
M 111	Blyth	MHC/SRMH	UK	829	ATC 131	Mollendo	AOR	Peru	559
PG 111	Bonny Serrano	PBF	Philippines	564	DD 131	Setoyuki	DDGHM	Japan	403
VL 111	Vapper	PB	Estonia	211	H 131	Scott	AGSH	UK	834
112	Harbin	DDGHM	China	126	PO 131	Capitán de Navio Sebastian José Holzinger	PSOH	Mexico	484
112	Typfoon	PTFG	Bulgaria	82	R 131	Norrköping	PTFG	Sweden	720
DD 112	Makinami	DDGHM	Japan	399	132	Hefei	DDGM	China	127
M 112	T 43 class	MSO	Albania	3	132	El Kobayat	LCT	Libya	457
M 112	Shoreham	MHC/SRMH	UK	829	132	Voron	PCM	Russian Federation	649
PG 112	Bienvenido Salting	PBF	Philippines	564	A 132	Diligence	ARH	UK	837
PM 112	Quitasueno	PGF	Colombia	155	DD 132	Asayuki	DDGHM	Japan	403
PO 112	Francisco Zarco	PG/PGH	Mexico	486	PO 132	Capitán De Navio Blas Godinez	PSOH	Mexico	484
113	Smerch	PTFG	Bulgaria	82	133	Chongoing	DDGM	China	127
113	Qingdao	DDGHM	China	126	L 133	Betano	LCH/LSM	Australia	30
113	Al Isar	MSO	Libya	457	PO 133	Brigadier José Mariá de la Vega	PSOH	Mexico	484
113	Menzhinsky	FFHM	Russian Federation	647	134	Zunyi	DDGM	China	127
DD 113	Sazanami	DDGHM	Japan	399	134	Ibn Harissa	LSTH	Libya	456
L 113	Audemer	RCL	UK	843	134	Krechet	PCM	Russia	649
M 113	T 301 class	MSI	Albania	2	F 134	Laksamana Hang Nadim	FSGM	Malaysia	467
P 113	Yarhisar	PBO	Turkey	783	PO 134	General Felipe B Berriozábal	PSOH	Mexico	484
PM 113	José Maria Garcia y Toledo	PB	Colombia	155	F 135	Laksamana Tun Abdul Jamil	FSGM	Malaysia	467
PO 113	Ignacio L Vallarta	PG/PGH	Mexico	486	A 135	Argus	HSS/APCR	UK	838
114	Grumete Perez	YFB	Chile	113	PF 135	Riohacha	PBR	Colombia	154
DD 114	Suzunami	DDGHM	Japan	399	136	Hangzhou	DDGHM	China	123
M 114	T 301class	MSI	Albania	2	136	Berkut	PCM	Russian Federation	649
P 114	Akhisar	PBO	Turkey	783	F 136	Laksamana Muhammad Amin	FSGM	Malaysia	467
PG 114	Salvador Abcede	PBF	Philippines	564	PF 136	Leticia	PBR	Colombia	154
PM 114	Juan Nepomuceno Eslava	PB	Colombia	155	137	Fuzhou	DDGHM	China	123
PO 114	Jesus Gonzalez Ortega	PG/PGH	Mexico	486	137	Derbent	PGG	Russian Federation	650
115	Emil Racovita	AGOR/AGI	Romania	598	F 137	Laksamana Tan Pusmah	FSGM	Malaysia	467
115	Ivan Lednev	AK	Russian Federation	652	PF 137	Aranca	PBR	Colombia	154
PG 115	Ramon Aguirre	PBF	Philippines	564	138	MPK 130	FFLM	Russian Federation	625
PM 115	Tecim Jaime E Cárdenas Gomez	PB	Colombia	155	139	Kizljar	PGG	Russian Federation	650
116	Pisagua	AKSL	Chile	113	A 140	Tornado	YDT/YPT	UK	840
PG 116	Nicolas Mahusay	PBF	Philippines	564	P 140	Rajshahi	PB	Pakistan	544
117	Ras al Fulaijah	MSO	Libya	457	PG 140	Emilo Aguinaldo	PBO	Philippines	564
PO 117	Mariano Matamoros	PG/PGH	Mexico	486	141	Vyborg	PGG	Russian Federation	650
119	Minsk	LSTM	Russian Federation	633	DDH 141	Haruna	DDHM	Japan	404
119	Nikolay Starshinov	AK	Russian Federation	652	DT 141	Paita	LSTH	Peru	558
U 120	Skadovsk	PB	Ukraine	798	P 141	Mubarraz	PGGFM	UAE	806
121	Vahakari	AKSL	Finland	219	PG 141	Antonio Luna	PBO	Philippines	564
121	Moskva	CGHM	Russian Federation	617	PM 141	Cabo Corrientes	PB	Colombia	154
DDK 121	Yuugumo	DDM/DDK	Japan	405	PO 141	Justo Sierra Mendez	PSOH	Mexico	484
PO 121	Cadete Virgilio Uribe	PSOH	Mexico	485	142	Aldan	PBO	Russian Federation	648
DD 122	Hatsuyuki	DDGHM	Japan	403	142	Novocherkassk	LSTM	Russian Federation	632
PO 122	Teniente José Azueta	PSOH	Mexico	485	A 142	Tormentor	YDT/YPT	UK	840
123	Ras al Massad	MSO	Libya	457	DDH 142	Hiei	DDHM	Japan	404
ARB 123	Guardian Rios	ATS	Peru	560	DT 142	Pisco	LSTH	Peru	558
DD 123	Shirayuki	DDGHM	Japan	403	P 142	Makasib	PGGFM	UAE	806
NL 123	Sarucabey	LSTH/ML	Turkey	786	PM 142	Cabo Manglares	PB	Colombia	154
PO 123	Capitan de Fragata Pedro Sáinz de Baranda	PSOH	Mexico	485	R 142	Ystad	PTFG	Sweden	720
124	Tarantul class	FSGM	Yemen	933	143	Almaz	PGG	Russian Federation	650
DD 124	Mineyuki	DDGHM	Japan	403	143	Sergey Sudetsky	AK	Russian Federation	652
NL 124	Karamürselbey	LSTH/ML	Turkey	786	DT 143	Callao	LSTH	Peru	558
PO 124	Comodoro Carlos Castillo Bretón	PSOH	Mexico	485	DDH 143	Shirane	DDHM	Japan	400
125	Ras Al Hani	MSO	Libya	457	PM 143	Cabo Tiburon	PB	Colombia	154
125	BDK 105	LSTM	Russian Federation	632	PO 143	Guillermo Prieto	PSOH	Mexico	484
DD 125	Sawayuki	DDGHM	Japan	403	DDH 144	Kurama	DDHM	Japan	400
NL 125	Osman Gazi	LSTH	Turkey	785	DT 144	Eten	LSTH	Peru	558
PO 125	Vicealmirante Othón P Blanco	PSOH	Mexico	485	PM 144	Cabo de la Vella	PB	Colombia	154
126	Huangfen class	PTGF	Yemen	934	PO 144	Matias Romero	PSOH	Mexico	484
DD 126	Hamayuki	DDGHM	Japan	403	F 145	Amatola	FSGHM	South Africa	681
L 126	Balikpapan	LCH/LSM	Australia	30	A 146	Waterman	AWT	UK	840
PO 126	Contralmirante Angel Ortiz Monasterio	PSOH	Mexico	485	F 146	Isandlwana	FSGHM	South Africa	681
127	Minsk	LSTM	Russian Federation	632	F 147	Spioenkop	FSGHM	South Africa	681
127	Huangfen class	PTFG	Yemen	934	148	Orsk	LSTM	Russian Federation	633
DD 127	Isoyuki	DDGHM	Japan	403	F 148	Mendi	FFGHM	South Africa	681
L 127	Brunei	LCH/LSM	Australia	30	149	Kuban	PCM	Russian Federation	648
128	Huangfen class	PTFG	Yemen	934	150	Aisberg	PGH	Russian Federation	648
DD 128	Haruyuki	DDGHM	Japan	403	150	Anzac	FFGHM	Australia	26
L 128	Labuan	LCH/LSM	Australia	30	150	Saratov	LSTM	Russian Federation	633
129	MPK 139	FFLM	Russian Federation	625	151	Arunta	FFGHM	Australia	26
DD 129	Yamayuki	DDGHM	Japan	403	151	Azov	LSTM	Russian Federation	632
L 129	Tarakan	LCH/LSM	Australia	30	P 151	Ban Yas	PGGF	UAE	806
PB 129	Merjen	WPB	Turkmenistan	794	PO 151	Durango	PSOH	Mexico	485
PF 129	Capitán Jaime Rook	PBR	Colombia	156	152	Warramunga	FFGHM	Australia	26
130	Ibn al Idrisi	LCT	Libya	457	152	Mutiara	AGSH	Malaysia	472
130	Kondor	PCM	Russian Federation	649	152	Nikolay Filchenkov	LSTM	Russian Federation	633
130	Korolev	LSTM	Russian Federation	632	152	Kobchik	PCM	Russian Federation	649
DD 130	Matsuyuki	DDGHM	Japan	403	ATP 152	Talara	AOT	Peru	559
H 130	Roebuck	AGS	UK	834	M 152	Podgora	MHSC	Serbia and Montenegro	668
L 130	Wewak	LCH/LSM	Australia	30	P 152	Marban	PGGF	UAE	806
PF 130	Manuela Saenz	PBR	Colombia	156	PO 152	Sonora	PSOH	Mexico	485
U 130	Hetman Sagaidachny	FFHM	Ukraine	796	153	Stuart	FFGHM	Australia	26
131	Nanjing	DDGM	China	127					

Number	Ship's name	Type	Country	Page	Number	Ship's name	Type	Country	Page
153	Perantau	AGS	Malaysia	472	L 173	Chios	LSTH	Greece	291
ATP 153	Lobitos	AOT	Peru	559	174	Terengganu	FSGHM	Malaysia	466
BH 153	Quindio	ABU	Colombia	157	174	Učka	PCM	Serbia and	667
DD 153	Yuugiri	DDGHM	Japan	402				Montenegro	
M 153	Blitvenica	MHSC	Serbia and	668	DDG 174	Kirishima	DDGHM	Japan	398
			Montenegro		L 174	Samos	LSTH	Greece	291
P 153	Rodqm	PGGF	UAE	806	175	Kelantan	FSGHM	Malaysia	466
PO 153	Guanajuato	PSOH	Mexico	485	AEH 175	Carrillo	AGSC/EH	Peru	558
U 153	Priluki	PGGK	Ukraine	798	DDG 175	Myoukou	DDGHM	Japan	398
154	Parramatta	FFGHM	Australia	26	L 175	Ikaria	LSTH	Greece	291
154	Berezina	AORH	Russian Federation	639	SSV 175	Odograf	AGIM	Russian Federation	634
154	Vasiliy Suntzov	AK	Russian Federation	652	176	Selangor	FSGHM	Malaysia	466
DD 154	Amagiri	DDGHM	Japan	402	176	Rahova	PGR	Romania	596
PO 154	Veracruz	PSOH	Mexico	485	176	Vyacheslav Denisov	AK	Russian Federation	652
P 154	Shaheen	PGGF	UAE	806	AH 176	Melo	AGSC/EH	Peru	558
U 154	Kahovka	PGGK	Ukraine	798	DDG 176	Choukai	DDGHM	Japan	398
155	Ballarat	FFGHM	Australia	26	L 176	Lesbos	LSTH	Greece	291
155	MPK 56	FFLM	Russian Federation	625	177	Opanez	PGR	Romania	596
BO 155	Providencia	AGOR	Colombia	157	L 177	Rodos	LSTH	Greece	291
DD 155	Hamagiri	DDGHM	Japan	402	SFP 177	Akademik Isanin	AGS	Russian Federation	637
P 155	Sagar	PGGF	UAE	806	178	Smardan	PGR	Romania	596
U 155	Pridneprovye	FSGM	Ukraine	798	178	Kosmaj	PCM	Serbia and	667
156	Toowoomba	FFGHM	Australia	26				Montenegro	
156	Orel	FFHM	Russian Federation	647	L 178	Naxos	LCU	Greece	292
156	Yamal	LSTM	Russian Federation	632	P 178	Ekpe	PGF	Nigeria	522
BO 156	Malpelo	AGOR	Colombia	157	179	Posada	PG	Romania	596
D 156	Nazim	DD	Pakistan	546	L 179	Paros	LCU	Greece	292
DD 156	Setogiri	DDGHM	Japan	402	P 179	Damisa	PGF	Nigeria	522
P 156	Tarif	PGGF	UAE	806	180	Rovine	PG	Romania	596
157	Perth	FFGHM	Australia	26	L 180	Kefallinia	LCUJ	Greece	291
ATP 157	Supe	AOT	Peru	559	P 180	Agu	PGF	Nigeria	522
DD 157	Sawagiri	DDGHM	Japan	402	D 181	Tariq	FFHM/FFGH	Pakistan	540
P 157	Larkana	PB	Pakistan	544	L 181	Ithaki	LCUJ	Greece	291
158	Tsesar Kunikov	LSTM	Russian Federation	632	PRF 181	Tenerife	PBR	Colombia	156
158	Yung Chuan	MSC	Taiwan	744	182	Korshun	PCM	Russian Federation	649
DD 158	Umigiri	DDGHM	Japan	402	D 182	Babur	FFHM/FFGH	Pakistan	540
160	Musytari	PSOH	Malaysia	469	L 182	Kerkira	LCUJ	Greece	291
BE 160	Gloria	AXS	Colombia	159	P 182	Ayam	PGGF	Nigeria	522
161	Changsha	DDGM	China	127	PRF 182	Tarapaca	PBR	Colombia	156
161	Marikh	PSOH	Malaysia	469	183	Volga	PGH	Russian Federation	648
161	Muray Jib	FSGHM	UAE	805	D 183	Khaibar	FFHM/FFGH	Pakistan	540
BL 161	Cartagena De Indias	AGP	Colombia	158	L 183	Zakynthos	LCUJ	Greece	291
FMB 161	Osa II class	PTFG	Eritrea	208	SFP 183	Akademik Seminikhin	AGS	Russian Federation	637
PO 161	Oaxaca	PSOH	Mexico	485	184	Mikhail Konovalov	AK	Russian Federation	652
162	Nanning	DDGM	China	127	D 184	Badr	FFHM/FFGH	Pakistan	540
162	Yung Fu	MSC	Taiwan	744	185	Sakhalin	PBO	Russian Federation	648
162	Das	FSGHM	UAE	805	A 185	Salmoor	ARSD	UK	839
BL 162	Buenaventura	AGP	Colombia	158	D 185	Tippu Sultan	FFHM/FFGH	Pakistan	540
P 162	Spejaren	PCGF	Sweden	721	D 186	Shahjahan	FFHM/FFGH	Pakistan	540
PO 162	Baja California	PSOH	Mexico	485	A 187	Salmaid	ARSD	UK	839
163	Nanchang	DDGM	China	127	T-AO 187	Henry J Kaiser	AOH	USA	898
M 163	Muhafiz	MHSC	Pakistan	544	188	Zborul	FSG	Romania	596
PO 163	Oaxaca class	PSOH	Mexico	485	T-AO 188	Joshua Humphries	AOH	USA	898
164	Guilin	DDGM	China	127	189	Pescarusul	FSG	Romania	596
164	MPK 7	FFFLM	Russian Federation	625	T-AO 189	John Lenthall	AOH	USA	898
M 164	Mujahid	MHSC	Pakistan	544	190	Lastunul	FSG	Romania	596
PO 164	Oaxaca class	PSOH	Mexico	485	190	MPK 14	FFLM	Russian Federation	625
165	Zhanjiang	DDG	China	122	191	Chung Cheng	LSD	Taiwan	741
166	Zhuhai	DDG	China	122	LSD 193	Shiu Hai	LSDH	Taiwan	742
M 166	Munsif	MHSC	Pakistan	544	T-AO 193	Walter S Diehl	AOH	USA	898
P 166	Tirfing	PCGF	Sweden	721	T-AO 194	John Ericsson	AOH	USA	898
167	Shenzhen	DDGHM	China	125	L 195	Serifos	LCU	Greece	292
167	Yung Ren	MSC	Taiwan	744	O 195	Westralia	AOR/AOT	Australia	34
L 167	Ios	LCU	Greece	292	T-AG 195	Hayes	AGE	USA	899
168	Guangzhou	DDGHM	China	124	T-AO 195	Leroy Grumman	AOH	USA	898
168	Yung Sui	MSC	Taiwan	744	196	MPK 59	FFLM	Russian Federation	625
DDG 168	Tachikaze	DDGM	Japan	404	P 196	Andromeda	PT	Greece	289
L 168	Sikinos	LCU	Greece	292	T-AO 196	Kanawha	AOH	USA	898
T-ATF 168	Catawba	ATF	USA	898	T-AO 197	Pecos	AOH	USA	898
169	Wuhan	DDGHM	China	124	198	Kamchatka	PBO	Russian Federation	648
DDG 169	Asakaze	DDGM	Japan	404	P 198	Kyknos	PT	Greece	289
L 169	Irakleia	LCU	Greece	292	T-AO 198	Big Horn	AOH	USA	898
SSV 169	Tavriya	AGIM	Russian Federation	634	199	MPK 194	FFLM	Russian Federation	625
T-ATF 169	Navajo	ATF	USA	898	P 199	Pigasos	PT	Greece	289
170	Lanzhou	DDGHM	China	125	T-AO 199	Tippecanoe	AOH	USA	898
170	Neva	PGH	Russian Federation	648	200	Yan Lon Aung	YDT	Myanmar	502
DDG 170	Sawakaze	DDGM	Japan	404	200	Perekop	AX	Russian Federation	637
L 170	Folegrandos	LCU	Greece	292	T-AO 200	Guadelupe	AOH	USA	898
T-ATF 170	Mohawk	ATF	USA	898	U 200	Lutsk	FFLM	Ukraine	797
171	Kedah	FSGHM	Malaysia	466	201	Kula	PB	Fiji	214
171	MPK 113	FFLM	Russian Federation	625	201	Iveria	PB	Georgia	260
A 171	Endurance	AGOBH	UK	833	201	Chung Hai	LST	Taiwan	742
AH 171	Carrasco	AGSC/EH	Peru	558	201	Natya class	MSO	Yemen	935
DDG 171	Hatakaze	DDGHM	Japan	400	A 201	Orion	AGIH	Sweden	725
T-ATF 171	Sioux	ATF	USA	898	F 201	Nicolas Bravo	FFH	Mexico	483
172	Quest	AGORH	Canada	93	P 201	Ruposhi Bangla	PB	Bangladesh	49
172	Pahang	FSGHM	Malaysia	466	P 201	Neiafu	PB	Tonga	768
172	Primorye	PBO	Russian Federation	648	SSV 201	Priazove	AGIM	Russian Federation	634
AH 172	Stiglich	AGSC/A	Peru	559	T-AO 201	Patuxent	AOH	USA	898
DDG 172	Shimakaze	DDGHM	Japan	400	UAM 201	Creoula	AXS	Portugal	589
T-ATF 172	Apache	ATF	USA	898	202	Kikau	PB	Fiji	214
173	Perak	FSGHM	Malaysia	466	202	Kutaisi	PB	Georgia	259
DDG 173	Kongou	DDGHM	Japan	398	202	Smeul	PTF	Romania	597

Number	Ship's name	Type	Country	Page	Number	Ship's name	Type	Country	Page
F 202	Hermenegildo Galeana	FFH	Mexico	482	216	Al Ula	YFU	Saudi Arabia	661
P 202	Pangai	PB	Tonga	768	216	Chung Kuang	LST	Taiwan	742
PC 202	Matias De Cordova	PB	Mexico	487	F 216	Schleswig-Holstein	FFGHM	Germany	266
T-AO 202	Yukon	AOH	USA	898	PC 216	Francisco J Mugica	PB	Mexico	487
U 202	Ternopil	FFLM	Ukraine	797	217	Bunbury	PB	Australia	30
203	Fremantle	PB	Australia	30	217	Chung Suo	LST	Taiwan	742
203	Kiro	PB	Fiji	214	F 217	Bayern	FFGHM	Germany	266
203	Mestia	PB	Georgia	260	218	Maryut	AOTL	Egypt	202
LDG 203	Bacamarte	LCU	Portugal	588	218	MPK 224	FFLM	Russian Federation	624
P 203	Savea	PB	Tonga	768	218	Afif	YFU	Saudi Arabia	661
T-AO 203	Laramie	AOH	USA	898	218	Chung Chi	LST	Taiwan	742
204	Warrnambool	PB	Australia	30	F 218	Mecklenburg-Vorpommern	FFGHM	Germany	266
204	Vijelia	PTF	Romania	597					
BH 204	El Idrissi	AGS	Algeria	7	PC 218	Jose Maria Del Castillo Velasco	PB	Mexico	487
T-AO 204	Rappahannock	AOH	USA	898					
205	Townsville	PB	Australia	30	F 219	Sachsen	FFGHM	Germany	270
205	Chung Chien	LST	Taiwan	742	220	Al Nil	AOTL	Egypt	202
U 205	Chernigiv	FFLM	Ukraine	797	F 220	Hamburg	FFGHM	Germany	270
206	Kapitan 1st Rank Dimiter Dobrev	ADG/AX	Bulgaria	85	PC 220	Jose Natividad Macias	PB	Mexico	487
					221	Jupiter	ATS	Bulgaria	85
206	Wollongong	PB	Australia	30	221	Regele Ferdinand	FFHM	Romania	595
PC 206	Ignacio López Rayón	PB	Mexico	487	221	Sobat	AFL	Sudan	712
U 206	Vinnitsa	FFLM	Ukraine	797	221	Chung Chuan	LST	Taiwan	742
207	Launceston	PB	Australia	30	F 221	Hessen	FFGHM	Germany	270
F 207	Bremen	FFGHM	Germany	268	P 221	Kaman	PGGF	Iran	352
L 207	Endurance	LPDM	Singapore	676	222	Regina Maria	FFHM	Romania	595
P 207	Utique	PB	Tunisia	771	222	Dinder	AFL	Sudan	712
PC 207	Manuel Crescencio Rejon	PB	Mexico	487	222	MPK 213	FFLM	Russian Federation	624
					P 222	Zoubin	PGGF	Iran	352
208	Whyalla	PB	Australia	30	P 223	Khadang	PGGF	Iran	352
208	Sevan	AKH/AGF	Russian Federation	638	PC 223	Tamaulipas	PB	Mexico	487
208	Chung Shun	LST	Taiwan	742	224	Al Furat	AOTL	Egypt	202
F 208	Niedersachsen	FFGHM	Germany	268	P 224	Peykan	PGGF	Iran	352
L 208	Resolution	LPDM	Singapore	676	PC 224	Yucatan	PB	Mexico	487
P 208	Separacion	PG	Dominican Republic	183	PC 225	Tabasco	PB	Mexico	487
P 208	Jerba	PB	Tunisia	771	226	Chung Chih	LST	Taiwan	742
PC 208	Juan Antonio De La Fuente	PB	Mexico	487	DE 226	Ishikari	FFG/DE	Japan	405
					P 226	Falakhon	PGGF	Iran	352
SSV 208	Kurily	AGIM	Russian Federation	634	PC 226	Cochimie	PB	Mexico	487
U 208	Khmelnitsky	PCM	Ukraine	798	227	Chung Ming	LST	Taiwan	742
209	Ipswich	PB	Australia	30	DE 227	Yuubari	FFG/DE	Japan	405
209	Vulcanul	PTF	Romania	597	P 227	Shamshir	PGGF	Iran	352
F 209	Rheinland-Pfalz	FFGHM	Germany	268	DE 228	Yuubetsu	FFG/DE	Japan	405
L 209	Persistence	LPDM	Singapore	676	P 228	Toxotis	PT	Greece	289
P 209	Kuriat	PB	Tunisia	771	P 228	Gorz	PGGF	Iran	352
PC 209	Leon Guzman	PB	Mexico	487	PC 228	Puebla	PB	Mexico	487
210	Cessnock	PB	Australia	30	A 229	Colonel Templer	AGOR	UK	840
210	Smolny	AX	Russian Federation	637	DE 229	Abukuma	FFGM	Japan	405
F 210	Emden	FFGHM	Germany	268	F 229	Lancaster	FFGHM	UK	824
F 210	Mussa Ben Nussair	FSG	Iraq	358	P 229	Tolmi	PGM	Greece	290
L 210	Endeavour	LPDM	Singapore	676	P 229	Gardouneh	PGGF	Iran	352
M 210	Thaleia	MHC/MSC	Greece	292	230	Shaladein	ARL	Egypt	202
PC 210	Ignacio Ramirez	PB	Mexico	487	230	Chung Pang	LST	Taiwan	742
211	Bendigo	PB	Australia	30	A 230	Admiral Pitka	AGFH/AGE/FFLH	Estonia	210
211	Parvin	PC	Iran	352					
F 211	Köln	FFGHM	Germany	268	DE 230	Jintsu	FFGM/DE	Japan	406
F 211	Ignacio Allende	FFHM	Mexico	482	P 230	Ormi	PGM	Greece	290
M 211	Alkyon	MSC	Greece	293	P 230	Khanjar	PGGF	Iran	352
P 211	Meghna	PB	Bangladesh	50	PC 230	Leona Vicario	PB	Mexico	487
PC 211	Ignacio Mariscal	PB	Mexico	487	231	Halaib	AEL	Egypt	202
212	Atabarah	AOTL	Egypt	202	231	Chung Yeh	LST	Taiwan	742
212	Gawler	PB	Australia	30	DE 231	Ooyodo	FFGM/DE	Japan	406
212	Bahram	PC	Iran	352	F 231	Argyll	FFGHM	UK	824
212	Yamal	AKH/AGF	Russian Federation	638	P 231	Neyzeh	PGGF	Iran	352
212	Al Qiaq	YFU	Saudi Arabia	661	PC 231	Josefa Ortiz De Dominguez	PB	Mexico	487
A 212	Ägir	YDT/AGF	Sweden	727					
F 212	Karlsruhe	FFGHM	Germany	268	SSV 231	Vassily Tatischev	AGIM	Russian Federation	634
F 212	Tariq Ibn Ziad	FSG	Iraq	358	232	Kalmykia	FFLM	Russian Federation	624
F 212	Al Hani	FFGM	Libya	455	232	Chung Ho	LSTH	Taiwan	742
F 212	Mariano Abasolo	FFH	Mexico	482	DE 232	Sendai	FFGM/DE	Japan	406
P 212	Jamuna	PB	Bangladesh	50	P 232	Tabarzin	PGGF	Iran	352
PC 212	Heriberto Jara Corona	PB	Mexico	487	233	Chung Ping	LSTH	Taiwan	742
213	Geraldton	PB	Australia	30	DE 233	Chikuma	FFGM/DE	Japan	406
213	Nahid	PC	Iran	352	DE 234	Tone	FFGM/DE	Japan	406
A 213	Nordanö	YDT	Sweden	727	F 234	Iron Duke	FFGHM	UK	824
F 213	Augsburg	FFGHM	Germany	268	F 235	Monmouth	FFGHM	UK	824
F 213	Al Qirdabiyah	FFGM	Libya	455	F 236	Montrose	FFGHM	UK	834
F 213	Guadaloupe Victoria	FFH	Mexico	482	F 237	Westminster	FFGHM	UK	834
M 213	Klio	MSC	Greece	293	F 238	Northumberland	FFGHM	UK	824
214	Dubbo	PB	Australia	30	F 239	Richmond	FFGHM	UK	824
214	Akdu	AOTL	Egypt	202	240	Kaszub	FSM	Poland	572
214	Al Sulayel	YFU	Saudi Arabia	661	F 240	Yavuz	FFGHM	Turkey	779
A 214	Belos III	ARSH	Sweden	728	M 240	Aidon	MSC	Greece	293
F 214	Lübeck	FFGHM	Germany	268	DBM 241	Krk	LCT/ML	Serbia and Montenegro	668
F 214	Francisco Javier Mina	FFH	Mexico	482					
M 214	Avra	MSC	Greece	293	F 241	Turgutreis	FFGHM	Turkey	779
PC 214	Colima	PB	Mexico	487	M 241	Kichli	MSC	Greece	293
215	Geelong	PB	Australia	30	PC 241	Démocrata	PBO	Mexico	486
F 215	Brandenburg	FFGHM	Germany	266	F 242	Fatih	FFGHM	Turkey	779
PC 215	Jose Joaquin Fernandez De Lizardi	PB	Mexico	487	M 242	Kissa	MSC	Greece	293
					243	MPK 227	FFLM	Russian Federation	624
216	Gladstone	PB	Australia	30	F 243	Yildirim	FFGHM	Turkey	779
216	Ayeda 3	AOTL	Egypt	202	244	Baskortostan	FFLM	Russian Federation	624

Number	Ship's name	Type	Country	Page	Number	Ship's name	Type	Country	Page
F 244	Barbaros	FFGHM	Turkey	778	P 284	Clyde	PSO	UK	832
245	MPK 105	FFLM	Russian Federation	624	P 284	Scimitar	PBF	UK	833
A 245	Leeuwin	AGS	Australia	32	P 285	Sabre	PBF	UK	833
F 245	Orucreis	FFGHM	Turkey	778	P 286	Diopos Antoniou	PB	Greece	291
A 246	Melville	AGS	Australia	32	P 287	Kelefstis Stamou	PB	Greece	291
F 246	Salihreis	FFGHM	Turkey	778	288	Mircea	AXS	Romania	598
LD 246	Morrosquillo	LCU	Colombia	157	296	Electronica	AGI	Romania	598
A 247	Pelikanan	YPT	Sweden	728	298	Magnetica	ADG	Romania	598
F 247	Kemalreis	FFGHM	Turkey	778	Y 298	Bandicoot	MSCD/YTB	Australia	31
M 247	Dafni	MSC	Greece	293	Y 299	Wallaroo	MSCD/YTB	Australia	31
A 248	Pingvinen	YPT	Sweden	728	Y 300	Barsø	PB	Denmark	177
LD 248	Bahí Honda	LCU	Colombia	157	301	Polnochny A class	LSM	Egypt	200
M 248	Pleias	MSC	Greece	293	301	Denden	LST	Eritrea	209
LD 249	Bahí Portete	LCU	Colombia	157	301	Batumi	PBF	Georgia	259
F 250	Muavenet	FFGH	Turkey	781	301	Hamzeh	YDT	Iran	356
251	Wodnik	AXTH	Poland	577	301	Teanoai	PB	Kiribati	428
A 251	Achilles	ATA	Sweden	729	301	MPK 67	FFLM	Russian Federation	624
LD 251	Bahí Solano	LCU	Colombia	157	A 301	Drakensberg	AORH	South Africa	684
LD 252	Bahí Cupica	LCU	Colombia	157	F 301	Bergen	FFGM	Norway	525
253	Iskra	AXS	Poland	578	MSO 301	Yaeyama	MSO	Japan	409
A 253	Hermes	YTM	Sweden	729	P 301	Inttisar	PB	Kuwait	449
C 253	Stalwart	PB	St Kitts and Nevis	653	P 301	Panquiaco	PB	Panama	547
F 253	Zafer	FFGH	Turkey	781	P 301	Bizerte	PBOM	Tunisia	771
LD 253	Bahí Utria	LCU	Colombia	157	P 301	Kozlu	PBO	Turkey	784
LD 254	Bahí Málaga	LCU	Colombia	157	Y 301	Drejø	PB	Denmark	177
F 255	Karadeniz	FFGH	Turkey	781	302	Atiya	AORL	Bulgaria	84
F 256	Ege	FFGH	Turkey	781	302	Tbilisi	PGGK	Georgia	259
258	MPK 216	FFLM	Russian Federation	624	302	Okba	PG	Morocco	484
P 258	Leeds Castle	PSOH	UK	833	ABH 302	Morona	AGSC/AH	Peru	559
260	Admiral Petre Barbuneanu	FS	Romania	595	F 302	Trondheim	FFGM	Norway	525
F 260	Braunschweig	FSGHM	Germany	272	MSO 302	Tsushima	MSO	Japan	409
M 260	Edincik	MHC	Turkey	784	P 302	Aman	PB	Kuwait	449
261	Kopernik	AGS	Poland	577	P 302	Ligia Elena	PB	Panama	547
F 261	Magdeburg	FSGHM	Germany	272	P 302	Horria	PBOM	Tunisia	771
GC 261	El Mourafik	ARL	Algeria	7	P 302	Kuşadasi	PBO	Turkey	784
M 261	Edremit	MHC	Turkey	784	Y 302	Romsø	PB	Denmark	177
262	Nawigator	AGI	Poland	577	303	Polnochny A class	LSM	Egypt	200
A 262	Skredsvik	YDT/AGFH	Sweden	727	303	Saku	PB	Fiji	214
F 262	Erfurt	FSGHM	Germany	272	303	Triki	PG	Morocco	494
F 262	Zulfiquar	FFH	Pakistan	542	MSO 303	Hachijyo	MSO	Japan	409
M 262	Enez	MHC	Turkey	784	P 303	Maimon	PB	Kuwait	449
263	Hydrograf	AGI	Poland	577	P 303	Naos	PB	Panama	547
263	Vice Admiral Eugeniu Rosca	FS	Romania	595	W 303	Svalbard	WPSOH	Norway	531
A 263	Galö	MCS	Sweden	727	Y 303	Samsø	PB	Denmark	177
F 263	Oldenburg	FSGHM	Germany	272	304	Saqa	PB	Fiji	214
M 263	Erdek	MHC	Turkey	784	304	El Khattabi	PGG	Morocco	494
264	Contre Admiral Eustatiu Sebastian	FSH	Romania	596	304	MPK 192	FFLM	Russian Federation	624
A 264	Trossö	AGP	Sweden	727	F 304	Narvik	FFGM	Norway	525
F264	Ludwigshafen	FSGHM	Germany	272	OR 304	Success	AORH	Australia	35
M 264	Erdemli	MHC	Turkey	784	P 304	Mobark	PB	Kuwait	449
265	Heweliusz	AGS	Poland	576	P 304	Monastir	PBOM	Tunisia	771
265	Admiral Horia Macellariu	FSH	Romania	596	Y 304	Thurø	PB	Denmark	177
A 265	Visborg	AKH	Sweden	727	305	Polnochny A class	LSM	Egypt	200
M 265	Alanya	MHSC	Turkey	784	305	Commandant Boutouba	PGG	Morocco	494
P 265	Dumbarton Castle	PSOH	UK	833	P 305	Al Shaheed	PB	Kuwait	449
266	Arctowski	AGS	Poland	576	P 305	Escudo De Veraguas	PB	Panama	547
M 266	Amasra	MHSC	Turkey	784	P 305	AG 5	ABU	Turkey	790
P 266	Machitis	PGG	Greece	291	Y 305	Vejrø	PB	Denmark	177
M 267	Ayvalik	MHSC	Turkey	784	306	Commandant El Harty	PGG	Morocco	494
P 267	Nikiforos	PGG	Greece	291	P 306	Bayan	PB	Kuwait	449
M 268	Akçakoca	MHSC	Turkey	784	P 306	AG 6	ABU	Turkey	790
P 268	Aittitos	PGG	Greece	291	Y 306	Farø	PB	Denmark	177
A 269	Grey Rover	AORLH	UK	836	307	Commandant Azougghar	PGG	Morocco	494
M 269	Anamur	MHSC	Turkey	784	A 307	Thetis	ANL	Greece	294
P 269	Krateos	PGG	Greece	291	P 307	Dasman	PB	Kuwait	449
M 270	Akçay	MHSC	Turkey	784	Y 307	Laesø	PB	Denmark	177
A 271	Gold Rover	AORLH	UK	836	308	El Hahiq	PBO	Morocco	494
PC 271	Cabo Corrientes	PB	Mexico	486	308	MPK 99	FFLM	Russian Federation	624
272	Generał Kazimierz Pułaski	FFGHM	Poland	571	P 308	Subahi	PB	Kuwait	449
PC 272	Cabo Corzo	PB	Mexico	486	Y 308	Rømø	PB	Denmark	177
273	Generał Tadeusz Kosciuszko	FFGHM	Poland	571	309	El Tawfiq	PBO	Morocco	494
A 273	Black Rover	AORLH	UK	836	P309	Jaberi	PB	Kuwait	449
PC 273	Cabo Catoche	PB	Mexico	486	310	L V Rabhi	PBO	Morocco	494
274	Vice Admiral Constantin Balescu	ML/MCS	Romania	597	F 310	Fridtjof Nansen	FFGHM	Norway	526
280	Iroquois	DDGHM	Canada	89	P 310	Saad	PB	Kuwait	449
281	Piast	ARS	Poland	579	U 310	Zhovti Vody	MSO	Ukraine	799
281	Constanta	AETLM	Romania	598	311	Errachiq	PBO	Morocco	494
P 281	Tyne	PSO	UK	832	311	Kazanets	FFLM	Russian Federation	624
PC 281	Punta Morro	PB	Mexico	486	311	Prabparapak	PTFG	Thailand	757
282	Athabaskan	DDGHM	Canada	89	F 311	Roald Amundsen	FFGHM	Norway	526
282	Lech	ARS	Poland	579	M 311	Wambola	MHC	Estonia	210
P 282	Severn	PSO	UK	832	P 311	Ahmadi	PB	Kuwait	449
PC 282	Punta Mastun	PB	Mexico	486	P 311	Bishkhali	PB	Bangladesh	50
283	Algonquin	DDGHM	Canada	89	P 311	Weeraya	PB	Sri Lanka	707
283	Midia	AETLM	Romania	598	U 311	Cherkasy	MSO	Ukraine	799
P 283	Mersey	PSO	UK	832	312	El Akid	PBO	Morocco	494
					312	Hanhak Sattru	PTFG	Thailand	757
					F 312	Otto Sverdrup	FFGHM	Norway	526
					M 312	Sulev	MHC	Estonia	210
					P 312	Padma	PB	Bangladesh	49
					P 312	Naif	PB	Kuwait	449
					313	El Maher	PBO	Morocco	494

Number	Ship's name	Type	Country	Page	Number	Ship's name	Type	Country	Page
313	Suphairin	PTFG	Thailand	757	341	Samadikun	FF	Indonesia	334
F 313	Helge Ingstad	FFGHM	Norway	526	341	Mei Chin	LSM	Taiwan	742
P 313	Surma	PB	Bangladesh	49	M 341	Karmøy	MHCM/MSCM	Norway	529
P 313	Thafir	PB	Kuwait	449	P 341	Marti	PGGF	Turkey	783
P 313-1	Fath	PTFG	Iran	351	P 341	Udara	PB	Sri Lanka	708
P 313-2	Nasr	PTFG	Iran	351	342	El Mourakeb	PG	Algeria	6
P 313-3	Saf	PTFG	Iran	351	342	Martadinata	FF	Indonesia	334
P 313-4	Ra'd	PTFG	Iran	351	M 342	Maløy	MHCM/MSCM	Norway	529
P 313-5	Fajr	PTFG	Iran	351	P 342	Tayfun	PGGF	Turkey	783
P 313-6	Shams	PTFG	Iran	351	343	El Kechef	PG	Algeria	6
P 313-7	Me'raj	PTFG	Iran	351	A 343	Sleipner	AKR	Sweden	726
P 313-8	Falaq	PTFG	Iran	351	M 343	Hinnøy	MHCM/MSCM	Norway	529
P 313-9	Hadid	PTFG	Iran	351	P 343	Volkan	PGGF	Turkey	783
P 313-10	Qadr	PTFG	Iran	351	344	El Moutarid	PG	Algeria	6
314	El Majid	PBO	Morocco	494	A 344	Loke	AKL	Sweden	728
F 314	Thor Heyerdahl	FFGHM	Norway	526	P 344	Rüzgar	PGGF	Turkey	783
P 314	Karnaphuli	PC	Bangladesh	49	345	El Rassed	PG	Algeria	6
P 314	Marzoug	PB	Kuwait	449	P 345	Poyraz	PGGF	Turkey	783
315	El Bachir	PBO	Morocco	494	346	El Djari	PG	Algeria	6
C 315	Late	LCM	Tonga	769	P 346	Gurbet	PGGF	Turkey	783
P 315	Tista	PC	Bangladesh	49	347	El Saher	PG	Algeria	6
P 315	Mash'noor	PB	Kuwait	449	347	Mei Sung	LSM	Taiwan	742
P 315	Jagatha	PB	Sri Lanka	707	P 347	Firtina	PGGF	Turkey	783
316	El Hamiss	PBO	Morocco	494	348	El Moukadem	PG	Algeria	6
P 316	Wadah	PB	Kuwait	449	P 348	Yildiz	PGGF	Turkey	783
P 316	Abeetha II	PB	Sri Lanka	707	349	Kebir class	PG	Algeria	6
317	El Karib	PBO	Morocco	494	P 349	Karayel	PGGF	Turkey	783
P 317	Taroub	PB	Kuwait	449	350	El Kanass	PG	Algeria	6
P 317	Edithara II	PB	Sri Lanka	707	350	MPK 214	FFLM	Russian Federation	625
318	Raïs Bargach	PSO	Morocco	495	M 350	Alta	MHCM/MSCM	Norway	529
P 318	Wickramaa II	PB	Sri Lanka	707	351	Djebel Chenoua	FSG	Algeria	5
319	Raïs Britel	PSO	Morocco	495	351	Ahmad Yani	FFGHM	Indonesia	333
320	Raïs Charkaoui	PSO	Morocco	495	351	Al Jouf	WPBF	Saudi Arabia	662
W 320	Nordkapp	WPSOH	Norway	531	M 351	Otra	MHCM/MSCM	Norway	529
321	Pauk II class	FSM	Cuba	168	P 351	Parakramabahu	PC	Sri Lanka	708
321	Raïs Maaninou	PSO	Morocco	495	352	El Chihab	FSG	Algeria	5
321	Ratcharit	PGGF	Thailand	757	352	Slamet Riyadi	FFGHM	Indonesia	333
P 321	Denizkusu	PTGF	Turkey	782	352	Turaif	WPBF	Saudi Arabia	662
W 321	Senja	WPSOH	Norway	531	353	Al Kirch	FSG	Algeria	5
322	Raïs Al Mounastiri	PSO	Morocco	495	353	Yos Sudarso	FFGHM	Indonesia	333
322	Witthayakhom	PGGF	Thailand	757	353	Hail	WPBF	Saudi Arabia	662
A 322	Heros	YTM	Sweden	729	353	Mei Ping	LSM	Taiwan	742
P 322	Ranarisi	PB	Sri Lanka	708	354	Oswald Siahann	FFGHM	Indonesia	333
P 322	Atmaca	PTGF	Turkey	782	354	MPK 221	FFLM	Russian Federation	625
W 322	Andenes	WPSOH	Norway	531	354	Najran	PBF	Saudi Arabia	662
323	MPK 64	FFLM	Russian Federation	625	F 354	Niels Juel	FFGM	Denmark	172
323	Udomdet	PGGF	Thailand	757	355	Abdul Halim	FFGHM	Indonesia	333
P 323	Sahin	PTGF	Turkey	782		Perdanakusuma			
A 324	Protea	AGSH	South Africa	683	F 355	Olfert Fischer	FFGM	Denmark	172
A 324	Hera	YTM	Sweden	729	356	Karel Satsuitubun	FFGHM	Indonesia	333
P 324	Kartal	PTGF	Turkey	782	356	Mei Lo	LSM	Taiwan	742
P 326	Pelikan	PTGF	Turkey	782	F 356	Peter Tordenskiold	FFGM	Denmark	172
P 327	Albatros	PTGF	Turkey	782	F 357	Thetis	FFH	Denmark	174
P 328	Simsek	PTGF	Turkey	782	F 358	Triton	FFH	Denmark	174
P 329	Kasirga	PTGF	Turkey	782	P 358	Hessa	AXL	Norway	530
330	Halifax	FFGHM	Canada	90	F 359	Vaedderen	FFHM	Denmark	174
F 330	Vasco da Gama	FFGHM	Portugal	584	P 359	Vigra	AXL	Norway	530
P 330	Ranajaya	PB	Sri Lanka	708	F 360	Hvidbjørnen	FFHM	Denmark	174
P 330	Kiliç	PGGF	Turkey	782	U 360	Genichesk	MHC	Ukraine	800
U 330	Melitopol	MHSC	Ukraine	799	361	Fatahillah	FFG/FFGH	Indonesia	335
331	Vancouver	FFGHM	Canada	90	362	Malahayati	FFG/FFGH	Indonesia	335
331	Sri Gaya	AP	Malaysia	471	362	MPK 17	FFLM	Russian Federation	625
331	Chon Buri	PG	Thailand	758	363	Nala	FFG/FFGH	Indonesia	335
F 331	Alvares Cabral	FFGHM	Portugal	584	364	Ki Hajar Dewantara	FFGH/FFT	Indonesia	334
P 331	Ranadeera	PB	Sri Lanka	708	A 367	Newton	AG	UK	839
P 331	Kalkan	PGGF	Turkey	782	A 368	Warden	YFRT	UK	841
332	Ville de Québec	FFGHM	Canada	90	369	MPK 191	FFLM	Russian Federation	625
332	Sri Tiga	AP	Malaysia	471	P 370	Rio Minho	PBR	Portugal	587
332	MPK 107	FFLM	Russian Federation	625	371	Kapitan Patimura	FS	Indonesia	336
332	Songkhla	PG	Thailand	758	HQ 371	Tarantul class	FSGM	Vietnam	929
F 332	Corte Real	FFGHM	Portugal	584	M 371	Ohue	MHSC	Nigeria	522
P 332	Ranawickrama	PB	Sri Lanka	708	372	Untung Suropati	FS	Indonesia	336
P 332	Mizrak	PGGF	Turkey	782	HQ 372	Tarantul class	FSGM	Vietnam	929
333	Toronto	FFGHM	Canada	90	M 372	Maraba	MHSC	Nigeria	522
333	Phuket	PG	Thailand	758	373	Nuku	FS	Indonesia	336
P 333	Tufan	PGGF	Turkey	782	HQ 373	Tarantul class	FSGM	Vietnam	929
334	Regina	FFGHM	Canada	90	374	Lambung Mangkurat	FS	Indonesia	336
P 334	Meltem	PGGF	Turkey	782	A 374	Prometheus	AORH	Greece	294
335	Calgary	FFGHM	Canada	90	HQ 374	Tarantul class	FSGM	Vietnam	929
P 335	Imbat	PGGF	Turkey	782	375	Cut Nyak Dien	FS	Indonesia	336
336	Montreal	FFGHM	Canada	90	375	MPK 82	FFLM	Russian Federation	625
P 336	Zipkin	PGGF	Turkey	782	A 375	Zeus	AOTL	Greece	294
337	Fredericton	FFGHM	Canada	90	376	Sultan Thaha Syaifuddin	FS	Indonesia	336
P 337	Atak	PGGF	Turkey	782	A 376	Orion	AOTL	Greece	294
338	Winnipeg	FFGHM	Canada	90	377	Sutanto	FS	Indonesia	336
P 338	Bora	PGGF	Turkey	782	378	Sutedi Senoputra	FS	Indonesia	336
339	Charlottetown	FFGHM	Canada	90	379	Wiratno	FS	Indonesia	336
340	St John's	FFGHM	Canada	90	380	Memet Sastrawiria	FS	Indonesia	336
M 340	Oksøy	MHCM/MSCM	Norway	529	381	Tjiptadi	FS	Indonesia	336
P 340	Dogan	PGGF	Turkey	783	HQ 381	BPS 500 class	FSGM	Vietnam	929
P 340	Prathpa	PB	Sri Lanka	708	382	Hasan Basri	FS	Indonesia	336
341	El Yadekh	PG	Algeria	6	383	Iman Bonjol	FS	Indonesia	336
341	Ottawa	FFGHM	Canada	90	384	Pati Unus	FS	Indonesia	336

Number	Ship's name	Type	Country	Page	Number	Ship's name	Type	Country	Page
385	Teuku Umar	FS	Indonesia	336	AOE 422	Towada	AOE/AORH	Japan	411
A 385	Fort Rosalie	AFSH	UK	837	423	Grom	FSGM	Poland	573
386	Silas Papare	FS	Indonesia	336	423	Smerch	FSGM	Russian Federation	627
A 386	Fort Austin	AFSH	UK	837	A 423	Heraklis	YTB	Greece	296
Y 386	Agdlek	PB	Denmark	175	AOE 423	Tokiwa	AOE/AORH	Japan	411
A 387	Fort Victoria	AORH	UK	837	424	Al Kharj	MHC	Saudi Arabia	660
Y 387	Agpa	PB	Denmark	175	A 424	Iason	YTB	Greece	296
A 388	Fort George	AORH	UK	837	AOE 424	Hamana	AOE/AORH	Japan	411
Y 388	Tulugaq	PB	Denmark	175	U 424	Artemivsk	ACV/LCUJM	Ukraine	799
A 389	Wave Knight	AORH	UK	835	A 425	Odisseus	YTB	Greece	296
390	MPK 222	FFLM	Russian Federation	625	AOE 425	Mashuu	AOE/AORH	Japan	410
A 390	Wave Ruler	AORH	UK	835	AOE 426	Oumi	AOE/AORH	Japan	410
396	MPK 28	FFLM	Russian Federation	625	430	Al Nour	PC	Egypt	200
400	Vitse Admiral Kulakov	DDGHM	Russian Federation	620	431	Kharg	AORH	Iran	357
U 400	Rivne	LST	Ukraine	799	431	Świnoujście	PTFGM	Poland	574
401	Admiral Branimir Ormanov	AGS	Bulgaria	84	431	Tapi	FS	Thailand	754
					A 431	Ahti	YDT	Estonia	210
401	Lieutenant Malghagh	LCU	Morocco	495	432	Khirirat	FS	Thailand	754
401	Ho Chi	LCU	Taiwan	743	433	Al Hady	PC	Egypt	200
401	Rade Končar	PTFG	Serbia and Montenegro	667	433	Władysławowo	PTFGM	Poland	574
					433	Makut Rajakumarn	FFH/AX	Thailand	763
A 401	Independencia	PBO	Panama	549	434	Gornik	FSGM	Poland	573
L 401	Al Soumood	LCU	Kuwait	450	434	Marshal Ushakov	DDGHM	Russian Federation	621
L 401	Ertuğrul	LST	Turkey	786	435	Hutnik	FSGM	Poland	572
P 401	Cassiopea	PSOH	Italy	382	436	Al Wakil	PC	Egypt	200
U 401	Kirovograd	LSM	Ukraine	798	436	Metalowiec	FSGM	Poland	573
402	Daoud Ben Aicha	LSMH	Morocco	495	437	Rolnik	FSGM	Poland	573
402	Ho Huei	LCU	Taiwan	743	439	Al Hakim	PC	Egypt	200
A 402	Manzanillo	AP	Mexico	490	441	Yan Sit Aung	PC	Myanmar	499
A 402	Flamenco	YO	Panama	549	441	Rattanakosin	FSGM	Thailand	753
L 402	Al Tahaddy	LCU	Kuwait	450	442	Yan Myat Aung	PC	Myanmar	499
L 402	Serdar	LST	Turkey	786	442	Al Salam	PC	Egypt	200
P 402	Libra	PSOH	Italy	382	442	Sukothai	FSGM	Thailand	753
U 402	Konstantin Olshansky	LST	Ukraine	799	443	Yan Nyein Aung	PC	Myanmar	499
403	Jalbout	LCU	Kuwait	450	444	Yan Khwin Aung	PC	Myanmar	499
403	Ahmed Es Sakali	LSMH	Morocco	495	445	Yan Ye Aung	PC	Myanmar	499
403	Ho Yao	LCU	Taiwan	743	445	Al Jabbar	PC	Egypt	200
ASR 403	Chihaya	ASRH	Japan	411	446	Yan Min Aung	PC	Myanmar	499
P 403	Spica	PSOH	Italy	382	447	Yan Paing Aung	PC	Myanmar	499
404	Abou Abdallah El Ayachi	LSMH	Morocco	495	448	Yan Win Aung	PC	Myanmar	499
					448	Al Qader	PC	Egypt	200
404	Hasan Zafirović-Laca	PTFG	Serbia and Montenegro	667	449	Yan Aye Aung	PC	Myanmar	499
					450	Yan Zwe Aung	PC	Myanmar	499
P 404	Vega	PSOH	Italy	382	450	Razliv	FSGM	Russian Federation	627
405	El Aigh	AKS	Morocco	496	F 450	Elli	FFGHM	Greece	287
AS 405	Chiyoda	AS/ASRH	Japan	411	451	Al Rafa	PC	Egypt	200
P 405	Esploratore	PB	Italy	382	F 451	Limnos	FFGHM	Greece	287
406	Ho Chao	LCU	Taiwan	743	F 452	Hydra	FFGHM	Greece	286
406	Ante Banina	PTFG	Serbia and Montenegro	667	F 453	Spetsai	FFGHM	Greece	286
					C 454	Prestol	PG	Dominican Republic	184
P 406	Sentinella	PB	Italy	382	F 454	Psara	FFGHM	Greece	286
407	Sidi Mohammed Ben Abdallah	LSTH	Morocco	495	455	Chao Phraya	FFG	Thailand	752
					F 455	Salamis	FFGHM	Greece	286
P 407	Vedetta	PB	Italy	382	456	Bangpakong	FFG	Thailand	752
408	Dakhla	AKS	Morocco	496	C 456	Almirante Juan Alexandro Acosta	PBO	Dominican Republic	183
P 408	Staffetta	PB	Italy	382	457	Kraburi	FFGH	Thailand	752
409	Moroz	FSGM	Russian Federation	627	C 457	Almirante Didiez Burgos	PBO	Dominican Republic	183
P 409	Sirio	PSOH	Italy	381					
P 410	Orione	PSOH	Italy	381	458	Saiburi	FFGH	Thailand	752
411	Kangan	AWT	Iran	356	F 459	Adrias	FFGHM	Greece	287
A 411	Rio Papaloapan	LSTH	Mexico	490	F 460	Aegeon	FFGHM	Greece	287
C 411	Tobruk	AX	Libya	457	461	Phuttha Yotfa Chulalok	FFGHM	Thailand	753
P 411	Shaheed Daulat	PC	Bangladesh	48	F 461	Navarinon	FFGHM	Greece	287
P 411	Ennasr	PB	Mauritania	477	462	Phuttha Loetia Naphalai	FFGHM	Thailand	753
412	Taheri	AWT	Iran	356	F 462	Kountouriotis	FFGHM	Greece	287
A 412	Usumacinta	LSTH	Mexico	490	F 463	Bouboulina	FFGHM	Greece	287
MSC 412	Addriyah	MHSC	Saudi Arabia	660	MST 463	Uraga	MSTH/ML	Japan	408
P 412	Shaheed Farid	PC	Bangladesh	48	A 464	Axios	ARL/AOTL	Greece	294
413	Pin Klao	FFT	Thailand	762	F 464	Kanaris	FFGHM	Greece	287
P 413	Shaheed Mohibullah	PC	Bangladesh	48	MST 464	Bungo	MSTH/MLH	Japan	408
MSC 414	Al Quysumah	MHSC	Saudi Arabia	660	F 465	Themistocles	FFGHM	Greece	287
P 414	Shaheed Aktheruddin	PC	Bangladesh	48	F 466	Nikiforos Fokas	FFGHM	Greece	287
A 415	Evros	AEL	Greece	295	A 470	Aliakmon	ARL/AOTL	Greece	294
M 415	Olev	MSI	Estonia	210	471	Polnochny B class	LSM	Algeria	6
416	Tariq Ibn Ziyad	FSGM	Libya	456	471	Maga	PTG	Myanmar	498
A 416	Ouranos	AOTL	Greece	294	F 471	Antonio Enes	FSH	Portugal	586
M 416	Vaindlo	MSI	Estonia	210	472	Kalaat Beni Hammad	LSLH	Algeria	6
MSC 416	Al Wadeeah	MHSC	Saudi Arabia	660	472	Saittra	PTG	Myanmar	498
A 417	Hyperion	AOTL	Greece	294	473	Kalaat Beni Rached	LSTH	Algeria	6
MSC 418	Safwa	MHSC	Saudi Arabia	660	473	Duwa	PTG	Myanmar	498
SSV 418	Ekvator	AGI/AGIM	Russian Federation	635	474	Zeyda	PTG	Myanmar	498
A 419	Pandora	AP	Greece	295	A 474	Pytheas	AGOR	Greece	293
420	Al Jawf	MHC	Saudi Arabia	660	475	Houxin class	PTG	Myanmar	498
A 420	Pandrosos	AP	Greece	295	F 475	João Coutinho	FSH	Portugal	586
U 420	Donetsk	ACV/LCUJ	Ukraine	799	476	Houxin class	PTG	Myanmar	498
421	Bandar Abbas	AORLH	Iran	357	A 476	Strabon	AGSC	Greece	293
421	Orkan	FSGM	Poland	573	F 476	Jacinto Candido	FSH	Portugal	586
421	Naresuan	FFGHM	Thailand	751	F 477	General Pereira d'Eça	FSH	Portugal	586
C 421	Ardent	PB	St Kitts and Nevis	653	A 478	Naftilos	AGS	Greece	293
422	Boushehr	AORLH	Iran	357	A 479	I Karavoyiannos Theophilopoulos	ABUH	Greece	295
422	Piorun	FSGM	Poland	573					
422	Shaqra	MHC	Saudi Arabia	660	F 480	Comandante João Belo	FF	Portugal	585
422	Taksin	FFGHM	Thailand	751					

Number	Ship's name	Type	Country	Page	Number	Ship's name	Type	Country	Page
481	Ho Shun	LCU	Taiwan	743	512	Wuxi	FFG	China	130
A 481	St Lykoudis	ABUH	Greece	295	512	Teluk Semangka	LSTH	Indonesia	340
F 481	Comandante Hermenegildo Capelo	FF	Portugal	585	512	Larak	LSLH	Iran	355
					A 512	Shahayak	YR	Bangladesh	51
ARC 482	Muroto	ARC	Japan	411	A 512	Mosel	ARLHM	Germany	275
F 483	Comandante Sacadura Cabral	FF	Portugal	585	HQ 512	Polnochny class	LCM	Vietnam	931
					SSV 512	Kildin	AGI/AGIM	Russian Federation	635
484	Ho Chung	LCU	Taiwan	743	513	Huayin	FFG	China	130
F 486	Baptista de Andrade	FSH	Portugal	585	513	Sinai	MSO	Egypt	201
F 487	João Roby	FSH	Portugal	585	513	Teluk Penju	LSTH	Indonesia	340
488	Ho Shan	LCU	Taiwan	743	513	Tonb	LSLH	Iran	355
F 488	Afonso Cerqueira	FSH	Portugal	585	513	Al Zuara	PTFG	Libya	456
489	Ho Chuan	LCU	Taiwan	743	513	Al Farouq	PGGF	Saudi Arabia	659
490	Ho Seng	LCU	Taiwan	743	A 513	Shahjalal	AG	Bangladesh	51
F 490	Gaziantep	FFGHM	Turkey	780	A 513	Rhein	ARLHM	Germany	275
P 490	Comandante Cigala Fulgosi	PSOH	Italy	381	HQ 513	Polnochny class	LCM	Vietnam	931
					514	Zhenjiang	FFG	China	130
491	Ho Meng	LCU	Taiwan	743	514	Teluk Mandar	LSTH	Indonesia	340
F 491	Giresun	FFGHM	Turkey	780	514	Lavan	LSLH	Iran	355
P 491	Comandante Borsini	PSOH	Italy	381	A 514	Werra	ARLHM	Germany	275
492	Ho Mou	LCU	Taiwan	743	M 514	Silifke	MSC	Turkey	785
F 492	Gemlik	FFGHM	Turkey	780	515	Xiamen	FFG	China	130
P 492	Comandante Bettica	PSOH	Italy	381	515	Teluk Sampit	LSTH	Indonesia	340
493	Ho Shou	LCU	Taiwan	743	515	Al Ruha	PTFG	Libya	456
F 493	Gelibolu	FFGHM	Turkey	780	515	Abdul Aziz	PGGF	Saudi Arabia	659
P 493	Comandante Foscari	PSOH	Italy	381	A 515	Khan Jahan Ali	AOT	Bangladesh	51
494	Ho Chun	LCU	Taiwan	743	A 515	Main	ARLHM	Germany	275
F 494	Gökçeada	FFGHM	Turkey	780	M 515	Saros	MSC	Turkey	785
495	Ho Yung	LCU	Taiwan	743	516	Jiujiang	FFG	China	130
F 495	Gediz	FFGHM	Turkey	780	516	Assiyut	MSO	Egypt	201
F 496	Gokova	FFGHM	Turkey	780	516	Teluk Banten	LSTH	Indonesia	340
F 497	Göksu	FFGHM	Turkey	780	A 516	Iman Gazzali	AOTL	Bangladesh	51
A 498	Lana	AGS	Nigeria	523	A 516	Donau	ARLHM	Germany	275
Y 498	Mario Marino	YDT	Italy	382	LT 516	Kalinga Apayao	LST	Philippines	565
Y 499	Alcide Pedretti	YDT	Italy	382	M 516	Sigacik	MSC	Turkey	785
500	Grozavu	ATA	Romania	599	517	Nanping	FFG	China	130
F 500	Bozcaada	FFGM	Turkey	777	517	Teluk Ende	LSTH	Indonesia	340
M 500	Foça	MSI	Turkey	785	517	Faisal	PGGF	Saudi Arabia	659
U 500	Donbas	AGF/AR	Ukraine	800	517	Hsin Lung	AOTL	Taiwan	745
501	Teluk Langsa	LST	Indonesia	340	M 517	Sapanca	MSC	Turkey	785
501	Eilat	FSGHM	Israel	363	518	Jian	FFG	China	130
501	Lieutenant Colonel Errhamani	FFGM	Morocco	493	518	Sharaba	PGGF	Libya	456
					M 518	Sariyer	MSC	Turkey	785
501	Hercules	ATA	Romania	599	519	Changzhi	FFG	China	130
501	La Galité	PGGF	Tunisia	771	519	Kahlid	PGGF	Saudi Arabia	659
A 501	Kyanwa	PBO	Nigeria	522	520	Rassvet	FSGM	Russian Federation	627
F 501	Bodrum	FFGM	Turkey	777	A 520	Sagres	AXS	Portugal	588
HQ 501	Tran Khanh Du	LST	Vietnam	931	M 520	Karamürsel	MSC	Turkey	784
LT 501	Laguna	LST	Philippines	565	SSV 520	Meridian	AGIM	Russian Federation	634
M 501	Fethiye	MSI	Turkey	785	521	Jiaxin	FFGHM	China	129
502	Teluk Bajur	LST	Indonesia	340	521	Hai	AGOR	China	147
502	Lahav	FSGHM	Israel	363	521	Amyr	PGGF	Saudi Arabia	659
502	Tunis	PGGF	Tunisia	771	521	Sattahip	PG	Thailand	758
A 502	Ologbo	PBO	Nigeria	522	A 521	Schultz Xavier	ABU	Portugal	589
F 502	Bandirma	FFGM	Turkey	777	M 521	Kerempe	MSC	Turkey	784
HQ 502	Qui Nonh	LST	Vietnam	931	P 521	Vigilante	PB	Cape Verde	103
M 502	Fatsa	MSI	Turkey	785	522	Lianyungang	FFGHM	China	129
503	Teluk Amboina	LST	Indonesia	340	522	Wahag	PGGF	Libya	456
503	Hanit	FSGHM	Israel	363	522	Klongyai	PG	Thailand	758
503	Carthage	PGGF	Tunisia	771	A 522	D Carlos I	AGS	Portugal	588
A 503	Nwamba	PBO	Nigeria	522	M 522	Kilimli	MSC	Turkey	784
F 503	Beykoz	FFGM	Turkey	777	523	Sanming	FFGHM	China	129
HQ 503	Vung Tau	LST	Vietnam	931	523	Al Fikah	PTFG	Libya	456
M 503	Finike	MSI	Turkey	785	523	Tariq	PGGF	Saudi Arabia	659
504	Teluk Kau	LST	Indonesia	340	523	Takbai	PG	Thailand	758
A 504	Obula	PBO	Nigeria	522	A 523	Almirante Gago Coutinho	AGS	Portugal	588
F 504	Bartin	FFGM	Turkey	777					
LT 504	Lanao del Norte	LST	Philippines	565	524	Putian	FFGHM	China	129
505	Uragan	FSGM	Russian Federation	627	524	Shehab	PGGF	Libya	456
F 505	Bafra	FFGM	Turkey	777	524	Yuen Feng	AKM	Taiwan	744
J 506	Yongxingdao	ASRH	China	142	524	Kantang	PG	Thailand	758
506	Dauriya	AKH/AGF	Russian Federation	638	525	Maanshan	FFGHM	China	128
SSV 506	Nakhoda	AGI/AGIM	Russian Federation	635	525	Al Mathur	PTFG	Libya	456
507	Daqahliya	MSO	Egypt	201	525	Oqbah	PGGF	Saudi Arabia	659
507	Chang Bai	AOTL	Taiwan	745	525	Wu Kang	AKM	Taiwan	744
508	Teluk Tomini	LST	Indonesia	340	525	Thepha	PG	Thailand	758
509	Chang De	FFG	China	130	526	Wenzhou	FFGHM	China	128
509	Teluk Ratai	LST	Indonesia	340	526	Nakat	FSG	Russian Federation	627
AOR 509	Protecteur	AORH	Canada	94	526	Hsin Kang	AKM	Taiwan	744
510	Shaoxing	FFG	China	130	526	Taimuang	PG	Thailand	758
510	Teluk Saleh	LST	Indonesia	340	527	Jiangwei class	FFGHM	China	129
AOR 510	Preserver	AORH	Canada	94	527	Abu Obaidah	PGGF	Saudi Arabia	644
U 510	Slavutich	AGFHM	Ukraine	800	528	Jinagwei II class	FFGHM	China	129
511	Nantong	FFG	China	130	528	Shouaiai	PGGF	Libya	456
511	Teluk Bone	LST	Indonesia	340	530	Giza	MSO	Egypt	201
511	Hengam	LSLH	Iran	355	530	Wu Yi	AOEHM	Taiwan	744
511	Kontradmiral X Czernicki	AKHM/AP/HM/AGI	Poland	578	A 530	Horten	ASH/AGP	Norway	530
					A 530	M 10 class hovercraft	UCAC	Sri Lanka	711
511	Al Siddiq	PGGF	Saudi Arabia	659					
A 511	Shaheed Ruhul Amin	AX	Bangladesh	49	P 530	Trabzon	PBO/AGI	Turkey	783
A 511	Elbe	ARLHM	Germany	275	531	Teluk Gilimanuk	LSM	Indonesia	341
HQ 511	Polnochny class	LCM	Vietnam	931	531	Najin class	FFG	Korea, North	430
U 511	Simferopol	AGS	Ukraine	801	531	Al Bitar	PTFG	Libya	456

Number	Ship's name	Type	Country	Page	Number	Ship's name	Type	Country	Page
531	Khamronsin	FS	Thailand	754	555	Zhaotong	FFG	China	130
P 531	Terme	PBO/AGI	Turkey	783	555	Geyzer	FSGM	Russian Federation	627
532	Teluk Celukan Bawang	LSM	Indonesia	341	ATF 555	Ta Fung	ATF/ARS	Taiwan	745
					F 555	Driade	FSM	Italy	378
532	Shoula	PGGF	Libya	456	P 555	Støren	PGGM/MHC/ MLC/AGSC	Denmark	176
532	Tulcea	AOT	Romania	599					
532	Thayanchon	FS	Thailand	754	F 556	Chimera	FSM	Italy	378
533	Ningbo	FFG	China	130	P 556	Svaerdfisken	PGGM/MHC/ MLC/AGSC	Denmark	176
533	Aswan	MSO	Egypt	201					
533	Teluk Cendrawasih	LSM	Indonesia	341	557	Jishou	FFG	China	130
533	Al Sadad	PTFG	Libya	456	F 557	Fenice	FSM	Italy	378
533	Tusha	FSGM	Russian Federation	627	P 557	Glenten	PGGM/MHC/ MLC/AGSC	Denmark	176
533	Longlom	FS	Thailand	754					
A 533	Norge	YAC	Norway	530	558	Zigong	FFG	China	130
534	Jinhua	FFG	China	130	F 558	Sibilla	FSM	Italy	378
534	Teluk Berau	LSM	Indonesia	341	P 558	Gribben	PGGM/MHC/ MLC/AGSC	Denmark	176
534	Shafak	PGGF	Libya	456					
535	Huangshi	FFG	China	132	559	Kangding	FFG	China	130
535	Teluk Peleng	LSM	Indonesia	341	A 559	Sleipner	AKS	Denmark	178
535	Aysberg	FSGM	Russian Federation	627	P 559	Lommen	PGGM/MHC/ MLC/AGSC	Denmark	176
A 535	Valkyrien	AKS/ATA	Norway	530					
SSV 535	Kareliya	AGIM	Russian Federation	634	560	Dongguan	FFG	China	130
536	Wuhu	FFG	China	132	560	Won San	MLH	Korea, South	444
536	Qena	MSO	Egypt	201	560	Zyb	FSGM	Russian Federation	627
536	Teluk Sibolga	LSM	Indonesia	341	D 560	Luigi Durand de La Penne	DDGHM	Italy	373
537	Zhoushan	FFG	China	132					
537	Teluk Manado	LSM	Indonesia	341	P 560	Ravnen	PGGM/MHC/ MLC/AGSC	Denmark	176
538	Teluk Hading	LSM	Indonesia	341					
538	Rad	PGGF	Libya	456	561	Shantou	FFG	China	130
539	Anqing	FFGHM	China	128	561	Multatuli	AGFH	Indonesia	343
539	Sohag	MSO	Egypt	201	561	Kang Kyeong	MHSC	Korea, South	444
539	Teluk Parigi	LSM	Indonesia	341	D 561	Francesco Mimbelli	DDGHM	Italy	373
540	Huainan	FFGHM	China	128	P 561	Skaden	PGGM/MHC/ MLC/AGSC	Denmark	176
540	Teluk Lampung	LSM	Indonesia	341					
A 540	Dannebrog	YAC	Denmark	180	562	Jiangmen	FFG	China	130
A 540	Hansaya	LCP	Sri Lanka	711	562	Kang Jin	MHSC	Korea, South	444
U 540	Chigirin	AXL	Ukraine	802	P 562	Viben	PGGM/MHC/ MLC/AGSC	Denmark	176
541	Huaibei	FFGHM	China	128					
541	Teluk Jakarta	LSM	Indonesia	341	563	Zhaoqing	FFG	China	130
541	Hua Hin	PSO	Thailand	757	563	Ko Ryeong	MHSC	Korea, South	444
P 541	Aboubekr Ben Amer	PBO	Mauritania	477	ATF 563	Ta Tai	ATF/ARS	Taiwan	745
U 541	Smila	AXL	Ukraine	802	P 563	Søløven	PGGM/MHC/ MLC/AGSC	Denmark	176
542	Tongling	FFGHM	China	128					
542	Dat Assawari	MHC	Egypt	201	564	Yichang	FFGHM	China	129
542	Teluk Sangkulirang	LSM	Indonesia	341	564	Admiral Tributs	DDGHM	Russian Federation	620
542	Laheeb	PGGF	Libya	456	565	Yulin	FFGHM	China	129
542	Klaeng	PSO	Thailand	757	565	Kim Po	MHSC	Korea, South	444
U 542	Darnicha	AXL	Ukraine	802	566	Yuxi	FFGHM	China	129
543	Dandong	FFG	China	130	566	Ko Chang	MHSC	Korea, South	444
543	Teluk Cirebon	AKL/ARL	Indonesia	344	567	Jiangwei II class	FFGHM	China	129
543	Marshal Shaposhnikov	DDGHM	Russian Federation	620	567	Kum Wha	MHSC	Korea, South	444
					568	Wenzhou	FFGHM	China	128
543	Si Racha	PSO	Thailand	757	570	Sonya class	MSC/MH	Cuba	168
544	Siping	FFGH	China	133	570	Passat	FSGM	Russian Federation	627
544	Teluk Sabang	AKL/ARL	Indonesia	344	A 570	Taskizak	AOTL	Turkey	787
545	Linfen	FFG	China	130	F 570	Maestrale	FFGHM	Italy	376
545	Navarin	MHC	Egypt	201	571	Yang Yang	MSC/MHC	Korea, South	444
548	Huainan	FFGHM	China	128	A 571	Albay Hakki Burak	AOT	Turkey	788
548	Burullus	MHC	Egypt	201	F 571	Grecale	FFGHM	Italy	376
548	Admiral Panteleyev	DDGHM	Russian Federation	620	SSV 571	Belomore	AGIM	Russian Federation	634
D 550	Ardito	DDGHM	Italy	374	572	Ongjin	MSC/MHC	Korea, South	444
LC 550	Bacolod City	LSVH	Philippines	565	572	Admiral Vinogradov	DDGHM	Russian Federation	620
P 550	Flyvefisken	PGGM/MHC/ MLC/AGSC	Denmark	176	A 572	Yuzbasi Ihsan Tolunay	AOT	Turkey	788
					F 572	Libeccio	FFGHM	Italy	376
551	Maoming	FFG	China	130	A 573	Binbaşi Saadettin Gürçan	AORL	Turkey	788
551	Liven	FSGM	Russian Federation	627					
A 551	Danbjørn	AGB	Denmark	181	F 573	Scirocco	FFGHM	Italy	376
ATF 551	Ta Wan	ATF/ARS	Taiwan	745	F 574	Aliseo	FFGHM	Italy	376
B 551	Voum-Legleita	PBO	Mauritania	478	575	Taicang	AORH	China	141
C 551	Giuseppe Garibaldi	CVGM	Italy	370	F 575	Euro	FFGHM	Italy	376
D 551	Audace	DDGHM	Italy	374	A 576	Degirmendere	ATA	Turkey	791
F 551	Minerva	FSM	Italy	378	F 576	Espero	FFGHM	Italy	376
LC 551	Dagupan City	LSVH	Philippines	565	A 577	Sokullu Mehmet Pasa	AG/AX	Turkey	787
P 551	Hajen	PGGM/MHC/ MLC/AGSC	Denmark	176	F 577	Zeffiro	FFGHM	Italy	376
					578	Sonya class	MSC/MH	Cuba	168
552	Yibin	FFG	China	130	A 578	Darica	ATR	Turkey	791
552	Ta Hu	ARS	Taiwan	745	A 579	Cezayirli Gazi Hasan Pasa	AG/AX	Turkey	787
A 552	Isbjørn	AGB	Denmark	181					
C 552	Cavour	CV	Italy	372	580	Dore	LCU	Indonesia	341
F 552	Urania	FSM	Italy	378	A 580	Akar	AORH	Turkey	788
P 552	Havkatten	PGGM/MHC/ MLC/AGSC	Denmark	176	A 581	Çinar	AWT	Turkey	788
					F 581	Carabiniere	AGEHM	Italy	384
553	Shaoguan	FFG	China	130	582	Kupang	LCU	Indonesia	341
A 553	Thorbjørn	AGB/AGS	Denmark	181	A 582	Kemer	AGS	Turkey	784
ATF 553	Ta Han	ATF/ARS	Taiwan	745	F 582	Artigliere	FFGHM	Italy	377
F 553	Danaide	FSM	Italy	378	583	Dili	LCU	Indonesia	341
P 553	Laxen	PGGM/MHC/ MLC/AGSC	Denmark	176	A 583	Agradoot	AGS	Bangladesh	51
					F 583	Aviere	FFGHM	Italy	377
554	Anshun	FFG	China	130	584	Nusuntara	LCU	Indonesia	341
ATF 554	Ta Kang	ATF/ARS	Taiwan	745	F 584	Bersagliere	FFGHM	Italy	377
F 554	Sfinge	FSM	Italy	378	A 585	Akin	ASR	Turkey	789
P 554	Makrelen	PGGM/MHC/ MLC/AGSC	Denmark	176	F 585	Granatiere	FFGHM	Italy	377
					A 587	Gazal	ATF	Turkey	790

Number	Ship's name	Type	Country	Page	Number	Ship's name	Type	Country	Page
A 588	Çanadarli	AGS	Turkey	787	620	Bespokoiny	DDGHM	Russian Federation	621
A 589	Isin	ARS	Turkey	789	620	Shtyl	FSGM	Russian Federation	627
590	Meteor	FSGM	Russian Federation	627	A 620	Jules Verne	ADH	France	250
591	Najin Class	FFG	Korea, North	430	D 620	Forbin	DDGHM	France	235
A 592	Karadeniz Ereglisi	AKS	Turkey	789	P 620	Sayura	PSO	Sri Lanka	707
A 593	Eceabat	AWT	Turkey	789	621	Mandau	PTFG	Indonesia	338
A 594	Çubuklu	AGS	Turkey	787	621	Flaming	MHCM	Poland	575
A 595	Yarbay Kudret Güngör	AORH	Turkey	788	621	Thalang	MCS	Thailand	759
A 596	Ulubat	YW	Turkey	788	D 621	Chevalier Paul	DDGHM	France	235
A 597	Van	YW	Turkey	788	622	Rencong	PTFG	Indonesia	338
A 598	Sögüt	AWT	Turkey	788	M 622	Pluton	MCD	France	247
A 599	Çesme	AGS	Turkey	787	623	Badik	PTFG	Indonesia	338
A 600	Kavak	AWT	Turkey	788	623	Mewa	MHCM	Poland	575
601	23 of July	PGGF	Egypt	198	624	Keris	PTFG	Indonesia	338
601	Lung Chiang	PGGF	Taiwan	741	624	Czajka	MHCM	Poland	575
601	Ras El Blais	PBO	Tunisia	773	625	TR 25	MSC	Poland	576
A 601	Monge	AGE	France	249	626	TR 26	MSC	Poland	576
P 601	Jayasagara	PB	Sri Lanka	707	630	Goplo	MHC	Poland	575
U 601	Alchevsk	AGS	Ukraine	801	A 630	Marne	AORHM	France	249
602	6 of October	PGGF	Egypt	198	631	Gardno	MHSC	Poland	575
602	Sui Chang	PGGF	Taiwan	741	631	Bang Rachan	MHSC	Thailand	760
602	Ras Ajdir	PBO	Tunisia	773	A 631	Somme	AORHM	France	249
U 602	Moma class	AGS	Ukraine	801	632	Bukowo	MHSC	Poland	575
603	Aiyar Lulin	LCU	Myanmar	501	632	Nongsarai	MHSC	Thailand	760
603	21 of October	PGGF	Egypt	198	633	Dabie	MHC	Poland	575
603	Jin Chiang	PCG	Taiwan	740	633	Lat Ya	MHSC	Thailand	760
603	Ras el Edrak	PBO	Tunisia	773	A 633	Taape	AG/ATS/YDT/ YPC/YPT	France	250
D 603	Duquesne	DDGM	France	235					
U 603	Biya class	AGS	Ukraine	801	634	Jamno	MHC	Poland	575
604	Aiyar Mai	LCU	Myanmar	501	634	Tha Din Daeng	MHSC	Thailand	760
604	18 of June	PGGF	Egypt	198	A 634	Rari	AFL	France	250
604	Ras El Manoura	PBO	Tunisia	773	635	Mielno	MHC	Poland	575
605	Aiyar Maung	LCU	Myanmar	501	A 635	Revi	AFL	France	250
605	25 of April	PGGF	Egypt	198	636	Wicko	MHC	Poland	575
605	Admiral Levchenko	DDGHM	Russian Federation	620	A 636	Maito	YTM	France	254
605	Andromache	PB	Seychelles	670	637	Resko	MHC	Poland	575
605	Tan Chiang	PCG	Taiwan	740	A 637	Maroa	YTM	France	254
605	Ras Enghela	PBO	Tunisia	773	638	Sarbsko	MHC	Poland	575
606	Aiyar Minthamee	LCU	Myanmar	501	A 638	Manini	YTM	France	254
606	Hsin Chiang	PCG	Taiwan	740	Y 638	Lardier	YTM	France	254
606	Ras Ifrikia	PBO	Tunisia	773	639	Necko	MHC	Poland	575
607	Aiyar Minthar	LCU	Myanmar	501	Y 639	Giens	YTM	France	254
607	Feng Chiang	PCG	Taiwan	740	640	Naklo	MHC	Poland	575
A 607	Meuse	AORHM	France	249	D 640	Georges Leygues	DDGHM	France	232
608	Tseng Chiang	PCG	Taiwan	740	Y 640	Mengam	YTM	France	254
A 608	Var	AORHM	France	249	641	Druzno	MHC	Poland	575
609	Kao Chiang	PCG	Taiwan	740	A 641	Esterel	YTM	France	253
610	Sechelt	YDT	Canada	93	D 641	Dupleix	DDGHM	France	232
610	Safaga	MSI	Egypt	201	M 641	Éridan	MHC	France	247
610	Nastoychivy	DDGHM	Russian Federation	621	Y 641	Balaguier	YTM	France	254
610	Svyazist	MSOM	Russian Federation	630	642	Hancza	MHSC	Poland	575
610	Jing Chiang	PCG	Taiwan	740	642	Natya class	MSC/AGORM	Syria	733
D 610	Tourville	DDGHM	France	236	A 642	Luberon	YTM	France	253
611	Sikanni	YTT/YPT	Canada	93	D 642	Montcalm	DDGHM	France	232
611	Tral class	FS	Korea, North	431	M 642	Cassiopée	MHC	France	247
611	Mohammed V	FFGHM	Morocco	492	Y 642	Taillat	YTM	France	254
611	Hsian Chiang	PCG	Taiwan	740	643	Mamry	MHSCM	Poland	575
611	Phosamton	AXL	Thailand	763	D 643	Jean de Vienne	DDGHM	France	233
M 611	Vulcain	MCD	France	247	M 643	Andromède	MHC	France	247
P 611	Tawheed	PC	Bangladesh	48	Y 643	Nividic	YTM	France	254
612	Sooke	YDT	Canada	93	644	Wigry	MHSCM	Poland	575
612	Tral class	FS	Korea, North	431	D 644	Primauguet	DDGHM	France	232
612	Hassan II	FFGHM	Morocco	492	M 644	Pégase	MHC	France	247
612	Badr	FSG	Saudi Arabia	659	645	Sniardwy	MHSCM	Poland	575
612	Tsi Chiang	PCG	Taiwan	740	D 645	La Motte-Picquet	DDGHM	France	232
612	Bangkeo	MSC	Thailand	760	M 645	Orion	MHC	France	247
D 612	De Grasse	DDGHM	France	236	646	Wdzydze	MHSCM	Poland	575
P 612	Tawfiq	PC	Bangladesh	48	D 646	Latouche-Treville	DDGHM	France	232
613	Stikine	YTT/YPT	Canada	93	M 646	Croix du Sud	MHC	France	247
613	Abu el Ghoson	MSI	Egypt	201	M 647	Aigle	MHC	France	247
613	Tral class	FS	Korea, North	421	Y 647	Le Four	YTM	France	254
613	Donchedi	MSC	Thailand	760	M 648	Lyre	MHC	France	247
A 613	Achéron	MCD	France	247	A 649	L'Etoile	AXS	France	253
P 613	Tamjeed	PC	Bangladesh	48	M 649	Persée	MHC	France	247
614	Tral class	FS	Korea, North	431	Y 649	Port Cros	YTM	France	254
614	Al Yarmook	FSG	Saudi Arabia	659	650	Admiral Chabanenko	DDGHM	Russian Federation	619
614	Po Chiang	PCG	Taiwan	740	A 650	La Belle Poule	AXS	France	253
D 614	Cassard	DDGHM	France	234	M 650	Sagittaire	MHC	France	247
M 614	Styx	MCD	France	247	651	Singa	PBO	Indonesia	338
P 614	Tanveer	PC	Bangladesh	48	M 651	Verseau	MHC	France	247
615	Bora	PGGJM	Russian Federation	628	A 652	Mutin	AXS	France	253
615	Chan Chiang	PCG	Taiwan	740	M 652	Céphée	MHC	France	247
A 615	Loire	AGH/AR	France	250	653	Ajak	PBO	Indonesia	338
D 615	Jean Bart	DDGHM	France	234	M 653	Capricorne	MHC	France	247
616	Samum	PGGJM	Russian Federation	628	661	Letuchy	FFM	Russian Federation	623
616	Hitteen	FSG	Saudi Arabia	659	A 664	Malabar	ATA	France	253
D 616	Forbin	DDGHM	France	235	MSC 666	Ogishima	MHSC	Japan	409
617	Mirazh	FSGM	Russian Federation	627	MSC 668	Yurishima	MHSC	Japan	409
617	Chu Chiang	PCG	Taiwan	740	A 669	Tenace	ATA	France	253
AD 617	Yakal	ARL	Philippines	565	MSC 669	Hikoshima	MHSC	Japan	409
D 617	Chevalier Paul	DDGHM	France	235	670	Ramadan	PGGF	Egypt	199
618	Tabuk	FSG	Saudi Arabia	659	MSC 670	Awashima	MHSC	Japan	409
619	Severomorsk	DDGHM	Russian Federation	620	671	Tral class	FS	Korea, North	431

Number	Ship's name	Type	Country	Page	Number	Ship's name	Type	Country	Page
671	Un Bong	LST	Korea, South	443	704	Hofouf	FFGHM	Saudi Arabia	658
MSC 671	Sakushima	MHSC	Japan	409	705	Whitehorse	MCDV	Canada	92
P 671	Glaive	PB	France	256	U 705	Kremenets	ATA	Ukraine	802
672	Khyber	PGGF	Egypt	199	706	Yellowknife	MCDV	Canada	92
MSC 672	Uwajima	MHSC	Japan	409	706	Abha	FFGHM	Saudi Arabia	658
P 672	Épée	PB	France	256	U 706	Izyaslav	YTM	Ukraine	802
MSC 673	Ieshima	MHSC	Japan	409	707	Goose Bay	MCDV	Canada	92
674	El Kadessaya	PGGF	Egypt	199	708	Moncton	MCDV	Canada	92
MSC 674	Tsukishima	MHSC	Japan	409	708	Taif	FFGHM	Saudi Arabia	658
P 674	D'Entrecasteaux	AG/AX	France	250	709	Saskatoon	MCDV	Canada	92
A 675	Fréhel	YTM	France	254	710	Brandon	MCDV	Canada	92
MSC 675	Maejima	MHSC	Japan	409	F 710	La Fayette	FFGHM	France	237
P 675	Arago	PBO	France	245	711	Summerside	MCDV	Canada	92
676	El Yarmouk	PGGF	Egypt	199	711	Pulau Rengat	MHSC	Indonesia	342
A 676	Saire	YTM	France	254	711	Podchorazy	AXL	Poland	578
676	Wee Bong	LST	Korea, South	443	F 711	Surcouf	FFGHM	France	237
MSC 676	Kumejima	MHSC	Japan	409	P 711	Barkat	PC	Bangladesh	49
P 676	Flamant	PBO	France	246	712	Pulau Rupat	MHSC	Indonesia	342
A 677	Armen	YTM	France	254	712	Neustrashimy	FFHM	Russian Federation	622
677	Su Yong	LST	Korea, South	443	712	Chang	LST	Thailand	761
MSC 677	Makishima	MHSC	Japan	409	F 712	Courbet	FFGHM	France	237
P 677	Cormoran	PBO	France	246	P 712	Salam	PB	Bangladesh	49
678	Badr	PGGF	Egypt	199	713	Kerch	CGHM	Russian Federation	618
678	Admiral Kharlamov	DDGHM	Russian Federation	620	713	Pangan	LST	Thailand	761
A 678	La Houssaye	YTM	France	254	F 713	Aconit	FFGHM	France	237
678	Buk Han	LST	Korea, South	443	P 713	Sangu	PBO	Bangladesh	48
MSC 678	Tobishima	MHSC	Japan	409	714	Lanta	LST	Thailand	761
P 678	Pluvier	PBO	France	246	F 714	Guépratte	FFGHM	France	237
A 679	Kéréon	YTM	France	254	P 714	Turag	PBO	Bangladesh	48
MSC 679	Yugeshima	MHSC	Japan	409	715	Bystry	DDGHM	Russian Federation	621
P 679	Grèbe	PBO	France	246	715	Prathong	LST	Thailand	761
680	Hettein	PGGF	Egypt	199	718	MT 265	MSO	Russian Federation	630
A 680	Sicié	YTM	France	254	P 720	Géranium	PB	France	256
MSC 680	Nagashima	MHSC	Japan	409	721	Pulau Rote	MSC	Indonesia	341
P 680	Sterne	PBO	France	246	721	Sichang	LSTH	Thailand	760
A 681	Taunoa	YTM	France	254	A 721	Khadem	ATA	Bangladesh	52
681	Kojoon Bong	LSTH	Korea, South	443	P 721	Jonquille	PB	France	256
MSC 681	Sugashima	MSC	Japan	409	722	Vaarlahti	AKSL	Finland	219
P 681	Albatros	PSO	France	245	722	Pulau Raas	MSC	Indonesia	341
682	Biro Bong	LSTH	Korea, South	443	722	Al Munjed	ARS	Libya	457
A 682	Rascas	YTM	France	254	722	MDK 113	ACV/LCUJ	Russian Federation	633
MSC 682	Notojima	MSC	Japan	409	722	Surin	LSTH	Thailand	760
P 682	L'Audacieuse	PBO	France	246	A 722	Sebak	YTM	Bangladesh	52
683	Hyangro Bong	LSTH	Korea, South	443	P722	Violette	PB	France	256
MSC 683	Tsunoshima	MSC	Japan	409	723	Väno	AKSL	Finland	219
P 683	La Boudeuse	PBO	France	246	723	Pulau Romang	MSC	Indonesia	341
MSC 684	Naoshima	MSC	Japan	409	A 723	Rupsha	YTM	Bangladesh	52
P 684	La Capricieuse	PBO	France	246	P 723	Jasmin	PB	France	256
685	Seongin Bong	LSTH	Korea, South	443	724	Pulau Rimau	MSC	Indonesia	341
MSC 685	Toyoshima	MSC	Japan	409	A 724	Shibsha	YTM	Bangladesh	52
P 685	La Fougueuse	PBO	France	246	MCL 724	Hahajima	MCSD	Japan	409
MSC 686	Ukushima	MSC	Japan	409	725	Pulau Rondo	MSC	Indonesia	341
P 686	La Glorieuse	PBO	France	246	MCL 725	Kamishima	MCSD	Japan	409
687	Marshal Vasilevsky	DDGHM	Russian Federation	620	726	Pulau Rusa	MSC	Indonesia	341
MSC 687	Izushima	MSC	Japan	409	727	Pulau Rangsang	MSC	Indonesia	341
P 687	La Gracieuse	PBO	France	246	728	Pulau Raibu	MSC	Indonesia	341
688	Haijiu class	PC	China	135	729	Pulau Rempang	MSC	Indonesia	341
MSC 688	Aishima	MSC	Japan	409	730	MDK 89	ACV/LCUJ	Russian Federation	633
P 688	La Moqueuse	PBO	France	246	F 730	Floréal	FFGHM	France	239
MSC 689	Aoshima	MSC	Japan	409	731	Neukrotimy	FFM	Russian Federation	623
P 689	La Railleuse	PBO	France	246	731	Kut	LSM	Thailand	761
MSC 690	Miyajima	MSC	Japan	409	F 731	Prairial	FFGHM	France	239
P 690	La Rieuse	PBO	France	246	F 732	Nivôse	FFGHM	France	239
691	Tatarstan	FFGM	Russian Federation	624	F 733	Ventôse	FFGHM	France	239
MSC 691	Shishijima	MSC	Japan	409	F 734	Vendémiaire	FFGHM	France	239
P 691	La Tapageuse	PBO	France	246	F 735	Germinal	FFGHM	France	239
Y 692	Telenn Mor	ABU	France	251	738	MT 264	MSOM	Russian Federation	630
A 693	Acharné	YTM	France	254	741	Prab	LCM	Thailand	761
A 695	Bélier	YTB	France	253	Y 741	Elfe	YFB	France	251
A 696	Buffle	YTB	France	253	742	Satakut	LCM	Thailand	761
697	Haijiu class	PC	China	135	A 743	Denti	AETL	France	248
A 697	Bison	YTB	France	253	747	DKA-67	LCU	Russian Federation	633
700	Kingston	MCDV	Canada	92	A 748	Léopard	AXL	France	252
A 700	Khaireddine	AGS	Tunisia	772	A 749	Panthère	AXL	France	252
U 700	Netisin	YDT	Ukraine	801	A 750	Jaguar	AXL	France	252
Y 700	Néréide	YFB	France	251	751	Houxin class	PGG	China	135
701	Sirius	LSM	Bulgaria	84	751	Dong Hae	FS	Korea, South	439
701	Glace Bay	MCDV	Canada	92	A 751	Lynx	AXL	France	252
701	Pulau Rani	MSO	Indonesia	341	752	Houxin class	PGG	China	135
701	Stupinets	PGGK	Russian Federation	629	752	Su Won	FS	Korea, South	439
A 701	N N O Salammbo	AGOR/AX	Tunisia	772	A 752	Guépard	AXL	France	252
P 701	Nandimithra	PGG	Sri Lanka	708	753	Houxin class	PGG	China	135
Y 701	Ondine	YFB	France	251	753	Kang Reung	FS	Korea, South	439
702	Antares	LSM	Bulgaria	84	A 753	Chacal	AXL	France	252
702	Nanaimo	MCDV	Canada	92	754	Houxin class	PGG	China	135
702	Abu Al Barakat Al Barbari	AGOR	Morocco	496	754	Bezboyaznennyy	DDGHM	Russian Federation	621
					A 754	Tigre	AXL	France	252
702	Pylky	FFM	Russian Federation	623	755	Houxin class	PGG	China	135
702	Madina	FFGHM	Saudi Arabia	658	755	An Yang	FS	Korea, South	439
P 702	Suranimala	PGG	Sri Lanka	708	A 755	Lion	AXL	France	252
Y 702	Naiade	YFB	France	251	756	Houxin class	PGG	China	135
703	Edmonton	MCDV	Canada	92	756	Po Hang	FS/FSG	Korea, South	439
704	Shawinigan	MCDV	Canada	92	U 756	Sudak	AWT	Ukraine	800

Number	Ship's name	Type	Country	Page	Number	Ship's name	Type	Country	Page
757	Houxin class	PGG	China	135	A 790	Coralline	YDT	France	252
757	Kun San	FS/FSG	Korea, South	439	F 790	Lieutenant de Vaisseau Lavallée	FFGM	France	238
A 757	D'Entrecasteaux	AG/AX	France	250	A 791	Lapérouse	AGS	France	249
758	Houxin class	PGG	China	135	F 791	Commandant l'Herminier	FFGM	France	238
758	Kyong Ju	FS/FSG	Korea, South	439	A 792	Borda	AGS	France	249
A 758	Beautemps-Beaupré	AGOR	France	248	F 792	Premier Maitre L'Her	FFGM	France	238
759	Houxin class	PGG	China	135	A 793	Laplace	AGS	France	249
759	Mok Po	FS/FSG	Korea, South	439	F 793	Commandant Blaison	FFGM	France	238
A 759	Dupuy de Lôme	AGIH	France	248	F 794	Enseigne de Vaisseau Jacoubet	FFGM	France	238
760	Houxin class	PGG	China	135	F 795	Commandant Ducuing	FFGM	France	238
A 760	Alize	YDT	France	250	795	MDK 108	ACV/LCUJM	Russian Federation	633
761	Kim Chon	FS/FSG	Korea, South	439	F 796	Commandant Birot	FFGM	France	238
761	Mataphon	LCM/LCVP/ LCP	Thailand	761	F 797	Commandant Bouan	FFGM	France	238
P 761	Kara	PB	Togo	768	799	Hylje	YPC	Finland	222
762	Chung Ju	FS/FSG	Korea, South	439	L 800	Rotterdam	LPD	Netherlands	512
762	V Gumanenko	MHOM	Russian Federation	631	801	Rais Hamidou	PTGM	Algeria	5
762	Rawi	LCM/LCVP/ LCP	Thailand	761	801	Pandrong	PBO	Indonesia	338
L 762	Lachs	LCU	Germany	274	801	Survey Ship	AGS	Myanmar	501
P 762	Mono	PB	Togo	768	801	Ladny	FFM	Russian Federation	623
763	Jin Ju	FS/FSG	Korea, South	439	801	Te Mataili	PB	Tuvalu	794
763	Adang	LCM/LCVP/ LCP	Thailand	761	A 801	Pelikaan	AP	Netherlands	513
764	Houxin class	PGG	China	135	L 801	Johan De Witt	LPD	Netherlands	512
764	Phetra	LCM/LCVP/ LCP	Thailand	761	802	Salah Rais	PTGM	Algeria	5
765	Houxin class	PGG	China	135	802	(ex-Changi)	AGS	Myanmar	501
765	Yo Su	FS/FSG	Korea, South	439	802	Sura	PBO	Indonesia	338
765	Kolam	LCM/LCVP/ LCP	Thailand	761	A 802	Snellius	AGSH	Netherlands	513
L 765	Schlei	LCU	Germany	274	F 802	De Zeven Provincien	FFGHM	Netherlands	507
766	Houxin class	PGG	China	135	803	Rais Ali	PTGM	Algeria	5
766	Jin Hae	FS/FSG	Korea, South	439	803	Todak	PBO	Indonesia	338
766	Talibong	LCM/LCVP/ LCP	Thailand	761	A 803	Luymes	AGSH	Netherlands	513
767	Houxin class	PGG	China	135	F 803	Tromp	FFGHM	Netherlands	507
767	Sun Chon	FS/FSG	Korea, South	439	804	Hiu	PBO	Indonesia	338
768	Houxin class	PGG	China	135	F 804	De Ruyter	FFGHM	Netherlands	507
768	Yee Ree	FS/FSG	Korea, South	439	805	Layang	PBO	Indonesia	338
A 768	Élan	AG/ATS/YDT/ YPC/YPT	France	250	F 805	Evertsen	FFGHM	Netherlands	507
769	Houxin class	PGG	China	135	806	Lemadang	PBO	Indonesia	338
769	Won Ju	FS/FSG	Korea, South	439	806	Motorist	MSOM	Russian Federation	630
770	Yangjiang	PTG	China	135	808	Pytlivy	FFM	Russian Federation	623
770	Yevgeniy Kocheshkov	ACV/LCUJM	Russian Federation	633	810	Smetlivy	DDGM	Russian Federation	618
770	Valentin Pikul	MSOM	Russian Federation	630	811	Kakap	PBOH	Indonesia	338
A 770	Glycine	AXL	France	253	811	Grunwald	LST	Poland	576
M 770	Antares	MHI	France	247	811	Chanthara	AGS	Thailand	764
771	Shunde	PTG	China	135	U 811	Balta	ADG	Ukraine	800
771	Kampela 1	LCU/AKSL	Finland	219	812	Kerapu	PBOH	Indonesia	338
771	An Dong	FS/FSG	Korea, South	439	812	Al Riyadh	FFGHM	Saudi Arabia	657
771	Thong Kaeo	LCU	Thailand	761	812	Suk	AGOR	Thailand	763
A 771	Eglantine	AXL	France	253	P 812	Nirbhoy	PC	Bangladesh	48
M 771	Altair	MHI	France	247	813	Tongkol	PBOH	Indonesia	338
772	Nanhai	PTG	China	135	F 813	Witte de With	FFGM	Netherlands	510
772	Kampela 2	LCU/AKSL	Finland	219	814	Barakuda	PBOH	Indonesia	338
772	Chon An	FS/FSG	Korea, South	439	814	Makkah	FFGHM	Saudi Arabia	657
772	Thong Lang	LCU	Thailand	761	816	Al Dammam	FFGHM	Saudi Arabia	657
M 772	Aldebaran	MHI	France	247	L 820	Yunnan class	LCU	Sri Lanka	711
773	Panyu	PTG	China	135	821	Misairutei-Ichi-Go	PTGK	Japan	410
773	Song Nam	FS/FSG	Korea, South	439	821	Lublin	LST/ML	Poland	576
773	Wang Nok	LCU	Thailand	761	821	Suriya	ABU	Thailand	765
P 773	Njambuur	PBO	Senegal	663	L 821	Yunnan class	LCU	Sri Lanka	711
774	Houjian class	PT5	China	135	822	Misairutei-Ni-Go	PTGK	Japan	410
774	Wang Nai	LCU	Thailand	761	822	Gniezno	LST/ML	Poland	576
A 774	Chevreuil	AG/ATS/YDT/ YPC/YPT	France	250	823	Misairutei-San-Go	PTGK	Japan	410
					823	Soho class	FFGH	Korea, North	430
775	Houjian class	PTG	China	135	823	Krakow	LST/ML	Poland	576
775	Bu Chon	FS/FSG	Korea, South	439	824	Hayabusa	PGGF	Japan	410
A 775	Gazelle	AG/ATS/YDT/ YPC/YPT	France	250	824	Poznan	LST/ML	Poland	576
					SSV 824	Liman	AGI/AGIM	Russian Federation	635
776	Houjian class	PTG	China	135	825	Wakataka	PGGF	Japan	410
776	Jae Chon	FS/FSG	Korea, South	439	825	Torun	LST/ML	Poland	576
A 776	Isard	AG/ATS/YDT/ YPC/YPT	France	250	826	Ootaka	PGGF	Japan	410
					827	Kumataka	PGGF	Japan	410
777	Porkkala	MLI	Finland	216	F 827	Karel Doorman	FFGHM	Netherlands	508
777	Dae Chon	FS/FSG	Korea, South	439	828	Umitaka	PGGF	Japan	410
778	Sok Cho	FS/FSG	Korea, South	439	F 828	Van Speijk	FFGHM	Netherlands	508
778	Burny	DDGHM	Russian Federation	621	829	Shirataka	PGGF	Japan	410
779	Yong Ju	FS/FSG	Korea, South	439	F 829	Willem van der Zaan	FFGHM	Netherlands	508
781	Nam Won	FS/FSG	Korea, South	439	F 830	Tjerk Hiddes	FFGHM	Netherlands	508
781	Man Nok	LCU	Thailand	761	U 830	Korets	ATA	Ukraine	802
782	Kwan Myong	FS/FSG	Korea, South	439	831	Chula	AORL	Thailand	764
782	Mordoviya	ACV/LCUJM	Russian Federation	633	F 831	Van Amstel	FFGHM	Netherlands	508
782	Man Klang	LCU	Thailand	761	U 831	Kover	ATA/YTM	Ukraine	802
U 782	Sokal	AXL	Ukraine	802	832	Samui	YO	Thailand	764
783	Sin Hung	FS/FSG	Korea, South	439	A 832	Zuiderkruis	AORH	Netherlands	513
783	Man Nai	LCU	Thailand	761	833	Prong	YO	Thailand	764
785	Kong Ju	FS/FSG	Korea, South	439	F 833	Van Nes	FFGHM	Netherlands	508
A 785	Thétis	MCD/BEGM	France	249	834	Proet	YO	Thailand	764
F 789	Lieutenant de Vaisseau Le Hénaff	FFGM	France	238	F 834	Van Galen	FFGHM	Netherlands	508
					835	Samed	YO	Thailand	764
					A 836	Amsterdam	AORH	Netherlands	514
					L 836	Ranavijaya	LCM	Sri Lanka	711
					L 839	Ranagaja	LCM	Sri Lanka	711
					PG 840	Conrado Yap	PBF	Philippines	565

Number	Ship's name	Type	Country	Page
841	Beidiao	AGI	China	146
841	Karabane	LCT	Senegal	664
841	Chuang	YW	Thailand	764
842	Chik	YO	Thailand	764
PG 842	Tedorico Dominado Jr	PBF	Philippines	565
PG 843	Cosme Acosta	PBF	Philippines	565
PG 844	José Artiaga Jr	PBF	Philippines	565
PG 846	Nicanor Jimenez	PBF	Philippines	565
847	Sibarau	PB	Indonesia	339
PG 847	Leopoldo Regis	PBF	Philippines	565
848	Siliman	PB	Indonesia	338
PG 848	Leon Tadina	PBF	Philippines	565
PG 849	Loreto Danipog	PBF	Philippines	565
851	Dongdiao	AGM/AGI	China	146
851	KD 11	LCU	Poland	576
A 851	Cerberus	YDT	Netherlands	514
HQ 851	Yurka class	MSO	Vietnam	932
PG 851	Apollo Tiano	PBF	Philippines	565
852	KD 12	LCU	Poland	576
A 852	Argus	YDT	Netherlands	514
U 852	Shostka	ABU	Ukraine	801
853	KD 13	LCU	Poland	576
853	Rin	YTB	Thailand	765
A 853	Nautilus	YDT	Netherlands	514
M 853	Haarlem	MHC	Netherlands	510
PG 853	Sulpicio Hernandez	PBF	Philippines	565
854	Rang	YTB	Thailand	765
A 854	Hydra	YDT	Netherlands	514
855	Kontradmiral Vlasov	MSOM	Russian Federation	630
855	Samaesan	YTR	Thailand	765
856	Raet	YTR	Thailand	765
M 856	Maasluis	MHC	Netherlands	510
857	Sigalu	PB	Indonesia	338
M 857	Makkum	MHC	Netherlands	510
858	Silea	PB	Indonesia	338
M 858	Middelburg	MHC	Netherlands	510
859	Siribua	PB	Indonesia	338
M 859	Hellevoetsluis	MHC	Netherlands	510
M 860	Schiedam	MHC	Netherlands	510
Y 860	Schwedeneck	AG	Germany	278
861	Changxingdao	ASRH	China	142
861	Kled Keo	AK	Thailand	765
HQ 861	Sonya class	MHSC	Vietnam	932
M 861	Urk	MHC	Netherlands	510
Y 861	Kronsort	AG	Germany	278
862	Chongmingdao	ASRH	China	142
862	Siada	PB	Indonesia	338
HQ 862	Sonya class	MHSC	Vietnam	932
M 862	Zierikzee	MHC	Netherlands	510
Y 862	Helmsand	AG	Germany	278
863	Sikuda	PB	Indonesia	338
HQ 863	Sonya class	MHSC	Vietnam	932
M 863	Vlaardingen	MHC	Netherlands	510
Y 863	Stollergrund	AG	Germany	278
864	Sigurot	PB	Indonesia	338
HQ 864	Sonya class	MHSC	Vietnam	932
M 864	Willemstad	MHC	Netherlands	510
Y 864	Mittelgrund	AG	Germany	278
Y 865	Kalkgrund	AG	Germany	278
Y 866	Breitgrund	AG	Germany	278
871	Similan	AORH	Thailand	764
A 874	Linge	YTM	Netherlands	515
875	Pyhäranta	MLI	Finland	216
A 875	Regge	YTM	Netherlands	515
876	Pansio	MLI	Finland	216
A 876	Hunze	YTM	Netherlands	515
877	Kampela 3	LCU/AKSL	Finland	219
A 877	Rotte	YTM	Netherlands	515
A 878	Gouwe	YTM	Netherlands	515
879	Valas	AKSL	Finland	219
L 880	Shakthi	LSM	Sri Lanka	711
882	Fengcang	AORH	China	141
883	Dayun class	AKH	China	144
884	Dayun class	AKH	China	144
885	Nancang	AORH	China	142
HQ 885	Yurka class	MSO	Vietnam	932
886	Fuchi	AORH	China	142
891	Dahua class	AGOR/AGE	China	146
Y 891	Altmark	APB	Germany	276
Y 895	Wische	APB	Germany	276
899	Halli	YPC	Finland	222
A 900	Mercuur	ASL/YTT	Netherlands	513
L 900	Shah Amanat	LSL	Bangladesh	52
901	Mourad Rais	FFLM	Algeria	4
901	Sharm el Sheikh	FFGHM	Egypt	195
901	Balikpapan	AOTL	Indonesia	343
901	A Zheleznyakov	MHOM	Russian Federation	631
L 901	Shah Poran	LCU	Bangladesh	52
902	Rais Kellich	FFLM	Algeria	4
902	Sambu	AOTL	Indonesia	343
902	Boraida	AORH	Saudi Arabia	661
A 902	Van Kinsbergen	AXL	Netherlands	514
L 902	Shah Makhdum	LCU	Bangladesh	52

Number	Ship's name	Type	Country	Page
P 902	Liberation	PBR	Belgium	55
903	Rais Korfou	FFLM	Algeria	4
903	Arun	AORLH	Indonesia	343
904	Yunbou	AORH	Saudi Arabia	661
906	Toushka	FFGHM	Egypt	195
908	Yuting I class	LSTH	China	138
909	Yuting I class	LSTH	China	138
910	Yuting I class	LSTH	China	138
F 910	Wielingen	FFGM	Belgium	54
911	Yuting II class	LSTH	China	139
911	Mubarak	FFGHM	Egypt	195
911	Sorong	AOTL	Indonesia	343
911	Chakri Naruebet	CVM	Thailand	750
F 911	Westdiep	FFGM	Belgium	54
P 911	Madhumati	PSO	Bangladesh	47
912	Yuting II class	LSTH	China	139
912	Turbinist	MSOM	Russian Federation	630
912	Chien Yang	DDGHM	Taiwan	736
P 912	Kapatakhaya	PBO	Bangladesh	48
913	Yuting II class	LSTH	China	139
913	Kovrovets	MSOM	Russian Federation	630
P 913	Karatoa	PBO	Bangladesh	48
P 914	Gomati	PBO	Bangladesh	48
M 915	Aster	MHC	Belgium	55
916	Taba	FFGHM	Egypt	195
M 916	Bellis	MHC	Belgium	55
M 917	Crocus	MHC	Belgium	55
VM 917	Al Manoud	YDT	Libya	448
919	Snayper	MSOM	Russian Federation	630
920	Dazhi	AS	China	143
921	El Fateh	AXT	Egypt	203
921	Jaya Wijaya	ARL	Indonesia	344
921	Liao Yang	DDGHM	Taiwan	736
M 921	Lobelia	MHC	Belgium	55
922	Rakata	AT/PBO	Indonesia	345
923	Soputan	ATF	Indonesia	345
923	Shen Yang	DDGHM	Taiwan	736
M 923	Narcis	MHC	Belgium	55
M 924	Primula	MHC	Belgium	55
925	Te Yang	DDGHM	Taiwan	736
927	Yukan class	LST	China	139
927	Yun Yang	DDGHM	Taiwan	736
928	Yukan class	LST	China	139
928	Chen Yang	DDGHM	Taiwan	736
929	Yukan class	LST	China	139
929	Shao Yang	DDGHM	Taiwan	736
930	Yukan class	LST	China	139
930	Legky	FFM	Russian Federation	623
931	Yukan class	LST	China	139
931	Tariq	AXT	Egypt	203
931	Burujulasad	AGORH	Indonesia	343
932	Yukan class	LST	China	139
932	Dewa Kembar	AGSH	Indonesia	342
932	Chin Yang	FFGH	Taiwan	737
933	Yukan class	LST	China	139
933	Jalanidhi	AGOR	Indonesia	343
933	Fong Yang	FFGH	Taiwan	737
934	Yuting I class	LSTH	China	138
934	Lampo Batang	YTM	Indonesia	345
934	Feng Yang	FFGH	Taiwan	737
935	Yuting I class	LSTH	China	138
935	Tambora	YTM	Indonesia	345
935	Lan Yang	FFGH	Taiwan	737
936	Yuting I class	LSTH	China	138
936	Bromo	YTM	Indonesia	345
936	Hae Yang	FFGH	Taiwan	737
937	Yuting I class	LSTH	China	138
937	Zharky	FFM	Russian Federation	623
937	Hwai Yang	FFGH	Taiwan	737
938	Ning Yang	FFGH	Taiwan	737
939	Yuting I class	LSTH	China	138
939	Yi Yang	FFGH	Taiwan	737
940	Yuting I class	LSTH	China	138
F 941	Abu Qir	FFGM	Egypt	196
F 946	El Suez	FFGM	Egypt	196
F 947	Krasnoperekopsk	ATA/YTM	Ukraine	802
951	Najim al Zaffer	FFG	Egypt	197
951	Ulsan	FFG	Korea, South	440
952	Nusa Telu	AKL	Indonesia	344
952	Seoul	FFG	Korea, South	440
953	Chung Nam	FFG	Korea, South	440
U 953	Dubno	ATA/YTM	Ukraine	802
955	Zadorny	FFM	Russian Federation	623
955	Masan	FFG	Korea, South	440
956	El Nasser	FFG	Egypt	197
956	Kyong Buk	FFG	Korea, South	440
957	Chon Nam	FFG	Korea, South	440
A 958	Zenobe Gramme	AXS	Belgium	57
958	Che Ju	FFG	Korea, South	440
959	Teluk Mentawai	AKL	Indonesia	344
959	Pusan	FFG	Korea, South	440
960	Karimata	AKL	Indonesia	344
A 960	Godetia	AGFH	Belgium	56

Number	Ship's name	Type	Country	Page	Number	Ship's name	Type	Country	Page
P 960	Skjold	PTGFM	Norway	528	1110	Tien Tan	FFGHM	Taiwan	738
961	Damyat	FFGH	Egypt	196	1131	Mondolkiri	PTF	Cambodia	86
961	Waigeo	AKL	Indonesia	344	1134	Ratanakiri	PTF	Cambodia	86
961	Chung Ju	FFG	Korea, South	440	P 1140	Cacine	PBO	Portugal	587
P 961	Storm	PTGMF	Norway	528	M 1142	Umzimkulu	MHC	South Africa	683
A 962	Belgica	AGOR/PBO	Belgium	56	P 1144	Quanza	PBO	Portugal	587
P 962	Skudd	PTGMF	Norway	528	P 1146	Zaire	PBO	Portugal	587
A 963	Stern	AGFH	Belgium	56	P 1150	Argos	PBR	Portugal	587
DD 963	Spruance	DDGHM	USA	872	P 1151	Dragão	PBR	Portugal	587
P 963	Steil	PTGMF	Norway	528	P 1152	Escorpião	PBR	Portugal	587
P 964	Glimt	PTGMF	Norway	528	P 1153	Cassiopeia	PBR	Portugal	587
P965	Gnist	PTGMF	Norway	528	P 1154	Hidra	PBR	Portugal	587
966	Rasheed	FFGH	Egypt	196	P 1155	Centauro	PBR	Portugal	587
966	Kosar	PGGK	Russian Federation	629	P 1156	Orion	PBR	Portugal	587
HQ 966	Truong	AKL	Vietnam	932	P 1157	Pégaso	PBR	Portugal	587
971	Kwanggaeto The Great	DDGHM	Korea, South	438	P 1158	Sagitario	PBR	Portugal	587
972	Tanjung Dalpele	LPD/APCR	Indonesia	340	P 1161	Save	PBO	Portugal	587
972	Euljimundok	DDGHM	Korea, South	438	P 1165	Aguia	PBR	Portugal	587
973	Yangmanchun	DDGHM	Korea, South	438	P 1167	Cisne	PBR	Portugal	587
975	Chungmugong Yi Sun-shin	DDGHM	Korea, South	437	1202	Kang Ding	FFGHM	Taiwan	739
					1203	Si Ning	FFGHM	Taiwan	739
976	Moonmu Daewang	DDGHM	Korea, South	437	1205	Kun Ming	FFGHM	Taiwan	739
977	Daejoyoung	DDGHM	Korea, South	437	1206	Di Hua	FFGHM	Taiwan	739
DD 985	Cushing	DDGHM	USA	872	1207	Wu Chang	FFGHM	Taiwan	739
P 986	Hauk	PTGM	Norway	527	1208	Chen Te	FFGHM	Taiwan	739
DD 987	O'Bannon	DDGHM	USA	872	M 1212	Umhloti	MHC	South Africa	683
P 987	Ørn	PTGM	Norway	527	M 1213	Umgeni	MHC	South Africa	683
P 988	Terne	PTGM	Norway	527	M 1223	Kapa	MSCD	South Africa	683
P 989	Tjeld	PTGM	Norway	527	M 1225	Tekwini	MSCD	South Africa	683
990	Yudeng class	LSM	China	138	1301	Yung Feng	MHC	Taiwan	744
P 990	Skarv	PTGM	Norway	527	1302	Yung Chia	MHC	Taiwan	744
991	Yuting I class	LSTH	China	138	1303	Yung Ting	MHC	Taiwan	744
P 991	Teist	PTGM	Norway	527	1305	Yung Shun	MHC	Taiwan	744
992	Yuting II class	LSTH	China	139	1306	Yung Yang	MSO	Taiwan	743
P 992	Jo	PTGM	Norway	527	1307	Yung Tzu	MSO	Taiwan	743
993	Yuting II class	LSTH	China	139	1308	Yung Ku	MSO	Taiwan	743
P 993	Lom	PTGM	Norway	527	1309	Yung Teh	MSO	Taiwan	743
994	Yuting II class	LSTH	China	139	LST 1312	Ambe	LST	Nigeria	522
P 994	Stegg	PTGM	Norway	527	A 1401	Eisvogel	AGB	Germany	280
995	Yuting II class	LSTH	China	139	A 1409	Wilhelm Pullwer	YAG	Germany	278
P 995	Falk	PTGM	Norway	527	A 1411	Berlin	AORH	Germany	275
996	Yuting II class	LSTH	China	139	A 1412	Frankfurt am Main	AORH	Germany	275
P 996	Ravn	PTGM	Norway	527	A 1425	Ammersee	AOL	Germany	276
997	Yuting II class	LSTH	China	139	A 1426	Tegernsee	AOL	Germany	276
P 997	Gribb	PTGM	Norway	527	A 1435	Westerwald	AEL	Germany	276
P 998	Geir	PTGM	Norway	527	A 1437	Planet	AGE	Germany	278
P 999	Erle	PTGM	Norway	527	A 1439	Baltrum	ATS/YDT	Germany	280
P 1026	Dehshat	PTFG	Pakistan	544	A 1440	Juist	ATS/YDT	Germany	280
P 1029	Jalalat	PTG	Pakistan	543	A 1441	Langeoog	ATS/YDT	Germany	280
P 1030	Shujaat	PTG	Pakistan	543	A 1442	Spessart	AOL	Germany	276
GC 1051	Kukulkán	PB	Guatemala	299	A 1443	Rhön	AOL	Germany	276
M 1052	Mühlhausen	MCD	Germany	275	A 1451	Wangerooge	ATS/YDT	Germany	280
M 1058	Fulda	MHC	Germany	274	A 1452	Spiekeroog	ATS/YDT	Germany	280
P 1011	Titas	PTF	Bangladesh	49	A 1456	Alliance	AGOR	NATO	504
P 1012	Kusiyara	PTF	Bangladesh	49	A 1458	Fehmarn	ATR	Germany	279
P 1013	Chitra	PTF	Bangladesh	49	FNH 1491	Punta Caxinas	LCU	Honduras	303
P 1014	Dhansiri	PTF	Bangladesh	49	M 1499	Umkomaas	MHC	South Africa	683
1026	Essequibo	PBO	Guyana	301	1503	Sri Indera Sakti	AORLH/ AETL/AX	Malaysia	471
D 1051	Al Gaffa	YDT	UAE	808					
M 1059	Weilheim	MHC	Germany	274	1504	Mahawangsa	AORLH/ AETL/AX	Malaysia	471
1060	Barkat	PBO	Pakistan	546					
M 1060	Weiden	MHC	Germany	274	1505	Sri Inderapura	LSTH	Malaysia	471
1061	Rehmat	PBO	Pakistan	546	P 1565	Isaac Dyobha	PGG	South Africa	682
M 1061	Rottweil	MHC	Germany	274	P 1567	Galeshewe	PGG	South Africa	682
1062	Nusrat	PBO	Pakistan	546	P 1569	Makhanda	PGG	South Africa	682
M 1062	Sulzbach-Rosenberg	MHC	Germany	274	A 1600	Iskenderun	AK	Turkey	788
1063	Vehdat	PBO	Pakistan	546	1601	Ta Kuan	AGOR	Taiwan	744
M 1063	Bad Bevensen	MHC	Germany	274	1603	Alacalufe	WPB	Chile	114
M 1064	Grömitz	MHC	Germany	274	1604	Hallef	WPB	Chile	114
M 1065	Dillingen	MHC	Germany	274	1608	Fresia	PB	Chile	110
M 1066	Frankenthal	MHC	Germany	274	1609	Aysen	WPB	Chile	114
P 1066	Sabqat	PB	Pakistan	546	1610	Corral	WPB	Chile	114
M 1067	Bad Rappenau	MHC	Germany	274	1611	Concepcion	WPB	Chile	114
M 1068	Datteln	MHC	Germany	274	1612	Caldera	WPB	Chile	114
P 1068	Rafaqat	PB	Pakistan	546	1613	San Antonio	WPB	Chile	114
M 1069	Homburg	MHC	Germany	274	1614	Antofagasta	WPB	Chile	114
M 1090	Pegnitz	MHCD	Germany	275	1615	Arica	WPB	Chile	114
M 1091	Kulmbach	MHC	Germany	274	1616	Coquimbo	WPB	Chile	114
M 1092	Hameln	MHCD	Germany	275	1617	Puerto Natales	WPB	Chile	114
M 1093	Auerbach	MHCD	Germany	275	1618	Valparaiso	WPB	Chile	114
M 1094	Ensdorf	MHCD	Germany	275	1619	Punta Arenas	WPB	Chile	114
M 1095	Überherrn	MHC	Germany	274	1620	Talcahuano	WPB	Chile	114
M 1096	Passau	MHC	Germany	274	1621	Quintero	WPB	Chile	114
M 1097	Laboe	MHC	Germany	274	1622	Chloe	WPB	Chile	114
M 1098	Siegburg	MHCD	Germany	275	1623	Puerto Montt	WPB	Chile	114
M 1099	Herten	MHC	Germany	274	1624	Iquique	WPB	Chile	114
1101	Cheng Kung	FFGHM	Taiwan	738	Y 1643	Bottsand	YPC	Germany	277
1103	Cheng Ho	FFGHM	Taiwan	738	Y 1644	Eversand	YPC	Germany	277
1105	Chi Kuang	FFGHM	Taiwan	738	1801	Chi The	DDGHM	Taiwan	737
1106	Yueh Fei	FFGHM	Taiwan	738	1802	Ming Teh	DDGHM	Taiwan	737
1107	Tzu-I	FFGHM	Taiwan	738	1803	Tong The	DDGHM	Taiwan	737
1108	Pan Chao	FFGHM	Taiwan	738	1805	Wu Teh	DDGHM	Taiwan	737
1109	Chang Chien	FFGHM	Taiwan	738	LCU 2001	Yusotei-Ichi-Go	LCU	Japan	408

Number	Ship's name	Type	Country	Page
LCU 2002	Yusotei-Ni-Go	LCU	Japan	408
LCAC 2101-6	Eakusshontei (1-6) Goo	LCAC	Japan	408
L 3004	Sir Bedivere	LSLH	UK	838
L 3005	Sir Galahad	LSLH	UK	839
L 3006	Largs Bay	LSD	UK	838
L 3007	Lyme Bay	LSD	UK	838
L 3008	Mounts Bay	LSD	UK	838
L 3009	Cardigan Bay	LSD	UK	838
P 3100	Mamba	PB	Kenya	428
P 3126	Nyayo	PGGF	Kenya	427
P 3127	Umoja	PGGF	Kenya	427
P 3130	Shujaa	PB	Kenya	428
P 3131	Shupavu	PB	Kenya	428
3144	Sri Sabah	PB	Malaysia	470
3145	Sri Sarawak	PB	Malaysia	470
3146	Sri Negri Sembilan	PB	Malaysia	470
3147	Sri Melaka	PB	Malaysia	470
P 3301	Ardhana	PB	UAE	806
P 3302	Zurara	PB	UAE	806
P 3303	Murban	PB	UAE	806
P 3304	Al Ghullan	PB	UAE	806
P 3305	Radoom	PB	UAE	806
P 3306	Ghanadhah	PB	UAE	806
3501	Perdana	PTFG	Malaysia	468
3501	Ilocos Norte	PB	Philippines	567
A 3501	Annad	YTB	UAE	808
3502	Serang	PTFG	Malaysia	468
3502	Neuva Vizcaya	PB	Philippines	567
3503	Ganas	PTFG	Malaysia	468
3503	Romblon	PB	Philippines	567
3504	Ganyang	PTFG	Malaysia	468
3504	Davao Del Norte	PB	Philippines	567
3505	Jerong	PB	Malaysia	470
L 3505	Sir Tristram	LSLH	UK	838
3506	Todak	PB	Malaysia	470
3507	Paus	PB	Malaysia	470
3508	Yu	PB	Malaysia	470
TV 3508	Kashima	AXH/TV	Japan	413
3509	Baung	PB	Malaysia	470
3510	Pari	PB	Malaysia	470
3511	Handalan	PTFG	Malaysia	469
3512	Perkasa	PTFG	Malaysia	469
3513	Pendekar	PTFG	Malaysia	469
TV 3513	Shimayuki	AXGHM/TV	Japan	413
3514	Gempita	PTFG	Malaysia	469
TV 3515	Yamagiri	AX/TV	Japan	414
TV 3516	Asagiri	AX/TV	Japan	414
P 3553	Moa	PB	New Zealand	518
P 3554	Kiwi	PB	New Zealand	518
P 3555	Wakakura	PB	New Zealand	518
P 3556	Hinau	PB	New Zealand	518
P 3711	Um Almaradim	PBM	Kuwait	448
P 3713	Ouha	PBM	Kuwait	448
P 3715	Failaka	PBM	Kuwait	448
P 3717	Maskan	PBM	Kuwait	448
P 3719	Al-Ahmadi	PBM	Kuwait	448
P 3721	Alfahaheel	PBM	Kuwait	448
P 3723	Al Yarmouk	PBM	Kuwait	448
P 3725	Garoh	PBM	Kuwait	448
LST 4001	Oosumi	LPD/LSTH	Japan	407
LST 4002	Shimokita	LPD/LSTH	Japan	407
LST 4003	Kunisaki	LPD/LSTH	Japan	407
LSU 4171	Yura	LSU/LCU	Japan	408
LSU 4172	Noto	LSU/LCU	Japan	408
ATS 4202	Kurobe	AVM/TV	Japan	414
ATS 4203	Tenryu	AVMH/TV	Japan	414
AMS 4301	Hiuchi	YTT	Japan	412
AMS 4302	Suou	YTT	Japan	412
AMS 4303	Amakusa	YTT	Japan	412
P 4505	Al Sanbouk	PGGF	Kuwait	448
AGB 5002	Shirase	AGBH	Japan	414
AGS 5102	Futami	AGS	Japan	413
AGS 5103	Suma	AGS	Japan	412
AGS 5104	Wakasa	AGS	Japan	413
AGS 5105	Nichinan	AGS	Japan	412
AOS 5201	Hibiki	AGOSH	Japan	412
AOS 5202	Harima	AGOSH	Japan	412
A 5203	Andromeda	AGSC	Portugal	588
A 5205	Auriga	AGSC	Portugal	588
A 5210	Bérrio	AORLH	Portugal	589
A 5303	Ammiraglio Magnaghi	AGSH	Italy	384
A 5304	Aretusa	AGS	Italy	384
A 5305	Galatea	AGS	Italy	384
A 5309	Anteo	ARSH	Italy	386
A 5311	Palinuro	AXS	Italy	387
A 5312	Amerigo Vespucci	AXS	Italy	387
A 5315	Raffaele Rossetti	AG/AGOR	Italy	384
A 5318	Prometeo	ATR	Italy	388
A 5319	Ciclope	ATR	Italy	388
A 5320	Vincenzo Martellotta	AG/AGE	Italy	384
A 5324	Titano	ATR	Italy	388
A 5325	Polifemo	ATR	Italy	388
A 5326	Etna	AORH	Italy	385
A 5327	Stromboli	AORH	Italy	385
A 5328	Gigante	ATR	Italy	388
A 5329	Vesuvio	AORH	Italy	385
A 5330	Saturno	ATR	Italy	388
A 5340	Elettra	AGORH/AGE/AGI	Italy	383
A 5347	Gorgona	AKL	Italy	386
A 5348	Tremiti	AKL	Italy	386
A 5349	Caprera	AKL	Italy	386
A 5351	Pantellaria	AKL	Italy	386
A 5352	Lipari	AKL	Italy	386
A 5353	Capri	AKL	Italy	386
A 5359	Bormida	AWT	Italy	385
A 5364	Ponza	ABU	Italy	386
A 5365	Tenace	ATR	Italy	388
A 5366	Levanzo	ABU	Italy	386
A 5367	Tavolara	ABU	Italy	386
A 5368	Palmaria	ABU	Italy	386
A 5376	Ticino	AWT	Italy	386
A 5377	Tirso	AWT	Italy	386
A 5379	Astice	AXL	Italy	388
A 5380	Mitilo	AXL	Italy	388
A 5382	Porpora	AXL	Italy	388
A 5383	Procida	ABU	Italy	386
A 5384	Alpino	MCSH	Italy	385
A 5390	Leonardo	AGOR (C)	NATO	505
S 5509	Qaruh	AGH	Kuwait	451
M 5550	Lerici	MHSC	Italy	380
M 5551	Sapri	MHSC	Italy	380
M 5552	Milazzo	MHSC	Italy	380
M 5553	Vieste	MHSC	Italy	380
M 5554	Gaeta	MHSC	Italy	380
M 5555	Termoli	MHSC	Italy	380
M 5556	Alghero	MHSC	Italy	380
M 5557	Numana	MHSC	Italy	380
M 5558	Crotone	MHSC	Italy	380
M 5559	Viareggio	MHSC	Italy	380
M 5560	Chioggia	MHSC	Italy	380
M 5561	Rimini	MHSC	Italy	380
P 5702	Istiqlal	PGGF	Kuwait	449
ASE 6101	Kurihama	ASE/AGE	Japan	413
ASE 6102	Asuka	AGEH	Japan	413
P 6113	Geier	PGGFM	Germany	273
P 6118	Seeadler	PGGFM	Germany	273
P 6119	Habicht	PGGFM	Germany	273
P 6120	Kormoran	PGGFM	Germany	273
P 6121	Gepard	PGGFM	Germany	273
P 6122	Puma	PGGFM	Germany	273
P 6123	Hermelin	PGGFM	Germany	273
P 6124	Nerz	PGGFM	Germany	273
P 6125	Zobel	PGGFM	Germany	273
P 6126	Frettchen	PGGFM	Germany	273
P 6127	Dachs	PGGFM	Germany	273
P 6128	Ozelot	PGGFM	Germany	273
P 6129	Wiesel	PGGFM	Germany	273
P 6130	Hyäne	PGGFM	Germany	273
P 8111	Durbar	PTFG	Bangladesh	48
P 8112	Duranta	PTFG	Bangladesh	48
P 8113	Durvedya	PTFG	Bangladesh	48
P 8114	Durdam	PTFG	Bangladesh	48
P 8125	Durdharsha	PTFG	Bangladesh	47
P 8126	Durdanta	PTFG	Bangladesh	47
P 8128	Dordanda	PTFG	Bangladesh	47
P 8131	Anirban	PTFG	Bangladesh	47
P 8141	Uttal	PTFG	Bangladesh	48
Y 8050	Urania	AXS	Netherlands	515
Y 8760	Patria	AOTL	Netherlands	514
L 9011	Foudre	LSDH/TCD	France	243
L 9012	Siroco	LSDH/TCD	France	243
L 9013	Mistral	LHDM/BPC	France	242
L 9014	Tonnerre	LHDM/BPC	France	242
L 9021	Ouragan	LSDH /TCD	France	244
L 9022	Orage	LSDH/ TCD	France	244
L 9031	Francis Garnier	LSTH	France	244
L 9032	Dumont D'Urville	LSTH	France	244
L 9033	Jacques Cartier	LSTH	France	244
L 9034	La Grandière	LSTH	France	244
L 9051	Sabre	LCT	France	245
L 9052	Dague	LCT	France	245
L 9061	Rapière	LCT	France	245
L 9062	Hallebarde	LCT	France	245
L 9077	Bougainville	AGIH	France	248
L 9090	Gapeau	LSL	France	251
L 9892	San Giorgio	LPD	Italy	383
L 9893	San Marco	LPD	Italy	383
L 9894	San Giusto	LPD	Italy	383

WORLD NAVIES

A — Z

Albania

Country Overview

After being governed by a communist regime since 1946, democratic elections in the Republic of Albania took place in 1991 although since then there have been periods of instability. Situated in western part of the Balkan Peninsula, the country has an area of 11,100 square miles and is bordered to the north by Serbia and Montenegro and to the south by Greece. There is a coastline of 195 n miles with the Adriatic Sea on which Durrës and Vlorë are the principal ports. The capital and largest city is Tirana. Territorial waters (12 n miles) are claimed but an EEZ has not been claimed. Italy provides strong operational, training and administrative support. Joint Coast Guard and Customs patrols are mounted within territorial waters while other personnel training is conducted in Italy. Turkey has established a 300-strong detachment at Vlorë to provide support and training, Greece assists with the support of navigational aids and the US also provides training.

Headquarters Appointments

Commander of the Navy:
   Rear Admiral Robert Bali

Personnel

2005:

(a)   1,500 approximately (60 per cent draftees)
(b)   military service

Bases

HQ: Durrës
Districts: Durrës (1st), Vlorë (2nd).
Bases: Shengyin, Himarë, Saranda, Vlorë.

PATROL FORCES

Notes: (1) Pennant numbers beginning with '1' indicate units from the Durrës district. Those beginning with '2' are from the Vlorë district.
(2) There are six inshore patrol craft of 12-15 m length.

5 HUCHUAN (TYPE 025/026) CLASS
(FAST ATTACK HYDROFOIL—TORPEDO) (PTK)

S 101-104   109

Displacement, tons: 39 standard; 45 full load
Dimensions, feet (metres): 71.5 × 20.7 × 11.8 (hullborne) *(21.8 × 6.3 × 3.6)*
Main machinery: 3 Type M 50F diesels; 3,300 hp(m) *(2.4 MW)* sustained; 3 shafts
Speed, knots: 50 foilborne. **Range, n miles:** 500 at 30 kt
Complement: 11
Guns: 4—14.5 mm (2 twin) MGs.
Torpedoes: 2—21 in *(533 mm)* tubes; Yu-1; 9.2 km *(5 n miles)* at 39 kt; warhead 400 kg.
Radars: Surface search/fire control: Skin Head; E-band.

Comment: Built in Shanghai and transferred from China as follows; six in 1968, 15 in 1969, two in 1970, seven in 1971, two in June 1974. Not all have foils but those that do have them forward while the stern planes on the surface. One escaped to Italy in May 1991 and was seized by the Italian authorities but handed back in October 1991. These are the last survivors of 10 that escaped to Italy in early 1997 and were returned in batches in 1998. Seldom seen at sea.
UPDATED

HUCHUAN CLASS (old numbers)   *6/2000, Massimo Annati, ITN /* 0104158

1 SHANGHAI II CLASS (FAST ATTACK CRAFT—GUN) (PC)

P 115

Displacement, tons: 113 standard; 134 full load
Dimensions, feet (metres): 127.3 × 17.7 × 5.6 *(38.8 × 5.4 × 1.7)*
Main machinery: 2 Type L-12V-180 diesels; 2,400 hp(m) *(1.76 MW)* (forward)
   2 Type 12-D-6 diesels; 1,820 hp(m) *(1.34 MW)* (aft); 4 shafts
Speed, knots: 30. **Range, n miles:** 700 at 16.5 kt
Complement: 34
Guns: 4 China 37 mm/63 (2 twin); 180 rds/min to 8.5 km *(4.6 n miles)*; weight of shell 1.42 kg.
   4 USSR 25 mm/60 (2 twin); 270 rds/min to 3 km *(1.6 n miles)*; weight of shell 0.34 kg.
Torpedoes: 2—21 in *(533 mm)* tubes; Yu-1; 9.2 km *(5 n miles)* at 39 kt; warhead 400 kg.
Depth charges: 2 projectors; 8 depth charges in lieu of torpedo tubes.
Mines: Rails can be fitted; probably only 10 mines.
Radars: Surface search/fire control: Skin Head; I-band.
Sonars: Hull-mounted set probably fitted.

Comment: Four transferred from China in mid-1974 and two in 1975. One ship escaped to Italy in early 1997, returned in early 1998 and was reported repaired in 2000. Has torpedo tubes on the stern taken from deleted Huchuan class. Seldom seen at sea.
VERIFIED

SHANGHAI II (China colours)   *6/1992 /* 0081445

2 PO 2 CLASS (COASTAL PATROL CRAFT) (PB)

A 120   A 212

Displacement, tons: 56 full load
Dimensions, feet (metres): 70.5 × 11.5 × 3.3 *(21.5 × 3.5 × 1)*
Main machinery: 1 Type 3-D-12 diesel; 300 hp(m) *(220 kW)* sustained; 1 shaft
Speed, knots: 12
Complement: 8
Guns: 2—12.7 mm MGs. At least one of the class has a twin 25 mm/60.
Radars: Surface search: I-band.

Comment: Two survive from a total of 11 transferred from USSR 1957-60. Previous minesweeping gear has been removed and the craft are used for utility roles. All escaped to Italy in early 1997 and returned, two in early 1998 and one in late 1998. Two others were towed back as being beyond repair. One other A 451 was sunk in a collision with an Italian corvette in March 1997. Seldom seen at sea.
VERIFIED

PO 2 (old number)   *7/1992, Terje Nilsen /* 0056447

MINE WARFARE FORCES

2 T 301 CLASS (MINESWEEPERS—INSHORE) (MSI)

M 113-114

Displacement, tons: 146 standard; 170 full load
Dimensions, feet (metres): 124.6 × 18.7 × 5.2 *(38 × 5.7 × 1.6)*
Main machinery: 3—6-cyl diesels; 900 hp(m) *(661 kW)*; 3 shafts
Speed, knots: 14. **Range, n miles:** 2,200 at 9 kt
Complement: 25
Guns: 2—37 mm/63; 160 rds/min to 8.5 km *(5 n miles)*; weight of shell 0.7 kg.
   4—14.5 mm (2 twin) MGs.
Mines: Mine rails fitted for 18.

Comment: Transferred from the USSR—two in 1957, two in 1959 and two in 1960. Those deleted have been cannibalised for spares. Three escaped to Italy in early 1997, were repaired and returned in 1998. Two remain operational but are seldom seen at sea.
VERIFIED

T 301 (old number)   *1991*

2 T 43 CLASS (MINESWEEPERS—OCEAN) (MSO)

M 111-112

Displacement, tons: 500 standard; 580 full load
Dimensions, feet (metres): 190.2 × 27.6 × 6.9 *(58 × 8.4 × 2.1)*
Main machinery: 2 Kolomna Type 9-D-8 diesels; 2,000 hp(m) *(1.47 MW)* sustained; 2 shafts
Speed, knots: 15. **Range, n miles:** 3,000 at 10 kt; 2,000 at 14 kt
Complement: 65
Guns: 4—37 mm/63 (2 twin); 160 rds/min to 9 km *(5 n miles)*; weight of shell 0.7 kg.
8—12.7 mm MGs.
Depth charges: 2 projectors.
Mines: 16.
Radars: Air/surface search: Ball End; E/F-band.
Navigation: Furuno; I-band.
Sonars: Stag Ear; hull-mounted set probably fitted.

Comment: Transferred from USSR in 1960. All escaped to Italy in early 1997 and were returned in 1998. M 111 refitted in Italy in 2002 and M 112 is expected to follow.

VERIFIED

T 43 class   5/1996 / 0056449

AUXILIARIES

Notes: In addition there are a survey ship of 20 tons *(A 110)*, an old ex-USSR Shalanda class tender *Marinza*, a 'Poluchat' torpedo recovery craft *(A 110)*, a water-barge, two tugs and a floating dock *(Vlorë)*.

1 LCT 3 CLASS (REPAIR SHIP) (ARL)

A 223 (ex-*MOC 1203*)

Displacement, tons: 640 full load
Dimensions, feet (metres): 192 × 31 × 7 *(58.6 × 9.5 × 2.1)*
Main machinery: 2 diesels; 1,000 hp *(746 kW)*; 2 shafts
Speed, knots: 8
Complement: 24

Comment: 1943 built LCT converted in Italian use as a repair craft. Refitted in Italy, transferred in 1999 and used for moored technical support. To be decommissioned in late 2005 following improvements to naval base facilities.

UPDATED

LCT 3 (Italian colours)   10/1998, Diego Quevedo / 0017507

COAST GUARD (ROJA BREGDETARE)

Note: (1) Coast Guard vessel pennant numbers are prefixed by the letter 'R'.
(2) A 'Nyryat 1' diving tender (R 218) was transferred from the Navy in 2003.

4 TYPE 227 INSHORE PATROL CRAFT (PBR)

R 123 (ex-*CP 229*)   R 225 (ex-*CP 234*)
R 124 (ex-*CP 235*)   R 226 (ex-*CP 236*)

Displacement, tons: 16 full load
Dimensions, feet (metres): 44.0 × 15.7 × 4.3 *(13.4 × 4.8 × 1.3)*
Main machinery: 2 AIFO 8281-SRM diesels; 1,770 hp *(1.32 MW)*; 2 shafts
Speed, knots: 24. **Range, n miles:** 400 at 24 kt
Complement: 5
Radars: Surface search: I-band.

Comment: Wooden construction. Built in Italy 1966-69. Transferred from Italian Coast Guard to Albanian Coast Guard in 2002.

UPDATED

2 COASTAL PATROL CRAFT (PB)

R 117   R 217

Displacement, tons: 18 full load
Dimensions, feet (metres): 45.6 × 13 × 3 *(13.9 × 4 × 0.9)*
Main machinery: 2 diesels; 1,300 hp *(942 kW)*; 2 waterjets
Speed, knots: 34. **Range, n miles:** 200 at 30 kt
Complement: 4
Guns: 2—12.7 mm MGs.
Radars: Surface search: Raytheon; I-band.

Comment: Transferred from the US on 27 February 1999. Reported operational.

VERIFIED

CPC (US colours)   6/1994, PBI / 0056448

3 SEA SPECTRE Mk III (PB)

R 118   R 215   R 216

Displacement, tons: 41 full load
Dimensions, feet (metres): 65 × 18 × 5.9 *(19.8 × 5.5 × 1.8)*
Main machinery: 3 Detroit 8V-71 diesels; 690 hp *(515 kW)* sustained; 3 shafts
Speed, knots: 28. **Range, n miles:** 450 at 25 kt
Complement: 9
Guns: 2—25 mm. 2—12.7 mm MGs.
Radars: Surface search: Raytheon; I-band.

Comment: Transferred from the US on 27 February 1999.

VERIFIED

SEA SPECTRE (US colours)   4/1991, Giorgio Arra / 0056446

7 TYPE 2010 INSHORE PATROL CRAFT (PBR)

R 125 (ex-CP 2008)   R 128 (ex-CP 2034)   R 228 (ex-CP 2023)
R 126 (ex-CP 2020)   R 224 (ex-CP 2010)
R 127 (ex-CP 2021)   R 227 (ex-CP 2007)

Displacement, tons: 15 full load
Dimensions, feet (metres): 41.0 × 11.8 × 3.6 *(12.5 × 3.6 × 1.1)*
Main machinery: 2 AIFO diesels; 1,072 hp *(800 kW)*; 2 shafts
Speed, knots: 24. **Range, n miles:** 533 at 20 kt
Complement: 5
Radars: Surface search: I-band.

Comment: Former harbour launches built in Italy in the 1970s. GRP construction. One transferred from Italian Coast Guard to Albanian Coast Guard in 2002 and a further six in 2004.

UPDATED

1 TYPE 303 COASTAL PATROL CRAFT (PB)

R 122 (ex-*CP 303*)

Displacement, tons: 20 full load
Dimensions, feet (metres): 44.0 × 12.5 × 3.6 *(13.4 × 3.8 × 1.1)*
Main machinery: 2 GM6V53 diesels; 730 hp *(544 kW)*; 2 shafts
Speed, knots: 13. **Range, n miles:** 350 at 13 kt
Complement: 5
Radars: Surface search: I-band.

Comment: Built in US in 1965. Transferred from Italian Coast Guard to Albanian Coast Guard in 2002.

VERIFIED

Algeria
MARINE DE LA REPUBLIQUE ALGERIENNE

Country Overview

Formerly a French colony, the People's Democratic Republic of Algeria gained independence in 1963. Situated in north Africa, it has an area of 919,595 square miles and is bordered to the east by Tunisia and Libya, to the south by Niger, Mali, and Mauritania and to the west by Morocco. It has a 540 n mile coastline with the Mediterranean. The capital, largest city and principal port is Algiers. Territorial seas (12 n miles) and Fishery zones (32/52 n miles) have been claimed but an EEZ has not been claimed.

Headquarters Appointments

Commander of the Navy:
  Lieutenant General Mohand Taha Yala
Inspector General of the Navy:
  General Major Abdelmadjid Taright

Personnel

(a)   2005: 7,500 (500 officers) (Navy) (includes at least 600 naval infantry); 500 (Coast Guard)
(b)   Voluntary service

Bases

Algiers (1st Region), Mers-el-Kebir (2nd Region), Jijel (3rd Region), Annaba (CG HQ)

Coast Defence

Four batteries of truck-mounted SS-C-3 Styx twin launchers. Permanent sites at Algiers, Mers-el-Kebir and Jijel linked by radar.

SUBMARINES

Notes: One decommissioned Romeo class is used for training.

2 KILO CLASS (PROJECT 877E) (SSK)

Name	No	Builders	Laid down	Launched	Commissioned
RAIS HADJ MUBAREK	012	Admiralty Yard, Leningrad	1985	1986	Oct 1987
EL HADJ SLIMANE	013	Admiralty Yard, Leningrad	1985	1987	Jan 1988

Displacement, tons: 2,325 surfaced; 3,076 dived
Dimensions, feet (metres): 238.2 × 32.5 × 21.7
  (72.6 × 9.9 × 6.6)
Main machinery: Diesel-electric; 2 diesels; 3,650 hp(m) *(2.68 MW)*; 2 generators; 1 motor; 5,900 hp(m) *(4.34 MW)*; 1 shaft; 2 auxiliary MT-168 motors; 204 hp(m) *(150 kW)*; 1 economic speed motor; 130 hp(m) *(95 kW)*
Speed, knots: 17 dived; 10 surfaced; 9 snorting
Range, n miles: 6,000 at 7 kt snorting; 400 at 3 kt dived
Complement: 52 (13 officers)

Torpedoes: 6—21 in *(533 mm)* tubes. Combination of Russian TEST-71ME; anti-submarine active/passive homing to 15 km *(8.2 n miles)* at 40 kt; warhead 205 kg and 53-65; anti-surface ship passive wake homing to 19 km *(10.3 n miles)* at 45 kt; warhead 300 kg. Total of 18 weapons.
Mines: 24 in lieu of torpedoes.
Countermeasures: ESM: Brick Pulp; radar warning.
Weapons control: MVU 110 TFCS.
Radars: Surface search: Snoop Tray; I-band.
Sonars: MGK 400 Shark Teeth/Shark Fin; hull-mounted; passive/active search and attack; medium frequency.
  MG 519 Mouse Roar; active attack; high frequency.

Programmes: New construction hulls, delivered as replacements for the Romeo class.

RAIS HADJ MUBAREK
3/1996 / 0056450

Structure: Diving depth, 790 ft *(240 m)*. 9,700 kW h batteries. Pressure hull 169.9 ft *(51.8 m)*. May be fitted with SA-N-5/8 portable SAM launcher.

Operational: One in refit at St Petersburg from June 1993, returned to service in May 1995. Second in refit in late 1993 and back in March 1996. Both are active. **UPDATED**

FRIGATES

3 MOURAD RAIS (KONI) CLASS (PROJECT 1159.2) (FFLM)

Name	No	Builders	Commissioned
MOURAD RAIS	901	Zelenodolsk Shipyard	20 Dec 1980
RAIS KELLICH	902	Zelenodolsk Shipyard	24 Mar 1982
RAIS KORFOU	903	Zelenodolsk Shipyard	3 Jan 1985

Displacement, tons: 1,440 standard; 1,900 full load
Dimensions, feet (metres): 316.3 × 41.3 × 11.5
  (96.4 × 12.6 × 3.5)
Main machinery: CODAG; 1 SGW, Nikolayev, M8B gas turbine (centre shaft); 18,000 hp(m) *(13.25 MW)* sustained; 2 Russki B-68 diesels; 15,820 hp(m) *(11.63 MW)* sustained; 3 shafts
Speed, knots: 27 gas; 22 diesel. **Range, n miles:** 1,800 at 14 kt
Complement: 130

Missiles: SAM: SA-N-4 Gecko twin launcher ❶; semi-active radar homing to 15 km *(8 n miles)* at 2.5 Mach; height envelope 9-3,048 m *(29.5-10,000 ft)*; warhead 50 kg; 20 missiles. Some anti-surface capability.

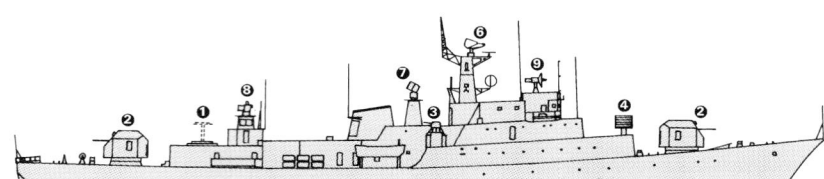

MOURAD RAIS
(Scale 1 : 900), Ian Sturton / 0567433

MOURAD RAIS
8/2004*, Schaeffer/Marsan / 1044063

RAIS KORFOU

8/2004 *, B Prézelin / 1044062

Guns: 4—3 in *(76 mm)*/60 (2 twin) ❷; 90 rds/min to 15 km *(8 n miles)*; weight of shell 6.8 kg.
4—30 mm/65 (2 twin) ❸; 500 rds/min to 5 km *(2.7 n miles)*; weight of shell 0.54 kg.
A/S mortars: 2—12-barrelled RBU 6000 ❹; range 6,000 m; warhead 31 kg.
Torpedoes: 4—533 mm (2 twin) (in 903 only) ❺.
Depth charges: 2 racks.
Mines: Rails; capacity 22.

Countermeasures: Decoys: 2 PK 16 chaff launchers.
ESM: Watch Dog. Cross Loop D/F.
Weapons control: 3P-60 UE.
Radars: Air/surface search: Pozitiv-ME1.2 ❻; I-band. (Strut Curve); E/F-band (in 901 and 902).
Navigation: Don 2; I-band.
Fire Control: Drum tilt ❼; H/I-band (for search/acquisition/FC).
Pop Group ❽; F/H/I-band (for missile control).
Hawk screech (901 and 902) ❾; I-band.

IFF: High Pole B. 2 Square Head.
Sonars: Hercules (MG 322) hull-mounted; active search and attack; medium frequency.

Programmes: New construction ships built in USSR with hull numbers 5, 7 and 10 in sequence. Others of the class built for Cuba, Yugoslavia, East Germany and Libya. Interest was shown in ex-GDR ships in 1991 but sale was rejected by the German government.
Modernisation: New generators fitted 1992-94. *Rais Korfou* in refit at Kronstadt from 1997 to November 2000. The refit included replacement of Strut Curve radar, removal of Hawk screech fire-control radar, fitting of torpedo tubes and a new electronic suite. The contract for refit of a second ship and a 'Nanuchka' is expected in 2005.
Structure: The deckhouse aft in Type II Konis houses air conditioning machinery.
Operational: All have been used for Training cruises.
UPDATED

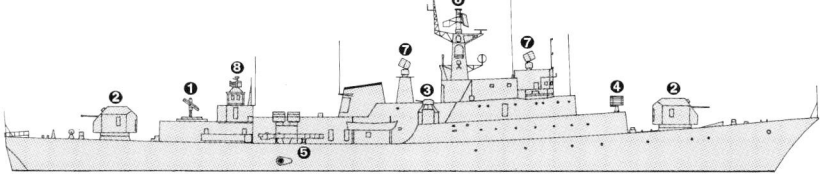

RAIS KORFOU

(Scale 1 : 900), Ian Sturton / 0104159

CORVETTES

3 NANUCHKA II (BURYA) CLASS (PROJECT 1234) (MISSILE CORVETTES) (PTGM)

Name	No	Builders	Commissioned
RAIS HAMIDOU	801	Petrovsky, Leningrad	4 July 1980
SALAH RAIS	802	Petrovsky, Leningrad	9 Feb 1981
RAIS ALI	803	Petrovsky, Leningrad	8 May 1982

Displacement, tons: 660 full load
Dimensions, feet (metres): 194.5 × 38.7 × 8.5 *(59.3 × 11.8 × 2.6)*
Main machinery: 6 M 504 diesels; 26,112 hp(m) *(19.2 MW)*; 3 shafts
Speed, knots: 33. **Range, n miles:** 2,500 at 12 kt; 900 at 31 kt
Complement: 42 (7 officers)

Missiles: SSM: 16 Zvezda SS-N-25 (in 802) (4 quad) (Kh 35E Uran); active radar homing to 130 km *(70.2 n miles)* at 0.9 Mach; sea skimmer.
4 SS-N-2C (in 801 and 803); active radar or IR homing to 46 km *(25 n miles)* at 0.9 Mach; warhead 513 kg.
SAM: SA-N-4 Gecko twin launcher; semi-active radar homing to 15 km *(8 n miles)* at 2.5 Mach; height envelope 9-3,048 m *(29.5-10,000 ft)*; warhead 50 kg; 20 missiles. Some anti-surface capability.
Guns: 2—57 mm/80 (twin); 120 rds/min to 6 km *(3.3 n miles)*; weight of shell 2.8 kg.
1—30 mm/65 AK 630 (in 802); 6 barrels per mounting; 3,000 rds/min combined to 2 km.
Countermeasures: Decoys: 2 PK 16 16-barrelled chaff launchers.
ESM: Bell Tap. Cross Loop; D/F.
Radars: Surface search: Square Tie (Radome) (801 and 803); I-band. Pozitiv-ME1.2 (802); I-band.
Navigation: Don 2; I-band.
Fire control: Pop Group; F/H/I-band (SA-N-4). Muff Cob or Drum Tilt (802); G/H-band. Plank Shave; E-band (SS-N-25).
IFF: Two Square Head. High Pole.

Programmes: Delivered as new construction.
Modernisation: *Salah Rais* refitted at Kronstadt 1997 to November 2000 with refurbished diesels, a replacement SSM system and electronic suite. The contract for refit of a second ship and a 'Koni' is expected in 2005.
UPDATED

RAIS HAMIDOU

5/2002, Globke Collection / 0529554

SALAH RAIS (with SS-N-25)

11/2000, German Navy / 0077834

3 + (1) DJEBEL CHENOUA (C 58) CLASS (PROJECT 802) (FSG)

Name	No	Builders	Launched	Commissioned
DJEBEL CHENOUA	351	ECRN, Mers-el-Kebir	3 Feb 1985	Nov 1988
EL CHIHAB	352	ECRN, Mers-el-Kebir	Feb 1990	June 1995
AL KIRCH	353	ECRN, Mers-el-Kebir	July 2000	2002

Displacement, tons: 496 standard; 540 full load
Dimensions, feet (metres): 191.6 × 27.9 × 8.5 *(58.4 × 8.5 × 2.6)*
Main machinery: 3 MTU 20V 538 TB92 diesels; 12,800 hp(m) *(9.4 MW)*; 3 shafts
Speed, knots: 31
Complement: 52 (6 officers)

Missiles: SSM: 4 China C 802 (CSS-N-8 Saccade) (2 twin); active radar homing to 120 km *(66 n miles)* at 0.9 Mach; warhead 165 kg.
Guns: 1 Russian 3 in *(76 mm)*/60; 90 rds/min to 15 km *(8 n miles)*; weight of shell 6.8 kg.
2—30 mm/65 (1 twin); 500 rds/min to 5 km *(2.7 n miles)*; weight of shell 0.54 kg.
Weapons control: Optronic director.
Radars: Surface search: Racal Decca 1226; I-band.

Programmes: Ordered July 1983. Project 802 built with Bulgarian assistance. First one completed trials in 1988. Work on the second of class was suspended in 1992 due to shipyard debt problems but the ship completed in 1995. Main guns were fitted at a later date. Construction of a fourth ship is reported to be under consideration.
Modernisation: First two ships fitted with main armaments comprising SSMs and 76 mm guns.
Structure: Hull size suggests association with Bazán Cormoran class.
UPDATED

DJEBEL CHENOUA

7/2002 / 0534071

LAND-BASED MARITIME AIRCRAFT

Numbers/Type: 2 Beechcraft Super King Air 200T.
Operational speed: 282 kt *(523 km/h)*.
Service ceiling: 35,000 ft *(10,670 m)*.
Range: 2,030 n miles *(3,756 km)*.
Role/Weapon systems: Operated by air force for close-range EEZ operations. Sensors: Weather radar only. Weapons: Unarmed.
VERIFIED

Numbers/Type: 3 Fokker F27-400/600.
Operational speed: 250 kt *(463 km/h)*.
Service ceiling: 25,000 ft *(7,620 m)*.
Range: 2,700 n miles *(5,000 km)*.
Role/Weapon systems: Visual reconnaissance duties in support of EEZ, particularly offshore platforms. Sensors: Weather radar and visual means only. Weapons: Limited armament.
VERIFIED

PATROL FORCES

9 OSA II CLASS (PROJECT 205) (FAST ATTACK CRAFT—MISSILE) (PTGF)

644-652

Displacement, tons: 245 full load
Dimensions, feet (metres): 126.6 × 24.9 × 8.8 *(38.6 × 7.6 × 2.7)*
Main machinery: 3 Type M 504 diesels; 10,800 hp(m) *(7.94 MW)* sustained; 3 shafts
Speed, knots: 37. **Range, n miles:** 500 at 35 kt
Complement: 30

Missiles: SSM: 4 SS-N-2B; active radar or IR homing to 46 km *(25 n miles)* at 0.9 Mach; warhead 513 kg.
Guns: 4—30 mm/65 (2 twin); 500 rds/min to 5 km *(2.7 n miles)*; weight of shell 0.54 kg.
Radars: Surface search: Square Tie; I-band.
Fire control: Drum Tilt; H/I-band.
IFF: 2 Square Head. High Pole B.

Programmes: Osa II transferred 1976-77 (four), fifth in September 1978, sixth in December 1978, next pair in 1979 and one from the Black Sea on 7 December 1981.
Modernisation: Plans to re-engine were reported as starting in late 1992 but there has been no confirmation.
Operational: At least six Osa IIs are active.

UPDATED

OSA 652　　　　　　　　　　　　　　　　　　　　　　　　　　　1989

13 KEBIR CLASS (FAST ATTACK CRAFT—GUN) (PG)

EL YADEKH 341	EL RASSED 345	— 349	— 357
EL MOURAKEB 342	EL DJARI 346	EL KANASS 350	— 358
EL KECHEF 343	EL SAHER 347	— 356	
EL MOUTARID 344	EL MOUKADEM 348		

Displacement, tons: 166 standard; 200 full load
Dimensions, feet (metres): 123 × 22.6 × 5.6 *(37.5 × 6.9 × 1.7)*
Main machinery: 2 MTU 12V 538 TB92 diesels; 5,110 hp(m) *(3.8 MW)*; 2 shafts (see *Structure*)
Speed, knots: 27. **Range, n miles:** 3,300 at 12 kt; 2,600 at 15 kt
Complement: 27 (3 officers)

Guns: 1 OTO Melara 3 in *(76 mm)*/62 compact (341-342); 85 rds/min to 16 km *(9 n miles)* anti-surface; 12 km *(6.5 n miles)* anti-aircraft; weight of shell 6 kg.
4 USSR 25 mm/60 (2 twin) (remainder); 270 rds/min to 3 km *(1.6 n miles)*; weight of shell 0.34 kg.
2 USSR 14.5 mm (twin) (in first five).
Weapons control: Lawrence Scott optronic director (in 341 and 342).
Radars: Surface search: Racal Decca 1226; I-band.

Programmes: Design and first pair ordered from Brooke Marine in June 1981. First left for Algeria without armament in September 1982, second arrived Algiers 12 June 1983. A further seven were then assembled or built at ECRN, Mers-el-Kebir. Of these, 346 commissioned 10 November 1985 and 347-349 delivered by 1993. After a delay two further craft were completed; 350 in late 1997 followed by 354 in 1998. 356-358 have since been added and original plans for a class of 15 may be achieved.
Structure: Same hull as Barbados *Trident*. There are some variations in armament.
Operational: Six of the class have been transferred to the Coast Guard.

UPDATED

KEBIR 350 (with 25 mm guns)　　　　　　　　　　　　　6/1998 / 0017511

EL MOURAKEB and EL YADEKH　　　　　　　　　5/1990 / 0056452

AMPHIBIOUS FORCES

2 LANDING SHIPS (LOGISTIC) (LSTH)

Name	No	Builders	Commissioned
KALAAT BENI HAMMAD	472	Brooke Marine, Lowestoft	Apr 1984
KALAAT BENI RACHED	473	Vosper Thornycroft, Woolston	Oct 1984

Displacement, tons: 2,450 full load
Dimensions, feet (metres): 305 × 50.9 × 8.1 *(93 × 15.5 × 2.5)*
Main machinery: 2 MTU 16V 1163 TB82 diesels; 8,880 hp(m) *(6.5 MW)* sustained; 2 shafts
Speed, knots: 15. **Range, n miles:** 3,000 at 12 kt
Complement: 81
Military lift: 240 troops; 7 MBTs and 380 tons other cargo; 2 ton crane with athwartships travel

Guns: 2 Breda 40 mm/70 (twin); 300 rds/min to 12.5 km *(6.8 n miles)*; weight of shell 0.96 kg.
4 USSR 25 mm/60 (2 twin); 270 rds/min to 3 km *(1.6 n miles)*; weight of shell 0.34 kg.
Countermeasures: Decoys: Wallop Barricade double layer chaff launchers.
ESM: Racal Cutlass; intercept.
ECM: Racal Cygnus; jammer.
Weapons control: CSEE Naja optronic.
Radars: Navigation: Racal Decca TM 1226; I-band.
Fire control: Marconi S 800; J-band.

Helicopters: Platform only for one Sea King.

Programmes: First ordered in June 1981, and launched 18 May 1983; second ordered 18 October 1982 and launched 15 May 1984. Similar hulls to Omani *Nasr Al Bahr*.
Structure: These ships have a through tank deck closed by bow and stern ramps. The forward ramp is of two sections measuring length 18 m (when extended) × 5 m breadth, and the single section stern ramp measures 4.3 × 5 m with the addition of 1.1 m finger flaps. Both hatches can support a 60 ton tank and are winch operated. In addition, side access doors are provided on each side forward. The tank deck side bulkheads extend 2.25 m above the upper deck between the forecastle and the forward end of the superstructure, and provide two hatch openings to the tank deck below. Additional 25 mm guns have been fitted either side of the bridge.
Operational: Both are reported active.

UPDATED

KALAAT BENI HAMMAD　　　　　　8/2004 *, B Prézelin / 1044061

1 POLNOCHNY B CLASS (PROJECT 771) (LSM)

471

Displacement, tons: 760 standard; 834 full load
Dimensions, feet (metres): 246.1 × 31.5 × 7.5 *(75 × 9.6 × 2.3)*
Main machinery: 2 Kolomna Type 40-D diesels; 4,400 hp(m) *(3.2 MW)* sustained; 2 shafts
Speed, knots: 18. **Range, n miles:** 1,000 at 18 kt
Complement: 42
Military lift: 180 troops; 350 tons including up to 6 tanks
Guns: 2—30 mm/65 (twin) AK 230; 500 rds/min to 5 km *(2.7 n miles)*; weight of shell 0.54 kg.
2—140 mm 18-tubed rocket launchers.
Radars: Navigation: Don 2; I-band.
Fire control: Drum Tilt; H/I-band.
IFF: Square Head. High Pole B.

Comment: Class built in Poland 1968-70. Transferred from USSR in August 1976. Tank deck covers 237 m². Operational and employed on training tasks.

UPDATED

POLNOCHNY 471　　　　　　　　1990, van Ginderen Collection

MINE WARFARE FORCES

Notes: (1) The Coast Guard support ship *El Mourafik* may have a minelaying capability.
(2) Two MCMV are expected to be out to tender in due course.

SURVEY SHIPS

2 SURVEY CRAFT (YFS)

RAS TARA　　　　　　　　　　　　　　ALIDADE

Comment: *Ras Tara* is of 16 tons displacement, built in 1980 and has a crew of four. *Alidade* is of 20 tons, built in 1983 and has a crew of eight.

UPDATED

1 SURVEY SHIP (AGS)

EL IDRISSI BH 204 (ex-A 673)

Displacement, tons: 540 full load
Complement: 28 (6 officers)

Comment: Built by Matsukara, Japan and delivered 17 April 1980. Based at Algiers.

VERIFIED

EL IDRISSI   9/1990 / 0056453

AUXILIARIES

Notes: A training ship was put out to tender in 1998 but there have been no reports of a contract.

1 POLUCHAT I CLASS (PROJECT 638) (YPT)

A 641

Displacement, tons: 70 standard; 100 full load
Dimensions, feet (metres): 97.1 × 19 × 4.8 *(29.6 × 5.8 × 1.5)*
Main machinery: 2 Type M 50F diesels; 2,200 hp(m) *(1.6 MW)* sustained; 2 shafts
Speed, knots: 20. Range, n miles: 1,500 at 10 kt
Complement: 15

Comment: Transferred from USSR in early 1970s. Has been used for SAR.

VERIFIED

POLUCHAT   1989

TUGS

Notes: There are a number of harbour tugs of about 265 tons. These include *Kader* A 210, *El Chadid* A 211 and *Mazafran* 1-4 Y 206-209.

MAZAFRAN 4   6/1994 / 0056454

COAST GUARD

Notes: (1) Six Kebir class were transferred from the Navy for Coast Guard duties but may have naval crews.
(2) There are also up to 12 small fishery protection vessels in the GC 301 series.

1 SUPPORT SHIP (WARL)

EL MOURAFIK GC 261

Displacement, tons: 600 full load
Dimensions, feet (metres): 193.6 × 27.6 × 6.9 *(59 × 8.4 × 2.1)*
Main machinery: 2 diesels; 2,200 hp(m) *(1.6 MW)*; 2 shafts
Speed, knots: 14
Complement: 54
Guns: 2—12.7 mm MGs.
Radars: Surface search: I-band.

Comment: Delivered by transporter ship from China in April 1990. The design appears to be a derivative of the T43 minesweeper but with a stern gantry. May have a minelaying capability. Based at Algiers.

VERIFIED

EL MOURAFIK   6/2003, B Lemachko / 0569789

7 EL MOUDERRIB (CHUI-E) CLASS (AXL)

EL MOUDERRIB I-VII GC 251-257

Displacement, tons: 388 full load
Dimensions, feet (metres): 192.8 × 23.6 × 7.2 *(58.8 × 7.2 × 2.2)*
Main machinery: 3 PCR/Kolomna diesels; 6,600 hp(m) *(4.92 MW)*; 3 shafts
Speed, knots: 24. Range, n miles: 1,400 at 15 kt
Complement: 42 including 25 trainees
Guns: 4 China 14.5 mm (2 twin).
Radars: Surface search: Type 756; I-band.

Comment: Two delivered by transporter ship from China in April 1990 and described as training vessels. Two more acquired in January 1991, the last three in July 1991. Hainan class hull with modified propulsion and superstructure similar to some Chinese paramilitary vessels. Used for training when boats are carried aft in place of the second 14.5 mm gun.

VERIFIED

EL MOUDERRIB I   6/1992, Diego Quevedo / 0056456

4 BAGLIETTO TYPE 20 (PBF)

EL HAMIL GC 325   EL ASSAD GC 326   MARKHAD GC 327   ETAIR GC 328

Displacement, tons: 44 full load
Dimensions, feet (metres): 66.9 × 17.1 × 5.5 *(20.4 × 5.2 × 1.7)*
Main machinery: 2 CRM 18DS diesels; 2,660 hp(m) *(2 MW)*; 2 shafts
Speed, knots: 36. Range, n miles: 445 at 20 kt
Complement: 11 (3 officers)
Guns: 1 Oerlikon 20 mm.

Comment: The first pair delivered by Baglietto, Varazze in August 1976 and the remainder in pairs at two monthly intervals. Fitted with radar and optical fire control. Four others of the class cannibalised for spares.

UPDATED

BAGLIETTO 20 GC CRAFT   1978, Baglietto

4 EL MOUNKID CLASS (SAR)

EL MOUNKID I GC 231	EL MOUNKID III GC 233
EL MOUNKID II GC 232	EL MOUNKID IV GC234

Comment: First three delivered by transporter ship from China which arrived in Algiers in April 1990, a fourth followed a year later. Used for SAR.

VERIFIED

CUSTOMS

Notes: The Customs service is a paramilitary organisation employing a number of patrol craft armed with small MGs. These include *Bouzagza, Djurdjura, Hodna, Aures* and *Hoggar*. The first three are P 1200 class 39 ton craft capable of 33 kt. The next pair are P 802 class. They were built by Watercraft, Shoreham and delivered in November 1985.

GC 231-233   *1991 /* 0056457

Angola
MARINHA DE GUERRA

Country Overview

Formerly known as Portuguese West Africa, the Republic of Angola became independent in 1975 but has been ravaged by civil war ever since. With an area of 481,354 square miles it has borders to the south with Namibia, to the east with Zambia and to the north and east with the Democratic Republic of the Congo which separates a small exclave, Cabinda, from the rest of the country. Angola has a coastline with the south Atlantic Ocean of some 864 n miles. The capital, largest city and principal port is

Luanda. Territorial seas (12 n miles) and a fisheries zone (200 n miles) are claimed. A 200 n mile Exclusive Economic Zone (EEZ) has been claimed but the limits have not been published. In 2004, there were no operational vessels in the Navy.

Personnel

(a)   2005: 890
(b)   Voluntary service

Bases

Luanda, Lobito, Namibe. (There are other good harbours available on the 1,000 mile coastline.) Naval HQ at Luanda on Ila de Luanda is in an old fort, as is Namibe.

Anguilla

Country Overview

British dependency since 1971 following secession from associated state of St Kitts-Nevis-Anguilla. With an area of 35 square miles, the island is situated at the northern end of the Leeward Islands in the Lesser Antilles and bordered by the Caribbean to the west and Atlantic to the east. Territorial seas (3 n miles) and a fishery zone (200 n miles) are claimed.

Headquarters Appointments

Commissioner of Police:
  Keithly Benjamin

Personnel

2005: 64

POLICE

1 HALMATIC M160 CLASS (INSHORE PATROL CRAFT) (PB)

DOLPHIN

Displacement, tons: 18 light
Dimensions, feet (metres): 52.5 × 15.4 × 4.6 *(16 × 4.7 × 1.4)*
Main machinery: 2 MAN V10 diesels; 820 hp *(610 kW)* sustained; 2 shafts
Speed, knots: 34. **Range, n miles:** 575 at 23 kt
Complement: 8
Guns: 1—12.7 mm MG.
Radars: Surface search: JRC 2254; I-band.

Comment: Built by Halmatic and delivered 22 December 1989. Identical craft to Qatar. GRP hull. Rigid inflatable boat launched by gravity davit. Returned to service on 30 August 2004 after refit.

UPDATED

1 BOSTON WHALER (INSHORE PATROL CRAFT) (PB)

LAPWING

Displacement, tons: 2.2 full load
Dimensions, feet (metres): 27 × 10 × 1.5 *(8.2 × 3 × 0.5)*
Main machinery: 2 Johnson outboards; 300 hp *(225 kW)*
Speed, knots: 38
Complement: 4

Comment: Delivered in 1990 and re-engined in 1992.

UPDATED

DOLPHIN   *8/1999, Anguilla Police /* 0056459

LAPWING   *8/1999, Anguilla Police /* 0056460

Antigua and Barbuda

Country Overview

Independent since 1981, the British monarch, represented by a governor-general, is head of state. Situated at the southern end of the Leeward Islands in the Lesser Antilles chain, the country comprises Antigua (108 square miles), Barbuda to the north and uninhabited Redonda to the southwest. The capital, largest town, and main port is St John's. An archipelagic state, territorial seas (12 n miles) and a fishery zone (200 n miles) are claimed.

A 200 n mile Exclusive Economic Zone (EEZ) has also been claimed but the limits are not defined. The Antigua Barbuda Defence Force (ABDF) took over the Coast Guard on 1 May 1995.

Headquarters Appointments

Commanding Officer, Coast Guard:
    Lieutenant Auden Nicholas

Personnel

2005: 50 (3 officers)

Bases

HQ: Deepwater Harbour, St Johns
Maintenance: Camp Blizzard

COAST GUARD

Notes: (1) In addition there is a Hurricane RIB, *CG 081* with a speed of 35 kt and two Boston Whalers, *CG 071-2*, with speeds of 30 kt. All were acquired in 1988/90.
(2) A 920 Zodiac RHIB, CG 091, was donated by the US government in 2003. It is capable of over 40 kt.

CG 081   *9/2004*, ABDFCG /* 0587691

CG 091   *9/2004*, ABDFCG /* 0587690

1 SWIFT 65 ft CLASS (PB)

Name	No	Builders	Commissioned
LIBERTA	P 01	Swiftships, Morgan City	30 Apr 1984

Displacement, tons: 36 full load
Dimensions, feet (metres): 65.5 × 18.4 × 5 *(20 × 5.6 × 1.5)*
Main machinery: 2 Detroit Diesel 12V-71TA diesels; 840 hp *(616 kW)* sustained; 2 shafts
Speed, knots: 22. **Range, n miles:** 250 at 18 kt
Complement: 9
Guns: 1—12.7 mm MG. 2—7.62 mm MGs.
Radars: Surface search: Furuno; I-band.

Comment: Ordered in November 1983. Aluminium construction. Funded by US. Refitted in 2001.
                                                   VERIFIED

LIBERTA   *5/2003 /* 0568341

1 DAUNTLESS CLASS (PB)

Name	No	Builders	Commissioned
PALMETTO	P 02	SeaArk Marine, Monticello	7 July 1995

Displacement, tons: 11 full load
Dimensions, feet (metres): 40 × 14 × 4.3 *(12.2 × 4.3 × 1.3)*
Main machinery: 2 Caterpillar 3208TA diesels; 870 hp *(650 kW)* sustained; 2 shafts
Speed, knots: 27. **Range, n miles:** 600 at 18 kt
Complement: 4
Guns: 1—7.62 mm MG.
Radars: Surface search: Raytheon R40; I-band.

Comment: Funded by USA. Similar craft delivered to several Caribbean countries in 1994-98.
                                                   UPDATED

PALMETTO   *9/2004*, ABDFCG /* 0587689

Argentina
ARMADA ARGENTINA

Country Overview

The Argentine Republic is situated in southern South America. With an area of 1,068,302 square miles it has borders to the north with Bolivia and Paraguay, to the east with Brazil and Uruguay and to the south and west with Chile. The country includes the Tierra del Fuego territory which comprises the eastern half of the Isla Grande de Tierra del Fuego and a number of adjacent islands to the east, including Isla de los Estados. It also claims sovereignty of the Falkland Islands. The capital, largest city and principal port is Buenos Aires. There are further ports at La Plata, Bahia Blanca, Comodoro Rivadavia and a river port at Rosario. There are some 5,940 n miles of navigable internal waterways. Territorial Seas (12 n miles) are claimed. An EEZ (200 n miles) is claimed but its limits are only partly defined by boundary agreements.

Headquarters Appointments

Chief of Naval General Staff:
    Vice Admiral Jorge Omar Godoy
Deputy Chief of Naval Staff:
    Rear Admiral Ernesto Telmo Juan Gaudiero
Naval Operations Commander:
    Rear Admiral Eduardo Luis Avilés

Senior Appointments

Commander Fleet:
    Vice Admiral Alejandro Daniel Giromini
Commander, Marine Infantry:
    Rear Admiral Carlos Alberto Comadira
Commander Naval Aviation:
    Captain Luis Alberto de Vincenti
Commander, Naval Area Austral:
    Rear Admiral Gustavo Adolfo Trama
Commander, Atlantic:
    Captain Francisco Antonio Galia
Commander, Naval Area Fluvial:
    Captain Guillermo Estevez

Organisation

Naval Area Austral covers coastal area from latitude 46° to 60° south.
Naval Area Atlantic covers coastal area from latitude 36° 18′ to 46° south. Naval Area Fluvial includes the rivers Paraná, Uruguay and Plate. Naval Area Antarctica is in force when *Almirante Irizar* deploys.

Personnel

2005: 16,200 (2,390 officers)

Special Forces

Consists of tactical divers who operate from submarines and other naval units, and amphibious commandos who are trained in parachuting and behind the lines operations. Both groups consist of about 150.

Bases

Buenos Aires (Dársena Norte): Some naval training.
Rio Santiago (La Plata): Schools.
Mar del Plata: Submarine base plus three frigates.
Puerto Belgrano: Main naval base, schools. Fleet Marine Force.
Ushuaia, Deseado, Dársena Sur, Zárate, Caleta Paula; Small naval bases.

Coast Guard (Prefectura Naval Argentina)

In January 1992 the Coast Guard was limited to operations inside 12 mile territorial seas but this legislation was then cancelled in favour of the previous 200 mile operating zone. In May 1994 a Community Protection Secretariat was formed to include the Coast Guard, Border Guard and Federal Police. In June 1996 the Coast Guard and Border Guard were placed under the Interior Ministry.

Prefix to Ships' Names

ARA (Armada Republica Argentina)

Naval Aviation

Personnel: 2,500
The Naval Air Command is at Puerto Belgrano.
1st Naval Air Wing (Punta Indio Naval Air Base): 1st Naval Attack Squadron with Aeromacchi MB-326; Naval Photographic Reconnaissance Group with Beech 200s. Naval Aviation School with Beech T-34s and Turbo Mentor.
2nd Naval Air Wing (Comandante Espora Naval Air Base): ASW Squadron with Grumman S-2T Trackers; 2nd Naval Helicopter Squadron with Agusta/Sikorsky SH-3H and AS-61D Sea Kings; 2nd Naval Attack Squadron with Super Etendards; 1st Naval Helicopter Squadron with Alouette III and Fennecs; 3rd Naval Helicopter Squadron with Bell UH-1Hs.

3rd Naval Air Wing (Almirante Zar Naval Air Base, Trelew): 6th Naval Reconnaissance and Surveillance Squadron with Lockheed P-3B Orions, Lockheed Electra L-188E, Beech 200s and Pilatus PC-6B.
52 Logistic Support Flight (Almirante Izar Naval Air Base): Fokker F-28s.

Marine Corps

Personnel: 2,800
2nd Marine Infantry Battalion (Puerto Belgrano)
3rd Marine Infantry Battalion (Zarate)
4th Marine Infantry Battalion (Ushuaia)
5th Marine Infantry Battalion (Training) (Rio Grande)

Marine Field Artillery Battalion (Puerto Belgrano)
Command and Logistics Support Battalion (Puerto Belgrano)
Amphibious Vehicles Battalion (Puerto Belgrano)
Communications Battalion (Puerto Belgrano)
Marine A/A Battalion (Puerto Belgrano)
Amphibious Engineers Company (Puerto Belgrano)
Amphibious Commandos Group (Puerto Belgrano)
There are Marine Security Battalions at Naval Bases in Buenos Aires and Puerto Belgrano.
There are Marine Security Companies at Naval Bases in Mar del Plata, Trelew, Ushuaia, Punta Indio and Zarate.

Strength of the Fleet

Type	Active (Reserve)	Building
Patrol Submarines	3	—
Destroyers	6	—
Frigates	8	1
Patrol Ships	7	—
Fast Attack Craft (Gun/Missile)	2	—
Coastal Patrol Craft	5	—
Survey/Oceanographic Ships	4	—
Survey Launches	1	—
Transports/Tankers	11	—
Training Ships	4	—

DELETIONS

Destroyers

2003   *Santisima Trinidad*

Mine Warfare Forces

2002   *Chaco, Formosa*

PENNANT LIST

Submarines

S 31	Salta
S 41	Santa Cruz
S 42	San Juan

Destroyers

B 52	Hercules
D 10	Almirante Brown
D 11	La Argentina
D 12	Heroina
D 13	Sarandi

Frigates

31	Drummond
32	Guerrico
33	Granville

41	Espora
42	Rosales
43	Spiro
44	Parker
45	Robinson
46	Gomez Roca

Patrol Forces

A 1	Comandante General Irigoyen
A 2	Teniente Olivieri
A 3	Francisco de Gurruchaga
A 6	Suboficial Castillo
A 9	Alferez Sobral
P 20	Murature
P 21	King
P 61	Baradero
P 62	Barranqueras
P 63	Clorinda

P 64	Concepción del Uruguay
P 65	Punta Mogotes
P 66	Rio Santiago
P 85	Intrepida
P 86	Indomita

Auxiliaries

B 1	Patagonia
B 3	Canal Beagle
B 4	Bahia San Blas
B 5	Cabo de Hornos
B 8	Astra Federico
B 9	Astra Valentina
B 13	Ingeniero Julio Krause
Q 2	Libertad
Q 5	Almirante Irizar
Q 11	Comodoro Rivadavia
Q 15	Cormoran

Q 20	Puerto Deseado
Q 61	Ciudad de Zarate
Q 62	Ciudad de Rosario
Q 63	Punta Alta
Q 73	Itati
Q 74	Fortuna I
Q 75	Fortuna II
R 2	Querandi
R 3	Tehuelche
R 5	Mocovi
R 6	Calchaqui
R 7	Ona
R 8	Toba
R 10	Chulupi
R 12	Mataco
R 16	Capayan
R 18	Chiquilyan
R 19	Morcoyan

SUBMARINES

Notes: Cosmos and Havas underwater chariots in service. Cosmos types are capable of carrying limpet or ground mines.

1 SALTA (209) (TYPE 1200) CLASS (SSK)

Name	No	Builders	Laid down	Launched	Commissioned
SALTA	S 31	Howaldtswerke, Kiel	30 Apr 1970	9 Nov 1972	7 Mar 1974

Displacement, tons: 1,140 surfaced; 1,248 dived
Dimensions, feet (metres): 183.4 × 20.5 × 17.9 *(55.9 × 6.3 × 5.5)*
Main machinery: Diesel-electric; 4 MTU 12V 493 AZ80 diesels; 2,400 hp(m) *(1.76 MW)* sustained; 4 alternators; 1.7 MW; 1 motor; 4,600 hp(m) *(3.36 MW)*; 1 shaft
Speed, knots: 10 surfaced; 22 dived; 11 snorting
Range, n miles: 6,000 at 8 kt surfaced; 230 at 8 kt; 400 at 4 kt dived
Complement: 31 (5 officers)

Torpedoes: 8—21 in *(533 mm)* bow tubes. 14 AEG SST 4 Mod 1; wire-guided; active/passive homing to 12/28 km *(6.5/15 n miles)* at 35/23 kt; warhead 260 kg or US Mk 37; wire-guided; active/passive homing to 8 km *(4.4 n miles)* at 24 kt; warhead 150 kg. Swim-out discharge.
Mines: Capable of carrying ground mines.
Countermeasures: ESM: DR 2000; radar warning.
Weapons control: Signaal M8 digital; computer-based; up to 3 targets engaged simultaneously.
Radars: Navigation: Thomson-CSF Calypso II.
Sonars: Atlas Elektronik CSU 3 (AN 526/AN 5039/41); active/passive search and attack; medium frequency.
Thomson Sintra DUUX 2C and DUUG 1D; passive ranging.

Programmes: Ordered in 1968. Built in sections by Howaldtswerke Deutsche Werft AG, Kiel from the IK 68 design of Ingenieurkontor, Lübeck. Sections were shipped to Argentina for assembly at Tandanor, Buenos Aires.
Modernisation: *Salta* completed a mid-life modernisation at the Domecq Garcia Shipyard. New engines, weapons and electrical systems fitted and the ship was relaunched on 4 October 1994 and recommissioned in May 1995. New batteries were installed at Domecq Garcia in 2004.
Structure: Diving depth, 250 m *(820 ft)*.
Operational: Operational and based at Mar del Plata. Second of class *(San Luis)* cannibalised for spares in 1997.

UPDATED

SALTA

2 SANTA CRUZ (TR 1700) CLASS (SSK)

Name	No	Builders	Laid down	Launched	Commissioned
SANTA CRUZ	S 41	Thyssen Nordseewerke	6 Dec 1980	28 Sep 1982	18 Oct 1984
SAN JUAN	S 42	Thyssen Nordseewerke	18 Mar 1982	20 June 1983	19 Nov 1985

Displacement, tons: 2,116 surfaced; 2,264 dived
Dimensions, feet (metres): 216.5 × 23.9 × 21.3
  (66 × 7.3 × 6.5)
Main machinery: Diesel-electric; 4 MTU 16V 6,720 hp diesels;
  6,720 hp(m) *(4.94 MW)* sustained; 4 alternators; 4.4 MW; 1
  Siemens Type 1HR4525 + 1HR 4525 4-circuit DC motor;
  6.6 MW; 1 shaft
Speed, knots: 15 surfaced; 12 snorting; 26 dived
Range, n miles: 12,000 at 8 kt surfaced; 20 at 25 kt; 460 at 6 kt
  dived
Complement: 29 (5 officers)

Torpedoes: 6—21 in *(533 mm)* bow tubes. 22 AEG SST 4; wire-
  guided; active/passive homing to 12/28 km *(6.5/15 n miles)*
  at 35/23 kt; warhead 260 kg; automatic reload in 50 seconds
  or US Mk 37; wire-guided; active/passive homing to 8 km

(4.4 n miles) at 24 kt; warhead 150 kg. Swim-out discharge.
  Mk 48 to replace Mk 37 in due course.
Mines: Capable of carrying 34 ground mines.
Countermeasures: ESM: Sea Sentry III; radar warning.
Weapons control: Signaal Sinbads; can handle 5 targets and 3
  torpedoes simultaneously.
Radars: Navigation: Thomson-CSF Calypso IV; I-band.
Sonars: Atlas Elektronik CSU 3/4; active/passive search and
  attack; medium frequency.
  Thomson Sintra DUUX 5; passive ranging.

Programmes: Contract signed 30 November 1977 with Thyssen
  Nordseewerke for two submarines to be built at Emden with
  parts and overseeing for four more boats to be built in
  Argentina by Astilleros Domecq Garcia, Buenos Aires. In early
  1996, S 44 was 30 per cent complete but no further work

was being done. In 2004, studies into the completion of S 43
  (Santa Fe) were conducted. The boat has been reported as 70
  per cent complete. The dockyard was sold in February 1996.
  Equipment for numbers five and six also used for spares.
Modernisation: *Santa Cruz* underwent mid-life update in Brazil
  between September 1999 and 2002. Refit includes new
  main motors and sonar upgrade. *San Juan* is to start a refit at
  Domecq Garcia in 2005.
Structure: Diving depth, 270 m *(890 ft)*.
Operational: Maximum endurance is 70 days. Both can be used
  for Commando insertion operations. They are based at Mar
  del Plata.

UPDATED

SAN JUAN   5/2004 *, A E Galarce / 1044065

SANTA CRUZ   7/2004 *, A E Galarce / 1044064

DESTROYERS

1 HERCULES (TYPE 42) CLASS (DDGHM)

Name	No	Builders	Laid down	Launched	Commissioned
HERCULES	B 52 (ex-D 1, ex-28)	Vickers, Barrow	16 June 1971	24 Oct 1972	12 July 1976

Displacement, tons: 3,150 standard; 4,100 full load
Dimensions, feet (metres): 412 × 47 × 19 (screws)
  (125.6 × 14.3 × 5.8)
Flight deck, feet (metres): 85.3 × 42.66 *(26 × 13)*
Main machinery: COGOG; 2 RR Olympus TM3B gas turbines;
  50,000 hp *(37.3 MW)* sustained
  2 RR Tyne RM1A gas-turbines; 9,900 hp *(7.4 MW)* sustained;
  2 shafts; cp props
Speed, knots: 29; 18 (Tynes). **Range, n miles:** 4,000 at 18 kt
Complement: 280

Missiles: SAM: British Aerospace Sea Dart Mk 30 twin launcher
  ❶; semi-active radar homing to 40 km *(21.5 n miles)* at 2
  Mach; height envelope 100-18,300 m *(328-60,042 ft)*; 22
  missiles; limited anti-ship capability.
Guns: 1 Vickers 4.5 in *(115 mm)/55 Mk 8 automatic ❷; 25 rds/
  min to 22 km *(12 n miles)*; weight of shell 21 kg; also fires
  chaff and illuminants.
  2 Oerlikon 20 mm Mk 7 ❸. 4—12.7 mm MGs.
Countermeasures: Decoys; Graseby towed torpedo decoy.
  Knebworth Corvus 8-tubed trainable launchers for chaff.
ESM: Racal RDL 257; radar intercept.
ECM: Racal RCM 2; jammer.
Combat data systems: Plessey-Ferranti ADAWS-4; Link 10.
Radars: Air search: Marconi Type 965P with double AKE2 array
  and 1010/1011 IFF ❹; A-band.
Surface search: Marconi Type 992Q ❺; E/F-band.
Navigation, HDWS and helicopter control: Kelvin Hughes Type
  1006; I-band.
Fire control: Marconi Type 909 ❻; I/J-band (for Sea Dart missile
  control).
Sonars: Graseby Type 184M; hull-mounted; active search and
  attack; medium frequency 6-9 kHz.
  Kelvin Hughes Type 162M classification set; sideways looking;
  active; high frequency.

Helicopters: 2 Sea King ❼.

Programmes: Contract signed 18 May 1970 between the
  Argentine government and Vickers Ltd.

HERCULES   (Scale 1 : 1,200), Ian Sturton / 0528400

HERCULES   6/2001, Argentine Navy / 0130745

Modernisation: Combat Data System has been improved with
  local modifications. Refitted in Chile from November 1999 to
  July 2000 to make flight deck and hanger Sea King capable.
  Further modifications included removal of MM38 launchers
  to be replaced by assault boats and possible adaptation of the
  Sea Dart magazine to accommodate 150 marines. Reports in
  2004 suggested that MM38 and torpedo tubes may be re-

installed but this has not been confirmed. The deletion of
  Santisima Trinidad has not been confirmed but she is unlikely
  to be refitted.
Operational: Based at Puerto Belgrano. SAM and Type 909 fire-
  control radars are probably non-operational. Officially
  described as a Fast Amphibious Transport.

UPDATED

4 ALMIRANTE BROWN (MEKO 360) CLASS (DDGHM)

Name	No	Builders	Laid down	Launched	Commissioned
ALMIRANTE BROWN	D 10	Blohm + Voss, Hamburg	8 Sep 1980	28 Mar 1981	26 Jan 1983
LA ARGENTINA	D 11	Blohm + Voss, Hamburg	30 Mar 1981	25 Sep 1981	4 May 1983
HEROINA	D 12	Blohm + Voss, Hamburg	24 Aug 1981	17 Feb 1982	31 Oct 1983
SARANDI	D 13	Blohm + Voss, Hamburg	9 Mar 1982	31 Aug 1982	16 Apr 1984

Displacement, tons: 2,900 standard; 3,630 full load
Dimensions, feet (metres): 413.1 × 46 × 19 (screws)
  (125.9 × 14 × 5.8)
Main machinery: COGOG; 2 RR Olympus TM3B gas turbines;
  50,000 hp *(37.4 MW)* sustained
  2 RR Tyne RM1C gas turbines; 9,900 hp *(7.4 MW)* sustained;
  2 shafts; cp props
Speed, knots: 30.5; 20.5 cruising. **Range, n miles:** 4,500 at
  18 kt
Complement: 200 (26 officers)

Missiles: SSM: 8 Aerospatiale MM 40 Exocet (2 quad) launchers
  ❶; inertial cruise; active radar homing to 70 km *(40 n miles)*;
  warhead 165 kg; sea-skimmer.
  SAM: Selenia/Elsag Albatros octuple launcher ❷; 24 Aspide;
  semi-active homing to 13 km *(7 n miles)* at 2.5 Mach; height
  envelope 15-5,000 m *(49.2-16,405 ft)*; warhead 30 kg.
Guns: 1 OTO Melara 5 in *(127 mm)*/54 automatic ❸; 45 rds/min
  to 23 km *(12.42 n miles)* anti-surface; 7 km *(3.6 n miles)* anti-
  aircraft; weight of shell 32 kg; also fires chaff and illuminants.
  8 Breda/Bofors 40 mm/70 (4 twin) ❹; 300 rds/min to
  12.6 km *(6.8 n miles)* anti-surface; 4 km *(2.2 n miles)* anti-
  aircraft; weight of shell 0.96 kg; 2 Oerlikon 20 mm.
Torpedoes: 6—324 mm ILAS 3 (2 triple) tubes ❺. Whitehead
  A 244; anti-submarine; active/passive homing to 7 km *(3.8 n
  miles)* at 33 kt; warhead 34 kg (shaped charge); 18 reloads.
Countermeasures: Decoys: CSEE Dagaie double mounting;
  Graseby G1738 towed torpedo decoy system.
  2 Breda 105 mm SCLAR chaff rocket launchers; 20 tubes per
  launcher; can be trained and elevated; chaff to 5 km *(2.7 n
  miles)*; illuminants to 12 km *(6.6 n miles)*.
ESM/ECM: Sphinx/Scimitar.
Combat data systems: Signaal SEWACO; Link 10/11. SATCOMs
  can be fitted.
Weapons control: 2 Signaal LIROD radar/optronic systems ❻
  each controlling 2 twin 40 mm mounts; Signaal WM25
  FCS ❼.
Radars: Air/surface search: Signaal DA08A ❽; F-band; range
  204 km *(110 n miles)* for 2 m² target.
  Surface search: Signaal ZW06 ❾; I-band.
  Navigation: Decca 1226; I-band.
  Fire control: Signaal STIR ❿; I/J/K-band; range 140 km *(76 n
  miles)* for 1 m² target.
Sonars: Atlas Elektronik 80 (DSQS-21BZ); hull-mounted; active
  search and attack; medium frequency.

Helicopters: AS 555 Fennec ⓫.

Programmes: Six were originally ordered in 1978, but later
  restricted to four when Meko 140 frigates were ordered in
  1979.
Modernisation: Block II Exocet MM 40 may be fitted when funds
  are available.
Operational: *Almirante Brown* took part in allied Gulf operations
  in late 1990. Fennec helicopters delivered in 1996 provide
  over the horizon targeting for SSMs and have the potential to
  improve ASW capability. All are active and form 2nd
  Destroyer Squadron based at Puerto Belgrano. All can be used
  as Flagships. While half-life refits are a high priority, they are
  not believed to be fully funded.

UPDATED

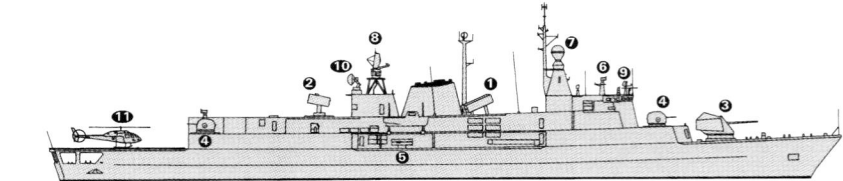

ALMIRANTE BROWN

(Scale 1 : 1,200), Ian Sturton / 0569252

LA ARGENTINA

11/2002, Mario R V Carneiro / 0528303

ALMIRANTE BROWN

7/2001, A E Galarce / 0130730

FRIGATES

6 ESPORA (MEKO 140) CLASS (FFGH)

Name	No	Builders	Laid down	Launched	Commissioned
ESPORA	41	AFNE, Rio Santiago	3 Oct 1980	23 Jan 1982	5 July 1985
ROSALES	42	AFNE, Rio Santiago	1 July 1981	4 Mar 1983	14 Nov 1986
SPIRO	43	AFNE, Rio Santiago	4 Jan 1982	24 June 1983	24 Nov 1987
PARKER	44	AFNE, Rio Santiago	2 Aug 1982	31 Mar 1984	17 Apr 1990
ROBINSON	45	AFNE, Rio Santiago	8 June 1983	15 Feb 1985	28 Aug 2000
GOMEZ ROCA	46	AFNE, Rio Santiago	1 Dec 1983	14 Nov 1986	20 May 2004

Displacement, tons: 1,470 standard; 1,836 full load
Dimensions, feet (metres): 299.1 × 36.4 × 11.2
  (91.2 × 11.1 × 3.4)
Main machinery: 2 SEMT-Pielstick 16 PC2-5 V 400 diesels;
  20,400 hp(m) *(15 MW)* sustained; 2 shafts
Speed, knots: 27. **Range, n miles:** 4,000 at 18 kt
Complement: 93 (11 officers)

Missiles: SSM: 4 Aerospatiale MM 38 Exocet ❶ inertial cruise;
  active radar homing to 42 km *(23 n miles)*; warhead 165 kg;
  sea-skimmer.
Guns: 1 OTO Melara 3 in *(76 mm)*/62 compact ❷; 85 rds/min to
  16 km *(8.7 n miles)* anti-surface; 12 km *(6.5 n miles)* anti-
  aircraft; weight of shell 6 kg; also fires chaff and illuminants.
  4 Breda 40 mm/70 (2 twin) ❸; 300 rds/min to 12.5 km *(6.8 n
  miles)*; weight of shell 0.96 kg; ready ammunition 736 (or
  444) using AP tracer, impact or proximity fuzing.
  2—12.7 mm MGs.
Torpedoes: 6—324 mm ILAS 3 (2 triple) tubes ❹. Whitehead A
  244/S; anti-submarine; active/passive homing to 7 km *(3.8 n
  miles)* at 33 kt; warhead 34 kg (shaped charge).
Countermeasures: Decoys: CSEE Dagaie double mounting; 10
  or 6 replaceable containers; trainable; chaff to 12 km *(6.5 n
  miles)*; illuminants to 4 km *(2.2 n miles)*; decoys in H- to
  J-bands.
ESM: Racal RQN-3B; radar warning.
ECM: Racal TQN-2X; jammer.
Combat data systems: Signaal SEWACO.
Weapons control: Signaal WM22/41 integrated system;
  1 LIROD 8 optronic director ❺ (plus 2 sights-1 on each bridge
  wing).
Radars: Air/surface search: Signaal DA05 ❻; E/F-band; range
  137 km *(75 n miles)* for 2 m² target.
Navigation: Decca TM 1226; I-band.
Fire control: Signaal WM28 ❼; I/J-band; range 46 km
  (25 n miles).
IFF: Mk 10.
Sonars: Atlas Elektronik ASO 4; hull-mounted; active search and
  attack; medium frequency.

Helicopters: 1 SA 319B Alouette III or AS 555 Fennec ❽ (in 44-
  46).

Programmes: A contract was signed with Blohm + Voss on
  1 August 1979 for this group of ships which are scaled down
  Meko 360s. All have been fabricated in AFNE, Rio Santiago.
  The last pair were to have been scrapped, but on 8 May
  1997 a decision was taken to complete them some 14 years
  after each was first launched. A formal restart ceremony was
  held on 18 July 1997 and *Robinson* became operational in
  2001. *Gomez Roca* was completed in 2004.
Modernisation: Plans to fit MM 40 Exocet from Meko 360. Flight
  deck extensions for AS 555 helicopters. *Robinson* and *Gomez
  Roca* may have different EW suite.
Structure: *Parker* fitted with a telescopic hangar which is
  planned to be retrofitted in first three and is being built into the
  last pair. Fitted with stabilisers.
Operational: Mostly used for offshore patrol and fishery
  protection duties but *Spiro* and *Rosales* sent to the Gulf in
  1990-91. Form 2nd Frigate Squadron based at Puerto
  Belgrano.

UPDATED

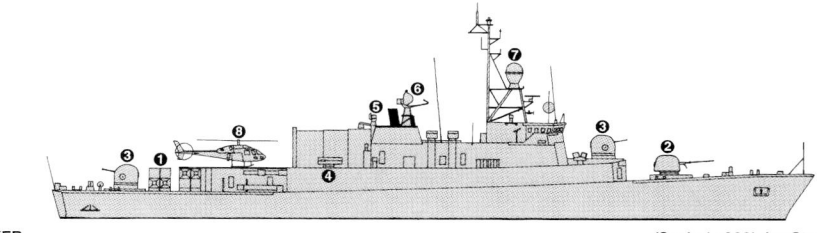

PARKER   *(Scale 1 : 900), Ian Sturton /* 0012007

ROSALES   *8/2004*, A E Galarce /* 1044068

PARKER
5/2004*, A E Galarce / 1044067

GOMEZ ROCA   *5/2004*, A E Galarce /* 1044069

3 DRUMMOND (TYPE A 69) CLASS (FFG)

Name	No	Builders	Laid down	Launched	Commissioned
DRUMMOND (ex-*Good Hope*, ex-*Lieutenant de Vaisseau le Hénaff* F 789)	31	Lorient Naval Dockyard	12 Mar 1976	5 Mar 1977	Mar 1978
GUERRICO (ex-*Transvaal*, ex-*Commandant l'Herminier* F 791)	32	Lorient Naval Dockyard	1 Oct 1976	13 Sep 1977	Oct 1978
GRANVILLE	33	Lorient Naval Dockyard	1 Dec 1978	28 June 1980	22 June 1981

Displacement, tons: 950 standard; 1,170 full load
Dimensions, feet (metres): 262.5 × 33.8 × 9.8; 18 (sonar)
   (80 × 10.3 × 3; 5.5)
Main machinery: 2 SEMT-Pielstick 12 PC2.2 V 400 diesels;
   12,000 hp(m) *(8.82 MW)* sustained; 2 shafts; LIPS cp props
Speed, knots: 23
Range, n miles: 4,500 at 15 kt; 3,000 at 18 kt
Complement: 93 (10 officers)

Missiles: SSM: 4 Aerospatiale MM 38 Exocet (2 twin) launchers
   ❶; inertial cruise; active radar homing to 42 km *(23 n miles)*;
   warhead 165 kg; sea-skimmer.
Guns: 1 Creusot-Loire 3.9 in *(100 mm)*/55 Mod 1953 ❷; 80°
   elevation; 60 rds/min to 17 km *(9 n miles)* anti-surface; 8 km
   (4.4 n miles) anti-aircraft; weight of shell 13.5 kg.
   2 Breda 40 mm/70 (twin) ❸; 300 rds/min to 12.5 km *(6.8 n
   miles)*; weight of shell 0.96 kg; ready ammunition 736 (or
   444) using AP tracer, impact or proximity fuzing.
   2 Oerlikon 20 mm ❹. 2—12.7 mm MGs.
Torpedoes: 6—324 mm Mk 32 (2 triple) tubes ❺. Whitehead A
   244; anti-submarine; active/passive homing to 7 km *(3.8 n
   miles)* at 33 kt; warhead 34 kg.
Countermeasures: Decoys: CSEE Dagaie double mounting; 10
   or 6 replaceable containers; trainable; chaff to 12 km *(6.5 n
   miles)*; illuminants to 4 km *(2.2 n miles)*; decoys in H- to
   J-bands or Corvus sextuple launchers for chaff.
ESM: DR 2000/DALIA 500; radar warning.
ECM: Thomson-CSF Alligator; jammer.
Combat data systems: MINIACO.
Weapons control: Thomson-CSF Vega system. CSEE Panda Mk
   2 optical director ❻. Naja optronic director (for 40 mm guns).
Radars: Air/surface search: Thomson-CSF DRBV 51A ❼ with
   UPX12 IFF; G-band.
   Navigation: Decca 1226; I-band.
   Fire control: Thomson-CSF DRBC 32E ❽; I/J-band (for 100 mm
   gun).
Sonars: Thomson Sintra Diodon; hull-mounted; active search
   and attack.

Programmes: The first pair was originally built for the French
   Navy and sold to the South African Navy in 1976 while under
   construction. As a result of a UN embargo on arms sales to
   South Africa this sale was cancelled. Purchased by Argentina
   in Autumn 1978. Both arrived in Argentina 2 November 1978
   (third ship being ordered some time later) and all have proved

very popular ships in the Argentine Navy. The transfer of a
further three of the class from the French Navy is very unlikely.
Modernisation: *Drummond* has had her armament updated to
the same standard as the other two, replacing the Bofors
40/60. It is reported that a SENIT combat data system may
have been installed but this is not confirmed.

Operational: Endurance, 15 days. Very economical in fuel
consumption. Assisted in UN operations off Haiti in 1994. All
based at Mar del Plata.

UPDATED

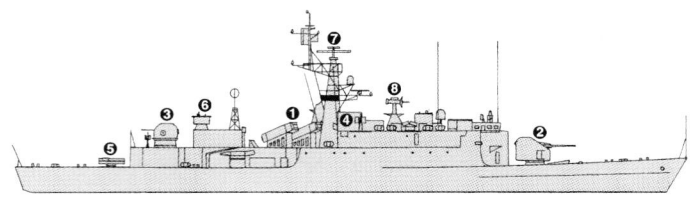

GRANVILLE   *(Scale 1 : 900), Ian Sturton*

DRUMMOND   *5/2004*, A E Galarce* / 1044066

GRANVILLE   *12/2002, A E Galarce* / 0529818

SHIPBORNE AIRCRAFT

Numbers/Type: 5 Aerospatiale SA 319B Alouette III.
Operational speed: 113 kt *(210 km/h)*.
Service ceiling: 10,500 ft *(3,200 m)*.
Range: 290 n miles *(540 km)*.
Role/Weapon systems: ASW Helicopter; used for liaison in peacetime; wartime role includes
   commando assault and ASW/ASVW. Sensors: Nose-mounted search radar. Weapons: ASW;
   2 × Mk 44 torpedoes. ASV: 2 × AS12 missiles.

UPDATED

Numbers/Type: 4 Aerospatiale AS 555 Fennec.
Operational speed: 121 kt *(225 km/h)*.
Service ceiling: 13,125 ft *(4,000 m)*.
Range: 389 n miles *(722 km)*.
Role/Weapon systems: Principal role OTHT with potential ASW capability. Delivered in 1996.
   More are wanted. Sensors: Bendix RDR 1500 radar; Mk 3 MAD. Weapons: ASW; 2 × A 244
   torpedoes or 4 depth bombs may be fitted.

UPDATED

ALOUETTE III   *7/2004*, A E Galarce* / 1044070

FENNEC   *7/2004*, A E Galarce* / 1044071

Numbers/Type: 2/5 Agusta-Sikorsky ASH-3H/AS-61D Sea King.
Operational speed: 120 kt *(222 km/h).*
Service ceiling: 12,205 ft *(3,720 m).*
Range: 630 n miles *(1,165 km).*
Role/Weapon systems: ASW Helicopter; with limited surface search capability. Can operate from *Hercules.* Sensors: APS-705 search radar, Bendix AQS 18 sonar. Weapons: ASW; up to 4 × A 244 torpedoes or 4 × depth bombs. ASV: 1 AM 39 Exocet ASM.
VERIFIED

SEA KING *8/2002, A E Galarce /* 0529816

Numbers/Type: 7 Bell UH-1H.
Operational speed: 110 kt *(204 km/h).*
Service ceiling: 15,000 ft *(4,570 m).*
Range: 250 n miles *(463 km).*
Role/Weapon systems: First pair acquired for the Marines in 1999. Six more in 2000. Sensors: none. Weapons: 2—7.62 mm MGs.
VERIFIED

UH-1H *3/2001 /* 0126382

LAND-BASED MARITIME AIRCRAFT

Notes: (1) In addition there are three Fokker F28 for Logistic Support; one Pilatus PC-6B for reconnaissance and nine Beech T-34 Turbo Mentor training aircraft. The four Lockheed Electra L-188 are no longer in service.
(2) Thirty-six ex-US Navy A4M Skyhawk with radar APG-66 acquired by the Air Force by July 1998. First 18 delivered in crates in 1995-96 and remainder modernised before delivery in 1997-98.
(3) Acquisition of second-hand Mirage 2000 aircraft is reported to be under consideration.

Numbers/Type: 5 + (6) Dassault-Breguet Super Etendard.
Operational speed: Mach 1.
Service ceiling: 44,950 ft *(13,700 m).*
Range: 920 n miles *(1,700 km).*
Role/Weapon systems: Strike Fighter with anti-shipping ability. In the past have flown from US or Brazilian aircraft carriers. Five aircraft are operational out of a total of eleven. Strike, air defence and ASV roles. Hi-lo-hi combat radius 460 n miles *(850 km).* Sensors: Thomson-CSF Agave multimode radar, ECM. Weapons: Strike; 2.1 tons of 'iron' bombs. ASVW; 1 AM 39 Exocet or 1 × Martin Pescador missiles. Self-defence; 2 × Magic AAMs. Standard; 2 × 30 mm cannon.
VERIFIED

SUPER ETENDARD *7/1999, A E Galarce /* 0056474

Numbers/Type: 5 Grumman S-2ET Tracker.
Operational speed: 130 kt *(241 km/h).*
Service ceiling: 25,000 ft *(7,620 m).*
Range: 1,350 n miles *(2,500 km).*
Role/Weapon systems: Used for MR and EEZ patrol. One shipped to Israel in 1989 for Garrett turboprop installation. Prototype for fleet conversion in Argentina when completed in 2000. Sensors: EL/M-2022 search radar up to 32 sonobuoys, ALD-2B or AES 210/E ESM, echo-ranging depth charges. Weapons: ASW; A 244 torpedoes, bombs and depth charges.
VERIFIED

S-2 TRACKER (landing on São Paulo) *5/2002, Walter Lastra/Fuerzas Navales /* 0528430

Numbers/Type: 6 Aermacchi MB-326GB.
Operational speed: 468 kt *(867 km/h).*
Service ceiling: 47,000 ft *(14,325 m).*
Range: 1,320 n miles *(2,446 km).*
Role/Weapon systems: Light Attack; supplements anti-shipping/strike; also has training role. Weapons: ASV; 1.8 tons of 'iron' bombs. Strike; 6 × rockets. Recce; underwing camera pod.
UPDATED

AERMACCHI 326 *4/2004* * /* 0051048

Numbers/Type: 4 Beechcraft B 200T Cormoran.
Operational speed: 260 kt *(482 km/h).*
Service ceiling: 31,000 ft *(9,448 m).*
Range: 2,000 n miles *(3,705 km).*
Role/Weapon systems: Multipurpose converted to Cormoran version for maritime patrol. There are three other unconverted aircraft. Sensors: Search radar. Weapons: Unarmed.
UPDATED

BEECH CORMORAN *5/2004* * /* 0570789

Numbers/Type: 6 Lockheed P-3B Orion.
Operational speed: 410 kt *(760 km/h).*
Service ceiling: 28,300 ft *(8,625 m).*
Range: 4,000 m *(7,410 km).*
Role/Weapon systems: Two acquired in 1997 from US; four more in 1998, and two for spares in 1999. Sensors: APS-115 radar; ESM. Weapons: ASW equipment may be carried in due course. ASV weapons may be fitted in due course. Electronic equipment from L-188E may be fitted in one aircraft.
VERIFIED

ORION *6/2002, Argentine Navy /* 0528429

PATROL FORCES

Notes: Project PAM (Patrulleros de Alta Mar) is for up to ten 1,200 ton offshore patrol vessels. With a length of up to 80 m, the ships are to have diesel propulsion and to be armed with a 40 mm gun. An order for the first two is expected in 2005. Collaboration with Chile is a possibility.

3 CHEROKEE CLASS (PATROL SHIPS) (PSO)

Name	No	Builders	Commissioned
COMANDANTE GENERAL IRIGOYEN	A 1	Charleston	10 Mar 1945
(ex-*Cahuilla*)		SB and DD Co	
FRANCISCO DE GURRUCHAGA	A 3	Charleston	16 June 1945
(ex-*Luiseno* ATF 156)		SB and DD Co	
SUBOFICIAL CASTILLO	A 6	United Engineering Co,	3 Aug 1944
(ex-*Takelma* ATF 113)		Alameda	

Displacement, tons: 1,235 standard; 1,731 full load
Dimensions, feet (metres): 205 × 38.5 × 17 *(62.5 × 11.7 × 5.2)*
Main machinery: Diesel-electric; 4 GM 12—278 diesels; 4,400 hp *(3.28 MW)*; 4 generators; 1 motor; 3,000 hp *(2.24 MW)*; 1 shaft
Speed, knots: 16. **Range, n miles:** 6,500 at 15 kt; 15,000 at 8 kt
Complement: 85
Guns: 4 Bofors 40/60 (2 twin). 2 Oerlikon 20 mm; 4 Bofors 40 mm/60 (2 twin) (A 1); 2 Bofors 40 mm/60 (A 3); 1 Bofors 40 mm/60 (A 6); 2 Oerlikon 20 mm/70 (A 1); 4 Oerlikon 20 mm (A 3, A 6); 2—12.7 mm MGs (A 6).
Radars: Surface search: Racal Decca 626; I-band.
Navigation: Racal Decca 1230; I-band.

Comment: Fitted with powerful pumps and other salvage equipment. *Comandante General Irigoyen* transferred by the US at San Diego, California, on 9 July 1961. Classified as a tug until 1966 when she was rerated as patrol ship. *Francisco De Gurruchaga* transferred on 24 July 1975 by sale, *Suboficial Castillo* on 30 September 1993 by grant aid. Armament has been reduced. All operational and based at Mar del Plata. ***UPDATED***

SUBOFICIAL CASTILLO   *7/2004*, A E Galarce* / 1044072

2 KING CLASS (PATROL SHIPS) (AX)

Name	No	Builders	Launched	Commissioned
MURATURE	P 20	Base Nav Rio Santiago	5 July 1943	12 Apr 1945
KING	P 21	Base Nav Rio Santiago	2 Nov 1943	28 July 1946

Displacement, tons: 913 standard; 1,000 normal; 1,032 full load
Dimensions, feet (metres): 252.7 × 29.5 × 13.1 *(77 × 9 × 4)*
Main machinery: 2 Werkspoor diesels; 2,500 hp(m) *(1.8 MW)*; 2 shafts
Speed, knots: 18. **Range, n miles:** 9,000 at 12 kt
Complement: 130
Guns: 3 Vickers 4 in *(105 mm)*/45; 16 rds/min to 19 km *(10 n miles)*; weight of shell 16 kg. 4 Bofors 40 mm/60 (1 twin, 2 single); 120 rds/min/barrel to 10 km *(5.5 n miles)*; weight of shell 0.89 kg. 5—12.7 mm MGs.
Radars: Surface search: Racal Decca 1226; I-band.

Comment: Named after Captain John King, an Irish follower of Admiral Brown, who distinguished himself in the war with Brazil, 1826-28; and Captain Jose Murature, who performed conspicuous service against the Paraguayans at the Battle of Cuevas in 1865. *King* laid down June 1938. *Murature* March 1940. Both used for cadet training. ***VERIFIED***

KING   *7/2003, A E Galarce* / 0572404

1 OLIVIERI CLASS (PATROL SHIP) (PBO)

Name	No	Builders	Commissioned
TENIENTE OLIVIERI (ex-*Marsea 10*)	A 2	Quality SB, Louisiana	1981

Displacement, tons: 1,640 full load
Dimensions, feet (metres): 184.8 × 40 × 14 *(56.3 × 12.2 × 4.3)*
Main machinery: 2 GM/EMD 16-645 E6; 3,230 hp *(2.4 MW)* sustained; 2 shafts; bow thruster
Speed, knots: 14. **Range, n miles:** 2,800 at 10 kt
Complement: 15 (4 officers)
Guns: 2—12.7 mm MGs.

Comment: Built by Quality Shipyards, New Orleans, as an oilfield support ship but rated as an Aviso. Acquired from US Maritime Administration 15 November 1987. Capable of carrying 600 tons of stores and 800 tons of liquids. Based at Puerto Belgrano. ***VERIFIED***

TENIENTE OLIVIERI   *3/2000* / 0104168

1 SOTOYOMO CLASS (PATROL SHIP) (PBO)

Name	No	Builders	Commissioned
ALFEREZ SOBRAL (ex-*Salish* ATA 187)	A 9	Levingstone, Orange	9 Sep 1944

Displacement, tons: 800 full load
Dimensions, feet (metres): 143 × 33.9 × 13 *(43.6 × 10.3 × 4)*
Main machinery: Diesel-electric; 2 GM 12-278A diesels; 2,200 hp *(1.64 MW)*; 2 generators; 1 motor; 1,500 hp *(1.12 MW)*; 1 shaft
Speed, knots: 12.5. **Range, n miles:** 16,500 at 8 kt
Complement: 49
Guns: 1 Bofors 40 mm/60. 2 Oerlikon 20 mm.
Radars: Surface search: Decca 1226; I-band.

Comment: Former US ocean tug transferred on 10 February 1972. Paid off in 1987 but back in service by 1996. Armament has been reduced. ***UPDATED***

SOTOYOMO CLASS   *10/1998* / 0017531

4 BARADERO (DABUR) CLASS (COASTAL PATROL CRAFT) (PB)

Name	No	Builders	Commissioned
BARADERO	P 61	Israel Aircraft Industries	1978
BARRANQUERAS	P 62	Israel Aircraft Industries	1978
CLORINDA	P 63	Israel Aircraft Industries	1978
CONCEPCIÓN DEL URUGUAY	P 64	Israel Aircraft Industries	1978

Displacement, tons: 33.7 standard; 39 full load
Dimensions, feet (metres): 64.9 × 18 × 5.8 *(19.8 × 5.5 × 1.8)*
Main machinery: 2 GM 12V-71TA diesels; 840 hp *(627 kW)* sustained; 2 shafts
Speed, knots: 19. **Range, n miles:** 450 at 13 kt
Complement: 9
Guns: 2 Oerlikon 20 mm. 4—12.7 mm MGs.
Depth charges: 2 portable rails.
Radars: Navigation: Decca 101; I-band.

Comment: Of all-aluminium construction. Employed in 1991 and 1992 as part of the UN Central American peacekeeping force. Based at Ushuaia. ***UPDATED***

BARADERO CLASS   *12/2000*, Eric Grove* / 1044073

2 INTREPIDA CLASS (TYPE TNC 45)
(FAST ATTACK CRAFT—GUN/MISSILE) (PGGF)

Name	No	Builders	Launched	Commissioned
INTREPIDA	P 85	Lürssen, Bremen	2 Dec 1973	20 July 1974
INDOMITA	P 86	Lürssen, Bremen	8 Apr 1974	12 Dec 1974

Displacement, tons: 268 full load
Dimensions, feet (metres): 147.3 × 24.3 × 7.9 *(44.9 × 7.4 × 2.4)*
Main machinery: 4 MTU MD 16V 538 TB90 diesels; 12,000 hp(m) *(8.82 MW)*; 4 shafts
Speed, knots: 38. **Range, n miles:** 1,450 at 20 kt
Complement: 39 (5 officers)
Missiles: SSM: 2 Aerospatiale Exocet mm 38; *(Intrepida)*; active radar homing to 42 km *(23 n miles)*; warhead 165 kg.
Guns: 1 OTO Melara 3 in *(76 mm)*/62 compact; 85 rds/min to 16 km *(9 n miles)* anti-surface; 12 km *(6.5 n miles)* anti-aircraft; weight of shell 6 kg.
1 or 2 Bofors 40 mm/70; 330 rds/min to 12 km *(6.5 n miles)* anti-surface; 4 km *(2.2 n miles)* anti-aircraft; weight of shell 0.89 kg.
2 Oerlikon 81 mm rocket launchers for illuminants.
Torpedoes: 2—21 in *(533 mm)* launchers. AEG SST-4; wire-guided; active/passive homing to 28 km *(15 n miles)* at 23 kt; warhead 250 kg.
Countermeasures: ESM: Racal RDL 1; radar warning.
Weapons control: Signaal WM22 optronic for guns/missiles. Signaal M11 for torpedo guidance and control.
Radars: Surface search: Decca 626; I-band.

Comment: These two vessels were ordered in 1970. Both are painted with a brown/green camouflage. Camouflage netting can also be fitted. Exocet SSM fitted vice the forward of the two Bofors guns in *Intrepida* in 1998. No indication of second of class being similarly refitted.
VERIFIED

INTREPIDA *6/2001, Argentine Navy /* 0130735

INTREPIDA (with camouflage netting) *3/2001 /* 0126381

2 POINT CLASS (PB)

Name	No	Builders	Commissioned
PUNTA MOGOTES (ex-*Point Hobart*)	P 65 (ex-82377)	J Martinac, Tacoma	13 July 1970
RIO SANTIAGO (ex-*Point Carrew*)	P 66 (ex-82374)	USCG Yard, Curtis Bay	18 May 1970

Displacement, tons: 67 full load
Dimensions, feet (metres): 83 × 17.2 × 15.8 *(25.3 × 5.2 × 1.8)*
Main machinery: 2 Caterpillar diesels; 1,600 hp *(1.19 MW)*; 2 shafts
Speed, knots: 22. **Range, n miles:** 1,200 at 8 kt
Complement: 10
Guns: 2—12.7 mm MGs.
Radars: Surface search: Raytheon SPS 64; I-band.

Comment: *Punta Mogotes* transferred from US Coast Guard on 8 July 1999 and is based at Mar del Plata. *Rio Santiago* transferred 22 August 2000.
UPDATED

PUNTA MOGOTES *7/2004 *, A E Galarce /* 1044074

MINE WARFARE FORCES

Notes: Options for the replacement of the deleted Ton class MCMV are under consideration.

AMPHIBIOUS FORCES

Notes: (1) Procurement of amphibious ships is a high priority. Acquisition of ex-French LSDs *Ouragan* and *Orage* when they leave service in 2006/07 is a possibility.
(2) Marine Corps acquired two Guardian craft in October 1999 and two more in February 2000. Powered by twin 150 hp Johnson outboards. Carry 1—12.7 mm MG and 4—7.62 mm MGs, Raytheon radar.
(3) A new class of indigenously built LCVPs is to start entering service in 2005.

GUARDIAN 35 *5/2004 *, A E Galarce /* 1044075

4 LCM 6 CLASS (LCM) and 16 LCVPs

EDM 1, 2, 3, 4	EDVP 30-37	+ 8

Displacement, tons: 56 full load
Dimensions, feet (metres): 56 × 14 × 3.9 *(17.1 × 4.3 × 1.2)*
Main machinery: 2 Gray 64 HN9 diesels; 330 hp *(246 kW)* sustained; 2 shafts
Speed, knots: 11. **Range, n miles:** 130 at 10 kt
Military lift: 30 tons
Guns: 2—12.7 mm MGs.

Comment: Details given are for the LCMs acquired from the US in June 1971. The LCVPs are split between those acquired from the USA in 1970 and a smaller variant built locally since 1971.
VERIFIED

SURVEY AND RESEARCH SHIPS

Notes: (1) There are also two Fisheries Research Ships employed by the government. These are *Oca Balda* and *Eduardo Holmberg*.
(2) Two 10 m hydrographic launches, *Monte Blanco* and *Kualchink* entered service in 2004.

1 SURVEY SHIP (AGOB)

Name	No	Builders	Commissioned
PUERTO DESEADO	Q 20 (ex-Q 8)	Astarsa, San Fernando	26 Feb 1979

Displacement, tons: 2,133 standard; 2,400 full load
Dimensions, feet (metres): 251.9 × 51.8 × 21.3 *(76.8 × 15.8 × 6.5)*
Main machinery: 2 Fiat-GMT diesels; 3,600 hp(m) *(2.65 MW)*; 1 shaft
Speed, knots: 15. **Range, n miles:** 12,000 at 12 kt
Complement: 61 (12 officers) plus 20 scientists
Radars: Navigation: Decca 1629; I-band.

Comment: Laid down on 17 March 1976 for Consejo Nacional de Investigaciones Tecnicas y Scientificas. Launched on 4 December 1976. For survey work fitted with: four Hewlett-Packard 2108-A, gravimeter, magnetometer, seismic systems, high-frequency sonar, geological laboratory. Omega and NAVSAT equipped. Painted with an orange hull in late 1996 for Antarctic deployments.
VERIFIED

PUERTO DESEADO *3/2000 /* 0104171

PUERTO DESEADO *12/2002, A E Galarce /* 0529812

1 RESEARCH SHIP (AGOR)

Name	No	Builders	Commissioned
COMODORO RIVADAVIA	Q 11	Mestrina, Tigre	6 Dec 1974

Displacement, tons: 820 full load
Dimensions, feet (metres): 171.2 × 28.9 × 8.5 *(52.2 × 8.8 × 2.6)*
Main machinery: 2 Stork Werkspoor RHO-218K diesels; 1,160 hp(m) *(853 kW)*; 2 shafts
Speed, knots: 12. **Range, n miles:** 6,000 at 12 kt
Complement: 34 (8 officers)

Comment: Laid down on 17 July 1971 and launched on 2 December 1972. Used for research.
VERIFIED

COMODORO RIVADAVIA   *3/2001* / 0126380

1 SURVEY CRAFT (AGSC)

Name	No	Builders	Commissioned
CORMORAN	Q 15	AFNE, Rio Santiago	20 Feb 1964

Displacement, tons: 102 full load
Dimensions, feet (metres): 83 × 16.4 × 5.9 *(25.3 × 5 × 1.8)*
Main machinery: 2 GM 6-71 diesels; 440 hp(m) *(323 kW)*; 2 shafts
Speed, knots: 11
Complement: 19 (3 officers)
Radars: Navigation: Decca TM1226; I-band.

Comment: Launched 10 August 1963. Classified as a coastal launch.
VERIFIED

CORMORAN   *5/2003, A E Galarce* / 0572406

TRAINING SHIPS

Notes: There are also three small yachts: *Itati* (Q 73), *Fortuna I* (Q 74) and *Fortuna II* (Q 75) plus a 25 ton yawl *Tijuca* acquired in 1993. *Fortuna III* was commissioned in 2004.

1 SAIL TRAINING SHIP (AXS)

Name	No	Builders	Commissioned
LIBERTAD	Q 2	AFNE, Rio Santiago	28 May 1963

Displacement, tons: 3,025 standard; 3,765 full load
Dimensions, feet (metres): 262 wl; 301 oa × 45.3 × 21.8 *(79.9; 91.7 × 13.8 × 6.6)*
Main machinery: 2 Sulzer diesels; 2,400 hp(m) *(1.76 MW)*; 2 shafts
Speed, knots: 13.5 under power. **Range, n miles:** 12,000 at 8 kt
Complement: 200 crew plus 150 cadets
Guns: 4 Hotchkiss 47 mm saluting guns.
Radars: Navigation: Decca; I-band.

Comment: Launched 30 May 1956. She set record for crossing the North Atlantic under sail in 1966. Sail area, 26,835 m². Based at Puerto Belgrano. Undergoing mid-life refit 2004-06.
UPDATED

LIBERTAD   *8/2003, Derek Fox* / 0569147

AUXILIARIES

Notes: A project for the acquisition of an Antarctic support vessel has been initiated.

1 DURANCE CLASS (AORH)

Name	No	Builders	Launched	Commissioned
PATAGONIA (ex-*Durance*)	B 1 (ex-A 629)	Brest Naval Dockyard	6 Sep 1975	1 Dec 1976

Displacement, tons: 17,900 full load
Dimensions, feet (metres): 515.9 × 69.5 × 38.5 *(157.3 × 21.2 × 10.8)*
Main machinery: 2 SEMT-Pielstick 16 PC2.5 V 400 diesels; 20,800 hp(m) *(15.3 MW)* sustained; 2 shafts; LIPS cp props
Speed, knots: 19. **Range, n miles:** 9,000 at 15 kt
Complement: 164 (10 officers) plus 29 spare
Cargo capacity: 9,000 tons fuel; 500 tons Avcat; 130 distilled water; 170 victuals; 150 munitions; 50 naval stores
Guns: 2 Bofors 40 mm/60. 4—12.7 mm MGs.
Countermeasures: ESM/ECM.
Radars: Navigation: 2 Racal Decca 1226; I-band.
Helicopters: 1 Alouette III.

Comment: Acquired from France on 12 July 1999 having been in reserve for two years. Entered Argentine Navy service in July 2000 after short refit.
VERIFIED

PATAGONIA   *5/2000, A E Galarce* / 0104175

3 CHARTERED SHIPS (AKS/AOTL)

Name	No	Builders	Commissioned	Chartered
INGENIERO JULIO KRAUSE	B 13	Astarsa, Tigre	1981	1992
ASTRA FEDERICO	B 8	Astarsa, Tigre	1981	1992
ASTRA VALENTINA	B 9	Astarsa, Tigre	1982	1992

Comment: Taken over by the Navy but also used for commercial trading. *Krause* is a 10,000 ton oiler which is capable of replenishment at sea. The other pair are 30,000 ton cargo ships. All have civilian crews.
VERIFIED

ASTRA VALENTINA   *5/2000, Hartmut Ehlers* / 0104174

3 RED CLASS (BUOY TENDERS) (ABU)

Name	No	Builders	Launched
PUNTA ALTA (ex-*Red Birch*)	Q 63 (ex-WLM 687)	CG Yard, Maryland	19 Feb 1965
CIUDAD DE ZARATE (ex-*Red Cedar*)	Q 61 (ex-WLM 688)	CG Yard, Maryland	1 Aug 1970
CIUDAD DE ROSARIO (ex-*Red Wood*)	Q 62 (ex-WLM 685)	CG Yard, Maryland	4 Apr 1964

Displacement, tons: 525 full load
Dimensions, feet (metres): 161.1 × 33 × 6 *(49.1 × 10.1 × 1.8)*
Main machinery: 2 Caterpillar D398 diesels; 1,800 hp *(1.34 MW)*; 2 shafts; cp props; bow thruster
Speed, knots: 12. **Range, n miles:** 2,248 at 11 kt
Complement: 31 (6 officers)
Guns: 2—12.7 mm MGs.

Comment: Ex-USCG buoy tenders. First one transferred on 10 June 1998 and recommissioned on 17 November 1998. Two more transferred 30 July 1999. Strengthened hull for light ice breaking. Equipped with a 10 ton boom. *Punta Alta* used as supply ship in the southern archipelago. The other pair are used as river supply ships.
VERIFIED

CIUDAD DE ZARATE   *7/2003, A E Galarce* / 0572407

3 COSTA SUR CLASS (TRANSPORT) (AKS)

Name	No	Builders	Commissioned
CANAL BEAGLE	B 3	Astillero Principe y Menghi SA	29 Apr 1978
BAHIA SAN BLAS	B 4	Astillero Principe y Menghi SA	27 Nov 1978
CABO DE HORNOS	B 5	Astillero Principe y Menghi SA	28 June 1979
(ex-Bahia Camarones)			

Measurement, tons: 5,800 dwt; 4,600 gross
Dimensions, feet (metres): 390.3 × 57.4 × 21 *(119 × 17.5 × 6.4)*
Main machinery: 2 AFNE-Sulzer diesels; 6,400 hp(m) *(4.7 MW)*; 2 shafts
Speed, knots: 16.5
Complement: 40

Comment: Ordered December 1975. Laid down 10 January 1977, 11 April 1977 and 29 April 1978. Launched 19 October 1977, 29 April 1978 and 4 November 1978. Used to supply offshore research installations in Naval Area South. One operated in the Gulf in 1991. *Bahia San Blas* painted grey in 1998 indicating an active naval role in amphibious support operations. Capable of carrying up to eight LCVPs on deck. 132 troops can be accommodated in containers. The ship is to be fitted with a helicopter deck.
VERIFIED

BAHIA SAN BLAS (LCVPs embarked)   5/2000, A E Galarce / 0104176

CANAL BEAGLE   8/1999, P Marsan / 0081446

2 FLOATING DOCKS

Number	Dimensions, feet (metres)	Capacity, tons
Y 1 (ex-ARD 23)	492 × 88.6 × 56 *(150 × 27 × 17.1)*	3,500
3	215.8 × 46 × 45.5 *(65.8 × 14 × 13.7)*	750

Comment: First one is at Mar del Plata naval base, the second at Puerto Belgrano. The ex-USN ARD was transferred 8 September 1993 by grant aid. All other docks have been sold.
VERIFIED

ICEBREAKERS

1 SUPPORT SHIP (AGB/AGOB)

Name	No	Builders	Launched	Commissioned
ALMIRANTE IRIZAR	Q 5	Wärtsilä, Helsinki	3 Feb 1978	15 Dec 1978

Displacement, tons: 14,900 full load
Dimensions, feet (metres): 398.1 × 82 × 31.2 *(121.3 × 25 × 9.5)*
Main machinery: Diesel-electric; 4 Wärtsilä-SEMT-Pielstick 8 PC2.5 L diesels; 18,720 hp(m) *(13.77 MW)* sustained; 4 generators; 2 Stromberg motors; 16,200 hp(m) *(11.9 MW)*; 2 shafts
Speed, knots: 17
Complement: 135 ship's company plus 45 passengers
Radars: Air/surface search: Plessey AWS 2; E/F-band.
Navigation: 2 Decca; I-band.
Helicopters: 2 ASH-3H Sea King.

Comment: Fitted for landing craft with two 16 ton cranes, fin stabilisers, Wärtsilä bubbling system and a 60 ton towing winch. RAST helicopter securing system. Red hull with white upperworks and red funnel. Designed for Antarctic support operations and able to remain in polar regions throughout the Winter with 210 people aboard. Used as a transport to South Georgia in December 1981 and as a hospital ship during the Falklands war April to June 1982. Has been used as a Patagonian supply ship, and for other activities associated with the Navy in the region. The ship is to undergo a major refit which is planned to be completed in late 2004. 40 mm guns have been removed.
UPDATED

ALMIRANTE IRIZAR   6/2004*, A E Galarce / 1044076

TUGS

11 TUGS (YTB/YTL)

QUERANDI R 2	CALCHAQUI R 6	CHULUPI R 10	CHIQUILYAN R 18
TEHUELCHE R 3	ONA R 7	MATACO R 12	MORCOYAN R 19
MOCOVI R 5	TOBA R 8	CAPAYAN R 16	

Comment: R 2-3 and R 7-8 and R 12 are coastal tugs of about 250 tons. The remainder are harbour tugs transferred from the USA.
VERIFIED

MATACO   5/2000, A E Galarce / 0104177

PREFECTURA NAVAL ARGENTINA — (COAST GUARD)

Headquarters Appointments

Commander:
Prefecto General Carlos Edgardo Fernández
Vice Commander:
Prefecto General Ricardo Rodriguez

Personnel

2005: 11,900 (1,600 officers)

Tasks

Under the General Organisation Act the PNA is charged with:

(a) Enforcement of Federal Laws on the high seas and waters subject to the Argentine Republic.
(b) Enforcement of environmental protection laws in Federal waters.
(c) Safety of ships in EEZ. Search and Rescue.
(d) Security of waterfront facilities and vessels in port.

(e) Operation of certain Navaids.
(f) Operation of some Pilot Services.
(g) Management and operation of Aviation Service; Coastguard Vessels; Salvage, Fire and Anti-Pollution Service; Yachtmaster School; National Diving School; several Fire Brigades and Anti-Narcotics Department.
(h) Operation of some Customs activities.

Organisation

Formed in 10 districts; High Parana River, Upper Parana and Paraguay Rivers, Lower Parana River, Upper Uruguay River, Lower Uruguay River, Delta, River Plate, Northern Argentine Sea, Southern Argentine Sea, Lakes and Comahue.

History

The Spanish authorities in South America established similar organisations to those in Spain. In 1756 the Captainship of the Port came into being in Buenos Aires—in 1810 the Ship Registry office was added to this title. On 29 October 1896 the title of Capitania General de Puertos was established by Act of Congress, the beginning of the PNA. Today, as a security and safety force, it has responsibilities throughout the rivers of Argentina, the ports and harbours as well as within territorial waters out to the 200 mile EEZ. An attempt was made in January 1992 to restrict operations to a 12 mile limit but the legislation was cancelled.

Identity markings

Two unequal blue stripes with, superimposed, crossed white anchors followed by the title Prefectura Naval.

Strength of Prefectura

Patrol Ships	6
Large Patrol Craft	4
Coastal Patrol Craft	63
Training Ships	4
Pilot Stations	1
Pilot and Patrol Craft	5

PATROL FORCES (PC)

Notes: In addition to the ships and craft listed below the PNA operates 400 craft, including floating cranes, runabouts and inflatables of all types.

1 PATROL SHIP (WPSO)

Name	No	Builders	Commissioned
DELFIN	GC 13	Ijsselwerf, Netherlands	14 May 1957

Displacement, tons: 700 standard; 1,000 full load
Dimensions, feet (metres): 193.5 × 29.8 × 13.8 *(59 × 9.1 × 4.2)*
Main machinery: 2 MAN diesels; 2,300 hp(m) *(1.69 MW)*; 2 shafts
Speed, knots: 15. **Range, n miles:** 6,720 at 10 kt
Complement: 27
Guns: 1 Oerlikon 20 mm. 2—12.7 mm Browning MGs.
Radars: Navigation: Decca; I-band.

Comment: Whaler acquired for PNA in 1969. Commissioned 23 January 1970.

VERIFIED

DELFIN *7/2003, A E Galarce /* 0572409

2 LYNCH CLASS (LARGE PATROL CRAFT) (WPB)

Name	No	Builders	Commissioned
LYNCH	GC 21	AFNE, Rio Santiago	20 May 1964
TOLL	GC 22	AFNE, Rio Santiago	7 July 1966

Displacement, tons: 100 standard; 117 full load
Dimensions, feet (metres): 98.4 × 21 × 6.9 *(30 × 6.4 × 2.1)*
Main machinery: 2 MTU Maybach diesels; 2,700 hp(m) *(1.98 MW)*; 2 shafts
Speed, knots: 22. **Range, n miles:** 2,000
Complement: 14 (3 officers)
Guns: 1 Oerlikon 20 mm (can be carried). 1—7.62 mm MG.
Radars: Surface search: Decca; I-band.

VERIFIED

LYNCH *1/1997, Prefectura Nava) /* 0012018

5 HALCON (TYPE B 119) CLASS (WPSO)

Name	No	Builders	Commissioned
MANTILLA	GC 24	Bazán, El Ferrol	20 Dec 1982
AZOPARDO	GC 25	Bazán, El Ferrol	28 Apr 1983
THOMPSON	GC 26	Bazán, El Ferrol	20 June 1983
PREFECTO FIQUE	GC 27	Bazán, El Ferrol	29 July 1983
PREFECTO DERBES	GC 28	Bazán, El Ferrol	20 Nov 1983

Displacement, tons: 910 standard; 1,084 full load
Dimensions, feet (metres): 219.9 × 34.4 × 13.8 *(67 × 10.5 × 4.2)*
Main machinery: 2 Bazán-MTU 16V 956 TB91 diesels; 7,500 hp(m) *(5.52 MW)* sustained; 2 shafts
Speed, knots: 20. **Range, n miles:** 5,000 at 18 kt
Complement: 33 (10 officers)
Guns: 1 Breda 40 mm/70; 300 rds/min to 12.5 km *(7 n miles)*; weight of shell 0.96 kg.
2—12.7 mm MGs.
Radars: Navigation: Decca 1226 ARPA; I-band.
Helicopters: Platform for 1 Dauphin 2.

Comment: Ordered in 1979 from Bazán, El Ferrol, Spain. All have Magnavox MX 1102 SATNAV. Hospital with four beds. Carry one rigid rescue craft *(6 m)* with a 90 hp MWM diesel powering a Hamilton water-jet and a capacity for 12 and two inflatable craft *(4.1 m)* with Evinrude outboard. Refits of these ships are to start in 2005.

UPDATED

MANTILLA *12/2000, A E Galarce /* 0130732

1 LARGE PATROL CRAFT (WAX)

Name	No	Builders	Commissioned
MANDUBI	GC 43	Base Naval Rio Santiago	1940

Displacement, tons: 270 full load
Dimensions, feet (metres): 108.9 × 20.7 × 6.2 *(33.2 × 6.3 × 1.9)*
Main machinery: 2 MAN G6V-23.5/33 diesels; 500 hp(m) *(367 kW)*; 1 shaft
Speed, knots: 14. **Range, n miles:** 800 at 14 kt; 3,400 at 10 kt
Complement: 12
Guns: 2—12.7 mm Browning MGs.
Radars: Surface search: Decca; I-band.

Comment: Since 1986 has acted as training craft for PNA Cadets School carrying 20 cadets.

VERIFIED

MANDUBI *8/1994, Mario Diaz /* 0056488

1 RIVER PATROL SHIP (WARS)

Name	No	Builders	Commissioned
TONINA	GC 47	SANYM SA San Fernando, Argentina	30 June 1978

Displacement, tons: 103 standard; 153 full load
Dimensions, feet (metres): 83.8 × 21.3 × 10.1 *(25.5 × 6.5 × 3.3)*
Main machinery: 2 GM 16V-71TA diesels; 1,000 hp *(746 kW)* sustained; 2 shafts
Speed, knots: 10. **Range, n miles:** 2,800 at 10 kt
Complement: 11 (3 officers)
Guns: 1 Oerlikon 20 mm.
Radars: Navigation: Decca 1226; I-band.

Comment: Served as training ship for PNA Cadets School until 1986. Now acts as salvage ship with salvage pumps and recompression chamber. Capable of operating divers and underwater swimmers. Also used as a patrol ship.

VERIFIED

TONINA *1/1998, Hartmut Ehlers /* 0017541

18 MAR DEL PLATA CLASS (COASTAL PATROL CRAFT) (WPB)

MAR DEL PLATA GC 64	RIO DE LA PLATA GC 70	INGENIERO WHITE GC 76
MARTIN GARCIA GC 65	LA PLATA GC 71	GOLFO SAN MATIAS GC 77
RIO LUJAN GC 66	BUENOS AIRES GC 72	MADRYN GC 78
RIO URUGUAY GC 67	CABO CORRIENTES GC 73	RIO DESEADO GC 79
RIO PARAGUAY GC 68	RIO QUEQUEN GC 74	USHUAIA GC 80
RIO PARANA GC 69	BAHIA BLANCA GC 75	CANAL DE BEAGLE GC 81

Displacement, tons: 81 full load
Dimensions, feet (metres): 91.8 × 17.4 × 5.2 *(28 × 5.3 × 1.6)*
Main machinery: 2 MTU 8V-331-TC92 diesels; 1,770 hp(m) *(1.3 MW)* sustained; 2 shafts
Speed, knots: 22. **Range, n miles:** 1,200 at 12 kt; 780 at 18 kt
Complement: 14 (3 officers)
Guns: 1 Oerlikon 20 mm. 2—12.7 mm Browning MGs.
Radars: Navigation: Decca 1226; I-band.

Comment: Ordered 24 November 1978 from Blohm + Voss to a Z-28 design. First delivered in June 1979 and then at monthly intervals. Steel hulls. GC 82 and 83 were captured by the British Forces in 1982.

UPDATED

RIO DESEADO *7/2004*, A E Galarce /* 1044077

1 COASTAL PATROL CRAFT (WPB)

Name	No	Builders	Commissioned
DORADO	GC 101	Base Naval, Rio Santiago	17 Dec 1939

Displacement, tons: 43 full load
Dimensions, feet (metres): 69.5 × 14.1 × 4.9 *(21.2 × 4.3 × 1.5)*
Main machinery: 2 GM 6071-6A diesels; 360 hp *(268 kW)*; 1 shaft
Speed, knots: 12. **Range, n miles:** 1,550
Complement: 7 (1 officer)
Radars: Navigation: Furuno; I-band.

VERIFIED

DORADO   *12/1999, R O Rivero /* 0056490

35 SMALL PATROL CRAFT (WPB)

ESTRELLEMAR GC 48	**SALMON** GC 54	**ORCA** GC 60	**ROBALDO** GC 92
REMORA GC 49	**BIGUA** GC 55	**PINGUINO** GC 61	**CAMARON** GC 93
CONGRIO GC 50	**FOCA** GC 56	**MEDUSA** GC 88	**GAVIOTA** GC 94
MERO GC 51	**TIBURON** GC 57	**PERCA** GC 89	**ABADEJO** GC 95
MARSOPA GC 52	**MELVA** GC 58	**CALAMAR** GC 90	**GC 102-114**
PETREL GC 53	**LENGUADO** GC 59	**HIPOCAMPO** GC 91	

Displacement, tons: 15 full load
Dimensions, feet (metres): 41 × 11.8 × 3.6 *(12.5 × 3.6 × 1.1)*
Main machinery: 2 GM diesels; 514 hp *(383 kW)*; 2 shafts
Speed, knots: 20. **Range, n miles:** 400 at 18 kt
Complement: 3
Guns: 12.7 mm Browning MG.
Radars: Navigation; I-band.

Comment: First delivered September 1978. First 14 built by Cadenazzi, Tigre 1977-79, most of the remainder by Ast Belen de Escobar 1984-86. *GC 102-114* are slightly smaller.

UPDATED

PERCA   *7/2002, A E Galarce /* 0529810

1 BAZAN TYPE (WPBF)

SUREL GC 142

Displacement, tons: 14.5 full load
Dimensions, feet (metres): 39 × 12.4 × 2.2 *(11.9 × 3.8 × 0.7)*
Main machinery: 2 MAN D2848 LXE diesels; 1,360 hp(m) *(1 MW)* sustained; 2 Hamilton 362 waterjets
Speed, knots: 38. **Range, n miles:** 300 at 25 kt
Complement: 4
Guns: 1—12.7 mm MG.
Radars: Navigation: Furuno; I-band.

Comment: Acquired in 1997 from Bazán, San Fernando. Similar to Spanish Bazán 39 class for Spanish Maritime Police. Plans to acquire further craft were not fulfilled.

UPDATED

SUREL   *12/2001, A E Galarce /* 0529809

10 ALUCAT 1050 CLASS (WPB)

CORMORAN GC 137	**SURUBI** GC 143	**HUALA** GC 146	**MANDURUYU** GC 148
CISNE GC 138	**BOGA** GC 144	**PACU** GC 147	**CORVINA** GC 149
PEJERREY GC 139	**SABALO** GC 145		

Displacement, tons: 9 full load
Dimensions, feet (metres): 37.7 × 12.5 × 2 *(11.5 × 3.8 × 0.6)*
Main machinery: 2 Volvo 61 ALD; 577 hp(m) *(424 kW)*; 2 Hamilton 273 waterjets
Speed, knots: 18
Complement: 4
Radars: Navigation: Furuno 12/24; I-band.

Comment: First six delivered in September 1994. Seven more ordered in 1999 but delivery not confirmed. *UPDATED*

HUALA   *4/2000, Hartmut Ehlers /* 0104180

33 ALUCAT 850 CLASS (WPB)

GC 152-184 (ex-LS 9201-9233)

Displacement, tons: 7 full load
Dimensions, feet (metres): 30.2 × 10.8 × 2 *(9.2 × 3.3 × 0.6)*
Main machinery: 2 Volvo TAMD 41B; 400 hp(m) *(294 kW)*; 2 waterjets
Speed, knots: 26
Complement: 4
Radars: Navigation: Furuno; I-band.

Comment: Alucat 850 class built by Damen. First six delivered in 1995, six more in February 1996, five more in December 1996 and five in December 1997. Five more ordered in 1999. *UPDATED*

ALUCAT 850 (old number)   *11/1995, Prefectura Naval /* 0012021

4 TRAINING SHIPS (WAXL/WAXS)

ESPERANZA	ADHARA II	TALITA II	DR BERNARDO HOUSSAY (ex-*El Austral*)

Displacement, tons: 33.5 standard
Dimensions, feet (metres): 62.3 × 14.1 × 8.9 *(19 × 4.3 × 2.7)*
Main machinery: 1 VM diesel; 90 hp(m) *(66 kW)*; 1 shaft
Speed, knots: 6; 15 sailing
Complement: 6 plus 6 cadets

Comment: Details given are for *Esperanza* built by Ast Central de la PNA. Launched and commissioned 20 December 1968 as a sail training ship. The 30 ton training craft *Adhara II* and *Talita II* are of similar dimensions. *Dr Bernardo Houssay* is a Danish-built ketch built in 1930. Displacement 460 tons and has a crew of 25 (five officers). Acquired by the PNA in 1996. *VERIFIED*

TALITA II   *6/1998, Prefectura Naval /* 0017545

DR BERNARDO HOUSSAY   *5/2000, Harald Carstens /* 0104181

6 SERVICE CRAFT (YTL/YTR)

PUERTO BUENOS AIRES SI 4 — SB 5 CANAL COSTANERO SB 9
— SB 3 CANAL EMILIO MITRE SB 8 — SB 10

Comment: *Canal Emilio Mitre* is a small tug of 53 tons full load, it has a speed of 10 kt and was built by Damen Shipyard, Netherlands in 1982.

UPDATED

PILOT VESSELS

1 PILOT STATION (WAGH/AHH)

Name	No	Builders	Commissioned
RECALADA (ex-*Rio Limay*)	DF 15	Astillero Astarsa	30 May 1972

Displacement, tons: 10,070 full load
Dimensions, feet (metres): 482.3 × 65.6 × 28 *(147 × 20 × 8.5)*
Speed, knots: 13
Complement: 28 (3 officers)

Comment: Commissioned as a Coast Guard ship 24 December 1991. Painted red with a white superstructure. Has a helicopter deck forward and a 20 bed hospital. After an extensive conversion and refit the ship replaced *Lago Lacar* in 1995. *VERIFIED*

RECALADA *8/1994, Marcelo Campodonico / 0056494*

22 PILOT CRAFT (PB)

ALUMINE GC 118 (ex-SP 14) MUSTERS GC 126 (ex-SP 26) HESS (ex-*Huechulafquen*) GC 135
TRAFUL GC 119 (ex-SP 15) COLHUE GC 129 (ex-SP 16) (ex-SP 34)
LACAR GC 120 (ex-SP 24) MARIA L PENDO GC 130 COLHUE HUAPI GC 136
MASCARDI GC 122 (ex-SP 17) (ex-SP 18) (ex-SP 33)
FONTANA GC 121 (ex-SP 32) ROCA GC 131 (ex-SP 28) YEHUIN GC 140 (ex-SP 30, ex-SP 35)
VIEDNA GC 123 (ex-SP 20) PUELO GC 132 (ex-SP 29) QUILLEN GC 141 (ex-SP 27)
SAN MARTIN GC 124 FUTALAUFQUEN GC 133 FAGNANO GC 150 (ex-SP 23)
(ex-SP 21) (ex-SP 30) NAHUEL HUAPI GC 151 (ex-SP 19)
BUENOS AIRES GC 125 FALKNER GC 134 (ex-SP 31) CARDIEL — (ex-SP 25)
(ex-SP 22)

(All names preceded by **LAGO**)

Comment: There are five different types of named pilot and patrol craft. SP 14-15 of 33.7 tons built in 1981; SP 16-18 of 47 tons built since 1981; SP 19-23 of 51 tons built since 1981; SP 25-27 of 20 tons built in 1981; SP 28-30 of 16.5 m built in 1983; SP 31-35 of 7 tons built in 1986-1991. Most built by Damen SY, Netherlands. The last one built by Astillero Mestrina, Tigre. No armament. Six were transferred to patrol duties in 1993 and 11 more in 1995 and all have GC numbers. SP 19, 23, 25, 27 and 30 are the pilot craft. *UPDATED*

HESS *12/1997, Hartmut Ehlers / 0017546*

LAND-BASED MARITIME AIRCRAFT

Notes: In addition to the aircraft listed, there are two Piper Warrior II/Archer II training aircraft and five Schweizer 300C training helicopters.

Numbers/Type: 2/3 CASA C-212 S 68 Aviocar/CASA C-212 A 68 Aviocar.
Operational speed: 190 kt *(353 km/h).*
Service ceiling: 24,000 ft *(7,315 m).*
Range: 1,650 n miles *(3,055 km).*
Role/Weapon systems: Two S 68 acquired in 1989, three A 68 in 1990. Medium-range reconnaissance and coastal surveillance duties in EEZ. Sensors: Bendix RDS 32 surface search radar. Omega Global GNS-500. Weapons: ASW; can carry torpedoes, depth bombs or mines. ASV; 2 × rockets or machine gun pods not normally fitted. *VERIFIED*

CASA C-212 *6/2002, CASA/EADS / 0528295*

Numbers/Type: 1 Aerospatiale SA 330 Super Puma.
Operational speed: 151 kt *(279 km/h).*
Service ceiling: 15,090 ft *(4,600 m).*
Range: 335 n miles *(620 km).*
Role/Weapon systems: Support and SAR helicopter for patrol work. Updated in France in 1996. Sensors: Omera search radar. Weapons: Can carry pintle-mounted machine guns but is usually unarmed. *VERIFIED*

SUPER PUMA *11/1996, Luis O Zunino / 0056495*

Numbers/Type: 3 Aerospatiale AS 365 Dauphin 2.
Operational speed: 150 kt *(278 km/h).*
Service ceiling: 15,000 ft *(4,575 m).*
Range: 410 n miles *(758 km).*
Role/Weapon systems: Acquired in 1995-96 to replace the Super Puma during the latter's update but have been retained. Sensors: Agrion search radar. Weapons: Unarmed. *VERIFIED*

DAUPHIN 2 *10/1996, Prefectura Naval / 0012022*

Australia

Country Overview

The Commonwealth of Australia comprises the island continent and the island of Tasmania which are separated by the Bass Strait. The British monarch, represented by a governor-general, is head of state. With an overall area of 2,966,151 square miles, it has a 13,910 n mile coastline with the Pacific (Coral and Tasman Seas) and Indian Oceans, the Timor Sea, Arafura Sea and the Torres Strait. External dependencies are the Australian Antarctic Territory, Christmas Island, the Territory of the Cocos Islands, the Territory of Heard Island and McDonald Islands, Norfolk Island, the Ashmore and Cartier Islands and the Coral Sea Islands Territory. Canberra is the capital while Sydney is the largest city and a major port. There are further ports at Melbourne, Fremantle, Newcastle, Port Kembla, Geelong, Brisbane, Gladstone, Port Hedland and Port Walcott. Territorial Seas (12 n miles) are claimed. An EEZ (200 n miles) is also claimed.

Headquarters Appointments

Chief of Navy:
    Vice Admiral C A Ritchie, AO

Maritime Commander, Australia:
    Rear Admiral R C Moffatt, AM
Commander Navy Systems Command:
    Commodore G J Geraghty
Commodore Flotillas:
    Commodore D R Thomas AM, CSC

Senior Appointments

Vice Chief of the Defence Force:
    Vice Admiral R E Shalders, AO, CSC
Head of Maritime Systems Division:
    Commodore T B Ruting AM, CSC
Head Defence Personnel Executive:
    Rear Admiral B L Adams, AO
Commander Australian Theatre:
    Rear Admiral M F Bonser, CSC
Director General Coast Watch:
    Rear Admiral R H Crane, CSM

Diplomatic Representation

Head Australian Defence Staff, Washington:
    Rear Admiral R W Gates
Head Australian Defence Staff, London:
    Brigadier V Williams
Defence Attaché in Washington:
    Commodore J R Stapleton, AM
Defence Adviser in Kuala Lumpur:
    Captain D L Garnock, CSC
Defence Attaché in Paris:
    Captain J B Dudley, CSC
Defence Attaché in Bangkok:
    Captain B A Fraser
Defence Attaché in Paris:
    Captain W R Haynes
Naval Attaché in Jakarta:
    Captain J B Dudley, CSC
Naval Adviser in London:
    Captain V S Jones
Defence Attaché in Phnom Penh:
    Captain T R Jenkinson
Defence Attaché in Singapore:
    Captain M T Jerrett

Personnel

(a) 2005: 13,155 officers and ratings
(b) 7,015 (2,495 active, 4,520 standby)

RAN Reserve

The Naval Reserve is integrated into the Permanent Force. Personnel are either Active Reservists with regular commitments or Inactive Reservists with periodic or contingent duty. The missions undertaken by the Reserve include Coordination and Guidance of Psychology, Public Relations, Intelligence, Diving and patrol boat/landing craft operations. In addition, members of the Ready Reserve (a component of the Active Reserve) are shadow posted to selected major fleet units.

Shore Establishments

Canberra: Navy Headquarters, Navy Systems Command Headquarters, *Harman* (Communications, Administration). Sydney: Maritime Headquarters, Fleet Base East (Garden Island), *Waterhen* (Mine Warfare and Clearance Diving), *Watson* (Warfare Training), *Penguin* (Diving, Hospital), *Kuttabul* (Administration). Wollongong Hydrographic Headquarters. Jervis Bay Area: *Albatross* (Air Station), *Creswell* (Leadership and Management Training and Fleet Support), Jervis Bay Range Facility.

Cockburn Sound (WA): Fleet Base West, *Stirling* (Administration and Maintenance Support, Submarines, Communications). Darwin: Minor warship base, *Coonawarra* (Administration). Cairns: *Cairns* (Administration), Minor Warship Base. Adelaide: Regional Naval Headquarters, South Australia. Brisbane: Regional Naval Headquarters, South Queensland. Hobart: Regional Naval Headquarters, Tasmania.

Fleet Deployment

Fleet Base East (and other Sydney bases): 3 FFG, 1 FFH, 1 AOR, 2 LPA, 1 LSH, 1 ASR, 6 MHC, 3 MSA.
Fleet Base West: 6 SS, 3 FFG, 3 FFH, 1 AO.
Darwin Naval Base: 10 PTF, 1 LCH.
Cairns: 5 PTF, 4 LCH, 4 AGS.

Fleet Air Arm (see *Shipborne Aircraft* section).

Squadron	Aircraft
723	Squirrel AS 350B, Utility, FFG embarked flights, SAR
	HS 748, Fixed-wing, EW operations and training
	Bell 206B, survey support
805	Seasprite SH-2G
817	Sea King Mk 50, Utility
816	Seahawk S-70B-2, ASW, ASST

Prefix to Ships' Names

HMAS. Her Majesty's Australian Ship

Strength of the Fleet

Type	Active	Building (Projected)
Patrol Submarines	6	—
Destroyers	—	(3)
Frigates (FFG)	6	—
Frigates (FF)	5	3
Minehunters (Coastal)	6	—
Minesweepers (Auxiliary)	3	—
Large Patrol Craft	15	—
Amphibious Heavy Lift Ship	1	—
Amphibious Transports	2	—
Landing Craft	10	—
Survey Ships	6	—
Replenishment Ships	2	—
Training Ships	1	—

DELETIONS

Mine Warfare Forces

2003 *Brolga*

PENNANT LIST

Submarines

73	Collins
74	Farncomb
75	Waller
76	Dechaineux
77	Sheean
78	Rankin

Frigates

01	Adelaide
02	Canberra
03	Sydney
04	Darwin
05	Melbourne
06	Newcastle
150	Anzac
151	Arunta
152	Warramunga
153	Stuart

154	Parramatta
155	Ballarat
156	Toowoomba (bldg)
157	Perth (bldg)

Mine Warfare Forces

M 82	Huon
M 83	Hawkesbury
M 84	Norman
M 85	Gascoyne
M 86	Diamantina
M 87	Yarra
Y 298	Bandicoot
Y 299	Wallaroo

Patrol Forces

83	Armidale
203	Fremantle
204	Warrnambool

205	Townsville
206	Wollongong
207	Launceston
208	Whyalla
209	Ipswich
210	Cessnock
211	Bendigo
212	Gawler
213	Geraldton
214	Dubbo
215	Geelong
216	Gladstone
217	Bunbury

Amphibious Forces

L 50	Tobruk
L 51	Kanimbla
L 52	Manoora
L 126	Balikpapan
L 127	Brunei

L 128	Labuan
L 129	Tarakan
L 130	Wewak
L 133	Betano

Survey Ships

A 01	Paluma
A 02	Mermaid
A 03	Shepparton
A 04	Benalla
A 245	Leeuwin
A 246	Melville

Auxiliaries

O 195	Westralia
OR 304	Success

SUBMARINES

Note: Stirling 4V-275R (75 kW) engines supplied for AIP trials ashore.

DECHAINEUX   *2/2004*, *John Mortimer* / 1121117

For details of the latest updates to *Jane's Fighting Ships* online and to discover the additional information available exclusively to online subscribers please visit
jfs.janes.com

6 COLLINS CLASS (SSK)

Name	No	Builders	Laid down	Launched	Commissioned
COLLINS	73	Australian Submarine Corp, Adelaide	14 Feb 1990	28 Aug 1993	27 July 1996
FARNCOMB	74	Australian Submarine Corp, Adelaide	1 Mar 1991	15 Dec 1995	31 Jan 1998
WALLER	75	Australian Submarine Corp, Adelaide	19 Mar 1992	14 Mar 1997	10 July 1999
DECHAINEUX	76	Australian Submarine Corp, Adelaide	4 Mar 1993	12 Mar 1998	23 Feb 2001
SHEEAN	77	Australian Submarine Corp, Adelaide	17 Feb 1994	1 May 1999	23 Feb 2001
RANKIN	78	Australian Submarine Corp, Adelaide	12 May 1995	7 Nov 2001	29 Mar 2003

Displacement, tons: 3,051 surfaced; 3,353 dived
Dimensions, feet (metres): 255.2 × 25.6 × 23
  (77.8 × 7.8 × 7)
Main machinery: Diesel-electric; 3 Hedemora/Garden Island
  Type V18B/14 diesels; 6,020 hp *(4.42 MW)*; 3 Jeumont
  Schneider generators; 4.2 MW; 1 Jeumont Schneider motor;
  7,344 hp(m) *(5.4 MW)*; 1 shaft; 1 MacTaggart Scott DM
  43006 hydraulic motor for emergency propulsion
Speed, knots: 10 surfaced; 10 snorting; 20 dived
Range, n miles: 9,000 at 10 kt (snort); 11,500 at 10 kt
  (surfaced)
  400 at 4 kt (dived)
Complement: 45 (8 officers)

Missiles: SSM: McDonnell Douglas Sub Harpoon; active radar
  homing to 130 km *(70 n miles)* at 0.9 Mach; warhead 227 kg.
Torpedoes: 6—21 in *(533 mm)* fwd tubes. Gould Mk 48 Mod 4;
  dual purpose; wire-guided; active/passive homing to 38 km
  (21 n miles) at 55 kt or 50 km *(27 n miles)* at 40 kt; warhead
  267 kg. Air turbine pump discharge. Total of 22 weapons
  including Mk 48 and Sub Harpoon.
Mines: 44 in lieu of torpedoes.
Countermeasures: Decoys: 2 SSE.
ESM: Condor CS-5600; intercept and warning.
Weapons control: Boeing/Rockwell integrated system. Link 11.
Radars: Navigation: Kelvin Hughes Type 1007; I-band.
Sonars: Thomson Sintra Scylla active/passive bow array and
  passive flank, intercept and ranging arrays.
  GEC-Marconi Kariwara (first pair) or Thomson Marconi Narama
  or Allied Signal TB 23; retractable passive towed array.

Programmes: Contract signed on 3 June 1987 for construction
  of six Swedish-designed Kockums Type 471. Fabrication work
  started in June 1989; bow and midships (escape tower)
  sections of the first submarines built in Sweden.

WALLER
4/2003, Mick Prendergast / 0569124

Structure: Stirling air independent propulsion (AIP) has been
  tested on a shore rig. Scylla is an updated Eledone sonar suite.
  Diving depth, 300 m *(984 ft)*. Anechoic tiles are fitted during
  build to all but *Collins* which is retrofitted. Pilkington Optronics
  CK 43 search and CH 93 attack periscopes fitted. Plans for an
  external mine belt have been abandoned.
Modernisation: The Replacement Combat System AN-BYG 1 is
  based on Raytheon's CCS Mk 2. The shore facilities version is
  to be established in mid-2005 and the first seagoing system in
  Waller in 2006. The other boats will follow by 2010.
  Meanwhile, following trials in *Collins* to improve the
  performance of the current combat system, the systems in
  Dechaineux and *Sheean* have been augmented. The
  remaining boats will not receive this upgrade. In parallel,
  significant improvements to noise signature have been
  achieved following modifications to propellers and casing
  sections and improvements to the hydraulics system and

engine reliability. These have been made to *Dechaineux*,
Sheean, Collins, Rankin and *Farncomb. Waller* is to be similarly
refitted at next scheduled docking period. Collaborative
development of the US Mk 48 ADCAP torpedo is being
progressed following the signature of a 'Statement of
Principles' agreement which promotes interoperability with
US Navy. Three boats have been fitted with the Condor CS
5600 ESM system with the remainder to follow in due course.
Collins has received a set of modifications to facilitate the
deployment and recovery of special forces.
Operational: *Dechaineux* and *Sheean* have achieved interim
operational capability but full operational capability will not be
achieved until the Replacement Combat System has been
installed. All submarines are based at Fleet Base West with
one or two deploying regularly to the east coast.

UPDATED

DECHAINEUX
2/2004*, Chris Sattler / 1042096

RANKIN
9/2003, Chris Sattler / 0569122

DESTROYERS

Notes: Project Sea 4000 is for the procurement of at least three area air-defence destroyers. Following Phase 1B, it was announced in August 2004 that the core of the combat system is to be the US Aegis system. Phase 1C, the selection of a designer and concept platform design is to be completed in mid-2005. In Phase 2, a fully costed design will be developed and an alliance established between the ship design company, Combat System Systems Engineering (CSSE) company, ship builder and Department of Defence. Following Requests for Proposals for the builder and CSSE in October 2004, announcements of successful bids is expected by mid-2005. Phase 3, ship construction, is to start in 2007. Delivery of the first ship is planned for 2013 with subsequent units to follow in 2015 and 2017. There is potential linkage between this programme and Australian participation in the US Theatre Ballistic Missile Defence programme.

FRIGATES

6 ADELAIDE (OLIVER HAZARD PERRY) CLASS (FFGHM)

Name	No	Builders	Laid down	Launched	Commissioned
ADELAIDE	01	Todd Pacific Shipyard Corporation, Seattle, USA	29 July 1977	21 June 1978	15 Nov 1980
CANBERRA	02	Todd Pacific Shipyard Corporation, Seattle, USA	1 Mar 1978	1 Dec 1978	21 Mar 1981
SYDNEY	03	Todd Pacific Shipyard Corporation, Seattle, USA	16 Jan 1980	26 Sep 1980	29 Jan 1983
DARWIN	04	Todd Pacific Shipyard Corporation, Seattle, USA	3 July 1981	26 Mar 1982	21 July 1984
MELBOURNE	05	Australian Marine Eng (Consolidated), Williamstown	12 July 1985	5 May 1989	15 Feb 1992
NEWCASTLE	06	Australian Marine Eng (Consolidated), Williamstown	21 July 1989	21 Feb 1992	11 Dec 1993

Displacement, tons: 4,100 full load
Dimensions, feet (metres): 453 × 45 × 24.5 (sonar); 14.8 (keel) *(138.1 × 13.7 × 7.5; 4.5)*
Main machinery: 2 GE LM 2500 gas turbines; 41,000 hp *(30.6 MW)* sustained; 1 shaft; cp prop; 2 auxiliary electric retractable propulsors fwd; 650 hp *(484 kW)*
Speed, knots: 29 (4 on propulsors). **Range, n miles:** 4,500 at 20 kt
Complement: 184 (15 officers) plus aircrew

Missiles: SSM: 8 McDonnell Douglas Harpoon; active radar homing to 130 km *(70 n miles)* at 0.9 Mach; warhead 227 kg.
SAM: GDC Pomona Standard SM-1MR; Mk 13 Mod 4 launcher for both SAM and SSM systems ❶; command guidance; semi-active radar homing to 46 km *(25 n miles)* at 2 Mach; height 45.7-18,288 m *(150-60,000 ft)*; 40 missiles (combined SSM and SAM). 32 Raytheon ESSM (8 quad forward) *(Sydney)*; semi-active radar homing to 18.5 km *(10 n miles)*.
Guns: 1 OTO Melara 3 in *(76 mm)*/62 US Mk 75 compact ❷; 85 rds/min to 16 km *(9 n miles)* anti-surface; 12 km *(6.5 n miles)* anti-aircraft; weight of shell 6 kg. Guns for 05 and 06 manufactured in Australia.
1 General Electric/GDC 20 mm Mk 15 Vulcan Phalanx ❸; anti-missile system with 6 barrels; 4,500 rds/min combined to 1.5 km.
Up to 6—12.7 mm MGs.
Torpedoes: 6—324 mm Mk 32 (2 triple) tubes ❹. Honeywell Mk 46 Mod 5; anti-submarine; active/passive homing to 11 km *(5.9 n miles)* at 40 kt; warhead 44 kg.
Countermeasures: Decoys: 2 Loral Hycor SRBOC Mk 36 chaff and IR decoy launchers; fixed 6-barrelled system; range 1-4 km. 4 BAe Nulka quad expendable decoy launchers.
SLQ-25; towed torpedo decoy or TMS Sea Defender torpedo countermeasures (from 2005).
ESM/ECM: Raytheon SLQ-32C ❺; intercept and jammer. Elbit EA-2118 jammer. Rafael C-Pearl (from 2005).
Combat data systems: NCDS using NTDS consoles and UYK 7 and UYK 43 computers; augmented as part of upgrade by ADACS from 2003. OE-2 SATCOM; Link 11.
Weapons control: Sperry Mk 92 Mod 2 (Mod 12 from 2005) gun and missile control (Signaal derivative). Radamec 2500 optronic director with TV, laser and IR imager.
Radars: Air search: Raytheon SPS-49 ❻; C-band.
Surface search/navigation: ISC Cardion SPS-55 ❼; I-band.
Fire control: Lockheed SPG-60 ❽; I/J-band; range 110 km *(60 n miles)*; Doppler search and tracking.
Sperry Mk 92 ❾; I/J-band.
IFF: AIMS Mk XII.
Tacan: URN 25.
Sonars: Raytheon SQS-56 or EMI/Honeywell Mulloka (05 and 06); hull-mounted; active; medium frequency. To be replaced by Thomson Marconi TMS 4131 medium frequency sonar from 2005. Petrel (TMS 5424) high frequency mine-avoidance from 2005, Albatros (TMS 4350) towed-array torpedo-warning system from 2005.

Helicopters: 2 Sikorsky S-70B-2 Seahawks ❿ or 1 Seahawk and 1 Squirrel.

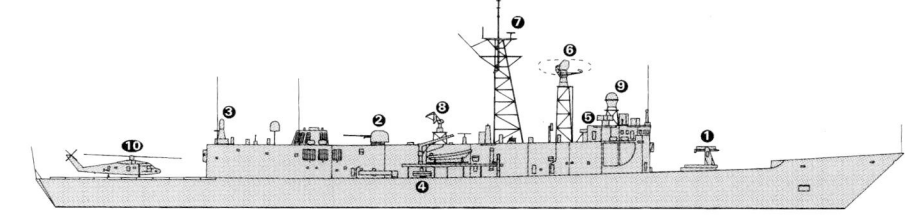

NEWCASTLE
(Scale 1 : 1,200), Ian Sturton / 0569258

NEWCASTLE
7/2004*, Michael Nitz / 1042113

Programmes: US numbers: *Adelaide* FFG 17; *Canberra* FFG 18; *Sydney* FFG 35; *Darwin* FFG 44.
Modernisation: *Adelaide* in November 1989, *Sydney* February 1989 and *Canberra* December 1991 completed a 12 month Helicopter Modification Programme to allow operation of Seahawk helicopters. The modification, fitted to *Darwin* during construction, involved angling the transom (increasing the ship's overall length by 8 ft) and fitting the RAST helo recovery system. *Melbourne* and *Newcastle* fitted during construction which also included longitudinal strengthening and buoyancy upgrades. The FFG Upgrade Programme (FFG UP) has commenced under Project Sea 1390 with *Sydney* being the first of four FFGs to have been upgraded at Garden Island in 2004. *Melbourne* is to be modernised in 2005. *Adelaide* and *Canberra* are not to be modernised. The work package includes upgrade of SPS 49 radar and the Mk 92 weapons control system and removal of NCDS and migration of its functionality to ADACS. Link 16 and improved ESM and decoy systems are also to be fitted. The upgraded FFGs are to be fitted with Standard SM-2 missiles to replace SM-1 missiles by 2009. An eight-cell Mk 41 VLS system for 32 ESSM is also to be fitted in the foredeck. Other upgrade work includes replacement of the hull-mounted sonar with TMS 4131 and the fitting of TMS 4350 torpedo defence system and TMS 5424 mine-avoidance sonar. Diesel generators and air compressors are also to be replaced.
Operational: *Adelaide, Darwin* and *Canberra* based at Fleet Base West. The remainder are based at Fleet Base East. For operational tasks ships are fitted with enhanced communications, electro-optical sights, rigid inflatable boats and portable RAM panels. All ships are fighter control capable. *Canberra* to be decommissioned in November 2005 and *Adelaide* in September 2006.

UPDATED

SYDNEY
2/2005*, Chris Sattler / 1042396

6 + 2 ANZAC (MEKO 200) CLASS (FFGHM)

Name	No	Builders	Laid down	Launched	Commissioned
ANZAC	150	Transfield, Williamstown	5 Nov 1993	16 Sep 1994	18 May 1996
ARUNTA (ex-Arrernte)	151	Transfield, Williamstown	22 July 1995	28 June 1996	12 Dec 1998
WARRAMUNGA (ex-Warumungu)	152	Tenix Defence Systems, Williamstown	26 July 1997	23 May 1998	31 Mar 2001
STUART	153	Tenix Defence Systems, Williamstown	25 July 1998	17 Apr 1999	17 Aug 2002
PARRAMATTA	154	Tenix Defence Systems, Williamstown	4 June 1999	17 June 2000	4 Oct 2003
BALLARAT	155	Tenix Defence Systems, Williamstown	4 Aug 2000	25 May 2002	26 June 2004
TOOWOOMBA	156	Tenix Defence Systems, Williamstown	26 July 2002	16 May 2003	Sep 2005
PERTH	157	Tenix Defence Systems, Williamstown	24 July 2003	20 Mar 2004	July 2006

Displacement, tons: 3,600 full load
Dimensions, feet (metres): 387.1 oa; 357.6 wl × 48.6 × 14.3
 (118; 109 × 14.8 × 4.35)
Main machinery: CODOG: 1 GE LM 2500 gas turbine;
 30,172 hp *(22.5 MW)* sustained; 2 MTU 12V 1163 TB83
 diesels; 8,840 hp(m) *(6.5 MW)* sustained; 2 shafts; cp props
Speed, knots: 27. **Range, n miles:** 6,000 at 18 kt
Complement: 174 (24 officers)

Missiles: SSM: 8 McDonnell Douglas Harpoon (152).
 SAM: Raytheon Sea Sparrow RIM-7NP (150, 151 only);
 Lockheed Martin Mk 41 Mod 5 octuple vertical launcher ❶;
 semi-active radar homing to 14.6 km *(8 n miles)* at 2.5 Mach;
 warhead 39 kg. 8 missiles total. Quadpack Evolved Sea
 Sparrow for 32 missiles in 152 onwards.
Guns: 1 United Defense 5 in (127 mm)/54/62 Mk 45 Mod 2 ❷;
 20 rds/min to 23 km *(12.6 n miles)*; weight of shell 32 kg.
 4—12.7 mm MGs.
Torpedoes: 6—324 mm (2 triple) Mk 32 Mod 5 tubes ❸. Mk 46
 Mod 5; anti-submarine; active/passive homing to 11 km *(5.9 n
 miles)* at 40 kt; warhead 44 kg.
Countermeasures: Decoys: G & D Aircraft SRBOC Mk 36 Mod 1
 decoy launchers ❹ for SRBOC/NATO Sea Gnat. 4 BAe Nulka
 quad expendable decoy launchers.
 FEL SLQ-25A towed torpedo decoy.
 ESM: Racal modified Sceptre A; radar intercept (being replaced
 by Thales Centaur). Telefunken PST-1720 Telegon 10; comms
 intercept.
Combat data systems: CelsiusTech 9LV 453 Mk 3 (Mk 3E in
 157). Link 11.
Weapons control: CelsiusTech 9LV 453 optronic director with
 Raytheon CW Mk 73 Mod 1 (for RIM-7NP) or CEA SSCWI (for
 ESSM).
Radars: Air search: Raytheon SPS-49(V)8 ANZ ❺; C-band.
 Air/surface search: Ericsson Sea Giraffe ❻; G/H-band.
 Navigation: Atlas Elektronik 9600 ARPA; I-band.
 Fire control: CelsiusTech 9LV 453 ❼; J-band.
 IFF: Cossor AIMS Mk XII.
Sonars: Thomson Sintra Spherion B Mod 5; hull-mounted; active
 search and attack; medium frequency. Provision for towed
 array; low frequency active/passive may be fitted; Petrel mine
 avoidance sonar.

Helicopters: 1 S-70B-2 Seahawk ❽ or SH-2G Seasprite.

Programmes: Contract signed with Australian Marine
 Engineering Consolidated (now Tenix Defence) on
 10 November 1989 to build eight Blohm + Voss designed
 MEKO 200 ANZAC frigates for Australia and two for New
 Zealand. First ship started construction 27 March 1992.
 Modules are constructed at Whangarei and shipped to
 Williamstown for assembly. The second and fourth ships of
 the class were delivered to New Zealand.

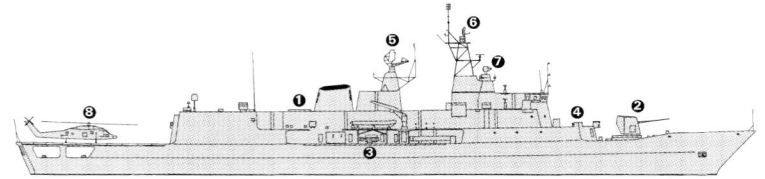

ANZAC
(Scale 1 : 1,200), Ian Sturton / 0569259

PARRAMATTA
2/2004 *, Bob Fildes / 1042097

Modernisation: Evolved Seasparrow missile (ESSM) was
 incorporated in 152, the first ship in the world to be so fitted
 (first missile launched 21 January 2003). Subsequent ships
 are similarly armed and the first two ships are also to be
 retrofitted from 2005. 157 is to be the first of class to be fitted
 with the 9LV Mk 3E combat management system, which is
 planned to be retrofitted in the remainder of the class
 between 2007-10. Other components of the ASMD upgrade
 project include installation of infra-red search and track and
 improvements to the fire-control radar. The potential of CEA-
 FAR active phased array multifunction radar was assessed
 during sea trials in 2004 although a decision on whether to
 proceed with wider installation has not yet been made.
 Harpoon has been installed on 152 and, following First of
 Class trials, is to be fitted to the rest of the class. Petrel mine

avoidance sonar is to enter service in 2005 while the MU 90
 torpedo is planned to replace the Mk 46.
Structure: 'Space and weight' reserved for a CIWS, an additional
 octuple VLS, second channel of fire for VLS, towed array
 sonar, offboard active ECM, extended ESM frequency
 coverage, Helo datalink and SATCOM. Stealth features are
 incorporated in the design. All-steel construction. Fin
 stabilisers. Indal RAST helicopter recovery system. Repairs to
 bilge keel cracks have been made to 150, 151 and 152 while
 design changes have been implemented for ships in build.
Operational: The SH-2G is to be ASM (Penguin) missile fitted
 once accepted. Two RHIBs are carried. F 150, F 151 and F 152
 based at Perth and F 153, F 154 and F 155 at Sydney.
UPDATED

ARUNTA
10/2004 *, Hachiro Nakai / 1042105

BALLARAT   *6/2004*, Chris Sattler* / 1042114

STUART   *2/2004*, Bob Fildes* / 1042098

PARRAMATTA   *7/2004*, Michael Nitz* / 1042115

SHIPBORNE AIRCRAFT

Notes: (1) Five Bell 206B Kiowa utility helicopters transferred to the Army in 2001.
(2) Up to 12 troop lift helicopters are planned to enter Army service from 2007. Fully navalised, they are to be capable of operating from *Kanimbla* and *Manoora* and from future amphibious ships. Consideration is also likely to be given to replacement of the Sea King utility helicopter (life-extended to 2008) and of Seahawk with the same air-frame. This might impact on plans for latter's mid-life upgrade programme.

Numbers/Type: 11 Kaman Seasprite SH-2G(A).
Operational speed: 130 kt *(241 km/h)*.
Service ceiling: 10,000 ft *(3,048 m)*
Range: 350 n miles *(650 km)*.
Role/Weapon systems: Contract placed in June 1997 for eleven aircraft for Anzac frigates. Refurbished USN aircraft being delivered from 2003-2005. First provisionally accepted in 2003 with remainder to follow in 2004. Sensors: Telephonics APS 143(V)3 radar; Raytheon AAQ-27 FLIR; AAR 54/AES 210/LWS 20 ESM; ALE 47 chaff and IR flares; Link 11, SATCOM. Weapons: ASW; 2 Mk 46 (to be replaced by MU 90) torpedoes. ASV: 2 Penguin Mk 2 Mod 7; 1—7.62 mm MG.

UPDATED

SH-2G(A)   *6/2002, Royal Australian Navy /* 0528405

Numbers/Type: 7 Westland Sea King HAS 50/50A.
Operational speed: 125 kt *(230 km/h)*.
Service ceiling: 14,500 ft *(4,400 m)*.
Range: 490 n miles *(908 km)*.
Role/Weapon systems: Utility helicopter; embarked periodically for operations from *Success*, *Tobruk* and the LPAs. Life extension to 2008 completed in November 1996 for six aircraft. One more acquired from UK in 1996 and upgraded to 50LEP (Mk 50) standard. Sensors: AW 391(A) radar. Weapons: MAG 58 7.62 mm MG.

UPDATED

SEA KING   *6/2004*, Chris Sattler /* 1042111

Numbers/Type: 16 Sikorsky S-70B-2 Seahawk.
Operational speed: 135 kt *(250 km/h)*.
Service ceiling: 10,000 ft *(3,050 m)*.
Range: 600 n miles *(1,110 km)*.
Role/Weapon systems: Seahawk SH-60F derivative aircraft designed by Sikorsky to meet RAN specifications for ASW and ASST operations. Eight assembled by ASTA in Victoria. Helicopters embarked in FFG-7 and used temporarily in ANZAC frigates. Upgrades from 2004 to include Raytheon AAQ 27 FLIR, Tracor ALE 47 countermeasures and Elisra AES 210 ESM. Sensors: MEL Surface surveillance radar, CDC Sonobuoy Processor and Barra Side Processor, and CAE Magnetic Anomaly Detector Set controlled by a versatile Tactical Display/Management System. Weapons: ASW; two Mk 46 Mod 5 (to be replaced by MU 90) torpedoes. ASV; one Mag 58 MG, possibly ASM after 2005.

UPDATED

SEAHAWK   *2/2003, Paul Jackson /* 0552766

Numbers/Type: 13 Aerospatiale AS 350B Squirrel.
Operational speed: 125 kt *(232 km/h)*.
Service ceiling: 10,000 ft *(3,050 m)*.
Range: 275 n miles *(510 km)*.
Role/Weapon systems: Support helicopter for utility tasks and training duties. Regularly embarked at sea. Sensors: None. Weapons: ASV; two Mag 58 MGs.

VERIFIED

SQUIRREL   *2/2003, Paul Jackson /* 0552765

LAND-BASED MARITIME AIRCRAFT

Notes: Replacement of the P3-C maritime patrol aircraft fleet from about 2013 is being taken forward under Project Air 7000. Interoperability with the US is a major factor and almost certainly points to selection of the Boeing 737 MMA. Such a choice would also offer significant logistical advantages as both the AEW and maritime patrol aircraft would then be based on a common airframe. In the meantime, procurement of high altitude, long endurance unmanned aerial vehicles for broad area surveillance over maritime and land environments is planned. Likely candidates include the Northrop Grumman RQ-4B Global Hawk and General Atomics Mariner.

Numbers/Type: 4 Boeing 737 AEW&C 'Wedgetail'.
Operational speed: to be confirmed.
Service ceiling: 41,000 ft *(12,500 m)*.
Range: to be confirmed.
Role/Weapon systems: Contract for four aircraft (adaptation of Boeing Business Jet) with options for further two signed on 20 December 2000. Delivery of first two aircraft scheduled in November 2006, second pair in 2007. IOC to be achieved late 2008. AAR capable. Sensors: Details unconfirmed but likely to include Northrop Grumman ESSD L-band multirole electronically scanned array (MESA) radar (fuselage mounted); electronic warfare self-protection (EWSP) system (including IR countermeasures, chaff and flares); Links 11 and 16; Satcom.

VERIFIED

BOEING WEDGETAIL   *7/2004*, Boeing /* 0566617

Numbers/Type: 17/4/15 General Dynamics F-111C/RF-111C/F-111G.
Operational speed: 793 kt *(1,469 km/h)*.
Service ceiling: 60,000 ft *(18,290 m)*.
Range: 2,540 n miles *(4,700 km)*.
Role/Weapon systems: Air Force operates the F-111 for anti-shipping strike and its small force of RF-111 for coastline surveillance duties using EW/ESM and photographic equipment underwing. Sensors: GE AN/APG-144, podded EW. Weapons: ASV; 4 × Harpoon missiles. Strike; 4 × Snakeye bombs. Self-defence; 2 × AIM-9P.

VERIFIED

F-111C   *2/2003, Paul Jackson /* 0552764

Numbers/Type: 18 Lockheed P-3C/AP-3C Orion.
Operational speed: 410 kt *(760 km/h)*.
Service ceiling: 28,300 ft *(8,625 m)*.
Range: 4,000 n miles *(7,410 km)*.
Role/Weapon systems: Operated by Air Force for long-range ocean surveillance and ASW. Three more aircraft (plus one for spare parts) without armament or sensors acquired for training. Six aircraft upgraded to AP-3C standard with remaining twelve to be delivered by late 2004. Sensors: Elta EL/M-2022A(V)3 radar, AQS-901 processor, AQS-81 MAD FLIR systems Star Safire electro-optic system, ECM, Elta/IAI, ALR 2001 ESM, 80 × BARRA sonobuoys. Weapons: ASW; 8 × Mk 46 (Mod 5 after upgrade) torpedoes, Mk 25 mines, 8 × Mk 54 depth bombs. ASV; up to 6 AGM-84A/C Harpoon.

VERIFIED

Numbers/Type: 68 McDonnell Douglas F/A-18 Hornet.
Operational speed: 1,032 kt *(1,910 km/h)*.
Service ceiling: 50,000 ft *(15,240 m)*.
Range: 1,000 n miles *(1,829 km)*.
Role/Weapon systems: Air defence and strike aircraft operated by Air Force but with fleet defence and anti-shipping secondary roles. Sensors: APG-65 attack radar, AAS-38 FLIR/ALR-67 radar warning receiver. Weapons: ASV; 4 × Harpoon missiles. Strike; 1 × 20 mm cannon, up to 7.7 tons of 'iron' bombs. Fleet defence; 4 × AIM-7 Sparrow and 4 × AIM-9L Sidewinder.

VERIFIED

ORION AP-3C   *4/2002, RAAF /* 0522152

AMPHIBIOUS FORCES

Notes: Replacements for the current Amphibious capability are being procured under Joint Project 2048. The Australian government announced in November 2003 that *Tobruk* and one of the LPA amphibious transports *(Kanimbla* and *Manoora)* are to be replaced by two helicopter capable amphibious ships (LHD) of approximately 20,000 tons in 2010 and 2013 respectively. The second LPA is to be replaced by a 'strategic sealift' capability in 2016. Short listed designs are those based on the French Mistral class BPC (LHD) and the Spanish Izar Strategic Projection Ship and a contract is expected in 2006. Replacement of the watercraft capability, represented by the current LCH, LCM 8 and LCVP and other ship-to-shore assets required to integrate with the new LHDs, will also be delivered under JP 2048. LCHs and LCM 8s will be decommissioned as the new capability enters service around the 2010-2013 timeframe.

2 KANIMBLA (NEWPORT) CLASS (LCCH/LLP)

Name	No	Builders	Laid down	Launched	Commissioned	Recommissioned
KANIMBLA (ex-*Saginaw*)	L 51 (ex-1188)	National Steel & Shipbuilding	24 May 1969	7 Feb 1970	23 Jan 1971	29 Aug 1994
MANOORA (ex-*Fairfax County*)	L 52 (ex-1193)	National Steel & Shipbuilding	28 Mar 1970	19 Dec 1970	16 Oct 1971	25 Nov 1994

Displacement, tons: 4,975 light; 8,450 full load
Dimensions, feet (metres): 552 × 69.5 × 17.5 (aft) *(168.2 × 21.2 × 5.3)*
Main machinery: 6 ALCO 16-251 diesels; 16,500 hp *(12.3 MW)* sustained; 2 shafts; cp props; bow thruster
Speed, knots: 20. **Range, n miles:** 14,000 at 15 kt
Complement: 213 (12 officers)
Military lift: 450 troops (25 officers); 229 lane-metres of vehicles; 2 LCM 8; 250 tons aviation fuel

Guns: 1 General Electric/General Dynamics 20 mm Vulcan Phalanx Mk 15 can be fitted **①** 4—12.7 mm MGs. Fitted for but not with army-operated RBS 70 launchers. To receive 2 Typhoon 25 mm guns in 2004-05.
Countermeasures: 2 SRBOC Mk 36 chaff and IR launchers.
Radars: Surface search: Kelvin Hughes 1007 **②**; I-band.
Navigation: Kelvin Hughes **③**; I-band.

Helicopters: 4 Army Black Hawks or 3 Sea Kings or 1 Chinook.

Programmes: Acquired by sale from USA on 25 August and 27 September 1994.
Modernisation: Conversion contract let to Forgacs Shipbuilding, Newcastle in May 1995. Both ships modified by fitting a hangar to take four Black Hawk helicopters, to incorporate a third landing spot forward, to increase aviation fuel capacity and to dispense with the bow landing ramp. The after flight deck is Chinook capable. A stern gate to the tank deck is retained. Two Army LCM 8 class are carried on the deck forward of the bridge and handled by a 70 ton crane. A classroom and improved medical facilities are installed. Installation of communications and command support system to support a deployable JTFHQ was undertaken in both ships in 2001.
Operational: Both based at Sydney. To be replaced in 2013 and 2016.

UPDATED

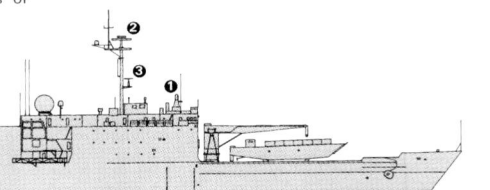

MANOORA   *(Scale 1 : 1,500), Ian Sturton /* 0569257

MANOORA   *1/2002, D Pawlenko, RAN /* 0528406

KANIMBLA   *11/2001, Royal Australian Navy /* 0528407

KANIMBLA   *6/2003, A Sharma /* 0569146

1 HEAVY LIFT SHIP (LSLH)

Name	No	Builders	Laid down	Launched	Commissioned
TOBRUK	L 50	Carrington Slipways Pty Ltd	7 Feb 1978	1 Mar 1980	23 Apr 1981

Displacement, tons: 3,300 standard; 5,700 full load
Dimensions, feet (metres): 417 × 60 × 16
(127 × 18.3 × 4.9)
Main machinery: 2 Mirrlees Blackstone KDMR8 diesels;
9,600 hp *(7.2 MW)*; 2 shafts
Speed, knots: 18. **Range, n miles:** 8,000 at 15 kt
Complement: 148 (13 officers)
Military lift: 314 troops (prolonged embarkation); 1,300 tons
cargo or 330 lane-metres of vehicles; 70 tons capacity
derrick; 2—4.25 ton cranes; 2 LCVP; 2 LCM 8

Guns: 2—12.7 mm MGs. To be fitted with 2 Typhoon 25 mm
guns in 2004-05.
Radars: Surface search: Kelvin Hughes Type 1006; I-band.
Navigation: Kelvin Hughes 1007; I-band.

Helicopters: Platform for one Sea King. Second Chinook capable
spot on forward flight deck (clear of cargo).

Structure: The design is an update of the British Sir Bedivere
class and provides facilities for the operation of helicopters,
landing craft, amphibians for ship-to-shore movement. A
special feature is the ship's heavy lift derrick system for
handling heavy loads. Able to embark a squadron of Leopard
tanks plus a number of wheeled vehicles and artillery in
addition to its troop lift. Bow and stern ramps are fitted. Two
LCM 8 carried on deck and two LCVPs at davits.
Operational: A basic communications fit enables participation in
amphibious operations but not in command role. Based at
Sydney. To be replaced in 2010. **UPDATED**

TOBRUK

TOBRUK   4/2004*, D Pawlenko, RAN / 1042095

6 LANDING CRAFT (HEAVY) (LCH/LSM)

Name	No	Builders	Commissioned
BALIKPAPAN	L 126	Walkers Ltd, Queensland	8 Dec 1971
BRUNEI	L 127	Walkers Ltd, Queensland	5 Jan 1973
LABUAN	L 128	Walkers Ltd, Queensland	9 Mar 1973
TARAKAN	L 129	Walkers Ltd, Queensland	15 June 1973
WEWAK	L 130	Walkers Ltd, Queensland	10 Aug 1973
BETANO	L 133	Walkers Ltd, Queensland	8 Feb 1974

Displacement, tons: 358 light; 509 full load
Dimensions, feet (metres): 146 × 33 × 6.5 *(44.5 × 10.1 × 2)*
Main machinery: 2 GM 6-71 diesels; 348 hp *(260 kW)* sustained; 2 shafts
Speed, knots: 10. **Range, n miles:** 3,000 at 10 kt
Complement: 16 (2 officers)
Military lift: 3 medium tanks or equivalent
Guns: 2—12.7 mm MGs.
Radars: Navigation: Racal Decca Bridgemaster; I-band.

Comment: Originally this class was ordered for the Army but only *Balikpapan* saw Army service
until being commissioned into the Navy on 27 September 1974. The remainder were built for
the Navy. *Balikpapan* based at Darwin. The remainder are based at Cairns. All have been given a
life extension refit, which started with *Wewak* in 2000, and completed with *Brunei* in 2002, for
retention until at least 2008. *Buna* and *Salamaua* transferred to Papua New Guinea Defence
Force in November 1974.

UPDATED

WEWAK   4/2004*, Chris Sattler / 1042103

4 LANDING CRAFT (LIGHT) (LCVP)

T 4-7

Displacement, tons: 6.5 full load
Dimensions, feet (metres): 43.3 × 11.5 × 2.3 *(13.2 × 3.5 × 0.7)*
Main machinery: 2 Volvo Penta Sterndrives; 400 hp(m) *(294 kW)*
Speed, knots: 22; 15 (fully laden)
Complement: 3
Military lift: 4.5 tons cargo or 1 Land Rover or 36 troops

Comment: Prototype built by Geraldton, Western Australia. Trials conducted in late 1992. Three
more delivered in July 1993. Two for *Tobruk*, one for *Success* (T 7) and one spare attached to
the shore base, *Penguin*.

VERIFIED

T5   8/1999, van Ginderen Collection / 0104188

PATROL FORCES

15 FREMANTLE CLASS (LARGE PATROL CRAFT) (PB)

Name	No	Builders	Commissioned
FREMANTLE	203	Brooke Marine, Lowestoft	17 Mar 1980
WARRNAMBOOL	204	NQEA Australia, Cairns	14 Mar 1981
TOWNSVILLE	205	NQEA Australia, Cairns	18 July 1981
WOLLONGONG	206	NQEA Australia, Cairns	28 Nov 1981
LAUNCESTON	207	NQEA Australia, Cairns	1 Mar 1982
WHYALLA	208	NQEA Australia, Cairns	3 July 1982
IPSWICH	209	NQEA Australia, Cairns	13 Nov 1982
CESSNOCK	210	NQEA Australia, Cairns	5 Mar 1983
BENDIGO	211	NQEA Australia, Cairns	28 May 1983
GAWLER	212	NQEA Australia, Cairns	27 Aug 1983
GERALDTON	213	NQEA Australia, Cairns	10 Dec 1983
DUBBO	214	NQEA Australia, Cairns	10 Mar 1984
GEELONG	215	NQEA Australia, Cairns	2 June 1984
GLADSTONE	216	NQEA Australia, Cairns	8 Sep 1984
BUNBURY	217	NQEA Australia, Cairns	15 Dec 1984

Displacement, tons: 245 full load
Dimensions, feet (metres): 137.1 × 23.3 × 5.9 *(41.8 × 7.1 × 1.8)*
Main machinery: 2 MTU 16V 538 TB91 diesels; 6,140 hp(m) *(4.5 MW)* sustained; 2 shafts
Speed, knots: 30. **Range, n miles:** 1,450 at 30 kt
Complement: 24 (4 officers)

Guns: 1 Bofors AN 4—40 mm/60; 120 rds/min to 10 km *(5.5 n miles)*. The 40 mm mountings
were designed by Australian Government Ordnance Factory and although the guns are of older
manufacture, this mounting gives greater accuracy particularly in heavy weather.
1—81 mm mortar. 3—12.7 mm MGs.
Countermeasures: ESM: AWA Defence Industries Type 133 PRISM.
Radars: Navigation: Kelvin Hughes Type 1006; I-band.

Programmes: The decision to buy these patrol craft was announced in September 1977. The
design is by Brooke Marine, Lowestoft which built the lead ship.
Modernisation: Original 15 year life was extended to 19 years and is now extended again. ESM
added in 1994-95. The cruise diesel on a centre line shaft has been deleted.
Operational: Bases: Darwin: P 203, P 204, P 206, P 207, P 210, P 212, P 213, P 214, P 215,
P 217. Cairns: P 205, P 208, P 209, P 211, P 216. To be replaced by Armidale class.
Decommissioning plan: *Cessnock* (June 2005); *Whyalla* (August 2005); *Warrnambool*
(November 2005); *Bunbury* (December 2005); *Wollongong* and *Dubbo* (January 2006);
Geraldton (April 2006); *Fremantle* (May 2006); *Gawler* and *Geelong* (June 2006); *Launceston*
and *Bendigo* (August 2006); *Townsville* and *Ipswich* (December 2006); *Gladstone* (February
2007).

UPDATED

GERALDTON   1/2003*, Chris Sattler / 1042104

1 + 11 ARMIDALE CLASS (PATROL CRAFT) (PB)

Name	No	Builders	Commissioned
ARMIDALE	83	Austal Ships, Fremantle	Apr 2005
BATHURST	—	Austal Ships, Fremantle	Oct 2005
BUNDABERG	—	Austal Ships, Fremantle	Oct 2005
ALBANY	—	Austal Ships, Fremantle	2006
PIRIE	—	Austal Ships, Fremantle	2006
MAITLAND	—	Austal Ships, Fremantle	2006
ARARAT	—	Austal Ships, Fremantle	2006
LAUNCESTON	—	Austal Ships, Fremantle	2007
LARRAKIA	—	Austal Ships, Fremantle	2007
WOLLONGONG	—	Austal Ships, Fremantle	2007
CHILDERS	—	Austal Ships, Fremantle	2007
BROOME	—	Austal Ships, Fremantle	2008

Displacement, tons: 270
Dimensions, feet (metres): 184.6 × 29.5 × 10.0 *(56.8 × 9.0 × 3.0)*
Main machinery: 2 MTU 4000 16V diesels; 6,225 hp *(4.64 MW)*; 2 shafts
Speed, knots: 25. **Range, n miles:** 3,000 at 12 kt
Complement: 21
Guns: 1—25 mm Rafael M242 Bushmaster. 2—12.7 mm MGs.

Comment: Austal Ships in conjunction with Defence Maritime Services contracted on 17 December 2003 to supply patrol boats to replace the Fremantle class under Project Sea 1444. The craft are to be of monohull design and to be capable of carrying two RHIBs. First steel cut on 5 May 2004 with delivery of the first of class due in 2005 followed by the whole class by 2008. DMS will provide through-life logistics and maintenance support over 15 years.The craft are named after Australian cities and towns. Eight of the craft are to be based at Darwin, Northern Territory, and the other four at Cairns, Queensland.

UPDATED

DIAMANTINA   *6/2004*, Chris Sattler /* 1042110

2 MINESWEEPERS AUXILIARY (TUGS) (MSCD/YTB)

BANDICOOT (ex-*Grenville VII*) Y 298   WALLAROO (ex-*Grenville V*) Y 299

Displacement, tons: 412 full load
Dimensions, feet (metres): 95.8 × 28 × 11.3 *(29.6 × 8.5 × 3.4)*
Main machinery: 2 Stork Werkspoor diesels; 2,400 hp(m) *(1.76 MW)*; 2 shafts
Speed, knots: 11. **Range, n miles:** 6,300 at 10 kt
Complement: 10
Radars: Navigation: Furuno 7040D; I-band.

Comment: Built in Singapore 1982 and operated by Maritime (PTE) Ltd. Purchased by the RAN and refurbished prior to delivery 11 August 1990. Used for minesweeping trials towing large AMASS influence and mechanical sweeps. No side scan sonar. Also used as berthing tugs. Bollard pull, 30 tons.

UPDATED

ARMIDALE   *1/2005*, Malcolm Back, RAN /* 1042395

MINE WARFARE FORCES

Notes: A new class of six minesweepers is planned. They are to use locally developed sweeps.

6 HUON (GAETA) CLASS (MINEHUNTERS—COASTAL) (MHC)

Name	No	Builders	Launched	Commissioned
HUON	82	Intermarine/ADI, Newcastle	25 July 1997	15 May 1999
HAWKESBURY	83	ADI, Newcastle	24 Apr 1998	12 Feb 2000
NORMAN	84	ADI, Newcastle	3 May 1999	26 Aug 2000
GASCOYNE	85	ADI, Newcastle	11 Mar 2000	2 June 2001
DIAMANTINA	86	ADI, Newcastle	2 Dec 2000	4 May 2002
YARRA	87	ADI, Newcastle	19 Jan 2002	1 Mar 2003

Displacement, tons: 720 full load
Dimensions, feet (metres): 172.2 × 32.5 × 9.8 *(52.5 × 9.9 × 3.0)*
Main machinery: 1 Fincantieri GMT diesel; 1,986 hp(m) *(1.46 MW)*; 1 shaft; LIPS cp prop; 3 Isotta Fraschini 1300 diesels; 1,440 hp(m) *(1,058 kW)*; 3 electrohydraulic motors; 506 hp(m) *(372 kW)*; Riva Calzoni retractable/rotatable APUs
Speed, knots: 14 diesel; 6 APUs. **Range, n miles:** 1,600 at 12 kt
Complement: 38 (6 officers) plus 11 spare

Guns: 1 MSI DS 30B 30 mm/75. 650 rds/min to 10 km *(5.4 n miles)* anti-surface; 3 km *(1.6 n miles)* anti-aircraft; weight of shell 0.36 kg.
Countermeasures: MCM systems: 2 Bofors SUTEC Double-Eagle Mk 2 mine disposal vehicles with DAMDIC charges; ADI double Oropesa mechanical sweep and capable of towing the Australian developed Mini-Dyad influence sweep.
Decoys: 2 MEL Aviation Super Barricade; chaff launchers.
ESM: AWADI Prism.
Combat data systems: GEC-Marconi Nautis 2M with Link 11 receive only.
Weapons control: Radamec 1400N optronic surveillance system.
Radars: Navigation: Kelvin Hughes 1007; I-band.
Sonars: GEC-Marconi Type 2093; VDS; VLF-VHF multifunction with five arrays; mine search and classification.

Programmes: The Force Structure Review of May 1991 recommended the acquisition of coastal minehunters of proven design. These ships would be required to operate in deeper and more exposed waters, to achieve lower transit times and remain on station longer than the two inshore minehunters which have now been paid off. A contract was signed with Australian Defence Industries (ADI) on 12 August 1994 to build six Intermarine designed Gaeta class derivatives. The hull of the first ship was constructed at Intermarine's Sarzana Shipyard in Italy and arrived in Australia as deck cargo on 31 August 1995 for fitting out in Newcastle, where the remaining five ships are being built at ADI's Throsby Basin. Local content for this project is about 69 per cent.
Structure: Monocoque GRP construction. A recompression chamber, one RIB and an inflatable diving boat are carried to support a six-man diving team.
Operational: This class which is named after Australian rivers, is based at HMAS *Waterhen* in Sydney.

UPDATED

WALLAROO   *10/2004*, Ian Edwards /* 1042394

3 MINESWEEPING DRONES (MSD)

MSD 02-04

Dimensions, feet (metres): 24 × 9.2 × 2 *(7.3 × 2.8 × 0.6)*
Main machinery: 2 Yamaha outboards; 300 hp(m) *(221 kW)*
Speed, knots: 45; 8 (sweeping)

Comment: Built by Hamil Haven in 1991-92. Remote-controlled drones. GRP hulls made by Hydrofield. Used for sweeping ahead of the MSA craft. Differential GPS navigation system with Syledis Vega back-up.

VERIFIED

MSD 03   *11/1992, John Mortimer*

SURVEY SHIPS (HYDROGRAPHIC SURVEY)

Notes: In addition to the ships listed below there are four civilian survey vessels; *Icebird, Franklin, Rig Seismic* and *Lady Franklin*. Also an arctic supply ship *Aurora Australis* started operating in the Antarctic in 1990; this vessel carries 70 scientists and has a helicopter hangar.

AURORA AUSTRALIS   *2/2003, Brian Morrison* / 0569131

2 LEEUWIN CLASS (AGS)

Name	No	Builders	Launched	Commissioned
LEEUWIN	A 245	NQEA, Cairns	19 July 1997	27 May 2000
MELVILLE	A 246	NQEA, Cairns	23 June 1998	27 May 2000

Displacement, tons: 2,170 full load
Dimensions, feet (metres): 233.6 × 49.9 × 14.1 *(71.2 × 15.2 × 4.3)*
Main machinery: Diesel-electric; 4 GEC Alsthom 6RK 215 diesel generators; 4,290 hp *(3.2 MW)* sustained; 2 Alsthom motors; 1.94 MW; 2 shafts; 1 Schottel bow thruster
Speed, knots: 14. **Range, n miles:** 18,000 at 9 kt
Complement: 60 (8 officers)
Radars: Navigation: STN Atlas 9600 ARPA; I-band.
Sonars: C-Tech CMAS 36/39; hull mounted; high frequency active.
Helicopters: 1 AS 350B (not permanently embarked)

Comment: Contract awarded 2 April 1996 to North Queensland Engineers & Agents (NQEA). Fitted with Atlas Fansweep-20 multibeam echo sounder and one AD 25 single beam echo sounder. Also fitted with Klein 2000 towed light-weight sidescan sonar. The ships carry three SMBs, two light utility boats and one RHIB. All repainted grey. Based at Cairns. *VERIFIED*

MELVILLE   *11/2003, John Mortimer* / 0569143

4 PALUMA CLASS (AGSC)

Name	No	Builders	Commissioned
PALUMA	A 01	Eglo, Adelaide	27 Feb 1989
MERMAID	A 02	Eglo, Adelaide	4 Dec 1989
SHEPPARTON	A 03	Eglo, Adelaide	24 Jan 1990
BENALLA	A 04	Eglo, Adelaide	20 Mar 1990

Displacement, tons: 320 full load
Dimensions, feet (metres): 118.9 × 42.0 × 8.6 *(36.6 × 12.8 × 2.65)*
Main machinery: 2 Detroit 12V-92TA diesels; 1,100 hp *(820 kW)* sustained; 2 shafts
Speed, knots: 11. **Range, n miles:** 3,600 at 11 kt
Complement: 14 (3 officers)
Radars: Navigation: Kelvin Hughes 1007; I-band
Sonars: Skipper S113; hull-mounted; active; high frequency. ELAC LAZ 72; hull-mounted side scan; active; high frequency.

Comment: Catamaran design based on Prince class ro-ro passenger ferries. Steel hulls and aluminium superstructure. Contract signed in November 1987. Also fitted with two ELAC LAX 4700 dual-frequency echo sounders and GEONAV data logging and processing system utilising Teramodel data display. All ships based at Cairns and operate in pairs when undertaking survey operations. *VERIFIED*

PALUMA   *6/2002, Royal Australian Navy* / 0528414

9 SURVEY MOTOR BOATS (YGS)

FANTOME 1005	TOM THUMB 1009	CASUARINA 1012
MEDA 1006	JOHN GOWLLAND 1010	CONDER 1021
DUYFKEN 1008	GEOGRAPHE 1011	WYATT EARP ASV 01

Dimensions, feet (metres): 35.1 × 9.5 × 5.6 *(10.7 × 2.9 × 1.7)*
Main machinery: 2 Volvo Penta AQAD-41A diesel stern drives; 400 hp(m) *(294 kW)*; 2 props
Speed, knots: 24. **Range, n miles:** 300 at 12 kt
Complement: 4 (1 officer)
Radars: Navigation: JRC; I-band.

Comment: Six survey motor boats built by Pro Marine, Victoria between October 1992 and 1993. Two additional SMBs (CAS and GEO) were built in 1997 to supplement the new AGSs. One craft has been taken out of service. The remaining seven are equipped with an Atlas Fansweep-20 multibeam echo sounder and one AD 15 single beam echo sounder, as well as a KLEIN 2000 towed lightweight side scan sonar. All collected data is fed to the Hydrographic Survey System provided by STN Atlas. Five of the class are allocated to the Leeuwin class hydrographic ships, two to the hydrographic school at HMAS *Penguin*. In addition, *Wyatt Earp* is a 9 m craft fitted for Antarctic service and allocated to the Hydrographic Office (Wollongong) Detached Survey Unit. It is fitted with the ODOM Hydrotrac Single Beam Echo Sounder and GEONAV/Terramodel. *Conder* has been built as a prototype replacement SMB. *UPDATED*

GEOGRAPHE   *6/2002, Royal Australian Navy* / 0528409

RESCUE VEHICLES

1 RESCUE SUBMERSIBLE (DSRV)

REMORA

Displacement, tons: 16.5
Dimensions, feet (metres): 19.7 × 7.9 × 13.4 (with skirt); 7.9 (without skirt) *(6.0 × 2.4 × 4.1; 2.4)*
Main machinery: 2 electric motors; 150 hp *(112 kW)*; 4 axial thrusters; 4 vertical thrusters; 2 transverse thrusters
Speed, knots: 3 dived
Complement: 1 operator and 6 survivors

Comment: Manufactured in 1995 by Can Dive Marine Services, Canada for Australian Submarine Corporation and subsequently in 2001 wholly owned by the RAN, *Remora* is operated and maintained (at 12 hours notice) by Fraser Diving, West Australia. Capable of operating to depths in excess of 500 m in a current of 3 kt, it can evacuate six personnel at a time and transfer them under pressure of up to 5 Bar directly to two 36-man decompression chambers for medical and hyperbaric treatment. A Remotely Operated Vehicle (ROV), *Remora* is flown and powered from the surface giving it unlimited endurance (emergency life support onboard is 240 man-hours). It is launchable from a craft of opportunity in up to sea state 5 using a Launch And Recovery System (LARS) that is part of the deployable suite. The skirt on the vehicle can be remotely manipulated to achieve mating angles up to 60°.
   Communications are by fibre-optic cable. The entire suite of *Remora*, LARS and all associated equipment can be fitted into ISO containers to facilitate rapid worldwide deployment. The USN replacement system, SRDRS, is based on the Remora system and is to enter service in 2005. *UPDATED*

REMORA   *6/2002, K Bristow, RAN* / 0528408

TUGS

Notes: In addition the two MSCD are used as tugs. Details under Mine Warfare Forces.

7 HARBOUR TUGS (YTL)

TAMMAR DT 2601	BRONZEWING HTS 501 (152)	MOLLYMAWK HTS 504 (154)
QUOKKA DT 1801	CURRAWONG HTS 502 (153)	SEAHORSE CHUDITCH
SEAHORSE QUENDA		

Comment: *Tammar* has a bollard pull of 35 tons and is based at *Stirling*; *Quokka* bollard pull 8 tons, is based at Darwin. The three HTS vessels have a bollard pull of 5 tons. Run as part of the commercial support programme from 1997. *Seahorse Chuditch* and *Seahorse Quenda* were built in Malaysia and delivered in 2003. 23 m long they have a bollard pull of 16 tons.
UPDATED

SEAHORSE QUENDA   7/2004*, John Mortimer / 1042393

QUOKKA   8/2003, John Mortimer / 0569142

TRAINING SHIPS

Notes: In addition to *Young Endeavour* and *Salthorse* there are five Fleet class yachts. Of 36.1 ft *(11 m)*. GRP yachts named *Charlotte of Cerberus, Friendship of Leeuwin, Scarborough of Cerberus, Lady Penrhyn of Nirimba* and *Alexander of Creswell*. The names are a combination of Australia's first colonising fleet and the training base to which each yacht is allocated.

1 SAIL TRAINING SHIP (AXS)

Name	Builders	Launched	Commissioned
YOUNG ENDEAVOUR	Brooke Yachts, Lowestoft	2 June 1987	25 Jan 1988

Displacement, tons: 239 full load
Dimensions, feet (metres): 144 × 26 × 13 *(44 × 7.8 × 4)*
Main machinery: 2 Perkins V8 diesels; 334 hp *(294 kW)*; 2 shafts
Speed, knots: 14 sail; 10 diesel. **Range, n miles:** 2,500 at 7 kt
Complement: 33 (9 RAN, 24 youth)

Comment: Built to Lloyds 100 AI LMC yacht classification by Brooke Yachts, Lowestoft. Sail area 707.1 m². Presented to Australia by UK Government as a bicentennial gift. Operated by RAN on behalf of the Young Endeavour Youth Scheme.
UPDATED

YOUNG ENDEAVOUR   1/2004*, Chris Sattler / 1042109

1 SAIL TRAINING SHIP (AXS)

SALTHORSE

Displacement, tons: 32 full load
Dimensions, feet (metres): 65.0 × 16.7 × 7.5 *(19.8 × 5.1 × 2.3)*
Main machinery: 2 Ford Lehman diesel; 120 hp *(89 kW)*
Speed, knots: 8. **Range, n miles:** 1,400 at 6 kt
Complement: 1 JRC JMA-2253; I-band

Comment: Ketch with steel hull and aluminium masts. Acquired in 1999 for officer training at HMAS *Creswell*.
VERIFIED

SALTHORSE   6/2002, Royal Australian Navy / 0528411

1 TRAINING SHIP (AXL)

Name	No	Builders	Launched
SEAHORSE MERCATOR	—	Tenix Shipbuilding, Henderson WA	15 Oct 1998

Displacement, tons: 165 full load
Dimensions, feet (metres): 103.3 × 26.9 × 7.9 *(31.5 × 8.2 × 2.4)*
Main machinery: 2 Caterpillar 3412 diesels; 2 shafts
Speed, knots: 16. **Range, n miles:** 2,700 at 10 kt
Complement: 8 plus 18 trainees

Comment: Operated by Defence Maritime Services as a Navigation training ship based at Sydney. Similar to Pacific class patrol craft.
VERIFIED

SEAHORSE MERCATOR   10/2003, Chris Sattler / 0572411

AUXILIARIES

Notes: (1) *Success* is due to be replaced in about 2015.
(2) Since 1998 all other support vessels have been contracted to the Defence Maritime Services. These craft have blue hulls and buff superstructures, and are chartered as required.
(3) In addition to the vessels listed there are some 24 workboats (AWB and NWB numbers), a VIP launch *Tresco II* and an admiral's barge *Green Parrot*.

2 TRIALS AND SAFETY VESSELS (ASR)

Name	No	Builders	Commissioned
SEAHORSE STANDARD (ex-*British Viking*)	—	Marystown Shipyard, Newfoundland	1980
SEAHORSE SPIRIT (ex-*British Magnus*)	—	Marystown Shipyard, Newfoundland	1980

Measurement, tons: 2,090 grt; 1,635 dwt
Dimensions, feet (metres): 236.2 × 52.5 × 17.4 *(72 × 16 × 5.3)*
Main machinery: 2 MLW-ALCO Model 251 V-12 diesels; 5,480 hp(m) *(4.03 MW)*; 1 shaft; cp prop; 2 stern and 2 bow thrusters
Speed, knots: 9
Complement: 20 plus 44 spare

Comment: Acquired 2 December 1998 by Defence Maritime Services to support RAN trials in Western and Southern Australian waters. Dynamic Positioning system. These ships are also used for weapon recovery and can embark the 'Remora' submarine rescue suite.
VERIFIED

SEAHORSE STANDARD   12/2002, G Hainsworth, RAN / 0569128

0 + 1 SIRIUS CLASS (REPLENISHMENT TANKER) (AORH)

Name	No	Builders	Laid down	Launched	Commissioned
SIRIUS (ex-*Delos*)	—	Hyundai Mipo Dockyard, Korea	-	2004	2006

Measurement, tons: 37,000 dwt
Dimensions, feet (metres): 598.9 × 89.6 × 36.7 *(182.55 × 27.3 × 11.2)*
Main machinery: 1 Burmeister & Wain diesel; 1 shaft
Speed, knots: 15
Complement: To be announced
Guns: 1—25 mm Rafael M242 Bushmaster (fitted for).
Helicopter: 1 medium.

Comment: New ship of double-hulled construction procured in June 2004 as replacement, following refit and conversion for military use, for *Westralia*. Modifications are likely to include the installation of replenishment-at-sea equipment, a flight deck (with one landing spot) and hangar aft and changes to accommodation and habitability arrangements. The ship is also likely to be fitted for but not with a 25 mm gun and possibly a 20 mm Vulcan Phalanx. A contract for the refit and initial logistic support of the ship is expected in 2005. The first RAN ship to carry the name *Sirius*; she is named after the flagship of the First Fleet.

NEW ENTRY

SEAHORSE HORIZON   *12/2004 *, Ian Edwards /* 1042392

SIRIUS (before conversion)   *7/2004 *, Royal Australian Navy /* 0566629

1 TRIALS AND SAFETY VESSEL (ASR)

Name	No	Builders	Commissioned
SEAHORSE HORIZON (ex-*Protector*, ex-*Blue*, *Nabilla*, ex-*Osprey*)	— (ex-ASR 241)	Stirling Marine Services, WA	1984

Displacement, tons: 670 full load
Dimensions, feet (metres): 140.1 × 31.2 × 9.8 *(42.7 × 9.5 × 3)*
Main machinery: 2 Detroit 12V-92TA diesels; 2,440 hp *(1.82 MW)* sustained; 2 Heimdal cp props
Speed, knots: 11.5. **Range, n miles:** 10,000 at 11 kt
Complement: 6 civilian or 9 navy (for training)
Radars: Navigation: JRC 310; I-band. Decca RM 970BT; I-band.
Sonars: Klein; side scan; high frequency.
Helicopters: Platform for 1 light.

Comment: A former National Safety Council of Australia vessel commissioned into the Navy in November 1990. Used to support contractor's sea trials of the Collins class submarines, and for mine warfare trials and diving operations. LIPS dynamic positioning, two ROVs and a recompression chamber. Helicopter deck and a submersible were removed in 1992. Based at Jervis Bay. Decommissioned in early 1998 and run as part of the commercial support programme. Also used for junior officer training.

UPDATED

3 FISH CLASS (TORPEDO RECOVERY VESSELS) (YPT)

TUNA TRV 801	TREVALLY TRV 802	TAILOR TRV 803

Displacement, tons: 91.6 full load
Dimensions, feet (metres): 88.5 × 20.9 × 4.5 *(27 × 6.4 × 1.4)*
Main machinery: 3 GM diesels; 890 hp *(664 kW)*; 3 shafts
Speed, knots: 13
Complement: 9
Radars: Navigation: I-band.

Comment: All built at Williamstown completed between January 1970 and April 1971. Can transport eight torpedoes. Based at Jervis Bay, Sydney and Fleet Base West respectively. Run as part of the commercial support programme from 1997. Blue hulls and buff superstructures.

UPDATED

TREVALLY   *4/2004 *, Chris Sattler /* 1042101

1 LEAF CLASS (UNDER WAY REPLENISHMENT TANKER) (AORH/AOT)

Name	No	Builder	Laid down	Launched	Commissioned
WESTRALIA (ex-*Appleleaf*, ex-*Hudson Cavalier*)	O 195 (ex-A 79)	Cammell Laird, Birkenhead	1974	24 July 1975	Nov 1979

Displacement, tons: 40,870 full load
Measurement, tons: 20,761 gross; 10,851 net; 33,595 dwt
Dimensions, feet (metres): 560 × 85 × 38.9 *(170.7 × 25.9 × 11.9)*
Main machinery: 2 SEMT-Pielstick 14 PC2.2 V 400 diesels; 14,000 hp(m) *(10.3 MW)* sustained; 1 shaft
Speed, knots: 16 (11 on 1 engine). **Range, n miles:** 7,260 at 15 kt
Complement: 89 (8 officers) plus 9 spare berths

Cargo capacity: 20,000 tons dieso; 3,000 tons aviation fuel; 1,500 tons water
Countermeasures: ESM: Matilda; radar warning.
Radars: Navigation: 2 Kelvin Hughes; 1007 ARPA (I-band) and Radpak (E/F-band).

Comment: Part of an order by the Hudson Fuel and Shipping Co which was subsequently cancelled. Leased by the RN from 1979 until transferred on 9 October 1989 on a five year lease to the RAN, arriving in Fremantle 20 December 1989. Purchased in 1994. Has three 3 ton cranes and two 5 ton derricks. Hospital facilities. Two beam replenishment stations. Stern refuelling restored in 1995. Based at *Stirling*. Also modified to provide a large Vertrep platform aft. Lifeboats have been replaced by liferafts. To be replaced in 2006 by *Sirius*.

UPDATED

WESTRALIA   *10/2004 *, Hachiro Nakai /* 1042106

1 DURANCE CLASS (UNDERWAY REPLENISHMENT TANKER) (AORH)

Name	No	Builders	Laid down	Launched	Commissioned
SUCCESS	OR 304	Cockatoo Dockyard, Sydney	9 Aug 1980	3 Mar 1984	19 Feb 1986

Displacement, tons: 17,933 full load
Dimensions, feet (metres): 515.7 × 69.5 × 30.6
(157.2 × 21.2 × 8.6)
Main machinery: 2 SEMT-Pielstick 16 PC2.5 V 400 diesels;
20,800 hp(m) *(15.3 MW)* sustained; 2 shafts; LIPS cp props
Speed, knots: 20. **Range, n miles:** 8,616 at 15 kt
Complement: 237 (25 officers)
Cargo capacity: 10,200 tons: 8,707 dieso; 975 Avcat; 116
distilled water; 57 victuals; 250 munitions including SM1
missiles and Mk 46 torpedoes; 95 naval stores and spares

Guns: Provision for 2 Vulcan Phalanx Mk 15 CIWS. 4—12.7 mm
MGs. Due to receive Rafael Typhoon 25 mm guns in
2004/05.
Radars: Navigation. 2 Kelvin Hughes Type 1006; I-band.
Helicopters: 1 AS 350B Squirrel, Sea King or Seahawk.

Comment: Based on French Durance class design.
Replenishment at sea from four beam positions (two having
heavy transfer capability) and vertrep. One LCVP is carried on
the starboard side aft. Hangar modified to take Sea Kings.
Phalanx guns fitted aft in 1997.

UPDATED SUCCESS

6/2004*, Chris Sattler / 1042108

4 SELF-PROPELLED LIGHTERS (WFL/AOTL)

WARRIGAL 333 (ex-WFL 8001)   WOMBAT 332 (ex-WFL 8003)
WALLABY 331 (ex-WFL 8002)   WYULDA 334 (ex-WFL 8004)

Displacement, tons: 265 light; 1,206 full load
Dimensions, feet (metres): 124.6 × 33.5 × 12.5 *(38 × 10.2 × 3.8)*
Main machinery: 2 Harbourmaster outdrives (1 fwd, 1 aft)
Speed, knots: 8
Cargo capacity: 560 tons dieso and 200 tons water

Comment: First three were laid down at Williamstown in 1978. The fourth, for HMAS *Stirling,* was
ordered in 1981 from Williamstown Dockyard. Used for water/fuel transport. Steel hulls with
twin, swivelling, outboard propellers. Based at Jervis Bay and Stirling (WFL 8001, 8004), other
pair at Sydney. Run as part of the commercial support operation from 1997.

UPDATED

WALLABY

10/2004*, Ian Edwards / 1042391

4 DIVING TENDERS (YDT/PB)

SEAL 2001   MALU BAIZAM 2003   SHARK 2004   DUGONG

Displacement, tons: 22 full load
Dimensions, feet (metres): 65.5 × 18.5 × 4.6 *(20 × 5.6 × 1.4)*
Main machinery: 2 MTU 8V 183 diesels; 2 shafts
Speed, knots: 26. **Range, n miles:** 450 at 20 kt
Complement: 6 plus 16 divers

Comment: Built by Geraldton Boat Builders, Western Australia and completed in August 1993.
Carry 2 tons of diving equipment to support 24 hour diving operations in depths of 54 m. *Shark*
based at *Stirling, Seal* at *Waterhen* and *Dugong* at Sydney, *Malu Baizam* is used for patrol duties
at Thursday Island. *Porpoise* grounded in 1995 and was assessed as being beyond economical
repair. Replacement built in 1996. Run as part of the commercial support operation from 1997.
Sister craft *Coral Snake* is operated by the Army.

VERIFIED

SEAL

6/2003, Mick Prendergast / 0569127

3 WATTLE CLASS STORES LIGHTERS (YE)

WATTLE CSL 01   BORONIA CSL 02   TELOPEA CSL 03

Displacement, tons: 147 full load
Dimensions, feet (metres): 79.4 × 32.8 × 5.4 *(24.2 × 10.0 × 1.66)*
Main machinery: 2 Caterpillar D333C diesels; 600 hp *(447kW)*
Speed, knots: 8. **Range, n miles:** 320 at 8 kt
Radars: 1 JRC JMA-2253; I-band.

Comment: Built by Cockatoo DY, Sydney and delivered in 1972. Employed to transport
ammunition and stores. Equipped with 3-ton electric crane. CSL 02 and 03 based at Sydney and
CSL 01 at Darwin.

UPDATED

ARMY

Notes: (1) Operated by Royal Australian Army Corps of Transport. Personnel: About 300 as
required.
(2) In addition to the craft listed below there are 159 assault boats 16.4 ft *(5 m)* in length and
capable of 30 kt. Can carry 12 troops or 1,200 kg of equipment. Also there are 12 ex-US Army
LARC-V amphibious wheeled lighters for service with *Manoora* and *Kanimbla.*

6 AMPHIBIOUS WATERCRAFT (LCM)

AB 2000-2005

Displacement, tons: 135 full load
Dimensions, feet (metres): 83.3 × 24.9 × 3.3 *(25.4 × 7.6 × 1.0)*
Main machinery: 2 Detroit 6062 diesels; 2 Doen waterjets
Speed, knots: 11. **Range, n miles:** 720 at 10 kt
Complement: To be announced
Guns: 2—12.7 mm MGs.

Comment: Contract signed with ADI in June 2002 to provide watercraft to operate in conjunction
with the LPAs. Two are to be carried by each ship. Of aluminium construction, they have
through-deck, roll-on/roll-off design and bow and stern ramps. With 65 tonne cargo capacity,
the craft can carry one Leopard tank or five armoured vehicles. An innovative feature is a
pontoon system to mate with LPA stern ramp to facilitate vehicle transfer.

UPDATED

TELOPEA

6/2003*, Mick Prendergast / 1042100

AB 2000

6/2004*, ADI / 1042099

14 LCM 8 CLASS

AB 1050, 1051, 1053, 1056, 1058-1067

Displacement, tons: 107 full load
Dimensions, feet (metres): 73.5 × 21 × 5.2 *(22.4 × 6.4 × 1.6)*
Main machinery: 2 8V92GM diesels; 720 hp *(547 kW)*; 2 shafts
Speed, knots: 11. **Range, n miles:** 290 at 10 kt
Complement: 4
Military lift: 55 tons
Guns: 2—12.7 mm MGs.

Comment: Built by North Queensland Engineers, Cairns and Dillinghams, Fremantle to US design. Based at Townsville and Darwin. *AB 1057* transferred to Tonga 1982, *AB 1052* and *AB 1054* sold to civilian use in 1992. All upgraded to Mod 2 standard by late 1999 with new engines and with endurance increased.

VERIFIED

AB 1056   *10/2002, John Mortimer /* 0528383

2 SAFCOL CRAFT

CORAL SNAKE AM 1353   RED VIPER

Displacement, tons: 22 full load
Dimensions, feet (metres): 65.5 × 20.0 × 4.6 *(20 × 6.1 × 1.4)*
Main machinery: 2 General Motors Detroit 8V92 diesels; 1,800 hp *(1.34 MW)*
Speed, knots: 28. **Range, n miles:** 350 at 25 kt
Complement: 3

Comment: Sister to Seal class built at Geraldton Boat Builders. *Coral Snake* delivered in 1994 and *Red Viper* in 1996. Used as Special Action Forces Craft Offshore Large (SAFCOL) to support dives and transport of stores and personnel.

VERIFIED

RED VIPER   *8/2003, Chris Sattler /* 0569126

9 EXPRESS SHARK CAT CLASS (PB)

AM 237-244   AM 428

Comment: Built by NoosaCat, Queensland and delivered by 1995. Trailer transportable. Similar craft in service with Navy and Police. Multihulls 30.8 ft *(9.4 m)* in length overall with twin Johnson outboards; 450 hp *(336 kW)* total power output, giving 40 kt maximum speed.

VERIFIED

AM 243   *11/1997, van Ginderen Collection /* 0012946

NON-NAVAL PATROL CRAFT

Notes: (1) In addition to the commercial support craft already listed, various State and Federal agencies, including some fishery departments, have built offshore patrol craft up to 25 m and 26 kt.
(2) Cocos Island patrol carried out by *Sir Zelman Cowan* of 47.9 × 14 ft *(14.6 × 4.3 m)* with two Cummins diesels; 20 kt, range 400 n miles at 17 kt, complement 13 (3 officers). Operated by West Australian Department of Harbours and Lights.
(3) All previously listed RAAF craft have been sold for civilian use.

4 SHARK CAT 800 CLASS (WORKBOATS) (YFL)

0801   0802   0803   0805

Displacement, tons: 13.7 full load
Dimensions, feet (metres): 27.4 × 9.2 × 3.3 *(8.35 × 2.8 × 1.0)*
Main machinery: 2 Mercury outboard engines
Speed, knots: 30
Complement: 1 plus 11 passengers

Comment: Built by Shark Cat, Noosaville, Queensland and delivered in 1980s. GRP construction. Used for target-towing, naval police and range clearance duties. Based at Sydney with the exception of *0805* which is based at HMAS *Creswell*.

UPDATED

SHARK CAT 0803   *6/2004*, Chris Sattler /* 1042107

4 NOOSACAT 930 WORKBOATS (YFL)

0901-0904

Dimensions, feet (metres): 30.5 × 11.5 × 2.3 *(9.3 × 3.5 × 0.7)*
Main machinery: 2 Volvo Penta ADQ41DP diesels; 2 props
Speed, knots: 30. **Range, n miles:** 240 at 20 kt

Comment: Built by Noosacat, Queensland and delivered in 1994. GRP hulled craft for general purpose stores and personnel transport. *0903* and *0904* based at Sydney, *0902* at HMAS *Creswell* and 0901 at HMAS *Cerberus*.

VERIFIED

NOOSACAT 0904   *6/2002, Royal Australian Navy /* 0528412

10 STEBER CLASS WORKBOATS (YFL/YDT)

NGPWB 01-10

Displacement, tons: 13.7 full load
Dimensions, feet (metres): 43.3 × 15.4 × 4.4 *(13.2 × 4.7 × 1.3)*
Main machinery: 2 diesels (01-06). 1 diesel (07-10)
Speed, knots: 25 (01-06). 20 (07-10)

Comment: Built by Steber craft and delivered in 1997. GRP hulled craft for general purpose stores and personnel transport and for use as diving tenders. Most have radars 01, 02, 07 and 08 based at Sydney, 03 at HMAS *Cresswell*, 04 and 09 at Fleet Base West and 06 at HMAS *Cerberus*.

VERIFIED

STEBER CRAFT   *6/2002, Royal Australian Navy /* 0528413

CUSTOMS

Notes: The Australian Customs Service operates a number of inshore vessels between 4.3 m and 6.4 m in length. All larger vessels have been sold and the Bay class are the only sea-going vessels.

8 BAY CLASS (PB)

ROEBUCK BAY ACV 10	**CORIO BAY** ACV 50
HOLDFAST BAY ACV 20	**ARNHEM BAY** ACV 60
BOTANY BAY ACV 30	**DAME ROMA MITCHELL** ACV 70
HERVEY BAY ACV 40	**STORM BAY** ACV 80

Displacement, tons: 134
Dimensions, feet (metres): 125.3 × 23.6 × 7.9 *(38.2 × 7.2 × 2.4)*
Main machinery: 2 MTU 16V 2000M 70 diesels; 2,856 hp(m) *(2.1 MW)* sustained; 2 shafts.
   1 Vosper Thornycroft bow thruster
Speed, knots: 24. **Range, n miles:** 1,000 at 20 kt
Complement: 12
Radars: Surface search: Racal Decca; E/F- and I-band.
Sonars: Wesmar SS 390E dipping sonar.

Comment: Built by Austal Ships and delivered from February 1999 to August 2000. The craft carry two RIBs capable of 35 kt.

UPDATED   ARNHEM BAY   *2/2004*, Chris Sattler /* 1042102

Austria
ÖSTERREICHISCHE MARINE

Country Overview

A landlocked central European country, the Republic of Austria has an area of 83,859 square miles and is bordered by the Czech Republic, Slovakia, Hungary, Slovenia, Italy, Switzerland, Liechtenstein and Germany. Vienna is the country's capital and largest city. The principal river is the Danube whose Austrian tributaries include the Inn, Traun, Enns and Ybbs. Among other important rivers are the Mur and Mürz. Additionally, there are numerous lakes including Bodensee (Lake Constance) and Neusiedler.

Commanding NCO

WO II Anton Sgarz

Diplomatic Representation

Defence Attaché in London:
   Brigadier W Plasche

Personnel

(a)   2005: 24 (cadre personnel and national service), plus a small shipyard unit
(b)   8 months' national service

Bases

Marinekaserne Tegetthof, Wien-Kuchelau (under command of Austrian School of Military Engineering)

PATROL FORCES

1 RIVER PATROL CRAFT (PBR)

Name	No	Builders	Commissioned
OBERST BRECHT	A 601	Korneuberg Werft AG	14 Jan 1958

Displacement, tons: 10 full load
Dimensions, feet (metres): 40.3 × 8.2 × 2.5 *(12.3 × 2.5 × 0.75)*
Main machinery: 2 MAN 6-cyl diesels; 290 hp(m) *(213 kW)*; 2 shafts
Speed, knots: 18
Complement: 5
Guns: 1—12.7 mm MG. 1—84 mm PAR 66 Carl Gustav AT mortar.

Comment: Refit in 2004 included overhaul of hull and electrical equipment, a redesigned superstructure and new paint scheme.

UPDATED

1 RIVER PATROL CRAFT (PBR)

Name	No	Builders	Launched	Commissioned
NIEDERÖSTERREICH	A 604	Korneuberg Werft AG	26 July 1969	16 Apr 1970

Displacement, tons: 78 full load
Dimensions, feet (metres): 96.8 × 17.8 × 3.6 *(29.4 × 5.4 × 1.1)*
Main machinery: 2 MWM V16 diesels; 1,640 hp(m) *(1.2 MW)*; 2 shafts
Speed, knots: 22
Complement: 9 (1 officer)
Guns: 1—20 mm Oerlikon SPz Mk 66. 1—12.7 mm MG. 1—7.62 mm MG. 1—84 mm PAR 66 'Carl Gustav' AT mortar.

Comment: Fully welded. Only one built of a projected class of 12. Main machinery and electrical equipment overhauled in 2000.

VERIFIED

OBERST BRECHT   *7/2004*, HBF /* 0589831

NIEDERÖSTERREICH   *7/2001, HBF /* 0114504

Azerbaijan

Country Overview

Formerly part of the USSR, the Republic of Azerbaijan declared its independence in 1991. Situated in the Transcaucasia region of western Asia, the country, which includes the disputed region of Nagorno-Karabakh, has an area of 33,400 square miles and is bordered to the north by Russia and Georgia and to the south with Iran. Armenia to the west includes the exclave of Nakhichevan. Azerbaijan has a coastline of 398 n miles with the Caspian Sea on which Baku, the capital and largest city, is the principal port. Maritime claims in the Caspian Sea have yet to be resolved. Coast Guard formed in July 1992 with ships transferred from the Russian Caspian Flotilla and Border Guard. Operational control and maintenance was assumed by Russia 1995-99 but since then, the Azeri Navy has taken back full responsibility. During 2003 there were increasing signs of a drive to improve effectiveness, reflecting heightened tensions in the Caspian Sea.

Headquarters Appointments

Commander of Navy:
   Rear Admiral Sahin Sultanov

Personnel

2005: 2,200

Bases

Baku

PATROL FORCES

Notes: An Osa II class (without SSM) and a Svetlyak are reported non-operational.

1 TURK CLASS (PB)

ARAZ (ex-AB 34)

Displacement, tons: 170 full load
Dimensions, feet (metres): 132 × 21 × 5.5 (40.2 × 6.4 × 1.7)
Main machinery: 4 SACM-AGO V16CSHR diesels; 9,600 hp(m) (7.06 MW); 2 cruise diesels; 300 hp(m) (220 kW); 2 shafts
Speed, knots: 22
Complement: 31 (3 officers)
Guns: 1 or 2 Bofors 40 mm/70.
  1 Oerlikon 20 mm (if only 1—40 mm fitted). 2—12.7 mm MGs.
Depth charges: 1 rack.
Radars: Surface search: Racal Decca; I-band.

Comment: Ex-AB 34 transferred from Turkey July 2000.

VERIFIED

TURK CLASS (Turkish colours)   *11/1998, Selim San /* 0050287

1 POINT CLASS (PB)

Name	No	Builders	Commissioned
— (ex-*Point Brower*)	S-201 (ex-82372)	USCG Yard, Curtis Bay	21 Apr 1970

Displacement, tons: 67 full load
Dimensions, feet (metres): 83 × 17.2 × 5.8 (25.3 × 5.3 × 1.8)
Main machinery: 2 Caterpillar diesels; 1,600 hp (1.19 MW); 2 shafts
Speed, knots: 22. Range, n miles: 1,200 at 8 kt
Complement: 10
Guns: 2—12.7 mm MGs.
Radars: Surface search: Hughes/Furuno SPS-73; I-band.

Comment: Transferred from US Coast Guard on 28 February 2003.

NEW ENTRY

POINT CLASS (Jamaica colours)   *10/1999, JDFCG /* 0080127

2 STENKA (PROJECT 205P) CLASS (PBF)

— (ex-AK 234)   — (ex-AK 374)

Displacement, tons: 253 full load
Dimensions, feet (metres): 129.3 × 25.9 × 8.2 (39.4 × 7.9 × 2.5)
Main machinery: 3 diesels; 14,100 hp(m) (10.36 MW); 3 shafts
Speed, knots: 37. Range, n miles: 2,300 at 14 kt
Complement: 25
Guns: 4—30 mm/65 (2 twin) AK 230.
Radars: Surface search: Pot Drum; H/I-band.
Fire control: Drum Tilt; H/I-band.
Navigation: Palm Frond; I-band.

Comment: Ex-Russian craft built in the 1970s. Sonar and torpedo tubes removed.

NEW ENTRY

STENKA CLASS (Georgia colours)   *7/1999 /* 0038747

1 ZHUK (GRIF) CLASS (PROJECT 1400M) (PB)

137 (ex-AK 55)

Displacement, tons: 39 full load
Dimensions, feet (metres): 78.7 × 16.4 × 3.9 (24 × 5 × 1.2)
Main machinery: 2 Type M 401B diesels; 2,200 hp(m) (1.6 MW) sustained; 2 shafts
Speed, knots: 30. Range, n miles: 1,100 at 15 kt
Complement: 13
Guns: 2—14.5 mm (twin). 1—12.7 mm MG.
Radars: Surface search: Spin Trough; I-band.

Comment: Ex-Russian craft built in the 1970s.

NEW ENTRY

ZHUK CLASS (Ukraine colours)   *7/2000, Hartmut Ehlers /* 0106655

MINE WARFARE FORCES

Notes: Three Sonya class are reported non-operational.

2 YEVGENYA CLASS (PROJECT 1258) (MINEHUNTERS) (MHC)

237   +1

Displacement, tons: 77 standard; 90 full load
Dimensions, feet (metres): 80.7 × 18 × 4.9 (24.6 × 5.5 × 1.5)
Main machinery: 2 Type 3-D-12 diesels; 600 hp(m) (440 kW) sustained; 2 shafts
Speed, knots: 11. Range, n miles: 300 at 10 kt
Complement: 10
Guns: 2—14.5 mm (twin) MGs.
Countermeasures: Minehunting gear is lowered on a crane at the stern.
Radars: Navigation: Don 2; I-band.
Sonars: MG 7 lifted over the stern.

Comment: Ex-Russian craft built in the 1970s.

NEW ENTRY

YEVGENYA (Russian colours)   *1991*

AUXILIARIES

Notes: A variety of auxiliary craft is reported to be in Azerbaijan service although operational status has not been confirmed. Vessels include a Petrushka class training vessel, a Shelon class torpedo recovery craft, an Emba class cable ship and four survey ships (one Kamenka, one Finik, one Vadim Popov and one Valeryan Uryvayev).

AMPHIBIOUS FORCES

Notes: (1) Two Vydra class are reported non-operational.
(2) A T4 LCM has also been reported.

1 POLNOCHNY B CLASS (PROJECT 771) (LSM)

309 (ex-MDK 107)

Displacement, tons: 760 standard; 834 full load
Dimensions, feet (metres): 246.1 × 31.5 × 7.5 (75 × 9.6 × 2.3)
Main machinery: 2 Kolomna Type 40-D diesels; 4,400 hp(m) (3.2 MW) sustained; 2 shafts
Speed, knots: 18. Range, n miles: 1,000 at 18 kt
Complement: 42
Military lift: 180 troops; 350 tons including up to 6 tanks
Guns: 2—30 mm/65 (twin) AK 230; 500 rds/min to 5 km (2.7 n miles); weight of shell 0.54 kg.
  2—140 mm 18-tubed rocket launchers.
Radars: Navigation: Don 2; I-band.
Fire control: Drum Tilt; H/I-band.
IFF: Square Head. High Pole B.

Comment: Built in Poland 1968-70. Tank deck covers 237 m².

NEW ENTRY

2 POLNOCHNY A (PROJECT 770) CLASS (LSM)

291 (ex-MDK 36)   **380** (ex-MDC 37)

Displacement, tons: 800 full load
Dimensions, feet (metres): 239.5 × 27.9 × 5.8 *(73 × 8.5 × 1.8)*
Main machinery: 2 Kolomna Type 40-D diesels; 4,400 hp(m) *(3.2 MW)* sustained; 2 shafts
Speed, knots: 19. **Range, n miles:** 1,000 at 18 kt
Complement: 40
Military lift: 6 tanks; 350 tons
Guns: 2 USSR 30 mm/65 (twin); 500 rds/min to 5 km *(2.7 n miles)*; weight of shell 0.54 kg.
    2—140 mm rocket launchers; 18 barrels to 9 km *(4.9 n miles)*.
Radars: Surface search: Decca; I-band.
Fire control: Drum Tilt; H/I-band.

Comment: Built at Northern Shipyard, Gdansk in the late 1960s.

NEW ENTRY   POLNOCHNY CLASS (Egyptian colours)   *10/2000, F Sadek / 0103742*

Bahamas

Country Overview

The Commonwealth of the Bahamas gained independence in 1971; the British monarch, represented by a governor-general, is head of state. Situated in the west Atlantic Ocean, it comprises about 700 islands and islets, and nearly 2,400 cays and rocks which stretch between Florida and Hispaniola. About 30 of the islands are inhabited. The capital, Nassau, is on New Providence Island which contains more than half of the total population. Grand Bahama, the most northerly of the group, is the second major island. An archipelagic regime, territorial seas (12 n miles) and a fishery zone (200 n miles) are claimed. A 200 n mile Exclusive Economic Zone (EEZ) has been claimed but the limits are not defined.

Headquarters Appointments

Commander Royal Bahamas Defence Force:
    Commodore Davy F Rolle
Squadron Commanding Officer:
    Commander Albert Armbrister

Bases

HMBS *Coral Harbour* (New Providence Island)
HMBS *Matthew Town* (Great Inagua Island)

Personnel

2005: 922

Prefix to Ships' Names

HMBS (Her Majesty's Bahamian Ship)

PATROL FORCES

2 BAHAMAS CLASS (PBO)

Name	No	Builders	Commissioned
BAHAMAS	P 60	Moss Point Marine, Escatawpa	27 Jan 2000
NASSAU	P 61	Moss Point Marine, Escatawpa	27 Jan 2000

Displacement, tons: 375 full load
Dimensions, feet (metres): 198.8 × 29.2 × 8.5 *(60.6 × 8.9 × 2.6)*
Main machinery: 3 Caterpillar 3516B diesels; 6,600 hp(m) *(4.85 MW)*; 3 shafts
Speed, knots: 24. **Range, n miles:** 3,000 at 10 kt
Complement: 35 plus 28 spare
Guns: 1 Bushmaster 25 mm. 3—12.7 mm MGs.
Radars: Surface search/Navigation: Decca Bridgemaster Type 656-14/CAB; I-band.

Comment: Order placed 14 March 1997 with Halter Marine Group. Aluminium superstructures fabricated at Equitable Shipyards while hulls built at Moss Point. The design is an adapted Vosper International Europatrol 250 with a RIB and launching crane at the stern. Based at Nassau.

UPDATED

NASSAU   *11/2001, Martin Mokrus / 0130420*

1 CHALLENGER CLASS (PB)

P 41

Displacement, tons: 8 full load
Dimensions, feet (metres): 27 × 5.5 × 1 *(8.2 × 1.7 × 0.3)*
Main machinery: 2 Evinrude outboards; 450 hp *(330 kW)*
Speed, knots: 26
Complement: 4
Guns: 1—7.62 mm MG.

Comment: Built by Boston Whaler Edgewater, Florida and delivered in September 1995.

VERIFIED

P 41   *9/1996, RBDF / 0056530*

3 PROTECTOR CLASS (PB)

Name	No	Builders	Commissioned
YELLOW ELDER	P 03	Fairey Marine, Cowes	20 Nov 1986
PORT NELSON	P 04	Fairey Marine, Cowes	20 Nov 1986
SAMANA	P 05	Fairey Marine, Cowes	20 Nov 1986

Displacement, tons: 110 standard; 180 full load
Dimensions, feet (metres): 108.3 × 22 × 6.9 *(33 × 6.7 × 2.1)*
Main machinery: 3 Detroit 16V-149TI diesels; 3,483 hp *(2.6 MW)* sustained; 3 shafts
Speed, knots: 30. **Range, n miles:** 300 at 24 kt; 600 at 14 kt on 1 engine
Complement: 20 (3 officers) plus 5 spare
Guns: 1 Rheinmetall 20 mm. 3—7.62 mm MGs.
Radars: Surface search: Furuno; I-band.

Comment: Ordered December 1984. Steel hulls. One RIB is carried and can be launched by a trainable crane. Based at Coral Harbour.

UPDATED

SAMANA   *4/1996, RBDF / 0056527*

1 ELEUTHERA (KEITH NELSON) CLASS (PB)

Name	No	Builders	Commissioned
INAGUA	P 27	Vosper Thornycroft	10 Dec 1979

Displacement, tons: 30 standard; 37 full load
Dimensions, feet (metres): 60 × 15.8 × 4.6 *(18.3 × 4.8 × 1.4)*
Main machinery: 2 Caterpillar 3408BTA diesels; 1,070 hp *(800 kW)* sustained; 2 shafts
Speed, knots: 20. **Range, n miles:** 650 at 16 kt
Complement: 11
Guns: 3—7.62 mm MGs.
Radars: Surface search: Furuno; I-band.

Comment: The survivor of a class of five. Light machine guns mounted in sockets either side of the bridge. One more is used as a museum. Main engine replaced in 1990.

VERIFIED

INAGUA   *6/1998, RBDF / 0017574*

2 DAUNTLESS CLASS (INSHORE PATROL CRAFT) (PB)

P 42   P 43

Displacement, tons: 11 full load
Dimensions, feet (metres): 40.4 × 14 × 4.3 *(12.3 × 4.3 × 1.3)*
Main machinery: 2 Caterpillar 3208TA diesels; 870 hp *(650 kW)* sustained; 2 shafts
Speed, knots: 25. **Range, n miles:** 600 at 18 kt
Complement: 5
Guns: 2—7.62 mm MGs.
Radars: Surface search: Furuno 1761; I-band.

Comment: Built by SeaArk Marine, Monticello, Arkansas and delivered in January 1996. Used primarily for medium-range search and rescue missions. Based at Coral Harbour.
UPDATED

P 43   6/1999, *RBDF* / 0081453

4 BOSTON WHALERS (PBF)

P 110-111   P 112-113

Displacement, tons: 1.5 full load
Dimensions, feet (metres): 20 × 7.2 × 1.1 *(6.1 × 2.2 × 0.4)*
Main machinery: 2 Evinrude outboards; 180 hp *(134 kW)* (P 110-111); 2 Mariner outboards; 150 hp *(120 kW)* (P 112-113)
Speed, knots: 45 *(P 110-111)*; 38 *(P 112-113)*
Complement: 3

Comment: P 110 and 111 are Impact designs commissioned 25 September 1995. P 112 and 113 are Wahoo types commissioned 23 October 1995.
VERIFIED

P 113   6/1999, *RBDF* / 0081454

Bahrain

Country Overview

Formerly under British control from 1861, Bahrain gained its independence in 1971. Situated in the southern Gulf, with which it has a coastline of 87 n miles, the country comprises a group of 33 islands between the Qatar Peninsula to the east and Saudi Arabia to the west. The principal islands include Bahrain (217 square miles), Al Muharraq; Umm an Na'san; Sitrah; Jiddah and the Hawar group. The capital, largest city and principal port is Manama. Territorial seas (12 n miles) are claimed. An EEZ has not been claimed.

Headquarters Appointments

Chief of Staff:
    Major General Shaikh Abdullah Bin Salman Bin Khalid Al Khalifa
Commander of Navy:
    Lieutenant Colonel Yusuf Ahmad Malullah
Director of Coast Guard:
    Colonel Abdul Ghaffar Abdul Aziz Mohammed

Personnel

(a)   2005: 1,300 (Navy), 770 (Coast Guard 240 seagoing)
(b)   Voluntary service

Bases

Mina Sulman (Navy)
Bandar-Dar (CG HQ)
Muharraq (CG base)

Coast Guard

This unit is under the direction of the Ministry of the Interior.

Prefix to Ships' Names

BRNS (Bahrain Royal Navy Ship)

FRIGATES

1 + (1) OLIVER HAZARD PERRY CLASS (FFGHM)

Name	No	Builders	Laid down	Launched	Commissioned	Recommissioned
SABHA (ex-*Jack Williams*)	90 (ex-FFG 24)	Bath Iron Works	25 Feb 1980	30 Aug 1980	19 Sep 1981	25 Feb 1997

Displacement, tons: 2,750 light; 3,638 full load
Dimensions, feet (metres): 445 × 45 × 14.8; 24.5 (sonar) *(135.6 × 13.7 × 4.5; 7.5)*
Main machinery: 2 GE LM 2500 gas turbines; 41,000 hp *(30.59 MW)* sustained; 1 shaft; cp prop
    2 auxiliary retractable props; 650 hp *(484 kW)*
Speed, knots: 29. **Range, n miles:** 4,500 at 20 kt
Complement: 206 (13 officers) including 19 aircrew

Missiles: SSM: 4 McDonnell Douglas Harpoon; active radar homing to 130 km *(70 n miles)* at 0.9 Mach; warhead 227 kg.
    SAM: 36 GDC Standard SM-1MR; command guidance; semi-active radar homing to 46 km *(25 n miles)* at 2 Mach.
    1 Mk 13 Mod 4 launcher for both SSM and SAM missiles ❶.
Guns: 1 OTO Melara 3 in *(76 mm)*/62 Mk 75 ❷; 85 rds/min to 16 km *(8.7 n miles)* anti-surface; 12 km *(6.6 n miles)* anti-aircraft; weight of shell 6 kg.
    1 General Electric/General Dynamics 20 mm/76 6-barrelled Mk 15 Vulcan Phalanx ❸; 3,000 rds/min (4,500 in Block 1) combined to 1.5 km.
    4—12.7 mm MGs.
Torpedoes: 6—324 mm Mk 32 Mod 7 (2 triple) tubes ❹. 24 Honeywell Mk 46; anti-submarine; active/passive homing to 11 km *(5.9 n miles)* at 40 kt; warhead 44 kg.
Countermeasures: Decoys: 2 Loral Hycor SRBOC 6-barrelled fixed Mk 36 ❺; IR flares and chaff to 4 km *(2.2 n miles)*.
    SLQ-25 Nixie; torpedo decoy.
ESM/ECM: SLQ-32(V)2 ❻; radar warning. Sidekick modification adds jammer and deception system.
Combat data systems: NTDS with Link 14. INMARSAT.
Weapons control: SWG-1 Harpoon LCS. Mk 92 (Mod 4). The Mk 92 is the US version of the Signaal WM28 system. Mk 13 weapon direction system. 2 Mk 24 optical directors.
Radars: Air search: Raytheon SPS-49(V)4 ❼; C-band; range 457 km *(250 n miles)*.
    Surface search: ISC Cardion SPS-55 ❽; I-band.
    Fire control: Lockheed STIR (modified SPG-60) ❾; I/J-band; range 110 km *(60 n miles)*.
    Sperry Mk 92 (Signaal WM28) ❿; I/J-band.
Tacan: URN 25.
Sonars: Raytheon SQS-56; hull-mounted; active search and attack; medium frequency.

Helicopters: 1 Eurocopter BO 105 ⓫. Space for 2 SH-2G.

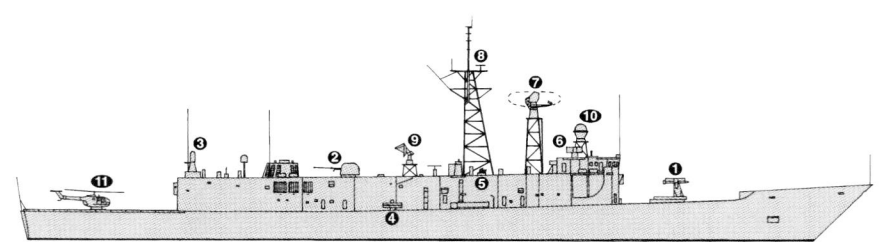

SABHA   *(Scale 1 : 1,200), Ian Sturton* / 0056532

SABHA   *4/2000, Guy Toremans* / 0104200

Programmes: *Sabha* transferred from the US by grant 18 September 1996. Arrived in the Gulf in June 1997 for a work-up and training period. Transfer of a second ship is a possibility.

Structure: Apart from the removal of the US SATCOM aerials there are no visible changes from US service.
Operational: A transfer of helicopters is required if the ASW potential of the ship is to be realised.   *UPDATED*

SABHA

6/2003, A Sharma / 0568881

CORVETTES

2 AL MANAMA (MGB 62) CLASS (FSGH)

Name	No	Builders	Commissioned
AL MANAMA	50	Lürssen	14 Dec 1987
AL MUHARRAQ	51	Lürssen	3 Feb 1988

Displacement, tons: 632 full load
Dimensions, feet (metres): 206.7 × 30.5 × 9.5
  (63 × 9.3 × 2.9)
Main machinery: 4 MTU 20V 538 TB92 diesels; 12,820 hp(m)
  (9.42 MW) sustained; 4 shafts
Speed, knots: 32. **Range, n miles:** 4,000 at 16 kt
Complement: 43 (7 officers)

Missiles: SSM: 4 Aerospatiale MM 40 Exocet launchers (2 twin)
  ❶; inertial cruise; active radar homing to 70 km *(40 n miles)* at
  0.9 Mach; warhead 165 kg; sea-skimmer.
Guns: 1 OTO Melara 3 in *(76 mm)*/62 compact ❷; 85 rds/min to
  16 km *(8.7 n miles)* anti-surface; 12 km *(6.5 n miles)* anti-
  aircraft; weight of shell 6 kg.
  2 Breda 40 mm/70 (twin) ❸; 300 rds/min to 12.5 km *(6.8 n
  miles)*; weight of shell 0.96 kg.
  2—7.62 mm MGs.
Countermeasures: Decoys: CSEE Dagaie ❹; chaff and IR flares.
  ESM/ECM: Racal Decca Cutlass/Cygnus ❺; intercept and
  jammer.
Weapons control: CSEE Panda Mk 2 optical director. Philips
  TV/IR optronic director ❻.
Radars: Air/surface search: Philips Sea Giraffe 50 HC ❼; G-band.
  Navigation: Racal Decca 1226; I-band.
  Fire control: Philips 9LV 331 ❽; J-band.

Helicopters: 1 Eurocopter BO 105 ❾.

Programmes: Ordered February 1984.
Modernisation: Upgrade planned to include a SAM self-defence
  system.
Structure: Similar to Singapore and UAE designs. Steel hull,
  aluminium superstructure. Fitted with a helicopter platform
  which incorporates a lift to lower the aircraft into the hangar.
Operational: Planned SA 365F helicopters were not acquired.

UPDATED

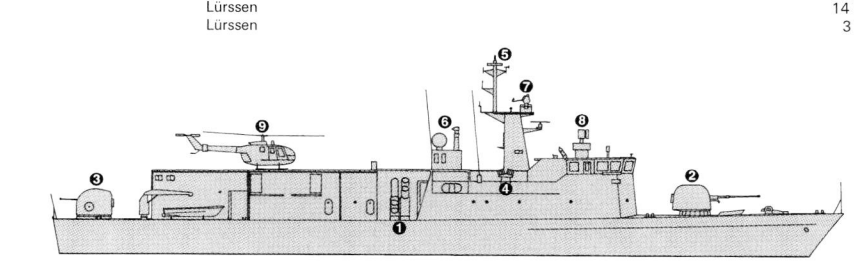

AL MANAMA

(Scale 1 : 600), Ian Sturton / 0104201

AL MANAMA

3/1999, Maritime Photographic / 0056536

AL MANAMA

11/2001, Royal Australian Navy / 0526836

PATROL FORCES

4 AHMAD EL FATEH (TNC 45) CLASS
(FAST ATTACK CRAFT–MISSILE) (PGGF)

Name	No	Builders	Commissioned
AHMAD EL FATEH	20	Lürssen	5 Feb 1984
AL JABIRI	21	Lürssen	3 May 1984
ABDUL RAHMAN AL FADEL	22	Lürssen	10 Sep 1986
AL TAWEELAH	23	Lürssen	25 Mar 1989

Displacement, tons: 228 half load; 259 full load
Dimensions, feet (metres): 147.3 × 22.9 × 8.2 *(44.9 × 7 × 2.5)*
Main machinery: 4 MTU 16V 538 TB92 diesels; 13,640 hp(m) *(10 MW)* sustained; 4 shafts
Speed, knots: 40. **Range, n miles:** 1,600 at 16 kt
Complement: 36 (6 officers)

Missiles: SSM: 4 Aerospatiale MM 40 Exocet (2 twin); inertial cruise; active radar homing to 70 km *(40 n miles)* at 0.9 Mach; warhead 165 kg; sea-skimmer.
Guns: 1 OTO Melara 3 in *(76 mm)*/62; dual purpose; 85 rds/min to 16 km *(8.7 n miles)* anti-surface; 12 km *(6.5 n miles)* anti-aircraft; weight of shell 6 kg.
2 Breda 40 mm/70 (twin); 300 rds/min to 12.5 km *(6.8 n miles)*; weight of shell 0.96 kg.
3—7.62 mm MGs.
Countermeasures: Decoys: CSEE Dagaie launcher; trainable mounting; 10 containers firing chaff decoys and IR flares.
ESM: Thales Sealion.
ECM: Racal Cygnus (not in 20 and 21); jammer.
Weapons control: 1 Panda optical director for 40 mm guns.
Radars: Air/surface search: Philips Sea Giraffe 50 HC; G-band.
Fire control: Philips 9LV 226/231; J-band.
Navigation: Racal Decca 1226; I-band.

Programmes: First pair ordered in 1979, second pair in 1985. Similar craft in service with Ecuador, Kuwait and UAE navies.
Structure: Only the second pair have the communication radome on the after superstructure.
Operational: Refits from 2000 by Lürssen at Abu Dhabi.   ***UPDATED***

AHMAD EL FATEH   *4/2003, A Sharma /* 0568844

AL TAWEELAH   *4/2000, Guy Toremans /* 0104203

2 AL RIFFA (FPB 38) CLASS (FAST ATTACK CRAFT–GUN) (PB)

Name	No	Builders	Commissioned
AL RIFFA	10	Lürssen	3 Mar 1982
HAWAR	11	Lürssen	3 Mar 1982

Displacement, tons: 188 half load; 205 full load
Dimensions, feet (metres): 126.3 × 22.9 × 7.2 *(38.5 × 7 × 2.2)*
Main machinery: 2 MTU 16V 538 TB92 diesels; 6,810 hp(m) *(5 MW)* sustained; 2 shafts
Speed, knots: 32. **Range, n miles:** 1,100 at 16 kt
Complement: 27 (3 officers)
Guns: 2 Breda 40 mm/70 (twin); dual purpose; 300 rds/min to 12 km *(6.5 n miles)* anti-surface; 4 km *(2.2 n miles)*; weight of shell 0.96 kg.
1—57 mm Starshell rocket launcher.
Mines: Mine rails fitted.
Countermeasures: Decoys: 1 Wallop Barricade chaff launcher.
ESM: Racal RDL-2 ABC; radar warning.
Weapons control: CSEE Lynx optical director with Philips 9LV 126 optronic system.
Radars: Surface search: Philips 9GR 600; I-band.
Navigation: Racal Decca 1226; I-band.

Comment: Ordered in 1979. *Al Riffa* launched April 1981. *Hawar* launched July 1981.   ***UPDATED***

HAWAR   *6/2003, A Sharma /* 0568880

2 AL JARIM (FPB 20) CLASS (FAST ATTACK CRAFT–GUN) (PB)

Name	No	Builders	Commissioned
AL JARIM	30	Swiftships, Morgan City	9 Feb 1982
AL JASRAH	31	Swiftships, Morgan City	26 Feb 1982

Displacement, tons: 33 full load
Dimensions, feet (metres): 63 × 18.4 × 6.5 *(19.2 × 5.6 × 2)*
Main machinery: 2 Detroit 12V-71TA diesels; 840 hp(m) *(627 kW)* sustained; 2 shafts
Speed, knots: 30. **Range, n miles:** 1,200 at 18 kt
Guns: 1 Oerlikon GAM-BO1 20 mm.
Radars: Surface search: Decca 110; I-band.

Comment: Aluminium hulls.

VERIFIED

AL JARIM   *5/2003, A Sharma /* 0568879

SHIPBORNE AIRCRAFT

Notes: SH-2G helicopters may be acquired for the frigate in due course.

Numbers/Type: 2 Eurocopter BO 105.
Operational speed: 113 kt *(210 km/h)*.
Service ceiling: 9,845 ft *(3,000 m)*.
Range: 407 n miles *(754 km)*.
Role/Weapon systems: Acquired in August 1994 as the first aircraft of a Naval Air Arm. Sensors: Bendix RDR 1500B radar. Weapons: Unarmed.

UPDATED

BO 105   *6/1995 /* 0056541

AUXILIARIES

Notes: There are also two RTK Medevac boats and one Diving Boat (512).

1 AJEERA CLASS (SUPPLY SHIPS) (YFU)

Name	No	Builders	Commissioned
AJEERA	41	Swiftships, Morgan City	21 Oct 1982

Displacement, tons: 420 full load
Dimensions, feet (metres): 129.9 × 36.1 × 5.9 *(39.6 × 11 × 1.8)*
Main machinery: 2 Detroit 16V-71 diesels; 811 hp *(605 kW)* sustained; 2 shafts
Speed, knots: 13. **Range, n miles:** 1,500 at 10 kt
Complement: 21
Guns: 2—12.7 mm MGs.
Radars: Navigation: Racal Decca; I-band.

Comment: Used as general purpose cargo ships and can carry up to 200 tons of fuel and water. Built to an LCU design with a bow ramp and 15 ton crane.

VERIFIED

AJEERA   *4/2003, A Sharma /* 0568843

4 LCU 1466 CLASS (LCU)

MASHTAN 42	**RUBODH** 43	**SUWAD** 44	**JARADAH** 45

Displacement, tons: 360 full load
Dimensions, feet (metres): 119 × 34 × 6 *(36.3 × 10.4 × 1.8)*
Main machinery: 3 Gray Marine 64 YTL diesels; 675 hp *(504 kW)*; 3 shafts
Speed, knots: 8. **Range, n miles:** 800 at 8 kt
Complement: 15
Cargo capacity: 167 tons
Guns: 2—12.7 mm MGs.
Radars: Navigation: Racal Decca; I-band.

Comment: Transferred from US in 1991.

VERIFIED

RUBODH   *4/2003, A Sharma /* 0568842

1 PERSONNEL TRANSPORT CRAFT (YFL)

TIGHATLIB 46

Comment: Details not confirmed.

UPDATED

TIGHATLIB   *6/2003, A Sharma /* 0568877

1 LOADMASTER II CLASS (LCU)

AL ZUBARA (ex-*Sabha*) 40

Displacement, tons: 150 full load
Dimensions, feet (metres): 73.8 × 24.6 × 3.9 *(22.5 × 7.5 × 1.2)*
Main machinery: 2 General Motors 8V92N diesels; 780 hp *(575 kW)*; 2 props
Speed, knots: 6
Radars: Navigation: I-band.

Comment: Built by Fairey Marine Cowes, UK and entered service in 1981.

NEW ENTRY

AL ZUBARA   *4/2004*, Guy Toremans /* 0580522

COAST GUARD

Notes: (1) In addition to the craft listed below about 14 small open fibreglass boats are used for patrol duties.
(2) Procurement of new patrol craft was expected in 2004 but has not been confirmed.

1 WASP 30 METRE CLASS (WPB)

AL MUHARRAQ

Displacement, tons: 90 standard; 103 full load
Dimensions, feet (metres): 98.5 × 21 × 5.5 *(30 × 6.4 × 1.6)*
Main machinery: 2 Detroit 16V-149TI diesels; 2,322 hp *(1.73 MW)* sustained; 2 shafts
Speed, knots: 25. **Range, n miles:** 500 at 22 kt
Complement: 9
Guns: 2—7.62 mm MGs.
   1 Hughes chain 7.62 mm.
Radars: Surface search: Racal Decca; I-band.

Comment: Ordered from Souters, Cowes, Isle of Wight in 1984. Laid down November 1984, launched 12 August 1985, shipped 21 October 1985. GRP hull.

VERIFIED

AL MUHARRAQ   *4/2003, A Sharma /* 0568841

4 HALMATIC 20 METRE CLASS (WPB)

DERA'A 2, 6, 7 and **8**

Displacement, tons: 31.5 full load
Dimensions, feet (metres): 65.9 × 19.4 × 5.1 *(20.1 × 5.9 × 1.5)*
Main machinery: 2 Detroit 12V-71TA diesels; 840 hp *(626 kW)* sustained; 2 shafts
Speed, knots: 25. **Range, n miles:** 500 at 20 kt
Complement: 7
Guns: 2—7.62 mm MGs.

Comment: Three delivered in late 1991, the last in early 1992. GRP hulls.

VERIFIED

DERA'A 2   *7/2003, A Sharma /* 0568878

2 WASP 20 METRE CLASS (WPB)

DERA'A 4 and **5**

Displacement, tons: 36.3 full load
Dimensions, feet (metres): 65.6 × 16.4 × 4.9 *(20 × 5 × 1.5)*
Main machinery: 2 Detroit 12V-71TA diesels; 840 hp *(626 kW)* sustained; 2 shafts
Speed, knots: 24.5. **Range, n miles:** 500 at 20 kt
Complement: 8
Guns: 2—7.62 mm MGs.
Radars: Surface search: Racal Decca; I-band.

Comment: Built by Souters, Cowes, Isle of Wight. Delivered 1983. GRP hulls.

VERIFIED

DERA'A 4   *6/2000, Bahrain Coast Guard /* 0104206

For details of the latest updates to *Jane's Fighting Ships* online and to discover the additional information available exclusively to online subscribers please visit
jfs.janes.com

6 HALMATIC 160 CLASS (WPB)

SAIF 5, 6, 7, 8, 9 and 10

Displacement, tons: 17 full load
Dimensions, feet (metres): 47.2 × 12.8 × 3.9 *(14.4 × 3.9 × 1.2)*
Main machinery: 2 Detroit 6V-92TA diesels; 520 hp *(388 kW)* sustained; 2 shafts
Speed, knots: 20. **Range, n miles:** 500 at 20 kt
Complement: 4
Guns: 1—7.62 mm MG.
Radars: Surface search: Furuno; I-band.

Comment: Built by Halmatic, UK, and delivered in 1990-91. GRP hulls.

UPDATED

SAIF 5   *4/2001, Guy Toremans /* 0114683

4 FAIREY SWORD CLASS (WPB)

SAIF 1, 2, 3 and 4

Displacement, tons: 15
Dimensions, feet (metres): 44.9 × 13.4 × 4.3 *(13.7 × 4.1 × 1.3)*
Main machinery: 2 GM 8V-71 diesels; 590 hp *(440 kW)* sustained; 2 shafts
Speed, knots: 22
Complement: 6
Radars: Navigation: Furuno; I-band.

Comment: Purchased in 1980. Built by Fairey Marine Ltd.

VERIFIED

SAIF 3   *11/1999, Bahrain Coast Guard /* 0056543

2 HAWAR CLASS (PB)

HAWAR 1   HAWAR 2

Displacement, tons: 10.5 full load
Dimensions, feet (metres): 40.7 × 13.0 × 2.3 *(12.4 × 4.0 × 0.7)*
Main machinery: 2 Cummins 6CTA8.3 diesels
Speed, knots: 30
Guns: 1—7.62 mm MG.

Comment: Entered service in 2003.

VERIFIED

HAWAR 1   *6/2003, John Fidler /* 0567903

3 WASP 11 METRE CLASS (WPB)

SAHAM 1   SAHAM 2   SAHAM 3

Displacement, tons: 7 full load
Dimensions, feet (metres): 36.1 × 10.5 × 2.6 *(11 × 3.2 × 0.8)*
Main machinery: 2 Yamaha outboards; 400 hp(m) *(294 kW)*
Speed, knots: 25. **Range, n miles:** 125 at 20 kt
Complement: 3
Radars: Navigation: Koden; I-band.

Comment: Built by Souters, Cowes in 1983. *Saham 2* re-engined with outboards and recommissioned in August 1997. *Saham 1* similarly back in service in 1998. *Saham 3* awaits upgrade.

VERIFIED

SAHAM 2   *10/1997, Bahrain Coast Guard /* 0012061

1 SUPPORT CRAFT (YAG)

SAFRA 3

Displacement, tons: 165 full load
Dimensions, feet (metres): 85 × 25.9 × 5.2 *(25.9 × 7.9 × 1.6)*
Main machinery: 2 Detroit 16V-92TA diesels; 1,380 hp *(1.03 MW)*; 2 shafts
Speed, knots: 13. **Range, n miles:** 700 at 12 kt
Complement: 6
Radars: Navigation: Racal Decca; I-band.

Comment: Built by Halmatic, Havant and delivered in early 1992. Logistic support work boat equipped for towing and firefighting. Can carry 15 tons.

VERIFIED

SAFRA 3   *4/2003, A Sharma /* 0568840

1 LANDING CRAFT (LCM)

SAFRA 2

Displacement, tons: 150 full load
Measurement, tons: 90 dwt
Dimensions, feet (metres): 73.9 × 24.9 × 4 *(22.5 × 7.5 × 1.2)*
Main machinery: 2 Detroit 12V-71 diesels; 680 hp *(508 kW)* sustained; 2 shafts
Speed, knots: 8
Complement: 8
Radars: Navigation: Furuno; I-band.

Comment: Built by Fairey Marine and delivered in 1981. Based at Bandar-Dar.

VERIFIED

SAFRA 2   *4/2003, A Sharma /* 0568839

Bangladesh

Country Overview

The People's Republic of Bangladesh, formerly East Pakistan, proclaimed independence in 1971. Situated in south Asia and with an area of 55,598 square miles, most of its land border is with India (cutting off north-east India from the rest). There is a short border with Myanmar to the south-east. Its 313 n mile coastline is with the Bay of Bengal on which the principal port of Chittagong is situated. The capital and largest city is Dhaka. Territorial waters (12 n miles) are claimed. An EEZ (200 n miles) has been claimed but the limits have not been defined.

Headquarters Appointments

Chief of Naval Staff:
Rear Admiral Shah Iqbal Mujtaba
Assistant Chief of Naval Staff (Operations):
Commodore S J Nizam
Assistant Chief of Naval Staff (Personnel):
Commodore Mohammad Emdadul Islam
Assistant Chief of Naval Staff (Logistics):
Commodore Mohammad Hasan Ali Khan
Assistant Chief of Naval Staff (Materials):
Commodore M A Haque

Senior Appointments

Naval Administrative Authority, Dhaka:
Commodore Mohammad Hasan Ali Khan
Commodore Commanding BN Flotilla:
Commodore M Atiqur Rahman

Commodore Commanding Chittagong:
Commodore M S Kabir
Commodore Commanding Khulna:
Captain M M Rahman
Director General Coast Guard:
Captain Sarkar Mohammad Humayun Kabir

Bases

Chittagong (BNS *Issa Khan*, BN Dockyard, Naval Stores Depot, Chittagong, BNS *Ulka*, Bangladesh. Naval Academy, BNS *Patenga*, BNS *Bhatiary*, Naval Units *Cox's Bazar, Chanua* and *St Martins*), Kaptai (BNS *Shaheed Moazzam*).
Dhaka (NHQ, BNS *Haji Mohsin* and Naval Unit *Pagla*).
Khulna (BNS *Titumir*, BNS *Mongla*, BNS *Upasham*, Forward Bases *Khepupara* and *Hiron Point*.

Personnel

(a)   2005: 11,658 (1,071 officers)
(b)   Voluntary service

Coast Guard

Formed on 19 December 1995 with two ships on loan from the Navy. Bases at Chittagong (East Zone) and Khulna (West Zone). Personnel 330 (41 officers). Colours thick red and thin blue diagonal stripes on hull with COAST GUARD on ships side.

Strength of the Fleet

Type	Active	Building
Frigates	5	—
Fast Attack Craft (Missile)	9	—
Fast Attack Craft (Torpedo)	8	—
Fast Attack Craft (Gun)	13	—
Large Patrol Craft	7	—
Coastal Patrol Craft	9	—
Riverine Patrol Craft	5	—
Minesweepers	4	—
Training Ships	1	—
Repair Ship	1	—
Tankers	2	—
Survey Craft	4	—

Prefix to Ships' Names

Navy: BNS
Coast Guard: CGS

PENNANT LIST

Frigates

F 15	Abu Bakr
F 16	Umar Farooq
F 17	Ali Haider
F 18	Osman
F 25	DW 2000H

Patrol Forces

P 111	Pabna (CG)
P 112	Noakhali (CG)
P 113	Patuakhali (CG)
P 114	Rangamati (CG)
P 115	Bogra (CG)
P 201	Ruposhi Bangla (CG)
P 211	Meghna
P 212	Jamuna
P 311	Bishkhali
P 312	Padma
P 313	Surma
P 314	Karnaphuli
P 315	Tista
P 411	Shaheed Daulat
P 412	Shaheed Farid

P 413	Shaheed Mohibullah
P 414	Shaheed Akteruddin
P 611	Tawheed (CG)
P 612	Tawfiq
P 613	Tamjeed
P 614	Tanveer
P 711	Barkat
P 712	Salam
P 713	Sangu
P 714	Turag
P 811	Nirbhoy
P 911	Madhumati
P 912	Kapatakhaya
P 913	Karatoa
P 914	Gomati
P 1011	Titas
P 1012	Kusiyara
P 1013	Chitra
P 1014	Dhansiri
P 8111	Durbar
P 8112	Duranta
P 8113	Durvedya
P 8114	Durdam
P 8125	Durdharsha
P 8126	Durdanta

P 8128	Dordanda
P 8131	Anirban
P 8141	Uttal
P 8221	TB 1
P 8222	TB 2
P 8223	TB 3
P 8224	TB 4
P 8235	TB 35
P 8236	TB 36
P 8237	TB 37
P 8238	TB 38

Mine Warfare Forces

M 91	Sagar
M 95	Shapla
M 96	Saikat
M 97	Surovi
M 98	Shaibal

Auxiliaries

A 511	Shaheed Ruhul Amin
A 512	Shahayak
A 513	Shahjalal

A 515	Khan Jahan Ali
A 516	Imam Gazzali
A 581	Darshak
A 582	Tallashi
A 583	Agradoot
A 584	LCT-101
A 585	LCT-102
A 587	LCT-104
A 711	Sundarban
A 721	Khadem
A 722	Sebak
A 723	Rupsha
A 724	Shibsha
A 731	Balaban
L 900	Shah Amanat
L 901	Shah Paran
L 902	Shah Makhdum

SUBMARINES

Notes: Plans to acquire a submarine service by 2012 were announced by the Defence Minister in April 2004.

FRIGATES

Notes: Replacement of the Salisbury and Leopard class frigates is under consideration although timescales have not been announced.

1 SALISBURY CLASS (TYPE 61) (FF)

Name	No	Builders	Laid down	Launched	Commissioned
UMAR FAROOQ (ex-*Llandaff*)	F 16	Hawthorn Leslie Ltd	27 Aug 1953	30 Nov 1955	11 Apr 1958

Displacement, tons: 2,170 standard; 2,408 full load
Dimensions, feet (metres): 339.8 × 40 × 15.5 (screws) *(103.6 × 12.2 × 4.7)*
Main machinery: 8 16 VTS ASR 1 diesels; 14,400 hp *(10.7 MW)* sustained; 2 shafts
Speed, knots: 24. **Range, n miles:** 2,300 at 24 kt; 7,500 at 16 kt
Complement: 237 (14 officers)

Guns: 2 Vickers 4.5 in *(115 mm)*/45 (twin) Mk 6 ❶; dual purpose; 20 rds/min to 19 km *(10 n miles)* anti-surface; 6 km *(3.3 n miles)* anti-aircraft; weight of shell 25 kg.
2 Bofors 40 mm/60 Mk 9 ❷; 120 rds/min to 3 km *(1.6 n miles)* anti-aircraft; 10 km *(5.5 n miles)* maximum.
A/S mortars: 1 triple-barrelled Squid Mk 4 ❸; fires pattern of 3 depth charges to 300 m ahead of ship.
Countermeasures: Decoys: Corvus chaff launchers.

Weapons control: 1 Mk 6M gun director.
Radars: Air search: Marconi Type 965 with double AKE 2 array ❹; A-band.
Air/surface search: Plessey Type 993 ❺; E/F-band.
Heightfinder: Type 278M ❻; E-band.
Surface search: Decca Type 978 ❼; I-band.
Navigation: Decca Type 978; I-band.
Fire control: Type 275 ❽; F-band.
Sonars: Type 174; hull-mounted; active search; medium frequency.
Graseby Type 170B; hull-mounted; active attack; 15 kHz.

Programmes: Transferred from UK at Royal Albert Dock, London 10 December 1976.
Operational: The radar Type 982 aerial is still retained on the after mast but the set is non-operational. The ship has been modified as a training ship and is expected to remain in service for some years.

UPDATED

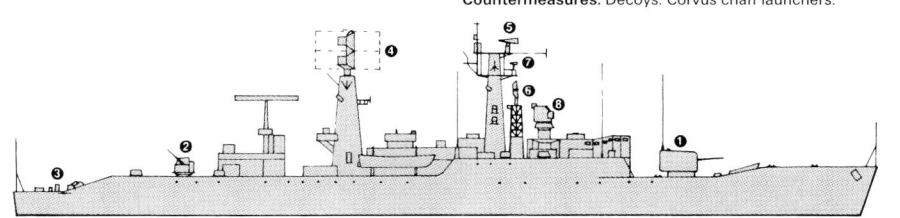

UMAR FAROOQ   (Scale 1 : 900), Ian Sturton

1 MODIFIED ULSAN CLASS (FFGH)

Name	No	Builders	Laid down	Launched	Commissioned
—	F 25	Daewoo Heavy Industries	12 May 1999	29 Aug 2000	20 June 2001

Displacement, tons: 2,170 standard; 2,370 full load
Dimensions, feet (metres): 340.3 × 41 × 12.5
(103.7 × 12.5 × 3.8)
Main machinery: CODAD: 4 SEMT-Pielstick 12V PA6V280 STC
diesels; 22,501 hp *(16.78 MW)* sustained; 2 shafts
Speed, knots: 25. **Range, n miles:** 4,000 at 18 kt
Complement: 186 (16 officers)

Missiles: SSM: 4 Otomat Mk 2 ❶; command guidance; active
radar homing to 180 km *(97.2 n miles)*, at 0.9 Mach; warhead
210 kg; sea-skimmer.
Guns: 1 Otobreda 3 in *(76 mm)*/62 Super Rapid ❷; 120 rds/min
to 16 km *(8.7 n miles)*; weight of shell 6 kg.
4 Otobreda 40 mm/70 (2 twin) compact ❸; 300 rds/min to
12.5 km *(6.8 n miles)*; weight of shell 0.96 kg.
Torpedoes: 6-324 mm B-515 (2 triple) tubes ❹; Whitehead
A244S; anti-submarine; active/passive homing to 7 km *(3.8 n
miles)*; warhead 34 kg (shaped charge).
Countermeasures: Decoys: 2 Super Barricade launchers ❺.
ESM: Racal Cutlass 242; intercept.
ECM: Racal Scorpion; jammer.
Combat data systems: Thales TACTICOS.
Weapons control: Signaal Mirador optronic director ❻.

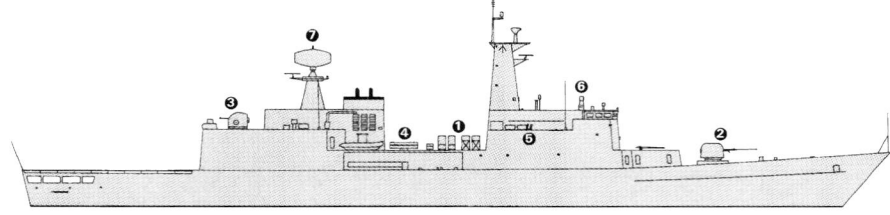

DW 2000H   (Scale 1 : 900), Ian Sturton / 0130076

Radars: Air search: Signaal DA08 ❼; F-band.
Fire control: Signaal Lirod Mk 2; K-band.
Navigation: 2 KH-1007; I-band.
Sonars: STN Atlas ASO 90; hull-mounted; active search; medium
frequency.

Helicopters: Hangar and platform for operation of 'Lynx' sized
helicopter.

Programmes: Modified Ulsan class ordered from Daewoo in
March 1998. Arrived at Chittagong on 16 June 2001.
Operational: The ship was decommissioned on 13 February
2002 for design modification, warranty repairs and capability
upgrades. Expected to be recommissioned on completion of
these programmes. Until then the ship is to be known by the
builder's designation DW 2000H.

UPDATED

DW 2000H   6/2001, Daewoo / 0094449   DW 2000H   6/2001 / 0111271

1 OSMAN (JIANGHU I) CLASS (TYPE 053 H1) (FFG)

Name	No	Builders	Laid down	Launched	Commissioned
OSMAN (ex-*Xiangtan*)	F 18 (ex-556)	Hudong Shipyard, Shanghai	1986	Dec 1988	4 Nov 1989

Displacement, tons: 1,425 standard; 1,702 full load
Dimensions, feet (metres): 338.6 × 35.4 × 10.2
(103.2 × 10.7 × 3.1)
Main machinery: 2 Type 12 E 390V diesels; 16,000 hp(m)
(11.9 MW) sustained; 2 shafts
Speed, knots: 26. **Range, n miles:** 2,700 at 18 kt
Complement: 300 (27 officers)

Missiles: SSM: 4 Hai Ying 2 (2 twin) launchers ❶; active radar or
IR homing to 80 km *(43.2 n miles)* at 0.9 Mach; warhead
513 kg.

Guns: 4 China 3.9 in *(100 mm)*/56 (2 twin) ❷; 18 rds/min to
22 km *(12 n miles)*; weight of shell 15.9 kg.
8 China 37 mm/76 (4 twin) ❸; 180 rds/min to 8.5 km *(4.6 n
miles)* anti-aircraft; weight of shell 1.42 kg.
A/S mortars: 2 RBU 1200 5-tubed fixed launchers ❹; range
1,200 m; warhead 34 kg.
Depth charges: 2 BMB-2 projectors; 2 racks.
Mines: Can carry up to 60.

Countermeasures: Decoys: 2 Loral Hycor SRBOC Mk 36
6-barrelled chaff launchers.
ESM: Watchdog; radar warning.
Weapons control: Wok Won director (752A) ❺.
Radars: Air/surface search: MX 902 Eye Shield (922-1) ❻;
G-band.
Surface search/fire control: Square Tie (254) ❼; I-band.
Navigation: Fin Curve (352); I-band.
IFF: High Pole A.
Sonars: Echo Type 5; hull-mounted; active search and attack;
medium frequency.

Programmes: Transferred 26 September 1989 from China,
arrived Bangladesh 8 October 1989. Second order expected
in 1991 was cancelled.
Structure: This is a Jianghu Type I (version 4) hull with twin
100 mm guns (vice the 57 mm in the ships sold to Egypt),
Wok Won fire-control system and a rounded funnel.
Operational: Damaged in collision with a merchant ship in
August 1991. One 37 mm mounting uprooted and SSM and
RBU mountings misaligned. Repaired in 1992-93.

VERIFIED

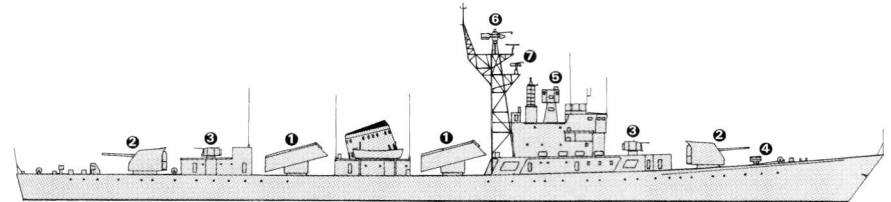

OSMAN   (Scale 1 : 900), Ian Sturton / 0130383

OSMAN   10/2003, Hartmut Ehlers / 0569148

2 LEOPARD CLASS (TYPE 41) (FF/FFT)

Name	No	Builders		Laid down	Launched	Commissioned
ABU BAKR (ex-*Lynx*)	F 15	John Brown & Co Ltd, Clydebank		13 Aug 1953	12 Jan 1955	14 Mar 1957
ALI HAIDER (ex-*Jaguar*)	F 17	Wm Denny & Bros Ltd, Dumbarton		2 Nov 1953	30 July 1957	12 Dec 1959

Displacement, tons: 2,300 standard; 2,520 full load
Dimensions, feet (metres): 339.8 × 40 × 15.5 (screws) *(103.6 × 12.2 × 4.7)*
Main machinery: 8 16 VTS ASR 1 diesels; 14,400 hp *(10.7 MW)* sustained; 2 shafts; F 17 fitted with cp props
Speed, knots: 24. **Range, n miles:** 2,300 at full power; 7,500 at 16 kt
Complement: 235 (15 officers)

Guns: 4 Vickers 4.5 in *(115 mm)*/45 (2 twin) Mk 6 ❶; dual purpose; 20 rds/min to 19 km *(10 n miles)* anti-surface; 6 km *(3.3 n miles)* anti-aircraft; weight of shell 25 kg.
1 Bofors 40 mm/60 Mk 9 ❷; 120 rds/min to 3 km *(1.6 n miles)* anti-aircraft; 10 km *(5.5 n miles)*.
2—7.62 mm MGs.

Countermeasures: Decoys: Corvus chaff launchers.
ESM: Radar warning.
Weapons control: Mk 6M gun director.
Radars: Air search: Marconi Type 965 with single AKE 1 array ❸; A-band.
Air/surface search: Plessey Type 993 ❹; E/F-band.
Navigation: Decca Type 978; Kelvin Hughes 1007; I-band.
Fire control: Type 275 ❺; F-band.

Programmes: *Ali Haider* transferred from UK 16 July 1978 and *Abu Bakr* on 12 March 1982. *Ali Haider* refitted at Vosper Thornycroft August-October 1978. *Abu Bakr* extensively refitted in 1982.
Structure: All welded. Fitted with stabilisers. Sonars removed while still in service with RN. Fuel tanks have a water compensation system to improve stability.
Operational: Both to remain in service until replacements have been acquired.

UPDATED

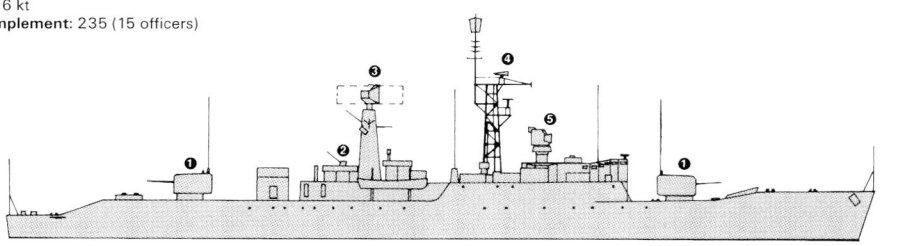

ABU BAKR   (Scale 1 : 900), Ian Sturton

ALI HAIDER   2/2001, Michael Nitz / 0529082

PATROL FORCES

Notes: Plans to acquire an offshore patrol vessel, four missile-firing craft and six further patrol boats were announced by the Defence Minister in April 2004.

1 MADHUMATI (SEA DRAGON) CLASS
(LARGE PATROL CRAFT) (PSO)

Name	No	Builders	Commissioned
MADHUMATI	P 911	Hyundai, Ulsan	18 Feb 1998

Displacement, tons: 635 full load
Dimensions, feet (metres): 199.5 × 26.2 × 8.9 *(60.8 × 8 × 2.7)*
Main machinery: 2 SEMT-Pielstick 12 PA6 diesels; 9,600 hp(m) *(7.08 MW)* sustained; 2 shafts
Speed, knots: 24. **Range, n miles:** 6,000 at 15 kt
Complement: 43 (7 officers)
Guns: 1 Bofors 57 mm/70 Mk 1; 220 rds/min to 17 km *(9.3 n miles)*; weight of shell 2.4 kg. 1 Bofors 40 mm/70. 2 Oerlikon 20 mm.
Weapons control: Optronic director.
Radars: Surface search: Kelvin Hughes KH 1007; I-band.
Navigation: GEM Electronics SPN 753B; I-band.

Comment: Ordered in 1995 and delivered in October 1997. Very similar to the South Korean Coast Guard vessels, but with improved fire-control equipment. Vosper stabilisers.

VERIFIED

4 DURDHARSHA (HUANGFEN) CLASS (TYPE 021)
(FAST ATTACK CRAFT—MISSILE) (PTFG)

DURDHARSHA	P 8125	DORDANDA	P 8128
DURDANTA	P 8126	ANIRBAN	P 8131

Displacement, tons: 171 standard; 205 full load
Dimensions, feet (metres): 126.6 × 24.9 × 8.9 *(38.6 × 7.6 × 2.7)*
Main machinery: 3 diesels; 12,000 hp(m) *(8.8 MW)*; 3 shafts
Speed, knots: 35. **Range, n miles:** 800 at 30 kt
Complement: 35 (5 officers)
Missiles: SSM: 4 HY-2; active radar or IR homing to 80 km *(43.2 n miles)* at 0.9 Mach; warhead 513 kg.
Guns: 4 USSR 30 mm/65 (2 twin).
Radars: Surface search: Square Tie; I-band.
Fire control: Rice Lamp; H/I-band.
IFF: High Pole A.

Comment: Built in China. First four commissioned in Bangladesh Navy on 10 November 1988. Chinese equivalent of the Soviet Osa class which started building in 1985. All damaged in April 1991 typhoon but recovered and repaired (*Durnibar* was converted to a patrol craft). A fifth vessel *Anirban* was delivered in June 1992. Original main machinery replaced.

VERIFIED

MADHUMATI   2/1998, Bangladesh Navy / 0017589

DORDANDA   6/2003, Bangladesh Navy / 0572413

6 ISLAND CLASS (COASTAL PATROL CRAFT/TRAINING CRAFT)
(PBO/AX)

Name	No	Builders	Commissioned	Recommissioned
SHAHEED RUHUL AMIN	A 511	Hall Russell, Aberdeen	15 Oct 1976	1994
(ex-*Jersey*)	(ex-P 295)			
KAPATAKHAYA	P 912	Hall Russell, Aberdeen	14 July 1977	4 May 2003
(ex-*Shetland*)	(ex-P 298)			
KARATOA	P 913	Hall Russell, Aberdeen	6 Oct 1979	4 May 2003
(ex-*Alderney*)	(ex-P 278)			
GOMATI	P 914	Hall Russell, Aberdeen	1 June 1979	3 Oct 2004
(ex-*Anglesey*)	(ex-P 277)			
SANGU	P 713	Hall Russell, Aberdeen	28 Oct 1977	3 Oct 2004
(ex-*Guernsey*)	(ex-P 297)			
TURAG	P 714	Hall Russell, Aberdeen	3 Mar 1978	3 Oct 2004
(ex-*Lindisfarne*)	(ex-P 300)			

Displacement, tons: 925 standard; 1,260 full load
Dimensions, feet (metres): 176 wl; 195.3 oa × 36 × 15 *(53.7; 59.5 × 11 × 4.5)*
Main machinery: 2 Ruston 12RKC diesels; 5,640 hp *(4.21 MW)* sustained; 1 shaft; cp prop
Speed, knots: 16.5. **Range, n miles:** 7,000 at 12 kt
Complement: 39
Guns: 1 Bofors 40 mm/60 Mk 3. 2 FN 7.62 mm MGs.
Countermeasures: ESM: Orange Crop; intercept.
Combat data systems: Racal CANE DEA-1 action data automation.
Radars: Navigation: Kelvin Hughes Type 1006; I-band.

Comment: *Shaheed Ruhul Amin* transferred as a training craft in 1993. Five further former UK Island class acquired as patrol craft. *Kapatakhaya* transferred 29 July 2002, *Karatoa* 31 October 2002, *Gomati* on 12 September 2003 and *Sangu* and *Turag* on 29 January 2004.
UPDATED

TURAG *3/2004 *, Derek Fox /* 1042116

5 DURBAR (HEGU) CLASS (TYPE 024)
(FAST ATTACK CRAFT—MISSILE) (PTFG)

DURBAR P 8111	DURVEDYA P 8113	UTTAL P 8141
DURANTA P 8112	DURDAM P 8114	

Displacement, tons: 68 standard; 79.2 full load
Dimensions, feet (metres): 88.6 × 20.7 × 4.3 *(27 × 6.3 × 1.3)*
Main machinery: 4 Type L-12V-180B diesels; 4,800 hp(m) *(3.57 MW)*; 4 shafts
Speed, knots: 37.5. **Range, n miles:** 400 at 30 kt
Complement: 17 (4 officers)
Missiles: SSM: 2 SY-1; active radar or IR homing to 45 km *(24.3 n miles)* at 0.9 Mach; warhead 513 kg.
Guns: 2—25 mm/80 (twin); 270 rds/min to 3 km *(1.6 n miles)*; weight of shell 0.34 kg.
Radars: Surface search: Square Tie; I-band.

Comment: Built in China. First pair commissioned in Bangladesh Navy on 6 April 1983, second pair on 10 November 1983. Two badly damaged in April 1991 typhoon but were repaired. *Uttal* was delivered in June 1992. Missiles are seldom embarked. All have been refitted with new versions of original engines.
VERIFIED

UTTAL *3/1998 /* 0017590

4 HUCHUAN CLASS (TYPE 026)
(FAST ATTACK CRAFT—TORPEDO) (PTK)

TB 35 P 8235	TB 36 P 8236	TB 37 P 8237	TB 38 P 8238

Displacement, tons: 46 full load
Dimensions, feet (metres): 73.8 × 16.4 × 6.9 (foil) *(22.5 × 5 × 2.1)*
Main machinery: 3 Type L-12V-180 diesels; 3,600 hp(m) *(2.64 MW)*; 3 shafts
Speed, knots: 50. **Range, n miles:** 500 at 30 kt
Complement: 23 (3 officers)
Guns: 4 China 14.5 mm (2 twin); 600 rds/min to 7 km *(3.8 km)*.
Torpedoes: 2—21 in *(533 mm)* China YU-1; anti-ship; to 9.2 km *(5 n miles)* at 39 kt or 3.7 km *(2.1 n miles)* at 51 kt; warhead 400 kg.
Radars: Surface search: China Type 753; I-band.

Comment: Chinese Huchuan class. Two damaged in April 1991 typhoon but were repaired. All reported operational.
VERIFIED

TB 38 *6/2003, Bangladesh Navy /* 0572415

1 DURJOY (HAINAN) CLASS
(TYPE 037) (LARGE PATROL CRAFT) (PC)

NIRBHOY P 812

Displacement, tons: 375 standard; 392 full load
Dimensions, feet (metres): 192.8 × 23.6 × 7.2 *(58.8 × 7.2 × 2.2)*
Main machinery: 4 diesels; 4,000 hp(m) *(2.94 MW)* sustained; 4 shafts
Speed, knots: 30.5. **Range, n miles:** 1,300 at 15 kt
Complement: 70
Guns: 4 China 57 mm/70 (2 twin); 120 rds/min to 12 km *(6.5 n miles)*; weight of shell 6.31 kg. 4—25 mm/60 (2 twin); 270 rds/min to 3 km *(1.6 n miles)* anti-aircraft.
A/S mortars: 4 RBU 1200 fixed 5-barrelled launchers; range 1,200 m; warhead 34 kg.
Depth charges: 2 racks; 2 throwers. 18 DCs.
Mines: Fitted with rails for 12 mines.
Radars: Surface search: Pot Head; I-band.
IFF: High Pole.
Sonars: Tamir II; hull-mounted; short-range attack; high frequency.

Comment: Transferred from China and commissioned 1 December 1985. Forms part of Escort Squadron 81 at Chittagong. *Durjoy* damaged beyond repair by cyclone in 1991. *Nirbhoy* refitted with new main machinery.
VERIFIED

NIRBHOY *6/2003, Bangladesh Navy /* 0572414

8 SHAHEED (SHANGHAI II) (TYPE 062) CLASS
(FAST ATTACK CRAFT—GUN) (PC)

SHAHEED DAULAT P 411	SHAHEED MOHIBULLAH P 413	TAWHEED P 611	TAMJEED P 613
SHAHEED FARID P 412	SHAHEED AKTHERUDDIN P 414	TAWFIQ P 612	TANVEER P 614

Displacement, tons: 113 standard; 134 full load
Dimensions, feet (metres): 127.3 × 17.7 × 5.6 *(38.8 × 5.4 × 1.7)*
Main machinery: 4 Type L 12-180 diesels; 4,400 hp(m) *(3.2 MW)* sustained; 4 shafts
Speed, knots: 30. **Range, n miles:** 800 at 16.5 kt
Complement: 36 (4 officers)
Guns: 4—37 mm/63 (2 twin); 180 rds/min to 8.5 km *(4.6 n miles)*; weight of shell 1.4 kg. 4—25 mm/80 (2 twin); 270 rds/min to 3 km *(1.6 n miles)* anti-aircraft.
Depth charges: 2 throwers; 8 charges.
Mines: 10 can be carried.
Radars: Surface search: Skin Head/Pot Head; E-band.
Sonars: Hull-mounted; active; short range; high frequency. Some reported to have VDS.

Comment: Transferred from China March 1982. Different engine arrangement from Chinese craft. P 411-414 form Patrol Squadron 41 based at Khulna. P 611-614 form Patrol Squadron 61 based at Chittagong. P 611 operated by the Coast Guard.
VERIFIED

TAMJEED *3/1998 /* 0017591

4 SEA DOLPHIN CLASS (FAST ATTACK CLASS—GUN) (PTF)

TITAS P 1011   **KUSIYARA** P 1012   **CHITRA** P 1013   **DHANSIRI** P 1014

Displacement, tons: 143 full load
Dimensions, feet (metres): 107.9 × 22.6 × 7.9 *(32.9 × 6.9 × 2.4)*
Main machinery: 2 MTU MD 16V 538 TB90 diesels; 4,500 hp(m) *(3.35 MW)* sustained; 2 shafts
Speed, knots: 37. **Range, n miles:** 600 at 20 kt
Complement: 28 (4 officers)
Guns: 1—40 mm.
    2—30 mm (1 twin).
    2—20 mm.
Weapons control: Optical director.
Radars: Surface search: Raytheon 1645; I-band.

Comment: Built by Korea SEC in the 1980s and transferred from South Korea as a gift. First pair (P 1011, 1012) recommissioned on 27 May 2000 and second pair (P 1013, 1014) on 3 October 2004. All form 101 Patrol Squadron based at Chittagong.

UPDATED

TITAS   *6/2001, Bangladesh Navy /* 0529005

1 RUPOSHI BANGLA CLASS (COASTAL PATROL CRAFT) (PB)

Name	No	Builders	Launched	Commissioned
RUPOSHI BANGLA	P 201	Hong Leong-Lürssen	28 June 1999	23 Jan 2000

Displacement, tons: 195 full load
Dimensions, feet (metres): 126.3 × 23 × 13.5 *(38.5 × 7 × 4.1)*
Main machinery: 2 Paxman 12VP 185 diesels; 6,729 hp(m) *(4.95 MW)* sustained; 2 shafts
Speed, knots: 27 (5 officers)
Complement: 30
Guns: 1 Oto Melara 25 mm KBA. 2—7.62 mm MGs.
Radars: Surface search: Furuno; I-band.

Comment: Ordered in June 1998 and laid down 11 August 1998. Based on the PZ design for the Malaysian Police. Operated by the Coast Guard.

VERIFIED

RUPOSHI BANGLA   *10/1999, Hong Leong-Lürssen /* 0064625

1 COASTAL PATROL CRAFT (PB)

Name	No	Builders	Commissioned
SALAM (ex-*Durnibar*)	P 712 (ex-*P 8127*)	Khulna Shipyard	19 Mar 2002

Displacement, tons: 185 standard; 216 full load
Dimensions, feet (metres): 126.6 × 24.9 × 8.9 *(38.6 × 7.6 × 2.7)*
Main machinery: 2 Paxman 12V 185 diesels; 4,800 hp *(3.6 MW)* sustained; 2 shafts
Speed, knots: 24. **Range, n miles:** 3,460 at 13 kt
Complement: 27 (5 officers)
Guns: 1 Bofors 40 mm/60; 120 rds/min 20 3 km (1.6 n miles).
    2 GCM AO2 30 mm (twin).
Radars: Surface search: Furuno HR 2010; E/F-band.
Navigation: Anritsu; I-band.

Comment: Former Huangfen class missile craft transferred from China in 1988. Sunk in River Kamaphuli in 1991 during cyclone and later recovered. Renovated and converted to patrol craft role and recommissioned in 2002.

VERIFIED

1 HAIZHUI (TYPE 062/1) CLASS (COASTAL PATROL CRAFT) (PC)

BARKAT P 711

Displacement, tons: 139 full load
Dimensions, feet (metres): 134.5 × 17.4 × 5.9 *(40.9 × 5.3 × 1.8)*
Main machinery: 4 Chinese L12-180A diesels; 4,800 hp(m) *(35.3 MW)*; 4 shafts
Speed, knots: 28. **Range, n miles:** 750 at 17 kt
Complement: 43 (4 officers)
Guns: 4 China 37 mm/63 (2 twin); 180 rds/min to 8.5 km *(4.6 n miles)*; weight of shell 1.42 kg.
    4 China 25 mm/80 (2 twin).
Depth charges: 2 rails.
Radars: Surface search: Anitsu 726; I-band.
Sonars: Stag Ear; active; high frequency.

Comment: Acquired from China in 1995. This is the Shanghai III, the larger and slower version of the Shanghai II which in Chinese service has anti-submarine mortars. An inclined pole mast and platform behind the bridge are distinguishing features.

VERIFIED

BARKAT   *3/1998 /* 0017592

2 KARNAPHULI (KRALJEVICA) CLASS
(LARGE PATROL CRAFT) (PC)

Name	No	Builders	Commissioned
KARNAPHULI (ex-*PBR 502*)	P 314	Yugoslavia	1956
TISTA (ex-*PBR 505*)	P 315	Yugoslavia	1956

Displacement, tons: 195 standard; 245 full load
Dimensions, feet (metres): 141.4 × 20.7 × 5.7 *(43.1 × 6.3 × 1.8)*
Main machinery: 2 Paxman 12V P185 (P 314); 2 MTU 12V 396 TE84 (P 315); 2 shafts
Speed, knots: 24. **Range, n miles:** 1,500 at 12 kt
Complement: 44 (4 officers)
Guns: 2 Bofors 40 mm/70. 2 Oerlikon 20 mm. 2—128 mm rocket launchers (5 barrels per mounting).
Depth charges: 2 racks; 2 Mk 6 projectors.
Radars: Surface search: Decca 1229; I-band.
Sonars: QCU 2; hull-mounted; active; high frequency.

Comment: Transferred and commissioned 6 June 1975. *Karnaphuli* re-engined in 1995, *Tista* in 1998.

VERIFIED

TISTA   *6/1999, Bangladesh Navy /* 0056550

2 AKSHAY CLASS (COASTAL PATROL CRAFT) (PB)

Name	No	Builders	Commissioned
PADMA (ex-*Akshay*)	P 312	Hooghly D & E Co, Calcutta	Jan 1962
SURMA (ex-*Ajay*)	P 313	Hooghly D & E Co, Calcutta	Apr 1962

Displacement, tons: 120 standard; 150 full load
Dimensions, feet (metres): 117.2 × 20 × 5.5 *(35.7 × 6.1 × 1.7)*
Main machinery: 2 Paxman YHAXM diesels; 1,100 hp *(820 kW)*; 2 shafts
Speed, knots: 18. **Range, n miles:** 500 at 12 kt
Complement: 35 (3 officers)
Guns: 4 or 8 Oerlikon 20 mm 1 or (2 quad). 2 Bofors 40 mm/60 (twin) *(Surma)*.
Radars: Surface search: Racal Decca; I-band.

Comment: Transferred from India and commissioned 12 April 1973 and 26 July 1974 respectively. *Surma* has a 40 mm gun aft vice the second quad 20 mm.

VERIFIED

PADMA   *6/1997, Bangladesh Navy /* 0012065

2 MEGHNA CLASS (COASTAL PATROL CRAFT) (PB)

Name	No	Builders	Launched
MEGHNA	P 211	Vosper Private, Singapore	19 Jan 1984
JAMUNA	P 212	Vosper Private, Singapore	19 Mar 1984

Displacement, tons: 410 full load
Dimensions, feet (metres): 152.5 × 24.6 × 6.6 *(46.5 × 7.5 × 2)*
Main machinery: 2 Paxman Valenta 12CM diesels; 5,000 hp *(3.73 MW)* sustained; 2 shafts
Speed, knots: 20. **Range, n miles:** 2,000 at 16 kt
Complement: 47 (3 officers)
Guns: 1 Bofors 57 mm/70 Mk 1; 200 rds/min to 17 km *(9.3 n miles)*; weight of shell 2.4 kg.
 1 Bofors 40 mm/70; 300 rds/min to 12 km *(6.5 n miles)*; weight of shell 0.96 kg.
 2—7.62 mm MGs; launchers for illuminants on the 57 mm gun.
Weapons control: Selenia NA 18 B optronic system.
Radars: Surface search: Decca 1229; I-band.

Comment: Built for EEZ work under the Ministry of Agriculture. Both completed late 1984. Both
 damaged in April 1991 typhoon but have been repaired. P 212 damaged by container ship at
 Chittagong in September 2003. **VERIFIED**

MEGHNA
6/2003, Bangladesh Navy / 0572416

4 TYPE 123K (CHINESE P4) CLASS
(FAST ATTACK CRAFT—TORPEDO) (PTL)

TB 1 P 8221   **TB 2** P 8222   **TB 3** P 8223   **TB 4** P 8224

Displacement, tons: 25 full load
Dimensions, feet (metres): 62.3 × 10.8 × 3.3 *(19 × 3.3 × 1)*
Main machinery: 2 Type L-12V-180 diesels; 2,400 hp(m) *(1.76 MW)*; 2 shafts
Speed, knots: 50. **Range, n miles:** 410 at 30 kt
Complement: 12 (1 officer)
Guns: 2—14.5 mm (twin) MG.
Torpedoes: 2—17.7 in *(450 mm)*; anti-ship.

Comment: Transferred from China 6 April 1983. Three reported to be operational. **VERIFIED**

TB 4
6/2003, Bangladesh Navy / 0572417

1 RIVER CLASS (COASTAL PATROL CRAFT) (PB)

Name	No	Builders	Commissioned
BISHKHALI (ex-*Jessore*)	P 311	Brooke Marine Ltd	20 May 1965

Displacement, tons: 115 standard; 143 full load
Dimensions, feet (metres): 107 × 20 × 6.9 *(32.6 × 6.1 × 2.1)*
Main machinery: 2 MTU 12V 538 TB90 diesels; 4,500 hp(m) *(3.3 MW)* sustained; 2 shafts
Speed, knots: 24
Complement: 30
Guns: 2 Breda 40 mm/70; 300 rds/min to 12.5 km *(6.8 n miles)*; weight of shell 0.96 kg.
Radars: Surface search: Racal Decca; I-band.

Comment: PNS *Jessore*, which was sunk during the 1971 war, was salvaged and extensively
 repaired at Khulna Shipyard and recommissioned as *Bishkhali* on 23 November 1978.
 VERIFIED

BISHKHALI
6/1996, Bangladesh Navy / 0056554

5 PABNA CLASS (RIVERINE PATROL CRAFT) (PBR)

Name	No	Builders	Commissioned
PABNA	P 111	DEW Narayangonj, Dhaka	12 June 1972
NOAKHALI	P 112	DEW Narayangonj, Dhaka	8 July 1972
PATUAKHALI	P 113	DEW Narayangonj, Dhaka	7 Nov 1974
RANGAMATI	P 114	DEW Narayangonj, Dhaka	11 Feb 1977
BOGRA	P 115	DEW Narayangonj, Dhaka	15 July 1977

Displacement, tons: 69.5 full load
Dimensions, feet (metres): 75 × 20 × 3.5 *(22.9 × 6.1 × 1.1)*
Main machinery: 2 Cummins diesels; 2 shafts
Speed, knots: 10.8. **Range, n miles:** 700 at 8 kt
Complement: 33 (3 officers)
Guns: 1 Bofors 40 mm/60 or Oerlikon 20 mm.

Comment: The first indigenous naval craft built in Bangladesh. Form River Patrol Squadron 11 at
 Mongla. All operated by the Coast Guard from 2003. **VERIFIED**

PABNA
6/2003, Bangladesh Navy / 0572418

MINE WARFARE FORCES

4 SHAPLA (RIVER) CLASS (MINESWEEPERS/PATROL CRAFT/
SURVEY SHIPS) (MHSC/PBO/AGS)

Name	No	Builders	Commissioned
SHAPLA (ex-*Waveney*)	M 95	Richards, Lowestoft	12 July 1984
SAIKAT (ex-*Carron*)	M 96	Richards, Great Yarmouth	30 Sep 1984
SUROVI (ex-*Dovey*)	M 97	Richards, Great Yarmouth	30 Mar 1985
SHAIBAL (ex-*Helford*)	M 98	Richards, Great Yarmouth	7 June 1985

Displacement, tons: 890 full load
Dimensions, feet (metres): 156 × 34.5 × 9.5 *(47.5 × 10.5 × 2.9)*
Main machinery: 2 Ruston 6RKC diesels; 3,100 hp *(2.3 MW)* sustained; 2 shafts; cp props
Speed, knots: 14. **Range, n miles:** 4,500 at 10 kt
Complement: 30 (7 officers)
Guns: 1 Bofors 40 mm/60 Mk 3.
Radars: Navigation: 2 Racal Decca TM 1226C; I-band.

Comment: These ships are four of a class of 12 of which seven are in service with Brazil.
 Transferred from the UK on 3 October 1994 and recommissioned on 27 April 1995. Steel
 hulled for deep-armed team sweeping with wire sweeps, and intended for use both as
 minesweepers and as patrol craft. Fitted with Racal Integrated Minehunting System. *Shaibal*
 converted for hydrographic survey duties but retains minesweeping gear. Fitted with echo
 sounders, side-scan sonar and a laboratory.
 UPDATED

SUROVI
3/1998 / 0017593

1 SAGAR (T 43) CLASS (MINESWEEPER) (MSO)

Name	No	Builders	Commissioned
SAGAR	M 91	Wuhan Shipyard	27 Apr 1995

Displacement, tons: 520 standard; 590 full load
Dimensions, feet (metres): 196.8 × 27.6 × 6.9 *(60 × 8.8 × 2.3)*
Main machinery: 2 CXZ MAN B&W Type 9L 20-27 diesels; 2,400 hp *(1.8 MW)* sustained; 2 shafts;
 cp props
Speed, knots: 14. **Range, n miles:** 3,000 at 10 kt
Complement: 70 (10 officers)

Guns: 4 China 37 mm/63 (2 twin); 180 rds/min to 8.5 km *(4.6 n miles)*; weight of shell 1.42 kg.
 4—25 mm/60 (2 twin); 270 rds/min to 3 km *(1.6 n miles)*.
 4 China 14.5 mm/93 (2 twin); 600 rds/min to 7 km *(3.8 n miles)*.
Depth charges: 2 BMB-2 projectors; 20 depth charges.
Mines: Can carry 12-16.
Countermeasures: MCMV; MPT-1 paravanes; MPT-3 mechanical sweep; acoustic and magnetic
 gear.
Radars: Surface search: Fin Curve; I-band.
Sonars: Celcius Tech CMAS 36/39; active high frequency mine detection.

Programmes: Ordered from China in 1993.
Modernisation: New sonar fitted in 1998.
Structure: Based on Type 010G minesweeper design.
Operational: Used mostly as a patrol ship.
 VERIFIED

SAGAR
3/1998 / 0017594

SURVEY AND RESEARCH SHIPS

1 SURVEY SHIP (AGS)

Name	No	Builders	Commissioned
AGRADOOT (ex-*Kodan*)	A 583	Khulna Shipyard	19 Mar 2002

Displacement, tons: 687 full load
Dimensions, feet (metres): 157.0 × 25.6 × 11.5 *(47.8 × 7.8 × 3.5)*
Main machinery: 2 Baudouin diesels
Speed, knots: 12.5
Complement: 70 (8 officers)
Guns: 1 Oerlikon 20 mm.
Radars: Furuno HR 2110. Kelvin Hughes HR-3000A.

Comment: Former Thai trawler converted into a Survey vessel by Khulna shipyard. Fitted with two dual frequency digital hydrographic echo sounders, side-scan sonar and laboratories. Carries a survey launch.

UPDATED

AGRADOOT 6/2003, Bangladesh Navy / 0572419

AUXILIARIES

Notes: Floating Dock A 711 (*Sundarban*) acquired from Brodogradiliste Joso Lozovina-Mosor, Trogir, Yugoslavia in 1980; capacity 3,500 tons. Has a complement of 85 (5 officers). Floating crane A 731 (*Balaban*) is self-propelled at 9 kt and has a lift of 70 tons; built at Khulna Shipyard and commissioned 18 May 1988, she has a complement of 29 (two officers).

1 TANKER (AOTL)

KHAN JAHAN ALI A 515

Displacement, tons: 2,900 full load
Measurement, tons: 1,343 gross
Dimensions, feet (metres): 250.8 × 37.5 × 18.4 *(76.4 × 11.4 × 5.6)*
Main machinery: 1 diesel; 1,350 hp(m) *(992 kW)*; 1 shaft
Speed, knots: 12
Complement: 26 (3 officers)
Cargo capacity: 1,500 tons
Guns: 2 Oerlikon 20 mm.

Comment: Completed in Japan in 1983. Can carry out stern replenishment at sea but is seldom used in this role.

VERIFIED

KHAN JAHAN ALI 3/1998 / 0017595

1 TANKER (AOTL)

IMAN GAZZALI A 516

Displacement, tons: 213 full load
Dimensions, feet (metres): 146.8 × 23 × 11.2 *(44.8 × 7 × 3.4)*
Main machinery: 1 Cummins diesel; 1 shaft
Speed, knots: 8
Complement: 30 (2 officers)

Comment: An oil tanker of some 600,000 litres capacity acquired in 1996.

VERIFIED

IMAN GAZZALI 6/1999, Bangladesh Navy / 0056556

1 REPAIR SHIP (YR)

SHAHAYAK A 512

Displacement, tons: 477 full load
Dimensions, feet (metres): 146.6 × 26.2 × 6.6 *(44.7 × 8 × 2)*
Main machinery: 1 Cummins 12 VTS 6 diesel; 425 hp *(317 kW)*; 1 shaft
Speed, knots: 11.5. **Range, n miles:** 3,800 at 11.5 kt
Complement: 45 (1 officer)
Guns: 1 Oerlikon 20 mm.

Comment: Re-engined and modernised at Khulna Shipyard and commissioned in 1978 to act as repair vessel.

VERIFIED

SHAHAYAK 6/1996, Bangladesh Navy / 0056557

1 TENDER (AG)

SHAHJALAL A 513

Displacement, tons: 600 full load
Dimensions, feet (metres): 131.8 × 29.7 × 12.6 *(40.2 × 9.1 × 3.8)*
Main machinery: 1 V 16-cyl type diesel; 1 shaft
Speed, knots: 12. **Range, n miles:** 7,000 at 12 kt
Complement: 55 (3 officers)
Guns: 2 Oerlikon 20 mm.

Comment: Ex-Thai fishing vessel SMS *Gold 4*. Probably built in Tokyo. Commissioned on 15 January 1987 and used as a diving/salvage tender.

VERIFIED

SHAHJALAL 6/1996, Bangladesh Navy / 0056558

1 HARBOUR TENDER (YAG)

SANKET

Displacement, tons: 80 full load
Dimensions, feet (metres): 96.5 × 20 × 5.9 *(29.4 × 6.1 × 1.8)*
Main machinery: 2 Deutz diesels; 2,400 hp(m) *(1.76 MW)*; 2 shafts
Speed, knots: 16. **Range, n miles:** 1,000 at 16 kt
Complement: 16 (1 officer)
Guns: 1 Oerlikon 20 mm.

Comment: A former MFV taken over in 1989 and used as a utility harbour craft. No pennant number has been allocated. A second vessel of this type *Shamikha* is in civilian service.

VERIFIED

SANKET 3/1996 / 0056559

1 LANDING CRAFT LOGISTIC (LSL)

SHAH AMANAT L 900

Displacement, tons: 366 full load
Dimensions, feet (metres): 154.2 × 34.1 × 8 *(47 × 10.4 × 2.4)*
Main machinery: 2 Caterpillar D 343 diesels; 730 hp *(544 kW)* sustained; 2 shafts
Speed, knots: 9.5
Complement: 31 (3 officers)
Military lift: 150 tons
Guns: 2—12.7 mm MGs.

Comment: Australian civil vessel confiscated by the Navy while engaged in smuggling in 1988. Transferred to the Navy and commissioned in 1990.

VERIFIED

SHAH AMANAT   *6/1996, Bangladesh Navy* / 0056562

2 LCU 1512 CLASS (LCU)

SHAH PORAN (ex-*Cerro Gordo*) L 901   **SHAH MAKHDUM** (ex-*Cadgel*) L 902

Displacement, tons: 375 full load
Dimensions, feet (metres): 134.9 × 29 × 6.1 *(41.1 × 8.8 × 1.9)*
Main machinery: 4 Detroit 6-71 diesels; 696 hp *(508 kW)* sustained; 2 shafts
Speed, knots: 11. **Range, n miles:** 1,200 at 8 kt
Complement: 14 (2 officers)
Military lift: 170 tons
Guns: 2—12.7 mm MGs.
Radars: Navigation: LN 66; I-band.

Comment: Ex-US Army landing craft transferred in April 1991 and commissioned 16 May 1992 after refit.

VERIFIED

SHAH MAKHDUM   *6/1996, Bangladesh Navy* / 0056563

5 YUCH'IN CLASS (TYPE 068/069) (LCU/LCP)

DARSHAK A 581	**TALLASHI** A 582
LCT 101 A 584	**LCT 102** A 585 **LCT 104** A 587

Displacement, tons: 85 full load
Dimensions, feet (metres): 81.2 × 17.1 × 4.3 *(24.8 × 5.2 × 1.3)*
Main machinery: 2 Type 12V 150 diesels; 600 hp(m) *(440 kW)*; 2 shafts
Speed, knots: 11.5. **Range, n miles:** 450 at 11.5 kt
Complement: 23
Military lift: Up to 150 troops *(L 101-104)*
Guns: 4 China 14.5 mm (2 twin) MGs can be carried.

Comment: Named craft transferred from China in 1983 fitted with survey equipment and used as inshore survey craft. Second pair transferred 4 May 1986; third pair 1 July 1986. Probably built in the late 1960s. Two badly damaged in April 1991 typhoon and LCT 103 was scrapped.

UPDATED

LCT 101   *2/1992, Bangladesh Navy* / 0056561

TALLASHI (survey)   *6/2003, Bangladesh Navy* / 0572421

3 LCVP

L 011	L 012	L 013

Displacement, tons: 83 full load
Dimensions, feet (metres): 69.9 × 17.1 × 4.9 *(21.3 × 5.2 × 1.5)*
Main machinery: 2 Cummins diesels; 730 hp *(544 kW)*; 2 shafts
Speed, knots: 12
Complement: 10 (1 officer)

Comment: First two built at Khulna Shipyard and *013* at DEW Narayangong; all completed in 1984.

VERIFIED

L 011   *6/1996, Bangladesh Navy* / 0056564

TUGS

1 HUJIU CLASS (OCEAN TUG) (ATA)

KHADEM A 721

Displacement, tons: 1,472 full load
Dimensions, feet (metres): 197.5 × 38 × 16.1 *(60.2 × 11.6 × 4.9)*
Main machinery: 2 LVP 24 diesels; 1,800 hp(m) *(1.32 MW)*; 2 shafts
Speed, knots: 14. **Range, n miles:** 7,200 at 14 kt
Complement: 56 (7 officers)
Guns: 2—12.7 mm MGs.
Radars: Navigation: China Type 756; I-band.

Comment: Commissioned 6 May 1984 after transfer from China.

VERIFIED

KHADEM   *6/1996, Bangladesh Navy* / 0056565

3 COASTAL TUGS (YTM)

SEBAK A 722	**RUPSHA** A 723	**SHIBSHA** A 724

Displacement, tons: 330 full load
Dimensions, feet (metres): 99.9 × 28.1 × 1.6 *(30.0 × 8.4 × 3.5)*
Main machinery: 2 Caterpillar 12V 3512B diesels; 2,700 hp *(2.0 MW)*; 2 shafts
Speed, knots: 12. **Range, n miles:** 1,800 at 12 kt
Complement: 23 (3 officers)
Guns: 2—7.62 mm MGs (fitted for).

Comment: Details are for *Rupsha* and *Shibsha* built to a Damen Stan Tug 3008 design by Khulna Shipyard. Construction started in 2001, completed in 2003 and commissioned on 3 October 2004. *Sebak* built in Narayangang Dockyard in 1993 and commissioned on 23 December 1993.

UPDATED

SHIBSHA   *6/2003, Bangladesh Navy* / 0572422

Barbados

Country Overview

Barbados gained independence in 1966; the British monarch, represented by a governor-general, is head of state. The easternmost island of the Windward Islands of the Lesser Antilles chain, it consists of a single island of 166 square miles. The capital, largest town and principal port is Bridgetown, located on the southwestern coast. Territorial seas (12 n miles) are claimed. A 200 n mile Exclusive Economic Zone (EEZ) has also been claimed but the limits are not defined. A Coast Guard was formed in 1973 and became the naval arm of the Barbados Defence Force in 1979.

Headquarters Appointments

Chief of Staff, Barbados Defence Force:
  Colonel H D Maynard
Commanding Officer Coast Guard Squadron:
  Lieutenant Commander D A Dowridge

Personnel

(a)   2005: 96 (11 officers)
(b)   Voluntary service

Bases

Bridgetown (HMBS *Willoughby Fort*)

Prefix to Ships' Names

HMBS

PATROL FORCES

1 KEBIR CLASS (LARGE PATROL CRAFT) (PB)

Name	No	Builders	Launched	Commissioned
TRIDENT	P 01	Brooke Marine	14 Apr 1981	Nov 1981

Displacement, tons: 155.5 standard; 190 full load
Dimensions, feet (metres): 123 × 22.6 × 5.6 *(37.5 × 6.9 × 1.7)*
Main machinery: 2 Paxman Valenta 12CM diesels; 5,000 hp *(3.73 MW)* sustained; 2 shafts
Speed, knots: 29. **Range, n miles:** 3,000 at 12 kt
Complement: 28
Guns: 2—12.7 mm MGs. 2—12.7 mm MGs.
Radars: Surface search: Racal Decca Bridgemaster; I-band.

Comment: Refitted by Bender Shipyard in 1990 when the old guns were removed. Refitted again by Cable Marine in 1998 after a main engine seized. Same hull as Algerian Kebir class.
VERIFIED

5 INSHORE PATROL CRAFT (PB)

Comment: Two Boston Whaler *(P 08* and *P 09)* 22 ft craft; speed 25 kt; commissioned early 1989. One Zodiac Hurricane 24 ft, speed 25 kt commissioned July 1995. A third *(PO 10)* Boston Whaler acquired in late 1996. A Zodiac 920 RHIB was donated by the US in 2004. All used for law enforcement.
UPDATED

2 DAUNTLESS CLASS (INSHORE PATROL CRAFT) (PB)

Name	No	Builders	Commissioned
ENDEAVOUR	P 04	SeaArk Marine, Monticello	Dec 1997
EXCELLENCE	P 05	SeaArk Marine, Monticello	Apr 1999

Displacement, tons: 11 full load
Dimensions, feet (metres): 40 × 14 × 4.3 *(12.2 × 4.3 × 1.3)*
Main machinery: 2 Caterpillar 3208TA diesels; 870 hp *(650 kW)*; 2 shafts
Speed, knots: 27. **Range, n miles:** 600 at 18 kt
Complement: 4
Guns: 1—12.7 mm MG. 1—7.62 mm MG.
Radars: Surface search: Raytheon R40; I-band.

Comment: Aluminium construction. One further craft is sought.
VERIFIED

ENDEAVOUR   12/2002, *Judy Ross* / 0528291

TRIDENT   12/2002, *Judy Ross* / 0528292

Belgium

Country Overview

The Kingdom of Belgium is situated in north-western Europe. With an area of 11,787 square miles, it is bordered to the north by the Netherlands and to the south by France. It has a 35 n mile coastline with the North Sea. The capital and largest city is Brussels while the principal port is Antwerp, which is accessible via the Schelde and Meuse estuaries, which lie within the Netherlands. Antwerp is also connected to an extensive canal system. Territorial seas (12 n miles) are claimed and an EEZ has also been claimed.

Headquarters Appointments

Commander, Maritime Command:
  Rear Admiral W Goethals
Deputy Commander, Maritime Command:
  Captain E Verbrugghe

Diplomatic Representation

Defence Attaché in London:
  Colonel D De Cock

Personnel

(a)   2005: 2,566
(b)   Voluntary service

Bases

Zeebrugge: Frigates, MCMV, Reserve Units, Training Ships, Logistics, Diving Centre. Mine Warfare Operational Sea Test centre (MOST).
Oostende: Belgium-Netherlands Mine-warfare school (EGUERMIN).
Koksijde: Naval aviation.
Brugge: Naval training centre.

DELETIONS

Frigates

2004   *Wandelaar* (to Bulgaria)

Mine Warfare Forces

2004   *Myosotis*

Auxiliaries

2003   *Zinnia* (reserve)

FRIGATES

Notes: Plans to acquire MultiPurpose Escort Vessels (MPEVs) of 4,000 to 6,000 tons have been abandoned in favour of modernisation of the two remaining Wielingen class and acquisition of two second-hand frigates.

2 WIELINGEN CLASS (TYPE E-71) (FFGM)

Name	No	Builders	Laid down	Launched	Commissioned
WIELINGEN	F 910	Boelwerf, Temse	5 Mar 1974	30 Mar 1976	20 Jan 1978
WESTDIEP	F 911	Cockerill, Hoboken	2 Sep 1974	8 Dec 1975	20 Jan 1978

Displacement, tons: 1,940 light; 2,430 full load
Dimensions, feet (metres): 349 × 40.3 × 18.4
  (106.4 × 12.3 × 5.6)
Main machinery: CODOG; 1 RR Olympus TM3B gas-turbine;
  25,440 hp (19 MW) sustained; 2 Cockerill 240 CO V 12
  diesels; 6,000 hp(m) (4.4 MW); 2 shafts; LIPS; cp props
Speed, knots: 26; 15 on 1 diesel; 20 on 2 diesels
Range, n miles: 4,500 at 18 kt; 6,000 at 15 kt
Complement: 159 (13 officers)

Missiles: SSM: 4 Aerospatiale MM 38 Exocet (2 twin) launchers
  ❶; inertial cruise; active radar homing to 42 km (23 n miles) at
  0.9 Mach; warhead 165 kg; sea-skimmer.
  SAM: Raytheon Sea Sparrow RIM-7P; Mk 29 octuple launcher **❷**;
  semi-active radar homing to 14.6 km (8 n miles) at 2.5 Mach;
  warhead 39 kg.
Guns: 1 Creusot-Loire 3.9 in (100 mm)/55 Mod 68 **❸**; 80 rds/
  min to 17 km (9 n miles) anti-surface; 8 km (4.4 n miles) anti-
  aircraft; weight of shell 13.5 kg.
Torpedoes: 2—21 in (533 mm) launchers. ECAN L5 Mod 4; anti-
  submarine; active/passive homing to 9.5 km (5 n miles) at
  35 kt; warhead 150 kg; depth to 550 m (1,800 ft).

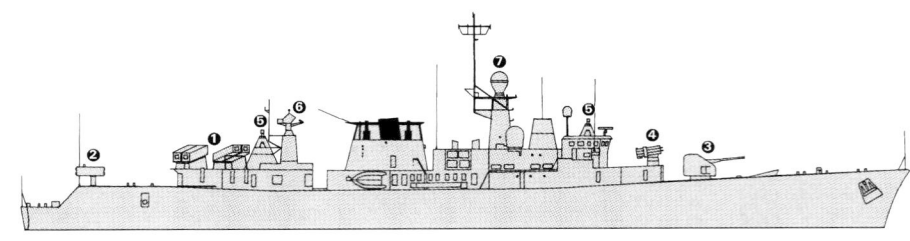

WESTDIEP
(Scale 1 : 900), Ian Sturton / 1047857

A/S Mortars: 1 Creusot-Loire 375 mm 6-barrelled trainable
  launcher **❹**; Bofors rockets to 1,600 m; warhead 107 kg.
Countermeasures: Decoys: 2 Tracor MBA SRBOC 6-barrelled
  Mk 36 launchers; chaff (Mk 214 Seagnat) and IR flares to
  4 km (2.2 n miles).
  Nixie SLQ-25; towed anti-torpedo decoy.
  ESM: Argos AR 900; intercept.

Combat data systems: Signaal SEWACO IV action data
  automation; Link 11. SATCOM.
Weapons control: Sagem Vigy 105 optronic director **❺**.
Radars: Air/surface search: Signaal DA05 **❻**; E/F-band.
  Surface search/fire control: Signaal WM25 **❼**; I/J-band.
  Navigation: Signaal Scout; I/J-band.
  IFF: Mk XII.
Sonars: Computing Devices Canada SQS 510; hull-mounted;
  active search and attack; medium frequency.

Programmes: This compact, well-armed class of frigate was
  designed by the Belgian Navy and built in Belgian yards. The
  programme was approved on 23 June 1971 and an order
  placed in October 1973.
Modernisation: A rolling programme of upgrades started in
  1996 and is planned to be completed in 2005. Sea Sparrow
  has been updated from 7M to 7P. WM25 radar has been
  modified to include improved ECCM and MTI capabilities and
  a new navigation radar and sonar have been fitted. New
  optronic director, IFF and communications facilities have also
  been installed. Finally, new diesel engines (2 ABC 12V DZC
  (2,915 kW each)) and alternators are to be fitted. Hull
  modernisation has included measures to reduce RCS. A
  further modernisation package is expected.
Structure: Fully air conditioned. Fin stabilisers fitted.
Operational: Based at Zeebrugge.

UPDATED

WESTDIEP
11/2004 *, B Prézelin / 1047871

WIELINGEN
4/2003, B Prézelin / 0568850

WESTDIEP
11/2004 *, B Prézelin / 1047872

MINE WARFARE FORCES

6 FLOWER CLASS (TRIPARTITE)
(MINEHUNTERS—COASTAL) (MHC/AEL)

Name	No	Builders	Launched	Commissioned
ASTER	M 915	Beliard, Ostend	6 June 1985	17 Dec 1985
BELLIS	M 916	Beliard, Ostend	14 Feb 1986	14 Aug 1986
CROCUS	M 917	Beliard, Ostend	6 Aug 1986	5 Feb 1987
LOBELIA	M 921	Beliard, Ostend	6 Jan 1988	9 May 1989
NARCIS	M 923	Beliard, Ostend	30 Mar 1990	27 Sep 1990
PRIMULA	M 924	Beliard, Ostend	17 Dec 1990	29 May 1991

Displacement, tons: 562 standard; 595 full load
Dimensions, feet (metres): 168.9 × 29.2 × 8.2 *(51.5 × 8.9 × 2.5)*
Main machinery: 1 Stork Wärtsilä A-RUB 215W-12 diesel; 1,860 hp(m) *(1.37 MW)* sustained; 1 shaft; LIPS cp prop; 2 motors; 240 hp(m) *(176 kW)*; 2 active rudders; 2 bow thrusters
Speed, knots: 15. **Range, n miles:** 3,000 at 12 kt
Complement: 46 (5 officers)

Guns: 1 DCN 20 mm/20; 720 rds/min to 10 km *(5.5 n miles)*. 2—12.7 mm MGs.
Countermeasures: MCM: 2 PAP 104 remote-controlled mine locators; 39 charges. Mechanical sweep gear (medium depth).
Radars: Navigation: Racal Decca 1229; I-band.
Sonars: Thomson Sintra DUBM 21B; hull-mounted; active minehunting; 100 kHz ± 10 kHz.

Programmes: Developed in co-operation with France and the Netherlands. A 'ship factory' for the hulls was built at Ostend and the hulls were towed to Rupelmonde for fitting out. Each country built its own hulls but France provided all MCM gear and electronics, Belgium electrical installation and the Netherlands the engine room equipment.
Modernisation: Propulsion system upgrade completed in 1999 for all of the class. Capability upgrade to extend service life of six ships to 2020 is planned. Modifications include an MCM command and control system, an Integrated Mine Countermeasures System (comprising hull-mounted and self-propelled variable-depth sonar (installed in Double Eagle Mk III Mod 1 ROV)) and a Mine-Identification and Disposal System (MIDS) based on the STN Atlas Seafox. The equipment will be first installed in HNMS *Hellevoetsluis*. BNS *Lobelia* is to be the first Belgium ship to be upgraded from 2005.
Structure: GRP hull fitted with active tank stabilisation, full NBC protection and air conditioning. Has automatic pilot and buoy tracking.
Operational: A 5 ton container can be carried, stored for varying tasks-HQ support, research, patrol, extended diving, drone control. The ship's company varies from 33 to 46 depending on the assigned task. Six divers are carried when minehunting. All of the class are based at Zeebrugge.
Sales: Three of the class paid off for sale in July 1993 and were bought by France in 1997.

UPDATED

LOBELIA   6/2004*, J Ciślak / 1044083

PRIMULA   3/2004*, Derek Fox / 1044082

PATROL FORCES

Notes: Three 7 m RIC were acquired in May 1994 from RIBTEC, Swanwick.

1 GRIFFON 2000 TDX CLASS (HOVERCRAFT) (UCAC)

BARBARA   A 999

Displacement, tons: 3.5 full load
Dimensions, feet (metres): 38.4 × 19.4 *(11.7 × 5.9)*
Main machinery: 1 Deutz BFBL diesel; 350 hp(m) *(235 kW)*
Speed, knots: 35. **Range, n miles:** 450 at 35 kt
Complement: 3
Radars: Navigation: Furuno 7010 D; I-band.

Comment: Built by Griffon Hovercraft, Southampton and delivered in 1995. Aluminium hull. Owned and operated by the Belgian Army for drone recovery and artillery range patrol.

VERIFIED

BARBARA   7/2003, A A de Kruijf / 0567879

1 RIVER PATROL CRAFT (PBR/YFLB)

Name	No	Builders	Launched	Commissioned
LIBERATION	P 902	Hitzler, Regensburg	29 July 1954	4 Aug 1954

Displacement, tons: 45 full load
Dimensions, feet (metres): 85.5 × 13.1 × 3.2 *(26.1 × 4 × 1)*
Main machinery: 2 MWM diesels; 440 hp(m) *(323 kW)*; 2 shafts
Speed, knots: 19
Complement: 7
Guns: 2—12.7 mm MGs.
Radars: Navigation: Racal Decca; I-band.

Comment: Laid down 12 March 1954. Paid off 12 June 1987 but put back in active service 15 September 1989 after repairs. Last of a class of 10 used for patrol and personnel transport. Replacement planned when funds are available.

VERIFIED

LIBERATION   7/2003, A A de Kruijf / 0568846

SHIPBORNE AIRCRAFT

Numbers/Type: 3 Aerospatiale SA 316B Alouette III.
Operational speed: 113 kt *(210 km/h)*.
Service ceiling: 10,500 ft *(3,200 m)*.
Range: 290 n miles *(540 km)*.
Role/Weapon systems: CG helicopter; used for close-range search and rescue and support for commando forces. Sensors: Carries Thomson-CSF search radar. Weapons: Unarmed. It is planned to upgrade these aircraft with new navigation and communications systems.

VERIFIED

ALOUETTE III   7/2000, van Ginderen Collection / 0104215

ALOUETTE III   7/2001, van Ginderen Collection / 0114692

LAND-BASED MARITIME AIRCRAFT

Numbers/Type: 5 Westland Sea King Mk 48.
Operational speed: 140 kt *(260 km/h).*
Service ceiling: 10,500 ft *(3,200 m).*
Range: 630 n miles *(1,165 km).*
Role/Weapon systems: SAR helicopter; operated by air force; used for surface search and combat rescue tasks. Upgraded in 1995 with new radar, FLIR and GPS. Sensors: Bendix RDR 1500B search radar. FLIR 2000F. Weapons: Unarmed.

VERIFIED

SEA KING   *7/2001, van Ginderen Collection /* 0114691

AUXILIARIES

Notes: It is planned to acquire a Command and Support Ship (MCS) to replace BNS *Godetia* in about 2012.

1 COMMAND AND SUPPORT SHIP (AGFH)

Name	No	Builders	Launched	Commissioned
GODETIA	A 960	Boelwerf, Temse	7 Dec 1965	23 May 1966

Displacement, tons: 2,000 standard; 2,260 full load
Dimensions, feet (metres): 301 × 46 × 11.5 *(91.8 × 14 × 3.5)*
Main machinery: 4 ACEC-MAN diesels; 5,400 hp(m) *(3.97 MW);* 2 shafts; cp props
Speed, knots: 19. **Range, n miles:** 8,700 at 12.5 kt
Complement: 105 (8 officers)
Guns: 6—12.7 mm MGs.
Radars: Surface search: Racal Decca 1229; I-band.
Helicopters: 1 Alouette III.

Comment: Laid down 15 February 1965. Rated as Command and Logistic Support Ship. Refit (1979-80) and mid-life conversion (1981-82) included helicopter deck and replacement cranes. Refitted again in 1992. Minesweeping cables fitted either side of helo deck have been removed. Can also serve as a Royal Yacht. To be replaced by new ship in about 2012.

UPDATED

GODETIA   *5/2002, W Sartori /* 0533277

1 SUPPORT SHIP (AGFH)

Name	No	Builders	Commissioned
STERN (ex-*KBV 171*)	A 963	Karlskronavarvet	3 Sep 1980

Displacement, tons: 375 full load
Dimensions, feet (metres): 164 × 27.9 × 7.9 *(50 × 8.5 × 2.4)*
Main machinery: 2 Hedemora V16A diesels; 4,480 hp(m) *(3.28 MW)* sustained; 2 shafts; cp props
Speed, knots: 18. **Range, n miles:** 3,000 at 12 kt
Complement: 13
Radars: Navigation: 2 Kelvin Hughes; E/F- and I-band.
Helicopters: Platform for 1 light.

Comment: Transferred from Swedish Coast Guard on 6 October 1998. GRP hull indentical to Landsort class. In Swedish service the ship carried a 20 mm gun, and had a Subsea sonar. Used for fishery protection and SAR duties.

UPDATED

STERN   *7/2003*, M Declerck /* 1044078

TUGS

2 COASTAL TUGS (YTM)

VALCKE (ex-*Steenbank,* ex-*Astroloog*) A 950   ALBATROS (ex-*Westgat*) A 996

Displacement, tons: 183 full load
Dimensions, feet (metres): 99.7 × 24.9 × 11.8 *(30.4 × 7.6 × 3.6)*
Main machinery: Diesel-electric; 2 Deutz diesel generators; 1,240 hp(m) *(911 kW);* 1 shaft; 1 bow thruster
Speed, knots: 11
Complement: 8

Comment: Details given are for A 950 which was launched in 1960. A 996 is 206 tons and was launched in 1967.

UPDATED

ALBATROS   *7/2000, van Ginderen Collection /* 0104220

3 HARBOUR TUGS (YTL)

WESP A 952   ZEEMEEUW A 954   MIER A 955

Displacement, tons: 195 full load
Dimensions, feet (metres): 86.5 × 24.7 × 10.7 *(26.23 × 7.5 × 3.25)*
Main machinery: 2 ABC 6 MDUS diesels; 1,000 hp *(746 kW)*
Speed, knots: 11
Complement: 4

Comment: Details given are for A 952 and A 955. A 954 is 146 tons.

UPDATED

MIER   *7/2004*, Manuel Declerck /* 1047870

SURVEY SHIPS

Notes: In addition to *Belgica* there are five small civilian manned survey craft: *Ter Streep, Scheldewacht II, De Parel II, Veremans* and *Prosper.*

1 SURVEY SHIP (AGOR/PBO)

Name	No	Builders	Launched	Commissioned
BELGICA	A 962	Boelwerf, Temse	6 Jan 1984	5 July 1984

Displacement, tons: 1,085 full load
Dimensions, feet (metres): 167 × 32.8 × 14.4 *(50.9 × 10 × 4.4)*
Main machinery: 1 ABC 6M DZC diesel; 1,600 hp(m) *(1.18 MW)* sustained; 1 Kort nozzle prop
Speed, knots: 13.5. **Range, n miles:** 5,000 at 12 kt
Complement: 26 (11 civilian)
Radars: Navigation: Racal Decca 1229; I-band.

Comment: Ordered 1 December 1982. Laid down 17 October 1983. Used for hydrography, oceanography, meteorology and fishery control. Marisat fitted. Based at Zeebrugge. Painted white.

UPDATED

BELGICA   *6/2004*, B Prézelin /* 1044079

TRAINING SHIPS

1 SAIL TRAINING VESSEL (AXS)

Name	No	Builders	Commissioned
ZENOBE GRAMME	A 958	Boel and Zonen, Temse	27 Dec 1961

Displacement, tons: 149 full load
Dimensions, feet (metres): 92 × 22.5 × 7 *(28 × 6.8 × 2.1)*
Main machinery: 1 MWM diesel; 200 hp(m) *(147 kW)*; 1 shaft
Speed, knots: 10
Complement: 14 (2 officers)
Radars: Navigation: Racal Decca; I-band.

Comment: Auxiliary sail ketch. Laid down 7 October 1960 and launched 23 October 1961. Designed for scientific research but now only used as a training ship.

UPDATED

ZENOBE GRAMME
7/2003 *, P Marsan / 1044080

Belize

Country Overview

Formerly known as British Honduras, Belize became an independent state in 1981. The British monarch, represented by a governor-general, is head of state. With an area of 8,867 square miles, it has borders with Mexico to the north and Guatemala to the west; its 208 n mile coastline is on the Caribbean Sea and fringed by numerous coral barrier reefs and cays. The capital city is Belmopan while the largest city and major port is Belize City. Territorial seas (12 n miles) are claimed. A 200 n mile Exclusive Economic Zone (EEZ) has been claimed but the limits are not defined. Transformation of the Defence Force Maritime Wing into a Coast Guard awaits legislation.

Headquarters Appointments

Commanding Officer Defence Force Maritime Wing:
Major J F Teck

Personnel

(a)   2005: 45 (2 officers)
(b)   The Maritime Wing of the Belize Defence Force comprises volunteers from the Army.

Bases

Ladyville, Hunting Cay, Calabash Cay (planned)

Maritime Patrol

Two Pilatus Britten-Norman Defenders are used for maritime surveillance.

PATROL FORCES

Notes: Transformation of the Maritime Wing into a Coast Guard awaits legislation. Current assets include:

(a)   Two Halmatic 22 ft RIBs with twin Yamaha 115 hp outboards. Names *Stingray Commando* and *Blue Marlin Ranger*.
(b)   Two Pelikan 35 ft craft with twin Yamaha 200 hp outboards. Built at Bradleys Boatyard in 1996 and called *Ocean Sentinel* and *Reef Sniper*.
(c)   Six Colombian 32 ft skiffs with twin Yamaha 200 hp outboards, confiscated and commissioned in service 1995-97.
(d)   One 36 ft skiff.

PELIKAN CRAFT
6/2001, Belize Defence Force / 0109932

Benin
FORCES NAVALES

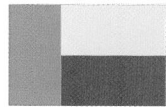

Country Overview

Formerly part of French West Africa, the republic gained full independence in 1960 as the Republic of Dahomey; it was renamed The Republic of Benin in 1975. With an area of 43,484 square miles it has borders to the east with Nigeria and to the west with Togo. Benin has a short coastline of 65 n miles with the Gulf of Guinea. The capital is Porto-Novo while Cotonou is the largest city and principal port. Benin has not claimed an Exclusive Economic Zone (EEZ) but is one of a few coastal states which claims a 200 n miles territorial sea. The naval force was established in 1978.

Headquarters Appointments

Commander Navy:
Capitaine de Vaisseau Soulémane Marcos

Aircraft

Dornier Do 128 and a DHC-6 Twin Otter reconnaissance aircraft are used for surveillance.

Bases

Cotonou

Personnel

2005: 220 (30 officers)

PATROL FORCES

Notes: There are two French-built 6 m river patrol craft with hydrojet propulsion.

2 CHINESE 27 METRE CLASS (PATROL CRAFT) (PB)

MATELOT BRICE KPOMASSE 798   **LA SOTA** 799

Displacement, tons: 55
Dimensions, feet (metres): 88.6 × 13.1 × 3.9 *(27 × 4 × 1.2)*
Main machinery: 2 diesels; 1,000 hp *(746 kW)*
Complement: 13
Guns: 4—14.5 mm (2 twin) MGs.
Radars: Navigation: I-band.

Comment: Understood to have been transferred from China in 2000.

VERIFIED

KPOMASSE and SOTA
2001, Benin Navy / 0114348

Bermuda

Country Overview

A British self-governing dependency, a Governor, appointed by the British Crown, is responsible for external affairs, internal security, defence, and the police. Situated in the north Atlantic Ocean some 650 n miles southeast of Cape Hatteras, the country consists of six principal islands, of which the largest is 14 miles long, linked by bridges and a causeway; there are some 150 other small islands, islets, and rocks, of which about 20 are inhabited. Hamilton is the capital, chief port and largest town. Territorial seas (12 n miles) and an Exclusive Economic Zone (EEZ) (200 n miles) are claimed.

Headquarters Appointments

Commanding Officer:
Inspector Mark Bothello

Bases

Hamilton

POLICE

Notes: In addition to patrol craft, three tugs, *Powerful, Faithful* and *Refit* are operated by the Department of Marine and Port Services.

1 PATROL CRAFT (PBI)

BLUE HERON

Comment: Donated by the US Drug Enforcement Agency in May 1996 to replace the original craft of the same name. 46 ft *(14 m)* in length and fitted with a Furuno radar. Complement six. The craft is used by the Joint Marine Interdiction Team to patrol inshore waters to intercept drug runners. Replacement by a 16 m craft is expected by 2005.

UPDATED

HERON II   *6/1997, Bermuda Police /* 0012079

2 SAR CRAFT (SAR)

RESCUE I   RESCUE II

Comment: *Rescue I* replaced the craft of the same name in November 1998 and *Rescue II* replaced the craft of the same name in 2001. Both are Halmatic 24 ft Arctic RIBs with twin 200 hp Yamaha outboards and a complement of three.

VERIFIED

BLUE HERON   *5/1996, Bermuda Police /* 0056583

4 PATROL CRAFT (PBI)

HERON I   HERON II   HERON III   HERON IV

Comment: *Heron I*, delivered in July 1997 to replace the previous craft of the same name, and *Heron III* delivered in June 1992 are 22 ft Boston Whalers fitted with twin Yamaha 225 hp and twin Yamaha 115 hp outboards, respectively. *Heron II* delivered in August 1996 to replace the previous craft of the same name, is a 27 ft Boston Whaler with twin Yamaha 250 hp(m) outboard engines. *Heron IV*, delivered in 2001, is a further 22 ft Boston Whaler with twin 115 hp outboards.

VERIFIED

HALMATIC ARCTIC RIB   *2001, Bermuda Police /* 0109933

Bolivia
ARMADA BOLIVIANA

Country Overview

The Republic of Bolivia is one of two landlocked countries in South America; Paraguay is the other. With an area of 424,165 square miles, it has borders to the north and east with Brazil, to the southeast with Paraguay, to the south with Argentina, and to the west with Chile and Peru. It has a 211 n mile shoreline with Lake Titicaca. The constitutional capital is Sucre while the administrative capital and seat of government is La Paz which is connected by railway to the Chilean port of Antofagasta.

The Bolivian Navy was founded in 1963 and received its present name in 1982. Its purpose is to patrol some 10,000 miles in three geographical areas. The Amazon basin includes the rivers Ichilo, Mamore, Itenez, Yacuma, Orthon, Abuna, Beni and Madre de Dios. The central basin comprises Lake Titicaca while the Del Plata basin includes the rivers Paraguay and Bermejo. Most advanced training is carried out in Argentina and Peru.

Headquarters Appointments

Commandant General of the Navy:
Admiral Luis Alberto Aranda Granados
Chief of the Naval Staff:
Vice Admiral Marco Antonio Justiniano Escalante

Personnel

(a)   2005: 6,659 (including Marines)
(b)   12 months' selective military service

Organisation

The country is divided into six naval districts, three naval areas and a Fuerza de Tareas Especiales.
1st Naval District (Beni) (HQ Riberalta). River Beni.
2nd Naval District (Mamore) (HQ Trinidad). Rivers Ichilo and Mamore.
3rd Naval District (Madera) (HQ Puerto Guayamerin). Rivers Madera and Itenez.
4th Naval District (Titicaca) (HQ San Pedro de Tiquina). Lake Titicaca.
5th Naval District (Santa Cruz de la Sierra) (HQ Puerto Quijarro). River Paraguay.
6th Naval District (Pando) (HQ Cobija). Rivers Acre, Madre dos Dios and Tahuamanu.
1st Naval Area (Cochabamba) (Puerto Villarroel). Naval yard and oil transport.

Organisation — *continued*

2nd Naval Area (Santa Cruz). Support duties.
3rd Naval Area (Bermejo).
4th Naval Area (La Paz).

Fuerza de Tareas Especiales consists of five task groups (based at Guayamerin, Cobija, Riberalta, Puerto Suarez and Copacabana) to provide support in counter-drug operations.

Marine Corps

The Bolivian Navy has seven marine corps battalions (BIM I-VII) located in the Naval Districts.

Prefix to Ships' Names

ARB

PATROL FORCES

Notes: There are 28 Rodman transport craft of 6-16 m. All were delivered in 1998.

3 RIVER PATROL CRAFT (PBR)

CAPITÁN PALOMEQUE PR 221　　**ANTOFAGASTA** PR 302　　**GENERAL BANZER** PR 301

Displacement, tons: 8 full load
Dimensions, feet (metres): 42.7 × 10.5 × 1.6 *(13 × 3.2 × 0.5)*
Main machinery: 2 diesels; 2 shafts
Speed, knots: 27
Complement: 4
Guns: 1—7.62 mm MG.

Comment: Details given are for *Capitán Palomeque* acquired in 1993. The others are similar in appearance and all are less than ten years old. Operate in the 2nd and 3rd Districts.
VERIFIED

CAPITÁN PALOMEQUE　　　　　　　　　　*1996, Bolivian Navy /* 0056585

1 SANTA CRUZ CLASS (PBR)

SANTA CRUZ DE LA SIERRA PR 501

Displacement, tons: 46 full load
Dimensions, feet (metres): 68.9 × 19 × 3.9 *(21 × 5.8 × 1.2)*
Main machinery: 2 Detroit diesels; 2 shafts
Speed, knots: 20. **Range, n miles:** 800 at 16 kt
Complement: 10
Guns: 2—12.7 mm MGs.
Radars: Surface search: Furuno; I-band.

Comment: Built by Hope Shipyards, Louisiana, in 1985. Used both as a patrol craft and supply ship. Operates in the 5th District on the river Paraguay.
VERIFIED

SANTA CRUZ DE LA SIERRA (old number)　　　*1996, Bolivian Navy /* 0056584

8 RIVER PATROL CRAFT (PBR)

PAZ ZAMORA LP 101　　**MARISCAL DE ZAPITA** LP 409　　**GUAQUI** LA 414
RAIDER LP 351　　**CAPITÁN BRETEL** LP 410　　**INDEPENDENCIA** LP 416
GENERAL BEJAR LP 406　　**TENIENTE SOLIZ** LP 411

Displacement, tons: 5 full load
Dimensions, feet (metres): 42.3 × 12.7 × 3.3 *(12.9 × 3.9 × 1)*
Main machinery: 2 diesels; 2 shafts
Speed, knots: 15
Complement: 5
Guns: 1—12.7 mm MG.
Radars: Surface search: Raytheon; I-band.

Comment: Details given are for *Capitán Bretel, Teniente Soliz* and *Guaqui* which is used as a logistic craft. The remainder are Boston Whaler types. All operate in the 4th District except *Paz Zamora* (1st) and *Raider* (5th).
UPDATED

CAPITÁN BRETEL alongside TENIENTE SOLIZ　　*1996, Bolivian Navy /* 0056587

42 RIVER PATROL CRAFT (PBR)

LP 01-42

Comment: Thirty-two Piranas were delivered from 1992-96. Fitted with one 12.7 mm MG and has twin outboards. Ten more craft delivered by the US 1998-99.
UPDATED

PIRANA Mk II　　　　　　　　　　　　　*1996, Bolivian Navy /* 0056588

AUXILIARIES

Notes: (1) Approximately 30 Rodman craft are used for transport and logistic support. A mixture of 17 m, 11 m, 8 m and 6 m were commissioned on 11 February 1999.
(2) *Guayamerin* (TNTB-01) is an LCM used as a transport vessel on Lake Titicaca. Built in Bolivia she was commissioned on 22 July 1998.
(3) A dredge *Pirai II* (FNDR-01) was commissioned on 11 August 2001.

11 RIVER TRANSPORTS (YFL)

ALMIRANTE GRAU M 101　　　　**INGENIERO GUMUCIO** M 341
COMANDANTE ARANDIA M 103　　**JORGE VILLARROEL** M 342
GERMAN BUSCH M 107　　　　　**COATI** M 401
LIBERTADOR M 223　　　　　　**COBIJA** M 402
TRINIDAD M 224　　　　　　　**SUAREZ ARANA** M 501
RIO GUAPORÉ M 301

Displacement, tons: 70 full load
Dimensions, feet (metres): 78.7 × 21.3 × 4.6 *(24 × 6.5 × 1.4)*
Speed, knots: 12. **Range, n miles:** 500 at 12 kt
Complement: 11
Radars: Navigation: Raytheon; I-band.

Comment: Details given are for *Ingeniero Gumucio* which is a troop transport and supply ship. The remainder are craft of various types, some acquired from China.
UPDATED

INGENIERO GUMUCIO　　　　　　　　　*1996, Bolivian Navy /* 0056589

6 LOGISTIC VESSELS (YAG)

JOSE MANUEL PANDO TNR 01	**JULIO OLMOS** TNR 05
NICOLAS SUAREZ TNR 02	**HORACIO UGARTECHE** TNBTL-06
MAX PAREDES TNR 04	**THAMES CRESPO** TNR 07

Comment: TNR-01 is a tug. The remainder are pusher/lighter combinations. There are eight
lighters TNBTP-02A, -02B, -04A, -04B, -05A, -06A, -06B and -07A.   ***VERIFIED***

MAX PAREDES   6/2000, Bolivian Navy / 0104222

1 TRAINING VESSEL (YXT)

BUQUE ESCUELA NAVAL MILITAR

Displacement, tons: 80 full load
Dimensions, feet (metres): 117.3 × 29.5 × 3.9 *(35.7 × 9.0 × 1.2)*
Main machinery: 2 diesels; 1,300 hp *(969 kW)*
Speed, knots: 18
Complement: 15 plus 50 trainees

Comment: Catamaran design. Launched at Tiquina, Lake Titicaca on 9 May 2001. Commissioned
on 24 April 2004.   ***UPDATED***

2 HOSPITAL SHIPS

Name	No	Tonnage
JULIAN APAZA	TNBH 401	150
XAVIER PINTO TELLERIA	TNBH 01	—

Comment: *Julian Apaza* given by the US; assembled in 1972 and based at Lake Titicaca. *Telleria*
was built in 1997 and is based at Puerto Villarod.   ***VERIFIED***

TELLERIA   6/2000, Bolivian Navy / 0104223

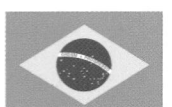

Brazil
MARINHA DO BRASIL

Country Overview

The Federal Republic of Brazil is the largest country in South
America. With an area of 3,286,500 square miles it has borders
to the north with Colombia, Venezuela, Guyana, Suriname and
French Guiana, to the south with Uruguay and to the west with
Argentina, Paraguay, Bolivia, and Peru. It has a coastline of
4,045 n miles with the south Atlantic Ocean. There are some
23,220 n miles of internal waterways that consist primarily of
the Amazon and its tributaries; the river is navigable by ocean-
going ships from its mouth to Iquitos in Peru. The capital is
Brasilia while the largest city is São Paulo. The principal ports are
the former capital, Rio de Janeiro, Santos, Paranaguá, Recife, and
Vitória. Manaus is an important river port. Territorial seas (12 n
miles) are claimed. An EEZ (200 n miles) is claimed and its limits
have been partly defined by boundary agreements.

Headquarters Appointments

Commander of the Navy:
    Admiral Roberto de Guimarães Carvalho
Chief of Naval Staff:
    Admiral Rayder Alencar da Silveira
Chief of Naval Operations:
    Mauro Magalhães de Souza Pinto
Commandant General Brazilian Marine Corps:
    Admiral (Marine Corps) Marcelo Gaya Cardoso Tosta
General Director of Personnel:
    Admiral Julio Soares de Moura Neto
General Director of Material:
    Admiral Euclides Duncan Janot de Matos
General Secretary of Navy:
    Admiral Kleber Luciano de Assis
Vice Chief of Naval Staff:
    Vice Admiral Gerson Carvalho Ravanelli

Senior Officers

Commander-in-Chief, Fleet:
    Vice Admiral Carlos Augusto Vasconcelos Saraiva Ribeiro
Commander, Fleet Marine Force:
    Vice Admiral (Marine Corps) Álvaro Augusto Dias Monteiro
Commander, I Naval District:
    Vice Admiral José Eduardo Pimental de Oliveira
Commander, II Naval District:
    Vice Admiral Álvaro Luiz Pinto
Commander, III Naval District:
    Vice Admiral Afonso Barbosa
Commander, IV Naval District:
    Vice Admiral Marcus Vinicius Oliveira dos Santos
Commander, V Naval District:
    Vice Admiral Luiz Umberto de Mendonça
Commander, VI Naval District:
    Rear Admiral Carlos Augusto de Sousa
Commander, VII Naval District:
    Rear Admiral Newton Cardoso
Commander, VIII Naval District
    Vice Admiral Marcelio Carmo de Castro Pereira
Commander, Occidental Amazonia Naval Command:
    Rear Admiral Marcus Vinícius Iório Hollanda

Diplomatic Representation

Naval Attaché in USA and Canada:
    Rear Admiral Edison Lawrence Mariath Dantas
Naval Attaché in United Kingdom, Sweden and Norway:
    Captain Paulo Fontes da Rocha Vianna
Naval Attaché in Uruguay:
    Captain Hamilton Jorge da Gama Henrique
Naval Attaché in France and Belgium:
    Captain Carlos Alberto Pêgas
Naval Attaché in Italy:
    Captain Moacyr Cavichiolo Filho

Diplomatic Representation — *continued*

Naval Attaché in Germany and Netherlands:
    Commander Ernesto Martins Tavares de Souza
Naval Attaché in South Africa and Mozambique:
    Captain José Carlos Mathias
Naval Attaché in Bolivia:
    Captain (Marine Corps) Augusto Cesar Lobato Pousada
Naval Attaché in Argentina:
    Captain Julio Cesar da Costa Fonseca
Naval Attaché in Venezuela:
    Captain (Marine Corps) Jorge de Oliveira Carlos
Naval Attaché in Peru:
    Captain (Marine Corps) Alexandre José Barreto da Mattos
Naval Attaché in Portugal:
    Captain Girlano Bezerra Santiago Freitas
Naval Attaché in Chile:
    Captain Marco Antônio Soares Garrido
Defence Attaché in Japan and Indonesia:
    Captain Carlos Alberto Auffinger
Defence Attaché in China and South Korea:
    Captain Bernardo Augusto Cunha de Hollanda
Defence Attaché in Russia:
    Captain Sergio Luiz Coutinho

Personnel

(a)   2005: 36,700 (5,900 officers) Navy; (including 1,300
      naval air)
      14,600 (690 officers) Marines
(b)   One year's national service

Bases

Arsenal de Marinha do Rio de Janeiro – Rio de Janeiro (Naval
shipyard with three dry docks and one floating dock with graving
docks of up to 70,000 tons capacity)
Base Naval do Rio de Janeiro – Rio de Janeiro (Main Naval Base
with two dry docks)
Base Almirante Castro e Silva – Rio de Janeiro (Naval Base for
submarines)
Base Naval de Aratu – Bahia (Naval Base and repair yard with one
dry dock and synchrolift)
Base Naval de Val-de-Cães – Pará (Naval River and repair yard
with one dry dock)
Base Naval de Natal – Rio Grande do Norte (Small Naval Base and
repair yard with one floating dock)
Base Fluvial de Ladário – Mato Grosso do Sul (Small Naval River
Base and repair yard with one dry dock)
Base Aérea Naval de São Pedro d'Aldeia – Rio de Janeiro (Naval
Air Station)
Estação Naval do Rio Negro – Amazonas (Small Naval River
Station and repair yard with one floating dock)
Estação Naval do Rio Grande – Rio Grande do Sul (Small Naval
Station and repair yard)

Organisation

Naval Districts as follows:
I Naval District (HQ Rio de Janeiro)
II Naval District (HQ Salvador)
III Naval District (HQ Natal)
IV Naval District (HQ Belém)
V Naval District (HQ Rio Grande)
VI Naval District (HQ Ladário)
VII Naval District (HQ Brasilia)
VIII Naval District (HQ São Paulo)
Comando Naval da Amazonia Ocidental (HQ Manaus)

Prefix to Ships' Names

These vary, indicating the type of ship for example, N Ae =
Aircraft Carrier; CT = Destroyer.

Naval Aviation

Squadrons: São Pedro da Aldeira; HA-1 Super Lynx; HS-1 Sea
King; HI-1 JetRanger; HU-1 Ecureuil 1 and 2; HU-2 Super Puma/
Cougar; VF 1 Skyhawk AF1.
    Manaus; HU-3 Ecureuil.
    Ladário; HU-4 Jet Ranger.
    Rio Grande; HU-5 Ecureuil.

Pennant Numbers

As a result of cuts in Officer numbers, ships are sometimes
formally decommissioned from the Navy and lose their pennant
numbers. They are then retained in service as tenders to Naval
establishments and are commanded by Warrant Officers.

Marines (Corpo de Fuzileiros Navais)

Headquarters at Fort São José, Rio de Janeiro
Divisão Anfibia: 3 Infantry Battalions (Riachuelo, Humaita and
Paissandu), 1 Artillery Battalion, 1 HQ Company, 1 Air Defence
Battery, 1 Tank Company.
Tropa de Reforço: 1 Engineer Battalion, 1 Amphib Vehicles
Battalion, 1 Logistic Battalion.
Special Forces Battalion (Tonelero).
Grupamentos Regionais: One security group in each naval district
and command (Rio de Janeiro, Salvador, Natal, Belém, Rio
Grande, Ladário, Manaus, Brasilia).

Strength of the Fleet

Type	Active	Building (Planned)
Submarines (Patrol)	4	1
Aircraft Carrier	1	—
Frigates	10	—
Corvettes	4	1
Patrol Forces	35	—
LSD/LST	3	—
Minesweepers (Coastal)	6	—
Survey and Research Ships	13	—
Buoy Tenders	17	—
S/M Rescue Ship	1	—
Tankers	2	—
Hospital Ships	3	—
Training Ships	8	—

DELETIONS

Frigates

2002   *Paraíba, Paraná*
2004   *Dodsworth, Pernambuco* (both reserve)

Patrol Forces

2002   *Bahiana, Purus*
2003   *Solimóes* (museum)
2004   *Angostura, Caboclo*

Survey Ships and Tenders

2002   *Barão de Teffé*
2003   *Almirante Câmara*
2004   *Caravelas, Itacurassá, Nogueira da Gama*

Auxiliaries

2002   *Custódio de Mello*

PENNANT LIST

Submarines		Amphibious Forces			
S 30	Tupi	G 28	Mattoso Maia		
S 31	Tamoio	G 30	Ceará		
S 32	Timbira	G 31	Rio de Janeiro		
S 33	Tapajó	L 10	Guarapari		
S 34	Tikuna (bldg)	L 11	Tambaú		
		L 12	Camboriú		

Patrol Forces	
V 15	Imperial Marinheiro
P 10	Piratini
P 11	Pirajá
P 12	Pampeiro
P 13	Parati
P 14	Penedo
P 15	Poti
P 20	Pedro Teixeira
P 21	Raposo Tavares
P 30	Roraima
P 31	Rondônia
P 32	Amapá
P 40	Grajaú
P 41	Guaiba
P 42	Graúna
P 43	Goiana
P 44	Guajará
P 45	Guaporé
P 46	Gurupá
P 47	Gurupi
P 48	Guanabara
P 49	Guarujá
P 50	Guaratuba

P 51	Gravataí
P 60	Bracui
P 61	Benevente
P 62	Bocaina
P 63	Babitonga

Mine Warfare Forces	
M 15	Aratú
M 16	Anhatomirim
M 17	Atalaia
M 18	Araçatuba
M 19	Abrolhos
M 20	Albardão

Survey Ships and Tenders	
H 18	Comandante Varella
H 19	Tenente Castelo
H 20	Comandante Manhães
H 21	Sirius
H 25	Tenente Boanerges
H 26	Faroleiro Mário Seixas
H 31	Argus
H 34	Almirante Graça Aranha
H 35	Amorim do Valle
H 36	Taurus
H 37	Garnier Sampaio
H 40	Antares
H 44	Ary Rongel
SSN-4 03	Paraibano
SSN-4 04	Rio Branco
BHMN 03	Camocin

Auxiliaries	
G 15	Paraguassú
G 17	Potengi
G 21	Ary Parreiras
G 23	Almirante Gastao Motta
G 27	Marajo
G 40	Atlântico Sud
K 11	Felinto Perry
R 21	Tritão
R 22	Tridente
R 23	Triunfo
R 24	Almirante Guilhem
R 25	Almirante Guillobel
R 26	Trindade
U 10	Aspirante Nascimento
U 11	Guarda Marinha Jensen
U 12	Guarda Marinha Brito
U 16	Doutor Montenegro
U 17	Parnaiba
U 18	Oswaldo Cruz
U 19	Carlos Chagas
U 20	Cisne Branco
U 27	Brasil
U 29	Piraim

Aircraft Carriers	
A 12	São Paulo

Destroyers/Frigates	
D 27	Pará
F 40	Niteroi
F 41	Defensora
F 42	Constituição
F 43	Liberal
F 44	Independência
F 45	União
F 46	Greenhalgh
F 48	Bosisio
F 49	Rademaker

Corvettes	
V 30	Inhaúma
V 31	Jaceguay
V 32	Julio de Noronha
V 33	Frontin
V 34	Barroso (bldg)

SUBMARINES

Notes: (1) Plans for the construction of nuclear-powered submarines continue although the programme has been constrained by lack of funding. The prototype nuclear reactor IPEN/MB-1 built at Aramar, Iperó, São Paulo is expected to be in service in 2016. An uranium enrichment plant was inaugurated at Iperó in April 1988. Published plans of the prototype SSN (SNAC-2) are of a boat of about 2,825 tons with power plant developing 48 MW for a speed of 28 kt. Entry into service is not expected before 2025.
(2) Design of a new SSK, designated S-MB-10 is in progress. Five boats of about 2,500 tons, 67 m length and 8 m beam are planned. The construction programme has not been published although it is thought that building of the first class is to begin in 2007 and all five are required by 2018.

0 + 1 (1) TIKUNA CLASS (SSK)

Name	No	Builders	Laid down	Launched	Commissioned
TIKUNA (ex-*Tocantins*)	S 34	Arsenal de Marinha, Rio de Janeiro	11 June 1996	Mar 2005	2006

Displacement, tons: 1,490 surfaced; 1,620 dived
Dimensions, feet (metres): 200.2 × 20.3 × 18
(61 × 6.2 × 5.5)
Main machinery: Diesel-electric; 4 MTU 12V 396 diesels; 3,760 hp(m) *(2.76 MW)*; 4 alternators; 1 motor; 1 shaft
Speed, knots: 11 surfaced/snorting; 22 dived
Range, n miles: 11,000 at 8 kt surfaced; 400 at 4 kt dived
Complement: 36 (7 officers)

Torpedoes: 8—21 in *(533 mm)* bow tubes. Bofors Torpedo 2000; wire guided active/passive homing to 50 km *(27 n miles)* at 20-50 kt; warhead 250 kg. Swim-out discharge. IPqM designed A/S torpedoes may also be carried; 18 km *(9.7 n miles)* at 45 kt. Total of 16 torpedoes.
Mines: 32 IPqM/Consub MCF-01/100 carried in lieu of torpedoes.
Countermeasures: ESM: Thomson-CSF DR-4000; intercept.
Weapons control: STN Atlas Electronik ISUS 83-13; 2 Kollmorgen Mod 76 periscopes.

Radars: Navigation: Terma Scanter; I-band.
Sonars: Atlas Elektronik CSU-83/1; hull-mounted; passive/active search and attack; medium frequency.

Programmes: Planned intermediate stage between Tupi class and the first SSN. Designed by the Naval Engineering Directorate. Contract effective with HDW in October 1995. A second of class, possibly to be named *Tapuia* is likely to be built. Construction is expected to start in 2006 and delivery in 2012.
Structure: Improved Tupi design similar to Turkish Gur class. Diving depth, 300 m *(985 ft)*. Very high-capacity batteries with GRP lead-acid cells by Microlite. More powerful engines than *Tupi*. Fitted with two Kollmorgen Mod 76 non-penetrative optronic masts.
Operational: Endurance, 60 days.

UPDATED

TIKUNA
6/2004*, Brazilian Navy / 1044088

4 TUPI CLASS (209 TYPE 1400) (SSK)

Name	No	Builders	Laid down	Launched	Commissioned
TUPI	S 30	Howaldtswerke-Deutsche Werft, Kiel	8 Mar 1985	28 Apr 1987	6 May 1989
TAMOIO	S 31	Arsenal de Marinha, Rio de Janeiro	15 July 1986	18 Nov 1993	12 Dec 1994
TIMBIRA	S 32	Arsenal de Marinha, Rio de Janeiro	15 Sep 1987	5 Jan 1996	16 Dec 1996
TAPAJÓ	S 33	Arsenal de Marinha, Rio de Janeiro	6 Mar 1996	5 June 1998	16 Nov 1999

Displacement, tons: 1,453 surfaced; 1,590 dived
Dimensions, feet (metres): 200.8 × 20.3 × 18
(61.2 × 6.2 × 5.5)
Main machinery: Diesel-electric; 4 MTU 12V 493 AZ80 GA31L diesels; 2,400 hp(m) *(1.76 MW)*; 4 alternators; 1.7 MW; 1 Siemens motor; 4,600 hp(m) *(3.36 MW)* sustained; 1 shaft
Speed, knots: 11 surfaced/snorting; 21.5 dived
Range, n miles: 8,200 at 8 kt surfaced; 400 at 4 kt dived
Complement: 36 (7 officers)

Torpedoes: 8—21 in *(533 mm)* bow tubes. 16 Marconi Mk 24 Tigerfish Mod 1 or 2; wire-guided; active homing to 13 km *(7 n miles)* at 35 kt; passive homing to 29 km *(15.7 n miles)* at 24 kt; warhead 134 kg. IPqM anti-submarine torpedoes may also be carried; range 18 km *(9.7 n miles)* at 45 kt. Swim-out discharge.
Countermeasures: ESM: Thomson-CSF DR-4000; radar warning.
Weapons control: Ferranti KAFS-A10 action data automation. 2 Kollmorgen Mod 76 periscopes.
Radars: Navigation: Terma Scanter; I-band.
Sonars: Atlas Elektronik CSU-83/1; hull-mounted; passive/active search and attack; medium frequency.

Programmes: Contract signed with Howaldtswerke in February 1984. Financial negotiations were completed with the West German Government in October 1984. Original plans included building four in Brazil followed by two improved Tupis for a total of six. In the end only three were constructed in Brazil.
Modernisation: A programme to upgrade auxiliary machinery, sonars, weapon control, countermeasures and navigation systems was announced in 2003.
Structure: Hull constructed of HY 80 steel. Single hull. Diving depth, 250 m *(820 ft)*. Equipped with Sperry Mk 29 Mod 3 SINS.
Operational: Based at Niteroi, Rio de Janeiro.

UPDATED

TUPI

6/2003, Mario R V Carneiro / 0569159

AIRCRAFT CARRIERS

1 CLEMENCEAU CLASS (CVM)

Name	No	Builders	Laid down	Launched	Commissioned
SÃO PAULO (ex-Foch)	A 12 (ex-R 99)	Chantiers de l'Atlantique, St. Nazaire	15 Feb 1957	23 July 1960	15 July 1963

Displacement, tons: 27,307 standard; 33,673 full load
Dimensions, feet (metres): 869.4 oa; 780.8 pp × 104.1 hull (168 oa) × 28.2 *(265; 238 × 31.7 (51.2) × 8.6)*
Flight deck, feet (metres): 850 × 154 *(259 × 47)*
Main machinery: 6 boilers; 640 psi *(45 kg/cm²)*; 840°F *(450°C)*; 2 GEC Alsthom turbines; 126,000 hp(m) *(93 MW)*; 2 shafts
Speed, knots: 30. **Range, n miles:** 7,000 at 18 kt; 4,800 at 24 kt; 3,500 at full power
Complement: 1,220 (80 officers); 358 (80 officers) aircrew

Missiles: SAM: 2 AESN Albatros Mk 2 (8 cells, 2 reloads); Aspide 2000 missiles; semi-active homing to 21 km (11 n miles) at 2.5 Mach; warhead 30 kg (to be fitted).
Guns: 2 Bofors SAK 40 mm/L 70-600 Mk 3 Sea Trinity; 330 rds/min to 4 km *(2.2 n miles)* (to be fitted).
5—12.7 mm MGs.
Countermeasures: 2 CSEE AMBL 2A Sagai (10 barrelled trainable launchers); chaff and IR flares.
Combat data systems: To be fitted with IPqM/Elebra SICONTA Mk 1 tactical system; Links YB and 14.
Radars: Air search: Thomson-CSF DRBV 23B ❶; D-band.
Air/surface search: Thomson-CSF DRBV 15 ❷; E/F-band.
Height finder: 2 DRBI 10 ❸; E/F-band.
Navigation: Racal Decca 1226; I-band.
Fire control: 2 AESN Orion RTN 30X; I/J-band (to be fitted).
Tacan: NRBP-2B.
Landing approach control: NRBA 51 ❹; I-band.

Fixed-wing aircraft: 15-18 A-4 Skyhawks.
Helicopters: 4-6 Agusta SH-3A/D Sea Kings; 3 Aerospatiale UH-12/13; 2 UH-14 Cougar.

Programmes: Acquired from France on 15 November 2000 and following modifications in Brest, arrived in Brazil in February 2001.
Modernisation: A foldable mini ski-jump has been fitted to both catapults. The jet deflectors are enlarged (this implies reducing the area of the forward lift). Crotale and Sadral systems disembarked before transfer. Aspide SAM may be fitted. An Umkhonto VLS system is also reported to be under consideration. Refit in 2003 included re-tubing of boilers and refurbishment of catapults.
Structure: Flight deck, island superstructure and bridges, hull (over machinery spaces and magazines) are all armour plated. There are three bridges: Flag, Command and Aviation.
Two Mitchell-Brown steam catapults; Mk BS 5; able to launch 20 ton aircraft at 110 kt. The flight deck is angled at 8°. Two lifts 52.5 × 36 ft *(16 × 10.97 m)* one of which is on the starboard deck edge. Dimensions of the hangar are 590.6 × 78.7 × 23 ft *(180 × 24 × 7 m)*.

SÃO PAULO · 9/2003, S C Neto/Mario R V Carneiro / 0569158

Operational: Oil fuel capacity is 3,720 tons. The aircraft complement for the helicopter carrier role includes between 30 and 40 with a mixture of Sea Kings, Super Puma, Super-Lynx, Ecureuil and Jet Ranger III.   **UPDATED**

SÃO PAULO · · · · · · · · · · · · · · · 2/2001, Mario R V Carneiro / 0059751

SÃO PAULO · · · · · · · · · · · · · · · 2/2001, Mario R V Carneiro / 0059752

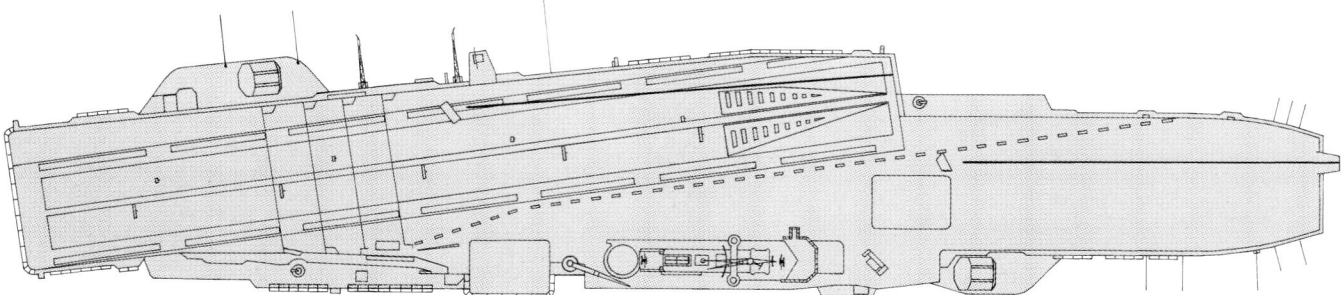

SÃO PAULO · (Scale 1 : 1,500), Ian Sturton / 0529159

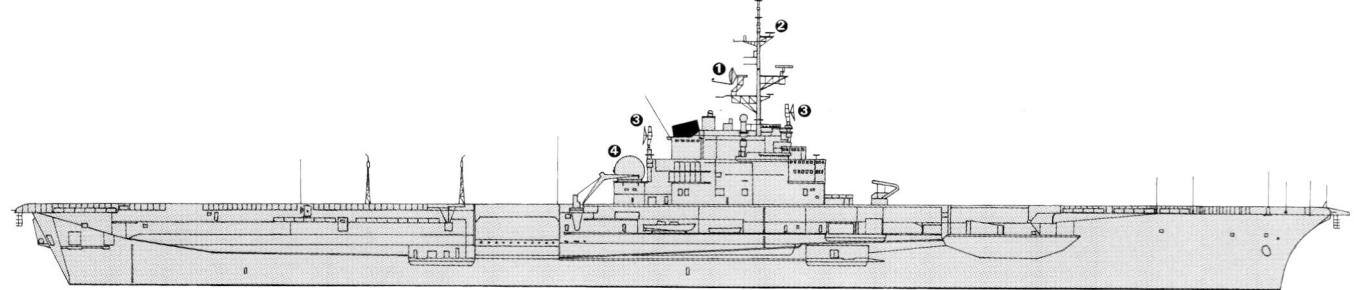

SÃO PAULO · (Scale 1 : 1,500), Ian Sturton / 0130381

SÃO PAULO

2/2001, Mario R V Carneiro / 0059750

FRIGATES

3 BROADSWORD CLASS (TYPE 22) (FFGHM)

Name	No	Builders	Laid down	Launched	Commissioned	Recommissioned
GREENHALGH (ex-*Broadsword*)	F 46 (ex-F 88)	Yarrow Shipbuilders, Glasgow	7 Feb 1975	12 May 1976	3 May 1979	30 June 1995
BOSISIO (ex-*Brazen*)	F 48 (ex-F 91)	Yarrow Shipbuilders, Glasgow	18 Aug 1978	4 Mar 1980	2 July 1982	31 Aug 1996
RADEMAKER (ex-*Battleaxe*)	F 49 (ex-F 89)	Yarrow Shipbuilders, Glasgow	4 Feb 1976	18 May 1977	28 Mar 1980	30 Apr 1997

Displacement, tons: 3,500 standard; 4,731 full load
Dimensions, feet (metres): 430 oa; 410 wl × 48.5 × 19.9 (screws) *(131.2; 125 × 14.8 × 6)*
Main machinery: COGOG; 2 RR Olympus TM3B gas turbines; 50,000 hp *(37.3 MW)* sustained; 2 RR Tyne RM1C gas turbines; 9,900 hp *(7.4 MW)* sustained; 2 shafts; cp props
Speed, knots: 30; 18 on Tynes
Range, n miles: 4,500 at 18 kt on Tynes
Complement: 239 (17 officers)

Missiles: SSM: 4 Aerospatiale MM 38 Exocet (F 48, F 49) **❶**; inertial cruise; active radar homing to 42 km *(23 n miles)* at 0.9 Mach; warhead 165 kg; sea-skimmer. 4 Aerospatiale MM 40 (F 46); inertial cruise; active radar homing to 70 km *(40 n miles)* at 0.9 Mach; warhead 165 kg.

SAM: 2 British Aerospace 6-barrelled Seawolf GWS 25 Mod 4 **❷**; command line of sight (CLOS) TV/radar tracking to 5 km *(2.7 n miles)* at 2+ Mach; warhead 14 kg; 32 rounds.
Guns: 2 Bofors SAK 40 mm/L 70-350 A-3 **❸** (F 46 only); 300 rds/min to 12 km *(6.5 n miles)*
2 Oerlikon BMARC 20 mm GAM-BO1; 1,000 rds/min to 2 km.
Torpedoes: 6—324 mm Plessey STWS Mk 2 (2 triple) tubes **❹**. Honeywell Mk-46 Mod 5; active/passive homing to 11 km *(5.9 n miles)* at 40 kt; warhead 44 kg.
Countermeasures: Decoys: 4 Loral Hycor SRBOC Mk 36; 6-barrelled fixed launchers **❺**; for chaff.
Graseby Type 182; towed torpedo decoy.
ESM: MEL UAA-2; intercept.
ECM: Type 670; jammers.

Combat data systems: CAAIS; Link YB being fitted. Inmarsat.
Weapons control: GWS 25 Mod 4 (for SAM); GWS 50 (Exocet).
Radars: Air/surface search: Marconi Type 967/968 **❻**; D/E-band.
Navigation: Kelvin Hughes Type 1006; I-band.
Fire control: Two Marconi Type 910 **❼**; I/Ku-band (for Seawolf).
Sonars: Plessey Type 2050; hull-mounted; search and attack; medium frequency.

Helicopters: 2 Westland Super Lynx AH-11A **❽**.

Programmes: Contract signed on 18 November 1994 to transfer four Batch I Type 22 frigates from the UK, one in 1995, two in 1996 and one in 1997. It is not planned to buy more Type 22s.
Modernisation: Plans to fit a single 57 mm gun on the bow were shelved in favour of a 40 mm gun on each beam. These guns are being taken from the Niteroi class. The modernisation programme has been suspended due to lack of funds. F 46 reported to have been upgraded with Exocet MM 40 in late 2003.
Structure: Accommodation modified in UK service to take 65 officers under training.
Operational: Primary role is ASW. Form part of Second Frigate Squadron at Niteroi, Rio de Janeiro. F 47 placed in reserve in 2004.

UPDATED

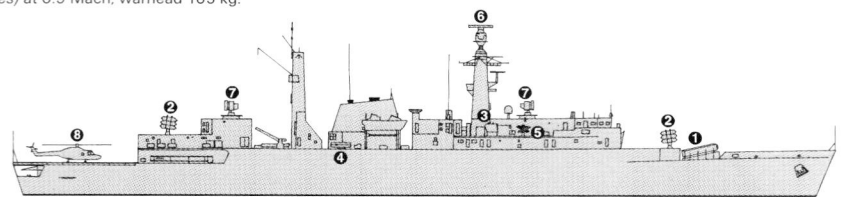

GREENHALGH   (Scale 1 : 1,200), Ian Sturton / 0012084

RADEMAKER   4/2000, Mario R V Carneiro / 0104224

GREENHALGH   2/2001, Mario R V Carneiro / 0130425

6 NITERÓI CLASS (FFGHM)

Name	No	Builders	Laid down	Launched	Commissioned
NITERÓI	F 40	Vosper Thornycroft Ltd	8 June 1972	8 Feb 1974	20 Nov 1976
DEFENSORA	F 41	Vosper Thornycroft Ltd	14 Dec 1972	27 Mar 1975	5 Mar 1977
CONSTITUIÇÃO	F 42	Vosper Thornycroft Ltd	13 Mar 1974	15 Apr 1976	31 Mar 1978
LIBERAL	F 43	Vosper Thornycroft Ltd	2 May 1975	7 Feb 1977	18 Nov 1978
INDEPENDÊNCIA	F 44	Arsenal de Marinha, Rio de Janeiro	11 June 1972	2 Sep 1974	3 Sep 1979
UNIÃO	F 45	Arsenal de Marinha, Rio de Janeiro	11 June 1972	14 Mar 1975	12 Sep 1980

Displacement, tons: 3,200 standard; 3,707 full load
Dimensions, feet (metres): 424 × 44.2 × 18.2 (sonar)
(129.2 × 13.5 × 5.5)
Main machinery: CODOG; 2 RR Olympus TM3B gas turbines;
50,880 hp *(37.9 MW)* sustained; 4 MTU 20V 1163 TB 93
diesels; 20,128 hp(m) *(14.8 MW)* sustained; 2 shafts; cp
props
Speed, knots: 30 gas; 22 diesels. **Range, n miles:** 5,300 at 17 kt
on 2 diesels; 4,200 at 19 kt on 4 diesels; 1,300 at 28 kt on gas
Complement: 209 (22 officers)

Missiles: SSM: 4 Aerospatiale MM 40 Exocet (2 twin) launchers
❶; inertial cruise; active radar homing to 70 km *(40 n miles)* at
0.9 Mach; warhead 165 kg; sea-skimmer.
SAM: AESN Albatros (8 cell, 2 reloads) ❷; Aspide 2000; semi-
active radar homing to 21 km *(11 n miles)* at 2.5 Mach.
Guns: 1 Vickers 4.5 in *(115 mm)*/55 Mk 8 ❸; 25 rds/min to
22 km *(12 n miles)* anti-surface; 6 km *(3.2 n miles)* anti-
aircraft; weight of shell 30 kg.
2 Bofors SAK 40 mm/L 70-600 Mk 3 Sea Trinity ❹; 330 rds/
min to 4 km *(2.2 n miles)*.
Torpedoes: 6—324 mm Plessey STWS-1 (2 triple) tubes ❺.
Honeywell Mk 46 Mod 5; anti-submarine; active/passive
homing to 11 km *(5.9 n miles)* at 40 kt; warhead 44 kg.
A/S mortars: 1 Bofors 375 mm trainable rocket launcher (twin-
tube) ❻; automatic loading; range 1,600 m.
Countermeasures: Decoys: 4 IPqM/Elebra MDLS octuples chaff
launchers ❼.
ESM: Racal Cutlass B-1B; intercept.
ECM: Racal Cygnus or IPqM/Elebra ET/SLQ-2X; jammer.
Combat data systems: IPqM/Elebra Siconta II. Link YB.

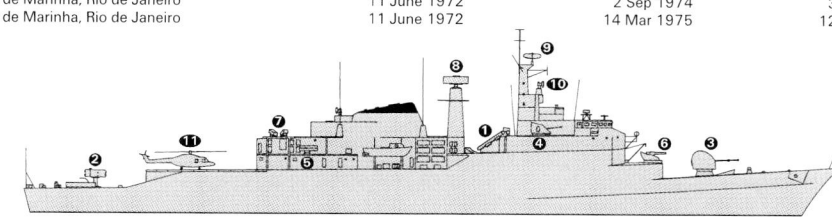

LIBERAL (after modernisation)   *(Scale 1 : 1,200), Ian Sturton /* 0529160

Weapons control: Saab/Combitech EDS-400/10B optronic
director. WSA 401. FCS.
Radars: Air/surface search: AESN RAN 20 S (3L) ❽; D-band.
Surface search: Terma Scenter MiP ❾; I-band.
Fire control: 2 AESN RTN 30X ❿; I/J-band.
Navigation: Furuno FR-1942 Mk 2; I-band.
Sonars: EDO 997F; hull-mounted; active search and attack;
medium frequency.
EDO 700E VDS (F 40 and 41); active search and attack;
medium frequency.

Helicopters: 1 Westland Super Lynx AH-11A ⓫.

Programmes: A contract announced on 29 September 1970
was signed between the Brazilian Government and Vosper
Thornycroft for the design and building of six Vosper
Thornycroft Mark 10 frigates. Seventh ship with differing
armament was ordered from Navyard, Rio de Janeiro in June
1981 and is used as a training ship.

Modernisation: The modernisation plan first signed in March
1995 included replacing Seacat by Aspide, Plessey AWS
2 radar by Alenia RAN 20S, RTN 10X by RTN 30X,
ZW06 radar by Terma Scanter, new 40 mm mountings, new
EW equipment, combat data system and hull-mounted sonar.
Ikara removed. Work is being done by Elebra. *Liberal*
completed 2001. *Defensora* (2002), *Independência* (2004)
and *Niterói* (2004). *Constituição* and *União* are due to be
completed in 2005.
Structure: Originally F 40, 41, 44 and 45 were of the A/S
configuration. F 42 and 43 general purpose design. Fitted
with retractable stabilisers.
Operational: Endurance, 45 days' stores, 60 days' provisions.
The helicopter has Sea Skua ASM. All are based at Niterói and
form the First Frigate Squadron.

UPDATED

LIBERAL (after modernisation)   *2/2001, Mario R V Carneiro /* 0130422

NITERÓI (Seacat and Ikara removed)   *10/2002, Mario R V Carneiro /* 0528957

For details of the latest updates to *Jane's Fighting Ships* online and to discover the
additional information available exclusively to online subscribers please visit
jfs.janes.com

1 PARÁ (GARCIA) CLASS (FFHM)

Name	No	Builders	Laid down	Launched	Commissioned	Recommissioned
PARÁ (ex-*Albert David*)	D 27 (ex-FF 1050)	Lockheed SB & Construction Co	29 Apr 1964	19 Dec 1964	19 Oct 1968	18 Sep 1989

Displacement, tons: 2,620 standard; 3,560 full load
Dimensions, feet (metres): 414.5 × 44.2 × 24 sonar; 14.5 keel *(126.3 × 13.5 × 7.3; 4.4)*
Main machinery: 2 Foster-Wheeler boilers; 1,200 psi *(83.4 kg/ cm²)*; 950°F *(510°C)*; 1 Westinghouse or GE turbine; 35,000 hp *(26 MW)*; 1 shaft
Speed, knots: 27.5. **Range, n miles:** 4,000 at 20 kt
Complement: 286 (18 officers) + 25 spare

Missiles: A/S: Honeywell ASROC Mk 116 Mod 3 octuple launcher ❶; inertial guidance to 1.6-10 km *(1-5.4 n miles)*; payload Mk 46 torpedo. *Pará* has automatic ASROC reload system.

Guns: 2 USN 5 in *(127 mm)*/38 Mk 30 ❷; 15 rds/min to 17 km *(9.3 n miles)*; weight of shell 25 kg.
2—12.7 mm MGs.
Torpedoes: 6—324 mm Mk 32 (2 triple) tubes ❸. 14 Honeywell Mk 46 Mod 5; anti-submarine; active/passive homing to 11 km *(5.9 n miles)* at 40 kt; warhead 44 kg.
Countermeasures: Decoys: 2 Loral Hycor Mk 33 RBOC 4 barrelled chaff launchers. T-Mk 6 Fanfare; torpedo decoy system. Prairie/Masker; hull/blade rate noise suppression.
ESM: WLR-1; WLR-6; radar warning.
ECM: ULQ-6; jammer.
Combat data systems: IPqH/Elebra Mini-Siconta.

Weapons control: Mk 56 GFCS *(127 mm)*. Mk 114 ASW FCS. SATCOM.
Radars: Air search: Lockheed SPS-40B ❹; B-band; range 320 km *(175 n miles)*.
Surface search: Raytheon SPS-10C ❺; G-band.
Navigation: Marconi LN66; I-band.
Fire control: General Electric Mk 35 ❻; I/J-band.
Tacan: SRN 15. IFF: UPX XII.
Sonars: SQS-26B; bow-mounted; active search and attack; medium frequency.

Helicopters: Westland Super Lynx AH-11 ❼.

Programmes: Last remaining ship of three transferred from USA in 1989. Classified as Destroyer in the Brazilian Navy. The lease was renewed in 1994.
Structure: Enlarged hangar capable of taking a helicopter the size of a Sea King. In USN service had the flight deck area converted to take SQR-15 towed array which was removed on transfer.
Operational: Part of Second Escort Squadron at Niterói, Rio de Janeiro. Operational readiness is proving difficult to maintain. D 28 and D 29 decommissioned and D 30 placed in reserve.
UPDATED

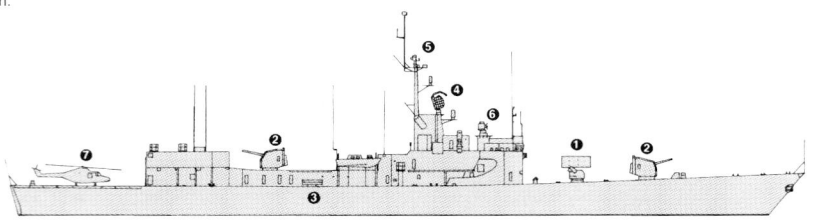

PARÁ

(Scale 1 : 1,200), Ian Sturton

PARÁ CLASS

2/2001, Mario R V Carneiro / 0130424

CORVETTES

0 + 1 BARROSO CLASS (FSGH)

Name	No	Builders	Laid down	Launched	Commissioned
BARROSO	V 34	Arsenal de Marinha, Rio de Janeiro	21 Dec 1994	20 Dec 2002	Apr 2008

Displacement, tons: 1,785 standard; 2,350 full load
Dimensions, feet (metres): 339.3 × 37.4 × 13.0; 17.4 (sonar) *(103.4 × 11.4 × 3.95; 5.3)*
Main machinery: CODOG; 1 GE LM 2500 gas turbine; 27,500 hp *(20.52 MW)* sustained; 2 MTU 20V 1163 TB83 diesels; 11,780 hp(m) *(8.67 MW)* sustained; 2 shafts; Kamewa cp props
Speed, knots: 29. **Range, n miles:** 4,000 at 15 kt
Complement: 160 (15 officers)

Missiles: SSM: 4 Aerospatiale MM 40 Exocet ❶; inertial cruise; active radar homing to 70 km *(40 n miles)* at 0.90 Mach; warhead 165 kg; sea-skimmer.
Guns: 1 Vickers 4.5 in *(115 mm)* Mk 8 ❷; 55° elevation; 25 rds/min to 22 km *(12 n miles)* anti-surface; 6 km *(3.3 n miles)* anti-aircraft; weight of shell 21 kg.
1 Bofors SAK Sea Trinity CIWS 40 mm/70 Mk 3 ❸; 330 rds/min to 4 km *(2.2 n miles)*; anti-aircraft; 2.5 km *(1.4 n miles)*

anti-missile; weight of shell 0.96 kg; with '3P' improved ammunition.
2—12.7 mm MGs.
Torpedoes: 6—324 mm Mk 32 (2 triple) tubes ❹; Honeywell Mk 46 Mod 5; anti-submarine; active/passive homing to 11 km *(5.9 n miles)* at 40 kt; warhead 44 kg.

Countermeasures: Decoys: 2 IPqM octuple chaff launchers ❺.
ESM: IPqM/Elebra ET/SLQ-1A ❻; radar warning.
ECM: IPqM/Elebra ET/SLQ-2 ❼; jammer.
Combat data systems: IPqM/Esca Siconta Mk III with Link YB.
Weapons control: Saab/Combitech EOS-400 FCS with optronic director ❽; two OFDLSE optical directors ❾.
Radars: Surface search: AESN RAN-20S ❿; F-band.
Navigation: Terma Scanter; I-band.
Fire control: AESN RTN-30-X ⓫; I/J-band (for Albatross and guns).
Sonars: EDO 997(F); hull-mounted; active; medium frequency.

Helicopters: 1 AH-11A Westland Super Lynx ⓬.

Programmes: Ordered in 1994 as a follow-on to the Inhauma programme. The building programme has been beset by funding difficulties and although a class of six ships is projected by 2018, construction of furthers units has not started.
Structure: The hull is some 4.2 m longer than the Inhauma class to improve sea-keeping qualities and allow extra space in the engine room. The design allows the use of containerised equipment to aid modernisation. Efforts have been made to incorporate stealth technology. Vosper stabilisers.
UPDATED

BARROSO

(Scale 1 : 900), Ian Sturton

BARROSO

5/2003, A E Galarce / 0572423

4 INHAÚMA CLASS (FSGH)

Name	No	Builders	Laid down	Launched	Commissioned
INHAÚMA	V 30	Arsenal de Marinha, Rio de Janeiro	23 Sep 1983	13 Dec 1986	12 Dec 1989
JACEGUAY	V 31	Arsenal de Marinha, Rio de Janeiro	15 Oct 1984	8 June 1987	2 Apr 1991
JULIO DE NORONHA	V 32	Verolme, Angra dos Reis	8 Dec 1986	15 Dec 1989	27 Oct 1992
FRONTIN	V 33	Verolme, Angra dos Reis	14 May 1987	6 Feb 1992	11 Mar 1994

Displacement, tons: 1,600 standard; 1,970 full load
Dimensions, feet (metres): 314.2 × 37.4 × 12.1; 17.4 (sonar)
(95.8 × 11.4 × 3.7; 5.3)
Main machinery: CODOG; 1 GE LM 2500 gas turbine; 27,500 hp
(20.52 MW) sustained; 2 MTU 16V 396 TB94 diesels;
5,800 hp(m) *(4.26 MW)* sustained; 2 shafts; Kamewa cp
props
Speed, knots: 27. **Range, n miles:** 4,000 at 15 kt
Complement: 133 (20 officers)

Missiles: SSM: 4 Aerospatiale MM 40 Exocet ❶; inertial cruise;
active radar homing to 70 km *(40 n miles)* at 0.9 Mach;
warhead 165 kg; sea-skimmer.
Guns: 1 Vickers 4.5 in *(115 mm)* Mk 8 ❷; 55° elevation; 25 rds/
min to 22 km *(12 n miles)* anti-surface; 6 km *(3.3 n miles)* anti-
aircraft, weight of shell 21 kg.
2 Bofors 40 mm/70 ❸; 300 rds/min to 12 km *(6.5 n miles)*
anti-surface; 4 km *(2.2 n miles)* anti-aircraft; weight of shell
0.96 kg.
Torpedoes: 6—324 mm Mk 32 (2 triple) tubes ❹. Honeywell Mk
46 Mod 5; anti-submarine; active/passive homing to 11 km
(5.9 n miles) at 40 kt; warhead 44 kg.
Countermeasures: Decoys: 2 Plessey Shield chaff launchers ❺;
fires chaff and IR flares in distraction, decoy or centroid
patterns.
ESM/ECM: Racal Cygnus B1 radar intercept ❻ and IPqM SDR-7
or Elebra SLQ-1; jammer ❼.

Combat data systems: Ferranti CAAIS 450/WSA 421; Link YB.
Weapons control: Saab EOS-400 FCS with optronic director ❽
and two OFDLSE optical ❾ directors.
Radars: Surface search: Plessey AWS 4 ❿; E/F-band.
Navigation: Kelvin Hughes Type 1007; I/J-band.
Fire control: Selenia Orion RTN 10X ⓫; I/J-band.
Sonars: Atlas Elektronik DSQS-21C; hull-mounted; active;
medium frequency.

Helicopters: 1 Westland Super Lynx ⓬ or UH-12/13 Ecureuil.

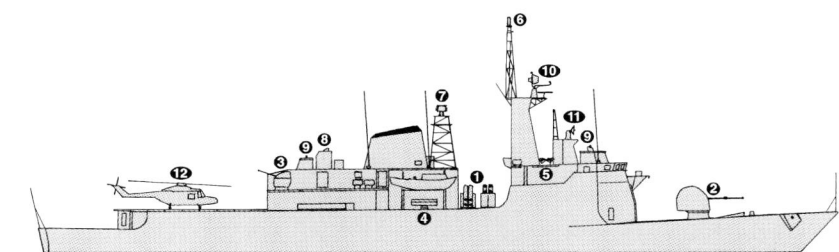

INHAÚMA

(Scale 1 : 900), Ian Sturton / 0017617

Programmes: Designed by Brazilian Naval Design Office with
advice from West German private Marine Technik design
company. Signature of final contract on 1 October 1981. First
pair ordered on 15 February 1982 and second pair 9 January
1986. In mid-1986 the government approved, in principle,
construction of a total of 16 ships but this was reduced to four.
Modernisation: Plans to fit Simbad SAM have been shelved.
Operational: Form part of First Escort Squadron based at Niterói,
Rio de Janeiro.

UPDATED

FRONTIN

2/2001, Mario R V Carneiro / 0130426

JACEGUAY

4/2000, Mario R V Carneiro / 0104227

SHIPBORNE AIRCRAFT (FRONT LINE)

Notes: It is planned to acquire up to three AEW aircraft. Options include Grumman S-2 Trackers.

Numbers/Type: 3/20 McDonnell Douglas AF-1 /AF-1A Skyhawk.
Operational speed: 560 kt *(1,040 km/h)*.
Service ceiling: 45,000 ft *(13,780 m)*.
Range: 1,060 n miles *(1,965 km)*.
Role/Weapon systems: Acquired from Kuwait Air Force in September 1998 to restore carrier
fixed wing flying. Sensors: APQ 145B radar; ESM/ECM. Weapons: AAM; 4 MAA-1 or 4 AIM 9H;
2 Colt 20 mm cannon; ASVW; bombs and rocket pods.

VERIFIED

AF-1   *10/2001, S C Neto/Mario R V Carneiro / 0569157*

Numbers/Type: 4/3/3/3 Agusta SH-3D/Sikorsky SH-3D/SH-3G/SH-3H.
Operational speed: 125 kt *(230 km/h)*.
Service ceiling: 12,200 ft *(3,720 m)*.
Range: 400 n miles *(740 km)*.
Role/Weapon systems: ASW helicopter; carrierborne and shore-based for medium-range ASW,
ASVW and SAR. Sixteen delivered between 1970 and 1997. Three have been lost. Seven are
reported operational. Sensors: SMA APS-705(V)II or APS-24 search radar; Bendix AQS 13F or
AQS 18(V) dipping sonar. Weapons: ASW; up to 4 × Mk 46 torpedoes, or 4 Mk II depth bombs.
ASVW; 2 × AM 39 Exocet missiles.

UPDATED

SH-3B   *2/2001, Mario R V Carneiro / 0130466*

Numbers/Type: 2/5 Aerospatiale UH-14 (AS 332F1 Super Puma)/(AS 532 SC Cougar).
Operational speed: 100 kt *(182 km/h)*.
Service ceiling: 20,000 ft *(6,100 m)*.
Range: 345 n miles *(635 km)*.
Role/Weapon systems: SAR, troop transport and ASVW. Sensors: Bendix RDR-1400C search
radar. Weapons: None.

VERIFIED

UH-14   *6/2003, S C Neto/Mario R V Carneiro / 0569155*

Numbers/Type: 12 AH-IIA Westland Super Lynx.
Operational speed: 125 kt *(232 km/h)*.
Service ceiling: 12,000 ft *(3,650 m)*.
Range: 160 n miles *(296 km)*.
Role/Weapon systems: ASW/ASV roles. First batch upgraded in 1994-97 to Super Lynx standard
with Mk 3 radar and Racal Kestrel EW suite. Sensors: Sea Spray Mk 1/Mk 3 radar; Racal MIR 2
ESM. Weapons: ASW; 2 × Mk 46 torpedoes, or Mk II depth bombs. ASV; 4 × BAe/Ferranti Sea
Skua missiles.

VERIFIED

AH-IIA   *6/2002, Mario R V Carneiro / 0528959*

Numbers/Type: 18 Aerospatiale UH-12 Esquilo (AS-350BA Ecureuil).
Operational speed: 140 kt *(260 km/h)*.
Service ceiling: 10,000 ft *(3,050 m)*.
Range: 240 n miles *(445 km)*.
Role/Weapon systems: Support helicopters for Fleet liaison and Marine Corps transportation.
Sensors: None. Weapons: 2 × axial 7.62 mm MGs or 1 × lateral MG or 1 × rocket pod.

VERIFIED

UH-12   *12/2002, Mario R V Carneiro / 0569156*

Numbers/Type: 8 Aerospatiale UH-13 Esquilo (AS 355F2 Ecureuil 2).
Operational speed: 121 kt *(224 km/h)*.
Service ceiling: 11,150 ft *(3,400 m)*.
Range: 240 n miles *(445 km)*.
Role/Weapon systems: SAR, liaison and utility in support of Marine Corps. Seven more ordered in
1998. Sensors: Search radar. Weapons: 2 × axial 7.62 mm MGs or 1 × lateral MG or 1 × rocket
pod.

VERIFIED

UH-13   *8/2002, Mario R V Carneiro / 0528966*

Numbers/Type: 19 IH-6B (Bell JetRanger III).
Operational speed: 115 kt *(213 km/h)*.
Service ceiling: 20,000 ft *(6,100 m)*.
Range: 368 n miles *(682 km)*.
Role/Weapon systems: Utility and training helicopters. Sensors: None. Weapons: 2 × 7.62 mm
MGs or 2 × rocket pods.

VERIFIED

IH-6B   *12/2002, Mario R V Carneiro / 0569154*

LAND-BASED MARITIME AIRCRAFT (FRONT LINE)

Numbers/Type: 9 Lockheed P-3BR Orion.
Operational speed: 411 kt *(761 km/h).*
Service ceiling: 28,300 ft *(8,625 m).*
Range: 4,000 m *(7,410 km).*
Role/Weapon systems: Twelve P-3 A/B acquired by the Air Force from the US Navy in 2002. Eight to be upgraded to P-3BR standard by EADS/CASA from 2003. First aircraft to be operational in 2005. The aircraft are to be fitted with CASA Fully Integrated Tactical System (FITS) mission suite. The remaining four aircraft are to be used for spare parts.

VERIFIED

P-3BR   *2002, Brazilian Navy /* 0536045

Numbers/Type: 10/9 Bandeirante P-95A /P-95B (EMB-111(B)).
Operational speed: 194 kt *(360 km/h).*
Service ceiling: 25,500 ft *(7,770 m).*
Range: 1,590 n miles *(2,945 km).*
Role/Weapon systems: Air Force operated for coastal surveillance role by four squadrons. Sensors: MEL search radar, searchlight pod on starboard wing, EFIS-74 (electronic flight instrumentation) and Collins APS-65 (autopilot); ESM Thomson-CSF DR2000A/Dalia 1000A Mk II, GPS (Trimble). Weapons: 4 or 6 × 127 mm rockets, or up to 28 × 70 mm rockets.

UPDATED

EMB-111   *6/1995 * /* 0503428

Numbers/Type: 53 A-1 (Embraer/Alenia/Aermacchi) AMX.
Operational speed: 493 kt *(914 km/h).*
Service ceiling: 42,650 ft *(13,000 m).*
Range: 1,800 n miles *(3,336 km).*
Role/Weapon systems: Air Force operated for strike, reconnaissance and anti-shipping attack; shore-based for fleet air defence and ASV primary roles; operated by 3rd/10th Group at Santa Maria Air Base (KS) and Santa Cruz Air Base. Sensors: Tecnasa/SMA SCP-01 Scipio radar. ECM suite/ESM flares and chaffs; GPS and IFF. Weapons: Strike; up to 3,800 kg of 'IRON' bombs; Self-defence; AAM; 2 × MAA-1 Piranha or 2 × AIM-9 Sidewinder missiles; 2 DEFA 30 mm cannon.

UPDATED

AMX   *6/1998 * /* 0013614

Numbers/Type: 8 Tucano AT-27 (EMB-312A).
Operational speed: 247 kt *(457 km/h).*
Service ceiling: 32,570 ft *(9,936 m).*
Range: 995 n miles *(1,844 km).*
Role/Weapon systems: Air Force operated for liaison and attack by 2 ELO. Sensors: None. Weapons: 6 or 8 × 127 mm rockets or bombs and 1 × 7.62 mm MG pod in each wing.

VERIFIED

PATROL FORCES

Notes: (1) Four 1,000 ton offshore patrol craft and five 500 ton patrol craft are planned. Construction is to start in 2005 and delivery is to be completed by 2014.
(2) There are plans to acquire seven river patrol ships (three for Amazon flotilla and four for Mato Grosso flotilla).
(3) There are 115 LAEP series Instruction and Support craft. 24 are 10 m long and 91 are 7 m long.
(4) There are 182 LPN series River patrol craft of 3 to 15 m length.
(5) There are ten Swift class patrol boats operated by port authorities.

12 GRAJAÚ CLASS (LARGE PATROL CRAFT) (PBO)

Name	No	Builders	Launched	Commissioned
GRAJAÚ	P 40	Arsenal de Marinha	21 May 1993	1 Dec 1993
GUAIBA	P 41	Arsenal de Marinha	10 Dec 1993	12 Sep 1994
GRAÚNA	P 42	Estaleiro Mauá, Niteroi	10 Nov 1993	15 Aug 1994
GOIANA	P 43	Estaleiro Mauá, Niteroi	26 Jan 1994	26 Feb 1997
GUAJARÁ	P 44	Peenewerft, Germany	24 Oct 1994	28 Apr 1995
GUAPORÉ	P 45	Peenewerft, Germany	23 Jan 1995	29 Aug 1995
GURUPÁ	P 46	Peenewerft, Germany	11 May 1995	8 Dec 1995
GURUPI	P 47	Peenewerft, Germany	6 Sep 1995	23 Apr 1996
GUANABARA	P 48	Inace, Fortalesa	5 Nov 1997	9 July 1999
GUARUJÁ	P 49	Inace, Fortalesa	24 Apr 1998	25 Nov 1999
GUARATUBA	P 50	Peenewerft, Germany	16 June 1999	1 Dec 1999
GRAVATAÍ	P 51	Peenewerft, Germany	26 Aug 1999	17 Feb 2000

Displacement, tons: 263 full load
Dimensions, feet (metres): 152.6 × 24.6 × 7.5 *(46.5 × 7.5 × 2.3)*
Main machinery: 2 MTU 16V 396 TB94 diesels; 5,800 hp(m) *(4.26 MW)* sustained; 2 shafts
Speed, knots: 26. **Range, n miles:** 2,200 at 12 kt
Complement: 29 (4 officers)
Guns: 1 Bofors 40 mm/70. 2 Oerlikon 20 mm (P 40-44). 2 Oerlikon BMARC 20 mm GAM-BO1 (P 45-51).
Weapons control: Radamec 1000N optronic director may be fitted in due course.
Radars: Surface search: Racal Decca 1290A; I-band.

Comment: Two ordered in late 1987 to a Vosper QAF design similar to Bangladesh Meghna class. Technology transfer in February 1988 and construction started in July 1988 for the first pair; second pair started construction in September 1990. Class name changed in 1993 when the first four were renumbered to reflect revised delivery dates. Building problems are also reflected in the replacing of the order for the third pair with Peenewerft in November 1993 and the fourth pair in August 1994. Two more ordered from Inace in September 1996 and from Peenewerft in 1998. Used for patrol duties and diver support. Carry one RIB and telescopic launching crane.

UPDATED

GUAPORÉ   *2/2001, Mario R V Carneiro /* 0130468

1 IMPERIAL MARINHEIRO CLASS
(COASTAL PATROL SHIPS) (PG/ATR)

Name	No	Builders	Commissioned
IMPERIAL MARINHEIRO	V 15	Smit, Kinderdijk, Netherlands	8 June 1955

Displacement, tons: 911 standard; 1,025 full load
Dimensions, feet (metres): 184 × 30.5 × 11.7 *(56 × 9.3 × 3.6)*
Main machinery: 2 Sulzer 6TD36 diesels; 2,160 hp(m) *(1.59 MW)*; 2 shafts
Speed, knots: 16
Complement: 64 (6 officers)
Guns: 1—3 in *(76 mm)*/50 Mk 33; 50 rds/min to 12.8 km *(6.9 n miles)*; weight of shell 6 kg. 2 or 4 Oerlikon 20 mm.
Radars: Surface search: Racal Decca; I-band.

Comment: Fleet tugs classed as corvettes. Equipped for firefighting. *Imperial Marinheiro* has acted as a submarine support ship but gave up the role in 1990. V 21 and V 23 withdrawn from service in 2002, V 24 in 2003 and V 19 and V 20 in 2004.

UPDATED

IMPERIAL MARINHEIRO CLASS   *2/2000, van Ginderen Collection /* 0104229

2 PEDRO TEIXEIRA CLASS (RIVER PATROL SHIPS) (PBR)

Name	No	Builders	Launched	Commissioned
PEDRO TEIXEIRA	P 20	Arsenal de Marinha	14 Oct 1970	17 Dec 1973
RAPOSO TAVARES	P 21	Arsenal de Marinha	11 June 1972	17 Dec 1973

Displacement, tons: 690 standard
Dimensions, feet (metres): 208.7 × 31.8 × 5.6 *(63.6 × 9.7 × 1.7)*
Main machinery: 4 MAN V6 V16/18 TL diesels; 3,840 hp(m) *(2.82 MW)*; 2 shafts
Speed, knots: 16. **Range, n miles:** 6,800 at 13 kt
Complement: 60 (6 officers)
Guns: 1 Bofors 40 mm/60; 300 rds/min to 12 km *(6.5 n miles)*.
6—12.7 mm MGs. 2—81 mm Mk 2 mortars.
Radars: Surface search: 2 Racal Decca; I-band.
Helicopters: 1 Bell JetRanger or UH-12 Esquilo.

Comment: Built in Rio de Janeiro. Belong to Amazon Flotilla. Can carry two armed LCVPs and 85 marines in deck accommodation. Both ships to be re-engined. *VERIFIED*

PEDRO TEIXEIRA   *6/1997, Brazilian Navy /* 0012091

3 RORAIMA CLASS (RIVER PATROL SHIPS) (PBR)

Name	No	Builders	Launched	Commissioned
RORAIMA	P 30	Maclaren, Niteroi	2 Nov 1972	21 Feb 1975
RONDÔNIA	P 31	Maclaren, Niteroi	10 Jan 1973	3 Dec 1975
AMAPÁ	P 32	Maclaren, Niteroi	9 Mar 1973	12 Jan 1976

Displacement, tons: 340 standard; 365 full load
Dimensions, feet (metres): 151.9 × 27.9 × 4.6 *(46.3 × 8.5 × 1.4)*
Main machinery: 2 MAN V6 V16/18TL diesels; 1,920 hp(m) *(1.41 MW)*; 2 shafts
Speed, knots: 14. **Range, n miles:** 3,000 at 12 kt
Complement: 48 (5 officers)
Guns: 1 Bofors 40 mm/60; 300 rds/min to 12 km *(6.5 n miles)*.
2 Oerlikon 20 mm. 2—81 mm mortars. 6—12.7 mm MGs.
Radars: Surface search: 2 Racal Decca; I-band.

Comment: Carry two armed LCVPs. Belong to Amazon Flotilla. P 32 re-engined with Volvo engines and P 30 and P 31 to be similarly refitted. *VERIFIED*

RORAIMA   *6/1998, Brazilian Navy /* 0017623

4 BRACUI (RIVER) CLASS (COASTAL PATROL CRAFT) (PBO)

Name	No	Builders	Commissioned
BRACUI (ex-*Itchen*)	P 60 (ex-M 2009)	Richards, Lowestoft	12 Oct 1985
BENEVENTE (ex-*Blackwater*)	P 61 (ex-M 2008)	Richards, Great Yarmouth	5 July 1985
BOCAINA (ex-*Spey*)	P 62 (ex-M 2013)	Richards, Lowestoft	4 Apr 1986
BABITONGA (ex-*Arun*)	P 63 (ex-M 2014)	Richards, Lowestoft	29 Aug 1986

Displacement, tons: 770 standard; 890 full load
Dimensions, feet (metres): 156 × 34.5 × 9.5 *(47.5 × 10.5 × 2.9)*
Main machinery: 2 Ruston 6 RKC diesels; 3,100 hp(m) *(2.3 MW)* sustained; 2 shafts
Speed, knots: 14. **Range, n miles:** 4,500 at 10 kt
Complement: 32 (4 officers)
Guns: 1 Bofors 40 mm/60.
2—7.62 mm MGs.
Mines: Rails for up to 20.
Radars: Surface search: 2 Racal Decca TM 1226C; I-band.

Comment: Second batch of ex-UK River class minesweepers transferred in 1998. These four were converted as patrol craft in UK service. Recommissioned 6 April, 10 July, 10 July and 9 September respectively. Three others transferred in 1995 are listed as Survey Ships and as a Buoy Tender. *VERIFIED*

BOCAINA   *7/1998, Maritime Photographic /* 0056608

6 PIRATINI CLASS (COASTAL PATROL CRAFT) (PB)

Name	No	Builders	Commissioned
PIRATINI (ex-PGM 109)	P 10	Arsenal de Marinha, Rio de Janeiro	30 Nov 1970
PIRAJÁ (ex-PGM 110)	P 11	Arsenal de Marinha, Rio de Janeiro	8 Mar 1971
PAMPEIRO (ex-PGM 118)	P 12	Arsenal de Marinha, Rio de Janeiro	16 June 1971
PARATI (ex-PGM 119)	P 13	Arsenal de Marinha, Rio de Janeiro	29 July 1971
PENEDO (ex-PGM 120)	P 14	Arsenal de Marinha, Rio de Janeiro	30 Sep 1971
POTI (ex-PGM 121)	P 15	Arsenal de Marinha, Rio de Janeiro	29 Oct 1971

Displacement, tons: 105 standard; 146 full load
Dimensions, feet (metres): 95 × 19 × 6.5 *(29 × 5.8 × 2)*
Main machinery: 4 Cummins VT-12M diesels; 1,100 hp *(820 kW)*; 2 shafts
Speed, knots: 17. **Range, n miles:** 1,700 at 12 kt
Complement: 16 (2 officers)
Guns: 1 Oerlikon 20 mm. 2—12.7 mm MGs.
Radars: Surface search: Racal Decca 1070; I-band.
Navigation: Furuno 3600; I-band.

Comment: Built under offshore agreement with the USA and similar to the US Cape class. 81 mm mortar removed in 1988. Carries an inflatable launch. P 10, P 11, P 14 and P 15 are based at Ladário Fluvial Base, Mato Grosso, the other two at Amazonas. *VERIFIED*

POTI   *6/1998, Brazilian Navy /* 0017624

4 TRACKER II (LPaN-21) CLASS (COASTAL PATROL CRAFT) (PB)

RIO CPRJ-05 (ex-P 8002)   TIMBIRA CPAOR-17 (ex-P 8005)
MUCURIPE CPCE-03 (ex-P 8004)   ESPADARTE CPSP-07 (ex-P 8003)

Displacement, tons: 31 standard; 45 full load
Dimensions, feet (metres): 68.6 × 17 × 4.8 *(20.9 × 5.2 × 1.5)*
Main machinery: 2 MTU 8V 396 TB83 diesels; 2,100 hp(m) *(1.54 MW)* sustained; 2 shafts
Speed, knots: 25. **Range, n miles:** 600 at 15 kt
Complement: 8 (2 officers)
Guns: 2—12.7 mm MGs.
Radars: Surface search: Racal Decca RM 1070A; I-band.

Comment: Ordered in February 1987 to a Fairey design and built at Estaleiro Shipyard, Porto Alegre. National input is 60 per cent. First of class completed building 22 February 1990. All entered service in May 1991. Employed as Police patrol boats. *UPDATED*

RIO   *10/2003, Gomel/Marsan /* 0569150

1 PARNAIBA CLASS (RIVER MONITOR) (PGRH)

Name	No	Builders	Commissioned
PARNAIBA	U 17 (ex-P 2)	Arsenal de Marinha, Rio de Janeiro	6 Nov 1938

Displacement, tons: 620 standard; 720 full load
Dimensions, feet (metres): 180.5 × 33.3 × 5.1 *(55 × 10.1 × 1.6)*
Main machinery: 2 diesels; 2 shafts
Speed, knots: 12. **Range, n miles:** 1,350 at 10 kt
Complement: 74 (6 officers)
Guns: 1 US 76 mm. 2 Bofors 40 mm/70. 6 Oerlikon 20 mm.
Radars: Surface search: Racal Decca; I-band.
Navigation: Furuno 3600; I-band.
Helicopters: Platform for one Esquilo.

Comment: Laid down 11 June 1936. Launched 2 September 1937. In Mato Grosso Flotilla. Re-armed with new guns in 1960. 3 in *(76 mm)* side armour and partial deck protection. Refitted in 1995/96 with improved armament, and with diesel engines replacing the steam reciprocating propulsion plant. Converted again in 1998 with Bofors 40 mm/70 guns taken from Niterói class frigates and a helo deck at the stern. Facilities to refuel and re-arm a UH-12 helicopter. Recommissioned 6 May 1999. *VERIFIED*

PARNAIBA   *5/2000, Hartmut Ehlers /* 0087859

AMPHIBIOUS FORCES

Notes: (1) Replacement of the two Ceará class LSDs is under consideration. Four LCU and eight LCM are also required.
(2) There are 8 EDVP II class landing craft of 13 tons built by BFL, Ladario and capable of carrying 3.7 tons or 37 troops at 9 kt. These are based at Ladario.
(3) There are 32 RIBs for special operations.

1 NEWPORT CLASS (LSTH)

Name	No	Builders	Laid down	Launched	Commissioned	Recommissioned
MATTOSO MAIA (ex-*Cayuga*)	G 28 (ex-LST 1186)	National Steel & Shipbuilding Co	28 Sep 1968	12 July 1969	8 Aug 1970	30 Aug 1994

Displacement, tons: 4,975 light; 8,750 full load
Dimensions, feet (metres): 522.3 (hull) × 69.5 × 17.5 (aft)
(159.2 × 21.2 × 5.3)
Main machinery: 6 ALCO 16-251 diesels; 16,500 hp *(12.3 MW)*
sustained; 2 shafts; cp props; bow thruster
Speed, knots: 20. **Range, n miles:** 14,250 at 14 kt
Complement: 257 (20 officers)
Military lift: 351 (33 officers); 500 tons vehicles; 3 LCVPs and 1
LCPL on davits

Guns: 1 General Electric/General Dynamics 20 mm Vulcan
Phalanx Mk 15. 8—12.7 mm MGs.
Radars: Surface search: Raytheon SPS-10F; G-band.
Navigation: Raytheon SPS-64(V)6 and Furuno FR 2120; I-band.

Helicopters: Platform only.

Programmes: Transferred from the USN by lease 26 August
1994, arriving in Brazil in late October. Purchased outright on
19 September 2000.
Structure: The ramp is supported by twin derrick arms. A stern
gate to the tank deck permits unloading of amphibious
tractors into the water, or unloading of other vehicles into an
LCU or onto a pier. Vehicle stowage covers 19,000 sq ft.
Length over derrick arms is 562 ft *(171.3 m)*; full load draught
is 11.5 ft forward and 17.5 ft aft.

UPDATED

MATTOSO MAIA

6/2003, S C Neto/Mario R V Carneiro / 0569153

2 CEARÁ (THOMASTON) CLASS (LSDH)

Name	No	Builders	Laid down	Launched	Commissioned	Recommissioned
CEARÁ (ex-*Hermitage*)	G 30 (ex-LSD 34)	Ingalls, Pascagoula	11 Apr 1955	12 June 1956	14 Dec 1956	28 Nov 1989
RIO DE JANEIRO (ex-*Alamo*)	G 31 (ex-LSD 33)	Ingalls, Pascagoula	11 Oct 1954	20 Jan 1956	24 Aug 1956	21 Nov 1990

Displacement, tons: 6,880 light; 12,150 full load
Dimensions, feet (metres): 510 × 84 × 19
(155.5 × 25.6 × 5.8)
Main machinery: 2 Babcock & Wilcox boilers; 580 psi *(40.8 kg/
cm²)*; 2 GE turbines; 24,000 hp *(17.9 MW)*; 2 shafts
Speed, knots: 22.5. **Range, n miles:** 10,000 at 18 kt
Complement: 345 (20 officers)
Military lift: 340 troops; 21 LCM 6s or 3 LCUs and 6 LCMs or 50
LVTs; 30 LVTs on upper deck

Guns: 6 USN 3 in *(76 mm)*/50 (3 twin) Mk 33; 50 rds/min to
12.8 km *(7 n miles)*; weight of shell 6 kg.
4—12.7 mm MGs.
Radars: Surface search: Raytheon SPS-10F; G-band.
Air/Surface search: Plessey AWS-2 (G 30); E/F-band.
Navigation: Raytheon CRP 3100; I-band.
Helicopters: Platform for Super Puma.

Programmes: The original plan to build a 4,500 ton LST was
overtaken by the acquisition of these two LSDs from the US
initially on a lease and finally by purchase on 24 January
2001.
Structure: Has two 50 ton capacity cranes and a docking well of
391 × 48 ft *(119.2 × 14.6 m)*. SATCOM fitted. Phalanx guns
and SRBOC chaff launchers removed before transfer. Air
search radars removed.

VERIFIED

RIO DE JANEIRO

4/2000, Hartmut Ehlers / 0104231

3 LCU 1610 CLASS (EDCG/LCU)

Name	No	Builders	Commissioned
GUARAPARI	L 10 (ex-GED 10)	Arsenal de Marinha, Rio de Janeiro	27 Mar 1978
TAMBAÚ	L 11 (ex-GED 11)	Arsenal de Marinha, Rio de Janeiro	27 Mar 1978
CAMBORIÚ	L 12 (ex-GED 12)	Arsenal de Marinha, Rio de Janeiro	6 Jan 1981

Displacement, tons: 390 full load
Dimensions, feet (metres): 134.5 × 27.6 × 6.6 *(41 × 8.4 × 2.0)*
Main machinery: 2 GM 12V-71 diesels; 874 hp *(650 kW)* sustained; 2 shafts; cp props
Speed, knots: 11. **Range, n miles:** 1,200 at 8 kt
Complement: 14 (2 officers)
Military lift: 172 tons
Guns: 3—12.7 mm MGs.
Radars: Navigation: Furuno 3600; I-band.

Comment: Original pennant numbers restored in 2004. Based at Niteroi.

UPDATED

CAMBORIÚ
6/2001, Brazilian Navy / 0130473

3 EDVM 17 CLASS (LCM)

301-303

Displacement, tons: 55 full load
Dimensions, feet (metres): 55.8 × 14.4 × 3.9 *(17 × 4.4 × 1.2)*
Main machinery: 2 Saab Scania diesels; 470 hp(m) *(345 kW)*; 2 shafts
Speed, knots: 9
Complement: 3
Military lift: 80 troops plus 31 tons equipment

Comment: LCM 6 type acquired from the US.

UPDATED

301   *1985, Ronaldo S Olive*

5 EDVM 25 CLASS (LCM)

801-805

Displacement, tons: 61 standard; 130 full load
Dimensions, feet (metres): 71 × 21 × 4.8 *(21.7 × 6.4 × 1.5)*
Main machinery: 2 Detroit diesels; 400 hp *(294 kW)* sustained; 2 shafts
Speed, knots: 9. **Range, n miles:** 95 at 9 kt
Complement: 5
Military lift: 150 troops plus 72 tons equipment

Comment: First of class launched 18 January 1994 by AMRJ, remainder by Inace. LCM 8 type. Based at Niteroi.

UPDATED

801   *6/2001, Brazilian Navy /* 0130472

MINE WARFARE FORCES

6 ARATU (SCHÜTZE) CLASS (MINESWEEPERS—COASTAL) (MSC)

Name	No	Builders	Commissioned
ARATU	M 15	Abeking & Rasmussen, Lemwerder	5 May 1971
ANHATOMIRIM	M 16	Abeking & Rasmussen, Lemwerder	30 Nov 1971
ATALAIA	M 17	Abeking & Rasmussen, Lemwerder	13 Dec 1972
ARAÇATUBA	M 18	Abeking & Rasmussen, Lemwerder	13 Dec 1972
ABROLHOS	M 19	Abeking & Rasmussen, Lemwerder	25 Feb 1976
ALBARDÃO	M 20	Abeking & Rasmussen, Lemwerder	25 Feb 1976

Displacement, tons: 230 standard; 280 full load
Dimensions, feet (metres): 154.9 × 23.6 × 6.9 *(47.2 × 7.2 × 2.1)*
Main machinery: 2 MTU Maybach diesels; 4,500 hp(m) *(3.3 MW)*; 2 shafts; 2 Escher-Weiss cp props
Speed, knots: 24. **Range, n miles:** 710 at 20 kt
Complement: 32 (4 officers)
Guns: 1 Bofors 40 mm/70.
Radars: Surface search: Bridge Master E; I-band.
Navigation: Furuno FR 1831; I-band.

Comment: Wooden hulled. First four ordered in April 1969 and last pair in November 1973. Same design as the now deleted German Schütze class. Can carry out wire, magnetic and acoustic sweeping. A life-extension refit programme started in 2001. M 15 completed in 2002 and the remainder of the class are to be completed by 2006. Modifications include replacement of the surface search radar, communications upgrade and hull preservation measures. Based at Aratu, Bahia.

UPDATED

ABROLHOS   *3/1998, Brazilian Navy /* 0017625

SURVEY AND RESEARCH SHIPS

Notes: (1) Survey ships are painted white except for those operating in the Antarctic which have red hulls.
(2) There are also 21 buoy tenders of between 15 and 26 m: nine LB-15, two LB-17 (*Lufanda* and *Piracema*), four LB-19, two LB-23 and four LB-26.

1 POLAR RESEARCH SHIP (AGOBH)

Name	No	Builders	Commissioned
ARY RONGEL (ex-*Polar Queen*)	H 44	Eides, Norway	22 Jan 1981

Displacement, tons: 3,628 full load
Dimensions, feet (metres): 247 × 42.7 × 17.4 *(75.3 × 13 × 5.3)*
Main machinery: 2 MAK 6M-453 diesels; 4,500 hp(m) *(3.3 MW)*; 1 shaft; cp prop; 2 bow thrusters; 1 stern thruster
Speed, knots: 14.5. **Range, n miles:** 17,000 at 12 kt
Complement: 70 (19 officers) + 22 scientists
Radars: Navigation: Sperry; I-band Racal-Decca; I/J-band
Cargo capacity: 2,400 m³
Helicopters: Platform for Ecureuil 2.

Comment: Acquired by sale 19 April 1994. Ice-strengthened hull fitted with Simrad Albatross dynamic positioning system.

VERIFIED

ARY RONGEL   *6/2002, Carlos Veras, Brazilian Navy /* 0572424

1 RESEARCH SHIP (AGS)

Name	No	Builders	Commissioned
ANTARES (ex-M/V *Lady Harrison*)	H 40	Mjellem and Karlsen A/S, Bergen	Aug 1984

Displacement, tons: 1,076 standard; 1,248 full load
Dimensions, feet (metres): 180.3 × 33.8 × 14.1 *(55 × 10.3 × 4.3)*
Main machinery: 1 Burmeister & Wain Alpha diesel; 1,860 hp(m) *(1.37 MW)*; 1 shaft; bow thruster
Speed, knots: 13.5. **Range, n miles:** 10,000 at 12 kt
Complement: 58 (12 officers) + 12
Radars: Navigation: 2 Racal Decca; I-band.

Comment: Research vessel acquired from Racal Energy Resources. Used for seismographic survey. Recommissioned 6 June 1988.

VERIFIED

ANTARES   *4/2000, Hartmut Ehlers /* 0104233

1 SIRIUS CLASS (SURVEY SHIP) (AGSH)

Name	No	Builders	Launched	Commissioned
SIRIUS	H 21	Ishikawajima Co Ltd, Tokyo	30 July 1957	17 Jan 1958

Displacement, tons: 1,463 standard; 1,741 full load
Dimensions, feet (metres): 255.7 × 39.3 × 12.2 *(78 × 12.1 × 3.7)*
Main machinery: 2 Sulzer 7T6-36 diesels; 2,700 hp(m) *(1.98 MW)*; 2 shafts; cp props
Speed, knots: 15.7. **Range, n miles:** 12,000 at 11 kt
Complement: 116 (16 officers) plus 14 scientists
Radars: Navigation: Racal Decca TM 1226C; I-band.
Helicopters: 1 Bell JetRanger or UH-12.

Comment: Laid down 1955-56. Special surveying apparatus, echo-sounders, Raydist equipment, sounding machines installed, and landing craft (LCVP), jeep, and survey launches carried. All living and working spaces are air conditioned.
VERIFIED

SIRIUS *8/1999* / 0056615

2 AMORIM DO VALLE (RIVER) CLASS (SURVEY SHIPS) (AGS)

Name	No	Commissioned
AMORIM DO VALLE (ex-*Humber*)	H 35 (ex-M 2007)	7 June 1985
TAURUS (ex-*Helmsdale*/*Jorge Leite*)	H 36 (ex-M 2010)	1 Mar 1986

Displacement, tons: 890 full load
Dimensions, feet (metres): 156 × 34.5 × 9.5 *(47.5 × 10.5 × 2.9)*
Main machinery: 2 Ruston 6RKC diesels; 3,100 hp *(2.3 MW)* sustained; 2 shafts
Speed, knots: 14. **Range, n miles:** 4,500 at 10 kt
Complement: 36 (4 officers)
Radars: Navigation: 2 Racal Decca TM 1226C; I-band.

Comment: H 35 and H 36 were two of the three ships transferred from the UK on 31 January 1995. The contract was signed on 18 November 1994. Steel hulled for deep-armed team sweeping with wire sweeps. All minesweeping gear and the 40 mm gun removed on transfer. Used as hydrographic ships. H 35 has a stern gantry and second crane amidships for oceanographic research. Four others of the class transferred in 1998 are listed under Patrol Forces. The class is also in service with the Bangladesh Navy.
VERIFIED

AMORIM DO VALLE *6/2001, Brazilian Navy* / 0130471

1 ARGUS CLASS (SURVEY SHIP) (AGSH)

Name	No	Builders	Launched	Commissioned
ARGUS	H 31	Arsenal de Marinha, Rio de Janeiro	6 Dec 1957	29 Jan 1959

Displacement, tons: 250 standard; 343 full load
Dimensions, feet (metres): 146.7 × 21.3 × 9.2 *(44.7 × 6.5 × 2.8)*
Main machinery: 2 Caterpillar D 379 diesels; 1,098 hp *(818 kW)* sustained; 2 shafts
Speed, knots: 15. **Range, n miles:** 3,000 at 15 kt
Complement: 34 (6 officers)
Guns: 2 Oerlikon 20 mm (removed).
Radars: Navigation: 2 Racal Decca 1226C; I-band.
Helicopters: Platform for one medium.

UPDATED

ARGUS *6/2004*, Brazilian Navy* / 1044087

1 LIGHTHOUSE TENDER (ABUH)

Name	No	Builders	Launched	Commissioned
ALMIRANTE GRAÇA ARANHA	H 34	Ebin, Niteroi	23 May 1974	9 Sep 1976

Displacement, tons: 2,440 full load
Dimensions, feet (metres): 245.3 × 42.6 × 13.8 *(74.8 × 13 × 4.2)*
Main machinery: 1 diesel; 2,440 hp(m) *(1.8 MW)*; 1 shaft; bow thruster
Speed, knots: 13
Complement: 80 (13 officers)
Radars: Navigation: 2 Racal Decca; I-band.
Helicopters: 1 Bell JetRanger.

Comment: Laid down in 1971. Fitted with telescopic hangar, 10 ton crane, two landing craft, GP launch and two Land Rovers. Omega navigation system.
VERIFIED

ALMIRANTE GRAÇA ARANHA *4/2000, Hartmut Ehlers* / 0104234

1 GARNIER SAMPAIO (RIVER) CLASS (BUOY TENDER) (ABU)

Name	No	Builders	Commissioned
GARNIER SAMPAIO (ex-*Ribble*)	H 37 (ex-M 2012)	Richards, Great Yarmouth	19 Feb 1986

Displacement, tons: 890 full load
Dimensions, feet (metres): 156 × 34.5 × 9.5 *(47.5 × 10.5 × 2.9)*
Main machinery: 2 Ruston 6RKC diesels; 3,100 hp *(2.3 MW)* sustained; 2 shafts
Speed, knots: 14. **Range, n miles:** 4,500 at 10 kt
Complement: 36 (6 officers)
Radars: Navigation: 2 Racal Decca TM 1226C; I-band.

Comment: One of three ships transferred from UK in 1995. Steel hulled for deep-armed team sweeping with wire sweeps. All minesweeping gear and the 40 mm gun removed on transfer and used as light buoy tender. Four others of the class transferred in 1998 are listed under Patrol Forces. The class is also in service with the Bangladesh Navy.
UPDATED

GARNER SAMPAIO *6/2002, Brazilian Navy* / 0529149

4 BUOY TENDERS (ABU)

Name	No	Builders	Commissioned
COMANDANTE VARELLA	H 18	Arsenal de Marinha, Rio de Janeiro	20 May 1982
TENENTE CASTELO	H 19	Estanave, Manaus	15 Aug 1984
COMANDANTE MANHÃES	H 20	Estanave, Manaus	15 Dec 1983
TENENTE BOANERGES	H 25	Estanave, Manaus	29 Mar 1985

Displacement, tons: 420 full load
Dimensions, feet (metres): 123 × 28.2 × 8.5 *(37.5 × 8.6 × 2.6)*
Main machinery: 2—8-cyl diesels; 1,300 hp(m) *(955 kW)*; 2 shafts
Speed, knots: 12. **Range, n miles:** 2,880 at 9 kt
Complement: 28 (2 officers)
Radars: Navigation: Racal Decca; I-band.

Comment: Dual-purpose minelayers. *Tenente Castelo* is based at Santana, *Tenente Boanerges* at Salvador.
UPDATED

COMANDANTE VARELLA *1/2000, van Ginderen Collection* / 0104235

1 BUOY TENDER (ABU)

FAROLEIRO MÁRIO SEIXAS (ex-*Mestre Jerânimo*) H 26

Displacement, tons: 294 full load
Dimensions, feet (metres): 116.4 × 21.8 × 11.8 *(35.5 × 6.6 × 3.6)*
Main machinery: 2 Scania DSI 14 MO3 diesels, 2 shafts
Speed, knots: 10
Complement: 18 (2 officers)
Radars: Navigation: Furuno FR 1831; I-band.

Comment: Former fishing vessel built in Vigo, Spain. Acquired by Brazilian Navy in 1979 and rebuilt as a buoy tender. Commissioned 31 January 1984.

VERIFIED

FAROLEIRO MÁRIO SEIXAS   6/2002, Brazilian Navy / 0529148

10 LB 20 CLASS BUOY TENDERS (ABU)

ACHERNAR CPSP 02	**CAPELLA** CPES 03	**RIGEL** SSN-5 06 (ex-SSN 409)
ALDEBARAN SSN-2 01	**DENÉBOLA** SSN-4 02	**VEGA** SSN-4 01 (ex-SSN 506)
BETELGEUSE CPSC 05	**FOMALHAUT** CPPR 05	**POLLUX** CAMR 11
(ex-SUL 03)	**REGULUS** SSN 4201 (ex-SSN 4204)	

Displacement, tons: 102
Dimensions, feet (metres): 65 × 19.7 × 5.9 *(19.8 × 6 × 1.8)*
Main machinery: 2 Cummins NT 855M diesels; 720 hp(m) *(530 kW)*; 2 shafts
Speed, knots: 10. **Range, n miles:** 1,000 at 10 kt
Complement: 6
Radars: Navigation: Furuno; I-band.

Comment: Built by Damen, Gorichen and assembled by Wilson, Sao Paolo. First one commissioned 20 December 1995 and the last on 29 December 1997. The pennant numbers correspond to naval facilities in which they are stationed.

UPDATED

RIGEL   2/2000, van Ginderen Collection / 0104236

3 SURVEY LAUNCHES (YGS)

PARAIBANO SSN-4 03 (ex-H 11)   **CAMOCIM** BHMN 03 (ex-H 16)
RIO BRANCO SSN-4 04 (ex-H 12)

Displacement, tons: 32 standard; 50 full load
Dimensions, feet (metres): 52.5 × 15.1 × 4.3 *(16 × 4.6 × 1.3)*
Main machinery: 2 GM diesels; 330 hp *(246 kW)*; 2 shafts
Speed, knots: 11. **Range, n miles:** 600 at 11 kt
Complement: 10 (1 officer)
Radars: Navigation: Racal Decca 110; I-band.

Comment: Built by Bormann, Rio de Janeiro and commissioned 1969-72. Majority work in Amazon Flotilla. Wooden hulls. All decommissioned in 1991 but retained in service as support to naval establishments and reclassified AvHi (inshore survey craft). Three decommissioned in 2004.

UPDATED

PARAIBANO (old number)   1985, Brazilian Navy

1 OCEAN SURVEY VESSEL (AGSC)

Name	No	Builders	Commissioned
SUBOFICIAL OLIVEIRA	CAMR 02 (ex-DHN 02, ex-U 15)	Inace	22 May 1981

Displacement, tons: 170 full load
Dimensions, feet (metres): 116.4 × 22 × 15.7 *(35.5 × 6.7 × 4.8)*
Main machinery: 2 diesels; 740 hp(m) *(544 kW)*; 2 shafts
Speed, knots: 8. **Range, n miles:** 1,400 at 8 kt
Complement: 10 (2 officers)
Radars: Navigation: Racal Decca 110; I-band.

Comment: Commissioned at Fortaleza for Naval Research Institute. Decommissioned in 1991 but retained in service as an AvPqOc (ocean survey craft). Based at Niterói.

VERIFIED

SUBOFICIAL OLIVEIRA (old number)   1990, Brazilian Navy / 0056618

TRAINING SHIPS

Notes: (1) There are 10 small sail training ships.
(2) One new training vessel *Braz de Aguiar* (ex-*Calha Norte*) has been reported.

1 MODIFIED NITERÓI CLASS (AXH)

Name	No	Builders	Commissioned
BRASIL	U 27	Arsenal de Marinha, Rio de Janeiro	21 Aug 1986

Displacement, tons: 2,548 light; 3,729 full load
Dimensions, feet (metres): 430.7 × 44.3 × 13.8 *(131.3 × 13.5 × 4.2)*
Main machinery: 2 Pielstick/Ishikawajima (Brazil) 6 PC2.5 L 400 diesels; 7,020 hp(m) *(5.17 MW)* sustained; 2 shafts
Speed, knots: 18. **Range, n miles:** 7,000 at 15 kt
Complement: 218 (27 officers) plus 201 midshipmen
Guns: 2 Bofors 40 mm/70. 4 saluting guns.
Countermeasures: Decoys: 2 CBV 50.8 mm flare launchers.
ESM: Racal RDL-2 ABC; radar intercept.
Weapons control: Saab Scania TVT 300 optronic director.
Radars: Surface search: Racal Decca RMS 1230C; E/F-band.
Navigation: Racal Decca TM 1226C and TMS 1230; I-band.
Helicopters: Platform for 1 Sea King.

Comment: A modification of the Vosper Thornycroft Mk 10 Frigate design ordered in June 1981. Laid down 18 September 1981, launched 23 September 1983. Designed to carry midshipmen and other trainees from the Naval and Merchant Marine Academies. Minimum electronics as required for training. There are four 51 mm launchers for flares and other illuminants.

UPDATED

BRASIL   8/2004*, Harald Carstens / 1044085

3 NASCIMENTO CLASS (AXL)

Name	No	Builders	Commissioned
ASPIRANTE NASCIMENTO	U 10	Ebrasa, Santa Catarina	13 Dec 1980
GUARDA MARINHA JENSEN	U 11	Ebrasa, Santa Catarina	22 July 1981
GUARDA MARINHA BRITO	U 12	Ebrasa, Santa Catarina	22 July 1981

Displacement, tons: 108.5 standard; 136 full load
Dimensions, feet (metres): 91.8 × 21.3 × 5.9 *(28 × 6.5 × 1.8)*
Main machinery: 2 Mercedes Benz OM-352A diesels; 650 hp(m) *(478 kW)*; 2 shafts
Speed, knots: 10. **Range, n miles:** 700 at 10 kt
Complement: 6 (2 officers) + 10 midshipmen
Guns: 1—12.7 mm MG.
Radars: Navigation: Racal Decca; I-band.

Comment: Can carry 10 trainees overnight. All of the class are attached to the Naval Academy at Rio de Janeiro. *VERIFIED*

GUARDA MARINHA JENSEN 5/2003, A E Galarce / 0572425

3 ROSCA FINA CLASS (AXL)

ROSCA FINA CN 31 **VOGA PICADA** CN 32 **LEVA ARRIBA** CN 33

Displacement, tons: 50 full load
Dimensions, feet (metres): 61 × 15.4 × 3.9 *(18.6 × 4.7 × 1.2)*
Main machinery: 1 diesel; 650 hp(m) *(477 kW)*; 1 shaft
Speed, knots: 11. **Range, n miles:** 200
Complement: 5 plus trainees
Radars: Navigation: Racal Decca 110; I-band.

Comment: Built by Carbrasmar, Rio de Janeiro. All commissioned 21 February 1984. All attached to the Naval college at Angra dos Reis. *UPDATED*

VOGA PICADA (old number) 1984, Brazilian Navy

1 SAIL TRAINING SHIP (AXS)

Name	No	Builders	Launched	Commissioned
CISNE BRANCO	U 20	Damen Shipyards, Gorinchem	4 Aug 1999	28 Feb 2000

Displacement, tons: 1,038 full load
Dimensions, feet (metres): 249.3 × 34.4 × 15.7 *(76 × 10.5 × 4.8)*
Main machinery: 1 Caterpillar 3508B DI-TA diesel; 1,015 hp(m) *(746 kW)* sustained; 1 shaft; Berg cp prop; bow thruster; 408 hp(m) *(300 kW)*
Speed, knots: 17 (sail); 11 (diesel)
Complement: 41 (10 officers) + 30 midshipmen
Radars: Navigation: Furuno FR 1510 Mk 3; I-band.

Comment: Ordered in 1998. Maximum sail area 2,195 m². *VERIFIED*

CISNE BRANCO 4/2000, Mario R V Carneiro / 0104238

AUXILIARIES

Notes: Future procurement plans include two support transport ships to replace *Ary Parreiras* and *Marajó* 2005-08.

1 TRANSPORT SHIP (AKS)

Name	No	Builders	Commissioned
ATLÂNTICO SUD (ex-*Lloyd Atlântico*)	G 40	Ishibras, Rio de Janeiro	2001

Displacement, tons: 28,977 full load
Dimensions, feet (metres): 617 × 99.1 × 36.7 *(188 × 30.2 × 11.2)*
Main machinery: Sulzer/Ishibras 6RTA76 diesel; 16,560 hp *(12.35 MW)* sustained; 1 shaft
Speed, knots: 17.5. **Range, n miles:** 17,500 at 17 kt
Complement: 25 (6 officers)
Radars: Navigation: 2—I-band.
Helicopters: Platform and refuelling facilities being installed.

Comment: Former merchant ship completed in 1986 and transferred in 2001. Capacity for 30 containers. Two cranes capable of lifting 30 tons. Based at Niteroi. *VERIFIED*

ATLÂNTICO SUD 6/2001, Brazilian Navy / 0130470

1 SUBMARINE RESCUE SHIP (ASRH)

Name	No	Builders	Commissioned
FELINTO PERRY	K 11	Stord Verft, Norway	Dec 1979
(ex-*Holger Dane*, ex-*Wildrake*)			

Displacement, tons: 1,380 standard; 3,850 full load
Dimensions, feet (metres): 256.6 × 57.4 × 15.1 *(78.2 × 17.5 × 4.6)*
Main machinery: Diesel-electric; 2 BMK KVG B12 and 2 KVG B16 diesels; 11,400 hp(m) *(8.4 MW)*; 2 motors; 7,000 hp(m) *(5.15 MW)*; 2 shafts; cp props; 2 bow thrusters; 2 stern thrusters
Speed, knots: 14.5
Complement: 65 (9 officers)
Radars: Navigation: 2 Raytheon; I-band.
Helicopters: Platform only.

Comment: Former oilfield support ship acquired 28 December 1988. Has an octagonal heliport (62.5 ft diameter) above the bridge. Equipped with a moonpool for saturation diving, and rescue and recompression chambers as the submarine rescue ship. Dynamic positioning system. Based at Niteroi, Rio de Janeiro. *VERIFIED*

FELINTO PERRY 2/2003, Mario R V Carneiro / 0569152

1 RIVER TRANSPORT SHIP (AP)

Name	No	Builders	Commissioned
PARAGUASSÚ (ex-*Garapuava*)	G 15	Amsterdam Drydock	1951

Displacement, tons: 285 full load
Dimensions, feet (metres): 131.2 × 23 × 6.6 *(40 × 7 × 2)*
Main machinery: 3 diesels; 2,505 hp(m) *(1.84 MW)*; 1 shaft
Speed, knots: 13. **Range, n miles:** 2,500 at 10 kt
Complement: 43 (4 officers)
Military lift: 178 troops
Guns: 6—7.62 mm MGs.
Radars: Navigation: Furuno 3600; I-band.

Comment: Passenger ship converted into a troop carrier in 1957 and acquired on 20 June 1972. *VERIFIED*

PARAGUASSÚ 5/2000, Hartmut Ehlers / 0104240

1 BARROSO PEREIRA CLASS (TRANSPORT) (AKSH)

Name	No	Builders	Commissioned
ARY PARREIRAS	G 21	Ishikawajima, Tokyo	6 Mar 1957

Displacement, tons: 4,800 standard; 7,300 full load
Measurement, tons: 4,200 dwt; 4,879 gross (Panama)
Dimensions, feet (metres): 362 pp; 391.8 oa × 52.5 × 20.5 *(110.4; 119.5 × 16 × 6.3)*
Main machinery: 2 Ishikawajima boilers and turbines; 4,800 hp(m) *(3.53 MW)*; 2 shafts
Speed, knots: 15
Complement: 159 (15 officers)
Military lift: 1,972 troops (overload); 497 troops (normal)
Cargo capacity: 425 m³ refrigerated cargo space; 4,000 tons
Guns: 2—3 in *(76 mm)* Mk 33; 50 rds/min to 12.8 km *(6.9 n miles)* anti-aircraft; weight of shell 6 kg.
2 or 4 Oerlikon 20 mm.
Radars: Navigation: Two Racal Decca; I-band.
Helicopters: Platform for one medium.

Comment: Transport and cargo vessel. Helicopter landing platform aft. Medical, hospital and dental facilities. Working and living quarters are mechanically ventilated with partial air conditioning. Refrigerated cargo space 15,500 cu ft. Operates commercially from time to time. To be decommissioned in 2005.

UPDATED

BARROSO PEREIRA CLASS   *4/2000, Hartmut Ehlers /* 0104239

1 RIVER TRANSPORT (YFBH)

Name	No	Builders	Commissioned
PIRAIM (ex-*Guaicuru*)	U 29	Estaleiro SNBP, Mato Grosso	10 Mar 1982

Displacement, tons: 91.5 full load
Dimensions, feet (metres): 82.0 × 18.0 × 3.2 *(25.0 × 5.5 × 0.97)*
Main machinery: 2 MWM diesels; 400 hp(m) *(294 kW)*; 2 shafts
Speed, knots: 7. **Range, n miles:** 700 at 7 kt
Complement: 17 (2 officers)
Guns: 4—7.62 mm MG.
Radars: Navigation: Furuno 3600; I-band.
Helicopters: Platform for UH-12.

Comment: Used as a logistics support ship for the Mato Grosso Flotilla. Can carry 2 platoons of marines and 2 rigid inflatable boats.

VERIFIED

PIRAIM   *6/1998, Brazilian Navy /* 0017635

1 HOSPITAL SHIP (AH)

Name	No	Builders	Commissioned
DOUTOR MONTENEGRO	U 16	CONAVE Shipyard, Manaus	17 May 2000

Displacement, tons: 347 full load
Dimensions, feet (metres): 137.8 × 36 × 7.9 *(42 × 11 × 2.4)*
Main machinery: 2 diesels; 600 hp(m) *(448 kW)*; 2 shafts
Speed, knots: 10
Complement: 50 (8 officers) plus 11 (8 doctors/dentists)
Radars: Navigation: Furuno 1942 Mk 2.

Comment: U 16 was built in January 1997 and belonged to the government of the Acre state before transfer to the Brazilian Navy. The ship has two wards, a pediatric ICU, an operating theatre, an X-ray room, a dentist office, a lab for clinical analysis, a trauma room and a pharmacy.

UPDATED

DOUTOR MONTENEGRO   *6/2002, Brazilian Navy /* 0529147

2 HOSPITAL SHIPS (AHH)

Name	No	Builders	Commissioned
OSWALDO CRUZ	U 18	Arsenal de Marinha, Rio de Janeiro	29 May 1984
CARLOS CHAGAS	U 19	Arsenal de Marinha, Rio de Janeiro	7 Dec 1984

Displacement, tons: 500 full load
Dimensions, feet (metres): 154.2 × 26.9 × 5.9 *(47.2 × 8.5 × 1.8)*
Main machinery: 2 Volvo diesels; 714 hp(m) *(525 kW)*; 2 shafts
Speed, knots: 12. **Range, n miles:** 4,000 at 9 kt
Complement: 27 (5 officers) plus 21 medical (6 doctors/dentists)
Radars: Navigation: Racal Decca; I-band.
Helicopters: 1 Helibras HB-350B.

Comment: *Oswaldo Cruz* launched 11 July 1983, and *Carlos Chagas* 16 April 1984. Has two sick bays, dental surgery, a laboratory, two clinics and X-ray centre. The design is a development of the Roraima class with which they operate in the Amazon Flotilla. Since 1992 both ships painted grey with dark green crosses on the hull.

UPDATED

OSWALDO CRUZ   *6/2004*, Brazilian Navy /* 1044086

1 REPLENISHMENT TANKER (AOR)

Name	No	Builders	Commissioned
ALMIRANTE GASTÃO MOTTA	G 23	Ishibras, Rio de Janeiro	26 Nov 1991

Displacement, tons: 10,320 full load
Dimensions, feet (metres): 442.9 × 62.3 × 24.6 *(135 × 19 × 7.5)*
Main machinery: Diesel-electric; 2 Wärtsilä 12V32 diesel generators; 11,700 hp(m) *(8.57 MW)* sustained; 1 motor; 1 shaft; Kamewa cp prop
Speed, knots: 20. **Range, n miles:** 9,000 at 15 kt
Complement: 121 (13 officers)
Cargo capacity: 5,000 tons liquid; 200 tons dry
Guns: 2—12.7 mm MGs.

Comment: Ordered March 1987. Laid down 11 December 1989 and launched 1 June 1990. Fitted for abeam and stern refuelling.

VERIFIED

ALMIRANTE GASTÃO MOTTA   *3/2002, Robert Pabst /* 0528968

4 RIO PARDO CLASSES (YFB)

RIO PARDO BNAJ 08 (ex-*U 40*)		RIO CHUI CIAW 14 (ex-*U 42*)
RIO NEGRO BNRJ 07 (ex-*U 41*)		RIO OIAPOQUE BNRJ 09 (ex-*U 43*)

Displacement, tons: 150 full load
Dimensions, feet (metres): 120 × 21.3 × 6.2 *(36.6 × 6.5 × 1.9)*
Main machinery: 2 Sulzer 6TD24; 900 hp(m) *(661 kW)*; 2 shafts
Speed, knots: 14. **Range, n miles:** 700 at 14 kt
Complement: 10
Radars: Navigation: Racal Decca 110; I-band.

Comment: Can carry 600 passengers. Built by Inconav de Niterói in 1975-76. Pennant numbers removed in 1989.

UPDATED

1 REPLENISHMENT TANKER (AOR)

Name	No	Builders	Launched	Commissioned
MARAJO	G 27	Ishikawajima do Brasil	31 Jan 1968	8 Jan 1969

Displacement, tons: 10,500 full load
Dimensions, feet (metres): 440.7 × 63.3 × 24 *(134.4 × 19.3 × 7.3)*
Main machinery: 1 Sulzer GRD 68 diesel; 8,000 hp(m) *(5.88 MW)*; 1 shaft
Speed, knots: 13. **Range, n miles:** 9,200 at 13 kt
Complement: 80 (13 officers)
Cargo capacity: 6,600 tons fuel

Comment: Fitted for abeam replenishment with two stations on each side. Was to have been replaced by *Gastão Motta* but is to be retained in service until 2008. **UPDATED**

MARAJO *1/1999 /* 0056623

1 RIVER TENDER (AG)

Name	No	Builders	Commissioned
POTENGI	G 17	Papendrecht, Netherlands	28 June 1938

Displacement, tons: 600 full load
Dimensions, feet (metres): 178.8 × 24.5 × 6 *(54.5 × 7.5 × 1.8)*
Main machinery: 2 diesels; 550 hp(m) *(404 kW)*; 2 shafts
Speed, knots: 10. **Range, n miles:** 600 at 8 kt
Complement: 19 (2 officers)
Cargo capacity: 450 tons dieso and avcat
Guns: 4—7.62 mm MGs.
Radars: Navigation: Furuno 3600; I-band.

Comment: Launched 16 March 1938. Employed in the Mato Grosso Flotilla on river service. Converted to logistic support ship and recommissioned 6 May 1999. **VERIFIED**

POTENGI *5/2000, Hartmut Ehlers /* 0104241

1 TORPEDO RECOVERY VESSEL (YPT)

Name	No	Builders	Commissioned
ALMIRANTE HESS	BACS 01 (ex-U 30)	Inace, Fortaleza	2 Dec 1983

Displacement, tons: 91 full load
Dimensions, feet (metres): 77.4 × 19.7 × 6.6 *(23.6 × 6 × 2)*
Main machinery: 2 diesels; 2 shafts
Speed, knots: 13
Complement: 14
Radars: Navigation: Racal Decca 110; I-band.

Comment: Attached to the Submarine Naval Base. Can transport up to four torpedoes. Decommissioned in 1991 but retained in service as an AvPpCo (coast support craft). BACS (Base Almirante Castro y Silva). **UPDATED**

ALMIRANTE HESS *6/1997, Brazilian Navy /* 0012096

4 FLOATING DOCKS

CIDADE DE NATAL (ex-G 27, ex-AFDL 39)
ALMIRANTE SCHIECK
ALFONSO PENA (ex-ARD 14)

ALMIRANTE JERONIMO GONÇALVES
(ex-G 26, ex-*Goiaz* AFDL 4)

Comment: The first two are floating docks loaned to Brazil by US Navy in the mid-1960s and purchased 11 February 1980. Ship lifts of 2,800 tons and 1,000 tons respectively. *Cidade de Natal* based at Natal and *Almirante Jeronimo Gonçalves* at Manaus. *Almirante Schieck* of 3,600 tons displacement was built by Arsenal de Marinha, Rio de Janeiro and commissioned 12 October 1989. *Alfonso Pena* acquired from US and based at Val-de-Caes (Para). **UPDATED**

TUGS

Notes: (1) In addition to the vessels listed below there are two harbour tugs: *Olga* (CASOP 01) and *Alves Barbosa* (AMRJ 11).
(2) There are plans to procure six ocean tugs from 2007-14. These are also to serve as offshore patrol ships.

2 ALMIRANTE GUILHEM CLASS (FLEET OCEAN TUGS) (ATF)

Name	No	Builders	Commissioned
ALMIRANTE GUILHEM (ex-*Superpesa 4*)	R 24	Sumitomo, Uraga	1976
ALMIRANTE GUILLOBEL (ex-*Superpesa 5*)	R 25	Sumitomo, Uraga	1976

Displacement, tons: 2,400 full load
Dimensions, feet (metres): 207 × 44 × 14.8 *(63.2 × 13.4 × 4.5)*
Main machinery: 2 GM EMD 20-645F7B diesels; 7,120 hp *(5.31 MW)* sustained; 2 shafts; cp props; bow thruster
Speed, knots: 14. **Range, n miles:** 10,000 at 10 kt
Complement: 40 (4 officers)
Guns: 2 Oerlikon 20 mm (not always carried)
Radars: Navigation: Racal Decca; I-band. Furuno; I-band.

Comment: Originally built as civilian tugs. Bollard pull, 84 tons. Commissioned into the Navy 22 January 1981. **VERIFIED**

ALMIRANTE GUILLOBEL *5/2003, A E Galarce /* 0572426

3 TRITÃO CLASS (FLEET OCEAN TUGS) (ATA)

Name	No	Builders	Commissioned
TRITÃO (ex-*Sarandi*)	R 21	Estanave, Manaus	19 Feb 1987
TRIDENTE (ex-*Sambaiba*)	R 22	Estanave, Manaus	8 Oct 1987
TRIUNFO (ex-*Sorocaba*)	R 23	Estanave, Manaus	5 July 1986

Displacement, tons: 1,680 full load
Dimensions, feet (metres): 181.8 × 38.1 × 11.2 *(55.4 × 11.6 × 3.4)*
Main machinery: 2 Vilares-Burmeister and Wain Alpha diesels; 2,480 hp(m) *(1.82 MW)*; 2 shafts; bow thruster
Speed, knots: 12
Complement: 43 (6 officers)
Guns: 2 Oerlikon 20 mm.
Radars: Navigation: 2 Racal Decca; I-band.

Comment: Offshore supply vessels acquired from National Oil Company of Brazil and converted for naval use. Assumed names of previous three ships of Sotoyomo class. Fitted to act both as tugs and patrol vessels. Bollard pull, 23.5 tons. Firefighting capability. Endurance, 45 days. **VERIFIED**

TRIUNFO *10/2003, Gomel/Marsan /* 0572427

1 TARGET TOWING TUG (ATA)

Name	No	Builders	Commissioned
TRINDADE (ex-*Nobistor*)	R 26 (ex-U 16)	J G Hitzler, Lavenburg	1969

Displacement, tons: 590 light; 1,308 full load
Dimensions, feet (metres): 176.1 × 36.1 × 11.1 *(53.7 × 11 × 3.4)*
Main machinery: 2 MWM diesels; 2,740 hp(m) *(2 MW)* sustained; 2 shafts
Speed, knots: 12.7 kt
Complement: 22 (2 officers)
Guns: 2—12.7 mm MGs.
Radars: Navigation: Furuno 1830; I-band.

Comment: Ex-Panamanian tug seized for smuggling in 1989 and commissioned in the Navy 31 January 1990. Used for target towing. **VERIFIED**

TRINDADE (old number) *1990, Mário R V Carneiro /* 0056625

8 COASTAL TUGS (YTB)

COMANDANTE MARROIG	CABO SCHRAM	VALENTE
BNRJ 03 (ex-R 15)	BNVC 01 (ex-R 18)	BNRJ 18
COMANDANTE DIDIER	INTRÉPIDO	IMPÁVIDO
BNRJ 04 (ex-R 16)	BNRJ 16	BNRJ 19
TENENTE MAGALHÃES	ARROJADO	
BNA 06 (ex-R 17)	BNRJ 17	

Comment: BNRJ 16-19 are Stan Tug 2207s of 200 tons with a bollard pull of 22.5 tons, built in 1992. BNRJ 03-04 and BNA 06 are 115 tons and built in 1981.

UPDATED

ARROJADO
2/2003, Mario R V Carneiro / 0569151

Brunei

ANGKATAN TENTERA LAUT DIRAJA BRUNEI

Country Overview

Formerly a British dependency, the Nation of Brunei is a sultanate that gained full independence in 1984. Situated on the northern coast of the island of Borneo, the country has a total area of 2,226 square miles and is bordered and divided into two halves by the Malaysian state of Sarawak. It has an 87 n mile coastline with the South China Sea. The capital and largest town is Bandar Seri Begawan which also has port facilities. There are further ports at Kuala Belait and Muara. Territorial seas (3 n miles) and an EEZ (200 n mile) are claimed.

Headquarters Appointments

Commander of the Navy:
   Colonel Joharie Bin Haji Matusin
Fleet Commander:
   Lieutenant Colonel Haji Saied Hussain

Bases

Muara

Personnel

(a)   2005: 747 (58 officers)
      This total, which includes the River Division, is planned to rise to 1,200 but will require concomitant enhancements to training infrastructure.
(b)   Voluntary service

Prefix to Ships' Names

KDB (Kapal Di-Raja Brunei)

CORVETTES

3 BRUNEI CLASS (FSGH)

Name	No	Builders	Laid down	Launched	Commissioned
NAKHODA RAGAM	28	BAE System Marine (Scotstoun)	16 Mar 1999	13 Jan 2001	2005
BENDAHARA SAKAM	29	BAE System Marine (Scotstoun)	15 Nov 1999	23 June 2001	2005
JERAMBAK	30	BAE System Marine (Scotstoun)	5 Apr 2000	22 June 2002	2005

Displacement, tons: 1,940 full load
Dimensions, feet (metres): 311.7 oa; 294.9 wl × 42 × 11.8 *(95; 89.9 × 12.8 × 3.6)*
Main machinery: CODAD; 4 MAN 20 RK270 diesels; 2 shafts; cp props
Speed, knots: 30. **Range, n miles:** 5,000 at 12 kt
Complement: 79 plus 24 spare

Missiles: SSM: 8 MBDA Exocet MM 40 Block II **❶**; active radar homing to 70 km *(40 n miles)* at 0.9 Mach.
SAM: BAe 16 cell VLS **❷**. BAe Sea Wolf; Command Line Of Sight (CLOS) radar/TV tracking to 6 km *(3.3 n miles)* at 2.5 Mach; warhead 14 kg; 16 missiles.
Guns: Otobreda 76 mm Super Rapid **❸**. 120 rds/min to 16 km *(8.7 n miles)*; weight of shell 6 kg.
2 MSI 30 mm/75. 650 rds/min to 10 km *(5.4 n miles)* **❹**.
Torpedoes: 6 Marconi 324 mm (2 triple) tubes **❺**.
Countermeasures: Decoys: 2 Super Barricade chaff launchers **❻**.
ECM: Thales Scorpion; jammer.
ESM: Thales Cutlass 242; intercept.
CESM: Falcon DS 300; intercept.
Combat data systems: Nautis Mk 2 with Link Y.
Weapons control: Radamec 2500 optronic director **❼**.
Radars: Air/surface search: Plessey AWS 9 **❽**; E/F-band.
Surface search: Kelvin Hughes 1007 **❾**; I-band.
Fire control: 2 Marconi 1802 **❿**; I/J-band.
Sonars: Thomson Marconi 4130C1; hull mounted.

Helicopters: Platform for 1 medium.

Programmes: Tenders requested on 28 April 1995. Yarrow Shipbuilders selected in August 1995. Detailed design done in 1996 with final contract signed 14 January 1998. Long-term support contract signed with BAE Systems in May 2002.
Structure: Scaled down version of Malaysian Lekiu class. Facilities to land and refuel S-70A and Bell 212 helicopters.
Operational: Sea trials of first of class began in January 2002. Training for all three crews provided by Flagship Training. *Jerambak* conducted acceptance trials in late 2004. Formal acceptance of all three ships is expected in 2005.

UPDATED

AKHODA RAGAM
(Scale 1 : 900), Ian Sturton / 0526842

BENDAHARA SAKAM
4/2004, John Brodie /* 1044352

NAKHODA RAGAM (on trials)
6/2002, H M Steele / 0533228

PATROL FORCES

Notes: There are also up to 15 Rigid Raider assault boats operated by the River Division for infantry battalions. These boats are armed with 1—7.62 mm MG.

3 WASPADA CLASS (FAST ATTACK CRAFT—MISSILE) (PTG)

Name	No	Builders	Launched	Commissioned
WASPADA	P 02	Vosper (Singapore)	3 Aug 1977	2 Aug 1978
PEJUANG	P 03	Vosper (Singapore)	15 Mar 1978	25 Mar 1979
SETERIA	P 04	Vosper (Singapore)	22 June 1978	22 June 1979

Displacement, tons: 206 full load
Dimensions, feet (metres): 121 × 23.5 × 6 *(36.9 × 7.2 × 1.8)*
Main machinery: 2 MTU 20V 538 TB91 diesels; 7,680 hp(m) *(5.63 MW)* sustained; 2 shafts
Speed, knots: 32. **Range, n miles:** 1,200 at 14 kt
Complement: 24 (4 officers)

Missiles: SSM: 2 Aerospatiale MM 38 Exocet; inertial cruise; active radar homing to 42 km *(23 n miles)* at 0.9 Mach; warhead 165 kg.
Guns: 2 Oerlikon 30 mm GCM-B01 (twin); 650 rds/min to 10 km *(5.5 n miles);* weight of shell 1 kg. 2—7.62 mm MGs. 2 MOD(N) 2 in launchers for illuminants.
Countermeasures: ESM: Decca RDL; radar warning.
Weapons control: Sea Archer system with Sperry Co-ordinate Calculator and 1412A digital computer. Radamec 2500 optronic director.
Radars: Surface search: Kelvin Hughes Type 1007; I-band.

Modernisation: Started in 1988 and included improved gun fire control and ESM equipment. Further improvements in 1998-2000 included Type 1007 radar and a Radamec 2500 optronic director.
Structure: Welded steel hull with aluminium alloy superstructure. *Waspada* has an enclosed upper bridge for training purposes.
Operational: Reported active. **UPDATED**

PEJUANG (with optronic director) *7/2000 /* 0104244

SETERIA *5/1998 /* 0017639

3 PERWIRA CLASS (COASTAL PATROL CRAFT) (PB)

Name	No	Builders	Launched	Commissioned
PERWIRA	P 14	Vosper (Singapore)	5 May 1974	9 Sep 1974
PEMBURU	P 15	Vosper (Singapore)	30 Jan 1975	17 June 1975
PENYERANG	P 16	Vosper (Singapore)	20 Mar 1975	24 June 1975

Displacement, tons: 38 full load
Dimensions, feet (metres): 71 × 20 × 5 *(21.7 × 6.1 × 1.2)*
Main machinery: 2 MTU MB 12V 331 TC81 diesels; 2,450 hp(m) *(1.8 MW)* sustained; 2 shafts
Speed, knots: 32. **Range, n miles:** 600 at 22 kt; 1,000 at 16 kt
Complement: 14 (2 officers)
Guns: 2 Oerlikon/BMARC 20 mm GAM-B01; 800 rds/min to 2 km; weight of shell 0.24 kg. 2—7.62 mm MGs.
Radars: Surface search: Racal Decca RM 1290; I-band.

Comment: Of all-wooden construction on laminated frames. Fitted with enclosed bridges-modified July 1976. A high speed RIB is launched from a stern ramp. New guns fitted in mid-1980s. P 15 may not be operational. **VERIFIED**

PENYERANG *6/1999, Royal Brunei Armed Forces /* 0056628

AUXILIARIES

2 TERABAN CLASS (LCU)

Name	No	Builders	Commissioned
TERABAN	33	Transfield, Perth	8 Nov 1996
SERASA	34	Transfield, Perth	8 Nov 1996

Displacement, tons: 220 full load
Dimensions, feet (metres): 119.8 × 26.2 × 4.9 *(36.5 × 8 × 1.5)*
Main machinery: 2 diesels; 2 shafts
Speed, knots: 12
Complement: 12
Military lift: 100 tons
Radars: Navigation: Racal; I-band.

Comment: Ordered in November 1995 and delivered in December 1996. Used as utility transports. Bow and side ramps are fitted. Reported active.
VERIFIED

TERABAN *5/1998, John Mortimer /* 0056629

2 CHEVERTON LOADMASTERS (YFU)

Name	No	Builders	Commissioned
DAMUAN	L 31	Cheverton Ltd, Isle of Wight	May 1976
PUNI	L 32	Cheverton Ltd, Isle of Wight	Feb 1977

Displacement, tons: 60; 64 *(Puni)* standard
Dimensions, feet (metres): 65 × 20 × 3.6 *(19.8 × 6.1 × 1.1)* (length 74.8 *(22.8)* Puni)
Main machinery: 2 Detroit 6-71 diesels; 442 hp *(305 kW)* sustained; 2 shafts
Speed, knots: 9. **Range, n miles:** 1,000 at 9 kt
Complement: 8
Military lift: 32 tons
Radars: Navigation: Racal Decca RM 1216; I-band.

VERIFIED

DAMUAN *6/1997, Royal Brunei Armed Forces /* 0012141

LAND-BASED MARITIME AIRCRAFT

Notes: There are also five BO-105, two S-70A and ten Bell 212 utility helicopters.

Numbers/Type: 3 CASA/IPTN CN-235 MPA.
Operational speed: 240 kt *(445 km/h).*
Service ceiling: 26,600 ft *(8,110 m).*
Range: 669 n miles *(1,240 km).*
Role/Weapon systems: Long-range maritime patrol for surface surveillance and ASW. Sensors: Search radar: Litton AN/APS 504(V)5; MAD; acoustic processors; sonobuoys. Weapons: Mk 46 torpedoes.

VERIFIED

CN-235 *3/2003, CASA/EADS /* 0569788

POLICE

Notes: In addition to the vessels listed below there are two Rotork type *Behagia* 07 and *Selamat* 10 and four River Patrol Craft *Aman* 01, *Damai* 02, *Sentosa* 04 and *Sejahtera* 06.

PDB 15   *3/1999, John Webber /* 0056631

7 INSHORE PATROL CRAFT

PDB 11-15   PDB 63   PDB 68

Displacement, tons: 20 full load
Dimensions, feet (metres): 47.7 × 13.9 × 3.9 *(14.5 × 4.2 × 1.2)*
Main machinery: 2 MAN D 2840 LE diesels; 1,040 hp(m) *(764 kW)* sustained; 2 shafts
Speed, knots: 30. **Range, n miles:** 310 at 22 kt
Complement: 7
Guns: 1—7.62 mm MG.
Radars: Surface search: Furuno; I-band.

Comment: Built by Singapore SBEC. First three handed over in October 1987, second pair in 1988, last two in 1996. Aluminium hulls.
VERIFIED

3 BENDEHARU CLASS (PB)

BENDEHARU P 21   MAHARAJALELA P 22   KEMAINDERA P 23

Displacement, tons: 68 full load
Dimensions, feet (metres): 93.5 × 17.8 × 5.6 *(28.5 × 5.4 × 1.7)*
Main machinery: 2 MTU diesels; 2,260 hp *(1.7 MW)*; 2 shafts
Speed, knots: 29
Guns: 1—12.7 mm MG.
Radars: Navigation: I-band.

Comment: Constructed by PT Pal, Surabaya, and entered service in 1991.
NEW ENTRY

Bulgaria
VOENNOMORSKI SILI

Country Overview

Situated in the Balkan Peninsula, the Republic of Bulgaria has an area of 42,823 square miles and is bordered to the north by Romania and to the south by Turkey and Greece. The River Danube forms much of the northern border. Bulgaria has a coastline of 191 n miles with the Black Sea on which Varna and Burgas are the principal ports. The capital is Sofia. Territorial waters (12 n miles) are claimed. An Exclusive Economic Zone (EEZ) was declared in 1987 but the precise limits have yet to be fully agreed and defined.

Headquarters Appointments

Commander of the Navy and Chief of Staff:
    Rear Admiral Neiko Petrov Atanasov

Diplomatic Representation

Defence Attaché, London:
    Major General Luben Pandev

Organisation

Four squadrons: Submarine, Surface, MCMV and Auxiliary, with Headquarters at Varna and Burgas. There is also a Border Guard Unit.

Personnel

(a)   2005: 4,140 (695 officers)
(b)   12 months' national service
(c)   Reserves 10,000

Bases

Varna; Naval HQ (North Zone), Naval Base, Air Station
Burgas: Naval HQ (South Zone)
Sozopol, Atiya, Balchik, Vidin (Danube); Naval Bases
Higher Naval School *(Nikola Yonkov Vaptsarov)* at Varna.

Coast Defence

One battalion with six truck-mounted SS-C-3 Styx twin launchers. Two Army regiments of coastal artillery with 100 mm and 130 mm guns.

SUBMARINES

1 ROMEO CLASS (PROJECT 633) (SS)

SLAVA 84

Displacement, tons: 1,475 surfaced; 1,830 dived
Dimensions, feet (metres): 251.3 × 22 × 16.1 *(76.6 × 6.7 × 4.9)*
Main machinery: Diesel-electric; 2 Type 37-D diesels; 4,000 hp (m) *(2.94 MW)*; 2 motors; 2,700 hp(m) *(1.98 MW)*; 2 creep motors; 2 shafts
Speed, knots: 16 surfaced; 13 dived.
Range, n miles: 9,000 at 9 kt surfaced
Complement: 54

Torpedoes: 8—21 in *(533 mm)* tubes (6 bow, 2 stern). 14 SAET-60; passive homing to 15 km *(8.1 n miles)* at 40 kt; warhead 400 kg.
Mines: Can carry up to 28 in lieu of torpedoes.
Countermeasures: ESM: Stop Light; radar warning.
Radars: Surface search: Snoop Plate; I-band.
Sonars: Hull-mounted; active/passive search and attack; high frequency.

Programmes: Built in 1961. Transferred from the USSR in 1986. An order for two Kilo class was subsequently cancelled.
Operational: Restricted to diving to about 50 m *(165 ft)*. Based at Varna. Attempts have been made to keep this last boat operational by cannibalising others of the class and operational status is doubtful.
VERIFIED

SLAVA   *6/1997 /* 0012100

FRIGATES

Notes: A Letter of Intent was signed in late 2004 between the governments of Belgium and Bulgaria for the acquisition of the frigate *Wandelaar* for the Bulgarian Navy. Subject to satisfactory trials and a final contract, transfer is expected to take place in late 2005.

1 KONI CLASS (PROJECT 1159) (FFLM)

SMELI (ex-*Delfin*) 11

Displacement, tons: 1,440 standard; 1,900 full load
Dimensions, feet (metres): 316.3 × 41.3 × 11.5
(96.4 × 12.6 × 3.5)
Main machinery: CODAG; 1 SGW, Nikolayev M8B gas turbine (centre shaft); 18,000 hp(m) *(13.25 MW)* sustained; 2 Russki B-68 diesels; 15,820 hp(m) *(11.63 MW)* sustained; 3 shafts
Speed, knots: 27 gas; 22 diesel. **Range, n miles:** 1,800 at 14 kt
Complement: 110

Missiles: SAM: SA-N-4 Gecko twin launcher ❶; semi-active radar homing to 15 km *(8 n miles)* at 2.5 Mach; warhead 50 kg; altitude 9.1-3,048 m *(30-10,000 ft)*; 20 missiles.
Guns: 4—3 in *(76 mm)*/60 (2 twin) ❷; 60 rds/min to 15 km *(8 n miles)*; weight of shell 7 kg.
4—30 mm/65 (2 twin) ❸; 500 rds/min to 5 km *(2.7 n miles)*; weight of shell 0.54 kg.
A/S mortars: 2 RBU 6000 12-tubed trainable ❹; range 6,000 m; warhead 31 kg.
Depth charges: 2 racks.
Mines: Capacity for 22.
Countermeasures: Decoys: 2 PK 16 chaff launchers.
ESM: 2 Watch Dog; radar warning.

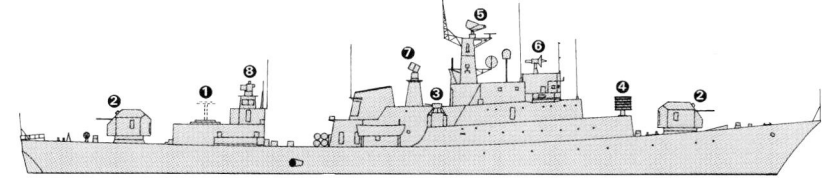

SMELI *(Scale 1 : 900)*, Ian Sturton / 0114505

Radars: Air search: Strut Curve ❺; F-band; range 110 km *(60 n miles)* for 2 m² target.
Surface search: Don 2; I-band.
Fire control: Hawk Screech ❻; I-band (for 76 mm). Drum Tilt ❼; H/I-band (for 30 mm). Pop Group ❽; F/H/I-band (for SA-N-4).
IFF: High Pole B.
Sonars: Hercules (MG 322); hull-mounted; active search and attack; medium frequency.

Programmes: First reported in the Black Sea in 1976. Type I retained by the USSR for training foreign crews but transferred in February 1990 when the Koni programme terminated. Others of the class acquired by the former East German Navy (now deleted), Yugoslavia, Algeria, Cuba and Libya.
Modernisation: Marisat fitted in 1996. Reported to be RAS capable. Communications upgrade planned to achieve NATO interoperability.
Operational: Based at Varna.

UPDATED

SMELI 6/2004*, C D Yaylali / 0587693

LAND-BASED MARITIME AIRCRAFT (FRONT LINE)

Notes: (1) A number of Air Force MiG-23s have AS-7 Kerry ASMs.
(2) Three Hormone B helicopters are non-operational.

Numbers/Type: 3 Mil Mi-14PL 'Haze A'.
Operational speed: 120 kt *(222 km/h)*.
Service ceiling: 15,000 ft *(4,570 m)*.
Range: 240 n miles *(445 km)*.
Role/Weapon systems: Primary role as inshore/coastal ASW and Fleet support helicopter; one converted as transport. Based at Asparukhovo airport. Sensors: Search radar, MAD, sonobuoys, dipping sonar. Weapons: ASW; up to 2 × torpedoes, or mines, or depth bombs.
VERIFIED

MULNIYA 7/2000, van Ginderen Collection / 0104245

CORVETTES

Notes: A plan to procure a new class of multirole corvettes is understood to be a high priority within a wider modernisation programme to meet NATO requirements. A class of six ships is planned of which the first is expected to be built by a strategic shipbuilding partner with a further five to be built at the Navy Maintenance and Repair Shipyard near Varna. Talks have been conducted with prospective partners.

1 TARANTUL II CLASS (PROJECT 1241.1M) (FSGM)

MULNIYA 101

Displacement, tons: 385 standard; 455 full load
Dimensions, feet (metres): 184.1 × 37.7 × 8.2 *(56.1 × 11.5 × 2.5)*
Main machinery: COGAG; 2 Nikolayev Type DR 77 gas turbines; 16,016 hp(m) *(11.77 MW)* sustained; 2 Nikolayev Type DR 76 gas turbines with reversible gearboxes; 4,993 hp(m) *(3.67 MW)* sustained; 2 shafts
Speed, knots: 36 on 4 turbines. **Range, n miles:** 400 at 36 kt; 2,000 at 20 kt
Complement: 34 (5 officers)

Missiles: SSM: 4 Raduga SS-N-2C Styx (2 twin) launchers; active radar or IR homing to 83 km *(45 n miles)* at 0.9 Mach; warhead 513 kg; sea-skimmer.
SAM: SA-N-5 Grail quad launcher; manual aiming; IR homing to 6 km *(3.2 n miles)* at 1.5 Mach; altitude 2,500 m *(8,000 ft)*; warhead 1.5 kg.
Guns: 1—3 in *(76 mm)*/60; 120 rds/min to 15 km *(8.1 n miles)*; weight of shell 7 kg.
2—30 mm/65; 6 barrels per mounting; 3,000 rds/min to 2 km.
Countermeasures: Decoys: 2 PK 16 chaff launchers.
ESM: 2 Half Hat; intercept.
Weapons control: Hood Wink optronic director. Band Stand datalink for SSM.
Radars: Air/surface search: Plank Shave; E-band.
Navigation: Kivach; I-band.
Fire control: Bass Tilt; H/I-band.
IFF: Square Head. High Pole.

Programmes: Built at Volodarski, Rybinsk. Transferred from USSR in December 1989. Name means Thunderbolt.
Operational: Based at Sozopol.

VERIFIED

4 LETYASHTI (POTI) CLASS (PROJECT 204) (FS)

LETYASHTI 41	BDITELNI 42	BEZSTRASHNI 43	KHRABRI 44

Displacement, tons: 545 full load
Dimensions, feet (metres): 196.8 × 26.2 × 6.6 *(60 × 8 × 2)*
Main machinery: CODAG; 2 gas-turbines; 30,000 hp(m) *(22.4 MW)*; 2 Type M 503A diesels; 5,350 hp(m) *(3.91 MW)* sustained; 2 shafts
Speed, knots: 32. **Range, n miles:** 3,000 at 18 kt; 500 at 37 kt
Complement: 80

Guns: 2 USSR 57 mm/80 (twin); 120 rds/min to 6 km *(3 n miles)*; weight of shell 2.8 kg.
Torpedoes: 4—16 in *(406 mm)* tubes. SAET-40; anti-submarine; active/passive homing to 10 km *(5.4 n miles)* at 30 kt; warhead 100 kg.
A/S mortars: 2 RBU 6000 12-tubed trainable launchers; range 6,000 m; warhead 31 kg.
Countermeasures: ESM: 2 Watch Dog; radar warning.
Radars: Air search: Strut Curve; F-band.
Surface search: Don; I-band.
Fire control: Muff Cob; G/H-band.
IFF: Square Head. High Pole.
Sonars: Hull-mounted; active search and attack; high frequency.

Programmes: Series built at Zelenodolsk between 1961 and 1968. Three transferred from USSR December 1975, the fourth at the end of 1986 and the last two in 1990. Two deleted in 1993. Names: 41 Flying, 42 Vigilant, 43 Fearless and 44 Gallant.
Operational: Based at Atiya and probably non-operational.
VERIFIED

LETYASHTI 6/1996, Bulgarian Navy

2 RESHITELNI (PAUK I) (PROJECT 1241P) CLASS (FSM)

RESHITELNI 13   BODRI 14

Displacement, tons: 440 full load
Dimensions, feet (metres): 195.2 × 33.5 × 10.8 *(59.5 × 10.2 × 3.3)*
Main machinery: 2 Type 521 diesels; 16,180 hp(m) *(11.9 MW)* sustained; 2 shafts
Speed, knots: 32. **Range, n miles:** 2,200 at 14 kt
Complement: 38

Missiles: SAM: SA-N-5 Grail quad launcher; manual aiming; IR homing to 6 km *(3.2 n miles)* at
   1.5 Mach; altitude to 2,500 m *(8,000 ft)*; warhead 1.5 kg; 8 missiles.
Guns: 1—3 in *(76 mm)*/60; 120 rds/min to 15 km *(8 n miles)*; weight of shell 7 kg.
   1—30 mm/65; 6 barrels; 3,000 rds/min combined to 2 km.
Torpedoes: 4—16 in *(406 mm)* tubes. Type 40; anti-submarine; active/passive homing up to 15 km
   (8 n miles) at up to 40 kt; warhead 100-150 kg.
A/S mortars: 2 RBU 1200 5-tubed fixed; range 1,200 m; warhead 34 kg.
Depth charges: 2 racks (12).
Countermeasures: Decoys: 2 PK 16 chaff launchers.
ESM: 3 Brick Plug; intercept.
Radars: Air/surface search: Peel Cone; E-band.
Surface search: Spin Trough; I-band.
Fire control: Bass Tilt; H/I-band.
Sonars: Foal Tail VDS (mounted on transom); active attack; high frequency.

Programmes: *Reshitelni* transferred from USSR in September 1989, *Bodri* in December 1990.
Operational: Based at Varna. *Reshitelni* is non-operational with propulsion problems.
                                                                      VERIFIED

BODRI   *9/2002, C D Yaylali* / 0533316

PATROL FORCES

Notes: Customs craft operate on the Danube. Vessels include three Boston Whalers donated by
the US and RIBs given by the UK in 1992-93.

9 ZHUK (PROJECT 1400M) CLASS
(COASTAL PATROL CRAFT) (PB)

511-513   521-523   531-533

Displacement, tons: 39 full load
Dimensions, feet (metres): 78.7 × 16.4 × 3.9 *(24 × 5 × 1.2)*
Main machinery: 2 Type M 401B diesels; 2,200 hp(m) *(1.6 MW)* sustained; 2 shafts
Speed, knots: 30. **Range, n miles:** 1,100 at 15 kt
Complement: 11 (3 officers)
Guns: 4 USSR 14.5 mm (2 twin) MGs.
Radars: Surface search: Spin Trough; I-band.

Comment: Transferred from USSR 1980-81. Belong to the Border Police under the Minister of the
Interior and have 'Border Guard' insignia on the ships side. Based at Atiya and at Varna.
                                                                      VERIFIED

ZHUK 512 (and others)   *6/1996, Bulgarian Navy*

2 NEUSTADT CLASS (PB)

SOZOPOL (ex-*Rosenheim*) 525 (ex-BG 18)   NESEBAR (ex-*Neustadt*) 526 (ex-BG 11)

Displacement, tons: 218 full load
Dimensions, feet (metres): 127.1 × 23 × 5 *(38.5 × 7 × 2.2)*
Main machinery: 2 MTU MD diesels; 6,000 hp(m) *(4.41 MW)*; 1 MWM diesel; 685 hp(m)
   (500 kW); 3 shafts
Speed, knots: 30. **Range, n miles:** 450 at 27 kt
Complement: 17
Guns: 2—7.62 mm MGs.
Radars: Surface search: Selenia ARP 1645; I-band.
Navigation: Racal Decca Bridgemaster MA 180/4; I-band.

Comment: Built in 1970 by Lürssen, Vegesack. 525 transferred from German Border Guard in June
2002 and *526* on 16 April 2004. Operated by the Border Police.
                                                                      UPDATED

NESEBAR   *5/2004*, Martin Mokrus* / 0587692

6 OSA (PROJECT 205) CLASS
(FAST ATTACK CRAFT—MISSILE) (PTFG)

URAGON 102   SVETKAVITSA 111
BURYA 103 (Osa I)   TYPHOON 112 (Osa 1)
GRUM 104   SMERCH 113

Displacement, tons: 245 full load; 210 (Osa I)
Dimensions, feet (metres): 126.6 × 24.9 × 8.8 *(38.6 × 7.6 × 2.7)*
Main machinery: 3 Type M 504 diesels; 10,800 hp(m) *(7.94 MW)* sustained; 3 shafts (Osa II)
   3 Type 503A diesels; 8,025 hp(m) *(5.9 MW)* sustained; 3 shafts (Osa I)
Speed, knots: 37 (Osa II); 35 (Osa I). **Range, n miles:** 500 at 35 kt
Complement: 26 (3 officers)
Missiles: SSM: 4 SS-N-2A/B Styx; active radar/IR homing to 46 km *(25 n miles)* at 0.9 Mach;
   warhead 513 kg. SS-N-2A in Osa I.
Guns: 4 USSR 30 mm/65 (2 twin); 500 rds/min to 5 km *(2.7 n miles)*; weight of shell 0.54 kg.
Radars: Surface search/fire control: Square Tie; I-band.
Fire control: Drum Tilt; H/I-band.
IFF: High Pole. Square Head.

Comment: Four Osa IIs built between 1965 and 1970, and transferred from USSR between 1977
and 1982. Two Osa Is transferred in 1972 and survived longer than expected. Names: 102
Hurricane, 103 Storm, 104 Thunder, 111 Lightning, 112 Typhoon and 113 Tornado. All based at
Sozopol and seldom go to sea.
                                                                      VERIFIED

GRUM   *6/2002, A Sheldon-Duplaix* / 0524968

BURYA   *6/2002, A Sheldon-Duplaix* / 0524970

2 COASTAL PATROL CRAFT (PB)

BOURGAS   VARNA

Displacement, tons: 50 standard
Dimensions, feet (metres): 68.9 × 19.0 × 4.6 *(21.0 × 5.8 × 1.4)*
Main machinery: 2 Deutz MWM TBD 616 diesels; 2,970 hp(m) *(2.2 MW)*; 2 shafts
Speed, knots: 30

Comment: Contract awarded in November 2002 to Lürssen, Berne-Bardenfleth. Delivery made in
2003. Operated by Border Police.
                                                                      UPDATED

MINE WARFARE FORCES

Notes: Six Vydra class (see *Amphibious Forces*) converted to minelayers in 1992-93. Some are in reserve.

4 BRIZ (SONYA) (PROJECT 12650) CLASS
(MINESWEEPERS—COASTAL) (MSC)

BRIZ 61	SHKVAL 62	PRIBOY 63	SHTORM 64

Displacement, tons: 450 full load
Dimensions, feet (metres): 157.4 × 28.9 × 6.6 *(48 × 8.8 × 2)*
Main machinery: 2 Kolomna Type 9-D-8 diesels; 2,000 hp(m) *(1.47 MW)* sustained; 2 shafts
Speed, knots: 15. **Range, n miles:** 1,500 at 14 kt
Complement: 43 (5 officers)
Guns: 2 USSR 30 mm/65 (twin); 500 rds/min to 5 km *(2.7 n miles)*; weight of shell 0.54 kg.
    2 USSR 25 mm/80 (twin); 270 rds/min to 3 km *(1.6 n miles)*; weight of shell 0.34 kg.
Mines: 5.
Radars: Surface search/navigation: Kivach; I-band.
IFF: Two Square Head. High Pole B.
Sonars: MG 69/79; hull-mounted; active minehunting; high frequency.

Comment: Wooden hulled ships transferred from USSR in 1981-84. Based at Atiya.   ***VERIFIED***

PRIBOY   *7/2000* / 0114506

4 ISCAR (VANYA) (PROJECT 257D) CLASS
(MINESWEEPERS—COASTAL) (MSC)

ISKAR 31	ZIBAR 32	DOBROTICH 33	EVSTATI VINAROV 34

Displacement, tons: 245 full load
Dimensions, feet (metres): 131.2 × 23.9 × 5.9 *(40 × 7.3 × 1.8)*
Main machinery: 2 M 870 diesels; 2,502 hp(m) *(1.84 MW)*; 2 shafts; cp props
Speed, knots: 16. **Range, n miles:** 2,400 at 10 kt
Complement: 36
Guns: 2 USSR 30 mm/65 (twin); 500 rds/min to 5 km *(2.7 n miles)*; weight of shell 0.54 kg.
Mines: Can carry 8.
Radars: Surface search: Don 2; I-band.
Sonars: MG 69/79; hull-mounted; active minehunting; high frequency.

Comment: Built 1961 to 1973. Transferred from the USSR—two in 1970, two in 1971 and two in 1985. Can act as minehunters. Two paid off in 1992, but back in service in 1994 and then finally scrapped in 1995. Based at Varna.   ***VERIFIED***

EVSTATI VINAROV   *8/2000* / 0114508

4 YEVGENYA (PROJECT 1258) CLASS
(MINESWEEPERS—COASTAL) (MSC)

65	66	67	68

Displacement, tons: 77 standard; 90 full load
Dimensions, feet (metres): 80.4 × 18 × 4.6 *(24.5 × 5.5 × 1.4)*
Main machinery: 2 Type 3-D-12 diesels; 600 hp(m) *(440 kW)* sustained; 2 shafts
Speed, knots: 11. **Range, n miles:** 300 at 10 kt
Complement: 10 (1 officer)
Guns: 2—25 mm/80 (twin).
Mines: 8 racks.
Radars: Surface search: Spin Trough; I-band.
IFF: High Pole.
Sonars: MG-7 lifted over stern; active; high frequency.

Comment: GRP hulls built at Kolpino. Transferred from USSR 1977. *65* and *66* based at Varna and *67* and *68* at Atiya.   ***VERIFIED***

YEVGENYA 66   *6/1996, Bulgarian Navy*

2 PO 2 (PROJECT 501) CLASS
(MINESWEEPERS—INSHORE) (MSB)

57	58

Displacement, tons: 56 full load
Dimensions, feet (metres): 70.5 × 11.5 × 3.3 *(21.5 × 3.5 × 1)*
Main machinery: 1 Type 3-D-12 diesel; 300 hp(m) *(220 kW)* sustained; 2 shafts
Speed, knots: 12
Complement: 8

Comment: Built in Bulgaria. First units completed in early 1950s and last in early 1960s. Originally a class of 24 and these two are the last to survive. Occasionally carry a 12.7 mm MG, when used for patrol duties. Both based at Balchik.   ***VERIFIED***

PO 2 58   *7/2000, van Ginderen Collection* / 0104252

6 OLYA (PROJECT 1259) CLASS
(MINESWEEPERS—INSHORE) (MSB)

51	52	53	54	55	56

Displacement, tons: 64 full load
Dimensions, feet (metres): 84.6 × 14.9 × 3.3 *(25.8 × 4.5 × 1)*
Main machinery: 2 Type 3D 6S11/235 diesels; 471 hp(m) *(346 kW)* sustained; 2 shafts
Speed, knots: 12. **Range (miles):** 300 at 10 kt
Complement: 15
Guns: 2—12.7 mm MGs (twin).
Radars: Navigation: Pechora; I-band.

Comment: First five built between 1988 and 1992 in Bulgaria to the Russian Olya design. *56* completed in 1996. Minesweeping equipment includes AT-6, SZMT-1 and 3 PKT-2 systems. *55* based at Varna, the remainder at Balchik.   ***VERIFIED***

OLYA 52   *7/2000, van Ginderen Collection* / 0104250

AMPHIBIOUS FORCES

2 POLNOCHNY A (PROJECT 770) CLASS (LSM)

SIRIUS (ex-*Ivan Zagubanski*) 701   ANTARES 702

Displacement, tons: 750 standard; 800 full load
Dimensions, feet (metres): 239.5 × 27.9 × 5.8 *(73 × 8.5 × 1.8)*
Main machinery: 2 Kolomna Type 40-D diesels; 4,400 hp(m) *(3.2 MW)* sustained; 2 shafts
Speed, knots: 19. **Range, n miles:** 1,000 at 18 kt
Complement: 40
Military lift: 350 tons including 6 tanks; 180 troops
Guns: 2 USSR 30 mm (twin). 2—140 mm 18-barrelled rocket launchers.
Radars: Navigation: Spin Trough; I-band.

Comment: Built 1963 to 1968. Transferred from USSR 1986-87. Not fitted either with the SA-N-5 Grail SAM system or with Drum Tilt fire-control radars. Plans to convert them to minelayers have been shelved and both are now used as transports. Based at Atiya.   *VERIFIED*

ANTARES   *7/2000, van Ginderen Collection /* 0104251

7 VYDRA (PROJECT 106K) CLASS (LCU)

205   703-707   712

Displacement, tons: 425 standard; 550 full load
Dimensions, feet (metres): 179.7 × 25.3 × 6.6 *(54.8 × 7.7 × 2)*
Main machinery: 2 Type 3-D-12 diesels; 600 hp(m) *(440 kW)* sustained; 2 shafts
Speed, knots: 12. **Range, n miles:** 2,500 at 10 kt
Complement: 20
Military lift: 200 tons or 100 troops or 3 MBTs
Radars: Navigation: Don 2; I-band.
IFF: High Pole.

Comment: Built 1963 to 1969. Ten transferred from the USSR in 1970, the remainder built in Bulgaria between 1974 and 1978. In 1992-93 *703-707* and *712* converted to be used as minelayers. Many deleted. *205* based at Varna, the remainder at Atiya.   *VERIFIED*

VYDRA 706 (and others)   *7/1995, Alexander Mladenov /* 0056636

SURVEY SHIPS

1 MOMA (PROJECT 861) CLASS (AGS)

ADMIRAL BRANIMIR ORMANOV 401

Displacement, tons: 1,580 full load
Dimensions, feet (metres): 240.5 × 36.8 × 12.8 *(73.3 × 11.2 × 3.9)*
Main machinery: 2 Zgoda-Sulzer 6TD48 diesels; 3,300 hp(m) *(2.43 MW)* sustained; 2 shafts; cp props
Speed, knots: 17. **Range, n miles:** 9,000 at 12 kt
Complement: 37 (5 officers)
Radars: Navigation: 2 Don-2; I-band.

Comment: Built at Northern Shipyard, Gdansk, Poland in 1977. Based at Varna. Two others of the class belonging to Russia were refitted in Bulgaria in 1995-96.   *VERIFIED*

ADMIRAL BRANIMIR ORMANOV   *7/2000, van Ginderen Collection /* 0104253

2 COASTAL SURVEY VESSELS (PROJECT 612) (AGSC)

231   331

Displacement, tons: 114 full load
Dimensions, feet (metres): 87.6 × 19 × 4.9 *(26.7 × 5.8 × 1.5)*
Main machinery: 2 Type 3-D-12 diesels; 600 hp(m) *(440 kW)* sustained; 2 shafts
Speed, knots: 12. **Range, n miles:** 600 at 10 kt
Complement: 9 (2 officers)
Radars: Navigation: I-band.

Comment: Built in Bulgaria in 1986 and 1988 respectively. Can carry 2 tons of equipment. *231* is based at Varna and *331* at Atiya.   *VERIFIED*

AGSC 331   *6/1996, Bulgarian Navy*

AUXILIARIES

1 SUPPORT TANKER (AOTL)

203

Displacement, tons: 1,250 full load
Dimensions, feet (metres): 181.8 × 36.1 × 11.5 *(55.4 × 11 × 3.5)*
Main machinery: 2 Sulzer 6AL-20-24 diesels; 1,500 hp(m) *(1.1 MW)*; 2 shafts
Speed, knots: 12. **Range, n miles:** 1,000 at 8 kt
Complement: 23
Cargo capacity: 650 tons fuel
Guns: 2 ZU-23-2F Wrobel 23 mm (twin).
Radars: Navigation: I-band.

Comment: Laid down 1989, launched 1993 and completed in 1994 at Burgas Shipyards, Burgas. Based at Varna.   *VERIFIED*

203   *1/1998 /* 0017648

1 MESAR CLASS (PROJECT 102) (SUPPORT TANKER) (AORL)

ATIYA 302

Displacement, tons: 3,240 full load
Dimensions, feet (metres): 319.8 × 45.6 × 16.4 *(97.5 × 13.9 × 5)*
Main machinery: 2 diesels; 12,000 hp(m) *(8.82 MW)*; 2 shafts
Speed, knots: 18. **Range, n miles:** 12,000 at 15 kt
Complement: 32 (6 officers)
Cargo capacity: 1,593 tons
Guns: 4 USSR 30 mm/65 (2 twin).
Radars: Navigation: 2 Don 2; I-band.

Comment: Built in Bulgaria in 1987. Abeam fuelling to port and astern fuelling. Mount 1.5 ton crane amidships. Also carries dry stores. Based at Atiya.   *VERIFIED*

ATIYA   *7/2002, S Breyer /* 0568845

1 DIVING TENDER (PROJECT 245) (YDT)

223

Displacement, tons: 112 full load
Dimensions, feet (metres): 91.5 × 17.1 × 7.2 *(27.9 × 5.2 × 2.2)*
Main machinery: Diesel-electric; 2 MCK 83-4 diesel generators; 1 motor; 300 hp(m) *(220 kW)*;
1 shaft
Speed, knots: 10. **Range, n miles:** 400 at 10 kt
Complement: 6 + 7 divers
Radars: Navigation: Don 2; I-band.

Comment: Built in Bulgaria in mid-1980s. A twin 12.7 mm MG can be fitted. Capable of bell diving
to 60 m. Based at Varna.
VERIFIED

YDT 223 *6/1998, S Breyer collection /* 0017650

1 BEREZA (PROJECT 130) CLASS (ADG/AX)

KAPITAN 1st RANK DIMITRI DOBREV 206

Displacement, tons: 2,051 full load
Dimensions, feet (metres): 228 × 45.3 × 13.1 *(69.5 × 13.8 × 4)*
Main machinery: 2 Zgoda-Sulzer 8 AL 25/30 diesels; 2,925 hp(m) *(2.16 MW)* sustained; 2 shafts;
cp props
Speed, knots: 13. **Range, n miles:** 1,000 at 13 kt
Complement: 48
Radars: Navigation: Kivach; I-band.

Comment: New construction built in Poland and transferred July 1988. Used as a degaussing ship.
Fitted with an NBC citadel and upper deck wash-down system. The ship has three laboratories.
Has also been used as a training ship. Based at Varna.
UPDATED

KAPITAN 1st RANK DIMITRI DOBREV *6/2004 *, Giorgio Ghiglione /* 0587694

1 TYPE 700 SALVAGE TUG (ATS)

JUPITER 221

Displacement, tons: 792 full load
Dimensions, feet (metres): 146.6 × 35.1 × 12.7 *(44.7 × 10.7 × 3.9)*
Main machinery: 2—12 KVD 21 diesels; 1,760 hp(m) *(1.3 MW)*; 2 shafts
Speed, knots: 12.5. **Range, n miles:** 3,000 at 12 kt
Complement: 39 (6 officers)
Guns: 4—25 mm/70 (2 twin) automatic (can be carried).
Radars: Navigation: I-band.

Comment: Built at Peenewerft Shipyard and completed 20 March 1964. Bollard pull, 16 tons.
Based at Varna.
VERIFIED

JUPITER *8/1996 /* 0056638

6 AUXILIARIES

| KALIAKRA | OLEV BLAGOEV 421 | 224 | 312 | 313 | 321 |

Comment: *Olev Blagoev* is a survey vessel converted to a training ship. *224* and *321* are firefighting
vessels. *312* and *313* are tugs. *Kaliakra* is a 380 ton barquentine used for sail training.
VERIFIED

OLEV BLAGOEV *6/2003, Schaeffer/Marsan /* 0567877

224 *7/2000, van Ginderen Collection /* 0104254

1 SALVAGE SHIP (ARS)

Name	No	Builders	Commissioned
— (ex-*Proteo*, ex-*Perseo*)	224 (ex-A 5310)	Cantieri Navali Riuniti, Ancona	24 Aug 1951

Displacement, tons: 1,865 standard, 2,147 full load
Dimensions, feet (metres): 248 × 38 × 21 *(75.6 × 11.6 × 6.4)*
Main machinery: 2 Fiat diesels; 4,800 hp(m) *(3.53 MW)*; 1 shaft
Speed, knots: 16. **Range, n miles:** 7,500 at 13 kt
Complement: 122 (8 officers)
Radars: Navigation: SMA-748; I-band.

Comment: Transferred to Bulgaria on 3 June 2004 having been decommissioned from the Italian
Navy in 2002. Originally laid down in 1943, construction was suspended until restarted in 1949.
Details are those of the ship when in Italian service.
NEW ENTRY

224 *6/2004 *, Giorgio Ghiglione /* 0580523

Cambodia

Country Overview

Formerly a French protectorate, the south-east Asian Kingdom of
Cambodia was ravaged by the Vietnam War and then by the
Khmer Rouge regime before relative stability followed the
nation's first multiparty elections in 1993. With an overall land
area of 69,898 square miles, the country is bordered to the north
by Thailand and Laos and to the east by Vietnam. There is a 239 n
mile coastline with the Gulf of Thailand. The capital and largest
city is Phnom Penh while the principal port is Kompong Som.
There are extensive inland waterways. Territorial seas (12 n
miles) are claimed. An EEZ (200 n miles) is claimed but the limits
have not been fully defined.

Headquarters Appointments

Commander of Navy:
Vice Admiral Ung Som Khan

Personnel

2005: 2,800 (780 officers) including marines

Bases

Ream (ocean), Phnom Penh (river), Kompongson (civil)

Organisation

Ocean Division has nine battalions and the River Division seven
battalions. Command HQ is at Phnom Penh.

PATROL FORCES

Notes: (1) There are also about 170 motorised and manual canoes.
(2) Three new patrol craft are required.

2 KAOH CLASS (RIVER PATROL CRAFT) (PBR)

KAOH CHHLAM 1105   KAOH RONG 1106

Displacement, tons: 44 full load
Dimensions, feet (metres): 76.4 × 20 × 3.9 *(23.3 × 6.1 × 1.2)*
Main machinery: 2 Deutz/MWM TBD 616 V16 diesels; 2,992 hp(m) *(2.2 MW)*; 2 shafts
Speed, knots: 34. **Range, n miles:** 400 at 30 kt
Complement: 13 (3 officers)
Guns: 2—14.5 mm MG (twin). 2—12.7 mm MGs.
Radars: Surface search: Racal Decca Bridgemaster; I-band.

Comment: Ordered from Hong Leong Shipyard, Butterworth to a German design in 1995 and delivered 20 January 1997. Aluminium construction.

VERIFIED

KAOH CHHLAM   *1/1997, Hong Leong Shipyard /* 0056667

2 MODIFIED STENKA CLASS (PROJECT 205P)
(FAST ATTACK CRAFT—PATROL) (PBF)

MONDOLKIRI 1131   RATANAKIRI 1134

Displacement, tons: 211 standard; 253 full load
Dimensions, feet (metres): 129.3 × 25.9 × 8.2 *(39.4 × 7.9 × 2.5)*
Main machinery: 3 Caterpillar diesels; 14,000 hp(m) *(10.29 MW)*; 3 shafts
Speed, knots: 37. **Range, n miles:** 800 at 24 kt; 500 at 35 kt
Complement: 25 (5 officers)
Guns: 2—23 mm/87 (twin). 1 Bofors 40 mm/70.
Radars: Surface search: Racal Decca Bridgemaster; I-band.
Fire control: Muff Cob; G/H-band.
Navigation: Racal Decca; I-band.
IFF: High Pole. 2 Square Head.

Comment: Four transferred from USSR in November 1987. Export model without torpedo tubes and sonar. One pair were modernised in Hong Leong Shipyard, Butterworth, from early 1995 to April 1996. New engines, guns and radars were fitted. The second pair similarly refitted by August 1997. By late 1998 only two were operational although it was reported in 2000 that a third may have undergone a further refit. Pennant numbers were changed for UN operations but changed back again in November 1993.

VERIFIED

MONDOLKIRI   *8/1997, Hong Leong Shipyard /* 0056666

Cameroon
MARINE NATIONALE RÉPUBLIQUE

Country Overview

The Republic of Cameroon became a unitary republic in 1972 and replaced the federation of East Cameroon (formerly French Cameroons) and West Cameroon (formerly part of British Cameroons). With an area of 183,569 square miles, the country has borders to the west with Nigeria and to the south with Gabon and Equatorial Guinea. It has a 217 n mile coastline with Atlantic Ocean on the Bight of Bonny. The capital is Yaoundé while Douala is the principal port which also serves adjacent landlocked states. Kribi is the country's second port. Cameroon is

the only coastal state to claim territorial seas of 50 n miles. It has not been declared an Exclusive Economic Zone (EEZ) and claims to jurisdiction would be complicated by the offshore islands of Bioko (Equatorial Guinea), São Tomé and Principe.

Headquarters Appointments

Chief of Naval Staff:
    Commander Guillaume Ngouah Ngally

Personnel

2005: 1,250

Bases

Douala (HQ), Limbe, Kribi
Construction of new maintenance facilities at Douala started in 2003.

PATROL FORCES

Notes: (1) Ten Rodman 6.5 m craft were delivered in 2000. All have speeds in excess of 25 kt.
(2) There are some eight Simmoneau 10 m craft in service and a further 15 are reported to have been ordered from Raidco Marine, Lorient.

1 BIZERTE (TYPE PR 48) CLASS (LARGE PATROL CRAFT) (PBO)

Name	No	Builders	Commissioned
L'AUDACIEUX	P 103	SFCN, Villeneuve-La-Garenne	11 May 1976

Displacement, tons: 250 full load
Dimensions, feet (metres): 157.5 × 23.3 × 7.5 *(48 × 7.1 × 2.3)*
Main machinery: 2 SACM 195 V12 CZSHR diesels; 6,000 hp(m) *(4.41 MW)* sustained; 2 shafts; cp props
Speed, knots: 23. **Range, n miles:** 2,000 at 16 kt
Complement: 25 (4 officers)
Guns: 2 Bofors 40 mm/70; 300 rds/min to 12.8 km *(7 n miles)*; weight of shell 0.96 kg.

Comment: *L'Audacieux* ordered in September 1974. Laid down on 10 February 1975, launched on 31 October 1975. Similar to Bizerte class in Tunisia. Operational status doubtful and not reported at sea since 1995. Fitted for SS 12M missiles but these are not embarked.

VERIFIED

BIZERTE CLASS (Tunisian colours)   *1993, van Ginderen Collection /* 0056668

1 BAKASSI (TYPE P 48S) CLASS
(OFFSHORE PATROL CRAFT) (PBO)

Name	No	Builders	Launched	Commissioned
BAKASSI	P 104	SFCN, Villeneuve-La-Garenne	22 Oct 1982	9 Jan 1984

Displacement, tons: 308 full load
Dimensions, feet (metres): 172.5 × 23.6 × 7.9 *(52.6 × 7.2 × 2.4)*
Main machinery: 2 SACM 195 V16 CZSHR diesels; 8,000 hp(m) *(5.88 MW)* sustained; 2 shafts
Speed, knots: 25. **Range, n miles:** 2,000 at 16 kt
Complement: 39 (6 officers)
Guns: 2 Bofors 40 mm/70; 300 rds/min to 12.8 km *(7 n miles)*; weight of shell 0.96 kg.
Weapons control: 2 Naja optronic systems. Racal Decca Cane 100 command system.
Radars: Navigation/surface search: 2 Furuno; I-band.

Comment: Ordered January 1981. Laid down 16 December 1981. Six month major refit by Raidco Marine (Lorient) in 1999. This included removing the Exocet missile system and EW equipment, and fitting new propellers and a funnel aft of the mainmast to replace the waterline exhausts. New radars were also installed. Two RIBs are carried.

UPDATED

BAKASSI   *7/1999, H M Steele /* 0121304

1 COASTAL PATROL CRAFT (PB)

QUARTIER MAÎTRE ALFRED MOTTO

Displacement, tons: 96 full load
Dimensions, feet (metres): 95.4 × 20.3 × 6.3 *(29.1 × 6.2 × 1.9)*
Main machinery: 2 Baudouin diesels; 1,290 hp(m) *(948 kW)*; 2 shafts
Speed, knots: 14
Complement: 17 (2 officers)
Guns: 2—7.62 mm MGs.
Radars: Surface search: I-band.

Comment: Built at Libreville, Gabon in 1974. Discarded as a derelict hulk in 1990 but refurbished and brought back into service with assistance from the French Navy in 1995-96.
VERIFIED

QUARTIER MAÎTRE ALFRED MOTTO　　　　*2/1996, French Navy /* 0056670

2 SWIFT PBR CLASS (RIVER PATROL CRAFT) (PBR)

PR 001　　　　　　　　　　PR 005

Displacement, tons: 12 full load
Dimensions, feet (metres): 38 × 12.5 × 3.2 *(11.6 × 3.8 × 1)*
Main machinery: 2 Stewart and Stevenson 6V-92TA diesels; 520 hp *(388 kW)* sustained; 2 shafts
Speed, knots: 32. **Range, n miles:** 210 at 20 kt
Complement: 4
Guns: 2—12.7 mm MGs. 2—7.62 mm MGs.

Comment: Built by Swiftships and supplied under the US Military Assistance Programme. First 10 delivered in March 1987, second 10 in September 1987 and the remainder in March 1988. These last survivors are used by the gendarmerie. Several others have been cannibalised for spares.
UPDATED

PBR class　　　　　　　　　　*4/1992 /* 0056671

2 RODMAN 101 (COASTAL PATROL CRAFT) (PB)

AKWAYAFE P 106　　　　　　　JABANNE P 107

Displacement, tons: 63 full load
Dimensions, feet (metres): 98.4 × 19 × 5.9 *(30 × 5.8 × 1.8)*
Main machinery: 2 diesels; 2,800 hp(m) *(2.06 MW)*; 2 shafts
Speed, knots: 26. **Range, n miles:** 800 at 18 kt
Complement: 9
Guns: 2—12.7 mm MGs.
Radars: Surface search: Furuno; I-band.

Comment: Delivered in late 2000.
VERIFIED

AKWAYAFE　　　　　　*7/2001, Adolfo Ortigueira Gil /* 0524974

4 RODMAN 46 CLASS (PB)

IDABATO VS 201　　ISONGO VS 202　　MOUANCO VS 203　　CAMPO VS 204

Displacement, tons: 12.5
Dimensions, feet (metres): 45.9 × 12.5 × 2.9 *(14.0 × 3.8 × 0.9)*
Main machinery: 2 Caterpillar diesels; 900 hp *(671 kW)*; 2 Hamilton waterjets
Speed, knots: 30
Complement: 4

Comment: GRP hull. Built in 2000 by Rodman, Vigo. Pennant numbers have not been confirmed.
UPDATED

AMPHIBIOUS FORCES

2 YUNNAN CLASS (TYPE 067) (LCU)

DEBUNDSHA　　　　　　　KOMBO A JANEA

Displacement, tons: 135 full load
Dimensions, feet (metres): 93.8 × 17.7 × 4.9 *(28.6 × 5.4 × 1.5)*
Main machinery: 2 diesels; 600 hp(m) *(441 kW)*; 2 shafts
Speed, knots: 12. **Range, n miles:** 500 at 10 kt
Complement: 22 (2 officers)
Military lift: 46 tons
Guns: 2—14.5 mm (1 twin) MGs.
Radars: Surface search: Fuji; I-band.

Comment: Acquired from China in 2002.
NEW ENTRY

Canada

Country Overview

Canada is the world's second-largest country. The British monarch, represented by a governor-general, is head of state. With an area of 3,849,652 square miles, it occupies most of northern North America and is bordered to the south by the United States and to the west by the US state of Alaska. It has a coastline of 131,647 n miles with the Pacific, Arctic and Atlantic Oceans and with Baffin Bay and the Davis Strait. Numerous coastal islands include the Arctic Archipelago to the north, Newfoundland, Cape Breton, Prince Edward, and Anticosti to the east and Vancouver Island and the Queen Charlotte Islands to the west. Hudson Bay contains Southampton Island and many smaller islands. The 2,035 n mile St Lawrence-Great Lakes navigation system enables ocean-going vessels to sail between the Atlantic Ocean and the Great Lakes via the St Lawrence Seaway (opened 1959). Ottawa is the capital while Toronto is the largest city and a leading port. Other major ports include Vancouver, Montreal, Halifax, Sept-Îles, Port-Cartier, Quebec, Saint John (New Brunswick), Thunder Bay, Prince Rupert, and Hamilton. Territorial seas (12 n miles) are claimed. A 200 n mile EEZ has been claimed but the limits have only been partly defined by boundary agreements.

Headquarters Appointments

Deputy Chief of Defence Staff:
　Vice Admiral G R Maddison, CMM, MSC, CD
Assistant Chief of Maritime Staff:
　Commodore J R Sylvester, CD
Director General Maritime Personnel and Readiness:
　Commodore J A D Rouleau, CD
Director General Maritime Force Development:
　Captain K J Pickford, CD

Flag Officers

Chief of Maritime Staff:
　Vice Admiral B MacLean, CMM, CD
Commander, Maritime Forces, Atlantic:
　Rear Admiral D G McNeil, CD
Commander, Maritime Forces, Pacific:
　Rear Admiral J Y Forcier, CD
Commander, Canadian Fleet Atlantic:
　Commodore T H W Pile, OMM, CD
Commander, Canadian Fleet Pacific:
　Commodore R Girouard, OMM, CD
Commander, Naval Reserves:
　Commodore R Blakely

Diplomatic Representation

Defence Attaché, Washington:
　Rear Admiral I D Mack, OMM, CD
Naval Attaché, Washington:
　Captain S C Bertrand, CD
Naval Adviser, London:
　Captain E P Webster, OMM, CD
Defence Adviser, Canberra:
　Captain R R Town, CD
Defence Attaché, Tokyo:
　Captain S E King, OMM, CD
Naval Attaché, Paris:
　Commander C Gauthier, CD
Naval Attaché, Warsaw:
　Captain K M Carlé, CD
Defence Attaché, Berlin:
　Captain S D Andrews, XD
Defence Attaché, The Hague:
　Commander K A Heemskerk
Defence Attaché, Bogotá:
　Commander B R Struthers

Establishment

The Royal Canadian Navy (RCN) was officially established on 4 May 1910, when Royal Assent was given to the Naval Service Act. On 1 February 1968 the Canadian Forces Reorganisation Act unified the three branches of the Canadian Forces and the title 'Royal Canadian Navy' was dropped.

Personnel

2005: 9,000 (Regular), 3,447 (Reserves)

Prefix to Ships' Names

HMCS

Bases

Halifax and Esquimalt

Fleet Deployment

Atlantic:
Canadian Fleet Atlantic (destroyers, frigates, AOR)
Operations Group Five (maritime warfare forces, submarines and coastal defence districts)

Pacific:
Canadian Fleet Pacific (destroyers, frigates, AOR)
Operations Group Four (maritime warfare forces and coastal defence districts)

Maritime Air Components (MAC)

Commander MAC (Atlantic)-based in Halifax
Commander MAC (Pacific)-based in Esquimalt

Squadron/ Unit	Base	Aircraft	Function
MP 404	Greenwood, NS	Aurora/	LRMP/
		Arcturus	Training
MP 405	Greenwood, NS	Aurora	LRMP
HT 406	Shearwater, NS	Sea King	Training
MP 407	Comox, BC	Aurora	LRMP
MP 415	Greenwood, NS	Aurora	LRMP
MH 423	Shearwater, NS	Sea King	General
MH 443	Victoria, BC	Sea King	General
HOTEF	Shearwater, NS	Sea King	Test
MPEU	Greenwood, NS	Aurora	Test

Notes

(a) Detachments from 423 and 443 meet ships' requirements in Atlantic and Pacific Fleets respectively. Sea King helicopters are now classified as General Purpose vice the former ASW designation.

(b) 413 Squadron based in Greenwood, NS, and 442 Squadron based in Comox, BC, are two maritime search and rescue squadrons under the command of 1 Canadian Air Division (CAD).

(c) Combat training support provided by commercial contract from March 2002.

Strength of the Fleet

Type	Active	Building
Submarines	2	2
Destroyers	3	—
Frigates	12	—
Mine Warfare Forces	14	6
Survey Ships	1	—
Support Ships	2	3

DELETIONS

Destroyers		Mine Warfare Forces	
2004	*Huron*	2002	*YDT 10*

PENNANT LIST

Submarines

876	Victoria
877	Windsor
878	Corner Brook
879	Chicoutimi

Destroyers

280	Iroquois
282	Athabaskan
283	Algonquin

Frigates

330	Halifax
331	Vancouver
332	Ville de Québec
333	Toronto
334	Regina
335	Calgary
336	Montreal
337	Fredericton
338	Winnipeg
339	Charlottetown
340	St John's
341	Ottawa

Mine Warfare Forces

700	Kingston
701	Glace Bay
702	Nanaimo
703	Edmonton
704	Shawinigan
705	Whitehorse
706	Yellowknife
707	Goose Bay
708	Moncton
709	Saskatoon
710	Brandon
711	Summerside

Auxiliaries

172	Quest
509	Protecteur
510	Preserver
610	Sechelt
611	Sikanni
612	Sooke
613	Stikine

SUBMARINES

4 VICTORIA (UPHOLDER) CLASS (TYPE 2400) (SSK)

Name	No	Builders	Start date	Launched	Commissioned	Recommissioned
VICTORIA (ex-*Unseen*)	876 (ex-S 41)	Cammell Laird, Birkenhead	Jan 1986	14 Nov 1989	7 June 1991	2 Dec 2000
WINDSOR (ex-*Unicorn*)	877 (ex-S 43)	Cammell Laird, Birkenhead (VSEL)	Feb 1989	16 Apr 1992	25 June 1993	4 Oct 2003
CORNER BROOK (ex-*Ursula*)	878 (ex-S 42)	Cammell Laird, Birkenhead (VSEL)	Aug 1987	28 Feb 1991	8 May 1992	29 June 2003
CHICOUTIMI (ex-*Upholder*)	879 (ex-S 40)	Vickers Shipbuilding and Engineering, Barrow	Nov 1983	2 Dec 1986	9 June 1990	2 Oct 2004

Displacement, tons: 2,168 surfaced; 2,455 dived
Dimensions, feet (metres): 230.6 × 25 × 17.7
  (70.3 × 7.6 × 5.5)
Main machinery: Diesel-electric; 2 Paxman Valenta 16SZ diesels; 3,620 hp *(2.7 MW)* sustained; 2 GEC alternators; 2.8 MW; 1 GEC motor; 5,400 hp *(4 MW)*; 1 shaft
Speed, knots: 12 surfaced; 20 dived; 12 snorting
Range, n miles: 8,000 at 8 kt snorting
Complement: 47 (7 officers) plus 5 spare

Torpedoes: 6—21 in *(533 mm)* bow tubes. 18 Gould Mk 48 Mod 4; dual purpose; active/passive homing to 50 km *(27 n miles)*/38 km *(21 n miles)* at 40/55 kt; warhead 267 kg. Air turbine pump discharge.
Countermeasures: Decoys: 2 SSE launchers.
ESM: AR 900; intercept.
Weapons control: Lockheed Martin SFCS.
Radars: Navigation: Kelvin Hughes Type 1007; I-band.
  Furuno (portable); I-band.
Sonars: Thomson Sintra Type 2040; hull-mounted; passive search and intercept; medium frequency.
  BAE Type 2007; flank array; passive; low frequency.
  Thales Type 2046; towed array; passive very low frequency.
  Thales Type 2019; passive/active range and intercept (PARIS).

Programmes: First ordered 2 November 1983. Further three ordered on 2 January 1986. Laid up after post Cold War defence cuts in 1994 and acquired from the UK on 6 April 1998. Refitted at Vickers, Barrow, for delivery from June 2000.
Modernisation: The Canadianisation Work Period (CWP) includes the installation of Mk 48 torpedoes and its associated fire-control system and new communications and ESM systems. AIP is under consideration for the future but is unlikely to be installed until at least 2010. There are also plans to replace the Mk 48 torpedo and to procure an anti-aircraft missile by 2009.
Structure: Single-skinned NQ1 high tensile steel hull, tear dropped shape 9:1 ratio, five man lock-out chamber in fin. Fitted with elastomeric acoustic tiles. Diving depth, greater than 200 m *(650 ft)*. Fitted with Pilkington Optronics CK 35 search and CH 85 attack optronic periscopes.
Operational: Reactivation of the four submarines took longer than originally planned. *Victoria* arrived in Canada in October 2000 and, following an extended CWP during which repairs to exhaust valves and replacement of seawater valves were also made, transferred to Esquimault, British Columbia in August 2003. The other boats are based at Halifax, NS. *Windsor* was accepted in 2002 and completed CWP in 2003. *Corner Brook* arrived in mid-2003 and is to operate as a training boat before undergoing CWP. She is expected to become operational in 2006. *Chicoutimi* suffered a serious fire on 5 October 2004 while on passage to Canada. Extensive repairs are to be carried out in Canada after being ship-lifted in January 2005. Mid-life refits for the whole class and the possible transfer of a second submarine to the Pacific are under consideration.

UPDATED

VICTORIA

10/2000, CDF / 0104257

WINDSOR
6/2004*, Ships of the World / 1042121

VICTORIA

10/2000, CDF / 0094514

DESTROYERS

Notes: A Single Combatant Class (SCC), to replace both the Iroquois class and the Halifax class, is under consideration. The first vessels of this class would have an Air Defence role while follow-on vessels would be general purpose.

3 IROQUOIS CLASS (DDGHM)

Name	No	Builders	Laid down	Launched	Commissioned
IROQUOIS	280	Marine Industries Ltd, Sorel	15 Jan 1969	28 Nov 1970	29 July 1972
ATHABASKAN	282	Davie Shipbuilding, Lauzon	1 June 1969	27 Nov 1970	30 Sep 1972
ALGONQUIN	283	Davie Shipbuilding, Lauzon	1 Sep 1969	23 Apr 1971	3 Nov 1973

Displacement, tons: 5,300 full load

Dimensions, feet (metres): 398 wl; 426 oa × 50 × 15.5 keel/21.5 screws *(121.4; 129.8 × 15.2 × 4.7/6.6)*

Main machinery: COGOG; 2 Pratt & Whitney FT4A2 gas turbines; 50,000 hp *(37 MW)*; 2 GM Allison 570-KF gas turbines; 12,700 hp *(9.5 MW)* sustained; 2 shafts; LIPS cp props

Speed, knots: 27. **Range, n miles:** 4,500 at 15 kt (cruise turbines)

Complement: 255 (23 officers) plus 30 (9 officers) aircrew

Missiles: SAM: 1 Martin Marietta Mk 41 VLS ❶ for 29 GDC Standard SM-2MR Block III; command/inertial guidance; semi-active radar homing to 167 km *(90 n miles)* at Mach 2.

Guns: 1 OTO Melara 3 in *(76 mm)*/62 Super Rapid ❷; 120 rds/min to 16 km *(8.7 n miles)*; weight of shell 6 kg. 6—12.7 mm MGs.

1 GE/GDC 20 mm/76 6-barrelled Vulcan Phalanx Mk 15 ❸; 3,000 rds/min combined to 1.5 km.

Torpedoes: 6—324 mm Mk 32 (2 triple) tubes ❹. Honeywell Mk 46 Mod 5; anti-submarine; active/passive homing to 11 km *(5.9 n miles)* at 40 kt; warhead 44 kg.

Countermeasures: Decoys: 4 Plessey Shield Mk 2 6-tubed fixed launchers ❺. P 8 chaff or P 6 IR flares.

BAe Nulka offboard decoys in quad pack launchers.

SLQ-25 Nixie; torpedo decoy.

ESM: MEL SLQ-501 Canews ❻; radar warning.

ECM: BAe Nulka.

Combat data systems: SHINPADS, automated data handling with UYQ-504 and UYK-507 processors. Links 11, 14 and 16. GCCS-M and Marconi Matra SHF SATCOM ❼ (DDG 280 and 282).

Weapons control: Signaal LIROD 8 ❽ optronic director. UYS-503(V) sonobuoy processor.

Radars: Air search: Signaal SPQ-502 (LW08) ❾; D-band. **Surface search:** Signaal SPQ-501 (DA08) ❿; E/F-band. **Fire control:** 2 Signaal SPG-501 (STIR 1.8) ⓫; I/J-band.

Navigation: 2 Raytheon Pathfinder; I-band.

Koden MD 373 *(Iroquois* only, on hangar roof); I-band.

Tacan: URN 26.

Sonars: General Dynamics SQS-510; combined VDS and hull-mounted; active search and attack. 2 sets.

Helicopters: 2 CH-124A Sea King ASW ⓬.

Modernisation: A contract for the Tribal Class Update and Modernisation Project (TRUMP) was awarded to Litton Systems Canada Limited in June 1986. The equipment reflected the changing role of the ship and replaced systems that did not meet the air defence requirement. *Algonquin* completed modernisation in October 1991, followed by *Iroquois* in May 1992. *Athabaskan* in August 1994 and *Huron* on 17 January 1995. Sonar upgraded from 1998. Nulka system replaced ULQ-6 in 1999. JMCIS has been fitted vice JOTS II, with SHF SATCOM in DDG 280 and 282, and 283 in due course. Wescan 14 optronics to be fitted in all. Vulcan Phalanx to be upgraded to Block 1B from 2003.

Structure: These ships are also fitted with a landing deck equipped with double hauldown and Beartrap, pre-wetting system to counter NBC conditions, enclosed citadel and bridge control of machinery. The flume type anti-roll tanks have been replaced during modernisation with a water displaced fuel system. Design weight limit has been reached.

Operational: Helicopters can carry 12.7 mm MGs and ESM/FLIR instead of ASW gear. After three years at extended notice, *Huron* was decommissioned in 2004.

UPDATED

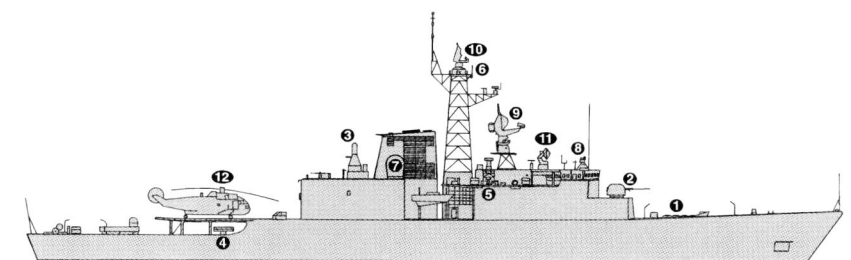

IROQUOIS

(Scale 1 : 1,200), Ian Sturton / 0056677

ALGONQUIN

7/2004*, Michael Nitz / 1042129

ALGONQUIN
6/2001, Bob Fildes / 0126373

FRIGATES

12 HALIFAX CLASS (FFGHM)

Name	No	Builders	Laid down	Launched	Commissioned
HALIFAX	330	Saint John SB Ltd, New Brunswick	19 Mar 1987	30 Apr 1988	29 June 1992
VANCOUVER	331	Saint John SB Ltd, New Brunswick	19 May 1988	8 July 1989	23 Aug 1993
VILLE DE QUÉBEC	332	Marine Industries Ltd, Sorel	17 Jan 1989	16 May 1991	14 July 1994
TORONTO	333	Saint John SB Ltd, New Brunswick	24 Apr 1989	18 Dec 1990	29 July 1993
REGINA	334	Marine Industries Ltd, Sorel	6 Oct 1989	25 Oct 1991	30 Sep 1994
CALGARY	335	Marine Industries Ltd, Sorel	15 June 1991	28 Aug 1992	12 May 1995
MONTREAL	336	Saint John SB Ltd, New Brunswick	8 Feb 1991	28 Feb 1992	21 July 1994
FREDERICTON	337	Saint John SB Ltd, New Brunswick	25 Apr 1992	13 Mar 1993	10 Sep 1994
WINNIPEG	338	Saint John SB Ltd, New Brunswick	19 Mar 1993	5 Dec 1993	23 June 1995
CHARLOTTETOWN	339	Saint John SB Ltd, New Brunswick	5 Dec 1993	10 July 1994	9 Sep 1995
ST JOHN'S	340	Saint John SB Ltd, New Brunswick	24 Aug 1994	12 Feb 1995	26 June 1996
OTTAWA	341	Saint John SB Ltd, New Brunswick	29 Apr 1995	22 Nov 1995	28 Sep 1996

Displacement, tons: 4,770 full load
Dimensions, feet (metres): 441.9 oa; 408.5 pp × 53.8 × 16.4;
    23.3 (screws) *(134.7; 124.5 × 16.4 × 5; 7.1)*
Main machinery: CODOG; 2 GE LM 2500 gas turbines;
    47,494 hp *(35.43 MW)* sustained
    1 SEMT-Pielstick 20 PA6 V 280 diesel; 8,800 hp(m)
    (6.48 MW) sustained; 2 shafts; cp props
Speed, knots: 29
Range, n miles: 9,500 at 13 kt (diesel); 3,930 at 18 kt (gas)
Complement: 198 (17 officers) plus 17 (8 officers) aircrew

Missiles: SSM: 8 McDonnell Douglas Harpoon Block 1C (2 quad)
    launchers ❶; active radar homing to 130 km *(70 n miles)* at
    0.9 Mach; warhead 227 kg.
SAM: 2 Raytheon Sea Sparrow RIM-7P Mk 48 octuple vertical
    launchers ❷; semi-active radar homing to 14.6 km *(8 n miles)*
    at 2.5 Mach; warhead 39 kg; 16 missiles.
Guns: 1 Bofors 57 mm/70 Mk 2 ❸; 220 rds/min to 17 km *(9 n
    miles)*; weight of shell 2.4 kg.
    1 GE/GDC 20 mm Vulcan Phalanx Mk 15 Mod 1 ❹; anti-
    missile; 3,000 rds/min (6 barrels combined) to 1.5 km.
    8—12.7 mm MGs.
Torpedoes: 4—324 mm Mk 32 Mod 9 (2 twin) tubes ❺. 24
    Honeywell Mk 46 Mod 5; anti-submarine; active/passive
    homing to 11 km *(5.9 n miles)* at 40 kt; warhead 44 kg.
Countermeasures: Decoys: 4 Plessey Shield Mk 2 decoy
    launchers ❻; sextuple mountings; fires P8 chaff and P6 IR
    flares in distraction, decoy or centroid modes.
    Nixie SLQ-25; towed acoustic decoy.
ESM: MEL/Lockheed Canews SLQ-501 ❼; radar intercept; (1-
    18 GHz). SRD 502; intercept. Sea Search AN/ULR 501.
ECM: MEL/Lockheed Ramses SLQ-503 ❽; jammer.
Combat data systems: UYC-501 SHINPADS action data
    automation with UYQ-504 and UYK-505 or 507 (336-341)
    processors. Links 11 and 14.
Weapons control: AHWCS for Harpoon. CDC UYS-503(V);
    sonobuoy processing system.
Radars: Air search: Raytheon SPS-49(V)5 ❾; C-band.
Air/surface search: Ericsson Sea Giraffe HC 150 ❿; G/H-band.
Fire control: Two Signaal SPG-503 (STIR 1.8) ⓫; K/I-band.
Navigation: Sperry Mk 340 being replaced by Kelvin Hughes
    1007; I-band.
Tacan: URN 25. IFF Mk XII.
Sonars: Westinghouse SQS-510; hull-mounted; active search
    and attack; medium frequency.
    General Dynamics SQR-501 CANTASS towed array (uses part
    of Martin Marietta SQR-19 TACTASS).

Helicopters: 1 CH-124A ASW or 1 CH-124B Heltas Sea King ⓬.

Programmes: On 29 June 1983 Saint John Shipbuilding Ltd
    won the competition for the first six of a new class of patrol
    frigates. Combat system design and integration was
    subcontracted to Loral Canada (formerly Paramax, a
    subsidiary of Unisys). Three ships were subcontracted to
    Marine Industries Ltd in Lauzon and Sorel. On 18 December
    1987 six additional ships of the same design were ordered
    from Saint John SB Ltd.
Modernisation: The Frigate Life Extension (FELEX) programme is
    to subsume all maintenance, sustainment and stand-alone
    projects planned to ensure the continued operation of the
    class for the duration of its life. In general, combat system

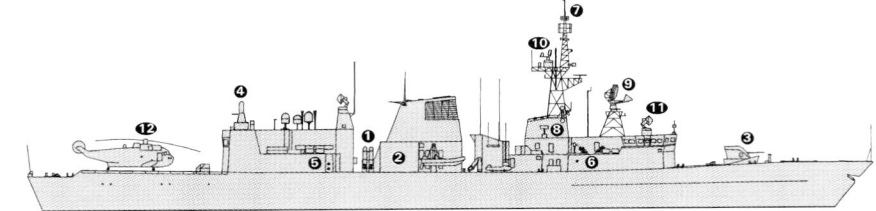

HALIFAX   *(Scale 1 : 1,200), Ian Sturton /* 0528399

CHARLOTTETOWN   *6/2004*, Harald Carstens /* 1042117

enhancements are to reflect increasing emphasis on littoral
operations in a joint and coalition context and the integration
of the Cyclone helicopter. Key components are the upgrade or
replacement of the combat data and communication systems,
including the fitting of Link 22. Projects already underway
include modifications to receive Evolved Sea Sparrow (ESSM)
(from 2004), the upgrade of Vulcan Phalanx to Block 1B (from
2003) and the fitting of Wescam 14PS-MAR optronics (from
2004). Other enhancements under consideration include
upgrade of the Bofors 57 mm Mk 2 gun to Mk 3 standard and
Harpoon to Block II. SPS-49, Sea Giraffe and STIR 1.8 radars
are also likely to be upgraded or replaced. Trials of the Sirius
infra-red search-and-track (IRST) took place in 2003 although

procurement decisions have not been taken. Of the EW suite,
SLQ-501 is likely to be modernised while SLQ-503 and
Plessey Shield decoy system are likely to be replaced. ASW
projects include improvement of torpedo defence and the
integration of active and passive sensors.
Structure: Much effort has gone into stealth technology. Gas
turbine engines are raft mounted. Dresball IR suppression is
fitted. Indal RAST helicopter handling system.
Operational: Problems on first of class trials included higher than
designed radiated noise levels which were reported as speed
associated. These have been rectified and the ships are stable
and quiet in all sea conditions. *Vancouver, Regina, Calgary,
Winnipeg* and *Ottawa* are Pacific based.   **UPDATED**

TORONTO   *4/2003*, Declerck & Steeghers /* 1042118

REGINA

7/2004*, Michael Nitz / 1042128

ST JOHN'S

10/2004*, Maritime Photographic / 1042127

OTTAWA

6/2002, John Chaney / 0529806

SHIPBORNE AIRCRAFT

Notes: The Maritime Helicopter Project to procure up to 28 new aircraft, to replace the Sea King, was given government approval in mid-2000. The Sikorsky H-92 Superhawk, to be designated Cyclone, selected on 22 July 2004. Contracts to cover acquisition, modification of the Halifax class destroyers and 20 years in-service support are expected in 2005 with delivery of the first aircraft planned in 2008. Final retirement of the Sea Kings is not expected until 2012.

Numbers/Type: 23/6 Sikorsky CH-124A ASW Sea King/CH-124B Helicopter Towed Array Support (HELTAS) Sea King.
Operational speed: 110 kt *(203 km/h)*.
Service ceiling: 10,000 ft *(3,030 m)*.
Range: 380 n miles *(705 km)*.
Role/Weapon systems: ASW, surface surveillance and support, convertible for carriage of six troops; deployed from shore or from three classes of ships (Halifax class FFG (1 aircraft), Iroquois class DDG (2 aircraft) and 'Protecteur' AOR (3 aircraft)); Sensors: CH-124A/B: APS-903 radar, ASN-123 mission computer, GPS, ARA-5 direction finder, APX-77A IFF, HF/VHF/UHF comms (with secure voice capability), ALQ-144 IR countermeasures (fitted for but not with). CH-124A: AQS-502 dipping sonar, ARR-52A sono receiver and ARR-1047 OTPI. CH-124B: UYS-503 sono processor, ARR-75 sono receiver, ASQ-504 MAD. Weapons: Two Mk 46 torpedoes and C6 light machine gun for both aircraft types.   *VERIFIED*

SEA KING   *4/1999, Winter & Findler /* 0056683

LAND-BASED MARITIME AIRCRAFT (FRONT LINE)

Numbers/Type: 18/3 Lockheed CP-140 Aurora/CP-140A Arcturus.
Operational speed: 405 kt *(750 km/h)*.
Service ceiling: 34,000 ft *(9,930 m)*.
Range: 4,000 n miles *(7,410 km)*.
Role/Weapon systems: Aurora operated for long-range maritime surveillance over Atlantic, Pacific and Arctic Oceans; roles include ASW/ASV and SAR; Arcturus for unarmed Arctic patrol, maritime surveillance, SAR and training. Arcturus fitted with same equipment as Aurora but without the ASW fit. Incremental modernisation programme scheduled 2000-2009. Contract for update of navigation and flight instruments awarded to BAE Systems in late 2000 and to MacDonald Dettwiler in January 2003 for replacement of AN/APS 506 radar. Aurora sensors: APS-506 radar, IFF, ALR-502 ESM, ECM, FLIR OR 5008 (to be replaced by L-3 Wescam MX-20), ASQ-502 MAD, OL 5004 acoustic processor. Weapons: 8 Mk 46 Mod 5 torpedoes. Arcturus sensors: APS-507 radar, IFF.

UPDATED

AURORA   *7/2003, Paul Jackson /* 0569161

MINE WARFARE FORCES

Notes: The Remote Minehunting System Technology Demonstrator (RMS-TD) Project was completed in early 2003. The prime contractor, MacDonald, Dettwiler and Associates, successfully demonstrated the technology for a remote-controlled semi-submersible drone (based on the DORADO developed by International Submarine Engineering Limited) to tow a sensor suite within a variable depth towfish capable of minehunting down to 200 m. The projected operational system would be deployable from a ship (displacing more than 900 tons), transportable by air and operable from a shore position. A Remote Minehunting and Disposal System (RMDS) is under active consideration by the Canadian Navy.

12 KINGSTON CLASS (MCDV)

Name	No	Builders	Laid down	Launched	Commissioned
KINGSTON	700	Halifax Shipyards	15 Dec 1994	12 Aug 1995	21 Sep 1996
GLACE BAY	701	Halifax Shipyards	28 Apr 1995	22 Jan 1996	26 Oct 1996
NANAIMO	702	Halifax Shipyards	11 Aug 1995	17 May 1996	10 May 1997
EDMONTON	703	Halifax Shipyards	8 Dec 1995	16 Aug 1996	21 June 1997
SHAWINIGAN	704	Halifax Shipyards	26 Apr 1996	15 Nov 1996	14 June 1997
WHITEHORSE	705	Halifax Shipyards	26 July 1996	24 Feb 1997	17 Apr 1998
YELLOWKNIFE	706	Halifax Shipyards	7 Nov 1996	5 June 1997	18 Apr 1998
GOOSE BAY	707	Halifax Shipyards	22 Feb 1997	4 Sep 1997	26 July 1998
MONCTON	708	Halifax Shipyards	31 May 1997	5 Dec 1997	12 July 1998
SASKATOON	709	Halifax Shipyards	5 Sep 1997	30 Mar 1998	21 Nov 1998
BRANDON	710	Halifax Shipyards	6 Dec 1997	3 Sep 1998	5 June 1999
SUMMERSIDE	711	Halifax Shipyards	28 Mar 1998	4 Oct 1998	18 July 1999

Displacement, tons: 962 full load
Dimensions, feet (metres): 181.4 × 37.1 × 11.2 *(55.3 × 11.3 × 3.4)*
Main machinery: Diesel-electric; 4 Wärtsilä UD 23V12 diesels; 4 Jeumont ANR-53-50 alternators; 7.2 MW; 2 Jeumont CI 560L motors; 3,000 hp(m) *(2.2 MW)*; 2 LIPS Z drive azimuth thrusters
Speed, knots: 15; 10 sweeping. **Range, n miles:** 5,000 at 8 kt
Complement: 31 (Patrol); 37 (MCM)

Guns: 1 Bofors 40 mm/60 Mk 5C. 2—12.7 mm MGs.
Countermeasures: MCM: 1 of 4 modular payloads: (a) Indal Technologies SLQ 38 deep mechanical minesweeping system; (b) MDA Ltd AN/SQS 511 Route Survey System

(RSS); (c) ISE Ltd TB 25 Bottom Object Inspection Vehicle (BOIV) System; (d) Fullerton and Sherwood Containerised Diving System. The RSS and BOIV systems can be carried at the same time.
Radars: Surface search: Kelvin Hughes 6000; E/F-band.
Navigation: Kelvin Hughes; I-band.
Sonars: AN/SQS 511 towed side scan; high frequency active; minehunting.

Programmes: Contract awarded to Fenco MacLaren on 15 May 1992. Halifax Shipyards is owned by Saint John Shipbuilding. Known as Maritime Coastal Defence Vessels (MCDV) combining MCM with general patrol duties.

Structure: MacDonald Dettwiler combat systems integration, MCM systems and integrated logistics support. Modular payloads comprising two MMS, four Route Survey and one ISE Trail Blazer 25 ROV. The Z drives can be rotated through 360°. Options for diving and minehunting equipment are being considered.
Operational: Predominantly manned by reservists. Six on each coast (700, 701, 704, 707, 708 and 711 Atlantic, remainder Pacific). One ship per coast is kept at extended readiness on a rotational basis.

UPDATED

NANAIMO   *10/2004*, Frank Findler /* 1042120

2 MCM DIVING TENDERS (YDT/YAG)

YDT 11   GRANBY YDT 12

Displacement, tons: 110
Dimensions, feet (metres): 99 × 20 × 8.5 *(27.3 × 6.2 × 2.6)*
Main machinery: Diesel; 228 hp (170 kW); 1 shaft
Speed, knots: 11
Complement: 13 (2 officers)
Radars: Navigation: Racal Decca; I-band.
Sonar: fitted for AN/SQQ 505(V); side scan.

Comment: Fitted with a two-compartment recompression chamber and for 100 m surface supplied diving and underwater maintenance. Both ships are due for replacement by 2010. ***VERIFIED***

GRANBY   *11/1995, CDF /* 0056682

TRAINING SHIPS

0 + 6 ORCA CLASS (TRAINING SHIPS) (AXL)

Measurement, tons: 165 full load
Dimensions, feet (metres): 103.3 × 26.9 × 7.9 *(31.5 × 8.2 × 2.4)*
Main machinery: 2 diesels; 2 shafts
Speed, knots: 16. **Range, n miles:** 2,700 at 10 kt
Complement: 8 plus 18 trainees

Comment: Contract awarded on 8 November 2004 for the construction of six (with the option for a further two) training vessels. Based on the Australian *Seahorse Mercator* design (on which dimensions are based), delivery of the first ship is expected in mid-2006, with the sixth and final vessel to be delivered in late 2008. ***NEW ENTRY***

ORCA CLASS (artist's impression)   *1/2005*, CDF /* 1042397

1 SAIL TRAINING SHIP (AXS)

Name	No	Builders	Launched
ORIOLE	YAC 3	Owens	4 June 1921

Displacement, tons: 92 full load
Dimensions, feet (metres): 102 × 19 × 9 *(31.1 × 5.8 × 2.7)*
Main machinery: 1 Cummins diesel; 165 hp *(123 kW)*; 1 shaft
Speed, knots: 8
Complement: 6 (1 officer) plus 18 trainees

Comment: Commissioned in the Navy in 1948 and based at Esquimalt. Sail area (with spinnaker) 11,000 sq ft. Height of mainmast 94 ft *(28.7 m)*, mizzen 55.2 ft *(16.8 m)*. ***VERIFIED***

ORIOLE   *12/1997, van Ginderen Collection /* 0017662

SURVEY AND RESEARCH SHIPS

1 RESEARCH SHIP (AGORH)

Name	No	Builders	Launched	Commissioned
QUEST	AGOR 172	Burrard, Vancouver	9 July 1968	21 Aug 1969

Displacement, tons: 2,130 full load
Dimensions, feet (metres): 235 × 42 × 15.5 *(71.6 × 12.8 × 4.6)*
Main machinery: Diesel-electric; 4 Fairbanks-Morse 38D8-1/8-9 diesel generators; 4.37 MW sustained; 2 GE motors; 2 shafts; cp props
Speed, knots: 16. **Range, n miles:** 10,000 at 12 kt
Complement: 55
Helicopters: Platform only.

Comment: Built for the Naval Research Establishment of the Defence Research Board for acoustic, hydrographic and general oceanographic work. Capable of operating in heavy ice in the company of an icebreaker. Launched on 9 July 1968. Based at Halifax and does line array acoustic research in the straits of the northern archipelago. Mid-life update in 1997-99 included new communications and navigation equipment and improved noise insulation. ***VERIFIED***

QUEST   *6/2000, CDF /* 0104261

AUXILIARIES

0 + 3 JOINT SUPPORT SHIPS (AFSH/AGFH/APCR)

Displacement, tons: 28,000 full load
Dimensions, feet (metres): 656.2 × 105 × 27.9 *(200 × 32 × 8.5)*
Main machinery: To be decided. Podded propulsion under investigation
Speed, knots: 21. **Range, n miles:** 10,800 at 15 kt
Complement: 165 + 210
Cargo capacity: 8,000 tons fuel; 500 tons aviation fuel; 300 tons ammunition and 230 tons of drinking water; combination of up to 120 × 20 ft containers and vehicles/equipment occupying 2,500 lane-metres (1,500 covered) of cargo space
Guns: Self-defence system.
Countermeasures: Likely to include active and passive missile and torpedo self-defence systems.
Radars: Air/surface search: E/F-band.
Navigation: I-band.
Helicopters: 4 medium.

Comment: Following an announcement by the Canadian government on 14 April 2004, it is planned to acquire three multirole ships, combining afloat support, sealift and shore support functions, to replace the current AORs between 2011 and 2013. There will be two RAS stations on each side, a modular onboard hospital that could include up to 60 beds and facilities for a Joint Task Force Headquarters. The double-hulled vessels will be capable of navigation in first-year ice (up to 0.7 m thick) while the hangar will be capable of providing second-line helicopter servicing. A flexible cargo system will consist of a container handling system, ramps, cranes and a capability to transfer cargo (single unit loads of up to 30 tonnes) from the ship to a destination ashore using landing craft in Sea State 1. A well-dock is unlikely to be included. ***UPDATED***

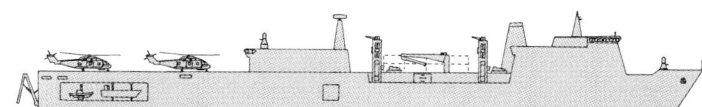

JOINT SUPPORT SHIP   *(not to scale), Ian Sturton /* 0569250

4 SECHELT CLASS (YTT/YPT/YDT)

Name	No	Builders	Commissioned
SECHELT	YDT 610	West Coast Manly	10 Nov 1990
SIKANNI	YPT 611	West Coast Manly	10 Nov 1990
SOOKE	YDT 612	West Coast Manly	10 Nov 1990
STIKINE	YPT 613	West Coast Manly	10 Nov 1990

Displacement, tons: 290 full load
Dimensions, feet (metres): 108.5 × 27.8 × 7.8 *(33.1 × 8.5 × 2.4)*
Main machinery: 2 Caterpillar 3412T diesels; 1,080 hp *(806 kW)* sustained; 2 shafts
Speed, knots: 12.5
Complement: 4 or 12 (610 and 612)
Sonar: fitted for AN/SQQ 505(V); side scan.

Comment: *Sikanni* and *Stikine* based at the Nanoose Bay Maritime Experimental and Test Range. *Sechelt* and *Sooke* converted to diving tenders in 1997 with a 6 place recompression chamber embarked. Diving operations supported to 80 m. *Sechelt* based at Halifax, Novia Scotia, *Sooke* at Esquimault, British Columbia. ***VERIFIED***

SOOKE (with containerised diving system)   *6/2002, CDF /* 0528415

2 PROTECTEUR CLASS (AORH)

Name	No	Builders	Laid down	Launched	Commissioned
PROTECTEUR	AOR 509	St John Dry Dock Co, NB	17 Oct 1967	18 July 1968	30 Aug 1969
PRESERVER	AOR 510	St John Dry Dock Co, NB	17 Oct 1967	29 May 1969	30 July 1970

Displacement, tons: 8,380 light; 24,700 full load
Dimensions, feet (metres): 564 × 76 × 33.3
(171.9 × 23.2 × 10.1)
Main machinery: 2 Babcock & Wilcox boilers; 1 GE Canada turbine; 21,000 hp *(15.7 MW)*; 1 shaft; bow thruster
Speed, knots: 21
Range, n miles: 4,100 at 20 kt; 7,500 at 11.5 kt
Complement: 365 (27 officers) including 45 aircrew
Cargo capacity: 14,590 tons fuel; 400 tons aviation fuel; 1,048 tons dry cargo; 1,250 tons ammunition; 2 cranes (15 ton lift)

Guns: 2 GE/GDC 20 mm/76 6-barrelled Vulcan Phalanx Mk 15. 6—12.7 mm MGs.

Countermeasures: Decoys: 6 Loral Hycor SRBOC chaff launchers.
ESM: Racal Kestrel SLQ-504; radar warning.
Combat data systems: EDO Link 11; SATCOM WSC-3(V).
Radars: Surface search: Norden SPS-502 with Mk XII IFF.
Navigation: Racal Decca 1630 and 1629; I-band.
Tacan: URN 20.
Helicopters: 3 CH-124A ASW or CH-124B Heltas Sea King.

Comment: Four replenishment positions. Both have been used as Flagships and troop carriers. They can carry anti-submarine helicopters, military vehicles and bulk equipment for sealift purposes; also four LCVPs. For the Gulf deployment in 1991, the 76 mm gun was remounted, two Vulcan Phalanx and two

Bofors 40/60 guns were fitted, four Plessey Shield chaff launchers and ESM equipment were provided for *Protecteur*. Additionally, all helicopters carried 12.7 mm MGs and ESM/ FLIR equipment instead of ASW gear. Bofors, 76 mm guns and hull mounted sonars later removed from both ships and are unlikely to be fitted again. *Protecteur* transferred to the Pacific Fleet November 1992. The refit of *Preserver* leaves Atlantic forces without a supply ship for most of 2004.
UPDATED

PRESERVER   *4/2003*, Declerck & Steeghers* / 1042119

TUGS AND TENDERS

13 COASTAL TUGS (YTB/YTL)

GLENDYNE YTB 640	**LAWRENCEVILLE** YTL 590	**FIREBIRD** YTR 561
GLENDALE YTB 641	**PARKSVILLE** YTL 591	**FIREBRAND** YTR 562
GLENEVIS YTB 642	**LISTERVILLE** YTL 592	**TILLICUM** YTM 555
GLENBROOK YTB 643	**MERRICKVILLE** YTL 593	
GLENSIDE YTB 644	**MARYSVILLE** YTL 594	

Comment: Glen class are 255 ton tugs built in the mid-1970s. Ville class are 70 ton tugs built in 1974. The two YTRs are firefighting craft of 130 tons. The YTM is a 140 ton tug. ***VERIFIED***

GLENEVIS   *6/2001, CDF* / 0126371

6 DIVING SUPPORT VESSELS (YDT)

FORTUNE	RESOLUTE	TONNERRE
ABALONE	DUNGENESS	SCULPIN

Displacement, tons: 2.2 full load
Dimensions, feet (metres): 39 × 12.5 × 2.3 *(11.9 × 3.8 × 0.7)*
Main machinery: 2 Caterpillar 3126TA diesels; 740 hp(m) *(548 kW)*; 2 WMC 357 waterjets
Speed, knots: 36. **Range, n miles:** 600 at 29 kt
Complement: 3 plus 14 divers

Comment: Built by Celtic Shipyards and delivered in early 1997. Landing craft bows for launching unmanned submersibles. Bollard pull 6,560 lb. *Fortune, Resolute* and *Tonnerre* based at Halifax, Nova Scotia, and the remainder at Esquimault, British Columbia. ***VERIFIED***

DSVs   *6/1997, CDF* / 0012131

COAST GUARD

Administration

Commissioner Canadian Coast Guard/Assistant Deputy Minister Marine:
John Adams

Establishment

In January 1962 all ships owned and operated by the Federal Department of Transport, with the exception of pilotage and canal craft, were amalgamated into the Canadian Coast Guard Fleet. On 1 April 1995 ships of the Fisheries and Oceans department merged with the Coast Guard under the direction of the Minister of Fisheries and Oceans. Its headquarters are in Ottawa while field operations are administered from five regional offices located in Vancouver, British Columbia (Pacific Region); Winnipeg, Manitoba (Central & Arctic Region); Quebec, Quebec (Quebec Region); Dartmouth, Nova Scotia (Maritimes Region); and St John's, Newfoundland and Labrador (Newfoundland and Labrador Region).

Missions

The Canadian Coast Guard carries out the following missions:

(a) Icebreaking and Escort. Icebreaking and escort of commercial ships is carried out in waters off the Atlantic seaboard, in the Gulf of St Lawrence, St Lawrence River and the Great Lakes in Winter and in Arctic waters in Summer.
(b) Aids to Navigation. Installation, supply and maintenance of fixed and floating aids to navigation in Canadian waters.
(c) Organise and provide icebreaker escort to commercial shipping in support of the annual Northern Sealift which supplies bases and settlements in the Canadian Arctic, Hudson Bay and Foxe Basin.
(d) Provide and operate a wide range of marine search and rescue vessels.
(e) Provide and operate Hydrographic survey, Oceanographic and Fishery Research vessels.
(f) Carry out Fishery Patrol and enforcement of fishery regulations.

Shipborne Aircraft

A total of 26 helicopters can be embarked in ships with aircraft facilities. These include six Bell 206, four Bell 212 and 16 MBB BO 105s. A long range Sikorsky S-61N is based ashore. All have Coast Guard markings.

Small Craft

In addition to the ships listed there are numerous lifeboats, surfboats, self-propelled barges and other small craft which are carried on board the larger vessels. Also excluded are shore-based work boats, floating oil spill boats, oil slick-lickers or any of the small boats which are available for use at the various Canadian Coast Guard Bases and lighthouse stations.

DELETIONS

2002	*Tofino, Shippegan, Comox Post, R B Young*
2003	*Westfort, Parizeau, CG 045*
2004	*Navicula, Souris, CG 141*

HEAVY GULF ICEBREAKERS

1 GULF CLASS (Type 1300)

Name	Builders	Launched	Commissioned
LOUIS S ST LAURENT	Canadian Vickers Ltd, Montreal	3 Dec 1966	Oct 1969

Displacement, tons: 14,500 full load
Measurement, tons: 11,441 grt; 5,370 net
Dimensions, feet (metres): 392.7 × 80.1 × 32.2 *(119.7 × 24.4 × 9.8)*
Main machinery: Diesel-electric; 5 Krupp MaK 16 M 453C diesels; 39,400 hp(m) *(28.96 MW)*; 5 Siemens alternators; 3 GE motors; 27,000 hp(m) *(19.85 MW)*; 3 shafts; bow thruster
Speed, knots: 18. **Range, n miles:** 23,000 at 17 kt
Complement: 47 (13 officers) plus 38 scientists
Radars: Navigation: 3 Kelvin Hughes; I-band.
Helicopters: 2 BO 105 CBS.

Comment: Larger than any of the former Coast Guard icebreakers. Two 49.2 ft *(15 m)* landing craft embarked. Mid-life modernisation July 1988 to early 1993 included replacing main engines with a diesel-electric system, adding a more efficient *Henry Larsen* type icebreaking bow (adds 8 m to length) with an air bubbler system and improving helicopter facilities with a fixed hangar. In addition the complement was reduced. Based in the Maritimes Region at Dartmouth, NS. On 22 August 1994 became the first Canadian ship to reach the North Pole, in company with USCG *Polar Sea*.

VERIFIED

LOUIS S ST LAURENT *6/1998, Harald Carstens* / 0017665

LOUIS S ST LAURENT *6/1998, Harald Carstens* / 0056691

MEDIUM GULF/RIVER ICEBREAKERS

3 R CLASS (Type 1200)

Name	Builders	Launched	Commissioned
PIERRE RADISSON	Burrard, Vancouver	3 June 1977	June 1978
AMUNDSEN	Burrard, Vancouver	10 Mar 1978	Mar 1979
(ex-*Sir John Franklin*)			
DES GROSEILLIERS	Port Weller, Ontario	20 Feb 1982	Aug 1982

Displacement, tons: 6,400 standard; 8,180 (7,594, *Des Groseilliers*) full load
Measurement, tons: 5,910 gross; 1,678 net
Dimensions, feet (metres): 322 × 64 × 23.6 *(98.1 × 19.5 × 7.2)*
Main machinery: Diesel-electric; 6 Montreal Loco 251V-16F diesels; 17,580 hp *(13.1 MW)*; 6 GEC generators; 11.1 MW sustained; 2 motors; 13,600 hp *(10.14 MW)*; 2 shafts; bow thruster
Speed, knots: 16. **Range, n miles:** 15,000 at 13.5 kt
Complement: 38 (12 officers)
Radars: Navigation: Sperry; E/F- and I-band.
Helicopters: 1 Bell 212.

Comment: Based in the Quebec Region at Quebec. *Amundsen* underwent a major refit in 2003 to convert her to an Arctic research role.

VERIFIED

AMUNDSEN *6/2003, P Dionne* / 0572428

1 MODIFIED R CLASS (Type 1200)

Name	Builders	Launched	Commissioned
HENRY LARSEN	Versatile Pacific SY, Vancouver, BC	3 Jan 1987	29 June 1988

Displacement, tons: 5,798 light; 8,290 full load
Measurement, tons: 6,172 gross; 1,756 net
Dimensions, feet (metres): 327.3 × 64.6 × 24 *(99.8 × 19.7 × 7.3)*
Main machinery: Diesel-electric; 3 Wärtsilä Vasa 16V32 diesel generators; 17.13 MW/60 Hz sustained; 3 motors; 16,320 hp(m) *(12 MW)*; 3 shafts
Speed, knots: 16. **Range, n miles:** 15,000 at 13.5 kt
Complement: 52 (15 officers) plus 20 spare berths
Radars: Navigation: Racal Decca Bridgemaster; I-band.
Helicopters: 1 Bell 212.

Comment: Contract date 25 May 1984, laid down 23 August 1985. Although similar in many ways to the R class she has a different hull form particularly at the bow and a very different propulsion system. Fitted with Wärtsilä air bubbling system. Based at St John's in the Newfoundland and Labrador Region. Engine room fire in 1998 put her out of commission for some time.

UPDATED

HENRY LARSEN *3/1999, Canadian Coast Guard* / 0056707

MAJOR NAVAIDS TENDERS/LIGHT ICEBREAKERS

6 MARTHA L BLACK CLASS (Type 1100)

Name	Builders	Commissioned
MARTHA L BLACK	Versatile Pacific, Vancouver, BC	30 Apr 1986
GEORGE R PEARKES	Versatile Pacific, Vancouver, BC	17 Apr 1986
EDWARD CORNWALLIS	Marine Industries Ltd, Tracy, Quebec	14 Aug 1986
SIR WILLIAM ALEXANDER	Marine Industries Ltd, Tracy, Quebec	13 Feb 1987
SIR WILFRID LAURIER	Canadian Shipbuilding Ltd, Ontario	15 Nov 1986
ANN HARVEY	Halifax Industries Ltd, Halifax, NS	29 June 1987

Displacement, tons: 4,662 full load
Measurement, tons: 3,818 *(Martha L Black)*; 3,809 *(George R Pearkes)*; 3,812 *(Sir Wilfrid Laurier)*; 3,727 *(Edward Cornwallis* and *Sir William Alexander)*; 3,823 *(Ann Harvey)* gross
Dimensions, feet (metres): 272.2 × 53.1 × 18.9 *(83 × 16.2 × 5.8)*
Main machinery: Diesel-electric; 3 Bombardier/Alco 12V-251 diesels; 8,019 hp *(6 MW)* sustained; 3 Canadian GE generators; 6 MW; 2 Canadian GE motors; 7,040 hp *(5.25 MW)*; 2 shafts; bow thrusters
Speed, knots: 15.5. **Range, n miles:** 6,500 at 15 kt
Complement: 25 (10 officers)
Radars: Navigation: Racal Decca Bridgemaster; I-band.
Helicopters: 1 light type, such as Bell 206L.

Comment: *Black* based in the Quebec Region at Quebec, *Cornwallis* and *Alexander* in the Maritimes Region at Dartmouth, *Ann Harvey* and *Pearkes* in the Newfoundland and Labrador Region at St Johns and *Laurier* in the Pacific Region at Victoria. The feasibility of converting *Cornwallis* to a survey ship was investigated but not taken forward.

UPDATED

GEORGE R PEARKES *4/1996, van Ginderen Collection* / 0056692

SIR WILLIAM ALEXANDER *8/1998, M B MacKay* / 0017668

1 GRIFFON CLASS (Type 1100)

Name	Builders	Commissioned
GRIFFON	Davie Shipbuilding, Lauzon	Dec 1970

Displacement, tons: 3,096 full load
Measurement, tons: 2,212 gross; 752 net
Dimensions, feet (metres): 233.9 × 49 × 15.5 *(71.3 × 14.9 × 4.7)*
Main machinery: Diesel-electric; 4 Fairbanks-Morse 38D8-1/8-12 diesel generators; 5.8 MW sustained; 2 motors; 3,982 hp(m) *(2.97 MW)*; 2 shafts
Speed, knots: 14. **Range, n miles:** 5,500 at 10 kt
Complement: 25 (9 officers)
Radars: Navigation: 2 Kelvin Hughes; I-band.
Helicopters: Platform for 1 light type, such as Bell 206L.

Comment: Based in the Central and Arctic Region at Prescott, Ontario.

VERIFIED

GRIFFON   *7/1998, van Ginderen Collection /* 0017669

HEAVY ICEBREAKER/SUPPLY TUG

1 TERRY FOX CLASS (Type 1200)

Name	Builders	Launched	Commissioned
TERRY FOX	Burrard Yarrow, Vancouver	1982	1983

Displacement, tons: 7,100 full load
Measurement, tons: 4,233 gross; 1,955 net
Dimensions, feet (metres): 288.7 × 58.7 × 27.2 *(88 × 17.9 × 8.3)*
Main machinery: 4 Werkspoor 8-cyl 4SA diesels; 23,200 hp(m) *(17 MW)*; 2 shafts; cp props; bow and stern thrusters
Speed, knots: 16. **Range, n miles:** 1,920 at 15 kt
Complement: 23 (10 officers)
Radars: Navigation: 2 Racal Decca ARPA; 1 Furuno 1411; E/F- and I-bands.

Comment: Initially leased for two years from Gulf Canada Resources during the completion of *Louis S St Laurent* conversion but has now been retained. Commissioned in Coast Guard colours 1 November 1991 and purchased 1 November 1993. Based in the Maritimes Region at Dartmouth.

VERIFIED

TERRY FOX   *7/1997, M B MacKay /* 0012133

MAJOR NAVAIDS TENDERS/LIGHT

1 J E BERNIER CLASS (Type 1100)

Name	Builders	Commissioned
J E BERNIER	Davie Shipbuilding, Lauzon	Aug 1967

Displacement, tons: 3,096 full load
Measurement, tons: 2,457 gross; 705 net
Dimensions, feet (metres): 218.9 × 49 × 16 *(66.7 × 14.9 × 4.9)*
Main machinery: Diesel-electric; 4 Fairbanks-Morse 4SA 8-cyl diesels; 5,600 hp *(4.12 MW)*; 4 generators; 3.46 MW; 2 motors; 4,250 hp *(3.13 MW)*; 2 shafts
Speed, knots: 13.5. **Range, n miles:** 4,000 at 11 kt
Complement: 21 (9 officers)
Radars: Navigation: 2 Kelvin Hughes; I-band.
Helicopters: 1 Bell 206L/L-1.

Comment: Based in Newfoundland and Labrador Region at St Johns. Laid up in preparation for possible decommissioning in 2006.

UPDATED

J E BERNIER   *7/1996, van Ginderen Collection /* 0056693

MEDIUM NAVAIDS TENDERS/LIGHT ICEBREAKERS

2 SAMUEL RISLEY CLASS (Type 1050)

Name	Builders	Commissioned
SAMUEL RISLEY	Vito Construction Ltd, Delta, BC	4 July 1985
EARL GREY	Pictou Shipyards Ltd, Pictou, NS	30 May 1986

Displacement, tons: 2,935 full load
Measurement, tons: 1,988 gross *(Grey)*; 1,967 gross *(Risley)*; 642 net *(Grey)*; 649.5 net *(Risley)*
Dimensions, feet (metres): 228.7 × 44.9 × 19 *(69.7 × 13.7 × 5.8)*
Main machinery: Diesel-electric; 4 Wärtsilä 4SA 12-cyl diesels; 8,644 hp(m) *(6.4 MW)* *(Samuel Risley)*; 4 Deutz 4SA 9-cyl diesels; 8,836 hp(m) *(6.5 MW)* *(Earl Grey)*; 2 shafts; cp props
Speed, knots: 13. **Range, n miles:** 18,000 at 12 kt
Complement: 22
Radars: Navigation: 2 Racal Decca; I-band.

Comment: *Risley* based in the Central and Arctic Region at Pary Sound, Ontario, *Grey* in the Maritimes Region at Charlottetown, PEI.

UPDATED

SAMUEL RISLEY   *4/1993, Canadian Coast Guard /* 0056694

MEDIUM NAVAIDS TENDERS/ICE STRENGTHENED

2 PROVO WALLIS CLASS (Type 1000)

Name	Builders	Commissioned
BARTLETT	Marine Industries, Sorel	Dec 1969
PROVO WALLIS	Marine Industries, Sorel	Oct 1969

Displacement, tons: 1,620 full load *(Bartlett)*
Measurement, tons: 1,317 gross; 491 net
Dimensions, feet (metres): 189.3; 209 *(Provo Wallis)* × 42.5 × 15.4 *(57.7; 63.7 × 13 × 4.7)*
Main machinery: 2 National Gas 6-cyl diesels; 2,100 hp *(1.55 MW)*; 2 shafts; LIPS cp props
Speed, knots: 12.5. **Range, n miles:** 3,300 at 11 kt
Complement: 24 (9 officers)
Radars: Navigation: 2 Kelvin Hughes; I-band.

Comment: *Bartlett* based in Pacific Region at Victoria, *Provo Wallis* in the Maritimes Region at Saint Johns, New Brunswick. *Bartlett* was modernised in 1988 and *Provo Wallis* completed one year modernisation at Marystown, Newfoundland at the end of 1990. Work included lengthening the hull by 6 m, installing new equipment and improving accommodation.

VERIFIED

BARTLETT   *5/1999, Hartmut Ehlers /* 0056695

0 + (3) TYPE 1000

Displacement, tons: 2,013 full load
Dimensions, feet (metres): 213.3 × 45.9 × 11.8 *(65 × 14 × 3.6)*
Main machinery: 2 diesels; 2 shafts
Speed, knots: 14. **Range, n miles:** 6,000 at 12 kt
Complement: 34

Comment: Designated as a shallow draft, multitaskable utility vessel. The long-term plan is to order three first followed by four more to replace all existing Type 1000 ships. Construction dates have not been decided.

VERIFIED

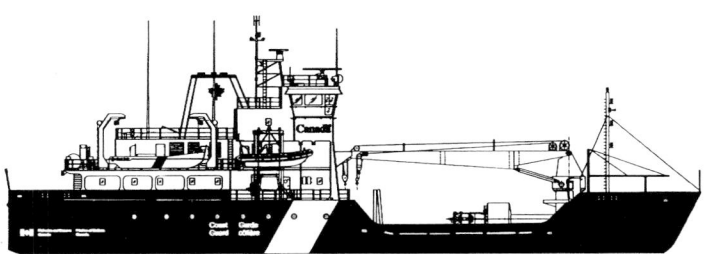

TYPE 1000   *1998, Canadian Coast Guard /* 0017670

1 TRACY CLASS (Type 1000)

Name	Builders	Commissioned
TRACY	Port Weller Drydocks, Ontario	17 Apr 1968

Displacement, tons: 1,300 full load
Measurement, tons: 963 gross; 290 net
Dimensions, feet (metres): 181.1 × 38 × 12.1 *(55.2 × 11.6 × 3.7)*
Main machinery: Diesel-electric; 2 Fairbanks-Morse 38D8-1/8-8 diesel generators; 1.94 MW sustained; 2 motors; 2,000 hp *(1.49 MW)*; 2 shafts
Speed, knots: 13. **Range, n miles:** 5,000 at 11 kt
Complement: 23 (8 officers)
Radars: Navigation: Kelvin Hughes; I-band.

Comment: Based in Quebec Region at Sorel.

VERIFIED

TRACY   *4/1999, Canadian Coast Guard /* 0056716

1 SIMCOE CLASS (Type 1000)

Name	Builders	Commissioned
SIMCOE	Canadian Vickers Ltd, Montreal	Oct 1962

Displacement, tons: 1,390 full load
Measurement, tons: 961 gross; 361 net
Dimensions, feet (metres): 179.5 × 38 × 12.5 *(54.7 × 11.6 × 3.8)*
Main machinery: Diesel-electric; 2 Paxman 4SA 12-cyl diesels; 3,000 hp *(2.24 MW)*; 2 motors; 2,000 hp *(1.49 MW)*; 2 shafts
Speed, knots: 13. **Range, n miles:** 5,000 at 10 kt
Complement: 27 (10 officers)
Radars: Navigation: 2 Kelvin Hughes; I-band.

Comment: Based in Central and Arctic Region at Prescott, Ontario. Modernised in 1988.

VERIFIED

SIMCOE   *4/1999, Canadian Coast Guard /* 0056712

SMALL NAVAIDS TENDERS/ICE STRENGTHENED

1 NAMAO CLASS (Type 900)

Name	Builders	Commissioned
NAMAO	Riverton Boat Works, Manitoba	1975

Displacement, tons: 380 full load
Measurement, tons: 318 gross; 107 net
Dimensions, feet (metres): 110 × 28 × 7 *(33.5 × 8.5 × 2.1)*
Main machinery: 2 Detroit 12V-71 diesels; 1,350 hp *(1 MW)*; 2 shafts
Speed, knots: 12. **Range, n miles:** 2,000 at 12 kt
Complement: 11 (4 officers)
Radars: Navigation: Racal Decca; I-band.

Comment: Based in the Central and Arctic Region on Lake Winnipeg at Selkirk. Planned to decommission in 1996 but kept in service until at least 2005.

UPDATED

NAMAO   *4/1999, Canadian Coast Guard /* 0056710

SMALL NAVAIDS TENDERS

7 COVE ISLAND CLASS (Type 800)

Name	Builders	Commissioned
COVE ISLE	Canadian D and D, Kingston, Ontario	1980
GULL ISLE	Canadian D and D, Kingston, Ontario	1980
TSEKOA II	Allied Shipbuilders, Vancouver	1984
PARTRIDGE ISLAND	Breton Industries, Port Hawkesbury, NS	31 Oct 1985
ILE DES BARQUES	Breton Industries, Port Hawkesbury, NS	26 Nov 1985
ILE SAINT-OURS	Breton Industries, Port Hawkesbury, NS	15 May 1986
CARIBOU ISLE	Breton Industries, Port Hawkesbury, NS	16 June 1986

Displacement, tons: 138 full load
Measurement, tons: 92 gross; 36 net
Dimensions, feet (metres): 75.5 × 19.7 × 4.4 *(23 × 6 × 1.4)*
Main machinery: 2 Detroit 8V-92 diesels; 475 hp *(354 kW)*; 2 shafts
Speed, knots: 11. **Range, n miles:** 1,800 at 11 kt
Complement: 5
Radars: Navigation: Sperry 1270; I-band.

Comment: Details given are for the last four. *Cove Isle* and *Gull Isle* are 3 m shorter in length; *Tsekoa II* is 3.7 m longer. *Cove Isle* and *Gull Isle* are based in the Central and Arctic Region at Parry Sound and Amherstburg respectively. *Tsekoa II* is based in the Pacific at Victoria. *Partridge Island* and *Ile des Barques* based in the Maritimes Region at Dartmouth, Nova Scotia, *Caribou Isle* in the Central and Arctic Region at Sault Ste Marie, Ontario, and *Ile Saint-Ours* in the Quebec Region at Sorel. Can carry 20 tons of stores.

UPDATED

ILE SAINT-OURS   *9/1994, van Ginderen Collection /* 0056696

SPECIAL RIVER NAVAIDS TENDERS

1 NAHIDIK CLASS (Type 700)

Name	Builders	Commissioned
NAHIDIK	Allied Shipbuilders Ltd, N Vancouver	1974

Displacement, tons: 1,125 full load
Measurement, tons: 856 gross; 392 net
Dimensions, feet (metres): 175.2 × 49.9 × 6.6 *(53.4 × 15.2 × 2)*
Main machinery: 2 Detroit diesels; 4,290 hp *(3.2 MW)*; 2 shafts
Speed, knots: 14. **Range, n miles:** 5,000 at 10 kt
Complement: 15

Comment: Based in Central and Arctic Region at Hay River, North West Territories.

UPDATED

NAHIDIK   *6/2004*, Canadian Coast Guard* / 1042126

1 DUMIT CLASS (Type 700)

Name	Builders	Commissioned
DUMIT	Allied Shipbuilders Ltd, N Vancouver	July 1979

Displacement, tons: 629 full load
Measurement, tons: 569 gross; 176 net
Dimensions, feet (metres): 160.1 × 40 × 5.2 *(48.8 × 12.2 × 1.6)*
Main machinery: 2 Caterpillar 3512TA; 2,420 hp *(1.8 MW)* sustained; 2 shafts
Speed, knots: 12. **Range, n miles:** 8,500 at 10 kt
Complement: 10

Comment: Similar to *Eckaloo*. Based in Central and Arctic Region at Hay River, North West Territories.

VERIFIED

DUMIT   *7/1996, Canadian Coast Guard* / 0017671

1 TEMBAH CLASS (Type 700)

Name	Builders	Commissioned
TEMBAH	Allied Shipbuilders Ltd, N Vancouver	Oct 1963

Measurement, tons: 189 gross; 58 net
Dimensions, feet (metres): 123 × 25.9 × 3 *(37.5 × 7.9 × 0.9)*
Main machinery: 2 Cummins diesels; 500 hp *(373 kW)*; 2 shafts
Speed, knots: 12. **Range:** 1,300 at 10 kt
Complement: 9

Comment: Based in Central and Arctic Region at Hay River, North West Territories.

VERIFIED

TEMBAH   *4/1999, Canadian Coast Guard* / 0056714

1 ECKALOO CLASS (Type 700)

Name	Builders	Commissioned
ECKALOO	Vancouver SY Ltd	31 Aug 1988

Displacement, tons: 534 full load
Measurement, tons: 661 gross; 213 net
Dimensions, feet (metres): 160.8 × 44 × 4 *(49 × 13.4 × 1.2)*
Main machinery: 2 Caterpillar 3512TA; 2,420 hp *(1.8 MW)* sustained; 2 shafts
Speed, knots: 13. **Range:** 2,500 at 12 kt
Complement: 10
Helicopters: Platform for 1 Bell 206L/L-1.

Comment: Replaced vessel of the same name. Similar design to *Dumit*. Based in Central and Arctic Region at Hay River, North West Territories.
VERIFIED

ECKALOO   *9/1994, van Ginderen Collection* / 0056697

OFFSHORE MULTITASK PATROL CUTTERS

1 SIR WILFRED GRENFELL (Type 600)

Name	Builders	Commissioned
SIR WILFRED GRENFELL	Marystown SY, Newfoundland	1987

Displacement, tons: 3,753 full load
Measurement, tons: 2,403 gross; 664.5 net
Dimensions, feet (metres): 224.7 × 49.2 × 16.4 *(68.5 × 15 × 5)*
Main machinery: 4 Deutz 4SA (2—16-cyl, 2—9-cyl) diesels; 12,862 hp(m) *(9.46 MW)*; 2 shafts; cp props
Speed, knots: 16. **Range:** 11,000 at 14 kt
Complement: 20

Comment: Built on speculation in 1984-85. Modified to include an 85 tonne towing winch and additional SAR accommodation and equipment. Ice strengthened hull. Based in the Newfoundland Region and Labrador at St John's.
VERIFIED

SIR WILFRED GRENFELL   *8/1997, M B MacKay* / 0012137

1 LEONARD J COWLEY CLASS (Type 600)

Name	Builders	Commissioned
LEONARD J COWLEY	Manly Shipyard, RivTow Ind, Vancouver BC	June 1985

Displacement, tons: 2,080 full load
Measurement, tons: 2,244 grt; 655 net
Dimensions, feet (metres): 236.2 × 45.9 × 16.1 *(72 × 14 × 4.9)*
Main machinery: 2 Wärtsilä Nohab F 312A diesels; 2,325 hp(m) *(1.71 MW)*; 1 shaft; bow thruster
Speed, knots: 12. **Range, n miles:** 12,000 at 12 kt
Complement: 20
Guns: 2—12.7 mm MGs.
Radars: Surface search: Sperry 340; E/F-band.
Navigation: Sperry ARPA; I-band.
Helicopters: Capability for 1 light.

Comment: Based in Newfoundland and Labrador Region at St John's.
VERIFIED

LEONARD J COWLEY   *9/1996, D Maginley* / 0056698

2 CAPE ROGER CLASS (Type 600)

Name	Builders	Commissioned
CYGNUS	Marystown SY, Newfoundland	May 1981
CAPE ROGER	Ferguson Industries, Pictou NS	Aug 1977

Displacement, tons: 1,465 full load
Measurement, tons: 1,255 grt; 357 net
Dimensions, feet (metres): 205 × 40 × 13 *(62.5 × 12.2 × 4.1)*
Main machinery: 2 Wärtsilä Nohab F 212V diesels, 4,461 hp(m) *(3.28 MW)*; 1 shaft; bow thruster
Speed, knots: 13. **Range, n miles:** 10,000 at 12 kt
Complement: 19
Guns: 2—12.7 mm MGs.
Helicopters: Capability for 1 light.

Comment: *Cygnus* based in Maritimes Region at Dartmouth and *Cape Roger* in Newfoundland and Labrador Region at St John's. Half-life refits completed in 1995-97.

VERIFIED

CYGNUS   *9/1999, Canadian Coast Guard* / 0056704

INTERMEDIATE MULTITASK PATROL CUTTERS

1 TANU CLASS (Type 500)

Name	Builders	Commissioned
TANU	Yarrows Ltd, Victoria BC	Sep 1968

Displacement, tons: 925 full load
Measurement, tons: 746 grt; 203 net
Dimensions, feet (metres): 164.3 × 3.2 × 15.1 *(50.1 × 9.8 × 4.6)*
Main machinery: 2 Fairbanks-Morse diesels; 2,624 hp *(1.96 MW)*; 1 shaft
Speed, knots: 11. **Range, n miles:** 4,000 at 11 kt
Complement: 18 plus 16 spare
Guns: 2—12.7 mm MGs.

Comment: Based in Pacific Region at Patricia Bay.

UPDATED

TANU   *7/2004*, M K Mitchell* / 1042125

1 LOUISBOURG CLASS (Type 500)

Name	Builders	Commissioned
LOUISBOURG	Breton Industries, Port Hawkesbury, NS	1977

Displacement, tons: 460 full load
Measurement, tons: 295 grt; 65 net
Dimensions, feet (metres): 125 × 27.2 × 8.5 *(38.1 × 8.3 × 2.6)*
Main machinery: 2 MTU 12V 538 TB91 diesels; 4,600 hp(m) *(3.38 MW)*; 2 shafts
Speed, knots: 13. **Range, n miles:** 6,200 at 12 kt
Complement: 14
Guns: 2—12.7 mm MGs.

Comment: Based in the Quebec Region at Gaspé.

UPDATED

LOUISBOURG   *9/1999, Canadian Coast Guard* / 0056708

1 QUÉBÉCOIS CLASS

Name	Builders	Commissioned
E P LE QUÉBÉCOIS	Les Chantiers Maritimes, Paspebiac, Quebec	1968

Measurement, tons: 186 gross; 32 net
Dimensions, feet (metres): 78.1 × 23.3 × ? *(28.3 × 7.1 × ?)*
Main machinery: 1 Caterpillar 3509 diesel; 509 hp *(380 kW)*; 1 shaft
Speed, knots: 11. **Range, n miles:** 2,800 at 9 kt
Complement: 8 (4 officers)

Comment: Based at Sept Îles, Quebec. Refitted in 1994.

VERIFIED

E P LE QUÉBÉCOIS   *6/2002, Canadian Coast Guard* / 0529823

1 ARROW POST CLASS

Name	Builders	Commissioned
ARROW POST	Hike Metal Products, Wheatley, Ontario	1991

Measurement, tons: 228 gross; 93.1 net
Dimensions, feet (metres): 94.8 × 28.9 × ? *(28.9 × 8.8 × ?)*
Main machinery: 1 Caterpillar 3512 diesel; 711 hp *(954 kW)*; 1 shaft
Speed, knots: 12. **Range, n miles:** 2,800 at 11 kt
Complement: 6 (3 officers). 6 additional

Comment: Based in Pacific Region at Prince Rupert, British Columbia.

UPDATED

ARROW POST   *6/2004*, M K Mitchell* / 1042124

1 GORDON REID CLASS (Type 500)

Name	Builders	Commissioned
GORDON REID	Versatile Pacific, Vancouver	Oct 1990

Measurement, tons: 836 gross; 247 net
Dimensions, feet (metres): 163.9 × 36.1 × 13.1 *(49.9 × 11 × 4)*
Main machinery: 4 Deutz SBV-6M-628 diesels; 2,475 hp(m) *(1.82 MW)* sustained; 2 shafts; bow thruster; 400 hp *(294 kW)*
Speed, knots: 15. **Range, n miles:** 2,500 at 15 kt
Complement: 14 plus 8 spare

Comment: Designed for long-range patrols along the British Columbian coast out to 200 mile limit. Has a stern ramp for launching Zodiac Hurricane 733 rigid inflatables in up to Sea State 6. The Zodiac has a speed of 50 kt and is radar equipped. Based in the Pacific Region at Victoria.

UPDATED

GORDON REID   *6/2004*, M K Mitchell* / 1042122

SMALL MULTITASK CUTTERS

5 CUTTERS (Type 400)

Name	Builders	Commissioned
ADVENT	Alloy Manufacturing, Qu	1972
POINT HENRY	Breton Industrial and Machinery	1980
ISLE ROUGE	Breton Industrial and Machinery	1980
POINT RACE	Pt Hawkesbury, NS	1982
CAPE HURD	Pt Hawkesbury, NS	1982

Displacement, tons: 97 full load
Measurement, tons: 57 gross; 14 net
Dimensions, feet (metres): 70.8 × 18 × 5.6 *(21.6 × 5.5 × 1.7)*
Main machinery: 2 MTU 8V 396 TC82 diesels; 1,740 hp(m) *(1.28 MW)* sustained; 2 shafts
Speed, knots: 20. **Range, n miles:** 950 at 12 kt
Complement: 5

Comment: Aluminium alloy hulls. *Point Henry* and *Point Race* based in Pacific Region at Prince Rupert and Campbell River respectively; *Cape Hurd* and *Advent* in Central and Arctic Region at Goderich and Cobourg respectively; *Isle Rouge* in the Quebec Region at Tadoussac. *Advent* is slightly larger at 23.5 m and has a top speed of 16 kt.

UPDATED

POINT RACE   *6/2001, Canadian Coast Guard /* 0126356

2 POST CLASS

Name	Builders	Commissioned
ATLIN POST	Philbrooks Shipyard Ltd, Sidney, BC	1975
KITIMAT II	Philbrooks Shipyard Ltd, Sidney, BC	1974

Measurement, tons: 57 gross; 15 net
Dimensions, feet (metres): 65.0 × 17.1 × ? *(19.8 × 5.2 × ?)*
Main machinery: 2 General Motors V12-71 diesels; 800 hp *(596 kW)*; 2 shafts
Speed, knots: 15. **Range, n miles:** 400 at 12 kt
Complement: 4 (3 officers)

Comment: Atlin Post based at Patricia Bay, British Columbia, and Kitimat II at Prince Rupert, British Columbia.

VERIFIED

ATLIN POST   *6/2001, Canadian Coast Guard /* 0126355

1 CUMELLA CLASS

Name	Builders	Commissioned
CUMELLA	A F Theriault & Son, Meteghan, NS	1983

Measurement, tons: 80 gross; 19 net
Dimensions, feet (metres): 76.1 × 15.7 × ? *(23.2 × 4.8 × ?)*
Main machinery: 2 General Motors V6-24L diesels; 1,680 hp *(1.25 MW)*; 2 shafts
Speed, knots: 15. **Range, n miles:** 600 at 12 kt
Complement: 4 (2 officers)

Comment: Based in Maritimes Region at Grand Manaan, New Brunswick.

VERIFIED

CUMELLA   *6/2001, Canadian Coast Guard /* 0126354

SMALL SAR CUTTERS/ICE STRENGTHENED

1 CUTTER (Type 200)

Name	Builders	Commissioned
HARP	Georgetown SY, PEI	12 Dec 1986

Displacement, tons: 225 full load
Measurement, tons: 179 gross; 69 net
Dimensions, feet (metres): 76.1 × 24.9 × 8.2 *(23.2 × 7.6 × 2.5)*
Main machinery: 2 Caterpillar 3408 diesels; 850 hp *(634 kW)*; 2 Kort nozzle props
Speed, knots: 10. **Range, n miles:** 500 at 10 kt
Complement: 7 plus 10 spare berths
Radars: Navigation: Sperry Mk 1270; I-band.

Comment: Ordered 26 April 1985. Ice strengthened hull. Based in Newfoundland and Labrador Region at St Anthony.

UPDATED

TYPE 200 CUTTER   *3/1999, Canadian Coast Guard /* 0056706

SMALL SAR UTILITY CRAFT

Notes: There are also at least 15 Inshore Rescue boats with CG numbers.

7 SAR CRAFT (Type 100)

Name	Builders	Commissioned
CG 119	Eastern Equipment, Montreal	1973
SORA	Eastern Equipment, Montreal	1982
BITTERN	Canadian Dredge, Kingston	1982
MALLARD	Matsumoto Shipyard, Vancouver, BC	Feb 1986
SKUA	Matsumoto Shipyard, Vancouver, BC	Mar 1986
OSPREY	Matsumoto Shipyard, Vancouver, BC	May 1986
STERNE	Matsumoto Shipyard, Vancouver, BC	Mar 1987

Measurement, tons: 15 gross
Dimensions, feet (metres): 40.8 × 13.2 × 4.2 *(12.4 × 4.1 × 1.3)*
Main machinery: 2 Mitsubishi diesels; 637 hp *(475 kW)*; 2 shafts
Speed, knots: 26. **Range, n miles:** 200 at 16 kt
Complement: 3

Comment: CG 119, *Sora* and *Bittern* based in Central and Arctic Region at Prescott, Amherstburg and Kingston respectively; *Sterne* is based in Quebec Region at Quebec and *Mallard, Skua* and *Osprey* in the Pacific Region at Powell River, Ganges and Kitsilano respectively. CG 119 is structurally different to and slower than the remainder.

UPDATED

CG 119   *1990, van Ginderen Collection*

MULTITASK LIFEBOATS

4 LIFEBOATS (Type 300)

Name	Builders	Commissioned
KESTREL	Chantier Maritimes Du St Laurent, Quebec	1969
CG 106 (ex-*Port Handy*, ex-*Bull Harbour*)	McKay Cormack Ltd, Victoria, BC	1970
TOBERMORY	Georgetown SY, PEI	1974
CGR 100	Hurricane Rescue Craft, Richmond, BC	1986

Measurement, tons: 10 gross
Dimensions, feet (metres): 44.1 × 12.7 × 3.4 *(13.5 × 3.9 × 1)*
Main machinery: 2 diesels; 485 hp *(362 kW)*; 2 shafts
Speed, knots: 12.5; 26 *(CGR 100)*
Complement: 3 or 4

Comment: CG 106 and *Kestrel* based in Pacific Region at Port Hardy and French Creek respectively; and *Tobermory* is in Central and Arctic Region at Tobermory. *CGR 100* is a self-righting Medina lifeboat and has a speed of 26 kt. She is based at Port Weller, Ontario. Three others, CG 117, CG 118 and *Cap Goeland*, serve as training vessels at the Canadian Coast Guard College. *Kestrel*, CG 106 and Tobermory are to be decommissioned in 2005 and CGR 100 in 2006.

UPDATED

KESTREL   7/2003, Chris Sattler / 0569163

10 LIFEBOATS (Type 300A)

Name	Builders	Commissioned
BICKERTON	Halmatic, Havant	Aug 1989
SPINDRIFT	Georgetown, PEI	Oct 1993
SPRAY	Industrie Raymond, Quebec	Sep 1994
COURTENAY BAY (ex-*Spume*)	Industrie Raymond, Quebec	Oct 1994
W JACKMAN (ex-*Cap Aux Meules*)	Industrie Raymond, Quebec	Sep 1995
W G GEORGE	Industrie Raymond, Quebec	Sep 1995
CAP AUX MEULES	Hike Metal Products Ltd, Ontario	Oct 1996
CLARKS HARBOUR	Hike Metal Products Ltd, Ontario	Sep 1996
SAMBRO	Hike Metal Products Ltd, Ontario	Jan 1997
WESTPORT	Hike Metal Products Ltd, Ontario	May 1997

Measurement, tons: 34 gross
Dimensions, feet (metres): 52 × 17.5 × 4.6 *(15.9 × 5.3 × 1.5)*
Main machinery: 2 Caterpillar 3408BTA diesels; 1,070 hp *(786 kW)* sustained; 2 shafts
Speed, knots: 16-20. **Range:** 100-150 m
Complement: 5
Radars: Navigation: Furuno; I-band.

Comment: Seven based in Martimes Region, two in Newfoundland and Labrador Region, one in Quebec Region. *Bickerton* has GRP hull, remainder aluminium.

VERIFIED

CLARKS HARBOUR   8/1996, Kathy Johnson / 0056702

28 + 3 LIFEBOATS (Type 300B)

Name	Builders	Commissioned
THUNDER CAPE	Metalcraft Marine, Kingston	Aug 2000
CAPE SUTIL	Metalcraft Marine, Kingston	Dec 1998
CAPE CALVERT	Metalcraft Marine, Kingston	Aug 1999
CAPE ST JAMES	Metalcraft Marine, Kingston	Nov 1999
CAPE MERCY	Metalcraft Marine, Kingston	Dec 2000
CAPE LAMBTON	Metalcraft Marine, Kingston	July 2001
CAPE STORM	Metalcraft Marine, Kingston	Nov 2002
CAPE FOX	Victoria Shipyard Co Ltd, Victoria, BC	May 2003
CAPE NORMAN	Victoria Shipyard Co Ltd, Victoria, BC	May 2003
CAP DE RABAST	Victoria Shipyard Co Ltd, Victoria, BC	Aug 2003
CAP ROZIER	Victoria Shipyard Co Ltd, Victoria, BC	Aug 2003
CAPE MUDGE	Victoria Shipyard Co Ltd, Victoria, BC	Nov 2003
CAPE FAREWELL	Victoria Shipyard Co Ltd, Victoria, BC	Nov 2003
CAPE COCKBURN	Victoria Shipyard Co Ltd, Victoria, BC	Jan 2004
CAPE SPRY	Victoria Shipyard Co Ltd, Victoria, BC	Apr 2004
CAP NORD	Victoria Shipyard Co Ltd, Victoria, BC	Apr 2004
CAP BRETON	Victoria Shipyard Co Ltd, Victoria, BC	Apr 2004
CAPE MCKAY	Victoria Shipyard Co Ltd, Victoria, BC	June 2004
CAPE CHAILLON	Victoria Shipyard Co Ltd, Victoria, BC	Oct 2004
CAPE PROVIDENCE	Victoria Shipyard Co Ltd, Victoria, BC	Oct 2004
CAPE COMMODORE	Victoria Shipyard Co Ltd, Victoria, BC	Oct 2004
CAPE ANN	Victoria Shipyard Co Ltd, Victoria, BC	Nov 2004
CAPE CAUTION	Victoria Shipyard Co Ltd, Victoria, BC	Dec 2004
CAPE DISCOVERY	Victoria Shipyard Co Ltd, Victoria, BC	Jan 2005
CAPE HEARNE	Victoria Shipyard Co Ltd, Victoria, BC	Feb 2005
CAPE DUNDAS	Victoria Shipyard Co Ltd, Victoria, BC	Mar 2005
CAP TOURMENTE	Victoria Shipyard Co Ltd, Victoria, BC	Apr 2005
CAP D'ESPOIR	Victoria Shipyard Co Ltd, Victoria, BC	June 2005
CAP PERCÉ	Victoria Shipyard Co Ltd, Victoria, BC	Aug 2005
CAPE EDENSAW	Victoria Shipyard Co Ltd, Victoria, BC	Sep 2005
CAPE KUPER	Victoria Shipyard Co Ltd, Victoria, BC	Oct 2005

Measurement, tons: 33.8 gross
Dimensions, feet (metres): 47.9 × 14 × 4.5 *(14.6 × 4.27 × 1.37)*
Main machinery: 2 Caterpillar 3196 diesels; 905 hp *(675 kW)* sustained; 2 shafts
Speed, knots: 22-25. **Range, n miles:** 200 n miles
Complement: 4
Radars: Navigation: Furuno 1942; I-band.

Comment: Multitask medium endurance lifeboat. A total of 31 craft to be built.

UPDATED

THUNDER CAPE   2000, Canadian Coast Guard / 0104265

HOVERCRAFT

1 API-88/200 TYPE

Name	Builders	Commissioned
WABAN-AKI	Westland Aerospace	15 July 1987

Displacement, tons: 47.6 light
Dimensions, feet (metres): 80.4 × 36.7 × 19.6 *(24.5 × 11.2 × 6.6)* (height on cushion)
Main machinery: 4 Deutz diesels; 2,394 hp(m) *(1.76 MW)*
Speed, knots: 50; 35 cruising
Complement: 3
Cargo capacity: 12 tons

Comment: *Waban-Aki* is based at Trois Rivières and capable of year round operation as a Navaid Tender for flood control operations in the St Lawrence. Fitted with a hydraulic crane. The name means People of the Dawn.

UPDATED

WABAN-AKI   4/1999, Canadian Coast Guard / 0056717

2 AP. I-88/400 TYPE

SIPU MUIN   SIYAY

Displacement, tons: 69 full load
Dimensions, feet (metres): 93.5 × 39.4 *(28.5 × 12)*
Main machinery: 4 Caterpillar 3412 TTA diesels; 3,650 hp(m) *(2.68 MW)* sustained
Speed, knots: 50; 35 cruising
Complement: 4
Cargo capacity: 22.6 tons

Comment: Contract awarded to GKN Westland in May 1996. Built at Hike Metal Products, Wheatley, Ontario and completed in August and December 1998 respectively. Well-deck size 8.2 × 4.6 m. There is a 5,000 kg load crane. *Sipu Muin* is based at Trois Rivières and the second at Sea Island, BC. **UPDATED**

SIPU MUIN   *5/1998, Canada Coast Guard /* 0017672

1 AP.I-88/100S TYPE

PENAC (ex-*Liv Viking*)

Displacement, tons: 45.5 full load
Dimensions, feet (metres): 80.4 × 39.0 *(24.5 × 11.9)*
Main machinery: 2 Deutz BF 12L513 diesels; 1,050 hp(m) *(785 kW)*. 2 MTU 12V 183TB32 diesels; 1,640 hp(m) *(1.25 MW)* sustained
Speed, knots: 50; 35 cruising
Complement: 7
Cargo capacity: 5.3 tons

Comment: Built by Hoverworks Ltd, Isle of Wight, UK in 1984. Procured by Canadian Coast Guard in 2004. Based in Vancouver, BC. **NEW ENTRY**

PENAC   *6/2004*, Canadian Coast Guard /* 1042123

ROYAL CANADIAN MOUNTED POLICE

Notes: The Marine Branch of the Royal Canadian Mounted Police is responsible for enforcement of Customs, Immigration, Shipping and Drug regulations as well as for standard policing duties in areas that are difficult to access by land. Of five 17-19 m catamaran-design patrol craft, *Inkster, Nadon, Higgitt* and *Lindsay* are stationed on the West Coast while *Simmonds* is based in Newfoundland on the East Coast. In addition there are some 377 smaller craft for use on inland waterways.

FISHERY RESEARCH SHIPS

11 RESEARCH SHIPS

Name	Commissioned	Based	Measurement, tons
ALFRED NEEDLER	Aug 1982	Dartmouth, NS	925 grt
WILFRED TEMPLEMAN	Mar 1982	St John's, NL	925 grt
W E RICKER	Dec 1978	Nanaimo, BC	1,040 grt
(ex-*Callistratus*)			
TELEOST	1996	St John's, NL	
PANDALUS III	1986	St Andrew's, NB	13 grt
SHAMOOK	1975	St John's, NL	187 grt
OPILIO	1989	Shippagan, NB	74 grt
SHARK	1971	Burlington, ON	19 grt
CALANUS II	1991	Rimouski, QC	160 grt
J L HART	1974	St Andrew's, NB	93 grt
NEOCALIGUS	2001	Nanaimbo, QC	98 grt

Comment: First four are classified as Offshore Fishery Research vessels, remainder as Inshore Fishery Research vessels. **UPDATED**

TELEOST   *4/1999, Canadian Coast Guard /* 0056713

SURVEY AND RESEARCH SHIPS

7 RESEARCH SHIPS

Name	Commissioned	Based	Displacement, tons
MATTHEW	1990	Dartmouth, NS	950
F C G SMITH	1986	Quebec, QC	300
HUDSON	1963	Dartmouth, NS	3,740
JOHN P TULLY	1985	Patricia Bay, BC	1,800
VECTOR	1967	Patricia Bay, BC	520
LIMNOS	1968	Burlington, ON	—
FREDERICK G CREED	1988	Rimouski, QC	81

Comment: The one ship in the Central and Arctic Region is employed on Limnology, and the remainder on Oceanographic Research. *Hudson* and *Tully* are classified as Offshore vessels, and the remainder are Coastal. *Smith* and *Creed* are multihulled. *R B Young* is in reserve, and *Tully* and *Vector* only operate for part of the year. **UPDATED**

F C G SMITH   *7/1998, C D Maginley /* 0017673

Cape Verde

Country Overview

A former Portuguese colony, the Republic of Cape Verde became independent in 1975. Situated in the Atlantic Ocean some 335 n miles due west of the western point of Africa, it consists of ten islands and five islets, which are divided into the northerly windward (Barlavento) and southerly leeward (Sotavento) groups. The windward group includes Santo Antão, São Vicente, Santa Luzia, São Nicolau, Sal and Boa Vista; the leeward group includes São Tiago, Brava, Fogo and Maio. Mindelo, on São Vicente, is the principal port and economic centre while Praia on São Tiago is the capital and largest town. An archipelagic state, territorial seas (12 n miles) are claimed. A 200 n mile Exclusive Economic Zone (EEZ) has been claimed but the limits are not fully defined.

Personnel

2005: 50

Bases

Praia, main naval base.
Porto Grande (Isle of São Vicente), naval repair yard.

Maritime Aircraft

One EMB-111 and one Dornier Do 328 are used for maritime surveillance.

PATROL FORCES

Notes: (1) One Zhuk class may still be in service but seldom goes to sea.
(2) There is a 25 m patrol boat *Tainha* P 262.

TAINHA   *6/2004* * / 1044089

1 KONDOR I CLASS (COASTAL PATROL CRAFT) (PBO)

Name	No	Builders	Commissioned
VIGILANTE (ex-*Kühlungsborn*)	P 521 (ex-BG 32, ex-GS 07)	Peenewerft, Wolgast	1970

Displacement, tons: 377 full load
Dimensions, feet (metres): 170.3 × 23.3 × 7.2 *(51.9 × 7.1 × 2.2)*
Main machinery: 2 Russki/Kolomna Type 40DM diesels; 4,408 hp(m) *(3.24 MW)* sustained; 2 shafts; cp props
Speed, knots: 20. **Range, n miles:** 1,800 at 15 kt
Complement: 25 (3 officers)
Guns: 2—25 mm (twin).
Radars: Surface search: Racal Decca; I-band.

Comment: Former GDR minesweeper taken over by the German Coast Guard, and then acquired by Cape Verde in September 1998. Armament is uncertain. Reported refitted in Germany in 1999-2000. ***VERIFIED***

KONDOR I (Malta colours)   *6/1997, Robert Pabst* / 0017674

1 ESPADARTE CLASS (PETERSON Mk 4 TYPE) (COASTAL PATROL CRAFT) (PB)

Name	No	Builders	Commissioned
ESPADARTE	P 151	Peterson Builders Inc	19 Aug 1993

Displacement, tons: 22 full load
Dimensions, feet (metres): 51.3 × 14.8 × 4.3 *(15.6 × 4.5 × 1.3)*
Main machinery: 2 Detroit 6V-92TA diesels; 520 hp *(388 kW)* sustained; 2 shafts
Speed, knots: 24. **Range, n miles:** 500 at 20 kt
Complement: 6 (2 officers)
Guns: 2—12.7 mm MGs (twin). 2—7.62 mm MGs.
Radars: Surface search: Raytheon; I-band.

Comment: Ordered from Peterson Builders Inc, under FMS programme on 25 September 1992. Option on three more not taken up. Aluminium hulls. The 12.7 mm mounting is aft with the smaller guns on the bridge roof. ***VERIFIED***

Mk 4 CPC (US colours)   *11/1993, Peterson Builders* / 0081500

Cayman Islands

Country Overview

A British dependency since 1962, the island group is situated south of Cuba in the Caribbean Sea. It comprises three islands: Grand Cayman, containing the capital George Town, Little Cayman and Cayman Brac, located about 80 miles northeast of Grand Cayman. Territorial seas (12 n miles) and a Fishery Zone (200 n miles) are claimed. A governor, appointed by the British Crown, is responsible for external affairs, internal security, defence and the police. The Marine section is a division of the Royal Cayman Islands Police (RCIP) and UK Customs Drugs Task Force. Its roles are Maritime Drug Interdiction, SAR, Safety, Conservation and Fishery Protection.

Headquarters Appointments

Commander Royal Cayman Islands Police (Marine):
   Bruce D Smith

Personnel

2005: 15 (mixture of police and customs)

Bases

Grand Cayman (main), Little Cayman, Cayman Brac.

POLICE

Notes: A Concept pursuit craft, *Derry's Pride,* with twin 225 hp Johnson outboards is based at Grand Cayman together with *Intrepid,* an 'Eduardono' Colombian craft, and *Typhoon,* a 24 ft RIB. Two Boston Whalers, *Lima 1* and *Miss Molly,* are based at Little Cayman and Cayman Brac respectively.

DERRY'S PRIDE   *6/2001, RCIP* / 0121307

LIMA 1   *6/2001, RCIP* / 0121306

1 DAUNTLESS CLASS (PB)

CAYMAN PROTECTOR

Displacement, tons: 17 full load
Dimensions, feet (metres): 47.9 × 14.1 × 3.3 *(14.6 × 4.3 × 1)*
Main machinery: 2 Caterpillar 3208TA diesels; 720 hp(m) *(529 kW)* sustained; 2 shafts
Speed, knots: 26. **Range, n miles:** 400 at 20 kt
Complement: 11
Guns: 2—7.62 mm MGs.
Radars: Surface search: Raytheon R40; I-band.

Comment: Built by SeaArk Marine, Monticello and acquired in July 1994. Based at Grand Cayman. ***VERIFIED***

CAYMAN PROTECTOR   *6/2001, RCIP* / 0121305

For details of the latest updates to *Jane's Fighting Ships* online and to discover the additional information available exclusively to online subscribers please visit
jfs.janes.com

Chile
ARMADA DE CHILE

Country Overview

The Republic of Chile is situated in western South America. With an area of 292,135 square miles it has borders to the north with Peru and to the east with Bolivia and Argentina. Off the 2,305 n mile coastline with the Pacific Ocean lie the Chonos Archipelago, Wellington Island and the western portion of Tierra del Fuego. Chilean islands in the south Pacific include the Juan Fernández Islands, Easter Island, and Sala y Gómez. The capital, largest city and principal port is Santiago. There are further ports at Talcahuano, Tomé, Antofagasta, San Antonio, Arica, Iquique, Coquimbo, San Vicente, Puerto Montt, and Punta Arenas. Territorial seas (12 n miles) and an EEZ (200 n miles) are claimed.

Headquarters Appointments

Commander-in-Chief:
    Admiral Miguel Angel Vergara Villalobos
Naval Operations Command:
    Vice Admiral Rodolfo Codina Díaz
Chief of the Naval Staff:
    Vice Admiral Oscar Manzano Soko
Director General, Naval Personnel:
    Vice Admiral Eduardo García Domínguez
Director General, Naval Services:
    Vice Admiral Juan Illanes Laso
Director General Maritime Territory and Merchant Marine:
    Vice Admiral Francisco Martínez Villarroel
Flag Officer, Fleet:
    Rear Admiral Gerardo Covacevich Castex
Flag Officer, Submarines:
    Rear Admiral Alejandro Herrmann Hartung
Flag Officer, 1st Naval Zone:
    Rear Admiral Gudelio Mondaca Oyarzun
Flag Officer, 2nd Naval Zone:
    Rear Admiral Daniel Arellano Walbaum
Flag Officer, 3rd Naval Zone:
    Rear Admiral Edmundo Gonzales Robles
Flag Officer, 4th Naval Zone:
    Rear Admiral Percy Richter Silberstein

Diplomatic Representation

Naval Attaché in Ottawa:
    Captain Kenneth Pugh Olavarría
Naval Attaché in Beijing:
    Captain Sergio Cabezas Ferrari
Naval Attaché in London:
    Captain Charles Le-May Vizcaya
Naval Attaché in Washington:
    Rear Admiral Roberto Carvajal Gacitúa
Naval Attaché in Paris:
    Captain Matías Purcell Echeverría
Naval Attaché in Buenos Aires:
    Captain Julian Elorrieta Grimalt

Diplomatic Representation — *continued*

Naval Attaché in Seoul:
    Captain Roggero Cozzi Paredes
Naval Attaché in Lima:
    Captain Humberto Ramirez Navarro
Naval Attaché in Madrid:
    Captain Gastón Massa Barros
Naval Attaché in Brasilia:
    Captain Luis Catalan Cruz
Naval Attaché in Quito:
    Captain Alejandro Campos Calvo
Naval Attaché in Panama:
    to be announced

Personnel

(a)   2005: 19,829 (1,988 officers)
(b)   4,500 Marines
(c)   2 years' national service (1,300)

Command Organisation

1st Naval Zone. HQ at Valparaiso. From 26° 00′ S to 34° 09′ S.
2nd Naval Zone. HQ at Talcahuano. From 34° 09′ S to 46° 00′ S.
3rd Naval Zone. HQ at Punta Arenas. From 46° 00′ S to South Pole.
4th Naval Zone. HQ at Iquique. From 18° 21′ S to 26° 00′S.
Coast Guard is fully integrated with the Navy.

Naval Air Stations and Organisation

Having won the battle to own all military aircraft flying over the sea, a fixed-wing squadron of about 20 CASA/ENAER Halcón is envisaged when finances permit.
Viña del Mar (Valparaiso); *Almirante Von Schroeders* (Punta Arenas); *Guardiamarina Zañartu* (Puerto Williams).
Four Squadrons: VP1: EMB-111, P-3A
    HA1: NAS 332C Cougar
    VC1: EMB-110, CASA 212, PC-7
    HU1: Bell 206B, BO 105C
    VP1: Mod Skymaster

Infanteria de Marina

Organisation: 4 detachments each comprising Amphibious Warfare, Coast Defence and Local Security. Also embarked are detachments of commandos, engineering units and a logistic battalion.
1st Marine Infantry Detachment 'Patricio Lynch'. At Iquique.
2nd Marine Infantry Detachment 'Miller'. At Viña del Mar.
3rd Marine Infantry Detachment 'Sargento Aldea'. At Talcahuano.

Infanteria de Marina — *continued*

4th Marine Infantry Detachment 'Cochrane'. At Punta Arenas.
51 Commando Group. At Valparaiso.
Some embarked units, commando and engineering units and a logistics battalion.

Bases

Valparaiso. Main naval base, schools, repair yard. HQ 1st Naval Zone. Air station.
Talcahuano. Naval base, schools, major repair yard (two dry docks, three floating docks), two floating cranes. HQ 2nd Naval Zone. Submarine base.
Punta Arenas. Naval base. Dockyard with slipway having building and repair facilities. HQ 3rd Naval Zone. Air station.
Iquique. Small naval base. HQ 4th Naval Zone.
Puerto Montt. Small naval base.
Puerto Williams (Beagle Channel). Small naval base. Air station.
Dawson Island (Magellan Straits). Small naval base.

Strength of the Fleet (including Coast Guard)

Type	Active	Building
Patrol Submarines	3	1
Destroyers	2	—
Frigates	4	4
Landing Ships (Tank)	3	—
Landing Craft	2	—
Fast Attack Craft (Missile)	7	—
Large Patrol Craft	6	—
Coastal Patrol Craft	40	—
Survey Ships	3	—
Training Ships	1	—
Transports	1	—
Tankers	1	—
Tenders	5	—

DELETIONS

Destroyers

2003   *Blanco Encalada* (old)

Patrol Forces

2002   *Tegualda*
2005   *Guacolda*

PENNANT LIST

Notes: From 1997 pennant numbers have been painted on major warship hulls.

Submarines

20	Thomson
21	Simpson
22	O'Higgins
23	Carrera

Destroyers

11	Prat
12	Cochrane

Frigates

06	Condell
07	Lynch
08	Ministro Zenteno
19	Almirante Williams

Patrol Forces

30	Casma
31	Chipana
34	Angamos
36	Riquelme
37	Orella
38	Serrano
39	Uribe
73	Isaza
74	Morel
77	Cabrales
78	Sibbald
1601	Ona (CG)
1602	Yagan (CG)
1603	Alacalufe (CG)
1604	Hallef (CG)

1608	Fresia
1609	Aysen (CG)
1610	Corral (CG)
1611	Concepcion (CG)
1612	Caldera (CG)
1613	San Antonio (CG)
1614	Antofagasta (CG)
1615	Arica (CG)
1616	Coquimbo [1616] (CG)
1617	Natales (CG)
1618	Valparaiso (CG)
1619	Punta Arenas (CG)
1620	Talcahuano (CG)
1621	Quintero (CG)
1622	Chiloe (CG)
1623	Puerto Montt (CG)
1624	Iquique
1814	Diaz
1815	Bolados
1816	Salinas
1817	Tellez
1818	Bravo
1819	Campos
1820	Machado
1821	Johnson
1822	Troncoso
1823	Hudson
1901	Maule (CG)
1902	Rapel (CG)
1903	Aconcagua (CG)
1904	Lauca (CG)
1905	Isluga (CG)
1906	Loa (CG)
1907	Maullín (CG)
1908	Copiapó (CG)
1909	Cau-Cau (CG)
1910	Pudeto (CG)
1911	Robinson Crusoe (CG)

Survey Ships

46	Contre-almirante Oscar Viel Toro
60	Vidal Gormaz
63	George Slight Marshall

Training Ships

43	Esmeralda

Amphibious Forces

90	Elicura
92	Rancagua
93	Valdivia
94	Orompello
95	Chacabuco

Auxiliaries

41	Aquiles
42	Merino
53	Araucano
71	Micalvi
72	Ortiz
YFB 114	Grumete Perez
116	Pisagua

Tugs/Supply Ships

ATF 66	Galvarino
ATF 67	Lautaro
ATF 68	Leucoton

SUBMARINES

Notes: There are some Swimmer Delivery Vehicles French Havas Mk 8 in service. This is the two-man version.

1 + 1 SCORPENE CLASS (SSK)

Name	No	Builders	Laid down	Launched	Commissioned
O'HIGGINS	22	DCN Cherbourg/IZAR	18 Nov 1999	1 Nov 2003	2004
CARRERA	23	IZAR, Cartagena/DCN	Nov 2000	24 Nov 2004	2006

Displacement, tons: 1,668 dived
Dimensions, feet (metres): 217.8 × 20.3 × 19
  (66.4 × 6.2 × 5.8)
Main machinery: Diesel electric; 4 MTU 16V 396 SE84 diesels;
  2,992 hp(m) *(2.2 MW)*; 1 Jeumont Schneider motor;
  3,808 hp(m) *(2.8 MW)*; 1 shaft
Speed, knots: 20 dived; 12 surfaced
Range, n miles: 550 at 4 kt dived; 6,500 at 8 kt surfaced
Complement: 31 (6 officers)

Torpedoes: 6—21 in *(533 mm)* tubes. 18 Whitehead Black Shark
  184 Mod 3 torpedoes.
Countermeasures: ESM; Argos AR 900; intercept.
Weapons control: UDS International SUBTICS.
Radars: Navigation: Sagem; I-band.
Sonars: Hull mounted; active/passive search and attack,
  medium frequency.

Programmes: Project Neptune. Contract awarded to DCN and
  Bazán on 18 December 1997 and became effective in April
  1998. The bows of both boats were built at Cherbourg and
  the sterns at Cartagena. First steel cut for the first of class on
  22 July 1998 and final assembly by DCN began on 15
  November 2002 when the stern arrived at Cherbourg. Final
  assembly of the second of class began on 22 March 2004
  when the bow arrived at Cartagena.
Structure: Sagem provides the APS attack periscope and an
  SMS optronic search periscope. SISDEF is fitting a datalink
  terminal. Diving depth more than 300 m *(984 ft)*. AIP is not
  being fitted.
Operational: Sea trials of *O'Higgins* started at Lorient in early
  2004.

UPDATED

O'HIGGINS   *9/2004*, B Prézelin /* 1044092

CARRERA
11/2004*, Diego Quevedo / 1121118

2 THOMSON (TYPE 209) CLASS (TYPE 1300) (SSK)

Name	No	Builders	Laid down	Launched	Commissioned
THOMSON	20	Howaldtswerke	1 Nov 1980	28 Oct 1982	31 Aug 1984
SIMPSON	21	Howaldtswerke	15 Feb 1982	29 July 1983	18 Sep 1984

Displacement, tons: 1,260 surfaced; 1,390 dived
Dimensions, feet (metres): 195.2 × 20.3 × 18
  (59.5 × 6.2 × 5.5)
Main machinery: Diesel-electric; 4 MTU 12V 493 AZ80 GA31L
  diesels; 2,400 hp(m) *(1.76 MW)* sustained; 4 Piller
  alternators; 1.7 MW; 1 Siemens motor; 4,600 hp(m)
  (3.38 MW) sustained; 1 shaft
Speed, knots: 11 surfaced; 21.5 dived
Range, n miles: 400 at 4 kt dived; 16 at 21.5 kt dived; 8,200 at
  8 kt snorkel
Complement: 32 (5 officers)

Torpedoes: 8—21 in *(533 mm)* bow tubes. 14 AEG SUT Mod 1;
  wire-guided; active homing to 12 km *(6.5 n miles)* at 35 kt;

passive homing to 28 km *(15 n miles)* at 23 kt; warhead
  250 kg.
Countermeasures: ESM: Thomson-CSF DR 2000U; radar
  warning.
Radars: Surface search: Thomson-CSF Calypso II; I-band.
Sonars: Atlas Elektronik CSU 3; hull-mounted; active/passive
  search and attack; medium frequency.

Programmes: Ordered from Howaldtswerke, Kiel in 1980.
Modernisation: *Thomson* refit completed at Talcahuano in late
  1990, *Simpson* in 1991. Refit duration about 10 months
  each. A major programme to upgrade and extend the service
  life of both boats to 2025 has been initiated. The work is to
  include the fitting of a UDS Subtics combat management

system and a new fire-control system. Torpedo tubes are to be
  upgraded to enable the Whitehead Black Shark torpedoes and
  anti-ship missiles to be fired while platform improvements are
  likely to include a new engine-control system and battery set.
  Work on *Simpson* is to start in 2005 and is to complete in
  2008. Modernisation of *Thomson* is to run approximately 18
  months behind. The CSU 90 sonars from the 'Oberons' may
  have been transferred.
Structure: Fin and associated masts lengthened by 50 cm to
  cope with wave size off Chilean coast.

UPDATED

THOMSON   *7/2001, Maritime Photographic /* 0121308

DESTROYERS

2 PRAT (COUNTY) CLASS (DDGHM)

Name	No	Builders	Laid down	Launched	Commissioned
CAPITÁN PRAT (ex-*Norfolk*)	11	Swan Hunter, Wallsend	15 Mar 1966	16 Nov 1967	7 Mar 1970
ALMIRANTE COCHRANE (ex-*Antrim*)	12	Fairfield SB & Eng Co Ltd, Govan	20 Jan 1966	19 Oct 1967	14 July 1970

Displacement, tons: 6,200 standard; 6,800 full load
Dimensions, feet (metres): 520.5 × 54 × 20.5
 (158.7 × 16.5 × 6.3)
Main machinery: COSAG; 2 Babcock & Wilcox boilers; 700 psi
 (49.2 kg/cm²); 950°F *(510°C)*; 2 AEI steam turbines;
 30,000 hp *(22.4 MW)*; 4 English Electric G6 gas turbines;
 30,000 hp *(22.4 MW)*; 2 shafts
Speed, knots: 28. **Range, n miles:** 3,500 at 28 kt
Complement: 470 (36 officers)

Missiles: SSM: 4 Aerospatiale MM 38 Exocet ❶; inertial cruise;
 active radar homing to 42 km *(23 n miles)* at 0.9 Mach;
 warhead 165 kg; sea-skimmer.
 2 octuple IAI/Rafael Barak I ❷ command line of sight radar or
 optical guidance to 10 km *(5.5 n miles)* at 2 Mach; warhead
 22 kg.
Guns: 2 Vickers 4.5 in *(115 mm)* Mk 6 semi-automatic (twin) ❸;
 20 rds/min to 19 km *(10.3 n miles)* anti-surface; 6 km *(3.2 n
 miles)* anti-aircraft; weight of shell 25 kg.
 2 or 4 Oerlikon 20 mm Mk 9 ❹; 800 rds/min to 2 km.
Torpedoes: 6—324 mm Mk 32 (2 triple) tubes ❺; Honeywell Mk
 46 Mod 2; active/passive homing to 11 km *(5.9 n miles)* at
 40 kt; warhead 44 kg.
Countermeasures: ESM: Elisra 9003, intercept. Elta IR sensor.
 ECM: Type 667; jammer.
 Decoys: SLQ-25 Nixi; noisemaker.
Combat data systems: Sisdef Imagen SP 100 with datalink.
 SATCOM.
Weapons control: Gunnery MRS 3 system.
Radars: Air search: Marconi Type 966 ❻; A-band.
 Elta LM 2228S ❼; E/F-band (for Barak).
 Surface search: Marconi Type 992 Q or R ❽; E/F-band; range
 55 km *(30 n miles)*.
 Navigation: Decca Type 978/1006; I-band.
 Fire control: Plessey Type 903 ❾; I-band (for Guns).
 Two Elta EL/M-2221GM ❿; I/J/K-band (for Barak).
Sonars: Kelvin Hughes Type 162 M; hull-mounted; sideways
 looking classification; high-frequency.
 Graseby Type 184 M; hull-mounted; active search and attack;
 medium range; 7 to 9 kHz.

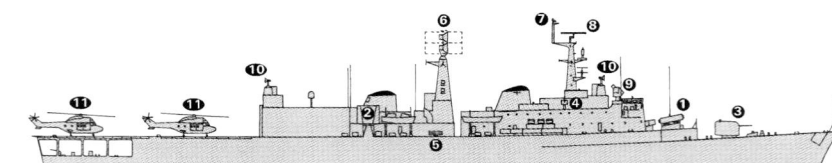

ALMIRANTE COCHRANE *(Scale 1 : 1,500), Ian Sturton /* 0529544

ALMIRANTE COCHRANE *7/2001, Maritime Photographic /* 0121309

Helicopters: 2 (1 in *Prat*) NAS-332SC Cougar ⓫.

Programmes: Transferred from UK 6 April 1982 *(Prat)* and 22
 June 1984 *(Cochrane)*. Extensive refits carried out after
 transfer. The titles Almirante and Capitán are sometimes
 omitted from the ships' names.

Modernisation: *Cochrane* converted to carry two Super Puma
 helicopters and completed in May 1994. *Prat* completed refit
 in February 2001 without full flight deck extension. All of the
 class fitted with the Israeli Barak I and new communications,
 optronic directors and EW equipment. Imagen combat data
 system fitted. Indal Assist helo recovery system also fitted and
 magazine stowage increased.
Structure: In *Cochrane*, greatly enlarged flight deck (617 m²)
 continued right aft to accommodate two large helicopters
 simultaneously, making them effectively flush-decked. The
 hangar has also been completely rebuilt (dimensions
 16.9 × 11.7 m) and the foremast extended. Indal ASIST helo
 handling system. Chaft launchers have been removed.
Operational: *Latorre* paid off in 1998 and *Blanco Encalada* in
 2003. These ships expected to be decommissioned as ex-
 Netherlands ships enter service.

UPDATED

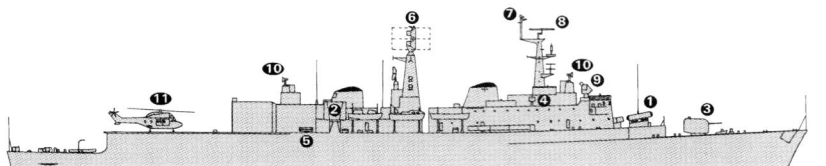

CAPITÁN PRAT *(Scale 1 : 1,500), Ian Sturton /* 0529543

CAPITÁN PRAT *7/2001, Maritime Photographic /* 0121311

FRIGATES

Notes: Following cancellation of the 'Tridente' frigate programme in early 2002, the 'Fragata' programme was initiated in February 2003. This too was superseded by the decision in February 2004 to acquire four ex-Netherlands frigates. The acquisition of further second-hand ships, possibly from UK, is expected to achieve total of eight to ten major surface combatants.

0 + 2 JACOB VAN HEEMSKERCK CLASS (FFGM)

Name	No	Builders	Laid down	Launched	Commissioned
LATORRE (ex-*Jacob van Heemskerck*)	(ex-F 812)	Koninklijke Maatschappij De Schelde, Flushing	21 Jan 1981	5 Nov 1983	15 Jan 1986
— (ex-*Witte de With*)	(ex-F 813)	Koninklijke Maatschappij De Schelde, Flushing	15 Dec 1981	25 Aug 1984	17 Sep 1986

Displacement, tons: 3,750 full load approx
Dimensions, feet (metres): 428 × 47.9 × 14.1 (20.3 screws) *(130.5 × 14.6 × 4.3 (6.2))*
Main machinery: COGOG; 2 RR Olympus TM3B gas turbines; 50,880 hp *(37.9 MW)* sustained
2 RR Tyne RM1C gas turbines; 9,900 hp *(7.4 MW)* sustained; 2 shafts; LIPS cp props
Speed, knots: 30. **Range, n miles:** 4,700 at 16 kt on Tynes
Complement: 197 (23 officers)

Missiles: SSM: 8 McDonnell Douglas Harpoon (2 quad) launchers ❶; active radar homing to 130 km *(70 n miles)* at 0.9 Mach; warhead 227 kg.

SAM: 40 GDC Pomona Standard SM-1MR; Block IV; Mk 13 Mod 1 launcher ❷; command guidance; semi-active radar homing to 46 km *(25 n miles)* at 2 Mach.
Raytheon Sea Sparrow Mk 29 octuple launcher ❸; semi-active radar homing to 14.6 km *(8 n miles)* at 2.5 Mach; warhead 39 kg; 24 missiles.
Guns: 1 Signaal SGE-30 Goalkeeper ❹ with General Electric 30 mm 7-barrelled; 4,200 rds/min combined to 2 km.
2 Oerlikon 20 mm.
Torpedoes: 4—324 mm US Mk 32 (2 twin) tubes ❺. Honeywell Mk 46 Mod 5; anti-submarine; active/passive homing to 11 km *(5.9 n miles)* at 40 kt; warhead 44 kg.

Countermeasures: Decoys: 2 Loral Hycor Mk 36 SRBOC 6-tubed fixed quad launchers ❻; IR flares and chaff to 4 km *(2.2 n miles)*.
ESM/ECM: Ramses; intercept and jammer.
Combat data systems: Signaal SEWACO VI action data automation; Link 11. SHF SATCOM ❼. JMCIS.
Radars: Air search: Signaal LW08 ❽; D-band; range 264 km *(145 n miles)* for 2 m² target.
Air/surface search: Signaal Smart; 3D ❾; F-band.
Surface search: Signaal Scout ❿; I-band.
Fire control: 2 Signaal STIR 240 ⓫; I/J/K-band; range 140 km *(76 n miles)* for 1 m² target.
Signaal STIR 180 ⓬; I/J/K-band.
Sonars: Westinghouse SQS-509; hull-mounted; active search and attack; medium frequency.

Programmes: Contract signed on 26 March 2004 for the acquisition of two air-defence frigates. Ex-*Jacob van Heemskerck* to be transferred in December 2005 and ex-*Witte de With* (possibly to be named *Capitán Prat* in June 2006. 200 SM-1 missiles also reported acquired.
Operational: Command facilities for a task group commander and his staff.

NEW ENTRY

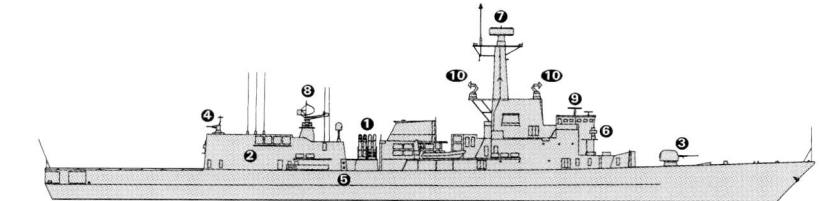

JACOB VAN HEEMSKERCK CLASS   *(Scale 1 : 1,200), Ian Sturton* / 0114748

0 + 2 KAREL DOORMAN CLASS (FFGHM)

Name	No	Builders	Laid down	Launched	Commissioned
ALMIRANTE RIVEROS (ex-*Tjerk Hiddes*)	— (ex-F 830)	Koninklijke Maatschappij De Schelde, Flushing	28 Oct 1986	9 Dec 1989	3 Dec 1992
ALMIRANTE BLANCO ENCALADA (ex-*Abraham van der Hulst*)	— (ex-F 832)	Koninklijke Maatschappij De Schelde, Flushing	8 Feb 1989	7 Sep 1991	15 Dec 1993

Displacement, tons: 3,320 full load
Dimensions, feet (metres): 401.2 oa; 374.7 wl × 47.2 × 14.1 *(122.3; 114.2 × 14.4 × 4.3)*
Flight deck, feet (metres): 72.2 × 47.2 *(22 × 14.4)*
Main machinery: CODOG; 2 RR Spey SM1C; 33,800 hp *(25.2 MW)* sustained (early ships of the class will initially only have SM1A gas generators and 30,800 hp *(23 MW)* sustained available; 2 Stork-Wärtsilä 12SW280 diesels; 9,790 hp(m) *(7.2 MW)* sustained; 2 shafts; LIPS cp props
Speed, knots: 30 (Speys); 21 (diesels)
Range, n miles: 5,000 at 18 kt
Complement: 156 (16 officers) (accommodation for 163)

Missiles: SSM: 8 McDonnell Douglas Harpoon Block 1C (2 quad) launchers ❶; active radar homing to 130 km *(70 n miles)* at 0.9 Mach; warhead 227 kg (to be confirmed).
SAM: Raytheon Sea Sparrow Mk 48 vertical launchers ❷; semi-active radar homing to 14.6 km *(8 n miles)* at 2.5 Mach; warhead 39 kg; 16 missiles. Canisters mounted on port side of hangar.
Guns: 1—3 in *(76 mm)*/62 OTO Melara compact Mk 100 ❸; 100 rds/min to 16 km *(8.6 n miles)* anti-surface; 12 km *(6.5 n miles)* anti-aircraft; weight of shell 6 kg. This is the version with an improved rate of fire.
1 Signaal SGE-30 Goalkeeper with General Electric 30 mm 7-barrelled ❹; 4,200 rds/min combined to 2 km.
2 Oerlikon 20 mm; 800 rds/min to 2 km.
Torpedoes: 4—324 mm US Mk 32 Mod 9 (2 twin) tubes (mounted inside the after superstructure) ❺. Honeywell Mk 46 Mod 5; anti-submarine; active/passive homing to 11 km *(5.9 n miles)* at 40 kt; warhead 44 kg.

Countermeasures: Decoys: 2 Loral Hycor SRBOC 6-tubed fixed Mk 36 quad launchers; IR flares and chaff to 4 km *(2.2 n miles)*.
SLQ-25 Nixie towed torpedo decoy.
ESM/ECM: Argo APECS II (includes AR 700 ESM) ❻; intercept and jammers.
Combat data systems: Signaal SEWACO VIIB action data automation; Link 11. WSC-6 twin aerials.
Weapons control: Signaal IRSCAN infra-red detector (fitted in F 829 for trials and may be retrofitted in all in due course). Signaal VESTA helo transponder.
Radars: Air/surface search: Signaal SMART ❼; 3D; F-band.
Air search: Signaal LW08 ❽; D-band.
Surface search: Signaal Scout ❾; I-band.
Navigation: Racal Decca 1226; I-band.
Fire control: 2 Signaal STIR ❿; I/J/K-band; range 140 km *(76 n miles)* for 1 m² target.

Sonars: Signaal PHS-36; hull-mounted; active search and attack; medium frequency.
Thomson Sintra Anaconda DSBV 61; towed array; passive low frequency. LFAS may be fitted in due course.

Helicopters: 1 NAS 332SC Cougar.

Programmes: Contract signed on 26 March 2004 for the acquisition of two frigates. Ex-*Abraham van der Hulst* to be transferred in November 2005 and ex-*Tjerk Hiddes* to be transferred in April 2007. The transfer of Harpoon missiles has not been confirmed.
Structure: The VLS SAM is similar to Canadian Halifax and Greek MEKO classes.

NEW ENTRY

KAREL DOORMAN CLASS   *(Scale 1 : 1,200), Ian Sturton* / 1044090

1 BROADSWORD CLASS (TYPE 22) (FFHM)

Name	No	Builders	Laid down	Launched	Commissioned
ALMIRANTE WILLIAMS (ex-*Sheffield*)	19 (ex-F 96)	Swan Hunter Shipbuilders, Wallsend-on-Tyne	29 Mar 1984	26 Mar 1986	26 July 1988

Displacement, tons: 4,100 standard; 4,800 full load
Dimensions, feet (metres): 480.5 × 48.5 × 21 *(146.5 × 14.8 × 6.4)*
Main machinery: COGOG; 2 RR Olympus TM3B gas turbines; 50,000 hp *(37.3 MW)* sustained; 2 RR Tyne RM1C gas turbines; 9,900 hp *(7.4 MW)*; 2 shafts; cp props
Speed, knots: 30; 18 on Tynes.
Complement: 273 (30 officers) (accommodation for 296)

Missiles: SAM: 2 British Aerospace Seawolf GWS 25 Mod 3; Command Line Of Sight (CLOS) with 2 channel radar tracking to 5 km *(2.7 n miles)* at 2+ Mach; warhead 14 kg.
Guns: Medium calibre (to be announced).
Torpedoes: to be announced.
Countermeasures: to be announced.
Combat data systems: CACS 1.
Weapons control: to be announced.
Radars: Air/Surface search: Marconi Type 967/968; D/E-band.
Surface search: Racal Decca Type 2008; E/F-band.
Navigation: Kelvin Hughes Type 1008; I-band.
Fire control: 2 Marconi Type 911; I-Ku-band (for Seawolf).
Sonars: Ferranti/Thomson Sintra Type 2050; hull-mounted; active search and attack.

Helicopters: 1 NAS 332SC Cougar.

Programmes: Originally successors to the UK Leander class, these ships entered RN service in 1987 but were withdrawn, half-way through their ships' lives, as a result of the 1998 UK Defence Review. Agreement for transfer to Chile ratified by the Chilean government in April 2003.

ALMIRANTE WILLIAMS   *9/2003, B Sullivan* / 0567435

Modernisation: The ship is to undergo a modernisation programme which is to start in late 2006 at ASMAR-Talcahuano Yard. This will include installation of a new combat data system, surface-to-surface missiles, and a medium calibre gun. The latter may be 76 mm or 114 m *(4.5 in)* Mk 8 acquired from Brazil. Further upgrades are under consideration.

Structure: Broadsword Batch 2 ships were stretched versions of Batch 1. The flight deck may be modified to allow operation of the Cougar replacement helicopter.
Operational: The ship entered service on 5 September 2003 when it replaced *Almirante Blanco Encalada*.

UPDATED

3 LEANDER CLASS (FFGHM)

Name	No	Builders	Laid down	Launched	Commissioned
ALMIRANTE CONDELL	06	Yarrow & Co, Scotstoun	5 June 1971	12 June 1972	21 Dec 1973
ALMIRANTE LYNCH	07	Yarrow & Co, Scotstoun	6 Dec 1971	6 Dec 1972	25 May 1974
MINISTRO ZENTENO (ex-*Achilles*)	08 (ex-F 12)	Yarrow & Co, Scotstoun	1 Dec 1967	21 Nov 1968	9 July 1970

Displacement, tons: 2,500 standard; 2,962 full load
Dimensions, feet (metres): 372 oa; 360 wl × 43 × 18 (screws)
(113.4; 109.7 × 13.1 × 5.5)
Main machinery: 2 Babcock & Wilcox boilers; 550 psi *(38.7 kg/cm²)*; 850°F *(450°C)*; 2 White/English Electric turbines; 30,000 hp *(22.4 MW)*; 2 shafts
Speed, knots: 27. **Range, n miles:** 4,500 at 12 kt
Complement: 263 (20 officers)

Missiles: SSM: 4 Aerospatiale MM 40 Exocet or MM 38 (08) ❶; inertial cruise; active radar homing to 70 km *(40 n miles)* (MM 40) or 42 km *(23 n miles)* (MM 38) at 0.9 Mach; warhead 165 kg; sea-skimmer.
SAM: Short Brothers Seacat GWS 22 quad launcher (06); optical/radar guidance to 5 km *(2.7 n miles)*; warhead 10 kg; 16 reloads.
Guns: 2 Vickers 4.5 in *(115 mm)*/45 Mk 6 (twin) semi-automatic ❸; 20 rds/min to 19 km *(10 n miles)* anti-surface; 6 km *(3.2 n miles)* anti-aircraft; weight of shell 25 kg.
4 Oerlikon 20 mm Mk 9 (2 twin) ❹; 800 rds/min to 2 km.
1 GE/GD 20 mm/76 Mk 15 Vulcan Phalanx (07) ❺; 6 barrels per mounting; 3,000 rds/min combined to 1.5 km.
Torpedoes: 6—324 mm Mk 32 (2 triple) tubes ❻. Honeywell Mk 46 Mod 2; active/passive homing to 11 km *(5.9 n miles)* at 40 kt; warhead 44 kg. To be replaced by Murene in due course.
Countermeasures: Decoys: 2 Corvus 8-barrelled trainable chaff rocket launchers ❼; distraction or centroid patterns to 1 km. Wallop Barricade double layer chaff launchers.
ESM/ECM: Elta EW system; intercept and jammer.
Combat data systems: Sisdef Imagen SP 100 includes datalink. Link 11 receive (06 and 07).
Weapons control: Maiten-1/CH for gunnery. GWS 22 system for Seacat.
Radars: Air search: Marconi Type 965/966 ❽; A-band.
Surface search: Marconi Type 992 Q or Plessey Type 994 (08) ❾; E/F-band.
Navigation: Kelvin Hughes Type 1006; I-band.
Fire control: Plessey Type 903 ❿; I-band (for guns).
Plessey Type 904 (06); I-band (for Seacat).
Sonars: Graseby Type 184 M/P; hull-mounted; active search and attack; medium frequency (6/9 kHz).
Graseby Type 170 B; hull-mounted; active attack; high frequency (15 kHz).
Kelvin Hughes Type 162 M; hull-mounted; sideways-looking classification; high frequency.

Helicopters: 1 Bell 206B (08) ⓬ or Cougar ⓭.

Programmes: First two ordered from Yarrow & Co Ltd, Scotstoun in the late 1960s. Third ship purchased from UK in September 1990.
Modernisation: The first major modernisations of *Lynch* (1989) and *Condell* (1993) were undertaken by ASMAR, Talcahuano. Upgrades included enlargement of the hangar and flight deck to operate a Cougar helicopter, the fitting of the Indal Assist helicopter recovery system, mounting of two twin MM 40 Exocet launchers on each side of the hangar (instead of the MM 38 aft) and moving the torpedo tubes down one deck. Other modifications include a new combat data system, improvements to the fire-control radars and the installation of Israeli EW systems. *Ministro Zenteno* was modified in 1997 with MM 38 launchers and a new foremast. The hangar was not enlarged but the flight deck can operate a Cougar helicopter. The Seacat system has subsequently been removed and a Satcom aerial installed in its place. A second major modernisation of *Lynch* was completed in 2002. Upgrades included complete overhaul of propulsion and machinery systems, improvements to habitability, replacement of the Seacat system with a Phalanx 20 mm CIWS and modernisation of the communications systems. *Condell* may be similarly refitted to extend service life to about 2012.
Operational: A fourth of class paid off in 1998.

UPDATED

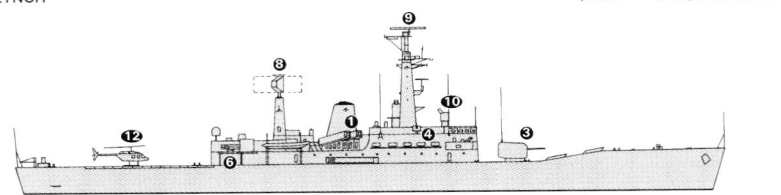

ALMIRANTE LYNCH
(Scale 1 : 1,200), Ian Sturton / 0534087

MINISTRO ZENTENO
(Scale 1 : 1,200), Ian Sturton / 1044091

MINISTRO ZENTENO
6/2003, Chilean Navy / 0569806

ALMIRANTE CONDELL
7/2001, Maritime Photographic / 0121313

ALMIRANTE LYNCH
7/2002, Chris Sattler / 0534045

SHIPBORNE AIRCRAFT

Notes: Replacement of the four Cougar helicopters by a specialist ASW helicopter is under consideration. Ex-USN SH-60 Seahawks are reported to be the preferred option.

Numbers/Type: 4 Nurtanio (Aerospatiale) NAS 332C Cougar.
Operational speed: 151 kt *(279 km/h)*.
Service ceiling: 15,090 ft *(4,600 m)*.
Range: 335 n miles *(620 km)*.
Role/Weapon systems: ASV/ASW helicopters for DLG conversions; surface search and SAR secondary roles. Sensors: Thomson-CSF Varam radar and Thomson Sintra HS-312 dipping sonar. DR 2000 ESM. Weapons: ASW; 2 × Alliant Mk 46 Mod 2 torpedoes or depth bombs. ASV; 1 or 2 × Aerospatiale AM 39 Exocet missiles. ***UPDATED***

COUGAR   *7/2001, Maritime Photographic /* 0121314

Numbers/Type: 4 MBB BO 105C.
Operational speed: 113 kt *(210 km/h)*.
Service ceiling: 9,845 ft *(3,000 m)*.
Range: 407 n miles *(754 km)*.
Role/Weapon systems: Coastal patrol helicopter for patrol, training and liaison duties; SAR as secondary role. Sensors: Bendix search radar. Weapons: Unarmed. ***VERIFIED***

BO 105C   *11/2001, Freddie Philips /* 0534054

Numbers/Type: 4 Bell 206B JetRanger.
Operational speed: 115 kt *(213 km/h)*.
Service ceiling: 13,500 ft *(4,115 m)*.
Range: 368 n miles *(682 km)*.
Role/Weapon systems: Some tasks and training carried out by torpedo-armed liaison helicopter; emergency war role for ASW. To be replaced by Bell 412. Weapons: ASW; 1 × Mk 46 torpedo or 2 depth bombs. ***VERIFIED***

JETRANGER   *7/2001, Maritime Photographic /* 0121315

Numbers/Type: 2 Bell Textron 412.
Operational speed: 122 kt *(226 km/h)*.
Service ceiling: 6,300 ft *(1,920 m)*.
Range: 500 n miles *(744 km)*.
Role/Weapon systems: Multipurpose aircraft to replace Bell 206B in training, SAR and surveillance roles. Further aircraft are likely to follow. ***VERIFIED***

BELL 412   *4/2002, Mario R V Carniero /* 0534103

LAND-BASED MARITIME AIRCRAFT (FRONT LINE)

Notes: (1) In addition there are two EMB-110, and three Casa Aviocar 212/300 support aircraft. (2) The Air Force has one Boeing 707 converted for AEW duties.

Numbers/Type: 6 Embraer EMB-111 Bandeirante.
Operational speed: 194 kt *(360 km/h)*.
Service ceiling: 25,500 ft *(7,770 m)*.
Range: 1,590 n miles *(2,945 km)*.
Role/Weapon systems: Designated EMB-111N for peacetime EEZ and wartime MR. Sensors: Eaton-AIL AN/APS-128 search radar, Thomson-CSF DR 2000 ESM, searchlight. Weapons: Strike; 6 × 127 mm or 28 × 70 mm rockets. ***UPDATED***

Numbers/Type: 7 Pilatus PC-7 Turbo-Trainer.
Operational speed: 270 kt *(500 km/h)*.
Service ceiling: 32,000 ft *(9,755 m)*.
Range: 1,420 n miles *(2,630 km)*.
Role/Weapon systems: Training includes simulated attacks to exercise ships' AA defences; emergency war role for strike operations. Sensors: None. Weapons: 4 × 127 mm or similar rockets and machine gun pods. ***VERIFIED***

Numbers/Type: 4 Lockheed P-3A Orion.
Operational speed: 410 kt *(760 km/h)*.
Service ceiling: 28,300 ft *(8,625 m)*.
Range: 4,000 n miles *(7,410 km)*.
Role/Weapon systems: Long-range MR for surveillance and SAR. First one delivered from USA in March 1993 followed by seven more of which one has been modified for transport, two are in reserve and two are used for spares. Sensors: Three aircraft upgraded with new radar, ESM and FLIR. APS-115 radar. Weapons: Weapon systems removed but to be replaced in due course including ASMs. ***UPDATED***

ORION   *6/2003, Chilean Navy /* 0569794

Numbers/Type: 8 Cessna 0-2A Skymaster.
Operational speed: 130 kt *(241 km/h)*.
Service ceiling: 5,000 ft *(1,524 m)*.
Range: 550 n miles *(1,019 km)*.
Role/Weapon systems: Maritime coastal patrol and training acquired in 1998/99. Sensors: None. Weapons: May be equipped with 4 weapons stations in due course. ***UPDATED***

SKYMASTER   *6/1999, Chilean Navy /* 0056723

PATROL FORCES

10 GRUMETE DIAZ (DABUR) CLASS
(COASTAL PATROL CRAFT) (PB)

DIAZ 1814	BRAVO 1818	JOHNSON 1821
BOLADOS 1815	CAMPOS 1819	TRONCOSO 1822
SALINAS 1816	MACHADO 1820	HUDSON 1823
TELLEZ 1817		

Displacement, tons: 39 full load
Dimensions, feet (metres): 64.9 × 18 × 5.9 *(19.8 × 5.5 × 1.8)*
Main machinery: 2 Detroit 12V 71TA diesels; 840 hp *(627 kW)* sustained; 2 shafts
Speed, knots: 19. **Range, n miles:** 450 at 13 kt
Complement: 8 (2 officers)
Guns: 2 Oerlikon 20 mm.
Radars: Surface search: Racal Decca Super 101 Mk 3; I-band.

Comment: All have LPC numbers and Grumete precedes the ships' names. First six transferred from Israel and commissioned 3 January 1991. Second batch of four more transferred and commissioned 17 March 1995. A fast inflatable boat is carried on the stern. Five deployed in 4th Naval Zone (Iquique) and five in 2nd Naval Zone and operate in the Chiloé area. All underwent life extension refits in 2001-02 at Valparaiso and Puerto Montt. Service lives end by 2012. ***VERIFIED***

HUDSON   *7/2001, Maritime Photographic /* 0121321

3 CASMA (SAAR 4) CLASS
(FAST ATTACK CRAFT—MISSILE) (PGG)

Name	No	Builders	Commissioned
CASMA (ex-*Romah*)	LM 30	Haifa Shipyard	Mar 1974
CHIPANA (ex-*Keshet*)	LM 31	Haifa Shipyard	Oct 1973
ANGAMOS (ex-*Reshef*)	LM 34	Haifa Shipyard	Apr 1973

Displacement, tons: 415 standard; 450 full load
Dimensions, feet (metres): 190.7 × 24.9 × 9.2 *(58.1 × 7.6 × 2.8)*
Main machinery: 4 MTU 16V 396 diesels; 13,029 hp(m) *(9.58 MW)* (30 and 31) ; 4 MTU 16V 596 TB91 diesels; 15,000 hp(m) *(11.3 MW)* (34); 4 shafts
Speed, knots: 32. **Range, n miles:** 1,650 at 30 kt; 3,700 at 18 kt
Complement: 46 (8 officers)

Missiles: SSM: 4 IAI Gabriel I or II; radar or optical guidance; semi-active radar homing to 20 km *(10.8 n miles)* (I) or 36 km *(20 n miles)* (II); at 0.7 Mach; warhead 75 kg HE.
Guns: 2 OTO Melara 3 in *(76 mm)*/62 compact; 85 rds/min to 16 km *(8.7 n miles)* anti-surface; 12 km *(6.5 n miles)* anti-aircraft; weight of shell 6 kg.
2 Oerlikon 20 mm; 800 rds/min to 2 km.
2—12.7 mm MGs.
Countermeasures: Decoys: 4 Rafael LRCR chaff decoy launchers.
ESM: Elta Electronics MN-53; intercept.
ECM: Elta Rattler; jammer.
Radars: Surface search: Elta EL-2208C; E/F-band.
Navigation: Raytheon 20X; I-band.
Fire control: Selenia Orion RTN 10X; I/J-band.

Programmes: One transferred from Israel December 1979 and second in January 1981. Two more acquired from Israel 1 June 1997 but one (ex-*Tarshish*) was cannibalised for spares in 1998.
Modernisation: New engines fitted in the first pair in 2000. Weapons control systems have been upgraded in LM 30 and LM 34. Similar refit of LM 31 completed by 2003.
Operational: All operate in Third Naval Zone (Beagle Channel).

UPDATED

CASMA   *9/2000, MTU* / 0094035

4 RIQUELME (TIGER) CLASS (TYPE 148)
(FAST ATTACK CRAFT—MISSILE) (PGG)

Name	No	Builders	Commissioned
RIQUELME (ex-*Wolf*)	LM 36 (ex-P 6149)	CMN Cherbourg	26 Feb 1974
ORELLA (ex-*Elster*)	LM 37 (ex-P 6154)	CMN Cherbourg	14 Nov 1974
SERRANO (ex-*Tiger*)	LM 38 (ex-P 6141)	CMN Cherbourg	30 Oct 1972
URIBE (ex-*Luchs*)	LM 39 (ex-P 6143)	CMN Cherbourg	9 Apr 1973

Displacement, tons: 234 standard; 265 full load
Dimensions, feet (metres): 154.2 × 23 × 8.9 *(47 × 7 × 2.7)*
Main machinery: 4 MTU 16V 396 diesels; 13,029 hp(m) *(9.58 MW)* sustained; 4 shafts
Speed, knots: 31. **Range, n miles:** 570 at 30 kt; 1,600 at 15 kt
Complement: 30 (4 officers)

Missiles: SSM: 4 Aerospatiale MM 38 Exocet (2 twin) launchers; inertial cruise; active radar homing to 42 km *(23 n miles)* at 0.9 Mach; warhead 165 kg; sea-skimmer.
Guns: 1 OTO Melara 3 in *(76 mm)*/62 compact; 85 rds/min to 16 km *(8.6 n miles)* anti-surface; 12 km *(6.5 n miles)* anti-aircraft; weight of shell 6 kg.
1 Bofors 40 mm/70; 330 rds/min to 12 km *(6.5 n miles)* anti-surface; 4 km *(2.2 n miles)* anti-aircraft; weight of shell 0.96 kg; fitted with GRP dome (1984).
2—12.7 mm MGs.
Mines: Laying capability.
Countermeasures: Decoys: Wolke chaff launcher.
Combat data systems: PALIS and Link 11.
Weapons control: CSEE Panda optical director. Thomson-CSF Vega PCET system, controlling missiles and guns.
Radars: Air/surface search: Thomson-CSF Triton; G-band; range 33 km *(18 n miles)* for 2 m² target.
Navigation: SMA 3 RM 20; I-band; range 73 km *(40 n miles)*.
Fire control: Thomson-CSF Castor; I/J-band.

Programmes: First pair transferred from Germany on 27 August 1997 and sailed in a transport ship on 2 September 1997. Four more transferred on 22 September 1998 and sailed 11 October. These four were all damaged during a storm in transit, and the two best were taken into service, with the other pair *(Pelikan* and *Kranich)* being used for spares. The ship names have prefixed ranks but these are not used.
Modernisation: New engines fitted in 2000. Speed reduced to 31 kt.
Structure: Similar to Combattante II craft. EW equipment was removed prior to transfer.
Operational: Operate in 4th Naval Zone (Iquique). Exocet missiles were not part of the transfer but have been acquired separately.

UPDATED

URIBE   *7/2001, Maritime Photographic* / 0121316

SERRANO   *7/2001, Maritime Photographic* / 0121317

6 MICALVI CLASS (LARGE PATROL CRAFT) (PB/AEM)

Name	No	Builders	Launched	Commissioned
MICALVI	PSG 71	ASMAR, Talcahuano	12 Sep 1992	30 Mar 1993
ORTIZ	PSG 72	ASMAR, Talcahuano	23 July 1993	15 Dec 1993
ISAZA	PSG 73	ASMAR, Talcahuano	7 Jan 1994	31 May 1994
MOREL	PSG 74	ASMAR, Talcahuano	21 Apr 1994	11 Aug 1994
CABRALES	PSG 77	ASMAR, Talcahuano	4 Apr 1996	29 June 1996
SIBBALD	PSG 78	ASMAR, Talcahuano	5 June 1996	29 Aug 1996

Displacement, tons: 518 full load
Dimensions, feet (metres): 139.4 × 27.9 × 9.5 *(42.5 × 8.5 × 2.9)*
Main machinery: 2 Caterpillar 3512 TA diesels; 2,560 hp(m) *(1.88 MW)* sustained; 2 shafts
Speed, knots: 15. **Range, n miles:** 4,200 at 12 kt
Complement: 23 (5 officers) plus 10 spare
Guns: 1 Bofors 40 mm/60. 2 Oerlikon 20 mm.
Radars: Surface search: Racal Decca; I-band.

Comment: First four built under design project Taitao. Last pair built for export but bought by the Navy. Multipurpose patrol vessels with a secondary mission of transport and servicing navigational aids. Provision for bow thruster, sonar and mine rails. Can carry 35 tons cargo in holds and 18 tons in containers. Crane lift of 2.5 tons. The ships' names all have prefixed ranks but these are not used. *Micalvi* and *Ortiz* were classified as missile tenders in 1999 but reclassified as patrol craft in 2004. *Cabrales* is also used as a survey vessel.

UPDATED

MICALVI   *11/2001, Freddie Philips* / 0534131

ORTIZ   *7/2001, Maritime Photographic* / 0534129

1 GUACOLDA CLASS (COASTAL PATROL CRAFT) (PB)

Name	No	Builders	Commissioned
FRESIA	1608 (ex-81)	Bazán, San Fernando	9 Dec 1965

Displacement, tons: 134 full load
Dimensions, feet (metres): 118.1 × 18.4 × 7.2 *(36 × 5.6 × 2.2)*
Main machinery: 2 Caterpillar diesels; 3,200 hp(m) *(2.35 MW)* sustained; 2 shafts
Speed, knots: 22. **Range, n miles:** 1,500 at 15 kt
Complement: 20
Guns: 1 Bofors 40 mm/70.
Radars: Navigation: Decca 505; I-band.

Comment: Built to West German Lürssen design from 1963 to 1966. Launched 1964. By mid-1998 had been converted to coastal patrol craft with torpedo tubes and after gun removed. Two deleted 2001-02 and a third in 2005.

UPDATED

FRESIA   *7/2001, Maritime Photographic* / 0121322

SURVEY SHIPS

1 TYPE 1200 CLASS (AGS/AGOBH)

Name	No	Builders	Commissioned
CONTRE-ALMIRANTE OSCAR VIEL TORO	AP 46	Canadian Vickers, Montreal	Oct 1960
(ex-*Norman McLeod Rogers*)			

Displacement, tons: 6,320 full load
Measurement, tons: 4,179 gross; 1,847 net
Dimensions, feet (metres): 294.9 × 62.5 × 20 *(89.9 × 19.1 × 6.1)*
Main machinery: 4 Fairbanks-Morse 38D8-1/8-12 diesels; 8,496 hp *(6.34 MW)* sustained; 4 GE generators; 4.8 MW; 2 Ruston RK3CZ diesels; 7,250 hp *(5.6 MW)* sustained; 2 GE generators; 2.76 MW; 2 GE motors; 12,000 hp *(8.95 MW)*; 2 shafts
Speed, knots: 15. **Range, n miles:** 12,000 at 12 kt
Complement: 33
Guns: 2 Oerlikon 20 mm.
Helicopters: 1 BO 105C.

Comment: Acquired from the Canadian Coast Guard on 16 February 1995. The ship was formerly based on the west coast at Victoria, BC, and was laid up in 1993. Has replaced the deleted *Piloto Pardo* as the Antarctic patrol and survey ship.   **UPDATED**

CONTRE-ALMIRANTE OSCAR VIEL TORO   *6/2004*, Chilean Navy / 1044093*

1 BUOY TENDER (ABU)

Name	No	Builders	Commissioned
GEORGE SLIGHT MARSHALL (ex-*M V Vigilant*)	BRS 63	Netherlands	July 1978

Displacement, tons: 816 full load
Dimensions, feet (metres): 173.9 × 36.7 × 11.5 *(53 × 11.2 × 3.5)*
Main machinery: 2 Ruston 6AP230 diesels; 1,360 hp *(1 MW)*; 2 shafts; bow thruster
Speed, knots: 12
Complement: 20
Guns: 2 Oerlikon 20 mm.

Comment: Acquired from the UK Mersey Harbour Board and recommissioned 5 February 1997. Carries a 15 ton derrick.   **VERIFIED**

GEORGE SLIGHT MARSHALL   *1/1999, van Ginderen Collection / 0050081*

1 ROBERT D CONRAD CLASS (AGOR)

Name	No	Builders	Commissioned
VIDAL GORMAZ (ex-*Thomas Washington*)	60 (ex-AGOR 10)	Marinette Marine, WI	27 Sep 1965

Displacement, tons: 1,370 full load
Dimensions, feet (metres): 208.9 × 40 × 15.3 *(63.7 × 12.2 × 4.7)*
Main machinery: Diesel-electric; 2 Cummins diesel generators; 1 motor; 1,000 hp *(746 kW)*; 1 shaft
Speed, knots: 13.5. **Range, n miles:** 12,000 at 12 kt
Complement: 41 (9 officers, 15 scientists)
Guns: 2 Oerlikon 20 mm.
Radars: Navigation: TM 1660/12S; I-band.

Comment: Transferred from US on 28 September 1992. This is the first class of ships designed and built by the US Navy for oceanographic research. Fitted with instrumentation and laboratories to measure gravity and magnetism, water temperature, sound transmission in water, and the profile of the ocean floor. Special features include 10 ton capacity boom and winches for handling over-the-side equipment; 620 hp gas turbine (housed in funnel structure) for providing 'quiet' power when conducting experiments; can propel the ship at 6.5 kt. Ships of this class are in service with several other navies.   **VERIFIED**

VIDAL GORMAZ   *7/2001, Maritime Photographic / 0121325*

AMPHIBIOUS FORCES

1 NEWPORT CLASS (LSTH)

Name	No	Builders	Laid down	Launched	Commissioned	Recommissioned
VALDIVIA (ex-*San Bernardino*)	93 (ex-LST 1189)	National Steel & Shipbuilding Co	12 July 1969	28 Mar 1970	27 Mar 1971	30 Sep 1995

Displacement, tons: 4,975 light; 8,450 full load
Dimensions, feet (metres): 522.3 (hull) × 69.5 × 17.5 (aft) *(159.2 × 21.2 × 5.3)*
Main machinery: 6 ALCO 16-251 diesels; 16,500 hp *(12.3 MW)* sustained; 2 shafts; cp props; bow thruster
Speed, knots: 20. **Range, n miles:** 14,250 at 14 kt
Complement: 257 (13 officers)
Military lift: 400 troops; 500 tons vehicles; 3 LCVPs and 1 LCPL on davits
Radars: Surface search: Raytheon SPS-67; G-band. Navigation: Marconi LN66; I/J-band.
Helicopters: Platform only.

Programmes: Transferred from the US by lease on 30 September 1995. A second of class was offered but not accepted due to its poor condition.
Structure: The hull form required to achieve 20 kt would not permit bow doors, thus these ships unload by a 112 ft ramp over their bow. The ramp is supported by twin derrick arms. A ramp just forward of the superstructure connects the lower tank deck with the main deck and a vehicle passage through the superstructure provides access to the parking area amidships. A stern gate to the tank deck permits unloading of amphibious tractors into the water, or unloading of other vehicles into an LCU or on to a pier. Vehicle stowage covers 19,000 sq ft. Length over derrick arms is 562 ft *(171.3 m)*; full load draught is 11.5 ft forward and 17.5 ft aft. Bow thruster fitted to hold position offshore while unloading amphibious tractors.
Operational: Damaged by grounding in mid-1997, but subsequently repaired. Vulcan Phalanx removed in 2002 and fitted in *Almirante Lynch*.   **UPDATED**

VALDIVIA   *1/1999, van Ginderen Collection / 0056726*

2 MAIPO (BATRAL) CLASS (LSTH)

Name	No	Builders	Launched	Commissioned
RANCAGUA	92	ASMAR, Talcahuano	6 Mar 1982	8 Aug 1983
CHACABUCO	95 (ex-93)	ASMAR, Talcahuano	16 July 1985	15 Apr 1986

Displacement, tons: 873 standard; 1,409 full load
Dimensions, feet (metres): 260.4 × 42.7 × 13 × 2.5 (79.4 × 13 × 2.5)
Main machinery: 2 Caterpillar diesels; 4,012 hp(m) (2.95 MW) sustained; 2 shafts; cp props
Speed, knots: 16. **Range, n miles:** 3,500 at 13 kt
Complement: 43 (5 officers)
Military lift: 180 troops; 12 vehicles; 350 tons
Guns: 2 Bofors 40 mm/60. 1 Oerlikon 20 mm. 2—81 mm mortars.
Radars: Navigation: Decca 1229; I/J-band.
Helicopters: Platform for 1 Bell 206B or BO 105C.

Comment: First laid down in 1980 to standard French design with French equipment. Have 40 ton bow ramps and vehicle stowage above and below deck. Both ships underwent life-extension refits in 2002-03.

UPDATED

CHACABUCO   *12/2004*, Globke Collection* / 1047869

2 ELICURA CLASS (LSM)

Name	No	Builders	Commissioned
ELICURA	90	Talcahuano	10 Dec 1968
OROMPELLO	94	Dade Dry Dock Co, MI	15 Sep 1964

Displacement, tons: 290 light; 750 full load
Dimensions, feet (metres): 145 × 34 × 12.8 (44.2 × 10.4 × 3.9)
Main machinery: 2 Cummins VT-17-700M diesels; 900 hp (660 kW); 2 shafts
Speed, knots: 10.5. **Range, n miles:** 2,900 at 9 kt
Complement: 20
Military lift: 350 tons
Guns: 3 Oerlikon 20 mm (can be carried).
Radars: Navigation: Raytheon 1500B; I/J-band.

Comment: Two of similar class operated by Chilean Shipping Co. Oil fuel, 77 tons.   *VERIFIED*

ELICURA   *10/2001, Freddie Philips* / 0534132

TRAINING SHIPS

1 SAIL TRAINING SHIP (AXS)

Name	No	Builders	Commissioned
ESMERALDA (ex-Don Juan de Austria)	43	Bazán, Cadiz	15 June 1954

Displacement, tons: 3,420 standard; 3,754 full load
Dimensions, feet (metres): 269.2 pp; 360 oa × 44.6 × 23 (82; 109.8 × 13.1 × 7)
Main machinery: 1 Burmeister & Wain diesel; 1,400 hp(m) (1.03 MW); 1 shaft
Speed, knots: 11. **Range, n miles:** 8,000 at 8 kt
Complement: 271 plus 80 cadets
Guns: 2 Hotchkiss saluting guns.

Comment: Four-masted schooner originally intended for the Spanish Navy. Near sister ship of *Juan Sebastian de Elcano* in the Spanish Navy. Refitted Saldanha Bay, South Africa, 1977. Sail area, 26,910 sq ft.   *VERIFIED*

ESMERALDA   *10/2002, Guy Toremans* / 0534133

AUXILIARIES

1 TRANSPORT SHIP (APH)

Name	No	Builders	Launched	Commissioned
AQUILES	AP 41	ASMAR, Talcahuano	4 Dec 1987	15 July 1988

Displacement, tons: 2,767 light; 4,550 full load
Dimensions, feet (metres): 337.8 × 55.8 × 18 (103 × 17 × 5.5 (max))
Main machinery: 2 Krupp MaK 8 M 453B diesels; 7,080 hp(m) (5.10 MW) sustained; 1 shaft; bow thruster
Speed, knots: 18
Complement: 80
Military lift: 250 troops
Helicopters: Platform for up to Cougar size.

Comment: Ordered 4 October 1985. Can be converted rapidly to act as hospital ship.

VERIFIED

AQUILES   *7/2001, Maritime Photographic* / 0121327

1 ÄLVSBORG CLASS (SUPPORT SHIP) (AGP/ASH)

Name	No	Builders	Launched	Commissioned
MERINO (ex-Älvsborg)	42 (ex-A 234, ex-M 02)	Karlskronavarvet	11 Nov 1969	6 Apr 1971

Displacement, tons: 2,660 full load
Dimensions, feet (metres): 303.1 × 48.2 × 13.2 (92.4 × 14.7 × 4)
Main machinery: 2 Nohab-Polar 112 VS diesels; 4,200 hp(m) (3.1 MW); 1 shaft; cp prop; bow thruster; 350 hp(m) (257 kW)
Speed, knots: 16
Complement: 52 (accommodation for 205)
Guns: 3 Bofors 40 mm/70 SAK 48.
Countermeasures: Decoys: 2 Philax chaff/IR launchers.
Radars: Surface search: Raytheon; E/F-band.
Fire control: Philips 9LV 200 Mk 2; I/J-band.
Navigation: Terma Scanter 009; I-band.
Helicopters: Platform for 1 medium.

Comment: Ordered in 1968 as a minelayer. Transferred from the Swedish Navy in November 1996, having been paid off in 1995. Recommissioned 7 February 1997. Originally designed as a minelayer with a capacity of 300 mines. Converted to act as a general support ship with improved accommodation and workshops. Acts as a depot ship for submarines and attack craft. The full name is *Almirante José Toribio Merino Castro*.

VERIFIED

MERINO   *7/2001, Maritime Photographic* / 0121328

1 REPLENISHMENT SHIP (AOR)

Name	No	Builders	Commissioned
ARAUCANO	AO 53	Burmeister & Wain, Copenhagen	10 Jan 1967

Displacement, tons: 23,000 full load
Dimensions, feet (metres): 497.6 × 74.9 × 28.8 (151.7 × 22.8 × 8.8)
Main machinery: 1 Burmeister & Wain Type 62 VT 2BF140 diesel; 10,800 hp(m) (7.94 MW); 1 shaft
Speed, knots: 17. **Range, n miles:** 12,000 at 15.5 kt
Complement: 130 (14 officers)
Cargo capacity: 21,126 m³ liquid; 1,444 m³ dry
Guns: 4 Bofors 40 mm/60 (2 twin).
Radars: Navigation: Racal Decca; I-band.

Comment: Launched on 21 June 1966. Single-hulled design.

UPDATED

ARAUCANO   *7/2001, Chilean Navy* / 0121329

1 HARBOUR TRANSPORT (YFB)

Name	No	Builders	Commissioned
GRUMETE PEREZ	YFB 114	ASMAR, Talcahuano	12 Dec 1975

Displacement, tons: 165 full load
Dimensions, feet (metres): 80 × 22 × 8.5 *(24.4 × 6.7 × 2.6)*
Main machinery: 1 diesel; 370 hp(m) *(272 kW)*; 1 shaft
Speed, knots: 10
Complement: 6
Guns: 1 Oerlikon 20 mm can be carried.
Radars: Navigation: Furuno; I-band.

Comment: Transferred to Seaman's School as harbour transport. Modified fishing boat design.
VERIFIED

GRUMETE PEREZ   *8/1997, Chilean Navy /* 0012168

1 SUPPLY SHIP (AKSL)

Name	No	Builders	Commissioned
PISAGUA	116	SIMAR, Santiago	11 July 1995

Displacement, tons: 195 full load
Dimensions, feet (metres): 73.2 × 19.7 × 4.9 *(22.3 × 6 × 1.5)*
Main machinery: 1 diesel; 1 shaft
Speed, knots: 8. **Range, n miles:** 500 at 8 kt
Cargo capacity: 50 tons
Radars: Navigation: Furuno; I-band.

Comment: LCU design operated by the Seaman's School, Quiriquina Island as a general purpose stores ship.
VERIFIED

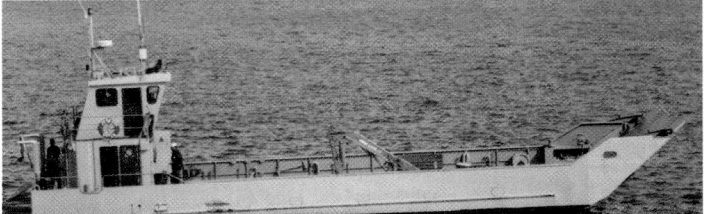

PISAGUA   *8/1997, Chilean Navy /* 0012169

3 FLOATING DOCKS (YFD)

Name	No	Lift	Commissioned
INGENIERO MERY (ex-*ARD 25*)	131	3,000 tons	1944 (1973)
MUTILLA (ex-*ARD 32*)	132	3,000 tons	1944 (1960)
TALCAHUANO (ex-*ARD 5*)	133	3,000 tons	1944 (1999)

Comment: There is also a Floating Dock *Marinero Gutierrez* with a 1,200 ton lift. Built in 1991.
VERIFIED

TUGS

Notes: Small harbour tugs *Reyes, Cortés* (both 100 tons and built in 1960) and *Galvez* (built in 1975), and the small personnel transport *Buzo Sobenes* BRT 112 are also in commission.

BUZO SOBENES   *7/1997, Chilean Navy /* 0012170

1 SMIT LLOYD CLASS (TUG/SUPPLY VESSEL) (AFL/ATF)

Name	No	Builders	Commissioned
LEUCOTON (ex-*Smit Lloyd* 44)	ATF 68	de Waal, Zaltbommel	1972

Displacement, tons: 1,750 full load
Dimensions, feet (metres): 174.2 × 39.4 × 14.4 *(53.1 × 12 × 4.4)*
Main machinery: 2 Burmeister & Wain Alpha diesels; 4,000 hp(m) *(2.94 MW)*; 2 shafts
Speed, knots: 13
Complement: 12
Guns: 2 Bofors 40 mm/60.
Radars: Surface search: E/F-band.

Comment: Acquired in February 1991. Modified at Punta Arenas and now used mainly as a supply ship.
VERIFIED

LEUCOTON   *11/2001, Freddie Philips /* 0534134

2 VERITAS CLASS (TUG/SUPPLY VESSELS) (ATF)

Name	No	Builders	Commissioned
GALVARINO (ex-*Maersk Traveller*)	ATF 66	Aukra Bruk, Aukra	1974
LAUTARO (ex-*Maersk Tender*)	ATF 67	Aukra Bruk, Aukra	1973

Displacement, tons: 941 light; 2,380 full load
Dimensions, feet (metres): 191.3 × 41.4 × 12.8 *(58.3 × 12.6 × 3.9)*
Main machinery: 2 Krupp MaK 8 M 453AK diesels; 6,400 hp(m) *(4.7 MW)*; 2 shafts; cp props; bow thruster
Speed, knots: 14
Complement: 11 plus 12 spare berths
Cargo capacity: 1,400 tons
Guns: 1 Bofors 40 mm/70 can be carried.
Radars: Navigation: Terma Pilot 7T-48; Furuno FR 240; I-band.

Comment: First one delivered from Maersk and commissioned into Navy 26 January 1988. Third one delivered in 1991. Bollard pull, 70 tonnes; towing winch, 100 tons. Fully air conditioned. Designed for towing large semi-submersible platform in extreme weather conditions. Ice strengthened.
VERIFIED

GALVARINO   *7/2001, Maritime Photographic /* 0121330

COAST GUARD

Note: (1) Project Danubio IV is for the procurement of up to four offshore patrol vessels to be built at ASMAR Talcahuano Shipyard. Proposals were presented on 8 July 2004 by Vosper Thornycroft, Damen Shipyards, Fincantieri, Fassmer GmbH and Kvaerner Masa Marine, Canada. Selection of the winning bid and award of a contract for the initial two ships is to be made in December 2004. Construction is to start in 2005. A ship of 1,500-1,800 tons is expected.
(2) There are also large numbers of harbour and SAR craft.

6 TYPE 44 CLASS (WPB)

PELLUHUE LSR 1703	CHACAO LSR 1705	GUAITECA LSR 1707
ARAUCO LSR 1704	QUEITAO LSR 1706	CURAUMILA LSR 1708

Displacement, tons: 18 full load
Dimensions, feet (metres): 44 × 12.8 × 3.6 *(13.5 × 3.9 × 1.1)*
Main machinery: 2 Detroit 6V-38 diesels; 185 hp *(136 kW)*; 2 shafts
Speed, knots: 14. **Range, n miles:** 215 at 10 kt
Complement: 3

Comment: Acquired from the US and recommissioned on 31 May 2001.
VERIFIED

TYPE 44 (Uruguay Colours)   *5/2000, Hartmut Ehlers /* 0105801

18 PROTECTOR CLASS (WPB)

ALACALUFE LEP 1603	COQUIMBO LSG 1616
HALLEF LEP 1604	PUERTO NATALES LSG 1617
AYSEN LSG 1609	VALPARAÍSO LSG 1618
CORRAL LSG 1610	PUNTA ARENAS LSG 1619
CONCEPCION LSG 1611	TALCAHUANO LSG 1620
CALDERA LSG 1612	QUINTERO LSG 1621
SAN ANTONIO LSG 1613	CHILOÉ LSG 1622
ANTOFAGASTA LSG 1614	PUERTO MONTT LSG 1623
ARICA LSG 1615	IQUIQUE LSG 1624

Displacement, tons: 120 full load
Dimensions, feet (metres): 107.3 × 22 × 6.6 *(33.1 × 6.6 × 2)*
Main machinery: 2 MTU diesels; 5,200 hp(m) *(3.82 MW)*; 2 shafts
Speed, knots: 22. **Range, n miles:** 800 at 16 kt
Complement: 10 (2 officers)
Guns: 1 Hornicon.50.

Comment: All built under licence from FBM at ASMAR, Talcahuano, in conjunction with FBM Marine. There are minor differences between LEP 1603-4 and the rest. First commissioned 24 June 1989 and last on 10 March 2004. A class of 19 (Project Danube) is envisaged. All conduct coastal patrols between Arica and Puerto Williams.

UPDATED

ARICA   *12/2004*, Globke Collection /* 1047868

ALACALUFE   *6/2003, Chilean Navy /* 0569805

2 COASTAL PATROL CRAFT (WPB)

ONA LEP 1601   YAGAN LEP 1602

Displacement, tons: 79 full load
Dimensions, feet (metres): 80.7 × 17.4 × 9.5 *(24.6 × 5.3 × 2.9)*
Main machinery: 2 MTU 8V 331 TC82 diesels; 1,300 hp(m) *(960 kW)* sustained; 2 shafts
Speed, knots: 22
Complement: 5
Guns: 2—12.7 mm MGs.

Comment: Built by Asenav and commissioned in 1980.   *VERIFIED*

YAGAN   *6/2003, Chilean Navy /* 0569804

11 INSHORE PATROL CRAFT (WPB)

MAULE LPM 1901	ISLUGA LPM 1905	CAU-CAU LPM 1909
RAPEL LPM 1902	LOA LPM 1906	PUDETO LPM 1910
ACONCAGUA LPM 1903	MAULLÍN LPM 1907	ROBINSON CRUSOE LPM 1911
LAUCA LPM 1904	COPIAPÓ LPM 1908	

Displacement, tons: 14 full load
Dimensions, feet (metres): 43.3 × 11.5 × 3.5 *(13.2 × 3.5 × 1.1)*
Main machinery: 2 MTU 6V 331 TC82 diesels; 1,300 hp(m) *(960 kW)* sustained; 2 shafts
Speed, knots: 18
Guns: 1—12.7 mm MG.
Radars: Surface search: I-band.

Comment: LPM 1901-1910 ordered in August 1981. Completed by Asenav 1982-83. LPM 1911 is a smaller 12 m craft built by Ast Sitecna, Puerto Montt, and commissioned 19 July 2000.

UPDATED

ACONCAGUA   *12/2004*, Globke Collection /* 1047867

18 RODMAN 800 CLASS (WPB)

PM 2031-2048

Dimensions, feet (metres): 29.2 × 9.8 × 3.6 *(8.9 × 3 × 0.8)*
Main machinery: 2 Volvo diesels; 300 hp(m) *(220 kW)*; 2 shafts
Speed, knots: 28. **Range, n miles:** 150 at 25 kt
Complement: 3
Guns: 1—12.7 mm MG.

Comment: Built by Rodman Polyships, Vigo and all delivered by 17 May 1996.   *VERIFIED*

PM 2034   *7/2001, Maritime Photographic /* 0121331

China
PEOPLE'S LIBERATION ARMY NAVY (PLAN)

Country Overview

The People's Republic of China, proclaimed on 1 October 1949, is the world's third-largest country by area (3,695,000 square miles) and the largest by population. It is bordered to the north by Kyrgyzstan, Kazakhstan, Mongolia and Russia, to the south by Vietnam, Laos, Myanmar, India, Bhutan, Nepal and North Korea and to the west by Pakistan, Afghanistan and Tajikistan. It has a 7,830 n mile coastline with the Yellow, East China and South China seas. There are more than 3,400 offshore islands of which Hainan is the largest. Sovereignty over Taiwan, still formally a province of China, is also claimed. Ownership of some or all of the Spratly Islands is disputed between China, Brunei, Taiwan, Vietnam, Malaysia and the Philippines. The principal ports are Shanghai (largest city), Fuzhou, Qingdao, Tianjin, Guangzhou and Hangzhou which is linked to the capital Beijing by the Grand Canal. Overall there are 54,000 n miles of navigable inland waterways including the Yangtze River on which the port of Wuhan is situated. Territorial seas (12 n miles) are claimed. A 200 n mile EEZ has also been claimed but the limits have not been defined.

Headquarters Appointments

Commander-in-Chief of the Navy:
  Admiral Zhang Dingfa
Political Commissar of the Navy:
  Admiral Hu Yianlin
Deputy Commanders-in-Chief of the Navy:
  Vice Admiral Shen Binyi
  Vice Admiral Wang Shouye
  Vice Admiral Wang Yucheng
  Vice Admiral Jin Mao
Chief of Naval Staff:
  Vice Admiral Zhao Xingfa

Fleet Commanders

North Sea Fleet:
  Vice Admiral Zhang Zhannan
East Sea Fleet:
  Vice Admiral Zhao Guojun
South Sea Fleet:
  Vice Admiral Gu Wengen

Personnel

(a) 2005: 250,000 officers and men, including 25,000 naval air force, 8-10,000 marines (28,000 in time of war) and 28,000 for coastal defence
(b) 2 years' national service for sailors afloat; 3 years for those in shore service. Some stay on for up to 15 years. 41,000 conscripts

Operational Numbers

Because numbers of vessels are kept in operational reserve, the Chinese version of the order of battle tends to show fewer ships than are counted by Western observers.

Organisation

Each of the North, East and South Sea Fleets has two submarine divisions, three DD/FF divisions and one MCMV division. The North also has one Amphibious Division, and the other Fleets have two each. The South has two Marine Infantry Brigades.

Bases

North Sea Fleet. Major bases: Qingdao (HQ), Huludao, Jianggezhuang, Guzhen Bay, Lushun, Xiaopingdao. Minor bases: Weihai Wei, Qingshan, Luda, Lianyungang, Ling Shan, Ta Ku Shan, Changshandao, Liuzhuang, Dayuanjiadun, Dalian
East Sea Fleet. Major bases: Ningbo (HQ), Zhoushan, Shanghai, Daxie, Fujan. Minor bases: Zhenjiangguan, Wusong, Xinxiang, Wenzhou, Sanduao, Xiamen, Xingxiang, Quandou, Wen Zhou SE, Wuhan, Dinghai, Jiaotou
South Sea Fleet. Major bases: Zhanjiang (HQ), Yulin, Huangfu, Hong Kong, Guangzhou (Canton). Minor bases: Haikou, Shantou, Humen, Kuanchuang, Tsun, Kuan Chung, Mawai, Beihai, Ping Tan, San Chou Shih, Tang-Chiah Huan, Longmen, Bailong, Dongcun, Baimajing, Xiachuandao, Yuchi

Coast Defence

A large number of HY-2 (CSSC-3) and HY-3 (CSSC-301) SSMs in 20 semi-fixed armoured sites. 35 Coastal Artillery regiments.

Equipment Procurement

Although often listed under the name of the designer, equipment has not necessarily been supplied direct from the parent company. It may have been acquired from a third party or by reverse engineering.

Training

The main training centres are:

Dalian: First Surface Vessel Academy
Guangzhou (Canton): Second Surface Vessel Academy
Qingdao: Submarine Academy
Wuhan: Engineering College
Nanjing: Naval Staff College, Medical School, Electronic Engineering College
Yan Tai: Aviation Engineering College
Tianjin: Logistic School

Marines

There are two brigades based at Heieu and subordinate to the Navy. Each has three Infantry regiments and one Artillery regiment.

Naval Air Force

With 25,000 officers and men and over 800 aircraft, this is a considerable naval air force primarily land-based. There is a total of eight Divisions with 27 Regiments split between the three Fleets. Some aircraft are laid up unrepaired.

Air bases include:

North Sea Fleet: Dalian, Qingdao, Jinxi, Jiyuan, Laiyang, Jiaoxian, Xingtai, Laishan, Anyang, Changzhi, Liangxiang and Shan Hai Guan
East Sea Fleet: Danyang, Daishan, Shanghai, Ningbo, Luqiao, and Shitangqiao
South Sea Fleet: Foluo, Haikou, Lingshui, Sanya, Guiping, Jialaishi and Lingling

Strength of the Fleet

Type	Active (Reserve)	Building (Planned)
SSBN	1	2 (2)
SSB	1	—
SSN	4	2 (3)
SSG	16	10 (1)
Patrol Submarines	41 (10)	—
Destroyers	25	4
Frigates	45	3
Fast Attack Craft (Missile)	40	—
Fast Attack Craft (Gun)	35	—
Fast Attack Craft (Patrol)	118	—
Patrol Craft	17	2
Minesweepers (Ocean)	18 (26)	1
Mine Warfare Drones	4 (42)	—
Minelayer	1	—
Hovercraft	10	—
LSTs	26	2
LSMs	47	—
LCMs-LCUs	148	—
Training Ships	2	—
Troop Transports (AP/AH)	6	—
Submarine Support Ships	11	—
Salvage and Repair Ships	3	1
Supply Ships	68+	—
Fleet Replenishment Ships	4	2
Icebreakers	4	

DELETIONS

Submarines

2003   *Han* 401

Patrol Forces

2003   14 'Houkou', 15 'Huchuan'

PENNANT LIST

Submarines

406	Xia

Destroyers

105	Jinan
106	Xian
107	Yinchuan
108	Xining
109	Kaifeng
110	Dalian
112	Harbin
113	Qingdao
131	Nanjing
132	Hefei
133	Chongqing
134	Zunyi
136	Hangzhou
137	Fuzhou
161	Changsha
162	Nanning
163	Nanchang
164	Guilin
165	Zhanjiang
166	Zhuhai
167	Shenzhen
168	Guangzhou
169	Wuhan
170	Lanzhou
171	—

Frigates

509	Chang De
510	Shaoxing
511	Nantong
512	Wuxi
513	Huayin
514	Zhenjiang
515	Xiamen
516	Jiujiang
517	Nanping
518	Jian
519	Changzhi
521	Jiaxing
522	Lianyungang
523	Sanming
524	Putian
525	Maanshan
533	Ningbo
534	Jinhua
535	Huangshi
536	Wuhu
537	Zhoushan
539	Anqing
540	Huainan
541	Huaibei
542	Tongling
543	Dandong
544	Siping
545	Linfen
551	Maoming
552	Yibin
553	Shaoguan
554	Anshun
555	Zhaotong
557	Jishou
558	Zigong
559	Kangding
560	Dongguan
561	Shantou
562	Jiangmen
563	Zhaoqing
564	Yichang
565	Yulin
566	Yuxi
568	Wenzhou

Principal Auxiliaries

81	Zhenghe
82	Shichang
920	Dazhi
121	Changxingdao
302	Chongmingdao
506	Yongxingdao
891	Dagushan
575	Taicang
882	Fengcang
885	Nancang
886	Fuchi

For details of the latest updates to *Jane's Fighting Ships* online and to discover the additional information available exclusively to online subscribers please visit

jfs.janes.com

SUBMARINES
Strategic Missile Submarines

0 + 2 (2) TYPE 094 CLASS (SSBN)

Name	No	Builders	Laid down	Launched	Commissioned
—	—	Huludao Shipyard	2001	28 July 2004	2008
—	—	Huludao Shipyard	2003	2006	2010

Displacement, tons: 8,000
Dimensions, feet (metres): 433.1 × 36 × 7.5
(132.0 × 11.0 × 7.5)
Main machinery: Nuclear: 2 PWR; 150 MW; 2 turbines; 1 shaft
Speed, knots: To be announced
Complement: 140

Missiles: SLBM: 12 JL-2 (CSS-NX-5); 3-stage solid-fuel rocket; stellar inertial guidance to over 8,000 km *(4,320 n miles)*; single nuclear warhead of 1 MT or 3-8 MIRV of smaller yield. CEP 300 m approx.

Torpedoes: 6—21 in (533 mm tubes).
Countermeasures: Decoys: ESM.
Radars: Surface search.
Sonars: Hull mounted passive/active; flank and towed arrays.

Programmes: The in-service date for the first boat is expected to be 2008 but deployment of the system will be dependent on successful testing of the missile. Further units are expected, probably at two-year intervals.
Structure: Details of both the boat and the SLBM are speculative. Likely to be based on the Type 093 SSN design which in turn is believed to be derived from the Russian Victor III design. The dimensions of the hull assume the incorporation of a 25 m 'missile plug' of 12 tubes for the 42 ton JL-2 missiles.
Operational: Likely to be based at Jianggezhuang. The long range of the missile may prompt a change in operating concept to a 'bastion' patrol approach.

NEW ENTRY

1 XIA CLASS (TYPE 092) (SSBN)

Name	No	Builders	Laid down	Launched	Commissioned
XIA	406	Huludao Shipyard	1978	30 Apr 1981	1987

Displacement, tons: 6,500 dived
Dimensions, feet (metres): 393.6 × 33 × 26.2
(120 × 10 × 8)
Main machinery: Nuclear; turbo-electric; 1 PWR; 90 MW; 1 shaft
Speed, knots: 22 dived
Complement: 140

Missiles: SLBM: 12 JL-1 (CSS-N-3); inertial guidance to 2,150 km *(1,160 n miles)*; warhead single nuclear 250 kT.
Torpedoes: 6—21 in *(533 mm)* bow tubes. Yu-3 (SET-65E); active/passive homing to 15 km *(8.1 n miles)* at 40 kt; warhead 205 kg.

Countermeasures: ESM: Type 921-A; radar warning.
Radars: Surface search: Snoop Tray; I-band.
Sonars: Trout Cheek; hull-mounted; active/passive search and attack; medium frequency.

Programmes: A second of class was reported launched in 1982 and an unconfirmed report suggests that one of the two was lost in an accident in 1985.
Modernisation: Started major update in late 1995 at Huludao, thought to include fitting improved JL-1A missile with increased range but this has not been confirmed.
Structure: Diving depth 300 m *(985 ft)*.

Operational: First test launch of the JL-1 missile took place on 30 April 1982 from a submerged pontoon near Huludao (Yellow Sea). Second launched on 12 October 1982, from the Golf class trials submarine. The first firing from *Xia* was in 1985 and was unsuccessful (delaying final acceptance into service of the submarine) and it was not until 27 September 1988 that a satisfactory launch took place. Based in the North Sea Fleet at Jianggezhuang. Following a refit which completed in late 1998, was reported to be operational again in 2003 although firing of a JL-1 missile has not been reported.

UPDATED

XIA

2002, Ships of the World / 0529138

1 GOLF CLASS (TYPE 031) (SSB)

200

Displacement, tons: 2,350 surfaced; 2,950 dived
Dimensions, feet (metres): 319.9 × 28.2 × 21.7
(97.5 × 8.6 × 6.6)
Main machinery: Diesel-electric; 3 Type 37-D diesels; 6,000 hp (m) *(4.41 MW)*; 3 motors; 5,500 hp(m) *(4 MW)*; 3 shafts
Speed, knots: 17 surfaced; 13 dived
Range, n miles: 6,000 surfaced at 15 kt
Complement: 86 (12 officers)

Missiles: SLBM: 1 JL-2 (CSS-NX-5); 2 or 3-stage solid fuel; inertial guidance to 8,000 km *(4,320 n miles)*; warhead 3 or 4 MIRV nuclear 90 kT or single nuclear 250 kT or 650 kT.
Torpedoes: 10—21 in *(533 mm)* tubes (6 bow, 4 stern). 12 Type Yu-4 (SAET-60); passive homing to 15 km *(8.1 n miles)* at 40 kt; warhead 400 kg.
Radars: Navigation: Snoop Plate; I-band.
Sonars: Pike Jaw; hull-mounted; active/passive search; medium frequency.

Programmes: Ballistic missile submarine similar but not identical to the deleted USSR Golf class. Built at Dalian and launched in September 1966.
Modernisation: Refitted in 1995 to take the JL-2 missile.
Operational: This was the trials submarine for the JL-1 ballistic missile which was successfully launched to 1,800 km in October 1982. Continues to be available as a trials platform for the successor missile JL-2. Based in the North Sea Fleet.

UPDATED

GOLF 200

2002, Ships of the World / 0529137

Attack Submarines (SSN)

0 + 2 (3) TYPE 093 (SSN)

Name	No	Builders	Laid down	Launched	Commissioned
—	—	Huludao Shipyard	1994	24 Dec 2002	2005
—	—	Huludao Shipyard	2000	2003	2005

Displacement, tons: 6,000 dived
Dimensions, feet (metres): 351 × 36 × 24.6
  (107 × 11 × 7.5)

Main machinery: Nuclear: 2 PWR; 150 MW; 2 turbines; 1 shaft
Speed, knots: 30 dived
Complement: 100

Missiles: SLCM; SSM.
Torpedoes: 6—21 in *(533 mm tubes).*
Countermeasures: Decoys: ESM.
Radars: Surface search.
Sonars: Hull mounted passive/active; flank and towed arrays.

Programmes: Designed in conjunction with Russian experts. Prefabrication started in late 1994 and the first launch took place in late 2002. The in-service date of the first of class is expected to be 2006 with a second boat to follow in 2007. Construction of a third boat may have started but has not been confirmed. Five boats of the class are expected.
Structure: Details given are speculative, based on the Russian Victor III design from which this submarine is reported to be derived.

TYPE 093

1997, US Navy / 0012178

UPDATED

4 HAN CLASS (TYPE 091) (SSN)

No	Builders	Laid down	Launched	Commissioned
402	Huludao Shipyard	1974	1977	Jan 1980
403	Huludao Shipyard	1980	1983	21 Sep 1984
404	Huludao Shipyard	1984	1987	Nov 1988
405	Huludao Shipyard	1987	8 Apr 1990	Dec 1990

Displacement, tons: 4,500 surfaced; 5,550 dived
Dimensions, feet (metres): 321.5; 347.8 *(403 onwards)* × 32.8 × 24.2
  (98; 106 × 10 × 7.4)

Main machinery: Nuclear; turbo-electric; 1 PWR; 90 MW; 1 shaft
Speed, knots: 25 dived; 12 surfaced
Complement: 75

Missiles: SSM: YJ-801Q (C-801); inertial cruise; active radar homing to 40 km *(22 n miles)* at 0.9 Mach; warhead 165 kg; sea-skimmer may be carried.
Torpedoes: 6—21 in *(533 mm)* bow tubes; combination of Yu-3 (SET-65E); active/passive homing to 15 km *(8.1 n miles)* at 40 kt; warhead 205 kg and Yu-1 (Type 53-51) to 9.2 km *(5 n miles)* at 39 kt or 3.7 km *(2 n miles)* at 51 kt; warhead 400 kg. 20 weapons.
Mines: 36 in lieu of torpedoes.
Countermeasures: ESM: Type 921-A; radar warning.
Radars: Surface search: Snoop Tray; I-band.
Sonars: Trout Cheek; hull-mounted; active/passive search and attack; medium frequency.
  DUUX-5; passive ranging and intercept; low frequency.

Programmes: First of this class delayed by problems with the power plant. Although completed in 1974 she was not fully operational until the 1980s.
Modernisation: The basic Russian ESM equipment was replaced by a French design. A French intercept sonar set has been fitted.
Structure: From *403* onwards the hull has been extended by some 8 m although this was not to accommodate missile tubes as previously reported. SSMs may be fired from the torpedo tubes. Diving depth 300 m *(985 ft).*
Operational: In North Sea Fleet based at Jianggezhuang. *403* and *404* started mid-life refits in 1998 which completed in early 2000. *405* started mid-life refit in 2000 and was reported completed in 2002. Torpedoes are a combination of older straight running and more modern Russian homing types. The first of class *401* was reported to have been decommissioned in 2003 and it is expected that others will follow as the Type 093 enter service.

UPDATED

HAN 404
5/1996, Ships of the World

HAN 402

1990

Patrol Submarines

Notes: An unknown number of midget submarines are reported building.

1 + 1 YUAN CLASS (TYPE 041) (SSK)

Name	No	Builders	Laid down	Launched	Commissioned
—	—	Wuhan Shipyard	—	31 May 2004	2006
—	—	Wuhan Shipyard	—	Dec 2004	2006

Displacement, tons: To be announced
Dimensions, feet (metres): 236.2 × 27.5 × ?
  (72.0 × 8.4 × ?)
Main machinery: To be announced
Speed, knots: To be announced
Complement: To be announced

Missiles: Anti-ship (possibly Klub or indigenous missile).
Torpedoes: 6—21 in *(533 mm)* bow tubes.
Countermeasures: To be announced.
Weapons control: To be announced.
Radars: To be announced.
Sonars: To be announced.

Programmes: A new class of submarine which came as a surprise when the first of class was launched in May 2004. A second of class launched in December 2004 suggests that further submarines can be expected.
Structure: There are few details at present but the design appears to exhibit some features of the Song class, although it appears to be shorter and broader, and possibly also of the Russian AMUR 1650 class. The design of the fin is similar to that of the former while a chin-mounted sonar and a distinctive 'hump' on top of a teardrop shaped hull are both characteristics of the latter. It is possible therefore that the boat is of single-hulled construction. Fitted with a seven-bladed propeller. The AMUR 1650 is offered with AIP but it is not known whether an AIP system has been incorporated.

NEW ENTRY

YUAN CLASS   6/2004* / 0583294

9 + 3 SONG CLASS (TYPE 039/039G) (SSG)

No	Builders	Laid down	Launched	Commissioned
320	Wuhan Shipyard	1991	25 May 1994	June 1999
321	Wuhan Shipyard	1995	11 Nov 1999	Apr 2001
322	Wuhan Shipyard	1996	28 June 2000	Dec 2001
323	Wuhan Shipyard	1998	May 2002	Nov 2003
324	Wuhan Shipyard	1999	28 Nov 2002	Dec 2003
325	Wuhan Shipyard	2001	3 Dec 2002	2004
314	Wuhan Shipyard	2001	19 May 2003	2004
—	Wuhan Shipyard	2002	29 Sep 2003	2004
—	Wuhan Shipyard	2002	2004	2005
—	Wuhan Shipyard	2002	July 2004	2005
—	Jiangnan Shipyard	2002	Aug 2004	2005
—	Wuhan Shipyard	2003	Sep 2004	2006

Displacement, tons: 1,700 surfaced; 2,250 dived
Dimensions, feet (metres): 246 × 24.6 × 17.5
  (74.9 × 7.5 × 5.3)
Main machinery: Diesel-electric; 4 MTU 16V 396 SE; 6,092 hp
  (m) *(4.48 MW)* diesels; 4 alternators; 1 motor; 1 shaft
Speed, knots: 15 surfaced; 22 dived
Complement: 60 (10 officers)

Missiles: SSM: YJ-801Q (C-801); radar active homing to 40 km
  (22 n miles) at 0.9 Mach; warhead 165 kg.
Torpedoes: 6—21 in *(533 mm)* tubes. Combination of Yu-4
  (SAET-60); passive homing to 15 km *(8.1 n miles)* at 40 kt;
  warhead 400 kg and Yu-1 (Type 53-51) to 9.2 km *(5 n miles)*
  at 39 kt or 3.7 km *(2.1 n miles)* at 51 kt; warhead 400 kg.
Mines: In lieu of torpedoes.
Countermeasures: ESM: Type 921-A; radar warning.
Radars: Surface search: I-band.
Sonars: Bow-mounted; passive/active search and attack;
  medium frequency.
  Flank array; passive search; low frequency.

Programmes: First of class (Type 039) started sea trials in August 1995 as a result of which substantial modifications were made. Second of class (Type 039G) trials started in early 2000 and third in early 2001. Fourth commissioned in 2003 while fifth and sixth were conducting trials in late 2003. Construction of the seventh hull is understood to have started in 2001 and of the eighth, ninth and tenth hulls in 2002. Construction of the eleventh of class at Jiangnan Shipyard suggested that a new building line may have been started but this has not been confirmed. The twelfth boat is under construction at Wuhan. It is unclear how the Song class programme will be affected by the Yuan class.
Structure: Comparable in size to Ming class but with a single skew propeller and an integrated spherical bow sonar. The forward hydroplanes are mounted below the bridge, which is on a step lower than the part of the fin that contains the masts in earlier boats. The fin is of a different shape (no cutaway) in later boats. Some of the details are speculative and the latest

SONG CLASS   4/2004*, Ships of the World / 1042142

SONG CLASS   6/2004* / 1042169

hulls of the class may have benefited from experience gained with the Kilos. The diesel engines are likely to be reverse engineered. Sonars are reported to be of French design.

Operational: The YJ-82 is the submarine launched version of the C-801 and is fired from torpedo tubes. Reports of an anti-submarine CY-1 air flight weapon are not confirmed.

UPDATED

SONG CLASS   4/2004*, Ships of the World / 1042155

海军 324 艇
NAVY SHIP 324

SONG 324

11/2004*, Ships of the World / 1042159

5 + 7 KILO CLASS (PROJECT 877EKM/636) (SSK)

No	Builders	Launched	Commissioned
364 (ex-B 171)	Nizhny Novgorod		1995
365 (ex-B 177)	Nizhny Novgorod	31 Mar 1985	1995
366	Admiralty, St Petersburg	24 Apr 1997	1998
367	Admiralty, St Petersburg	18 June 1998	1999
—	Nizhny Novgorod	17 May 2004	2005
—	Admiralty, St Petersburg	27 May 2004	2005
—	Admiralty, St Petersburg	19 Aug 2004	2005
—	Admiralty, St Petersburg	2005	2005
—	Admiralty, St Petersburg	2005	2005
—	Admiralty, St Petersburg	2005	2005
—	Severodvinsk Shipyard	2005	2006
—	Severodvinsk Shipyard	2005	2006

Displacement, tons: 2,325 surfaced; 3,076 dived
Dimensions, feet (metres): 238.2; 242.1 (Project 636) × 32.5 × 21.7 *(72.6; 73.8 × 9.9 × 6.6)*
Main machinery: Diesel-electric; 2 diesels; 3,650 hp(m) *(2.68 MW)*; 2 generators; 1 motor; 5,900 hp(m) *(4.34 MW)*; 1 shaft; 2 auxiliary motors; 204 hp(m) *(150 kW)*; 1 economic speed motor; 130 hp(m) *(95 kW)*
Speed, knots: 17 dived; 10 surfaced
Complement: 52 (13 officers)

Missiles: SLCM: Novator Alfa Klub SS-N-27 (3M-54E1); active radar homing to 180 km *(97.2 n miles)* at 0.7 Mach (cruise) and 2.5 Mach (attack); warhead 450 kg.
Torpedoes: 6—21 in *(533 mm)* tubes. 18 torpedoes. Combination of TEST 71/96; wire-guided; active/passive homing to 15 km *(8.1 n miles)* at 40 kt; warhead 205 kg and 53-65; passive wake homing to 19 km *(10.3 n miles)* at 45 kt; warhead 300 kg.
Mines: 24 in lieu of torpedoes.
Countermeasures: ESM: Squid Head or Brick Pulp; radar warning.
Weapons control: MVU-119 EM Murena TFCS.
Radars: Surface search: Snoop Tray; I-band.
Sonars: Shark Teeth; hull-mounted; passive/active search and attack; medium frequency.
Mouse Roar; hull-mounted; active attack; high frequency.

Programmes: Four of the class were ordered in mid-1993. The first two are Project 877 hulls built for a former Warsaw Pact country and subsequently cancelled. The first one departed the Baltic in December 1994 and arrived by transporter ship in February 1995. The second was delivered by the same method in November 1995. The third and fourth are of the newer Project 636 design. The first of these two left the Baltic by transporter in November 1997 and arrived in January 1998. The second followed in December 1998 arriving on 1 February 1999. A contract for a further eight 636 or 636M variants armed with SS-N-27 was signed on 3 May 2002. The

KILO   *6/2001, Ships of the World* / 0126367

first of these was originally laid down at Nizhny Novgorod for the Russian Navy in about 1991 but was never completed due to lack of funding. She is likely to be the last submarine to be built at the shipyard. Five of the boats are being built by Admiralty Yard, St Petersburg. Two were launched in 2004, the first of which was delivered in late 2004. Three further boats are expected to be launched in 2005. The remaining two boats are being built at Severodvinsk where they were laid down on 29 May 2003. The programme is to be completed by 2007.
Modernisation: The first four submarines are to be refitted in Russian shipyards. Upgrades are likely to include installation of the Klub (3M54) (SS-N-27) anti-ship missile system.
Structure: Latest export version of the elderly Kilo design and has better weapon systems co-ordination and improved accommodation than the earlier ships of the class. Double-hull

construction with six watertight compartments. Normal diving depth is 240 m with 300 m available in emergency. At least two torpedo tubes can fire wire-guided weapons. An SA-N-8 SAM launcher may be fitted on top of the fin. Some modifications have been carried out after arrival in China including a possible new ESM. SSMs may be fitted in due course.
Operational: Based at Xiangshan in the East Sea Fleet. The torpedoes are far more advanced than those previously available to China. The first pair were reported initially as having propulsion/battery problems, due to Chinese cost cutting in the initial fitting of equipment.
Opinion: This is a key programme which, in addition to meeting an operational need, involves the acquisition of technology.
UPDATED

KILO

20 MING CLASS (TYPE 035) (SS)

342	352	353	354	356	357	358	359	360
361	362	363	305	306	307	308	310	311
312	313							

Displacement, tons: 1,584 surfaced; 2,113 dived
Dimensions, feet (metres): 249.3 × 24.9 × 16.7
(76 × 7.6 × 5.1)
Main machinery: Diesel-electric; 2 diesels; 5,200 hp(m)
(3.82 MW); 2 shafts
Speed, knots: 15 surfaced; 18 dived; 10 snorting
Range, n miles: 8,000 at 8 kt snorting; 330 at 4 kt dived
Complement: 57 (10 officers)

Torpedoes: 8—21 in *(533 mm)* (6 fwd, 2 aft) tubes. Combination of Yu-4 (SAET-60); passive homing to 15 km *(8.1 n miles)* at 40 kt; warhead 400 kg, and Yu-1 (53-51) to 9.2 km *(5 n miles)* at 39 kt or 3.7 km *(2.1 n miles)* at 51 kt; warhead 400 kg; 16 weapons.
Mines: 32 in lieu of torpedoes.

Radars: Surface search: Snoop Tray; I-band.
Sonars: Pike Jaw; hull-mounted; active/passive search and attack; medium frequency.
DUUX 5; passive ranging and intercept; low frequency.

Programmes: First three completed between 1971 and 1979 one of which was scrapped after a fire and another *(232)* has been decommissioned. These were Type ES5C/D. Building resumed at Wuhan Shipyard in 1987 at the rate of one per year to a modified design ES5E. The programme was thought to have ended with hull number 14 *(363)* launched in May 1996, but *305* was launched in June 1997 followed by *306* in September 1997, *307* in May 1998, *308* in October 1998, *310* in June 2000, *311* in September 2000, *312* in May 2001 and *313* in April 2002. The expected launch of a further boat

in 2003 did not take place and, in view of the 'Kilo' programme, this programme has probably been discontinued.
Structure: Diving depth, 300 m *(985 ft)*. Only the later models have the DUUX 5 sonar. Hull 20 is reported to have a 2 m extension to its machinery space.
Operational: Thirteen are based in the North Sea Fleet at Lushun, Qingdao and Xiapingdao. From *305* onwards, based in the South Sea Fleet. Fitted with Magnavox SATNAV. The entire crew of *361* (70 officers and men) killed in an accident in April 2003. The cause of the accident is believed to have been carbon monoxide or chlorine poisoning. After repairs at Dalian, the submarine became operational again in 2004.

UPDATED

MING CLASS

7/2004*, Ian Edwards / 1042410

1 MODIFIED ROMEO CLASS (PROJECT 033G) (SSG)

351

Displacement, tons: 1,650 surfaced; 2,100 dived
Dimensions, feet (metres): 251.3 × 22 × 17.1
(76.6 × 6.7 × 5.2)
Main machinery: Diesel-electric; 2 Type 37-D diesels; 4,000 hp (m) *(2.94 MW)*; 2 motors; 2,700 hp(m) *(1.98 MW)*; 2 creep motors; 2 shafts
Speed, knots: 13 dived; 15 surfaced; 10 snorting
Complement: 54 (10 officers)

Missiles: SSM: 6 YJ-1 (Eagle Strike) (C-801); three launchers either side of fin; inertial cruise; active radar homing to 40 km *(22 n miles)* at 0.9 Mach; warhead 165 kg; sea-skimmer. May be replaced by C-802 in due course.
Torpedoes: 8—21 in *(533 mm)* (6 bow, 2 stern) tubes.
Mines: 28 in lieu of torpedoes.
Radars: Surface search: Snoop Plate and Snoop Tray; I-band.

Sonars: Hercules or Pike Jaw; hull-mounted; active/passive search and attack; medium frequency.

Programmes: This design, designated ES5G, is a modified Romeo (Wuhan) rebuilt as a trials SSM platform.
Structure: The six missile tubes are built into the casing abreast the fin and elevate to fire. To provide target acquisition an additional radar mast (Snoop Tray) is mounted between the two periscopes.
Operational: Has to surface to fire missiles although trials are reported to include an encapsulated missile which is launched from a torpedo tube while dived. Based in the North Sea Fleet and reported still doing trials in 1999. Clearly there is no intention to fit this type of missile tube in other classes, and the submarine may soon be scrapped.

VERIFIED

MOD ROMEO 351
6/1987, Xinhua

21 (+ 10 RESERVE) ROMEO CLASS (PROJECT 033) (SS)

Displacement, tons: 1,475 surfaced; 1,830 dived
Dimensions, feet (metres): 251.3 × 22 × 17.1
(76.6 × 6.7 × 5.2)
Main machinery: Diesel-electric; 2 Type 37-D diesels; 4,000 hp (m) *(2.94 MW)*; 2 motors; 2,700 hp(m) *(1.98 MW)*; 2 creep motors; 2 shafts
Speed, knots: 15.2 surfaced; 13 dived; 10 snorting
Range, n miles: 9,000 at 9 kt surfaced
Complement: 54 (10 officers)

Torpedoes: 8—21 in *(533 mm)* (6 bow, 2 stern) tubes. Combination of Yu-4 (SAET-60); passive homing to 15 km *(8.1 n miles)* at 40 kt; warhead 400 kg and Yu-1 (53-51) to 9.2 km *(5 n miles)* at 39 kt or 3.7 km *(2.1 n miles)* at 55 kt; warhead 400 kg. 14 weapons.

Mines: 28 in lieu of torpedoes.
Radars: Surface search: Snoop Plate or Snoop Tray; I-band.
Sonars: Hercules or Tamir 5; hull-mounted; active/passive search and attack; high frequency.
Thomson Sintra DUUX 5 intercept in some of the class.

Programmes: The first boats of this class were built at Jiangnan SY, Shanghai in mid-1962 with Wuhan being used later. The basic Romeo class design has evolved from the Type 031 (ES3B). Construction stopped around 1987 with the resumption of the Ming class programme. A total of 84 was built.
Modernisation: Battery refits are being done and the more modern boats have French passive ranging sonar.

Structure: Diving depth, 300 m *(984 ft)*. There are probably some dimensional variations between newer and older ships of the class.
Operational: Operational numbers are declining as these obsolete submarines are being scrapped. Few are now kept in reserve. With the exception of some more modern boats, ASW capability is virtually non-existent. The submarines are split between the three Fleets.
Sales: Seven to North Korea in 1973-75. Two to Egypt in 1982, two in 1984. All new construction.

UPDATED

ROMEO 272

3/1995, van Ginderen Collection

AIRCRAFT CARRIERS

Notes: After six years of discussions and negotiations about a new aircraft carrier, the Russian Nevskoye Design Bureau was given a contract in about 1994 to design an aircraft carrier based on Chinese requirements. There have been a number of unconfirmed reports of ships under construction and considerable speculation has arisen over the acquisition of the ex-Russian carrier *Varyag* which finally left the Black Sea on 2 November 2001 and arrived at Dalian on 4 March 2002. While a carrier capability is a high priority, such a complex and expensive project is not likely to be achieved in the short term and may involve several stages (for example, Amphibious Ships) first. The former Russian aircraft carrier *Minsk* is a tourist attaction at Shenzhen.

DESTROYERS

0 + 2 TYPE 051C CLASS (DDGHM)

Name	No	Builders	Laid down	Launched	Commissioned
—	115	Dalian Shipyard	2002	Dec 2004	2006
—	—	Dalian Shipyard	2003	2005	2007

Displacement, tons: 7,000 full load
Dimensions, feet (metres): 508.5 × 55.8 × 19.7
(155.0 × 17.0 × 6.0)
Main machinery: To be announced
Speed, knots: To be announced
Complement: To be announced

Missiles: SSM: 8 (to be confirmed) C-803 (YJ-83/CSS-N-8 Saccade) 2 quad; active radar homing to 120 km *(66 n miles)* at 0.9 Mach; warhead 165 kg; sea skimmer.
SAM: 6 (2 forward, 4 aft) SA-N-6 Grumble (Fort) revolving vertical launchers; 8 rounds per launcher; command guidance; semi-active radar homing to 100 km *(54 n miles)*; warhead 90 kg; altitude 27,432 m *(90,000 ft)*. 48 missiles.
Guns: 1—3.9 in *(100 mm)/56*; 60-80 rds/min to 17 km *(9.3 n miles)*; weight of shell 13.5 kg.
2 Type 730 30 mm 7 barrels per mounting; 4,200 rds/min combined to 1.5 km.
Torpedoes: To be announced.
A/S mortars: To be announced.
Countermeasures: To be announced.
Combat data systems: To be announced. SATCOM.
Weapons control: Datalink for SSM.
Radars: Air search: Top Plate; 3D; E-band.
Air/Surface search: Type 364 Seagull C; G-band.
Fire control: To be announced.
Navigation: To be announced.
Sonars: Bow mounted, to be announced.

Helicopters: 1 Harbin Zhi-9A Haitun or Kamov KA-28 Helix.

Programmes: The requirement for these ships arose from a need to address AAW deficiencies. It may predate the Luyang

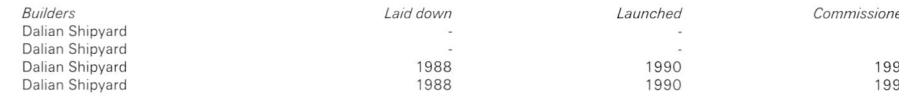

DDG 115   *1/2005* / 1042172*

programmes and could have been delayed by procurement of the SAM system.
Structure: Design appears to be based on the Type 051B/Luhai DDG but to be less stealthy than the Luyang classes although it has been assumed that the dimensions, which have not been confirmed, are similar. Two VLS launchers are reported

to be installed in the platform in front of the bridge and four ahead of the helicopter hangar. The number of SSMs is likely to be limited to eight due to lack of space between the forward funnel and aft mast.
Operational: Likely to be based in the North Sea Fleet.
NEW ENTRY

4 LUDA CLASS (TYPE 051DT/051G/051G II) (DDG)

Name	No	Builders	Laid down	Launched	Commissioned
KAIFENG	109	Dalian Shipyard	-	-	-
DALIAN	110	Dalian Shipyard	-	-	-
ZHANJIANG	165	Dalian Shipyard	1988	1990	1991
ZHUHAI	166 (168 out of area)	Dalian Shipyard	1988	1990	1991

Displacement, tons: 3,250 standard; 3,730 full load
Dimensions, feet (metres): 433.1 × 42 × 15.3
(132 × 12.8 × 4.7)
Main machinery: 2 boilers; 2 turbines; 72,000 hp(m) *(53 MW)*; 2 shafts
Speed, knots: 32. **Range, n miles:** 2,970 at 18 kt
Complement: 280 (45 officers)

Missiles: SSM: 16 C 801A (YJ 81/CSS-N-4) (Sardine) **❶**; active radar homing to 95 km *(51 n miles)* at 0.9 Mach; warhead 165 kg; sea-skimmer.
SAM: 1 HQ-7 (Crotale) octuple launcher **❷**; line of sight guidance to 13 km *(7 n miles)* at 2.4 Mach; warhead 14 kg.
Guns: 2 USSR 5.1 in *(130 mm)* (109, 110) **❸**; 17 rds/min to 29 km *(16 n miles)*; weight of shell 33.4 kg.
4—3.9 in *(100 mm)/56* (2 twin) (165, 166) **❹**; 18 rds/min to 22 km *(12 n miles)*; weight of shell 15 kg.
6 China 57 mm/63 (3 twin) (109,110) **❺**; 120 rds/min to 12 km *(6.5 n miles)*; weight of shell 6.31 kg.
6 China 37 mm/63 Type 76A (3 twin) (165, 166) **❻**; 180 rds/min to 8.5 km *(4.6 n miles)*; weight of shell 1.42 kg.
Torpedoes: 6—324 mm Whitehead B515 (2 triple tubes) **❼**; Yu-2 (Mk 46 Mod 1); active/passive homing to 11 km *(5.9 n miles)* at 40 kt; warhead 44 kg.
A/S mortars: 2 FQF 2500 12-tubed fixed launchers **❽**; 120 rockets; range 1,200 m; warhead 34 kg. Similar in design to the RBU 1200.
Countermeasures: Decoys: 2 Type 946; 15 barrelled 100 mm chaff launchers.
ESM: Type 825; intercept.
ECM: Type 981; jammer.
Combat data systems: Thomson-CSF Tavitac with Vega FCS (109); ZKJ-1 (110); ZKJ 4A (165); ZKJ 4B (166).
Radars: Air search: Type 517 Knife Rest **❾**; A-band.
Surface search: Type 363 Sea Tiger (109). Type 354 Eye Shield (165, 166) **❿**; E/F-band.
Navigation: Racal Decca 1290; I-band.
Fire control: Type 344 (MR 34) (165, 166) **⓫**; I-band (for SSM and 100 mm).
Type 343 Sun Visor (109, 110) **⓬**; I-band.
Type 347G Rice Lamp **⓭**; I-band (for 57/37 mm).
Type 345 (MR 35) **⓮**; I/J-band (for Crotale).
IFF: High Pole.
Sonars: DUBV 23 (165, 166); hull-mounted; active search and attack; medium frequency.

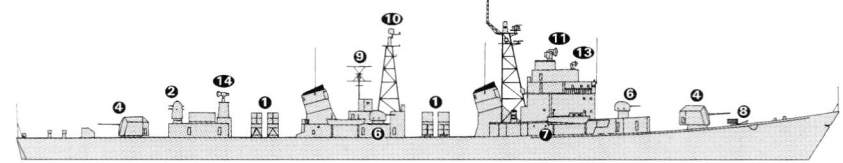

ZHANJIANG   *(Scale 1 : 1,200), Ian Sturton / 0572402*

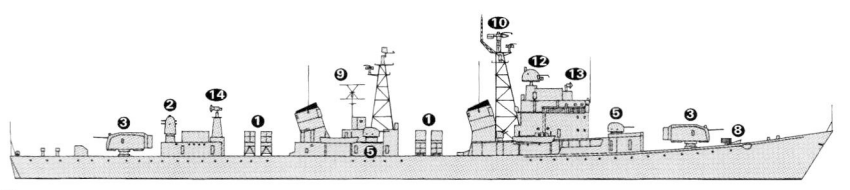

KAIFENG   *(Scale 1 : 1,200), Ian Sturton / 0126350*

KAIFENG (with C-801)   *1/2000, Ships of the World / 0103654*

Programmes: Updated Luda designs sometimes known collectively as the Luda III class.
Modernisation: 109 redesignated Type 051DT after being fitted with Tavitac, Sea Tiger radar and HQ-7 (Crotale). In 1999, she was further modified to receive 16 C0801A missiles, Type 825 ESM, Type 981 ECM and Type 946 chaff launchers. 110

subsequently modernised with ZKJ-1 command system and an otherwise similar configuration as 109. 166 underwent extensive modernisation 2001-03. Principal enhancements include the replacement of YJ-1 by four quadruple YJ-81 missiles, the installation of an octuple HQ-7 SAM launcher in place of the aft (X turret) 37 mm gun and the replacement of

the 130 mm guns with twin 100 mm guns fore and aft. 165 is reported to have undergone a similar upgrade.
Structure: The VDS sonar is a copy of DUBV 43.
Operational: South Seas Fleet based at Zhanjiang. 166 has used different pennant number on foreign deployments.
UPDATED

2 + 2 (2) SOVREMENNY CLASS (PROJECT 956E/956EM) (DDGHM)

Name	No	Builders	Laid down	Launched	Commissioned
HANGZHOU (ex-*Vazhny*, ex-*Yekaterinbugr*)	136 (ex-698)	North Yard, St Petersburg	4 Nov 1988	23 May 1994	25 Dec 1999
FUZHOU (ex-*Alexandr Nevsky*)	137	North Yard, St Petersburg	22 Feb 1989	16 Apr 1999	16 Jan 2001
—	—	North Yard, St Petersburg	27 June 2002	27 Apr 2004	2005
—	—	North Yard, St Petersburg	2003	23 July 2004	2006

Displacement, tons: 7,940 full load
Dimensions, feet (metres): 511.8 × 56.8 × 21.3
  (156 × 17.3 × 6.5)
Main machinery: 4 KVN boilers; 2 GTZA-674 turbines; 99,500 hp(m) *(73.13 MW)* sustained; 2 shafts; bow thruster
Speed, knots: 32. **Range, n miles:** 2,400 at 32 kt; 4,000 at 14 kt
Complement: 296 (25 officers) plus 60 spare

Missiles: SSM: 8 Raduga SS-N-22 Sunburn (Moskit 3M-80E) (2 quad) launchers ❶; active/passive radar homing to 160 km *(87 n miles)* at 2.5 (4.5 for attack) Mach; warhead 300 kg; sea-skimmer.
SAM: 2 SA-N-7 Gadfly (Uragan) ❷ 9M38M1 Smerch; command/semi-active radar and IR homing to 25 km *(13.5 n miles)* at 3 Mach; warhead 70 kg; altitude 15-14,020 m *(50-46,000 ft)*; 44 missiles. Multiple channels of fire.
Guns: 4—130 mm/70 (2 twin) AK 130 ❸; 35-45 rds/min to 29.5 km *(16 n miles)*; weight of shell 33.4 kg.
  4—30 mm/65 AK 630 ❹; 6 barrels per mounting; 3,000 rds/min combined to 2 km.
Torpedoes: 4—21 in *(533 mm)* (2 twin) tubes ❺.
A/S mortars: 2 RBU 1000 6-barrelled ❻; range 1,000 m; warhead 55 kg; 120 rockets carried. Torpedo countermeasure.
Mines: Mine rails for up to 40.
Countermeasures: Decoys: 8 PK 10 and 2 PK 2 chaff launchers. ESM/ECM: 4 Foot Ball. 6 Half Cup laser warner.
Weapons control: 1 China optronic director and laser rangefinder ❼. Band Stand ❽ datalink for SS-N-22. Bell Nest, 2 Light Bulb and 2 Tee Pump datalinks.
Radars: Air search: Top Plate ❾; 3D; E-band.
  Surface search: 3 Palm Frond ❿; I-band.
  Fire control: 6 Front Dome ⓫; F-band (for SA-N-7). Kite Screech ⓬; H/I/K-band (for 130 mm guns). 2 Bass Tilt ⓭; H/I-band (for 30 mm guns).
Sonars: Bull Horn (Platina) and Whale Tongue; hull-mounted; active search and attack; medium frequency.

Helicopters: 1 Harbin Zhi-9C Haitun ⓮ or Kamov Ka-28 Helix.

Programmes: After prolonged negotiations, a contract was signed in September 1996 for two uncompleted Russian Sovremenny class destroyers. These were hulls 18 and 19. Progress was held up for a time because China wanted KA-28 helicopters included, and the Russians demanded extra payment for the aircraft. Deleted Russian units of the class may have been cannibalised for some equipment. A contract for the procurement of two more ships was signed on 3 January 2002. The keel of the first modified Sovremenny class was laid down on 27 June 2002. An option for two further ships was also agreed.
Structure: These are the first Chinese warships to have a data system link. The optronic director is probably a Chinese

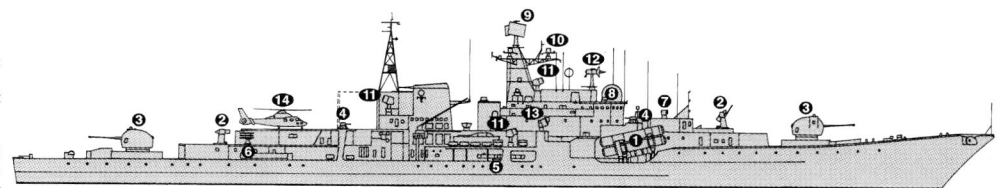

HANGZHOU

(Scale 1 : 1,200), Ian Sturton / 0103652

FUZHOU

12/2000, van Ginderen Collection / 0126288

version of Squeeze Box. The modified 'Sovremenny' are to include variations in weapon fit including replacement of the AK 630 system with 'Kashtan' (with associated Cross Dome target indication radar) CIWS and a reduction to one forward AK 130 turret. The flight deck is to be extended. Two single-armed launchers for SA-N-7 are to be retained. An uprated SS-N-22 system with 200 km range is also to be fitted.

Operational: First one arrived Dinghai on 16 February 2000 and second in February 2001. SS-N-22 test fired on 15 September 2001. Both based in the East Sea Fleet.
Opinion: The main role of these ships is anti-surface warfare although they also possess a good AAW capability. Together with the new AAW destroyers, they represent a step-change in Chinese naval capabilities.

UPDATED

FUZHOU

12/2000, Maritime Photographic / 0105533

HANGZHOU

2/2004, Ships of the World /* 1042158

2 LUYANG I (TYPE 052B) CLASS (DDGHM)

Name	No	Builders	Laid down	Launched	Commissioned
GUANGZHOU	168	Jiangnan Shipyard, Shanghai	2001	25 May 2002	18 July 2004
WUHAN	169	Jiangnan Shipyard, Shanghai	2001	9 Sep 2002	18 July 2004

Displacement, tons: 7,000 full load
Dimensions, feet (metres): 508.5 × 55.8 × 19.7
(155 × 17 × 6)
Main machinery: CODOG: 2 Ukraine DA80 gas turbines;
48,600 hp(m) *(35.7 MW)*; 2 diesels; 8,840 hp(m) *(6.5 MW)*; 2
shafts; cp props
Speed, knots: 29. **Range, n miles:** 4,500 at 15 kt
Complement: 280 (40 officers)

Missiles: SSM: 16 C-802 (YJ-82/C SS-N-8 Saccade) 4 quad ❶;
active radar homing to 120 km *(66 n miles)* at 0.9 Mach;
warhead 165 kg; sea skimmer.
SAM: SA-N-12 Grizzly (Shtil-1) 9M38M2 ❷; command/semi-
active radar and IR homing to 35 km *(18.9 n miles)* at 3 Mach;
warhead 70 kg; 2 magazines (forward and aft). 48 missiles.
Guns: 1—3.9 in *(100 mm)*/56 ❸; 60-80 rds/min to 17 km *(9.3 n
miles)*; weight of shell 13.5 kg.
2—30 mm Type 730 ❹; 7 barrels per mounting; 4,200 rds/
min combined to 1.5 km.
A/S mortars: 4 multiple rocket launchers (possibly multirole) ❺.
Countermeasures: Decoys: 4—18 tube 100 mm launchers.
ESM: SRW 210A.
ECM: Type 984 (I-band jammer). Type 985 (E/F-band jammer).
Combat data systems. To be announced. SATCOM.
Weapons control: Band Stand ❻ datalink (for C-802).
Radars: Air search: Top Plate ❼; E/F-band.
Air/Surface search: Type 364 Seagull C ❽; G-band.
Fire control: 4 Front Dome (Orekh) ❾; F-band (for SA-N-12).
Type 344 (MR 34) ❿; I-band (for SSM and 100 mm).
Type 347G Rice Lamp; I-band (for Type 730).
Navigation: To be announced.
Sonars: Bow mounted. To be announced.

Helicopters: 1 Harbin Zhi-9A Haitun or Kamov KA-28 Helix ⓫.

Programmes: Construction of new multirole destroyers with
medium-range air defence capability started in 2001.
Structure: Based on 'Luhai' design but with more advanced
stealth features. The aft superstructure contains the hangar
on the port side and aft missile magazine to starboard. Details
of both the hull and its weapon systems are speculative.
Operational: Names are unconfirmed. Sea trials of 168 began on
24 July 2003. Likely to be based in the South Sea Fleet.
UPDATED

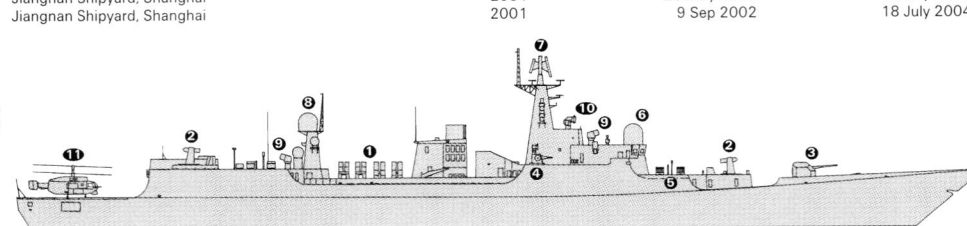

GUANGZHOU
(Scale 1 : 1,200), Ian Sturton / 1042087

GUANGZHOU
6/2004* / 1042168

WUHAN
7/2004* / 0583664

GUANGZHOU
11/2004*, Ships of the World / 1042157

2 LUYANG II (TYPE 052C) CLASS (DDGHM)

Name	No	Builders	Laid down	Launched	Commissioned
LANZHOU	170	Jiangnan Shipyard, Shanghai	June 2002	29 Apr 2003	18 July 2004
HAIKOU	171	Jiangnan Shipyard, Shanghai	Nov 2002	29 Oct 2003	2005

Displacement, tons: 7,000 full load
Dimensions, feet (metres): 508.5 × 55.8 × 19.7
   (155 × 17 × 6)
Main machinery: CODOG: 2 Ukraine DA80 gas turbines;
   48,600 hp(m) *(35.7 MW)*; 2 diesels; 8,840 hp(m) *(6.5 MW)*; 2
   shafts; cp props
Speed, knots: 29. **Range, n miles:** 4,500 at 15 kt
Complement: 280 (40 officers)

Missiles: SSM: 8 C-803 (YJ-83) **❶** 2 quad; active radar homing to
   160 km *(86 n miles)* at 1.3 Mach; warhead 165 kg; sea
   skimmer.
   SAM: HHQ-9 **❷**; 8 vertical revolving sextuple launchers (6
   forward, 2 aft); command guidance; semi-active radar homing
   to 100 km *(54 n miles)* at 3 Mach; warhead 90 kg; 48
   missiles.
Guns: 1—3.9 in *(100 mm)*/56 **❸**; 60-80 rds/min to 17 km *(9.3 n
   miles)*; weight of shell 13.5 kg.
   2—30 mm Type 730 **❹**; 7 barrels per mounting; 4,200 rds/
   min combined to 1.5 km.
Countermeasures: To be announced.
Combat data systems. To be announced. SATCOM.
Weapons control: Band Stand **❺** datalink for C-803.

Radars: Air search: Type 517 Knife Rest **❻**; A-band.
   Air search/fire control: Type 346 phased arrays **❼**; 3D; F-band.
   Air/Surface search: Type 364 Seagull C **❽**; G-band
   Fire control: Type 344 (MR 34) **❾**; I-band (for SSM and 100 mm).
      Type 347G Rice Lamp; I-band (for Type 730).
   Navigation: To be announced.
Sonars: Bow mounted. To be announced.

Helicopters: 2 Harbin Zhi-9A Haitun or Kamov KA-28 Helix **❿**.

Programmes: The second phase of the destroyer construction
   programme which introduces the long-range HHQ-9 missile
   system into service.
Structure: Appears to share the same basic hull design as the
   Type 052B destroyers which in turn are based on the Luhai
   class. As well as incorporating stealth features, the design
   includes a taller forward superstructure in which the four
   phased array antennas are installed. The helicopter hangar is
   on the port side of the aft superstructure. Details are
   speculative and firm details of both the SAM and SSM
   systems are yet to be confirmed. The CIWS systems are on
   raised platforms forward and on top of the hangar.

Operational: Likely to be based in the South Sea Fleet. Name
   unconfirmed.   *UPDATED*

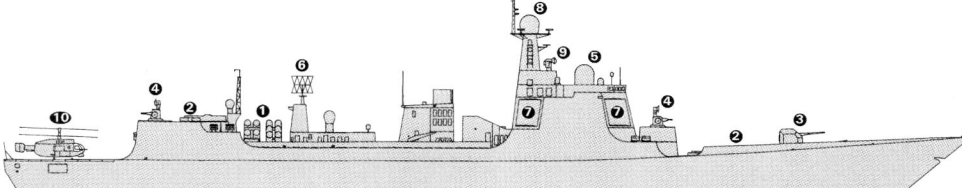

LANZHOU   *(Scale 1 : 1,200), Ian Sturton* / 1042086

LANZHOU   7/2004 * / 0583665

LANZHOU   *12/2004 *, Ships of the World* / 1042156

1 LUHAI CLASS (TYPE 051B) (DDGHM)

Name	No	Builders	Laid down	Launched	Commissioned
SHENZHEN	167	Dalian Shipyard	July 1996	16 Oct 1997	4 Jan 1999

Displacement, tons: 6,000 full load
Dimensions, feet (metres): 505 × 52.5 × 19.7
   (154 × 16 × 6)
Main machinery: CODOG: 2 Ukraine gas turbines; 48,600 hp(m)
   (35.7 MW); 2 MTU 12V 1163 TB 83 diesels; 8,840 hp(m)
   (6.5 MW) sustained; 2 shafts; cp props
Speed, knots: 29. **Range, n miles:** 4,500 at 14 kt
Complement: 250 (42 officers)

Missiles: SSM: 16 C-802 (YJ-82/CSS-N-8 Saccade) **❶**; active
   radar homing to 120 km *(66 n miles)* at 0.9 Mach; warhead
   165 kg; sea skimmer.
   SAM: 1 HQ-7 (Crotale) octuple launcher **❷**; CSA-N-4 line of sight
   guidance to 13 km *(7 n miles)* at 2.4 Mach; warhead 14 kg.
   Possible reloading hatch aft of the HQ-7 launcher.

Guns: 2—3.9 in *(100 mm)*/56 (twin) **❸**; 18 rds/min to 22 km
   (12 n miles); weight of shell 15 kg.
   8—37 mm/63 Type 76A (4 twin) **❹**; 180 rds/min to 8.5 km
   (4.6 n miles) anti-aircraft; weight of shell 1.42 kg.
Torpedoes: 6—324 mm B515 (2 triple) tubes **❺** Yu-2/5/6;
   active/passive homing to 11 km *(5.9 n miles)* at 40 kt;
   warhead 44 kg.
Countermeasures: Decoys: 2 Type 946 15-tube 100 mm chaff
   launchers **❻**.
   2 Type 947 10-tube 130 mm chaff launchers.
   ESM: Type 826.
   ECM: Type 984; I-band jammer; Type 985; E/F-band jammer.
Combat data systems: Thomson-CSF Tavitac; SATCOM.

Weapons control: 2 GDG 776 optronic directors.
Radars: Air search: Type 517 Knife Rest **❼**; A-band.
   Air search: Type 381C Rice Shield **❽**; E/F-band.
   Air/surface search: Type 360 Seagull S **❾**; E/F-band.
   Fire control: Type 344 (MR 34) **❿**; I-band (for SSM and
      100 mm).
      2 Type 347G Rice Lamp **⓫**; I-band (for 37 mm).
      Type 345 (MR 35) **⓬**; I/J-band (for HQ-7).
   Navigation: Racal/Decca 1290; I-band.
Sonars: DUBV-23; hull mounted; active search and attack;
   medium frequency.

Helicopters: 2 Harbin Zhi-9C Haitun **⓭** or Kamov Ka-28 Helix.

Programmes: Follow-on from the Luhu class. Although the only
   ship of its class, it would appear to be the baseline design for
   the Type 052B and 052C destroyers.
Structure: Apart from the second funnel and octuple SSM
   launchers, there are broad similarities with the smaller Luhu.
   Anti-aircraft guns are all mounted aft allowing more space in
   front of the bridge which seems to show a reloading hatch for
   HQ-7.
Operational: Based at Zhanjiang in South Sea Fleet. Out of area
   deployment to Europe in 2001.

                                              UPDATED

SHENZHEN   *(Scale 1 : 1,200), Ian Sturton* / 0569249

SHENZHEN   *10/2001, Derek Fox* / 0126287

2 LUHU (TYPE 052) CLASS (DDGHM)

Name	No	Builders	Laid down	Launched	Commissioned
HARBIN	112	Jiangnan Shipyard, Shanghai	Nov 1990	Oct 1991	July 1994
QINGDAO	113	Jiangnan Shipyard, Shanghai	Jan 1993	Oct 1993	Mar 1996

Displacement, tons: 4,600 full load
Measurement, tons: 472.4 × 52.5 × 16.7
(144 × 16 × 5.1)
Main machinery: CODOG: 2 GE LM 2500 gas turbines (112);
55,000 hp *(41 MW)* sustained or 2 Ukraine gas turbines (113)
48,600 hp(m) *(35.7 MW)*; 2 MTU 12V 1163 TB83 diesels;
8,840 hp(m) *(6.5 MW)* sustained; 2 shafts; cp props
Speed, knots: 31
Range, n miles: 5,000 at 15 kt
Complement: 266 (38 officers)

Missiles: SSM: 16 C-802 (YJ 82/CSS-N-8) Saccade ❶; active
radar homing to 120 km *(66 n miles)* at 0.9 Mach; warhead
165 kg; sea-skimmer.
SAM: 1 HQ-7 (Crotale) octuple launcher ❷; CSA-4; line of sight
guidance to 13 km *(7 n miles)* at 2.4 Mach; warhead 14 kg. 32
missiles.
Guns: 2—3.9 in *(100 mm)/56* (twin) ❸; 18 rds/min to 22 km
(12 n miles); weight of shell 15 kg.
8—37 mm/63 Type 76A (4 twin) ❹; 180 rds/min to 8.5 km
(4.6 n miles) anti-aircraft; weight of shell 1.42 kg.
Torpedoes: 6—324 mm Whitehead B515 (2 triple) tubes ❺. Yu-2
(Mk 46 Mod 1); active/passive homing to 11 km *(5.9 n miles)*
at 40 kt; warhead 44 kg.
A/S mortars: 2 FQF 2500 ❻ 12-tubed fixed launchers; range
1,200 m; warhead 34 kg. 120 rockets.
Countermeasures: Decoys: 2 Type 946; 15 barrelled 100 mm
chaff launchers.
ESM: Rapids (112). Type 826 (113).
ECM: Ramses (113). Type 984 (113).
Combat data systems: Thomson-CSF Tavitac (112). ZKJ-6 (113);
action data automation. SATCOM. Link W.
Weapons control: 2 GDG-775 optronic directors ❼.
Radars: Air search: Type 518 (navalised REL-1) ❽; D-band.
Air/surface search: Type 363S Sea Tiger (113) ❾; E/F-band.
Type 360 Seagull S (112); E/F-band.
Surface search: Type 362 (ESR 1) ❿; I-band.
Navigation: Racal Decca 1290; I-band.
Fire control: Type 344 (MR 34) ⓫; I-band (for SSM and
100 mm).
Two Type 347G Rice Lamp ⓬; I-band (for 37 mm).
Type 345 (MR 35) ⓭; I/J-band (for Crotale).
Sonars: DUBV-23; Hull-mounted; active search and attack;
medium frequency.
DUBV-43 VDS; active attack; medium frequency.

Helicopters: 2 Harbin Zhi-9C Haitun ⓮.

HARBIN   *(Scale 1 : 1,200), Ian Sturton /* 0569255

HARBIN (after refit)   *2/2003, Y Chang /* 0531594

Programmes: Class of two ordered in 1985 but delayed by
priority being given to export orders for Thailand.
Modernisation: *Harbin* completed refit in early 2003. It has been
fitted with a new low radar profile 100 mm gun turret.
Qingdao reported to have started refit in 2004.
Structure: The most notable features are the SAM launcher,
improved radar and fire-control systems and a modern
100 mm gun. Gas turbines for the second of class came from
the Ukraine. The HQ-7 launcher is a Chinese copy of Crotale.

DCN Samahe 110N helo handling system. *Harbin* has a dome-
shaped radome on the superstructure while *Qingdao* has
cylindrical antennae in the same position. Both are likely to be
ECM systems.
Operational: First of class based in North Sea Fleet at Guzhen
Bay, second in the East Sea Fleet at Jianggezhuang. *Qingdao*
deployed to the Pacific and South America in 2002.

UPDATED

HARBIN   *11/2004*, Ships of the World /* 1042144

QINGDAO   *5/1998 /* 0017715

QINGDAO

5/1998, Sattler/Steele / 0056747

12 LUDA (TYPES 051/051D/051Z) CLASS (DDGM/DDGHM)

Name	No	Name	No	Name	No
JINAN	105	NANJING	131	CHANGSHA	161
XIAN	106	HEFEI	132	NANNING	162
YINCHUAN	107	CHONGQING	133	NANCHANG	163
XINING	108	ZUNYI	134	GUILIN	164

Displacement, tons: 3,250 standard; 3,670 full load
Dimensions, feet (metres): 433.1 × 42 × 15.1
(132 × 12.8 × 4.6)
Main machinery: 2 or 4 boilers; 2 turbines; 72,000 hp(m)
(53 MW); 2 shafts
Speed, knots: 32. **Range, n miles:** 2,970 at 18 kt
Complement: 280 (45 officers)

Missiles: SSM: 6 HY-2 (C-201) (CSS-C-3A Seersucker) (2 triple)
launchers ❶; active radar or IR homing to 95 km *(51 n miles)*
at 0.9 Mach; warhead 513 kg.
Guns: 4 USSR 5.1 in *(130 mm)*/58 (2 twin) ❷; 17 rds/min to
29 km *(16 n miles)*; weight of shell 33.4 kg.
8 China 57 mm/70 (4 twin); 120 rds/min to 12 km *(6.5 n
miles)*; weight of shell 6.31 kg or 8 China 37 mm/63 (4 twin)
❸; 180 rds/min to 8.5 km *(4.6 n miles)*; weight of shell
1.42 kg.
8 USSR 25 mm/60 (4 twin) ❹; 270 rds/min to 3 km *(1.6 n
miles)* anti-aircraft; weight of shell 0.34 kg.
Torpedoes: 6—324 mm Whitehead B515 (2 triple tubes) (fitted
in some); Yu-2 (Mk 46 Mod 1); active/passive homing to
11 km *(5.9 n miles)* at 40 kt; warhead 44 kg.
A/S mortars: 2 FQF 2500 12-tubed fixed launchers ❺; 120
rockets; range 1,200 m; warhead 34 kg. Similar in design to
the RBU 1200.
Depth charges: 2 or 4 BMB projectors; 2 or 4 racks.
Mines: 38.
Combat data systems: ZKJ-1 (132).
Radars: Air search: Type 515 Bean Sticks ❻; A-band.
Type 381 Rice Screen ❼ (132); 3D; G-band. Similar to Hughes
SPS-39A.
Surface search: Type 354 Eye Shield ❽; G-band.
Square Tie (not in all); I-band.
Navigation: Fin Curve or Racal Decca 1290; I-band.
Fire control: Wasp Head (also known as Wok Won) or Type 343
Sun Visor B (series 2) ❾; I-band.
2 Type 347G Rice Lamp ❿; I-band.
IFF: High Pole.
Sonars: Pegas 2M and Tamir 2; hull-mounted; active search and
attack; high frequency.

Helicopters: 2 Harbin Z-9C (Dauphin) (105).

Programmes: The first Chinese-designed destroyers of such a
capability to be built. First of class completed in 1971. 105 to

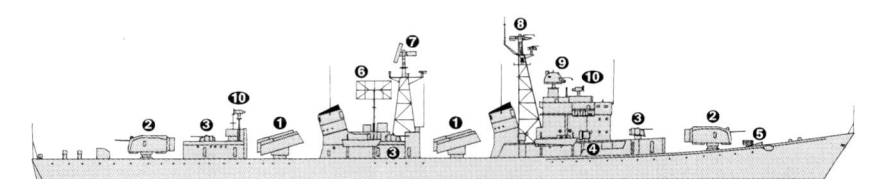

HEFEI

(Scale 1 : 1,200), Ian Sturton / 0056749

HEFEI

6/1999, Ships of the World / 0056752

108 built at Luda; 131 to 134 at Shanghai and 161 to 164 at
Dalian. Similar to the deleted USSR Kotlin class. The
programme was much retarded after 1971 by drastic cuts in
the defence budget. In early 1977 building of series two of this
class was put in hand and includes those after 108, with the
latest 164 completed in April 1990. The order of completion
was 105, 160 (scrapped), 106, 161, 107, 162, 131, 108, 132,
163, 133, 134 and 164.
Modernisation: Equipment varies considerably from ship to ship.
The original Type 051 ships are 105, 106, 107, 131, 161 and
162. Of these 105 was converted in 1987 to act as a trials
ship; a twin helicopter hangar and deck replaced the after
armament. Type 051D ships are 108, 133, 134, 163 and 164.
132 is a command ship (Type 051Z) fitted with ZKJ-1

command system and Rice Screen (Type 381A) 3-D radar
although this has been temporarily replaced by a SATCOM
terminal.
Structure: Electronics vary in later ships. Some ships have
57 mm guns, others 37 mm. 105 may have Alcatel
'Safecopter' landing aid. SAM is fitted in *Kaifeng* and *Dalian* in
X gun position.
Operational: Capable of foreign deployment, although
command and control is limited. Underway refuelling is
practised. Deployment; 105 series in North Sea Fleet at Yuchi;
131 series in East Sea Fleet at Dalian; 161 series in South Sea
Fleet at Zhanjiang. 160 was damaged by an explosion in
1978, and was scrapped.

UPDATED

JINAN (with helicopter deck)

1/1994 / 0056753

FRIGATES

2 + 1 JIANGKAI (TYPE 054) CLASS (FFGHM)

Name	No	Builders	Laid down	Launched	Commissioned
MAANSHAN	525	Hudong Shipyard, Shanghai	Dec 2001	11 Sep 2003	18 July 2004
WENZHOU	526	Huangpu Shipyard, Guangzhou	Feb 2002	Nov 2003	June 2005
—	—	Huangpu Shipyard, Guangzhou	2003	2005	2006

Displacement, tons: 3,500 standard; 3,900 full load
Dimensions, feet (metres): 433.2 × 49.2 × 16.4
(132.0 × 15.0 × 5.0)
Main machinery: CODAD; 4 SEMT-Pielstick diesels; 2 shafts
Speed, knots: 27. **Range, n miles:** 3,800 at 18 kt
Complement: 190

Missiles: SSM: 8 C-802 (YJ-82/CSS-N-8 Saccade) ❶; active
radar homing to 120 km *(66 n miles)* at 0.9 Mach; warhead
165 kg; sea skimmer.
SAM: 1 HQ-7 (Crotale) ❷; CSA-N-4 line-of-sight guidance to
13 km *(7 n miles)* at 2.4 Mach; warhead 14 kg.
Guns: 1—3.9 in *(100 mm)*/56 ❸; 18 rds/min to 22 km *(12 n
miles)*; weight of shell 15 kg.
4—300 mm/65 AK 630 ❹; 6 barrels per mounting;
3,000 rds/min combined to 2km.
Torpedoes: 6—324 mm B515 (2 triple) tubes; Yu-2/6/7; active/
passive homing to 11 km *(5.9 n miles)* at 40 kt; warhead
44 kg.
Countermeasures: to be announced.
Combat data systems: to be announced.
Radars: Air/surface search: Type 363S Sea Tiger ❺; E/F-band.
Surface search: Type 364 Seagull C ❻; G-band.
Fire control: Type 344 (MR 34) ❼; I-band (for SSM and 100 mm).
Type 345 (MR 35) ❽; I/J-band (for HQ-7).
Type 347G Rice Lamp ❾ (for AK 630).
Sonars: to be announced.

Helicopters: 1 Harbin Zhi-9C Haitun ❿.

Programmes: The first two vessels of a new general-purpose
frigate class to follow the Jiangwei II class and to replace the
Jianghu class. A third ship believed to be under construction

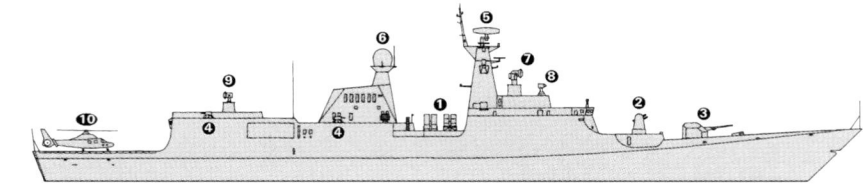

MAANSHAN
(Scale 1 : 1,200), Ian Sturton / 1042084

MAANSHAN
7/2004* / 0583666

at Huangpu Shipyard and further ships (possibly more
advanced variants) are expected at Hudong.

Structure: A new design incorporating stealth features.
UPDATED

4 JIANGWEI I (TYPE 053 H2G) CLASS (FFGHM)

Name	No	Builders	Laid down	Launched	Commissioned
ANQING	539	Hudong Shipyard, Shanghai	Nov 1990	July 1991	Dec 1991
HUAINAN	540 (548 out of area)	Hudong Shipyard, Shanghai	Jan 1991	Oct 1991	July 1992
HUAIBEI	541	Hudong Shipyard, Shanghai	July 1992	Apr 1993	Aug 1993
TONGLING	542	Hudong Shipyard, Shanghai	Dec 1992	Sep 1993	Apr 1994

Displacement, tons: 2,250 full load
Dimensions, feet (metres): 366.5 × 40.7 × 15.7
(111.7 × 12.4 × 4.8)
Main machinery: 2 Type 18E 390 diesels; 24,000 hp(m)
(17.65 MW) sustained; 2 shafts
Speed, knots: 27. **Range, n miles:** 4,000 at 18 kt
Complement: 170

Missiles: SSM: 6 YJ-1 (Eagle Strike) (C-801) (CSS-N-4 Sardine) or
C-802 (2 triple) launchers ❶; active radar homing to 40 km
(22 n miles) or 120 km *(66 n miles)* (C-802) at 0.9 Mach;
warhead 165 kg; sea-skimmer.
SAM: 1 HQ-61 sextuple launcher ❷; RF 61 (CSA-N-2); semi-active
radar homing to 10 km *(5.5 n miles)* at 2 Mach. Similar to Sea
Sparrow. May be replaced in due course.
Guns: 2 China 3.9 in *(100 mm)*/56 (twin) ❸; 18 rds/min to
22 km *(12 n miles)*; weight of shell 15.9 kg.
8 China 37 mm/63 Type 76A (4 twin) ❹; 180 rds/min to
8.5 km *(4.6 n miles)* anti-aircraft; weight of shell 1.42 kg.
A/S mortars: 2 Type 87 ❺ 6-tubed launchers.
Countermeasures: Decoys: 2 China Type 945 26-barrelled chaff
launchers ❻.
ESM: RWD8; intercept.
ECM: NJ81-3; jammer. Similar to Scimitar.

HUAIBEI
(Scale 1 : 900), Ian Sturton / 0130723

Radars: Air search: Type 517 Knife Rest ❼; A-band.
Air/surface search: Type 360 Seagull S ❽; E/F-band.
Fire control: Type 343 (Wok Won) (Wasp Head) ❾; I-band.
Fog Lamp ❿; I/J-band (for SAM).
Type 347G Rice Lamp ⓫; I/J-band (for 37 mm).
Navigation: Racal Decca 1290 and China Type 360; I-band.
Sonars: Echo Type 5; hull-mounted; active search and attack;
medium frequency.

Helicopters: 2 Harbin Z-9C (Dauphin) ⓬.

Programmes: Programme started in 1988. First one conducted
sea trials in late 1991. Four of the class built before the design
moved on to the Jiangwei II.
Modernisation: SAM system has been unsatisfactory and may
be replaced in due course.
Structure: The sextuple launcher is a multiple launch SAM
system using the CSA-N-2 missile.
Operational: All based in the East Sea Fleet at Dinghai.
UPDATED

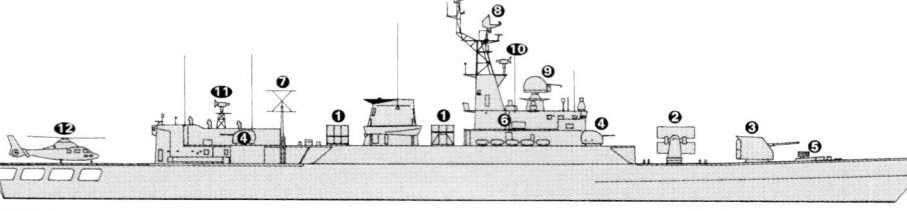

HUAIBEI
4/2000, Ships of the World / 0103659

8 + 2 JIANGWEI II (TYPE 053H3) CLASS (FFGHM)

Name	No	Builders	Laid down	Launched	Commissioned
JIAXIN	521 (ex-597)	Hudong Shipyard, Shanghai	Oct 1996	10 Aug 1997	Nov 1998
LIANYUNGANG	522	Hudong Shipyard, Shanghai	Dec 1996	8 Aug 1997	Feb 1999
SANMING	523	Hudong Shipyard, Shanghai	June 1997	10 Aug 1998	Oct 1999
PUTIAN	524	Hudong Shipyard, Shanghai	Dec 1997	Dec 1998	Nov 1999
YICHANG	564	Huangpu Shipyard, Guangzhou	Dec 1997	Oct 1998	Dec 1999
YULIN	565	Huangpu Shipyard, Guangzhou	May 1998	Apr 1999	Mar 2000
YUXI	566	Hudong Shipyard, Shanghai	May 2000	Jan 2001	Mar 2002
—	567	Huangpu Shipyard, Guangzhou	Mar 2001	Aug 2001	Sep 2002
—	527	Huangpu Shipyard, Guangzhou	2003	1 Oct 2004	2005
—	528	Hudong Shipyard, Shanghai	2003	30 May 2004	2005

Displacement, tons: 2,250 full load
Dimensions, feet (metres): 366.5 × 40.7 × 15.7
(111.7 × 12.4 × 4.8)
Main machinery: 2 Type 18E 390 diesels; 24,000 hp(m)
(17.65 MW) sustained; 2 shafts
Speed, knots: 27. **Range, n miles:** 4,000 at 18 kt
Complement: 170

Missiles: SSM: 8 YJ-1 (Eagle Strike) (C-801) (CSS-N-4 Sardine) or
C-802 (2 quad) launchers ❶; active radar homing to 40 km
(22 n miles) or 120 km *(66 n miles)* (C-802) at 0.9 Mach;
warhead 165 kg; sea-skimmer.
SAM: 1 HQ-7 (Crotale) octuple launcher ❷; CSA-N-4 line of sight
guidance to 13 km *(7 n miles)* at 2.4 Mach; warhead 14 kg.
Guns: 2 China 3.9 in *(100 mm)*/56 (twin) ❸; 18 rds/min to
22 km *(12 n miles)*; weight of shell 15.9 kg.
8 China 37 mm/63 Type 76A (4 twin) ❹; 180 rds/min to
8.5 km *(4.6 n miles)* anti-aircraft; weight of shell 1.42 kg.
A/S mortars: 2 RBU 1200 ❺; 5-tubed fixed launchers; range
1,200 m; warhead 34 kg.
Countermeasures: Decoys: 2 SRBOC Mk 36 6-barrelled chaff
launchers ❻; 2 China 26-barrelled chaff launchers ❼.
ESM: SR-210; intercept.
ECM: 981-3 noise jammer. RWD-8 deception jammer.
Combat data systems: ZKJ 3C.
Weapons control: JM-83H optronic director.
Radars: Air search: Type 517 Knife Rest ❽; A-band.
Air/surface search: Type 360 Seagull S ❾; E/F-band.
Fire control: Type 343G ❿; I-band (for SSM and 100 mm).
Type 345 (MR 35) ⓫; I/J-band (for SAM).
Type 347G Rice Lamp ⓬; I/J-band (for 37 mm).
Navigation: 2 RM-1290; I-band.
Sonars: Echo Type 5; hull-mounted; active search and attack;
medium frequency.

Helicopters: 2 Harbin Z-9C (Dauphin) ⓭.

Programmes: Follow-on to the Jiangwei class, building some
four years later. The building programme appeared to have
been terminated after eight ships but reports indicate that two
further ships are under construction. Further units are likely.
Structure: An improved SAM system, updated fire-control radars
and a redistribution of the after anti-aircraft guns are the

obvious differences from the original Jiangwei. New Type 99
turret fitted in 567 and to be retro-fitted to the remainder of
the class.
Operational: 521-524 assigned to East Sea Fleet and 564-566
to South Sea Fleet.

Sales: Possibly two for Pakistan in due course, but with some
changes in equipment fitted.

UPDATED

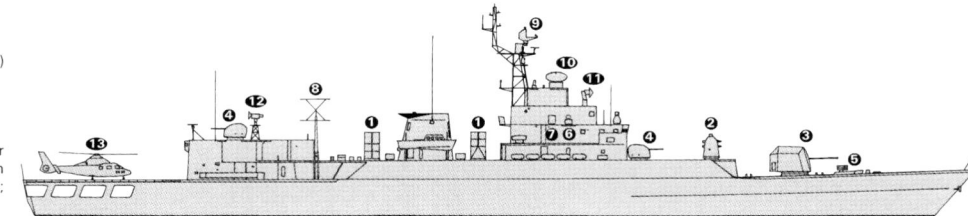

JIANGWEI II　　　　　　　　　　　　　　　　　*(Scale 1 : 900), Ian Sturton /* 0569248

JIANGWEI 527　　　　　　　　　　　　　　　　　　　*10/2004* /* 1042167

LIANYUNGANG　　　　　　　　　　　　　　　*3/2004*, L-G Nilsson /* 1042148

YICHANG　　　　　　　　　　　　　　　　*10/2001, John Mortimer /* 0126229

27 JIANGHU I (TYPE 053H) CLASS (FFG)

Name	No	Name	No	Name	No	Name	No
CHANG DE	509	JIUJIANG	516	LINFEN	545	ZIGONG	558
SHAOXING	510	NANPING	517	MAOMING	551	KANGDING	559
NANTONG	511	JIAN	518	YIBIN	552	DONGGUAN	560
WUXI	512	CHANGZHI	519	SHAOGUAN	553	SHANTOU	561
HUAYIN	513	NINGBO	533	ANSHUN	554	JIANGMEN	562
ZHENJIANG	514	JINHUA	534	ZHAOTONG	555	ZHAOQING	563
XIAMEN	515	DANDONG	543	JISHOU	557		

Displacement, tons: 1,425 standard; 1,702 full load
Dimensions, feet (metres): 338.5 × 35.4 × 10.2
(103.2 × 10.8 × 3.1)
Main machinery: 2 Type 12E 390V diesels; 14,400 hp(m)
(10.6 MW) sustained; 2 shafts
Speed, knots: 26. **Range, n miles:** 4,000 at 15 kt; 2,700 at 18 kt
Complement: 200 (30 officers)

Missiles: SSM: 4 HY-2 (C-201) (CSSC-3 Seersucker) (2 twin)
launchers ❶; active radar or IR homing to 80 km *(43.2 n miles)*
at 0.9 Mach; warhead 513 kg.
Guns: 2 or 4 China 3.9 in *(100 mm)*/56 (2 single ❷ or 2 twin ❸);
18 rds/min to 22 km *(12 n miles)*; weight of shell 15.9 kg.
12 China 37 mm/63 (6 twin) ❹ (8 (4 twin), in some); 180 rds/
min to 8.5 km *(4.6 n miles)* anti-aircraft; weight of shell
1.42 kg.
A/S mortars: 2 RBU 1200 5-tubed fixed launchers (4 in some) ❺;
range 1,200 m; warhead 34 kg.
Depth charges: 2 BMB-2 projectors; 2 racks (in some).
Mines: Can carry up to 60.
Countermeasures: Decoys: 2 RBOC Mk 33 6-barrelled chaff
launchers or 2 China 26-barrelled launchers.
ESM: Jug Pair or Watchdog; radar warning.
Weapons control: Wok Won director (in some) ❻.
Radars: Air search: Type 517 Knife Rest ❼; A-band.
Air/surface search: Type 354 Eye Shield (MX 902) ❽; G-band.
Type (unknown) ❾; I-band.
Surface search/fire control: Type 352 Square Tie ❿; I-band.
Navigation: Don 2 or Fin Curve or Racal Decca; I-band.
Fire control: Type 347G Rice Lamp (in some) ⓫; I/J-band.
Type 343 (Wok Won) (Wasp Head) (in some) ⓬; I-band.
IFF: High Pole A. Yard Rake or Square Head.
Sonars: Echo Type 5; hull-mounted; active search and attack;
medium frequency.

Programmes: Pennant numbers changed in 1979. All built in
Shanghai starting in the mid-1970s at the Hudong, Jiangnan
and Huangpu shipyards. Ships were completed in the
following order: 515, 516, 517, 511, 512, 513, 514, 518, 509,
510, 519, 520, 551, 552, 533, 534, two for Egypt, 543, 553,
554, 555, 556 (to Bangladesh), 557, 544, 558, 560,
561, 559, 562 and 563. The last of class 563 completed in
February 1996. Reports that construction had restarted in
1997 were incorrect.
Modernisation: Equipment varies considerably from ship to ship.
The Type 053H ships are 509-519 and 551 and 552. These
are equipped with SY-1 or SY-2 SSM, a single 100 mm gun
and SJD-3 sonar. Type 053H1 ships are 533, 534, 543, 544,
553, 554, 555 and 557. These are similar to Type 053H but
are equipped with twin 100 mm guns and SJD-5 (Echo 5)
sonar. Type 053H1G ships are 558-563. These are similar to
Type 053H1 but are equipped with 37 mm enclosed gun
mounts. A larger bridge structure suggests a possible CIC
compartment. The designation of the Air/Surface search
radar in Type 053H1G is not yet known but it bears similarities
to the I-band MR-36A which has been promoted as a
replacement for Type 352 'Square Tie'. 516 appears to have
been modified for a shore bombardment role having been
fitted with seven 122 mm MLRs.
Structure: All of the class have the same hull dimensions.
Previously reported Type numbers have been superseded by
the following designations:
  Type I has at least five versions. Version 1 has an oval funnel
  and square bridge wings; version 2 a square funnel with
  bevelled bridge face; version 3 an octagonal funnel; version 4
  reverts back to the oval funnel and version 5 has a distinctive
  fluting arrangement with cowls on the funnel, as well as
  gunhouses on the 37 mm guns. Some have bow bulwarks.
  Type II. See separate entry.
  Types III and IV. See separate entry.
Operational: 520 paid off in 1993. Ten are based in the Eastern
Fleet, three in the North and the remainder in the South.
Sales: Two have been transferred to Egypt, one in September
1984, the other in March 1985, and one, *Xiangtan* 556, to
Bangladesh in November 1989.

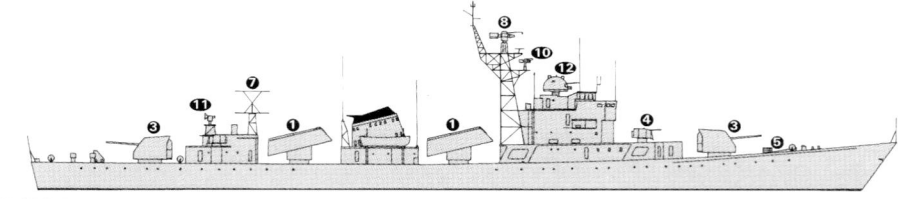

ZHENJIANG (single 100 mm gun)   *(Scale 1 : 900), Ian Sturton /* 0529151

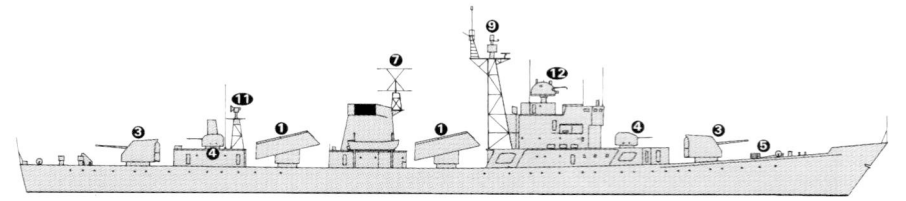

NINGBO (Rice Lamp FC radar)   *(Scale 1 : 900), Ian Sturton /* 0130728

DONGGUAN (37 mm gunhouses)   *(Scale 1 : 900), Ian Sturton /* 0130727

UPDATED   KANGDING   *1/1999, 92 Wing RAAF /* 0056763

ZHENJIANG   *5/2004* */* 1042134

JIUJIANG
7/2004*, Ian Edwards / 1042409

WUXI
5/2004* / 1042132

SHANTOU
9/2000 / 0103662

3 JIANGHU III and IV (TYPE 053 H2) CLASS (FFG)

HUANGSHI 535 (Type III)　　　　　　**WUHU** 536 (Type III)　　　　　　**ZHOUSHAN** 537 (Type IV)

Displacement, tons: 1,924 full load
Dimensions, feet (metres): 338.5 × 35.4 × 10.2
　(103.2 × 10.8 × 3.1)
Main machinery: 2 Type 18E 390V diesels; 14,400 hp(m)
　(10.6 MW) sustained; 2 shafts
Speed, knots: 26
Range, n miles: 4,000 at 15 kt; 2,700 at 18 kt
Complement: 200 (30 officers)

Missiles: SSM: 8 YJ-1 (Eagle Strike) (C-801) (CSS-N-4 Sardine) ❶;
　active radar homing to 40 km *(22 n miles)* at 0.9 Mach;
　warhead 165 kg. Type IV is fitted with C-802 (CSS-N-8
　Saccade) with an extended range to 120 km *(66 n miles)*.
Guns: 4 China 3.9 in *(100 mm)*/56 (2 twin) ❷; 18 rds/min to
　22 km *(12 n miles)*; weight of shell 15.9 kg.
　8 China 37 mm/63 (4 twin) ❸; 180 rds/min to 8.5 km *(4.6 n
　miles)* anti-aircraft; weight of shell 1.42 kg.
A/S mortars: 2 RBU 1200 5-tubed fixed launchers ❹; range
　1,200 m; warhead 34 kg.
Depth charges: 2 BMB-2 projectors; 2 racks.
Mines: Can carry up to 60.
Countermeasures: Decoys: 2 China 26-barrelled chaff
　launchers.
ESM: Elettronica Newton; radar warning.
ECM: Elettronica 929 (Type 981); jammer.
Combat data systems: ZKJ-3.

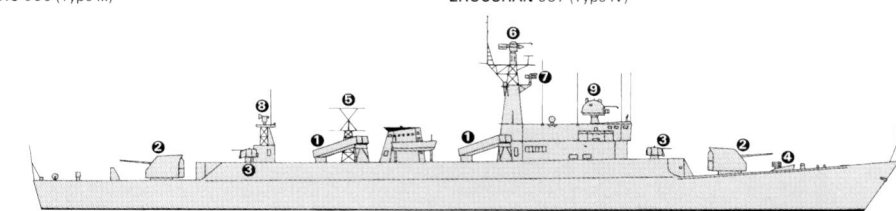

ZHOUSHAN

(Scale 1 : 900), Ian Sturton / 0130726

Radars: Air search: Type 517 Knife Rest ❺; A-band.
Air/surface search: Type 354 Eye Shield (MX 902) ❻; G-band.
Surface search/fire control: Type 352 Square Tie ❼; I-band.
Navigation: Fin Curve; I-band.
Fire control: Type 347G Rice Lamp ❽; I/J-band.
　Type 343G (Wok Won) (Wasp Head) ❾; I-band.
IFF: High Pole A. Square Head.
Sonars: Echo Type 5; hull-mounted; active search and attack;
　medium frequency.

Programmes: These ships are Jianghu hulls 27, 28 and 30 and
　are referred to as New Missile Frigates. *Huangshi*
　commissioned 14 December 1986, *Wuhu* in 1987, and

Zhoushan completed in 1989. They were the first Chinese
warships to be equipped with a computerised combat system.
A fourth of class 538 was reported in 1991 but may have been
confused with one of the Thai ships.
Structure: The main deck is higher in the midships section and
the lower part of the mast is solid. Type IV has an improved
SSM missile which is probably the turbojet C-802. The
arrangement of the launchers is side by side, as opposed to
the staggered pairings in Type III. These are the first all-
enclosed, air conditioned ships built in China.
Operational: Based in East Sea Fleet at Dinghai.
Sales: Four modified Type III to Thailand in 1991-92.

UPDATED

ZHOUSHAN

10/1992, Ships of the World / 0056766

HUANGSHI

2/2001, Ships of the World / 0126362

1 JIANGHU II (TYPE 053) CLASS (FFGH)

Name	No	Builders	Laid down	Launched	Commissioned
SIPING	544	Hudong Shipyard, Shanghai	1984	Sep 1985	Nov 1986

Displacement, tons: 1,550 standard; 1,865 full load
Dimensions, feet (metres): 338.5 × 35.4 × 10.2
(103.2 × 10.8 × 3.1)
Main machinery: 2 Type 12E 390V diesels; 14,400 hp(m)
(10.6 MW) sustained; 2 shafts
Speed, knots: 26. **Range, n miles:** 4,000 at 15 kt; 2,700 at 18 kt
Complement: 185 (30 officers)

Missiles: SSM: 2 HY-2 (C-201) (CSSC-3 Seersucker) (twin)
launchers ❶; active radar or IR homing to 80 km *(43.2 n miles)*
at 0.9 Mach; warhead 513 kg.
Guns: 1 Creusot-Loire 3.9 in *(100 mm)*/55 ❷; 60-80 rds/min to
17 km *(9.3 n miles)*; weight of shell 13.5 kg.
8 China 37 mm/63 (4 twin) ❸; 180 rds/min to 8.5 km *(4.6 n
miles)* anti-aircraft; weight of shell 1.42 kg.
Torpedoes: 6—324 mm ILAS (2 triple) tubes ❹. Yu-2 (Mk 46
Mod 1) active/passive homing to 11 km *(5.9 n miles)* at 40 kt;
warhead 44 kg.
A/S mortars: 2 RBU 1200 5-tubed fixed launchers ❺; range
1,200 m; warhead 34 kg.
Countermeasures: Decoys: 2 SRBOC Mk 33 6-barrelled chaff
launchers or 2 China 26-barrelled launchers.
ESM: Jug Pair or Watchdog; radar warning.
Weapons control: CSEE Naja optronic director for 100 mm gun.
Radars: Air/surface search: Type 354 Eye Shield (MX 902) ❻;
G-band.
Surface search/fire control: Type 352 Square Tie ❼; I-band.
Navigation: Don 2 or Fin Curve; I-band.
IFF: High Pole A. Yard Rake or Square Head.
Sonars: Echo Type 5; hull-mounted; active search and attack;
medium frequency.

Helicopters: Harbin Z-9C (Dauphin) ❽.

Programmes: Built as a standard Jianghu I and then converted
before being commissioned.
Structure: The after part of the ship has been rebuilt to take a
hangar and flight deck for a single helicopter. Alcatel

SIPING *(Scale 1 : 900), Ian Sturton / 0572397*

SIPING *6/2003 / 0569166*

'Safecopter' landing aid. This ship also has a French 100 mm
gun and optronic director, and Italian triple torpedo tubes
mounted on the quarterdeck.
Operational: Based in North Sea Fleet at Guzhen Bay.

Opinion: More of the class were expected to be converted, but
this may have been a one-off helicopter trials ship for the Luhu
and Jiangwei designs.

UPDATED

SHIPBORNE AIRCRAFT

Numbers/Type: 20 Changhe Z-8 Super Frelon.
Operational speed: 134 kt *(248 km/h)*.
Service ceiling: 10,000 ft *(3,100 m)*.
Range: 440 n miles *(815 km)*.
Role/Weapon systems: ASW helicopter; SA 321G delivered from France in 1977 but
supplemented by locally built Zhi-8, of which the first operational aircraft was delivered in late
1991. Thomson Sintra HS-12 in four SA 321Gs for SSBN escort role. Sensors: HS-12 dipping
sonar and processor, some have French-built search radar. Weapons: ASW; Whitehead A244 or
Yu-2 (Mk 46 Mod 1) torpedo. ASV: C-802K ASM.

UPDATED

Z-8 *9/2002 *, Paul Jackson / 0525833*

Numbers/Type: 25 Hai Z-9C Haitun (Dauphin 2).
Operational speed: 140 kt *(260 km/h)*.
Service ceiling: 15,000 ft *(4,575 m)*.
Range: 410 n miles *(758 km)*.
Role/Weapon systems: China has an option to continue building. Sensors: Thomson-CSF Agrion;
HS-12 dipping sonar; Crouzet MAD. Weapons: ASV; up to four locally built radar-guided anti-ship
missiles and Whitehead A244 torpedoes or Yu-2 (Mk 46 Mod 1).

UPDATED

Z-9C *4/2002 * / 0106480*

Numbers/Type: 6/4 Kamov Ka 28PL/Ka 28PS Helix A.
Operational speed: 135 kt *(250 km/h)*.
Service ceiling: 19,685 ft *(6,000 m)*.
Range: 432 n miles *(800 km)*.
Role/Weapon systems: First pair are (Ka 28PL) ASW helicopters acquired in 1997 for evaluation.
Four more ASW versions and four (Ka 28PS) for SAR delivered in late 1999. Sensors: Splash
Drop radar; VGS-3 dipping sonar; MAD; ESM. Weapons: three torpedoes or depth bombs or
mines.

UPDATED

Ka-28 *6/2004 * / 1042165*

LAND-BASED MARITIME AIRCRAFT (FRONT LINE)

Notes: (1) Surveillance and targeting is a high priority. Up to three KJ-2000 (similar to Mainstay)
are in air force service and a national programme is understood to be under development.
(2) In addition to those listed there are about 170 training and transport aircraft.

Numbers/Type: 18 XAC JH-7.
Operational speed: 653 kt *(1,210 km/h)*.
Service ceiling: 51,180 ft *(15,600 m)*.
Range: 891 n miles *(1,650 km)*.
Role/Weapon systems: All-weather dual seat 'Fencer' type attack fighter first revealed in 1988
and being redesigned for export market (FBC-1). A second batch of JH-7A is expected. Further
domestic orders not thought likely. Sensors: Letri JL-10A Shen-Ying pulse Doppler fire-control
radar capable of tracking four targets to 29 n miles *(54 km)* in look-down mode simultaneously.
Weapons: AAM; PL-5b, PL-7 and 23 mm gun. ASM; Two C-801 or C-802 anti-ship missiles;
C-701 anti-ship missile and 500 kg LGBs. AS-17 (Kh-31) may be fitted in due course.

UPDATED

JH-7 *5/2003 / 0114641*

Numbers/Type: 48 Sukhoi Su-30 MKK Flanker.
Operational speed: 1,345 kt *(2,500 km/h).*
Service ceiling: 59,000 ft *(18,000 m).*
Range: 2,160 n miles *(4,000 km).*
Role/Weapon systems: 24 delivered in 2004 and delivery of a further 24 expected. The air force operates at least 130 of the similar Su-27 which also might be used for fleet air-defence. Sensors: Doppler radar. Weapons: One 30 mm cannon; 10 AAMs. Kh-35 anti-ship missiles may be fitted to some aircraft in due course.
UPDATED

Su-27 *5/2003 /* 0114638

Numbers/Type: 4 Harbin SH-5.
Operational speed: 243 kt *(450 km/h).*
Service ceiling: 23,000 ft *(7,000 m).*
Range: 2,563 n miles *(4,750 km).*
Role/Weapon systems: Multipurpose amphibian introduced into service in 1986. Final total of about 20 planned with ASW and avionics upgrade. Sensors: Doppler radar; MAD; sonobuoys. Weapons: ASV; four C 101, two gun turret, bombs. ASW; Yu-2 (Mk 46 Mod 1) torpedoes, mines, depth bombs.
UPDATED

SH-5 *6/2004 * /* 1042160

Numbers/Type: 3 SAC Y-8X (Cub).
Operational speed: 351 kt *(650 km/h).*
Service ceiling: 34,120 ft *(10,400 m).*
Range: 3,020 n miles *(5,600 km).*
Role/Weapon systems: Maritime patrol version of An-12 Cub transport; first flown 1985. There are reported to be two Y-8J variants equipped with Searchwater radar in a dropped nose radome. In addition a Y-8DZ Elint variant was reported undergoing trials in 2004. Sensors: Litton APSO-504(V)3 search radar in undernose radome. Two Litton LTN 72R INS and Omega/Loran. Weapons: No weapons carried.
UPDATED

Y-8X *7/1997 /* 0012195

Numbers/Type: 100 Harbin H-5 (Il-28 Beagle).
Operational speed: 487 kt *(902 km/h).*
Service ceiling: 40,350 ft *(12,300 m).*
Range: 1,175 n miles *(2,180 km).*
Role/Weapon systems: Overwater strike aircraft with ASW/ASVW roles. Numbers are doubtful as some have been phased out and others moved into second line roles such as target towing and ECM training. Weapons: ASW; two torpedoes or four depth bombs. ASVW; one torpedo + mines. Standard; four 23 mm cannon.
UPDATED

H-5 (Romanian colours) *2002, Lindsay Peacock /* 0524583

Numbers/Type: 50/50/20 SAC J-8-I Finback A/J-8-II Finback B/J-8-IV Finback D.
Operational speed: 701 kt *(1,300 km/h).*
Service ceiling: 65,620 ft *(20,000 m).*
Range: 1,187 n miles *(2,200 km).*
Role/Weapon systems: Dual role, all-weather fighter introduced into service in 1990 and production continues. There are at least 450 more in service with the Air Force. Weapons: 23 mm twin-barrel cannon; PL-2/7 AAM; ASM. PL-2 has some ASM capability. *UPDATED*

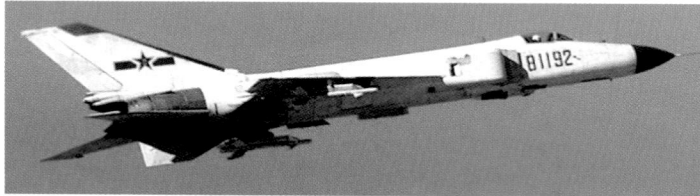

J-8-II *4/2002 * /* 0116938

Numbers/Type: 40 Nanchang A-5 (Fantan-A).
Operational speed: 643 kt *(1,190 km/h).*
Service ceiling: 52,500 ft *(16,000 m).*
Range: 650 n miles *(1,188 km).*
Role/Weapon systems: Strike aircraft developed from Shenyang J-6; operated in the beachhead and coastal shipping attack role. A-5M version adapted to carry two torpedoes or C-801 ASM. Weapons: Two 23 mm cannon, two cluster bombs, one or two air-to-air missiles. Capable of carrying 1 ton warload.
UPDATED

FANTAN-A *6/2002, Ships of the World /* 0554726

Numbers/Type: 200 Shenyang J-6 (MiG-19 Farmer).
Operational speed: 831 kt *(1,540 km/h).*
Service ceiling: 58,725 ft *(17,900 m).*
Range: 1,187 n miles *(2,200 km).*
Role/Weapon systems: Strike fighter for Fleet air defence and anti-shipping strike. Weapons: Fleet air defence role; four AA-1 ('Alkali') beam-riding missiles. Attack; some 1,000 kg of underwing bombs or depth charges, PL-2 missile has anti-ship capability.
UPDATED

Numbers/Type: 30/1 XAC H-6D/H-6X (Tu-16 Badger).
Operational speed: 535 kt *(992 km/h).*
Service ceiling: 40,350 ft *(12,300 m).*
Range: 2,605 n miles *(4,800 km).*
Role/Weapon systems: Three regiments of H-6D bomber and maritime reconnaissance aircraft. Some converted as tankers. H-6s now believed to be out of service and deliveries of new version H-6X, armed with ASM, have begun. Sensors: Search/attack radar; ECM. Weapons: ASV; two underwing anti-shipping missiles of local manufacture, including C-801. Up to five 23 mm cannon; bombs.
UPDATED

H-6X *6/2004 * /* 1042161

Numbers/Type: 200 CAC J-7.
Operational speed: 1,175 kt *(2,175 km/h).*
Service ceiling: 61,680 ft *(18,800 m).*
Range: 804 n miles *(1,490 km).*
Role/Weapon systems: Land-based Fleet air defence fighter with limited strike role against enemy shipping or beachhead. Most are J-7 but also some J-7E in service. Sensors: Search attack radar, some ECM. Weapons: ASV; 500 kg bombs or 36 rockets. Standard; two 30 mm cannon. AD; two 'Atoll' AAMs.
UPDATED

J-7E *6/2002, Ships of the World /* 0554725

PATROL FORCES

Notes: (1) Many patrol craft carry the HN-5 shoulder-launched Chinese version of the SA-N-5 SAM.
(2) More Patrol Craft are listed under Paramilitary vessels at the end of the Chinese section.

3 FAST ATTACK CRAFT—MISSILE (PGGF)

2208-2210

Displacement, tons: 220 full load
Dimensions, feet (metres): 139.8 × 40.0 × 4.9 *(42.6 × 12.2 × 1.5)*
Main machinery: 2 diesels; 6,865 hp *(5.1 MW)*; 2 waterjet propulsors
Speed, knots: 36
Complement: 12
Missiles: 4 SSM.
Guns: 1—30 mm/65 AK 630; 6 barrels; 3,000 rds/min combined to 2 km; 12 missiles.
Weapons control: Optronic director.
Radars: Surface search: Type 362 (ESR-1); I-band.
Navigation: I-band.

Comment: A new coastal patrol craft, the first of which was launched at Qiuxin Shipyard, Shanghai in April 2004. The design is believed to be based on a 42 m hull developed by AMD Marine Consulting, Sydney. This was further progressed by its joint venture company in Guangzhou, Sea Bus International (SBI), into a patrol boat configuration which was selected by the Chinese Navy after a five-year investigation into various platform contenders. The craft has a wave-piercing catamaran hull form and a centre bow. Likely to be of aluminium alloy construction, the design clearly incorporates RCS reduction measures. Following extensive first of class trials, two further craft have been constructed and it is likely that full production has started. Large numbers are expected to replace the ageing patrol boat inventory. Dimensions are based on the original AMD design and other details are speculative.

NEW ENTRY

2210   11/2004 * / 1042171

2208   6/2004 * / 0589733

16 HOUXIN (TYPE 037/1G) CLASS
(FAST ATTACK CRAFT-MISSILE) (PTG)

751-760   **764-769**

Displacement, tons: 478 full load
Dimensions, feet (metres): 203.4 × 23.6 × 7.5 *(62.8 × 7.2 × 2.4)*
Main machinery: 4 China PR 230ZC diesels; 4,000 hp(m) *(2.94 MW)*; 4 shafts
Speed, knots: 28. Range, n miles: 750 at 18 kt
Complement: 71

Missiles: SSM: 4 YJ-1 (Eagle Strike) (C-801) (CSS-N-4 Sardine) (2 twin); active radar homing to 40 km *(22 n miles)* at 0.9 Mach; warhead 165 kg; sea-skimmer. C-802 in due course.
Guns: 4—37 mm/63 (Type 76A) (2 twin); 180 rds/min to 8.5 km *(4.6 n miles)* anti-aircraft; weight of shell 1.42 kg
4—14.5 mm (Type 69) (2 twin); 600 rds/min to 7 km *(3.8 n miles)*.
Countermeasures: ESM/ECM: Intercept and jammer.
Radars: Surface search: Square Tie; I-band.
Fire control: Rice Lamp; I-band.
Navigation: Anritsu Type 723; I-band.

Programmes: First seen in 1991 and built at the rate of up to three per year at Qiuxin and Huangpu Shipyards to replace the Houku class and for export. Building may have stopped in mid-1999.
Structure: This is a missile armed version of the Hainan class. There are some variations in the bridge superstructure in later ships of the class.
Operational: Split between the East and South Sea Fleets.
Sales: Two to Burma in December 1995, two in July 1996 and two in late 1997.

UPDATED

HOUXIN 758   3/2003, Bob Fildes / 0569184

2 HAIJIU CLASS (LARGE PATROL CRAFT) (PC)

688   **697**

Displacement, tons: 490 full load
Dimensions, feet (metres): 210 × 23.6 × 7.2 *(64 × 7.2 × 2.2)*
Main machinery: 4 diesels; 8,800 hp(m) *(6.47 MW)*; 4 shafts
Speed, knots: 28. Range, n miles: 750 at 18 kt
Complement: 72
Guns: 4 China 57 mm/70 (2 twin); 120 rds/min to 12 km *(6.5 n miles)*; weight of shell 6.31 kg.
2 USSR 30 mm/65 (1 twin); 500 rds/min to 5 km *(2.7 n miles)* anti-aircraft; weight of shell 0.54 kg.
A/S mortars: 4 RBU 1200 5-tubed fixed launchers; range 1,200 m; warhead 34 kg.
Depth charges: 2 rails.
Radars: Surface search: Pot Head; I-band.
Fire control: Round Ball; I-band.
Sonars: Stag Ear or Thomson Sintra SS 12.

Comment: A lengthened version of the Hainan class probably used as a prototype for the Houxin class. *697* seen with a Thomson Sintra SS 12 VDS Sonar. Both in East Sea Fleet. Two others have been scrapped.

VERIFIED

HAIJIU CLASS   4/1990, John Mapletoft / 0056768

7 HOUJIAN (OR HUANG) (TYPE 037/2) CLASS
(FAST ATTACK CRAFT—MISSILE) (PTG)

Name	No	Builders	Launched	Commissioned
YANGJIANG	770	Huangpu Shipyard	Jan 1991	May 1991
SHUNDE	771	Huangpu Shipyard	July 1994	Feb 1995
NANHAI	772	Huangpu Shipyard	Feb 1995	Apr 1995
PANYU	773	Huangpu Shipyard	May 1995	July 1995
—	774	Huangpu Shipyard	Sep 1998	Feb 1999
—	775	Huangpu Shipyard	Apr 1999	Nov 1999
—	776	Huangpu Shipyard	—	2000

Displacement, tons: 520 standard
Dimensions, feet (metres): 214.6 × 27.6 × 7.9 *(65.4 × 8.4 × 2.4)*
Main machinery: 3 SEMT-Pielstick 12 PA6 280 diesels; 15,840 hp(m) *(11.7 MW)* sustained; 3 shafts
Speed, knots: 32. Range, n miles: 1,800 at 18 kt
Complement: 75

Missiles: SSM: 6 YJ-1 (Eagle Strike) (C-801) (CSS-N-4 Sardine) (2 triple); inertial cruise; active radar homing to 40 km *(22 n miles)* at 0.9 Mach; warhead 165 kg or C-802 (CSS-N-8 Saccade); range 120 km *(66 n miles)*.
Guns: 2—37 mm/63 (twin) Type 76A; 180 rds/min to 8.5 km *(4.6 n miles)* anti-aircraft; weight of shell 1.42 kg.
4—30 mm/65 (2 twin) Type 69; 500 rds/min to 5 km *(2.7 n miles)*; weight of shell 0.54 kg.
Countermeasures: Decoys: 2 Type 945G 26-barrelled launcher.
ESM: Type 928; intercept.
Weapons control: Type JM-83 optronic director.
Radars: Surface search: Type 381C Rice Shield; G-band.
Fire control: Type 347G Rice Lamp; I-band.
Navigation: Type 765; I-band.

Programmes: First of class laid down in 1989 and built in a very short time. Sometimes called the Huang class.
Modernisation: Some may be fitted with Type 363 search radar and Type 344 (MR 34) fire-control radar rather than Type 347G.
Operational: Based in South Sea Fleet at Hong Kong from mid-1997. One possibly sunk in late 1997.

UPDATED

YANGJIANG   12/2002 *, Ian Edwards / 1042408

98 HAINAN (TYPE 037) CLASS
(FAST ATTACK CRAFT—PATROL) (PC)

Nos 275-285, 290, 302, 305, 609, 610, 618-622, 626-629, 636-643, 646-681, 683-687, 689-692, 695-699, 701, 707, 723-733, 740-742

Displacement, tons: 375 standard; 392 full load
Dimensions, feet (metres): 192.8 × 23.6 × 7.2 *(58.8 × 7.2 × 2.2)*
Main machinery: 4 PCR/Kolomna Type 9-D-8 diesels; 4,000 hp(m) *(2.94 MW)* sustained; 4 shafts
Speed, knots: 30.5. **Range, n miles:** 1,300 at 15 kt
Complement: 78

Missiles: Can be fitted with 4 YJ-1 launchers in lieu of the after 57 mm gun.
Guns: 4 China 57 mm/70 (2 twin); 120 rds/min to 12 km *(6.5 n miles)*; weight of shell 6.31 kg.
 4 USSR 25 mm/60 (2 twin); 270 rds/min to 3 km *(1.6 n miles)* anti-aircraft; weight of shell 0.34 kg.
A/S mortars: 4 RBU 1200 5-tubed fixed launchers; range 1,200 m; warhead 34 kg.
Depth charges: 2 BMB-2 projectors; 2 racks. 18 DCs.
Mines: Rails fitted for 12.
Radars: Surface search: Pot Head or Skin Head; E/F-band.
IFF: High Pole.
Sonars: Stag Ear; hull-mounted; active search and attack; high frequency.
 Thomson Sintra SS 12 (in some); VDS.

Programmes: A larger Chinese-built version of the former Soviet SO 1. Low freeboard. Programme started 1963-64 and continued with new hulls replacing the first ships of the class. There are at least six variants with minor differences.
Structure: Later ships have a tripod or solid foremast in place of a pole and a short stub mainmast. Two trials SS 12 sonars fitted in 1987.
Operational: Divided between the three Fleets.
Sales: Two to Bangladesh, one in 1982 and one in 1985; eight to Egypt in 1983-84; six to North Korea 1975-78; four to Pakistan, two in 1976 and two in 1980; six to Burma in 1991 and four in 1993.

UPDATED

HAINAN 685   *5/2004* / 1042138*

HAINAN 686   *3/2001, Ships of the World / 0126358*

14 HUANGFEN (TYPE 021) (OSA I TYPE) and 1 HOLA CLASS
(FAST ATTACK CRAFT—MISSILE) (PTGF)

3100 series   6100 series   7100 series

Displacement, tons: 171 standard; 205 full load
Dimensions, feet (metres): 126.6 × 24.9 × 8.9 *(38.6 × 7.6 × 2.7)*
Main machinery: 3 Type 42-160 diesels; 12,000 hp(m) *(8.8 MW)* sustained; 3 shafts
Speed, knots: 35. **Range, n miles:** 800 at 30 kt
Complement: 28

Missiles: SSM: 4 HY-2 (CSS-N-3 Seersucker) (2 twin) launchers; active radar or IR homing to 80 km *(43.2 n miles)* at 0.9 Mach; warhead 513 kg.
Guns: 4 USSR 25 mm/60 (2 twin); 270 rds/min to 3 km *(1.6 n miles)* anti-aircraft.
 Replaced in some by 4 USSR 30 mm/65 (2 twin) AK 230.
Radars: Surface search: Square Tie; I-band.
Fire control: Round Ball or Rice Lamp; H/I-band.
IFF: 2 Square Head; High Pole A.

Programmes: First reported in 1985.
Structure: The only Hola class has a radome aft, four launchers, no guns, slightly larger dimensions (137.8 ft *(42 m)* long) and a folding mast. This radome is also fitted in others which carry 30 mm guns. Pennant numbers: Hola, 5100 and the remainder 3100/7100 series.
Operational: China credits this class with a speed of 39 kt. Split between the Fleets. Numbers continue to be reduced.
Sales: Four to North Korea, 1980; four to Pakistan, 1984; four to Bangladesh, 1988; and one more in 1992. Three of a variant were transferred to Yemen in June 1995, delivery having been delayed by the Yemen civil war. A variant called the Houdong class has been built for Iran. Five delivered to Iran in September 1994, five more in March 1996.

VERIFIED

HUANGFEN 6120   *3/2002, Ships of the World / 0529118*

20 HAIQING (TYPE 037/1) CLASS
(FAST ATTACK CRAFT—PATROL) (PC)

710-717, 743-744, 761-763, 786-792

Displacement, tons: 478 full load
Dimensions, feet (metres): 206 × 23.6 × 7.9 *(62.8 × 7.2 × 2.4)*
Main machinery: 4 Chinese PR 230ZC diesels; 4,000 hp(m) *(2.94 MW)* sustained; 4 shafts
Speed, knots: 28. **Range, n miles:** 1,300 at 15 kt
Complement: 71

Guns: 4 China 37 mm/63 (2 twin) Type 76. 4 China 14.5 mm (2 twin) Type 69.
A/S mortars: 2 Type 87 6-tubed launchers.
Radars: Surface search: Anritsu RA 723; I-band.
Sonars: Hull mounted; active search and attack; medium frequency Thomson Sintra SS 12; VDS.

Programmes: Starting building at Qiuxin Shipyard in 1992 and replaced the Hainan class programme. First one completed in November 1993. Production continued at Qingdao, Chongqing and Huangpu as well as Qiuxin.
Structure: Based on the Hainan class, but the large A/S mortars suggest a predominantly ASW role, and this may explain the rapid building rate.
Operational: In service in all three Fleets. Some pennant numbers may have changed.
Sales: One to Sri Lanka in December 1995.

VERIFIED

HAIQING 713   *6/2003, Ships of the World / 0569183*

15 HAIZHUI/SHANGHAI III (TYPE 062/1) CLASS
(COASTAL PATROL CRAFT) (PC)

1203	2327	4339	4341	4345-4348
1204	2329	4340	4342	+3

Displacement, tons: 170 full load
Dimensions, feet (metres): 134.5 × 17.4 × 5.9 *(41 × 5.3 × 1.8)*
Main machinery: 4 Chinese L12-180A diesels; 4,400 hp(m) *(3.22 MW)* sustained; 4 shafts
Speed, knots: 25
Range, n miles: 750 at 17 kt
Complement: 43

Guns: 4 China 37 mm/63 (2 twin); 180 rds/min to 8.5 km *(4.6 n miles)*; weight of shell 1.42 kg.
 4 China 14.5 mm (2 twin) Type 69 or 4 China 25 mm (2 twin).
Radars: Surface search: Pot Head or Anritsu 726; I-band.
Sonars: Stag Ear; hull-mounted; active search; high frequency (in some).

Programmes: First seen in 1992 and built for Chinese use and for export. Sometimes referred to as Shanghai III class when not fitted with ASW equipment.
Structure: Lengthened Shanghai II hull. Inclined pole mast and a pronounced step at the back of the bridge superstructure are recognition features. Much reduced top speed. Some may be equipped with RBU 1200 launchers in place of other armament.
Operational: Based in the North and East Sea Fleets.
Sales: Three of a variant to Tunisia in 1994, three to Sri Lanka in August 1995, three more in May 1996 and three more in August 1998. One to Bangladesh in mid-1996. One to Sierra Leone in 1997.

UPDATED

HAIZHUI 4339   *5/2000, M Declerck / 0103665*

HAIZHUI 1204   *4/2004* / 1042139*

35 SHANGHAI II (TYPE 062) CLASS
(FAST ATTACK CRAFT—GUN) (PC)

Displacement, tons: 113 standard; 134 full load
Dimensions, feet (metres): 127.3 × 17.7 × 5.6 *(38.8 × 5.4 × 1.7)*
Main machinery: 2 Type L-12V-180 diesels; 2,400 hp(m) *(1.76 MW)* (forward); 2 Type 12-D-6 diesels; 1,820 hp(m) *(1.34 MW)* (aft); 4 shafts
Speed, knots: 30. **Range, n miles:** 700 at 16.5 kt on 1 engine
Complement: 38

Guns: 4 China 37 mm/63 (2 twin); 180 rds/min to 8.5 km *(4.6 n miles)*; weight of shell 1.42 kg.
   4 USSR 25 mm/60 (2 twin); 270 rds/min to 3 km *(1.6 n miles)* anti-aircraft; weight of shell 0.34 kg.
   Some are fitted with a twin 57 mm/70, some have a twin 75 mm Type 56 recoilless rifle mounted forward and some have a twin 14.5 mm MG.
Depth charges: 2 projectors; 8 weapons.
Mines: Mine rails can be fitted for 10 mines.
Radars: Surface search: Skin Head ; E/F-band or Pot Head; I-band.
IFF: High Pole.
Sonars: Hull-mounted active sonar or VDS in some.

Programmes: Construction began in 1961 and continued at Shanghai and other yards at rate of about 10 a year for 30 years before being replaced by the Type 062/1G Haizhui class.
Structure: The five versions of this class vary slightly in the outline of their bridges. A few of the class have been reported as fitted with RBU 1200 anti-submarine mortars.
Operational: Evenly divided between the three Fleets. Reported but not confirmed that up to 20 have been converted to sweep mines. Numbers continue to decline.
Sales: Eight to North Vietnam in May 1966, plus Romanian craft of indigenous construction. Seven to Tanzania in 1970-71, six to Guinea, 12 to North Korea, 12 to Pakistan, five to Sri Lanka in 1972, two to Tunisia in 1977, six to Albania, eight to Bangladesh in 1980-82, three to Congo, four to Egypt in 1984, three to Sri Lanka in 1991, two to Tanzania in 1992. Many of the earlier craft have since been deleted. ***VERIFIED***

SHANGHAI II (Sri Lankan colours)   *1992* / 0012772

4 HARBOUR PATROL CRAFT (PBI)

Displacement, tons: 80 full load
Dimensions, feet (metres): 82 × 13.3 × 4.5 *(25 × 4.1 × 1.4)*
Main machinery: 2 diesels; 2 shafts
Speed, knots: 28
Guns: 2—14.5 mm (twin).
Radars: Surface search: I-band.

Comment: Four new patrol craft arrived at Hong Kong on 1 July 1997. There may be more of the class, which are similar to some of the paramilitary patrol craft, but much faster. ***VERIFIED***

HARBOUR PATROL CRAFT   *6/1999, Ships of the World* / 0056772

MINE WARFARE FORCES

Notes: There are also some 50 auxiliary minesweepers of various types including trawlers and motor-driven junks. Up to 20 Shanghai II class, known as the Fushun class, may be used.

0 + 1 MINE COUNTERMEASURES VESSEL (MCMV)

Name	No	Builders	Launched	Commissioned
—	804	Qiuxin Shipyard, Shanghai	Apr 2004	2005

Displacement, tons: 575 full load
Dimensions, feet (metres): 180.4 × 30.5 × 8.5 *(55.0 × 9.3 × 2.6)*
Main machinery: To be announced
Speed, knots: To be announced
Complement: To be announced
Guns: 8—37 mm (4 twin).
Countermeasures: To be announced.
Combat data system: To be announced.
Radars: To be announced.
Sonars: To be announced.

Comment: A new class of mine-countermeasures vessel which is likely to be a successor to the T43 class. Little is known about the capabilities of the vessel. There are 10-15 units expected. ***NEW ENTRY***

804   *1/2005** / 1042170

14 (+ 26 RESERVE) T 43 CLASS (TYPE 010)
(MINESWEEPERS—OCEAN) (MSO)

830	831	832	+11

Displacement, tons: 520 standard; 590 full load
Dimensions, feet (metres): 196.8 × 27.6 × 6.9 *(60 × 8.8 × 2.3)*
Main machinery: 2 PCR/Kolomna Type 9-D-8 diesels; 2,000 hp(m) *(1.47 MW)*; 2 shafts
Speed, knots: 14. **Range, n miles:** 3,000 at 10 kt
Complement: 70 (10 officers)

Guns: 2 or 4 China 37 mm/63 (1 or 2 twin) (3 of the class have a 65 mm/52 forward instead of one twin 37 mm/63); dual purpose; 180 rds/min to 8.5 km *(4.6 n miles)*; weight of shell 1.42 kg.
   4 USSR 25 mm/60 (2 twin); 270 rds/min to 3 km *(1.6 n miles)*.
   4 China 14.5 mm/93 (2 twin); 600 rds/min to 7 km *(3.8 n miles)*.
   Some also carry 1—85 mm/52 Mk 90K; 18 rds/min to 15 km *(8 n miles)*; weight of shell 9.6 kg.
Depth charges: 2 BMB-2 projectors; 20 depth charges.
Mines: Can carry 12-16.
Countermeasures: MCMV; MPT-1 paravanes; MPT-3 mechanical sweep; acoustic and magnetic gear.
Radars: Surface search: Fin Curve or Type 756; F-band.
IFF: High Pole or Yard Rake.
Sonars: Tamir II; hull-mounted; active search and attack; high frequency.

Programmes: Started building in 1956 and continued intermittently until the late 1980s at Wuhan and at Guangzhou.
Structure: Based on the USSR T 43s, some of which transferred in the mid-1950s but have all now been deleted.
Operational: Ten in North Sea Fleet, nine in East Sea Fleet and eight in South Sea Fleet. Some are used as patrol ships with sweep gear removed. Three units reported as having a 65 mm/52 gun forward. Thirteen of the class are in reserve.
Sales: One to Bangladesh in 1995.

UPDATED

T 43 831   *5/2004** / 1042137

T 43 830   *3/2004**, L-G Nilsson / 1042145

1 WOLEI CLASS (MINELAYER) (ML/MST)

814

Displacement, tons: 3,100 full load
Dimensions, feet (metres): 307.7 × 47.2 × 13.1 *(93.8 × 14.4 × 4)*
Main machinery: 4 diesels; 6,400 hp(m) *(4.7 MW)*; 2 shafts
Speed, knots: 18. **Range, n miles:** 7,000 at 14 kt
Complement: 180
Guns: 2 China 57 mm/50 (twin).
  6 China 37 mm/63 (3 twin); 180 rds/min to 8.5 km *(4.6 n miles)*; weight of shell 1.42 kg.
Mines: 300.
Radars: Surface search. Fire control. Navigation.

Comment: Built at Dalian Shipyard and completed successful sea trials in 1988. Resembles the
  deleted Japanese Souya class and may be used as a support ship as well as a minelayer. Based
  in the North Sea Fleet.

UPDATED

WOLEI 814   *6/2002* / 0529145

4 WOSAO (TYPE 082) CLASS (MINESWEEPER—COASTAL) (MSC)

800-803

Displacement, tons: 320 full load
Dimensions, feet (metres): 147 × 22.3 × 7.5 *(44.8 × 6.8 × 2.3)*
Main machinery: 4 M 50 diesels; 4,400 hp(m) *(3.23 MW)*; 4 shafts
Speed, knots: 25. **Range, n miles:** 500 at 15 kt
Complement: 40 (6 officers)
Guns: 4 China 25 mm/60 (2 twin); 270 rds/min to 3 km *(1.6 n miles)*.
Mines: 6.
Countermeasures: Acoustic, magnetic and mechanical sweeps.
Radars: Navigation: China Type 753; I-band.
Sonars: Hull-mounted; active minehunting.

Comment: Building started in 1986. First of class commissioned in 1988 but second, with
  modified bridge structure, not seen until 1997. There are further craft but numbers have not
  been confirmed. Steel hull with low magnetic properties. Equipped with mechanical (Type 316),
  magnetic (Type 317), acoustic (Type 318) and infrasonic (Type 319) sweeps. Based in the East
  Sea Fleet. Pennant numbers changed.

UPDATED

WOSAO 800-803   *3/2004**, L-G Nilsson / 1042146

4 (+ 42 RESERVE) FUTI CLASS (TYPE 312)
(DRONE MINESWEEPERS) (MSD)

Displacement, tons: 47 standard
Dimensions, feet (metres): 68.6 × 12.8 × 6.9 *(20.9 × 3.9 × 2.1)*
Main machinery: Diesel-electric; 1 Type 12V 150C diesel generator; 300 hp(m) *(220 kW)*; 1
  motor; cp prop
Speed, knots: 12. **Range, n miles:** 144 at 12 kt
Complement: 3

Comment: A large number of these craft, similar to the German Troikas, has been built since the
  early 1970s. Fitted to carry out magnetic and acoustic sweeping under remote control up to
  5 km *(2.7 n miles)* from shore control station. Most are kept in reserve.

VERIFIED

DRONE Type 312   *1988, CSSC* / 0056775

AMPHIBIOUS FORCES

Notes: (1) In addition to the ships listed below there are up to 500 minor LCM/LCVP types used to
transport stores and personnel.
(2) Eight Yuchai class (USSR T 4 design) and ten T4 LCMs are still in reserve in the South Sea Fleet.
(3) A 20 m WIG (wing-in-ground effect) craft assembled at Shanghai and completed in late 1997.
Resembles Russian Volga II passenger ferry and may enter naval service if it proves to be reliable.

1 YUDENG (TYPE 073) CLASS (LSM)

No	Builders	Launched	Commissioned
990	Zhonghua Shipyard	Mar 1991	Aug 1994

Displacement, tons: 1,850 full load
Dimensions, feet (metres): 285.4 × 42.7 × 12.5 *(87 × 13 × 3.8)*
Main machinery: 2 diesels; 2 shafts
Speed, knots: 14
Complement: 35
Military lift: 500 troops; 9 tanks
Guns: 2 China 57 mm/50 (twin). 4—25 mm (2 twin).
Radars: Navigation: China Type 753; I-band.

Comment: The only one of the class. Based in the South Sea Fleet. Production may have been for
  export or the design was overtaken by the smaller Wuhu-A class.

VERIFIED

YUDENG 990   *1999, China Shipbuilding Ltd* / 0105535

10 YUTING I (TYPE 072 IV) CLASS (LSTH)

No	Builders	Launched	Commissioned
991	Zhonghua Shipyard, Shanghai	Sep 1991	Sep 1992
934	Zhonghua Shipyard, Shanghai	Apr 1995	Sep 1995
935	Zhonghua Shipyard, Shanghai	July 1995	Dec 1995
936	Zhonghua Shipyard, Shanghai	Dec 1995	May 1996
937	Zhonghua Shipyard, Shanghai	Apr 1996	Aug 1996
908 (ex-938)	Zhonghua Shipyard, Shanghai	Aug 1996	Jan 1997
909 (ex-939)	Zhonghua Shipyard, Shanghai	Nov 1999	Apr 2000
910	Zhonghua Shipyard, Shanghai	May 2000	Dec 2001
939	Zhonghua Shipyard, Shanghai	Apr 2001	Aug 2001
940	Zhonghua Shipyard, Shanghai	Dec 2001	Apr 2002

Displacement, tons: 3,770 standard; 4,800 full load
Dimensions, feet (metres): 393.7 × 52.5 × 10.5 *(120 × 16 × 3.2)*
Main machinery: 2 diesels; 2 shafts
Speed, knots: 17. **Range, n miles:** 3,000 at 14 kt
Complement: 120
Military lift: 250 troops; 10 tanks; 4 LCVP
Guns: 6 China 37 mm/63 (3 twin); 180 rds/min to 8.5 km *(4.6 n miles)*; weight of shell 1.42 kg.
Radars: Navigation: 2 China Type 753; I-band.
Helicopters: Platform for 2 medium.

Comment: To augment amphibious lift capabilities and provide helicopter lift. Bow and bridge
  structures are very similar to the Yukan class but there is a large helicopter deck. *934-937* and
  991 based in South Sea Fleet. *908-910* based in East Sea Fleet.

UPDATED

YUTING 909   *5/2004** / 1042136

YUTING 908   *11/2004*, Ships of the World* / 1042143

9 YUTING II CLASS (LSTH)

Name	Builders	Launched	Commissioned
913	Zhonghua Shipyard, Shanghai	Mar 2003	Oct 2003
911	Dalian Shipyard	May 2003	2003
992	Wuhan Shipyard	June 2003	2003
993	Zhonghua Shipyard, Shanghai	July 2003	Jan 2004
912	Dalian Shipyard	Sep 2003	2004
994	Wuhan Shipyard	2004	2004
995	Zhonghua Shipyard, Shanghai	2004	2004
996	Dalian Shipyard	2004	2004
997	Wuhan Shipyard	2004	2004

Displacement, tons: 3,770 standard; 4,800 full load
Dimensions, feet (metres): 393.7 × 53.8 × 10.5 *(120 × 16.4 × 3.2)*
Main machinery: 2 diesels; 2 shafts
Speed, knots: 17. **Range, n miles:** 3,000 at 14 kt
Complement: 120
Military lift: 250 troops; 10 tanks; 4 LCVP
Guns: To be announced.
Radars: Navigation: 2 China Type 753; I-band.
Helicopters: Platform for 2 medium.

Comment: Details are speculative but reported to be an improved version of the Yuting I class with similar dimensions. Design differences include modifications to the stern, including the ramp and a taller funnel. A tunnel in the centre of the superstructure connects the main and after decks. With construction underway at three shipyards, this appears to be a very active programme. At least ten are expected.

UPDATED

YUTING II 913   *3/2004 *, L-G Nilsson /* 0580524

YUTING II 995   *4/2004 *, Ships of the World /* 1042147

7 YUKAN (TYPE 072) CLASS (LST)

927	928	929	930	931	932	933

Displacement, tons: 3,110 standard; 4,170 full load
Dimensions, feet (metres): 393.6 × 50 × 9.5 *(120 × 15.3 × 2.9)*
Main machinery: 2 Type 12E 390 diesels; 14,400 hp(m) *(10.6 MW)* sustained; 2 shafts
Speed, knots: 18. **Range, n miles:** 3,000 at 14 kt
Complement: 109
Military lift: 200 troops; 10 tanks; 2 LCVP; total of 500 tons
Guns: 2 China 57 mm/50 (1 twin); 120 rds/min to 12 km *(6.5 n miles)*; weight of shell 6.31 kg.
4, 6 or 8—37 mm (2, 3 or 4 twin); 180 rds/min to 8.5 km *(4.6 n miles)*; weight of shell 1.42 kg.
4—25 mm/60 (2 twin) (some also have 4—25 mm (2 twin) mountings amidships above the tank deck); 270 rds/min to 3 km *(1.6 n miles)*.
Radars: Navigation: 2 China Type 753; I-band.

Comment: First completed in 1980 at Wuhan Shipyard. Building appeared to terminate in November 1995. Bow and stern ramps fitted. Carry two LCVPs. Bow ramp maximum load 50 tons, stern ramp 20 tons. Five based in the East and two in South Sea Fleets.

UPDATED

YUKAN 930   *5/2004 * /* 1042133

YUKAN 930   *3/2001, Ships of the World /* 0126363

22 YULIANG (TYPE 079) CLASS (LSM)

Displacement, tons: 1,100 full load
Dimensions, feet (metres): 206.7 × 32.8 × 7.9 *(63 × 10 × 2.4)*
Main machinery: 2 diesels; 2 shafts
Speed, knots: 14
Complement: 60
Military lift: 3 tanks
Guns: 4—25 mm/60 (2 twin); 270 rds/min to 3 km *(1.6 n miles)*.
2 BM 21 MRL rocket launchers; range about 9 km *(5 n miles)*.
Radars: Navigation: Fin Curve; I-band.

Comment: Production started in 1980 in three or four smaller shipyards. Numbers have been overestimated in the past and production stopped in favour of Yuhai class. Four in the North Sea Fleet, remainder based in the South Sea Fleet.

UPDATED

YULIANG 1122   *5/2004 * /* 1042135

10 YUNSHU CLASS (LSM)

No	Builders	Launched	Commissioned
946	Hudong Zhonghua Shipyard, Shanghai	June 2003	2004
947	Qingdao Naval Dockyard	July 2003	2004
—	Lushun Shipyard	July 2003	2004
—	Wuhu Shipyard	July 2003	2004
—	Lushun Shipyard	Oct 2003	2004
—	Hudong Zhonghua Shipyard, Shanghai	Dec 2003	2004
949	Lushun Shipyard	Feb 2004	2004
—	Wuhu Shipyard	2004	2004
—	Qingdao Naval Dockyard	2004	2004
—	Hudong Zhonghua Shipyard, Shanghai	Mar 2004	2004

Displacement, tons: 1,460 standard; 1,850 full load
Dimensions, feet (metres): 285.4 × 41.3 × 7.4 *(87.0 × 12.6 × 2.25)*
Main machinery: 2 diesels; 2 shafts
Speed, knots: 17. **Range, n miles:** 1,500 at 14 kt
Complement: 70
Military lift: 6 tanks or 12 trucks or 250 tons dry stores
Guns: 2—57 mm.
Radars: Navigation: I-band.

Comment: A new class of LSM, based on the Yudeng class, under construction at Zhonghua, Wuhu, Qingdao and Lushun. Series production at four shipyards suggests that a large class is planned.

NEW ENTRY

YUNSHU 949   *3/2004 *, L-G Nilsson /* 0580525

YUNSHU 949   *6/2004 * /* 1042162

8 YUBEI CLASS (LCU)

No	Builders	Launched	Commissioned
3128	Qingdao Naval Dockyard	Sep 2003	2004
3315	Zhanjiang Shipyard North	2003	2004
3232	Shanghai Shipyard International	Sep 2003	2004
3129	Qingdao Naval Dockyard	Dec 2003	2004
3316	Dinghai Naval Dockyard	Sep 2003	2004
—	Dinghai Naval Dockyard	Nov 2003	2004
3318	Dinghai Naval Dockyard	Jan 2004	2004
—	Qingdao Naval Dockyard	2004	2004

Displacement, tons: To be announced
Dimensions, feet (metres): 213.2 × 36.1 × 88.6 *(65.0 × 11.0 × 2.7)*
Main machinery: 2 diesels; 2 shafts
Speed, knots: To be announced
Complement: To be announced
Military lift: 10 tanks; 150 troops
Guns: To be announced.
Radars: To be announced.

Comment: A new class of LCU under construction at Qingdao, Zhanjiang, Shanghai and Dinghai. Series production at four shipyards suggests that a large class is planned.
NEW ENTRY

YUBEI 3315   *8/2003* / 1042164*

YUBEI 3232   *7/2004*, Ian Edwards / 1042407*

13 YUHAI (TYPE 074) (WUHU-A) CLASS (LSM)

481	6562	7579	7595	+9

Displacement, tons: 799 full load
Dimensions, feet (metres): 191.6 × 34.1 × 8.9 *(58.4 × 10.4 × 2.7)*
Main machinery: 2 MAN-8L 20/27 diesels; 4,900 hp(m) *(3.6 MW)*; 2 shafts
Speed, knots: 14
Complement: 56
Military lift: 2 tanks; 250 troops
Guns: 2—25 mm/80 (1 twin).
Radars: Navigation: I-band.

Comment: First one completed in Wuhu Shipyard in 1995. One sold to Sri Lanka in December 1995. Three based in the North, four in the East and six in the South Sea Fleet. Further units are not expected.
UPDATED

YUHAI 7595   *12/2002*, Ian Edwards / 1042406*

120 YUNNAN CLASS (TYPE 067) (LCU)

Displacement, tons: 135 full load
Dimensions, feet (metres): 93.8 × 17.7 × 4.9 *(28.6 × 5.4 × 1.5)*
Main machinery: 2 diesels; 600 hp(m) *(441 kW)*; 2 shafts
Speed, knots: 12. **Range, n miles:** 500 at 10 kt
Complement: 12
Military lift: 46 tons
Guns: 4—14.5 mm (2 twin) MGs.
Radars: Navigation: Fuji; I-band.

Comment: Built in China 1968-72 although a continuing programme was reported in 1982. Pennant numbers in 3000 series (3313, 3321, 3344 seen). 5000 series (5526 seen) and 7000 series (7566 and 7568 seen). The majority of the operational hulls are based in the South Sea Fleet. One to Sri Lanka in 1991 and a second in 1995. Estimation of numbers is difficult but most are believed to be in reserve or in non-naval service. Some may have 12.7 mm MGs. Twelve in the East Sea Fleet, remainder in the South.
VERIFIED

YUNNAN 3003   *8/2000, Hachiro Nakai / 0103675*

1 YUDAO CLASS (TYPE 073) (LSM)

965

Displacement, tons: 1,650 full load
Dimensions, feet (metres): 253.9 × 34.1 × 9.8 *(77.4 × 10.4 × 3)*
Speed, knots: 18. **Range, n miles:** 1,000 at 16 kt
Complement: 60
Guns: 4—25 mm/60 (2 twin); 270 rds/min to 3 km *(1.6 n miles)*.
Radars: Navigation: Fin Curve; I-band.

Comment: First entered service in early 1980s. *965* is the only one left and is in the East Fleet.
VERIFIED

YUDAO 965   *6/1995 / 0056781*

20 YUCH'IN (TYPE 068/069) CLASS (LCM)

Displacement, tons: 58 standard; 85 full load
Dimensions, feet (metres): 81.2 × 17.1 × 4.3 *(24.8 × 5.2 × 1.3)*
Main machinery: 2 Type 12V 150C diesels; 600 hp(m) *(441 kW)*; 2 shafts
Speed, knots: 11.5. **Range, n miles:** 450 at 11.5 kt
Complement: 12
Military lift: Up to 150 troops
Guns: 4—14.5 mm (2 twin) MGs.

Comment: Built in Shanghai 1962-72. Smaller version of Yunnan class with a shorter tank deck and longer poop deck. Primarily intended for personnel transport. Based in South Sea Fleet. Six sold to Bangladesh and two to Tanzania in 1995.
VERIFIED

YUCH'IN 4507   *5/2000, van Ginderen Collection / 0103674*

10 JINGSAH II CLASS (HOVERCRAFT) (UCAC)

452	+9

Displacement, tons: 70
Dimensions, feet (metres): 72.2 × 26.2 *(22 × 8)*
Main machinery: 2 propulsion motors; 2 lift motors
Speed, knots: 55
Military lift: 15 tons
Guns: 4—14.5 mm (2 twin) MGs.

Comment: The prototype was built at Dagu in 1979. This may now have been scrapped and been superseded by this improved version which has a bow door for disembarkation. Numbers are uncertain and may be conditional on progress with WIG craft.
VERIFIED

JINGSAH II   *1993, Ships of the World / 0056783*

TRAINING SHIPS

1 SHICHANG CLASS (HSS/AHH)

Name	No	Builders	Launched	Commissioned
SHICHANG	82	Qiuxin, Shanghai	Apr 1996	27 Jan 1997

Displacement, tons: 10,000 full load
Dimensions, feet (metres): 393.7 × 59.1 × 23 *(120 × 18 × 7)*
Main machinery: 2 diesels; 2 shafts
Speed, knots: 17.5. **Range, n miles:** 8,000 at 17 kt
Complement: 170 plus 200 trainees
Military lift: 300 containers
Helicopters: 2 Zhi-9A Haitun.

Comment: China's first air training ship described officially as a defence mobilisation vessel which can be used for civilian freight, for helicopter or navigation training, or as a hospital ship. The vessel looks like a scaled down version of the UK *Argus* with the bridge superstructure forward and an after funnel on the starboard side of the flight deck. There are two landing spots. Based in the South Sea Fleet and deployed to Australia in mid-1998.

VERIFIED

SHICHANG   *5/1998, Sattler/Steele /* 0017738

SHICHANG   *5/1998, RAN /* 0017739

1 DAXIN CLASS (AXH)

Name	No	Builders	Launched	Commissioned
ZHENGHE	81	Qiuxin, Shanghai	12 July 1986	27 Apr 1987

Displacement, tons: 5,470 full load
Dimensions, feet (metres): 426.5 × 52.5 × 15.7 *(130.0 × 16.0 × 4.8)*
Main machinery: 2 6PC2-5L diesels; 7,800 hp(m) *(5.73 MW)*; 2 shafts
Speed, knots: 15. **Range, n miles:** 5,000 at 15 kt
Complement: 170 plus 30 instructors plus 200 Midshipmen
Guns: 4 China 57 mm/70 (2 twin). 4—30 mm AK 230 (2 twin). 4—12.7 mm MGs.
A/S mortars: 2 FQF 2500 fixed 12-tubed launchers; range 1,200 m; warhead 34 kg.
Radars: Air/surface search: Eye Shield; E-band.
Surface search: China Type 756; I-band.
Navigation: Racal Decca 1290; I-band.
Fire control: Round Ball; I-band.
Sonars: Echo Type 5; hull-mounted; active; high frequency.
Helicopters: Platform only.

Comment: Resembles a small cruise liner. Subordinate to the Naval Academy and replaced *Huian*. Based in the North Sea Fleet.

UPDATED

ZHENGHE   *9/2000, B Lemachko /* 0126258

AUXILIARIES

Notes: (1) There is a water-tanker with similar characteristics to the Fuzhou class with pennant number 1101.
(2) There is a water tanker of unknown dimensions with pennant number 1102.

1102   *9/2001*, Ian Edwards /* 1042405

1101   *7/2003*, Ian Edwards /* 1042404

2 FUQING CLASS (REPLENISHMENT SHIPS) (AORH)

TAICANG 575	FENGCANG (ex-*Dongyun*) 882 (ex-615)	

Displacement, tons: 7,500 standard; 21,750 full load
Dimensions, feet (metres): 552 × 71.5 × 30.8 *(168.2 × 21.8 × 9.4)*
Main machinery: 1 Sulzer 8RL B66 diesel; 15,000 hp(m) *(11 MW)* sustained; 1 shaft
Speed, knots: 18. **Range, n miles:** 18,000 at 14 kt
Complement: 130 (24 officers)
Cargo capacity: 10,550 tons fuel; 1,000 tons dieso; 200 tons feed water; 200 tons drinking water; 4 small cranes
Guns: 8—37 mm (4 twin) (fitted for but not with).
Radars: Navigation: Fin Curve or Racal Decca 1290; I-band.
Helicopters: Platform for 1 medium.

Comment: Operational in late 1979. This is the first class of ships built for underway replenishment in the Chinese Navy. Helicopter platform but no hangar. Both built at Dalian. Two liquid replenishment positions each side with one solid replenishment position each side by the funnel. A third of the class *Hongcang* (X 950) was converted to merchant use in 1989 and renamed *Hai Lang*, registered at Dalian. A fourth (X 350) was sold to Pakistan in 1987. One based in the North and one in the East. Both ships deployed out of area in 2001 and *Fengcang* appears to have a command role.

UPDATED

FENGCANG   *3/2004*, L-G Nilsson /* 1042154

TAICANG   *10/2001, Mick Prendergast /* 0126257

1 NANYUN CLASS (REPLENISHMENT SHIP) (AORH)

Name	No	Builders	Launched	Commissioned
NANCANG (ex-*Vladimir Peregudov*)	885 (ex-953)	Kherson/Dalian	Apr 1992	2 June 1996

Displacement, tons: 37,000 full load
Measurement, tons: 28,750 dwt
Dimensions, feet (metres): 586.9 × 83 × 36.1 *(178.9 × 25.3 × 11)*
Main machinery: 1 B&W diesel; 11,600 hp(m) *(8.53 MW)*; 1 shaft
Speed, knots: 16
Complement: 125
Cargo capacity: 9,630 tons fuel
Helicopters: 1 Super Frelon.

Comment: Sometimes referred to as Fusu class. One of a class of 11 built at Kherson Shipyard, Crimea. Laid down in January 1989. Sailed from Ukraine to Dalian Shipyard in 1993. Completed fitting out in China and joined the South Sea Fleet. RAS rigs on both sides and stern refuelling. Similar to Indian *Jyoti* but with better helicopter facilities. Deployed out of area with DDG 167 in 2000.

VERIFIED

NANCANG   8/2000, Robert Pabst / 0103677

NANCANG   5/1998, Sattler/Steele / 0017742

6 QIONGSHA CLASS (4 AP + 2 AH)

Y 830	Y 831	Y 832	Y 833	Y 834	Y 835

Displacement, tons: 2,150 full load
Dimensions, feet (metres): 282.1 × 44.3 × 13.1 *(86 × 13.5 × 4)*
Main machinery: 3 SKL 8 NVD 48 A-2U diesels; 3,960 hp(m) *(2.91 MW)* sustained; 3 shafts
Speed, knots: 16
Complement: 59
Military lift: 400 troops; 350 tons cargo
Guns: 8 China 14.5mm/93 (4 twin); 600 rds/min to 7 km *(3.8 n miles)*.
Radars: Navigation: Fin Curve; I-band.

Comment: Personnel attack transports begun about 1980. Previous numbers of this class were overestimated. All South Sea Fleet. Has four sets of davits, light cargo booms serving forward and aft. No helicopter pad. Twin funnels. Carries a number of LCAs. *Y 833* and *Y 834* converted to Hospital Ships (AH) and painted white.

VERIFIED

QIONGSHA 831   2/1999 / 0056784

3 DAJIANG CLASS (SUBMARINE SUPPORT SHIPS) (ASRH)

CHANGXINGDAO 861 (ex-J 121)   CHONGMINGDAO 862 (ex-J 302)   YONGXINGDAO J 506

Displacement, tons: 11,975 full load
Dimensions, feet (metres): 511.7 × 67.2 × 22.3 *(156 × 20.5 × 6.8)*
Main machinery: 2 MAN K9Z60/105E diesels; 9,000 hp(m) *(6.6 MW)*; 2 shafts
Speed, knots: 20
Complement: 308
Guns: Light MGs. Can carry 6—37 mm (3 twin).
Radars: Surface search: Eye Shield; E-band.
Navigation: 2 Fin Curve; I-band.
Helicopters: 2 Aerospatiale SA 321G Super Frelon.

Comment: Submarine support and salvage ships built at Shanghai. First launched in mid-1973, operational in 1976. *Yongxingdao* has a smoke deflector on funnel. Provision for DSRV on forward well-deck aft of launching crane. A fourth and fifth of the class are listed under *Research Ships*. Foremast on *Yongxingdao* suggests long-range communications capability, possibly for submarine command. One based in each Fleet.

VERIFIED

CHONGMINGDAO (old number)   8/2000, Hachiro Nakai / 0103678

YONGXINGDAO   6/1996 / 0012206

2 FUCHI CLASS (REPLENISHMENT SHIPS) (AORH)

Name	No	Builders	Laid down	Launched	Commissioned
FUCHI	886	Hudong Shipyard, Shanghai	2002	29 Mar 2003	2004
—	887	Guangzhou Shipyard	—	June 2003	2004

Displacement, tons: 23,000 full load
Dimensions, feet (metres): 584.0 × 82.0 × 29.5 *(178.0 × 25.0 × 9.0)*
Main machinery: 2 SEMT-Pielstick diesels; 24,000 hp *(17.9 MW)*; 2 shafts
Speed, knots: 19. **Range, n miles:** 10,000 at 14 kt
Complement: 130

Cargo capacity: 10,500 tons fuel, 250 tons of water, 680 tons of ammunition and stores
Guns: 8—37 mm (4 twin).
Radars: To be announced.
Helicopters: Platform for 1 medium.

Comment: New ships which bear a marked resemblance to Type R22T Similan class tanker built for Thailand in 1996. Fitted with two RAS stations (one liquids, one solids) on each side. In view of the Chinese Navy's increasing requirement for underway replenishment, further ships are likely.

UPDATED

887   6/2003 * / 1042163

FUCHI   6/2004 * / 0583295

1 DAZHI CLASS (SUBMARINE SUPPORT SHIP) (AS)

DAZHI 920

Displacement, tons: 5,600 full load
Dimensions, feet (metres): 350 × 50 × 20 *(106.7 × 15.3 × 6.1)*
Main machinery: Diesel-electric; 2 diesel generators; 3,500 hp(m) *(2.57 MW)*; 2 shafts
Speed, knots: 14. **Range, n miles:** 6,000 at 14 kt
Complement: 290
Cargo capacity: 500 tons dieso
Guns: 4 China 37 mm/63 (2 twin). 4—25 mm/60 (2 twin).
Radars: Navigation: Fin Curve; I-band.

Comment: Built at Hudong, Shanghai 1963-65. Has four electrohydraulic cranes. Carries large stock of torpedoes and stores. Based in East Sea Fleet but ship has not been seen at sea in recent years and may have been withdrawn from service. ***VERIFIED***

DAZHI *(not to scale)*

1 DADONG and 1 DADAO CLASSES (SALVAGE SHIPS) (ARS)

304 +1

Displacement, tons: 1,500 full load
Dimensions, feet (metres): 269 × 36.1 × 8.9 *(82 × 11 × 2.7)*
Main machinery: 2 diesels; 7,400 hp(m) *(5.44 MW)*; 2 shafts
Speed, knots: 18
Complement: 150
Guns: 4—25 mm/80 (2 twin).
Radars: Navigation: Type 756; F-band.

Comment: *304* reported to have been built at Hudong. Has a large and conspicuous crane aft. Principal role is wreck location and salvage. A similar ship called the Dadao class was launched in January 1986. This vessel is slightly larger (84 × 12.4 m) and is civilian manned. Both ships are in the East Sea Fleet. ***VERIFIED***

DADAO class *1989, Gilbert Gyssels*

2 DAZHOU CLASS (SUBMARINE TENDERS) (ASL)

502 504

Displacement, tons: 1,100 full load
Dimensions, feet (metres): 259.2 × 31.2 × 8.5 *(79 × 9.5 × 2.6)*
Main machinery: 2 diesels; 2 shafts
Speed, knots: 18
Complement: 130
Guns: 2 China 37 mm/63 (twin). 4—14.5 mm/93 (2 twin).
Radars: Navigation: Fin Curve; I-band.

Comment: Built in 1976-77. One in South Sea Fleet, one in the North and both have been used as AGIs. ***VERIFIED***

DAZHOU 504 *12/1990, DTM*

5 DALANG CLASS (SUBMARINE SUPPORT SHIPS) (AS)

503 122 911 428 332

Displacement, tons: 3,700 standard; 4,200 full load
Dimensions, feet (metres): 367 × 47.9 × 14.1 *(111.9 × 14.6 × 4.3)*
Main machinery: 2 diesels; 4,000 hp(m) *(2.94 MW)*; 2 shafts
Speed, knots: 16. **Range, n miles:** 8,000 at 14 kt
Complement: 180
Guns: 2—25 mm/80 (1 twin) or 2—14.5 mm/93 (1 twin).
Radars: Navigation: Fin Curve; I-band.

Comment: Details given are for the first two built at Guangzhou Shipyard. *503* commissioned November 1975, *122* in 1986. *911* built at Wuhu Shipyard, commissioning in late 1986, and *428* was launched in June 1996. Sometimes called Dalang I and Dalang II classes. Have been used as AGIs. Upper deck modifications (which may include a decompression chamber) have been incorporated in *332*. ***UPDATED***

DALANG 911 *3/2004*, L-G Nilsson / 1042151*

DALANG 332 *5/2000, M Declerck / 0103679*

2 DSRV (SALVAGE SUBMARINES) (DSRV)

Displacement, tons: 35 full load
Dimensions, feet (metres): 48.9 × 8.5 × 8.5 *(14.9 × 2.6 × 2.6)*
Main machinery: 2 silver-zinc batteries; 1 mortar; 1 shaft
Speed, knots: 4. **Range, n miles:** 40 at 2 kt
Complement: 3

Comment: First tested in 1986 and can be carried on large salvage ships. Capable of 'wet' rescue at 200 m and of diving to 600 m. Capacity for six survivors. Underwater TV, high-frequency active sonar and a manipulator arm are all fitted. Life support duration is 1,728 man-hours. An upgrade of submarine rescue capabilities may be planned following attendance at international conferences in 2001 and talks with industry. Up to three modern DSRV may be required. ***VERIFIED***

DSRV *1991, CSSC / 0056786*

2 YANTAI CLASS SUPPLY SHIPS (AK)

800 801

Displacement, tons: 3,330 full load
Dimensions, feet (metres): 393.6 × 50 × 9.8 *(120 × 15.3 × 3)*
Main machinery: 2 diesels; 9,600 hp(m) *(7.06 MW)*; 2 shafts
Speed, knots: 17. **Range, n miles:** 3,000 at 16 kt
Complement: 100
Guns: 2 China 37 mm/63 (twin).
Radars: Navigation: Type 756; I-band.

Comment: First seen in 1992. Appears to be based on a landing ship design but without a bow door. Fitted with cargo-handling cranes fore and aft. A ship with pennant number 938 has been reported unloading missile containers but it is not known whether this is an additional ship. Based in South Sea Fleet. ***VERIFIED***

YANTAI 800 *6/1996 / 0056789*

For details of the latest updates to *Jane's Fighting Ships* online and to discover the additional information available exclusively to online subscribers please visit
jfs.janes.com

2 DAYUN (TYPE 904) CLASS SUPPLY SHIPS (AKH)

883 (ex-951)   884 (ex-952)

Displacement, tons: 8,500 full load
Dimensions, feet (metres): 407.5 × 42 × 12.5 *(124.2 × 12.8 × 3.8)*
Main machinery: 2 diesels; 9,000 hp(m) *(6.6 MW)*; 2 shafts
Speed, knots: 22
Complement: 240
Guns: 4—37 mm/63 (2 twin). 4—25 mm/80 (2 twin).
Radars: Navigation: 2 Type 756; I-band.
Helicopters: 2 SA 321 Super Frelon.

Comment: First of class completed at Hudong Shipyard in March 1992, second in August 1992. Four landing craft are embarked. Both based in South Sea Fleet. A reported third of class was in fact the first of the larger Nanyun class.

UPDATED

DAYUN 884 (old pennant number)   *12/1998, Ships of the World* / 0056788

13 DANLIN CLASS SUPPLY SHIPS (AK/AOT)

| 531 | 591 | 592 | 594 | 794 | +3 |
| 827 | 834 | 835 | 972 | 975 | |

Displacement, tons: 1,290 full load
Dimensions, feet (metres): 198.5 × 29.5 × 13.1 *(60.5 × 9 × 4)*
Main machinery: 1 USSR/PRC Type 6DRN 30/50 diesel; 750 hp(m) *(551 kW)*; 1 shaft
Speed, knots: 15
Complement: 35
Cargo capacity: 750-800 tons
Guns: 4—25 mm/80 (2 twin). 4—14.5 mm (2 twin).
Radars: Navigation: Fin Curve or Skin Head; I-band.

Comment: Built in China in early 1960-62. The six AKs have refrigerated stores capability and serve in the South Sea Fleet. The seven AOTs are split between the Fleets. Not all are armed.

VERIFIED

DANLIN 794   *5/1992, Henry Dodds* / 0056790

3 DANDAO CLASS (AK/AOT)

599   802   803

Displacement, tons: 1,600 full load
Dimensions, feet (metres): 215.6 × 41 × 13 *(65.7 × 12.5 × 4)*
Main machinery: 1 diesel; 1 shaft
Speed, knots: 12
Complement: 40
Guns: 4 China 37 mm/63 (2 twin). 4 China 14.5 mm/93 (2 twin).
Radars: Navigation: Fin Curve; I-band.

Comment: Built in the late 1970s. Similar to the Danlin class. Two in the North and one in the East Sea Fleet.

VERIFIED

DANDAO 802   *5/2000, van Ginderen Collection* / 0126255

5 HONGQI CLASS (AK)

| 443 | 528 | 755 | 756 | 771 |

Displacement, tons: 1,950 full load
Dimensions, feet (metres): 203.4 × 39.4 × 14.4 *(62 × 12 × 4.4)*
Main machinery: 1 diesel; 1 shaft
Speed, knots: 14. **Range, n miles:** 2,500 at 11 kt
Complement: 35
Guns: 4 China 25/80 (2 twin).

Comment: Used to support offshore military garrisons. A further ship, L 202, appears to be similar but carries no armament. Others of this type in civilian use. Three in the North, two in the East Sea Fleet.

VERIFIED

HONGQI 755   *3/2003, Bob Fildes* / 0569175

10 LEIZHOU CLASS (AWT/AOT)

| 728 | 755 | 793 | 826 | 973 |
| 736 | 792 | 823 | 828 | 974 |

Displacement, tons: 900 full load
Dimensions, feet (metres): 173.9 × 32.2 × 10.5 *(53 × 9.8 × 3.2)*
Main machinery: 1 diesel; 500 hp(m) *(367 kW)*; 1 shaft
Speed, knots: 12. **Range, n miles:** 1,200 at 10 kt
Complement: 25-30
Cargo capacity: 450 tons
Guns: 4—14.5 mm/93 (2 twin).
Radars: 2 navigation; I-band.

Comment: Built in late 1960s at Qingdao and Wudong. Split between the Fleets. Some have been converted to carry water, others carry oil. Many deleted or in civilian use.

UPDATED

LEIZHOU 755   *7/2004*, Ian Edwards* / 1042403

16 FULIN CLASS (REPLENISHMENT SHIPS) (AOT)

560	589	620	629
563	606	623	630
582	607	625	632
583	609	628	633

Displacement, tons: 2,300 standard
Dimensions, feet (metres): 216.5 × 42.6 × 13.1 *(66 × 13 × 4)*
Main machinery: 1 diesel; 600 hp(m) *(441 kW)*; 1 shaft
Speed, knots: 10. **Range, n miles:** 1,500 at 8 kt
Complement: 30
Guns: 4—14.5 mm/93 (2 twin).
Radars: Navigation: Fin Curve; I-band.

Comment: A total of 20 of these ships built at Hudong, Shanghai, beginning 1972. Naval ships painted grey. Both in South Sea Fleet. Many others of the class are civilian but may carry pennant numbers.

UPDATED

FULIN 632   *3/2004*, L-G Nilsson* / 1042150

2 SHENGLI CLASS (AOT)

620 621

Displacement, tons: 3,300 standard; 4,950 full load
Dimensions, feet (metres): 331.4 × 45.3 × 18 *(101 × 13.8 × 5.5)*
Main machinery: 1 6 ESDZ 43/82B diesel; 2,600 hp(m) *(1.91 MW)*; 1 shaft
Speed, knots: 14. **Range, n miles:** 2,400 at 11 kt
Complement: 48
Cargo capacity: 3,400 tons dieso
Guns: 2—37 mm/63 (twin). 4—25 mm/80 (2 twin).
Radars: Navigation: Fin Curve; I-band.

Comment: Built at Hudong SY, Shanghai in late 1970s. Others of the class in commercial service.
UPDATED

SHENGLI 621 *7/2004*, Ian Edwards /* 1042402

3 JINYOU CLASS (AOT)

622 625 675

Displacement, tons: 4,800 full load
Dimensions, feet (metres): 324.8 × 104.3 × 187.0 *(99.0 × 31.8 × 5.7)*
Main machinery: 1 SEMT-Pielstick 8PC2.2L diesel; 3,000 hp *(2.24 MW)*; 1 shaft
Speed, knots: 15. **Range, n miles:** 4,000 at 10 kt
Complement: 40
Radars: Navigation: I-band.

Comment: Built by Kanashashi Shipyard, Japan and entered service 1989-90.
NEW ENTRY

JINYOU 625 *6/2004*, Ian Edwards /* 1042401

9 FUZHOU CLASS (AOT/AWT)

Displacement, tons: 2,100 full load
Dimensions, feet (metres): 208.3 × 41.3 × 12.5 *(63.5 × 12.6 × 3.8)*
Main machinery: 1 diesel; 600 hp(m) *(441 kW)*; 1 shaft
Speed, knots: 11
Complement: 35
Cargo capacity: 600 tons
Guns: 4—25 mm/80 (2 twin). 4—14.5 mm/93 (2 twin).
Radars: Navigation: Fin Curve; I-band.

Comment: Built 1964-70. At least 18 others of the class are civilian and used as transport oilers but may carry pennant numbers. Not all are armed.
VERIFIED

FUZHOU 1104 *4/2003, Bob Fildes /* 0569173

5 GUANGZHOU CLASS (AOTL/AWTL)

412 555 558 645 +1

Displacement, tons: 530 full load
Dimensions, feet (metres): 160.8 × 24.6 × 9.8 *(49 × 7.5 × 3)*
Main machinery: 1 diesel; 1 shaft
Speed, knots: 10
Complement: 19
Guns: 4—14.5 mm/93 (2 twin).

Comment: Coastal tankers built in the 1970s and 1980s. At least 18 others of the class are civilian but may carry pennant numbers.
UPDATED

GUANGZHOU 645 *7/2004*, Ian Edwards /* 1042400

7 YANNAN CLASS (BUOY TENDER) (ABU)

124 263 463 982 983 B-22 B-25

Displacement, tons: 1,750 standard
Dimensions, feet (metres): 237.2 × 38.7 × 13.1 *(72.3 × 11.8 × 4)*
Main machinery: 2 diesels; 2,640 hp(m) *(1.94 MW)*; 2 shafts
Speed, knots: 12
Complement: 95
Radars: Navigation: Fin Curve; I-band.

Comment: Built 1978-79; commissioned 1980.
UPDATED

YANNAN B-25 *3/2004*, L-G Nilsson /* 1042153

4 YEN PAI CLASS (ADG)

735 736 746 863

Displacement, tons: 746 standard
Dimensions, feet (metres): 213.3 × 29.5 × 8.5 *(65 × 9 × 2.6)*
Main machinery: Diesel-electric; 2 12VE 230ZC diesels; 2,200 hp(m) *(1.62 MW)*; 2 ZDH-99/57 motors; 2 shafts
Speed, knots: 16. **Range, n miles:** 800 at 15 kt
Complement: 55
Guns: 4—37 mm/63 (2 twin). 4—25 mm/80 (2 twin).
Radars: Navigation: Type 756; I-band.

Comment: Enlarged version of T 43 MSF with larger bridge and funnel amidships. Reels on quarterdeck for degaussing function. Not all the guns are embarked.
UPDATED

YEN PAI 736 *3/2004*, L-G Nilsson /* 1042152

SURVEY AND RESEARCH SHIPS

Notes: (1) In addition to the naval ships shown in this section there are large numbers of civilian marine survey ships. The majority belong to the **National Marine Bureau** and have funnel markings of a red star with light blue wave patterns on either side. There are about 37 ships with names *Zhong Guo Hai Jian* or *Xiang Yang Hong* followed by a pennant number. The **National Land Resources Department** has two Geological Survey Squadrons and these ships have a red star and light blue ring on a white or yellow background. The **State Education Department** Science section owns ships with funnel markings of yellow and blue lines either side of a circular blue design. Also there are a few nationalised companies such as the **China Marine Oil Company** which have a band of light blue round the top of the funnel.
(2) There is a large number of ocean surveillance fishing trawlers. These sometimes engage in fishing activities and are not easily distinguishable from civilian fishing vessels.

XIANG YANG HONG 14 (National Marine Bureau) *4/2004*, Ships of the World /* 1042141

AGI 201 (converted trawler)   6/1997, A Sharma / 0017746

ZHONG GUO HAI JIAN 52 (National Marine Bureau)   7/2004*, Ian Edwards / 1042399

FENDOU SHIHAO (National Land Resources)   6/1999, Ships of the World / 0056795

DONG FANG HONG 2 (State Education Department)   4/2004*, Ships of the World / 1042130

HAI YING 12 HAO (China Marine Oil Company)   6/1997, A Sharma / 0006690

1 DAHUA CLASS (AGOR/AGE)

891 (ex-970, ex-909)

Displacement, tons: 6,000 full load
Dimensions, feet (metres): 433.1 × 58.1 × 23 *(132 × 17.7 × 7)*
Main machinery: 2 diesels; 2 shafts
Speed, knots: 20
Complement: 80

Comment: Launched on 9 March 1997 with pennant number 909 at Zhonghua, and completed in August 1997 with new pennant number. There is a helicopter deck aft. This is a key unit which has been involved in a number of trials including those for the HQ-9 phased array radar. Large cylindrical launch tubes have been fitted midships.
VERIFIED

891   6/2003 / 0572429

4 SPACE EVENT SHIPS (AGMH/AGI)

YUAN WANG 1	YUAN WANG 2	YUAN WANG 3	YUAN WANG 4

Displacement, tons: 17,100 standard; 18,400 full load
Dimensions, feet (metres): 610.2 × 74.1 × 24.6 *(186 × 22.6 × 7.5)*
Main machinery: 1 Sulzer diesel; 17,400 hp(m) *(12.78 MW)*; 1 shaft
Speed, knots: 20. **Range, n miles:** 18,000 at 20 kt
Complement: 470

Comment: Built by Shanghai Jiangnan Yard. First two commissioned in 1979, the third in April 1995. The fourth is a former survey ship of 11,000 tons. Have helicopter platform but no hangar. Extensive communications, SATNAV and meteorological equipment fitted in the first pair in Jiangnan SY in 1986-87. Both refitted in 1991-92. Based in the East Sea Fleet, but all belong to the National Marine Bureau.
VERIFIED

YUAN WANG 3   10/2003, Robert Pabst / 0569171

YUAN WANG 4   12/2003, M Back, RAN / 0569178

1 SPACE EVENT SHIP (AGM/AGI)

DONGDIAO 851 (ex-232)

Displacement, tons: 6,000 full load
Dimensions, feet (metres): 426.5 × 53.8 × 21.3 *(130 × 16.4 × 6.5)*
Main machinery: 2 diesels; 2 shafts
Speed, knots: 20
Complement: 250
Guns: 1—37 mm. 2—14.5 mm.

Comment: First seen fitting out in 1999. A larger version of Dadie class with extensive space monitoring equipment. In service in March 2000.
VERIFIED

DONGDIAO (old number)   9/2002, Ships of the World / 0569177

1 DADIE CLASS (AGI)

BEIDIAO 841

Displacement, tons: 2,550 full load
Dimensions, feet (metres): 308.4 × 37.1 × 13.1 *(94 × 11.3 × 4)*
Main machinery: 2 diesels; 2 shafts
Speed, knots: 17
Complement: 170 (18 officers)
Guns: 4—14.5 mm (2 twin)
Radars: Navigation: 2 Type 753; I-band.

Comment: Built at Wuhan shipyard, Wuchang and commissioned in 1986. North Sea Fleet and seen regularly in Sea of Japan and East China Sea.
VERIFIED

BEIDIAO   10/1997 / 0017748

2 KAN CLASS (AGOR)

101   102

Displacement, tons: 1,100 full load
Dimensions, feet (metres): 225 × 22.5 × 9 *(68.6 × 6.9 × 2.7)*
Main machinery: 2 diesels; 2 shafts
Speed, knots: 18
Complement: 150
Radars: Navigation: Fin Curve; I-band.

Comment: Details given are for *102* which is believed built in 1985-87, possibly at Shanghai. Large open stern area. Aft main deck area covered and may have cable reel system. *101* is similar but slightly larger and may have been built in 1965 as an ASR. Operate in East China Sea and Sea of Japan.
VERIFIED

KAN 101   *5/2000, van Ginderen Collection* / 0103684

1 BIN HAI CLASS (AGOR)

HAI 521

Displacement, tons: 550 full load
Dimensions, feet (metres): 164 × 32.8 × 11.5 *(50 × 10 × 3.5)*
Main machinery: 2 Niigata Type 6M26KHHS diesels; 1,600 hp(m) *(1.18 MW)*; 2 shafts; bow thruster
Speed, knots: 14. **Range, n miles:** 5,000 at 11 kt
Complement: 15 (7 officers) plus 25 scientists
Radars: Navigation: Japanese AR-M31; I-band.

Comment: A purpose-built research ship built by Niigata Engineering Co, Niigata (Japan) in 1974-75. Launched 10 March 1975. Commissioned July 1975. First operated by the China National Machinery Export-Import Corporation on oceanographic duties. Operates on East and South China research projects but based in North Sea Fleet. For small vessel, has cruiser stern with raked bow and small funnel well aft. Capability to operate single DSRV and the Chinese Navy has a number of Japanese-built KSWB-300 submersibles. Painted white. This ship may belong to the China Marine Oil Company and further vessels may be in service.
UPDATED

1 SHUGUANG CLASS (ex-T-43) (AGOR/AGS)

203

Displacement, tons: 500 standard; 570 full load
Dimensions, feet (metres): 190.3 × 28.9 × 11.5 *(58 × 8.8 × 3.5)*
Main machinery: 2 PRC/Kolomna Type 9-D-8 diesels; 2,000 hp(m) *(1.47 MW)* sustained; 2 shafts
Speed, knots: 15. **Range, n miles:** 5,300 at 8 kt
Complement: 55-60

Comment: Converted from ex-Soviet T-43 minesweeper in late 1960s. Painted white. This last survivor is based in the North Sea Fleet.
VERIFIED

SHUGUANG 203   *10/1997, van Ginderen Collection* / 0012980

1 GANZHU CLASS (AGS)

420

Displacement, tons: 1,000 full load
Dimensions, feet (metres): 213.2 × 29.5 × 9.7 *(65 × 9 × 3)*
Main machinery: 4 diesels; 4,400 hp(m) *(3.23 MW)*; 2 shafts
Speed, knots: 20
Complement: 125
Guns: 4—37 mm/63 (2 twin); 8—14.5 mm (4 twin).

Comment: Built at Zhujiang in 1973-75. Long refit in 1996 for up to two years.
VERIFIED

GANZHU 420   *8/1998* / 0056802

5 YENLAI CLASS (AGS)

226   227   420   427   943

Displacement, tons: 1,040 full load
Dimensions, feet (metres): 241.8 × 32.1 × 9.7 *(73.7 × 9.8 × 3)*
Main machinery: 2 PRC/Kolomna Type 9-D-8 diesels; 2,000 hp(m) *(1.47 MW)* sustained; 2 shafts
Speed, knots: 16. **Range, n miles:** 4,000 at 14 kt
Complement: 25
Guns: 4 China 37 mm/63 (2 twin). 4—25 mm/80 (2 twin).
Radars: Navigation: Fin Curve; I-band.

Comment: Built at Zhonghua Shipyard, Shanghai in early 1970s. Carries four survey motor boats.
UPDATED

YENLAI 226   *7/2004*, Ships of the World* / 1042131

ICEBREAKERS

1 YANBING (MOD YANHA) CLASS (AGB/AGI)

723

Displacement, tons: 4,420 full load
Dimensions, feet (metres): 334.6 × 56 × 19.5 *(102 × 17.1 × 5.9)*
Main machinery: Diesel-electric; 2 diesel generators; 2 motors; 2 shafts
Speed, knots: 17
Complement: 95
Guns: 8—37 mm/63 Type 61/74 (4 twin).
Radars: Navigation: 2 Fin Curve; I-band.

Comment: Enlarged version of Yanha class icebreaker, built in 1982, with greater displacement, longer and wider hull, added deck level and curved upper funnel. In October 1990, painted white while operating in Sea of Japan. Used as an AGI in the North Sea Fleet.
VERIFIED

YANBING 723   *12/2001, Ships of the World* / 0529115

3 YANHA CLASS (AGB/AGI)

519   721   722

Displacement, tons: 3,200 full load
Dimensions, feet (metres): 290 × 53 × 17 *(88.4 × 16.2 × 5.2)*
Main machinery: Diesel-electric; 2 diesel generators; 1 motor; 1 shaft
Speed, knots: 17.5
Complement: 90
Guns: 8—37 mm/63 Type 61/74 (4 twin). 4—25 mm/80 Type 61.
Radars: Navigation: Fin Curve; I-band.

Comment: *721* and *722* built in 1969-70. *519* commissioned in 1989. Used as AGIs in the North Sea Fleet.
VERIFIED

519   *10/1991, G Jacobs*

TUGS

Notes: The vessels below represent a cross-section of the craft available.

4 TUZHONG CLASS (ATF)

154	710	830	890

Displacement, tons: 3,600 full load
Dimensions, feet (metres): 278.5 × 46 × 18 (84.9 × 14 × 5.5)
Main machinery: 2 10 ESDZ 43/82B diesels; 8,600 hp(m) (6.32 MW); 2 shafts
Speed, knots: 18.5
Complement: 120
Radars: Navigation: Fin Curve; I-band.

Comment: Built in late 1970s. Can be fitted with twin 37 mm AA armament and at least one of the class (710) has been fitted with a Square Tie radar. 35 ton towing winch. One in each Fleet and one in reserve.

VERIFIED

TUZHONG   11/1996, A Sharma / 0012228

1 DAOZHA CLASS (ATF)

Displacement, tons: 4,000 full load
Dimensions, feet (metres): 275.6 × 41.3 × 17.7 (84 × 12.6 × 5.4)
Main machinery: 2 diesels; 8,600 hp(m) (6.32 MW); 2 shafts
Speed, knots: 18
Complement: 125

Comment: Built in 1993-94 probably as a follow-on to the Tuzhong class. Based in South Sea Fleet.
VERIFIED

DAOZHA   9/1993, Hachiro Nakai

17 GROMOVOY CLASS (ATF)

149	156	166	167	680	683	684	716	802
809	811	813	814	817	822	824	827	

Displacement, tons: 795 standard; 890 full load
Dimensions, feet (metres): 149.9 × 31.2 × 15.1 (45.7 × 9.5 × 4.6)
Main machinery: 2 diesels; 1,300 hp(m) (956 kW); 2 shafts
Speed, knots: 11
Range, n miles: 7,000 at 7 kt
Complement: 25-30 (varies)
Guns: 4—14.5 mm (2 twin) or 12.7 mm (2 twin) MGs.
Radars: Navigation: Fin Curve or OKI X-NE-12 (Japanese); I-band.

Comment: Built at Luda Shipyard and Shanghai International, 1958-62. Four in North Sea Fleet, nine in East Sea Fleet and four in South Sea Fleet.
VERIFIED

GROMOVOY 802   5/1992, Henry Dodds / 0056805

10 HUJIU CLASS (ATF)

147	155	622	711	717	837	842	843	875	877

Displacement, tons: 1,470 full load
Dimensions, feet (metres): 197.5 × 38.1 × 14.4 (60.2 × 11.6 × 4.4)
Main machinery: 2 LVP 24 diesels; 1,800 hp(m) (1.32 MW); 2 shafts
Speed, knots: 15. **Range, n miles:** 7,200 at 14 kt
Complement: 56
Radars: Navigation: Fin Curve or Type 756; I-band.

Comment: Built at Wuhu in 1980s. One sold to Bangladesh in 1984 and a second in 1995. Three based in the North and East, three in the South Sea Fleet.
UPDATED

HUJIU 877   3/2004*, L-G Nilsson / 1042149

19 ROSLAVL CLASS (ATA/ARS)

153	159	161-164	168	518	604	613
618	646	707	852-854	862	863	867

Displacement, tons: 670 full load
Dimensions, feet (metres): 149.9 × 31 × 15.1 (45.7 × 9.5 × 4.6)
Main machinery: Diesel-electric; 2 diesel generators; 1,200 hp(m) (882 kW); 1 motor; 1 shaft
Speed, knots: 12. **Range, n miles:** 6,000 at 11 kt
Complement: 28
Guns: 4—14.5 mm (2 twin) MGs.

Comment: Built in China in mid-1960s to the USSR design. One carries diving bell and submarine rescue gear on stern and is classified as ARS. Split evenly between the fleets.
UPDATED

ROSLAVL 854   9/2002*, Ian Edwards / 1042398

MARITIME MILITIA (MBDF)

Notes: (1) China has four regular paramilitary maritime Security Forces: the Customs Service (Hai Guan); the maritime section of the Public Security Bureau (Hai Gong); the maritime command (Gong Bian) of the Border Security Force (which is itself a part of the PLA-subordinated People's Armed Police); and the Border Defence (Bian Jian).

These four organisations patrol extensively with a variety of vessels. In recent years the better disciplined and centrally controlled Hai Guan has received a significant number of new vessels, many of them with offshore capabilities. A number of Haitun helicopters are also in service.

There have been many reports of Chinese paramilitary vessels committing acts of piracy in the South China Sea, particularly Gong Bian vessels. Gong Bian and Hai Guan patrol vessels have been operating as far as the coasts of Luzon and Taiwan.

(2) Types of vessels vary from Huxins, Shanghai IIs and Huludaos to a number of other designs spread across all forces. For example Huxin and Huludao classes can show the markings of all four services.

(3) From December 1999 pennant numbers have been standardised to show the vessels' legitimate operating area. This is an attempt to crack down on illegal activities by making it easier for merchant ships to report violations to the Maritime Police (Hai Gong), who have taken overall responsibility.

a. 海关   HAI GUAN (HOI KWAN) – CUSTOMS

b. 海公   HAI GONG (HOI KUNG) – MARITIME POLICE

c. 公边   GONG BIAN (KUNG BIN) – BORDER SECURITY

d. 边检   BIAN JIAN (PIN KAM) – BORDER DEFENCE

BORDER SECURITY FORCE MARITIME COMMAND (GONG BIAN)

HUXIN CLASS (PB)

Displacement, tons: 165 full load
Dimensions, feet (metres): 91.9 × 13.8 × 5.2 *(28 × 4.2 × 1.6)*
Main machinery: 2 diesels; 1,000 hp(m) *(735 kW)*; 2 shafts
Speed, knots: 17. **Range, n miles:** 400 at 10 kt
Complement: 26
Guns: 2 China 14.5 mm/93 (twin).
Radars: Surface search: Skin Head; I-band.

Comment: This is a class of modified Huangpu design with a greater freeboard and a slightly larger displacement. First seen in 1989 and now in series production. Huxin 178 is a modified command vessel with a forward superstructure extension.
VERIFIED

HUXIN   *4/2003, Bob Fildes /* 0569179

COASTAL PATROL CRAFT (NEW) (PB)

Displacement, tons: 58 full load
Dimensions, feet (metres): 73.8 × 15.7 × 5.2 *(22.5 × 4.8 × 1.6)*
Main machinery: 2 diesels; 1,600 hp(m) *(1.18 MW)*; 2 shafts
Speed, knots: 22. **Range, n miles:** 850 at 11 kt
Complement: 13
Guns: 2—14.5 mm (twin).
Radars: Surface search: I-band.

Comment: Large numbers of this type in all Fleet areas. Sometimes involved in piracy and other illegal activities, although whether as official policy or as a result of private enterprise is unknown. Armaments vary.
UPDATED

GONG BIAN 4401   *6/1999 /* 0056807

GONG BIAN 4407   *6/1997 /* 0017754

COASTAL PATROL CRAFT (OLD) (PB)

Displacement, tons: 82 full load
Dimensions, feet (metres): 82 × 13.5 × 4.6 *(25 × 4.1 × 1.4)*
Main machinery: 2 diesels; 900 hp(m) *(662 kW)*; 2 shafts
Speed, knots: 14. **Range, n miles:** 900 at 11 kt
Complement: 12
Guns: 4—14.5 mm/93 2 (twin).
Radars: Surface search: Fin Curve; I-band.

Comment: Large numbers of this type still extensively used although numbers are declining in favour of Huxin and the newer CPC design.
VERIFIED

GONG BIAN 1301   *3/1995, van Ginderen Collection /* 0056808

STEALTH CRAFT (PBF)

Comment: Since 1996 large numbers of low profile stealth craft have been active in the South Sea areas, and have been reported as far away as the Philippines. Sizes vary from 30 to 60 m in length and many are capable of speeds in excess of 30 kt. Most are paramilitary vessels but some may be privately owned.
VERIFIED

STEALTH   *8/1996 /* 0012232

INSHORE PATROL CRAFT (PBI)

Displacement, tons: 32 full load
Dimensions, feet (metres): 62 × 13.1 × 3.6 *(18.9 × 4 × 1.1)*
Main machinery: 2 diesels; 900 hp(m) *(662 kW)*; 2 shafts
Speed, knots: 15
Complement: 5
Guns: 1—12.7 mm MG.

Comment: Details given are for the standard small patrol craft. In addition there are a number of speedboats confiscated from smugglers and used for interception duties.
VERIFIED

GONG BIAN 3110   *4/1998 /* 0017755

GONG BIAN SPEEDBOAT   *2/1995, T Hollingsbee /* 0056809

CUSTOMS (HAI GUAN) AND PUBLIC SECURITY BUREAU (HAI GONG) AND BORDER DEFENCE (BIAN JIAN)

Notes: A new class of 20-24 Qui-M class offshore patrol craft is reported to have entered service. Armed with twin 30 mm guns, a distinguishing feature is a stern ramp to facilitate the handling of high-speed interceptor craft. At 100 m length, they are substantially larger than previous Customs vessels and, despite appearances, there has been some speculation as to whether these craft are manned by naval personnel.

HULUDAO CLASS (TYPE 206)
(FAST ATTACK CRAFT—PATROL) (PC)

Displacement, tons: 180 full load
Dimensions, feet (metres): 147.6 × 21 × 5.6 *(45 × 6.4 × 1.7)*
Main machinery: 3 MWM TBD604BV12 diesels; 5,204 hp(m) *(3.82 MW)* sustained; 3 shafts
Speed, knots: 29. **Range, n miles:** 1,000 at 15 kt
Complement: 24 (6 officers)
Guns: 6 China 14.5 mm Type 82 (3 twin); 600 rds/min to 7 km *(3.8 n miles)*; weight of shell 1.42 kg.

Comment: EEZ patrol craft first seen at Wuxi Shipyard in 1988. The craft is sometimes referred to as the Wuting class.
VERIFIED

HAI GONG HULUDAO   *6/1995 /* 0056810

7 TYPE P 58E (COMMAND SHIPS) (AGF)

901-907

Displacement, tons: 435 full load
Dimensions, feet (metres): 190.3 × 24.9 × 7.5 *(58 × 7.6 × 2.3)*
Main machinery: 4 MTU diesels; 8,720 hp(m) *(6.4 MW)* sustained; 4 shafts
Speed, knots: 27. **Range, n miles:** 1,500 at 12 kt
Complement: 50
Guns: 2 China 14.5 mm/93 (twin) MGs.
Radars: Surface search: I-band.

Comment: First one built at Guangzhou in 1990, last one in 1998. Less well armed but similar to those in service with Pakistan's MSA. Used as command ships.

VERIFIED

HAI GUAN 901   *1993, T Hollingsbee* / 0056811

42 COASTAL PATROL CRAFT (NEW) (PB)

801-842

Displacement, tons: 98 full load
Dimensions, feet (metres): 101.7 × 15.4 × 4.6 *(31 × 4.7 × 1.4)*
Main machinery: 2 diesels; 2 shafts
Speed, knots: 32
Complement: 15
Guns: 2 China 14.5 mm/93 (twin).
Radars: Surface search: Racal Decca ARPA; I-band.

Comment: Building in Shanghai at about six a year since 1992. More may follow.

UPDATED

HAI GUAN 812   *1993, T Hollingsbee*

COASTAL PATROL CRAFT (OLD) (PB)

Comment: Shanghai type hull but with a different superstructure. Two twin 14.5 mm MGs. Being phased out and replaced by the 800 series of patrol craft.

VERIFIED

HAI GUAN 62   *6/1995* / 0056812

2 COMBATBOAT 90E (PBF)

Displacement, tons: 9 full load
Dimensions, feet (metres): 39 × 9.5 × 2.3 *(11.9 × 2.9 × 0.7)*
Main machinery: 1 Scania AB DSI 14 diesel; 398 hp(m) *(293 kW)*; waterjet
Speed, knots: 40
Complement: 2

Comment: Two delivered to Hai Guan in April 1997. This is the transport version of the Swedish raiding craft and can lift two tons of stores or 6-10 troops.

VERIFIED

COMBATBOAT 90E (Swedish colours)   *5/1999, Per Körnefeldt* / 0056813

COAST GUARD

Notes: The China Coast Guard (Maritime Safety Administration), part of the Ministry of Communications, was established in 1998 and is responsible for safety at sea, security and pollution control in Chinese offshore waters, ports and inland rivers. The agency reportedly operates some 150 vessels which are painted white with a large diagonal red stripe and four thin blue stripes.

1 + 2 HAIXUN CLASS (PBOH)

HAIXUN 21

Displacement, tons: 1,500 full load
Dimensions, feet (metres): 305.8 × 40.0 × 17.7 *(93.2 × 12.2 × 5.4)*
Main machinery: 2 diesels; 2 shafts
Speed, knots: 22
Radars: Navigation.
Helicopters: Platform for one medium.

Comment: The first of possibly three patrol ships. Conducted joint exercises with the Japanese Coast Guard in May 2004.

NEW ENTRY

HAIXUN 21   *5/2004*, Hachiro Nakai* / 0589002

Colombia
ARMADA DE LA REPUBLICA

Country Overview

The Republic of Colombia is the only South American country that fronts both the Caribbean Sea and the Pacific Ocean with coastlines of 950 n miles and 782 n miles respectively. With an area of 440,831 square miles, it is bordered to the north by Panama, to the east by Venezuela and Brazil and to the south by Peru and Ecuador. The capital and largest city is Bogotá. Buenaventura and Tumaco are the main Pacific ports while Cartagena, Santa Marta and Barranquilla, which is near the mouth of the principal river and transport artery, the Magdalena, are on the Caribbean side. Territorial seas (12 n miles) are claimed but while it has claimed a 200 n mile EEZ, its limits have not been fully defined.

Headquarters Appointments

Commander of the Navy:
   Admiral Mauricio Soto Gomez
Deputy Commander of the Navy:
   Vice Admiral Alonso Navarro Dallos
Inspector General:
   Vice Admiral René Moreno Moreno
Chief of Naval Staff:
   Rear Admiral Rengifo Sánchez
Chief of Naval Operations:
   Rear Admiral Fernando Eliás Román Campos
Commander Caribbean Force:
   Rear Admiral Enrique Barrera Hurtado
Commander Pacific Force:
   Rear Admiral Alvaro Echandia Duran
Commander South Force:
   Captain Juan Prieto Vasquez
Chief of Logistics:
   Rear Admiral Jaime Parra Cifuentes
Commander Marine Corps:
   Rear Admiral Luis Fernando Yance Villamil

Personnel

(a)   2005: 7,501 (Navy); 9,000 (Marines); 200 (Coast Guard);
      100 (Aircrew)
(b)   2 years' national service (few conscripts in the Navy)

Organisation

Caribbean Force Command: HQ at Cartagena.
Pacific Force Command: HQ at Bahia Malaga.
Naval Force South: HQ at Puerto Leguízamo.
Riverine Brigade: HQ at Bogotá, DC.
Coast Guard: HQ at Bogotá.

Bases

ARC Bolivar, Cartagena, Main naval base (floating dock, 1 slipway), schools.
ARC Bahía Málaga: Major Pacific base.
ARC Barranquilla: Naval training base.
ARC Puerto Leguízamo: Putumayo River base.
ARC Leticia: Minor River base.
Puerto López: Minor River base.
Puerto Carreño: Minor River base.
Barrancabermeja: Minor River base.
San Andrés y Providencia: Specific Command

Marine Corps

Organisation: First Brigade (Sincelejo):
31, 32, 33 Battalions (Sincelejo)
No. 3 Battalion (Malaga)
No. 21 MP Battalion (Cartagena de Indias)
No. 5 Battalion (Corozal)
No. 43 Training Battalion, (Coveñas)
No. 23 MP Battalion (Coveñas)
No. 41 Training Battalion (Coveñas)
Specials Forces Battalion (Cartagena de Indias)
Nos. 31-33 Battalions (Sincelejo)
Second Brigade (Buenaventura).
No. 6 Battalion (Bahía Solano)
No. 40 training Battalion (Tumaco)
No. 2 Battalion (Tumaco)
Riverine Brigade (Bogotá DC)
Five battalions at Turbo (20 Bn), Yati (30 Bn), Puerto Carreno (40 Bn), Puerto Leguizamo (60 Bn) and Puerto Inirida (50 Bn).

Strength of the Fleet

Type	Active
Patrol Submarines	2
Midget Submarines	2
Frigates	4
Patrol Ships and Fast Attack Craft (Gun)	12
Coast Patrol Craft	45
Amphibious Forces	8
River Patrol Craft	32
River Patrol Craft Support	11
River Assault Boats	150
Survey Vessels	8
Auxiliaries	27
Training Ships	4

Prefix to Ships' Names

ARC (Armada Republica de Colombia)

Dimar

Maritime authority in charge of hydrography and navigational aids.

Coast Guard and Customs (DIAN)

The Coast Guard was established in 1979 but then gave way to the Customs Service before being re-established in January 1992 under the control of the Navy. Headquarters at Bogotá. Main bases are Cartagena, Buenaventura y Turbo and Valle. Ships have a red and yellow diagonal stripe on the hull and patrol craft have a PM number. Customs craft were absorbed into the Coast Guard but by 1995 were again independent as part of the DIAN (Direccion de Impuestos y Aduanas Nacionales). Customs craft have Aduana written on the ship's side, a thick and two thin diagonal stripes and have AN numbers.

PENNANT LIST

Submarines

SO 28	Pijao
SO 29	Tayrona
ST 20	Intrépido
ST 21	Indomable

Frigates

FL 51	Almirante Padilla
FL 52	Caldas
FL 53	Antioquia
FL 54	Independiente

Patrol Forces

PO 41	Espartana
PO 42	Capitán Pablo José de Porto
PO 43	Capitán Jorge Enrique Marques Duran
PO 44	Valle del Cauca
PM 102	Rafael del Castillo y Rada
PM 103	TN José María Palas
PM 104	CN Medardo Monzon Coronado
PM 105	S2 Jaime Gómez Castro
PM 106	S2 Juan Nepomuceno Peña
PM 112	Quitasueño
PM 113	José María García y Toledo
PM 114	Juan Nepomuceno Eslava
PM 115	TECIM Jaime E Cárdenas Gomez
PM 141	Cabo Corrientes
PM 142	Cabo Manglares
PM 143	Cabo Tiburon
PM 144	Cabo de la Vella
PG 401	Altair
PG 402	Castor
PG 403	Pollux
PG 404	Vega
PB 421	Antares
PB 422	Capricornio
PB 423	Acuario
PB 424	Piscis
PB 425	Aries
PB 426	Tauro
PB 427	Géminis
PB 428	Deneb
PB 429	Rigel
PB 430	Júpiter
PB 431	Leo
PB 432	Aldebarán
PB 433	Neptuno
PB 434	Spica
PB 435	Denebola
PB 436	Libra
PB 437	Escorpión
PB 438	Alpheraz
PB 439	Bellatrix
PB 440	Canopus
PB 441	Procycom
PB 442	Tulcán
PB 443	Halley
PB 444	Hooker Bay
PB 445	Isla Bolívar
PB 446	Capella
PC 451	Andrómeda
PC 452	Casiopea
PC 453	Centauro
PC 454	Dragón
PC 455	Vela
PC 456	Polaris
PC 457	Fenix
PC 458	Regulus
PC 459	Aquila
PC 460	Perseus
PC 461	Ramadan
PC 462	Apolo
PC 463	Zeus
PC 464	Sagitario
PC 465	Lince
PF 121	Diligente
PF 122	Juan Lucio
PF 123	Alfonso Vargas
PF 124	Fritz Hagale
PF 125	Vengadora
PF 126	Humberto Cortez
PF 128	Carlos Galindo
PF 129	Capitán Jaime Rook
PF 130	Manuela Saenz
PF 135	Riohacha
PF 136	Leticia
PF 137	Arauca
PRF 176	Río Magdalena
PRF 177	Río Cauca
PRF 178	Río Atrato
PRF 179	Río Sinú
PRF 180	Río San Jorge
PRF 181	Tenerife
PRF 182	Tarapaca
PRF 183	Mompox
PRF 184	Orocué
PRF 185	Calamar
PRF 186	Magangue
PRF 187	Monclart
PRF 188	Caucaya
PRF 189	Mitú
PRF 190	Río Putumayo
PRF 191	Río Caquetá
PRF 192	Río Orinoco
PRF 193	Río Orteguaza
PRF 194	Río Vichada
PRF 195	Río Guaviare
PRF 320-322	

Amphibious Forces

LD 240	Bahía Zapzurro
LD 246	Morrosquillo
LD 248	Bahía Honda
LD 249	Bahía Portete
LD 251	Bahía Solano
LD 252	Bahía Cupica
LD 253	Bahía Utría
LD 254	Bahía Málaga

Auxiliaries

BL 161	Cartagena de Indias
BL 162	Buenaventura
TM 501	Bocachica
TM 502	Arturus
TM 503	Pedro David Salas
TM 504	Sirius
TM 506	Tolú
TM 507	Calima
TM 508	Bahí Santa Catalina
TM 509	Móvil I
TM 510	Móvil II
TM 511	Renacer del Pacifico
TM 512	Jhonny Cay
TM 513	Punta Evans
TB 542	Playa Blanca
TB 543	Tierra Bomba
TB 544	Bell Salter
TB 545	Maldonado
TB 546	Orion
TB 547	Pegasso
TB 548	Almirante I
TB 549	Almirante II
TB 550	Ara
TB 551	Valerosa
TB 552	Luchadora
TB 554	Orca
DF 170	Mayor Jaime Arias Arango
NF 601	Filigonio Hichamón
NF 602	SSIM Manuel Antonio Moyar
NF 603	Igaraparaná
NF 604	SSIM Julio Correa Hernández
NF 605	Manacacías
NF 606	Cotuhe
NF 607	SSCIM Senen Alberto Araujo
NF 608	CPCIM Guillermo Londoño Vargas
NF 609	Ariarí
NF 610	Mario Villegas
NF 611	Tony Pastrana Contreras
NF 131	Socorro
NF 132	Hernando Gutiérrez

Survey Vessels

BO 155	Providencia
BO 156	Malpelo
BH 153	Quindio
BB 31	Gorgona
BB 32	Capitán Binney
BB 33	Abadía Médez
BB 34	Ciénaga de Mayorquin
BB 35	Isla Palma

Training Ships

BE 160	Gloria
YT 230	Comodoro
YT 231	Tridente
YT 232	Cristina

Tugs

RM 75	Andagoya
RM 76	Josué Alvarez
RB 78	Portete
RB 79	Maldonado
RB 80	Ciénaga de San Juan
RF 81	Capitán Castro
RF 83	Joves Fiallo
RF 84	Capitán Alvaro Ruiz
RF 85	Miguel Silva
RF 86	Capitán Rigoberto Giraldo
RF 87	Vladimir Valek
RF 88	Teniente Luis Bernal
RF 91	TN Alejandro Baldomero Salgado
RF 92	Carlos Rodríguez
RF 93	Sejeri
RF 94	Ciudad de Puerto López
RF 96	Inirida

SUBMARINES

Notes: There are three Swimmer Delivery Vehicles: *Defensora*, *Poderosa* and *Protectora*.

2 PIJAO (209 TYPE 1200) CLASS (SS)

Name	No	Builders	Laid down	Launched	Commissioned
PIJAO	SO 28	Howaldtswerke, Kiel	1 Apr 1972	10 Apr 1974	18 Apr 1975
TAYRONA	SO 29	Howaldtswerke, Kiel	1 May 1972	16 July 1974	16 July 1975

Displacement, tons: 1,180 surfaced; 1,285 dived
Dimensions, feet (metres): 183.4 × 20.5 × 17.9
 (55.9 × 6.3 × 5.4)
Main machinery: Diesel-electric; 4 MTU 12V 493 AZ80 diesels;
 2,400 hp(m) *(1.76 MW)* sustained; 4 AEG alternators;
 1.7 MW; 1 Siemens motor; 4,600 hp(m) *(3.38 MW)*
 sustained; 1 shaft
Speed, knots: 22 dived; 11 surfaced
Range, n miles: 8,000 at 8 kt surfaced; 4,000 at 4 kt dived
Complement: 34 (7 officers)

Torpedoes: 8—21 in *(533 mm)* bow tubes. 14 AEG SUT; dual
 purpose; wire-guided; active/passive homing to 12 km *(6.5 n
 miles)* at 35 kt; 28 km *(15 n miles)* at 23 kt; warhead 250 kg.
 Swim-out discharge.
Countermeasures: ESM: Thomson-CSF DR 2000; intercept.
Weapons control: Signaal M8/24 TFCS.
Radars: Surface search: Thomson-CSF Calypso II; I-band.
Sonars: Krupp Atlas PSU 83-55; hull-mounted; active/passive
 search and attack; medium frequency.
 Atlas Elektronik PRS 3-4; passive ranging; integral with CSU 3.

Programmes: Ordered in 1971. Both refitted by HDW at Kiel;
 Pijao completed refit in July 1990 and *Tayrona* in September
 1991. Main batteries were replaced.
Structure: Single-hulled. Diving depth, 820 ft *(250 m)*.
Operational: Refitted 1999-2002. Both boats employed on
 counter-drug operations.

UPDATED

PIJAO
2000, Colombian Navy / 0103689

2 MIDGET SUBMARINES (SSW)

Name	No	Builders	Launched	Commissioned
INTRÉPIDO	ST 20	Cosmos, Livorno	1 Jan 1972	17 Apr 1973
INDOMABLE	ST 21	Cosmos, Livorno	1 Jan 1972	17 Apr 1973

Displacement, tons: 58 surfaced; 70 dived
Dimensions, feet (metres): 75.5 × 13.1 *(23 × 4)*
Main machinery: Diesel-electric; 1 diesel; 1 motor; 300 hp(m)
 (221 kW); 1 shaft
Speed, knots: 11 surfaced; 6 dived.
Range, n miles: 1,200 surfaced; 60 dived
Complement: 4
Mines: 6 Mk 21 with 300 kg warhead. 8 Mk 11 with 50 kg
 warhead.

Comment: They can carry eight swimmers with 2 tons of
 explosive as well as two swimmer delivery vehicles (SDVs).
 Built by Cosmos, Livorno and commissioned at 40 tons, but
 subsequently enlarged in the early 1980s. Listed by the Navy
 as 'Tactical Submarines'.

VERIFIED

INTRÉPIDO
2000, Colombian Navy / 0103690

FRIGATES

4 ALMIRANTE PADILLA CLASS (TYPE FS 1500) (FLGHM)

Name	No	Builders	Laid down	Launched	Commissioned
ALMIRANTE PADILLA	FL 51	Howaldtswerke, Kiel	17 Mar 1981	6 Jan 1982	31 Oct 1983
CALDAS	FL 52	Howaldtswerke, Kiel	14 June 1981	23 Apr 1982	14 Feb 1984
ANTIOQUIA	FL 53	Howaldtswerke, Kiel	22 June 1981	28 Aug 1982	30 Apr 1984
INDEPENDIENTE	FL 54	Howaldtswerke, Kiel	22 June 1981	21 Jan 1983	24 July 1984

Displacement, tons: 1,500 standard; 2,100 full load
Dimensions, feet (metres): 325.1 × 37.1 × 12.1
 (99.1 × 11.3 × 3.7)
Main machinery: 4 MTU 20V 1163 TB92 diesels; 23,400 hp(m)
 (17.2 MW) sustained; 2 shafts; cp props
Speed, knots: 27; 18 on 2 diesels
Range, n miles: 7,000 at 14 kt; 5,000 at 18 kt
Complement: 94

Missiles: SSM: 8 Aerospatiale MM 40 Exocet ❶; inertial cruise;
 active radar homing to 70 km *(40 n miles)* at 0.9 Mach;
 warhead 165 kg; sea-skimmer.
 SAM: ❷ 2 Matra Simbad twin launchers; Mistral; IR homing to
 4 km *(2.2 n miles)*; warhead 3 kg; anti-sea-skimmer.
Guns: 1 OTO Melara 3 in *(76 mm)*/62 compact ❸; 85 rds/min to
 16 km *(8.7 n miles)*; weight of shell 6 kg.
 2 Breda 40 mm/70 (twin) ❹; 300 rds/min to 12.5 km *(6.8 n
 miles)* anti-surface; weight of shell 0.96 kg.
Torpedoes: 6—324 mm ILAS 3 (2 triple) tubes ❺; Whitehead
 A244S; anti-submarine; active/passive homing to 7 km *(3.8 n
 miles)*; warhead 38 kg (shaped charge).
Countermeasures: Decoys: 1 CSEE Dagaie double mounting; IR
 flares and chaff decoys (H- to J-band).
 ESM: Argo AC672; radar warning.
 ECM: Racal Scimitar; jammer.
Combat data systems: Thomson-CSF TAVITAC action data
 automation. Possibly Link Y fitted.
Weapons control: 2 Canopus optronic directors. Thomson-CSF
 Vega II GFCS.

ALMIRANTE PADILLA
(Scale 1 : 900), Ian Sturton / 0056815

CALDAS
5/2003 / 0568909

ANTIOQUIA   *5/2003* / 0587696*

Radars: Air/surface search: Thomson-CSF Sea Tiger ❻;
  E/F-band; range 110 km *(60 n miles)* for 2 m² target.
Navigation: Furuno; I-band.
Fire control: Castor II B ❼; I/J-band; range 15 km *(8 n miles)* for
  1 m² target.
IFF: Mk 10.
Sonars: Atlas Elektronik ASO 4-2; hull-mounted; active attack;
  medium frequency.

Helicopters: 1 MBB BO 105 CB ❽ or 1 Bell 412.

Programmes: Order for four Type FS 1500 placed late 1980.
  Reclassified as light frigates in 1999. Similar to Malaysian
  Kasturi class frigates.
Modernisation: Mistral SAM system reported to have been
  fitted. Helicopter deck lengthened by 2 m to take Bell 412

aircraft. There have also been minor modifications to ship
systems and superstructure.

UPDATED

SHIPBORNE AIRCRAFT

Numbers/Type: 2 MBB BO 105CB.
Operational speed: 113 kt *(210 km/h)*.
Service ceiling: 9,854 ft *(3,000 m)*.
Range: 407 n miles *(754 km)*.
Role/Weapon systems: Surface search and limited ASW helicopter. Sensors: Search/weather
  radar. Weapons: ASW; provision to carry depth bombs. ASV; light attack role with machine gun
  pods.

VERIFIED

BO 105   *2000, Colombian Navy / 0103692*

Numbers/Type: 2 Eurocopter AS 555 Fennec.
Operational speed: 121 kt *(225 km/h)*.
Service ceiling: 13,125 ft *(4,000 m)*.
Range: 389 n miles *(722 km)*.
Role/Weapon systems: OTHT capability for surface-to-surface role. Also used for logistic support.
  More are being acquired. Sensors: Bendix RDR 1500B radar. Weapons: Torpedoes may be fitted
  in due course.

VERIFIED

Numbers/Type: 4 Bell 412.
Operational speed: 122 kt *(226 km/h)*.
Service ceiling: 10,000 ft *(3,300 m)*.
Range: 500 n miles *(744 km)*.
Role/Weapon systems: Multipurpose used mostly for surveillance, troop transport and logistic
  support. Sensors: Weather radar. Weapons: ASV 7.62 mm MG can be carried.

VERIFIED

BELL 412   *6/1999, Colombian Navy / 0056820*

LAND-BASED MARITIME AIRCRAFT

Notes: The Navy operates the following fixed-wing aircraft for maritime surveillance and transport:
four RC690, two PA-31, one Cessna 206, one Beech B-350, one Gavillan 358, one Gulfstream I and
two PA-28 Cherokee. There are also one R22, one Bell 212 and five Bell Huey II helicopters for
training and transport.

Numbers/Type: 2 Casa CN-235 200.
Operational speed: 210 kt *(384 km/h)*.
Service ceiling: 24,000 ft *(7,315 m)*.
Range: 2,000 n miles *(3,218 km)*.
Role/Weapon systems: EEZ surveillance. Delivered in 2003. Sensors: Search radar Bendix APS
  504(V)5; FLIR. Weapons: Unarmed.

UPDATED

CN-235   *6/2003*, CASA / 0587695*

AS 555   *6/2000, Colombian Navy / 0103693*

PATROL FORCES

Notes: At least three Orca class 12 m fast intercept craft, capable of 40 kt, are reported to be in service.

1 RELIANCE CLASS (PSOH)

Name	No	Builders	Commissioned
VALLE DEL CAUCA	PO 44	Coast Guard Yard, Baltimore	8 Dec 1967
(ex-*Durable*)	(ex-WMEC 628)		

Displacement, tons: 1,129 full load
Dimensions, feet (metres): 210.5 × 34 × 10.5 *(64.2 × 10.4 × 3.2)*
Main machinery: 2 Alco 16V-251 diesels; 6,480 hp *(4.83 MW)* sustained; 2 shafts; LIPS cp props
Speed, knots: 18. **Range, n miles:** 6,100 at 14 kt; 2,700 at 18 kt
Complement: 75 (12 officers)
Guns: 1 Boeing 25 mm/87 Mk 38 Bushmaster; 200 rds/min to 6.8 km *(3.4 n miles)*. 2—12.7 mm MGs.
Radars: Surface search: Hughes/Furuno SPS-73; I-band.
Helicopters: Platform for one medium.

Comment: Transferred to Colombia on 4 September 2003. During 34 years in USCG service, underwent Major Maintenance Availability (MMA) in 1989. The exhausts for main engines, ship service generators and boilers were run in a vertical funnel which reduced flight deck size. Capable of towing ships up to 10,000 tons. Based in the Pacific.

UPDATED

RELIANCE CLASS (USCG colours)   *10/2002, M Mazumdar /* 0530032

2 LAZAGA CLASS (FAST ATTACK CRAFT—GUN) (PBO)

Name	No	Builders	Commissioned
CAPITÁN PABLO JOSÉ DE PORTO (ex-*Recalde*)	PO 42 (ex-PM 116, ex-P 06)	Bazán, La Carraca	17 Dec 1977
CAPITÁN JORGE ENRIQUE MARQUEZ DURAN (ex-*Cadarso*)	PO 43 (ex-PM 117, ex-P 03)	Bazán, La Carraca	10 July 1976

Displacement, tons: 393 full load
Dimensions, feet (metres): 190.6 × 24.9 × 8.5 *(58.1 × 7.6 × 2.6)*
Main machinery: 2 MTU/Bazán 16V 956 TB 91 diesels; 7,500 hp(m) *(5.5 MW)* sustained; 2 shafts
Speed, knots: 26. **Range, n miles:** 2,400 at 15 kt
Complement: 40 (4 officers)
Guns: 1 Breda 40 mm/70. 1 Oerlikon 20 mm L85. 1—12.7 mm MG.
Weapons control: CSEE optical director.
Radars: Surface search: Furuno; E/F-band.
Navigation: Furuno; I-band.

Comment: Paid off from the Spanish Navy in 1993 and put into reserve. Acquired by Colombia in March 1997 for extensive refurbishment at Bazán, San Fernando. Recommissioned 25 April 1998 and 25 June 1998 respectively. Radars have been changed and the 76 mm gun replaced by a 20 mm cannon. These ships may be used to carry troops. Four more of the class are available and more may be acquired in due course.

VERIFIED

CAPITÁN JORGE ENRIQUE MARQUEZ DURAN   *6/2001, Maritime Photographic /* 0114510

1 CORMORAN CLASS (FAST ATTACK CRAFT—GUN) (PBO)

Name	No	Builders	Commissioned
ESPARTANA (ex-*Cormoran*)	PO 41	Bazán, San Fernando	27 Oct 1989

Displacement, tons: 358 full load
Dimensions, feet (metres): 185.7 × 24.7 × 6.5 *(56.6 × 7.5 × 2)*
Main machinery: 3 MTU-Bazán 16V 956 TB91 diesels; 11,250 hp(m) *(8.27 MW)* sustained; 3 shafts
Speed, knots: 32. **Range, n miles:** 2,500 at 15 kt
Complement: 31 (5 officers)
Guns: 1 Bofors 40/70 SP 48. 1 Oerlikon 20 mm.
Weapons control: Alcor C optronic director.
Radars: Surface search: Raytheon; I-band.

Comment: Built with overseas sales in mind, this ship was launched in October 1985, but from 1989 served in the Spanish Navy until April 1994 when she was laid up at Cartagena. Transferred in September 1995, she was then refitted at Cadiz, before sailing for Colombia in mid-1996. Based at San Andres Island and belongs to the Coast Guard.

VERIFIED

ESPARTANA   *10/1996, Colombian Navy /* 0056822

4 POINT CLASS (PB)

Name	No	Builders	Commissioned
CABO CORRIENTES (ex-*Point Warde*)	PM 141 (ex-82368)	J M Martinac, Tacoma	14 Aug 1967
CABO MANGLARES (ex-*Point Wells*)	PM 142 (ex-82343)	USCG Yard, Curtis Bay	20 Nov 1963
CABO TIBURON (ex-*Point Estero*)	PM 143 (ex-82344)	USCG Yard, Curtis Bay	11 Dec 1963
CABO DE LA VELLA (ex-*Point Sal*)	PM 144 (ex-82352)	J M Martinac, Tacoma	5 Dec 1966

Displacement, tons: 66; 69 full load
Dimensions, feet (metres): 83 × 17.2 × 5.8 *(25.3 × 5.2 × 1.8)*
Main machinery: 2 Caterpillar 3412 diesels; 1,600 hp *(1.19 MW)*; 2 shafts
Speed, knots: 23.5. **Range, n miles:** 1,500 at 8 kt
Complement: 10 (1 officer)
Guns: 2—12.7 mm MGs.
Radars: Surface search: Hughes/Furuno SPS-73; I-band.

Comment: Steel hulled craft with aluminium superstructure built in United States 1960-70. *Cabo Corrientes* transferred on 29 June 2000 followed by *Cabo Manglares* on 13 October 2000. *Cabo Tiburon* and *Cabo de la Vella* transferred on 8 February 2001 and 29 May 2001 respectively.

VERIFIED

CABO CORRIENTES   *2000, Colombian Navy /* 0103695

3 ARAUCA CLASS (RIVER GUNBOATS) (PBR)

Name	No	Builders	Commissioned
RIOHACHA	PF 135 (ex-35)	Union Industrial de Barranquilla	6 Sep 1956
LETICIA	PF 136 (ex-36)	Union Industrial de Barranquilla	6 Sep 1956
ARAUCA	PF 137 (ex-37)	Union Industrial de Barranquilla	6 Sep 1956

Displacement, tons: 275 full load
Dimensions, feet (metres): 163.5 × 27.2 × 8.9 *(49.9 × 8.3 × 2.7)*
Main machinery: 2 Caterpillar diesels; 916 hp *(683 kW)*; 2 shafts
Speed, knots: 14. **Range, n miles:** 1,890 at 14 kt
Complement: 43; 39 plus 6 orderlies *(Leticia)*
Guns: 2 USN 3 in *(76 mm)*/50 Mk 26. 4 Oerlikon 20 mm (not *Leticia*).

Comment: Launched in 1955. *Leticia* can be equipped as a hospital ship with six beds. Based in Naval Force South.

UPDATED

ARAUCA   *1991, Colombian Navy /* 0056821

2 JOSÉ MARIA PALAS (SWIFT 110) CLASS
(LARGE PATROL CRAFT) (PB)

Name	No	Builders	Commissioned
JOSÉ MARIA PALAS	PM 103 (ex-GC 103)	Swiftships Inc, Berwick	Sep 1989
MEDARDO MONZON CORONADO	PM 104 (ex-GC 104)	Swiftships Inc, Berwick	July 1990

Displacement, tons: 99 full load
Dimensions, feet (metres): 109.9 × 24.6 × 6.6 *(33.5 × 7.5 × 2)*
Main machinery: 4 Detroit 12V-71TI diesels; 2,400 hp *(1.79 MW)*; 4 shafts
Speed, knots: 25. **Range, n miles:** 2,250 at 15 kt
Complement: 19 (3 officers)
Guns: 1 Bofors 40 mm/70. 1—12.7 mm MG. 2—7.62 mm MGs.
Radars: Surface search: Furuno FR 8100D; I-band.

Comment: Acquired under US FMS programme. These ships belong to the Coast Guard.
VERIFIED

JOSÉ MARIA PALAS   *1/1996, van Ginderen Collection* / 0056824

1 ASHEVILLE CLASS (FAST ATTACK CRAFT—GUN) (PGF)

Name	No	Builders	Commissioned
QUITASUEÑO (ex-*Tacoma*)	PM 112	Tacoma Boat Building	14 July 1969

Displacement, tons: 225 standard; 245 full load
Dimensions, feet (metres): 164.5 × 23.8 × 9.5 *(50.1 × 7.3 × 2.9)*
Main machinery: CODOG; 2 Cummins VT12-875M diesels; 1,450 hp *(1.08 MW)*; 1 GE LM 1500 gas turbine; 13,300 hp *(9.92 MW)*; 2 shafts; cp props
Speed, knots: 40. **Range, n miles:** 1,700 at 16 kt on diesels; 325 at 37 kt
Complement: 24
Guns: 1 US 3 in *(76 mm)*/50 Mk 34; 50 rds/min to 12.8 km *(7 n miles)*; weight of shell 6 kg.
1 Bofors 40 mm/56; 160 rds/min to 11 km *(5.9 n miles)* anti-aircraft; weight of shell 0.96 kg.
2—12.7 mm (twin) MGs.
Radars: Surface search: Raytheon 3100; I-band.

Comment: Transferred from US by lease 16 May 1983 and recommissioned 6 September 1983 and by sale August 1989. Fire-control system removed. Unreliable propulsion system prevented further transfers of this class and it is unlikely the gas turbine is operational, which reduces the top speed to 16 kt. Belongs to the Coast Guard.
UPDATED

QUITASUEÑO   *2000, Colombian Navy* / 0103696

2 TOLEDO CLASS (LARGE PATROL CRAFT) (PB)

Name	No	Builders	Commissioned
JOSÉ MARIA GARCIA Y TOLEDO	PM 113	Bender Marine, Mobile	15 July 1994
JUAN NEPOMUCENO ESLAVA	PM 114	Bender Marine, Mobile	25 May 1994

Displacement, tons: 142 full load
Dimensions, feet (metres): 116 × 24.9 × 7 *(35.4 × 7.6 × 2.1)*
Main machinery: 2 MTU 12V 396 TE94 diesels; 8,240 hp(m) *(6.1 MW)*; 2 shafts
Speed, knots: 25. **Range, n miles:** 1,200 at 15 kt
Complement: 25 (5 officers)
Guns: 1 Bushmaster 25 mm/87 Mk 96. 2—12.7 mm MGs.
Radars: Surface search: Furuno FR 151OD; I-band.

Comment: Acquired under US FMS programme. These ships belong to the Coast Guard.
VERIFIED

JUAN NEPOMUCENO ESLAVA   *6/2001, Maritime Photographic* / 0114511

2 RAFAEL DEL CASTILLO Y RADA (SWIFT 105) CLASS
(LARGE PATROL CRAFT) (PB)

Name	No	Builders	Commissioned
RAFAEL DEL CASTILLO Y RADA	PM 102 (ex-GC 102, ex-AN 202)	Swiftships Inc, Berwick	28 Feb 1983
TECIM JAIME E CÁRDENAS GOMEZ (ex-*Olaya Herrera*)	PM 115 (ex-AN 21, ex-AN 201)	Swiftships Inc, Berwick	16 Oct 1981

Displacement, tons: 115 full load
Dimensions, feet (metres): 105 × 22 × 7 *(31.5 × 6.7 × 2.1)*
Main machinery: 4 MTU 12V 331 TC92 diesels; 5,320 hp(m) *(3.97 MW)* sustained; 4 shafts
Speed, knots: 25. **Range, n miles:** 1,200 at 18 kt
Complement: 19 (3 officers)
Guns: 1 Bofors 40 mm/60 Mk 3 (PM 102). 2—12.7 mm MGs.
Weapons control: 1 COAR optronic director.
Radars: Surface search: Raytheon; I-band.

Comment: Delivered for the Customs service. PM 102 is part of the Coast Guard. PM 115 was paid off, but returned unarmed as part of the resurrected Customs service until being transferred back to the Coast Guard in 1997.
VERIFIED

RAFAEL DEL CASTILLO Y RADA   *6/1999, Colombian Navy* / 0056826

2 JAIME GÓMEZ (Mk III PB) CLASS
(COASTAL PATROL CRAFT) (PB)

Name	No	Builders	Commissioned
JAIME GÓMEZ CASTRO	PM 105 (ex-GC 105)	Peterson Builders	1975
JUAN NEPOMUCENO PEÑA	PM 106 (ex-GC 106)	Peterson Builders	1977

Displacement, tons: 34 full load
Dimensions, feet (metres): 64.9 × 18 × 5.1 *(19.8 × 5.5 × 1.6)*
Main machinery: 3 Detroit 8V-71 diesels; 690 hp *(515 kW)* sustained; 3 shafts
Speed, knots: 28. **Range, n miles:** 450 at 26 kt
Complement: 7 (1 officer)
Guns: 2-12.7 mm MGs. 2-7.62 mm MGs. 1 Mk 19 grenade launcher.
Radars: Surface search: 2 Furuno FR 1510D; I-band.

Comment: Acquired from the USA. Recommissioned in December 1989 and February 1990 respectively. Original 40 mm and 20 mm guns replaced by lighter armament. Both based at Leticia, Rio Amazonas, under coast guard control.
UPDATED

JAIME GÓMEZ CASTRO   *2000, Colombian Navy* / 0103697

3 SWIFTSHIPS CLASS (RIVER PATROL CRAFT) (PBR)

PRF 320-322

Displacement, tons: 17 full load
Dimensions, feet (metres): 45.5 × 11.8 × 1.8 *(13.9 × 3.6 × 0.6)*
Main machinery: 2 Detroit 6V-92TA diesels; 900 hp *(671 kW)*; 2 Hamilton water-jets
Speed, knots: 22. **Range, n miles:** 600 at 22 kt
Complement: 4
Guns: 2 M2HB 12.7 mm MGs; 2 M60D 7.62 mm MGs.
Radars: Surface search: Raytheon 40; I-band.

Comment: Acquired in 2000. Hard chine modified V hull form. Can carry up to eight troops.
NEW ENTRY

SWIFTSHIP CLASS   *6/2001, Ecuador Coast Guard* / 0114516

2 ROTORK 412 CRAFT (RIVER PATROL CRAFT) (PBR)

CAPITÁN JAIME ROOK PF 129 (ex-PM 107)   **MANUELA SAENZ** PF 130 (ex-PM 108)

Displacement, tons: 9 full load
Dimensions, feet (metres): 41.7 × 10.5 × 2.3 *(12.7 × 3.2 × 0.7)*
Main machinery: 2 Caterpillar diesels; 240 hp *(179 kW)*; 2 shafts
Speed, knots: 25
Complement: 4
Military lift: 4 tons or 8 marines
Guns: 1—12.7 mm MG. 2—7.62 mm MGs.
Radars: Surface search: Raytheon; I-band.

Comment: Acquired in 1989-90. Capable of transporting eight fully equipped marines but used as
  river patrol craft.

VERIFIED

CAPITÁN JAIME ROOK   *1990, Colombian Navy /* 0056828

9 TENERIFE CLASS (RIVER PATROL CRAFT) (PBR)

TENERIFE PRF 181   **OROCUÉ** PRF 184   **MONCLART** PRF 187
TARAPACA PRF 182   **CALAMAR** PRF 185   **CAUCAYA** PRF 188
MOMPOX PRF 183   **MAGANGUE** PRF 186   **MITÚ** PRF 189

Displacement, tons: 12 full load
Dimensions, feet (metres): 40.7 × 9.5 × 2 *(12.4 × 2.9 × 0.6)*
Main machinery: 2 Caterpillar 3208 TA diesels; 850 hp *(634 kW)* sustained; 2 shafts
Speed, knots: 29. **Range, n miles:** 530 at 15 kt
Complement: 5 plus 12 troops
Guns: 3—12.7 mm MGs (1 twin, 1 single). 1 Mk 19 grenade launcher. 1—7.62mm MGs.
Radars: Surface search: Raytheon 1900; I-band.

Comment: Built by Bender Marine, Mobile, Alabama. Acquired in October 1993 for anti-narcotics
  patrols. Aluminium hulls. Can be transported by aircraft.

VERIFIED

MITÚ   *2000, Colombian Navy /* 0103698

13 INSHORE PATROL CRAFT (PBI)

ALTAIR PG 401   **NEPTUNO** PB 433   **APOLO** PC 462
CASTOR PG 402   **HALLEY** PB 443   **ZEUS** PC 463
POLLUX PG 403   **HOOKER BAY** PB 444   **SAGITARIO** PC 464
VEGA PG 404   **ISLA BOLIVAR** PB 445   **LINCE** PC 465
JÚPITER PB 430

Comment: All are of about 10 tons. PG 401-404 (Altair class) have a speed of 10 kt and are armed
  with 2—7.62 mm MGs. The remainder (Bay class and Sea Ark class) have outboard engines and
  are capable of speeds in excess of 30 kt.

VERIFIED

VEGA   *6/1999, Colombian Navy /* 0056830

11 ANDRÓMEDA CLASS (INSHORE PATROL CRAFT) (PBI)

ANDRÓMEDA PC 451   **VELA** PC 455   **AQUILA** PC 459
CASIOPEA PC 452   **POLARIS** PC 456   **PERSEUS** PC 460
CENTAURO PC 453   **FENIX** PC 457   **RAMADAN** PC 461
DRAGÓN PC 454   **REGULUS** PC 458

VERIFIED

ANDROMEDA   *2000, Colombian Navy /* 0103699

11 RIO CLASS (RIVER PATROL CRAFT) (PBR)

RIO MAGDALENA PRF 176   **RIO SAN JORGE** PRF 180   **RIO ORTEGUAZA** PRF 193
RIO CAUCA PRF 177   **RIO PUTUMAYO** PRF 190   **RIO VICHADA** PRF 194
RIO ATRATO PRF 178   **RIO CAQUETÁ** PRF 191   **RIO GUAVIARE** PRF 195
RIO SINÚ PRF 179   **RIO ORINOCO** PRF 192

Displacement, tons: 7 full load
Dimensions, feet (metres): 31 × 11.1 × 2 *(9.8 × 3.5 × 0.6)*
Main machinery: 2 Detroit 6V-53 diesels; 296 hp *(221 kW)* sustained; 2 water-jets
Speed, knots: 24
Range, n miles: 150 at 22 kt
Complement: 4
Guns: 2—12.7 mm (twin) MGs. 1—7.62 mm MG. 1—60 mm mortar.
Radars: Surface search: Raytheon 1900; I-band.

Comment: Acquired in 1989-90. Ex-US PBR Mk II built by Uniflite in 1970. All recommissioned in
  September 1990. GRP hulls.

VERIFIED

RIO MAGDALENA   *2000, Colombian Navy /* 0103700

7 RIVER PATROL CRAFT (PBR)

DILIGENTE   **ALFONSO VARGAS**   **VENGADORA**   **CARLOS GALINDO**
  PF 121 (ex-LR 121)   PF 123   PF 125 (ex-LR 125)   PF 128
JUAN LUCIO   **FRITZ HAGALE**   **HUMBERTO CORTEZ**
  PF 122   PF 124   PF 126

Comment: All between 31 and 40 tons. Various designs and ages, but all are armed with two
  12.7 mm MGs and most have 7.62 mm MGs as well.

VERIFIED

VENGADORA (old number)   *2000, Colombian Navy /* 0103701

21 DELFIN CLASS (INSHORE PATROL CRAFT) (PBI)

ANTARES PB 421	GÉMINIS PB 427	SPICA PB 434	BELLATRIX PB 439
CAPRICORNIO PB 422	DENEB PB 428	DENEBOLA PB 435	CANOPUS PB 440
ACUARIO PB 423	RIGEL PB 429	LIBRA PB 436	PROCYON PB 441
PISCIS PB 424	LEO PB 431	ESCORPIÓN PB 437	TULCÁN PB 442
ARIES PB 425	ALDEBARÁN PB 432	ALPHERAZ PB 438	CAPELLA PB 446
TAURO PB 426			

Displacement, tons: 5.4 full load
Dimensions, feet (metres): 25.9 × 8.5 × 3.1 *(7.9 × 2.6 × 0.9)*
Main machinery: 2 Evinrude outboards; 400 hp *(294 kW)*
Speed, knots: 40
Complement: 4
Guns: 1—12.7 mm MG. 2—7.62 mm MGs.
Radars: Surface search: Raytheon; I-band.

Comment: First two built by Mako Marine, Miami and delivered in December 1992. Remainder acquired locally from 1993-94.
UPDATED

DELFIN CLASS   *6/2001, Maritime Photographic /* 0114512

AMPHIBIOUS FORCES

150 RIVER ASSAULT BOATS (RAB) (PBR)

Comment: These are 6.8 m river assault boats acquired from Boston Whaler for use by Marines. Armed with 1—12.7 mm and 2—7.62 mm MGs. 14 patrol units each operate with one Rio or Tenerife class and three Pirañas. Capable of 25 to 30 kt depending on load. Some have been damaged beyond repair. There are also about 110 small river assault boats.
VERIFIED

RAB   *2000, Colombian Navy /* 0103703

1 LCM 8

BAHÍA ZAPZURRO LD 240

Displacement, tons: 125 full load
Dimensions, feet (metres): 71.9 × 20.7 × 9.9 *(21.9 × 6.3 × 3)*
Main machinery: 1 diesel; 285 hp *(213 kW)*; 1 shaft
Speed, knots: 12
Complement: 5
Military lift: 60 tons or 150 troops

Comment: Transferred in 1993.
VERIFIED

BAHÍA ZAPZURRO   *6/1999, Colombian Navy /* 0056832

7 MORROSQUILLO (LCU 1466A) CLASS (LCU)

MORROSQUILLO LD 246	BAHÍA SOLANO LD 251	BAHÍA UTRIA LD 253
BAHÍA HONDA LD 248	BAHÍA CUPICA LD 252	BAHÍA MALAGA LD 254
BAHÍA PORTETE LD 249		

Displacement, tons: 347 full load
Dimensions, feet (metres): 119 × 34 × 6 *(36.3 × 10.4 × 1.8)*
Main machinery: 3 Detroit 6-71 diesels; 522 hp *(389 kW)* sustained; 3 shafts
Speed, knots: 7. **Range, n miles:** 700 at 7 kt
Complement: 14
Cargo capacity: 167 tons or 300 troops
Guns: 2—12.7 mm MGs.
Radars: Navigation: Raytheon; I-band.

Comment: Former US Army craft built in 1954 and transferred in 1991 and 1992 with new engines. Used as inshore transports. Speed quoted is fully laden. Numbers split between each coast.
VERIFIED

MORROSQUILLO   *1/1993 /* 0056833

SURVEY SHIPS

Notes: There are also four small buoy tenders: *Capitán Binney* BB 32, *Abadía Médez* BB 33, *Ciénaga de Mayorquin* BB 34, and *Isla Palma* BB 35.

2 PROVIDENCIA CLASS (AGOR)

Name	No	Builders	Commissioned
PROVIDENCIA	BO 155	Martin Jansen SY, Leer	24 July 1981
MALPELO	BO 156	Martin Jansen SY, Leer	24 July 1981

Displacement, tons: 1,157 full load
Dimensions, feet (metres): 164.3 × 32.8 × 13.1 *(50.3 × 10 × 4)*
Main machinery: 2 MAN-Augsburg diesels; 1,570 hp(m) *(1.15 MW)*; 1 Kort nozzle prop; bow thruster
Speed, knots: 13. **Range, n miles:** 15,000 at 12 kt
Complement: 48 (5 officers) plus 6 scientists
Radars: Navigation: Raytheon; I-band.

Comment: Both launched in January 1981. *Malpelo* employed on fishery research and *Providencia* on geophysical research. Both are operated by DIMAR, the naval authority in charge of hydrographic, pilotage, navigational and ports services. Painted white.
VERIFIED

MALPELO   *2000, Colombian Navy /* 0103704

1 BUOY TENDER (ABU)

Name	No	Builders	Commissioned
QUINDIO (ex-YFR 443)	BH 153	Niagara SB Corporation	11 Nov 1943

Displacement, tons: 600 full load
Dimensions, feet (metres): 131 × 29.8 × 9 *(40 × 9.1 × 2.7)*
Main machinery: 2 Union diesels; 600 hp *(448 kW)*; 2 shafts
Speed, knots: 10
Complement: 17 (2 officers)

Comment: Transport ship transferred by lease from the US in July 1964 and by sale on 31 March 1979. Used as a buoy tender.
VERIFIED

QUINDIO   *2000, Colombian Navy /* 0103705

1 SURVEY SHIP (AGSC)

Name	No	Builders	Commissioned
GORGONA	BB 31 (ex-BO 154, ex-BO 161, ex-FB 161)	Lidingoverken, Sweden	28 May 1954

Displacement, tons: 574 full load
Dimensions, feet (metres): 135 × 29.5 × 9.3 *(41.2 × 9 × 2.8)*
Main machinery: 2 Wärtsilä Nohab diesels; 910 hp(m) *(669 kW)*; 2 shafts
Speed, knots: 13
Complement: 45 (2 officers)

Comment: Paid off in 1982 but after a complete overhaul at Cartagena naval base was back in service in late 1992.

VERIFIED

GORGONA (old number)   *1993, Colombian Navy /* 0056835

AUXILIARIES

2 LUNEBURG CLASS (TYPE 701) (SUPPORT SHIPS) (AGP)

Name	No	Builders	Commissioned
CARTAGENA DE INDIAS (ex-*Luneburg*)	BL 161 (ex-A 1411)	Flensburger	31 Jan 1966
BUENAVENTURA (ex-*Nienburg*)	BL 162 (ex-A 1416)	Bremer Vulcan	1 Aug 1968

Displacement, tons: 3,483 full load
Dimensions, feet (metres): 341.2 × 43.3 × 13.8 *(104 × 13.2 × 4.2)*
Main machinery: 2 MTU MD 16V 538 TB90 diesels; 6,000 hp(m) *(4.1 MW)* sustained; 2 shafts; cp props; bow thruster
Speed, knots: 16. **Range, n miles:** 3,200 at 14 kt
Complement: 70 (9 officers)
Cargo capacity: 1,100 tons
Guns: 4 Bofors 40 mm/70 (2 twin).
Radars: Navigation: I-band.

Comment: BL 161 paid off from the German Navy in 1994. Taken in hand for refit by HDW, Kiel in August 1997. Recommissioned on 2 November 1997. Guns were cocooned in German service. The ship acts as a depot ship for patrol craft. BL 162 paid off and was transferred the same day on 27 March 1998. She is now based at Cartagena.

VERIFIED

BUENAVENTURA   *5/1998, Michael Nitz /* 0056836

2 RIVER SUPPORT CRAFT (YAG)

Name	No	Builders	Commissioned
HERNANDO GUTIÉRREZ	NF 132 (ex-BD 35, ex-TF 52)	Ast Naval, Cartagena	1955
SOCORRO (ex-*Alberto Gomez*)	NF 131 (ex-BD 33, ex-TF 53)	Ast Naval, Cartagena	1956

Displacement, tons: 190 full load
Dimensions, feet (metres): 98.4 × 18 × 3.9 *(30 × 5.5 × 1.2)*
Main machinery: 2 Lister 8KB FRAPIL diesels; 260 hp *(194 kW)*; 2 shafts
Speed, knots: 6. **Range, n miles:** 650 at 9 kt
Complement: 20 plus berths for 48 troops and medical staff
Guns: 2—12.7 mm MGs.

Comment: River transports. Named after Army officers. *Socorro* was converted in July 1967 into a floating surgery. *Hernando Gutierrez* was converted into a dispensary ship in 1970. Both used as support river patrol craft.

VERIFIED

HERNANDO GUTIÉRREZ   *1/2002, van Ginderen Collection /* 0533239

11 RIVER SUPPORT CRAFT (YDT/YAG)

FILIGONIO HICHAMÓN NF 601 (ex-NF 141)
SSIM MANUEL A MOYAR NF 602 (ex-NF 144)
IGARAPARANÁ NF 603 (ex-RR 92, LR 92)
SSIM JULIO CORREA HERNÁNDEZ NF 604 (ex-NF 143)
MANACACÍAS NF 605 (ex-RR 95, LR 95)
COTUHE NF 606 (ex-RR 98)
SSCIM SENEN ALBERTO ARAUJO NF 607 (ex-NF 147)
CPCIM GUILLERMO LONDOÑO VARGAS NF 608 (ex-NF 146)
ARIARÍ NF 609 (ex-PF-127, RR 97)
MARIO VILLEGAS NF 610
TONY PASTRANA CONTRERAS NF 611 (ex-NF 149)

Displacement, tons: 260
Dimensions, feet (metres): 109.6 × 31.2 × 3.1 *(33.4 × 9.5 × 0.95)*
Main machinery: Diesels
Speed, knots: 9
Complement: 18 plus 82 troops
Guns: 8—12.7 mm MGs.
Helicopters: Platform (NF 610, 611) for 1 small.

Comment: Details are for NF 607, 608, 610 and 611 which were built by COTECMAR, Cartagena de Indias and delivered 2000-2004. Six further are projected. The remainder have various characteristics and are deployed as river patrol craft, command and support ships.

UPDATED

TONY PASTRANA CONTRERAS   *3/2004*, Colombian Navy /* 0563761

12 TRANSPORTS

BOCACHICA TM 501
ARTURUS TM 502
PEDRO DAVID SALAS TM 503 (ex-TM 101)
SIRIUS TM 504 (ex-TM 62)
TOLÚ TM 506
CALIMA TM 507 (ex-TM 49)
BAHÍA SANTA CATALINA TM 508
MÓVIL I TM 509
MÓVIL II TM 510
RENACER DEL PACIFICO TM 511
JOHNNY CAY TM 512
PUNTA EVANS TM 513

Comment: Small supply ships of various characteristics from 300 tons (TM 506) to 3 tons (TM 508-513). The others are mostly about 30 tons with a speed of 10 kt.

VERIFIED

CALIMA (old number)   *6/1999, Colombian Navy /* 0056838

12 BAY SUPPORT CRAFT

PLAYA BLANCA TB 542
TIERRA BOMBA TB 543
BELL SALTER TB 544
MALDONADO TB 545
ORION TB 546
PEGASSO TB 547
ALMIRANTE I TB 548
ALMIRANTE II TB 549
ARA TB 550
VALEROSA TB 551
LUCHADORA TB 552
ORCA TB 554

Comment: Mostly small craft of less than 10 tons. The largest is TB 544 which is 87 tons and has previously been listed as an Admiral's Yacht.

VERIFIED

BELL SALTER   *6/1999, Colombian Navy /* 0056840

1 FLOATING DOCK (ASL)

MAYOR JAIME ARIAS ARANGO DF 170 (ex-DF 41, ex-170)

Comment: Capacity of 165 tons, length 140 ft *(42.7 m)*, displacement 700 tons. Used as a non-self-propelled depot ship for the midget submarines. **VERIFIED**

MAYOR JAIME ARIAS ARANGO   *6/2001, Maritime Photographic /* 0114513

TUGS

17 TUGS (YTL)

ANDAGOYA RM 75	CAPITAN RIGOBERTO GIRALDO RF 86
JOSUÉ ALVAREZ RB 76	VLADIMIR VALEK RF 87
PORTETE RB 78	TENIENTE LUIS BERNAL RF 88
MALDONADO RB 79	TENIENTE ALEJANDRO BALDOMERO SALGADO RF 91
CIENAGA DE SAN JUAN RB 80	CARLOS RODRIGUEZ RF 92
CAPITÁN CASTRO RF 81	SEJERI RF 93
JOVES FIALLO RF 83	CIUDAD DE PUERTO LÓPEZ RF 94
CAPITÁN ALVARO RUIZ RF 84	INIRIDA RF 96
MIGUEL SILVA RF 85	

Comment: River craft of various types described as 'Remolcador Bahia (RB), Fluvial (RF) or Mar (RM)'. Used for transport and ferry duties in harbours and rivers. RM 75 and RM 76 are harbour tugs. **UPDATED**

JOSUÉ ALVAREZ   *6/1999, Colombian Navy /* 0056841

TRAINING SHIPS

Notes: There are also three sail training yachts *Comodoro* YT 230, *Tridente* YT 231 and *Cristina* YT 232.

1 SAIL TRAINING SHIP (AXS)

Name	No	Builders	Launched	Commissioned
GLORIA	BE 160	AT Celaya, Bilbao	6 Sep 1966	16 May 1969

Displacement, tons: 1,250 full load
Dimensions, feet (metres): 249.3 oa; 211.9 wl; × 34.8 × 21.7 *(76; 64.6 × 10.6 × 6.6)*
Main machinery: 1 auxiliary diesel; 530 hp(m) *(389 kW)*; 1 shaft
Speed, knots: 10.5
Complement: 51 (10 officers) plus 88 trainees

Comment: Sail training ship. Barque rigged. Hull is entirely welded. Sail area, 1,675 sq yds *(1,400 sq m)*. Endurance, 60 days. Similar to Ecuador, Mexico and Venezuelan vessels. **VERIFIED**

GLORIA   *6/2000, Adolfo Ortigueira Gil /* 0533240

Comoros

Country Overview

A former French territory, the Federal Islamic Republic of the Comoros declared independence in 1975. The islands are situated at the northern entrance to the Mozambique Channel, between the African mainland and the island of Madagascar. There are three islands: Njazidja (formerly known as Grande Comore), Mwali (Mohéli), and Nzwani (Anjouan). A fourth island in the archipelago, Mayotte (Mahoré), is formally claimed by Comoros but chose to remain a French dependency. The largest town, capital and principal port is Moroni on southwestern Njazidja. This archipelagic state claims 12 n miles of territorial seas. A 200 n mile Exclusive Economic Zone (EEZ) has been claimed but the limits are not fully defined.

Bases

Moroni.

PATROL FORCES

2 YAMAYURI CLASS (PBI)

Name	No	Builders	Commissioned
KARTHALA	—	Ishihara Dockyard Co Ltd	Oct 1981
NTRINGUI	—	Ishihara Dockyard Co Ltd	Oct 1981

Displacement, tons: 26.5 standard; 41 full load
Dimensions, feet (metres): 59 × 14.1 × 3.6 *(18 × 4.3 × 1.1)*
Main machinery: 2 Nissan RD10TA06 diesels; 900 hp(m) *(661 kW)* maximum; 2 shafts
Speed, knots: 20
Complement: 6
Guns: 2—12.7 mm (twin) MGs.
Radars: Surface search: FRA 10; I-band.

Comment: These two patrol vessels of the MSA type (steel-hulled), supplied under Japanese government co-operation plan. Used for fishery protection services. Due to be replaced. **VERIFIED**

KARTHALA
10/1981, Ishihara DY / 0056842

Democratic Republic of Congo

Country Overview

Formerly known as the Belgian Congo until it became independent in 1960, the Democratic Republic of the Congo was known as Zaire from 1971-97. With an area of 905,568 square miles, it has borders to the north with the Republic of the Congo. A 22 n mile coastline with the Atlantic Ocean separates Angola, to the south, from its Cabinda province. The capital and largest city is Kinshasa (formerly Léopoldville) while the principal ports are Matadi and Boma, on the lower Congo, and Banana, at its mouth. Territorial seas (12 n miles) are claimed. An EEZ has reportedly been claimed but the details have not been published. A cease fire in the civil war was declared in September 1999 although some fighting continued until January 2001. In July 2003, the Transitional National Government was established as part of the evolving peace process.

While some Shanghai II class survived the war, all are reported derelict. Some barges and small patrol craft have been mounted with guns.

Headquarters Appointments

Chief of the Navy:
  Major General Dieudonne Amuli Bahigwa

Personnel

(a)   2005: 1,000 (70 officers)
(b)   Voluntary service

Organisation

There are four commands which came under the Army in 1997: Matadi (coastal), Kinshasa (riverine), Kalémié (Lake Tanganyika) and Goma (Lake Kivu). Kalémié was in rebel hands by mid-1998 and Moba by mid-1999.

Bases

Matadi, Kinshasa, Kalémié (Lake Tanganyika), Goma.

Congo-Brazzaville

Country Overview

Formerly known as the Middle Congo, part of a French colony, the Republic of Congo gained independence in 1960. An unstable political period followed, culminating in civil war between 1997 and 2000 when a Transitional Council was created. A new constitution was approved by referendum in 2002. With an area of 132,000 square miles, it is situated in west-central Africa and has borders to the north with Cameroon and the Central African Republic, to the south-west with Angola

(Cabinda enclave) and to the west with Gabon. The River Congo, a major transport artery, provides the southern and much of the eastern border with the Democratic Republic of Congo (formerly Zaire). It has a 91 n mile coastline with the Atlantic Ocean. Brazzaville is the capital and largest city while Pointe Noire is the principal port and centre of the offshore oil industry. Congo has not claimed an EEZ but is one of a few coastal states which claims a 200 n mile territorial sea. The navy consists mainly of riverine craft but acquisition of offshore patrol vessels to protect offshore resources is a possibility.

Headquarters Appointments

Chief of the Navy:
  Capitaine de Vaisseau Fulgort Ongobo

Bases

Pointe Noire, Brazzaville, Impfondo.

Cook Islands

Country Overview

The Cook Islands are a South Pacific island group which became self-governing in 1965; defence and external affairs remain the responsibility of the New Zealand government. Situated some 2,430 n miles south of Hawaii, they comprise two groups of widely scattered islands. The Southern Group includes Rarotonga, Aitutaki, Atiu, Mangaia, Mauke, Mitiaro, Manuae and Takutea. The Northern Group is composed of low-lying coral islands and includes Pukapuka, Tongareva (also called Penrhyn), Manihiki, Palmerston, Rakahanga, Suwarrow and Nassau.

The port of Avarua on the island of Rarotonga is the administrative centre. Territorial seas (12 n miles) are claimed. An Exclusive Economic Zone (EEZ) (200 n miles) is claimed but limits have not been fully defined by boundary agreements.

Headquarters Appointments

Maritime Commander:
  Chief Inspector Garth Henderson
Maritime Surveillance Adviser:
  Lieutenant Commander M Parsons, RAN

Bases

Avatiu Wharf, Rarotonga

PATROL FORCES

1 PACIFIC CLASS
(LARGE PATROL CRAFT) (PB)

Name	Builders	Commissioned
TE KUKUPA	Australian Shipbuilding Industries	1 Sep 1989

Displacement, tons: 162 full load
Dimensions, feet (metres): 103.3 × 26.6 × 6.9
  (31.5 × 8.1 × 2.1)
Main machinery: 2 Caterpillar 3516TA diesels; 2,820 hp
  (2.1 MW) sustained; 2 shafts
Speed, knots: 20. **Range, n miles:** 2,500 at 12 kt
Complement: 17 (3 officers)
Radars: Surface search: Furuno 1011; I-band.

Comment: Laid down 16 May 1988 and launched 27 January 1989. Cost, training and support provided by Australia under defence co-operation. Acceptance date was 9 March 1989 but the handover was deferred another six months because of the change in local government. Has Furuno D/F equipment, SATNAV and a Stressl seaboat with a 40 hp outboard engine. A half-life refit was conducted in 1997 and, following the announcement by the Australian government to extend the Pacific Patrol Boat programme to a 30 year ship life, *Te Kukupa* will be due for a life extension refit in 2006.
                                         UPDATED

TE KUKUPA
6/1995, van Ginderen Collection

Costa Rica
SERVICIO NACIONAL GUARDACOSTAS

Country Overview

The Republic of Costa Rica is an independent Central American State which lies between Nicaragua to the north and Panama to the south-east. With an area of 19,652 square miles, it has a 584 n mile coastline with the North Pacific Ocean and of 112 n miles with the Caribbean. The uninhabited Cocos Island, about 290 n miles southwest of Burrica Point, is also under Costa Rican sovereignty. The country's capital is San José while other important cities are the Caribbean port of Limón and the Pacific port of Puntarenas. Territorial seas (12 n miles) are claimed. While a 200 n mile EEZ has been claimed, the limits have only been partly defined by boundary agreements.

Personnel

(a) 2005: 350 officers and men
(b) Voluntary service

Bases

Pacific: Golfito, Punta Arenas, Cuajiniquil, Quepos.
Atlantic: Limon, Moin.

PATROL FORCES

Notes: Three Boston Whalers, *Tauro* (20-1), *Villa Mar* (20-2) and *Cocori* (22-1) are operational. The first of six Costa Rican-built Apex RIBs, *Escorpion* (24-1), entered service in 2001.

APEX RIB   *5/2001, Julio Montes* / 0109935

1 CAPE CLASS (LARGE PATROL CRAFT) (PB)

Name	No	Builders	Commissioned
ASTRONAUTA FRANKLIN CHANG (ex-*Cape Henlopen*)	95-1	Coast Guard Yard, Curtis Bay	5 Dec 1958

Displacement, tons: 98 standard; 148 full load
Dimensions, feet (metres): 94.8 × 20.3 × 6.6 *(28.9 × 6.2 × 2)*
Main machinery: 2 Detroit 16V-149TI diesels; 2,070 hp *(1.54 MW)* sustained; 2 shafts
Speed, knots: 20. **Range, n miles:** 2,500 at 10 kt
Complement: 14 (1 officer)
Guns: 2—12.7 mm MGs.
Radars: Surface search: Raytheon SPS-64(V)1; I-band.

Comment: Transferred from US Coast Guard 28 September 1989 after a refit by Bender SB and Repair Co. Painted white in 1989. Based at Punta Arenas.
VERIFIED

ASTRONAUTA FRANKLIN CHANG   *6/1989, Bender Shipbuilding* / 0056843

1 SWIFT 105 ft CLASS (FAST PATROL CRAFT) (PB)

Name	No	Builders	Commissioned
ISLA DEL COCO	105-1 (ex-1055)	Swiftships, Morgan City	Feb 1978

Displacement, tons: 118 full load
Dimensions, feet (metres): 105 × 23.3 × 7.2 *(32 × 7.1 × 2.2)*
Main machinery: 3 MTU 12V 1163 TC92 diesels; 10,530 hp(m) *(7.74 MW)*; 3 shafts
Speed, knots: 33. **Range, n miles:** 1,200 at 18 kt; 2,000 at 12 kt
Complement: 17 (3 officers)
Guns: 1—12.7 mm MG. 4—7.62 mm (2 twin) MGs. 1—60 mm mortar.
Radars: Navigation: Furuno; I-band.

Comment: Aluminium construction. Refitted in 1985-86 under FMS funding. The twin MGs are fitted abaft the bridge and the mortar is on the stern. Based at Punta Arenas.
VERIFIED

ISLA DEL COCO (old number)   *2/1989* / 0056844

3 POINT CLASS (COASTAL PATROL CRAFT) (PB)

Name	No	Builders	Commissioned
SANTAMARIA (ex-*Point Camden*)	82-2 (ex-82373)	J Martinac, Tacoma	4 May 1970
JUAN RAFAEL MORA (ex-*Point Chico*)	82-3 (ex-82339)	US Coast Guard Yard, Curtis Bay	29 Oct 1962
PANCHA CARRASCO (ex-*Point Bridge*)	82-4 (ex-82338)	US Coast Guard Yard, Curtis Bay	10 Oct 1962

Displacement, tons: 67 full load
Dimensions, feet (metres): 83 × 17.2 × 5.8 *(25.3 × 5.2 × 1.8)*
Main machinery: 2 Caterpillar 3412 diesels; 1,600 hp *(1.19 MW)*; 2 shafts
Speed, knots: 23. **Range, n miles:** 1,200 at 8 kt
Complement: 10
Guns: 2—12.7 mm MGs.
Radars: Navigation: Raytheon SPS-64/Hughes SPS-73; I-band.

Comment: First transferred from USCG on 15 December 1999. A second transferred on 22 June 2001 and third on 28 September 2001.
VERIFIED

SANTAMARIA   *2/2000, Julio Montes* / 0109937

1 SWIFT 36 ft CLASS (INSHORE PATROL CRAFT) (PB)

PUERTO QUEPOS (ex-*Telamanca*) 36-1

Displacement, tons: 11 full load
Dimensions, feet (metres): 36 × 10 × 2.6 *(11 × 3.1 × 0.8)*
Main machinery: 2 Detroit diesels; 500 hp *(373 kW)*; 2 shafts
Speed, knots: 24. **Range, n miles:** 250 at 18 kt
Complement: 4 (1 officer)
Guns: 1—12.7 mm MG. 1—60 mm mortar.
Radars: Navigation: Raytheon 1900; I-band.

Comment: Built by Swiftships, Morgan City and completed in March 1986.
VERIFIED

PUERTO QUEPOS   *2/2000, Julio Montes* / 0109936

3 SWIFT 65 ft CLASS (COASTAL PATROL CRAFT) (PB)

CABO BLANCO 65-3   PUNTA BURICA 65-4   ISLA UVITA 65-6

Displacement, tons: 35 full load
Dimensions, feet (metres): 65.5 × 18.4 × 6.6 *(20 × 5.6 × 2)*
Main machinery: 2 MTU 8V 331 TC92 diesels; 1,770 hp(m) *(1.3 MW)*; 2 shafts
Speed, knots: 23. **Range, n miles:** 500 at 18 kt
Complement: 7 (2 officers)
Guns: 1—12.7 mm MG. 4—7.62 mm (2 twin) MGs. 1—60 mm mortar.
Radars: Navigation: Furuno; I-band.

Comment: Built by Swiftships, Morgan City in 1979. Refitted 1985-86 under FMS funding. 65-3 is
based at Limon.

UPDATED

1 SWIFT 42 ft CLASS (INSHORE PATROL CRAFT) (PB)

PRIMERA DAMA 42-1

Displacement, tons: 11 full load
Dimensions, feet (metres): 42 × 14.1 × 3 *(12.8 × 4.3 × 0.9)*
Main machinery: 2 Detroit 8V-92TA diesels; 700 hp *(522 kW)*; 2 shafts
Speed, knots: 33. **Range, n miles:** 300 at 30 kt; 450 at 18 kt
Complement: 4 (1 officer)

Comment: Probably ex-*Donna Margarita* (ex-*Puntarena*), completed in 1986 and formerly used as
a hospital ship.

VERIFIED

CABO BLANCO   *11/2003*, A A de Kruijf /* 0587697

Côte d'Ivoire
MARINE CÔTE D'IVOIRE

Country Overview

Formerly a French colony, The Republic of Côte d'Ivoire gained
full independence in 1960. Located in west Africa, the country
has an area of 133,425 square miles and a 281 n mile coastline
with the Gulf of Guinea. It is bordered to the east by Ghana and to
the west by Liberia and Guinea. The capital is Yamoussoukro
while the former capital, Abidjan, is the largest city, principal port
and commercial centre. A further port at San Pedro is linked to
Mali by rail. Territorial seas (12 n miles) are claimed. A 200 n mile
EEZ has been claimed but the limits have not been defined by
boundary agreements.

Following the rebellion of September 2002, a Government of
National Conciliation has restored a level of stability although
internal tensions continue. While the navy remains unchanged,
operational effectiveness is likely to have suffered.

Headquarters Appointments

Chief of Naval Staff:
  Capitaine de Frigate Diomande Megna

Bases

Use made of ports at Locodjo (Abidjan), Sassandra, Tabouand
San-Pédro

Personnel

2005: 930 (75 officers)

PATROL FORCES

Notes: Two Rodman 890 craft delivered in late 1997 for the Police.

1 PATRA CLASS (LARGE PATROL CRAFT) (PBO)

Name	No	Builders	Launched	Commissioned
L'INTRÉPIDE	—	Auroux, Arcachon	21 July 1978	6 Oct 1978

Displacement, tons: 147.5 full load
Dimensions, feet (metres): 132.5 × 19.4 × 5.2 *(40.4 × 5.9 × 1.6)*
Main machinery: 2 SACM AGO 195 V12 CZSHR diesels; 4,340 hp(m) *(3.19 MW)* sustained;
  2 shafts; cp props
Speed, knots: 26. **Range, n miles:** 1,750 at 10 kt; 750 at 20 kt
Complement: 19 (2 officers)
Guns: 1 Breda 40 mm/70. 1 Oerlikon 20 mm. 2—7.62 mm MGs.
Radars: Surface search: Racal Decca 1226; I-band.

Comment: Of similar design to French Patra class. Laid down 7 July 1977. Patrol endurance of five
days. SS-12M missiles are no longer carried. Sister ship *L'Ardent* decommissioned in 2003 to
provide spares.

UPDATED

L'ARDENT   *3/1994 /* 0080123

AFFAIRES MARITIMES

2 RODMAN 890 (PBR)

AMOUGNA AF 003   MONSEKELA AF 004

Dimensions, feet (metres): 29.2 × 9.8 × 3.6 *(8.9 × 3 × 0.8)*
Main machinery: 2 Volvo diesels; 300 hp(m) *(220 kW)*; 2 shafts
Speed, knots: 28. **Range, n miles:** 150 at 25 kt
Complement: 3
Guns: 1—7.62 mm MG.
Radars: Surface search: I-band.

Comment: Two craft delivered by Rodman in 1997. Employed on Fishery Protection duties.

NEW ENTRY

AMOUGNA   *6/1997*, Rodman /* 0583296

AUXILIARIES

Notes: (1) There are also some Rotork 412 craft supplied in 1980. Some are naval, some civilian.
(2) Two French harbour tugs *Merisier* and *Meronnior* were acquired in September 1999.
(3) A Yunnan class LCM *Atchan* may still be in limited service.

2 CTM (LCM)

ABY (ex-*CTM 15*)   TIAGHA (ex-*CTM 16*)

Displacement, tons: 150 full load
Dimensions, feet (metres): 78 × 21 × 4.2 *(23.8 × 6.4 × 1.3)*
Main machinery: 2 Poyaud 520 V8 diesels; 225 hp(m) *(165 kW)*; 2 shafts
Speed, knots: 9.5. **Range, n miles:** 350 at 8 kt
Complement: 6
Military lift: 48 tons

Comment: Transferred from France in March 1999. Built in about 1968. Bow ramps are fitted.

VERIFIED

CTM (French colours)   *6/1995 /* 0012960

Croatia
HRVATSKA RATNA MORNARICA

Country Overview

Formerly a constituent republic of the Federal Republic of Yugoslavia, Croatia declared its independence in 1991. With an area of 21,829 square miles, it is situated in south-east Europe in the Balkan Peninsula and bordered to the north by Slovenia and Hungary, to the east and south by Bosnia and Herzegovina and to the east by Serbia and Montenegro. There is a coastline of 3,127 n miles with the Adriatic Sea on which Dubrovnik, Split, Ploče and Rijeka are the principal ports. The capital and largest city is Zagreb. Territorial waters (12 n miles) are claimed but an EEZ has not been claimed.

Headquarters Appointments

Commander of the Navy:
    Rear Admiral Zdravko Kardum
Deputy Commander of the Navy:
    Commodore Zdenko Simićić
Chief of Staff, Navy HQ:
    Captain Ante Urlić
Commander of the Fleet:
    Commodore Ivica Tolić

Personnel

(a)   2005: 1,900 (620 officers)
(b)   Reserve: 8,000

General

The Navy was established on 12 September 1991. Ships captured from the Yugoslav Federation form the bulk of the Fleet. The main task is the protection and defence of territorial waters.

Bases and Organisation

Headquarters: Lora-Split.
Main base: Split.
Minor bases: Sibenik, Pula, Ploče.
River Patrol Flotillas: Osijek (Drava) and Sisak (Sava).
There are two coastal command sectors: North and South Adriatic. Radar surveillance stations and coastal batteries are established on key islands and peninsulas. All the bases and naval installations of the former federal Navy were taken over with the exception of those in the Gulf of Kotor.

Coast Defence

Three mobile RBS 15 batteries on trucks. Total of 10 coastal artillery batteries. Jadran command system for coastal defence using Italian built and US radars installed in 2003.

Naval Infantry

Headquarters in Split. All in reserve.

DELETIONS

Submarines

2003   *Velebit*

Amphibious Forces

2003   DJC 101, DJC 102, DSM 110

SUBMARINES

Notes: The Una class submarine *Velebit* is not operational.

2 R-2 MALA CLASS (TWO-MAN SWIMMER DELIVERY VEHICLES) (LDW)

Displacement, tons: 1.4
Dimensions, feet (metres): 16.1 × 4.6 × 4.3
    (4.9 × 1.4 × 1.3)
Main machinery: 1 motor; 4.7 hp(m) *(3.5 kW)*; 1 shaft
Speed, knots: 4.4
Range, n miles: 18 at 4.4 kt; 23 at 3.7 kt
Complement: 2
Mines: 250 kg of limpet mines.

Comment: Free-flood craft with the main motor, battery, navigation pod and electronic equipment housed in separate watertight cylinders. Instrumentation includes aircraft type gyrocompass, magnetic compass, depth gauge (with 0 to 100 m scale), echo-sounder, sonar and two searchlights. Constructed of light aluminium and plexiglass, it is fitted with fore and after-hydroplanes, the tail being a conventional cruciform with a single rudder abaft the screw. Large perspex windows give a good all-round view. Operating depth, 60 m *(196.9 ft)*, maximum. Two reported sold to Syria and one to Sweden.

Notes: There is also an R-1 craft which is 3.7 m long and capable of 2.8 kt down to 50 m. It has a range of 4 n miles. There may also be some locally built SDVs.

VERIFIED   R-2

2/2002, RH-Alan / 0528428

R-1

2/2002, RH-Alan / 0528427

CORVETTES

2 KRALJ (TYPE R-03) CLASS (FSG)

Name	No	Builders	Launched	Commissioned
KRALJ PETAR KRESIMIR IV	RTOP 11	Kraljevica Shipyard	21 Mar 1992	7 July 1992
KRALJ DMITAR ZVONIMIR	RTOP 12	Kraljevica Shipyard	30 Mar 2001	June 2002

Displacement, tons: 385 (11), 401 (12) full load
Dimensions, feet (metres): 175.9 × 27.9 × 7.5
(53.6 × 8.5 × 2.3)
Main machinery: 3 M 504B-2 diesels; 12,500 hp(m) *(9.2 MW)*
sustained; 3 shafts
Speed, knots: 36. **Range, n miles:** 1,700 at 18 kt
Complement: 29 (11), 30 (12) (5 officers)

Missiles: SSM: 4 or 8 Saab RBS 15B (2 or 4 twin) ❶; active radar
homing to 70 km *(37.8 n miles)* at 0.8 Mach; warhead 83 kg.
Guns: 1 Bofors 57 mm/70 (RTOP 11) ❷; 200 rds/min to 17 km
(9.3 n miles); weight of shell 2.4 kg. Launchers for illuminants
on side of mounting.
1—30 mm/65 AK 630M ❸; 6 barrels; 3,000 rds/min
combined to 4 km.
Mines: 4 AIM-70 magnetic or 6 SAG-1 acoustic in lieu of SSMs.
Countermeasures: Decoys: 2 Wallop Barricade chaff/IR
launchers.
Weapons control: PEAB 9LV 249 Mk 2 director.
Kolonka for AK 630M.
Radars: Surface search: Racal BT 502 ❹; E/F-band.
Fire control: PEAB 9LV 249 Mk 2 ❺; I/J-band.
Navigation: Racal 1290A; I-band.
Sonars: RIZ PP10M; hull-mounted; active search; high frequency.

Programmes: The building of this class (formerly called Kobra by
NATO) was officially announced as 'suspended' in 1989 but
was restarted in 1991. Designated as a missile Gunboat.
Structure: Derived from the Koncar class with a stretched hull
and a new superstructure. Either missiles or mines may be
carried. The second of class is 0.6 m longer than the first ship
and incorporates modifications to the bridge structure.
Operational: Based at Split.

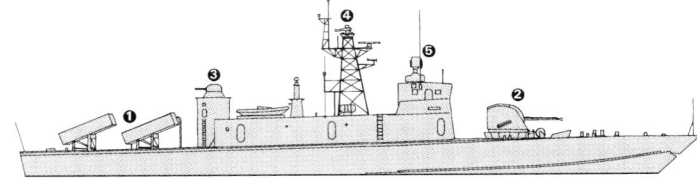

KRALJ DMITAR ZVONIMIR *(Scale 1 : 600), Ian Sturton /* 1044094

UPDATED KRALJ PETAR KRESIMIR IV *6/2003, Agencija Alan /* 0569185

KRALJ DMITAR ZVONIMIR *2/2002, RH-Alan /* 0528425

KRALJ PETAR KRESIMIR IV *2/2002, Hrvatski Vosnik /* 0528426

PATROL FORCES

Notes: A requirement for up to ten offshore patrol vessels is unlikely to be realised in the current financial climate.

1 KONČAR (TYPE R-02) CLASS
(FAST ATTACK CRAFT—MISSILE) (PTGF)

Name	No	Builders	Launched	Commissioned
ŠIBENIK (ex-*Vlado Četković*)	RTOP 21 (ex-402)	Tito SY, Kraljevica	20 Aug 1977	Mar 1978

Displacement, tons: 260 full load
Dimensions, feet (metres): 147.6 × 27.6 × 8.5 *(45 × 8.4 × 2.6)*
Main machinery: CODAG; 2 RR Proteus 52-M558 gas turbines; 7,200 hp *(5.37 MW)* sustained; 2 MTU 16V 538 TB91 diesels; 7,200 hp(m) *(5.29 MW)* sustained; 4 shafts; cp props
Speed, knots: 38; 23 (diesels). **Range, n miles:** 500 at 35 kt; 880 at 23 kt (diesels)
Complement: 29 (5 officers)

Missiles: SSM: 4 Saab RBS 15B; active radar homing to 70 km *(37.8 n miles)* at 0.8 Mach; warhead 83 kg.
Guns: 1 Bofors 57 mm/70; 200 rds/min to 17 km *(9.3 n miles)*; weight of shell 2.4 kg. 128 mm rocket launcher for illuminants.
1-30 mm/65 AK 630M; 6 barrels; 3,000 rds/min to 4 km.
Countermeasures: Decoys: 2 Wallop Barricade double layer chaff launchers.
Weapons control: PEAB 9LV 202 GFCS.
Radars: Surface search: Decca 1226; I-band.
Fire control: Philips TAB; I/J-band.

Programmes: Type name, Raketna Topovnjaca. Recommissioned into the Croatian Navy on 28 September 1991. Others of the class serve with the Yugoslav Navy.
Modernisation: The original Styx missiles have been replaced by RBS 15 and the after 57 mm gun by a 30 mm AK 630. Fire-control radar was updated in 1994.
Structure: Aluminium superstructure. Designed by the Naval Shipping Institute in Zagreb based on Swedish Spica class with bridge amidships like Malaysian boats.
Operational: Based at Split. Reported operational.

VERIFIED

ŠIBENIK   *7/1997, Dario Vuljanić /* 0012240

1 RLM-301 CLASS (RIVER PATROL CRAFT) (PBR)

SLAVONAC PB 91

Displacement, tons: 48 full load
Dimensions, feet (metres): 63.6 × 14.4 × 3.3 *(19.4 × 4.4 × 1)*
Main machinery: 2 Torpedo B 536 diesels; 280 hp(m) *(206 kW)*; 2 shafts
Speed, knots: 12
Complement: 9
Guns: 1 Bofors 40 mm/70. 4—14.5 mm (quad) MGs; 2—12.7 mm MG.
Radars: Surface search: Racal Decca; I-band.

Comment: Former minesweeper launched in 1952 at Mačvanska, Mitrovica. Used as a river patrol vessel. Based at Sisak.

VERIFIED

SLAVONAC   *12/1996, Croatian Navy /* 0017764

4 RIVER PATROL CRAFT (PBR)

BREKI	PB 92	VUKOVAR 91	DOMAGOJ

Displacement, tons: 48 full load
Dimensions, feet (metres): 59.4 × 13.4 × 3.4 *(19.5 × 4.4 × 3.4)*
Main machinery: 2 Torpedo B 538 diesels; 345 hp(m) *(254 kW)*; 2 shafts
Speed, knots: 8
Guns: 1 Bofors 40 mm/70. 4—14.5 mm (quad) MGs. 2—12.7 mm MGs.

Comment: Details given are for *PB 92* which is based at Sisak. *Breki* is of similar size but unarmed while *Vukovar 91* and *Domagoj* are of 30 tons and capable of 6 kt. *Vukovar 91* has a 40 mm/70 gun. The latter three craft are based at Osijek.

VERIFIED

PB 92   *12/1996, Croatian Navy /* 0017765

4 MIRNA (TYPE 140) CLASS
(FAST ATTACK CRAFT—PATROL) (PCM)

Name	No	Builders	Launched
NOVIGRAD (ex-*Biokovo*)	OB 61 (ex-171)	Kraljevica Shipyard	18 Dec 1980
ŠOLTA (ex-*Mukos*)	OB 62 (ex-176)	Kraljevica Shipyard	11 Nov 1982
CAVTAT (ex-*Vrlika*, ex-*Cer*)	OB 63 (ex-180)	Kraljevica Shipyard	27 Sep 1984
HRVATSKA KOSTAJNICA (ex-*Durmitor*)	OB 64 (ex-181)	Kraljevica Shipyard	10 Jan 1985

Displacement, tons: 142 full load
Dimensions, feet (metres): 106.9 × 22 × 7.5 *(32.6 × 6.7 × 2.3)*
Main machinery: 2 SEMT-Pielstick 12 PA4 200 VGDS diesels; 5,292 hp(m) *(3.89 MW)* sustained; 2 shafts
Speed, knots: 25. **Range, n miles:** 600 at 24 kt
Complement: 19 (3 officers)
Missiles: SAM: 1 SA-N-5 Grail quad mounting; manual aiming; IR homing to 6 km *(3.2 n miles)* at 1.5 Mach; altitude to 2,500 m *(8,000 ft)*; warhead 1.5 kg.
Guns: 1 Bofors 40 mm/70. 4 Hispano 20 mm (quad) Type M75. 2—128 mm illuminant launchers.
Depth charges: 8 DCs.
Countermeasures: Decoys: chaff launcher (PB 62).
Radars: Surface search: Racal Decca 1216C; I-band.
Sonars: Simrad SQS-3D/SF; active high frequency.

Comment: An electric outboard motor has been removed. Two were captured after sustaining heavy damage, one by a missile and the other by a torpedo fired from the island of Brač. Both fully repaired and all four are operational, although some may be deleted due to budget constraints.

UPDATED

HRVATSKA KOSTAJNICA   *6/2000, N A Sifferlinger /* 0103706

For details of the latest updates to *Jane's Fighting Ships* online and to discover the additional information available exclusively to online subscribers please visit
jfs.janes.com

AMPHIBIOUS FORCES

2 CETINA (SILBA) CLASS (LCT/ML)

Name	No	Builders	Launched	Commissioned
CETINA	DBM 81	Brodosplit, Split	18 July 1992	19 Feb 1993
KRKA	DBV 82	Brodosplit, Split	17 Sep 1994	9 Mar 1995

Displacement, tons: 880 full load
Dimensions, feet (metres): 163.1 oa; 144 wl × 33.5 × 10.5 *(49.7; 43.9 × 10.2 × 3.2)*
Main machinery: 2 Alpha 10V23L-VO diesels; 3,100 hp(m) *(2.28 MW)* sustained; 2 shafts; cp props
Speed, knots: 12. **Range, n miles:** 1,200 at 12 kt
Complement: 25 (3 officers)
Military lift: 460 tons or 6 medium tanks or 7 APCs or 4—130 mm guns plus towing vehicles or 300 troops with equipment
Missiles: SAM: 1 SA-N-5 Grail quad mounting *(Cetina)*.
Guns: 4—30 mm/65 (2 twin) AK 230 *(Cetina)*.
2 *(Krka)* Hispano 20 mm M71.
Mines: 94 *(Krka)* or 70 *(Cetina)* SAG-1.
Radars: Surface search: Racal Decca 1290A; I-band.

Comment: Ro-ro design with bow and stern ramps. *Krka*'s 40 mm gun is mounted at the bow; *Cetina*'s two 30 mm guns are either side of the bridge. Can be used for minelaying, transporting weapons or equipment and personnel. *Krka* is being used as a water carrier. Both are operational and based at Split.

UPDATED

KRKA   4/2001 / 0528298

CETINA   4/2001 / 0528299

3 TYPE 21 (LCVP)

DJB 103, 104, 107

Displacement, tons: 38 full load
Dimensions, feet (metres): 69.9 × 14.1 × 5.2 *(21.3 × 4.3 × 1.1)*
Main machinery: 1 (2 in *103*) MTU 12V 331 TC81 diesel; 1,450 hp(m) *(1.07 MW)*; 1 shaft (2 waterjets in *103*)
Speed, knots: 21. **Range, n miles:** 320 at 18 kt
Complement: 6
Military lift: 6 tons or 40 troops
Guns: 1—20 mm M71. 1—30 mm grenade launcher.
Radars: Navigation: Decca 1213; I-band.

Comment: Built at Greben Shipyard 1987-88. *DJB 103* upgraded with new main machinery in 1991.

UPDATED

DJB 103   5/1997, Dario Vuljanić / 0012246

1 TYPE 22 (LCVPF)

DJC 106 (ex-624)

Displacement, tons: 42 full load
Dimensions, feet (metres): 73.2 × 15.7 × 3.3 *(22.3 × 4.8 × 1)*
Main machinery: 2 MTU MWM 604 TDV8 diesels; 1,740 hp(m) *(1.28 MW)*; 2 waterjets
Speed, knots: 35. **Range, n miles:** 320 at 22 kt
Complement: 8
Military lift: 40 troops or 15 tons cargo
Guns: 2 Hispano 20 mm. 1—30 mm grenade launcher.
Radars: Navigation: Decca 150; I-band.

Comment: Built at Greben Shipyard in 1987 of polyester and glass fibre.

VERIFIED

DJC 106   8/1998, N A Sifferlinger / 0038489

MINE WARFARE FORCES

0 + 1 MPMB CLASS (MINEHUNTER—INSHORE) (MHI)

Name	No	Builders	Launched
—	—	Greben, Vela Luka	—

Displacement, tons: 173 full load
Dimensions, feet (metres): 84.3 × 22.3 × 8.5 *(25.7 × 6.8 × 2.6)*
Main machinery: 2 MTU 8V 183TE62 diesels; 993 hp(m) *(730 kW)*; 2 Holland Roerpropeler stern azimuth thrusters; bow thruster; 190 hp(m) *(140 kW)*
Speed, knots: 11. **Range, n miles:** 1,000 at 9 kt
Complement: 14 (3 officers)
Guns: 1—20 mm M71.
Countermeasures: Minehunting: 1 ECA38 PAP 104; 1 Super Sea Rover (Benthos); Minesweeping: MDL3 mechanical sweep.
Radars: Navigation: Kelvin Hughes 5000 ARPA, NINAS Mod.
Sonars: Reson mine avoidance; active; high frequency.
Klein 2000 side scan; active for route survey; high frequency.

Comment: Ordered in 1995. The ship has a trawler appearance with a gun on the forecastle and a hydraulic crane on the sweep deck. GRP hull. Due to a shortage of funds, building had stopped by late 1999 but, since then, the contract has been updated and it is planned to complete the ship in late 2005.

VERIFIED

TRAINING SHIPS

1 MOMA (PROJECT 861) CLASS (AX)

Name	No	Builders	Commissioned
ANDRIJA MOHOROVIČIČ	BS 72 (ex-PH 33)	Northern Shipyard, Gdansk	1972

Displacement, tons: 1,514 full load
Dimensions, feet (metres): 240.5 × 36.7 × 12.8 *(73.3 × 11.2 × 3.9)*
Main machinery: 2 Zgoda-Sulzer 6TD48 diesels; 3,300 hp(m) *(2.4 MW)* sustained; 2 shafts; cp props
Speed, knots: 17. **Range, n miles:** 9,000 at 11 kt
Complement: 27 (4 officers)
Radars: Navigation: Racal Decca BT 502; I-band.

Comment: Built in 1971 for the Yugoslav Navy as a survey vessel. Based at Split. Has a 5 ton crane and carries a launch. Used as the Naval Academy training ship.

UPDATED

ANDRIJA MOHOROVIČIČ   4/2001 / 0528297

AUXILIARIES

Notes: In addition there are two harbour tugs *LR-71* and *LR-73*, two diving tenders *BRM-81* and *BRM-83*, auxiliary transport ship *PDS-713*, five harbour transport boats *BMT-1/5*, and two yachts *Učka* (ex-*Podgorka*) and *Jadranka* (ex-civilian *Smile*).

LR 73   10/2004*, E & M Laursen / 1047866

1 SPASILAC CLASS (ASR)

Name	No	Builders	Commissioned
FAUST VRANČIČ (ex-Spasilac)	BS 73 (ex-PS 12)	Tito Shipyard, Belgrade	10 Sep 1976

Displacement, tons: 1,590 full load
Dimensions, feet (metres): 182 × 39.4 × 12.5 *(55.5 × 12 × 3.8)*
Main machinery: 2 diesels; 4,340 hp(m) *(3.19 MW)*; 2 shafts; Kort nozzle props; bow thruster
Speed, knots: 13. **Range, n miles:** 4,000 at 12 kt
Complement: 28 (4 officers)
Cargo capacity: 350 tons fuel; 300 tons deck cargo
Radars: Navigation: Kelvin Hughes Nucleus 5000R; I-band.

Comment: Fitted for firefighting and fully equipped for salvage work. Decompression chamber and can support a German built manned rescue submersible. Can be fitted with two quadruple M 75 and two single M 71 20 mm guns. Reported operational and based at Split.

UPDATED

FAUST VRANČIČ   *6/2000, N A Sifferlinger /* 0103708

1 PT 71 TYPE (TRANSPORT) (AKL)

PT 71 (ex-*Meduza*)

Displacement, tons: 710 full load
Dimensions, feet (metres): 152.2 × 23.6 × 17.1 *(46.4 × 7.2 × 5.2)*
Main machinery: 1 Burmeister & Wain diesel; 930 hp(m) *(684 kW)*; 1 shaft
Speed, knots: 10
Complement: 16 (2 officers)
Guns: 1 Bofors 40 mm/60. 2 Hispano 20 mm M71 can be carried.
Radars: Navigation: Racal Decca 1216A; I-band.

Comment: Built in 1953. Reported operational.

VERIFIED

PT 71   *6/1996 /* 0056850

MINISTRY OF INTERIOR

Notes: (1) Plans to form a Coast Guard were shelved in favour of a strengthened Ministry of Interior maritime force to police inshore waters. These vessels are in five types:
Type 1: 3—24 m craft capable of 30 kt; P-1 *(Srd)*, P-2 *(Marino)*, P-101 *(Sveti Mihovil)*
Type 2: 6—13 m craft capable of 23 kt; P-11 to P-16
Type 3: 6—11 m craft capable of 23 kt; P-111 to P-116
Type 4: 2—14 m craft capable of 30 kt; P-201 and P-202
Type 5: Numerous small craft under 10 m; RIB or inflatable construction
(2) In addition there are civilian registered base port craft with PU (Pula), SB (Sibenic), ST (Split) and so on markings.

P-112   *5/2004*, Martin Mokrus /* 1044095

P-201   *3/2003, Agencija Alan /* 0569186

SVETI MIHOVIL   *3/2003, Agencija Alan /* 0569187

Cuba
MARINA DE GUERRA REVOLUCIONARIA

Country Overview

The Republic of Cuba is an independent republic located in the Caribbean Sea with which it has a 2,020 n mile coastline. The most westerly of the Greater Antilles group, the country comprises two main islands, Cuba (40,519 square miles) and Isla de la Juventud (849 square miles), and more than 1,600 small coral cays and islets. To the west, Cuba commands the approaches to the Gulf of Mexico; the Straits of Florida and the Yucatán Channel separate the country from Florida and Mexico respectively. To the east, the Windward Passage separates the island from Hispaniola (Haiti and the Dominican Republic). Jamaica lies to the south and the Bahamas to the north-east. Havana is the capital, largest city and principal port. Territorial seas (6 n miles) are claimed. A 200 n mile EEZ has been claimed but the limits have not been defined.
    The Navy is in a parlous state and has no capability to sustain operations beyond territorial waters. The Naval Academy is at Punta Santa Ana.

Headquarters Appointments

Chief of Naval Staff:
    Vice Admiral Pedro Perez Betancourt

Personnel

2005: 2,000 (approximately) (including 500 marines)

Command Organisation

Western Naval District (HQ Cabanas).
Eastern Naval District (HQ Holguin).

Naval Aviation

Four Kamov Ka-28 and 14 Mi-14PL Haze A have been reported but operational status is not known.

Coast Defence

Truck mounted SS-N-2B Styx.

Bases

Cabanas, Nicaro, Cienfuegos, Havana, Santiago de Cuba, Banes.

DELETIONS

Notes: Some vessels have been disposed of. Others are decaying alongside in harbour.

CORVETTES

1 PAUK II CLASS (PROJECT 1241PE) (FSM)

321

Displacement, tons: 440 full load
Dimensions, feet (metres): 191.9 × 33.5 × 11.2 *(58.5 × 10.2 × 3.4)*
Main machinery: 2 Type M 521 diesels; 16,184 hp(m) *(11.9 MW)* sustained; 2 shafts
Speed, knots: 32. **Range, n miles:** 2,400 at 14 kt
Complement: 32
Missiles: SAM: SA-N-5 quad launcher; manual aiming, IR homing to 10 km *(5.4 n miles)* at 1.5 Mach; warhead 1.1 kg.
Guns: 1 USSR 76 mm/60; 120 rds/min to 7 km *(3.8 n miles)*; weight of shell 16 kg.
1—30 mm/65; 6 barrels; 3,000 rds/min combined to 2 km. 4—25 mm (2 twin).
A/S mortars: 2 RBU 1200 5-tubed fixed; range 1,200 m; warhead 34 kg.
Countermeasures: 2 PK 16 chaff launchers.
Radars: Air/surface search: Positive E; E/F-band.
Navigation: Pechora; I-band.
Fire control: Bass Tilt; H/I-band.
Sonars: Rat Tail; VDS (on transom); attack; high frequency.

Comment: Built at Yaroslav Shipyard in the USSR and transferred in May 1990. Similar to the ships built for India. Has a longer superstructure than the Pauk I and electronics with a radome similar to the Parchim II class. Torpedo tubes removed. Two twin 25 mm guns fitted on the stern. Based at Havana. Operational status doubtful. *UPDATED*

PAUK II (Indian colours) *2/1998 /* 0052339

PATROL FORCES

6 OSA II CLASS (PROJECT 205)
(FAST ATTACK CRAFT—MISSILE) (PTGF)

| 261 | 262 | 267 | 268 | 271 | 274 |

Displacement, tons: 171 standard; 245 full load
Dimensions, feet (metres): 126.6 × 24.9 × 8.8 *(38.6 × 7.6 × 2.7)*
Main machinery: 3 Type M 504 diesels; 10,800 hp(m) *(7.94 MW)* sustained; 3 shafts
Speed, knots: 37. **Range, n miles:** 500 at 35 kt
Complement: 30
Missiles: SSM: 4 SS-N-2B Styx; active radar or IR homing to 46 km *(25 n miles)* at 0.9 Mach; warhead 513 kg.
Guns: 4—30 mm/65 (2 twin); 500 rds/min to 5 km *(2.7 n miles)*; weight of shell 0.54 kg.
Radars: Surface search: Square Tie; I-band.
Fire control: Drum Tilt; H/I-band.
IFF: Square Head. High Pole B.

Comment: One Osa II delivered in mid-1976, one in January 1977 and one in March 1978. Further two delivered in December 1978, one in April 1979, one in October 1979, two from Black Sea November 1981, four in February 1982. While a few may be seagoing, most have been cannibalised for spares and all have had their missiles disembarked for use in shore batteries. One was sunk as a tourist attraction in 1998. Based at Nicaro and Cabanas. *UPDATED*

OSA II (Bulgarian colours) *8/1998, E & M Laursen /* 0017645

MINE WARFARE FORCES

2 SONYA CLASS (PROJECT 1265)
(MINESWEEPERS/HUNTERS) (MSC/MH)

| 570 | 578 |

Displacement, tons: 450 full load
Dimensions, feet (metres): 157.4 × 28.9 × 6.6 *(48 × 8.8 × 2)*
Main machinery: 2 Kolomna Type 9-D-8 diesels; 2,000 hp(m) *(1.47 MW)* sustained; 2 shafts
Speed, knots: 15. **Range, n miles:** 3,000 at 10 kt
Complement: 43
Guns: 2—30 mm/65 (twin); 500 rds/min to 5 km *(2.7 n miles)*; weight of shell 0.54 kg.
2—25 mm/80 (twin); 270 rds/min to 3 km *(1.6 n miles)*.
Mines: Can carry 8.
Radars: Navigation: Don 2; I-band.
IFF: 2 Square Head. High Pole B.
Sonars: MG 69/79; hull-mounted; active minehunting; high frequency.

Comment: Transferred from USSR in January and December 1985. Two others are non-operational and these two have not been reported at sea since 1999. *UPDATED*

SONYA (Russian colours) *5/1990 /* 0056851

3 YEVGENYA CLASS (PROJECT 1258) (MINEHUNTERS) (MHC)

501, 510, 511

Displacement, tons: 77 standard; 90 full load
Dimensions, feet (metres): 80.7 × 18 × 4.9 *(24.6 × 5.5 × 1.5)*
Main machinery: 2 Type 3-D-12 diesels; 600 hp(m) *(440 kW)* sustained; 2 shafts
Speed, knots: 11. **Range, n miles:** 300 at 10 kt
Complement: 10
Guns: 2—14.5 mm (twin) MGs.
Countermeasures: Minehunting gear is lowered on a crane at the stern.
Radars: Navigation: Don 2; I-band.
Sonars: MG 7 lifted over the stern.

Comment: First pair transferred from USSR in November 1977, one in September 1978, two in November 1979, two in December 1980, two from the Baltic on 10 December 1981, one in October 1982 and four on 1 September 1984. There are two squadrons, based at Cabanas and Nicaro although these last three are the only seaworthy units. *VERIFIED*

YEVGENYA (Russian colours) *1991*

AUXILIARIES

Notes: In addition there are two other vessels: *Siboney* H 101 of 535 tons and used for cadet training, and a buoy tender *Taino* H 102 of 1,123 tons. Neither are active.

1 PELYM (PROJECT 1799) CLASS (AXT)

40

Displacement, tons: 1,050 full load
Dimensions, feet (metres): 210.3 × 38.4 × 11.5 *(64.1 ×11.7 × 3.5)*
Main machinery: 1 diesel; 1,540 hp *(1.1 MW)*; 1 shaft
Speed, knots: 13.5. **Range, n miles:** 1,000 at 13 kt
Complement: 40
Radars: Navigation: Don; I-band.

Comment: Transferred from the USSR in 1982 equipped as deperming vessel. Deperming gear since deleted and now employed as a training ship. Based at Havana. *VERIFIED*

PELYM (Russian colours) *9/1998, T Gander /* 0050067

1 BIYA (PROJECT 871) CLASS (ABU)

GUAMA H 103

Displacement, tons: 766 full load
Dimensions, feet (metres): 180.4 × 32.1 × 8.5 *(55 × 9.8 × 2.6)*
Main machinery: 2 diesels; 1,200 hp(m) *(882 kW)*; 2 shafts; cp props
Speed, knots: 13. **Range, n miles:** 4,700+ at 11 kt
Complement: 29 (7 officers)
Radars: Navigation: Don 2; I-band.

Comment: Has laboratory facilities, one survey launch and a 5 ton crane. Built in Poland and acquired from USSR in November 1980. Subordinate to Institute of Hydrography. Last deployed in 1993, but is used locally as a buoy tender and is based at Havana.
VERIFIED

STENKA *1990 / 0056852*

BIYA (Russian colours)   *10/1993, van Ginderen Collection*

BORDER GUARD

Notes: A 5,000 strong force which operates under the Ministry of the Interior at a higher state of readiness than the Navy. Pennant numbers painted in red.

2 STENKA (TARANTUL) CLASS
(PROJECT 205P) (FAST ATTACK CRAFT—PATROL) (PB)

801   816

Displacement, tons: 211 standard; 253 full load
Dimensions, feet (metres): 129.3 × 25.9 × 8.2 *(39.4 × 7.9 × 2.5)*
Main machinery: 3 M 583A diesels; 12,172 hp(m) *(8.95 MW)*; 3 shafts
Speed, knots: 34. **Range, n miles:** 2,250 at 14 kt
Complement: 25 (5 officers)
Guns: 4—30 mm/65 (2 twin) AK 230; 500 rds/min to 5 km *(2.7 n miles)*; weight of shell 0.54 kg.
Radars: Surface search: Pot Drum; H/I-band.
Fire control: Muff Cob; G/H-band.
IFF: High Pole. Square Head.

Comment: Similar to class operated by Russian border guard with torpedo tubes and sonar removed. Transferred from USSR in February 1985 (two) and August 1985 (one). These two reported to be operational.
VERIFIED

18 ZHUK (GRIF) CLASS (PROJECT 1400M)
(COASTAL PATROL CRAFT) (PB)

Displacement, tons: 39 full load
Dimensions, feet (metres): 78.7 × 16.4 × 3.9 *(24 × 5 × 1.2)*
Main machinery: 2 Type M 401B diesels; 2,200 hp(m) *(1.6 MW)* sustained; 2 shafts
Speed, knots: 30. **Range, n miles:** 1,100 at 15 kt
Complement: 11 (3 officers)
Guns: 4—14.5 mm (2 twin) MGs.
Radars: Surface search: Spin Trough; I-band.

Comment: A total of 40 acquired since 1971. Last batch of two arrived December 1989. Some transferred to Nicaragua. The total has been reduced to allow for wastage. In some of the class the after gun has been removed. Most of the remaining vessels are still active.
VERIFIED

ZHUK (Yemen colours)   *11/1989 / 0056853*

Cyprus

Country Overview

Formerly a British colony, the Republic of Cyprus gained independence in 1960. The United Kingdom retained sovereignty over two military bases on the south coast. The total area of the country is 3,572 square miles but, since 1974, the northern third of the country has been occupied by Turkish troops and has formed, de facto, a separate (not UN recognised) state called the Turkish Republic of Northern Cyprus. Situated in the eastern Mediterranean Sea, with which it has a 351 n mile coastline, the island lies west of Syria and south of Turkey. Nicosia is the capital and largest city while Limassol and Larnaca are the principal ports. Territorial seas (12 n miles) are claimed. An EEZ has not been claimed.

Headquarters Appointments

Commander Navy Command of the National Guard:
   Captain Fotis Kotsis
Director Operations:
   Commander Constantinos Fitiris

General

Raif Denktas KKTC 101, *Erenköy* KKTC 02, KKTC 104 and two KAAN 15 craft, KKTC SG11 and SG12 are patrol craft permanently based at Kyrenia (Girne) flying the North Cyprus flag. For details of these vessels see Turkey Coast Guard section.

Bases

Limassol
Zyyi

Coast Defence

Twenty-four Exocet MM 40 Block 2. Truck-mounted in batteries of four.

PATROL FORCES

Notes: (1) The expected transfer of two or more Combattante II class fast attack craft from Greece did not take place in 2003 but remains under review.
(2) There are also three launches and a number of RIBs in use by the Underwater Diving section of the Navy.

1 MODIFIED PATRA CLASS (PBM)

Name	No	Builders	Commissioned
SALAMIS	P 01	Chantiers de l'Esterel	24 May 1983

Displacement, tons: 92 full load
Dimensions, feet (metres): 105.3 × 21.3 × 5.9 *(32.1 × 6.5 × 1.8)*
Main machinery: 2 SACM 195 CZSHRY12 diesels; 4,680 hp(m) *(3.44 MW)* sustained; 2 shafts
Speed, knots: 30. **Range, n miles:** 1,200 at 15 kt
Complement: 22
Missiles: SAM: 2 Matra Simbad twin launchers; Mistral; IR homing to 4 km *(2.2 n miles)*; warhead 3 kg.
Guns: 1—40 mm. 1 Oerlikon Breda 25 mm.
1 Rheinmetall Wegmann 20 mm. 2—12.7 mm MGs.
Radars: Surface search: Decca 1226; I-band.

Comment: Laid down in December 1981 for Naval Command of National Guard.
UPDATED

SALAMIS   *10/1999, E & M Laursen / 0056854*

1 DILOS CLASS (COASTAL PATROL CRAFT) (PBM)

Name	No	Builders	Commissioned
KYRENIA (ex-*Knossos*)	P 02 (ex-P 268)	Hellenic Shipyards, Skaramanga	1979

Displacement, tons: 92 full load
Dimensions, feet (metres): 95.1 × 16.2 × 5.6 *(29 × 5 × 1.7)*
Main machinery: 2 MTU 12V 331 TC81 diesels; 2,700 hp(m) *(1.97 MW)* sustained; 2 shafts
Speed, knots: 26. **Range, n miles:** 1,600 at 24 kt
Complement: 17 (4 officers)
Missiles: 2 Matra Simbad twin launchers; Mistral; IR homing to 4 km *(2.2 n miles)*; warhead 3 kg
Guns: 1 Oerlikon Breda 25 mm.
Radars: Surface search: Racal Decca 914C; I-band.

Comment: Ordered in May 1976 to a design by Abeking & Rasmussen. Transferred from Greece in March 2000 and used mainly for SAR. Others of the class are in service in Georgia, and with the Hellenic Coast Guard and Customs services.   **VERIFIED**

KYRENIA (Greek colours)   *6/1998, E M Cornish* / 0052296

2 RODMAN 55HJ CLASS (PBF)

PANAGOS   AGATHOS

Displacement, tons: 15.7 full load
Dimensions, feet (metres): 57.1 × 12.5 × 2.3 *(17.4 × 3.8 × 0.7)*
Main machinery: 2 MAN diesels; 2 waterjets
Speed, knots: 48. **Range, n miles:** 300 at 35 kt
Complement: 7
Guns: 1—12.7 mm MG. 2—7.62 mm MGs.
Radars: Surface search: Furuno; I-band.

Comment: GRP hulls built by Rodman, Vigo and commissioned on 8 June 2002.   **UPDATED**

PANAGOS   *9/2002*, van Ginderen Collection* / 1044096

2 + (2) VITTORIA CLASS (COASTAL PATROL CRAFT) (PB)

Name	No	Builders	Commissioned
COMMANDER TSOMAKIS	P 03	Cantiere Navale Vittoria, Adria	Aug 2004
COMMANDER GEORGIU	P 04	Cantiere Navale Vittoria, Adria	Aug 2004

Displacement, tons: 95 full load
Dimensions, feet (metres): 88.9 × 13.4 × 1.0 *(27.6 × 4.1 × 0.3)*
Main machinery: 2 MTU diesels; 2 shafts
Speed, knots: 45. **Range, n miles:** 800 at 35 kt
Complement: 12
Guns: 1—30 mm. 2—12.7 mm MGs.

Comment: Built by Cantiere Navale Vittoria, Italy. Two further craft are expected, possibly for the Police force.   **NEW ENTRY**

LAND-BASED MARITIME AIRCRAFT

Notes: There are also three Bell 206 utility helicopters.

Numbers/Type: 1 Pilatus Britten-Norman Maritime Defender BN-2A.
Operational speed: 150 kt *(280 km/h).*
Service ceiling: 18,900 ft *(5,760 m).*
Range: 1,500 n miles *(2,775 km).*
Role/Weapon systems: Operated around southern coastline of Cyprus to prevent smuggling and terrorist activity. Sensors: Search radar, searchlight mounted on wings. Weapons: ASV; various machine gun pods and rockets.   **VERIFIED**

POLICE

Notes: (1) In addition there are five Fletcher Malibu speed boats, *Astrapi I-V,* of 5.3 m with 200 hp engines built in Cyprus in 1986.
(2) Personnel numbers are approximately 330 Maritime Police.

5 SAB 12 TYPE (PB)

KARPASIA PL 11 (ex-G 50/GS 10)   DIONYSOS PL 14 (ex-G 55/GS 12)
ILARION PL 12 (ex-G 52/GS 25)   AKAMAS PL 15 (ex-G 57/GS 28)
KOURION PL 13 (ex-G 54/GS 27)

Displacement, tons: 14 full load
Dimensions, feet (metres): 41.3 × 13.1 × 3.6 *(12.6 × 4 × 1.1)*
Main machinery: 2 Volvo Penta diesels; 539 hp(m) *(396 kW)*; 2 shafts
Speed, knots: 16. **Range, n miles:** 300 at 15 kt
Complement: 5
Guns: 1—7.62 mm MG.
Radars: Surface search: Raytheon; I-band.

Comment: Built in 1979 by Veb Yachwerft, Berlin. Harbour patrol craft of the former GDR MAB 12 class transferred in December 1992. New radars fitted.   **VERIFIED**

ILARION   *10/1999, E & M Laursen* / 0056858

1 SHALDAG CLASS (PBF)

Name	No	Builders	Commissioned
ODYSSEUS	PV 22	Israel Shipyards	4 Sep 1997

Displacement, tons: 56 full load
Dimensions, feet (metres): 81.4 × 19.7 × 3.9 *(24.8 × 6 × 1.2)*
Main machinery: 2 MTU 12V 396TE diesels; 3,560 hp(m) *(2.62 MW)* sustained; 2 Kamewa waterjets
Speed, knots: 45. **Range, n miles:** 850 at 16 kt
Complement: 15
Guns: 1 Oerlikon 20 mm; 2—7.62 mm MGs.
Weapons control: Optronic director.
Radars: Surface search: Raytheon; I-band.

Comment: Similar to craft in service with Sri Lankan Navy.   **UPDATED**

PV 22   *10/1999, E & M Laursen* / 0056855

2 POSEIDON CLASS (PBF)

Name	No	Builders	Commissioned
POSEIDON	PV 20	Brodotehnika SY, Belgrade	21 Nov 1991
EVAGORAS	PV 21	Brodotehnika SY, Belgrade	21 Nov 1991

Displacement, tons: 58 full load
Dimensions, feet (metres): 80.7 × 18.7 × 3.9 *(24.6 × 5.7 × 1.2)*
Main machinery: 2 MTU 12V 396 TE94 diesels; 3,560 hp(m) *(2.62 MW)* sustained; 2 Kamewa 56 water-jets
Speed, knots: 42. **Range, n miles:** 600 at 20 kt
Complement: 9
Guns: 1 Breda KVA 25 mm; ISBRS rocket launcher. 2—12.7 mm MGs.
Radars: Surface search: JRC; I-band.

Comment: Designated as FAC-23 Jets. Aluminium construction. New radars fitted.   **VERIFIED**

POSEIDON   *9/2002, van Ginderen Collection* / 0569188

Denmark
DEN KONGELIGE DANSKE MARINE

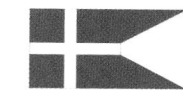

Country Overview

The Kingdom of Denmark is a constitutional monarchy. The southernmost of the Scandinavian countries, it comprises most of the Jutland peninsula and more than 400 islands, the principal of which are Sjaelland (the largest), Fyn, Lolland, Falster, Langeland and Møn. The island of Bornholm lies in the Baltic about 70 n miles east of Sjaelland. With an area of 16,639 square miles, the country is bordered to the south by Germany. Its 1,825 n mile coastline is with the North Sea to the west, the Skagerrak to the north and the Kattegatt, which is linked to the Baltic Sea by the Øresund, to the east. The capital, largest city and principal port is Copenhagen. There are further ports at Århus, Odense and Ålborg. Territorial seas (12 n miles) are claimed. It has claimed a 200 n mile EEZ for the mainland and 200 n mile Fishery Zones for the external territories of the Faroes and Greenland.

Headquarters Appointments

Admiral Fleet:
  Rear Admiral K B Jensen
Inspector Naval Home Guard:
  Captain K R Andersen

Diplomatic Representation

Defence Attaché, Washington and Ottawa:
  Brigadier J F Autzen
Defence Attaché, London, Dublin and The Hague:
  Captain N A K Olsen
Defence Attaché, Paris:
  Colonel C J D Dirksen
Defence Attaché, Berlin and Prague:
  Colonel P O Topp
Defence Attaché, Moscow and Minsk:
  Brigadier S B Bojesen
Defence Attaché, Warsaw:
  Colonel P E Tranberg
Defence Attaché, Vilnius:
  Commander Senior Grade C V Rasmussen
Defence Attaché, Riga:
  Lieutenant Colonel O C Grüner
Defence Attaché, Kiev:
  Lieutenant Colonel C Mathiesen
Defence Attaché, Helsinki and Tallin:
  Lieutenant Colonel E K Praestegaard

Personnel

(a) 2005: 3,770 (874 officers) including 450 national service
    Reserves: 4,000.
    Naval Home Guard: 4,800.
(b) 9 months' national service

Bases

Korsør (Corvettes, Stanflex), Frederikshavn (Submarines, Minelayers, MCMV, Fishery Protection Ships), Copenhagen, Grønnedal (Greenland)

Naval Air Arm

Naval helicopters owned and operated by Navy in naval squadron based at Karup, Jutland. All servicing and maintenance by Air Force. LRMP are flown by the Air Force.

Naval Home Guard

Established in 1952 as a separate service under the operational control of the navy. Duties include surveillance, harbour control, search and rescue and the guarding of naval installations ashore. Following the Defence Agreement 2004, the service is to play a greater role in home defence and further tasks include environmental survey, pollution control and support of the police and customs services.

Coast Defence

Ten radar stations and a number of coast watching stations. The two mobile Harpoon batteries were decommissioned in late 2003.

Command and Control

The Royal Danish Navy, on behalf of the Ministry of Defence, runs and maintains the icebreakers. Likewise, the Navy runs and maintains two environmental protection divisions based in Copenhagen and Korsør respectively. Responsibility for environmental survey, protection and pollution fighting in maritime areas around Denmark is executed by the Royal Danish Navy. Survey ships are run by the Farvandsvæsenet Nautisk Afdeling (Administration of Navigation and Hydrography) under the Ministry of Defence, and the Directorate of Fisheries has four rescue vessels.

Appearance

Ships are painted in six different colours as follows:
Grey: frigates, corvettes and patrol frigates.
Black: submarines.
Orange: survey vessels.
White: Royal Yacht and the sail training yawls.
Black/yellow: service vessels, tugs and ferryboats.

Strength of the Fleet

Type	Active	Building (Projected)
Frigates	7	(2-4)
Large Patrol Craft	26	(8-9)
Coastal Patrol Craft	2	—
Naval Home Guard	29	1
Minehunters and Drones	10	(8)
Support Ships	2	—
Transport Ship	1	—
Icebreakers	3	—
Royal Yacht	1	—

Prefix to Ships' Names

HDMS

DELETIONS

Submarines

2003 *Narhvalen, Nordkaperen*
2004 *Tumleren, Saelen, Springeren, Kronborg*

Mine Warfare Forces

2003 *Fyen*
2004 *Lindormen, Lossen, Møen*

Auxiliaries

2003 *Hugin, Elbjørn, Nordjylland, Jens Vaever*
2004 *Munin, Mimer*

PENNANT LIST

Frigates

F 354	Niels Juel	
F 355	Olfert Fischer	
F 356	Peter Tordenskiold	
F 357	Thetis	
F 358	Triton	
F 359	Vaedderen	
F 360	Hvidbjørnen	

Patrol Forces

P 550	Flyvefisken	
P 551	Hajen	
P 552	Havkatten	

P 553	Laxen	Y 304	Thurø	A 559	Sleipner
P 554	Makrelen	Y 305	Vejrø	A 560	Gunnar Thorson
P 555	Støren	Y 306	Farø	A 561	Gunnar Seidenfaden
P 556	Svaerdfisken	Y 307	Laesø	A 562	Mette Miljø
P 557	Glenten	Y 308	Rømø	A 563	Marie Miljø
P 558	Gribben	Y 386	Agdlek	L 16	Absalon
P 559	Lommen	Y 387	Agpa	L 17	Esbern Snare
P 560	Ravnen	Y 388	Tulugaq	Y 101	Svanen
P 561	Skaden			Y 102	Thyra
P 562	Viben	**Auxiliaries**			
P 563	Søløven				
Y 300	Barsø	A 540	Dannebrog		
Y 301	Drejø	A 551	Danbjørn		
Y 302	Romsø	A 552	Isbjørn		
Y 303	Samsø	A 553	Thorbjørn		

SUBMARINES

Notes: Following approval of the 2005-09 Defence Plan by the Danish parliament on 10 June 2004 the submarine service has been disestablished. *Tumleren, Saelen* and *Springeren* have been decommissioned and may be sold or scrapped. *Kronborg* has been returned to Sweden. Denmark has withdrawn from the Viking project.

FRIGATES

0 + 3 PATROL SHIPS (FFHM)

Name	No	Builders	Laid down	Launched	Commissioned
—	—	Odense Shipyard, Lindø	2007	2009	2010
—	—	Odense Shipyard, Lindø	2008	2010	2011
—	—	Odense Shipyard, Lindø	2009	2011	2012

Displacement, tons: 5,000
Dimensions, feet (metres): 449.6 × 64.0 × 20.7
  (137.0 × 19.5 × 6.3)
Main machinery: To be announced
Speed, knots: 26

Missiles: SSM: McDonnell Douglas Block II
  SAM: Standard SM-2 Block III. Raytheon ESSM. 4 octuple Lockheed Martin Mk 41 Vertical Launch Systems.
Guns: 1—5 in *(127 mm).*
Countermeasures: ESM. Decoys.
Combat data systems: To be announced.
Weapons control: To be announced.
Radars: Air search: SMART 1; 3D; D-band.
  Surface search: To be announced.
  Navigation: To be announced.
Sonars: Hull-mounted and towed.

Helicopters: 1 EH-101 or 2 Lynx.

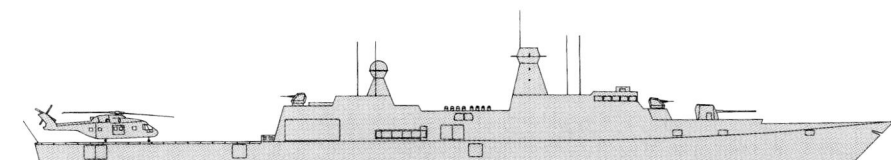

PATROL SHIP   *(Scale 1 : 1,200), Ian Sturton /* 1047858

Programmes: Contract in October 2001 for detailed design work. Following the Defense Agreement of 10 June 2004, the number of ships has been raised from two to one batch of three ships. Long-lead items ordered from mid-2004. Details of weapons and sensors are illustrative and have not been confirmed. The design has the flexibility to accommodate further weapons such as Tactical Tomahawk.

Structure: A variant of the Flexible Support Ships to be built to DNV Navy standards. Design evolved from the Thetis class and based on Stanflex concept. Large equipment items can be deployed in up to six container positions.
Operational: To replace Niels Juel class, ships to have general purpose role and to be able to act as command platform.
UPDATED

3 NIELS JUEL CLASS (FFGM)

Name	No	Builders	Laid down	Launched	Commissioned
NIELS JUEL	F 354	Aalborg Vaerft	20 Oct 1976	17 Feb 1978	26 Aug 1980
OLFERT FISCHER	F 355	Aalborg Vaerft	6 Dec 1978	10 May 1979	16 Oct 1981
PETER TORDENSKIOLD	F 356	Aalborg Vaerft	3 Dec 1979	30 Apr 1980	2 Apr 1982

Displacement, tons: 1,320 full load
Dimensions, feet (metres): 275.5 × 33.8 × 10.2
(84 × 10.3 × 3.1)
Main machinery: CODOG; 1 GE LM 2500 gas turbine;
24,600 hp *(18.35 MW)* sustained; 1 MTU 20 V 956 TB82
diesel; 5,210 hp(m) *(3.83 MW)* sustained; 2 shafts
Speed, knots: 28, gas; 20, diesel
Range, n miles: 2,500 at 18 kt
Complement: 94 (15 officers)

Missiles: SSM: 8 McDonnell Douglas Harpoon (2 quad)
launchers ❶; active radar homing to 130 km *(70 n miles)* at
0.9 Mach; warhead 227 kg.
SAM: 12 (2 sextuple) Raytheon Sea Sparrow Mk 48 Mod 3 VLS
(12 missiles) or Mk 56 Mod O VLS (24 missiles) modular
launchers ❷; semi-active radar homing to 14.6 km *(8 n miles)*
at 2.5 Mach; warhead 39 kg; 12 missiles.
4 Stinger mountings (2 twin) ❸.
Guns: 1 OTO Melara 3 in *(76 mm)*/62 compact ❹; 85 rds/min to
16 km *(8.7 n miles)* anti-surface; 12 km *(6.6 n miles)* anti-
aircraft; SAPOMER round weight 12.7 kg.
4—12.7 mm MGs.
Depth charges: 1 rack.
Countermeasures: Decoys: 2 DL-12T Sea Gnat 12-barrelled
chaff launchers ❺.

Combat data systems: CelciusTech 9LV Mk 3. Link 11.
SATCOMs (can be fitted forward or aft of the funnel).
Weapons control: Philips 9LV 200 Mk 3 GFCS with TV tracker.
Raytheon Mk 91 Mod 1 MFCS with two directors. Harpoon to
1A(V) standard.
Radars: Air search: DASA TRS-3D ❻; G/H-band.
Surface search: Philips 9GR 600 ❼; I-band.
Fire control: 2 Mk 95 ❽; I/J-band (for SAM).
Philips 9LV 200 Mk 1 Rakel 203C ❾; J-band (for guns and
SSM).
Navigation: Terma Scanter Mil; I-band.
Sonars: Plessey PMS 26; hull-mounted; active search and attack;
10 kHz.

Programmes: YARD Glasgow designed the class to Danish
order.
Modernisation: Mid-life update from 1996, including a NATO
Sea Sparrow VLS, and new communications. Air search radar
replaced by TST TRS-3D. Improved combat data system fitted.
F 356 completed in May 1998, F 354 in April 1999, F 355 in
December 2001. Stinger SAM mounted each side of the
funnel.
Operational: Normally only one sextuple SAM launcher is
carried, but the second set can be embarked in a few hours.
UPDATED

PETER TORDENSKIOLD

(Scale 1 : 900), Ian Sturton / 1047859

PETER TORDENSKIOLD

9/2004 *, Harald Carstens / 1044116

OLFERT FISCHER

9/2004 *, M Nitz / 1044119

PETER TORDENSKIOLD
9/2003*, Harald Carstens / 1044113

NIELS JUEL
9/2004*, Harald Carstens / 1044114

PETER TORDENSKIOLD
2/2004*, L-G Nilsson / 1044115

4 THETIS CLASS (FFHM)

Name	No	Builders	Laid down	Launched	Commissioned
THETIS	F 357	Svenborg Vaerft	10 Oct 1988	14 July 1989	1 July 1991
TRITON	F 358	Svenborg Vaerft	27 June 1989	16 Mar 1990	2 Dec 1991
VAEDDEREN	F 359	Svenborg Vaerft	19 Mar 1990	21 Dec 1990	9 June 1992
HVIDBJØRNEN	F 360	Svenborg Vaerft	2 Jan 1991	11 Oct 1991	30 Nov 1992

Displacement, tons: 2,600 standard; 3,500 full load
Dimensions, feet (metres): 369.1 oa; 327.4 wl × 47.2 × 19.7
(112.5; 99.8 × 14.4 × 6.0)
Main machinery: 3 MAN/Burmeister & Wain Alpha 12V 28/
32A diesels; 10,800 hp(m) *(7.94 MW)* sustained; 1 shaft;
Kamewa cp prop; bow and azimuth thrusters; 880 hp(m)
(647 kW), 1,100 hp(m) *(800 kW)*
Speed, knots: 20; 8 on thrusters
Range, n miles: 8,500 at 15.5 kt
Complement: 60 (12 officers) plus 30 spare berths

Missiles: SAM: 4 Stinger mountings (2 twin) on hangar roof near
mast.
Guns: 1 OTO Melara 3 in *(76 mm)*/62; Super Rapid ❶; dual
purpose; 120 rds/min to 16 km *(8.7 n miles)*; SAPOMER
round weight 12.7 kg.
2—12.7 mm MGs.
Depth charges: 2 Rails (door in stern).
Countermeasures: Decoys: 2 Sea Gnat DL-12T 12-barrelled
launchers for chaff and IR flares.
ESM: Racal Sabre; intercept.
Combat data systems: Terma TDS; SATCOM ❷.
Weapons control: Bofors 9LV 200 Mk 3 director. FSI Safire
surveillance director ❸.
Radars: Air/surface search: Plessey AWS 6 ❹; G-band.
Surface search: Furuno 2135; E/F-band.
Navigation: Furuno 2115; I-band.
Fire control: CelsiusTech 9LV Mk 3 ❺; I/J-band.
Sonars: Thomson Sintra TSM 2640 Salmon; VDS; active search
and attack; medium frequency.
C-Teck; hull-mounted; active search; medium frequency.

Helicopters: 1 Westland Lynx Mk 90B ❻.

Programmes: Preliminary study by YARD in 1986 led to Dwinger
Marine Consultants being awarded a contract for a detailed
design completed in mid-1987. All four ordered in October
1987.
Modernisation: There are plans for a new air search radar and
SAM in due course.
Structure: The hull is some 30 m longer than the Hvidbjørnen
class to improve sea-keeping qualities and allow considerable
extra space for additional armament. The design allows the
use of containerised equipment to be shipped depending on
role and there is some commonality with the Flex 300 ships.
The hull is ice strengthened to enable penetration of 1 m thick
ice and efforts have been made to incorporate stealth
technology, for instance by putting anchor equipment,
bollards and winches below the upper deck. There is a double
skin up to 2 m below the waterline. A rigid inflatable boarding

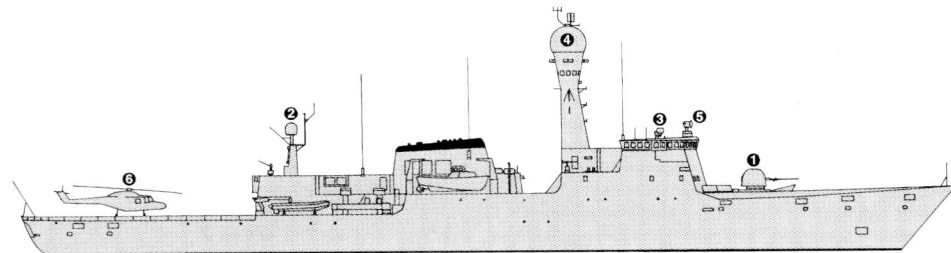

THETIS

(Scale 1 : 900), Ian Sturton / 0012258

THETIS

*6/2004 *, P Froud / 1121261*

craft plumbed by a hydraulic crane is fitted alongside the fixed
hangar. The bridge and ops room are combined. *Thetis* was
modified in the stern for seismological survey. Since these
operations have terminated, the stern has been remodified to
facilitate the ability to act as a command ship and to conduct
training.
Operational: Primary role is fishery protection. *UPDATED*

THETIS

*6/2004 *, J Cislak / 1044118*

THETIS

*9/2003 *, Harald Carstens / 1044112*

SHIPBORNE AIRCRAFT

Notes: The Defence Agreement of 10 June 2004 provided for the procurement of four maritime helicopters. These might be further Merlin aircraft in addition to the 14 utility variants being procured for the Danish Air Force.

Numbers/Type: 8 Westland Lynx Mk 90B.
Operational speed: 125 kt *(232 km/h)*.
Service ceiling: 12,500 ft *(3,810 m)*.
Range: 320 n miles *(593 km)*.
Role/Weapon systems: Shipborne helicopter for EEZ and surface search tasks. All being upgraded to Super Lynx standard with first delivered November 2000. Sensors: Ferranti Seaspray; Racal Kestrel ESM; FLIR 2000. Weapons: Unarmed.

UPDATED

LYNX   2/2004 *, Per Körnefeldt / 1044107

LAND-BASED MARITIME AIRCRAFT

Notes: Fourteen Agusta Westland EH 101 helicopters are being procured for SAR and troop transport missions. They are to be operated by the Air Force.

Numbers/Type: 3 Challenger 604.
Operational speed: 470 kt *(870 km/h)*.
Service ceiling: 41,000 ft *(12,497 m)*.
Range: 3,769 n miles *(6,980 km)*.
Role/Weapon systems: Maritime reconnaissance for EEZ patrol in the Baltic and off Greenland. Sensors: Terma SLAR radar; IR/UV scanner. Weapons: unarmed.

VERIFIED

CHALLENGER 604   6/2001, Royal Danish Navy / 0114705

Numbers/Type: 8 Sikorsky S-61A-1 Sea King.
Operational speed: 118 kt *(219 km/h)*.
Service ceiling: 14,700 ft *(4,480 m)*.
Range: 542 n miles *(1,005 km)*.
Role/Weapon systems: Land-based SAR helicopter for combat rescue and surface search. Sensors: Bendix weather radar; GEC Avionics FLIR. Weapons: unarmed.

VERIFIED

SEA KING   5/1999, H M Steele / 0056867

PATROL FORCES

Notes: *Lunden* Y 343 is a 20 m trawler type vessel. Previously thought to have been decommissioned, she remains in service as a target-towing vessel.

2 VTS CLASS (COASTAL PATROL CRAFT) (PB)

VTS 3   VTS 4

Displacement, tons: 34 full load
Dimensions, feet (metres): 55.8 × 16.1 × 6.9 *(17 × 4.9 × 2.1)*
Main machinery: 2 MWM TBD 616 V12 diesels; 979 hp(m) *(720 kW)*; 2 waterjets
Speed, knots: 33. **Range, n miles:** 300 at 30 kt
Complement: 3
Guns: 1—7.62 mm MG can be carried.
Radars: Surface search: Furuno FR 1505 Mk 2; I-band.
Navigation: Furuno M1831; I-band.

Comment: Built by Mulder & Rijke, Netherlands. Completed in 1997 and 1998 to replace Botved type.

VERIFIED

VTS 3   6/1999, Royal Danish Navy / 0056874

0 + 2 (1) ARCTIC PATROL SHIPS (PGBH)

Name	No	Builders	Laid down	Launched	Commissioned
—	—	Karstensens Skibsvaerft, Skagen	2006	2007	2008
—	—	Karstensens Skibsvaerft, Skagen	2006	2007	2008

Displacement, tons: 1,680
Dimensions, feet (metres): 231.0 × 45.9 × 16.7 *(70.4 × 14.0 × 5.1)*
Main machinery: To be announced
Speed, knots: 18
Missiles: SAM: Vertical Launch System for 12 Evolved Sea Sparrow (ESSM).
Guns: 1 OTO Melara 3 in *(76 mm)*/62 Super Rapid; 120 rds/min to 16 km *(8.7 n miles)*; SAPOMER round weight 12.7 kg.
Countermeasures: ESM. Decoys.
Combat data systems: To be announced.
Radars: Surface/air search: To be announced.
Fire control: To be announced.
Navigation: To be announced.
Sonars: To be announced.
Helicopters: 1 medium.

Programmes: Contract in December 2004 to achieve an in-service date of 2008. Option for a third unit.
Structure: To be built to DNV Navy standards using Stanflex concept. Large equipment items can be deployed in up to four container positions.
Operational: To replace Agdlek class.

NEW ENTRY

3 AGDLEK CLASS (LARGE PATROL CRAFT) (PB)

Name	No	Builders	Commissioned
AGDLEK	Y 386	Svendborg Vaerft	12 Mar 1974
AGPA	Y 387	Svendborg Vaerft	14 May 1974
TULUGAQ	Y 388	Svendborg Vaerft	26 June 1979

Displacement, tons: 394 full load
Dimensions, feet (metres): 103 × 25.3 × 11.2 *(31.4 × 7.7 × 3.4)*
Main machinery: 1 Burmeister & Wain Alpha A08-26 VO diesel; 800 hp(m) *(588 kW)*; 1 shaft
Speed, knots: 12
Complement: 14 (3 officers)
Guns: 2—12.7 mm MGs.
Radars: Surface search: Furuno 2135; E/F-band.
Navigation: Furuno 1510; I-band.

Comment: Designed for service off Greenland. Ice strengthened. SATCOM fitted. To be replaced from from 2007 by new 70 m Arctic patrol vessels.

UPDATED

TULUGAQ   10/2004 *, Per Körnefeldt / 1044106

0 + 6 SF MK II CLASS (LARGE PATROL CRAFT) (PB)

Displacement, tons: 276 full load
Dimensions, feet (metres): 141.1 × 26.9 × 7.2 *(43.0 × 8.2 ×2.2)*
Main machinery: 2 diesels; 2,700 hp(m) *(2 MW)*; 1 cp prop
Speed, knots: 18
Complement: 9 (accommodation for 15)

Comment: GRP vessels to replace the Ø class. Ordered from Faaborg Vaerft, Denmark, the hull, superstructure and machinery are to be built at Kockums, Karlskrona. To be delivered 2006-07.

NEW ENTRY

14 FLYVEFISKEN CLASS (LARGE PATROL/ATTACK CRAFT AND MINEHUNTERS/LAYERS) (PGGM/MHCD/MLC/AGSC)

Name	No	Builders	Commissioned
FLYVEFISKEN	P 550	Danyard A/S, Aalborg	19 Dec 1989
HAJEN	P 551	Danyard A/S, Aalborg	19 July 1990
HAVKATTEN	P 552	Danyard A/S, Aalborg	1 Nov 1990
LAXEN	P 553	Danyard A/S, Aalborg	22 Mar 1991
MAKRELEN	P 554	Danyard A/S, Aalborg	1 Oct 1991
STØREN	P 555	Danyard A/S, Aalborg	24 Apr 1992
SVAERDFISKEN	P 556	Danyard A/S, Aalborg	1 Feb 1993
GLENTEN	P 557	Danyard A/S, Aalborg	29 Apr 1993
GRIBBEN	P 558	Danyard A/S, Aalborg	1 July 1993
LOMMEN	P 559	Danyard A/S, Aalborg	21 Jan 1994
RAVNEN	P 560	Danyard A/S, Aalborg	17 Oct 1994
SKADEN	P 561	Danyard A/S, Aalborg	10 Apr 1995
VIBEN	P 562	Danyard A/S, Aalborg	15 Jan 1996
SØLØVEN	P 563	Danyard A/S, Aalborg	28 May 1996

Displacement, tons: 480 full load
Dimensions, feet (metres): 177.2 × 29.5 × 8.2 (54 × 9 × 2.5)
Main machinery: CODAG; 1 GE LM 500 gas turbine (centre shaft); 5,450 hp (4.1 MW) sustained; 2 MTU 16V 396 TB94 diesels (outer shafts); 5,800 hp(m) (4.26 MW) sustained; 3 shafts; cp props on outer shafts; bow thruster. Auxiliary propulsion by hydraulic motors on outer gearboxes; hydraulic pumps driven by 1 GM 12V-71 diesel; 500 hp (375 kW)
Speed, knots: 30; 20 on diesels; 10 on hydraulic propulsion. **Range, n miles:** 2,400 at 18 kt
Complement: 19-29 (depending on role) (4 officers)

Missiles: SSM: 8 McDonnell Douglas Harpoon; active radar homing to 130 km (70 n miles) at 0.9 Mach; warhead 227 kg. Attack role only. Block II from 2004 gives land attack option.
SAM: Raytheon Sea Sparrow Mk 48 Mod 3 VLS (6 missiles) or Mk 56 Mod 0 VLS (12 missiles); semi-active radar homing to 14.6 km (8 n miles) at 2.5 Mach; warhead 32 kg. 12 missiles from 2002. Fitted for Attack, MCM and Minelaying roles. In MCM role one Stinger twin launcher can be fitted instead of Sea Sparrow.
Guns: 1 OTO Melara 3 in (76 mm)/62 Super Rapid; dual purpose; 120 rds/min to 16 km (8.7 n miles); SAPOMER round weight 12.7 kg.
2—12.7 mm MGs.
Torpedoes: 2—21 in (533 mm) tubes; FFV Type 613; wire-guided passive homing to 15 km (8.2 n miles) at 45 kt; warhead 240 kg. Eurotorp Mu 90 Impact from 2001.
Depth charges: 4.
Mines: 60. Minelaying role only.
Countermeasures: MCMV: Ibis 43 minehunting system with Thomson Sintra 2061 tactical system and 2054 side scan sonar towed by MSF class drones (see *Mine Warfare Forces* section). Bofors Double Eagle ROV Mk II. Minehunting role only.
Decoys: 2 Sea Gnat 130 mm DL-6T 6-barrelled launcher for chaff and IR flares.
ESM: Racal Sabre; radar warning.
Combat data systems: Terma/CelsiusTech TDS. Link 11 being fitted.
Weapons control: CelsiusTech 9LV Mk 3 optronic director. Harpoon to 1A(V) standard or AHWCS with Block II.
Radars: Air/surface search: Plessey AWS 6 (550-556); G-band; or Telefunken SystemTechnik TRS-3D (557-563); G/H-band.
Surface search: Terma Scanter Mil; I-band.
Navigation: Furuno; I-band.
Fire control: CelsiusTech 9LV 200 Mk 3; Ku/J-band.

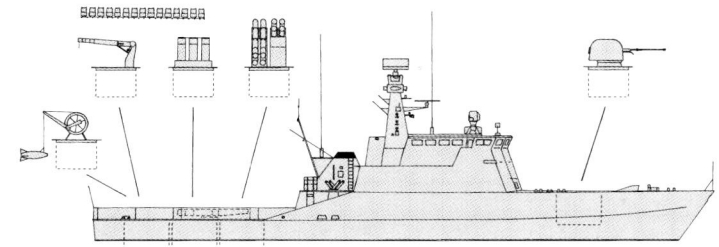

FLYVEFISKEN (composite fit)   (Scale 1 : 600), Ian Sturton / 0103718

GLENTEN   5/2004*, Per Körnefeldt / 1044110

Sonars: CelsiusTech CTS-36/39; hull-mounted; active search; high frequency.
Thomson Sintra TSM 2640 Salmon; VDS; medium frequency. For ASW only.

Programmes: Standard Flex 300 replaced Daphne class (seaward defence craft), Søløven class (fast attack craft torpedo), and Sund (MCM) class. First batch of seven with option on a further nine contracted with Danyard on 27 July 1985. Second batch of six ordered 14 June 1990 and last one authorised in 1993 to a total of 14, two less than originally planned.
Modernisation: Mk 48 Mod 3 SAM launchers replaced by Mk 56 launchers from 2002. Block II Harpoon with GPS from 2004. Link 11 being fitted. *Lommen* was used as test-bed for C-Flex, a major C[3]I upgrade which is likely to be installed in combat-role ships. Thales Albatross thermal imagers may be fitted.
Structure: GRP sandwich hulls. Four positions prepared to plug in armament and equipment containers in combinations meeting the requirements of the various roles. Torpedo tubes and minerails detachable. Combat data system modular with standard consoles of which three to six are embarked depending on the role. SAV control aerials are mounted on the bridge. TRS-3D radar fitted in last seven.
Operational: Following an operational review of the class, the original concept, to be able to re-role by the interchange of mission-specific containers for different taskings (ASUW, ASW, MCM and Patrol) has been abandoned. Under a revised concept of employment, the class is to be reduced to ten ships. Of these, four ships are to be permanently roled for MCM, four for a combat role (ASW or ASUW) and two (*Gribben* and *Søløven*) for Patrol duties. The remaining four ships *Svaerdfisken*, *Flyvefisken*, *Hajen* and *Lommen* are to be decommissioned by 2009. Gas turbines are not fitted in MCM ships.

UPDATED

HAVKATTEN   9/2003*, Harald Carstens / 1044108

VIBEN   2/2004*, L-G Nilsson / 1044111

9 Ø CLASS (LARGE PATROL CRAFT) (PB)

Name	No	Builders	Commissioned
BARSØ	Y 300	Svendborg Vaerft	13 June 1969
DREJØ	Y 301	Svendborg Vaerft	1 July 1969
ROMSØ	Y 302	Svendborg Vaerft	21 July 1969
SAMSØ	Y 303	Svendborg Vaerft	15 Aug 1969
THURØ	Y 304	Svendborg Vaerft	12 Sep 1969
VEJRØ	Y 305	Svendborg Vaerft	17 Oct 1969
FARØ	Y 306	Svendborg Vaerft	17 May 1973
LAESØ	Y 307	Svendborg Vaerft	23 July 1973
ROMØ	Y 308	Svendborg Vaerft	3 Sep 1973

Displacement, tons: 155 full load
Dimensions, feet (metres): 84 × 19.7 × 9.2 *(25.6 × 6 × 2.8)*
Main machinery: 1 diesel; 385 hp(m) *(283 kW)*; 1 shaft
Speed, knots: 11
Complement: 9 (2 officers)
Guns: 2—12.7 mm MG.
Radars: Navigation: Furuno 1510; I-band.

Comment: Rated as patrol cutters. *Laesø* acts as diver support ship with a recompression chamber and towed acoustic array. The last three have a wheelhouse which extends over the full beam. To be replaced by six new 43 m patrol craft from 2006.

UPDATED

SAMSØ 8/2001, E & M Laursen / 0533295

APOLLO 5/2004*, E & M Laursen / 1121263

ANDROMEDA 7/2004*, Frank Findler / 1121262

NAVAL HOME GUARD

18 MHV 800 CLASS (COASTAL PATROL CRAFT) (PB)

Name	No	Builders	Commissioned
ALDEBARAN	MHV 801	Soby Shipyard	9 July 1992
CARINA	MHV 802	Soby Shipyard	30 Sep 1992
ARIES	MHV 803	Soby Shipyard	30 Mar 1993
ANDROMEDA	MHV 804	Soby Shipyard	30 Sep 1993
GEMINI	MHV 805	Soby Shipyard	28 Feb 1994
DUBHE	MHV 806	Soby Shipyard	1 July 1994
JUPITER	MHV 807	Soby Shipyard	30 Nov 1994
LYRA	MHV 808	Soby Shipyard	30 May 1995
ANTARES	MHV 809	Soby Shipyard	30 Nov 1995
LUNA	MHV 810	Soby Shipyard	30 May 1996
APOLLO	MHV 811	Soby Shipyard	30 Nov 1996
HERCULES	MHV 812	Soby Shipyard	28 May 1997
BAUNEN	MHV 813	Soby Shipyard	17 Dec 1997
BUDSTIKKEN	MHV 814	Soby Shipyard	30 Aug 1998
KUREREN	MHV 815	Soby Shipyard	30 May 1999
PATRIOTEN	MHV 816	Soby Shipyard	25 Feb 2000
PARTISAN	MHV 817	Soby Shipyard	29 Nov 2000
SABOTØREN	MHV 818	Soby Shipyard	13 Oct 2001

Displacement, tons: 83 full load
Dimensions, feet (metres): 77.8 × 18.4 × 6.6 *(23.7 × 5.6 × 2)*
Main machinery: 2 Saab Scania DSI-14 diesels; 900 hp(m) *(661 kW)*; 2 shafts
Speed, knots: 13. **Range, n miles:** 990 at 11 kt
Complement: 8 + 4 spare
Guns: 2—7.62 mm MGs. 2—12.7 mm MGs (can be fitted).
Radars: Navigation: Furuno 1505; I-band.

Comment: First six ordered in April 1991, second six in July 1992, six more in 1997. Steel hulls with a moderate ice capability.

UPDATED

PATRIOTEN 8/2004*, Per Körnefeldt / 1044105

3 + 8 MHV 900 CLASS (COASTAL PATROL CRAFT) (PB)

Name	No	Builders	Commissioned
ENØ	MHV 901	Søby Shipyard	Oct 2003
MANØ	MHV 902	Søby Shipyard	Apr 2004
HJORTØ	MHV 903	Søby Shipyard	Jan 2005
LYØ	MHV 904	Søby Shipyard	Nov 2005

Displacement, tons: 95 full load
Dimensions, feet (metres): 89.3 × 18.7 × 8.2 *(27.2 × 5.7 × 2.5)*
Main machinery: 2 Saab Scania DI 16V8 diesels; 980 hp(m) *(730 kW)*; 2 shafts
Speed, knots: 13
Complement: 10
Guns: 2—7.62 mm MGs.

Comment: Similar to but 3.5 m longer than the MHV 800 class. Steel construction. Eleven vessels ordered.

UPDATED

BUDSTIKKEN 7/2004*, Frank Findler / 1121264

ENØ 10/2003, Per Körnefeldt / 0567436

6 MHV 90 CLASS (COASTAL PATROL CRAFT) (PB)

BOPA MHV 90	**HOLGER DANSKE** MHV 92	**RINGEN** MHV 94
BRIGADEN MHV 91	**HVIDSTEN** MHV 93	**SPEDITØREN** MHV 95

Displacement, tons: 85 full load
Dimensions, feet (metres): 64.9 × 18.7 × 8.2 *(19.8 × 5.7 × 2.5)*
Main machinery: 1 Burmeister & Wain diesel; 400 hp(m) *(294 kW)*; 1 shaft
Speed, knots: 11
Complement: 12
Guns: 2—7.62 mm MGs.
Radars: Navigation: Furuno 1505; I-band.

Comment: Built between 1973 and 1975. New radars fitted.

UPDATED

HVIDSTEN   *5/2004*, Per Körnefeldt /* 1044101

2 MHV 70 CLASS (COASTAL PATROL CRAFT) (PB)

SATURN MHV 70   **SCORPIUS** MHV 71

Displacement, tons: 125 full load
Dimensions, feet (metres): 64 × 16.7 × 8.2 *(19.5 × 5.1 × 2.5)*
Main machinery: 1 diesel; 200 hp(m) *(147 kW)*; 1 shaft
Speed, knots: 10
Complement: 12
Guns: 2—7.62 mm MGs.
Radars: Navigation: Raytheon RM 1290S; I-band.

Comment: Patrol boats and training craft for the Naval Home Guard. Built in the Royal Dockyard, Copenhagen and commissioned in 1958. Formerly designated DMH, but allocated MHV numbers in 1969.

UPDATED

SCORPIUS   *10/2003, Per Körnefeldt /* 0567455

MINE WARFARE FORCES

Notes: See also Flyvefisken class under *Patrol Forces.*

4 MSF CLASS (MRD)

MSF 1-4

Displacement, tons: 125 full load
Dimensions, feet (metres): 86.9 × 23 × 6.9 *(26.5 × 7 × 2.1)*
Main machinery: 2 Scania DSI 14 diesels; 1,000 hp(m) *(736 kW)*; 2 Schottel waterjets or 2 Schottel azimuth thrusters
Speed, knots: 12
Complement: 4
Combat data systems: IN-SNEC/INFOCOM.
Radars: Navigation: Raytheon 40 or Terma; I-band.
Sonars: Thomson Marconi STS 2054 side scan active; high frequency.

Comment: MSF (Minor Standard Vessel). Ordered in January 1997 from Danyard, Aalborg, and five delivered June 1998 to January 1999. Used primarily as MCM drones although built as multipurpose platform (with one Stanflex container position). Fitted with containerised MCM gear for working in conjunction with Flyvefisken class minehunters. GRP hulls. IN-SNEC is a high data rate sonar/TV link. INFOCOM is a low data rate command link. Plans for further craft are under consideration. One transferred to Sweden in 2001.

UPDATED

MSF 3   *6/2004*, Frank Findler /* 1044102

6 SAV CLASS (MINEHUNTER—DRONES) (MSD)

MRD 1 (ex-*MRF 1*)   **MRD 2** (ex-*MRF 2*)   **MRD 3-6**

Displacement, tons: 32 full load
Dimensions, feet (metres): 59.7 × 15.6 × 3.9 *(18.2 × 4.8 × 1.2)*
Main machinery: 2 Detroit diesels; 350 hp(m) *(257 kW)*; 2 Schottel waterjet propulsors
Speed, knots: 12
Complement: 4
Combat data systems: Terma link to Flyvefisken class (in MCMV configuration).
Radars: Navigation: Furuno; I-band.
Sonars: Thomson Sintra TSM 2054 side scan; active minehunting; high frequency.

Comment: Built by Danyard with GRP hulls. First one completed in March 1991, second in December 1991. Four more ordered in mid-1994 and delivered in 1996. The vessels are robot drones (or Surface Auxiliary Vessels (SAV)) operated in pairs by the Flyvefisken class in MCMV configuration. Hull is based on the Hugin class TRVs with low noise propulsion. The towfish with side scan sonar is lowered and raised from the stern-mounted gantry. The first two craft have slightly different funnel designs.

VERIFIED

MRD 4   *7/2002, E & M Laursen /* 0526826

AUXILIARIES

Notes: The Mobile Base (MOBA) and Mobile Logistics (M-LOG) units have been consolidated into one M-LOG detachment. It consists of some 30 vehicles (fuel-trucks, torpedo and helicopter handling facilities, communications, stores, provisions and workshops).

1 TRANSPORT SHIP (AKS)

Name	No	Builders	Commissioned
SLEIPNER	A 559	Åbenrå Vaerft og A/S	18 July 1986

Displacement, tons: 465 full load
Dimensions, feet (metres): 119.6 × 24.9 × 8.8 *(36.5 × 7.6 × 2.7)*
Main machinery: 1 Callesen diesel; 575 hp(m) *(423 kW)*; 1 shaft
Speed, knots: 11. **Range, n miles:** 2,400 at 11 kt
Complement: 7 (1 officer)
Cargo capacity: 150 tons
Radars: Navigation: Furuno 2115; I-band.

VERIFIED

SLEIPNER   *6/2003, J Ciślak /* 0567440

2 ABSALON CLASS (FLEXIBLE SUPPORT SHIPS) (AGF/AKR/AH)

Name	No	Builders	Laid down	Launched	Commissioned
ABSALON	L 16	Odense Shipyard, Lindø	28 Nov 2003	25 Feb 2004	1 July 2004
ESBERN SNARE	L 17	Odense Shipyard, Lindø	2004	24 June 2004	25 Feb 2005

Displacement, tons: 6,300 full load
Dimensions, feet (metres): 449.6 × 64.0 × 20.7
 (137.0 × 19.5 × 6.3)
Main machinery: CODAD. 2 MTU 8000 diesels; 22,300 hp
 (16.63 MW); 2 shafts; CP propellers; bow thruster
Speed, knots: 23
Complement: 100 + 70 staff

Missiles: SSM: 2 modules for 16 Harpoon Block II missiles.
 SAM: 3 modules for 36 RIM-162 Evolved Sea Sparrow (ESSM).
Guns: United Defense 5 in *(127 mm)*/62 Mk 45 Mod 4; ex-171
 ERGM 40 rds/min to 116 km *(63 n miles)*; weight of shell
 50 kg.
Countermeasures: Decoys.
Combat data systems: Terma C-Flex.
Radars: Air/Surface search: Thales SMART-S 3D; E/F-band.
 Fire control: SaabTech Ceros 200 Mk 3; I/J-band.
 Navigation: to be announced.
Sonars: Hull mounted.
Helicopter: 2 EH 101.

MULTIROLE SUPPORT SHIP

(Scale 1 : 1,500), Ian Sturton / 0569800

Programmes: Contract on 16 October 2001 for detailed design and construction of two multirole support ships. Construction of first of class started on 30 April 2003.
Structure: Built to DNV Navy standards using Stanflex concept, large equipment items can be deployed in up to five container positions. Ro-Ro ramp aft gives access to 900 m² of multipurpose deck (vehicles, logistics, ammunition, up to 34 TEU containers).

Operational: To be capable of acting as a command platform, transporting up to 200 personnel and equipment, provision of joint logistic support, acting as a hospital ship and minelaying. While likely to be optimised for self-defence, there is the flexibility to adopt an offensive role if required. Initial sea trials of L 16 in 2004. First of class trials are to continue until 2007.

UPDATED

ABSALON

6/2004*, Royal Danish Navy / 1044117

ABSALON

10/2004*, Per Körnefeldt / 1044104

ABSALON

10/2004*, Per Körnefeldt / 1044103

1 ROYAL YACHT (YAC)

Name	No	Builders	Commissioned
DANNEBROG	A 540	R Dockyard, Copenhagen	20 May 1932

Displacement, tons: 1,130 full load
Dimensions, feet (metres): 246 × 34 × 12.1 *(75 × 10.4 × 3.7)*
Main machinery: 2 Burmeister & Wain Alpha T23L-KVO diesels; 1,800 hp(m) *(1.32 MW)*; 2 shafts; cp props
Speed, knots: 14
Complement: 54 (12 officers)
Guns: 2—40 mm saluting guns.
Radars: Navigation: Furuno 2115; I-band.

Comment: Laid down 2 January 1931, launched on 10 October 1931. Major refit 1980 included new engines and electrical gear. Marisat fitted in 1992.

VERIFIED

DANNEBROG   *7/2002, E & M Laursen* / 0526825

4 LCP CLASS (COASTAL PATROL CRAFT) (PB)

LCP 1-4

Displacement, tons: 6.5 full load
Dimensions, feet (metres): 39.0 × 9.5 × 2.3 *(11.9 × 2.9 × 0.7)*
Main machinery: 1 Scania DSI 14 V8 diesel; 625 hp *(465 kW)*; 1 Kamewa water-jet
Speed, knots: 38
Complement: 3
Guns: 1—12.7 mm or 7.62 mm MG.

Comment: Based on the Swedish Combatboat 90E, these craft were developed as a joint venture between Forsvarets Materielverk and Storebro by whom the craft were constructed and completed in 2004. To be used as fast landing craft from the Absalon class support ships, they can carry 10 fully equipped soldiers or four stretchers.

NEW ENTRY

LCP 3   *5/2004*, E & M Laursen* / 1044097

2 POLLUTION CONTROL CRAFT (YPC)

MILJØ 101 and **102**

Displacement, tons: 16 full load
Dimensions, feet (metres): 53.8 × 14.4 × 7.1 *(16.2 × 4.2 × 2.2)*
Main machinery: 1 MWM TBD232V12 diesel; 454 hp(m) *(334 kW)* sustained; 1 shaft
Speed, knots: 15. **Range, n miles:** 350 at 8 kt
Complement: 3 (1 officer)

Comment: Built by Ejvinds Plastikbodevaerft, Svendborg. Carry derricks and booms for framing oil slicks and dispersant fluids. Naval manned. Delivered 1 November and 1 December 1977.

VERIFIED

MILJØ 101   *6/1999, Royal Danish Navy* / 0056890

0 + 6 SF MK I CLASS (MULTIROLE CRAFT) (MSD/AXL/AGSC)

Displacement, tons: 138 full load
Dimensions, feet (metres): 94.8 × 21.0 × 6.6 *(28.9 × 6.4 × 2.0)*
Main machinery: 2 diesels; 1,005 hp(m) *(750 kW)*; 2 azimuth thrusters
Speed, knots: 13
Complement: 3 (accommodation for 9)

Comment: Multirole GRP vessels constructed by Danish Yacht A/S, Skagen. First two vessels to be inshore survey craft launches and to be followed by two training vessels and two MCM drones. To be delivered between 2005 and 2007.

NEW ENTRY

4 RESCUE VESSELS (PBO)

NORDSØEN	VESTKYSTEN	HAVØRNEN	VIBEN

Measurement, tons: 594 gwt *(Nordsøen)*; 657 gwt *(Vestkysten)*; 188 gwt *(Havørnen)*; 23 gwt *(Viben)*
Dimensions, feet (metres): 174.6 × 33.8 × 10.8 *(53.2 × 10.3 × 3.3) (Nordsøen)*
163.7 × 32.8 × 13.8 *(49.9 × 10 × 4.2) (Vestkysten)*
101.4 × 21.6 × ? *(30.9 × 6.6 × ?) (Havørnen)*
56.4 × 11.8 × 5.2 *(17.2 × 3.6 × 1.6) (Viben)*

Comment: Non-naval ships operated by the Ministry of Food and Fisheries. *Nordsøen* and *Vestkysten* operate primarily in the North Sea and Kattegat area, *Havørnen* in the Baltic Sea around Bornholm and *Viben* in shallow waters. Capable of 14-18 kt.

VERIFIED

NORDSØEN   *1/1999, Harald Carstens* / 0056889

2 OIL POLLUTION CRAFT (YPC/ABU)

GUNNAR THORSON A 560	GUNNAR SEIDENFADEN A 561

Displacement, tons: 750 full load
Dimensions, feet (metres): 183.7 × 40.3 × 12.8 *(56 × 12.3 × 3.9)*
Main machinery: 2 Burmeister and Wain Alpha 8V23L-VO diesels; 2,320 hp(m) *(1.7 MW)*; 2 shafts; cp props; bow thruster
Speed, knots: 12.5
Complement: 16 (7 officers)

Comment: Built by Ørnskov Stålskibsvaerft, Frederikshavn. Delivered 8 May and 2 July 1981 respectively. *G Thorson* at Copenhagen, *G Seidenfaden* at Korsør. Carry firefighting equipment. Large hydraulic crane fitted in 1988 for the secondary task of buoy tending. Orange painted hulls.

UPDATED

GUNNAR SEIDENFADEN   *5/2002, L-G Nilsson* / 0526823

GUNNAR THORSON   *7/1998, M Declerck* / 0017790

2 SEA TRUCKS (AKL)

METTE MILJØ A 562   **MARIE MILJØ** A 563

Displacement, tons: 157 full load
Dimensions, feet (metres): 97.7 × 26.2 × 5.2 *(29.8 × 8 × 1.6)*
Main machinery: 2 Grenaa diesels; 660 hp(m) *(485 kW)*; 2 shafts
Speed, knots: 10
Complement: 9 (1 officer)

Comment: Built by Carl B Hoffmann A/S, Esbjerg and Søren Larsen & Sønners Skibsvaerft A/S, Nykøbing Mors. Delivered 22 February 1980. Have orange and yellow superstructure.
UPDATED

METTE MILJØ   *9/2002, Per Körnefeldt / 0526824*

1 RESEARCH SHIP (AGE)

DANA

Displacement, tons: 3,700 full load
Dimensions, feet (metres): 257.5 × 48.6 × 19.7 *(78.5 × 14.8 × 6)*
Main machinery: 2 Burmeister and Wain Alpha 16V23-LU diesels; 4,960 hp(m) *(3.65 MW)*; 1 shaft
     cp prop; bow and stern thrusters
Speed, knots: 15. **Range, n miles:** 8,000 at 14 kt
Complement: 27 plus 12 scientists

Comment: Built by Dannebrog, Aarhus in 1982. Used mostly for Fisheries survey and research. Has an ice-strengthened hull and three 6 ton cranes.
VERIFIED

DANA   *6/2002, Royal Danish Navy / 0533223*

2 ARVAK CLASS (HARBOUR TUGS) (YTL)

ARVAK Y 344   **ALSIN** Y 345

Displacement, tons: 79 full load
Dimensions, feet (metres): 52.5 × 21.7 × 8.2 *(16.0 × 6.6 × 2.5)*
Main machinery: 1 MTU 12V 183TE62 diesel; 737 hp(m) *(550 kW)*
Speed, knots: 10

Comment: Built by Hvide Sande Skibs & Baadebyggeri and delivered on 18 November 2002. In service at Korsør and Frederikshavn. Fitten with Stanflex container position aft to facilitate transport of containerised stores and equipment between naval bases.
UPDATED

ARVAK   *10/2004*, Per Körnefeldt / 1044098*

ICEBREAKERS

Notes: Icebreakers, are controlled by the Navy but have a combined naval and civilian crew. Maintenance is done at Frederikshavn in Summer. Surveying is no longer conducted by these vessels.

1 THORBJØRN CLASS (AGB/AGS)

Name	No	Builders	Commissioned
THORBJØRN	A 553	Svendborg Vaerft	June 1981

Displacement, tons: 2,344 full load
Dimensions, feet (metres): 221.4 × 50.2 × 15.4 *(67.5 × 15.3 × 4.7)*
Main machinery: Diesel-electric; 4 Burmeister & Wain Alpha 16U28L-VO diesels; 6,800 hp(m) *(5 MW)*; 2 motors; 2 shafts
Speed, knots: 16.5. **Range, n miles:** 22,000 at 16 kt
Complement: 22 (7 officers)

Comment: No bow thruster. Side rolling tanks. Fitted for surveying duties in non-ice periods.
VERIFIED

THORBJØRN   *6/2002, Royal Danish Navy / 0533297*

2 DANBJØRN CLASS (AGB)

Name	No	Builders	Commissioned
DANBJØRN	A 551	Lindø Vaerft, Odense	1965
ISBJØRN	A 552	Lindø Vaerft, Odense	1966

Displacement, tons: 3,685 full load
Dimensions, feet (metres): 252 × 56 × 20 *(76.8 × 17.1 × 6.1)*
Main machinery: Diesel-electric; 6 Burmeister and Wain 12-26MT-40V diesels; 10,500 hp(m) *(7.72 MW)*; 8 motors; 5,240 hp(m) *(38.5 MW)*; 4 shafts
Speed, knots: 14. **Range, n miles:** 11,500 at 14 kt
Complement: 25 (9 officers)

Comment: Two of the four propellers are positioned forward, two aft.
UPDATED

DANBJØRN   *10/2004*, Per Körnefeldt / 1044099*

TRAINING SHIPS

Notes: There are two small Sail Training Ships, *Svanen* Y 101 and *Thyra* Y 102. Of 32 tons they have a sail area of 480 m² and an auxiliary diesel of 72 hp(m) *(53 kW)*. Built in 1960 by Molich yacht builders, Hundested. Used to train midshimen before attending the naval academy.

THYRA   *6/2004*, Frank Findler / 1044100*

SURVEY SHIPS

6 SURVEY LAUNCHES (YGS)

SKA 11-16

Displacement, tons: 52 full load
Dimensions, feet (metres): 65.6 × 17.1 × 6.9 *(20 × 5.2 × 2.1)*
Main machinery: 1 GM diesel; 540 hp *(403 kW)*; 1 shaft
Speed, knots: 12
Complement: 6 (1 officer)
Radars: Navigation: Furuno; I-band.

Comment: GRP hulls. Built 1981-84 by Rantsausminde. *SKA 11* and *SKA 12* have strengthened hulls and are permanently deployed to Naval Station Grønnedal (Greenland) for surveying of Greenland waters. Multibeam echo sounders are fitted. *SKA 13* and *14* have been modified for other tasks at the Naval Bases. *SKA 15* and *16* are fitted for surveying of Danish waters. The survey launches can work alone, in pairs or in conjunction with Flyvefisken class vessels. All have red hulls and white superstructures. **VERIFIED**

SKA 16   *7/1997, E & M Laursen* / 0012275

Djibouti

Country Overview

Formerly the French territory of French Somaliland and later the Afars and the Issas, Djibouti became independent in 1977. With an area of 8,957 square miles and a coastline of 170 n miles, the country is situated in a strategic position on the Bab el Mandeb, the strait that links the Red Sea with the Gulf of Aden. It is bordered to the north by Eritrea, to the west by Ethiopia and to the south by Somalia. The largest town and capital is also called Djibouti whose port serves as an international transhipment and refuelling centre. It also provides Ethiopia with its only rail link to the sea. Territorial seas (12 n miles) are claimed. A 200 n mile Exclusive Economic Zone (EEZ) has been claimed but the limits are not fully defined.

Personnel

2005: 125

Bases

Djibouti

French Navy

The permanent French naval contingent usually includes up to three frigates and a repair ship.

PATROL FORCES

Notes: (1) One Zhuk and one Boghammar (*Dorra* P 15) patrol craft were transferred from Ethiopia in 1996. Neither is operational.
(2) Up to six RIBs are in use. Zodiac and Avon types.
(3) One LCM (ex-CTM 14) transferred from France in 1999.

1 SAWARI CLASS (INSHORE PATROL CRAFT) (PBR)

P 13

Comment: Acquired from Iraq in 1989. Can be armed with MGs and rocket launchers. Outboard engines give speeds up to 25 kt in calm conditions. Four further craft are no longer operational. **UPDATED**

1 PLASCOA CLASS (COASTAL PATROL CRAFT) (PB)

Name	No	Builders	Commissioned
MONT ARREH	P 11	Plascoa, Cannes	16 Feb 1986

Displacement, tons: 35 full load
Dimensions, feet (metres): 75.5 × 18 × 4.9 *(23 × 5.5 × 1.5)*
Main machinery: 2 SACM Poyaud V12-520 M25 diesels; 1,700 hp(m) *(1.25 MW)*; 2 shafts
Speed, knots: 25. **Range, n miles:** 750 at 12 kt
Complement: 15
Guns: 1 Giat 20 mm. 1—12.7 mm MG.
Radars: Navigation: Decca 36; I-band.

Comment: Ordered in October 1984 and transferred as a gift from France. GRP hulls. Refitted in 1988 and 1994. *Moussa Ali* decommissioned in 2001. **UPDATED**

MONT ARREH   *1986, Plascoa* / 0056896

2 BATTALION 17 (PBF)

P 16   P 17

Displacement, tons: 35.5 full load
Dimensions, feet (metres): 55.9 × 17 × 5.2 *(17.05 × 5.2 × 1.6)*
Main machinery: 2 MTU 12V 183 TE 92 diesels
Speed, knots: 35.2. **Range, n miles:** 680 at 30 kt
Complement: 8
Guns: 2—14.5 mm MGs (1 twin).
Radars: Surface search: Raytheon; I-band.

Comment: Built by Harena Boat Yard at Assab, Eritrea and delivered in 2001. Five similar craft in service in Eritrea. **NEW ENTRY**

Dominica

Country Overview

Formerly a British colony, the Commonwealth of Dominica became an independent republic in 1978. With an area of 290 sq miles and coastline of 80 n miles, it is the largest and most northerly of the Windward Islands in the Lesser Antilles chain and is situated in the Caribbean Sea between the French possessions of Guadeloupe to the north and Martinique to the south. The capital, major town, and port is Roseau. Territorial seas (12 n miles) are claimed. A 200 n mile Exclusive Economic Zone (EEZ) has been claimed but the limits are not fully defined.

Headquarters Appointments

Head of Coast Guard:
   Inspector O Frederick

Personnel

2005: 32

Bases

Roseau

COAST GUARD

1 DAUNTLESS CLASS (PB)

Name	No	Builders	Commissioned
UKALE	D 05	SeaArk Marine	8 Nov 1995

Displacement, tons: 11 full load
Dimensions, feet (metres): 40 × 14 × 4.3 *(12.2 × 4.3 × 1.3)*
Main machinery: 2 Caterpillar 3208TA diesels; 870 hp *(650 kW)* sustained; 2 shafts
Speed, knots: 27. **Range, n miles:** 600 at 18 kt
Complement: 6
Guns: 1—7.62 mm MG (can be carried).
Radars: Surface search: Raytheon; I-band.

Comment: Similar to craft delivered by the US to many Caribbean coast guards under FMS. **VERIFIED**

UKALE   *11/1995, SeaArk* / 0056897

1 SWIFT 65 ft CLASS (PB)

Name	No	Builders	Commissioned
MELVILLE	D 4	Swiftships, Morgan City	1 May 1984

Displacement, tons: 33 full load
Dimensions, feet (metres): 64.9 × 18.4 × 6.6 *(19.8 × 5.6 × 2)*
Main machinery: 2 Detroit 12V-71TA diesels; 840 hp *(616 kW)* sustained; 2 shafts
Speed, knots: 23. **Range, n miles:** 250 at 18 kt
Complement: 10
Guns: 1—7.62 mm MG.
Radars: Surface search: Furuno; I/J-band.

Comment: Donated by US government. Similar craft supplied to Antigua and St Lucia.

VERIFIED

MELVILLE *11/1993, Maritime Photographic*

3 PATROL CRAFT (PBR)

VIGILANCE OBSERVER RESCUER

Displacement, tons: 2.4 full load
Dimensions, feet (metres): 27 × 8.4 × 1 *(8.2 × 2.6 × 0.3)*
Main machinery: 1 Evinrude outboard; 225 hp *(168 kW)* sustained or 2 Johnson outboards *(Rescuer)*; 280 hp *(205 kW)*
Speed, knots: 28 or 45 *(Rescuer)*
Complement: 3

Comment: First two are Boston Whalers acquired in 1988. *Rescuer* is of similar size but is an RHIB acquired in 1994.

VERIFIED

OBSERVER *11/1993, Maritime Photographic*

Dominican Republic
MARINA DE GUERRA

Country Overview

The Dominican Republic is an independent state whose constitution was promulgated in 1966. With an area of 18,816 square miles, it occupies the eastern two thirds of the island of Hispaniola, which it shares with Haiti to the west. There are also a number of adjacent islands, notably Beata and Saona. It has a 697 n mile coastline and is bordered to the north by the Atlantic Ocean, to the east by the Mona Passage, which separates it from Puerto Rico, and to the south by the Caribbean Sea. Santo Domingo is the capital, largest city and principal port. Territorial seas (6 n miles) are claimed. A 200 n mile EEZ has been claimed but the limits have not been defined by boundary agreements.

Headquarters Appointments

Chief of Naval Staff:
  Vice Admiral Victor F Garcia Alecont
Vice Chief of Naval Staff:
  Rear Admiral Juan Thomas Diaz Polanco

Personnel

(a) 2005: 3,800 officers and men (including naval infantry)
(b) Selective military service

Bases

27 de Febrero, Santo Domingo: HQ of CNS, Naval School. Supply base.
Las Calderas, Las Calderas, Baní: Naval dockyard, 700 ton synchrolift. Training centre. Supply base.
Haina: Dockyard facility. Supply base.
Puerto Plata. Small naval base.

DELETIONS

Note: *Melia* still flies an ensign as a museum ship.

PATROL FORCES

Notes: (1) Two Super Dvora class may be acquired from Israel in 2005.
(2) One Eduardoño class patrol craft is reported to have been transferred from the US in 2003.

1 COHOES CLASS (PG)

Name	No	Builders	Commissioned
SEPARACION (ex-*Passaconaway* AN 86)	P 208	Marine SB Co	27 Apr 1945

Displacement, tons: 855 full load
Dimensions, feet (metres): 162.3 × 33.8 × 11.7 *(49.5 × 10.3 × 3.6)*
Main machinery: Diesel-electric; 2 Busch-Sulzer BS-539 diesels; 1,500 hp(m) *(1.1 MW)*; 2 generators; 1 motor; 1 shaft
Speed, knots: 12
Complement: 64 (5 officers)
Guns: 2—3 in *(76 mm)*/50 Mk 26. 3 Oerlikon 20 mm.
Radars: Surface search: Raytheon SPS-64; I-band.

Comment: Ex-netlayer laid up in reserve in US in 1963. Transferred by sale on 29 September 1976. Now used for patrol duties. Modified in 1980 with the removal of the bow horns. P 209 has been decommissioned.

UPDATED

SEPARACION *5/1999, A Sheldon-Duplaix / 0056898*

2 BALSAM CLASS (PBO/WMEC)

Name	No	Builders	Commissioned
ALMIRANTE JUAN ALEXANDRO ACOSTA (ex-*Citrus*)	C 456 (ex-WMEC 300)	Marine Iron, Duluth	30 May 1943
ALMIRANTE DIDIEZ BURGOS (ex-*Buttonwood*)	C 457 (ex-WLB 306)	Duluth Shipyard, Minnesota	24 Sep 1943

Displacement, tons: 1,034 full load
Dimensions, feet (metres): 180 × 37 × 12 *(54.9 × 11.3 × 3.8)*
Main machinery: Diesel-electric; 2 Cooper Bessemer diesels; 1,402 hp *(1.06 MW)*; 2 motors; 1,200 hp *(895 kW)*; 1 shaft; bow thruster
Speed, knots: 13
Complement: 54 (4 officers)
Guns: 1—4 in; 2—20 mm (456). 2—20 mm; 2—12.7 MGs (457).
Radars: Surface search: Raytheon SPS-64(V)1; I-band.

Comment: C 456 built as a buoy tender but served as a US Coast Guard cutter from 1979 to 1994. Transferred by gift on 16 September 1995 and recommissioned in January 1996 after a short refit. C 457 transferred from US Coast Guard on 30 June 2001.

VERIFIED

ALMIRANTE JUAN ALEXANDRO ACOSTA *8/2002, A Sheldon-Duplaix / 0534105*

For details of the latest updates to *Jane's Fighting Ships* online and to discover the additional information available exclusively to online subscribers please visit
jfs.janes.com

1 ADMIRABLE CLASS (PG)

Name	No	Builders	Launched
PRESTOL (ex-*Separacion*, ex-*Skirmish* MSF 303)	C 454	Associated SB	16 Aug 1943

Displacement, tons: 650 standard; 905 full load
Dimensions, feet (metres): 184.5 × 33 × 14.4 *(56.3 × 10.1 × 4.4)*
Main machinery: 2 Cooper-Bessemer GSB8 diesels; 1,710 hp *(1.28 MW)*; 2 shafts
Speed, knots: 15. **Range, n miles:** 4,300 at 10 kt
Complement: 90 (8 officers)
Guns: 1—3 in *(76 mm)*/50 Mk 26. 2 Bofors 40 mm/60. 6 Oerlikon 20 mm.
Radars: Surface search: Raytheon SPS-64(V)9; I-band.

Comment: Former US fleet minesweeper. Purchased on 13 January 1965. Sweep-gear removed. Classified as Cañoneros.

VERIFIED

PRESTOL　　　　　　　　　　　*6/1997, A Sheldon-Duplaix /* 0012278

1 SOTOYOMO CLASS (PG/ATA)

Name	No	Builders	Commissioned
ENRIQUILLO (ex-*Stallion* ATA 193)	RM 22	Levington SB Co, Orange, TX	26 Feb 1945

Displacement, tons: 534 standard; 860 full load
Dimensions, feet (metres): 143 × 33.9 × 13 *(43.6 × 10.3 × 4)*
Main machinery: Diesel-electric; 2 GM 12-278A diesels; 2,200 hp *(1.64 MW)*; 2 generators; 1 motor; 1,500 hp *(1.12 MW)*; 1 shaft
Speed, knots: 13. **Range, n miles:** 8,000 at 10kt
Complement: 45
Guns: 1 US 3 in *(76 mm)*/50 Mk 26. 2 Oerlikon 20 mm.
Radars: Surface search: Raytheon SPS-5D; G/H-band.

Comment: Leased from US 30 October 1980, renewed 15 June 1992 and approved for transfer 10 June 1997.

VERIFIED

ENRIQUILLO　　　　　　　　　　*8/2002, A Sheldon-Duplaix /* 0534084

1 PGM 71 CLASS (LARGE PATROL CRAFT) (PB)

Name	No	Builders	Commissioned
BETELGEUSE (ex-*PGM* 77)	GC 102	Peterson, USA	1966

Displacement, tons: 130 standard; 145 full load
Dimensions, feet (metres): 101.5 × 21 × 5 *(30.9 × 6.4 × 1.5)*
Main machinery: 2 Caterpillar D 348 diesels; 1,450 hp *(1.08 MW)* sustained; 2 shafts
Speed, knots: 21. **Range, n miles:** 1,500 at 10 kt
Complement: 20 (3 officers)
Guns: 1 Oerlikon 20 mm. 2—12.7 mm MGs.
Radars: Surface search: Raytheon; I-band.

Comment: Built in the USA and transferred to the Dominican Republic under the Military Aid Programme on 14 January 1966. Re-engined in 1980.

VERIFIED

BETELGEUSE (alongside ENRIQUILLO)　　*3/2001, A Sheldon-Duplaix /* 0114349

2 CANOPUS (SWIFTSHIPS 110 ft) CLASS
(LARGE PATROL CRAFT) (PB)

Name	No	Builders	Commissioned
CRISTOBAL COLON (ex-*Canopus*)	GC 107	Swiftships, Morgan City	June 1984
ORION	GC 109	Swiftships, Morgan City	Aug 1984

Displacement, tons: 93.5 full load
Dimensions, feet (metres): 109.9 × 23.9 × 5.9 *(33.5 × 7.3 × 1.8)*
Main machinery: 3 Detroit 12V-92TA diesels; 1,020 hp *(760 kW)* sustained; 3 shafts
Speed, knots: 23. **Range, n miles:** 1,500 at 12 kt
Complement: 19 (3 officers)
Guns: 1—20 mm or 2—12.7 mm MGs.
Radars: Surface search: Raytheon; I-band.

Comment: Built of aluminium. GC 107 completely rebuilt and reconditioned by Swiftships in 2003. GC 109 was similarly refitted in 2004.

UPDATED

CRISTOBAL COLON　　　　　　　*1/2004*, Swiftships /* 0587700

4 BELLATRIX CLASS (COASTAL PATROL CRAFT) (PB)

Name	No	Builders	Commissioned
PROCION	GC 103	Sewart Seacraft Inc, Berwick, LA	1967
ALDEBARÁN	GC 104	Sewart Seacraft Inc, Berwick, LA	1972
BELLATRIX	GC 106	Sewart Seacraft Inc, Berwick, LA	1967
CAPELLA	GC 108	Sewart Seacraft Inc, Berwick, LA	1968

Displacement, tons: 60 full load
Dimensions, feet (metres): 85 × 18 × 5 *(25.9 × 5.5 × 1.5)*
Main machinery: 2 GM 16V-71 diesels; 811 hp *(605 kW)* sustained; 2 shafts
Speed, knots: 18.7. **Range, n miles:** 800 at 15 kt
Complement: 12
Guns: 3—12.7 mm MGs.
Radars: Surface search: Raytheon SPS-64; I-band.

Comment: Transferred to the Dominican Navy by the US. *Procion* was taken out of service in 1995 but returned in 1997 after a long refit. GC 103 and GC 106 completely rebuilt and reconditioned by Swiftships, Morgan City, in 2003. GC 104 and GC 108 were similarly refitted in 2004.

UPDATED

BELLATRIX　　　　　　　　　　*6/2004*, A Sheldon-Duplaix /* 0587699

CAPELLA　　　　　　　　　　　*8/2002, A Sheldon-Duplaix /* 0534085

3 POINT CLASS (PB)

Name	No	Builders	Commissioned
ARIES (ex-*Point Batan*)	101 (ex-82340)	CG Yard, Maryland	21 Nov 1962
ANTARES (ex-*Point Martin*)	105 (ex-82379)	J Martinac, Tacoma	20 Aug 1970
SIRIUS (ex-*Point Spencer*)	110 (ex-82349)	J Martinac, Tacoma	25 Oct 1966

Displacement, tons: 67 full load
Dimensions, feet (metres): 83 × 17.2 × 5.8 *(25.3 × 5.2 × 1.8)*
Main machinery: 2 Caterpillar diesels; 1,600 hp *(1.19 MW)*; 2 shafts
Speed, knots: 22. **Range, n miles:** 1,200 at 8 kt
Complement: 10
Guns: 2—12.7 mm MGs.
Radars: Surface search: Hughes/Furuno SPS-73; I-band.

Comment: *Aries* and *Antares* transferred from US Coast Guard 1 October 1999. *Sirius* transferred 12 December 2000. These ships are widely spread throughout the Caribbean navies.
UPDATED

SIRIUS 6/2004*, *A Sheldon-Duplaix* / 0587698

2 SWIFTSHIPS 35M CLASS (LARGE PATROL CRAFT) (PB)

Name	No	Builders	Commissioned
ALTAIR	112	Swiftships, Morgan City	Oct 2003
ARCTURUS	114	Swiftships, Morgan City	Mar 2004

Displacement, tons: 95 standard
Dimensions, feet (metres): 115.1 × 24.0 × 5.0 *(35.1 × 7.3 × 1.5)*
Main machinery: 3 CAT 3412 diesels; 3,000 hp *(2.2 MW)*; 3 Hamilton HM 651 waterjets
Speed, knots: 25. **Range, n miles:** To be announced
Complement: To be announced
Guns: 1—25 mm. 2—12.7 mm MGs.

Comment: Two new craft ordered from Swiftships, Morgan City, LA as part of wider programme to increase capability to conduct counter-smuggling and drug-trafficking operations. Fitted with launching ramp for 4.7 m RIB.
UPDATED

ALTAIR 12/2003, *A Sheldon-Duplaix* / 0569189

AUXILIARIES

Notes: There are also two dredgers manned by the Navy.

DREDGER 10/1998, *A Sheldon-Duplaix* / 0056902

1 HARBOUR TANKER (AOTL)

Name	No	Builders	Commissioned
CAPITÁN BEOTEGUI (ex-*YO 215*)	BT 5	Ira S Bushey, Brooklyn	17 Dec 1945

Displacement, tons: 422 light; 1,400 full load
Dimensions, feet (metres): 174 × 32.9 × 13.3 *(53.1 × 10 × 4.1)*
Main machinery: 1 Union diesel; 525 hp *(392 kW)*; 1 shaft
Speed, knots: 8
Complement: 23
Cargo capacity: 6,570 barrels
Guns: 2 Oerlikon 20 mm.

Comment: Former US self-propelled fuel oil barge. Lent by the USA in April 1964. Lease renewed 31 December 1980 and again 5 August 1992 and approved for transfer 10 June 1997.
VERIFIED

CAPITÁN BEOTEGUI 1/1998, *M Mokrus* / 0017794

2 WHITE SUMAC CLASS (ABU)

Name	No	Builders	Commissioned
TORTUGUERO (ex-*White Pine*)	BA 1 (ex-WLM 547)	Erie Concrete and Steel, Erie	11 July 1944
CAPOTILLO (ex-*White Sumac*)	BA 2 (ex-WLM 540)	Niagara Shipbuilding Corporation, Buffalo	1943

Displacement, tons: 485 full load
Dimensions, feet (metres): 133 × 31 × 9 *(40.5 × 9.5 × 2.7)*
Main machinery: 2 Caterpillar diesels; 600 hp *(448 kW)*; 2 shafts
Speed, knots: 9
Complement: 24

Comment: BA 1 transferred from US Coast Guard in 1999 and BA 2 on 20 September 2002. Fitted with a 10 ton capacity boom.
UPDATED

TORTUGUERO 12/1999, *A Sheldon-Duplaix* / 0056903

2 FLOATING DOCKS (YFD)

ENDEAVOR DF 1 (ex-AFDL 1) DF 2 (ex-AFDM 2)

Comment: DF 1 lift, 1,000 tons. Commissioned in 1943. Transferred from US on loan 8 March 1986 and approved for transfer 10 June 1997. DF 2 lift 12,000 tons. Commissioned in 1942. Transferred from US in 1999.
VERIFIED

TRAINING SHIPS

Notes: In addition to those listed below there are various tenders mostly acquired 1986-88: *Cojinoa* BA 01, *Bonito* BA 02, *Beata* BA 14, *Albacora* BA 18, *Salinas* BA 19, *Carey* BA 20.

2 SAIL TRAINING SHIPS (AXS)

Name	No	Builders	Commissioned
RAMBO (ex-*Jurel*)	BA 15	Ast Navales Dominicanos	1975
NUBE DEL MAR	BA 7	Ast Navales Dominicanos	1979

Displacement, tons: 24
Dimensions, feet (metres): 45 × 13 × 6.6 *(13.7 × 4 × 1.9)*
Main machinery: 1 GM diesel; 101 hp *(75 kW)*; 1 shaft
Speed, knots: 9
Complement: 4
Guns: 1—7.62 mm MG.

Comment: *Rambo* is an auxiliary sailing craft with a sail area of 750 sq ft and a cargo capacity of 7 tons. There may be more of this class. *Nube del Mar* is an auxiliary yacht used for sail training at the Naval School and is slightly smaller.
VERIFIED

TUGS

5 COASTAL/HARBOUR TUGS (YTM/YTL)

HERCULES (ex-*R 2*) RP 12   BOHECHIO (ex-*YTL 600*) RP 16
GUACANAGARIX (ex-*R 5*) RP 13   CAYACCA RP 19
OCOA LPD 303

Displacement, tons: 200 full load
Dimensions, feet (metres): 70 × 15.6 × 9 *(21.4 × 4.8 × 2.7)*
Main machinery: 1 Caterpillar diesel; 500 hp *(373 kW)*; 1 shaft
Speed, knots:
Complement: 8

Comment: Details given are for RP 12 and 13 built in 1960. RP 16 and 19 are small harbour tugs of about 70 tons. *Ocoa* is an LCU type used as a tug.

VERIFIED   HERCULES   *5/1999, A Sheldon-Duplaix* / 0056904

East Timor

Country Overview

The Democratic Republic of Timor-Leste (also known as East Timor) has an area of 7,400 square miles and lies in the eastern part of Timor island, the largest and easternmost of the Lesser Sunda Islands in the Malay Archipelago. Originally settled in the early 16th century, the Portuguese and Dutch competed for influence until boundaries became established. Dutch Timor, in the west, later became part of the Republic of Indonesia in 1950. Portuguese Timor, comprising the region of Dili, in the east, and the small area of Oecussi in the north-west, was annexed by Indonesia in 1975. Following an armed conflict and two and a

half years of UN administration, East Timor gained independence on 20 May 2002 and became a UN member on 27 September 2002. A successor UN mission (UNMISET) is expected to provide assistance until 20 May 2005. The capital, principal city and port is Dili. Maritime claims are not known.

The East Timor Defence Force (FALINTIL) is being trained by teams from Australia, New Zealand, Portugal and United States. The role of the Naval Component of FALINTIL will be to conduct Fishery Protection duties in the East Timorese EEZ and to safeguard the only direct access to the enclave of Oecussi which is by sea.

Headquarters Appointments

Commander in Chief Defence Forces:
  Brigadier General Taur Matan Ruak

Personnel

2005: 150 (under training)

Bases

Hera Harbour

PATROL FORCES

2 ALBATROZ CLASS (RIVER PATROL CRAFT) (PB)

Name	No	Builders	Commissioned
OECUSSI (ex-*Açor*)	P 101 (ex-P 1163)	Arsenal do Alfeite	9 Dec 1974
ATAURO (ex-*Albatroz*)	P 102 (ex-P 1162)	Arsenal do Alfeite	9 Dec 1974

Displacement, tons: 45 full load
Dimensions, feet (metres): 77.4 × 18.4 × 5.2 *(23.6 × 5.6 × 1.6)*
Main machinery: 2 Cummins diesels; 1,100 hp *(820 kW)*; 2 shafts
Speed, knots: 20. **Range, n miles:** 2,500 at 12 kt
Complement: 8 (1 officer)
Guns: 1 Oerlikon 20 mm/65. 2—12.7 mm MGs.
Radars: Surface search: Decca RM 316P; I-band.

Comment: Transferred by Portugal in 2001 to establish the Naval Component of the ETDF.

UPDATED

ALBATROZ class
(Portuguese colours) *10/1994, van Ginderen Collection* / 0081608

Ecuador
ARMADA DE GUERRA

Country Overview

The Republic of Ecuador is situated in northwestern South America. With an area of 105,037 square miles it straddles the equator and has borders to the north with Colombia and to the south with Peru. It has a coastline of 1,210 n miles with the Pacific Ocean. The country also includes the Galápagos Islands about 520 n miles west of the mainland. The capital is Quito while Guayaquil is the principal port and commercial centre. Ecuador has not claimed an EEZ but is one of a few coastal states which claims a 200 n mile territorial sea.

Headquarters Appointments

Commander-in-Chief of the Navy:
  Vice Admiral Renán Sanchez Coba
Chief of Naval Staff:
  Vice Admiral Victor Hugo Rosero Barba
Chief of Naval Operations:
  Rear Admiral Manuel Zapater Ramos
Chief of Naval Materiel:
  Rear Admiral Luis Flores Cazañas

Diplomatic Representation

Naval Attaché in Rome:
  Captain Livio Espinoza Espinoza
Naval Attaché in London and Paris:
  Captain Marcos Salinas Haro
Naval Attaché in Washington:
  Captain Aland Molestina Malta

Personnel

(a)   2005: 4,200 (including 1,500 marines and 250 naval aviation)
(b)   1 year's selective national service

Prefix to Ships' Names

BAE (Buque de Armada de Ecuador)

Bases

Guayaquil (main naval base), Jaramijo, Salinas. San Lorenzo, Galapagos Islands. Guayaquil air base.

Establishments

The Naval Academy and Merchant Navy Academy in Salinas; Naval War College in Guayaquil.

Naval Infantry

A force of marines is based at Guayaquil, Esmeraldas San Lorenzo, Galapagos and Jaramijo.

Coast Guard

Small force formed in 1980. Hull markings include diagonal thick and thin red stripes on the hull.

PENNANT LIST

Submarines		Patrol Forces		RB 75	Iliniza	LG 33	10 de Agosto
				RB 76	Altar	LG 34	3 de Noviembre
S 101	Shyri	LM 21	Quito	RB 78	Quilotoa	LG 35	5 de Agosto
S 102	Huancavilca	LM 23	Guayaquil			LG 36	27 de Febrero
		LM 24	Cuenca	Auxiliaries		LG 37	9 de Octubre
						LG 38	27 de Octubre
Frigates		Amphibious Forces		TR 62	Calicuchima	LG 41	Rio Puyango
				TR 63	Atahualpa	LG 42	Rio Mataje
FM 01	Presidente Eloy Alfaro	TR 61	Hualcopo	TR 64	Quisquis	LG 43	Rio Zarumilla
FM 02	Moran Valverde			TR 65	Taurus	LG 44	Rio Chone
		Survey/Research Vessels		BE 91	Guayas	LG 45	Rio Daule
Corvettes				DF 81	Rio Amazonas	LG 46	Rio Babahoyo
		BI 91	Orion	DF 82	Rio Napo	LG 47	Rio Esmeraldas
CM 11	Esmeraldas	LH 94	Rigel	UT 111	Isla la Plata	LG 48	Rio Santiago
CM 12	Manabi			UT 112	Isla Puná		
CM 13	Los Rios	Tugs					
CM 14	El Oro			Coast Guard			
CM 15	Los Galápagos	RA 70	Chimborazo				
CM 16	Loja	RB 72	Sangay	LG 31	25 de Julio		
		RB 73	Cotopaxi	LG 32	24 de Mayo		

SUBMARINES

2 TYPE 209 CLASS (TYPE 1300) (SSK)

Name	No	Builders	Laid down	Launched	Commissioned
SHYRI	S 101 (ex-S 11)	Howaldtswerke, Kiel	5 Aug 1974	6 Oct 1976	5 Nov 1977
HUANCAVILCA	S 102 (ex-S 12)	Howaldtswerke, Kiel	2 Jan 1975	15 Mar 1977	16 Mar 1978

Displacement, tons: 1,285 surfaced; 1,390 dived
Dimensions, feet (metres): 195.1 × 20.5 × 17.9 *(59.5 × 6.3 × 5.4)*
Main machinery: Diesel-electric; 4 MTU 12V 493 AZ80 GA31L diesels; 2,400 hp(m) *(1.76 MW)* sustained; 4 Siemens alternators; 1.7 MW; 1 Siemens motor; 4,600 hp(m) *(3.38 MW)* sustained; 1 shaft
Speed, knots: 11 surfaced/snorting; 21.5 dived
Complement: 33 (5 officers)

Torpedoes: 8—21 in *(533 mm)* bow tubes. 14 AEG SUT; dual purpose; wire-guided; active/passive homing to 28 km *(15 n miles)* at 23 kt; 12 km *(6.5 n miles)* at 35 kt; warhead 250 kg.
Countermeasures: ESM: Thomson-CSF DR 2000U; intercept.
Weapons control: Signaal M8 Mod 24.
Radars: Surface search: Thomson-CSF Calypso; I-band.
Sonars: Atlas Elektronik CSU 3; hull-mounted; active/passive search and attack; medium frequency.
Thomson Sintra DUUX 2; passive ranging.

Programmes: Ordered in March 1974. *Shyri* underwent major refit in West Germany in 1983; *Huancavilca* in 1984. Second refits by ASMAR, Chile starting with *Shyri* in 1999/2000.
Operational: Based at Guayaquil. *Shyri* badly damaged by fire on 2 February 2003 and return to service is unlikely. *Huancavilca* also reported non-operational.

UPDATED

SHYRI   *6/1998 /* 0017796

TYPE 209   *6/2001, Maritime Photographic /* 0114670

FRIGATES

2 LEANDER CLASS (FFGHM)

Name	No	Builders	Laid down	Launched	Commissioned
PRESIDENTE ELOY ALFARO (ex-*Penelope*)	FM 01 (ex-F 127)	Vickers Armstrong, Newcastle	14 Mar 1961	17 Aug 1962	31 Oct 1963
MORAN VALVERDE (ex-*Danae*)	FM 02 (ex-F 47)	HM Dockyard, Devonport	16 Dec 1964	31 Oct 1965	7 Sep 1967

Displacement, tons: 2,450 standard; 3,200 full load
Dimensions, feet (metres): 360 wl; 372 oa × 41 × 14.8 (keel); 19 (screws) *(109.7; 113.4 × 12.5 × 4.5; 5.8)*
Main machinery: 2 Babcock & Wilcox boilers; 38.7 kg/cm²; 850°F *(450°C)*; 2 English Electric/White turbines; 30,000 hp *(22.4 MW)*; 2 shafts
Speed, knots: 28. **Range, n miles:** 4,000 at 15 kt
Complement: 248 (20 officers)

Missiles: SSM: 4 Aerospatiale MM 38 Exocet ❶; inertial cruise; active radar homing to 42 km *(23 n miles)* at 0.9 Mach; warhead 165 kg.
SAM: 3 twin Matra Simbad launchers ❷ for Mistral; IR homing to 4 km *(2.2 n miles)*; warhead 3 kg.

Guns: 2 Bofors 40 mm/60 Mk 9 ❸; 120 rds/min to 10 km *(5.4 n miles)* anti-surface; 3 km *(1.6 n miles)* anti-aircraft; weight of shell 0.89 kg.
2 Oerlikon/BMARC 20 mm GAM-BO1 can be fitted midships or aft.
Torpedoes: 6—324 mm ILAS-3 (2 triple) tubes ❹ Whitehead A 244; anti-submarine; pattern running to 7 km *(3.8 n miles)* at 33 kt; warhead 34 kg shaped charge.
Countermeasures: Decoys: Graseby Type 182; towed torpedo decoy.
4 Mk 36 Mod 2 SRBOC; chaff and IR launchers ❺.
ESM: ELISRA NS-9010; intercept.
ECM: Type 667/668; jammer.

Combat data systems: SISDEF. Link Y.
Radars: Air search: Marconi Type 966 ❻; A-band.
Surface search: Plessey Type 994 ❼; E/F-band.
Navigation: Kelvin Hughes Type 1006; I-band.
Fire control: Selenia ❽ I/J-band.
Sonars: Kelvin Hughes Type 162M; hull-mounted; bottom classification; 50 kHz.
Graseby Type 184P; hull-mounted; active search and attack; 7-9 kHz.

Helicopters: 1 Bell 206B ❾.

Programmes: Both ships acquired from UK 25 April 1991 and sailed for Ecuador after working up in July and August respectively.
Modernisation: Chilean SISDEF command system has been installed. Simbad launchers have replaced Seacat, and the torpedo tubes restored. ASMAR is investigating replacing the steam turbines by diesel engines.
Structure: These are Batch 2 Exocet conversions completed in 1980 and 1982. SRBOC chaff launchers fitted after transfer.
Operational: Both ships are reported to have propulsion and electrical problems and have not been to sea for several years..

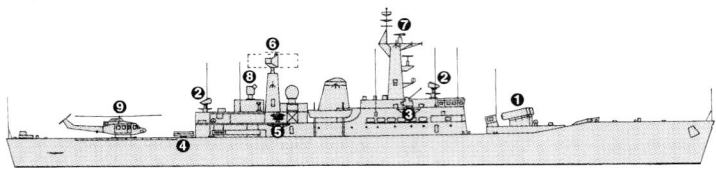

PRESIDENTE ELOY ALFARO

(Scale 1 : 1,200), Ian Sturton / 0056906

UPDATED

PRESIDENTE ELOY ALFARO

6/1998 / 0017799

PRESIDENTE ELOY ALFARO

6/1998 / 0017798

CORVETTES

6 ESMERALDAS CLASS (FSGHM)

Name	No	Builders	Laid down	Launched	Commissioned
ESMERALDAS	CM 11	Fincantieri Muggiano	27 Sep 1979	1 Oct 1980	7 Aug 1982
MANABI	CM 12	Fincantieri Ancona	19 Feb 1980	9 Feb 1981	21 June 1983
LOS RIOS	CM 13	Fincantieri Muggiano	5 Dec 1979	27 Feb 1981	9 Oct 1983
EL ORO	CM 14	Fincantieri Ancona	20 Mar 1980	9 Feb 1981	11 Dec 1983
LOS GALAPÁGOS	CM 15	Fincantieri Muggiano	4 Dec 1980	4 July 1981	26 May 1984
LOJA	CM 16	Fincantieri Ancona	25 Mar 1981	27 Feb 1982	26 May 1984

Displacement, tons: 685 full load
Dimensions, feet (metres): 204.4 × 30.5 × 8
  (62.3 × 9.3 × 2.5)
Main machinery: 4 MTU 20V 956 TB92 diesels; 22,140 hp(m)
  (16.27 MW) sustained; 4 shafts
Speed, knots: 37. **Range, n miles:** 4,400 at 14 kt
Complement: 51

Missiles: SSM: 6 Aerospatiale MM 40 Exocet (2 triple) launchers
  ❶; inertial cruise; active radar homing to 70 km *(40 n miles)* at
  0.9 Mach; warhead 165 kg; sea-skimmer.
SAM: Selenia Elsag Albatros quad launcher ❷; Aspide; semi-
  active radar homing to 13 km *(7 n miles)* at 2.5 Mach; height
  envelope 15-5,000 m *(49.2-16,405 ft)*; warhead 30 kg.
Guns: 1 OTO Melara 3 in *(76 mm)*/62 compact ❸; 85 rds/min to
  16 km *(8.7 n miles)*; weight of shell 6 kg.
  2 Breda 40 mm/70 (twin) ❹; 300 rds/min to 12.5 km *(6.8 n
  miles)* anti-surface; weight of shell 0.96 kg.
Torpedoes: 6—324 mm ILAS-3 (2 triple) tubes ❺; Whitehead
  Motofides A244; anti-submarine; self-adaptive patterns to
  7 km *(3.8 n miles)* at 33 kt; warhead 34 kg shaped charge.
  Not fitted in all.
Countermeasures: Decoys: 1 Breda 105 mm SCLAR launcher;
  chaff to 5 km *(2.7 n miles)*; illuminants to 12 km *(6.6 n miles)*.
ESM/ECM: Elettronika Gamma ED; radar intercept and jammer.
Combat data systems: Selenia IPN 10 action data automation.
  Link Y.
Weapons control: 2 Selenia NA21 with C03 directors.

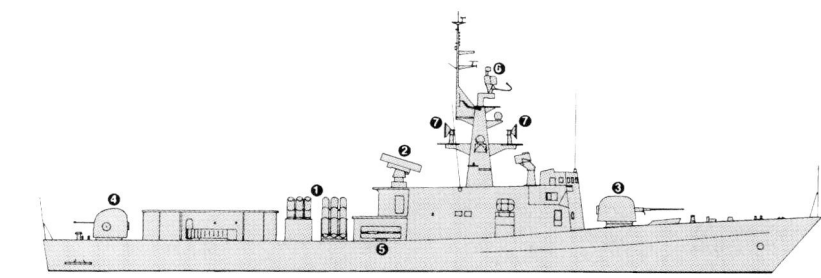

ESMERALDAS   *(Scale 1 : 600), Ian Sturton*

Radars: Air/surface search: Selenia RAN 10S ❻; E/F-band; range
  155 km *(85 n miles)*.
Navigation: SMA 3 RM 20; I-band.
Fire control: 2 Selenia Orion 10X ❼; I/J-band; range 40 km *(22 n
  miles)*.
Sonars: Thomson Sintra Diodon; hull-mounted; active search
  and attack; 11, 12 or 13 kHz.

Helicopters: Platform for 1 Bell 206B.

Programmes: Ordered in 1979.
Modernisation: Contracts for updating command and weapons
  control systems placed in 1993-94 but no funds are available.
Operational: Torpedo tubes removed from two of the class to
  refit in frigates. Two vessels are in reserve and the operational
  status of the remainder is doubtful due to lack of spares and
  replacements.

UPDATED

EL ORO   *2/2000 /* 0103731

MANABI   *6/2002, Ecuador Navy /* 0533898

SHIPBORNE AIRCRAFT

Numbers/Type: 2 Bell 230T.
Operational speed: 145 kt *(269 km/h)*.
Service ceiling: 10,300 ft *(3,140 m)*.
Range: 307 n miles *(568 km)*.
Role/Weapon systems: Support helicopter for afloat reconnaissance and SAR. Navalised Bell 230s acquired in 1995. Sensors: None. Weapons: None.

UPDATED

BELL 230   *6/2003, Ecuador Navy /* 0568886

Numbers/Type: 5 Bell 206 Jet Ranger.
Operational speed: 115 kt *(213 km/h)*.
Service ceiling: 13,500 ft *(4,115 m)*.
Range: 368 n miles *(682 km)*.
Role/Weapon systems: Support helicopter for afloat reconnaissance and SAR. Sensors: None. Weapons: None.

NEW ENTRY

BELL 206   *5/2004*, Paul Jackson /* 0569619

LAND-BASED MARITIME AIRCRAFT (FRONT LINE)

Notes: The Navy operates one Airtech CN-235M-100 transport aircraft, four ENAER T-35 Pillan training aircraft, one Beech 300 King Air and one Cessna Citation.

Numbers/Type: 3 Beech Super King Air 200T.
Operational speed: 250 kt *(450 km/h)*.
Service ceiling: 30,000 ft *(9,090 m)*.
Range: 2,030 n miles *(3,756 km)*.
Role/Weapon systems: Maritime reconnaissance and drug interdiction. Sensors: Weather radar. Weapons: Unarmed.

UPDATED

Numbers/Type: 1 Casa CN-235 Persuader.
Operational speed: 210 kt *(384 km/h)*.
Service ceiling: 24,000 ft *(7,315 m)*.
Range: 2,000 n miles *(3,218 km)*.
Role/Weapon systems: EEZ surveillance. To be delivered in 2004. Weapons and sensors to be confirmed.

VERIFIED

PERSUADER (Irish colours)   *6/1994 /* 0080057

Numbers/Type: 2 Beech King Air B-200 CATPASS 250.
Operational speed: 239 kt *(443 km/h)*.
Service ceiling: 9,144 m *(30,000 ft)*.
Range: 2,000 n miles *(3,218 km)*.
Role/Weapon systems: Maritime Patrol aircraft delivered in January and June 1997. Sensors: Not confirmed but CATPASS conversion includes bottom-mounted sea search radar and radome/ FLIR and ESM.

VERIFIED

CATPASS 250MP   *6/1999, Ecuador Navy /* 0054061

PATROL FORCES

3 QUITO (LÜRSSEN 45) CLASS
(FAST ATTACK CRAFT—MISSILE) (PGGF)

Name	No	Builders	Launched	Commissioned
QUITO	LM 21	Lürssen, Vegesack	20 Nov 1975	13 July 1976
GUAYAQUIL	LM 23	Lürssen, Vegesack	5 Apr 1976	22 Dec 1977
CUENCA	LM 24	Lürssen, Vegesack	6 Dec 1976	17 July 1977

Displacement, tons: 255
Dimensions, feet (metres): 147.6 × 23 × 8.1 *(45 × 7 × 2.5)*
Main machinery: 4 MTU 16V 396 diesels; 13,600 hp(m) *(10 MW)* sustained; 4 shafts
Speed, knots: 40. **Range, n miles:** 700 at 40 kt; 1,800 at 16 kt
Complement: 35

Missiles: SSM: 4 Aerospatiale MM 38 Exocet; inertial cruise; active radar homing to 42 km *(23 n miles)* at 0.9 Mach; warhead 165 kg; sea-skimmer.
Guns: 1 OTO Melara 3 in *(76 mm)*/62 compact; 85 rds/min to 16 km *(8.7 n miles)*; weight of shell 6 kg.
2 Oerlikon 35 mm/90 (twin); 550 rds/min to 6 km *(3.3 n miles)*; weight of shell 1.55 kg.
Countermeasures: ESM: ELISRA NS-9010; intercept.
Weapons control: Thomson-CSF Vega system.
Radars: Air/surface search: Thomson-CSF Triton; G-band; range 33 km *(18 n miles)* for 2 m² target.
Fire control: Thomson-CSF Pollux; I/J-band; range 31 km *(17 n miles)* for 2 m² target.
Navigation: Racal Decca 1226; I-band.

Modernisation: New engines fitted during refits in 1994-95 at Guayaquil.
Operational: *Quito* may be laid up.

UPDATED

CUENCA   *2/2000 /* 0103732

AMPHIBIOUS FORCES

1 512-1152 CLASS (LST)

Name	No	Builders	Commissioned
HUALCOPO	TR 61	Chicago Bridge and	9 June 1945
(ex-*Summit County* LST 1146)	(ex-T 61)	Iron Co	

Displacement, tons: 1,747 standard; 3,610 full load
Dimensions, feet (metres): 328 × 50 × 14 *(100 × 16.1 × 4.3)*
Main machinery: 2 GM 12-567A diesels; 1,800 hp *(1.34 MW)*; 2 shafts
Speed, knots: 11.6. **Range, n miles:** 7,200 at 10 kt
Complement: 120 (15 officers)
Military lift: 147 troops
Guns: 8 Bofors 40 mm. 2 Oerlikon 20 mm.
Radars: Navigation: I-band.

Comment: Purchased from US on 14 February 1977. Commissioned in November 1977 after extensive refit. Plans for replacement not yet realised. The ship had a bad fire in July 1998.

UPDATED

HUALCOPO   *2/2004*, Judy Ross /* 0587701

SURVEY AND RESEARCH SHIPS

1 SURVEY CRAFT (YFS)

Name	No	Builders	Commissioned
RIGEL	LH 94 (ex-LH 92)	Halter Marine	1975

Displacement, tons: 50 full load
Dimensions, feet (metres): 64.5 × 17.1 × 3.6 *(19.7 × 5.2 × 1.1)*
Main machinery: 2 diesels; 2 shafts
Speed, knots: 10
Complement: 10 (2 officers)

Comment: Used for inshore oceanographic work.
VERIFIED

1 SURVEY SHIP (YGS)

Name	No	Builders	Commissioned
ORION (ex-*Dometer*)	BI 91 (ex-HI 91, ex-HI 92)	Ishikawajima, Tokyo	10 Nov 1982

Measurement, tons: 1,105 gross
Dimensions, feet (metres): 210.6 pp × 35.1 × 11.8 *(64.2 × 10.7 × 3.6)*
Main machinery: Diesel-electric; 3 Detroit 16V-92TA diesel generators; 2,070 hp *(1.54 MW)* sustained; 2 motors; 1,900 hp *(1.42 MW)*; 1 shaft
Speed, knots: 12.6. **Range, n miles:** 6,000 at 12 kt
Complement: 45 (6 officers) plus 14 civilians
Radars: Navigation: 2 Decca 1226; I-band.

Comment: Research vessel for oceanographic, hydrographic and meteorological work.
VERIFIED

ORION *6/2001, Maritime Photographic /* 0114523

TRAINING SHIPS

1 SAIL TRAINING SHIP (AXS)

Name	No	Builders	Commissioned
GUAYAS	BE 91 (ex-BE 01)	Ast Celaya, Spain	23 July 1977

Measurement, tons: 234 dwt; 934 gross
Dimensions, feet (metres): 264 × 33.5 × 13.4 *(80 × 10.2 × 4.2)*
Main machinery: 1 GM 12V-149T diesel; 875 hp *(652 kW)* sustained; 1 shaft
Speed, knots: 11.3
Complement: 50 plus 80 trainees

Comment: Three masted. Launched 23 September 1976. Has accommodation for 180. Similar to ships in service with Colombia, Mexico and Venezuela. *VERIFIED*

GUAYAS *7/2000, van Ginderen Collection /* 0106847

AUXILIARIES

1 YW CLASS (WATER TANKER) (AWT)

Name	No	Builders	Commissioned
ATAHUALPA (ex-*YW 131*)	TR 63	Leatham D Smith SB Co	17 Sep 1945

Displacement, tons: 460 light; 1,481 full load
Dimensions, feet (metres): 174 × 32 × 15 *(53.1 × 9.8 × 4.6)*
Main machinery: 1 GM 8V-278A diesel; 640 hp *(477 kW)*; 2 shafts
Speed, knots: 8
Complement: 25 (5 officers)
Cargo capacity: 930 tons

Comment: Acquired from the US on 2 May 1963. Purchased on 1 December 1977. Paid off in 1988 but back in service in 1990 to provide water for the Galapagos Islands.
VERIFIED

ATAHUALPA *6/2001, Ecuador Navy /* 0114520

1 OIL TANKER (AOTL)

Name	No	Builders	Commissioned
TAURUS	TR 65 (ex-T 66)	Astinave, Guayaquil	1985

Measurement, tons: 1,175 dwt; 1,110 gross
Dimensions, feet (metres): 174.2 × 36 × 14.4 *(53.1 × 11 × 4.4)*
Main machinery: 2 GM diesels; 1,050 hp *(783 kW)*; 1 shaft
Speed, knots: 11
Complement: 20

Comment: Acquired for the Navy in 1987.
VERIFIED

TAURUS *6/2003, Ecuador Navy /* 0568887

1 ARMAMENT STORES CARRIER (AETL)

Name	No	Builders	Commissioned
CALICUCHIMA (ex-*Throsk*)	TR 62 (ex-A 379)	Cleland SB Co, Wallsend	20 Sep 1977

Displacement, tons: 2,184 full load
Dimensions, feet (metres): 231.2 × 39 × 15 *(70.5 × 11.9 × 4.6)*
Main machinery: 2 Mirrlees-Blackstone diesels; 3,000 hp *(2.2 MW)*; 1 shaft
Speed, knots: 14.5. **Range, n miles:** 4,000 at 11 kt
Complement: 29 (5 officers)
Cargo capacity: 785 tons
Radars: Navigation: Decca 926; I-band.

Comment: Acquired from the UK in November 1991. Recommissioned 24 March 1992.
VERIFIED

CALICUCHIMA *6/2001, Ecuador Navy /* 0114521

2 ARD 12 CLASS (FLOATING DOCKS) (YFD)

Name	No	Builders	Commissioned
RIO AMAZONAS (ex-*ARD 17*)	DF 81 (ex-DF 121)	USA	1944
RIO NAPO (ex-*ARD 24*)	DF 82	USA	1944

Dimensions, feet (metres): 492 × 81 × 17.7 *(150 × 24.7 × 5.4)*

Comment: *Amazonas* leased from US in 1961 and bought outright in 1982; *Napo* bought in 1988. Suitable for docking ships up to 3,200 tons.
VERIFIED

1 WATER CLASS (WATER TANKER) (AWT)

Name	No	Builders	Commissioned
QUISQUIS (ex-*Waterside*)	TR 64 (ex-Y 20)	Drypool Engineering, Hull	1968

Measurement, tons: 519 gross
Dimensions, feet (metres): 131.5 × 25.7 × 11.7 *(40.1 × 7.7 × 3.5)*
Main machinery: 1 Lister-Blackstone ERS-8-MCR diesel; 660 hp *(492 kW)*; 1 shaft
Speed, knots: 10. **Range, n miles:** 1,585 at 9 kt
Complement: 20 (4 officers)
Cargo capacity: 150 tons
Radars: Navigation: Furuno; I-band.

Comment: Acquired from the UK in November 1991.   *VERIFIED*

QUISQUIS   *2/1992, A J Moorey* / 0056909

TUGS

5 HARBOUR TUGS (YTM/YTL)

SANGAY RB 72   ILINIZA RB 75   QUILOTOA RB 78
COTOPAXI RB 73   ALTAR RB 76

Comment: Mostly built in the 1950s and 1960s.   *VERIFIED*

1 CHEROKEE CLASS (ATF)

Name	No	Builders	Commissioned
CHIMBORAZO (ex-*Chowanoc* ATF 100)	RA 70 (ex-R 710, ex-R 71, ex-R 105)	Charleston SB & DD Co	21 Feb 1945

Displacement, tons: 1,235 standard; 1,640 full load
Dimensions, feet (metres): 205 × 38.5 × 17 *(62.5 × 11.7 × 5.2)*
Main machinery: Diesel-electric; 4 Busch-Sulzer BS-539 diesels; 4 generators; 1 motor; 3,000 hp *(2.24 MW)*; 1 shaft
Speed, knots: 16.5. **Range, n miles:** 7,000 at 15 kt
Complement: 85
Guns: 1—3 in *(76 mm)*. 2 Bofors 40 mm. 2 Oerlikon 20 mm (not all fitted).

Comment: Launched 20 August 1943 and transferred 1 October 1977.   *VERIFIED*

CHIMBORAZO   *6/2001, Maritime Photographic* / 0114524

COAST GUARD

Notes: In addition to the vessels listed below, there are up to 40 river patrol launches operated by both the Coast Guard and the Army.

0 + 3 VIGILANTE CLASS (OFFSHORE PATROL CRAFT) (PBO)

Displacement, tons: 300
Dimensions, feet (metres): 147.7 × 32.1 × 8.0 *(45.0 × 9.8 × 2.5)*
Main machinery: 2 MTU 16V 4000 M90; 1 MTU 12V 4000 M80; 3 shafts
Speed, knots: 25. **Range, n miles:** 3,000 at 12 kt
Complement: 27 (5 officers)
Guns: To be announced.
Radars: Navigation: I-band.

Comment: Contract for three craft for the coast Guard let to FBM Babcock Marine in partnership with Astilleros de Murueta, Spain, on 4 March 2004. The steel-hulled craft, to be built in Spain, is based on the FBM Marine Protector 45 class. Propulsion arrangements allow for the use of two main engines or a smaller central engine for loiter. A 5 m interception craft is carried on the aft work deck. To replace the 10 de Agosto class.   *NEW ENTRY*

VIGILANTE CLASS (artist's impression)   *3/2004*, FBM Babcock Marine* / 0587702

2 MANTA CLASS (LARGE PATROL CRAFT) (WPBF)

Name	No	Builders	Commissioned
9 DE OCTUBRE (ex-*Manta*)	LG 37 (ex-LM 25)	Lürssen, Vegesack	11 June 1971
27 DE OCTUBRE (ex-*Nuevo Rocafuerte*)	LG 38 (ex-LM 27)	Lürssen, Vegesack	23 June 1971

Displacement, tons: 119 standard; 134 full load
Dimensions, feet (metres): 119.4 × 19.1 × 6 *(36.4 × 5.8 × 1.8)*
Main machinery: 3 Mercedes-Benz diesels; 9,000 hp(m) *(6.61 MW)*; 3 shafts
Speed, knots: 42. **Range, n miles:** 700 at 30 kt; 1,500 at 15 kt
Complement: 19
Radars: Navigation: I-band.

Structure: Similar design to the Chilean Guacolda class with an extra diesel, 3 kt faster.
Operational: A third of class sank in September 1998 after a collision with a tug. Transferred from the Navy in 2000.   *VERIFIED*

9 DE OCTUBRE   *6/2001, Ecuador Coast Guard* / 0114527

2 ESPADA CLASS (LARGE PATROL CRAFT) (WPB)

Name	No	Builders	Commissioned
5 DE AGOSTO	LG 35	Moss Point Marine, Escatawpa	May 1991
27 DE FEBRERO	LG 36	Moss Point Marine, Escatawpa	Nov 1991

Displacement, tons: 190 full load
Dimensions, feet (metres): 112 × 22.5 × 7 *(34.1 × 6.9 × 2.1)*
Main machinery: 2 Detroit 16V-149TI diesels; 2,322 hp *(1.73 MW)* sustained; 1 Detroit 16V-92TA; 690 hp *(514 kW)* sustained; 3 shafts
Speed, knots: 27. **Range, n miles:** 1,500 at 14 kt
Complement: 19 (5 officers)
Guns: 1—20 mm GAM-BO1. 2—12.7 mm MGs.
Radars: Surface search: Racal Decca; I-band.

Comment: Built under FMS programme. Steel hulls and aluminium superstructure. Accommodation is air conditioned. Carry a 10-man RIB and launching crane on the stern.   *VERIFIED*

5 DE AGOSTO   *6/2002, Ecuador Coast Guard* / 0533896

2 SWIFTSHIPS CLASS (RIVER PATROL CRAFT) (WPBR)

Name	No	Builders	Commissioned
RIO ESMERALDAS (ex-*9 de Octubre*)	LG 47 (ex-LG 37)	Swiftships, Morgan City	1 Oct 1992
RIO SANTIAGO (ex-*27 de Octubre*)	LG 48 (ex-LG 38)	Swiftships, Morgan City	1 Oct 1992

Displacement, tons: 17 full load
Dimensions, feet (metres): 45.5 × 11.8 × 1.8 *(13.9 × 3.6 × 0.6)*
Main machinery: 2 Detroit 6V-92TA diesels; 900 hp *(671 kW)*; 2 Hamilton water-jets
Speed, knots: 22. **Range, n miles:** 600 at 22 kt
Complement: 4
Guns: 2 M2HB 12.7 mm MGs; 2 M60D 7.62 mm MGs.
Radars: Surface search: Raytheon 40; I-band.

Comment: Transferred from US under MAP to the Navy and thence to the Coast Guard. Hard chine modified V hull form. Can carry up to eight troops. Used as command craft for river flotillas.   *VERIFIED*

RIO ESMERALDAS   *6/2001, Ecuador Coast Guard* / 0114516

1 POINT CLASS (COASTAL PATROL CRAFT) (WPB)

Name	No	Builders	Commissioned
24 DE MAYO (ex-*Point Richmond*)	LG 32 (ex-82370)	CG Yard, Curtis Bay	25 Aug 1967

Displacement, tons: 66 full load
Dimensions, feet (metres): 83 × 17.2 × 5.8 *(25.3 × 5.2 × 1.8)*
Main machinery: 2 Caterpillar 3412 diesels; 1,600 hp *(1.19 MW)*; 2 shafts
Speed, knots: 23. **Range, n miles:** 1,500 at 8 kt
Complement: 10
Guns: 2-12.7 mm MGs.
Radars: Navigation: Raytheon SPS 64(V)1; I-band.

Comment: Transferred from US Coast Guard on 22 August 1997.

VERIFIED

24 DE MAYO   *6/2001, Ecuador Coast Guard* / 0114515

6 RIO PUYANGO CLASS (RIVER PATROL CRAFT) (WPBR)

Name	No	Builders	Commissioned
RIO PUYANGO	LG 41 (ex-LGC 40)	Halter Marine, New Orleans	15 June 1986
RIO MATAJE	LG 42 (ex-LGC 41)	Halter Marine, New Orleans	15 June 1986
RIO ZARUMILLA	LG 43 (ex-LGC 42)	Astinave, Guayaquil	11 Mar 1988
RIO CHONE	LG 44 (ex-LGC 43)	Astinave, Guayaquil	11 Mar 1988
RIO DAULE	LG 45 (ex-LGC 44)	Astinave, Guayaquil	17 June 1988
RIO BABAHOYO	LG 46 (ex-LGC 45)	Astinave, Guayaquil	17 June 1988

Displacement, tons: 17
Dimensions, feet (metres): 44 × 13.5 × 3.5 *(13.4 × 4.1 × 1.1)*
Main machinery: 2 Detroit 8V-71 diesels; 460 hp *(343 kW)* sustained; 2 shafts
Speed, knots: 26. **Range, n miles:** 500 at 18 kt
Complement: 5 (1 officer)
Guns: 1—12.7 mm MG. 2—7.62 mm MGs.
Radars: Surface search: Furuno 2400; I-band.

Comment: Two delivered by Halter Marine in June 1986. Four more ordered in February 1987; assembled under licence at Astinave shipyard, Guayaquil. Used mainly for drug interdiction and all are very active.

VERIFIED

RIO BABAHOYO   *6/2002, Ecuador Coast Guard* / 0533895

1 PGM-71 CLASS (LARGE PATROL CRAFT) (WPB)

Name	No	Builders	Commissioned
25 DE JULIO (ex-*Quito*)	LG 31 (ex-LGC 31, ex-LC 71)	Peterson, USA	30 Nov 1965

Displacement, tons: 130 standard; 146 full load
Dimensions, feet (metres): 101.5 × 21 × 5 *(30.9 × 6.4 × 1.5)*
Main machinery: 4 MTU diesels; 3,520 hp(m) *(2.59 MW)*; 2 shafts
Speed, knots: 21. **Range, n miles:** 1,000 at 12 kt
Complement: 15
Guns: 1 Oerlikon 20 mm. 2—12.7 mm MGs.
Radars: Surface search: Raytheon; I-band.

Comment: Transferred from US to the Navy under MAP on 30 November 1965 and then to the Coast Guard in 1980. Paid off into reserve in 1983 and deleted from the order of battle. Refitted with new engines in 1988-89. Second of class deleted in 1997.

UPDATED

25 DE JULIO   *6/2002, Ecuador Coast Guard* / 0533894

2 10 DE AGOSTO CLASS (LARGE PATROL CRAFT) (WPB)

Name	No	Builders	Commissioned
10 DE AGOSTO	LG-33 (ex-LGC-33)	Bremen, Germany	1954
3 DE NOVIEMBRE	LG-34 (ex-LGC-34)	Bremen, Germany	1955

Displacement, tons: 35 standard; 45 full load
Dimensions, feet (metres): 76.75 × 15.7 × 4.6 *(23.4 × 4.8 × 1.4)*
Main machinery: 2 Detroit diesels
Speed, knots: 12. **Range, n miles:** 450 at 12 kt
Complement: 10
Radars: Surface search: Raytheon; I-band.
Guns: 2 Ametralladora .30.

Comment: Transferred from Coopno-Coopin to the coast guard on 12 January 1992 and 4 June 1992. To be replaced by Vigilante class.

UPDATED

10 DE AGOSTO   *6/2001, Ecuador Coast Guard* / 0114518

6 PIRAÑA CLASS (RIVER PATROL CRAFT) (WPBR)

LG-51-56

Main machinery: 2 outboard motors; 300 hp *(224 kW)*
Speed, knots: 35
Complement: 6
Guns: 1 Ametralladora MAG 7.62 mm.

Comment: Built by Astinave and commissioned 1994-95.

VERIFIED

LG 54   *6/2001, Ecuador Coast Guard* / 0114517

2 NAPO CLASS (PBF)

LG 59-60

Main machinery: 2 outboard motors; 300 hp *(224 kW)*
Speed, knots: 40
Complement: 6
Guns: 1 Ametralladora MAG 7.62 mm MG.

Comment: Built by Astinave, Guaquil. Entered service in 2002.

VERIFIED

LG 59   *6/2003, Ecuador Coast Guard* / 0568885

2 RINKER CLASS (PBF)

LG 57-58

Main machinery: 2 outboard motors; 300 hp *(224 kW)*
Speed, knots: 40
Complement: 5
Guns: 1 Ametralladora MAG 7.62 mm MG.

Comment: Built in US. Entered service in 2002.

VERIFIED

RINKER CLASS
6/2003, Ecuador Coast Guard / 0568884

Egypt

Country Overview

The Arab Republic of Egypt declared independence in 1952. The country was united with Syria as the United Arab Republic 1958-61. Located in north-eastern Africa and the Sinai Peninsula, the country has an area of 385,229 square miles and is bordered to the east by Israel, to the south by Sudan and to the west by Libya. It has a 1,323 n mile coastline with the Mediterranean and Red Seas. Cairo is the capital and largest city while Alexandria is the principal port. Port Said and Port Suez are at the northern and southern ends of the 88 n mile long Suez Canal respectively. Territorial seas (12 n miles) are claimed. An EEZ (200 n miles) has been claimed but the limits have not been defined.

Headquarters Appointments

Commander in Chief, Navy:
  Vice Admiral Tamer Abdel Alim
Chief of Naval Staff:
  Rear Admiral Tarek Ahmad Moneim
Chief of Operations:
  Rear Admiral Mohab Mohammed Hessen Mamesh
Chief of Armaments:
  Rear Admiral Fayez Yousif Noubar

Personnel

(a) 2005: 18,500 officers and men, including 2,000 Coast Guard and 10,000 conscripts (Reserves of 14,000)
(b) 1 to 3 years' national service (depending on educational qualifications)

Bases

Alexandria (HQ), Port Said, Mersa Matru, Abu Qir, Suez. Safaqa and Hurghada on the Red Sea.
Naval Academy: Abu Qir.

Coast Defence

There are three batteries of Border Guard Otomat truck-mounted SSMs (two twin launchers each) with targeting by Plessey radars (fixed) and Thomson-CSF radars (mobile). Two Artillery brigades, under naval co-operative control, are armed with 100, 130 and 152 mm guns.

Prefix to Ships' Name

ENS

Maritime Air

Although the Navy has no air arm the Air Force has a number of E-2Cs, ASW Sea Kings and Gazelles with an ASM capability (see *Land-based Maritime Aircraft* section). The Sea Kings and Seasprite helicopters are controlled by the Anti-Submarine Brigade, based at Alexandria, and have some naval aircrew.

Strength of the Fleet

Type	Active	Building (Projected)
Submarines (Patrol)	4	—
Frigates	10	(2)
Fast Attack Craft (Missile)	23	(5)
Fast Attack Craft (Gun)	10	—
Fast Attack Craft (Patrol)	8	—
LSMs/LST	3	(2)
LCUs	9	—
Minesweepers (Ocean)	7	—
Minehunters (Coastal)	3	—
Route Survey Vessels	2	—

PENNANT LIST

Frigates			Al Salam		Assiyut		Shaladein
		442	Al Salam	516	Assiyut	230	Shaladein
901	Sharm el Sheikh	445	Al Jabbar	530	Giza	231	Halaib
906	Toushka	448	Al Qader	533	Aswan	103	Al Maks
911	Mubarak	451	Al Rafa	536	Qena	105	Al Agami
916	Taba	601	23 of July	539	Sohag	107	Al Antar
951	Najim al Zaffer	602	6 of October	542	Dat Assawari	109	Al Dekheila
956	El Nasser	603	21 of October	545	Navarin	111	Al Iskandarani
961	Damyat	604	18 of June	548	Burullus		
966	Rasheed	605	25 of April	610	Safaga	**Training Ships**	
F 941	Abu Qir	670	Ramadan	613	Abu el Ghoson		
F 946	El Suez	672	Khyber			P 91	Al Kousser
		674	El Kadessaya			921	El Fateh
		676	El Yarmouk	**Auxiliaries**		931	Tariq
Patrol Forces		678	Badr				
		680	Hettein	212	Atabarah		
430	Al Nour			214	Akdu		
433	Al Hady	**Mine Warfare Forces**		216	Ayeda 3		
436	Al Wakil			218	Maryut		
439	Al Hakim	507	Daqahliya	220	Al Nil		
		513	Sinai	224	Al Furat		

SUBMARINES

Notes: (1) The Egyptian government signed a Letter of Intent in mid-2000 to purchase two new submarines from an industry team led by Litton Ingalls under FMS funding arrangements. The submarines will be to the RDM 'Moray' design. A contract was not signed in 2002 and the project has probably been abandoned. An alternative solution is the procurement of second-hand submarines and preliminary negotiations for the acquisition of Type 206A boats from Germany are reported to have started in December 2004.
(2) Some two-man Swimmer Delivery Vehicles (SDVs) of Italian CF2 FX 100 design are in service.

4 IMPROVED ROMEO CLASS (PROJECT 033) (SSK)

849   852   855   858

Displacement, tons: 1,475 surfaced; 1,830 dived
Dimensions, feet (metres): 251.3 × 22 × 16.1
(76.6 × 6.7 × 4.9)
Main machinery: Diesel-electric; 2 Type 37-D diesels; 4,000 hp (m) *(2.94 MW)*; 2 motors; 2,700 hp(m) *(1.98 MW)*; 2 creep motors; 2 shafts
Speed, knots: 16 surfaced; 13 dived
Range, n miles: 9,000 at 9 kt surfaced
Complement: 54 (8 officers)

Missiles: SSM: McDonnell Douglas Sub Harpoon; active radar homing to 130 km *(70 n miles)* at 0.9 Mach; warhead 227 kg.
Torpedoes: 8—21 in *(533 mm)* tubes (6 bow, 2 stern). 14 Alliant Mk 37F Mod 2; wire-guided; active/passive homing to 18 km *(9.7 n miles)* at 32 kt; warhead 148 kg.
Mines: 28 in lieu of torpedoes.

Countermeasures: ESM: Argo Phoenix AR-700-S5; radar warning.
Weapons control: Singer Librascope Mk 2. Datalink.
Radars: Surface search: I-band.
Sonars: Atlas Elektronik CSU 83; bow-mounted; active/passive; medium frequency.
  Loral; hull-mounted; active attack; high frequency.

Programmes: Two transferred from China 22 March 1982. Second pair arrived from China 3 January 1984, commissioned 21 May 1984.
Modernisation: In early 1988 a five year contract was signed with Tacoma, Washington to retrofit Harpoon, and Mk 37 wire-guided torpedoes; weapon systems improvements to include Loral active sonar, Atlas Elektronik passive sonar and fire-control system. New air conditioning was also installed.

The US Congress did not give approval to start work until July 1989 and then Tacoma went bankrupt and the work was not taken over by Loral/Lockheed Martin until April 1992. Towed communications wire and GPS are fitted. Kollmorgen 76 and 86 periscopes. Plans to fit optronic masts have not been confirmed. Plans to install an inertial navigation system were announced in 2003.
Operational: *855* was the first to complete modernisation and the remainder completed by mid-1996. All four are reported to have completed machinery overhauls in the last few years and are based at Alexandria. None have been reported active and operational status is doubtful. The ex-USSR submarines of this class have paid off but are still alongside at Alexandria.

VERIFIED

ROMEO 855

3/2000, van Ginderen Collection / 0103733

FRIGATES

Notes: The possible acquisition of two Koni class frigates, one as spares, from Serbia and Montenegro was reportedly discussed in 2004.

4 OLIVER HAZARD PERRY CLASS (FFGHM)

Name	No	Builders	Laid down	Launched	Commissioned
MUBARAK (ex-*Copeland*)	911 (ex-FFG 25)	Todd Shipyards, San Pedro	24 Oct 1979	26 July 1980	7 Aug 1982
TABA (ex-*Gallery*)	916 (ex-FFG 26)	Bath Iron Works	17 May 1980	20 Dec 1980	5 Dec 1981
SHARM EL SHEIKH (ex-*Fahrion*)	901 (ex-FFG 22)	Todd Shipyards, Seattle	1 Dec 1978	24 Aug 1979	16 Jan 1982
TOUSHKA (ex-*Lewis B Puller*)	906 (ex-FFG 23)	Todd Shipyards, San Pedro	23 May 1979	15 Mar 1980	17 Apr 1982

Displacement, tons: 2,750 light; 3,638 full load
Dimensions, feet (metres): 445 × 45 × 14.8; 24.5 (sonar) *(135.6 × 13.7 × 4.5; 7.5)*
Main machinery: 2 GE LM 2500 gas turbines; 41,000 hp *(30.59 MW)* sustained; 1 shaft; cp prop
2 auxiliary retractable props; 650 hp *(484 kW)*
Speed, knots: 29. **Range, n miles:** 4,500 at 20 kt
Complement: 206 (13 officers) including 19 aircrew

Missiles: SSM: 4 McDonnell Douglas Harpoon; active radar homing to 130 km *(70 n miles)* at 0.9 Mach; warhead 227 kg.
SAM: 36 GDC Standard SM-1MR; command guidance; semi-active radar homing to 46 km *(25 n miles)* at 2 Mach.
1 Mk 13 Mod 4 launcher for both SSM and SAM missiles ❶.
Guns: 1 OTO Melara 3 in *(76 mm)*/62 Mk 75 ❷; 85 rds/min to 16 km *(8.7 n miles)* anti-surface; 12 km *(6.6 n miles)* anti-aircraft; weight of shell 6 kg.
1 General Electric/General Dynamics 20 mm/76 6-barrelled Mk 15 Vulcan Phalanx ❸; 3,000 rds/min combined to 1.5 km.
4—12.7 mm MGs.
Torpedoes: 6—324 mm Mk 32 (2 triple) tubes ❹. 24 Alliant Mk 46 Mod 5; anti-submarine; active/passive homing to 11 km *(5.9 n miles)* at 40 kt; warhead 44 kg.
Countermeasures: Decoys: 2 Loral Hycor SRBOC 6-barrelled fixed Mk 36 ❺; IR flares and chaff to 4 km *(2.2 n miles)*.
T-Mk-6 Fanfare/SLQ-25 Nixie; torpedo decoy.
ESM/ECM: Raytheon SLQ-32 ❻ radar warning.
Combat data systems: NTDS with Link Y.
Weapons control: SWG-1 Harpoon LCS. Mk 92 (Mod 4). Mk 13 weapon direction system. 2 Mk 24 optical directors.
Radars: Air search: Raytheon SPS-49(V)4 ❼; C/D-band.
Surface search: ISC Cardion SPS-55 ❽; I-band.
Fire control: Lockheed STIR (modified SPG-60) ❾; I/J-band; range 110 km *(60 n miles)*.
Sperry Mk 92 (Signaal WM28) ❿; I/J-band.
Navigation: Furuno; I-band ⓫; JRC; I-band.
Tacan: URN 25. IFF Mk XII AIMS UPX-29.
Sonars: Raytheon SQS-56; hull-mounted; active search and attack; medium frequency.

Helicopters: 2 Kaman SH-2G Seasprite ⓬.

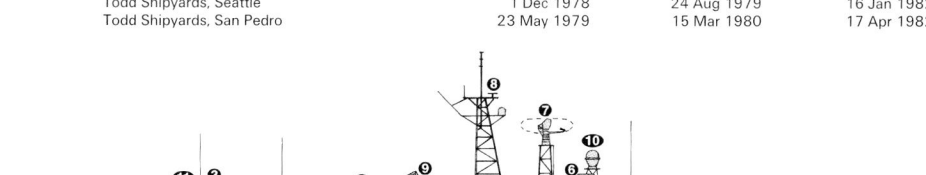

MUBARAK *(Scale 1 : 1,200), Ian Sturton /* 0103734

TABA *1/2001, van Ginderen Collection /* 0130478

Programmes: First one acquired from US on 18 September 1996, second on 28 September 1996, third on 31 March 1998, and fourth on 30 September 1998.
Modernisation: JRC radar fitted on hangar roof.

Operational: First pair arrived in Egypt in mid-1997 after working up, third in late 1998 and fourth in 1999. All reported active, at least one in the Red Sea.

UPDATED

TABA *6/2000, Guy Toremans /* 0103736

2 KNOX CLASS (FFGH)

Name	No	Builders	Laid down	Launched	Commissioned	Recommissioned
DAMYAT (ex-Jesse L Brown)	961 (ex-FF 1089)	Avondale Shipyard	8 Apr 1971	18 Mar 1972	17 Feb 1973	1 Oct 1994
RASHEED (ex-Moinester)	966 (ex-FF 1097)	Avondale Shipyard	25 Aug 1972	12 May 1973	2 Nov 1974	1 Oct 1994

Displacement, tons: 3,011 standard; 4,260 full load
Dimensions, feet (metres): 439.6 × 46.8 × 15; 24.8 (sonar)
   (134 × 14.3 × 4.6; 7.8)
Main machinery: 2 Combustion Engineering/Babcock & Wilcox
   boilers; 1,200 psi (84.4 kg/cm²); 950°F (510°C); 1 turbine;
   35,000 hp (26 MW); 1 shaft
Speed, knots: 27. **Range, n miles:** 4,000 at 22 kt on 1 boiler
Complement: 288 (17 officers)

Missiles: SSM: 8 McDonnell Douglas Harpoon; active radar
   homing to 130 km (70 n miles) at 0.9 Mach; warhead 227 kg.
   A/S: Honeywell ASROC Mk 16 octuple launcher with reload
   system (has 2 cells modified to fire Harpoon) ❶; inertial
   guidance to 1.6-10 km (1-5.4 n miles); payload Mk 46.
Guns: 1 FMC 5 in (127 mm)/54 Mk 42 Mod 9 ❷; 20-40 rds/min
   to 24 km (13 n miles) anti-surface; 14 km (7.7 n miles) anti-
   aircraft; weight of shell 32 kg.
   1 General Electric/General Dynamics 20 mm/76 6-barrelled
   Mk 15 Vulcan Phalanx ❸; 3,000 rds/min combined to 1.5 km.
Torpedoes: 4—324 mm Mk 32 (2 twin) fixed tubes ❹. 22 Alliant
   Mk 46 Mod 5; anti-submarine; active/passive homing to
   11 km (5.9 n miles) at 40 kt; warhead 44 kg.
Countermeasures: Decoys: 2 Loral Hycor SRBOC 6-barrelled
   fixed Mk 36 ❺; IR flares and chaff to 4 km (2.2 n miles). T Mk 6
   Fanfare/SLQ-25 Nixie; torpedo decoy. Prairie Masker hull and
   blade rate noise suppression.
ESM/ECM: Elettronica ❻ intercept and jammer.
Combat data systems: FFISTS mini NTDS with Link Y.
Weapons control: SWG-1A Harpoon LCS. Mk 68 GFCS. Mk 114
   ASW FCS. Mk 1 target designation system.
Radars: Air search: Lockheed SPS-40B ❼; B-band; range 320 km
   (175 n miles).
   Surface search: Raytheon SPS-10 or Norden SPS-67 ❽; G-band.
   Navigation: Marconi LN66; I-band.
   Fire control: Western Electric SPG-53A/D/F ❾; I/J-band.
   Tacan: SRN 15.
Sonars: EDO/General Electric SQS-26 CX; bow-mounted; active
   search and attack; medium frequency.

Helicopters: 1 Kaman SH-2G Seasprite ❿.

Programmes: Lease agreed from USA in mid-1993 and signed
   27 July 1994 when both ships sailed for Egypt. Two others
   were transferred for spares in 1996. Ships of this class have
   been transferred to Greece, Taiwan, Turkey and Thailand.
Modernisation: Vulcan Phalanx fitted in the mid-1980s. There
   are plans to fit quadruple Harpoon launchers and possibly to
   remove the ASROC launcher. EW suite replaced.

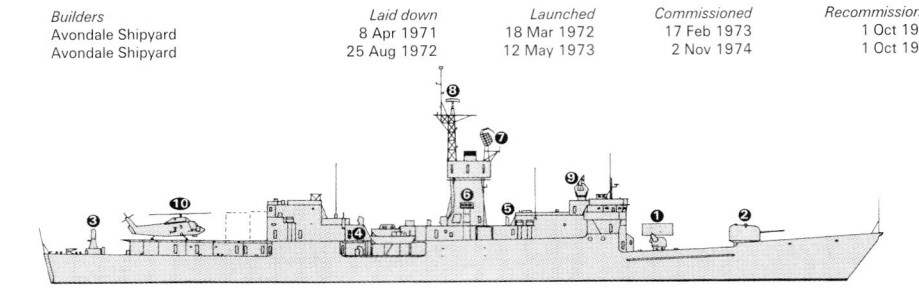

DAMYAT   (Scale 1 : 1,200), Ian Sturton

DAMYAT   3/2000, M Declerck / 0103737

Structure: Four torpedo tubes are fixed in the midship
   superstructure, two to a side, angled out at 45°. A lightweight
   anchor is fitted on the port side and an 8,000 lb anchor fits in
   to the after section of the sonar dome.

Operational: These ships have had boiler problems in Egyptian
   service, and refits are planned with US assistance if and when
   funds become available.

                                                           VERIFIED

2 DESCUBIERTA CLASS (FFGM)

Name	No	Builders	Laid down	Launched	Commissioned
EL SUEZ (ex-Serviola)	F 946	Bazán, Ferrol	28 Feb 1979	20 Dec 1979	27 Oct 1984
ABU QIR (ex-Centinela)	F 941	Bazán, Ferrol	31 Oct 1978	6 Oct 1979	21 May 1984

Displacement, tons: 1,233 standard; 1,479 full load
Dimensions, feet (metres): 291.3 × 34 × 12.5
   (88.8 × 10.4 × 3.8)
Main machinery: 4 MTU-Bazán 16V 956 TB91 diesels;
   15,000 hp(m) (11 MW) sustained; 2 shafts; cp props
Speed, knots: 25.5; 28 trials. **Range, n miles:** 4,000 at 18 kt
Complement: 116 (10 officers)

Missiles: SSM: 8 McDonnell Douglas Harpoon (2 quad)
   launchers ❶; active radar homing to 130 km (70 n miles) at
   0.9 Mach; warhead 227 kg.
   SAM: Selenia Elsag Albatros octuple launcher ❷; 24 Aspide;
   semi-active radar homing to 13 km (7 n miles) at 2.5 Mach;
   height envelope 15-5,000 m (49.2-16,405 ft); warhead
   30 kg.
Guns: 1 OTO Melara 3 in (76 mm)/62 compact ❸; 85 rds/min to
   16 km (8.7 n miles); weight of shell 6 kg.
   2 Bofors 40 mm/70 ❹; 300 rds/min to 12.5 km (6.8 n miles);
   weight of shell 0.96 kg.
Torpedoes: 6—324 mm Mk 32 (2 triple) tubes ❺. MUSL
   Stingray; anti-submarine; active/passive homing to 11 km
   (5.9 n miles) at 45 kt; warhead 35 kg (shaped charge); depth
   to 750 m (2,460 ft).
A/S mortars: 1 Bofors 375 mm twin-barrelled trainable launcher
   ❻; automatic loading; range 1,600 or 3,600 m depending on
   type of rocket.

Countermeasures: ESM/ECM: Elettronica SpA Beta; intercept
   and jammer.
   Prairie Masker; acoustic signature suppression.
Combat data systems: Signaal SEWACO action data
   automation. Link Y.
Radars: Air/surface search: Signaal DA05 ❼; E/F-band; range
   137 km (75 n miles) for 2 m² target.
   Navigation: Signaal ZW06; I-band.
   Fire control: Signaal WM25 ❽; I/J-band.
Sonars: Raytheon 1160B; hull-mounted; active search and
   attack; medium frequency.
   Raytheon 1167 ❾; VDS; active search; 12-7.5 kHz.

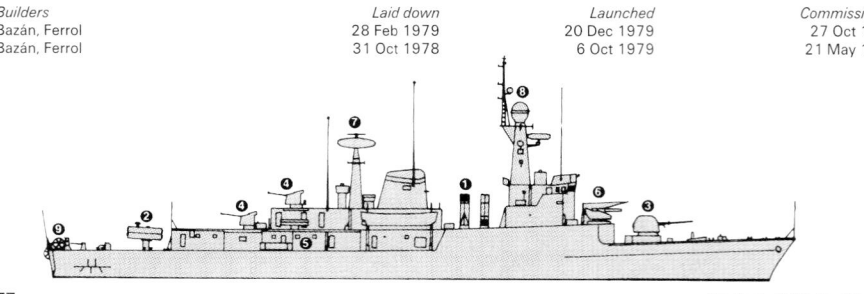

EL SUEZ   (Scale 1 : 900), Ian Sturton

Programmes: Ordered September 1982 from Bazán, Spain. The
   two Spanish ships Centinela and Serviola were sold to Egypt
   prior to completion and transferred after completion at Ferrol
   and modification at Cartagena. El Suez completed 28
   February 1984 and Abu Qir on 31 July 1984.
Modernisation: The combat data system, air search and fire-
   control radars were updated in 1995-96.
Operational: Stabilisers fitted. Modern noise insulation of main
   and auxiliary machinery. Both are active.

                                                           VERIFIED

EL SUEZ   10/1999 / 0085001

2 JIANGHU I CLASS (FFG)

Name	No	Builders	Commissioned
NAJIM AL ZAFFER	951	Hudong, Shanghai	27 Oct 1984
EL NASSER	956	Hudong, Shanghai	16 Apr 1985

Displacement, tons: 1,425 standard; 1,702 full load
Dimensions, feet (metres): 338.5 × 35.4 × 10.2
  (103.2 × 10.8 × 3.1)
Main machinery: 2 Type 12 E 390V diesels; 14,400 hp(m)
  (10.6 MW) sustained; 2 shafts
Speed, knots: 26. **Range, n miles:** 4,000 at 15 kt
Complement: 195

Missiles: SSM: 4 Hai Ying 2 (Flying Dragon) (2 twin) ❶; active
  radar or passive IR homing to 80 km *(43.2 n miles)* at 0.9
  Mach; warhead 513 kg.
Guns: 4 China 57 mm/70 (2 twin) ❷; 120 rds/min to 12 km
  (6.5 n miles); weight of shell 6.31 kg.
  12 China 37 mm/63 (6 twin) ❸; 180 rds/min to 8.5 km *(4.6 n
  miles)*; weight of shell 1.42 kg.
A/S mortars: 2 RBU 1200 5-tubed fixed launchers ❹; range
  1,200 m; warhead 34 kg.
Depth charges: 4 projectors.
Mines: Up to 60.
Countermeasures: ESM/ECM: Elettronica SpA Beta or Litton
  Triton; intercept and jammer.
Radars: Air search: Type 765 ❺; A-band.
  Surface search: Eye Shield ❻; G-band.
  Surface search/gun direction: Square Tie; I-band.
  Fire control: Fog Lamp.
  Navigation: Decca RM 1290A; I-band.
Sonars: China Type E5; hull-mounted; active search and attack;
  high frequency.

Programmes: Ordered from China in 1982. This is a Jianghu I
  class modified with 57 mm guns vice the standard 100 mm.
  These were the 17th and 18th hulls of the class.
Modernisation: Combat data system to be fitted together with
  CSEE Naja optronic fire-control directors. There are also plans,
  confirmed in October 1994, to remove the after
  superstructure and guns and build a flight deck for an SH-2G

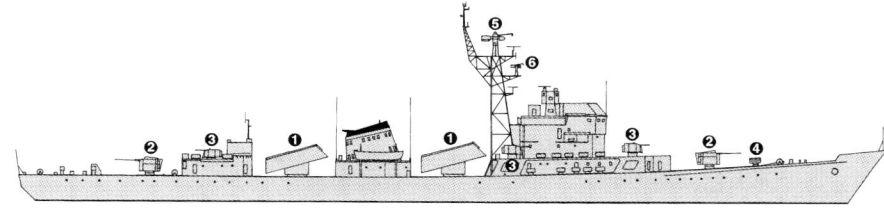

NAJIM AL ZAFFER
(Scale 1 : 900), Ian Sturton / 0056914

EL NASSER
6/2002 / 0528335

Seasprite helicopter. Although a refit programme is reported
to have been proposed by China, there is still no sign yet of
work being done.

Structure: The funnel is the rounded version of the Jianghu class.
Operational: Both ships are very active in the Red Sea.
VERIFIED

NAJIM AL ZAFFER

5/2001, Schaeffer/Marsan / 0130480

SHIPBORNE AIRCRAFT

Numbers/Type: 10 Kaman SH-2G(E) Seasprite.
Operational speed: 130 kt *(241 km/h).*
Service ceiling: 22,500 ft *(6,860 m).*
Range: 367 n miles *(679 km).*
Role/Weapon systems: Total of 10 upgraded SH-2F aircraft transferred under FMS by September 1998. New engines and avionics. A further avionics upgrade was reportedly under consideration in 2004. Sensors: LN66/HP radar; ALR-66 ESM; ALE-39 ECM; ARN-118 Tacan; Ocean Systems AQS-18A dipping sonar. Possible mine detection optronic sensor. Weapons: 2 × Mk 46 torpedoes or a depth bomb.

UPDATED

SEASPRITE   *1995, Kaman /* 0056915

LAND-BASED MARITIME AIRCRAFT (FRONT LINE)

Notes: There are also 2/4 Westland Commando Mk 2B/2E helicopters. Some refitted in 1997/98.

Numbers/Type: 9 Aerospatiale SA 342L Gazelle.
Operational speed: 142 kt *(264 km/h).*
Service ceiling: 14,105 ft *(4,300 m).*
Range: 407 n miles *(755 km).*
Role/Weapon systems: Air Force helicopter for coastal anti-shipping strike, particularly against FAC and insurgents. Sensors: SFIM sight. Weapons: ASV; 2 × AS-12 wire-guided missiles.

VERIFIED

Numbers/Type: 6 Grumman E-2C Hawkeye.
Operational speed: 323 kt *(598 km/h).*
Service ceiling: 37,000 ft *(11,278 m).*
Range: 1,540 n miles *(2,852 km).*
Role/Weapon systems: Air Force airborne early warning and control tasks; capable of handling up to 30 tracks over water or land. Sixth aircraft ordered in June 2001. Sensors: APS-138 search/warning radar being replaced by APS-145 from October 2002 as part of major upgrade programme. The first upgraded aircraft delivered in February 2003 and second in early 2004; various ESM/ECM systems. Weapons: Unarmed.

UPDATED

HAWKEYE 2000   *3/2003*, Northrop Grumman /* 0530203

Numbers/Type: 5 Westland Sea King Mk 47.
Operational speed: 112 kt *(208 km/h).*
Service ceiling: 14,700 ft *(4,480 m).*
Range: 664 n miles *(1,230 km).*
Role/Weapon systems: Air Force helicopter for ASW and surface search; secondary role as SAR helicopter. Airframe and engine refurbishment in 1990 for first five. Four more are in reserve and out of service. Sensors: MEL search radar. Weapons: ASW; 4 × Mk 46 or Stingray torpedoes or depth bombs. ASV: Otomat.

VERIFIED

Numbers/Type: 2 Beechcraft 1900C.
Operational speed: 267 kt *(495 km/h).*
Service ceiling: 25,000 ft *(7,620 m).*
Range: 1,569 n miles *(2,907 km).*
Role/Weapon systems: Two (of six) Air Force aircraft acquired in 1988 and used for maritime surveillance. Sensors: Litton search radar; Motorola multimode SLAMMR radar; Singer S-3075 ESM; Datalink Y. Weapons: Unarmed.

VERIFIED

PATROL FORCES

Notes: Following responses to an ITT issued in 1999, the Egyptian Navy placed an order in January 2001 for four Fast Attack Craft (Missile). The 60 m, diesel-powered 'Ambassador III' craft, were to have been built by Halter Marine under United States' FMS funding arrangements. Halter Marine has since been taken over by Singapore Technologies Engineering and there were reports of renewed interest in the project in 2004.

5 TIGER CLASS (TYPE 148)
(FAST ATTACK CRAFT—MISSILE) (PGGF)

Name	No	Builders	Commissioned
23 OF JULY (ex-*Alk*)	601 (ex-P 6155)	CMN, Cherbourg	7 Jan 1975
6 OF OCTOBER (ex-*Fuchs*)	602 (ex-P 6146)	CMN, Cherbourg	17 Oct 1973
21 OF OCTOBER (ex-*Löwe*)	603 (ex-P 6148)	CMN, Cherbourg	9 Jan 1974
18 OF JUNE (ex-*Dommel*)	604 (ex-P 6156)	CMN, Cherbourg	12 Feb 1975
25 OF APRIL (ex-*Weihe*)	605 (ex-P 6157)	CMN, Cherbourg	3 Apr 1975

Displacement, tons: 234 standard; 265 full load
Dimensions, feet (metres): 154.2 × 23 × 8.9 *(47 × 7 × 2.7)*
Main machinery: 4 MTU MD 16V 538 TB90 diesels; 12,000 hp(m) *(8.82 MW)* sustained; 4 shafts
Speed, knots: 36. **Range, n miles:** 570 at 30 kt; 1,600 at 15 kt
Complement: 30 (4 officers)

Missiles: SSM: 4 Aerospatiale MM 38 Exocet (2 twin) launchers; inertial cruise; active radar homing to 42 km *(23 n miles)* at 0.9 Mach; warhead 165 kg; sea-skimmer.
Guns: 1 OTO Melara 3 in *(76 mm)*/62 compact; 85 rds/min to 16 km *(8.6 n miles)* anti-surface; 12 km *(6.5 n miles)* anti-aircraft; weight of shell 6 kg.
1 Bofors 40 mm/70; 330 rds/min to 12 km *(6.5 n miles)* anti-surface; 4 km *(2.2 n miles)* anti-aircraft; weight of shell 0.96 kg; fitted with GRP dome (1984) (see *Modernisation*).
Mines: Laying capability.
Countermeasures: Decoys: Wolke chaff launcher. Hot Dog IR launcher.
Combat data systems: PALIS and Link 11.
Weapons control: CSEE Panda optical director. Thomson-CSF Vega PCET system, controlling missiles and guns.
Radars: Air/surface search: Thomson-CSF Triton; G-band; range 33 km *(18 n miles)* for 2 m² target.
Navigation: SMA 3 RM 20; I-band; range 73 km *(40 n miles).*
Fire control: Thomson-CSF Castor; I/J-band.

Programmes: 601 transferred from Germany in July 2002 and the remainder in March 2003. Weapons and sensors have also been transferred with the possible exception of EW equipment.
Modernisation: Triton search and Castor fire-control radars fitted to the whole class.
Structure: Steel-hulled craft. Similar to Combattante II craft.

VERIFIED

25 OF APRIL   *4/2003, Michael Nitz /* 0552771

21 OF OCTOBER   *4/2003, Michael Nitz /* 0552773

18 OF JUNE   *4/2003, Michael Nitz /* 0552772

6 RAMADAN CLASS (FAST ATTACK CRAFT—MISSILE) (PGGF)

Name	No	Builders	Launched	Commissioned
RAMADAN	670	Vosper Thornycroft	6 Sep 1979	20 July 1981
KHYBER	672	Vosper Thornycroft	31 Jan 1980	15 Sep 1981
EL KADESSAYA	674	Vosper Thornycroft	19 Feb 1980	6 Apr 1982
EL YARMOUK	676	Vosper Thornycroft	12 June 1980	18 May 1982
BADR	678	Vosper Thornycroft	17 June 1981	17 June 1982
HETTEIN	680	Vosper Thornycroft	25 Nov 1980	28 Oct 1982

Displacement, tons: 307 full load
Dimensions, feet (metres): 170.6 × 25 × 7.5 *(52 × 7.6 × 2.3)*
Main machinery: 4 MTU 20V 538 TB91 diesels; 15,360 hp(m) *(11.29 MW)* sustained; 4 shafts
Speed, knots: 40. **Range, n miles:** 1,600 at 18 kt
Complement: 30 (4 officers)

Missiles: SSM: 4 OTO Melara/Matra Otomat Mk 2; active radar homing to 160 km *(86.4 n miles)* at 0.9 Mach; warhead 210 kg.
Guns: 1 OTO Melara 3 in *(76 mm)* compact; 85 rds/min to 16 km *(8.7 n miles)*; weight of shell 6 kg.
2 Breda 40 mm/70 (twin); 300 rds/min to 12.5 km *(6.8 n miles)* anti-surface; weight of shell 0.96 kg.
Countermeasures: Decoys: 4 Protean fixed launchers each with 4 magazines containing 36 chaff decoy and IR flare grenades.
ESM: Racal Cutlass; radar intercept.
ECM: Racal Cygnus; jammer.
Combat data systems: Ferranti CAAIS action data automation.
Weapons control: Marconi Sapphire System with 2 radar/TV and 2 optical directors.
Radars: Air/surface search: Marconi S 820; E/F-band; range 73 km *(40 n miles)*.
Navigation: Marconi S 810; I-band.
Fire control: 2 Marconi ST 802; I-band.

Programmes: The contract was carried out at the Porchester yard of Vosper Thornycroft Ltd with some hulls built at Portsmouth Old Yard, being towed to Porchester for fitting out.
Modernisation: Contracts for the modernisation of these craft was let in 2001. Alenia Marconi Systems is to upgrade the Otomat missiles to Mk 2, renovate the S 820 and ST 802 radars and replace the CAAIS combat system by NAUTIS 3. Work being carried out 2002-2007.
Operational: Portable SAM SA-N-5 sometimes carried.

UPDATED

EL KADESSAYA *3/2000, van Ginderen Collection /* 0103739

3 + (5) OSA I (PROJECT 205) CLASS (FAST ATTACK CRAFT—MISSILE) (PTFG)

631	633	643

Displacement, tons: 171 standard; 210 full load
Dimensions, feet (metres): 126.6 × 24.9 × 8.9 *(38.6 × 7.6 × 2.7)*
Main machinery: 3 MTU diesels; 12,000 hp(m) *(8.82 MW)*; 3 shafts
Speed, knots: 35. **Range, n miles:** 400 at 34 kt
Complement: 30

Missiles: SSM: 4 SS-N-2A Styx; active radar or IR homing to 46 km *(25 n miles)* at 0.9 Mach; altitude preset up to 300 m *(984.3 ft)*; warhead 513 kg.
SAM: SA-N-5 Grail; manual aiming; IR homing to 6 km *(3.2 n miles)* at 1.5 Mach; altitude to 2,500 m *(8,000 ft)*; warhead 1.5 kg.
Guns: 4 USSR 30 mm/65 (2 twin); 500 rds/min to 5 km *(2.7 n miles)* anti-aircraft; weight of shell 0.54 kg.
2—12.7 mm MGs.
Countermeasures: ESM: Thomson-CSF DR 875; radar warning.
ECM: Racal; jammer.
Radars: Air/surface search: Kelvin Hughes; I-band.
Navigation: Racal Decca 916; I-band.
Fire control: Drum Tilt; H/I-band.
IFF: High Pole. Square Head.

Programmes: Thirteen reported to have been delivered to Egypt by the Soviet Navy in 1966-68 but some were sunk in war with Israel, October 1973. Four of the remainder were derelict in 1989 but one more was back in service in 1991 and two more in 1993 and the last of the four in 1995. Five vessels may be acquired from Serbia and Montenegro.
Modernisation: Refitted with MTU diesels, two machine guns, improved radars and EW equipment.
Operational: Three more *637* and *639* and *641* are laid up.

UPDATED

OSA 633 *1986*

5 OCTOBER CLASS (FAST ATTACK CRAFT—MISSILE) (PTFG)

781	783	787	789	791

Displacement, tons: 82 full load
Dimensions, feet (metres): 84 × 20 × 5 *(25.5 × 6.1 × 1.3)*
Main machinery: 4 CRM 12 D/SS diesels; 5,000 hp(m) *(3.67 MW)* sustained; 4 shafts
Speed, knots: 38. **Range, n miles:** 400 at 30 kt
Complement: 20

Missiles: SSM: 2 OTO Melara/Matra Otomat Mk 2; active radar homing to 160 km *(86.4 n miles)* at 0.9 Mach; warhead 210 kg; can be carried.
Guns: 4 BMARC/Oerlikon 30 mm/75 (2 twin); 650 rds/min to 10 km *(5.5 n miles)* anti-surface; 3 km *(1.6 n miles)* anti-aircraft; weight of shell 1 kg and 0.36 kg mixed.
Countermeasures: Decoys: 2 Protean fixed launchers each with 4 magazines containing 36 chaff decoy and IR flare grenades.
ESM: Racal Cutlass; radar warning.
Weapons control: Marconi Sapphire radar/TV system.
Radars: Air/surface search: Marconi S 810; range 48 km *(25 n miles)*.
Fire control: Marconi/ST 802; I-band.

Programmes: Built in Alexandria 1975-76. Hull of same design as USSR Komar class. Refitted by Vosper Thornycroft, completed 1979-81. *791* was washed overboard on return trip, recovered and returned to Portsmouth for refit. Left UK after repairs on 12 August 1982. Probably Link fitted.
Modernisation: Alenia Marconi systems to upgrade Otomat missiles to Mk 2 between 2002-2007.
Operational: *791* reported non-operational and *785* is laid up.

UPDATED

OCTOBER 783 *2/2004 * /* 1044123

4 HEGU CLASS (FAST ATTACK CRAFT—MISSILE) (PTFG)

609	611	613	615

Displacement, tons: 68 standard; 79.2 full load
Dimensions, feet (metres): 88.6 × 20.7 × 4.3 *(27 × 6.3 × 1.3)*
Main machinery: 4 Type L-12V-180 diesels; 4,800 hp(m) *(3.53 MW)*; 4 shafts
Speed, knots: 37.5. **Range, n miles:** 400 at 30 kt
Complement: 17 (2 officers)

Missiles: SSM: 2 SY-1; active radar or passive IR homing to 40 km *(22 n miles)* at 0.9 Mach; warhead 513 kg.
Guns: 2—23 mm (twin); locally constructed to fit 25 mm mounting.
Countermeasures: ESM: Litton Triton; radar intercept.
Radars: Surface search/fire control: Square Tie; I-band or Decca; I-band.
IFF: High Pole A.

Programmes: Acquired from China and commissioned in Egypt on 27 October 1984. The Hegu is the Chinese version of the deleted Komar.
Modernisation: ESM fitted in 1995-96.
Operational: *619* and *617* are reported laid up.

VERIFIED

HEGU 609 *7/1995 /* 0050086

HEGU 609 *3/2000 /* 0103740

6 SHERSHEN CLASS (FAST ATTACK CRAFT—GUN) (PTFM)

| 751 | 753 | 755 | 757 | 759 | 761 |

Displacement, tons: 145 standard; 170 full load
Dimensions, feet (metres): 113.8 × 22 × 4.9 *(34.7 × 6.7 × 1.5)*
Main machinery: 3 Type M 503A diesels; 8,025 hp(m) *(5.9 MW)* sustained; 3 shafts
Speed, knots: 45. **Range, n miles:** 850 at 30 kt
Complement: 23

Missiles: SAM: SA-N-5 Grail *(755-761)*; manual aiming; IR homing to 6 km *(3.2 n miles)* at 1.5
  Mach; warhead 1.5 kg.
Guns: 4 USSR 30 mm/65 (2 twin); 500 rds/min to 5 km *(2.7 n miles)*; weight of shell 0.54 kg.
  2 USSR 122 mm rocket launchers (*755-761* in lieu of torpedo tubes); 20 barrels per launcher;
  range 9 km *(5 n miles)*.
Depth charges: 12.
Countermeasures: ESM: Thomson-CSF DR 875; radar warning.
Radars: Surface search: Pot Drum; H/I-band.
Fire control: Drum Tilt, H/I-band (in some).
IFF: High Pole.

Programmes: Five delivered from USSR in 1967 and two more in 1968. One deleted. *753*
  completed an extensive refit at Ismailia in 1987; *751* in 1988.
Structure: The last four have had their torpedo tubes removed to make way for multiple BM21
  rocket launchers and one SA-N-5 Grail, which are not always carried. Some have Drum Tilt
  radars removed. The first two have also had their torpedo tubes removed but these may be
  replaced.
Operational: Based at Alexandria, Port Said and Mersa Matru. All are active.
VERIFIED

SHERSHEN 759   *5/2003, A Sharma /* 0552770

8 HAINAN CLASS (FAST ATTACK CRAFT—PATROL) (PC)

| AL NOUR 430 | AL HADY 433 | AL WAKIL 436 | AL HAKIM 439 |
| AL SALAM 442 | AL JABBAR 445 | AL QADER 448 | AL RAFA 451 |

Displacement, tons: 375 standard; 392 full load
Dimensions, feet (metres): 192.8 × 23.6 × 7.2 *(58.8 × 7.2 × 2.2)*
Main machinery: 4 PRC/Kolomna Type 9-D-8 diesels; 4,000 hp *(2.94 MW)* sustained; 4 shafts
Speed, knots: 30.5. **Range, n miles:** 1,300 at 15 kt
Complement: 69

Guns: 4 China 57 mm/70 (2 twin); 120 rds/min to 12 km *(6.5 n miles)*; weight of shell 6.31 kg.
  4—23 mm (2 twin); locally constructed to fit the 25 mm mountings.
Torpedoes: 6—324 mm (2 triple) tubes (in two of the class). Mk 44 or MUSL Stingray.
A/S mortars: 4 RBU 1200 fixed 5-tubed launchers; range 1,200 m; warhead 34 kg.
Depth charges: 2 projectors; 2 racks. 18 DCs.
Mines: Rails fitted. 12 mines.
Radars: Surface search: Pot Head or Skin Head; I-band.
Navigation: Decca; I-band.
IFF: High Pole.
Sonars: Stag Ear; hull-mounted; active search and attack; high frequency.

Programmes: First pair transferred from China in October 1983, next three in February 1984
  (commissioned 21 May 1984) and last three late 1984.
Modernisation: Two fitted with torpedo tubes and with Singer Librascope fire control. No sign of
  the remainder being similarly equipped. New sonar reported being fitted.
Operational: Based at Alexandria. *436, 439* and *451* reported not operational.
VERIFIED

AL SALAM   *3/2000 /* 0103741

4 SHANGHAI II CLASS (FAST ATTACK CRAFT—GUN) (PB)

| 793 | 795 | 797 | 799 |

Displacement, tons: 113 standard; 131 full load
Dimensions, feet (metres): 127.3 × 17.7 × 5.6 *(38.8 × 5.4 × 1.7)*
Main machinery: 2 Type L12-180 diesels; 2,400 hp(m) *(1.76 MW)* (forward); 2 Type L12-180Z
  diesels; 1,820 hp(m) *(1.34 MW)* (aft); 4 shafts
Speed, knots: 30. **Range, n miles:** 700 at 16.5 kt
Complement: 34

Guns: 4 China 37 mm/63 (2 twin); 180 rds/min to 8.5 km *(4.6 n miles)*; weight of shell 1.42 kg.
  4—23 mm (2 twin); locally constructed to fit the 25 mm mountings.
Mines: Rails can be fitted for 10 mines.
Countermeasures: ESM: Thomson-CSF; radar warning.
Radars: Surface search: Decca; I-band.
IFF: High Pole.

Programmes: Transferred from China in 1984.
Operational: Three based at Suez and one *(799)* at Mersa Matru. *795* refitted in 1998.
VERIFIED

SHANGHAI 797   *6/1997, J W Currie /* 0012295

AMPHIBIOUS FORCES

Notes: (1) Acquisition of LSTs is a high priority.
(2) Ro-Ro ferries are chartered for amphibious exercises.
(3) Rigid Raiders with Johnson outboards are also in service.
(4) Three small hovercraft similar to Slingsby SAH 2200 reported to be in service.

3 POLNOCHNY A (PROJECT 770) CLASS (LSM)

| 301 | 303 | 305 |

Displacement, tons: 800 full load
Dimensions, feet (metres): 239.5 × 27.9 × 5.8 *(73 × 8.5 × 1.8)*
Main machinery: 2 Kolomna Type 40-D diesels; 4,400 hp(m) *(3.2 MW)* sustained; 2 shafts
Speed, knots: 19. **Range, n miles:** 1,000 at 18 kt
Complement: 40
Military lift: 6 tanks; 350 tons
Guns: 2 USSR 30 mm/65 (twin); 500 rds/min to 5 km *(2.7 n miles)*; weight of shell 0.54 kg.
  2—140 mm rocket launchers; 18 barrels to 9 km *(4.9 n miles)*.
Radars: Surface search: Decca; I-band.
Fire control: Drum Tilt; H/I-band.

Comment: Built at Northern Shipyard, Gdansk and transferred from USSR 1973-74. All used for
  Gulf logistic support in 1990-91. SA-N-5 may be carried. Radar updated. All are active.
VERIFIED

POLNOCHNY 303   *10/2000, F Sadek /* 0103742

4 SEAFOX TYPE (SWIMMER DELIVERY CRAFT) (LDW)

| 21 | 23 | 27 | 30 |

Displacement, tons: 11.3 full load
Dimensions, feet (metres): 36.1 × 9.8 × 2.6 *(11 × 3 × 0.8)*
Main machinery: 2 GM 6V-92TA diesels; 520 hp *(388 kW)* sustained; 2 shafts
Speed, knots: 30. **Range, n miles:** 200 at 20 kt
Complement: 3
Guns: 2—12.7 mm MGs. 2—7.62 mm MGs.
Radars: Surface search: LN66; I-band.

Comment: Ordered from Uniflite, Washington in 1982. GRP construction painted black. There is a
  strong underwater team in the Egyptian Navy which is also known to use commercial two-man
  underwater chariots. Based at Abu Qir and all took part in the 1998 Fleet review. Six others are
  in various states of repair. RIBs are also in service.
UPDATED

SEAFOX   *1999 /* 0056917

9 VYDRA CLASS (LCU)

330	332	334	336	338	340	342	344	346

Displacement, tons: 425 standard; 600 full load
Dimensions, feet (metres): 179.7 × 25.3 × 6.6 *(54.8 × 7.7 × 2)*
Main machinery: 2 Type 3-D-12 diesels; 600 hp(m) *(440 kW)* sustained; 2 shafts
Speed, knots: 11. **Range, n miles:** 2,500 at 10 kt
Complement: 20
Military lift: 200 troops; 250 tons.
Guns: 2 or 4—37 mm/63 (1 or 2 twin) (may be fitted).
Radars: Navigation: Decca; I-band.

Comment: Built in late 1960s, transferred from USSR 1968-69. For a period after the Israeli war of October 1973 several were fitted with rocket launchers and two 37 or 40 mm guns, some of which have now been removed. All still in service.

UPDATED

VYDRA 346   *10/2002*, F Sadek /* 1044122

MINE WARFARE FORCES

4 YURKA CLASS (MINESWEEPERS—OCEAN) (MSO)

GIZA 530	ASWAN 533	QENA 536	SOHAG 539

Displacement, tons: 540 full load
Dimensions, feet (metres): 171.9 × 30.8 × 8.5 *(52.4 × 9.4 × 2.6)*
Main machinery: 2 Type M 503 diesels; 5,350 hp(m) *(3.91 MW)* sustained; 2 shafts
Speed, knots: 17. **Range, n miles:** 1,500 at 12 kt
Complement: 45
Guns: 4 USSR 30 mm/65 (2 twin); 500 rds/min to 5 km *(2.7 n miles)*; weight of shell 0.54 kg.
Mines: Can lay 10.
Radars: Navigation: Don; I-band.
Sonars: Stag Ear; hull-mounted; active search; high frequency.

Comment: Steel-hulled minesweepers transferred from the USSR in 1969. Built 1963-69. Egyptian Yurka class do not carry Drum Tilt radar and have a number of ship's-side scuttles. The plan to equip them with VDS sonar has been shelved. At least one operates an ROV.

UPDATED

SOHAG   *10/2001*, F Sadek /* 1044121

2 SWIFTSHIPS TYPE (ROUTE SURVEY VESSELS) (MSI)

Name	No	Builders	Commissioned
SAFAGA	610 (ex-RSV 1)	Swiftships	1 Oct 1994
ABU EL GHOSON	613 (ex-RSV 2)	Swiftships	1 Oct 1994

Displacement, tons: 165 full load
Dimensions, feet (metres): 90 × 24.8 × 8 *(27.4 × 7.6 × 2.4)*
Main machinery: 2 MTU 12V 183 TA61 diesels; 928 hp(m) *(682 kW)*; 2 shafts; bow thruster; 60 hp(m) *(44 kW)*
Speed, knots: 12. **Range, n miles:** 1,500 at 10 kt
Complement: 16 (2 officers)
Guns: 1—12.7 mm MG.
Radars: Navigation: Furuno 2020; I-band.
Sonars: EG & G side scan; active; high frequency.

Comment: Route survey vessels ordered from Swiftships in November 1990 and delivered in September 1993. Two more are planned to be built in Egyptian yards in due course. Unisys improved SYQ-12 command system. Provision for both shallow and deep towed bodies. The names have been taken from the obsolete K 8 class.

VERIFIED

SAFAGA   *3/2000 /* 0103744

3 SWIFTSHIPS TYPE (COASTAL MINEHUNTERS) (MHC)

Name	No	Builders	Launched	Commissioned
DAT ASSAWARI	542 (ex-CMH 1)	Swiftships, Morgan City	4 Oct 1993	13 July 1997
NAVARIN	545 (ex-CMH 2)	Swiftships, Morgan City	13 Nov 1993	13 July 1997
BURULLUS	548 (ex-CMH 3)	Swiftships, Morgan City	4 Dec 1993	13 July 1997

Displacement, tons: 203 full load
Dimensions, feet (metres): 111 × 27 × 8 *(33.8 × 8.2 × 2.3)*
Main machinery: 2 MTU 12V 183 TE61 diesels; 1,068 hp(m) *(786 kW)*; 2 Schottel steerable props; 1 White Gill thruster; 300 hp *(224 kW)*
Speed, knots: 12.4. **Range, n miles:** 2,000 at 10 kt
Complement: 25 (5 officers)
Guns: 2—12.7 mm MGs.
Radars: Navigation: Sperry; I-band.
Sonars: Thoray/Thomson Sintra TSM 2022; hull-mounted; active minehunting; high frequency.

Comment: MCM vessels with GRP hulls ordered from Swiftships in December 1990 with FMS funding. First one acceptance trials in June 1994 and completion in August. Fitted with a Unisys command data handling system which is an improved version of SYQ-12. GPS and line of sight navigation system. Dynamic positioning. A side scan sonar body and Gaymarine Pluto ROV can be streamed from a deck crane. Portable decompression chamber carried. Two delivered 29 November 1995 and the third in April 1996. All were finally commissioned after delays caused by problems with the minehunting equipment.

VERIFIED

DAT ASSAWARI   *3/2000, van Ginderen Collection /* 0088754

3 T 43 CLASS (MINESWEEPERS—OCEAN) (MSO)

DAQAHLIYA 507	SINAI 513	ASSIYUT 516

Displacement, tons: 580 full load
Dimensions, feet (metres): 190.2 × 27.6 × 6.9 *(58 × 8.4 × 2.1)*
Main machinery: 2 Kolomna Type 9-D-8 diesels; 2,000 hp(m) *(1.47 MW)* sustained; 2 shafts
Speed, knots: 15. **Range, n miles:** 3,000 at 10 kt
Complement: 65
Guns: 4—37 mm/63 (2 twin); 160 rds/min to 9 km *(5 n miles)*; weight of shell 0.7 kg.
8—12.7 mm (4 twin) MGs.
Mines: Can carry 20.
Radars: Navigation: Don 2; I-band.
Sonars: Stag Ear; hull-mounted; active search; high frequency.

Comment: Delivered in the early 1970s from the USSR. Others of the class have been sunk or used as targets or cannibalised for spares. The plan to fit them with VDS sonars and ROVs has been shelved.

VERIFIED

DAQAHLIYA   *3/2000 /* 0103745

AUXILIARIES

Notes: (1) There are also two survey launches *Misaha 1* and *2* with a crew of 14. Both were commissioned in 1991.
(2) A small barge *Amira Rama* was donated to the Navy in 1987 and is used as lighthouse tender.
(3) There are two supply ships with pennant numbers 113 and 115.
(4) *El Hurreya*, a 6,000 ton transport ship, was launched at Alexandria on 27 January 2004.

SUPPLY SHIP 115   3/2004*, Bob Fildes / 1044120

6 TOPLIVO 2 CLASS (TANKERS) (AOTL/AWTL)

ATABARAH 212	AKDU 214	AYEDA 3 216
MARYUT 218	AL NIL 220	AL FURAT 224

Displacement, tons: 1,029 full load
Dimensions, feet (metres): 176.2 × 31.8 × 10.5 *(53.7 × 9.7 × 3.2)*
Main machinery: 1 6DR 30/50-5 diesel; 600 hp(m) *(441 kW)*; 1 shaft
Speed, knots: 10. **Range, n miles:** 400 at 7 kt
Complement: 16
Cargo capacity: 500 tons diesel or water (211-215)
Radars: Navigation: Spin Trough; I-band.

Comment: Built in Alexandria in 1972-77 to a USSR design. Another of the class 217 is laid up.
VERIFIED

ATABARAH   6/1999 / 0080651

1 LÜNEBURG CLASS (TYPE 701) (SUPPORT SHIP) (ARL)

Name	No	Builders	Commissioned
SHALADEIN (ex-*Glücksburg*)	230 (ex-A 1414)	Bremer Vulkan/Flensburger Schiffbau	9 July 1968

Displacement, tons: 3,709 full load
Dimensions, feet (metres): 374.9 × 43.3 × 13.8 *(114.3 × 13.2 × 4.2)*
Main machinery: 2 MTU MD 16V 538 TB90 diesels; 6,000 hp(m) *(4.1 MW)* sustained; 2 shafts; cp props; bow thruster
Speed, knots: 17. **Range, n miles:** 3,200 at 14 kt
Complement: 71 (9 officers)
Cargo capacity: 1,100 tons
Guns: 4 Bofors 40 mm/70 (2 twin).
Countermeasures: Decoys: 2 Breda 105 mm SCLAR chaff launchers.

Comment: Transferred from Germany in early 2003 to act as support ship, including missile maintenance, of Type 148 patrol craft.
UPDATED

SHALADEIN   4/2003, Frank Findler / 0552746

1 WESTERWALD CLASS (TYPE 760) (AMMUNITION TRANSPORT) (AEL)

Name	No	Builders	Commissioned
HALAIB (ex-*Odenwald*)	231 (ex-A 1436)	Orenstein and Koppel, Lübeck	23 Mar 1967

Displacement, tons: 3,460 standard; 4,042 full load
Dimensions, feet (metres): 344.4 × 46 × 15.1 *(105 × 14 × 4.6)*
Main machinery: 2 MTU MD 16V 538 TB90 diesels; 6,000 hp(m) *(4.1 MW)* sustained; 2 shafts; cp props; bow thruster
Speed, knots: 17. **Range, n miles:** 3,500 at 17 kt
Complement: 31
Cargo capacity: 1,080 tons ammunition
Guns: 2 Bofors 40 mm.
Countermeasures: Decoys: 2 Breda SCLAR 105 mm chaff launchers are carried in A 1436.
Radars: Navigation: Kelvin Hughes; I-band.

Comment: Transferred from Germany in early 2003.
UPDATED

HALAIB   4/2003, Frank Findler / 0552745

1 POLUCHAT 1 CLASS (YPT)

P 937

Displacement, tons: 100 full load
Dimensions, feet (metres): 97.1 × 19 × 4.8 *(29.6 × 5.8 × 1.5)*
Main machinery: 2 Type M 50 diesels; 2,200 hp(m) *(1.6 MW)* sustained; 2 shafts
Speed, knots: 20. **Range, n miles:** 1,500 at 10 kt
Complement: 15
Radars: Surface search: Spin Trough; I-band.

Comment: Used as Torpedo Recovery Vessel. Unarmed.
VERIFIED

POLUCHAT   3/2000 / 0103779

2 NYRYAT I (PROJECT 522) CLASS (DIVING TENDERS) (YDT)

P 001   P 002

Displacement, tons: 116 full load
Dimensions, feet (metres): 93.8 × 17.1 × 5.6 *(28.6 × 5.2 × 1.7)*
Main machinery: 1 diesel; 450 hp(m) *(331 kW)* sustained; 1 shaft
Speed, knots: 12.5. **Range, n miles:** 1,500 at 10 kt
Complement: 15
Radars: Surface search: Spin Trough; I-band.

Comment: Transferred in 1964.
VERIFIED

TRAINING SHIPS

Notes: *Al Kousser* P 91 is a 1,000 ton vessel belonging to the Naval Academy. *Intishat* is a 500 ton training ship. Pennant number 160 is a USSR Sekstan class used as a cadet training ship. Two YSB training craft acquired from the US in 1989. A 3,300 ton training ship *Aida IV* presented by Japan in 1988 for delivery in March 1992 belongs to the Arab Maritime Transport Academy.

1 PRESIDENTIAL YACHT (YAC/AX)

Name		Builders	Commissioned
EL HORRIYA (ex-*Mahroussa*)		Samuda, Poplar	1865

Displacement, tons: 4,560 full load
Dimensions, feet (metres): 479 × 42.6 × 17.4 *(146 × 13 × 5.3)*
Main machinery: 3 boilers; 3 turbines; 5,500 hp *(4.1 MW)*; 3 shafts
Speed, knots: 16
Complement: 160

Comment: Became a museum in 1987 but was reactivated in 1992. Used as a training ship as well as a Presidential Yacht.
VERIFIED

EL HORRIYA   3/2000, van Ginderen Collection / 0103782

1 Z CLASS (AXT)

Name	No
EL FATEH (ex-*Zenith*, ex-*Wessex*)	921

Displacement, tons: 1,730 standard; 2,575 full load
Dimensions, feet (metres): 362.8 × 35.7 × 16
(*110.6 × 10.9 × 4.9*)
Main machinery: 2 Admiralty boilers; 2 Parsons turbines;
40,000 hp *(30 MW)*; 2 shafts
Speed, knots: 24. **Range, n miles:** 2,800 at 20 kt
Complement: 186

Missiles: SAM: 2 SA-N-5 mountings.
Guns: 4 Vickers 4.5 in *(115 mm)*/45 hand-loaded Mk 5
mounting; 50° elevation; 14 rds/min to 17 km *(9.3 n miles)*;

Builders
Wm Denny & Bros, Dumbarton

weight of shell 25 kg.
8 China 37 mm/63 (4 twin); 180 rds/min to 8.5 km *(4.6 n
miles)*; weight of shell 1.42 kg.
2 Bofors 40 mm/60 (twin).
Torpedoes: 8—21 in *(533 mm)* (2 quad) tubes.
Depth charges: 4 projectors.
Weapons control: Fly 4 director.
Radars: Air/surface search: Marconi SNW 10; D-band.
Navigation: Racal Decca 916; I-band.
Fire control: Marconi Type 275; F-band.

Laid down	Launched	Commissioned
19 May 1942	5 June 1944	22 Dec 1944

Programmes: Purchased from the UK in 1955.
Modernisation: Bofors replaced by Chinese 37 mm guns.
Sonars removed. Boilers renewed in 1993 and SA-N-5
mountings fitted.
Operational: Used primarily for harbour training, and the
intention is to keep the ship in service. Last seen at sea in
1994. The last survivor of its class, the ship may be preserved
as a museum.

VERIFIED

EL FATEH
4/1994, van Ginderen Collection / 0017808

1 BLACK SWAN CLASS (AXT)

Name	No	Builders	Laid down	Launched	Commissioned
TARIQ (ex-*Malek Farouk*, ex-*Whimbrel*)	931	Yarrows, Glasgow	31 Oct 1941	25 Aug 1941	13 Jan 1943

Displacement, tons: 1,925 full load
Dimensions, feet (metres): 299 × 38.5 × 11.5
(*81.2 × 11.7 × 3.5*)
Main machinery: 2 Admiralty boilers; 2 Parsons geared turbines;
3,600 hp *(2.69 MW)*; 2 shafts
Speed, knots: 18
Complement: 180

Guns: 6 Vickers 4 in *(102 mm)*/45 (3 twin) Mk 19; 16 rds/min to
19.5 km *(10.5 n miles)*; weight of shell 15.9 kg.
4—37 mm (2 twin). 4—12.7 mm MGs.
Depth charges: 4 projectors; 2 racks.
Radars: Surface search: 2 Decca; I-band.

Programmes: Transferred from UK in November 1949.
Structure: Still has the original class appearance with some
minor modifications to the armament.
Operational: Relegated for a time in the mid-1980s to an
accommodation ship and offered as part of a deal involving
the acquisition of two Oberon class submarines in 1989.
When this project was cancelled, the ship resumed service
as a training platform and was described as 'running like a train'
in 1993. Since then there has been limited activity but the ship
was reported at sea again in late 1997. Although seagoing
days are probably over, the ship still has a limited training role
and may be preserved as a museum.

VERIFIED

TARIQ
3/2000 / 0103738

TUGS

Notes: There are also four Coast Guard harbour tugs built by Damen in 1982. Names *Khoufou*,
Khafra, *Ramses* and *Kreir*. Two other harbour tugs were delivered in 1998. Names *Ajmi* and *Jihad*.

5 OKHTENSKY CLASS (ATA)

AL MAKS 103	**AL ANTAR** 107	**AL ISKANDARANI** 111
AL AGAMI 105	**AL DEKHEILA** 109	

Displacement, tons: 930 full load
Dimensions, feet (metres): 156.1 × 34 × 13.4 *(47.6 × 10.4 × 4.1)*
Main machinery: Diesel-electric; 2 BM diesel generators; 1 motor; 1,500 hp(m) *(1.1 MW)*; 1 shaft
Speed, knots: 13. **Range, n miles:** 6,000 at 13 kt
Complement: 38

Comment: Two transferred from USSR in 1966, others assembled at Alexandria. Replacements
are needed. 113 may have been deleted.

VERIFIED

AL AGAMI
3/2000, M Declerck / 0103780

For details of the latest updates to *Jane's Fighting Ships* online and to discover the
additional information available exclusively to online subscribers please visit

jfs.janes.com

COAST GUARD

Notes: (1) The Coast Guard is controlled by the Navy.
(2) There are four obsolete P 6 craft; pennant numbers 222, 246, 253 and 201.
(3) There is also a minimum of four ex-USN Bollinger type harbour security craft of 3.9 tons capable of 22 kt. Twin diesel engines. Carry a 7.62 mm MG.
(4) There is an unknown number of RIBs for inshore patrols.

9 TYPE 83 CLASS (LARGE PATROL CRAFT) (WPB)

46-54

Displacement, tons: 85 full load
Dimensions, feet (metres): 83.7 × 21.3 × 5.6 *(25.5 × 6.5 × 1.7)*
Main machinery: 2 diesels; 2 shafts
Speed, knots: 24
Complement: 12
Guns: 4—23 mm (2 twin). 1 Oerlikon 20 mm.
Radars: Surface search: Furuno; I-band.

Comment: Two of this class commissioned 13 July 1997. Built locally, these craft are similar to the Swiftships 93 ft class. Numbers uncertain but at least three are operational.
UPDATED

TYPE 83 CLASS   10/1995 / 0056923

21 TIMSAH CLASS (LARGE PATROL CRAFT) (WPB)

01-02   04-22

Displacement, tons: 106 full load
Dimensions, feet (metres): 101.8 × 17 × 4.8 *(30.5 × 5.2 × 1.5)*
Main machinery: 2 MTU 8V 331 TC92 diesels; 1,770 hp *(1.3 MW)* sustained; 2 shafts *(01-06)*; 2 MTU 12V 331 TC92 diesels; 2,660 hp(m) *(1.96 MW)* sustained; 2 shafts *(07-19)*
Speed, knots: 25. **Range, n miles:** 600 at 18 kt
Complement: 13
Guns: 2 Oerlikon 30 mm (twin) or 2 Oerlikon 20 mm.
Radars: Surface search: Racal Decca; I-band.

Comment: First three Timsah I completed December 1981, second three Timsah I December 1982 at Timsah SY, Ismailia. These all have funnels but there appear to be minor structural differences. *03* sunk in late 1993. Further six Timsah II ordered in January 1985 and completed in 1988-89 with a different type of engine and with waterline exhaust vice a funnel. Last of this batch in service in 1992, followed by ten more by 1999.
VERIFIED

TIMSAH 17   4/2002, A Sharma / 0528333

TIMSAH 02 (with funnel)   6/1995, Ships of the World / 0056922

TIMSAH 01   7/1992, F Sadek / 0528332

9 SWIFTSHIPS 93 ft CLASS (LARGE PATROL CRAFT) (WPB)

35-43

Displacement, tons: 102 full load
Dimensions, feet (metres): 93.2 × 18.7 × 4.9 *(28.4 × 5.7 × 1.5)*
Main machinery: 2 MTU 12V 331 TC92 diesels; 2,660 hp(m) *(1.96 MW)* sustained; 2 shafts
Speed, knots: 27. **Range, n miles:** 900 at 12 kt
Complement: 14 (2 officers)
Guns: 4—23 mm (2 twin); 1 Oerlikon 20 mm or 1—14.5 mm MG.
Radars: Surface search: Furuno; I-band.

Comment: Ordered November 1983. First three built in US, remainder assembled by Osman Shipyard, Ismailia. First four commissioned 16 April 1985, five more in 1986. Armament upgraded with 23 mm guns fitted forward in some of the class.
UPDATED

SWIFTSHIPS 343   2/2003, A Sharma / 0569931

6 CRESTITALIA MV 70 CLASS
(COASTAL PATROL CRAFT) (WPBF)

Displacement, tons: 36 full load
Dimensions, feet (metres): 68.9 × 17.4 × 3 *(21 × 5.3 × 0.9)*
Main machinery: 2 MTU 12V 331 TC92 diesels; 2,660 hp(m) *(1.96 MW)* sustained; 2 shafts
Speed, knots: 35. **Range, n miles:** 500 at 32 kt
Complement: 10 (1 officer)
Guns: 2 Oerlikon 30 mm A32 (twin). 1 Oerlikon 20 mm.
Radars: Surface search: Racal Decca; I-band.

Comment: Ordered 1980-GRP hulls. Naval manned but still employed on Coast Guard duties.
VERIFIED

CRESTITALIA 70 ft   1980, Crestitalia

12 SEA SPECTRE PB MK III CLASS
(COASTAL PATROL CRAFT) (WPB)

Displacement, tons: 37 full load
Dimensions, feet (metres): 64.9 × 18 × 5.9 *(19.8 × 5.5 × 1.8)*
Main machinery: 3 GM 8V-71TI diesels; 1,800 hp *(1.3 MW)*; 3 shafts
Speed, knots: 29. **Range, n miles:** 450 at 25 kt
Complement: 9 (1 officer)
Guns: 2—12.7 mm MGs.
Radars: Surface search: Raytheon; I-band.

Comment: PB Mk III type built by Peterson, Sturgeon Bay and delivered in 1980-81. Used for Customs duties.
VERIFIED

SPECTRE   1981, Peterson Builders / 0056924

9 PETERSON TYPE (COASTAL PATROL CRAFT) (WPB)

71-79

Displacement, tons: 18 full load
Dimensions, feet (metres): 45.6 × 13 × 3 *(13.9 × 4 × 0.9)*
Main machinery: 2 MTU 8V 183 TE92 diesels; 1,314 hp(m) *(966 kW)* sustained; Hamilton 362 water-jets
Speed, knots: 34. **Range, n miles:** 200 at 30 kt
Complement: 4
Guns: 2—12.7 mm MGs.
Radars: Surface search: Raytheon; I-band.

Comment: Built by Peterson Shipbuilders, Sturgeon Bay and delivered between June and October 1994 under FMS. Replaced Bertram type and used as pilot boats.
VERIFIED

PETERSON 72 (US colours)   6/1994, PBI / 0056925

5 NISR CLASS (LARGE PATROL CRAFT) (WPB)

THAR 701	NUR 703	NISR 713	NIMR 719	AL BAHR

Displacement, tons: 110 full load
Dimensions, feet (metres): 102 × 18 × 4.9 *(31 × 5.2 × 1.5)*
Main machinery: 2 Maybach diesels; 3,000 hp(m) *(2.2 MW)*; 2 shafts
Speed, knots: 24
Complement: 15
Guns: 2 or 4—23 mm (twin). 1 BM 21 122 mm 8-barrelled rocket launcher.
Radars: Surface search: Racal Decca 1230; I-band.

Comment: Built by Castro, Port Said on P6 hulls. First three launched in May 1963. Two more completed 1983. The rocket launcher and after 23 mm guns are interchangeable. 701 and 703 were refitted in 1998. Naval manned but employed on Coast Guard duties.
VERIFIED

3 PETERSON TYPE (COASTAL PATROL CRAFT) (WPBF)

80-82

Displacement, tons: 20 full load
Dimensions, feet (metres): 51 × 12 × 3 *(15.5 × 3.7 × 0.9)*
Main machinery: 2 MTU diesels; 2,266 hp(m) *(1.66 MW)*; Hamilton 391 water-jets
Speed, knots: 45. **Range, n miles:** 320 at 30 kt
Complement: 5
Guns: 2—12.7 mm MGs.
Radars: Surface search: Raytheon; I-band.

Comment: Built by Peterson Shipbuilders, Sturgeon Bay and delivered between October and December 1996 under FMS. Aluminium construction. Used mostly as pilot boats.
VERIFIED

PETERSON 81   3/2000 / 0103781

29 DC 35 TYPE (YFL)

Displacement, tons: 4 full load
Dimensions, feet (metres): 35.1 × 11.5 × 2.6 *(10.7 × 3.5 × 0.8)*
Main machinery: 2 Perkins T6-354 diesels; 390 hp *(287 kW)*; 2 shafts
Speed, knots: 25
Complement: 4

Comment: Built by Dawncraft, Wroxham, UK, from 1977. Harbour launches. One destroyed in September 1994. About half are laid up at Port Said.
VERIFIED

DC 35   8/1994, F Sadek / 0056927

El Salvador
FUERZA NAVAL DE EL SALVADOR

Country Overview

The Republic of El Salvador is an independent Central American State whose current constitution was established in 1983. With an area of 8,124 square miles, it has a 166 n mile coastline with the Pacific Ocean and is bounded to the north by Honduras and to the west by Guatemala. The country's capital is San Salvador while Acajutla, La Libertad and La Unión are the principal ports. El Salvador has not claimed an Exclusive Economic Zone (EEZ) but is one of a few coastal states which claims a 200 n mile territorial sea.

Senior Officer

Commander of the Navy:
    Captain Marco Antonio Palacios Luna

Personnel

(a)   2005: 877 (including 133 naval infantry)
(b)   Voluntary service

Bases

Acajutla, La Libertad, El Triunfo y La Union

Air Bases

El Tamarindo Air Station is reported to have been improved to enable the Third Air Brigade to provide air support to naval patrols. The US may donate fixed-wing aircraft and helicopters to assist in this task.

PATROL FORCES

Notes: There are two high-speed RHIBs donated by Taiwan and US.

3 CAMCRAFT TYPE (COASTAL PATROL CRAFT) (PB)

PM 6-8 (ex-*CG 6-8*)

Displacement, tons: 100 full load
Dimensions, feet (metres): 100 × 21 × 4.9 *(30.5 × 6.4 × 1.5)*
Main machinery: 3 Detroit 12V-71TA diesels; 1,260 hp *(939 kW)* sustained; 3 shafts
Speed, knots: 25. **Range, n miles:** 780 at 24 kt
Complement: 10
Guns: 1—20 mm Oerlikon or 1—12.7 mm MG. 2—7.62 mm MGs. 1—81 mm mortar.
Radars: Surface search: Furuno; I-band.

Comment: Aluminium hulled. Delivered 24 October, 8 November and 3 December 1975. Refitted in 1986 at Lantana Boatyard. Sometimes carry a combined 12.7 mm MG/81 mm mortar mounting in the stern. New radars fitted in 1995. Difficult to maintain and may be replaced by ASMAR Protector class.
VERIFIED

PM 6 (old number)   6/2000, Julio Montes / 0103783

1 POINT CLASS (PB)

Name	No	Builders	Commissioned
— (ex-*Point Stuart*)	PM 12 (ex-GC 12, ex-82358)	J Martinac, Tacoma	17 Mar 1967

Displacement, tons: 67 full load
Dimensions, feet (metres): 83 × 17.2 × 5.8 *(25.3 × 5.2 × 1.8)*
Main machinery: 2 Caterpillar diesels; 1,600 hp *(1.19 MW)*; 2 shafts
Speed, knots: 22. **Range, n miles:** 1,200 at 8 kt
Complement: 10
Guns: 2—12.7 mm MGs.
Radars: Surface search: Hughes/Furuno SPS-73; I-band.

Comment: Transferred from US Coast Guard on 27 April 2001.   *VERIFIED*

PM 12   *11/2001, Julio Montes /* 0130481

1 SWIFTSHIPS 77 ft CLASS (COASTAL PATROL CRAFT) (PB)

PM 11 (ex-GC 11)

Displacement, tons: 48 full load
Dimensions, feet (metres): 77.1 × 20 × 4.9 *(23.5 × 6.1 × 1.5)*
Main machinery: 3 Detroit 12V-71TA diesels; 1,260 hp *(939 kW)* sustained; 3 shafts
Speed, knots: 26
Complement: 7
Guns: 2—12.7 mm MGs. Aft MG combined with 81 mm mortar.
Radars: Surface search: Furuno; I-band.

Comment: Aluminium hull. Delivered by Swiftships, Morgan City 6 May 1985.   *VERIFIED*

PM 11 (old number)   *9/2000, Von Santos /* 0103784

1 SWIFTSHIPS 65 ft CLASS (COASTAL PATROL CRAFT) (PB)

PM 10 (ex-*GC 10*)

Displacement, tons: 36 full load
Dimensions, feet (metres): 65.6 × 18.3 × 5 *(20 × 6 × 1.5)*
Main machinery: 2 Detroit 12V-71TA diesels; 840 hp *(626 kW)* sustained; 2 shafts
Speed, knots: 23. **Range, n miles:** 600 at 18 kt
Complement: 6
Guns: 1 Oerlikon 20 mm. 1 or 2—12.7 mm MGs. 1—81 mm mortar.
Radars: Surface search: Furuno; I-band.

Comment: Aluminium hull. Delivered by Swiftships, Morgan City 14 June 1984. Was laid up for a time in 1989-90 but became operational again in 1991. Refitted in 1996.   *VERIFIED*

PM 10   *6/2003, El Salvador Navy /* 0568340

4 TYPE 44 CLASS (PBI)

PRM 01-04

Displacement, tons: 18 full load
Dimensions, feet (metres): 44 × 12.8 × 3.6 *(13.5 × 3.9 × 1.1)*
Main machinery: 2 Detroit 6V-38 diesels; 185 hp *(136 kW)*; 2 shafts
Speed, knots: 14. **Range, n miles:** 215 at 10 kt
Complement: 3

Comment: Ex-USCG craft similar to those transferred to Uruguay.

VERIFIED

PRM 04   *11/2001, Julio Montes /* 0130482

6 PIRANHA CLASS (RIVER PATROL CRAFT) (PBR)

PF 01-06 (ex-LOF 021-026) series

Displacement, tons: 8.2 full load
Dimensions, feet (metres): 36 × 10.1 × 1.6 *(11 × 3.1 × 0.5)*
Main machinery: 2 Caterpillar 3208TA diesels; 680 hp *(507 kW)* sustained; 2 shafts
Speed, knots: 26
Complement: 5
Guns: 2—12.7 mm (twin) MGs. 2—7.62 mm (twin) MGs.
Radars: Surface search: Furuno 3600; I-band.

Comment: Riverine craft with Kevlar hulls used by the Naval Infantry. Completed in March 1987 by Lantana Boatyard, Florida. Same type supplied to Honduras. Five craft reported operational.
VERIFIED

PF 06   *6/2003, El Salvador Navy /* 0568339

9 PROTECTOR CLASS (RIVER PATROL CRAFT) (PBR)

PC 01-09

Displacement, tons: 9 full load
Dimensions, feet (metres): 40.4 × 13.4 × 1.4 *(12.3 × 4 × 0.4)*
Main machinery: 2 Caterpillar 3208TA diesels; 680 hp *(507 kW)* sustained; 2 shafts
Speed, knots: 28. **Range, n miles:** 350 at 20 kt
Complement: 4
Guns: 2—12.7 mm MGs. 2—7.62 mm MGs.
Radars: Surface search: Furuno 3600; I-band.

Comment: Ordered in December 1987 from SeaArk Marine (ex-MonArk). Four delivered in December 1988 and four in February and March 1989. Eight reported operational and one in maintenance.
VERIFIED

PC 09   *11/2001, Julio Montes /* 0130483

8 AIR PATROL BOATS (PBI)

PFR 01-08

Comment: Purchased in Miami for SAR on inland waters.

VERIFIED

PFR 05   5/2003, El Salvador Navy / 0568338

2 MERCOUGAR INTERCEPT CRAFT (PBR)

PA 01-02

Comment: Two remaining of five 40 ft craft delivered by Mercougar in 1988. Powered by two Ford Merlin diesels; 600 hp *(448 kW)* giving speeds of up to 40 kt and range of 556 km *(300 n miles)*. Radar fitted.

VERIFIED

PA 02   5/2001, Julio Montes / 0109938

1 BALSAM CLASS (AGP)

Name	No	Builders	Commissioned
MANUEL JOSÉ ARCE (ex-*Madrona*)	BL 01 (ex-WLB 302)	Zenith Dredge, Duluth, MN	30 May 1943

Displacement, tons: 1,034 full load
Dimensions, feet (metres): 180 × 37 × 12 *(54.9 × 11.3 × 3.8)*
Main machinery: Diesel electric; 2 diesels; 1,402 hp *(1.06 MW)*; 1 motor; 1,200 hp *(895 kW)*; 1 shaft; bow thruster
Speed, knots: 13. **Range, n miles:** 8,000 at 12 kt
Complement: 53
Guns: 2—12.7 mm MGs.
Radars: Navigation: Raytheon SPS-64(V)1.

Comment: Transferred from the US Coast Guard on 14 June 2002. Used as a mother ship for coastal patrol craft.

VERIFIED

AUXILIARIES

3 LCM

BD 02 (ex-LD 02)   BD 04-05 (ex-LD 04-05)

Displacement, tons: 45 full load
Dimensions, feet (metres): 64.7 × 14 × 5 *(21.5 × 4.6 × 1.6)*
Main machinery: 2 Detroit 12V 71TA diesels; 840 hp *(626 kW)* sustained; 2 shafts
Speed, knots: 15
Complement: 6
Guns: 2—12.7 mm MGs. 2—7.62 mm MGs.
Radars: Navigation: Furuno; I-band.

Comment: First one delivered by SeaArk Marine in January 1987, second pair in May 1996.

VERIFIED

BD 04   6/2003, El Salvador Navy / 0568336

POLICE

Notes: Ten jet-skis are reported to have been delivered in 2002 for SAR.

11 RODMAN 890 (PBR)

L-01-01 to 01-11

Displacement, tons: 3.1 full load
Dimensions, feet (metres): 29.2 × 9.8 × 3.6 *(8.9 × 3 × 0.8)*
Main machinery: 2 Volvo diesels; 300 hp(m) *(220 kW)*; 2 shafts
Speed, knots: 28. **Range, n miles:** 150 at 25 kt
Complement: 3
Guns: 1—7.62 mm MG.
Radars: Surface search: I-band.

Comment: Eleven craft delivered by Rodman in 1998. Operational availability is reported to be constrained by lack of spares.

VERIFIED

RODMAN 890   6/1998, Rodman / 0576109

ARCE   6/2003, El Salvador Navy / 0568337

Equatorial Guinea

Country Overview

The Republic of Equatorial Guinea became independent in 1968 as a federation of the two former Spanish provinces of Fernando Po and Río Muni. It became a unitary state in 1973. Located in west Africa, the country has an overall area of 10,831 square miles and includes a mainland section which is bordered to the north by Cameroon and to the east and south by Gabon. It has a

160 n mile coastline with the Gulf of Guinea in which lie the islands of Bioko (formerly Fernando Po), Annobón, Corisco, Elobey Grande and Elobey Chico. The administrative capital on the mainland is Bata while Malabo, on the north coast of Bioko, is capital of the republic, largest city and prinicpal port. Territorial waters (12 n miles) are claimed. A 200 n mile Exclusive Economic Zone (EEZ) has been claimed but the boundaries have not been agreed.

Personnel

2005: 120 officers and men

Bases

Malabo, Bata.

PATROL FORCES

Notes: The Lantana 68 class *Isla de Bioko* and 20 m patrol craft *Riowele* are believed to be non-operational.

1 DAPHNE CLASS (PB)

Name	No	Builders	Commissioned
URECA (ex-*Nymfen*)	P 31 (ex-P 535)	Royal Dockyard, Copenhagen	4 Oct 1963

Displacement, tons: 170 full load
Dimensions, feet (metres): 121 × 20 × 6.5 *(36.9 × 6.1 × 2.0)*
Main machinery: 3 diesels; 3 shafts
Speed, knots: 20
Complement: 23
Guns: 2—14.5 mm.
Radars: Navigation: Furuno; I-band.

Comment: Acquired in 1999.

NEW ENTRY

2 ZHUK (GRIF) CLASS (PROJECT 1400M) (PB)

MIGUEL ELA EDJODJOMO LP 039   HIPOLITO MICHA LP 041

Displacement, tons: 39 full load
Dimensions, feet (metres): 78.7 × 16.4 × 3.9 *(24 × 5 × 1.2)*
Main machinery: 2 diesels; 2 shafts
Speed, knots: 30. Range, n miles: 1,100 at 15 kt
Complement: 13 (1 officer)
Guns: 2—14.5 mm (twin, fwd) MGs. 1—12.7 mm (aft) MG.
Radars: Surface search: Furuno; I-band.

Comment: Reported to have been transferred from Ukraine in 2000.

NEW ENTRY

2 KALKAN (PROJECT 50030) M CLASS (INSHORE PATROL CRAFT) (PBR)

GASPAR OBIANG ESONO LP 043   FERNANDO NUARA ENGONDA LP 045

Displacement, tons: 8.5 full load
Dimensions, feet (metres): 38.1 × 10.8 × 2.0 *(11.6 × 3.3 × 0.6)*
Main machinery: 1 Type 475K diesel; 496 hp *(370 kW)*; 1 waterjet
Speed, knots: 34
Complement: 2

Comment: Built by Morye Feodosiya and reportedly acquired in 2001.

NEW ENTRY

KALKAN CLASS   6/2003 *, Morye / 0572655

Eritrea

Country Overview

A British protectorate from 1941, The State of Eritrea was federated with Ethiopia in 1952 and incorporated as a province in 1962. The following war of liberation culminated in independence in 1993. The country is situated on the southwest shore of the Red Sea with which it has a 621 n mile coastline with an area of 46,842 square miles, it is bordered to the north by Sudan, to the west by Ethiopia and to the south by Djibouti. The largest town and capital is Asmara and the principal port is Massawa. There are no claims to maritime jurisdiction over territorial seas or Exclusive Economic Zone (EEZ).

All vessels of the former Ethiopian Navy were put up for sale at Djibouti from 16 September 1996. All were either taken over by Eritrea, sold to civilian firms or scrapped.

Headquarters Appointments

Commander Eritrean Navy:
  Major General Romedan Awliai
Chief of Staff:
  Brigadier General Tekie Russom

Personnel

2005:   1,100 including 500 conscripts

Bases

Massawa, Dahlak.

PATROL FORCES

Notes: There are also about 50 rigid raiding craft.

1 OSA II (PROJECT 205) CLASS (FAST ATTACK CRAFT—MISSILE) (PTFG)

FMB 161

Displacement, tons: 245 full load
Dimensions, feet (metres): 126.6 × 24.9 × 8.8 *(38.6 × 7.6 × 2.7)*
Main machinery: 3 Type M 504 diesels; 10,800 hp(m) *(7.94 MW)* sustained; 3 shafts
Speed, knots: 37. Range, n miles: 800 at 30 kt
Complement: 30

Missiles: SSM: 4 SS-N-2B Styx; active radar or IR homing to 46 km *(25 n miles)* at 0.9 Mach; warhead 513 kg.
Guns: 4—30 mm/65 (2 twin); 500 rds/min to 5 km *(2.7 n miles)* anti-aircraft; weight of shell 0.54 kg.
Radars: Surface search: Square Tie; I-band.
Fire control: Drum Tilt; H/I-band.
IFF: Square Head. High Pole B.

Programmes: Acquired from USSR on 13 January 1981. The rest of the class has been sunk or scuttled. *FMB 161* was refitted from October 1994 to January 1995 and taken over by Eritrea in 1997. A second of class *FMB 163* was acquired for spares. New missiles were reported to have been ordered in 2002 but operational status is doubtful.

UPDATED

FMB 161
1/1998 / 0017824

4 SUPER DVORA CLASS (FAST ATTACK CRAFT—GUN) (PTF)

P101   P 102   P 103   P 104

Displacement, tons: 58 full load
Dimensions, feet (metres): 82 × 18.7 × 3 *(25 × 5.7 × 0.9)*
Main machinery: 2 MTU 8V 396 TE 94 diesels; 3,046 hp(m) *(2.24 MW)*; 2 shafts; ASD 14 surface
  drives
Speed, knots: 40. **Range, n miles:** 1,200 at 17 kt
Complement: 10 (1 officer)
Guns: 2—23 mm (twin). 2—12 mm MGs.
Depth charges: 1 rail.
Weapons control: Optronic sight.
Radars: Surface search: Raytheon; I-band.

Comment: Built by Israel Aircraft Industries and delivered from July 1993 to a modified Super
  Dvora design. The original order may have been for six of the class. All are based at Massawa
  and all are active.

VERIFIED

SUPER DVORA P 104   *6/2000, Eritrean Navy /* 0103787

3 SWIFTSHIPS 105 ft CLASS (LARGE PATROL CRAFT) (PB)

P 151   P 152   P 153

Displacement, tons: 118 full load
Dimensions, feet (metres): 105 × 23.6 × 6.5 *(32 × 7.2 × 2)*
Main machinery: 2 MTU MD 16V 538 TB90 diesels; 6,000 hp(m) *(4.41 MW)* sustained; 2 shafts
Speed, knots: 30. **Range, n miles:** 1,200 at 18 kt
Complement: 21
Guns: 4 Emerlec 30 mm (2 twin) *(P 151)*; 600 rds/min to 6 km *(3.3 n miles)*; weight of shell
  0.35 kg.
  4—23 mm/60 (2 twin) *(P 152/153)*. 2—12.7 mm (twin).
Radars: Surface search: Decca RM 916; I-band.

Comment: Six ordered in 1976 of which four were delivered in April 1977 before the cessation of
  US arms sales to Ethiopia. Built by Swiftships, Louisiana. One deserted to Somalia and served in
  that Navy for a time. Based at Massawa and in reasonable condition. All are active.

VERIFIED

P 153   *1/1998 /* 0017825

5 BATTALION 17 (PBF)

P 084   P 085   P 086   P 087   P 088

Displacement, tons: 35.5 full load
Dimensions, feet (metres): 55.9 × 17 × 5.2 *(17.05 × 5.2 × 1.6)*
Main machinery: 2 MTU 12V 183 TE 92 diesels
Speed, knots: 35.2. **Range, n miles:** 680 at 30 kt
Complement: 8
Guns: 2—14.5 mm MGs (1 twin).
Radars: Surface search: Raytheon; I-band.

Comment: Built by Harena Boat Yard at Assab, Eritrea. Five craft delivered in 2000 with possible
  further orders since then.

VERIFIED

P 086   *6/2000, Eritrean Navy /* 0103788

AMPHIBIOUS FORCES

Notes: Two obsolete ex-USSR T4 LCUs are in harbour service at Massawa.

1 ASHDOD CLASS (LST)

P 63 (ex-302)

Displacement, tons: 400 standard; 730 full load
Dimensions, feet (metres): 205.5 × 32.8 × 5.8 *(62.7 × 10 × 1.8)*
Main machinery: 3 MWM diesels; 1,900 hp(m) *(1.4 MW)*; 3 shafts
Speed, knots: 10.5
Complement: 20
Guns: 2—23 mm (1 twin). 2—12.7 mm MGs.

Comment: Former Ethiopian commercial LST acquired from Israel in 1993, taken over by Eritrea in
  1997 and subsequently transferred to the Navy. Reported operational.

UPDATED

P 63 (Israeli pennant number)   *1995, Eritrean Navy /* 0103789

1 CHAMO CLASS (LST)

DENDEN 301

Displacement, tons: 884 full load
Dimensions, feet (metres): 197.5 × 39.3 × 4.7 *(60.2 × 12 × 1.44)*
Main machinery: 2 MTU 6V 396 TB 63; 1,350 hp(m) *(1 MW)*; 2 shafts
Speed, knots: 10
Complement: 23
Guns: 2—23 mm (1 twin); 2—12.7 mm MGs.

Comment: German built former Ethiopian commercial LST taken over by Eritrea in 1997 and
  subsequently transferred to the Navy. Reported operational.

VERIFIED

Estonia
EESTI MEREVÄGI

Country Overview

The Republic of Estonia regained independence in 1991 after 51
years as a Soviet republic. Situated in northeastern Europe, the
country includes more than 1,500 islands, the largest of which
are Saaremaa and Hiiumaa. With an area of 17,462 square miles
it has borders to the east with Russia and to the south with Latvia.
It has a 750 n mile coastline with the Baltic Sea and Gulf of
Finland. Tallinn is the capital, largest city and principal port.
Territorial seas (12 n miles) are claimed but while it has claimed a
200 n mile Exclusive Economic Zone (EEZ), its limits have not
been fully defined by boundary agreements.

The Navy was founded in 1918 and re-established on 22 April
1994. The Border Guard comes under the Ministry of Internal
Affairs and is responsible for SAR and Pollution Prevention.

Headquarters Appointments

Commander in Chief Estonian Defence Forces:
  Vice Admiral Tarmo Kõuts
Acting Commander of the Navy and Chief of Staff:
  Lieutenant Commander Ahti Piirimägi

Personnel

(a) 2005: 270 (70 officers)
(b) 8-11 months' national service
(c) Border Guard: 300

Bases

Major: Miinisadam (Tallinn)
Minor: Kopli (Tallinn) (Border Guard)

FRIGATES

1 MODIFIED HVIDBJØRNEN CLASS (FFLH/AGFH/AGE)

Name	No	Builders	Laid down	Launched	Commissioned
ADMIRAL PITKA (ex-Beskytteren)	A 230 (ex-F 340)	Aalborg Vaerft	11 Dec 1974	29 May 1975	27 Feb 1976

Displacement, tons: 1,970 full load
Dimensions, feet (metres): 245 × 40 × 17.4
   (74.7 × 12.2 × 5.3)
Main machinery: 3 MAN/Burmeister & Wain Alpha diesels;
   7,440 hp(m) (5.47 MW); 1 shaft; cp prop
Speed, knots: 18. **Range, n miles:** 4,500 at 16 kt on 2 engines;
   6,000 at 13 kt on 1 engine
Complement: 43 (9 officers)

Guns: 1 USN 3 in (76 mm)/50; Mk 22.
Countermeasures: ESM: Racal Cutlass; radar warning.
Radars: Navigation: 2 Litton Decca E; I-band.

Helicopters: Platform for 1 Lynx type.

Programmes: Transferred by gift from Denmark in July 2000
   and formally recommissioned on 21 November 2000.
Structure: Strengthened for ice operations.
Operational: Flagship of the Estonian Navy, its primary role is as
   a Command and Support ship and its secondary role is as a
   research ship. The vessel was refitted prior to being
   transferred. Modifications included the replacement of the
   military radars with Litton Marine radars, and the removal of
   PMS 26 sonar.

UPDATED

ADMIRAL PITKA

6/2002, Frank Findler / 0524988

MINE WARFARE FORCES

2 FRAUENLOB (TYPE 394) CLASS (MSI)

Name	No	Builders	Commissioned
OLEV (ex-Diana)	M 415 (ex-M 2664)	Krogerwerft, Rendsburg	21 Sep 1967
VAINDLO (ex-Undine)	M 416 (ex-M 2662)	Krogerwerft, Rendsburg	20 Mar 1967

Displacement, tons: 246 full load
Dimensions, feet (metres): 124.6 × 26.9 × 6.6 (38 × 8.2 × 2)
Main machinery: 2 MTU MB 12V 493 TY70 diesels; 2,200 hp(m) (1.62 MW) sustained; 2 shafts
Speed, knots: 14. **Range, n miles:** 400 at 12 kt
Complement: 23 (5 officers)
Guns: 1 Bofors 40 mm/70.
Mines: Laying capability.
Radars: Navigation: Atlas Elektronik; I-band.

Comment: *Olev* transferred in June 1997 and *Vaindlo*, which replaced *Kalev*, on 8 October
   2002 having paid off from the German Navy in 1995. Capable of influence and mechanical
   minesweeping. *UPDATED*

VAINDLO

4/2003, Per Körnefeldt / 0561498

OLEV

6/2000, Findler & Winter / 0103792

2 LINDAU (TYPE 331) CLASS (MHC)

Name	No	Builders	Commissioned
WAMBOLA (ex-Cuxhaven)	M 311 (ex-M 1078)	Burmeister, Bremen	11 Mar 1959
SULEV (ex-Lindau)	M 312 (ex-M 1072)	Burmeister, Bremen	24 Apr 1958

Displacement, tons: 463 full load
Dimensions, feet (metres): 154.5 × 27.2 × 9.8 (9.2 Troika) (47.1 × 8.3 × 3) (2.8)
Main machinery: 2 MTU MD diesels; 4,000 hp(m) (2.94 MW); 2 shafts
Speed, knots: 16.5. **Range, n miles:** 850 at 16.5 kt
Complement: 37 (6 officers)
Guns: 1 Bofors 40 mm/70; 330 rds/min to 12 km (6.5 n miles); weight of shell 0.96 kg.
Radars: Navigation: Raytheon SPS 64; I-band.
Sonars: Plessey 193M; minehunting; high frequency (100/300 kHz).

Comment: *Wambola* transferred from Germany on 23 March 2000 and *Sulev* on 29 September
   2000. Two PAP 104 ROVs were included in the transfer. *UPDATED*

SULEV

5/2004*, Per Körnefeldt / 0589735

WAMBOLA

6/2004*, Frank Findler / 0589734

AUXILIARIES

1 MAAGEN CLASS (YDT)

Name	No	Builders	Commissioned
AHTI (ex-Mallemukken)	A 431 (ex-Y 385)	Helsingor Dockyard	19 May 1960

Displacement, tons: 190 full load
Dimensions, feet (metres): 88.6 × 23.6 × 9.5 (27 × 7.2 × 2.9)
Main machinery: 1 diesel; 385 hp(m) (283 kW); 1 shaft
Speed, knots: 10
Complement: 11
Guns: 2—12.7 mm MGs.
Radars: Surface search: Pechora; I-band.
   Navigation: Skanter 009; I-band.
Sonars: Sidescan.

Comment: Handed over at Tallinn on 29 March 1994, having decommissioned from the Danish
   Navy in 1992. Serves as a diving tender and for route surveillance. *VERIFIED*

AHTI

6/2003, Hartmut Ehlers / 0561497

BORDER GUARD (EESTI PIIRIVALVE)

Notes: (1) *Director General:* Colonel Harry Hein
(2) The letters PV are visible on the national flag which is defaced with green and yellow markings.
(3) Three vessels are used for anti-pollution duties. *Triin* (PVL-200) (ex-*Bester*) and *Reet* (PVL-201) (ex-*EVA-200*) are both 34 m vessels which entered Border Guard service in May 2001. *Kati* (PVL-202) (ex-*KBV-003*) is a 40 m vessel transferred from Sweden in May 2002.
(4) PVL-110 is a Slavyanka class LCM acquired in 1997 and used as a harbour utility craft.
(5) The Border Guard Aviation Group was formed in February 1993 and includes two L-410 maritime patrol aircraft and two Mi-8 helicopters.

REET   6/2003, Hartmut Ehlers / 0561496

PVL-110   6/2003, Hartmut Ehlers / 0561495

L-410   7/2004*, Paul Jackson / 0589739

1 BALSAM CLASS (AGF)

Name	No	Builders	Commissioned
VALVAS (ex-*Bittersweet*)	PVL 109 (ex-*WLB 389*)	Duluth Shipyard, Minnesota	11 May 1944

Displacement, tons: 1,034 full load
Dimensions, feet (metres): 180 × 37 × 12 *(54.9 × 11.3 × 3.8)*
Main machinery: Diesel electric; 2 diesels; 1,402 hp *(1.06 MW)*; 1 motor; 1,200 hp *(895 kW)*; 1 shaft; bow thruster
Speed, knots: 13. **Range, n miles:** 8,000 at 12 kt
Complement: 53
Guns: 2—25 mm/L80 (1 twin). 2—12.7 mm MGs.
Radars: Navigation: Raytheon SPS-64(V)1.

Comment: Transferred from the US Coast Guard and recommissioned as a Border Guard Headquarters ship on 5 September 1997.

VERIFIED

VALVAS   6/2003, Hartmut Ehlers / 0561492

1 SILMÄ CLASS (LARGE PATROL CRAFT) (PBO)

Name	No	Builders	Commissioned
KOU (ex-*Silmä*)	PVL 107	Laivateollisuus, Turku	19 Aug 1963

Displacement, tons: 530 full load
Dimensions, feet (metres): 158.5 × 27.2 × 14.1 *(48.3 × 8.3 × 4.3)*
Main machinery: 1 Werkspoor diesel; 1,800 hp(m) *(1.32 MW)*; 1 shaft
Speed, knots: 15
Complement: 10
Guns: 2—25 mm/80 (twin).
Radars: Surface search: I-band.
Sonars: Simrad SS105; active scanning; 14 kHz.

Comment: Transferred from Finland Frontier Guard in January 1995.

VERIFIED

KOU   6/2000, Per Körnefeldt / 0103793

1 VIIMA CLASS (COASTAL PATROL CRAFT) (PB)

Name	No	Builders	Commissioned
MARU (ex-*Viima*)	PVL 106	Laivateollisuus, Turku	12 Oct 1964

Displacement, tons: 134 full load
Dimensions, feet (metres): 117.1 × 21.7 × 7.5 *(35.7 × 6.6 × 2.3)*
Main machinery: 3 MTU MB diesels; 4,050 hp(m) *(2.98 MW)*; 3 shafts; cp props
Speed, knots: 23
Complement: 9
Guns: 2—25 mm/L 80 (1 twin). 2—14.5 mm MGs (twin). 1—7.62 mm MG.
Radars: Surface search: I-band.

Comment: Acquired from Finland Frontier Guard in January 1995.

UPDATED

MARU   6/2003*, Hartmut Ehlers / 0589736

1 PIKKER II CLASS (COASTAL PATROL CRAFT) (PB)

Name	No	Builders	Commissioned
VAPPER	PVL 111	Baltic Ship Repairers, Tallinn	1 June 2000

Displacement, tons: 117 full load
Dimensions, feet (metres): 103 × 19.7 × 5.9 *(31.4 × 6.0 × 1.8)*
Main machinery: 2 Deutz TBD 620 V12 diesels; 4,087 hp(m) *(3.1 MW)*; 2 shafts
Speed, knots: 27
Complement: 7
Guns: 2—25 mm (1 twin). 1—14.5 mm.

Comment: Launched in April 2000. Carries one RIB for SAR and inspection.

VERIFIED

VAPPER   8/2000 / 0114351

1 PIKKER I CLASS (COASTAL PATROL CRAFT) (PB)

Name	No	Builders	Launched	Commissioned
PIKKER	PVL 103	Talinn	23 Dec 1995	Apr 1996

Displacement, tons: 90 full load
Dimensions, feet (metres): 98.4 × 19 × 4.9 *(30 × 5.8 × 1.5)*
Main machinery: 2 12YH 18/20 diesels; 2,700 hp(m) *(1.98 MW)* sustained; 2 shafts
Speed, knots: 23
Complement: 5
Guns: 2—14.5 mm MGs (twin). 2—7.62 mm MGs.
Radars: Surface search: I-band.

Comment: Carries an RIB with a hydraulic launch crane aft. Intended to be the first of a series of 10 of which the other nine may be funded in due course.

VERIFIED

PIKKER   *6/2003, Hartmut Ehlers /* 0561494

1 STORM CLASS (PB)

Name	No	Builders	Launched
TORM (ex-*Arg*)	PVL 105 (ex-*P968*)	Bergens Mek, Verksteder	24 May 1966

Displacement, tons: 100 standard; 135 full load
Dimensions, feet (metres): 120 × 20 × 5 *(36.5 × 6.1 × 1.5)*
Main machinery: 2 MTU MB 16V 538 TB90 diesels; 6,000 hp(m) *(4.41 MW)* sustained; 2 shafts
Speed, knots: 32. **Range, n miles:** 800 at 25 kt
Complement: 8
Guns: 2—25 mm/80 (twin). 2—14.5 mm MGs (twin).
Radars: Surface search: Racal Decca TM 1226; I-band.

Comment: Built in 1966 and paid off from the Norwegian Navy in 1991. Transferred 16 December 1994 stripped of all weapons and associated sensors. Rearmed in 1995 with light guns. No further transfers are expected.

VERIFIED

TORM   *6/1999, Estonian Border Guard /* 0056948

3 KBV 236 CLASS (PB)

PVK 001 (ex-*KBV 257*)   PVK 002 (ex-*KBV 259*)   PVK 003 (ex-*KBV 246*)

Displacement, tons: 17 full load
Dimensions, feet (metres): 63 × 13.1 × 4.3 *(19.2 × 4 × 1.3)*
Main machinery: 2 Volvo Penta TAMD120A diesels; 700 hp(m) *(515 kW)*; 2 shafts
Speed, knots: 22
Complement: 5
Guns: 1—7.62 mm MG.

Comment: Transferred on 4 April 1992, 20 October 1993 and 6 December 1993. Former Swedish Coast Guard vessel built in 1970. Similar craft to Latvia and Lithuania.

VERIFIED

PVK 003   *8/1995, Erki Holm /* 0056949

11 INSHORE PATROL CRAFT (PBI)

PVK 006, 008, 010-013, 016-017, 020-021, 025

Comment: *PVK 010* is a 15 m patrol craft built in 1997, *PVK 011* was commissioned in 1999, *PVK 017* (ex-*EVA 203*) is a 44 ton MFV type of vessel built in Finland in 1963. *PVK 018* (ex-*EVA 204*) is a 22 kt craft built in Finland in 1993 and *PVK 008* and *013* are 13.7 ton icebreaking launches acquired from Finland and based on Lake Peipus. There is also a Jet Combi 10 power boat based on Lake Peipus. Further craft under 12 m have numbers *PVK 004, 006, 012, 016, 020-021*. *PVK 025* is an ex-Swedish craft (KBV 275) acquired in January 1997.

UPDATED

PVK 010   *6/2002, Baltic Ship Repairers /* 0526817

1 GRIFFON 2000 TDX Mk II (HOVERCRAFT) (UCAC)

PVH 1

Displacement, tons: 6.8 full load
Dimensions, feet (metres): 36.1 × 15.1 *(11 × 4.6)*
Main machinery: 1 Deutz BF8L 513 diesel; 320 hp *(293 kW)* sustained
Speed, knots: 33. **Range, n miles:** 300 at 25 kt
Complement: 2
Military lift: 16 troops or 2 tons
Guns: 1—7.62 mm MG.

Comment: Similar to craft supplied to Finland. Acquired in 1999.

VERIFIED

PVH 1   *9/1999, Nick Hall /* 0103794

MARITIME ADMINISTRATION
(EESTI VEETEDE AMET (EVA))

Notes: The Maritime Administration (EVA) was re-established in 1990 and is responsible for hydrographic work, aids to navigation, ice-breaking and control of shipping. The main base is at Tallin. Ships are painted with a blue hull and white superstructure and are as follows:
Tarmo, icebreaker built in 1963 and acquired from Finland in 1992. Fleet flagship.
EVA 010, port control launch built in Finland in 1991
EVA 017, port control launch built in Finland in 1995
EVA 019, port control launch built in Estonia in 1997
EVA 300 (ex-*Tormilind*), hydrographic ship built in Russia in 1983
EVA 303 (ex-*Kaater*), buoy ship built in Poland in 1988
EVA 305, hydrographic launch built in Russia in 1979
EVA 308 (ex-GS-108-93), buoy ship built in Poland in 1968
EVA 309 (ex-BGK-117-93), buoy ship built in Russia in 1967
EVA 316 (ex-*Lonna*), buoy ship built in Finland in 1980
EVA 317-318, buoy ships built in Finland in 1994
EVA 319, buoy ship built in Finland in 1996
EVA 320, hydrographic ship built in Finland in 1997
EVA 321, buoy ship built in Estonia in 1999
EVA 322, launch built in Finland in 1997
EVA 323, launch built in Finland in 1994
EVA 324, workboat built in Japan in 1996
EVA 325, hydrographic ship built in Finland in 2002

TARMO   *6/2003, Hartmut Ehlers /* 0561491

EVA-318   *6/2003*, Hartmut Ehlers /* 0589737   EVA-308   *6/2003*, Hartmut Ehlers /* 0589738

Falkland Islands

Country Overview

The Falkland Islands are a self-governing British dependency administered by a Governor and a legislative council. Situated in the south Atlantic Ocean 323 n miles northeast of Cape Horn, approximately 200 islands are divided into two main groups on the east and west by the narrow Falkland Sound. The two largest islands are West Falkland Island (2,090 square miles) and East Falkland Island (2,610 square miles) on which the capital, largest town and principal port, Stanley, is situated. Territorial waters (12 n miles) are claimed as is a 200 n mile fishery zone.

Maritime Aircraft

There are two Pilatus Britten-Norman Defender unarmed maritime surveillance aircraft.

PATROL FORCES

1 FISHERY PATROL SHIP (PBO)

SIGMA

Measurement, tons: 1,467 grt
Dimensions, feet (metres): 196.92 × 36.2 × 20.7 *(60 × 11.03 × 6.3)*
Main machinery: 1 Wärtsilä 8R32 diesel; 3,680 hp *(2.74 MW)*; 1 shaft; cp prop; 1 Brunvoll bow thruster 500 hp *(373 kW)*
Speed, knots: 14.5
Complement: 12
Radars: Surface search: Atlas 8600; E/F-band.
Navigation: 2 Furuno 2115; I-band.

Comment: Stern trawler built in Norway in 1972. Converted in 1982 for seismic work and again in 2000 for Fishery Protection. Chartered from Sigma Marine Ltd in August 2000 and extended in 2003 until August 2006. Carries two RIBs capable of over 30 kt.

UPDATED

1 FISHERY PATROL SHIP (PSO)

DORADA

Measurement, tons: 2,360 grt
Dimensions, feet (metres): 249.3 × 47.9 × 20 *(76 × 14.6 × 6.1)*
Main machinery: 2 Sulzer-Cegielski diesels; 1—2,333 hp *(1,740 kW)*; 1—1,167 hp *(870 kW)*; 1 shaft; cp prop; 1 ABB bow thruster; 340 hp *(250 kW)*
Speed, knots: 15.5. **Range, n miles:** 14,000 at 12 kt
Complement: 15
Guns: 1 Oerlikon Mk VII A 20 mm.
Radars: Surface search/navigation: 2 Furuno 2100; I-band. 1 Furuno 1500 Mk 3; I-band.

Comment: Stern trawler built in Poland in 1991 and refitted in New Zealand. On long-term charter from Dorada Marine until January 2008. Has a red hull and white superstructure. Carries two RIBs capable of over 30 kt. Took part in multinational operation in 2003 to arrest Uruguayan fishing vessel.

UPDATED

SIGMA   *6/2002, Falkland Islands Fisheries /* 0137784   DORADA   *6/2002, Falkland Islands Fisheries /* 0137785

Faroe Islands

Country Overview

The Faroe Islands are a self-governing island group that is an integral part of Denmark which retains control of foreign relations. Located in the North Atlantic Ocean, about midway between the Shetland Islands and Iceland, there are 18 islands, of which the most important are Østerø, Suderø, Sandø, Vagø, Bordø and Strømø, on which the capital and principal port, Tórshavn, is situated. Territorial waters (12 n miles) are claimed. A 200 n mile fishery zone has also been claimed although the limits have only been partly defined by boundary agreements.

The Coast Guard and Fisheries come under the Landsstyri which is the islands' local government. Vessels work closely with the Danish Navy.

Headquarters Appointments

Head of Coast Guard:
   Captain Elmar Hojgaard

Personnel

2005: 60

Bases

Tórshavn (Isle of Streymoy)

COAST GUARD

1 PATROL SHIP (PBO)

TJALDRID

Displacement, tons: 650 full load
Dimensions, feet (metres): 146 × 33.1 × 10.5 *(44.5 × 10.1 × 3.2)*
Main machinery: 2 MWM diesels; 2,400 hp(m) *(1.76 MW)*; 2 shafts
Speed, knots: 14.5
Complement: 18 plus 4 divers
Guns: 1 Oerlikon 20 mm can be carried.
Radars: Surface search: Raytheon TM/TCPA; I-band.

Comment: Originally a commercial tug built in 1976 by Svolvaer, Verksted and acquired by the local government in 1987. The old 57 mm gun has been replaced. A decompression chamber can be carried.

VERIFIED

1 PATROL SHIP (PSO)

BRIMIL

Displacement, tons: 2,000 full load
Dimensions, feet (metres): 208.71 × 41.3 × 14.1 *(63.6 × 12.6 × 4.3)*
Main machinery: 2 Bergen diesels; 5,452 hp *(4.06 MW)*.
Speed, knots: 17
Complement: 12 with accommodation for 30 including 3 divers
Radars: Surface search: 2 Furuno.

Comment: Built for Faroese government as a patrol vessel by Myclebust Mek. Verksted, Norway. Entered service in April 2001.

UPDATED

TJALDRID   12/1999, Faroes Coast Guard / 0080652

BRIMIL   7/2003*, Martin Mokrus / 0589001

Fiji

Country Overview

A former British colony, the Republic of Fiji gained independence in 1970. Part of Melanesia, it is situated in the south Pacific Ocean some 972 n miles north of New Zealand and comprises more than 300 islands and islets, 100 of which are inhabited. The largest and most important of these are Viti Levu and Vanua Levu, which together contain more than 85 per cent of the total land area. To the southeast lie Taveuni, Kandavu, Koro and the Lau group while to the northwest lie Rotuma and the Yasawa group. The capital, largest town and principal port is Suva. An archipelagic state, territorial seas (12 n miles) are claimed. An Exclusive Economic Zone (EEZ) (200 n miles) is also claimed but limits have yet to be fully defined by boundary agreements.

Headquarters Appointments

Commander, Navy:
  Commander Villiamo Naupoto, RFN

Personnel

2005: 300

Prefix to Ships' Names

RFNS (Republic of Fiji naval ship)

Bases

RFNS *Viti*, at Togalevu (Training).
RFNS *Stanley Brown*.
Operation base at Walu Bay, Suva.
Forward base at Lautoka.

DELETIONS

Patrol Forces

2003   *Vai, Ogo*

PATROL FORCES

3 PACIFIC CLASS (LARGE PATROL CRAFT) (PB)

Name	No	Builders	Commissioned
KULA	201	Transfield Shipbuilding	28 May 1994
KIKAU	202	Transfield Shipbuilding	27 May 1995
KIRO	203	Transfield Shipbuilding	14 Oct 1995

Displacement, tons: 162 full load
Dimensions, feet (metres): 103.3 × 26.6 × 6.9 *(31.5 × 8.1 × 2.1)*
Main machinery: 2 Caterpillar 3516TA diesels; 2,820 hp *(2.09 MW)* sustained; 2 shafts
Speed, knots: 20. **Range, n miles:** 2,500 at 12 kt
Complement: 17 (4 officers)
Guns: 1—20 mm Oerlikon. 2—12.7 mm MGs.
Radars: Surface search: Furuno; I-band.

Comment: Ordered in December 1992. These are hulls 17, 19 and 20 of the class offered by the Australian government under Defence Co-operation Programme. *Kikau* underwent a half-life refit at Gladstone in 2001 followed by *Kula* and *Kiro* in 2002. Following the decision by the Australian government to extend the Pacific Patrol Boat project until 2025, life extension refits will be required in 2010, 2011 and 2011 respectively.

UPDATED

2 VAI (DABUR) CLASS (COASTAL PATROL CRAFT) (PB)

SAKU 303   SAQA 304

Displacement, tons: 39 full load
Dimensions, feet (metres): 64.9 × 18 × 5.8 *(19.8 × 5.5 × 1.8)*
Main machinery: 4 GM 12V-71TA diesels; 1,680 hp *(1.25 MW)* sustained; 2 shafts
Speed, knots: 19. **Range, n miles:** 450 at 13 kt
Complement: 9 (2 officers)
Guns: 2—20 mm Oerlikon. 2—12.7 mm MGs.
Radars: Surface search: Racal Decca Super 101 Mk 3; I-band.

Comment: Built in mid-1970s by Israel Aircraft Industries and transferred from Israel 22 November 1991. ASW equipment is not fitted. Reported as being no longer required by the Navy and may be used by other government departments.

VERIFIED

KIRO   9/1998, van Ginderen Collection / 0017831

SAQA   6/1995 / 0056954

2 COASTAL PATROL CRAFT (PB)

Name	No	Builders	Recommissioned
LEVUKA	101	Beaux's Bay Craft, Louisiana	22 Oct 1987
LAUTOKA	102	Beaux's Bay Craft, Louisiana	28 Oct 1987

Displacement, tons: 97 full load
Dimensions, feet (metres): 110 × 24 × 5 *(33.8 × 7.4 × 1.5)*
Main machinery: 4 GM 12V-71TA diesels; 1,680 hp *(1.25 MW)* sustained; 4 shafts
Speed, knots: 12
Complement: 12 (2 officers)
Guns: 1—12.7 mm MG.
Radars: Surface search: Racal Decca; I-band.

Comment: Built in 1979-80 as oil rig support craft. Purchased in September 1987. All aluminium construction.

VERIFIED

LAUTOKA

8/1996, Fiji Navy / 0056953

Finland
SUOMEN MERIVOIMAT

Country Overview

The Republic of Finland is situated in northern Europe. Nearly one third of the country lies north of the Arctic Circle. With an area of 130,559 square miles, which includes some 60,000 lakes, it has borders to the north with Norway and to the east with Russia. It has a 510 n mile coastline with the Baltic Sea and Gulf of Finland. The Ahvenanmaa archipelago (Åland Islands), consisting of some 6,500 islands, lies southwest of the mainland. Helsinki is the capital, largest city and principal port. Territorial Seas and a Fishing Zone, both of 12 n miles, have been claimed but not an EEZ.

Headquarters Appointments

Chief of Finnish Defence Forces:
   Admiral Julhani Kaskeala
Commander-in-Chief Finnish Navy:
   Commodore Hans Holström
Chief of Staff FNHQ:
   Commodore Hanno Strang

Diplomatic Representation

Defence Attaché in London:
   Lieutenant Colonel Juhani Karjomaa
Defence Attaché in Moscow:
   Captain Jukka Hellberg

Personnel

(a)   2005: 2,256 regulars
(b)   3,650 conscripts (6-12 months' national service)

Fleet Organisation

Naval Headquarters is to be relocated from Helsinki to Turku by 2008.
Gulf of Finland Naval Command; main base Upinniemi, Helsinki.
Archipelago Sea Naval Command; main base at Pansio, near Turku.
Kotka Coastal Command at Kotka.
Uusimaa Jaeger Brigade at Tammisaari.
Not all ships are fully manned all the time but all are rotated on a regular basis.

Coast Defence

Coastal Artillery and naval infantry troops. RBS 15 truck-mounted quadruple SSM launchers. 155 mm, 130 mm and 100 mm fixed and mobile guns.

Frontier Guard

All Frontier Guard vessels come under the Ministry of the Interior. The ships have dark green hulls with a thick red diagonal stripe superimposed by a thin white stripe. Superstructure is painted grey. Personnel numbers: 600.

Icebreakers

Icebreakers work for the Board of Navigation.

DELETIONS

Patrol Forces

2004   *Tuuli*

PENNANT LIST

Patrol Forces		05	Uusimaa	235	Hirsala	826	Isku
		21-26	Kuha 21-26	237	Hila	830	Högsåra
50	Kiisla	521-527	Kiiski 1-7	238	Haruna	831	Kallanpää
51	Kurki	777	Porkkala	241	Askeri	836	Houtskär
60	Helsinki	875	Pyhäranta	334	Hankoniemi	874	Kala 4
61	Turku	876	Pansio	511	Jymy	877	Kampela 3
62	Oulu			512	Raju	879	Valas
63	Kotka			531	Syöksy	894	Alskär
70	Rauma	**Auxiliaries**		541	Vinha	899	Halli
71	Raahe			722	Vaarlahti	993	Torsö
72	Porvoo	56	Kajava	723	Vänö		
73	Naantali	57	Lokki	730	Haukipää		
80	Hamina	92	Putsaari	731	Hakuni		
81	Tornio (bldg)	96	Pikkala	739	Hästö		
82	— (bldg)	98	Mursu	751	Lohi		
		99	Kustaanmiekka	752	Lohm		
Mine Warfare Forces		121	Vahakari	771	Kampela 1		
		133	Havouri	772	Kampela 2		
01	Pohjanmaa	176	Kala 6	792	Träskö		
02	Hämeenmaa	232	Hauki	799	Hylje		

PATROL FORCES

4 RAUMA CLASS (FAST ATTACK CRAFT—MISSILE) (PTGM)

Name	No	Builders	Commissioned
RAUMA	70	Hollming, Rauma	18 Oct 1990
RAAHE	71	Hollming, Rauma	20 Aug 1991
PORVOO	72	Finnyards, Rauma	27 Apr 1992
NAANTALI	73	Finnyards, Rauma	23 June 1992

Displacement, tons: 215 standard; 248 full load
Dimensions, feet (metres): 157.5 × 26.2 × 4.5 *(48 × 8 × 1.5)*
Main machinery: 2 MTU 16V 538 TB93 diesels; 7,510 hp(m) *(5.52 MW)* sustained; 2 Riva Calzoni IRC 115 water-jets
Speed, knots: 30
Complement: 19 (5 officers)

Missiles: SSM: 6 Saab RBS 15SF (could embark 8); active radar homing to 150 km *(80 n miles)* at 0.8 Mach; warhead 200 kg.
SAM: 1 sextuple launcher; Matra Mistral; IR homing to 4 km *(2.2 n miles)*; warhead 3 kg.
Guns: 1 Bofors 40 mm/70; 300 rds/min to 12 km *(6.6 n miles)*; weight of shell 0.96 kg.
   6—103 mm rails for rocket illuminants. 2—12.7 mm MGs.
   2 Sako 23 mm/87 (twin); can be fitted instead of Mistral launcher.
A/S mortars: 4 Saab Elma LLS-920 9-tubed launchers; range 300 m; warhead 4.2 kg shaped charge.
Depth charges: 1 rail.
Countermeasures: Decoys: Philax chaff and IR flares.
ESM: MEL Matilda; radar intercept.
Weapons control: Bofors Electronic 9LV Mk 3 optronic director with TV camera; infra-red and laser telemetry.

PORVOO

6/2001, Harald Carstens / 0114728

Radars: Surface search: 9GA 208; I-band.
Fire control: Bofors Electronic 9LV 225; J-band.
Navigation: Raytheon ARPA; I-band.
Sonars: Simrad Subsea Toadfish sonar; search and attack; active high frequency.
   Finnyards Sonac/PTA towed array; low frequency.

Programmes: Ordered 27 August 1987.
Structure: Developed from Helsinki class. Hull and superstructure of light alloy. SAM and 23 mm guns are interchangeable within the same Sako barbette which has replaced the ZU mounting.
Operational: Primary function is the anti-ship role but there is some ASW capability. Mine rails can be fitted in place of the missile launchers. Towed array cable is 78 m with 24 hydrophones and can be used at speeds between 3 and 12 kt.

UPDATED

2 + 2 HAMINA CLASS (FAST ATTACK CRAFT—MISSILE) (PTGM)

Name	No	Builders	Commissioned
HAMINA	80 (ex-74)	Aker Finnyards, Rauma	24 Aug 1998
TORNIO	81	Aker Finnyards, Rauma	12 May 2003
—	82	Aker Finnyards, Rauma	Sep 2005
—	—	Aker Finnyards, Rauma	June 2006

Displacement, tons: 270 full load
Dimensions, feet (metres): 164 × 26.2 × 6.2 *(50.8 × 8.3 × 2)*
Main machinery: 2 MTU 16V 538 TB93 diesels; 7,510 hp(m) *(5.52 MW)* sustained; 2 Kamewa 90SII waterjets
Speed, knots: 32. **Range, n miles:** 500 at 30 kt
Complement: 21 (5 officers)

Missiles: SSM: 4 Saab RBS 15SF; active radar homing to 150 km *(80 n miles)* at 0.8 Mach; warhead 200 kg.
SAM: 1 sextuple launcher; Matra Mistral; IR homing to 4 km *(2.2 n miles)*; warhead 3 kg.
Guns: Bofors 40 mm/70; 300 rds/min to 12 km *(6.6 n miles)*; weight of shell 0.96 kg.
6—103 mm rails for rocket illuminants. 2—12.7 mm MGs.
2 Sako 23 mm/87 (twin); can be fitted instead of Mistral launcher.
2—12.7 mm MGs.
A/S mortars: 4 Saab Elma LLS-920 9-tubed launchers; range 300 m; warhead 4.2 kg shaped charge.
Depth charges: 1 rail.
Countermeasures: Decoys: Philax chaff and IR flares.
Smoke system: Lacroix ATOS system.
ESM: MEL Matilda; radar intercept.
Combat data systems: Signaal TACTICOS (trialled in 2000).
Weapons control: Saab Dynamics EOS-400 optronic director.
Radars: Air search: DASA TRS-3D; C-band.
Surface search: Signaal Scout; I-band.
Fire control: Bofors Electronic 9LV 225; J-band.
Navigation: Raytheon ARPA; I-band.
Sonars: Simrad Subsea Toadfish sonar; search and attack; active high frequency.
Finnyards Sonac/PTA towed array; low frequency.

Programmes: First ordered on 31 December 1996. Second ordered in February 2000. A third of class was ordered on 3 December 2003 and a fourth on 15 February 2005 for delivery in mid-2006.
Modernisation: The weapon/sensor fit is likely to include the EADS ANCS 2000 combat data system, the Umkhonto point defence missile system, a Bofors 57 mm/70 Mk 3 gun, the SaabTech Systems Ceros fire-control system and the SAGEM Electro-Optical Multifunction System (EOMS). These are expected to be included in the second of class on build and to be retro-fitted in *Hamina*. This will entail removal of the Tacticos command system, the Mistral SAM system, the 9LV 225 fire-control system and EOS-400 optronic director. A new mine (Seamine 2000) is also under development.
Structure: A continuation of the Rauma design with greater use of composite materials. Aluminium hull. Signature reduction is aided by RAM coatings on the superstructure, submerged engine exhausts, upper deck pre-wetting, resilient mountings for all machinery, waterjet propulsion and conductive sealings on doors and hatches to prevent electromagnetic leakage.
Operational: To be command ships of 'Squadron 2000'.

UPDATED

TORNIO   8/2003 *, Finnish Navy / 0587711

4 HELSINKI CLASS (FAST ATTACK CRAFT—MISSILE) (PTGM)

Name	No	Builders	Commissioned
HELSINKI	60	Wärtsilä, Helsinki	1 Sep 1981
TURKU	61	Wärtsilä, Helsinki	3 June 1985
OULU	62	Wärtsilä, Helsinki	1 Oct 1985
KOTKA	63	Wärtsilä, Helsinki	16 June 1986

Displacement, tons: 280 standard; 300 full load
Dimensions, feet (metres): 147.6 × 29.2 × 9.9 *(45 × 8.9 × 3)*
Main machinery: 3 MTU 16V 538 TB92 diesels; 10,230 hp(m) *(7.52 MW)* sustained; 3 shafts
Speed, knots: 30
Complement: 30

Missiles: SSM: 8 Saab RBS 15; inertial guidance; active radar homing to 70 km *(37.8 n miles)* at 0.8 Mach; warhead 150 kg; sea-skimmer.
SAM: 2 sextuple launchers; Matra Mistral; IR homing to 4 km *(2.2 n miles)*; warhead 3 kg.
Guns: 1 Bofors 57 mm/70; 200 rds/min to 17 km *(9.3 n miles)*; weight of shell 2.4 kg.
6—103 mm rails for rocket illuminants.
4 Sako 23 mm/87 (2 twin); can be fitted in place of Mistral launcher.
Depth charges: 2 rails.
Countermeasures: Decoys: Philax chaff and IR flare launcher.
ESM: Argo; radar intercept.
Weapons control: Saab EOS 400 optronic director.
Radars: Surface search: 9GA 208; I-band.
Fire control: Philips 9LV 225; J-band.
Navigation: Raytheon ARPA; I-band.
Sonars: Simrad Marine SS 304; high-resolution active scanning.
Finnyards Sonac/PTA towed array; low frequency.

Programmes: *Helsinki* was launched 5 November 1980. Next three ordered to a revised design on 13 January 1983.
Modernisation: *Helsinki's* bridge and armament have been modified and are now the same as the other three of the class. A Sako barbette can take either twin 23 mm guns or a Sadral SAM launcher. The Sako mounting has replaced the original ZU version. A planned upgrade for all four ships has been cancelled.
Structure: The light armament can be altered to suit the planned role. Missile racks can also be replaced by mine rails. Hull and superstructure of light alloy.
Operational: All four ships are to be decommissioned by 2009.

UPDATED

HELSINKI   6/2003 *Bram Plokker / 0587713

2 KIISLA CLASS (COASTAL PATROL CRAFT) (PB)

Name	No	Builders	Commissioned
KIISLA	50	Hollming, Rauma	25 May 1987
KURKI	51	Hollming, Rauma	Nov 1990

Displacement, tons: 270 full load
Dimensions, feet (metres): 158.5 × 28.9 × 7.2 *(48.3 × 8.8 × 2.2)*
Main machinery: 2 MTU 16V 538 TB93 diesels; 7,510 hp(m) *(6.9 MW)* sustained; 2 Kamewa 90 waterjets
Speed, knots: 25
Complement: 10
Guns: 2 USSR 23 mm/60 (twin) or 1 Madsen 20 mm.
Weapons control: Radamec 2100 optronic director.
Sonars: Simrad SS304 hull-mounted and VDS; active search; high frequency.

Comment: First ordered on 23 November 1984 and second on 22 November 1988. Plans for two further craft were cancelled. The design allows for rapid conversion to attack craft, ASW craft, minelayer, minesweeper or minehunter. A central telescopic crane over the engine room casing is used to launch a 5.7 m rigid inflatable sea boat. A fire monitor is mounted in the bows. The Kamewa steerable water-jets extend the overall hull length by 2 m. Transferred from the Frontier Guard in 2004.

UPDATED

KIISLA   6/2004 *, Finnish Navy / 0587710

MINE WARFARE FORCES

Notes: Development of a new MCM squadron is in progress. Following proposals for three new ships and for a Mine Warfare Data Centre from European prime contractors in late 2004, a contract is expected to follow in early 2006. Ship deliveries are expected in 2009, 2010 and 2011.

3 PANSIO CLASS (MINELAYERS—LCU TYPE) (MLI)

Name	No	Builders	Commissioned
PANSIO	876 (ex-576)	Olkiluoto Shipyard	25 Sep 1991
PYHÄRANTA	875 (ex-575, ex-475)	Olkiluoto Shipyard	26 May 1992
PORKKALA	777	Olkiluoto Shipyard	29 Oct 1992

Displacement, tons: 450 standard
Dimensions, feet (metres): 144.3 oa; 128.6 wl × 32.8 × 6.6 *(44; 39.2 × 10 × 2)*
Main machinery: 2 MTU 12V 183 TE62 diesels; 1,500 hp(m) *(1.1 MW)*; 2 shafts; bow thruster
Speed, knots: 10
Complement: 12
Guns: 2 ZU 23 mm/87 (twin). 1—12.7 mm MG.
Mines: 50.
Radars: Navigation: Raytheon ARPA; I-band.

Comment: Ordered in May 1990. Used for inshore minelaying and transport with a capacity of 100 tons. Ice strengthened with ramps in bow and stern. Has a 15 ton crane fitted aft.

VERIFIED

PANSIO   6/2001, Finnish Navy / 0114730

2 HÄMEENMAA CLASS (MINELAYERS) (ML)

Name	No	Builders	Laid down	Launched	Commissioned
HÄMEENMAA	02	Finnyards, Rauma	2 Apr 1991	11 Nov 1991	15 Apr 1992
UUSIMAA	05	Finnyards, Rauma	12 Nov 1991	June 1992	2 Dec 1992

Displacement, tons: 1,330 full load
Dimensions, feet (metres): 252.6 oa; 228.3 wl × 38.1 × 9.8
  (77; 69.6 × 11.6 × 3)
Main machinery: 2 Wärtsilä 16V22 diesels; 6,300 hp(m)
  (4.64 MW) sustained; 2 Kamewa cp props; bow thruster;
  247 hp(m) *(184 kW)*
Speed, knots: 19
Complement: 70

Missiles: SAM: 1 sextuple launcher; Matra Mistral; IR homing to
  4 km *(2.2 n miles)*; warhead 3 kg.
Guns: 2 Bofors 40 mm/70; 300 rds/min to 12 km *(6.6 n miles)*;
  weight of shell 0.96 kg.
  4 or 6 Sako 23 mm/87 (2 or 3 twin) (the third mounting is
  interchangeable with Mistral launcher).
A/S mortars: 2 RBU 1200 fixed 5-tubed launchers; range
  1,200 m; warhead 34 kg.
Depth charges: 2 racks for 16 DCs.
Mines: 4 rails for 100-150.
Countermeasures: Decoys: 2 ML/Wallop Superbarricade
  multichaff and IR launchers.
  ESM: MEL Matilda; intercept.
Weapons control: Radamec System 2400 optronic director;
  2 Galileo optical directors.
Radars: Surface search and Navigation: 3 Selesmar ARPA;
  I-band.
Sonars: Simrad; hull-mounted; active mine detection; high
  frequency.

Programmes: First one ordered 29 December 1989 after the
  original order in July from Wärtsilä had been cancelled.
  Second ordered 13 February 1991.
Modernisation: A mid-life upgrade is planned to be ordered in
  2005. Modernisation, to be undertaken in 2007-08, is likely to
  include EADS ANCS 2000 combat data system, EADS TRS
  3D radar, Sagem EOMS and Umkhonto point defence missile
  system.
Structure: Steel hull and alloy superstructure. Ice strengthened
  (Ice class 1A) and capable of breaking up to 40 mm ice.
  Ramps in bow and stern. The Mistral launcher is mounted at
  the stern. SAM system can be replaced by a third twin 23 mm
  mounting within the same barbette.
Operational: Dual role as a transport and support ship.

UPDATED

UUSIMAA
6/2001, Findler & Winter / 0114718

UUSIMAA
6/2001, van Ginderen Collection / 0114727

1 MINELAYER (ML)

Name	No	Builders	Laid down	Launched	Commissioned
POHJANMAA	01	Wärtsilä, Helsinki	4 May 1978	28 Aug 1978	8 June 1979

Displacement, tons: 1,000 standard; 1,100 full load
Dimensions, feet (metres): 255.8 × 37.7 × 9.8
  (78.2 × 11.6 × 3)
Main machinery: 2 Wärtsilä Vasa 16V22 diesels; 6,300 hp(m)
  (4.64 MW) sustained; 2 shafts; cp props; bow thruster
Speed, knots: 19. **Range, n miles:** 3,500 at 15 kt
Complement: 90

Missiles: SAM: 2 sextuple launchers; Matra Mistral; IR homing to
  4 km *(2.2 n miles)*; warhead 3 kg.
Guns: 1 Bofors 57 mm/70; 200 rds/min to 17 km *(9.3 n miles)*;
  weight of shell 2.4 kg.
  6—103 mm launchers for illuminants fitted to the mounting.

2 Bofors 40 mm/70; 300 rds/min to 12 km *(6.6 n miles)*;
  weight of shell 0.96 kg.
  4 Sako 23 mm/87 (2 twin). 2—12.7 mm MGs.
A/S mortars: 2 RBU 1200 fixed 5-tubed launchers; range
  1,200 m; warhead 34 kg.
Depth charges: 2 rails.
Mines: 120 including UK Stonefish.
Countermeasures: Decoys: Philax chaff and IR flare launcher.
  ESM: Argo; radar intercept.
Radars: Air search: Signaal DA05; E/F-band.
  Fire control: Phillips 9LV 220; J-band.
  Navigation: I-band.

Sonars: Simrad; hull-mounted; active search and attack; high
  frequency.
  Bottom classification; search; high frequency.

Programmes: Design completed 1976. Ordered late 1977.
Modernisation: In 1992 the forward 23 mm guns were replaced
  by 12.7 mm MGs. Major refit in 1996-98 to replace the main
  gun, improve air defences and minelaying capability. The
  SAM mounting is interchangeable with 23 mm guns.
Operational: Also serves as training ship. Carries 70 trainees
  accommodated in Portakabins on the mine deck. Helicopter
  area on quarterdeck but no hangar.

UPDATED

POHJANMAA
9/2004*, B Sullivan / 0587712

6 KUHA CLASS (MINESWEEPERS—INSHORE) (MSI)

Name	No	Builders	Commissioned
KUHA 21-26	21-26	Laivateollisuus, Turku	1974-75

Displacement, tons: 90 full load
Dimensions, feet (metres): 87.2; 104 × 22.7 × 6.6 *(26.6; 31.7 × 6.9 × 2)*
Main machinery: 2 Cummins MT-380M diesels; 600 hp(m) *(448 kW)*; 1 shaft; cp prop; active rudder
Speed, knots: 12
Complement: 15 (3 officers)
Guns: 2 ZU 23 mm/60 (twin). 1—12.7 mm MG.
Radars: Navigation: Decca; I-band.
Sonars: Reson Seabat 6012 mine avoidance; active high frequency.

Comment: All ordered 1972. First one completed 28 June 1974, and last on 13 November 1975. Fitted for magnetic, acoustic and pressure-mine clearance. Hulls are of GRP. May carry a Pluto ROV. Four of the class were lengthened in 1997/98 and remaining two by 2000 to take a new minesweeping control system, and new magnetic and acoustic sweeps. New sonars installed.
VERIFIED

KUHA 21   *6/2001, Finnish Navy* / 0114724

7 KIISKI CLASS (MINESWEEPERS—INSHORE) (MSI)

Name	No	Builders	Commissioned
KIISKI 1-7	521-527	Fiskars, Turku	1983-84

Displacement, tons: 20 full load
Dimensions, feet (metres): 49.9 × 13.4 × 3.3 *(15.2 × 4.1 × 1.2)*
Main machinery: 2 Valmet 611 CSMP diesels; 340 hp(m) *(250 kW)*; 2 Hamilton water-jets
Speed, knots: 11. **Range, n miles:** 260 at 11 kt
Complement: 4

Comment: Ordered January 1983. All completed by 24 May 1984. GRP hull. Built to be used with Kuha class for unmanned teleguided sweeping, but this was not successful and they are now used for manned magnetic and acoustic sweeping operations with crew of four.
VERIFIED

KIISKI 5   *6/2001, Finnish Navy* / 0114725

ICEBREAKERS

Notes: Operation of Finnish icebreakers was transferred from the Finnish Maritime Administration to the new state-owned shipping enterprise, Finstaship, in January 2004. All the ships are based at Helsinki.

2 KARHU 2 CLASS (AGBH)

OTSO   KONTIO

Measurement, tons: 9,200 dwt
Dimensions, feet (metres): 324.7 × 79.4 × 26.2 *(99 × 24.2 × 8)*
Main machinery: Diesel-electric; 4 Wärtsilä Vasa 16V32 diesel generators; 22.84 MW 60 Hz sustained; 2 motors; 17,700 hp(m) *(13 MW)*; 2 shafts; 2 thrusters
Speed, knots: 18.5
Complement: 28
Helicopters: 1 light.

Comment: First ordered from Wärtsilä 29 March 1984, completed 30 January 1986. Second ordered 29 November 1985, delivered 29 January 1987. Fitted with Wärtsilä bubbler system. One other transferred to Estonia in 1993.
UPDATED

KONTIO   *3/2002*, Finstaship* / 0587709

2 URHO CLASS (AGBH)

URHO   SISU

Displacement, tons: 7,800 *Urho* (7,900, *Sisu*) standard; 9,500 full load
Dimensions, feet (metres): 349.7 × 78.1 × 27.2 *(106.6 × 23.8 × 8.3)*
Main machinery: Diesel-electric; 5 Wärtsilä-SEMT-Pielstick diesel generators; 25,000 hp(m) *(18.37 MW)*; 4 motors; 22,000 hp(m) *(16.2 MW)*; 4 shafts (2 fwd, 2 aft (cp props))
Speed, knots: 18
Complement: 47
Helicopters: 1 light.

Comment: Built by Wärtsilä and commissioned on 5 March 1975 and 28 January 1976 respectively. Fitted with two screws aft, taking 60 per cent of available power and two forward, taking the remainder. Similar to Swedish Atle class.
UPDATED

SISU   *1/2003*, Finstaship* / 0587708

2 TARMO CLASS (AGBH)

APU   VOIMA

Displacement, tons: 4,890 full load
Dimensions, feet (metres): 283.8 × 69.9 × 23.9 *(86.5 × 21.3 × 7.3)*
Main machinery: Diesel-electric; 4 Wärtsilä-Sulzer diesel generators; 12,000 hp(m) *(8.82 MW)*; 4 shafts (2 screws fwd, 2 aft)
Speed, knots: 17. **Range, n miles:** 7,000 at 17 kt
Complement: 45-55
Helicopters: 1 light.

Comment: Both built by Wärtsilä. *Voima* commissioned in 1954 and modernised 1978-79. *Apu* commissioned in 1970.
UPDATED

APU   *1/2003*, Finstaship* / 0587707

2 FENNICA CLASS (AGBH)

FENNICA   NORDICA

Measurement, tons: 1,650 (Winter); 3,900 (Arctic); 4,800 (Summer) dwt
Dimensions, feet (metres): 380.5 × 85.3 × 27.6 *(116 × 26 × 8.4)*
Main machinery: Diesel-electric; 2 Wärtsilä Vasa 16V32D/ABB Strömberg diesel generators; 12 MW; 2 Wärtsilä Vasa 12V32D/ABB Strömberg diesel generators; 9 MW; 2 ABB Strömberg motors; 2 Aquamaster US ARC 1 nozzles; 20,400 hp(m) *(15 MW)*; 3 Brunvoll bow thrusters; 6,120 hp(m) *(4.5 MW)*
Speed, knots: 16
Complement: 16 + 80 passengers
Radars: Navigation: 2 Selemar; I-band.
Helicopters: 1 light.

Comment: First of class ordered in October 1991, second in May 1992, from Finnyards, Rauma. *Fennica* launched 10 September 1992 and completed 15 March 1993. *Nordica* launched July 1993 and completed January 1994. Bollard pull 230 tons. Capable of 8 kt at 0.8 m level ice and continuous slow speed at 1.8 m arctic level ice. 115 ton A frame and two deck cranes of 15 and 5 tons each. Combination of azimuth propulsion units and bow thrusters gives full dynamic positioning capability.
UPDATED

FENNICA   *6/2004*, Finstaship* / 0587706

1 BOTNICA CLASS (AGBH)

BOTNICA

Measurement, tons: 6,370 grt
Dimensions, feet (metres): 317.2 × 78.7 × 27.9 *(96.7 × 24.0 × 8.5)*
Main machinery: Diesel-electric; 6 twin Caterpillar 3512B V12 diesels; 16,100 hp *(12 MW)*; 2 Azipod propulsors
Speed, knots: 15
Complement: 25
Radars: Navigation: I-band.
Helicopters: 1 light.

Comment: Combined icebreaker, tug and supply vessel built at Aker Finnyards to Det Norske Veritas class standards and delivered in 1998. During Summer, it is chartered for servicing and intervention work on oil and gas wells in the North Sea. There is a 6.5 × 6.5 m moonpool to enable underwater servicing work.

NEW ENTRY

BOTNICA *6/2004 *, Finstaship /* 0587705

TRAINING SHIPS

2 LOKKI CLASS (AX)

Name	No	Builders	Commissioned
LOKKI	57	Valmet/Lavateollisuus	28 Aug 1986
KAJAVA	56	Valmet/Lavateollisuus	3 Oct 1981

Displacement, tons: 59 *(Lokki)*; 64
Dimensions, feet (metres): 87.9 × 18 × 6.2 *(26.8 × 5.5 × 1.9)*
87.9 × 17.1 × 8.5 *(26.8 × 5.2 × 2.1) (Lokki)*
Main machinery: 2 MTU 8V 396 TB82 diesels; 1,740 hp(m) *(1.28 MW)* sustained *(Lokki)*
2 MTU 8V 396 TB84 diesels; 2,100 hp(m) *(1.54 MW)* sustained; 2 shafts
Speed, knots: 25
Complement: 6
Guns: 2 ZU 23 mm/60 can be carried.
Sonars: Simrad SS 242; hull-mounted; active search; high frequency.

Comment: Transferred from the Frontier Guard to the Navy in 1999 and used as training vessels. Built in light metal alloy. *Lokki* has a V-shaped hull. A third of class to Lithuania in 1997 and a fourth to Latvia in 2001.

UPDATED

LOKKI *6/2001, Finnish Navy /* 0114723

3 TRAINING SHIPS (AX)

681   683   685

Comment: Naval Academy training ships.

VERIFIED

681 *5/1997, N A Sifferlinger /* 0012319

AUXILIARIES

1 KEMIO CLASS (COMMAND SHIP) (AGF/AGI)

KUSTAANMIEKKA *(ex-Valvoja III)* 99

Displacement, tons: 340 full load
Dimensions, feet (metres): 118.1 × 29.5 × 9.8 *(36 × 9 × 3)*
Main machinery: 1 Burmeister & Wain diesel; 670 hp(m) *(492 kW)*; 1 shaft
Speed, knots: 11
Complement: 10
Guns: 2—12.7 mm MGs (not always carried).

Comment: Completed in 1963. Former buoy tender transferred from Board of Navigation and converted by Hollming, Rauma in 1989. Bofors 40 mm gun removed in 1988. A ship of the same class transferred to Estonia in 1992.

UPDATED

KUSTAANMIEKKA *6/2001 *, Finnish Navy /* 0587704

5 VALAS CLASS (GP TRANSPORTS) (AKSL)

VALAS 879   **MURSU** 98   **VAHAKARI** 121   **VAARLAHTI** 722   **VÄNÖ** 723

Displacement, tons: 285 full load
Dimensions, feet (metres): 100.4 × 26.5 × 10.4 *(30.6 × 8.1 × 3.2)*
Main machinery: 1 Wärtsilä Vasa 8V22 diesel; 1,576 hp(m) *(1.16 MW)* sustained; 1 shaft
Speed, knots: 12
Complement: 11
Military lift: 35 tons or 150 troops
Guns: 2—23 mm/60 (twin). 1—12.7 mm MG.
Mines: 28 can be carried.
Radars: Navigation: Decca 1226; I-band.

Comment: Completed 1979-80. *Mursu* acts as a diving tender. Funnel is offset to starboard. Can be used as minelayers or transport/cargo carriers and are capable of breaking thin ice.

VERIFIED

VALAS (old number) *7/1998, van Ginderen Collection /* 0069878

3 KAMPELA CLASS (LCU TRANSPORTS) (LCU/AKSL)

Name	No	Builders	Commissioned
KAMPELA 1	771	Enso Gutzeit	29 July 1976
KAMPELA 2	772	Enso Gutzeit	21 Oct 1976
KAMPELA 3	877	Finnmekano	23 Oct 1979

Displacement, tons: 90 light; 260 full load
Dimensions, feet (metres): 106.6 × 26.2 × 4.9 *(32.5 × 8 × 1.5)*
Main machinery: 2 Scania diesels; 460 hp(m) *(338 kW)*; 2 shafts
Speed, knots: 9
Complement: 10
Guns: 2 or 4 ZU 23 mm/60 (1 or 2 twin).
Mines: About 20 can be carried.

Comment: Can be used as amphibious craft, transports, minelayers or for shore support. Armament can be changed to suit role.

VERIFIED

KAMPELA 2 (old number) *6/2000, Finnish Navy /* 0103800

2 KALA CLASS (LCU TRANSPORTS) (LCU/AKSL)

KALA 4 874   KALA 6 176

Displacement, tons: 60 light; 200 full load
Dimensions, feet (metres): 88.6 × 26.2 × 6 (27 × 8 × 1.8)
Main machinery: 2 Valmet diesels; 360 hp(m) (265 kW); 2 shafts
Speed, knots: 9
Complement: 10
Guns: 2 Oerlikon 20 mm (not in all).
Mines: 34.
Radars: Navigation: Decca 1226; I-band.

Comment: Completed between 1956 and 4 December 1959 (Kala 6). Can be used as coastal transports, amphibious craft, minelayers or for shore support. Armament can be changed to suit role.
VERIFIED

KALA 6   6/2001, Finnish Navy / 0114729

6 HAUKI CLASS (TRANSPORTS) (AKSL)

| HAVOURI 133 | HIRSALA 235 | HAKUNI 731 |
| HAUKI 232 | HANKONIEMI 334 | HOUTSKÄR 836 |

Displacement, tons: 45 full load
Dimensions, feet (metres): 47.6 × 15.1 × 7.2 (14.5 × 4.6 × 2.2)
Main machinery: 2 Valmet 611 CSM diesels; 586 hp(m) (431 kW); 1 shaft
Speed, knots: 12
Complement: 4
Cargo capacity: 6 tons or 40 passengers
Radars: Navigation: I-band.

Comment: Completed 1979. Ice strengthened; two serve isolated island defences. Four converted in 1988 as tenders to the Marine War College, but from 1990 back in service as light transports.
UPDATED

HOUTSKÄR (old number)   9/1997*, Finnish Navy / 0587703

4 HILA CLASS (TRANSPORTS) (AKSL)

HILA 237   HARUNA 238   HÄSTÖ 739   HÖGSÅRA 830

Displacement, tons: 50 full load
Dimensions, feet (metres): 49.2 × 13.1 × 5.9 (15 × 4 × 1.8)
Main machinery: 2 diesels; 416 hp(m) (306 kW); 2 shafts
Speed, knots: 12
Complement: 4

Comment: Ordered from Kotkan Telakka in August 1990. Second pair completed in 1994. Ice strengthened.
VERIFIED

HILA   6/2000, Finnish Navy / 0103801

1 TRIALS SHIP (MLI)

Name	No	Builders	Launched	Commissioned
ISKU	826 (ex-829, ex-16)	Reposaaron Konepaja	4 Dec 1969	1970

Displacement, tons: 180 full load
Dimensions, feet (metres): 108.5 × 28.5 × 5.9 (33 × 8.7 × 1.8)
Main machinery: 4 Type M 50 diesels; 4,400 hp(m) (3.3 MW) sustained; 4 shafts
Speed, knots: 18
Complement: 25
Radars: Navigation: Raytheon ARPA; I-band.

Comment: Formerly a missile experimental craft, now used for various equipment trials. Modernised in 1989-90 by Uusikaupunki Shipyard and lengthened by 7 m. Can quickly be converted to a minelayer.
UPDATED

ISKU   6/2000, Finnish Navy / 0103798

2 LOHI CLASS (LCU TRANSPORTS) (LCU)

LOHI 751 (ex-351)   LOHM 752

Displacement, tons: 38 full load
Dimensions, feet (metres): 65.6 × 19.7 × 3 (20 × 6 × 0.9)
Main machinery: 2 WMB diesels; 1,200 hp(m) (882 kW); 2 water-jets
Speed, knots: 20. Range, n miles: 240 at 20 kt
Complement: 4
Guns: 2 ZU 23 mm/60 (twin). 1—14.5 mm MG.

Comment: Commissioned September 1984. Used as troop carriers and for light cargo. Guns not always carried.
UPDATED

LOHI   6/2000, Finnish Navy / 0103802

1 TRANSPORT and COMMAND LAUNCH (YFB)

ASKERI   241

Displacement, tons: 25 full load
Dimensions, feet (metres): 52.6 × 14.5 × 4.5 (16 × 4.4 × 1.4)
Main machinery: 2 Volvo Penta diesels; 1,100 hp(m) (808 kW); 2 shafts
Speed, knots: 22
Complement: 6
Radars: Surface search: I-band.
Navigation: Raytheon; I-band.

Comment: Completed in 1992. Closely resembles Spanish PVC II class.
VERIFIED

COMMAND LAUNCH   6/2000, Finnish Navy / 0103804

7 VIHURI CLASS (COMMAND LAUNCHES) (YFB)

JYMY 511	SYÖKSY 531	TRÄSKÖ 792	ALSKÄR 894
RAJU 512	VINHA 541	TORSÖ 993	

Displacement, tons: 13 full load
Dimensions, feet (metres): 42.7 × 13.1 × 3 *(13 × 4 × 0.9)*
Main machinery: 2 diesels; 772 hp(m) *(567 kW)*; 2 water-jets
Speed, knots: 30
Complement: 6
Radars: Surface search: I-band.

Comment: First of class *Vihuri* delivered in 1988, the next five in 1991 and the last pair in 1993. *Träskö, Torsö* and *Alskär* act as fast transports. The remainder are command launches for Navy squadrons. *Vihuri* was destroyed by fire in late 1991.

VERIFIED

VINHA · 5/1993, van Ginderen Collection / 0069883

30 MERIUISKO CLASS (LCP)

U 201-211	U 301-312	U 400 series

Displacement, tons: 10 full load
Dimensions, feet (metres): 36 × 11.5 × 2.9 *(11 × 3.5 × 0.9)*
Main machinery: 2 Volvo TAMD70E diesels; 418 hp(m) *(307 kW)* sustained; 2 Hamilton waterjets
Speed, knots: 36; 30 full load
Complement: 3
Military lift: 48 troops
Radars: Navigation (U 401 series): I-band.

Comment: First batch of 11 completed by Alumina Varvet from 1983 to 1986. A further four ordered in 1989. Constructed of light alloy. Fitted with small bow ramp. Two of the class equipped with cable handling system for boom defence work. Batch one has smaller cabins.

UPDATED

U 304 · 6/2000, Finnish Army / 0103805

6 + (30) JURMO CLASS (LCP)

U 601-606

Displacement, tons: 10 full load
Dimensions, feet (metres): 43.6 × 11.5 × 2.0 *(13.3 × 3.5 × 0.6)*
Main machinery: 2 Caterpillar diesels; 2 FF-jet 375 waterjets
Speed, knots: 30+
Complement: 2
Military lift: 21 troops with equipment or 2.5 tons cargo
Guns: 1—12.7 mm MG.
Radars: Navigation: I-band.

Comment: Developed from Meriusko class for troop carrying role. Prototype built by Alutech Ltd and delivered in 1999. Three further in 2002. Further orders projected. Cargo hatch of composite material to provide armoured protection.

UPDATED

U 603 · 8/2002, E & M Laursen / 0534066

23 RAIDING CRAFT (LCVP)

Displacement, tons: 3 full load
Dimensions, feet (metres): 26.2 × 6.9 × 1 *(8 × 2.1 × 0.3)*
Main machinery: 1 Yanmar 4LHA-STE diesel; 240 hp *(179 kW)*; 1 RR FF-jet 240 waterjet
Speed, knots: 30
Complement: 1
Military lift: 9 troops with equipment

Comment: First batch of 23 units ordered in February 2001. Based on Swedish Gruppbåt and built by Alutech Ltd. Delivered late 2001.

VERIFIED

RAIDING CRAFT · 6/2001, Finnish Navy / 0114721

1 CABLE SHIP (ANL)

PUTSAARI 92

Displacement, tons: 430 full load
Dimensions, feet (metres): 149.5 × 28.6 × 8.2 *(45.6 × 8.7 × 2.5)*
Main machinery: 1 Wärtsilä diesel; 510 hp(m) *(375 kW)*; 1 shaft; active rudder; bow thruster
Speed, knots: 10
Complement: 20

Comment: Built by Rauma-Repola, Rauma, launched on 15 December 1965 and commissioned in 1966. Modernised by Wärtsilä in 1987. Fitted with two 10 ton cable winches. Strengthened for ice operations.

VERIFIED

PUTSAARI · 6/2001, Finnish Navy / 0114720

1 SUPPORT CRAFT (YFB)

PIKKALA (ex-*Fenno*) 96

Displacement, tons: 66 full load
Dimensions, feet (metres): 75.5 × 14.4 × 6.6 *(23 × 4.4 × 2)*
Main machinery: 1 Valmet diesel; 177 hp(m) *(130 kW)*; 1 shaft
Speed, knots: 10
Complement: 5

Comment: Used for utility and transport roles at Helsinki. Commissioned in June 1946 at Turhu.

VERIFIED

PIKKALA · 6/2000, Finnish Navy / 0103806

2 POLLUTION CONTROL VESSELS (YPC)

HYLJE 799   **HALLI** 899

Displacement, tons: 1,500 *(Hylje)*; 1,600 *(Halli)* full load
Dimensions, feet (metres): 164; 198.5 *(Halli)* × 41 × 9.8 *(50; 60.5 × 12.5 × 3)*
Main machinery: 2 Saab diesels; 680 hp(m) *(500 kW)*; 2 shafts; active rudders; bow thruster *(Hylje)*
2 Wärtsilä diesels; 2,650 hp(m) *(19.47 MW)*; 2 shafts; active rudders *(Halli)*
Speed, knots: 7 *(Hylje)*; 13 *(Halli)*

Comment: Painted grey. Strengthened for ice. Owned by Ministry of Environment, civilian-manned but operated by Navy from Turku. *Hylje* commissioned 3 June 1981, *Halli* in January 1987. Capacity is about 550 m³ *(Hylje)* and 1,400 m³ *(Halli)* of contaminated seawater. The ships have slightly different superstructure lines aft.

VERIFIED

HALLI   *6/2000, Finnish Navy /* 0103812

TUGS

2 HARBOUR TUGS (YTM)

HAUKIPÄÄ 730   **KALLANPÄÄ** 831

Displacement, tons: 38 full load
Dimensions, feet (metres): 45.9 × 16.4 × 7.5 *(14 × 5 × 2.3)*
Main machinery: 2 diesels; 360 hp(m) *(265 kW)*; 2 shafts
Speed, knots: 9
Complement: 2

Comment: Delivered by Teijon Telakka Oy in December 1985. Similar to Hauki class. Also used as utility craft.

VERIFIED

HAUKIPÄÄ (old number)   *6/2000, Finnish Navy /* 0103813

FRONTIER GUARD

1 IMPROVED TURSAS CLASS
(OFFSHORE PATROL VESSEL) (WPBO)

MERIKARHU

Displacement, tons: 1,100 full load
Dimensions, feet (metres): 189.6 × 36.1 × 15.1 *(57.8 × 11 × 4.6)*
Main machinery: 2 Wärtsilä Vasa 8R26 diesels; 3,808 hp(m) *(2.8 MW)* sustained; 1 shaft; cp prop; bow and stern thrusters
Speed, knots: 15. **Range, n miles:** 2,000 at 15 kt
Complement: 30
Guns: 2—23 mm/87 (twin) can be carried.
Radars: Surface search. Navigation.

Comment: Ordered 17 June 1993 from Finnyards, and completed 28 October 1994. Capable of 5 kt in 50 cm of ice. Used as an all-weather patrol ship in the Baltic, capable of Command, SAR, tug work with 30 ton bollard pull, and environmental pollution cleaning up. Carries an RIB launched from a hydraulic crane.

VERIFIED

MERIKARHU   *6/2001, Bram Plokker /* 0114733

2 TURSAS CLASS (OFFSHORE PATROL VESSELS) (WPBO)

TURSAS   **UISKO**

Displacement, tons: 730 full load
Dimensions, feet (metres): 160.8 × 34.1 × 13.1 *(49 × 10.4 × 4)*
Main machinery: 2 Wärtsilä Vasa 8R22 diesels; 3,152 hp(m) *(2.32 MW)* sustained; 2 shafts
Speed, knots: 16
Complement: 32
Guns: 2 Sako 23 mm/60 (twin).
Sonars: Simrad SS105; active scanning; 14 kHz.

Comment: First ordered from Rauma-Repola on 21 December 1984, launched 31 January 1986 and delivered 6 June 1986. Second ordered 20 March 1986, launched 19 June 1986 and delivered 27 January 1987. Operate as offshore patrol craft and can act as salvage tugs. Ice strengthened.

VERIFIED

UISKO   *5/1994, van Ginderen Collection*

1 IMPROVED VALPAS CLASS
(OFFSHORE PATROL VESSEL) (WPBO)

TURVA

Displacement, tons: 550 full load
Dimensions, feet (metres): 159.1 × 28 × 12.8 *(48.5 × 8.6 × 3.9)*
Main machinery: 2 Wärtsilä diesels; 2,000 hp(m) *(1.47 MW)*; 1 shaft
Speed, knots: 15
Complement: 23
Guns: 1 Oerlikon 20 mm.
Sonars: Simrad SS105; active scanning; 14 kHz.

Comment: Built by Laivateollisuus, Turku and commissioned 15 December 1977. Armament changed in 1992.

VERIFIED

TURVA   *6/1993, van Ginderen Collection /* 0069889

3 TELKKÄ CLASS (WPBO)

TELKKÄ	TAVI	TIIRA

Displacement, tons: 400 full load
Dimensions, feet (metres): 160.8 × 24.6 × 11.8 *(49 × 7.5 × 3.6)*
Main machinery: 2 diesels; 6,120 hp(m) *(4.5 MW)*; 2 shafts
Speed, knots: 20
Complement: 17
Guns: 1—20 mm.
Sonars: Sonac PTA; towed array; low frequency.

Comment: *Telkkä* entered service in July 1999, *Tavi* in 2003 and *Tiira* on 27 May 2004.
UPDATED

TAVI   5/2003, J Ciślak / 0568851

4 SLINGSBY SAH 2200 (HOVERCRAFT) (UCAC)

Displacement, tons: 5.5 full load
Dimensions, feet (metres): 34.8 × 13.8 *(10.6 × 4.2)*
Main machinery: 1 Cummins 6CTA-8-3M-1 diesel; 300 hp *(224 kW)*
Speed, knots: 40. **Range, n miles:** 400 at 30 kt
Complement: 2
Military lift: 2.2 tons or 12 troops
Guns: 1—12.7 mm MG.
Radars: Navigation: Raytheon R41; I-band.

Comment: First one acquired from Slingsby Amphibious Hovercraft Company in March 1993. Three more ordered in February 1998 and delivered in late 1999.
VERIFIED

SLINGSBY 2200   6/1993, Slingsby / 0069892

3 GRIFFON 2000 TDX(M) (HOVERCRAFT) (UCAC)

Displacement, tons: 6.8 full load
Dimensions, feet (metres): 36.1 × 15.1 *(11 × 4.6)*
Main machinery: 1 Deutz BF8L513 diesel; 320 hp *(239 kW)* sustained
Speed, knots: 33. **Range, n miles:** 300 at 25 kt
Complement: 2
Military lift: 16 troops or 2 tons
Guns: 1—7.62 mm MG.
Radars: Navigation: I-band.

Comment: First two acquired from Griffon, UK and commissioned 1 December 1994; third one bought in June 1995. Can be embarked in an LCU. Speed indicated is at Sea State 3 with a full load. Similar to those in service with the UK Navy.
VERIFIED

GRIFFON 2000   6/1994, P Felstead / 0080653

39 INSHORE PATROL CRAFT AND TENDERS (PB)

Class	Total	Tonnage	Speed	Commissioned
RV-37	7	20	12	1978-85
RV-90	10	25	12	1992-96
PV-11	14	10	28	1984-90

VERIFIED

RV-11 class   6/1993 / 0069894

LAND-BASED MARITIME AIRCRAFT

Numbers/Type: 2 Agusta AB 412 Griffon.
Operational speed: 122 kt *(226 km/h).*
Service ceiling: 17,000 ft *(5,180 m).*
Range: 354 n miles *(656 km).*
Role/Weapon systems: Operated by Coast Guard/Frontier force for patrol and SAR. Sensors: Radar and FLIR. Weapons: Unarmed at present but mountings for machine guns.
VERIFIED

Numbers/Type: 3 Eurocopter AS 332L1 Super Puma.
Operational speed: 130 kt *(240 km/h).*
Service ceiling: 15,090 ft *(4,600 m).*
Range: 672 n miles *(1,245 km).*
Role/Weapon systems: Coastal patrol, surveillance and SAR helicopters. Sensors: Surveillance radar, FLIR, tactical navigation systems and SAR equipment. Weapons: Unarmed.
VERIFIED

Numbers/Type: 2 Agusta AB 206B JetRanger.
Operational speed: 116 kt *(215 km/h).*
Service ceiling: 13,500 ft *(4,120 m).*
Range: 364 n miles *(674 km).*
Role/Weapon systems: Coastal patrol and inshore surveillance helicopters. Sensors: Visual means only. FLIR may be fitted in due course. Weapons: Unarmed.
VERIFIED

AB 206 (Swedish colours)   6/2000, Andreas Karlsson, Swedish Defence Image / 0106563

Numbers/Type: 2 Dornier Do 228-212.
Operational speed: 223 kt *(413 km/h).*
Service ceiling: 29,600 ft *(9,020 m).*
Range: 939 n miles *(1,740 km).*
Role/Weapon systems: Maritime surveillance, SAR and pollution control. Acquired in 1995. Sensors: GEC-Marconi Seaspray radar; Terma Side scan radar; FLIR/TV, SLAR and IR/UV scanner. Weapons: Unarmed.
VERIFIED

DORNIER 228 (German colours)   9/2002, Frank Findler / 0528878

For details of the latest updates to *Jane's Fighting Ships* online and to discover the additional information available exclusively to online subscribers please visit
jfs.janes.com

France
MARINE NATIONALE

Country Overview

The French Republic, which includes the island of Corsica, is situated in western Europe. With an area of 210,026 square miles, the mainland is bordered to the north by Belgium, Luxembourg and Germany, to the south-east by Switzerland and Italy and to the south-west by Spain. It has a 1,852 n mile coastline with the Atlantic Ocean, Mediterranean Sea, North Sea and English Channel. Overseas departments are French Guiana, Martinique, Guadeloupe and Réunion. Dependencies include St Pierre and Miquelon, Mayotte, New Caledonia, French Polynesia, the French Southern and Antarctic Territories, and Wallis and Futuna Islands. The capital and largest city is Paris while the principal ports are Marseille, Le Havre, Dunkirk, St Nazaire and Rouen. Strasbourg is a port on the Rhine. Territorial seas (12 n miles) are claimed. An EEZ (200 n miles) has also been claimed but not all the large number of boundaries have been defined by agreements.

Headquarters Appointments

Chief of the Naval Staff:
  Amiral Jean Louis Battet
Inspector General of the Navy:
  Admiral Jean Moulin
Director of Personnel:
  Vice-Amiral Philippe Sauter
Major General of the Navy:
  Vice-Amiral d'Escadre Alain Oudot de Dainville

Senior Appointments

C-in-C Atlantic Theatre (CECLANT):
  Vice-Amiral d'Escadre Laurent Merer
C-in-C Mediterranean Theatre (CECMED):
  Vice-Amiral d'Escadre Jean-Marie Van Huffel
Flag Officer, French Forces Polynesia (ALPACI):
  Contre-Amiral Patrick Giaume
Flag Officer, Naval Forces Indian Ocean (ALINDIEN):
  Vice-Amiral Xavier Rolin
Flag Officer, Cherbourg:
  Contre-Amiral Edouard Guillaud
Flag Officer, Submarines (ALFOST):
  Vice-Amiral Pierre-François Forissier
Flag Officer, Naval Action Force (ALFAN):
  Vice-Amiral d'Escadre Alain Dumontet
Deputy Flag Officer, Naval Action Force (TOULON):
  Contre-Amiral Jacques Mazars
Deputy Flag Officer, Naval Action Force (BREST):
  Contre-Amiral Jean-Pierre Teule
Flag Officer Mine Warfare Force (ALMINES):
  Capitaine de Vaisseau Yves-Marie Marechal
Flag Officer Naval Aviation (ALAVIA):
  Vice-Amiral Jean-Pierre Tiffou
Flag Officer Lorient and Commandant Marines (Alfusco):
  Contre-Amiral Jean-Paul Cabrières

Diplomatic Representation

Defence Attaché in London:
  Contre-Amiral Jean-Pierre Tiffou
Naval Attaché in London:
  Capitaine de Vaisseau Henri-François Piot
Defence and Naval Attaché in Riyad:
  Contre-Amiral Bertand Vibert
Head of French Military Delegation to the European Union:
  Vice-Amiral d'Escadre Edouard Mac-Grath
Head of French Military Mission to SACLANT:
  Contre-Amiral Xavier Paitard
Head of French Military Mission to CINCSOUTH:
  Contre-Amiral Jean-Pierre Vadet
Naval Attaché in Washington:
  Capitaine de Vaisseau Philippe Alquier

Personnel

(a) 2005: 43,195 (4,832 officers)
(b) military service discontinued November 2001
(c) 2005: civilians in direct support: 10,064
(d) 2005: Active reserve: 6,500

Bases

Brest: Main Atlantic base. SSBN base
Toulon: Mediterranean Command base
Cherbourg: Channel base
Bayonne: Landes firing range
Small bases at Papeete (Tahiti), Fort-de-France (Martinique), Nouméa (New Caledonia), Degrad-des-Cannes (French Guiana), Port-des-Galets (La Réunion).

Shipyards (Naval)

Cherbourg: Submarines and Fast Attack Craft (private shipyard)
Brest: Major warships and refitting
Lorient: Destroyers and Frigates, MCMVs, Patrol Craft
Toulon: Major refits

Dates

Armement pour essais: After launching when the ship is sufficiently advanced to allow a crew to live on board, and the commanding officer has joined. From this date the ship hoists the French flag and is ready to undertake her first harbour trials.
Armement définitif: On this date the ship has received her full complement and is able to undergo sea trials.
Clôture d'armement: Trials are completed and the ship is now able to undertake her first endurance cruise.
Croisière de longue durée or *traversée de longue durée:* The endurance cruise follows the *clôture d'armement* and lasts until the ship is accepted with all systems fully operational.
Admission au service actif: Commissioning date.

Reserve

A ship in 'Reserve Normale' has no complement but is available at short notice. 'Reserve Speciale' means that a refit will be required before the ship can go to sea again. 'Condamnation' is the state before being broken up or sold; at this stage a Q number is allocated.

Prefix to Ships' Names

FS is used in NATO communications but is not official.

Strength of the Fleet

Type	Active (Reserve)	Building (Projected)
Submarines (SSBN)	4	1
Submarines (SSN)	6	(6)
Aircraft Carriers	1	(1)
Helicopter Carrier	1	—
Destroyers	12	2 (2)
Frigates	20	— (17)
Public Service Force	7	—
Patrol Craft	10	—
LSDs	4	2
LST/LCT	8	—
LCMs	17	—
Route Survey Vessels	3	—
Minehunters	13	—
Diving Tenders	4	—
Survey/Research Ships	6	2
Tankers (AOR)	4	—
Maintenance Ships	2	(2)
Supply Tenders	7	—
Transports	9	—
Training Ships	16	—

DELETIONS

Submarines

2005 *Indomptable*

Frigates

2002 *Second Maître Le Bihan* (to Turkey)

Patrol Forces

2003 *Stellis*

Amphibious Forces

2002 LCM 1058
2004 *Champlain*

Auxiliaries

2002 *Ile d'Oleron, Rhin*
2003 *Garonne*
2004 *Mouette*

Tugs

2002 *Le Fort, Le Travailleur*

Fleet Air Arm Bases

Notes: In addition to the following squadrons, there are two other squadrons operating with mixed Air Force and Navy crews on behalf of both services:
• Helicopter Squadron EH-1/67 ''Pyrénées'', based at Cazaux AFB, for the combat SAR role, operating seven specialised Aerospatiale SA-330 Puma helicopters. A Eurocopter AS-532 Cougar Mk 2 Resco was delivered in September 1999 for testing in order to define the equipment suite for the three other machines ordered. All four new Resco helicopters should be operational in 2003. Pumas and Cougars regularly embark on *Charles de Gaulle.*
• Training Squadron EAT-319, based at Avord AFB, with Embraer 121 Xingu light transport (some coming from the Navy) for pilot basic training.

Embarked Squadrons

Base/Squadron No	Aircraft	Task
Lann Bihoué/4F	E-2C Hawkeye	AEW
Landivisiau/11F	Super Étendard	Assault, Recce
Landivisiau/12F	Rafale M	Air Defence
Landivisiau/17F	Super Étendard	Assault, Recce
Hyères/31F	Lynx	ASW
Lanvéoc-Poulmic/34F	Lynx	ASW
Hyères/36F	Panther	Surveillance

Support Squadrons

Base/Squadron No	Aircraft	Task
Hyères/CEPA/10S	Various	Research, trials
Lanvéoc-Poulmic/22S (detachments on ships)	Alouette III	Support Atlantic Region
Lanvéoc-Poulmic/32F (detachment at Hyères)	Super Frelon,	Transport, SAR
Hyères/35F (detachments at various locations and ships)	Dauphin 2, Alouette III	Surveillance, SAR, Carrier-borne SAR
Landivisiau/57S	Falcon 10 MER	Support, Training

Maritime Patrol Squadrons

Base/Squadron No	Aircraft	Task
Nîmes-Garons/21F	Atlantique Mk 2	MP
Lann Bihoué/23F	Atlantique Mk 2	MP
Lann Bihoué/24F	Falcon 50M/Xingu Gardian	Surveillance, SAR
Faaa (Papeete)/25F (detachments at Nouméa, New Caledonia and Fort-de-France, Antilles)	Gardian/Nord 262E/Xingu	Surveillance, SAR
Nîmes-Garons/28F	Nord 262E/Xingu	Surveillance, SAR, Flying School, liaison

Training Squadrons

Base/Squadron No	Aircraft	Task
Lanvéoc-Poulmic/EIP/50S	MS 880 Rallye/CAP 10	Initial Flying School, Recreational

---

Approximate Fleet Dispositions 1 May 2005

		Channel	Atlantic	Mediterranean	Indian Ocean (1)	Pacific	Antilles F. Guiana
Carriers	FAN	—	1 (hel)	1	—	—	—
SSBN	FOST	—	4	—	—	—	—
SSN	FOST	—	—	6	—	—	—
DDG/DDH	FAN	—	5	7	—	—	—
FFG	FAN	—	5	9	2	2	1
MCMV (incl tenders)	FAN	1	14	5	—	—	—
Patrol Forces (2)	FAN/GM	3	6	1	5	6	5
LPD/LSD	FAN	—	—	4	—	—	—
LST/LCT	FAN	—	—	3	2	2	1
AOR	FAN	—	—	3	1	—	—

FAN = Force d'Action Navale (HQ at Toulon). All surface ships based at Toulon, Brest or overseas.
FOST = Force Océanique Stratégique (HQ at Brest). SSBNs based at l'Île Longue near Brest. All SSNs based at Toulon.
GM = Gendarmerie Maritime
Notes:
(1) Plus one or two DDG/DDH/FFG regularly deployed from Toulon.
(2) Patrol forces include vessels manned by the Navy and major craft from the Gendarmerie Maritime.

PENNANT LIST

Submarines

S 601	Rubis
S 602	Saphir
S 603	Casabianca
S 604	Émeraude
S 605	Améthyste
S 606	Perle
S 615	L'Inflexible
S 616	Le Triomphant
S 617	Le Téméraire
S 618	Le Vigilant
S 619	Le Terrible (bldg)

Aircraft and Helicopter Carriers

R 91	Charles de Gaulle
R 97	Jeanne d'Arc

Destroyers

D 603	Duquesne
D 610	Tourville
D 612	De Grasse
D 614	Cassard
D 615	Jean Bart
D 620	Forbin (bldg)
D 621	Chevalier Paul (bldg)
D 640	Georges Leygues
D 641	Dupleix
D 642	Montcalm
D 643	Jean de Vienne
D 644	Primauguet
D 645	La Motte-Picquet
D 646	Latouche-Tréville

Frigates

F 710	La Fayette
F 711	Surcouf
F 712	Courbet
F 713	Aconit
F 714	Guépratte
F 730	Floréal
F 731	Prairial
F 732	Nivôse
F 733	Ventôse
F 734	Vendémiaire
F 735	Germinal
F 789	Lieutenant de Vaisseau le Hénaff
F 790	Lieutenant de Vaisseau Lavallée
F 791	Commandant l'Herminier
F 792	Premier Maître l'Her
F 793	Commandant Blaison
F 794	Enseigne de Vaisseau Jacoubet
F 795	Commandant Ducuing
F 796	Commandant Birot
F 797	Commandant Bouan

Mine Warfare Forces

M 611	Vulcain
M 614	Styx
M 622	Pluton
M 641	Éridan
M 642	Cassiopée
M 643	Andromède
M 644	Pégase
M 645	Orion
M 646	Croix du Sud
M 647	Aigle
M 648	Lyre
M 649	Persée
M 650	Sagittaire
M 651	Verseau
M 652	Céphée
M 653	Capricorne
M 770	Antarès
M 771	Altaïr
M 772	Aldébaran

Patrol Forces

P 601	Elorn
P 602	Verdon
P 603	Adour
P 604	Scarpe
P 671	Glaive (GM)
P 672	Épée (GM)
P 675	Arago
P 676	Flamant
P 677	Cormoran
P 678	Pluvier
P 679	Grèbe
P 680	Sterne
P 681	Albatros
P 682	L'Audacieuse
P 683	La Boudeuse
P 684	La Capricieuse
P 685	La Fougueuse
P 686	La Glorieuse
P 687	La Gracieuse
P 688	La Moqueuse
P 689	La Railleuse
P 690	La Rieuse
P 691	La Tapageuse
P 703	Lilas (GM)
P 704	Bégonia (GM)
P 705	Pivoine (GM)
P 706	Nymphéa (GM)
P 707	MDLC Robet (GM)
P 708	Gendarme Perez (GM)
P 709	MDLC Richard (GM)
P 710	General Delfosse (GM)
P 711	Gentiane (GM)
P 712	Fuschia (GM)
P 713	Capitaine Moulié (GM)
P 714	Lieut Jamet (GM)

P 715	Bellis (GM)
P 716	MDLC Jacques (GM)
P 717	Lavande (GM)
P 720	Géranium (GM)
P 721	Jonquille (GM)
P 722	Violette (GM)
P 723	Jasmin (GM)
P 740	Fulmar (GM)
P 755	Miri (GM)
P 760	Pétulante (GM)
P 761	Mimosa (GM)
P 764	—
P 776	Sténia (GM)
P 778	Réséda (GM)
P 789	Melia (GM)
P 790	Vétiver (GM)
P 791	Hortensia (GM)

GM = Gendarmerie Maritime

Amphibious Forces

L 9011	Foudre
L 9012	Siroco
L 9013	Mistral (bldg)
L 9014	Tonnerre (bldg)
L 9021	Ouragan
L 9022	Orage
L 9031	Francis Garnier
L 9032	Dumont D'Urville
L 9033	Jacques Cartier
L 9034	La Grandière
L 9051	Sabre
L 9052	Dague
L 9061	Rapière
L 9062	Hallebarde
L 9090	Gapeau

Major Auxiliaries Survey and Support Ships

A 601	Monge
A 607	Meuse
A 608	Var
A 613	Achéron
A 615	Loire
A 620	Jules Verne
A 630	Marne
A 631	Somme
A 633	Taape
A 634	Rari
A 635	Revi
A 636	Maito
A 637	Maroa
A 638	Manini
A 649	L'Étoile
A 641	Esterel
A 642	Lubéron

A 650	La Belle Poule
A 652	Mutin
A 664	Malabar
A 669	Tenace
A 675	Fréhel
A 676	Saire
A 677	Armen
A 678	La Houssaye
A 679	Kéréon
A 680	Sicié
A 681	Taunoa
A 682	Rascas
A 693	Acharné
A 695	Bélier
A 696	Buffle
A 697	Bison
A 712	Athos
A 713	Aramis
A 722	Poséidon
A 743	Denti
A 748	Léopard
A 749	Panthère
A 750	Jaguar
A 751	Lynx
A 752	Guépard
A 753	Chacal
A 754	Tigre
A 755	Lion
A 758	Beautemps-Beaupré
A 759	Dupuy de Lôme (bldg)
A 760	Alize (bldg)
A 768	Élan
A 770	Glycine
A 771	Églantine
A 774	Chevreuil
A 775	Gazelle
A 776	Isard
A 785	Thétis
A 791	Lapérouse
A 792	Borda
A 793	Laplace
L 9077	Bougainville
P 674	D'Entrecasteaux
Y 613	Faune
Y 638	Lardier
Y 639	Giens
Y 640	Mengam
Y 641	Balaguier
Y 642	Taillat
Y 643	Nividic
Y 647	Le Four
Y 649	Port Cros
Y 662	Dryade
Y 692	Telenn Mor
Y 700	Nereide
Y 701	Ondine
Y 702	Naiade
Y 741	Elfe
Y 750	La Persévérante

SUBMARINES
Strategic Missile Submarines (SSBN/SNLE)

1 L'INFLEXIBLE M4 CLASS (SSBN/SNLE)

Name	No	Builders	Laid down	Launched	Commissioned
L'INFLEXIBLE	S 615	DCN, Cherbourg	27 Mar 1980	23 June 1982	1 Apr 1985

Displacement, tons: 8,080 surfaced; 8,920 dived
Dimensions, feet (metres): 422.1 × 34.8 × 32.8
(128.7 × 10.6 × 10)
Main machinery: Nuclear; turbo-electric; 1 PWR; 2 turbo-alternators; 1 Jeumont Schneider motor; 16,000 hp(m) *(11.76 MW)*; twin SEMT-Pielstick/Jeumont Schneider 8 PA4 V 185 SM diesel-electric auxiliary propulsion; 9.9 MW; 1 emergency motor; 1 shaft
Speed, knots: 20 dived
Range, n miles: 5,000 at 4 kt on auxiliary propulsion only
Complement: 135 (15 officers) (2 crews)

Missiles: SLBM: 16 Aerospatiale M45/TN 75; 3-stage solid fuel rockets; inertial guidance to 6,000 km *(3,240 n miles)*; thermonuclear warhead with 6 MRV each of 100 kT.
SSM: Aerospatiale SM 39 Exocet; launched from 21 in *(533 mm)* torpedo tubes; inertial cruise; active radar homing to 50 km *(27 n miles)* at 0.9 Mach; warhead 165 kg.
Torpedoes: 4—21 in *(533 mm)* tubes. ECAN L5 Mod 3; dual purpose; active/passive homing to 9.5 km *(5.1 n miles)* at 35 kt; warhead 150 kg; depth to 550 m *(1,800 ft)*; and ECAN F17 Mod 2; wire-guided; active/passive homing to 20 km *(10.8 n miles)* at 40 kt; warhead 250 kg; depth 600 m *(1,970 ft)*; total of 18 torpedoes and SSM carried in a mixed load.
Countermeasures: ESM: Thomson-CSF ARUR 12/DR 3000U; intercept.
Weapons control: SAD (Système d'Armes de Dissuasion) strategic data system (for SLBMs); SAT (Système d'Armes Tactique) tactical data system and DLA 1A weapon control system (for SSM and torpedoes).
Radars: Navigation: Thomson-CSF DRUA 33; I-band.
Sonars: Thomson Sintra DSUX 21B 'multifunction' passive bow and flank arrays.
DUUX 5; passive ranging and intercept; low frequency.
DSUV 61B; towed array; very low frequency.

Programmes: With the paying off of *Le Redoutable* in December 1991, the remaining submarines of the class became known as L'Inflexible class SNLE M4.

L'INFLEXIBLE
7/2004*, H M Steele / 1042262

L'INFLEXIBLE
7/2004*, B Prézelin / 1042173

Modernisation: A successful test firing of an M 45 missile, containing components of the next generation M 51 missile, was conducted on 18 April 2001. *L'Inflexible* to be refitted in 2005.
Structure: Diving depth, 250 m *(820 ft)* approx.

Operational: *L'Inflexible* is planned to pay off by 2010 and *L'Indomptable* in 2005. *Le Tonnant* paid off to reserve in December 1999.

UPDATED

3 + 1 LE TRIOMPHANT CLASS (SSBN/SNLE-NG)

Name	No	Builders	Laid down	Launched	Commissioned
LE TRIOMPHANT	S 616	DCN, Cherbourg	9 June 1989	13 July 1993	21 Mar 1997
LE TÉMÉRAIRE	S 617	DCN, Cherbourg	18 Dec 1993	8 Aug 1997	23 Dec 1999
LE VIGILANT	S 618	DCN, Cherbourg	1997	12 Apr 2003	26 Nov 2004
LE TERRIBLE	S 619	DCN, Cherbourg	Nov 2002	2008	July 2010

Displacement, tons: 12,640 surfaced; 14,335 dived
Dimensions, feet (metres): 453 × 41; 55.8 (aft planes) × 41
(138 × 12.5; 17 × 12.5)
Main machinery: Nuclear; turbo-electric; 1 PWR Type K15 (enlarged CAS 48); 150 MW; 2 turbo-alternators; 1 motor; 41,500 hp(m) *(30.5 MW)*; diesel-electric auxiliary propulsion; 2 SEMT-Pielstick 8 PA4 V 200 SM diesels; 900 kW; 1 emergency motor; 1 shaft; pump jet propulsor
Speed, knots: 25 dived
Complement: 111 (15 officers) (2 crews)

Missiles: SLBM: 16 Aerospatiale M45/TN 75; 3-stage solid fuel rockets; inertial guidance to 6,000 km *(3,240 n miles)*; thermonuclear warhead with 6 MRV each of 100 kT. To be replaced by M51.1/TN 75 which has a planned range of 8,000 km *(4,300 n miles)* and 6 MRVs (to be fitted first in S 619 in 2010) and from 2015 by M51.2 (to be fitted first in S 618) with the new TNO (Tête Nucléaire Océanique) warhead.
SSM: Aerospatiale SM 39 Exocet; launched from 21 in *(533 mm)* torpedo tubes; inertial cruise; active radar homing to 50 km *(27 n miles)* at 0.9 Mach; warhead 165 kg.
Torpedoes: 4—21 in *(533 mm)* tubes. ECAN L5 Mod 3; dual purpose; active/passive homing to 9.5 km *(5.1 n miles)* at 35 kt; warhead 150 kg; depth to 550 m *(1,800 ft)*; total of 18 torpedoes and SSM carried in a mixed load.
Countermeasures: ESM: Thomson-CSF ARUR 13/DR 3000U; intercept.
Weapons control: SAD (Système d'Armes de Dissuasion) strategic data system (for SLBMs) SAD M5I will be fitted in S 619; SAT (Système d'Armes Tactique) tactical data system and DLA 4A weapon control system (for SSM and torpedoes). SYCOBS to be fitted in S 619.
Radars: Search: Dassault; I-band.
Sonars: Thomson Sintra DMUX 80 'multifunction' passive bow and flank arrays. DUUX 5; passive ranging and intercept; low frequency.
DSUV 61; towed array; very low frequency.

LE VIGILANT
4/2004 *, H M Steele / 1042260

Programmes: *Le Triomphant* ordered 10 March 1986. *Le Téméraire* ordered 18 October 1989. *Le Vigilant* ordered 27 May 1993 with first steel cut 9 December 1993. Hull transferred to dock on 12 April 2003 and sea trials began in January 2004. *Le Terrible* ordered 28 July 2000 and first steel cut 24 October 2000. Class of six originally planned, but reduced to four after the end of the Cold War. Sous-marins Nucléaires Lanceurs d'Engins-Nouvelle Génération (SNLE-NG).

Modernisation: Development of the M5 missile discontinued in favour of the less expensive M51 which is planned to equip S 619 in 2010 and the first three submarines between 2010 and 2015. Warhead TN O is to replace TN 75 by 2015.

Structure: Built of HLES 100 steel capable of withstanding pressures of more than 100 kg/mm². Diving depth 500 m *(1,640 ft)*. Height from keel to top of fin is 21.3 m *(69.9 ft)*. Plans to lengthen the hull in later ships of the class have been shelved.

Operational: First sea cruise of *Le Triomphant* 16 July to 22 August 1995. First submerged M45 launch on 14 February 1995, second on 19 September 1996. *Le Téméraire* official trials started April 1998, first submerged M 45 launch 4 May 1999. *Le Triomphant* completed 30 month refit in October 2004 and conducted test launch of M45 missile on 2 February 2005. *Le Téméraire* is to start refit in late 2006. *Le Vigilant* started sea trials on 1 April 2004. Based at Brest.   **UPDATED**

LE VIGILANT
5/2004 *, B Prézelin / 1042175

LE TÉMÉRAIRE
6/2002, French Navy / 0529140

→ OUR BUSINESS, YOUR BENEFIT.

It's no accident that we count 30 navies among our satisfied customers. Or that international sales represent over 25 percent of our total revenues. We're a world leader in prime contracting and system integration with expertise spanning every aspect of naval defence from design concept to fully-integrated warships. Projects and achievements include, among all, the Charles-de-Gaulle, new-generation SSBNs, SSKs with or without AIP, Horizon anti-air frigates, stealth surface vessels, torpedoes, and through-life support for navies all over the world. DCN – men and women committed to better naval systems and customer satisfaction.

DCN

EXPERTS IN NAVAL SYSTEMS

www.dcn.fr

Attack Submarines (SSN/SNA)

Notes: The Agosta class submarine *Ouessant* is to be re-introduced into service in 2005 to support the sales of Scorpene class boats to Malaysia.

0 + 6 BARRACUDA CLASS (SSN)

Name	No	Builders		Laid down	Launched	Commissioned
—	—	DCN, Cherbourg		2006	2011	2013

Displacement, tons: 4,500 surfaced
Dimensions, feet (metres): 311.7 × ? × ?
(95.0 × ? × ?)
Main machinery: Nuclear; turbo-electric; 1 PWR K-15; 1 shaft
Speed, knots: 25 dived
Complement: 60 (8 officers)

Missiles: SLCM: MBDA Scalp land-attack missile launched in capsule from 21 in torpedo tubes; inertial cruise and tercom, electro-optic homing to 400 km *(215 n miles)* at 0.9 Mach; warhead 400 kg.
SSM: Aerospatiale SM 39 Exocet launched from 21 in *(533 mm)* torpedo tubes; inertial cruise; active radar homing to 50 km *(27 n miles)* at 0.9 Mach; warhead 165 kg.
Torpedoes: 4—21 in *(533 mm)* bow tubes. Whitehead Black Shark 184 Mod 3 torpedoes. Total of 24 torpedoes/missiles in mixed load.
Mines: In lieu of torpedoes.
Countermeasures: ESM.
Combat data systems: SYCOBS.
Radars: Surface search: I-band.
Sonars: Bow sonar, wide aperture flank array and a reelable thin-line towed array.

Programmes: Studies for a new generation SSN (Project Barracuda) funded under the 1997-2002 budget. Programme launched on 14 October 1998. DCN (submarine design authority) expected to present detailed commercial and technical offer in June 2005 for design, development and initial production. A contract for long lead items is expected in 2005 with a main contract, covering production of an initial three boats and a 10-year integrated logistic support package, likely to follow in early 2006. First of class expected to start building at Cherbourg in 2006 with subsequent boats to be built at a rate of one every 20 months to replace Rubis class submarines. The last one is to be delivered in 2022.
Structure: Much of the technology will emanate from the Le Triomphant design as well as new features developed for the

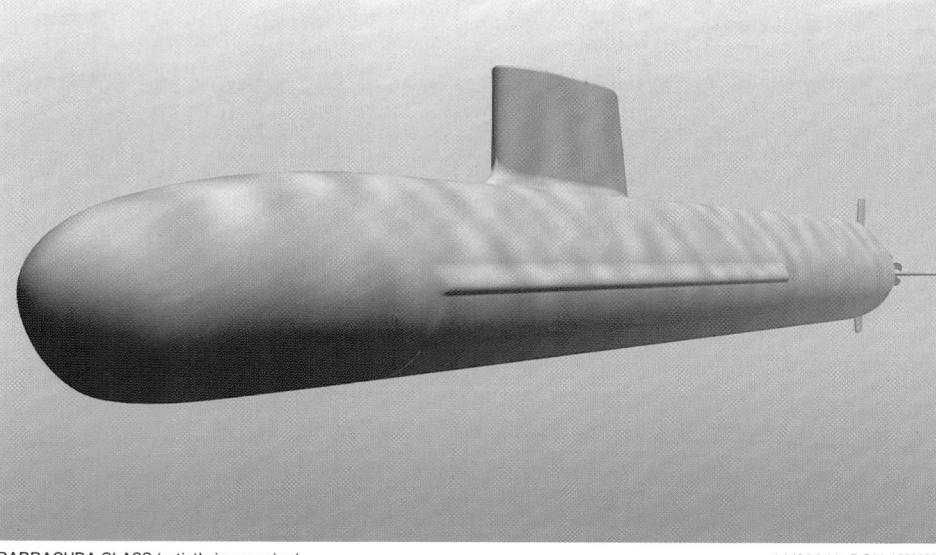

BARRACUDA CLASS (artist's impression)   *11/2004*, DCN /* 0590253

Scorpene design. A high level of automation is planned to reduce complement to 60. Diving depth is over 350 m. A hybrid propulsion system will use electric propulsion at cruise speeds and turbo-mechanical propulsion for higher speeds.

Operational: Sea trials for the first of class are scheduled for 2011 and entry into service in 2013.
NEW ENTRY

6 RUBIS AMÉTHYSTE CLASS (SSN/SNA)

Name	No	Builders	Laid down	Launched	Commissioned
RUBIS	S 601	Cherbourg Naval Dockyard	11 Dec 1976	7 July 1979	23 Feb 1983
SAPHIR	S 602	Cherbourg Naval Dockyard	1 Sep 1979	1 Sep 1981	6 July 1984
CASABIANCA	S 603	Cherbourg Naval Dockyard	19 Sep 1979	22 Dec 1984	13 May 1987
ÉMERAUDE	S 604	Cherbourg Naval Dockyard	4 Mar 1981	12 Apr 1986	15 Sep 1988
AMÉTHYSTE	S 605	Cherbourg Naval Dockyard	31 Oct 1983	14 May 1988	20 Mar 1992
PERLE	S 606	Cherbourg Naval Dockyard	27 Mar 1987	22 Sep 1990	7 July 1993

Displacement, tons: 2,410 surfaced; 2,670 dived
Dimensions, feet (metres): 241.5 × 24.9 × 21
(73.6 × 7.6 × 6.4)
Main machinery: Nuclear; turbo-electric; 1 PWR CAS 48; 48 MW; 2 turbo-alternators; 1 motor; 9,500 hp(m) *(7 MW)*; SEMT-Pielstick/Jeumont Schneider 8 PA4 V 185 SM diesel-electric auxiliary propulsion; 450 kW; 1 emergency motor; 1 shaft
Speed, knots: 25
Complement: 70 (10 officers) (2 crews)

Missiles: SSM: Aerospatiale SM 39 Exocet; launched from 21 in *(533 mm)* torpedo tubes; inertial cruise; active radar homing to 50 km *(27 n miles)* at 0.9 Mach; warhead 165 kg.
Torpedoes: 4—21 in *(533 mm)* tubes. ECAN F17 Mod 2; wire-guided; active/passive homing to 20 km *(10.8 n miles)* at 40 kt; warhead 250 kg; depth 600 m *(1,970 ft)*. Total of 14 torpedoes and missiles carried in a mixed load.
Mines: Up to 32 FG 29 in lieu of torpedoes.
Countermeasures: ESM: Thomson-CSF ARUR 13/DR 3000U; intercept.
Combat data systems: TIT (Traitement des Informations Tactiques) data system; OPSMER command support system; Syracuse 2 SATCOM. Link 11 (receive only).
Weapons control: LAT (Lancement des Armes Tactiques) system.
Radars: Navigation: Kelvin Hughes 1007; I-band.
Sonars: Thomson Sintra DMUX 20 multifunction; passive search; low frequency.
DSUV 62C; towed passive array; very low frequency.
DSUV 22; listening suite.

Programmes: The programme was terminated early by defence economies with the seventh of class *Turquoise* and eighth of class *Diamant* being cancelled.
Modernisation: Between 1989 and 1995 the first four of this class converted under operation Améthyste (AMÉlioration Tactique HYdrodynamique Silence Transmission Ecoute) to bring them to the same standard of ASW (included new sonars) efficiency as *Améthyste* and *Perle* rather than that required for the original anti-surface ship role. Two F17 torpedoes can be guided simultaneously against separate targets. *Saphir* recommissioned 1 July 1991, *Rubis* in February 1993; *Casabianca* in June 1994 and *Émeraude* in March 1996. A new radar added on a telescopic mast.
Structure: Diving depth, greater than 300 m *(984 ft)*. There has been a marked reduction in the size of the reactor compared with the L'Inflexible class. On completion of the modernisation programme, all six of the class are virtually identical.
Operational: All operational SSNs are assigned to Escadrille des Sous-Marins nucléaires d'attaque (ESNA) based at Toulon but frequently deploy to the Atlantic or overseas. Endurance rated at 45 days, limited by amount of food carried. *Rubis* collided with a tanker on 17 July 1993 and has undergone extensive repairs. *Émeraude* had a bad steam leak on 30 March 1994

ÉMERAUDE   *3/2004*, B Prézelin /* 1042174

CASABIANCA   *8/2004*, B Prézelin /* 1042177

AMÉTHYSTE   *5/2004*, H M Steele /* 1042261

which caused casualties amongst the crew. *Saphir* undertook a refit/refuel in September 2000 following reactor problems. The submarine returned to service in late 2001. Refits

scheduled for *Améthyste* (2005), *Saphir* (2006-07) and *Rubis* (2007-08). Service life of all boats extended to 30 years.
UPDATED

HELICOPTER CARRIERS

1 JEANNE D'ARC CLASS (CVHG)

Name	No	Builders	Laid down	Launched	Commissioned
JEANNE D'ARC (ex-*La Résolue*)	R 97	Brest Naval Dockyard	7 July 1960	30 Sep 1961	16 July 1964

Displacement, tons: 10,575 standard; 13,270 full load
Dimensions, feet (metres): 597.1 × 78.7 hull × 24.6 *(182 × 24 × 7.5)*
Flight deck, feet (metres): 203.4 × 68.9 *(62 × 21)*
Main machinery: 4 boilers; 640 psi *(45 kg/cm²)*; 840°F *(450°C)*; 2 Rateau-Bretagne turbines; 40,000 hp(m) *(29.4 MW)*; 2 shafts
Speed, knots: 26.5. **Range, n miles:** 7,500 at 15 kt
Complement: 506 (33 officers) plus 13 instructors and 150 cadets

Missiles: SSM: 6 Aerospatiale MM 38 Exocet (2 triple) ❶; inertial cruise; active radar homing to 42 km *(23 n miles)* at 0.9 Mach; warhead 165 kg; sea-skimmer.
Guns: 2 DCN 3.9 in *(100 mm)*/55 Mod 64 CADAM automatic ❷; 80 rds/min to 17 km *(9 n miles)* anti-surface; 8 km *(4.4 n miles)* anti-aircraft; weight of shell 13.5 kg.
4—12.7 mm MGs.
Countermeasures: Decoys: 2 CSEE/VSEL Syllex 8-barrelled trainable launchers for chaff (may not be fitted).
ESM: Thomson-CSF ARBR 16/ARBX 10; intercept.
Weapons control: 3 C T Analogiques; 2 Sagem DMAa optical sights. SATCOM ❸.
Radars: Air search: Thomson-CSF DRBV 22D ❹; D-band; range 366 km *(200 n miles)*.
Air/surface search: DRBV 51 ❺; G-band.
Navigation: 2 DRBN 34A (Racal-Decca); I-band.
Fire control: 2 Thomson-CSF DRBC 32A; I-band.
Tacan: SRN-6.
Sonars: Thomson Sintra DUBV 24C; hull-mounted; active search; medium frequency; 5 kHz.

Helicopters: 2 Pumas and 2 Gazelles from the Army and 2 Navy Alouette III for annual training cruises. Up to 8 Super Frelon or 10 mixed heavy/light aircraft in war time.

Modernisation: Long refits in the summers of 1989 and 1990 have allowed equipment to be updated to enable the ship to continue well into the next century. SENIT 2 combat data system was to have been fitted but this was cancelled as a cost-saving measure. Two 100 mm guns have been removed from quarterdeck. A refit is planned to start in 2006.
Structure: Flight deck lift has a capacity of 12 tons. Some of the hangar space is used to accommodate officers under training. The ship is almost entirely air conditioned. Carries two LCVPs. Topmast can be removed for passing under bridges or other obstructions.
Operational: Used for training officer cadets. After rapid modification, she could be used as a commando ship, helicopter carrier or troop transport with commando equipment and a battalion of 700 men. Flagship of the Training Squadron for an Autumn/Spring cruise with Summer refit. Army helicopters Super Puma/Cougar and Gazelle are embarked during training cruises. Thirty-three training cruises completed by April 1997 when the ship was docked for extensive propulsion machinery repairs which completed in July 1998. Service life has been extended to at least 2010.

UPDATED

JEANNE D'ARC *(Scale 1 : 1,500), Ian Sturton* / 0529162

JEANNE D'ARC *1/2004*, Robert Pabst* / 1042183

JEANNE D'ARC *6/2004*, B Prézelin* / 1042182

AIRCRAFT CARRIERS

0 + (1) FUTURE AIRCRAFT CARRIER CLASS (CV)

Name	No	Builders	Laid down	Launched	Commissioned
—	—	—	2009	2011	2014

Displacement, tons: 60,000 full load
Dimensions, feet (metres): 931.7 × 236.2 × ? *(284.0 × 72.0 × ?)*
Flight deck, feet (metres): To be announced
Main machinery: Integrated Full Electric Propulsion using gas turbines and/or diesels
Speed, knots: 27. **Range, n miles:** To be announced
Complement: 900 approx plus aircrew

Missiles: SAM: ASTER 15.
Guns: To be announced.
Countermeasures: To be announced.
Combat data systems: To be announced.
Weapons control: To be announced.
Radars: Air search: To be announced.
Surface search: To be announced.
Navigation: To be announced.
Fire control: To be announced.
Tacan: To be announced.

Fixed-wing aircraft: Up to 35. A typical mix might include 32 Rafale M and three E-2C Hawkeye.
Helicopters: Up to five NH 90.

Programmes: A second aircraft carrier (PA2) is planned under the 2003-08 Defence Programming Law. The ship is planned to enter service in 2014 before *Charles de Gaulle* undergoes a refuel/refit in 2015. It was announced on 13 February 2004 that the ship is to be built in co-operation with the UK carrier programme. DCN and Thales are to establish a joint venture company with a joint board to manage the project at industry level. Concept design and engineering studies, undertaken by a DCN/Thales team, are expected to be completed in mid-2005 although some specific engineering work, notably on the generation of steam for aircraft catapults, is to continue thereafter. System design work should start in late 2005 and a main contract for detailed design and build is expected in late 2006.
Structure: The so-called Romeo design, with an all-electric power and propulsion architecture, provides the baseline for further iterative development of the design and a reference against which industrial co-operation with UK suppliers can proceed. The ship is to be equipped with two 90 m C-13 steam catapults.

NEW ENTRY

PA 2 *10/2004*, DCN* / 1042254

1 CHARLES DE GAULLE CLASS (CVNM/PAN)

Name	No	Builders	Laid down	Launched	Commissioned
CHARLES DE GAULLE	R 91	DCN, Brest	14 Apr 1989	7 May 1994	18 May 2001

Displacement, tons: 36,600 standard; 42,000 full load

Dimensions, feet (metres): 857.7 oa; 780.8 wl × 211.3 oa; 103.3 wl × 30.9 *(261.5; 238 × 64.4; 31.5 × 9.4)*

Flight deck, feet (metres): 857.7 × 211.3 *(261.5 × 64.4)*

Main machinery: Nuclear; 2 PWR Type K15; 300 MW; 2 GEC Alsthom turbines; 82,000 hp(m) *(61 MW)* sustained; 2 shafts

Speed, knots: 27.

Complement: 1,256 ship's company (94 officers) plus 610 aircrew plus 42 flag staff (accommodation for 1,950) (plus temporary 800 marines)

Missiles: SAM: EUROSAAM SAAM/F system with 4 (2 port, 2 starboard) DCN Sylver A43 octuple VLS launchers ❶; Aerospatiale ASTER 15; anti-missile system (operational from mid-2002) with inertial guidance and mid-course update; active radar homing at 4.5 Mach to 15 km *(8.1 n miles)*; warhead 13 kg. 32 weapons.
2 Matra Sadral PDMS sextuple launchers ❷; Mistral; IR homing to 4 km *(2.2 n miles)*; warhead 3 kg; anti-sea-skimmer; able to engage targets down to 10 ft above sea level.

Guns: 4 Giat 20F2 20 mm; 720 rds/min to 8 km *(4.3 n miles)*; weight of shell: 0.25 kg.

Countermeasures: Decoys: 4 CSEE Sagaie AMBL-2A 10-barrelled trainable launchers ❸; medium range; chaff to 8 km *(4.3 n miles)*; IR flares to 3 km *(1.6 n miles)*. Dassault LAD offboard decoys. SLAT torpedo decoys from 2006.
ESM: Thomson-CSF ARBR 21; intercept. 1 SAT DIBV 2A Vampir MB; (IRST) ❹.
ECM: 2 ARBB 33B ❺; jammers.

Combat data systems: SENIT 8; Links 11, 14 and 16. Syracuse 2 and FLEETSATCOM ❻. AIDCOMER and MCCIS command support systems.

Weapons control: 2 DIBC 2A (Sagem VIGY-105) optronic directors.

Radars: Air search: Thomson-CSF DRBJ 11B ❼; 3D; E/F-band; range 366 km *(200 n miles)* for aircraft.
Thomson-CSF DRBV 26D Jupiter ❽; D-band; range 183 km *(100 n miles)* for 2 m² target.
Air/surface search: Thomson-CSF DRBV 15C Sea Tiger Mk 2 ❾; E/F-band; range 110 km *(60 n miles)* for 2 m² target.
Navigation: Two Racal 1229 (DRBN 34A) ❿; I-band.
Fire control: Thomson-CSF Arabel 3D ⓫; I/J-band (for SAAM); range 70 km *(38 n miles)* for 2 m² target.
Tacan: NRBP 20A ⓬.

Sonars: To include SLAT torpedo attack warning.

Fixed-wing aircraft: 20 Super Étendard, 2 E-2C Hawkeye. 12 Rafale F1.

Helicopters: 2 AS 565 Panther or 2 AS 322 Cougar (AF) or 2 Super Frelon plus 2 Dauphin SAR.

Programmes: On 23 September 1980 the Defence Council decided to build two nuclear-propelled carriers to replace *Clemenceau* in 1996 and *Foch* some years later. First of class ordered 4 February 1986, first metal cut 24 November 1987. Hull floated for technical trials on 19 December 1992, and back in dock on 8 January 1993. A 19.8 m *(65 ft)* long one-twelfth scale model was used for hydrodynamic trials. Building programme delayed three years due to defence budget cuts.

Modernisation: From October 1999 to March 2000 modifications included additional radiation shielding, and lengthening of angled flight deck by 4.4 m. A 43 launchers to be replaced by A 50 (for ASTER 15 and ASTER 30) in due course.

Structure: Two lifts 62.3 × 41 ft *(19 × 12.5 m)* of 36 tons capacity. Hangar for 20-25 aircraft; dimensions 454.4 × 96.5 × 20 ft *(138.5 × 29.4 × 6.1 m)*. Angled deck 8.5° and 655.7 ft *(200 m)* overall length. Catapults: 2 USN Type C13-3; length 246 ft *(75 m)* for Super Étendards and up to 23 tonne aircraft. Enhanced weight capability of flight deck

CHARLES DE GAULLE

6/2002, French Navy / 0529143

CHARLES DE GAULLE

6/2004*, B Prézelin / 1042176

to allow operation of AEW aircraft. Island placed well forward so that both lifts can be protected from the weather. CSEE Dallas (Deck Approach and Landing Laser System) fitted, later to be replaced by MLS system. Active fin stabilisers. Bunkerage of 3,000 cum of avgas and 1,500 cum dieso.

Operational: Seven years continuous steaming at 25 kt available before refuelling (same reactors as *Le Triomphant*). Both reactors self-sustaining by 10 June 1998. Sea trials started 26 January 1999 and after handover to the Navy on

28 September 2000, continued until 9 November 2000 when a large section of the port propeller was lost while steaming at high speed in the west Atlantic. Trials resumed on 26 March 2001 with spare propellers from decommissioned *Clemenceau* (speed limited to 23 kt). Deployed Indian Ocean, December 2001 to June 2002, successful ASTER 15 firing on 30 October 2002. New propellers are expected to be installed during the 2006 refuel/refit.

UPDATED

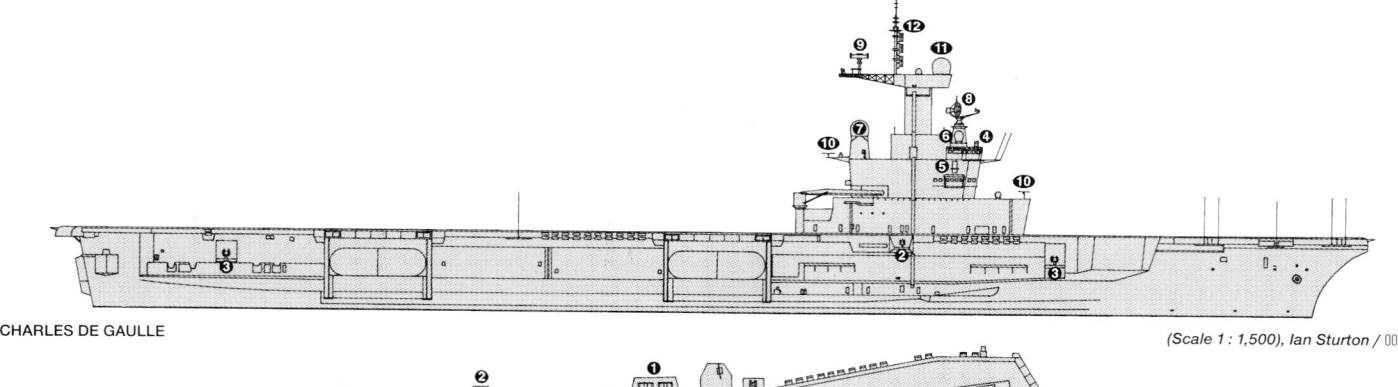

CHARLES DE GAULLE

(Scale 1 : 1,500), Ian Sturton / 0069903

CHARLES DE GAULLE

(Scale 1 : 1,500), Ian Sturton / 0104438

CHARLES DE GAULLE

5/2001, Ships of the World / 0130447

CHARLES DE GAULLE

5/2001, Ships of the World / 0130446

DESTROYERS

7 GEORGES LEYGUES CLASS (TYPE F 70 (ASW)) (DDGHM)

Name	No	Builders	Laid down	Launched	Commissioned
GEORGES LEYGUES	D 640	Brest Naval Dockyard	16 Sep 1974	17 Dec 1976	10 Dec 1979
DUPLEIX	D 641	Brest Naval Dockyard	17 Oct 1975	2 Dec 1978	13 June 1981
MONTCALM	D 642	Brest Naval Dockyard	5 Dec 1975	31 May 1980	28 May 1982
JEAN DE VIENNE	D 643	Brest Naval Dockyard	26 Oct 1979	17 Nov 1981	25 May 1984
PRIMAUGUET	D 644	Brest Naval Dockyard	17 Nov 1981	17 Mar 1984	5 Nov 1986
LA MOTTE-PICQUET	D 645	Brest Naval Dockyard/Lorient	12 Feb 1982	6 Feb 1985	18 Feb 1988
LATOUCHE-TRÉVILLE	D 646	Brest Naval Dockyard/Lorient	15 Feb 1984	19 Mar 1988	16 July 1990

Displacement, tons: 3,880 standard; 4,830 (D 640-643); 4,750 (D 644-646) full load

Dimensions, feet (metres): 455.9 × 45.9 × 18.7; 19.35 (D 640-643)
(139 × 14 × 5.7; 5.9)

Main machinery: CODOG; 2 RR Olympus TM3B gas turbines; 46,200 hp *(34.5 MW)* sustained; 2 SEMT-Pielstick 16 PA6 V280 diesels; 12,800 hp(m) *(9.41 MW)* sustained; 2 shafts; LIPS cp props

Speed, knots: 30; 21 on diesels

Range, n miles: 8,500 at 18 kt on diesels; 2,500 at 28 kt

Complement: 235 (20 officers) (D 644-646) 216 (18 officers) (D 641-643); 183 (18 officers) (D 640)

Missiles: SSM: 4 Aerospatiale MM 40 Exocet (MM 38 in D 640) ❶; inertial cruise; active radar homing to 42 km *(23 n miles)* at 0.9 Mach; active radar homing to 70 km *(40 n miles)* at 0.9 Mach (MM 40); warhead 165 kg; sea-skimmer. 4 additional Exocet MM 40 missiles can be carried as a warload (D 641-646).

SAM: Thomson-CSF Crotale Naval EDIR octuple launcher ❷; command line of sight guidance; radar/IR homing to 13 km *(7 n miles)* at 2.4 Mach; warhead 14 kg; 26 missiles.
2 Matra Simbad twin launchers mounted in lieu of 20 mm guns (D 644-646); 2 Matra Sadral sextuple launchers being fitted to D 640-643; Mistral; IR homing to 4 km *(2.2 n miles)*; warhead 3 kg.

Guns: 1 DCN/Creusot-Loire 3.9 in *(100 mm)*/55 Mod 68 CADAM automatic ❸; dual purpose; 78 rds/min to 17 km *(9 n miles)* anti-surface; 8 km *(4.4 n miles)* anti-aircraft; weight of shell 13.5 kg.
2 Breda/Mauser 30 mm (D 641-643) ❹. 800 rds/min to 3 km; weight of shell 0.37 kg.
2 Oerlikon 20 mm ❺; 720 rds/min to 10 km *(5.5 n miles)*.
4 M2HB 12.7 mm MGs (D 640, D 644-646).

Torpedoes: 2 fixed launchers. Eurotorp MU 90; anti-submarine; active/passive homing to 12 km *(6.5 n miles)* at 50 kt; warhead 50 kg; depth to 1,000 m *(3,300 ft)*.

Countermeasures: Decoys: 2 CSEE Dagaie Mk 1 or 2 10-barrelled double trainable launcher ❻; chaff and IR flares; H- to J-band. Dassault LAD offboard decoys.

ESM: ARBR 17 ❼; radar warning. Sagem DIBV 2A Vampir MB IRST (D 641-643).

ECM: ARBB 32B or Dassault ARBB 36A (D 641-646); jammer.

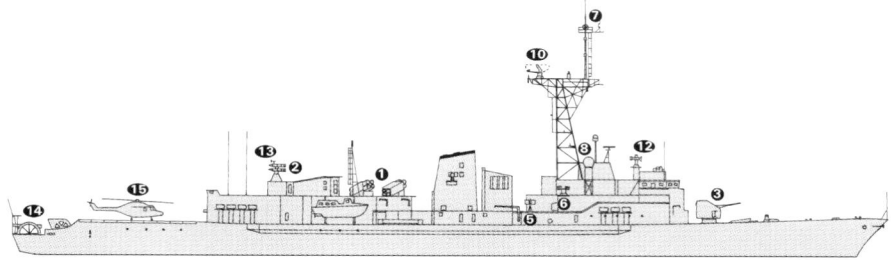

LA MOTTE-PICQUET   *(Scale 1 : 1,200), Ian Sturton / 0569907*

Combat data systems: SENIT 4 (D 640); STIDAV based on SENIT 8 added (D 641-646) action data automation; Links 11 and 14. Syracuse 2 SATCOM ❽. OPSMER command support system.

Weapons control: Thomson-CSF Vega (D 640-643) and DCN CTMS (D 644-646) optronic/radar systems. SAT Murène IR tracker being added to CTMS and Vega systems. CSEE Panda optical director. 2 Sagem VIGY-105 optronic systems (for 30 mm guns) fitted 1995-97. DLT L4 (D 640-643) and DLT L5 (D 644-646) torpedo control system. OPS-100F acoustic processor being fitted.

Radars: Air search: Thomson-CSF DRBV 26A (D 640-643) ❾; D-band; range 182 km *(100 n miles)* for 2 m² target.
Air/surface search: Thales DRBV 15A (D 640, D 641, D 645); Thales DRBV 15B (D 642-644, D 646) ❿; E/F-band.
Navigation: 2 DRBN-34 (Decca 1226); I-band (1 for close-range helicopter control).
Fire control: Thomson-CSF Vega with DRBC 32E (D 640-643) ⓫; I-band; DRBC 33A (D 644-646) ⓬; I-band. Castor 2 ⓭; I-band (for SAM).

Sonars: Thomson Sintra DUBV 23D (DUBV 24C in D 644-646); bow-mounted; active search and attack; 5 kHz.
DUBV 43B (43C in D 643-646) ⓮; VDS; search; medium frequency; paired with DUBV 23D/24; tows at 24 kt down to 200 m *(650 ft)*, (700 m *(3,000 ft)* for 43C). Length of tow 600 m *(2,000 ft)*; being upgraded to 43C.

DSBV 61B (in D 644 onward); passive linear towed array; very low frequency; 365 m *(1,200 ft)*. ATBF 2 lightweight towed array may be fitted in due course.

Helicopters: 2 Lynx Mk 4 ⓯ (except D 640).

Programmes: First three were in the 1971-76 new construction programme, fourth in 1978 estimates, fifth in 1980 estimates, sixth in 1981 estimates, seventh in 1983 estimates. D 645 and 646 were towed from Brest to Lorient for completion. Service lives: *Georges Leygues*, 2009; *Dupleix*, 2011; *Montcalm*, 2012; *Jean de Vienne*, 2014; *Primauguet*, 2015; *La Motte-Picquet*, 2016; *Latouche-Tréville*, 2017. Re-rated F 70 'frégates anti-sous-marines (FASM)' (ex-C 70) on 6 June 1988.

Modernisation: Air defence upgrade (Opération Amélioration Autodéfense Antimissiles, OP3A completed) for six of the class, *Jean de Vienne* (1996), *La Motte-Picquet* (1997), *Latouche-Tréville* (1998), *Primauget* (1999), *Dupleix* (1999) and *Montcalm* (2000). Large command structure fitted above the bridge, two Matra Sadral sextuple launchers, two Breda/Mauser 30 mm gun mounts controlled by Sagem Vigy 105 optronic sights, Vampir MB IRST and ARBB 36 jammers (replacing ARBB 32). ASW modernisation of the last six of the class planned but likely to be cancelled in view of plans for multimission frigates. Plans to fit Milas ASW missiles have been shelved. *Georges Leygues* hangar converted for training role in 1999 and crew reduced. Rolling six month refit programme started with *Primauguet* in April 2002, followed by *George Leygues*, *Dupleix*, *Montcalm*, *La Motte-Picquet*, *Latouche-Tréville* and *Jean de Vienne* which completed in June 2004. Work included hull strengthening, replacement of DRBV 51 with DRBV 15 and replacement of L5 torpedoes with MU 90. *La Motte-Picquet* to be refitted in 2005.

Structure: Bridge raised one deck in the last three of the class. Inmarsat aerial can be fitted forward of the funnel or between the Syracuse domes. 210 tons of ballast embarked to improve stability.

Operational: *Primauguet* and *Latouche-Tréville* allocated to ALFAN BREST. *Georges Leygues* acts as a tender to *Jeanne d'Arc* for training cruises.

UPDATED

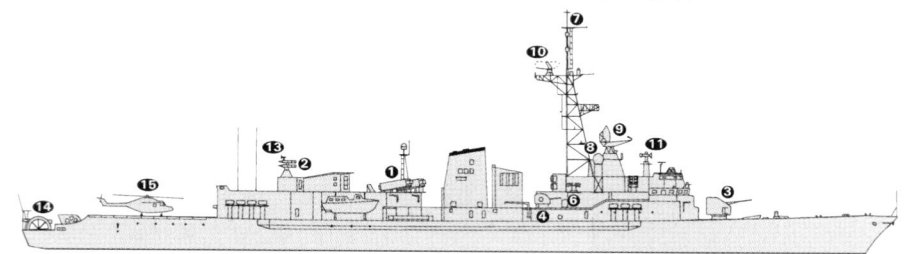

DUPLEIX   *(Scale 1 : 1,200), Ian Sturton / 0569908*

GEORGES LEYGUES   *12/2004*, B Prézelin / 1042179*

LA MOTTE-PICQUET
8/2004*, B Prézelin / 1042180

DUPLEIX
10/2004*, Schaeffer/Marsan / 1042181

LATOUCHE-TRÉVILLE
4/2004*, D Pawlenko, RAN / 1042257

2 CASSARD CLASS (TYPE F 70 (A/A)) (DDGHM)

Name	No	Builders	Laid down	Launched	Commissioned
CASSARD	D 614	Lorient Naval Dockyard	3 Sep 1982	6 Feb 1985	28 July 1988
JEAN BART	D 615	Lorient Naval Dockyard	12 Mar 1986	19 Mar 1988	21 Sep 1991

Displacement, tons: 4,230 standard; 5,000 full load
Dimensions, feet (metres): 455.9 × 45.9 × 21.3 (sonar)
  (139 × 14 × 6.5)
Main machinery: 4 SEMT-Pielstick 18 PA6 V 280 BTC diesels;
  43,200 hp(m) (31.75 MW) sustained; 2 shafts
Speed, knots: 29.5. **Range, n miles:** 8,000 at 17 kt.
Complement: 245 (25 officers) accommodation for 253

Missiles: SSM: 8 Aerospatiale MM 40 Exocet ❶; inertial cruise;
  active radar homing to 70 km (40 n miles) at 0.9 Mach;
  warhead 165 kg; sea-skimmer.
SAM: 40 GDC Pomona Standard SM-1MR; Mk 13 Mod 5
  launcher ❷; semi-active radar homing to 46 km (25 n miles) at
  2 Mach; height envelope 45-18,288 m (150-60,000 ft).
  Launchers taken from T 47 (DDG) ships.
  2 Matra Sadral PDMS sextuple launchers ❸; 39 Mistral; IR
  homing to 4 km (2.2 n miles); warhead 3 kg; anti-sea-
  skimmer; able to engage targets down to 10 ft above sea
  level.

Guns: 1 DCN/Creusot-Loire 3.9 in (100 mm)/55 Mod 68 CADAM
  automatic ❹; 80 rds/min to 17 km (9 n miles) anti-surface;
  8 km (4.4 n miles) anti-aircraft; weight of shell 13.5 kg.
  2 Oerlikon 20 mm ❺; 720 rds/min to 10 km (5.5 n miles).
  4—12.7 mm MGs.
Torpedoes: 2 fixed launchers model KD 59E ❻. 10 ECAN L5 Mod
  4; anti-submarine; active/passive homing to 9.5 km (5.1 n
  miles) at 35 kt; warhead 150 kg; depth to 550 m (1,800 ft).
Countermeasures: Decoys: 2 CSEE AMBL 1B Dagaie ❼ and 2
  AMBL 2A Sagaie 10-barrelled trainable launchers ❽; fires a
  combination of chaff and IR flares. Dassault LAD offboard
  decoys. Nixie; towed torpedo decoy.
ESM: Thomson-CSF ARBR 17B ❾; radar warning. DIBV 1A
  Vampir ❿; IR detector (integrated with search radar for
  active/passive tracking in all weathers). Saigon radio
  intercept at masthead.
ECM: Thomson-CSF ARBB 33; jammer; H-, I- and J-bands.
Combat data systems: SENIT 68; Links 11, 14 and 16. Syracuse
  2 SATCOM ⓫. OPSMER command support system.

Weapons control: DCN CTMS optronic/radar system with DIBC
  1A Piranha II IR/TV tracker; CSEE Najir optronic secondary
  director.
Radars: Air search: Thomson-CSF DRBJ 11B ⓬; 3D; E/F-band;
  range 366 km (200 n miles).
Air/surface search: Thomson-CSF DRBV 26C ⓭; D-band.
Navigation: 2 Racal DRBN 34A; I-band (1 for close-range
  helicopter control ⓮).
Fire control: Thomson-CSF DRBC 33A ⓯; I-band (for guns).
  2 Raytheon SPG-51C ⓰; G/I-band (for missiles).
Sonars: Thomson Sintra DUBA 25A (D 614); DUBA 25C (D 615);
  hull-mounted; active search and attack; medium frequency.

Helicopters: 1 AS 565SA Panther ⓱.

Programmes: The building programme was considerably slowed
  down by finance problems and doubts about the increasingly
  obsolescent Standard SM 1 missile system. Re-rated F 70
  (ex-C 70) on 6 June 1988, officially 'frégates anti-aériennes
  (FAA)'.
Modernisation: DRBJ 15 radar initially fitted in Cassard but this
  was replaced in 1992 by DRBJ 11. Panther has replaced Lynx
  helicopter. Cassard refitted 2000-2001. Upgrade included
  hull strengthening, fitting of new propellers and SENIT 68
  combat direction system (SENIT 6 core augmented by SENIT
  8 data-link processing component (for Link 16 and data
  forwarding). Jean Bart similarly refitted October 2002 to
  September 2003. Plans to fit ASTER 30 have been
  abandoned. Cassard to be refitted 2005.
Structure: Samahe 210 helicopter handling system.
Operational: Helicopter used for third party targeting for the
  SSM. Both ships are based at Toulon. Service lives: Cassard,
  2013; Jean Bart, 2015. To be replaced by second batch of
  Horizon class from 2012.

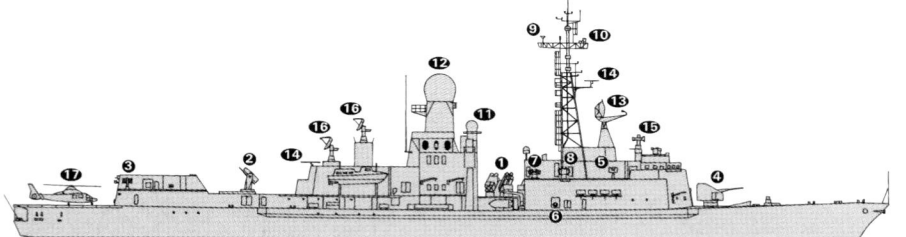

CASSARD

(Scale 1 : 1,200), Ian Sturton / 0569909

UPDATED

CASSARD

*6/2004 *, Derek Fox / 1042178*

JEAN BART

*8/2004 *, B Prézelin / 1042196*

0 + 2 (2) FORBIN (HORIZON) CLASS (DDGHM)

Name	No	Builders	Laid down	Launched	Commissioned
FORBIN	D 620	DCN, Lorient	Jan 2004	10 Mar 2005	Dec 2006
CHEVALIER PAUL	D 621	DCN, Lorient	2004	July 2006	Mar 2008

Displacement, tons: 6,700 full load
Dimensions, feet (metres): 502.0; 465.0 × 66.6 × 15.7
(153.0; 141.7 × 20.3 × 4.8)
Main machinery: CODOG: 2 Fiat/GE LM 2500 gas turbines;
58,500 hp *(43 MW)*; 2 SEMT-Pielstick 12PA 6STC; 11,700 hp
(m) *(8.6 MW)*; 2 shafts; cp props; bow thruster
Speed, knots: 29 (18 on diesels).
Range, n miles: 7,000 at 18 kt
Complement: 190 (accommodation for 230 with up to 20 per
cent female crew

Missiles: SSM: 8 Aerospatiale Matra Exocet MM 40 Block III
❶; inertial cruise; active radar homing to 70 km *(40 n miles)* at
0.9 Mach; warhead 165 kg; sea-skimmer.
SAM: EUROPAAMS PAAMS with DCN Sylver A50 VLS ❷ for
Aerospatiale Matra Aster 15 and Aster 30; 48 cells (six
octuple launcher modules); range (Aster 30): 120 km *(65 n
miles)* against large aircraft.
2 Aerospatiale Matra Sadral PDMS sextuple launchers ❸ for
Mistral SR SAMs; IR homing to 6 km; warhead 3 kg; anti-sea-
skimmer; able to engage targets down to 10 ft above sea
level.
Guns: 2 Otobreda 76 mm/62 Super Rapid ❹. 2 Giat 20 mm ❺.
Torpedoes: 2 EUROTORP TLS fixed launchers ❻. Up to 24
Eurotorp Mu 90 Impact torpedoes; active/passive homing to
15 km *(8 n miles)* at 29/50 kt.
Countermeasures: SIGEN EW suite comprising 2 EADS NGDS
multifunction decoy launchers ❼, radar warning equipment, a
high-power jammer ❽ and an ESM/ECM support aid. SLAT
torpedo defence system.
Combat data systems: EUROSYSNAV; 2 Link 11 (Link 22 in the
future) and Link 16; OPSMER or SIC 21 follow-on command
support system; Syracuse SATCOM ❾.

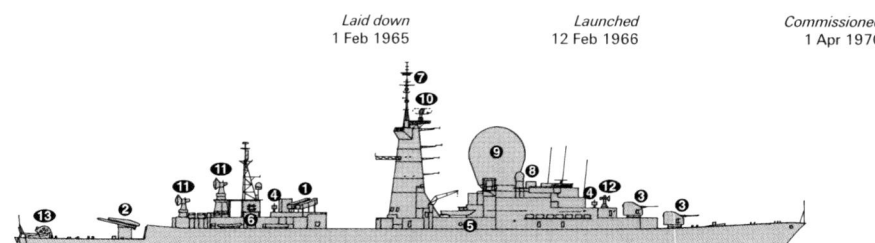

FORBIN   *(Scale 1 : 1,200), Ian Sturton /* 1042411

Weapons control: Sagem Vampir optronic director ❿.
Radars: Air/surface search: Thomson-CSF/Marconi DRBV 27
(S 1850M) Astral ⓫; L-band.
Surveillance/fire control: Alenia Marconi EMPAR ⓬; C-band;
multifunction.
Surface search: 2 SPN 753 ⓭; I-band.
Fire control: Alenia Marconi NA 25 ⓮; J-band.
Sonars: TUS-WASS 4110CL; hull-mounted; active search and
attack; medium frequency.

Helicopters: 1 Eurocopter NH90 ⓯.

Programmes: Classified as 'Frégates antiaériennes' (FAA).
Initially a three-nation project with Italy and UK. Joint project
office established in 1993. After UK withdrew in April 1999,

an agreement was signed on 7 September 1999 between
France and Italy to continue. Following a French/Italian MoU
on 22 September 2000 to build four destroyers, the French
government ordered two ships to be built by DCN Lorient and
delivered in December 2006 and April 2008. They are
planned to replace *Suffren* and *Duquesne*. First steel cut for
Forbin on 8 April 2002 and for *Chevalier Paul* on 1 December
2003. A second pair is planned to replace *Cassard* and *Jean
Bart* and the third of class is expected to be ordered by 2008
to meet an in-service date of 2012.
Structure: Details given are subject to change. Space available
for two additional missile launcher modules, possibly with
Sylver A70 VLS.

UPDATED

1 SUFFREN CLASS (DDGM)

Name	No	Builders	Laid down	Launched	Commissioned
DUQUESNE	D 603	Brest Naval Dockyard	1 Feb 1965	12 Feb 1966	1 Apr 1970

Displacement, tons: 5,335 standard; 6,780 full load
Dimensions, feet (metres): 517.1 × 50.9 × 23.8
(157.6 × 15.5 × 7.25)
Main machinery: 4 boilers; 640 psi *(45 kg/cm²)*; 842°F *(450°C)*;
2 Rateau turbines; 72,500 hp(m) *(53 MW)*; 2 shafts
Speed, knots: 34. **Range, n miles:** 5,100 at 18 kt; 2,400 at 29 kt
Complement: 355 (23 officers)

Missiles: SSM: 4 Aerospatiale MM 38 Exocet ❶; inertial cruise;
active radar homing to 42 km *(23 n miles)* at 0.9 Mach;
warhead 165 kg; sea-skimmer.
SAM: ECAN Ruelle Masurca twin launcher ❷; Mk 2 Mod 3 semi-
active radar homers; range 55 km *(30 n miles)*; warhead
98 kg; 48 missiles.
Guns: 2 DCN/Creusot-Loire 3.9 in *(100 mm)*/55 Mod 1964
CADAM automatic ❸; 80 rds/min to 17 km *(9 n miles)* anti-
surface; 8 km *(4.4 n miles)* anti-aircraft; weight of shell
13.5 kg.
4 or 6 Oerlikon 20 mm ❹; 720 rds/min to 10 km *(5.5 n miles)*.
2—12.7 mm MGs.
Torpedoes: 4 launchers (2 each side) ❺. 10 ECAN L5; anti-
submarine; active/passive homing to 9.5 km *(5.1 n miles)* at
35 kt; warhead 150 kg; depth to 550 m *(1,800 ft)*.
Countermeasures: Decoys: 2 CSEE Sagaie 10-barrelled
trainable launchers; chaff to 8 km *(4.4 n miles)* and IR flares to
3 km *(1.6 n miles)*. 2 Dagaie launchers ❻. Dassault LAD
offboard decoys.
ESM: ARBR 17 ❼; intercept.
ECM: ARBB 33; jammer.
Combat data systems: SENIT 2 action data automation; Links 11
and 14. Syracuse 2 SATCOM ❽. OPSMER command support
system. Marisat.

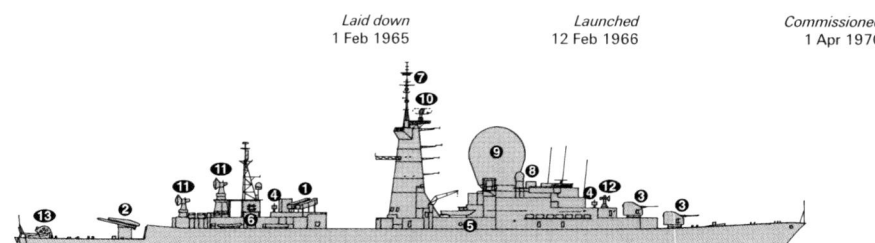

DUQUESNE   *(Scale 1 : 1,500), Ian Sturton /* 0069916

Weapons control: DCN CTMS radar/optronic control system
with SAT DIBC 1A Piranha IR and TV tracker. 2 Sagem DMA
optical directors.
Radars: Air search (radome): DRBI 23 ❾; D-band.
Air/surface search: DRBV 15A ❿; E/F-band.
Navigation: Racal Decca 1229 (DRBN 34A); I-band.
Fire control: 2 Thomson-CSF DRBR 51 ⓫; G/I-band (for
Masurca).
Thomson-CSF DRBC 33A ⓬; I-band (for guns).
Tacan: URN 20.
Sonars: Thomson Sintra DUBV 23; hull-mounted; active search
and attack; 5 kHz.
DUBV 43 ⓭; VDS; medium frequency 5 kHz; tows at up to
24 kt at 200 m *(656 ft)*.

Programmes: Ordered under the 1960 programme. Service life
2007.
Modernisation: MM 38 Exocet fitted in 1977; Masurca
modernised in 1984-85 with new computers. DRBV 15A

radars replaced DRBV 50. Major refit from June 1990 to
March 1991: modernisation of the DRBI-23 radar; new
computers for the SENIT combat data system; new CTMS fire-
control system for 100 mm guns fitted (with DRBC-33A radar,
TV camera and DIBC-1A Piranha IR tracker). New ESM/ECM
suite: ARBR 17 radar interceptor, ARBB 33 jammer and
Sagaie decoy launchers. Two 20 mm guns fitted either side of
DRBC 33A.
Structure: Equipped with gyro-controlled stabilisers operating
three pairs of non-retractable fins. NBC citadel fitted during
modernisation. Air conditioning of accommodation and
operational areas.
Operational: Based at Toulon. Officially frégate lance-missiles
(FLM). In refit from September 1998 to July 1999. Malafon
removed having been non-operational since 1997. To be
replaced by *Chevalier Paul* in 2008. *Suffren* decommissioned
on 2 April 2001.

UPDATED

DUQUESNE   *8/2004*, B Prézelin /* 1042186

For details of the latest updates to *Jane's Fighting Ships* online and to discover the
additional information available exclusively to online subscribers please visit
jfs.janes.com

2 TOURVILLE CLASS (TYPE F 67) (DDGHM)

Name	No	Builders	Laid down	Launched	Commissioned
TOURVILLE	D 610	Lorient Naval Dockyard	16 Mar 1970	13 May 1972	21 June 1974
DE GRASSE	D 612	Lorient Naval Dockyard	14 June 1972	30 Nov 1974	1 Oct 1977

Displacement, tons: 4,650 standard; 6,100 full load
Dimensions, feet (metres): 501.6 × 51.8 × 21.6
(152.8 × 15.8 × 6.6)
Main machinery: 4 boilers; 640 psi *(45 kg/cm²)*; 840°F *(450°C)*;
2 Rateau turbines; 58,000 hp(m) *(43 MW)*; 2 shafts
Speed, knots: 32. **Range, n miles:** 5,000 at 18 kt
Complement: 301 (26 officers)

Missiles: SSM: 6 Aerospatiale MM 38 Exocet **❶**; inertial cruise;
active radar homing to 42 km *(23 n miles)* at 0.9 Mach;
warhead 165 kg; sea-skimmer.
SAM: Thomson-CSF Crotale Naval EDIR octuple launcher **❷**;
command line of sight guidance; radar/IR homing to 13 km
(7 n miles) at 2.4 Mach; warhead 14 kg.
Guns: 2 DCN/Creusot-Loire 3.9 in *(100 mm)*/55 Mod 68
CADAM automatic **❸**; dual purpose; 80 rds/min to 17 km *(9 n
miles)* anti-surface; 8 km *(4.4 n miles)* anti-aircraft; weight of
shell 13.5 kg.
2 Giat 20 mm **❹**.
4—12.7 mm MGs.
Torpedoes: 2 launchers **❺**. 10 ECAN L5; anti-submarine; active/
passive homing to 9.5 km *(5.1 n miles)* at 35 kt; warhead
150 kg; depth to 550 m *(1,800 ft)*. Honeywell Mk 46 or
Eurotorp Mu 90 Impact torpedoes for helicopters.
Countermeasures: Decoys: 2 CSEE/VSEL Syllex 8-barrelled
trainable launcher (to be replaced by 2 Dagaie systems) **❻**;
chaff to 1 km in centroid and distraction patterns.
ESM: ARBR 16; radar warning.
ECM: ARBB 32; jammer.
Combat data systems: SENIT 3 action data automation; Links 11
and 14. Syracuse 2 SATCOM **❼**. OPSMER command support
system. Inmarsat.
Weapons control: SENIT 3 radar/TV tracker (possibly SAT
Murène in due course). 2 Sagem DMAa optical directors.
OPS-100F acoustic processor.
Radars: Air search: DRBV 26A **❽**; D-band; range 182 km *(100 n
miles)* for 2 m² target.
Air/surface search: Thomson-CSF DRBV 51D **❾**; G-band; range
29 km *(16 n miles)*.
Navigation: 2 DRBN 34 **❿** (Racal Decca Type 1226); I-band (1
for helicopter control).
Fire control: Thomson-CSF DRBC 32D **⓫**; I-band.
Crotale **⓬**; J-band (for SAM).
Sonars: Thomson Sintra DUBV 23; bow-mounted; active search
and attack; medium frequency.
Thomson Sintra DSBX 1A (ATBF) VDS **⓭**; active 1 kHz
transmitter and 5 kHz transceiver in same 10 tonne towed
body.
Thomson Sintra DSBV 62C; passive linear towed array; very
low frequency.

Helicopters: 2 Lynx Mk 4 **⓮**.

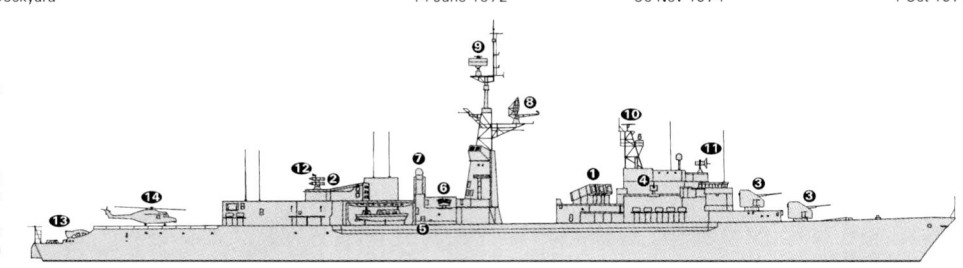

TOURVILLE *(Scale 1 : 1,200), Ian Sturton* / 0569912

TOURVILLE *6/2004*, B Prézelin* / 1042185

Programmes: Originally rated as corvettes but reclassified as
'frégates anti-sous-marins (FASM)' on 8 July 1971 and given D
pennant numbers.
Modernisation: Major communications and combat data
systems updates. The SLASM ASW combat suite installed in
Tourville from March 1994 to April 1995, *De Grasse* from May
1995 to September 1996. This included new signal
processing for the bow sonar, plus LF and MF towed active
sonar with separate towed passive array including torpedo

warning. Acoustic processor for helo borne sonobuoys. Milas
ASW missile cancelled. Passive towed arrays fitted in 1990.
Malafon removed from *Tourville* in 1994 and *De Grasse* in
1996. *De Grasse* refitted at Brest in 2003.
Operational: Assigned to ALFAN BREST. Helicopters are now
used primarily in the ASW role with sonar or sonobuoy
dispenser, and ASW weapons. Service lives: *Tourville* 2009;
De Grasse 2010.

UPDATED

TOURVILLE *7/2004*, H M Steele* / 1042259

DE GRASSE *4/2004*, B Prézelin* / 1042188

FRIGATES

5 LA FAYETTE CLASS (FFGHM)

Name	No	Builders	Laid down	Launched	Commissioned
LA FAYETTE	F 710	DCN, Lorient	15 Dec 1990	13 June 1992	23 Mar 1996
SURCOUF	F 711	DCN, Lorient	3 July 1992	3 July 1993	7 Feb 1997
COURBET	F 712	DCN, Lorient	15 Sep 1993	12 Mar 1994	1 Apr 1997
ACONIT (ex-*Jauréguiberry*)	F 713	DCN, Lorient	1 Aug 1996	8 June 1997	3 June 1999
GUÉPRATTE	F 714	DCN, Lorient	1 Oct 1998	3 Mar 1999	27 Oct 2001

Displacement, tons: 3,750 full load

Dimensions, feet (metres): 407.5 oa; 377.3 pp × 50.5 × 19.0 (screws) *(124.2; 115 × 15.4 × 5.8)*

Main machinery: CODAD; 4 SEMT-Pielstick 12 PA6 V 280 STC diesels; 21,107 hp(m) *(15.52 MW)* sustained; 2 shafts; LIPS cp props; bow thruster

Speed, knots: 25. **Range, n miles:** 7,000 at 15 kt; 9,000 at 12 kt

Complement: 140 (13 officers) plus 12 aircrew plus 25 Marines

Missiles: SSM: 8 Aerospatiale MM 40 Block 2 Exocet **❶**; inertial cruise; active radar homing to 70 km *(40 n miles)* at 0.9 Mach; warhead 165 kg; sea-skimmer.
SAM: Thomson-CSF Crotale Naval CN 2 (EDIR with V3 in *La Fayette* to be retrofitted with CN 2 in due course) octuple launcher (eighteen missiles in magazine) **❷**; VT 1; command line of sight guidance; radar/IR homing to 13 km *(7 n miles)* at 3.5 Mach; warhead 14 kg. 24 missiles. Space for 2 × 8 cell VLS **❸**.

Guns: 1 DCN 3.9 in *(100 mm)*/55 TR **❹**; 80 rds/min to 17 km *(9 n miles)*; weight of shell 13.5 kg.
2 Giat 20F2 20 mm **❺**; 720 rds/min to 10 km *(5.5 n miles)*.

Countermeasures: Decoys: 2 CSEE Dagaie Mk 2 **❻** 10-barrelled trainable launchers; chaff and IR flares.
ESM: Thomson-CSF ARBR 21 (DR 3000-S) **❼**; radar intercept. ARBG 2 Maigret; comms intercept.
DIBV 10 Vampir **❽**; IR detector (can be fitted).
ECM: Dassault ARBB 33; jammer (can be fitted).

Combat data systems: Thomson-CSF TAVITAC 2000. Links 11 and 14. Syracuse 2 SATCOM **❾**. OPSMER command support system. INMARSAT.

Weapons control: Thomson-CSF CTM radar/IR system. Sagem TDS 90 VIGY optronic system.

Radars: Air/surface search: Thomson-CSF Sea Tiger Mk 2 (DRBV 15C) **❿**; E/F-band; range 110 km *(60 n miles)* for 2 m² target.
Navigation: 2 Racal Decca 1229 (DRBN 34A) **⓫**; I-band. One set for helicopter control.
Fire control: Thomson-CSF Castor 2J **⓬**; J-band; range 17 km *(9.2 n miles)* for 1 m² target.
Crotale **⓭**; J-band (for SAM).

Helicopters: 1 Aerospatiale AS 565MA Panther **⓯** or platform for 1 Super Frelon. NH90 in due course.

Programmes: Originally described as 'Frégates Légères' but this was changed in 1992 to 'Frégates type La Fayette'. First three ordered 25 July 1988; three more 24 September 1992 but the last of these was cancelled in May 1996. The construction timetable was delayed by several months because of funding problems.

Structure: Constructed from high-tensile steel with a double skin from waterline to upperdeck. 10 mm plating protects vital spaces. External equipment and upper deck fittings are concealed or placed in low positions. Superstructure inclined at 10° to vertical to reduce REA. Extensive use of radar absorbent paint. DCN Samahe helicopter handling system. 'Raidco 700' fast assault craft fitted to F 712 and others in due course. The design includes potential to install new and/or replace old weapon systems in the future. Two octuple VLS launchers for ASTER 15 missiles could be installed forward of the bridge while there is space for the associated Arabel radar and also for SLAT anti-torpedo system. Plans to fit a sonar have been dropped and, in view of the FMM frigate programme, it is unlikely that an ASW variant of the ship will be developed.

Operational: *La Fayette* started sea trials 27 September 1993, *Surcouf* 4 July 1994, *Courbet* 14 September 1995, *Aconit* 14 April 1998 and *Guépratte* on 16 January 2001. These frigates are designed for out of area operations on overseas stations and the first three are assigned to FAN for the Indian Ocean. Super Frelon helicopters can land on the flight deck. NH 90 prototype trials in *Courbet* in 1998. The ship can launch inflatable boats from a hatch in the stern which hinges upwards. The Vampir IR detector and ARBB 33 jammer are fitted 'for but not with'. *Courbet* to be refitted 2005.

Sales: Three of an improved design to Saudi Arabia, six for Taiwan, and six for Singapore.

UPDATED

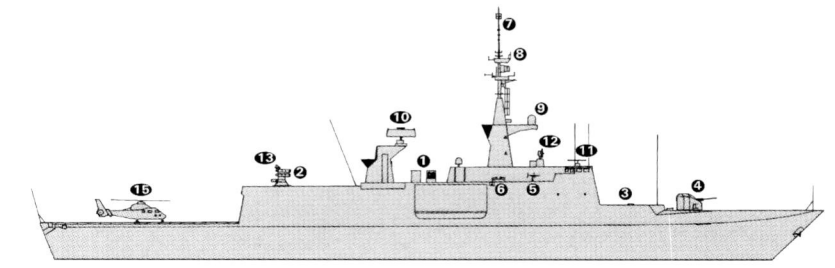

LA FAYETTE *(Scale 1 : 1,200), Ian Sturton /* 0569910

LA FAYETTE *1/2004*, B Prézelin /* 1042189

GUÉPRATTE *1/2004*, B Prézelin /* 1042191

COURBET *8/2004*, B Prézelin /* 1042190

9 D'ESTIENNE D'ORVES (TYPE A 69) CLASS (FFGM)

Name	No	Builders	Laid down	Launched	Commissioned
LIEUTENANT DE VAISSEAU LE HÉNAFF	F 789	Lorient Naval Dockyard	21 Mar 1977	16 Sep 1978	13 Feb 1980
LIEUTENANT DE VAISSEAU LAVALLÉE	F 790	Lorient Naval Dockyard	30 Nov 1977	11 May 1979	8 Oct 1980
COMMANDANT L'HERMINIER	F 791	Lorient Naval Dockyard	29 May 1979	7 Mar 1981	19 Jan 1986
PREMIER MAÎTRE L'HER	F 792	Lorient Naval Dockyard	15 Dec 1978	28 June 1980	5 Dec 1981
COMMANDANT BLAISON	F 793	Lorient Naval Dockyard	15 Nov 1979	7 Mar 1981	26 Apr 1982
ENSEIGNE DE VAISSEAU JACOUBET	F 794	Lorient Naval Dockyard	8 July 1980	28 Sep 1981	23 Oct 1982
COMMANDANT DUCUING	F 795	Lorient Naval Dockyard	1 Oct 1980	26 Sep 1981	17 Mar 1983
COMMANDANT BIROT	F 796	Lorient Naval Dockyard	23 Mar 1981	22 May 1982	14 Mar 1984
COMMANDANT BOUAN	F 797	Lorient Naval Dockyard	12 Oct 1981	23 Apr 1983	1 Nov 1984

Displacement, tons: 1,175 standard; 1,250 (F 789-791), 1,290 (F 792-793), 1,330 (F 794-797) full load

Dimensions, feet (metres): 264.1 × 33.8 × 18 (sonar) *(80.5 × 10.3 × 5.5)*

Main machinery: 2 SEMT-Pielstick 12 PC2 V 400 diesels; 12,000 hp(m) *(8.82 MW)*; 2 shafts; LIPS cp props 2 SEMT-Pielstick 12 PA6 V 280 BTC diesels; 14,400 hp(m) *(10.6 MW)* sustained; 2 shafts; LIPS cp props *(Commandant L'Herminier)*

Speed, knots: 24. **Range, n miles:** 4,500 at 15 kt

Complement: 90 (7 officers) plus 18 marines (in some)

Missiles: SSM: 4 Aerospatiale MM 40 (MM 38 in F 789-791) Exocet ❶; inertial cruise; active radar homing to 70 km *(40 n miles)* (or 42 km *(23 n miles)*) at 0.9 Mach (MM 40); warhead 165 kg; sea-skimmer; active radar homing to 42 km *(23 n miles)* at 0.9 Mach (MM 38).

SAM: Matra Simbad twin launcher for Mistral ❷; IR homing to 4 km *(2.2 n miles)*; warhead 3 kg.

Guns: 1 DCN/Creusot-Loire 3.9 in *(100 mm)*/55 Mod 68 CADAM automatic ❸; 80 rds/min to 17 km *(9 n miles)* anti-surface; 8 km *(4.4 n miles)* anti-aircraft; weight of shell 13.5 kg.

2 Giat 20 mm ❹; 720 rds/min to 10 km *(5.5 n miles)*. 4—12.7 mm MGs.

Torpedoes: 4 fixed tubes ❺. ECAN L5; dual purpose; active/passive homing to 9.5 km *(5.1 n miles)* at 35 kt; warhead 150 kg; depth to 550 m *(1,800 ft)*.

A/S mortars: 1 Creusot-Loire 375 mm Mk 54 6-tubed trainable launcher (F 789, F 790, F 791); range 1,600 m; warhead 107 kg. Removed from others.

Countermeasures: Decoys: 2 CSEE Dagaie 10-barrelled trainable launchers ❻; chaff and IR flares; H- to J-band. Nixie torpedo decoy.

ESM: ARBR 16; radar warning.

Combat data systems: Syracuse 2 SATCOM (F 792, F 793, F 794, F 795, F 796, F 797) ❼. OPSMER command support system with Link 11 (receive only) in MM 40 ships. INMARSAT (F 789, F 790, F 791).

Weapons control: Thomson-CSF Vega system; CSEE Panda optical secondary director.

Radars: Air/surface search: Thomson-CSF DRBV 51A ❽; G-band. Navigation: Racal Decca 1290 (DRBN 34); I-band. Fire control: Thomson-CSF DRBC 32E ❾; I-band.

Sonars: Thomson Sintra DUBA 25; hull-mounted; search and attack; medium frequency.

Programmes: Classified as 'Avisos'.

Modernisation: In 1985 *Commandant L'Herminier*, F 791, fitted with 12PA6 BTC Diesels Rapides as trial for Type F 70. Most have dual MM 38/MM 40 ITL (Installation de Tir Légère) capability. Weapon fit depends on deployment and operational requirement. Those without ITL are fitted with ITS (Installation de Tir Standard). Syracuse 2 SATCOM fitted in F 792-797, vice the A/S mortar, and accommodation provided for commandos. Matra Simbad launchers have been fitted aft of the A/S mortar/Syracuse SATCOM for operations. Fast raiding craft fitted to *Commandant Birot* and to others in due course.

Operational: Endurance, 30 days and primarily intended for coastal A/S operations. Also available for overseas patrols. All assigned to FAN with six ships based at Brest and remaining three at Toulon. F 789 to decommission in 2012 and F 790 in 2013 with remainder to follow by 2018.

Sales: The original *Lieutenant de Vaisseau Le Hénaff* and *Commandant l'Herminier* sold to South Africa in 1976 while under construction. As a result of the UN embargo on arms sales to South Africa, they were sold to Argentina in September 1978 followed by a third, specially built. Six ships were sold to Turkey in October 2000. All delivered by July 2002 after refit at Brest. The last one, *Second Maître Le Bihan*, decommissioned from the French Navy on 26 June 2002. No further sales are planned.

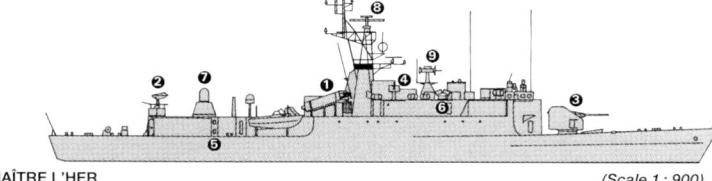

PREMIER MAÎTRE L'HER

(Scale 1 : 900), Ian Sturton / 0535887

PREMIER MAÎTRE L'HER

*10/2004 *, B Sullivan* / 1042258

UPDATED LIEUTENANT DE VAISSEAU LAVALLÉE

*3/2004 *, Derek Fox* / 1042187

COMMANDANT DUCUING

*8/2004 *, B Prézelin* / 1042184

6 FLORÉAL CLASS (FFGHM)

Name	No	Builders	Laid down	Launched	Commissioned
FLORÉAL	F 730	Chantiers de l'Atlantique, St Nazaire	2 Apr 1990	6 Oct 1990	27 May 1992
PRAIRIAL	F 731	Chantiers de l'Atlantique, St Nazaire	11 Sep 1990	23 Mar 1991	20 May 1992
NIVÔSE	F 732	Chantiers de l'Atlantique, St Nazaire	16 Jan 1991	10 Aug 1991	16 Oct 1992
VENTÔSE	F 733	Chantiers de l'Atlantique, St Nazaire	28 June 1991	14 Mar 1992	5 May 1993
VENDÉMIAIRE	F 734	Chantiers de l'Atlantique, St Nazaire	17 Jan 1992	29 Aug 1992	20 Oct 1993
GERMINAL	F 735	Chantiers de l'Atlantique, St Nazaire	17 Aug 1992	13 Mar 1993	17 May 1994

Displacement, tons: 2,600 standard; 2,950 full load
Dimensions, feet (metres): 306.8 × 45.9 × 14.1
 (93.5 × 14 × 4.3)
Main machinery: CODAD; 4 SEMT-Pielstick 6 PA6 L 280 diesels;
 8,820 hp(m) *(6.5 MW)* sustained; 2 shafts; LIPS cp props;
 bow thruster; 340 hp(m) *(250 kW)*
Speed, knots: 20. **Range, n miles:** 10,000 at 15 kt
Complement: 86 (10 officers) (including aircrew) plus 24
 Marines + 13 spare

Missiles: SSM: 2 Aerospatiale MM 38 Exocet ❶; inertial cruise;
 active radar homing to 42 km *(23 n miles)* at 0.9 Mach;
 warhead 165 kg; sea-skimmer.
SAM: 1 or 2 Matra Simbad twin launchers can replace 20 mm
 guns or Dagaie launcher.
Guns: 1 DCN 3.9 in *(100 mm)*/55 Mod 68 CADAM ❷; 80 rds/
 min to 17 km *(9 n miles)*; weight of shell 13.5 kg.
 2 Giat 20 F2 20 mm ❸; 720 rds/min to 10 km *(5.5 n miles)*.
Countermeasures: Decoys: 1 or 2 CSEE Dagaie Mk II; 10-
 barrelled trainable launchers ❹; chaff and IR flares.
ESM: Thomson-CSF ARBR 17 ❺; radar intercept.
 ARBG 1 Saigon; comms intercept (F 730 and F 733).
Weapons control: CSEE Najir optronic director ❻. Syracuse 2
 SATCOM (F 730 and F 733) INMARSAT ❼ (F 731, F 732,
 F 734, F 735).
Radars: Air/surface search: Thomson-CSF Mars DRBV 21A ❽;
 D-band.
Navigation: 2 Racal Decca 1229 (DRBN 34A); I-band (1 for
 helicopter control ❾).

Helicopters: 1 AS 565MA Panther or platform for 1 AS 332F
 Super Puma ❿.

Programmes: Officially described as 'Frégates de Surveillance'
 or 'Ocean capable patrol vessel' and designed to operate in
 the offshore zone in low-intensity operations. First two
 ordered on 20 January 1989; built at Chantiers de
 l'Atlantique, St Nazaire, with weapon systems fitted by DCAN
 Lorient. Second pair ordered 9 January 1990; third pair in
 January 1991. Named after the months of the Revolutionary
 calendar.
Structure: Built to merchant passenger marine standards with
 stabilisers and air conditioning. New funnel design improves
 air flow over the flight deck. Has one freight bunker aft for
 about 100 tons cargo. Second-hand Exocet MM 38 has been
 fitted instead of planned MM 40.
Operational: Endurance, 50 days. Range proved to be better
 than expected during sea trials. Able to operate a helicopter
 up to Sea State 5. Stations as follows: *Ventose* in Antilles,
 Germinal at Brest, *Prairial* in Tahiti. *Floréal* and *Nivôse* at Le
 Réunion and *Vendémiaire* at Noumea (New Caledonia).
 Floréal refitted in floating dry-dock at Papeete in 2003.
Sales: Two delivered to Morocco in 2002. **UPDATED**

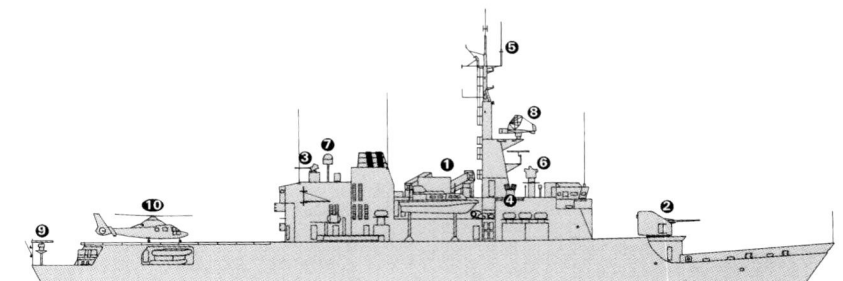

PRAIRIAL *(Scale 1 : 900), Ian Sturton* / 0529161

GERMINAL *8/2004*, B Prézelin* / 1042241

GERMINAL *9/2001, Derek Fox* / 0130505

0 + 8 (9) MULTIMISSION FRIGATES (FFGHM)

Displacement, tons: 5,700 full load (approx)
Dimensions, feet (metres): 459.3 × 62.3 × 16.4
(140.0 × 19.0 × 5.0)
Main machinery: CODLOG/CODLAG; 1 gas turbine; 2 motors
Speed, knots: 27. **Range, n miles:** 6,000 at 18 kt
Complement: 108

Missiles: SLCM: 16 cell Sylver A70 VLS for MBDA Scalp-Naval.
SAM: 16 cell Sylver A43 VLS for Aster 15.
SSM: 8 Exocet MM 40 Block 3.
Guns: 1 OTO 127 mm/64 ER (F-AVT). 1—OTO 76 mm SR (F-ASM). 2—20 mm.
Torpedoes: 2 Eurotorp TLS launchers; Eurotorp MU 90 torpedoes.
Countermeasures: Decoys: 2 EADS NGDS 12-barrelled chaff, IR and anti-torpedo decoy launchers.
ESM/ECM. SLAT torpedo defence system.
Combat data systems: Horizon class derivative. Links 11 and 16.
Weapons control: 1 optronic FCS.
Radars: Air/surface search: Thales Herakles 3-D multifunction; E/F-band.
Surface search: To be decided.
Navigation: To be decided.
Fire control: Alenia Marconi NA 25; J-band.

Sonars: Thomson Marconi 4110CL; hull mounted (bow dome); active search and attack. Thales Captas active/passive towed array (F-ASM).

Helicopters: 1 NH-90. ASW aircraft in ASW variant. Transport aircraft and tactical UAV in land attack variant.

Programmes: Agreement reached on 7 November 2002 for a 27 ship collaborative programme with Italy. The French requirement is for 17 'Frégates Européenne multimissions'

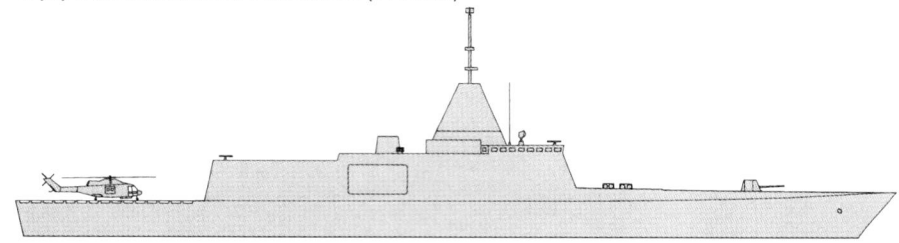

FREMM *(Scale 1 : 1,200), Ian Sturton /* 1042092

(FREMM) with common hull and machinery in two variants. Eight ASW ships (F-ASM) are to replace the Tourville and Georges Leygues classes while nine general purpose ships (FMM/F-AVT), with emphasis on land-attack capabilities, are to replace the A 69 Avisos and supplement the La Fayette class frigates. Eight ships are planned in the 2003-08 Defence Programming Law. This first batch is to comprise six F-ASM and two F-AVT. The order for construction is expected in 2005 for delivery of the first of class in 2009. ***UPDATED***

SHIPBORNE AIRCRAFT

Notes: (1) Three (out of a total of 13 for all three services) Eurocopter 725 helicopters for maritime counter-terrorism were ordered for delivery in 2004.
(2) The French Navy and the DGA (Armaments Directorate) have launched two drone demonstration programmes: a rotary wing naval tactical drone, DMT (Drone Maritime Tactique), to be operated from frigates (notably the F-AVT land attack variant of the multimission frigate) up to Sea State 6; a first demonstrator, Hetel, made its maiden flight in December 2002 and is planned for sea trials in 2005 (to enter service by 2009); the second programme concerns a fixed-wing long endurance drone, DELE (Drone Embarqué Longue Endurance), to be operated from aircraft carriers and from Mistral class LHDs; it should have an endurance of 12 hours and be able to operate up to 60 n miles away from carrying ship; sea trials expected in 2006.

Numbers/Type: 10 Dassault Aviation ACM Rafale M.
Operational speed: Mach 2.
Service ceiling: 50,000 ft *(15,240 m).*
Range: 1,800 n miles *(3,335 km).*
Role/Weapon systems: Total procurement of up to 60 Rafale M single-seaters (air superiority and ground/surface attack). First of two Rafale M naval prototypes (single seaters) flown 12 December 1992. First deck trials in *Foch* in 1993. First production Rafale M flown 7 July 1999 and assigned to development trials. Second aircraft delivered to the Navy 19 July 2000 and eight more delivered by 2002 to form Flottille 12F. Further deliveries of six (2006), ten (2007) and one (2008). All aircraft at standard F1 for air defence; standard F2 (from 2005) for ground attack; standard F3 (from 2007) for nuclear strike, air to surface and recce capabilities. All aircraft to be brought to this standard later. Sensors: Thales/Dassault Electronique RBE 2 multirole radar; Thales/Dassault Electronique/Matra SPECTRA EW/IR countermeasure suite, MIDSCO MIDS-LVT voice/data (Link 16), Thales/Sagem OSF optronic surveillance and target acquisition device (Standard F2); Thales Reco NG optronic reconnaissance pod with data link (two-seater F3s). Weapons: Giat M 791 30 mm cannon; up to eight AAMs (air-defence role), including MBDA Magic 2 short range and MICA EM medium range AAMs (standard F1), also MICA IR (standard F2); MBDA Apache stand-off weapon dispenser; Scalp/EG stand-off precision guided ASM; Sagem AASM general-purpose precision bomb; MBDA ASMP-A nuclear ASM and anti-ship missile (standard F3). Up to 8 tons of military load, on 13 hardpoints. ***UPDATED***

RAFALE M *7/2003, B Prézelin /* 0569995

Numbers/Type: 51 Dassault-Bréguet Super Étendard.
Operational speed: Mach 1 (approximately).
Service ceiling: 45,000 ft *(13,700 m).*
Range: 920 n miles *(1,682 km).*
Role/Weapon systems: Carrierborne strike fighter with nuclear strike capabilities and limited air defence role; tactical recce role to be added. All aircraft still in inventory modernised 1994-1999 to Standard F3. Standard F4 for all the fleet from mid-2000 to early 2005; tactical recce role added; standard 5 under development to enter service in 2005; service life extended to 2011. Sensors: Dassault Electronique Anémone radar, DRAX (standard 3) or Thales-Detexis Sherloc-F ESM (standard 4), SAGEM UAT 90 computer, Thomson-CSF Barracuda jammer, Phimat chaff dispenser, Alkan IR decoy dispenser; Thales Optrosys photo/optronic chassis (with Omera 40 panoramic camera and SDS-250 digital camera) in a ventral bay (Standard F4); Thales Atlis 2 FLIR/laser pod designator (Standard F3 and F4) and from late 2002 Thales Damocles day/night FLIR/designator (Standard F4 and F5). Weapons: air defence and self protection: two Matra BAe Dynamic Magic 2 short range AAMs and two DEFA 30 mm cannon; nuclear strike: one Aerospatiale ASMP nuclear ASM; air-to-surface: one Aerospatiale AM 39 Exocet anti-ship missile; air-to-ground: bombs, and (standard 3 aircraft) laser guided bombs (Paveway) or one Aerospatiale AS 30L laser guided missile. 7 hardpoints (from standard 4). ***UPDATED***

SUPER ÉTENDARD *6/2004*, B Prézelin /* 1042240

Numbers/Type: 3 Grumman E-2C Hawkeye Group 2.
Operational speed: 320 kt *(593 km/h).*
Service ceiling: 37,000 ft *(11,278 m).*
Range: 1,540 n miles *(2,852 km).*
Role/Weapon systems: Used for AEW, and direction of AD and strike operations. First pair ordered in May 1995 and delivered in April and December 1998 respectively. Third delivered in December 2003. All aircraft to be upgraded within five years. Sensors: APS-145 radar, ESM, ALR-73 PDS, ALQ-108 airborne tactical data system with Links 11 and 16. Weapons: Unarmed. ***UPDATED***

HAWKEYE *6/2004*, B Prézelin /* 1042242

Numbers/Type: 6/3/15 Eurocopter (Aerospatiale) SA 365F Dauphin 2/SA 365N Dauphin 2/AS 565MA Panther.
Operational speed: 165 kt *(305 km/h).*
Service ceiling: 16,700 ft *(5,100 m).*
Range: 483 n miles *(895 km).*
Role/Weapon systems: New-built SA 365F Dauphin 2s acquired to replace Alouette IIIs for carrierborne SAR. They feature the same ORB-32 radars as Panthers. SA 365Ns are second-hand helicopters purchased for SAR, general surveillance and public service roles from various locations in metropolitan France. They do not have any radar. Fifteen AS 565 Panthers purchased in several batches to operate from Cassard class DDGs, La Fayette and Floréal class frigates. 16th aircraft acquired from the Armée de l'Air (French Air Force). All Panthers to be modernised to Standard 2 before 2015 with new avionics, comprehensive countermeasures suite (laser, radar and missile warning systems, decoy dispenser), FLIR and, eventually, lightweight anti-ship missiles. Service life to 2025 (AS 565MA). Sensors: (AS 565MA and SA 365F) Thales ORB-32 radar and (AS 565MA) Thales Chlio FLIR on some helicopters (all fitted for); Titus tactical situation management aid (with encrypted data link). Weapons: (AS 565MA) provision for internally mounted 7.62 mm MG. ***UPDATED***

DAUPHIN 2 *6/2001, M Declerck /* 0132013

PANTHER *5/2001 /* 0130503

Numbers/Type: 27 NH Industries NH 90 NFH.
Operational speed: 157 kt *(291 km/h).*
Service ceiling: 13,940 ft *(4,250 m).*
Range: 621 n miles *(1,150 km).*
Role/Weapon systems: Total of 27 NH-90 ordered 30 June 2000 for the French Navy in two variants: 13 NHS support helicopters with secondary ASuW role; 14 NHC combat helicopters for ASW and ASuW. One to be delivered in 2005 and a further six by 2008; all 14 NHC 2009-15; and six last NHS 2016-18. Sensors: both variants: Thales ENR surveillance radar; Sagem OLOSP tactical FLIR; MBDA Saphir decoy dispenser; Link 11; NHC: TUS FLASH dipping sonar, and UMS 2000-TSM 8203 sonobuoy processing system. Weapons: ASM (NHC and NHS); 2 MU 90 Impact torpedoes (NHC).   *UPDATED*

NH 90   *3/2004*, NHI /* 0062373

Numbers/Type: 9 Aerospatiale SA 321G Super Frelon.
Operational speed: 160 kt *(296 km/h).*
Service ceiling: 10,170 ft *(3,100 m).*
Range: 420 n miles *(778 km).* 594 n miles *(1,100 km)* with auxiliary tank.
Role/Weapon systems: Formerly ASW helicopter; used for assault and support tasks embarked on carriers and LSDs; radar updated; provision for 27 passengers. Service life extended to 2008 to allow replacement by NH 90. Sensors: Omera ORB search radar. Thales Chlio FLIR fitted to some. Weapons: Provision for 20 mm gun.   *UPDATED*

SUPER FRELON   *6/2004*, B Prézelin /* 1042243

Numbers/Type: 31 Westland Lynx Mk 4 (FN).
Operational speed: 125 kt *(232 km/h).*
Service ceiling: 12,500 ft *(3,810 m).*
Range: 320 n miles *(593 km).*
Role/Weapon systems: Sole French ASW helicopter, all now of the Mk 4 variant; embarked in destroyers and deployed on training tasks. Service life to 2015. Sensors: Omera 31 search radar, Alcatel (DUAV 4) dipping sonar, sonobuoys, Sextant Avionique MAD. Thales Chlio FLIR. Weapons: ASW; two Mk 46 Mod 1 (or Mu 90 Impact in due course) torpedoes, or depth charges. ASV: 1—7.62 mm MG.   *UPDATED*

LYNX   *4/2004*, B Prézelin /* 1042244

Numbers/Type: 30 Aerospatiale SA 319B Alouette III.
Operational speed: 113 kt *(210 km/h).*
Service ceiling: 10,500 ft *(3,200 m).*
Range: 290 n miles *(540 km).*
Role/Weapon systems: General purpose helicopter; replaced by Lynx for ASW; now used for trials, surveillance and training tasks. Sensors: Some radar. Weapons: Unarmed.   *UPDATED*

ALOUETTE III   *7/2004*, B Prézelin /* 1042246

LAND-BASED MARITIME AIRCRAFT (FRONT LINE)

Numbers/Type: 5 Dassault-Aviation Falcon 20H/Gardian.
Operational speed: 470 kt *(870 km/h).*
Service ceiling: 45,000 ft *(13,715 m).*
Range: 2,425 n miles *(4,490 km).*
Role/Weapon systems: Assigned to Flotilla 25F based at Tahiti with permanent detachments at Tontouta (New Caledonia) and Martinique. Maritime reconnaissance role. Sensors: Thomson-CSF Varan radar, Omega navigation, ECM/ESM pods. Weapons: Unarmed.   *VERIFIED*

Numbers/Type: 4 Dassault Falcon 50M.
Operational speed: 370 kt *(685 km/h).*
Service ceiling: 49,000 ft *(14,930 m).*
Range: 3,300 n miles *(6,100 km).*
Role/Weapon systems: Maritime reconnaissance and SAR roles in the Atlantic and overseas stations (replaced deleted Atlantic Mk 1). First aircraft delivered in December 1999 (for Opeval), second in March 2000, third in March 2001; fourth and last one late 2002. Allocated to Flotille 24F (Lann-Bihoué). Sensors: Thales/DASA Ocean Master 100(V) search radar, Thales Chlio FLIR, Inmarsat C. Weapons: Unarmed (two SAR chains). Endurance: six hours 30 minutes at 100 n miles *(185 km)* from base, four hours at 500 n miles *(926 km)* or one hour at 1,200 n miles *(2,222 km).*   *UPDATED*

FALCON 50M   *7/2004*, B Prézelin /* 1042245

Numbers/Type: Boeing E-3F Sentry AWACS.
Operational speed: 460 kt *(853 km/h).*
Service ceiling: 30,000 ft *(9,145 m).*
Range: 870 n miles *(1,610 km).*
Role/Weapon systems: Air defence early warning aircraft with secondary role to provide coastal AEW for the Fleet; 6 hours endurance at the range given above. Sensors: Westinghouse APY-2 surveillance radar, Bendix weather radar, Mk XII IFF, Yellow Gate, ESM, ECM. Weapons: Unarmed. Operated by the Air Force.   *VERIFIED*

E-3F   *6/2002, Armée de l'Air /* 0118289

Numbers/Type: 28 Dassault Aviation Atlantique Mk 2.
Operational speed: 355 kt *(658 km/h).*
Service ceiling: 32,800 ft *(10,000 m).*
Range: 8 hours patrol at 1,000 n miles from base; 4 hours patrol at 1,500 n miles from base.
Role/Weapon systems: Maritime reconnaissance. ASW, ASV, COMINT/ELINT roles. Last one delivered in January 1998. Six aircraft are in long-term storage. Sensors: Thomson-CSF Iguane radar, ARAR 13 ESM, ECM, FLIR, MAD, sonobuoys (with DSAX-1 Thomson-CSF Sadang processing equipment). Link 11 (being fitted in all). COMINT/ELINT equipment optional. Integrated sensor/weapon system built around a CIMSA 15/125X computer. Weapons: Two AM 39 Exocet ASMs in ventral bay, or up to eight lightweight torpedoes (Mk 46 and later Mu 90), or depth charges, mines or bombs. Limited modernisation programme planned to adapt aircraft to Mu 90 torpedoes. More extensive modernisation planned for 2008-2015.   *UPDATED*

ATLANTIQUE II   *8/2003, Frank Findler /* 0569941

Numbers/Type: 6 Dassault-Aviation Falcon 10MER.
Operational speed: 492 kt *(912 km/h).*
Service ceiling: 35,500 ft *(10,670 m).*
Range: 1,920 n miles *(3,560 km).*
Role/Weapon systems: Primary aircrew/ECM training role but also has overwater surveillance role. Sensors: Search radar. Weapons: Unarmed. Allocated to Flottille 57S (Landivisiau).   *UPDATED*

FALCON 10MER   *7/2003, Paul Jackson /* 0569996

Numbers/Type: 15 Aerospatiale N262E.
Operational speed: 226 kt *(420 km/h).*
Service ceiling: 26,900 ft *(8,200 m).*
Role/Weapon systems: Crew training and EEZ surveillance role. All allocated to Flotilla 28F for surveillance, SAR and Flying School. Modified N262A aircraft. Sensors: Omera ORB 32 radar; photo pod. Weapons: Unarmed. Target towing capability.   *VERIFIED*

AMPHIBIOUS FORCES

0 + 2 MISTRAL CLASS (AMPHIBIOUS ASSAULT SHIPS) (LHDM/BPC)

Name	No	Builders	Laid down	Launched	Commissioned
MISTRAL	L 9013	DCN Brest	10 July 2003	6 Oct 2004	Dec 2005
TONNERRE	L 9014	DCN Brest	2004	2005	Dec 2006

Displacement, tons: 16,500 standard; 21,500 full load; 32,300 flooded

Dimensions, feet (metres): 653 × 105 × 20.3 (199 × 32 × 6.2)

Flight deck, feet (metres): 653 × 105 (199 × 32)

Main machinery: Electric propulsion: 4 diesel generators provide total of 20.8 MW for propulsion and services. Total of 20,400 hp(m) *(15 MW)*; 2 Alstom Mermaid podded propulsors trainable through 360°; bow thruster

Speed, knots: 19. **Range, n miles:** 11,000 at 15 kt

Complement: 160 (up to 900 in austerity conditions)

Military lift: 450 troops and 60 armoured vehicles/8 helicopters or 16 helicopters or 230 vehicles. 4 CTM (LCU) or 2 LCACs.

Missiles: SAM: Simbad twin PDMS launchers for Matra BAE Dynamics Mistral; IR homing to 6 km (3.2 n miles); warhead 3 kg; anti-sea-skimmer.

Guns: 2—30 mm. 4—12.7 mm MGs.

Countermeasures: ESM: ARBR 21; intercept.

Combat data systems: SENIT combat data system (SENIT 8 derivative) SIC 21 command support system for joint operations; space available for afloat CJTF command; Syracuse Satcom.

Weapons control: 2 optronic systems.

Radars: Air/surface search: Thales MRR; 3-D; G-band. Navigation: 2 Racal-Decca Bridgemaster (DRBN 38A); I-band.

Helicopters: Up to 16 NH90 or Super Puma/Cougars transport/assault helicopters or Eurocopter Tigre attack helicopters.

Programmes: Designated BPC (Bâtiment de Projection et de Commandement, support and command ship for force projection), ex-NTCD (new LHDs); to replace *Ouragan* and *Orage*. Design and definition phase launched 12 November 1999; building contract notified 22 December 2000; ordered from DCN (prime contractor) and Alstom Marine-Chantiers de l'Atlantique. Forward sections built at St Nazaire, and middle and aft blocks at Brest where final construction and outfitting is taking place. Sixty per cent of the aft section subcontracted to Stocznia Remontowa, Gdansk, and shipped to Brest by barge. First steel cut for *Mistral* 9 July 2002 and *Tonnerre* 13 December 2002. New landing craft planned to operate with *Mistral* and *Tonnerre*.

Structure: Built to merchant marine standards. Flight deck has 6 spots for 16 ton medium/heavy helicopters. Well dock 57.5 × 15.4 × 8.2 m; one 1,800 m² hangar for helicopters or vehicles (with 2 lifts) and one 1,000 m² hangar for vehicles only (1 lift); up to 1,000 tons load on vehicle deck. Hospital: 69 beds; additional modular field hospital may be embarked for humanitarian missions. Other modular facilities could also be embarked according to missions.

Operational: Roles: forward presence, force projection, logistic support for deployed force (ashore or at sea), humanitarian aid, disaster relief, command ship for combined operations. Endurance: 45 days. Sea trials of *Mistral* began in February 2005. **UPDATED**

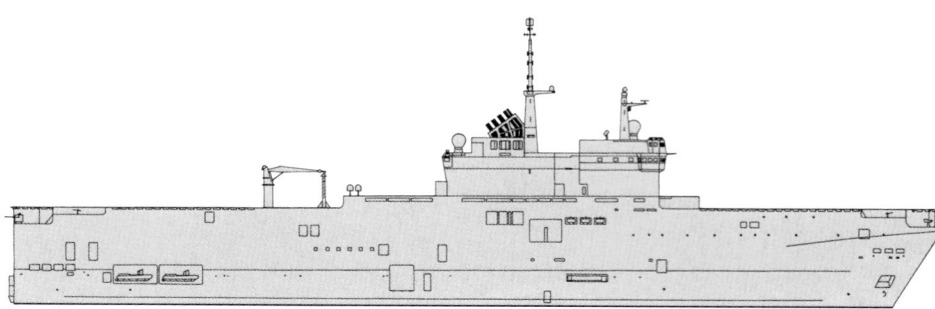

MISTRAL

(Scale 1 : 1,200), Ian Sturton / 1042093

MISTRAL

10/2004*, J Y Robert / 1042247

MISTRAL

10/2004*, B Prézelin / 1042221

MISTRAL

10/2004*, B Prézelin / 1042220

2 FOUDRE CLASS (LANDING SHIPS DOCK) (LSDH/TCD 90)

Name	No	Builders	Laid down	Launched	Commissioned
FOUDRE	L 9011	DCN, Brest	26 Mar 1986	19 Nov 1988	7 Dec 1990
SIROCO	L 9012	DCN, Brest	2 Oct 1994	14 Dec 1996	21 Dec 1998

Displacement, tons: 8,190 (L 9011), 8,230 (L 9012) light; 12,400 full load; 17,200 flooded

Dimensions, feet (metres): 551 × 77.1 × 17 (30.2 flooded) *(168 × 23.5 × 5.2 (9.2))*

Main machinery: 2 SEMT-Pielstick 16 PC2.5 V 400 diesels; 20,800 hp(m) *(15.3 MW)* sustained; 2 shafts; LIPS cp props; bow thruster; 1,000 hp(m) *(735 kW)*

Speed, knots: 21. **Range, n miles:** 11,000 at 15 kt

Complement: 215 (17 officers)

Military lift: 470 (up to 2,000 for 3 days) troops plus 1,880 tons load; 1 EDIC/CDIC plus 4 CTMs (typical) or 2 CDIC or 10 CTMs; 150 vehicles

Missiles: SAM: 2 Matra Simbad twin launchers ❶; Mistral; IR homing to 4 km *(2.2 n miles)*; warhead 3 kg.

Guns: 3 Breda/Mauser 30 mm/70 ❷ 4—12.7 mm MGs.

Countermeasures: ECM: 2 Thales ARBB 36 jammers. SLQ-25 Nixie towed torpedo decoy.

Combat data systems: SENIT 8/01 (Siroco); Syracuse SATCOM ❸. OPSMER command support system. Link 11 (receive only).

Weapons control: 2 Sagem VIGY-105 optronic systems (for 30 mm guns).

Radars: Air/surface search: Thomson-CSF DRBV 21A Mars ❹; D-band.
Surface search: Racal Decca 2459 (Foudre); I-band.
Navigation: 2 Racal-Decca DRBN 34A (L 9012) or Racal-Decca 1229 (L 9011); I-band (1 for helo control) ❺.

Helicopters: 4 AS 332F Super Puma ❻ or 2 Super Frelon.

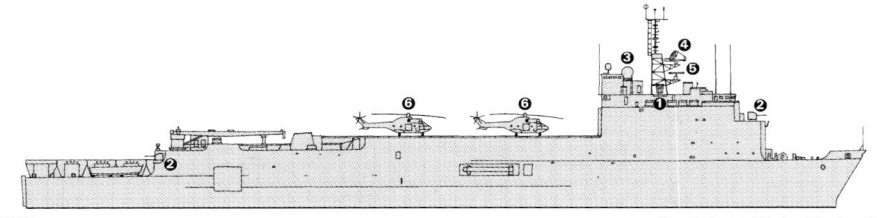

SIROCO   (Scale 1 : 1,500), Ian Sturton / 0529157

Programmes: First ordered 5 November 1984, second 11 April 1994. Transports de Chalands de Débarquement (TCD).

Modernisation: Sadral SAM replaced by two lightweight Simbad SAMs either side of bridge. New air search radar. 30 mm guns to replace 40 mm and 20 mm in *Foudre* and fitted on build in *Siroco*. Sagem optronic fire control fitted in 1997. ESM/ECM to be acquired.

Structure: Designed to take a mechanised regiment of the Rapid Action Force and act as a logistic support ship. Extensive command (OPSMER and other systems) and hospital facilities (500 m²) include two operating suites and 47 beds. Modular field hospital may be embarked on *Siroco*. Well dock of 122 × 14 m *(1,640 m²)* which can be used to dock a 400 tons ship. Crane of 37 tons and lift of 52 tons *(Foudre)* or 38 tons *(Siroco)*. Flight deck of *Foudre* 1,450 m² with two landing

spots (one fitted with landing grid and SAMAHE helo handling system). Additional landing spot on the (removable) well rolling cover. *Siroco* landing deck extended aft up to the lift to give a 1,740 m² area. Flume stabilisation fitted in 1993 to *Foudre*.

Operational: Two landing spots on flight deck plus one on deck well rolling cover. Can operate Super Frelons or Super Pumas. Could carry up to 1,600 troops in emergency. Endurance, 30 days (with 700 persons aboard). Assigned to FAN and based at Toulon. Typical loads: one CDIC, four CTM, 10 AMX 10RC armoured cars and 50 vehicles or total of 180 to 200 vehicles (without landing craft). *Siroco* deployed to East Timor in 1999.

UPDATED

FOUDRE   8/2004*, B Prézelin / 1042218

SIROCO   6/2002, French Navy / 0529144

2 OURAGAN CLASS (LANDING SHIPS DOCK) (LSDH/TCD)

Name	No	Builders	Laid down	Launched	Commissioned
OURAGAN	L 9021	Brest Naval Dockyard	June 1962	9 Nov 1963	1 June 1965
ORAGE	L 9022	Brest Naval Dockyard	June 1966	22 Apr 1967	1 Apr 1968

Displacement, tons: 5,800 light; 8,500 full load; 14,400 when fully docked down
Dimensions, feet (metres): 488.9 × 75.4 × 17.7 (28.5 flooded) *(149 × 23 × 5.4 (8.7))*
Main machinery: 2 SEMT-Pielstick diesels; 8,640 hp(m) *(6.35 MW) (Ouragan)*; 9,400 hp(m) *(6.91 MW) (Orage)*; 2 shafts; LIPS cp props
Speed, knots: 17. **Range, n miles:** 9,000 at 15 kt
Complement: 205 (12 officers)
Military lift: 343 troops (plus 129 short haul only); 2 LCTs (EDIC) with 11 light tanks each or 8 loaded CTMs; logistic load 1,500 tons; 2 cranes (35 tons each)

Missiles: SAM: 2 Matra Simbad twin launchers; Mistral; IR homing to 4 km *(2.2 n miles)*; warhead 3 kg; anti-sea-skimmer.
Guns: 2 Breda/Mauser 30 mm/70. 4—12.7 mm MGs.
Weapons control: 2 Sagem VIGY-105 optronic systems.
Radars: Air/surface search: Thomson-CSF DRBV 51A; G-band.
Navigation: 2 Racal Decca DRBN 34A; I-band.

Helicopters: 4 SA 321G Super Frelon or 10 SA 319B Alouette III.

Modernisation: Simbad SAM and new search radars fitted in 1993. Sagem VIGY-105 optronic fire-control system and 30 mm guns are being fitted to replace the 40 mm. *Ouragan* sonar has been removed. *Orage* has an enclosed Flag bridge.
Structure: Normal helicopter platform for operating three Super Frelon or 10 Alouette III plus a portable platform for a further one Super Frelon or three Alouette III. Bridge is on the starboard side. Three LCVPs can also be carried. Extensive

OURAGAN   8/2004*, B Prézelin / 1042218

workshops. Flight deck is 900 m²; docking well 120 × 13.2 m with 3 m of water. Two 35 ton cranes.
Operational: Typical loads-18 Super Frelon or 80 Alouette III helicopters or 120 AMX 10 APCs, or 84 DUKWs or 340 Jeeps or 12—50 ton barges. A 400 ton ship can be docked. Command facilities for directing amphibious and helicopter operations. Both ships assigned to FAN and based in Toulon. Typical loads: one CDIC, four CTM, 10 AMX 10RC armoured cars and 21 vehicles or total of 150 to 170 vehicles (without landing craft). Likely to be decommissioned in 2006 as *Mistral* and *Tonnerre* enter service.

UPDATED

ORAGE   4/2003, B Prézelin / 0569949

4 BATRAL TYPE
(LIGHT TRANSPORTS and LANDING SHIPS) (LSTH)

Name	No	Builders	Commissioned
FRANCIS GARNIER	L 9031	Brest Naval Dockyard	21 June 1974
DUMONT D'URVILLE	L 9032	Français de l'Ouest	5 Feb 1983
JACQUES CARTIER	L 9033	Français de l'Ouest	28 Sep 1983
LA GRANDIÈRE	L 9034	Français de l'Ouest	20 Jan 1987

Displacement, tons: 750 standard; 1,580 full load
Dimensions, feet (metres): 262.4 × 42.6 × 7.9 *(80 × 13 × 2.4)*
Main machinery: 2 SACM AGO 195 V12 diesels; 3,600 hp(m) *(2.65 MW)* sustained; 2 shafts; cp props
Speed, knots: 14.5. **Range, n miles:** 4,500 at 13 kt
Complement: 52 (5 officers)
Military lift: 180 troops; 12 vehicles; 350 tons load; 10 ton crane

Missiles: SAM: 2 Matra Simbad twin launchers (may be fitted).
Guns: 2 Bofors 40 mm/60 (L 9031). 2 Giat 20F2 20 mm (L 9032-L 9034). 2—12.7 mm MGs.
Radars: Navigation: DRBN 32; I-band.

Helicopters: Platform for Lynx or Panther.

Programmes: Classified as Batral 3F. Bâtiments d'Assaut et de TRAnsport Légers (BATRAL). First two launched 17 November 1973. *Dumont D'Urville* floated out 27 November 1981. *Jacques Cartier* launched 28 April 1982 and *La Grandière* 15 December 1985. *F Garnier* refitted at Brest 2000.

FRANCIS GARNIER   5/2004*, B Prézelin / 1042217

Structure: 40 ton bow ramp; stowage for vehicles above and below decks. One LCVP and one LCPS carried. Helicopter landing platform. Last three of class have bridge one deck higher, a larger helicopter platform and a crane replaces the boom on the cargo deck.
Operational: Deployment: *F Garnier*, Martinique; *D D'Urville*, Papeete; *J Cartier*, New Caledonia; *La Grandière*, Indian Ocean. Service lives of *F Garnier* (2005) and *J Cartier* (2008) extended. *Champlain* placed in reserve in Martinique 2004.
Sales: Ships of this class built for Chile, Gabon, Ivory Coast and Morocco. *La Grandière* was also built for Gabon under Clause 29 arrangements but funds were not available.

UPDATED

2 EDIC 700 CLASS (LCT)

Name	No	Builders	Commissioned
SABRE	L 9051	SFCN, Villeneuve la Garenne	13 June 1987
DAGUE	L 9052	SFCN, Villeneuve la Garenne	19 Dec 1987

Displacement, tons: 736 full load
Dimensions, feet (metres): 193.6 × 38.1 × 5.8 *(59 × 11.6 × 1.7)*
Main machinery: 2 SACM Uni Diesel UD 30 V12 M1 diesels; 1,200 hp(m) *(882 kW)* sustained; 2 shafts
Speed, knots: 12. **Range, n miles:** 1,800 at 12 kt
Complement: 17
Military lift: 350 tons
Guns: 2 Giat 20F2 20 mm. 2—12.7 mm MGs.
Radars: Navigation: Racal Decca 1229; I-band.

Comment: Ordered 10 March 1986. Given names on 29 April 1999. Rated as Engins de Débarquement d'Infanterie et Chars (EDIC III). Based at Toulon (L 9051) and Djibouti (L 9052). L 9051 refitted in 2004. ***UPDATED***

SABRE *11/2004*, B Prézelin /* 1042216

2 CDIC CLASS (LCT)

Name	No	Builders	Commissioned
RAPIÈRE	L 9061	SFCN, Villeneuve la Garenne	28 July 1988
HALLEBARDE	L 9062	SFCN, Villeneuve la Garenne	2 Mar 1989

Displacement, tons: 380 light; 750 full load
Dimensions, feet (metres): 194.9 × 39 × 5.9 *(59.4 × 11.9 × 1.8)*
Main machinery: 2 SACM Uni Diesel UD 30 V12 M1 diesels; 1,200 hp(m) *(882 kW)* sustained; 2 shafts
Speed, knots: 10.5. **Range, n miles:** 1,000 at 10 kt
Complement: 12 (1 officer)
Military lift: 336 tons
Guns: 2 Giat 20F2 20 mm. 2—12.7 mm MGs.
Radars: Navigation: Racal Decca 1229; I-band.

Comment: CDIC (Chaland de Débarquement d'Infanterie et de Chars) built to work with Foudre class. The wheelhouse can be lowered to facilitate docking manoeuvres in the LPDs. Assigned to FAN at Toulon. Given names on 21 July 1997. One other transferred to Senegal in February 1999. ***UPDATED***

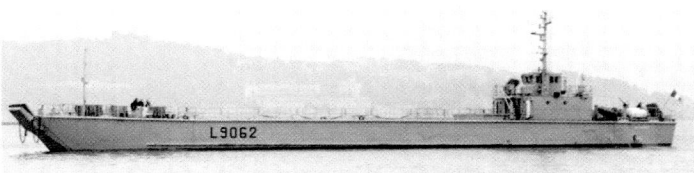

HALLEBARDE *3/2004*, B Prézelin /* 1042215

15 CTMs (LCM)

CTM 17-31

Displacement, tons: 59 standard; 150 full load
Dimensions, feet (metres): 78 × 21 × 4.2 *(23.8 × 6.4 × 1.3)*
Main machinery: 2 Poyaud V8520NS diesels; 450 hp(m) *(331 kW)*; 2 shafts
Speed, knots: 9.5. **Range, n miles:** 380 at 8 kt
Complement: 4 + 200 passengers
Military lift: 90 tons (maximum); 48 tons (normal)
Guns: 2—12.7 mm MGs.
Radars: Navigation: I-band.

Comment: First series of 16 built 1966-70 and all but *CTM 3* have been deleted. Second series *(CTM 17-31)* built at CMN, Cherbourg 1982-92. All have a bow ramp but the second series has a different shaped pilot-house. Chalands de Transport de Matériel (CTM). CTM 17 based at Lorient, CTM 18 at Mayotte, CTM 21, 24, 25 at Djibouti, CTM 26 at Dakar and ten at Toulon. Six others of the class are based at La Rochelle and operated by the French Army Transport Corps. They include L 14-16 and L 924-925. ***UPDATED***

CTM 31 *9/2004*, Schaeffer/Marsan /* 1042227

L 924 (Army) *9/1997, van Ginderen Collection /* 0012352

PATROL FORCES

Notes: 'Sauvegarde Maritime' is the organisation that encompasses the surveillance and traffic control of all maritime approaches around continental France and overseas territories. It also includes pollution control. Although all naval ships could participate in surveillance tasks, specialised vessels include the OPVs manned by the navy, patrol vessels and patrol craft of the 'Gendarmerie Maritime', French Customs and 'Affaires Maritimes'. Also merchant support vessels on long-term charter (see Government Maritime Forces). All these ships, including specialised naval ships, will display blue/white/red stripes on hull sides. Naval patrol ships (OPVs) are sometimes referred to as 'Patrouilleurs Spécialisés de Service Public' (PSSP, Public Service Special Patrol Vessel).

1 LAPÉROUSE CLASS (PBO)

Name	No	Builders	Launched	Commissioned
ARAGO	P 675 (ex-A 795)	Lorient Naval Dockyard	9 Sep 1990	9 July 1991

Displacement, tons: 980 full load
Dimensions, feet (metres): 193.5 × 35.8 × 11.9 *(59 × 10.9 × 3.6)*
Main machinery: 2 Wärtsilä UD 30 V12 M6D diesels; 2,500 hp(m) *(1.84 MW)*; 2 cp props; bow thruster; auxiliary electric motor; 220 hp *(160 kW)*
Speed, knots: 15. **Range, n miles:** 5,200 at 12 kt
Complement: 30 (3 officers)
Guns: 2—12.7 mm MGs.
Radars: Navigation: Decca 1226; I-band.

Comment: Ex-survey ship converted for patrol duties. Based at Toulon. ***UPDATED***

ARAGO *7/2004*, Schaeffer/Marsan /* 1042237

1 TRAWLER TYPE (PSO)

Name	No	Builders	Commissioned
ALBATROS (ex-*Névé*)	P 681	Ch de la Seine Maritime	1967

Displacement, tons: 1,940 standard; 2,800 full load
Dimensions, feet (metres): 278.1 × 44.3 × 18.4 *(84.8 × 13.5 × 5.6)*
Main machinery: Diesel-electric; 2 SACM UD 33 V12 S4 diesel generators; 3,050 hp(m) *(2.24 MW)* sustained; 2 motors; 2,200 hp(m) *(1.62 MW)*; 1 shaft
Speed, knots: 15. **Range, n miles:** 14,700 at 14 kt
Complement: 49 (7 officers) plus 15 passengers
Guns: 1 Bofors 40 mm/60. 2—12.7 mm MGs.
Countermeasures: ESM; ARBR 16 radar detector and ARUR 10B radio interceptor.
Radars: Surface search: 2 DRBN 38A; I-band.

Comment: Former trawler bought in April 1983 from Compagnie Nav. Caennaise for conversion into a patrol ship. Commissioned 19 May 1984. Conducts patrols from Réunion to Kerguelen, Crozet, St Paul and Amsterdam Islands with occasional deployments to South Pacific. Vertrep facilities. Can carry 200 tons cargo, and has 4 tonne telescopic crane. Hospital with six berths and operating room. Major refit in Lorient from June 1990 to March 1991 included new diesel-electric propulsion. A further major overhaul was undertaken August 2001-April 2002. Service life: 2015. ***UPDATED***

ALBATROS *4/2002, B Prézelin /* 0528841

10 P 400 CLASS (LARGE PATROL CRAFT) (PBO)

Name	No	Builders	Commissioned
L'AUDACIEUSE	P 682	CMN, Cherbourg	18 Sep 1986
LA BOUDEUSE	P 683	CMN, Cherbourg	15 Jan 1987
LA CAPRICIEUSE	P 684	CMN, Cherbourg	13 Mar 1987
LA FOUGUEUSE	P 685	CMN, Cherbourg	13 Mar 1987
LA GLORIEUSE	P 686	CMN, Cherbourg	18 Apr 1987
LA GRACIEUSE	P 687	CMN, Cherbourg	17 July 1987
LA MOQUEUSE	P 688	CMN, Cherbourg	18 Apr 1987
LA RAILLEUSE	P 689	CMN, Cherbourg	16 May 1987
LA RIEUSE	P 690	CMN, Cherbourg	13 June 1987
LA TAPAGEUSE	P 691	CMN, Cherbourg	11 Feb 1988

Displacement, tons: 406 standard; 477 full load
Dimensions, feet (metres): 178.6 × 26.2 × 8.5 *(54.5 × 8 × 2.5)*
Main machinery: 2 SEMT-Pielstick 16 PA4 200 VGDS diesels; 8,000 hp(m) *(5.88 MW)* sustained; 2 shafts
Speed, knots: 24.5. **Range, n miles:** 4,200 at 15 kt
Complement: 26 (3 officers) plus 20 passengers
Guns: 1 Bofors 40 mm/60; 1 Giat 20F2 20 mm; 2—7.62 mm MGs.
Radars: Surface search: DRBN 38A; I-band.

Programmes: First six ordered in May 1982, with further four in March 1984. The original propulsion system was unsatisfactory. Modifications were ordered and construction slowed. This class relieved the Patra fast patrol craft which have all transferred to the Gendarmerie.
Structure: Steel hull and superstructure protected by an upper deck bulwark. Design modified from original missile craft configuration. Now capable of transporting personnel with appropriate store rooms. Of more robust construction than previously planned and used as overseas transports. Can be converted for missile armament (MM 38) with dockyard assistance and Sadral PDMS has been considered. *L'Audacieuse* has done trials with a VDS-12 sonar. Twin funnels replaced the unsatisfactory submerged diesel exhausts in 1990-91. P 682 fitted with new propellers in 2003. If successful the rest of the class will be fitted.
Modernisation: A modernisation programme started in 2002. P 682, 683, 689, 690 and 691 have been refitted. The remainder are to be completed by 2006.
Operational: Deployments: Antilles; P 685, French Guiana; P 682, 684, 687. Nouméa; P 686, 688. La Réunion; P 683 and 690. Tahiti; P 689 and P 691. Endurance, 15 days with 45 people aboard. All are receiving a six month refit in Lorient which means some switching of deployments.
Sales: To Gabon and Oman.

UPDATED

L'AUDACIEUSE   7/2002, Per Körnefeldt / 0528830

LA GLORIEUSE   2/2003, Chris Sattler / 0569954

3 FLAMANT (OPV 54) CLASS (PBO)

Name	No	Builders	Launched	Commissioned
FLAMANT	P 676	CMN, Cherbourg	24 Apr 1995	18 Dec 1997
CORMORAN	P 677	Leroux & Lotz, Lorient	15 May 1995	29 Oct 1997
PLUVIER	P 678	CMN, Cherbourg	2 Dec 1996	18 Dec 1997

Displacement, tons: 314 standard; 390 full load
Dimensions, feet (metres): 179.8 × 32.8 × 9.2 *(54.8 × 10 × 2.8)*
Main machinery: CODAD; 2 Deutz/MWM 16V TBD 620 diesels and 2 MWM 12V TBD 234 diesels; 7,230 hp(m) *(5.32 MW)* sustained; 2 shafts; LIPS cp props
Speed, knots: 23. **Range, n miles:** 4,500 at 14 kt
Complement: 19 (3 officers)
Guns: 2—12.7 mm MGs.
Radars: Surface search: 1 Racal Decca Bridgemaster 250 (DRBN 38A); I-band.
Navigation: DRBN 38B; I-band.

Comment: Authorised in July 1992 and ordered in August 1993 to a Serter Deep V design. Has a stern door for a 7 m EDL 700 fast assault craft or a Zodiac Hurricane RIB, capable of 30 kt. Two passive stabilisation tanks are fitted, and a remotely operated water-jet gun for firefighting. Deck area of 12 × 9 m for Vertrep. Similar to craft built for Mauritania in 1994. Hulls of all three ships to be strengthened by DCN Brest by late 2004. *Flamant* and *Pluvier* based at Cherbourg, *Cormoran* at Brest.

UPDATED

CORMORAN   7/2004 *, Derek Fox / 1042234

1 GRÈBE CLASS (PBO)

Name	No	Builders	Commissioned
GRÈBE	P 679	SFCN, Villeneuve La Garenne	6 Apr 1991

Displacement, tons: 300 standard; 410 full load
Dimensions, feet (metres): 170.6 × 32.2 × 9 *(52 × 9.8 × 2.8)*
Main machinery: 2 Wärtsilä UD 33 V12 M6D diesels; 4,410 hp(m) *(3.24 MW)*; diesel-electric auxiliary propulsion; 245 hp(m) *(180 kW)*; 2 shafts; cp props
Speed, knots: 23; 7.5 on auxiliary propulsion. **Range, n miles:** 4,500 at 12 kt
Complement: 19 (4 officers); accommodation for 24; 2 crews
Guns: 2—12.7 mm MGs.
Radars: Navigation: Racal Decca; I-band.

Comment: Type Espadon 50 ordered 17 July 1988 and launched 16 November 1989. Serter 'Deep V' hull; stern ramp for craft handling. Large deck area (8 × 8 m) for Vertrep operations. Pollution control equipment and remotely operated water-jet gun for firefighting. Based at Toulon from November 1997.

UPDATED

GRÈBE   6/2004 *, Schaeffer/Marsan / 1042235

1 STERNE CLASS (PBO)

Name	No	Builders	Commissioned
STERNE	P 680	La Perrière, Lorient	20 Oct 1980

Displacement, tons: 250 standard; 380 full load
Dimensions, feet (metres): 160.7 × 24.6 × 9.2 *(49 × 7.5 × 2.8)*
Main machinery: 2 SACM 195 V12 CZSHR diesels; 4,340 hp(m) *(3.19 MW)* sustained; electrohydraulic auxiliary propulsion on starboard shaft; 150 hp(m) *(110 kW)*; 2 shafts
Speed, knots: 20; 6 on auxiliary propulsion. **Range, n miles:** 4,900 at 12 kt; 1,500 at 20 kt
Complement: 18 (3 officers); 2 crews
Guns: 2—12.7 mm MGs.
Radars: Navigation: Racal Decca; I-band.

Comment: *Sterne* was the first ship for the FSMC. Has active tank stabilisation. Launched 31 October 1979 and completed 18 July 1980 for the 'Affaires Maritimes' but then transferred and is now manned and operated by the Navy from Brest.

UPDATED

STERNE   11/2004 *, B Prézelin / 1042236

MINE WARFARE FORCES

13 ÉRIDAN (TRIPARTITE) CLASS (MINEHUNTERS) (MHC)

Name	No	Laid down	Launched	Commissioned
ÉRIDAN	M 641	20 Dec 1977	2 Feb 1979	16 Apr 1984
CASSIOPÉE	M 642	26 Mar 1979	26 Sep 1981	5 May 1984
ANDROMÈDE	M 643	6 Mar 1980	22 May 1982	18 Oct 1984
PÉGASE	M 644	22 Dec 1980	23 Apr 1983	30 May 1985
ORION	M 645	17 Aug 1981	6 Feb 1985	14 Jan 1986
CROIX DU SUD	M 646	22 Apr 1982	6 Feb 1985	14 Nov 1986
AIGLE	M 647	2 Dec 1982	8 Mar 1986	1 July 1987
LYRE	M 648	13 Oct 1983	14 Nov 1986	16 Dec 1987
PERSÉE	M 649	30 Oct 1984	19 Apr 1988	4 Nov 1988
SAGITTAIRE	M 650	1 Feb 1993	14 Jan 1995	2 Apr 1996
VERSEAU (ex-Iris)	M 651	20 May 1985	21 June 1987	6 Oct 1988
CÉPHÉE (ex-Fuchsia)	M 652	28 Oct 1985	23 Oct 1987	18 Feb 1988
CAPRICORNE (ex-Dianthus)	M 653	17 Apr 1985	26 Feb 1987	14 Aug 1987

Displacement, tons: 562 standard; 605 full load
Dimensions, feet (metres): 168.9 × 29.2 × 8.2 *(51.5 × 8.9 × 2.5)*
Main machinery: 1 Stork Wärtsilä A-RUB 215X-12 diesel; 1,860 hp(m) *(1.37 MW)* sustained;
 1 shaft; LIPS cp prop
 Auxiliary propulsion; 2 motors; 240 hp(m) *(179 kW)*; 2 active rudders; 2 bow thrusters
Speed, knots: 15; 7 on auxiliary propulsion. **Range, n miles:** 3,000 at 12 kt
Complement: 49 (5 officers)

Guns: 1 Giat 20F2 20 mm; 1—12.7 mm MG.
Countermeasures: MCM: 2 PAP 104 ROVs; OD3 mechanical sweep gear. AP-4 acoustic sweep.
 Double Eagle ROV from 2001.
Combat data systems: TSM 2061 being fitted.
Radars: Navigation: Racal Decca 1229; I-band.
Sonars: Thomson Sintra DUBM 21B or 21D (M 650); being replaced from 2001 by two TUS 2022 Mk III
 sonars (one hull-mounted and one PVDS on Bofors Double Eagle Mk II UUV; dual frequency.

Programmes: All built in Lorient. Belgium, France and the Netherlands each agreed to build 15 (10 in
 Belgium with option on five more). Subsequently the French programme was cut to 10. Belgium
 provided all the electrical installations, France all the minehunting gear and some electronics and the
 Netherlands the propulsion systems. Replacement for the last of class (sold to Pakistan) was ordered
 in January 1992. Three Belgian ships of the class acquired between March and August 1997 after
 being in reserve since 1990.
Modernisation: A modernisation programme started in 2001. Nine had been upgraded by 2004 and
 the programme is to end when M 649 is completed in August 2005. Modernisation includes
 replacement of sonar by TSM 2022 Mk 3, fitting of a Bofors Double Eagle Mk 2 ROV, a new tactical
 data system and upgrade of radar and comms.
Structure: GRP hull. Equipment includes: autopilot and hovering; automatic radar navigation;
 navigation aids by Loran and Syledis; Evec data system.
Operational: Minehunting, minesweeping, patrol, training, directing ship for unmanned minesweeping, HQ ship for diving operations and pollution control. Prepacked 5 ton modules of
 equipment embarked for separate tasks. M 645, 651 and 653 based at Toulon, remainder at Brest.
Sales: The original tenth ship of the class, completed in 1989, was transferred to Pakistan
 24 September 1992 as part of an order for three; the second built in Lorient, the third in Karachi.

UPDATED

VERSEAU   8/2004 *, B Prézelin / 1042233

AIGLE   6/2004 *, Frank Findler / 1042232

SAGITTAIRE   3/2004 *, Derek Fox / 1042231

3 ANTARÈS (BRS) CLASS (ROUTE SURVEY VESSELS) (MHI)

Name	No	Builders	Commissioned
ANTARÈS	M 770	Socarenam, Boulogne	15 Dec 1993
ALTAÏR	M 771	Socarenam, Boulogne	9 July 1994
ALDÉBARAN	M 772	Socarenam, Boulogne	10 Mar 1995

Displacement, tons: 250 standard; 340 full load
Dimensions, feet (metres): 92.8 × 25.3 × 13.1 *(28.3 × 7.7 × 4)*
Main machinery: 1 Baudouin 12P15-2SR diesel; 800 hp(m) *(590 kW)*; 1 shaft; cp prop; bow
 thruster
Speed, knots: 12. **Range, n miles:** 3,600 at 10 kt
Complement: 25 (1 officer)
Guns: 1—12.7 mm MG.
Radars: Navigation: 1 Racal-Decca Bridgemaster C 180; I-band.
Sonars: 2 Thomson Sintra DUBM 41B; towed side scan; active search; high frequency.

Comment: Has replaced the Aggressive class for route survey at Brest. BRS Bâtiments
 Remorqueurs de Sonars. Trawler type similar to Glycine class (see *Training Ships* section). The
 DUBM 41B towed bodies have been taken from the older MSOs. A mechanical sweep is also
 carried. There are two 4.5 ton hydraulic cranes. Original dual navigation training role has been
 lost.

UPDATED

ALTAIR   6/2004 *, M Declerck / 1042230

4 MCM DIVING TENDERS (MCD)

Name	No	Builders	Launched	Commissioned
VULCAIN	M 611	La Perrière, Lorient	17 Jan 1986	11 Oct 1986
PLUTON	M 622	La Perrière, Lorient	13 May 1986	12 Dec 1986
ACHÉRON	A 613	CMN, Cherbourg	9 Nov 1986	17 June 1987
STYX	M 614	CMN, Cherbourg	3 Mar 1987	22 July 1987

Displacement, tons: 375 standard; 505 full load
Dimensions, feet (metres): 136.5 × 24.6 × 12.5 *(41.6 × 7.5 × 3.8)*
Main machinery: 2 SACM MGO 175 V16 ASHR diesels; 2,200 hp(m) *(1.62 MW)*; 2 shafts;
 bow thruster; 70 hp(m) *(51 kW)*
Speed, knots: 13.7. **Range, n miles:** 2,800 at 13 kt; 7,400 at 9 kt
Complement: 14 (1 officer) plus 12 divers
Guns: 1—12.7 mm MG. 2—7.62 mm MGs.
Radars: Navigation: Decca 1226; I-band.

Comment: First pair ordered in December 1984. Second pair ordered July 1985. Designed to act
 as support ships for clearance divers. (Bâtiments Bases pour Plongeurs Démineurs – BBPD).
 Vulcain based at Cherbourg, *Pluton* at Toulon, *Achéron* at Toulon as a diving school tender and
 Styx at Brest. Modified Chamois (BSR) class design. 5 ton hydraulic crane.

UPDATED

PLUTON   1/2004 *, Schaeffer/Marsan / 1042229

SURVEY AND RESEARCH SHIPS

Notes: (1) These ships are painted white. A total of about 100 officers and technicians with
oceanographic and hydrographic training is employed in addition to the ships' companies listed
here. They occupy the extra billets marked as 'scientists'.
(2) In addition to the ships listed below there is a civilian-manned 25 m trawler *L'Aventurière II*
(launched July 1986) operated by GESMA, Brest for underwater research which comes under
DCN.
(3) Two survey launches, *Matthew* and *Hunter* were built in 1980.

7 TYPE VH 8 SURVEY LAUNCHES (YGS)

Displacement, tons: 4
Dimensions, feet (metres): 25.9 × 7.9 × 1.6 *(7.9 × 2.4 × 0.5)*
Main machinery: Volvo Penta Aquamatic Duotrop diesel; Z-drive; 1 shaft; 237 hp(m) *(174 kW)*
Speed, knots: 17. **Range, n miles:** 109
Complement: 6

Comment: Built by Fr. Frassmer GmbH & Co (Germany); first craft delivered October 2002 and
 based at Toulon since July 2003. To be carried by *Beautemps-Beaupré* and Lapérouse-class
 survey vessels. Vedette hydrographique de 8m (VH 8). Fitted with two echo sounders, one
 multipath echo sounder (Simrad EM 3200), side-scan towed sonar and towed magnetometer.

UPDATED

1 BEAUTEMPS-BEAUPRÉ CLASS (BHO HYDROGRAPHIC AND OCEANOGRAPHIC SURVEY SHIP) (AGOR)

Name	No	Builders	Laid down	Launched	Commissioned
BEAUTEMPS-BEAUPRÉ	A 758	Alstom Marine, Lorient	17 July 2001	26 Apr 2002	13 Dec 2003

Displacement, tons: 2,125 standard; 3,330 full load
Dimensions, feet (metres): 264.5 × 48.9 × 23.0 *(80.6 × 14.9 × 7)*
Main machinery: Diesel-electric; four 1,500 hp(m) *(1.1 MW)* Mitsubishi diesels; 1 electric motor; 1 shaft; 3,000 hp(m) *(2.2 MW)*.
2 active rudders 300 hp(m) *(220 kW)* each; bow thruster 600 hp(m) *(440 kW)*.
Speed, knots: 14
Complement: 26 (5 officers) (two crews) plus 25 to 30 scientists
Guns: 2—7.62 mm MGs.
Radars: Navigation: 2 Kongsberg; I-band.
Sonars: EG & G side looking towed sonar; Kongsberg/Simrad EM 120 deep multipath echo sounder (12 kHz); Kongsberg/Simrad EA 600 deep echo sounder (12 kHz); Kongsberg/Simrad EM 1002S shallow waters multipath echo-sounder (95 kHz); Kongsberg/Simrad EA 400-210 shallow waters echo sounder (33 kHz); Kongsber/Simrad SBP 120 (3 to 7 kHz) narrow beam and SHOM 9 TR 109 (3.5 kHz) wide beam sediment echo sounders. Bodenseewerk KSS31 gravimeter; Thales SMM II magnetometer; acoustic current profiler. Most sensor transducers mounted on a removable chassis fixed underneath the hull. Oceanographic buoys; Sippican Mk 21.

Comment: Contracted to Alstom-Leroux Naval 13 March 2001. Derived from the civilian research ship *Thalassa* built in 1995 by Leroux & Lotz (now part of Alstom Marine) for the French government civilian agency IFREMER. 95 per cent funded by the MoD and 5 per cent by the Ministry of civilian research on behalf of IFREMER that will use the ship 10 days per year. First steel cut 17 July 2001. Started builder sea trials 17 October 2002 and official acceptance trials late December. Two 7.85 m survey launches. 10 tonne stern gantry and 10 tonne crane; up to 5 shelters can be shipped and bolted on the deck to increase lab surfaces; up to 4 vehicles can be stored in the hold. Endurance 45 days. Bâtiment hydrographique et océangraphique (BHO, hydrographic and oceanographic survey ship). ***UPDATED***

BEAUTEMPS-BEAUPRÉ   *7/2003, B Prézelin* / 0569985

0 + 1 POURQUOIS PAS ? CLASS
(OCEANOGRAPHIC SURVEY SHIP (AGOR)

Name	No	Builders	Laid down	Launched	Commissioned
POURQUOIS PAS ?	—	Alstom Marine, St Nazaire	2003	14 Oct 2004	2005

Displacement, tons: 6,600
Dimensions, feet (metres): 353.0 × 65.6 × 22.6 *(107.6 × 20 × 6.9)*
Main machinery: Diesel-electric; four diesel generators; two electric motors; 4,350 hp(m) *(3.2 MW)*; 2 shafts
Speed, knots: 14.5. **Range, n miles:** 16,000 at 11 kt
Complement: 35 + 40 scientists
Radars: Navigation: I-band.
Sonars: Reason Seabat 7111 (100 kHz) and Seabat 7150 (12/24 kHz) multipath echo sounders; Simrad EA 600 (12/38/200 kHz) deep echo sounder; RDI Ocean Surveyor current profiler (38/150 kHz); Eramer/Triton Elics sediment echo sounder (2-8 kHz); most sensor transducers mounted on a removable chassis fixed underneath the hull; also optional towed sonars.

Comment: Contract awarded 17 December 2002 to Alstom Marine. Funded 55 per cent by the Ministry of Research and Education, and 45 per cent by the MoD which will use the ship 150 days per year; civilian manned (operated by Genavir on behalf of IFREMER research agency — see Government Maritime Forces), with navy specialists when operating for military campaigns. First steel cut 1 September 2003. Trials in February 2005 and delivery in March 2005. Optional additional labs in containers; helo deck. Able to operate the *Nautile* mini sub, the *Victor 6000* ROV or the future NATO Submarine Rescue System (NSRS); can embark up to three navy VH & survey launches (two under davits); stern gantry to handle equipments up to 22 tonnes. Space allocated to embark up to 20 20 ft containers. Endurance 60 days. *Pourquois Pas ?* (Why not?) is the name given by the famous explorer and oceanographer Jean-Baptiste Charcot (1867-1936) to several of his research vessels. ***UPDATED***

POURQUOI PAS ?   *10/2004*, B Prézelin* / 1042228

1 RESEARCH SHIP (AETL)

Name	No	Builders	Launched	Commissioned
DENTI	A 743	DCAN Toulon	7 Oct 1975	15 July 1976

Displacement, tons: 190 full load
Dimensions, feet (metres): 113.8 × 21.6 × 7.5 *(34.7 × 6.6 × 2.3)*
Main machinery: 2 Baudouin DP8 diesels; 960 hp(m) *(706 kW)*; 2 shafts; cp props
Speed, knots: 12. **Range, n miles:** 800 at 12 kt
Complement: 6 (2 officers) plus 6 scientists
Radars: Navigation: Decca; I-band.

Comment: Employed on ammunition trials for DCN off Toulon. Service life extended to 2006 but no replacement planned. ***UPDATED***

DENTI   *9/2004*, Schaeffer/Marsan* / 1042194

0 + 1 DUPUY DE LÔME INTELLIGENCE COLLECTION SHIP (AGIH)

Name	No	Builders	Laid down	Launched	Commissioned
DUPUY DE LÔME	A 759	Royal Niestern Sander, Delfzijl	1 Dec 2002	27 Mar 2004	Mar 2006

Displacement, tons: 3,100 standard; 4,000 full load
Dimensions, feet (metres): 333.8 × 51.7 × 16.1 *(101.8 × 15.8 × 4.9)*
Main machinery: 2 MaK 9M25 diesels; 8,130 hp *(6.1 MW)*; 2 shafts
Speed, knots: 16. **Range, n miles:** To be announced
Complement: 32 + 78 specialists
Guns: 2—12.7 mm MGs.

Programmes: Programme initiated 29 October 2001. Contract awarded 14 January 2002 to Thales Naval France (for the mission system) and Compagnie Nationale de Navigation to procure and maintain the vessel for initial five year period. Installation of the MINREM mission system started at Toulon in January 2005. The ship is to replace *Bougainville*.
Structure: The ship is to have a design life of 30 years and is to be fitted with a flight deck and underway replenishment facilities.
Operational: To be fitted with both COMINT and ELINT equipment. The ship is to be available for 350 days a year. ***UPDATED***

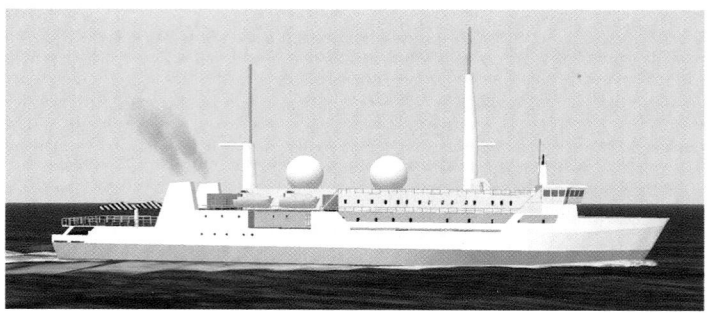

DUPUY DE LÔME (artist's impression)   *2002, Thales* / 0096691

1 BOUGAINVILLE CLASS (AGIH)

Name	No	Builders	Launched	Commissioned
BOUGAINVILLE	L 9077	Chantier Dubigeon, Nantes	3 Oct 1986	25 Mar 1988

Displacement, tons: 3,310 standard; 4,870 full load
Dimensions, feet (metres): 372.3; 344.4 wl × 55.8 × 14.1 *(113.5; 105 × 17 × 4.3)*
Flight deck, feet (metres): 85.3 × 55.8 *(26 × 17)*
Main machinery: 2 SACM AGO 195 V12 RVR diesels; 4,810 hp(m) *(3.6 MW)* sustained; 2 shafts; LIPS cp props; bow thruster; 400 hp(m) *(294 kW)*
Speed, knots: 15. **Range, n miles:** 6,000 at 12 kt
Complement: 53 (5 officers) plus 10 staff
Military lift: 500 troops for 8 days; 1,180 tons cargo; 2 LCU in support or 10 LCP plus 2 LCM for amphibious role
Missiles: SAM: 2 Matra Simbad twin launchers (may be fitted).
Guns: 2—12.7 mm MGs.
Radars: Navigation: 2 Decca 1226; I-band.

Helicopters: Platform for 2 Super Frelon.

Programmes: Ordered November 1984 for the Direction du Centre d'Experimentations Nucléaires (DIRCEN). As Chantier Dubigeon closed down after her launch she was completed by Chantiers de l'Atlantique of the Alsthom group. Bâtiment de Transport et de Soutien (BTS).
Modernisation: Conversion 30 November 1998, Syracuse II SATCOM and communications intercept equipment fitted.
Structure: Well size is 78 × 10.2 m *(256 × 33.5 ft)*. It can receive tugs and one BSR or two CTMs, a supply tender of the Chamois class, containers, mixed bulk cargo. Has extensive repair workshops and repair facilities for helicopters. Can act as mobile crew accommodation and has medical facilities. Storerooms for spare parts, victuals and ammunition. Hull to civilian standards. Carries a 37 ton crane.
Operational: Returned to France from Papeete in November 1998. Can dock a 500 ton ship. Based at Toulon since November 1999 and has replaced *Berry* as an AGI. To be replaced by *Dupuy de Lôme* in 2006 and is planned to replace *Loire* as MCM support ship. ***UPDATED***

BOUGAINVILLE   *3/2004*, B Prézelin* / 1042238

3 LAPÉROUSE (BH2) CLASS (AGS)

Name	No	Builders	Launched	Commissioned
LAPÉROUSE	A 791	Lorient Naval Dockyard	14 Nov 1986	20 Apr 1988
BORDA	A 792	Lorient Naval Dockyard	14 Nov 1986	16 June 1988
LAPLACE	A 793	Lorient Naval Dockyard	9 Nov 1988	5 Oct 1989

Displacement, tons: 970 standard; 1,100 full load
Dimensions, feet (metres): 193.5 × 35.8 × 13.8 *(59 × 10.9 × 4.2)*
Main machinery: 2 Wärtsilä UD 30 V12 M6D diesels; 2,500 hp(m) *(1.84 MW)*; 2 cp props; auxiliary propulsion; electric motor and 160 hp(m) *(120 kW)* bow thruster
Speed, knots: 15. **Range, n miles:** 6,000 at 12 kt
Complement: 31 (3 officers) plus 11-18 scientists
Guns: 2—7.5 mm MGs
Radars: Navigation: Decca Bridgemaster (DRBN 38A) (A 791, 792); Furuno (A 793); I-band.
Sonars: Thomson Sintra DUBM 42 or DUBM 21C (A 791); active search; high frequency.
EG & G towed sidescan sonar.
Kongsberg/Simrad EM 1002 S shallow water multipath echo sounder (95 kHz); Thales SMM II magnetometer; sediment echo sounder

Comment: Ordered under 1982 and 1986 estimates, first two on 24 July 1984, third 22 January 1986 and fourth *(Arago* — converted in 2002 to patrol craft) on 12 April 1988. BH2 (Bâtiments Hydrographiques de 2e classe). To carry 2-3 new VH 8 survey launches. Based at Brest.
UPDATED

BORDA
*11/2003 *, P Marsan /* 1042239

1 RESEARCH SHIP (AGMH)

Name	No	Builders	Launched	Commissioned
MONGE	A 601	Chantiers de l'Atlantique, St Nazaire	6 Oct 1990	5 Nov 1992

Displacement, tons: 17,160 standard; 21,040 full load
Dimensions, feet (metres): 740.1 × 81.4 × 25.3 *(225.6 × 24.8 × 7.7)*
Main machinery: 2 SEMT-Pielstick 8 PC2.5 L 400 diesels; 10,400 hp(m) *(7.65 MW)* sustained; 1 shaft; LIPS cp props; bow thruster; 1,360 hp(m) *(1 MW)*
Speed, knots: 15.8. **Range, n miles:** 15,000 at 15 kt
Complement: 120 (10 officers) plus 100 military and civilian technicians
Guns: 2 Giat F2 20 mm. 2—12.7 mm MGs.
Combat data systems: Tavitac 2000 for trials.
Radars: Air search: Thomson-CSF DRBV 15C; E/F-band.
Missile tracking: Thomson-CSF Stratus; L-band; 2 Gascogne; 2 Armor; Savoie; 5 Antarès.
Navigation: Two Racal Decca (DRBN 34A) (one for helo control); I-band.
Helicopters: 2 Super Frelon or Alouette III.

Comment: Ordered 25 November 1988. Rated as a BEM (Bâtiment d'Essais et de Mesures). Laid down 26 March 1990, and launched 6 October 1990. She has 14 telemetry antennas; optronic tracking unit; LIDAR; Syracuse SATCOM. Flume tank stabilisation restricts the ship to a maximum of 9° roll at slow speed in Sea State 6. Flagship of the Trials Squadron. Used for space surveillance by the French Space Agency (CNES) and for M 45 and M 51 ballistic missile tests.
UPDATED

MONGE
*4/2004 *, H M Steele /* 1042256

1 LAPÉROUSE CLASS (MCD/BEGM)

Name	No	Builders	Launched	Commissioned
THÉTIS (ex-*Nereide*)	A 785	Lorient Naval Dockyard	14 Dec 1986	9 Nov 1988

Displacement, tons: 883 standard; 1,015 full load
Dimensions, feet (metres): 183.4 × 35.8 × 12.5 *(55.9 × 10.9 × 3.8)*
Main machinery: 2 Uni Diesel UD 30 V16 M4 diesels; 2,710 hp(m) *(1.99 MW)* sustained; 1 shaft; cp prop
Speed, knots: 15. **Range, n miles:** 6,000 at 12 kt
Complement: 38 (2 officers) plus 7 passengers
Guns: 2—12.7 mm MGs.
Radars: Navigation: Racal Decca Bridgemaster (DRBN 38A); I-band.
Sonars: VDS; Thomson Sintra DUBM 42 and DUBM 60A; active search; high frequency.

Comment: Same hull as Lapérouse class. Classified as Bâtiment Experimental Guerre de Mines (BEGM). Operated by the Centre d'Études, d'Instruction et d'Entraînement de la Guerre des Mines (CETIEGM) in Brest. Launched 19 March 1988. Renamed to avoid confusion with Y 700. Equipped to conduct trials on all underwater weapons and sensors for mine warfare. Can lay mines. Can support six divers. Fitted with the Thomson Sintra Lagadmor mine warfare combat system designed for the cancelled Narvik class. Also used for experiments with Propelled Variable Depth Sonar system.
UPDATED

THÉTIS
*12/2004 *, B Prézelin /* 1042195

AUXILIARIES

4 DURANCE CLASS (UNDERWAY REPLENISHMENT TANKERS) (AORHM)

Name	No	Builders	Laid down	Launched	Commissioned
MEUSE	A 607	Brest Naval Dockyard	2 June 1977	2 Dec 1978	21 Nov 1980
VAR	A 608	Brest Naval Dockyard	8 May 1979	1 June 1981	29 Jan 1983
MARNE	A 630	Brest Naval Dockyard	4 Aug 1982	2 Feb 1985	16 Jan 1987
SOMME	A 631	Normed, la Seyne	3 May 1985	3 Oct 1987	7 Mar 1990

Displacement, tons: 7,600 (A 607); 7,800 (others) standard; 17,900 (A 607); 18,500 (others) full load
Dimensions, feet (metres): 515.9 × 69.5 × 38.5 *(157.3 × 21.2 × 10.8)*
Main machinery: 2 SEMT-Pielstick 16 PC2.5 V 400 diesels; 20,800 hp(m) *(15.3 MW)* sustained; 2 shafts; LIPS cp props
Speed, knots: 19. **Range, n miles:** 9,000 at 15 kt
Complement: 164 (10 officers) plus 29 spare

Cargo capacity: 5,000 tons FFO; 3,200 diesel; 1,800 TR5 Avcat; 130 distilled water; 170 victuals; 150 munitions; 50 naval stores *(Meuse).* 5,090 tons FFO; 3,310 diesel; 1,090 TR5 Avcat; 260 distilled water; 180 munitions; 15 stores *(Var, Somme and Marne)*

Missiles: SAM: 3 (1 in A 607) Matra Simbad twin launchers; Mistral; IR homing to 4 km *(2.2 n miles)*; warhead 3 kg.

Guns: 1—40 mm. 2—20 mm (A 607). 4—12.7 mm MGs.
Countermeasures: ESM/ECM.
Combat data systems: Syracuse 2 SATCOM. OPSMER command support system (fitted for BCR ships).
Radars: Navigation: 2 Racal Decca Bridgemaster (DRBN 38A); I-band.

Helicopters: 1 Dauphin or Lynx Mk 4.

Programmes: One classed as Pétroliers Ravitailleurs d'Escadres (PRE). Three classed as Bâtiments de Commandement et de Ravitaillement (BCR; Command and Replenishment Ships).
Modernisation: EW equipment fitted to improve air defences under the 3A programme in 1996-99. Simbad SAM may be carried at bridge deck level.
Structure: Four beam transfer positions and two astern, two of the beam positions having heavy transfer capability. *Var, Marne* and *Somme* differ from *Meuse* in several respects. The bridge extends further aft, boats are located either side of the funnel and a crane is located between the gantries. Also fitted with Syracuse SATCOM.
Operational: *Var, Marne* and *Somme* are designed to carry a Maritime Zone staff or Commander of a Logistic Formation and a commando unit of up to 45 men. Capable of accommodating 250 men. Assigned to FAN with one of the three BCR ships deployed to the Indian Ocean as a Flagship. To be replaced after 2010 by new ships.
Sales: One to Australia built locally; two of similar but smaller design to Saudi Arabia. One to Argentina in July 1999.
UPDATED

MEUSE
*8/2004 *, B Prézelin /* 1042193

1 MAINTENANCE and REPAIR SHIP (ADH)

Name	No	Builders	Launched	Commissioned
JULES VERNE (ex-*Achéron*)	A 620	Brest Naval Dockyard	30 May 1970	17 Sep 1976

Displacement, tons: 7,815 standard; 10,250 full load
Dimensions, feet (metres): 482.2 × 70.5 × 21.3 *(147 × 21.5 × 6.5)*
Main machinery: 2 SEMT-Pielstick 12 PC2.2 V 400 diesels; 13,600 hp(m) *(10.1 MW)* sustained; 1 shaft
Speed, knots: 19. **Range, n miles:** 9,500 at 18 kt
Complement: 132 plus 135 for support
Guns: 2 Bofors 40 mm/60. 4—12.7 mm MGs.
Radars: Navigation: 1 DRBN 34A; 1 DRBN 38A; I-band.
Helicopters: 3 SA 319B Alouette III.

Comment: Ordered in 1961 budget, originally as an Armament Supply Ship. Role and design changed whilst building-now rated as Engineering and Electrical Maintenance Ship. Also equipped with 16-bed hospital. Serves in Indian Ocean, providing general support for all ships. Carries stocks of torpedoes and ammunition. Refit in France November 1988/June 1989 and another refit at Brest from January to June 1995. Based at Toulon from December 1997 and assigned to FAN. Refitted in 1998 after a collision with *Var*. Super Frelon and Cougar helicopters can be landed on the flight deck. Service life 2012. **UPDATED**

JULES VERNE *8/2004*, B Prézelin /* 1042192

1 SUPPORT/TRAINING SHIP (AG/AX)

Name	No	Builders	Commissioned
D'ENTRECASTEAUX	P 674 (ex-A 757)	Brest Naval Dockyard	8 Oct 1971

Displacement, tons: 1,925 standard; 2,450 full load
Dimensions, feet (metres): 292 × 42.7 × 14.4 *(89 × 13 × 4.4)*
Main machinery: Diesel-electric; 2 diesel generators; 2,720 hp(m) *(2 MW)*; 2 motors; 2 shafts; LIPS cp props; auxiliary propulsion; 2 Schottel trainable and retractable props
Speed, knots: 15. **Range, n miles:** 10,000 at 12 kt
Complement: 55 (5 officers) plus 50 passengers
Radars: Navigation: 1 Racal Decca 1226; 1 DRBN 38A; I-band.
Helicopters: 1 SA 319B Alouette III.

Comment: This ship was originally designed for oceanographic surveys. Telescopic hangar. Based at Brest from September 1995. After being replaced by *Beautemps Beaupré* in her survey role, refitted in 2004 to undertake support role (including pollution control) and training until at least 2006. **UPDATED**

D'ENTRECASTEAUX *11/2004*, B Prézelin /* 1042204

1 RHIN CLASS (SUPPORT SHIP) (AGH/AR)

Name	No	Builders	Commissioned
LOIRE	A 615	Lorient Naval Dockyard	17 Oct 1967

Displacement, tons: 2,050 standard; 2,445 full load
Dimensions, feet (metres): 333.0 × 45.3 × 12.5 *(101.5 × 13.8 × 3.8)*
Main machinery: 2 SEMT-Pielstick 12 PA4 V 185VG diesels; 4,000 hp(m) *(2.94 MW)*; 1 shaft
Speed, knots: 16.5. **Range, n miles:** 13,000 at 13 kt
Complement: 156 (12 officers)
Guns: 3 Bofors 40 mm/60. 3—12.7 mm MGs.
Radars: Navigation: 2 Racal Decca 1226; I-band.
Helicopters: 1-3 SA 310B Alouette III *(Loire)*.

Comment: Has a 5 ton crane and carries two LCPs. Used for minesweeper support at Brest. To decommission in 2008 and may be replaced by *Bougainville*. **UPDATED**

LOIRE *6/2004*, Michael Nitz /* 1042255

5 CHAMOIS CLASS (SUPPLY TENDERS) (AG/ATS/YDT/YPC/YPT)

Name	No	Builders	Commissioned
TAAPE	A 633	La Perrière, Lorient	2 Nov 1983
ÉLAN	A 768	La Perrière, Lorient	7 Apr 1978
CHEVREUIL	A 774	La Perrière, Lorient	7 Oct 1977
GAZELLE	A 775	La Perrière, Lorient	13 Jan 1978
ISARD	A 776	La Perrière, Lorient	15 Dec 1978

Displacement, tons: 315 light; 505 full load
Dimensions, feet (metres): 136.1 × 24.6 × 10.5 *(41.5 × 7.5 × 3.2)*
Main machinery: 2 SACM AGO 175 V16 diesels; 2,700 hp(m) *(1.98 MW)*; 2 shafts; cp props; bow thruster
Speed, knots: 14.2. **Range, n miles:** 6,000 at 12 kt
Complement: 12 plus 12 spare berths
Radars: Navigation: Racal Decca 1226; I-band.

Comment: Similar to the standard fish oil rig support ships. Can act as tugs, oil pollution vessels, salvage craft (two 30 ton and two 5 ton winches), coastal and harbour controlled minelaying, torpedo recovery, diving tenders and a variety of other tasks. Bollard pull 25 tons. Can carry 100 tons of stores on deck or 125 tons of fuel and 40 tons of water or 65 tons of fuel and 120 tons of water. *Taape* ordered in March 1982 from La Perrière-of improved design but basically similar with bridge one deck higher. *Elan* based at Cherbourg, remainder at Toulon. *Isard* serves as a special diving support ship with an extra deckhouse. To be replaced in this role by *Alize* in 2005. Two paid off so far, one of which transferred to Madagascar in May 1996. To be replaced by eight Bâtiments de Soutien et d'Assistance Hauturiers (BSAH) from 2008. **UPDATED**

TAAPE *8/2004*, Schaeffer/Marsan /* 1042203

0 + 1 ALIZE CLASS (DIVING TENDER) (YDT)

ALIZE A 760

Displacement, tons: 1,100 standard; 1,500 full load
Dimensions, feet (metres): 196.8 × 42.6 × 13.1 *(60.0 × 13.0 × 4.0)*
Main machinery: 2 diesels; 2,800 hp *(2.1 MW)*; 2 shafts; bow thruster
Speed, knots: 14. **Range, n miles:** 7,500 at 12 kt
Complement: 17 (3 officers) plus 30 passengers
Guns: 2—12.7 mm MGs.
Radars: Navigation: Racal Decca Bridgemaster (DRBN 38A); I-band.

Comment: Under construction at SOCARENAM, Boulogne. Laid down January 2004 for completion in December 2005. To replace *Isard* in diving support role. **NEW ENTRY**

2 RR 4000 TYPE (SUPPLY TENDERS) (AFL)

Name	No	Builders	Commissioned
RARI	A 634	Breheret, Conéron	21 Feb 1985
REVI	A 635	Breheret, Conéron	9 Mar 1985

Displacement, tons: 900 light; 1,577 full load
Dimensions, feet (metres): 167.3 × 41.3 × 13.1 *(51 × 12.6 × 4)*
Main machinery: 2 SACM AGO 195 V12 diesels; 4,410 hp(m) *(3.24 MW)*; 2 shafts; cp props; 2 bow thrusters
Speed, knots: 14.5. **Range, n miles:** 6,000 at 12 kt
Complement: 22 plus 18 passengers
Guns: 2—7.62 mm MGs.
Radars: Navigation: Racal Decca Bridgemaster (DRBN 38A); I-band.

Comment: Two 'remorqueurs ravitailleurs' built for le Centre d'Expérimentation du Pacifique. Can carry 400 tons of cargo on deck. Bollard pull 47 tons. *Revi* based at Papeete and *Rari* at Brest. **UPDATED**

RARI *6/2004*, B Prézelin /* 1042202

For details of the latest updates to *Jane's Fighting Ships* online and to discover the additional information available exclusively to online subscribers please visit

jfs.janes.com

1 TRANSPORT LANDING SHIP (LSL)

Name	No	Builders	Commissioned
GAPEAU	L 9090	Chantier Serra, la Seyne	2 Oct 1987

Displacement, tons: 563 standard; 1,090 full load
Dimensions, feet (metres): 216.5 × 41.0 × 11.2 *(66 × 12.5 × 3.4)*
Main machinery: 2 diesels; 550 hp(m) *(404 kW)*; 2 shafts
Speed, knots: 11. **Range, n miles:** 1,900 at 10 kt
Complement: 6 + 30 scientists
Cargo capacity: 460 tons
Radars: Navigation: Racal Decca 1226 and Furuno FRS 1000; I-band.

Comment: Supply ship with bow doors. Operates for Centre d'Essais de la Mediterranée, Levant Island (missile range).

UPDATED

GAPEAU 5/2003, *Per Körnefeldt* / 0569982

1 MOORING VESSEL (ABU)

TELENN MOR Y 692

Comment: 392 tons with 450 hp(m) *(335 kW)* diesel. Commissioned on 16 January 1986 and based at Brest. *UPDATED*

TELENN MOR 7/2004*, *B Prézelin* / 1042201

6 ARIEL CLASS (TRANSPORTS) (YFB)

FAUNE Y 613	**NEREIDE** Y 700	**NAIADE** Y 702
DRYADE Y 662	**ONDINE** Y 701	**ELFE** Y 741

Displacement, tons: 195 standard; 225 full load
Dimensions, feet (metres): 132.8 × 24.5 × 10.8 *(40.5 × 7.5 × 3.3)*
Main machinery: 2 SACM MGO or Poyaud diesels; 1,640 hp(m) *(1.21 MW)* or 1,730 hp(m) *(1.27 MW)*; 2 shafts
Speed, knots: 15.3. **Range, n miles:** 940 at 14 kt
Complement: 9
Radars: Navigation: Racal Decca 1226; I-band.

Comment: All built 1964-80 by Société Française de Construction Naval (ex-Franco-Belge) except for *Nereide*, *Ondine* and *Naiade* by DCAN Brest. Can carry 400 passengers (250 seated). To be decommissioned 2005-06. Based at Brest. *UPDATED*

NAIADE 10/2004*, *J Y Robert* / 1042252

1 LA PRUDENTE CLASS BUOY TENDER (ABU)

Name	No	Builders	Launched	Commissioned
LA PERSÉVÉRANTE	Y 750	AC de la Rochelle shipyard	14 May 1968	3 Mar 1970

Displacement, tons: 446 tons standard; 626 full load
Dimensions, feet (metres): 142.7 × 32.8 × 9.2 *(43.5 × 10 × 2.8)*
Main machinery: Diesel-electric; 2 diesels; 600 hp *(440 kW)*. 1 motor; 1 shaft
Speed, knots: 10. **Range, n miles:** 4,000 at 10 kt
Complement: 6
Radars: Decca RM 914; I-band.

Comment: Last survivor of three. *La Fidele* sank off Cherbourg on 30 April 1997 after an explosion while she was destroying old ammunition; *La Prudente* paid off in late 2000. *La Perséverante* was to have been decommissioned in late 2001 but her service life has been extended to 2008. Operates in Toulon area with a reduced crew. Capacity of forward gantry: 25 tonnes. *NEW ENTRY*

LA PERSÉVÉRANTE 2/2004*, *B Prézelin* / 1042200

1 DIVING TENDER (YDT)

POSÉIDON A 722

Displacement, tons: 220 full load
Dimensions, feet (metres): 132.9 × 23.6 × 7.3 *(40.5 × 7.2 × 2.2)*
Main machinery: 1 diesel; 600 hp(m) *(441 kW)*; 1 shaft
Speed, knots: 13
Complement: 15 (1 officer) plus 27 swimmers
Radars: Navigation: Racal Decca 1226; I-band.

Comment: Base ship for assault swimmers at Toulon. Built in St Malo and completed 6 August 1975. *VERIFIED*

POSÉIDON 9/2002, *Schaeffer/Marsan* / 0528854

15 HARBOUR CRAFT (YFL/YP/YTR)

Y 754	**Y 765**	**Y 777**	**Y 783-786**	**Y 789**
Y 762-763	**Y 772** (ex-P 772)	**Y 779-781**	**Y 787**	

Displacement, tons: 14.5 standard; 19.5 full load
Dimensions, feet (metres): 47.9 × 15.1 × 3.3 *(14.9 × 4.6 × 0.9)*
Main machinery: 2 Baudouin diesels; 1,000-750 hp(m) *(735-551 kW)*; 2 shafts
Speed, knots: 25-17. **Range, n miles:** 400 at 11 kt
Complement: 4

Comment: Y 772 built in 1975, the remainder between 1988 and 1994. Details are for VPIL 14 class. Y 754 and Y 786 are VSTP 14 class based at Brest. Y 762 and 765 are VSTA 14 class based at Toulon and Brest respectively. Y 763 is a VSC 14 class based at Mayotte. Y 779-781 are VPIL class (pilot craft) based at Cherbourg, Toulon and Brest respectively. Y 783-785 are VIR 14 class (firefighting) based at Brest, Toulon and Cherbourg respectively. Y 787 is a VTP 14 class based at Nouméa. Y 772 is a VPIL 14 class based at Toulon. Y 789 is a VSTP 13 class based at Papeete and Y 777 is a VSR class used for radiological monitoring craft and based at Brest. Y 776 is in reserve. *UPDATED*

Y 777 12/2004*, *B Prézelin* / 1042198

Y 783 10/2004*, *J Y Robert* / 1042253

2 RANGE SUPPORT VESSELS (YFRT)

ATHOS A 712   ARAMIS A 713

Displacement, tons: 89 standard; 108 full load
Dimensions, feet (metres): 105.3 × 21.3 × 6.2 *(32.1 × 6.5 × 1.9)*
Main machinery: 2 SACM UD 33V12 M5 diesels; 3,950 hp(m) *(2.94 MW)*; 2 shafts
Speed, knots: 28. **Range, n miles:** 1,500 at 15 kt
Complement: 13 plus 6 passengers
Guns: 1—12.7 mm MG.
Radars: Navigation: Racal Decca 1226 (A 712); Furuno (A 713); I-band.

Comment: Built by Chantiers Navals de l'Esterel for Missile Trials Centre of Les Landes (CEL). Based at Bayonne, forming Groupe des Vedettes de l'Adour. A 712 commissioned 20 November 1979 and A 713 on 9 September 1980. Classified as Range Safety Craft from July 1995. *Athos* completed refit at Cherbourg in April 2003.

UPDATED

ATHOS   *9/1983, van Ginderen Collection*

10 DIVING TENDERS (YDT)

CORALLINE A 790	LISERON Y 793	GENÊT Y 796
DIONÉE Y 790	MAGNOLIA Y 794	GIROFLÉE Y 797
MYOSOTIS Y 791	AJONC Y 795	ACANTHE Y 798
GARDÉNIA Y 792		

Displacement, tons: 49 full load
Dimensions, feet (metres): 68.9 × 16.1 × 5.2 *(21.0 × 4.9 × 1.6)*
Main machinery: 2 diesels; 264 hp(m) *(194 kW)*; 2 shafts
Speed, knots: 13
Complement: 4 plus 14 divers

Comment: Diving tenders built at Lorient. First one delivered in February 1990. *Coralline* is used for radioactive monitoring in Cherbourg. *Y 794* and *Y 798* based at Cherbourg. *Y 790, Y 791, Y 792, Y 795* and *Y 797* based at Toulon. *Y 793* and *Y 796* based at Brest. Rated as 'Vedettes d'Instruction Plongée (VIP)', divers training craft, and 'Vedettes d'Intervention Plongeurs-Démineurs (VIPD)', clearance diving team support craft. *UPDATED*

AJONC   *2/2004*, Schaeffer/Marsan /* 1042199

2 PHAÉTON CLASS (TOWED ARRAY TENDERS) (YAG)

PHAÉTON Y 656   MACHAON Y 657

Displacement, tons: 72 full load
Dimensions, feet (metres): 63.0 × 22.3 × 3.9 *(19.2 × 6.8 × 1.2)*
Main machinery: 1 diesel; 660 hp(m) *(485 kW)*; waterjet
Speed, knots: 9
Complement: 4

Comment: 18.6 m catamarans built in 1993-94 at Brest. Water-jet propulsion, speed 8 kt. Hydraulic crane and winch to handle submarine towed arrays. *Phaéton* based at Toulon, *Machaon* at Brest. *UPDATED*

MACHAON   *7/2004*, B Prézelin /* 1042197

42 HARBOUR SUPPORT CRAFT

Comment: There are 11 oil barges (CICGH), one of which is of 1,200 tonnes and the rest between 100 and 800 tonnes, eight 400 tonne oily bilge barges (CIEM), three anti-pollution barges (800 tonne BAPM, and two 400 tonne CIEP), and seven water barges (CIE, 120 to 400 tonnes). Some self-propelled. Also 12 self-propelled YFUs (CHA 27-38), and two 15 m Sea Truck craft.

UPDATED

CHA 30   *10/1999, van Ginderen Collection /* 0069961

1 FLOATING DOCK and 5 FLOATING CRANES

Comment: The dock is of 3,800 tons capacity, built at Brest in 1975. Based at Papeete. 150 × 33 m. The cranes have lifts up to 15 tons and are self-propelled. Three at Toulon and one each at Brest and Cherbourg.

UPDATED

TRAINING SHIPS

Notes: In addition there are two naval school tenders of 100 tons, *Chimère* Y 706 and *Farfadet* Y 711, built in 1971, and an ex-fishing vessel built in 1932, *La Grande Hermine* A 653 which works for the navigation school.

CHIMÈRE   *7/2002, Per Körnefeldt /* 0528859

8 LÉOPARD CLASS (AXL)

Name	No	Builders	Commissioned
LÉOPARD	A 748	ACM, St Malo	4 Dec 1982
PANTHÈRE	A 749	ACM, St Malo	4 Dec 1982
JAGUAR	A 750	ACM, St Malo	18 Dec 1982
LYNX	A 751	La Perrière, Lorient	18 Dec 1982
GUÉPARD	A 752	ACM, St Malo	1 July 1983
CHACAL	A 753	ACM, St Malo	10 Sep 1983
TIGRE	A 754	La Perrière, Lorient	1 July 1983
LION	A 755	La Perrière, Lorient	10 Sep 1983

Displacement, tons: 335 standard; 463 full load
Dimensions, feet (metres): 141 × 27.1 × 10.5 *(43 × 8.3 × 3.2)*
Main machinery: 2 SACM MGO 175 V16 ASHR diesels; 2,200 hp(m) *(1.62 MW)*; 2 shafts
Speed, knots: 15. **Range, n miles:** 4,100 at 12 kt
Complement: 14 plus 21 trainees
Guns: 2—12.7 mm MGs.
Radars: Navigation: Racal Decca 1226; I-band.

Comment: First four ordered May 1980. Further four ordered April 1981. Form 20ème Divec (Training division) for shiphandling training and occasional EEZ patrols.

UPDATED

CHACAL   *2/2004*, B Prézelin /* 1042214

2 GLYCINE CLASS (AXL)

Name	No	Builders	Commissioned
GLYCINE	A 770	Socarenam, Boulogne	11 Apr 1992
EGLANTINE	A 771	Socarenam, Boulogne	9 Sep 1992

Displacement, tons: 250 standard; 295 full load
Dimensions, feet (metres): 92.8 × 25.3 × 12.5 *(28.3 × 7.7 × 3.8)*
Main machinery: 1 Baudouin 12P15-2SR diesel; 800 hp(m) *(588 kW)*; 1 shaft; cp prop
Speed, knots: 10. **Range, n miles:** 3,600 at 10 kt
Complement: 10 + 16 trainees
Radars: Navigation: 4 Furuno; I-band.

Comment: Trawler type. Three more built in 1995-96 as route survey craft (included under *Mine Warfare Forces* section). ***UPDATED***

GLYCINE *8/2004*, B Prézelin /* 1042213

2 LA BELLE POULE CLASS (AXS)

L'ÉTOILE A 649   LA BELLE POULE A 650

Displacement, tons: 275 full load
Dimensions, feet (metres): 127 × 24.3 × 12.1 *(37.5 × 7.4 × 3.7)*
Main machinery: 1 Sulzer diesel; 300 hp(m) *(220 kW)*; 1 shaft
Speed, knots: 9 (diesel)
Complement: 20 (1 officer) plus 20 trainees

Comment: Auxiliary sail vessels. Built by Chantiers de Normandie (Fécamp) and launched 7 July 1932 and 8 February 1932 respectively. Accommodation for three officers, 30 cadets, five petty officers, 12 men. Sail area 450 m². Attached to Naval School. A 649 major overhaul in 1994. ***UPDATED***

LA BELLE POULE *9/2004*, B Prézelin /* 1042212

1 SAIL TRAINING SHIP (AXS)

Name	No	Builders	Launched
MUTIN	A 652	Chaffeteau, Les Sables d'Olonne	18 Mar 1927

Displacement, tons: 57 full load
Dimensions, feet (metres): 108.3 × 21 × 11.2 *(33 × 6.4 × 3.4)*
Main machinery: 1 diesel; 112 hp(m) *(82 kW)*; 1 auxiliary prop
Speed, knots: 6 (diesel). **Range, n miles:** 860 at 6 kt
Complement: 12 + 6 trainees

Comment: Attached to the Navigation School. Has a sail area of 312 m². This is the oldest ship in the French Navy. Used by the SOE during the Second World War. ***UPDATED***

MUTIN *9/2004*, B Prézelin /* 1042211

TUGS

2 TYPE RP 50 (COASTAL/HARBOUR TUGS) (YTM)

ESTEREL A 641 (ex-Y 601)   LUBERON A 642 (ex-Y 602)

Displacement, tons: 510 standard; 670 full load
Dimensions, feet (metres): 119.1 oa; 116.5 wl × 36.1 × 16.4 *(36.3; 35.5 × 11 × 5)*
Main machinery: 2 ABC 8 DZ 1000. 179 diesels; 2 Voith-Schneider 28 GII propulsors; 5,120 hp(m) *(3,812 kW)*.
Speed, knots: 14. **Range, n miles:** 1,500 at 12 kt
Complement: 8

Comment: Ordered 15 December 2000; built by SOCARENAM, Boulogne. *Esterel* delivered 27 March 2002 and *Lubéron* 4 July 2002. Based at Toulon to assist *Charles de Gaulle* in harbour. Bollard pull 52 tonnes; 1,350 kN towing winch; fire fighting equipment; 20 cubic metre tank for pollution control dispersal agent. Classified as 'Remorqueurs portuaires de 50 tonnes de traction' (RP 50, 50 tonne bollard pull harbour tugs). ***UPDATED***

ESTEREL *8/2004*, B Prézelin /* 1042210

2 OCEAN TUGS (ATA)

MALABAR A 664   TENACE A 669

Displacement, tons: 1,080 light; 1,454 full load
Dimensions, feet (metres): 167.3 × 37.8 × 18.6 *(51 × 11.5 × 5.7)*
Main machinery: 2 Krupp MaK 9 M 452 AK diesels; 4,600 hp(m) *(3.38 MW)*; 1 shaft; Kort nozzle
Speed, knots: 15. **Range, n miles:** 9,500 at 13 kt
Complement: 56 (2 officers)
Radars: Navigation: Racal Decca RM 1226; I-band.
Racal Decca 060; I-band.

Comment: *Malabar* and *Tenace* built by J. Oelkers, Hamburg. *Tenace* commissioned 15 November 1973, and *Malabar* on 3 February 1976. Based at Brest. Carry firefighting equipment. Bollard pull, 60 tons. One of the class to Turkey in 1999. To be replaced by BSAH by 2010. ***UPDATED***

MALABAR *7/2004*, B Prézelin /* 1042209

3 BÉLIER CLASS (YTB)

BÉLIER A 695   BUFFLE A 696   BISON A 697

Displacement, tons: 356 light; 500 full load
Dimensions, feet (metres): 104.3 × 30.2 × 13.8 *(31.8 × 9.2 × 4.2)*
Main machinery: 2 SACM AGO 195 V8 CSHR diesels; 2,600 hp(m) *(1.91 MW)*; 2 Voith-Schneider props
Speed, knots: 11
Complement: 12

Comment: Built by DCN at Cherbourg. *Bélier* commissioned 10 July 1980, *Buffle* on 19 July 1980, *Bison* on 16 April 1981. A 695 and 697 based at Toulon and A 696 at Brest. Bollard pull, 25 tons. ***UPDATED***

BÉLIER *8/2004*, B Prézelin /* 1042208

16 + 6 FRÉHEL CLASS (COASTAL TUGS) (YTM)

FRÉHEL A 675	KÉRÉON (ex-*Sicie*) A 679	LARDIER Y 638	TAILLAT Y 642
SAIRE A 676	SICIÉ A 680	GIENS Y 639	NIVIDIC Y 643
ARMEN A 677	TAUNOA A 681	MENGAM Y 640	LE FOUR Y 647
LA HOUSSAYE A 678	RASCAS A 682	BALAGUIER Y 641	PORT CROS Y 649

Displacement, tons: 220 standard; 295 full load
Dimensions, feet (metres): 82 × 27.6 × 11.2 *(25 × 8.4 × 3.4)*
Main machinery: 2 SACM-Wärtsilä UD 30 V12 M3 diesels; 2 Voith-Schneider propulsors; 1,280 hp (m) *(941 kW)*; 1,360 hp(m) *(1 MW)* in later vessels
Speed, knots: 11. **Range, n miles:** 800 at 10 kt
Complement: 8 (coastal); 5 (harbour)
Radars: 1 Racal Decca; I-band.

Comment: Building at Lorient Naval et Industries shipyard (formerly Chantiers et Ateliers de la Perrière, now part of Leroux et Lotz) and at Boulogne by SOCARENAM. *Fréhel* in service 23 May 1989, based at Cherbourg, *Saire* 6 October 1989 at Cherbourg, *Armen* 6 December 1991 at Brest, *La Houssaye* 30 October 1992 at Lorient, *Kereon* 5 December 1992 at Brest. *Mengam* 6 October 1994 at Brest and *Sicié* 6 October 1994 at Toulon, *Giens* 2 December 1994 at Toulon, *Lardier* 12 March 1995 at Toulon, *Balaguier* 8 July 1995 at Toulon, *Taillat* 18 October 1995 at Toulon. *Taunoa* completed 9 March 1996 at Brest, *Nividic* on 13 December 1996 at Brest, *Port Cros* on 21 June 1997 at Brest, *Le Four* on 13 March 1998 at Brest and *Rascas* on 12 December 2003. Bollard pull 12 tons. Type RPC 12 coastal tugs, with 'A' pennant numbers and a crew of eight. Type RP12 harbour tugs with 'Y' pennant numbers and a crew of five. A further order for six craft is expected to bring the total class number to 23 units.

UPDATED

LA HOUSSAYE　　　　　　　　　　　　10/2004*, J Y Robert / 1042251

37 + 4 HARBOUR TUGS (YTL/YTR)

MÉSANGE Y 621	PAPAYER Y 740	P 13-24
BONITE Y 630	AIGUIÈRE Y 745	P 26-38
ROUGET Y 634	EMBRUN Y 746	P 101-104
MARTINET Y 636	P 6	

Comment: All between 65 and 100 tons. Those without names are pusher tugs. At least seven others in reserve. Older 700 hp and 250 hp tugs are being scrapped and replaced by harbour pushers (with Voith-Schneider propellors) or RPC 12 tugs. One to Senegal in 1998, two to Ivory Coast in 1999. four new 90 ton RP 10 class tugs launched on 4 November 2004 for entry into service in 2005. To be called *Morse, Otarie, Loutre* and *Phoque* with pennant numbers Y 771-774.

UPDATED

MÉSANGE　　　　　　　　　　　　6/2002, Schaeffer/Marsan / 0528864

PHOQUE　　　　　　　　　　　　11/2004*, P Marsan / 1042207

3 MAROA CLASS (YTM)

MAITO A 636	MAROA A 637	MANINI A 638

Displacement, tons: 228 standard; 280 full load
Dimensions, feet (metres): 90.5 × 27.2 × 11.5 *(27.6 × 8.3 × 3.5)*
Main machinery: 2 SACM-Wärtsilä UD 30 L6 M6 diesels; 1,280 hp(m) *(940 kW)*; 2 Voith-Schneider propulsors
Speed, knots: 11. **Range, n miles:** 1,200 at 11 kt
Complement: 8 + 2 spare
Radar: Navigation: Racal-Decca 1226; I-band

Comment: Built by SFCN and Villeneuve La Garenne (A 638) and formerly used at the CEP Nuclear Test Range. *Maito* commissioned 25 July 1984 and is based at Martinique. *Maroa* (commissioned 28 July 1984) and *Manini* (commissioned 12 September 1985) are both based at Papeete, Tahiti. Bollard pull, 12 tons. Fire-fighting water cannon.

VERIFIED

MAITO　　　　　　　　　　　　3/2003, A Sheldon-Duplaix / 0569975

1 ACTIF CLASS (YTM)

ACHARNÉ A 693

Displacement, tons: 293 full load
Dimensions, feet (metres): 89.9 × 24.6 × 14.8 *(27.4 × 7.5 × 4.5)*
Main machinery: 1 SACM MGO diesel; 1,050 hp(m) *(773 kW)* or 1,450 hp(m) *(1.07 MW)* (later ships); 1 shaft
Speed, knots: 11. **Range, n miles:** 2,400 at 10 kt
Complement: 15

Comment: Last of 12 coastal tugs commissioned 5 July 1974. Bollard pull 13 tons. Not to be replaced before 2010.

UPDATED

ACHARNÉ　　　　　　　　　　　　6/2003, Schaeffer/Marsan / 0569968

GOVERNMENT MARITIME FORCES

Notes: (1) From 29 April 2003, all ships and craft involved in maritime constabulary operations and public service bear a new livery of blue/white/red on their hull sides. This has also been applied to some naval vessels and to all those belonging to the Gendarmerie Maritime, Douanes Français and Affaires Maritimes. It is also to be applied to chartered vessels involved in pollution control and general support.
(2) All navy and government maritime forces share a common maritime picture network (Spatio.nav).

AUXILIARIES

Notes: (1) Support vessels are designated Bâtiment de Soutien de Haute Mer (BHSM). An Invitation to Tender has been issued for the charter of another ship to replace *Carangue*.
(2) It is planned to replace *Alcyon, Ailette,* the ocean-going tugs *Malabar* and *Tenace,* and the Chamois class tenders by up to eight 65 m Bâtiments de Soutien et d'Assistance Hauturiers (BSAH) from 2008. These new vessels are to have a bollard pull of 100 tonnes, to be capable of pollution control and to be fitted with undersea equipment. Some might be procured by the Navy and others chartered from civilian companies.
(3) In addition to the vessels listed below, the UK tug *Anglian Monarch* (1,480 grt) is chartered from Klyne Tugs Ltd, under a share agreement with the UK Maritime and Coast Guard Agency. Based at Dover.
(4) A contract was renewed with Abeilles International in July 2002 for emergency use of various harbour tugs.
(5) *Langevin* is chartered for submarine associated trials.

0 + 2 ULSTEIN UT 515 (SALVAGE AND RESCUE TUGS) (ARS)

Displacement, tons: 3,200 standard; 4,000 full load
Dimensions, feet (metres): 262.5 × 54.1 × 21.3 *(80.0 × 16.5 × 6.5)*
Main machinery: 4 MaK 8M32C diesels; 21,700 hp(m) *(16 MW)*; 2 shafts
Speed, knots: 19.5
Complement: 7
Radars: Navigation: I-band.

Comment: Contract awarded in November 2003 to Abeilles International for the procurement and operation (over eight years) of two Ulstein 515 salvage tugs, classified as Remorques d'Intervention, d'Assistance et de Sauvetage (RIAS). To be equipped with extensive fire-fighting and pollution control equipment. Vessels built by Myklebust, Norway for delivery in early 2005 and commissioning in April 2005. To be based at Brest and Cherbourg.   ***UPDATED***

1 ULSTEIN UT 710 (SALVAGE AND RESCUE TUG) (ARS)

ARGONAUTE (ex-*Island Patriot*)

Displacement, tons: 2,371 standard; 4,420 full load
Dimensions, feet (metres): 226.0 × 50.8 × 19.3 *(68.9 × 15.5 × 5.9)*
Main machinery: 2 Rolls Royce Bergen BRM-9 diesels; 10,800 hp(m) *(8.1 MW)*; 2 shafts
Speed, knots: 15.5. **Range, n miles:** 19,000 at 10 kt
Radars: Navigation: I-band.

Comment: Built by Aker-Brevik Construction AS, Norway, the ship was launched on 7 July 2003 and entered service with Island Offshore on 12 December 2003. Chartered by the French government in 2004. Based at Brest.   ***NEW ENTRY***

ARGONAUTE   *11/2004*, B Prézelin /* 1042206

2 ABEILLE FLANDRE CLASS (SALVAGE TUGS) (ARS)

ABEILLE FLANDRE (ex-*Neptun Suecia*)   ABEILLE LANGEDOC (ex-*Neptun Gothia*)

Displacement, tons: 3,800 full load
Dimensions, feet (metres): 207.7 × 48.2 × 22.6 *(63.4 × 14.7 × 6.9)*
Main machinery: 4 MaK 8M453AK diesels; 12,800 hp *(9.5 MW)*; 2 cp props; bow thruster
Speed, knots: 17
Complement: 12
Radars: Navigation: 1 Racal Decca Bridgemaster; 1 Racal Decca RMS 2080; I-band.

Comment: Built by Ulstein Hatho A/S, Norway and entered service in 1978 and 1979. On long-term charter from Abeilles International. Bollard pull 160 tons. *Abeille Flandre* is based at Toulon. *Abeille Langedoc* is currently based at Cherbourg but will move to La Pollice, near La Rochelle, in 2005.   ***UPDATED***

ABEILLE FLANDRE   *2/2004, B Prézelin /* 0573523

1 CARANGUE CLASS (SALVAGE AND RESCUE TUG) (ARS)

CARANGUE (ex-*Pilot Fish*, ex-*Smit Lloyd 119*, ex-*Maersk Handler*)

Displacement, tons: 2,500 full load
Dimensions, feet (metres): 211.3 × 45.3 × 16.7 *(64.4 × 13.8 × 5.1)*
Main machinery: 2 Nohars-Nohab F2116V-D diesels; 7,050 hp *(5.2 MW)*; 2 cp props; bow thruster
Speed, knots: 16. **Range, n miles:** 21,000 at 10 kt
Complement: 8
Radars: Navigation: I-band.

Comment: Built by Samsung SB, Koje, South Korea and entered service in 1980. On long-term charter from Abeilles International. Equipped with two fire-pumps, two water cannons and anti-pollution equipment. Based at Toulon.   ***UPDATED***

CARANGUE   *8/2004*, Schaeffer/Marsan /* 1042205

2 ALCYON CLASS (BUOY TENDERS) (ABU)

ALCYON (ex-*Bahram*)   AILETTE (ex-*Cyrus*)

Displacement, tons: 1,900 full load
Dimensions, feet (metres): 173.9 × 42.7 × 14.8 *(53.0 × 13.0 × 4.5)*
Main machinery: 2 Bergens-Normo KVMB-12 diesels; 5,200 hp *(3.9 MW)*; 2 cp props; 2 bow thrusters
Speed, knots: 14.5
Complement: 7
Radars: Navigation: Racal-Decca Bridgemaster and Furuno FR 1830; I-band.

Comment: Built by A & C de la Manche, Dieppe and entered service in 1981 and 1982. On long-term charter from SURF (Groupe Bourbon). Former oil-field supply vessels both modernised in 2002-03 at Brest to install dynamic positioning system and improve pollution control capabilities. Also fitted with a buoy-handling crane. Capable of operating ROVs or TRANSREC 250 sea-skimming system. Deck capacity 480 tons and bollard pull of 64 tons. *Alcyon* based at Brest and *Aillette* at Toulon.   ***UPDATED***

ALCYON   *10/2004*, J Y Robert /* 1042250

1 AQUITAINE EXPLORER CLASS (SALVAGE VESSEL) (ARS)

AQUITAINE EXPLORER (ex-*Abeille Supporter*, ex-*Seaway Hawk*, ex-*Seaway Devon*)

Displacement, tons: 2,500 full load
Dimensions, feet (metres): 208.7 × 44.0 × 18.9 *(63.6 × 13.4 × 5.75)*
Main machinery: 2 MaK 12M453AK diesels; 8,800 hp *(5.9 MW)*; 2 cp props; bow and stern lateral thrusters
Speed, knots: 14
Radars: Navigation: I-band.

Comment: Built by Aukra Bruk, Norway and entered service in 1975. Acquired in 1982 by DGA (Armaments Directorate) for support of undersea activities. Operated by Abeilles International. Bollard pull 100 tons. Also used in support of pollution control. Based at Bayonne.   ***UPDATED***

RESEARCH SHIPS

Note: Several government agencies use research vessels for various purposes. Most of them are operated by GENAVIR on their behalf. Main agency is IFREMER (Institut Français de Recherche pour l'Exploitation de la Mer) that operates four large ocean-going vessels; *Thalassa* (1996, 3,022 tons), *L'Atalante* (1989, 3,550 tons), *Le Suroît* (1975, modernised 1999, 1,132 tons) and *Nadir* (1974, 1857 tons); and three coastal operations vessels: *L'Europe* (1993, 264 tons catamaran), *Thalia* (1978, 135 grt, trawler type) and *Gwen Drez* (1976, 249 tons, trawler type). *Pourquoi pas?* (5,800 tons) is to replace *Nadir* from 2005. She is partially funded by the MoD, under a general agreement signed in July 2002 for the co-operation between IFREMER and the hydrographic and oceanographic service of the Navy. IRD (Institut de Recherche pour le Développement, ex-ORSTOM) operates in the Pacific two research vessels: *Antéa* (1995, 421 grt, catamaran) and *Alis* (1987, 198 grt UMS, trawler type), and two smaller craft. INSU (Institut National des Sciences de l'Univers) operates five coastal vessels (12.5 to 24.9 m) along the French coasts. TAAF (Administration des Terres Australes et Antarctiques Françaises) uses *Marion Dufresne* (1995, 9,403 GRT UMS, 120 m long), a large support ship for Antarctic operations also fitted for scientific research work, *L'Astrolabe*, (ex-*Austral Fish*, 1986, 1,370 grt) and *La Curieuse* (1989, 150 grt UMS, trawler type).

POLICE (GENDARMERIE MARITIME)

Note: The Gendarmerie Maritime is a force of 1,152 officers and men belonging to the Gendarmerie Nationale (constabulary force) but acting under the operational control of the Marine Nationale. The Gendarmes are tasked to protect naval bases and establishments ashore, but also operate 7 large patrol craft (32 to 40 m) and 24 smaller (10 to 24 m) for constabulary tasks in metropolitan France and overseas territories. All ships are procured and maintained on the Navy budget. They display the indication 'Gendarmerie' or Gendarmerie Maritime.

1 STELLIS CLASS (COASTAL PATROL CRAFT) (PB)

Name	No	Builders	Commissioned
STÉNIA	P 776	DCN, Lorient	1 Mar 1993

Displacement, tons: 60 full load
Dimensions, feet (metres): 81.7 × 20 × 5.6 *(24.9 × 6.1 × 1.7)*
Main machinery: 3 diesels; 2,500 hp(m) *(1.8 MW)*; 2 shafts; 1 waterjet
Speed, knots: 28; 10 (water-jet only). **Range, n miles:** 700 at 22 kt
Complement: 8
Guns: 1—12.7 mm MG. 2—7.62 mm MGs.
Radars: Navigation: Racal Decca 1226; I-band.

Comment: Based at Kourou in French Guiana. *Stellis* reduced to reserve status in February 2003. GRP hulls.   ***UPDATED***

STELLIS CLASS   *2000, French Navy /* 0104484

2 PATRA CLASS (COASTAL PATROL CRAFT) (PB)

Name	No	Builders	Commissioned
GLAIVE	P 671	Auroux, Arcachon	2 Apr 1977
ÉPÉE	P 672	CMN, Cherbourg	9 Oct 1976

Displacement, tons: 115 standard; 147.5 full load
Dimensions, feet (metres): 132.5 × 19.4 × 5.2 *(40.4 × 5.9 × 1.6)*
Main machinery: 2 SACM AGO 195 V12 diesels; 4,410 hp(m) *(3.24 MW)*; 2 shafts; cp props
Speed, knots: 26. **Range, n miles:** 1,750 at 10 kt; 750 at 20 kt
Complement: 18 (1 officer)
Guns: 1 Bofors 40 mm/60. 2—7.5 mm MGs.
Radars: Surface search: Racal Decca 1226; I-band.

Comment: *Glaive* based at Cherbourg and *Épée* at Brest. Service life extended to 2006. ***UPDATED***

GLAIVE   7/2003, B Prézelin / 0569969

4 GERANIUM CLASS (PB)

Name	No	Builders	Commissioned
GÉRANIUM	P 720	DCN, Lorient	19 Feb 1997
JONQUILLE	P 721	Chantiers Guy Couach Plascoa	15 Nov 1997
VIOLETTE	P 722	DCN, Lorient	4 Dec 1997
JASMIN	P 723	Chantiers Guy Couach Plascoa	15 Nov 1997

Displacement, tons: 80 standard; 96 full load
Dimensions, feet (metres): 105.7 × 20 × 6.2 *(32.2 × 6.1 × 1.9)*
Main machinery: 2 Deutz/MWM TBD 516 V16; 1 Deutz/MWM TBD 516 V12; 3,960 hp *(2.95 MW)*; 2 shafts; 1 Hamilton 422 water-jet
Speed, knots: 30. **Range, n miles:** 1,200 at 15 kt
Complement: 15 (2 officers)
Guns: 1—12.7 mm MG. 1—7.62 mm MG.
Radars: Navigation: Racal-Decca CH 180/6; E/F-band.

Comment: There are some minor differences between the DCN (details shown) and the Plascoa craft. *Géranium* based at Cherbourg; *Jonquille* at Réunion Island; *Violette* at Pointe-à-Pitre, Guadeloupe; *Jasmin* at Papeete, Tahiti. Two similar craft built for Affaires Maritimes. ***UPDATED***

GÉRANIUM   9/2003, Marian Ferrette / 0569970

8 VSC 14 CLASS (PB)

MIRI P 755 (ex-Y 755)	— P 764	VÉTIVER P 790
PÉTULANTE P 760	RÉSÉDA P 778	HORTENSIA P 791
MIMOSA P 761	MELIA P 789	

Displacement, tons: 21 full load
Dimensions, feet (metres): 47.9 × 15.1 × 3.3 *(14.6 × 4.6 × 1)*
Main machinery: 2 Baudouin 12 F11 SM diesels; 800 hp(m) *(588 kW)*; 2 shafts
Speed, knots: 20. **Range, n miles:** 360 at 18 kt
Complement: 7
Guns: 1—12.7 mm MG. 1—7.5 mm MG.
Radars: Navigation: Furuno; I-band.

Comment: Type V14 SC. Built 1985-87 except P 791 in 1990. Similar to naval tenders with Y pennant numbers. P 790 based at Mayotte, P 760 at Nouméa, P 761 at Ajaccio, P 778 and P 791 at Brest and P 789 at Toulon. P 755 (ex-Y 755) has replaced *Stellis* at Degrad-des-Cannes, French Guiana. P 764 is in reserve. ***UPDATED***

HORTENSIA   10/2004*, J Y Robert / 1042249

1 FULMAR CLASS (COASTAL PATROL CRAFT) (PB)

FULMAR (ex-*Jonathan*) P 740

Displacement, tons: 680 full load
Dimensions, feet (metres): 120.7 × 27.9 × 15.4 *(36.8 × 8.5 × 4.7)*
Main machinery: 1 diesel; 1,200 hp(m) *(882 kW)*; 1 shaft. Bow thruster
Speed, knots: 13. **Range, n miles:** 3,500 at 12 kt
Complement: 9 (1 officer)
Guns: 1—12.7 mm MG.
Radars: Surface search: 2 Furuno; I-band.

Comment: Former trawler built in 1990, acquired in October 1996 and converted for patrol duties by April 1997. Recommissioned 28 October 1997 and is based at St Pierre for western Atlantic Fishery Protection duties. ***UPDATED***

FULMAR   2000, French Navy / 0104486

8 + 16 TYPE VCSM (PATROL CRAFT) (PB)

ÉLORN P 601	DUMBÉA P 606	ODET P 611	TRIEUX P 616	ABER-WRACH P 621
VERDON P 602	YSER P 607	MAURY P 612	VÉSUBIE P 617	ESTÉRON P 622
ADOUR P 603	ARGENS P 608	CHARENTE P 613	ESCAUT P 618	MAHURY P 623
SCARPE P 604	HÉRAULT P 609	TECH P 614	HUVEAUNE P 619	ORGANABO P 624
VERTONNE P 605	GRAVONA P 610	PANFELD P 615	SÈVRE P 620	

Displacement, tons: 40
Dimensions, feet (metres): 65.6 × 16.4 × 4.9 *(20.0 × 5.0 × 1.5)*
Main machinery: 2 MAN V12 diesels; 2 shafts; 2,000 hp(m) *(1,470 kW)*
Speed, knots: 25. **Range, n miles:** 530 at 15 kt
Complement: 5
Guns: 1—7.62 mm MG.
Radars: Navigation: Furuno; I-band.

Comment: Designated 'Vedette Côtière de Surveillance Maritime' (VCSM), coastal surveillance craft. Raidco RPB 20. Ordered in two batches of 11 on 6 Dec 2001 and 6 June 2002. Built at l'Herbaudière by Raidco Marine with the co-operation of Chantiers Beneteau. Bear names of rivers. First of class (P 601) entered service on 20 June 2003 followed by P 602, P 603 and P 604 in 2003 and P 605, P 607, P 608 and P 610 in 2004. The remainder are to enter service by 2009. To replace VSC 14 and VSC 10 craft. GRP hull and superstructure. One 4.9 m RIB fitted aft on an inclined ramp. Also fitted with water-cannon. ***UPDATED***

VERDON   8/2003*, A Campanera i Rovira / 1042248

13 VSC 10 CLASS (PB)

LILAS P 703	MDLC RICHARD P 709	BELLIS P 715
BÉGONIA P 704	GENERAL DELFOSSE P 710	MDLS JACQUES P 716
PIVOINE P 705	GENTIANE P 711	LAVANDE P 717
MDLC ROBET P 707	CAPITAINE MOULIÉ P 713	
GENDARME PEREZ P 708	LIEUT JAMET P 714	

Comment: 10 m craft capable of 25 kt built 1985-95. Based at Dunkirk, Rochefort, Saint-Raphaël, Boulogne, Saint-Malo, Sète, Marseilles, Sables d'Olonnes, Port Vendres, Dieppe and Pornichet. Some to be replaced by VCSM craft as they enter service. ***UPDATED***

MDLC ROBET   7/2004*, B Prézelin / 1042226

CUSTOMS (DOUANES FRANÇAISES)

Notes: The French customs service has a number of tasks not normally associated with such an organisation. In addition to the usual duties of dealing with ships entering either its coastal area or ports it also has certain responsibilities for rescue at sea, control of navigation, fishery protection and pollution protection. Operated by about 650 personnel, the fleet comprises 12 large patrol vessels (28 to 35 m), 16 patrol boats (15 to 27 m) and 27 smaller craft. The larger vessels include DF 48 *Arafenua* (105 tons), DF 41 *Avel Gwalarn* (67 tons), DF 42 *Suroît* (67 tons), DF 31 *Alizé* (64 tons), DF 37 *Vent d'Aval* (64 tons), DF 43 *Haize Hegoa* (64 tons), DF 44 *Mervent* (64 tons), DF 45 *Vent d'Autan* (64 tons), DF 46 *Avel Sterenn* (64 tons), DF 47 *Lissero* (64 tons), DF 36 *Kan Avel* (64 tons) and DF 40 *Vent d'Amont* (61 tons). All vessels have DF numbers painted on the bow. There are also four Cessna 404 Titan, and 13 Reims-Cessna F406 patrol aircraft (2 equipped for pollution control) and six AS 350B1 Ecureuil helicopters. A further F 406, configured for pollution control, was delivered late 2002. A 28 m patrol craft is being built by Chantier Naval Guy Couach.

THEMIS   *9/2004*, B Prézelin /* 1042224

MERVENT   *7/2004*, B Prézelin /* 1042225

MAUVE   *9/2004*, Schaeffer/Marsan /* 1042223

AFFAIRES MARITIMES

Notes: A force administered and funded by the Ministry of Transport, these vessels are operated by the Préfectures Maritimes. Their duties comprise fishery protection, pollution control, navigation safety and SAR. The vessels are unarmed and manned by civilians and can be identified by a grey/blue hull with blue/red stripes, PM pennant numbers and 'Affaires Maritimes' written on the superstructure. The fleet comprises six large patrol vessels (28 to 46 m), 27 patrol boats (8 to 17 m) and 37 minor craft and service craft (mostly buoy tenders). The larger patrol vessels include PM 41 *Themis* (400 tons), PM 40 *Iris* (230 tons), PM 30 *Gabian* (76 tons), PM 31 *Origan* (70 tons), PM 29 *Mauve* (65 tons) and PM 32 *Armoise* (91 tons). Recent acquisitions include four FPB 50 Mk II patrol boats built by OCEA, Les Stables d'Olonne. These are 22 ton, 16 m craft capable of 25 kt, the first of which, PM 101 *Calisto*, was delivered in December 2000. A further 12 m craft is being built by the same shipyard. Service craft can be identified by 'Phares & Balises' written on the superstructure. Larger vessels include: *Armorique* (500 tons), *Hauts de France* (450 tons), *Provence* (326 tons), *Chef de Caux* (128 tons), *Louis Henin* (73 tons) and *Le Kahouanne* (73 tons).

HAUTS DE FRANCE   *5/2004*, Schaeffer/Marsan /* 1042222

Gabon
MARINE GABONAISE

Country Overview

A former French colony, the Gabonese Republic achieved independence in 1960. Located astride the Equator, the country has an area of 103,347 square miles and has borders to the north with Cameroon and Equatorial Guinea and to the east and south with Congo. It has a 480 n mile coastline with the Atlantic Ocean. The capital, largest city and principal port is Libreville and there is a further port at Port-Gentil. Territorial seas (12 n miles) are claimed. A 200 n mile Exclusive Economic Zone (EEZ) has been claimed but the limits are not defined; jurisdiction is complicated by the offshore islands of Isla de Annobon (Equatorial Guinea) and São Tomé and Principe.

Bases

Port Gentil, Mayumba

Personnel

2005: 600 (65 officers)

PATROL FORCES

2 P 400 CLASS (LARGE PATROL CRAFT) (PBO)

Name	No	Builders	Commissioned
GÉNÉRAL d'ARMÉE BA-OUMAR	P 07	CMN, Cherbourg	27 June 1988
COLONEL DJOUE-DABANY	P 08	CMN, Cherbourg	14 Sep 1990

Displacement, tons: 446 full load
Dimensions, feet (metres): 179 × 26.2 × 8.5 *(54.6 × 8 × 2.5)*
Main machinery: 2 Wärtsilä UD 33 V16 diesels; 8,000 hp(m) *(5.88 MW)* sustained; 2 shafts; cp props
Speed, knots: 24. **Range, n miles:** 4,200 at 15 kt
Complement: 32 (4 officers)
Military lift: 20 troops
Guns: 1 Bofors 57 mm/70 SAK 57 Mk 2 (P 07); 220 rds/min to 17 km *(9 n miles)*; weight of shell 2.4 kg. Not in P 08 which has a second Oerlikon 20 mm.
2 Giat F2 20 mm (twin) (P 08).
Weapons control: CSEE Naja optronic director (P 07).
Radars: Surface search: Racal Decca 1226C; I-band.

Programmes: Contract signed May 1985 with CMN Cherbourg. First laid down 2 July 1986, launched 18 December 1987 and arrived in Gabon 6 August 1988 for a local christening ceremony. Second ordered in February 1989 and launched 29 March 1990.
Structure: There is space on the quarterdeck for two MM 40 Exocet surface-to-surface missiles. These craft are similar to the French vessels but with different engines. *Djoue-Dabany* had twin funnels fitted in 1992, similar to French P 400 class conversions.   *VERIFIED*

COLONEL DJOUE-DABANY   *2000, Gabon Navy /* 0104491

1 PATRA CLASS (FAST ATTACK CRAFT—MISSILE) (PTM)

Name	No	Builders	Commissioned
GÉNÉRAL NAZAIRE BOULINGUI	P 10	Chantiers Naval de l'Estérel	7 Aug 1978
(ex-*Président Omar Bongo*)			

Displacement, tons: 160 full load
Dimensions, feet (metres): 138 × 25.3 × 6.5 *(42 × 7.7 × 1.9)*
Main machinery: 3 SACM 195 V12 CSHR diesels; 5,400 hp(m) *(3.97 MW)* sustained; 3 shafts
Speed, knots: 32. **Range, n miles:** 1,500 at 15 kt
Complement: 20 (3 officers)
Missiles: SSM: 4 Aerospatiale SS 12M; wire-guided to 5.5 km *(3 n miles)* subsonic; warhead 30 kg.
Guns: 1 Bofors 40 mm/60. 1 DCN 20 mm.
Radars: Surface search: Racal Decca RM1226; I-band.

Comment: Re-activated in 2000.   *VERIFIED*

GÉNÉRAL NAZAIRE BOULINGUI   *2000, Gabon Navy /* 0104492

LAND-BASED MARITIME AIRCRAFT

Numbers/Type: 1 Embraer EMB-111 Bandeirante.
Operational speed: 194 kt *(360 km/h).*
Service ceiling: 25,500 ft *(7,770 m).*
Range: 1,590 n miles *(2,945 km).*
Role/Weapon systems: Air Force coastal surveillance and EEZ protection tasks are primary roles. Sensors: APS-128 search radar, limited ECM, searchlight. Weapons: ASV; 8 × 127 mm rockets or 28 × 70 mm rockets.   ***VERIFIED***

AMPHIBIOUS FORCES

1 BATRAL TYPE (LSTH)

Name	No	Builders	Launched	Commissioned
PRESIDENT EL HADJ OMAR BONGO	L 05	Français de l'Ouest, Rouen	16 Apr 1984	26 Nov 1984

Displacement, tons: 770 standard; 1,336 full load
Dimensions, feet (metres): 262.4 × 42.6 × 7.9 *(80 × 13 × 2.4)*
Main machinery: 2 SACM Type 195 V12 CSHR diesels; 3,600 hp(m) *(2.65 MW)*; 2 shafts; cp props
Speed, knots: 16. **Range, n miles:** 4,500 at 13 kt
Complement: 39
Military lift: 188 troops; 12 vehicles; 350 tons cargo
Guns: 1 Bofors 40 mm/60; 300 rds/min to 12 km *(6.5 n miles)*; weight of shell 0.89 kg. 2—81 mm mortars. 2 Browning 12.7 mm MGs. 1—7.62 mm MG.
Radars: Surface search: Racal Decca 1226; I-band.
Helicopters: Capable of operating up to SA 330 Puma size.

Comment: Sister to French *La Grandière.* Carries one LCVP and one LCP. Started refit by Denel, Cape Town in April 1996, and returned to service in 1997 with bow doors welded shut. Completed repair and cleaning at Abidjan during 2000.   ***VERIFIED***

PRESIDENT EL HADJ OMAR BONGO   *6/1993, Gabon Navy* / 0069977

2 SEA TRUCKS (LCVP)

Comment: Built by Tanguy Marine, Le Havre in 1985. One of 12.2 m with two Volvo Penta 165 hp (m) *(121 kW)* engines and one of 10.2 m with one engine.   ***VERIFIED***

POLICE

Notes: The Police have about 12—6.8 m LCVPs and Simonneau 11 m patrol craft, 10 of which were delivered in 1989.

SIMONNEAU SM 360   *1989, Simonneau Marine* / 0069978

Gambia

Country Overview

The Republic of Gambia was a British protectorate until 1965 when it gained independence. With an area of 4,361 square miles, it has a short 43 n mile coastline with the Atlantic Ocean but is otherwise completely surrounded by Senegal. The two countries united in 1981 to form the confederation of Senegambia but this collapsed in 1989 when the countries reverted to being separate states. The capital, largest city and principal port is Banjul (formerly Bathurst). Territorial seas (12 n miles) and a 200 mile fishing zone are claimed. The patrol craft came under 3 Marine Company of the National Army until 1996 when a navy was established.

Headquarters Appointments

Commander, Navy:
   Lieutenant Commander M B Sarr

Personnel

(a)   2005: 150
(b)   Voluntary service

Bases

Banjul

PATROL FORCES

Notes: Acquisition of new patrol craft is under consideration.

2 PATROL CRAFT (PB)

FATIMAH   SULAYMAN JUNKUNG

Displacement, tons: 25
Dimensions, feet (metres): 52.8 × 14.8 × 5.3 *(16.1 × 4.5 × 1.6)*
Main machinery: Caterpillar diesel; 800 hp *(596 kW)*
Speed, knots: 40
Radars: Surface search: Furuno; I-band.

Comment: Procured from Taiwan in 1999.   ***VERIFIED***

FATIMAH   *2000, Gambian Navy* / 0104493

1 PETERSON Mk 4 CLASS (PB)

Name	No	Builders	Commissioned
BOLONG KANTA	P 14	Peterson Builders, Sturgeon Bay	15 Oct 1993

Displacement, tons: 24 full load
Dimensions, feet (metres): 50.9 × 14.8 × 4.3 *(15.5 × 4.5 × 1.3)*
Main machinery: 2 Detroit 6V-92A diesels; 520 hp *(388 kW)* sustained; 2 shafts
Speed, knots: 24. **Range, n miles:** 500 at 20 kt
Complement: 6
Guns: 2—12.7 mm MGs.
Radars: Surface search: Raytheon R41X; I-band.

Comment: Reported seaworthy. Similar craft in service in Egypt, Cape Verde and Senegal.   ***VERIFIED***

PETERSON Mk 4 (Senegal colours)   *1/1998* / 0050096

1 FAIREY MARINE LANCE CLASS
(COASTAL PATROL CRAFT) (PB)

Name	No	Builders	Commissioned
SEA DOG	P 11	Fairey Marine, Cowes	28 Oct 1976

Displacement, tons: 17 full load
Dimensions, feet (metres): 48.7 × 15.3 × 4.3 *(14.8 × 4.7 × 1.3)*
Main machinery: 2 GM 8V-71TA diesels; 650 hp *(485 kW)* sustained; 2 shafts
Speed, knots: 24. **Range, n miles:** 500 at 16 kt
Complement: 9
Guns: 2—7.62 mm MGs (not carried).
Radars: Surface search: Racal Decca 110; I-band.

Comment: Delivered 28 October 1976. Unarmed and used for training and personnel transfer. Still seaworthy.

VERIFIED

SEA DOG
1/1990, E Grove / 0069980

Georgia

Country Overview

Formerly part of the USSR, the Republic of Georgia declared independence in 1991. Situated in the Transcaucasia region of western Asia, the country has an area of 26,900 square miles and is bordered to the north by Russia and to the south by Turkey, Armenia and Azerbaijan. It has a coastline of 167 n miles with the Black Sea on which Poti and Batumi are the principal ports. T'bilisi is the capital and largest city. The country includes two autonomous republics, Abkhazia and Adzharia, and one autonomous region, South Ossetia. USSR legislation appears still to apply to maritime claims. Territorial waters (12 n miles) are claimed, as is an EEZ (200 n miles) although the limits of the latter are not defined. Naval and Coast Guard Forces (part of the Border Guard) formed 7 July 1993. While merger of the two forces has been considered, they are likely to remain different commands.

Headquarters Appointments

Commander of the Navy:
    Rear Admiral Gennady Khaidarov
Commander of the Border Guard:
    Major General Chkheidze

Personnel

2005: 3,100

Bases

Poti (HQ), Batumi.

PATROL FORCES

Notes: In addition to the vessels listed below, the following vessels are on the Navy List:
(1) A former fishing vessel *Gantiadi* (016) is used as a patrol craft and tender. It is armed with two 23 mm guns and 2—12.7 mm MGs.
(2) Three 'Aist' (Project 1398) class patrol launches (10, 12, 14). 14 is active with the Hydrographic Service and has a blue hull.
(3) A 'Nyryat' (DHK-81) and 'Flamingo' (DHK-82) are active with the Hydrographic service and are civilian manned.
(4) A Project 371U patrol launch *Gali* (04).

1 TURK CLASS (PB)

Name	No	Builders	Commissioned
KUTAISI (ex-*AB 30*)	202 (ex-*P 130*)	Haliç Shipyard	21 Feb 1969

Displacement, tons: 170 full load
Dimensions, feet (metres): 132 × 21 × 5.5 *(40.2 × 6.4 × 1.7)*
Main machinery: 4 SACM-AGO V16 CSHR diesels; 9,600 hp(m) *(7.06 MW)*; 2 cruise diesels; 300 hp(m) *(220 kW)*; 2 shafts
Speed, knots: 22
Complement: 31
Guns: 1 Bofors 40 mm/60. 2 ZSU 23 mm. 2—12.7 mm MGs.
Radars: Surface search: Racal Decca; I-band.

Comment: Transferrred from Turkish Navy on 5 December 1998 to the Navy. May retain its active sonar and ASW rocket launcher but this is unlikely.

VERIFIED

KUTAISI
10/2002, Hartmut Ehlers / 0552754

1 STENKA (PROJECT 205P) CLASS (PBF)

BATUMI (ex-*PSKR 638*) 301 (ex-*648*)

Displacement, tons: 253 full load
Dimensions, feet (metres): 129.3 × 25.9 × 8.2 *(39.4 × 7.9 × 2.5)*
Main machinery: 3 diesels; 14,100 hp(m) *(10.36 MW)*; 3 shafts
Speed, knots: 37. **Range, n miles:** 2,300 at 14 kt
Complement: 25
Guns: 2—37 mm/L 68.
Radars: Surface search: Pot Drum; H/I-band.
Fire control: Drum Tilt; H/I-band.
Navigation: Palm Frond; I-band.

Comment: Transferred from Ukraine in 1998 in a disarmed state. In a poor state of repair at Balaklava and operational status is doubtful. **UPDATED**

BATUMI
9/2004*, Hartmut Ehlers / 1044124

1 MATKA (PROJECT 206MP) CLASS (PGGK)

TBILISI (ex-*Konotop*) 302 (ex-R-15)

Displacement, tons: 225 standard; 260 full load
Dimensions, feet (metres): 129.9 × 24.9 (41 over foils) × 6.9 (13.1 over foils) *(39.6 × 7.6 (12.5) × 2.1 (4))*
Main machinery: 3 Type M 504 diesels; 10,800 hp(m) *(7.94 MW)* sustained; 3 shafts
Speed, knots: 40. **Range, n miles:** 600 at 35 kt foilborne; 1,500 at 14 kt hullborne
Complement: 33
Missiles: SSM: 2 SS-N-2C/D Styx; active radar or IR homing to 83 km *(45 n miles)* at 0.9 Mach; warhead 513 kg; sea-skimmer at end of run.
Guns: 1—3 in *(76 mm)*/60; 120 rds/min to 15 km *(8 n miles)*; weight of shell 7 kg.
    1—30 mm/65 AK 630; 6 barrels per mounting; 3,000 rds/min to 2 km.
Countermeasures: Decoys: 2 PK 16 chaff launchers.
ESM: Clay Brick; intercept.
Weapons control: Hood Wink optronic directors.
Radars: Air/surface search: Plank Shave; E-band.
Navigation: SRN-207; I-band.
Fire control: Bass Tilt; H/I-band.

Comment: Acquired from Ukraine in 1999 with full armament. Based at Poti.

VERIFIED

TBILISI
10/2002, Hartmut Ehlers / 0552755

1 POLUCHAT 1 CLASS (PB)

AKHMETA 102

Displacement, tons: 100 full load
Dimensions, feet (metres): 97.1 × 19 × 4.8 *(29.6 × 5.8 × 1.5)*
Main machinery: 2 Type M 50 diesels; 2,200 hp(m) *(1.6 MW)* sustained; 2 shafts
Speed, knots: 20. **Range, n miles:** 1,500 at 10 kt
Complement: 15
Guns: 1—37 mm/L 68. 1—140 mm 17 round rocket launcher.
Radars: Surface search: Spin Trough; I-band.

Comment: Acquired in a disarmed state from commercial sources in Ukraine. Refitted at Metallist Ship Repair Yard, Balaklava 2000-2002.

VERIFIED

P 102   *8/2000, Hartmut Ehlers /* 0104494

1 PROJECT 360 PATROL CRAFT (PB)

TSKALTUBO (ex-*Mercuriy*) 101

Displacement, tons: 58; 70 full load
Dimensions, feet (metres): 88.6 × 21.3 × 4.6 *(27 × 6.5 × 1.4)*
Main machinery: 4 diesels; 4,800 hp *(3.58 MW)*; 4 shafts
Speed, knots: 38
Guns: 1—37 mm/L 68.
Radars: 1 navigation.

Comment: Former Black Sea Fleet Flag Officers' Yacht built by Almaz, St Petersburg. Acquired from commercial sources in Ukraine in 1997 and used as a patrol boat.

VERIFIED

TSKALTUBO   *10/2002, Hartmut Ehlers /* 0552756

2 DILOS CLASS (PB)

Name	No	Builders	Commissioned
IVERIA (ex-*Lindos*)	201 (ex-P 269)	Hellenic Shipyard, Skaramanga	1978
MESTIA (ex-*Dilos*)	203 (ex-P 267)	Hellenic Shipyard, Skaramanga	1978

Displacement, tons: 86 full load
Dimensions, feet (metres): 95.1 × 16.2 × 5.6 *(29 × 5 × 1.7)*
Main machinery: 2 MTU 12V 331 TC92 diesels; 2,660 hp(m) *(1.96 MW)* sustained; 2 shafts
Speed, knots: 27. **Range, n miles:** 1,600 at 24 kt
Complement: 15
Guns: 4—23 mm ZSU (2 twin).
Radars: Surface search: Racal Decca 1226C; I-band.

Comment: First one transferred from the Greek Navy in February 1998, second in September 1999. Reported to have been refitted in Greece in 2004. *UPDATED*

IVERIA and MESTIA   *10/2002, Hartmut Ehlers /* 0552757

1 ANNINOS (LA COMBATTANTE II) CLASS
(FAST ATTACK CRAFT—MISSILE) (PTFG)

Name	No	Builders	Commissioned
— (ex-*Ypoploiarchos Batsis*, ex-*Calypso*)	— (ex-P 17)	CMN Cherbourg	Dec 1971

Displacement, tons: 234 standard; 255 full load
Dimensions, feet (metres): 154.2 × 23.3 × 8.2 *(47 × 7.1 × 2.5)*
Main machinery: 4 MTU MD 16V 538 TB90 diesels; 12,000 hp(m) *(8.82 MW)* sustained; 4 shafts
Speed, knots: 36.5. **Range, n miles:** 850 at 25 kt
Complement: 40 (4 officers)

Guns: 4 Oerlikon 35 mm/90 (2 twin); 550 rds/min to 6 km *(3.2 n miles)* anti-surface; 5 km *(2.7 n miles)* anti-aircraft; weight of shell 1.55 kg.
Torpedoes: 2—21 in *(533 mm)* tubes. AEG SST-4; wire-guided; active homing to 12 km *(6.5 n miles)* at 35 kt; passive homing to 28 km *(15 n miles)* at 23 kt; warhead 250 kg.
Countermeasures: ESM: Thomson-CSF DR 2000S; intercept.
Weapons control: Thomson-CSF Vega system.
Radars: Surface search: Thomson-CSF Triton; G-band.
Navigation: Decca 1226C; I-band.
Fire control: Thomson-CSF Pollux; I/J-band.
IFF: Plessey Mk 10.

Comment: Transferred from Greece on 23 April 2004. Exocet SSM removed. It is likely that torpedoes and EW equipment have also been removed although this has not been confirmed.
UPDATED

ANNINOS CLASS (with Exocet)   *7/2000, Amintika Themata /* 0104558

AMPHIBIOUS FORCES

2 VYDRA (PROJECT 106K) CLASS (LCU)

GURIA 001   **ATIA** 002

Displacement, tons: 425 standard; 550 full load
Dimensions, feet (metres): 179.7 × 25.3 × 6.6 *(54.8 × 7.7 × 2)*
Main machinery: 2 Type 3-D-12 diesels; 600 hp(m) *(440 kW)* sustained; 2 shafts
Speed, knots: 12. **Range, n miles:** 2,500 at 10 kt
Complement: 20
Military lift: 200 tons or 100 troops or 3 MBTs
Guns: 4—23 mm ZSU (2 twin).
Radars: Navigation: Don 2; I-band.
IFF: High Pole.

Comment: Built at Burgas Shipyard 1974-75. Transferred to Georgia on 6 July 2001.
VERIFIED

ATIA   *10/2002, Hartmut Ehlers /* 0552747

2 ONDATRA (PROJECT 1176) CLASS (LCU)

MDK 01   **MDK 02**

Displacement, tons: 145 full load
Dimensions, feet (metres): 78.7 × 16.4 × 4.9 *(24 × 5 × 1.5)*
Main machinery: 1 diesel; 300 hp(m) *(221 kW)*; 1 shaft
Speed, knots: 10. **Range, n miles:** 500 at 5 kt
Complement: 4
Military lift: 1 MBT
Guns: 2—12.7 mm MGs.
Radars: Navigation: Spin Trough; I-band.

Comment: Transferred from USSR January 1983. Has a tank deck of 45 × 13 ft. MDK 02 awaiting refit at Poti.
VERIFIED

MDK 02   *10/2002, Hartmut Ehlers /* 0552748

LAND-BASED MARITIME AIRCRAFT

Notes: While there is no naval air arm, two Mi-14 helicopters were reported delivered in April 2004 after five years undergoing refitting in Ukraine. They are believed to be for patrol and SAR duties and to be unarmed.

COAST GUARD

Notes: In addition to the vessels listed below, the following vessels are on the Coast Guard list:
(1) A former fishing vessel *P 101*.
(2) A patrol launch *P 105*.
(3) Nine Aist (Project 1398) class patrol launches *(P 0111-0115, 0212, 702-703)*.
(4) One Strizh (Project 1390) class launch *P 0116*.
(5) One 44 m tug *Poti* (ex-*Zorro*) acquired from Ukraine in 1999 for salvage purposes.

1 LINDAU (TYPE 331) CLASS (WPBO)

Name	No	Builders	Commissioned
AYETY (ex-*Minden*)	P 22 (ex-M1085)	Burmester, Bremen	22 Jan 1960

Displacement, tons: 463 full load
Dimensions, feet (metres): 154.5 × 27.2 × 9.8 *(47.1 × 8.3 × 2.8)*
Main machinery: 2 MTU MD diesels; 4,000 hp(m) *(2.94 MW)*; 2 shafts
Speed, knots: 16. **Range, n miles:** 850 at 16 kt
Complement: 43
Guns: 1 Bofors 40 mm/70. 2—12.7 mm MGs.
Radars: Surface search: Atlas Elektronik TRS; I-band.

Comment: Paid off from German Navy in 1997 and transferred 15 November 1998 to the Coast Guard. Former minehunter refitted as a patrol craft in Germany before transfer. ***VERIFIED***

AYETY   *10/2002, Hartmut Ehlers* / 0552749

1 STENKA (TARANTUL) CLASS (PROJECT 205P) (WPBF)

GIORGI TORELI (ex-*Anastasiya*, ex-*PSKR 629*) P 21

Displacement, tons: 253 full load
Dimensions, feet (metres): 129.3 × 25.9 × 8.2 *(39.4 × 7.9 × 2.5)*
Main machinery: 3 Type M 517 or M 583 diesels; 14,100 hp(m) *(10.36 MW)*; 3 shafts
Speed, knots: 37. **Range, n miles:** 500 at 35 kt; 1,540 at 14 kt
Complement: 30 (5 officers)
Guns: 2—37 mm/L 68.
Radars: Navigation: Palm Frond; I-band.

Comment: Acquired in a disarmed state from commercial sources in Ukraine. ***VERIFIED***

GIORGI TORELI   *10/2002, Hartmut Ehlers* / 0552750

2 DAUNTLESS CLASS (WPB)

P 106 (ex-P 208)   P 209

Displacement, tons: 11 full load
Dimensions, feet (metres): 40 × 14 × 4.3 *(12.2 × 4.3 × 1.3)*
Main machinery: 2 Caterpillar 3208TA diesels; 870 hp *(650 kW)*; 2 shafts
Speed, knots: 27. **Range, n miles:** 600 at 18 kt
Complement: 5
Guns: 1—12.7 mm MG.
Radars: Surface search: Raytheon; I-band.

Comment: Aluminium construction. Acquired in July 1999 from SeaArk Marine. ***VERIFIED***

P 209   *10/2002, Hartmut Ehlers* / 0552751

2 POINT CLASS (WPB)

Name	No	Builders	Commissioned
TSOTNE DADIANI (ex-*Point Countess*)	P 210 (ex-82335)	USCG Yard, Curtis Bay	8 Aug 1962
GENERAL MAZNIASHVILI (ex-*Point Baker*)	P 211 (ex-82342)	USCG Yard, Curtis Bay	30 Oct 1963

Displacement, tons: 66; 69 full load
Dimensions, feet (metres): 2 Caterpillar 3412 diesels; 1,600 hp *(1.19 MW)*; 2 shafts
Speed, knots: 23.5. **Range, n miles:** 1,500 at 8 kt
Complement: 10 (1 officer)
Guns: 2—12.7 mm MGs.
Radars: Surface search: Hughes/Furuno SPS-73; I-band.

Comment: Steel hulled craft with aluminium superstructure. First transferred from United States in June 2000 and second on 12 February 2002. ***UPDATED***

GENERAL MAZNIASHVILI   *10/2002*, Hartmut Ehlers* / 0589740

8 ZHUK CLASS (WPB)

P 102-104   P 203-207

Displacement, tons: 25 full load
Dimensions, feet (metres): 49 × 14.8 × 2.4 *(14.9 × 4.5 × 0.7)*
Main machinery: 2 GM diesels; 450 hp *(335 kW)*; 2 shafts
Speed, knots: 12. **Range, n miles:** 200 at 12 kt
Complement: 7
Guns: 2—23 mm (1 twin) *(P 204, P 205)*.
   2—12.7 mm MGs *(P 203)*.

Comment: P 102-104 constructed at Batumi 1997-99. P 203 transferred from Ukraine in April 1997. P 204-205 acquired from Ukraine and P 206-207 transferred from Georgian Navy in 1998. ***VERIFIED***

P 203   *10/2002, Hartmut Ehlers* / 0552753

Germany
DEUTSCHE MARINE

Country Overview

The Federal Republic of Germany (FRG) is situated in central Europe. The country was re-unified in 1990 when the German Democratic Republic became part of the FRG. With an area of 137,823 square miles, it is bordered to the north by Denmark, to the east by Poland and the Czech Republic, to the south by Austria and Switzerland and to the west by France, Luxembourg, Belgium and the Netherlands. It has a 1,290 n mile coastline with the North and Baltic Seas which are linked by the Kiel Canal. The capital and largest city is Berlin. North Sea ports include Hamburg, Wilhelmshaven, Bremen, Nordenham and Emden, while the main Baltic ports are Lübeck, Wismar, Rostock and Stralsund. The Rhine is the principal inland waterway on which Duisburg is the largest port. Territorial seas (12 n miles) are claimed. An EEZ (200 n miles) has also been claimed.

Headquarters Appointments

Chief of Naval Staff:
Vice Admiral Lutz Feldt
Chief of Staff:
Rear Admiral Jörg Auer

Commander-in-Chief

Commander-in-Chief, Fleet:
Vice Admiral Wolfgang Nolting
Deputy Commander-in-Chief, Fleet:
Rear Admiral Gottfried Hoch

Diplomatic Representation

Defence Attaché in Washington:
Rear Admiral H von Puttkamer
Naval Attaché in London:
Captain P Monte
Naval Attaché in Washington:
Captain L Helmeich
Naval Attaché in Paris:
Captain H Walz
Naval Attaché in Moscow:
Captain K Seemann
Defence Attaché in Buenos Aires:
Captain T G Hering
Defence Attaché in Brasilia:
Captain H J Liedke
Defence Attaché in Pretoria:
Captain G Kiehnle
Defence Attaché in Jakarta:
Commander G Eschle
Naval and Military Attaché in Rome:
Commander Dr F Ganseuer
Defence Attaché in Tokyo:
Captain R Wallner
Defence Attaché in Kuala-Lumpur:
Commander U Becker
Defence Attaché in Oslo:
Commander V Brügmann
Defence Attaché in Stockholm:
Commander G Bruch
Naval Attaché in Madrid:
Commander B Wehner
Naval Attaché in Ankara:
Commander B Schmidt
Naval Attaché in Kiev:
Commander H-H Schneider
Defence Attaché in Abu Dhabi:
Commander A Ostermann

Defence Attaché in Helsinki:
Commander R Rusch
Defence Attaché in Tunis:
Commander R Wittek
Defence Attaché in Copenhagen:
Commander U Arnold
Defence Attaché in Lisbon:
Commander G Kleinert
Naval Attaché in Peking:
Commander C Klenke
Naval Attaché in Tel Aviv:
Commander G Lintner
Naval Attaché in Santiago:
Commander Marusche
Defence Attaché in Canberra:
Commander F J Birkel
Defence Attaché in Lima:
Commander W Wirtz

Personnel

(a) 2005: 21,600 (5,017 officers) (including naval air arm) plus 5,600 conscripts
(b) 9 months' national service

Fleet Disposition

Submarine Flotilla (Eckernförde)
1st and 3rd Squadrons; Type 206A and Type 205

Destroyer Flotilla (Wilhelmshaven)
1st Destroyer Squadron; Lütjens class
1st Frigate Squadron Sachsen class
2nd and 4th Frigate Squadron; Bremen class
6th Frigate Squadron; Brandenburg class

Mine Warfare Flotilla (Kappeln)

FPB Flotilla
2nd and 7th Squadrons (Warnemünde) Type 143

Bases

C-in-C Fleet: Glücksburg. Flag Officer Naval Command: Rostock.
Baltic: Kiel, Olpenitz (all mine warfare forces in due course), Flensburg*, Neustadt*, Warnemünde (all patrol craft in due course). Eckernförde (all submarines in due course).
North Sea: Wilhelmshaven.
Naval Arsenal: Wilhelmshaven, Kiel.
Training (other than in Bases above): Bremerhaven, Glückstadt, List/Sylt, Plön, Grossenbrode*, Parow.

The administration of the bases is vested in the Naval Support Command at Wilhelmshaven. Those marked with an asterisk are to close by 2003 although Neustadt may retain some minor support facilities.

Naval Air Arm

MFlg Fltl (Flotille der Marineflieger)
MFG 2 (Fighter-Bomber and Reconnaissance Wing at Eggebek)
PA 200 Tornado
MFG 3 'Graf Zeppelin' (LRMP Wing at Nordholz)
Breguet Atlantic of which 4 converted for Sigint, Sea Lynx (landbased for embarkation and maintenance). Dornier Do 228 (for pollution control)
MFG 5 (SAR and Liaison Wing at Kiel). To be evacuated and returned to civil use in due course.
Sea King Mk 41

Strength of the Fleet

Type	Active	Building (Projected)
Submarines—Patrol	13	3 (2)
Frigates	14	1 (4)
Corvettes	—	5
Fast Attack Craft—Missile	14	—
LCM/LCU	6	—
Minehunters	17	—
Minesweepers—Coastal	5	—
Minesweepers—Drones	18	1 (3)
Tenders	6	—
Replenishment Tankers	6	—
Ammunition Transports	1	—
Tugs—Icebreaking	1	—
AGIs	3	—
Sail Training Ships	2	—

Prefix to Ships' Names

Prefix FGS is used in communications.

Hydrographic Service

This service, under the direction of the Ministry of Transport, is civilian-manned with HQ at Hamburg. Survey ships are listed at the end of the section.

DELETIONS

Submarines

2003	U 11
2004	U 28

Destroyers

2002	*Mölders*
2003	*Lütjens*

Patrol Forces

2002	*Alk, Dommel, Fuchs, Löwe, Weihe* (all to Egypt)
2004	*Falke, Kondor*
2005	*Albatros, Bussard, Sperber* (to Tunisia), *Greif* (to Tunisia)

Mine Warfare Vessels

2002	*Frauenlob, Gefion, Loreley*

Survey and Research Ships

2004	*Planet* (old)

Auxiliaries

2002	*Nordeney* (to Uruguay), *Neuwerk, TF 6*
2003	*Freiburg, Westensee, Bant*
2004	*Neuende*
2005	*Meersberg*

Coast Guard

2002	*Bad Bramstedt* (old)

PENNANT LIST

Submarines

S 171	U 22
S 172	U 23
S 173	U 24
S 174	U 25
S 175	U 26
S 178	U 29
S 179	U 30
S 181	U 31
S 182	U 32
S 183	U 33 (bldg)
S 184	U 34 (bldg)
S 191	U 12
S 194	U 15
S 195	U 16
S 196	U 17
S 197	U 18

Frigates

F 207	Bremen
F 208	Niedersachsen
F 209	Rheinland-Pfalz
F 210	Emden
F 211	Köln
F 212	Karlsruhe
F 213	Augsburg
F 214	Lübeck
F 215	Brandenburg
F 216	Schleswig-Holstein
F 217	Bayern
F 218	Mecklenburg-Vorpommern
F 219	Sachsen (bldg)
F 220	Hamburg (bldg)
F 221	Hessen (bldg)

Patrol Forces

P 6113	S 63 Geier
P 6118	S 68 Seeadler
P 6119	S 69 Habicht
P 6120	S 70 Kormoran
P 6121	S 71 Gepard
P 6122	S 72 Puma
P 6123	S 73 Hermelin
P 6124	S 74 Nerz
P 6125	S 75 Zobel
P 6126	S 76 Frettchen
P 6127	S 77 Dachs
P 6128	S 78 Ozelot
P 6129	S 79 Wiesel
P 6130	S 80 Hyäne

Mine Warfare Forces

M 1052	Mühlhausen
M 1058	Fulda
M 1059	Weilheim
M 1060	Weiden
M 1061	Rottweil
M 1062	Sulzbach-Rosenberg
M 1063	Bad Bevensen
M 1064	Grömitz
M 1065	Dillingen
M 1066	Frankenthal
M 1067	Bad Rappenau
M 1068	Datteln
M 1069	Homburg
M 1090	Pegnitz
M 1091	Kulmbach
M 1092	Hameln
M 1093	Auerbach
M 1094	Ensdorf
M 1095	Überherrn
M 1096	Passau
M 1097	Laboe
M 1098	Siegburg
M 1099	Herten

Amphibious Forces

L 762	Lachs
L 765	Schlei

Auxiliaries

A 50	Alster
A 52	Oste
A 53	Oker
A 60	Gorch Fock
A 511	Elbe
A 512	Mosel
A 513	Rhein
A 514	Werra
A 515	Main
A 516	Donau
A 1401	Eisvogel
A 1409	Wilhelm Pullwer
A 1411	Berlin
A 1412	Frankfurt Am Main
A 1425	Ammersee
A 1426	Tegernsee
A 1435	Westerwald
A 1437	Planet
A 1439	Baltrum
A 1440	Juist
A 1441	Langeoog
A 1442	Spessart
A 1443	Rhön
A 1451	Wangerooge
A 1452	Spiekeroog
A 1458	Fehmarn
Y 811	Knurrhahn
Y 812	Lütje Hörn
Y 814	Knechtsand
Y 815	Scharhörn
Y 816	Vogelsand
Y 817	Nordstrand
Y 819	Langeness
Y 834	Nordwind
Y 835	Todendorf
Y 836	Putlos
Y 837	Baumholder
Y 838	Bergen
Y 839	Munster
Y 842	Schwimmdock A
Y 855	TF 5
Y 860	Schwedeneck
Y 861	Kronsort
Y 862	Helmsand
Y 863	Stollergrund
Y 864	Mittelgrund
Y 865	Kalkgrund
Y 866	Breitgrund
Y 875	Hiev
Y 876	Griep
Y 891	Altmark
Y 895	Wische
Y 1643	Bottsand
Y 1644	Eversand
Y 1656	Wustrow
Y 1658	Drankse
Y 1671	AK 1
Y 1672	AK 3
Y 1674	AM 6
Y 1675	AM 8
Y 1676	MA 2
Y 1677	MA 3
Y 1678	MA 1
Y 1679	AM 7
Y 1683	AK 6
Y 1685	Aschau
Y 1686	AK 2
Y 1687	Borby
Y 1689	Bums

SUBMARINES

1 + 3 (2) TYPE 212A (SSK)

Name	No	Builders	Laid down	Launched	Commissioned
U 31	S 181	HDW, Kiel	Feb 2000	20 Mar 2002	15 Sep 2004
U 32	S 182	TNSW, Emden	Jan 2002	4 Dec 2003	May 2005
U 33	S 183	HDW, Kiel	Oct 2002	13 Sep 2004	Jan 2006
U 34	S 184	TNSW, Emden	June 2003	July 2005	June 2006

Displacement, tons: 1,450 surfaced; 1,830 dived
Dimensions, feet (metres): 183.4 × 23 × 19.7
(55.9 × 7 × 6)
Main machinery: Diesel-electric; 1 MTU 16V 396 diesel; 4,243 hp(m) *(3.12 MW)*; 1 alternator; 1 Siemens Permasyn motor; 3,875 hp(m) *(2.85 MW)*; 1 shaft; 9 Siemens/HDW PEM fuel cell (AIP) modules; 306 kW; sodium sulphide high-energy batteries
Speed, knots: 20 dived; 12 surfaced
Range, n miles: 8,000 at 8 kt surfaced
Complement: 27 (8 officers)

Torpedoes: 6—21 in *(533 mm)* bow tubes; water ram discharge; STN (formerly AEG) DM 2A4. Total 12 weapons.
Countermeasures: Decoys: TAU 2000 (C 303) torpedo countermeasures.
ESM: DASA FL 1800U; radar warning.
Weapons control: Kongsberg MSI-90U weapons control system.
Radars: Navigation: Kelvin Hughes 1007; I-band.
Sonars: STN Atlas Elektronik DBQS-40; passive ranging and intercept; FAS-3 flank and passive towed array.
STN Atlas Elektronik MOA 3070 or Allied Signal ELAK; mine detection; active; high frequency.

Programmes: Design phase first completed in 1992 by ARGE 212 (HDW/TNSW) in conjunction with IKL. Authorisation for the first four of the class was given on 6 July 1994, but the first steel cut was delayed to 1 July 1998 because of modifications needed to achieve commonality with the Italian

U 32   7/2004 *, Frank Findler / 0587716

Navy, which is building two identical hulls. Changes included greater diving depth and improved habitability. HDW Kiel, and TNSW Emden, are sharing the work with the forward half being built by HDW and the stern by TNSW with final assembly alternating between the shipyards. A second batch is likely to be confined to two boats to enter service from 2012.
Structure: Equipped with a hybrid fuel cell/battery propulsion based on the Siemens PEM fuel cell technology. The submarine is designed with a partial double hull which has a larger diameter forward. This is joined to the after end by a short conical section which houses the fuel cell plant. Two LOX tanks and hydrogen stored in metal cylinders are carried around the circumference of the smaller hull section. Zeiss search and attack periscopes.
Operational: Maximum speed on AIP is 8 kt without use of main battery. All to be based at Eckenförde as part of the First Submarine Squadron. Sea trials of *U 32* began on 16 June 2004.
Sales: Two identical submarines are being built in Italy. Four Type 214 submarines are being built for Greece and three for South Korea.

UPDATED

U 31   4/2003, Michael Nitz / 0552768

1 TYPE 205 (SSC)

Name	No	Builders	Laid down	Launched	Commissioned
U 12	S 191	Howaldtswerke, Kiel	1 Sep 1966	10 Sep 1968	14 Jan 1969

Displacement, tons: 419 surfaced; 450 dived
Dimensions, feet (metres): 144 × 15.1 × 14.1
(43.9 × 4.6 × 4.3)
Main machinery: Diesel-electric; 2 MTU 12V 493 AZ80 GA 31L diesels; 1,200 hp(m) *(882 kW)* sustained; 2 alternators; 810 kW; 1 Siemens motor; 1,800 hp(m) *(1.32 MW)* sustained; 1 shaft
Speed, knots: 10 surfaced; 17 dived
Complement: 22 (4 officers)

Torpedoes: 8—21 in *(533 mm)* tubes.
Countermeasures: ESM: Radar warning.
Weapons control: Signaal Mk 8.
Radars: Surface search: Thomson-CSF Calypso II; I-band.
Sonars: Atlas Elektronik SRS M1H; passive/active search and attack; high frequency.

Programmes: Built in floating docks. Last survivor of first submarine class designed and built by Germany after the Second World War.
Structure: Diving depth, 159 m *(490 ft)*. Hulls of steel alloys with non-magnetic properties. Acts as a sonar trials platform.
Operational: Based at Eckernförde. To be decommissioned in December 2006.

UPDATED

U 12
4/2002, Michael Nitz / 0529112

11 TYPE 206A (SSK)

Name	No	Builders	Laid down	Launched	Commissioned
U 15	S 194	Howaldtswerke, Kiel	1 June 1970	15 June 1972	17 July 1974
U 16	S 195	Rheinstahl Nordseewerke, Emden	1 Nov 1970	29 Aug 1972	9 Nov 1973
U 17	S 196	Howaldtswerke, Kiel	1 Oct 1970	10 Oct 1972	28 Nov 1973
U 18	S 197	Rheinstahl Nordseewerke, Emden	1 Apr 1971	31 Oct 1972	19 Dec 1973
U 22	S 171	Rheinstahl Nordseewerke, Emden	18 Nov 1971	27 Mar 1973	26 July 1974
U 23	S 172	Rheinstahl Nordseewerke, Emden	5 Mar 1973	25 May 1974	2 May 1975
U 24	S 173	Rheinstahl Nordseewerke, Emden	20 Mar 1972	26 June 1973	16 Oct 1974
U 25	S 174	Howaldtswerke, Kiel	1 July 1971	23 May 1973	14 June 1974
U 26	S 175	Rheinstahl Nordseewerke, Emden	14 July 1972	20 Nov 1973	13 Mar 1975
U 29	S 178	Howaldtswerke, Kiel	10 Jan 1972	5 Nov 1973	27 Nov 1974
U 30	S 179	Rheinstahl Nordseewerke, Emden	5 Dec 1972	26 Mar 1974	13 Mar 1975

Displacement, tons: 450 surfaced; 498 dived
Dimensions, feet (metres): 159.4 × 15.1 × 14.8
(48.6 × 4.6 × 4.5)
Main machinery: Diesel-electric; 2 MTU 12V 493 AZ80 GA31L
diesels; 1,200 hp(m) *(882 kW)* sustained; 2 alternators;
810 kW; 1 Siemens motor; 1,800 hp(m) *(1.32 MW)*
sustained; 1 shaft
Speed, knots: 10 surfaced; 17 dived
Range, n miles: 4,500 at 5 kt surfaced
Complement: 22 (4 officers)

Torpedoes: 8—21 in *(533 mm)* bow tubes. STN Atlas DM 2A3;
wire-guided; active homing to 13 km *(7 n miles)* at 35 kt;
passive homing to 28 km *(15 n miles)* at 23 kt; warhead
260 kg.

Mines: GRP container secured outside hull each side. Each
container holds 12 mines, carried in addition to the normal
torpedo or mine armament (16 in place of torpedoes).
Countermeasures: ESM: Thomson-CSF DR 2000U with THORN
EMI Sarie 2; intercept.
Weapons control: SLW 83 (TFCS).
Radars: Surface search: Thomson-CSF Calypso II; I-band.
Sonars: Atlas Elektronik DBQS-21D; passive/active search and
attack; medium frequency.
Thomson Sintra DUUX 2; passive ranging.

Programmes: Authorised on 7 June 1969.
Modernisation: Mid-life conversion of the class was a very
extensive one, including the installation of new sensors (sonar
DBQS-21D with training simulator STU-5), periscopes,

weapon control system (LEWA), ESM, weapons (torpedo
Seeal), GPS navigation, and a comprehensive refitting of the
propulsion system, as well as habitability improvements.
Conversion work was shared between Thyssen Nordseewerke
(U 23, 30, 22, 24, 15, 26) at Emden and HDW *(U 29, 16, 25,
17, 18)* at Kiel. The work started in mid-1987 and completed in
February 1992.
Structure: Hulls are built of high-tensile non-magnetic steel.
Operational: First and third squadrons based at Eckernförde.
U 26 to be decommissioned in December 2005 and *U 22* and
U 29 in December 2006.
Sales: Two unmodernised (Type 206) were to have been
acquired by Indonesia but the sale was cancelled in late 1998.
UPDATED

U 16

6/2004 *, *B Sullivan* / 0587754

U 24

8/2004 *, *E & M Laursen* / 1044261

U 29

4/2003, Martin Mokrus / 0570654

U 15

3/2003, A Sheldon-Duplaix / 0570645

U 29

10/2003, Curt Borgenstam / 0570646

FRIGATES

Notes: The design phase for the next generation Type 125 frigate is expected to be completed by early 2006 with a contract to follow in the same year. Four ships are required. Various financing options including bank loans and leasing are under consideration. The first ship is to enter service in 2014.

4 BRANDENBURG CLASS (TYPE 123) (FFGHM)

Name	No	Builders	Laid down	Launched	Commissioned
BRANDENBURG	F 215	Blohm + Voss, Hamburg	11 Feb 1992	28 Aug 1992	14 Oct 1994
SCHLESWIG-HOLSTEIN	F 216	Howaldtswerke, Kiel	1 July 1993	8 June 1994	2 Nov 1995
BAYERN	F 217	Thyssen Nordseewerke, Emden	16 Dec 1993	30 June 1994	15 June 1996
MECKLENBURG-VORPOMMERN	F 218	Bremer Vulkan/Thyssen Nordseewerke	23 Nov 1993	8 July 1995	6 Dec 1996

Displacement, tons: 4,900 full load
Dimensions, feet (metres): 455.7 oa; 416.3 wl × 54.8 × 22.3 *(138.9; 126.9 × 16.7 × 6.8)*
Main machinery: CODOG; 2 GE 7LM2500SA-ML gas turbines; 51,000 hp *(38 MW)* sustained; 2 MTU 20V 956 TB92 diesels; 11,070 hp(m) *(8.14 MW)* sustained; 2 shafts; Escher Weiss; cp props
Speed, knots: 29; 18 on diesels. **Range, n miles:** 4,000 at 18 kt
Complement: 199 (27 officers) plus 19 aircrew

Missiles: SSM: 4 Aerospatiale MM 38 Exocet (2 twin) ❶ (from Type 101A); inertial cruise; active radar homing to 42 km *(23 n miles)* at 0.9 Mach; warhead 165 kg; sea-skimmer.
SAM: Martin Marietta VLS Mk 41 Mod 3 ❷ for 16 NATO Sea Sparrow; semi-active radar homing to 14.6 km *(8 n miles)* at 2.5 Mach; warhead 39 kg.
2 RAM 21 cell Mk 49 launchers ❸; passive IR/anti-radiation homing to 9.6 km *(5.2 n miles)* at 2 Mach; warhead 9.1 kg; 32 missiles.
Guns: 1 OTO Melara 3 in *(76 mm)*/62 Mk 75 ❹; 85 rds/min to 16 km *(8.6 n miles)* anti-surface; 12 km *(6.5 n miles)* anti-aircraft; weight of shell 6 kg.
2 Rheinmetall 20 mm Rh 202 to be replaced by Mauser 27 mm.
Torpedoes: 4—324 mm Mk 32 Mod 9 (2 twin) tubes ❺; anti-submarine. Honeywell Mk 46 Mod 2; anti-submarine; active/passive homing to 11 km *(5.9 n miles)* at 40 kt; warhead 44 kg. To be replaced by Eurotorp Mu 90 Impact in due course.
Countermeasures: Decoys: 2 Breda SCLAR ❻. Chaff and IR flares.
ESM/ECM: TST FL 1800S Stage II; intercept and jammers.
Combat data systems: Atlas Elektronik/Paramax SATIR action data automation with Unisys UYK 43 computer; Link 11. Link 16. Matra Marconi SCOT 3 SATCOM ❼.
Weapons control: Signaal MWCS. 2 optical sights. STN Atlas Elektronic WBA optronic sensor.
Radars: Air search: Signaal LW08 ❽; D-band.
Air/Surface search: Signaal SMART; 3D; F-band.
Fire control: 2 Signaal STIR 180 trackers ❿.
Navigation: 2 Raytheon Raypath; I-band.
Sonars: Atlas Elektronik DSQS-23BZ; hull-mounted; active search and attack; medium frequency.
Towed array (provision only); active; low frequency.

Helicopters: 2 Westland Sea Lynx Mk 88 or 88A ⓫.

BRANDENBURG   *(Scale 1 : 1,200), Ian Sturton /* 0069988

MECKLENBURG-VORPOMMERN   *10/2004*, B Sullivan /* 0587753

Programmes: Formerly Deutschland class. Four ordered 28 June 1989. Developed by Blohm + Voss whose design was selected in October 1988. Replaced deleted Hamburg class.
Modernisation: Noise-reduction measures have been taken in F 215 and are to be carried out in the other three ships. LFASS (low frequency active sonar system) towed array to be fitted from 2005. SCOT 3 SATCOM and STN optronic sensor fitted from 1998.

Structure: The design is a mixture of MEKO and improved serviceability Type 122 having the same propulsion as the Type 122. Contemporary stealth features. All steel. Fin stabilisers. Space allocated for a Task Group Commander and Staff.
Operational: 6th Frigate Squadron based at Wilhelmshaven. One RIB is carried for boarding operations.

UPDATED

MECKLENBURG-VORPOMMERN   *9/2004*, John Brodie /* 0587720

BAYERN   *5/2004*, Frank Findler /* 0587719

SCHLESWIG-HOLSTEIN

8/2001, Michael Nitz / 0529111

BRANDENBURG

4/2002, John Brodie / 0528871

BAYERN

11/2004*, John Brodie / 1044259

BRANDENBURG

3/2004*, B Prézelin / 0587718

8 BREMEN CLASS (TYPE 122) (FFGHM)

Name	No	Builders	Laid down	Launched	Commissioned
BREMEN	F 207	Bremer Vulkan	9 July 1979	27 Sep 1979	7 May 1982
NIEDERSACHSEN	F 208	AG Weser/Bremer Vulkan	9 Nov 1979	9 June 1980	15 Oct 1982
RHEINLAND-PFALZ	F 209	Blohm + Voss/Bremer Vulkan	29 Sep 1979	3 Sep 1980	9 May 1983
EMDEN	F 210	Thyssen Nordseewerke, Emden/Bremer Vulkan	23 June 1980	17 Dec 1980	7 Oct 1983
KÖLN	F 211	Blohm + Voss/Bremer Vulkan	16 June 1980	29 May 1981	19 Oct 1984
KARLSRUHE	F 212	Howaldtswerke, Kiel/Bremer Vulkan	10 Mar 1981	8 Jan 1982	19 Apr 1984
AUGSBURG	F 213	Bremer Vulkan	4 Apr 1987	17 Sep 1987	3 Oct 1989
LÜBECK	F 214	Thyssen Nordseewerke, Emden/Bremer Vulkan	1 June 1987	15 Oct 1987	19 Mar 1990

Displacement, tons: 3,680 full load
Dimensions, feet (metres): 426.4 × 47.6 × 21.3
(130 × 14.5 × 6.5)
Main machinery: CODOG; 2 GE LM 2500 gas turbines;
51,000 hp *(38 MW)* sustained; 2 MTU 20V 956 TB92 diesels;
11,070 hp(m) *(8.14 MW)* sustained; 2 shafts; cp props
Speed, knots: 30; 20 on diesels. **Range, n miles:** 4,000 at 18 kt
Complement: 219 (26 officers)

Missiles: SSM: 8 McDonnell Douglas Harpoon (2 quad)
launchers ❶; active radar homing to 130 km *(70 n miles)* at
0.9 Mach; warhead 227 kg.
SAM: 16 Raytheon NATO Sea Sparrow RIM-7M; Mk 29 octuple
launcher ❷; semi-active radar homing to 14.6 km *(8 n miles)* at
2.5 Mach; warhead 39 kg.
2 GDC RAM 21 cell ❸; passive IR/anti-radiation homing to
9.6 km *(5.2 n miles)* at 2 Mach; warhead 9.1 kg.
Guns: 1 OTO Melara 3 in *(76 mm)*/62 Mk 75 ❹; 85 rds/min to
16 km *(8.6 n miles)* anti-surface; 12 km *(6.5 n miles)* anti-
aircraft; weight of shell 6 kg.
2 Rheinmetall 20 mm Rh 202, to be replaced by Mauser
27 mm.
Torpedoes: 4—324 mm Mk 32 (2 twin) tubes ❺. 8 Honeywell
Mk 46 Mod 2; anti-submarine; active/passive homing to
11 km *(5.9 n miles)* at 40 kt; warhead 44 kg. To be replaced by
Eurotorp Mu 90.
Countermeasures: Decoys: 4 Loral Hycor SRBOC ❻ 6-barrelled
fixed Mk 36; chaff and IR flares to 4 km *(2.2 n miles)*.
SLQ-25 Nixie; towed torpedo decoy. Prairie bubble noise
reduction.
ESM/ECM: TST FL 1800 Stage II ❼; intercept and jammer.
Combat data systems: SATIR action data automation; Link 11;
Matra Marconi SCOT 1A SATCOM ❽ (3 sets for the class).
Weapons control: Signaal WM25/STIR. STN Atlas Elektronic
WBA optronic sensor.
Radars: Air/surface search: DASA TRS-3D/32 ❾; C-band.
Navigation: SMA 3 RM 20; I-band.
Fire control: Signaal WM25 ❿; I/J-band.
Signaal STIR ⓫; I/J/K-band; range 140 km *(76 n miles)* for
1 m² target.
Sonars: Atlas Elektronik DSQS-21BZ (BO); hull-mounted; active
search and attack; medium frequency.

Helicopters: 2 Westland Sea Lynx Mk 88 or 88A ⓬.

Programmes: Approval given in early 1976 for first six of this
class, a modification of the Netherlands Kortenaer class.
Replaced the deleted Fletcher and Köln classes. Equipment
ordered February 1986 after order placed 6 December 1985
for last pair. Hulls and some engines provided in the five
building yards. Ships were then towed to the prime contractor
Bremer Vulkan where weapon systems and electronics were
fitted and trials conducted. The three names for F 210-212
were changed from the names of Länder to take the well
known town names of the Köln class as they were paid off.
Modernisation: RAM fitted from 1993-1996: Updated EW fit
from 1994. 20 mm guns, taken from Type 520 LCUs, fitted aft
of the bridge on each side. TRS-3D/32 radar has replaced DA
08 in all ships. STN optronic sensor fitted from 1998. 27 mm
guns to replace 20 mm in due course. A further modernisation
plan is under consideration.
Operational: Form 2nd and 4th Frigate Squadrons. Three
containerised SCOT 1A terminals acquired in 1988 and when
fitted are mounted on the hangar roof.   ***UPDATED***

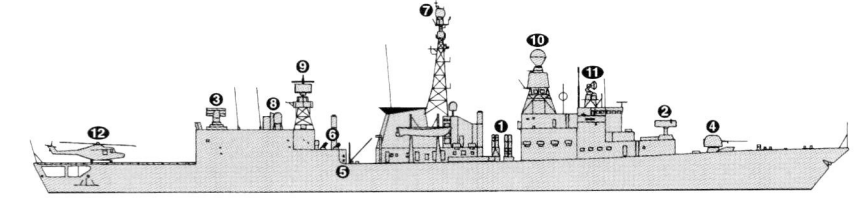

EMDEN   *(Scale 1 : 1,200), Ian Sturton /* 0012400

BREMEN   *3/2004*, B Prézelin /* 0587752

LÜBECK   *3/2004*, Frank Findler /* 0587751

RHEINLAND-PFALZ   *4/2004*, John Brodie /* 0587750

For details of the latest updates to *Jane's Fighting Ships* online and to discover the
additional information available exclusively to online subscribers please visit
jfs.janes.com

German Quality at Sea

2 + 1 SACHSEN CLASS (TYPE 124) (FFGHM)

Name	No	Builders	Laid down	Launched	Commissioned
SACHSEN	F 219	Blohm + Voss, Hamburg	1 Feb 1999	1 Dec 1999	4 Nov 2004
HAMBURG	F 220	Howaldtswerke, Kiel	1 Sep 2000	16 Aug 2002	Dec 2004
HESSEN	F 221	Thyssen Nordseewerke, Emden	14 Sep 2002	27 June 2003	Dec 2005

Displacement, tons: 5,600 full load
Dimensions, feet (metres): 469.2 oa; 433.7 wl × 57.1 × 14.4
  (143; 132.2 × 17.4 × 4.4)
Main machinery: CODAG; 1 GE LM 2500 gas turbine; 31,514 hp
  (23.5 MW); 2 MTU 20V 1163 TB 93 diesels; 20,128 hp(m)
  (14.8 MW); 2 shafts; cp props
Speed, knots: 29. **Range, n miles:** 4,000 at 18 kt
Complement: 255 (39 officers)

Missiles: SSM: 8 Harpoon ❶ 2 (quad).
  SAM: Mk 41 VLS (32 cells) ❷ 24 GDC Standard SM-2 (Block IIIA);
  command/inertial guidance; semi-active radar homing to
  167 km (90 n miles) at 2 Mach. 32 Evolved Sea Sparrow RIM
  162B; semi-active radar homing to 18 km (9.7 n miles) at 3.6
  Mach; warhead 39 kg.
  2 RAM launchers ❸. 21 rounds per launcher; passive IR/anti-
  radiation homing to 9.6 km (5.2 n miles) at 2 Mach; warhead
  9.1 kg.
Guns: 1 Otobreda 76 mm/62 IROF ❹; 85 rds/min to 16 km
  (8.6 n miles) anti-surface; 12 km (6.5 n miles) anti-aircraft;
  weight of shell 6 kg.
  2 Mauser 27 mm ❺.
Torpedoes: 6—324 mm (2 triple) Mk 32 Mod 7 tubes ❻.
  Eurotorp Mu 90 Impact.
Countermeasures: Decoys: 6 SRBOC 130 mm chaff launchers
  ❼.
ESM/ECM: DASA FI 1800S-II; intercept ❽ and jammer.
Combat data systems: SEWACO FD; Link 11/16.
Weapons control: MSP optronic director ❾.
Radars: Air search: SMART L ❿ 3D; D-band.
  Air/surface search: Signaal APAR phased array ⓫; I/J-band.
  Surface search: ⓬.
  Navigation: 2 sets ⓭; I-band.
  IFF: Mk XII.
Sonars: Atlas DSQS-21B (Mod); bow-mounted; active search;
  medium frequency.

Helicopters: 2 NH90 NFH ⓮ or 2 Lynx 88A.

Programmes: Type 124 air defence ships built to replace the
  Lütjens class. A collaborative design with the Netherlands. A
  Memorandum of Understanding (MoU) was signed in October

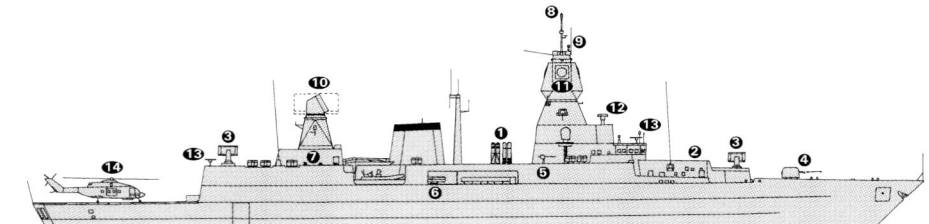

SACHSEN   (Scale 1 : 1,200), Ian Sturton / 0529152

HAMBURG   4/2004*, Michael Winter / 0587749

1993 between Blohm + Voss, Royal Schelde and Bazán
shipyards. A contract to build three ships was authorised on
12 June 1996 with an option on a fourth. *Sachsen* started sea
trials in August 2001 and *Hamburg* on 12 January 2004.
Structure: Based on the Type 123 hull with improved stealth
features. MBB-FHS helo handling system.

Operational: Successful sea-firings of Standard SM-2 and ESSM
conducted at USN range off southern California in July/
August 2004.

UPDATED

SACHSEN   5/2004*, Harald Carstens / 0587748

SACHSEN   6/2004*, Michael Nitz / 1044272

Simply the best for the navies

It's EADS.

EADS Defence and Communications Systems develops and manufactures system solutions to provide world-class combat systems.

Combat system management is achieved through modern object-orientated real-time software modules designed in an open architecture. These modules are platform-independent with a Plug and Fight Philosophy to meet all specific program requirements.

EADS is well qualified as:
- Combat Systems Designer
- Combat Systems Integrator
- Combat Management System Supplier
- System Prime Contractor

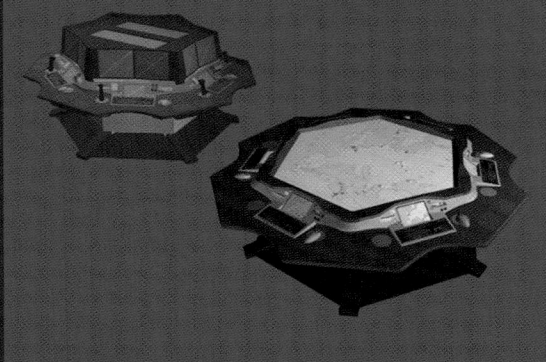

Our latest product, **Advanced Naval Combat System (ANCS™)**, is a well-proven Combat Management System and is based on our extensive experience combined with leading-edge technology.

ANCS™ is managed and operated through the next generation of operator console OMADA™ designed for the network operations environment.

Defence and Communications Systems
The EADS Systems House

EADS
Defence and Communications Systems
89070 Ulm/Germany
Phone   + 49 (0) 7 31 . 3 92 - 73 69
Fax   + 49 (0) 7 31 . 3 92 - 29 13
contact.info@eads.com

www.eads.com

CORVETTES

0 + 5 BRAUNSCHWEIG (K130) CLASS (FSGHM)

Name	No	Builders	Laid down	Launched	Commissioned
BRAUNSCHWEIG	F 260	Blohm + Voss, Hamburg	2005	2006	May 2007
MAGDEBURG	F 261	Lürssen, Vegesack	2005	2006	Nov 2007
ERFURT	F 262	Thyssen Nordseewerke, Emden	2006	2007	Apr 2008
OLDENBURG	F 263	Blohm + Voss, Hamburg	2006	2007	Aug 2008
LUDWIGSHAFEN	F 264	Lürssen, Vegesack	2006	2007	Nov 2008

Displacement, tons: 1,662 full load
Dimensions, feet (metres): 289.8 × 43.4 × 15.7 *(88.3 × 13.2 × 4.8)*
Main machinery: 2 diesels; total of 19,850 hp(m) *(14.8 MW)*; 2 shafts
Speed, knots: 26. **Range, n miles:** 2,500 at 15 kt
Complement: 50 + 15 spare
Missiles: SSM: 4 Saab RBS-15 Mk 3 ❶; active radar homing to 200 km *(108 n miles)* at 0.9 Mach; warhead 200 kg.
SAM: 2 RAM 21 cell Mk 49 launchers ❷.
Guns: 1 Otobreda 76 mm/62 ❸; 2 Mauser 27 mm. ❹.
Countermeasures: Decoys: 2 MASS ❺; softkill launchers.
ESM: EADS SPS-N-5000; intercept.
ECM: DASA SPN/KJS 5000; jammer.
Combat data systems: SEWACO; Link 11/16.
Weapons control: 2 Signaal Mirador optronic directors. ❻.
Radars: Air/surface search: DASA TRS-3D ❼; C-band.
Surface search: E/F-band.
Fire control: I/J-band.
Helicopters: 1 medium (NH 90) and 2 VTOL drones.

Programmes: Invitations to tender accepted at the end of 1998. Blohm + Voss selected as consortium leader 18 July 2000. Consortium includes Thyssen Nordseewerke and Lürssen. Batch of five ships ordered on 14 December 2001 and first steel cut for the first of class on 19 July 2004. There will be no further ships of this class.
Structure: Measures to reduce radar, IR (underwater exhaust system) and noise signatures have been included in the design. The Polyphem SAM consists of eight vertical launchers recessed in the flight deck.

UPDATED

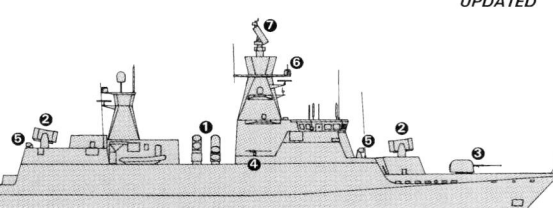

K130   *(Scale 1 : 900), Ian Sturton* / 0529158

K130 (model)   *9/2002, Michael Nitz* / 0529080

SHIPBORNE AIRCRAFT

Notes: (1) NH 90s are required in due course to replace Sea Kings and Lynx aircraft.
(2) Plans to procure a vertical take-off and landing UAV for the K 130 corvette have been abandoned.

Numbers/Type: 15/7 Westland Sea Lynx Mk 88/Super Lynx Mk 88A.
Operational speed: 125 kt *(232 km/h)*.
Service ceiling: 12,500 ft *(3,010 m)*.
Range: 320 n miles *(593 km)*.
Role/Weapon systems: Shipborne ASW/ASV role. Seven Mk 88A ordered in 1996 for delivery in 1999. Mid-life upgrade for the Mk 88 aircraft to Super Lynx standard placed in mid-1998 and completed by 2003. Sensors: Ferranti Sea Spray Mk 1 or GEC Marine Sea Spray 3000 and FLIR (Mk 88A) radar and Bendix AQS-18 dipping sonar. Weapons: ASW; up to two Mk 46 Mod 2 (or Eurotorp Mu 90 Impact in due course) torpedoes or depth charges. ASV; BAe Sea Skua (Mk 88A).

UPDATED

LYNX Mk 88A   *6/2004*, B Prézelin* / 0587747

LAND-BASED MARITIME AIRCRAFT (FRONT LINE)

Note: The Maritime Patrol Replacement programme (MPA-R), a collaborative programme between Germany and Italy, has been abandoned. It was announced jointly by the German and Netherlands defence ministries on 20 July 2004 that Germany is to procure eight P-3C Orions, modernised to full Update II standard, for delivery between November 2005 and March 2006. Training began in August 2004 and the aircraft, which will replace the Atlantic MPAs, will be based at MFG 3, Nordholz.

Numbers/Type: 4 Dornier Do 228-212.
Operational speed: 156 kt *(290 km/h)*.
Service ceiling: 20,700 ft *(6,300 m)*.
Range: 667 n miles *(1,235 km)*.
Role/Weapon systems: Liaison and transport. Two converted for pollution control. Sensors: Weather radar; converted aircraft also has SLAR, IR/UR scanner, microwave radiometer, LLL TV camera and data downlink. Weapons: Unarmed.

VERIFIED

DORNIER 228   *8/2003, Frank Findler* / 0570631

Numbers/Type: 14 Breguet Atlantic 1.
Operational speed: 355 kt *(658 km/h)*.
Service ceiling: 32,800 ft *(10,000 m)*.
Range: 4,850 n miles *(8,990 km)*.
Role/Weapon systems: Long-range/endurance MR tasks carried out in North and Baltic Seas, also Atlantic Ocean; four aircraft allocated to Elint/SIGINT tasks. 16 upgraded by Raytheon with GPS, FLIR, new ESM and radar. To be replaced by P-3C Orions by March 2006. Sensors: APS-134 radar, Loral ESM, MAD, sonobuoys. Weapons: ASW; eight torpedoes (Mk 46 or Mu 90 Impact in due course) or mines or depth bombs.

UPDATED

ATLANTIC   *8/2003, Frank Findler* / 0570632

Numbers/Type: 50 Panavia Tornado IDS.
Operational speed: Mach 2.2.
Service ceiling: 80,000 ft *(24,385 m)*.
Range: 1,500 n miles *(2,780 km)*.
Role/Weapon systems: Swing-wing strike and recce; shore-based for fleet air defence and ASV strike primary roles; update with Kormoran 2 and Texas Instruments HARM. To be transferred to the Air Force in 2005. Sensors: Texas Instruments nav/attack system, MBB/Alenia multisensor recce pod. Weapons: ASV; four Kormoran 2 missiles. Fleet AD; two 27 mm cannon, four AIM-9L Sidewinder.

UPDATED

TORNADO   *9/2004*, Frank Findler* / 1044258

Numbers/Type: 22 Westland Sea King Mk 41 KWS.
Operational speed: 140 kt *(260 km/h)*.
Service ceiling: 10,500 ft *(3,200 m)*.
Range: 630 n miles *(1,165 km)*.
Role/Weapon systems: Role change from primary combat rescue helicopter to ASV started in 1988 with new camouflage appearance and an update programme by MBB with BAe/Ferranti support which completed in 1995. Sensors: Ferranti Sea Spray Mk 3 radar, ALR 68 ESM, chaff and flare dispenser. Weapons: ASV; four BAe Sea Skua missiles.

UPDATED

SEA KING   *9/2004*, Frank Findler* / 1044266

PATROL FORCES

Notes: Vessels in this section have an 'S' number as part of their name as well as a 'P' pennant number. The 'S' number is shown in the Pennant List at the front of this country.

4 ALBATROS CLASS (TYPE 143B)
(FAST ATTACK CRAFT—MISSILE) (PGGFM)

Name	No	Builders	Commissioned
GEIER	P 6113	Lürssen, Vegesack	2 June 1976
SEEADLER	P 6118	Lürssen, Vegesack	28 Mar 1977
HABICHT	P 6119	Kröger, Rendsburg	23 Dec 1977
KORMORAN	P 6120	Lürssen, Vegesack	29 July 1977

Displacement, tons: 398 full load
Dimensions, feet (metres): 189 × 25.6 × 8.5 *(57.6 × 7.8 × 2.6)*
Main machinery: 4 MTU 16V 956 TB91 diesels; 17,700 hp(m) *(13 MW)* sustained; 4 shafts
Speed, knots: 40. **Range, n miles:** 1,300 at 30 kt
Complement: 40 (4 officers)

Missiles: SSM: 4 Aerospatiale MM 38 Exocet (2 twin) launchers; inertial cruise; active radar homing to 42 km *(23 n miles)* at 0.9 Mach; warhead 165 kg; sea-skimmer.
Guns: 2 OTO Melara 3 in *(76 mm)*/62 compact; 85 rds/min to 16 km *(8.6 n miles)* anti-surface; 12 km *(6.5 n miles)* anti-aircraft; weight of shell 6 kg.
2—12.7 mm MGs (may be fitted).
Torpedoes: 2—21 in *(533 mm)* aft tubes. AEG Seeal; wire-guided; active homing to 13 km *(7 n miles)* at 35 kt; passive homing to 28 km *(15 n miles)* at 23 kt; warhead 260 kg.
Countermeasures: Decoys: Buck-Wegmann Hot Dog/Silver Dog; IR/chaff dispenser.
ESM/ECM: Racal Octopus (Cutlass intercept, Scorpion jammer).
Combat data systems: AEG/Signaal command and fire-control system; Link 11.
Weapons control: ORG7/3 optronics GFCS. STN Atlas WBA optronic sensor to be fitted.
Radars: Surface search/fire control: Signaal WM27; I/J-band.
Navigation: SMA 3 RM 20; I-band.

Programmes: AEG-Telefunken main contractor. Ordered in 1972.
Modernisation: *Habicht* started trials with RAM-ASDM mounting in 1983. Plans for major modernisation were reduced to fitting a new EW system, Racal Octopus, which completed in 1995 and a Signaal update to the command system which completed in 1999. 12.7 mm MGs may be fitted for deployments.
Structure: Wooden hulled craft.
Operational: Form 2nd Squadron based on tender *Donau* at Warnemünde. Last remaining craft to be decommissioned by December 2005. All to be sold to Tunisia without MM 38 Exocet.

UPDATED

10 GEPARD CLASS (TYPE 143 A)
(FAST ATTACK CRAFT—MISSILE) (PGGFM)

Name	No	Builders	Launched	Commissioned
GEPARD	P 6121	AEG/Lürssen	25 Sep 1981	13 Dec 1982
PUMA	P 6122	AEG/Lürssen	8 Feb 1982	24 Feb 1983
HERMELIN	P 6123	AEG/Kröger	8 Dec 1981	5 May 1983
NERZ	P 6124	AEG/Lürssen	18 Aug 1982	14 July 1983
ZOBEL	P 6125	AEG/Kröger	30 June 1982	25 Sep 1983
FRETTCHEN	P 6126	AEG/Lürssen	26 Jan 1983	15 Dec 1983
DACHS	P 6127	AEG/Lürssen	14 Dec 1982	22 Mar 1984
OZELOT	P 6128	AEG/Lürssen	7 June 1983	3 May 1984
WIESEL	P 6129	AEG/Lürssen	8 Aug 1983	12 July 1984
HYÄNE	P 6130	AEG/Lürssen	5 Oct 1983	13 Nov 1984

Displacement, tons: 391 full load
Dimensions, feet (metres): 190 × 25.6 × 8.5 *(57.6 × 7.8 × 2.6)*
Main machinery: 4 MTU MA 16V 956 SB80 diesels; 13,200 hp(m) *(9.7 MW)* sustained; 4 shafts
Speed, knots: 40. **Range, n miles:** 2,600 at 16 kt; 600 at 33 kt
Complement: 34 (4 officers)

Missiles: SSM: 4 Aerospatiale MM 38 Exocet; inertial cruise; active radar homing to 42 km *(23 n miles)* at 0.9 Mach; warhead 165 kg; sea-skimmer.
SAM: GDC RAM 21 cell point defence system; passive IR/anti-radiation homing to 9.6 km *(5.2 n miles)* at 2 Mach; warhead 9.1 kg.
Guns: 1 OTO Melara 3 in *(76 mm)*/62 compact; 85 rds/min to 16 km *(8.6 n miles)* anti-surface; 12 km *(6.5 n miles)* anti-aircraft; weight of shell 6 kg.
Mines: Can lay mines.
Countermeasures: Decoys: Buck-Wegmann Hot Dog/Silver Dog; IR/chaff dispenser.
ESM/ECM: Dasa FL 1800 Mk 2; radar intercept and jammer.
Combat data systems: AEG AGIS with Signaal update; Link 11.
Weapons control: STN Atlas WBA optronic sensor being fitted.
Radars: Surface search/fire control: Signaal WM27; I/J-band; range 46 km *(25 n miles)*.
Navigation: SMA 3 RM 20; I-band.

Programmes: Ordered mid-1978 from AEG-Telefunken with subcontracting to Lürssen (P 6121, 6122, 6124-6128) and Kröger (P 6123, 6129, 6130).
Modernisation: Updated EW fit in 1994-95. RAM fitted in *Puma* in 1992, and to the rest from 1993-98. Combat data system update completed in 1999. Improved EW aerials being fitted from 1999.
Structure: Wooden hulls on aluminium frames.
Operational: Form 7th Squadron based on the tender *Elbe* at Warnemünde.

UPDATED

HABICHT 6/2004*, Frank Findler / 0587746

GEIER 6/2003, Frank Findler / 0570635

SEEADLER 6/2004*, B Sullivan / 0587714

FRETTCHEN 6/2004*, Frank Findler / 0587745

HYÄNE 6/2004*, Frank Findler / 0587744

DACHS 6/2004*, B Sullivan / 0587715

AMPHIBIOUS FORCES

Notes: Procurement of multirole amphibious shipping is unlikely due to budget restrictions.

4 TYPE 521 (LCM)

SARDELLE LCM 14	KRABBE LCM 23	MUSCHEL LCM 25	KORALLE LCM 26

Displacement, tons: 168 full load
Dimensions, feet (metres): 77.4 × 20.9 × 4.9 *(23.6 × 6.4 × 1.5)*
Main machinery: 2 MWM 8-cyl diesels; 685 hp(m) *(503 kW)*; 2 shafts
Speed, knots: 10.5
Complement: 7
Military lift: 60 tons or 50 troops
Radars: Navigation: Atlas Elektronik; I-band.

Comment: Built by Rheinwerft, Walsam. Completed in 1964-67 and later placed in reserve. All are rated as 'floating equipment' without permanent crews. The design is similar to US LCM 8. LCM 1-11 sold to Greece in April 1991. All but three of the class paid off in 1993-94 but six were brought back into service in 1995 and one paid off again in 1996. Four paid off in 2000. In addition LCM 17 is used by the Bremerhaven Fire Brigade. A new class of LCM is being designed but is not yet funded.   *UPDATED*

SARDELLE   *7/2004*, Frank Findler /* 1044265

2 TYPE 520 (LCU)

LACHS L 762	SCHLEI L 765

Displacement, tons: 430 full load
Dimensions, feet (metres): 131.2 × 28.9 × 7.2 *(40 × 8.8 × 2.2)*
Main machinery: 2 MWM 12-cyl diesels; 1,020 hp(m) *(750 kW)*; 2 shafts
Speed, knots: 11
Complement: 17
Military lift: 150 tons

Comment: Similar to the US LCU (Landing Craft Utility) type. Provided with bow and stern ramp. Built by Howaldtswerke, Hamburg, 1965-66. Two sold to Greece in November 1989 and six more in 1992. Based at Olpenitz with Minesweeper Squadron 3. Guns have been removed.
VERIFIED

LACHS   *6/2003, Frank Findler /* 0570640

MINE WARFARE FORCES

1 + (3) MINESWEEPING DRONES (MSD)

Displacement, tons: 267 full load
Dimensions, feet (metres): 130.6 × 19.2 × 6.9 *(39.8 × 5.86 × 2.1)*
Main machinery: To be decided
Complement: Passage crew

Comment: The requirement for a minesweeping capability is expected to be fulfilled by unmanned minesweeping drones which may be developed in collaboration with the Netherlands. A number of designs, including monohull, SWATH and catamaran are under consideration. With an expected endurance of six days they would be controlled from the upgraded Alkmaar class minehunters which would also programme the target signature simulation system. The requirement to survive explosions so triggered poses a considerable design challenge. Due to budget constraints, procurement has been delayed until 2007 to meet an earliest in-service date of 2010. In the meantime, the upgraded Alkmaar class will be capable of controlling the German 'Seehund' drones. A prototype drone 'Explorer' started sea trials in February 2004 and is to be handed over to the navy on completion.   *UPDATED*

EXPLORER   *6/2004*, Frank Findler /* 0587726

12 FRANKENTHAL CLASS (TYPE 332)
(MINEHUNTERS—COASTAL) (MHC)

Name	No	Builders	Launched	Commissioned
FRANKENTHAL	M 1066	Lürssenwerft	6 Feb 1992	16 Dec 1992
WEIDEN	M 1060	Abeking & Rasmussen	14 May 1992	30 Mar 1993
ROTTWEIL	M 1061	Krögerwerft	12 Mar 1992	7 July 1993
BAD BEVENSEN	M 1063	Lürssenwerft	21 Jan 1993	9 Dec 1993
BAD RAPPENAU	M 1067	Abeking & Rasmussen	3 June 1993	19 Apr 1994
GRÖMITZ	M 1064	Krögerwerft	29 Apr 1993	23 Aug 1994
DATTELN	M 1068	Lürssenwerft	27 Jan 1994	8 Dec 1994
DILLINGEN	M 1065	Abeking & Rasmussen	26 May 1994	25 Apr 1995
HOMBURG	M 1069	Krögerwerft	21 Apr 1994	26 Sep 1995
SULZBACH-ROSENBERG	M 1062	Lürssenwerft	27 Apr 1995	23 Jan 1996
FULDA	M 1058	Abeking & Rasmussen	29 Sep 1997	16 June 1998
WEILHEIM	M 1059	Lürssenwerft	26 Feb 1998	3 Dec 1998

Displacement, tons: 650 full load
Dimensions, feet (metres): 178.8 × 30.2 × 8.5 *(54.5 × 9.2 × 2.6)*
Main machinery: 2 MTU 16V 396 TB84 diesels; 5,550 hp(m) *(4.08 MW)* sustained; 2 shafts; cp props; 1 motor (minehunting)
Speed, knots: 18
Complement: 37 (5 officers)

Missiles: SAM: 2 Stinger quad launchers.
Guns: 1 Bofors 40 mm/70; being replaced by Mauser 27 mm.
Combat data systems: STN MWS 80-4.
Radars: Navigation: Raytheon SPS-64; I-band.
Sonars: Atlas Elektronik DSQS-11M; hull-mounted; high frequency.

Programmes: First 10 ordered in September 1988 with STN Systemtechnik Nord as main contractor. M 1066 laid down at Lürssen 6 December 1989. Last pair ordered 16 October 1995.
Structure: Same hull, similar superstructure and high standardisation as Type 343. Built of amagnetic steel. Two STN Systemtechnik Nord Pinguin-B3 drones with sonar, TV cameras and two countermining charges, but not Troika control and minelaying capabilities.
Sales: Six of the class being built for Turkey from late 1999.   *UPDATED*

SULZBACH-ROSENBERG   *5/2004*, Per Körnefeldt /* 0587743

5 KULMBACH CLASS (TYPE 333)
(MINEHUNTERS—COASTAL) (MHC)

Name	No	Builders	Launched	Commissioned
ÜBERHERRN	M 1095	Abeking & Rasmussen	30 Aug 1988	19 Sep 1989
LABOE	M 1097	Krögerwerft	13 Sep 1988	7 Dec 1989
KULMBACH	M 1091	Abeking & Rasmussen	15 June 1989	24 Apr 1990
PASSAU	M 1096	Abeking & Rasmussen	1 Mar 1990	18 Dec 1990
HERTEN	M 1099	Krögerwerft	22 Dec 1989	26 Feb 1991

Displacement, tons: 635 full load
Dimensions, feet (metres): 178.5 × 30.2 × 8.2 *(54.4 × 9.2 × 2.5)*
Main machinery: 2 MTU 16V 538 TB91 diesels; 6,140 hp(m) *(4.5 MW)* sustained; 2 shafts; cp props
Speed, knots: 18
Complement: 37 (4 officers)

Missiles: SAM: 2 Stinger quad launchers.
Guns: 2 Mauser 27 mm.
Mines: 60.
Countermeasures: Decoys: 2 Silver Dog chaff rocket launchers.
ESM: Thomson-CSF DR 2000; radar warning.
Combat data systems: PALIS with Link 11.
Radars: Surface Search/fire control: Signaal WM20/2; I/J-band.
Navigation: Raytheon SPS-64; I-band.
Sonars: Atlas Elektronik DSQS-11M; hull-mounted; high frequency.

Programmes: On 3 January 1985 an STN Systemtechnik Nord-headed consortium was awarded the order. The German designation of 'Schnelles Minenkampfboot' was changed in 1989 to 'Schnelles Minensuchboot'. After modernisation redesignated 'Minenjagdboote'.
Modernisation: Five ships of Hameln class converted to minehunters 1999-2001 and redesignated Kulmbach class (Type 333). Eight to ten disposable ROV Sea Fox I are carried for inspection and up to 30 Sea Fox C for mine disposal. It has a range of 500 m at 6 kt and uses a shaped charge.
Structure: Ships built of amagnetic steel adapted from submarine construction. Signaal M 20 System removed from the deleted Zobel class fast attack craft. PALIS active link.
UPDATED

HERTEN   *5/2004*, Per Körnefeldt /* 0587742

5 ENSDORF CLASS (TYPE 352)
(MINESWEEPERS—COASTAL) (MHCD)

Name	No	Builders	Launched	Commissioned
HAMELN	M 1092	Lürssenwerft	15 Mar 1988	29 June 1989
PEGNITZ	M 1090	Lürssenwerft	13 Mar 1989	9 Mar 1990
SIEGBURG	M 1098	Krögerwerft	14 Apr 1989	17 July 1990
ENSDORF	M 1094	Lürssenwerft	8 Dec 1989	25 Sep 1990
AUERBACH	M 1093	Lürssenwerft	18 June 1990	7 May 1991

Displacement, tons: 635 full load
Dimensions, feet (metres): 178.5 × 30.2 × 8.2 *(54.4 × 9.2 × 2.5)*
Main machinery: 2 MTU 16V 538 TB91 diesels; 6,140 hp(m) *(4.5 MW)* sustained; 2 shafts; cp props
Speed, knots: 18
Complement: 37 (4 officers)

Missiles: SAM: 2 Stinger quad launchers.
Guns: 2 Mauser 27 mm.
Mines: 60.
Countermeasures: Decoys: 2 Silver Dog chaff rocket launchers.
ESM: Thomson-CSF DR 2000; radar warning.
Combat data systems: PALIS with Link 11. STN C2 remote-control system for minesweeping drone Seehund.
Radars: Surface Search/fire control: Signaal WM20/2; I/J-band.
Navigation: Raytheon SPS-64; I-band.
Sonars: STN ADS DSQS 15A mine-avoidance; active high frequency.

Programmes: On 3 January 1985 an STN Systemtechnik Nord-headed consortium was awarded the order. The German designation of 'Schnelles Minenkampfboot' was changed in 1989 to 'Schnelles Minensuchboot'. After modernisation redesignated 'Hohlstablenkboote'.
Modernisation: Five ships of Hameln class converted to minesweepers 2000-2001 with capability to control up to four remotely controlled minesweeping drones (Seehund). RoV Sea Fox C for mine disposal. Double oropesa system for mechanical sweeping.
Structure: Ships built of amagnetic steel adapted from submarine construction. Signaal M 20 System removed from the deleted Zobel class fast attack craft. PALIS active link. ***UPDATED***

SIEGBURG *4/2004*, Harald Carstens /* 0587741

18 SEEHUND (MINESWEEPERS—DRONES) (MSD)

SEEHUND 1-18

Displacement, tons: 99 full load
Dimensions, feet (metres): 88.5 × 15 × 4.5 *(26.9 × 4.6 × 1.4)*
Main machinery: 1 Deutz MWM D602 diesel; 446 hp(m) *(328 kW)*; 1 shaft
Speed, knots: 10. **Range, n miles:** 520 at 9 kt
Complement: 3 (passage crew)

Comment: Built by MaK, Kiel and Blohm + Voss, Hamburg between August 1980 and May 1982. Modernised in conjunction with the 352 conversion programme 2000-2001. ***UPDATED***

SEEHUND 6 *7/2003*, Martin Mokrus /* 0587740

1 DIVER SUPPORT SHIP (TYPE 742) (MCD)

Name	No	Builders	Commissioned
MÜHLHAUSEN	M 1052	Burmester, Bremen	21 Dec 1967
(ex-*Walther von Ledebur*)	(ex-A 1410, ex-Y 841)		

Displacement, tons: 775 standard; 825 full load
Dimensions, feet (metres): 206.6 × 34.8 × 8.9 *(63 × 10.6 × 2.7)*
Main machinery: 2 Maybach MTU 16-cyl diesels; 5,200 hp(m) *(3.82 MW)*; 2 shafts
Speed, knots: 19
Complement: 11 plus 10 trials party
Radars: Navigation: I-band.

Comment: Wooden hulled vessel. Launched on 30 June 1966 as a prototype minesweeper but completed as a trials ship. Paid off in April 1994 but reactivated as a diver support ship. Recommissioned in its new role 6 April 1995. ***VERIFIED***

MÜHLHAUSEN *6/2003, B Prézelin /* 0570628

AUXILIARIES

2 BERLIN CLASS (TYPE 702) (AORH)

Name	No	Builders	Launched	Commissioned
BERLIN	A 1411	Flensburger	30 Apr 1999	11 Apr 2001
FRANKFURT AM MAIN	A 1412	Flensburger	5 Jan 2001	27 May 2002

Displacement, tons: 20,240 full load
Dimensions, feet (metres): 569.9 oa; 527.6 wl × 78.7 × 24.3 *(173.7; 160.8 × 24 × 7.4)*
Main machinery: 2 MAN 12V 32/40 diesels; 14,388 hp(m) *(10.58 MW)* sustained; 2 shafts; cp props; bow thruster; 1,000 hp(m) *(735 kW)*
Speed, knots: 20
Complement: 139 (12 officers) plus 94 for embarked staff
Cargo capacity: 9,540 tons fuel; 450 tons water; 280 tons cargo; 160 tons ammunition
Missiles: SAM: 2 RAM launchers fitted for but not with.
Guns: 4 Mauser 27 mm.
Radars: Navigation and aircraft control; I-band.
Helicopters: 2 NH90 or Sea Kings.

Comment: First ship ordered 15 October 1997, and second 3 July 1998. Hulls built by FSG, superstructure by Kröger and electronics by Lürssen. MBB-FHS helo handling system. Two RAS beam stations and stern refuelling. Two portable SAM launchers are carried. EW equipment may be fitted. These ships are designed to support UN type operations abroad. Twenty-six containers can be mounted in two layers on the upper deck. This could include a containerised hospital unit for 50. A 1411 based at Wilhelmshaven and A 1412 at Kiel. A second batch of two ships is planned to start entering service in 2011. ***UPDATED***

FRANKFURT AM MAIN *7/2003, B Sullivan /* 0570660

BERLIN *5/2004*, John Brodie /* 0587739

6 ELBE CLASS (TYPE 404) (TENDERS) (ARLHM)

Name	No	Builders	Launched	Commissioned
ELBE	A 511	Bremer Vulkan	24 June 1992	28 Jan 1993
MOSEL	A 512	Bremer Vulkan	22 Apr 1993	22 July 1993
RHEIN	A 513	Flensburger Schiffbau	11 Mar 1993	22 Sep 1993
WERRA	A 514	Flensburger Schiffbau	17 June 1993	9 Dec 1993
MAIN	A 515	Lürssen/Krögerwerft	15 June 1993	23 June 1994
DONAU	A 516	Lürssen/Krögerwerft	24 Mar 1994	22 Nov 1994

Displacement, tons: 3,586 full load
Dimensions, feet (metres): 329.7 oa; 285.4 wl × 50.9 × 13.5 *(100.5; 87 × 15.5 × 4.1)*
Main machinery: 1 Deutz MWM 8V 12M 628 diesel; 3,335 hp(m) *(2.45 MW)*; 1 shaft; bow thruster
Speed, knots: 15. **Range, n miles:** 2,000 at 15 kt
Complement: 40 (4 officers) plus 12 squadron staff plus 38 maintainers
Cargo capacity: 450 tons fuel; 150 tons water; 11 tons luboil; 130 tons ammunition
Missiles: SAM: 2 Stinger (Fliegerfaust 2) quad launchers.
Guns: 2 Rheinmetall 20 mm or Mauser 27 mm.
Radars: Navigation: I-band.
Helicopters: Platform for 1 Sea King.

Comment: Funds released in November 1990 for the construction of six ships to replace the Rhein class. Containers for maintenance and repairs, spare parts and supplies for fast attack craft and minesweepers. Waste disposal capacity: 270 m³ liquids, 60 m³ solids. The use of the Darss class (all sold in 1991) was investigated as an alternative but rejected on the grounds of higher long-term costs because of the age of the ships. Allocated as follows: *Elbe* to 7th Squadron FPBs, *Mosel* to 5th Squadron MSC, *Rhein* to 6th Squadron MSC, *Werra* to 1st Squadron MSC, *Donau* to 2nd Squadron FPBs. *Main* is to undergo conversion to submarine depot ship from November 2005 to November 2006. 20 mm guns are fitted at the break of the forecastle. Converted with helicopter refuelling facilities from July 1996 to July 1997. ***UPDATED***

RHEIN *5/2004*, Frank Findler /* 0587738

2 REPLENISHMENT TANKERS (TYPE 704) (AOL)

Name	No	Builders	Commissioned
SPESSART (ex-*Okapi*)	A 1442	Kröger, Rendsburg	1974
RHÖN (ex-*Okene*)	A 1443	Kröger, Rendsburg	1974

Displacement, tons: 14,169 full load
Measurement, tons: 6,103 grt; 10,800 dwt
Dimensions, feet (metres): 427.1 × 63.3 × 26.9 *(130.2 × 19.3 × 8.2)*
Main machinery: 1 MaK 12-cyl diesel; 8,000 hp(m) *(5.88 MW)*; 1 shaft; cp prop
Speed, knots: 16. **Range, n miles:** 7,400 at 16 kt
Complement: 42
Cargo capacity: 11,000 m³ fuel; 400 m³ water
Radars: Navigation: I-band.

Comment: Completed for Terkol Group as tankers. Acquired in 1976 for conversion *(Spessart* at Bremerhaven, *Rhön* at Kröger*)*. The former commissioned for naval service on 5 September 1977 and the latter on 23 September 1977. Has two portable SAM positions. Civilian manned.
UPDATED

RHÖN   *3/2004*, Frank Findler /* 0587737

2 WALCHENSEE CLASS (TYPE 703)
(REPLENISHMENT TANKERS) (AOL)

Name	No	Builders	Commissioned
AMMERSEE	A 1425	Lindenau, Kiel	2 Mar 1967
TEGERNSEE	A 1426	Lindenau, Kiel	23 Mar 1967

Displacement, tons: 2,191 full load
Dimensions, feet (metres): 243.4 × 36.7 × 14.8 *(74.2 × 11.2 × 4.5)*
Main machinery: 2 MWM 12-cyl diesels; 1,370 hp(m) *(1 MW)*; 1 Kamewa prop
Speed, knots: 12.6. **Range, n miles:** 3,250 at 12 kt
Complement: 21
Radars: Navigation: Kelvin Hughes; I-band.

Comment: Civilian manned.
UPDATED

AMMERSEE   *8/2004*, Martin Mokrus /* 0587736

1 KNURRHAHN CLASS (TYPE 730) (APB)

Name	No	Builders	Commissioned
KNURRHAHN	Y 811	Sietas, Hamburg	Nov 1989

Displacement, tons: 1,424 full load
Dimensions, feet (metres): 157.5 × 45.9 × 5.9 *(48 × 14 × 1.8)*

Comment: Accommodation for 200 people.
VERIFIED

KNURRHAHN   *7/2003, Frank Findler /* 0570616

1 WESTERWALD CLASS (TYPE 760)
(AMMUNITION TRANSPORT) (AEL)

Name	No	Builders	Commissioned
WESTERWALD	A 1435	Orenstein and Koppel, Lübeck	11 Feb 1967

Displacement, tons: 3,460 standard; 4,042 full load
Dimensions, feet (metres): 344.4 × 46 × 15.1 *(105 × 14 × 4.6)*
Main machinery: 2 MTU MD 16V 538 TB90 diesels; 6,000 hp(m) *(4.1 MW)* sustained; 2 shafts; cp props; bow thruster
Speed, knots: 17. **Range, n miles:** 3,500 at 17 kt
Complement: 31
Cargo capacity: 1,080 tons ammunition
Guns: 2 Bofors 40 mm (cocooned).
Countermeasures: Decoys: 2 Breda SCLAR 105 mm chaff launchers are carried in A 1436.
Radars: Navigation: Kelvin Hughes; I-band.

Comment: Based at Wilhelmshaven. Civilian manned.
VERIFIED

WESTERWALD   *8/2003, Martin Mokrus /* 0570617

2 OHRE CLASS (ACCOMMODATION SHIPS) (APB)

ALTMARK Y 891 (ex-H 11)   **WISCHE** (ex-*Harz*) Y 895 (ex-H 31)

Displacement, tons: 1,320 full load
Dimensions, feet (metres): 231 × 39.4 × 5 *(70.4 × 12 × 1.6)*

Comment: Ex-GDR Type 162 built by Peenewerft, Wolgast. One hydraulic 8 ton crane fitted. First commissioned 1985. Classified as 'Schwimmende Stuetzpunkte'. Propulsion and armament has been removed and they are used as non-self-propelled accommodation ships for crews of vessels in refit. Civilian manned. Both modernised at Wilhelmshaven and to remain in service until further notice. Two others paid off in 2000 later than expected.
VERIFIED

ALTMARK   *9/2003, Frank Findler /* 0570618

1 BATTERY CHARGING CRAFT (TYPE 718) (YAG)

LP 3

Displacement, tons: 234 full load
Dimensions, feet (metres): 90.6 × 23 × 5.2 *(27.6 × 7.0 × 1.6)*
Main machinery: 1 MTU MB diesel; 250 hp(m) *(184 kW)*; 1 shaft
Speed, knots: 9
Complement: 6

Comment: Built in 1964. Has diesel charging generators for submarine batteries.
VERIFIED

LP 3   *6/2001, van Ginderen Collection /* 0130313

7 LAUNCHES (TYPE 946/945) (YFL)

AK 1 Y 1671	**MA 3** Y 1677	**ASCHAU** Y 1685
AK 3 Y 1672	**MA 1** Y 1678	**BORBY** Y 1687
MA 2 Y 1676		

Dimensions, feet (metres): 39.4 × 12.8 × 6.2 *(12.0 × 3.9 × 1.9)*
Main machinery: 1 MAN D2540MTE diesel; 366 hp(m) *(269 kW)*; 1 shaft

Comment: Built by Hans Boost, Trier. All completed in 1985 except *MA 1* and *Aschau* which are larger at 16.2 m and completed in 1992. AK prefix indicates Kiel, and MA Wilhelmshaven.
UPDATED

MA 2   *4/2004*, Frank Findler /* 0587735

5 LAUNCHES (TYPES 743, 744, 744A, 1344) (YFL)

AM 7 Y 1679	**AK 2** Y 1686	**A 41**
AM 8 Y 1675	**AK 6** Y 1683	

Dimensions, feet (metres): 52.5 × 13.1 × 3.9 *(16 × 4 × 1.2)* approx
Main machinery: 1 or 2 diesels

Comment: For personnel transport and trials work. Types 744 *(AK 6)* and 744A *(AK 2)* are radio calibration craft. AM prefix indicates Eckernförde, and AK Kiel. *A 41* is a former GDR tug (Type 1344) used as a diving boat at Warnemünde. AM 6 and AM 7 to decommission in 2005 and AM 8 in 2006.
UPDATED

AK 6   *6/2004*, Frank Findler /* 0587734

23 PERSONNEL TENDERS (TYPES 934 and GDR 407) (YFL)

V 3-21	**B 11**	**B 33**	**B 34**	**B 83**

Comment: *V 3-V 21* built in 1987-88 by Hatecke. The B series are ex-GDR craft built by Yachtwerft, Berlin.
UPDATED

V 15   *7/2004*, Frank Findler /* 0587733

5 RANGE SAFETY CRAFT (TYPE 905) (YFRT)

Name	No	Builders	Commissioned
TODENDORF	Y 835	Lürssen, Vegesack	25 Nov 1993
PUTLOS	Y 836	Lürssen, Vegesack	24 Feb 1994
BAUMHOLDER	Y 837	Lürssen, Vegesack	30 Mar 1994
BERGEN	Y 838	Lürssen, Vegesack	19 May 1994
MUNSTER	Y 839	Lürssen, Vegesack	14 July 1994

Displacement, tons: 126 full load
Dimensions, feet (metres): 91.2 × 19.7 × 4.6 *(27.8 × 6 × 1.4)*
Main machinery: 2 KHD TBD 234 diesels; 2,054 hp(m) *(1.51 MW)*; 2 shafts
Speed, knots: 16
Complement: 15

Comment: Replaced previous Types 369 and 909 craft. Funded by the Army and manned by the Navy.
UPDATED

BERGEN   *6/2004*, Harald Carstens /* 0587732

2 OIL RECOVERY SHIPS (TYPE 738) (YPC)

Name	No	Builders	Commissioned
BOTTSAND	Y 1643	Lühring, Brake	24 Jan 1985
EVERSAND	Y 1644	Lühring, Brake	11 June 1988

Measurement, tons: 500 gross; 650 dwt
Dimensions, feet (metres): 151.9 × 39.4 (137.8, bow opened) × 10.2 *(46.3 × 12 (42) × 3.1)*
Main machinery: 1 Deutz BA12M816 diesel; 1,000 hp(m) *(759 kW)* sustained; 2 shafts
Speed, knots: 10
Complement: 6

Comment: Built with two hulls which are connected with a hinge in the stern. During pollution clearance the bow is opened. Ordered by Ministry of Transport but taken over by West German Navy. Normally used as tank cleaning vessels and harbour oilers. Civilian manned. *Bottsand* based at Warnemünde, *Eversand* at Wilhelmshaven. A third of class *Thor* belongs to the Ministry of Transport.
UPDATED

BOTTSAND   *6/2004*, Frank Findler /* 0587731

4 FLOATING DOCKS (TYPES 712-715) and 2 CRANES (TYPE 711)

SCHWIMMDOCK 3 Y 842	**HIEV** Y 875
DRUCKDOCK (DOCK C)	**GRIEP** Y 876
DOCKS A and 2	

Comment: Schwimmdock 3 is 8,000 tons while Dock C is used for submarine pressure tests. Docks A and 2 are 1,000 tons. Y 875 and Y 876 are self-propelled floating cranes with a 100 ton crane.
VERIFIED

DOCK C   *4/2003, Frank Findler /* 0570624

1 TORPEDO RECOVERY VESSEL (TYPE 430A) (YPT)

TF 5 Y 855

Comment: Built in 1966 of approximately 56 tons. Provided with stern ramp for torpedo recovery. Two to Greece in 1989 and two more in 1991.   *UPDATED*

TF 5   *9/2004*, Frank Findler /* 1044264

INTELLIGENCE VESSELS

3 OSTE CLASS (TYPE 423) (AGI)

Name	No	Builders	Commissioned
ALSTER	A 50	Schiffsbaugesellschaft, Flensburg	5 Oct 1989
OSTE	A 52	Schiffsbaugesellschaft, Flensburg	30 June 1988
OKER	A 53	Schiffsbaugesellschaft, Flensburg	10 Nov 1988

Displacement, tons: 3,200 full load
Dimensions, feet (metres): 273.9 × 47.9 × 13.8 *(83.5 × 14.6 × 4.2)*
Main machinery: 2 Deutz-MWM BV16M728 diesels; 8,980 hp(m) *(6.6 MW)* sustained; 2 shafts; 2 motors (for slow speed)
Speed, knots: 21 (diesels); 8 (motors)
Complement: 36 plus 40 specialists or 51 plus 36 specialists
Missiles: SAM: 2 Stinger launchers.
Guns: 2—12.7 mm Mauser MGs.

Comment: Ordered in March 1985 and December 1986 and have replaced the Radar Trials Ships of the same name (old *Oker* and *Alster* transferred to Greece and Turkey respectively). *Oste* launched 15 May 1987, *Oker* 24 September 1987, *Alster* 4 November 1988. Carry Atlas Elektronik passive sonar and optical ELAM and electronic surveillance equipment. Particular attention has been given to accommodation standards. Reduced to one crew only for each ship in 1994. Fitted for but not with light armaments.   *UPDATED*

OKER   *6/2004*, E & M Laursen /* 1044263

SURVEY AND RESEARCH SHIPS

Notes: A 12 ton midget submarine *Narwal* was recommissioned in April 1996 for research. Originally built by Krupp Atlas as an SDV.

1 TYPE 751 (AGE)

Name	No	Builders	Commissioned
PLANET	A 1437	Thyssen Nordseewerke, Emden	2004

Displacement, tons: 3,500 full load
Dimensions, feet (metres): 239.5 × 89.26 × 22.3 *(73 × 27.2 × 6.8)*
Main machinery: Diesel electric; 2 permanent magnet motors; 6,034 hp(m) *(4.5 MW)*; 2 shafts
Speed, knots: 15. **Range, n miles:** 5,000 at 15 kt
Complement: 25 plus 20 trials personnel

Comment: Ex-Type 752 SWATH design which is to replace the old *Planet*. The roles of the ship will include both research and trials. It will be run by Forschungsanstalt für Wasserschall und Geophysik (FWG) in Kiel and Wehrtechmische Dienststelle (WTD 71) in Eckenförde. First authorised in April 1998 and contract placed with TNSW, Emden. After a delay of over two years, firm order finally made in December 2000. Launched on 12 August 2003, the ship has a sonar well, torpedo tubes and can carry five 20 ft containers.   *UPDATED*

PLANET   *8/2003, Harald Carstens /* 0558663

3 SCHWEDENECK CLASS (TYPE 748) (MULTIPURPOSE) (AG)

Name	No	Builders	Commissioned
SCHWEDENECK	Y 860	Krögerwerft, Rendsburg	20 Oct 1987
KRONSORT	Y 861	Elsflether Werft	2 Dec 1987
HELMSAND	Y 862	Krögerwerft, Rendsburg	4 Mar 1988

Displacement, tons: 1,018 full load
Dimensions, feet (metres): 185.3 × 35.4 × 17 *(56.5 × 10.8 × 5.2)*
Main machinery: Diesel-electric; 3 MTU 6V 396 TB53 diesel generators; 1,485 kW 60 Hz sustained; 1 motor; 1 shaft
Speed, knots: 13. **Range, n miles:** 2,400 at 13 kt
Complement: 13 plus 10 trials parties
Radars: Navigation: 2 Raytheon; I-band.

Comment: Order for first three placed in mid-1985. One more was planned after 1995 to replace *Mühlhausen* (ex-*Walther von Ledebur*) but was not funded. Based at Eckernförde.   *UPDATED*

HELMSAND   *7/2004*, Frank Findler /* 1044262

4 STOLLERGRUND CLASS (TYPE 745) (MULTIPURPOSE) (AG)

Name	No	Builders	Commissioned
STOLLERGRUND	Y 863	Krögerwerft	31 May 1989
MITTELGRUND	Y 864	Elsflether Werft	23 Aug 1989
KALKGRUND	Y 865	Krögerwerft	23 Nov 1989
BREITGRUND	Y 866	Elsflether Werft	19 Dec 1989

Displacement, tons: 450 full load
Dimensions, feet (metres): 126.6 × 30.2 × 10.5 *(38.6 × 9.2 × 3.2)*
Main machinery: 1 Deutz-MWM BV6M628 diesel; 1,690 hp(m) *(1.24 MW)* sustained; 1 shaft; bow thruster
Speed, knots: 12. **Range, n miles:** 1,000 at 12 kt
Complement: 7 plus 6 trials personnel

Comment: Five ordered from Lürssen in November 1987; two subcontracted to Elsflether. Equipment includes two I-band radars and an intercept sonar. The first four are based at the Armed Forces Technical Centre, Eckernförde.   *VERIFIED*

KALKGRUND   *2/2002, Findler & Winter /* 0528895

1 TRIALS SHIP (TYPE 741) (YAG)

Name	No	Builders	Commissioned
WILHELM PULLWER	A 1409 (ex-Y 838)	Schürenstadt, Bardenfleth	22 Dec 1967

Displacement, tons: 160 full load
Dimensions, feet (metres): 103.3 × 24.6 × 7.2 *(31.5 × 7.5 × 2.2)*
Main machinery: 2 MTU MB diesels; 700 hp(m) *(514 kW)*; 2 Voith-Schneider props
Speed, knots: 12.5
Complement: 17

Comment: Wooden hulled trials ship for barrage systems.   *UPDATED*

WILHELM PULLWER   *9/2004*, Hartmut Ehlers /* 1044260

1 TRIALS BOAT (TYPE 740) (YAG)

Name	No	Builders	Commissioned
BUMS	Y 1689	Howaldtswerke, Kiel	16 Feb 1970

Dimensions, feet (metres): 86.6 × 22.3 × 4.9 *(26.4 × 6.8 × 1.5)*

Comment: Single diesel engine. Has a 3 ton crane. Based at Eckernförde. To be decommissioned in 2006.

VERIFIED

BUMS
8/1997, N Sifferlinger / 0012437

TRAINING SHIPS

Notes: In addition to the two listed below there are 54 other sail training vessels (Types 910-915).

1 SAIL TRAINING SHIP (AXS)

Name	No	Builders	Commissioned
GORCH FOCK	A 60	Blohm + Voss, Hamburg	17 Dec 1958

Displacement, tons: 2,006 full load
Dimensions, feet (metres): 293 × 39.2 × 16.1 *(89.3 × 12 × 4.9)*
Main machinery: Auxiliary 1 Deutz MWM BV6M628 diesel; 1,690 hp(m) *(1.24 MW)* sustained; 1 shaft; Kamewa cp prop
Speed, knots: 11 power; 15 sail. **Range, n miles:** 1,990 at 10 kt
Complement: 206 (10 officers, 140 cadets)

Comment: Sail training ship of the improved Horst Wessel type. Barque rig. Launched on 23 August 1958. Sail area, 21,141 sq ft. Major modernisation in 1985 at Howaldtswerke. Second major refit in 1991 at Motorenwerke, Bremerhaven included a new propulsion engine and three diesel generators, which increased displacement. Third major refit at Elsfleth-Werft in 2000-2001 included modernisation of electrical distribution system.

UPDATED

GORCH FOCK
6/2004*, Frank Findler / 0587730

1 SAIL TRAINING CRAFT (AXSL)

Name	No	Builders	Commissioned
NORDWIND	Y 834 (ex-W 43)	—	1944

Displacement, tons: 110
Dimensions, feet (metres): 78.8 × 21.3 × 8.5 *(24 × 6.5 × 2.6)*
Main machinery: 1 Demag diesel; 150 hp(m) *(110 kW)*; 1 shaft
Speed, knots: 8. **Range, n miles:** 1,200 at 7 kt
Complement: 10

Comment: Ketch rigged. Sail area, 2,037.5 sq ft. Ex-Second World War patrol craft. Taken over from Border Guard in 1956.

VERIFIED

NORDWIND
6/2003, Frank Findler / 0570602

TUGS

8 HARBOUR TUGS (TYPES 725, 724, 660) (YTM)

Name	No	Builders	Commissioned
VOGELSAND	Y 816	Orenstein und Koppel, Lübeck	14 Apr 1987
NORDSTRAND	Y 817	Orenstein und Koppel, Lübeck	20 Jan 1987
LANGENESS	Y 819	Orenstein und Koppel, Lübeck	5 Mar 1987
LÜTJE HÖRN	Y 812	Husumer Schiffswerft	31 May 1990
KNECHTSAND	Y 814	Husumer Schiffswerft	16 Nov 1990
SCHARHÖRN	Y 815	Husumer Schiffswerft	1 Oct 1990
WUSTROW (ex-*Zander*)	Y 1656	VEB Yachtwerft, Berlin	25 May 1989
DRANKSE (ex-*Kormoran*)	Y 1658	VEB Yachtwerft, Berlin	12 Dec 1989

Displacement, tons: 445 full load
Dimensions, feet (metres): 99.3 × 29.8 × 8.5 *(30.3 × 9.1 × 2.6)*
Main machinery: 2 Deutz MWM BV6M628 diesels; 3,360 hp(m) *(2.47 MW)* sustained; 2 Voith-Schneider props
Speed, knots: 12
Complement: 10

Comment: Details given are for the Type 725 (Y 812-819) which have a bollard pull of 23 tons. Y 1656 and Y 1658 are Type 660 former GDR vessels of 320 tons. Y 823 to Greece in 1998.

UPDATED

LÜTJE HÖRN
6/2004*, Frank Findler / 0587729

DRANSKE
4/2004*, Martin Mokrus / 0587728

1 HELGOLAND CLASS (TYPE 720B) (ATR)

Name	No	Builders	Commissioned
FEHMARN	A 1458	Unterweser, Bremerhaven	1 Feb 1967

Displacement, tons: 1,310 standard; 1,643 full load
Dimensions, feet (metres): 223.1 × 41.7 × 14.4 *(68 × 12.7 × 4.4)*
Main machinery: Diesel-electric; 4 MWM 12-cyl diesel generators; 2 motors; 3,300 hp(m) *(2.43 MW)*; 2 shafts
Speed, knots: 17. **Range, n miles:** 6,400 at 16 kt
Complement: 34
Mines: Laying capacity.
Radars: Navigation: Raytheon; I-band.
Sonars: High definition, hull-mounted for wreck search.

Comment: Launched on 9 April 1965. Carry firefighting equipment and has an ice-strengthened hull. Employed as safety ship for the submarine training group. Twin 40 mm guns removed. One of the class to Uruguay in 1998.

VERIFIED

FEHMARN
1/2003, Diego Quevedo / 0570604

5 WANGEROOGE CLASS (3 TYPE 722 and 3 TYPE 754) (ATS/YDT)

Name	No	Builders	Commissioned
WANGEROOGE	A 1451	Schichau, Bremerhaven	9 Apr 1968
SPIEKEROOG	A 1452	Schichau, Bremerhaven	14 Aug 1968
BALTRUM	A 1439	Schichau, Bremerhaven	8 Oct 1968
JUIST	A 1440	Schichau, Bremerhaven	1 Oct 1971
LANGEOOG	A 1441	Schichau, Bremerhaven	14 Aug 1968

Displacement, tons: 854 standard; 1,024 full load
Dimensions, feet (metres): 170.6 × 39.4 × 12.8 *(52 × 12.1 × 3.9)*
Main machinery: Diesel-electric; 4 MWM 16-cyl diesel generators; 2 motors; 2,400 hp(m) *(1.76 MW)*; 2 shafts
Speed, knots: 14. **Range, n miles:** 5,000 at 10 kt
Complement: 24 plus 33 trainees (A 1439-1441)
Guns: 1 Bofors 40 mm/70 (cocooned in some, not fitted in all).

Comment: First two are salvage tugs with firefighting equipment and ice-strengthened hulls. *Wangerooge* sometimes used for pilot training and *Spiekeroog* as submarine safety ship. The other three were converted 1974-78 to training ships with *Baltrum* and *Juist* being used as diving training vessels at Neustadt, with recompression chambers and civilian crews. A 1455 sold to Uruguay in 2002.

UPDATED

LANGEOOG   3/2004*, Martin Mokrus / 0587727

ICEBREAKERS

1 EISVOGEL CLASS (TYPE 721) (AGB)

Name	No	Builders	Commissioned
EISVOGEL	A 1401	J G Hitzler, Lauenburg	11 Mar 1961

Displacement, tons: 640 full load
Dimensions, feet (metres): 125.3 × 31.2 × 15.1 *(38.2 × 9.5 × 4.6)*
Main machinery: 2 Maybach 12-cyl diesels; 2,400 hp(m) *(1.76 MW)*; 2 Kamewa cp props
Speed, knots: 13. **Range, n miles:** 2,000 at 12 kt
Complement: 16
Radars: Navigation: Kelvin Hughes; I-band.

Comment: Launched on 28 April 1960. Icebreaking tug of limited capability. Civilian manned. Fitted for but not with one Bofors 40 mm/70.

VERIFIED

EISVOGEL   7/2003, Frank Findler / 0570606

ARMY

Notes: Four companies of River Engineers are located along the River Rhine at Krefeld, Koblenz, Neuwied and Wiesbaden. Each company is provided with Bodan Landing Craft. There are also large numbers of M-Boot type work boats. All other landing craft and patrol boats have been paid off.

12 BODAN CLASS (RIVER LANDING CRAFT) (LCM)

Displacement, tons: 148 full load
Dimensions, feet (metres): 98.4 × 19 *(30 × 5.8)* (loading area)
Main machinery: 4 diesels; 596 hp(m) *(438 kW)*; 4 Schottel props
Speed, knots: 6
Guns: 1 Oerlikon 20 mm.

Comment: Built of 12 pontoons, provided with bow and stern ramp. Can carry 90 tons. These are the only Army LCMs still in service.

VERIFIED

BODAN   6/1992, Horst Dehnst

1 RIVER TUG (YTL)

Dimensions, feet (metres): 91.8 × 19.4 × 3.9 *(28 × 5.9 × 1.2)*
Main machinery: 2 KHD SBF 12M716 diesels; 760 hp(m) *(559 kW)*; 2 shafts
Speed, knots: 11
Complement: 7
Guns: 2—7.62 mm MGs.

Comment: The only survivor of a class of four.

VERIFIED

RIVER TUG   4/1995, van Ginderen Collection / 0056994

COAST GUARD (KÜSTENWACHE)

Notes: The Coast Guard was formed on 1 July 1974 and is a loose affiliation of the forces of several organisations including: the Border Guard (Sea) (Bundesgrenzschutz-See); Fishery Protection (Fischereischutz); Maritime Police (Wasserschutzpolizei); Water and Navigation Board (Schiffahrtspolizei); Customs (Zoll). These organisations have responsibility for the operation and maintenance of their own craft but all have the inscription *Küstenwache* on the side.

BORDER GUARD (Bundesgrenzschutz—See)

Notes: (1) The force consists of about 600 men. Headquarters at Neustadt and bases at Warnemunde and Cuxhaven. There are three Flotillas; one each at Neustadt, Cuxhaven and Warnemunde.
(2) The force is augmented by a maritime section of the anti-terrorist force GSG 9.
(3) Craft have dark blue hulls and white superstructures with a black, red and yellow diagonal stripe and the inscription Küstenwache painted on the ship's side.
(4) There is a total of some 60 helicopters including 13 Eurocopter EC 155, 9 EC 135, 13 Bell UH-1D, 8 Bell 212, 17 BO-105 and a number of AS 330 Puma.
(5) All 40 mm guns removed in 1997.

3 BAD BRAMSTEDT CLASS (WPSO)

Name	No	Builders	Commissioned
BAD BRAMSTEDT	BG 24	Abeking and Rasmussen, Lemwerder	8 Nov 2002
BAYREUTH	BG 25	Abeking and Rasmussen, Lemwerder	2 May 2003
ESCHWEGE	BG 26	Abeking and Rasmussen, Lemwerder	18 Dec 2003

Displacement, tons: 800 standard
Dimensions, feet (metres): 216.3 × 34.8 × 10.5 *(65.9 × 10.6 × 3.2)*
Main machinery: 1 MTU 16V 1163 diesel; 7,000 hp(m) *(5.2 MW)*; 1 shaft; fixed propeller
Speed, knots: 21.5
Complement: 14 + 10 in temporary accommodation
Radars: Surface search: I-band.
Navigation: I-band.
Helicopters: Platform for 1 light.

Comment: Contract awarded in 2000 to Prime Contractor Abeking and Rasmussen for three craft to replace six ships of Neustadt class. Hulls constructed by Yantar, Kaliningrad and completed at Lemwerder. Steel hull with aluminium superstructure. The Russian Federal Border Guard Sprut class offshore patrol vessels is based on this design.

UPDATED

BAD BRAMSTEDT   6/2004*, Michael Nitz / 1044271

1 BREDSTEDT CLASS (TYPE PB 60) (WPSO)

Name	No	Builders	Commissioned
BREDSTEDT	BG 21	Elsflether Werft	24 May 1989

Displacement, tons: 673 full load
Dimensions, feet (metres): 214.6 × 30.2 × 10.5 *(65.4 × 9.2 × 3.2)*
Main machinery: 1 MTU 20V 1163 TB93 diesel; 8,325 hp(m) *(6.12 MW)* sustained; 1 shaft; bow thruster; 1 auxiliary diesel generator; 1 motor
Speed, knots: 25 (12 on motor). **Range, n miles:** 2,000 at 25 kt; 7,000 at 10 kt
Complement: 17 plus 4 spare
Guns: 1—40 mm MGs.
Radars: Surface search: Racal AC 2690 BT; I-band.
Navigation: 2 Racal ARPA; I-band.
Helicopters: Platform for 1 light.

Comment: Ordered 27 November 1987, laid down 3 March 1988 and launched 18 December 1988. An Avon Searider rigid inflatable craft can be lowered by a stern ramp. A second RIB on the port side is launched by crane. Based at Cuxhaven.

VERIFIED

BREDSTEDT   *6/2002, Michael Nitz /* 0529097

2 SASSNITZ CLASS (TYPE PB 50 ex-TYPE 153) (WPBO)

Name	No	Builders	Commissioned
NEUSTRELITZ (ex-*Sassnitz*)	BG 22 (ex-P 6165, ex-591)	Peenewerft, Wolgast	31 July 1990
BAD DÜBEN (ex-*Binz*)	BG 23 (ex-593)	Peenewerft, Wolgast	23 Dec 1990

Displacement, tons: 369 full load
Dimensions, feet (metres): 160.4 oa; 147.6 wl × 28.5 × 7.2 *(48.9; 45 × 8.7 × 2.2)*
Main machinery: 2 MTU 12V 595 TE90 diesels; 8,800 hp(m) *(6.48 MW)* sustained; 2 shafts
Speed, knots: 25. **Range, n miles:** 2,400 at 20 kt
Complement: 33 (7 officers)
Guns: 2—7.62 mm MGs.
Radars: Surface search: Racal AC 2690 BT; I-band (BG 22 and 23).
Navigation: Racal ARPA; I-band (BG 22 and 23).

Comment: Ex-GDR designated Balcom 10 and seen for the first time in the Baltic in August 1988. The original intention was to build up to 50 for the USSR, Poland and the GDR. In 1991 the first three were transferred to the Border Guard, based at Neustadt. *Neustrelitz* fitted with German engines and electronics in 1992-93 and accommodation improved. *Bad Düben* similarly modified at Peenewerft in 1995-96. The original design had the SS-N-25 SSM and three engines. The third of class, *Sellin*, had been on loan to WTD 71 (weapons trials) at Eckernförde but was sold in 1999.

VERIFIED

BAD DÜBEN   *4/2003, Frank Findler /* 0570608

3 BREMSE CLASS (TYPE GB 23) (WPB)

UCKERMARK BG 62 (ex-G 34, ex-GS 23)   BÖRDE BG 64 (ex-G 35, ex-GS 50)
ALTMARK BG 63 (ex-G 21, ex-GS 21)

Displacement, tons: 42 full load
Dimensions, feet (metres): 74.1 × 15.4 × 3.6 *(22.6 × 4.7 × 1.1)*
Main machinery: 2 DM 6VD 18/5 AL-1 diesels; 1,020 hp(m) *(750 kW)*; 2 shafts
Speed, knots: 14
Complement: 6
Radars: Navigation: TSR 333; I-band.

Comment: Built in 1971-72 for the ex-GDR GBK. Five of the class sold to Tunisia, two to Malta and two to Jordan, all in 1992. BG 62 was based on the Danube for WEU embargo operations in 1994-96. All belong to BGSAMT-Rostock.

VERIFIED

BÖRDE   *8/2002, Frank Findler /* 0528897

4 SCHWEDT CLASS (WPBR)

SCHWEDT BG 42   FRANKFURT/ODER BG 43
KUSTRIN-KIEZ BG 41   AURITH BG 44

Displacement, tons: 6 full load
Dimensions, feet (metres): 33.5 × 10.5 × 2.6 *(10.2 × 3.2 × 0.8)*
Main machinery: 2 Volvo Penta TAMD 42 WJ; 462 hp(m) *(340 kW)*; 2 Hamilton 211 waterjets
Speed, knots: 32. **Range, n miles:** 200 at 25 km
Complement: 3
Guns: 1—7.62 mm MG.
Radars: Navigation: I-band.

Comment: River patrol craft which belong to the BGSAMT-Frankfurt/Oder since 1994.

VERIFIED

FRANKFURT/ODER   *12/1998, BGSAMT /* 0056996

4 TYPE SAB 12 (WPB)

VOGTLAND BG 51 (ex-G 56, ex-GS 17)   SPREEWALD BG 53 (ex-G 51, ex-GS 16)
RHÖN BG 52 (ex-G 53, ex-GS 26)   ODERBRUCH BG 54

Displacement, tons: 14 full load
Dimensions, feet (metres): 41.3 × 13.1 × 3.6 *(12.6 × 4 × 1.1)*
Main machinery: 2 Volvo Penta diesels; 539 hp(m) *(396 kW)*; 2 shafts
Speed, knots: 16
Complement: 5

Comment: Ex-GDR MAB 12 craft based at Karnin, Stralsund and Frankfurt/Oder. Five sold to Cyprus in 1992. Belong to BGSAMT-Rostock.

VERIFIED

RHÖN   *6/1998, Hartmut Ehlers /* 0052272

2 EUROPA CLASS (WPBR)

EUROPA 1   EUROPA 2

Displacement, tons: 10
Dimensions, feet (metres): 47.2 × 12.5 × 3.1 *(14.4 × 3.8 × 0.9)*
Main machinery: 2 MAN diesels; 240 hp(m) *(180 kW)*
Speed, knots: 22
Radars: Kelvin Hughes; I-band

Comment: River patrol craft.

UPDATED

FISHERY PROTECTION SHIPS (Fischereischutz)

Notes: Operated by Ministry of Food and Agriculture.

3 PATROL SHIPS

MEERKATZE of 2,250 tons and 15 kt. Completed December 1977
SEEFALKE of 2,400 tons gross and 20 kt. Completed August 1981
SEEADLER of 1,600 tons gross (approximately) and 19 kt. Completed 2000

Comment: Fishery Protection Ships. Black hulls with grey superstructure and black, red and yellow diagonal stripes.

VERIFIED

SEEADLER   *5/2001, Michael Nitz /* 0130380

MARITIME POLICE (Wasserschutzpolizei)

Notes: (1) Under the control of regional governments. Most have Küstenwache markings but colours vary from region to region.
(2) There are 13 seaward patrol craft: *WSP 1, 4, 5* and *7, Bremen 2, 3* and *9, Helgoland, Sylt, Fehmarn, Eider, Kieper, Falshöft, Bürgermeister Brauer* and *Bürgermeister Weichmann.*
(3) Harbour craft include *Stegnitz, Greif, Schwansen, Vossbrook, Brunswick, Trave, Wagrien, Bussard* and *Habicht.*

EIDER   *6/2004*, Frank Findler /* 0587725

FALSHÖFT   *6/2004*, Frank Findler /* 0587724

CUSTOMS (Zoll)

Notes: (1) Operated by Ministry of Finance with a total of over 100 craft. Green hulls with grey superstructure and sometimes carry machine guns. Some have Küstenwache markings.
(2) Seaward patrol craft include *Hamburg, Bremerhaven, Schleswig-Holstein, Emden, Kniepsand, Priwall, Glückstadt, Oldenburg, Hohwacht, Hiddensee, Riigen* and *Kalkgrund.*

HAMBURG   *8/2004*, Michael Nitz /* 1044269

HOHWACHT   *6/2004*, Frank Findler /* 0587723

WATER AND NAVIGATION BOARD (Schiffahrtspolizei)

Notes: (1) Comes under the Ministry of Transport. Most ships have black hulls with black/red/yellow stripes. Some have Küstenwache markings.
(2) Two icebreakers: *Max Waldeck* and *Stephan Jantzen* (ex-GDR).
(3) Nine buoy tenders: *Walter Körte, Kurt Burkowitz, Otto Treplin, Gustav Meyer, Bruno Illing, Konrad Meisel, Barsemeister Brehme, J G Repsold, Buk* (ex-GDR).
(4) Six oil recovery ships: *Scharhörn, Oland, Nordsee, Mellum, Kiel, Neuwerk.*
(5) Seven SKB 64 and 601 types (ex-GDR). *Golwitz, Ranzow, Landtief, Grasort, Gellen, Darsser Ort, Arkona, Vogelsand.*
(6) One launch: *Friedrich Voss.*

BUK   *5/2004*, Michael Nitz /* 1044270

MELLUM   *8/2004*, Harald Carstens /* 0587717

CIVILIAN SURVEY AND RESEARCH SHIPS

Notes: The following ships operate for the Bundesamt für Seeschiffahrt und Hydrographie (BSH), either under the Ministry of Transport or the Ministry of Research and Technology (*Polarstern, Meteor, Poseidon, Sonne* and *Alkor*).

KOMET (survey and research) 1,590 tons completed by Krögerwerft in October 1998.
ATAIR (survey), **DENEB** (survey), **WEGA** (survey) 1,050 tons, diesel-electric, 11.5 kt. Complement 16 plus 6 scientists. Built by Krögerwerft and Peenewerft *(Deneb),* completed 3 August 1987, 24 November 1994 and 26 October 1990 respectively.
METEOR (research) 3,500 tons, diesel-electric, 14 kt, range 10,000 n miles. Complement 33 plus 29 research staff. Completed by Schlichting, Travemünde 15 March 1986
GAUSS (survey and research) 1,813 grt, completed 6 May 1980 by Schlichting, speed 13.5 kt. Complement 19 plus 12 scientists. Modernised 1985
WALTHER HERWIG III 2,400 tons. Completed 1993
CAPELLA 455 tons. Completed by Fassmerwerft in 2003.
POLARSTERN (polar research) 10,878 grt. Completed 1982
SONNE (research) 1,200 grt. Completed by Rickmerswerft 1990.
HEINCKE, ALKOR 1,200 tons. Completed 1990
SENCKENBERG 185 tons. Completed in 1977.

POLARSTERN   *1/2003, Robert Pabst /* 0569998

ALKOR   *5/2003, Martin Mokrus* / 0570648

CAPELLA   *2/2004*, Martin Mokrus* / 0587721

METEOR
11/2003*, Martin Mokrus / 0587722

Ghana

Country Overview

Formerly a British colony known as the Gold Coast, Ghana gained independence in 1957. Located in west Africa, the country has an area of 92,100 square miles and a 292 n mile coastline with the Gulf of Guinea. It is bordered to the east by Togo and to the west by Ivory Coast. The capital and largest city is Accra which has links to a deep-water port at Tema. There is a second port at Sekondi-Takoradi. Territorial seas (12 n miles) are claimed. A 200 n mile Exclusive Economic Zone (EEZ) has been claimed but the limits are not defined.

Headquarters Appointments

Commander, Navy:
   Rear Admiral J K Gbena
Western Naval Command:
   Commodore C B Puplampu
Eastern Naval Command:
   Commodore F Daley

Personnel

(a)   2005: 1,214 (132 officers)
(b)   Voluntary service

Bases

Burma Camp, Accra (Headquarters)
Sekondi (Western Naval Command)
Tema (near Accra) (Eastern Naval Command)

Maritime Aircraft

Four Defender aircraft are available for maritime surveillance but only one is used.

PATROL FORCES

Note: A 20 m patrol craft *David Hansen* has also been reported.

2 BALSAM CLASS (PBO)

Name	No	Builders	Commissioned
ANZONE (ex-*Woodrush*)	P 30 (ex-WLB 407)	Duluth Shipyard, Minnesota	22 Sep 1944
BONSU (ex-*Sweetbrier*)	P 31 (ex-WLB 405)	Duluth Shipyard, Minnesota	26 July 1944

Displacement, tons: 1,034 full load
Dimensions, feet (metres): 180 × 37 × 12 *(54.9 × 11.3 × 3.8)*
Main machinery: Diesel electric; 2 diesels; 1,402 hp *(1.06 MW)*; 1 motor; 1,200 hp *(895 kW)*; 1 shaft; bow thruster
Speed, knots: 13. **Range, n miles:** 8,000 at 12 kt
Complement: 53
Guns: 2—12.7 mm MGs.
Radars: Navigation: Raytheon SPS-64(V)1.

Comment: *Anzone* transferred from the US Coast Guard on 4 May 2001 and *Bonsu* on 27 August 2001. **VERIFIED**

2 LÜRSSEN PB 57 CLASS (FAST ATTACK CRAFT—GUN) (PG)

Name	No	Builders	Commissioned
ACHIMOTA	P 28	Lürssen, Vegesack	27 Mar 1981
YOGAGA	P 29	Lürssen, Vegesack	27 Mar 1981

Displacement, tons: 389 full load
Dimensions, feet (metres): 190.6 × 25 × 9.2 *(58.1 × 7.6 × 2.8)*
Main machinery: 3 MTU 16V 538 TB91 diesels; 9,210 hp(m) *(6.78 MW)* sustained; 3 shafts
Speed, knots: 30
Complement: 55 (5 officers)
Guns: 1 OTO Melara 3 in *(76 mm)* compact; 85 rds/min to 16 km *(8.6 n miles)* anti-surface; 12 km *(6.5 n miles)* anti-air; weight of shell 6 kg; 250 rounds.
   1 Breda 40 mm/70; 300 rds/min to 12.5 km *(6.8 n miles)* anti-surface; weight of shell 0.96 kg.
Weapons control: LIOD optronic director.
Radars: Surface search/fire control: Thomson-CSF Canopus A; I/J-band.
Navigation: Decca TM 1226C; I-band.

Comment: Ordered in 1977. *Yogaga* completed a major overhaul at Swan Hunter's Wallsend, Tyneside yard 8 May 1989. *Achimota* started a similar refit at CMN Cherbourg in May 1991 and was joined by *Yogaga* for repairs in late 1991. Both completed by August 1992. Employed on Fishery Protection duties. Planned refit for *Yogaga* at Sekondi in 2000 did not take place. **UPDATED**

ANZONE   *5/2002* / 0533317

ACHIMOTA   *12/2001* / 0137789

2 LÜRSSEN FPB 45 CLASS (FAST ATTACK CRAFT—GUN) (PBO)

Name	No	Builders	Commissioned
DZATA	P 26	Lürssen, Vegesack	25 July 1980
SEBO	P 27	Lürssen, Vegesack	25 July 1980

Displacement, tons: 269 full load
Dimensions, feet (metres): 147.3 × 23 × 8.9 *(44.9 × 7 × 2.7)*
Main machinery: 2 MTU 16V 538 TB91 diesels; 6,140 hp(m) *(4.5 MW)* sustained; 2 shafts
Speed, knots: 27. **Range, n miles:** 1,800 at 16 kt; 700 at 25 kt
Complement: 45 (5 officers)
Guns: 2 Bofors 40 mm/70; 300 rds/min to 12.5 km *(6.8 n miles)*; weight of shell 0.96 kg.
Radars: Surface search: Decca TM 1226C; I-band.

Comment: Ordered in 1976. *Dzata* completed a major overhaul at Swan Hunter's Wallsend, Tyneside yard on 8 May 1989. *Sebo* started a similar refit at CMN Cherbourg in May 1991 which completed in August 1992. Employed in Fishery Protection role. *Dzata* refitted in 2000 at Sekondi.

UPDATED

DZATA
5/2002 / 0533318

Greece
HELLENIC NAVY

Country Overview

The Hellenic Republic is situated in south-eastern Europe and occupies the southernmost part of the Balkan Peninsula. It includes more than 3,000 islands, most of which are in the Aegean Sea. With an area of 50,949 square miles, it has borders to north-west with Albania, to the north with the Former Yugoslav Republic of Macedonia and with Bulgaria and to the north-east with Turkey. It has a 7,387 n mile coastline with the Aegean, Mediterranean and Ionian Seas. The capital and largest city is Athens whose seaport, Piraeus, is also the largest. Other major ports include Thessaloníki, Patras and Iráklion. Territorial seas (6 n miles) are claimed but an EEZ is not claimed.

Headquarters Appointments

Chief of the Hellenic Navy:
   Vice Admiral A Antoniades
Deputy Chief of Staff:
   Rear Admiral A Sourvinos
Commander, Navy Training Command:
   Rear Admiral N Spilianakis
Commander, Navy Logistics Command:
   Rear Admiral N Diamantopoulos
Inspector General:
   Rear Admiral I Nanos

Fleet Command

Commander of the Fleet:
   Vice Admiral P Hinofotis
Deputy Commander of the Fleet:
   Rear Admiral E Korovesis

Diplomatic Representation

Naval Attaché in Ankara:
   Commander K Tsovos
Naval Attaché in Berlin:
   Captain N Krioneritis
Naval Attaché in Cairo:
   Captain D Protonotarios
Naval Attaché in London:
   Captain K Kyriakidis
Naval Attaché in Paris:
   Commander J Paulopoylos
Naval Attaché in Washington:
   Captain G Dimitriadis
Naval Attaché in Madrid:
   Captain J Ntouniadakis
Naval Attaché in Tel Aviv:
   Commander B Tsoytsias

Personnel

(a)   2005: 20,940 (3,800 officers) including 4,396 conscripts
(b)   12 months' national service

Bases

Salamis and Suda Bay

Naval Commands

Commander of the Fleet has under his flag all combatant ships. Navy Logistic Command is responsible for the bases at Salamis and Suda Bay, the Supply Centre and all auxiliary ships. Navy Training Command is in charge of the Naval Officers' Academy, Petty Officers' School, two training centres and two training ships.

Naval Districts

Aegean, Ionian and Northern Greece

Naval Aviation

Alouette III helicopters (Training).
AB 212ASW helicopters (No 1 Squadron).
S-70B-6 Seahawk (No 2 Squadron).
P-3B Orions are operated under naval command by mixed Air Force and Navy crews.

Strength of the Fleet

Type	Active	Building (Planned)
Patrol Submarines	8	4
Frigates	14	—
Corvettes	5	—
Fast Attack Craft—Missile	16	4
Fast Attack Craft—Torpedo	8	—
Offshore Patrol Craft	8	—
Coastal Patrol Craft	4	—
LST/LSD/LSM	5	—
LCU/LCM	16	—
Hovercraft	3	1
Minesweepers—Coastal	13	—
Survey and Research Ships	4	—
Support Ships	2	—
Training Ships	3	—
Tankers	5	—
Auxiliary Transports	2	—
Ammunition Ship	1	—

Prefix to Ships' Names

HS (Hellenic Ship)

DELETIONS

Notes: Some of the deleted ships are in unmaintained reserve in anchorages.

Destroyers

2002	*Formion, Themistocles*
2003	*Nearchos*
2004	*Kimon*

Frigates

2001	*Thrace*
2002	*Epirus*

Patrol Forces

2002	*Antipoploiarchos Anninos, Ypoploiarchos Arliotis*
2003	*Ypoploiarchos Konidis*
2004	*Ypoploiarchos Batsis (to Georgia), Carteria*

Amphibious Forces

2001	*Ypoploiarchos Roussen, Kos*
2003	*Inouse, Kythera, Milos*

Mine Warfare Forces

2002	*Amvrakia*

Survey and Research Ships

2003	*Hermis*

Auxiliaries, Training and Survey Ships

2003	*Ariadne*
2004	*Aris, Arethusa*

PENNANT LIST

Submarines

S 110	Glavkos
S 111	Nereus
S 112	Triton
S 113	Proteus
S 116	Poseidon
S 117	Amphitrite
S 118	Okeanos
S 119	Pontos
S 120	Papanikolis (bldg)
S 121	Matrozos (bldg)
S 122	Pipinos (bldg)
S 123	Katsonis (bldg)

Frigates

F 450	Elli
F 451	Limnos
F 452	Hydra
F 453	Spetsai
F 454	Psara
F 455	Salamis
F 459	Adrias
F 460	Aegeon
F 461	Navarinon
F 462	Kountouriotis
F 463	Bouboulina
F 464	Kanaris
F 465	Themistocles
F 466	Nikiforos Fokas

Corvettes

P 62	Niki
P 63	Doxa
P 64	Eleftheria
P 65	Carteria
P 66	Agon

Patrol Forces

P 18	Armatolos
P 19	Navmachos
P 20	Anthyploiarchos Laskos
P 21	Plotarchis Blessas
P 22	Ypoploiarchos Mikonios
P 23	Ypoploiarchos Troupakis
P 24	Simeoforos Kavaloudis
P 26	Ypoploiarchos Degiannis

P 27	Simeoforos Xenos
P 28	Simeoforos Simitzopoulos
P 29	Simeoforos Starakis
P 50	Hesperos
P 53	Kyklon
P 54	Lelaps
P 56	Tyfon
P 57	Pyrpolitis
P 61	Polemistis
P 67	Ypoploiarchos Roussen
P 68	Ypoploiarchos Daniolos
P 69	Ypoploiarchos Kristallidis (bldg)
P 70	Ypoploiarchos Grigoro Poulos (bldg)
P 71	Anthypoploiarchos Ritsos (bldg)
P 72	Ypoploiarchos Votsis
P 73	Anthyploiarchos Pezopoulos
P 74	Plotarchis Vlahavas
P 75	Plotarchis Maridakis
P 76	Ypoploiarchos Tournas
P 77	Plotarchis Sakipis
P 196	Andromeda
P 198	Kyknos
P 199	Pigasos
P 228	Toxotis
P 229	Tolmi

P 230	Ormi
P 266	Machitis
P 267	Nikiforos
P 268	Aittitos
P 269	Krateos (bldg)
P 286	Diopos Antoniou
P 287	Kelefstis Stamou

Amphibious Forces

L 167	Ios
L 168	Sikinos
L 169	Irakleia
L 170	Folegandros
L 173	Chios
L 174	Samos
L 175	Ikaria
L 176	Lesbos
L 177	Rodos
L 178	Naxos
L 179	Paros
L 180	Kefallinia
L 181	Ithaki
L 183	Zakynthos
L 195	Serifos

PENNANT LIST

Minesweepers/Hunters		A 307	Thetis	A 425	Odisseus	A 464	Axios
		A 359	Ostria	A 426	Cyclops	A 466	Trichonis
M 60	Erato	A 374	Prometheus (bldg)	A 427	Danaos	A 467	Doirani
M 61	Evniki	A 375	Zeus	A 428	Nestor	A 468	Kalliroe
M 62	Evropi	A 376	Orion	A 429	Perseus	A 469	Stimfalia
M 63	Kallisto	A 407	Antaios	A 430	Pelops	A 470	Aliakmon
M 210	Thaleia	A 408	Atlas	A 432	Gigas	A 474	Pytheas
M 211	Alkyon	A 409	Acchileus	A 433	Kerkini	A 476	Strabon
M 213	Klio	A 410	Atromitos	A 434	Prespa	A 478	Naftilos
M 214	Avra	A 411	Adamastos	A 435	Kekrops	A 479	I Karavoyiannos Theophilopoulos
M 240	Aidon	A 412	Aias	A 436	Minos	A 481	St Lykoudis
M 241	Kichli	A 413	Pilefs	A 437	Pelias		
M 242	Kissa	A 415	Evros	A 438	Aegeus		
M 247	Dafni	A 416	Ouranos	A 439	Atrefs		
M 248	Pleias	A 417	Hyperion	A 440	Diomidis		
		A 419	Pandora	A 441	Theseus		
Auxiliaries, Training and Survey Ships		A 420	Pandrosos	A 460	Evrotas		
		A 422	Kadmos	A 461	Arachthos		
A 233	Maistros	A 423	Heraklis	A 462	Strymon		
A 234	Sorokos	A 424	Iason	A 463	Nestos		

SUBMARINES

0 + 4 PAPANIKOLIS (TYPE 214) CLASS (SSK)

Name	No	Builders	Laid down	Launched	Commissioned
PAPANIKOLIS	S 120	Howaldtswerke, Kiel	27 Feb 2001	22 Apr 2004	Oct 2005
MATROZOS	S 121	Hellenic Shipyards, Skaramanga	Feb 2003	Nov 2005	July 2008
PIPINOS	S 122	Hellenic Shipyards, Skaramanga	Feb 2004	Nov 2006	July 2009
KATSONIS	S 123	Hellenic Shipyards, Skaramanga	2005	2007	July 2010

Displacement, tons: 1,700 (surfaced); 1,800 (dived)
Dimensions, feet (metres): 213.3 × 20.7 × 21.6
 (65 × 6.3 × 6.6)
Main machinery: 2 MTU 16V 396 diesels; 5,600 hp(m)
 (4.17 MW); 1 Siemens Permasyn motor; 1 shaft; 2 HDW PEM
 fuel cells; 240 kW
Speed, knots: 20 dived; 11 surfaced
Complement: 27 (5 officers)

Missiles: SSM: Sub Harpoon.
Torpedoes: 8—21 in *(533 mm)* bow tubes; four (SUT, SST-4,
 DM2A4) fitted for Sub Harpoon discharge; STN Atlas
 torpedoes; total of 16 weapons.
Countermeasures: Decoys: Circe torpedo countermeasures.
 ESM. Elbit TIMNEX II.
Weapons control: STN Atlas. ISUS – 90.
Radars: Surface search: Thales Sphynx; I-band.
Sonars: Bow, flank and towed arrays (fitted for but not with).

Programmes: Decision taken on 24 July 1998 and announced
 on 9 October to order three HDW designed submarines with
 an option for a fourth. The first of class is being built at Kiel and
 subsequent hulls at Hellenic. First steel for S 121 was cut on
 15 October 2002. Contracts to build signed 15 February
 2000 and the fourth was ordered in 2002.
Structure: Details given are mainly for the Type 214 Air-
 Independent Propulsion (AIP) submarines as advertised by
 HDW although there are likely to be changes. Diving depth
 400 m *(1,300 ft)*. Zeiss optronic mast and SATCOM to be
 fitted.
Operational: *Papanikolis* started initial sea trials on 2 February
 2005. **UPDATED**

PAPANIKOLIS   *2/2005*, Michael Nitz /* 1043491

8 GLAVKOS CLASS (209 TYPES 1100 and 1200) (SSK)

Name	No	Builders	Laid down	Launched	Commissioned
GLAVKOS	S 110	Howaldtswerke, Kiel	1 Sep 1968	15 Sep 1970	6 Sep 1971
NEREUS	S 111	Howaldtswerke, Kiel	15 Jan 1969	7 June 1971	10 Feb 1972
TRITON	S 112	Howaldtswerke, Kiel	1 June 1969	14 Oct 1971	8 Aug 1972
PROTEUS	S 113	Howaldtswerke, Kiel	1 Oct 1969	1 Feb 1972	8 Aug 1972
POSEIDON	S 116	Howaldtswerke, Kiel	15 Jan 1976	21 Mar 1978	22 Mar 1979
AMPHITRITE	S 117	Howaldtswerke, Kiel	26 Apr 1976	14 June 1978	14 Sep 1979
OKEANOS	S 118	Howaldtswerke, Kiel	1 Oct 1976	16 Nov 1978	15 Nov 1979
PONTOS	S 119	Howaldtswerke, Kiel	25 Jan 1977	21 Mar 1979	29 Apr 1980

Displacement, tons: 1,125 surface; 1,235 dived (S 110-113)
 1,200 surfaced; 1,285 dived (S 116-119)
Dimensions, feet (metres): 179.5 × 20.3 × 18.5
 (54.4 × 6.2 × 5.6) (S 110-113)
 183.4 × 20.3 × 18.8 *(55.9 × 6.2 × 5.7)* (S 116-119)
Main machinery: Diesel-electric; 4 MTU 12V 493 AZ80 diesels;
 2,400 hp(m) *(1.76 MW)* sustained; 4 Siemens alternators;
 1.7 MW; 1 Siemens motor; 4,600 hp(m) *(3.38 MW)*
 sustained; 1 shaft
Speed, knots: 11 surfaced; 21.5 dived
Complement: 31 (6 officers)

Missiles: McDonnell Douglas Sub Harpoon; active radar homing
 to 130 km *(70 n miles)* at 0.9 Mach; warhead 258 kg. Can be
 discharged from 4 tubes only (S 110-113).
Torpedoes: 8—21 in *(533 mm)* bow tubes. 14 AEG SUT Mod 0;
 wire-guided; active/passive homing to 12 km *(6.5 n miles)* at
 35 kt; warhead 250 kg. Swim-out discharge.
Countermeasures: ESM: Argos AR-700-S5; radar warning (S
 110-113).
 Thomson Arial DR 2000; radar warning (S 116-119).
Weapons control: Signaal Sinbads (S 116-S 119). Unisys/
 Kanaris with UYK-44 computers (S 110-113).
Radars: Surface search: Thomson-CSF Calypso II (S 116-119).
 Thomson MILNAV (S 110-113); I-band.
Sonars: Atlas Elektronik CSU 83-90 (DBQS-21); (S 110-113);
 Atlas Elektronik CSU 3-4 (S 116-119); hull-mounted; active/
 passive search and attack; medium frequency.
 Atlas Elektronik PRS-3-4; passive ranging.

Programmes: Designed by Ingenieurkontor, Lübeck for
 construction by Howaldtswerke, Kiel and sale by Ferrostaal,
 Essen all acting as a consortium.

NEREUS   *7/2000, Michael Nitz /* 0104553

Modernisation: Contract signed 5 May 1989 with HDW and
 Ferrostaal to implement a Neptune I update programme to
 bring first four up to an improved standard and along the same
 lines as the German S 206A class. Included Sub Harpoon,
 flank array sonar, Unisys FCS, Sperry Mk 29 Mod 3 inertial
 navigation system, GPS and Argos ESM. *Triton* completed
 refit at Kiel in May 1993, *Proteus* at Salamis in December
 1995, *Glavkos* in November 1997, and *Nereus* in March
 2000. A contract signed 31 May 2002 with Hellenic
 Shipyards (main sub-contractor HDW) for a Neptune II
 modernisation programme for three boats (plus one option) of
 S 116-119. The programme is to start in 2004 and to be
 completed in 2012. The hulls are to receive a 'plug-in'

extension of 6.5 m to incorporate AIP (Siemens PEM fuel cell
 system). In addition an STN Atlas ISUS-90 combat
 management system, flank array sonar, electro-optic mast,
 SATCOM, Link II and Sub Harpoon are to be fitted. The
 conversion is expected to take up to three years for each boat.
Structure: A single-hull design with two ballast tanks and
 forward and after trim tanks. Fitted with snort and remote
 machinery control. The single screw is slow revving. Very
 high-capacity batteries with GRP lead-acid cells and battery
 cooling by Wilh Hagen and VARTA. Diving depth, 250 m
 (820 ft). Fitted with two periscopes.
Operational: Endurance, 50 days. A mining capability is reported
 but not confirmed. **UPDATED**

FRIGATES

Notes: Procurement of air-defence capable frigates is a high priority and a new programme is expected to be initiated by 2005.

4 HYDRA CLASS (MEKO 200HN) (FFGHM)

Name	No	Builders	Laid down	Launched	Commissioned
HYDRA	F 452	Blohm + Voss, Hamburg	17 Dec 1990	25 June 1991	12 Nov 1992
SPETSAI	F 453	Hellenic Shipyards, Skaramanga	11 Aug 1992	9 Dec 1993	24 Oct 1996
PSARA	F 454	Hellenic Shipyards, Skaramanga	12 Dec 1993	20 Dec 1994	30 Apr 1998
SALAMIS	F 455	Hellenic Shipyards, Skaramanga	20 Dec 1994	15 May 1997	16 Dec 1998

Displacement, tons: 2,710 light; 3,350 full load
Dimensions, feet (metres): 383.9; 357.6 (wl) × 48.6 × 19.7
(117; 109 × 14.8 × 6)
Main machinery: CODOG; 2 GE LM 2500 gas turbines;
60,000 hp *(44.76 MW)* sustained; 2 MTU 20V 956 TB82
diesels; 10,420 hp(m) *(7.66 MW)* sustained; 2 shafts; cp
props
Speed, knots: 31 gas; 20 diesel. **Range, n miles:** 4,100 at 16 kt
Complement: 173 (22 officers) plus 16 flag staff

Missiles: SSM: 8 McDonnell Douglas Harpoon Block 1C; 2 quad
launchers ❶; active radar homing to 130 km *(70 n miles)* at
0.9 Mach; warhead 227 kg.
SAM: Raytheon NATO Sea Sparrow Mk 48 Mod 2 vertical
launcher ❷; 16 missiles; semi-active radar homing to 14.6 km
(8 n miles) at 2.5 Mach; warhead 39 kg.
Guns: 1 FMC 5 in *(127 mm)*/54 Mk 45 Mod 2A ❸ 20 rds/min to
24 km *(13 n miles)* anti-surface; 14 km *(7.7 n miles)* anti-
aircraft; weight of shell 32 kg.
2 GD/GE Vulcan Phalanx 20 mm Mk 15 Mod 12 ❹; 6 barrels
per mounting; 3,000 rds/min combined to 1.5 km.
Torpedoes: 6—324 mm Mk 32 Mod 5 (2 triple) tubes ❺.
Honeywell Mk 46 Mod 5; anti-submarine; active/passive
homing to 11 km *(5.9 n miles)* at 40 kt; warhead 44 kg.
Countermeasures: Decoys: 4 Mk 36 Mod 2 SRBOC chaff
launchers ❻.
SLQ-25 Nixie; torpedo decoy.
ESM: Argo AR 700; Telegon 10; intercept.
ECM: Argo APECS II; jammer.
Combat data systems: Signaal STACOS Mod 2; Links 11 and 14.
Weapons control: 2 Signaal Mk 73 Mod 1 (for SAM). Vesta Helo
transponder with datalink for OTHT. SAR-8 IR search. SWG 1
A(V) Harpoon LCS.
Radars: Air search: Signaal MW08 ❼; 3D; F/G-band.
Air Surface search: Signaal/Magnavox; DA08 ❽; G-band.
Navigation: Racal Decca 2690 BT; ARPA; I-band.
Fire Control: 2 Signaal STIR ❾; I/J/K-band.
IFF: Mk XII Mod 4.
Sonars: Raytheon SQS-56/DE 1160; hull-mounted and VDS.

HYDRA

(Scale 1 : 1,200), Ian Sturton / 0052282

SPETSAI

8/2004*, B Sullivan / 1044267

Helicopters: 1 Sikorsky S-70B-6 Aegean Hawk ❿.

Programmes: Decision to buy four Meko 200 Mod 3HN
announced on 18 April 1988. West German government
'offset' of tanks and aircraft went with the sale, and the
electronics and some of the weapon systems secured through
US FMS credits. The first ship ordered 10 February 1989 built
by Blohm + Voss, Hamburg and the remainder ordered
10 May 1989 at Hellenic Shipyards, Skaramanga. Programme

was delayed by financial problems at Hellenic Shipyards in
1992 and some of the prefabrication of *Spetsai* was done in
Hamburg.
Modernisation: A mid-life upgrade programme is to be initiated
by 2010.
Structure: The design follows the Portuguese Vasco da Gama
class. All steel fin stabilisers.
Operational: Aegean Hawk carried from 1995.

UPDATED

SALAMIS

1/2004*, B Prézelin / 0587766

PSARA

10/2003, Camil Busquets i Vilanova / 0569190

10 ELLI (KORTENAER) CLASS (FFGHM)

Name	No	Builders	Laid down	Launched	Commissioned
ELLI (ex-*Pieter Florisz*)	F 450 (ex-F 812)	Koninklijke Maatschappij de Schelde, Flushing	1 July 1977	15 Dec 1979	10 Oct 1981
LIMNOS (ex-*Witte de With*)	F 451 (ex-F 813)	Koninklijke Maatschappij de Schelde, Flushing	13 June 1978	27 Oct 1979	18 Sep 1982
AEGEON (ex-*Banckert*)	F 460 (ex-F 810)	Koninklijke Maatschappij de Schelde, Flushing	25 Feb 1976	13 July 1978	29 Oct 1980
ADRIAS (ex-*Callenburgh*)	F 459 (ex-F 808)	Koninklijke Maatschappij de Schelde, Flushing	30 June 1975	12 Mar 1977	26 July 1979
NAVARINON (ex-*Van Kinsbergen*)	F 461 (ex-F 809)	Koninklijke Maatschappij de Schelde, Flushing	2 Sep 1975	16 Apr 1977	24 Apr 1980
KOUNTOURIOTIS (ex-*Kortenaer*)	F 462 (ex-F 807)	Koninklijke Maatschappij de Schelde, Flushing	8 Apr 1975	18 Dec 1976	26 Oct 1978
BOUBOULINA (ex-*Pieter Florisz*, ex-*Willem van der Zaan*)	F 463 (ex-F 826)	Koninklijke Maatschappij de Schelde, Flushing	21 Jan 1981	8 May 1982	1 Oct 1983
KANARIS (ex-*Jan van Brakel*)	F 464 (ex-F-825)	Koninklijke Maatschappij de Schelde, Flushing	16 Nov 1979	16 May 1981	14 Apr 1983
THEMISTOCLES (ex-*Philips Van Almonde*)	F 465 (ex-F-823)	Dok en Werfmaatschappij-Fijenoord	3 Oct 1977	11 Aug 1979	2 Dec 1981
NIKIFOROS FOKAS (ex-*Bloys van Treslong*)	F 466 (ex- F 824)	Dok en Werfmaatschappij-Fijenoord	27 Apr 1978	15 Nov 1980	25 Nov 1982

Displacement, tons: 3,050 standard; 3,630 full load
Dimensions, feet (metres): 428 × 47.9 × 20.3 (screws)
 (130.5 × 14.6 × 6.2)
Main machinery: COGOG; 2 RR Olympus TM3B gas turbines;
 50,880 hp *(39.7 MW)* sustained; 2 RR Tyne RM1C gas
 turbines; 9,900 hp *(7.4 MW)* sustained; 2 shafts; LIPS cp
 props
Speed, knots: 30. **Range, n miles:** 4,700 at 16 kt
Complement: 176 (17 officers)

Missiles: SSM: 8 McDonnell Douglas Harpoon (2 quad)
 launchers ❶; active radar homing to 130 km *(70 n miles)* at
 0.9 Mach; warhead 227 kg.
 SAM: Raytheon NATO Sea Sparrow ❷; 24 missiles; semi-active
 radar homing to 14.6 km *(8 n miles)* at 2.5 Mach; warhead
 39 kg.
 Portable Redeye; shoulder-launched; short range.
Guns: 1 (459-462) or 2 (450, 451) OTO Melara 3 in *(76 mm)*/62
 compact ❸; 85 rds/min to 16 km *(8.6 n miles)* anti-surface;
 12 km *(6.5 n miles)* anti-aircraft; weight of shell 6 kg.
 1 or 2 (450, 451) GE/GD Vulcan Phalanx 20 mm Mk 15
 6-barrelled ❹; 3,000 rds/min combined to 1.5 km. Not fitted
 in 463-466.
Torpedoes: 4—324 mm Mk 32 (2 twin) tubes ❺. 16 Honeywell
 Mk 46 Mod 5; anti-submarine; active/passive homing to
 11 km *(5.9 n miles)* at 40 kt; warhead 44 kg. Can be fitted.
Countermeasures: Decoys: 2 Loral Hycor Mk 36 SRBOC chaff
 launchers.
 ESM: Elettronika Sphinx and MEL Scimitar; intercept.
 ECM: ELT 715; jammer.
Combat data systems: Signaal SEWACO II action data
 automation; Links 10, 11 and 14.
Radars: Air search: Signaal LW08 ❻; D-band; range 264 km
 (145 n miles) for 2 m² target.
 Surface search: Signaal ZW06 ❼; I-band.
 Fire control: Signaal WM25 ❽; I/J-band; range 46 km *(25 n
 miles)*.
 Signaal STIR ❾; I/J/K-band; range 39 km *(21 n miles)* for 1 m²
 target.
Sonars: Canadian Westinghouse SQS-505; hull-mounted; active
 search and attack; 7 kHz.

Helicopters: 2 AB 212ASW ❿.

Programmes: A contract was signed with the Netherlands on
 15 September 1980 for the purchase of one of the Kortenaer
 class building for the Netherlands' Navy, and an option on a
 second of class, which was taken up 7 June 1981. A second
 contract, signed on 9 November 1992, transferred three more
 of the class. Recommissioning dates for the second batch
 were *Aegeon* 14 May 1993, *Adrias* 30 March 1994 and
 Navarinon 1 March 1995. *Kountouriotis*, the sixth ship to
 transfer, recommissioned on 15 December 1997, *Bouboulina*,
 the seventh, on 14 December 2001, *Kanaris*, the eighth, on
 29 November 2002, *Themistocles*, the ninth, on 24 October
 2003 and *Nikiforos Fokas*, the tenth, on 17 December 2003.
Modernisation: The original plan was to fit one Phalanx CIWS in
 place of the after 76 mm gun but for Gulf deployments in
 1990-91 the gun was retained in 450 and 451 and two
 Phalanx fitted on the deck above the torpedo tubes. Corvus
 chaff launchers replaced by SRBOC (fitted either side of the
 bridge). The second batch of four ships were to be similarly
 modified but the original plan of one Phalanx vice the after
 76 mm has been used as a cheaper alternative. Mid-life
 modernisation programme (MLM) is planned for the first six
 ships of the class to extend life to 2020. There is an option for
 two further ships at a later date. To be undertaken by Hellenic
 Shipyards with Thales Nederland acting as main sub-
 contractor, the MLM is to include replacement of the combat
 data system with Tacticos, replacement of ZW06 surface
 search radar with Scout, improvements to the tracking
 performance of LW08 and WM25/STIR and installation of the
 Mirador optronic director. Upgrades to the EW capability are
 to include EDO CS-3701 ESM receiver and upgrade of SRBOC.
 Upgrade of the Sea Sparrow system to RIM 162 ESSM has
 been postponed indefinitely. F 462 is the first to be
 modernised and will be completed with F 460 in 2009.
Structure: Hangar is 2 m longer than in Netherlands' ships to
 accommodate AB 212ASW helicopters.

UPDATED

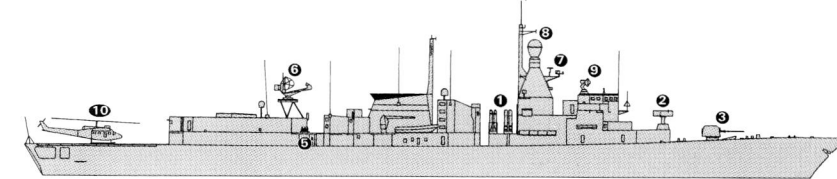

KANARIS   *(Scale 1 : 1,200), Ian Sturton / 1044255*

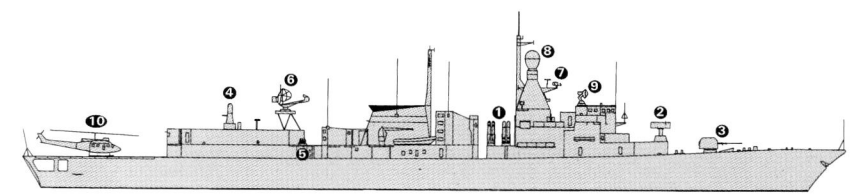

ADRIAS   *(Scale 1 : 1,200), Ian Sturton / 0126345*

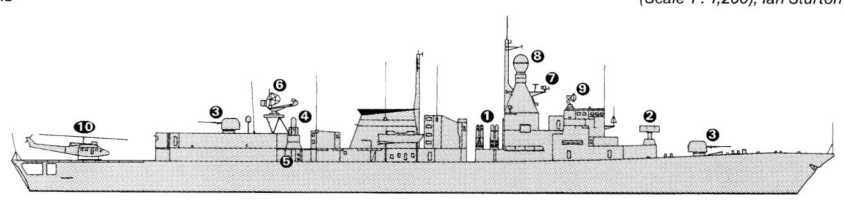

ELLI   *(Scale 1 : 1,200), Ian Sturton / 0126346*

ELLI   *5/2002, A Sharma / 0525893*

THEMISTOCLES   *2/2004*, John Brodie / 0587765*

ADRIAS   *10/2003*, John Brodie / 0587764*

CORVETTES

Notes: The corvette programme was cancelled in 2004.

4 NIKI (THETIS) (TYPE 420) CLASS (GUNBOATS) (FS)

Name	No	Commissioned	Recommissioned
NIKI (ex-*Thetis*)	P 62 (ex-P 6052)	1 July 1961	6 Sep 1991
DOXA (ex-*Najade*)	P 63 (ex-P 6054)	12 May 1962	6 Sep 1991
ELEFTHERIA (ex-*Triton*)	P 64 (ex-P 6055)	10 Nov 1962	7 Sep 1992
AGON (ex-*Andreia*, ex-*Theseus*)	P 66 (ex-P 6056)	15 Aug 1963	8 Nov 1993

Displacement, tons: 575 standard; 732 full load
Dimensions, feet (metres): 229.7 × 26.9 × 8.6 *(70 × 8.2 × 2.7)*
Main machinery: 2 MAN V84V diesels; 6,800 hp(m) *(5 MW)*; 2 shafts
Speed, knots: 19.5. **Range, n miles:** 2,760 at 15 kt
Complement: 48 (4 officers)

Guns: 4 Breda 40 mm/70 (2 twin); 300 rds/min to 12.5 km *(6.7 n miles)*; weight of shell 0.96 kg.
   2 Rheinmetall 20 mm.
Torpedoes: 6—324 mm Mk 32 (2 triple) tubes; 4 Honeywell Mk 46 Mod 5; active/passive homing
   to 11 km *(5.9 n miles)* at 40 kt; warhead 44 kg.
Depth charges: 2 rails.
Countermeasures: ESM: Thomson-CSF DR 2000S; intercept.
Weapons control: Signaal Mk 9 TFCS.
Radars: Surface search: Thomson-CSF TRS 3001; E/F-band.
   Navigation: Decca BM-E; I-band.
Sonars: Atlas Elektronik ELAC 1 BV; hull-mounted; active search and attack; high frequency.

Programmes: All built by Rolandwerft, Bremen, and transferred from Germany.
Modernisation: The A/S mortars have been replaced by a second 40 mm gun and single torpedo
   tubes by triple mountings. Upgrades started in 2000 and completed in 2002 included new
   diesel generators, two Rheinmetall 20 mm guns to replace the MGs and a new navigation suite.
Structure: *Doxa* has a deckhouse before bridge for sick bay.

UPDATED

DOXA   *7/2002, Ptisi /* 0525866

SHIPBORNE AIRCRAFT

Notes: There are also two Alouette IIIs used for SAR and training.

Numbers/Type: 11 Sikorsky S-70B-6 Aegean Hawk.
Operational speed: 135 kt *(250 km/h)*.
Service ceiling: 10,000 ft *(3,050 m)*.
Range: 600 n miles *(1,110 km)*.
Role/Weapon systems: Five ordered 17 August 1991. First one delivered 14 October 1994,
   remainder in July 1995. The option was taken up on three more of which one was delivered in
   1997, and two more in 1998. Three further more modern aircraft ordered June 2000, all of
   which have been delivered (differences are indicated in brackets). All of the original eight aircraft
   are to be similarly upgraded. Sensors: Telephonica APS 143(V)3 search radar and AAQ-22 (or
   AAS 44) FLIR, AlliedSignal AQS 18(V)3 (or Ocean Systems HELRAS) dipping sonar, MAD, Litton
   ALR 606(V)2 (or LR 100) ESM, Litton ASN 150(V) tactical data system with CD22 or Link 11.
   Weapons: ASV; Kongsberg Penguin Mk 2 Mod 7, two AS 12 (or four AGM-114K Hellfire). ASW;
   two (or three) Mk 46 torpedoes.

UPDATED

AEGEAN HAWK   *10/2001, Diego Quevedo /* 0126292

Numbers/Type: 8/2 Agusta AB 212ASW/212EW.
Operational speed: 106 kt *(196 km/h)*.
Service ceiling: 14,200 ft *(4,330 m)*.
Range: 230 n miles *(425 km)*.
Role/Weapon systems: Shipborne ASW, Elint and surface search role from escorts. Sensors:
   Selenia APS-705 radar, ESM/ECM (Elint version), AlliedSignal AQS-18 dipping sonar (ASW
   version). Weapons: ASV; two AS 12. ASW; two Mk 46 or two A244/S homing torpedoes.

VERIFIED

AB 212ASW   *6/2003, Adolfo Ortigueira Gil /* 0568866

LAND-BASED MARITIME AIRCRAFT

Notes: (1) A squadron of Air Force Mirage 2000 EG fighters is assigned to the naval strike role
using Exocet AM 39 ASMs.
(2) Replacement or upgrade of the six P-3B Orions is under consideration.

Numbers/Type: 6 Lockheed P-3B Orion.
Operational speed: 410 kt *(760 km/h)*.
Service ceiling: 28,300 ft *(8,625 m)*.
Range: 4,000 n miles *(7,410 km)*.
Role/Weapon systems: Four P-3A transferred from the USN in 1992-93 as part of the Defence Co-
   operation. Four P-3B acquired in 1996 plus two more P-3A. Two more P-3B in 1997. The six P-3B
   are operational; two P-3A are used for ground training only and the remainder for spares.
   Sensors: APS 115 radar; sonobuoys; ESM. Weapons: ASW; Mk 46 torpedoes, depth bombs and
   mines.

VERIFIED

ORION   *6/1997, Hellenic Navy /* 0012468

PATROL FORCES

Notes: Eight coastal patrol craft ordered on 24 September 2002 from Motomarine Shipyards. The
first was planned to enter service in 2003. Eight further craft ordered by the Hellenic Coast Guard
and delivered in 2004.

4 HESPEROS (JAGUAR) CLASS
(FAST ATTACK CRAFT—TORPEDO) (PTF)

Name	No	Builders	Commissioned
HESPEROS (ex-*Seeadler* P 6068)	P 50	Lürssen, Vegesack	29 Aug 1958
KYKLON (ex-*Greif* P 6071)	P 53	Lürssen, Vegesack	3 Mar 1959
LELAPS (ex-*Kondor* P 6070)	P 54	Lürssen, Vegesack	24 Feb 1959
TYFON (ex-*Geier* P 6073)	P 56	Lürssen, Vegesack	3 June 1959

Displacement, tons: 160 standard; 190 full load
Dimensions, feet (metres): 139.4 × 23.6 × 7.9 *(42.5 × 7.2 × 2.4)*
Main machinery: 4 MTU MD 16V 538 TB90 diesels; 12,000 hp(m) *(8.82 MW)* sustained; 4 shafts
Speed, knots: 42. **Range, n miles:** 500 at 40 kt; 1,000 at 32 kt
Complement: 39
Guns: 2 Bofors 40 mm/70; 300 rds/min to 12 km *(6.5 n miles)* anti-surface; 4 km *(2.2 n miles)*
   anti-aircraft; weight of shell 2.4 kg.
Torpedoes: 4—21 in *(533 mm)* tubes. Mk.8; anti-surface; straight-running to 4.6 km *(2.5 n miles)* at
   45 kt; warhead 350 kg.
Mines: 2 in lieu of each torpedo.
Radars: Surface search: Decca 1226; I-band.

Comment: Transferred from Germany 1976-77. P 53 and P 56 commissioned in Hellenic Navy
   12 December 1976. P 50 and P 54 on 24 March 1977, P 52 and P 55 on 22 May 1977. Three
   others (ex-*Albatros*, ex-*Bussard*, and ex-*Sperber*) transferred at same time for spares.

UPDATED

LELAPS   *10/1999, van Ginderen Collection /* 0104559

4 NASTY CLASS (FAST ATTACK CRAFT—TORPEDO) (PT)

Name	No	Builders	Commissioned
ANDROMEDA	P 196	Mandal, Norway	Nov 1966
KYKNOS	P 198	Mandal, Norway	Feb 1967
PIGASOS	P 199	Mandal, Norway	Apr 1967
TOXOTIS	P 228	Mandal, Norway	May 1967

Displacement, tons: 72 full load
Dimensions, feet (metres): 80.4 × 24.6 × 6.9 *(24.5 × 7.5 × 2.1)*
Main machinery: 2 MTU 12V 331 TC92 diesels; 2,660 hp(m) *(1.96 MW)* sustained; 2 shafts
Speed, knots: 25. **Range, n miles:** 676 at 17 kt
Complement: 20
Guns: 1 Bofors 40 mm/70. 1 Rheinmetall 20 mm.
Torpedoes: 4—21 in *(533 mm)* tubes. Mk.14 and Mk.23; anti-surface; straight running to 4.2 km *(2.2 n miles)* at 45 kt; warhead 292 kg.
Radars: Surface search: Decca 1226; I-band.

Comment: Six of the class acquired from Norway in 1967 and paid off into reserve in the early 1980s. Four re-engined and brought back into service in 1988. These craft continue to be active although top speed has been markedly reduced. Torpedo tubes have been reported removed from P 198 and P 199.

UPDATED

PIGASOS
7/2004*, C D Yaylali / 0587756

1 + 4 ROUSSEN (SUPER VITA) CLASS
(FAST ATTACK CRAFT—MISSILE) (PGGM)

Name	No	Builders	Commissioned
YPOPLOIARCHOS ROUSSEN	P 67	Elefsis Shipyard	July 2004
YPOPLOIARCHOS DANIOLOS	P 68	Elefsis Shipyard	Nov 2004
YPOPLOIARCHOS KRISTALLIDIS	P 69	Elefsis Shipyard	Nov 2005
YPOPLOIARCHOS GRIGORO POULOS	P 70	Elefsis Shipyard	Oct 2006
ANTHYPOPLOIARCHOS RITSOS	P 71	Elefsis Shipyard	Mar 2007

Displacement, tons: 580 full load
Dimensions, feet (metres): 203.1 × 31.2 × 8.5 *(61.9 × 9.5 × 2.6)*
Main machinery: 4 MTU 16V 595 TE 90 diesels; 4 shafts
Speed, knots: 34. **Range, n miles:** 1,800 at 12 kt
Complement: 45

Missiles: SSM: MBDA Exocet MM 40 Block 2 ❶; inertial cruise; active radar homing to 70 km *(40 n miles)* at 0.9 Mach; warhead 165 kg; sea skimmer.
SAM: RAM ❷. Mk 31 Mod 1 launcher with 21 missiles.
Guns: 1 Otobreda 76 mm/62 Super Rapid ❸; 120 rds/min to 16 km *(8.7 n miles)*; weight of shell 6 kg.
2 Otobreda 30 mm ❹.
Countermeasures: Decoys: 2 Loral Hycor Mk 36 SRBOC chaff launchers ❺.
ESM: Argo AR 900 ❻; intercept.
Combat data systems: Signaal Tacticos. Link 11.
Weapons control: Signaal Mirador optronic director ❼.
Radars: Air/surface search: Thomson-CSF MW 08 ❽; G-band.
Surface search: Signaal Scout Mk 2 LPI; I-band.
Navigation: Litton Marine Bridgemaster; I-band.
Fire control: Signaal Sting ❾; I/J-band.
IFF: Mk XII.

Programmes: Design selected 21 September 1999 based on Vosper Thornycroft Vita corvettes in service in Qatar. Contract signed 7 January 2000 for the building of first three vessels which started in March 2000. *Roussen* launched on 13 November 2002 and conducted sea trials in late 2003. *Daniolos* launched on 8 July 2003 and *Kristallidis* on 5 April 2004. Contract for a further two ships signed on 23 August 2003 for delivery in 2006 and 2007.

UPDATED

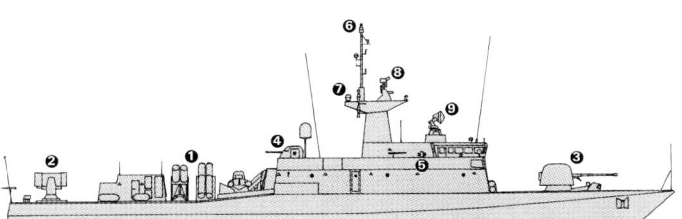

YPOPLOIARCHOS ROUSSEN
(Scale 1 : 900), Ian Sturton / 0126344

DANIOLOS
7/2003, TLV / 0549881

9 LASKOS (*LA COMBATTANTE III*) CLASS
(FAST ATTACK CRAFT—MISSILE) (PGGF/PGG)

Name	No	Builders	Commissioned
ANTHYPOPLOIARCHOS LASKOS	P 20	CMN Cherbourg	20 Apr 1977
PLOTARCHIS BLESSAS	P 21	CMN Cherbourg	7 July 1977
YPOPLOIARCHOS MIKONIOS	P 22	CMN Cherbourg	10 Feb 1978
YPOPLOIARCHOS TROUPAKIS	P 23	CMN Cherbourg	8 Nov 1977
SIMEOFOROS KAVALOUDIS	P 24	Hellenic Shipyards, Skaramanga	14 July 1980
YPOPLOIARCHOS DEGIANNIS	P 26	Hellenic Shipyards, Skaramanga	Dec 1980
SIMEOFOROS XENOS	P 27	Hellenic Shipyards, Skaramanga	31 Mar 1981
SIMEOFOROS SIMITZOPOULOS	P 28	Hellenic Shipyards, Skaramanga	June 1981
SIMEOFOROS STARAKIS	P 29	Hellenic Shipyards, Skaramanga	12 Oct 1981

Displacement, tons: 359 standard; 425 full load (P 20-23)
329 standard; 429 full load (P 24-29)
Dimensions, feet (metres): 184 × 26.2 × 7 *(56.2 × 8 × 2.1)*
Main machinery: 4 MTU 20V 538 TB92 diesels; 17,060 hp(m) *(12.54 MW)* sustained; 4 shafts (P 20-23)
4 MTU 20V 538 TB91 diesels; 15,360 hp(m) *(11.29 MW)* sustained; 4 shafts (P 24-29)
Speed, knots: 36 (P 20-23); 32.5 (P 24-29). **Range, n miles:** 700 at 32 kt; 2,700 at 15 kt
Complement: 42 (5 officers)

Missiles: SSM: 4 Aerospatiale MM 38 Exocet (P 20-P 23); inertial cruise; active radar homing to 42 km *(23 n miles)* at 0.9 Mach; warhead 165 kg.
6 Kongsberg Penguin Mk 2 Mod 3 (P 24-P 29); inertial/IR homing to 27 km *(15 n miles)* at 0.8 Mach; warhead 120 kg.
Guns: 2 OTO Melara 3 in *(76 mm)*/62 compact; 85 rds/min to 16 km *(8.6 n miles)* anti-surface; 12 km *(6.5 n miles)* anti-aircraft; weight of shell 6 kg.
4 Emerson Electric 30 mm (2 twin); multipurpose; 1,200 rds/min combined to 6 km *(3.2 n miles)*; weight of shell 0.35 kg.
Torpedoes: 2—21 in *(533 mm)* aft tubes. AEG SST-4; anti-surface; wire-guided; active homing to 12 km *(6.5 n miles)* at 35 kt; passive homing to 28 km *(15 n miles)* at 23 kt; warhead 250 kg.
Countermeasures: Decoys: Wegmann chaff launchers.
ESM: Thomson-CSF DR 2000S; intercept.
Weapons control: 2 CSEE Panda optical directors for 30 mm guns. Thomson-CSF Vega I or II system (P 20-P 23). NFT PFCS-2 (P 24-P 29).
Radars: Surface search: Thomson-CSF Triton; G-band; range 33 km *(18 n miles)* for 2 m² target.
Navigation: Decca 1226C; I-band.
Fire control: Thomson-CSF Castor II; I/J-band; range 31 km *(17 n miles)* for 2 m² target.
Thomson-CSF Pollux; I/J-band; range 31 km *(17 n miles)* for 2 m² target.

Programmes: First four ordered in September 1974. Second group of six ordered 1978.
Modernisation: P 24-29 upgraded to fire Penguin Mk 2 Mod 3 missiles. Thales DR 3000 ESM to be fitted to P21, P23 and P29 and the remainder in due course. A contract for the upgrade of P 20-23 was signed on 31 October 2003.
Structure: First four fitted with SSM Exocet; remainder have Penguin.
Operational: P 25 sunk after collision with a ferry in November 1996.

UPDATED

SIMEOFOROS STARAKIS (with Penguin)
9/1998, van Ginderen Collection / 0052292

YPOPLOIARCHOS TROUPAKIS (with Exocet)
7/2002, Ptisi / 0525867

SIMEOFOROS SIMITZOPOULOS (with Penguin)
9/2000, A Sharma / 0126333

6 VOTSIS (LA COMBATTANTE IIA) (TYPE 148) CLASS
(FAST ATTACK CRAFT—MISSILE) (PGGF)

Name	No	Builders	Commissioned
YPOPLOIARCHOS VOTSIS (ex-*Iltis*)	P 72 (ex-P 51)	CMN, Cherbourg	8 Jan 1973
ANTHYPOPLOIARCHOS PEZOPOULOS (ex-*Storch*)	P 73 (ex-P 30)	CMN, Cherbourg	17 July 1974
PLOTARCHIS VLAHAVAS (ex-*Marder*)	P 74	CMN, Cherbourg	14 June 1973
PLOTARCHIS MARIDAKIS (ex-*Häher*)	P 75	CMN, Cherbourg	12 June 1974
YPOPLOIARCHOS TOURNAS (ex-*Leopard*)	P 76	CMN, Cherbourg	21 Aug 1973
PLOTARCHIS SAKIPIS (ex-*Jaguar*)	P 77	CMN, Cherbourg	13 Nov 1973

Displacement, tons: 265 full load
Dimensions, feet (metres): 154.2 × 23 × 8.9 *(47 × 7 × 2.7)*
Main machinery: 4 MTU MD 16V 538 TB90 diesels; 12,000 hp(m) *(8.82 MW)* sustained; 4 shafts
Speed, knots: 36. **Range, n miles:** 570 at 30 kt; 1,600 at 15 kt
Complement: 30 (4 officers)

Missiles: SSM: 4 Aerospatiale MM 38 Exocet (2 twin) launchers (P 72-73); inertial cruise; active radar homing to 42 km *(23 n miles)* at 0.9 Mach; warhead 165 kg; sea-skimmer.
4 McDonnell Douglas Harpoon (2 twin) launchers (P 74-75); active radar homing to 130 km *(70 n miles)* at 0.9 Mach; warhead 227 kg.
Guns: 1 OTO Melara 3 in *(76 mm)*/62 compact; 85 rds/min to 16 km *(8.6 n miles)* anti-surface; 12 km *(6.5 n miles)* anti-aircraft; weight of shell 6 kg.
1 Bofors 40 mm/70; 330 rds/min to 12 km *(6.5 n miles)* anti-surface; 4 km *(2.2 n miles)* anti-aircraft; weight of shell 0.96 kg; fitted with GRP dome (1984).
Mines: Laying capability.
Countermeasures: Decoys: Wolke chaff launcher.
ESM: Thomson-CSF DR 2000S; intercept.
Combat data systems: PALIS and Link 11.
Weapons control: CSEE Panda optical director. Thomson-CSF Vega PCET system, controlling missiles and guns.
Radars: Air/surface search: Thomson-CSF Triton; G-band; range 33 km *(18 n miles)* for 2 m² target.
Navigation: SMA 3 RM 20; I-band.
Fire control: Thomson-CSF Castor; I/J-band.

Programmes: First pair transferred from Germany in September 1993 and recommissioned 17 February 1994. Two more transferred 16 March 1995 and recommissioned 30 June 1995. Third pair transferred from Germany and recommissioned on 27 October 2000.
Modernisation: Mid-life updates in 1980s. P 74-75 fitted with Harpoon. New ESM fitted after transfer. P 76-77 to be modernised at Lamda Shipyards in 2003-2004.
Structure: Steel hulls. Similar to Combattante II class. Spray rails have been fitted to improve hydrodynamic performance.

UPDATED

PLOTARCHIS MARIDAKIS   7/2002, Ptisi / 0525870

2 ARMATOLOS (OSPREY 55) CLASS
(LARGE PATROL CRAFT) (PGG)

Name	No	Builders	Commissioned
ARMATOLOS	P 18	Hellenic Shipyards, Skaramanga	27 Mar 1990
NAVMACHOS	P 19	Hellenic Shipyards, Skaramanga	15 July 1990

Displacement, tons: 555 full load
Dimensions, feet (metres): 179.8; 166.7 (wl) × 34.4 × 8.5 *(54.8; 50.8 × 10.5 × 2.6)*
Main machinery: 2 MTU 16V 1163 TB63 diesels; 10,000 hp(m) *(7.3 MW)* sustained; 2 shafts; Kamewa cp props
Speed, knots: 25. **Range, n miles:** 500 at 25 kt, 2,800 at 12 kt
Complement: 36 plus 25 troops
Missiles: SSM: 4 McDonnell Douglas Harpoon (can be fitted).
Guns: 1 OTO Melara 3 in *(76 mm)*/62 compact; 85 rds/min to 16 km *(8.6 n miles)* anti-surface; 12 km *(6.6 n miles)* anti-aircraft; weight of shell 6 kg.
1 Bofors 40 mm/70.
Mines: Rails.
Countermeasures: Decoys: 2 chaff launchers.
ESM: Thomson-CSF DR 2000S; intercept.
Weapons control: Selenia Elsag NA 21.
Radars: Surface search: Thomson-CSF Triton; G-band.
Fire control: Selenia RTNX; I/J-band.

Comment: Built in co-operation with Danyard A/S. Ordered in March 1988. First one laid down 8 May 1989 and launched 19 December 1989. Second laid down 9 November 1989 and launched 16 May 1990. Armament is of modular design and therefore can be changed. 76 mm guns replaced the Bofors 40 mm in 1995, after being taken from decommissioned Gearing class destroyers. Options on more of the class were shelved in favour of the Hellenic 56 design.

UPDATED

NAVMACHOS   7/2002, Ptisi / 0525871

2 PYRPOLITIS (HELLENIC 56) CLASS (BATCH 1)
(LARGE PATROL CRAFT) (PGG)

Name	No	Builders	Commissioned
PYRPOLITIS	P 57	Hellenic Shipyard, Skaramanga	4 May 1993
POLEMISTIS	P 61	Hellenic Shipyard, Skaramanga	16 June 1994

Displacement, tons: 555 full load
Dimensions, feet (metres): 185.4 × 32.8 × 8.9 *(56.5 × 10 × 2.7)*
Main machinery: 2 Wärtsilä Nohab 16V25 diesels; 9,200 hp(m) *(6.76 MW)* sustained; 2 shafts
Speed, knots: 24. **Range, n miles:** 2,470 at 15 kt; 900 at 24 kt
Complement: 36 (6 officers) plus 25 spare
Missiles: SSM: 4 McDonnell Douglas Harpoon (can be fitted).
Guns: 1 OTO Melara 3 in *(76 mm)*/62 compact; 85 rds/min to 16 km *(8.6 n miles)* anti-surface; 12 km *(6.6 n miles)* anti-aircraft; weight of shell 6 kg.
1 Bofors 40 mm/70. 2 Rheinmetall 20 mm.
Mines: 2 rails.
Countermeasures: ESM: Thomson-CSF DR 2000S; intercept.
Weapons control: Selenia Elsag NA 21.
Radars: Surface search: Thomson-CSF Triton; I-band.

Comment: First pair ordered 20 February 1990. This is a design by the Hellenic Navy which uses the modular concept so that weapons and sensors can be changed as required. Appearance is similar to Osprey 55 class. *Pyrpolitis* launched 16 September 1992, *Polemistis* 21 June 1993. Completion delayed by the shipyard's financial problems. Alternative guns and Harpoon SSM can be fitted. 25 fully equipped troops can be carried. Engines are resiliently mounted.

UPDATED

POLEMISTIS   8/2000, van Ginderen Collection / 0104560

POLEMISTIS   5/2004*, Martin Mokrus / 0587755

2 TOLMI (ASHEVILLE) CLASS (COASTAL PATROL CRAFT) (PGM)

Name	No	Builders	Commissioned
TOLMI (ex-*Green Bay*)	P 229	Peterson, Wisconsin	5 Dec 1969
ORMI (ex-*Beacon*)	P 230	Peterson, Wisconsin	21 Nov 1969

Displacement, tons: 225 standard; 245 full load
Dimensions, feet (metres): 164.5 × 23.8 × 9.5 *(50.1 × 7.3 × 2.9)*
Main machinery: 2 MTU 12V 596 TE94 diesels; 4,500 hp *(3.3 MW)*; 2 shafts
Speed, knots: 20. **Range, n miles:** 1,700 at 16 kt
Complement: 24 (3 officers)
Missiles: SSM: 4 Aerospatiale SS 12M; wire-guided to 5.5 km *(3 n miles)* subsonic; warhead 30 kg.
Guns: 1 USN 3 in *(76 mm)*/50 Mk 34; 50 rds/min to 12.8 km *(7 n miles)*; weight of shell 6 kg.
1 Bofors 40 mm/70 Mk 10. 4—12.7 mm (2 twin) MGs.
Weapons control: Mk 63 GFCS.
Radars: Surface search: Sperry SPS-53; I/J-band.
Fire control: Western Electric SPG-50; I/J-band.

Comment: Transferred from the USA in mid-1990 after a refit and recommissioned 18 June 1991. Both were in reserve from April 1977 having originally been built for the Cuban crisis. Similar craft in Turkish, Colombian and South Korean navies. Original gas-turbine propulsion engine was removed prior to transfer and both craft reported re-engined in 2004.

UPDATED

TOLMI   9/2001, A Sharma / 0126331

4 MACHITIS CLASS (LARGE PATROL CRAFT) (PGG)

Name	No	Builders	Commissioned
MACHITIS	P 266	Hellenic Shipyards, Skaramanga	29 Oct 2003
NIKIFOROS	P 267	Hellenic Shipyards, Skaramanga	30 Mar 2004
AITTITOS	P 268	Hellenic Shipyards, Skaramanga	5 Aug 2004
KRATEOS	P 269	Hellenic Shipyards, Skaramanga	Nov 2004

Displacement, tons: 575 full load
Dimensions, feet (metres): 185.4 × 32.8 × 8.9 *(56.5 × 10 × 2.7)*
Main machinery: 2 Wärtsilä Nohab 16V25 diesels; 9,200 hp(m) *(6.76 MW)* sustained; 2 shafts
Speed, knots: 24. **Range, n miles:** 2,000 at 15 kt; 900 at 24 kt
Complement: 36 (6 officers) plus 37 spare
Missiles: SSM: 4 McDonnell Douglas Harpoon (can be fitted).
Guns: 1 Otobreda 3 in *(76 mm)*/62 compact; 85 rds/min to 16km *(8.6 n miles)* anti-surface; 12 km *(6.6 n miles)* anti-aircraft; weight of shell 6 kg.
  1 Otobreda 40 mm/70-520R. 2 Rheinmetall 20 mm.
Mines: 2 rails.
Countermeasures: ESM: Thomson-CSF DR 3000.
Combat data systems: TACTICOS with Link 11.
Weapons control: LIROD Mk 2 optronic tracker.
Radars: Air/surface search: Signaal variant; E/F-band.
Navigation: Bridgemaster I-band.

Comment: Contract to build four improved Pyrpolitis class given to Hellenic Shipyard on 21 December 1999. Building started in February 2000 *Machitis* launched in June 2002, *Nikiforos* on 13 December 2002, *Aittitos* on 26 February 2003 and *Krateos* on 30 October 2003. An option for a fifth vessel is unlikely to be exercised. **UPDATED**

MACHITIS (at launch) 6/2002, *T L Valmas* / 0521912

2 COASTAL PATROL CRAFT (PB)

Name	No	Builders	Commissioned
DIOPOS ANTONIOU	P 286	Ch N de l'Esterel	4 Dec 1975
KELEFSTIS STAMOU	P 287	Ch N de l'Esterel	28 July 1975

Displacement, tons: 115 full load
Dimensions, feet (metres): 105 × 19 × 5.3 *(32 × 5.8 × 1.6)*
Main machinery: 2 MTU 12V 331 TC81 diesels; 2,610 hp(m) *(1.92 MW)* sustained; 2 shafts
Speed, knots: 30. **Range, n miles:** 1,500 at 15 kt
Complement: 17
Missiles: SSM: 4 Aerospatiale SS 12M; wire-guided to 5.5 km *(3 n miles)* subsonic; warhead 30 kg.
Guns: 1 Rheinmetall 20 mm. 1—12.7 mm MG.
Radars: Surface search: Decca 1226; I-band.

Comment: Originally ordered for Cyprus, later transferred to Greece. Wooden hulls. Fast RIB carried on the stern. **VERIFIED**

KELEFSTIS STAMOU 6/2002, *C D Yaylali* / 0525890

15 FAST INTERCEPT CRAFT (HSIC)

SAP 1-9   + 6

Displacement, tons: 6.3 full load
Dimensions, feet (metres): 42 × ? × ? *(12.8 × ? × ?)*
Speed, knots: 60+
Complement: 4
Guns: 1—40 mm Mk 19 grenade launcher.
  1—12.7 mm MG. 2—7.62 mm MGs.

Comment: Details are for *SAP 7-9*, donated by Angelopoulos family to Hellenic Navy for use by special forces. Rigid Hull Inflatable Boat with removable synthetic armour panels. Built by Italian shipyard Fabio Buzzi. Six smaller craft *(SAP 1-6)* are also in service and there are at least six further similar craft. **UPDATED**

FAST INTERCEPT CRAFT 6/2001, *Elias Daloumis* / 0094361

AMPHIBIOUS FORCES

Notes: There is a number of paid off LSTs and LSMs in unmaintained reserve at Salamis.

5 JASON CLASS (LSTH)

Name	No	Builders	Launched	Commissioned
CHIOS	L 173	Eleusis Shipyard	16 Dec 1988	30 May 1996
SAMOS	L 174	Eleusis Shipyard	6 Apr 1989	20 May 1994
LESBOS	L 176	Eleusis Shipyard	5 July 1990	25 Feb 1999
IKARIA	L 175	Eleusis Shipyard	22 Oct 1998	6 Oct 1999
RODOS	L 177	Eleusis Shipyard	6 Oct 1999	30 May 2000

Displacement, tons: 4,400 full load
Dimensions, feet (metres): 380.5 × 50.2 × 11.3 *(116 × 15.3 × 3.4)*
Main machinery: 2 Wärtsilä Nohab 16V25 diesels; 9,200 hp(m) *(6.76 MW)* sustained; 2 shafts
Speed, knots: 16
Military lift: 300 troops plus vehicles; 4 LCVPs
Guns: 1 OTO Melara 76 mm/62 Mod 9 compact; 100 rds/min to 16 km *(8.6 n miles)* anti-surface; 12 km *(6.5 n miles)* anti-aircraft; weight of shell 6 kg.
  2 Breda 40 mm/70; 300 rds/min to 12 km *(6.5 n miles)*; weight of shell 0.96 kg.
  4 Rheinmetall 20 mm (2 twin).
Weapons control: 1 CSEE Panda optical director. Thomson-CSF Canopus GFCS.
Radars: Thomson-CSF Triton; G-band.
Fire control: Thomson-CSF Pollux; I/J-band.
Navigation: Kelvin Hughes Type 1007; I-band.
Helicopters: Platform for one medium.

Comment: Contract for construction of five LSTs by Eleusis Shipyard signed 15 May 1986. Bow and stern ramps, drive through design. First laid down 18 April 1987, second in September 1987, third in May 1988, fourth April 1989 and fifth November 1989. Completion of all five and in particular the last three, severely delayed by shipyard financial problems which were later overcome, following privatisation. Combat data system is a refurbished German system. **VERIFIED**

SAMOS 9/2003, *Schaeffer/Marsan* / 0568865

LESBOS 8/2001, *A Campanera i Rovira* / 0126318

3 + 1 (1) POMORNIK (ZUBR) (PROJECT 1232) HOVERCRAFT (LCUJ)

Name	No	Builders	Commissioned
KEFALLINIA	L 180 (ex-717)	Almaz, St Petersburg	22 Jan 2001
ITHAKI	L 181 (ex-U 421)	Morye Shipyard, Ukraine	2 Mar 2001
ZAKYNTHOS	L 183	Almaz, St Petersburg	5 Oct 2001
KERKIRA	L 182	Almaz, St Petersburg	2005

Displacement, tons: 550 full load
Dimensions, feet (metres): 189 × 84 *(57.6 × 25.6)*
Main machinery: 5 Type NK-12MV gas-turbines; 2 for lift, 23,672 hp(m) *(17.4 MW)* nominal; 3 for drive, 35,508 hp(m) *(26.1 MW)* nominal
Speed, knots: 60. **Range, n miles:** 300 at 55 kt
Complement: 27 (4 officers)
Military lift: 3 MBT or 10 APC plus 230 troops (total 130 tons)
Guns: 2—30 mm/65 AK 630; 6 barrels per mounting.
  2 retractable 122 mm rocket launchers.
Mines: 2 rails can be carried for 80.
Countermeasures: ESM: intercept.
Weapons control: Optronic director.
Radars: Air/surface search: Cross Dome; I-band.
Fire control: Bass Tilt; H/I-band.

Comment: Two ordered from Russia and two from Ukraine on 24 January 2000. First delivered late December 2000, the remainder in 2001. L 180 was second-hand, L 181 was completion of a half-built vessel and L 183 was new build. The second Ukrainian ship was not accepted into service and a replacement (L 182) was ordered from Russia on 30 September 2002 and launched on 24 June 2004. An order for a fifth craft from Almaz has also been reported. **UPDATED**

KEFALLINIA 1/2001, *T L Valmas* / 0034713

7 TYPE 520 (LCU)

NAXOS (ex-*Renke*) L 178	**IOS** (ex-*Barbe*) L 167	**IRAKLEIA** (ex-*Forelle*) L 169
PAROS (ex-*Salm*) L 179	**SIKINOS** (ex-*Dorsch*) L 168	**FOLEGANDROS** (ex-*Delphin*) L 170
SERIFOS (ex-*Rochen*) L 195		

Displacement, tons: 430 full load
Dimensions, feet (metres): 131.2 × 28.9 × 7.2 *(40 × 8.8 × 2.2)*
Main machinery: 2 MWM 12-cyl diesels; 1,020 hp(m) *(750 kW)*; 2 shafts
Speed, knots: 11. **Range, n miles:** 1,200 at 11 kt
Complement: 17
Military lift: 150 tons
Guns: 2 Rheinmetall 20 mm (not all fitted).
Radars: Navigation: Kelvin Hughes; I-band.

Comment: First two transferred from Germany 16 November 1989, remainder on 31 January 1992. Built by HDW, Hamburg in 1966. Bow and stern ramps similar to US Type. One other (ex-*Murane*) used for spares. L 178 and L 195 modified to act as auxiliary transports in 2002.
UPDATED

FOLEGANDROS   *11/1999* / 0081940

IOS   *9/1997, van Ginderen Collection* / 0052299

9 TYPE 521 (LCM)

Displacement, tons: 168 full load
Dimensions, feet (metres): 77.4 × 20.9 × 4.9 *(23.6 × 6.4 × 1.5)*
Main machinery: 2 MWM 8-cyl diesels; 685 hp(m) *(503 kW)*; 2 shafts
Speed, knots: 10.5. **Range, n miles:** 700 at 10 kt
Complement: 7
Military lift: 60 tons or 50 troops

Comment: Built in 1964-67 but spent much of their time in reserve. Transferred from Germany in April 1991 and numbered between ABM 20-30.
VERIFIED

TYPE 521   *9/1994, van Ginderen Collection*

59 LANDING CRAFT

Displacement, tons: 56 full load
Dimensions, feet (metres): 56 × 14.4 × 3.9 *(17 × 4.4 × 1.2)*
Main machinery: 2 Gray Marine 64 HN9 diesels; 330 hp *(264 kW)*; 2 shafts
Speed, knots: 10. **Range, n miles:** 130 at 10 kt
Military lift: 30 tons

Comment: Details given are for the 11 LCMs transferred from the USA in 1956-58. Twenty-nine LCVPs were also transferred from the USA 1956-71 and the remainder (12 LCPs and 7 LCAs) were built in Greece from 1977.
VERIFIED

MINE WARFARE FORCES

2 HUNT CLASS (MHSC)

Name	No	Builders	Commissioned
EVROPI (ex-*Bicester*)	M 62 (ex-M 36)	Vosper Thornycroft	20 Mar 1986
KALLISTO (ex-*Berkeley*)	M 63 (ex-M 40)	Vosper Thornycroft	14 Jan 1988

Displacement, tons: 750 full load
Dimensions, feet (metres): 197 × 34.1 × 10.5 *(60 × 10.4 × 3.2)*
Main machinery: 2 MTU diesels; 1,900 hp *(1.42 MW)*; 1 Deltic Type 9-55B diesel for pulse generator and auxiliary drive; 780 hp *(582 kW)*; 2 shafts; bow thruster
Speed, knots: 15 diesels; 8 hydraulic drive. **Range, n miles:** 1,500 at 12 kt
Complement: 50 (8 officers)
Guns: 1 DES/MSI DS 30B 30 mm/75; 650 rds/min to 10 km *(5.4 n miles)* anti-surface; 3 km *(1.6 n miles)* anti-aircraft; weight of shell 0.36 kg.
Countermeasures: MCM: 2 PAP 104 remotely controlled submersibles, MS 14 magnetic loop, Sperry MSSA Mk   1 Towed Acoustic Generator and conventional Mk 8 Oropesa sweeps.
ESM: MEL Matilda UAR 1.
Combat data systems: CAAIS DBA 4 action data automation.
Radars: Navigation: Kelvin Hughes Type 1006; I-band.
Sonars: Plessey 193M Mod 1; hull-mounted; minehunting; 100/300 kHz.
Mil Cross mine avoidance sonar; hull-mounted; active; high frequency.
Type 2059 to track PAP 104.

Comment: First one transferred from UK 31 July 2000, second one 28 February 2001. Main machinery replaced by MTU units between May 2004 and January 2005.
UPDATED

KALLISTO   *3/2001, W Sartori* / 0126329

3 ADJUTANT CLASS
(MINESWEEPERS/HUNTERS—COASTAL) (MHC/MSC)

ERATO (ex-*Castagno* M 5504) M 60	**EVNIKI** (ex-*Gelso* M 5509) M 61
THALEIA (ex-*Blankenberge* M 923, ex-*MSC 170*) M 210	

Displacement, tons: 330 standard; 402 full load
Dimensions, feet (metres): 145 × 27.9 × 8 *(44.2 × 8.5 × 2.4)*
Main machinery: 2 GM 8-268A diesels; 880 hp *(656 kW)*; 2 shafts
Speed, knots: 14. **Range, n miles:** 2,500 at 10 kt
Complement: 38 (4 officers)
Guns: 1 Oerlikon 20 mm.
Radars: Navigation: Decca or SMA 3RM 20R; I-band.
Sonars: SQQ-14 or UQS-1D; active; high frequency.

Comment: M 210 last of four originally supplied by the USA to Belgium under MDAP; built in 1954 in the USA by Consolidated SB Corp, Morris Heights. Subsequently returned to the USA and simultaneously transferred to Greece on 26 September 1969. M 60 and M 61 bought from Italy on 10 October 1995. Both are classed as minehunters.
UPDATED

EVNIKI   *3/2002, Schaeffer/Marsan* / 0525889

8 ALKYON (MSC 294) CLASS
(MINESWEEPERS—COASTAL) (MSC)

Name	No	Builders	Commissioned
ALKYON (ex-MSC 319)	M 211	Peterson Builders	3 Dec 1968
KLIO (ex-Argo, ex-MSC 317)	M 213	Peterson Builders	7 Aug 1968
AVRA (ex-MSC 318)	M 214	Peterson Builders	3 Oct 1968
AIDON (ex-MSC 314)	M 240	Peterson Builders	22 June 1967
KICHLI (ex-MSC 308)	M 241	Peterson Builders	14 July 1964
KISSA (ex-MSC 309)	M 242	Peterson Builders	1 Sep 1964
DAFNI (ex-MSC 307)	M 247	Peterson Builders	23 Sep 1964
PLEIAS (ex-MSC 310)	M 248	Peterson Builders	13 Oct 1964

Displacement, tons: 320 standard; 370 full load
Dimensions, feet (metres): 144 × 28 × 8.2 (43.3 × 8.5 × 2.5)
Main machinery: 2 GM-268A diesels; 1,760 hp (1.3 MW); 2 shafts
Speed, knots: 13. **Range, n miles:** 2,500 at 10 kt
Complement: 39 (4 officers)
Guns: 2 Oerlikon 20 mm (twin).
Radars: Navigation: Decca; I-band.
Sonars: UQS-1D; active; high frequency.

Comment: Built in the USA for Greece, wooden hulls. Modernisation programme from 1990 to 1995 with replacement main engines and navigation radar. New sonar under consideration but unlikely to be funded.

UPDATED

KISSA
2/2004*, Schaeffer/Marsan / 0587757

SURVEY AND RESEARCH SHIPS

1 SURVEY SHIP (AGS)

Name	No	Builders	Commissioned
NAFTILOS	A 478	Annastadiades Tsortanides, Perama	3 Apr 1976

Displacement, tons: 1,470 full load
Dimensions, feet (metres): 207 × 38 × 13.8 (63.1 × 11.6 × 4.2)
Main machinery: 2 Burmeister & Wain SS28LM diesels; 2,640 hp(m) (1.94 MW); 2 shafts
Speed, knots: 15
Complement: 74 (8 officers)

Comment: Launched 19 November 1975. Of similar design to the two lighthouse tenders.

VERIFIED

NAFTILOS
9/1999, van Ginderen Collection / 0079497

1 RESEARCH SHIP (AGOR)

Name	No	Builders	Commissioned
PYTHEAS	A 474	Annastadiades Tsortanides, Perama	15 Dec 1983

Displacement, tons: 670 standard; 840 full load
Dimensions, feet (metres): 164.7 × 31.5 × 21.6 (50.2 × 9.6 × 6.6)
Main machinery: 2 Detroit 12V-92TA diesels; 1,020 hp (760 kW) sustained; 2 shafts
Speed, knots: 14
Complement: 58 (8 officers)

Comment: *Pytheas* ordered in May 1982. Launched 19 September 1983. A similar ship, *Aegeon*, was constructed to Navy specification in 1984 but now belongs to the Maritime Research Institute.

VERIFIED

PYTHEAS
6/2000, Hellenic Navy / 0104566

1 RESEARCH AND TRAINING CRAFT (AXSL)

OLYMPIAS

Dimensions, feet (metres): 121.4 × 17.1 × 4.9 (37 × 5.2 × 1.5)
Main machinery: 170 oars (85 each side in three rows)
Speed, knots: 8
Complement: 180

Comment: Construction started in 1985 and completed in 1987. Made of Oregon pine. Built for historic research and as a reminder of the naval hegemony of ancient Greeks. Part of the Hellenic Navy. Refit in 1992-93.

VERIFIED

OLYMPIAS
6/1996, Hellenic Navy / 0079500

1 SURVEY SHIP (AGSC)

Name	No	Builders	Commissioned
STRABON	A 476	Emanuil-Maliris, Perama	27 Feb 1989

Displacement, tons: 252 full load
Dimensions, feet (metres): 107.3 × 20 × 8.2 (32.7 × 6.1 × 2.5)
Main machinery: 1 MAN D2842LE; 571 hp(m) (420 kW) sustained; 1 shaft
Speed, knots: 12.5
Complement: 20 (2 officers)

Comment: Ordered in 1987, launched September 1988. Used as coastal survey vessel.

VERIFIED

STRABON
6/2000, Hellenic Navy / 0104567

TRAINING SHIPS

3 SAIL TRAINING CRAFT (AXS)

MAISTROS A 233   SOROKOS A 234   OSTRIA A 359

Displacement, tons: 12 full load (A 233 and 234)
Dimensions, feet (metres): 48.6 × 12.8 × 6.9 (14.8 × 3.9 × 2.1)

Comment: Sail training ships acquired in 1983-84 (A 233-234) and 1989 (A 359). A 359 is slightly smaller at 12.1 × 3.6 m.

VERIFIED

AUXILIARIES

2 FLOATING DOCKS and 5 FLOATING CRANES

Comment: One floating dock is 45 m (147.6 ft) in length and has a 6,000 ton lift. Built at Eleusis with Swedish assistance and launched 5 May 1988; delivered 1989. The second is the ex-US AFDM 2 transferred in 1999. This dock was built in 1942 and has a 12,000 ton lift. The cranes were all built in Greece.

VERIFIED

1 ETNA CLASS (AORH)

Name	No	Builders	Commissioned
PROMETHEUS	A 374	Elefsis Shipyard	4 July 2003

Displacement, tons: 13,400 full load
Dimensions, feet (metres): 480.6 × 68.9 × 24.3 (146.5 × 21 × 7.4)
Flight deck, feet (metres): 91.9 × 68.9 (28 × 21)
Main machinery: 2 Sulzer 12 ZAV 40S diesels; 22,400 hp(m) (16.46 MW) sustained; 2 shafts; cp props ; bow thruster
Speed, knots: 21. **Range, n miles:** 7,600 at 18 kt
Complement: 137 plus 119 spare including flag staff
Cargo capacity: 6,350 tons gas oil; 1,200 tons JP5; 2,100 m³ ammunition and stores
Missiles: SAM: 2 Stinger mountings.
Guns: 4—20 mm guns.
Countermeasures: SLQ-25 Nixie; torpedo decoy.
Radars: Surface search: Raytheon SPS-10D; G-band.
Navigation: GEM LD-1825; I-band.
Helicopters: Aegean Hawk or AB 212.

Comment: Ordered in August 1999 from Fincantieri and from Elefsis on 7 January 2000. First steel cut July 2000, launched 18 February 2002. Almost identical to the Italian Etna class. Two CIWS are to be fitted, one on the hangar roof and forward of the bridge. There is one RAS station on each side and one astern station.

VERIFIED

PROMETHEUS
6/2003, *Hellenic Navy* / 0567904

2 LÜNEBURG (TYPE 701) CLASS (SUPPORT SHIPS) (ARL/AOTL)

Name	No	Builders	Commissioned	Recommissioned
AXIOS	A 464	Bremer Vulcan	9 July 1968	30 Sep 1991
(ex-*Coburg*)	(ex-A 1412)			
ALIAKMON	A 470	Blohm + Voss	30 July 1968	19 Oct 1994
(ex-*Saarburg*)	(ex-A 1415)			

Displacement, tons: 3,709 full load
Dimensions, feet (metres): 374.9 × 43.3 × 13.8 (114.3 × 13.2 × 4.2)
Main machinery: 2 MTU MD 16V 538 TB90 diesels; 6,000 hp(m) (4.41 MW) sustained; 2 shafts; cp props; bow thruster
Speed, knots: 17. **Range, n miles:** 3,200 at 14 kt
Complement: 71
Cargo capacity: 1,400 tons fuel ; 200 tons ammunition; 130 tons water
Guns: 4 Bofors 40 mm/70 (2 twin); 300 rds/min to 12 km (6.5 n miles); weight of shell 0.96 kg.
Radars: Navigation: Decca; I-band.

Comment: Both ships converted to Fleet oilers by Hellenic Shipyards. Contract signed 21 December 1999. *Axios* completed September 2000 and *Aliakmon* in December 2002.

VERIFIED

ALIAKMON
8/2002, *A A de Kruijf* / 0525886

6 WATER TANKERS (AWT)

KERKINI (ex-German *FW 3*) A 433	TRICHONIS (ex-German *FW 6*) A 466	KALLIROE A 468
PRESPA A 434	DOIRANI A 467	STIMFALIA A 469

Comment: All built between 1964 and 1990. Capacity, 600 tons except A 433 and A 466 which can carry 300 tons and A 469 which can carry 1,000 tons. Three in reserve. *Stimfalia* is similar to *Ouranos*. A 433 damaged in collision on 15 April 2002.

VERIFIED

KALLIROE
9/2001, *A Sharma* / 0126325

4 OURANOS CLASS (AOTL)

Name	No	Builders	Commissioned
OURANOS	A 416	Kinosoura Shipyard	27 Jan 1977
HYPERION	A 417	Kinosoura Shipyard	27 Apr 1977
ZEUS	A 375 (ex-A 490)	Hellenic Shipyards	21 Feb 1989
ORION	A 376	Hellenic Shipyards	5 May 1989

Displacement, tons: 2,100 full load
Dimensions, feet (metres): 219.8; 198.2 (wl) × 32.8 × 13.8 (67; 60.4 × 10 × 4.2)
Main machinery: 1 MAN-Burmeister & Wain 12V 20/27 diesel; 1,632 hp(m) (1.2 MW) sustained; 1 shaft
Speed, knots: 11
Complement: 28
Cargo capacity: 1,300 tons oil or petrol
Guns: 2 Rheinmetall 20 mm.

Comment: First two are oil tankers. The others were ordered from Hellenic Shipyards, Skaramanga in December 1986 and are used as petrol tankers. There are some minor superstructure differences between the first two and the last two which have a forward crane instead of kingposts.

VERIFIED

HYPERION
9/2001, *A Sharma* / 0126326

ZEUS
5/2001, *E & M Laursen* / 0130738

1 NETLAYER (ANL)

Name	No	Builders	Commissioned
THETIS (ex-*AN 103*)	A 307	Kröger, Rendsburg	Apr 1960

Displacement, tons: 680 standard; 805 full load
Dimensions, feet (metres): 169.5 × 33.5 × 11.8 (51.7 × 10.2 × 3.6)
Main machinery: Diesel-electric; 1 MAN GTV-40/60 diesel generator; 1 motor; 1,470 hp(m) (1.08 MW); 1 shaft
Speed, knots: 12. **Range, n miles:** 6,500 at 10 kt
Complement: 48 (5 officers)
Guns: 1 Bofors 40 mm/60. 3 Rheinmetall 20 mm.
Radars: Navigation: Decca; I-band.

Comment: US offshore order. Launched in 1959. Some guns not always embarked.

VERIFIED

THETIS
9/1998, *A Sharma* / 0052305

1 AMMUNITION SHIP (AEL)

Name	No	Builders	Commissioned
EVROS (ex-*Schwarzwald*, ex-*Amaltheé*)	A 415	Ch Dubigeon Nantes	7 June 1956

Displacement, tons: 2,400 full load
Measurement, tons: 1,667 gross
Dimensions, feet (metres): 263.1 × 39 × 15.1 *(80.2 × 11.9 × 4.6)*
Main machinery: 1 Sulzer 6SD60 diesel; 3,000 hp(m) *(2.2 MW)*; 1 shaft
Speed, knots: 15. **Range, n miles:** 4,500 at 15 kt
Guns: 4 Bofors 40 mm/60.

Comment: Bought by FDR from Société Navale Caënnaise in February 1960. Transferred to Greece 6 June 1976.

VERIFIED

EVROS   *1987, Hellenic Navy*

2 AUXILIARY TRANSPORTS (AP)

Name	No	Builders	Commissioned
PANDORA	A 419	Perama Shipyard	26 Oct 1973
PANDROSOS	A 420	Perama Shipyard	1 Dec 1973

Displacement, tons: 390 full load
Dimensions, feet (metres): 153.5 × 27.2 × 6.2 *(46.8 × 8.3 × 1.9)*
Main machinery: 2 diesels; 2 shafts
Speed, knots: 12
Military lift: 500 troops
Radars: Navigation: Racal Decca; I-band.

Comment: Launched 1972 and 1973.

UPDATED

PANDROSOS   *3/2004*, Bob Fildes / 1044257*

4 TYPE 430A (TORPEDO RECOVERY VESSELS) (YPT)

EVROTAS (ex-*TF 106*) A 460 (ex-Y 872)   STRYMON (ex-*TF 107*) A 462 (ex-Y 873)
ARACHTHOS (ex-*TF 108*) A 461 (ex-Y 874)   NESTOS (ex-*TF 4*) A 463 (ex-Y 854)

Comment: First two acquired from Germany on 16 November 1989, second pair on 5 March 1991. Of about 56 tons with stern ramps for torpedo recovery. Built in 1966. A 461 ran aground on 20 June 2002.

VERIFIED

TYPE 430A (German colours)   *6/1998, Michael Nitz / 0052255*

2 LIGHTHOUSE TENDERS (ABUH)

Name	No	Builders	Commissioned
I KARAVOYIANNOS THEOPHILOPOULOS	A 479	Perama Shipyard	17 Mar 1976
ST LYKOUDIS	A 481	Perama Shipyard	2 Jan 1976

Displacement, tons: 1,450 full load
Dimensions, feet (metres): 207.3 × 38 × 13.1 *(63.2 × 11.6 × 4)*
Main machinery: 1 Deutz MWM TBD5008UD diesel; 2,400 hp(m) *(1.76 MW)*; 1 shaft
Speed, knots: 15
Complement: 40
Radars: Navigation: Racal Decca; I-band.
Helicopters: Platform for 1 light.

Comment: Similar to *Naftilos*, the survey ship.

VERIFIED

I KARAVOYIANNOS THEOPHILOPOULOS   *1/1999, van Ginderen Collection / 0064674*

TUGS

21 HARBOUR TUGS (YTM/YTL)

Name	No	Commissioned
ANTAIOS (ex-*Busy* YTM 2012)	A 407	1947
ATLAS (ex-*Mediator*)	A 408	1944
ACCHILEUS (ex-*Confident*)	A 409	1947
ATROMITOS	A 410	1968
ADAMASTOS	A 411	1968
AIAS (ex-*Ankachak* YTM 767)	A 412	1972
PILEFS (ex-German)	A 413	1991
KADMOS (ex-US)	A 422	1989
CYCLOPS	A 426	1947
DANAOS (ex-US)	A 427	1989
NESTOR (ex-US)	A 428	1989
PERSEUS	A 429	1989
PELOPS	A 430	1989
GIGAS	A 432	1961
KEKROPS	A 435	1989
MINOS (ex-German)	A 436	1991
PELIAS (ex-German)	A 437	1991
AEGEUS (ex-German)	A 438	1991
ATREFS (ex-*Ellerbek*)	A 439	1971
DIOMIDIS (ex-*Neuwerk*)	A 440	1963
THESEUS (ex-*Heppens*)	A 441	2000

Comment: Some may be armed.

UPDATED

NESTOR   *9/1998, M Declerck / 0064675*

PELIAS   *1/2002, M Declerck / 0525884*

3 COASTAL TUGS (YTB)

HERAKLIS A 423	IASON A 424	ODISSEUS A 425

Displacement, tons: 345 full load
Dimensions, feet (metres): 98.5 × 26 × 11.3 *(30 × 7.9 × 3.4)*
Main machinery: 1 Deutz MWM diesel; 1,200 hp(m) *(882 kW)*; 1 shaft
Speed, knots: 12

Comment: Laid down 1977 at Perama Shipyard. Commissioned 6 April, 6 March and 28 June 1978 respectively.

VERIFIED

IASON (with *Gigas*)
8/1997, A Sharma / 0012482

COAST GUARD (Limenikon Soma)

Senior Officers

Commander-in-Chief:
Vice Admiral C Delimichalis
Deputy Commanders-in-Chief:
Rear Admiral Aggelopoulos Pelopidas
Rear Admiral George Papachristodoulou
Inspector General:
Rear Admiral D Trigalos

Personnel

2005: 4,000 (1,055 officers)

Bases

HQ: Piraeus
Main bases: Piraeus, Eleusis, Thessalonika, Volos, Patra, Corfu, Rhodes, Mytilene, Heraklion (Crete), Chios, Kavala, Chalcis, Igoumenitsa, Rafina
Minor bases: Every port and island of Greece

Ships and Craft

In general very similar in appearance to naval ships, being painted grey. Since 1990 pennant numbers have been painted white and on both sides of the hull they carry a blue and white band with two crossed anchors. From 1993 ships have been given grey hulls and white superstructures.

Pennant Numbers

OPV:	010-090	FPB:	101-199	FPO:	210-299

General

This force consists of about 150 patrol craft and anti-pollution vessels including 24 inflatables for the 48 man Underwater Missions Squad and 12 anti-pollution vessels. Administration in peacetime is by the Ministry of Merchant Marine. In wartime it would be transferred to naval command.
Officers are trained at the Naval Academy and ratings at two special schools.
The pennant numbers are all preceded as in the accompanying photographs by Greek 'Lambda Sigma' for Limenikon Soma.

Duties

The policing of all Greek harbours, coasts and territorial waters, navigational safety, SAR operations, anti-pollution surveillance and operations, supervision of port authorities, merchant navy training, inspection of Greek merchant ships worldwide.

Coast Guard Air Service

In October 1981 the Coast Guard acquired two Cessna Cutlass 172 RG aircraft and in July 1988 two Socata TB 20s. Maintenance and training by the Air Force. Based at Dekelia air base. Four Eurocopter Super Pumas AS 322C1 ordered in August 1998. First pair delivered in December 1999, second pair in May 2000. Being operated by mixed Air Force and Coast Guard crews. Bendix radar fitted. Three Reims Cessna Vigilant maritime patrol aircraft ordered in July 1999. First (F 406) delivered on 7 March 2001 and the other two in 2002. Six AS 365N3 Dauphin 2 helicopters were delivered in mid-2004.

Notes: Three 8 m Boston Whalers were donated by the US government on 26 June 2004.

7 DILOS CLASS (WPB)

LS 010	LS 020	LS 030	LS 040	+ 3

Displacement, tons: 86 full load
Dimensions, feet (metres): 95.1 × 16.2 × 5.6 *(29 × 5 × 1.7)*
Main machinery: 2 MTU 12V 331 TC92 diesels; 2,660 hp(m) *(1.96 MW)* sustained; 2 shafts
Speed, knots: 27. **Range, n miles:** 1,600 at 24 kt
Complement: 18
Guns: 2 Rheinmetall 20 mm.
Radars: Surface search: Racal Decca 1226C; I-band.

Comment: Same Abeking and Rasmussen design as the three naval craft and built at Hellenic Shipyards in the early 1980s. Three former Customs craft transferred to the Coast Guard in 2004.

UPDATED

LS 010
6/2002, C D Yaylali / 0525872

3 + (1) SAAR 4 CLASS (LARGE PATROL CRAFT) (PB)

FOURNOI LS 060	RO LS 070	A G EFSTRATIOS LS 080

Displacement, tons: 415 standard; 450 full load
Dimensions, feet (metres): 190.6 × 25 × 8 *(58.0 × 7.8 × 2.4)*
Main machinery: 4 MTU 16V956 TB91 diesels; 15,000 hp(m) *(11.03 MW)* sustained; 4 shafts
Speed, knots: 32. **Range, n miles:** 1,650 at 30 kt; 4,000 at 17.5 kt
Complement: 30
Guns: 1—30 mm. 2—12.7 mm MGs.
Weapons control: Rafael DAFCO.
Radars: Air/Surface search: SIGNAAL variant; E/F-band.
Navigation: Bridgemaster; I-band.

Comment: Three vessels ordered in November 2002. The first two built at Israel Shipyards while the third assembled at Hellenic Shipyards, Skaramanga. The first vessel delivered 23 December 2003, the second in February 2004 and the third in April 2004. A fourth vessel may be ordered. Armament is to be fitted at a later date.

NEW ENTRY

A G EFSTRATIOS
7/2004*, C D Yaylali / 0583668

A G EFSTRATIOS
7/2004*, C D Yaylali / 0583669

39 COLVIC CRAFT (WPB)

LS 114-119	LS 121-123	LS 125-128	LS 133	LS 137-161

Displacement, tons: 24 full load
Dimensions, feet (metres): 54.1 × 15.4 × 4.6 *(16.5 × 4.7 × 1.4)*
Main machinery: 2 MAN D2840 LE 401 diesels; 1,644 hp(m) *(1.21 MW)* sustained; 2 shafts
Speed, knots: 34. **Range, n miles:** 500 at 25 kt
Complement: 5 (1 officer)
Guns: 1—12.7 mm MG. 1—7.62 mm MG.
Radars: Surface search: Raytheon; I-band.

Comment: Ordered from Colvic Craft, Colchester in 1993. Shipped to Motomarine, Glifada for engine and electronics installation. First 12 completed in mid-1994. The remainder delivered at about 12 per year from 1995. GRP hulls with a stern platform for recovery of divers. Later craft have a higher superstructure sited further forward. Two have been lost in accidents.

UPDATED

LS 149
5/2003, A A de Kruijf / 0568864

1 VOSPER EUROPATROL 250 Mk 1 (PBF)

LS 050

Displacement, tons: 240 full load
Dimensions, feet (metres): 155.2 × 24.6 × 7.9 *(47.3 × 7.5 × 2.4)*
Main machinery: 3 GEC/Paxman Valenta 16CM diesels; 13,328 hp(m) *(9.8 MW)*; 3 shafts
Speed, knots: 40. **Range, n miles:** 2,000 at 16 kt
Complement: 21
Radars: Surface search: Racal Decca; I-band.

Comment: Ordered from McTay Marine, Bromborough in July 1993 and completed in November 1994. This is a Vosper International design with a steel hull and aluminium superstructure. Replenishment at sea facilities are provided by light jackstay and the ship carries a 45 kt RIB with water-jet propulsion. A continuous patrol speed of 4 kt is achievable using the centre shaft. Air conditioned accommodation. Similar craft built for the Bahamas. Fitted for a 40 mm gun but this is not carried. Transferred to the Coast Guard in 2004.

UPDATED

LS 050   *5/2004*, Martin Mokrus* / 0587762

4 INTERMARINE CRAFT (WPB)

LS 129-132

Displacement, tons: 25 full load
Dimensions, feet (metres): 53.8 × 14.8 × 7.5 *(16.4 × 4.5 × 2.3)*
Main machinery: 2 MAN diesels; 2,000 hp(m) *(1.47 MW)* sustained; 2 shafts
Speed, knots: 36

Comment: Constructed by Intermarine, La Spezia and delivered 1996-97.

NEW ENTRY

3 COMBATBOAT 90H (WPBF)

LS 134-136

Displacement, tons: 19 full load
Dimensions, feet (metres): 52.2 × 12.5 × 2.6 *(15.9 × 3.8 × 0.8)*
Main machinery: 2 Volvo Penta TAMD 163P diesels; 1,500 hp(m) *(1.1 MW)*; 2 waterjets
Speed, knots: 45. **Range, n miles:** 240 at 30 kt
Complement: 3
Guns: 3—12.7 mm MGs.
Radars: Surface search: I-band.

Comment: Built by Dockstavarvet in Sweden and delivered 6 July 1998. Same design as Swedish naval craft but with more powerful engines. GRP construction with armoured protection for cockpit.

UPDATED

LS 136   *7/2004*, A Campanera i Rovira* / 0587761

16 OL 44 CLASS (WPB)

LS 55	LS 95	LS 103	LS 109
LS 65	LS 97	LS 106	LS 110
LS 84-88	LS 101	LS 107	LS 112

Displacement, tons: 14 full load
Dimensions, feet (metres): 44.9 × 14.4 × 2 *(13.7 × 4.4 × 0.6)*
Main machinery: 2 diesels; 630 hp(m) *(463 kW)*; 2 shafts
Speed, knots: 23
Complement: 4
Guns: 1—7.62 mm MG.
Radars: Surface search: JRC; I-band.

Comment: Built by Olympic Marine. GRP hulls.

VERIFIED

LS 101   *5/2000, van Ginderen Collection* / 0104571

8 MOTORMARINE CRAFT (WPB)

LS 601-608

Displacement, tons: To be announced
Dimensions, feet (metres): To be announced
Main machinery: To be announced
Speed, knots: To be announced

Comment: Constructed by Motormarine, Koropi, Greece. *LS 601* delivered November 2003 and remainder by August 2004.

NEW ENTRY

LS 601   *5/2004*, Martin Mokrus* / 0587758

16 LS 51 CLASS (WPB)

LS 51	LS 52	LS 155-157	+ 11

Displacement, tons: 13 full load
Dimensions, feet (metres): 44 × 11.5 × 3.3 *(13.4 × 3.5 × 1)*
Main machinery: 2 diesels; 630 hp(m) *(463 kW)*; 2 shafts
Speed, knots: 25. **Range, n miles:** 400 at 18 kt
Complement: 4
Guns: 1—7.62 mm MG.
Radars: Surface search: Racal Decca; I-band.

Comment: Built by Olympic Marine. GRP hulls.

VERIFIED

60 COASTAL CRAFT and 22 CRISS CRAFT

Comment: Included in the total are 20 of 8.2 m, 17 of 7.9 m, 26 of 5.8 m and 19 ex-US Criss craft. In addition the Coast Guard operates 24 Inflatable craft, and 10 SAR craft *(LS 509-518)*.

UPDATED

LS 214   *7/2004*, C D Yaylali* / 0587760

LS 130   *10/2002, E & M Laursen* / 0533891

4 POLLUTION CONTROL SHIPS (YPC)

LS 401   **LS 413-415**

Displacement, tons: 230 full load
Dimensions, feet (metres): 95.1 × 20.3 × 8.2 *(29 × 6.2 × 2.5)*
Main machinery: 2 CAT 3512 DITA diesels; 2,560 hp(m) *(1.88 MW)* sustained; 2 shafts
Speed, knots: 15. **Range, n miles:** 500 at 13 kt
Complement: 12
Radars: Navigation: Furuno; I-band.

Comment: Details given are for *LS 413-415*. Built by Astilleros Gondan, Spain in collaboration with Motomarine. Delivered in 1993-94. *LS 401* is an older pollution control ship.   **UPDATED**

LS 401
5/2004 *, Martin Mokrus / 0587759

CUSTOMS

Notes: The Customs service also operates large numbers of coastal and inshore patrol. The craft have a distinctive Alpha Lambda (A/) on the hull and are sometimes armed with 7.62 mm MGs.

AL 20   *6/2002, C D Yaylali* / 0525874

Grenada

Country Overview

Grenada gained independence in 1974; the British monarch, represented by a governor-general, is the head of state. The southernmost of the Windward Islands in the Lesser Antilles chain, the country comprises the island of Grenada (311 square miles) and some of the southern Grenadines including Carriacou and Petit Martinique. The capital, largest town, and main port is St George's. Territorial seas (12 n miles) are claimed. A 200 n mile Exclusive Economic Zone (EEZ) has been claimed but the limits are not defined. The Coast Guard craft are operated under the direction of the Commissioner of Police.

Headquarters Appointments

Coast Guard Commander:
    Assistant Superintendent Osmond Griffith

Personnel

2005: 30

Bases

Prickly Bay

COAST GUARD

Notes: A 920 Zodiac RHIB, donated by the US government, entered service in 2004.

1 GUARDIAN CLASS (COASTAL PATROL CRAFT) (PB)

Name	No	Builders	Commissioned
TYRREL BAY	PB 01	Lantana, Florida	21 Nov 1984

Displacement, tons: 90 full load
Dimensions, feet (metres): 105 × 20.6 × 7 *(32 × 6.3 × 2.1)*
Main machinery: 3 Detroit 12V-71TA diesels; 1,260 hp *(939 kW)* sustained; 3 shafts
Speed, knots: 24. **Range, n miles:** 1,500 at 18 kt
Complement: 15 (2 officers)
Guns: 2—12.7 mm MGs. 2—7.62 mm MGs.
Radars: Surface search: Furuno 1411 Mk II; I-band.

Comment: Similar to Jamaican and Honduras vessels. Refit in 1995/96.
                                                        VERIFIED

TYRREL BAY   *11/1990, Bob Hanlon* / 0064681

1 DAUNTLESS CLASS (PB)

Name	No	Builders	Commissioned
LEVERA	PB 02	SeaArk Marine	8 Sep 1995

Displacement, tons: 11 full load
Dimensions, feet (metres): 40 × 14 × 4.3 *(12.2 × 4.3 × 1.3)*
Main machinery: 2 Caterpillar 3208TA diesels; 870 hp *(650 kW)* sustained; 2 shafts
Speed, knots: 27. **Range, n miles:** 600 at 18 kt
Complement: 5
Guns: 1—7.62 mm MG.
Radars: Surface search: Raytheon R40X; I-band.

Comment: One of many of this type, provided by the US, throughout the Caribbean navies.
                                                        VERIFIED

LEVERA   *9/1995, SeaArk Marine* / 0064683

2 BOSTON WHALERS (PB)

Displacement, tons: 1.3 full load
Dimensions, feet (metres): 22.3 × 7.4 × 1.2 *(6.7 × 2.3 × 0.4)*
Main machinery: 2 outboards; 240 hp *(179 kW)*
Speed, knots: 40+
Complement: 4
Guns: 1—12.7 mm MG.

Comment: Acquired in 1988-89.
                                                        VERIFIED

BOSTON WHALER   *11/1990, Bob Hanlon* / 0064682

Guatemala

Country Overview

The Republic of Guatemala is situated in Central America between Mexico to the north, Belize to the east and Honduras and El Salvador to the south-east. With an area of 42,042 square miles, it has an 83 n mile coastline with the Caribbean and a 133 n mile coastline with the Pacific Ocean. The capital city is Guatemala City while the principal Caribbean ports are Puerto Barrios and Santo Tomás de Castilla and Pacific ports are Puerto Quetzal, San José and Champerico. Territorial seas (12 n miles) are claimed. A 200 n mile EEZ has been claimed but the limits are not defined.

Headquarters Appointments

Commander Caribbean Naval Region:
    To be confirmed
Commander Pacific Naval Region:
    To be confirmed

Personnel

(a)   2005: 1,250 (130 officers) including 650 Marines (2 battalions) (mostly volunteers)
(b)   2¼ years' national service

Bases

Santo Tomás de Castillas (BANATLAN); Sipacate and Puerto Quetzal (BANAPAC)

PATROL FORCES

Notes: (1) There is also a naval manned Ferry *15 de Enero* (T 691) and a 69 ft launch *Orca* which was built locally in 1996/97.
(2) Three sail training craft, *Mendieta*, *Margarita* and *Ostuncalco* are based at Santo Thomás de Castilla.

1 BROADSWORD CLASS (COASTAL PATROL CRAFT) (PB)

Name	No	Builder	Commissioned
KUKULKÁN	GC 1051 (ex-P 1051)	Halter Marine	4 Aug 1976

Displacement, tons: 90.5 standard; 110 full load
Dimensions, feet (metres): 105 × 20.4 × 6.3 *(32 × 6.2 × 1.9)*
Main machinery: 2 Detroit 8V 92TA Model 91; 1,300 hp *(970 kW)*; 2 shafts
Speed, knots: 22. **Range, n miles:** 1,150 at 20 kt
Complement: 20 (5 officers)
Guns: 2 Oerlikon GAM/204 GK 20 mm. 2—7.62 mm MGs.
Radars: Surface search: Furuno; I-band.

Comment: As the flagship she used to rotate between Pacific and Atlantic bases every two years but has remained in the Pacific since 1989. Rearmed with 20 mm guns in 1989. These were replaced by GAM guns in 1990-91 when the ship received a new radar. Refitted again in 1996 with new engines.
VERIFIED

KUKULKÁN   *2/1998* / 0052314

2 SEWART CLASS (COASTAL PATROL CRAFT) (PB)

Name	No	Builders	Commissioned
UTATLAN	GC 851 (ex-P 851)	Sewart, Louisiana	May 1967
SUBTENIENTE OSORIO SARAVIA	GC 852 (ex-P 852)	Sewart, Louisiana	Nov 1972

Displacement, tons: 54 full load
Dimensions, feet (metres): 85 × 18.7 × 7.2 *(25.9 × 5.7 × 2.2)*
Main machinery: 2 Detroit 8V 92TA Model 91; 1,300 hp *(970 kW)*; 2 shafts
Speed, knots: 22. **Range, n miles:** 400 at 12 kt
Complement: 17 (4 officers)
Guns: 1 Oerlikon GAM/204 GK 20 mm. 2—7.62 mm MGs.
Radars: Surface search: Furuno; I-band.

Comment: Aluminium superstructure. Both rearmed with 20 mm guns, and 75 mm recoilless removed in 1990. P 851 is based in the Atlantic; P 852 in the Pacific. Refitted in 1995-96 with new engines.
VERIFIED

UTATLAN   *12/1999* / 0104573

6 CUTLASS CLASS
(5 COASTAL PATROL CRAFT AND 1 SURVEY CRAFT) (PB)

Name	No	Builders	Commissioned
TECUN UMAN	GC 651 (ex-P 651)	Halter Marine	26 Nov 1971
KAIBIL BALAN	GC 652 (ex-P 652)	Halter Marine	8 Feb 1972
AZUMANCHE	GC 653 (ex-P 653)	Halter Marine	8 Feb 1972
TZACOL	GC 654 (ex-P 654)	Halter Marine	10 Mar 1976
BITOL	GC 655 (ex-P 655)	Halter Marine	4 Aug 1976
GUCUMAZ	GC-H-656 (ex-BH 656)	Halter Marine	15 May 1981

Displacement, tons: 45 full load
Dimensions, feet (metres): 64.5 × 17 × 3 *(19.7 × 5.2 × 0.9)*
Main machinery: 2 Detroit 8V 92TA Model 91 diesels; 1,300 hp *(970 kW)*; 2 shafts
Speed, knots: 25. **Range, n miles:** 400 at 15 kt
Complement: 10 (2 officers)
Guns: 2 Oerlikon GAM/204 GK 20 mm. 2 or 3—12.7 mm MGs.
Radars: Surface search: Furuno; I-band.

Comment: First five rearmed with 20 mm guns in 1991. P 651, 654 and 655 are in the Atlantic, remainder in the Pacific. Aluminium hulls. *Gucumaz* was used as a survey craft but by 1996 was again serving as a patrol craft with three MGs. 654 and 656 refitted in 1994-95, remainder in 1995-97. New engines fitted.
UPDATED

GUCUMAZ   *2/1999* / 0064684

BITOL   *11/2003*, A A de Kruijf / 0587767

1 DAUNTLESS CLASS (PB)

IXIMCHE

Displacement, tons: 11 full load
Dimensions, feet (metres): 40 × 12.66 × 2.3 *(12.19 × 3.86 × 0.69)*
Main machinery: 2 Caterpillar 3208TA diesels; 850 hp *(635 kW)*; 2 shafts
Speed, knots: 28. **Range, n miles:** 400 at 22 kt
Complement: 5
Guns: 1—7.62 mm MG.
Radars: Surface search: Raytheon R40X; I-band.

Comment: Built by SeaArk, Monticello, of aluminium construction. Donated by US government as foreign aid in 1997.
UPDATED

DAUNTLESS CLASS (Cayman Islands colours)   *6/2001*, RCIS / 0121305

6 VIGILANTE CLASS (PBI)

GC 271-276

Displacement, tons: 3.5 full load
Dimensions, feet (metres): 26.6 × 10 × 1.8 *(8.1 × 3 × 0.5)*
Main machinery: 2 Evinrude outboards; 600 hp *(448 kW)*
Speed, knots: 40+
Complement: 4
Guns: 1—12.7 mm MG.
Radars: Surface search: Furuno; I-band.

Comment: Ordered in 1993 from Boston Whaler. Delivered in 1994 and divided three to each coast.
VERIFIED

GC 275   *12/1999* / 0104574

20 RIVER PATROL CRAFT (PBR)

Group A	Group B	Group C	Group D
DENEB	LAGO DE ATITLAN	CHOCHAB	MERO
SIRIUS	MAZATENANGO	ALIOTH	SARDINA
PROCYON	RETALHULEU	MIRFA	PAMPANA
VEGA	ESCUINTLA	SCHEDAR	MAVRO-I
POLLUX		COMAMEFA	
SPICA			
STELLA MARIS			

Comment: Group A are wooden hull craft with a speed of 19 kt. Group B have aluminium hulls and a speed of 28 kt. Group C are probably of Israeli design and Group D are commercial craft caught smuggling and confiscated. All can be armed with 7.62 mm MGs and are used by Marine battalions as well as the Navy.
VERIFIED

CHOCHAB AND COMAMEFA   *2/1996, Julio Montes* / 0064686

Guinea

Country Overview

A former French colony, The Republic of Guinea became independent in 1958. Located in west Africa, the country has an area of 94,926 square miles, a 173 n mile coastline with the Atlantic Ocean and includes the Iles de Los. It is bordered to the north by Guinea-Bissau and Senegal and to the south by Liberia and Sierra Leone. The capital, largest city and principal port is Conakry. Territorial seas (12 n miles) are claimed. A 200 n mile Exclusive Economic Zone (EEZ) has been claimed but the limits have not been formally agreed. Fishery Protection may be provided by civilian contractors.

Personnel

(a)   2005: 400 officers and men
(b)   2 years' conscript service

Bases

Conakry, Kakanda

Notes: A number of craft, including two Bogomol, two Stinger and two Swiftships *(Vigilante* and *Intrepide)* are laid up alongside. Some of these might be resurrected to combat piracy problems in the region. A Damen 13 m patrol boat, *Matakang,* is reported to have been delivered in 1999.

Guinea-Bissau

Country Overview

A former Portuguese colony, The Republic of Guinea-Bissau gained independence in 1974. Located in west Africa, the country has an area of 13,948 square miles, a 189 n mile coastline with the Atlantic Ocean and includes about 60 offshore islands, among them the Bijagós (Bissagos) Islands. It is bordered to the north by Senegal and to the south by Guinea. The capital, largest city and principal port is Bissau. Other ports include Cacheu and Bolama. Territorial seas (12 n miles) are claimed. A 200 n mile Exclusive Economic Zone (EEZ) has been claimed and has been partially defined by boundary agreements.

Personnel

(a)   2005: 310 officers and men
(b)   Voluntary service

Base

Bissau

Maritime Aircraft

A Cessna 337 patrol aircraft is used for offshore surveillance, when serviceable.

PATROL FORCES

Notes: One Rodman R 800 8.7 m patrol craft with a speed of 28 kt acquired in 1999.

2 ALFEITE TYPE (COASTAL PATROL CRAFT) (PC)

Name	No	Builders	Commissioned
CACINE	LF 01	Arsenal do Alfeite	9 Mar 1994
CACHEU	LF 02	Arsenal do Alfeite	9 Mar 1994

Displacement, tons: 55 full load
Dimensions, feet (metres): 64.6 × 19 × 10.6 *(19.7 × 5.8 × 3.2)*
Main machinery: 3 MTU 12V 183 TE92 diesels; 3,000 hp(m) *(2.2 MW)* maximum; 3 Hamilton MH 521 water-jets
Speed, knots: 28
Complement: 9 (1 officer)
Radars: Navigation: Furuno FR 2010; I-band.

Comment: Ordered from Portugal in 1991. GRP hulls. Used for fishery protection patrols and customs duties. They are the only vessels regularly reported at sea.
VERIFIED

CACHEU
3/1994, Arsenal do Alfeite / 0064688

Guyana

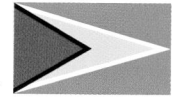

Country Overview

Formerly known as British Guiana, the Cooperative Republic of Guyana became an independent state in 1966. With an area of 83,000 square miles it has borders to the east with Suriname, to the west with Venezuela and to the south with Brazil; its 248 n mile coastline is on the Atlantic Ocean. The capital, largest city and chief port is Georgetown. Territorial seas (12 n miles) and a fisheries zone (200 n miles) are claimed. A 200 n mile Exclusive Economic Zone (EEZ) has also been claimed but the limits are not defined. Rebuilding of the Coast Guard started in 2001.

Headquarters Appointments

Commanding Officer, Coast Guard:
    Commander Terrence Pile

Personnel

(a)   2005: 30 plus 160 reserves
(b)   Voluntary service

Bases

Georgetown, Benab (Corentyne)

PATROL FORCES

4 TYPE 44 CLASS (WPB)

BARRACUDA	HYMANA	PIRAI	TIRAPUKA

Displacement, tons: 18 full load
Dimensions, feet (metres): 44 × 12.8 × 3.6 *(13.5 × 3.9 × 1.1)*
Main machinery: 2 Detroit 6V-38 diesels; 185 hp *(136 kW)*; 2 shafts
Speed, knots: 14. **Range, n miles:** 215 at 10 kt
Complement: 3

Comment: Acquired from the US and recommissioned on 9 August 2003.

NEW ENTRY

TYPE 44 (Uruguay Colours)   *5/2000, Hartmut Ehlers /* 0105801

1 RIVER CLASS (COASTAL PATROL CRAFT) (PBO)

Name	No	Builders	Commissioned
ESSEQUIBO (ex-*Orwell*)	1026 (ex-M 2011)	Richards, Great Yarmouth	27 Nov 1985

Displacement, tons: 890 full load
Dimensions, feet (metres): 156 × 34.5 × 9.5 *(47.5 × 10.5 × 2.9)*
Main machinery: 2 Ruston 6 RKC diesels; 3,100 hp(m) *(2.3 MW)* sustained; 2 shafts
Speed, knots: 14. **Range, n miles:** 4,500 at 10 kt
Complement: 32 (4 officers)
Guns: 1 Bofors 40 mm/60.
    2—7.62 mm MGs.
Radars: Surface search: 2 Racal Decca TM 1226C; I-band.

Comment: Ex-UK River class previously employed as patrol ship and then officers' training ship. Transferred on 22 June 2001.

VERIFIED

ESSEQUIBO   *7/2001, Derek Fox /* 0114272

Honduras
FUERZA NAVAL REPUBLICA

Country Overview

The Republic of Honduras is one of the largest Central American republics. With an area of 43,433 square miles, it is situated between El Salvador and Guatemala to the west and Nicaragua to the south and east. It has a 350 n mile coastline with the Caribbean and a 93 n mile coastline with the Pacific Ocean. The capital and largest city is Tegucigalpa while the principal Caribbean ports are La Ceiba and Puerto Cortés and Pacific port is Amapala. Territorial seas (12 n miles) are claimed. A 200 n mile EEZ is claimed and has been partly defined by boundary agreements.

Headquarters Appointments

Commanding Officer, General HQ:
    Capitan de Navio D E M N Rolando Gonzalez Flores

Personnel

2005: 1,000 including 350 marines

Bases

Tegucigalpa (General HQ)
Puerto Cortés, Puerto Castilla (Atlantic HQ), Amapala (Pacific HQ), La Ceiba, Puerto Trujillo

---

PATROL FORCES

Notes: (1) In addition there may be three Piranha river craft still in limited service. (2) Five 23 m catamarans reported to have been ordered in 2004.

3 SWIFT 105 ft CLASS (FAST ATTACK CRAFT—GUN) (PB)

GUAYMURAS FNH 101	HONDURAS FNH 102	HIBUERAS FNH 103

Displacement, tons: 111 full load
Dimensions, feet (metres): 105 × 23.6 × 7 *(32 × 7.2 × 2.1)*
Main machinery: 2 MTU 16V 538 TB90 diesels; 6,000 hp(m) *(4.4 MW)* sustained; 2 shafts
Speed, knots: 30. **Range, n miles:** 1,200 at 18 kt
Complement: 17 (3 officers)
Guns: 6 Hispano-Suiza 20 mm (2 triple). 2—12.7 mm MGs.
Weapons control: Kollmorgen 350 optronic director.
Radars: Surface search: Furuno; I-band.

Comment: First delivered by Swiftships, Morgan City in April 1977 and last two in March 1980. Aluminium hulls. Armament changed 1996-98.

UPDATED

HONDURAS
4/1991 / 0064689

1 GUARDIAN CLASS (COASTAL PATROL CRAFT) (PB)

TEGUCIGALPA FNH 104 (ex-FNH 107)

Displacement, tons: 94 full load
Dimensions, feet (metres): 106 × 20.6 × 7 *(32.3 × 6.3 × 2.1)*
Main machinery: 3 Detroit 16V-92TA diesels; 2,070 hp *(1.54 MW)* sustained; 3 shafts
Speed, knots: 30. **Range, n miles:** 1,500 at 18 kt
Complement: 17 (3 officers)
Guns: 1 General Electric Sea Vulcan 20 mm Gatling.
  3 Hispano Suiza 20 mm (1 triple). 2—12.7 mm MGs.
Weapons control: Kollmorgen 350 optronic director.
Radars: Surface search: Furuno; I-band.

Comment: Delivered by Lantana Boatyard, Florida August 1986. Second of class, *Copan*, no longer
  in service. A third of the class, completed in May 1984, became the Jamaican *Paul Bogle*.
  Aluminium hulls.
VERIFIED

GUARDIAN CLASS   *7/1986, Giorgio Arra*

6 SWIFT 65 ft CLASS (COASTAL PATROL CRAFT) (PB)

NACAOME (ex-*Aguan*, ex-*Gral*) FNH 651   **ULUA** FNH 654
GOASCORAN (ex-*General J T Cabanas*) FNH 652   **CHOLUTECA** FNH 655
PATUCA FNH 653   **RIO COCO** FNH 656

Displacement, tons: 33 full load
Dimensions, feet (metres): 69.9 × 17.1 × 5.2 *(21.3 × 5.2 × 1.6)*
Main machinery: 2 GM 12V-71TA diesels; 840 hp *(627 kW)* sustained; 2 shafts (FNH 651-2)
  2 MTU 8V 396 TB93 diesels; 2,180 hp(m) *(1.6 MW)* sustained; 2 shafts (FNH 653-5)
Speed, knots: 25 (FNH 651-2); 36 (FNH 653-5). **Range, n miles:** 2,000 at 22 kt (FNH 651-2)
Complement: 9 (2 officers)
Guns: 1 Oerlikon 20 mm. 2—12.7 mm (twin) MGs. 2—7.62 MGs.
Radars: Surface search: Racal Decca; I-band.

Comment: First pair built by Swiftships, Morgan City originally for Haiti. Contract cancelled and
  Honduras bought the two that had been completed in 1973-74. Delivered in 1977. Last four
  ordered in 1979 and delivered 1980.
VERIFIED

PATUCA   *5/1993 / 0064690*

1 SWIFT 85 ft CLASS (COASTAL PATROL CRAFT) (PB)

CHAMELECON (ex-*Rio Kuringwas*) FNH 8501

Displacement, tons: 60 full load
Dimensions, feet (metres): 85 × 20 × 5 *(25.9 × 6.1 × 1.8)*
Main machinery: 2 Detroit diesels; 2 shafts
Speed, knots: 25
Complement: 10 (2 officers)
Radars: Surface search: Racal/Decca; I-band.

Comment: Built by Swiftships, Morgan City in about 1967 for Nicaragua from where it was
  transferred in 1979.
UPDATED

CHAMELECON   *2000, Honduran Navy / 0105811*

5 OUTRAGE CLASS (RIVER PATROL CRAFT) (PBR)

Displacement, tons: 2.2 full load
Dimensions, feet (metres): 24.9 × 7.9 × 1.3 *(7.6 × 2.4 × 0.4)*
Main machinery: 2 Evinrude outboards; 300 hp *(224 kW)*
Speed, knots: 30. **Range, n miles:** 200 at 30 kt
Complement: 4
Guns: 1—12.7 mm MG. 2—7.62 mm MGs.
Radars: Navigation: Furuno 3600; I-band.

Comment: Built by Boston Whaler in 1982. Seven deleted so far. Radar is sometimes embarked.
VERIFIED

OUTRAGE   *10/1997, Julio Montes / 0012491*

15 RIVER CRAFT (PBR)

Comment: Acquired from Taiwan in 1996. Nine based at Castilla, three at Cortes and three at
  Amapala. Single Mercury outboard engine. Carry a 7.62 mm MG. Three sunk in 1998.
VERIFIED

PBR   *10/1997, R Torrento / 0012492*

AUXILIARIES

Notes: In addition there are two ex-US LCM 8 (*Warunta* FNH 7401, *Tansin* FNH 7402) transferred in 1987. Both are used as transport vessels.

LCM 8   *2000, Honduran Navy* / 0105812

1 LANDING CRAFT (LCU)

PUNTA CAXINAS FNH 1491

Displacement, tons: 625 full load
Dimensions, feet (metres): 149 × 33 × 6.5 *(45.4 × 10 × 2)*
Main machinery: 3 Caterpillar 3412 diesels; 1,821 hp *(1.4 MW)* sustained; 3 shafts
Speed, knots: 14. **Range, n miles:** 3,500 at 12 kt
Complement: 18 (3 officers)
Cargo capacity: 100 tons equipment or 50,000 gallons dieso plus 4 standard containers
Radars: Navigation: Furuno 3600; I-band.

Comment: Ordered in 1986 from Lantana, Florida, and commissioned in May 1988.

VERIFIED

PUNTA CAXINAS   *6/1997* / 0064691

Hong Kong
POLICE MARINE REGION

Country Overview

Formerly a British colony, the Hong Kong Special Administrative Region of China reverted to Chinese sovereignty on 30 June 1997. While China has assumed responsibility for foreign affairs and defence, the territory is to maintain its own legal, social, and economic systems until at least 2047. Hong Kong comprises three main regions, Hong Kong Island (29 sq miles), Kowloon Peninsula and Stonecutters Island (6 sq miles) and the New Territories (380 sq miles). As with the remainder of China, territorial seas (12 n miles) are claimed. An EEZ (200 n mile) is also claimed but the limits have not been defined by boundary agreements. The role of the Marine Police is to maintain the integrity of the sea boundary and territorial waters of Hong Kong, enforce the laws of Hong Kong in territorial waters, prevent illegal immigration by sea, SAR in territorial and adjacent waters, and casualty evacuation.

Headquarters Appointments

Regional Commander:
  Au Hok-Lam
Deputy Regional Commander:
  P C Burbidge-King

Organisation

Marine Police Regional HQ, Sai Wan Ho
Bases at Ma Liu Shui, Tui Min Hoi, Tai Lam Chung, Aberdeen, Sai Wan Ho

Personnel

(a)   2005: 2,600
(b)   Voluntary service

DELETIONS

2002   *PL 61, PL 66, PL 6-8*
2004   *PL 63, PL 65*

POLICE

Notes: The naming of craft has been discontinued.

2 SURVEILLANCE BARGES (YAG)

PB 1-2

Displacement, tons: 227
Dimensions, feet (metres): 98.4 × 42.6 × 2.6 *(30.0 × 13 × 0.8)*
Main machinery: 2 Onan 75 MDGDB
Complement: 10
Radars: Surface search: Decca; I-band.

Comment: Steel-hulled barges constructed by Bonny Fair Ltd and delivered in June 2002. Permanently moored in Deep Bay.   *UPDATED*

PB 1   *6/2004*, Hong Kong Police* / 0589752

1 TRAINING VESSEL (WAX)

PL 3

Displacement, tons: 420 full load
Dimensions, feet (metres): 131.2 × 28.2 × 10.5 *(40 × 8.6 × 3.2)*
Main machinery: 2 Caterpillar 3512TA diesels; 2,420 hp *(1.81 MW)* sustained; 2 shafts
Speed, knots: 14. **Range, n miles:** 1,500 at 14 kt
Complement: 7
Radars: Surface search: 2 Racal Decca ARPA C342/8; I-band.

Comment: Built by Hong Kong SY in 27 July 1987 and commissioned 1 February 1988. Steel hull. Racal ARPA and GPS Electronic Chart system. 12.7 mm MGs removed in mid-1997. Can carry up to 30 armed police for short periods. Former command vessel converted to a training role.   *UPDATED*

PL 3   *6/2004*, Hong Kong Police* / 0589753

4 + 2 KEKA CLASS (PATROL CRAFT) (WPB)

PL 60-63

Displacement, tons: 105
Dimensions, feet (metres): 98.4 × 20.6 × 5.6 *(30.0 × 6.3 × 1.7)*
Main machinery: 2 MTU 12V-396 TE 84 diesels
Speed, knots: 25. **Range, n miles:** 360 at 15 kt
Complement: 14
Radars: Surface search: Decca; I-band.

Comment: Aluminium-hulled craft built by Cheoy Lee Shipyards Ltd to replace Damen Mk1 class patrol craft. Delivered in 2002 and 2004. Two further craft expected by 2005.   *UPDATED*

PL 62   *6/2004*, Hong Kong Police* / 0589751

14 DAMEN Mk III CLASS (PATROL CRAFT) (WPB)

PL 70-80 **PL 82-84**

Displacement, tons: 95 full load
Dimensions, feet (metres): 87 × 19 × 6 *(26.5 × 5.8 × 1.8)*
Main machinery: 2 MTU 12V 396 TC82 diesels; 2,610 hp(m) *(1.92 MW)* sustained; 2 shafts
1 Mercedes-Benz OM 424A 12V diesel; 341 hp(m) *(251 kW)* sustained; 1 Kamewa 45 water-jet
Speed, knots: 26 on 3 diesels; 8 on water-jet and cruising diesel. **Range, n miles:** 600 at 14 kt
Complement: 14
Radars: Surface search: Racal Decca; I-band.

Comment: Steel-hulled craft constructed by Chung Wah SB & Eng Co Ltd, 1984-85. 12.7 mm MGs removed in mid-1997.

UPDATED

PL 76   *6/2004*, Hong Kong Police /* 0589750

6 PROTECTOR (ASI 315) CLASS
(COMMAND/PATROL CRAFT) (WPB)

PL 51-56

Displacement, tons: 170 full load
Dimensions, feet (metres): 107 × 26.9 × 5.2 *(32.6 × 8.2 × 1.6)*
Main machinery: 2 Caterpillar 3516TA diesels; 4,400 hp *(3.28 MW)* sustained; 2 shafts; 1 Caterpillar 3412TA; 1,860 hp *(1.24 MW)* sustained; Hamilton jet (centreline); 764 hp *(570 kW)*
Speed, knots: 30. **Range, n miles:** 600 at 18 kt
Complement: 19
Weapons control: GEC V3901 optronic director.
Radars: Surface search: Racal Decca; I-band.

Comment: Built by Australian Shipbuilding Industries and completed in 1993. As well as patrol work, the craft provide command platforms for Divisional commanders. 12.7 mm guns removed in 1996 and the optronic director is used for surveillance only.

UPDATED

PL 52   *6/2004*, Hong Kong Police /* 0589749

2 DAMEN Mk I CLASS (PATROL CRAFT) (WPB)

PL 67-68

Displacement, tons: 86 full load
Dimensions, feet (metres): 85.9 × 19.4 × 5.5 *(26.2 × 5.9 × 1.7)*
Main machinery: 2 MTU 12V 396 TC82 diesels; 2,610 hp(m) *(1.92 MW)* sustained; 2 shafts
1 MAN D2566 diesel; 195 hp(m) *(143 kW)*
Speed, knots: 23 MTU; 6 MAN. **Range, n miles:** 600 at 14 kt
Complement: 14
Radars: Surface search: Racal Decca; I-band.

Comment: Designed by Damen SY, Netherlands. Steel-hulled craft built by Chung Wah SB & Eng Co Ltd between 1980 and 1981. 12.7 mm MGs removed in mid-1997. Contract signed with Cheoy Lee Shipyards Ltd in August 2000 to replace original six craft. Two decommissioned in 2002 and two in 2004.

UPDATED

PL 67   *6/2004*, Hong Kong Police /* 0589748

7 HARBOUR PATROL CRAFT (WPB)

PL 11-17

Displacement, tons: 36 full load
Dimensions, feet (metres): 52.5 × 15.1 × 4.9 *(16 × 4.6 × 1.5)*
Main machinery: 2 Cummins NTA-855-M diesels; 700 hp *(522 kW)* sustained; 2 shafts
Speed, knots: 12
Complement: 6
Radars: Surface search: Racal Decca; I-band.

Comment: Built by Chung Wah SB & Eng Co Ltd in 1986-87.

UPDATED

PL 13   *6/2004*, Hong Kong Police /* 0589747

5 SEA STALKER 1500 CLASS (INTERCEPTOR CRAFT) (HSIC)

PL 85-89

Displacement, tons: 7.5 full load
Dimensions, feet (metres): 48.6 × 9.5 × 2.6 *(14.8 × 2.9 × 0.8)*
Main machinery: 3 Innovation Marine Sledge Hammers; 1,500 hp(m) *(1.1 MW)*; 3 shafts
Speed, knots: 60; 45 in Sea State 3
Complement: 5
Radars: Surface search: Raytheon; I-band.

Comment: Built by Damen, Gorinchem in 1999. Used by the Small Boat Division.

UPDATED

PL 86   *6/2004*, Hong Kong Police /* 0589746

4 SEASPRAY CLASS (LOGISTIC CRAFT) (YFB)

PL 46-49

Displacement, tons: 10.7 full load
Dimensions, feet (metres): 37.4 × 13.8 × 3.9 *(11.4 × 4.2 × 1.2)*
Main machinery: 2 Caterpillar 3208TA diesels; 550 hp *(410 kW)* sustained; 2 shafts
Speed, knots: 32
Complement: 4
Radars: Navigation: Koden; I-band.

Comment: Built by Seaspray Boats, Fremantle in 1992. Catamaran hulls capable of carrying 16 police officers.

UPDATED

PL 47   *6/2004*, Hong Kong Police /* 0589745

11 SEASPRAY CLASS (INSHORE PATROL CRAFT) (WPB)

PL 22-32

Displacement, tons: 8.7 full load
Dimensions, feet (metres): 32.5 × 13.8 × 4.3 *(9.9 × 4.2 × 1.3)*
Main machinery: 2 Caterpillar 3208TA diesels; 680 hp *(508 kW)*; 2 shafts
Speed, knots: 35
Complement: 4
Radars: Surface search: Koden; I-band.

Comment: Built by Seaspray Boats, Fremantle in 1992-93.

UPDATED

PL 22   6/2004*, Hong Kong Police / 0589744

9 INSHORE PATROL CRAFT (WPB)

PL 20-21   PL 90-92   PL 93-96

Displacement, tons: 4.5
Dimensions, feet (metres): 27 × 9.2 × 1.6 *(8.3 × 2.8 × 0.5)*
Main machinery: 2 outboards; 540 hp *(403 kW)*
Speed, knots: 40+
Complement: 4
Radars: Surface search: Koden; I-band.

Comment: Details given are for *PL 20-21* which are Sharkcat class of catamaran construction, commissioned in October 1988. *PL 90-92* are Boston Whaler Guardians with 2 Johnson 115 hp outboards, and *PL 93-96* are Boston Whaler Vigilantes with 2 Johnson 250 hp outboards. The Whalers were all delivered in 1997 and are capable of speeds in excess of 33 kt.

UPDATED

PL 93   6/2004*, Hong Kong Police / 0589743

6 CHEOY LEE CLASS (INSHORE PATROL CRAFT) (WPB)

PL 40-45

Displacement, tons: 15
Dimensions, feet (metres): 42.9 × 13 × 2.3 *(13.07 × 3.96 × 0.7)*
Main machinery: 2 MAN D2842LE403 diesels; 720 hp *(537 kW)* sustained; 2 Hamilton water-jets
Speed, knots: 35
Complement: 4
Radars: Surface search: Bridgemaster E 180; I-band.

Comment: Based upon a design from Peterson Shipbuilders, these shallow draft vessels were constructed by Cheoy Lee Shipyards Ltd and delivered in 2000.

UPDATED

PL 43   6/2004*, Hong Kong Police / 0589742

8 HIGH SPEED INTERCEPTORS (HSIC)

PV 30-37

Displacement, tons: 2.7 full load
Dimensions, feet (metres): 28.3 × 8.7 × 2.4 *(8.5 × 2.6 × 0.7)*
Main machinery: 2 Mercury outboards; 500 hp *(373 kW)*
Speed, knots: 51
Complement: 3

Comment: Built by Queensland Ships in 1997.

UPDATED

PV 36   6/2004*, Hong Kong Police / 0589741

CUSTOMS

Notes: The Marine Enforcement Group is based at Stonecutters Island. There are five Sector Command launches. Three Damen 26 m craft were completed in 1986 by Chung Wah SB & Eng Co Ltd, Kowloon. In all essentials these craft are sisters of the 14 operated by the Hong Kong Police with the exception of the latter's slow speed waterjet. Names: *Sea Glory* (Customs 6), *Sea Guardian* (Customs 5), *Sea Leader* (Customs 2). Two 32 m launches, *Sea Reliance* (Customs 8) and *Sea Fidelity* (Customs 9) were commissioned in October 2000. With a gross tonnage of 125 tonnes, the craft have a maximum speed of 28 kt. Equipped with a 'sea-rider' they are also fitted with night vision aids and narcotics and explosives scanning devices. There are also two shallow water launches, eight inflatable boats and four high speed pursuit craft.

SEA FIDELITY   3/2003, Bob Fildes / 0568853

LAND-BASED MARITIME AIRCRAFT

Notes: (1) All aircraft belong to the Government Flying Service based at Hong Kong International Airport.
(2) There are also two BAe Jetstream J 41 aircraft used for transport and SAR.

Numbers/Type: 3 Eurocopter AS 332 L2 Super Puma.
Operational speed: 130 kt *(240 km/h)*.
Service ceiling: 15,090 ft *(4,600 m)*.
Range: 672 n miles *(1,245 km)*.
Role/Weapon systems: SAR/coastal surveillance, Medevac and transport. Sensors: radar, Spectrolab searchlight. Weapons: Unarmed. Medical equipment and up to six stretchers. Ordered on 17 September 1999. The aircraft entered service in April 2002.

VERIFIED

Numbers/Type: 4 Eurocopter EC 155B1.
Operational speed: 140 kt *(260 km/h)*.
Service ceiling: 16,760 ft *(5,110 m)*.
Range: 432 n miles *(800 km)*.
Role/Weapon systems: SAR, Medevac, VIP transport; enlarged variant of 'Dauphin'. Sensors: Radar, searchlight, siren, loudspeaker. Weapons: Unarmed. Two stretchers. Ordered on 17 September 1999; aircraft delivered in late 2002.

UPDATED

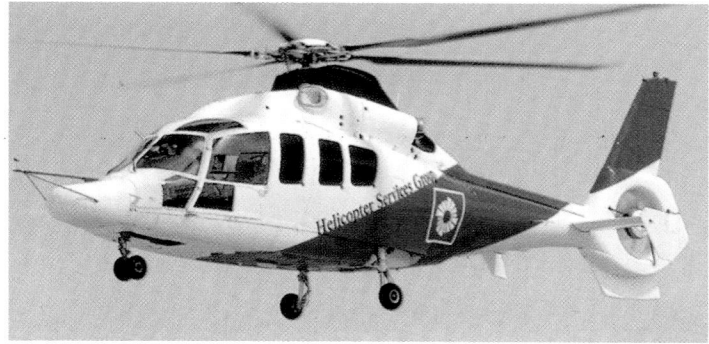

EC 155 (Norwegian markings)   4/2004* / 0015402

Hungary

Country Overview

A landlocked central European country, the Republic of Hungary has an area of 35,919 square miles and is bordered by Slovakia, Ukraine, Romania, Yugoslavia, Croatia, Slovenia and Austria. Budapest is the country's capital and largest city. The country is divided into two general regions by the principal river, the Danube, which flows for 145 n miles north-south through the centre of the country and serves as a major artery of the transport system.

Diplomatic Representation

Defence Attaché in London:
Lieutenant Colonel József Gulyás

Personnel

(a) 2005: 36 officers and men
(b) 6 months' national service

Bases

Budapest.

MINE WARFARE FORCES

3 NESTIN CLASS (RIVER MINESWEEPERS) (MSR)

ÓBUDA AM 22   **DUNAÚJVÁROS** AM 31   **DUNAFOLDVAR** AM 32

Displacement, tons: 72.3 full load
Dimensions, feet (metres): 88.6 × 21.3 × 3.9 *(27 × 6.5 × 1.2)*
Main machinery: 2 Torpedo 12-cyl diesels; 520 hp(m) *(382 kW)*; 2 shafts
Speed, knots: 15. **Range, n miles:** 860 at 11 kt
Complement: 17 (1 officer)
Guns: 6 Hispano 20 mm (1 quad M75 fwd, 2 single M70 aft).
Mines: 24 ground mines.
Radars: Navigation: Decca 101; I-band.

Comment: Built by Brodotehnika, Belgrade in 1979-80. Full magnetic/acoustic and wire sweeping capabilities. Kram minesweeping system employs a towed sweep at 200 m. The ships form the first 'Honved' Ordnance Disposal and Warship Regiment.

VERIFIED

ÓBUDA   *10/1998, Hungary Maritime Wing /* 0064703

Iceland
LANDHELGISGAESLAN

Country Overview

An island republic, the Republic of Iceland lies just south of the Arctic Circle in the North Atlantic Ocean about 162 n miles southeast of Greenland and 432 n miles northwest of Scotland. With an area of 39,769 square miles, the country has a 2,695 n mile coastline. Reykjavík is the capital, largest city and principal port. Territorial waters (12 n miles) are claimed. A 200 n mile Exclusive Economic Zone (EEZ) has also been claimed although the limits are not fully defined by boundary agreements. The Coast Guard Service deals with fishery protection, salvage, rescue, hydrographic research, surveying and lighthouse duties.

Headquarters Appointments

Director of Coast Guard:
Hafsteinn Hafsteinsson

Personnel

2005: 128 officers and men

Colours

Since 1990 vessels have been marked with red, white and blue diagonal stripes on the ships' side and the Coast Guard name (Landhelgisgaeslan).

Bases

Reykjavík

Research Ships

A number of government Research Ships bearing RE pennant numbers operate off Iceland.

Maritime Aircraft

Maritime aircraft include a Fokker Friendship plus AS 332 Super Puma, Dauphin 2 and Ecureuil helicopters

COAST GUARD

Notes: Plans to replace *Odinn* are under consideration.

2 AEGIR CLASS (PSOH)

Name	No	Builders	Commissioned
AEGIR	—	Aalborg Vaerft, Denmark	1968
TYR	—	Dannebrog Vaerft, Denmark	15 Mar 1975

Displacement, tons: 1,200 (1,300 *Tyr*) standard; 1,500 full load
Dimensions, feet (metres): 231.3 × 33 × 14.8 *(70.5 × 10 × 4.6)*
Main machinery: 2 MAN/Burmeister & Wain 8L 40/54 diesels; 13,200 hp(m) *(9.68 MW)* sustained; 2 shafts; cp props
Speed, knots: 19 *(Aegir)*; 20 *(Tyr).* **Range, n miles:** 9,000 at 18 kt
Complement: 19
Guns: 1 Bofors 40 mm/60 Mk 3.
Radars: Surface search: Sperry; E/F-band.
Navigation: Furuno; I-band.
Sonars: Hull-mounted; active search; high frequency *(Tyr).*
Helicopters: 1 Dauphin 2.

Comment: Similar ships but *Tyr* has a slightly improved design and *Aegir* has no sonar. The hangar is between the funnels. In 1994 a large crane was fitted on the starboard side at the forward end of the flight deck. In 1997 the helicopter deck was extended and a radome fitted on the top of the tower.

VERIFIED

1 BALDUR CLASS (AGS)

Name	No	Builders	Commissioned
BALDUR	—	Vélsmiöja Seyöisfjaröar	8 May 1991

Displacement, tons: 54 full load
Dimensions, feet (metres): 65.6 × 17.1 × 5.6 *(20 × 5.2 × 1.7)*
Main machinery: 2 Caterpillar 3406TA diesels; 640 hp *(480 kW)*; 2 shafts
Speed, knots: 12
Complement: 5
Radars: Navigation: Furuno; I-band.

Comment: Built in an Icelandic Shipyard. Used for survey work.

VERIFIED

AEGIR   *9/1997, Iceland Coast Guard /* 0012501

BALDUR   *1993, Iceland Coast Guard /* 0064705

1 ODINN CLASS (PSOH)

Name	No	Builders	Commissioned
ODINN	—	Aalborg Vaerft, Denmark	Jan 1960

Displacement, tons: 1,200 full load
Dimensions, feet (metres): 210 × 33 × 13 *(64 × 10 × 4)*
Main machinery: 2 MAN/Burmeister & Wain diesels; 5,700 hp(m) *(4.19 MW)*; 2 shafts
Speed, knots: 18. **Range, n miles:** 9,500 at 17 kt
Complement: 19
Guns: 1 Bofors 40 mm/60 Mk 3.
Radars: Surface search: Sperry; E/F-band.
Navigation: Furuno; I-band.

Comment: Refitted in Denmark in late 1975 by Aarhus Flydedock AS with a hangar and helicopter deck which was later adapted in 1989 for the operation of RHIB inspection craft; a crane was fitted at the starboard forward end of the flight deck. The original 57 mm gun was replaced in 1990.

VERIFIED   ODINN   *2/2002, L-G Nilsson / 0561502*

India

Country Overview

The Republic of India is a federal democracy which gained independence in 1947. It consists of the entire Indian peninsula and parts of the Asian mainland. With an area of 1,269,219 square miles, it is bordered to the north by Pakistan, Tibet, Nepal, China, and Bhutan and to the east by Burma and Bangladesh, which almost separates north-east India from the rest of the country. The status of Jammu and Kashmir is disputed with Pakistan. It has a 4,104 n mile coastline with the Arabian Sea, the Gulf of Mannar (which separates it from Sri Lanka) and the Bay of Bengal. The capital is New Delhi while the largest city is Mumbai. The principal ports include Mumbai, Calcutta, Madras and Vishakapatnam. Territorial waters (12 n miles) are claimed. A 200 n mile EEZ has been claimed although the limits have only been partly defined by boundary agreements.

Headquarters Appointments

Chief of Naval Staff:
    Admiral Arun Prakash, PVSM, AVSM, VrC, VSM
Vice Chief of Naval Staff:
    Vice Admiral Yashwant Prasad, PVSM, AVSM, VSM
Deputy Chief of Naval Staff:
    Vice Admiral Sureesh Mehta, AVSM
Chief of Personnel:
    Vice Admiral Venkat Bharathan, AVSM, VSM
Chief of Material:
    Vice Admiral Parvesh Jaitly, PVSM, AVSM, VSM

Senior Appointments

Flag Officer Commanding Western Naval Command:
    Vice Admiral Madanjit Singh, PVSM, AVSM
Flag Officer Commanding Eastern Naval Command:
    Vice Admiral Om Prakash Bansal, PVSM, AVSM, VSM
Flag Officer Commanding Southern Naval Command:
    Vice Admiral Satish Chandra Suresh Bangara
Commander-in-Chief, Andaman and Nicobar:
    Lieutenant General Bhupendera Singh Thakur, AVSM
Flag Officer Commanding Western Fleet:
    Rear Admiral Pratap Singh Byce
Flag Officer Commanding Eastern Fleet:
    Rear Admiral Sunil K Damle
Flag Officer, Naval Aviation and Goa Area (at Goa):
    Rear Admiral Shekar Sinha
Flag Officer, Submarines (Vishakapatnam):
    Rear Admiral Krishan Nair Sushil, NM
Flag Officer, Sea Training:
    Rear Admiral I K Saluja

Personnel

(a)   2005: 56,500 (7,500 officers) (including 5,000 Naval Air Arm and 1,000 Marines)
(b)   Voluntary service
(c)   The Marine Commando Force was formed in 1986.

Naval Air Arm

Squadron	Aircraft	Role
300 (Goa)	Sea Harrier FRS. Mk 51	Fighter/Strike
310 (Goa)	Dornier 228	MRMP
312 (Madras)	Tu-142M 'Bear F'	LRMP/ASW
315 (Goa)	Il-38 May	LRMP/ASW
318 (Goa)	Dornier 228	MRMP
321 (Mumbai)	HAL Chetak	Utility/SAR
330 (Kochi)	Sea King Mk 42B	ASW

Naval Air Arm — *continued*

Squadron	Aircraft	Role
333 (ships) (Goa)	Kamov Ka-28 'Helix'	ASW
336 (Kochi)	Sea King Mk 42A/42B	ASW
339 (Mumbai)	Ka-28 'Helix'	ASW/ASVW
550 (Goa)	Dornier 228, PBN Defender, Deepak	Training
551 A (Goa)	Kiran Mk I/II	Training
551 B (Goa)	Sea Harrier T Mk 60	Training
561 (Kochi)	HAL Chetak, Hughes 300	Training

Air Stations

Name	Location	Role
INS *Kunjali*	Mumbai	Helicopters
INS *Garuda*	Willingdon Island, Kochi	Helicopters
INS *Hansa*	Goa	HQ Flag Officer Naval Air Stations, LRMP, Strike/Fighter
	Karwar	Fleet Support
INS *Utkrosh*	Port Blair, Andaman Isles	Maritime Patrol Maritime Patrol Maritime Patrol
INS *Dega*	Vishakapatnam	Fleet support and maritime patrol
INS *Rajali*	Arakonam	LRMP, Helo Training
INS *Ramnad*	Bangalore	LRMP Naval Air Technical School

Prefix to Ships' Names

INS

Bases and Establishments

Note: Bombay is now referred to as Mumbai, Cochin as Kochi and Madras as Chennai.

New Delhi, Integrated HQ of Ministry of Defence (Navy).
Mumbai, C-in-C **Western Command**, barracks and main Dockyard; with one 'Carrier' dock. Submarine base (INS *Vajrabahu*). Supply school (INS *Hamla*). The region includes Mazagon and Goa shipyards.
Vishakapatnam, C-in-C **Eastern Command**, submarine base (INS *Virbahu*), submarine school (INS *Satyavahana*) and major dockyard built with Soviet support and being extended. Naval Air Station (INS *Dega*). Marine Gas Turbine maintenance facility (INS *Eksila*). New entry training (INS *Chilka*). At Thirunelveli is the submarine VLF W/T station completed in September 1986. Facilities at Chennai and Calcutta. The region includes Hindustan and Garden Reach shipyards.
Kochi, C-in-C **Southern Command**, Naval Air Station, and professional schools (INS *Venduruthy*) (all naval training comes under Southern Command). Ship repair yard. Gunnery Training establishment (INS *Dronacharya*).
There are also limited support facilities including a floating dock at Port Blair in the Andaman Islands.
Goa is HQ Flag Officer Naval Aviation.
Karwar (near Goa) has been selected as the site for a new naval base; first phase due for completion in 2005. Alongside berthing for Aircraft Carriers and a naval air station are planned. At Lakshadweep in the Laccadive Islands there is a patrol craft base.
Shipbuilding: Mumbai (submarines, destroyers, frigates,

Bases and Establishments — *continued*

corvettes); Calcutta (frigates, corvettes, LSTs, auxiliaries); Goa (patrol craft, LCU, MCMV facility planned). Vishakapatnam (corvettes, patrol craft).

Marine Commando Force (MCF)

The MCF was formed in 1987. Known as MARCOS, elements are based in the three regional commands. The force is trained in counter-terrorism operations.

Coast Defence

Truck-mounted SS-3-C Styx missiles. At least seven fixed sites. UAV Squadrons: to be based at Porbandar, Port Blair and Kochi.

Strength of the Fleet

Type	Active	Building (Projected)
Attack Submarine (SSN)	—	2
Patrol Submarines	16	(6)
Aircraft Carriers	1	2
Destroyers	8	3
Frigates	13	4 (3)
Corvettes	25	6
Patrol Ships	6	—
Patrol Craft	14	6
LST	2	1 (2)
LCM	11	—
Minesweepers—Ocean	12	—
Minesweepers—Inshore	2	—
Minehunters	—	(8)
Research and Survey Ships	11	—
Training Ships	4	—
Submarine Tender	1	—
Diving Support/Rescue Ship	1	—
Replenishment Tankers	3	(1)
Transport Ships	3	—
Support Tankers	6	—
Water Carriers	2	—
Ocean Tugs	1	—

DELETIONS

Submarines

2002   *Karanj*

Patrol Forces

2002   *Amini, Chamak, Chatak, Makar, Mesh*
2003   *Anjadip*
2004   *Sindhudurg*

Amphibious Forces

2002   L 32, L33

Mine Warfare Forces

2003   *Mulki, Malvan*
2004   *Mangrol*

Auxiliaries

2002   *Astravahini, A 71*

PENNANT LIST

Submarines

S 40	Vela
S 42	Vagli
S 44	Shishumar
S 45	Shankush
S 46	Shalki
S 47	Shankul
S 55	Sindhughosh
S 56	Sindhudhvaj
S 57	Sindhuraj
S 58	Sindhuvir
S 59	Sindhuratna
S 60	Sindhukesari
S 61	Sindhukirti
S 62	Sindhuvijay
S 63	Sindhurakshak
S 64	Sindhushastra

Aircraft Carriers

R 22	Viraat

Destroyers

D 51	Rajput
D 52	Rana
D 53	Ranjit
D 54	Ranvir
D 55	Ranvijay
D 60	Mysore
D 61	Delhi
D 62	Mumbai

Frigates

F 20	Godavari
F 21	Gomati
F 22	Ganga
F 31	Brahmaputra
F 32	Betwa
F 34	Himgiri
F 35	Udaygiri
F 36	Dunagiri
F 37	Beas (bldg)

Frigates — *continued*		K 47	Vinash	M 83	Mahé	A 74	Sagardhwani
		K 48	Vidyut	M 86	Malpe	A 75	Tarangini
F 40	Talwar	K 83	Nashak			A 86	Tir
F 41	Taragiri	K 91	Pralaya	**Amphibious Forces**		J 14	Nirupak
F 42	Vindhyagiri	K 92	Prabal			J 15	Investigator
F 43	Trishul	K 98	Prahar	L 17	Sharabh	J 16	Jamuna
F 44	Tabar			L 18	Cheetah	J 17	Sutlej
F 46	Krishna (training)	**Patrol Forces**		L 19	Mahish	J 18	Sandhayak
				L 20	Magar	J 19	Nirdeshak
Corvettes		P 50	Sukanya	L 21	Guldar	J 21	Darshak
		P 51	Subhadra	L 22	Kumbhir	J 22	Sarvekshak
P 33	Abhay	P 52	Suvarna	L 23	Gharial	J 33	Meen
P 34	Ajay	P 53	Savitri	L 34	—	J 34	Mithun
P 35	Akshay	P 55	Sharada	L 35	—		
P 36	Agray	P 56	Sujata	L 36	—	**Seaward Defence Forces**	
P 44	Kirpan			L 37	—		
P 46	Kuthar	**Mine Warfare Forces**		L 38	—	T 58	—
P 47	Khanjar			L 39	—	T 59	—
P 49	Khukri	M 61	Pondicherry			T 61	Trinkat
P 61	Kora	M 62	Porbandar	**Auxiliaries and Survey Ships**		T 62	Tillan Chang
P 62	Kirch	M 63	Bedi			T 63	Tarasa
P 63	Kulish	M 64	Bhavnagar	—	Nicobar	T 64	Tarmugli
P 64	Karmukh	M 65	Alleppey	—	Andamans	T 65	Bangaram
K 40	Veer	M 66	Ratnagiri	A 15	Nireekshak	T 66	Bitra
K 41	Nirbhik	M 67	Karwar	A 53	Matanga	J 33	Meen
K 42	Nipat	M 68	Cannanore	A 54	Amba	J 34	Mithun
K 43	Nishank	M 69	Cuddalore	A 57	Shakti		
K 44	Nirghat	M 70	Kakinada	A 58	Jyoti		
K 45	Vibhuti	M 71	Kozhikode	A 59	Aditya		
K 46	Vipul	M 72	Konkan	A 72	Torpedo Recovery Vessel		

SUBMARINES

Notes: (1) The Advanced Technology Vessel (ATV) project was initiated in the 1980s. In addition to traditional SSN/SSGN functions, the boat is likely to have a strategic role and, to this end, may also be capable of deploying nuclear-tipped cruise or ballistic missiles in addition to torpedo-tube launched conventional anti-ship and land-attack missiles. Prithvi III, a navalised version of the short-range ballistic missile, was successfully launched at Balasore on 27 October 2004. Currently lead by Vice Admiral Bhasin, the ATV project has facilities in Delhi, Hyderabad, Vishakapatnam and Kalpakkam

(where the PWR reactors are being tested). Companies in support of the project are reported to be Larsen and Toubro at Hazira, Mazagon Dock Ltd and Bharat Electronics. It is believed that the submarine is a development of a Russian design, derived either from the Project 885 Severodvinsk class SSGN or more probably from the Victor/Akula class generation. The nuclear propulsion system is understood to be an Indo-Russian PWR although reports that it may be a Russian supplied VM-5 PWR have also circulated. In parallel, negotiations for the lease-purchase of one or two nuclear-powered submarines from Russia

are believed to be under consideration. Uncertainty remains, however, over key programme dates and given the complexities involved, it is unlikely that the ATV will enter service before 2012. Up to five of the class are expected by 2025.
(2) India has been reported to be interested in the Russian Amur 1650 class SSK. This may also be known as Project 78.
(3) India operates up to 11 Cosmos CE2F/FX100 swimmer delivery vehicles, delivered in 1991.

0 + (6) SCORPENE CLASS (SSK)

Displacement, tons: 1,668 dived
Dimensions, feet (metres): 217.8 × 20.3 × 19 *(66.4 × 6.2 × 5.8)*
Main machinery: Diesel electric; 4 MTU 16V 396 SE84 diesels; 2,992 hp(m) *(2.2 MW)*; 1 Jeumont Schneider motor; 3,808 hp(m) *(2.8 MW)*; 1 shaft
Speed, knots: 20 dived; 12 surfaced
Range, n miles: 550 at 4 kt dived; 6,500 at 8 kt surfaced
Complement: 31 (6 officers)

Torpedoes: 6—21 in *(533 mm)* tubes.
Countermeasures: ESM.
Weapons control: UDS International SUBTICS.
Radars: Navigation: Sagem; I-band.
Sonars: Hull mounted; active/passive search and attack, medium frequency.

Programmes: Project 75. Negotiations for construction of six submarines continued in 2004 and may be concluded in 2005. Options include building one or two boats in France followed by the remainder under licence at Mazagon Dock Ltd, Mumbai or building of all six in India. Delivery is expected to be at the rate of one per year from about 2010. Details are speculative and based on those being built for Chile. AIP is believed to be a requirement.
Structure: Diving depth more than 300 m *(984 ft)*.

UPDATED

SCORPENE (computer graphic)
1998, DCN / 0017689

2 FOXTROT (PROJECT 641) CLASS (SS)

Name	*No*	*Builders*	*Commissioned*
VELA	S 40	Sudomekh, Leningrad	Aug 1973
VAGLI	S 42	Sudomekh, Leningrad	Aug 1974

Displacement, tons: 1,952 surfaced; 2,475 dived
Dimensions, feet (metres): 299.5 × 24.6 × 19.7 *(91.3 × 7.5 × 6)*
Main machinery: Diesel-electric; 3 Type 37-D diesels; 6,000 hp (m) *(4.4 MW)*; 3 motors (1 × 2,700 and 2 × 1,350); 5,400 hp (m) *(3.97 MW)*; 3 shafts; 1 auxiliary motor; 140 hp(m) *(103 kW)*
Speed, knots: 16 surfaced; 15 dived
Range, n miles: 20,000 at 8 kt surfaced; 380 at 2 kt dived
Complement: 75 (8 officers)

Torpedoes: 10—21 in *(533 mm)* (6 fwd, 4 aft) tubes. 22 SET-65E/SAET-60; active/passive homing to 15 km *(8.1 n miles)* at 40 kt; warhead 205 kg.
Mines: 44 in lieu of torpedoes.
Countermeasures: ESM: Stop Light; radar warning.
Radars: Surface search: Snoop Tray; I-band.
Sonars: Herkules/Fenik; bow-mounted; passive search and attack; medium frequency.

Structure: Diving depth 250 m *(820 ft)*, reducing with age.
Operational: Survivors of an original eight of the class. *Vela* completed refit in 2004. Form 8th Submarine Squadron at Vishakapatnam.

UPDATED

FOXTROT
2/2001, Guy Toremans / 0105814

4 SHISHUMAR (209) CLASS (TYPE 1500) (SSK)

Name	No	Builders	Laid down	Launched	Commissioned
SHISHUMAR	S 44	Howaldtswerke, Kiel	1 May 1982	13 Dec 1984	22 Sep 1986
SHANKUSH	S 45	Howaldtswerke, Kiel	1 Sep 1982	11 May 1984	20 Nov 1986
SHALKI	S 46	Mazagon Dock Ltd, Mumbai	5 June 1984	30 Sep 1989	7 Feb 1992
SHANKUL	S 47	Mazagon Dock Ltd, Mumbai	3 Sep 1989	21 Mar 1992	28 May 1994

Displacement, tons: 1,450 standard; 1,660 surfaced; 1,850 dived

Dimensions, feet (metres): 211.2 × 21.3 × 19.7 *(64.4 × 6.5 × 6)*

Main machinery: Diesel-electric; 4 MTU 12V 493 AZ80 GA31L diesels; 2,400 hp(m) *(1.76 MW)* sustained; 4 Siemens alternators; 1.8 MW; 1 Siemens motor; 4,600 hp(m) *(3.38 MW)* sustained; 1 shaft

Speed, knots: 11 surfaced; 22 dived

Range, n miles: 8,000 snorting at 8 kt; 13,000 surfaced at 10 kt

Complement: 40 (8 officers)

Missiles: SSM: *S 48* and *S 49* only.

Torpedoes: 8—21 in *(533 mm)* tubes. 14 AEG SUT Mod 1; wire-guided; active/passive homing to 28 km *(15.3 n miles)* at 23 kt; 12 km *(6.6 n miles)* at 35 kt; warhead 250 kg.

Mines: External 'strap-on' type for 24 mines.

Countermeasures: Decoys: C 303 acoustic decoys. ESM: Argo Phoenix II AR 700 or Kollmorgen Sea Sentry; radar warning.

Weapons control: Singer Librascope Mk 1.

Radars: Surface search: Thomson-CSF Calypso; I-band.

Sonars: Atlas Elektronik CSU 83; active/passive search and attack; medium frequency. TSM 2272 to be fitted. Thomson Sintra DUUX-5; passive ranging and intercept.

Programmes: Howaldtswerke concluded an agreement with the Indian Navy on 11 December 1981. This was in four basic parts: the building in West Germany of two Type 1500 submarines; the supply of 'packages' for the building of two more boats at Mazagon, Mumbai; training of various groups of specialists for the design and construction of the Mazagon pair; logistic services during the trials and early part of the commissions as well as consultation services in Mumbai. In 1984 it was announced that a further two submarines would

SHALKI *2/2001, Sattler/Steele /* 0121373

be built at Mazagon for a total of six but this was overtaken by events in 1987-88 and the agreement with HDW terminated at four. This was reconsidered in 1992 and again in 1997. Government approval was given in mid-1999 for the construction of further submarines under Project 75.

Modernisation: Thomson Sintra Eledone sonars may be fitted in due course. Trials for integration of indigenous Panchendriya ATAS developed by NPOL are in progress in Karanj.

Structure: The Type 1500 has a central bulkhead and an IKL designed integrated escape sphere which can carry the full crew of up to 40 men, has an oxygen supply for 8 hours, and can withstand pressures at least as great as those that can be withstood by the submarine's pressure hull. Diving depth 260 m *(853 ft)*.

Operational: Form 10th Submarine Squadron based at Mumbai. *Shishumar* mid-life refit started in 1999 and had been completed by 2001. Refits of the other three are also due.

UPDATED

SHISHUMAR *2/2001, Guy Toremans /* 0105813

10 SINDHUGHOSH (KILO) (PROJECT 877EM/8773) CLASS (SSK)

Name	No	Builders	Commissioned
SINDHUGHOSH	S 55	Sudomekh, Leningrad	30 Apr 1986
SINDHUDHVAJ	S 56	Sudomekh, Leningrad	12 June 1987
SINDHURAJ	S 57	Sudomekh, Leningrad	20 Oct 1987
SINDHUVIR	S 58	Sudomekh, Leningrad	16 May 1988
SINDHURATNA	S 59	Sudomekh, Leningrad	19 Nov 1988
SINDHUKESARI	S 60	Sudomekh, Leningrad	19 Dec 1988
SINDHUKIRTI	S 61	Sudomekh, Leningrad	9 Dec 1990
SINDHUVIJAY	S 62	Sudomekh, Leningrad	17 Dec 1990
SINDHURAKSHAK	S 63	Sudomekh, St Petersburg	24 Dec 1997
SINDHUSHASTRA	S 64	Sudomekh, St Petersburg	19 July 2000

Displacement, tons: 2,325 surfaced; 3,076 dived

Dimensions, feet (metres): 238.2 × 32.5 × 21.7 *(72.6 × 9.9 × 6.6)*

Main machinery: Diesel-electric; 2 Model 4-2AA-42M diesels; 3,650 hp(m) *(2.68 MW)*; 2 generators; 1 motor; 5,900 hp(m) *(4.34 MW)*; 1 shaft; 2 MT-168 auxiliary motors; 204 hp(m) *(150 kW)*; 1 economic speed motor; 130 hp(m) *(95 kW)*

Speed, knots: 10 surfaced; 17 dived; 9 snorting

Range, n miles: 6,000 at 7 kt snorting; 400 at 3 kt dived

Complement: 52 (13 officers)

Missiles: SLCM: Novator Alfa Klub SS-N-27 (3M-54E1) (S 55, S 57, 59, 60 and 64); active radar homing to 180 km *(97.2 n miles)* at 0.7 Mach (cruise) and 2.5 Mach (attack); warhead 450 kg.
SAM: SA-N-8 portable launcher; IR homing to 3.2 n miles *(6 km)*.

Torpedoes: 6—21 in *(533 mm)* tubes. Combination of Type 53-65; passive wake homing to 19 km *(10.3 n miles)* at 45 kt; warhead 305 kg and TEST 71/96; anti-submarine; active/passive homing to 15 km *(8.1 n miles)* at 40 kt or 20 km *(10.8 n miles)* at 25 kt; warhead 220 kg. Total of 18 weapons. Wire-guided on 2 tubes.

Mines: 24 DM-1 in lieu of torpedoes.

Countermeasures: ESM: Squid Head; radar warning.

Weapons control: Uzel MVU-119EM TFCS.

Radars: Navigation: Snoop Tray; I-band.

Sonars: Shark Teeth/Shark Fin; MGK-400; hull-mounted; active/passive search and attack; medium frequency.
Mouse Roar; MG-519; hull-mounted; active search; high frequency.

Programmes: The Kilo class was launched in the former Soviet Navy in 1979 and although India was the first country to acquire one they have since been transferred to Algeria, Poland, Romania, Iran and China. Because of the slowness of the S 209 programme, the original order in 1983 for six Kilo class expanded to 10 but was then cut back again to eight. Two further orders were confirmed in May 1997. S 63 was a spare Type 877 hull built for the Russian Navy, but never purchased. S 64 is a Type 8773 and is fitted for SLCM. She was launched on 14 October 1999. Plans to manufacture the class under licence in India have probably been abandoned in favour of an AMUR 1650 programme.

Modernisation: An engine change is probable during major refits in Russia which started in 1997. A tube launched SLCM capability is also part of the refit. A German designed main battery with a five year life has replaced Russian batteries in all of the class. Battery cooling has been improved.

Structure: Diving depth, 300 m *(985 ft)*. Reported that from *Sindhuvir* onwards these submarines have an SA-N-8 SAM capability. The launcher is shoulder held and stowed in the fin for use when the submarine is surfaced. Two torpedo tubes can fire wire-guided torpedoes and four tubes have automatic reloading. Anechoic tiles are fitted on casings and fins.

Operational: First four form the 11th Submarine Squadron. Based at Vishakapatnam and the remainder of the 12th Squadron based at Mumbai. *Sindhuvir* completed major refit at Severodvinsk from May 1997 to July 1999. *Sindhuraj* and *Sindhukesari* completed similar refits at Admiralty Yard, St Petersburg from May 1999 to November 2001. *Sindhuratna* completed a two-year refit at Severodvinsk in 2002. *Sindhughosh*, following refit work at Vishakapatnam from 1999, started modernisation at Severodvinsk in September 2002 which is expected to be completed in 2005. She is likely to become the fifth boat to be fitted with SS-N-27 although land-attack cruise missiles are another possibility. *Sindhudhvaj* is to start a two-year refit at Severodvinsk in May 2005. All of these may be based in the Western Fleet in future.

UPDATED

SINDHURATNA *10/2002, Diego Quevedo /* 0529550

SINDHURAJ *11/2001, Michael Nitz /* 0534072

AIRCRAFT CARRIERS

0 + 1 MODIFIED KIEV CLASS (PROJECT 1143.4) (CVGM)

Name	Builders	Laid down	Launched	Commissioned
— (ex-*Admiral Gorshkov*, ex-*Baku*)	Nikolayev South	17 Feb 1978	1 Apr 1982	11 Jan 1987

Displacement, tons: 45,400 full load
Dimensions, feet (metres): 928.5 oa; 818.6 wl × 167.3 oa;
  107.3 wl ×32.8 *(283; 249.5 × 51; 32.7 × 10)*
Main machinery: 8 KWG4 boilers; 4 GTZA 674 turbines;
  200,000 hp(m) *(147 MW)*; 4 shafts
Speed, knots: 28. **Range, n miles:** 13,800 at 18 kt
Complement: 1,200 plus aircrew

Missiles: SAM/Guns: 6 CADS-N-1 (Kortik/Kashtan) (3M 87)
Countermeasures: Decoys: 2 PK2 chaff launchers; 2 towed
  torpedo decoys.
ESM/ECM: Bharat intercept and jammers.
Combat data systems: Lesorub 11434.
Radars: Air search: Plate Steer.
Surface search: 2 Strut Pair.
Navigation: Aircraft control.
Sonars: Horse Jaw (MG 355) ; hull-mounted; active search;
  medium frequency.

Fixed-wing aircraft: Up to 24 MiG-29K or Rafale.
Helicopters: 6 Helix 27/28/31 or Westland Sea King 42A/42B.

Programmes: First offered for sale to India by Russia in 1994. By
  1999 the proposal was to gift the ship as long as India pays for
  the refit. Following a Government to Government agreement
  on 4 October 2000 and protracted negotiations, contract
  signed on 20 January 2004 for a five-year refit at a cost
  estimated to be US$625 million. Most of the work is to be
  undertaken at Severodvinsk with final outfitting to be
  undertaken at Cochin.
Modernisation: New propulsion, power and air conditioning
  systems to be fitted. All the original Russian weapons systems
  removed and to be replaced by six Kashtan SAM/gun

ex-GORSHKOV (artist's impression)

10/2004*, Nevskoye Design Bureau / 1042276

systems. The flight deck is to be converted to a STOBAR
configuration with a 14° ski-jump.
Structure: The ship has a 198 m angled deck with three arrestor
  wires. Flight deck lifts are 19.2 × 10.3 m and 18.5 × 4.7 m,
and can lift 30 tons and 20 tons respectively. The hangar is
  130 × 22.5 m.
Operational: The ship is expected to become operational in
  2008 and to be based at Karwar. ***UPDATED***

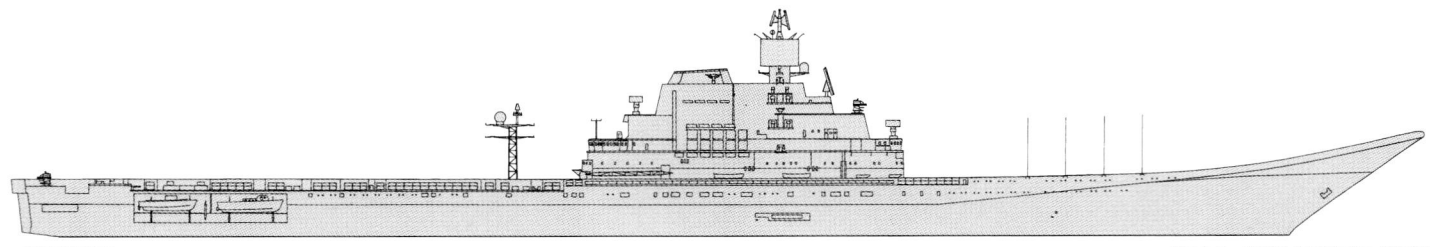

ex-GORSHKOV

(Scale 1 : 1,500), Ian Sturton / 1042091

0 + 1 AIR DEFENCE SHIP CLASS (PROJECT 71) (CVM)

Name	No	Builders	Laid down	Launched	Commissioned
—	—	Cochin Shipyard Ltd	2005	2009	2013

Displacement, tons: 38,000 full load
Dimensions, feet (metres): 826.8 oa; 770.1 wl × 186.3 oa;
  106.6 wl × 25.6
  (252.3; 235 × 56.8; 32.5 × 7.8)
Main machinery: 4 gas turbines; 2 shafts
Speed, knots: 32
Complement: 1,350
Missiles: SAM.
Guns: CIWS.
Radars: Air search; surface search; fire control.

Fixed-wing aircraft: 16 MiG 29s.

Helicopters: 20 Helix and Sea King and ALH.

Programmes: The plan announced in 1989 was to build two
  new aircraft carriers. The indigenously-built Air Defence Ship
  (ADS) is to replace the former *Vikrant* while the Russian carrier
  ex-*Admiral Gorshkov* is to replace *Viraat* in 2008. A number of
  international companies including DCN, IZAR and Fincantieri
  are believed to have been involved in conceptual and design
  work of the ADS and it is understood that the shipbuilder,
  Cochin Shipyard Ltd (CSL), has sub-contracted specialist 'task
  forces' to collaborate in building the ship. Two contracts
  signed in mid-2004 with Fincantieri to finalise the ADS design
and its ancillary propulsion systems and main power plants.
  After construction begins in 2005, Fincantieri is likely to
  provide further assistance during the vessel's construction,
  tests and sea trials.
Structure: All details are still speculative and the diagrams show
  an indicative design including a short take off (with ski jump)
  and arrested recovery (STOBAR) system. The ADS is likely to
  have a similar propulsion system (four General Electric LM
  2500 gas turbines) as the *Cavour* being built for the Italian
  Navy. ***UPDATED***

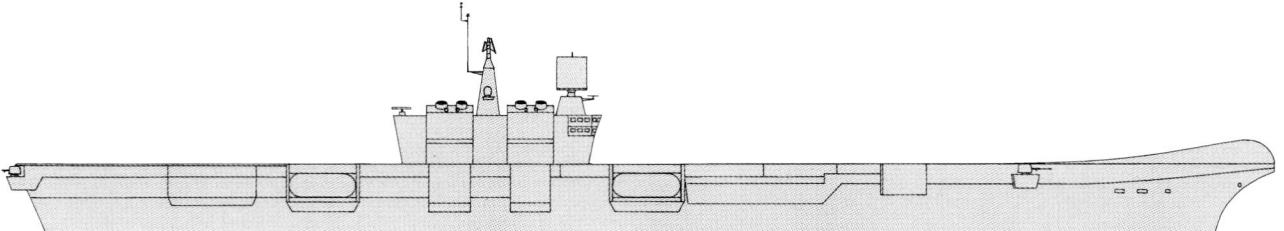

PROJECT 71

(Scale 1 : 1,500), Ian Sturton (via M Mazumdar) / 0529540

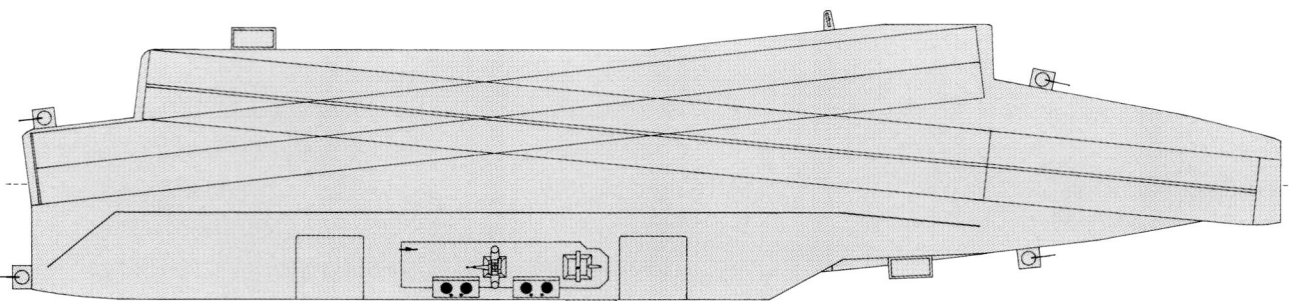

PROJECT 71

(Scale 1 : 1,500), Ian Sturton (via M Mazumdar) / 0529542

1 HERMES CLASS (CVM)

Name	No	Builders	Laid down	Launched	Commissioned
VIRAAT (ex-*Hermes*)	R 22	Vickers Shipbuilding Ltd, Barrow-in-Furness	21 June 1944	16 Feb 1953	18 Nov 1959

Displacement, tons: 23,900 standard; 28,700 full load
Dimensions, feet (metres): 685 wl; 744.3 oa × 90; 160 oa × 28.5 *(208.8; 226.9 × 27.4; 48.8 × 8.7)*
Main machinery: 4 Admiralty boilers; 400 psi *(28 kg/cm²)*; 700°F *(370°C)*; 2 Parsons geared turbines; 76,000 hp *(57 MW)*; 2 shafts
Speed, knots: 28
Complement: 1,350 (143 officers)

Missiles: SAM/Guns: 2 Octuple IAI/Rafael Barak VLS, command line of sight radar or optical guidance to 10 km *(5.5 n miles)* at 2 Mach; warhead 22 kg.
Guns: 2 Bofors 40 mm ❶. 4—30 mm.
2 USSR 30 mm AK 230 Gatlings ❷.
Countermeasures: Decoys: 2 Knebworth Corvus chaff launchers ❸.
ESM: Bharat Ajanta; intercept ❹.
Combat data systems: CAAIS action data automation. SATCOM ❺.
Radars: Air search: Bharat RAWL-2 (PLN 517) ❻; D-band.
Air/surface search: Bharat RAWS (PFN 513) ❼; E/F-band.
Fire control: IAI/Elta EL/M-2221; Ka-band.
Navigation: Bharat Rashmi; I-band.
Tacan: FT 13-S/M.
Sonars: Graseby Type 184M; hull-mounted; active search and attack; 6-9 kHz.

Fixed-wing aircraft: 12 Sea Harriers FRS Mk 51 ❽ (capacity for 30).
Helicopters: 7 Sea King Mk 42B/C ❾ ASW/ASV/Vertrep and Ka-27 Helix. Ka-31 Helix.

Programmes: Purchased in May 1986 from the UK, thence to an extensive refit in Devonport Dockyard. Life extension of at least 10 years. Commissioned in Indian Navy 20 May 1987.
Modernisation: UK refit included new fire-control equipment, navigation radars and deck landing aids. Boilers were converted to take distillate fuel and the ship was given improved NBC protection. New search radar in 1995. In 1996 single 40 mm guns fitted on starboard bow and forward of the island; AK 230 Gatlings on the sponsons previously occupied by Seacat SAM. Further modernisation in 1999-2001 refit, improved indigenous RAWL 02 (Mk II) and Rashmi radars for CCA/navigation, EW equipment and new communications systems. A further refit, completed in December 2004, included installation of Barak CIWS.
Structure: Fitted with 12° ski jump. Reinforced flight deck (0.75 in); 1 to 2 in of armour over magazines and machinery spaces. Four LCVP on after davits. Magazine capacity includes 80 lightweight torpedoes.
Operational: The Sea Harrier complement is normally no more than 12 aircraft leaving room for a greater mix of Sea King and Helix helicopters. Based at Mumbai.

UPDATED

VIRAAT

2/2001, Guy Toremans / 0105815

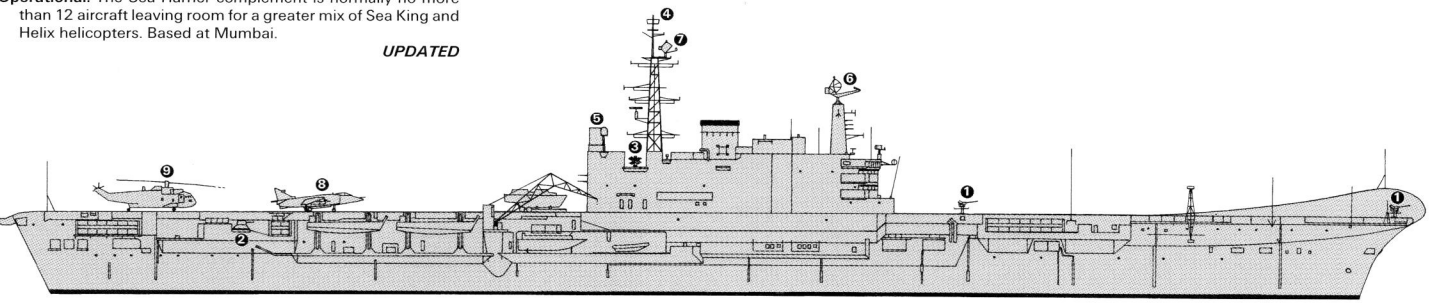

VIRAAT

(Scale 1 : 1,200), Ian Sturton / 0126209

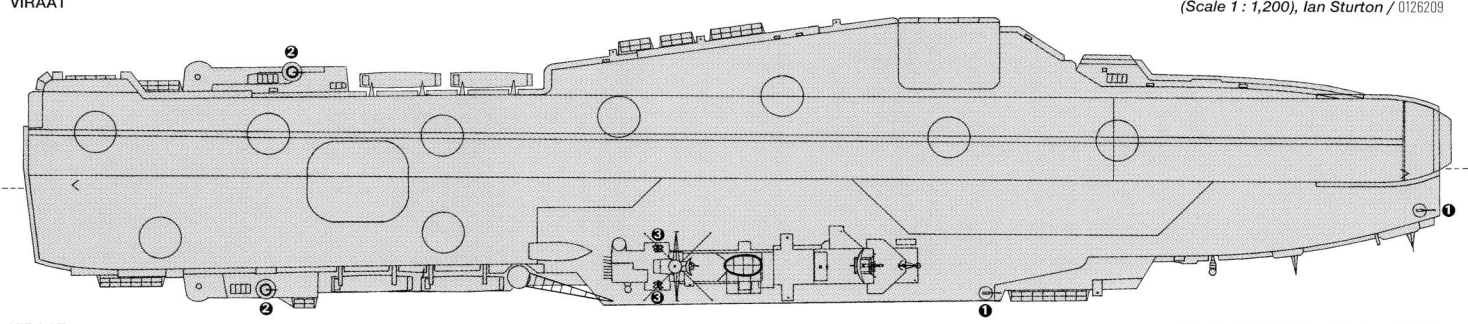

VIRAAT

(Scale 1 : 1,200), Ian Sturton / 0126210

VIRAAT

2/2001, Michael Nitz / 0534073

DESTROYERS

3 DELHI CLASS (PROJECT 15) (DDGHM)

Name	No	Builders	Laid down	Launched	Commissioned
DELHI	D 61	Mazagon Dock Ltd, Mumbai	14 Nov 1987	1 Feb 1991	15 Nov 1997
MYSORE	D 60	Mazagon Dock Ltd, Mumbai	2 Feb 1991	4 June 1993	2 June 1999
MUMBAI	D 62	Mazagon Dock Ltd, Mumbai	14 Dec 1992	20 Mar 1995	22 Jan 2001

Displacement, tons: 6,700 full load
Dimensions, feet (metres): 534.8 × 55.8 × 21.3
 (163 × 17 × 6.5)
Main machinery: 4 Zorya/Mashprockt DT-59 gas turbines;
 82,820 hp(m) *(61.7 MW)*; 2 shafts; cp props
Speed, knots: 32. **Range, n miles:** 4,500 at 18 kt
Complement: 360 (40 officers)

Missiles: SSM: 16 Zvezda SS-N-25 (4 quad) (KH 35E Uran) ❶
 active radar homing to 130 km *(70.2 n miles)* at 0.9 Mach;
 warhead 145 kg; sea skimmer.
SAM: 2 SA-N-7 Gadfly (Kashmir/Uragan) ❷ command, semi-
 active radar and IR homing to 25 km *(13.5 n miles)* at 3 Mach;
 warhead 70 kg. Total of 48 missiles.
 2 Octuple IAI/Rafael Barak VLS (D 61) ❸; command line of
 sight radar or optical guidance to 10 km *(5.5 n miles)* at 2
 Mach; warhead 22 kg.
Guns: 1 USSR 3.9 in *(100 mm)*/59 ❹. AK 100; 60 rds/min to
 15 km *(8.2 n miles)*; weight of shell 16 kg.
 4 (2 in D 61) USSR 30 mm/65 ❺ AK 630; 6 barrels per
 mounting; 3,000 rds/min combined to 2 km.
Torpedoes: 5 PTA 21 in *(533 mm)* (quin) tubes ❻. Combination
 of SET 65E; anti-submarine; active/passive homing to 15 km
 (8.1 n miles) at 40 kt; warhead 205 kg and Type 53-65;
 passive wake homing to 19 km *(10.3 n miles)* at 45 kt;
 warhead 305 kg.
A/S mortars: 2 RBU 6000 ❼; 12 tubed trainable; range 6,000 m;
 warhead 31 kg.
Depth charges: 2 rails.
Countermeasures: Decoys: 2 PK2 chaff launchers ❽. Towed
 torpedo decoy.
ESM: Bharat Ajanta Mk 2; intercept.
ECM: Elettronica TQN-2; jammer.

Combat data systems: Bharat IPN Shikari (IPN 10).
Radars: Air search: Bharat/Signaal RAWL (LW08) ❾; D-band.
Air/Surface search: Half Plate ❿; E-band.
Fire control: 6 Front Dome ⓫; G-band (for SAM); Kite Screech
 ⓬; I/J-band (for 100 mm); 2 Bass Tilt (MR-123) (D 60 and D
 62) ⓭; I/J-band (for AK 630); Plank Shave (Granit Garpun B)
 ⓮ (for SSM); I/J-band.
Navigation: 3 Nyada MR-212/201; I-band.
Sonars: Bharat HUMVAD; hull-mounted; active search; medium
 frequency.
 Bharat HUMSA; hull-mounted; medium frequency (D 62)
 Indal/Garden Reach Model 15-750 VDS.
 Thales ATAS; active towed array (D 62)

Helicopters: 2 Westland Sea Kings Mk 42B ⓯ or 2 Hindustan
 Aeronautics ALH.

Programmes: Built with Russian Severnoye Design Bureau
 assistance. *Delhi* ordered in March 1986. Programme is called

Project 15. Much delay was caused by the breakdown in the
central control of Russian export equipment.
Structure: The design is described as a 'stretched *Rajput*' with
 some *Godavari* features. A combination of Russian and Indian
 weapon systems fitted. Missile blast deflectors indicate an
 original intention to fit SS-N-22 Sunburn. Samahé helo
 handling system. Forward funnel offset to port and after
 funnel to starboard.
Modernisation: Barak has replaced the forward AK-630
 mountings in D 61. The two Bass Tilt radars have also been
 replaced. D 60 and D 62 are to be similarly refitted. SS-N-25
 may be replaced by Brahmos.
Operational: After nine years being built, the first of class started
 sea trials in February 1997, second of class in September
 1998. Based at Mumbai. Have Flag facilities.

UPDATED

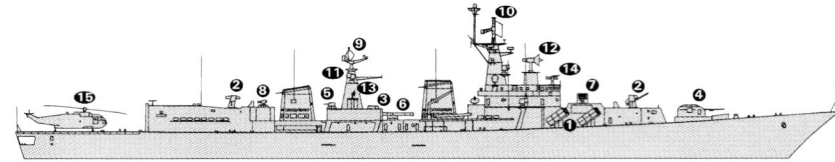

DELHI   *(Scale 1 : 1,500), Ian Sturton /* 0572398

DELHI (before being fitted with Barak)   *2/2001, Sattler/Steele /* 0121369

MYSORE   *10/2002, Chris Sattler /* 0534075

MUMBAI

2/2001, *Michael Nitz* / 0534074

MYSORE

2/2001, *Sattler/Steele* / 0121371

5 RAJPUT (KASHIN II) CLASS (PROJECT 61ME) (DDGHM)

Name	No	Builders	Laid down	Launched	Commissioned
RAJPUT (ex-*Nadezhniy*)	D 51	Nikolayev North (61 Kommuna)	11 Sep 1976	17 Sep 1977	4 May 1980
RANA (ex-*Gubitelyniyy*)	D 52	Nikolayev North (61 Kommuna)	29 Nov 1976	27 Sep 1978	19 Feb 1982
RANJIT (ex-*Lovkiyy*)	D 53	Nikolayev North (61 Kommuna)	29 June 1977	16 June 1979	24 Nov 1983
RANVIR (ex-*Tverdyy*)	D 54	Nikolayev North (61 Kommuna)	24 Oct 1981	12 Mar 1983	21 Apr 1986
RANVIJAY (ex-*Tolkoviyy*)	D 55	Nikolayev North (61 Kommuna)	19 Mar 1982	1 Feb 1986	21 Dec 1987

Displacement, tons: 3,950 standard; 4,974 full load
Dimensions, feet (metres): 480.5 × 51.8 × 15.7
(146.5 × 15.8 × 4.8)
Main machinery: COGAG; 4 Ukraine gas turbines; 72,000 hp(m)
(53 MW); 2 shafts
Speed, knots: 35. **Range, n miles:** 4,500 at 18 kt; 2,600 at 30 kt
Complement: 320 (35 officers)

Missiles: SSM: 4 SS-N-2D Mod 2 Styx ❶; IR homing to 83 km
(45 n miles) at 0.9 Mach; warhead 513 kg; sea-skimmer.
6 Brahmos PJ-10 (to be confirmed); active/passive radar
terminal homing to 290 km *(157 n miles)* at 2.6 Mach;
warhead 200 kg.
SAM: 2 SA-N-1 Goa twin launchers ❷; command guidance to
31.5 km *(17 n miles)* at 2 Mach; height 91-22,860 m *(300-
75,000 ft)*; warhead 60 kg; 44 missiles. Some SSM capability.
Guns: 2—3 in *(76 mm)*/60 (twin, fwd) ❸; 90 rds/min to 15 km
(8 n miles); weight of shell 6.8 kg.
8—30 mm/65 (4 twin) AK 230 *(Rajput, Rana* and *Ranjit)* ❹;
500 rds/min to 5 km *(2.7 n miles)*; weight of shell 0.54 kg.
4—30 mm/65 ADG 630 (6 barrels per mounting) *(Ranvir* and
Ranvijay); 3,000 rds/min combined to 2 km.
Torpedoes: 5—21 in *(533 mm)* (quin) tubes ❺. Combination of
SET-65E; anti-submarine; active/passive homing to 15 km
(8.1 n miles) at 40 kt; warhead 205 kg and Type 53-65;
passive wake homing to 19 km *(10.3 n miles)* at 45 kt;
warhead 305 kg.
A/S mortars: 2 RBU 6000 12-tubed trainable ❻; range 6,000 m;
warhead 31 kg.
Countermeasures: 4 PK 16 chaff launchers for radar decoy and
distraction.
ESM: 2 Bell Squat/Bell Shroud (last pair); Bell Clout/Bell Slam/
Bell Tap (first three); intercept.
ECM: 2 Top Hat; jammers.
Radars: Air search: Big Net A ❼; C-band; range 183 km *(100 n
miles)* for 2 m² target.
Air/surface search: Head Net C ❽; 3D; E-band.
Navigation: 2 Don Kay; I-band.
Fire control: 2 Peel Group ❾; H/I-band; range 73 km *(40 n miles)*
for 2 m² target.
Owl Screech ❿; G-band.
2 Drum Tilt ⓫ or 2 Bass Tilt *(Ranvir* and *Ranvijay)*; H/I-band.
IFF: 2 High Pole B.
Sonars: Vycheda MG 311; hull-mounted; active search and
attack; medium frequency.
Mare Tail VDS; active search; medium frequency.

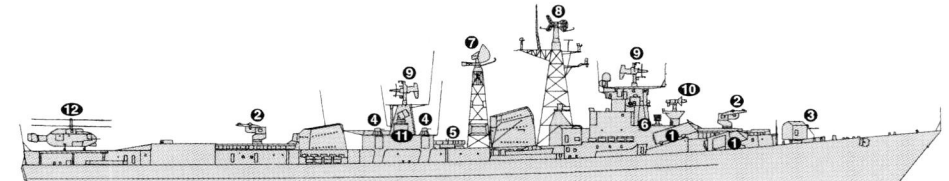

RANA
(Scale 1 : 1,200), Ian Sturton

RANJIT
10/2004*, Toshiyuki Hanta / 1042266

Helicopters: 1 Ka-28 Helix ⓬.

Programmes: First batch of three ordered in the mid-1970s.
Ranvir was the first of the second batch ordered on
20 December 1982.
Modernisation: New EW equipment installed on all ships refitted
since 1993. SS-N-2D may be replaced by Brahmos (PJ-10)
cruise missile from 2005. Third test-launch of Brahmos
conducted from D 51 on 12 February 2003. Brahmos
launchers have replaced forward SS-N-2D launchers. Barak
may replace SA-N-1.

Structure: All built as new construction for India at Nikolayev
with considerable modifications to the Kashin design.
Helicopter hangar, which is reached by a lift from the flight
deck, replaces after 76 mm twin mount and the SS-N-2D
launchers are sited forward of the bridge. *Ranvir* and *Ranvijay*
differ from previous ships in class by being fitted with
ADGM-630 30 mm guns and two Bass Tilt fire-control radars.
It is possible that an Italian combat data system compatible
with Selenia IPN-10 is installed. Inmarsat fitted.
Operational: First three based at Vishakapatnam, last pair at
Mumbai. **UPDATED**

RANVIJAY
10/2004*, Ships of the World / 1042265

RANA
2/2001, Michael Nitz / 0534076

0 + 3 PROJECT 15A CLASS (DDGHM)

Name	No	Builders	Laid down	Launched	Commissioned
—	—	Mazagon Dock Ltd, Mumbai	26 Sep 2003	2005	2007
—	—	Mazagon Dock Ltd, Mumbai	2005	2007	2009
—	—	Mazagon Dock Ltd, Mumbai	2006	2008	2010

Displacement, tons: 7,000 full load
Dimensions, feet (metres): 534.8 × 55.8 × 21.3
(163 × 17 × 6.5)
Main machinery: 4 Zorya/Mashprockt DT-59 gas turbines;
82,820 hp(m) *(61.7 MW)*; 2 shafts; cp props
Speed, knots: 32. **Range, n miles:** 4,500 at 18 kt
Complement: 360 (40 officers)

Missiles: SSM: 16 Brahmos PJ-10 (2 octuple VLS) ❶; active/
passive radar homing to 290 km *(157 n miles)* at 2.6 Mach;
warhead 200 kg; sea skimmer in terminal phase.
SAM: SA-N-12 Grizzly (Shtil-1) (9M317ME) ❷; command/semi-
active radar and IR homing to 35 km *(18.9 n miles)* at 3 Mach;
warhead 70 kg. 2 × 24-cell VLS launchers (forward and aft);
total of 48 missiles.
SAM/Guns: 2 CADS-N-1 (Kashtan) (may replace AK 630); each
has twin 30 mm Gatling combined with 8 SA-N-11 (Grisson)
and Hot Flash/Hot Spot radar/optronic director. Laser beam
guidance for missiles to 8 km *(4.4 n miles)*; warhead 9 kg;
9,000 rds/min (combined) to 1.5 km for guns.
Guns: 1—3.9 in *(100 mm)*/59 A 190E ❸; 60 rds/min to 21.5 km
(11.6 n miles); weight of shell 16 kg.
2—30 mm/AK 630 ❹; 6 barrels per mounting; 3,000 rds/min
combined to 2 km.
Torpedoes: 5 PTA 21 in *(533 mm)* (quin) tubes ❺. Combination
of SET 65E; anti-submarine; active/passive homing to 15 km
(8.1 n miles) at 40 kt; warhead 205 kg and Type 53-65;

passive wake homing to 19 km *(10.3 n miles)* at 45 kt;
warhead 305 kg.
A/S mortars: 2 RBU 6000 ❻; 12 tubed trainable; range 6,000 m;
warhead 31 kg.
Countermeasures: Decoys: 2 PK2 chaff launchers ❼. Towed
torpedo decoy.
ESM: Bharat Ajanta Mk 2; intercept.
ECM: Elettronica TQN-2; jammer.
Combat data systems: To be announced.
Radars: Air search: Bharat RAWL (LW08) ❽; D-band.
Air/surface search: Top Plate (Fregat M2EM) ❾; 3D; E/F-band.
Fire control: 6 Front Dome (MR 90); F-band (for SAM) ❿; Kite
Screech; I/J-band (for 100 mm); Plank Shave (Granit Garpun
B) (for SSM) ⓫; I/J-band.
Navigation: Kelvin Hughes Nucleus 6000; E/F-band. 2 Nyada
MR-212/201; I-band.

Sonars: Bharat HUMSA; hull-mounted; medium frequency.
Towed array (to be confirmed).

Helicopters: 2 Westland Sea Kings Mk 42B ⓬ or 2 Hindustan
Aeronautics ALH.

Programmes: The first of three modified Delhi class was laid
down in 2003.
Structure: Designed by the Indian Naval Design Bureau, the
design appears to be a development of the Delhi class
incorporating some features of both the Talwar and Project 17
frigates.

NEW ENTRY

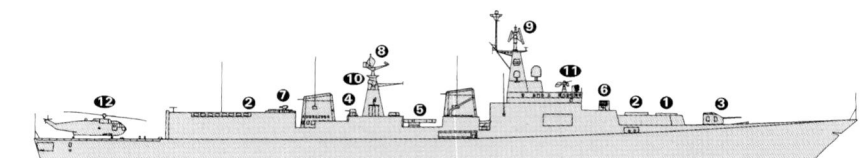

PROJECT 15A

(Scale 1 : 1,500), Ian Sturton / 1042090

FRIGATES

3 GODAVARI CLASS (PROJECT 16) (FFGHM)

Name	No	Builders	Laid down	Launched	Commissioned
GODAVARI	F 20	Mazagon Dock Ltd, Mumbai	2 June 1978	15 May 1980	10 Dec 1983
GOMATI	F 21	Mazagon Dock Ltd, Mumbai	1981	19 Mar 1984	16 Apr 1988
GANGA	F 22	Mazagon Dock Ltd, Mumbai	1980	21 Oct 1981	30 Dec 1985

Displacement, tons: 4,209 full load
Dimensions, feet (metres): 414.9 × 47.6 × 14.8 (29.5 sonar)
(126.5 × 14.5 × 4.5 (9))
Main machinery: 2 Babcock & Wilcox boilers; 550 psi *(38.7 kg/
cm²)*; 850°F *(450°C)*; 2 turbines; 30,000 hp *(22.4 MW)*;
2 shafts
Speed, knots: 28. **Range, n miles:** 4,500 at 12 kt
Complement: 313 (40 officers including 13 aircrew)

Missiles: SSM: 4 SS-N-2D Styx ❶; active radar (Mod 1) or IR (Mod
2) homing to 83 km *(45 n miles)* at 0.9 Mach; warhead
513 kg; sea-skimmer at end of run. Indian designation.
SAM: SA-N-4 Gecko twin launcher (F 20, F 21) ❷; semi-active
radar homing to 15 km *(8 n miles)* at 2.5 Mach; height 9.1-
3,048 m *(130-10,000 ft)*; warhead 50 kg; limited surface-to-
surface capability; 20 missiles.
1 Octuple IAI/Rafael Barak VLS (F 22); command line of sight
radar or optical guidance to 10 km *(5.5 n miles)* at 2 Mach;
warhead 22 kg.
Guns: 2—57 mm/70 (twin) ❸; 120 rds/min to 8 km *(4.4 n
miles)*; weight of shell 2.8 kg.
8—30 mm/65 (4 twin) AK 230 ❹; 500 rds/min to 5 km *(2.7 n
miles)*; weight of shell 0.54 kg.
2—7.63 mm MGs.
Torpedoes: 6—324 mm ILAS 3 (2 triple) tubes ❺. Whitehead
A244S; anti-submarine; active/passive homing to 7 km *(3.8 n
miles)* at 33 kt; warhead 34 kg (shaped charge). *Godavari* has
tube modifications for the Indian NST 58 version of A244S.
Countermeasures: Decoys: 2 chaff launchers (Super Barricade).
Graseby G738 towed torpedo decoy.
ESM/ECM: Selenia INS-3 (Bharat Ajanta and Elettronica TQN-2);
intercept and jammer.
Combat data systems: Selenia IPN-10 action data automation.
Inmarsat communications (JRC) ❻.
Weapons control: MR 301 MFCS. MR 103 GFCS.
Radars: Air search: Signaal LW08 ❼; D-band.
Air/surface search: Head Net C ❽; 3D; E/F-band.
Navigation/helo control: 2 Signaal ZW06 ❾; or Don Kay; I-band.
Fire control: 2 Drum Tilt ❿; H/I-band (for 30 mm).
Pop Group ⓫; F/H/I-band (for SA-N-4).
Muff Cob ⓬; G/H-band (for 57 mm).
Sonars: Bharat APSOH; hull-mounted; active panoramic search
and attack; medium frequency.
Fathoms Oceanic VDS.

Thomson Sintra DSBV 62 (in *Ganga*); passive towed array; very
low frequency.
Type 162M; bottom classification; high frequency.

Helicopters: 2 Sea King or 1 Sea King and 1 Chetak ⓭.

Modernisation: Barak launchers have replaced SA-N-4 in F 22
and are to be fitted to F 20 and F 21.
Structure: A further modification of the original Leander design
with an indigenous content of 72 per cent and a larger hull.

Poor welding is noticeable in *Godavari*. *Gomati* is the first
Indian ship to have digital electronics in her combat data
system.
Operational: French Samahé helicopter handling equipment is
fitted. Usually only one helo is carried with more than one
crew. These ships have a unique mixture of Russian, Western
and Indian weapon systems which has inevitably led to some
equipment compatibility problems.

UPDATED

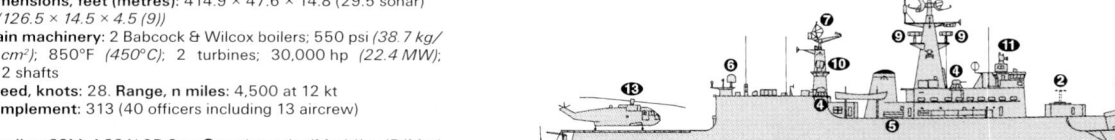

GODAVARI

(Scale 1 : 1,200), Ian Sturton / 0052330

GOMATI

2/2001, Michael Nitz / 0534079

GANGA

2/2001, Michael Nitz / 0534077

3 + (3) TALWAR (PROJECT 1135.6) CLASS (FFGHM)

Name	No	Builders	Laid Down	Launched	Commissioned
TALWAR	F 40	Northern Shipyard, St Petersburg	10 Mar 1999	12 May 2000	18 June 2003
TRISHUL	F 43	Northern Shipyard, St Petersburg	24 Sep 1999	24 Nov 2000	25 June 2003
TABAR	F 44	Northern Shipyard, St Petersburg	26 May 2000	25 May 2001	19 Apr 2004

Displacement, tons: 3,620 standard; 4,035 full load
Dimensions, feet (metres): 409.6 × 49.9 × 15.1
 (124.8 × 15.2 × 4.6)
Main machinery: COGAG; 2 Zorya DN-59 gas turbines; 43,448 hp/m *(34.2 MW)*; 2 Zorya UGT 6000 gas turbines; 16,628 hp(m) *(12.4 MW)*; 2 shafts; fixed propellers
Speed, knots: 32. **Range, n miles:** 4,850 at 14 kt; 1,600 at 30 kt

Complement: 180 (18 officers)

Missiles: SSM: 8 SS-N-27 Novator Alfa Klub (3M-54E1) ❶ active radar homing to 180 km *(97.2 n miles)* at 0.7 Mach (cruise) and 2.5 Mach (attack); warhead 450 kg. VLS silo.
SAM: SA-N-7 Gadfly (Kashmir/Uragan) single launcher ❷ command, semi-active radar and IR homing to 25 km *(13.5 n miles)* at 3 Mach; warhead 70 kg. 24 9M 317 missiles.

SAM/Guns: 2 CADS-N-1 (Kashtan) ❸ each has twin 30 mm Gatling combined with 8 SA-N-11 (Grisson) and Hot Flash/Hot Spot radar/optronic director. Laser beam guidance for missiles to 8 km *(4.4 n miles)* warhead 9 kg; 9,000 rds/min (combined) to 1.5 km for guns.
Guns: 1—3.9 in *(100 mm)*/59 A 190E ❹; 60 rds/min to 21.5 km *(11.6 n miles)*; weight of shell 16 kg.
Torpedoes: 4 PTA-53 21 in *(533 mm)* (2 twin) fired launchers ❺.
A/S mortars: 1 RBU 6000 12-barrelled launcher ❻ range 6 km; warhead 31 kg.
Countermeasures: Decoys: 2 PK 2 chaff launchers (to be fitted).
ESM: Bharat Ajanta; intercept.
ECM: ASOR 11356; jammer.
Combat data systems: Trebovaniye-M.
Radars: Air search: Top Plate (Fregat-M2EM) ❼ 3D; E/F-band.
Air/surface search: Cross Dome (Positiv-E) ❽; E/F-band.
Fire control: 4 Front Dome (MR-90) ❾; F-band (for SA-N-7). Plank Shave (Garpun-B) ❿; I/J-band (for SSM); Ratep 5P-10E Puma ⓫; I-band (for 100 mm gun).
Navigation: Kelvin Hughes Nucleus 6000 ⓬; E/F-band. 2 Nyada MR 212/201 (Palm Frond) ⓭; I-band.
Sonars: HUMSA; hull mounted; active/passive medium frequency.
VDS (may be fitted in future).

Helicopters: 1 Ka-28/Ka-31 Helix ⓮ or ALH.

Programmes: Contract placed in 1997 and confirmed 21 July 1998 for three modified Krivak IIIs. Mutual interference difficulties reportedly delayed entry into service of first of class by one year. There is an option for a further three ships.
Structure: The first surface unit to be fitted with SS-N-27 missile. This is likely to be replaced by the Brahmos missile.
 UPDATED

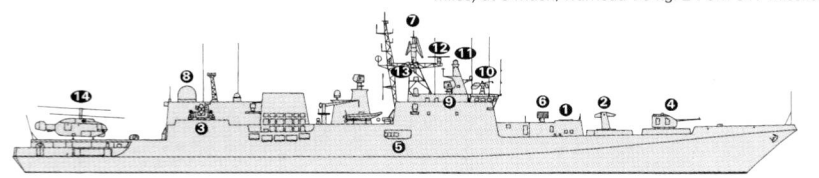

TALWAR
(Scale 1 : 1,200), Ian Sturton / 0569246

TALWAR
7/2003, H M Steele / 0554931

TRISHUL
7/2003, Michael Nitz / 0569191

TABAR
5/2004*, Harald Carstens / 1042274

0 + 3 SHIVALIK (PROJECT 17) CLASS (FFGHM)

Name	No	Builders	Laid Down	Launched	Commissioned
SHIVALIK	—	Mazagon Dock Ltd, Mumbai	11 July 2001	18 Apr 2003	Dec 2005
SATPURA	—	Mazagon Dock Ltd, Mumbai	Oct 2002	4 June 2004	2006
SAHYADI	—	Mazagon Dock Ltd, Mumbai	Sep 2003	2005	2007

Displacement, tons: 4,600 standard; 4,900 full load
Dimensions, feet (metres): 469.3 × 55.5 × 17.4
(143.0 × 16.9 × 5.3)
Main machinery: CODOG; 2 GE LM 2,500 gas turbines;
44,000 hp *(32.8 MW)*; 2 SEMT-Pielstick PA6 STC diesels;
15,200 hp *(11.3 MW)*; 2 cp propellers.
Speed, knots: 32. **Range, n miles:** 4,500 at 18 kt; 1,600 at 30 kt
Complement: 250 (25 officers)

Missiles: SSM: 8 SS-N-27 Novator Alfa Klub (3M-54E1) ❶ active
radar homing to 180 km *(97.2 n miles)* at 0.7 Mach (cruise)
and 2.5 Mach (attack); warhead 450 kg. VLS silo.

SAM: SA-N-7 Gadfly (Kashmir/Uragan) single launcher 6 ❷
command, semi-active radar and IR homing to 25 km *(13.5 n
miles)* at 3 Mach; warhead 70 kg. 24 9M38M1 missiles.
SAM/Guns: 2 CADS-N-1 (Kashtan) ❸ each has twin 30 mm
Gatling combined with 8 SA-N-11 (Grisson) and Hot Flash/Hot
Spot radar/optronic director. Laser beam guidance for
missiles to 8 km *(4.4 n miles)* warhead 9 kg; 9,000 rds/min
(combined) to 1.5 km for guns.
Guns: 1 OTO Melara 3 in *(76 mm)*/62 Super Rapid ❹; 120 rds/
min to 16 km *(8.7 n miles)*; weight of shell 6 kg.
Torpedoes: 6—324 mm ILAS 3 (2 triple) ❺.
A/S mortars: 2 RBU 6000 12-barrelled launcher ❻ range 6 km;
warhead 31 kg.

Countermeasures: Decoys: 2 PK 2 chaff launchers.
ESM: Bharat Ajanta; intercept.
ECM: ASOR 11356; jammer.
Combat data systems: Bharat (IPN-10).
Radars: Air search: Bharat RAWL-02 ❼; E/F-band.
Air/surface search: Top Plate (Fregat-M2EM) ❽ 3D; D/E-band.
Fire control: 2 BEL Shikari (based on Contraves Seaguard) ❾ (for
76 mm); I/K-bands.
1 Bharat Aparna (modified Plank Shave/Garpun B) ❿ (for
SSMs); I/J-bands.
4 Front Dome (MR 90) ⓫ (for SA-N-7); F-band.
Navigation: 1 BEL Rashmi; I-band.
Sonars: Bharat HUMSA; hull-mounted; active search and attack;
medium frequency.
VDS; active search; medium frequency.

Helicopters: 1 Sea King Mk 42B ⓬.

Programmes: Three Project 17 ships approved in June 1999 and
construction of the first of class began in 2001. The second
and third units are planned to be delivered at one year
intervals, an unprecedented rate of construction. A class of 12
ships is envisaged.
Structure: An enlarged and modified version of the Talwar class,
the aft section resembles the Delhi class. Signature reduction
(IR and RCS) features are believed to be incorporated. Details
are speculative. **UPDATED**

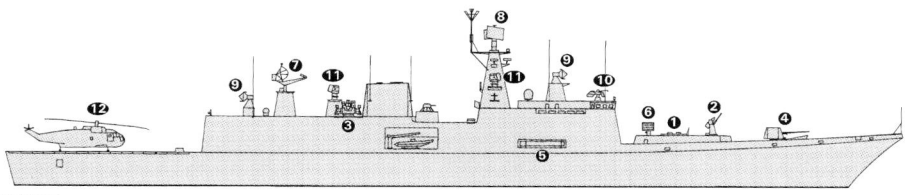

SHIVALIK   *(Scale 1 : 1,200), Ian Sturton /* 0569247

2 + 1 BRAHMAPUTRA CLASS (PROJECT 16A) (FFGHM)

Name	No	Builders	Laid down	Launched	Commissioned
BRAHMAPUTRA	F 31	Garden Reach SY, Calcutta	1989	29 Jan 1994	14 Apr 2000
BETWA	F 32	Garden Reach SY, Calcutta	22 Aug 1994	26 Feb 1998	7 July 2004
BEAS	F 37	Garden Reach SY, Calcutta	26 Feb 1998	2002	2005

Displacement, tons: 4,450 full load
Dimensions, feet (metres): 414.9 × 47.6 × 14.8 (29.5 sonar)
(126.5 × 14.5 × 4.5 (9))
Main machinery: 2 boilers; 550 psi *(38.7 kg/cm²)*; 850°F
(450°C); 2 Bhopal turbines; 30,000 hp *(22.4 MW)*; 2 shafts
Speed, knots: 27. **Range, n miles:** 4,500 at 12 kt
Complement: 351 (31 officers and 13 aircrew)

Missiles: SSM: 16 SS-N-25 (4 quad) (KH-35E Uran) ❶; active
radar homing to 130 km *(70.2 n miles)* at 0.9 Mach; warhead
145 kg; sea skimmer.
SAM: 1 Octuple IAI/Rafael Barak VLS ❷ (to be fitted); command
line of sight radar or optical guidance to 10 km *(5.5 n miles)* at
2 Mach; warhead 22 kg.
Guns: OTO Melara 76 mm/62 ❸; 85 rds/min to 16 km *(8.6 n
miles)* weight of shell 6 kg.
4—30 mm/65 AK 630 ❹; 6 barrels per mounting; 3,000 rds/
min combined to 2 km.
Torpedoes: 6—324 mm ILAS 3 (2 triple) tubes ❺. Whitehead
A244S; anti-submarine; active/passive homing to 7 km *(3.8 n
miles)* at 33 kt; warhead 34 kg (shaped charge).
Countermeasures: Decoys: 2 chaff launchers (Super Barricade
in due course). Graseby G738 towed torpedo decoy.
ESM/ECM: Selenia INS-3 (Bharat Ajanta and Elettronica TQN-2)
❻; intercept and jammer.
Combat data systems: Selenia IPN-10 action data automation.
Inmarsat communications (JRC).
Weapons control: MR 103 GFCS.
Radars: Air search: Signaal LW08/Bharat RAWL-02 (PLN 517)
❼; D-band.
Air/surface search: Bharat RAWS-03 (using DA 08 antenna)
(PFN 513) ❽; E/F-band.
Navigation/helo control: Decca Bridgemaster; I-band. BEL
Rashmi (PIN 524) (using ZW 06 antenna); I-band.
Fire control: 2 BEL Shikari (based on Contraves Seaguard) ❾ (for
76 mm and Ak 630); I/K-bands.
Bharat Aparna (modified Plank Shave/Garpun B) ❿ (for
SSM); I/J-band.
Selenia RAN (for SAM); I-band.
Sonars: Bharat HUMSA (APSOH); hull-mounted; active
panoramic search and attack; medium frequency.
Thales towed array.

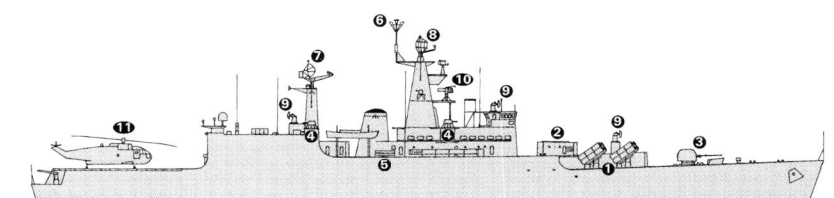

BRAHMAPUTRA   *(Scale 1 : 1,200), Ian Sturton /* 0121334

BRAHMAPUTRA   *2/2001 /* 0126189

Helicopters: 2 Sea King or 1 Sea King and 1 Chetak ⓫.

Programmes: Project 16A. Progress has been very slow.
Structure: The main difference is the replacement of the
Godavari SS-N-2 by SS-N-25. Following the cancellation of the

Trishul SAM programme, Barak is to be fitted in its place. Gun
armament has also improved.
Operational: *Betwa* started sea trials in late 2003 and is to be
commissioned in 2004. *Beas* is also likely to be commissioned
in 2005. **UPDATED**

BRAHMAPUTRA   *2/2001, Sattler/Steele /* 0121366

5 NILGIRI (LEANDER) CLASS (FFH)

Name	No	Builders	Laid down	Launched	Commissioned
HIMGIRI	F 34	Mazagon Dock Ltd, Mumbai	4 Nov 1968	6 May 1970	23 Nov 1974
DUNAGIRI	F 36	Mazagon Dock Ltd, Mumbai	25 Jan 1973	9 Mar 1974	5 May 1977
UDAYGIRI	F 35	Mazagon Dock Ltd, Mumbai	14 Sep 1970	24 Oct 1972	18 Feb 1976
TARAGIRI	F 41	Mazagon Dock Ltd, Mumbai	15 Oct 1975	25 Oct 1976	16 May 1980
VINDHYAGIRI	F 42	Mazagon Dock Ltd, Mumbai	5 Nov 1976	12 Nov 1977	8 July 1981

Displacement, tons: 2,962 full load (F 34-F 36). 3,039 full load (F 41-F 42)

Dimensions, feet (metres): 372 × 36.1 (F 34-F 36). 44.3 (F 41 and F 42) × 18 (113.5 × 11 (F 34-F 36); 13.5 (F 41 and F 42) × 5.5

Main machinery: 2 Babcock & Wilcox boilers; 550 psi *(38.7 kg/cm²)*; 850°F *(450°C)*; 2 turbines; 30,000 hp *(22.4 MW)*; 2 shafts

Speed, knots: 27; 28 (F 41 and F 42)

Range, n miles: 4,500 at 12 kt

Complement: 267 (17 officers). 300 (20 officers) (F 41 and F 42)

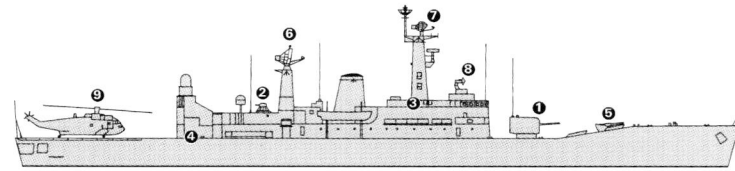

VINDHYAGIRI   *(Scale 1 : 1,200), Ian Sturton /* 1042089

Guns: 2 Vickers 4.5 in *(114 mm)*/45 (twin) Mk 6 ❶; 20 rds/min to 19 km *(10.4 n miles)* anti-surface; 6 km *(3.3 n miles)* anti-aircraft; weight of shell 25 kg.
4—30 mm/65 (2 twin) AK 230 ❷; 500 rds/min to 5 km *(2.7 n miles)*; weight of shell 0.54 kg.
2 Oerlikon 20 mm/70 ❸; 800 rds/min to 2 km.

Torpedoes: 6—324 mm ILAS 3 (2 triple) tubes (F 41 and F 42) ❹. Whitehead A244S or Indian NST 58 version; anti-submarine; active/passive homing to 7 km *(3.8 n miles)* at 33 kt; warhead 34 kg (shaped charge).

A/S mortars: 1 Bofors 375 mm twin-tubed launcher (F 41 and F 42) ❺; range 1,600 m.
1 Limbo Mk 10 triple-tubed launcher (remainder); range 1,000 m; warhead 92 kg.

Countermeasures: Decoys: Graseby 738; towed torpedo decoy.
ESM: Bharat Ajanta; intercept. FH5 Telegon D/F.
ECM: Racal Cutlass; jammer.
Combat data systems: Signaal DS-22.

Radars: Air search: Signaal LW04 ❻; D-band.
Air/Surface search: Signaal DA 05 ❼; E/F-band.
Navigation: Signaal ZW 06; I-band.
Fire control: Signaal M 45 ❽; I/J-band.
IFF: Type 944; 954M.

Sonars: Westinghouse SQS-505; Graseby 750 (APSOH fitted in *Himgiri* as trials ship); hull-mounted; active search and attack; medium frequency. Type 170; active attack; high frequency. Westinghouse VDS (first two only); active; medium frequency. Thomson Diodon VDS in F 41 and F 42.

Helicopters: 1 Chetak or 1 Sea King Mk 42 (in *Taragiri* and *Vindhyagiri*) ❾.

Programmes: The first major warships built in Indian yards to a UK design with a 60 per cent indigenous component. An ex-UK Leander class was acquired in 1995 and is listed under Training Ships.

Modernisation: The VDS arrays are installed inside towed bodies built by Fathom Oceanology Ltd of Canada. The transducer elements in both cases are identical. AK 230 guns have replaced the obsolete Seacat.

Structure: In the first two the hangar was provided with telescopic extension to take the Alouette III helicopter while in the last pair, a much-changed design, the Mk 10 Mortar has been removed as well as VDS and the aircraft space increased to make way for a Sea King helicopter with a telescopic hangar and Canadian Beartrap haul-down gear. In these two an open deck has been left below the flight deck for handling mooring gear and there is a cut-down to the stern.

Operational: *Vindhyagiri* and *Taragiri* have more powerful engines than the remainder.

UPDATED

VINDHYAGIRI   *11/2003* / 1042277

DUNAGIRI
2/2001, Michael Nitz / 0534080

SHIPBORNE AIRCRAFT

Numbers/Type: 15/2 British Aerospace Sea Harrier FRS. Mk 51/Mk 60.
Operational speed: 640 kt *(1,186 km/h)*.
Service ceiling: 51,200 ft *(15,600 m)*.
Range: 800 n miles *(1,480 km)*.
Role/Weapon systems: Fleet air defence, strike and reconnaissance STOVL fighter. Three more acquired from UK in 1999 to make good losses. Of total numbers, only about one third are operational. Sensors: Ferranti Blue Fox air interception radar, limited ECM/RWR (Elta 8420 in due course). Weapons: Air defence; two Magic AAMs (possibly ASRAAM in due course), two 30 mm Aden cannon. Plans for a mid-life upgrade have been abandoned. Avionics are to be improved to extend life of aircraft to 2008.

UPDATED

Numbers/Type: 12/4 MIG 29K Fulcrum/MIG 29UB.
Operational speed: 750 kt *(1,400 km/h)*.
Service ceiling: 57,000 ft *(17,400 m)*.
Range: 1,400 n miles *(2,600 km)*.
Role/Weapon systems: All-weather single-seat fighter with attack capability, optimised for ski-jump take off, is to be main weapon of *Admiral Gorshkov* aircraft carrier. Initial order for 12 aircraft and four trainers to be delivered from 2007 with an option to acquire a further 30 aircraft by 2015. Sensors: Pulse Doppler look down/shoot down radar (Slot Back) able to track 10 targets simultaneously. Weapons: AAM; R77. ASM: CH-31A/P anti-ship and anti-radar. Conventional bombs: KAB-500 Kr. 30 mm cannon.

UPDATED

SEA HARRIER   *1994, Indian Navy /* 0012970

MiG-29   *7/2004 *, Paul Jackson /* 0572477

Numbers/Type: 2/20/5 Westland Sea King Mk 42A/42B/42C.
Operational speed: 112 kt *(208 km/h).*
Service ceiling: 11,500 ft *(3,500 m).*
Range: 664 n miles *(1,230 km).*
Role/Weapon systems: Mk 42A has primary ASW and 42B primary ASV capability; Mk 42C for commando assault/vertrep. Not all aircraft are operational. Sensors: MEL Super Searcher radar, Thomson Sintra H/S-12 dipping sonar (Mk 42A and B), AQS 902B acoustic processor (Mk 42B); Marconi Hermes ESM (Mk 42B); Bendix weather radar (Mk 42C). Weapons: ASW; 2 Whitehead A244S or USSR APR-2 torpedoes; Mk 11 depth bombs, mines (Mk 42B only). ASV; two Sea Eagle (Mk 42B only). Unarmed (Mk 42C).

VERIFIED

SEA KING 42B   *8/2002, Arjun Sarup /* 0569194

Numbers/Type: 12 Kamov Ka-28 Helix A.
Operational speed: 110 kt *(204 km/h).*
Service ceiling: 12,000 ft *(3,660 m).*
Range: 270 n miles *(500 km).*
Role/Weapon systems: ASW helicopter embarked in large escorts. Has replaced Ka-25. Sensors: Splash Drop search radar; VGS-3 dipping sonar, sonobuoys. Weapons: ASW; two Whitehead A244S or USSR APR-2 torpedoes or four depth bombs.

VERIFIED

HELIX A   *8/2002, Arjun Sarup /* 0569192

Numbers/Type: 9 Kamov Ka-31 Helix B.
Operational speed: 119 kt *(220 km/h).*
Service ceiling: 11,480 ft *(3,500 m).*
Range: 325 n miles *(600 km).*
Role/Weapon systems: AEW helicopter. First two delivered late 2002 with remainder in 2003. Radar antenna folds beneath fuselage. Sensors: OKO E-80/M radar.

UPDATED

Ka-31 (AEW)   *6/2004* /* 1042278

Numbers/Type: 23 Aerospatiale (HAL) SA 319B Chetak (Alouette III).
Operational speed: 113 kt *(210 km/h).*
Service ceiling: 10,500 ft *(3,200 m).*
Range: 290 n miles *(540 km).*
Role/Weapon systems: Several helicopter roles performed including embarked ASW and carrier-based SAR, utility and support to commando forces. 15 aircraft are operated by Coast Guard. Weapons: ASW; two Whitehead A244S torpedoes.

VERIFIED

CHETAK   *2/2001, Wingman Aviation /* 0102181

Numbers/Type: 12 HAL Dhruv.
Operational speed: 156 kt *(290 km/h).*
Service ceiling: 9,850 ft *(3,000 m).*
Range: 216 n miles *(400 km).*
Role/Weapon systems: Formerly known as Advanced Light Helicopter (ALH), full production was delayed by thrust and vibration problems which have now been overcome. The naval variant started trials in March 1995 and the first two were delivered in 2003. Sensors: Dipping sonar, ECM. Weapons: ASW; torpedoes, depth charges. ASV; Sea Eagle ASM.

VERIFIED

Dhruv   *2/2001, HAL /* 0095088

LAND-BASED MARITIME AIRCRAFT (FRONT LINE)

Notes: The requirement for future maritime reconnaissance aircraft may be met by joint development (with an international partner) of an indigenously built aircraft and/or by procurement of second-hand aircraft from Russia or from the US. A total of up to 40 aircraft over 20 years are said to be required.

Numbers/Type: 2 Fokker F27 Friendship.
Operational speed: 250 kt *(463 km/h).*
Service ceiling: 29,500 ft *(8,990 m).*
Range: 2,700 n miles *(5,000 km).*
Role/Weapon systems: Operated by coast guard for long-range patrol. Search radar only. Unarmed.

VERIFIED

Numbers/Type: 15 Dornier 228.
Operational speed: 200 kt *(370 km/h).*
Service ceiling: 28,000 ft *(8,535 m).*
Range: 940 n miles *(1,740 km).*
Role/Weapon systems: Coastal surveillance and EEZ protection duties for Navy and Coast Guard. Sensors: MEL Marec or THORN EMI Super Marec search radar with FLIR, cameras and searchlight. Weapons: Unarmed, but may carry anti-ship missiles in due course.

UPDATED

DORNIER 228   *12/2000 /* 0121342

Numbers/Type: 3 Ilyushin Il-38 (May).
Operational speed: 347 kt *(645 km/h).*
Service ceiling: 32,800 ft *(10,000 m).*
Range: 3,887 n miles *(7,200 km).*
Role/Weapon systems: Shore-based long-range ASW reconnaissance into Indian Ocean. Following the loss of two aircraft in a mid-air collision in 2002, two replacement aircraft were donated by Russia. All five aircraft being upgraded to Il-38SD standard with improved avionics, ASM and ASW capabilities. First delivery 2004. Sensors: Leninets Sea Dragon/Novella radar, MAD, sonobuoys, ESM. Weapons: ASW; various torpedoes, mines and depth bombs.

VERIFIED

MAY   *2/2001, Wingman Aviation /* 0121338

Numbers/Type: 3 Pilatus Britten-Norman Maritime Defender.
Operational speed: 150 kt *(280 km/h).*
Service ceiling: 18,900 ft *(5,760 m).*
Range: 1,500 n miles *(2,775 km).*
Role/Weapon systems: Coastal and short-range reconnaissance tasks undertaken in support of Navy (6) and Coast Guard. Six upgraded with turboprop engines 1996-97. Sensors: Search radar, camera. Weapons: Unarmed.

UPDATED

DEFENDER   *2/2001, Wingman Aviation /* 0121339

Numbers/Type: 8 Tupolev Tu-142M (Bear F).
Operational speed: 500 kt *(925 km/h).*
Service ceiling: 45,000 ft *(13,720 m).*
Range: 6,775 n miles *(12,550 km).*
Role/Weapon systems: First entered service in April 1988 for long-range surface surveillance and ASW. Air Force manned. Planned acquisition of further eight in 2001 not confirmed. Sensors: Wet Eye search and attack radars, MAD, cameras. 75 active and passive sonobuoys. Weapons: ASW; 12 torpedoes, depth bombs. ASV; two 23 mm cannon. Avionics, ASM (possibly SS-N-25) and ASW package upgraded in mid-life update from 2001.

VERIFIED

BEAR F   *2/2001 /* 0121345

Numbers/Type: 8 SEPECAT/HAL Jaguar International.
Operational speed: 917 kt *(1,699 km/h)* (max).
Service ceiling: 36,000 ft *(11,000 m).*
Range: 760 n miles *(1,408 km).*
Role/Weapon systems: A maritime strike squadron. Air Force operated. Sensors: Thomson-CSF Agave radar. Weapons: ASV; 2 BAe Sea Eagle; 2 DEFA 30 mm cannon or up to 8—1,000 lb bombs. Can carry 2 Magic AAM overwing.

VERIFIED

JAGUAR   *2/2001, Wingman Aviation /* 0121340

CORVETTES

Note: Project 28 is a programme for six new 90 m corvettes. These are expected to have an ASW role and to be of the order of 2,000 tons. The keel of the first of class is expected to be laid in 2005.

4 ABHAY (PROJECT 1241 PE) (PAUK II) CLASS (FSM)

Name	No	Builders	Commissioned
ABHAY	P 33	Volodarski	10 Mar 1989
AJAY	P 34	Volodarski	24 Jan 1990
AKSHAY	P 35	Volodarski	10 Dec 1990
AGRAY	P 36	Volodarski	30 Jan 1991

Displacement, tons: 485 full load
Dimensions, feet (metres): 191.9 × 33.5 × 11.2 *(58.5 × 10.2 × 3.4)*
Main machinery: 2 Type M 521 diesels; 16,184 hp(m) *(11.9 MW)* sustained; 2 shafts
Speed, knots: 28. **Range, n miles:** 2,400 at 14 kt
Complement: 32 (6 officers)

Missiles: SAM: SA-N-5/8 Grail quad launcher; manual aiming, IR homing to 6 km *(3.2 n miles)* at 1.5 Mach; warhead 1.5 kg.
Guns: 1 USSR 3 in *(76 mm)*/60; 120 rds/min to 15 km *(8 n miles)*; weight of shell 7 kg.
1—30 mm/65 AK 630; 6 barrels; 3,000 rds/min combined to 2 km.
Torpedoes: 4—21 in *(533 mm)* (2 twin) tubes. SET-65E; active/passive homing to 15 km *(8.1 n miles)* at 40 kt; warhead 205 kg.
A/S mortars: 2 RBU 1200 5-tubed fixed; range 1,200 m; warhead 34 kg.
Countermeasures: 2 PK 16 chaff launchers.
Radars: Air/Surface search: Cross Dome; E/F-band.
Navigation: Pechora; I-band.
Fire Control: Bass Tilt; H/I-band.
Sonars: Rat Tail VDS (on transom); attack; high frequency.

Programmes: Modified Pauk II class built in the USSR at Volodarski, Rybinsk for export. Original order in late 1983 but completion of the first delayed by lack of funds and the order for the others was not reinstated until 1987. Names associated with former coastal patrol craft.
Modernisation: There are plans to re-engine all four ships.
Structure: Has a longer superstructure than the Pauk I, larger torpedo tubes and improved electronics.
Operational: Classified as ASW ships. Comprise 23rd Patrol Boat Squadron based at Mumbai.   *UPDATED*

ABHAY   *2/2001, Michael Nitz /* 0534082

4 KORA CLASS (PROJECT 25A) (FSGHM)

Name	No	Builders	Laid down	Launched	Commissioned
KORA	P 61	Garden Reach SY, Calcutta	10 Jan 1990	23 Sep 1992	10 Aug 1998
KIRCH	P 62	Garden Reach SY, Calcutta/Mazagon Dock	31 Jan 1990	28 Sep 1995	22 Jan 2001
KULISH	P 63	Garden Reach SY, Calcutta	4 Oct 1995	18 Aug 1997	20 Aug 2001
KARMUKH	P 64	Garden Reach SY, Calcutta/Mazagon Dock	27 Aug 1997	6 Apr 2000	4 Feb 2004

Displacement, tons: 1,460 full load
Dimensions, feet (metres): 298.9 × 34.4 × 14.8 *(91.1 × 10.5 × 4.5)*
Main machinery: 2 SEMT-Pielstick/Kirloskar 18 PA6 V 280 diesels; 14,400 hp(m) *(10.58 MW)* sustained; 2 shafts; LIPS cp props
Speed, knots: 25. **Range, n miles:** 4,000 at 16 kt
Complement: 134 (14 officers)

Missiles: SSM: 16 Zvezda SS-N-25 (4 quad) (Kh 35E Uran) ❶; active radar homing to 130 km *(70.2 n miles)* at 0.9 Mach; warhead 145 kg; sea skimmer.
SAM: 2 SA-N-5 Grail ❷; manual aiming; IR homing to 6 km *(3.2 n miles)* at 1.5 Mach; altitude to 2,500 m *(8,000 ft)*; warhead 1.5 kg.
Guns: 1 USSR 3 in *(76 mm)*/60 AK 176 (P 61) ❸; 90 rds/min to 12 km *(6.4 n miles)*; weight of shell 7 kg. 1 Otobreda 76 mm/62 (P 62, P 63 and P 64).
2—30 mm/65 AK 630 ❹; 6 barrels per mounting; 3,000 rds/min to 2 km.
Countermeasures: Decoys: 2 PK 10 chaff launchers ❺. 2 BEL TOTED; towed torpedo decoys.
ESM: Bharat Ajanta P Mk II intercept ❻.

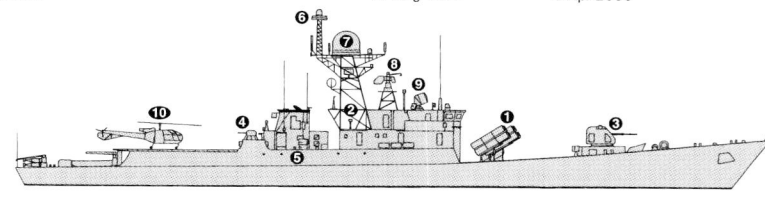

KORA

(Scale 1 : 900), Ian Sturton / 0064715

Combat data systems: Bharat Vympal IPN-10.
Radars: Air search: Cross Dome ❼; E/F-band; range 130 km *(70 n miles)*.
Air/surface search: Plank Shave (Granit Harpun B) ❽; I/J-band.
Fire control: Bass Tilt (P 61) ❾; H/I-band; BEL Lynx (P62-64); I-band.
Navigation: Bharat 1245; I-band.
IFF: Square Head.

Helicopters: Platform only ❿ for Chetak (to be replaced by Hindustan Aeronautics ALH in due course).

Programmes: First pair ordered in April 1990 and second pair in October 1994. Programme slowed by delays in provision of Russian equipment and it is not clear whether further vessels are to be built.
Structure: Very similar to the original Khukri class except that SS-N-25 has replaced SS-N-2. Trishul SAM may replace the 76 mm gun in future. Stabilisers fitted.
Operational: Sea trials for *Kirch* and *Kulish* probably took place in 2000. All 16 SS-N-25 can be fired in one salvo.

UPDATED

KULISH

1/2004*, Ships of the World / 1042275

KIRCH

3/2004*, Bob Fildes / 1042273

4 KHUKRI CLASS (PROJECT 25) (FSGHM)

Name	No	Builders	Laid down	Launched	Commissioned
KHUKRI	P 49	Mazagon Dock Ltd, Mumbai	27 Sep 1985	3 Dec 1986	23 Aug 1989
KUTHAR	P 46	Mazagon Dock Ltd, Mumbai	13 Sep 1986	15 Apr 1989	7 June 1990
KIRPAN	P 44	Garden Reach SY, Calcutta	15 Nov 1985	16 Aug 1988	12 Jan 1991
KHANJAR	P 47	Garden Reach SY, Calcutta	15 Nov 1985	16 Aug 1988	22 Oct 1991

Displacement, tons: 1,423 full load
Dimensions, feet (metres): 298.9 × 34.4 × 13.1 *(91.1 × 10.5 × 4)*
Main machinery: 2 SEMT-Pielstick/Kirloskar 18 PA6 V 280 diesels; 14,400 hp(m) *(10.58 MW)* sustained; 2 shafts; LIPS cp props
Speed, knots: 24. **Range, n miles:** 4,000 at 16 kt
Complement: 112 (12 officers)

Missiles: SSM: 4 SS-N-2D Mod 1 Styx (2 twin) launchers ❶; IR homing to 83 km *(45 n miles)* at 0.9 Mach; warhead 513 kg.
SAM: SA-N-5 Grail ❷; manual aiming; IR homing to 6 km *(3.2 n miles)* at 1.5 Mach; altitude to 2,500 m *(8,000 ft)*; warhead 1.5 kg.
Guns: 1 USSR 3 in *(76 mm)*/60 AK 176 ❸; 120 rds/min to 12 km *(6.4 n miles)*; weight of shell 7 kg.
2—30 mm/65 AK 630 ❹; 6 barrels per mounting; 3,000 rds/min to 2 km.
Countermeasures: Decoys: 2 PK 16 chaff launchers ❺. NPOL; towed torpedo decoy.
ESM: Bharat Ajanta P; intercept.
Combat data systems: Selenia IPN-10 *(Khukri)*; Bharat Vympal IPN-10 (remainder).
Radars: Air search: Cross Dome ❻; E/F-band; range 130 km *(70 n miles)*.
Air/surface search: Plank Shave ❼; I-band.
Fire control: Bass Tilt ❽; H/I-band.
Navigation: Bharat 1245; I-band.

Helicopters: Platform only ❾ for Chetak (to be replaced by HAL Dhruv in due course).

Programmes: First two ordered December 1983, two in 1985. The diesels were assembled in India under licence by Kirloskar. Indigenous content of the whole ship is about 65 per cent.
Structure: The reported plan was to make the first four ASW ships, and the remainder anti-aircraft or general purpose. However *Khukri* has neither torpedo tubes nor a sonar (apart from an Atlas Elektronik echo-sounder), so if the plan is correct these ships will rely on an ALH helicopter which has dunking sonar and ASW torpedoes and depth charges. All have fin stabilisers and full air conditioning.
Operational: All based at Vishakapatnam. The advanced light helicopter (ALH) to have Sea Eagle SSM, torpedoes and dipping sonar. **UPDATED**

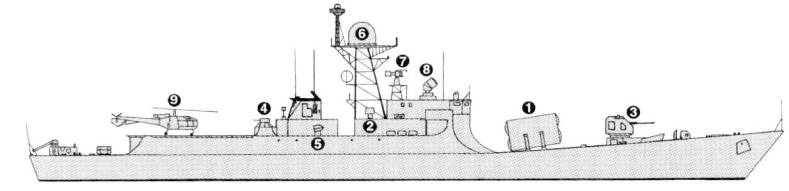

KHUKRI

(Scale 1 : 900), Ian Sturton / 0064713

KHANJAR

10/1998, John Mortimer / 0052335

KUTHAR

3/1996 / 0012974

KIRPAN

2/2001, Michael Nitz / 0534081

13 VEER (TARANTUL I) CLASS (PROJECT 1241RE) (FSGM)

Name	No	Builders	Laid down	Launched	Commissioned
VEER	K 40	Volodarski, Rybinsk	1984	Oct 1986	26 Mar 1987
NIRBHIK	K 41	Volodarski, Rybinsk	1985	Oct 1987	21 Dec 1987
NIPAT	K 42	Volodarski, Rybinsk	1986	Nov 1988	5 Dec 1988
NISHANK	K 43	Volodarski, Rybinsk	1987	June 1989	2 Sep 1989
NIRGHAT	K 44	Volodarski, Rybinsk	1988	Mar 1990	4 June 1990
VIBHUTI	K 45	Mazagon Dock, Mumbai	28 Sep 1987	26 Apr 1990	3 June 1991
VIPUL	K 46	Mazagon Dock, Mumbai	29 Feb 1988	3 Jan 1991	16 Mar 1992
VINASH	K 47	Goa Shipyard	30 Jan 1989	24 Jan 1992	20 Nov 1993
VIDYUT	K 48	Goa Shipyard	27 May 1990	12 Dec 1992	16 Jan 1995
NASHAK	K 83	Mazagon Dock, Mumbai	21 Jan 1991	12 Nov 1993	29 Dec 1994
PRAHAR	K 98	Goa Shipyard	28 Aug 1992	26 Aug 1995	1 Mar 1997
PRABAL	K 92	Mazagon Dock, Mumbai	31 Aug 1998	28 Sep 2000	11 Apr 2002
PRALAYA	K 91	Goa Shipyard	14 Nov 1998	14 Dec 2000	18 Dec 2002

Displacement, tons: 455 full load; 477 full load (K 92 and K 91)
Dimensions, feet (metres): 184.1 × 37.7 × 8.2
(56.1 × 11.5 × 2.5)
Main machinery: COGAG (M15E); 2 Nikolayev Type DR 77 (DS 71 in K 92) gas turbines; 16,016 hp(m) *(11.77 MW)* sustained; 2 Nikolayev Type DR 76 gas turbines with reversible gearboxes; 4,993 hp(m) *(3.67 MW)* sustained; 2 shafts
Speed, knots: 36. **Range, n miles:** 2,000 at 20 kt; 400 at 36 kt
Complement: 41 (5 officers)

Missiles: SSM: 4 SS-N-2D Mod 1 Styx; IR homing to 83 km *(45 n miles)* at 0.9 Mach; warhead 513 kg; sea-skimmer at end of run. 16 (4 quad) SS-N-25 (Kh 35 Uran) in K 91 and K 92; active radar homing to 130 km *(70.2 n miles)* at 0.9 Mach; warhead 145 kg; sea skimmer.

SAM: SA-N-5 Grail quad launcher; manual aiming; IR homing to 6 km *(3.2 n miles)* at 1.5 Mach; warhead 1.5 kg.
Guns: 1 USSR 3 in *(76 mm)*/60; 120 rds/min to 15 km *(8 n miles)*; weight of shell 7 kg.
1 OTO Melara 3 in *(76 mm)*/62 Super Rapid (K 91 and K 92); 120 rds/min to 16 km (8.7 n miles); weight of shell 6 kg.
2—30 mm/65 AK 630; 6 barrels per mounting; 3,000 rds/min combined to 2 km. 2—7.62 mm MGs.
Countermeasures: Decoys: PK 16 chaff launcher.
ESM: Bharat Ajanta P Mk II; intercept.
Weapons control: Hood Wink optronic director.
Radars: Air/surface search: Plank Shave; E-band.
Cross Dome (K 91 and K 92); E/F-band.
Navigation: Mius; I-band.
Fire control: Bass tilt: H/I-band.

BEL Lynx (K 91 and K 92) (for guns); I-band; Bharat Aparna (modified Plank Shave/Harpun B) (for SSM); I/J-band.
IFF: Salt Pot, Square Head A.

Programmes: First five are USSR Tarantul I class built for export. Six further of the same type built in India. Two further craft, armed with the SS-N-25 missile were delivered in 2002. It is not clear whether there are to be further vessels.
Structure: K 92 and K 91 are to a modified design to accommodate the SS-N-25 missile. Principal differences are the bridge and mast configurations.
Operational: All form the 22nd Missile Vessel Squadron at Mumbai.

UPDATED

PRABAL

12/2002, Kapil Chandni / 0529546

PRAHAR

2/2001, Michael Nitz / 0534061

PATROL FORCES

6 SUKANYA CLASS (PSOH)

Name	No	Builders	Launched	Commissioned
SUKANYA	P 50	Korea Tacoma, Masan	1989	31 Aug 1989
SUBHADRA	P 51	Korea Tacoma, Masan	1989	25 Jan 1990
SUVARNA	P 52	Korea Tacoma, Masan	22 Aug 1990	4 Apr 1991
SAVITRI	P 53	Hindustan SY, Vishakapatnam	23 Aug 1989	27 Nov 1990
SHARADA	P 55	Hindustan SY, Vishakapatnam	22 Aug 1990	27 Oct 1991
SUJATA	P 56	Hindustan SY, Vishakapatnam	25 Oct 1991	3 Nov 1993

Displacement, tons: 1,890 full load
Dimensions, feet (metres): 331.7 oa; 315 wl × 37.7 × 14.4 *(101.1; 96 × 11.5 × 4.4)*
Main machinery: 2 SEMT-Pielstick 16 PA6 V 280 diesels; 12,800 hp(m) *(9.41 MW)* sustained; 2 shafts
Speed, knots: 21. **Range, n miles:** 5,800 at 15 kt
Complement: 140 (15 officers)
Guns: 1 Bofors 40 mm/60. 4—12.7 mm MGs.
A/S mortars: 4 RBU 2500 16-tubed trainable launchers; range 2,500 m; warhead 21 kg. Two launchers fitted in forward section.
Radars: Surface search: Racal Decca 2459; I-band.
Navigation: Bharat 1245; I-band.

Helicopters: 1 Chetak.

Programmes: First three ordered in March 1987 from Korea Tacoma to an Ulsan class design. Second four ordered in August 1987. The Korean-built ships commissioned at Masan and then sailed for India where the armament was fitted. Three others of a modified design have been built for the Coast Guard. P 54 transferred to Sri Lanka December 2000.
Structure: Lightly armed and able to 'stage' helicopters, they are fitted out for offshore patrol work only but have the capacity to be much more heavily armed. Fin stabilisers fitted. Firefighting pump on hangar roof aft.

Operational: These ships are used for harbour defence, protection of offshore installations and patrol of the EEZ. Potential for role change is considerable. *Subhadra* modified in early 2000 to test fire Dhanush (naval version of Prithvi) SRBM from her flight deck. The missile was first successfully fired on 20 September 2001 and, on 7 November 2004, a 350 km range missile was reportedly fired in the Bay of Bengal. First three based at Mumbai, last pair at Kochi, P 53 at Vishakapatnam.

UPDATED

SUJATA
3/2004*, Bob Fildes / 1042272

5 + 2 SUPER DVORA Mk II CLASS (PBF)

No	Builders	Commissioned
T 80	IAI, Ramta	24 June 1998
T 81	IAI, Ramta	14 June 1999
T 82	Goa Shipyard Ltd	9 Oct 2003
T 83	Goa Shipyard Ltd	14 Jan 2004
T 84	Goa Shipyard Ltd	19 Apr 2004
T 85	Goa Shipyard Ltd	2005
T 86	Goa Shipyard Ltd	2005

Displacement, tons: 60 full load
Dimensions, feet (metres): 82 × 18.4 × 4.9 *(25 × 5.6 × 1.5)*
Main machinery: 2 MTU 12V 396 TE94 diesels; 4,570 hp(m) *(3.36 MW)*; 2 Arneson ASD 16 surface drives
Speed, knots: 50. **Range, n miles:** 700 at 42 kt
Complement: 10 (1 officer)
Guns: 1—20 mm. 1—12.7 mm MG.
Weapons control: Elop MSIS optronic director.
Radars: Surface search: Koden; I-band.

Comment: Collaborative programme involving IAI, Ramta, Israel and Goa Shipyard Ltd. First pair ordered 2 December 1996 and built at Ramta. Three indigenously built craft delivered by Spring 2004 with a further two reported to be under construction. Further orders are expected.
UPDATED

SUPER DVORA
2/2001 / 0126188

2 SDB Mk 3 CLASS (LARGE PATROL CRAFT) (PB)

T 58-59

Displacement, tons: 210 full load
Dimensions, feet (metres): 124 × 24.6 × 6.2 *(37.8 × 7.5 × 1.9)*
Main machinery: 2 MTU 16V 538 TB92 diesels; 6,820 hp(m) *(5 MW)* sustained; 2 shafts
Speed, knots: 30
Complement: 32
Guns: 2 Bofors 40 mm/60; 120 rds/min to 10 km *(5.5 n miles)*; weight of shell 0.89 kg.
Radars: Surface search: Bharat 1245; I-band.

Comment: Built at Garden Reach and Goa and completed 1984-86. Employed as seaward defence forces.
UPDATED

SDB MK 3 CLASS
6/2004*, Indian Navy / 1042279

4 + 4 SDB MK 5 (TRINKAT) CLASS (LARGE PATROL CRAFT) (PBO)

TRINKAT T 61	TARASA T 63	BANGARAM T 65
TILLAN CHANG T 62	TARMUGLI T 64	BITRA T 66

Displacement, tons: 260 full load
Dimensions, feet (metres): 151.0 × 24.6 × 8.2 *(46.0 × 7.5 × 2.5)*
Main machinery: 2 MTU 16V 538 TB92 diesels; 6,820 hp(m) *(5 MW)* sustained; 2 shafts
Speed, knots: 30. **Range, n miles:** 2,000 at 12 kt
Complement: 34 (4 officers)
Guns: 1 Medak 30 mm 2A42
Radars: Surface search: Bharat 1245; I-band.

Comment: The first four built at Garden Reach and commissioned between September 2000 and March 2002. Four of a modified design are under construction. T 65 was launched on 11 December 2004 and T 66 on 14 December 2004. Both were commissioned in November 2005. T 67 and T 68 are to be launched in 2005 for commissioning in 2006.
UPDATED

TARASA
5/2002 / 0534083

AMPHIBIOUS FORCES

2 + 1 (2) MAGAR CLASS (LSTH)

Name	No	Builders	Launched	Commissioned
MAGAR	L 20	Garden Reach	7 Nov 1984	15 July 1987
GHARIAL	L 23	Hindustan/Garden Reach	1 Apr 1991	14 Feb 1997
SHARDUL	—	Hindustan/Garden Reach	3 Apr 2004	2006

Displacement, tons: 5,655 full load
Dimensions, feet (metres): 409.4 oa; 393.7 wl × 57.4 × 13.1 *(124.8; 120 × 17.5 × 4)*
Main machinery: 2 SEMT-Pielstick 12 PA6 V280 diesels; 8,560 hp(m) *(6.29 MW)* sustained; 2 shafts
Speed, knots: 15. **Range, n miles:** 3,000 at 14 kt
Complement: 136 (16 officers)
Military lift: 15 tanks plus 8 APC plus 500 troops
Guns: 4 Bofors 40 mm/60. 2—122 mm multibarrel rocket launchers at the bow.
Countermeasures: ESM: Bharat Ajanta; intercept.
Radars: Navigation: Bharat; I-band.
Helicopters: 1 Sea King 42C; platform for 2.

Comment: Based on the *Sir Lancelot* design. *Magar* was built entirely at Garden Reach. *Gharial* ordered in 1985. Built at Hindustan Shipyard but fitted out at Garden Reach. Internal design differs from *Magar*. Carries four LCVPs on davits. Bow door. Can beach on gradients 1 in 40 or more. *Magar* refitted in 1995. Both based at Vishakapatnam. A third of class with major design changes is expected to be commissioned in 2006. Two further ships are to be acquired. **UPDATED**

MAGAR

2/2001, Guy Toremans / 0121348

6 Mk 2/3 LANDING CRAFT (LSM)

VASCO DA GAMA L 34	— L 36	MIDHUR L 38	
— L 35	— L 37	MANGALA L 39	

Displacement, tons: 500 full load
Dimensions, feet (metres): 188.6 oa; 174.5 pp × 26.9 × 5.2 *(57.5; 53.2 × 8.2 × 1.6)*
Main machinery: 3 Kirloskar-MAN V8V 17.5/22 AMAL diesels; 1,686 hp(m) *(1.24 MW)*; 3 shafts
Speed, knots: 11. **Range, n miles:** 1,000 at 8 kt
Complement: 167
Military lift: 250 tons; 2 PT 76 or 2 APC. 120 troops.
Guns: 2 Bofors 40 mm/60 (aft).
Mines: Can be embarked.
Radars: Navigation: Decca 1229; I-band.

Comment: L 34 and 35 are Mk 2 craft built by Hooghly D and E Co. The remaining Mk 3 craft were built at Goa Shipyard. First craft commissioned 28 January 1980 and the last one commissioned 25 March 1987. L 36-39 have a considerably modified superstructure and a higher bulwark on the cargo deck. **UPDATED**

L 36

2/1999, 92 Wing RAAF / 0064719

L 34

3/1995 / 0064720

5 POLNOCHNY C (PROJECT 773 I) and D CLASS (PROJECT 773 IM) (LSM/LSMH)

SHARABH L 17	CHEETAH L 18	MAHISH L 19
GULDAR L 21	KUMBHIR L 22	

Displacement, tons: 1,150 (C class); 1,190 (D class) full load
Dimensions, feet (metres): 266.7; 275.3 (D class) × 31.8 × 7.9 *(81.3; 83.9 × 9.7 × 2.4)*
Main machinery: 2 Kolomna Type 40-D diesels; 4,400 hp(m) *(3.2 MW)* sustained; 2 shafts
Speed, knots: 16. **Range, n miles:** 3,000 at 12 kt
Complement: 60 (6 officers)
Military lift: 160 troops; 5 MBT or 5 APC or 5 AA guns or 8 trucks
Guns: 4—30 mm (2 twin) Ak 230. 2—140 mm 18-tubed rocket launchers.
Radars: Navigation: Don 2 or Krivach (SRN 745); I-band.
Fire control: Drum Tilt; H/I-band (in D class).
Helicopters: Platform only (in D class).

Comment: A original class of eight built in two batches by Naval Shipyard, Gdynia. *Sharabh* transferred in February 1976, *Cheetah* in February 1985, *Mahish* in July 1985, *Guldar* in March 1986 and *Kumbhir* in November 1986. The last four are Polnochny Ds with the flight deck forward of the bridge and different radars. All are being restricted operationally through lack of spares, but all are seaworthy. Drum Tilt radar removed from some ships. Four Polnochny Ds (L 18-22) form 5th landing Ship Squadron based at Port Blair. **UPDATED**

MAHISH

2/2001, Guy Toremans / 0121353

MINE WARFARE FORCES

Notes: Procurement of up to eight minehunters has been approved. It is anticipated that the ships will be to a foreign design and of GRP construction. Building is expected to take place at Goa Shipyards. The ships are likely to be equipped with a minehunting sonar and with a remote-control mine-disposal system. Requests for proposals are expected to be sought from predominantly European manufacturers. Building of the first of class is due to start in 2008.

12 PONDICHERRY (NATYA I) CLASS (PROJECT 266M) (MINESWEEPERS—OCEAN) (MSO)

Name	No	Builders	Commissioned
PONDICHERRY	M 61	Isora, Leningrad	2 Feb 1978
PORBANDAR	M 62	Isora, Leningrad	19 Dec 1978
BEDI	M 63	Isora, Leningrad	27 Apr 1979
BHAVNAGAR	M 64	Isora, Leningrad	27 Apr 1979
ALLEPPEY	M 65	Isora, Leningrad	10 June 1980
RATNAGIRI	M 66	Isora, Leningrad	10 June 1980
KARWAR	M 67	Isora, Leningrad	14 July 1986
CANNANORE	M 68	Isora, Leningrad	17 Dec 1987
CUDDALORE	M 69	Isora, Leningrad	29 Oct 1987
KAKINADA	M 70	Isora, Leningrad	23 Dec 1986
KOZHIKODE	M 71	Isora, Leningrad	19 Dec 1988
KONKAN	M 72	Isora, Leningrad	8 Oct 1988

Displacement, tons: 804 full load
Dimensions, feet (metres): 200.1 × 33.5 × 10.8 *(61 × 10.2 × 3)*
Main machinery: 2 Type 504 diesels; 5,000 hp(m) *(3.67 MW)* sustained; 2 shafts; cp props
Speed, knots: 16. **Range, n miles:** 3,000 at 12 kt
Complement: 82 (10 officers)

Guns: 4—30 mm/65 (2 twin); 500 rds/min to 5 km *(2.7 n miles)*; weight of shell 0.54 kg.
4—25 mm/70 (2 twin); 270 rds/min to 3 km *(1.6 n miles)*.
A/S mortars: 2 RBU 1200 5-tubed fixed; range 1,200 m; warhead 34 kg.
Mines: Can carry 10.
Countermeasures: MCM: 1 GKT-2 contact sweep; 1 AT-2 acoustic sweep; 1 TEM-3 magnetic sweep.
Radars: Navigation: Don 2; I-band.
Fire control: Drum Tilt; H/I-band.
IFF: 2 Square Head. High Pole B.
Sonars: MG 69/79; hull-mounted; active mine detection; high frequency.

Programmes: Built for export. Last six were delivered out of pennant number order.
Structure: Steel hulls but do not have stern ramp as in Russian class.
Operational: Some are fitted with two quad SA-N-5 systems. *Pondicherry* was painted white and used as the Presidential yacht for the Indian Fleet Review by President R Venkataramen on 15 February 1989; she reverted to her normal role and colour on completion. One serves as an AGI. First six form 19th MCM Squadron based at Mumbai and second batch form 21st MCM Squadron based at Vishakapatnam. **UPDATED**

KARWAR

4/2004*, John Mortimer / 1042271

2 MAHÉ (YEVGENYA) CLASS (PROJECT 1258)
(MINESWEEPERS—INSHORE) (MSI)

MAHÉ M 83   **MALPE** M 86

Displacement, tons: 90 full load
Dimensions, feet (metres): 80.7 × 18 × 4.9 *(24.6 × 5.5 × 1.5)*
Main machinery: 2 Type 3-D-12 diesels; 600 hp(m) *(440 kW)* sustained; 2 shafts
Speed, knots: 11. **Range, n miles:** 300 at 10 kt
Complement: 10 (1 officer)
Guns: 2 USSR 25 mm/80 (twin).
Radars: Navigation: Spin Trough; I-band.
Sonars: MG 7 small transducer streamed over the stern on a crane.

Comment: First batch commissioned 16 May 1983 and second batch on 10 May 1984. A mid-1960s design with GRP hulls built at Kolpino. Form 20th MCM Squadron based at Cochin. *Mulki, Malvan, Magdala* and *Mangrol* have decommissioned.   ***UPDATED***

MAHÉ CLASS   *6/1994 /* 0064722

SURVEY AND RESEARCH SHIPS

Notes: The National Institute of Oceanography operates several research and survey ships including *Sagar Kanya, Samudra Manthan, Sagar Sampada, Samudra Sarvekshak, Samudra Nidhi* and *Samudra Sandhari.*

8 SANDHAYAK CLASS (SURVEY SHIPS) (AGSH)

Name	No	Builders	Launched	Commissioned
SANDHAYAK	J 18	Garden Reach, Calcutta	6 Apr 1977	1 Mar 1981
NIRDESHAK	J 19	Garden Reach, Calcutta	16 Nov 1978	4 Oct 1982
NIRUPAK	J 14	Garden Reach, Calcutta	10 July 1981	14 Aug 1985
INVESTIGATOR	J 15	Garden Reach, Calcutta	8 Aug 1987	11 Jan 1990
JAMUNA	J 16	Garden Reach, Calcutta	4 Sep 1989	31 Aug 1991
SUTLEJ	J 17	Garden Reach, Calcutta	1 Dec 1991	19 Feb 1993
DARSHAK	J 21	Goa Shipyard	3 Mar 1999	28 Apr 2001
SARVEKSHAK	J 22	Goa Shipyard	24 Nov 1999	14 Jan 2002

Displacement, tons: 1,929 full load
Dimensions, feet (metres): 288 × 42 × 11.1 *(87.8 × 12.8 × 3.4)*
Main machinery: 2 GRSE/MAN 66V 30/45 ATL diesels; 7,720 hp(m) *(5.67 MW)* sustained; 2 shafts; active rudders
Speed, knots: 16. **Range, n miles:** 6,000 at 14 kt; 14,000 at 10 kt
Complement: 178 (18 officers) plus 30 scientists
Guns: 1 or 2 Bofors 40 mm/60.
Countermeasures: ESM: Telegon IV HF D/F.
Radars: Navigation: Racal Decca 1629; I-band.
Helicopters: 1 Chetak.

Comment: Telescopic hangar. Fitted with three echo-sounders, side scan sonar, extensively equipped laboratories, and carries four GRP survey launches on davits amidships. Painted white with yellow funnels. An active rudder with a DC motor gives speeds of up to 5 kt. First three based at Vishakapatnam and have been used as troop transports. *Investigator* is at Mumbai and *Jamuna* and *Sutlej* at Kochi. The last pair were laid down in May and August 1995 and have a secondary role as casualty holding ships.   ***VERIFIED***

DARSHAK   *4/2002, Giorgio Ghiglione /* 0534057

1 SAGARDHWANI CLASS (RESEARCH SHIP) (AGORH)

Name	No	Builders	Commissioned
SAGARDHWANI	A 74	Garden Reach, Calcutta	30 July 1994

Displacement, tons: 2,050 full load
Dimensions, feet (metres): 279.2 × 42 × 12.1 *(85.1 × 12.8 × 3.7)*
Main machinery: 2 GRSE/MAN 66V 30/45 ATL diesels; 3,860 hp(m) *(2.84 MW)* sustained; 2 shafts; 2 auxiliary thrusters
Speed, knots: 16. **Range, n miles:** 6,000 at 16 kt
Complement: 80 (10 officers) plus 16 scientists
Radars: Navigation: Racal Decca 1629; I-band.
Helicopters: Platform for Alouette III.

Comment: Marine Acoustic Research Ship (MARS) launched in May 1991. The hull and main machinery are very similar to the Sandhayak class survey ships, but there are marked superstructure differences with the bridge positioned amidships and a helicopter platform forward. Aft there are two large cranes and a gantry for deploying and recovering research equipment. The vessel is designed to carry out acoustic and geological research and special attention has been paid to noise reduction. The ship is painted white except for the lift equipment and two boats which are orange. Employed in advanced torpedo trials and missile range support. Based at Kochi.   ***VERIFIED***

SAGARDHWANI   *2/2001, Michael Nitz /* 0534058

2 MAKAR CLASS (SURVEY SHIPS) (AGS)

MEEN J 33   **MITHUN** J 34

Displacement, tons: 210 full load
Dimensions, feet (metres): 123 × 24.6 × 6.2 *(37.5 × 7.5 × 1.9)*
Main machinery: 2 diesels; 1,124 hp(m) *(826 kW)*; 2 shafts
Speed, knots: 12. **Range, n miles:** 1,500 at 12 kt
Complement: 36 (4 officers)
Guns: 1 Bofors 40 mm/60.
Radars: Navigation: Decca 1629; I-band.

Comment: Launched at Goa in 1981-82. Similar hulls to deleted SDB Mk 2 class but with much smaller engines. Employed as seaward defence forces.   ***VERIFIED***

MEEN   *4/1992 /* 0064723

TRAINING SHIPS

1 TIR CLASS (TRAINING SHIP) (AXH)

Name	No	Builders	Launched	Commissioned
TIR	A 86	Mazagon Dock Ltd, Bombay	15 Apr 1983	21 Feb 1986

Displacement, tons: 3,200 full load
Dimensions, feet (metres): 347.4 × 43.3 × 15.7 *(105.9 × 13.2 × 4.8)*
Main machinery: 2 Crossley-Pielstick 8 PC2 V Mk 2 diesels; 7,072 hp(m) *(5.2 MW)* sustained; 2 shafts
Speed, knots: 18. **Range, n miles:** 6,000 at 12 kt
Complement: 239 (35 officers) plus 120 cadets
Guns: 2 Bofors 40 mm/60 (twin) with launchers for illuminants. 4 saluting guns.
Countermeasures: ESM: Telegon IV D/F.
Radars: Navigation: Bharat/Decca 1245; I-band.
Helicopters: Platform for Alouette III.

Comment: Second of class reported ordered May 1986 but was cancelled as an economy measure. Built to commercial standards, Decca collision avoidance plot and SATNAV. Can carry up to 120 cadets and 20 instructors. Based at Kochi.   ***VERIFIED***

TIR   *2/2001, Michael Nitz /* 0534059

1 LEANDER (BATCH 3A) CLASS (AXH)

Name	No	Builders	Commissioned
KRISHNA (ex-*Andromeda*)	F 46 (ex-F 57)	Portsmouth Dockyard	2 Dec 1968

Displacement, tons: 2,960 full load
Dimensions, feet (metres): 372 × 43 × 18 (screws) *(113.4 × 13.1 × 5.5)*
Main machinery: 2 Babcock & Wilcox boilers; 550 psi *(38.7 kg/cm²)*; 850°F *(454°C)*; 2 White/English Electric turbines; 30,000 hp *(22.4 MW)*; 2 shafts
Speed, knots: 28. **Range, n miles:** 4,000 at 15 kt
Complement: 260 (19 officers)
Guns: 2 Bofors 40 mm/60. 2 Oerlikon 20 mm.
Radars: Air/surface search: Marconi Type 968; D/E-band.
Navigation: Kelvin Hughes Type 1006; I-band.
Helicopters: 1 Chetak.

Comment: Laid down 25 May 1966 and launched 24 May 1967. Acquired from the UK in April 1995 having paid off in June 1993 to a state of extended readiness. Refitted by DML, Devonport, before recommissioning 22 August 1995. The original 114 mm gun turret, Seacat SAM and ASW Limbo mortar were removed in 1979-80 when Exocet SSM, Seawolf SAM, STWS torpedo tubes and facilities for a Lynx helicopter were fitted. Acquired for training purposes to supplement the *Tir*. Armament has been reduced to the minimum required for the training role, and now includes 40 mm guns on either side, aft of the funnel. Based at Kochi.

VERIFIED

KRISHNA
8/1995, H M Steele / 0064724

2 SAIL TRAINING SHIPS (AXS)

VARUNA   TARANGINI A 75

Displacement, tons: 420 full load
Dimensions, feet (metres): 177.2 × 27.9 × 13.1 *(54 × 8.5 × 4)*
Main machinery: 2 diesels; 640 hp(m) *(470 kW)*; 2 shafts; LIPS props
Speed, knots: 10 (diesels)
Complement: 15 (6 officers) plus 45 cadets

Comment: *Varuna* completed in April 1981 by Alcock-Ashdown, Bhavnagar. Can carry 26 cadets. Details given are for *Tarangini* which is based on a Lord Nelson design by Colin Mudie of Lymington and has been built by Goa Shipyard. Launched on 23 December 1995, and completed in December 1997. Three masted barque, square rigged on forward and main mast and 'fore and aft' rigged on mizzen mast. Based at Mumbai.

VERIFIED

TARANGINI
2/2001, Guy Toremans / 0121359

AUXILIARIES

Notes: (1) There is also a small hospital ship *Lakshadweep* of 865 tons and a crew of 35 including 16 medics.
(2) *Ambika* is a 1,000 ton oiler commissioned in 1995. Built by Hindustan Shipyard, it is based at Vishakhapatnam.

1 UGRA CLASS (SUBMARINE TENDER) (ASH)

Name	No	Builders	Launched	Commissioned
AMBA	A 54	Nikolayev Shipyard	18 Jan 1968	28 Dec 1968

Displacement, tons: 6,750 standard; 9,650 full load
Dimensions, feet (metres): 462.6 × 57.7 × 23 *(141 × 17.6 × 7)*
Main machinery: Diesel-electric; 4 Kolomna Type 2-D-42 diesel generators; 2 motors; 8,000 hp(m) *(5.88 MW)*; 2 shafts
Speed, knots: 17. **Range, n miles:** 21,000 at 10 kt
Complement: 400
Guns: 4 USSR 3 in *(76 mm)*/60 (2 twin).
Radars: Air/surface search: Slim Net; E/F-band.
Fire control: 2 Hawk Screech; I-band.
Navigation: Don 2; I-band.
IFF: 2 Square Head. High Pole A.

Comment: Acquired from the USSR in 1968. Provision for helicopter. Can accommodate 750. Two cranes, one of 6 tons and one of 10 tons. Differs from others of the class by having 76 mm guns. After extensive repairs, the ship is now deployed on the east coast.

VERIFIED

AMBA
2/1998 / 0052348

1 + (1) JYOTI CLASS (REPLENISHMENT TANKER) (AORH)

Name	No	Builders	Launched	Commissioned
JYOTI	A 58	Admiralty Yard, St Petersburg	8 Dec 1995	20 July 1996

Displacement, tons: 35,900 full load
Dimensions, feet (metres): 587.3 × 72.2 × 26.2 *(179 × 22 × 8)*
Main machinery: 1 Burmeister & Wain diesel; 10,948 hp(m) *(8.05 MW)*; 1 shaft
Speed, knots: 15. **Range, n miles:** 12,000 at 15 kt
Complement: 92 (16 officers)
Cargo capacity: 25,040 tons diesel
Radars: Navigation: I-band.
Helicopters: Platform for 1 medium.

Comment: This was the third of a class of merchant tankers, modified for naval use for the Indian Navy and acquired in 1995. The ship was laid down in September 1993. Based at Mumbai where she arrived in November 1996. There are two replenishment positions on each side and stern refuelling is an option. Similar ship sold to China and two others are in commercial service. Procurement of another ship is reported to be under consideration.

UPDATED

JYOTI
10/2004*, Hachiro Nakai / 1042267

For details of the latest updates to *Jane's Fighting Ships* online and to discover the additional information available exclusively to online subscribers please visit
jfs.janes.com

1 DEEPAK CLASS (REPLENISHMENT TANKER) (AORH)

Name	No	Builders	Commissioned
SHAKTI	A 57	Bremer-Vulkan	31 Dec 1975

Displacement, tons: 6,785 light; 15,828 full load
Measurement, tons: 12,013 gross
Dimensions, feet (metres): 552.4 × 75.5 × 30 *(168.4 × 23 × 9.2)*
Main machinery: 2 Babcock & Wilcox boilers; 1 BV/BBC steam turbine; 16,500 hp(m) *(12.13 MW)*; 1 shaft
Speed, knots: 18.5. **Range, n miles:** 5,500 at 16 kt
Complement: 169
Cargo capacity: 1,280 tons diesel; 12,624 tons FFO; 1,495 tons avcat; 812 tons FW
Guns: 4 Bofors 40 mm/60. 2 Oerlikon 20 mm can be carried.
Countermeasures: ESM: Telegon IV HF D/F.
Radars: Navigation: 2 Decca 1226; I-band.
Helicopters: 1 Chetak.

Comment: Automatic tensioning fitted to replenishment gear. Heavy and light jackstays. Stern fuelling as well as alongside. DG fitted. Based at Mumbai. **VERIFIED**

SHAKTI   6/2000 / 0104589

1 ADITYA CLASS
(REPLENISHMENT AND REPAIR SHIP) (AORH/AS)

Name	No	Builders	Launched	Commissioned
ADITYA (ex-*Rajaba Gan Palan*)	A 59	Garden Reach, Calcutta	15 Nov 1993	3 Apr 2000

Displacement, tons: 24,600 full load
Measurement, tons: 17,000 dwt
Dimensions, feet (metres): 564.3 × 75.5 × 29.9 *(172 × 23 × 9.1)*
Main machinery: 2 MAN/Burmeister & Wain 16V 40/45 diesels; 23,936 hp(m) *(17.59 MW)* sustained; 1 shaft
Speed, knots: 20. **Range, n miles:** 10,000 at 16 kt
Complement: 156 (16 officers) + 6 aircrew
Cargo capacity: 14,200 m³ diesel and avcat; 2,250 m³ water; 2,170 m³ ammunition and stores
Guns: 3 Bofors 40 mm/60.
Helicopters: 1 Chetak.

Comment: Ordered in July 1987 to a Bremer-Vulkan design. Lengthened version of Deepak class but with a multipurpose workshop. Four RAS stations alongside. Fully air conditioned. Building progress was very slow and sea trials were curtailed by propulsion problems during 1999. Ship has the capability to carry a Sea King 42B or KA 28 helicopter. **VERIFIED**

ADITYA   2/2001, Guy Toremans / 0121357

1 DIVING SUPPORT SHIP (ASR)

Name	No	Builders	Commissioned
NIREEKSHAK	A 15	Mazagon Dock Ltd, Bombay	8 June 1989

Displacement, tons: 2,160 full load
Dimensions, feet (metres): 231.3 × 57.4 × 16.4 *(70.5 × 17.5 × 5)*
Main machinery: 2 Bergen KRM-8 diesels; 4,410 hp(m) *(3.24 MW)* sustained; 2 shafts; cp props; 2 bow thrusters; 2 stern thrusters; 990 hp(m) *(727 kW)*
Speed, knots: 12
Complement: 63 (15 officers)

Comment: Laid down in August 1982 and launched January 1984. Acquired on lease with an option for purchase which was taken up in March 1995, and the ship was recommissioned on 15 September 1995. The vessel was built for offshore support operations but has been modified for naval requirements. Two DSRV, capable of taking 12 men to 300 m, are carried together with two six-man recompression chambers and one three-man bell. Kongsberg ADP-503 Mk II. Dynamic positioning system. The ship is used for submarine SAR. Based at Mumbai. **VERIFIED**

NIREEKSHAK   1991 / 0064727

3 TRANSPORT SHIPS (APH)

Name	No	Builders	Launched
NICOBAR	—	Szczecin Shipyard, Poland	12 Apr 1990
ANDAMANS (ex-*Nancowry*)	—	Szczecin Shipyard, Poland	5 Oct 1990
SWARAJ DEEP	—	Vishakhapatnam	1997

Displacement, tons: 19,000 full load
Measurement, tons: 14,176 grt
Dimensions, feet (metres): 515.1 × 68.9 × 22 *(157 × 21 × 6.7)*
Main machinery: 2 Cegielski-Burmeister am Wain 6L35MC diesels; 72,000 hp *(5.3 MW)*; 2 shafts; bow thruster
Speed, knots: 16
Complement: 160
Cargo capacity: 1,200 troops
Helicopters: Platform for 1 medium.

Comment: The first two ships designed and built in Poland. *Nicobar* delivered to the Shipping Corporation of India (which operated the ship for the Andaman and Nicobar Islands Administration) on 5 June 1991 and subsequently acquired for use by the Indian Navy in April 1998. *Andamans* delivered to the Shipping Corporation of India on 31 March 1992 and acquired for use by the Indian Navy in April 2000. Both ships used to trans-ship stores and personnel to the Andaman and Nicobar Islands. They have large davits capable of operating LCVPs. *Swaraj Deep* is of a similar design. **UPDATED**

SWARAJ DEEP   6/2004 *, M Mazumdar / 1042280

6 SUPPORT TANKERS (AOTL)

POSHAK	PURAN	PUSHPA	PRADHAYAK	PURAK	PALAN

Comment: First two built at Mazagon Dock Ltd, Bombay. *Poshak* completed April 1982, and *Puran* in November 1988. *Pushpa* (capacity 650 tons) built at Goa Shipyard and completed in 1990. *Pradhayak*, *Purak* and *Palan* built at Rajabagan Shipyard, Bombay, the first two in 1977 and *Palan* in May 1986. Cargo capacities vary. Civilian manned. **VERIFIED**

PUSHPA   1990, Goa Shipyard / 0064728

2 WATER CARRIERS (AWT)

AMBUDA	COCHIN

Comment: First laid down Rajabagan Shipyard 18 January 1977. Second built at Mazagon Dock Ltd, Bombay. Civilian manned. **VERIFIED**

AMBUDA   4/1992 / 0064729

1 TORPEDO RECOVERY VESSEL (YPT)

A 72

Displacement, tons: 110 full load
Dimensions, feet (metres): 93.5 × 20 × 4.6 *(28.5 × 6.1 × 1.4)*
Main machinery: 2 Kirloskar V12 diesels; 720 hp(m) *(529 kW)*; 2 shafts
Speed, knots: 11
Complement: 13

Comment: Completed in 1981 at Goa Shipyard. Based at Vishakapatnam.

VERIFIED

A 72   *2/1989, G Jacobs*

3 DIVING TENDERS (YDT)

Displacement, tons: 36 full load
Dimensions, feet (metres): 48.9 × 14.4 × 3.9 *(14.9 × 4.4 × 1.2)*
Main machinery: 2 diesels; 130 hp(m) *(96 kW)*; 2 shafts
Speed, knots: 12

Comment: Built at Cleback Yard. First completed 1979; second and third in 1984.

VERIFIED

YDT   *9/1996 /* 0012531

TUGS

1 TUG (OCEAN) (ATA/ATR)

MATANGA A 53

Measurement, tons: 1,313 grt
Dimensions, feet (metres): 222.4 × 40.4 × 13.1 *(67.8 × 12.3 × 4)*
Main machinery: 2 GRSE/MAN G7V diesels; 3,920 hp(m) *(2.88 MW)*; 2 shafts
Speed, knots: 15. **Range, n miles:** 4,000 at 15 kt
Complement: 78 (8 officers)
Guns: 1 Bofors 40 mm/60.
Radars: Navigation: I-band.

Comment: Built by Garden Reach SY. *Matanga* launched 29 October 1977. Bollard pull of 40 tons and capable of towing a 20,000 ton ship at 8 kt. Carries a divers' decompression chamber and other salvage equipment.

UPDATED

MATANGA   *2/2001, Michael Nitz /* 0143309

14 HARBOUR TUGS (YTM/YTL)

AGARAL	BC DUTT	RAJAJI	ANAND	BHIM
SHAMBU SINGH	BALSHIL	MADAN SINGH	BAJARANG	AJRAL
ARJUN	TARAFDAAR	BALRAM	GAJ A 51	

Measurement, tons: 216 grt
Dimensions, feet (metres): 96.1 × 27.9 × 8.5 *(29.3 × 8.5 × 2.6)*
Main machinery: 2 SEMT-Pielstick 8 PA4 V 200 diesels; 3,200 hp(m) *(2.35 MW)*; 2 shafts
Speed, knots: 11
Complement: 12

Comment: First three built by Mazagon Dock Ltd, Bombay in 1973-74. Five more delivered in 1988-89, and four more in 1991 from Mazagon Dock Ltd, Goa. *Gaj* is a 25 ton bollard pull tug built by Hindustan Shipyard and commissioned on 10 October 2002. Details given are for *Balram* and *Bajrang*; *Rajaji* is of comparable size built in 1982; *Bhim*, *Balshil* and *Ajral* were built by Tebma Shipyard, Chennai, the others are of varying types.

UPDATED

MADAN SINGH   *2/2001, Sattler/Steele /* 0121356

COAST GUARD

Senior Appointments

Director General:
  Vice Admiral Arun Kumar Singh
Deputy Director General:
  Inspector General P Paleri, TM

Personnel

2005:   5,393 (717 officers)

General

The Coast Guard was constituted as an independent paramilitary service on 19 August 1978. It functions under the Ministry of Defence.

Responsibilities include:

(a)   Ensuring the safety and protection of artificial islands, offshore terminals and other installations in the Maritime Zones.

(b)   Measures for the safety of life and property at sea including assistance to mariners in distress.
(c)   Measures to preserve and protect the marine environment and control marine pollution.
(d)   Assisting the Customs and other authorities in anti-smuggling operations.
(e)   Enforcing the provisions of enactments in force in the Maritime Zones.
(f)   Protection of fishermen and assistance to them at sea while in distress.

Bases

The Headquarters of the Coast Guard is located in Delhi with Regional Headquarters in Mumbai, Chennai and Port Blair. West Coast District Headquarters at Mumbai, New Mangalore, Goa, Porbandar, Kochi. East Coast District/Headquarters at Vishakapatnam, Chennai, Paradip and Haldia. Andaman and Nicobar District Headquarters at Campbell Bay and Diglipur. Stations at Vadinar, Mandapam, Okha and Tuticorin.

Aviation

Air Squadrons at Daman CGAS 750 (11 Dorniers 228); Kochi CGAS 747 (2 Dornier); Chennai CGAS 744 (7 Dorniers 228); Kolkatta CGAS 700 (2 Dornier 228); Port Blair CGAS 745 (2 Dornier 228); Daman CGAS 841 (4 Chetaks); Mumbai CGAS 842 (3 Chetaks); Goa CGAS 800 (4 Chetaks); Chennai CGAS 848 (3 Chetaks); Port Blair (1 Chetak). Vajra flight (1 Chetak), Veera flight (1 Chetak) and CGEFU Goa (3 ALH).

DELETIONS

2002   *C-05*

PATROL FORCES

0 + 1 ADVANCED OFFSHORE PATROL VESSEL (WPSOH)

Name	No	Builders	Laid down	Launched	Commissioned
—	—	Goa Shipyard	17 July 2004	2005	2007

Displacement, tons: 2,230 full load
Dimensions, feet (metres): 344.5 × 42.3 × 11.8 *(105.0 × 12.9 × 3.6)*
Main machinery: 2 SEMT-Pielstick 20 PA6B stc diesels; 20,900 hp(m) *(15.58 MW)*; 2 shafts; cp props
Speed, knots: 24. **Range, n miles:** 6,500 at 12 kt
Complement: 126 (18 officers)
Guns: 1–3 in *(76 mm)*.
Radars: Surface search: To be announced.
Navigation: To be announced.
Helicopters: 1 HAL Dhruv.

Comment: Designed and built under ABS and IRS classification by Goa Shipyard for patrol and SAR operations, pollution control and firefighting. This will be the largest Coast Guard vessel when completed and further orders are expected.
NEW ENTRY

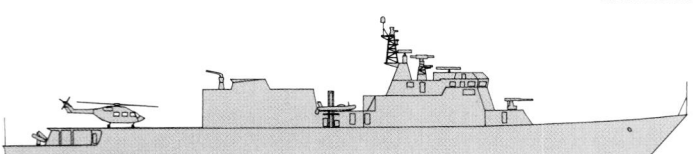

ADVANCED OPV   *(Scale 1 : 900), Ian Sturton / 1042088*

0 + 3 OFFSHORE PATROL VESSELS (UT 517 CLASS) (WPSOH)

Name	No	Builders	Laid down	Launched	Commissioned
—	—	ABG Shipyard, Surat	2004	2005	2006
—	—	ABG Shipyard, Surat	2005	2006	2007
—	—	ABG Shipyard, Surat	2006	2007	2008

Displacement, tons: 3,300 full load
Dimensions, feet (metres): 308.4 × 50.9 × 14.8
 (94.0 × 15.5 × 4.5)
Main machinery: 2 Bergen B32 diesels; 8,050 hp *(6.0 MW)*; 2 shafts; cp props. 1 Ulstein Aquamaster bow thruster; 1,185 hp *(883 kW)*
Speed, knots: 20. **Range, n miles:** 6,000 at 14 kt
Complement: To be announced
Guns: To be announced.
Radars: Navigation: To be announced.
Helicopters: Platform for 1 medium.

Comment: Rolls-Royce UT 517 design selected on 25 October 2004 for three environmental protection ships. The ships are to feature a range of Rolls-Royce propulsion, steering and motion control equipment and are similar to those selected for use by the French Navy and Norwegian Coast Guard. The ships are to be capable of deploying a boom system to contain oil spillages while additional tasks are to include surveillance and law enforcement, anti-smuggling and fishery protection, search and rescue, collecting data, and assistance with salvage and fire fighting.
NEW ENTRY

UT 517 CLASS (artist's impression)   *10/2004 *, Rolls-Royce / 1042264*

9 VIKRAM CLASS (OFFSHORE PATROL VESSELS) (WPSOH)

Name	No	Builders	Launched	Commissioned
VIKRAM	33	Mazagon Dock, Mumbai	26 Sep 1981	19 Dec 1983
VIJAYA	34	Mazagon Dock, Mumbai	5 June 1982	12 Apr 1985
VEERA	35	Mazagon Dock, Mumbai	30 June 1984	3 May 1986
VARUNA	36	Mazagon Dock, Mumbai	28 Jan 1986	27 Feb 1988
VAJRA	37	Mazagon Dock, Mumbai	3 Jan 1987	22 Dec 1988
VIVEK	38	Mazagon Dock, Mumbai	5 Nov 1987	19 Aug 1989
VIGRAHA	39	Mazagon Dock, Mumbai	27 Sep 1988	12 Apr 1990
VARAD	40	Goa Shipyard	3 Sep 1989	19 July 1990
VARAHA	41	Goa Shipyard	5 Nov 1990	11 Mar 1992

Displacement, tons: 1,224 full load
Dimensions, feet (metres): 243.1 × 37.4 × 10.5 *(74.1 × 11.4 × 3.2)*
Main machinery: 2 SEMT-Pielstick 16 PA6 V 280 diesels; 12,800 hp(m) *(9.41 MW)* sustained; 2 shafts; cp props
Speed, knots: 22. **Range, n miles:** 4,250 at 12 kt
Complement: 96 (11 officers)
Guns: 1 Bofors 40 mm/60.
Weapons control: Lynx optical sights.
Radars: Navigation: 2 Decca 1226; I-band.
Helicopters: 1 HAL (Aerospatiale) Chetak.

Comment: Owes something to a NEVESBU (Netherlands) design, being a stretched version of its 750 ton offshore patrol vessels. Ordered in 1979. Fin stabilisers. Diving equipment. 4.5 ton deck crane. External firefighting pumps. Has one GRP boat and two inflatable craft. This class is considered too small for its required task and hence the need for the larger Samar class.
VERIFIED

VIJAYA   *2/2001, Guy Toremans / 0121354*

8 PRIYADARSHINI CLASS (COASTAL PATROL CRAFT) (WPBO)

Name	No	Builders	Commissioned
PRIYADARSHINI	221	Garden Reach, Calcutta	25 May 1992
RAZIA SULTANA	222	Goa Shipyard	18 Nov 1992
ANNIE BESANT	223	Goa Shipyard	7 Dec 1992
KAMLA DEVI	224	Goa Shipyard	20 May 1992
AMRIT KAUR	225	Goa Shipyard	20 Mar 1993
KANAK LATA BAURA	226	Garden Reach, Calcutta	27 Mar 1997
BHIKAJI CAMA	227	Garden Reach, Calcutta	24 Sep 1997
SUCHETA KRIPALANI	228	Garden Reach, Calcutta	16 Mar 1998

Displacement, tons: 306 full load
Dimensions, feet (metres): 150.9 × 24.6 × 6.2 *(46.0 × 7.5 × 1.9)*
Main machinery: 2 MTU 12V 538 diesels; 4,025 hp(m) *(2.96 MW)* sustained; 2 shafts
Speed, knots: 23. **Range, n miles:** 2,400 at 12 kt
Complement: 34 (7 officers)
Guns: 1 Bofors 40 mm/60.
 2—7.62 mm MGs.
Radars: Surface search: Racal Decca 1226 or BEL 1245/6X (221-225); I-band.

Comment: A development of the Tara Bai class. *Razia Sultana* (222), previously thought to have been lost at sea, remains in commission.
UPDATED

KAMLA DEVI   *3/2004 *, Bob Fildes / 1042270*

4 SAMAR CLASS (OFFSHORE PATROL VESSELS) (WPSOH)

Name	No	Builders	Laid down	Launched	Commissioned
SAMAR	42	Goa Shipyard	1990	26 Aug 1992	14 Feb 1996
SANGRAM	43	Goa Shipyard	1992	18 Mar 1995	29 Mar 1997
SARANG	44	Goa Shipyard	1993	8 Mar 1997	21 June 1999
SAGAR	45	Goa Shipyard	1999	14 Dec 2001	3 Nov 2003

Displacement, tons: 2,005 full load
Dimensions, feet (metres): 334.6 oa; 315 wl × 37.7 × 11.5
 (102; 96 × 11.5 × 3.5)
Main machinery: 2 SEMT-Pielstick 16 PA6 V 280 diesels; 12,800 hp(m) *(9.41 MW)* sustained; 2 shafts; LIPS cp props
Speed, knots: 22. **Range, n miles:** 7,000 at 15 kt
Complement: 124 (12 officers)

Guns: 1 OTO Melara 3 in *(76 mm)*/62; 60 rds/min to 16 km *(8.7 n miles)*; weight of shell 6 kg.
 2—12.7 mm MGs.
Weapons control: BEL/Radamec optronic 2400 director.
Radars: Surface search: Decca 2459; F/I-band.
Navigation: BEL 1245; I-band.

Helicopters: 1 Chetak.

Programmes: First three ordered in April 1991. Fourth of class ordered 1999.
Structure: Similar to the Navy's Sukanya class but more heavily armed and carrying a helicopter capable of transporting a Marine contingent. Telescopic hangar.
VERIFIED

SANGRAM   *9/2003, Hachiro Nakai / 0572431*

6 TARA BAI CLASS (COASTAL PATROL CRAFT) (WPBO)

Name	No	Builders	Commissioned
TARA BAI	71	Singapore SBEC	26 June 1987
AHALYA BAI	72	Singapore SBEC	9 Sep 1987
LAKSHMI BAI	73	Garden Reach, Calcutta	20 Mar 1989
AKKA DEVI	74	Garden Reach, Calcutta	9 Aug 1989
NAIKI DEVI	75	Garden Reach, Calcutta	19 Mar 1990
GANGA DEVI	76	Garden Reach, Calcutta	19 Nov 1990

Displacement, tons: 195 full load
Dimensions, feet (metres): 147.3 × 23.0 × 6.23 *(44.9 × 7.0 × 1.9)*
Main machinery: 2 MTU 12V 538 diesels; 4,025 hp(m) *(2.96 MW)* sustained; 2 shafts
Speed, knots: 26. **Range, n miles:** 2,400 at 12 kt
Complement: 34 (7 officers)
Guns: 1 Bofors 40 mm/60.
2—7.6 mm MGs.
Radars: Surface search: Racal Decca 1226 or BEL 1245/6X (221-225); I-band.

Comment: Two ordered in June 1986 with license to build further four in India. These were laid down in 1987. *UPDATED*

AKKA DEVI   *6/2000*, Indian Navy* / 1042263

2 + 5 SAROJINI NAIDU CLASS (WPBO)

Name	No	Builders	Commissioned
SAROJANI NAIDU	229	Goa Shipyard	11 Nov 2002
DURGABAI DESHMUKH	230	Goa Shipyard	30 Apr 2003

Displacement, tons: 260 full load
Dimensions, feet (metres): 157.8 × 24.6 × 6.6 *(48.1 × 7.5 × 2)*
Main machinery: 3 MTU-F 16V4000 M90 diesels; total of 10,942 hp(m) *(8.2 MW)* sustained; 3 Kamewa 71SII waterjets
Speed, knots: 35
Complement: 35
Guns: 1—30 mm.
2—12.7 mm MGs.
Radars: Surface search: to be announced.

Comment: A new class of patrol ship designed and developed by Goa Shipyard. Following the initial delivery of two vessels, an order for a further five was made in 2004. The first of these is to be delivered in March 2006 and the remainder at three month intervals. *UPDATED*

SAROJINI NAIDU   *11/2002, Indian Coast Guard* / 0530081

7 JIJA BAI MOD 1 CLASS (TYPE 956)
(COASTAL PATROL CRAFT) (WPBO)

Name	No	Builders	Commissioned
JIJA BAI	64	Sumidagawa, Tokyo	22 Feb 1984
CHAND BIBI	65	Sumidagawa, Tokyo	22 Feb 1984
KITTUR CHENNAMMA	66	Sumidagawa, Tokyo	21 Oct 1983
RANI JINDAN	67	Sumidagawa, Tokyo	21 Oct 1983
HABBAH KHATUN	68	Garden Reach, Calcutta	27 Apr 1985
RAMADEVI	69	Garden Reach, Calcutta	3 Aug 1985
AVVAIYYAR	70	Garden Reach, Calcutta	19 Oct 1985

Displacement, tons: 181 full load
Dimensions, feet (metres): 144.3 × 24.3 × 7.5 *(44 × 7.4 × 2.3)*
Main machinery: 2 MTU 12V 538 TB82 diesels; 4,025 hp(m) *(2.96 MW)* sustained; 2 shafts
Speed, knots: 25. **Range, n miles:** 2,375 at 14 kt
Complement: 34 (7 officers)
Guns: 1 Bofors 40 mm/60. 2—7.62 mm MGs.
Radars: Surface search: Racal Decca 1226; I-band.

Comment: All were ordered in 1981 and are similar to those in service with the Philippines Coast Guard. *VERIFIED*

RANI JINDAN   *6/1996, Indian Coast Guard* / 0064732

2 SDB Mk 2 RAJ CLASS (COASTAL PATROL CRAFT) (WPB)

Name	No	Builders	Commissioned
RAJKIRAN	59	Garden Reach, Calcutta	29 Mar 1984
RAJKAMAL	61	Garden Reach, Calcutta	19 Sep 1986

Displacement, tons: 203 full load
Dimensions, feet (metres): 123 × 24.6 × 5.9 *(37.5 × 7.5 × 1.8)*
Main machinery: 2 MTU 12V 538 diesels; 4,025 hp(m) *(2.96 MW)* sustained; 2 shafts
Speed, knots: 29. **Range, n miles:** 1,400 at 14 kt
Complement: 28 (4 officers)
Guns: 1 Bofors 40/60 mm.
Radars: Surface search: Racal Decca; I-band.

Comment: Earlier vessels of this class belonged to the Navy but have been scrapped. *VERIFIED*

RAJKAMAL   *6/2000, Indian Coast Guard* / 0104591

1 SWALLOW 65 CLASS (WPB)

C 63

Displacement, tons: 32 full load
Dimensions, feet (metres): 65.6 × 15.4 × 5 *(20 × 4.7 × 1.5)*
Main machinery: 2 Detroit 12V-71TA diesels; 840 hp *(627 kW)* sustained; 2 shafts
Speed, knots: 20. **Range, n miles:** 400 at 20 kt
Complement: 9 (1 officer)
Guns: 1—7.62 mm MG.
Radars: Navigation: I-band.

Comment: Built by Swallow Craft Co, Pusan, South Korea in early 1980s. Last remaining craft in service. *VERIFIED*

C 63   *1982, Swallow Craft* / 0012534

11 INSHORE PATROL CRAFT (WPB)

C 131-138   C 140   C 141-142

Displacement, tons: 49 full load
Dimensions, feet (metres): 68.2 × 19 × 5.9 *(20.8 × 5.8 × 1.8)*
Main machinery: 2 Deutz MWM TBD234V12 diesels; 1,646 hp(m) *(1.21 MW)* sustained; 1 Deutz MWM TBD234V8 diesel; 550 hp(m) *(404 kW)* sustained; 3 Hamilton 402 water-jets
Speed, knots: 25. **Range, n miles:** 600 at 15 kt
Complement: 8 (1 officer)
Guns: 1—12.7 mm MG.
Radars: Navigation: Furuno; I-band.

Comment: Ordered from Anderson Marine, Goa in September 1990 to a P-2000 design by Amgram, similar to British Archer class. GRP hull. Official description is 'Interceptor Boats'. All built at Goa. Commissioned: *C 131-132* on 16 November 1993, *C 133-134* on 20 May 1994, *C 135-136* on 16 February 1995, *C 137-138* on 4 September 1996, and *C 139* on 16 October 1997. *C 139* was leased to Mauritius in 2001. Two further craft (C 141-142) of unknown type were commissioned on 8 February 2002. An interceptor boat, C 140, was commissioned on 15 November 2003. *UPDATED*

C 36 (old number)   *2/2001, Sattler/Steele* / 0121360

C 141   *6/2004*, Kapil Chandni* / 1042268

6 GRIFFON 8000 TD(M) CLASS HOVERCRAFT (UCAC)

H 181-186

Displacement, tons: 18.2; 24.6 full load
Dimensions, feet (metres): 69.5 × 36.1 × 1 *(21.15 × 11 × 0.32)*
Main machinery: 2 MTU 12V 183 TB 32 V12 diesels; 800 hp *(596 kW)*
Speed, knots: 50. **Range, n miles:** 400 at 45 kt
Complement: 13 (2 officers)
Guns: 1—12.7 mm MG.
Radar: Raytheon R-80; I-band.

Comment: Six hovercraft were ordered from GRSE Calcutta in May 1999 for construction in technical collaboration with Griffon UK. The first craft H 181 was commissioned on 18 September 2000, four further in 2001, and the final one on 21 March 2002. ***UPDATED***

H 185
6/2004*, Kapil Chandni / 1042269

Indonesia
TENTARA NASIONAL

Country Overview

The Republic of Indonesia gained full independence from the Netherlands in 1949. Straddling the equator, the country comprises more than 13,670 islands, of which some 6,000 are inhabited. The major islands include Sumatra, Java, Sulawesi (Celebes), southern Borneo (Kalimantan) and western New Guinea (Irian Jaya). Smaller islands include Madura, western Timor, Lombok, Sumbawa, Flores, and Bali. The Moluccas and Lesser Sunda Islands are the largest island groups. The coastline of 29,550 n miles is with the South China Sea, the Celebes Sea, the Pacific Ocean and the Indian Ocean. The total land area is 741,903 square miles. The capital, largest city and principal port is Jakarta (Java). Further main ports are at Surabaya (Java), Medan (Sumatra) and Ujung Pandang (Sulawesi). An archipelagic state, territorial seas (12 n miles) are claimed. A 200 n mile EEZ has also been claimed but the limits are only partly defined by boundary agreements.

Headquarters Appointments

Chief of the Naval Staff:
   Admiral Bernard Kent Sondakh
Deputy Chief of the Naval Staff:
   Vice Admiral Sahrani Kasnadi
Inspector General of the Navy:
   Major General Sudarsono Kadsi

Fleet Command

Commander-in-Chief Western Fleet (Jakarta):
   Rear Admiral Mualimin Santoso
Commander-in-Chief Eastern Fleet (Surabaya):
   Rear Admiral Wayan Argawa
Commandant of Navy Marine Corps:
   Major General Suharto
Commander, Military Sealift Command:
   Rear Admiral Rudolf Ompusunggu

Fleet Command — *continued*

Commander, Naval Training Command:
   Rear Admiral Stanny Fofied

Personnel

2005:
a)   56,000 (including 20,000 Marine Commando Corps and 1,000 Naval Air Arm)
b)   Selective national service

Bases

Tanjung Priok (North Jakarta), Ujung (Surabaya), Sabang, Belawan (North Sumatera), Ujung Pandang (South Sulawesi), Balikpapan (East Kalimantan), Jayapura (Irian Jaya), Tanjung Pinang, Bitung (North Sulawesi), Teluk Ratai (South Sumatera), Banjarmasin, South Kalmantan. Naval Air Base at Juanda (Surabaya), Biak (Irian Jaya), Pekan Baru, Sam Ratulangi (North Sulawesi), Sabang, Natuna, P Aru.

Command Structure

Eastern Command (Surabaya)
Western Command (Jakarta)
Training Command
Military Sea Communications Command (Maritime Security Agency)
Military Sealift Command (Logistic Support)

Marine Corps

Reorganisation in March 2001 created the 1st Marine Corps Group (1st, 3rd and 5th battalions) based at Surabaya and the Independent Marine Corps Brigade (2nd, 4th and 6th battalions) based in Jakarta. A new formation (7th, 8th and 9th battalions) is to be based at Teluk Ratai, Sumatra. Equipment includes amphibious tanks, field artillery and anti-aircraft missiles and guns. There are plans to expand the Corps to 22,800 by 2009. Further reorganisation is expected to include relocation of the eastern command from Surabaya to Makassar and the central command from Jakarta to Surabaya.

Strength of the Fleet

Type	Active	Building (Projected)
Patrol Submarines	2	2
Frigates	9	—
Corvettes	19	3 (1)
Fast Attack Craft—Missile	4	—
Large Patrol Craft	21	—
Patrol craft	5	(9)
LPD	1	(4)
LST/LSM	26	—
MCMV	12	—
Survey and Research Ships	9	—
Command Ship	1	—
Repair Ship	1	—
Replenishment Tankers	2	—
Coastal Tankers	3	—
Support Ships	7	—
Transports	1	—
Sail Training Ships	2	—

Prefix to Ships' Names

KRI (Kapal di Republik Indonesia)

PENNANT LIST

Submarines

401	Cakra
402	Nanggala

Frigates

341	Samadikun
342	Martadinata
351	Ahmad Yani
352	Slamet Riyadi
353	Yos Sudarso
354	Oswald Siahann
355	Abdul Halim Perdanakusuma
356	Karel Satsuitubun
364	Ki Hajar Dewantara

Corvettes

361	Fatahillah
362	Malahayati
363	Nala
371	Kapitan Patimura
372	Untung Suropati
373	Nuku
374	Lambung Mangkurat
375	Cut Nyak Dien
376	Sultan Thaha Syaifuddin
377	Sutanto
378	Sutedi Senoputra
379	Wiratno
380	Memet Sastrawiria
381	Tjiptadi
382	Hasan Basri
383	Iman Bonjol
384	Pati Unus
385	Teuku Umar
386	Silas Papare

Patrol Forces

621	Mandau
622	Rencong
623	Badik
624	Keris
651	Singa
653	Ajak
801	Pandrong
802	Sura
803	Todak
804	Hiu
805	Layang (bldg)
806	Lemadang (bldg)
811	Kakap
812	Kerapu
813	Tongkol
814	Barakuda
847	Sibarau
848	Siliman
857	Sigalu
858	Silea
859	Siribua
862	Siada
863	Sikuda
864	Sigurot
—	Cucuk

Amphibious Forces

501	Teluk Langsa
502	Teluk Bajur
503	Teluk Amboina
504	Teluk Kau
508	Teluk Tomini
509	Teluk Ratai
510	Teluk Saleh
511	Teluk Bone
512	Teluk Semangka
513	Teluk Penju
514	Teluk Mandar
515	Teluk Sampit
516	Teluk Banten
517	Teluk Ende
531	Teluk Gilimanuk
532	Teluk Celukan Bawang
533	Teluk Cendrawasih
534	Teluk Berau
535	Teluk Peleng
536	Teluk Sibolga
537	Teluk Manado
538	Teluk Hading
539	Teluk Parigi
540	Teluk Lampung
541	Teluk Jakarta
542	Teluk Sangkulirang
580	Dore
582	Kupang
583	Dili
584	Nusantara

Survey Ships

KAL-IV-02	Baruna Jaya I
KAL-IV-03	Baruna Jaya II
KAL-IV-04	Baruna Jaya III
KAL-IV-05	Baryna Jaya IV
931	Burujulasad
932	Dewa Kembar
933	Jalanidhi

Mine Warfare Forces

701	Pulau Rani
711	Pulau Rengat
712	Pulau Rupat
721	Pulau Rote
722	Pulau Raas
723	Pulau Romang
724	Pulau Rimau
725	Pulau Rondo
726	Pulau Rusa
727	Pulau Rangsang
728	Pulau Raibu
729	Pulau Rempang

Auxiliaries

543	Teluk Cirebon
544	Teluk Sabang
561	Multatuli
901	Balikpapan
902	Sambu
903	Arun
906	Sungai Gerong
911	Sorong
921	Jaya Wijaya
922	Rakata
923	Soputan
934	Lampo Batang
935	Tambora
936	Bromo
952	Nusa Telu
959	Teluk Mentawai
960	Karimata
961	Waigeo
972	Tanjung Oisina

SUBMARINES

Notes: Two ex-German Type 206 submarines were taken over on 25 September 1997 with plans to refit them, followed by three others. Funds ran out in June 1998 and the whole project was then cancelled. New plans to acquire two submarines from South Korea were announced in October 2003. Delivery from 2008 is a reported requirement and this may point to the Chang Bogo class. Funding is reported to be provided through a counter-trade agreement involving CN-235 aircraft to South Korea.

2 CAKRA (209) CLASS (1300 TYPE) (SSK)

Name	No	Builders	Laid down	Launched	Commissioned
CAKRA	401	Howaldtswerke, Kiel	25 Nov 1977	10 Sep 1980	19 Mar 1981
NANGGALA	402	Howaldtswerke, Kiel	14 Mar 1978	10 Sep 1980	6 July 1981

Displacement, tons: 1,285 surfaced; 1,390 dived
Dimensions, feet (metres): 195.2 × 20.3 × 17.9 *(59.5 × 6.2 × 5.4)*
Main machinery: Diesel-electric; 4 MTU 12V 493 AZ80 GA31L diesels; 2,400 hp(e) *(1.76 MW)* sustained; 4 Siemens alternators; 1.7 MW; 1 Siemens motor; 4,600 hp(m) *(3.38 MW)* sustained; 1 shaft
Speed, knots: 11 surfaced; 21.5 dived

Range, n miles: 8,200 at 8 kt
Complement: 34 (6 officers)

Torpedoes: 8—21 in *(533 mm)* bow tubes. 14 AEG SUT Mod 0; dual purpose; wire-guided; active/passive homing to 12 km *(6.5 n miles)* at 35 kt; 28 km *(15 n miles)* at 23 kt; warhead 250 kg.
Countermeasures: ESM: Thomson-CSF DR 2000U; radar warning.
Weapons control: Signaal Sinbad system.

Radars: Surface search: Thomson-CSF Calypso; I-band.
Sonars: Atlas Elektronik CSU 3-2; active/passive search and attack; medium frequency. PRS-3/4; (integral with CSU) passive ranging.

Programmes: Ordered on 2 April 1977. Designed by Ingenieurkontor, Lübeck for construction by Howaldtswerke, Kiel and sale by Ferrostaal, Essen—all acting as a consortium.
Modernisation: Major refits at HDW spanning three years from 1986 to 1989. These refits were expensive and lengthy and may have discouraged further orders at that time. *Cakra* refitted again at Surabaya from 1993 completing in April 1997, including replacement batteries and updated Sinbad TFCS. *Nanggala* received a similar refit from October 1997 to mid-1999. *Cakra* began a refit at Daewoo Shipyard, South Korea in 2004. Work is likely to include new batteries, overhaul of engines and modernisation of the combat system and is to be completed in 2006. A similar refit of *Nanggala* is also under consideration.
Structure: Have high-capacity batteries with GRP lead-acid cells and battery cooling supplied by Wilhelm Hagen AG. Diving depth, 240 m *(790 ft)*.
Operational: Endurance, 50 days. Operational status of both boats is doubtful until refits have been completed. ***UPDATED***

NANGGALA   *8/1999, van Ginderen Collection / 0080001*

FRIGATES

6 AHMAD YANI (VAN SPEIJK) CLASS (FFGHM)

Name	No	Builders	Laid down	Launched	Commissioned
AHMAD YANI (ex-*Tjerk Hiddes*)	351	Nederlandse Dok en Scheepsbouw Mij, Amsterdam	1 June 1964	17 Dec 1965	16 Aug 1967
SLAMET RIYADI (ex-*Van Speijk*)	352	Nederlandse Dok en Scheepsbouw Mij, Amsterdam	1 Oct 1963	5 Mar 1965	14 Feb 1967
YOS SUDARSO (ex-*Van Galen*)	353	Koninklijke Maatschappij de Schelde, Flushing	25 July 1963	19 June 1965	1 Mar 1967
OSWALD SIAHAAN (ex-*Van Nes*)	354	Koninklijke Maatschappij de Schelde, Flushing	25 July 1963	26 Mar 1966	9 Aug 1967
ABDUL HALIM PERDANAKUSUMA (ex-*Evertsen*)	355	Koninklijke Maatschappij de Schelde, Flushing	6 July 1965	18 June 1966	21 Dec 1967
KAREL SATSUITUBUN (ex-*Isaac Sweers*)	356	Nederlandse Dok en Scheepsbouw Mij, Amsterdam	5 May 1965	10 Mar 1967	15 May 1968

Displacement, tons: 2,225 standard; 2,835 full load
Dimensions, feet (metres): 372 × 41 × 13.8 *(113.4 × 12.5 × 4.2)*
Main machinery: 2 Babcock & Wilcox boilers; 550 psi *(38.7 kg/cm²)*; 850°F *(450°C)*; 2 Werkspoor/English Electric turbines; 30,000 hp *(22.4 MW)*; 2 shafts
Speed, knots: 28.5. **Range, n miles:** 4,500 at 12 kt
Complement: 180

Missiles: SSM: 8 McDonnell Douglas Harpoon ❶; active radar homing to 130 km *(70 n miles)* at 0.9 Mach; warhead 227 kg.
SAM: 2 Short Brothers Seacat quad launchers.
Being replaced by 2 Matra Simbad twin launchers for Mistral ❷; IR homing to 4 km *(2.2 n miles)*; warhead 3 kg.
Guns: 1 OTO Melara 3 in *(76 mm)/*62 compact ❸; 85 rds/min to 16 km *(8.7 n miles)* anti-surface; 12 km *(6.6 n miles)* anti-aircraft; weight of shell 6 kg. 4—12.7 mm MGs.
Torpedoes: 6—324 mm Mk 32 (2 triple) tubes ❹. Honeywell Mk 46; anti-submarine; active/passive homing to 11 km *(5.9 n miles)* at 40 kt; warhead 44 kg.
Countermeasures: Decoys: 2 Knebworth Corvus 8-tubed trainable; radar distraction or centroid chaff to 1 km.
ESM: UA 8/9; UA 13 (355 and 356); radar warning. FH5 D/F.
Combat data systems: SEWACO V action data automation and Daisy data processing.
Weapons control: Signaal LIOD optronic director. Mk 2 fitted in 354, 353 and 356. SWG-1A Harpoon LCS.
Radars: Air search: Signaal LW03 ❺; D-band; range 219 km *(120 n miles)* for 2 m² target.
Air/surface search: Signaal DA05 ❻; E/F-band; range 137 km *(75 n miles)* for 2 m² target.
Navigation: Racal Decca 1229; I-band.
Fire control: Signaal M 45 ❼; I/J-band (for 76 mm gun and SSM). 2 Signaal M 44; I/J-band (for Seacat) (being removed).
Sonars: Signaal CWE 610; hull-mounted; active search and attack; medium frequency. VDS; medium frequency.

Helicopters: 1 Westland Wasp ❽ or NBO-105C.

Programmes: On 11 February 1986 agreement signed with the Netherlands for transfer of two of this class with an option on two more. Transfer dates: *Tjerk Hiddes*, 31 October 1986; *Van Speijk*, 1 November 1986; *Van Galen*, 2 November 1987; *Van Nes*, 31 October 1988. Contract of sale for the last two of the class signed 13 May 1989. *Evertsen* transferred 1 November 1989 and *Isaac Sweers* 1 November 1990. Ships provided with all spare parts but not towed arrays or helicopters.
Modernisation: This class underwent mid-life modernisation at Rykswerf Den Helder from 1976. This included replacement of 4.5 in turret by 76 mm, A/S mortar by torpedo tubes, new electronics and electrics, updating combat data system, improved communications, extensive automation with reduction in complement, enlarged hangar for Lynx and improved habitability. Harpoon for first two only initially because there was no FMS funding for the others. However the USN then provided sufficient SWG 1A panels for all of the class to be retrofitted with Harpoon missiles. LIOD optronic directors Mk 2 fitted in 354, 353 and 356 in 1996-97. Seacat which is non-operational is being replaced by Simbad twin launchers when funds are available. Plans to recondition all six ships were announced by the Chief of Naval Staff on 31 May 2002 but there have been no firm indications that work has started.
Operational: Operational availability has been drastically reduced by propulsion problems. Harpoon missiles are reported to be time-expired. ***UPDATED***

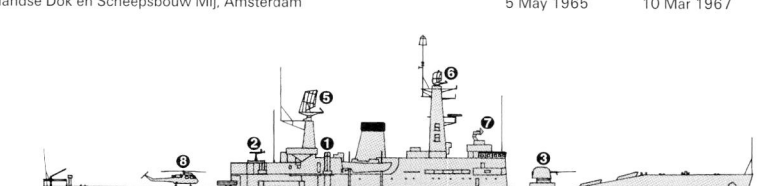

AHMAD YANI (with Simbad)   *(Scale 1 : 1,200), Ian Sturton / 0012535*

KAREL SATSUITUBUN   *10/2004*, D Pawlenko, RAN / 1044131*

KAREL SATSUITUBUN   *11/2004*, Chris Gee / 1047873*

1 KI HAJAR DEWANTARA CLASS (FFGH/FFT)

Name	No	Builders	Laid down	Launched	Commissioned
KI HAJAR DEWANTARA	364	Split SY, Yugoslavia	11 May 1979	11 Oct 1980	31 Oct 1981

Displacement, tons: 2,050 full load

Dimensions, feet (metres): 317.3 × 36.7 × 15.7
(96.7 × 11.2 × 4.8)

Main machinery: CODOG; 1 RR Olympus TM3B gas turbine;
24,525 hp *(18.3 MW)* sustained; 2 MTU 16V 956 TB92
diesels; 11,070 hp(m) *(8.14 MW)* sustained; 2 shafts; cp props

Speed, knots: 26 gas; 20 diesels

Range, n miles: 4,000 at 18 kt; 1,150 at 25 kt

Complement: 76 (11 officers) plus 14 instructors and 100 cadets

Missiles: SSM: 4 Aerospatiale MM 38 Exocet ❶; inertial cruise;
active radar homing to 42 km *(23 n miles)* at 0.9 Mach;
warhead 165 kg; sea-skimmer.

Guns: 1 Bofors 57 mm/70 ❷; 200 rds/min to 17 km *(9.3 n
miles)*; weight of shell 2.4 kg.
2 Rheinmetall 20 mm ❸.

Torpedoes: 2—21 in *(533 mm)* tubes ❹. AEG SUT; dual
purpose; wire-guided; active/passive homing to 28 km *(15 n
miles)* at 23 kt; 12 km *(6.5 n miles)* at 35 kt; warhead
250 kg.

Depth charges: 1 projector/mortar.

Countermeasures: Decoys: 2—128 mm twin-tubed flare
launchers.

ESM: MEL Susie; radar intercept.

Combat data systems: Signaal SEWACO-RI action data
automation.

Radars: Surface search: Racal Decca 1229 ❺; I-band.

Fire control: Signaal WM28 ❻; I/J-band.

Sonars: Signaal PHS-32; hull-mounted; active search and attack;
medium frequency.

Helicopters: Platform ❼ for 1 NBO-105 helicopter.

Programmes: First ordered 14 March 1978 from Split SY,
Yugoslavia where the hull was built and engines fitted.
Armament and electronics fitted in the Netherlands and
Indonesia.

Structure: For the training role there is a classroom and
additional wheelhouse, navigation and radio rooms. Torpedo
tubes are fixed in the stern transom. Two LCVP-type ship's
boats are carried.

Operational: Used for training and troop transport. War roles
include escort, ASW and troop transport.

UPDATED

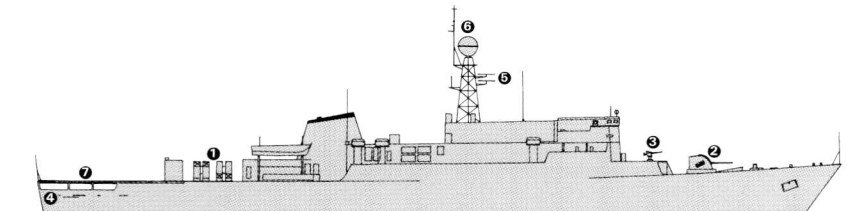

KI HAJAR DEWANTARA   *(Scale 1 : 900), Ian Sturton*

KI HAJAR DEWANTARA   *12/1992 /* 0080005

2 SAMADIKUN (CLAUD JONES) CLASS (FF)

Name	No	Builders	Laid down	Launched	Commissioned
SAMADIKUN (ex-*John R Perry* DE 1034)	341	Avondale Marine Ways	1 Oct 1957	29 July 1958	5 May 1959
MARTADINATA (ex-*Charles Berry* DE 1035)	342	American SB Co, Toledo, OH	29 Oct 1958	17 Mar 1959	25 Nov 1959

Displacement, tons: 1,720 standard; 1,968 full load

Dimensions, feet (metres): 310 × 38.7 × 18
(95 × 11.8 × 5.5)

Main machinery: 2 Fairbanks-Morse 38TD 8-1/8-12 diesels (not
in 343); 7,000 hp *(5.2 MW)* sustained; 1 shaft

Speed, knots: 22. **Range, n miles:** 3,000 at 18 kt

Complement: 171 (12 officers)

Guns: 1 or 2 US 3 in *(76 mm)*/50 Mk 34; 50 rds/min to 12.8 km
(7 n miles); weight of shell 6 kg.
2 USSR 37 mm/63 (twin); 160 rds/min to 9 km *(5 n miles)*;
weight of shell 0.7 kg.

Torpedoes: 6—324 mm Mk 32 (2 triple) tubes. Honeywell Mk
46; anti-submarine; active/passive homing to 11 km *(5.9 n
miles)* at 40 kt; warhead 44 kg.

Depth charges: 2 DC throwers.

Countermeasures: ESM: WLR-1C (except *Samadikun*); radar
warning.

Weapons control: Mk 70 Mod 2 for guns.

Radars: Air search: Westinghouse SPS-6E; D-band; range
146 km *(80 n miles)* (for fighter).
Surface search: Raytheon SPS-5D; G/H-band; range 37 km *(20 n
miles)*.
Raytheon SPS-4 *(Ngurahrai)*; G/H-band.
Navigation: Racal Decca 1226; I-band.
Fire control: Lockheed SPG-52; K-band.

Sonars: EDO *(Samadikun)*; SQS-45V *(Martadinata)*; SQS-39V
(Monginsidi); SQS-42V *(Ngurahrai)*; hull-mounted; active
search and attack; medium/high frequency.

Programmes: *Samadikun* transferred from USA 20 February
1973; *Martadinata*, 31 January 1974; *Monginsidi* and
Ngurahrai, 16 December 1974. All refitted at Subic Bay
1979-82.

Modernisation: The Hedgehog A/S mortars have been removed,
as have the 25 mm guns. Some have a second 76 mm gun
vice the 37 mm.

Operational: It was planned that the Van Speijk class would
replace these ships. Two have been deleted and the
operational status of these remaining two is doubtful.

UPDATED

SAMADIKUN   *10/2001, Chris Sattler /* 0121379

CORVETTES

3 FATAHILLAH CLASS (FFG/FFGH)

Name	No	Builders	Laid down	Launched	Commissioned
FATAHILLAH	361	Wilton Fijenoord, Schiedam	31 Jan 1977	22 Dec 1977	16 July 1979
MALAHAYATI	362	Wilton Fijenoord, Schiedam	28 July 1977	19 June 1978	21 Mar 1980
NALA	363	Wilton Fijenoord, Schiedam	27 Jan 1978	11 Jan 1979	4 Aug 1980

Displacement, tons: 1,200 standard; 1,450 full load
Dimensions, feet (metres): 276 × 36.4 × 10.7
(84 × 11.1 × 3.3)
Main machinery: CODOG; 1 RR Olympus TM3B gas turbine;
25,440 hp *(19 MW)* sustained; 2 MTU 20V 956 TB92 diesels;
11,070 hp(m) *(8.14 MW)* sustained; 2 shafts; LIPS cp props
Speed, knots: 30. **Range, n miles:** 4,250 at 16 kt
Complement: 89 (11 officers)

Missiles: SSM: 4 Aerospatiale MM 38 Exocet ❶; inertial cruise;
active radar homing to 42 km *(23 n miles)* at 0.9 Mach;
warhead 165 kg; sea-skimmer.
Guns: 1 Bofors 4.7 in *(120 mm)*/46 ❷; 80 rds/min to 18.5 km
(10 n miles); weight of shell 21 kg.
1 or 2 Bofors 40 mm/70 (2 in *Nala*) ❸; 300 rds/min to 12 km
(6.6 n miles); weight of shell 0.96 kg.
2 Rheinmetall 20 mm; 1,000 rds/min to 2 km anti-aircraft;
weight of shell 0.24 kg.
Torpedoes: 6—324 mm Mk 32 or ILAS 3 (2 triple) tubes
(none in *Nala*) ❹. 12 Mk 46 (or A244S); anti-submarine;
active/passive homing to 11 km *(5.9 n miles)* at 40 kt;
warhead 44 kg.
A/S mortars: 1 Bofors 375 mm twin-barrelled trainable ❺; 54
Erika; range 1,600 m and Nelli; range 3,600 m.

Countermeasures: Decoys: 2 Knebworth Corvus 8-tubed
trainable chaff launchers ❻; radar distraction or centroid
modes to 1 km. 1 T-Mk 6; torpedo decoy.
ESM: MEL Susie 1 (UAA-1); radar intercept.
Combat data systems: Signaal SEWACO-RI action data
automation.
Weapons control: Signaal LIROD optronic director.
Radars: Air/surface search: Signaal DA05 ❼; E/F-band; range
137 km *(75 n miles)* for 2 m² target.
Surface search: Racal Decca AC 1229 ❽; I-band.
Fire control: Signaal WM28 ❾; I/J-band; range 46 km *(25 n
miles)*.

Sonars: Signaal PHS-32; hull-mounted; active search and attack;
medium frequency.

Helicopters: 1 Westland Wasp (*Nala* only) ❿.

Programmes: Ordered August 1975. Officially rated as
Corvettes.
Structure: NEVESBU design. *Nala* is fitted with a folding hangar/
landing deck.
Operational: These ships are the busiest of the larger warships.
Three successful Exocet (locally modified after life-expiry)
firings conducted on 25 August 2002. **UPDATED**

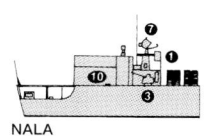

NALA   FATAHILLAH   *(Scale 1 : 1,200), Ian Sturton /* 0126692/0121374

NALA   *6/2000, van Ginderen Collection /* 0104593

FATAHILLAH   *11/2004*, Chris Gee /* 1047876

16 KAPITAN PATIMURA (PARCHIM I) CLASS (PROJECT 1331) (FS)

Name	No	Builders	Commissioned	Recommissioned
KAPITAN PATIMURA (ex-*Prenzlau*)	371 (ex-231)	Peenewerft, Wolgast	11 May 1983	23 Sep 1993
UNTUNG SUROPATI (ex-*Ribnitz*)	372 (ex-233)	Peenewerft, Wolgast	29 Oct 1983	23 Sep 1993
NUKU (ex-*Waren*)	373 (ex-224)	Peenewerft, Wolgast	23 Nov 1982	15 Dec 1993
LAMBUNG MANGKURAT (ex-*Angermünde*)	374 (ex-214)	Peenewerft, Wolgast	26 July 1985	12 July 1994
CUT NYAK DIEN (ex-*Lübz*)	375 (ex-P 6169, ex-221)	Peenewerft, Wolgast	12 Feb 1982	25 Feb 1994
SULTAN THAHA SYAIFUDDIN (ex-*Bad Doberan*)	376 (ex-222)	Peenewerft, Wolgast	30 June 1982	25 Feb 1995
SUTANTO (ex-*Wismar*)	377 (ex-P 6170, ex-241)	Peenewerft, Wolgast	9 July 1981	10 Mar 1995
SUTEDI SENOPUTRA (ex-*Parchim*)	378 (ex-242)	Peenewerft, Wolgast	9 Apr 1981	19 Sep 1994
WIRATNO (ex-*Perleberg*)	379 (ex-243)	Peenewerft, Wolgast	19 Sep 1981	19 Sep 1994
MEMET SASTRAWIRIA (ex-*Bützow*)	380 (ex-244)	Peenewerft, Wolgast	30 Dec 1981	2 June 1995
TJIPTADI (ex-*Bergen*)	381 (ex-213)	Peenewerft, Wolgast	1 Feb 1985	10 May 1996
HASAN BASRI (ex-*Güstrow*)	382 (ex-223)	Peenewerft, Wolgast	10 Nov 1982	10 May 1996
IMAN BONJOL (ex-*Teterow*)	383 (ex-P 6168, ex-234)	Peenewerft, Wolgast	27 Jan 1984	26 Apr 1994
PATI UNUS (ex-*Ludwiglust*)	384 (ex-232)	Peenewerft, Wolgast	4 July 1983	21 July 1995
TEUKU UMAR (ex-*Grevesmühlen*)	385 (ex-212)	Peenewerft, Wolgast	21 Sep 1984	27 Oct 1996
SILAS PAPARE (ex-*Gadebusch*)	386 (ex-P 6167, ex-211)	Peenewerft, Wolgast	31 Aug 1984	27 Oct 1996

Displacement, tons: 769 standard
Dimensions, feet (metres): 246.7 × 32.2 × 11.5
  (75.2 × 9.8 × 3.5)
Main machinery: 1 Zvezda M 504A diesel; 4,700 hp *(3.5 MW)*
  for centreline cp prop
  2 Deutz TBD 620 V16 diesels (372, 373, 374, 377, 378, 381);
  6,000 hp *(4.5 MW)*
  or 2 MTU 16V 4000 M 90 diesels (371, 379, 380, 382, 383
  and 386); 7,300 hp *(5.4 MW)*
  or 2 CAT 3516B diesels (355, 376, 384, 385); 5,200 hp
  (3.9 MW); 2 outboard shafts
Speed, knots: 24. **Range, n miles:** 1,750 at 18 kt
Complement: 64 (9 officers)

Missiles: SAM: SA-N-5/8 launchers fitted in some. May be replaced
  by twin Simbad launchers.
Guns: 2 USSR 57 mm/80 (twin) ❶ automatic; 120 rds/min to
  12 km *(6.4 n miles)*; weight of shell 2.8 kg.
  2—30 mm (twin) ❷; 500 rds/min to 5 km *(2.7 n miles)* anti-
  aircraft; weight of shell 0.54 kg.
Torpedoes: 4—400 mm tubes ❸.
A/S mortars: 2 RBU 6000 12-barrelled trainable launchers ❹;
  automatic loading; range 6,000 m; warhead 31 kg.
Depth charges: 2 racks.
Mines: Mine rails fitted.
Countermeasures: Decoys: 2 PK 16 chaff rocket launchers.
  ESM: 2 Watch Dog; radar warning.
Radars: Air/surface search: Strut Curve ❺; F-band; range 110 km
  (60 n miles) for 2 m² target.
  Navigation: TSR 333; I-band.
  Fire control: Muff Cob ❻; G/H-band.
  IFF: High Pole B.
Sonars: MG 332T; hull-mounted; active search and attack; high
  frequency.
  Elk Tail; VDS system on starboard side (in some hulls).

Programmes: Ex-GDR ships mostly paid off in 1991. Formally
  transferred on 4 January 1993 and became Indonesian ships on
  25 August 1993. First three arrived Indonesia in November 1993.
Modernisation: All refitted prior to sailing for Indonesia. Range
  increased and air conditioning added to accommodation. SAM
  launchers can be carried. A re-engining programme is underway
  and is to be completed in 2005.
Structure: Basically very similar to Russian Grisha class but with a
  higher freeboard and different armament.
Operational: Some pennant numbers begin with 8 rather than 3.

UPDATED

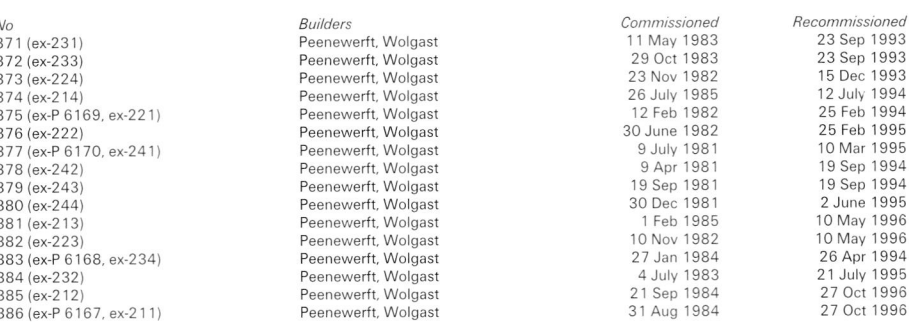

KAPITAN PATIMURA   *(Scale 1 : 600), Ian Sturton*

UNTUNG SUROPATI   *5/2000, M Declerck* / 0104596

0 + 2 (2) SIGMA CLASS (CORVETTES) (FS)

Name	No	Builders	Laid down	Launched	Commissioned
—	—	Royal Schelde, Vlissengen	2005	2007	2008
—	—	Royal Schelde, Vlissengen	2005	2007	2008

Displacement, tons: 1,650 full load
Dimensions, feet (metres): 296.9 × 35.9 × 11.1
  (90.5 × 11.0 × 3.4)
Main machinery: 2 diesels; 21,725 hp *(16.2 MW)*; 2 shafts; cp
  props
Speed, knots: 28. **Range, n miles:** 4,000 at 18 kt
Complement: To be announced

Missiles: SAM: 2 quadruple Tetral launchers ❶; MBDA Mistral; IR
  homing to 4 km *(2.2 n miles)*; warhead 3 kg.
  SSM: 4 MBDA MM 40 Exocet Block II ❷; inertial cruise; active
  radar homing to 70 km *(40 n miles)* at 0.9 Mach; warhead
  165 kg; sea-skimmer.
Guns: 1 OTO Melara 3 in *(76 mm)*/62 Super Rapid ❸; 120 rds/
  min to 16 km *(8.7 n miles)*; weight of shell 6 kg. 2 Giat
  20 mm ❹.
Torpedoes: 6—324 mm (2 triple) tubes.
Combat data systems: Tacticos including Link Y.
Weapons control: LIROD Mk 2 optronic tracker.
Radars: Surface search: Thales MW 08; G-band.
  Navigation: To be announced.
Sonars: Hull mounted, to be announced.

Helicopters: To be announced.

Programmes: Contract for the construction of two corvettes,
  both to be built in the Netherlands, signed on 7 January 2004.
  The role of the ships is to conduct coastal security operations.
  Delivery of the first ship is expected in 2008. The second ship
  is to follow four months later. There is an option to build two
  further craft in Indonesia.

NEW ENTRY

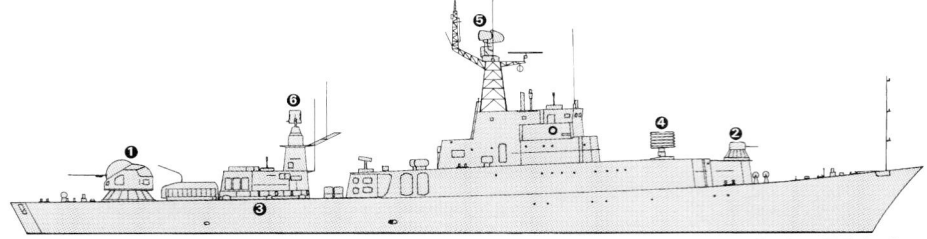

SIGMA CORVETTE   *(Scale 1 : 900), Ian Sturton* / 1044125

SIGMA CORVETTE (artist's impression)
1/2004*, Schelde Naval Shipbuilding / 0563344

0 + 1 (1) PT PAL CORVETTE (FS)

Name	No	Builders	Laid down	Launched	Commissioned
—	—	PT Pal, Surabaya	2005	2006	2007

Displacement, tons: 1,500 full load
Dimensions, feet (metres): 263.8 × 40.0 × 11.4 *(80.4 × 12.2 × 3.46)*
Main machinery: 2 diesels; 19,850 hp *(14.8 MW)*; 2 shafts; cp props
Speed, knots: 25. **Range, n miles:** 3,500 at 14 kt
Complement: To be announced

Guns: To be announced.
Combat data systems: To be announced.
Weapons control: To be announced.
Radars: Surface search: To be announced.
Navigation: To be announced.

Helicopters: To be announced.

Programmes: A programme for the construction of corvettes at the PT PAL Shipyard in Surabaya was officially launched on 8 October 2004 when first steel was cut. Designed in collaboration with Orrizonte Sistemi Navali S.P.A (Fincantieri) and Italian Navy Corvette, the design is probably a development of the Comandante class offshore patrol vessels on which outline details are based. Further orders are expected.

NEW ENTRY

SHIPBORNE AIRCRAFT

Notes: (1) One NB 412 helicopter acquired in August 1996. A total of four is reported in service.
(2) Two Mi-17 medium lift helicopters for the Indonesian Marine Corps were reportedly acquired from Russia in 2002.

Numbers/Type: 6 Nurtanio (MBB) NBO 105C.
Operational speed: 113 kt *(210 km/h)*.
Service ceiling: 9,845 ft *(3,000 m)*.
Range: 407 n miles *(754 km)*.
Role/Weapon systems: Surveillance/support aircraft. A further three for SAR. Sensors: Thomson-CSF AMASCOS surveillance system; Chlio FLIR. Weapons: Unarmed.

VERIFIED

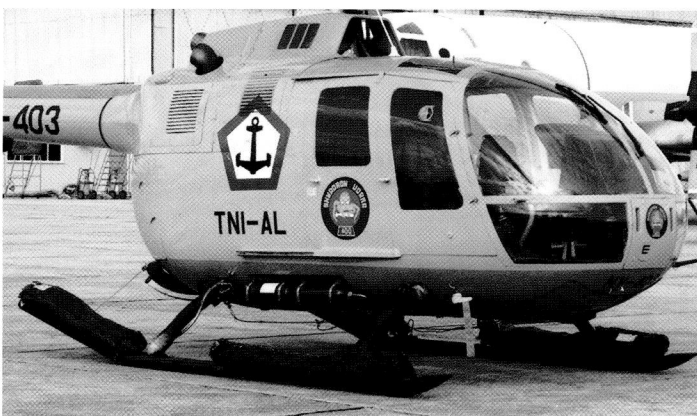

NBO 105C   11/1990 / 0080007

Numbers/Type: 3 Nurtanio (Aerospatiale) NAS-332 Super Puma.
Operational speed: 151 kt *(279 km/h)*.
Service ceiling: 15,090 ft *(4,600 m)*.
Range: 335 n miles *(620 km)*.
Role/Weapon systems: ASW and assault operations with secondary role in utility and SAR; ASVW development possible with Exocet or similar. Sensors: Thomson-CSF Omera radar and Alcatel dipping sonar in some. Weapons: ASW; two Mk 46 torpedoes or depth bombs.

VERIFIED

SUPER PUMA (French colours)   6/1994 / 0080008

Numbers/Type: 3 Westland Wasp (HAS Mk 1).
Operational speed: 96 kt *(177 km/h)*.
Service ceiling: 12,200 ft *(3,720 m)*.
Range: 263 n miles *(488 km)*.
Role/Weapon systems: Shipborne ASW helicopter weapons carrier and reconnaissance; SAR and utility as secondary roles. Probably grounded. Preferred replacement is Westland Navy Lynx. Sensors: None. Weapons: ASW; two Mk 44 or one Mk 46 torpedoes, depth bombs or mines.

VERIFIED

WASP (NZ colours)   4/1997, Maritime Photographic / 0012537

LAND-BASED MARITIME AIRCRAFT (FRONT LINE)

Notes: 15 IPTN NC-212 MPA/Elint aircraft reported delivered since 1984. A further four have been ordered.

Numbers/Type: 3 Boeing 737-200 Surveiller.
Operational speed: 462 kt *(856 km/h)*.
Service ceiling: 50,000 ft *(15,240 m)*.
Range: 2,530 n miles *(4,688 km)*.
Role/Weapon systems: Land based for long-range maritime surveillance roles. Air Force manned. Sensors upgraded in 1993-94 to include IFF. Sensors: Motorola APS-135(v) SLAM MR radar, Thomson-CSF Oceanmaster radar. Weapons: Unarmed.

UPDATED

BOEING 737   9/2003 *, Boeing / 0560018

Numbers/Type: 29 GAF Searchmaster Nomad B and 6 GAF Searchmaster Nomad L.
Operational speed: 168 kt *(311 km/h)*.
Service ceiling: 21,000 ft *(6,400 m)*.
Range: 730 n miles *(1,352 km)*.
Role/Weapon systems: Nomad type built in Australia. Short-range maritime patrol, EEZ protection and anti-smuggler duties. 20 more acquired from Australian Army in August 1997 for use in maritime role. Not all are operational and NC-212 replacements are planned. Sensors: Nose-mounted search radar. Weapons: Unarmed.

VERIFIED

Numbers/Type: 8 Northrop F-5E Tiger II.
Operational speed: 940 kt *(1,740 km/h)*.
Service ceiling: 51,800 ft *(15,790 m)*.
Range: 300 n miles *(556 km)*.
Role/Weapon systems: Fleet air defence and strike fighter, formed 'naval co-operation unit'. Planned to be replaced by BAe Hawk 200 in due course. Sensors: AI radar. Weapons: AD; two AIM-9 Sidewinder, two 20 mm cannon. Strike; 3,175 tons of underwing stores.

UPDATED

Numbers/Type: 9 IPTN (CASA) CN-235.
Operational speed: 240 kt *(445 km/h)*.
Service ceiling: 26,600 ft *(8,110 m)*.
Range: 669 n miles *(1,240 km)*.
Role/Weapon systems: Six surveillance aircraft operated by Air Force. Two surveillance and one utility aircraft operated by Navy. Sensors: Thomson-CSF Oceanmaster radar. Chlio FLIR. Weapons: ASV; may have Exocet AM 39.

UPDATED

CN-235   6/2004 *, CASA/EADS / 1044130

PATROL FORCES

Note: A new 40 m missile craft has been reported under construction and may enter service in 2005.

4 DAGGER CLASS (FAST ATTACK CRAFT—MISSILE) (PTFG)

Name	No	Builders	Commissioned
MANDAU	621	Korea Tacoma, Masan	20 July 1979
RENCONG	622	Korea Tacoma, Masan	20 July 1979
BADIK	623	Korea Tacoma, Masan	Feb 1980
KERIS	624	Korea Tacoma, Masan	Feb 1980

Displacement, tons: 270 full load
Dimensions, feet (metres): 164.7 × 23.9 × 7.5 *(50.2 × 7.3 × 2.3)*
Main machinery: CODOG; 1 GE LM 2500 gas turbine; 23,000 hp *(17.16 MW)* sustained; 2 MTU 12V 331 TC81 diesels; 2,240 hp(m) *(1.65 MW)* sustained; 2 shafts; cp props
Speed, knots: 41 gas; 17 diesel. **Range, n miles:** 2,000 at 17 kt
Complement: 43 (7 officers)
Missiles: SSM: 4 Aerospatiale MM 38 Exocet; inertial cruise; active radar homing to 42 km *(23 n miles)* at 0.9 Mach; warhead 165 kg; sea-skimmer.
Guns: 1 Bofors 57 mm/70 Mk 1; 200 rds/min to 17 km *(9.3 n miles)*; weight of shell 2.4 kg. Launchers for illuminants on each side.
1 Bofors 40 mm/70; 300 rds/min to 12 km *(6.6 n miles)*; weight of shell 0.96 kg.
2 Rheinmetall 20 mm.
Countermeasures: ESM: Thomson-CSF DR 2000S (in 623 and 624); radar intercept.
Weapons control: Selenia NA-18 optronic director.
Radars: Surface search: Racal Decca 1226; I-band.
Fire control: Signaal WM28; I/J-band.

Programmes: PSMM Mk 5 type craft ordered in 1975.
Structure: Shorter in length and smaller displacement than South Korean units. *Mandau* has a different shaped mast with a tripod base.

UPDATED

RENCONG   10/1998 / 0052358

4 TODAK (PB 57) CLASS (NAV V)
(LARGE PATROL CRAFT) (PBO)

Name	No	Builders	Commissioned
TODAK	803	PT Pal Surabaya	4 May 2000
HIU	804	PT Pal Surabaya	Sep 2000
LAYANG	805	PT Pal Surabaya	10 July 2002
LEMADANG (ex-*Dorang*)	806	PT Pal Surabaya	Aug 2004

Displacement, tons: 447 full load
Dimensions, feet (metres): 190.6 × 25 × 9.2 *(58.1 × 7.6 × 2.8)*
Main machinery: 2 MTU 16V 956 TB92 diesels; 8,850 hp(m) *(6.5 MW)* sustained; 2 shafts
Speed, knots: 27. **Range, n miles:** 6,100 at 15 kt; 2,200 at 27 kt
Complement: 53
Guns: 1 Bofors SAK 57 mm/70 Mk 2; 220 rds/min to 14 km *(7.6 n miles)*; weight of shell 2.4 kg.
1 Bofors SAK 40 mm/70; 300 rds/min to 12 km *(6.6 n miles)*; weight of shell 0.96 kg.
2 Rheinmetall 20 mm.
Countermeasures: Decoys: CSEE Dagaie chaff launchers.
ESM: Thomson-CSF DR 3000 S1; intercept.
Combat data systems: TACTICOS type.
Weapons control: Signaal LIOD 73 Ri Mk 2 optronic director.
Radars: Surface search: Signaal Scout variant; G-band.
Fire control: Signaal LIROD Mk 2; K-band.
Navigation: Kelvin Hughes KH 1007; I-band.

Comment: Ordered in mid-1993 from PT Pal Surabaya. Weapon systems ordered in November 1994. Much improved combat data system is fitted. The after gun was intended to be a second 57 mm but this was changed to a 40 mm.

UPDATED

TODAK   10/2003, Hartmut Ehlers / 0569200

4 KAKAP (PB 57) CLASS (NAV III and IV)
(LARGE PATROL CRAFT) (PBOH)

Name	No	Builders	Commissioned
KAKAP	811	Lürssen/PT Pal Surabaya	29 June 1988
KERAPU	812	Lürssen/PT Pal Surabaya	5 Apr 1989
TONGKOL	813	PT Pal Surabaya	Dec 1993
BARAKUDA (ex-*Bervang*)	814	PT Pal Surabaya	Aug 1995

Displacement, tons: 423 full load
Dimensions, feet (metres): 190.6 × 25 × 9.2 *(58.1 × 7.6 × 2.8)*
Main machinery: 2 MTU 16V 956 TB92 diesels; 8,850 hp(m) *(6.5 MW)* sustained; 2 shafts
Speed, knots: 28. **Range, n miles:** 6,100 at 15 kt; 2,200 at 27 kt
Complement: 49 plus 8 spare berths
Guns: 1 Bofors 40 mm/70; 240 rds/min to 12.6 km *(6.8 n miles)*; weight of shell 0.96 kg.
2—12.7 mm MGs.
Countermeasures: ESM: Thomson-CSF DR 3000 S1; intercept.
Radars: Surface search: Racal Decca 2459; I-band.
Navigation: KH 1007; I-band.
Helicopters: Platform for 1 NBO-105.

Comment: Ordered in 1982. First pair shipped from West Germany and completed at PT Pal Surabaya. Second pair assembled at Surabaya taking longer than expected to complete. The first three are NAV III SAR and Customs versions and by comparison with NAV I are very lightly armed and have a 13 × 7.1 m helicopter deck in place of the after guns and torpedo tubes. Vosper Thornycroft fin stabilisers are fitted. Can be used for Patrol purposes as well as SAR, and can transport two rifle platoons. There is also a fast seaboat with launching crane at the stern and two water guns for firefighting. The single NAV IV version has some minor variations and is used as Presidential Yacht manned by a special unit.

UPDATED

TONGKOL   2/2001, Sattler/Steele / 0121380

BARAKUDA (NAV IV)   8/1995, van Ginderen Collection / 0080012

8 SIBARAU (ATTACK) CLASS (LARGE PATROL CRAFT) (PB)

Name	No	Builders	Commissioned
SIBARAU (ex-*Bandolier*)	847	Walkers, Australia	14 Dec 1968
SILIMAN (ex-*Archer*)	848	Walkers, Australia	15 May 1968
SIGALU (ex-*Barricade*)	857	Walkers, Australia	26 Oct 1968
SILEA (ex-*Acute*)	858	Evans Deakin	24 Apr 1968
SIRIBUA (ex-*Bombard*)	859	Walkers, Australia	5 Nov 1968
SIADA (ex-*Barbette*)	862	Walkers, Australia	16 Aug 1968
SIKUDA (ex-*Attack*)	863	Evans Deakin	17 Nov 1967
SIGUROT (ex-*Assail*)	864	Evans Deakin	12 July 1968

Displacement, tons: 146 full load
Dimensions, feet (metres): 107.5 × 20 × 7.3 *(32.8 × 6.1 × 2.2)*
Main machinery: 2 Paxman 16YJCM diesels; 4,000 hp *(2.98 MW)* sustained; 2 shafts
Speed, knots: 21. **Range, n miles:** 1,220 at 13 kt
Complement: 19 (3 officers)
Guns: 1 Bofors 40 mm/60. 1—12.5 mm MG.
Countermeasures: ESM: DASA Telegon VIII; intercept.
Radars: Surface search: Decca 916; I-band.

Comment: Transferred from Australia after refit—*Bandolier* 16 November 1973, *Archer* in 1974, *Barricade* March 1982, *Acute* 6 May 1983, *Bombard* September 1983, *Attack* 22 February 1985 (recommissioned 24 May 1985), *Barbette* February 1985, *Assail* February 1986. All carry rocket/flare launchers. Two similar craft with pennant numbers 860 and 861 were built locally in 1982/83 but have not been reported for some years.

UPDATED

SIGALU   4/1999 / 0080013

4 SINGA (PB 57) CLASS (NAV I and II)
(LARGE PATROL CRAFT) (PBO)

Name	No	Builders	Commissioned
SINGA	651	Lürssen/PT Pal Surabaya	Apr 1988
AJAK	653	Lürssen/PT Pal Surabaya	5 Apr 1989
PANDRONG	801	PT Pal Surabaya	1992
SURA	802	PT Pal Surabaya	1993

Displacement, tons: 447 full load (NAV I); 428 full load (NAV II)
Dimensions, feet (metres): 190.6 × 25 × 9.2 *(58.1 × 7.6 × 2.8)*
Main machinery: 2 MTU 16V 956 TB92 diesels; 8,850 hp(m) *(6.5 MW)* sustained; 2 shafts
Speed, knots: 27. **Range, n miles:** 6,100 at 15 kt; 2,200 at 27 kt
Complement: 42 (6 officers)
Guns: 1 Bofors SAK 57 mm/70 Mk 2; 220 rds/min to 14 km *(7.6 n miles)*; weight of shell 2.4 kg.
1 Bofors SAK 40 mm/70; 300 rds/min to 12 km *(6.6 n miles)*; weight of shell 0.96 kg.
2 Rheinmetall 20 mm.
Torpedoes: 2-21 in *(533 mm)* Toro tubes (651 and 653). AEG SUT; anti-submarine; wire-guided; active/passive homing to 12 km *(6.6 n miles)* at 35 kt; 28 km *(15 n miles)* at 23 kt warhead 250 kg.
Countermeasures: Decoys: CSEE Dagaie single trainable launcher; automatic dispenser for IR flares and chaff; H/J-band.
ESM: Thomson-CSF DR 2000 S3 with Dalia analyser; intercept. DASA Telegon VIII D/F.
Weapons control: Signaal LIOD 73 Ri optronic director. Signaal WM22 72 Ri WCS (651 and 653).
Radars: Surface search: Racal Decca 2459; I-band; Signaal Scout; H/I-band (801 and 802).
Fire control: Signaal WM22; I/J-band (651 and 653).
Sonars: Signaal PMS 32 (NAV I); active search and attack; medium frequency.

Comment: Class ordered from Lürssen in 1982. First launched and shipped incomplete to PT Pal Surabaya for fitting out in January 1984. Second shipped July 1984. The first two are NAV I ASW versions with torpedo tubes and sonars. The second pair are NAV II AAW versions with an augmented gun armament, an improved surveillance and fire-control radar, but without torpedo tubes and sonars and completed later than expected in 1992-93. Vosper Thornycroft fin stabilisers are fitted.

UPDATED

SINGA (NAV I)   *5/1999, G Toremans /* 0080009

AJAK (NAV I)   *5/1998, John Mortimer /* 0052359

SURA   *5/2000, M Declerck /* 0104597

1 PATROL CRAFT (PB)

Name	No	Builders	Launched	Commissioned
CUCUK (ex-*Jupiter*)	— (ex-A 102)	Singapore SBEC	3 Apr 1990	19 Aug 1991

Displacement, tons: 170 full load
Dimensions, feet (metres): 117.5 × 23.3 × 7.5 *(35.8 × 7.1 × 2.3)*
Main machinery: 2 Deutz MWM TBD234V12 diesels; 1,360 hp(m) *(1 MW)* sustained; 2 shafts; bow thruster
Speed, knots: 14. **Range, n miles:** 200 at 14 kt
Complement: 33 (5 officers)
Guns: 1 Oerlikon 20 mm GAM-BO1. 4—12.7 mm MGs.
Radars: Navigation: Racal Decca; I-band.

Comment: Designed as an underwater search and salvage craft, decommissioned from the Singapore Navy and transferred on 21 March 2002. Deployed as a patrol craft.

VERIFIED

CUCUK (Singapore colours)   *6/1994, van Ginderen Collection /* 0084281

13 PC-36 PATROL CRAFT (PB)

Name	No	Builders	Commissioned
KOBRA	867	Fasharkan, Mentigi	31 Mar 2003
ANAKONDA	868	Fasharkan, Jakarta	31 Mar 2003
PATOLA	869	PT Pelindo, Tanjung Pinang	Oct 2003
BOA	807	Fasharkan, Mentigi	6 Aug 2004
WELANG	808	Fasharkan, Mentigi	6 Aug 2004
TALIWANGSA	870	Fasharkan, Manokwari	6 Aug 2004
SULUH PARI	809	Fasharkan, Mentigi	20 Jan 2005
KATON	810	Fasharkan, Mentigi	20 Jan 2005
SANCA	815	Fasharkan, Manokwari	20 Jan 2005
WARAKAS	816	Fasharkan, Jakarta	20 Jan 2005
PANANA	817	Fasharkan, Makassar	20 Jan 2005
KALAKAE	818	Fasharkan, Makassar	20 Jan 2005
TEDONG NAGA	819	Fasharkan, Jakarta	20 Jan 2005

Displacement, tons: 90 full load
Dimensions, feet (metres): 118.1 × 23.0 × 4.4 *(36 × 7.0 × 1.35)*
Main machinery: 3 MAN D2842 LE 410 diesels; 3,300 hp *(2.46 MW)*; or 3 Caterpillar 3412E diesels; 3,600 hp *(2.7 MW)*
Speed, knots: 38
Complement: 18
Guns: 1—20 mm. 1—12.7 mm MG.
Radars: Navigation: I-band.

Comment: *Kobra* was the prototype vessel first demonstrated in late 2002. Glass fibre hull. Some are known as KAL-35 and others as KAL-36 craft. There are some differences in armament and superstructure, some being fitted with a stern ramp for RIB. *Patola* funded by Bali province and others may have been similarly procured. Constructed by variety of shipbuilders and operated by the Indonesian Navy. Further craft are expected.

UPDATED

BOA   *7/2004*, EPA/Bagus/Indahono /* 0584211

AMPHIBIOUS FORCES

Notes: (1) This section includes some vessels of the Military Sealift Command-Kolinlamil.
(2) *Tanjung Kambani* 971 is a converted Ro-Ro ferry which is reported to have a military lift of one battalion and four LCUs. Super Pumas can be operated from a large helicopter deck. Delivered in mid-2000, there may be further vessels.

1 + 4 MULTIROLE VESSEL (LPD/APCR)

Name	No	Builders	Laid down	Launched	Commissioned
TANJUNG DALPELE	972	Daesun Shipbuilders, Pusan	2002	17 May 2003	Sep 2003

Displacement, tons: 11,400
Dimensions, feet (metres): 400.00 × 72.2 × 22.0 *(122.0 × 22.0 × 6.7)*
Main machinery: To be announced
Speed, knots: 15. **Range, n miles:** 8,600 at 12 kt
Complement: To be announced
Military lift: To be announced.
Guns: 1 Bofors 40 mm. 2—20 mm.
Radars: Navigation: 2—I-band.

Helicopters: 2 SH-2G Super Seasprites.

Programmes: Officially designated a Multipurpose Hospital Ship. Following delivery of the first vessel in mid-2003, a contract for a further four vessels, to be delivered 2007-08, was finalised in December 2004. One of these is to have command facilities.
Structure: Has a docking well, capable of accommodating two LCU-23M, stern and side ramps and hospital facilities.

NEW ENTRY

TANJUNG DALPELE   *6/2004*, Daesun /* 1047875

7 LST 1-511 and 512-1152 CLASSES (LST)

Name	No	Builders	Commissioned
TELUK LANGSA (ex-*LST 1128*)	501	Chicago Bridge	9 Mar 1945
TELUK BAJUR (ex-*LST 616*)	502	Chicago Bridge	29 May 1944
TELUK KAU (ex-*LST 652*)	504	Chicago Bridge	1 Jan 1945
TELUK TOMINI (ex-*Inagua Crest*, ex-*Brunei*, ex-*Bledsoe County, LST 356*)	508	Charleston, NY	22 Dec 1942
TELUK RATAI (ex-*Inagua Shipper*, ex-*Presque Isle*, APB 44, ex-*LST 678*, ex-*Teluk Sindoro*)	509	American Bridge, PA	30 June 1944
TELUK SALEH (ex-*Clark County, LST 601*)	510	Chicago Bridge	25 Mar 1944
TELUK BONE (ex-*Iredell County, LST 839*)	511	American Bridge, PA	6 Dec 1944

Displacement, tons: 1,653 standard; 4,080 full load
Dimensions, feet (metres): 328 × 50 × 14 *(100 × 15.2 × 4.3)*
Main machinery: 2 GM 12-567A diesels; 1,800 hp *(1.34 MW)*; 2 shafts
Speed, knots: 11.6. **Range, n miles:** 11,000 at 10 kt
Complement: 119 (accommodation for 266)
Military lift: 2,100 tons
Guns: 7—40 mm. 2—20 mm *(Teluk Langsa)*. 8—37 mm (remainder).
Radars: Surface search: SPS-21 *(Teluk Tomini, Teluk Sindoro)*.
SPS-53 *(Teluk Saleh, Teluk Bone)*. SO-1 *(Teluk Kau)*. SO-6 *(Teluk Langsa)*.

Comment: *Teluk Bajur, Teluk Saleh* and *Teluk Bone* transferred from USA in June 1961 (and purchased 22 February 1979). *Teluk Kau* and *Teluk Langsa* in July 1970. These ships are used as transports and stores carriers. It was anticipated that they would decay in reserve Fleet anchorages once the Frosch class were in service, but all remain active. *Bajur* and *Tomini* serve with the Military Sealift Command.

UPDATED

TELUK BONE   *8/1995, John Mortimer /* 0080014

6 TACOMA TYPE (LSTH)

Name	No	Builders	Commissioned
TELUK SEMANGKA	512	Korea-Tacoma, Masan	20 Jan 1981
TELUK PENJU	513	Korea-Tacoma, Masan	20 Jan 1981
TELUK MANDAR	514	Korea-Tacoma, Masan	July 1981
TELUK SAMPIT	515	Korea-Tacoma, Masan	June 1981
TELUK BANTEN	516	Korea-Tacoma, Masan	May 1982
TELUK ENDE	517	Korea-Tacoma, Masan	2 Sep 1982

Displacement, tons: 3,750 full load
Dimensions, feet (metres): 328 × 47.2 × 13.8 *(100 × 14.4 × 4.2)*
Main machinery: 2 diesels; 12,800 hp(m) *(9.41 MW)* sustained; 2 shafts
Speed, knots: 15. **Range, n miles:** 7,500 at 13 kt
Complement: 90 (13 officers)
Military lift: 1,800 tons (including 17 MBTs); 2 LCVPs; 200 troops
Guns: 2 or 3 Bofors 40 mm/70. 2 Rheinmetall 20 mm.
Radars: Surface search: Raytheon; E/F-band *(Teluk Banten* and *Teluk Ende)*.
Navigation: Racal Decca; I-band.
Helicopters: 1 Westland Wasp; 3 NAS-332 Super Pumas can be carried in last pair.

Comment: First four ordered in June 1979, last pair June 1981. No hangar in *Teluk Semangka* and *Teluk Mandar*. Two hangars in *Teluk Ende*. The last pair differ in silhouette having drowned exhausts in place of funnels and having their LCVPs carried forward of the bridge. They also have only two 40 mm guns and an additional radar fitted above the bridge. Battalion of marines can be embarked if no tanks are carried. *Teluk Ende* and *Teluk Banten* act as Command ships, the former also able to serve as a hospital ship.

UPDATED

TELUK BANTEN (command ship)   *8/1995, van Ginderen Collection /* 0080016

TELUK SEMANGKA (no hangar)   *8/1995, van Ginderen Collection /* 0080015

TELUK ENDE   *5/2001 /* 0126190

1 LST

Name	No	Builders	Commissioned
TELUK AMBOINA	503	Sasebo, Japan	June 1961

Displacement, tons: 2,378 standard; 4,200 full load
Dimensions, feet (metres): 327 × 50 × 15 *(99.7 × 15.3 × 4.6)*
Main machinery: 2 MAN V6V 22/30 diesels; 3,425 hp(m) *(2.52 MW)*; 2 shafts
Speed, knots: 13.1. **Range, n miles:** 4,000 at 13.1 kt
Complement: 88
Military lift: 212 troops; 2,100 tons; 4 LCVP on davits
Guns: 6—37 mm; anti-aircraft.

Comment: Launched on 17 March 1961 and transferred from Japan in June 1961. A faster copy of US LST 511 class with 30 ton crane forward of bridge. Serves with the Military Sealift Command.

UPDATED

TELUK AMBOINA   *8/1995, van Ginderen Collection /* 0080018

12 FROSCH I CLASS (TYPE 108) (LSM)

Name	No	Commissioned	Recommissioned
TELUK GILIMANUK	531 (ex-611)	12 Nov 1976	12 July 1994
(ex-*Hoyerswerda*)			
TELUK CELUKAN BAWANG	532 (ex-632)	1 Dec 1976	25 Feb 1994
(ex-*Hagenow*)			
TELUK CENDRAWASIH	533 (ex-613)	2 Feb 1977	9 Dec 1994
(ex-*Frankfurt/Oder*)			
TELUK BERAU	534 (ex-634)	28 May 1977	10 Mar 1995
(ex-*Eberswalde-Finow*)			
TELUK PELENG (ex-*Lübben*)	535 (ex-631)	15 Mar 1978	23 Sep 1993
TELUK SIBOLGA (ex-*Schwerin*)	536 (ex-612)	19 Oct 1977	15 Dec 1993
TELUK MANADO	537 (ex-633)	28 Dec 1977	2 June 1995
(ex-*Neubrandenburg*)			
TELUK HADING (ex-*Cottbus*)	538 (ex-614)	26 May 1978	12 July 1994
TELUK PARIGI (ex-*Anklam*)	539 (ex-635)	14 July 1978	21 July 1995
TELUK LAMPUNG (ex-*Schwedt*)	540 (ex-636)	7 Sep 1979	26 Apr 1994
TELUK JAKARTA	541 (ex-615)	4 Jan 1979	19 Sep 1994
(ex-*Eisenhüttenstadt*)			
TELUK SANGKULIRANG	542 (ex-616)	4 Jan 1979	9 Dec 1994
(ex-*Grimmen*)			

Displacement, tons: 1,950 full load
Dimensions, feet (metres): 321.5 × 36.4 × 9.2 *(98 × 11.1 × 2.8)*
Main machinery: 2 diesels; 5,000 hp(m) *(3.68 MW)*; 2 shafts
Speed, knots: 18
Complement: 46
Military lift: 600 tons
Guns: 1—40 mm/60. 2—37 mm/63 (1 twin). 4—25 mm (2 twin).
Mines: Can lay 40 mines through stern doors.
Countermeasures: Decoys: 2 PK 16 chaff launchers.
Radars: Air/surface search: Strut Curve; F-band.
Navigation: TSR 333; I-band.

Comment: All built by Peenewerft, Wolgast. Former GDR ships transferred from Germany on 25 August 1993. Demilitarised with all guns removed, but 37 mm guns have replaced the original 57 mm and 30 mm twin guns. All refitted in Germany prior to sailing. First two arrived Indonesia in late 1993, remainder throughout 1994 and 1995. *Teluk Lampung* damaged by heavy seas during transit in June 1994 but was repaired.

UPDATED

TELUK BERAU   *8/1995, van Ginderen Collection /* 0075855

TELUK PELENG   *6/2000, M Declerck /* 0104598

54 LANDING CRAFT (LCU)

DORE 580   KUPANG 582   DILI 583   NUSANTARA 584   +50

Displacement, tons: 400 full load
Dimensions, feet (metres): 140.7 × 29.9 × 4.6 *(42.9 × 9.1 × 1.4)*
Main machinery: 4 diesels; 2 shafts
Speed, knots: 12. Range, n miles: 700 at 11 kt
Complement: 17
Military lift: 200 tons

Comment: Details given are for LCUs 582-584 built at Naval Training Centre, Surabaya in 1978-80. Military Sealift Command. LCU 580 is a smaller ship at 275 tons and built in 1968. About 20 LCM 6 type and 30 LCVPs are also in service.

VERIFIED

LCVP   *8/1995, van Ginderen Collection /* 0080020

MINE WARFARE FORCES

9 KONDOR II (TYPE 89) CLASS
(MINESWEEPERS—COASTAL) (MSC)

Name	No	Builders	Commissioned
PULAU ROTE (ex-*Wolgast*)	721 (ex-V 811)	Peenewerft, Wolgast	1 June 1971
PULAU RAAS (ex-*Hettstedt*)	722 (ex-353)	Peenewerft, Wolgast	22 Dec 1971
PULAU ROMANG (ex-*Pritzwalk*)	723 (ex-325)	Peenewerft, Wolgast	26 June 1972
PULAU RIMAU (ex-*Bitterfeld*)	724 (ex-332, ex-M 2672)	Peenewerft, Wolgast	7 Aug 1972
PULAU RONDO (ex-*Zerbst*)	725 (ex-335)	Peenewerft, Wolgast	30 Sep 1972
PULAU RUSA (ex-*Oranienburg*)	726 (ex-341)	Peenewerft, Wolgast	1 Nov 1972
PULAU RANGSANG (ex-*Jüterbog*)	727 (ex-342)	Peenewerft, Wolgast	7 Apr 1973
PULAU RAIBU (ex-*Sömmerda*)	728 (ex-311, ex-M 2670)	Peenewerft, Wolgast	9 Aug 1973
PULAU REMPANG (ex-*Grimma*)	729 (ex-336)	Peenewerft, Wolgast	10 Nov 1973

Displacement, tons: 310 full load
Dimensions, feet (metres): 186 × 24.6 × 7.9 *(56.7 × 7.5 × 2.4)*
Main machinery: 2 Russki Kolomna Type 40-DM diesels; 4,408 hp(m) *(3.24 MW)* sustained; 2 shafts; cp props
Speed, knots: 17. Range, n miles: 2,000 at 14 kt
Complement: 31 (6 officers)
Guns: 6—25 mm/80 (3 twin).
Mines: 2 rails.
Radars: Navigation: TSR 333; I-band.
Sonars: Bendix AQS 17 VDS; minehunting; active; high frequency (in some).

Comment: Former GDR minesweepers transferred from Germany in Russian dockship *Trans-Shelf* arriving 22 October 1993. MCM is secondary role with EEZ patrol taking priority. ADI Dyads can be embarked for MCM. *Pulau Rondo* was used for trials. **UPDATED**

PULAU RIMAU   *4/2004*, Chris Sattler /* 1044128

PALAU RUSA   *8/1995, van Ginderen Collection /* 0080021

PULAU RONDO   *6/2001, John Mortimer /* 0121376

1 T 43 (PROJECT 244) CLASS (MINESWEEPER—OCEAN) (MSO)

PULAU RANI 701

Displacement, tons: 580 full load
Dimensions, feet (metres): 190.2 × 27.6 × 6.9 *(58 × 8.4 × 2.1)*
Main machinery: 2 Kolomna 9-D-8 diesels; 2,000 hp(m) *(1.6 MW)* sustained; 2 shafts
Speed, knots: 15. Range, n miles: 3,000 at 10 kt
Complement: 55
Guns: 4—37 mm/63 (2 twin). 8—12.7 mm (4 twin) MGs.
Depth charges: 2 projectors.
Radars: Navigation: Decca 110; I-band.
Sonars: Stag Ear; hull-mounted; active search and attack; high frequency.

Comment: Transferred from USSR in 1964. Mostly used as patrol craft. Second of class sunk in collision in May 2000. **UPDATED**

PULAU RANI   *1983, P D Jones*

2 PULAU RENGAT (TRIPARTITE) CLASS (MHSC)

Name	No	Builders	Launched	Commissioned
PULAU RENGAT	711	van der Giessen-de Noord	23 July 1987	26 Mar 1988
PULAU RUPAT	712	van der Giessen-de Noord	27 Aug 1987	26 Mar 1988

Displacement, tons: 502 standard; 568 full load
Dimensions, feet (metres): 168.9 × 29.2 × 8.2 *(51.5 × 8.9 × 2.5)*
Main machinery: 2 MTU 12V 396 TC82 diesels; 2,610 hp(m) *(1.92 MW)* sustained; 1 shaft; LIPS cp prop; auxiliary propulsion; 3 Turbomeca gas-turbine generators; 2 motors; 2,400 hp(m) *(1.76 MW)*; 2 retractable Schottel propulsors; 2 bow thrusters; 150 hp(m) *(110 kW)*
Speed, knots: 15; 7 auxiliary propulsion. **Range, n miles:** 3,000 at 12 kt
Complement: 46 plus 4 spare berths

Guns: 2 Rheinmetall 20 mm. Matra Simbad SAM launcher may be added for patrol duties or a third 20 mm gun.
Countermeasures: MCM: OD3 Oropesa mechanical sweep gear; Fiskars F-82 magnetic and SA Marine AS 203 acoustic sweeps; Ibis V minehunting system; 2 PAP 104 Mk 4 mine disposal systems.
Combat data systems: Signaal SEWACO-RI action data automation.
Radars: Navigation: Racal Decca AC 1229C; I-band.
Sonars: Thomson Sintra TSM 2022; active minehunting; high frequency.

Programmes: First ordered on 29 March 1985, laid down 22 July 1985, second ordered 30 August 1985 and laid down 15 December 1985. More were to have been built in Indonesia up to a total of 12 but this programme was cancelled by lack of funds.
Structure: There are differences in design between these ships and the European Tripartites, apart from their propulsion. Deckhouses and general layout are different as they are required to act as minehunters, minesweepers and patrol ships. Hull construction is GRP shock-proven.
Operational: Endurance, 15 days. Automatic operations, navigation and recording systems, Thomson-CSF Naviplot TSM 2060 tactical display. A 5 ton container can be shipped, stored for varying tasks—research; patrol; extended diving; drone control.
UPDATED

PULAU RUPAT   *3/2004*, Chris Sattler /* 1044129

SURVEY AND RESEARCH SHIPS

2 SURVEY SHIPS (AGS)

BARUNA JAYA VII   BARUNA JAYA VIII

Comment: Survey vessels reported ordered in November 1995.
VERIFIED

BARUNA JAYA VII   *9/1998, S Tattam, RAN /* 0050681

BARUNA JAYA VIII   *9/1998, Maritime Photographic /* 0044067

4 RESEARCH SHIPS (AGS/AGOR)

Name	No	Builders	Commissioned
BARUNA JAYA I	KAL-IV-02	CMN, Cherbourg	10 Aug 1989
BARUNA JAYA II	KAL-IV-03	CMN, Cherbourg	25 Sep 1989
BARUNA JAYA III	KAL-IV-04	CMN, Cherbourg	3 Jan 1990
BARUNA JAYA IV	KAL-IV-05	CMN, Cherbourg	2 Nov 1995

Displacement, tons: 1,180 (1,425 IV) full load
Dimensions, feet (metres): 198.2 × 39.7 × 13.8 *(60.4 × 12.1 × 4.2)*
Main machinery: 2 Niigata/SEMT-Pielstick 5 PA5 L 255 diesels; 2,990 hp(m) *(2.2 MW)* sustained; 1 shaft; cp prop; bow thruster
Speed, knots: 14. **Range, n miles:** 7,500 at 12 kt
Complement: 37 (8 officers) plus 26 scientists

Comment: First three ordered from La Manche, Dieppe in February 1985 by the office of Technology, Ministry of Industry and Research. Badly delayed by the closing down of the original shipbuilders (ACM, Dieppe) and construction taken over by CMN at Cherbourg. Fourth of class ordered in 1993 to a slightly enlarged design and with a more enclosed superstructure. *Baruna Jaya 1* is employed on hydrography, the second on oceanography and the third combines both tasks. *Baruna Jaya IV* is operated by the Agency responsible for developing new technology. All are part of the Naval Auxiliary Service.
VERIFIED

BARUNA JAYA II   *4/1998, John Mortimer /* 0052362

BARUNA JAYA IV   *11/1995, van Ginderen Collection /* 0080023

1 HECLA CLASS (SURVEY SHIP) (AGSH)

Name	No	Builders	Commissioned
DEWA KEMBAR (ex-*Hydra*)	932	Yarrow and Co, Blythswood	5 May 1966

Displacement, tons: 1,915 light; 2,733 full load
Dimensions, feet (metres): 260.1 × 49.1 × 15.4 *(79.3 × 15 × 4.7)*
Main machinery: Diesel-electric; 3 Paxman 12YJCZ diesels; 3,780 hp *(2.82 MW)*; 3 generators; 1 motor; 2,000 hp(m) *(1.49 MW)*; 1 shaft; bow thruster
Speed, knots: 14. **Range, n miles:** 12,000 at 11 kt
Complement: 123 (14 officers)
Guns: 2—12.7 mm MGs.
Radars: Navigation: Kelvin Hughes Type 1006; I-band.
Helicopters: 1 Westland Wasp.

Comment: Transferred from UK 18 April 1986 for refit. Commissioned in Indonesian Navy 10 September 1986. SATCOM fitted. Two survey launches on davits.
VERIFIED

DEWA KEMBAR   *11/1997, van Ginderen Collection /* 0012542

1 RESEARCH SHIP (AGORH)

Name	No	Builders	Commissioned
BURUJULASAD	931	Schlichting, Lübeck-Travemünde	1967

Displacement, tons: 2,165 full load
Dimensions, feet (metres): 269.5 × 37.4 × 11.5 *(82.2 × 11.4 × 3.5)*
Main machinery: 4 MAN V6V 22/30 diesels; 6,850 hp(m) *(5.03 MW)*; 2 shafts
Speed, knots: 19.1. **Range, n miles:** 14,500 at 15 kt
Complement: 108 (15 officers) plus 28 scientists
Guns: 4—12.7 mm (2 twin) MGs.
Radars: Surface search: Decca TM 262; I-band.
Helicopters: 1 Bell 47J.

Comment: *Burujulasad* was launched in August 1965; her equipment includes laboratories for oceanic and meteorological research and a cartographic room. Carries one LCVP and three surveying motor boats. A 37 mm gun was added in 1992 but by 1998 had been removed again.
VERIFIED

BURUJULASAD   *4/1998, John Mortimer* / 0052361

1 RESEARCH SHIP (AGOR)

Name	No	Builders	Commissioned
JALANIDHI	933	Sasebo Heavy Industries	12 Jan 1963

Displacement, tons: 985 full load
Dimensions, feet (metres): 176.8 × 31.2 × 14.1 *(53.9 × 9.5 × 4.3)*
Main machinery: 1 MAN G6V 30/42 diesel; 1,000 hp(m) *(735 kW)*; 1 shaft
Speed, knots: 11.5. **Range, n miles:** 7,200 at 10 kt
Complement: 87 (13 officers) plus 26 scientists
Radars: Navigation: Nikkon Denko; I-band. Furuno; I-band.

Comment: Launched in 1962. Oceanographic research ship with hydromet facilities and weather balloons. 3 ton boom aft. Operated by the Navy for the Hydrographic Office. *VERIFIED*

JALANIDHI   *8/1995, van Ginderen Collection* / 0080024

AUXILIARIES

Notes: (1) The Don class depot ship *Ratulangi* 400 is in use as a floating workshop at Surabaya naval base, but is not seaworthy.
(2) There is also a small oiler *Sungai Gerong* 906.

1 REPLENISHMENT TANKER (AOTL)

Name	No	Builders	Commissioned
SORONG	911	Trogir SY, Yugoslavia	Apr 1965

Displacement, tons: 8,400 full load
Dimensions, feet (metres): 367.4 × 50.5 × 21.6 *(112 × 15.4 × 6.6)*
Main machinery: 1 diesel; 1 shaft
Speed, knots: 15
Complement: 110
Cargo capacity: 4,200 tons fuel; 300 tons water
Guns: 4—12.7 mm (2 twin) MGs.
Radars: Navigation: Don; I-band.

Comment: Has limited underway replenishment facilities on both sides and stern refuelling.
VERIFIED

SORONG   *8/1995, van Ginderen Collection* / 0080026

1 COMMAND SHIP (AGFH)

Name	No	Builders	Launched	Commissioned
MULTATULI	561	Ishikawajima-Harima	15 May 1961	Aug 1961

Displacement, tons: 3,220 standard; 6,741 full load
Dimensions, feet (metres): 365.3 × 52.5 × 23 *(111.4 × 16 × 7)*
Main machinery: 1 Burmeister & Wain diesel; 5,500 hp(m) *(4.04 MW)*; 1 shaft
Speed, knots: 18.5. **Range, n miles:** 6,000 at 16 kt
Complement: 135
Guns: 6 USSR 37 mm/63 (2 twin, 2 single); 160 rds/min to 9 km *(5 n miles)*; weight of shell 0.7 kg.
8—12.7 mm MGs.
Radars: Surface search: Ball End; E/F-band.
Navigation: I-band.
Helicopters: 1 Bell 47J.

Comment: Built as a submarine tender. Original after 76 mm mounting replaced by helicopter deck with a hangar added in 1998. Living and working spaces air conditioned. Capacity for replenishment at sea (fuel oil, fresh water, provisions, ammunition, naval stores and personnel). Medical and hospital facilities. Used as fleet flagship (Eastern Force) and is fitted with ICS-3 communications.
VERIFIED

MULTATULI   *8/1995, van Ginderen Collection* / 0080025

1 ROVER CLASS (REPLENISHMENT TANKER) (AORLH)

Name	No	Builders	Commissioned
ARUN (ex-*Green Rover*)	903	Swan Hunter, Tyneside	15 Aug 1969

Displacement, tons: 4,700 light; 11,522 full load
Dimensions, feet (metres): 461 × 63 × 24 *(140.6 × 19.2 × 7.3)*
Main machinery: 2 SEMT-Pielstick 16 PA4 diesels; 15,360 hp(m) *(11.46 MW)*; 1 shaft; Kamewa cp prop; bow thruster
Speed, knots: 19. **Range, n miles:** 15,000 at 15 kt
Complement: 49 (16 officers)
Cargo capacity: 6,600 tons fuel
Guns: 2 Bofors 40 mm/60. 2 Oerlikon 20 mm.
Radars: Navigation: Kelvin Hughes Type 1006; I-band.
Helicopters: Platform for Super Puma.

Comment: Transferred from UK in September 1992 after a refit. Small fleet tanker designed to replenish ships at sea with fuel, fresh water, limited dry cargo and refrigerated stores under all conditions while under way. No hangar but helicopter landing platform is served by a stores lift, to enable stores to be transferred at sea by 'vertical lift'. Capable of HIFR. Used as the Flagship for the Training Commander.
UPDATED

ARUN   *10/2004*, Chris Gee* / 1047874

2 KHOBI CLASS (COASTAL TANKERS) (AOTL)

BALIKPAPAN 901		SAMBU 902

Displacement, tons: 1,525 full load
Dimensions, feet (metres): 206.6 × 33 × 14.8 *(63 × 10.1 × 4.5)*
Main machinery: 2 diesels; 1,600 hp(m) *(1.18 MW)*; 2 shafts
Speed, knots: 13. **Range, n miles:** 2,500 at 12 kt
Complement: 37 (4 officers)
Cargo capacity: 550 tons dieso
Guns: 4—14.5 mm (2 twin) MGs. 2—12.7 mm MGs.
Radars: Navigation: Neptun; I-band.

Comment: *Balikpapan* and *Sambu* are Japanese copies of the Khobi class built in the 1960s.
VERIFIED

SAMBU   *8/1995, van Ginderen Collection* / 0080028

1 ACHELOUS CLASS (REPAIR SHIP) (ARL)

Name	No	Builders	Commissioned
JAYA WIJAYA (ex-*Askari*, ex-*ARL 30*, ex-*LST 1131*)	921	Chicago Bridge and Iron Co	15 Mar 1945

Displacement, tons: 4,325 full load
Dimensions, feet (metres): 328 × 50 × 14 *(100 × 15.3 × 4.3)*
Main machinery: 2 GM 12-567A diesels; 1,800 hp *(1.34 MW)*; 2 shafts
Speed, knots: 12. **Range, n miles:** 17,000 at 7 kt
Complement: 180 (11 officers)
Cargo capacity: 300 tons; 60 ton crane
Guns: 8 Bofors 40 mm/56 (2 quad); 160 rds/min to 11 km *(5.9 n miles)*; weight of shell 0.9 kg.
Radars: Air/surface search: Sperry SPS-53; I/J-band.
Navigation: Raytheon 1900; I/J-band.
IFF: UPX 12B.

Comment: In reserve from 1956-66. She was recommissioned and reached Vietnam in 1967 to support River Assault Flotilla One. She was used by the US Navy and Vietnamese Navy working up the Mekong in support of the Cambodian operations in May 1970. Transferred from US on lease to Indonesia at Guam on 31 August 1971 and purchased 22 February 1979. Bow doors welded shut. Carries two LCVPs.
VERIFIED

JAYA WIJAYA   9/1998, 92 Wing RAAF

2 FROSCH II CLASS (TYPE 109) (SUPPORT SHIPS) (AKL/ARL)

Name	No	Builders	Commissioned
TELUK CIREBON (ex-*Nordperd*)	543 (ex-E 171)	Peenewerft, Wolgast	3 Oct 1979
TELUK SABANG (ex-*Südperd*)	544 (ex-E 172)	Peenewerft, Wolgast	26 Feb 1980

Displacement, tons: 1,700 full load
Dimensions, feet (metres): 297.6 × 36.4 × 9.2 *(90.7 × 11.1 × 2.8)*
Main machinery: 2 diesels; 4,408 hp(m) *(3.24 MW)* sustained; 2 shafts
Speed, knots: 18
Cargo capacity: 650 tons
Guns: 4—37 mm/63 (2 twin). 4—25 mm (2 twin).
Countermeasures: Decoys: 2 PK 16 chaff launchers.
Radars: Air/surface search: Strut Curve; F-band.
Navigation: I-band.

Comment: Ex-GDR ships disarmed and transferred from Germany 25 August 1993. 5 ton crane amidships. In GDR service these ships had two twin 57 mm and two twin 25 mm guns plus Muff Cob fire-control radar. Both refitted at Rostock and recommissioned 25 April 1995. 37 mm guns fitted after transfer. Rocket launchers are mounted forward of the bridge.
VERIFIED

TELUK SABANG   5/1995, Frank Behling / 0075856

4 SUPPORT SHIPS (AKL)

NUSA TELU 952	TELUK MENTAWAI 959	KARIMATA 960	WAIGEO 961

Displacement, tons: 2,400 full load
Dimensions, feet (metres): 258.4 × 35.4 × 15.1 *(78.8 × 10.8 × 4.6)*
Main machinery: 1 MAN diesel; 1,000 hp(m) *(735 kW)*; 1 shaft
Speed, knots: 12. **Range, n miles:** 3,000 at 11 kt
Complement: 26
Cargo capacity: 875 tons dry; 11 tons liquid
Guns: 4—14.5 mm (2 twin) MGs.
Radars: Navigation: Spin Trough; I-band.

Comment: 959-961 are Tisza class. Built in Hungary. Transferred in 1963-64. Military Sealift Command since 1978. 952 is much smaller ship of 1950s vintage.
VERIFIED

NUSA TELU   8/1995, van Ginderen Collection / 0080029

TRAINING SHIPS

1 SAIL TRAINING SHIP (AXS)

Name	No	Builders	Commissioned
DEWARUCI	—	HC Stülcken & Sohn, Hamburg	9 July 1953

Displacement, tons: 810 standard; 1,500 full load
Dimensions, feet (metres): 136.2 pp; 191.2 oa × 31.2 × 13.9 *(41.5; 58.3 × 9.5 × 4.2)*
Main machinery: 1 MAN diesel; 600 hp(m) *(441 kW)*; 1 shaft
Speed, knots: 10.5
Complement: 110 (includes 78 midshipmen)

Comment: Barquentine of steel construction. Sail area, 1,305 sq yards *(1,091 sq m)*. Launched on 24 January 1953.
VERIFIED

DEWARUCI   7/2003, B Prézelin / 0569199

1 SAIL TRAINING SHIP (AXS)

Name	Builders	Launched	Commissioned
ARUNG SAMUDERA (ex-*Adventurer*)	Hendrik Oosterbroek, Tauranga	July 1991	9 Jan 1996

Measurement, tons: 96 grt
Dimensions, feet (metres): 128 oa; 103.7 wl × 21.3 × 8.5 *(39; 31.6 × 6.5 × 2.6)*
Main machinery: 2 Ford 2725E diesels; 292 hp *(218 kW)*; 2 shafts
Speed, knots: 10 (diesels)
Complement: 20 (includes trainees)

Comment: Three masted schooner acquired from New Zealand. Sail area 433.8 m².
VERIFIED

ARUNG SAMUDERA   5/2000, A Campanera i Rovira / 0104601

TUGS

Notes: Two BIMA VIII class of 423 tons completed in 1991 are not naval. Names *Merapi* and *Merbabu*.

1 CHEROKEE CLASS (AT/PBO)

Name	No	Builders	Commissioned
RAKATA (ex-*Menominee* ATF 73)	922	United Engineering, Alameda	25 Sep 1942

Displacement, tons: 1,235 standard; 1,640 full load
Dimensions, feet (metres): 205 × 38.5 × 17 *(62.5 × 11.7 × 5.2)*
Main machinery: Diesel-electric; 4 GM 12-278 diesels; 4,400 hp *(3.28 MW)*; 4 generators; 1 motor; 3,000 hp *(2.24 MW)*; 1 shaft
Speed, knots: 15. **Range, n miles:** 6,500 at 15 kt
Complement: 67
Guns: 1 US 3 in *(76 mm)*/50. 2 Bofors 40 mm/60 aft. 4—25 mm (2 twin) (bridge wings).
Radars: Surface search: Racal Decca; I-band.

Comment: Launched on 14 February 1942. Transferred from US at San Diego in March 1961. Used mostly as a patrol ship.

VERIFIED

RAKATA   8/1995, John Mortimer / 0080030

3 HARBOUR TUGS (YTM)

Name	No	Builders	Commissioned
LAMPO BATANG	934	Ishikawajima-Harima	Sep 1961
TAMBORA (Army)	935	Ishikawajima-Harima	June 1961
BROMO	936	Ishikawajima-Harima	Aug 1961

Comment: All of 250 tons displacement. There are a number of other naval tugs in the major ports.

VERIFIED

1 NFI CLASS (ATF)

Name	No	Builders	Commissioned
SOPUTAN	923	Dae Sun SB & Eng, Busan	11 Aug 1995

Measurement, tons: 1,279 grt
Dimensions, feet (metres): 217.2 × 39 × 17.1 *(66.2 × 11.9 × 5.2)*
Main machinery: Diesel-electric; 4 SEMT-Pielstick diesel generators; 1 motor; 12,240 hp(m) *(9 MW)*; 1 shaft; bow thruster
Speed, knots: 13.5
Complement: 42
Radars: Navigation: Racal Decca; I-band.

Comment: Ocean Cruiser class NFI. Bollard pull 120 tons.

VERIFIED

SOPUTAN   8/1995, van Ginderen Collection / 0080031

CUSTOMS

Notes: Identified by BC (Tax and Customs) preceding the pennant number.

14 COASTAL PATROL CRAFT (WPB)

BC 2001-2007   BC 3001-3007

Displacement, tons: 70.3 full load
Dimensions, feet (metres): 93.5 × 17.7 × 5.5 *(28.5 × 5.4 × 1.7)*
Main machinery: 2 MTU 12V 331 TC92 diesels; 2,660 hp(m) *(1.96 MW)* sustained; 2 shafts
Speed, knots: 28-34
Complement: 19
Guns: 1—20 mm or 1—12.7 mm MG.

Comment: Built CMN Cherbourg. Delivered in 1980 and 1981.

VERIFIED

BC 2007   1/1990, 92 Wing RAAF

10 LÜRSSEN VSV 15 CLASS (WHSIC)

BC 1601-1610

Displacement, tons: 11 full load
Dimensions, feet (metres): 52.5 × 9.2 × 3.3 *(16 × 2.8 × 1)*
Main machinery: 2 MTU diesels; 600 hp(m) *(441 kW)*; 2 shafts
Speed, knots: 50. **Range, n miles:** 750 at 30 kt
Complement: 5 (1 officer)
Guns: 1—7.62 mm MG.

Comment: Built in Germany and delivered between November 1998 and June 1999.

VERIFIED

BC 1608   5/1999, Lürssen / 0080032

48 LÜRSSEN 28 METRE TYPE (WPB)

BC 4001-3, 5001-3, 6001-24, 7001-6, 8001-6, 9001-6

Displacement, tons: 68 full load
Dimensions, feet (metres): 91.8 × 17.7 × 5.9 *(28 × 5.4 × 1.8)*
Main machinery: 2 Deutz diesels; 2,720 hp(m) *(2 MW)*; or 2 MTU diesels; 2,260 hp(m) *(1.66 MW)*; 2 shafts
Speed, knots: 30. **Range, n miles:** 1,100 at 15 kt; 860 at 28 kt
Complement: 19 (6 officers)
Guns: 1—12.7 mm MG.

Comment: Lürssen design, some built by Fulton Marine and Scheepswerven van Langebrugge of Belgium, some by Lürssen Vegesack and some by PT Pal Surabaya (which also assembled most of them). Programme started in 1980. Some of these craft are operated by the Navy, the Police and the Maritime Security Agency.

VERIFIED

BC 7001   5/2000, van Ginderen Collection / 0104602

5 LÜRSSEN NEW 28 METRE TYPE (WHSIC)

BC 10001-10002   BC 20001-20003

Displacement, tons: 85 full load
Dimensions, feet (metres): 92.5 × 21.7 × 4.6 *(28.2 × 6.6 × 1.4)*
Main machinery: 2 MTU 16V 396 TE94 diesels; 2,955 hp(m) *(2.14 MW)* sustained; 2 shafts
Speed, knots: 40. **Range, n miles:** 1,100 at 30 kt
Complement: 11 (3 officers)
Guns: 2—7.62 mm MGs.
Radars: Surface search: Furuno FR 8731; I-band.

Comment: First pair built in Germany and delivered between May 1999 and November 1999. Last three built by PT Pal Surabaya and delivered between September 1999 and November 1999. Aluminium construction. *VERIFIED*

BC 10001   *5/1999, Lürssen /* 0080034

BC 20001   *9/1999, PT Pal /* 0075857

COAST AND SEAWARD DEFENCE COMMAND

Notes: (1) Established in 1978 as the Maritime Security Agency to control the 200 mile EEZ and to maintain navigational aids. Comes under the Military Sea Communications Agency. Some craft have blue hulls with a diagonal thick white and thin red stripe plus KPLP on the superstructure. In addition to the craft listed there are large numbers of small harbour boats.
(2) There are also a number of civilian manned vessels used for transport and servicing navigational aids.

5 KUJANG CLASS (WPB)

KUJANG 201   PARANG 202   CELURIT 203   CUNDRIK 204   BELATI 205

Displacement, tons: 162 full load
Dimensions, feet (metres): 125.6 × 19.6 × 6.8 *(38.3 × 6 × 2.1)*
Main machinery: 2 AGO SACM 195 V12 CZSHR diesels; 4,410 hp(m) *(3.24 MW)*; 2 shafts
Speed, knots: 28. **Range, n miles:** 1,500 at 18 kt
Complement: 18
Guns: 1—12.7 mm MG.

Comment: Built by SFCN, Villeneuve la Garenne. Completed April 1981 *(Kujang* and *Parang)*, August 1981 *(Celurit)*, October 1981 *(Cundrik)*, December 1981 *(Belati)*. Pennant numbers are preceded by PAT. *VERIFIED*

CUNDRIK   *11/1998, van Ginderen Collection /* 0052366

4 GOLOK CLASS (WSAR)

GOLOK 206   **PANAN** 207   **PEDANG** 208   **KAPAK** 209

Displacement, tons: 190 full load
Dimensions, feet (metres): 123 pp × 23.6 × 6.6 *(37.5 × 7.2 × 2)*
Main machinery: 2 MTU 16V 652 TB91 diesels; 4,610 hp(m) *(3.39 MW)* sustained; 2 shafts
Speed, knots: 25. **Range, n miles:** 1,500 at 18 kt
Complement: 18
Guns: 1 Rheinmetall 20 mm.

Comment: All launched 5 November 1981. First pair completed 12 March 1982. Last pair completed 12 May 1982. Built by Deutsche Industrie Werke, Berlin. Fitted out by Schlichting, Travemünde. Used for SAR and have medical facilities. Pennant numbers preceded by PAT. *VERIFIED*

KAPAK   *11/1998, van Ginderen Collection /* 0052367

15 HARBOUR PATROL CRAFT (WPB)

PAT 01-15

Displacement, tons: 12 full load
Dimensions, feet (metres): 40 × 14.1 × 3.3 *(12.2 × 4.3 × 1)*
Main machinery: 1 Renault diesel; 260 hp(m) *(191 kW)*; 1 shaft
Speed, knots: 14
Complement: 4
Guns: 1—7.62 mm MG.

Comment: First six built at Tanjung Priok Shipyard 1978-79. Four more of a similar design built in 1993-94 by Mahalaya Utama Shipyard and delivered from 1995. *VERIFIED*

HARBOUR PATROL CRAFT   *11/1998, van Ginderen Collection /* 0052368

NAVAL AUXILIARY SERVICE

Notes: This is a paramilitary force of non-commissioned craft. They have KAL pennant numbers. About 24 vessels operate in the eastern Fleet and 47 in the western Fleet, and three belong to the Naval Academy. In addition, the Baruna Jaya ships listed under Survey Ships are also part of the NAS.

KAL   *4/1999 /* 0080035

65 KAL KANGEAN CLASS (COASTAL PATROL CRAFT) (WPB)

Displacement, tons: 44.7 full load
Dimensions, feet (metres): 80.4 × 14.1 × 3.3 *(24.5 × 4.3 × 1)*
Main machinery: 2 diesels; 2 shafts
Speed, knots: 18
Guns: 2 USSR 25 mm/80 (twin). 2 USSR 14.5 mm (twin) MGs.

Comment: Ordered from Tanjung Uban Navy Yard in about 1984 and completed between 1987 and 1996. Numbers are uncertain. Have four figure pennant numbers in the 1101 series.
VERIFIED

KAL KANGEAN 1112   *10/1998, Trevor Brown*

6 CARPENTARIA CLASS (COASTAL PATROL CRAFT) (WPB)

201-206

Displacement, tons: 27 full load
Dimensions, feet (metres): 51.5 × 15.7 × 4.3 *(15.7 × 4.8 × 1.3)*
Main machinery: 2 MTU 8V 331 TC92 diesels; 1,770 hp(m) *(1.3 MW)* sustained; 2 shafts
Speed, knots: 29. **Range, n miles:** 950 at 18 kt
Complement: 10
Guns: 2—12.7 mm MGs.
Radars: Surface search: Decca; I-band.

Comment: Built 1976-77 by Hawker de Havilland, Australia. Endurance, four to five days. Transferred from the Navy in the mid-1980s to the Police and now with the Naval Auxiliary Service.
VERIFIED

CARPENTARIA 203   *8/1995, van Ginderen Collection / 0080036*

ARMY

Notes: The Army (ADRI) craft have mostly been transferred to the Military Sealift Command (Logistic Support).

27 LANDING CRAFT LOGISTICS (LCL)

ADRI XXXII-LVIII

Displacement, tons: 580 full load
Dimensions, feet (metres): 137.8 × 35.1 × 5.9 *(42 × 10.7 × 1.8)*
Main machinery: 2 Detroit 6-71 diesels; 348 hp(m) *(260 kW)* sustained; 2 shafts
Speed, knots: 10. **Range, n miles:** 1,500 at 10 kt
Complement: 15
Military lift: 122 tons equipment

Comment: A variety of LCL built in Tanjung Priok Shipyard 1979-82. Details are for *Adri XL.* XXXI sank in February 1993.
VERIFIED

ADRI XXXIII   *10/1999, David Boey / 0080037*

POLICE

Notes: The police operate about 85 craft of varying sizes including 14 Bango class of 194 tons and 32 Hamilton water-jet craft of 7.9 m, 234 hp giving a speed of 28 kt. Lürssen type (619-623) are identical to Customs craft.

POLICE 622   *8/1995 / 0080038*

POLICE 620   *3/1997, A Sharma / 0569198*

2 OFFSHORE PATROL CRAFT (PBO)

Name	No	Builders	Commissioned
BISMA	520	Astilleros Gondan, Castropol	May 2003
BALADEWA	521	Astilleros Gondan, Castropol	June 2003

Dimensions, feet (metres): 200.2 × 32.5 × 8.5 *(61.0 × 9.9 × 2.6)*
Main machinery: 2 MTU 12V 595TE 90 diesels; 8,700 hp *(6.5 MW)*
Speed, knots: 22. **Range, n miles:** 3,500 at 12 kt
Helicopters: Platform for one medium.

Comment: Primary role Search and Rescue.
UPDATED

BISMA   *5/2003, Astilleros Gondan / 0569201*

For details of the latest updates to *Jane's Fighting Ships* online and to discover the additional information available exclusively to online subscribers please visit
jfs.janes.com

Iran

Country Overview

Formerly a constitutional monarchy ruled by a shah, The Islamic Republic of Iran was established in 1979. With an area of 636,296 square miles, it is situated in the Middle East and is bordered to the north by Armenia, Azerbaijan and Turkmenistan, to the west by Iraq and Turkey and to the east by Afghanistan and Pakistan. It has a 1,318 n mile coastline with the Gulf, the Gulf of Oman and the Caspian Sea. The capital and largest city is Tehran. The principal Caspian ports are Bandar-e Anzali and Bandar-e Torkeman while those in the Gulf include the oil-shipping facilities on Kharg Island, Khorramshahr, Bandar-Khomeini and Bandar-Abbas on the strategic Strait of Hormuz. Territorial Seas (12 n miles) are claimed. An EEZ (200 n miles) has been claimed but the limits have not been defined.

Headquarters Appointments

Commander of Navy:
   Rear Admiral Abbas Mohtaj
Head of Naval Equipment:
   Rear Admiral Mohammed Hossein Shafii
Head of IRCG(N) (Sepah):
   Rear Admiral Morteza Saffari

Personnel

2005: 18,000 Navy (including 2,000 Naval Air and Marines), 20,000 IRGCN

Bases

Persian Gulf: Bandar Abbas (MHQ and 1st Naval District), Boushehr (2nd Naval District and also a Dockyard), Kharg Island, Qeshm Island, Bandar Lengeh
Indian Ocean: Chah Bahar (Bandar Beheshti) (3rd Naval District and forward base)
Caspian Sea: Bandar Anzali (4th Naval District)
Pasdaran: Al Farsiyah, Halileh, Sirri, Abu Musa, Larak

Coast Defence

Three Navy and one IRGCN brigades with many fixed installations and command posts. Approximately 100 truck-mounted C 802 and 80 CSSC-3 (Seersucker) Chinese SSMs in at least four sites.

Mines

Stocks of up to 3,000 mines are reported including Chinese EM 52 rising mines.

Strength of the Fleet

Type	Active	Building
Submarines	3	—
Mini Submarines	1	(2)
Frigates	3	—
Corvettes	2	3
Fast Attack Craft—Missile	20	—
Large Patrol Craft	8	—
Coastal Patrol Craft	123+	12
Landing Ships (Logistic)	7	—
Landing Ships (Tank)	6	—
Hovercraft	6	—
Minesweepers—Coastal	3	—
Replenishment Ship	1	—
Supply Ships	1 (1)	—
Support Ships	7	—
Water Tankers	4	—
Tenders	13	—

Prefix to Ships' Names

IS

PENNANT LIST

Submarines		Patrol Forces			P 232	Tabarzin	Amphibious Warfare Forces and Auxiliaries	
					P 313-1	Fath		
901	Tareq	202	Azadi		P 313-2	Nasr	21	Hejaz
902	Noor	203	Mehran		P 313-3	Saf	22	Karabala
903	Yunes	211	Parvin		P 313-4	Ra'd	23	Amir
		212	Bahram		P 313-5	Fajr	24	Farsi
		213	Nahid		P 313-6	Shams	25	Sardasht
Frigates		P 221	Kaman		P 313-7	Me'raj	26	Sab Sahel
		P 222	Zoubin		P 313-8	Falaq	101	Fouque
71	Alvand	P 223	Khadang		P 313-9	Hadid	411	Kangan
72	Alborz	P 226	Falakhon		P 313-10	Qadr	412	Taheri
73	Sabalan	P 227	Shamshir				421	Bandar Abbas
		P 228	Gorz				422	Boushehr
		P 229	Gardouneh				431	Kharg
Corvettes		P 230	Khanjar				511	Hengam
		P 231	Neyzeh		**Mine Warfare Forces**		512	Larak
81	Bayandor						513	Tonb
82	Naghdi				301	Hamzeh	514	Lavan

SUBMARINES

3 KILO CLASS (PROJECT 877 EKM) (SSK)

Name	No	Builders	Laid down	Launched	Commissioned
TAREQ	901	Admiralty Yard, St Petersburg	1988	1991	21 Nov 1992
NOOR	902	Admiralty Yard, St Petersburg	1989	1992	6 June 1993
YUNES	903	Admiralty Yard, St Petersburg	1990	1993	25 Nov 1996

Displacement, tons: 2,356 surfaced; 3,076 dived
Dimensions, feet (metres): 238.2 × 32.5 × 21.7 *(72.6 × 9.9 × 6.6)*
Main machinery: Diesel-electric; 2 diesels; 3,650 hp(m) *(2.68 MW)*; 2 generators; 1 motor; 5,500 hp(m) *(4.05 MW)*; 1 economic speed motor; 130 hp(m) *(95 kW)*; 1 shaft; 2 auxiliary propulsion motors; 204 hp(m) *(150 kW)*
Speed, knots: 17 dived; 10 surfaced; 9 snorting
Range, n miles: 6,000 at 7 kt snorting; 400 at 3 kt dived
Complement: 53 (12 officers)

Torpedoes: 6—21 in *(533 mm)* tubes; combination of TEST-71/96; wire-guided active/passive homing to 15 km *(8.1 n miles)* at 40 kt; warhead 220 kg and 53-65; passive wake homing to 19 km *(10.3 n miles)* at 45 kt; warhead 350 kg. Total of 18 weapons.
Mines: 24 in lieu of torpedoes.
Countermeasures: ESM; Squid Head; radar warning. Quad Loop D/F.

Weapons control: MVU-119EM Murena TFCS.
Radars: Surface search: Snoop Tray MRP-25; I-band.
Sonars: Sharks Teeth MGK-400; hull-mounted; passive/active search and attack; medium frequency.
   Mouse Roar MG-519; active attack; high frequency.

Programmes: Contract signed in 1988 for three of the class. The first submarine to be transferred sailed from the Baltic in October 1992 flying the Russian flag and with a predominantly Russian crew. The second sailed in June 1993. The third completed in 1994 but delivery delayed by funding problems. She arrived in Iran in mid-January 1997.
Modernisation: Chinese YJ-1 or Russian Novator Alfa SSMs may be fitted in due course.
Structure: Diving depth, 240 m *(787 ft)* normal. Has a 9,700 kW/h battery. SA-N-10 SAM system may be fitted, but this is not confirmed.
Operational: Based at Bandar Abbas but planned to move to Chah Bahar (Bandar Beheshti) on the northern shore of the

Gulf of Oman. So far a jetty has been extended to facilitate operations. Training is being done with assistance from Russia. There have been no indications of *Tareq* or *Noor* returning to Russia for long overdue refits although this remains a possibility. Problems with battery cooling and air conditioning were reported to have been resolved using Indian batteries. It is unlikely that refits will be conducted in India as has been speculated.
Opinion: The northern Gulf of Oman and the few deep water parts of the Persian Gulf are notoriously difficult areas for anti-submarine warfare. These submarines will be vulnerable to attack when alongside in harbour but pose a severe threat to merchant shipping either with torpedoes or mines.
UPDATED

YUNES

6/1999 / 0080039

1 + 2 COASTAL SUBMARINES (SSC)

Displacement, tons: 120 (approx) surfaced
Dimensions, feet (metres): 98.4 × ? × ? *(30.0 × ? × ?)*
Main machinery: To be announced
Speed, knots: To be announced
Complement: 32
Torpedoes: To be announced.
Sonars: To be announced.

Programmes: Little is known about this submarine whose existence was noted in February 2004. Dimensions are approximate. It is reported that perhaps two further boats have been constructed or are in manufacture. If this boat has been indigenously built, as is claimed, this would represent a significant technological development. The submarines are known as *Qadir 1, 2* and *3* and are likely to be employed in shallow areas of the Gulf such as the Strait of Hormuz.
NEW ENTRY

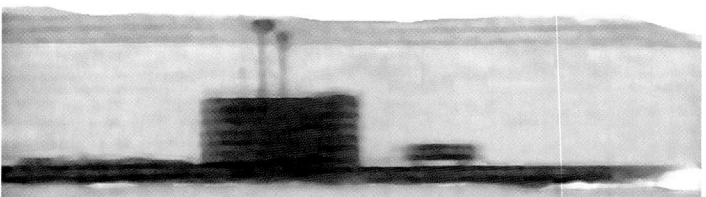

COASTAL SUBMARINE   *2/2004 * / 1044353*

AL SABEHAT 15   *8/2000 / 0104860*

1 SWIMMER DELIVERY VEHICLE (LDW)

Comment: On 29 August 2000, the first Iranian-built Swimmer Delivery Vehicle (SDV) *Al Sabehat 15* was launched at Bandar Abbas. The 8 m craft can accommodate a two-man crew and has the capability to carry three additional divers. It is well suited to coastal reconnaissance, Special Forces insertion/extraction and mining (it can carry 14 limpet mines) of ports and anchorages but not to open water operations. The absence of further deliveries suggests that first of class difficulties have yet to be overcome.

VERIFIED

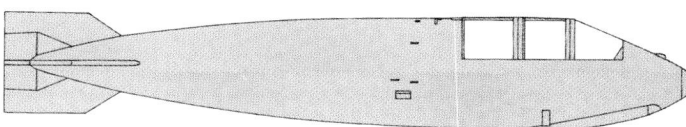

AL SABEHAT 15   *(not to scale), Ian Sturton / 0104859*

CORVETTES

Notes: A new corvette, known as *Mowj,* of about 1,200 tons was launched in October 2002 and is being completed at Bandar Abbas although there have been no reports of building progress. Further units are expected. The design appears to be a development of the Vosper Mk 5 and weapons are likely to include SSM (C-802?) and a medium calibre gun.

2 BAYANDOR (PF 103) CLASS (FS)

Name	No	Builders	Laid down	Launched	Commissioned
BAYANDOR (ex-US *PF 103*)	81	Levingstone Shipbuilding Co, Orange, TX	20 Aug 1962	7 July 1963	18 May 1964
NAGHDI (ex-US *PF 104*)	82	Levingstone Shipbuilding Co, Orange, TX	12 Sep 1962	10 Oct 1963	22 July 1964

Displacement, tons: 900 standard; 1,135 full load
Dimensions, feet (metres): 275.6 × 33.1 × 10.2 *(84 × 10.1 × 3.1)*
Main machinery: 2 Fairbanks-Morse 38TD8-1/8-9 diesels; 5,250 hp *(3.92 MW)* sustained; 2 shafts
Speed, knots: 20. **Range, n miles:** 2,400 at 18 kt; 4,800 at 12 kt
Complement: 140

Guns: 2 US 3 in *(76 mm)*/50 Mk 34 ❶; 50 rds/min to 12.8 km *(7 n miles)*; weight of shell 6 kg.
1 Bofors 40 mm/60 (twin) ❷; 120 rds/min to 10 km *(5.5 n miles)*; weight of shell 0.89 kg.
2 Oerlikon GAM-BO1 20 mm ❸. 2—12.7 mm MGs.
Weapons control: Mk 63 for 76 mm gun. Mk 51 Mod 2 for 40 mm guns.
Radars: Air/surface search: Westinghouse SPS-6C ❹; D-band; range 146 km *(80 n miles)* (for fighter).
Surface search: Racal Decca ❺; I-band.
Navigation: Raytheon 1650 ❻; I/J-band.
Fire control: Western Electric Mk 36 ❼; I/J-band.
IFF: UPX-12B.

Sonars: EDO SQS-17A; hull-mounted; active attack; high frequency.

Programmes: Transferred from the USA to Iran under the Mutual Assistance programme in 1964.
Modernisation: *Naghdi* change of engines and reconstruction of accommodation completed in mid-1988. 23 mm gun and depth charge racks replaced by 20 mm guns in 1990.

Operational: *Milanian* and *Khanamuie* sunk in 1982 during war with Iraq. Both remaining ships are very active. Sonars may have been removed.
VERIFIED

BAYANDOR   *(Scale 1 : 900), Ian Sturton*

BAYANDOR   *2/1998 / 0052372*

FRIGATES

3 ALVAND (VOSPER Mk 5) CLASS (FFG)

Name	No	Builders	Laid down	Launched	Commissioned
ALVAND (ex-*Saam*)	71	Vosper Thornycroft, Woolston	22 May 1967	25 July 1968	20 May 1971
ALBORZ (ex-*Zaal*)	72	Vickers, Barrow	3 Mar 1968	25 July 1969	1 Mar 1971
SABALAN (ex-*Rostam*)	73	Vickers, Newcastle & Barrow	10 Dec 1967	4 Mar 1969	28 Feb 1972

Displacement, tons: 1,350 full load
Dimensions, feet (metres): 310 × 36.4 × 14.1 (screws)
    (94.5 × 11.1 × 4.3)
Main machinery: CODOG; 2 RR Olympus TM2A gas turbines;
    40,000 hp *(29.8 MW)* sustained; 2 Paxman 16YJCM diesels;
    3,800 hp *(2.83 MW)* sustained; 2 shafts; cp props
Speed, knots: 39 gas; 18 diesel
Range, n miles: 3,650 at 18 kt; 550 at 36 kt
Complement: 125 (accommodation for 146)

Missiles: SSM: 4 China C-802 (2 twin) ❶; active radar homing to
    120 km *(66 n miles)* at 0.9 Mach; warhead 165 kg; sea-
    skimmer.
Guns: 1 Vickers 4.5 in *(114 mm)*/55 Mk 8 ❷; 25 rds/min to
    22 km *(12 n miles)* anti-surface; 6 km *(3.3 n miles)* anti-
    aircraft; weight of shell 21 kg.
    2 Oerlikon 35 mm/90 (twin) ❸; 550 rds/min to 6 km *(3.3 n
    miles)*; weight of shell 1.55 kg.
    3 Oerlikon GAM-BO1 20 mm ❹. 2—12.7 mm MGs.
Torpedoes: 6—324 mm Mk 32 (2 triple) tubes (71).
A/S mortars: 1—3-tubed Limbo Mk 10 (72 and 73) ❺; automatic
    loading; range 1,000 m; warhead 92 kg.
Countermeasures: Decoys: 2 UK Mk 5 rocket flare launchers.
    ESM: Decca RDL 2AC; radar warning. Racal FH 5-HF/DF.
Radars: Air/surface search: Plessey AWS 1 ❻; E/F-band; range
    110 km *(60 n miles)*.
    Surface search: Racal Decca 1226 ❼; I-band.
    Navigation: Decca 629; I-band.

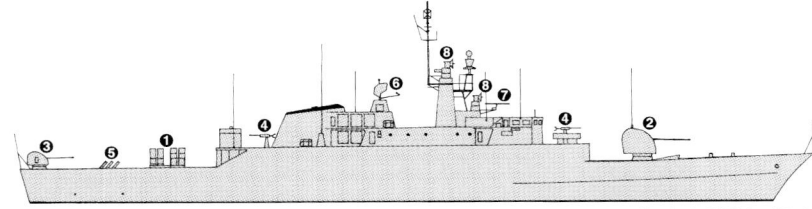

ALBORZ   *(Scale 1 : 900), Ian Sturton /* 0012550

Fire control: Contraves Sea Hunter ❽; I/J-band.
    IFF: UK Mk 10.
Sonars: Graseby 174; hull-mounted; active search; medium/high
    frequency.
    Graseby 170; hull-mounted; active attack; high frequency.

Programmes: Ordered on 25 August 1966.
Modernisation: Major refits including replacement of 4.5 in Mk
    5 gun by Mk 8 completed 1977. Modifications in 1988
    included replacing Seacat with a 23 mm gun and boat davits
    with minor armaments. By mid-1991 the 23 mm and both
    boats had been replaced by GAM-BO1 20 mm guns and the
    SSM launcher had effectively become a twin launcher. In
    1996/97 two of the class had the Sea Killer SSM replaced by

C-802 launchers and a new communications mast fitted
between the two fire-control radars. The third has been
similarly modified. *Sabalan* appears to be fitted with Rice
Screen air/surface search radar. Torpedo tubes which
replaced the mortars in *Alvand* were probably taken from
decommissioned Babr class.
Structure: Air conditioned throughout. Fitted with Vosper
stabilisers.
Operational: *Sahand* sunk by USN on 18 April 1988. *Sabalan* had
her back broken by a laser-guided bomb in the same skirmish
but was out of dock by the end of 1990 and was operational
again in late 1991. ASW mortars probably unserviceable. All
are active.

UPDATED

ALVAND   *1/2002 /* 0569203

SABALAN   *2/1998 /* 0052371

SHIPBORNE AIRCRAFT

Numbers/Type: 6 Agusta AB 204ASW/212.
Operational speed: 104 kt *(193 km/h).*
Service ceiling: 11,500 ft *(3,505 m).*
Range: 332 n miles *(615 km).*
Role/Weapon systems: Mainly engaged in ASV operations in defence of oil installations. Numbers are uncertain. Sensors: APS 705 search radar, dipping sonar (if carried). Weapons: ASW; two China YU-2 torpedoes. ASV; two AS 12 missiles. *UPDATED*

AB 212 (Spanish colours)   *3/2002*, A Campanera i Rovira /* 0529019

Numbers/Type: 8 Agusta-Sikorsky ASH-3D Sea King.
Operational speed: 120 kt *(222 km/h).*
Service ceiling: 12,200 ft *(3,720 m).*
Range: 630 n miles *(1,165 km).*
Role/Weapon systems: Shore-based ASW helicopter to defend major port and oil installations. Can be embarked in *Kharg.* Sensors: Selenia search radar, dipping sonar. Weapons: ASW; four A244/S torpedoes or depth bombs. ASV; trials of an anti-ship missile 'Fajr-e-Darya' are reported to have taken place. Capabilities not known but could be a development of Sea Killer. *UPDATED*

SEA KING   *3/1997 /* 0012549

LAND-BASED MARITIME AIRCRAFT (FRONT LINE)

Notes: (1) The Air Force also has up to six F-4 Phantoms equipped with C 80IK ASMs for the maritime role.
(2) Four F-27 Fokker Friendship aircraft are used in a utility MPA role.
(3) Five Dornier 228 are also in service but are reported not to be very active.
(4) The Iranian Air Force operates some 14 (plus 18 ex-Iraqi) Su-24 Fencer ground attack, some of which may be 'marinised' for an anti-ship role.
(5) An-140 transport aircraft are under licensed production at Esfahan. The first aircraft flew in January 2001. These provide a potential airframe for the replacement of the ageing P3F fleet.

Numbers/Type: 6 Sikorsky RH/MH-53D Sea Stallion.
Operational speed: 125 kt *(232 km/h).*
Service ceiling: 11,100 ft *(3,385 m).*
Range: 405 n miles *(750 km).*
Role/Weapon systems: Surface search helicopter which could be used for mine clearance but so far has only been used for Logistic purposes. Can be carried on Hengam class flight deck. Sensors: Weather radar. Weapons: Unarmed. *UPDATED*

Numbers/Type: 5 Lockheed C-130H-MP Hercules.
Operational speed: 325 kt *(602 km/h).*
Service ceiling: 33,000 ft *(10,060 m).*
Range: 4,250 n miles *(7,876 km).*
Role/Weapon systems: Long-range maritime reconnaissance role by Air Force which has a total of 23 of these aircraft. Sensors: Search/weather radar. Weapons: Unarmed. *UPDATED*

Numbers/Type: 3 Lockheed P-3F Orion.
Operational speed: 410 kt *(760 km/h).*
Service ceiling: 28,300 ft *(8,625 m).*
Range: 4,000 n miles *(7,410 km).*
Role/Weapon systems: Air Force manned. One of the remaining aircraft can be used for early warning and control duties for strikes. Replacements are being sought. Sensors: Search radar, sonobuoys. Weapons: ASW; various weapons can be carried. ASV; C-802 SSM. *UPDATED*

P3F   *12/2001, A Sharma /* 0528307

PATROL FORCES

Notes: Two new classes of semi-submersible torpedo boat have been reported. Three 'Taedong-C' and two 'Taedong-B' were delivered from North Korea on the Iranian freighter *Iran Meead* on 8 December 2002. Little is known about them although it is possible that they are derivatives of high-speed infiltration craft reported to be in service in North Korea. These are believed to be 12.8 m in length and to have a top surface speed of about 45 kt. They can submerge to a depth of 3 m using a snort mast.

10 THONDOR (HOUDONG) CLASS
(FAST ATTACK CRAFT—MISSILE) (PTFG)

FATH	P 313-1	FAJR	P 313-5	FALAQ	P 313-8
NASR	P 313-2	SHAMS	P 313-6	HADID	P 313-9
SAF	P 313-3	ME'RAJ	P 313-7	QADR	P 313-10
RA'D	P 313-4				

Displacement, tons: 171 standard; 205 full load
Dimensions, feet (metres): 126.6 × 22.3 × 8.9 *(38.6 × 6.8 × 2.7)*
Main machinery: 3 diesels; 8,025 hp(m) *(7.94 MW)* sustained; 3 shafts
Speed, knots: 35. **Range, n miles:** 800 at 30 kt
Complement: 28 (3 officers)

Missiles: SSM: 4 China C-802; active radar homing to 120 km *(66 n miles)* at 0.9 Mach; warhead 165 kg; sea-skimmer.
Guns: 2—30 mm/65 (twin) AK 230. 2—23 mm/87 (twin).
Radars: Surface search: China SR-47A; I-band.
Navigation: China RM 1070A; I-band.
Fire control: Rice Lamp Type 341; I/J-band.

Programmes: Negotiations for sale started in 1991 but were held up by arguments over choice of missile. Built at Zhanjiang Shipyard. First five delivered in September 1994 by transporter

SAF   *10/1997 /* 0052375

vessel, second batch in March 1996. Original pennant numbers 301-310. More may be built in Iran under licence.
Structure: The hull is a shortened version of the Chinese Huangfen (Osa 1) class but the superstructure has a lattice mast to support two I-band radars and there is a separate director plinth for the fire-control system. A twin 23 mm gun is fitted aft of the mast.
Operational: Manned by the Pasdaran. *VERIFIED*

ME'RAJ   *6/1998 /* 0052376

11 + (2) KAMAN (COMBATTANTE II) CLASS
(FAST ATTACK CRAFT—MISSILE) (PGGF)

Name	No	Builders	Commissioned
KAMAN	P 221	CMN, Cherbourg	12 Aug 1977
ZOUBIN	P 222	CMN, Cherbourg	12 Sep 1977
KHADANG	P 223	CMN, Cherbourg	15 Mar 1978
PEYKAN	P 224		2004
FALAKHON	P 226	CMN, Cherbourg	31 Mar 1978
SHAMSHIR	P 227	CMN, Cherbourg	31 Mar 1978
GORZ	P 228	CMN, Cherbourg	22 Aug 1978
GARDOUNEH	P 229	CMN, Cherbourg	11 Sep 1978
KHANJAR	P 230	CMN, Cherbourg	1 Aug 1981
NEYZEH	P 231	CMN, Cherbourg	1 Aug 1981
TABARZIN	P 232	CMN, Cherbourg	1 Aug 1981

Displacement, tons: 249 standard; 275 full load
Dimensions, feet (metres): 154.2 × 23.3 × 6.2 *(47 × 7.1 × 1.9)*
Main machinery: 4 MTU 16V 538 TB91 diesels; 12,280 hp(m) *(9.03 MW)* sustained; 4 shafts
Speed, knots: 37.5. **Range, n miles:** 2,000 at 15 kt; 700 at 33.7 kt
Complement: 31

Missiles: SSM: 2 or 4 China C-802 (1 or 2 twin); active radar homing to 120 km *(66 n miles)* at 0.9 Mach; warhead 165 kg; sea-skimmer or 4 McDonnell Douglas Harpoon (2 twin); active radar homing to 40 km *(22 n miles)* at 0.9 Mach; warhead 165 kg; sea-skimmer or Standard SM1-MR box launchers *(Gorz)*.
Guns: 1 OTO Melara 3 in *(76 mm)*/62 compact; 85 rds/min to 16 km *(8.7 n miles)* anti-surface; 12 km *(6.6 n miles)* anti-aircraft; weight of shell 6 kg; 320 rounds.
1 Breda Bofors 40 mm/70; 300 rds/min to 12 km *(6.6 n miles)*; weight of shell 0.96 kg; 900 rounds. Some have a 23 mm or 20 mm gun in place of the 40 mm.
2—12.7 mm MGs.
Countermeasures: ESM: Thomson-CSF TMV 433 Dalia; radar intercept.
ECM: Thomson-CSF Alligator; jammer.
Radars: Surface search/fire control: Signaal WM28; I/J-band.
Navigation: Racal Decca 1226; I-band.
IFF: UPZ-27N/APX-72.

Programmes: Twelve ordered in February 1974. The transfer of the last three craft was delayed by the French Government after the Iranian revolution. On 12 July 1981 France decided to hand them over. This took place on 1 August, on 2 August they sailed and soon after *Tabarzin* was seized by a pro-Royalist group off Cadiz. After the latter surrendered to the French in Toulon further problems were prevented by sending all three to Iran in a merchant ship. Further indigenously built craft have been developed for operations in the Caspian Sea. Known as the SINA 1 programme, the first vessel was launched on 29 September 2003 and a further two are reported to be under construction. *Peykan* has the same name and pennant number of a vessel sunk in 1980 and is reported to have similar capabilities.
Modernisation: Most of the class fitted with C-802 SSM in 1996-98. *Gorz* has been used for trials, first with Harpoon, and now with SM 1 launchers taken from the deleted Sumner class destroyers.
Structure: Portable SA-7 launchers may be embarked in some.
Operational: The original *Peykan* was sunk in 1980 by Iraq; *Joshan* in April 1988 by the USN.

UPDATED

GARDOUNEH (with Harpoon)   *11/2001, Royal Australian Navy /* 0528433

GORZ (with SM1)   *12/2002 /* 0569204

KHANJAR   *6/1998 /* 0052374

3 PARVIN (PGM-71) CLASS (LARGE PATROL CRAFT) (PC)

Name	No	Builders	Commissioned
PARVIN (ex-*PGM 103*)	211	Peterson Builders Inc	1967
BAHRAM (ex-*PGM 112*)	212	Peterson Builders Inc	1969
NAHID (ex-*PGM 122*)	213	Peterson Builders Inc	1970

Displacement, tons: 98 standard; 148 full load
Dimensions, feet (metres): 101 × 21.3 × 8.3 *(30.8 × 6.5 × 2.5)*
Main machinery: 8 GM 6-71 diesels; 2,040 hp *(1.52 MW)* sustained; 2 shafts
Speed, knots: 22. **Range, n miles:** 1,140 at 17 kt
Complement: 20
Guns: 1 Bofors 40 mm/60. 1 GAM-BO1 20 mm. 2—12.7 mm MGs.
Depth charges: 4 racks (8 US Mk 6).
Radars: Surface search: I-band.
Sonars: SQS-17B; hull-mounted active attack; high frequency.

Comment: The heavier 40 mm gun is mounted aft and the 20 mm forward to compensate for the large SQS-17B sonar dome under the bows. Mousetrap A/S mortar removed. Beginning to be difficult to maintain in an operational state.

VERIFIED

PARVIN   *1/2002, A Sharma /* 0528306

2 KAIVAN (CAPE) CLASS (LARGE PATROL CRAFT) (PB)

AZADI 202   **MEHRAN** 203

Displacement, tons: 98 standard; 148 full load
Dimensions, feet (metres): 95 × 20.2 × 6.6 *(28.9 × 6.2 × 2)*
Main machinery: 24 Cummins NYHMS-1200 diesels; 2,120 hp *(1.58 MW)*; 2 shafts
Speed, knots: 21. **Range, n miles:** 2,324 at 8 kt
Complement: 15
Guns: 1 Bofors 40 mm/60. 2 USSR 23 mm/80 (twin). 2—12.7 mm MGs.

Comment: Three patrol craft originally built by the US Coast Guard, Curtis Bay, Maryland in the 1950s were withdrawn from Iranian service in approximately 1995. In light of reports that at least two of the craft have been refitted, it is assumed that they have been recommissioned. Details are as for the craft in 1994 but it is likely that machinery and armament may now be different.

VERIFIED

6 MIG-S-2600 CLASS (PBF)

Displacement, tons: 85 full load
Dimensions, feet (metres): 80 × 20.3 × 4.6 *(26.2 × 6.2 × 1.4)*
Main machinery: 4 diesels; 4,000 hp(m) *(2.94 MW)*; 4 shafts
Speed, knots: 35
Complement: 12
Guns: 2—23 mm/80 (twin). 1—12-barrelled 107 mm MRL.
Radars: Surface search: I-band.

Comment: Numbers are uncertain. Built by Joolaee Marine Industries, Tehran, to a similar specification as the North Korean Chaho class but with a different superstructure and a raised mast to give an improved radar horizon. Pasdaran manned.

VERIFIED

MIG-S-2600   *1996, Joolaee Marine Industries*

9 US Mk III CLASS (COASTAL PATROL CRAFT) (PB)

Displacement, tons: 41.6 full load
Dimensions, feet (metres): 65 × 18.1 × 6 *(19.8 × 5.5 × 1.8)*
Main machinery: 3 GM 8V-71TI diesels; 690 hp *(515 kW)* sustained; 3 shafts
Speed, knots: 30. **Range, n miles:** 500 at 28 kt
Complement: 8
Guns: 1—20 mm GAM-BO1. 1—12.7 mm MG.
Radars: Surface search: RCA LN66; I-band.

Comment: Twenty ordered from Marinette Marine Corporation, Wisconsin, USA; the first delivered in December 1975 and the last in December 1976. A further 50 were ordered in 1976 to be shipped out and completed in Iran. It is not known how many were finally assembled. Six lost in the Gulf War, others have been scrapped. These last nine are based at Boushehr and Bandar Abbas. Continue to be active.   ***VERIFIED***

US Mk III   *5/1999* / 0080041

10 MIG-G-1900 CLASS (COASTAL PATROL CRAFT) (PBF)

Displacement, tons: 30 full load
Dimensions, feet (metres): 64 × 13.8 × 3 *(19.5 × 4.2 × 0.9)*
Main machinery: 2 MWM TBD 234 V12 diesels; 1,646 hp(m) *(1.21 MW)*; 2 shafts
Speed, knots: 36
Complement: 8
Guns: 2—23 mm/80 (twin).
Radars: Surface search: I-band.

Comment: Building in Iran to a modified US Mk II design. Numbers uncertain. Pasdaran craft.   ***VERIFIED***

MIG-G-1900   *1992, Iranian Marine Industries* / 0080042

20 MIG-S-1800 CLASS (COASTAL PATROL CRAFT) (PB)

Displacement, tons: 60 full load
Dimensions, feet (metres): 61.3 × 18.9 × 3.4 *(18.7 × 5.8 × 1.1)*
Main machinery: 2 MWM TBD 234 V12 diesels; 1,646 hp(m) *(1.21 MW)*; 2 shafts
Speed, knots: 18
Complement: 10
Guns: 1 Oerlikon 20 mm. 2—7.62 mm MGs.
Radars: Surface search: I-band.

Comment: Assembled in Iran as general purpose patrol craft. Numbers uncertain. Pasdaran craft.   ***VERIFIED***

MIG-S-1800   *1996, Joolaee Marine Industries*

10 PEYKAAP CLASS (INSHORE PATROL CRAFT) (PTF)

Displacement, tons: 7 approx
Dimensions, feet (metres): 49.2 × 9.8 × 2.3 *(15 × 3 × 0.7)*
Speed, knots: 50 approx
Torpedoes: 2 lightweight

Comment: Up to ten of this class in service with the Pasdaran. Built in North Korea, six craft were reported to have been delivered on 8 December 2002 on the Iranian freighter *Iran Meead*. An apparently stealthy craft whose unusual armament of lightweight torpedoes suggest a ship-disabling role.   ***VERIFIED***

PEYKAAP CLASS   *6/2002, Royal Australian Navy* / 0528431

10 TIR CLASS (INSHORE PATROL CRAFT) (PTF)

Displacement, tons: 30 approx
Dimensions, feet (metres): 65.6 × 16.4 × 3.3 *(20 × 5 × 1)*
Speed, knots: 50 approx
Torpedoes: 2 unknown

Comment: Up to ten of this class in service with the Pasdaran. Built in North Korea, two craft were reported to have been delivered on 8 December 2002 on the Iranian freighter *Iran Meead*. Anti-surface ship role.   ***VERIFIED***

TIR CLASS   *6/2002, Royal Australian Navy* / 0528434

6 US Mk II CLASS (COASTAL PATROL CRAFT) (PB)

Displacement, tons: 22.9 full load
Dimensions, feet (metres): 49.9 × 15.1 × 4.3 *(15.2 × 4.6 × 1.3)*
Main machinery: 2 GM 8V-71TI diesels; 460 hp *(343 kW)* sustained; 2 shafts
Speed, knots: 28. **Range, n miles:** 750 at 26 kt
Complement: 8
Guns: 2—12.7 mm MGs.
Radars: Surface search: SPS-6; I-band.

Comment: Twenty-six ordered from Peterson, USA in 1976-77. Six were for the Navy and the remainder for the Imperial Gendarmerie. All were built in association with Arvandan Maritime Corporation, Abadan. The six naval units operate in the Caspian Sea. Of the remaining 20, six were delivered complete and the others were only 65 per cent assembled on arrival in Iran. Some were lost when the Iraqi Army captured Koramshahr. Others have been lost at sea. Numbers uncertain.   ***VERIFIED***

US Mk II   *3/1996* / 0080043

30 PBI TYPE (COASTAL PATROL CRAFT) (PBM)

Displacement, tons: 20.1 full load
Dimensions, feet (metres): 50 × 15 × 4 *(15.2 × 4.6 × 1.2)*
Main machinery: 2 GM 8V-71TI diesels; 460 hp *(343 kW)* sustained; 2 shafts
Speed, knots: 28. **Range, n miles:** 750 at 26 kt
Complement: 5 (1 officer)
Missiles: SSM: Tigercat; range 6 km *(3.2 n miles).*
Guns: 2—12.7 mm MGs.
Radars: Surface search: I-band.

Comment: Ordered by Iranian Arvandan Maritime Company. First 19 completed by Petersons and remainder shipped as kits for completion in Iran. The SSM is crude and unguided. Numbers are approximate.

VERIFIED

PBI   *4/1995* / 0080044

6 + 4 CHINA CAT CLASS (PTGF)

Displacement, tons: 19 full load
Dimensions, feet (metres): 45.9 × 13.1 × 3.4 *(14 × 4 × 1)*
Main machinery: 2 diesels; 2 shafts
Speed, knots: 50
Complement: 10
Missiles: SSM: 4 FL-10 (2 twin) launchers.
Guns: 2 China 25 mm (twin).
Weapons control: Optronic director.
Radars: Surface search: I-band.

Comment: Prototype reported delivered late 2000 and commissioned in February 2001. A further five have been reported operational and a class of at least ten is expected. Catamaran hull. Type of missile has not been confirmed although this may be the new short-range Kowsar missile.

UPDATED

CHINA CAT   *2001, China State Shipbuilding Corporation* / 0096378

30 BOGHAMMAR CRAFT (PBF)

Displacement, tons: 6.4 full load
Dimensions, feet (metres): 41.2 × 8.6 × 2.3 *(13 × 2.7 × 0.7)*
Main machinery: 2 Seatek 6-4V-9 diesels; 1,160 hp *(853 kW)*; 2 shafts
Speed, knots: 46. **Range, n miles:** 500 at 40 kt
Complement: 5/6
Guns: 3—12.7 mm MGs. 1 RPG-7 rocket launcher or 106 mm recoilless rifle. 1—12-barrelled 107 mm rocket launcher (MRL).
Radars: Surface search: I-band.

Comment: Ordered in 1983 and completed in 1984-85 for Customs Service. Total of 51 delivered. Used extensively by the Pasdaran. Maximum payload 450 kg. Speed is dependent on load carried. They can be transported by Amphibious Lift Ships and can operate from bases at Farsi, Sirri and Abu Musa Islands with a main base at Bandar Abbas. Re-engined with Seatek diesels from 1991. There are also a further 10—11 m craft with similar characteristics. Known as TORAGH boats and manned by the Pasdaran and the Navy. Numbers approximate.

VERIFIED

20 MIG-G-0800 (INSHORE PATROL CRAFT) (PBF)

Displacement, tons: 1.3 full load
Dimensions, feet (metres): 22.3 × 7.4 × 1.2 *(6.7 × 2.3 × 0.4)*
Main machinery: 2 outboards; 240 hp *(179 kW)*
Speed, knots: 40+
Complement: 4
Guns: Various, but can include 1—12-barrelled 107 mm MRL or 1—12.7 mm MG.

Comment: Boston Whaler type craft based on a Watercraft (UK) design. Numerous indigenously constructed GRP hulls. Numbers uncertain. Manned by the Pasdaran and the Navy.

VERIFIED

MIG-G-0800   1988

RIVER ROADSTEAD PATROL AND HOVERCRAFT (PBR)

Comment: Numerous craft used by the Revolutionary Guard include:
Type 2: Dimensions, feet (metres): 22.0 × 7.2 *(6.7 × 2.2)*; single outboard engine; 1—12.7 mm MG.
Type 3: Dimensions, feet (metres): 16.4 × 5.2 *(5.0 × 1.6)*; single outboard engine; small arms.
Type 4: Dimensions, feet (metres): 13.1-26.2 × 7.9 *(4-8 × 1.6)*; two outboard engines; small arms.
Type 5: Dimensions, feet (metres): 24.6 × 9.2 *(7.5 × 2.8)*; assault craft.
Type 6: Dimensions, feet (metres): 30.9 × 11.8 *(9.4 × 3.6)*; single outboard engine; 1—12.7 mm MG.
Dhows: Dimensions, feet (metres): 77.1 × 20 *(23.5 × 6.1)*; single diesel engine; mine rails.
Yunus: Dimensions, feet (metres): 27.6 × 9.8 *(8.4 × 3)*; speed 32 kt.
Ashoora: Dimensions, feet (metres): 26.6 × 7.9 *(8.1 × 2.4)*; two outboards; speed 42 kt; 1—7.62 mm MG.
Jet Skis: RPGs.

VERIFIED

ASHOORA   *6/1994* / 0080045

JET SKI (with RPG)   *5/1999* / 0080046

Type 4   *5/1997* / 0012558

20 MIG-G-0900 CLASS (INSHORE PATROL CRAFT) (PBI)

Displacement, tons: 3.5 full load
Dimensions, feet (metres): 30.2 × 9.2 × 1.5 *(9.2 × 2.8 × 0.45)*
Main machinery: 2 Volvo Penta diesels; 1,260 hp *(940 kW)*
Speed, knots: 30
Complement: 3
Guns: 3—12.7 mm MGs. 1 RPG-7 rocket launcher or 106 mm recoilless rifle. 1—12-barrelled 107 mm rocket launcher (MRL).
Radars: Surface search: I-band.

Comment: Built by MiG, the unarmed variant has been produced in relatively large numbers since the mid-1990s. This approximate number of armed variant is believed to be in Pasdaran or naval service. *VERIFIED*

MIG-G-0900 *2000, MiG /* 0126375

AMPHIBIOUS FORCES

Notes: (1) Commercial LSLs have been built at Bandar Abbas. These include two 1,151 grt ships, *Chavoush* launched in December 1995 and *Chalak* in June 1996.
(2) There are an unknown number of small Wing-In-Ground (WIG) vehicles, possibly for operations in the Caspian Sea.

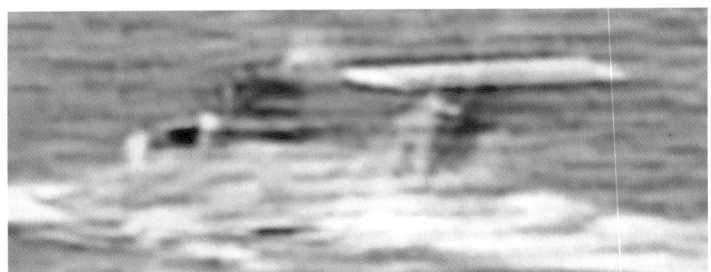

WIG *6/2004 * /* 1044357

4 HENGAM CLASS (LSLH)

Name	No	Builders	Commissioned
HENGAM	511	Yarrow (Shipbuilders) Ltd, Clyde	12 Aug 1974
LARAK	512	Yarrow (Shipbuilders) Ltd, Clyde	12 Nov 1974
TONB	513	Yarrow (Shipbuilders) Ltd, Clyde	21 Feb 1985
LAVAN	514	Yarrow (Shipbuilders) Ltd, Clyde	16 Jan 1985

Displacement, tons: 2,540 full load
Dimensions, feet (metres): 305 × 49 × 7.3 *(93 × 15 × 2.4)*
Main machinery: 4 Paxman 12YJCM diesels *(Hengam, Larak)*; 3,000 hp *(2.24 MW)* sustained; 2 shafts. 4 MTU 16V 652 TB81 diesels *(Tonb, Lavan)*; 4,600 hp(m) *(3.38 MW)* sustained; 2 shafts
Speed, knots: 14.5. **Range, n miles:** 4,000+ at 12 kt
Complement: 80
Military lift: Up to 9 tanks depending on size; 600 tons cargo; 227 troops; 10 ton crane

Guns: 4 Bofors 40 mm/60 *(Hengam and Larak)*. 8 USSR 23 mm/80 (4 twin) *(Tonb and Lavan)*. 2—12.7 mm MGs.
1 BM-21 multiple rocket launcher.
Countermeasures: Decoys: 2 UK Mk 5 rocket flare launchers.
Radars: Navigation: Racal Decca 1229; I-band.
IFF: SSR 1520 *(Hengam and Larak)*.
Tacan: URN 25.
Helicopters: Can embark 1 Sikorsky MH-53D.

Programmes: Named after islands in the Gulf. First two ordered 25 July 1972. Four more ordered 20 July 1977. The material for the last two ships of the second order had been ordered by Yarrows when the order was cancelled in early 1979. *Tonb* carried out trials in October 1984 followed by *Lavan* later in the year and both were released by the UK in 1985 as 'Hospital Ships'.
Structure: Smaller than British *Sir Lancelot* design with no through tank deck. Rocket launcher mounted in the bows.
Operational: Two LCVPs and a number of small landing craft can be carried. Can act as Depot Ships for MCMV and small craft and have been used to ferry Pasdaran small craft around the Gulf. *UPDATED*

TONB *6/2004 * /* 1044354

LARAK *3/1997, Maritime Photographic /* 0052378

3 IRAN HORMUZ 24 CLASS (LST)

FARSI 24   SARDASHT 25   SAB SAHEL 26

Displacement, tons: 2,014 full load
Dimensions, feet (metres): 239.8 × 46.6 × 8.2 *(73.1 × 14.2 × 2.5)*
Main machinery: 2 Daihatsu 6DLM-22 diesels; 2,400 hp(m) *(1.76 MW)*; 2 shafts
Speed, knots: 12
Complement: 30 plus 110 berths
Military lift: 9 tanks, 140 troops

Comment: Built at Inchon, South Korea in 1985-86 and as with the Iran Hormuz 21 class officially classed as Merchant Ships. Have been used to support Pasdaran activities. *VERIFIED*

IRAN HORMUZ 24 *5/1999 /* 0080048

3 FOUQUE (MIG-S-3700) CLASS (LSL)

FOUQUE 101   102   103

Displacement, tons: 276 full load
Dimensions, feet (metres): 121.4 × 26.2 × 4.9 *(37 × 8 × 1.5)*
Main machinery: 2 MWM TBD 234 V8 diesels; 879 hp(m) *(646 kW)*; 2 shafts
Speed, knots: 10. **Range, n miles:** 400 at 10 kt
Complement: 8
Military lift: 140 tons of vehicles

Comment: *Fouque* assembled in Iran by Martyr Darvishi Marine, Bandar Abbas. Launched in June 1998. Others of the class are in commercial service and more can be taken over by the Navy if required. Two others for the Navy were launched in September 1995. *VERIFIED*

FOUQUE *1994, Iranian Marine Industries /* 0080049

3 IRAN HORMUZ 21 CLASS (LST)

HEJAZ 21	KARABALA 22	AMIR 23

Displacement, tons: 1,280 full load
Measurement, tons: 750 dwt
Dimensions, feet (metres): 213.3 × 39.4 × 8.5 *(65 × 12 × 2.6)*
Main machinery: 2 MAN V12V-12.5/14 or 2 MWM TBD 604 V12 diesels; 1,460 hp(m) *(1.07 MW)*;
   2 shafts
Speed, knots: 9
Complement: 12
Military lift: 600 tons

Comment: Officially ordered for 'civilian use' and built by Ravenstein, Netherlands in 1984-85. 21 and
   22 are manned by the Pasdaran. A local version is assembled as the MIG-S-5000 for commercial use.
   One was launched in mid-1995 at Boushehr and a second in 1997.
   VERIFIED

6 WELLINGTON (BH.7) CLASS (HOVERCRAFT) (UCAC)

101-106

Displacement, tons: 53.8 full load
Dimensions, feet (metres): 78.3 × 45.6 × 5.6 (skirt) *(23.9 × 13.9 × 1.7)*
Main machinery: 1 RR Proteus 15 M/541 gas turbine; 4,250 hp *(3.17 MW)* sustained
Speed, knots: 70; 30 in Sea State 5 or more. **Range, n miles:** 620 at 66 kt
Guns: 2 Browning 12.7 mm MGs.
Radars: Surface search: Decca 1226; I-band.

Comment: First pair are British Hovercraft Corporation 7 Mk 4 commissioned in 1970-71 and the
   next four are Mk 5 craft commissioned in 1974-75. Mk 5 craft fitted for, but not with Standard
   missiles. Some refitted in UK in 1984. Can embark troops and vehicles or normal support
   cargoes. The Iranian Aircraft Manufacturing Industries (HESA) is reported to be able to maintain
   these craft in service.
   UPDATED

WELLINGTON   *6/1998, HESA* / 0033385

1 IRAN CLASS (HOVERCRAFT) (UCAC)

Displacement, tons: 10 full load
Dimensions, feet (metres): 48.4 × 25.3 × 15.9 *(14.8 × 7.7 × 4.8)*
Main machinery: 1 gas turbine
Speed, knots: 60

Comment: The first of a new Iran class was completed in March 2000 and is probably based on the
   old SRN-6 class on which the approximate dimensions are based. Reports suggest a military lift
   of 2 tons and 26 troops.
   NEW ENTRY

IRAN CLASS   *6/2004* / 1044355

AUXILIARIES

Notes: (1) There is also an inshore survey vessel *Abnegar*.
(2) Two 65 ton training vessels of Kialas-C-Qasem class are reported to have commissioned
mid-2000. There may be further craft. No other details are known.

2 FLOATING DOCKS

400 (ex-US *ARD 29*, ex-*FD 4*)   DOLPHIN

Dimensions, feet (metres): 487 × 80.2 × 32.5 *(149.9 × 24.7 × 10)* *(400)*
   786.9 × 172.1 × 58.4 *(240 × 52.5 × 17.8)* *(Dolphin)*

Comment: *400* is an ex-US ARD 12 class built by Pacific Bridge, California and transferred in 1977;
   lift 3,556 tons. *Dolphin* built by MAN-GHH Nordenham, West Germany and completed in
   November 1985; lift 28,000 tons.
   VERIFIED

1 MSC 268/292 CLASS (YDT)

Name	No	Builders	Commissioned
HAMZEH (ex-*Shahrokh*, ex-*MSC 276*)	301	Bellingham Shipyard	1960

Displacement, tons: 384 full load
Dimensions, feet (metres): 145.8 × 28 × 8.3 *(44.5 × 8.5 × 2.5)*
Main machinery: 4 GM 6-71 diesels; 696 hp *(519 kW)* sustained; 2 shafts
Speed, knots: 13. **Range, n miles:** 2,400 at 10 kt
Complement: 40 (6 officers)
Guns: 2 Oerlikon 20 mm (twin).
Radars: Surface search: Decca; I-band.

Comment: Originally class of four. Transferred from the USA under MAP in 1959-62. Wooden
   construction. *Hamzeh* in the Caspian Sea was reported scrapped, but has re-emerged as a
   diving tender with the new name of the former Royal Yacht. *Simorgh*, *Karkas* and *Shabaz* have
   been deleted.
   VERIFIED

MSC 268 (Spanish colours)   *4/1999, Diego Quevedo* / 0080047

4 KANGAN CLASS (WATER TANKERS) (AWT)

KANGAN 411	TAHERI 412	SHAHID MARJANI	AMIR

Displacement, tons: 12,000 full load
Measurement, tons: 9,430 dwt
Dimensions, feet (metres): 485.6 × 70.5 × 16.4 *(148 × 21.5 × 5)*
Main machinery: 1 MAN 7L52/55A diesel; 7,385 hp(m) *(5.43 MW)* sustained; 1 shaft
Speed, knots: 15
Complement: 14
Cargo capacity: 9,000 m³ of water
Guns: 2 USSR 23 mm/80 (twin). 2—12.7 mm MGs.
Radars: Navigation: Decca 1229; I-band.

Comment: The first two were built in Mazagon Dock, Bombay in 1978 and 1979. The second pair
   to a slightly modified design was acquired in 1991-92 but may be civilian manned. Some of the
   largest water tankers afloat and can be used to supply remote coastal towns and islands,
   although there have been few reports of activity in 2001. Accommodation is air conditioned. All
   have a 10 ton boom crane.
   VERIFIED

TAHERI   *5/1989*

12 HENDIJAN CLASS (TENDERS) (PBO)

HENDIJAN	GENAVEH	BAHREGAN (ex-*Geno*)	NAYBAND
KALAT	SIRIK	MOGAM	MACHAM
KONARAK	GAVATAR	ROSTANI	KORAMSHAHR

Displacement, tons: 460 full load
Dimensions, feet (metres): 166.7 × 28.1 × 11.5 *(50.8 × 8.6 × 3.5)*
Main machinery: 2 Mitsubishi S16MPTK diesels; 7,600 hp(m) *(5.15 MW)*; 2 shafts
Speed, knots: 25
Complement: 15 plus 90 passengers
Cargo capacity: 40 tons on deck; 95 m³ of liquid/solid cargo space
Guns: 1—20 mm (sometimes fitted in patrol craft). 2—12.7 mm MGs.
Radars: Navigation: Racal Decca or China RM 1070A; I-band.

Comment: First eight built by Damen, Netherlands 1988-91. Remainder built at Bandar Abbas
   under the MIG-S-4700 programme. Last pair launched on 25 November 1995. Reports of three
   more being built may be caused by confusion with new corvettes. Variously described in the
   Iranian press as 'frigates' or 'patrol ships', they are regularly used for coastal surveillance. One is
   used as a training ship. Pennant numbers in the 1400 series.
   UPDATED

HENDIJAN 1410   *6/2004* / 1044358

10 DAMEN 1550 (PILOT CRAFT)

Displacement, tons: 25 full load
Dimensions, feet (metres): 52.5 × 15.1 × 4.6 *(16 × 4.6 × 1.4)*
Main machinery: 2 MTU diesels; 2 shafts
Speed, knots: 19
Complement: 3
Radars: Navigation: Furuno; I-band.

Comment: Ordered from Damen, Gorinchen in February 1993. Steel hull and aluminium superstructure. Used primarily as pilot craft. ***VERIFIED***

DAMEN 1550 *1993, Damen Shipyards /* 0080051

2 FLEET SUPPLY SHIPS (AORLH)

Name	No	Builders	Commissioned
BANDAR ABBAS	421	C Lühring Yard, Brake, West Germany	Apr 1974
BOUSHEHR	422	C Lühring Yard, Brake, West Germany	Nov 1974

Displacement, tons: 4,673 full load
Measurement, tons: 3,250 dwt; 3,186 gross
Dimensions, feet (metres): 354.2 × 54.4 × 14.8 *(108 × 16.6 × 4.5)*
Main machinery: 2 MAN 6L 52/55 diesels; 12,060 hp(m) *(8.86 MW)* sustained; 2 shafts
Speed, knots: 20. **Range, n miles:** 3,500 at 16 kt
Complement: 59
Guns: 3 GAM-BO1 20 mm can be carried. 2—12.7 mm MGs.
Radars: Navigation: 2 Decca 1226; I-band.
Helicopters: 1 AB 212.

Comment: *Bandar Abbas* launched 11 August 1973, *Boushehr* launched 23 March 1974. Combined tankers and store-ships carrying victualling, armament and general stores. Telescopic hangar. Both carry 2 SA-7 portable SAM and 20 mm guns have replaced the former armament. *Bandar Abbas* damaged by an explosion in early 1999 but has been repaired. Deployment to the Caspian Sea has not taken place. ***UPDATED***

BANDAR ABBAS *6/2004* */* 1044356

7 DELVAR CLASS (SUPPORT SHIPS) (AEL/AKL/AWT)

CHARAK (AKL)	CHIROO (AKL)	DELVAR (AEL)	DILIM (AWT)
SOURU (AKL)	SIRJAN (AEL)	DAYER (AWT)	

Measurement, tons: 890 gross; 765 dwt
Dimensions, feet (metres): 210 × 34.4 × 10.9 *(64 × 10.5 × 3.3)*
Main machinery: 2 MAN G6V 23.5/33ATL diesels; 1,560 hp(m) *(1.15 MW)*; 2 shafts
Speed, knots: 11
Complement: 20
Guns: 1 GAM-BO1 20 mm. 2—12.7 mm MGs.
Radars: Navigation: Decca 1226; I-band.

Comment: All built by Karachi SY in 1980-82. *Delvar* and *Sirjan* are ammunition ships, *Dayer* and *Dilim* water carriers and the other three are general cargo ships. The water carriers have only one crane (against two on the other types), and have rounded sterns (as opposed to transoms). Re-armed. 424 and 483 have been reported. ***VERIFIED***

CHARAK *10/1997 /* 0012563

DELVAR CLASS 424 *5/2003, A Sharma /* 0569202

1 REPLENISHMENT SHIP (AORH)

Name	No	Builders	Commissioned
KHARG	431	Swan Hunter Ltd, Wallsend	5 Oct 1984

Displacement, tons: 11,064 light; 33,014 full load
Measurement, tons: 9,367 dwt; 18,582 gross
Dimensions, feet (metres): 679 × 86.9 × 30 *(207.2 × 26.5 × 9.2)*
Main machinery: 2 Babcock & Wilcox boilers; 2 Westinghouse turbines; 26,870 hp *(19.75 MW)*; 1 shaft
Speed, knots: 21.5
Complement: 248
Guns: 1 OTO Melara 76 mm/62 compact. 4 USSR 23 mm/80 (2 twin). 2—12.7 mm MGs.
Radars: Navigation: Decca 1229; I-band.
Tacan: URN 20.
Helicopters: 3 Sea Kings (twin hangar).

Comment: Ordered October 1974. Laid down 27 January 1976. Launched 3 February 1977. Ship handed over to Iranian crew on 25 April 1980 but remained in UK. In 1983 Iranian Government requested this ship's transfer. The UK Government delayed approval until January 1984. On 10 July 1984 began refit at Tyne Ship Repairers. Trials began 4 September 1984 and ship was then delivered without guns which were subsequently fitted. A design incorporating some of the features of the British OI class but carrying ammunition and dry stores in addition to fuel. Inmarsat fitted. ***VERIFIED***

KHARG *5/1997 /* 0052379

KHARG *6/1998 /* 0052380

TUGS

17 HARBOUR TUGS (YTB/YTM)

HAAMOON	ALBAN	ASLAM	DARYAVAND II
HIRMAND	SEFID-RUD	DEHLORAN	KHANDAG
MENAB	ATRAK	ILAM	ARVAND
HARI-RUD	ABAD	HANGAM	KARKHEH
ARAS			

Comment: All between 70 and 90 ft in length, built since 1984. ***VERIFIED***

Iraq

Country Overview

The Republic of Iraq was proclaimed in 1958 following a coup d'état. With an area of 168,754 square miles, it is situated in the Middle East and is bordered to the north by Turkey, to the east by Iran (with which it was at war 1980-88), to the west by Jordan and Syria and to the south by Saudi Arabia (with which it jointly administers the Neutral Zone) and Kuwait (which it invaded and occupied 1990-91 until expelled in the Gulf War 1991). It has a 31 n mile coastline with the Gulf. Baghdad is the capital and largest city. Al Basrah, located on the Shatt Al-Arab, is the only

port. Territorial Seas (12 n miles) are claimed. An EEZ has not been claimed.

In the wake of the US-led occupation in March-April 2003, Iraq remained under coalition control until 30 June 2004 when full authority was handed over to an Iraqi Interim Government. Direct elections are to be conducted by late 2005. All naval coastal defence units, surface ships and aircraft were destroyed or disabled during the war and are unlikely to be resurrected. An oiler *(Agnadeen)* at Alexandria could be reclaimed but this is unlikely to be a high priority. The Iraqi Coastal Defence Force (ICDF) was formally established at Umm Qasr on 30 September

2004 and a training programme aims to build a force of about 400 ex-Iraqi Navy personnel.

Headquarters Appointments

Commander, ICDF:
   Colonel Hameed Balafam

Bases

Al Basra (Navy HQ), Khor Az Zubayr, Umm Qasr.

CORVETTES

0 + 2 ASSAD CLASS (FSG)

Name	No	Builders	Laid down	Launched	Commissioned
MUSSA BEN NUSSAIR	F 210	Fincantieri, Muggiano	15 Jan 1982	22 Oct 1982	17 Sep 1986
TARIQ IBN ZIAD	F 212	Fincantieri, Muggiano	20 May 1982	8 July 1983	29 Oct 1986

Displacement, tons: 685 full load
Dimensions, feet (metres): 204.4 × 30.5 × 8.0
   (62.3 × 9.3 × 2.5)
Main machinery: 4 MTU 20V 956 TB92 diesels; 20,120 hp(m)
   (14.8 MW) sustained; 4 shafts
Speed, knots: 37. **Range, n miles:** 4,000 at 18 kt
Complement: 51 (without aircrew)

Missiles: SSM: 2 OTO Melara/Matra Teseo (fitted for).
SAM: 1 Selenia/Elsag Albatros launcher (4 cell-2 reloads)
   Aspide; semi-active radar homing to 13 km *(7 n miles)* at
   2.5 Mach; warhead 30 kg.
Guns: 1 OTO Melara 3 in *(76 mm)*/62 compact; 85 rds/min to
   16 km *(8.7 n miles)*; weight of shell 6 kg.

Countermeasures: Decoys: 2 Breda 105 mm 6-tubed fixed
   multipurpose launchers.
ESM: Selenia INX-3; intercept.
ECM: Selenia TQN-2; jammer.
Combat data systems: Selenia IPN-10.
Weapons control: 2 Selenia 21 (for SAN). Dardo (for guns).
Radars: Air/surface search: Selenia RAN 12L/X; D/I-band.
Fire control: 2 Selenia RTN 10X; I/J-band.
Navigation: SMA SPN-703; I-band.
Sonars: Kae ASO 84-41; hull-mounted; active search and attack.

Helicopters: 1 Agusta AB 212 type.

Programmes: Originally ordered in February 1981 and formally handed over in 1986 without weapon systems. Trials with Iraqi crews began in 1990 but final delivery was halted by the Iraqi invasion of Kuwait and subsequent UN sanctions. Thereafter the ships remained at La Spezia with a caretaker team of about 12. In December 2004, the Italian government announced plans to refurbish both ships for delivery in 2006. Details are as for original ships and may differ.
Structure: Similar to Malaysian Laksamana class which were originally ordered by Iraq but fitted with flight deck and telescopic hangar.
   NEW ENTRY

TARIQ IBN ZIAD

8/1995, Ships of the World / 0581387

PATROL FORCES

5 INSHORE PATROL CRAFT (PB)

P 101-105

Displacement, tons: To be announced
Dimensions, feet (metres): 88.9 × 9.2 × 5.9 *(27.1 × 2.8 × 1.8)*
Main machinery: 2 MTU 12V 396 TE742; 4,025 hp *(3 MW)*
Speed, knots: 32
Complement: 6
Guns: 1—7.62 mm MG.

Comment: Built by Wuhan Nanhu High-Speed Engineering Company and originally acquired in 2002. Maintained in dry-dock at Jebel Ali, UAE, until the first two were commissioned into the Iraqi Coastal Defence Force on 4 April 2004. Three further craft followed in May 2004. Acquisition and refit costs funded by the US. The craft are to be used initially for training of Iraqi personnel with a view to assuming greater responsibility for security in Iraqi waters. Based at Umm Qasr.
   NEW ENTRY

P 102

5/2004*, US Navy / 0580527

P 102

5/2004*, US Navy / 0580528

Ireland
AN SEIRBHIS CHABHLAIGH

Country Overview

The Republic of Ireland comprises about five sixths of the island of Ireland. Situated west of Great Britain, the country consists of the provinces of Leinster, Munster, Connaught and three counties of the province of Ulster. The remaining six counties of Ulster form Northern Ireland, a constituent part of the United Kingdom. With an area of 27,136 square miles, the country has a 783 n mile coastline with the Atlantic Ocean and Irish Sea. Dublin is the capital, largest city and principal port. There is another major port at Cork. Territorial waters (12 n miles) are claimed. A 200 n mile Fishery zone has also been claimed.

Headquarters Appointments

Flag Officer Commanding Naval Service:
  Commodore F Lynch

Bases

Haulbowline Island, Cork Harbour—Naval HQ, Base and Dockyard

Personnel

(a)   2005: 1,050 (135 officers)
(b)   Voluntary service
(c)   Reserves: 400 (one unit in each of the following cities: Dublin, Waterford, Cork and Limerick)

Fishery Protection

In late 2004 and early 2005, all ships were fitted with the new Lirguard system. This system incorporates the previously separate functions of the database, GIS database display, Vessel Monitoring System (VMS) and legislation browser. This system will be updated several times daily by satellite link from the Fisheries Monitoring Centre (FMC) at Haulbowline. These will provide a near real-time display and analysis tool of fishing activity to allow more intelligent and efficient use of the ships in the Fishery Protection role. Research is also continuing into the incorporation of VMS data with data from Earth Observation (EO) technology.

Prefix to Ships' Names

LÉ (Long Éirennach = Irish Ship)

PATROL FORCES

Notes: Replacement of the P-21 class with more capable vessels is under consideration.

1 EITHNE CLASS (PSOH)

Name	No	Builders	Laid down	Launched	Commissioned
EITHNE	P 31	Verolme, Cork	15 Dec 1982	19 Dec 1983	7 Dec 1984

Displacement, tons: 1,760 standard; 1,910 full load
Dimensions, feet (metres): 265 × 39.4 × 14.1
  (80.8 × 12 × 4.3)
Main machinery: 2 Ruston 12RKC diesels; 6,800 hp *(5.07 MW)* sustained; 2 shafts; cp props
Speed, knots: 20+; 19 normal. **Range, n miles:** 7,000 at 15 kt
Complement: 73 (10 officers) plus 8 (2 officers) aircrew

Guns: 1 Bofors 57 mm/70 Mk 1; 200 rds/min to 17 km *(9.3 n miles)*; weight of shell 2.4 kg.
  2 Rheinmetall 20 mm/20. 2—7.62 mm MGs.
  2 Wallop 57 mm launchers for illuminants.
Weapons control: Signaal LIOD director. 2 Signaal optical sights.
Radars: Air/surface search: Signaal DA05 Mk 4; E/F-band.
Surface search: Kelvin Hughes; E/F-band.
Navigation: 2 Kelvin Hughes; 6000A; I-band.
Tacan: MEL RRB transponder.

Helicopters: Platform only.

Programmes: Ordered 23 April 1982 from Verolme, Cork, this was the last ship to be built at this yard.

EITHNE   *6/2003, D Jones, Irish Navy /* 0568897

Structure: Fitted with retractable stabilisers. Closed circuit TV for flight deck operations. Satellite navigation and communications. CTD tactical displays.

Operational: Helicopter no longer operational. Two Delta 6.4 m RIBs fitted in addition to two 5.4 m RIBs in 2003. Long refit (SLEP) in 1998/99. **UPDATED**

2 ROISIN CLASS (PSO)

Name	No	Builders	Laid down	Launched	Commissioned
ROISIN	P 51	Appledore Shipbuilders, Bideford	Dec 1998	12 Aug 1999	15 Dec 1999
NIAMH	P 52	Appledore Shipbuilders, Bideford	June 2000	10 Feb 2001	18 Sep 2001

Displacement, tons: 1,700 full load
Dimensions, feet (metres): 258.7 × 45.9 × 12.8
  (78.9 × 14 × 3.9)
Main machinery: 2 Wärtsilä 16V26 diesels; 6,800 hp(m) *(5 MW)* sustained; 2 shafts; LIPS cp props; bow thruster; 462 hp(m) *(340 kW)*
Speed, knots: 23. **Range, n miles:** 6,000 at 15 kt

Complement: 44 (6 officers)
Guns: 1 OTO Melara 3 in *(76 mm)*/62; 85 rds/min to 16 km *(8.6 n miles)*; weight of shell 6 kg.
  2—12.7 mm MGs. 4—7.62 mm MGs.
Weapons control: Radamec 1500 optronic director.
Radars: Surface search: Kelvin Hughes; E/F-band.
Navigation: Kelvin Hughes; I-band.

Programmes: Contract for first ship signed on 16 December 1997 with 65 per cent of EU funding. Option on a second of class taken up on 6 April 2000.
Operational: Designated Large Patrol Vessel, the design is a modification of the Mauritius ship *Vigilant* but without the hangar or flight deck. Two Delta 6.5 m and one Avon 5.4 m RIBs are carried. CTD tactical displays. **UPDATED**

ROISIN   *6/2004*, Irish Navy /* 0589758

3 P 21 CLASS (OFFSHORE PATROL VESSELS) (PSO)

Name	No	Builders	Launched	Commissioned
EMER	P 21	Verolme, Cork	4 Aug 1977	16 Jan 1978
AOIFE	P 22	Verolme, Cork	12 Apr 1979	29 Nov 1979
AISLING	P 23	Verolme, Cork	3 Oct 1979	21 May 1980

Displacement, tons: 1,019.5
Dimensions, feet (metres): 213.7 × 34.4 × 14 *(65.2 × 10.5 × 4.4)*
Main machinery: 2 SEMT-Pielstick 6 PA6 L 280 diesels; 4,800 hp *(3.53 MW)*; 1 shaft; bow thruster *(Aoife and Aisling)*
Speed, knots: 17. **Range, n miles:** 4,000 at 17 kt; 6,750 at 12 kt
Complement: 47 (6 officers)

Guns: 1 Bofors 40 mm/60 Mk 22; 120 rds/min to 10 km *(5.4 n miles)*; weight of shell 0.89 kg.
2 GAM-B01 20 mm; 900 rds/min to 2 km.
2—12.7 mm MGs. 2—7.62 mm MGs.
Radars: Surface search: Kelvin Hughes I-band.
Navigation: Kelvin Hughes Nucleus 6000A; I-band.

Modernisation: New search radars were fitted in 1994-95. CTD tactical display fitted.
Structure: Stabilisers fitted. *Aoife* and *Aisling* are equipped with a bow thruster. Inmarsat SATCOM fitted.
Operational: *Emer* refitted in 1995, *Aoife* in 1996/97 and *Aisling* in 1997/98. Sonars have been removed.

UPDATED

AISLING   *6/2004*, Irish Navy /* 0589757

AOIFE   *6/2004*, Irish Navy /* 0589756

2 P 41 PEACOCK CLASS (COASTAL PATROL VESSELS) (PSO)

Name	No	Builders	Commissioned
ORLA (ex-*Swift*)	P 41	Hall Russell, Aberdeen	3 May 1985
CIARA (ex-*Swallow*)	P 42	Hall Russell, Aberdeen	17 Oct 1984

Displacement, tons: 712 full load
Dimensions, feet (metres): 204.1 × 32.8 × 8.9 *(62.6 × 10 × 2.7)*
Main machinery: 2 Crossley SEMT-Pielstick 18 PA6 V 280 diesels; 14,400 hp(m) *(10.58 MW)* sustained; 2 shafts; auxiliary drive; Schottel prop; 181 hp(m) *(133 kW)*
Speed, knots: 25. **Range, n miles:** 2,500 at 17 kt
Complement: 39 (5 officers)

Guns: 1—3 in (76 mm)/62 OTO Melara compact; 85 rds/min to 16 km *(8.6 n miles)*; weight of shell 6 kg.
2—12.7 mm MGs. 4—7.62 mm MGs.
Weapons control: Radamec 1500 optronic director (for 76 mm).
Radars: Surface search: Kelvin Hughes; I-band.
Navigation: Kelvin Hughes Nucleus 5000A/6000A; I-band.

Programmes: *Orla* launched 11 September 1984 and *Ciara* 31 March 1984. Both served in Hong Kong from mid-1985 until early 1988. Acquired from UK and commissioned 21 November 1988. Others of the class acquired by the Philippines in 1997.
Modernisation: New radars fitted in 1993. CTD tactical display fitted.
Structure: Have loiter drive. Displacement increased after building by the addition of more electronic equipment.

UPDATED

CIARA   *6/2004*, Irish Navy /* 0589755

SHIPBORNE AIRCRAFT

Numbers/Type: 2 Aerospatiale SA 365F Dauphin 2.
Operational speed: 140 kt *(260 km/h)*.
Service ceiling: 15,000 ft *(4,575 m)*.
Range: 410 n miles *(758 km)*.
Role/Weapon systems: No longer operated from *Eithne*; some shore land-based training by Army Air Corps and SAR. One aircraft lost in 2001. Sensors: Bendix RDR 1500 radar.
Weapons: Unarmed.   *UPDATED*

DAUPHIN 2   *6/2002, D Jones, Irish Navy /* 0525904

LAND-BASED MARITIME AIRCRAFT

Notes: Four civilian operated Sikorsky S-61 helicopters provide long-range SAR services.

Numbers/Type: 2 Casa CN-235 Persuader.
Operational speed: 210 kt *(384 km/h)*.
Service ceiling: 24,000 ft *(7,315 m)*.
Range: 2,000 n miles *(3,218 km)*.
Role/Weapon systems: EEZ surveillance. First one delivered in June 1992 but returned to Spain in 1995. Two more delivered in December 1994. Sensors: Search radar Bendix APS 504(V)5; FLIR.
Weapons: Unarmed.   *VERIFIED*

CN-235   *7/2003, Paul Jackson /* 0568896

AUXILIARIES

Notes: (1) In addition there are a number of mostly civilian manned auxiliaries including: *Seabhac* a small tug acquired in 1983; *Fainleog*, *David F* (built in 1962) and *Fiach Dubh* passenger craft, the last two taken over after lease in 1988 and the first in 1983; *Tailte* a Dufour 35 ft sail training yacht bought in 1979 and an elderly training yacht *Creidne*.
(2) *Granuaile* is an 80 m lighthouse tender with a helicopter flight deck forward operated by the Commissioners of Irish Lights. Launched on 14 August 1999 this ship replaced a previous vessel of the same name on 23 March 2000.

GRANUAILE   *3/2000, Commissioners of Irish Lights /* 0093593

Israel

HEYL HAYAM

Country Overview

Established in 1948, The State of Israel is situated on the eastern shore of the Mediterranean Sea and has borders to the north with Lebanon, to the north-east with Syria, to the east with Jordan and to the south-west with Egypt. It has coastlines with the Mediterranean (142 n miles) and with the Gulf of Aqaba (5 n miles) in the northern Red Sea. A land area of 8,463 square miles includes East Jerusalem and other territory (including Gaza Strip, the West Bank region of Jordan, the Golan Heights area of south-western Syria) annexed in 1967. Jerusalem is the largest city but, although claimed as the capital, is not so recognised by the United Nations. Many nations maintain embassies at Tel Aviv. Haifa is the principal port. Territorial seas (12 n miles) are claimed but an EEZ is not claimed.

Headquarters Appointments

Commander-in-Chief:
  Vice Admiral David Ben-Ba'ashat

General

Less than 5 per cent of Israeli defence budget is allocated to the Navy.

Personnel

(a)   2005: 6,500 (880 officers) of whom 2,500 are conscripts. Includes a Naval Commando of 300
(b)   3 years' national service for Jews and Druzes

Notes: An additional 5,000 Reserves available on mobilisation.

Bases

Haifa, Ashdod, Eilat
(The repair base at Eilat has a synchrolift)

Coast Defence

There are ten integrated coastal radar stations.

Prefix to Ships' Names

INS (Israeli Naval Ship)

SUBMARINES

Notes: (1) The two decommissioned Gal class submarines were reported to have started a refit in Germany on 18 December 2003 although it is unclear whether the boats are for sale or to be reactivated.
(2) Negotiations for the construction of two German-built Type 212 submarines, incorporating air-independent propulsion, have reportedly taken place although funding arrangements are yet to be finalised.

GAL CLASS

4/2004*, Martin Mokrus / 0580526

3 DOLPHIN (TYPE 800) CLASS (SSK)

Name	No	Builders	Laid down	Launched	Commissioned
DOLPHIN	—	Howaldtswerke/Thyssen Nordseewerke	7 Oct 1994	12 Apr 1996	37 July 1999
LEVIATHAN	—	Howaldtswerke/Thyssen Nordseewerke	13 Apr 1995	25 Apr 1997	15 Nov 1999
TEKUMA	—	Howaldtswerke/Thyssen Nordseewerke	12 Dec 1996	26 June 1998	25 July 2000

Displacement, tons: 1,640 surfaced; 1,900 dived
Dimensions, feet (metres): 188 × 22.3 × 20.3
  (57.3 × 6.8 × 6.2)
Main machinery: 3 MTU 16V 396 SE 84 diesels; 4,243 hp(m)
  (3.12 MW) sustained; 3 alternators; 2.91 MW; 1 Siemens
  motor; 3,875 hp(m) *(2.85 MW)* sustained; 1 shaft
Speed, knots: 20 dived; 11 snorting
Range, n miles: 8,000 at 8 kt surfaced; 420 at 8 kt dived
Complement: 30 (6 officers)

Missiles: SSM: Sub Harpoon; UGM-84C; active radar or GPS
  homing to 130 km *(70 n miles)* at 0.9 Mach; warhead 227 kg.
  SAM: Fitted for Triten anti-helicopter system.
Torpedoes: 4—25.6 in *(650 mm)* and 6—21 in *(533 mm)* bow
  tubes. STN Atlas DM2A4 Seehecht; wire-guided active
  homing to 13 km *(7 n miles)* at 35 kt; passive homing to
  28 km *(15 n miles)* at 23 kt; warhead 260 kg. Total of 16
  torpedoes and 5 SSMs. The four 650 mm tubes may be for
  SDVs, but could carry torpedoes if liners are fitted.
Mines: In lieu of torpedoes.
Countermeasures: ESM: Elbit Timnex 4CH(V)2; intercept.
Weapons control: STN/Atlas Elektronik ISUS 90-1 TCS.
Radars: Surface search: Elta; I-band.
Sonars: Atlas Elektronik CSU 90; hull-mounted; passive/active
  search and attack.
  Atlas Elektronik PRS-3; passive ranging.
  FAS-3; flank array; passive search.

Programmes: In mid-1988 Ingalls Shipbuilding Division of Litton
  Corporation was chosen as the prime contractor for two IKL-
  designed Dolphin class submarines to be built in West
  Germany with FMS funds by HDW in conjunction with
  Thyssen Nordseewerke. Funds approved in July 1989 with an
  effective contract date of January 1990 but the project was
  cancelled in November 1990 due to pressures on defence
  funds. After the Gulf War in April 1991 the contract was
  resurrected, this time with German funding for two
  submarines with an option on a third taken up in July 1994.
Modernisation: Installation of air-independent propulsion is
  under consideration.
Structure: Diving depth, 350 m *(1,150 ft)*. Similar to German
  Type 212 in design but with a 'wet and dry' compartment for
  underwater swimmers. Two Kollmorgen periscopes. Probably
  fitted for Triten anti-helicopter SAM system.
Operational: Endurance, 30 days. Used for interdiction,
  surveillance and special boat operations. Development of a
  submarine-launched cruise missile would complete the final
  part of a triad of nuclear deterrents. However, while Israel
  probably has the expertise and technology to deploy SLCM,

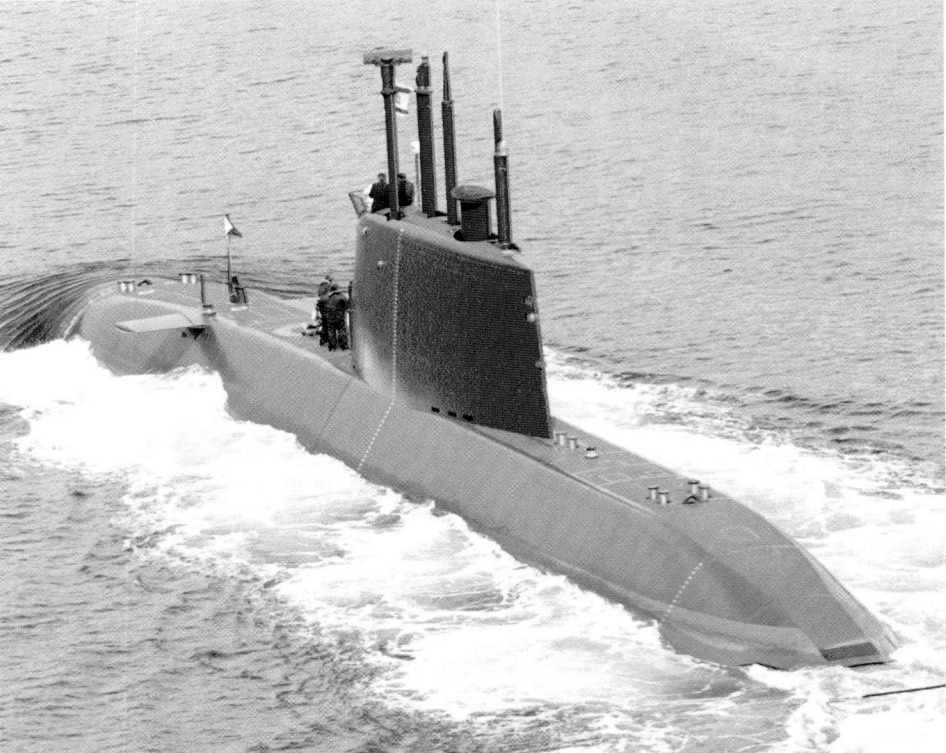

DOLPHIN   *6/1999, Michael Nitz /* 0080058

little information exists to confirm or deny such a programme.
Adaptation of the indigenous Delilah and Popeye groups of
missles is a possible option although encapsulation of the
missile would pose a significant challenge. Painted blue/

green to aid concealment in the eastern Mediterranean. Some
other NT 37E torpedoes are embarked until full Seehecht
outfits are available.   ***UPDATED***

TEKUMA   *9/2000, Michael Nitz /* 0104865

CORVETTES

Notes: A Request for Information was issued in September 2003 for the acquisition of up to three multimission corvettes. While this programme was temporarily superseded in 2004 by a proposal to procure a 13,000 ton amphibious ship, it has re-emerged as the priority due to budget realities. Options include a new design, a development of the USCG Offshore Patrol Cutter design or development of the existing SAAR 5 class.

3 EILAT (SAAR 5) CLASS (FSGHM)

Name	No	Builders	Laid down	Launched	Commissioned
EILAT	501	Ingalls, Pascagoula	24 Feb 1992	9 Feb 1993	24 May 1994
LAHAV	502	Ingalls, Pascagoula	25 Sep 1992	20 Aug 1993	23 Sep 1994
HANIT	503	Ingalls, Pascagoula	5 Apr 1993	4 Mar 1994	7 Feb 1995

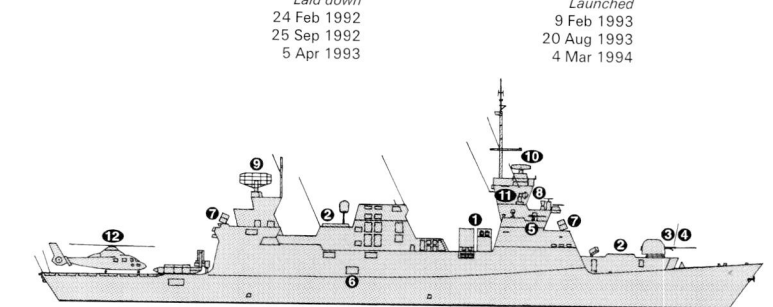

EILAT

(Scale 1 : 900), Ian Sturton / 0012570

Displacement, tons: 1,075 standard; 1,295 full load
Dimensions, feet (metres): 279.0 × 39.0 × 10.5
(85.0 × 11.9 × 3.2)
Main machinery: CODOG; 1 GE LM 2500 gas turbine; 30,000 hp *(22.38 MW)* sustained; 2 MTU 12V 1163 TB82 diesels; 6,600 hp(m) *(4.86 MW)* sustained; 2 shafts; Kamewa cp props
Speed, knots: 33 gas; 20 diesels. **Range, n miles:** 3,500 at 17 kt
Complement: 64 (16 officers) plus 10 (4 officers) aircrew

Missiles: SSM: 8 McDonnell Douglas Harpoon (2 quad) launchers ❶; active radar homing to 130 km *(70 n miles)* at 0.9 Mach; warhead 227 kg.
SAM: 2 Israeli Industries Barak I (vertical launch) ❷; 2 × 32 cells; command line of sight radar or optical guidance to 10 km *(5.5 n miles)* at 2 Mach; warhead 22 kg (see *Operational*).
Guns: OTO Melara 3 in *(76 mm)*/62 compact ❸; 85 rds/min to 16 km *(8.7 n miles)*; weight of shell 6 kg.
The main gun is interchangeable with a Bofors 57 mm gun or Vulcan Phalanx CIWS ❹.
2 Sea Vulcan 20 mm CIWS ❺; range 1 km.
Torpedoes: 6—324 mm Mk 32 (2 triple) tubes ❻. Honeywell Mk 46; anti-submarine; active/passive homing to 11 km *(5.9 n miles)* at 40 kt; warhead 44 kg. Mounted in the superstructure.
Countermeasures: Decoys: 3 Elbit/Deseaver 72-barrelled chaff and IR launchers ❼; Rafael ATC-1 towed torpedo decoy.
ESM: Elisra NS 9003; intercept. Tadiran NATACS.
ECM: 2 Rafael 1010; Elisra NS 9005; jammers.
Combat data systems: Elbit NTCCS using Elta EL/S-9000 computers. Reshet datalink.
Weapons control: 2 Elop MSIS optronic directors ❽.
Radars: Air search: Elta EL/M-2218S ❾; E/F-band.

Surface search: Cardion SPS-55 ❿; I-band.
Navigation: I-band.
Fire control: 3 Elta EL/M-2221 GM STGR ⓫; I/K/J-band.
Sonars: EDO Type 796 Mod 1; hull-mounted; search and attack; medium frequency.
Rafael towed array (fitted for).

Helicopters: 1 Dauphin SA 366G ⓬ or Sea Panther can be carried.

Programmes: A design by John J McMullen Associates Inc for Israeli Shipyards, Haifa in conjunction with Ingalls Shipbuilding Division of Litton Corporation which was authorised to act as main contractor using FMS funding. Contract awarded 8 February 1989. All delivered to Israel for combat system installation, first two completed in 1996 and last one in mid-1997. Major refits of these ships is reported to be under consideration. Plans to procure a further five new

ships (SAAR 5+) under similar FMS funding are now unlikely to be taken forward in view of the requirement for multimission ships. The option for a fourth SAAR 5 is not thought to have been taken up.
Structure: Steel hull and aluminium superstructure. Stealth features including resilient mounts for main machinery, funnel exhaust cooling, Radar Absorbent Material (RAM), NBC washdown and Prairie Masker Bubbler system. A secondary operations room is fitted aft. There are some Flag capabilities. Plans to carry Gabriel SSMs have been scrapped because of topweight problems. The planned third MSIS director has not yet been seen on the platform aft of the air search radar.
Operational: Endurance, 20 days. The main role is to counter threats in shipping routes. ICS-2 integrated communications system. The position of the satellite aerial suggests that the SAM after VLS launchers are not used. Barak has still to be installed, because of lack of funds. For the same reason the normal Harpoon load may be reduced to four. **UPDATED**

HANIT

12/2001, M Declerck / 0567460

HANIT

12/2001, M Declerck / 0533267

For details of the latest updates to *Jane's Fighting Ships* online and to discover the additional information available exclusively to online subscribers please visit

jfs.janes.com

PATROL FORCES

Notes: (1) There are about 12 'Firefish' type fast attack boats in service with Special Forces.
(2) A 50 ft *(15.2 m)* shallow draft Stealth craft has been built in a Vancouver Shipyard and delivered in late 1998. A second completed by Oregon Iron Works, Portland in 1999 and painted dark green. Two diesels giving 35 kt and a Rafael optronic surveillance system are included. Crew of five.
(3) The Saar 2 class are no longer operational.

8 HETZ (SAAR 4.5) CLASS
(FAST ATTACK CRAFT—MISSILE) (PGGM)

Name	Builders	Launched	Commissioned
ROMAT	Israel Shipyards, Haifa	30 Oct 1981	Oct 1981
KESHET	Israel Shipyards, Haifa	Oct 1982	Nov 1982
HETZ (ex-*Nirit*)	Israel Shipyards, Haifa	Oct 1990	Feb 1991
KIDON	Israel Shipyards, Haifa	1993	7 Feb 1994
TARSHISH	Israel Shipyards, Haifa	1995	June 1995
YAFFO	Israel Shipyards, Haifa	1998	1 July 1998
HEREV	Israel Shipyards, Haifa	2002	June 2002
SUFA	Israel Shipyards, Haifa	2002	Aug 2002

Displacement, tons: 488 full load
Dimensions, feet (metres): 202.4 × 24.9 × 8.2 *(61.7 × 7.6 × 2.5)*
Main machinery: 4 MTU 16V 538 TB93 or 4 MTU 16V 396 TE diesels; 16,600 hp(m) *(12.2 MW)*; 4 shafts
Speed, knots: 31. **Range, n miles:** 3,000 at 17 kt; 1,500 at 30 kt
Complement: 53

Missiles: SSM: 4 McDonnell Douglas Harpoon ❶; active radar homing to 130 km *(70 n miles)* at 0.9 Mach; warhead 227 kg.
6 IAI Gabriel II ❷; radar or optical guidance; semi-active radar plus anti-radiation homing to 36 km *(19.4 n miles)* at 0.7 Mach; warhead 75 kg.
SAM: Israeli Industries Barak I (vertical launch) ❸; 32 or 16 cells in 2- or 4—8 pack launchers; command line of sight radar or optical guidance to 10 km *(5.5 n miles)* at 2 Mach; warhead 22 kg. Most fitted for but not with.
Guns: 1 OTO Melara 3 in *(76 mm)*/62 ❹; 85 rds/min to 16 km *(8.7 n miles)*; weight of shell 6 kg.
2 Oerlikon 20 mm; 800 rds/min to 2 km.
1 Rafael Typhoon 25 mm (Herev).
1 General Electric/General Dynamics Vulcan Phalanx 6-barrelled 20 mm Mk 15 ❺; 3,000 rds/min combined to 1.5 km anti-missile.
2 or 4—12.7 mm (twin or quad) MGs.
Countermeasures: Decoys: Elbit/Deseaver 72-barrelled launchers for chaff and IR flares ❻.
ESM/ECM: Elisra NS 9003/5; intercept and jammer.
Combat data systems: IAI Reshet datalink.
Weapons control: Galileo OG 20 optical director; Elop MSIS optronic director ❼.
Radars: Air/surface search: Thomson-CSF TH-D 1040 Neptune ❽; G-band.
Fire control: 2 Elta EL/M-2221 GM STGR ❾; I/K/J-band.

Programmes: *Hetz* started construction in 1984 as the fifth of the SAAR 4 class but was not completed, as an economy measure. Taken in hand again in 1989 and fitted out as the

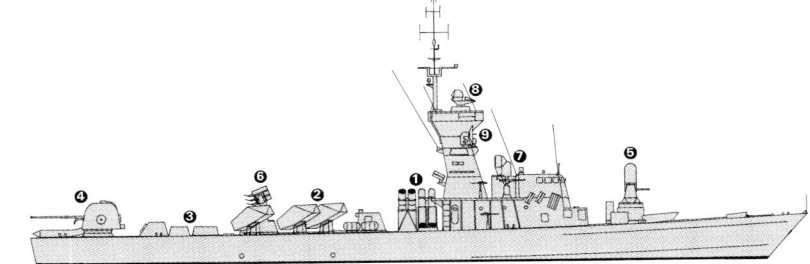

HETZ

(Scale 1 : 600), Ian Sturton / 0126347

HETZ CLASS

8/2000 / 0105824

trials ship for some of the systems installed in the Eilat class.
Modernisation: *Romat* and *Keshet* were modernised to same standard as *Hetz* in what was called the Nirit programme. The remaining craft were new build and some of these have been given names previously allocated to decommissioned/transferred SAAR 4s.
Structure: The CIWS is mounted in the eyes of the ship replacing the 40 mm gun. The eight pack Barak launchers are fully containerised and require no deck penetration or onboard maintenance. They are fitted aft in place of two of the Gabriel launchers. The fire-control system for Barak is fitted on the platform aft of the bridge on the port side.
Operational: Davits can be installed aft of the Gabriel missiles for special forces boats. Barak SAM is only operational in one of the class.

UPDATED

HEREV

6/2002, Israeli Navy / 0127285

15 DABUR CLASS (COASTAL PATROL CRAFT) (PC)

| 850 | 851 | 853 | 860-862 | 864 | 865 | 868 |
| 873 | 902 | 905 | 906 | 909 | 910 | |

Displacement, tons: 39 full load
Dimensions, feet (metres): 64.9 × 18 × 5.8 *(19.8 × 5.5 × 1.8)*
Main machinery: 2 GM 12V-71TA diesels; 840 hp *(627 kW)* sustained; 2 shafts
About 8 have more powerful GE engines.
Speed, knots: 19; 30 (GE engines). **Range, n miles:** 450 at 13 kt
Complement: 6/9 depending on armament

Guns: 2 Oerlikon 20 mm; 800 rds/min to 2 km.
2—12.7 mm MGs. Carl Gustav 84 mm portable rocket launchers.
Torpedoes: 2—324 mm tubes. Honeywell Mk 46; anti-submarine; active/passive homing to 11 km *(5.9 n miles)* at 40 kt; warhead 44 kg.
Depth charges: 2 racks in some.
Weapons control: Elop optronic director.
Radars: Surface search: Decca Super 101 Mk 3 or HDWS; I-band.
Sonars: Active search and attack; high frequency.

Programmes: Twelve built by Sewart Seacraft USA and remainder by Israel Aircraft Industries (RAMTA) between 1973 and 1977. Final total of 34. Likely to be phased out as new fast attack craft enter service.
Structure: Aluminium hull. Several variations in the armament. Up to eight of the class are fitted with more powerful General Electric engines to increase speed to 30 kt.
Operational: These craft have been designed for overland transport. Good rough weather performance. Portable rocket launchers are carried for anti-terrorist purposes. Not considered fast enough to cope with modern terrorist speedboats and some have been sold as Super Dvoras commissioned. Two based at Eilat, remainder at Ashdod.
Sales: Four to Argentina in 1978; four to Nicaragua in 1978 and three more in 1996; two to Sri Lanka in 1984; four to Fiji, six to Chile in 1991 and four more in 1995. Five also given to Lebanon Christian Militia in 1976 but these were returned.

VERIFIED

DABUR *12/1998* / 0075862

2 RESHEF (SAAR 4) CLASS
(FAST ATTACK CRAFT—MISSILE) (PTG)

Name	Builders	Launched	Commissioned
NITZHON	Israel Shipyards, Haifa	10 July 1978	Sep 1978
ATSMOUT	Israel Shipyards, Haifa	3 Dec 1978	Feb 1979

Displacement, tons: 415 standard; 450 full load
Dimensions, feet (metres): 190.6 × 25 × 8 *(58 × 7.8 × 2.4)*
Main machinery: 4 MTU/Bazán 16V 956 TB91 diesels; 15,000 hp(m) *(11.03 MW)* sustained; 4 shafts
Speed, knots: 32. **Range, n miles:** 1,650 at 30 kt; 4,000 at 17.5 kt
Complement: 45

Missiles: SSM: 2-4 McDonnell Douglas Harpoon (twin or quad) launchers; active radar homing to 130 km *(70 n miles)* at 0.9 Mach; warhead 227 kg.
4-6 Gabriel II; radar or TV optical guidance; semi-active radar plus anti-radiation homing to 36 km *(20 n miles)* at 0.7 Mach; warhead 75 kg.
Harpoons fitted with Israeli homing systems. The Gabriel II system carries a TV camera which can transmit a homing picture to the firing ship beyond the radar horizon. The missile fit currently varies in training boats-2 Harpoon, 5 Gabriel II.
Guns: 2 Oerlikon 20 mm; 800 rds/min to 2 km.
1 General Electric/General Dynamics Vulcan Phalanx 6-barrelled 20 mm Mk 15; 3,000 rds/min combined to 1.5 km anti-missile.
2—12.7 mm MGs.
Torpedoes: Tubes fitted in VDS fitted ship.
Countermeasures: Decoys: 1—45-tube, 4- or 6—24-tube, 4 single-tube chaff launchers.
ESM/ECM: Elisra NS 9003/5; intercept and jammer.
Combat data systems: IAI Reshet datalink.
Radars: Air/surface search: Thomson-CSF TH-D 1040 Neptune; G-band; range 33 km *(18 n miles)* for 2 m² target.
Fire control: Selenia Orion RTN 10X; I/J-band.
Sonars: EDO 780; VDS; fitted in one of the class.

Modernisation: Some of the class modernised to Nirit standards and transferred to the Saar 4.5 class. Gabriel III SSM did not go into production.
Operational: Operational status of last two remaining craft is doubtful. Replacement by SAAR 5 is a high priority.
Sales: Nine built for South Africa in Haifa and Durban. One transferred to Chile late 1979, one in February 1981, and two more in June 1997. Two transferred to Sri Lanka in 2000.

UPDATED

SAAR 4 (with VDS) *8/2000* / 0105825

SAAR 4 *10/1999* / 0075863

13 SUPER DVORA MK I and MK II CLASSES
(FAST ATTACK CRAFT—GUN) (PTFM)

811-819 (Mk I)	820-823 (Mk II)

Displacement, tons: 54 full load
Dimensions, feet (metres): 71 × 18 × 5.9 screws *(21.6 × 5.5 × 1.8)* (Mk I)
82 × 18.4 × 3.6 *(25 × 5.6 × 1.1)* (Mk II)
Main machinery: 2 Detroit 16V-92TA diesels; 1,380 hp *(1.03 MW)* sustained; 2 shafts (Mk I)
2 MTU 12V 396 TE94 diesels; 4,175 hp(m) *(3.07 MW)* sustained; 2 ASD 16 drives (Mk II)
Speed, knots: 36 or 46 (Mk II). **Range, n miles:** 1,200 at 17 kt
Complement: 10 (1 officer)

Missiles: SSM Hellfire; range 8 km *(4.3 n miles)*; can be carried.
Guns: 2 Oerlikon 20 mm/80 or Bushmaster 25 mm/87 Mk 96 or 3 Typhoon 12.7 mm (triple) MGs.
2—12.7 or 7.62 mm MGs. 1—84 mm rocket launcher.
Depth charges: 2 racks.
Weapons control: Elop MSIS optronic director.
Radars: Surface search: Raytheon; I-band.

Programmes: An improvement on the Dabur design ordered in March 1987 from Israel Aircraft Industries (RAMTA). First started trials in November 1988, and first two commissioned in June 1989. First 10 are Mk I. From 820 onwards the ships are fitted with more powerful engines for a higher top speed and surface drives which greatly reduce maximum draft. First Mk II commissioned in 1993.
Structure: All gun armament and improved speed and endurance compared with the prototype Dvora. SSM, depth charges, torpedoes or a 130 mm MRL can be fitted if required.
Operational: Two (Mk II) are based at Eilat, the remainder at Haifa. The 25 mm or 12.7 mm Gatling guns can be operated by joystick control from the bridge. Hellfire SSM is sometimes carried.
Sales: Six Mk I sold to Sri Lanka in 1988 and four to Eritrea in 1993. One Mk II to Sri Lanka in 1995 and three more in 1996. One to Slovenia in 1996 and a second in 1997. Two to India in 1997, with more building under licence in India.

UPDATED

SUPER DVORA Mk II 820 *1995, IAI* / 0080068

SUPER DVORA Mk I 819 *4/1996* / 0080069

SUPER DVORA Mk I 816 *8/2000* / 0105826

1 + 5 (4) SUPER DVORA MK III CLASS (PTFM)

Displacement, tons: 72 full load
Dimensions, feet (metres): 89.9 × 18.7 × 3.6 *(27.4 × 5.7 × 1.1)*
Main machinery: 2 MTU 12V 4000 diesels; 2 Arneson ASD16 surface drives
Speed, knots: 45. **Range, n miles:** 1,000 at cruising speed
Complement: 5
Guns: 1 Bushmaster 25 mm M242 chain gun. 1—20 mm. 2—7.62 mm MGs.
Weapons control: ELOP optronic director.
Radars: Surface search: I-band.

Comment: An order for six craft, with an option for a further four, was made with IAI-Ramta on 13 January 2002. First delivered in July 2004 with deliveries of the first batch to be completed by early 2006. *NEW ENTRY*

SUPER DVORA Mk III   *7/2004*, IDF* / 0590154

2 + (2) SHALDAG CLASS (FAST ATTACK CRAFT—GUN) (PBF)

Displacement, tons: 58 full load
Dimensions, feet (metres): 81.4 × 19.7 × 3.9 *(24.8 × 6 × 1.2)*
Main machinery: 2 Deutz 620 TB 16V or MTU 396 TE diesels; 5,000 hp(m) *(3.68 MW)*; 2 LIPS or MJP water-jets
Speed, knots: 50. **Range, n miles:** 700 at 32 kt
Complement: 10
Guns: 1 Rafael Typhoon 25 mm. 1—20 mm.
Weapons control: ELOP compass optronic director. Typhoon GFCS.
Radars: Surface search: MD 3220 Mk II; I-band.

Comment: Order in January 2002 for two craft, with option for two further hulls, made from Israel Shipyards, Haifa. Details reflect those in Sri Lankan service and are thus speculative. First delivery was due in late 2003. *VERIFIED*

SHALDAG CLASS (Sri Lankan colours)   *2/1996, Sri Lanka Navy* / 0080700

3 STINGRAY INTERCEPTOR CLASS (PBF)

Displacement, tons: 10.5 full load
Dimensions, feet (metres): 39.4 × 14.5 × 2.9 *(12 × 4.4 × 0.9)*
Main machinery: 2 Caterpillar marine diesels; 2 shafts
Speed, knots: 35. **Range, n miles:** 300 at cruising speed
Complement: 5
Radar: Surface search: I-band.

Comment: Catamaran design of GRP construction built by Stingray Marine of Durbanville, Western Cape and delivered in 1997 and 1998. *VERIFIED*

STINGRAY INTERCEPTOR   *2000, Stingray Marine* / 0104866

AMPHIBIOUS FORCES

Notes: Plans to acquire an assault ship (LPD) of about 13,000 tons have been suspended.

AUXILIARIES

Notes: (1) Two new construction landing ships are required by the Navy to transport troops. No funds available. A Newport class *Peoria* LST 1183 was authorised for lease from the US but this is unlikely to go ahead.
(2) A Ro-Ro ship *Queshet* is used for research and development. Built in Japan in 1979 and formerly used as a general purpose cargo ship.
(3) Two former merchant ships *Nir* and *Naharya* are used as alongside tenders in Haifa and Eilat respectively.
(4) A 19 m Alligator class semi-submersible craft was reported delivered in 1998. It is likely to be used for special forces operations.

1 ASHDOD CLASS (LCT)

Name	No	Builders	Commissioned
ASHDOD	61	Israel Shipyards, Haifa	1966

Displacement, tons: 400 standard; 730 full load
Dimensions, feet (metres): 205.5 × 32.8 × 5.8 *(62.7 × 10 × 1.8)*
Main machinery: 3 MWM diesels; 1,900 hp(m) *(1.4 MW)*; 3 shafts
Speed, knots: 10.5
Complement: 20
Guns: 2 Oerlikon 20 mm.

Comment: Used as a trials ship for Barak VLS. Based at Ashdod but refitted at Eilat in 1999. *VERIFIED*

ASHDOD   *3/1989* / 0080070

SHIPBORNE AIRCRAFT

Numbers/Type: 1 Aerospatiale SA 366G Dauphin.
Operational speed: 140 kt *(260 km/h)*.
Service ceiling: 15,000 ft *(4,575 m)*.
Range: 410 n miles *(758 km)*.
Role/Weapon systems: Air Force SAR/MR helicopters acquired in 1985 and primarily SAR/MR. Sensors: Israeli-designed radar/FLIR systems. Integrated Elop MSIS for OTHT. Weapons: Unarmed. *VERIFIED*

Numbers/Type: 8 Eurocopter AS 565SA Sea Panther.
Operational speed: 165 kt *(305 km/h)*.
Service ceiling: 16,700 ft *(5,100 m)*.
Range: 483 n miles *(895 km)*.
Role/Weapon systems: Built by American Eurocopter in Texas. Three delivered by October 1998 with one more in 1999. Sensors: Telephonics search radar; Elop MSIS for OTHT. Weapons: Unarmed. *UPDATED*

AS 565SB   *6/2002, Adolfo Ortigueira Gil* / 0567461

LAND-BASED MARITIME AIRCRAFT

Notes: (1) Army helicopters can be used including Cobras.
(2) Two C-130 aircraft used for maritime surveillance.

Numbers/Type: 17 Bell 212.
Operational speed: 100 kt *(185 km/h)*.
Service ceiling: 13,200 ft *(4,025 m)*.
Range: 224 n miles *(415 km)*.
Role/Weapon systems: SAR and coastal helicopter surveillance tasks undertaken. Sensors: IAI EW systems. Weapons: Unarmed except for self-defence machine guns. *VERIFIED*

Numbers/Type: 3 IAI 1124N Sea Scan.
Operational speed: 471 kt *(873 km/h)*.
Service ceiling: 45,000 ft *(13,725 m)*.
Range: 2,500 n miles *(4,633 km)*.
Role/Weapon systems: Air Force manned. Coastal surveillance tasks with long endurance; used for intelligence gathering. Sensors: Elta EL/M-2022 radar, IFF, MAD, Sonobuoys, and various EW systems of IAI manufacture. *UPDATED*

SEA SCAN   *6/1994*, R A Cooper* / 0503199

Italy
MARINA MILITARE

Country Overview

Italy is situated in southern Europe and comprises, in addition to the Italian mainland, the islands of Sardinia, Sicily, Elba and many smaller islands. Enclaves within mainland Italy are the independent countries of San Marino and Vatican City. With an area of 116,341 square miles, it is bordered to the north by France, Switzerland, Austria and Slovenia. It has a 2,700 n mile coastline with the Mediterranean, Ionian, Adriatic, Tyrrhenian Sea and Ligurian Seas. The capital and largest city is Rome while the principal ports are Genoa, Naples, Trieste, Taranto, Palermo and Venice. Territorial waters (12 n miles) are claimed but an EEZ has not been claimed.

Headquarters Appointments

Chief of Naval Staff:
  Admiral Giampaolo di Paola
Vice Chief of Naval Staff:
  Admiral Giovanni Vitaloni
Chief of Joint Military Intelligence:
  Vice Admiral Andrea Campregher
Chief of Procurement:
  Engineer Admiral Dino Nascetti
Chief of Technical Support:
  Engineer Admiral Giancarlo Cecchi
Chief of Naval Personnel:
  Vice Admiral Vincenzo del Vento

Flag Officers

Commander, Allied Naval Forces, Southern Europe (Naples):
  Admiral Ferdinando San Felice di Monteforte
Commander-in-Chief of Fleet (Rome):
  Admiral Bruno Branciforte
Commander, Tyrrhenian Sea (La Spezia):
  Admiral Quinto Gramellini
Commander, Ionian Sea (Taranto):
  Admiral Francesco Ricci
Commander, Adriatic Sea (Ancona):
  Vice Admiral Paolo Pagnottella
Commander, Sicily (Augusta):
  Vice Admiral Armando Molaschi
Commander, Sardinia (Cagliari):
  Vice Admiral Roberto Baggioni
Commander, High Seas Fleet (COMFORAL):
  Vice Admiral Andrea Toscano
Commander, Naval Group (COMGRUPNAV) (Taranto):
  Rear Admiral Salvatore Ruzzittu
Commander (1ˢᵗ Frigate Squadron) (Taranto):
  Captain Michele Lafortezza
Commander, Naval Group (2ⁿᵈ Frigate Squadron) (La Spezia):
  Captain Isidoro Fusco
Commander, MCM Forces (COMFORDRAG) (La Spezia):
  Rear Admiral Lorenzo Spagnuolo
Commander, Amphibious Force (COMFORSBARC) (Brindisi):
  Vice Admiral Sirio Lanfredini
Commander, Training Command (MARICENTADD) (Taranto):
  Vice Admiral Michele De Pinto
Commander, Coastal and Patrol Forces (COMFORPAT) (Augusta):
  Captain Francesco De Biase
Commander, Naval Air Arm (COMFORAER) (Rome):
  Rear Admiral Giuseppe de Giorgi
Commander Submarine Force (COMFORSUB) (Taranto):
  Captain Mario Caruso
Commander, Naval Special Forces (COMSUBIN) (La Spezia):
  Rear Admiral Roberto Paperini
Commander Coast Guard:
  Admiral Eugenio Sicurezza

Diplomatic Representation

Naval Attaché in Bonn:
  Captain Gianluca Turilli
Naval Attaché in Peking:
  Captain Gianfranco Cucchiaro
Naval Attaché in London:
  Rear Admiral Cristiano Bettini
Naval Attaché in Moscow:
  Captain Marco Scano
Naval Attaché in Paris:
  Captain Roberto Ive
Naval Attaché in Washington:
  Rear Admiral Raffaele Caruso

Bases

Regional Commands: La Spezia (Tyrrhenian Sea), Taranto (Ionian Sea), Ancona (Adriatic Sea), Augusta (Sicily), Cagliari (Sardinia). Main bases (Major Arsenals/Navy Shipyards): Taranto, La Spezia. Secondary base (Minor Arsenal/Navy Shipyard): Augusta. Minor bases: Brindisi.

Organisation

CINCNAV is responsible for all operational activities. There are six subordinate commands:
High-Sea Forces Command (COMFORAL) including all Major and Amphibious Ships. Based in Taranto with subordinated command COMGRUPNAVIT.
Patrol Forces Command (COMFORPAT) Corvettes and OPVs. Based in Augusta.
Naval Air Command. Based at Santa Rosa, Rome.
Submarine Force Command. Based at Taranto.
Mine Countermeasures Command. Based at La Spezia.
COMFORSBARC with San Marco Regiment, Carlotto (logistic) regiment and one assault boat group. Based at Brindisi.
Special Forces Command (COMSUBIN) Commandos and support craft. Based near La Spezia. Controlled directly by Chief of Naval Staff.

Prefix to Ships' Names

ITS (Italian Ship)

Strength of the Fleet

Type	Active	Building (Planned)
Submarines	6	1 (2)
Aircraft Carriers	1	1
Destroyers	4	2 (2)
Frigates	12	6 (4)
Corvettes	8	—
Offshore Patrol Vessels	10	—
Coastal Patrol Craft	4	—
LPD	3	—
Minehunters/sweepers	12	—
Survey/Research Ships	7	—
Replenishment Tankers	3	—
Coastal Tankers	11	—
Coastal Transports	6	—
Sail Training Ships	7	—
Training Ships	4	—
Lighthouse Tenders	5	—
Salvage Ships	1	—
Repair Ships	1	—

Personnel

(a) 2005: 36,837 (5,050 officers) including 1,550 naval air and 2,100 naval infantry (amphib)
(b) 10 months' national service (about 6,500 are conscripts). The Italian Armed Forces will become all professional from the end of 2006.

Naval Air Arm

1 LRMP squadron-Bréguet Atlantic (No 41, Catania); operated by Air Force under Navy command with air force/navy aircrews
2 SH-3D/H helicopter squadrons (1st and 3rd based at Luni and Catania respectively)
3 AB 212 helicopter squadrons (2nd, 4th and 5th based at Luni, Taranto and Catania respectively)
1 AV-8B Harrier II squadron at Grottaglie, Taranto (2 TAV-8B plus 16 AV-8B)
1 amphibious squadron at Taranto (AB 212 and SH-3D)

Naval Infantry and Army Amphibious Units

A Landing Force Command was established in 1998 including a collaborative Spanish/Italian amphibious brigade (SIAF). Landing Force Command is based at Brindisi and comprises the San Marco assault regiment (two assault battalions), the Carlotto support regiment (one logistic and one training battalion) and a Landing Craft Group. The Amphibious assault air squadron has eight modified SH-3D and seven modified AB-212 helicopters.
  The Italian Army operates an amphibious regiment named 'Serenissima' which is based at Venice. It is equipped with four LCM, six LCVP and 47 rigid raider and assault craft.

DELETIONS

Submarines

2002 *Marconi*
2004 *Fecia di Cossato*

Cruisers

2003 *Vittorio Veneto* (museum)

Frigates

2002 *Sagittario* (reserve)
2003 *Perseo* (reserve)

Auxiliaries

2002 *Proteo* (to Bulgaria), *Basento*, *Simeto* (to Tunisia), *Porto d'Ischia* (to Tunisia), *Riva Trigoso*
2003 *MOC 1204* (to Tunisia), *MOC 1201*
2004 *Basento* (reserve)

Training Ships

2001 *Aragosta*, *Polipo*

PENNANT LIST

Submarines

S 520	Leonardo da Vinci
S 522	Salvatore Pelosi
S 523	Giuliano Prini
S 524	Primo Longobardo
S 525	Gianfranco Gazzana Priaroggia
S 526	Salvatore Todaro
S 527	Scire (bldg)

Light Aircraft Carriers

C 551	Giuseppe Garibaldi
C 552	Cavour (bldg)

Destroyers

D 550	Ardito
D 551	Audace
D 560	Luigi Durand de la Penne
D 561	Francesco Mimbelli

Frigates

F 570	Maestrale
F 571	Grecale
F 572	Libeccio
F 573	Scirocco
F 574	Aliseo
F 575	Euro
F 576	Espero
F 577	Zeffiro
F 582	Artigliere
F 583	Aviere
F 584	Bersagliere
F 585	Granatiere

Corvettes

F 551	Minerva
F 552	Urania
F 553	Danaide
F 554	Sfinge
F 555	Driade
F 556	Chimera
F 557	Fenice
F 558	Sibilla

Patrol Forces

P 401	Cassiopea
P 402	Libra
P 403	Spica
P 404	Vega
P 405	Esploratore
P 406	Sentinella
P 407	Vedetta
P 408	Staffetta
P 409	Sirio
P 410	Orione
P 490	Comandante Cigala Fulgosi
P 491	Comandante Borsini
P 492	Comandante Bettica
P 493	Comandante Foscari

Minehunters

M 5550	Lerici
M 5551	Sapri
M 5552	Milazzo
M 5553	Vieste
M 5554	Gaeta
M 5555	Termoli
M 5556	Alghero

M 5557	Numana
M 5558	Crotone
M 5559	Viareggio
M 5560	Chioggia
M 5561	Rimini

Amphibious Forces

L 9892	San Giorgio
L 9893	San Marco
L 9894	San Giusto

Survey and Research Ships

A 5303	Ammiraglio Magnaghi
A 5304	Aretusa
A 5308	Galatea
A 5315	Raffaele Rossetti
A 5320	Vincenzo Martellotta
A 5340	Elettra
F 581	Carabiniere

Auxiliaries

A 5302	Caroly
A 5309	Anteo
A 5311	Palinuro
A 5312	Amerigo Vespucci
A 5313	Stella Polare
A 5316	Corsaro II
A 5318	Prometeo
A 5319	Ciclope
A 5322	Capricia
A 5323	Orsa Maggiore
A 5324	Titano
A 5325	Polifemo
A 5326	Etna

A 5327	Stromboli
A 5328	Gigante
A 5329	Vesuvio
A 5330	Saturno
A 5347	Gorgona
A 5348	Tremiti
A 5349	Caprera
A 5351	Pantelleria
A 5352	Lipari
A 5353	Capri
A 5359	Bormida
A 5364	Ponza
A 5365	Tenace
A 5366	Levanzo
A 5367	Tavolara
A 5368	Palmaria
A 5370-3	MCC 1101-4
A 5376	Ticino
A 5377	Tirso
A 5379	Astice
A 5380	Mitilo
A 5382	Porpora
A 5383	Procida
A 5384	Alpino
Y 413	Porto Fossone
Y 416	Porto Torres
Y 417	Porto Corsini
Y 421	Porto Empedocle
Y 422	Porto Pisano
Y 423	Porto Conte
Y 425	Porto Ferraio
Y 426	Porto Venere
Y 428	Porto Salvo
Y 498	Mario Marino
Y 499	Alcide Pedretti

SUBMARINES

1 + 1 (2) TYPE 212A (SSK)

Name	No	Builders	Laid down	Launched	Commissioned
SALVATORE TODARO	S 526	Fincantieri, Muggiano	Jan 2001	6 Nov 2003	June 2005
SCIRÈ	S 527	Fincantieri, Muggiano	Apr 2002	18 Dec 2004	May 2006

Displacement, tons: 1,450 surfaced; 1,830 dived
Dimensions, feet (metres): 183.4 × 23 × 19.7
  (55.9 × 7 × 6)
Main machinery: Diesel-electric; 1 MTU 16V 396 diesel;
  4,243 hp(m) *(3.12 MW)*; 1 alternator; 1 Siemens PEM motor;
  3,875 hp(m) *(2.85 MW)*; 1 shaft; Siemens/HDW PEM 9 fuel
  cell (AIP) modules; 306 kW
Speed, knots: 20 dived; 12 surfaced
Range, n miles: 8,000 at 8 kt surfaced
Complement: 27 (8 officers)

Torpedoes: 6—21 in *(533 mm)* bow tubes; water ram discharge;
  Whitehead A184 Mod 3. Total 12 weapons.
Mines: In lieu of torpedoes.

Countermeasures: Decoys: CIRCE Torpedo countermeasures.
  ESM: DASA FL 1800U; intercept.
Weapons control: Kongsberg MSI-90U TFCS.
Radars: Navigation: KH 1007; I-band.
Sonars: STN Atlas Elektronik DBQS-40; passive ranging and
  intercept; FAS-3 Flank and passive towed array.
  STN Atlas Moa 3070, mine detection, active, high frequency.

Programmes: German design phase first completed in 1992 by
  ARGE 212 (HDW/TNSW) in conjunction with IKL. MoU
  signed with Germany 22 April 1996 for a common design.
  First pair ordered from Fincantieri in August 1997. First
  steel cut for first of class 19 July 1999, and for second in July
  2000.

Structure: Equipped with a hybrid fuel cell/battery propulsion
  based on the Siemens PEM fuel cell technology. The
  submarine is designed with a partial double hull which has a
  larger diameter forward. This is joined to the after end by a
  short conical section which houses the fuel cell plant. Two
  LOX tanks and hydrogen stored in metal cylinders are carried
  around the circumference of the smaller hull section. Italian
  requirements included a greater diving depth, improved
  external communications, and better submerged escape
  facilities. The final design is identical to the German
  submarines. Fitted with Zeiss search and attack periscopes.
Operational: Dived speeds up to 8 kt are projected, without use
  of main battery. Sea trials of *Salvatore Todaro* started in June
  2004.   *UPDATED*

SALVATORE TODARO   *10/2004 *, Giorgio Ghiglione /* 1044359

1 SAURO (TYPE 1081) CLASS (SSK)

Name	No	Builders	Laid down	Launched	Commissioned
LEONARDO DA VINCI	S 520	Italcantieri, Monfalcone	8 June 1978	20 Oct 1979	23 Oct 1981

Displacement, tons: 1,456 surfaced; 1,631 dived
Dimensions, feet (metres): 210 × 22.5 × 18.9
  (63.9 × 6.8 × 5.7)
Main machinery: Diesel-electric; 3 Fincantieri GMT 210.16 NM
  diesels; 3,350 hp(m) *(2.46 MW)* sustained; 3 alternators;
  2.16 MW; 1 motor; 3,210 hp(m) *(2.36 MW)*; 1 shaft. AIP trial
  in *Fecia di Cossato*
Speed, knots: 11 surfaced; 19 dived; 12 snorting
Range, n miles: 11,000 surfaced at 11 kt; 250 dived at 4 kt
Complement: 49 (6 officers) plus 4 trainees

Torpedoes: 6—21 in *(533 mm)* bow tubes. 12 Whitehead A184
  Mod 3; dual purpose; wire-guided; active/passive homing to
  25 km *(13.7 n miles)* at 24 kt; 17 km *(9.2 n miles)* at 38 kt;
  warhead 250 kg. Swim-out discharge.
Countermeasures: ESM: Elettronica BLD 727; radar warning.
Weapons control: SMA BSN 716(V)1 SACTIS data processing
  and computer-based TMA. CCRG FCS.
Radars: Search/navigation: SMA BPS 704; I-band.
Sonars: Selenia Elsag IPD 70/S; linear passive array;
  200 Hz-7.5 kHz; active and UWT transducers in bow (15 kHz).

Programmes: Ordered 12 February 1976.
Modernisation: Modernised in 1993. New batteries have
  greater capacity, some auxiliary machinery replaced and
  habitability improved.
Structure: Diving depth, 300 m *(985 ft)* (max) and 250 m
  (820 ft) (normal). Periscopes: Barr & Stroud CK 31 search and
  CH 81 attack.
Operational: Endurance, 35 days. *Nazario Sauro* paid off in early
  2001, *Guglielmo Marconi* in 2002 and *Fecia di Cossato* in
  2004.   *UPDATED*

SAURO CLASS   *6/2001, H M Steele /* 0130519

4 IMPROVED SAURO CLASS (SSK)

Name	No
SALVATORE PELOSI	S 522
GIULIANO PRINI	S 523
PRIMO LONGOBARDO	S 524
GIANFRANCO GAZZANA PRIAROGGIA	S 525

Builders	Laid down	Launched	Commissioned
Fincantieri, Monfalcone	24 May 1984	29 Dec 1986	14 July 1988
Fincantieri, Monfalcone	30 May 1985	12 Dec 1987	11 Nov 1989
Fincantieri, Monfalcone	19 Dec 1991	20 June 1992	20 May 1994
Fincantieri, Monfalcone	12 Nov 1992	26 June 1993	12 Apr 1995

Displacement, tons: 1,476 (1,653, S 524-5) surfaced; 1,662 (1,862, S 524-5) dived

Dimensions, feet (metres): 211.2 (217.8 S 524-5) × 22.3 × 18.4 *(64.4 (66.4) × 6.8 × 5.6)*

Main machinery: Diesel-electric; 3 Fincantieri GMT 210.16 SM diesels; 3,672 hp(m) *(2.7 MW)* sustained; 3 generators; 2.16 MW; 1 motor; 3,128 hp(m) *(2.3 MW)*; 1 shaft

Speed, knots: 11 surfaced; 19 dived; 12 snorting

Range, n miles: 11,000 at 11 kt surfaced; 250 at 4 kt dived

Complement: 51 (7 officers)

Missiles: Capability to launch Harpoon or Exocet being considered (for S 524-5).

Torpedoes: 6—21 in *(533 mm)* bow tubes. 12 Whitehead A184 Mod 3; dual purpose; wire-guided; active/passive homing to 25 km *(13.7 n miles)* at 24 kt; 17 km *(9.2 n miles)* at 38 kt; warhead 250 kg. Swim-out discharge.

Countermeasures: ESM: Elettronica BLD-727; radar warning; 2 aerials-1 on a mast, second in search periscope.

Weapons control: STN Atlas ISUS 90-20.

Radars: Search/navigation: SMA BPS 704; I-band; also periscope radar for attack ranging.

Sonars: Selenia Elsag IPD 70/S; linear passive array; 200 Hz-7.5 kHz; active and UWT transducers in bow (15 kHz).

Programmes: The first two were ordered in March 1983 and the second pair in July 1988.

Modernisation: An upgrade programme included replacement of acoustic sensors, weapons control system (STN Atlas ISUS 90-20) and communications. Work on S 522, S 524 and S 525 has been completed and was finished in S 523 on 25 November 2004.

Structure: Pressure hull of HY 80 steel with a central bulkhead for escape purposes. Diving depth, 300 m *(985 ft)* (test) and 600 m *(1,970 ft)* (crushing). The second pair has a slightly longer hull to give space for SSMs.

Periscopes: Kollmorgen; S 76 Mod 322 with laser rangefinder and ESM-attack; S 76 Mod 323 with radar rangefinder and ESM-search. Wave contour snort head has a very low radar profile. The last pair have anechoic tiles.

Operational: Litton Italia PL 41 inertial navigation; Ferranti autopilot (in S 522-3) or Sepa autopilot (S 524-5) Omega and Transit. Endurance, 45 days.

UPDATED

PRIMO LONGOBARDO

6/2004*, Diego Quevedo / 1044360

PRIMO LONGOBARDO

9/2002, Giorgio Ghiglione / 0528329

AIRCRAFT CARRIERS

1 GARIBALDI CLASS (CVGM)

Name	No	Builders	Laid down	Launched	Commissioned
GIUSEPPE GARIBALDI	C 551	Italcantieri, Monfalcone	26 Mar 1981	4 June 1983	30 Sep 1985

Displacement, tons: 10,100 standard; 13,850 full load
Dimensions, feet (metres): 591 × 110.2 × 22
(180 × 33.4 × 6.7)
Flight deck, feet (metres): 570.2 × 99.7 *(173.8 × 30.4)*
Main machinery: COGAG; 4 Fiat/GE LM 2500 gas turbines;
81,000 hp *(60 MW)* sustained; 2 shafts
Speed, knots: 30. **Range, n miles:** 7,000 at 20 kt
Complement: 582 ship plus 230 air group (accommodation for
825 including Flag and staff)

Missiles: SAM: 2 Selenia Elsag Albatros octuple launchers ❶; 48
Aspide; semi-active radar homing to 13 km *(7 n miles)* at 2.5
Mach; height envelope 15-5,000 m *(49.2-16,405 ft)*; warhead
30 kg.
Guns: 6 Breda 40 mm/70 (3 twin) MB ❷; 300 rds/min to
12.5 km *(6.8 n miles)* anti-surface; 4 km *(2.2 n miles)* anti-
aircraft; weight of shell 0.96 kg.
Torpedoes: 6—324 mm B-515 (2 triple) tubes ❸. Honeywell Mk
46; anti-submarine; active/passive homing to 11 km *(5.9 n
miles)* at 40 kt; warhead 44 kg. Being replaced by new A 290.
Countermeasures: Decoys: SLQ-25 Nixie; noisemaker.
2 Breda SCLAR 105 mm 20-barrelled launchers; trains and
elevates; chaff to 5 km *(2.7 n miles)*; illuminants to 12 km
(6.6 n miles). SLAT in 2002.
ESM/ECM: Elettronica Nettuno SLQ-732; integrated intercept
and jamming system.
Combat data systems: IPN 20 (SADOC 2) action data
automation including Links 11 and 14. SATCOM ❹.
Weapons control: 3 Alenia NA 30E electro-optical back-up for
SAM. 3 Dardo NA21 for guns.
Radars: Long-range air search: Hughes SPS-52C ❺; 3D; E/F-
band; range 440 km *(240 n miles)*.

Air search: Selenia SPS-768 (RAN 3L) ❻; D-band; range 220 km
(120 n miles).
Air/surface search: Selenia SPS-774 (RAN 10S) ❼; E/F-band.
Surface search/target indication: SMA SPS-702 UPX; 718
beacon; I-band.
Navigation: ARPA SPN-753 G(V); I-band.
Fire control: 3 Selenia SPG-75 (RTN 30X) ❽; I/J-band; range
15 km *(8 n miles)* (for Albatros).
3 Selenia SPG-74 (RTN 20X) ❾; I/J-band; range 13 km *(7 n
miles)* (for Dardo).
CCA: Selenia SPN-728(V)1; I-band.
IFF: Mk XII. Tacan: SRN-15A.
Sonars: Raytheon DE 1160 LF; bow-mounted; active search;
medium frequency.

Fixed-wing aircraft: 15 AV-8B Harrier II.
Helicopters: 17 SH-3D Sea King or EH 101 Merlin helicopters
(12 in hangar, 6 on deck). The total capacity is either 15
Harriers or 17 helicopters, but this leaves no space for
movement. In practice a combination is embarked (see
Operational).

Programmes: Contract awarded 21 November 1977. The design
work completed February 1980. Started sea trials
3 December 1984.
Modernisation: A major C⁴I upgrade programme, completed in
September 2003, has given the ship a Maritime Component
Commander (MCC) capability. Improvements to the combat
data system include a MCC data system and Link 16.
SATCOM domes have replaced the TESEO launchers which
have been removed. SHF SATCOM has been installed in the
old positions of the chaff launchers while SCLAR-D chaff

launchers have been installed on new sponsons aft and below
the flight deck. Other work includes modernisation of the
ESM/ECM equipment, replacement of the DE 1150F sonar
with DMSS 2000 and the fitting of an electro-optic tracking
device on the bridge roof in lieu of SPN-728 radar which has
been removed. A further upgrade of air defence systems is
projected once *Andrea Doria* has commissioned in 2007
although the scope of this is to be decided.
Structure: Six decks with 13 vertical watertight bulkheads. Fitted
with 6.5° ski-jump and VSTOL operating equipment. Two
15 ton lifts 18 × 10 m *(59 × 32.8 ft)*. Hangar size
110 × 15 × 6 m *(361 × 49.2 × 19.7 ft)*. Hangar capacity is for
10 Harriers or 12 Sea Kings. Has a slightly narrower flight deck
than UK Invincible class. Two MEN class fast personnel
launches (capacity 250) can be embarked for amphibious
operations or disaster relief.
Operational: Fleet Flagship. Equipped for Joint Task Force
command and control. The long-standing dispute between
the Navy and the Air Force concerning the former's operation
of fixed-wing aircraft (dating back to pre-Second World War
legislation) was finally resolved by legislation passed on
29 January 1989. Embarked aircraft are operated by the Navy
with the Air Force providing evaluation and maintenance. The
carrier has operated in the assault role with seven SH-3D, four
AB 212 and Army helicopters including six AB 205, three A
129 and two CH-47. First operational Harriers embarked for
permanent duty in December 1994.

UPDATED

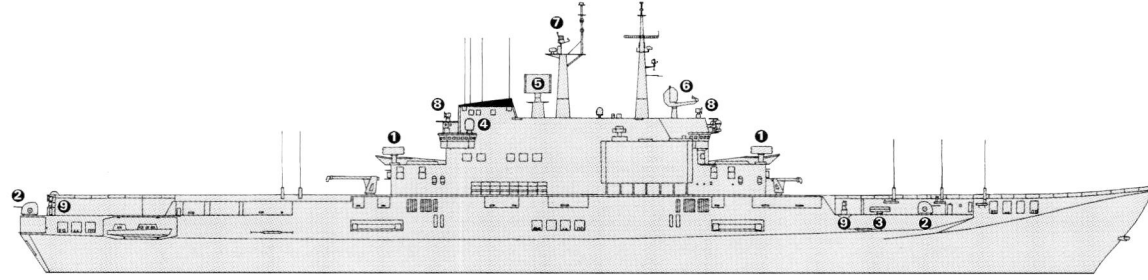

GIUSEPPE GARIBALDI
(Scale 1 : 1,200), Ian Sturton / 1043173

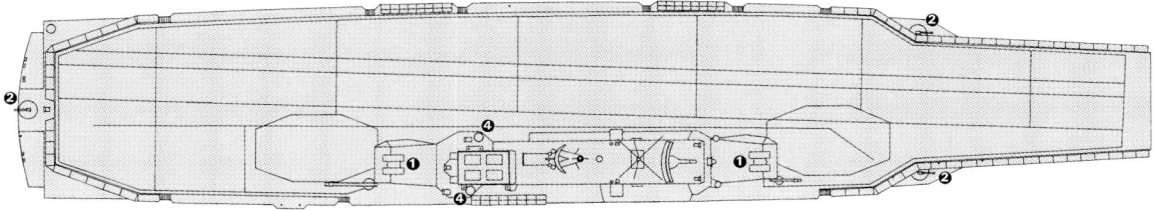

GIUSEPPE GARIBALDI
(Scale 1 : 1,200), Ian Sturton / 1043172

GIUSEPPE GARIBALDI (before modernisation)
5/2002, Ships of the World / 0528345

GIUSEPPE GARIBALDI

7/2004*, United States Navy / 1043185

GIUSEPPE GARIBALDI

8/2004*, Guy Toremans / 1044368

0 + 1 CAVOUR CLASS (CV)

Name	No	Builders	Laid down	Launched	Commissioned
CAVOUR (ex-Andrea Doria)	C 552	Fincantieri Muggiano/Riva Trigoso	17 July 2001	20 July 2004	2007

Displacement, tons: 27,100 full load
Dimensions, feet (metres): 772.9 oa; 707.3 wl × 128 oa;
96.8 wl × 24.6 *(235.6; 215.6 × 39; 29.5 × 7.5)*
Flight deck, feet (metres): 721.8 × 111.5 *(220 × 34)*
Main machinery: COGAG: 4 GE/Fiat LM 2500 gas turbines;
118,000 hp(m) *(88 MW)* sustained; 2 shafts; cp props; bow
and stern thrusters; 6—2.2 MW diesel generators and 2
motors
Speed, knots: 28. **Range, n miles:** 7,000 at 16 kt
Complement: 451 ship plus 203 air group plus 145 staff (CJTF
or CATF/CLF) plus 360 marines (90 additional marines for
short period). Total accommodation for 1,210
Military lift: (garage only): 100 wheeled vehicles or 60 armoured
vehicles or 24 MBTs (Ariete) or mixture

Missiles: 4 Sylver 8 cell VLS for Aster 15.
Guns: 2 OTO Melara 3 in *(76 mm)*/62 Super Rapid. 2 Otobreda
25 mm.
Countermeasures: Decoys: 2 Breda SCLAR-H 20-barrel
trainable chaff/decoy launchers.
TCM: 2 SLAT TCM launchers.

ESM: Radar and Comms intercept ❶.
ECM: Jammer.
Combat data systems: 'Horizon' derivative flag and command
support system. Links 11 and 16; provision for Link 22.
Satcom ❷.
Weapons control: FIAR SSAS optronic director.
Radars: Long-range air search: RAN-40L; D-band ❸.
Air search and missile guidance: EMPAR; G-band ❹.
CCA: SPN-41; J-band.
Surface search: SPS-791; E/F-band ❺.
Navigation: SPN-753G(V); I-band.
Tacan.
Sonars: WASS mine avoidance sonar (bow dome).

Fixed-wing aircraft: 8 AV-8B Harrier II or JSF.
Helicopters: 12 EH 101 (fitted also for AB 212, NH90 and
SH-3D).

Programmes: Following a study phase which included
significant changes to the initial configuration of the Nuova
Unità Maggiore (NUM) design, the Italian government placed
a contract with Fincantieri for the construction of a ship to
replace *Vittorio Veneto* in 2007. Capabilities include afloat
command, air and amphibious operations. The ship is being
constructed at Muggiano (bow) and Riva Trigoso (centre and
stern) and is to be joined, outfitted and tested at Muggiano. A
second contract, for the development and supply of the
combat system was signed with an AMS-led industrial group
in October 2002.
Structure: The flight deck features six helicopter take-off spots,
one spot for SAR, eight parking spots and a 12° ski jump. A
notional air group includes 12 EH-101 helicopters and 8 AV-8B
Harrier IIs. There is provision in the design to operate JSF and
UAVs. The hangar/garage can accommodate various
combinations of aircraft and vehicles (including MBT and
trucks). There are two 30 ton lifts, one forward of the island
and the other starboard side aft. Two Ro-Ro ramps are
positioned aft and starboard side. Two 15 ton and one 7 ton
lifts are fitted for ordnance and logistic needs respectively.
UPDATED

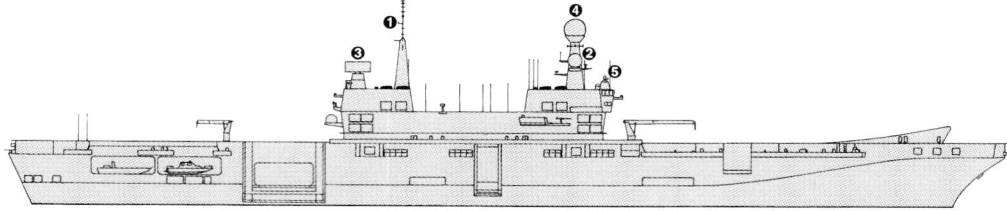

CAVOUR

(Scale 1 : 1,800), Ian Sturton / 0104868

CAVOUR (at launch)

7/2004*, Giorgio Ghiglione / 1044361

CAVOUR (artist's impression)

6/2002, Fincantieri / 0528403

DESTROYERS

2 DE LA PENNE (ex-ANIMOSO) CLASS (DDGHM)

Name	No	Builders	Laid down	Launched	Commissioned
LUIGI DURAND DE LA PENNE (ex-*Animoso*)	D 560	Fincantieri, Riva Trigoso/Muggiano	20 Jan 1988	29 Oct 1989	18 Mar 1993
FRANCESCO MIMBELLI (ex-*Ardimentoso*)	D 561	Fincantieri, Riva Trigoso/Muggiano	15 Nov 1989	13 Apr 1991	19 Oct 1993

Displacement, tons: 4,330 standard; 5,400 full load
Dimensions, feet (metres): 487.4 × 52.8 × 28.2 (sonar)
  (147.7 × 16.1 × 8.6)
Flight deck, feet (metres): 78.7 × 42.7 *(24 × 13)*
Main machinery: CODOG; 2 Fiat/GE LM 2500 gas turbines;
  54,000 hp *(40.3 MW)* sustained; 2 GMT BL 230.20 DVM
  diesels; 12,600 hp(m) *(9.3 MW)* sustained; 2 shafts; cp props
Speed, knots: 31 (21 on diesels). **Range, n miles:** 7,000 at 18 kt
Complement: 331 (25 officers)

Missiles: SSM: 4 or 8 OTO Melara/Matra Teseo Mk 2 (TG 2) (2 or
  4 twin) **①**; mid-course guidance; active radar homing to
  180 km *(98.4 n miles)* at 0.9 Mach; warhead 210 kg; sea-
  skimmer.
  Mk 3 with radar/IR homing to 300 km *(162 n miles)*; warhead
  160 kg in due course.
A/S: OTO Melara/Matra Milas launcher; inertial guidance with
  command update to 55 km *(29.8 n miles)* at 0.9 Mach;
  payload Mk 46 Mod 5 or Mu 90 torpedo; 4 weapons (see
  Modernisation).
SAM: 40 GDC Pomona Standard SM-1MR; Mk 13 Mod 4
  launcher **②**; command guidance; semi-active radar homing to
  46 km *(25 n miles)* at 2 Mach.
  Selenia Albatros Mk 2 octuple launcher for Aspide **③**; semi-
  active radar homing to 13 km *(7 n miles)* at 2.5 Mach; 16
  missiles. Automatic reloading.
Guns: 1 OTO Melara 5 in *(127 mm)*/54 **④**; 45 rds/min to 23 km
  (12.42 n miles); weight of shell 32 kg.
  3 OTO Melara 3 in *(76 mm)*/62 Super Rapid **⑤**; 120 rds/min
  to 16 km *(8.7 n miles)*; weight of shell 6 kg. 2—20 mm.
Torpedoes: 6—324 mm B-515 (2 triple) tubes **⑥**. Honeywell Mk
  46; anti-submarine; active/passive homing to 11 km *(5.9 n
  miles)* at 40 kt; warhead 44 kg. May be replaced by
  Whitehead Mu 90 in due course.
Countermeasures: Decoys: 2 CSEE Sagaie chaff launchers **⑦**. 1
  SLQ-25 Nixie anti-torpedo system.
ESM/ECM: Elettronica SLQ-732 Nettuno **⑧**; integrated intercept
  and jamming system. SLC 705.
Combat data systems: Selenia Elsag IPN 20 (SADOC 2); Links 11
  and 14. SATCOM.
Weapons control: 4 Dardo-E systems (3 channels for Aspide).
  Milas TFCS.
Radars: Long-range air search: Hughes SPS-52C; 3D **⑨**; E/F-
  band.
Air search: Selenia SPS-768 (RAN 3L) **⑩**; D-band.
Air/surface search: Selenia SPS-774 (RAN 10S) **⑪**; E/F-band.
Surface search: SMA SPS-702 **⑫**; I-band.
Fire control: 4 Selenia SPG-76 (RTN 30X) **⑬**; I/J-band (for
  Dardo).
  2 Raytheon SPG-51D **⑭**; G/I-band (for SAM).
Navigation: SMA SPN-748; I-band.
IFF: Mk X/XII. Tacan: SRN-15A.
Sonars: Raytheon DE 1164 LF-VDS; integrated bow and VDS;
  active search and attack; medium frequency (3.75 kHz (hull);
  7.5 kHz (VDS)).

Helicopters: 2 AB 212ASW **⑮**; SH-3D Sea King and EH 101
  Merlin capable.

Programmes: Order placed 9 March 1986 with Riva Trigoso. All
  ships built at Riva Trigoso are completed at Muggiano after
  launching. Names changed on 10 June 1992 to honour
  former naval heroes. Acceptance dates were delayed by
  reduction gear radiated noise problems which have been
  resolved.
Modernisation: Milas ASW launchers fitted by late 2004. New
  sonar dome fitted in D 560 in 2000 increased draft by 1.5 m.
  A major upgrade is planned to be undertaken in D 561
  (starting late 2005) and D 560 (starting in 2007). SPS-768 is
  to be replaced by AMS RAN-40L (also to be fitted in *Cavour*);

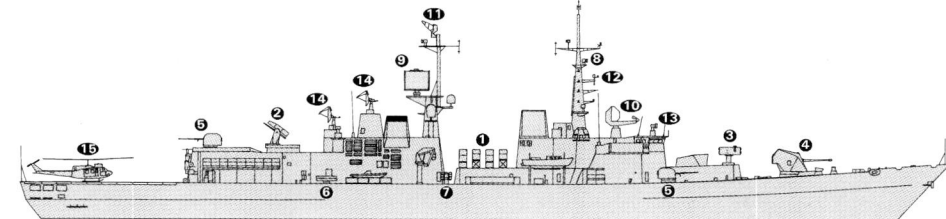

FRANCESCO MIMBELLI   *(Scale 1 : 1,200), Ian Sturton /* 0569913

LUIGI DURAND DE LA PENNE   *6/2004*, John Brodie /* 1044362

LUIGI DURAND DE LA PENNE   *5/2004*, B Sullivan /* 1044363

SPS-774 to be replaced by AMS RAN-21S; SPN-702 to be
replaced by SPN-763 ARPA; Dardo-E to be replaced by four
new fire-control systems (Dardo-F with RTN-30X); Sagem
IRST and new combat data system to be installed.
Structure: Kevlar armour fitted. Steel alloys used in
superstructure. Prairie Masker noise suppression system. The

127 mm guns are ex-Audace class B turrets. Fully stabilised.
Hangar is 18.5 m in length.
Operational: GPS and Meteosat receivers fitted. The three Super
Rapid 76 mm guns are used as a combined medium-range
anti-surface armament and CIWS against missiles.

UPDATED

FRANCESCO MIMBELLI   *5/2003, A Sharma /* 0570684

2 AUDACE CLASS (DDGHM)

Name	No	Builders	Laid down	Launched	Commissioned
ARDITO	D 550	Italcantieri, Castellammare	19 July 1968	27 Nov 1971	5 Dec 1972
AUDACE	D 551	Fincantieri, Riva Trigoso/Muggiano	27 Apr 1968	2 Oct 1971	16 Nov 1972

Displacement, tons: 3,600 standard; 4,400 full load
Dimensions, feet (metres): 448 × 46.6 × 15.1
(136.6 × 14.2 × 4.6)
Main machinery: 4 Foster-Wheeler boilers; 600 psi *(43 kg/cm²)*; 850°F *(450°C)*; 2 turbines; 73,000 hp(m) *(54 MW)*; 2 shafts
Speed, knots: 34
Range, n miles: 3,000 at 20 kt
Complement: 380 (30 officers)

Missiles: SSM: 8 OTO Melara/Matra Teseo Mk 2 (TG 2) (4 twin) ❶; mid-course guidance; active radar homing to 180 km *(98.4 n miles)* at 0.9 Mach; warhead 210 kg; sea-skimmer.
SAM: 40 GDC Pomona Standard SM-1MR; Mk 13 Mod 4 launcher ❷; command guidance; semi-active radar homing to 46 km *(25 n miles)* at 2 Mach; height envelope 45.7 to 18,288 m *(150 to 60,000 ft)*.
Selenia Albatros octuple launcher for Aspide ❸; semi-active radar homing to 13 km *(7 n miles)* at 2.5 Mach.

Guns: 1 OTO Melara 5 in *(127 mm)*/54 ❹; 45 rds/min to 23 km *(12.42 n miles)* anti-surface; 7 km *(3.8 n miles)* anti-aircraft; weight of shell 32 kg.
3 OTO Melara 3 in *(76 mm)*/62 Compact (*Ardito*) and 1 (*Ardito*) or 4 (*Audace*) Super Rapid ❺; 85 rds/min (Compact) or 120 rds/min (Super Rapid) to 16 km *(8.7 n miles)* anti-surface; 12 km *(6.6 n miles)* anti-aircraft; weight of shell 6 kg.
Torpedoes: 6—324 mm US Mk 32 (2 triple) tubes ❻. Honeywell Mk 46; anti-submarine; active/passive homing to 11 km *(5.9 n miles)* at 40 kt; warhead 44 kg. Transom tubes have been removed.
Countermeasures: Decoys: 2 Breda 105 mm SCLAR 20-barrelled trainable; chaff to 5 km *(2.7 n miles)*; illuminants to 12 km *(6.6 n miles)*. SLQ-25 Nixie; towed torpedo decoy.
ESM/ECM: Elettronica SLQ-732 Nettuno; integrated intercept and jammer.
Combat data systems: Selenia Elsag IPN-20 SADOC 2 action data automation; Links 11 and 14. SATCOM ❼.

Weapons control: 3 Dardo E FCS (3 channels for Aspide). Selenia NA 30 optronic director. Mk 74 Mod 13 MFCS.
Radars: Long-range air search: Hughes SPS-52C ❽; 3D; E/F-band.
Air search: Selenia SPS-768 (RAN 3L) ❾; D-band.
Air/surface search: Selenia SPS-774 (RAN 10S) ❿; E/F-band.
Surface search: SMA SPQ-2D ⓫; I-band.
Navigation: SMA SPN-748; I-band.
Fire control: 3 Selenia SPG-76 (RTN 30X) ⓬; I/J-band; range 40 km *(22 n miles)* (for Dardo E).
2 Raytheon SPG-51 ⓭; G/I-band (for Standard).
IFF: Mk XII.
Tacan: SRN-15A.
Sonars: CWE 610; hull-mounted; active search and attack; medium frequency.

Helicopters: 2 AB 212ASW ⓮ or EH 101 Merlin.

Programmes: It was announced in April 1966 that two new guided missile destroyers would be built. They were to be similar to, but an improvement in design on, that of the Impavido class (now deleted).
Modernisation: B gun has been replaced by Albatros PDMS. Stern torpedo tubes removed. *Audace* fitted with four and *Ardito* one Super Rapid guns vice the 76 mm Compacts. Plans to give *Ardito* three more by 1994 were shelved. *Ardito* completed modernisation in March 1988 and *Audace* in early 1991. Improved EW equipment also fitted.
Structure: Both fitted with stabilisers.
Operational: First deck landing trials of EH 101 helicopters were carried out on 14 May 1992.

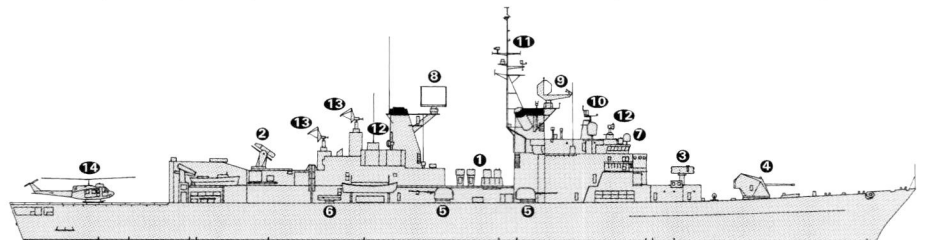

ARDITO

(Scale 1 : 1,200), Ian Sturton / 0569914

VERIFIED

ARDITO

6/2001, Giorgio Ghiglione / 0130514

AUDACE

7/2001, C D Yaylali / 0130361

0 + 2 (2) ANDREA DORIA (HORIZON) CLASS (DDGHM)

Name	No	Builders	Laid down	Launched	Commissioned
ANDREA DORIA (ex-*Carlo Bergamini*)	—	Fincantieri, Riva Trigoso/Muggiano	19 July 2002	2005	2007
CAIO DUILIO	—	Fincantieri, Riva Trigoso/Muggiano	19 Sep 2003	2007	2009

Displacement, tons: 6,700 full load
Dimensions, feet (metres): 494.1 oa; 464.9 wl × 57.4 × 16.7 *(150.6; 141.7 × 17.5 × 5.1)*
Main machinery: CODOG: 2 GE LM 2500 gas turbines; 55,750 hp(m) *(41 MW)*; 2 SEMT Pielstick 12 PA6B STC diesels; 11,700 hp(m) *(8.6 MW)*; 2 shafts; cp props
Speed, knots: 29. **Range, n miles:** 7,000 at 18 kt
Complement: 200 (35 officers)

Missiles: SSM: 8 (2 quad) Teseo Mk 2A ❶.
SAM: DCN Sylver VLS ❷ PAAMS (principal anti-air missile system); 48 cells for Aster 15 and Aster 30 weapons.
Guns: 3 Otobreda 76 mm/62 Super Rapid ❸.
2 Breda Oerlikon 25 mm/80 ❹.
Torpedoes: 2 fixed launchers ❺. Eurotorp Mu 90 Impact torpedoes.
Countermeasures: Decoys: 2 Otobreda SCLAR-H chaff/IR flare launchers ❻. SLAT torpedo defence system.
ESM/ECM. Elettronica JANEWS ❼.
Combat data systems: DCN/Alenia CMS; Link 16. Link 14 SATCOM ❽.
Weapons control: Sagem Vampir optronic director ❾.
Radars: Air/surface search: S 1850M ❿; D-band.
Surveillance/fire control: Alenia EMPAR ⓫; G-band; multifunction.
Surface search: Alenia RASS ⓬; E/F-band.
Fire control: 2 Alenia Marconi NA 25XP ⓭.
Navigation: Alenia SPN 753(V)4; I-band.

Sonars: Thomson Marconi 4110CL; hull-mounted; active search and attack; medium frequency.

Helicopters: 1 Augusta/Westland EH 101 Merlin ⓮ or NH-90.

Programmes: Three-nation project for a new air defence ship with Italy, France and UK. Joint project office established in 1993. Memorandum of Understanding for joint development signed 11 July 1994. After UK withdrew in April 1999, an agreement was signed on 7 September 1999 between France and Italy to continue. Following a preliminary agreement on 2 August 2000, a Memorandum of Understanding was signed by the French and Italian Defence Ministries on 22 September 2000 for the joint development of the 'Horizon' destroyer. A Horizon Joint Venture Company was created by DCN/Thomson-CSF and Fincantieri/Finmeccanica on 16 October 2000. The first batch of two vessels for each country was ordered on 27 October 2000.

VERIFIED

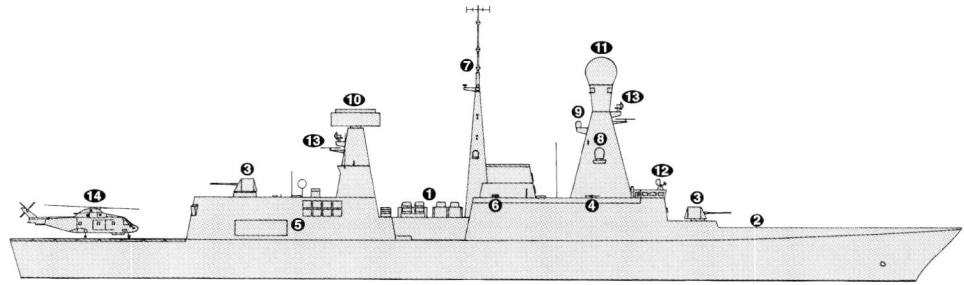

ANDREA DORIA   *(Scale 1 : 1,200), Ian Sturton /* 0104870

ANDREA DORIA (computer graphic)   *2000, ECPA /* 0104871

FRIGATES

0 + 6 (4) MULTIMISSION FRIGATES (FFGH)

Displacement, tons: 5,750 full load
Dimensions, feet (metres): 456.0 × 62.3 × 16.4 *(139.0 × 19.0 × 5.0)*
Main machinery: CODLOG/CODLAG; 1 Rolls Royce MT30 or General Electric LM 2500 gas turbine; 40,230 hp *(30 MW)*; 4 diesels; 11,270 hp *(8.4 MW)*; 2 motors; 5,900 hp *(4.4 MW)*; 2 shafts
Speed, knots: 28. **Range, n miles:** 6,000 at 15 kt
Complement: 123 (accommodation for 165)

Missiles: SLCM: to be decided.
SAM: 16 Sylver A43 cell VLS for Aster 15/30.
SSM: 4 (8 in GP variant) Teseo Mk 2 Block 4.
Guns: 1 OTO 127 mm/64ER (GP). 2 (1 GP) OTO 76 mm SR. 2—25 mm.
Torpedoes: 4 (2 twin) tubes; MU-90.
A/S mortars: 4 MILAS (ASW variant).
Countermeasures: Decoys: 2 Breda SCLAR-H 20-barrel trainable chaff/decoy launchers.
TCM: SLAT launchers.
ESM: Radar and Comms intercept.
ECM: jammer.
Combat data systems: Cavour derivative system.
Weapons control: IRST optronic director.
Radars: Air search: Alenia EMPAR; G-band.

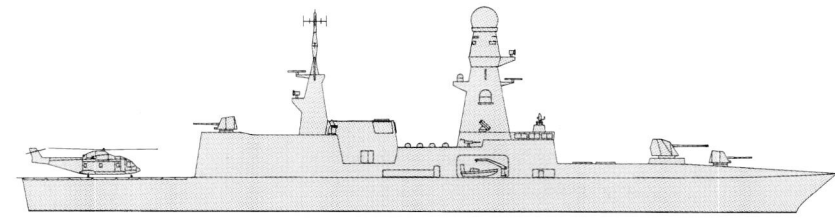

MULTIMISSION FRIGATE   *(Scale 1 : 1,200), Ian Sturton /* 0528393

Surface search: SPS 791; E/F-band.
Navigation: 2 (to be announced).
Fire control: Alenia Marconi NA-25XP; J-band.
Sonars: Thales UMS 4110CL; hull-mounted (bow dome). CAPTAS VDS (ASW variant). Mine avoidance sonar.

Helicopters: 2 NH 90 or 1 NH 90 plus 1 EH 101.

Programmes: A €5.68 billion frigate programme was approved by the Italian government in April 2002. Agreement was later reached on 7 November 2002 for a 27 ship collaborative programme with France. The programme is known as 'Fregate Europee Multi-Missione' (FREMM). The Italian requirement is for 10 frigates to replace the Lupo and Maestrale classes. There are to be two variants: four ASW ships and six for general purpose/land attack. The first batch of six ships (two ASW, four GP) are to be ordered in 2005 with first delivery to take place in 2010. All ten ships are to be in service by 2010.

UPDATED

8 MAESTRALE CLASS (FFGHM)

Name	No	Builders	Laid down	Launched	Commissioned
MAESTRALE	F 570	Fincantieri, Riva Trigoso	8 Mar 1978	2 Feb 1981	6 Mar 1982
GRECALE	F 571	Fincantieri, Muggiano	21 Mar 1979	12 Sep 1981	5 Feb 1983
LIBECCIO	F 572	Fincantieri, Riva Trigoso	1 Aug 1979	7 Sep 1981	5 Feb 1983
SCIROCCO	F 573	Fincantieri, Riva Trigoso	26 Feb 1980	17 Apr 1982	20 Sep 1983
ALISEO	F 574	Fincantieri, Riva Trigoso	10 Aug 1980	29 Oct 1982	7 Sep 1983
EURO	F 575	Fincantieri, Riva Trigoso	15 Apr 1981	25 Apr 1983	24 Jan 1984
ESPERO	F 576	Fincantieri, Riva Trigoso	29 July 1982	19 Nov 1983	4 May 1984
ZEFFIRO	F 577	Fincantieri, Riva Trigoso	15 Mar 1983	19 May 1984	4 May 1985

Displacement, tons: 2,500 standard; 3,200 full load
Dimensions, feet (metres): 405 × 42.5 × 15.1 *(122.7 × 12.9 × 4.6)*
Flight deck, feet (metres): 89 × 39 *(27 × 12)*
Main machinery: CODOG; 2 Fiat/GE LM 2500 gas turbines; 50,000 hp *(37.3 MW)* sustained; 2 GMT B 230.20 DVM diesels; 11,000 hp(m) *(8.1 MW)* sustained; 2 shafts; LIPS cp props
Speed, knots: 32 gas; 21 diesels. **Range, n miles:** 6,000 at 16 kt
Complement: 205 (16 officers)

Missiles: SSM: 4 OTO Melara Teseo Mk 2 (TG 2) ❶; mid-course guidance; active radar homing to 180 km *(98.4 n miles)*; warhead 210 kg; sea-skimmer. Mk 3 with radar/IR homing to 300 km *(162 n miles)*; warhead 160 kg in due course.
SAM: Selenia Albatros octuple launcher; 16 Aspide ❷; semi-active homing to 13 km *(7 n miles)* at 2.5 Mach; height envelope 15-5,000 m *(49.2-16,405 ft)*; warhead 30 kg.
Guns: 1 OTO Melara 5 in *(127 mm)*/54 automatic ❸; 45 rds/min to 23 km *(12.42 n miles)* anti-surface; 7 km *(3.8 n miles)* anti-aircraft; weight of shell 32 kg; fires chaff and illuminants.
4 Breda 40 mm/70 (2 twin) compact ❹; 300 rds/min to 12.5 km *(6.8 n miles)* anti-surface; 4 km *(2.2 n miles)* anti-aircraft; weight of shell 0.96 kg.
2 Oerlikon 20 mm fitted for Gulf deployments in 1990-91. 2 Breda Oerlikon 25 mm/90 (twin) tested in *Espero.*
Torpedoes: 6—324 mm US Mk 32 (2 triple) tubes ❺. Honeywell Mk 46; anti-submarine; active/passive homing to 11 km *(5.9 n miles)* at 40 kt; warhead 44 kg.
Countermeasures: Decoys: 2 Breda 105 mm SCLAR 20-tubed trainable chaff rocket launchers ❻; chaff to 5 km *(2.7 n miles)*; illuminants to 12 km *(6.6 n miles)*. 2 Dagaie chaff launchers.
SLQ-25; towed torpedo decoy. Prairie Masker; noise suppression system.
ESM: Elettronica SLR-4; intercept.
ECM: 2 SLQ-D; jammers.

Combat data systems: IPN 20 (SADOC 2) action data automation; Link 11. SATCOM ❼.
Weapons control: NA 30 for Albatros and 5 in guns. 2 Dardo for 40 mm guns.
Radars: Air/surface search: Selenia SPS-774 (RAN 10S) ❽; E/F-band.
Surface search: SMA SPS-702 ❾; I-band.
Navigation: SMA SPN-703; I-band.
Fire control: Selenia SPG-75 (RTN 30X) ❿; I/J-band (for Albatros and 12.7 mm gun).
2 Selenia SPG-74 (RTN 20X) ⓫; I/J-band; range 15 km *(8 n miles)* (for Dardo).
IFF: Mk XII.
Sonars: Raytheon DE 1164; hull-mounted; VDS; active/passive attack; medium frequency. VDS can be towed at up to 28 kt. Maximum depth 300 m. Modified to include mine detection active high frequency.

Helicopters: 2 AB 212ASW ⓬.

Programmes: First six ordered December 1976 and last pair in October 1980. All Riva Trigoso ships completed at Muggiano after launch.

Modernisation: Hull and VDS sonars modified from 1994 to give better shallow water performance and a mine detection capability. A major upgrade is planned to be undertaken in F 573 and F 577 (starting 2005) and F 572 and F 576 (starting 2006). SPS-774 to be replaced by AMS RAN-21S, SPN-703 to be replaced by SPN-753 ARPA, Dardo to be replaced by two new fire-control systems (Dardo-F with RTN-30X), Sagem IRST and new combat data system to be installed.
Structure: There has been a notable increase of 34 ft in length and 5 ft in beam over the Lupo class to provide for the fixed hangar and VDS, the result providing more comfortable accommodation but a small loss of top speed. Fitted with stabilisers.
Operational: A towed passive LF array may be attached to the VDS body. Aft A 184 torpedo tubes have been removed. F 572, F 573, F 576 and F 577 to remain in service until 2015-2018. F 570, F 571, F 574 and F 575 are to be decommissioned from 2010 as FREMM enter service.

UPDATED

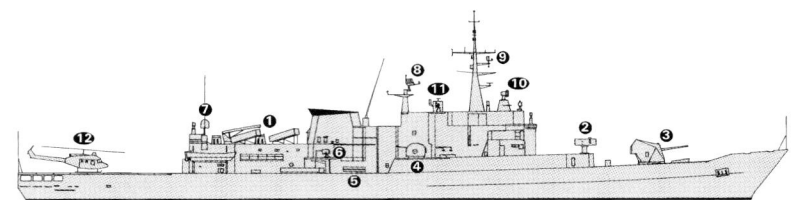

MAESTRALE   *(Scale 1 : 1,200), Ian Sturton* / 0569915

LIBECCIO   *10/2004*, Giorgio Ghiglione* / 1044365

ESPERO   *9/2003, B Prézelin* / 0570683

4 ARTIGLIERE (LUPO) CLASS (FLEET PATROL SHIPS) (FFGHM)

Name	No	Builders	Laid down	Launched	Commissioned
ARTIGLIERE (ex-*Hittin*)	F 582 (ex-F 14)	Fincantieri, Ancona	31 Mar 1982	27 July 1983	28 Oct 1994
AVIERE (ex-*Thi Qar*)	F 583 (ex-F 15)	Fincantieri, Ancona	3 Sep 1982	19 Dec 1984	4 Jan 1995
BERSAGLIERE (ex-*Al Yarmouk*)	F 584 (ex-F 17)	Fincantieri, Riva Trigoso	12 Mar 1984	18 Apr 1985	8 Nov 1995
GRANATIERE (ex-*Al Qadisiya*)	F 585 (ex-F 16)	Fincantieri, Ancona	1 Dec 1983	1 June 1985	20 Mar 1996

Displacement, tons: 2,208 standard; 2,525 full load
Dimensions, feet (metres): 371.3 × 37.1 × 12.1
  (113.2 × 11.3 × 3.7)
Main machinery: CODOG; 2 Fiat/GE LM 2500 gas turbines;
  50,000 hp *(37.3 MW)* sustained; 2 GMT BL 230.20 M
  diesels; 7,800 hp(m) *(5.7 MW)* sustained; 2 shafts; LIPS cp
  props
Speed, knots: 35 turbines; 21 diesels
Range, n miles: 5,000 at 15 kt on diesels
Complement: 177 (13 officers)

Missiles: SSM: 8 OTO Melara Teseo Mk 2 (TG 2) ❶; mid-course
  guidance; active radar homing to 180 km *(98.4 n miles)* at 0.9
  Mach; warhead 210 kg; sea-skimmer.
  SAM: Selenia Elsag Aspide octuple launcher ❷; semi-active radar
  homing to 14.6 km *(8 n miles)* at 2.5 Mach; warhead 39 kg. 8
  reloads.
Guns: 1 OTO Melara 5 in *(127 mm)*/54 ❸; 45 rds/min to 23 km
  (12.42 n miles) anti-surface; 7 km *(3.8 n miles)* anti-aircraft;
  weight of shell 32 kg.
  4 Breda 40 mm/70 (2 twin) compact ❹; 300 rds/min to
  12.5 km *(6.8 n miles)* anti-surface; 4 km *(2.2 n miles)* anti-
  aircraft; weight of shell 0.96 kg.
  2 Oerlikon 20 mm can be fitted.
Countermeasures: Decoys: 2 Breda 105 mm SCLAR 20-tubed
  trainable ❺; chaff to 5 km *(2.7 n miles)*; illuminants to 12 km
  (6.6 n miles).
  ESM/ECM: Selenia SLQ-747 (INS-3M); intercept and jammer.
Combat data systems: IPN 10 mini SADOC action data
  automation; Link 11. SATCOM.

Weapons control: 2 Elsag Mk 10 Argo with NA 21 directors for
  missiles and 5 in gun. 2 Dardo for 40 mm guns.
Radars: Air search: Selenia SPS-774 (RAN 10S) ❻; E/F-band.
  Surface search: Selenia SPQ-712 (RAN 12 L/X) ❼; I-band.
  Navigation: SMA SPN-703; I-band.
  Fire control: 2 Selenia SPG-70 (RTN 10X) ❽; I/J-band; range
    40 km *(22 n miles)* (for Argo).
    2 Selenia SPG-74 (RTN 20X) ❾; I/J-band; range 15 km *(8 n
    miles)* (for Dardo).
  IFF: Mk XII.

Helicopters: 1 AB 212 ❿.

Programmes: On 20 January 1992 it was decided to transfer the
  four ships built for Iraq to the Italian Navy. The original sale to
  Iraq was first delayed by payment problems and then
  cancelled in 1990 when UN embargoes were placed on
  military sales to Iraq. After several attempts by the Italian

Defence Committee to cancel the project, finance was finally
  authorised in July 1993.
Modernisation: The details given are for the ships as modernised
  for Italian service. All ASW equipment removed, new combat
  and communications systems to Italian standards and a major
  upgrading of damage control and accommodation facilities.
  F 584 was fitted with the lightweight 127 mm/54 gun in late
  2000 for trials which were successfully concluded in 2002.
Operational: The first two commissioned with only machinery,
  damage control and accommodation upgraded. The weapon
  systems' changes were made during 1995. The last pair
  entered service fully modified. Official designation is Fleet
  Patrol Ships. All based at Taranto.

VERIFIED

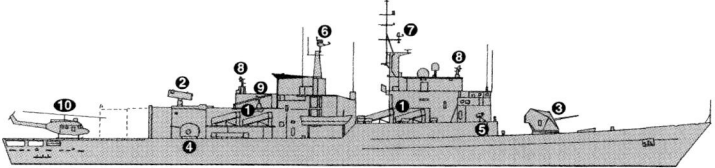

ARTIGLIERE

(Scale 1 : 1,200), Ian Sturton

ARTIGLIERE

2/2002, Giorgio Ghiglione / 0528350

BERSAGLIERE

9/2002, Giorgio Ghiglione / 0528349

CORVETTES

8 MINERVA CLASS (FSM)

Name	No	Builders	Laid down	Launched	Commissioned
MINERVA	F 551	Fincantieri, Riva Trigoso	11 Mar 1985	3 Apr 1986	10 June 1987
URANIA	F 552	Fincantieri, Riva Trigoso	4 Apr 1985	21 June 1986	1 June 1987
DANAIDE	F 553	Fincantieri, Muggiano	26 June 1985	18 Oct 1986	9 Sep 1987
SFINGE	F 554	Fincantieri, Muggiano	2 Sep 1986	16 May 1987	13 Feb 1988
DRIADE	F 555	Fincantieri, Riva Trigoso	18 Mar 1988	11 Mar 1989	19 Apr 1990
CHIMERA	F 556	Fincantieri, Riva Trigoso	21 Dec 1988	7 Apr 1990	15 Jan 1991
FENICE	F 557	Fincantieri, Riva Trigoso	6 Sep 1988	9 Sep 1989	11 Sep 1990
SIBILLA	F 558	Fincantieri, Muggiano	16 Oct 1989	15 Sep 1990	16 May 1991

Displacement, tons: 1,029 light; 1,285 full load
Dimensions, feet (metres): 284.1 × 34.5 × 10.5
(86.6 × 10.5 × 3.2)
Main machinery: 2 Fincantieri GMT BM 230.20 DVM diesels;
11,000 hp(m) *(8.1 MW)* sustained; 2 shafts; cp props
Speed, knots: 24. **Range, n miles:** 3,500 at 18 kt
Complement: 106 (8 officers)

Missiles: SSM: Fitted for but not with 4 or 6 Teseo Otomat
between the masts.
SAM: Selenia Elsag Albatros octuple launcher (F 555-558) ❶; 8
Aspide; semi-active radar homing to 13 km *(7 n miles)* at 2.5
Mach; height envelope 15-5,000 m *(49.2-16,405 ft)*; warhead
30 kg. Capacity for larger magazine.
Guns: 1 OTO Melara 3 in *(76 mm)*/62 Compact ❷; 85 rds/min to
16 km *(8.7 n miles)* anti-surface; 12 km *(6.6 n miles)* anti-
aircraft; weight of shell 6 kg.
Torpedoes: 6—324 mm Whitehead B 515 (2 triple) tubes (F 555-
558) ❸. Honeywell Mk 46; active/passive homing to 11 km
(5.9 n miles) at 40 kt; warhead 44 kg. Being replaced by
Whitehead Mu 90.
Countermeasures: Decoys: 2 Wallop Barricade double layer
launchers for chaff and IR flares. SLQ-25 Nixie; towed torpedo
decoy.

ESM/ECM: Selenia SLQ-747 intercept and jammer.
Combat data systems: Selenia IPN 10 Mini SADOC action data
automation; Link 11. SATCOM.
Weapons control: 1 Elsag Dardo E system. Selenia/Elsag NA
18L Pegaso optronic director ❹. Elmer TLC system.
Radars: Air/surface search: Selenia SPS-774 (RAN 10S) ❺; E/F-
band.
Navigation: SMA SPN-728(V)2 ❻; I-band.
Fire control: Selenia SPG-76 (RTN 30X) ❼; I/J-band (for Albatros
and gun).
Sonars: Raytheon/Elsag DE 1167; hull-mounted; active search
and attack; 7.5-12 kHz.

Programmes: First four ordered in November 1982, second four
in January 1987. A third four were planned, but this plan was
overtaken by the acquisition of the Artigliere class.
Structure: The funnels remodelled to reduce turbulence and IR
signature. Two fin stabilisers.
Operational: Omega transit fitted. Intended for a number of roles
including EEZ patrol, fishery protection and Commanding
Officers' training. SAM launchers and torpedo tubes removed
from first four units. All based at Augusta, Sicily.

VERIFIED

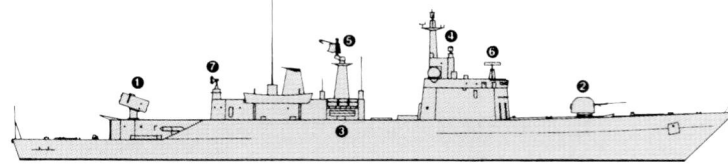

DRIADE

(Scale 1 : 900), Ian Sturton

DRIADE

4/2002, Schaeffer/Marsan / 0528348

SFINGE

9/2003, Giorgio Ghiglione / 0570674

SHIPBORNE AIRCRAFT

Numbers/Type: 15/2 AV-8B Harrier II Plus/TAV-8B Harrier II Plus..
Operational speed: 562 kt *(1,041 km/h).*
Service ceiling: 50,000 ft *(15,240 m).*
Range: 800 n miles *(1,480 km).*
Role/Weapon systems: Two trainers delivered in July 1991 plus 15 front-line aircraft from 1994 to December 1997. Sensors: Radar derived from Hughes APG-65, FLIR, ALQ-164 ESM. Weapons: Maverick ASM; AMRAAM AIM-120 AAM; bombs and 25 mm cannon.   ***VERIFIED***

HARRIER PLUS   *6/2003, Adolfo Ortigueira Gil /* 0570675

Numbers/Type: 16 Agusta/Westland EH 101 Merlin.
Operational speed: 160 kt *(296 km/h).*
Service ceiling: 15,000 ft *(4,572 m).*
Range: 550 n miles *(1,019 km).*
Role/Weapon systems: Primary anti-submarine role with secondary anti-surface and troop carrying capabilities. 16 ordered in October 1995 and approved in July 1997. Six delivered by mid-2002 and further 10 by June 2004. Total of eight for ASW/ASV, four for AEW and four utility/assault. Four further special operations aircraft ordered in 2002 for delivery 2005-06. Sensors: APS-784 (ASW/ASV version); Eliradar HEW-784 (AEW version) radar, Bendix dipping sonar, Galileo FLIR, ALR 735 ESM, ELT 156X ESM, Marconi RALM 1 decoys, Link 11, sonobuoy acoustic processor. Weapons: ASW; four Mk 46 or Mu 90 torpedoes. ASV; four Marte ASM capability for guidance of ship-launched SSM.   ***UPDATED***

MERLIN   *6/2004 *, Paul Jackson /* 0131742

Numbers/Type: 1 NH Industries NH 90 NFH.
Operational speed: 157 kt *(291 km/h).*
Service ceiling: 13,940 ft *(4,250 m).*
Range: 621 n miles *(1,150 km).*
Role/Weapon systems: Total of 56 NFH-90 ordered 30 June 2000 for the Italian Navy in two variants: 46 combat helicopters for ASW/ASV; 10 utility/assault helicopters. First aircraft to be delivered in September 2006. Sensors and weapons to be announced.   ***UPDATED***

NH 90   *3/2004 *, NHI /* 0062373

Numbers/Type: 42 Agusta-Bell 212.
Operational speed: 106 kt *(196 km/h).*
Service ceiling: 17,000 ft *(5,180 m).*
Range: 360 n miles *(667 km).*
Role/Weapon systems: ASW/ECM/Assault helicopter; mainly deployed to escorts, but also shore-based for ASW support duties and nine used for assault. Five are for EW. Sensors: Selenia APS 705 (APS 707 in five Artigliere class aircraft) search/attack radar, AQS-13B dipping sonar or GUFO (not in Artigliere aircraft) ESM/ECM. Weapons: ASW; two Mk 46 torpedoes. Assault aircraft have an armoured cabin, no sensors and are armed with two 7.62 mm MGs and two 70 mm MRLs.   ***VERIFIED***

AB-212   *6/2001, Adolfo Ortigueira Gil /* 0528387

Numbers/Type: 27 Agusta-Sikorsky SH-3D/H Sea King.
Operational speed: 120 kt *(222 km/h).*
Service ceiling: 12,200 ft *(3,720 m).*
Range: 630 n miles *(1,165 km).*
Role/Weapon systems: ASW helicopter; embarked in larger ASW ships, including CVL; also shore-based for medium ASV-ASW in Mediterranean Sea; nine are fitted for ASV, 12 with ASW and EW equipment, six transport/assault. To be replaced by EH-101. Sensors: Selenia APS 705 search radar, AQS-13B dipping sonar, sonobuoys. ESM/ECM. Weapons: ASW; four Mk 46 torpedoes. ASV; two Marte 2 missiles. Assault aircraft have armoured cabins, no sensors, and are armed with two 7.62 mm MGs.   ***UPDATED***

SEA KING   *6/2003, Adolfo Ortigueira Gil /* 0570676

LAND-BASED MARITIME AIRCRAFT

Notes: (1) The Maritime Patrol Replacement programme (MPA-R), a collaborative programme between Italy and Germany, has been abandoned. In the short term, it is likely that the 'Atlantic' MPAs will receive an upgrade to prolong service life to about 2012. Lease of P-3Cs from the US is also a possibility. Future replacement options include participation in the US Multimission Maritime Aircraft (MMA) programme.
(2) One Agusta A 109 transport helicopter procured in 2002 for liaison duties. Three further aircraft may be acquired.
(3) Two Piaggio P-180 transport aircraft procured in 2002 for liaison duties.

Numbers/Type: 18 Bréguet Atlantic 1.
Operational speed: 355 kt *(658 km/h).*
Service ceiling: 22,800 ft *(10,000 m).*
Range: 4,855 n miles *(8,995 km).*
Role/Weapon systems: Air Force shore-based for long-range MR and shipping surveillance; wartime role includes ASW support to helicopters. Sensors: Thomson-CSF Iguane radar, ECM/ESM, MAD, sonobuoys; Marconi ASQ-902 acoustic system. Weapons: ASW; nine torpedoes (including Mk 46 torpedoes) or depth bombs or mines.   ***UPDATED***

ATLANTIC   *7/2004 *, Paul Jackson /* 1044366

Numbers/Type: 16 Panavia Tornado IDS.
Operational speed: 2.2 Mach.
Service ceiling: 80,000 ft *(24,385 m).*
Range: 1,500 n miles *(2,780 km).*
Role/Weapon systems: Air Force swing wing strike and recce; part of a force of a total of 100 aircraft of which 16 are used for maritime operations based near Bari. Sensors: Texas Instruments nav/attack systems. Weapons: ASV; four Kormoran missiles; two 27 mm cannon. AD; four AIM-9L Sidewinder.   ***VERIFIED***

TORNADO IDS   *8/2001, C Hoyle/Jane's /* 0034970

MINE WARFARE FORCES

Notes: Plans for future MCM vessels have been shelved and replacement vessels are not expected before 2010.

12 LERICI/GAETA CLASS (MINEHUNTERS/SWEEPERS) (MHSC)

Name	No	Builders	Launched	Commissioned
LERICI	M 5550	Intermarine, Sarzana	3 Sep 1982	22 Mar 1985
SAPRI	M 5551	Intermarine, Sarzana	5 Apr 1984	4 June 1985
MILAZZO	M 5552	Intermarine, Sarzana	4 Jan 1985	6 Aug 1985
VIESTE	M 5553	Intermarine, Sarzana	18 Apr 1985	2 Dec 1985
GAETA	M 5554	Intermarine, Sarzana	28 July 1990	3 July 1992
TERMOLI	M 5555	Intermarine, Sarzana	15 Dec 1990	13 Nov 1992
ALGHERO	M 5556	Intermarine, Sarzana	11 May 1991	31 Mar 1993
NUMANA	M 5557	Intermarine, Sarzana	26 Oct 1991	30 July 1993
CROTONE	M 5558	Intermarine, Sarzana	11 Apr 1992	19 Jan 1994
VIAREGGIO	M 5559	Intermarine, Sarzana	3 Oct 1992	1 July 1994
CHIOGGIA	M 5560	Intermarine, Sarzana	9 May 1994	19 May 1996
RIMINI	M 5561	Intermarine, Sarzana	17 Sep 1994	26 Nov 1996

Displacement, tons: 620 (697, *Gaeta* onwards) full load
Dimensions, feet (metres): 164 (172.1 *Gaeta*) × 32.5 × 8.6 *(50 (52.5) × 9.9 × 2.6)*
Main machinery: 1 Fincantieri GMT BL 230.8 M diesel (passage); 1,985 hp(m) *(1.46 MW)* sustained; 1 shaft; LIPS cp prop; 3 Isotta Fraschini ID 36 SS 6V diesels (hunting); 1,481 hp(m) *(1.1 MW)* sustained; 3 hydraulic 360° rotating thrust props; 506 hp(m) *(372 kW)* (1 fwd, 2 aft)
Speed, knots: 14; 6 hunting. **Range, n miles:** 1,500 at 14 kt
Complement: 44 (4 officers) including 7 divers

Guns: 1 Oerlikon 20 mm/70.
Countermeasures: Minehunting: 1 Plutogigas and 1 Pluto standard RoV; 1 MIN Mk 2 and 1 Pluto Plus (*Gaeta* onwards); diving equipment and Galeazzi recompression chamber; Galeazzi Z1 two-man recompression chamber (*Gaeta* onwards).
Minesweeping: Oropesa Mk 4 wire sweep.

Combat data systems: Motorola MRS III/GPS Eagle precision navigation system with Datamat SMA SSN-714V(3) automatic plotting and radar indicator IP-7113. Datamat SMA SSN-714 V(2) (*Gaeta* onwards).
Radars: Navigation: SMA SPN-728V(3); I-band.
Sonars: FIAR SQQ-14(IT) VDS (lowered from keel forward of bridge); classification and route survey; high frequency.

Programmes: First four (Lerici class) ordered 7 January 1978 under Legge Navale. Next six ordered from Intermarine 30 April 1988 and two more in 1991. From No 5 onwards (Gaeta class) ships are 2 m longer and are of an improved design. Construction of Gaetas started in 1988. The last pair delayed by budget cuts but re-ordered on 17 September 1992.
Modernisation: Improvements to Gaeta class include a better minehunting sonar system which was backfitted to the Lerici class in 1991. Other Gaeta upgrades include a third hydraulic

system, improved electrical generators, Pluto Gigas ROV, a new type of recompression chamber, and a reduced magnetic signature. Plans to replace the guns with 25 mm have been shelved.
Structure: Of heavy GRP throughout hull, decks and bulkheads, with frames eliminated. All machinery is mounted on vibration dampers and main engines made of a magnetic material. Fitted with crane for launching RoVs and for diving operations.
Operational: Endurance, 12 days. For long passages passive roll-stabilising tanks can be used for extra fuel increasing range to 4,000 miles at 12 kt.
Sales: Four to Malaysia, two to Nigeria and two to Thailand. 12 of a modified design built by the US and six by Australia.
UPDATED

RIMINI

3/2004*, Giorgio Ghiglione / 1044373

GAETA

9/2004*, B Prézelin / 1044372

CHIOGGIA   *1/2003, Giorgio Ghiglione /* 0570673

PATROL FORCES

6 COMANDANTE CLASS PATROL VESSELS (PSOH)

Name	No	Builders	Launched	Commissioned
COMANDANTE CIGALA FULGOSI	P 490	Fincantieri, Riva Trigoso	7 Oct 2000	31 July 2001
COMANDANTE BORSINI	P 491	Fincantieri, Riva Trigoso	17 Feb 2001	4 Dec 2001
COMANDANTE BETTICA	P 492	Fincantieri, Riva Trigoso	23 June 2001	4 Apr 2002
COMANDANTE FOSCARI	P 493	Fincantieri, Riva Trigoso	24 Nov 2001	1 Aug 2002
SIRIO	P 409	Fincantieri, Riva Trigoso	11 May 2002	31 May 2003
ORIONE	P 410	Fincantieri, Riva Trigoso	24 July 2002	1 Aug 2003

Displacement, tons: 1,520 full load
Dimensions, feet (metres): 290.0 × 40 × 15.1 (screws)
(88.4 × 12.2 × 4.6)
Main machinery: 2 GM Trieste-Wärtsilä W18-V 26 XIV diesels;
17,600 hp(m) *(13.2 MW)*; 2 shafts; cp props; bow thruster
2 Wärtsilä 12V26X diesels (P 409-410); 11,585 hp
(8.64 MW); 2 shafts; cp props; bow thruster
Speed, knots: 26 (22 kt for P 409-410)
Range, n miles: 3,500 at 14 kt
Complement: 60 (5 officers)

Guns: 1 Otobreda 3 in *(76 mm)*/62 compact (P 490-493) **❶**.
2 Otobreda 25 mm/90 **❷**.
Countermeasures: Decoys: chaff launcher.
ESM/ECM: Selenia SLQ-747; intercept and jammer.
Combat data systems: AMS IPNS.
Weapons control: 1 optronic director **❸**.
Radars: Surface search **❹**: SPS 703; I/J-band.
Fire control **❺**: SPG 76 (RTN 30X); I/J-band.
Navigation: SPS 753 **❻**; I-band.

Helicopters: 1 AB 212 **❼** or NH90 in due course.

Programmes: Four (P 490-493) for the Navy, and two funded by
the Ministry of Transport, manned by the Navy, but equipped
with more simple command data systems for anti-pollution
and SAR tasks.
Structure: Stealth features include IR suppression and reduced
radar cross-section. P 493 has a superstructure of composite
material. P 409-410 have less powerful engines, no hangar,
no countermeasures and MGs vice 25 mm guns. *UPDATED*

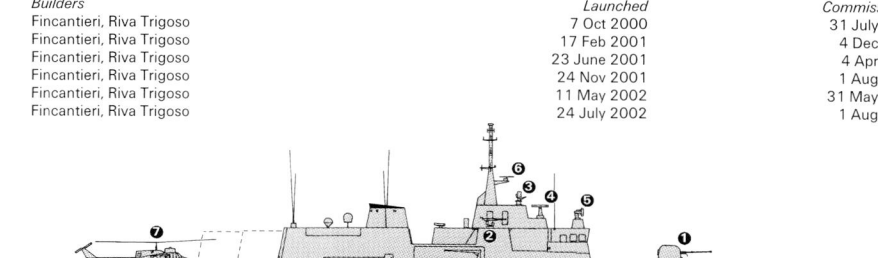

COMANDANTE FOSCARI   *(Scale 1 : 900), Ian Sturton /* 0589003

ORIONE   *7/2003, Giorgio Ghiglione /* 0570679

COMANDANTE BETTICA   *3/2004*, Giorgio Ghiglione /* 1044367

4 CASSIOPEA CLASS (OFFSHORE PATROL VESSELS) (PSOH)

Name	No	Builders	Laid down	Launched	Commissioned
CASSIOPEA	P 401	Fincantieri, Muggiano	16 Dec 1987	20 July 1988	6 July 1989
LIBRA	P 402	Fincantieri, Muggiano	17 Dec 1987	27 July 1988	28 Nov 1989
SPICA	P 403	Fincantieri, Muggiano	5 Sep 1988	27 May 1989	3 May 1990
VEGA	P 404	Fincantieri, Muggiano	20 June 1989	24 Feb 1990	25 Oct 1990

Displacement, tons: 1,002 standard; 1,475 full load
Dimensions, feet (metres): 261.8 × 38.7 × 11.5
(79.8 × 11.8 × 3.5)
Flight deck, feet (metres): 72.2 × 26.2 *(22 × 8)*
Main machinery: 2 Fincantieri/GMT BL 230.16 M diesels;
7,940 hp(m) *(5.84 MW)* sustained; 2 shafts; LIPS cp props
Speed, knots: 20. **Range, n miles:** 3,300 at 17 kt
Complement: 65 (5 officers)

Guns: 1 OTO Melara 3 in *(76 mm)*/62; 60 rds/min to 16 km
(8.7 n miles); weight of shell 6 kg. Breda Oerlikon 25 mm/90.
2—12.7 mm MGs.
Weapons control: Argo NA 10.
Radars: Surface search: SMA SPS-702(V)2; I-band.
Navigation: SMA SPN-748(V)2; I-band.
Fire control: Selenia SPG-70 (RTN 10X); I/J-band.

Helicopters: 1 AB 212ASW.

Programmes: Ordered in December 1986 for operations in EEZ.
Officially 'pattugliatori marittimi'. Funded by the Ministry of
Transport but all operated by the Navy.
Structure: Fitted for firefighting, rescue and supply tasks.
Telescopic hangar. The 20 mm guns were old stock taken

VEGA

3/2001, Giorgio Ghiglione / 0130329

from deleted Bergamini class and have been replaced by
25 mm guns. There is a 500 m³ tank for storing oil polluted
water.

Operational: All based at Augusta.

VERIFIED

4 ESPLORATORE CLASS (PB)

Name	No	Builders	Launched	Commissioned
ESPLORATORE	P 405	Coinaval, La Spezia	4 Nov 1996	26 June 1997
SENTINELLA	P 406	Coinaval, La Spezia	13 Nov 1997	10 July 1998
VEDETTA	P 407	Coinaval, La Spezia	11 Jan 1997	29 July 1999
STAFFETTA	P 408	Coinaval, La Spezia/ Oromare	Nov 2002	1 Mar 2005

Displacement, tons: 165 full load
Dimensions, feet (metres): 122 × 23.3 × 6.2 *(37.2 × 7.1 × 1.9)*
Main machinery: 2 Isotta Fraschini M1712T2 diesels; 3,810 hp(m) *(2.8 MW)*; 2 shafts
Speed, knots: 20. **Range, n miles:** 1,200 at 20 kt
Complement: 14 (2 officers)
Guns: 1 Oerlikon 20 mm/70. 2—7.62 mm MGs.
Weapons control: AESN Medusa optronic director to be fitted.
Radars: Surface search: 2 SPS-753B/C; F/I-band.

Comment: Ordered from Ortona Shipyard in December 1993 but the contract was then
transferred to Coinaval Yards, La Spezia in 1994 which caused inevitable delays and
construction did not start until 1995. An option on a fourth of class, was taken up in February
1998 but shipbuilding programme delayed launch until January 2001 when it was transferred
to Oromare shipyard for completion which has been delayed by financial problems. Based in
Red Sea for Multinational Force Observer (MFO) operations. To return to Adriatic on completion.

UPDATED

SENTINELLA

10/1999, Giorgio Ghiglione / 0080087

MEN 219 and 220

2000, M Annati / 0104881

2 PEDRETTI CLASS
(SPECIAL OPERATIONS SUPPORT CRAFT) (YDT)

Name	No	Builders	Commissioned
ALCIDE PEDRETTI	Y 499 (ex-MEN 213)	Crestitalia-Ameglia	23 Oct 1984
MARIO MARINO	Y 498 (ex-MEN 214)	Crestitalia-Ameglia	21 Dec 1984

Displacement, tons: 75.4 *(Alcide Pedretti)*, 69.5 *(Mario Marino)* full load
Dimensions, feet (metres): 86.6 × 22.6 × 3.3 *(26.4 × 6.9 × 1)*
Main machinery: 2 Isotta Fraschini ID 36 SS 12V diesels; 2,640 hp(m) *(1.94 MW)* sustained;
2 shafts
Speed, knots: 25. **Range, n miles:** 450 *(Alcide Pedretti)*, 250 *(Mario Marino)* at 23 kt
Complement: 8 (1 officer)
Radars: Navigation: I-band.

Comment: Both laid down 8 September 1983. For use by assault swimmers of COMSUBIN. Both
have decompression chambers. *Alcide Pedretti* has a floodable dock aft and is used for combat
swimmers and special operations, while *Mario Marino* is fitted for underwater work and rescue
missions. Based at Varignano, La Spezia. A similar but more heavily equipped vessel serves with
the UAE Navy.

VERIFIED

AMPHIBIOUS FORCES

Notes: (1) A Ro-Ro ship MV *Major* built in 1984 is on long term charter to the Army Mobility and
Transport Command. 6,830 tons displacement with 1,240 m of vehicle lanes. Can carry
3,955 tons of cargo.
(2) There are also 54 Rigid Raider Craft in service with Amphibious Forces.

9 MTM 217 CLASS (LCM)

MEN 217-222   MEN 227-228   MEN 551

Displacement, tons: 64.6 full load
Dimensions, feet (metres): 60.7 × 16.7 × 3 *(18.5 × 5.1 × 0.9)*
Main machinery: 2 Fiat diesels; 560 hp(m) *(412 kW)*; 2 shafts
Speed, knots: 9. **Range, n miles:** 300 at 9 kt
Complement: 3
Military lift: 30 tons

Comment: First six built at Muggiano, La Spezia by Fincantieri. Three completed 9 October 1987
for *San Giorgio*, three completed 8 March 1988 for *San Marco*. Three more ordered in March
1991 from Balzamo Shipyard and completed in 1993 for *San Giusto*. Others of this class are also
in service with the Army.

VERIFIED

ALCIDE PEDRETTI

10/1999, Giorgio Ghiglione / 0080088

For details of the latest updates to *Jane's Fighting Ships* online and to discover the
additional information available exclusively to online subscribers please visit
jfs.janes.com

3 SAN GIORGIO CLASS (LPD)

Name	No	Builders	Laid down	Launched	Commissioned
SAN GIORGIO	L 9892	Fincantieri, Riva Trigoso	27 June 1985	25 Feb 1987	9 Oct 1987
SAN MARCO	L 9893	Fincantieri, Riva Trigoso	28 June 1986	21 Oct 1987	18 Mar 1988
SAN GIUSTO	L 9894	Fincantieri, Riva Trigoso	30 Nov 1992	2 Dec 1993	9 Apr 1994

Displacement, tons: 6,687 standard; 7,960 (8,000 *San Giusto*) full load
Dimensions, feet (metres): 449.5 *(San Giusto)*; 437.2 × 67.3 × 17.4 *(137; 133.3 × 20.5 × 5.3)*
Flight deck, feet (metres): 328.1 × 67.3 *(100 × 20.5)*
Main machinery: 2 Fincantieri GMT A 420.12 diesels; 16,800 hp (m) *(12.35 MW)* sustained; 2 shafts; LIPS cp props; bow thruster
Speed, knots: 21. **Range, n miles:** 7,500 at 16 kt; 4,500 at 20 kt
Complement: 168 (12 officers); 167 (15 officers) *(San Giusto)*
Military lift: Battalion of 400 plus 30-36 APCs or 30 medium tanks. 2 LCMs in stern docking well. 2 LCVPs on upper deck *(San Marco)* or 3 LCVPs on sponsons *(San Giusto* and *San Giorgio)*. 1 LCPL

Guns: 1 OTO Melara 3 in *(76 mm)*/62 (Compact in *San Giusto)*; 60 rds/min to 16 km *(8.7 n miles)*; weight of shell 6 kg.
2 Breda Oerlikon 25 mm/90. 2—12.7 mm MGs.
Countermeasures: ESM: SLR 730; intercept.
ESM/ECM: SLQ-747 *(San Giusto)*.
Combat data systems: Selenia IPN 20 *(San Giusto)*. Marisat. SATCOM.
Weapons control: Elsag NA 10.
Radars: Surface search: SMA SPS-702; I-band.
Navigation: SMA SPN-748; I-band.
Fire control: Selenia SPG-70 (RTN 10X); I/J-band.

Helicopters: 3 SH-3D Sea King or EH 101 Merlin or 5 AB 212.

Programmes: *San Giorgio* ordered 26 November 1983, *San Marco* on 5 March 1984 and *San Giusto* 1 March 1991. Launching dates of the first two are slightly later than the 'official' launching ceremony because of poor weather and for the third because of industrial problems.
Modernisation: 25 mm guns replaced 20 mm from 1999. Modifications to *San Giorgio* include removal of the 76 mm gun, movement of LCVPs from davits to a new sponson, and lengthening and enlargement of the flight deck to allow two Merlin and two AB 212 to operate simultaneously on deck. Work completed in early 2003. Similar work on *San Marco* completed in March 2004. *San Giusto* has been fitted with an MCC data system to enable her to act as CJTF.
Structure: Aircraft carrier type flight deck with island to starboard. Following modernisation, *San Giorgio* and *San*

SAN GIORGIO   *8/2003, C D Yaylali* / 0570661

SAN GIORGIO   *2/2002, Italian Navy* / 0570697

Marco have four landing spots, a stern docking well (20.5 × 7 m), LCVPs on a port side sponson, a 30 ton lift and two 40 ton travelling cranes for LCMs. *San Giusto* is of similar design, but was 300 tons heavier on build to include extra accommodation and a slightly longer island. Bow doors and beaching capability removed from *San Marco* in refit.

Operational: *San Marco* was paid for by the Ministry of Civil Protection, is specially fitted for disaster relief but is run by the Navy. All are based at Brindisi and assigned to COMFORAL. One of the three ships carries out the annual Summer cruise for officer and petty officer cadets.   *UPDATED*

SAN GIUSTO   *9/2004*, John Brodie* / 1044369

17 MTP 96 CLASS (LCVP)

MDN 94-104	MDN 108-109	MDN 114-117

Displacement, tons: 14.3 full load
Dimensions, feet (metres): 44.9 × 12.5 × 2.3 *(13.7 × 3.8 × 0.7)*
Main machinery: 2 diesels; 700 hp(m) *(515 kW)*; 2 shafts or 2 water-jets
Speed, knots: 29 or 22. **Range, n miles:** 100 at 12 kt
Complement: 3

Comment: Built by Technomatic Ancona in 1985 (two), Technomatic Bari in 1987-88 (six) and Technoplast Venezia 1991-94 (nine). Can carry 45 men or 4.5 tons of cargo. These craft have Kevlar armour. The most recent versions have water-jet propulsion which gives a top speed of 29 kt (22 kt fully laden). This is being backfitted to all GRP LCVPs.
                                                 VERIFIED

MDN 101   *10/2001, Chris Sattler* / 0130326

SURVEY AND RESEARCH SHIPS

1 SURVEY SHIP (AGORH/AGE/AGI)

Name	No	Builders	Commissioned
ELETTRA	A 5340	Fincantieri, Muggiano	2 Apr 2003

Displacement, tons: 3,180 full load
Dimensions, feet (metres): 305.1 × 49.9 × 17.1 *(93 × 15.2 × 5.2)*
Main machinery: Diesel electric; 2 Wartsila CW 12V 200 diesel generators; 5,750 kVA. 2 ABB motors; 4,023 hp *(3 MW)*; 2 shafts; bow thruster
Speed, knots: 17. **Range, n miles:** 8,000 at 12 kt
Complement: 94 (12 officers)
Radars: Navigation: I-band.

Comment: Ordered on 1 December 1999; construction started in March 2000 and launch on 24 July 2002. The design is derived from that of the NATO *Alliance* but is equipped as an intelligence collector. The propulsion system, based on two multi permanent magnet electric motors, is the first of its type to be fitted in a surface vessel.   *UPDATED*

ELETTRA   *5/2003, Giorgio Ghiglione* / 0570671

1 SURVEY SHIP (AGSH)

Name	No	Builders	Commissioned
AMMIRAGLIO MAGNAGHI	A 5303	Fincantieri, Riva Trigoso	2 May 1975

Displacement, tons: 1,700 full load
Dimensions, feet (metres): 271.3 × 44.9 × 11.5 *(82.7 × 13.7 × 3.5)*
Main machinery: 2 GMT B 306 SS diesels; 3,000 hp(m) *(2.2 MW)*; 1 shaft; cp prop; auxiliary motor; 240 hp(m) *(176 kW)*; bow thruster
Speed, knots: 16. **Range, n miles:** 6,000 at 12 kt (1 diesel); 4,200 at 16 kt (2 diesels)
Complement: 148 (14 officers, 15 scientists)
Guns: 1 Breda 40 mm/70 (not fitted).
Radars: Navigation: SMA 3 RM 20; I-band.
Helicopters: Platform only.

Comment: Ordered under 1972 programme. Laid down 13 June 1973. Launched 11 October 1974. Full air conditioning, bridge engine controls, flume-type stabilisers. Equipped for oceanographical studies including laboratories and underwater TV. Two Qubit Trac V integrated navigation and logging systems and a Chart V data processing system installed in 1992 to augment the existing Trac 100-based HODAPS. Carries six surveying motor boats.

VERIFIED

AMMIRAGLIO MAGNAGHI   *3/2002, Giorgio Ghiglione /* 0528354

2 SURVEY SHIPS (AGS)

Name	No	Builders	Commissioned
ARETUSA	A 5304	Intermarine	10 Jan 2002
GALATEA	A 5308	Intermarine	10 Jan 2002

Displacement, tons: 415 full load
Dimensions, feet (metres): 128.6 × 41.3 × 8.2 *(39.2 × 12.6 × 2.5)*
Main machinery: Diesel electric; 2 Isotta Fraschini V170812 ME diesels; 2 ABB generators 1,904 hp(m) *(1.4 MW)*; 2 shafts; Schottel props; 2 bow thrusters
Speed, knots: 13. **Range, n miles:** 1,700 at 13 kt
Complement: 29 (4 officers)
Guns: 2—7.62 mm MGs.
Radars: 2 Navigation; I-band.

Comment: GRP catamaran design. Ordered in January 1998. *Aretusa* launched 8 May 2000 and *Galatea* 7 June 2000. Fitted with Kongsberg EA 500 single-beam echo sounder, towed sidescan sonar and dynamic positioning system.

UPDATED

GALATEA   *1/2004 *, Giorgio Ghiglione /* 1044371

1 RESEARCH SHIP (AG/AGOR)

Name	No	Builders	Launched	Commissioned
RAFFAELE ROSSETTI	A 5315	Picchiotti, Viareggio	12 July 1986	20 Dec 1986

Displacement, tons: 320 full load
Dimensions, feet (metres): 146.3 × 25.9 × 6.9 *(44.6 × 7.9 × 2.1)*
Main machinery: 2 Fincantieri Isotta Fraschini ID 36 N 6V diesels; 3,520 hp(m) *(2.55 kW)* sustained; 2 shafts; cp props; bow thruster
Speed, knots: 17.5. **Range, n miles:** 700 at 15 kt
Complement: 17 (2 officers)

Comment: Five different design torpedo tubes fitted for above and underwater testing and trials. Other equipment for research into communications, surface and air search as well as underwater weapons. There is a stern doorway which is partially submerged and the ship has a set of 96 batteries to allow 'silent' propulsion. Operated by the Permanent Commission for Experiments of War Materials at La Spezia.

VERIFIED

RAFFAELE ROSSETTI   *3/1998, Giorgio Ghiglione /* 0052416

1 RESEARCH SHIP (AG/AGE)

Name	No	Builders	Commissioned
VINCENZO MARTELLOTTA	A 5320	Picchiotti, Viareggio	22 Dec 1990

Displacement, tons: 340 full load
Dimensions, feet (metres): 146.3 × 25.9 × 7.5 *(44.6 × 7.9 × 2.3)*
Main machinery: 2 Fincantieri Isotta Fraschini ID 36 SS 16V diesels; 3,520 hp(m) *(2.59 MW)* sustained; 2 shafts; cp props; bow thruster
Speed, knots: 17. **Range, n miles:** 700 at 15 kt
Complement: 19 (2 officers)

Comment: Launched on 28 May 1988. Has one 21 in *(533 mm)* and three 12.75 in *(324 mm)* torpedo tubes and acoustic equipment to operate a 3-D tracking range for torpedoes or underwater vehicles. Like *Rossetti* she is operated by the Commission for Experiments at La Spezia.

VERIFIED

VINCENZO MARTELLOTTA   *9/2002, Giorgio Ghiglione /* 0528352

1 ALPINO CLASS (AGEHM)

Name	No	Builders	Commissioned
CARABINIERE (ex-*Climene*)	F 581	Fincantieri, Riva Trigoso	28 Apr 1968

Displacement, tons: 2,400 standard; 2,700 full load
Dimensions, feet (metres): 371.7 × 43.6 × 12.7 *(113.3 × 13.3 × 3.9)*
Main machinery: 4 Tosi OTV-320 diesels; 16,800 hp(m) *(12.35 MW)*; 2 shafts
Speed, knots: 20. **Range, n miles:** 3,500 at 18 kt
Complement: 158 (13 officers)

Missiles: SAM: Alenia/DCN 8-cell VLS for Aster 15 trials; radar homing to 30 km *(16.2 n miles)* at 4.5 Mach; warhead 13 kg.
A/S: OTO Melara/Matra Milas launcher; command guidance to 55 km *(29.8 n miles)* at Mach 0.9; payload Mu 90 or Mk 46 Mod 5 torpedo.
Guns: 1 OTO Melara 3 in *(76 mm)*/62; 60 rds/min to 16 km *(8.7 n miles)*; weight of shell 6 kg.
Torpedoes: 6—324 mm Whitehead (2 triple tubes).
Countermeasures: ESM/ECM: Selenia SLQ-747; integrated intercept and jammer.
Combat data systems: SADOC 3 for EMPAR trials.
Weapons control: 2 Argo 'O' for 3 in guns.
Radars: Air search: RCA SPS-12; D-band.
Surface search: SMA SPS-702(V)3; I-band.
Navigation: SMA SPN-748; I-band.
Fire control: 2 Selenia SPG-70 (RTN 10X); I/J-band.
Alenia/Marconi EMPAR SPY 790; for PAAMS; G-band; range 50 km *(27 n miles)* for 0.1 m² target.

Modernisation: Modified as a trials ship. B gun turret and the Whitehead mortar removed and, following a refit which completed in October 2000, fitted with an eight cell A43 vertical launch system aft of the forward gun and an upgrade of the communication system. The A43 VLS module was then replaced by an A50 module (for Aster 30) in late 2003. An oil rig style mast on the flight deck carries the EMPAR SPY 790 radar that was installed in 1995.
Operational: Since 1993, served as testbed for the Milas anti-submarine missile and EMPAR radar. EMPAR serves as surveillance, acquisition and missile guidance system for the SAAM/IT point defence system on which final qualification firings with Aster 15 missiles were conducted during 2002 and early 2003. SLAT soft-kill anti-torpedo system trials were conducted at the same time. Full PAAMS combat system trials (for the French, Italian and UK navies) began off Toulon from mid-2004. A live firing is to be completed during 2005.

UPDATED

CARABINIERE   *2/2004 *, Giorgio Ghiglione /* 1044370

AUXILIARIES

1 ETNA CLASS (REPLENISHMENT TANKER) (AORH)

Name	No	Builders	Laid down	Launched	Commissioned
ETNA	A 5326	Fincantieri, Riva Trigoso	4 July 1995	12 July 1997	29 Aug 1998

Displacement, tons: 13,400 full load
Dimensions, feet (metres): 480.6 × 68.9 × 24.3
 (146.5 × 21 × 7.4)
Flight deck, feet (metres): 91.9 × 68. 9 *(28 × 21)*
Main machinery: 2 Sulzer 12 ZAV 40S diesels; 22,400 hp(m)
 (16.46 MW) sustained; 2 shafts; bow thruster
Speed, knots: 21. **Range, n miles:** 7,600 at 18 kt

Complement: 162 (14 officers) plus 81 spare
Cargo capacity: 6,350 tons gas oil; 1,200 tons JP5; 2,100
 m³ ammunition and stores
Guns: 1 OTO Melara 76 mm/62. 2 Breda Oerlikon 25 mm/93.
Radars: Surface search: SMA SPS-702(V)3; I-band.
 Navigation: GEM SPN-753; I-band.
Helicopters: 1 EH 101 Merlin or SH-3D or 2 AB 212.

Comment: Details revised in 1992 for an order 29 July 1994.
 Construction authorised on 3 January 1995. The main gun is
 not fitted, and the specification includes a CIWS on the hangar
 roof. Two RAS stations on each side. A similar ship has been
 built for Greece. A major upgrade to C⁴I capability has given
 the ship a Maritime Component Commander capability.
 UPDATED

ETNA   *6/2001, M Declerck* / 0132012

1 ALPINO CLASS (MCSH)

Name	No	Builders	Commissioned
ALPINO (ex-*Circe*)	A 5384 (ex-F 580)	Fincantieri, Riva Trigoso	14 Jan 1968

Displacement, tons: 2,400 standard; 2,700 full load
Dimensions, feet (metres): 371.7 × 43.6 × 12.7 *(113.3 × 13.3 × 3.9)*
Main machinery: 4 Tosi OTV-320 diesels; 16,800 hp(m) *(12.35 MW)*; 2 shafts
Speed, knots: 20. **Range, n miles:** 3,500 at 18 kt
Complement: 155 (13 officers)
Guns: 3 OTO Melara 3 in *(76 mm)*/62; 60 rds/min to 16 km *(8. 7 n miles)*; weight of shell 6 kg.
 2—12.7 mm MGs.
Countermeasures: ESM/ECM: Selenia SLQ-747; integrated intercept and jammer.
Weapons control: 2 Argo 'O' for 3 in guns.
Radars: Air search: RCA SPS-12; D-band.
 Surface search: SMA SPS-702(V)3; I-band.
 Navigation: SMA SPN-748; I-band.
 Fire control: 2 Selenia SPG-70 (RTN 10X); I/J-band.
Helicopters: 1 AB 212ASW.

Modernisation: *Alpino* modified as an MCMV command and support ship 1996-97. Gun
 armament reduced, A/S mortar, torpedo tubes, sonars and chaff launchers removed. A 2.5 ton
 crane has been fitted and a recompression chamber embarked. Gas turbines removed.
Structure: Stabilisers fitted.
Operational: To remain in service to 2008.   ***UPDATED***

ALPINO   *4/2004*, Giorgio Ghiglione* / 1044374

2 STROMBOLI CLASS (REPLENISHMENT TANKERS) (AORH)

Name	No	Builders	Launched	Commissioned
STROMBOLI	A 5327	Fincantieri, Riva Trigoso	20 Feb 1975	20 Nov 1975
VESUVIO	A 5329	Fincantieri, Muggiano	4 June 1977	18 Nov 1978

Displacement, tons: 3,556 light; 8,706 full load
Dimensions, feet (metres): 423.1 × 59 × 21.3 *(129 × 18 × 6.5)*
Main machinery: 2 GMT C428 SS diesels; 9,600 hp(m) *(7.06 MW)*; 1 shaft; LIPS cp prop
Speed, knots: 18.5. **Range, n miles:** 5,080 at 18 kt
Complement: 131 (10 officers)
Cargo capacity: 3,000 tons FFO; 1,000 tons dieso; 400 tons JP5; 300 tons other stores
Guns: 1 OTO Melara 3 in *(76 mm)*/62.
 2 Oerlikon 20 mm/70.
Weapons control: Argo NA 10 system.
Radars: Surface search: SMA SPQ-2; I-band.
 Navigation: SMA SPN-748; I-band.
 Fire control: Selenia SPG-70 (RTN 10X); I/J-band.
Helicopters: Platform for 1 medium.

Comment: *Vesuvio* was the first large ship to be built at Muggiano (near La Spezia) since the
 Second World War and the first with funds under Legge Navale 1975. Beam and stern refuelling
 stations for fuel and stores. Also Vertrep. The two ships have different midships crane
 arrangements. Similar ship built for Iraq and laid up in Alexandria since 1986. 20 mm guns
 replaced by 25 mm from 2002.   ***VERIFIED***

VESUVIO   *6/2001, Giorgio Ghiglione* / 0130345

1 BORMIDA CLASS (WATER TANKER) (AWT)

BORMIDA (ex-*GGS 1011*) A 5359

Displacement, tons: 736 full load
Dimensions, feet (metres): 131.9 × 23.6 × 10.5 *(40.2 × 7.2 × 3.2)*
Main machinery: 1 diesel; 130 hp(m) *(95.6 kW)*; 1 shaft
Speed, knots: 7
Complement: 11 (1 officer)
Cargo capacity: 260 tons

Comment: Converted at La Spezia in 1974.   ***VERIFIED***

BORMIDA   *9/2002, Giorgio Ghiglione* / 0528367

4 MCC 1101 CLASS (WATER TANKERS) (AWT)

MCC 1101 A 5370	**MCC 1102** A 5371	**MCC 1103** A 5372	**MCC 1104** A 5373

Displacement, tons: 898 full load
Dimensions, feet (metres): 155.2 × 32.8 × 10.8 *(47.3 × 10 × 3.3)*
Main machinery: 2 Fincantieri Isotta Fraschini ID 36 SS 6V diesels; 1,320 hp(m) *(970 kW)*
 sustained; 2 shafts
Speed, knots: 13. **Range, n miles:** 1,500 at 12 kt
Complement: 12 (2 officers)
Cargo capacity: 550 tons
Radars: Navigation: SPN-753; I-band.

Comment: Built by Ferrari, La Spezia and completed one in 1986, two in May 1987, one in
 May 1988.   ***VERIFIED***

MCC 1103   *7/2003, Giorgio Ghiglione* / 0570663

2 SIMETO CLASS (WATER TANKERS) (AWT)

Name	No	Builders	Commissioned
TICINO	A 5376	Poli Shipyard, Pellestrina	10 June 1994
TIRSO	A 5377	Poli Shipyard, Pellestrina	12 Mar 1994

Displacement, tons: 1,858 full load; 1,968 *(Ticino* and *Tirso)* full load
Dimensions, feet (metres): 229 × 33.1 × 14.4 *(69.8 × 10.1 × 4.1)*
Main machinery: 2 GMT B 230.6 BL diesels; 2,530 hp(m) *(1.86 MW)* sustained; 2 shafts; cp props; bow thruster; 300 hp(m) *(220 kW)*
Speed, knots: 13. **Range, n miles:** 1,800 at 12 kt
Complement: 36 (3 officers)
Cargo capacity: 1,130 tons; 1,200 tons *(Ticino* and *Tirso)*
Guns: 1—20 mm/70. 2—7.62 mm MGs can be carried.
Radars: Navigation: 2 SPN-753B(V); I-band.

Comment: Guns are not normally carried. *Simeto* transferred to Tunisia on 30 June 2003.
VERIFIED

TIRSO   *2/2000, van Ginderen Collection /* 0104888

7 DEPOLI CLASS TANKERS (AOTL/AWT)

GGS 1012-1014   GRS/G 1010-1012   GRS/J 1013

Dimensions, feet (metres): 128.3 × 27.9 × 10.2 *(39.1 × 8.5 × 3.1)*
Main machinery: 2 diesels; 748 hp(m) *(550 kW)*; 2 shafts
Speed, knots: 11
Complement: 12
Cargo capacity: 500 m³ liquids
Radars: Navigation: I-band.

Comment: Built by DePoli and delivered between February 1990 and February 1991. The GGS series is for water, GRS/G for fuel and GRS/J for JP5.
VERIFIED

GGS 1012   *5/1999, A Sharma /* 0080101

6 MTC 1011 CLASS (RAMPED TRANSPORTS) (AKL)

Name	No	Builders	Commissioned
GORGONA (1011)	A 5347	CN Mario Marini	23 Dec 1986
TREMITI (1012)	A 5348	CN Mario Marini	2 Mar 1987
CAPRERA (1013)	A 5349	CN Mario Marini	10 Apr 1987
PANTELLERIA (1014)	A 5351	CN Mario Marini	26 May 1987
LIPARI (1015)	A 5352	CN Mario Marini	10 July 1987
CAPRI (1016)	A 5353	CN Mario Marini	16 Sep 1987

Displacement, tons: 631 full load
Dimensions, feet (metres): 186 × 32.8 × 8.2 *(56.7 × 10 × 2.5)*
Main machinery: 2 CRM 12D/SS diesels; 1,760 hp(m) *(1.29 MW)*; 2 shafts
Speed, knots: 14.5. **Range, n miles:** 1,500 at 14 kt
Complement: 32 (4 officers)
Guns: 1 Oerlikon 20 mm (fitted for). 2—7.62 mm MGs.
Radars: Navigation: SMA SPN-748; I-band.

Comment: As well as transporting stores, oil or water they can act as support ships for Light Forces, salvage ships or minelayers. Stern ramp fitted. 1011 based at La Spezia, 1012 at Ancona, 1013 at La Maddalena and 1014 at Taranto.
VERIFIED

CAPRI   *5/2001, Paolo Marsan /* 0130342

1 SALVAGE SHIP (ARSH)

Name	No	Builders	Launched	Commissioned
ANTEO	A 5309	C N Breda-Mestre	11 Nov 1978	31 July 1980

Displacement, tons: 3,200 full load
Dimensions, feet (metres): 322.8 × 51.8 × 16.7 *(98.4 × 15.8 × 5.1)*
Main machinery: 2 GMT A 230.12 diesels; 5,000 hp(m) *(3.68 MW)*; 2 motors; 6,000 hp(m) *(4.41 MW)*; 1 shaft; 2 bow thrusters; 1,000 hp(m) *(735 kW)*
Speed, knots: 20. **Range, n miles:** 4,000 at 14 kt
Complement: 121 (including salvage staff)
Guns: 2 Oerlikon 20 mm/70 fitted during deployments.
Radars: Surface search: SMA SPN-751; I-band.
Navigation: SMA SPN-748; I-band.
Helicopters: 1 AB 212.

Comment: Ordered mid-1977. Comprehensively fitted with flight deck and hangar, extensive salvage gear, including rescue bell and recompression chambers. Carries four lifeboats of various types. Three firefighting systems. Full towing equipment. Carries midget submarine, *Usel*, of 13.2 tons dived with dimensions 26.2 × 6.2 × 8.9 ft *(8 × 1.9 × 2.7 m)*. Carries two men and can dive to 600 m. Endurance, 120 hours at 5 kt. Also has a McCann rescue chamber.
VERIFIED

ANTEO   *5/2000, Giorgio Ghiglione /* 0104889

5 PONZA CLASS (LIGHTHOUSE TENDERS) (ABU)

Name	No	Builders	Commissioned
PONZA	A 5364	Morini Yard, Ancona	9 Dec 1988
LEVANZO	A 5366	Morini Yard, Ancona	24 Jan 1989
TAVOLARA	A 5367	Morini Yard, Ancona	12 Apr 1989
PALMARIA	A 5368	Morini Yard, Ancona	12 May 1989
PROCIDA	A 5383	Morini Yard, Ancona	14 Nov 1990

Displacement, tons: 608 full load
Dimensions, feet (metres): 186 × 35.4 × 8.2 *(56.7 × 10.8 × 2.5)*
Main machinery: 2 Fincantieri Isotta Fraschini ID 36 SS 8V diesels; 1,760 hp(m) *(1.29 MW)* sustained; 2 shafts; cp props; bow thruster; 120 hp(m) *(88 kW)*
Speed, knots: 14.5. **Range, n miles:** 1,500 at 14 kt
Complement: 34 (2 officers)
Guns: 2—7.62 mm MGs.
Radars: Navigation: SPN-732; I-band.

Comment: MTF 1304-1308. Similar to MTC 1011 class.
UPDATED

PALMARIA   *6/2004*, Giorgio Ghiglione /* 1044375

1 MEN 212 CLASS (YPT)

MEN 212

Displacement, tons: 32 full load
Dimensions, feet (metres): 58.4 × 16.7 × 3.3 *(17.8 × 5.1 × 1)*
Main machinery: 2 HP diesels; 1,380 hp(m) *(1.01 MW)*; 2 shafts
Speed, knots: 22. **Range, n miles:** 250 at 20 kt
Complement: 4
Radars: Navigation: SPN-732; I-band.

Comment: Torpedo Recovery Vessel completed in October 1983 by Crestitalia. GRP construction with a stern ramp. Capacity for up to three torpedoes.
VERIFIED

MEN 212   *8/2003, P Marsan /* 0570664

2 MEN 215 CLASS (YFU/YFB)

MEN 215   MEN 216

Displacement, tons: 82 full load
Dimensions, feet (metres): 89.6 × 23 × 3.6 *(27.3 × 7 × 1.1)*
Main machinery: 2 Isotta Fraschini ID 36 SS 12V diesels; 2,640 hp(m) *(1.94 MW)* sustained; 2 shafts
Speed, knots: 28. **Range, n miles:** 250 at 14 kt
Complement: 4
Radars: Navigation: SPN-732; I-band.

Comment: Fast personnel launches completed in June 1986 by Crestitalia. Can also be used for amphibious operations or disaster relief. One is based at La Spezia and one in Taranto, where they are used as local ferries. *VERIFIED*

MEN 216   *3/1998, Giorgio Ghiglione /* 0052424

HARBOUR CRAFT

Comment: There are large numbers of naval manned harbour craft with MDN, MCN, MBN and MEN numbers. *Argo* (ex-MEN 209) is being used as a Presidential yacht. There is also a ferry *Cheradi* Y 402 at Taranto. Craft with VF numbers are non-naval. *VERIFIED*

MCN 1634   *5/2001, L-G Nilsson /* 0130349

ARGO   *5/2000, Giorgio Ghiglione /* 0104891

19 FLOATING DOCKS

Number	Date	Capacity-tons	Number	Date	Capacity-tons
GO 1	1942	1,000	GO 18B	1920	600
GO 5	1893	100	GO 20	1935	1,600
GO 8	1904	3,800	GO 22-23	1935	1,000
GO 10	1900	2,000	GO 51	1971	2,000
GO 11	1920	2,700	GO 52-54	1988-93	6,000
GO 17	1917	500	GO 55-57	1995-96	850
GO 18A	1920	800	GO 58	1995	2,000

Comment: Stationed at La Spezia *(GO 52)*, Augusta *(GO 53)* and Taranto *(GO 54)*. *VERIFIED*

TRAINING SHIPS

Notes: (1) In addition to the ships listed the LPDs are used in a training role.
(2) There is a requirement for new training ships to replace the Aragosta class but the programme is not funded.

1 SAIL TRAINING SHIP (AXS)

Name	No	Builders	Commissioned
AMERIGO VESPUCCI	A 5312	Castellammare	15 May 1931

Displacement, tons: 3,543 standard; 4,146 full load
Dimensions, feet (metres): 229.5 pp; 270 oa hull; 330 oa bowsprit × 51 × 22 *(70; 82.4; 100 × 15.5 × 7)*
Main machinery: Diesel-electric; 2 Fiat B 306 ESS diesel generators; 2 Marelli motors; 2,000 hp (m) *(1.47 MW)*; 1 shaft
Speed, knots: 10. **Range, n miles:** 5,450 at 6.5 kt
Complement: 243 (13 officers)
Radars: Navigation: 2 SMA SPN-748; I-band.

Comment: Launched on 22 March 1930. Hull, masts and yards are of steel. Sail area, 22,604 sq ft. Extensively refitted at La Spezia Naval Dockyard in 1973 and again in 1984. Used for Naval Academy Summer cruise with up to 150 trainees. *UPDATED*

AMERIGO VESPUCCI   *7/2004*, Ships of the World /* 1044364

1 SAIL TRAINING SHIP (AXS)

Name	No	Builders	Commissioned
PALINURO (ex-*Commandant Louis Richard*)	A 5311	Ch Dubigeon, Nantes	1934

Displacement, tons: 1,042 standard; 1,450 full load
Measurement, tons: 858 gross
Dimensions, feet (metres): 193.5 × 32.8 × 15.7 *(59 × 10 × 4.8)*
Main machinery: 1 GMT A 230.6N diesel; 600 hp *(447 kW)*; 1 shaft
Speed, knots: 7.5. **Range, n miles:** 5,390 at 7.5 kt
Complement: 69 (6 officers)
Radars: Navigation: SPN-748; I-band.

Comment: Barquentine launched in 1934. Purchased in 1951. Rebuilt in 1954-55 and commissioned in Italian Navy on 1 July 1955. She was one of the last two French Grand Bank cod-fishing barquentines. Owned by the Armement Glâtre she was based at St Malo until bought by Italy. Used for seamanship basic training. *UPDATED*

PALINURO   *7/2004*, Diego Quevedo /* 1044376

5 SAIL TRAINING YACHTS (AXS)

Name	No	Builders	Commissioned
CAROLY	A 5302	Baglietto, Varazze	1948
STELLA POLARE	A 5313	Sangermani, Chiavari	8 Oct 1965
CORSARO II	A 5316	Costaguta, Voltri	5 Jan 1961
CAPRICIA	A 5322	Bengt-Plym	1963
ORSA MAGGIORE	A 5323	Tencara, Venezia	1994

Comment: The first three are sail training yachts between 40 and 60 tons with a crew including trainees of about 16. *Capricia* is a yawl of 55 tons and was donated by the Agnelli foundation as replacement for *Cristoforo Colombo II* which was not completed when the shipyard building her went bankrupt. *Capricia* commissioned in the Navy 23 May 1993. *Orsa Maggiore* is a ketch of 70 tons. *VERIFIED*

ORSA MAGGIORE   *7/2003, J Ciślak /* 0570669

3 ARAGOSTA (HAM) CLASS (AXL)

ASTICE A 5379	MITILO A 5380	PORPORA A 5382

Displacement, tons: 188 full load
Dimensions, feet (metres): 106 × 21 × 6 *(32.5 × 6.4 × 1.8)*
Main machinery: 2 Fiat-MTU 12V 493 TY7 diesels; 2,200 hp(m) *(1.62 MW)* sustained; 2 shafts
Speed, knots: 14. **Range, n miles:** 2,000 at 9 kt
Complement: 13 (2 officers)
Radars: Navigation: BX 732; I-band.

Comment: Builders: CRDA, Monfalcone: *Astice*. Picchiotti, Viareggio: *Mitilo*. Costaguta, Voltri: *Porpora*. Similar to the late UK Ham class. All constructed to the order of NATO in 1955-57. Designed armament of one 20 mm gun not mounted. Originally class of 20. Remaining three converted for training 1986. *Porpora* used by the Naval Academy. *Astice* has a modified bridge structure. Plans for transfer to Albania have been shelved. There are no funded replacement plans. ***VERIFIED***

ASTICE 7/2003, *Giorgio Ghiglione* / 0570667

TUGS

32 HARBOUR TUGS (YTM)

RP 101 Y 403 (1972)	RP 113 Y 463 (1978)	RP 125 Y 478 (1983)
RP 102 Y 404 (1972)	RP 114 Y 464 (1980)	RP 126 Y 479 (1983)
RP 103 Y 406 (1974)	RP 115 Y 465 (1980)	RP 127 Y 480 (1984)
RP 104 Y 407 (1974)	RP 116 Y 466 (1980)	RP 128 Y 481 (1984)
RP 105 Y 408 (1974)	RP 118 Y 468 (1980)	RP 129 Y 482 (1984)
RP 106 Y 410 (1974)	RP 119 Y 470 (1980)	RP 130 Y 483 (1985)
RP 108 Y 452 (1975)	RP 120 Y 471 (1980)	RP 131 Y 484 (1985)
RP 109 Y 456 (1975)	RP 121 Y 472 (1984)	RP 132 Y 485 (1985)
RP 110 Y 458 (1975)	RP 122 Y 473 (1981)	RP 133 Y 486 (1985)
RP 111 Y 460 (1975)	RP 123 Y 467 (1981)	RP 134 Y 487 (1985)
RP 112 Y 462 (1975)	RP 124 Y 471 (1981)	

Displacement, tons: 120 full load
Dimensions, feet (metres): 64.9 × 17.1 × 6.9 *(19.8 × 5.2 × 2.1)*
Main machinery: 1 Fiat diesel; 368 hp *(270 kW)*; 1 shaft
Speed, knots: 9.5

Comment: *RP 101-124* built by Visitini, Dorada 1972-81. *RP 125-134* are larger tugs as shown in details. ***UPDATED***

RP 129 11/2004*, *Declerck/Cracco* / 1043182

RP 109 3/2004*, *Giorgio Ghiglione* / 1044377

7 OCEAN TUGS (ATR)

PROMETEO A 5318	POLIFEMO A 5325	SATURNO A 5330
CICLOPE A 5319	GIGANTE A 5328	TENACE A 5365
TITANO A 5324		

Displacement, tons: 658 full load
Dimensions, feet (metres): 127.6 × 32.5 × 12.1 *(38.9 × 9.9 × 3.7)*
Main machinery: 2 GMT B 230.8 M diesels; 3,970 hp(m) *(2.02 MW)* sustained; 2 shafts; LIPS cp props
Speed, knots: 14.5. **Range, n miles:** 3,000 at 14 kt
Complement: 12
Radars: Navigation: SPN-748; I-band.

Comment: Details given are for all except A 5318. Built by CN Ferrari, La Spezia. Completed *Ciclope*, 5 September 1985; *Titano*, 7 December 1985; *Polifemo*, 21 April 1986; *Gigante*, 18 July 1986; *Saturno* 5 April 1988 and *Tenace* 9 July 1988. All fitted with firefighting equipment and two portable submersible pumps. Bollard pull 45 tons. *Prometeo* was completed 14 August 1975 and is slightly larger at 746 tons and has single engine propulsion. ***VERIFIED***

POLIFEMO 9/2002, *Martin Mokrus* / 0528363

9 COASTAL TUGS (YTB)

PORTO FOSSONE Y 413	PORTO EMPEDOCLE Y 421	PORTO FERRAIO Y 425
PORTO TORRES Y 416	PORTO PISANO Y 422	PORTO VENERE Y 426
PORTO CORSINI Y 417	PORTO CONTE Y 423	PORTO SALVO Y 428

Displacement, tons: 412 full load
Measurement, tons: 122 dwt
Dimensions, feet (metres): 106.3 × 27.9 × 12.8 *(32.4 × 8.5 × 3.9)*
Main machinery: 1 GMT B 230.8 M diesels; 1,600 hp(m) *(1.18 MW)* sustained; 1 shaft; cp prop
Speed, knots: 12.7. **Range, n miles:** 4,000 at 12 kt
Complement: 13
Radars: Navigation: GEM BX 132; I-band.

Comment: Details given are for all except Y 436 and 443. Six ordered from CN De Poli (Pellestrina) and further three from Ferbex (Naples) in 1986.
Delivery dates *Porto Salvo* (13 September 1985), *Porto Pisano* (22 October 1985), *Porto Ferraio* (20 July 1985), *Porto Conte* (21 November 1985), *Porto Empedocle* (19 March 1986), *Porto Venere* (16 May 1989), *Porto Fossone* (24 September 1990), *Porto Torres* (16 January 1991) and *Porto Corsini* (4 March 1991).
Fitted for firefighting and anti-pollution. Carry a 1 ton telescopic crane. Based at Taranto, La Spezia, Augusta and La Maddalena. *Porto d'Ischia* transferred to Tunisia in 2002 and *Riva Trigoso* decommissioned. ***VERIFIED***

PORTO TORRES 5/2001, *Giorgio Ghiglione* / 0130337

ARMY

Notes: The following units are operated by the 'Serenissima Amphibious Regiment' in the Venice Lagoons area. EIG means Italian Army Craft and is part of the hull number. Four LCM (EIG 28-31), 60 tons; two LCVP (EIG 26, 27), 13 tons; four recce craft (EIG 32, 33, 48, 49), 5 tons; two command craft (EIG 208, 210), 21.5 tons; one rescue tug (EIG 209), 45 tons; one inshore tanker (EIG 44), 95 tons; one ambulance and rescue craft (EIG 142) and about 70 minor craft (ferries, barges, river boats, rigid inflatable raiders).

ARMY CRAFT 7/1993, *van Ginderen Collection* / 0075865

GOVERNMENT MARITIME FORCES

Notes: Consideration has been given to combine all these forces into one Coast Guard.

CUSTOMS (SERVIZIO NAVALE GUARDIA DI FINANZA)

Notes: (1) This force is operated by the Ministry of Finance but in time of war would come under the command of the Marina Militare. It is divided into 16 naval stations, 30 operational sectors and 28 squadrons. Their task is to patrol ports, lakes and rivers. The total manpower is 5,400. Nearly all the larger craft are armed. There are 12 P-166 and 3 ATR 42 patrol aircraft plus 53 Hughes NH 500, 21 Agusta A 109 and 18 Agusta Bell AB 412 helicopters.
(2) In addition to the classes detailed below there are:

(a)   180 inshore patrol craft of between 27 and 3 tons and 23 to 64 kt, including seized smugglers' craft.
(b)   100 lake patrol craft.

(3) The replacement programme for the Meatini class has been shelved.
(4) Fourteen 13 m 2000 class are being constructed by Intermarine for delivery by December 2005.
(5) 33 units of the 12 m V600 class, capable of 54 kt are under construction.

ATR-42   *7/2003, Adolfo Ortigueira Gil* / 0570665

1 TRAINING SHIP (AX)

GIORGIO CINI

Displacement, tons: 800 full load
Dimensions, feet (metres): 172.2 × 32.8 × 9.5 *(54.0 × 10.0 × 2.9)*
Main machinery: 1 Fiat B306-SS diesel; 1,500 hp *(1.1 MW)*; 1 shaft
Speed, knots: 14. **Range, n miles:** 800 at 14 kt
Radars: Navigation: 2 BX-732; I-band.

Comment: Former merchant navy training ship acquired in 1981 for training role.
VERIFIED

GIORGIO CINI   *3/2002, Adolfo Ortigueira Gil* / 0570670

3 ANTONIO ZARA CLASS (PB)

Name	No	Builders	Commissioned
ANTONIO ZARA	P 01	Fincantieri, Muggiano	23 Feb 1990
GIUSEPPE VIZZARI	P 02	Fincantieri, Muggiano	27 Apr 1990
GIOVANNI DENARO	P 03	Fincantieri, Muggiano	20 Mar 1998

Displacement, tons: 340 full load
Dimensions, feet (metres): 167 × 24.6 × 6.2 *(51 × 7.5 × 1.9)*
Main machinery: 2 GMT BL 230.12 M diesels; 5,956 hp(m) *(4.38 MW)* sustained; 2 shafts
   4 MTU 16V 396 TB94 diesels; 13,029 hp(m) *(9.58 MW)* sustained; 2 shafts (P 03)
Speed, knots: 27; 35 (P 03). **Range, n miles:** 3,800 at 15 kt
Complement: 33 (3 officers)
Guns: 1 or 2 Breda 30 mm/70 (single or twin). 2—7.62 mm MGs.
Weapons control: Selenia Pegaso 2 AESN Medusa (P 03) optronic director.
Radars: Surface search: Gemant 2 ARPA and SPN 749; I-band.

Comment: Similar to the Ratcharit class built for Thailand in 1976-79. First pair ordered in August 1987. Third ordered in October 1995 with more powerful engines and with a modified armament of a single 30 mm gun with a Medusa optronic director. All are being fitted with an infra-red search and surveillance sensor (AMS SVIR).
UPDATED

ANTONIO ZARA   *6/2001, L-G Nilsson* / 0130338

5 + 2 MAZZEI CLASS (PB/YXT)

Name	No	Builders	Commissioned
MAZZEI	G 01	Intermarine, Sarzana	Apr 1998
VACCARO	G 02	Intermarine, Sarzana	May 1998
DI BARTOLO	G 03	Intermarine, Sarzana	Oct 2003
AVALLONE	G 04	Intermarine, Sarzana	Dec 2003
OLTRAMONTI	G 05	Intermarine, Sarzana	Apr 2004
BARBARISO	G 06	Intermarine, Sarzana	2005
PAOLINI	G 07	Intermarine, Sarzana	2005

Displacement, tons: 115 full load
Dimensions, feet (metres): 116.5 × 24.8 × 3.6 *(35.5 × 7.6 × 1.1)*
Main machinery: 2 MTU 16V 396 TB94 diesels; 5,800 hp(m) *(4.26 MW)* sustained; 2 shafts
Speed, knots: 35. **Range, n miles:** 700 at 18 kt
Complement: 17 plus 18 trainees
Guns: 1 Breda Mauser 30 mm/70. 2—7.62 mm MGs.
Weapons control: Elsag Medusa optronic director.
Radars: Surface search: GEM 3072A ARPA; I-band.
Navigation: GEM 1410; I-band.

Comment: Based on the Bigliani class but with an extended hull. G 01 and G 02 used as training ships. G 03-05 are patrol craft. All are being fitted with an infra-red search and surveillance sensor (AMS SVIR).
UPDATED

MAZZEI   *7/2001, Giorgio Ghiglione* / 0130339

41 MEATINI CLASS (PB)

G 13-66 series

Displacement, tons: 40 full load
Dimensions, feet (metres): 65.9 × 17.1 × 3.3 *(20.1 × 5.2 × 1)*
Main machinery: 2 CRM 18D/52 diesels; 2,500 hp(m) *(1.84 MW)*; 2 shafts
Speed, knots: 34. **Range, n miles:** 550 at 20 kt
Complement: 11 (1 officer)
Guns: 1—12.7 mm MG.
Radars: Surface search: 1 GEM 1210; I-band.

Comment: Fifty-six of the class built from 1970 to 1978. Numbers are reducing. Replacement by new craft is in progress.
UPDATED

DARIDA   *4/2004*, Giorgio Ghiglione* / 1044378

36 V 5000/6000 CLASS (FAST PATROL CRAFT) (HSIC)

V 5000-5020	V 5100	V 6000-6012	V 6100

Displacement, tons: 16 (V 6000), 27 (V 5000) full load
Dimensions, feet (metres): 53.8 × 9.2 × 2.6 *(16.4 × 2.8 × 0.8)*
Main machinery: 4 Seatek 6-4V-10D diesels; 2,856 hp(m) *(2.13 MW)* sustained; 4 surface-piercing propellers
   2 MTU 8V 396 TE94 diesels (V 5000)
Speed, knots: 70 (V 6000); 52 (V 5000)
Complement: 4
Radars: Surface search: I-band.

Comment: V 6003-6012 were delivered in 2002-03. V 6001-6002 are smaller prototype craft with three engines and a top speed of 64 kt.
VERIFIED

V 6001   *2/2000, Guardia di Finanzia* / 0104897

V 5006   *6/2001, Guardia di Finanzia* / 0130143

26 CORRUBIA CLASS (PBF)

CORRUBIA G 90	FAIS G 97	APRUZZI G 104	LETIZIA G 110
GIUDICE G 91	FELICIANI G 98	BALLALI G 105	MAZZARELLA G 111
ALBERTI G 92	GARZONI G 99	BOVIENZO G 106	NIOI G 112
ANGELINI G 93	LIPPI G 100	CARRECA G 107	PARTIPILO G 113
CAPPELLETTI G 94	LOMBARDI G 101	CONVERSANO G 108	PULEO G 114
CIORLIERI G 95	MICCOLI G 102	INZERILLI G 109	ZANNOTTI G 115
D'AMATO G 96	TREZZA G 103		

Displacement, tons: 92 full load
Dimensions, feet (metres): 87.9 × 24.9 × 3.9 *(26.8 × 7.6 × 1.2)*
Main machinery: 2 Isotta Fraschini ID 36 SS 16V diesels; 6,400 hp(m) *(4.7 MW)*; 2 shafts (G 90-91)
2 MTU 16V 396 TB94; 5,800 hp(m) *(4.26 MW)* sustained; 2 shafts (G 92-103)
Speed, knots: 43. **Range, n miles:** 700 at 20 kt
Complement: 12 (1 officer)
Guns: 1 Breda Mauser 30 mm/70 (G 90-103). 1 Astra 20 mm (G 104-115). 2—7.62 mm MGs.
Weapons control: Elsag Medusa optronic director.
Radars: Surface search: GEM 3072A ARPA; I-band.
Navigation: GEM 1210; I-band.

Comment: First two built by Cantieri del Golfo, Gaeta and delivered in 1990. Others built by Cantieri del Golfo (G 92-100), and Crestitalia (G 101-103), and Intermarine from 1995 onwards. G 115 completed in 1999. There are minor structural differences between the first series (G 90-91), the second series (G 92-103) and the third batch (G 104-115). All are being fitted with an infra-red search and surveillance sensor (AMS SVIR). *UPDATED*

MAZZARELLA
3/2002, Giorgio Ghiglione / 0528362

22 + 2 BIGLIANI CLASS (PB)

OTTONELLI G 78	SMALTO G 84	LAGANÀ G 116	LA SPINA G 122
BARLETTA G 79	FORTUNA G 85	SANNA G 117	SALONE G 123
BIGLIANI G 80	BUONOCORE G 86	INZUCCHI G 118	CAVATORTO G 124
CAVAGLIA G 81	SQUITIERI G 87	VITALI G 119	FUSCO G 125
GALIANO G 82	LA MALFA G 88	CALABRESE G 120	— G 126
MACCHI G 83	ROSATI G 89	URSO G 121	— G 127

Displacement, tons: 87 full load
Dimensions, feet (metres): 86.6 × 23 × 3.6 *(26.4 × 7 × 1.1)*
Main machinery: 2 MTU 16V 396 TB94 diesels; 6,850 hp(m) *(5.12 MW)* sustained; 2 shafts
Speed, knots: 42. **Range, n miles:** 770 at 18 kt
Complement: 12
Guns: 1 Breda Mauser 30 mm/80. 2—7.62 mm MGs.
Combat data systems: AMS IPNS.
Weapons control: Elsag Medusa Mk 4 optronic director.
Radars: Surface search: GEM 3072A ARPA; I-band.
Navigation: GEM 1410; I-band.

Comment: First eight built by Crestitalia and delivered from October 1987 to September 1992. Three more were ordered from Crestitalia/Intermarine in October 1994 and were delivered from December 1996 to April 1997. A fourth was delivered in late 1999. There are minor structural differences between Series I (G 80-81), Series II (G 82-87) and Series III (G 78-79, G 88-89). Twelve series IV (G 116-127) craft ordered from Intermarine, Sarzana, for delivery in 2004-06. These include Kevlar armour and are fitted with a remote-control Breda 12.7 mm gun and 40 mm grenade launcher. All are being fitted with an infra-red search and surveillance sensor (AMS SVIR). *UPDATED*

OTTONELLI
1/2000, Giorgio Ghiglione / 0075866

COAST GUARD (GUARDIA COSTIERA—CAPITANERIE DI PORTO)

Notes: This is a force which is affiliated with the Marina Militare under whose command it would be placed in an emergency. The Coast Guard denomination was given after the Sea Protection Law in 1988. The force is responsible for the Italian Maritime Rescue Co-ordination Centre (MRCC) in Rome and 13 sub-centres (MRSC). The SAR network consists of 109 stations, three air stations and one helicopter station. All vessels have a red diagonal stripe painted on the white hull and many are armed with 7.62 mm MGs. There are some 10,500 naval personnel including 1,200 officers of which about half are doing national service. Ranks are the same as the Navy. In addition to the Saettia class (detailed separately), the following craft are in service. All have the prefix CP (Capitaneria di Porto):
(a) SAR craft: *Giulio Ingianni* CP 409 (205 tons); *Antonio Scialoja* CP 406, *Michele Lolini* CP 407, *Mario Grabar* CP 408 (136 tons), *Oreste Cavallari* CP 401, *Renato Pennetti* CP 402, *Walter Fachin*

ANTONIO SCIALOJA
8/2004*, Paolo Marsan / 1044379

CP 285
7/2004*, Paolo Marsan / 1044380

CP 256
7/2002, Martin Mokrus / 0528359

CP 403, *Gaetano Magliano* CP 404 (100 tons), *Bruno Gregoretti* CP 312 (65 tons); *Dante Novaro* CP 313 (57 tons); CP 314-318 (45 tons).
(b) Fast Patrol craft: CP 265-292 (54 tons), CP 262 (30 tons), CP 246-253 (23 tons), CP 254-260 (22 tons), CP 454-456 (19.4 tons).
(c) Inshore Patrol craft: 408 craft of between 3 and 15 tons. CP 2201-2205 (15 tons), CP 2084-2103 (12 tons), CP 2001-2009, 2011-2015, 2017-2083, 2201-2205 (15 tons); CP 829-831, 836-838 (14.6 tons); CP 825-828, 832-835 (12.5 tons); CP 839, 862, 872-881, 884-889 (13.3 tons), CP 863, 871, 882-883, 890-892 (10 tons); CP 814-824 (12.5 tons); CP 801-813 (9 tons), CP 701-712 (6 tons), CP 512-523, 540-564 (7.5 tons); CP 6001-6022 (3.7 tons), CP 1001-1006 (5.4 tons), CP 601-605 (3 tons), CG 101 class, 64 CG 20 RHIB.
(d) Aircraft include 14 Piaggio P 166 DL3-SEM and three ATR 42MP maritime patrol and 12 Griffon AB-412-CP helicopters.
(e) CP 451 is a 1,278 ton training ship (ex-US ATF *Bannock*); *Barbara* CP 452 (190 tons) is a former naval Range Safety patrol craft which recommissioned in late 1999.
(f) CP 210 and CP 211 are airboats used for SAR in the Venice Lagoon area.

6 SAETTIA CLASS (SAR)

Name	No	Builders	Commissioned
SAETTIA	CP 901	Fincantieri, Muggiano	Dec 1985
UBALDO DICIOTTI	CP 902	Fincantieri, Muggiano	20 July 2002
LUIGI DATTILO	CP 903	Fincantieri, Muggiano	28 Nov 2002
MICHELE FIORILLO	CP 904	Fincantieri, Muggiano	7 Apr 2003
ANTONIO PELUSO	CP 905	Fincantieri, Muggiano	2 July 2003
ORAZIO CORSI	CP 906	Fincantieri, Muggiano	7 Feb 2004

Displacement, tons: 427 full load
Dimensions, feet (metres): 173.3 × 26.6 × 6.6 *(52.8 × 8.1 × 2.0)*
Main machinery: 4 Isotta Fraschini V1716T2MSD diesels; 12,660 hp *(9.44 MW)*; 4 cp props; bow thruster
Speed, knots: 29. **Range, n miles:** 1,800 at 18 kt
Complement: 30 (2 officers)
Guns: 1 Oerliken 20 mm/70.
Weapons control: Eurocontrol optronic sensor.
Radars: Surface search: SPN 753; I-band.

Comment: Details are for CP 902-906 which were ordered on 29 June 2000. CP 901 was built as an attack missile craft demonstrator by Fincantieri in 1984 and was later taken over by the Coast Guard on 20 July 1999. 30 tons lighter and with some structural differences, she is powered by 4 MTU 16V538 TB93 engines providing 17,598 hp and a top speed of 40 kt. She is armed with an Otobreda 25 mm gun. All the vessels form a 'Squadrilla' based at Messina, Sicily, whose role is fishery protection and immigration control. *UPDATED*

MICHELE FIORILLO
9/2004*, Adolfo Ortigueira Gil / 1044381

POLICE (SERVIZIO NAVALE CARABINIERI)

Notes: (1) The Carabinieri established its maritime force in 1969 and has some 600 personnel. There are 179 craft in service or building which operate in coastal waters within the 3 mile limit and in inshore waters. Craft currently in service include: 20—800 class of 28 tons; 2—700 class of 22 tons; 6—600 class of 12 tons; 27 N 500 class of 6 tons; 3 S 500 class of 18 tons; 65—200 class of 2 tons, 28 minor craft and 30 RHIBs.
Most are capable of 20 to 25 kt except the 800 class at 35 kt.
(2) There is also a Sea Police Force of the State. All craft have POLIZIA written on the side. Vessels include 37 Squalo class of 14 tons, 4 Nelson class of 11 tons, 7 Intermarine class of 8.4 tons, 37 Crestitalia class of 6 tons and 25 Aquamaster/Drago classes of 3 tons. Speeds vary between 23 and 45 kt.
(3) A programme to replace all coastal craft by 2012 is in progress.

803
4/2004*, Bob Fildes / 1044382

Jamaica

Country Overview

Jamaica gained independence in 1962; the British monarch, represented by a governor-general, is head of state. The island country (area 4,244 square miles), third-largest of the Greater Antilles, is situated south of Cuba and has a 552 n mile coastline with the Caribbean Sea. Kingston is the capital, largest town and principal port. An archipelagic state, territorial seas (12 n miles) are claimed. A 200 n mile Exclusive Economic Zone (EEZ) has been claimed but the limits are not defined.

Headquarters Appointments

Commanding Officer Coast Guard:
Commander Sydney R Innis, MVO

Aviation

Three fixed-wing aircraft (Cessna 2 1-M, Pilatus-Britten-Norman BN-2A and a Beech 100) are used for coastal patrol and seven helicopters (three Bell 412 and four Eurocopter AS 355N) are used for SAR.

Personnel

(a) 2005: 216 (18 officers) Regulars
(b) 55 (14 officers) Reserve Forces

Bases

Main: HMJS *Cagway*, Port Royal
Coastguard: Discovery Bay, Pedro Cays, Port Antonio, Montego Bay and Black River

COAST GUARD

Notes: There are also two Boston Whalers CG 091 and CG 092 built in 1992.

0 + 3 DAMEN STAN PATROL 4207 (PB)

Displacement, tons: 205
Dimensions, feet (metres): 140.4 × 23.3 × 8.3 *(42.8 × 7.11 × 2.52)*
Main machinery: 2 Caterpillar 3516B DI-TA; 5,600 hp *(4.17 MW);* 2 cp props
Speed, knots: 26
Complement: To be announced
Guns: To be announced.

Comment: Contract signed on 21 April 2004 with Damen Shipyard Gorinchem for construction of three Damen 4207 offshore patrol craft. The first to be delivered in late 2005 and following vessels at six month intervals. Details are based on those in UK Customs service.
NEW ENTRY

1 FORT CLASS (PB)

Name	No	Builders	Commissioned
FORT CHARLES	P 7	Sewart Seacraft Inc, Berwick	Sep 1974

Displacement, tons: 130 full load
Dimensions, feet (metres): 116 × 22 × 7 *(35.3 × 6.7 × 2.1)*
Main machinery: 2 MTU 16V 538 TB90 diesels; 6,000 hp(m) *(4.41 MW)* sustained; 2 shafts
Speed, knots: 32. **Range, n miles:** 1,200 at 18 kt
Complement: 16 (3 officers)
Guns: 1 Oerlikon 20 mm. 2—12.7 mm MGs.
Radars: Surface search: Sperry 4016; I-band.

Comment: Of all-aluminium construction, launched July 1974. Underwent refit at Jacksonville, Florida, in 1980-81 which included extensive modifications to the bow resulting in increased length. Refitted again in 1987-88. A further refit is not planned. Accommodation for 18 soldiers and may be used as 18-bed mobile hospital in an emergency.
UPDATED

4 DAUNTLESS CLASS (INSHORE PATROL CRAFT) (PB)

CG 121   CG 122   CG 123   CG 124

Displacement, tons: 11 full load
Dimensions, feet (metres): 40 × 14 × 4.3 *(12.2 × 4.3 × 1.3)*
Main machinery: 2 Caterpillar 3208TA diesels; 870 hp *(650 kW);* 2 shafts
Speed, knots: 27. **Range, n miles:** 600 at 18 kt
Complement: 5
Guns: 1—7.62 mm MG (can be carried).
Radars: Surface search: Raytheon 40X; I-band.

Comment: Delivered in September and November 1992, January 1993 and May 1994. Built by SeaArk Marine, Monticello. Aluminium construction. Craft of this class have been distributed throughout the Caribbean under FMS funding.
VERIFIED

CG 121
10/2000 / 0121383

3 FAST COASTAL INTERCEPTORS (PBF)

CG 131, 132, 133

Displacement, tons: 11 full load
Dimensions, feet (metres): 44 × 10.5 × 3 *(13.4 × 3.2 × 0.92)*
Main machinery: 2 Caterpillar 3196 diesels; 1,140 hp *(850 kW);* two twin disc waterjets
Speed, knots: 37. **Range, n miles:** 400 at 20 kt
Complement: 6
Guns: 1—7.62 mm M60 MG.
Radars: Surface search: Raytheon Pathfinder; I-band.

Comments: Aluminium construction. Built by Silver Ships, Mobile, Alabama. Funded by the US State Department, Narcotics Affairs Section. Delivered in March 2003.
VERIFIED

FORT CHARLES
6/1999, JDFCG / 0080125

CG 131
6/2003, JDFCG / 0568335

2 POINT CLASS (PB)

Name	No	Builders	Commissioned
SAVANNAH POINT (ex-*Point Nowell*)	CG 251 (ex-82363)	CG Yard, Maryland	1 June 1967
BELMONT POINT (ex-*Point Barnes*)	CG 252 (ex-82371)	J Martinac, Tacoma	21 Apr 1970

Displacement, tons: 67 full load
Dimensions, feet (metres): 83 × 17.2 × 5.8 *(25.3 × 5.3 × 1.8)*
Main machinery: 2 Caterpillar diesels; 1,600 hp *(1.19 MW)*; 2 shafts
Speed, knots: 22. **Range, n miles:** 1,200 at 8 kt
Complement: 10
Guns: 2—12.7 mm MGs.
Radars: Surface search: Hughes/Furuno SPS-73; I-band.

Comment: Transferred from US Coast Guard on 15 October 1999 and 21 January 2000 respectively. These ships are spread throughout the Caribbean navies. **VERIFIED**

SAVANNAH POINT 10/1999, JDFCG / 0080127

3 OFFSHORE PERFORMANCE TYPE (INSHORE PATROL CRAFT) (HSIC)

CG 101   CG 102   CG 103

Displacement, tons: 3 full load
Dimensions, feet (metres): 33 × 8 × 1.8 *(10.1 × 2.4 × 0.6)*
Main machinery: 2 Johnson OMC outboards; 450 hp *(336 kW)*
Speed, knots: 48
Complement: 3
Guns: 1—7.62 mm MG (can be carried).
Radars: Surface search: Raytheon 40X; I-band.

Comment: Delivered in April 1992. Built by Offshore Performance Marine, Miami. Used in the anti-narcotics role. **VERIFIED**

CG 102 5/1992, JDFCG

1 HERO CLASS (PB)

Name	No	Builders	Commissioned
PAUL BOGLE	P 8	Lantana Boatyard Inc, FL	17 Sep 1985

Displacement, tons: 93 full load
Dimensions, feet (metres): 105 × 20.6 × 7 *(32 × 6.3 × 2.1)*
Main machinery: 3 MTU 8V 396 TB93 diesels; 3,270 hp(m) *(2.4 MW)* sustained; 3 shafts
Speed, knots: 32
Complement: 20 (4 officers)
Guns: 1 Oerlikon 20 mm. 2—12.7 mm MGs.
Radars: Surface search: Furuno 2400; I-band.
Navigation: Sperry 4016; I-band.

Comment: Of all-aluminium construction, launched in 1984. *Paul Bogle* was originally intended for Honduras as the third of the Guardian class. Similar to patrol craft in Honduras and Grenada navies. Refitted in March 1998 at Network Marine, Louisiana and to be further refitted in 2004-05 by Damen Shipyards, Gorinchem. **UPDATED**

PAUL BOGLE 6/1999, JDFCG / 0080126

Japan
MARITIME SELF-DEFENCE FORCE (MSDF)
KAIJOH JIEI-TAI

Country Overview

Japan is a constitutional monarchy in East Asia that comprises four main islands: Hokkaido, Honshu, Shikoku and Kyushu. It also includes the Ryukyu Islands to the southwest and more than 1,000 lesser islands. The sovereignty of the South Kuril Islands (Etorofu, Kunashiri, Shikotan and the Habomai Group) is disputed with Russia. With an overall area of 145,850 square miles it has a coastline of 16,065 n miles, with the Pacific Ocean, Sea of Japan, the La Perouse Strait (which separates it from Sakhalin Island), Sea of Okhotsk, East China Sea and the Korea Strait (which separates it from South Korea). The capital and largest city is Tokyo while the principal ports are Yokohama, Osaka and Kobe. Territorial seas of 12 n miles (3 n miles in Korea Strait) are claimed. A 200 n mile EEZ has also been claimed but the limits have not been defined.

Headquarters Appointments

Chief of Staff, Maritime Self-Defence Force:
  Admiral Takashi Saitou
Commander-in-Chief, Self-Defence Fleet:
  Vice Admiral Eiichi Nakashima

Senior Appointments

Commander Fleet Escort Force:
  Vice Admiral Nobuharu Yasui
Commander Fleet Air Force:
  Vice Admiral Toru Takahashi
Commander Fleet Submarine Force:
  Vice Admiral Tsutomu Tamura

Diplomatic Representation

Defence (Naval) Attaché in London:
  Captain Takaki Mizuma

Personnel

2005: 45,842 (including Naval Air) plus 3,566 civilians

Fleet Escort Force (Yokosuka)
  Tachikaze (DDG 168) Flagship

Escort Flotilla 1 (Yokosuka)
  Shirane (DDH 143)
1st Destroyer Division (Y)
  Murasame (DD 101)
  Harusame (DD 102)
  Ikazuchi (DD 107)
5th Destroyer Division (Y)
  Takanami (DD 110)
  Oonami (DD 111)
61st Destroyer Division (Y)
  Hatakaze (DDG 171)
  Kirishima (DDG 174)

Escort Flotilla 2 (Sasebo)
  Kurama (DDH 144)
2nd Destroyer Division (S)
  Makinami (DD 112)
  Yuudachi (DD 103)
  Sawagiri (DD 157)
6th Destroyer Division (S)
  Kirisame (DD 104)
  Ariake (DD 109)
62nd Destroyer Division (S)
  Sawakaze (DDG 170)
  Kongou (DDG 173)

Escort Flotilla 3 (Maizuru)
  Haruna (DDH 141)
3rd Destroyer Division (M)
  Hamayuki (DD 126)
  Amagiri (DD 154)
7th Destroyer Division (O)
  Yuugiri (DD 153)
  Hamagiri (DD 155)
  Setogiri (DD 156)
63rd Destroyer Division (M)
  Shimakaze (DDG 172)
  Myoukou (DDG 175)

Escort Flotilla 4 (Kure)
  Hiei (DDH 142)
4th Destroyer Division (K)
  Inazuma (DD 105)
  Samidare (DD 106)
  Akebono (DD 108)
8th Destroyer Division (K)
  Umigiri (DD 158)
  Asagiri (DD 151)
64th Destroyer Division (S)
  Asakaze (DDG 169)
  Choukai (DDG 176)

Organisation of the Major Surface Units of Japan (MSDF)

In addition to the Escort Force, there are two Submarine Flotillas (Kure and Yokosuka), one Minesweeper Flotilla (Yokosuka), which are to be merged, and five District Flotillas (Yokosuka, Maizuru, Ohminato, Sasebo and Kure). The District Flotillas are comprised of three or five destroyers, a LST or a LSU and a number of MSC and patrol craft.

Bases

Naval-Yokosuka, Kure, Sasebo, Maizuru, Ohminato
Naval Air-Atsugi, Hachinohe, Iwakuni, Kanoya, Komatsujima, Naha, Ozuki, Ohminato, Ohmura, Shimofusa, Tateyama, Tokushima, Ioujima

Coast Defence

The Army controls 92 SSM-1 truck-mounted sextuple launchers.

Strength of the Fleet (31 March 2005)

Type	Active (Auxiliary)	Building (Projected)
Submarines	16 (2)	4 (1)
Destroyers	43	4 (1)
Frigates	9	—
Patrol Forces	9	—
LSTs	4	—
LCUs	4	—
LCACs	6	—
Landing Craft (LCM)	13	—
MCM Tenders/Controllers	4	—
Minesweepers—Ocean	3	—
Minesweepers—Coastal	26	3 (2)
Major Auxiliaries	33	(2)

New Construction Programme (Warships)

2003	1—7,700 ton DDG, 1—2,700 ton SS, 1—510 ton MSC.
2004	1—13,500 ton DDH, 1—2,900 ton SS, 1—570 ton MSC.
2005	1—2, 900 ton SS, 1—570 ton MSC, 2—980 ton AMS.

Naval Air Force

16 Air Patrol Sqns: P-3C, EP-3, OP-3C, HSS-2B, SH-60J
Six Air Training Sqns: P-3C, YS-11, TC-90, T-5, OH-6D, HSS-2B, SH-60J
One Air Training Support Squadron: U-36A, UP-3D, LC-90
One Transport Sqn: YS-11, LC-90
One MCM Sqn: MH-53E
Air Training Command (Shimofusa)
Air Wings at Kanoya (Wing 1), Hachinohe (Wing 2), Atsugi (Wing 4), Naha (Wing 5), Tateyama (Wing 21), Ohmura (Wing 22), Iwakuni (Wing 31)

DELETIONS and CONVERSIONS

Submarines

2003	*Okishio*
2004	*Akishio*
2005	*Takeshio*

Destroyers

2002	*Takatsuki*
2003	*Kikuzuki*
2005	*Asagiri* (converted)

Frigates

2003	*Noshiro*

Amphibious Forces

2002	*Satsuma*
2005	*Nemuro*

Mine Warfare Forces

2002	*Nuwajima, Etajima, Niijima, Hahajma* (converted)
2004	*Yakushima*
2005	*Himeshima, Moroshima*

Auxiliaries

2002	*Hayase, Ninoshima, Miyajima, ASU-85*
2003	*Aokumo*
2005	*Sagami*

Training Ships

2005	*Akigumo*

PENNANT LIST

Submarines—Patrol

SS 501	— (bldg)
SS 581	Yukishio
SS 582	Sachishio
SS 583	Harushio
SS 584	Natsushio
SS 585	Hayashio
SS 586	Arashio
SS 587	Wakashio
SS 588	Fuyushio
SS 590	Oyashio
SS 591	Michishio
SS 592	Uzushio
SS 593	Makishio
SS 594	Isoshio
SS 595	Narushio
SS 596	Kuroshio
SS 597	Takashio
SS 598	Yaeshio (bldg)
SS 599	— (bldg)

Submarines—Auxiliary

TSS 3601	Asashio
TSS 3604	Hamashio

Destroyers

DD 101	Murasame
DD 102	Harusame
DD 103	Yuudachi
DD 104	Kirisame
DD 105	Inazuma
DD 106	Samidare
DD 107	Ikazuchi
DD 108	Akebono
DD 109	Ariake
DD 110	Takanami
DD 111	Oonami
DD 112	Makinami
DD 113	Sazanami
DD 114	Suzunami (bldg)
DDK 121	Yuugumo
DD 122	Hatsuyuki
DD 123	Shirayuki
DD 124	Mineyuki
DD 125	Sawayuki
DD 126	Hamayuki
DD 127	Isoyuki
DD 128	Haruyuki
DD 129	Yamayuki
DD 130	Matsuyuki
DD 131	Setoyuki
DD 132	Asayuki
DDH 141	Haruna
DDH 142	Hiei
DDH 143	Shirane
DDH 144	Kurama
DDH 145	— (bldg)
DD 153	Yuugiri
DD 154	Amagiri
DD 155	Hamagiri
DD 156	Setogiri
DD 157	Sawagiri
DD 158	Umigiri
DDG 168	Tachikaze
DDG 169	Asakaze
DDG 170	Sawakaze
DDG 171	Hatakaze
DDG 172	Shimakaze
DDG 173	Kongou
DDG 174	Kirishima
DDG 175	Myoukou
DDG 176	Choukai
DDG 177	— (bldg)
DDG 178	— (bldg)

Frigates

DE 226	Ishikari
DE 227	Yuubari
DE 228	Yuubetsu
DE 229	Abukuma
DE 230	Jintsu
DE 231	Ooyodo
DE 232	Sendai
DE 233	Chikuma
DE 234	Tone

Patrol Forces

PG 821	PG 01
PG 822	PG 02
PG 823	PG 03
PG 824	Hayabusa
PG 825	Wakataka
PG 826	Ootaka
PG 827	Kumataka
PG 828	Umitaka
PG 829	Shirataka

Minehunters/Sweepers—Ocean

MSO 301	Yaeyama
MSO 302	Tsushima
MSO 303	Hachijyo

Minesweepers—Coastal

MSC 666	Ogishima
MSC 668	Yurishima
MSC 669	Hikoshima
MSC 670	Awashima
MSC 671	Sakushima
MSC 672	Uwajima
MSC 673	Ieshima
MSC 674	Tsukishima
MSC 675	Maejima
MSC 676	Kumejima
MSC 677	Makishima
MSC 678	Tobishima
MSC 679	Yugeshima
MSC 680	Nagashima
MSC 681	Sugashima
MSC 682	Notojima
MSC 683	Tsunoshima
MSC 684	Naoshima
MSC 685	Toyoshima
MSC 686	Ukushima
MSC 687	Izushima
MSC 688	Aishima
MSC 689	Aoshima
MSC 690	Miyajima
MSC 691	Shishijima (bldg)
MSC 692	— (bldg)

MCM Tenders/Control Ships

MCL 724	Hahajima
MCL 725	Kamishima

MST 463	Uraga
MST 464	Bungo

Amphibious Forces

LCU 2001	Yusotei-Ichi-Go
LCU 2002	Yusotei-Ni-Go
LST 4001	Oosumi
LST 4002	Shimokita
LST 4003	Kunisaki
LSU 4171	Yura
LSU 4172	Noto

Submarine Depot/Rescue Ships

AS 405	Chiyoda
ASR 403	Chihaya

Fleet Support Ships

AOE 422	Towada
AOE 423	Tokiwa
AOE 424	Hamana
AOE 425	Mashuu
AOE 426	Oumi

Training Ships

TV 3508	Kashima
TV 3513	Shimayuki
TV 3515	Yamagiri
TV 3516	Asagiri

Training Support Ships

ATS 4202	Kurobe
ATS 4203	Tenryu
AMS 4301	Hiuchi
AMS 4302	Suou
AMS 4303	Amakusa

Cable Repair Ship

ARC 482	Muroto

PENNANT LIST

Icebreakers		Survey and Research Ships		Ocean Surveillance Ships		Tenders	
AGB 5002	Shirase	AGS 5102	Futami	AOS 5201	Hibiki	ASY 91	Hashidate
		AGS 5103	Suma	AOS 5202	Harima	YDT 01-06	—
		AGS 5104	Wakasa				
		AGS 5105	Nichinan				
		ASE 6101	Kurihama				
		ASE 6102	Asuka				

SUBMARINES

0 + 2 IMPROVED OYASHIO CLASS (SSK)

Name	No	Builders	Laid down	Launched	Commissioned
—	—	—	Mar 2005	2007	Mar 2009
—	—	—	2006	2008	2010

Displacement, tons: 2,900 dived
Dimensions, feet (metres): 275.6 × ? × ?
   (84.0 × ? × ?)
Main machinery: Diesel-stirling-electric; 2 diesels; 4 Kockums Stirling AIP; 1 motor; 1 shaft
Speed, knots: 20 dived; 12 surfaced
Range, n miles: To be announced
Complement: To be announced

Missiles: SSM: McDonnell Douglas Sub-Harpoon; active radar homing to 130 km *(70 n miles)* at 0.9 Mach; warhead 227 kg.
Torpedoes: 6—21 in *(533 mm)* bow tubes. Japanese Type 89; wire-guided (option); active/passive homing to 50 km *(27 n miles)* at 40/55 kt; warhead 267 kg. Type 80 ASW. SSM and torpedoes (total unknown).
Countermeasures: To be announced.
Weapons control: To be announced.
Radars: Surface search: JRS ZPS 6; I-band.

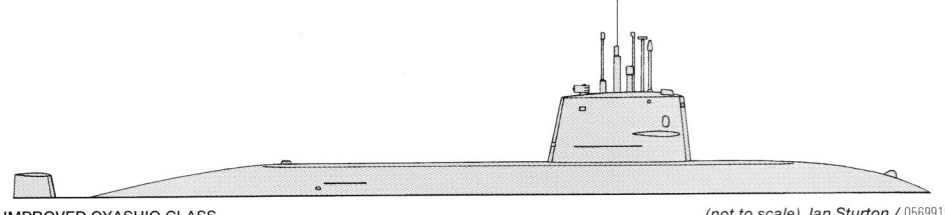

IMPROVED OYASHIO CLASS   *(not to scale), Ian Sturton* / 0569919

Sonars: Hughes/OKI ZQQ 5B/6; hull and flank arrays; active/ passive search and attack; medium/low frequency. Towed array.

Programmes: First of new class authorised in FY04 budget.
                                                       UPDATED

8 + 3 OYASHIO CLASS (SSK)

Name	No	Builders	Laid down	Launched	Commissioned
OYASHIO	SS 590	Kawasaki, Kobe	26 Jan 1994	15 Oct 1996	16 Mar 1998
MICHISHIO	SS 591	Mitsubishi, Kobe	16 Feb 1995	18 Sep 1997	10 Mar 1999
UZUSHIO	SS 592	Kawasaki, Kobe	6 Mar 1996	26 Nov 1998	9 Mar 2000
MAKISHIO	SS 593	Mitsubishi, Kobe	26 Mar 1997	22 Sep 1999	29 Mar 2001
ISOSHIO	SS 594	Kawasaki, Kobe	9 Mar 1998	27 Nov 2000	14 Mar 2002
NARUSHIO	SS 595	Mitsubishi, Kobe	2 Apr 1999	4 Oct 2001	3 Mar 2003
KUROSHIO	SS 596	Kawasaki, Kobe	27 Mar 2000	23 Oct 2002	8 Mar 2004
TAKASHIO	SS 597	Mitsubishi, Kobe	30 Jan 2001	1 Oct 2003	9 Mar 2005
YAESHIO	SS 598	Kawasaki, Kobe	15 Jan 2002	4 Nov 2004	Mar 2006
—	SS 599	Mitsubishi, Kobe	23 Jan 2003	Sep 2005	Mar 2007
—	SS 501	Kawasaki, Kobe	23 Feb 2004	2006	2008

Displacement, tons: 2,750 standard; 3,000 dived
Dimensions, feet (metres): 268 × 29.2 × 24.3
   (81.7 × 8.9 × 7.4)
Main machinery: Diesel-electric; 2 Kawasaki 12V25S diesels; 5,520 hp(m) *(4.1 MW)*; 2 Kawasaki alternators; 3.7 MW; 2 Toshiba motors; 7,750 hp(m) *(5.7 MW)*; 1 shaft
Speed, knots: 12 surfaced; 20 dived
Complement: 70 (10 officers)

Missiles: SSM: McDonnell Douglas Sub-Harpoon; active radar homing to 130 km *(70 n miles)* at 0.9 Mach; warhead 227 kg.
Torpedoes: 6—21 in *(533 mm)* tubes; Type 89; wire-guided; active/passive homing to 50 km *(27 n miles)*/38 km *(21 n miles)* at 40/55 kt; warhead 267 kg and Type 80 ASW. Total of 20 SSM and torpedoes.
Countermeasures: ESM: NZLR-1B; radar warning.
Weapons control: SMCS type TFCS.
Radars: Surface search: JRC ZPS 6; I-band.
Sonars: Hughes/Oki ZQQ 5B/6; hull and flank arrays; active/ passive search and attack; medium/low frequency. ZQR 1 (BQR 15) towed array; passive search; very low frequency.

TAKASHIO   *7/2004*, Hachiro Nakai* / 1044384

Programmes: First of a new class approved in the 1993 budget and then one a year up to FY03.
Structure: Fitted with large flank sonar arrays which are reported as the reason for the increase in displacement over the

Harushio class. Double hull sections forward and aft and anechoic tiles on the fin. A new type of deck casing and faired fin are other distinguishing features.
                                                       UPDATED

KUROSHIO   *3/2004*, Hachiro Nakai* / 1044383

7 HARUSHIO CLASS (SSK)

Name	No	Builders	Laid down	Launched	Commissioned
HARUSHIO	SS 583	Mitsubishi, Kobe	21 Apr 1987	26 July 1989	30 Nov 1990
NATSUSHIO	SS 584	Kawasaki, Kobe	8 Apr 1988	20 Mar 1990	20 Mar 1991
HAYASHIO	SS 585	Mitsubishi, Kobe	9 Dec 1988	17 Jan 1991	25 Mar 1992
ARASHIO	SS 586	Kawasaki, Kobe	8 Jan 1990	17 Mar 1992	17 Mar 1993
WAKASHIO	SS 587	Mitsubishi, Kobe	12 Dec 1990	22 Jan 1993	1 Mar 1994
FUYUSHIO	SS 588	Kawasaki, Kobe	12 Dec 1991	16 Feb 1994	7 Mar 1995
ASASHIO	TSS 3601 (ex-SS 589)	Mitsubishi, Kobe	24 Dec 1992	12 July 1995	12 Mar 1997

Displacement, tons: 2,450 (2,900, TSS 3601) standard; 2,750 (3,250, TSS 3601) dived
Dimensions, feet (metres): 252.6; 285.5 (TSS 3601) × 32.8 × 25.3 *(77; 87 × 10 × 7.7)*
Main machinery: Diesel-electric; 2 Kawasaki 12V25/25S diesels; 5,520 hp(m) *(4.1 MW)*; 2 Kawasaki alternators; 3.7 MW; 2 Fuji motors; 7,200 hp(m) *(5.3 MW)*; 1 shaft 4 Stirling engines (TSS 3601) Kockums V4-275R Mk 2; 348 hp *(260 kW)*
Speed, knots: 12 surfaced; 20 dived
Complement: 75 (10 officers); 71 (10 officers) (SS 589)

Missiles: SSM: McDonnell Douglas Sub-Harpoon; active radar homing to 130 km *(70 n miles)* at 0.9 Mach; warhead 227 kg.

Torpedoes: 6—21 in *(533 mm)* tubes. Japanese Type 89; wire-guided (option); active/passive homing to 50 km *(27 n miles)*/38 km *(21 n miles)* at 40/55 kt; warhead 267 kg; depth to 900 m, and Type 80 ASW. Total of 20 SSM and torpedoes.
Countermeasures: ESM: NZLR-1; radar warning.
Radars: Surface search: JRC ZPS 6; I-band.
Sonars: Hughes/Oki ZQQ 5B; hull-mounted; active/passive search and attack; medium/low frequency.
ZQR 1 towed array similar to BQR 15; passive search; very low frequency.

Programmes: First approved in 1986 estimates and then one per year until 1992.

Structure: The slight growth in all dimensions is a natural evolution from the Yuushio class and includes more noise reduction, towed sonar and wireless aerials, as well as anechoic coating. Double hull construction. *Asashio* had a slightly larger displacement on build and a small cutback in the crew as a result of greater systems automation for machinery and snorting control. The hull was extended in 2001 to accommodate an AIP module (Stirling engine) which was fitted by Mitsubishi, Kobe. Diving depth 350 m *(1,150 ft)*.
Operational: A remote periscope viewer is fitted in *Asashio*. *Asashio* is an experimental submarine for test of AIP propulsion.

UPDATED

FUYUSHIO   *6/2004*, Hachiro Nakai* / 1044385

HAYASHIO   *6/2004*, Hachiro Nakai* / 1121411

ASASHIO   *8/2004*, Hachiro Nakai* / 1044386

3 YUUSHIO CLASS (SSK/SSA)

Name	No	Builders	Laid down	Launched	Commissioned
HAMASHIO	TSS 3604 (ex-SS 578)	Kawasaki, Kobe	8 Apr 1982	1 Feb 1984	5 Mar 1985
YUKISHIO	SS 581	Mitsubishi, Kobe	11 Apr 1985	23 Jan 1987	11 Mar 1988
SACHISHIO	SS 582	Kawasaki, Kobe	11 Apr 1986	17 Feb 1988	24 Mar 1989

Displacement, tons: 2,250 (SS 578-582); 2,300 (SS 576) standard; 2,450 dived
Dimensions, feet (metres): 249.3 × 32.5 × 24.3 *(76 × 9.9 × 7.4)*
Main machinery: Diesel-electric; 2 Kawasaki V8V24/30ATL diesels; 6,800 hp(m) *(5 MW)*; 2 Fuji motors; 7,200 hp(m) *(5.3 MW)*; 1 shaft
Speed, knots: 12 surfaced; 20 dived
Complement: 75 (10 officers)

Missiles: SSM: McDonnell Douglas Sub-Harpoon; active radar homing to 130 km *(70 n miles)* at 0.9 Mach; warhead 227 kg.

Torpedoes: 6—21 in *(533 mm)* tubes amidships. Japanese Type 89; active/passive homing to 50 km *(27 n miles)*/38 km *(21 n miles)* at 40/55 kt; warhead 267 kg; depth to 900 m and Type 80 ASW. Total of 20 SSM and torpedoes.
Countermeasures: ESM: ZLR 5, 6; radar warning.
Radars: Surface search: JRC ZPS 5, 6; I-band.
Sonars: Hughes/Oki ZQQ 5 (modified BQS 4); bow-mounted; passive/active search and attack; medium/low frequency. ZQR 1 towed array similar to BQR 15 (in most of the class); passive search; very low frequency.

Programmes: First one approved in FY75, then one per year from 1977 until 1985.
Modernisation: Towed sonar array fitted in *Okishio* in 1987 and now backfitted to others in the class. ZQQ 5 retrofitted.
Structure: An enlarged version of the Uzushio class with improved diving depth to 275 m *(900 ft)*. Double hull construction. The towed array is stowed in a conduit on the starboard side of the casing.
Operational: *Hamashio* was assigned as a training submarine in March 2003 and completed its conversion on 30 June 2003.
UPDATED

SACHISIO   7/2003, Hachiro Nakai / 0570840

HAMASHIO   6/2003, Hachiro Nakai / 0576114

DESTROYERS

0 + 1 (1) FUTURE DESTROYER CLASS (DDHM)

Name	No	Builders	Laid down	Launched	Commissioned
—	145	IHI Marine United, Yokohama	2005	2007	2009

Displacement, tons: 13,500 standard
Dimensions, feet (metres): 639.9 × 105.0 × ?
(195.0 × 32.0 × ?)
Main machinery: COGAG; 4 LM 2500 gas turbines; 2 shafts
Speed, knots: 30. **Range, n miles:** 6,000 at 20 kt
Complement: 322 (+25 HQ staff)

Missiles: SAM: Raytheon Sea Sparrow RIM-7P; Lockheed Martin
 Marietta Mk 41 Mod 5 sixteen cell vertical launcher ❶; semi-
 active radar homing to 14.6 km *(8 n miles)* at 2.5 Mach;
 warhead 39 kg.
A/S: Vertical launch ASROC.
Guns: 2 GE 20 mm/76 Sea Vulcan 20 ❷; 3 barrels per
 mounting; 1,500 rds/min.
 2—20 mm. 2—12.7 mm MGs.

Torpedoes: 6—324 mm (2 triple) tubes ❸.
Countermeasures: Decoys: 4 Hycor Mk 137 sextuple RBOC
 chaff launchers ❹.
ESM/ECM: NOLQ-2.
Combat data system: to be announced.
Weapons control: FCS-3.
Radars: Air/Surface search: JRC OPS-28D; 3D; G-band.
 Fire control: Signaal Lirod Mk 2; K-band.
 Navigation: JRC OPS-20; I-band.
Sonars: Bow-mounted sonar. TACTAS.

Helicopters: 3 SH-60K, 1 EH-101 (MCH).

Programmes: Two new aviation capable ships to replace the
 Haruna class authorised in the FY01-05 and FY05-09
 programmes. The first authorised in the FY04 budget and the
 second is a candidate for the FY06 budget.
Structure: Broadly similar to the Spanish light carrier *Príncipe de
 Asturias* although not fitted with a ski jump and VSTOL
 capability. The flight deck has two lifts and four helicopter
 spots. The Mk 41 VLS launcher is situated on the starboard
 quarter.
Operational: To be capable of acting as Command Vessels to
 replace *Haruna* and *Hiei*.

UPDATED

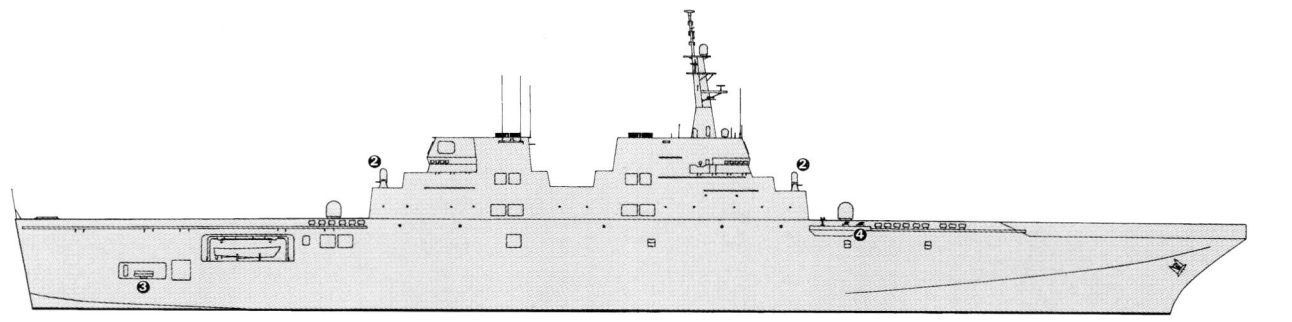

DDH

(Scale 1 : 1,200), Ian Sturton / 0569917

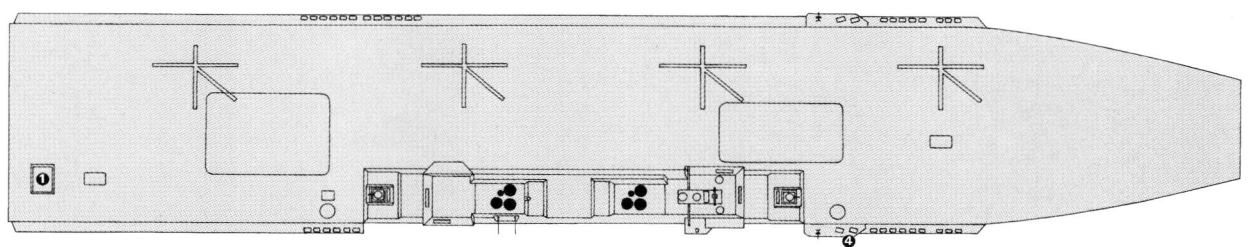

DDH

(Scale 1 : 1,200), Ian Sturton / 0569916

0 + 2 IMPROVED KONGOU CLASS (DDGHM)

Name	No	Builders	Laid down	Launched	Commissioned
—	DDG 177	Mitsubishi, Nagasaki	5 Apr 2004	Aug 2005	Mar 2007
—	DDG 178	Mitsubishi, Nagasaki	Apr 2005	Aug 2006	Mar 2008

Displacement, tons: 7,700 standard
Dimensions, feet (metres): 540.1 × 68.9 × 20.3
(164.9 × 21.0 × 6.2)
Main machinery: COGAG; 4 GE LM 2500 gas turbines;
 102,160 hp *(76.21 MW)* sustained; 2 shafts; cp props
Speed, knots: 30. **Range, n miles:** 4,500 at 20 kt
Complement: 309 (27 officers)

Missiles: SSM: 8 Mitsubishi Type 90 SSM-1B (2 quad) ❶.
 SAM: Standard SM-2MR. FMC Mk 41 VLS; 64 cells forward ❷ 32
 cells aft ❸.
A/S: Vertical launch ASROC.
Guns: 1 United States Mk 45 Mod 4 5 in *(127 mm)*/62 ❹.
 2 GE/GD 20 mm/76 Mk 15 Vulcan Phalanx Block IB ❺.

Torpedoes: 6—324 mm (2 triple) HOS 324 tubes ❻.
Countermeasures: Decoys: 4 Mk 36 SRBOC ❼ 6-barrelled Mk
 36 chaff launchers; Type 4 towed torpedo decoy.
ESM/ECM: NOLQ-2.
Radars: Air search: RCA SPY 1(V) ❽; 3D; F-band.
 Surface search: JRC OPS-28D ❾; G-band.
 Navigation: JRC OPS-20; I-band.
 Fire control: 3 SPG-62 ❿; 1 Mk 2/21 ⓫; I/J-band.
Sonars: SQQ-89(V) bow sonar.

Helicopters: 1 Mitsubishi/Sikorsky SH-60J ⓬.

Programmes: Two ships authorised in the FY01-05 programme.
 The first authorised in the FY02 budget and the second in
 FY03 budget.
Structure: The upgrade from the Kongou class includes one
 hangar for embarked helicopters. Vertical launchers are
 increased by three cells at each end.

UPDATED

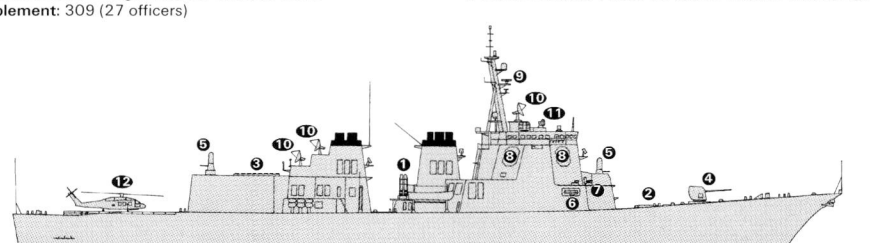

IMPROVED KONGOU

(Scale 1 : 1,500), Ian Sturton / 0529154

For details of the latest updates to *Jane's Fighting Ships* online and to discover the
additional information available exclusively to online subscribers please visit
jfs.janes.com

4 KONGOU CLASS (DDGHM)

Name	No	Builders	Laid down	Launched	Commissioned
KONGOU	DDG 173	Mitsubishi, Nagasaki	8 May 1990	26 Sep 1991	25 Mar 1993
KIRISHIMA	DDG 174	Mitsubishi, Nagasaki	7 Apr 1992	19 Aug 1993	16 Mar 1995
MYOUKOU	DDG 175	Mitsubishi, Nagasaki	8 Apr 1993	5 Oct 1994	14 Mar 1996
CHOUKAI	DDG 176	Ishikawajima Harima, Tokyo	29 May 1995	27 Aug 1996	20 Mar 1998

Displacement, tons: 7,250 standard; 9,485 full load
Dimensions, feet (metres): 528.2 × 68.9 × 20.3; 32.7 (sonar)
(161 × 21 × 6.2; 10)
Main machinery: COGAG; 4 GE LM 2500 gas turbines;
102,160 hp *(76.21 MW)* sustained; 2 shafts; cp props
Speed, knots: 30. **Range, n miles:** 4,500 at 20 kt
Complement: 300 (27 officers)

Missiles: SSM: 8 McDonnell Douglas Harpoon (2 quad) ❶
launchers; active radar homing to 130 km *(70 n miles)* at 0.9
Mach; warhead 227 kg.
SAM: GDC Pomona Standard SM-2MR (SM-3 in due course).
FMC Mk 41 VLS (29 cells) forward ❷. Martin Marietta Mk 41
VLS (61 cells) aft ❸; command/inertial guidance; semi-active
radar homing to 167 km *(90 n miles)* at 2 Mach. Total of 90
Standard and ASROC weapons.
A/S: Vertical launch ASROC; inertial guidance to 1.6-10 km *(1-5.4 n miles)*; payload Mk 46 Mod 5 Neartip.
Guns: 1 OTO Melara 5 in *(127 mm)*/54 Compatto ❹; 45 rds/min
to 23 km *(12.42 n miles)*; weight of shell 32 kg.
2 GE/GD 20 mm/76 Mk 15 Vulcan Phalanx Block IB ❺.
6 barrels per mounting; 3,000 rds/min combined to 1.5 km.
Torpedoes: 6—324 mm (2 triple) HOS 302 tubes ❻. Honeywell
Mk 46 Mod 5 Neartip; anti-submarine; active/passive homing
to 11 km *(5.9 n miles)* at 40 kt; warhead 44 kg.
Countermeasures: Decoys: 4 Mk 36 SRBOC ❼ 6-barrelled Mk
36 chaff launchers; Type 4 towed torpedo decoy.
ESM/ECM: Melko NOLQ 2; intercept/jammer.
Combat data systems: Aegis NTDS with Link 11. SATCOM
WSC-3/OE-82C ❽. OQR-1 helicopter datalink ❾.
Weapons control: 3 Mk 99 Mod 1 MFCS. Type 2-21 GFCS. Mk
116 Hitachi OYQ 102 (Mod 7 for ASW).

KONGOU   *(Scale 1 : 1,500), Ian Sturton* / 0130387

Radars: Air search: RCA SPY 1D ❿; 3D; F-band.
Surface search: JRC OPS-28D ⓫; G-band.
Navigation: JRC OPS-20; I-band.
Fire control: 3 SPG-62 ⓬; 1 Mk 2/21 ⓭; I/J-band.
IFF: UPX 29.
Sonars: Nec OQS 102 (SQS-53B/C) bow-mounted; active search
and attack.
Oki OQR 2 (SQR-19A (V)) TACTASS; towed array; passive; very
low frequency.

Helicopters: Platform ⓮ and fuelling facilities for SH-60J.

Programmes: Proposed in the FY87 programme; first one
accepted in FY88 estimates, second in FY90, third in FY91,
fourth in FY93. Designated as destroyers but these ships are
of cruiser size. The combination of cost and US Congressional
reluctance to release Aegis technology slowed the

programme down. The ships' names were last used by
battleships and cruisers of the Second World War era.
Modernisation: It is likely that these ships will be equipped with
Standard SM-3 anti-ballistic missiles as they become
available. Upgrade of the first ship is expected to start in
2006, with the others following at one year intervals.
Structure: This is an enlarged and improved version of the USN
Arleigh Burke with a lightweight version of the Aegis system.
There are two missile magazines. OQS 102 plus OQR 2 towed
array is the equivalent of SQQ-89. Prairie-Masker acoustic
suppression system.
Operational: As well as air defence of the Fleet, these ships
contribute to the air defences of mainland Japan. Standard
SM-3 Block 0 to be fitted in due course.

UPDATED

CHOUKAI   *8/2004 *, Hachiro Nakai* / 1044387

KIRISHIMA   *10/2003, Hachiro Nakai* / 0570841

4 + 1 TAKANAMI CLASS (DDGHM)

Name	No	Builders	Laid down	Launched	Commissioned
TAKANAMI	DD 110	IHI Marine United, Yokosuka (Uraga)	25 Apr 2000	26 July 2001	12 Mar 2003
OONAMI	DD 111	Mitsubishi, Nagasaki	17 May 2000	20 Sep 2001	13 Mar 2003
MAKINAMI	DD 112	IHI Marine United, Yokohama	7 July 2001	8 Aug 2002	18 Mar 2004
SAZANAMI	DD 113	Mitsubishi, Nagasaki	4 Apr 2002	29 Aug 2003	16 Feb 2005
SUZUNAMI	DD 114	IHI Marine United, Yokohama	24 Sep 2003	26 Aug 2004	Feb 2006

Displacement, tons: 4,650 standard; 5,300 full load
Dimensions, feet (metres): 495.4 × 57.1 × 17.4
  (151 × 17.4 × 5.3)
Main machinery: COGAG; 2 RR Spey SM1C gas turbines;
  41,630 hp *(31 MW)* sustained; 2 GE LM 2500 gas turbines;
  43,000 hp *(32.08 MW)* sustained; 2 shafts
Speed, knots: 30
Complement: 175

Missiles: SSM: 8 Mitsubishi Type 90 SSM-1B (2 quad) **❶**; active
  radar homing to 150 km *(81 n miles)* at 0.9 Mach; warhead
  225 kg.
SAM: Mk 41 VLS 32 cells **❷** Sea Sparrow RIM-7M (PIP); semi-
  active radar homing to 14.6 km *(8 n miles)* at 2.5 Mach;
  warhead 39 kg and VL ASROC; internal guidance to 1.6-
  10 km *(1-5.4 n miles)*; payload Mk 46 Mod 5 Neartip.
Guns: 1 Otobreda 5 in *(127 mm)*/54 **❸**; 45 rds/min to 23 km
  (12.42 n miles); weight of shell 32 kg.
  2 General Electric/General Dynamics 20 mm Phalanx Mk 15
  CIWS **❹**; 6 barrels per mounting; 3,000 rds/min combined to
  1.5 km.
Torpedoes: 6—324 mm HOS-302 (2 triple) tubes **❺** Mk 46 Mod
  5; anti-submarine; active/passive homing to 11 km *(5.9 n
  miles)* at 40 kt; warhead 44 kg.
Countermeasures: Decoys: 4 Mk 36 SRBOC chaff launchers **❻**.
  SLQ-25 Nixie towed torpedo decoy.
ESM/ECM: Nec NOLQ 2/3; intercept and jammer.

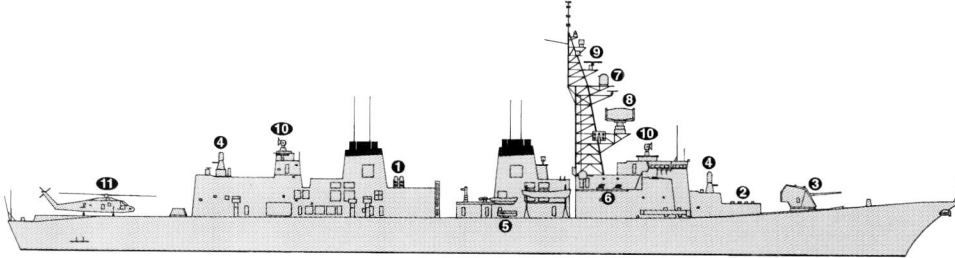

TAKANAMI CLASS

(Scale 1 : 1,200), Ian Sturton / 0080138

Combat data systems: OYQ-7 with Link 11. ORQ-1 helicopter
  datalink **❼**.
Weapons control: Hitachi OYQ-103 ASW control system.
Radars: Air search: Melco OPS-24B **❽**; 3D; D-band.
Surface search: JRC OPS-28D **❾**; G-band.
Fire control: Two FCS 2-3/B **❿**.
Navigation: OPS-20; I-band.
Sonars: OQS-5; Bow-mounted; active search and attack; low
  frequency.
  OQR-2; towed array; passive search; very low frequency.

Helicopters: 1 Mitsubishi/Sikorsky SH-60J **⓫**.

Programmes: First two approved in FY98, then one a year up to
  FY01.
Modernisation: Evolved Sea Sparrow (ESSM) (RIM-162) to be
  fitted in due course.
Structure: Murasame class modified to fit a Mk 41 VLS,
  improved missile fire control and new sonar.

UPDATED

TAKANAMI

4/2004*, Hachiro Nakai / 1044389

OONAMI

4/2004*, Hachiro Nakai / 1044390

2 HATAKAZE CLASS (DDGHM)

Name	No	Builders	Laid down	Launched	Commissioned
HATAKAZE	DDG 171	Mitsubishi, Nagasaki	20 May 1983	9 Nov 1984	27 Mar 1986
SHIMAKAZE	DDG 172	Mitsubishi, Nagasaki	30 Jan 1985	30 Jan 1987	23 Mar 1988

Displacement, tons: 4,600 (4,650, DDG 172) standard; 6,400 full load
Dimensions, feet (metres): 492 × 53.8 × 15.7 *(150 × 16.4 × 4.8)*
Main machinery: COGAG; 2 RR Olympus TM3B gas turbines; 49,400 hp *(36.8 MW)* sustained; 2 RR Spey SM1A gas turbines, 26,650 hp *(19.9 MW)* sustained; 2 shafts; Kamewa cp props
Speed, knots: 30
Complement: 260 (23 officers)

Missiles: SSM: 8 McDonnell Douglas Harpoon ❶; active radar homing to 130 km *(70 n miles)* at 0.9 Mach; warhead 227 kg.
SAM: 40 GDC Pomona Standard SM-1MR; Mk 13 Mod 4 launcher ❷; command guidance; semi-active radar homing to 46 km *(25 n miles)* at 2 Mach; height envelope 45-18,288 m *(150-60,000 ft)*.
A/S: Honeywell ASROC Mk 112 octuple launcher ❸; inertial guidance to 1.6-10 km *(1-5.4 n miles)* at 0.9 Mach; payload Mk 46 Mod 5 Neartip. Reload capability.
Guns: 2 FMC 5 in *(127 mm)*/54 Mk 42 automatic ❹; 20-40 rds/min to 24 km *(13 n miles)* anti-surface; 14 km *(7.6 n miles)* anti-aircraft; weight of shell 32 kg.
2 General Electric/General Dynamics 20 mm Phalanx Mk 15 CIWS ❺; 6 barrels per mounting; 3,000 rds/min combined to 1.5 km.
Torpedoes: 6—324 mm Type 68 or HOS 301 (2 triple) tubes ❻. Honeywell Mk 46 Mod 5 Neartip; anti-submarine; active/passive homing to 11 km *(5.9 n miles)* at 40 kt; warhead 44 kg.
Countermeasures: Decoys: 2 Loral Hycor SRBOC 6-barrelled Mk 36 chaff launchers; range 4 km *(2.2 n miles)*.
ESM/ECM: Melco NOLQ-1; intercept/jammer. Fujitsu OLR 9B; intercept.
Combat data systems: OYQ-4 Mod 1 action data automation; Link 11. SATCOM ❼.
Weapons control: Type 2-21C for 127 mm guns. General Electric Mk 74 Mod 13 for Standard.
Radars: Air search: Hughes SPS-52C ❽; 3D; E/F-band. Melco OPS-11C ❾; B-band.
Surface search: JRC OPS-28B ❿; G/H-band.
Fire control: 2 Raytheon SPG-51C ⓫; G-band.
Melco 2-21 ⓬; I/J-band. Type 2-12 ⓭; I-band.
Sonars: Nec OQS 4 Mod 1; bow-mounted; active search and attack; medium frequency.

Helicopters: Platform for 1 SH-60J Seahawk ⓮.

Programmes: DDG 171 provided for in 1981 programme. DDG 172 provided for in 1983 programme, ordered 29 March 1984.

UPDATED

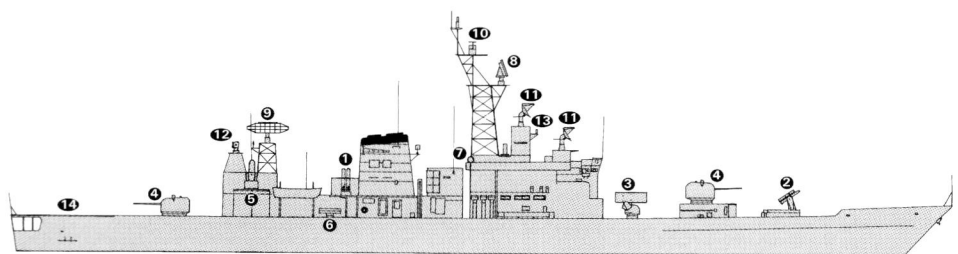

HATAKAZE
(Scale 1 : 1,200), Ian Sturton

SHIMAKAZE
4/2004*, Hachiro Nakai / 1044388

2 SHIRANE CLASS (DDHM)

Name	No	Builders	Laid down	Launched	Commissioned
SHIRANE	DDH 143	Ishikawajima Harima, Tokyo	25 Feb 1977	18 Sep 1978	17 Mar 1980
KURAMA	DDH 144	Ishikawajima Harima, Tokyo	17 Feb 1978	20 Sep 1979	27 Mar 1981

Displacement, tons: 5,200 standard; 7,500 full load
Dimensions, feet (metres): 521.5 × 57.5 × 17.5 *(159 × 17.5 × 5.3)*
Main machinery: 2 IHI boilers; 850 psi *(60 kg/cm²)*; 900°F *(480°C)*; 2 IHI turbines; 70,000 hp(m) *(51.5 MW)*; 2 shafts
Speed, knots: 31
Complement: 350; 360 (DDH 144) plus 20 staff

Missiles: SAM: Raytheon Sea Sparrow (RIM-7E (DDH 143); RIM-7M (DDH 144); Mk 25 launcher (DDH 143) ❶; Type 3 launcher (DDH 144); semi-active radar homing to 14.6 km *(8 n miles)* at 2.5 Mach; warhead 39 kg; 24 missiles.
A/S: Honeywell ASROC Mk 112 octuple launcher ❷; inertial guidance to 10 km *(5.4 n miles)* at 0.9 Mach; payload Mk 46 Mod 5 Neartip.
Guns: 2 FMC 5 in *(127 mm)*/54 Mk 42 automatic ❸; 20-40 rds/min to 24 km *(13 n miles)* anti-surface; 14 km *(7.6 n miles)* anti-aircraft; weight of shell 32 kg.
2 General Electric/General Dynamics 20 mm Phalanx Mk 15 CIWS ❹; 6 barrels per mounting; 3,000 rds/min combined to 1.5 km.
Torpedoes: 6—324 mm HOS 301 (2 triple) tubes ❺. Honeywell Mk 46 Mod 5 Neartip; anti-submarine; active/passive homing to 11 km *(5.9 n miles)* at 40 kt; warhead 44 kg.
Countermeasures: Decoys: 4 Mk 36 SRBOC chaff launchers. Prairie Masker; blade rate suppression system.
ESM/ECM: Melco NOLQ 1; intercept/jammer. Fujitsu OLR 9B; intercept.
Combat data systems: OYQ-3B; Links 11 and 14. SATCOM ❻.
Weapons control: Singer Mk 114 for ASROC and TFCS; Type 72-1A GFCS.
Radars: Air search: Nec OPS-12 ❼; 3D; D-band.
Surface search: JRC OPS-28 ❽; G-band.
Navigation: JRC OPS-20; I-band.
Fire control: Signaal WM25 (DDH 143) ❾; I/J-band; range 46 km *(25 n miles)*.
Type 2-12; I/J-band.
2 Type 72-1A FCS ❿; I/J-band.
Tacan: ORN-6C/6C-Y.
Sonars: EDO/Nec SQS-35(J); VDS; active/passive search; medium frequency.
Nec OQS 101; bow-mounted; low frequency.
EDO/Nec SQR-18A; towed array; passive; very low frequency.

Helicopters: 3 SH-60J Seahawk ⓫.

Programmes: One each in 1975 and 1976 programmes.
Modernisation: DDH 143 refit in 1989-90. Both fitted with CIWS and towed array sonars by mid-1990. DDH 144 upgraded with Type 3 launcher to fire RIM-7M during 2003-04 refit at Mitsubishi, Nagasaki.
Structure: Fitted with Vosper Thornycroft fin stabilisers. The after funnel is set to starboard and the forward one to port. The crane is on the starboard after corner of the hangar. Bear Trap helicopter hauldown gear.
Operational: Both ships carry SH-60J helicopters.

UPDATED

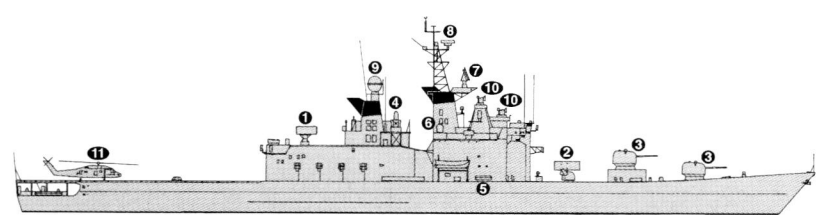

SHIRANE
(Scale 1 : 1,500), Ian Sturton / 0012639

KURAMA
9/2004*, Hachiro Nakai / 1044398

9 MURASAME CLASS (DDGHM)

Name	No	Builders	Laid down	Launched	Commissioned
MURASAME	DD 101	Ishikawajima Harima, Tokyo	18 Aug 1993	23 Aug 1994	12 Mar 1996
HARUSAME	DD 102	Mitsui, Tamano	11 Aug 1994	16 Oct 1995	24 Mar 1997
YUUDACHI	DD 103	Marine United (Sumitomo, Uraga)	18 Mar 1996	19 Aug 1997	4 Mar 1999
KIRISAME	DD 104	Mitsubishi, Nagasaki	3 Apr 1996	21 Aug 1997	18 Mar 1999
INAZUMA	DD 105	Mitsubishi, Nagasaki	8 May 1997	9 Sep 1998	15 Mar 2000
SAMIDARE	DD 106	Marine United (Ishikawajima Harima, Tokyo)	11 Sep 1997	24 Sep 1998	21 Mar 2000
IKAZUCHI	DD 107	Hitachi, Maizuru	25 Feb 1998	24 June 1999	14 Mar 2001
AKEBONO	DD 108	Marine United (Ishikawajima Harima, Tokyo)	29 Oct 1999	25 Sep 2000	19 Mar 2002
ARIAKE	DD 109	Mitsubishi, Nagasaki	18 May 1999	16 Oct 2000	6 Mar 2002

Displacement, tons: 4,550 standard; 5,100 full load
Dimensions, feet (metres): 495.4 × 57.1 × 17.1
   (151 × 17.4 × 5.2)
Main machinery: COGAG; 2 RR Spey SM1C gas turbines;
   27,600 hp *(20.6 MW)* sustained; 2 GE LM 2500 gas turbines;
   33,700 hp *(25.1 MW)* sustained; 2 shafts
Speed, knots: 30
Complement: 165

Missiles: SSM: 8 Type 90 SSM-1B ❶ (Harpoon); active radar
   homing to 130 km *(70 n miles)* at 0.9 Mach; warhead 227 kg.
SAM: Raytheon Mk 48 VLS 16 cells ❷ Sea Sparrow RIM-7M;
   semi-active radar homing to 14.6 km *(8 n miles)* at 2.5 Mach;
   warhead 39 kg.
A/S: Mk 41 VL ASROC 16 cells ❸. Total of 29 missiles can be
   carried.
Guns: 1 Otobreda 3 in *(76 mm)*/62 compact ❹; 85 rds/min to
   16 km *(8.6 n miles)* anti-surface; 12 km *(6.5 n miles)* anti-
   aircraft; weight of shell 6 kg.
   2 General Electric/General Dynamics 20 mm Phalanx Mk 15
   CIWS ❺; 6 barrels per mounting; 3,000 rds/min combined to
   1.5 km.
Torpedoes: 6—324 mm HOS 302 (2 triple) tubes ❻ Mk 46 Mod
   5; anti-submarine; active/passive homing to 11 km *(5.9 n
   miles)* at 40 kt; warhead 44 kg.
Countermeasures: Decoys: 4 Mk 36 SRBOC chaff launchers ❼.
   Type 4 towed torpedo decoy.
ESM/ECM: Nec NOLQ 3; intercept and jammer.
Combat data systems: OYQ-9B with Link 11. ORQ-1 helicopter
   datalink ❽.
Weapons control: Hitachi OYQ-103 ASW control system.
Radars: Air search: Melco OPS-24B ❾; 3D; D-band.
   Surface search: JRC OPS-28D ❿; G-band.
   Fire control: Two Type 2-31 ⓫.
   Navigation: OPS-20; I-band.
Sonars: Mitsubishi OQS-5; hull-mounted; active search and
   attack; low frequency.
   OQR-1 towed array; passive search; very low frequency.

Helicopters: 1 SH-60J Seahawk ⓬.

Programmes: First one approved in FY91 as an addition to the
   third Aegis-type destroyer. Second approved in FY92. Two

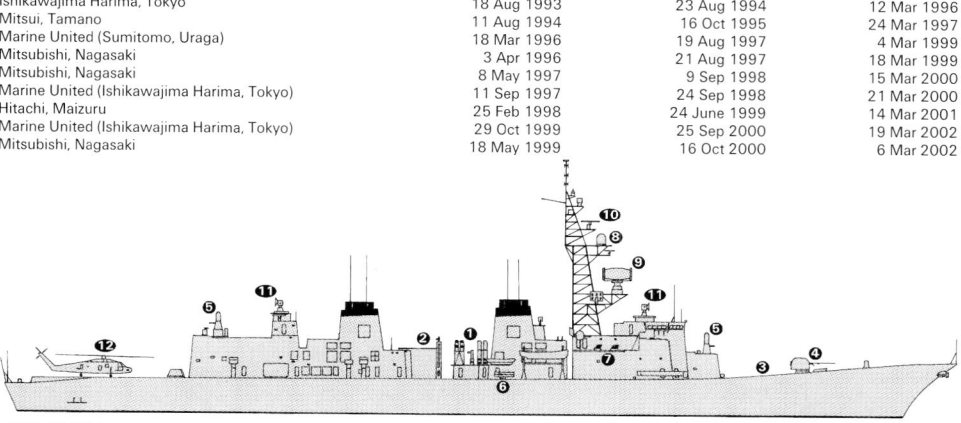

MURASAME *(Scale 1 : 1,200), Ian Sturton*

IKAZUCHI *6/2004, Hachiro Nakai / 1121410*

more approved in FY94, two in FY95, one in FY96 and two in
FY97. The programme was given added priority as the
Kongou class was reduced to four ships because of the cost of
Aegis.
Modernisation: Evolved Sea Sparrow (ESSM) to be fitted in due
   course.
Structure: More like a mini-Kongou than an enlarged Asagiri
   class, with VLS and a much reduced complement. Stealth

features are evident in sloping sides and rounded
superstructure. Indal RAST helicopter hauldown.
Operational: ASROC missiles are not carried. *Kirisame* deployed
   to Indian Ocean in November 2001 to provide non-combatant
   support to US forces.

UPDATED

MURASAME *10/2004*, Hachiro Nakai / 1044392*

YUUDACHI *9/2004*, Hachiro Nakai / 1044391*

6 ASAGIRI CLASS (DDGHM)

Name	No	Builders	Laid down	Launched	Commissioned
YUUGIRI	DD 153	Sumitomo, Uraga	25 Feb 1986	21 Sep 1987	28 Feb 1989
AMAGIRI	DD 154	Ishikawajima Harima, Tokyo	3 Mar 1986	9 Sep 1987	17 Mar 1989
HAMAGIRI	DD 155	Hitachi, Maizuru	20 Jan 1987	4 June 1988	31 Jan 1990
SETOGIRI	DD 156	Sumitomo, Uraga	9 Mar 1987	12 Sep 1988	14 Feb 1990
SAWAGIRI	DD 157	Mitsubishi, Nagasaki	14 Jan 1987	25 Nov 1988	6 Mar 1990
UMIGIRI	DD 158	Ishikawajima Harima, Tokyo	31 Oct 1988	9 Nov 1989	12 Mar 1991

Dimensions, feet (metres): 449.4 × 48 × 14.6
(137 × 14.6 × 4.5)
Main machinery: COGAG; 4 RR Spey SM1A gas turbines;
53,300 hp *(39.8 MW)* sustained; 2 shafts; cp props
Speed, knots: 30+
Complement: 220

Missiles: SSM: 8 McDonnell Douglas Harpoon (2 quad)
launchers ❶; active radar homing to 130 km *(70 n miles)* at
0.9 Mach; warhead 227 kg.
SAM: Raytheon Sea Sparrow Mk 29 (Type 3/3A) octuple
launcher ❷; semi-active radar homing to 14.6 km *(8 n miles)* at
2.5 Mach; warhead 39 kg; 20 missiles.
A/S: Honeywell ASROC Mk 112 octuple launcher ❸; inertial
guidance to 1.6-10 km *(1-5.4 n miles)* at 0.9 Mach; payload Mk
46 Mod 5 Neartip. Reload capability.
Guns: 1 Otobreda 3 in *(76 mm)*/62 compact ❹; 85 rds/min to
16 km *(8.6 n miles)* anti-surface; 12 km *(6.5 n miles)* anti-
aircraft; weight of shell 6 kg.
2 General Electric/General Dynamics 20 mm Phalanx Mk 15
CIWS ❺; 6 barrels per mounting; 3,000 rds/min combined to
1.5 km.
Torpedoes: 6—324 mm Type 68 (2 triple) HOS 301 tubes ❻.
Honeywell Mk 46 Mod 5 Neartip; anti-submarine; active/
passive homing to 11 km *(5.9 n miles)* at 40 kt; warhead
44 kg.
Countermeasures: Decoys: 2 Loral Hycor SRBOC 6-barrelled Mk
36 chaff launchers ❼; range 4 km *(2.2 n miles)*.
1 SLQ-25 Nixie or Type 4; towed torpedo decoy.
ESM: Nec NOLR 6C or NOLR 8 (DD 152) ❽; intercept.
ECM: Fujitsu OLT-3; jammer.
Combat data systems: OYQ-7B data automation; Link 11/14.
SATCOM. ORQ-1 helicopter datalink ❾ for SH-60J.

Radars: Air search: Melco OPS-14C (DD 151-154); D-band.
Melco OPS-24 (DD 155-158) ❿; 3D; D-band.
Surface search: JRC OPS-28C ⓫; G-band (DD 151, 152, 155-
158).
JRC OPS-28C-Y; G-band (DD 153-154).
Fire control: Type 2-22 (for guns) ⓬; Type 2-12E (for SAM) (DD
151-154); Type 2-12G (for SAM) ⓭ (DD 155-158).
Tacan: ORN-6D (URN 25).
Sonars: Mitsubishi OQS 4A (II); hull-mounted; active search and
attack; low frequency.
OQR-1; towed array; passive search; very low frequency.

Helicopters: 1 SH-60J Seahawk ⓮.

Programmes: DD 153-154 in 1984 estimates, DD 155-157 in
1985 and DD 158 in 1986.
Modernisation: The last four were fitted on build with improved
air search radar, updated fire-control radars and a helicopter
datalink. Plans to fit the first four may have been postponed.

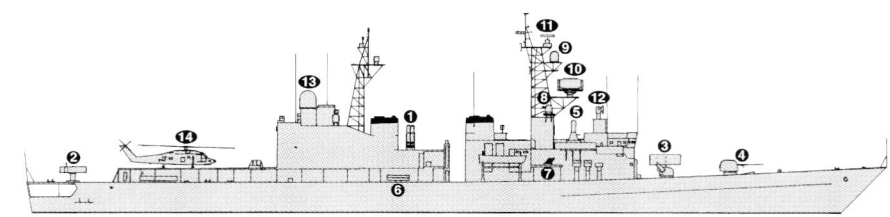

UMIGIRI   *(Scale 1 : 1,200), Ian Sturton /* 0012635

Umigiri also commissioned with a sonar towed array which
has been fitted to the rest of the class.

Structure: Because of the enhanced IR signature and damage to
electronic systems on the mainmast caused by after funnel
gases there have been modifications to help contain the
problem. The mainmast is now slightly higher than originally
designed and has been offset to port, more so in the last four
of the class. The forward funnel is also offset slightly to port
and the after funnel to the starboard side of the
superstructure. The hangar structure is asymmetrical
extending to the after funnel on the starboard side but only to
the mainmast to port. SATCOM is fitted at the after end of the
hangar roof.
Operational: Beartrap helicopter hauldown system. Sea Kings
have been phased out. *Sawagiri* deployed to Indian Ocean in
November 2001 to provide non-combatant support of US
forces. *Yamagiri* (D 152) converted to training ship on
18 March 2004 and *Asagiri* (D 151) on 16 February 2005.
UPDATED

HAMAGIRI   *4/2004*, Kazumasa Watanabe /* 1044393

SAWAGIRI   *6/2004*, Hachiro Nakai /* 1044394

11 HATSUYUKI CLASS (DDGHM)

Name	No	Builders	Laid down	Launched	Commissioned
HATSUYUKI	DD 122	Sumitomo, Uraga	14 Mar 1979	7 Nov 1980	23 Mar 1982
SHIRAYUKI	DD 123	Hitachi, Maizuru	3 Dec 1979	4 Aug 1981	8 Feb 1983
MINEYUKI	DD 124	Mitsubishi, Nagasaki	7 May 1981	19 Oct 1982	26 Jan 1984
SAWAYUKI	DD 125	Ishikawajima Harima, Tokyo	22 Apr 1981	21 June 1982	15 Feb 1984
HAMAYUKI	DD 126	Mitsui, Tamano	4 Feb 1981	27 May 1982	18 Nov 1983
ISOYUKI	DD 127	Ishikawajima Harima, Tokyo	20 Apr 1982	19 Sep 1983	23 Jan 1985
HARUYUKI	DD 128	Sumitomo, Uraga	11 Mar 1982	6 Sep 1983	14 Mar 1985
YAMAYUKI	DD 129	Hitachi, Maizuru	25 Feb 1983	10 July 1984	3 Dec 1985
MATSUYUKI	DD 130	Ishikawajima Harima, Tokyo	7 Apr 1983	25 Oct 1984	19 Mar 1986
SETOYUKI	DD 131	Mitsui, Tamano	26 Jan 1984	3 July 1985	11 Dec 1986
ASAYUKI	DD 132	Sumitomo, Uraga	22 Dec 1983	16 Oct 1985	20 Feb 1987

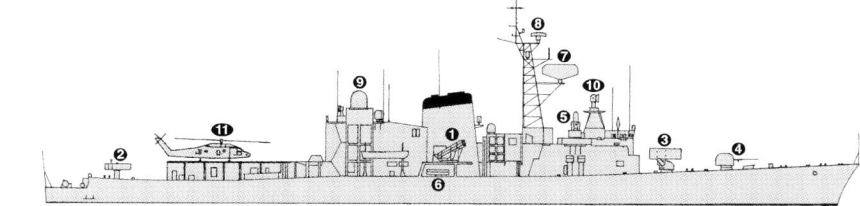

Displacement, tons: 2,950 (3,050 from DD 129 onwards) standard; 3,700 (3,800) full load
Dimensions, feet (metres): 426.4 × 44.6 × 13.8 (14.4 from 129 onwards) *(130 × 13.6 × 4.2) (4.4)*
Main machinery: COGOG; 2 Kawasaki-RR Olympus TM3B gas turbines; 49,400 hp *(36.8 MW)* sustained; 2 RR Type RM1C gas turbines; 9,900 hp *(7.4 MW)* sustained; 2 shafts; cp props
Speed, knots: 30; 19 cruise
Complement: 195 (200, DD 124 onwards)

Missiles: SSM: 8 McDonnell Douglas Harpoon (2 quad) launchers ❶; active radar homing to 130 km *(70 n miles)* at 0.9 Mach; warhead 227 kg.
SAM: Raytheon Sea Sparrow Mk 29 Type 3A launcher ❷; semi-active radar homing to 14.6 km *(8 n miles)* at 2.5 Mach; warhead 39 kg; 12 missiles.
A/S: Honeywell ASROC Mk 112 octuple launcher ❸; inertial guidance to 1.6-10 km *(1-5.4 n miles)* at 0.9 Mach; payload Mk 46 Mod 5 Neartip.
Guns: 1 Otobreda 3 in *(76 mm)*/62 compact ❹; 85 rds/min to 16 km *(8.6 n miles)* anti-surface; 12 km *(6.5 n miles)* anti-aircraft; weight of shell 6 kg.
2 General Electric/General Dynamics 20 mm Phalanx Mk 15 CIWS ❺; 6 barrels per mounting; 3,000 rds/min combined to 1.5 km.
Torpedoes: 6—324 mm Type 68 or HOS 301 (2 triple) tubes ❻. Honeywell Mk 46 Mod 5 Neartip; anti-submarine; active/passive homing to 11 km *(5.9 n miles)* at 40 kt; warhead 44 kg.
Countermeasures: Decoys: 2 Loral Hycor SRBOC 6-barrelled Mk 36 chaff launchers; range 4 km *(2.2 n miles)*.
ESM: Nec NOLR 6C; intercept.
ECM: Fujitsu OLT 3; jammer.
Combat data systems: OYQ-5B action data automation. SATCOM.
Radars: Air search: Melco OPS-14B ❼; D-band.
Surface search: JRC OPS-18-1 ❽; G-band.
Fire control: Type 2-12 A ❾; I/J-band (for SAM).
2 Type 2-21/21A ❿; I/J-band (for guns).
Tacan: ORN-6C-Y (DD 122, 125 and 132); ORN-6C (remainder).
Sonars: Nec OQS 4A (II) (SQS-23 type); bow-mounted; active search and attack; low frequency.
OQR 1 TACTASS (in some); passive; low frequency.

Helicopters: 1 SH-60J Seahawk ⓫.

SHIRAYUKI   *7/2004*, Hachiro Nakai /* 1044396

Modernisation: *Shirayuki* retrofitted with Phalanx in early 1992, and the rest of the class by 1996. *Matsuyuki* first to get sonar towed array in 1990 and *Hatsuyuki* in 1994; the others are being fitted. All of the class converted to carry Seahawk helicopters.

Structure: Fitted with fin stabilisers. Steel in place of aluminium alloy for bridge etc after DD 129 which increased displacement.
Operational: Canadian Beartrap helicopter landing aid. Improved ECM equipment in the last two of the class. Last of class *Shimayuki* converted to a training ship 18 March 1999.

UPDATED

YAMAYUKI   *6/2004*, Hachiro Nakai /* 1044395

ASAYUKI   *7/2004*, Hachiro Nakai /* 1121409

3 TACHIKAZE CLASS (DDGM)

Name	No	Builders	Laid down	Launched	Commissioned
TACHIKAZE	DDG 168	Mitsubishi, Nagasaki	19 June 1973	17 Dec 1974	26 Mar 1976
ASAKAZE	DDG 169	Mitsubishi, Nagasaki	27 May 1976	15 Oct 1977	27 Mar 1979
SAWAKAZE	DDG 170	Mitsubishi, Nagasaki	14 Sep 1979	4 June 1981	30 Mar 1983

Displacement, tons: 3,850 (3,950, DDG 170) standard; 5,500 full load

Dimensions, feet (metres): 469 × 47 × 15.1 *(143 × 14.3 × 4.6)*

Main machinery: 2 Mitsubishi boilers; 600 psi *(40 kg/cm²)*; 850°F *(454°C)*; 2 Mitsubishi turbines; 70,000 hp(m); *(51.5 MW)*; 2 shafts

Speed, knots: 32

Complement: 230-255

Missiles: SSM: 8 McDonnell Douglas Harpoon (DDG 170); active radar homing to 130 km *(70 n miles)* at 0.9 Mach; warhead 227 kg HE.
SAM: GDC Pomona Standard SM-1MR; Mk 13 Mod 1 or 4 launcher ❶; command guidance; semi-active radar homing to 46 km *(25 n miles)* at 2 Mach; height envelope 45-18,288 m *(150-60,000 ft)*; 40 missiles (SSM and SAM combined).

A/S: Honeywell ASROC Mk 112 octuple launcher ❷; inertial guidance to 1.6-10 km *(1-5.4 n miles)* at 0.9 Mach; payload Mk 46 Mod 5 Neartip. Reloads in DDG 170 only.

Guns: 1 or 2 FMC 5 in *(127 mm)*/54 Mk 42 automatic ❸; 20-40 rds/min to 24 km *(13 n miles)* anti-surface; 14 km *(7.6 n miles)* anti-aircraft; weight of shell 32 kg.
2 General Electric/General Dynamics 20 mm Phalanx CIWS Mk 15 ❹; 6 barrels per mounting; 3,000 rds/min combined to 1.5 km.

Torpedoes: 6—324 mm Type 68 or HOS 301 (2 triple) tubes ❺. Honeywell Mk 46 Mod 5 Neartip; anti-submarine; active/passive homing to 11 km *(5.9 n miles)* at 40 kt; warhead 44 kg.

Countermeasures: Decoys: 4 Loral Hycor SRBOC Mk 36 multibarrelled chaff launchers. SLQ-25 towed torpedo decoy.

ESM: Nec NOLR 6 (DDG 168); Nec NOLQ 1 (others); intercept.

ECM: Fujitsu OLT 3; jammer.

Combat data systems: OYQ-1B (DDG 168), OYQ-2B (DDG 169), OYQ-4 (DDG 170) action data automation; Links 11 and 14. SATCOM.

Weapons control: 2 Mk 74 Mod 13 missile control directors. US Mk 114 ASW control. GFCS-2-21 for gun (DDG 170). GFCS-72-1A for gun (others).

Radars: Air search: Melco OPS-11C ❻; B-band.
Hughes SPS-52B ❼ or 52C (DDG 170); 3D; E/F-band.
Surface search: JRC OPS-16D ❽; G-band (DDG 168).
JRC OPS-28 (DDG 170); G-band.
JRC OPS-18-3; G-band (DDG 169).
Fire control: 2 Raytheon SPG-51 ❾; G/I-band.
Type 2 FCS ❿; I/J-band.

IFF: NYPX-2.

Sonars: Nec OQS-3A (Type 66); bow-mounted; active search and attack; low frequency.

Modernisation: CIWS added to DDG 168 in 1983, DDG 169 and 170 in 1987. After gun removed to allow increased Flag accommodation in *Tachikaze* in 1998.

Operational: *Tachikaze* is the Escort Force Flagship.

UPDATED

ASAKAZE *(Scale 1 : 1,200), Ian Sturton*

TACHIKAZE *6/2004*, Hachiro Nakai /* 1044397

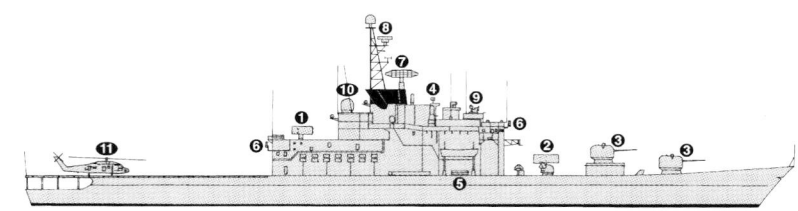

SAWAKAZE
4/2003, Hachiro Nakai / 0570722

2 HARUNA CLASS (DDHM)

Name	No	Builders	Laid down	Launched	Commissioned
HARUNA	DDH 141	Mitsubishi, Nagasaki	19 Mar 1970	1 Feb 1972	22 Feb 1973
HIEI	DDH 142	Ishikawajima Harima, Tokyo	8 Mar 1972	13 Aug 1973	27 Nov 1974

Displacement, tons: 4,950 (5,050, DDH 142) standard; 6,900 full load

Dimensions, feet (metres): 502 × 57.4 × 17.1 *(153 × 17.5 × 5.2)*

Main machinery: 2 Mitsubishi (DDH 141) or IHI (DDH 142) boilers; 850 psi *(60 kg/cm²)*; 900°F *(480°C)*; 2 Mitsubishi (DDH 141) or IHI (DDH 142) turbines; 70,000 hp *(51.5 MW)*; 2 shafts

Speed, knots: 31

Complement: 370 (360, DDH 141) (36 officers)

Missiles: SAM: Raytheon Sea Sparrow Mk 29 (Type 3A) octuple launcher ❶; semi-active radar homing to 14.6 km *(8 n miles)* at 2.5 Mach; warhead 39 kg; 24 missiles.

A/S: Honeywell ASROC Mk 112 octuple launcher ❷; inertial guidance to 1.6-10 km *(1-5.4 n miles)* at 0.9 Mach; payload Mk 46 Mod 5 Neartip.

Guns: 2 FMC 5 in *(127 mm)*/54 Mk 42 automatic ❸; 20-40 rds/min to 24 km *(13 n miles)* anti-surface; 14 km *(7.6 n miles)* anti-aircraft; weight of shell 32 kg.
2 General Electric/General Dynamics 20 mm Phalanx Mk 15 CIWS ❹; 6 barrels per mounting; 3,000 rds/min combined to 1.5 km.

Torpedoes: 6—324 mm HOS 301 (2 triple) tubes ❺. Honeywell Mk 46 Mod 5 Neartip; anti-submarine; active/passive homing to 11 km *(5.9 n miles)* at 40 kt; warhead 44 kg.

Countermeasures: Decoys: 4 Loral Hycor SRBOC Mk 36 multibarrelled chaff launchers.

ESM/ECM: Melco NOLQ 1; intercept/jammer. Fujitsu OLR 9; intercept.

Combat data systems: OYQ-7B action data automation; Links 11 and 14; US SATCOM ❻.

Weapons control: 2 Type 2-12 FCS (1 for guns, 1 for SAM).

Radars: Air search: Melco OPS-11C ❼; B-band.
Surface search: JRC OPS-28C/28C-Y ❽; G-band.
Fire control: 1 Type 1A ❾; I/J-band (guns).
1 Type 2-12 ❿; I/J-band (SAM).

Navigation: Koden OPN-11; I-band.

IFF: YPA-2. YPX-3.

Tacan: Nec ORN-6D/6C.

Sonars: Sangamo/Mitsubishi OQS 3; bow-mounted; active search and attack; low frequency with bottom bounce.

HARUNA *(Scale 1 : 1,500), Ian Sturton /* 0012641

HIEI *4/2004*, Hachiro Nakai /* 1044399

Helicopters: 3 SH-60J Seahawk ⓫.

Programmes: Ordered under the third five-year defence programme (from 1967-71). The ships' names were last used by capital ships of the Second World War era.

Modernisation: DDH 141 taken in hand from 31 March 1986 to 31 October 1987 for FRAM at Mitsubishi, Nagasaki; DDH 142 received FRAM from 31 August 1987 to 30 March 1989 at IHI, Tokyo; included Sea Sparrow, two CIWS and chaff launchers.

Structure: The funnel is offset slightly to port. Fitted with fin stabilisers. A heavy crane has been fitted on the top of the hangar, starboard side.

Operational: Fitted with Canadian Beartrap hauldown gear.

UPDATED

1 YAMAGUMO CLASS (DDM/DDK)

Name	No	Builders	Laid down	Launched	Commissioned
YUUGUMO	DDK 121	Sumitomo, Uraga	4 Feb 1976	31 May 1977	24 Mar 1978

Displacement, tons: 2,150 standard; 2,900 full load
Dimensions, feet (metres): 377.2 × 38.7 × 13.1
 (114.9 × 11.8 × 4)
Main machinery: 6 Mitsubishi 12UEV30/40N diesels;
 21,600 hp(m) (15.9 MW); 2 shafts
Speed, knots: 28. **Range, n miles:** 7,000 at 20 kt
Complement: 220 (19 officers)

Missiles: A/S: Honeywell ASROC Mk 112 octuple launcher **❶**;
 inertial guidance to 1.6-10 km (1-5.4 n miles) at 0.9 Mach;
 payload Mk 46 Mod 5 Neartip.
Guns: 4 USN 3 in (76 mm)/50 Mk 33 (2 twin) **❷**; 50 rds/min to
 12.8 km (6.9 n miles); weight of shell 6 kg.
Torpedoes: 6—324 mm Type 68 (2 triple) tubes **❸**. Honeywell
 Mk 46 Mod 5 Neartip; anti-submarine; active/passive homing
 to 11 km (5.9 n miles) at 40 kt; warhead 44 kg.
A/S mortars: 1 Bofors 375 mm Type 71 4-barrelled trainable
 rocket launcher **❹**; automatic loading; range 1.6 km.
Countermeasures: ESM: Nec NOLR 6; radar intercept.
Weapons control: Japanese GFCS-1 for 76 mm guns.

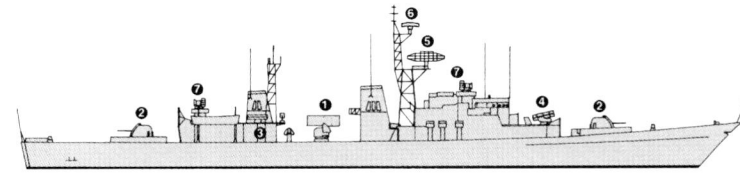

YUUGUMO

Radars: Air search: Melco OPS-11B/11C **❺**; B-band.
 Surface search: JRC OPS-18-3 **❻**; G-band.
 Fire control: 2 General Electric Mk 35 **❼**; I/J-band.
 IFF: YPA-2, YPX-3.
Sonars: Nec OQS 3A; hull-mounted; active search and attack;
 medium frequency.
 EDO SQS-35(J); VDS; active/passive search; medium
 frequency.

(Scale 1 : 1,200), Ian Sturton

Operational: The first two of the class were converted to Training
Ships on 20 June 1991 but were deleted in 1995 and the third
became a Submarine Support Ship on 18 October 1993,
before being deleted in 1998. DD 119 and DDK 120 were
converted to Training Ships on 18 March 1999 and 13 March
2000 respectively.
 UPDATED

YUUGUMO
 10/2003, Hachiro Nakai / 0570718

FRIGATES

Notes: The MSDF classifies these ships as Destroyer Escorts.

3 ISHIKARI/YUUBARI CLASS (FFG/DE)

Name	No	Builders	Laid down	Launched	Commissioned
ISHIKARI	DE 226	Mitsui, Tamano	17 May 1979	18 Mar 1980	28 Mar 1981
YUUBARI	DE 227	Sumitomo, Uraga	9 Feb 1981	22 Feb 1982	18 Mar 1983
YUUBETSU	DE 228	Hitachi, Maizuru	14 Jan 1982	25 Jan 1983	14 Feb 1984

Displacement, tons: 1,470 (1,290 (DE 226)) standard; 1,690
 (1,450 (DE 226)) full load
Dimensions, feet (metres): 298.5; 278.8 (DE 226) ×
 35.4 × 11.8 (91; 85 × 10.8 × 3.6)
Main machinery: CODOG; 1 Kawasaki/RR Olympus TM3B gas
 turbine; 24,700 hp (18.4 MW) sustained; 1 Mitsubishi/MAN
 6DRV diesel; 4,700 hp(m) (3.45 MW); 2 shafts; cp props
Speed, knots: 25
Complement: 95

Missiles: SSM: 8 McDonnell Douglas Harpoon (2 quad)
 launchers **❶**; active radar homing to 130 km (70 n miles) at
 0.9 Mach; warhead 227 kg.
Guns: 1 Otobreda 3 in (76 mm)/62 compact **❷**; 85 rds/min to
 16 km (8.6 n miles) anti-surface; 12 km (6.5 n miles) anti-
 aircraft; weight of shell 6 kg. 1 General Electric/General
 Dynamics 20 mm Phalanx (unlikely to be fitted) **❸**.
Torpedoes: 6—324 mm Type 68 (2 triple) tubes **❹**. Honeywell
 Mk 46 Mod 5 Neartip; anti-submarine; active/passive homing
 to 11 km (5.9 n miles) at 40 kt; warhead 44 kg.
A/S mortars: 1—375 mm Bofors Type 71 4 to 6-barrelled
 trainable rocket launcher **❺**; automatic loading; range 1.6 km.
Countermeasures: Decoys: 2 Loral Hycor SRBOC 6-barrelled Mk
 36 chaff launchers **❻**; range 4 km (2.2 n miles).
 ESM: Nec NOLR 6B **❼**; intercept.
Combat data systems: OYQ-5.
Weapons control: Type 2-21 system for 76 mm gun.
Radars: Surface search: JRC OPS-28B/28-1 **❽**; G-band.
 Navigation: Fujitsu OPS-19B; I-band.
 Fire control: Type 2-21 **❾**; I/J-band.
Sonars: Nec SQS-36J; hull-mounted; active/passive; medium
 frequency.

Programmes: The name *Yuubari* commemorates that of a light
 cruiser sunk in the Second World War.
Structure: *Yuubari* and *Yuubetsu* were slightly larger versions of
 Ishikari. The increased space for the same weapons systems
 as *Ishikari* has meant improved accommodation and an
 increase in fuel oil carried.
 VERIFIED

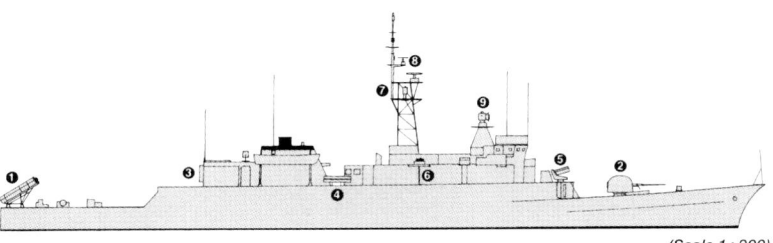

YUUBARI
 (Scale 1 : 900), Ian Sturton

ISHIKARI
 7/2003, Hachiro Nakai / 0570716

6 ABUKUMA CLASS (FFGM/DE)

Name	No	Builders	Laid down	Launched	Commissioned
ABUKUMA	DE 229	Mitsui, Tamano	17 Mar 1988	21 Dec 1988	12 Dec 1989
JINTSU	DE 230	Hitachi, Maizuru	14 Apr 1988	31 Jan 1989	28 Feb 1990
OOYODO	DE 231	Mitsui, Tamano	8 Mar 1989	19 Dec 1989	23 Jan 1991
SENDAI	DE 232	Sumitomo, Uraga	14 Apr 1989	26 Jan 1990	15 Mar 1991
CHIKUMA	DE 233	Hitachi, Maizuru	14 Feb 1991	25 Jan 1992	24 Jan 1993
TONE	DE 234	Sumitomo, Uraga	8 Feb 1991	6 Dec 1991	8 Feb 1993

Displacement, tons: 2,000 standard; 2,550 full load
Dimensions, feet (metres): 357.6 × 44 × 12.5
  (109 × 13.4 × 3.8)
Main machinery: CODOG; 2 RR Spey SM1A gas turbines;
  26,650 hp *(19.9 MW)* sustained; 2 Mitsubishi S12U-MTK
  diesels; 6,000 hp(m) *(4.4 MW)*; 2 shafts
Speed, knots: 27
Complement: 120

Missiles: SSM: 8 McDonnell Douglas Harpoon (2 quad)
  launchers ❶; active radar homing to 130 km *(70 n miles)* at
  0.9 Mach; warhead 227 kg.
A/S: Honeywell ASROC Mk 112 octuple launcher ❷; inertial
  guidance to 1.6-10 km *(1-5.4 n miles)* at 0.9 Mach; payload Mk
  46 Mod 5 Neartip.
Guns: 1 Otobreda 3 in *(76 mm)*/62 compact ❸; 85 rds/min to
  16 km *(8.6 n miles)* anti-surface; 12 km *(6.5 n miles)* anti-
  aircraft; weight of shell 6 kg.
  1 General Electric/General Dynamics 20 mm Phalanx CIWS
  Mk 15 ❹; 6 barrels per mounting; 3,000 rds/min combined to
  1.5 km.
Torpedoes: 6—324 mm HOS 301 (2 triple) tubes ❺. Honeywell
  Mk 46 Mod 5 Neartip; anti-submarine; active/passive homing
  to 11 km *(5.9 n miles)* at 40 kt; warhead 44 kg.
Countermeasures: Decoys: 2 Loral Hycor SRBOC 6-barrelled Mk
  36 chaff launchers.
ESM: Nec NOLR-8; intercept.
Combat data systems: OYQ-6. SATCOM.

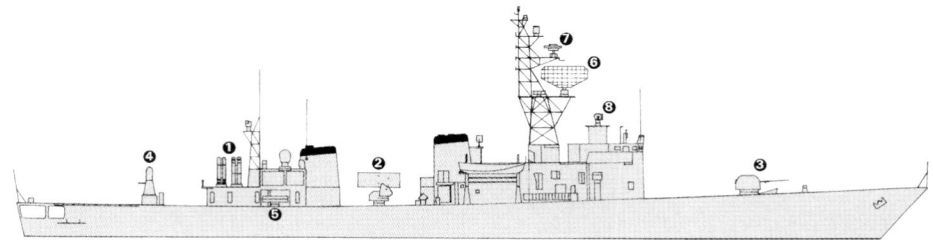

ABUKUMA *(Scale 1 : 900), Ian Sturton*

Weapons control: Type 2-21; GFCS.
Radars: Air search: Melco OPS-14C ❻; D-band.
Surface search: JRC OPS-28D (DE 233-234); JRS OPC-28C
  (remainder) ❼; G-band.
Fire control: Type 2-21 ❽.
Sonars: Hitachi OQS-8; hull-mounted; active search and attack;
  medium frequency.
  SQR-19A towed passive array in due course.

Programmes: First pair of this class approved in 1986 estimates,
  ordered March 1987; second pair in 1987 estimates, ordered
February 1988; last two in 1989 estimates, ordered
24 January 1989. The name of the first of class
commemorates that of a light cruiser which was sunk in the
battle of Leyte Gulf in October 1944.
Structure: Stealth features include non-vertical and rounded
  surfaces. German RAM PDMS may be fitted later, although
  this now seems unlikely, and space has been left for a towed
  sonar array. SATCOM fitted aft of the after funnel.

UPDATED

SENDAI *7/2004*, Hachiro Nakai / 1044400*

SHIPBORNE AIRCRAFT

Notes: Agusta Westland EH-101 selected on 5 June 2003 to replace MH-53E Sea Dragon
minesweeping aircraft and S-61A utility aircraft. Of 14 aircraft to be acquired, 11 are expected to
be in the minesweeping configuration. The first aircraft is to be built in the UK and the remainder in
Japan under licence.

Numbers/Type: 91/2 Sikorsky/Mitsubishi SH-60J/SH-60K (Seahawk).
Operational speed: 143 kt *(264 km/h).*
Service ceiling: 13,500 ft *(4,090 m).*
Range: 600 n miles *(1,110 km).*
Role/Weapon systems: ASW helicopter; started replacing HSS-2B in July 1991; built in Japan;
  prototypes fitted by Mitsubishi with Japanese avionics and mission equipment. Overall
  requirement for 103 aircraft. Half of force land-based. SH-60K are prototype aircraft. Sensors:
  Texas Instruments APS 124 search radar; sonobuoys plus datalink; Bendix AQS 18/Nippon
  HQS 103 dipping sonar, ECM, HLR 108 ESM. Weapons: ASW; two Mk 46 torpedoes or depth
  bombs.

UPDATED

S-61A *10/2003, Hachiro Nakai / 0570732*

LAND-BASED MARITIME AIRCRAFT (FRONT LINE)

Notes: Aircraft type names are not used by the MSDF.

Numbers/Type: 10 Sikorsky/Mitsubishi S-80M-1 (Sea Dragon) (MH53E).
Operational speed: 170 kt *(315 km/h).*
Service ceiling: 18,500 ft *(5,640 m).*
Range: 1,120 n miles *(2,000 km).*
Role/Weapon systems: Three-engined AMCM helicopter tows Mk 103, 104, 105 and 106 MCM
  sweep equipment; self-deployed. Weapons: Two 12.7 mm guns for mine disposal. *UPDATED*

SH-60J *7/2004*, Hachiro Nakai / 1044401*

Numbers/Type: 2 Mitsubishi-Sikorsky S-61A.
Operational speed: 144 kt *(267 km/h).*
Service ceiling: 14,700 ft *(4,280 m).*
Range: 542 n miles *(1,005 km).*
Role/Weapon systems: Support helicopter based on Sea King airframe. Deployed in ice-patrol
  ship *Shirase.*

VERIFIED

MH-53E *9/2002, Hachiro Nakai / 0529025*

Numbers/Type: 80/5/1/3/3 Lockheed/Kawasaki P-3C/EP-3/UP-3C/UP-3D/OP-3C.
Operational speed: 395 kt *(732 km/h).*
Service ceiling: 28,300 ft *(8,625 m).*
Range: 3,300 n miles *(6,100 km).*
Role/Weapon systems: Long-range MR/ASW and surface surveillance and attack. Most maritime surveillance is done by these aircraft. Four EW version EP-3. Sensors: APS-115 radar, ASQ-81 MAD, AQA 7 processor, Unisys CP 2044 computer, IFF, ECM, ALQ 78, ESM, ALR 66, sonobuoys. Weapons: ASW; eight Mk 46 torpedoes, depth bombs or mines, four underwing stations for Harpoon and ASM-1.   *UPDATED*

US-1A   *10/2004*, Hachiro Nakai /* 1044403

P-3C   *10/2004*, Hachiro Nakai /* 1044402

Numbers/Type: 7/1 Shinmeiwa US-1A Rescue/Shinmeiwa US-1A Kai.
Operational speed: 265 kt *(491 km/h).*
Service ceiling: 30,000 ft *(9,144 m).*
Range: 2,300 n miles *(4,260 km).*
Role/Weapon systems: Turboprop amphibian designed for maritime patrol and SAR missions. Crew of 12. Accommodation for 16 survivors or 12 stretchers. The US-1A Kai is undertaking trials with a view to entering service in 2007. A second test vehicle is expected by 2005. Sensors: Raytheon AN/APS-115-2 search radar.   *UPDATED*

US-1A KAI   *6/2004*, Hachiro Nakai /* 1044404

AMPHIBIOUS FORCES

3 OOSUMI CLASS (LPD/LSTH)

Name	No	Builders	Laid down	Launched	Commissioned
OOSUMI	LST 4001	Mitsui, Tamano	6 Dec 1995	18 Nov 1996	11 Mar 1998
SHIMOKITA	LST 4002	Mitsui, Tamano	30 Nov 1999	29 Nov 2000	12 Mar 2002
KUNISAKI	LST 4003	Universal, Maizuru	7 Sep 2000	13 Dec 2001	26 Feb 2003

Displacement, tons: 8,900 standard
Dimensions, feet (metres): 584 × 84.6 × 19.7 *(178 × 25.8 × 6)*
Flight deck, feet (metres): 426.5 × 75.5 *(130 × 23)*
Main machinery: 2 Mitsui 16V42MA diesels; 27,600 hp(m) *(20.29 MW)*; 2 shafts; 2 bow thrusters
Speed, knots: 22
Complement: 135
Military lift: 330 troops; 2 LCAC; 10 Type 90 tanks or 1,400 tons cargo

Guns: 2 GE/GD 20 mm Vulcan Phalanx Mk 15 ❶. 6 barrels per mounting; 3,000 rds/min combined to 1.5 km.
Countermeasures: ESM/ECM.
Radars: Air search: Mitsubishi OPS-14C ❷; C-band.
Surface search: JRC OPS-28D ❸; G-band.
Navigation: JRC OPS-20; I-band.

Helicopters: Platform for 2 CH-47J.

Programmes: A 5,500 ton LST was requested and not approved in the 1989 or 1990 estimates. The published design resembled the Italian San Giorgio with a large flight deck and a stern dock. No further action was taken for two years but the FY93 request included a larger ship showing the design of a USN LPH, although smaller in size. This vessel, with some modifications, was authorised in the 1993 estimates. A second of class approved in FY98 and third in FY99.
Structure: Through deck, flight deck and stern docking well make this more like a mini LHA than an LST, except that the ship is described as providing only 'platform and refuelling facilities for helicopters'.

UPDATED

OOSUMI   *(Scale 1 : 1,500), Ian Sturton /* 0012652

KUNISAKI
2/2003, Hachiro Nakai / 0570731

SHIMOKITA   *2/2004*, Hachiro Nakai /* 1044405

2 YURA CLASS (LSU/LCU)

Name	No	Builders	Commissioned
YURA	LSU 4171	Sasebo Heavy Industries	27 Mar 1981
NOTO	LSU 4172	Sasebo Heavy Industries	27 Mar 1981

Displacement, tons: 590 standard
Dimensions, feet (metres): 190.2 × 31.2 × 5.6 *(58 × 9.5 × 1.7)*
Main machinery: 2 Fuji 6L27.5XF diesels; 3,250 hp(m) *(2.39 MW)*; 2 shafts; cp props
Speed, knots: 12
Complement: 31
Military lift: 70 troops
Guns: 1 GE 20 mm/76 Sea Vulcan 20; 3 barrels per mounting; 1,500 rds/min combined to 4 km *(2.2 n miles)*.
Radars: Navigation: Fujitsu OPS-9B; I-band.

Comment: Both laid down 23 April 1980. 4171 launched 15 October 1980 and 4172 on 12 November 1980.

UPDATED

YURA   *6/2004*, Hachiro Nakai /* 1044406

2 YUSOTEI CLASS (LCU)

Name	No	Builders	Commissioned
YUSOTEI-ICHI-GO	LCU 2001	Sasebo Heavy Industries	17 Mar 1988
YUSOTEI-NI-GO	LCU 2002	Sasebo Heavy Industries	11 Mar 1992

Displacement, tons: 420 standard
Dimensions, feet (metres): 170.6 × 28.5 × 5.2 *(52 × 8.7 × 1.6)*
Main machinery: 2 Mitsubishi S6U-MTK diesels; 3,040 hp(m) *(2.23 MW)*; 2 shafts
Speed, knots: 12
Complement: 28
Guns: 1 GE 20 mm/76 Sea Vulcan; 3 barrels per mounting; 1,500 rds/min combined to 4 km *(2.2 n miles)*.
Radars: Navigation: OPS-19B/26; I-band.

Comment: First approved in 1986 estimates, laid down 11 May 1987, launched 9 October 1987. Second approved in FY90 estimates, laid down 17 May 1991, launched 7 October 1991; plans for a third have been scrapped. Official names are *LCU 01* and *LCU 02*.

UPDATED

YUSOTEI-ICHI-GO   *7/2004*, Hachiro Nakai /* 1044407

6 LANDING CRAFT AIR CUSHION (LCAC)

EAKUSSHONTEI – **(1-6)** – GOO LCAC 2101-2106

Displacement, tons: 89.3 light; 167 full load
Dimensions, feet (metres): 88 oa (on cushion) (81 between hard structures) × 47 beam (on cushion) (43 beam hard structure) × 2.9 draught (off cushion) *(26.8 (24.7) × 14.3 (13.1) × 0.9)*
Main machinery: 4 Avco-Lycoming TF-40B gas turbines; 2 for propulsion and 2 for lift; 16,000 hp *(12 MW)* sustained; 2 shrouded reversible-pitch airscrews (propulsion); 4 double entry fans, centrifugal or mixed flow (lift)
Speed, knots: 40 (loaded). **Range, n miles:** 300 at 35 kt; 200 at 40 kt
Complement: 5
Military lift: 24 troops; 1 MBT or 60-75 tons
Radars: Navigation: LN-66; I-band.

Comment: Built by Textron Marine, New Orleans for embarkation in LPDs. Approval for sale given by US on 8 April 1994. First one authorised in FY93, second in FY95, third and fourth in FY99 and fifth and sixth in FY00. Cargo space capacity is 1,809 sq ft.

UPDATED

EAKUSSHONTEI – 5 – GOO   *4/2004*, Kazumasa Watanabe /* 1044408

11 LCM TYPE (LCM)

YF 2075	2121	2124-25	2127-29	2132	2135	2138	2141

Displacement, tons: 25 standard
Dimensions, feet (metres): 55.8 × 14 × 2.3 *(17.0 × 4.3 × 0.7)*
Main machinery: 2 Isuzu E120-MF6R diesels; 480 hp(m) *(353 kW)*; 2 shafts
Speed, knots: 10. **Range, n miles:** 130 at 9 kt
Complement: 3
Military lift: 34 tons or 80 troops

Comment: Built in Japan. *YF 2127-29* commissioned in March 1992, *2132* in March 1993, *2135* in March 1995, *2138* in March 1996 and *2141* in March 1997. *YF 2150-51* are 50 ton vessels built by Yokohama Yacht and completed in March 2003. With a military lift of 100 tons they are capable of 16 kt.

UPDATED

YF 2135   *6/2004*, Hachiro Nakai /* 1044409

2 YF 2150 CLASS LCM (LCM)

YF 2150-51

Displacement, tons: 50 standard
Dimensions, feet (metres): 121.4 × 22.0 × 11.2 *(19.8 × 5.4 × 2.3)*
Main machinery: 2 Mitsubishi S12R-MTK diesels; 3,000 hp *(2.24 MW)*; 2 waterjets
Speed, knots: 16
Complement: 4
Military lift: 100 troops or 1 vehicle

Comment: Built in Japan by Universal, Keihin and commissioned on 19 March 2003.

VERIFIED

YF 2150   *3/2003, Universal /* 0570727

MINE WARFARE FORCES

2 URAGA CLASS (MINESWEEPER TENDERS) (MSTH/ML)

Name	No	Builders	Launched	Commissioned
URAGA	MST 463	Hitachi, Maizuru	22 May 1996	19 Mar 1997
BUNGO	MST 464	Mitsui, Tamano	24 Apr 1997	23 Mar 1998

Displacement, tons: 5,700 standard
Dimensions, feet (metres): 462.6 × 72.2 × 17.7 *(141 × 22 × 5.4)*
Main machinery: 2 Mitsui 12V42MA diesels; 19,500 hp(m) *(14.33 MW)*; 2 shafts
Speed, knots: 22
Complement: 160
Guns: 1 OTO Melara 3 in *(76 mm)*/62 compact; 85 rds/min to 16 km *(8.6 n miles)*; weight of shell 6 kg.
Mines: Laying capability; 4 rails (Type 3).
Radars: Fire control: Type 2-21; I/J-band.
Navigation: JRC OPS-39C; I-band.
Helicopters: Platform for 1 MH-53E.

Comment: First one authorised 15 February 1994 and laid down 19 May 1995; second authorised in FY95 and laid down 4 July 1996. Capable of laying mines, from four internal rails. Phalanx is planned to be fitted forward of the bridge and on the superstructure aft of the funnel.

UPDATED

BUNGO   *11/2004*, Hachiro Nakai /* 1044410

3 YAEYAMA CLASS (MINESWEEPERS—OCEAN) (MSO)

Name	No	Builders	Launched	Commissioned
YAEYAMA	MSO 301	Hitachi Zosen, Kanagawa	29 Aug 1991	16 Mar 1993
TSUSHIMA	MSO 302	Nippon Koukan, Tsurumi	20 Sep 1991	23 Mar 1993
HACHIJYO	MSO 303	Nippon Koukan, Tsurumi	15 Dec 1992	24 Mar 1994

Displacement, tons: 1,000 standard; 1,275 full load
Dimensions, feet (metres): 219.8 × 38.7 × 10.2 *(67 × 11.8 × 3.1)*
Main machinery: 2 Mitsubishi 6NMU-TA1 diesels; 2,400 hp(m) *(1.76 MW)*; 2 shafts; 1 hydrojet
  bow thruster; 350 hp(m) *(257 kW)*
Speed, knots: 14
Complement: 60
Guns: 1 JM-61 20 mm/76 Sea Vulcan; 3 barrels per mounting; 1,500 rds/min combined to 4 km
  (2.2 n miles).
Radars: Surface search: Fujitsu OPS-39B; I-band.
Sonars: Raytheon SQQ-32 VDS; high frequency; active.

Comment: First two approved in 1989 estimates, third in 1990. First laid down 30 August 1990,
  second 20 July 1990 and third 17 May 1991. Wooden hulls. Fitted with S 7 deep sea
  minehunting system, S 8 (SLQ-48) deep sea moored minesweeping equipment and ADI Dyad
  sweeps. Appears to be a derivative of the USN Avenger class. An integrated tactical system is
  fitted. Termination of the programme at three of the class suggests similar problems to US ships
  of the same class. *UPDATED*

IZUSHIMA   *5/2004*, Hachiro Nakai /* 1044412

TSUSHIMA   *9/2004*, Hachiro Nakai /* 1044411

2 NIIJIMA CLASS (DRONE CONTROL SHIPS) (MCSD)

Name	No	Builders	Commissioned
HAHAJIMA	MCL 724 (ex-MSC 660)	Nippon Koukan, Tsurumi	18 Dec 1984
KAMISHIMA	MCL 725 (ex-MSC 664)	Nippon Koukan, Tsurumi	16 Dec 1986

Displacement, tons: 440 standard; 510 full load
Dimensions, feet (metres): 180.4 × 30.8 × 7.9 *(55 × 9.4 × 2.4)*
Main machinery: 2 Mitsubishi 12ZC diesels; 1,440 hp(m) *(1.06 MW)*; 2 shafts
Speed, knots: 14
Complement: 28
Guns: 1 GE 20 mm/76 Sea Vulcan 20; 3 barrels per mounting; 1,500 rds/min combined to 4 km
  (2.2 n miles).
Radars: Surface search: Fujitsu OPS-9B; I-band.

Comment: Both converted to act as Minesweeper Control Ship (MCLs) and equipped to operate
  SAM remote controlled drones. All minesweeping gear removed. *VERIFIED*

14 HATSUSHIMA/UWAJIMA CLASS
(MINEHUNTERS/SWEEPERS—COASTAL) (MHSC)

Name	No	Builders	Commissioned
OGISHIMA	MSC 666	Hitachi, Kanagawa	19 Dec 1987
YURISHIMA	MSC 668	Nippon Koukan, Tsurumi	15 Dec 1988
HIKOSHIMA	MSC 669	Hitachi, Kanagawa	15 Dec 1988
AWASHIMA	MSC 670	Hitachi, Kanagawa	13 Dec 1989
SAKUSHIMA	MSC 671	Nippon Koukan, Tsurumi	13 Dec 1989
UWAJIMA	MSC 672	Nippon Koukan, Tsurumi	19 Dec 1990
IESHIMA	MSC 673	Hitachi, Kanagawa	19 Dec 1990
TSUKISHIMA	MSC 674	Hitachi, Kanagawa	17 Mar 1993
MAEJIMA	MSC 675	Hitachi, Kanagawa	15 Dec 1993
KUMEJIMA	MSC 676	Nippon Koukan, Tsurumi	12 Dec 1994
MAKISHIMA	MSC 677	Hitachi, Kanagawa	12 Dec 1994
TOBISHIMA	MSC 678	Nippon Koukan, Tsurumi	10 Mar 1995
YUGESHIMA	MSC 679	Hitachi, Kanagawa	11 Dec 1996
NAGASHIMA	MSC 680	Nippon Koukan, Tsurumi	25 Dec 1996

Displacement, tons: 440 (490, MSC 670 onwards) standard; 510 full load
Dimensions, feet (metres): 180.4 (190.3, MSC 670 onwards) × 30.8 × 8.2 (9.5)
  (55 (58.0) × 9.4 × 2.5 (2.9))
Main machinery: 2 Mitsubishi 6NMU-TAI diesels; 1,400 hp(m) *(1.03 MW)*; 2 shafts
Speed, knots: 14
Complement: 45; 40 (MSC 675 onwards)

Guns: 1 JM-61 20 mm/76 Sea Vulcan 20; 3 barrels per mounting; 1,500 rds/min combined to
  4 km *(2.2 n miles).*
Radars: Surface search: Fujitsu OPS-9 or OPS-39 (MSC 674 onwards); I-band.
Sonars: Nec/Hitachi ZQS 2B or ZQS 3 (MSC 672 onwards); hull-mounted; minehunting; high
  frequency.

Programmes: First ordered in 1976. Last two authorised in FY94. Because of the new sonar and
  mine detonating equipment vessels from MSC 672 onwards are known as the Uwajima class.
Structure: From MSC 670 onwards the hull is lengthened by 2.7 m in order to improve the
  sleeping accommodation from three tier to two tier bunks. Hulls are made of wood. The last pair
  has more powerful engines developing 1,800 hp(m) *(1.32 MW)*.
Operational: Fitted with S 4 (S 7 from MSC 672 onwards) mine detonating equipment, a remote-
  controlled counter-mine charge. Four clearance divers are carried. MSC 668, 669, 670 and 671
  formed the Minesweeper Squadron to deploy to the Gulf in 1991. Earlier vessels of the class
  converted to drone control or paid off at a rate of one or two a year. *UPDATED*

HAHAJIMA   *1/2003, Takeshi Oosaki /* 0573520

10 + 2 SUGASHIMA CLASS (MINESWEEPER (COASTAL)) (MSC)

Name	No	Builders	Launched	Commissioned
SUGASHIMA	MSC 681	NKK, Tsurumi	25 Aug 1997	16 Mar 1999
NOTOJIMA	MSC 682	Hitachi, Kanagawa	3 Sep 1997	16 Mar 1999
TSUNOSHIMA	MSC 683	Hitachi, Kanagawa	22 Oct 1998	13 Mar 2000
NAOSHIMA	MSC 684	NKK, Tsurumi	7 Oct 1999	16 Mar 2001
TOYOSHIMA	MSC 685	Hitachi, Kanagawa	13 Sep 2000	4 Mar 2002
UKUSHIMA	MSC 686	Universal, Keihin (Tsurumi)	17 Sep 2001	18 Mar 2003
IZUSHIMA	MSC 687	Universal, Keihin (Kanawaga)	31 Oct 2001	18 Mar 2003
AISHIMA	MSC 688	Universal, Keihin (Tsurumi)	8 Oct 2002	16 Feb 2004
AOSHIMA	MSC 689	Universal, Keihin (Kanawaga)	16 Sep 2003	9 Feb 2005
MIYAJIMA	MSC 690	Universal, Keihin (Tsurumi)	10 Oct 2003	9 Feb 2005
SHISHIJIMA	MSC 691	Universal, Keihin (Tsurumi)	29 Sep 2004	Feb 2006
—	MSC 692	Universal, Keihin (Tsurumi)	Aug 2005	Feb 2007

Displacement, tons: 510 standard
Dimensions, feet (metres): 177.2 × 30.8 × 9.8 *(54.0 × 9.4 × 3.0)*
Main machinery: 2 Mitsubishi 6 NMU-TA1 diesels; 1,800 hp(m) *(1.33 MW)*; 2 shafts; bow
  thrusters
Speed, knots: 14
Complement: 45
Guns: 1 JM-61 20 mm/76 Sea Vulcan; 3 barrels for mounting; 1,500 rds/min combined to 4 km
  (2 n miles).
Combat data systems: AMS/NEC Nautis-M type MCM control system.
Radars: Surface search: Fujitsu OPS-39B; I-band.
Sonars: THALES Hitachi GEC Type 2093 VDS; high frequency; active.

Comment: First pair authorised in FY95, third in FY96, fourth in FY97, fifth in FY98, sixth and
  seventh in FY99, eighth in FY00, ninth and tenth in FY01, eleventh in FY02 and twelfth in FY03.
  Hull is similar to *Uwajima* but the upper deck is extended aft to provide more stowage for mine
  disposal gear, and there are twin funnels. PAP 104 Mk 5 ROVs are carried and ADI Dyad
  minesweeping gear fitted. *UPDATED*

NAGASHIMA   *6/2004*, Hachiro Nakai /* 1044414

KUMEJIMA   *6/2004*, Hachiro Nakai /* 1044415

TOBISHIMA   *2/2003, Hachiro Nakai /* 0570737

0 + 2 570 TON CLASS (MINESWEEPERS—COASTAL) (MSC)

Displacement, tons: 570 standard
Dimensions, feet (metres): 187 × 32.1 × 9.8 *(57.0 × 9.8 × 3.0)*
Main machinery: 2 diesels; 2 shafts
Speed, knots: 14
Complement: To be announced
Guns: 1—20 mm Sea Vulcan.

Comment: First authorised in FY04 budget to enter service in 2008. Second authorised in FY05 budget. To be equipped with S-10 minesweeping and disposal system.   *UPDATED*

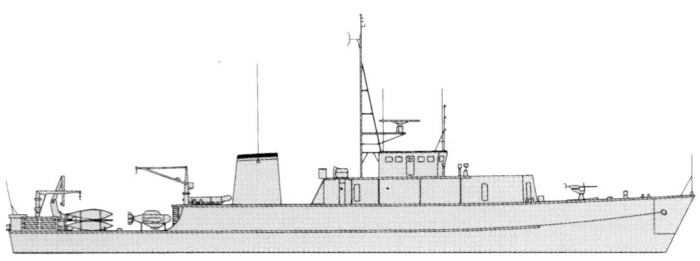

570 TON MSC   *(not to scale), Ian Sturton /* 0569918

6 SAM CLASS (MSD)

SAM 01-06

Displacement, tons: 20 full load
Dimensions, feet (metres): 59.1 × 20 × 5.2 *(18 × 6.1 × 1.6)*
Main machinery: 1 Volvo Penta TAMD 70D diesel; 210 hp(m) *(154 kW)*; 1 Schottel prop
Speed, knots: 8. **Range, n miles:** 330 at 8 kt

Comment: First pair acquired from Karlskronavarvet, Sweden in February 1998 followed by two more in December 1998 and two more in 2000. Remote controlled magnetic and acoustic catamaran sweepers operated by *Kamishima* and *Ogishima*.   *UPDATED*

SAM 02   *2/2000, Hachiro Nakai /* 0104934

PATROL FORCES

6 HAYABUSA CLASS (PGGF)

Name	No	Builders	Launched	Commissioned
HAYABUSA	824	Mitsubishi, Shimonoseki	13 June 2001	25 Mar 2002
WAKATAKA	825	Mitsubishi, Shimonoseki	13 Sep 2001	25 Mar 2002
OOTAKA	826	Mitsubishi, Shimonoseki	13 May 2002	24 Mar 2003
KUMATAKA	827	Mitsubishi, Shimonoseki	2 Aug 2002	24 Mar 2003
UMITAKA	828	Mitsubishi, Shimonoseki	21 May 2003	24 Mar 2004
SHIRATAKA	829	Mitsubishi, Shimonoseki	8 Aug 2003	24 Mar 2004

Displacement, tons: 200 standard
Dimensions, feet (metres): 164.4 × 27.6 × 13.8 *(50.1 × 8.4 × 4.2)*
Main machinery: 3 LM 500-G07 gas turbines 16,200 hp *(12.08 MW)*; 3 water jets
Speed, knots: 44
Complement: 18 (+3 staff)
Missiles: 4 Mitsubishi Type 90 SSM-1B; active radar homing to 130 km *(70 n miles)* at 0.9 Mach; warhead 227 kg.
Guns: 1 OTO Melara 3 in *(76 mm)*/62 compact; 85 rds/min to 16 km *(8.7 n miles)* anti-surface; 12 km *(6.6 n miles)* anti-aircraft; weight of shell 6 kg.
2—12.7 mm MGs.
Countermeasures: Decoys: chaff launchers.
ESM/ECM.
Radars: Surface search: OPS-18-3; G-band.
Fire control: Type 2-31.
Navigation: OPS-20; I-band.

Comment: First pair authorised in FY99 budget, second pair in FY00 and third pair in FY01. Single hull.   *UPDATED*

SHIRATAKA   *3/2004*, Hachiro Nakai /* 1044416

3 PG 01 (SPARVIERO) CLASS
(FAST ATTACK HYDROFOIL—MISSILE) (PTGK)

Name	No	Builders	Launched	Commissioned
MISAIRUTEI-ICHI-GO	821	Sumitomo, Uraga	17 July 1992	25 Mar 1993
MISAIRUTEI-NI-GO	822	Sumitomo, Uraga	17 July 1992	25 Mar 1993
MISAIRUTEI-SAN-GO	823	Sumitomo, Uraga	15 June 1994	13 Mar 1995

Displacement, tons: 50 standard
Dimensions, feet (metres): 71.5 × 22.9 × 4.6 *(21.8 × 7 × 1.4)* (hull)
80.7 × 23.1 × 14.4 *(24.6 × 7 × 4.4)* (foilborne)
Main machinery: 1 GE/IHI LM 500 gas turbine; 5,000 hp *(3.72 MW)* sustained; 1 pumpjet (foilborne); 1 diesel; 1 retractable prop (hullborne)
Speed, knots: 46; 8 (diesel). **Range, n miles:** 400 at 45 kt; 1,200 at 8 kt
Complement: 11 (3 officers)
Missiles: SSM: 4 Mitsubishi SSM-1B (derivative of land-based system); range 150 km *(81 n miles)*.
Guns: 1 GE 20 mm/76 Sea Vulcan; 3 barrels per mounting; 1,500 rds/min combined to 4 km *(2.2 n miles)*.
Countermeasures: Decoys: 2 Loral Hycor Mk 36 SRBOC chaff launchers.
ESM/ECM: intercept and jammer.
Combat data systems: Link 11.
Radars: Surface search: JRC OPS-28-2; G-band.

Comment: Classified as Guided Missile Patrol Boats. First two approved in FY90 and both laid down 25 March 1991. One more approved in FY92, laid down 8 March 1993. A fourth was asked for but not authorised in FY95 and the programme is now complete. Built with Italian assistance from Fincantieri. Planned to improve the Navy's interceptor capabilities, this was an ambitious choice of vessel bearing in mind the falling popularity of the hydrofoil in the few navies (US, Italy and Russia) that built them.   *UPDATED*

MISAIRUTEI-SAN-GO   *7/2003, Hachiro Nakai /* 0570739

AUXILIARIES

2 MASHUU CLASS
(FAST COMBAT SUPPORT SHIPS) (AOE/AORH)

Name	No	Builders	Laid down	Launched	Commissioned
MASHUU	AOE 425	Mitsui, Tamano	21 Jan 2002	5 Feb 2003	15 Mar 2004
OUMI	AOE 426	Universal, Maizuru	7 Feb 2003	19 Feb 2004	3 Mar 2005

Displacement, tons: 13,500 standard
Dimensions, feet (metres): 725 × 88.6 × 27,2 *(221 × 27 × 8.3)*
Main machinery: 2 Kawasaki RR Spey SM1C gas turbines; 40,000 hp *(29.8 MW)*; 2 shafts
Speed, knots: 24
Complement: 145
Guns: 2—20 mm CIWS (to be fitted).
Countermeasures: Decoys: 4 SRBOC Mk 36 chaff and IR launchers.

Comment: First ship approved in FY00 and second in FY01. Capacity for 30 containers. Cranes capable of lifting 15 tons. Three replenishment at sea positions on each side.   *UPDATED*

MASHUU   *3/2004*, Hachiro Nakai /* 1044417

6 300 TON CLASS (EOD TENDERS) (YDT)

YDT 01-06

Displacement, tons: 309 standard
Dimensions, feet (metres): 150.9 × 28.2 × 7.2 *(46 × 8.6 × 2.2)*
Main machinery: 2 Niigata 6NSDL diesels; 1,500 hp(m) *(1.1 MW)*; 2 shafts
Speed, knots: 15
Complement: 15 plus 15 divers

Comment: Built by Maehata Zousen. First pair approved in FY98, third in FY99, fourth in FY00, fifth and sixth in FY01. First two commissioned 24 March 2000, third on 21 March 2001, fourth in December 2001 and last two on 14 March 2003. Used as diving tenders.   *UPDATED*

YDT 04   *2/2004*, Hachiro Nakai /* 1044418

1 CHIYODA CLASS
(SUBMARINE TENDER DEPOT AND RESCUE SHIP) (AS/ASRH)

Name	No	Builders	Launched	Commissioned
CHIYODA	AS 405	Mitsui, Tamano	7 Dec 1983	27 Mar 1985

Displacement, tons: 3,650 standard; 4,450 full load
Dimensions, feet (metres): 370.6 × 57.7 × 15.1 *(113 × 17.6 × 4.6)*
Main machinery: 2 Mitsui 8L42M diesels; 10,540 hp(m) *(8.8 MW)*; 2 shafts; cp props; bow and stern thrusters
Speed, knots: 17
Complement: 120
Radars: Navigation: JRC OPS-18-1; G-band.
Helicopters: Platform for up to MH-53 size.

Comment: Laid down 19 January 1983. Carries a Deep Submergence Rescue Vehicle (DSRV), which is lowered and recovered through a centreline moonpool. The DSRV can mate to a decompression chamber. A personnel transfer capsule can also be deployed. Flagship Second Submarine Flotilla based at Yokosuka. ***UPDATED***

CHIYODA *1/2004*, Hachiro Nakai /* 1044421

1 CHIHAYA CLASS (SUBMARINE RESCUE SHIP) (ASRH)

Name	No	Builders	Launched	Commissioned
CHIHAYA	ASR 403	Mitsui, Tamano	8 Oct 1998	23 Mar 2000

Displacement, tons: 5,450 standard
Dimensions, feet (metres): 419.9 × 65.6 × 16.7 *(128 × 20 × 5.1)*
Main machinery: 2 Mitsui 12V 42M-A diesels; 19,500 hp(m) *(14.33 MW)*; 2 shafts; 2 bow and 2 stern thrusters
Speed, knots: 21
Complement: 125
Radars: Navigation: I-band.
Helicopters: Platform for up to MH-53 size.

Comment: Authorisation approved in the 1996 budget as a replacement for *Fushimi*. Laid down 13 October 1997. Fitted with a search sonar and carries a 40 ton DSRV. Also used as a hospital ship. ***UPDATED***

CHIHAYA *6/2004*, Hachiro Nakai /* 1044422

3 TOWADA CLASS
(FAST COMBAT SUPPORT SHIPS) (AOE/AORH)

Name	No	Builders	Launched	Commissioned
TOWADA	AOE 422	Hitachi, Maizuru	25 Mar 1986	24 Mar 1987
TOKIWA	AOE 423	Ishikawajima Harima, Tokyo	23 Mar 1989	12 Mar 1990
HAMANA	AOE 424	Hitachi, Maizuru	18 May 1989	29 Mar 1990

Displacement, tons: 8,150 standard; 15,850 full load
Dimensions, feet (metres): 547.8 × 72.2 × 26.9 *(167 × 22 × 8.2)*
Main machinery: 2 Mitsui 16V42MA diesels; 23,950 hp(m) *(17.6 MW)*; 2 shafts
Speed, knots: 22
Complement: 140
Cargo capacity: 5,700 tons
Countermeasures: Decoys: 2 chaff launchers can be fitted.
Radars: Surface search: JRC OPS-18-1/28C; G-band.
Helicopters: Platform for 1 Sea King size.

Comment: First approved under 1984 estimates, laid down 17 April 1985. Second and third of class in 1987 estimates. AOE 423 laid down 12 May 1988, and AOE 424 8 July 1988. Three replenishment at sea positions on each side (two fuel only, one stores). ***VERIFIED***

TOWADA *4/2003, Hachiro Nakai /* 0570746

34 HARBOUR TANKERS (YO/YW/YG)

Comment: There are: 17 of 490 tons (YO 14, 21-27, 29-31, 33-38); eight of 310 tons (YW 17-24); two of 290 tons (YO 12-13); seven of 270 tons (YO 28, 32 and YG 201-205). ***VERIFIED***

YW 23 *4/2003, Hachiro Nakai /* 0570748

2 FIREFIGHTING TENDERS (YTR)

YR 01-02

Displacement, tons: 60 standard
Dimensions, feet (metres): 82.0 × 18.0 × 3.6 *(25.0 × 5.5 × 1.1)*
Main machinery: 2 diesels; 2 shafts
Speed, knots: 19
Complement: 10

Comment: Built in Japan by Ishikawajima-Harima Heavy Industries. *YR 01* approved in FY99 budget and commissioned in 2001. *YR 02* approved in FY00 budget and commissioned in 2002. Fitted with three waterjets forward and a crane aft. ***VERIFIED***

YR 01 *9/2002, Takatoshi Okano /* 0570888

1 MUROTO CLASS (CABLE REPAIR SHIP) (ARC)

Name	No	Builders	Launched	Commissioned
MUROTO	ARC 482	Mitsubishi, Shimonoseki	25 July 1979	27 Mar 1980

Displacement, tons: 4,500 standard
Dimensions, feet (metres): 436.2 × 57.1 × 18.7 *(133 × 17.4 × 5.7)*
Main machinery: 4 Kawasaki-MAN V8V22/30ATL diesels; 8,000 hp(m) *(5.88 MW)*; 2 shafts; bow thruster
Speed, knots: 18
Complement: 135
Radars: Navigation: Fujitsu OPS-9B; I-band.

Comment: Ocean survey capability. Laid down 28 November 1978. Similar vessels in civilian use. ***VERIFIED***

MUROTO *1/2003, Hachiro Nakai /* 0535883

1 HASHIDATE CLASS (ASY/YAC)

Name	No	Builders	Launched	Commissioned
HASHIDATE	ASY 91	Hitachi, Kanagawa	26 July 1999	30 Nov 1999

Displacement, tons: 400 standard
Dimensions, feet (metres): 203.4 × 30.8 × 6.6 *(62 × 9.4 × 2.0)*
Main machinery: 2 Niigata 16V 16FX diesels; 5,500 hp(m) *(4.04 MW)*; 2 shafts
Speed, knots: 20. **Range, n miles:** 1,000 at 12 kt
Complement: 29 plus 130 passengers

Comment: Authorised in FY97 budget. Laid down 28 October 1998. Has replaced *Hiyodori* as a ceremonial yacht. Has facilities for disaster relief. Based at Yokosuka.

UPDATED

HASHIDATE   *10/2004*, Hachiro Nakai /* 1044423

3 HIUCHI CLASS (MULTIPURPOSE SUPPORT SHIPS) (YTT)

Name	No	Builders	Launched	Commissioned
HIUCHI	AMS 4301	NKK, Tsurumi	4 Sep 2001	27 Mar 2002
SUOU	AMS 4302	Universal, Keihin (Tsurumi)	25 Apr 2003	16 Mar 2004
AMAKUSA	AMS 4303	Universal, Keihin (Tsurumi)	6 Aug 2003	16 Mar 2004

Displacement, tons: 980 standard
Dimensions, feet (metres): 213.3 × 39.4 × 11.5 *(65 × 12 × 3.5)*
Main machinery: 2 Daihatsu 6 DKM-28 (L) diesels; 5,000 hp(m) *(3.67 MW)*; 2 shafts
Speed, knots: 15
Complement: 40

Comment: First authorised in FY99 and two more in FY01. Equipped for torpedo launch and recovery. Replaced ASU 81 class. Used as an ocean tug.

UPDATED

AMAKUSA   *4/2004*, Hachiro Nakai /* 1044424

RESCUE VEHICLES

2 RESCUE SUBMARINES (DSRV)

Displacement, tons: 40
Dimensions, feet (metres): 40.7 × 10.5 × 14.1 *(12.4 × 17.6 × 4.6)*
Main machinery: Electric; 30 hp *(22 kW)*; single shaft
Speed, knots: 4
Complement: 4

Comment: Rescue submersibles built by Kawasaki Heavy Industries, Kobe and delivered on 27 August 1999. Space for 12 people. Sonars are fitted on the bow, upper and lower casings for depth sounding and obstacle avoidance. Can be deployed in the submarine rescue ships *Chiyoda* (AS 405) and *Chihaya* (ASR 403).

UPDATED

DSRV   *10/2003, Hachiro Nakai /* 0570848

SURVEY AND RESEARCH SHIPS

Notes: (1) The SES trials ship *Merguro II* does not belong to the MSDF.
(2) Survey ships are also included in the Coast Guard section.

2 HIBIKI CLASS (OCEAN SURVEILLANCE SHIPS) (AGOSH)

Name	No	Builders	Launched	Commissioned
HIBIKI	AOS 5201	Mitsui, Tamano	27 July 1990	30 Jan 1991
HARIMA	AOS 5202	Mitsui, Tamano	11 Sep 1991	10 Mar 1992

Displacement, tons: 2,850 standard
Dimensions, feet (metres): 219.8 × 98.1 × 24.6 *(67 × 29.9 × 7.5)*
Main machinery: Diesel-electric; 4 Mitsubishi 6SU diesels; 6,700 hp(m) *(4.93 MW)*; 4 generators; 2 motors; 3,000 hp(m) *(2.2 MW)*; 2 shafts
Speed, knots: 11 (3 towing). **Range, n miles:** 3,800 at 10 kt
Complement: 40
Radars: Surface search: JRC OPS-18-1; G-band.
Navigation: Koden OPS-20; I-band.
Sonars: UQQ 2 SURTASS; passive surveillance.
Helicopters: Platform only.

Comment: First authorised 24 January 1989, laid down 28 November, second approved in FY90, laid down 26 December 1990. Auxiliary Ocean Surveillance (AOS) ships to a SWATH design similar to USN TAGOS-19 class. A data collection station is based at Yokosuka Bay using WSC-6 satellite data relay to the AOS.

VERIFIED

HARIMA   *10/2000, Hachiro Nakai /* 0104945

1 NICHINAN CLASS (SURVEY SHIP) (AGS)

Name	No	Builders	Launched	Commissioned
NICHINAN	AGS 5105	Mitsubishi, Shimonoseki	11 June 1998	24 Mar 1999

Displacement, tons: 3,300 standard
Dimensions, feet (metres): 364.2 × 55.8 × 14.8 *(111 × 17 × 4.5)*
Main machinery: Diesel-electric; 2 Mitsubishi S16U diesel generators; 2 motors; 5,800 hp(m) *(4.26 MW)*; 2 shafts
Speed, knots: 18
Complement: 80

Comment: Authorisation approved in FY96 to replace *Akashi*. Combination cable repair and hydrographic survey ship.

UPDATED

NICHINAN   *3/1999*, Hachiro Nakai /* 1044425

1 SUMA CLASS (AGS)

Name	No	Builders	Launched	Commissioned
SUMA	AGS 5103	Hitachi, Maizuru	1 Sep 1981	30 Mar 1982

Displacement, tons: 1,180 standard
Dimensions, feet (metres): 236.2 × 42 × 11.1 *(72 × 12.8 × 3.4)*
Main machinery: 2 Fuji 6L27.5XF diesels; 3,250 hp(m) *(2.39 MW)*; 2 shafts; cp props; bow thruster
Speed, knots: 15
Complement: 65

Comment: Laid down 24 September 1980. Carries an 11 m launch for surveying work.

VERIFIED

SUMA   *2/1999, Hachiro Nakai /* 0080180

2 FUTAMI CLASS (AGS)

Name	No	Builders	Launched	Commissioned
FUTAMI	AGS 5102	Mitsubishi, Shimonoseki	9 Aug 1978	27 Feb 1979
WAKASA	AGS 5104	Hitachi, Maizuru	21 May 1985	25 Feb 1986

Displacement, tons: 2,050 standard; 3,175 full load
Dimensions, feet (metres): 318.2 × 49.2 × 13.8 *(97 × 15 × 4.2)*
Main machinery: 2 Kawasaki-MAN V8V22/30ATL diesels; 4,000 hp(m) *(2.94 MW)* (AGS 5102);
   2 Fuji 8L27.5XF diesels; 3,250 hp(m) *(2.39 MW)* (AGS 5104); 2 shafts; cp props; bow thruster
Speed, knots: 16
Complement: 105
Radars: Navigation: JRC OPS-18-3; G-band.

Comment: AGS 5102 laid down 20 January 1978, AGS 5104 21 August 1984. Built to merchant marine design. Carry an RCV-225 remote-controlled rescue/underwater survey submarine. *Wakasa* has a slightly taller funnel.

UPDATED

FUTAMI   *10/2002*, Takeshi Oosaki /* 1044426

1 KURIHAMA CLASS (ASE/AGE)

Name	No	Builders	Launched	Commissioned
KURIHAMA	ASE 6101	Sasebo Heavy Industries	20 Sep 1979	8 Apr 1980

Displacement, tons: 950 standard
Dimensions, feet (metres): 223 × 37.9 × 9.8 (screws) *(68 × 11.6 × 3)*
Main machinery: 2 Fuji 6S30B diesels; 4,800 hp(m) *(3.5 MW)*; 2 shafts; 2 cp props; 2 auxiliary electric props; bow thruster
Speed, knots: 15
Complement: 40 plus 12 scientists
Radars: Navigation: Fujitsu OPS-9B; I-band.

Comment: Experimental ship built for the Technical Research and Development Institute and used for testing underwater weapons and sensors.

VERIFIED

KURIHAMA   *7/2002, Hachiro Nakai /* 0529049

1 ASUKA CLASS (AGEH)

Name	No	Builders	Launched	Commissioned
ASUKA	ASE 6102	Sumitomo, Uraga	21 June 1994	22 Mar 1995

Displacement, tons: 4,250 standard
Dimensions, feet (metres): 495.4 × 56.8 × 16.4 *(151 × 17.3 × 5)*
Main machinery: COGLAG; 2 IHI/GE LM 2500 gas turbines; 43,000 hp *(31.6 MW)*; 2 shafts; cp props
Speed, knots: 27
Complement: 70 plus 100 scientists
Missiles: SAM: 8 cell VLS.
Weapons control: Type 3 FCS.
Radars: Air search: SPY-1D type; E/F-band.
Air/surface search: Melco OPS-14C; D-band.
Surface search: JRC OPS-18-1; G-band.
Fire control: Type 3; I/J-band.
Sonars: Bow-mounted; active search; medium frequency.
   Towed passive/active array in due course.
Helicopters: 1 SH-60J Seahawk.

Comment: Included in the FY92 programme and laid down 21 April 1993. For experimental and weapon systems testing which started with the FCS 3 in 1996. The bow sonar dome extends aft to the bridge. The VLS system is on the forecastle. Surveillance and countermeasures systems are also being evaluated.

UPDATED

ASUKA   *4/2004*, Hachiro Nakai /* 1044427

TRAINING SHIPS

1 SHIMAYUKI CLASS (TRAINING SHIP) (AXGHM/TV)

Name	No	Builders	Commissioned
SHIMAYUKI	TV 3513 (ex-DD 133)	Mitsubishi, Nagasaki	17 Feb 1987

Displacement, tons: 3,050 standard; 3,800 full load
Dimensions, feet (metres): 426.4 × 44.6 × 14.4 *(130 × 13.6 × 4.4)*
Main machinery: COGOG; 2 Kawasaki-RR Olympus TM3B gas turbines; 49,400 hp *(36.8 MW)* sustained; 2 RR Type RM1C gas turbines; 9,900 hp *(7.4 MW)* sustained; 2 shafts; cp props
Speed, knots: 30; 19 cruise
Complement: 200

Missiles: SSM: 8 McDonnell Douglas Harpoon (2 quad) launchers; active radar homing to 130 km *(70 n miles)* at 0.9 Mach; warhead 227 kg.
SAM: Raytheon Sea Sparrow Mk 29 Type 3A launcher; semi-active radar homing to 14.6 km *(8 n miles)* at 2.5 Mach; warhead 39 kg; 12 missiles.
A/S: Honeywell ASROC Mk 112 octuple launcher; inertial guidance to 1.6-10 km *(1-5.4 n miles)* at 0.9 Mach; payload Mk 46 Mod 5 Neartip.
Guns: 1 OTO Melara 3 in *(76 mm)*/62 compact; 85 rds/min to 16 km *(8.6 n miles)* anti-surface; 12 km *(6.5 n miles)* anti-aircraft; weight of shell 6 kg.
   2 General Electric/General Dynamics 20 mm Phalanx Mk 15 CIWS; 6 barrels per mounting; 3,000 rds/min combined to 1.5 km.
Torpedoes: 6—324 mm Type 68 (2 triple) tubes. Honeywell Mk 46 Mod 5 Neartip; anti-submarine; active/passive homing to 11 km *(5.9 n miles)* at 40 kt; warhead 44 kg.
Countermeasures: Decoys: 2 Loral Hycor SRBOC 6-barrelled Mk 36 chaff launchers; range 4 km *(2.2 n miles)*.
ESM: NOLR 6C; intercept.
ECM: Fujitsu OLT 3; jammer.
Combat data systems: OYQ-5 action data automation; Link 14 (receive only). SATCOM.
Radars: Air search: Melco OPS-14B; D-band.
Surface search: JRC OPS-18-1; G-band.
Fire control: Type 2-12 A; I/J-band (for SAM).
   2 Type 2-21/21A; I/J-band (for guns).
Tacan: ORN-6C.
Sonars: Nec OQS 4A (II) (SQS-23 type); bow-mounted; active search and attack; low frequency.
Helicopters: 1 SH-60J Seahawk.

Comment: Converted to training ship in March 1999. Lecture room added to helicopter hangar.

VERIFIED

SHIMAYUKI   *6/2002, Mitsuhiro Kadota /* 0528911

1 KASHIMA CLASS (TRAINING SHIP) (AXH/TV)

Name	No	Builders	Launched	Commissioned
KASHIMA	TV 3508	Hitachi, Maizuru	23 Feb 1994	26 Jan 1995

Displacement, tons: 4,050 standard
Dimensions, feet (metres): 469.2 × 59.1 × 15.1 *(143 × 18 × 4.6)*
Main machinery: CODOG; 2 RR Spey SM1C gas-turbines; 26,650 hp *(19.9 MW)* sustained; 2 Mitsubishi S16U-MTK diesels; 8,000 hp(m) *(5.88 MW)*; 2 shafts
Speed, knots: 25. **Range, n miles:** 7,000 at 18 kt
Complement: 389 (includes 140 midshipmen)
Guns: 1 OTO Melara 76 mm/62. 2—40 mm saluting guns.
Torpedoes: 6—324 mm (2 triple) tubes.
Radars: Air/surface search: Melco OPS-14C; D-band.
Surface search: JRC OPS-18-1; D-band.
Navigation: Fujitsu OPS-20; I-band.
Fire control: Type 2-22; I/J-band.
Sonars: Hull-mounted; active search and attack; medium frequency.
Helicopters: Platform for 1 medium.

Comment: Approved in FY91 as a dedicated training ship but the project postponed to FY92 as a budget saving measure. Laid down 20 April 1993. The ship conducted a world tour in 1995.

UPDATED

KASHIMA   *6/2003, Mick Prendergast /* 0570890

2 ASAGIRI CLASS (TRAINING SHIPS) (AX/TV)

Name	No	Builders	Laid down	Launched	Commissioned
YAMAGIRI	TV 3515 (ex-DD 152)	Mitsui, Tamano	5 Feb 1986	8 Oct 1987	25 Jan 1989
ASAGIRI	TV 3516 (ex-DD 151)	Ishikawajima Harima, Tokyo	13 Feb 1985	19 Sep 1986	17 Mar 1988

Displacement, tons: 3,500 (3,550, DD 155-158) standard; 4,200 full load
Dimensions, feet (metres): 449.4 × 48 × 14.6 *(137 × 14.6 × 4.5)*
Main machinery: COGAG; 4 RR Spey SM1A gas turbines; 53,300 hp *(39.8 MW)* sustained; 2 shafts; cp props
Speed, knots: 30+
Complement: 220

Missiles: SSM: 8 McDonnell Douglas Harpoon (2 quad) launchers; active radar homing to 130 km *(70 n miles)* at 0.9 Mach; warhead 227 kg.
SAM: Raytheon Sea Sparrow Mk 29 (Type 3/3A) octuple launcher; semi-active radar homing to 14.6 km *(8 n miles)* at 2.5 Mach; warhead 39 kg; 20 missiles.
A/S: Honeywell ASROC Mk 112 octuple launcher; inertial guidance to 1.6-10 km *(1-5.4 n miles)* at 0.9 Mach; payload Mk 46 Mod 5 Neartip. Reload capability.
Guns: 1 Otobreda 3 in *(76 mm)*/62 compact; 85 rds/min to 16 km *(8.6 n miles)* anti-surface; 12 km *(6.5 n miles)* anti-aircraft; weight of shell 6 kg.
2 General Electric/General Dynamics 20 mm Phalanx Mk 15 CIWS; 6 barrels per mounting; 3,000 rds/min combined to 1.5 km.
Torpedoes: 6—324 mm Type 68 (2 triple) HOS 301 tubes. Honeywell Mk 46 Mod 5 Neartip; anti-submarine; active/passive homing to 11 km *(5.9 n miles)* at 40 kt; warhead 44 kg.
Countermeasures: Decoys: 2 Loral Hycor SRBOC 6-barrelled Mk 36 chaff launchers; range 4 km *(2.2 n miles)*.
1 SLQ-25 Nixie or Type 4; towed torpedo decoy.

YAMAGIRI

8/2004*, Takeshi Oosaki / 1044419

ESM: Nec NOLR 6C or NOLR 8 (DD 152); intercept.
ECM: Fujitsu OLT-3; jammer.
Combat data systems: OYQ-7B data automation; Link 11/14. SATCOM. ORQ-1 helicopter datalink for SH-60J.
Radars: Air search: Melco OPS-14C (DD 151-154); D-band. Melco OPS-24 (DD 155-158); 3D; D-band.
Surface search: JRC OPS-28C; G-band (DD 151, 152, 155-158). JRC OPS-28C-Y; G-band (DD 153-154).
Fire control: Type 2-22 (for guns). Type 2-12E (for SAM) (DD 151-154); Type 2-12G (for SAM) (DD 155-158).

Tacan: ORN-6D (URN 25).
Sonars: Mitsubishi OQS 4A (II); hull-mounted; active search and attack; low frequency.
OQR-1; towed array; passive search; very low frequency.

Helicopters: 1 SH-60J Seahawk.

Comment: TV 3515 converted to training ship on 18 March 2004 and TV 3516 on 16 February 2005.   *NEW ENTRY*

1 TENRYU CLASS (TRAINING SUPPORT SHIP) (AVHM/TV)

Name	No	Builders	Launched	Commissioned
TENRYU	ATS 4203	Sumitomo, Uraga	14 Apr 1999	17 Mar 2000

Displacement, tons: 2,450 standard
Dimensions, feet (metres): 347.8 × 54.1 × 13.5 *(106 × 16.5 × 4.1)*
Main machinery: 4 Niigata 8MG28H diesels; 12,500 hp(m) *(9.19 MW)* sustained; 2 shafts
Speed, knots: 22
Complement: 140
Guns: 1 OTO Melara 3 in *(76 mm)*/62 compact; 85 rds/min to 16 km *(8.6 n miles)*; weight of shell 6 kg.
Radars: Air/surface search: Melco OPS-14; D-band.
Navigation: Fujitsu OPS-19; I-band.
Fire control: Type 2-22; I/J-band.
Helicopters: 1 medium.

Comment: Authorised in 1997 budget as a replacement for *Azuma* and laid down 19 June 1998. Carries four BQM-34J drones and four Northrop Chukar III drones used for evaluating performance of ships SAM systems. Improved 'Kurobe' design.   *UPDATED*

TENRYU

6/2004*, Hachiro Nakai / 1044420

1 KUROBE CLASS (TRAINING SUPPORT SHIP) (AVM/TV)

Name	No	Builders	Commissioned
KUROBE	ATS 4202	Nippon Koukan, Tsurumi	23 Mar 1989

Displacement, tons: 2,270 standard; 3,200 full load
Dimensions, feet (metres): 331.4 × 54.1 × 13.1 *(101 × 16.5 × 4)*
Main machinery: 4 Fuji 8L27.5XF diesels; 8,700 hp(m) *(6.4 MW)*; 2 shafts; cp props
Speed, knots: 20
Complement: 156 (17 officers)
Guns: 1 FMC/OTO Melara 3 in *(76 mm)*/62 Mk 75; 85 rds/min to 16 km *(8.6 n miles)* anti-surface; 12 km *(6.5 n miles)* anti-aircraft; weight of shell 6 kg.
Radars: Air search: Melco OPS-14C; D-band.
Surface search: JRC OPS-18-1; G-band.
Fire control: Type 2-21; I/J-band.

Comment: Approved under 1986 estimates, laid down 31 July 1987, launched 23 May 1988. Carries four BQM-34AJ high-speed drones and four Northrop Chukar II drones with two stern launchers. Used for training crews in anti-aircraft operations and evaluating the effectiveness and capability of ships' anti-aircraft missile systems.   *VERIFIED*

KUROBE

4/2003, Hachiro Nakai / 0570744

1 TRAINING TENDER (YXT)

YTE 13

Displacement, tons: 170 standard
Dimensions, feet (metres): 115.0 × 24.2 × 5.6 *(35.3 × 7.4 × 1.72)*
Main machinery: 2 Yanmar 12 LAK ST2 diesels; 2,200 hp(m) *(1.16 MW)*; 2 shafts
Speed, knots: 16

Comment: Approved in FY00 budget and commissioned in 2002. Assigned to 1st Maritime Service School for cadet training.   *UPDATED*

YTE 13

3/2002, Takatoshi Okano / 0570887

ICEBREAKERS

Notes: It is planned to replace *Shirase* with a new Antarctic expedition ship by 2007. The ship is likely to be of the order of 12,500 tons.

1 SHIRASE CLASS (AGBH)

Name	No	Builders	Launched	Commissioned
SHIRASE	AGB 5002	Nippon Koukan, Tsurumi	11 Dec 1981	12 Nov 1982

Displacement, tons: 11,600 standard; 19,000 full load
Dimensions, feet (metres): 439.5 × 91.8 × 32.2 *(134 × 28 × 9.8)*
Main machinery: Diesel-electric; 6 Mitsui 12V42M diesels; 53,900 hp(m) *(39.6 MW)*; 6 generators; 6 motors; 30,000 hp(m) *(22 MW)*; 3 shafts
Speed, knots: 19. **Range, n miles:** 25,000 at 15 kt
Complement: 170 (37 officers) plus 60 scientists
Cargo capacity: 1,000 tons
Radars: Surface search: JRC OPS-18-1; G-band.
Navigation: OPS-22; I-band.
Tacan: ORN-6 (URN 25).
Helicopters: 2 Mitsubishi S-61A; 1 Kawasaki OH-6D.

Comment: Laid down 5 March 1981. Fully equipped for marine and atmospheric research. Stabilised. The dome covers a weather radar.   *VERIFIED*

SHIRASE

11/2001, M Back, RAN / 0529126

TUGS

20 + 1 COASTAL AND HARBOUR TUGS (YTM/YTB)

YT 51   YT 59-62   YT 80   YT 85   YT 91
YT 53-57   YT 75-77   YT 82-83   YT 87-88

Displacement, tons: 53 standard
Dimensions, feet (metres): 55.8 × 15.8 × 7.8 *(17.0 × 4.8 × 2.4)*
Main machinery: 2 Isuzu UM6SD1TCB diesels; 500 hp (373 kW); 2 shafts
Speed, knots: 8
Complement: 4

Comment: Details given are for 50 ton class (YT 75-77, YT 80, YT 85, YT 87-88 and YT 91). There are also four of 190 tons (YT 53, YT 55-57), two of 35 tons (YT 60-61), one of 30 tons (YT 62) and three of 29 tons (YT 51, YT 54, YT 59). A further 50 ton harbour tug is to enter service in 2005.

UPDATED

20 + 1 OCEAN TUGS (ATA/YT)

YT 58   YT 63-74   YT 78-79   YT 81   YT 84   YT 86   YT 89-90

Displacement, tons: 260 standard
Dimensions, feet (metres): 93 × 28 × 8.2 *(28.4 × 8.6 × 2.5)*
Main machinery: 2 Niigata 6L25B diesels; 1,800 hp(m) *(1.32 MW)*; 2 shafts
Speed, knots: 11
Complement: 10

Comment: *YT 58* entered service on 31 October 1978, *YT 63* on 27 September 1982, *YT 64* on 30 September 1983, *YT 65* on 20 September 1984, *YT 66* on 20 September 1985, *YT 67* on 4 September 1986, *YT 68* on 9 September 1987, *YT 69* on 16 September 1987, *YT 70* on 2 September 1988, *YT 71* on 28 July 1989, *YT 72* on 28 July 1990, *YT 73* on 31 July 1991, *YT 74* on 30 September 1991, *YT 78* in July 1994, *YT 79* on 29 September 1994, *YT 81* on 8 July 1996, *YT 84* on 30 September 1998, *YT 86* on 21 March 2000, *YT 89* and *90* on 16 March 2001. All built by Yokohama Yacht. A further 260 ton tug is expected to enter service in 2005.

UPDATED

YT 87   *9/2002, Takatoshi Okano /* 0570889

YT 90   *7/2003, Hachiro Nakai /* 0570749

COAST GUARD

KAIJYOU HOANCHOU

Headquarters Appointments

Commandant of the Coast Guard:
  Hiromi Ishikawa

Establishment

The Japan Coast Guard (Maritime Safety Agency before 1 April 2000) was established on 1 May 1948. Its five missions are Maintaining Peace and Security, Ensuring Maritime Traffic Safety, Maritime Search and Rescue, Environmental Protection and Enforcement and Co-operation with other national and international agencies. The HQ is at Tokyo, the Coast Guard Academy is at Kure and the Coast Guard School is at Maizuru.
The main operational branches are the Guard and Rescue, the Hydrographic and the Aids to Navigation Departments. Regional offices control the 11 districts with their location as follows (airbases in brackets): RMS 1-Otaru (Chitose, Hakodate, Kushiro); 2-Shiogama (Sendai); 3-Yokohama (Haneda); 4-Nagoya (Ise); 5-Kobe (Yao); 6-Hiroshima (Hiroshima); 7-Kitakyushu (Fukuoka); 8-Maizuru (Miho); 9-Niigata (Niigata); 10-Kagoshima (Kagoshima); 11-Naha (Naha, Ishigaki). This organisation includes, as well as the RMS HQ, 66 MS offices, 58 MS stations, 14 MS air stations, five district communication centres, seven traffic advisory service centres, four hydrographic observatories, 18 aids to navigation offices, one Special Rescue station, one Special Security station, one National Strike Team station and one Transnational Organised Crime Strike Force station.

Personnel

2005: 12,258 (2,630 officers)

Strength of the Fleet

Type	Active	Building
GUARD AND RESCUE SERVICE		
Patrol Vessels:		
Large with helicopter (PLH)	13	—
Large (PL)	39	5
Medium (PM)	40	—
Small (PS)	27	—
Firefighting Vessels (FL)	5	—
Patrol Craft:		
Patrol Craft (PC)	63	(1)
Patrol Craft (CL)	170	—
Firefighting Craft (FM)	4	—
Special Service Craft:		
Monitoring Craft (MS)	3	—
Guard Boats (GS)	2	—
Surveillance Craft (SS)	22	—
Oil Recovery Craft (OR)	5	—
Oil Skimming Craft (OS)	3	—
Oil Boom Craft (OX)	19	—

Type	Active	Building
HYDROGRAPHIC SERVICE		
Surveying Vessels:		
Large (HL)	5	—
Small (HS)	8	—
AIDS TO NAVIGATION SERVICE		
Aids to Navigation Research Vessel (LL)	1	—
Buoy Tenders:		
Large (LL)	3	—
Medium (LM)	1	—
Aids to Navigation Tenders:		
Medium (LM)	10	—
Small (LS)	35	—

DELETIONS

2002   *Murakumo, Yodo, CL 205, CL 210, CL 212, CL 216-217, CL 222-223, CL 227, CL 236, CL 243, LM 107, LS 148, LS 186, LS 209-210*
2003   *Kamishima, Yaeyama, Yamagiri, Kotobiki, Nachi, Kurushima (HS 35), Kegon, LS 211, LS 149, LS 154, LS 181*
2004   *Fuji, Miyake, Kabashima, Okushiri, SS 25, LS 155, LS 157-158, LS 160, LS 208*

LARGE PATROL VESSELS

1 SHIKISHIMA CLASS (PLH/PSOH)

Name	No	Builders	Laid down	Launched	Commissioned
SHIKISHIMA	PLH 31	Ishikawajima Harima, Tokyo	24 Aug 1990	27 June 1991	8 Apr 1992

Displacement, tons: 6,500 standard; 9,350 full load
Dimensions, feet (metres): 492.1 × 55.8 × 19.7 *(150 × 17 × 6)*
Main machinery: 2 SEMT-Pielstick 16 PC2.5 V 400; 20,800 hp (m) *(15.29 MW)*; 2 shafts; bow thruster
Speed, knots: 25. **Range, n miles:** 20,000 at 18 kt
Complement: 110 plus 30 aircrew
Guns: 4 Oerlikon 35 mm/90 Type GDM-C (2 twin); 1,100 rds/min to 6 km *(3.2 n miles)*; weight of shell 1.55 kg.
  2 JM-61 MB 20 mm Gatling.
Radars: Air/surface search: Melco Ops 14; D-band.
Surface search: JMA 1576; I-band.
Navigation: JMA 1596; I-band.
Helo control: JMA 3000; I-band.
Tacan: ORN-6 (URN 25).
Helicopters: 2 Bell 212 or 2 Super Puma.

Comment: Authorised in the FY89 programme in place of the third Mizuho class. Used to escort the plutonium transport ship. SATCOM fitted.

UPDATED   SHIKISHIMA

SHIKISHIMA   *5/2004*, Mitsuhiro Kadota /* 1044429

2 MIZUHO CLASS (PLH/PSOH)

Name	No	Builders	Launched	Commissioned
MIZUHO	PLH 21	Mitsubishi, Nagasaki	5 June 1985	19 Mar 1986
YASHIMA	PLH 22	Nippon Koukan, Tsurumi	20 Jan 1988	1 Dec 1988

Displacement, tons: 4,900 standard; 5,204 full load
Dimensions, feet (metres): 426.5 × 50.9 × 17.7 *(130 × 15.5 × 5.4)*
Main machinery: 2 SEMT-Pielstick 14 PC2.5 V 400 diesels; 18,200 hp(m) *(13.38 MW)* sustained; 2 shafts; cp props; bow thruster
Speed, knots: 23. **Range, n miles:** 8,500 at 22 kt
Complement: 100 plus 30 aircrew
Guns: 1 Oerlikon 35 mm/90; 550 rds/min to 6 km *(3.2 n miles)* anti-surface; 5 km *(2.7 n miles)* anti-aircraft; weight of shell 1.55 kg.
1 JM-61 MB 20 mm Gatling.
Radars: Surface search: JMA 8303; I-band.
Navigation and helo control: 2 JMA 3000; I-band.
Helicopters: 2 Fuji-Bell 212.

Comment: PLH 21 ordered under the FY83 programme laid down 27 August 1984. PLH 22 in 1986 estimates, laid down 3 October 1987. Two sets of fixed electric fin stabilisers that have a lift of 26 tons × 2 and reduce rolling by 90 per cent at 18 kt. Employed in search and rescue outside the 200 mile economic zone.

VERIFIED

MIZUHO   *5/2003, Hachiro Nakai /* 0570751

10 SOYA CLASS (PLH/PSOH)

Name	No	Builders	Commissioned
SOYA	PLH 01	Nippon Kokan, Tsurumi	22 Nov 1978
TSUGARU	PLH 02	IHI, Tokyo	17 Apr 1979
OOSUMI	PLH 03	Mitsui Tamano	18 Oct 1979
HAYATO (ex-*Uraga*)	PLH 04	Hitachi, Maizuru	5 Mar 1980
ZAO	PLH 05	Mitsubishi, Nagasaki	19 Mar 1982
CHIKUZEN	PLH 06	Kawasaki, Kobe	28 Sep 1983
SETTSU	PLH 07	Sumitomo, Oppama	27 Sep 1984
ECHIGO	PLH 08	Mitsui Tamano	28 Feb 1990
RYUKYU	PLH 09	Mitsubishi, Nagasaki	31 Mar 2000
DAISEN	PLH 10	Nippon Kokan, Tsurumi	1 Oct 2001

Displacement, tons: 3,200 normal; 3,744 full load
Dimensions, feet (metres): 323.4 × 51.2 × 17.1 *(98.6 × 15.6 × 5.2)* (PLH 01)
345.8 × 47.9 × 15.7 *(105.4 × 14.6 × 4.8)*
Main machinery: 2 SEMT-Pielstick 12 PC2.5 V 400 diesels; 15,604 hp(m) *(11.47 MW)* sustained; 2 shafts; cp props; bow thruster
Speed, knots: 21 (PLH 01); 22 (others). **Range, n miles:** 5,700 at 18 kt
Complement: 71 (PLH 01-04); 69 (others)
Guns: 1 Bofors 40 mm or Oerlikon 35 mm. 1 Oerlikon 20 mm (PLH 01, 02, 05-07) or 1—20 mm JM61MB Gatling gun.
Radars: Surface search: JMA 1576; I-band.
Navigation: JMA 1596; I-band.
Helo control: JMA 1596; I-band.
Helicopters: 1 Fuji-Bell 212.

Comment: PLH 01 has an icebreaking capability while the other ships are only ice strengthened. Fitted with both fin stabilisers and anti-rolling tanks of 70 tons capacity. The fixed electric hydraulic fins have a lift of 26 tons × 2 at 18 kt which reduces rolling by 90 per cent at that speed. At slow speed the reduction is 50 per cent, using the tanks. PLH 04 name changed on 27 March 1997. PLH 10 laid down 8 March 1999.

UPDATED

TSUGARU   *5/2004*, Mitsuhiro Kadota /* 1044430

1 IZU CLASS (PL/PSOH)

Name	No	Builders	Launched	Commissioned
IZU	PL 31	Kawasaki, Sakaide	7 Feb 1997	25 Sep 1997

Displacement, tons: 3,500 normal
Dimensions, feet (metres): 360.9 × 49.2 × 17.4 *(110 × 15 × 5.3)*
Main machinery: 2 diesels; 12,000 hp(m) *(8.82 MW)*; 2 shafts; bow thruster
Speed, knots: 20
Complement: 40 plus 70 spare
Guns: 1 Oerlikon 35 mm. 1 JM-61 MB 20 mm Gatling.
Radars: Surface search: I-band.
Navigation: I-band.
Helicopters: Platform for 1 Fuji-Bell 212.

Comment: Authorised in the FY95 programme. Laid down 22 March 1996. Replaced the former *Izu* in 1998, taking the same name and pennant number. Carries two launches.

UPDATED

IZU   *4/2002, Hachiro Nakai /* 0529052

1 MIURA CLASS (PL/PSOH)

Name	No	Builders	Launched	Commissioned
MIURA	PL 22	Sumitomo, Uraga	11 Mar 1998	28 Oct 1998

Displacement, tons: 3,000 normal
Dimensions, feet (metres): 377.3 × 45.9 × 15.7 *(115 × 14 × 4.8)*
Main machinery: 2 diesels; 8,000 hp(m) *(5.88 MW)*; 2 shafts; cp props
Speed, knots: 18
Complement: 40 plus 10 spare
Guns: 1 Oerlikon 35 mm. 1—20 mm JM 61-B Gatling.

Comment: Authorised in FY96 programme. Laid down in October 1996. Has replaced ship of the same name.

UPDATED

MIURA   *5/2003, Hachiro Nakai /* 0570753

1 KOJIMA CLASS (PL/PSOH)

Name	No	Builders	Commissioned
KOJIMA	PL 21	Hitachi, Maizuru	11 Mar 1993

Displacement, tons: 2,650 normal; 2,950 full load
Dimensions, feet (metres): 377.3 × 45.9 × 16.4 *(115 × 14 × 5)*
Main machinery: 2 diesels; 8,000 hp(m) *(5.9 MW)*; 2 shafts; cp props
Speed, knots: 18. **Range, n miles:** 7,000 at 15 kt
Complement: 118
Guns: 1 Oerlikon 35 mm/90. 1—20 mm JM-61B Gatling. 1—12.7 mm MG.
Radars: Navigation: Two JMA 1596; I-band.
Helicopters: Platform for 1 medium.

Comment: Authorised in the FY90 programme and ordered in March 1991. Laid down 7 November 1991, launched 10 September 1992. Training ship which has replaced the old ship of the same name and pennant number. SATCOM fitted.

UPDATED

KOJIMA   *1/2004*, Hachiro Nakai /* 1044428

1 NOJIMA CLASS (PL/PSOH)

Name	No	Builders	Commissioned
OKI (ex-*Nojima*)	PL 01	Ishikawajima Harima, Tokyo	21 Sep 1989

Displacement, tons: 1,500 normal
Dimensions, feet (metres): 285.4 × 34.4 × 11.5 *(87 × 10.5 × 3.5)*
Main machinery: 2 Fuji 8S40B diesels; 8,120 hp(m) *(5.97 MW)*; 2 shafts
Speed, knots: 19
Complement: 34
Guns: 1 Oerlikon 35 mm/90. 1—20 mm JM-61B Gatling.
Radars: Navigation: 2 JMA 1596; I-band.
Helicopters: Platform for 1 Bell 212.

Comment: Laid down 16 August 1988 and launched 30 May 1989. Equipped as surveillance and rescue command ship. SATCOM fitted. Name changed on 30 November 1997.

VERIFIED

OKI   *5/2002, Takatoshi Okano /* 0570886

27 SHIRETOKO CLASS (PL/PSO)

Name	No	Builders	Commissioned
SHIRETOKO	PL 101	Mitsui Tamano	8 Nov 1978
ESAN	PL 102	Sumitomo	16 Nov 1978
WAKASA	PL 103	Kawasaki, Kobe	29 Nov 1978
KII (ex-*Shimanto*, ex-*Yahiko*)	PL 104	Mitsubishi, Shimonoseki	16 Nov 1978
RISHIRI	PL 106	Shikoku	12 Sep 1979
MATSUSHIMA	PL 107	Tohoku	14 Sep 1979
IWAKI	PL 108	Naikai	10 Aug 1979
SHIKINE	PL 109	Usuki	20 Sep 1979
SURUGA	PL 110	Kurushima	28 Sep 1979
REBUN	PL 111	Narasaki	21 Nov 1979
CHOKAI	PL 112	Nihonkai	30 Nov 1979
NOJIMA (ex-*Ashizuri*)	PL 113	Sanoyasu	31 Oct 1979
TOSA (ex-*Oki*)	PL 114	Tsuneishi	16 Nov 1979
NOTO	PL 115	Miho	30 Nov 1979
YONAKUNI	PL 116	Hayashikane	31 Oct 1979
IWAMI (ex-*Kudaka*, ex-*Daisetsu*, ex-*Kurikoma*)	PL 117	Hakodate	31 Jan 1980
SHIMOKITA	PL 118	Ishikawajima, Kakoki	12 Mar 1980
SUZUKA	PL 119	Kanazashi	7 Mar 1980
KUNISAKI	PL 120	Kouyo	29 Feb 1980
GENKAI	PL 121	Oshima	31 Jan 1980
GOTO	PL 122	Onomichi	29 Feb 1980
KOSHIKI	PL 123	Kasado	25 Jan 1980
HATERUMA	PL 124	Osaka	12 Mar 1980
KATORI	PL 125	Tohoku	21 Oct 1980
KUNIGAMI	PL 126	Kanda	17 Oct 1980
ETOMO	PL 127	Naikai	17 Mar 1982
AMAGI (ex-*Mashu*)	PL 128	Shiikoku	12 Mar 1982

Displacement, tons: 974 normal; 1,360 full load
Dimensions, feet (metres): 255.8 × 31.5 × 10.5 *(78 × 9.6 × 3.2)*
Main machinery: 2 Fuji 8S40B; 8,120 hp(m) *(5.97 MW)*; or 2 Niigata 8MA40 diesels; 2 shafts; cp props
Speed, knots: 20. **Range, n miles:** 4,400 at 17 kt
Complement: 41
Guns: 1 Bofors 40 mm or 1 Oerlikon 35 mm or 1 JM-61 20 mm Gatling (PL 101). 1 Oerlikon 20 mm (PL 101-105, 127 and 128).
Radars: Surface search: JMA 1576; I-band.
Navigation: JMA 1596; I-band.

Comment: Average time from launch to commissioning was about four to five months. Designed for EEZ patrol duties. PL 117 changed her name on 1 April 1988, again 1 August 1994 and again on 1 October 2000. PL 128 changed 1 April 1997, and PL 104 on 28 September 1999 and again on 1 October 2004. PL 105 paid off on 20 October 2000 after being involved in a collision.

UPDATED

SHIMOKITA *5/2004*, Mitsuhiro Kadota /* 1044432

7 OJIKA CLASS (PL/PSOH)

Name	No	Builders	Launched	Commissioned
ERIMO (ex-*Ojika*)	PL 02	Mitsui, Tamano	23 Apr 1991	31 Oct 1991
KUDAKA	PL 03	Hakodate Dock	10 May 1994	25 Oct 1994
YAHIKO (ex-*Satsuma*)	PL 04	Sumitomo, Uraga	3 June 1995	26 Oct 1995
HAKATA	PL 05	Ishikawajima, Tokyo	6 July 1998	26 Nov 1998
KURIKOMA (ex-*Dejima*)	PL 06	Mitsui, Tamano	28 June 1999	29 Oct 1999
SATAUMA	PL 07	Kawasaki, Kobe	3 June 1999	29 Oct 1999
MOTOBU	PL 08	Sasebo Heavy Industries	5 June 2000	31 Oct 2000

Displacement, tons: 1,883 normal
Dimensions, feet (metres): 299.9 × 36.1 × 11.5 *(91.4 × 11 × 3.5)*
Main machinery: 2 Fuji 8S40B diesels; 7,000 hp(m) *(5.15 MW)*; 2 shafts; cp props; 2 bow thrusters
Speed, knots: 18. **Range, n miles:** 4,400 at 15 kt
Complement: 38
Guns: 1 Oerlikon 35 mm/90. 1—20 mm JM-61B Gatling.
Radars: Navigation: JMA 1596; I-band.
Helicopters: Platform for 1 Bell 212 or Super Puma.

Comment: Equipped as SAR command ships. SATCOM fitted. 30 ton bollard pull. Stern dock for RIB. PL 04 name changed 28 September 1999. PL 02 name changed 1 October 2000. PL 06 name changed 4 January 2005.

UPDATED

HAKATA *5/2004*, Hachiro Nakai /* 1044431

0 + 3 1,800 TON CLASS (PL/PSO)

Name	No	Builders	Commissioned
—	PL 51	Mitsubishi, Shimonoseki	Apr 2006
—	PL 52	Mitsubishi, Shimonoseki	Apr 2006
—	PL 53		Apr 2007

Displacement, tons: 1,800 standard
Dimensions, feet (metres): 362.6 × 42.7 × 19.7 *(95.0 × 13.0 × 6.0)*
Main machinery: 4 diesels; waterjet propulsion
Speed, knots: 30
Guns: 1—40 mm Bofors Mk 3. 1—20 mm JM61 Gatling.
Helicopters: Platform for one medium.

Programme: Two ships authorised in FY03 budget and a third in FY04 budget.

UPDATED

1 + 2 ASO CLASS (PL/PSO)

Name	No	Builders	Laid down	Launched	Commissioned
ASO	PL 41	Mitsubishi, Shimonseki	18 Dec 2003	28 Oct 2004	Apr 2005
—	PL 42	Universal, Keihin	—	—	Apr 2006
—	PL 43	Universal, Keihin	—	—	Apr 2006

Displacement, tons: 770 standard
Dimensions, feet (metres): 259.2 × 32.8 × 19.7 *(79.0 × 10.0 × 6.0)*
Main machinery: 4 diesels, waterjet propulsion
Speed, knots: 30
Guns: 1—40 mm.

Comment: PL 41 authorised in FY02 budget and *PL 42-43* in FY03 budget.

UPDATED

ASO *2/2005*, Hachiro Nakai /* 1121408

SHIPBORNE AIRCRAFT

Numbers/Type: 4 Aerospatiale AS 332L1 Super Puma.
Operational speed: 125 kt *(231 km/h)*.
Service ceiling: 15,090 ft *(4,600 m)*.
Range: 500 n miles *(926 km)*.
Role/Weapon systems: Medium lift, support and SAR. Sensors: Search radar. Weapons: Unarmed.

UPDATED

AS 332L *5/2004*, Hachiro Nakai /* 1044434

Numbers/Type: 3 Sikorsky S-76C.
Operational speed: 135 kt *(250 km/h)*.
Service ceiling: 11,800 ft *(3,505 m)*.
Range: 607 n miles *(1,125 km)*.
Role/Weapon systems: Utility aircraft acquired in 1994-98. One aircraft lost on 10 January 2005. Up to 20 required to replace Bell 212s. Sensors: Search radar. Weapons: Unarmed.

UPDATED

S-76C *5/2004*, Mitsuhiro Kadota /* 1044437

Numbers/Type: 26/8 Bell 212/412.
Operational speed: 103 kt *(191 km/h)*.
Service ceiling: 10,000 ft *(3,048 m)*.
Range: 412 n miles *(763 km)*.
Role/Weapon systems: Liaison, medium-range support and SAR. Sensors: Search radar. Weapons: Unarmed.

UPDATED

BELL 212   *7/2004*, Hachiro Nakai /* 1044435

BELL 412   *5/2004*, Mitsuhiro Kadota /* 1044436

LAND-BASED MARITIME AIRCRAFT (FRONT LINE)

Notes: There are also a Cessna U 206G and four Bell 206 B.

Numbers/Type: 7/2/10 Beech Super King Air 200T/B200T/350.
Operational speed: 200 kt *(370 km/h)*.
Service ceiling: 35,000 ft *(10,670 m)*.
Range: 1,460 n miles *(2,703 km)*.
Role/Weapon systems: Visual reconnaissance in support of EEZ. Two are trainers. Sensors: Weather/search radar. Weapons: Unarmed.

UPDATED

BEECH B200T   *7/2004*, Hachiro Nakai /* 1044433

Numbers/Type: 5 NAMC YS-11A.
Operational speed: 230 kt *(425 km/h)*.
Service ceiling: 21,600 ft *(6,580 m)*.
Range: 1,960 n miles *(3,629 km)*.
Role/Weapon systems: Maritime surveillance and associated tasks. Sensors: Weather/search radar. Weapons: Unarmed.

UPDATED

YS-11A   *5/2004*, Mitsuhiro Kadota /* 1044438

Numbers/Type: 2 Dassault Falcon 900.
Operational speed: 428 kt *(792 km/h)*.
Service ceiling: 51,000 ft *(15,544 m)*.
Range: 4,170 n miles *(7,722 km)*.
Role/Weapon systems: Maritime surveillance. Sensors: Weather/search radar. Weapons: Unarmed.

UPDATED

FALCON 900   *5/2004*, Mitsuhiro Kadota /* 1044439

Numbers/Type: 2 SAAB 340B.
Operational speed: 250 kt *(463 km/h)*.
Service ceiling: 25,000 ft *(7,620 m)*.
Range: 570 n miles *(1,056 km)*.
Role/Weapon systems: Patrol aircraft procured in 1997.

NEW ENTRY

SAAB 340B   *5/2004*, Mitsuhiro Kadota /* 1044440

MEDIUM PATROL VESSELS

14 TESHIO CLASS (PM/PSO)

Name	No	Builders	Commissioned
NATSUI (ex-*Teshio*)	PM 01	Shikoku	30 Sep 1980
KITAKAMI (ex-*Oirose*)	PM 02	Naikai	29 Aug 1980
ECHIZEN	PM 03	Usuki	30 Sep 1980
TOKACHI	PM 04	Narazaki	24 Mar 1981
HITACHI	PM 05	Tohoku	19 Mar 1981
OKITSU	PM 06	Usuki	17 Mar 1981
ISAZU	PM 07	Naikai	18 Feb 1982
CHITOSE	PM 08	Shikoku	15 Mar 1983
KUWANO	PM 09	Naikai	10 Mar 1983
SORACHI	PM 10	Tohoku	30 Aug 1984
YUBARI	PM 11	Usuki	28 Nov 1985
MOTOURA	PM 12	Shikoku	21 Nov 1986
KANO	PM 13	Naikai	13 Nov 1986
SENDAI	PM 14	Shikoku	1 June 1988

Displacement, tons: 630 normal; 670 full load
Dimensions, feet (metres): 222.4 × 25.9 × 6.6 *(67.8 × 7.9 × 2.7)*
Main machinery: 2 Fuji 6S32F or Arakata 6M31E diesels; 3,650 hp(m) *(2.69 MW)*; 2 shafts
Speed, knots: 18. **Range, n miles:** 3,200 at 16 kt
Complement: 33
Guns: 1 JN-61B 20 mm Gatling.
Radars: Navigation: 2 JMA 159B; I-band.

Comment: First three built under FY79 programme and second three under FY80, seventh under FY81, PM 08-09 under FY82, PM 10 under FY83, PM 11 under FY84, PM 12-13 under FY85, PM 14 under FY87. *Isazu* has an additional structure aft of the mainmast which is used as a classroom.

VERIFIED

ISAZU   *9/2003, Hachiro Nakai /* 0570760

2 TAKATORI CLASS (PM/PBO)

Name	No	Builders	Commissioned
TAKATORI	PM 89	Naikai	24 Mar 1978
KUMANO	PM 94	Namura	23 Feb 1979

Displacement, tons: 634 normal
Dimensions, feet (metres): 152.5 × 30.2 × 9.3 *(46.5 × 9.2 × 2.9)*
Main machinery: 2 Niigata 6M31EX diesels; 3,000 hp(m) *(2.21 MW)*; 2 shafts; cp props
Speed, knots: 15. **Range, n miles:** 700 at 14 kt
Complement: 34
Radars: Navigation: JMA 1596 and JMA 1576; I-band.

Comment: SAR vessels equipped for salvage and firefighting.   *UPDATED*

TAKATORI   *5/2004*, Mitsuhiro Kadota /* 1044441

17 BIHORO CLASS (350-M4 TYPE) (PM/PSO)

Name	No	Builders	Commissioned
BIHORO	PM 73	Tohoku	28 Feb 1974
KUMA	PM 74	Usuki	28 Feb 1974
ISHIKARI	PM 78	Tohoku	13 Mar 1976
ABUKUMA	PM 79	Tohoku	30 Jan 1976
ISUZU	PM 80	Naikai	10 Mar 1976
KIKUCHI	PM 81	Usuki	6 Feb 1976
KUZURYU	PM 82	Usuki	18 Mar 1976
HOROBETSU	PM 83	Tohoku	27 Jan 1977
SHIRAKAMI	PM 84	Tohoku	24 Mar 1977
MATSUURA (ex-*Sagami*)	PM 85	Naikai	30 Nov 1976
TONE	PM 86	Usuki	30 Nov 1976
MISASA (ex-*Yoshino*)	PM 87	Usuki	28 Jan 1977
KUROBE	PM 88	Shikoku	15 Feb 1977
CHIKUGO	PM 90	Naikai	27 Jan 1978
YAMAKUNI	PM 91	Usuki	26 Jan 1978
KATSURA	PM 92	Shikoku	15 Feb 1978
OOYODO (ex-*Shinano*)	PM 93	Tohoku	23 Feb 1978

Displacement, tons: 615 normal; 636 full load
Dimensions, feet (metres): 208 × 25.6 × 8.3 *(63.4 × 7.8 × 2.5)*
Main machinery: 2 Niigata 6M31EX diesels; 3,000 hp(m) *(2.21 MW)*; 2 shafts; cp props
Speed, knots: 18. **Range, n miles:** 3,200 at 16 kt
Complement: 34
Guns: 1 USN 20 mm/80 Mk 10.
Radars: Navigation: JMA 1596 and JMA 1576; I-band.

Comment: Average time from launch to commissioning, four months. PM 85 and 87 changed names 3 April 2000, PM 77 on 24 January 2001 and PM 93 on 1 April 2001.   *UPDATED*

BIHORO CLASS   *3/2003, Hachiro Nakai /* 0570762

4 AMAMI CLASS (PM/PBO)

Name	No	Builders	Commissioned
AMAMI	PM 95	Hitachi, Kanagawa	28 Sep 1992
KUROKAMI (ex-*Matsuura*)	PM 96	Hitachi, Kanagawa	24 Nov 1995
KUNASHIRI	PM 97	Mitsubishi, Shimonoseki	26 Aug 1998
MINABE	PM 98	Mitsubishi, Shimonoseki	26 Aug 1998

Displacement, tons: 230 normal
Dimensions, feet (metres): 183.7 × 24.6 × 6.6 *(56 × 7.5 × 2)*
Main machinery: 2 Fuji 8S40B diesels; 8,120 hp(m) *(5.97 MW)*; 2 shafts; cp props
Speed, knots: 25
Guns: 1—20 mm JM-61B Gatling.
Radars: Navigation: I-band.

Comment: First one authorised in the FY91 programme; laid down 22 October 1991. Second authorised in FY93 programme; laid down 7 October 1994. Last pair authorised in FY96 programme and both laid down 30 September 1997. Stern ramp for launching RIB. PM 96 changed name 3 April 2000. PM 95 damaged in incident with possible North Korean intelligence collection ship on 22 December 2001.   *UPDATED*

MINABE   *7/2004*, Hachiro Nakai /* 1044443

1 TESHIO CLASS (ICEBREAKER) (PM/AGOB)

Name	No	Builders	Commissioned
TESHIO	PM 15	Nippon Koukan, Tsurumi	19 Oct 1995

Displacement, tons: 550 normal
Dimensions, feet (metres): 180.4 × 34.8 × 12.8 *(55 × 10.6 × 3.9)*
Main machinery: 2 diesels; 3,600 hp(m) *(2.65 MW)*; 2 shafts; bow thruster
Speed, knots: 14.5
Complement: 35
Guns: 1—20 mm JM-61B Gatling.
Radars: Navigation: 2 sets; I-band.

Comment: Authorised in FY93; laid down 7 October 1994, launched 20 April 1995. Has an icebreaker bow.   *VERIFIED*

TESHIO   *6/2002, Japan Coast Guard /* 0570891

3 TOKARA CLASS (PM/PBO)

Name	No	Builders	Commissioned
TOKARA	PM 21	Universal, Keihin (Kanagawa)	12 Mar 2003
FUKUE	PM 22	Mitsubishi, Shimonoseki	12 Mar 2003
OIRASE	PM 23	Mitsui, Tamano	18 Mar 2004

Displacement, tons: 335 standard
Dimensions, feet (metres): 183.8 × 32.4 × 14.4 *(56.0 × 8.5 × 4.4)*
Main machinery: 3 diesels; 3 waterjets
Speed, knots: 30+
Guns: 1—20 mm Gatling gun. 1—12.7 mm MG

Comment: First two authorised in FY01 budget and launched on 4 and 10 December 2002. *Oirase* authorised in FY02 budget.   *UPDATED*

OIRASE   *5/2004*, Mitsuhiro Kadota /* 1044442

SMALL PATROL VESSELS

7 AKAGI CLASS (PS/PB)

Name	No	Builders	Commissioned
AKAGI	PS 101	Sumidagawa	26 Mar 1980
TSUKUBA	PS 102	Sumidagawa	24 Feb 1982
KONGOU	PS 103	Ishihara	16 Mar 1987
KATSURAGI	PS 104	Ishihara	24 Mar 1988
BIZAN (ex-*Hiromine*)	PS 105	Yokohama Yacht Co	24 Mar 1988
SHIZUKI	PS 106	Sumidagawa	24 Mar 1988
TAKACHIHO	PS 107	Sumidagawa	24 Mar 1988

Displacement, tons: 115 full load
Dimensions, feet (metres): 114.8 × 20.7 × 4.3 *(35 × 6.3 × 1.3)*
Main machinery: 2 Pielstick 16 PA4 V 185 diesels; 5,344 hp(m) *(3.93 MW)* sustained; 2 shafts
Speed, knots: 28. **Range, n miles:** 500 at 20 kt
Complement: 22
Guns: 1 Browning 12.7 mm MG.
Radars: Navigation: 1 set; I-band.

Comment: Carry a 25-man inflatable rescue craft. The last four were ordered on 31 August 1987 and commissioned less than seven months later. PS 105 name changed on 1 October 2004.   *UPDATED*

BIZAN   *2/2004*, Hachiro Nakai /* 1044444

12 MIHASHI and BANNA CLASS (PS/PBF)

Name	No	Builders	Commissioned
SHINZAN (ex-*Akiyoshi*, ex-*Mihashi*)	PS 01	Mitsubishi, Shimonoseki	9 Sep 1988
SAROMA	PS 02	Hitachi, Kanagawa	24 Nov 1989
INASA	PS 03	Mitsubishi, Shimonoseki	31 Jan 1990
KIRISHIMA	PS 04	Hitachi, Kanagawa	22 Mar 1991
KAMUI	PS 05	Mitsubishi, Shimonoseki	31 Jan 1994
BANNA (ex-*Bizan*)	PS 06	Hitachi, Kanagawa	31 Jan 1994
ASHITAKI	PS 07	Mitsui, Tamano	30 Sep 1994
KARIBA (ex-*Kurama*)	PS 08	Mitsubishi, Shimonoseki	29 Aug 1995
ARASE	PS 09	Mitsubishi, Shimonoseki	29 Jan 1997
SANBE	PS 10	Hitachi, Kanagawa	29 Jan 1997
MIZUKI	PS 11	Mitsui, Tamano	9 June 2000
KOUYA	PS 12	Universal, Keihin	18 Mar 2004

Displacement, tons: 195 normal
Dimensions, feet (metres): 141.1 × 24.6 × 5.6 *(43 × 7.5 × 1.7)*
Main machinery: 2 SEMT-Pielstick 16 PA4 V 200 VGA diesels; 7,072 hp(m) *(5.2 MW)*; 2 shafts
   1 SEMT-Pielstick 12 PA4 V 200 VGA diesel; 2,720 hp(m) *(2 MW)*; Kamewa 80 water-jet
Speed, knots: 35. **Range, n miles:** 650 at 34 kt
Complement: 34
Guns: 1—12.7 mm MG or 1—20 mm JM 61 Gatling (PS 01, 03, 08 and 11).
Radars: Navigation: Furuno; I-band.

Comment: Capable of 15 kt on the water-jet alone. PS 01 name changed 28 January 1997 and
   again on 24 January 2001, PS 06 on 17 April 1999. PS 11 authorised in FY98 programme and
   PS 12 in FY02 programme. PS 08 name changed on 29 March 2004.   **UPDATED**

KOUYA   *5/2004*, Mitsuhiro Kadota /* 1044445

2 TAKATSUKI CLASS (PS/PBF)

Name	No	Builders	Commissioned
TAKATSUKI	PS 108	Mitsubishi, Shimonoseki	23 Mar 1992
NOBARU	PS 109	Hitachi, Kanagawa	22 Mar 1993

Displacement, tons: 115 normal; 180 full load
Dimensions, feet (metres): 114.8 × 22 × 4.3 *(35 × 6.7 × 1.3)*
Main machinery: 2 MTU 16V 396 TB94 diesels; 5,200 hp(m) *(3.82 MW)*; 2 Kamewa 71 water-jets
Speed, knots: 35
Complement: 13
Guns: 1—12.7 mm MG.
Radars: Navigation: I-band.

Comment: First authorised in the FY91 programme, second in FY92. Aluminium hulls.
   VERIFIED

TAKATSUKI   *5/2001, Hachiro Nakai /* 0130246

6 TSURUUGI CLASS (PS/PBOF)

Name	No	Builders	Commissioned
TSURUUGI	PS 201	Hitachi, Kanagawa	15 Feb 2001
HOTAKA	PS 202	Mitsubishi, Shimonoseki	16 Mar 2001
NORIKURA	PS 203	Mitsui, Tamano	16 Mar 2001
KAIMON	PS 204	Mitsui, Tamano	21 Apr 2004
ASAMA	PS 205	Mitsui, Tamano	21 Apr 2004
HOUOU	PS 206	Mitsui, Tamano	Apr 2005

Displacement, tons: 220 standard
Dimensions, feet (metres): 164.1 × 26.2 × 13.1 *(50.0 × 8.0 × 4.0)*
Main machinery: 3 diesels; 3 waterjets
Speed, knots: 45+
Guns: 1—20 mm JM-61 RFS Gatling.

Comment: First three authorised in FY99 budget, fourth and fifth in FY02 budget and sixth in FY03
   budget.   **UPDATED**

HOTAKA   *5/2004*, Mitsuhiro Kadota /* 1044446

COASTAL PATROL CRAFT

4 + (1) YODO CLASS (PC/YTR)

Name	No	Builders	Launched	Commissioned
YODO	PC 51	Sumidagawa	2 Oct 2001	29 Mar 2002
KOTOBIKI	PC 52	Sumidagawa	23 Oct 2002	27 Mar 2003
NACHI	PC 53	Ishihara	29 Jan 2003	27 Mar 2003
NUNOBIKI	PC 54	Sumidagawa	4 Dec 2002	27 Mar 2003

Displacement, tons: 125 standard
Dimensions, feet (metres): 121.4 × 22.0 × 11.2 *(37.0 × 6.7 × 3.4)*
Main machinery: 2 diesels; 2 waterjets
Speed, knots: 25

Comment: The first authorised in FY00 budget, three more in FY01 budget and a fifth proposed in
   FY04 budget. Also equipped for firefighting and replaced firefighting vessel of the same name.
   UPDATED

YODO   *5/2004*, Mitsuhiro Kadota /* 1044447

21 MURAKUMO CLASS (PC/PB)

Name	No	Builders	Commissioned
KITAGUMO	PC 202	Hitachi, Kanagawa	17 Mar 1978
YUKIGUMO	PC 203	Hitachi, Kanagawa	27 Sep 1978
ASAGUMO	PC 204	Mitsubishi, Shimonoseki	21 Sep 1978
HAYAGUMO	PC 205	Mitsubishi, Shimonoseki	30 Jan 1979
AKIGUMO	PC 206	Hitachi, Kanagawa	28 Feb 1979
YAEGUMO	PC 207	Mitsubishi, Shimonoseki	16 Mar 1979
NATSUGUMO	PC 208	Hitachi, Kanagawa	22 Mar 1979
KAWAGIRI	PC 210	Hitachi, Kanagawa	27 July 1979
TOSAGIRI (ex-*Bizan*, ex-*Teruzuki*)	PC 211	Mitsubishi, Shimonoseki	26 June 1979
NATSUZUKI	PC 212	Mitsubishi, Shimonoseki	26 July 1979
MIYAZUKI	PC 213	Hitachi, Kanagawa	13 Mar 1980
NIJIGUMO	PC 214	Mitsubishi, Shimonoseki	29 Jan 1981
TATSUGUMO	PC 215	Mitsubishi, Shimonoseki	19 Mar 1981
ISEYUKI (ex-*Hamayuki*)	PC 216	Hitachi, Kanagawa	27 Feb 1981
ISONAMI	PC 217	Mitsubishi, Shimonoseki	19 Mar 1981
NAGOZUKI	PC 218	Hitachi, Kanagawa	29 Jan 1981
YAEZUKI	PC 219	Hitachi, Kanagawa	19 Mar 1981
YAMAYUKI	PC 220	Hitachi, Kanagawa	16 Feb 1982
KOMAYUKI	PC 221	Mitsubishi, Shimonoseki	10 Feb 1982
UMIGIRI	PC 222	Hitachi, Kanagawa	17 Feb 1983
ASAGIRI	PC 223	Mitsubishi, Shimonoseki	23 Feb 1983

Displacement, tons: 85 normal
Dimensions, feet (metres): 98.4 × 20.7 × 7.2 *(30 × 6.3 × 2.2)*
Main machinery: 2 Ikegai MTU MB 16V 652 SB70 diesels; 4,400 hp(m) *(3.23 MW)* sustained;
   2 shafts
Speed, knots: 30. **Range, n miles:** 350 at 28 kt
Complement: 13
Guns: 1—12.7 mm MG.
Radars: Navigation: I-band.

Comment: PC 211 name changed on 17 April 1999 and again on 1 October 2004. P 216 changed
   name on 22 February 2001.
   UPDATED

KAWAGIRI   *5/2004*, Mitsuhiro Kadota /* 1044450

UMIGIRI   *9/2002, Mitsuhiro Kadota /* 0528998

3 HAMAYUKI CLASS (PC/PBF)

Name	No	Builders	Commissioned
HAMAYUKI (ex-*Kagayuki*)	PC 105	Mitsubishi, Shimonoseki	24 Dec 1999
MURAKUMO	PC 106	Hitachi, Kanagawa	19 Aug 2002
IZUNAMI	PC 107	Mitsui, Tamano	18 Mar 2003

Displacement, tons: 100 standard
Dimensions, feet (metres): 105.0 × 21.3 × 10.8 *(32.0 × 6.5 × 3.3)*
Main machinery: 2 diesels; 5,200 hp(m) *(3.82 MW)*; 2 waterjets
Speed, knots: 36
Complement: 10
Guns: 1—12.7 mm MG.

Comment: Larger version of Asogiri class with waterjet propulsion and higher top speed. PC 105 changed name on 22 February 2001. PC 106 authorised in FY01 budget and PC 107 in FY01 extra budget.

UPDATED

IZUNAMI
5/2004*, Hachiro Nakai / 1044448

9 AKIZUKI CLASS (PC/SAR)

Name	No	Builders	Commissioned
URAYUKI	PC 72	Mitsubishi, Shimonoseki	31 May 1975
HATAGUMO	PC 75	Mitsubishi, Shimonoseki	21 Feb 1976
MAKIGUMO	PC 76	Mitsubishi, Shimonoseki	19 Mar 1976
HAMAZUKI	PC 77	Mitsubishi, Shimonoseki	29 Nov 1976
ISOZUKI	PC 78	Mitsubishi, Shimonoseki	18 Mar 1977
SHIMANAMI	PC 79	Mitsubishi, Shimonoseki	23 Dec 1977
YUZUKI	PC 80	Mitsubishi, Shimonoseki	22 Mar 1979
TAMANAMI (ex-*Hanayuki*)	PC 81	Mitsubishi, Shimonoseki	27 Mar 1981
AWAGIRI	PC 82	Mitsubishi, Shimonoseki	24 Mar 1983

Displacement, tons: 77 normal
Dimensions, feet (metres): 85.3 × 20.7 × 6.9 *(26 × 6.3 × 2.1)*
Main machinery: 3 Mitsubishi 12DM20MTK diesels; 3,000 hp(m) *(2.21 MW)*; 3 shafts
Speed, knots: 22. Range, n miles: 220 at 21.5 kt
Complement: 10
Radars: Navigation: FRA 10 Mk 2; I-band.

Comment: Aluminium hulls. Used mostly for SAR. Being paid off.

VERIFIED

AWAGIRI
10/2003, Hachiro Nakai / 0570723

1 MATSUNAMI CLASS (PC/PB)

Name	No	Builders	Commissioned
MATSUNAMI	PC 01	Mitsubishi, Shimonoseki	22 Feb 1995

Displacement, tons: 165 normal
Dimensions, feet (metres): 114.8 × 26.2 × 10.8 *(35 × 8 × 3.3)*
Main machinery: 2 diesels; 5,200 hp(m) *(3.82 MW)*; 2 water-jets
Speed, knots: 25
Complement: 30
Radars: Navigation: I-band.

Comment: Has replaced old craft of the same name. Laid down 10 May 1994. Used for patrol and for VIPs.

UPDATED

MATSUNAMI
5/2004*, Hachiro Nakai / 1044449

3 SHIMAGIRI CLASS (PC/PB)

Name	No	Builders	Commissioned
SHIMAGIRI	PC 83	Hitachi, Kanagawa	7 Feb 1985
OKINAMI (ex-*Setogiri*)	PC 84	Hitachi, Kanagawa	22 Mar 1985
HAYAGIRI	PC 85	Mitsubishi, Shimonoseki	22 Feb 1985

Displacement, tons: 51 normal
Dimensions, feet (metres): 75.5 × 17.4 × 6.2 *(23 × 5.3 × 1.9)*
Main machinery: 2 Ikegai 12V 175 RTC diesels; 3,000 hp(m) *(2.21 MW)*; 2 shafts
Speed, knots: 30
Complement: 10
Guns: 1—12.7 mm MG (not in all).
Radars: Navigation: FRA 10 Mk 2; I-band.

Comment: Aluminium hulls. PC 84 name changed 1 October 2000.

VERIFIED

HAYAGIRI
4/2003, Bob Fildes / 0570885

1 SHIKINAMI CLASS (PC/PB)

Name	No	Builders	Commissioned
ASOYUKI	PC 74	Hitachi, Kanagawa	16 June 1975

Displacement, tons: 46 normal
Dimensions, feet (metres): 69 × 17.4 × 3.3 *(21 × 5.3 × 1)*
Main machinery: 2 MTU MB 12V 493 TY7 diesels; 2,200 hp(m) *(1.62 MW)* sustained; 2 shafts
Speed, knots: 26. Range, n miles: 230 at 23.8 kt
Complement: 10
Radars: Navigation: MD 806; I-band.

Comment: Built completely of light alloy. PC 69 paid off 8 December 1999 and PC 70 on 19 October 2000.

UPDATED

ASOYUKI
5/2004*, Hirotoshi Yamamoto / 1044454

2 NATSUGIRI CLASS (PC/PB)

Name	No	Builders	Commissioned
NATSUGIRI	PC 86	Sumidagawa	29 Jan 1990
SUGANAMI	PC 87	Sumidagawa	29 Jan 1990

Displacement, tons: 68 normal
Dimensions, feet (metres): 88.6 × 18.4 × 3.9 *(27 × 5.6 × 1.2)*
Main machinery: 2 diesels; 3,000 hp(m) *(2.21 MW)*; 2 shafts
Speed, knots: 27
Complement: 10
Radars: Navigation: I-band.

Comment: Built under FY88 programme. Steel hulls.

UPDATED

SUGANAMI
5/2004*, Hachiro Nakai / 1044452

15 HAYANAMI CLASS (PC/PB/YTR)

Name	No	Builders	Commissioned
HAYANAMI	PC 11	Sumidagawa	25 Mar 1993
SETOGIRI (ex-*Shikinami*)	PC 12	Sumidagawa	24 Mar 1994
MIZUNAMI	PC 13	Ishihara	24 Mar 1994
IYONAMI	PC 14	Sumidagawa	30 June 1994
KURINAMI	PC 15	Sumidagawa	30 Jan 1995
HAMANAMI	PC 16	Sumidagawa	28 Mar 1996
SHINONOME	PC 17	Ishihara	29 Feb 1996
HARUNAMI	PC 18	Ishihara	28 Mar 1996
KIYOZUKI	PC 19	Sumidagawa	23 Feb 1996
AYANAMI	PC 20	Yokohama Yacht	28 Mar 1996
TOKINAMI	PC 21	Yokohama Yacht	28 Mar 1996
HAMAGUMO	PC 22	Sumidagawa	27 Aug 1999
AWANAMI	PC 23	Sumidagawa	27 Aug 1999
URANAMI	PC 24	Sumidagawa	24 Jan 2000
SHIKINAMI	PC 25	Ishihara	24 Oct 2000

Displacement, tons: 110 normal; 190 full load
Dimensions, feet (metres): 114.8 × 20.7 × 7.5 *(35 × 6.3 × 2.3)*
Main machinery: 2 diesels; 4,000 hp(m) *(2.94 MW)*; 2 shafts
Speed, knots: 25
Guns: 1—12.7 mm MG.
Radars: Navigation: I-band.

Comment: One more authorised in FY99 budget. From PC 22 onwards these craft are equipped for firefighting. PC 12 changed name 1 October 2000. *UPDATED*

HARUNAMI *5/2004*, Mitsuhiro Kadota / 1044453*

4 ASOGIRI CLASS (PC/PB)

Name	No	Builders	Commissioned
ASOGIRI	PC 101	Yokohama Yacht	19 Dec 1994
MUROZUKI	PC 102	Ishihara	27 July 1995
WAKAGUMO	PC 103	Ishihara	17 July 1996
NAOZUKI	PC 104	Sumidagawa	23 Jan 1997

Displacement, tons: 88 normal
Dimensions, feet (metres): 108.3 × 20.7 × 4.6 *(33 × 6.3 × 1.4)*
Main machinery: 2 diesels; 5,200 hp(m) *(3.82 MW)*; 2 shafts
Speed, knots: 30
Complement: 10
Guns: 1—12.7 mm MG.

Comment: First pair authorised in FY93 programme, third and fourth in FY95.
UPDATED

MUROZUKI *8/2001, Hachiro Nakai / 0130250*

195 COASTAL PATROL AND RESCUE CRAFT (CL/PB)

CL 01-04	CL 206-209	CL 213-215	CL 228	CL 237-242	GS 01-02
CL 11-134	CL 211	CL 226	CL 231-235	CL 244-264	SS 51-73

Comment: Some have firefighting capability. Built by Shigi, Ishihara, Sumidagawa, Yokohama Yacht Co and Yamaha. For coastal patrol and rescue duties. Built of high tensile steel. Fourteen CL 11 class authorised in FY01 budget.
UPDATED

SS 51 *6/2004*, Hachiro Nakai / 1044451*

CL 130 *5/2004*, Mitsuhiro Kadota / 1044455*

FIREFIGHTING VESSELS AND CRAFT

1 MODIFIED HIRYU CLASS (FL/YTR)

Name	No	Builders	Launched	Commissioned
HIRYU	FL 01	NKK, Tsurumi	5 Sep 1997	24 Dec 1997

Displacement, tons: 280 normal
Dimensions, feet (metres): 114.8 × 40 × 8.9 *(35 × 12.2 × 2.7)*
Main machinery: 2 diesels; 4,000 hp(m) *(2.94 MW)*; 2 shafts
Speed, knots: 14
Complement: 15

Comment: Authorised in FY96 programme. Catamaran design. Replaced ship of the same name and pennant number. *UPDATED*

HIRYU *5/2004*, Mitsuhiro Kadota / 1044457*

4 NUNOBIKI CLASS (FM/YTR)

Name	No	Builders	Commissioned
SHIRAITO	FM 04	Yokohama Yacht Co	25 Feb 1975
MINOO	FM 08	Sumidagawa	27 Jan 1978
RYUSEI	FM 09	Yokohama Yacht Co	24 Mar 1980
KIYOTAKI	FM 10	Sumidagawa	25 Mar 1981

Displacement, tons: 89 normal
Dimensions, feet (metres): 75.4 × 19.7 × 5.2 *(23 × 6 × 1.6)*
Main machinery: 1 MTU MB 12V 493 TY7 diesel; 1,100 hp(m) *(810 kW)* sustained; 1 shaft 2 Nissan diesels; 500 hp(m) *(515 kW)*; 3 shafts
Speed, knots: 14. **Range, n miles:** 180 at 13.5 kt
Complement: 12
Radars: Navigation: FRA 10; I-band.

Comment: Equipped for chemical firefighting. FM 01 paid off 31 October 2000 and FM 02 in 2002. FM 05, FM 06 and FM 07 paid off on 11 March 2003. *VERIFIED*

KIYOTAKI *10/2003, Hachiro Nakai / 0570706*

4 HIRYU CLASS (FL/YTR)

Name	No	Builders	Commissioned
SHORYU	FL 02	Nippon Kokan, Tsurumi	4 Mar 1970
NANRYU	FL 03	Nippon Kokan, Tsurumi	4 Mar 1971
KAIRYU	FL 04	Nippon Kokan, Tsurumi	18 Mar 1977
SUIRYU	FL 05	Yokohama Yacht Co	24 Mar 1978

Displacement, tons: 215 normal
Dimensions, feet (metres): 90.2 × 34.1 × 7.2 *(27.5 × 10.4 × 2.2)*
Main machinery: 2 Ikegai MTU MB 12V 493 TY7 diesels; 2,200 hp(m) *(1.62 MW)* sustained; 2 shafts
Speed, knots: 13.2. **Range, n miles:** 300 at 13 kt
Complement: 14

Comment: Catamaran type fire boats designed and built for firefighting services to large tankers.
UPDATED

TENYO   *11/2003, Hachiro Nakai* / 0570705

NANRYU   *7/2004 *, Hachiro Nakai* / 1044456

SURVEY SHIPS

1 SHOYO CLASS (AGS)

Name	No	Builders	Launched	Commissioned
SHOYO	HL 01	Mitsui, Tamano	23 June 1997	20 Mar 1998

Displacement, tons: 3,000 normal
Dimensions, feet (metres): 321.5 × 49.9 × 11.8 *(98 × 15.2 × 3.6)*
Main machinery: Diesel-electric; 2 diesels; 8,100 hp(m) *(5.95 MW)*; 2 motors; 5,712 hp(m) *(4.2 MW)*; 2 shafts; cp props
Speed, knots: 17
Complement: 60

Comment: Authorised in FY95 programme. Laid down 4 October 1996. Has replaced former *Shoyo*.
VERIFIED

SHOYO   *5/2003, Hachiro Nakai* / 0570707

1 TENYO CLASS (AGS)

Name	No	Builders	Commissioned
TENYO	HL 04	Sumitomo, Oppama	27 Nov 1986

Displacement, tons: 770 normal
Dimensions, feet (metres): 183.7 × 32.2 × 9.5 *(56 × 9.8 × 2.9)*
Main machinery: 2 Akasaka diesels; 1,300 hp(m) *(955 kW)*; 2 shafts
Speed, knots: 13. **Range, n miles:** 5,400 at 12 kt
Complement: 43 (18 officers)
Radars: Navigation: 2 JMA 1596; I-band.

Comment: Laid down 11 April 1986, launched 5 August 1986. Based at Tokyo.
VERIFIED

1 TAKUYO CLASS (AGS)

Name	No	Builders	Commissioned
TAKUYO	HL 02	Nippon Kokan, Tsurumi	31 Aug 1983

Displacement, tons: 3,000 normal
Dimensions, feet (metres): 314.9 × 46.6 × 15.1 *(96 × 14.2 × 4.6)*
Main machinery: 2 Fuji 6S40B diesels; 6,090 hp(m) *(4.47 MW)*; 2 shafts; cp props
Speed, knots: 17. **Range, n miles:** 12,000 at 16 kt
Complement: 60 (24 officers)
Radars: Navigation: 2 sets; I-band.

Comment: Laid down on 14 April 1982, launched on 24 March 1983. Based at Tokyo. Side scan sonar fitted. Two survey launches.
VERIFIED

TAKUYO   *2/1999, JMSA* / 0080210

7 HAMASHIO CLASS (YGS)

Name	No	Builders	Commissioned
HAMASHIO	HS 21	Yokohama Yacht	25 Mar 1991
ISOSHI	HS 22	Yokohama Yacht	25 Mar 1993
UZUSHIO	HS 23	Yokohama Yacht	22 Dec 1995
OKISHIO	HS 24	Ishihara	4 Mar 1999
ISESHIO	HS 25	Ishihara	10 Mar 1999
HAYASHIO	HS 26	Ishihara	10 Mar 1999
KURUSHIMA	HS 27	Nissui Marine	26 Mar 2003

Displacement, tons: 42 normal
Dimensions, feet (metres): 66.6 × 14.8 × 3.9 *(20.3 × 4.5 × 1.2)*
Main machinery: 3 diesels; 1,015 hp(m) *(746 kW)*; 3 shafts
Speed, knots: 15
Complement: 10
Radars: Navigation: I-band.

Comment: Survey launches. HS 27 authorised in FY01 extra budget.
UPDATED

UZUSHIO   *7/2004 *, Hachiro Nakai* / 1044460

For details of the latest updates to *Jane's Fighting Ships* online and to discover the additional information available exclusively to online subscribers please visit
jfs.janes.com

2 MEIYO CLASS (AGS)

Name	No	Builders	Commissioned
MEIYO	HL 03	Kawasaki, Kobe	24 Oct 1990
KAIYO	HL 05	Mitsubishi, Shimonoseki	7 Oct 1993

Displacement, tons: 550 normal
Dimensions, feet (metres): 196.9 × 34.4 × 10.2 *(60 × 10.5 × 3.1)*
Main machinery: 2 Daihatsu 6 DLM-24 diesels; 3,000 hp(m) *(2.2 MW)*; 2 shafts; bow thruster
Speed, knots: 15. **Range, n miles:** 5,280 at 11 kt
Complement: 25 + 13 scientists
Radars: Navigation: 2 sets; I-band.

Comment: *Meiyo* laid down 24 July 1989 and launched 29 June 1990; *Kaiyo* laid down 7 July 1992 and launched 26 April 1993. Have anti-roll tanks and resiliently mounted main machinery. Has a 12 kHz bottom contour sonar. A large survey launch is carried on the port side.
UPDATED

KAIYO   5/2004*, Hachiro Nakai / 1044461

AIDS TO NAVIGATION SERVICE

1 SUPPLY SHIP (AKSL)

Name	No	Builders	Commissioned
TSUSHIMA	LL 01	Mitsui, Tamano	9 Sep 1977

Displacement, tons: 1,950 normal
Dimensions, feet (metres): 246 × 41 × 13.8 *(75 × 12.5 × 4.2)*
Main machinery: 1 Fuji-Sulzer 8S40C diesel; 4,200 hp(m) *(3.09 MW)*; 1 shaft; cp prop; bow thruster
Speed, knots: 15.5. **Range, n miles:** 10,000 at 15 kt
Complement: 54

Comment: Lighthouse Supply Ship launched 7 April 1977. Fitted with tank stabilisers. Equipped with modern electronic instruments for carrying out research on electronic aids to navigation.
UPDATED

TSUSHIMA   5/2004*, Mitsuhiro Nakai / 1044458

3 HOKUTO CLASS (ABU)

Name	No	Builders	Commissioned
HOKUTO	LL 11	Sasebo	29 June 1979
KAIOU	LL 12	Sasebo	11 Mar 1980
GINGA	LL 13	Kawasaki, Kobe	18 Mar 1980

Displacement, tons: 700 normal
Dimensions, feet (metres): 180.4 × 34.8 × 8.7 *(55 × 10.6 × 2.7)*
Main machinery: 2 Asakasa MH23R diesels; 1,030 hp(m) *(757 kW)*; 2 shafts
Speed, knots: 12. **Range, n miles:** 3,900 at 12 kt
Complement: 31 (9 officers)

Comment: Used as buoy tenders.
UPDATED

HOKUTO   5/2004*, Hachiro Nakai / 1044459

1 MYOJO CLASS (ABU)

Name	No	Builders	Commissioned
MYOJO	LM 11	Nippon Kokan, Tsurumi	25 Mar 1974

Displacement, tons: 303 normal
Dimensions, feet (metres): 88.6 × 39.4 × 8.8 *(27 × 12 × 2.7)*
Main machinery: 2 Niigata 6M9 16HS diesels; 600 hp(m) *(441 kW)*; 2 shafts; cp props
Speed, knots: 10.5. **Range, n miles:** 1,360 at 10.5 kt
Complement: 18

Comment: Catamaran type buoy tender, this ship is employed in maintenance and position adjustment service to floating aids to navigation.
VERIFIED

MYOJO   9/2003, Hachiro Nakai / 0570713

AIDS TO NAVIGATION TENDERS

1 SUPPLY SHIP (AKSL)

Name	No	Builders	Commissioned
ZUIUN	LM 101	Usuki	27 July 1983

Displacement, tons: 370 normal
Dimensions, feet (metres): 146.3 × 24.6 × 7.2 *(44.6 × 7.5 ×2.2)*
Main machinery: 2 Mitsubishi-Asakasa MH23R diesels; 1,030 hp(m) *(757 kW)*; 2 shafts
Speed, knots: 13.5. **Range, n miles:** 1,000 at 13 kt
Complement: 20

Comment: Classed as a medium tender and used to service lighthouses. Can carry 85 tons of stores.
VERIFIED

ZUIUN   6/2003, Japan Coast Guard / 0570701

9 SUPPLY CRAFT (AKSL)

Name	No	Builders	Commissioned
TOKUUN	LM 114	Yokohama Yacht Co	23 Mar 1981
SHOUN	LM 201	Sumidagawa	26 Mar 1986
SEIUN	LM 202	Sumidagawa	22 Feb 1989
SEKIUN	LM 203	Ishihara	12 Mar 1991
HOUUN	LM 204	Ishihara	22 Feb 1991
REIUN	LM 205	Ishihara	28 Feb 1992
GENUN	LM 206	Wakamatsu	19 Mar 1996
AYABANE	LM 207	Ishihara	9 Mar 2000
KOUN	LM 208	Sumidagawa	16 Mar 2001

Displacement, tons: 58 full load
Dimensions, feet (metres): 75.5 × 19.7 × 3.3 *(23 × 6 × 1)*
Main machinery: 2 GM 12V-71TA diesels; 840 hp *(627 kW)* sustained; 2 shafts
Speed, knots: 14. **Range, n miles:** 250 at 14 kt
Complement: 9
Radars: Navigation: FRA 10 Mk III; I-band.
UPDATED

KOUN   7/2004*, Hachiro Nakai / 1044462

35 SMALL TENDERS (YAG)

LS 161	LS 187-195	LS 212-223
LS 164-170	LS 201	LS 231-235

Displacement, tons: 27 full load
Dimensions, feet (metres): 65 × 14.7 × 7.5 *(20 × 4.5 × 2.3)*
Main machinery: 2 diesels; 1,820 hp(m) *(1.34 MW)*; 2 shafts
Speed, knots: 25
Complement: 8

Comment: Details given are for *LS 231-233*. Others with varying characteristics.   **UPDATED**

LS 220   *7/2004*, Hachiro Nakai /* 1044463

ENVIRONMENT MONITORING CRAFT

Notes: In addition to those listed there are three oil skimmers OS 01-03.

3 SERVICE CRAFT (YPC)

Name	No	Builders	Commissioned
KINUGASA	MS 01	Ishihara, Takasago	31 Jan 1992
SAIKAI	MS 02	Ishihara, Takasago	4 Feb 1994
KATSUREN	MS 03	Sumidagawa	18 Dec 1997

Displacement, tons: 39 normal
Dimensions, feet (metres): 59.1 × 29.5 × 4.3 *(18 × 9 × 1.3)*
Main machinery: 2 diesels; 1,000 hp(m) *(735 kW)*; 2 shafts
Speed, knots: 15
Complement: 8

Comment: Details given are for *Kinugasa* which has a catamaran hull. *Saikai* and *Katsuren* are monohulls of 26 tons. Used for monitoring pollution.   **UPDATED**

SAIKAI   *8/2004*, Hachiro Nakai /* 1044464

KINUGASA   *7/2001, Mitsuhiro Kadota /* 0130213

5 SERVICE CRAFT (YAG)

SHIRASAGI OR 01	MIZUNANGI OR 03	ISOSHIGI OR 05
SHIRATORI OR 02	CHIDORI OR 04	

Displacement, tons: 153 normal
Dimensions, feet (metres): 72.3 × 21 × 2.6 *(22 × 6.4 × 0.9)*
Main machinery: 2 Nissan UD626 diesels; 360 hp(m) *(265 kW)*; 2 shafts
Speed, knots: 6. **Range, n miles:** 160 at 6 kt
Complement: 7

Comment: Completed by Sumidagawa (OR 01), Shigi (OR 02 and 04) and Ishihara (OR 03 and 05) between 31 January 1977 and 23 March 1979. Used for oil recovery.   **UPDATED**

SHIRASAGI   *5/2004*, Hachiro Nakai /* 1044465

Jordan

Country Overview

The Hashemite Kingdom of Jordan is situated in the Middle East. With an area of 34,445 square miles, it has borders to the north with Syria, to the east with Iraq, to the west with Israel and the West Bank and to the east and south with Saudi Arabia. It has a 14 n mile coastline with the Gulf of Aqaba (in the northern Red Sea) on which Aqaba, the only seaport, is situated. Amman is the capital and largest city. Territorial seas (3 n miles) are claimed but an Exclusive Economic Zone (EEZ) is not claimed.

Headquarters Appointments

Commander Naval Forces:
    Brigadier General Dari al-Zaban
Deputy Commander:
    Lieutenant Colonel Abdelkareem Fdoul

Organisation

The Royal Jordanian Naval Force comes under the Director of Operations at General Headquarters.

Bases

Dead Sea, Aqaba

Personnel

(a)   2005: 500 officers and men
(b)   Voluntary service

PATROL FORCES

Notes: In addition to the craft listed, there are also four 17 ft launches and four 14 ft GRP boats used by the Underwater Swimmer unit.

3 AL HUSSEIN (HAWK) CLASS (FAST ATTACK CRAFT—GUN) (PB)

AL HUSSEIN 101	AL HASSAN 102	KING ABDULLAH 103

Displacement, tons: 124 full load
Dimensions, feet (metres): 100 × 22.5 × 4.9 *(30.5 × 6.9 × 1.5)*
Main machinery: 2 MTU 16V 396 TB94 diesels; 5,800 hp(m) *(4.26 MW)* sustained; 2 shafts
Speed, knots: 32. **Range, n miles:** 750 at 15 kt; 1,500 at 11 kt
Complement: 16 (3 officers)
Guns: 1 Oerlikon GCM-A03 30 mm. 1 Oerlikon GAM-BO1 20 mm. 2—12.5 mm MGs.
Countermeasures: Decoys: 2 Wallop Stockade chaff launchers.
Combat data systems: Racal Cane 100.
Weapons control: Radamec Series 2000 optronic director for 30 mm gun.
Radars: Surface search: Kelvin Hughes 1007; I-band.

Comment: Ordered from Vosper Thornycroft in December 1987. GRP structure. First one on trials in May 1989 and completed December 1989. Second completed in March 1990 and the third in early 1991. All transported to Aqaba in September 1991.   **VERIFIED**

AL HUSSEIN   *6/2002 /* 0554724

4 FAYSAL CLASS (INSHORE PATROL CRAFT) (PB)

FAYSAL	HUSSEIN (ex-*Han*)	HASSAN (ex-*Hasayu*)	MUHAMMED

Displacement, tons: 8 full load
Dimensions, feet (metres): 38 × 13.1 × 1.6 *(11.6 × 4 × 0.5)*
Main machinery: 2 6M 8V715 diesels; 600 hp *(441 kW)*; 2 shafts
Speed, knots: 22. **Range, n miles:** 240 at 20 kt
Complement: 8
Guns: 1—12.7 mm MG. 1—7.62 mm MG.
Radars: Surface search: Decca; I-band.

Comment: Acquired from Bertram, Miami in 1974. GRP construction. Still operational and no replacements are planned yet.

UPDATED

MUHAMMED   *3/2004*, Bob Fildes /* 0587768

3 HASHIM (ROTORK) CLASS (PB)

HASHIM	FAISAL	HAMZA

Displacement, tons: 9 full load
Dimensions, feet (metres): 41.7 × 10.5 × 3 *(12.7 × 3.2 × 0.9)*
Main machinery: 2 Deutz diesels; 240 hp *(179 kW)*; 2 shafts
Speed, knots: 28
Complement: 5
Military lift: 30 troops
Guns: 1—12.7 mm MG. 1—7.62 mm MG.
Radars: Surface search: Furuno; I-band.

Comment: Delivered in late 1990 for patrolling the Dead Sea. Due to the annual decrease of water depth, the three craft were moved to Aqaba in 2000.

UPDATED

HAMZA   *3/2004*, Bob Fildes /* 0587769

Kazakhstan

Country Overview

Formerly part of the USSR, the Republic of Kazakhstan declared its independence in 1991. Situated in Central Asia, it has an area of 1,049,155 square miles and is bordered to the north and west with Russia, to the east with China and to the south with Kyrgyzstan, Uzbekistan, and Turkmenistan. It has a 755 n mile coastline with the Caspian Sea on which Aktau, the principal port, is situated. Astana became the capital city in 1995 while Almaty, the former capital, is the largest city. Maritime claims in the Caspian Sea are not clear. The naval Flotilla was inaugurated by President Nazarbayev in June 1998. The plan was to absorb about 30 per cent of the former USSR Caspian Flotilla, but many of these craft are derelict.

Headquarters Appointments

Commander, Navy:
    Rear Admiral Ratmir Komratov

Bases

Aktau (Caspian) (HQ)
Aralsk (Aral Sea), Bautino (Caspian)

Personnel

2005: 3,000

PATROL FORCES

Notes: (1) Plans to expand the Navy were announced by Rear Admiral Komratov in July 2003 although this is likely to take at least ten years, subject to funding.
(2) There is also an ex-trawler *Tyulen II* of 39 m with a single diesel of 578 hp(m) *(425 kW)* capable of 10 kt. Acquired in 1997.
(3) Six Customs cutters acquired from the UAE in 1998. At least one sunk in transit.
(4) Five Guardian class Boston Whalers delivered in November 1995 are unseaworthy.
(5) Plans to transfer three Yevgenya class from Russia appear to have been abandoned.

2 TURK CLASS (PB)

Name	No	Builders	Commissioned
— (ex-AB 32)	— (ex-P 132)	Haliç Shipyard	6 June 1969
— (ex-AB 26)	— (ex-P 126)	Haliç Shipyard	6 Feb 1970

Displacement, tons: 170 full load
Dimensions, feet (metres): 132 × 21 × 5.5 *(40.2 × 6.4 × 1.7)*
Main machinery: 4 SACM-AGO V16 CSHR diesels; 9,600 hp(m) *(7.06 MW)*; 2 cruise diesels; 300 hp(m) *(220 kW)*; 2 shafts
Speed, knots: 22
Complement: 31
Guns: 1 Bofors 40 mm/70. 1 Oerlikon 20 mm.
Radars: Surface search: Racal Decca; I-band.

Comment: Presented by the Turkish Navy on 3 July 1999 (AB 32) and 25 July 2001 (AB 26) at Geljuk. May have retained active sonar and ASW rocket launcher but this is unlikely.

UPDATED

TURK CLASS (Turkish colours)   *10/2000, Selim San /* 0106636

4 KW 15 (TYPE 369) CLASS (PB)

ALMATY (ex-*KW 15*) 2013 (ex-201)		ATYRAU (ex-*KW 17*) 2033 (ex-203)
AKTAU (ex-*KW 16*) 2023 (ex-202)		SCHAMBYL (ex-*KW 20*) 2043 (ex-204)

Displacement, tons: 70 full load
Dimensions, feet (metres): 93.5 × 15.4 × 4.9 *(28.9 × 4.7 × 1.5)*
Main machinery: 2 Mercedes-Benz diesels; 2,000 hp(m) *(1.47 MW)*; 2 shafts
Speed, knots: 25
Complement: 17
Guns: 2—20 mm can be fitted.
Radars: Surface search: Kelvin Hughes 14/9; I-band.

Comment: Transferred from Germany at Wilhelmshaven on 23 August 1996. Built in Germany 1952-53 and paid off in 1994, having been used for river patrols and later as range safety craft. Disarmed on transfer. Reported as being non-operational.

VERIFIED

ALMATY (old number)   *8/1996, Michael Nitz /* 0080219

1 ZHUK (PROJECT 1400) CLASS (PB)

BERKUT

Displacement, tons: 39 full load
Dimensions, feet (metres): 78.7 × 16.4 × 3.9 *(24 × 5 × 1.2)*
Main machinery: 2 Type M401B diesels; 2,200 hp(m) *(1.6 MW)* sustained; 2 shafts
Speed, knots: 30
Range, n miles: 1,100 at 15 kt
Complement: 11
Guns: 2—14.5 mm (twin); 1—12.7 mm MG.
Radars: Surface search: Spin Trough; I-band.

Comment: Built at the Zenith Shipyard, Uralsk, and commissioned 15 July 1998. Reports of a second craft have not been confirmed.

VERIFIED

DAUNTLESS *7/1996, SeaArk Marine /* 0080220

2 SAYGAK (PROJECT 1408) CLASS (PB)

Displacement, tons: 13 full load
Dimensions, feet (metres): 46.3 × 11.5 × 3 *(14.1 × 3.5 × 0.9)*
Main machinery: 1 diesel; 980 hp(m) *(720 kW)*; 1 water-jet
Speed, knots: 35. **Range, n miles:** 135 at 35 kt
Complement: 6
Guns: 2—7.62 mm MGs.
Radars: Surface search: I-band.

Comment: Russian-built small craft primarily found on the Amur river. Built in 1995 and acquired in early 1996.

VERIFIED

ZHUK (Russian colours) *11/1996, MoD Bonn /* 0019041

1 DAUNTLESS CLASS (PB)

ABAY

Displacement, tons: 11 full load
Dimensions, feet (metres): 42 × 14 × 4.3 *(12.8 × 4.3 × 1.3)*
Main machinery: 2 Detroit 8V-92TA diesels; 1,270 hp *(935 kW)*; 2 shafts
Speed, knots: 35. **Range, n miles:** 600 at 18 kt
Complement: 5
Guns: 1—12.7 mm MG. 2—7.62 mm MGs.
Radars: Surface search: Furuno; I-band.

Comment: Ordered under US funding in November 1995. Built by SeaArk, Monticello. Used to interdict the smuggling of nuclear materials across the Caspian Sea.

VERIFIED

SAYGAK (Russian colours) *7/1996, Hartmut Ehlers /* 0052520

Kenya

Country Overview

A former British colony, The Republic of Kenya gained independence in 1963. Located astride the Equator, the country has an area of 224,082 square miles and has borders to the north with Somalia and Ethiopia and to the south with Tanzania. It has a 292 n mile coastline with the Indian Ocean. The country includes almost all of Lake Turkana (Lake Rudolf) and a small portion of Lake Victoria. The capital and largest city is Nairobi and the main seaport is Mombasa. Kisumu is a port on Lake Victoria. Perhaps the first proponent of the Exclusive Economic Zone (EEZ) concept, Kenya claims a 200 n mile EEZ whose limits have been partly defined. Territorial seas (12 n miles) are claimed.

Headquarters Appointments

Commander, Navy:
  Major General Pastore Awitta
Fleet Commander:
  Colonel Joe Waswa

Personnel

(a) 2005: 1,250 plus 120 marines
(b) Voluntary service

Bases

Mombasa, Manda, Malindi, Lamu, Kisumu (Lake Victoria)

Coast Defence

There are nine Masura coastal radar stations spread along the coast. Each station has 30 ft fast boats to investigate contacts.

Customs/Police

There are some 14 Customs and Police patrol craft of between 12 and 14 m. Mostly built by Cheverton, Performance Workboats and Fassmer in the 1980s. One Cheverton 18 m craft acquired in early 1997.

PATROL FORCES

Notes: There are also five Spanish built inshore patrol craft of 16 m armed with 12.7 mm MGs and driven by twin 538 hp diesels for a speed of 16 kt. Acquired in 1995 and have pennant numbers P 943-947.

2 NYAYO CLASS (FAST ATTACK CRAFT—MISSILE) (PGGF)

Name	No	Builders	Launched	Commissioned
NYAYO	P 3126	Vosper Thornycroft	20 Aug 1986	23 July 1987
UMOJA	P 3127	Vosper Thornycroft	5 Mar 1987	16 Sep 1987

Displacement, tons: 310 light; 430 full load
Dimensions, feet (metres): 186 × 26.9 × 7.9 *(56.7 × 8.2 × 2.4)*
Main machinery: 4 Paxman Valenta 18CM diesels; 15,000 hp *(11.19 MW)* sustained; 4 shafts; 2 motors (slow speed patrol); 100 hp *(74.6 kW)*
Speed, knots: 40. **Range, n miles:** 2,000 at 18 kt
Complement: 40

Missiles: SSM: 4 OTO Melara/Matra Otomat Mk 2 (2 twin); active radar homing to 160 km *(86.4 n miles)* at 0.9 Mach; warhead 210 kg; sea-skimmer for last 4 km *(2.2 n miles)*.
Guns: 1 OTO Melara 3 in *(76 mm)*/62; 85 rds/min to 16 km *(8.7 n miles)* anti-surface; 12 km *(6.5 n miles)* anti-aircraft; weight of shell 6 kg.
   2 Oerlikon/BMARC 30 mm GCM-AO2 (twin); 650 rds/min to 10 km *(5.4 n miles)* anti-surface; 3 km *(1.6 n miles)* anti-aircraft; weight of shell 0.36 kg.
   2 Oerlikon/BMARC 20 mm A41A; 800 rds/min to 2 km; weight of shell 0.24 kg.
Countermeasures: Decoys: 2 Wallop Barricade 18-barrelled launchers; Stockade and Palisade rockets.
ESM: Racal Cutlass; radar warning.
ECM: Racal Cygnus; jammer.
Weapons control: CAAIS 450.
Radars: Surface search: Plessey AWS 4; E/F-band; range 101 km *(55 n miles)*.

NYAYO *2/2001, Sattler/Steele /* 0114357

Navigation: Decca AC 1226; I-band.
Fire control: Marconi ST802; I-band.

Programmes: Ordered in September 1984. Sailed in company from the UK, arriving at Mombasa 30 August 1988. Similar to Omani Province class.
Operational: First live Otomat firing in February 1989. RIB carried right aft. Form Squadron 86. Both ships awaiting refit.

VERIFIED

1 MAMBA CLASS (LARGE PATROL CRAFT) (PB)

Name	No	Builders	Commissioned
MAMBA	P 3100	Brooke Marine, Lowestoft	7 Feb 1974

Displacement, tons: 125 standard; 160 full load
Dimensions, feet (metres): 123 × 22.5 × 5.2 *(37.5 × 6.9 × 1.6)*
Main machinery: 2 Paxman 16YJCM diesels; 4,000 hp *(2.98 MW)* sustained; 2 shafts
Speed, knots: 25. **Range, n miles:** 3,300 at 13 kt
Complement: 25 (3 officers)
Missiles: SSM: 4 IAI Gabriel II.
Guns: 2 Oerlikon/BMARC 30 mm GCM-A02 (twin); 650 rds/min to 10 km *(5.4 n miles)* anti-surface; 3 km *(1.6 n miles)* anti-aircraft; weight of shell 0.36 kg.
Radars: Navigation: Decca AC 1226; I-band.
Fire control: Selenia RTN 10X; I/J-band; range 40 km *(22 n miles)*.

Programmes: Laid down 17 February 1972, launched 6 November 1973.
Modernisation: New missiles, gunnery equipment and optronic director fitted in 1982.
Operational: Refitted at Vosper Thornycroft 1989-90. Although still seagoing, operational capability is limited. Gabriel system non-operational.

VERIFIED

SHUJAA   *2/2001, Michael Nitz /* 0137788

AUXILIARIES

2 GALANA CLASS (LCM)

Name	No	Builders	Commissioned
GALANA	L 38	Astilleros Gondan, Spain	Feb 1994
TANA	L 39	Astilleros Gondan, Spain	Feb 1994

Displacement, tons: 1,400 full load
Dimensions, feet (metres): 208.3 × 43.6 × 7.9 *(63.5 × 13.3 × 2.4)*
Main machinery: 2 MTU/Bazán diesels; 2,700 hp(m) *(1.98 MW)* sustained; 2 shafts; bow thruster
Speed, knots: 12.5
Complement: 30
Radars: Navigation: Racal Decca; I-band.

Comment: Acquired by Galway Ltd for civilian use and taken over by the Navy for logistic support. The 4 m wide ramp is capable of taking 70 ton loads. Guns may be fitted in due course.

VERIFIED

MAMBA   *6/2002 /* 0533319

2 SHUPAVU CLASS (LARGE PATROL CRAFT) (PBO)

SHUJAA P 3130   SHUPAVU P 3131

Displacement, tons: 480 full load
Dimensions, feet (metres): 190.3 × 26.9 × 9.2 *(58 × 8.2 × 2.8)*
Main machinery: 2 diesels; 2 shafts
Speed, knots: 22
Complement: 24
Guns: 1 OTO Melara 3 in *(76 mm)*/62. 1 Breda 25 mm KBA (P 3130).
Weapons control: Breda optronic director.
Radars: Surface search: I-band.

Comment: Built to civilian standards at Astilleros Gondan, Castropol and delivered in 1997 when they were taken over by the Navy. Armament fitted in Kenya.

VERIFIED

TANA   *2/1999 /* 0052523

2 TENDERS

Dimensions, feet (metres): 60 × 15.7 × 4.9 *(18.3 × 4.8 × 1.5)*
Main machinery: 2 Caterpillar 3306B-DIT diesels; 880 hp(m) *(647 kW)*; 2 shafts
Speed, knots: 10. **Range, n miles:** 200 at 10 kt
Complement: 2 plus 136 passengers

Comment: Built by Souters, Cowes and delivered in 1998. Personnel tenders.   *VERIFIED*

Kiribati

Country Overview

The Republic of Kiribati, formerly the Gilbert Islands, is a south Pacific island group which gained independence in 1979 after the other part of the former British colony, the Ellice Islands, became independent as Tuvalu the previous year. Straddling the equator some 1,385 n miles southwest of Hawaii, it comprises from west to east Banaba (Ocean Island) and three detached island groups: the 16 Gilbert Islands, including Tarawa, on which the capital, Bairiki, is located, nine Phoenix Islands and eight of the 11 Line Islands. About 20 of the 34 islands are permanently inhabited. An archipelagic state, territorial seas (12 n miles) are claimed. An Exclusive Economic Zone (EEZ) (200 n miles) is also claimed but limits have not been fully defined by boundary agreements.

Headquarters Appointments

Head of Police Maritime Unit:
    Inspector John Mote

Bases

Tarawa

PATROL FORCES

1 PACIFIC CLASS (LARGE PATROL CRAFT) (PB)

Name	No	Builders	Commissioned
TEANOAI	301	Transfield Shipbuilding	22 Jan 1994

Displacement, tons: 165 full load
Dimensions, feet (metres): 103.3 × 26.6 × 6.9 *(31.5 × 8.1 × 2.1)*
Main machinery: 2 Caterpillar 3516TA diesels; 4,400 hp *(3.28 MW)* sustained; 2 shafts
Speed, knots: 18. **Range, n miles:** 2,500 at 12 kt
Complement: 18 (3 officers)
Guns: Can carry 1—12.7 mm MG but is unarmed.
Radars: Navigation: Furuno 1011; I-band.

Comment: The Pacific Patrol Boat programme was started by Australia in 1987. *Teanoai*, the 16th of the class, was handed over to Kiribati in 1994. The Australian government has announced that the programme will be extended so that all 22 boats will be able to operate for 30 years. *Teanoai* completed a half-life refit at Gladstone in 2001 and will require a life extension refit in 2010.

UPDATED

TEANOAI
6/1998, RAAF / 0052524

Korea, North
PEOPLE'S DEMOCRATIC REPUBLIC

Country Overview

The Democratic People's Republic of Korea (DPRK) was proclaimed in 1948 and occupies the northern part of the Korean peninsula. Located in north-eastern Asia and with an area of 46,540 square miles, it is bordered to the north by China and Russia and to the south by South Korea. It has a 1,350 n mile coastline with the Sea of Japan and the Yellow Sea. The capital and largest city is Pyóngyang while the principal ports are Nampo and Haeju on the west coast and Chojin and Wónsan on the east coast. Territorial seas (12 n miles) are claimed. A 200 n mile EEZ has also been claimed but the limits have not been defined. A source of tension at sea is the dispute concerning the status of the *Northern Limit Line* and a number of South Korean islands off the south-west coast of DPRK.

The North Korean Navy is principally a coastal force and is the lowest priority military service. Ships are allocated to East or West Fleet Command. The Navy is manpower intensive and most equipment is technologically outdated and incapable of bluewater operations. Nevertheless, considerable emphasis has been placed on high speed infiltration and assault craft and the ability to conduct unconventional operations. Fishing vessels are likely to be converted and/or commandeered for military use.

Headquarters Appointments

Commander of the Navy:
Admiral Kim Yun-Sim

Bases

East Fleet Command (HQ Toejo Dong).
East coast: Toejo (HQ), Mayang-do, Najin, Cha-ho (submarines).
Minor bases: Songjon Pando, Munchon-up, Mayang-do-ri, Mugye-po, Changjon, Puam-Dong.
West Fleet Command (HQ Nampo).
West coast: Nampo (HQ), Pipa-got (submarines), Sagon-ri.
Minor bases: Tasa-ri, Sohae-ri, Chodo, Sunwi-do, Pupo-ri, Koampo.
A number of these bases has underground berthing facilities.

Personnel

(a) 2005: 46,000 officers and other ranks
(b) 5 years' national service

Maritime Security Battalions

In addition to the Navy there is a Coastal and Port Security Police Force which would be subordinate to the Navy in war. It is reported that the strength of this force is 10-15 Chong-Jin patrol craft and 130 patrol boats of various types.

Coast Defence

Two Regiments (12-15 batteries) with six fixed and several mobile launchers of SSC-2 and SSC-3 missiles. Large numbers of 130 mm and 120 mm guns controlled from radar sites.

Strength of the Fleet

Type	Active
Submarines—Patrol	21
Submarines—Coastal	29
Submarines—Midgets	20
Frigates	3
Corvettes	4
Patrol Forces	400+
Amphibious Craft	129
Hovercraft (LCPA)	135
Minesweepers	24
Depot Ships for Midget Submarines	8
Survey Vessels	4

DELETIONS

Notes: The order of battle and fleet dispositions represent the best estimates that can be made based on incomplete information.

SUBMARINES

Notes: (1) There are four obsolete ex-Soviet Whiskey class based at Pipa-got used for training. Probably restricted to periscope depth when dived.
(2) Reports of a sea-based ballistic missile capability have not been substantiated. A surface-ship based system is considered more likely than a submarine-launched missile which would present considerable technical challenges.

21 ROMEO (PROJECT 033) CLASS (SS)

Displacement, tons: 1,475 surfaced; 1,830 dived
Dimensions, feet (metres): 251.3 × 22 × 17.1
(76.6 × 6.7 × 5.2)
Main machinery: Diesel-electric; 2 Type 37-D diesels; 4,000 hp (m) *(2.94 MW)*; 2 motors; 2,700 hp(m) *(1.98 MW)*; 2 creep motors; 2 shafts
Speed, knots: 15 surfaced; 13 dived
Range, n miles: 9,000 at 9 kt surfaced
Complement: 54 (10 officers)

Torpedoes: 8—21 in *(533 mm)* tubes (6 bow, 2 stern). 14 probably SAET-60; passive homing up to 15 km *(8.1 n miles)* at 40 kt; warhead 400 kg. Also some 53-56 may be carried.
Mines: 28 in lieu of torpedoes.
Countermeasures: ESM: China Type 921A Golf Ball (Stop Light); radar warning.
Radars: Surface search: Snoop Plate/Tray; I-band.
Sonars: Pike Jaw; hull-mounted; active.
Feniks; hull-mounted; passive.

Programmes: Two transferred from China 1973, two in 1974 and three in 1975. First three of class built at Sinpo and Mayang-do shipyards in 1976. Programme ran at about one every 14 months until 1995 when it stopped in favour of the Sang-O class. One reported sunk in February 1985.
Operational: Seventeen are stationed on east coast and have occasionally operated in Sea of Japan. Four ex-Chinese units are based on the west coast. By modern standards these are basic attack submarines with virtually no anti-submarine performance or potential and their operational status is doubtful.

UPDATED

ROMEO (China colours)

3/1995, van Ginderen Collection / 0080222

32 SANG-O CLASS (SSC)

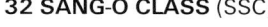

Displacement, tons: 256 surfaced; 277 dived
Dimensions, feet (metres): 116.5 × 12.5 × 12.1
(35.5 × 3.8 × 3.7)
Main machinery: 1 Russian diesel generator; 1 North Korean motor; 1 shaft; shrouded prop
Speed, knots: 7.6 surfaced; 7.2 snorting; 8.8 dived
Range, n miles: 2,700 at 7 kt
Complement: 19 (2 officers) plus 6 swimmers

Torpedoes: 2 or 4—21 in *(533 mm)* tubes (in some). Probably Russian Type 53-56.
Mines: 16 can be carried (in some).
Radars: Surface search: Furuno; I-band.
Sonars: Russian hull-mounted; passive/active search and attack.

Programmes: Started building in 1995 at Sinpo accelerating up to about four to six a year by 1996. Reported to have been building at about three a year from 1997. One reported delivered in 2002 and one in 2003 and production continues at one or two units per year.
Structure: A variation of a reverse engineered Yugoslav design. There are at least two types, one with torpedo tubes and one capable of carrying up to six external bottom mines. There is a single periscope and a VLF radio receiver in the fin. Rocket launchers and a 12.7 mm MG can be carried. Diving depth 180 m *(590 ft)*. A longer (39 m) variant submarine may replace older boats.
Operational: Used extensively for infiltration operations. The submarine can bottom, and swimmer disembarkation is reported as being normally exercised from periscope depth. One of the class grounded and was captured by South Korea on 18 September 1996. Some crew members may be replaced by special forces for short operations. 17 stationed on east coast.

UPDATED

SANG-O

9/1996 / 0080223

20 (+ 10 RESERVE) YUGO and P-4 CLASS (MIDGET SUBMARINES) (SSW)

Displacement, tons: 90 surfaced; 110 dived
Dimensions, feet (metres): 65.6 × 10.2 × 15.1
(20 × 3.1 × 4.6)
Main machinery: 2 diesels; 320 hp(m) *(236 kW)*; 1 shaft
Speed, knots: 12 surfaced; 8 dived
Range, n miles: 550 at 10 kt surfaced; 50 at 4 kt dived
Complement: 4 plus 6—7 divers
Torpedoes: 2—406 mm tubes.
Radars: Navigation: I-band.

Comment: Built at Yukdaeso-ri shipyard since early 1960s. More
than one design. Details given are for the latest type, at least
one of which has been exported to Iran, and have been
building since 1987 to a Yugoslavian design. Some have two
short external torpedo tubes and some have a snort mast. The
conning tower acts as a wet and dry compartment for divers.
There is a second and smaller propeller for slow speed
manoeuvring while dived. Two of the class are designated
P-4s and belong to the KWP. This type has two internal
torpedo tubes. Operate from eight merchant mother ships
(see *Auxiliaries*). Some have been lost in operations against
South Korea, the latest in June 1998. Numbers are reducing
in favour of the Sang-O class. Two exported to Vietnam in
June 1997. There are also about 50 two-man submersibles of
Italian design 4.9 × 1.4 m. Overall numbers are approximate
due to scrapping of older units.

VERIFIED

YUGO P-4
6/1998, Ships of the World / 0052525

FRIGATES

1 SOHO CLASS (FFGH)

Name	No	Builders	Laid down	Launched	Commissioned
—	823	Najin Shipyard	June 1980	Nov 1981	May 1982

Displacement, tons: 1,640 full load
Dimensions, feet (metres): 242.1 × 50.9 × 12.5
(73.8 × 15.5 × 3.8)
Main machinery: 2 diesels; 15,000 hp(m) *(11.03 MW)*; 2 shafts
Speed, knots: 23
Complement: 189 (17 officers)

Missiles: SSM: 4 CSS-N-2 ❶; active radar or IR homing to 46 km
(25 n miles) at 0.9 Mach; warhead 513 kg.
Guns: 1—3.9 in *(100 mm)*/56 ❷; 40° elevation; 15 rds/min to
16 km *(8.6 n miles)*; weight of shell 13.5 kg.
4—37 mm/63 (2 twin) ❸.
4—30 mm/65 (2 twin) ❹. 4—25 mm/60 (2 twin) ❺.
A/S mortars: 2 RBU 1200 5-tubed fixed launchers ❻; range
1,200 m; warhead 34 kg.
Countermeasures: ESM: China RW-23 Jug Pair (Watch Dog);
intercept.
Radars: Surface search: Square Tie ❼; I-band.
Fire control: Drum Tilt ❽; H/I-band.
Navigation: I-band.
Sonars: Stag Horn; hull-mounted; active search and attack; high
frequency.

Helicopters: Platform for 1 medium.

Programmes: Planned class of six but only one was ordered.
Structure: One of the largest warships built anywhere with a
twin hull design and a helicopter deck aft. Has a large central
superstructure to carry the heavy gun armament.

Operational: Reported operational but is probably very weather
limited like many catamaran designs. Based at Mugye-po on
the east coast.

VERIFIED

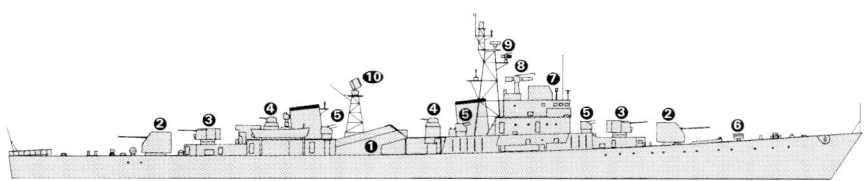

SOHO
(Scale 1 : 600), Ian Sturton

2 NAJIN CLASS (FFG)

531	591

Displacement, tons: 1,500 full load
Dimensions, feet (metres): 334.6 × 32.8 × 8.9
(102 × 10 × 2.7)
Main machinery: 3 SEMT-Pielstick Type 16 PA6 280 diesels;
18,000 hp(m) *(13.2 MW)*; 3 shafts
Speed, knots: 24. **Range, n miles:** 4,000 at 13 kt
Complement: 180 (16 officers)

Missiles: SSM: 2 CSS-N-1 ❶; active radar or IR homing to 46 km
(25 n miles) at 0.9 Mach; warhead 513 kg HE. Replaced
torpedo tubes on both ships.
Guns: 2—3.9 in *(100 mm)*/56 ❷; 40° elevation; 15 rds/min to
16 km *(8.6 n miles)*; weight of shell 13.5 kg.
4—57 mm/80 (2 twin) ❸; 120 rds/min to 6 km *(3.2 n miles)*;
weight of shell 2.8 kg.
12 or 4—30 mm/60 (6 or 2 twin) ❹ (see *Structure*).
12—25 mm (6 twin) ❺.
A/S mortars: 2 RBU 1200 5-tubed fixed launchers ❻; range
1,200 m; warhead 34 kg (not in *531*).
Depth charges: 2 projectors; 2 racks. 30 weapons.
Mines: 30 (estimated).
Countermeasures: Decoys: 6 chaff launchers.
ESM: China RW-23 Jug Pair (Watch Dog); intercept.
Weapons control: Optical director ❼.
Radars: Air search: Square Tie ❽; I-band.
Surface search: Pot Head ❾; I-band.
Navigation: Pot Drum; H/I-band.
Fire control: Drum Tilt ❿; H/I-band.
IFF: High Pole. Square Head.
Sonars: Stag Horn; hull-mounted; active search; high frequency.

Programmes: Built at Najin and Nampo shipyards. First
completed 1973, second 1975.
Structure: There is some resemblance to the ex-Soviet Kola
class, now deleted. The original torpedo tubes were replaced
by CSS-N-1 missile launchers in the mid-1980s and the RBU

NAJIN
(Scale 1 : 900), Ian Sturton

NAJIN 531
5/1993, JMSDF / 0080224

1200 mortars have been removed in at least one of the class.
Gun armaments differ, one having six twin 30 mm while the
other only has one twin 30 mm and six twin 25 mm.

Operational: One based on each coast but seldom seen at sea.
UPDATED

CORVETTES

4 SARIWON CLASS (FS) and 1 TRAL CLASS (FS)

671   611   612   613   614

Displacement, tons: 650; 580 (Tral); full load
Dimensions, feet (metres): 203.7 × 23.9 × 7.8
(62.1 × 7.3 × 2.4)
Main machinery: 2 diesels; 3,000 hp(m) *(2.21 MW)*; 2 shafts
Speed, knots: 16. **Range, n miles:** 2,700 at 16 kt
Complement: 60 (7 officers)

Guns: 1—85 mm/52 tank turret (Tral) ❶.
4—57 mm/80 (2 twin) (Sariwon).
2 or 4—37 mm/6 (single (Tral) ❷; 2 twin (Sariwon)).
16—14.5 mm ❸; 4 quad.
A/S mortars: 2 RBU 1200 5-tubed fixed launchers (Sariwon 513).
Depth charges: 2 rails.
Mines: 30.
Radars: Surface search: Pot Head or Don 2 ❹; I-band.
Navigation: Model 351; I-band.
IFF: Ski Pole.
Sonars: Stag Horn; hull-mounted; active; high frequency.

TRAL 671   *(Scale 1 : 600), Ian Sturton*

Programmes: Four Sariwon class built in Korea in the mid-1960s. The two Tral class were transferred from the USSR in the mid-1950s, were paid off in the early 1980s but one returned to service in the early 1990s.
Structure: The Sariwon design is based on the original USSR Fleet minelayer Tral or Fugas class in service in the mid-1930s. Minelaying rails are visible along the whole of upper deck aft of the bridge superstructure. Sariwon *671* is one of the original Tral class built in 1938 and restored to service. Some variations in gun armament and one Sariwon is reported as having sonar and an ASW armament.
Operational: All ships are based on the east coast (Changjon) and, despite their age, have been reported as seagoing. Pennant numbers 611-614 have been reported.   *UPDATED*

TRAL 671   *5/1993, JMSDF /* 0080225

PATROL FORCES

8 OSA I (PROJECT 205) and 4 HUANGFEN CLASSES
(FAST ATTACK CRAFT—MISSILE) (PTFG)

Displacement, tons: 171 standard; 210 full load
Dimensions, feet (metres): 126.6 × 24.9 × 8.9 *(38.6 × 7.6 × 2.7)*
Main machinery: 3 Type M 503A diesels; 8,025 hp(m) *(5.9 MW)* sustained; 3 shafts
Speed, knots: 35. **Range, n miles:** 800 at 30 kt
Complement: 30
Missiles: SSM: 4 SS-N-2A Styx; active radar or IR homing to 46 km *(25 n miles)* at 0.9 Mach; warhead 513 kg.
Guns: 4—30 mm/65 (2 twin) AK 230; 500 rds/min to 5 km *(2.7 n miles)*; weight of shell 0.54 kg.
Countermeasures: ESM: China BM/HZ 8610; intercept (Huangfen class).
Radars: Surface search: Square Tie; I-band.
Fire control: Drum Tilt; H/I-band (Osa I).
IFF: High Pole B. Square Head.

Programmes: Twelve Osa I class transferred from USSR in 1968 and four more in 1972-83. Eight remain of which four are based on each coast. Four Huangfen class acquired from China in 1980 and are based on the west coast.   *VERIFIED*

6 KOMAR CLASS and 6 SOHUNG CLASS
(FAST ATTACK CRAFT—MISSILE) (PTFG)

Displacement, tons: 75 standard; 85 full load
Dimensions, feet (metres): 84 × 24 × 5.9 *(25.6 × 7.3 × 1.8)* (Sohung)
Main machinery: 4 Type M 50 diesels; 4,400 hp(m) *(3.3 MW)* sustained; 4 shafts
Speed, knots: 40. **Range, n miles:** 400 at 30 kt
Complement: 19
Missiles: SSM: 2 SS-N-2A Styx or CSS-N-1; active radar or IR homing to 46 km *(25 n miles)* at 0.9 Mach; warhead 513 kg.
Guns: 2—25 mm/80 (twin); 270 rds/min to 3 km *(1.6 n miles)*; weight of shell 0.34 kg.
2—14.5 mm (twin) MGs.
Radars: Surface search: Square Tie; I-band.
IFF: Square Head.

Programmes: Ten Komar class transferred by USSR in 1968, six still in service but with wood hulls replaced by steel. The Sohung class is a North Korean copy of the Komar class, first built in 1980-81 and no longer in production. The 'Komars' and four 'Sohung' are based on the east coast.   *VERIFIED*

OSA I

KOMAR

10 SOJU CLASS (FAST ATTACK CRAFT—MISSILE) (PTG)

Displacement, tons: 265 full load
Dimensions, feet (metres): 139.4 × 24.6 × 5.6 *(42.5 × 7.5 × 1.7)*
Main machinery: 3 Type M 503A diesels; 8,025 hp(m) *(5.9 MW)* sustained; 3 shafts
Speed, knots: 34. **Range, n miles:** 600 at 30 kt
Complement: 32 (4 officers)
Missiles: SSM: 4 SS-N-2 Styx; active radar or IR homing to 46 km *(25 n miles)* at 0.9 Mach; warhead 513 kg.
Guns: 4—30 mm/65 (2 twin) AK 230; 500 rds/min to 5 km *(2.7 n miles)*; weight of shell 0.54 kg.
Countermeasures: ESM: China BM/HZ 8610; intercept.
Radars: Surface search: Square Tie; I-band.
Fire Control: Drum Tilt; H/I-band.

Comment: North Korean built and enlarged version of Osa class. First completed in 1981; built at about one per year at Nampo, Najin and Yongampo shipyards, but the programme terminated in 1996. Six based on the east coast and four on the west.
VERIFIED

6 HAINAN CLASS (LARGE PATROL CRAFT) (PC)

201-204   292-293

Displacement, tons: 375 standard; 392 full load
Dimensions, feet (metres): 192.8 × 23.6 × 6.6 *(58.8 × 7.2 × 2)*
Main machinery: 4 Kolomna/PCR Type 9-D-8 diesels; 4,000 hp(m) *(2.94 MW)*; 4 shafts
Speed, knots: 30.5. **Range, n miles:** 1,300 at 15 kt
Complement: 69
Guns: 4—57 mm/70 (2 twin); 120 rds/min to 8 km *(4.4 n miles)*; weight of shell 2.8 kg.
    4—25 mm/80 (2 twin); 270 rds/min to 3 km *(1.6 n miles)*; weight of shell 0.34 kg.
A/S mortars: 4 RBU 1200 5-tubed launchers; range 1,200 m; warhead 34 kg.
Depth charges: 2 projectors; 2 racks for 30 DCs.
Mines: Laying capability for 12.
Countermeasures: Decoys: 2 PK 16 chaff launchers.
ESM: China BM/HZ 8610; intercept.
Radars: Surface search: Pot Head (Model 351); I-band.
Sonars: Stag Ear; hull-mounted; active search and attack; high frequency.

Comment: Transferred from China in 1975 (two), 1976 (two), 1978 (two). All based on the west coast.
VERIFIED

HAINAN (China colours)   *4/1998* / 0080226

19 SO 1 CLASS (LARGE PATROL CRAFT) (PC)

Displacement, tons: 170 light; 215 normal
Dimensions, feet (metres): 137.8 × 19.7 × 5.9 *(42 × 6 × 1.8)*
Main machinery: 3 Kolomna Type 40-D diesels; 6,600 hp(m) *(4.85 MW)* sustained; 3 shafts
Speed, knots: 28. **Range, n miles:** 1,100 at 13 kt
Complement: 31
Guns: 1—85 mm/52; 18 rds/min to 15 km *(8 n miles)*; weight of shell 9.5 kg.
    2—37 mm/63 (twin); 160 rds/min to 9 km *(4.9 n miles)*; weight of shell 0.7 kg.
    4 or 6—25 mm/60 (2 or 3 twin); 270 rds/min to 3 km *(1.6 n miles)*; weight of shell 0.34 kg.
    4—14.5 mm/93 MGs.
A/S mortars: 4 RBU 1200 5-tubed launchers; range 1,200 m; warhead 34 kg.
Radars: Surface search: Pot Head (Model 351); I-band.
Navigation: Don 2; I-band.
IFF: Ski Pole or Dead Duck.
Sonars: Stag Ear; hull-mounted; active.

Comment: Eight transferred by the USSR in early 1960s, with RBU 1200 ASW rocket launchers and depth charges instead of the 85 mm and 37 mm guns. Remainder built in North Korea to modified design. Twelve are fitted out for ASW with sonar and depth charges; the other seven are used as gunboats. The majority are based on the east coast.
VERIFIED

SO 1 (USSR colours)   *1988*

12 SHANGHAI II CLASS (FAST ATTACK CRAFT—GUN) (PBT)

381-388   391-395

Displacement, tons: 113 standard; 131 full load
Dimensions, feet (metres): 126.3 × 17.7 × 5.6 *(38.5 × 5.4 × 1.7)*
Main machinery: 2 Type L12-180 diesels; 2,400 hp(m) *(1.76 MW)* (forward)
    2 Type 12-D-6 diesels; 1,820 hp(m) *(1.34 MW)* (aft); 4 shafts
Speed, knots: 30. **Range, n miles:** 700 at 16.5 kt
Complement: 34
Guns: 4—37 mm/63 (2 twin); 160 rds/min to 9 km *(4.9 n miles)*; weight of shell 0.7 kg.
    4—25 mm/60 (2 twin); 270 rds/min to 3 km *(1.6 n miles)*; weight of shell 0.34 kg.
    2—3 in *(76 mm)* recoilless rifles.
Depth charges: 8.
Mines: Rails can be fitted for 10 mines.
Countermeasures: ESM: China BM/HZ 8610; intercept.
Radars: Surface search: Pot Head (Model 351) or Skin Head; I-band.

Comment: Acquired from China since 1967. Based in the west fleet.
VERIFIED

SHANGHAI II   *1994* / 0080227

7 TAECHONG I CLASS and 5 TAECHONG II CLASS (LARGE PATROL CRAFT) (PC)

Displacement, tons: 385 standard; 410 full load (I); 425 full load (II)
Dimensions, feet (metres): 196.3 (I); 199.5 (II) × 23.6 × 6.6 *(59.8; 60.8 × 7.2 × 2)*
Main machinery: 4 Kolomna Type 40-D diesels; 8,800 hp(m) *(6.4 MW)* sustained; 4 shafts
Speed, knots: 25. **Range, n miles:** 2,000 at 12 kt
Complement: 80
Guns: 1—3.9 in *(100 mm)*/56 (Taechong II); 15 rds/min to 16 km *(8.6 n miles)*; weight of shell 13.5 kg or 1—85 mm/52.
    2—57 mm/70 (twin); 120 rds/min to 8 km *(4.4 n miles)*; weight of shell 2.8 kg.
    4—30 mm/65 (2 twin) (Taechong II). 2—25 mm/60 (twin) (Taechong I).
    16 or 4—14.5 mm MGs (4 quad (Taechong II); 2 twin (Taechong I)).
A/S mortars: 2 RBU 1200 5-tubed fixed launchers; range 1,200 m; warhead 34 kg.
Depth charges: 2 racks.
Radars: Surface search: Pot Head (Model 351); I-band.
Fire control: Drum Tilt; H/I-band.
IFF: High Pole A. Square Head.
Sonars: Stag Ear; hull-mounted; active attack; high frequency.

Comment: North Korean class of mid-1970s design, slightly larger than Hainan class. The first seven are Taechong I class. Taechong II built at about one per year at Najin shipyard up to 1995. They are slightly longer, and are heavily armed for units of this size. Based in both fleets.
VERIFIED

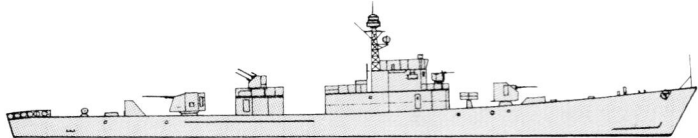

TAECHONG   *(not to scale)*

TAECHONG II (with *Najin*)   *1998*

6 CHONG-JU CLASS (LARGE PATROL CRAFT) (PC)

Displacement, tons: 205 full load
Dimensions, feet (metres): 138.8 × 23.6 × 6.9 *(42.3 × 7.2 × 2.1)*
Main machinery: 4 diesels; 4,406 hp(m) *(3.24 MW)*; 4 shafts
Speed, knots: 20. **Range, n miles:** 1,350 at 12 kt
Complement: 48 (7 officers)
Missiles: SSM: 4 CSS-N-1; active radar or IR homing to 46 km *(25 n miles)* at 0.9 Mach; warhead 513 kg. In three of the class.
Guns: 1—85 mm/52; 18 rds/min to 15 km *(8 n miles)*; weight of shell 9.5 kg.
    4—37 mm/63 (2 twin). 4—25 mm/60 (2 twin).
    4—14.5 mm/93 (2 twin) MGs.
A/S mortars: 2 RBU 1200; 5-tubed launchers; range 1,200 m; warhead 34 kg.
Radars: Surface search: Pot Head (Model 351); I-band.
Sonars: Stag Ear; hull-mounted; active attack; high frequency.

Comment: Built between 1975 and 1989. At least one has been converted to fire torpedoes and three others have CSS-N-1 missiles and resemble the Soju class. Based in both fleets.
VERIFIED

59 CHAHO CLASS (FAST ATTACK CRAFT—GUN) (PTF)

Displacement, tons: 82 full load
Dimensions, feet (metres): 85.3 × 19 × 6.6 *(26 × 5.8 × 2)*
Main machinery: 4 Type M 50 diesels; 4,400 hp(m) *(3.2 MW)* sustained; 4 shafts
Speed, knots: 37. **Range, n miles:** 1,300 at 18 kt
Complement: 16 (2 officers)
Guns: 1 BM 21 multiple rocket launcher. 2 USSR 23 mm/87 (twin). 2—14.5 mm (twin) MGs.
Radars: Surface search: Pot Head (Model 351); I-band.

Comment: Building in North Korea since 1974. Based on P 6 hull. Three transferred to Iran in April 1987. Still building and new hulls are replacing the old ones. 35 based in the east and 24 in the west.
VERIFIED

CHAHO (Iranian colours)   *4/1998*

54 CHONG-JIN CLASS (FAST ATTACK CRAFT—GUN) (PTF/PTK)

Displacement, tons: 80 full load
Dimensions, feet (metres): 85.3 × 19 × 5.9 (26 × 5.8 × 1.8)
Main machinery: 4 Type M 50 diesels; 4,400 hp(m) (3.2 MW) sustained; 4 shafts
Speed, knots: 36. **Range, n miles:** 450 at 30 kt
Complement: 17 (3 officers)
Guns: 1—85 mm/52; 18 rds/min to 15 km (8 n miles); weight of shell 9.5 kg.
 4 or 8—14.5 mm (2 or 4 twin) MGs.
Radars: Surface search: Skin Head; I-band.
IFF: High Pole B; Square Head.

Comment: Particulars similar to Chaho class of which this is an improved version. Building began
 about 1975. About one third reported to be a hydrofoil development. Up to 15 are operated by
 the Coastal Security Force. Based in both fleets. ***VERIFIED***

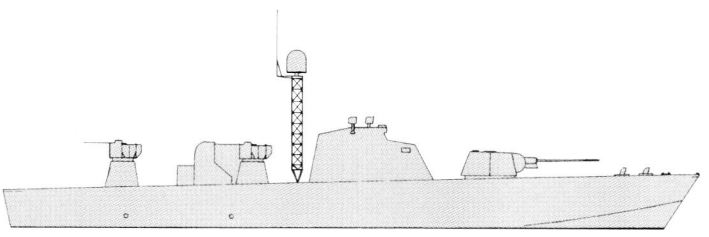

CHONG-JIN *(not to scale), Ian Sturton*

33 SINPO CLASS (FAST ATTACK CRAFT—TORPEDO) (PTF/PTK)

Displacement, tons: 64 standard; 73 full load
Dimensions, feet (metres): 85.3 × 20 × 4.9 (26 × 6.1 × 1.5)
Main machinery: 4 Type M 50 diesels; 4,400 hp(m) (3.2 MW) sustained; 4 shafts
Speed, knots: 45. **Range, n miles:** 450 at 30 kt; 600 at 15 kt
Complement: 15
Guns: 4—25 mm/80 (2 twin) (original). 2—37 mm (others). 6—14.5 mm MGs (Sinpo class).
Torpedoes: 2—21 in (533 mm) tubes (in some). Sinpo class has no tubes.
Depth charges: 8 in some.
Radars: Surface search: Skin Head; I-band (some have Furuno).
IFF: Dead Duck. High Pole.

Comment: Thirteen craft remain of the 27 P 6 class transferred from the USSR and 15 Shantou
 class transferred from China. Some of the P 6s have hydrofoils and one sank in June 1999. The
 Sinpo (or Sinnam) class are locally built versions of these craft of which 20 now remain. Based in
 both fleets. ***VERIFIED***

SINPO

P 6

142 KU SONG, SIN HUNG and MOD SIN HUNG CLASSES
(FAST ATTACK CRAFT—TORPEDO) (PTF/PTK)

Displacement, tons: 42 full load
Dimensions, feet (metres): 75.4 × 16.1 × 5.5 (23 × 4.9 × 1.7)
Main machinery: 2 Type M 50 diesels; 2,200 hp(m) (1.6 MW) sustained; 2 shafts
Speed, knots: 40; 50 (Mod Sin Hung). **Range, n miles:** 500 at 20 kt
Complement: 20 (3 officers)
Guns: 4—14.5 mm (2 twin) MGs.
Torpedoes: 2—18 in (457 mm) or 2—21 in (533 mm) tubes (not fitted in all).
Radars: Surface search: Skin Head; I-band.
IFF: Dead Duck.

Comment: Ku Song and Sin Hung built in North Korea between mid-1950s and 1970s. Frequently
 operated near South Korean border. A modified version of Sin Hung with hydrofoils built from
 1981-85. Fifty craft, previously thought to have been scrapped, are in various states of repair.
 Based in both fleets. ***UPDATED***

SIN HUNG (no torpedo tubes) *1991*

MODIFIED FISHING VESSELS
(COASTAL PATROL CRAFT) (PB/AGI)

Comment: Approximately 15 fishing vessels have been converted for naval use. Some act as patrol
 craft, others as AGIs. The vessel sunk by the Japanese Coast Guard on 22 December 2001
 carried a 14.5 mm machine-gun, two anti-air missile launchers and numerous small arms. The
 stern was fitted with outward opening doors. ***UPDATED***

MFV 801 *7/1991, G Jacobs*

Fishing Vessel (being salvaged) *9/2002, P A News / 0522267*

HIGH-SPEED AND SEMI-SUBMERSIBLE INFILTRATION CRAFT
(HSIC/PBF)

Displacement, tons: 5 full load
Dimensions, feet (metres): 30.5 × 8.2 × 3.1 (9.3 × 2.5 × 1)
Main machinery: 1 diesel; 260 hp(m) (191 kW); 1 shaft
Speed, knots: 35
Complement: 2
Guns: 1—7.62 mm MG.
Radars: Navigation: Furuno 701; I-band.

Comment: Up to a hundred built for Agent infiltration and covert operations. These craft have a
 very low radar cross-section and 'squat' at high speeds. High rate of attrition. A newer version
 was reported in 1998. This is 12.8 m in length and has a top speed of about 45 kt. It is reported
 to travel on the surface until submerging to a depth of 3 m using a snort mast. It has a dived
 speed of 4 kt. ***UPDATED***

HSIC *1991, J Bermudez*

15 TB 11PA AND 10 TB 40A CLASSES
(INSHORE PATROL CRAFT) (PBF)

Displacement, tons: 8 full load
Dimensions, feet (metres): 36.7 × 8.6 × 3.3 (11.2 × 2.7 × 1)
Main machinery: 2 diesels; 520 hp(m) (382 kW); 2 shafts
Speed, knots: 35. **Range, n miles:** 200 at 15 kt
Complement: 4
Guns: 1—7.62 mm MG.
Radars: Surface search: Furuno; I-band.

Comment: High-speed patrol boats. Reinforced fibreglass hull. Design closely resembles a number
 of UK/Western European commercial craft. Larger hull design, known as 'TB 40A' also built.
 Both classes being operated by the Coastal Security Force. ***VERIFIED***

AMPHIBIOUS FORCES

10 HANTAE CLASS (LSM)

Displacement, tons: 350 full load
Dimensions, feet (metres): 157.5 × 21.3 × 6.6 *(48 × 6.5 × 2)*
Main machinery: 2 diesels; 4,352 hp(m) *(3.2 MW)*; 2 shafts
Speed, knots: 18. **Range, n miles:** 2,000 at 12 kt
Complement: 36 (4 officers)
Military lift: 350 troops plus 3 MBTs
Guns: 8—25 mm/80 (4 twin).

Comment: Built in the early 1980s. Most are based on the east coast.

VERIFIED

96 NAMPO CLASS (LLP)

Displacement, tons: 75 full load
Dimensions, feet (metres): 85.3 × 19 × 5.6 *(26 × 5.8 × 1.7)*
Main machinery: 4 Type M 50 diesels; 4,400 hp(m) *(3.2 MW)* sustained; 4 shafts
Speed, knots: 36. **Range, n miles:** 450 at 30 kt
Complement: 19
Military lift: 35 troops
Guns: 4—14.5 mm (2 twin) MGs.
Radars: Surface search: Skin Head; I-band.

Comment: A class of assault landing craft. Similar to the Chong-Jin class but with a smaller forward gun mounting and with retractable ramp in bows. Building began about 1975. Several have been deleted due to damage. There are 18 of the original class and 73 of a modified version which have a covered-in deck. Most have bow doors welded shut. Four sold to Madagascar in 1979 but now deleted. The Nampo D is the latest version with a multihull design. The first of these entered service in 1997 and four further craft have followed. Based in both fleets.
VERIFIED

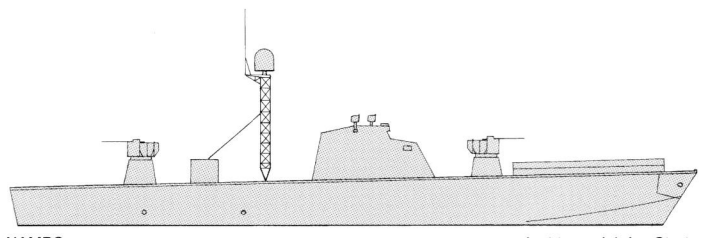

NAMPO   *(not to scale), Ian Sturton*

7 HANCHON CLASS (LCM) and 18 HUNGNAM CLASS (LCM)

Displacement, tons: 145 full load
Dimensions, feet (metres): 117.1 × 25.9 × 3.9 *(35.7 × 7.9 × 1.2)*
Main machinery: 2 Type 3-D-12 diesels; 600 hp(m) *(443 kW)* sustained; 2 shafts
Speed, knots: 10. **Range, n miles:** 600 at 6 kt
Complement: 15 (1 officer)
Military lift: 2 tanks or 300 troops
Guns: 2—14.5 mm/93 (twin) MG.
Radars: Surface search: Skin Head; I-band.

Comment: Details given for the Hanchons, built in the 1980s. The Hungnams are older and half the size. Based in both fleets.
VERIFIED

136 KONGBANG CLASS (HOVERCRAFT) (LCPA)

Comment: Three types: one Type I, 57 Type II and 78 are Type III. Length 25 m (I), 21 m (II) and 18 m (III). A series of high-speed air cushion landing craft first reported in 1987 and building continued until 1996 and then stopped. Use of air cushion technology is an adoption of commercial technology based on the SRN-6. Kongbang II has twin propellers and can carry up to 50 commandos at 50 kt. Kongbang III has a single propeller and can take about 40 troops at 40 kt. All are radar fitted. Some have Styx SSM missiles. Older craft are being replaced continuously in a high priority programme. Divided between both fleets.
VERIFIED

MINE WARFARE FORCES

19 YUKTO I CLASS and 5 YUKTO II CLASSES
(COASTAL MINESWEEPERS) (MSC)

Displacement, tons: 60 full load (I); 52 full load (II)
Dimensions, feet (metres): 78.7 × 13.1 × 5.6 *(24 × 4 × 1.7)* (Yukto I)
Main machinery: 2 diesels; 2 shafts
Speed, knots: 18
Complement: 22 (4 officers)
Guns: 1—37 mm/63 or 2—25 mm/80 (twin). 2—14.5 mm/93 (twin) MGs.
Mines: 2 rails for 4.
Radars: Surface search: Skin Head; I-band.

Comment: North Korean design built in the 1980s and replaced the obsolete ex-Soviet KN-14 class. Yukto IIs are 3 m shorter (at 21 m length) overall and have no after gun. Wooden construction. One completed mid-1996 but this may have been of a new design. Based in both fleets.
VERIFIED

SURVEY SHIPS

Notes: The Hydrographic Department has four survey ships but also uses a number of converted fishing vessels.

AUXILIARIES

Notes: (1) Trawlers operate as AGIs on the South Korean border where several have been sunk over the years. In addition many ocean-going commercial vessels are used for carrying weapons and ammunition worldwide in support of international terrorism.
(2) There are also eight ocean cargo ships adapted as mother ships for midget submarines. Their names are *Soo Gun-Ho, Dong Geon Ae Gook-Ho, Dong Hae-Ho, Choong Seong-Ho Number One, Choong Seong-Ho Number Two, Choong Seong-Ho Number Three, Hae Gum Gang-Ho* and the *Song Rim-Ho*.

1 KOWAN CLASS (ASR)

Displacement, tons: 2,010 full load
Dimensions, feet (metres): 275.6 × 46.9 × 12.8 *(84 × 14.3 × 3.9)*
Main machinery: 4 diesels; 8,160 hp(m) *(6 MW)*; 2 shafts
Speed, knots: 16
Complement: 150
Guns: 12—14.5 mm (6 twin) MGs.
Radars: Navigation: Furuno; I-band.

Comment: Used as a submarine rescue ship. Probable catamaran construction.
VERIFIED

Korea, South
REPUBLIC

Country Overview

The Republic of Korea was proclaimed in 1948 and occupies the southern part of the Korean peninsula. Located in northeastern Asia and with an area of 38,375 square miles, it is bordered to the north by North Korea. It has a 1,300 n miles coastline with the Sea of Japan, the Yellow Sea and the Korea Strait, which separates it from Japan. There are numerous offshore islands in the south and west, the largest of which is Cheju. A source of tension at sea is the dispute concerning the status of the *Northern Limit Line* and a number of South Korean islands off the southwest coast of DPRK. The capital and largest city is Seoul. The principal port is Pusan while others include Inchon, the major port on the Yellow Sea, Mokp'o and Kunsan. Territorial seas (12 n miles) are claimed. A 200 n mile EEZ has also been claimed but the limits have not been defined.

Headquarters Appointments

Chief of Naval Operations:
  Admiral Mun Jung Il
Commandant Marine Corps:
  Lieutenant General Kim In Sik
Vice Chief of Naval Operations:
  Vice Admiral Choi Ki Chul
Chief of Maritime Police:
  Commissioner General Lee Seng Jae

Operational Commands

Commander-in-Chief Fleet:
  Vice Admiral Kim, Sung Man
Commander First Fleet:
  Rear Admiral Ahn, Kee Suk
Commander Second Fleet:
  Rear Admiral Sur, Yang Won

Operational Commands — *continued*

Commander Third Fleet:
  Rear Admiral Park, In Young

Diplomatic Representation

Defence Attaché in London:
  Captain K Y Lee

Personnel

(a) 2005: Regulars: 33,000 (Navy) and 24,000 (Marines)
    Conscripts: 17,000 (Navy and Marines)
(b) 2¼ years' national service for conscripts
(c) Reserves: 9,000

Bases

Major: Chinhae (Fleet HQ and 3rd Fleet), Donghae (1st Fleet), Pyongtaek (2nd Fleet)
Minor: Cheju, Mokpo, Mukho, Pohang, Pusan
Aviation: Pohang (MPA base), Chinhae, Cheju
Marines: Pohang, Kimpo, Pengyongdo

A new base, to replace Chinhae as HQ of 3rd Fleet is to be constructed at Nansu, south of Pusan.

Coast Defence

Three batteries of Marines with truck-mounted quadruple Harpoon SSM launchers.

Pennant Numbers

Numbers ending in 4 are not used as they are unlucky.

Organisation

In 1986 the Navy was reorganised into three Fleets, each commanded by a Rear Admiral, whereas the Marines retained two Divisions and one brigade plus smaller and support units. From October 1973 the RoK Marine Force was placed directly under the RoK Navy command with a Vice Chief of Naval Operations for Marine Affairs replacing the Commandant of Marine Corps. The Marine Corps was re-established as an independent service on 1 November 1987.

1st Fleet: No 11, 12, 13 DD/FF Sqn; No 101, 102 Coastal Defence Sqn; 181, 191, 111, 121 Coastal Defence Units; 121st Minesweeper Sqn.
2nd Fleet: No 21, 22, 23 DD/FF Sqn; No 201, 202 Coastal Defence Sqn; 211, 212 Coastal Defence Units; 522nd Minesweeper Sqn.
3rd Fleet: 301, 302, 303 DD/FF Sqn; 304, 406th Coastal Defence Units.

Strength of the Fleet

Type	Active (Reserve)	Building (Proposed)
Submarines (Patrol)	9	3
Submarines (Midget)	11	—
Destroyers	7	5
Frigates	9	—
Corvettes	28	—
Fast Attack Craft—Missile	5	40
Fast Attack Craft—Patrol	83	—
Minehunters	—	(2)
Minesweepers	3	—
Minelayers	1	—
LPD	—	1
LSTs	8	—
LSMs	—	—
LCU/LCM/LCF	20	—
Logistic Support Ships	3	—

DELETIONS

Destroyers		Patrol Forces		Amphibious Forces		Mine Warfare Forces	
1999	*Taejon, Kwang Ju, Kwang Won*	1999	*Pae Ku 56-61*	1999	*Wol Mi, Ki Rin,* Mulkae 72-73, Mulkae 75-79, Mulkae 81-82, *Kae Bong, Solgae*	2001	*Kum San, Ko Hung, Kum Kok, Nam Yang, Ha Dong, Sam Kok, Yong Dong, Ok Cheon*

PENNANT LIST

Submarines

061	Chang Bogo
062	Yi Chon
063	Choi Muson
065	Park Wi
066	Lee Jongmu
067	Jung Woon
068	Lee Sunsin
069	Na Daeyong
071	Lee Eokgi

Destroyers

971	Kwanggaeto the Great
972	Euljimundok
973	Yangmanchun
975	Chungmugong Yi Sun-Shin
976	Moonmu Daewang (bldg)
977	Daejoyoung (bldg)

Frigates

951	Ulsan
952	Seoul
953	Chung Nam

955	Masan
956	Kyong Buk
957	Chon Nam
958	Che Ju
959	Pusan
961	Chung Ju

Corvettes

751	Dong Hae
752	Su Won
753	Kang Reung
755	An Yang
756	Po Hang
757	Kun San
758	Kyong Ju
759	Mok Po
761	Kim Chon
762	Chung Ju
763	Jin Ju
765	Yo Su
766	Jin Hae
767	Sun Chon
768	Yee Ree
769	Won Ju
771	An Dong

772	Chon An
773	Song Nam
775	Bu Chon
776	Jae Chon
777	Dae Chon
778	Sok Cho
779	Yong Ju
781	Nam Won
782	Kwan Myong
783	Sin Hung
785	Kong Ju

Mine Warfare Forces

560	Won San
561	Kang Kyeong
562	Kang Jin
563	Ko Ryeong
565	Kim Po
566	Ko Chang
567	Kum Wha
571	Yang Yang
572	Ongjin

Amphibious Forces

671	Un Bong
676	Wee Bong
677	Su Yong
678	Buk Han
681	Kojoon Bong
682	Biro Bong
683	Hyangro Bong
685	Seongin Bong

Auxiliaries

21	Cheong Hae Jin
27	Pyong Taek
28	Kwang Yang
57	Chun Jee
58	Dae Chung
59	Hwa Chun
AGS 11	Sunjin

SUBMARINES

Notes: (1) Plans to acquire three Russian Kilo (Project 636) class have been abandoned.
(2) The Type 214 programme may be followed by a larger 3,000 ton submarine class. Numbers and funding are yet to be finalised.

0 + 3 KSS-2 (TYPE 214) CLASS (SSK)

Name	No	Builders	Laid down	Launched	Commissioned
—	—	Hyundai, Ulsan	2003	2005	2007
—	—	Hyundai, Ulsan	2004	2006	2008
—	—	Hyundai, Ulsan	2005	2007	2009

Displacement, tons: 1,700 surfaced; 1,860 dived
Dimensions, feet (metres): 213.3 × 20.7 × 19.7
 (65 × 6.3 × 6)
Main machinery: 1 MTU 16V 396 diesel; 4,243 hp *(3.12 MW)*;
 1 Siemens Permasyn motor; 3,875 hp(m) *(2.85 MW)*; 1 shaft;
 9 Siemens/HDW PEM fuel cell (AIP) modules; 306 kW;
 sodium sulphide high-energy batteries
Speed, knots: 20 dived; 12 surfaced
Complement: 27 (5 officers)
Torpedoes: 8—21 in *(533 mm)* bow tubes.
Countermeasures: Decoys: ESM.
Weapons control: STN Atlas.
Radars: Surface search; I-band.
Sonars: Bow, flank and towed arrays.

Programmes: Decision taken in November 2000 to order three
 HDW designed Air Independent Propulsion (AIP) submarines.
 The boats are being built by Hyundai Heavy Industries with the
 German Submarine Corporation, led by HDW, providing
 construction plans, materials and other equipment. First steel
 cut for the first of class in November 2002.
Structure: The Type 214 is a synthesis of the proven Type 209
 design with AIP from the Type 212. South Korea is the second
 customer for the Type 214 after Greece. Details given are
 mainly for the Type 214 as advertised by HDW but changes
 may have been made. Diving depth 400 m.
UPDATED

TYPE 214

9/2002, Michael Nitz / 0529079

2 KSS-1 DOLGORAE CLASS and 9 DOLPHIN (COSMOS) CLASS (MIDGET SUBMARINES)

052-053 (Dolgorae)

Displacement, tons: 150 surfaced; 175 dived (Dolgorae); 70
 surfaced; 83 dived (Cosmos)
Dimensions, feet (metres): 82 × 6.9 *(25 × 2.1)* (Cosmos)
Main machinery: Diesel-electric; 1 diesel generator; 1 motor;
 1 shaft
Speed, knots: 9 surfaced; 6 dived
Complement: 6 + 8 swimmers
Torpedoes: 2—406 mm tubes (Dolgorae). 2—533 mm tubes
 (Cosmos).
Sonars: Atlas Elektronik; hull-mounted; passive search; high
 frequency.

Comment: Dolgorae class started entering service in 1983.
 Cosmos type used by Marines. Limited endurance, for use
 only in coastal waters. Fitted with Pilkington Optronics
 periscopes (CK 37 in Dolgorae and CK 41 in Cosmos).
 Numbers of each type confirmed but the Dolgorae class are
 being replaced by more Cosmos. All are based at Cheju Island.
VERIFIED

DOLGORAE
11/1985, G Jacobs

9 CHANG BOGO (TYPE 209) CLASS (1200) (SSK)

Name	No	Builders	Laid down	Launched	Commissioned
CHANG BOGO	061	HDW, Kiel	1989	18 June 1992	2 June 1993
YI CHON	062	Daewoo, Okpo	1990	14 Oct 1992	30 Apr 1994
CHOI MUSON	063	Daewoo, Okpo	1991	25 Aug 1993	27 Feb 1995
PARK WI	065	Daewoo, Okpo	1992	20 May 1994	3 Feb 1996
LEE JONGMU	066	Daewoo, Okpo	1993	17 Apr 1995	29 Aug 1996
JUNG WOON	067	Daewoo, Okpo	1994	7 May 1996	29 Aug 1997
LEE SUNSIN	068	Daewoo, Okpo	1995	21 May 1998	15 June 1999
NA DAEYONG	069	Daewoo, Okpo	1996	15 June 1999	Nov 2000
LEE EOKGI	071	Daewoo, Okpo	1997	26 May 2000	30 Nov 2001

Displacement, tons: 1,100 surfaced; 1,285 dived
Dimensions, feet (metres): 183.7 × 20.3 × 18
(56 × 6.2 × 5.5)
Main machinery: Diesel-electric; 4 MTU 12V 396 SE diesels;
3,800 hp(m) *(2.8 MW)* sustained; 4 alternators; 1 motor;
4,600 hp(m) *(3.38 MW)* sustained; 1 shaft
Speed, knots: 11 surfaced/snorting; 22 dived
Range, n miles: 7,500 at 8 kt surfaced
Complement: 33 (6 officers)

Missiles: SSM: McDonnell Douglas UGM-84B Sub Harpoon;
active radar homing to 130 km *(70 n miles)* at 0.9 Mach;
warhead 227 kg (fitted to at least three boats).
Torpedoes: 8—21 in *(533 mm)* bow tubes. 14 SystemTechnik
Nord (STN) SUT Mod 2; wire-guided; active/passive homing
to 12 km *(6.6 n miles)* at 35 kt or 28 km *(15.1 n miles)* at
23 kt; warhead 260 kg. Swim-out discharge.
Mines: 28 in lieu of torpedoes.
Countermeasures: ESM: Argo; radar warning.
Weapons control: Atlas Elektronik ISUS 83 TFCS.
Radars: Navigation: I-band.
Sonars: Atlas Elektronik CSU 83; hull-mounted; passive search
and attack; medium frequency.

Programmes: First three ordered in late 1987, one built at Kiel by
HDW, and two assembled at Okpo by Daewoo from material
packages transported from Germany. Second three ordered in
October 1989 and a further batch of three in January 1994.
Modernisation: Mid-life upgrade of all nine boats is under
consideration. It is envisaged that AIP propulsion and Sub-
Harpoon SSM may be fitted in stretched hulls and that the
refit programme will be conducted between 2005-10.
Structure: Type 1200 similar to those built for the Turkish Navy
with a heavy dependence on Atlas Elektronik sensors and STN
torpedoes. Diving depth 250 m *(820 ft)*. A passive towed
array may be fitted in due course.
Operational: An indigenous torpedo based on the Honeywell NP
37 may be available in due course. The class is split between
the three Fleets. Operations conducted off Hawaii from 1997
to improve operating standards.

UPDATED   NA DAEYONG   *6/2002, Ships of the World* / 0529129

CHANG BOGO   *7/2004*, Michael Nitz* / 1042341

CHANG BOGO   *7/2004*, Michael Nitz* / 1042342

DESTROYERS

0 + 3 KDX-3 CLASS (DDGHM)

Name	No	Builders	Laid down	Launched	Commissioned
—	—	Hyundai, Ulsan	11 Nov 2004	2006	2008
—	—	Hyundai, Ulsan	2007	2008	2010
—	—	Hyundai, Ulsan	2009	2010	2012

Displacement, tons: 7,000 standard
Dimensions, feet (metres): 544.6 × ? × ?
(166.0 × ? × ?)
Main machinery: To be announced
Speed, knots: 30

Missiles: SSM: 8 Harpoon.
SAM: Mk 41 VLS; 32 cells for Standard SM-2.
1 RAM system.

Guns: 1—5 in *(127 mm)*.
1 Goalkeeper system.
Combat data system: Aegis variant.
Weapons control: To be announced.
Radars: Air search: SPY 1D; 3D; F-band.
Air/Surface search: To be announced.
Fire control: To be announced.
Navigation: To be announced.
Sonars: To be announced.
Helicopters: To be announced.

Programmes: A further development of the KDX-2 destroyers but optimised for anti-air warfare. Lockheed Martin selected on 24 July 2002 to supply the combat data system and multifunction radar. Design details are speculative.
UPDATED

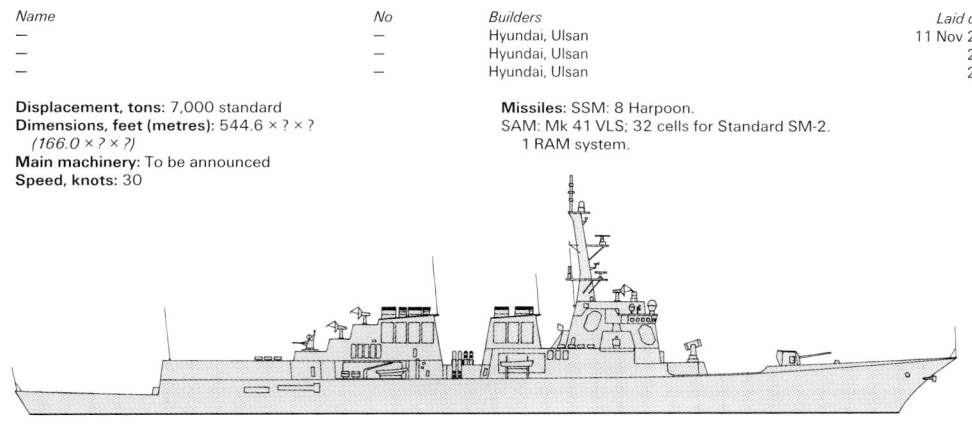

KDX-3   *(not to scale), Ian Sturton / 0569924*

2 + 1 KDX-2 CLASS (DDGHM)

Name	No	Builders	Laid down	Launched	Commissioned
CHUNGMUGONG YI SUN-SHIN	975	Daewoo, Okpo	2001	20 May 2002	Nov 2003
MOONMU DAEWANG	976	Hyundai, Ulsan	2002	11 Apr 2003	30 Sep 2004
DAEJOYOUNG	977	Daewoo, Okpo	2002	12 Nov 2003	2005

Displacement, tons: 4,800 full load
Dimensions, feet (metres): 506.6 × 55.5 × 14.1
(154.4 × 16.9 × 4.3)
Main machinery: CODOG; 2 GE LM 2500 gas turbines; 58,200 hp *(43.42 MW)* sustained; 2 MTU 20V 956 TB92 diesels; 8,000 hp(m) *(5.88 MW)*; 2 shafts
Speed, knots: 29. **Range, n miles:** 4,000 at 18 kt
Complement: 200 (18 officers)

Missiles: SSM: 8 Harpoon (Block 1C) (2 quad) **❶**.
SAM: Mk 41 Mod 2 VLS **❷** 32 cells for GDC Standard SM-2MR (Block IIIA); command/inertial guidance; semi-active radar homing to 167 km *(90 n miles)* at 2 Mach.
1 Raytheon RAM Mk 31 systems **❸**; 21 rounds per launcher; passive IR/anti-radiation homing to 9.6 km *(5.2 n miles)* at 2 Mach; warhead 9.1 kg.
A/S: ASROC VLS; inertial guidance 1.6-10 km *(1-5.4 n miles)* at 0.9 Mach; payload Mk 48.
Guns: 1 United Defense 5 in *(127 mm)*/62 Mk 45 Mod 4 **❹**; 20 rds/min to 23 km *(12.6 n miles)*; weight of shell 32 kg.
1 Signaal Goalkeeper 30 mm **❺**; 7 barrels per mounting; 4,200 rds/min to 1.5 km.
Countermeasures: 4 chaff launchers. ESM/ECM.
Combat data systems: BAeSema/Samsung KD COM-2; Link 11.
Weapons control: Marconi Mk 14 weapons direction system.
Radars: Air search: Raytheon SPS-49(V)5 **❻**; C/D-band.
Surface search: Signaal MW08 **❼**; G-band.
Fire control: 2 Signaal STIR 240 **❽**; I/J/K-band.
Sonars: DSQS-23; hull-mounted; active search; medium frequency. Daewoo Telecom towed array; passive low frequency.

Helicopters: 1 Westland Super Lynx Mk 99 **❾**.

Programmes: Approval for first three given in late 1996 but the final decision was not taken until 1998. Contract to design and build the first of class won by Daewoo in November 1999. Further ships are not expected.
Operational: Successful SM-2 firings conducted on the Pacific Missile Range Facility, off Hawaii, in mid-2004.
UPDATED

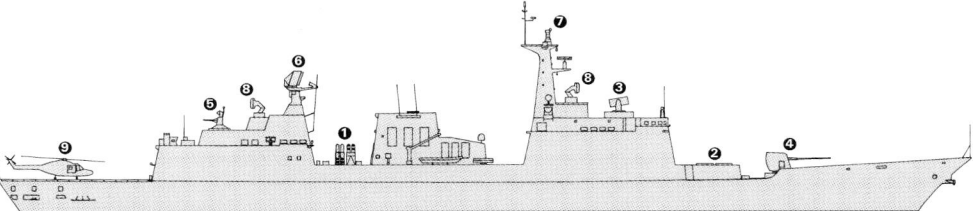

CHUNGMUGONG YI SUN-SHIN   *(Scale 1 : 1,200), Ian Sturton / 0569921*

CHUNGMUGONG YI SUN-SHIN   *7/2004*, Michael Nitz / 1042339*

CHUNGMUGONG YI SUN-SHIN   *7/2004*, Michael Nitz / 1042340*

3 KWANGGAETO THE GREAT (KDX-1) CLASS (DDGHM)

Name	No	Builders	Laid down	Launched	Commissioned
KWANGGAETO THE GREAT	971	Daewoo, Okpo	June 1995	28 Oct 1996	24 July 1998
EULJIMUNDOK	972	Daewoo, Okpo	Jan 1996	16 Oct 1997	20 June 1999
YANGMANCHUN	973	Daewoo, Okpo	Aug 1997	19 Oct 1998	29 June 2000

Displacement, tons: 3,855 full load
Dimensions, feet (metres): 444.2 × 46.6 × 13.8
  (135.4 × 14.2 × 4.2)
Main machinery: CODOG; 2 GE LM 2500 gas turbines;
  58,200 hp *(43.42 MW)* sustained; 2 MTU 20V 956 TB92
  diesels; 8,000 hp(m) *(5.88 MW)*; 2 shafts
Speed, knots: 30. **Range, n miles:** 4,000 at 18 kt
Complement: 170 (15 officers)

Missiles: SSM: 8 McDonnell Douglas Harpoon Block 1C (2 quad)
  launchers ❶; active radar homing to 130 km *(70 n miles)* at
  0.9 Mach; warhead 227 kg.
  SAM: Raytheon Sea Sparrow; Mk 48 Mod 2 VLS launcher ❷ for
  16 cells RIM-7P; semi-active radar homing to 14.6 km *(8 n
  miles)* at 2.5 Mach; warhead 39 kg.
Guns: 1 Otobreda 5 in *(127 mm)*/54 ❸; 45 rds/min to 23 km
  (12.4 n miles); weight of shell 32 kg.
  2 Signaal 30 mm Goalkeeper ❹; 7 barrels per mounting;
  4,200 rds/min combined to 2 km.
Torpedoes: 6—324 mm (2 triple) Mk 32 tubes ❺; Alliant
  Techsystems Mk 46 Mod 5; anti-submarine; active/passive
  homing to 11 km *(5.9 n miles)* at 40 kt; warhead 44 kg.
Countermeasures: Decoys: 4 CSEE Dagaie Mk 2 chaff launchers
  ❻. SLQ-25 Nixie towed torpedo decoy.
  ESM/ECM: Argo AR 700/APECS II ❼; intercept and jammer.
Combat data systems: BAeSEMA/Samsung SSCS Mk 7; Litton
  NTDS (Link 11). SATCOM ❽.
Radars: Air search: Raytheon SPS-49V5 ❾; C/D-band.
  Surface search: Signaal MW08 ❿; G-band.
  Fire control: 2 Signaal STIR 180 ⓫; I/J/K-band.
  Navigation: Daewoo DTR 92 (SPS 55M) ⓬; I-band,
  IFF: UPX-27.
Sonars: Atlas Elektronik DSQS-21BZ; hull-mounted active
  search; medium frequency.
  Daewoo Telecom towed array; passive low frequency.

Helicopters: 1 Westland Super Lynx ⓭.

Programmes: Project KDX-1. A much delayed programme. The
  first keel was to have been laid down at Daewoo in late 1992
  for completion in 1996, but definition studies extended to late
  1993, when contracts started to be signed for the weapon
  systems. First steel cut at Daewoo Okpo in April 1994.

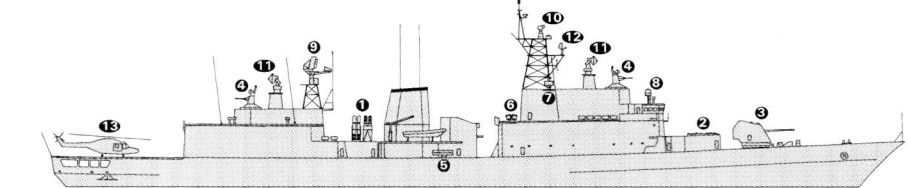

KWANGGAETO THE GREAT   *(Scale 1 : 1,200), Ian Sturton / 0572485*

EULJIMUNDOK   *7/2004*, Michael Nitz / 1042337*

Structure: Emphasis is on air defence but the design has taken
  so long to reach fulfilment that it has been overtaken by the
  KDX-2. McTaggart Scott Trigon 5 helo handling system.

Operational: The Goalkeepers are also to be used against
  close-in surface threats using FAPDS (Frangible Armour
  Penetrating Discarding Sabot). ***UPDATED***

YANGMANCHUN   *12/2003*, Bob Fildes / 1042335*

KWANGGAETO THE GREAT   *12/2002, Hachiro Nakai / 0529069*

CORVETTES

24 PO HANG CLASS (FS/FSG)

Name	No	Builders	Commissioned
PO HANG	756	Korea SEC, Pusan	Dec 1984
KUN SAN	757	Korea Tacoma	Dec 1984
KYONG JU	758	Hyundai, Ulsan	Nov 1986
MOK PO	759	Daewoo, Okpo	Aug 1986
KIM CHON	761	Korea SEC, Pusan	May 1985
CHUNG JU	762	Korea Tacoma	May 1985
JIN JU	763	Hyundai, Ulsan	June 1988
YO SU	765	Daewoo, Okpo	Nov 1988
JIN HAE	766	Korea SEC, Pusan	Feb 1989
SUN CHON	767	Korea Tacoma	June 1989
YEE REE	768	Hyundai, Ulsan	June 1989
WON JU	769	Daewoo, Okpo	Aug 1989
AN DONG	771	Korea SEC, Pusan	Nov 1989
CHON AN	772	Korea Tacoma	Nov 1989
SONG NAM	773	Daewoo, Okpo	May 1989
BU CHON	775	Hyundai, Ulsan	Apr 1989
JAE CHON	776	Korea SEC, Pusan	May 1989
DAE CHON	777	Korea Tacoma	Apr 1989
SOK CHO	778	Korea SEC, Pusan	Feb 1990
YONG JU	779	Hyundai, Ulsan	Mar 1990
NAM WON	781	Daewoo, Okpo	Apr 1990
KWAN MYONG	782	Korea Tacoma	July 1990
SIN HUNG	783	Korea SEC, Pusan	Mar 1993
KONG JU	785	Korea Tacoma	July 1993

Displacement, tons: 1,220 full load
Dimensions, feet (metres): 289.7 × 32.8 × 9.5 *(88.3 × 10 × 2.9)*
Main machinery: CODOG; 1 GE LM 2500 gas turbine; 26,820 hp *(20 MW)* sustained; 2 MTU 12V 956 TB82 diesels; 6,260 hp(m) *(4.6 MW)* sustained; 2 shafts; Kamewa cp props
Speed, knots: 32. **Range, n miles:** 4,000 at 15 kt (diesel)
Complement: 95 (10 officers)

Missiles: SSM: 2 Aerospatiale MM 38 Exocet (756-759) ❶; inertial cruise; active radar homing to 42 km *(23 n miles)* at 0.9 Mach; warhead 165 kg; sea-skimmer.
4 McDonnell Douglas Harpoon (2 twin) launchers ❷; active radar homing to 130 km *(70 n miles)* at 0.9 Mach; warhead 227 kg.
Guns: 1 or 2 OTO Melara 3 in *(76 mm)*/62 compact ❸; 85 rds/min to 16 km *(8.6 n miles)* anti-surface; 12 km *(6.5 n miles)* anti-aircraft; weight of shell 6 kg.
4 Emerson Electric 30 mm (2 twin) (756-759) ❹; 4 Breda 40 mm/70 (2 twin) (761 onwards) ❺.
Torpedoes: 6—324 mm Mk 32 (2 triple) tubes ❻. Honeywell Mk 46; anti-submarine; active/passive homing to 11 km *(5.9 n miles)* at 40 kt; warhead 44 kg.
Depth charges: 12 (761 onwards).
Countermeasures: Decoys: 4 MEL Protean fixed launchers; 36 grenades.
2 Loral Hycor SRBOC 6-barrelled Mk 36 launchers (in some); range 4 km *(2.2 n miles)*.
ESM/ECM: THORN EMI or NobelTech; intercept/jammer.
Combat data systems: Signaal Sewaco ZK (756-759); Ferranti WSA 423 (761 onwards).
Weapons control: Signaal Liod or Radamec 2400 (766 onwards) optronic director ❼.
Radars: Surface search: Marconi 1810 ❽ and/or Raytheon SPS-64 ❾; I-band.
Fire control: Signaal WM28 ❿; I/J-band; or Marconi 1802 ⓫; I/J-band.
Sonars: Signaal PHS-32; hull-mounted; active search and attack; medium frequency.

Programmes: First laid down early 1983. After early confusion, names, pennant numbers and shipbuilders are now correct. The programme terminated in 1993.
Modernisation: Harpoon has been installed in 769 and may also have been fitted to others.
Structure: The first four are Exocet fitted and have a different weapon systems arrangement. The remainder have an improved combat data system with Ferranti/Radamec/Marconi fire-control systems and radars as in the later versions of the Ulsan class. *UPDATED*

MOK PO
10/1998, John Mortimer / 0052541

4 DONG HAE CLASS (FS)

Name	No	Builders	Commissioned
DONG HAE	751	Korea SEC, Pusan	Aug 1982
SU WON	752	Korea Tacoma	Oct 1983
KANG REUNG	753	Hyundai, Ulsan	Nov 1983
AN YANG	755	Daewoo, Okpo	Dec 1983

Displacement, tons: 1,076 full load
Dimensions, feet (metres): 256.2 × 31.5 × 8.5 *(78.1 × 9.6 × 2.6)*
Main machinery: CODOG; 1 GE LM 2500 gas turbine; 26,820 hp *(20 MW)* sustained; 2 MTU 12V 956 TB82 diesels; 6,260 hp(m) *(4.6 MW)* sustained; 2 shafts; Kamewa cp props
Speed, knots: 31. **Range, n miles:** 4,000 at 15 kt (diesel)
Complement: 95 (10 officers)

Guns: 1 OTO Melara 3 in *(76 mm)*/62 compact ❶; 85 rds/min to 16 km *(8.6 n miles)*; weight of shell 6 kg.
4 Emerson Electric 30 mm (2 twin) ❷. 2 Bofors 40 mm/60 (twin) ❸.
Torpedoes: 6—324 mm Mk 32 (2 triple) tubes ❹. Honeywell Mk 46; anti-submarine; active/passive homing to 11 km *(5.9 n miles)* at 40 kt; warhead 44 kg.
Depth charges: 12.
Countermeasures: Decoys: 4 MEL Protean chaff launchers.
ESM/ECM: THORN EMI or NobelTech; intercept and jammer.
Combat data systems: Signaal Sewaco ZK.
Weapons control: Signaal Liod optronic director ❺.
Radars: Surface search: Raytheon SPS-64 ❻; I-band.
Fire control: Signaal WM28 ❼; I/J-band.
Sonars: Signaal PHS-32; hull-mounted; active search and attack; medium frequency.

Programmes: This was the first version of the corvette series, with four being ordered in 1980, one each from the four major warship building yards.
Structure: The design was too small for the variety of different weapons which were intended to be fitted for different types of warfare and was therefore discontinued in favour of the Po Hang class.

VERIFIED

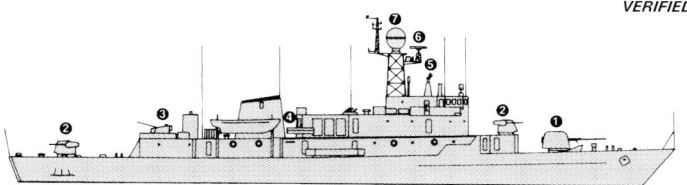
DONG HAE
(Scale 1 : 900), Ian Sturton

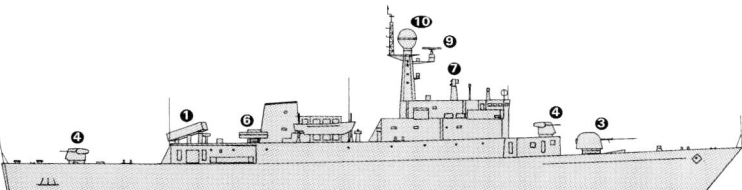
PO HANG
(Scale 1 : 900), Ian Sturton / 0572484

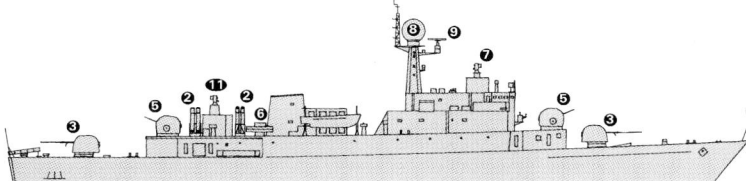
WON JU
(Scale 1 : 900), Ian Sturton / 0569920

SU WON
9/2000 / 0104998

WON JU
7/2002, Chris Sattler / 0528917

FRIGATES

9 ULSAN CLASS (FFG)

Name	No	Builders	Laid down	Launched	Commissioned
ULSAN	951	Hyundai, Ulsan	1979	8 Apr 1980	1 Jan 1981
SEOUL	952	Hyundai, Ulsan	1982	24 Apr 1984	30 June 1985
CHUNG NAM	953	Korean SEC, Pusan	1984	26 Oct 1984	1 June 1986
MASAN	955	Korea Tacoma	1983	26 Oct 1984	20 July 1985
KYONG BUK	956	Daewoo, Okpo	1984	15 Jan 1986	30 May 1986
CHON NAM	957	Hyundai, Ulsan	1986	19 Apr 1988	17 June 1989
CHE JU	958	Daewoo, Okpo	1986	3 May 1988	1 Jan 1990
PUSAN	959	Hyundai, Ulsan	1990	20 Feb 1992	1 Jan 1993
CHUNG JU	961	Daewoo, Okpo	1990	20 Mar 1992	1 June 1993

Displacement, tons: 1,496 light; 2,180 full load (2,300 for FF 957-961)
Dimensions, feet (metres): 334.6 × 37.7 × 11.5 *(102 × 11.5 × 3.5)*
Main machinery: CODOG; 2 GE LM 2500 gas turbines; 53,640 hp *(40 MW)* sustained; 2 MTU 16V 538 TB82 diesels; 5,940 hp(m) *(4.37 MW)* sustained; 2 shafts; cp props
Speed, knots: 34; 18 on diesels. **Range, n miles:** 4,000 at 15 kt
Complement: 150 (16 officers)

Missiles: SSM: 8 McDonnell Douglas Harpoon (4 twin) launchers ❶; active radar homing to 130 km *(70 n miles)* at 0.9 Mach; warhead 227 kg.
Guns: 2—3 in *(76 mm)*/62 OTO Melara compact ❷; 85 rds/min to 16 km *(8.6 n miles)* anti-surface; 12 km *(6.5 n miles)* anti-aircraft; weight of shell 6 kg.
8 Emerson Electric 30 mm (4 twin) (FF 951-955) ❸; 6 Breda 40 mm/70 (3 twin) (FF 956-961) ❹.
Torpedoes: 6—324 mm Mk 32 (2 triple) tubes ❺. Honeywell Mk 46 Mod 1; anti-submarine; active/passive homing to 11 km *(5.9 n miles)* at 40 kt; warhead 44 kg.
Depth charges: 12.
Countermeasures: Decoys: 4 Loral Hycor SRBOC 6-barrelled Mk 36 launchers ❻; range 4 km *(2.2 n miles)*.
SLQ-25 Nixie; towed torpedo decoy.
ESM: ULQ-11K; intercept.
Combat data systems: Samsung/Ferranti WSA 423 action data automation (FF 957-961). Litton systems retrofitted to others. Link 11 in three of the class. WSC-3 SATCOM (F 957).
Weapons control: 1 Signaal Liod optronic director (FF 951-956) ❼; 1 Radamec System 2400 optronic director (FF 957-961) ❽.
Radars: Air/surface search: Signaal DA05 ❾; E/F-band.
Surface search: Signaal ZW06 (FF 951-956) ❿; Marconi S 1810 (FF 957-961) ⓫; I-band.
Fire control: Signaal WM28 (FF 951-956) ⓬; Marconi ST 1802 (FF 957-961) ⓭; I/J-band.
Navigation: Raytheon SPS-10C (FF 957-961) ⓮; I-band.
Tacan: SRN 15.
Sonars: Raytheon DE 1167; hull-mounted; active search and attack; medium frequency.

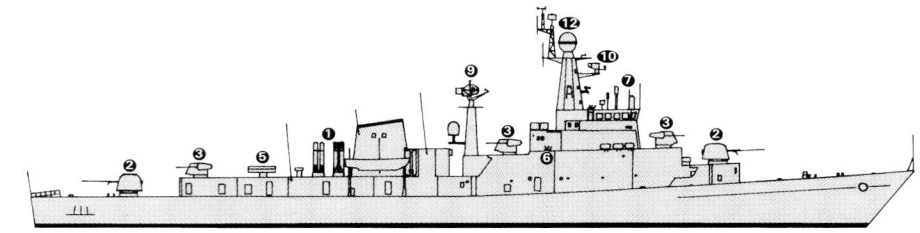

ULSAN   *(Scale 1 : 900), Ian Sturton*

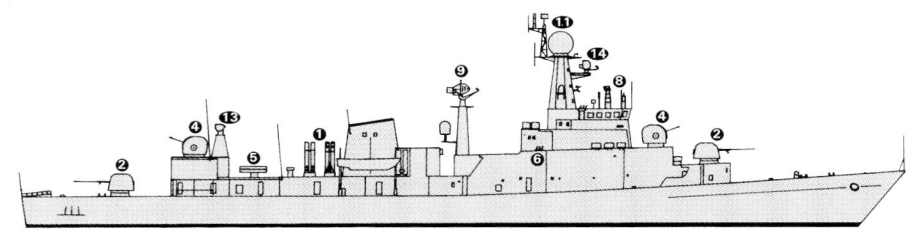

CHE JU   *(Scale 1 : 900), Ian Sturton*

Modernisation: New sonars fitted. WSC-3 SATCOM fitted in *Chon Nam*.
Structure: Steel hull with aluminium alloy superstructure. There are three versions. The first five ships are the same but *Kyong Buk* has the four Emerson Electric twin 30 mm guns replaced by three Breda twin 40 mm, and the last four of the class have a built-up gun platform aft and a different combination of surface search, target indication and navigation radars. Weapon systems integration caused earlier concern and a Ferranti combat data system has been installed in the last five; Litton Systems Link 11 fitted in three of the class.
Operational: *Che Ju* and *Chung Nam* conducted the first ever deployment of South Korean warships to Europe during a four month tour from September 1991 to January 1992. Trainees were embarked. Three of the class have a shore datalink and act as local area commanders to control attack craft carrying out coastal protection patrols.

VERIFIED

CHON NAM   *7/2000, Sattler/Steele* / 0104995

ULSAN   *8/2000, van Ginderen Collection* / 0104996

CHE JU   *10/2002, Guy Toremans* / 0528915

CHUNG JU   *2/2001, Ships of the World* / 0130106

CHUNG NAM   *9/2002, Hachiro Nakai* / 0529073

SHIPBORNE AIRCRAFT

Notes: A Request for Proposals for eight mine-hunting helicopters is expected in 2008 with deliveries in 2010-2011.

Numbers/Type: 17/13 Westland Super Lynx Mk 99/Mk 100.
Operational speed: 125 kt *(231 km/h)*.
Service ceiling: 12,000 ft *(3,660 m)*.
Range: 320 n miles *(593 km)*.
Role/Weapon systems: Six ASV helicopters delivered in 1991; 11 ASW versions acquired later and 13 more ordered in June 1997 and delivered in 1999/2000. Sensors: Ferranti Sea Spray Mk 3 radar and Racal ESM. Bendix AQS 18(V) dipping sonar and ASQ 504(V) MAD in ASW versions. Weapons: 4 BAe Sea Skua missiles. Mk 46 (Mod 5) torpedo (in ASW version). Sea Skua may be replaced in due course.
VERIFIED

SUPER LYNX   *1/2001, Mitsuhiro Kadota* / 0130104

Numbers/Type: 6 Aerospatiale SA 316B/SA 319B Alouette III.
Operational speed: 113 kt *(210 km/h)*.
Service ceiling: 10,500 ft *(3,200 m)*.
Range: 290 n miles *(540 km)*.
Role/Weapon systems: Marine support helicopter; operated by RoK Marine Corps. Sensors: None. Weapons: Unarmed.
VERIFIED

ALOUETTE III   *1990* / 0081162

LAND-BASED MARITIME AIRCRAFT (FRONT LINE)

Notes: (1) F-16 fighters are capable of firing Harpoon ASV missiles.
(2) There are also 10 UH-60 and 10 UH-1 utility helicopters.

Numbers/Type: 8 Grumman S-2A/F Tracker.
Operational speed: 130 kt *(241 km/h)*.
Service ceiling: 25,000 ft *(7,620 m)*.
Range: 1,350 n miles *(2,500 km)*.
Role/Weapon systems: Maritime surveillance and limited ASW operations; coastal surveillance and EEZ patrol. Sensors: Search radar, ECM. Weapons: ASW; torpedoes, depth bombs and mines. ASV; underwing 127 mm rockets. HARM missiles may be acquired.
VERIFIED

Numbers/Type: 8 Lockheed P-3C Orion Update III.
Operational speed: 411 kt *(761 km/h)*.
Service ceiling: 28,300 ft *(8,625 m)*.
Range: 4,000 n miles *(7,410 km)*.
Role/Weapon systems: Maritime patrol aircraft ordered in December 1990. First pair delivered April 1995, remainder April 1996. Funding constraints have stalled plans for a second squadron of eight aircraft. The Update III version is fitted with ASQ-212 tactical computer. Sensors: APS-134 or 137(V)6 search radar; AAS-36 IR. Weapons: four Harpoon ASM.
VERIFIED

Numbers/Type: 5 Rheims-Cessna F 406 Caravan II.
Operational speed: 229 kt *(424 km/h)*.
Service ceiling: 30,000 ft *(9,145 m)*.
Range: 1,153 m *(2,135 km)*.
Role/Weapon systems: Maritime surveillance version ordered in 1997 with first one delivered in mid-1999. Sensors: APS 134 radar; Litton FLIR. Weapons: none.
VERIFIED

F 406 (Australian colours)   *6/1997, Reims Aviation*

PATROL FORCES

83 SEA DOLPHIN/WILDCAT CLASS
(FAST ATTACK CRAFT—PATROL) (PBF/PTF)

PKM 212-375 series

Displacement, tons: 148 full load
Dimensions, feet (metres): 121.4 × 22.6 × 5.6 *(37 × 6.9 × 1.7)*
Main machinery: 2 MTU MD 16V 538 TB90 diesels; 6,000 hp(m) *(4.41 MW)* sustained; 2 shafts
Speed, knots: 37. **Range, n miles:** 600 at 20 kt
Complement: 31 (5 officers)
Guns: 2 Emerson Electric 30 mm (twin) or USN 3 in *(76 mm)*/50 or Bofors 40 mm/60. 2 GE/GD 20 mm Sea Vulcan Gatlings (in most).
2—12.7 mm MGs. Rocket launchers in lieu of after Gatling in some.
Weapons control: Optical director.
Radars: Surface search: Raytheon 1645; I-band.

Comment: Fifty-four Sea Dolphins built by Korea SEC, and 47 Wildcats by Korea Tacoma. First laid down 1978. The class has some gun armament variations and some minor superstructure changes in later ships of the class. These craft form the basis of the coastal patrol effort against incursions by North Korean amphibious units. Five sold to the Philippines in 1995, two transferred to Bangladesh in 2000. Some deleted so far, others are in reserve.
UPDATED

WILDCAT 253   *6/1999* / 0081163

SEA DOLPHIN 279   *6/2004** / 1042332

SEA DOLPHIN 372 (with 40 mm gun)   *10/1998, L Stephenson* / 0052551

0 + 40 PKM-X FAST ATTACK CRAFT (PGGF)

Displacement, tons: 300
Dimensions, feet (metres): 183.8 × ? × ? *(56.0 × ? × ?)*
Main machinery: To be announced
Missiles: SSM: 4 Harpoon (2 twin).
Guns: 1—3 in *(76 mm)*.
Radars: Air/Surface search: Thales MW 08; G-band.
Navigation: I-band.

Comment: A new class of patrol craft, to replace existing vessels, is reported to have started construction in 2003. Design and details are speculative but armament is believed to include Harpoon and a 76 mm gun.
UPDATED

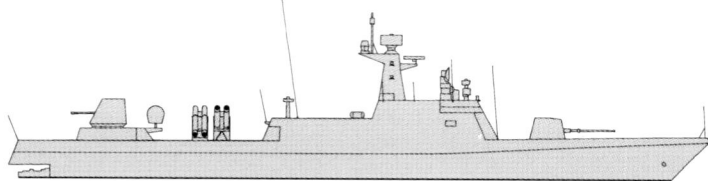

PKM-X (illustrative design)   *(Scale 1 : 600), Ian Sturton* / 0569923

AMPHIBIOUS FORCES

0 + 1 (1) AMPHIBIOUS TRANSPORT DOCK (LPD)

Name	No	Builders	Laid down	Launched	Commissioned
—	—	Hanjin Heavy Industries, Pusan	2003	Mar 2005	June 2007

Displacement, tons: 13,000 standard; 19,000 full load
Dimensions, feet (metres): 656.3 × 105.0 × 21.33 *(200.0 × 32.0 × 6.5)*
Main machinery: 4 SEMT Pielstick 16PC 2.5 STC diesels; 41,615 hp(m) *(30.6 MW)* sustained; 2 shafts
Speed, knots: 22
Complement: 400 ship plus 700
Military lift: 700 troops, 10 tanks and two air-cushion landing craft
Missiles: 1 Raytheon RAM system.
Guns: 2 TNNL Goalkeeper systems.
Combat data systems: To be based on Tacticos.
Radars: Air search: To be announced.
Surface search: Signaal MW 08; G-band.
Navigation: To be announced.
Helicopters: 10.

Programmes: The contract for an amphibious assault ship was placed with Hanjin Heavy Industries on 28 October 2002. An order for a second ship, to begin construction in 2006, is expected.
Structure: The design may include a 'ski-jump' on the flight deck.

UPDATED

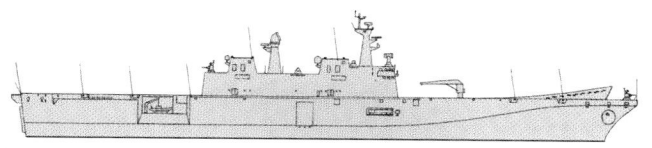

LPX
(Scale 1 : 2,400), Ian Sturton / 0569925

4 ALLIGATOR CLASS (LSTH)

Name	No	Builders	Launched	Commissioned
KOJOON BONG	681	Korea Tacoma, Masan	Sep 1992	June 1993
BIRO BONG	682	Korea Tacoma, Masan	Dec 1996	Nov 1997
HYANGRO BONG	683	Korea Tacoma, Masan	Oct 1998	Aug 1999
SEONGIN BONG	685	Korea Tacoma, Masan	Feb 1999	Nov 1999

Displacement, tons: 4,278 full load
Dimensions, feet (metres): 369.1 × 50.2 × 9.8 *(112.5 × 15.3 × 3)*
Main machinery: 2 SEMT-Pielstick 16 PA6 V 280; 12,800 hp(m) *(9.41 MW)* sustained; 2 shafts; cp props
Speed, knots: 16. **Range, n miles:** 4,500 at 12 kt
Complement: 169
Military lift: 200 troops; 15 MBT; 6—3 ton vehicles; 4 LCVPs.
Guns: 2 Breda 40 mm/70. 2 Vulcan 20 mm Gatlings.
Countermeasures: Decoys: 1 RBOC chaff launcher.
ESM: radar intercept.
Weapons control: Selenia NA 18. Optronic director. Daeyoung WCS-86.
Radars: Surface search: Raytheon SPS 64; E/F-band.
Navigation: Raytheon SPS 64; I-band.
Tacan: SRN 15.
Helicopters: Platform for 1 UH-60A.

Comment: First one ordered in June 1990 from Korea Tacoma, Masan but delayed by financial problems. Korea Tacoma is now Hanjin Heavy Industries. Design improvements include stern ramp for underway launching of LVTs, helicopter deck, and a lengthened bow ramp. Three more may be added in due course.

VERIFIED

KOJOON BONG
11/1997 / 0081164

KOJOON BONG
10/1998, John Mortimer / 0052552

4 LST 512-1152 CLASS (LST)

Name	No	Builders	Commissioned
UN BONG (ex-*LST 1010*)	671	Bethlehem Steel	25 Apr 1944
WEE BONG (ex-*Johnson County* LST 849)	676	American Bridge	16 Jan 1945
SU YONG (ex-*Kane County* LST 853)	677	Chicago Bridge	11 Dec 1945
BUK HAN (ex-*Lynn County* LST 900)	678	Dravo, Pittsburg	28 Dec 1944

Displacement, tons: 1,653 standard; 2,366 beaching; 4,080 full load
Dimensions, feet (metres): 328 × 50 × 14 (screws) *(100 × 15.2 × 4.3)*
Main machinery: 2 GM 12-567A diesels; 1,800 hp *(1.34 MW)*; 2 shafts
Speed, knots: 10
Complement: 80
Military lift: 2,100 tons including 20 tanks and 2 LCVPs
Guns: 8 Bofors 40 mm (2 twin, 4 single). 2 Oerlikon 20 mm.

Comment: Former US Navy tank landing ships. Transferred to South Korea between 1955 and 1959. All purchased 15 November 1974. Planned to be replaced by the Alligator class but four reported as still in service.

UPDATED

BUK HAN
10/1997 / 0081165

0 + 3 TSAPLYA (MURENA E) (PROJECT 12061) CLASS (ACV)

Displacement, tons: 149 full load
Dimensions, feet (metres): 103.7 × 47.6 × 5.2 *(31.6 × 14.5 × 1.6)*
Main machinery: 2 PR-77 gas turbines for lift and propulsion; 8,000 hp *(5.88 MW)*
Speed, knots: 50. **Range, n miles:** 500 at 50 kt
Complement: 11 (3 officers) + 100 troops
Guns: 2—30 mm AK 306M. 2—30 mm grenade launchers. 2—12.7 mm MGs.

Comment: Designed by Almaz, all to be built at Khabarovsk. First laid down on 26 April 2004 for delivery in mid-2005. The second and third are to follow in 2006 and 2007. Capable of carrying one medium tank or 130 troops.

NEW ENTRY

10 LCM 8 CLASS (LCM)

Displacement, tons: 115 full load
Dimensions, feet (metres): 74.5 × 21 × 4.6 *(22.7 × 6.4 × 1.4)*
Main machinery: 4 GM 6-71 diesels; 696 hp *(519 kW)* sustained; 2 shafts
Speed, knots: 11
Complement: 11
Military lift: 55 tons

Comment: Previously US Army craft. Transferred in September 1978.

VERIFIED

LCM 8
5/1995, David Jordan

MISCELLANEOUS LANDING CRAFT

Comment: A considerable number of US LCVP type built of GRP in South Korea. In addition there were plans to build up to 20 small hovercraft for special forces; first two reported building in 1994, and one seen on sea trials in May 1995. Also 56 combat support boats of 8 m were ordered from FBM Marine for assembly by Hanjin Heavy Industries.

VERIFIED

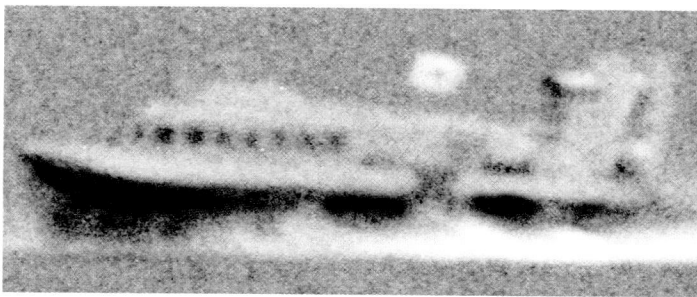

HOVERCRAFT
5/1995, David Jordan / 0081167

For details of the latest updates to *Jane's Fighting Ships* online and to discover the additional information available exclusively to online subscribers please visit
jfs.janes.com

MINE WARFARE FORCES

6 SWALLOW CLASS (MINEHUNTERS) (MHSC)

Name	No	Builders	Commissioned
KANG KYEONG	561	Kangnam Corporation	Dec 1986
KANG JIN	562	Kangnam Corporation	May 1991
KO RYEONG	563	Kangnam Corporation	Nov 1991
KIM PO	565	Kangnam Corporation	Apr 1993
KO CHANG	566	Kangnam Corporation	Oct 1993
KUM WHA	567	Kangnam Corporation	Apr 1994

Displacement, tons: 470 standard; 520 full load
Dimensions, feet (metres): 164 × 27.2 × 8.6 *(50 × 8.3 × 2.6)*
Main machinery: 2 MTU diesels; 2,040 hp(m) *(1.5 MW)* sustained; 2 Voith-Schneider props; bow thruster; 102 hp(m) *(75 kW)*
Speed, knots: 15. **Range, n miles:** 2,000 at 10 kt
Complement: 44 (5 officers) plus 4 divers
Guns: 1 Oerlikon 20 mm. 2—7.62 mm MGs.
Countermeasures: MCM: 2 Gaymarine Pluto remote-control submersibles (possibly to be replaced by Double Eagle).
Combat data systems: Racal MAINS 500.
Radars: Navigation: Raytheon SPS 64; I-band.
Sonars: GEC-Marconi 193M Mod 1 or Mod 3; minehunting; high frequency.

Comment: Built to a design developed independently by Kangnam Corporation but similar to the Italian Lerici class. GRP hull. Single sweep gear deployed at 8 kt. Decca/Racal plotting system. First delivered at the end of 1986 for trials. Two more with some modifications ordered in 1988, three more in 1990. *VERIFIED*

KANG JIN   *1/2000, van Ginderen Collection /* 0104999

1 WON SAN CLASS (MINELAYER) (MLH)

Name	No	Builders	Launched	Commissioned
WON SAN	560	Hyundai, Ulsan	Sep 1996	Sep 1997

Displacement, tons: 3,300 full load
Dimensions, feet (metres): 340.6 × 49.2 × 11.2 *(103.8 × 15 × 3.4)*
Main machinery: CODAD; 4 SEMT-Pielstick 12 PA6 diesels; 17,200 hp(m) *(12.64 MW)*; 2 shafts
Speed, knots: 22. **Range, n miles:** 4,500 at 15 kt
Complement: 160
Guns: 1 OTO Melara 3 in *(76 mm)*/62; 85 rds/min to 16 km *(8.6 n miles)*; weight of shell 6 kg. 2 Breda 40 mm/70.
Torpedoes: 6—324 mm Mk 32 (2 triple) launchers.
Mines: 2 stern launchers. Up to 300.
Countermeasures: Decoys: 2 chaff launchers. ESM/ECM.
Weapons control: Radamec optronic director.
Radars: Air/surface search: E/F-band.
Fire control: Marconi 1802; I/J-band.
Navigation: I-band.
Sonars: Bow-mounted; active search and attack; medium frequency.
Helicopters: Platform only.

Comment: Project design contract ordered October 1991 and completed July 1993 by Hyundai. Order to build given in October 1994. *UPDATED*

WON SAN   *3/2004*, Chris Sattler /* 1042336

2 YANG YANG CLASS (MSC/MHC)

Name	No	Builders	Commissioned
YANG YANG	571	Kangnam Corporation	Dec 1999
ONGJIN	572	Kangnam Corporation	2004

Displacement, tons: 880 full load
Dimensions, feet (metres): 195 × 34.4 × 9.8 *(59.4 × 10.5 × 3.0)*
Main machinery: 2 MTU diesels; 4,000 hp(m) *(2.98 MW)* sustained; 2 Voith-Schneider props; bow thruster; 134 hp(m) *(100 kW)*
Speed, knots: 15. **Range, n miles:** 3,000 at 12 kt
Complement: 56 (7 officers) plus 5 divers
Guns: 1—20 mm Sea Vulcan Gatling. 2—7.62 mm MGs.
Countermeasures: MCM: BAE Systems deep mechanical and combined influence sweep system. 2 Gayrobot Pluto GIGAS ROVs.
Combat data systems: Thomson Marconi TSM 2061 Mk 3.
Radars: Navigation: Raytheon; I-band.
Sonars: Thomson Marconi Type 2093 VDS; minehunting; active multifrequency.

Comment: The first one ordered in late 1995 and a second delivered in 2004. Further orders have not been confirmed. A large version of the Swallow class built to a design developed by Kangnam Corporation. GRP hull. The integrated navigation and dynamic positioning system developed by Kongsberg Simrad. *UPDATED*

ONGJIN   *8/2004*, John Mortimer /* 1042334

AUXILIARIES

Notes: The South Korean Navy also operates nine small harbour tugs (designated YTLs). These include one ex-US Navy craft and five ex-US Army craft. There are also approximately 35 small service craft in addition to the YO-type tankers listed and the harbour tugs. These craft include open lighters, floating cranes, diving tenders, dredgers, ferries, non self-propelled fuel barges, pontoon barges, and sludge removal barges; most are former US Navy craft.

1 CHEONG HAE JIN CLASS (ARS)

Name	No	Builders	Launched	Commissioned
CHEONG HAE JIN	21	Daewoo, Okpo	Oct 1995	30 Nov 1996

Displacement, tons: 4,300 full load
Dimensions, feet (metres): 337.3 × 53.8 × 15.1 *(102.8 × 16.4 × 4.6)*
Main machinery: Diesel-electric; 4 MAN Burmeister & Wain 16V 28/32 diesels; 11,800 hp(m) *(8.67 MW)*; 2 motors; 5,440 hp(m) *(4 MW)*; 2 shafts; cp props; 3 bow and 2 stern thrusters
Speed, knots: 18. **Range, n miles:** 9,500 at 15 kt
Complement: 130
Guns: 1 GE/GD 20 mm Vulcan Gatling (can be fitted). 6—12.7 mm MGs.
Radars: Navigation: I-band.
Sonars: Hull-mounted; active search; high frequency.
Helicopters: Platform for 1 light.

Comment: Ordered in 1992. Laid down December 1994. A multipurpose salvage and rescue ship which carries a 300 m ROV as well as two LCVPs on davits plus a diving bell for nine men and a decompression chamber. Two large hydraulic cranes fore and aft and one towing winch. There are also two salvage ships which belong to the Coast Guard. *VERIFIED*

CHEONG HAE JIN   *9/2003, Hartmut Ehlers /* 0570936

2 EDENTON CLASS (SALVAGE SHIPS) (ATS)

Name	No	Builders	Commissioned
PYONG TAEK (ex-*Beaufort*)	27	Brooke Marine, Lowestoft	22 Jan 1972
KWANG YANG (ex-*Brunswick*)	28	Brooke Marine, Lowestoft	19 Dec 1972

Displacement, tons: 2,929 full load
Dimensions, feet (metres): 282.6 × 50 × 15.1 *(86.1 × 15.2 × 4.6)*
Main machinery: 4 Paxman 12YJCM diesels; 6,000 hp *(4.48 MW)* sustained; 2 shafts; cp props; bow thruster
Speed, knots: 16. **Range, n miles:** 10,000 at 13 kt
Complement: 129 (7 officers)
Guns: 2 Oerlikon 20 mm Mk 68.
Radars: Navigation: Sperry SPS-53; I/J-band.

Comment: Transferred from USA on 29 August 1996. Capable of (1) ocean towing, (2) supporting diver operations to depths of 850 ft, (3) lifting submerged objects weighing as much as 600,000 lb from a depth of 120 ft by static tidal lift or 30,000 lb by dynamic lift, (4) fighting ship fires. Fitted with 10 ton capacity crane forward and 20 ton capacity crane aft. Both recommissioned 28 February 1997. *VERIFIED*

PYONG TAEK (US colours)   *12/1995, Giorgio Arra*

3 CHUN JEE CLASS (LOGISTIC SUPPORT SHIPS) (AORH)

Name	No	Builders	Launched	Commissioned
CHUN JEE	57	Hyundai, Ulsan	May 1990	Dec 1990
DAE CHUNG	58	Hyundai, Ulsan	Jan 1997	Nov 1997
HWA CHUN	59	Hyundai, Ulsan	July 1997	Mar 1998

Displacement, tons: 7,500 full load
Dimensions, feet (metres): 426.5 × 58.4 × 21.3 *(130 × 17.8 × 6.5)*
Main machinery: 2 SEMT-Pielstick 16 PA6 V 280 (AO 57) or 12 PC2.5 diesels; 12,800 hp(m) *(9.4 MW)* sustained; 2 shafts
Speed, knots: 20. **Range, n miles:** 4,500 at 15 kt
Cargo capacity: 4,200 tons liquids; 450 tons solids
Guns: 4 Emerlec 30 mm (2 twin) or 2 Breda 40 mm/70. 2 GE/GD 20 mm Vulcan Gatlings.
Radars: Navigation: 2 Racal Decca; I-band.
Helicopters: Platform for 1 medium.

Comment: *Chun Jee* laid down September 1989. Underway replenishment stations on both sides. Helicopter for Vertrep but no hangar. There are three 6 ton lifts. Possibly based on Italian Stromboli class. Second of class was to have followed on but was eventually ordered together with the third in May 1995, to a slightly different design. More may be built when funds are available.
UPDATED

DAE CHUNG   12/2003*, Bob Fildes / 1042333

CHUN JEE   10/2002, Chris Sattler / 0528920

1 TRIALS SUPPORT SHIP (AGE)

Name	No	Builders	Launched	Commissioned
SUNJIN	AGS 11	Hyundai, Ulsan	Nov 1992	Apr 1993

Displacement, tons: 320 full load
Dimensions, feet (metres): 113.2 × 49.2 × 12.1 *(34.5 × 15 × 3.7)*
Main machinery: 1 MTU 16V 396 TE74L diesel; 2,680 hp(m) *(2 MW)*; 1 shaft; cp prop; 2 bow thrusters
Speed, knots: 21. **Range, n miles:** 600 at 16 kt
Complement: 5 plus 20 scientists
Guns: 1—20 mm Gatling.
Radars: Navigation: I-band.

Comment: Experimental design built by Hyundai. Ordered June 1991, laid down June 1992. Aluminium SWATH hull with dynamic positioning system. Fitted with various trials equipment including an integrated navigation system and torpedo tracking pinger system. VDS and towed arrays. Used by the Defence Development Agency and civilian operated.
VERIFIED

SUNJIN   1993, Hyundai / 0081169

TUGS

Notes: In addition to the Edenton class ATS there are a further 10 harbour tugs and numerous port service auxiliaries.

HARBOUR TUG   10/1998, John Mortimer / 0052559

SURVEY SHIPS

17 SURVEY SHIPS (AGOR)

PUSAN 801	PUSAN 806	201-204
PUSAN 802	PUSAN 810	208-209
PUSAN 803	CH'UNGNAM 821	215-216
PUSAN 805	KANGWON 831	220

Comment: All ships are painted white with a distinctive yellow coloured crest on the funnel. Most were commissioned in the 1980s. The Hydrographic Service is responsible to the Ministry of Transport.
VERIFIED

204   8/1997 / 0012710

202   4/2000, M Declerck / 0105000

PUSAN 801   7/1997, van Ginderen Collection / 0012709

MARITIME POLICE

Notes: (1) The South Korean Maritime Police operates a number of small ships and several hundred craft including tugs and rescue craft, and has taken over the Coast Guard. The overall colour scheme on all units is a medium blue coloured hull with white superstructure and black block style pennant numbers, with the lettering 'POLICE' in both Korean and English on a prominent space on the superstructure. The colour scheme is further distinguished with a black funnel top where appropriate. Immediately below the funnel top is a thin white band separating a thick lightish green band with the Police logo superimposed. This logo is in gold with the blue and red colours of Korea in the centre. Bell 412 helicopters are being acquired.
(2) A 5,000 ton ship, based on *Tae Pung Yang II*, is reported to have been commissioned in 2001.
(3) Six new offshore patrol vessels are to be acquired by 2008. These will include three 3,000 ton helicopter-capable ships and three 1,500 ton ships.

1 DAEWOO TYPE (PSO)

SUMJINKANG PC 1006

Displacement, tons: 1,650 full load
Dimensions, feet (metres): 275.6 × 34.1 × 11.8 *(84 × 10.4 × 3.6)*
Main machinery: 2 Wärtsilä Nohab 16V25 diesels; 10,000 hp(m) *(7.35 MW)* sustained; 2 shafts
Speed, knots: 21. **Range, n miles:** 4,500 at 18 kt
Complement: 57 (7 officers)
Guns: 1—20 mm Sea Vulcan Gatling. 4—12.7 mm MGs.
Radars: Surface search: I-band.

Comment: Ordered in 1997 from Daewoo. Described as a multipurpose patrol ship this is the largest patrol vessel yet built for the Maritime Police. Launched 22 January 1999, and delivered 20 June 1999.
VERIFIED

SUMJINKANG   8/1999, Ships of the World / 0081173

3 MAZINGER CLASS (PSO)

PC 1001-1003

Displacement, tons: 1,200 full load
Dimensions, feet (metres): 264.1 × 32.2 × 11.5 *(80.5 × 9.8 × 3.2)*
Main machinery: 2 SEMT-Pielstick 12 PA6 V 280 diesels; 9,600 hp(m) *(7.08 MW)* sustained; 2 shafts
Speed, knots: 22. **Range, n miles:** 7,000 at 18 kt
Complement: 69 (11 officers)
Guns: 1 Bofors 40 mm/70. 4 Oerlikon 20 mm (2 twin).
Radars: Surface search: Raytheon; I-band.

Comment: Ordered 7 November 1980 from Korea Tacoma and Hyundai. *PC 1001* delivered 29 November 1981. *PC 1002* 31 August 1982 and *PC 1003* on 31 August 1983. All-welded mild steel construction. Used for offshore surveillance and general coast guard duties. *PC 1001* is the Coast Guard Command ship. Only three of this class were completed.
VERIFIED

MAZINGER (old colours)   1987, Korea Tacoma

1 HAN KANG CLASS (PG)

HAN KANG PC 1005

Displacement, tons: 1,180 full load
Dimensions, feet (metres): 289.7 × 32.8 × 9.5 *(88.3 × 10 × 2.9)*
Main machinery: CODOG; 1 GE LM 2500 gas turbine; 26,820 hp *(20 MW)* sustained; 2 MTU 12V 956 TB82 diesels; 6,260 hp(m) *(4.6 MW)* sustained; 3 shafts
Speed, knots: 32. **Range, n miles:** 4,000 at 15 kt
Complement: 72 (11 officers)
Guns: 1 OTO Melara 76/62 compact. 1 Bofors 40 mm/70. 2 GE/GD 20 mm Vulcan Gatlings.
Weapons control: Signaal LIOD optronic director.
Radars: Surface search: Raytheon SPS-64(V); I-band.
Fire control: Signaal WM28; I/J-band.

Comment: Built between May 1984 and December 1985 by Daewoo. Same hull as Po Hang class but much more lightly armed. Only one of the class was completed.
VERIFIED

HAN KANG   9/2000 / 0097740

6 SEA DRAGON/WHALE CLASS (PBO)

PC 501, 502, 503, 505, 506, 507

Displacement, tons: 640 full load
Dimensions, feet (metres): 199.5 × 26.2 × 8.9 *(60.8 × 8 × 2.7)*
Main machinery: 2 SEMT-Pielstick 12 PA6 V 280 diesels; 9,600 hp(m) *(7.08 MW)* sustained; 2 shafts
Speed, knots: 24. **Range, n miles:** 6,000 at 15 kt
Complement: 40 (7 officers)
Guns: 1 Bofors 40 mm/60. 2 Oerlikon 20 mm. 2 Browning 12.7 mm MGs.
Radars: Navigation: Two sets.

Comment: Delivered 1978-1982 by Hyundai, Korea and Korea Tacoma. Fitted with SATNAV. Welded steel hull. Armament varies between ships, one 76 mm gun can be mounted on the forecastle. Variant of this class built for Bangladesh and delivered in October 1997. *VERIFIED*

SEA DRAGON 507 (old colours)   1987, Korea Tacoma

6 430 TON CLASS (PBO)

300-303   402-403

Displacement, tons: 430 full load
Dimensions, feet (metres): 176.2 × 24.3 × 7.9 *(53.7 × 7.4 × 2.4)*
Main machinery: 2 MTU 16V 396 TB83 diesels; 1,990 hp(m) *(1.49 MW)*; 2 shafts; cp props
Speed, knots: 19. **Range, n miles:** 2,100 at 17 kt
Complement: 14
Guns: 2 GD/GE 20 mm Vulcan Gatlings. 4—12.7 mm MGs.
Radars: Surface search: Raytheon; I-band.

Comment: All built between 1990 and 1995 by Hyundai except 301 which was built by Daewoo. Multipurpose patrol ships.
VERIFIED

301   8/2000, van Ginderen Collection / 0097741

300   3/1996, D Swetnam / 0081172

23 SEA WOLF/SHARK CLASS (PBO)

| 207 | 255-259 | 265-269 | 275-277 |
| 251-253 | 261-263 | 271-273 | |

Displacement, tons: 310 full load
Dimensions, feet (metres): 158.1 × 23.3 × 8.2 *(48.2 × 7.1 × 2.5)*
Main machinery: 2 diesels; 7,320 hp(m) *(5.38 MW)*; 2 shafts
Speed, knots: 25. **Range, n miles:** 2,400 at 15 kt
Complement: 35 (3 officers)
Guns: 4 Oerlikon 20 mm (2 twin or 1 twin, 2 single). Some have a twin Bofors 40 mm/70 vice the twin Oerlikon. 2 Browning 12.7 mm MGs.
Radars: Surface search: I-band.

Comment: First four ordered in 1979-80 from Korea SEC (Sea Shark), Hyundai and Korea Tacoma (Sea Wolf). Programme terminated in 1988. Pennant numbers in 200 series up to 277.
VERIFIED

SEA WOLF 207   5/1997, van Ginderen Collection / 0012711

4 BUKHANSAN CLASS (PBO)

BUKHANSAN 278   CHULMASAN 279   P 281   P 282

Displacement, tons: 380 full load
Dimensions, feet (metres): 174.2 × 24 × 7.2 *(53.1 × 7.3 × 2.2)*
Main machinery: 2 MTU diesels; 8,300 hp(m) *(6.1 MW)* sustained; 2 shafts
Speed, knots: 28. **Range, n miles:** 2,500 at 15 kt
Complement: 35 (3 officers)
Guns: 1 Breda 40 mm/70. 1 GE/GD 20 mm Vulcan Gatling. 2—12.7 mm MGs.
Weapons control: Radamec optronic director.
Radars: Surface search: I-band.

Comment: Follow on to Sea Wolf class developed by Hyundai in 1987. Ordered in 1988 from Hyundai and Daewoo respectively. First pair in service in 1989, and second pair in 1990.
VERIFIED

CHULMASAN (old colours)   1989, Daewoo

5 HYUNDAI TYPE (PB)

| 105 | 113 | 118 | 121 | 125 |

Displacement, tons: 110 full load
Dimensions, feet (metres): 105.6 × 19.7 × 4.6 *(32.2 × 6 × 1.4)*
Main machinery: 2 diesels; 2 shafts
Speed, knots: 25
Complement: 19
Guns: 1 Rheinmetall 20 mm. 2—12.7 mm MGs.
Radars: Surface search: Furuno; I-band.

Comment: Ordered in 1996 and delivered from June 1997.

VERIFIED

HYUNDAI 113   *4/2000, M Declerck /* 0097742

INSHORE PATROL CRAFT (PBR)

Displacement, tons: 47 full load
Dimensions, feet (metres): 69.9 × 17.7 × 4.6 *(21.3 × 5.4 × 1.4)*
Main machinery: 2 diesels; 1,800 hp(m) *(1.32 MW)*; 2 shafts
Speed, knots: 22. **Range, n miles:** 400 at 12 kt
Complement: 11
Guns: 1 Rheinmetall 20 mm. 3—12.7 mm MGs.
Radars: Surface search: Furuno; I-band.

Comment: Details are for the largest design of patrol craft. There are numbers of this type of vessel used for inshore patrol work. All Police craft have P pennant numbers. Armaments vary. Customs craft have double numbers.

VERIFIED

P 71   *9/2000 /* 0097743

CUSTOM 71-810   *6/1996, D Swetnam*

1 SALVAGE SHIP (ARSH)

Name	No	Builders	Launched	Commissioned
TAE PUNG YANG I	3001	Hyundai, Ulsan	Oct 1991	18 Feb 1993

Displacement, tons: 3,200 standard; 4,300 full load
Dimensions, feet (metres): 343.5 × 49.2 × 17 *(104.7 × 15 × 5.2)*
Main machinery: 4 Ssangyoung MAN Burmeister & Wain 16V 28/32 diesels; 4,800 hp(m) *(3.53 MW)*; 2 shafts; cp props; bow and stern thrusters
Speed, knots: 21. **Range, n miles:** 8,500 at 15 kt
Complement: 121
Guns: 1 GD/GE 20 mm Vulcan Gatling. 6—12.7 mm MGs.
Radars: Navigation: I-band.
Helicopters: 1 light.

Comment: Laid down February 1991. Has a helicopter deck and hangar, an ROV capable of diving to 300 m and a firefighting capability. Dynamic positioning system. Can be used for cable laying. Operates for the Marine Police.

VERIFIED

TAE PUNG YANG 1   *1/2000, van Ginderen Collection /* 0097744

1 SALVAGE SHIP (ARSH)

Name	No	Builders	Commissioned
TAE PUNG YANG II	3002	Hyundai, Ulsan	Nov 1988

Displacement, tons: 3,900 standard
Dimensions, feet (metres): 362.5 × 50.5 × 16.1 *(110.5 × 15.4 × 4.9)*
Main machinery: 2 diesels; 2 shafts
Speed, knots: 18
Complement: 120
Guns: 2—20 mm Vulcan Gatlings. 6—12.7 mm MGs.
Radars: Surface search: I-band.
Helicopters: Platform for 1 large.

Comment: Ordered from Hyundai in mid-1996. Also used for SAR operations.

VERIFIED

TAE PUNG YANG II   *8/2000, Ships of the World /* 0097746

1 SALVAGE SHIP (ARSH)

Name	No	Builders	Commissioned
JAEMIN I	1501	Daewoo, Okpo	28 Dec 1992

Displacement, tons: 2,072 full load
Dimensions, feet (metres): 254.6 × 44.3 × 13.8 *(77.6 × 13.5 × 4.2)*
Main machinery: 2 MTU diesels; 8,000 hp(m) *(5.88 MW)*; 2 shafts; cp props
Speed, knots: 18. **Range, n miles:** 4,500 at 12 kt
Complement: 92
Guns: 1 GD/GE 20 mm Vulcan Gatling.
Radars: Navigation: I-band.

Comment: Ordered in 1990. Fitted with diving equipment and has a four point mooring system. Carries two LCVPs.

VERIFIED

JAEMIN I   *8/2000 /* 0097745

1 SALVAGE SHIP (ARS)

Name	No	Builders	Launched	Commissioned
JAEMIN II	1502	Hyundai, Ulsan	15 July 1995	Apr 1996

Displacement, tons: 2,500 full load
Dimensions, feet (metres): 288.7 × 47.6 × 15.1 *(88 × 14.5 × 4.6)*
Main machinery: 2 MTU diesels; 12,662 hp(m) *(9.31 MW)*; 2 shafts; Kamewa cp props; bow and stern thrusters
Speed, knots: 20. **Range, n miles:** 4,500 at 15 kt
Complement: 81
Guns: 1 GE/GD 20 mm Vulcan Gatling.
Radars: Navigation: I-band.

Comment: Ordered in December 1993 for Maritime Police. A general purpose salvage ship capable of towing, firefighting, supply or patrol duties.

VERIFIED

JAEMIN II   *8/1999, Ships of the World /* 0081176

1 SALVAGE SHIP (ARSH)

Name	No	Builders	Commissioned
JAEMIN III	1503	Hyundai, Ulsan	Nov 1998

Displacement, tons: 4,200 full load
Dimensions, feet (metres): 362.6 × 50.5 × 16 *(110.5 × 15.4 × 4.9)*
Main machinery: 2 diesels; 2 shafts
Speed, knots: 18
Complement: 120
Guns: 2 GE 20 mm Vulcan Gatlings. 6—12.7 mm MGs.
Radars: Navigation: I-band.

Comment: Ordered in 1996, from Hyundai, Ulsan. Large helicopter deck but no hangar.
VERIFIED

1503
8/2000, van Ginderen Collection / 0097747

Kuwait

Country Overview

Formerly a British protectorate, the Kingdom of Kuwait gained independence in 1961. Situated on the northwestern coast of the Gulf, it is bordered to the north by Iraq and to the south by Saudi Arabia. The country's total area, including the islands of Bubiyan, Warbah, and Faylakah, is 6,880 square miles. It has a 269 n mile coastline with the Gulf. The capital, largest city and principal port is Kuwait City. The country was annexed by Iraq from August 1990 to February 1991 when the country was liberated. Territorial seas (12 n miles) are claimed. An EEZ has not been claimed.

Headquarters Appointments

Commander of the Navy:
Brigadier Ahmed Yousuf Al Mulla

Personnel

2005: 2,700 (including 500 Coast Guard)

Aviation

The Air Force operates three Eurocopter AS 532C Cougar helicopters armed with Exocet AM 39 ASMs and 40 F/A-18 Hornets.

Bases

Navy: Ras Al Qalayah
Coast Guard: Shuwaikh, Umm Al-Hainan, Al-Bida

PATROL FORCES

Notes: There is a requirement for up to four 88 m OPVs armed with SSMs.

8 UM ALMARADIM (COMBATTANTE I) CLASS (PBM)

Name	No	Builders	Launched	Commissioned
UM ALMARADIM	P 3711	CMN, Cherbourg	27 Feb 1997	31 July 1998
OUHA	P 3713	CMN, Cherbourg	29 May 1997	31 July 1998
FAILAKA	P 3715	CMN, Cherbourg	29 Aug 1997	19 Dec 1998
MASKAN	P 3717	CMN, Cherbourg	6 Jan 1998	19 Dec 1998
AL-AHMADI	P 3719	CMN, Cherbourg	2 Apr 1998	1 July 1999
ALFAHAHEEL	P 3721	CMN, Cherbourg	16 June 1998	1 July 1999
AL-YARMOUK	P 3723	CMN, Cherbourg	3 Mar 1999	7 June 2000
GAROH	P 3725	CMN, Cherbourg	June 1999	7 June 2000

Displacement, tons: 245 full load
Dimensions, feet (metres): 137.8 oa; 121.4 wl × 26.9 × 6.2 *(42; 37 × 8.2 × 1.9)*
Main machinery: 2 MTU 16V 538 TB93 diesels; 4,000 hp(m) *(2.94 MW)*; 2 Kamewa waterjets
Speed, knots: 30. **Range, n miles:** 1,350 at 14 kt
Complement: 29 (5 officers)

Missiles: SSM: 4 BAe Sea Skua (2 twin). Semi-active radar homing to 15 km *(8.1 n miles)* at 0.9 Mach.
SAM: Sadral sextuple launcher fitted for only.
Guns: 1 Otobreda 40 mm/70; 120 rds/min to 12.5 km *(6.8 n miles)*; weight of shell 0.96 kg.
1 Giat 20 mm M 621. 2—12.7 mm MGs.
Countermeasures: Decoys: 2 Dagaie Mk 2 chaff launchers fitted for only.
ESM: Thomson-CSF DR 3000 S1; intercept.
Combat data systems: Thomson-CSF TAVITAC NT; Link Y.
Weapons control: CS Defence Najir Mk 2 optronic director.
Radars: Air/surface search: Thomson-CSF MRR; 3D; G-band.
Fire control: BAe Seaspray Mk 3; I/J-band (for SSM).
Navigation: Litton Marine 20V90; I-band.

Programmes: Contract signed with CMN Cherbourg on 27 March 1995. First steel cut 9 June 1995. Names are taken from former Kuwaiti patrol craft.
Structure: Late decisions were made on the missile system which has been fitted in the last pair on build and to the remainder from 2000. Provision is also made for Simbad SAM and Dagaie decoy launchers, which may be fitted later. Positions of smaller guns are uncertain.
Operational: Training done in France. The aim is to have 10 crews capable of manning the eight ships. First four arrived in the Gulf in mid-August 1999, second four arrived in mid-2000.
UPDATED

MASKAN
5/2003, A Sharma / 0559834

1 TNC 45 TYPE (FAST ATTACK CRAFT—MISSILE) (PGGF)

Name	No	Builders	Commissioned
AL SANBOUK	P 4505	Lürssen, Vegesack	26 Apr 1984

Displacement, tons: 255 full load
Dimensions, feet (metres): 147.3 × 23 × 7.5 *(44.9 × 7 × 2.3)*
Main machinery: 4 MTU 16V 538 TB92 diesels; 13,640 hp(m) *(10 MW)* sustained; 4 shafts
Speed, knots: 41. **Range, n miles:** 1,800 at 16 kt
Complement: 35 (5 officers)

Missiles: SSM: 4 Aerospatiale MM 40 Exocet; inertial cruise; active radar homing to 70 km *(40 n miles)* at 0.9 Mach; warhead 165 kg; sea-skimmer.
Guns: 1 OTO Melara 3 in *(76 mm)*/62 compact; 85 rds/min to 16 km *(8.6 n miles)* anti-surface; 12 km *(6.5 n miles)* anti-aircraft; weight of shell 6 kg.
2 Breda 40 mm/70 (twin); 300 rds/min to 12.5 km *(6.6 n miles)*; weight of shell 0.96 kg.
Countermeasures: Decoys: CSEE Dagaie; IR flares and chaff; H/J-band.
ESM: Racal Cutlass; intercept.
Weapons control: PEAB 9LV 228 system; Link Y; CSEE Lynx optical sight.
Radars: Air/surface search: Ericsson Sea Giraffe 50HC; G/H-band.
Fire control: Philips 9LV 200; J-band.
Navigation: Decca TM 1226C; I-band.

Programmes: Six ordered from Lürssen in 1980 and delivered in 1983-84.
Operational: *Al Sanbouk* escaped to Bahrain when the Iraqis invaded in August 1990, but the rest of this class was taken over by the Iraqi Navy, and either sunk or severely damaged by Allied forces in February 1991. The ship was refitted by Lürssen in 1995.
UPDATED

OUHA
3/2003, A Sharma / 0568871

AL SANBOUK
3/2003, A Sharma / 0568872

1 FPB 57 TYPE (FAST ATTACK CRAFT—MISSILE) (PGGF)

Name	No	Builders	Commissioned
ISTIQLAL	P 5702	Lürssen, Vegesack	9 Aug 1983

Displacement, tons: 410 full load
Dimensions, feet (metres): 190.6 × 24.9 × 8.9 *(58.1 × 7.6 × 2.7)*
Main machinery: 4 MTU 16V 956 TB91 diesels; 15,000 hp(m) *(11 MW)* sustained; 4 shafts
Speed, knots: 36. **Range, n miles:** 1,300 at 30 kt
Complement: 40 (5 officers)

Missiles: SSM: 4 Aerospatiale MM 40 Exocet; inertial cruise; active radar homing to 70 km *(40 n miles)* at 0.9 Mach; warhead 165 kg; sea-skimmer.
Guns: 1 OTO Melara 3 in *(76 mm)*/62 compact; 85 rds/min to 16 km *(8.6 n miles)* anti-surface; 12 km *(6.5 n miles)* anti-aircraft; weight of shell 6 kg.
2 Breda 40 mm/70 (twin); 300 rds/min to 12.5 km *(6.6 n miles)*; weight of shell 0.96 kg.
Mines: Fitted for minelaying.
Countermeasures: Decoys: CSEE Dagaie trainable mounting; automatic dispenser; IR flares and chaff; H/J-band.
ESM: Racal Cutlass; radar intercept.
ECM: Racal Cygnus; jammer.
Weapons control: PEAB 9LV 228 system; Link Y; CSEE Lynx optical sight.
Radars: Surface search: Marconi S 810 (after radome); I-band; range 43 km *(25 n miles)*.
Navigation: Decca TM 1226C; I-band.
Fire control: Philips 9LV 200; J-band.

Programmes: Two ordered from Lürssen in 1980.
Operational: *Istiqlal* escaped to Bahrain when the Iraqis invaded in August 1990. The second of this class was captured and sunk in February 1991. Having been laid up since 1997 *Istiqlal* was refitted at Lürssen 2003-2005.

UPDATED

ISTIQLAL   4/2005*, Michael Nitz / 1121416

COAST GUARD

PATROL FORCES

Headquarters Appointments

Director of Coast Guard:
   Lieutenant Colonel Sayid Al-Buaijan

4 INTTISAR (OPV 310) CLASS (PB)

Name	No	Builders	Commissioned
INTTISAR	P 301	Australian Shipbuilding Industries	20 Jan 1993
AMAN	P 302	Australian Shipbuilding Industries	20 Jan 1993
MAIMON	P 303	Australian Shipbuilding Industries	7 Aug 1993
MOBARK	P 304	Australian Shipbuilding Industries	7 Aug 1993

Displacement, tons: 150 full load
Dimensions, feet (metres): 103.3 oa; 88.9 wl × 21.3 × 6.6 *(31.5; 27.1 × 6.5 × 2)*
Main machinery: 2 MTU 16V 396 TB94 diesels; 5,800 hp(m) *(4.26 MW)* sustained; 2 shafts; 1 MTU 8V 183 TE62 diesel; 750 hp(m) *(550 kW)* maximum; 1 Hamilton 422 water-jet
Speed, knots: 28. **Range, n miles:** 300 at 28 kt
Complement: 11 (3 officers)
Guns: 1 Oerlikon 20 mm. 1—12.7 mm MG.
Radars: Surface search: 2 Racal Decca; I-band.

Comment: First two ordered from Australian Shipbuilding Industries in 1991. Second pair ordered in July 1992. Steel hulls, aluminium superstructure. The third engine drives a small waterjet to provide a loiter capability. Carries an RIB. Used by the Coast Guard.

VERIFIED

AMAN   1992, Australian Shipbuilding Industries / 0081178

---

9 + 1 SUBAHI CLASS (PB)

Name	No	Builders	Commissioned
SUBAHI	P 308	OCEA, St Nazaire	6 Aug 2003
JABERI	P 309	OCEA, St Nazaire	Dec 2003
SAAD	P 310	OCEA, St Nazaire	Feb 2004
AHMADI	P 311	OCEA, St Nazaire	Mar 2004
NAIF	P 312	OCEA, St Nazaire	May 2004
THAFIR	P 313	OCEA, St Nazaire	July 2004
MARZOUG	P 314	OCEA, St Nazaire	Sep 2004
MASH'NOOR	P 315	OCEA, St Nazaire	Jan 2005
WADAH	P 316	OCEA, St Nazaire	May 2005
TAROUB	P 317	OCEA, St Nazaire	Sep 2006

Displacement, tons: 116 full load
Dimensions, feet (metres): 115.5 × 22.3 × 4.0 *(35.2 × 6.8 × 1.2)*
Main machinery: 2 MTU 12V 4000 M70 diesels; 4,600 hp *(3.43 MW)*; 2 Kamewa waterjets
Speed, knots: 32. **Range, n miles:** 300 at 28 kt
Complement: 11 (3 officers)
Guns: 1 Oerlikon 20 mm. 2-12.7 mm MGs.
Radars: Sperry Bridgemaster E; I-band.

Comment: Built by OCEA, France based on Al Shaheed class design. Aluminium construction. Operated by the Coast Guard. A class of 10 is planned for delivery by 2005.

UPDATED

NAIF   4/2004*, B Prézelin / 0587771

3 INSHORE PATROL CRAFT (PBR)

KASSIR T 205	DASTOOR T 210	MAHROOS T 215

Displacement, tons: to be announced
Dimensions, feet (metres): 70.9 × 19.5 × 4.9 *(21.6 × 5.96 × 1.5)*
Main machinery: 2 MTU 12V 183 TE92 diesels; 1,800 hp *(1.45 MW)*; 2 shafts
Speed, knots: 25. **Range, n miles:** 325 at 25 kt
Complement: 3 + 41 passengers
Radars: Navigation: to be announced.

Comment: Order for three craft for the Coast Guard announced on 7 January 2003. Based on the 22 m craft in service with the New South Wales Police, the vessels were constructed by Austal Ships subsidiary, Image Marine and delivered in June 2004. Aluminium hull.

UPDATED

DASTOOR   6/2004*, Austal Ships / 0587772

3 AL SHAHEED CLASS (PB)

Name	No	Builders	Commissioned
AL SHAHEED	P 305	OCEA, Les Sables d'Olonne	July 1997
BAYAN	P 306	OCEA, Les Sables d'Olonne	Apr 1999
DASMAN	P 307	OCEA, Les Sables d'Olonne	2001

Displacement, tons: 104 full load
Dimensions, feet (metres): 109.3 × 23 × 4 *(33.3 × 7 × 1.2)*
Main machinery: 2 MTU 12V 396 TE94; 4,352 hp(m) *(3.2 MW)* sustained; 2 shafts
Speed, knots: 30. **Range, n miles:** 360 at 25 kt
Complement: 11 (3 officers)
Guns: 1 Oerlikon 20 mm. 2—12.7 mm MGs.
Radars: Surface search: Racal Decca 20V 90 TA; E/F-band.
Navigation: Racal Decca Bridgemaster ARPA; I-band.

Comment: Built by OCEA, France to FPB 100K design. Operated by the Coast Guard.

VERIFIED

AL SHAHEED   10/1997, Ships of the World / 0012718

33 AL-SHAALI TYPE (INSHORE PATROL CRAFT) (PBF)

Comment: Ten 10 m and 23 8.5 m patrol craft built by Al-Shaali Marine, Dubai, and delivered in June 1992. Also used by UAE Coast Guard. More Rapid Intervention patrol craft are to be acquired in due course.

UPDATED

12 MANTA CLASS (INSHORE PATROL CRAFT) (PBF)

1B 1501-1523 series

Displacement, tons: 10 full load
Dimensions, feet (metres): 45.9 × 12.5 × 2.3 *(14 × 3.8 × 0.7)*
Main machinery: 2 Caterpillar 3208 diesels; 810 hp(m) *(595 kW)* sustained; 2 shafts
Speed, knots: 40. **Range, n miles:** 180 at 35 kt
Complement: 4
Guns: 3 Herstal M2HB 12.7 mm MGs.
Radars: Surface search: Furuno; I-band.

Comment: Original craft ordered in September 1992 from Simonneau Marine and delivered in 1993. Aluminium construction. This version has two inboard engines. Pennant numbers are in odd number sequence. All the class reported to be inoperable due to technical problems. An underlying cause may be that the boats were fitted with inboard engines although designed for outboards.

VERIFIED

MANTA 1501   *11/1996* / 0012719

6 COUGAR ENFORCER 40 CLASS
(INSHORE PATROL CRAFT) (PBF)

Displacement, tons: 5.7 full load
Dimensions, feet (metres): 40 × 9 × 2.1 *(12.2 × 2.8 × 0.8)*
Main machinery: 2 Sabre 380 S diesels; 760 hp(m) *(559 kW)*; 2 Arneson ASD 8 surface drives; 2 shafts
Speed, knots: 45. **Range, n miles:** 250 at 35 kt
Complement: 4
Guns: 1—12.7 mm MG.
Radars: Surface search: Koden; I-band.

Comment: First one completed in July 1996 for the Coastguard by Cougar Marine, Warsash. The craft has a V monohull design.

VERIFIED

ENFORCER 40   *7/1996, Cougar Marine* / 0081179

17 COUGAR TYPE (INSHORE PATROL CRAFT) (PBF)

Comment: Three Cat 900 (32 ft) and six Predator 1100 (35 ft) all powered by two Yamaha outboards (400 hp(m) *(294 kW)*). Four Type 1200 (38 ft) and four Type 1300 (41 ft) all powered by two Sabre diesels (760 hp(m) *(559 kW)*). All based on the high-performance planing hull developed for racing, and acquired in 1991-92. Most have a 7.62 mm MG and a Kroden I-band radar. Used by the Coast Guard. Most have K numbers on the side.

UPDATED

COUGAR 1200   *1991, Cougar Marine* / 0081180

AUXILIARIES

Notes: (1) There are some unarmed craft with Sawahil numbers which are not naval.
(2) A 95 m ship of about 2,000 tons is required to act as a support ship for patrol vessels. It would also be equipped to undertake a training role. Revised bids were submitted in June 2001 but there have been no further developments.
(3) There is a logistic craft P 140.
(4) A landing craft has been ordered from Singapore Technologies Marine for delivery in November 2005.

P 140   *10/2002** / 0587770

1 LOADMASTER Mk 2 (LOGISTIC SUPPORT CRAFT) (LCU)

JALBOUT L 403

Displacement, tons: 420 full load
Dimensions, feet (metres): 108.3 × 33.5 × 5.7 *(33.0 × 10.2 ×1.75)*
Main machinery: 2 Caterpillar V12 diesels; 1,000 hp *(745 kW)*; 2 props
Speed, knots: 10
Complement: 7 (1 officer)
Radars: Navigation: I-band.

Comment: Built by Fairey Marine Cowes, UK and entered service in 1985. Captured by Iraqi forces in 1990 and subsequently recovered and reactivated in 1992.

VERIFIED

L 403   *5/2001* / 0525907

2 AL TAHADDY CLASS (LCU)

Name	No	Builders	Commissioned
AL SOUMOOD	L 401	Singapore SBEC	July 1994
AL TAHADDY	L 402	Singapore SBEC	July 1994

Displacement, tons: 215 full load
Dimensions, feet (metres): 141.1 × 32.8 × 6.2 *(43 × 10 × 1.9)*
Main machinery: 2 MTU diesels; 2 shafts
Speed, knots: 13
Complement: 12
Military lift: 80 tons
Radars: Navigation: Racal Decca; I-band.

Comment: Ordered in 1993 and launched on 15 April 1994. Multipurpose supply ships with cargo tanks for fuel, fresh water, refrigerated stores and containers on the main deck. Has 3 ton crane. Capable of beaching. Used by the Coast Guard.

VERIFIED

AL TAHADDY   *1/1999, Maritime Photographic* / 0053294

1 SUPPORT SHIP (AGH)

QARUH S 5509

Measurement, tons: 545 dwt
Dimensions, feet (metres): 181.8 × 31.5 × 6.6 *(55.4 × 9.6 × 2)*
Main machinery: 2 diesels; 2,400 hp(m) *(1.76 MW)*; 2 shafts
Speed, knots: 9
Complement: 40
Guns: 2—12.7 mm MGs.
Radars: Navigation: Racal Decca; I-band.

Comment: This is a Sawahil class oil rig replenishment and accommodation ship which was built in South Korea in 1986 and taken on by the Coast Guard in 1990. She escaped to Bahrain during the Iraqi invasion, and is back in service. High-level helicopter platform aft. Used as a utility transport. Refitted in 1996/97.

VERIFIED

QARUH
11/1997, Kuwait Navy / 0012721

Latvia
LATVIJAS JURAS SPEKI

Country Overview

The Republic of Latvia regained independence in 1991 after 51 years as a Soviet republic. Situated in northeastern Europe, the country has an area of 24,938 square miles and borders to the north with Estonia, east with Russia and to the south with Belarus and Lithuania. It has a 286 n mile coastline with the Baltic Sea. Riga is the capital, largest city and principal port. Territorial seas (12 n miles) are claimed but while it has claimed a 200 n mile Exclusive Economic Zone (EEZ), its limits have not been fully defined by boundary agreements.

Headquarters Appointments

Commander of the Navy:
  Captain Ilmars Lešinskis

Bases

Liepaja, Ventspils, Riga

Personnel

2005: 800 Navy (including Coast Guard)

Coastal Surveillance

Work began in 2002 on a maritime sea surveillance system which includes Swedish PS2-39 radars at Jurmalciens, Ventspils, Ovici and Kolka. These are to be linked data-link by late 2005 and will be part of a wider network including other Baltic and Scandinavian navies.

Coast Guard

These ships have a diagonal thick white and thin white line on the hull, and have KA numbers. They operate as part of the Navy.

PATROL FORCES

Notes: A programme for replacement of the Storm class patrol craft with new 50 m craft is under consideration.

4 STORM CLASS (PB)

Name	No	Builders	Commissioned
ZIBENS (ex-*Djerv*)	P 01 (ex-P 966)	Westermoen, Mandal	1966
LODE (ex-*Hvass*)	P 02 (ex-P 972)	Westermoen, Mandal	1966
LINGA (ex-*Gnist*)	P 03 (ex-P 979)	Bergens Mek Verksteder	1967
BULTA (ex-*Traust*)	P 04 (ex-P 973)	Bergens Mek Verksteder	1967

Displacement, tons: 135 full load
Dimensions, feet (metres): 120 × 20 × 5 *(36.5 × 6.1 × 1.5)*
Main machinery: 2 MTU MB 872A diesels; 7,200 hp(m) *(5.3 MW)* sustained; 2 shafts
Speed, knots: 32
Complement: 20 (4 officers)
Guns: 1 Bofors 40 mm/60 (P 04). 1 TAK Bofors 76 mm; 1 Bofors 40 mm/70 (P 02 and P 03).
Radars: Surface search: Racal Decca TM 1226; I-band.

Comment: P 04 disarmed and acquired from Norway on 13 December 1994 as a gun patrol craft. Recommissioned 1 February 1995 at Liepaja. 40 mm gun fitted aft in 1998. P 01, P 02 and P 03 transferred from Norway and recommissioned 11 June 2001. Service lives extended to 2010. Other craft given to Lithuania and Estonia.

UPDATED

LINGA
8/2004*, Guy Toremans / 0587775

ZIBENS
6/2004*, Harald Carstens / 0587776

MINE WARFARE FORCES

Notes: Replacement of the MCMV force is under consideration. This may include second-hand craft in the short-term and/or new ships by about 2012.

2 KONDOR II (TYPE 89.2) CLASS (MINESWEEPERS) (MSC)

Name	No	Builders	Commissioned
VIESTURS (ex-*Kamenz*)	M 01 (ex-351)	Peenewerft, Wolgast	24 July 1971
IMANTA (ex-*Röbel*)	M 02 (ex-324)	Peenewerft, Wolgast	1 Dec 1971

Displacement, tons: 410 full load
Dimensions, feet (metres): 186 × 24.6 × 7.9 *(56.7 × 7.5 × 2.4)*
Main machinery: 2 Type 40D diesels; 4,408 hp(m) *(3.24 MW)*; 2 shafts; cp props
Speed, knots: 17
Complement: 31 (6 officers)
Guns: 2 Wrobel ZU 23-2MR 23 mm (twin). 2 FK 20 20 mm.
Radars: Surface search: Racal Decca; I-band.
Sonars: Klein 2000 sidescan; 100/500 kHz.

Comment: Former GDR vessels transferred from Germany on 30 August 1993. Weapons and minesweeping equipment were removed on transfer. New guns have been fitted and ex-German Shultz class minesweeping gear was installed in 1997. Both recommissioned in April 1994. Based at Liepaja.

UPDATED

VIESTURS
6/2004*, Harald Carstens / 0587777

IMANTA
8/2004*, Guy Toremans / 0587774

1 LINDAU (TYPE 331) CLASS (MINEHUNTER) (MHC)

Name	No	Builders	Commissioned
NEMEJS (ex-*Völklingen*)	M 03 (ex-M 1087)	Burmester, Bremen	21 May 1960

Displacement, tons: 463 full load
Dimensions, feet (metres): 154.5 × 27.2 × 9.8 (9.2 Troika) *(47.1 × 8.3 × 3) (2.8)*
Main machinery: 2 MTU MD 871 UM/1D diesels; 4,000 hp(m) *(2.94 MW)*; 2 shafts
Speed, knots: 16.5. **Range, n miles:** 850 at 16.5 kt
Complement: 45 (9 officers)
Guns: 1 Bofors 40 mm/70.
Radars: Navigation: Raytheon, SPS 64; I-band.
Sonars: Plessey 193 m; minehunting; high frequency (100/300 kHz).

Comment: Acquired in June 1999 from Germany. Recommissioned 1 October 1999. Hull is of wooden construction. Converted to minehunter 1979. PAP 105 ROV fitted. Based at Lepaja. *Göttingen* transferred 24 January 2001 for spares.

UPDATED

NEMEJS   6/2004 *, W Sartori / 0587778

AUXILIARIES

1 VIDAR CLASS (MCCS/AG)

Name	No	Builders	Launched	Commissioned
VIRSAITIS (ex-*Vale*)	A 53 (ex-N 53)	Mjellem and Karlsen, Bergen	5 Aug 1977	10 Feb 1978

Displacement, tons: 1,500 standard; 1,673 full load
Dimensions, feet (metres): 212.6 × 39.4 × 13.1 *(64.8 × 12 × 4)*
Main machinery: 2 Wichmann 7AX diesels; 4,200 hp(m) *(3.1 MW)*; 2 shafts; auxiliary motor; 425 hp(m) *(312 kW)*; bow thruster
Speed, knots: 15
Complement: 50
Guns: 2 Bofors 40 mm/70; 300 rds/min to 12 km *(6.6 n miles)*; weight of shell 0.96 kg.
Weapons control: TVT optronic director.
Radars: Surface search: 2 Racal Decca TM 1226; I-band.
Sonars: Simrad; hull-mounted; search and attack; medium/high frequency.

Programmes: Decommissioned from Norwegian Navy in 2001 and transferred to Latvia on 27 January 2003.
Operational: Former minelayer modified to undertake mine countermeasures command and support roles. Additional tasks are likely to include training and support of diving operations.

UPDATED

VIRSAITIS   7/2004 *, M Ferrette / 0587779

1 GOLIAT CLASS (PROJECT 667R) (ATA)

PERKONS A 18 (ex-*H 18*)

Displacement, tons: 150 full load
Dimensions, feet (metres): 70.2 × 20 × 8.5 *(21.4 × 6.1 × 2.6)*
Main machinery: 1 Buckau-Wolf 8NVD diesel; 300 hp(m) *(221 kW)*; 1 shaft
Speed, knots: 9
Complement: 8 (2 officers)

Comment: Built at Gdynia in the 1960s and transferred from Poland 16 November 1993 at Liepaja.

VERIFIED

PERKONS   4/1995, Hartmut Ehlers

1 NYRYAT 1 CLASS (DIVING TENDER) (YDT)

LIDAKA (ex-*Gefests*) A 51 (ex-A 101)

Displacement, tons: 92 standard; 116 full load
Dimensions, feet (metres): 93.8 × 17.1 × 5.6 *(28.6 × 5.2 × 1.7)*
Main machinery: 1 6CSP 28/3C diesel; 450 hp(m) *(331 kW)*; 1 shaft
Speed, knots: 11. **Range, n miles:** 1,500 at 10 kt
Complement: 18 (3 officers)
Radars: Navigation: SNN-7; I-band.

Comment: Former SAR vessel acquired in 1992. Based at Liepaja.

VERIFIED

LIDAKA   9/1996, Hartmut Ehlers

1 LOGISTICS VESSEL (AKS/AXL)

Name	No	Builders	Commissioned
VARONIS (ex-*Buyskes*)	A 90 (ex-A 904)	Boele's Scheepswerven	9 Mar 1973

Displacement, tons: 967 standard; 1,033 full load
Dimensions, feet (metres): 196.6 × 36.4 × 12 *(60 × 11.1 × 3.7)*
Main machinery: Diesel-electric; 3 Paxman 12 RPH diesel generators; 2,100 hp *(1.57 MW)*; 1 motor; 1,400 hp(m) *(1.03 MW)*; 1 shaft
Speed, knots: 13.5. **Range, n miles:** 3,000 at 11.5 kt
Complement: 43 (6 officers)
Radars: Navigation: Racal Decca 1229; I-band.
Sonars: Side scanning and wreck search.

Comment: Originally designed and operated as a hydrographic vessel by the Royal Netherlands Navy from which she was decommissioned in 2003. Donated to Latvia on 8 November 2004 for use as a logistic and training vessel. Hydrographic launchers were not transferred and the ship is fitted with an inflatable boat.

NEW ENTRY

VARONIS (Netherlands colours)   5/2000 *, van Ginderen Collection / 0105152

COAST GUARD

1 RIBNADZOR-4 CLASS (WPB)

COMETA KA 03 (ex-KA 103)

Displacement, tons: 160 full load
Dimensions, feet (metres): 114.8 × 22 × 5.6 *(35 × 6.7 × 1.7)*
Main machinery: 1 40 DMM3 diesel; 2,200 hp(m) *(1.62 MW)*; 1 shaft
Speed, knots: 15
Complement: 17 (3 officers)
Guns: 1—12.7 mm MG.
Radars: Surface search: MIUS; I-band.

Comment: Ex-fishing vessel built in 1978 and converted in 1992. Recommissioned 5 May 1992 at Bolderaja. Refitted in 1998. Belongs to Coast Guard. Sister ship *Spulga* lost in 2000.

VERIFIED

COMETA   8/1994, E & M Laursen / 0081182

5 KBV 236 CLASS (WPB)

KRISTAPS KA 01 (ex-KBV 244)
GAISMA KA 06 (ex-KBV 249)
AUSMA KA 07 (ex-KBV 260)
SAULE KA 08 (ex-KBV 256)
KLINTS KA 09 (ex-KBV 250)

Displacement, tons: 17 full load
Dimensions, feet (metres): 63 × 13.1 × 4.3 *(19.2 × 4 × 1.3)*
Main machinery: 2 Volvo Penta TMD 100C diesels; 526 hp(m) *(387 kW)*; 2 shafts
Speed, knots: 20
Complement: 3 (1 officer)
Radars: Navigation: Raytheon or Furuno; I-band.

Comment: Former Swedish Coast Guard vessel built in 1964. First one recommissioned 5 March 1993, second pair 9 November 1993 and last pair 27 April 1994. KA 01, 06 and 09 are based at Bolderaja, 07 at Liepaja and 08 at Ventspils. Not all are identical. All belong to Coast Guard.

UPDATED

GAISMA   8/2004*, Guy Toremans / 0587773

SAULE   10/2000, van Ginderen Collection / 0114738

1 LOKKI CLASS (PB)

TIIRA

Displacement, tons: 76 full load
Dimensions, feet (metres): 87.9 × 18 × 6.2 *(26.8 × 5.5 × 1.9)*
Main machinery: 2 MTU 8V 396 TB82 diesels; 1,740 hp(m) *(1.28 MW)* sustained
    2 MTU 8V 396 TB84 diesels; 2,100 hp(m) *(1.54 MW)* sustained; 2 shafts
Speed, knots: 25
Complement: 6

Comment: Donated by Finland in 2001. Armament and sonar removed.

NEW ENTRY

LOKKI class (Finnish colours)   6/2001, Finnish Navy / 0114723

1 PATROL CRAFT (WPB)

ASTRA KA 14

Displacement, tons: 22 full load
Dimensions, feet (metres): 74.8 × 18.4 × 9.2 *(22.8 × 5.6 × 2.8)*
Main machinery: 3 Scania D91 1467M diesels; 1,850 hp *(1.38 MW)*
Speed, knots: 25. **Range, n miles:** 575 at 25 kt
Complement: 4 (1 officer)
Radars: Navigation: Furuno; I-band.

Comment: Built in Finland in 1996. Commissioned on 12 March 2001.

VERIFIED

ASTRA   4/2002, Guy Toremans / 0524992

3 HARBOUR PATROL CRAFT (WPB)

KA 10   KA 11   GRANATA KA 12

Displacement, tons: 9.6 full load *(KA 10-11)*; 5.4 full load *(KA 12)*
Dimensions, feet (metres): 41.3 × 10.5 × 2 *(12.6 × 3.2 × 0.6)*
Main machinery: 1 3D6C diesel; 150 hp(m) *(110 kW)*; 1 shaft
Speed, knots: 13
Complement: 2
Radars: Navigation: Furuno; I-band.

Comment: Former USSR craft. *KA 10* and *11* were Sverdlov class cruiser boats. KA 12 is an ex-Border Guard launch of 9 m with a 300 hp(m) *(220 kW)* engine and a speed of 18 kt. All acquired in 1993-94. All belong to Coast Guard.

VERIFIED

KA 11   9/1996, Hartmut Ehlers

1 VALPAS CLASS (OFFSHORE PATROL VESSEL) (WPBO)

VALPAS

Displacement, tons: 545 full load
Dimensions, feet (metres): 159.1 × 27.9 × 12.5 *(48.5 × 8.5 × 3.8)*
Main machinery: 1 Werkspoor diesel; 2,000 hp(m) *(1.47 MW)*; 1 shaft; cp prop
Speed, knots: 15
Complement: 18
Guns: 1 Oerlikon 20 mm.
Sonars: Simrad SS105; active scanning; 14 kHz.

Comment: An improvement on the *Silmä* design. Built by Laivateollisuus, Turku, and commissioned 21 July 1971. Ice strengthened. Donated by Finland on 25 September 2002 and operated by State Border Security Service.

VERIFIED

VALPAS   5/2003, J Ciślak / 0568321

Lebanon

Country Overview

The Lebanese Republic gained independence from France in 1946 but was devastated by civil war between 1975-1991. Situated on the eastern shore of the Mediterranean Sea, it has an area of 4,015 square miles and is bordered to the north and east by Syria and to the south by Israel. It has a 121 n mile coastline with the Mediterranean Sea. The capital, largest city and principal port is Beirut. Other important ports include Tripoli and Sidon. Territorial seas (12 n miles) are claimed but an EEZ is not claimed.

Headquarters Appointments

Navy Commander:
   Major General Georges Chehwan

Personnel

2005: 800 (45 officers)

Bases

Beirut (HQ), Jounieh

PATROL FORCES

2 FRENCH EDIC CLASS (LST)

Name	No	Builders	Commissioned
SOUR	21	SFCN, Villeneuve la Garonne	28 Mar 1985
DAMOUR	22	SFCN, Villeneuve la Garonne	28 Mar 1985

Displacement, tons: 670 full load
Dimensions, feet (metres): 193.5 × 39.2 × 4.2 *(59 × 12 × 1.3)*
Main machinery: 2 SACM MGO 175 V12 M1 diesels; 1,200 hp(m) *(882 kW)*; 2 shafts
Speed, knots: 10. Range, n miles: 1,800 at 9 kt
Complement: 20 (2 officers)
Military lift: 96 troops; 11 trucks or 8 APCs
Guns: 2 Oerlikon 20 mm. 1—81 mm mortar. 2—12.7 mm MGs. 1—7.62 mm MG.
Radars: Navigation: Decca; I-band.

Comment: Both were damaged in early 1990 but repaired in 1991 and are fully operational. Used by the Marine Regiment formed in 1997. *VERIFIED*

DAMOUR *2/2000 /* 0097755

2 TRACKER Mk 2 CLASS (COASTAL PATROL CRAFT) (PB)

SARAFAND (ex-*Swift*) 307   BATROUN (ex-*Safeguard*) 303

Displacement, tons: 31 full load
Dimensions, feet (metres): 63.3 × 16.4 × 4.9 *(19.3 × 5 × 1.5)*
Main machinery: 2 Detroit 12V-71TA diesels; 840 hp *(616 kW)* sustained; 2 shafts
Speed, knots: 25. Range, n miles: 650 at 20 kt
Complement: 13 (1 officer)
Guns: 3—12.7 mm MGs.
Radars: Surface search: Racal Decca 1216; I-band.

Comment: Two ex-UK Customs Craft first commissioned in 1979 and acquired by Lebanon in late 1993. Fitted with a twin 23 mm gun after transfer but this may have been replaced by a 12.7 mm MG. *VERIFIED*

BATROUN *6/1999, Lebanese Navy /* 0081188

5 ATTACKER CLASS (COASTAL PATROL CRAFT) (PB)

TRIPOLI (ex-*Attacker*) 301	BYBLOS (ex-*Chaser*) 304	SIDON (ex-*Striker*) 306
JOUNIEH (ex-*Fencer*) 302	BEIRUT (ex-*HunterII*) 305	

Displacement, tons: 38 full load
Dimensions, feet (metres): 65.6 × 17 × 4.9 *(20 × 5.2 × 1.5)*
Main machinery: 2 Detroit 12V-71TA diesels; 840 hp *(616 kW)* sustained; 2 shafts
Speed, knots: 21. Range, n miles: 650 at 14 kt
Complement: 13 (1 officer)
Guns: 3—12.7 mm MGs.
Radars: Surface search: Racal Decca 1216; I-band.

Comment: Built at Cowes and Southampton, and commissioned in March 1983. First three transferred from UK 17 July 1992 after serving as patrol craft for the British base in Cyprus. The other two were acquired in 1993. All are operational. *UPDATED*

SIDON *2/2000 /* 0097756

25 INSHORE PATROL CRAFT (PBR)

401	403-418	420-427

Displacement, tons: 6 full load
Dimensions, feet (metres): 26.9 × 8.2 × 2 *(8.2 × 2.5 × 0.6)*
Main machinery: 2 Sabre 212 diesels; 212 hp(m) *(156 kW)*; 2 waterjets
Speed, knots: 22. Range, n miles: 154 at 22 kt
Complement: 4
Guns: 3—5.56 mm MGs.

Comment: M-boot type used by the US Army on German rivers and 27 were transferred in January 1994. Called Combat Support Boats, there are 20 operational and five laid up. Two were decommissioned in 2002. *VERIFIED*

CSB *6/1998, Lebanese Navy /* 0052571

Libya

Country Overview

The Socialist People's Libyan Arab Jamahiriyah is situated in north Africa. With an area of 679,362 square miles, it has a 956 n mile coastline with the Mediterranean Sea and is bordered to the east by Egypt, to the south by Sudan, Chad and Niger and to the west by Algeria and Tunisia. The capital and largest city is Tripoli which, with Benghazi, is a principal port. Territorial seas (12 n miles) are claimed. An EEZ has not been claimed. The status of the Gulf of Sirte, which Libya claims as internal waters, is disputed by numerous states including USA, United Kingdom, France, Italy and Greece.

Headquarters Appointments

Chief of Staff Navy:
   Rear Admiral Muhammad al Shaybani Ahmad al Suwaihili
Deputy Chief of Staff Navy:
   Captain al-Din Mufti

Personnel

(a)   2005: 8,000 officers and ratings, including Coast Guard
(b)   Voluntary service

Bases

Naval HQ at Al Khums.
Operating Ports at Tripoli, Darnah (Derna) and Benghazi.
Naval bases at Al Khums and Tobruq.
Submarine base at Ras Hilal.
Naval air station at Al Girdabiyah.
Naval infantry battalion at Sidi Bilal.

Coast Defence

Batteries of truck-mounted SS-C-3 Styx missiles.

General

Specialist teams in unconventional warfare are a threat and most Libyan vessels can lay mines, but overall operational effectiveness is very low, not least because of poor maintenance and stores support. Sanctions imposed by the UN in April 1992 were reported as 'destroying' the Fleet. The situation improved in late 1995 when mostly Ukrainian technicians were hired on maintenance contracts. Further progress was reported in 1998 and the situation could improve following the lifting of UN sanctions on 12 September 2003. The EU arms embargo was lifted on 11 October 2004 although export licences are still required.

SUBMARINES

2 FOXTROT CLASS (PROJECT 641) (SS)

AL KHYBER 315   AL HUNAIN 316

Displacement, tons: 1,950 surfaced; 2,475 dived
Dimensions, feet (metres): 299.5 × 24.6 × 19.7
(91.3 × 7.5 × 6)
Main machinery: Diesel-electric; 3 Type 37-D diesels (1 × 2,700
and 2 × 1,350); 6,000 hp(m) *(4.4 MW)*; 3 motors; 5,400 hp
(m) *(3.97 MW)*; 3 shafts; 1 auxiliary motor; 140 hp(m)
(103 kW)
Speed, knots: 16 surfaced; 15 dived
Range, n miles: 20,000 at 8 kt surfaced; 380 at 2 kt dived
Complement: 75 (8 officers)

Torpedoes: 10—21 in *(533 mm)* (6 bow, 4 stern) tubes.
SAET-60; passive homing to 15 km *(8.1 n miles)* at 40 kt;
warhead 400 kg, and SET-65E; active/passive homing to
15 km *(8.1 n miles)* at 40 kt; warhead 205 kg or Type 53-56.
Total of 22 torpedoes.
Mines: 44 in place of torpedoes.
Countermeasures: ESM: Stop Light; radar warning.
Radars: Surface search: Snoop Tray; I-band.
Sonars: Herkules; hull-mounted; active; medium frequency.
Feniks; hull-mounted; passive.

Programmes: Six of the class originally transferred from USSR;
this last one in April 1982.
Operational: Libyan crews trained in the USSR and much of the
maintenance was done by Russian personnel. No routine
patrols have been seen since 1984 although both boats have
been reported to conduct surface patrols. One submarine was
reported to be in dry dock at Tripoli during 2003 but, in view of
the poor material state of these submarines, a return to full
operational capability remains highly unlikely.
UPDATED

FOXTROT   *6/1992, van Ginderen Collection /* 0081190

FRIGATES

Notes: The *Dat Assawari* F 211 is a training hulk alongside in Tripoli.

2 KONI (PROJECT 1159) CLASS (FFGM)

AL HANI F 212   AL QIRDABIYAH F 213

Displacement, tons: 1,440 standard; 1,900 full load
Dimensions, feet (metres): 316.3 × 41.3 × 11.5
(96.4 × 12.6 × 3.5)
Main machinery: CODAG; 1 SGW, Nikolayev, M8B gas turbine
(centre shaft); 18,000 hp(m) *(13.25 MW)* sustained; 2 Russki
B-68 diesels; 15,820 hp(m) *(11.63 MW)* sustained; 3 shafts
Speed, knots: 27 on gas; 22 on diesel.
Range, n miles: 1,800 at 14 kt
Complement: 120

Missiles: SSM: 4 Soviet SS-N-2C Styx (2 twin) launchers ❶;
active radar/IR homing to 83 km *(45 n miles)* at 0.9 Mach;
warhead 513 kg; sea-skimmer at end of run.
SAM: SA-N-4 Gecko twin launcher ❷; semi-active radar homing
to 15 km *(8 n miles)* at 2.5 Mach; altitude 9.1-3,048 m *(29.5-
10,000 ft)*; warhead 50 kg; 20 missiles.
Guns: 4 USSR 3 in *(76 mm)*/60 (2 twin) ❸; 60 rds/min to 15 km
(8 n miles) anti-surface; 14 km *(7.6 n miles)* anti-aircraft;
weight of shell 6.8 kg.
4 USSR 30 mm/65 (2 twin) automatic ❹; 500 rds/min to
2 km *(1.1 n miles)*; weight of shell 0.54 kg.
Torpedoes: 4—406 mm (2 twin) tubes amidships ❺. USET-95;
active/passive homing to 10 km *(5.5 n miles)* at 30 kt;
warhead 100 kg.
A/S mortars: 1 RBU 6000 12-tubed trainable launcher ❻;
automatic loading; range 6,000 m; warhead 31 kg.
Depth charges: 2 racks.

AL HANI   *(Scale 1 : 900), Ian Sturton*

Mines: Capacity for 20.
Countermeasures: Decoys: 2—16-barrelled chaff launchers.
Towed torpedo decoys.
ESM: 2 Watch Dog; radar warning.
Radars: Air search: Strut Curve ❼; F-band; range 110 km *(60 n
miles)* for 2 m² target.
Surface search: Plank Shave ❽; E/F-band.
Navigation: Don 2; I-band.
Fire control: Drum Tilt ❾; H/I-band (for 30 mm).
Hawk Screech ❿; I-band; range 27 km *(15 n miles)* (for
76 mm).
Pop Group ⓫; F/H/I-band (for SAM).
IFF: High Pole B. Square Head.
Sonars: Hercules (MG 322); hull-mounted; active search and
attack; medium frequency.

Programmes: Type III Konis built at Zelenodolsk and transferred
from the Black Sea. 212 commissioned 28 June 1986 and
213 on 24 October 1987.
Structure: SSMs mounted either side of small deckhouse on
forecastle behind gun. A deckhouse amidships contains air
conditioning machinery. Changes to the standard Koni
include SSM, four torpedo tubes, only one RBU 6000 and
Plank Shave surface search and target indication radar.
Camouflage paint applied in 1991.
Operational: One of the class fired an exercise Styx missile in
September 1999. 213 has been reported active but 212 may
be in reserve.
VERIFIED

AL HANI   *7/1999, van Ginderen Collection /* 0081191

CORVETTES

1 NANUCHKA II (PROJECT 1234) CLASS
(MISSILE CORVETTE) (FSGM)

TARIQ IBN ZIYAD (ex-*Ean Mara*) 416

Displacement, tons: 660 full load
Dimensions, feet (metres): 194.5 × 38.7 × 8.5 *(59.3 × 11.8 × 2.6)*
Main machinery: 6 M 504 diesels; 26,112 hp(m) *(19.2 MW)*; 3 shafts
Speed, knots: 33. **Range, n miles:** 2,500 at 12 kt; 900 at 31 kt
Complement: 42 (7 officers)

Missiles: SSM: 4 Soviet SS-N-2C Styx launchers; auto-pilot; active radar/IR homing to 83 km *(45 n miles)* at 0.9 Mach; warhead 513 kg HE; sea-skimmer at end of run.
SAM: SA-N-4 Gecko twin launcher; semi-active radar homing to 15 km *(8 n miles)* at 2.5 Mach; altitude 9.1-3,048 m *(29.5-10,000 ft)*; warhead 50 kg HE; 20 missiles.
Guns: 2 USSR 57 mm/80 (twin) automatic; 120 rds/min to 6 km *(3.2 n miles)*; weight of shell 2.8 kg.
Countermeasures: Decoys: 2 chaff 16-barrelled launchers.
ESM: Bell Tap; radar warning.
Radars: Surface search: Square Tie; I-band (Bandstand radome).
Navigation: Don 2; I-band.
Fire control: Muff Cob; G/H-band.
Pop Group; F/H/I-band (for SAM).

Programmes: First transferred from USSR in October 1981; second in February 1983; third in February 1984; fourth in September 1985.
Structure: Camouflage paint applied in 1991 but have been reported as having blue hulls since 1993.
Operational: *Ean Zaquit* (419) sunk on 24 March 1986. *Ean Mara* (416) severely damaged on 25 March 1986 by forces of the US Sixth Fleet; repaired in Leningrad and returned to Libya in early 1991 as the *Tariq Ibn Ziyad*. *Ean Al Gazala* (417) probably in reserve as a source of spares and *Ean Zarrah* (418) reported non-operational. Reports of refit plans are doubtful.
UPDATED

TARIQ IBN ZIYAD   *7/1991, van Ginderen Collection* / 0081192

PATROL FORCES

Notes: (1) More than 50 remote-control explosive craft acquired from Cyprus. Based on Q-Boats with Q-26 GRP hulls and speed of about 30 kt. Also reported that 15 31 ft craft delivered by Storebro, and 60 more built locally are similarly adapted. No reports of recent activity.
(2) There is also a Hamelin 37 m patrol craft *Al Ziffa* 206 based at Tripoli.

6 OSA II (PROJECT 205) CLASS
(FAST ATTACK CRAFT—MISSILE) (PTFG)

AL ZUARA 513	AL FIKAH 523	AL BITAR 531
AL RUHA 515	AL MATHUR 525	AL SADAD 533

Displacement, tons: 245 full load
Dimensions, feet (metres): 126.6 × 24.9 × 8.8 *(38.6 × 7.6 × 2.7)*
Main machinery: 3 Type M 504 diesels; 10,800 hp(m) *(7.94 MW)* sustained; 3 shafts
Speed, knots: 37. **Range, n miles:** 800 at 30 kt; 500 at 35 kt
Complement: 30

Missiles: SSM: 4 Soviet SS-N-2C Styx; active radar or IR homing to 83 km *(45 n miles)* at 0.9 Mach; warhead 513 kg HE; sea-skimmer at end of run.
Guns: 4 USSR 30 mm/65 (2 twin) automatic; 500 rds/min to 5 km *(2.7 n miles)*; weight of shell 0.54 kg.
Radars: Surface search: Square Tie; I-band; range 73 km *(45 n miles)*.
Fire control: Drum tilt; H/I-band.
IFF: 2 Square Head. High Pole.

Programmes: The first craft arrived from USSR in October 1976, four more in August-October 1977, a sixth in July 1978, three in September-October 1979, one in April 1980, one in May 1980 (521) and one in July 1980 (529).
Structure: Some painted with camouflage stripes in 1991 and some were given blue hulls in 1993.
Operational: Although there have been few sightings of these ships at sea, they remain more active than most of the Libyan fleet. One fired an exercise Styx missile in September 1999. Six further craft inactive alongside. *Al Sadad* 533 also reported non-operational. Based at Tobruk.
UPDATED

AL MATHUR   1993

---

8 COMBATTANTE II G CLASS
(FAST ATTACK CRAFT—MISSILE) (PGGF)

SHARABA (ex-*Beir Grassa*) 518	SHOULA (ex-*Beir Ktitat*) 532
WAHAG (ex-*Beir Gzir*) 522	SHAFAK (ex-*Beir Alkrarim*) 534
SHEHAB (ex-*Beir Gtifa*) 524	RAD (ex-*Beir Alkur*) 538
SHOUAIAI (ex-*Beir Algandula*) 528	LAHEEB (ex-*Beir Alkuefat*) 542

Displacement, tons: 311 full load
Dimensions, feet (metres): 160.7 × 23.3 × 6.6 *(49 × 7.1 × 2)*
Main machinery: 4 MTU 20V 538 TB91 diesels; 15,360 hp(m) *(11.29 MW)* sustained; 4 shafts
Speed, knots: 39. **Range, n miles:** 1,600 at 15 kt
Complement: 27

Missiles: SSM: 4 OTO Melara/Matra Otomat Mk 2 (TG1); active radar homing to 80 km *(43.2 n miles)* at 0.9 Mach; warhead 210 kg.
Guns: 1 OTO Melara 3 in *(76 mm)*/62 compact; 85 rds/min to 16 km *(8.6 n miles)* anti-surface; 12 km *(6.8 n miles)* anti-aircraft; weight of shell 6 kg.
2 Breda 40 mm/70 (twin); 300 or 450 rds/min to 12.5 km *(6.8 n miles)* anti-surface; 4 km *(2.2 n miles)* anti-aircraft; weight of shell 0.96 kg.
Weapons control: CSEE Panda director. Thomson-CSF Vega II system.
Radars: Surface search: Thomson-CSF Triton; G-band; range 33 km *(18 n miles)* for 2 m² target.
Fire control: Thomson-CSF Castor IIB; I-band; range 15 km *(8 n miles)* (associated with Vega fire-control system).

Programmes: Ordered from CMN Cherbourg in May 1977. 518 completed February 1982; 522 3 April 1982; 524 29 May 1982; 528 5 September 1982; 532 29 October 1982; 534 17 December 1982; 542 29 July 1983.
Structure: Steel hull with alloy superstructure.
Operational: *Waheed* (526) sunk on 24 March 1986 and one other severely damaged on 25 March 1986 by forces of the US Sixth Fleet. 524, 534 and 542 visited Malta in late 2001. *Shoula* (532) reported non-operational but continued activity suggests that the remaining craft are at least partly operational.
UPDATED

SHAFAK   1993

LAND-BASED MARITIME AIRCRAFT

Numbers/Type: 2/5 Aerospatiale SA 321 Frelon/Aerospatiale SA 324 Super Frelon.
Operational speed: 134 kt *(248 km/h)*.
Service ceiling: 10,000 ft *(3,050 m)*.
Range: 440 n miles *(815 km)*.
Role/Weapon systems: Obsolescent helicopter; Air Force manned but used for naval support tasks. Most are non-operational due to lack of spares. Sensors: None. Weapons: Fitted for Exocet AM 39.
VERIFIED

Numbers/Type: 5 Aerospatiale SA 316B Alouette III.
Operational speed: 113 kt *(210 km/h)*.
Service ceiling: 10,500 ft *(3,200 m)*.
Range: 290 n miles *(540 km)*.
Role/Weapon systems: Support helicopter. Probably non-operational. Another six are used by the Police. Sensors: None. Weapons: Unarmed.
VERIFIED

AMPHIBIOUS FORCES

Note: Three Polochny D class landing craft (112, 116 and 118) are in reserve and are unlikely to be restored to operational status.

1 PS 700 CLASS (LSTH)

Name	No	Builders	Commissioned
IBN HARISSA	134	CNI de la Mediterranée	10 Mar 1978

Displacement, tons: 2,800 full load
Dimensions, feet (metres): 326.4 × 51.2 × 7.9 *(99.5 × 15.6 × 2.4)*
Main machinery: 2 SEMT-Pielstick 16 PA4 V 185 diesels; 5,344 hp(m) *(3.93 MW)* sustained; 2 shafts; cp props
Speed, knots: 15.4. **Range, n miles:** 4,000 at 14 kt
Complement: 35
Military lift: 240 troops; 11 tanks
Guns: 6 Breda 40 mm/70 (3 twin). 1—81 mm mortar.
Weapons control: CSEE Panda director.
Radars: Air search: Thomson-CSF Triton; D-band.
Surface search: Decca 1226; I-band.
Helicopters: 1 Aerospatiale SA 316B Alouette III.

Comment: Laid down 18 April 1977, launched 18 October 1977. Remains active. The status of *Ibn Ouf* 132 is unclear.
UPDATED

IBN HARISSA   1981

3 TURKISH Ç 107 CLASS (LCT)

IBN AL IDRISI 130   IBN MARWAN 131   EL KOBAYAT 132

Displacement, tons: 280 standard; 600 full load
Dimensions, feet (metres): 183.7 × 37.8 × 3.6 *(56 × 11.6 × 1.1)*
Main machinery: 3 GM 6-71TI diesels; 930 hp *(694 kW)* maximum; 3 shafts
Speed, knots: 8.5 loaded; 10 max. Range, n miles: 600 at 10 kt
Complement: 15
Military lift: 100 troops; 350 tons including 5 tanks
Guns: 2—30 mm (twin).

Comment: First two transferred 7 December 1979 (ex-Turkish *C130* and *C131*) from Turkish fleet. Third of class reported in 1991. Previously reported numbers were much exaggerated. Not reported at sea in recent years and operational status of 132 is doubtful.
                                                                                    VERIFIED

TURKISH LCT (Turkish colours)   *10/1991, Harald Carstens*

2 SLINGSBY SAH 2200 (HOVERCRAFT) (UCAC)

Displacement, tons: 5.5 full load
Dimensions, feet (metres): 34.8 × 13.8 *(10.6 × 4.2)*
Main machinery: 1 diesel; 300 hp(m) *(224 kW)*
Speed, knots: 40. Range, n miles: 400 at 30 kt
Complement: 2
Military lift: 2.2 tons
Guns: 1—12.7 mm MG.
Radars: Surface search: I-band.

Comment: Ordered in September 1999 for delivery to Greece in mid-2000 and subsequently to Libya in 2001.
                                                                                    VERIFIED

TRAINING SHIPS

1 VOSPER CLASS (AX)

Name	No	Builders	Commissioned
TOBRUK	C 411	Vosper Thornycroft	20 Apr 1966

Displacement, tons: 500 full load
Dimensions, feet (metres): 177.3 × 28.5 × 13 *(54 × 8.7 × 4)*
Main machinery: 2 Paxman Ventura diesels; 3,800 hp *(2.83 MW)*; 2 shafts
Speed, knots: 18. Range, n miles: 2,900 at 14 kt
Complement: 63 (5 officers)
Guns: 1 Vickers 4 in *(102 mm)*/33 Mk 52; 2 Bofors 40 mm/60.
Radars: Surface search: Decca TM 1226C; I-band.

Comment: State apartments are included in the accommodation. The ship is used mostly for training.
                                                                                    VERIFIED

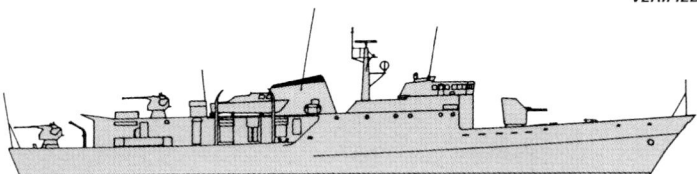

TOBRUK   *(Scale 1 : 600), Ian Sturton*

MINE WARFARE FORCES

5 NATYA (PROJECT 266ME) CLASS
(OCEAN MINESWEEPERS) (MSO)

AL ISAR (ex-*Ras El Gelais*) 113   RAS AL FULAIJAH 117   RAS AL HANI 125
AL TIYAR (ex-*Ras Hadad*) 111   RAS AL MASSAD 123

Displacement, tons: 804 full load
Dimensions, feet (metres): 200.1 × 33.5 × 10.8 *(61 × 10.2 × 3)*
Main machinery: 2 Type M 504 diesels; 5,000 hp(m) *(3.67 MW)* sustained; 2 shafts; cp props
Speed, knots: 16. Range, n miles: 3,000 at 12 kt
Complement: 67
Guns: 4 USSR 30 mm/65 (2 twin) automatic; 500 rds/min to 5 km *(2.7 n miles)*; weight of shell 0.54 kg.
    4 USSR 25 mm/60 (2 twin); 270 rds/min to 3 km *(1.6 n miles)*; weight of shell 0.34 kg.
A/S mortars: 2 RBU 1200 5-tubed fixed launchers; elevating; range 1,200 m; warhead 34 kg.
Mines: 10.
Countermeasures: MCM: 1 GKT-2 contact sweep; 1 AT-2 acoustic sweep; 1 TEM-3 magnetic sweep.
Radars: Surface search: Don 2; I-band.
Fire control: Drum Tilt; H/I-band.
IFF: 2 Square Head. 1 High Pole B.
Sonars: Hull-mounted; active search; high frequency.

Comment: Transferred from USSR between 1981 and 1986. At least one of the class painted in green striped camouflage in 1991. Others may have blue hulls. Capable of magnetic, acoustic and mechanical sweeping. Mostly used for coastal patrols and never observed minesweeping. *Ras Al Massad* has been used for training cruises. *Ras Al Hamman* (115), *Ras Al Qula* (119) and *Ras Al Madwar* are inactive.
                                                                                    UPDATED

NATYA   *2/1988*

AUXILIARIES

Note: *Zeltin* 711 is used as an alongside tender for patrol forces but is no longer capable of going to sea.

10 TRANSPORTS (AG/ML)

GARYOUNIS (ex-*Mashu*)	EL TEMSAH	DERNA	GHAT
GARNATA (ex-*Monte Granada*)	TOLETELA (ex-*Monte Toledo*)	RAHMA (ex-*Krol*)	LA GRAZIETTA
HANNA	GHARDIA		

Measurement, tons: 2,412 gross
Dimensions, feet (metres): 546.3 × 80.1 × 21.3 *(166.5 × 24.4 × 6.5)*
Main machinery: 2 SEMT-Pielstick diesels; 20,800 hp(m) *(15.29 MW)*; 2 shafts; bow thruster
Speed, knots: 20

Comment: Details are for *Garyounis*, a converted ro-ro passenger/car ferry used as a training vessel in 1989. In addition the 117 m *El Temsah* was refitted and another four of these vessels are of Ro-Ro design. All are in regular civilian service and *Garyounis* is also used by the military. All have minelaying potential.
                                                                                    VERIFIED

TOLETELA   *2000, Sky Photos /* 0130092

1 SPASILAC CLASS (SALVAGE SHIP) (ARS)

AL MUNJED (ex-*Zlatica*) 722

Displacement, tons: 1,590 full load
Dimensions, feet (metres): 182 × 39.4 × 14.1 *(55.5 × 12 × 4.3)*
Main machinery: 2 diesels; 4,340 hp(m) *(3.19 MW)*; 2 shafts; cp props; bow thruster
Speed, knots: 13. Range, n miles: 4,000 at 12 kt
Complement: 50
Guns: 4—12.7 mm MGs. Can also be fitted with 8—20 mm (2 quad) and 2—20 mm.
Radars: Surface search: Racal Decca; I-band.

Comment: Transferred from Yugoslavia in 1982. Fitted for firefighting, towing and submarine rescue-carries recompression chamber. Built at Tito SY, Belgrade. Used as the lead vessel for the 1998 training cruise, and is still operational.
                                                                                    VERIFIED

SPASILAC (Iraqi colours)   *1988, Peter Jones*

1 YELVA (PROJECT 535M) CLASS (DIVING TENDER) (YDT)

AL MANOUD VM 917

Displacement, tons: 300 full load
Dimensions, feet (metres): 134.2 × 26.2 × 6.6 *(40.9 × 8 × 2)*
Main machinery: 2 Type 3-D-12A diesels; 630 hp(m) *(463 kW)* sustained; 2 shafts
Speed, knots: 12.5
Complement: 30
Radars: Navigation: Spin trough; I-band.
IFF: High Pole.

Comment: Built in early 1970s. Transferred from USSR December 1977. Carries two 1.5 ton cranes and has a portable decompression chamber. Based at Tripoli and does not go to sea.
VERIFIED

YELVA (Russian colours)
7/1996, Hartmut Ehlers

2 FLOATING DOCKS

Comment: One of 5,000 tons capacity at Tripoli. One of 3,200 tons capacity acquired in April 1985.
VERIFIED

TUGS

7 COASTAL TUGS (YTB)

RAS EL HELAL A 31		**AL KERIAT**	A 33-35
AL AHWEIRIF A 32		**AL TABKAH**	

Comment: First four of 34.8 m built in Portugal in 1976-78. Last three of 26.6 m built in the Netherlands in 1979-80. All are in service.
VERIFIED

Lithuania
KARINĖS JŪRØ PAJĖGOS

Country Overview

The Republic of Lithuania regained independence in 1991 after 51 years as a Soviet republic. Situated in northeastern Europe, the country has an area of 25,175 square miles and borders to the north with Latvia, to the east and south with Belarus, to the southwest with Poland and the Russian exclave of Kaliningrad. It has a 58 n mile coastline with the Baltic Sea. Vilnius is the capital and largest city while Klaipeda is the principal port. Territorial seas (12 n miles) are claimed but while it has claimed a 200 n mile Exclusive Economic Zone (EEZ), its limits have not been fully defined by boundary agreements.

Headquarters Appointments

Commander of the Navy:
 Captain Kęstutis Macijauskas
Chief of Staff:
 Commander Oleg Marinic

Personnel

2005: 680

Bases

Klaipeda

State Border Police (Pakrančių Apsauga)

Coast Guard Force formed in late 1992. Name changed in 1996 to Border Police. Vessels have one thick and one thin diagonal yellow stripe on the hull.

FRIGATES

2 GRISHA III (ALBATROS) CLASS (PROJECT 1124M) (FFLM)

Name	No	Builders	Commissioned	Recommissioned
ŽEMAITIS	F 11 (ex-MPK 108)	Zelenodolsk Shipyard	1 Oct 1981	6 Nov 1992
AUKŠTAITIS	F 12 (ex-MPK 44)	Kiev Shipyard	15 Aug 1980	6 Nov 1992

Displacement, tons: 950 standard; 1,200 full load
Dimensions, feet (metres): 233.6 × 32.2 × 12.1 *(71.2 × 9.8 × 3.7)*
Main machinery: CODAG; 1 gas-turbine; 15,000 hp(m) *(11 MW)*; 2 diesels; 16,000 hp(m) *(11.8 MW)*; 3 shafts
Speed, knots: 30
Range, n miles: 2,500 at 14 kt diesels; 950 at 27 kt
Complement: 48 (5 officers)

Missiles: SAM: SA-N-4 Gecko twin launcher ❶; semi-active radar homing to 15 km *(8 n miles)* at 2.5 Mach; warhead 50 kg; altitude 9.1-3,048 m *(30-10,000 ft)*; 20 missiles.
Guns: 2—57 mm/80 (twin) ❷; 120 rds/min to 6 km *(3.3 n miles)*; weight of shell 2.8 kg.
 1—30 mm/65 ❸; 6 barrels; 3,000 rds/min combined to 2 km.
 1—12.7 mm MG.
A/S mortars: 2 RBU 6000 12-tubed trainable ❹; range 6,000 m; warhead 31 kg.
Depth charges: 2 racks (12).
Mines: Capacity for 18 in lieu of depth charges.
Countermeasures: Decoys: 1 PK-16 (F 11) chaff launcher.
ESM: 2 Watch Dog.
Radars: Air/surface search: Strut Curve ❺; F-band.
Navigation: Terma Scanter; I-band.
Fire control: Pop Group ❻; F/H/I-band (for SA-N-4). Bass Tilt ❼; H/I-band (for guns).
Sonars: Bull Nose; hull-mounted; active search and attack; high/ medium frequency.

Modernisation: Torpedo tubes removed from F 12 in 1996 and from F 11 in 1997.
UPDATED

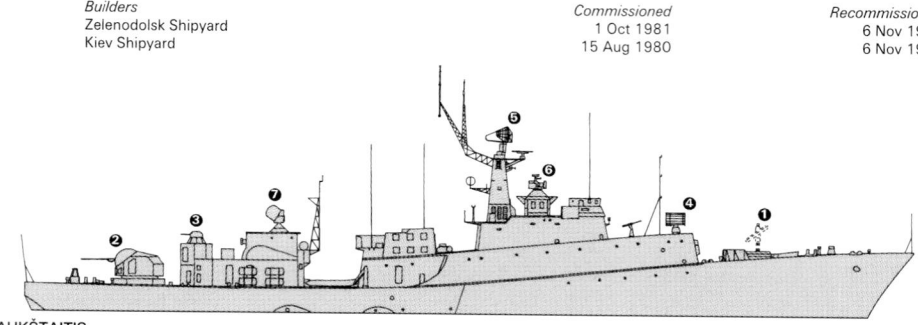

AUKŠTAITIS
(Scale 1 : 600), Ian Sturton / 0587562

AUKŠTAITIS
6/2003, J Cišlak / 0568320

ŽEMAITIS
6/2004*, Harald Carstens / 0589759

PATROL FORCES

3 STORM CLASS (PB)

Name	No	Builders	Commissioned
DZŪKAS (ex-Kjekk)	P 31 (ex-P 965)	Bergens Mek Verksteder	1966
SELIS (ex-Skudd)	P 32 (ex-P 967)	Bergens Mek Verksteder	1966
SKALVIS (ex-Steil)	P 33 (ex-P 969)	Westermoen, Mandal	1967

Displacement, tons: 138 full load
Dimensions, feet (metres): 120 × 20.3 × 5.9 (36.5 × 6.2 × 1.8)
Main machinery: 2 MTU MB 16V 538 TB90 diesels; 6,000 hp(m) (4.41 MW) sustained; 2 shafts
Speed, knots: 32. **Range, n miles:** 550 at 32 kt
Complement: 20 (3 officers)
Guns: 1 Bofors 3 in (76 mm)/50; 30° elevation; 30 rds/min to 13 km (7 n miles). Surface fire only; weight of shell 5.9 kg.
   1 Bofors 40 mm/70; 90° elevation; 300 rds/min to 12 km (6.6 n miles); weight of shell 0.96 kg.
Weapons control: TVT 300 optronic tracker.
Radars: Navigation: Furuno; I-band.

Comment: P 31 (ex-Glimt) disarmed and acquired from Norway on 12 December 1994 as a gun patrol craft. Re-armed in 1998. A further craft (ex-Kjekk) transferred in October 2001 and, following refit, commissioned as P 31 in August 2002. Ex-Glimt is laid up at Klaipeda. P 32 and P 33 transferred from Norway in June 2001. Others of the class given to Latvia and Estonia.
VERIFIED

SKALVIS   4/2002, Guy Toremans / 0524995

1 COASTAL PATROL CRAFT (PB/YFS)

HK 21 (ex-Vilnele)

Displacement, tons: 88 full load
Dimensions, feet (metres): 75.8 × 19 × 5.9 (23.1 × 5.8 × 1.8)
Main machinery: 2 diesels; 600 hp(m) (441 kW); 2 shafts
Speed, knots: 12
Complement: 7 (1 officer)

Comment: Acquired in 1992. Former pilot boat and tender to Vetra. Used as a hydrographic vessel.
UPDATED

HK 21   6/2004*, Lithuanian Navy / 0589761

MINE WARFARE FORCES

2 LINDAU (TYPE 331) CLASS (MINEHUNTER) (MHC)

Name	No	Builders	Commissioned
SŪDUVIS (ex-Koblenz)	M 52 (ex-M 1071)	Burmester, Bremen	8 July 1958
KURŠIS (ex-Marburg)	M 51 (ex-M 1080)	Burmester, Bremen	11 June 1959

Displacement, tons: 463 full load
Dimensions, feet (metres): 154.5 × 27.2 × 9.8 (9.2 Troika) (47.1 × 8.3 × 3) (2.8)
Main machinery: 2 MTU MD diesels; 4,000 hp(m) (2.94 MW); 2 shafts
Speed, knots: 16.5. **Range, n miles:** 850 at 16.5 kt
Complement: 43 (5 officers)
Guns: 1 Bofors 40 mm/70.
Radars: Navigation: Raytheon Mariner Pathfinder; I-band.
Sonars: Plessey 193 m; minehunting; high frequency (100/300 kHz).
   EdgeTech DF-1000 sidescan (M 51); high frequency (100/400 kHz).

Comment: M 52 acquired from Germany in June 1999 and recommissioned 2 December 1999. M 51 transferred in November 2000. Converted to minehunters in 1978. Hulls of wooden construction. Full minehunting equipment including PAP 104 ROVs transferred with the vessels.
UPDATED

KURŠIS   8/2004*, Guy Toremans / 1044132

AUXILIARIES

Notes: *Victoria* 245 is an ex-Swedish Coast Guard vessel now owned by the Fishery Inspection Service.

1 VALERIAN URYVAYEV CLASS (AGOR/AX)

VETRA (ex-Rudolf Samoylovich) A 41

Displacement, tons: 1,050 full load
Dimensions, feet (metres): 180.1 × 31.2 × 13.1 (54.9 × 9.5 × 4)
Main machinery: 1 Deutz diesel; 850 hp(m) (625 kW); 1 shaft
Speed, knots: 12
Complement: 34 (8 officers)
Guns: 2—12.7 mm MGs.
Radars: Navigation: Racal Decca RM 1290; I-band.

Comment: Built at Khabarovsk in early 1980s. Transferred from the Russian Navy in 1992 where she was used as a civilian oceanographic research vessel. Now used as the Flag ship for the Baltic States MCMV unit which includes *Olev* and *Kalev* (from Estonia) and *Viesturs* and *Imanta* (from Latvia). A second of class *Vejas* works for the Ministry of Environment.
UPDATED

VETRA   6/2004*, Marian Wright / 0589760

1 HARBOUR TUG (YTL)

H 22 (ex-A 330)

Displacement, tons: 35
Dimensions, feet (metres): 48 × 14.8 × 8.2 (14.65 × 4.5 × 2.5)
Main machinery: 1 Scania-Vabis DSI 11R82A diesel; 230 hp (171 kW)
Speed, knots: 9
Complement: 4
Radars: Navigation: Racal Decca; I-band.

Comment: Ex-Swedish *Atlas* transferred in 2000.
VERIFIED

H 22   6/2003, Hartmut Ehlers / 0561507

1 KUTTER CLASS (PB)

LOKYS (ex-Apollo) S 07

Displacement, tons: 35 full load
Dimensions, feet (metres): 60.4 × 17.1 × 7.5 (18.4 × 5.2 × 2.3)
Main machinery: 1 diesel; 165 hp(m) (121 kW); 1 shaft
Speed, knots: 9
Complement: 7
Guns: 2—7.62 mm MGs.
Radars: Surface search: Raytheon RM 1290S; I-band.

Comment: Built in the 1930s and served with the Danish Naval Home Guard. Transferred in July 1997. Manned by naval volunteer organisation.
VERIFIED

LOKYS   6/2003, Hartmut Ehlers / 0561505

STATE BORDER SECURITY SERVICE

1 LOKKI CLASS (PB)

KIHU   102 (ex-003)

Displacement, tons: 76 full load
Dimensions, feet (metres): 87.9 × 17 × 6.2 *(26.8 × 5.2 × 1.9)*
Main machinery: 2 MTU 8V 396 TB84 diesels; 2,120 hp(m) *(1.58 MW)* sustained; 2 shafts
Speed, knots: 25
Complement: 6
Radars: Navigation: Furuno FR 2010 and FCR 1411; I-band.

Comment: Armament and sonar removed on transfer. Donated by Finland in 1998.

VERIFIED

KIHU
6/2003, *Hartmut Ehlers* / 0561506

1 KBV 041 CLASS (PB)

MADELEINE 042 (ex-KBV 041)

Displacement, tons: 69 full load
Dimensions, feet (metres): 73.5 × 17.72 × 5.6 *(22.4 × 5.4 × 1.7)*
Main machinery: 2 diesels; 450 hp(m) *(331 kW)*; 2 shafts
Speed, knots: 10
Complement: 4
Radars: Navigation: Furuno FRS 1000C and FR 1510; I-band.

Comment: Class B sea truck transferred from the Swedish Coast Guard in April 1995. Used for pollution control in Swedish service but now used as patrol craft.

VERIFIED

MADELEINE
6/2003, *Hartmut Ehlers* / 0561504

1 KBV 101 CLASS (PB)

LILIAN 101 (ex-*KBV 101*)

Displacement, tons: 69 full load
Dimensions, feet (metres): 82 × 16.4 × 6.5 *(25 × 5 × 2)*
Main machinery: 2 Cummins KTA38-M diesels; 2,120 hp(m) *(1.56 MW)*; 2 shafts
Speed, knots: 18. Range, n miles: 1,000 at 15 kt
Complement: 5
Radars: Navigation: Furuno FR 2010 and FCR 1411; I-band.

Comment: Built in Sweden in 1969. Transferred from Swedish Coast Guard on 24 June 1996. Used in Swedish service as a salvage diving vessel and had a high frequency active hull-mounted sonar.

VERIFIED

LILIAN
6/2003, *Hartmut Ehlers* / 0561503

1 CHRISTINA (GRIFFON 2000 TD) CLASS HOVERCRAFT (UCAC)

CHRISTINA

Displacement, tons: 5 full load
Dimensions, feet (metres): 41.35 × 20 *(12.6 × 6.1)*
Main machinery: 1 Deutz BF8L diesel; 355 hp *(265 kW)*
Speed, knots: 35
Complement: 3
Radars: Furuno 1000C; I-band.

Comment: Built by Griffon UK and delivered in 2000. Similar to crafts supplied to Estonia and Finland.

VERIFIED

CHRISTINA
6/2001, *Lithuanian Navy* / 0114364

![Flag of Macedonia] # Macedonia, Former Yugoslav Republic of

Country Overview

The Former Yugoslav Republic of Macedonia declared its independence in 1991. A land-locked country with an area of 9,928 square miles, it is situated in south-eastern Europe and is bordered to the north by Serbia, to the east by Bulgaria, to the south by Greece and to the west by Albania. Parts of the borders with Albania and Greece pass through the two principal lakes, Ohrid and Prespa. The capital and largest city is Skopje.

PATROL FORCES

Notes: The Macedonian Lake Service (Ezerska sluzba – EZ) consists of about 400 soldiers and is nominally an independent arm of the Army although in practice it is almost integrated with Land Forces. In addition to the five ex-Yugoslavian Army patrol boats on Lake Ohrid, there are two further small craft on Lake Prespa.

5 BOTICA CLASS (TYPE 16) (RIVER PATROL CRAFT) (PBR)

Displacement, tons: 23 full load
Dimensions, feet (metres): 55.8 × 11.8 × 2.8 *(17.0 × 3.6 × 0.8)*
Main machinery: 2 diesels; 464 hp *(340 kW)*; 2 shafts
Speed, knots: 15. Range, n miles: 340 at 14 kt
Complement: 7
Military lift: 3 tons or 30 troops
Guns: 1 Oerlikon 20 mm. 2—7.62 mm MGs.
Radars: Surface search: Decca 110; I-band.

Comment: Former Yugoslavian craft which entered service in the 1970s.

NEW ENTRY

BOTICA CLASS
6/2003* / 1044466

Madagascar
MALAGASY REPUBLIC MARINE

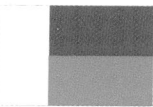

Country Overview

Formerly a French Protectorate, the Malagasy Republic became self-governing in 1958 and fully independent in 1960. It adopted the name Democratic Republic of Madagascar in 1975. Situated in the Indian Ocean and separated from the southeastern coast of Africa by the Mozambique Channel, it comprises Madagascar Island, the fourth largest island in the world, and several small islands. The country's total area is 226,658 square miles and it has a coastline of 2,608 n miles. Antananarivo is the capital while Toamasina is the principal commercial port. There are further ports at Antsiranana, Mahajanga and Toliara. Territorial seas (12 n miles) are claimed. An Exclusive Economic Zone (EEZ) has been claimed but boundaries have not been agreed.

Headquarters Appointments

Head of Navy:
    Rear Admiral Manny Ranaivonativo

Personnel

2005: 430 officers and men (including Marine Company of 120 men)

Bases

Antsiranana (main), Toamasina, Mahajanga, Toliary, Nosy-Be, Tolagnaro, Manakara.

PATROL FORCES

Notes: It was planned to order coastal patrol craft from South Korea in 2001 but this has not been confirmed.

AMPHIBIOUS FORCES

1 EDIC CLASS (LCT)

AINA VAO VAO (ex-*L9082*)

Displacement, tons: 250 standard; 670 full load
Dimensions, feet (metres): 193.5 × 39.2 × 4.5 *(59 × 12 × 1.3)*
Main machinery: 2 SACM MGO diesels; 1,000 hp(m) *(753 kW)*; 2 shafts
Speed, knots: 8. **Range, n miles:** 1,800 at 8 kt
Complement: 32 (3 officers)
Military lift: 250 tons
Guns: 2 Giat 20 mm.

Comment: Built in 1964 by Chantier Naval Franco-Belge. Transferred from France 28 September 1985 having been paid off by the French Navy in 1981. Repaired by the French Navy in 1996 and now back in service.
VERIFIED

AINA VAO VAO   6/1999, Madagascar Navy / 0081202

AUXILIARIES

Notes: (1) There are three Aigrette class harbour tugs, *Tourterelle*, was acquired from France in 1975 and *Engoulevent* and *Martin-Pêcheur* May 1996.
(2) There is also a 400 ton coastal tug *Trozona*.

1 CHAMOIS CLASS (SUPPLY TENDER) (AG)

MATSILO (ex-*Chamois*) (ex-A 767)

Displacement, tons: 495 full load
Dimensions, feet (metres): 136.1 × 24.6 × 10.5 *(41.5 × 7.5 × 3.2)*
Main machinery: 2 SACM AGO 175 V16 diesels; 2,700 hp(m) *(1.98 MW)*; 2 shafts; cp props; bow thruster
Speed, knots: 14. **Range, n miles:** 6,000 at 12 kt
Complement: 13 plus 7 spare
Cargo capacity: 100 tons cargo; 165 tons of fuel or water
Radars: Navigation: Racal Decca 1226; I-band.

Comment: Built by La Perrière, Lorient and commissioned in the French Navy 24 September 1976. Paid off in 1995 and transferred from France in May 1996. Can act as a tug (bollard pull 25 tons) or for SAR and supply tasks but is mostly used as a patrol craft. There are two 30 ton winches and up to 100 tons of stores can be carried on deck.
VERIFIED

MATSILO
6/1999, Madagascar Navy / 0081203

Malawi

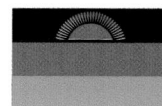

Country Overview

Formerly the British Protectorate of Nyasaland, the Republic of Malawi gained independence in 1964. A landlocked country situated in east Central Africa, it is bordered to the north by Tanzania, to the west by Zambia and to the south and east by Mozambique. The country's total area is 45,747 square miles, nearly a quarter of which is water. The principal lake is Lake Malawi (formerly Lake Nyasa), with which there is a shoreline of some 475 n miles. The largest city is the former capital Blantyre and the capital, since 1975, is Lilongwe. The naval base at Monkey Bay is situated on a peninsula at the south of the lake.

Headquarters Appointments

Commander of the Malawi Army Marine Unit:
    Lieutenant Colonel G A Ziyabu

Bases

Monkey Bay, Lake Malawi

Personnel

2005: 225

PATROL FORCES

Notes: One survey craft built in France in 1988 is operated on Lake Malawi by Department of Surveys.

1 ANTARES CLASS (PB)

KASUNGU (ex-*Chikala*) P 703

Displacement, tons: 41 full load
Dimensions, feet (metres): 68.9 × 16.1 × 4.9 *(21 × 4.9 × 1.5)*
Main machinery: 2 Poyaud 520 V12 M2 diesels; 1,300 hp(m) *(956 kW)*; 2 shafts
Speed, knots: 22. **Range, n miles:** 650 at 15 kt
Complement: 16
Guns: 1 MG 21 20 mm. 2—7.62 mm MGs.
Radars: Surface search: Decca; I-band.

Comment: Built in prefabricated sections by SFCN Villeneuve-la-Garenne and shipped to Malawi for assembly on 17 December 1984. Commissioned May 1985. Out of service and in need of refit if and when funds become available.
VERIFIED

KASUNGU
6/1996, Malawi Navy / 0012737

1 NAMACURRA CLASS (PB)

KANING'A (ex-*Y 1520*) P 704

Displacement, tons: 5 full load
Dimensions, feet (metres): 29.5 × 9 × 2.8 *(9 × 2.7 × 0.8)*
Main machinery: 2 BMW 3.3 outboards; 380 hp(m) *(279 kW)*
Speed, knots: 32. **Range, n miles:** 180 at 20 kt
Complement: 4
Guns: 1—12.7 mm MG. 2—7.62 mm MGs.
Radars: Surface search: Decca; I-band.

Comment: Donated by South Africa on 29 October 1988.

1 ROTORK CLASS (LCU)

CHIKOKO I L 702

Displacement, tons: 9 full load
Dimensions, feet (metres): 41.5 × 10.5 × 1.5 *(12.7 × 3.2 × 0.5)*
Main machinery: 2 Volvo diesels; 260 hp(m) *(191 kW)*; 2 shafts
Speed, knots: 24. **Range, n miles:** 3,000 at 15 kt
Complement: 8
Guns: 3—7.62 mm MGs.

Comment: Built by Rotork Marine. Needs a refit but no funds are available.

VERIFIED

VERIFIED

KANING'A
6/1997, Malawi Navy / 0012736

CHIKOKO I
6/1996, Malawi Navy / 0012738

Malaysia
TENTERA LAUT DIRAJA

Country Overview

The Federation of Malaysia was formed in 1963. Situated in south-east Asia, its two regions are separated by some 350 n miles of the South China Sea. Peninsular Malaysia (formerly West Malaysia) is bordered to the north by Thailand and to the south by Singapore (which left the federation in 1965) and includes 11 states occupying the southern half of the Malay Peninsula. To the east, the states (former British colonies) of Sabah and Sarawak (which surrounds the sultanate of Brunei) occupy the northern third of the island of Borneo, the remainder of which forms the Indonesian province of Kalimantan. With an overall land area of 127,320 square miles, Malaysia has a coastline of 2,527 n miles with the Strait of Malacca, the South China Sea, the Sulu and Celebes Seas. Kuala Lumpur is the capital and largest city while the principal ports are George Town, Klang, Tanjung Pelepas and Kuching. Territorial seas (12 n miles) are claimed. An EEZ (200 n miles) is claimed but the limits have not been fully defined.

Headquarters Appointments

Chief of Navy:
    Admiral Dato' Sri Mohd Anwar bin Mohd Nor
Deputy Chief of Navy:
    Vice Admiral Datuk Ilyas bin Hj Din
Fleet Commander:
    Vice Admiral Dato' Mohammad bin Nik
Commander Naval Area I (Kuantan):
    First Admiral Mohammad Nordin bin Ali
Commander Naval Area II (Sabah and Sarawak):
    First Admiral Datuk Abdul Aziz bin Hj Jaafar

Coastal Defence

Procurement of a coastal surveillance system is under consideration. About 15 sites are required and a short-list of potential contractors is expected by 2005.

Personnel

(a)   2005: 17,141 (2,013 officers)
(b)   Voluntary service: Royal Malaysian Navy Voluntary Reserve (RMNVR): Total, 3,362 (1,061 officers)
(c)   Bases for regular and reserve forces at Penang, Perak, Selangor, Kuala Lumpur, Pahang, Johor, Terengganu, Sabah den Sarawak.

Bases

(a)   Lumut Naval Base comprises Fleet Operations Command, HQ Fleet System, HQ Support, Naval Education and Training Centre, Naval Air Station – KD *Rajawali*, Diving Centre and Special Forces Training Centre
(b)   HQ Area 1 – Kuantan (West of 109E)
(c)   HQ Area 2 – Labuan (East of 109E). Comprises mainly two forward bases, KD *Sri Sandakan* and KD *Sri Tawau*
(d)   HQ Area 3 – Langkawi Island *
(e)   Sapanggar Naval Base – Kota Kinabalu*
(f)   Semporna Naval Base – Sabah

* Under construction to be completed 2005

Prefix to Ships' Names

The names of Malaysian warships are prefixed by KD (Kapal DiRaja meaning His Majesty's Ship).

Maritime Patrol Craft

There are large numbers of armed patrol craft belonging to the Police, Customs and Fisheries Departments. Details at the end of the section.

Strength of the Fleet

Type	Active	Building (Planned)
Submarines	—	2
Submarines Mini	—	3
Frigates	2	—
Corvettes	6	6
Offshore Patrol Vessels	2	6
Logistic Support Vessels	2	—
Fast Attack Craft—Missile	8	—
Fast Attack Craft—Gun	6	—
Patrol Craft	18	—
Minehunters	4	—
Survey Ships	2	1
LSTs	1	(1)
Training Ships	2	—

DELETIONS

Training Ships

2004   *Rahmat*

PENNANT LIST

Frigates			37	Badek
			38	Renchong
29	Jebat		39	Tombak
30	Lekiu		40	Lembing
			41	Serampang
Corvettes			42	Panah
			43	Kerambit
25	Kasturi		44	Beladau
26	Lekir		45	Kelewang
134	Laksamana Hang Nadim		46	Rentaka
135	Laksamana Tun Abdul Jamil		47	Sri Perlis
136	Laksamana Muhammad Amin		49	Sri Johor
137	Laksamana Tan Pusmah		160	Musytari
171	Kedah (bldg)		161	Marikh
172	Pahang (bldg)		3144	Sri Sabah
173	Perak (bldg)		3145	Sri Sarawak
174	Terengganu (bldg)		3146	Sri Negri Sembilan
175	Kelantan (bldg)		3147	Sri Melaka
176	Selangor (bldg)		3501	Perdana
			3502	Serang
Patrol Forces			3503	Ganas
			3504	Ganyang
34	Kris		3505	Jerong
36	Sundang		3506	Todak

3507	Paus	
3508	Yu	
3509	Baung	
3510	Pari	
3511	Handalan	
3512	Perkasa	
3513	Pendekar	
3514	Gempita	

Mine Warfare Forces		
11	Mahamiru	
12	Jerai	
13	Ledang	
14	Kinabalu	

Amphibious Forces		
1505	Sri Inderapura	

Training Ships		
76	Hang Tuah	
A 13	Tunas Samudera	

Survey Ships		
152	Mutiara	
153	Perantau	

Auxiliaries		
4	Penyu	
331	Sri Gaya	
332	Sri Tiga	
1503	Sri Indera Sakti	
1504	Mahawangsa	

SUBMARINES

Notes: (1) There are no plans to procure mini-submarines as has been previously reported.
(2) The French Agosta class submarine *Ouessant* is on loan to Malaysia to provide initial training which is to start in early 2005. The boat continues to belong to the French Navy and is based at Brest. The submarine may go to Malaysia on completion of training in 2009.

0 + 2 SCORPENE CLASS (SSK)

Name	No	Builders	Laid down	Launched	Commissioned
—	—	DCN, Cherbourg	2003	2006	2008
—	—	Izar, Cartagena	2004	2007	2009

Displacement, tons: 1,564 surfaced; 1,711 dived
Dimensions, feet (metres): *217.9 × 20.3 × 17.7 (66.4 × 6.2 × 5.4)*
Main machinery: Diesel electric; 2 SEMT-Pielstick 12 PA4 200 SM DS diesels; 1 Jeumont Industrie motor; 4,700 hp *(3.5 MW)*; 1 shaft
Speed, knots: 20.5 dived; 12 surfaced
Range, n miles: 360 at 4 kt dived; 6,000 at 8 kt surfaced
Complement: 31 (7 officers)

Missiles: SSM: Aerospatiale SM39 Exocet; launched from 21 in *(533 mm)* torpedo tubes; inertial cruise; active radar homing to 50 km *(27 n miles)* at 0.9 Mach; warhead 165 kg.
Torpedoes: 6—21 in *(533 mm)* tubes. Alenia Whitehead Black Shark torpedoes; wire (fibre-optic cable) guided; active and passive homing. Total of 18 weapons.
Countermeasures: ESM: Thales DR 3000; intercept.
Weapons control: UDS International SUBTICS.
Radars: Navigation: I-band.
Sonars: Hull mounted; active/passive search and attack, medium frequency.

Programmes: Contract for the construction of two submarines awarded to Armaris and IZAR on 5 June 2002. A four-year training programme aboard an Agosta-70 (ex-*Ouessant*) is included in the package. First steel cut for first of class 2 December 2003.
Structure: Similar in design to the Chilean boats. Diving depth more than 300 m *(984 ft)*. Option to retrofit AIP at a later date.
Operational: To be based at Sapanggar Naval Base, Sabah.
UPDATED

SCORPENE (computer graphic)   *1998, DCN /* 0017689

FRIGATES

Notes: One older frigate is listed under Training Ships.

JEBAT   *10/2001, Hartmut Ehlers /* 0132701

2 LEKIU CLASS (FFGHM)

Name	No	Builders	Laid down	Launched	Commissioned
JEBAT	29	Yarrow (Shipbuilders), Glasgow	Nov 1994	27 May 1995	20 Nov 1999
LEKIU	30	Yarrow (Shipbuilders), Glasgow	Mar 1994	3 Dec 1994	9 Oct 1999

Displacement, tons: 1,845 standard; 2,390 full load
Dimensions, feet (metres): 346 oa; 319.9 wl × 42 × 11.8
(105.5; 97.5 × 12.8 × 3.6)
Main machinery: CODAD; 4 MTU 20V 1163 TB93 diesels;
33,300 hp(m) *(24.5 MW)* sustained; 2 shafts; Kamewa cp
props
Speed, knots: 28. **Range, n miles:** 5,000 at 14 kt
Complement: 146 (18 officers)

Missiles: SSM: 8 Aerospatiale MM 40 Exocet Block II ❶; inertial
cruise; active radar homing to 70 km *(40 n miles)* at 0.9 Mach;
warhead 165 kg; sea-skimmer.
SAM: British Aerospace VLS Seawolf; 16 launchers ❷; command
line of sight (CLOS) radar/TV tracking to 6 km *(3.3 n miles)* at
2.5 Mach; warhead 14 kg.
Guns: 1 Bofors 57 mm/70 SAK Mk 2 ❸; 220 rds/min to 17 km
(9.3 n miles); weight of shell 2.4 kg.
2 MSI 30 mm/75 DS 30B ❹; 650 rds/min to 10 km *(5.4 n
miles)*; weight of shell 0.36 kg.
Torpedoes: 6 Whitehead B 515 324 mm (2 triple) tubes ❺; anti-
submarine; Marconi Stingray; active/passive homing to 11 km
(5.9 n miles) at 45 kt; warhead 35 kg (shaped charge).
Countermeasures: Decoys: 2 Super Barricade 12-barrelled
launchers for chaff ❻; Graseby Sea Siren torpedo decoy.
ESM: AEG Telefunken/Marconi Mentor; intercept.
ECM: MEL Scimitar; jammer.

Combat data systems: GEC-Marconi Nautis-F; Signaal Link Y Mk
2.
Weapons control: Radamec 2400 Optronic director ❼.
Thomson-CSF ITL 70 (for Exocet); GEC-Marconi Type V 3901
thermal imager.
Radars: Air search: Signaal DA08 ❽; E/F-band.
Surface search: Ericsson Sea Giraffe 150HC ❾; G/H-band.
Navigation: Racal Decca; I-band.
Fire control: 2 Marconi 1802 ❿; I/J-band.
Sonars: Thomson Sintra Spherion; hull-mounted active search
and attack; medium frequency.

Helicopters: 1 Westland Super Lynx ⓫.

LEKIU

(Scale 1 : 900), Ian Sturton / 0081204

Programmes: Contract announced 31 March 1992 for two ships
originally classed as corvettes but uprated to light frigates.
First steel cut in March 1993. *Jebat* is the senior ship and has
therefore taken the lower pennant number.
Structure: GEC Naval Systems Frigate 2000 design with a
modern combat data system and automated machinery
control.
Operational: Delivery dates were delayed by weapon system
integration problems but both arrived in Malaysia in early
2000. Form 23rd Frigate Squadron.

UPDATED

LEKIU

8/2001, John Mortimer / 0126290

JEBAT

2/2001, Sattler/Steele / 0126376

CORVETTES

2 KASTURI (TYPE FS 1500) CLASS (FSGH)

Name	No	Builders	Laid down	Launched	Commissioned
KASTURI	25	Howaldtswerke, Kiel	3 Jan 1983	14 May 1983	15 Aug 1984
LEKIR	26	Howaldtswerke, Kiel	3 Jan 1983	14 May 1983	15 Aug 1984

Displacement, tons: 1,500 standard; 1,850 full load
Dimensions, feet (metres): 319.1 × 37.1 × 11.5
(97.3 × 11.3 × 3.5)
Main machinery: 4 MTU 20V 1163 TB92 diesels; 23,400 hp(m)
(17.2 MW) sustained; 2 shafts
Speed, knots: 28; 18 on 2 diesels
Range, n miles: 3,000 at 18 kt; 5,000 at 14 kt
Complement: 124 (13 officers)

Missiles: SSM: 8 Aerospatiale MM 40 Exocet Block II ❶; inertial
cruise; active radar homing to 70 km *(40 n miles)* at 0.9 Mach;
warhead 165 kg; sea-skimmer.
Guns: 1 Creusot-Loire 3.9 in *(100 mm)*/55 Mk 2 compact ❷; 20/
45/90 rds/min to 17 km *(9.2 n miles)* anti-surface; 6 km
(3.2 n miles) anti-aircraft; weight of shell 13.5 kg.
1 Bofors 57 mm/70 ❸; 200 rds/min to 17 km *(9.2 n miles)*;
weight of shell 2.4 kg. Launchers for illuminants.
4 Emerson Electric 30 mm (2 twin) ❹; 1,200 rds/min
combined to 6 km *(3.2 n miles)*; weight of shell 0.35 kg.
A/S mortars: 1 Bofors 375 mm twin trainable launcher ❺;
automatic loading; range 3,600 m.
Countermeasures: Decoys: 2 CSEE Dagaie trainable systems;
replaceable containers for IR or chaff.
ESM: Rapids.
ECM: MEL Scimitar; jammer.
Combat data systems: Signaal Sewaco-MA. Link Y Mk 2.
Weapons control: 2 Signaal LIOD optronic directors for gunnery.

Radars: Air/surface search: Signaal DA08 ❻; F-band; range
204 km *(110 n miles)* for 2 m² target.
Navigation: Kelvin Hughes 1007; I-band.
Fire control: Signaal WM22 ❼; I/J-band.
IFF: US Mk 10.
Sonars: Atlas Elektronik DSQS-21C; hull-mounted; active search
and attack; medium frequency.

Helicopters: Platform for 1 medium ❽.

Programmes: First two ordered in February 1981. Fabrication
began early 1982. Rated as Corvettes even though they are
bigger ships than *Rahmat*.

Modernisation: Plans to fit telescopic hangars have been
shelved. Mid-life update planned to include CIWS, and
torpedo tubes vice the obsolete ASW mortar. Sewaco combat
data system upgraded in 1997-98. Exocet upgraded to MM
40 2001-02.
Structure: Near sisters to the Colombian ships with differing
armament.
Operational: *Kasturi* is the trials ship for 'smart ship' reduced
manning. Form 22nd Corvette Squadron.

UPDATED

KASTURI

(Scale 1 : 900), Ian Sturton

KASTURI

10/2002, Hachiro Nakai / 0526822

LEKIR

10/2001, Chris Sattler / 0126291

1 + 5 KEDAH (MEKO A 100) CLASS (FSGHM)

Name	No	Builders	Laid down	Launched	Commissioned
KEDAH	171	Blohm + Voss/Penang Shipbuilding	13 Nov 2001	21 Mar 2003	Nov 2004
PAHANG	172	Blohm + Voss/Penang Shipbuilding	21 Dec 2001	2 Oct 2003	Aug 2005
PERAK	173	Penang Shipbuilding, Lumut	Mar 2002	Mar 2005	Oct 2005
TERENGGANU	174	Penang Shipbuilding, Lumut	Aug 2004	Dec 2005	Apr 2006
KELANTAN	175	Penang Shipbuilding, Lumut	July 2005	Nov 2006	June 2007
SELANGOR	176	Penang Shipbuilding, Lumut	July 2006	Sep 2007	Feb 2008

Displacement, tons: 1,650 full load
Dimensions, feet (metres): 298.9 × 39.4 × 9.8
  (91.1 × 12.0 × 3.0)
Main machinery: 2 Caterpillar 3616 diesels; total of 14,617 hp
  (m) *(10.9 MW)* sustained; 2 shafts; cp propellors
Speed, knots: 22. **Range, n miles:** 6,050 at 12 kt
Complement: 68 (11 officers)

Missiles: Fitted for SSM (MM40) ❶ and SAM (RAM CIWS) ❷.
Guns: 1 Otobreda 3 in *(76 mm)*/62 ❸; Super Rapid; 120 rds/min
  to 16 km *(8.7 n miles)*; weight of shell 6 kg.
  1—30 mm Otobreda/Mauser ❹. 2—12.7 mm MGs.
Countermeasures: Decoys: RBOC chaff launcher.
Combat data systems: STN Atlas Cosys 110M1.
Weapons control: Contraves TMEO optronic director ❺.
Radars: Air/surface search: TRS-3D/16ES ❻; G-band.
Sonars: Fitted for.
Helicopters: Platform for medium helicopter.

Programmes: Following a delay of 19 months since initial signature, the Malaysian government, the Penang Shipbuilding Corporation and German Naval Group

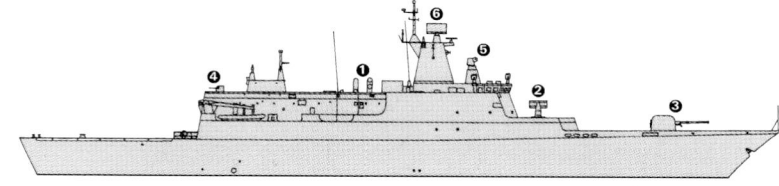

KEDAH

(Scale 1 : 900), Ian Sturton / 1044256

consortium (led by Blohm + Voss) reached final agreement in November 2000 for the supply of an initial batch of six vessels. The first two OPVs are being built in Germany for shipment to Malaysia and assembly and fitting out at Lumut. The first ship left Hamburg on 1 April 2003 and the second ship on 6 October 2003. Subsequent ships to be built in Malaysia. Delivery of the first ship to the Royal Malaysian Navy is expected in mid-2004 and all ships are to have been delivered by 2008.

Structure: Design based on Blohm + Voss MEKO A 100 including measures to reduce the radar and IR signatures. Space has been included for future enhancements which may include SSM, SAM, sonar and an EW suite.
Operational: Principal tasks are expected to be maritime surveillance and patrol duties in the Malaysian EEZ.

UPDATED

PAHANG

10/2003, Michael Nitz / 0568876

KEDAH

3/2003, Michael Nitz / 0552763

4 LAKSAMANA (ASSAD) CLASS (FSGM)

Name	No	Builders	Laid down	Launched	Commissioned
LAKSAMANA HANG NADIM (ex-*Khalid Ibn Al Walid*)	F 134 (ex-F 216)	Fincantieri, Breda, Mestre	3 June 1982	5 July 1983	28 July 1997
LAKSAMANA TUN ABDUL JAMIL (ex-*Saad Ibn Abi Waccade*)	F 135 (ex-F 218)	Fincantieri, Breda, Marghera	17 Sep 1982	2 Dec 1983	28 July 1997
LAKSAMANA MUHAMMAD AMIN (ex-*Abdulla Ben Abi Sarh*)	F 136 (ex-F 214)	Fincantieri, Breda, Mestre	22 Mar 1982	5 July 1983	31 July 1999
LAKSAMANA TAN PUSMAH (ex-*Salahi Ad Deen Alayoori*)	F 137 (ex-F 220)	Fincantieri, Breda, Marghera	17 Sep 1982	30 Mar 1984	31 July 1999

Displacement, tons: 705 full load
Dimensions, feet (metres): 204.4 × 30.5 × 8
(62.3 × 9.3 × 2.5)
Main machinery: 4 MTU 20V 956 TB92 diesels; 20,120 hp(m)
(14.8 MW) sustained; 4 shafts
Speed, knots: 36. **Range, n miles:** 2,300 at 18 kt
Complement: 47

Missiles: SSM: 6 OTO Melara/Matra Otomat Teseo Mk 2 (TG 2) (3 twin) **❶**; command guidance; active radar homing to 180 km *(98.4 n miles)* at 0.9 Mach; warhead 210 kg; sea-skimmer.
SAM: 1 Selenia/Elsag Albatros launcher **❷** (4 cell-2 reloads); Aspide; semi-active radar homing to 13 km *(7 n miles)* at 2.5 Mach; height envelope 15-5,000 m *(49.2-16,405 ft)*; warhead 30 kg.
Guns: 1 OTO Melara 3 in *(76 mm)*/62 Super Rapid **❸**; 120 rds/min to 16 km *(8.7 n miles)* anti-surface; 12 km *(6.6 n miles)* anti-aircraft; weight of shell 6 kg.
2 Breda 40 mm/70 (twin) **❹**; 300 rds/min to 12.5 km *(6.8 n miles)*; weight of shell 0.96 kg.
Torpedoes: 6—324 mm ILAS 3 (2 triple) tubes **❺**. Whitehead A244S; anti-submarine; active/passive homing to 7 km *(3.8 n miles)*; warhead 34 kg (shaped charge).
Countermeasures: Decoys: 2 Breda 105 mm 6-tubed multipurpose launchers; chaff to 5 km *(2.7 n miles)*; illuminants to 12 km *(6.6 n miles)*.
ESM: Selenia INS-3; intercept.
ECM: Selenia TQN-2; jammer.
Combat data systems: Selenia IPN 10 (136, 137); Alenia IPN-S (134, 135); Signaal/AESN Link Y Mk 2.

Weapons control: 2 Selenia NA 21; Dardo.
Radars: Air/surface search: Selenia RAN 12L/X **❻**; D/I-band; range 82 km *(45 n miles)*.
Navigation: Kelvin Hughes 1007; I-band.
Fire control: 2 Selenia RTN 10X **❼**; I/J-band; 1 Selenia RTN 20X **❽**; I/J-band.
Sonars: Atlas Elektronik ASO 84-41; hull-mounted; active search and attack.

Programmes: Ordered in February 1981 for the Iraqi Navy and fell foul of UN sanctions before they could either be paid for or delivered. Subsequently completed in 1988 and maintained by Fincantieri. Two near sister ships were paid for by Iraq and remain laid up in Italian ports. Contract signed on 26 October 1995, and confirmed on 26 July 1996, to transfer two of the class to the Malaysian Navy after refit at Muggiano and three

months training in Italy. Contract for two more signed on 20 February 1997 for conversion and delivery.
Modernisation: Super Rapid 76 mm gun, datalink, new navigation radar and GPS fitted in 1996. Bridge wings are extended to the after gun deck. Contract signed with Alenia Marconi on 11 April 2002 to upgrade command systems of F 134 and F 135 to IPN-S.
Structure: NBC citadel and full air conditioning fitted.
Operational: First pair arrived in Malaysia in September 1997. Second pair delayed by payment problems but arrived in September 1999. Constitute 24th Corvette Squadron.
Opinion: This was an unusual purchase because of the lack of equipment commonality with the rest of the Fleet.

UPDATED

LAKSAMANA HANG NADIM   *(Scale 1 : 600), Ian Sturton /* 0126348

LAKSAMANA TAN PUSMAH   *10/2003, Hartmut Ehlers /* 0567894

LAKSAMANA HANG NADIM   *5/2001 /* 0126341

SHIPBORNE AIRCRAFT

Notes: Sikorsky S-61A Nuri Army support helicopter can be embarked in two Logistic Support Ships and two OPVs.

NURI   6/1997 / 0081211

Numbers/Type: 6 GKN Westland Super Lynx.
Operational speed: 120 kt *(222 km/h)*.
Service ceiling: 10,000 ft *(3,048 m)*.
Range: 320 n miles *(593 km)*.
Role/Weapon systems: Ordered on 3 September 1999. All delivered in 2003. Surveillance. Sensors: Seaspray radar; ESM: Sky Guardian 2500; MST-S FLIR. Weapons: ASW; two A244S torpedoes. ASV; Sea Skua ASM; 2—12.7 mm MG.

VERIFIED

SUPER LYNX   9/2003, Hartmut Ehlers / 0567893

LAND-BASED MARITIME AIRCRAFT

Notes: The Air Force has eight F/A-18D fighter-bombers with Harpoon ASM, and 16 Hawk fighters with Sea Eagle ASM.

Numbers/Type: 6 Aerospatiale AS 555 Fennec.
Operational speed: 114 kt *(211 km/h)*.
Service ceiling: 10,000 ft *(3,050 m)*.
Range: 389 n miles *(722 km)*.
Role/Weapon systems: Unarmed aircraft ordered late 2001 for delivery in June 2004. Utility, SAR and training roles. Sensors: RDR 1500B radar; EWR 99 Fruit RWR; ARGOS 410-A5 FLIR. Weapons: 7.62 mm MG.

VERIFIED

FENNEC (on deck KD *Jebat*)   10/2001, Kathryn Shaw / 0096316

Numbers/Type: 4 Beechcraft B 200T Super King.
Operational speed: 282 kt *(523 km/h)*.
Service ceiling: 35,000 ft *(10,670 m)*.
Range: 2,030 n miles *(3,756 km)*.
Role/Weapon systems: Used for maritime surveillance. Acquired in 1994. Air Force operated. Sensors: Search radar. Weapons: Unarmed.

VERIFIED

SUPER KING   6/1993 / 0084007

PATROL FORCES

17 COMBATBOAT 90E (PBF)

TANGKIS   1
TEMPUR   11-15, 21-24, 32-34, 41-44

Displacement, tons: 19 full load
Dimensions, feet (metres): 52.2 × 12.5 × 2.6 *(15.9 × 3.8 × 0.8)*
Main machinery: 2 Volvo Penta TAMD 163P diesels; 1,500 hp(m) *(1.1 MW)*; 2 waterjets
Speed, knots: 45. **Range, n miles:** 240 at 30 kt
Complement: 3
Guns: 1—7.62 mm MG.
Radars: Surface search: I-band.

Comment: Ordered from Dockstavarvet in Sweden in April 1997. Have more powerful engines than the boats in Swedish service. Primary role is maritime law enforcement particularly on east coast of Sabah.

VERIFIED

TEMPUR 44   9/2003, Hartmut Ehlers / 0567892

4 PERDANA (LA COMBATTANTE II) CLASS
(FAST ATTACK CRAFT—MISSILE) (PTFG)

Name	No	Builders	Launched	Commissioned
PERDANA	3501	CMN, Cherbourg	31 May 1972	21 Dec 1972
SERANG	3502	CMN, Cherbourg	22 Dec 1971	31 Jan 1973
GANAS	3503	CMN, Cherbourg	26 Oct 1972	28 Feb 1973
GANYANG	3504	CMN, Cherbourg	16 Mar 1972	20 Mar 1973

Displacement, tons: 234 standard; 265 full load
Dimensions, feet (metres): 154.2 × 23.1 × 12.8 *(47 × 7 × 3.9)*
Main machinery: 4 MTU MB 870 diesels; 14,000 hp(m) *(10.3 MW)*; 4 shafts
Speed, knots: 36.5. **Range, n miles:** 800 at 25 kt; 1,800 at 15 kt
Complement: 30 (4 officers)

Missiles: SSM: 2 Aerospatiale MM 38 Exocet; inertial cruise; active radar homing to 42 km *(23 n miles)* at 0.9 Mach; warhead 165 kg; sea-skimmer. Not always carried.
Guns: 1 Bofors 57 mm/70; 200 rds/min to 17 km *(9.2 n miles)*; weight of shell 2.4 kg.
1 Bofors 40 mm/70; 300 rds/min to 12 km *(6.5 n miles)* anti-surface; 4 km *(2.2 n miles)* anti-aircraft; weight of shell 0.96 kg.
Countermeasures: Decoys: 4—57 mm chaff/flare launchers.
ESM: Thomson-CSF DR 3000; intercept.
Weapons control: Thomson-CSF Vega optical for guns.
Radars: Air/surface search: Thomson-CSF TH-D 1040 Triton; G-band; range 33 km *(18 n miles)* for 2 m² target.
Navigation: Kelvin Hughes 1007; I-band.
Fire control: Thomson-CSF Pollux; I/J-band; range 31 km *(17 n miles)* for 2 m² target.

Programmes: Left Cherbourg for Malaysia 2 May 1973.
Modernisation: There are plans to replace MM 38 with MM 40 or Teseo SSMs and to update radar and EW.
Structure: All of basic La Combattante II design with steel hulls and aluminium superstructure.
Operational: Form 1st Fast Attack Craft Squadron.

UPDATED

GANAS   10/2003, Hartmut Ehlers / 0567889

2 MUSYTARI CLASS (OFFSHORE PATROL VESSELS) (PSOH)

Name	No	Builders	Launched	Commissioned
MUSYTARI	160	Korea Shipbuilders, Pusan	20 July 1984	19 Dec 1985
MARIKH	161	Malaysia SB and E Co, Johore	21 Jan 1985	9 Apr 1987

Displacement, tons: 1,300 full load
Dimensions, feet (metres): 246 × 35.4 × 12.1
(75 × 10.8 × 3.7)
Main machinery: 2 SEMT-Pielstick diesels; 12,720 hp(m)
(9.35 MW); 2 shafts
Speed, knots: 22. **Range, n miles:** 5,000 at 15 kt
Complement: 76 (10 officers)
Guns: 1 Creusot-Loire 3.9 in *(100 mm)*/55 Mk 2 compact (161);
20/45/90 rds/min to 17 km *(9.2 n miles)* anti-surface; 6 km
(3.2 n miles) anti-aircraft; weight of shell 13.5 kg.
1—40 mm (160).
2 Emerson Electric 30 mm (twin); 1,200 rds/min combined to
6 km *(3.2 n miles)*; weight of shell 0.35 kg.
Countermeasures: ESM: Thales DR 3000; intercept.
Weapons control: PEAB 9LV 230 optronic system.
Radars: Air/surface search: Signaal DA05; E/F-band; range
137 km *(75 n miles)* for 2 m² target.
Navigation: Kelvin Hughes 1007; I-band.
Fire control: Philips 9LV; J-band.
Helicopters: Platform for 1 medium.

Programmes: Ordered in June 1983. Names translate to Jupiter
and Mars.
Structure: Flight deck suitable for Sikorsky S-61A Nuri army
support helicopter.
Operational: Form 16th Offshore Patrol Vessel Squadron based
at Kuantan. These ships may be transferred to the new
Maritime Enforcement Agency. **UPDATED**

MARIKH *10/2003, Hartmut Ehlers /* 0567891

MUSYTARI *10/2003, Hartmut Ehlers /* 0567890

4 HANDALAN (SPICA-M) CLASS (FAST ATTACK CRAFT—MISSILE) (PTFG)

Name	No	Builders	Commissioned
HANDALAN	3511	Karlskrona, Sweden	26 Oct 1979
PERKASA	3512	Karlskrona, Sweden	26 Oct 1979
PENDEKAR	3513	Karlskrona, Sweden	26 Oct 1979
GEMPITA	3514	Karlskrona, Sweden	26 Oct 1979

Displacement, tons: 240 full load
Dimensions, feet (metres): 142.6 × 23.3 × 7.4 (screws)
(43.6 × 7.1 × 2.4)
Main machinery: 3 MTU 16V 538 TB91 diesels; 9,180 hp(m)
(6.75 MW) sustained; 3 shafts
Speed, knots: 34.5. **Range, n miles:** 1,850 at 14 kt
Complement: 40 (6 officers)

Missiles: SSM: 4 Aerospatiale MM 38 Exocet; inertial cruise;
active radar homing to 42 km *(23 n miles)* at 0.9 Mach;
warhead 165 kg; sea-skimmer.
Guns: 1 Bofors 57 mm/70 Mk 1; 200 rds/min to 17 km *(9.2 n
miles)*; weight of shell 2.4 kg. Illuminant launchers.
1 Bofors 40 mm/70; 300 rds/min to 12 km *(6.5 n miles)* anti-
surface; 4 km *(2.2 n miles)* anti-aircraft; weight of shell
0.96 kg.
Countermeasures: ESM: Thales DR 3000; intercept.
Weapons control: 1 PEAB 9LV212 Mk 2 weapon control system
with TV tracking. LME anti-aircraft laser and TV rangefinder.
Radars: Surface search: Philips 9GR 600; I-band (agile
frequency).
Navigation: Kelvin Hughes 1007; I-band.
Fire control: Philips 9LV 212; J-band.

Programmes: Ordered 15 October 1976. All named in one
ceremony on 11 November 1978, arriving in Port Klang on 26
October 1979.
Modernisation: There are plans to replace the MM 38 with MM
40 or Teseo missiles and to update radar and EW.
Structure: Bridge further forward than in Swedish class to
accommodate Exocet. Plans to fit an ASW capability were
shelved and the sonar removed.
Operational: Form 2nd Fast Attack Craft Squadron.
UPDATED

GEMPITA *2/2001, Chris Sattler /* 0126278

18 31 METRE PATROL CRAFT (PB)

Name	No	Builders	Commissioned
SRI SABAH	3144	Vosper Ltd, Portsmouth	2 Sep 1964
SRI SARAWAK	3145	Vosper Ltd, Portsmouth	30 Sep 1964
SRI NEGRI SEMBILAN	3146	Vosper Ltd, Portsmouth	28 Sep 1964
SRI MELAKA	3147	Vosper Ltd, Portsmouth	2 Nov 1964
KRIS	34	Vosper Ltd, Portsmouth	1 Jan 1966
SUNDANG	36	Vosper Ltd, Portsmouth	29 Nov 1966
BADEK	37	Vosper Ltd, Portsmouth	15 Dec 1966
RENCHONG	38	Vosper Ltd, Portsmouth	17 Jan 1967
TOMBAK	39	Vosper Ltd, Portsmouth	2 Mar 1967
LEMBING	40*	Vosper Ltd, Portsmouth	12 Apr 1967
SERAMPANG	41	Vosper Ltd, Portsmouth	19 May 1967
PANAH	42	Vosper Ltd, Portsmouth	27 July 1967
KERAMBIT	43*	Vosper Ltd, Portsmouth	28 July 1967
BELADAU	44*	Vosper Ltd, Portsmouth	12 Sep 1967
KELEWANG	45	Vosper Ltd, Portsmouth	4 Oct 1967
RENTAKA	46	Vosper Ltd, Portsmouth	22 Sep 1967
SRI PERLIS	47*	Vosper Ltd, Portsmouth	24 Jan 1968
SRI JOHOR	49*	Vosper Ltd, Portsmouth	14 Feb 1968

* Training

Displacement, tons: 96 standard; 109 full load
Dimensions, feet (metres): 103 × 19.8 × 5.5 *(31.4 × 6 × 1.7)*
Main machinery: 2 Bristol Siddeley or MTU MD 655/18 diesels; 3,500 hp(m) *(2.57 MW)*; 2 shafts
Speed, knots: 27. **Range, n miles:** 1,400 (1,660 Sabah class) at 14 kt
Complement: 22 (3 officers)
Guns: 2 Bofors 40 mm/70. 2—7.62 mm MGs.
Radars: Surface search: Racal Decca Bridgemaster ARPA; I-band.

Comment: The four Sabah class were ordered in 1963 for delivery in 1964. The boats of the Kris class were ordered in 1965 for delivery between 1966 and 1968. All are of prefabricated steel construction and are fitted with air conditioning and Vosper roll damping equipment. The differences between the classes are minor, the later ones having improved radar, communications, evaporators and engines of MTU, as opposed to Bristol Siddeley construction. All have been refitted to extend their operational lives. Eight form the 13th Patrol Craft Squadron, based at Sandakan, and eight form the 14th Patrol Craft Squadron based at Kuantan. Two form the 12th Patrol Craft Squadron based at Lumut. These craft may be transferred to the new Maritime Enforcement Agency. Similar craft in service in Panama.

UPDATED

SUNDANG   *4/1997, Maritime Photographic* / 0012749

SERAMPANG   *11/2001, Maritime Photographic* / 0130744

6 JERONG CLASS (FAST ATTACK CRAFT—GUN) (PB)

Name	No	Builders	Commissioned
JERONG	3505	Hong Leong-Lürssen, Butterworth	27 Mar 1976
TODAK	3506	Hong Leong-Lürssen, Butterworth	16 June 1976
PAUS	3507	Hong Leong-Lürssen, Butterworth	16 Aug 1976
YU	3508	Hong Leong-Lürssen, Butterworth	15 Nov 1976
BAUNG	3509	Hong Leong-Lürssen, Butterworth	11 Jan 1977
PARI	3510	Hong Leong-Lürssen, Butterworth	23 Mar 1977

Displacement, tons: 244 full load
Dimensions, feet (metres): 147.3 × 23 × 8.3 *(44.9 × 7 × 2.5)*
Main machinery: 3 MTU MB 16V 538 TB90 diesels; 9,000 hp(m) *(6.6 MW)* sustained; 3 shafts
Speed, knots: 32. **Range, n miles:** 2,000 at 14 kt
Complement: 36 (4 officers)
Guns: 1 Bofors 57 mm/70 Mk 1. 200 rds/min to 17 km *(9.2 n miles)*; weight of shell 2.4 kg.
1 Bofors 40 mm/70.
Countermeasures: ESM: Thales DR 3000; intercept.
Radars: Surface search: Kelvin Hughes 1007; I-band.

Comment: Lürssen 45 type. Illuminant launchers on both gun mountings. Design of hull modification is reported to have been contracted.

UPDATED

PARI   *5/1990, John Mortimer* / 0081217

MINE WARFARE FORCES

4 MAHAMIRU (LERICI) CLASS (MINEHUNTERS) (MHC)

Name	No	Builders	Launched	Commissioned
MAHAMIRU	11	Intermarine, Italy	23 Feb 1984	11 Dec 1985
JERAI	12	Intermarine, Italy	5 Jan 1984	11 Dec 1985
LEDANG	13	Intermarine, Italy	14 July 1983	11 Dec 1985
KINABALU	14	Intermarine, Italy	19 Mar 1983	11 Dec 1985

Displacement, tons: 610 full load
Dimensions, feet (metres): 167.3 × 32.5 × 9.2 *(51 × 9.9 × 2.8)*
Main machinery: 2 MTU 12V 396 TC82 diesels (passage); 2,605 hp(m) *(1.91 MW)* sustained; 2 shafts; Kamewa cp props; 3 Fincantieri Isotta Fraschini ID 36 SS 6V diesels; 1,481 hp(m) *(1.09 MW)* sustained; 2 Riva Calzoni hydraulic thrust jets
Speed, knots: 16 diesels; 7 thrust jet. **Range, n miles:** 2,000 at 12 kt
Complement: 42 (5 officers)
Guns: 1 Bofors 40 mm/70; 300 rds/min to 12.5 km *(6.8 n miles)*; weight of shell 0.96 kg.
Countermeasures: Thomson-CSF IBIS II minehunting system; 2 improved PAP 104 ROVs. Oropesa 'O' MIS-4 mechanical sweep.
Radars: Navigation: Kelvin Hughes 1007; Thomson-CSF Tripartite III; I-band.
Sonars: Thomson Sintra TSM 2022 with Display 2060; minehunting; high frequency.

Comment: Ordered on 20 February 1981. All arrived in Malaysia on 26 March 1986. Heavy GRP construction without frames. Snach active tank stabilisers. Draeger Duocom decompression chamber. Slightly longer than Italian sisters. Endurance, 14 days. Upgrade of tactical data system completed in 2001; Minehunter Technical Display System (MTDS) installed by Altech Defence System, South Africa. Form the 26th Mine Countermeasures Squadron.

UPDATED

JERAI   *2/2004*, Bob Fildes* / 0587780

AMPHIBIOUS FORCES

33 LCM/LCP/LCU

LCM 1-5	LCP 1-15	RCP 1-9	LCU 1-4

Displacement, tons: 56 (LCM); 30 (LCU/RCP); 18.5 (LCVP) full load
Main machinery: 2 diesels; 330 hp *(246 kW)* (LCM); 400 hp *(298 kW)* (LCVP); 2 shafts
Speed, knots: 10 (LCM); 16 (LCVP); 17 (LCU/RCP)
Military lift: 30 tons (LCM); 35 troops (LCU/RCP/LCP)

Comment: LCMs and LCPs are Australian built and transferred 1965-70. LCMs have light armour on sides and some have gun turrets. RCPs and LCUs are Malaysian built and in service 1974-84. Transferred to the Army in 1993.

VERIFIED

LCM 5 (with gun turret)   *6/1995* / 0012753

1 NEWPORT CLASS (LSTH)

Name					
SRI INDERAPURA (ex-*Spartanburg County*)					

	No 1505 (ex-1192)	Builders National Steel, San Diego	Laid down 7 Feb 1970	Launched 11 Nov 1970	Commissioned 1 Sep 1971

Displacement, tons: 4,975 light; 8,450 full load
Dimensions, feet (metres): 522.3 (hull) × 69.5 × 17.5 (aft)
(159.2 × 21.2 × 5.3)
Main machinery: 6 ALCO 16-251 diesels; 16,500 hp *(12.3 MW)*
sustained; 2 shafts; cp props; bow thruster
Speed, knots: 20. **Range, n miles:** 14,250 at 14 kt
Complement: 257 (13 officers)
Military lift: 400 troops (20 officers); 500 tons vehicles; 3 LCVPs
and 1 LCPL on davits

Guns: 1 General Electric/General Dynamics 20 mm Vulcan
Phalanx Mk 15.
Radars: Surface search: Raytheon SPS-67; G-band.
Navigation: Marconi LN66; I/J-band.
Kelvin Hughes 1007; I-band.

Helicopters: Platform only.

Programmes: Transferred by sale from the USN 16 December
1994, arriving in Malaysia in June 1995. Second authorised
for transfer by lease in 1998 but this was not confirmed.
Structure: The hull form required to achieve 20 kt would not
permit bow doors, thus these ships unload by a 112 ft ramp
over their bow. The ramp is supported by twin derrick arms. A
ramp just forward of the superstructure connects the lower
tank deck with the main deck and a vehicle passage through
the superstructure provides access to the parking area
amidships. A stern gate to the tank deck permits unloading of
amphibious tractors into the water, or unloading of other

SRI INDERAPURA   *5/1995, Robert Pabst /* 0081219

vehicles into an LCU or onto a pier. Vehicle stowage covers
19,000 sq ft. Length over derrick arms is 562 ft *(171.3 m)*; full
load draught is 11.5 ft forward and 17.5 ft aft.

Operational: 3 in guns removed before transfer. Repeated refits
in Johore shipyard between late 1995 and 1998. Damaged by
fire on 15 December 2002 at Lumut.   ***UPDATED***

130 DAMEN ASSAULT CRAFT 540 (LCP)

Dimensions, feet (metres): 17.7 × 5.9 × 2 *(5.4 × 1.8 × 0.6)*
Main machinery: 1 outboard; 40 hp(m) *(29.4 kW)*
Speed, knots: 12
Military lift: 10 troops

Comment: First 65 built by Damen Gorinchem, Netherlands in 1986. Remainder built by
Limbungan Timor SY. Army assault craft. Manportable and similar to Singapore craft. Used by
the Army. Some have been deleted.

     VERIFIED

AUXILIARIES

2 LOGISTIC SUPPORT SHIPS (AOR/AE/AXH)

Name	No	Builders	Commissioned
SRI INDERA SAKTI	1503	Bremer Vulkan	24 Oct 1980
MAHAWANGSA	1504	Korea Tacoma	16 May 1983

Displacement, tons: 4,300 (1503); 4,900 (1504) full load
Dimensions, feet (metres): 328; 337.9 (1504) × 49.2 × 15.7 *(100; 103 × 15 × 4.8)*
Main machinery: 2 Deutz KHD SBV6M540 diesels; 5,865 hp(m) *(4.31 MW)*; 2 shafts; cp props;
bow thruster
Speed, knots: 16.5. **Range, n miles:** 4,000 at 14 kt
Complement: 136 (14 officers) plus 65 spare
Military lift: 17 tanks; 600 troops
Cargo capacity: 1,300 tons dieso; 200 tons fresh water (plus 48 tons/day distillers)

Guns: 2 Bofors 57 mm Mk 1 (1 only fwd in 1503). 2 Oerlikon 20 mm.
Countermeasures: ESM: Thales DR 3000; intercept.
Weapons control: 2 CSEE Naja optronic directors (1 only in 1503).
Radars: Navigation: Kelvin Hughes 1007; I-band.

Helicopters: 1 Sikorsky S-61A Nuri (army support) can be carried.

Programmes: Ordered in October 1979 and 1981 respectively.
Modernisation: 100 mm gun included in original design but used for OPVs.
Structure: Fitted with stabilising system, vehicle deck, embarkation ramps port and starboard,
recompression chamber and a stern anchor. Large operations room and a conference room are
provided. Transfer stations on either beam and aft, light jackstay on both sides and a 15 ton
crane for replenishment at sea. 1504 has additional capacity to transport ammunition and the
funnel has been removed to enlarge the flight deck which is also higher in the superstructure.
Operational: Used as training ships for cadets in addition to main roles of long-range support of
Patrol Forces and MCM vessels, command and communications and troop or ammunition
transport.

     VERIFIED

MAHAWANGSA   *10/2001, Chris Sattler /* 0126274

2 FAST TROOP VESSELS (AP)

SRI GAYA 331   SRI TIGA 332

Displacement, tons: 116.5 full load
Dimensions, feet (metres): 123.1 × 23.0 × 3.6 *(37.5 × 7.0 × 1.1)*
Main machinery: 4 MAN D2842 LE 408 diesels; 2,080 hp *(1.55 MW)*; 4 water-jets
Speed, knots: 25. **Range, n miles:** 540
Complement: 8
Military lift: 32 troops + stores
Radars: Navigation: Furuno; I-band.

Comment: Design based on Australian Wave Master fast-ferry monohull. Procured to transport
troops and stores particularly in Sabah and Sarawak waters. Built by Naval Dockyard, Lumut and
commissioned on 29 May 2001. Based at Labuan.

     VERIFIED

SRI TIGA   *10/2001, Chris Sattler /* 0126273

7 COASTAL SUPPLY SHIPS AND TANKERS (AOTL/AKSL)

LANG TIRAM	ENTERPRISE	KEPAH	LANG SIPUT
MELEBAN	JERNIH	TERIJAH	

Comment: Various auxiliaries mostly acquired in the early 1980s. There are also Sabah supply
ships identified by M numbers.

     VERIFIED

SRI INDERA SAKTI   *10/2001, Chris Sattler /* 0126275

SURVEY SHIPS

Notes: There is also an ex-Survey craft *Penyu* 4 of 465 tons commissioned in 1979. Used as a diving tender. Complement is 26 (two officers).

1 SURVEY VESSEL (AGSH)

Name	No	Builders	Commissioned
MUTIARA	152	Hong Leong-Lürssen, Butterworth	12 Jan 1978

Displacement, tons: 1,905 full load
Dimensions, feet (metres): 232.9 × 42.6 × 13.1 *(71 × 13 × 4)*
Main machinery: 2 Deutz SBA12M528 diesels; 4,000 hp(m) *(2.94 MW)*; 2 shafts
Speed, knots: 16. **Range, n miles:** 4,500 at 16 kt
Complement: 155 (14 officers)
Guns: 4 Oerlikon 20 mm (2 twin).
Radars: Navigation: 2 Racal Decca 1226/1229; I-band.
Helicopters: Platform only.

Comment: Ordered in early 1975. Carries satellite navigation, auto-data system and computerised fixing system. Davits for six survey launches.

VERIFIED

MUTIARA   *9/2003, Hartmut Ehlers /* 0567888

1 SURVEY VESSEL (AGS)

Name	No	Builders	Commissioned
PERANTAU	153	Hong Leong-Lürssen, Butterworth	12 Oct 1998

Displacement, tons: 1,996 full load
Dimensions, feet (metres): 222.4 × 43.6 × 13.1 *(67.8 × 13.3 × 4)*
Main machinery: 2 Deutz/MWM SBV8 M628 diesels; 4,787 hp(m) *(3.52 MW)*; 2 shafts; Berg cp props; Schottel bow thruster
Speed, knots: 16. **Range, n miles:** 6,000 at 10 kt
Complement: 94 (17 officers)
Radars: Navigation: STN Atlas; I-band.

Comment: Ordered from Krogerwerft in 1996. The ship is equipped with two survey launches and four multipurpose boats and has three winches and two cranes, including a hoist for a STN Atlas side scan sonar. Full range of hydrographic and mapping equipment embarked.

VERIFIED

PERANTAU   *10/2001, Hartmut Ehlers /* 0130737

PERANTAU   *9/2003, Hartmut Ehlers /* 0567884

TRAINING SHIPS

1 HANG TUAH (TYPE 41/61) CLASS (FFH/AX)

Name	No	Builders	Commissioned
HANG TUAH (ex-*Mermaid*)	76	Yarrow (Shipbuilders), Glasgow	16 May 1973

Displacement, tons: 2,300 standard; 2,520 full load
Dimensions, feet (metres): 339.3 × 40 × 16 (screws) *(103.5 × 12.2 × 4.9)*
Main machinery: 2 Stork Wärtsilä 12SW28 diesels; 9,928 hp(m) *(7.3 MW)* sustained; 2 shafts; cp props
Speed, knots: 24. **Range, n miles:** 4,800 at 15 kt
Complement: 210
Guns: 1 Bofors 57 mm/70 Mk 1; 200 rds/min to 17 km *(9.2 n miles)*; weight of shell 2.4 kg.
   2 Bofors 40 mm/70; 300 rds/min to 12 km *(6.5 n miles)* anti-surface; 4 km *(2.2 n miles)* anti-aircraft; weight of shell 0.96 kg.
Radars: Navigation: Kelvin Hughes 1007; I-band.
Helicopters: Platform for 1 medium.

Comment: Originally built for Ghana as a display ship for ex-President Nkrumah but put up for sale after his departure. She was launched without ceremony on 29 December 1966 and completed in 1968. Commissioned in Royal Navy 16 May 1973 and transferred to Royal Malaysian Navy May 1977. Refitted in 1991-92 to become a training ship. Main gun and main engines replaced in 1995-96. Sonars removed but Limbo mounting still fitted. There are no plans for further modifications.

UPDATED

HANG TUAH   *10/2001, Chris Sattler /* 0126271

1 SAIL TRAINING SHIP (AXS)

Name	No	Builders	Commissioned
TUNAS SAMUDERA	A 13	Brooke Yacht, Lowestoft	16 Oct 1989

Displacement, tons: 239 full load
Dimensions, feet (metres): 114.8 × 25.6 × 13.1 *(35 × 7.8 × 4)*
Main machinery: 2 Perkins diesels; 370 hp *(272 kW)*; 2 shafts
Speed, knots: 9
Complement: 10 plus 26 trainees
Radars: Navigation: Racal Decca; I-band.

Comment: Laid down 1 December 1988 and launched 4 August 1989. Two-masted brig manned by the Navy but used for training all sea services.

VERIFIED

TUNAS SAMUDERA   *10/2001, Chris Sattler /* 0126270

TUGS

12 HARBOUR TUGS (YTM/YTL)

TUNDA SATU 1-3	KETAM	TERITUP
SIPUT	BELAWKAS	KEMPONG
PENYU A4	SELAR	TEPURUK
SOTONG A 6		

VERIFIED

PENYU   *10/2003, Hartmut Ehlers /* 0567883

GOVERNMENT MARITIME FORCES

Notes: (1) The Malaysian Maritime Enforcement Agency (MMEA) was established in March 2005. The Agency is to operate patrol craft from the Police, Customs and Fishery Protection services and also some naval units.
(2) A fourth government agency, the Fire and Rescue Department, operates at least eight 10 m rescue craft. Helicopters include Mi-17 and Agusta Westland A 109.

POLICE

14 LANG HITAM CLASS (PBF)

LANG HITAM PZ 1	BELIAN PZ 6	HARIMAU AKAR PZ 12
LANG MALAM PZ 2	KURITA PZ 7	PERANGAN PZ 13
LANG LEBAH PZ 3	SERANGAN BATU PZ 8	MERSUJI PZ 14
LANG KUIK PZ 4	HARIMAU BINTANG PZ 9	ALU-ALU PZ 15
BALONG PZ 5	HARIMAU BELANG PZ 11	

Displacement, tons: 230 full load
Dimensions, feet (metres): 126.3 × 22.9 × 5.9 *(38.5 × 7 × 1.8)*
Main machinery: 2 MTU 20V 538 TB92 diesels; 8,360 hp(m) *(6.14 MW)* sustained; 2 shafts
Speed, knots: 35. **Range, n miles:** 1,200 at 15 kt
Complement: 38 (4 officers)
Guns: 1 Bofors 40 mm/70 (in a distinctive plastic turret).
1 Oerlikon 20 mm. 2 FN 7.62 mm MGs.
Radars: Navigation: Kelvin Hughes; I-band.

Comment: Ordered from Hong Leong-Lürssen, Butterworth, Malaysia in 1979. First delivered August 1980, last in April 1983. One deleted in 1994.

VERIFIED

LANG HITAM *9/2003, Hartmut Ehlers / 0567882*

6 BROOKE MARINE 29 METRE CLASS (PBF)

SANGITAN PX 28	DUNGUN PX 30	TUMPAT PX 32
SABAHAN PX 29	TIOMAN PX 31	SEGAMA PX 33

Displacement, tons: 114 full load
Dimensions, feet (metres): 95.1 × 19.7 × 5.6 *(29 × 6 × 1.7)*
Main machinery: 2 Paxman Valenta 6CM diesels; 2,250 hp *(1.68 MW)* sustained; 2 shafts
Speed, knots: 36. **Range, n miles:** 1,200 at 24 kt
Complement: 18 (4 officers)
Guns: 1 Oerlikon 20 mm. 2—7.62 mm MGs.

Comment: Ordered 1979 from Penang Shipbuilding Co. First delivery June 1981, last pair completed June 1982. Brooke Marine provided lead yard services.

VERIFIED

SANGITAN *1991, RM Police*

120 INSHORE/RIVER PATROL CRAFT (PBI/PBR)

Comment: Built in several batches and designs since 1964. Some are armed with 7.62 mm MGs. All have PX/PA/PC/PSC/PGR numbers. Included are 23 Simonneau SM 465 type (PC 6-28) built between January 1992 and mid-1993.

UPDATED

PSC 6 *10/2001, Chris Sattler / 0126267*

PC 6 (SIMONNEAU) *4/1997, Maritime Photographic / 0012763*

PA 20 *9/2003, Hartmut Ehlers / 0567887*

6 STAN PATROL 1500 CLASS (PBF)

Dimensions, feet (metres): 48.6 × 8.9 × 2.6 *(14.8 × 2.7 × 0.8)*
Main machinery: 4 diesels; 4,500 hp(m) *(33.1 MW)*; 4 shafts; LIPS props
Speed, knots: 55
Complement: 8
Guns: 2—12.7 mm MGs.

Comment: Built in Malaysia and completed in 1998/99. Details are not confirmed.

VERIFIED

CUSTOMS

Notes: In addition there are about 25 interceptor craft of 9 m, and 30 of 13.7 m and some inflatable chase boats.

KB 51 *2/2004*, Bob Fildes / 0587781*

027 *10/2001, Chris Sattler / 0126265*

10 PERANTAS FAST INTERCEPT CRAFT (PBF)

KB 71   +9

Displacement, tons: 16.2 full load
Dimensions, feet (metres): 54.1 × 12.8 × ? *(16.5 × 3.9 × ?)*
Main machinery: 2 MTU 12V183 TE94 diesels; 2,600 hp *(1.94 MW)*; 2 Kamewa waterjets
Speed, knots: 45. **Range, n miles:** 400 at 40 kt
Complement: 8

Comment: Built by Destination Marine Services, Port Klang. GRP hulls.

VERIFIED

KB 71 *9/2003, Hartmut Ehlers / 0567881*

4 PEMBANTERAS CLASS (PB)

Displacement, tons: 58 full load
Dimensions, feet (metres): 94.5 × 19.4 × 6.6 *(28.8 × 5.9 × 2)*
Main machinery: 2 Deutz SBA16M816C diesels; 3,140 hp(m) *(2.31 MW)*; 2 shafts
Speed, knots: 20
Complement: 8

Comment: Built at Limbungan Timor shipyard, Terengganu and completed in 1993.
VERIFIED

KA 45   9/2003, Hartmut Ehlers / 0567886

10 VOSPER 32 METRE PATROL CRAFT (PB)

JUANG K 33	— K 35	**BAYU** K 37	— K 39	— K 41
PULAI K 34	**PERAK** K 36	**HIJAU** K 38	**JERAI** K 40	— K 42

Displacement, tons: 143 full load
Dimensions, feet (metres): 106.2 × 23.6 × 5.9 *(32.4 × 7.2 × 1.8)*
Main machinery: 2 Paxman Valenta 16CM diesels; 6,650 hp *(5 MW)* sustained; 2 shafts
    1 Cummins diesel; 575 hp *(423 kW)*; 1 shaft
Speed, knots: 27; 8 on cruise diesel. **Range, n miles:** 2,000 at 8 kt
Complement: 26
Guns: 1 Oerlikon 20 mm. 2—7.62 mm MGs.
Radars: Surface search: Kelvin Hughes; I-band.

Comment: Ordered February 1981 from Malaysia Shipyard and Engineering Company with technical support from Vosper Thornycroft (Private) Ltd, Singapore. Two completed 1982, the remainder in 1983-84. Names are preceded by 'Bahtera'.
VERIFIED

JERAI   10/2003, Hartmut Ehlers / 0567885

FISHERIES DEPARTMENT

Notes: Patrol craft have distinctive thick blue and thin red diagonal bands on the hull and have been mistaken for a Coast Guard. All have P numbers. There is also a research vessel *K K Senangin II*.

PL 65   12/1999, Sattler/Steele / 0081229

P 204   12/1999, Sattler/Steele / 0081228

K K SENANGIN II   9/2003, Hartmut Ehlers / 0567880

Maldives

Country Overview

Formerly a British Protectorate, The Maldives gained independence in 1965 and a republic was established in 1968. Situated in the northern Indian Ocean, southwest of the southern tip of India, the country comprises a 468 n mile long chain of nearly 2,000 small coral islands that are grouped together into clusters of atolls. The capital and principal commercial centre is Malé and other populous atolls include Suvadiva and Tiladummati. An archipelagic state, territorial waters (12 n miles) are claimed. A 200 n mile Exclusive Economic Zone (EEZ) has been claimed although the limits have only been partly defined by boundary agreements.

Headquarters Appointments

Chief of Coast Guard:
    Colonel Moosa Ali Jaleel

Bases

Malé

Personnel

2005: 400

COAST GUARD

Notes: (1) All pennant numbers add up to seven.
(2) Two LSLs were ordered from Colombo Dockyard in March 1996. One was completed in January 2000. Displacing 33 tons, it is capable of 20 kt. Delivery of the second craft has not been confirmed.
(3) The ex-UK patrol craft *Kingfisher* was acquired by a civilian company in early 1997. It is painted white and is used as a survey ship.
(4) There are also four RIBs in service.

2 ISKANDHAR CLASS (PB)

ISKANDHAR   **GHAZEE**

Displacement, tons: 58 full load
Dimensions, feet (metres): 80.1 × 19 × 4.3 *(24.4 × 5.8 × 1.3)*
Main machinery: 2 Paxman diesels; 8,506 hp(m) *(6.26 MW)*; 2 shafts
Speed, knots: 30
Complement: 18
Guns: 2—12.7 mm MGs.
Radars: Surface search: I-band.

Comment: Ordered from Colombo Dockyard in 1997. First one delivered in 1999 and second reported delivered in 2002.
VERIFIED

GHAZEE   7/2003, A Sharma / 0561538

4 TRACKER II CLASS (PB)

KAANI 133 (ex-11)   **KUREDHI** 142 (ex-12)   **MIDHILI** 151 (ex-13)   **NIROLHU** 106 (ex-14)

Displacement, tons: 38 full load
Dimensions, feet (metres): 65.6 × 17.1 × 4.9 *(20 × 5.2 × 1.5)*
Main machinery: 2 Detroit 12V-71TA diesels; 840 hp *(627 kW)* sustained; 2 shafts
Speed, knots: 25. **Range, n miles:** 450 at 20 kt
Complement: 10
Guns: 1—12.7 mm MG. 1—7.62 mm MG.
Radars: Surface search: Kroden; I-band.

Comment: First one ordered June 1985 from Fairey Marinteknik and commissioned in April 1987. Three more acquired July 1987 ex-UK Customs craft. GRP hulls. Seven days normal endurance. Used for fishery protection and EEZ patrols.

VERIFIED

BUREVI   *7/2003, A Sharma /* 0561537

1 DAGGER CLASS (PB)

FUNA 124

Displacement, tons: 20 full load
Dimensions, feet (metres): 36.8 × 11.2 × 5 *(11.2 × 3.4 × 1.2)*
Main machinery: 2 Sabre diesels; 660 hp *(492 kW)*; 2 shafts
Speed, knots: 35
Complement: 6
Guns: 2—7.62 mm MGs.
Radars: Surface search: Furuno; I-band.

Comment: Built by Fairey Marine at Cowes, Isle of Wight and delivered in 1982.

VERIFIED

KAANI   *6/1996 /* 0081230

1 CHEVERTON CLASS (PB)

BUREVI 115 (ex-7)

Displacement, tons: 24 full load
Dimensions, feet (metres): 55.8 × 14.8 × 3.9 *(17 × 4.5 × 1.2)*
Main machinery: 2 Detroit 8V-71TI diesels; 850 hp *(634 kW)* sustained; 2 shafts
Speed, knots: 22. **Range, n miles:** 590 at 18 kt
Complement: 10
Guns: 1—12.7 mm MG. 1—7.62 mm MG.
Radars: Surface search: Kroden; I-band.

Comment: GRP hull and aluminium superstructure. Originally built for Kiribati and subsequently sold to Maldives in 1984. Has a GRP hull.

VERIFIED

FUNA   *6/1993, Maldives CG /* 0081232

Malta

Country Overview

Formerly a British colony, the Republic of Malta gained independence in 1964. Situated 45 n miles south of Sicily, the country comprises the islands of Malta (95 square miles), Gozo (26 square miles), Comino, Kemmunett, and Filfla. It has a 76 n mile coastline with the Mediterranean Sea. The capital, largest town and principal port is Valletta. Territorial seas (12 n miles) are claimed. A fishery conservation zone of 25 n miles is also claimed.

Headquarters Appointments

Officer Commanding Maritime Squadron:
  Major Martin Cauchi Inglott

General

A coastal patrol force of small craft was formed in 1971. It is manned by the 2nd Regiment of the Armed Forces of Malta and primarily employed as a Coast Guard.

Personnel

2005: 201 (11 officers)

PATROL FORCES

2 MARINE PROTECTOR CLASS (PB)

P 51   P 52

Displacement, tons: 91 full load
Dimensions, feet (metres): 86.9 × 19 × 5.2 *(26.5 × 5.8 × 1.6)*
Main machinery: 2 MTU 8V 396 TE94 diesels; 2,680 hp(m) *(1.97 MW)* sustained; 2 shafts
Speed, knots: 25. **Range, n miles:** 900 at 8 kt
Complement: 10 (1 officer)
Radars: Navigation: I-band.

Comment: *P 51* ordered on 30 July 2001 and delivered on 25 October 2002. *P 52* delivered on 7 July 2004. Built by Bollinger Shipyards to US Coast Guard specifications. The vessels are based on the hull of the Damen Stan Patrol 2600 in service with the Hong Kong police. Steel hull with GRP superstructure. A stern ramp is used for launching a 5.5 m RIB.

UPDATED

P 52
7/2004*, Armed Forces of Malta / 0589762

0 + 1 DICIOTTI CLASS (OFFSHORE PATROL VESSEL) (PBO)

P 61

Displacement, tons: 391
Dimensions, feet (metres): 175.2 × 26.6 × 17.7 *(53.4 × 8.1 × 5.4)*
Main machinery: 2 Isotto Fraschini V1716 T2 MSD diesels; 6,335 hp *(4.7 MW)*; 2 shafts
Speed, knots: 23. Range, n miles: 2,100 at 16 kt
Complement: 25 (4 officers)
Guns: 1 Otobreda 25 mm.
Radars: Surface search: E/F-band.
Navigation: I-band.
Helicopters: Platform for 1 medium.

Comment: Financed from the 4th Italo-Maltese Protocol, contract signed on 12 March 2004 with Fincantieri, Muggiano, Italy for the construction of one vessel for delivery in August 2005. The contract included a training and logistic support package. Design based on Diciotti (modified Saettia) class vessels in service with the Italian Coast Guard. Steel hull with helicopter deck and stern ramp for launching a 6.5 m RIB.

NEW ENTRY

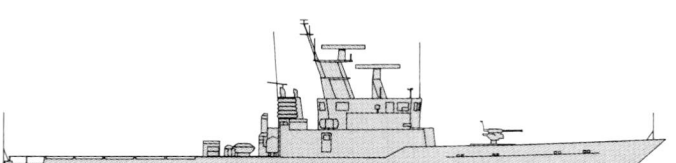

DICIOTTI CLASS   *(Scale 1 : 600), Ian Sturton /* 1044133

2 BREMSE CLASS (INSHORE PATROL CRAFT) (PBI)

P 32 (ex-*G 33/GS 20*)   P 33 (ex-*G 22/GS 22*)

Displacement, tons: 42 full load
Dimensions, feet (metres): 74.1 × 15.4 × 3.6 *(22.6 × 4.7 × 1.1)*
Main machinery: 2 Iveco diesels; 1,000 hp(m) *(745 kW)*; 2 shafts
Speed, knots: 17
Complement: 9
Guns: 1—12.7 mm MG.
Radars: Surface search: Racal 1290A; I-band.

Comment: Built in 1971-72 for the ex-GDR GBK. Transferred from Germany in mid-1992. Others of the class acquired by Tunisia. *P 32* started mid-life upgrade in 2004.

UPDATED

P 33   *11/2000, Lawrence Dalli /* 0114529

2 SWIFT CLASS (HARBOUR PATROL CRAFT) (YP)

P 23 (ex-*C 6823*)   P 24 (ex-*C 6824*)

Displacement, tons: 22.5 full load
Dimensions, feet (metres): 50 × 13 × 4.9 *(15.6 × 4 × 1.5)*
Main machinery: 2 GM 12V-71 diesels; 680 hp *(507 kW)* sustained; 2 shafts
Speed, knots: 25. Range, n miles: 400 at 18 kt
Complement: 6
Guns: 1—12.7 mm MG.
Radars: Surface search: Furuno 1040; I-band.

Comment: Built by Sewart Seacraft Ltd in 1967. Transferred from US in February 1971. Have an operational endurance of about 24 hours. Modernised in Malta in 1998/99.

UPDATED

P 24   *6/2001, Hartmut Ehlers /* 0524999

2 SUPERVITTORIA 800 CLASS (SAR)

MELITA I   MELITA II

Displacement, tons: 12.5 full load
Dimensions, feet (metres): 37.7 × 16.1 × 2.6 *(11.5 × 4.9 × 0.8)*
Main machinery: 2 Cummins 6CTA 8.3 DIAMONS; 840 hp(m) *(618 kW)*; 2 Kamewa FF310 waterjets
Speed, knots: 34. Range, n miles: 160 at 34 kt
Complement: 4
Radars: Surface search: Raytheon Pathfinder SL 70; I-band.

Comment: Built in 1998 by Vittoria Naval Shipyard, Italy, for the Civil Protection Department of Malta. Transferred to the Armed Forces of Malta (AFM) in May 1999 for search and rescue duties. Although still the property of the Civil Protection Department, the Melita I and II are operated and maintained by the Maritime Squadron of the AFM.

VERIFIED

MELITA II   *11/2000, Lawrence Dalli /* 0114532

LAND-BASED MARITIME AIRCRAFT

Notes: The Armed Forces of Malta operate two Britten-Norman BN-2B maritime patrol aircraft, five BAe Bulldog T. Mk 1 observation aircraft, two Nardi-MD NH 500HM, five SA.316B/D Alouette III and two AB-47G-2 helicopters.

NH 500   *4/2002, Adolfo Ortigueira Gil /* 0568875

ALOUETTE III   *2001, Pierre Gillard /* 0114533

BN ISLANDER   *2001, Douglas-John Falzon /* 0114534

Marshall Islands

Country Overview

The Republic of the Marshall Islands was a US-administered UN Trust territory from 1947 before becoming a self-governing republic in 1979. In 1986, a Compact of Free Association, delegating to the US the responsibility for defence and foreign affairs, came into effect. The country consists of some 1,200 atolls and reefs in the central Pacific. There are two main island groups: the Ratak and Ralik chains. Majuro is the capital island. Kwajalein is the largest atoll and is leased as a US missile test range. Bikini and Enewetak are former US nuclear test sites.

An archipelagic state, territorial seas (12 n miles) are claimed. An Exclusive Economic Zone (EEZ) (200 n miles) is also claimed but limits have not been fully defined.

Headquarters Appointments

Chief of Surveillance:
  Major Thomas Heine
Maritime Surveillance Adviser:
  Lieutenant Commander P McCarthy, RAN

Personnel

2005: 30

Bases

Majuro

PATROL FORCES

1 PACIFIC CLASS (LARGE PATROL CRAFT) (PB)

Name	No	Builders	Commissioned
LOMOR	03	Australian Shipbuilding Industries	29 June 1991

Displacement, tons: 162 full load
Dimensions, feet (metres): 103.3 × 26.6 × 6.9 *(31.5 × 8.1 × 2.1)*
Main machinery: 2 Caterpillar 3516TA diesels; 4,400 hp *(3.3 MW)* sustained; 2 shafts
Speed, knots: 20. **Range, n miles:** 2,500 at 12 kt
Complement: 17 (3 officers)
Guns: 1—12.7 mm MG.
Radars: Surface search: Furuno 8111; I-band.

Comment: The 14th craft to be built in this series for a number of Pacific Island coast guards. Ordered in 1989. Following the decision by the Australian government to extend the Pacific Patrol Boat project to a 30-year life for each boat, *Lomor* will undergo a life-extension refit in 2009 having undergone a half-life refit in 1999.

UPDATED

LOMOR
2001, Marshall Islands Sea Patrol Force / 0109942

Mauritania
MARINE MAURITANIENNE

Country Overview

A former French colony, The Islamic Republic of Mauritania gained full independence in 1960. With an area of 397,955 square miles, it is situated in northwestern Africa and has borders to the north with western Sahara and Algeria, to the east with Mali and to the south with Senegal. It has a 405 n mile coastline with the Atlantic Ocean. The capital and largest city is Nouakchott while Nouadhibou is the principal port. Territorial

seas (12 n miles) are claimed but while it has claimed a 200 n mile Exclusive Economic Zone (EEZ), its limits have not been defined by boundary agreements.

Headquarters Appointments

Commander of Navy:
  Colonel A Ould Yahya

Personnel

(a)   2005: 500 (40 officers) plus 200 marines
(b)   Voluntary service

Bases

Port Etienne, Nouadhibou
Port Friendship, Nouakchott

PATROL FORCES

Notes: An 18 m patrol vessel *Yacoub Ould Rajel* was donated by the European Union. Constructed by Raidco Marine, Lorient, it was delivered in 2000 and is used for fishery protection.

1 OPV 54 CLASS (PBO)

Name	No	Builders	Launched	Commissioned
ABOUBEKR BEN AMER	P 541	Leroux & Lotz, Lorient	17 Dec 1993	7 Apr 1994

Displacement, tons: 374 full load
Dimensions, feet (metres): 177.2 × 32.8 × 9.2 *(54 × 10 × 2.8)*
Main machinery: 2 MTU 16V 396 TE94 diesels; 5,712 hp(m) *(4.2 MW)* sustained; 2 auxiliary motors; 250 hp(m) *(184 kW)*; 2 shafts; cp props
Speed, knots: 23 (8 on motors). **Range, n miles:** 4,500 at 12 kt
Complement: 21 (3 officers)
Guns: 2—12.7 mm MGs.
Radars: Surface search: Racal Decca Bridgemaster 250; I-band.

Comment: Ordered in September 1992. This is the prototype to a Serter design of three similar craft built for the French Navy. Stern ramp for a 30 kt RIB. Option on a second of class not taken up. Refitted at Lorient 2001.

VERIFIED

1 PATRA CLASS (LARGE PATROL CRAFT) (PB)

Name	No	Builders	Commissioned
ENNASR (ex-*Le Dix Juillet*, ex-*Rapière*)	P 411	Auroux, Arcachon	14 May 1982

Displacement, tons: 147.5 full load
Dimensions, feet (metres): 132.5 × 19.4 × 5.2 *(40.4 × 5.9 × 1.6)*
Main machinery: 2 Wärtsilä UD 33 V12 diesels; 4,340 hp(m) *(3.2 MW)* sustained; 2 shafts
Speed, knots: 26.3. **Range, n miles:** 1,750 at 10 kt
Complement: 20 (2 officers)
Guns: 1 Bofors 40 mm/60. 1 Oerlikon 20 mm. 2—12.7 mm MGs.
Radars: Surface search: Racal/Decca 1226; I-band.

Comment: Originally built as a private venture by Auroux. Carried out trials with French crew as *Rapière*. Laid down 15 February 1980, launched 3 June 1981, commissioned for trials 1 November 1981. Transferred to Mauritania in 1982. Re-engined in 1993-94.

UPDATED

ABOUBEKR BEN AMER
7/2001, Peron/Marsan / 0137787

ENNASR
4/1998 / 0052598

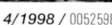

1 LARGE PATROL CRAFT (PBO)

Name	No	Builders	Commissioned
VOUM-LEGLEITA (ex-*Poseidon*)	B 551 (ex-A 12)	Bazán	8 Aug 1964

Displacement, tons: 1,069 full load
Dimensions, feet (metres): 183.5 × 32.8 × 13.1 *(55.9 × 10 × 4)*
Main machinery: 2 Sulzer diesels; 3,200 hp *(2.53 MW)*; 1 shaft; cp prop
Speed, knots: 15. **Range, n miles:** 4,640 at 14 kt
Complement: 60
Guns: 2 Oerlikon 20 mm.
Radars: Navigation: 2 Decca TM 626; I-band.

Comment: Ocean going tug transferred from Spain in January 2000, about a year later than planned. Used primarily as an OPV and for fishery protection.   **VERIFIED**

VOUM-LEGLEITA   *1/2000, Diego Quevedo /* 0081240

1 HUANGPU CLASS PB)

LIMAM EL HADRAMI P 601

Displacement, tons: 430 full load
Dimensions, feet (metres): 196.8 × 26.9 × 14.8 *(60.0 × 8.2 × 4.5)*
Main machinery: 3 MTU 12V 4000 diesels; 3 shafts
Speed, knots: 20
Guns: 4—37 mm.
Radars: Navigation: I-band.

Comment: Delivered from China on 20 April 2002.   **NEW ENTRY**

1 ARGUIN CLASS (PBO)

Name	Builders	Commissioned
ARGUIN	Fassmer Werft, Berne/Motzen, Germany	17 July 2000

Measurement, tons: 1,000 dwt
Dimensions, feet (metres): 178.8 × 35.8 × 14.8 *(54.5 × 10.9 × 4.5)*
Main machinery: 2 MaK 6M20 diesels; 2,735 hp *(2.04 MW)*; 1 shaft; cp prop
Speed, knots: 16.5. **Range, n miles:** 15,000 at 12 kt
Complement: 13

Comment: Ordered in 1998. Hull construction at Yantar, Kaliningrad. Steel hull and superstructure. Equipped with interception craft on centreline ramp in mother-daughter configuration.   **UPDATED**

ARGUIN   *7/2000, Fassmer Werft /* 1044268

4 MANDOVI CLASS (INSHORE PATROL CRAFT) (PB)

Displacement, tons: 15 full load
Dimensions, feet (metres): 49.2 × 11.8 × 2.6 *(15 × 3.6 × 0.8)*
Main machinery: 2 Deutz MWM TBD232V12 Marine diesels; 750 hp(m) *(551 kW)*; 2 Hamilton water-jets
Speed, knots: 24. **Range, n miles:** 250 at 14 kt
Complement: 8
Guns: 1—7.62 mm MG.
Radars: Navigation: Furuno FR 8030; I-band.

Comment: Built by Garden Reach, Calcutta and delivered from India in 1990. Same type acquired by Mauritius. Some may not be operational.   **VERIFIED**

LAND-BASED MARITIME AIRCRAFT

Notes: There are also two Cessna 337F.

Numbers/Type: 2 Piper Cheyenne II.
Operational speed: 283 kt *(524 km/h)*.
Service ceiling: 31,600 ft *(9,630 m)*.
Range: 1,510 n miles *(2,796 km)*.
Role/Weapon systems: Coastal surveillance and EEZ protection acquired 1981. Sensors: Bendix 1400 weather radar; cameras. Weapons: Unarmed.   **VERIFIED**

Mauritius

Country Overview

A former British colony, the Republic of Mauritius gained independence in 1968 and became a republic in 1992. Situated in the western Indian Ocean, east of Madagascar, it comprises the islands of Mauritius (720 square miles), Rodrigues (42 square miles), the Agalega islands to the north and the St Brandon Group (also known as the Cargados Carajos Shoals) to the northeast. The capital, largest town and principal port is Port Louis. Territorial seas (12 n miles) are claimed but, while it has declared a 200 n mile Exclusive Economic Zone (EEZ), the claim is complicated by disputes over the sovereignty of Tromelin Island (France) and Diego Garcia (UK).

Headquarters Appointments

Commandant National Coast Guard:
  M A Hampiholi

Bases

Port Louis (plus 24 manned CG stations)

Personnel

2005: 750 (including officers on deputation)

Maritime Aircraft

2 Dornier 228 (MPCG 01).
1 Britten-Norman BN-2-T Defender.

COAST GUARD

Notes: There are approximately 60 inshore craft (RHIBs, glass fibre boats and so on) in addition to those listed.

1 GUARDIAN CLASS (PSOH)

Name	No	Builders	Launched	Commissioned
VIGILANT	21	Talcahuano Yard, Chile	6 Dec 1995	10 May 1996

Displacement, tons: 1,650 full load
Dimensions, feet (metres): 246.1 × 45.9 × 12.8 *(75 × 14 × 3.9)*
Main machinery: 4 Caterpillar 3516 diesels; 11,530 hp *(8.6 MW)*; 2 shafts; cp props; bow thruster; 671 hp *(500 kW)*
Speed, knots: 22. **Range, n miles:** 6,500 at 19 kt
Complement: 57 (11 officers) plus 20 spare
Guns: 2 Bofors 40 mm/70 (1 twin). 2—12.7 mm MGs.
Radars: Surface search: Kelvin Hughes; I-band.
Helicopters: 1 light.

Comment: Contract signed with the Western Canada Marine Group in March 1994. Keel was laid in April 1994. All-steel construction. The ship can be operated by a crew of 18. Full helicopter facilities are included in the design which is based on a Canadian Fisheries vessel *Leonard J Cowley*. Not operational from 1998-2000 due to shaft problems.   **UPDATED**

VIGILANT
2/2001, Sattler/Steele / 0114366

1 SDB Mk 3 CLASS (PB)

GUARDIAN

Displacement, tons: 210 full load
Dimensions, feet (metres): 124 × 24.6 × 6.2 *(37.8 × 7.5 × 1.9)*
Main machinery: 2 MTU 16V 538 TB92 diesels; 6,820 hp(m) *(5 MW)* sustained; 2 shafts
Speed, knots: 21
Complement: 32
Guns: 2 Bofors 40 mm/60; 120 rds/min to 10 km *(5.5 n miles)*; weight of shell 0.89 kg.
Radars: Surface search: Furuno FK 1505 DA; I-band.

Comment: Transferred from Indian Navy in 1993. Built by Garden Reach, Calcutta in 1984.
UPDATED

GUARDIAN   7/2003, Arjun Sarup / 0568319

2 ZHUK (TYPE 1400M) CLASS (PB)

RESCUER   RETRIEVER

Displacement, tons: 39 full load
Dimensions, feet (metres): 78.7 × 16.4 × 3.9 *(24 × 5 × 1.2)*
Main machinery: 2 M 401B diesels; 2,200 hp(m) *(1.6 MW)* sustained; 2 shafts
Speed, knots: 30. **Range, n miles:** 1,100 at 15 kt
Complement: 14
Guns: 4—12.7 mm (2 twin) MGs.
Radars: Surface search: Spin Trough; I-band.

Comment: Acquired from the USSR on 3 December 1989.
UPDATED

RETRIEVER   7/2003, Arjun Sarup / 0568318

1 P-2000 CLASS (PB)

OBSERVER (ex-C 39)

Displacement, tons: 40 full load
Dimensions, feet (metres): 68.2 × 19 × 5.9 *(20.8 × 5.8 × 1.8)*
Main machinery: 2 Deutz MWM TBD234 V12 diesels; 1,646 hp(m) *(1.21 MW)* sustained; 1 Deutz MWM TBD234 V8 diesel; 550 hp(m) *(404 kW)* sustained; 3 Hamilton 402 waterjets
Speed, knots: 25. **Range, n miles:** 600 at 15 kt
Complement: 8 (1 officer)
Guns: 1—7.62 mm MG.
Radars: Navigation: Furuno; I-band.

Comment: Leased from the Indian Coast Guard in 2001. Originally commissioned in 1997, one of ten ordered from Anderson Marine, Goa in September 1990 to a P-2000 design by Amgram, similar to Archer class. GRP hull. Built at Goa.   *VERIFIED*

OBSERVER   4/2001, Arjun Sarup / 0568317

4 HEAVY DUTY BOATS (PBI)

HDB 01-04

Displacement, tons: 5
Dimensions, feet (metres): 29.25 × 11.5 × 1.5 *(8.9 × 3.5 × 0.45)*
Main machinery: 2 Johnson outboard motors; 400 hp
Speed, knots: 45. **Range, n miles:** 300 at 35 kt
Complement: 4 (plus 14 passengers)

Comment: An initial order of four boats supplied by M/S Praga Marine, India in 2000. Option for six additional boats.
UPDATED

HEAVY DUTY BOAT   2000, Mauritius Coast Guard / 0105127

8 KAY MARINE HEAVY DUTY BOATS (PBI)

HDB 5-12

Displacement, tons: ?
Dimensions, feet (metres): 29.0 × 10.5 × 1.5 *(8.85 × 3.21 × 0.45)*
Main machinery: 2 Suzuki (1 twin) outboard motors; 450 hp
Speed, knots: 40
Complement: 18 including passengers

Comment: Acquired from Kay Marine Malaysia in November 2002. Deep Vee monohull of aluminium construction.
UPDATED

KAY MARINE HDB 05   8/2003, Arjun Sarup / 0568316

4 HALMATIC HEAVY DUTY BOATS (PBI)

HDB 13-16

Displacement, tons: 6
Dimensions, feet (metres): 30.2 × 10.2 × 3.3 *(9.2 × 3.1 × 1.0)*
Main machinery: 2 Yamaha V6 outboard motors; 400 hp
Speed, knots: 35
Complement: 18 including passengers

Comment: Acquired from Halmatic Ltd UK in June 2003.
UPDATED

HALMATIC HDB 16   7/2003, Arjun Sarup / 0568315

2 ROVER PATROL BOATS (PBI)

ROVER 1   ROVER 2

Displacement, tons: 3.5
Dimensions, feet (metres): 21.3 × 6.5 × 1.6 *(6.5 × 2.0 × 0.5)*
Main machinery: 1 Mariner outboard motor; 90 hp

Comment: Donated by Australia in 1988-89.

VERIFIED

6 TORNADO VIKING 580 RHIB (PBI)

Dimensions, feet (metres): 18.7 × 8.5 × 2.5 *(5.7 × 2.6 × 0.75)*
Main machinery: 1 Yamaha outboard motor; 90 hp
Speed, knots: 35
Complement: 10 including passengers

Comment: Acquired in 2004.

NEW ENTRY

ROVER 2   4/2003, Arjun Sarup / 0568314

VIKING 580   6/2004 *, Mauritius Coast Guard / 0589763

Mexico
MARINA NACIONAL

Country Overview

The United Mexican States is a federal republic in North America. A total land area of 756,066 square miles includes a number of offshore islands. Bordered to the north by the United States and to the south by Belize and Guatemala, it has a 1,382 n mile coastline with the Caribbean and Gulf of Mexico and 3,656 n mile coastline with the Pacific Ocean. The capital and largest city is Mexico City while the principal ports are Acapulco (Pacific) and Veracruz (Gulf of Mexico). Territorial seas (12 n miles) are claimed. A 200 n mile EEZ has also been claimed but the limits have not been fully defined by boundary agreements.

Headquarters Appointments

Secretary of the Navy:
    Admiral Marco Antonio Peyrot Gonzalez
Under-Secretary of the Navy:
    Admiral Armando Sanchez Moreno
Inspector General of the Navy:
    Admiral Enrique Ramos Martinez
Chief of the Naval Staff:
    Vice Admiral Alberto Castro Rosas

Flag Officers

Commander in Chief, Gulf and Caribbean:
    Admiral Daniel Zamora Contreras
Commander in Chief, Pacific:
    Admiral Casimiro Armando Martinez Pretelin

Personnel

(a)   2005: 46,972 officers and men (including 946 Naval Air Force and 11,385 Marines)
(b)   Military service

Naval Bases and Commands

The Naval Command is split between the Pacific and Gulf areas each with a Commander-in-Chief with HQs at Manzanillo and Tuxpan respectively. Each area has three naval Regions which are further subdivided into Zones (9), Sectors (11) and Subsectors (7). There is a Central Naval Region that has an HQ in Mexico City.

Gulf Area
First Naval Region (HQ Tampico, Tamaulipas).
    I Naval Zone (HQ Tampico, Tamaulipas).
        Naval Subsector (Matamoros, Tamaulipas).
    III Naval Zone (HQ Veracruz, Veracruz).
        Naval Sector (HQ Tuxpan, Veracruz).
        Naval Sector (HQ Coatzacoalcos, Veracruz).
Third Naval Region (HQ Lerma, Campeche).
    V Naval Zone (HQ Cuidad del Carmen, Campeche).
        Naval Sector (HQ Lerma, Campeche).
            Naval Subsector (HQ Dos Bocas, Tabasco).
            Naval Subsector (HQ Frontera, Tabasco).
Fifth Naval Region (HQ Yukalpeten, Yucatan).
    VII Naval Zone (HQ Isla Mujeres, Quintana Roo).
        Naval Sector (HQ Chetumal, Quintana Roo).
        Naval Sector (HQ Yukalpeten, Yucatan).
            Naval Subsector (HQ Isla Cozumel, Quintana Roo).

Pacific Area
Second Naval Region (HQ Mazatlan, Sinaloa).
    II Naval Zone (HQ Ensenada, Baja California).
    IV Naval Zone (HQ Guaymas, Sonora).
        Naval Sector (HQ Mazatlan, Sinaloa).
        Naval Sector (HQ Topolobampo, Sinaloa).
        Naval Sector (HQ La Paz, Baja California Sur).
            Naval Subsector (HQ Puerto Cortes, Baja California Sur).
            Naval Subsector (HQ Puerto Peñasco, Sonora).
Fourth Naval Region (HQ Manzanillo, Colima).
    VI Naval Zone (HQ Lazaro Cardenas, Michoacán).
        Naval Sector (HQ Puerto Vallarta, Jalisco).
        Naval Sector (HQ Manzanillo, Colima).
            Naval Subsector (HQ San Blas, Nayarit).
Sixth Naval Region (HQ Acapulco, Guerrero).
    VIII Naval Zone (HQ Acapulco, Guerrero).
    X Naval Zone (HQ Salina Cruz, Oaxaca).
        Naval Sector (HQ Puerto Madero, Chiapas).
Central Naval Region (HQ Mexico, DF).

Naval Air Force

Six naval air bases at Mexico City, Veracruz, Campeche, Chetumal, Tapachula and La Paz; there are two Naval Air Stations at Guaymas and Tampico.

Marine Forces

There are two Amphibious Reaction Forces, each one based in Manzanillo and Tuxpan; one Parachute Battalion, two Infantry Battalions and one Presidential Guards Battalion in Mexico City.

Strength of the Fleet

Type	Active	Building
Destroyers	1	—
Frigates	7	—
Gunships	19	2
Large Patrol Craft	26	—
Coast Guard	11	—
Coastal and River Patrol Craft	60	12
Survey Ships	7	—
Support Ships	7	—
Tankers	2	—
Sail Training Ship	1	—

Names and Pennant Numbers

Many of the ship names and pennant numbers were changed in early 1994 and again in 2001. Destroyers and frigates are named after Aztec emperors and forerunners of the Independence War (1810-1825). Gunboats are named after naval and military heroes.

DELETIONS

Destroyers

2002   *Quetzalcoatl*

Patrol Forces

2003   *Andrés Quintana Roo, Manel Ramos Arizpe, José Maria Izazaga, Juan Bautista Morales, José Maria Mata, Pastor Rollaix, Luis Manue Rojas, Ignacio Zaagoza, Campeche, Margarita Maza de Juarez, Leandro Valle, Sebastian Lerdo de Tejada, Ignacio de la Llave, Ignacio Manuel Altamirano, Felipe Xicoténcatl, Juan Aldama*
2004   *Laguna de Tamiahua, Laguna de Lagartos, Lago de Patzcuaro, Benito Juarez*

Auxiliaries

2002   *Iguala*
2004   *Rio Panuco, Vicente Guerrero*

PENNANT LIST

Destroyers

D 102	Netzahualcoyotl
D 111	Comodoro Manuel Azueta

Frigates

F 201	Nicolas Bravo
F 202	Hermengildo Galeana
F 211	Ignacio Allende
F 212	Mariano Abasolo
F 213	Guadaloupe Victoria
F 214	Francisco Javier Mina

Patrol Forces

A 301	Huracán
A 302	Tormenta

PC 202	Cordova
PC 206	Ignacio López Rayón
PC 207	Manuel Crescencio Rejon
PC 208	Juan Antomio de la Fuente
PC 209	Leon Guzman
PC 210	Ignacio Ramirez
PC 211	Ignacio Mariscal
PC 212	Heriberto Jara Corona
PC 214	Colima
PC 215	José Joaquin Fernandez de Lizardi
PC 216	Francisco J Mugica
PC 218	José Maria del Castillo Velasco
PC 220	José Natividad Macias
PC 223	Tamaulipas
PC 224	Yucatan
PC 225	Tabasco
PC 226	Cochimie
PC 228	Puebla

PC 230	Leon Vicario
PC 231	Josefa Ortiz de Dominguez
PC 241	Démocrata
PC 271	Cabo Corrientes
PC 272	Cabo Corzo
PC 273	Cabo Catoche
PC 281	Punta Morro
PC 282	Punta Mastun
PI 1101	Polaris
PI 1102	Sirius
PI 1103	Capella
PI 1104	Canopus
PI 1105	Vega
PI 1106	Achernar
PI 1107	Rigel
PI 1108	Arcturus
PI 1109	Alpheratz
PI 1110	Procyon

PI 1111	Avior
PI 1112	Deneb
PI 1113	Formalhaut
PI 1114	Pollux
PI 1115	Regulus
PI 1116	Acrux
PI 1117	Spica
PI 1118	Hadar
PI 1119	Shaula
PI 1120	Mirfak
PI 1121	Ankaa
PI 1122	Bellatrix
PI 1123	Elnath
PI 1124	Alnilam
PI 1125	Peacock
PI 1126	Betelgeuse
PI 1127	Adhara
PI 1128	Alioth

Patrol Forces — *continued*		PO 102		PO 154	Veracruz	A 412	Usumacinta
		PO 103	Juan de la Barrera	PO 161	Oaxaca (bldg)	AMP 01	Huasteco
PI 1129	Rasalhague	PO 104	Mariano Escobedo	PO 162	Baja California (bldg)	AMP 02	Zapoteco
PI 1130	Nunki	PO 106	Manuel Doblado	PR 293	Laguna de Cuyutlan	ARE 01	Otomi
PI 1131	Hamal	PO 108	Santos Degollado	PR 294	Laguna de Alvarado	ARE 02	Yaqui
PI 1132	Suhail	PO 109	Juan Alvares	PR 295	Laguna de Catemaco	ARE 03	Seri
PI 1133	Dubhe	PO 110	Manuel Gutierrez Zamora	PR 310	Arrecife Rizo	ARE 04	Cora
PI 1134	Denebola	PO 112	Valentin Gomez Farias	PR 311	Arrecife Pajaros	ARE 05	Iztaccihuatl
PI 1135	Alkaid	PO 113	Francisco Zarco	PR 312	Arrecife de Emmedio	ARE 06	Popocatepetl
PI 1136	Alphecca	PO 114	Ignacio Vallarta	PR 313	Arrecife de Hornos	ARE 07	Citlaltepl
PI 1137	Eltanin	PO 117	Jesus Gonzalez Ortega	PR 322	Lago de Chapala	ARE 08	Xinantecatl
PI 1138	Kochab	PO 121	Mariano Matamoros			ARE 09	Matlalcueye
PI 1139	Enif	PO 122	Cadete Virgilio Uribe	**Survey and Research Ships**		ARE 10	Tlaloc
PI 1140	Schedar	PO 123	Teniente José Azueta			ATQ 01	Aguascalientes
PI 1141	Markab		Capitán de Fragata Pedro Sáinz de	BI 01	Alejandro de Humboldt	ATQ 02	Tlaxcala
PI 1142	Megrez	PO 124	Baranda	BI 02	Onjuku	ATR 01	Maya
PI 1143	Mizar	PO 125	Comodoro Carlos Castillo Bretón	BI 03	Altair	ATR 03	Tarasco
PI 1144	Phekda	PO 126	Vicealmirante Othón P Blanco	BI 04	Antares		
PI 1201	Isla Coronado		Contralmirante Angel Ortiz	BI 05	Rio Suchiate	**Training Ships**	
PI 1202	Isla Lobos		Monasterio	BI 06	Rio Ondo		
PI 1203	Isla Guadalupe	PO 131	Capitán de Navio Sebastian José	BI 07	Moctezuma II	BE 01	Cuauhtemoc
PI 1204	Isla Cozumel		Holzinger	BI 08	Arrecife Alacrán	BE 02	Aldebarán
PI 1301	Acuario	PO 132	Capitán de Navio Blas Godinez	BI 09	Arrecife Rizo		
PI 1302	Aguila	PO 133	Brigadier José Mariá de la Vega	BI 10	Arrecife Cabezo		
PI 1303	Aries	PO 134	General Felipe B Berriozábal	BI 11	Arrecife Anegada de Adentro		
PI 1304	Auriga	PO 141	Justo Sierra Mendez				
PI 1305	Cancer	PO 143	Guillermo Prieto	**Auxiliaries**			
PI 1306	Capricorno	PO 144	Matias Romero				
PI 1307	Centauro (bldg)	PO 151	Durango	A 402	Manzanillo		
PI 1308	Geminis (bldg)	PO 152	Sonora	A 411	Rio Papaloapan		
		PO 153	Guanajuato				

DESTROYERS

1 QUETZALCOATL (GEARING FRAM I) CLASS (DDH)

Name
NETZAHUALCOYOTL (ex-*Steinaker* DD 863)

No	*Builders*	*Laid down*	*Launched*	*Commissioned*
D 102 (ex-E 11, ex-E 04)	Bethlehem, Staten Island	1 Sep 1944	13 Feb 1945	26 May 1945

Displacement, tons: 3,030 standard; 3,690 full load
Dimensions, feet (metres): 390.2 × 41.9 × 15
   (118.7 × 12.5 × 4.6)
Main machinery: 4 Babcock & Wilcox boilers; 600 psi *(43.3 kg/cm²)*; 850°F *(454°C)*; 2 GE turbines; 60,000 hp *(45 MW)*;
   2 shafts
Speed, knots: 15. **Range, n miles:** 5,800 at 15 kt
Complement: 250

Guns: 4 USN 5 in *(127 mm)*/38 (2 twin) Mk 38 **❶**; 15 rds/min to
   17 km *(9.3 n miles)* anti-surface; 11 km *(5.9 n miles)* anti-
   aircraft; weight of shell 25 kg.
Countermeasures: ESM: WLR-1; radar warning.
Weapons control: Mk 37 GFCS. Mk 112 TFCS.
Radars: Air search: Lockheed SPS-40; B-band (E 10).
   Westinghouse SPS-29 **❷**; B/C-band (E 11).
   Surface search: Kelvin Hughes 17/9 **❸**; I-band.
   Navigation: Marconi LN66; I-band.
   Fire control: Western Electric Mk 12/22 **❹**; I/J-band.

Helicopters: 1 MBB BO 105 CB **❺**.

Programmes: Transferred from US by sale 24 February 1982.
Modernisation: A Bofors 57 mm gun was mounted between the
   torpedo tubes in B gun position in 1993 but removed in 2002.
   ASROC, torpedo tubes and sonar removed in 1996, and the
   flight deck slightly extended. New topmast and search radar
   also fitted in 1996.
Structure: The devices on top of the funnel are to reduce IR
   signature.
Operational: Top speed much reduced from the original 32 kt.
   Helicopter seldom carried. Pennant number changed in 2001.
   Based at Manzanillo.

UPDATED

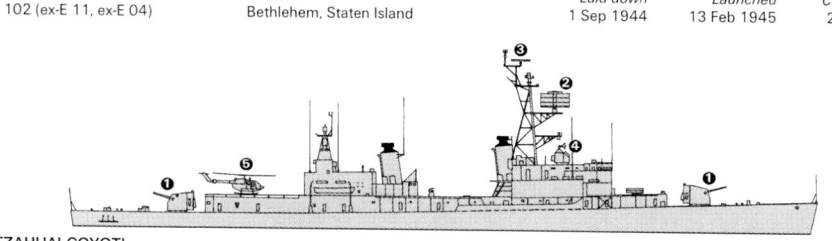

NETZAHUALCOYOTL

(Scale 1 : 1,200), Ian Sturton / 0587564

NETZAHUALCOYOTL

6/2004*, Mexican Navy / 0589779

FRIGATES

1 MANUEL AZUETA (EDSALL) CLASS (FF/AX)

Name
COMODORO MANUEL AZUETA (ex-*Hurst* DE 250)

No	*Builders*	*Laid down*	*Launched*	*Commissioned*
D 111 (ex-E 30, ex-A 06)	Brown SB Co, Houston, TX	27 Jan 1943	14 Apr 1943	30 Aug 1943

Displacement, tons: 1,400 standard; 1,850 full load
Dimensions, feet (metres): 302.7 × 36.6 × 13
   (92.3 × 11.3 × 4)
Main machinery: 4 Fairbanks-Morse 38D8-1/8-10 diesels;
   7,080 hp *(5.3 MW)* sustained; 2 shafts
Speed, knots: 12. **Range, n miles:** 13,000 at 12 kt
Complement: 216 (15 officers)

Guns: 2 USN 3 in *(76 mm)*/50; 20 rds/min to 12 km *(6.6 n
   miles)*; weight of shell 6 kg.
   8 Bofors 40 mm/60 (1 quad, 2 twin) Mk 2 and Mk 1; 120 rds/
   min to 10 km *(5.5 n miles)*; weight of shell 0.89 kg.
   2 Oerlikon 20 mm. 2—37 mm saluting guns.
Weapons control: Mk 52 (for 3 in); Mk 51 Mod 2 (for 40 mm).
Radars: Surface search: Kelvin Hughes Type 17; I-band.
   Navigation: Kelvin Hughes Type 14; I-band.

Programmes: Transferred from US 1 October 1973.
Modernisation: OTO Melara 76 mm gun fitted in 1995 but
   subsequently removed and US 3 in gun restored.
Operational: Employed as training ship and based at Tuxpan.
   A/S weapons and sensors removed. Speed much reduced.
   Pennant number changed in 2001.

UPDATED

COMODORO MANUEL AZUETA (old number)
10/1998, E & M Laursen / 0052604

4 ALLENDE (KNOX) CLASS (FFHM)

Name	No	Builders	Laid down	Launched	Commissioned
IGNACIO ALLENDE (ex-Stein)	F 211 (ex-E 50, ex-FF 1065)	Lockheed	1 June 1970	19 Dec 1970	8 Jan 1972
MARIANO ABASOLO (ex-Marvin Shields)	F 212 (ex-E 51, ex-FF 1066)	Todd Shipyards	12 Apr 1968	23 Oct 1969	10 Apr 1971
GUADALOUPE VICTORIA (ex-Pharris)	F 213 (ex-E 52, ex-FF 1094)	Avondale Shipyards	11 Feb 1972	16 Dec 1972	26 Jan 1974
FRANCISCO JAVIER MINA (ex-Whipple)	F 214 (ex-FF 1062)	Todd Shipyards	24 Apr 1967	12 Apr 1968	22 Aug 1970

Displacement, tons: 3,011 standard; 4,260 full load
Dimensions, feet (metres): 439.6 × 46.8 × 15; 24.8 (sonar)
  (134 × 14.3 × 4.6; 7.8)
Main machinery: 2 Combustion Engineering/Babcock & Wilcox
  boilers; 1,200 psi *(84.4 kg/cm²)*; 950°F *(510°C)*;
  1 Westinghouse turbine; 35,000 hp *(26 MW)*; 1 shaft
Speed, knots: 27. **Range, n miles:** 4,000 at 22 kt on 1 boiler
Complement: 288 (20 officers)

Missiles: SAM: 1 Mk 25 launcher for Sea Sparrow (in F 211) ❶
  (see *Structure*).
  SA-N-10; IR homing to 5 km *(2.7 n miles)* at 1.7 Mach; warhead
  1.5 kg.
A/S: Honeywell ASROC Mk 16 octuple launcher with reload
  system (has 2 cells modified to fire Harpoon) ❷; inertial
  guidance to 1.6-10 km *(1-5.4 n miles)*; payload Mk 46.
Guns: 1 FMC 5 in *(127 mm)*/54 Mk 42 Mod 9 ❸; 20-40 rds/min
  to 24 km *(13 n miles)* anti-surface; 14 km *(7.7 n miles)* anti-
  aircraft; weight of shell 32 kg.
Torpedoes: 4—324 mm Mk 32 (2 twin) fixed tubes ❹. 22
  Honeywell Mk 46; anti-submarine; active/passive homing to
  11 km *(5.9 n miles)* at 40 kt; warhead 44 kg.
Countermeasures: Decoys: 2 Loral Hycor SRBOC 6-barrelled
  fixed Mk 36 ❺; IR flares and chaff to 4 km *(2.2 n miles)*. T Mk-6
  Fanfare/SLQ-25 Nixie; torpedo decoy. Prairie Masker hull and
  blade rate noise suppression.
  ESM: SLQ-32(V)2 ❻; intercept.
Weapons control: Mk 68 Mod 3 GFCS. Mk 114 Mod 6 ASW FCS.
  Mk 1 target designation system. MMS target acquisition sight
  (for mines, small craft and low flying aircraft).

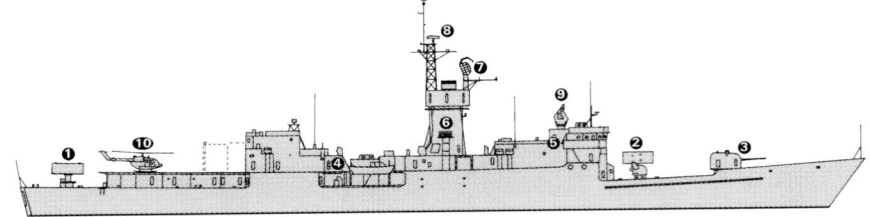

IGNACIO ALLENDE

(Scale 1 : 1,200), Ian Sturton / 0114668

Radars: Air search: Lockheed SPS-40B ❼; B-band.
  Surface search: Raytheon SPS-10 or Norden SPS-67 ❽; G-band.
  Navigation: Marconi LN66; I-band.
  Fire control: Western Electric SPG-53D/F ❾; I/J-band.
  Tacan: SRN 15.
Sonars: EDO/General Electric SQS-26CX; bow-mounted; active
  search and attack; medium frequency.

Helicopters: 1 BO 105 CB ❿.

Programmes: First pair decommissioned from USN in 1992/93.
  Both transferred on 29 January 1997 and arrived in Mexico
  16 August 1997. Both then underwent extensive refits,
  entering service on 23 November 1998. Third of class *(ex-*

Pharris) transferred 2 February 2000 and recommissioned on
  16 March 2000. The fourth ship *(ex-Whipple)* transferred in
  August 2001 and recommissioned on 1 November 2002.
Modernisation: To be fitted with SSM (Harpoon or Gabriel II).
Structure: Four Mk 32 torpedo tubes are fixed in the midships
  structure, two to a side, angled out at 45°. The original Knox
  class SAM launcher has been put back aft, in F 211 only.
Operational: In US service these ships had Harpoon SSM, but it is
  reported that these weapons are not carried. Pennant
  numbers changed in 2001. F 214 based at Manzanillo, the
  remainder at Tuxpan.

UPDATED

IGNACIO ALLENDE (old number)

11/1998, Mexican Navy / 0017679

MARIANO ABASOLO (old number)

4/1999, Mexican Navy / 0081243

2 BRAVO (BRONSTEIN) CLASS (FFH)

Name	No	Builders	Laid down	Launched	Commissioned
NICOLAS BRAVO (ex-*McCloy*)	F 201 (ex-E 40, ex-FF 1038)	Avondale Shipyards	15 Sep 1961	9 June 1962	21 Oct 1963
HERMENEGILDO GALEANA (ex-*Bronstein*)	F 202 (ex-E 42, ex-FF 1037)	Avondale Shipyards	16 May 1961	31 Mar 1962	16 June 1963

Displacement, tons: 2,360 standard; 2,650 full load
Dimensions, feet (metres): 371.5 × 40.5 × 13.5; 23 (sonar) *(113.2 × 12.3 × 4.1; 7)*
Main machinery: 2 Foster-Wheeler boilers; 1 De Laval geared turbine; 20,000 hp *(14.92 MW)*; 1 shaft
Speed, knots: 23.5. **Range, n miles:** 3,924 at 15 kt
Complement: 207 (17 officers)

Missiles: A/S: Honeywell ASROC Mk 112 octuple launcher ❶.
Guns: 2 USN 3 in *(76 mm)*/50 (twin) Mk 33 ❷; 50 rds/min to 12.8 km *(7 n miles)*; weight of shell 6 kg, or 1 Bofors 57 mm/70 Mk 2; 220 rds/min to 17 km *(9.3 n miles)*; weight of shell 2.4 kg.
Torpedoes: 6—324 mm US Mk 32 Mod 7 (2 triple) tubes ❸. 14 Honeywell Mk 46; anti-submarine; active/passive homing to 11 km *(5.9 n miles)* at 40 kt; warhead 44 kg.
Countermeasures: Decoys: 2 Loral Hycor 6-barrelled fixed Mk 33; IR flares and chaff to 4 km *(2.2 n miles)*. T-Mk 6 Fanfare; torpedo decoy system.
Weapons control: Mk 56 GFCS. Mk 114 ASW FCS. Mk 1 target designation system. Elsag NA 18 optronic director may be fitted.
Radars: Air search: Lockheed SPS-40D ❹; B-band; range 320 km *(175 n miles)*.
Surface search: Raytheon SPS-10F ❺; G-band.
Navigation: Marconi LN66; I-band.
Fire control: General Electric Mk 35 ❻; I/J-band.

Sonars: EDO/General Electric SQS-26 AXR; bow-mounted; active search and attack; medium frequency.

Helicopters: Platform and some facilities but no hangar.

Programmes: Transferred from the US to Mexico by sale 12 November 1993 having paid off in December 1990.
Modernisation: Bofors 57 mm SAK may be fitted to replace the Mk 33 gun, possibly with an Elsag NA 18 optronic director.

Structure: Position of stem anchor and portside anchor (just forward of gun mount) necessitated by large bow sonar dome. As built, a single 3 in (Mk 34) open mount was aft of the helicopter deck; removed for installation of towed sonar which has since been taken out.
Operational: ASROC is non-operational. Pennant numbers changed in 2001. Both based at Manzanillo.

UPDATED

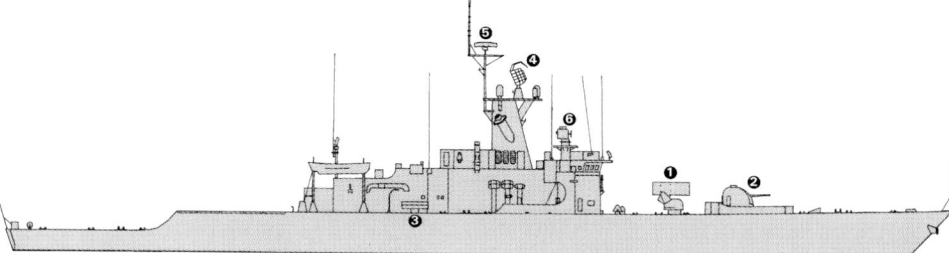

NICOLAS BRAVO

(Scale 1 : 900), Ian Sturton

HERMENEGILDO GALEANA

6/2004 *, Mexican Navy / 0589777

HERMENEGILDO GALEANA

6/2004 *, Mexican Navy / 0589778

SHIPBORNE AIRCRAFT

Numbers/Type: 11/6 MBB BO 105CB/MD 902 Explorer.
Operational speed: 113 kt *(210 km/h)*.
Service ceiling: 9,845 ft *(3,000 m)*.
Range: 407 n miles *(754 km)*.
Role/Weapon systems: Coastal patrol helicopter for patrol, fisheries protection and EEZ protection duties; SAR as secondary role. A modernisation programme was announced in October 2003 with the first upgraded aircraft to be delivered by early 2004. Sensors: Bendix search radar. Weapons: MGs or rocket pods. *UPDATED*

BO 105CB

9/1994, Mexican Navy / 0052606

Numbers/Type: 3 Eurocopter AS 555 AF Fennec.
Operational speed: 121 kt *(225 km/h)*.
Service ceiling: 13,120 ft *(4,000 m)*.
Range: 389 n miles *(722 km)*.
Role/Weapon systems: Patrol helicopter for EEZ protection and SAR. To be operated from Oaxaca class patrol ships. More may be acquired when funds are available. Sensors: Bendix 1500 search radar. Weapons: Can carry up to two torpedoes, rocket pods or an MG. *UPDATED*

AS 555 AF

9/1994, Mexican Navy / 0052607

LAND-BASED MARITIME AIRCRAFT (FRONT LINE)

Notes: (1) Transport aircraft used include six Antonov An-32B, 21 Mil Mi-8/17 and four Mil Mi-2 Hoplite.
(2) Training aircraft include seven Aeromacchi M-290TP Redigos, 14 Maule MX-7-180, ten Zlin Z242L, three Beech B55 Baron, one Robinson R22 Mariners and four MD 500 helicopters.
(3) Aircraft with a communications role include one Learjet 60, one Learjet 31A, one Learjet 25D, two Rockwell 306 Sabreliner and five Gulfstream 695.
(4) Patrol aircraft include one Piper, one Dash-8 200.
(5) A contract for the supply of two Eurocopter AS.565 helicopters was signed on 13 October 2003. Roles envisaged include coastal patrol and SAR and delivery is expected in 2005.

Mi-17

6/2004 *, Mexican Navy / 0589776

An-32B

6/2004 *, Mexican Navy / 0589775

SABRELINER

6/2004 *, Mexican Navy / 0589774

Numbers/Type: 3 Grumman E-2C Hawkeye.
Operational speed: 323 kt *(598 km/h).*
Service ceiling: 37,000 ft *(11,278 m).*
Range: 1,540 n miles *(2,852 km).*
Role/Weapon systems: Acquired from Israel in 2004 after refurbishment by Israel Aircraft Industries' (IAI's) Bedek Aviation Group. Equipment details are speculative. Sensors: ESM: ALR-73 PDS; Airborne tactical data system with Links 4A and 11; AN/APS-125 radar; Mk XII IFF. Weapons: Unarmed.

NEW ENTRY

Numbers/Type: 8 CASA C-212 Aviocar.
Operational speed: 190 kt *(353 km/h).*
Service ceiling: 24,000 ft *(7,315 m).*
Range: 1,650 n miles *(3,055 km).*
Role/Weapon systems: Acquired from 1987 and used for Maritime Surveillance. Two aircraft upgraded in Spain with EADS/CASA Integrated Tactical System (FITS) in 2003. The remainder are to be upgraded in Mexico. Sensors: Search radar; APS 504. Weapons: Unarmed.

UPDATED

E-2C

6/2004*, Mexican Navy / 0589773

C-212

6/2004*, Mexican Navy / 0589772

PATROL FORCES

4 HOLZINGER CLASS (GUNSHIPS) (PSOH)

Name	No	Builders	Launched	Commissioned
CAPITÁN DE NAVIO SEBASTIAN JOSÉ HOLZINGER (ex-*Uxmal*)	PO 131 (ex-C 01, ex-GA 01)	Tampico	1 June 1988	1 May 1991
CAPITÁN DE NAVIO BLAS GODINEZ (ex-*Mitla*)	PO 132 (ex-C 02, ex-GA 02)	Veracruz	22 Mar 1988	1 Nov 1991
BRIGADIER JOSÉ MARIÁ DE LA VEGA (ex-*Peten*)	PO 133 (ex-C 03, ex-GA 03)	Tampico	22 Mar 1992	16 Mar 1994
GENERAL FELIPE B BERRIOZÁBAL (ex-*Anahuac*)	PO 134 (ex-C 04, ex-GA 04)	Veracruz	21 Apr 1991	16 Mar 1994

Displacement, tons: 1,290 full load
Dimensions, feet (metres): 244.1 × 34.4 × 11.2
(74.4 × 10.5 × 3.4)
Main machinery: 2 MTU 20V 956 TB92 diesels; 11,700 hp(m)
(8.6 MW) sustained; 2 shafts
Speed, knots: 22. **Range, n miles:** 3,820 at 16 kt
Complement: 75 (11 officers)

Guns: 1 Bofors 57 mm/70 Mk 2; 220 rds/min to 17 km *(9.3 n miles);* weight of shell 2.4 kg or 2 Bofors 40 mm/60 (1 twin).
Combat data systems: Elsag 2 CSDA-10.
Weapons control: Elsag NA 18 optronic director.
Radars: Surface search: Raytheon SPS-64(V)6A; I-band.
Navigation: Kelvin Hughes Nucleus; I-band.

Helicopters: 1 MBB BO 105 CB.

Programmes: Originally four were ordered from Tampico and Veracruz. First laid down November 1983, second in 1984 but then there were delays caused by financial problems. Named after military heroes.
Structure: An improved variant of the Bazán Halcon (Uribe) class with a flight deck extended to the stern. PO 131 and 132 commissioned with a Bofors 40 mm/60 in lieu of the 57 mm and without the optronic director. PO 133 fitted with what appears to be a 3 in gun. The Navy describes this as a temporary arrangement.
Operational: Pennant numbers changed in 2001. PO 131 and PO 134 based at Tampico; PO 132 and PO 133 based at Ensenada. *UPDATED*

GENERAL FELIPE B BERRIOZÁBAL (old number)

9/2001 / 0533279

3 SIERRA CLASS (GUNSHIPS) (PSOH)

Name	No	Builders	Laid down	Launched	Commissioned
JUSTO SIERRA MENDEZ	PO 141 (ex-C 2001)	Tampico, Tamaulipas	19 Jan 1998	1 June 1998	1 June 1998
GUILLERMO PRIETO	PO 143 (ex-C 2003)	Tampico, Tamaulipas	1 June 1998	18 Sep 1999	17 Sep 1999
MATIAS ROMERO	PO 144 (ex-C 2004)	Salina Cruz, Oaxaco	23 July 1998	17 Sep 1999	17 Sep 1999

Displacement, tons: 1,344 full load
Dimensions, feet (metres): 231 × 34.4 × 9.3
(70.4 × 10.5 × 2.8)
Main machinery: 2 Caterpillar 3616 V16 diesels; 6,197 hp(m)
(4.55 MW); 2 shafts
Speed, knots: 18
Complement: 76 (10 officers)

Missiles: SA-N-10 (PO 144); IR homing to 5 km *(2.7 n miles)* at 1.7 Mach; warhead 1.5 kg.
Guns: 1 Bofors 57 mm/70 Mk 3; 220 rds/min to 17 km *(9.3 n miles);* weight of shell 2.4 kg.
Combat data systems: Alenia 2.
Weapons control: Saab EOS 450 optronic director.
Radars: Air/surface search: E/F-band.
Surface search: I-band.

Helicopters: 1 MD 902 Explorer.

Programmes: Follow on to the Holzinger class. Ordered in 1997.
Structure: Derived from the Holzinger class but with a markedly different superstructure. All ships carry 11 m interceptor craft capable of 50 kt.
Operational: All based at Lazaro. PO 142 badly damaged by fire in October 2003 and subsequently decommissioned.

UPDATED

JUSTO SIERRA MENDEZ
6/2004*, Mexican Navy / 0589771

4 DURANGO CLASS (GUNSHIPS) (PSOH)

Name	No	Builders	Laid down	Launched	Commissioned
DURANGO	PO 151	Tampico, Tamaulipas	18 Dec 1999	11 Sep 2000	11 Sep 2000
SONORA	PO 152	Salina Cruz, Oaxaco	14 Dec 1999	4 Sep 2000	4 Sep 2000
GUANAJUATO	PO 153	Tampico, Tamaulipas	2000	13 Dec 2001	13 Dec 2001
VERACRUZ	PO 154	Salina Cruz, Oaxaco	4 Sep 2000	17 Dec 2001	17 Dec 2001

Displacement, tons: 1,470 full load
Dimensions, feet (metres): 268 × 34.4 × 9.3
(81.8 × 10.5 × 2.8)
Main machinery: 2 Caterpillar 3616 V16 diesels; 6,197 hp(m)
(4.55 MW); 2 shafts
Speed, knots: 18
Complement: 76 (10 officers)

Guns: 1 Bofors 57 mm/70 Mk 3; 220 rds/min to 17 km *(9.3 n miles)*; weight of shell 2.4 kg.
Combat data systems: Alenia 2.
Weapons control: Saab EOS 450 optronic director.
Radars: Air/surface search: E/F-band.
Surface search: I-band.

Helicopters: 1 MD 902 Explorer.

Programmes: Follow on to the Sierra class. Ordered on 1 June 1998.
Structure: Derived from the Holzinger class but with a markedly different superstructure. Durango class slightly larger than the Sierra class. All ships carry 11 m interceptor craft capable of 50 kt.
Operational: All based at Coatzacoalcos.

UPDATED   DURANGO

6/2004, Mexican Navy / 0589770*

2 + 2 OAXACA CLASS (GUNSHIPS) (PSOH)

Name	No	Builders	Laid down	Launched	Commissioned
OAXACA	PO 161	Salina Cruz, Oaxaco	17 Dec 2001	11 Apr 2003	1 May 2003
BAJA CALIFORNIA	PO 162	Tampico, Tamaulpas	13 Dec 2001	21 May 2003	1 Apr 2003
—	PO 163	Salina Cruz, Oaxaco	11 Apr 2003	2004	2005
—	PO 164	Tampico, Tamaulpas	21 May 2003	2004	2005

Displacement, tons: 1,680
Dimensions, feet (metres): 282.2 × 34.4 × 9.3
(86.0 × 10.5 × 3.6)
Main machinery: 2 Caterpillar 3916 V16 diesels; 2 shafts
Speed, knots: 20
Complement: 77

Guns: 1 Oto Melara 3 in *(76 mm)*/62 compact; 85 rds/min to 16 km *(8.7 n miles)*; weight of shell 6 kg. 1 Oto Melara 25 mm.
Combat data systems: Alenia.
Radars: Surface search/navigation: Terma Scanter 2001; I-band.
Fire control: Alenia NA-25; I-band.
Helicopters: Eurocopter AS 565 Panther.

Structure: A further derivation of the basic Holzinger class and a slightly longer version of the Durango class. Capable of operating a helicopter and equipped with a fast 11 m interception boat capable of 50 kt. PO 161 based at Salina Cruz and PO 162 at Tampico.

UPDATED

6 URIBE CLASS (GUNSHIPS) (PSOH)

Name	No	Builders	Laid down	Launched	Commissioned
CADETE VIRGILIO URIBE	PO 121 (ex-C 11, ex-GH 01)	Bazán, San Fernando	1 July 1981	12 Nov 1981	1 Aug 1982
TENIENTE JOSÉ AZUETA	PO 122 (ex-C 12, ex-GH 02)	Bazán, San Fernando	7 Sep 1981	12 Dec 1981	23 Sep 1982
CAPITÁN de FRAGATA PEDRO SÁINZ de BARANDA	PO 123 (ex-C 13, ex-GH 03)	Bazán, San Fernando	22 Oct 1981	29 Jan 1982	1 May 1983
COMODORO CARLOS CASTILLO BRETÓN	PO 124 (ex-C 14, ex-GH 04)	Bazán, San Fernando	11 Nov 1981	26 Feb 1982	24 May 1983
VICEALMIRANTE OTHÓN P BLANCO	PO 125 (ex-C 15, ex-GH 05)	Bazán, San Fernando	18 Dec 1981	26 Mar 1982	24 Feb 1983
CONTRALMIRANTE ANGEL ORTIZ MONASTERIO	PO 126 (ex-C 16, ex-GH 06)	Bazán, San Fernando	30 Dec 1981	4 May 1982	24 Feb 1983

Displacement, tons: 988 full load
Dimensions, feet (metres): 219.9 × 34.4 × 11.5
(67 × 10.5 × 3.5)
Main machinery: 2 MTU-Bazán 16V 956 TB91 diesels; 7,500 hp (m) *(5.52 MW)* sustained; 2 shafts
Speed, knots: 22. **Range, n miles:** 5,000 at 13 kt
Complement: 46 (7 officers)

Guns: 1 Bofors 40 mm/70; 300 rds/min to 12.5 km *(6.7 n miles)*; weight of shell 0.96 kg.

Weapons control: Naja optronic director.
Radars: Surface search: Decca AC 1226; I-band.
Navigation: I-band.
Tacan: SRN 15.

Helicopters: 1 MBB BO 105 CB.

Programmes: Ordered in 1980 to a Halcon class design. Contracts for a further eight of the class have been shelved.

Pennant numbers changed in 1992. Named after naval heroes.
Structure: Flight deck extends to the stern. Similar ships built for Argentina.
Operational: Used for EEZ patrol. Pennant numbers changed in 2001. PO 121 and PO 123 based at Ensenada, PO 122 and PO 125 at Tampico and PO 124 and PO 126 at Lazaro.

UPDATED

COMODORO CARLOS CASTILLO BRETÓN

9/2002, B Sullivan / 0533280

For details of the latest updates to *Jane's Fighting Ships* online and to discover the additional information available exclusively to online subscribers please visit
jfs.janes.com

11 VALLE (AUK) CLASS (COAST GUARD) (PG/PGH)

JUAN DE LA BARRERA (ex-*Guillermo Prieto*, ex-*Symbol* MSF 123) PO 102 (ex-C 71, ex-G-02)
MARIANO ESCOBEDO (ex-*Champion* MSF 314) PO 103 (ex-C 72, ex-G-03)
MANUEL DOBLADO (ex-*Defense* MSF 317) PO 104 (ex-C 73, ex-G-05)
SANTOS DEGOLLADO (ex-*Gladiator* MSF 319) PO 106 (ex-C 75, ex-G-07)
JUAN N ALVARES (ex-*Ardent* MSF 340) PO 108 (ex-C 77, ex-G-09)
MANUEL GUTIERREZ ZAMORA (ex-*Roselle* MSF 379) PO 109 (ex-C 78, ex-G-10)
VALENTIN GOMEZ FARIAS (ex-*Starling* MSF 64) PO 110 (ex-C 79, ex-G-11)
FRANCISCO ZARCO (ex-*Threat* MSF 124) PO 112 (ex-C 81, ex-G-13)
IGNACIO L VALLARTA (ex-*Velocity* MSF 128) PO 113 (ex-C 82, ex-G-14)
JESUS GONZALEZ ORTEGA (ex-*Chief* MSF 315) PO 114 (ex-C 83, ex-G-15)
MARIANO MATAMOROS (ex-*Hermenegildo Galeana*, ex-*Sage* MSF 111) PO 117 (ex-C 86, ex-G-19)

Displacement, tons: 1,065 standard; 1,250 full load
Dimensions, feet (metres): 221.2 × 32.2 × 10.8 *(67.5 × 9.8 × 3.3)*
Main machinery: Diesel-electric; 2 Caterpillar diesels; 2 shafts
Speed, knots: 18. **Range, n miles:** 6,900 at 10 kt
Complement: 73 (9 officers)
Guns: 1 USN 3 in *(76 mm)*/50. 4 Bofors 40 mm/60 (2 twin).
   2—12.7 mm MGs (in some on quarterdeck).
Radars: Surface search: Kelvin Hughes 14/9 (in most); I-band.
Helicopters: Platform for 1 BO 105 (C 72, C 73 and C 79 only).

Comment: Transferred from US, six in February 1973, four in April 1973, nine in September 1973. Eight have since been deleted. Employed on Coast Guard duties. All built during Second World War. Variations are visible in the mid-ships section where some have a bulwark running from the break of the forecastle to the quarterdeck. Minesweeping gear removed. All ships re-engined 1999-2002. Some carry a Pirana 26 kt motor launch armed with 40 mm grenade launchers and 7.62 mm MGs. P 103, P 104 and P 110 have had helicopter flight decks installed aft. Plans to fit flight decks in the others have been shelved. PO 102, 103, 104, 106, 108, 112 and 113 based at Lazaro; PO 109 and 114 based at Tampico; PO 110 and 117 based at Ensenada.

UPDATED

MANUEL GUTIERREZ ZAMORA (old number)   11/2000 / 0105130

VALLE (AUK) CLASS   6/2003, *Mexican Navy* / 0567906

1 + 1 (4) CENTENARIO CLASS (PBO)

Name	No	Builders	Launched	Commissioned
DÉMOCRATA	PC 241 (ex-C 101)	Varadero, Guaymas	16 Oct 1997	12 Jan 1998

Displacement, tons: 450 full load
Dimensions, feet (metres): 172.2 × 29.5 × 8.5 *(52.5 × 9 × 2.6)*
Main machinery: 2 MTU 20V 956 TB92 diesels; 6,119 hp(m) *(4.5 MW)*; 2 shafts
Speed, knots: 30
Complement: 36 (13 officers)
Guns: 2 Bofors 40 mm/60 (twin).
Radars: Surface search: Racal Decca; E/F-band.

Comment: Based at Tuxpan. A second unit was reported under construction in August 2002 and further ships may be built if funds become available. A 50 kt Boston Whaler launch is carried at the stern.

UPDATED

DÉMOCRATA   6/2004*, *Mexican Navy* / 0589769

3 CAPE (PGM 71) CLASS (LARGE PATROL CRAFT) (PB)

Name	No	Builders	Recommissioned
CABO CORRIENTES	PC 271 (ex-P 42)	CG Yard, Curtis Bay	16 Mar 1990
(ex-*Jalisco*, ex-*Cape Carter*)			
CABO CORZO	PC 272 (ex-P 43)	CG Yard, Curtis Bay	21 Apr 1990
(ex-*Nayarit*, ex-*Cape Hedge*)			
CABO CATOCHE	PC 273 (ex-P 44	CG Yard, Curtis Bay	18 Mar 1991
(ex-*Cape Hattaras*)			

Displacement, tons: 98 standard; 148 full load
Dimensions, feet (metres): 95 × 20.2 × 6.6 *(28.9 × 6.2 × 2)*
Main machinery: 2 GM 16V-149TI diesels; 2,322 hp *(1.73 MW)* sustained; 2 shafts
Speed, knots: 20. **Range, n miles:** 2,500 at 10 kt
Complement: 14 (1 officer)
Guns: 1—20 mm. 2—12.7 mm MGs.
Radars: Navigation: Raytheon SPS-64; I-band.

Comment: All built in 1953; have been re-engined and extensively modernised. Transferred under the FMS programme, having paid off from the US Coast Guard. Pennant numbers changed in 2001. PC 271 and PC 272 based at Puerto Vallarta and PC 273 at Isla Cozumel.

UPDATED

CABO CORRIENTES (old number)   6/2001, *Mexican Navy* / 0114675

2 POINT CLASS (LARGE PATROL CRAFT) (PB)

Name	No	Builders	Recommissioned
PUNTA MORRO (ex-*Point Verde*)	PC 281 (ex-P 60, ex-P 45)	CG Yard, Curtis Bay	19 July 1991
PUNTA MASTUN (ex-*Point Herron*)	PC 282 (ex-P 61, ex-P 46)	CG Yard, Curtis Bay	19 July 1991

Displacement, tons: 67 full load
Dimensions, feet (metres): 83 × 17.2 × 5.8 *(25.3 × 5.2 × 1.8)*
Main machinery: 2 Caterpillar diesels; 1,600 hp *(1.19 MW)*; 2 shafts
Speed, knots: 12. **Range, n miles:** 1,500 at 8 kt
Complement: 10
Guns: 2—12.7 mm MGs (can be carried).
Radars: Surface search: Raytheon SPS-64; I-band.

Comment: Ex-US Coast Guard craft built in 1961. Steel hulls and aluminium superstructures. Speed much reduced from original 23 kt. Pennant numbers changed in 2001. Both based at Lerma.

UPDATED

PUNTA MASTUN (old number)   6/1994, *Mexican Navy* / 0081251

20 AZTECA CLASS (LARGE PATROL CRAFT) (PB)

Name	No	Builders	Commissioned
MATIAS DE CORDOVA	PC 202	Scott & Sons, Bowling	6 Jan 1974
(ex-*Guaycura*)	(ex-P 02)		
IGNACIO LÓPEZ RAYÓN	PC 206	Ailsa Shipbuilding Co Ltd	18 Apr 1975
(ex-*Tarahumara*)	(ex-P 06)		
MANUEL CRESCENCIO REJON	PC 207	Ailsa Shipbuilding Co Ltd	1 Dec 1975
(ex-*Tepehuan*)	(ex-P 07)		
JUAN ANTONIO DE LA FUENTE	PC 208	Ailsa Shipbuilding Co Ltd	28 Dec 1975
(ex-*Mexica*)	(ex-P 08)		
LEON GUZMAN	PC 209	Scott & Sons, Bowling	1 June 1975
(ex-*Zapoteca*)	(ex-P 09)		
IGNACIO RAMIREZ	PC 210	Ailsa Shipbuilding Co Ltd	1 June 1975
(ex-*Huastela*)	(ex-P 10)		
IGNACIO MARISCAL	PC 211	Ailsa Shipbuilding Co Ltd	25 Dec 1975
(ex-*Mazahua*)	(ex-P 11)		
HERIBERTO JARA CORONA	PC 212	Ailsa Shipbuilding Co Ltd	17 Nov 1975
(ex-*Huichol*)	(ex-P 12)		
COLIMA	PC 214	Scott & Sons, Bowling	1 July 1975
(ex-*Yacqui*)	(ex-P 14)		
JOSE JOAQUIN FERNANDEZ DE LIZARDI	PC 215	Ailsa Shipbuilding Co Ltd	1 June 1976
(ex-*Tlapaneco*)	(ex-P 15)		
FRANCISCO J MUGICA	PC 216	Ailsa Shipbuilding Co Ltd	1 June 1976
(ex-*Tarasco*)	(ex-P 16)		
JOSE MARIA DEL CASTILLO VELASCO	PC 218	Lamont & Co Ltd	1 Nov 1976
(ex-*Otomi*)	(ex-P 18)		
JOSE NATIVIDAD MACIAS	PC 220	Lamont & Co Ltd	29 Dec 1976
(ex-*Pimas*)	(ex-P 20)		
TAMAULIPAS	PC 223	Veracruz	18 May 1977
(ex-*Mazateco*)	(ex-P 23)		
YUCATAN	PC 224	Veracruz	1 May 1975
(ex-*Tolteca*)	(ex-P 24)		
TABASCO	PC 225	Veracruz	1 Dec 1978
(ex-*Maya*)	(ex-P 25)		
COCHIMIE	PC 226	Veracruz	1 Dec 1978
(ex-*Veracruz*)	(ex-P 26)		
PUEBLA	PC 228	Veracruz	1 Aug 1982
(ex-*Totonaca*)	(ex-P 28)		
LEONA VICARIO	PC 230	Salina Cruz	1 May 1977
(ex-*Olmeca*)	(ex-P 30)		
JOSEFA ORTIZ DE DOMINGUEZ	PC 231	Salina Cruz	1 June 1977
(ex-*Tlahuica*)	(ex-P 31)		

Displacement, tons: 148 full load
Dimensions, feet (metres): 112.7 × 28.3 × 7.2 *(34.4 × 8.7 × 2.2)*
Main machinery: 2 Paxman 12YJCM diesels; 3,000 hp *(2.24 MW)* sustained; 2 shafts
Speed, knots: 24. **Range, n miles:** 1,537 at 14 kt
Complement: 24 (2 officers)
Guns: 1 Bofors 40 mm/60; 300 rds/min to 12 km *(6.5 n miles)* anti-surface; 4 km *(2.2 n miles)* anti-aircraft; weight of shell 2.4 kg.
  1 Oerlikon 20 mm or 1—7.62 mm MG.
Radars: Surface search: Kelvin Hughes; I-band.

Comment: Ordered by Mexico on 27 March 1973 from Associated British Machine Tool Makers Ltd to a design by TT Boat Designs, Bembridge, Isle of Wight. The first 21 were modernised in 1987 in Mexico with spare parts and equipment supplied by ABMTM Marine Division who supervised the work which included engine refurbishment and the fitting of air conditioning. Names and pennant numbers changed in 2001. Based at: Veracruz (PC 202, 207, 223, 228); Yukaltepen (PC 224, 225, 226); Salina Cruz (PC 206, 209, 218, 220); Guaymas (PC 208, 210, 214); Mazatlan (PC 211, 216, 230, 231); Acapulco (PC 212, 215). ***UPDATED***

AZTECA CLASS   4/1999, van Ginderen Collection / 0081249

4 ISLA CLASS (FAST ATTACK CRAFT) (PBF)

Name	No	Builders	Commissioned
ISLA CORONADO	PI 1201 (ex-P 51)	Equitable Shipyards	1 Sep 1993
ISLA LOBOS	PI 1202 (ex-P 52)	Equitable Shipyards	1 Nov 1993
ISLA GUADALUPE	PI 1203 (ex-P 53)	Equitable Shipyards	1 Feb 1994
ISLA COZUMEL	PI 1204 (ex-P 54)	Equitable Shipyards	1 Apr 1994

Displacement, tons: 52 full load
Dimensions, feet (metres): 82 × 17.9 × 4 *(25 × 5.5 × 1.2)*
Main machinery: 3 Detroit diesels; 16,200 hp *(12.9 MW)*; 3 Arneson surface drives
Speed, knots: 50. **Range, n miles:** 1,200 at 30 kt
Complement: 9 (3 officers)
Guns: 1—12.7 mm MG. 2—7.62 mm MGs.
Radars: Surface search: Raytheon SPS 69; I-band.
Fire control: Thomson-CSF Agrion; J-band.

Comment: Built by the Trinity Marine Group to an XFPB (extra fast patrol boat) design. Deep Vee hulls with FRP/Kevlar construction. Similar craft built for US Navy. May be fitted with MM 15 SSMs in due course and armed with 40 mm or 20 mm guns. Pennant numbers changed 2001. Based at Topolobampo (PI 1201, 1202) and Guaymas (PI 1203, 1204). ***UPDATED***

ISLA CORONADO (old number)   6/2001, Mexican Navy / 0114674

3 LAGUNA (ex-POLIMAR) CLASS (RIVER PATROL CRAFT) (PB)

Name	No	Builders	Commissioned
CUYUTLAN (ex-*Polcinco*)	PR 293 (ex-P 73, ex-F 05)	Astilleros de Tampico	2 Sep 1959
ALVARADO (ex-*Polsiete*)	PR 294 (ex-P 75, ex-F 07)	—	1 Jan 1989
CATEMACO (ex-*Polocho*)	PR 295 (ex-P 76, ex-F 08)	—	15 Dec 1990

Displacement, tons: 155 full load
Dimensions, feet (metres): 85 × 14.1 × 6.6 *(25.9 × 4.3 × 2)*
Main machinery: 2 diesels; 456 hp *(335 kW)*; 2 shafts
Speed, knots: 8
Guns: 1—40 mm.
Radars: Surface search: Kelvin Hughes; I-band.

Comment: Steel construction. The last two were transferred from the USN on the dates shown and are similar in size. All have *Laguna de* in front of the names. Pennant numbers changed in 2001. ***UPDATED***

LAGUNA CLASS   6/2001, Mexican Navy / 0114673

44 POLARIS CLASS (COMBATBOAT 90 HMN) (PBF)

POLARIS PI 1101	DENEB PI 1112	ELNATH PI 1123	DENEBOLA PI 1134
SIRIUS PI 1102	FOMALHAUT PI 1113	ALNILÁN PI 1124	ALKAID PI 1135
CAPELLA PI 1103	POLLUX PI 1114	PEACOCK PI 1125	ALPHECCA PI 1136
CANOPUS PI 1104	RÉGULUS PI 1115	BETELGEUSE PI 1126	ELTANIN PI 1137
VEGA PI 1105	ACRUX PI 1116	ADHARA PI 1127	KOCHAB PI 1138
ACHERNAR PI 1106	SPICA PI 1117	ALIOTH PI 1128	ENIF PI 1139
RIGEL PI 1107	HADAR PI 1118	RASALHAGUE PI 1129	SCHEDAR PI 1140
ARCTURUS PI 1108	SHAULA PI 1119	NUNKI PI 1130	MARKAB PI 1141
ALPHERATZ PI 1109	MIRFAK PI 1120	HAMAL PI 1131	MEGREZ PI 1142
PROCYÓN PI 1110	ANKAA PI 1121	SUHAIL PI 1132	MIZAR PI 1143
AVIOR PI 1111	BELLATRIX PI 1122	DUBHE PI 1133	PHEKDA PI 1144

Displacement, tons: 19 full load
Dimensions, feet (metres): 52.2 × 12.5 × 2.6 *(15.9 × 3.8 × 0.8)*
Main machinery: 2 CAT 3406E diesels; 1,605 hp(m) *(1.18 MW)*; 2 waterjets
Speed, knots: 47. **Range, n miles:** 240 at 30 kt
Complement: 4
Guns: 1 Oto Melara 12.7 mm MG.
Radars: Surface search: Litton Decca Bridgemaster E; I-band.

Comment: All named after stars. First 12 ordered from Dockstavarvet, Sweden, on 15 April 1999, second batch of eight on 29 July 1999 and last batch of 20 on 1 February 2000. All delivered by 2001. A further batch of four delivered in 2004. This is the fastest version of this type built so far. These craft are in service with the Swedish and Norwegian navies and with paramilitary forces in Malaysia and China. Based at Ciudad del Carmen (1103, 1104, 1125, 1140); Cozumel (1105, 1106, 1143, 1144); El Mezquital (1101, 1102); Yucalpeten (1107, 1108); Isla Mujeres (1109, 1110); Tuxpan (1113, 1114); Chetumel (1128, 1129); Veracruz (1131, 1132); Ensenada (1111, 1112); Manzanillo (1115, 1116); Puerto Madero (1117, 1120); Topolobampo (1118, 1119); Mazatlan (1121, 1122); Puerto Cortes (1123, 1141); Puerto Vallarta (1124, 1136); Acapulco (1126, 1127); Guaymas (1130, 1139); Salina Cruz (1133, 1134); Puerto Penasco (1137, 1138); Isla Socorro (1135); Frontera (1142). ***UPDATED***

ACHERNAR   4/2002, Martin Mokrus / 0533282

1 LAGO CLASS (RIVER PATROL CRAFT) (PB)

Name	No	Builders	Commissioned
LAGO DE CHAPALA (ex-*AM 05*)	PR 322 (ex-P 81)	Vera Cruz	12 Oct 1959

Displacement, tons: 37 full load
Dimensions, feet (metres): 56.1 × 13.1 × 3.9 *(17.1 × 4 × 1.2)*
Main machinery: 1 diesel; 320 hp(m) *(235 kW)*; 1 shaft
Speed, knots: 5
Complement: 5
Radars: Surface search: Kelvin Hughes; I-band.

Comment: Steel construction. Three of original class now deleted and PR 323 was lost 1 September 2003. Pennant numbers changed in 2001. ***UPDATED***

LAGO CLASS   1/2003, Gomel/Marsan / 0568892

8 ACUARIO CLASS (COMBATBOAT 90HMN) (PBF)

ACUARIO PI 1301		CANCER PI 1305
AGUILA PI 1302		CAPRICORNO PI 1306
ARIES PI 1303		CENTAURO PI 1307
AURIGA PI 1304		GEMINIS PI 1308

Displacement, tons: 19 full load
Dimensions, feet (metres): 52.2 × 12.5 × 2.6 *(15.9 × 3.8 × 0.8)*
Main machinery: 2 CAT 3406E diesels; 1,605 hp(m) *(1.18 MW)*; 2 waterjets
Speed, knots: 47. **Range, n miles:** 240 at 30 kt
Complement: 4
Guns: 1 Oto Melara 12.7 mm MG.
Radars: Surface search: Litton Decca Bridgemaster E; I-band.

Comment: A further development of the Polaris class which are based on the Swedish Combatboat 90. 1301 and 1302 commissioned on 1 June 2004 and the remainder on 1 September 2004. All named after stars.

UPDATED

ACUARIO   *6/2004*, Mexican Navy /* 0589768

52 FAST PATROL CRAFT (PBF)

G 01-36	C 152-01	C 241-01	+14

Dimensions, feet (metres): 22.3 × 7.5 × 1 *(6.8 × 2.3 × 0.3)*
Main machinery: 2 Johnson outboards; 280 hp *(209 kW)*
Speed, knots: 40. **Range, n miles:** 190 at 40 kt
Complement: 2
Guns: 1 or 2—7.62 mm MGs.

Comment: Details are for the 36 G 01-36 Piraña class. Acquired in 1993/94. The 50 kt Interceptor class launch is carried in *Démocrata* and modified versions can be embarked in the gunships. More are to be acquired as OPVs enter service. There are also ten 29 ft Mako Marine craft, with twin Mercury outboards acquired in 1995. Five Sea Force 730 RIBs with Hamilton water-jets, also acquired in 1995-96.

UPDATED

INTERCEPTOR (mod)   *6/2004*, Mexican Navy /* 0589767

INTERCEPTOR (old number)   *7/1998, Mexican Navy /* 0052610

PIRAÑA CLASS   *9/2002, Julio Montes /* 0533285

2 ALIYA (SAAR 4.5) CLASS
(FAST ATTACK CRAFT—MISSILE) (PTG)

Name	No	Builders	Launched	Commissioned
HURACAN (ex-*Aliya*)	301	Israel Shipyards, Haifa	11 July 1980	Aug 1980
TORMENTA (ex-*Geoula*)	302	Israel Shipyards, Haifa	Oct 1980	31 Dec 1980

Displacement, tons: 498 full load
Dimensions, feet (metres): 202.4 × 24.9 × 8.2 *(61.7 × 7.6 × 2.5)*
Main machinery: 4 MTU/Bazán 16V 956 TB91 diesels; 15,000 hp(m) *(11.03 MW)* sustained; 4 shafts
Speed, knots: 31. **Range, n miles:** 3,000 at 17 kt; 1,500 at 30 kt
Complement: 53

Missiles: SSM: 6 IAI Gabriel II; radar or optical guidance; semi-active radar plus anti-radiation homing to 36 km *(19.4 n miles)* at 0.7 Mach; warhead 75 kg.
Guns: 2 Oerlikon 20 mm; 800 rds/min to 2 km.
    1 General Electric/General Dynamics Vulcan Phalanx 6-barrelled 20 mm Mk 15; 3,000 rds/min combined to 1.5 km anti-missile.
    4—12.7 mm (twin or quad) MGs.
Countermeasures: Decoys: 1—45-tube, 4—24-tube, 4 single-tube chaff launchers.
ESM/ECM: Elisra NS 9003/5; intercept and jammer.
Combat data systems: IAI Reshet datalink.
Radars: Air/surface search: Thomson-CSF TH-D 1040 Neptune; G-band.
Fire control: Selenia Orion RTN-10X; I/J-band.

Helicopters: To be announced.

Programmes: First two of the original class of five Saar 4.5s, before conversions from Saar 4s were started. Transferred to Mexico in July 2004.
Structure: The CIWS mounted in the eyes of the ship replaced a 40 mm gun.

NEW ENTRY

HURACAN   *7/2004*, Diego Quevedo /* 0583999

TORMENTA   *7/2004*, Diego Quevedo /* 0584000

SURVEY AND RESEARCH SHIPS

1 ONJUKU CLASS (SURVEY SHIP) (AGS)

Name	No	Builders	Commissioned
ONJUKU	BI 02 (ex-H 04)	Uchida Shipyard	10 Jan 1980

Displacement, tons: 494 full load
Dimensions, feet (metres): 121 × 26.2 × 11.5 *(36.9 × 8 × 3.5)*
Main machinery: 1 Yanmar 6UA-UT diesel; 700 hp(m) *(515 kW)*; 1 shaft
Speed, knots: 12. **Range, n miles:** 5,645 at 10.5 kt
Complement: 20 (4 officers)
Radars: Navigation: Furuno; I-band.
Sonars: Furuno; hull-mounted; high frequency active.

Comment: Launched 9 December 1977 in Japan. Sonar is a fish-finder type. New pennant number in 2001. Based at Veracruz.

UPDATED

ONJUKU (old number)   *4/1999, van Ginderen Collection /* 0081256

2 ROBERT D CONRAD CLASS (RESEARCH SHIPS) (AGOR)

Name	No	Builders	Commissioned
ALTAIR (ex-*James M Gilliss*)	BI 03 (ex-H 05, ex-AGOR 4)	Christy Corp, WI	5 Nov 1962
ANTARES (ex-*S P Lee*)	BI 04 (ex-H 06, ex-AG 192)	Defoe, Bay City	2 Dec 1962

Displacement, tons: 1,370 full load
Dimensions, feet (metres): 208.9 × 40 × 15.4 *(63.7 × 12.2 × 4.7)*
Main machinery: Diesel-electric; 2 Caterpillar diesel generators; 1,200 hp *(895 kW)*; 2 motors; 1,000 hp *(746 kW)*; 1 shaft; bow thruster
Speed, knots: 13.5. **Range, n miles:** 10,500 at 10 kt
Complement: 41 (12 officers) plus 15 scientists
Radars: Navigation: Raytheon 1025; Raytheon R4iY; I-band.

Comment: *Altair* leased from US 14 June 1983. Refitted and modernised in Mexico. Recommissioned 23 November 1984. Primarily used for oceanography. *Antares* served as an AGI with the USN until February 1974 when she transferred on loan to the Geological Survey. Acquired by sale and recommissioned on 1 December 1992. New pennant numbers in 2001. Based at Manzanillo (BI 03) and Tampico (BI 04).

UPDATED

ALTAIR   *8/2002*, Rahn collection /* 0589764

4 ARRECIFE (ex-OLMECA II) CLASS (SURVEY CRAFT) (YGS)

ALACRAN BI 08 (ex-PR 301)	CABEZO BI 10 (ex-PR 304)
RIZO BI 09 (ex-PR 310)	ANEGAGADA DE ADENTRO BI 11 (ex-PR 309)

Displacement, tons: 18 full load
Dimensions, feet (metres): 54.8 × 14.4 × 3.9 *(16.7 × 4.4 × 1.2)*
Main machinery: 2 Detroit 8V-92TA diesels; 700 hp *(562 kW)* sustained; 2 shafts
Speed, knots: 20. **Range, n miles:** 460 at 10 kt
Complement: 15 (2 officers)
Guns: 1—12.7 mm MG.
Radars: Navigation: Raytheon 1900; I-band.

Comment: Built at Acapulco and completed between 1982 and 1989. GRP hulls. Converted for inshore hydrographic duties in 2003. All have *Arrecife* in front of the names. Based at Manzanillo (BI 08, 09) and Veracruz (BI 10, 11).

UPDATED

1 SURVEY SHIP (AGS)

Name	No	Builders	Commissioned
RIO ONDO (ex-*Deer Island*)	BI 06 (ex-H 08, ex-A 26, ex-YAG 62)	Halter Marine	May 1962

Displacement, tons: 400 full load
Dimensions, feet (metres): 120.1 × 27.9 × 6.9 *(36.6 × 8.5 × 2.1)*
Main machinery: 2 diesels; 2 shafts
Speed, knots: 10. **Range, n miles:** 6,000 at 10 kt
Complement: 20

Comment: Acquired from US on 1 August 1996 and adapted for a support ship role in 1997. Converted to Survey Ship in 1999. Used in US service from 1983 as an acoustic research ship to test noise reduction equipment. Started life as an oil rig supply tug. New pennant number in 2001. Based at Coatzacoalcos.

UPDATED

RIO ONDO (old number)   *4/1999, M Declerck /* 0081258

1 HUMBOLDT CLASS (RESEARCH SHIP) (AGOR)

Name	No	Builders	Recommissioned
ALEJANDRO DE HUMBOLDT	BI 01 (ex-H 03)	JG Hitzler, Elbe	22 June 1987

Displacement, tons: 585 standard; 700 full load
Dimensions, feet (metres): 140.7 × 32 × 13.5 *(42.3 × 9.6 × 4.1)*
Main machinery: 2 diesels; 2 shafts
Speed, knots: 14
Complement: 20 (4 officers)
Radars: Navigation: Kelvin Hughes; I-band.

Comment: Former trawler built in Germany and launched in January 1970. Converted in 1982 to become a hydrographical and acoustic survey ship. Based at Manzanillo. New pennant number in 2001.

UPDATED

ALEJANDRO DE HUMBOLDT (old number)   *6/2001, Mexican Navy /* 0114671

1 SURVEY SHIP (AGSC)

MOCTEZUMA II BI 07 (ex-A-09)

Displacement, tons: 150 full load
Dimensions, feet (metres): 108.3 × 20.3 × 13.1 *(33.0 × 6.2 × 4.0)*
Main machinery: 1 Detroit diesel; 1 shaft
Complement: 17

Comment: Two-masted sailing vessel built in 1972 and taken over by the Navy on 6 December 1985. Based at Guaymas.

UPDATED

1 SUPPORT SHIP (AKS)

Name	No	Builders	Commissioned
RIO SUCHIATE (ex-*Monob 1*)	BI 05 (ex-A 27, ex-YAG 61, ex-YW 87)	Zenith Dredge Co	11 Nov 1943

Displacement, tons: 1,390 full load
Dimensions, feet (metres): 191.9 × 33.1 × 15.7 *(58.5 × 10.1 × 4.8)*
Main machinery: 1 Caterpillar D 398 diesel; 850 hp *(634 kW)*; 1 shaft
Speed, knots: 9. **Range, n miles:** 2,500 at 9 kt
Complement: 21

Comment: Acquired from US on 1 August 1996. The ship was converted from a water carrier to an acoustic research role in 1969, and had four laboratories in US service. Adapted to act also in support ship role in 1997. New pennant number in 2001. Based at Guaymas.

UPDATED

RIO SUCHIATE (US colours)   *7/1988, Giorgio Arra*

AMPHIBIOUS FORCES

2 NEWPORT CLASS (LSTH)

Name	No	Builders	Laid down	Launched	Commissioned	Recommissioned
RIO PAPALOAPAN (ex-*Sonora*, ex-*Newport*)	A 411 (ex-A-04, ex-LST-1179)	Philadelphia Naval Shipyard	1 Nov 1966	3 Feb 1968	7 June 1969	5 June 2001
USUMACINTA (ex-*Frederick*)	A 412 (ex-LST-1184)	National Steel & Shipbuilding Co	13 Apr 1968	8 Mar 1969	11 Apr 1970	1 Dec 2002

Displacement, tons: 4,975 light; 8,450 full load
Dimensions, feet (metres): 522.3 (hull) × 69.5 × 17.5 (aft) *(159.2 × 21.2 × 5.3)*
Main machinery: 6 General Motors 16-645-E5 diesels; 16,500 hp *(12.3 MW)* sustained; 2 shafts; cp props; bow thruster
Speed, knots: 20. **Range, n miles:** 14,250 at 14 kt
Complement: 257 (13 officers)

Military lift: 400 troops; 500 tons vehicles; 3 LCVPs and 1 LCPL on davits.
Guns: 4 USN 3 in *(76 mm)*/50 (A 411).
Radars: Surface search: Raytheon SPS-10F; G-band.
Navigation: Raytheon SPS-64; I-band.

Helicopters: Platform only.

Programmes: A-411 sold to Mexico by the US Navy on 18 January 2001. A-412 sold on 9 December 2002. Both ships employed in amphibious role rather than as transport ships as previously reported. A 411 based at Tampico and A 412 at Manzanillo.
UPDATED

AUXILIARIES

Notes: Procurement of up to two Hospital Ships is reported to be under consideration.

1 TRANSPORT SHIP (AP)

Name	No	Builders	Commissioned
MANZANILLO (ex-*Clearwater County*)	A 402 (ex-A 02)	Chicago Bridge & Iron Co	31 Mar 1944

Displacement, tons: 4,080 full load
Dimensions, feet (metres): 328 × 50 × 14 *(100 × 15.3 × 4.3)*
Main machinery: 2 GM 12-567A diesels; 1,800 hp *(1.34 MW)*; 2 shafts
Speed, knots: 11. **Range, n miles:** 6,000 at 11 kt
Complement: 250
Guns: 8 Bofors 40 mm (2 twin, 4 single).

Comment: Ex-US LST 452 class transferred and recommissioned on 1 July 1972. Deployed also as SAR and disaster relief ship. Based at Manzanillo.
UPDATED

TRANSPORT SHIP
7/1991, Harald Carstens / 0081259

1 LOGISTIC SUPPORT SHIP (AKS)

Name	No	Builders	Recommissioned
MAYA (ex-*Rio Nautla*)	ATR 01 (ex-A 20, ex-A 23)	Isla Gran Cayman, Ru	1 June 1988

Displacement, tons: 924 full load
Dimensions, feet (metres): 160.1 × 38.7 × 16.1 *(48.8 × 11.8 × 4.9)*
Main machinery: 1 MAN diesel; 1 shaft
Speed, knots: 12
Complement: 15 (8 officers)

Comment: First launched in 1962 and acquired for the Navy in 1988. Unarmed. New name and pennant number in 2001. Based at Mazatlan.
UPDATED

MAYA
6/2004 *, Mexican Navy / 0589766

1 LOGISTIC SUPPORT SHIP (AK)

Name	No	Builders	Recommissioned
TARASCO (ex-*Rio Lerma*, ex-*Sea Point*, ex-*Tricon*, ex-*Marika*, ex-*Arneb*)	ATR 03 (ex-A 22, A 25)	Solvesborg, Sweden	1 Mar 1990

Displacement, tons: 1,970 full load
Dimensions, feet (metres): 282.2 × 40.7 × 16.1 *(86 × 12.4 × 4.9)*
Main machinery: 1 Kloeckner Humboldt Deutz diesel; 2,100 hp(m) *(1.54 MW)*; 1 shaft
Speed, knots: 14
Complement: 35
Cargo capacity: 778 tons

Comment: Built in 1962 as a commercial ship and taken into the Navy in 1990. New name and pennant number in 2001. Based at Tampico.
UPDATED

TARASCO (old number)
1990, Mexican Navy / 0081260

2 HUASTECO CLASS (APH/AK/AH)

Name	No	Builders	Commissioned
HUASTECO (ex-*Rio Usumacinta*)	AMP 01 (ex-A 10, ex-A 21)	Tampico, Tampa	21 May 1986
ZAPOTECO (ex-*Rio Coatzacoalcos*)	AMP 02 (ex-A11, ex-A 22)	Salina Cruz	1 Sep 1986

Displacement, tons: 1,854 standard; 2,650 full load
Dimensions, feet (metres): 227 × 42 × 18.6 *(69.2 × 12.8 × 5.7)*
Main machinery: 1 GM-EMD diesel; 3,600 hp(m) *(2.65 MW)*; 1 shaft
Speed, knots: 14.5. **Range, n miles:** 5,500 at 14 kt
Complement: 85 plus 300 passengers
Guns: 1 Bofors 40/60 Mk 3.
Radars: Navigation: I-band.
Helicopters: Platform for 1 MBB BO 105C.

Comment: Used in a training role but can also serve as troop transports, supply or hospital ships. New names and pennant numbers in 2001. AMP 01 based at Tampico and AMP 02 at Manzanillo.
UPDATED

HUASTECO
7/2004 *, Diego Quevedo / 0589765

2 AGUASCALIENTES CLASS (YOG/YO)

Name	No	Builders	Recommissioned
AGUASCALIENTES (ex-*Las Choapas*)	ATQ 01 (ex-A 45, ex-A 03)	Geo H Mathis Co Ltd,	26 Nov 1964
TLAXCALA (ex-*Amatlan*)	ATQ 02 (ex-A 46, ex-A 04)	Geo Lawley & Son, Neponset, MA	26 Nov 1964

Displacement, tons: 895 standard; 1,480 full load
Dimensions, feet (metres): 159.2 × 32.9 × 13.3 *(48.6 × 10 × 4.1)*
Main machinery: 1 Fairbanks-Morse diesel; 500 hp *(373 kW)*; 1 shaft
Speed, knots: 6
Complement: 26 (5 officers)
Cargo capacity: 6,570 barrels
Guns: 1 Oerlikon 20 mm.

Comment: Former US self-propelled fuel oil barges built in 1943. Purchased in August 1964. New names and pennant numbers in 2001. ATQ 01 based at Puerto Cortes and ATQ 02 at Coatzacoalcos.
UPDATED

TLAXCALA (old number)
6/1994, Mexican Navy / 0081265

5 FLOATING DOCKS

ADI 01 (ex-US ARD 2)   **ADI 03** (ex-US AFDL 28)   — (ex-US ARD 31)
ADI 02 (ex-US ARD 15)   **ADI 04** (ex-US ARD 11)

Comment: ARD 2 (150 × 24.7 m) transferred 1963 and ARD 11 (same size) 1974 by sale. Lift 3,550 tons. Two 10 ton cranes and one 100 kW generator. ARD 15 has the same capacity and facilities-transferred 1971 by lease. AFDL 28 built in 1944, transferred 1973. Lift, 1,000 tons. ARD 30 transferred on 20 March 2001 and ARD 31 in 2004.

UPDATED

17 DREDGERS (YM)

BANDERAS ADR 01 (ex-D 01)	**CHACAGUA** ADR 10 (ex-D 24)
MAGDALENA ADR 02 (ex-D 02)	**COYUCA** ADR 11 (ex-D 25)
KINO ADR 03 (ex-D 03)	**FARRALLON** ADR 12 (ex-D 26)
YAVAROS ADR 04 (ex-D 04)	**CHAIREL** ADR 13 (ex-D 27)
CHAMELA ADR 05 (ex-D 05)	**SAN ANDRES** ADR 14 (ex-D 28)
TEPOCA ADR 06 (ex-D 06)	**SAN IGNACIO** ADR 15 (ex-D 29)
TODO SANTOS ADR 07 (ex-D 21)	**TERMINOS** ADR 16 (ex-D 30)
ASUNCION ADR 08 (ex-D 22)	**TECULAPA** ADR 17 (ex-D 31)
ALMEJAS ADR 09 (ex-D 23)	

Comment: Ships vary in size from 113 m *Kino* to 8 m *Terminos*. Most were taken over by the navy from the Transport Ministry in 1994.

VERIFIED

ALDEBARÁN (old number)   *4/1999, M Declerck /* 0081264

TRAINING SHIPS

1 SAIL TRAINING SHIP (AXS)

Name	No	Builders	Launched	Commissioned
CUAUHTÉMOC	BE 01 (ex-A 07)	Astilleros Talleres Calaya SA, Bilbao	9 Jan 1982	23 Sep 1982

Displacement, tons: 1,662 full load
Dimensions, feet (metres): 296.9 (bowsprit); 220.5 wl × 39.4 × 17.7 *(90.5; 67.2 × 12 × 5.4)*
Main machinery: 1 Detroit 12V-149T diesel; 1,125 hp *(839 kW)*; 1 shaft
Speed, knots: 17 sail; 7 diesel
Complement: 268 (20 officers, 90 midshipmen)
Guns: 2–65 mm Schneider Model 1902 saluting guns.

Comment: Has 2,368 m² of sail. Similar ships in Ecuador, Colombia and Venezuela. Based at Acapulco.

UPDATED

CUAUHTEMOC   *7/2003, J Ciślak /* 0567900

1 ADMIRABLE CLASS (SUPPORT SHIP) (AKS)

Name	No	Builders	Launched
ALDEBARÁN (ex-*DM 20*, ex-*Harlequin* AM 365, ex-ID-20)	BE 02 (ex-A 08, ex-N 02)	Willamette Iron & Steel	28 Sep 1945

Displacement, tons: 804 full load
Dimensions, feet (metres): 184.5 × 33 × 14.5 *(56.3 × 10.1 × 4.4)*
Main machinery: 4 Cooper-Bessemer GSB-8 diesels; 1,710 hp *(1.26 MW)*; 2 shafts
Speed, knots: 15. **Range, n miles:** 4,300 at 10 kt
Complement: 62 (12 officers)
Guns: 1–3 in *(76 mm)*/50 Mk 22. 4–7.62 mm MGs.
Radars: Navigation: Kelvin Hughes; I-band.

Comment: Acquired in 1962 and converted for survey work in 1978. Converted again as a training ship in 1999 and re-armed. Based at Veracruz.

UPDATED

TUGS

4 ABNAKI CLASS (ATF)

Name	No	Builders	Commissioned
OTOMI (ex-*Kukulkan*, ex-*Molala* ATF 106)	ARE 01 (ex-A 52, ex-A 17)	United Eng Co, Alameda, CA	29 Sep 1943
YAQUI (ex-*Ehacatl*, ex-*Abnaki* ATF 96)	ARE 02 (ex-A 53, ex-A 18)	Charleston SB and DD Co	15 Nov 1943
SERI (ex-*Tonatiuh*, ex-*Cocopa* ATF 101)	ARE 03 (ex-A 54, ex-A 19)	Charleston SB and DD Co	25 Mar 1944
CORA (ex-*Chac*, ex-*Hitchiti* ATF 103)	ARE 04 (ex-A 55, ex-A 20)	Charleston SB and DD Co	27 May 1944

Displacement, tons: 1,640 full load
Dimensions, feet (metres): 205 × 38.5 × 17 *(62.5 × 11.7 × 5.2)*
Main machinery: Diesel-electric; 4 Busch-Sulzer BS-539 diesels; 6,000 hp *(4.48 MW)*; 4 generators; 1 motor; 3,000 hp(m) *(2.24 MW)*; 1 shaft
Speed, knots: 10. **Range, n miles:** 6,500 at 10 kt
Complement: 75
Guns: 1 US 3 in *(76 mm)*/50 Mk 22.
Radars: Navigation: Marconi LN66; I-band.

Comment: *Otomi* transferred from US 27 September 1978, remainder 1 October 1978. All by sale. Speed reduced. Based at Tampico (ARE 01, ARE 03) and Manzanillo (ARE 02, ARE 04).

UPDATED

SERI   *6/2003, Mexican Navy /* 0567905

6 HARBOUR TUGS (YTL)

IZTACCIHUATL ARE 05 (ex-R-60)	**CITLALTEPL** ARE 07 (ex-R-62)	**MATLALCUEYE** ARE 09 (ex-R-64)
POPOCATEPTL ARE 06 (ex-R-61)	**XINANTECATL** ARE 08 (ex-R-63)	**TLALOC** ARE 10 (ex-R-65)

Displacement, tons: 140 full load
Dimensions, feet (metres): 73.8 × 22.3 × 9.8 *(22.5 × 6.8 × 3.0)*
Complement: 12

Comment: Details are for ARE 05 taken over by the Navy on 1 November 1994.

VERIFIED

Federated States of Micronesia

Country Overview

The Federated States of Micronesia was a US-administered UN Trust territory from 1947 before becoming a self-governing republic in 1979. In 1986, a Compact of Free Association, delegating to the US the responsibility for defence and foreign affairs, came into effect. Composed of the states of Pohnpei (location of capital, Palikir), Kosrae, Chuuk, and Yap, the country consists of 607 islands in the western Pacific Ocean which extend 1,566 n miles across the Caroline Islands archipelago.

Moen Island in Chuuk, is the largest community. Territorial seas (12 n miles) are claimed. An Exclusive Economic Zone (EEZ) (200 n miles) is also claimed but limits have not been fully defined.

Headquarters Appointments

Maritime Wing Commander:
  Commander Robert Maluweirang

Personnel

2005: 120

Bases

Kolonia (main base), Kosral, Moen, Takatik.

PATROL FORCES

3 PACIFIC CLASS (LARGE PATROL CRAFT) (PB)

Name	No	Builders	Commissioned
PALIKIR	FSM 01	Australian Shipbuilding Industries	28 Apr 1990
MICRONESIA	FSM 02	Australian Shipbuilding Industries	3 Nov 1990
INDEPENDENCE	FSM 05	Transfield	22 May 1997

Displacement, tons: 162 full load
Dimensions, feet (metres): 103.3 × 26.6 × 6.9 *(31.5 × 8.1 × 2.1)*
Main machinery: 2 Caterpillar 3516TA diesels; 4,400 hp *(3.28 MW)* sustained; 2 shafts
Speed, knots: 20. **Range, n miles:** 2,500 at 12 kt
Complement: 17 (3 officers)
Radars: Surface search: Furuno 1011; I-band.

Comment: First pair ordered in June 1989 from Australian Shipbuilding Industries. Training and support provided by Australia at Port Kolonia. Third of class negotiated with Transfield (former ASI) in 1997. Following the decision by the Australian government to extend the Pacific Patrol Boat programme to enable 30-year boat lives, *Palikir*, *Micronesia* and *Independence*, which underwent half-life refits in 1998, 1999 and 2003, will be due life-extension refits in 2007, 2008 and 2012 respectively. *UPDATED*

MICRONESIA   *11/1990, Royal Australian Navy*

2 CAPE CLASS (LARGE PATROL CRAFT) (PB)

Name	No	Builders	Commissioned
PALUWLAP (ex-*Cape Cross*)	FSM 03	Coast Guard Yard, Curtis Bay	20 Aug 1958
CONSTITUTION (ex-*Cape Corwin*)	FSM 04	Coast Guard Yard, Curtis Bay	14 Nov 1958

Displacement, tons: 148 full load
Dimensions, feet (metres): 95 × 20.2 × 6.6 *(28.9 × 6.2 × 2)*
Main machinery: 2 GM 16V-149TI diesels; 2,070 hp *(1.54 MW)* sustained; 2 shafts
Speed, knots: 20. **Range, n miles:** 2,500 at 10 kt
Complement: 14 (3 officers)
Guns: 2—12.7 mm MGs. 2—40 mm mortars.
Radars: Surface search: Raytheon SPS-64; I-band.

Comment: Two former Coast Guard ships transferred from US on loan in March and September 1991. Re-engined in 1982 and now restricted in speed. *VERIFIED*

CAPE class (USCG colours)   *1990*

Morocco
MARINE ROYALE MAROCAINE

Country Overview

Formerly divided into French and Spanish protectorates, the Kingdom of Morocco gained independence in 1956. Situated in north-western Africa, it has an area of 172,414 square miles and is bordered to the east by Algeria; it occupies 80 per cent of Western Sahara (formerly Spanish Sahara), the country to the south. Two Spanish exclaves, Ceuta and Melilla, are located on the Mediterranean coast. It has coastlines with Atlantic Ocean (756 n miles) and Mediterranean Sea (238 n miles). The capital is Rabat while Casablanca is the largest city and principal port. Other ports are at Tangier, Agadir, Kenitra, Mohammedia, and Safi. Territorial seas (12 n miles) are claimed. An EEZ (200 n mile) has also been claimed but its limits have not been fully defined.

Headquarters Appointments

Inspector of the Navy:
   Captain Mohammed Triki

Personnel

(a)   2005: 7,800 officers and ratings (including 1,500 Marines)
(b)   18 months' national service

Bases

Casablanca (HQ), Safi, Agadir, Kenitra, Tangier, Dakhla, Al Hoceima

Aviation

The Ministry of Fisheries operates 11 Pilatus Britten-Norman Defender maritime surveillance aircraft.

FRIGATES

2 FLOREAL CLASS (FFGHM)

Name	No	Builders	Laid down	Launched	Commissioned
MOHAMMED V	611	Chantiers de L'Atlantique, St Nazaire	June 1999	9 Mar 2001	12 Mar 2002
HASSAN II	612	Chantiers de L'Atlantique, St Nazaire	Dec 1999	11 Feb 2002	20 Dec 2002

Displacement, tons: 2,950 full load
Dimensions, feet (metres): 306.8 × 45.9 × 14.1 *(93.5 × 14 × 4.3)*
Main machinery: CODAD; 4 SEMT-Pielstick 6 PA6 L 280 diesels; 9,600 hp(m) *(7.06 MW)* sustained; 2 shafts; LIPS cp props; bow thruster; 340 hp(m) *(250 kW)*
Speed, knots: 20. **Range, n miles:** 10,000 at 15 kt
Complement: 89 (11 officers)

Missiles: SSM: 2 Aerospatiale MM 38 Exocet ❶.
SAM: 2 Matra Simbad twin launchers ❷ can replace 20 mm guns or Dagaie launcher.
Guns: 1 Otobreda 76 mm/62 ❸.
   2 Giat 20 F2 20 mm ❹.
Countermeasures: Decoys: 2 CSEE Dagaie Mk II ❺; 10-barrelled trainable launchers; chaff and IR flares.
ESM: Thomson-CSF ARBR 17 ❻; radar intercept.
Weapons control: CSEE Najir optronic director ❼.
Radars: Surface search/Fire control: Thales WM28 ❽; I/J-band.
Navigation: 2 Decca Bridgemaster E ❾; I-band (1 for helicopter control).

Helicopters: 1 Aerospatiale AS 565MA Panther.

Programmes: Contract signed with Alstom on 12 July 1999. 611 delivered on 12 March 2002 and 612 on 20 December 2002.
Structure: Constructed to DNV standards. Very similar to ships in French service with 76 mm in place of 100 mm gun. *UPDATED*

MOHAMMED V   *(Scale 1 : 900), Ian Sturton / 1044134*

HASSAN II
1/2004, B Prézelin / 0589780*

MOHAMMED V

8/2004*, B Prézelin / 0589781

1 MODIFIED DESCUBIERTA CLASS (FFGM)

Name	No	Builders	Laid down	Launched	Commissioned
LIEUTENANT COLONEL ERRHAMANI	501	Bazán, Cartagena	20 Mar 1979	26 Feb 1982	28 Mar 1983

Displacement, tons: 1,233 standard; 1,479 full load
Dimensions, feet (metres): 291.3 × 34 × 12.5
  (88.8 × 10.4 × 3.8)
Main machinery: 4 MTU-Bazán 16V 956 TB91 diesels;
  15,000 hp(m) *(11 MW)* sustained; 2 shafts; cp props
Speed, knots: 25.5. **Range, n miles:** 4,000 at 18 kt (1 engine)
Complement: 100

Missiles: SSM: 4 Aerospatiale MM 38 Exocet ❶; inertial cruise;
  active radar homing to 42 km *(23 n miles)* at 0.9 Mach;
  warhead 165 kg; sea-skimmer. Frequently not embarked.
SAM: Selenia/Elsag Albatros octuple launcher ❷; 24 Aspide;
  semi-active radar homing to 13 km *(8 n miles)* at 2.5 Mach;
  height envelope 15-5,000 m *(49.2-16,405 ft)*; warhead
  30 kg.
Guns: 1 OTO Melara 3 in *(76 mm)*/62 compact ❸; 85 rds/min to
  16 km *(8.6 n miles)* anti-surface; 12 km *(6.5 n miles)* anti-
  aircraft; weight of shell 6 kg.
  2 Breda Bofors 40 mm/70 ❹; 300 rds/min to 12.5 km *(6.7 n
  miles)*; weight of shell 0.96 kg.
Torpedoes: 6—324 mm Mk 32 (2 triple) tubes ❺. Honeywell Mk
  46 Mod 1; anti-submarine; active/passive homing to 11 km
  (5.9 n miles) at 40 kt; warhead 44 kg.
A/S mortars: 1 Bofors SR 375 mm twin trainable launcher ❻;
  range 3.6 km *(1.9 n miles)*; 24 rockets.
Countermeasures: Decoys: 2 CSEE Dagaie double trainable
  mounting; IR flares and chaff; H/J-band.
ESM/ECM: Elettronica ELT 715; intercept and jammer.
Combat data systems: Signaal SEWACO-MR action data
  automation. SATCOM.
Radars: Air/surface search: Signaal DA05 ❼; E/F-band (see
  Operational).
Surface search: Signaal ZW06 ❽; I-band.
Fire control: Signaal WM25/41 ❾; I/J-band; range 46 km *(25 n
  miles)*.
Sonars: Raytheon DE 1160 B; hull-mounted; active/passive;
  medium range; medium frequency.

Programmes: Ordered 7 June 1977.
Modernisation: New 40 mm guns fitted in 1995. Refit in Spain in
  1996.
Operational: The ship is fitted to carry Exocet but the missiles are
  seldom embarked. The air search radar was removed in 1998
  but reinstated in 1999.

UPDATED

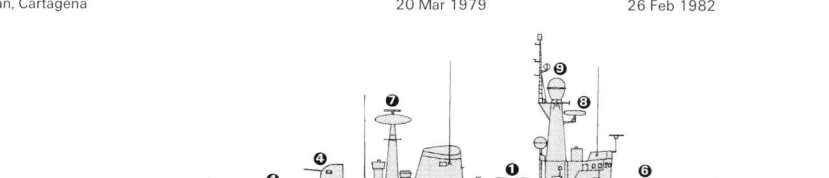

LIEUTENANT COLONEL ERRHAMANI
(Scale 1 : 900), Ian Sturton

LIEUTENANT COLONEL ERRHAMANI
5/2003, B Prézelin / 0568858

LIEUTENANT COLONEL ERRHAMANI
5/2003, B Prézelin / 0568859

SHIPBORNE AIRCRAFT

Numbers/Type: 3 Aerospatiale AS 565MB Panther.
Operational speed: 165 kt *(305 km/h)*.
Service ceiling: 16,700 ft *(5,100 m)*.
Range: 483 n miles *(895 km)*.
Role/Weapon systems: Procured from France for operation from Floréal class. Sensors: Thomson-CSF Varan radar. FLIR. Weapons: 7.62 mm MG.

UPDATED

PANTHER (French colours)   9/1998, M Declerck / 0052167

PATROL FORCES

4 LAZAGA CLASS (FAST ATTACK CRAFT—MISSILE) (PGG)

Name	No	Builders	Commissioned
EL KHATTABI	304	Bazán, San Fernando	26 July 1981
COMMANDANT BOUTOUBA	305	Bazán, San Fernando	2 Aug 1982
COMMANDANT EL HARTY	306	Bazán, San Fernando	20 Nov 1981
COMMANDANT AZOUGGARH	307	Bazán, San Fernando	25 Feb 1982

Displacement, tons: 425 full load
Dimensions, feet (metres): 190.6 × 24.9 × 8.9 *(58.1 × 7.6 × 2.7)*
Main machinery: 2 MTU-Bazán 16V 956 TB91 diesels; 7,500 hp(m) *(5.51 MW)* sustained; 2 shafts
Speed, knots: 30. **Range, n miles:** 3,000 at 15 kt
Complement: 41
Missiles: SSM: 4 Aerospatiale MM 38 Exocet; inertial cruise; active radar homing to 42 km *(23 n miles)* at 0.9 Mach; warhead 165 kg; sea-skimmer.
Guns: 1 OTO Melara 3 in *(76 mm)*/62 compact; 85 rds/min to 16 km *(8.6 n miles)* anti-surface; 12 km *(6.5 n miles)* anti-aircraft; weight of shell 6 kg.
1 Breda Bofors 40 mm/70; 300 rds/min to 12.5 km *(6.7 n miles)*; weight of shell 0.96 kg.
2 Oerlikon 20 mm/90 GAM-BO1; 800 rds/min to 2 km.
Weapons control: CSEE Panda optical director.
Radars: Surface search: Signaal ZW06; I-band; range 26 km *(14 n miles)*.
Fire control: Signaal WM25; I/J-band; range 46 km *(25 n miles)*.
Navigation: Furuno; I-band.

Comment: Ordered from Bazán, San Fernando (Cadiz), Spain 14 June 1977. New Bofors guns fitted aft in 1996/97. 76 mm gun removed from 305 in 1998.

UPDATED

EL KHATTABI   8/1997, Diego Quevedo / 0012784

COMMANDANT AZOUGGARH   9/2001 / 0525910

---

2 OKBA (PR 72) CLASS (LARGE PATROL CRAFT) (PG)

Name	No	Builders	Commissioned
OKBA	302	SFCN, Villeneuve la Garenne	16 Dec 1976
TRIKI	303	SFCN, Villeneuve la Garenne	12 July 1977

Displacement, tons: 375 standard; 445 full load
Dimensions, feet (metres): 188.8 × 25 × 7.1 *(57.5 × 7.6 × 2.1)*
Main machinery: 2 SACM AGO V16 ASHR diesels; 5,520 hp(m) *(4.1 MW)*; 2 shafts
Speed, knots: 20. **Range, n miles:** 2,500 at 16 kt
Complement: 53 (5 officers)
Guns: 1 OTO Melara 3 in *(76 mm)*/62 compact; 85 rds/min to 16 km *(8.6 n miles)* anti-surface; 12 km *(6.5 n miles)* anti-aircraft; weight of shell 6 kg.
1 Bofors 40 mm/70; 300 rds/min to 12.5 km *(6.7 n miles)*; weight of shell 0.96 kg.
Weapons control: 2 CSEE Panda optical directors.
Radars: Surface search: Racal Decca 1226; I-band.

Comment: Ordered June 1973. *Okba* launched 10 October 1975, *Triki* 1 February 1976. Can be Exocet fitted (with Vega control system). *Triki* refitted at Lorient 2002-03. Modifications included installation of a funnel and removal of two diesels and two shafts. Speed reduced to 20 kt. Similar refit for *Okba* in 2004.

UPDATED

TRIKI   6/2003*, B Prézelin / 0589787

4 OSPREY MK II CLASS (LARGE PATROL CRAFT) (PBO)

Name	No	Builders	Commissioned
EL HAHIQ	308	Danyard A/S, Frederickshaven	11 Nov 1987
EL TAWFIQ	309	Danyard A/S, Frederickshaven	31 Jan 1988
EL HAMISS	316	Danyard A/S, Frederickshaven	9 Aug 1990
EL KARIB	317	Danyard A/S, Frederickshaven	23 Sep 1990

Displacement, tons: 475 full load
Dimensions, feet (metres): 179.8 × 34 × 8.5 *(54.8 × 10.5 × 2.6)*
Main machinery: 2 MAN Burmeister & Wain Alpha 12V23/30-DVO diesels; 4,440 hp(m) *(3.23 MW)* sustained; 2 water-jets
Speed, knots: 22. **Range, n miles:** 4,500 at 16 kt
Complement: 15 plus 20 spare berths
Guns: 1 Bofors 40 mm/60. 2 Oerlikon 20 mm.
Radars: Surface search; Racal Decca; I-band.
Navigation: Racal Decca; I-band.

Comment: First two ordered in September 1986; two more on 30 January 1989. There is a stern ramp with a hinged cover for launching the inspection boat. Used for Fishery Protection duties.

UPDATED

EL HAHIQ   2/2004*, Adolfo Ortigueira Gil / 0589783

6 CORMORAN CLASS (LARGE PATROL CRAFT) (PBO)

Name	No	Builders	Launched	Commissioned
L V RABHI	310	Bázan, San Fernando	23 Sep 1987	16 Sep 1988
ERRACHIQ	311	Bázan, San Fernando	23 Sep 1987	16 Dec 1988
EL AKID	312	Bázan, San Fernando	29 Mar 1988	4 Apr 1989
EL MAHER	313	Bázan, San Fernando	29 Mar 1988	20 June 1989
EL MAJID	314	Bázan, San Fernando	21 Oct 1988	26 Sep 1989
EL BACHIR	315	Bázan, San Fernando	21 Oct 1988	19 Dec 1989

Displacement, tons: 425 full load
Dimensions, feet (metres): 190.6 × 24.9 × 8.9 *(58.1 × 7.6 × 2.7)*
Main machinery: 2 MTU-Bázan 16V 956 TB82 diesels; 8,340 hp(m) *(6.13 MW)* sustained; 2 shafts
Speed, knots: 22. **Range, n miles:** 6,100 at 12 kt
Complement: 36 (4 officers) plus 15 spare
Guns: 1 Bofors 40 mm/70. 2 Giat 20 mm.
Weapons control: CSEE Lynx optronic director.
Radars: Surface search: Racal Decca; I-band.

Comment: Three ordered from Bazán, Cadiz in October 1985 as a follow on to the Lazaga class of which these are a slower patrol version with a 10 day endurance. Option on three more taken up. Used for fishery protection.

UPDATED

EL AKID   10/1999, M Declerck / 0105136

5 RAÏS BARGACH CLASS (TYPE OPV 64) (PSO)

Name	No	Builders	Launched	Commissioned
RAÏS BARGACH	318	Leroux & Lotz, Lorient	9 Oct 1995	14 Dec 1995
RAÏS BRITEL	319	Leroux & Lotz, Lorient	19 Mar 1996	14 May 1996
RAÏS CHARKAOUI	320	Leroux & Lotz, Lorient	25 Sep 1996	10 Dec 1996
RAÏS MAANINOU	321	Leroux & Lotz, Lorient	7 Mar 1997	21 May 1997
RAÏS AL MOUNASTIRI	322	Leroux & Lotz, Lorient	15 Oct 1997	17 Dec 1997

Displacement, tons: 580 full load
Dimensions, feet (metres): 210 × 37.4 × 9.8 *(64 × 11.4 × 3)*
Main machinery: 2 Wärtsilä Nohab 25 V16 diesels; 10,000 hp(m) *(7.36 MW)* sustained; 2 Leroy auxiliary motors; 326 hp(m) *(240 kW)*; 2 shafts; cp props
Speed, knots: 24; 7 (on motors). **Range, n miles:** 4,000 at 12 kt
Complement: 24 (3 officers) + 30 spare
Guns: 1 Bofors 40 mm/60. 1 Oerlikon 20 mm. 4—14.5 mm MGs (2 twin).
Radars: Surface search: Racal Decca Bridgemaster; I-band.

Comment: First pair ordered to a Serter design from Leroux & Lotz, Lorient in December 1993, second pair in October 1994. Option on fifth taken up in 1996. There is a stern door for launching a 7 m RIB, a water gun for firefighting and two passive stabilisation tanks. This version of the OPV 64 does not have a helicopter deck and the armament is fitted after delivery. Manned by the Navy for the Fisheries Department. Based at Agadir.

UPDATED

RAS AL MOUNASTIRI   *9/2003, Schaeffer/Marsan /* 0568855

RAÏS CHARKAOUI   *4/2004*, B Prézelin /* 0589786

6 EL WACIL (P 32) CLASS (COASTAL PATROL CRAFT) (PB)

Name	No	Builders	Launched	Commissioned
EL WACIL	203	CMN, Cherbourg	12 June 1975	9 Oct 1975
EL JAIL	204	CMN, Cherbourg	10 Oct 1975	3 Dec 1975
EL MIKDAM	205	CMN, Cherbourg	1 Dec 1975	30 Jan 1976
EL KHAFIR	206	CMN, Cherbourg	21 Jan 1976	16 Apr 1976
EL HARIS	207	CMN, Cherbourg	31 Mar 1976	30 June 1976
EL ESSAHIR	208	CMN, Cherbourg	2 June 1976	16 July 1976

Displacement, tons: 74 light; 89 full load
Dimensions, feet (metres): 105 × 17.7 × 4.6 *(32 × 5.4 × 1.4)*
Main machinery: 2 SACM MGO 12V BZSHR diesels; 2,700 hp(m) *(1.98 MW)*; 2 shafts
Speed, knots: 28. **Range, n miles:** 1,500 at 15 kt
Complement: 17
Guns: 1 Oerlikon 20 mm.
Radars: Surface search: Decca; I-band.

Comment: Ordered in February 1974. In July 1985 a further four of this class were ordered from the same builders but for the Customs Service. Wooden hull sheathed in plastic.

UPDATED

EL JAIL   *9/2004*, S D Llosá /* 1044135

AMPHIBIOUS FORCES

1 EDIC CLASS (LCU)

Name	No	Builders	Commissioned
LIEUTENANT MALGHAGH	401	Chantiers Navals Franco-Belges	1965

Displacement, tons: 250 standard; 670 full load
Dimensions, feet (metres): 193.5 × 39.2 × 4.3 *(59 × 12 × 1.3)*
Main machinery: 2 SACM MGO diesels; 1,000 hp(m) *(735 kW)*; 2 shafts
Speed, knots: 8. **Range, n miles:** 1,800 at 8 kt
Complement: 16 (1 officer)
Military lift: 11 vehicles
Guns: 2 Oerlikon 20 mm. 1—120 mm mortar.
Radars: Navigation: Decca 1226; I-band.

Comment: Ordered early in 1963. Similar to the former French landing craft of the Edic type built at the same yard.   *UPDATED*

LIEUTENANT MALGHAGH   *2/1995, Diego Quevedo /* 0081277

1 NEWPORT CLASS (LSTH)

Name	No	Builders	Commissioned
SIDI MOHAMMED BEN ABDALLAH (ex-*Bristol County*)	407 (ex-1198)	National Steel, San Diego	5 Aug 1972

Displacement, tons: 4,975 light; 8,450 full load
Dimensions, feet (metres): 522.3 (hull) × 69.5 × 17.5 (aft) *(159.2 × 21.2 × 5.3)*
Main machinery: 6 ALCO 16-251 diesels; 16,500 hp *(12.3 MW)* sustained; 2 shafts; cp props; bow thruster
Speed, knots: 20. **Range, n miles:** 14,250 at 14 kt
Complement: 257 (13 officers)
Military lift: 400 troops (20 officers); 500 tons vehicles; 3 LCVPs and 1 LCPL on davits
Guns: 1 GE/GD 20 mm 6-barrelled Vulcan Phalanx Mk 15.
Radars: Surface search: Raytheon SPS-67; G-band.
Navigation: Marconi LN66; I/J-band.
Helicopters: Platform only.

Comment: Received from the US by grant transfer on 16 August 1994. Has replaced *Arrafiq*. The ship was non-operational by late 1995 and although back in service, has so far proved to be a poor bargain. The bow ramp is supported by twin derrick arms. A ramp just forward of the superstructure connects the lower tank deck with the main deck and a vehicle passage through the superstructure provides access to the parking area amidships. A stern gate to the tank deck permits unloading of amphibious tractors into the water, or unloading of other vehicles into an LCU or on to a pier. Vehicle stowage covers 19,000 sq ft. Length over derrick arms is 562 ft *(171.3 m)*; full load draught is 11.5 ft forward and 17.5 ft aft. Based at Casablanca.

UPDATED

SIDI MOHAMMED BEN ABDALLAH   *8/2003*, Diego Quevedo /* 0589785

3 BATRAL CLASS (LSMH)

Name	No	Builders	Commissioned
DAOUD BEN AICHA	402	Dubigeon, Normandie	28 May 1977
AHMED ES SAKALI	403	Dubigeon, Normandie	Sep 1977
ABOU ABDALLAH EL AYACHI	404	Dubigeon, Normandie	Mar 1978

Displacement, tons: 750 standard; 1,409 full load
Dimensions, feet (metres): 262.4 × 42.6 × 7.9 *(80 × 13 × 2.4)*
Main machinery: 2 SACM Type 195 V12 CSHR diesels; 3,600 hp(m) *(2.65 MW)* sustained; 2 shafts
Speed, knots: 16. **Range, n miles:** 4,500 at 13 kt
Complement: 47 (3 officers)
Military lift: 140 troops; 12 vehicles or 300 tons
Guns: 2 Bofors 40 mm/70. 2—81 mm mortars. 2—12.7 mm MGs.
Radars: Surface search: Thomson-CSF DRBN 32 (Racal Decca 1226); I-band.
Helicopters: Platform only.

Comment: Two ordered on 12 March 1975. Third ordered 19 August 1975. Of same type as the French *Champlain*. Vehicle-stowage above and below decks. *Daoud Ben Aicha* was refitted in Lorient by Leroux & Lotz in 1995 and *Abou Abdallah el Ayachi* in 1997.

UPDATED

ABOU ABDALLAH EL AYACHI   *10/2003*, Martin Mokrus /* 0589784

AUXILIARIES

Notes: (1) There is also a yacht, *Essaouira*, 60 tons, from Italy in 1967, used as a training vessel for watchkeepers.
(2) Bazán delivered a harbour pusher tug, similar to Spanish Y 171 class, in December 1993.
(3) There are two sail training craft *Al Massira* and *Boujdour*.
(4) There is a stern trawler used as a utility and diver support vessel (803 (ex-YFU 14)).

803   *9/2004 *, S D Llosá* / 1044141

1 LOGISTIC SUPPORT SHIP (AKS)

EL AIGH (ex-*Merc Nordia*) 405

Measurement, tons: 1,500 grt
Dimensions, feet (metres): 252.6 × 40 × 15.4 *(77 × 12.2 × 4.7)*
Main machinery: 1 Burmeister & Wain diesel; 1,250 hp(m) *(919 kW)*; 1 shaft
Speed, knots: 11
Complement: 25
Guns: 2—14.5 mm MGs.

Comment: Logistic support vessel with four 5 ton cranes. Former cargo ship with ice-strengthened bow built by Fredrickshavn Vaerft in 1973 and acquired in 1981.   **VERIFIED**

EL AIGH   *5/1994, M Declerck*

1 DAKHLA CLASS (LOGISTIC SUPPORT SHIP) (AKS)

Name	No	Builders	Launched	Commissioned
DAKHLA	408	Leroux & Lotz, Lorient	5 June 1997	1 Aug 1997

Displacement, tons: 2,160 full load
Dimensions, feet (metres): 226.4 × 37.7 × 13.8 *(69 × 11.5 × 4.2)*
Main machinery: 1 Wärtsilä Nohab 8V25 diesel; 2,300 hp(m) *(1.69 MW)* sustained; 1 shaft; cp prop
Speed, knots: 12. **Range, n miles:** 4,300 at 12 kt
Complement: 24 plus 22 spare
Cargo capacity: 800 tons
Guns: 2—12.7 mm MGs.
Radars: Navigation: 2 Racal Decca Bridgemaster ARPA; I-band.

Comment: Ordered from Leroux & Lotz, Nantes in 1995. Side entry for vehicles. One 15 ton crane. Based at Agadir.   **VERIFIED**

DAKHLA   *8/1997, Leroux & Lotz* / 0012789

SURVEY AND RESEARCH SHIPS

1 ROBERT D CONRAD CLASS (AGOR)

Name	No	Builders	Commissioned
ABU AL BARAKAT AL BARBARI (ex-*Bartlett*)	702 (ex-T-AGOR 13)	Northwest Marine Iron Works, Portland, OR	31 Mar 1969

Displacement, tons: 1,200 light; 1,370 full load
Dimensions, feet (metres): 208.9 × 40 × 15.3 *(63.7 × 12.2 × 4.7)*
Main machinery: Diesel-electric; 2 Caterpillar D 378 diesel generators; 1 motor; 1,000 hp *(746 kW)*; 1 shaft; bow thruster
Speed, knots: 13.5. **Range, n miles:** 12,000 at 12 kt
Complement: 41 (9 officers, 15 scientists)
Radars: Navigation: TM 1660/12S; I-band.

Comment: Leased from the USA on 26 July 1993. Fitted with instrumentation and laboratories to measure gravity and magnetism, water temperature, sound transmission in water, and the profile of the ocean floor. Special features include 10 ton capacity boom and winches for handling over-the-side equipment; bow thruster; 620 hp gas turbine (housed in funnel structure) for providing 'quiet' power when conducting experiments; can propel the ship at 6.5 kt.
Ships of this class are in service with Brazil, Mexico, Chile, Tunisia and Portugal. **UPDATED**

ABU EL BARAKAT AL BARABARI   *3/2001* / 0525909

CUSTOMS/COAST GUARD/POLICE

Notes: (1) The Coast Guard was created by Royal Decree on 9 September 1997. Responsibility for Search and Rescue conferred on the Ministére des Pêches Maritimes (MPM). Operational control is exercised from the National Rescue Service HQ at Rabat in co-ordination with the Merchant Marine HQ at Casablanca.
(2) There is a 17 m SAR craft *Al Fida* delivered in August 2002.
(3) There are four SAR craft: *Rif, Loukouss, Souss* and *Dghira*.

AL FIDA   *7/2004 *, S D Llosá* / 1044137

SOUSS   *7/1995 *, Zamacona* / 1044138

2 SAR CRAFT (SAR)

AL AMANE 2344   **AIT BAÂMRANE** 2345

Displacement, tons: 68 full load
Dimensions, feet (metres): 51.7 × 14.7 × 3.4 *(15.75 × 4.48 × 1.05)*
Main machinery: 2 Volvo D12; 1,300 hp *(970 kW)*; Hamilton waterjets
Speed, knots: 34
Complement: 4

Comment: Constructed by Auxnaval Shipbuilders, Spain and delivered in March 2003. Aluminium hull.   **UPDATED**

AL AMANE   *7/2003 *, Auxnaval* / 1044136

2 SAR CRAFT (SAR)

AL WHADA 12-64   SEBOU 12-65

Displacement, tons: 70 full load
Dimensions, feet (metres): 68.0 × 19.2 × 5.9 *(20.7 × 5.8 × 1.8)*
Main machinery: 2 MAN D2842 LE401 diesels; 2,000 hp *(1.49 MW)*; 2 shafts
Speed, knots: 20
Complement: 4

Comment: Constructed by Auxnaval, Asturias, Spain and delivered in 2004.

NEW ENTRY

AL WHADA   7/2004*, Auxnaval / 1044139

4 ERRAID (P 32) CLASS (COASTAL PATROL CRAFT) (WPB)

Name	No	Builders	Launched	Commissioned
ERRAID	209	CMN, Cherbourg	20 Dec 1987	18 Mar 1988
ERRACED	210	CMN, Cherbourg	21 Jan 1988	15 Apr 1988
EL KACED	211	CMN, Cherbourg	10 Mar 1988	17 May 1988
ESSAID	212	CMN, Cherbourg	19 May 1988	4 July 1988

Displacement, tons: 89 full load
Dimensions, feet (metres): 105 × 17.7 × 4.6 *(32 × 5.4 × 1.4)*
Main machinery: 2 SACM MGO 12V BZSHR diesels; 2,700 hp(m) *(1.98 MW)*; 2 shafts
Speed, knots: 28. **Range, n miles:** 1,500 at 15 kt
Complement: 17
Guns: 1 Oerlikon 20 mm.
Radars: Navigation: Decca; I-band.

Comment: Similar to the El Wacil class listed under Patrol Forces. Ordered in July 1985.

VERIFIED

EL KACED   6/1999 / 0081279

18 ARCOR 46 CLASS (COASTAL PATROL CRAFT) (WPB)

Displacement, tons: 15 full load
Dimensions, feet (metres): 47.6 × 13.8 × 4.3 *(14.5 × 4.2 × 1.3)*
Main machinery: 2 SACM UD18V8 M5D diesels; 1,010 hp(m) *(742 kW)* sustained; 2 shafts
Speed, knots: 32. **Range, n miles:** 300 at 20 kt
Complement: 6
Guns: 2 Browning 12.7 mm MGs.
Radars: Surface search: Furuno 701; I-band.

Comment: Ordered from Arcor, La Teste in June 1985. GRP hulls. Delivered in groups of three from April to September 1987. Used for patrolling the Mediterranean coastline.   *UPDATED*

ARCOR 46 CLASS   9/2004*, S D Llosá / 1044140

15 ARCOR 53 CLASS (COASTAL PATROL CRAFT) (WPBF)

Displacement, tons: 17 full load
Dimensions, feet (metres): 52.5 × 13 × 3.9 *(16 × 4 × 1.2)*
Main machinery: 2 Saab DSI-14 diesels; 1,250 hp(m) *(919 kW)*; 2 shafts
Speed, knots: 35. **Range, n miles:** 300 at 20 kt
Complement: 6
Guns: 1—12.7 mm MG.
Radars: Surface search: Furuno; I-band.

Comment: Ordered from Arcor, La Teste in 1990 for the Police Force. Delivered at one a month from October 1992.   *VERIFIED*

ARCOR 53 (Police)   10/1999, M Declerck / 0105138

3 SAR CRAFT (SAR)

HAOUZ   ASSA   TARIK

Displacement, tons: 40 full load
Dimensions, feet (metres): 63.6 × 15.7 × 4.3 *(19.4 × 4.8 × 1.3)*
Main machinery: 2 diesels; 1,400 hp(m) *(1.03 MW)*; 2 shafts
Speed, knots: 20
Complement: 6

Comment: Rescue craft built by Schweers, Bardenfleth and delivered in 1991.   *VERIFIED*

Mozambique
MARINHA MOÇAMBIQUE

Country Overview

The Republic of Mozambique gained independence from Portugal in 1975. Situated in south-eastern Africa, it has an area of 308,642 square miles and is bordered to the north by Tanzania, to the south by South Africa and Swaziland and to the west by Zimbabwe, Zambia, and Malawi. It has a 1,334 n mile coastline with the Mozambique Channel of the Indian Ocean. Maputo (formerly Lourenço Marques) is the capital, largest city and principal port. There is another major port at Beira. Territorial Seas (12 n miles) are claimed. A 200 n mile EEZ has also been claimed but the limits are not fully defined by boundary agreements.

All the Russian built Zhuks and Yevgenyas have sunk alongside or been sold. There are some motorboats operational on Lake Malawi. Restoration of the navy began with the donation by South Africa of two Namacurra class inshore patrol craft in late 2004.

Headquarters Appointments

Head of Navy:
Vice Admiral Pascoal Jose Nhalungo

Personnel

2005: 200

Bases

Maputo (Naval HQ); Nacala; Beira; Pemba (Porto Amelia); Metangula (Lake Malawi); Tete (River Zambesi); Inhambane.

Myanmar
TATMADAW YAY

Country Overview

The Union of Myanmar, also known as the Republic of Burma, gained independence in 1948. Situated in South East Asia, it has an area of 261,218 square miles, is bordered to the north-east by China, to the north-west by India and Bangladesh, and to the south-east by Laos and Thailand. It has a 1,042 n mile coastline with the Andaman Sea and the Bay of Bengal. Rangoon (Yangon) is the capital, largest city and principal port. Some 6,900 n miles of navigable inland waterways are important transport arteries. Territorial waters (12 n miles) are claimed. A 200 n mile EEZ has been claimed although the limits have only been partly defined by boundary agreements.

Headquarters Appointments

Commander in Chief:
Vice Admiral Kyi Min

Personnel

(a)   2005: 13,000 (including 800 naval infantry)
(b)   Voluntary service

Naval Infantry

Mostly deployed to Arakan and Tenasserim coastal regions and to the Irawaddy delta for counter-insurgency operations.

Bases

There are five regional commands with principal bases as indicated:

Ayeyarwady (Rangoon (Monkey Point) (HQ), Yadanabon, Great Coco Island)
Taninthayi (Mergui (HQ), Kathekyun, Pale Island, Thetkatan, Pearl Island, Lampi Island)
Panmawady (Hainggyi Island (HQ), Pathein)
Mawrawady (Moulmein (HQ))
Danyawady (Sittwe (HQ), Kyaukpyu, Thandwe)

The Headquarters of Training Command is in Rangoon. The main training depot is currently at Syriam (Thanlyin), but is in the process of being moved to Thilawa, near the mouth of the Hlaing (Rangoon) River.
The Pathein base will reportedly be moved to Bya Date Kyee Island, where an expanded airfield will permit the basing of air force equipment and personnel as well as navy. The Great Coco Island base is also being expanded through the construction of a large landing jetty to replace the existing small pier.

Organisations

Naval units are usually commanded directly from Rangoon, but operational control is occasionally delegated to regional commands.

CORVETTES

3 SINMALAIK CLASS (CORVETTES) (FS)

Name	No	Builders	Laid down	Launched	Commissioned
—	771	Sinmalaik Shipyard, Rangoon	1998	2000	2001
—	—	Sinmalaik Shipyard, Rangoon	1998	2001	2002
—	—	Sinmalaik Shipyard, Rangoon	1998	2001	2003

Displacement, tons: 1,200 full load
Dimensions, feet (metres): 252.6 × ? × ? *(77.0 × ? × ?)*

Main machinery: To be announced
Speed, knots: To be announced
Complement: To be announced

Guns: 1 OTO Breda 3 in *(76 mm)*/62 compact; 85 rds/min to 16 km *(8.7 n miles)* anti-surface; 12 km *(6.5 n miles)* anti-aircraft; weight of shell 6 kg.
  2 Breda 40 mm/70 (twin); 300 rds/min to 12.5 km *(6.8 n miles)*; weight of shell 0.96 kg.
Countermeasures: To be announced.
Radars: Surface search: To be announced.
Navigation: To be announced.
Fire control: To be announced.
Sonars: To be announced.

Helicopters: Platform for 1 medium.

Programmes: The programme to acquire ships to replace the now decommissioned PCE-827 and Admirable class corvettes was probably instituted in the 1990s. As frigates proved to be too expensive, three Chinese hulls are believed to have been acquired in about 1998 for fitting out at Sinmalaik Shipyard. There have been reports that Israeli electronic systems (radars and sonar) have been fitted. The details of the programme are speculative. Further vessels may be under construction.
Operational: There has been speculation that these vessels were to be armed with four C-801 anti-ship missiles but is unclear as to whether they have been fitted. The first ship conducted sea trials in 2001 when the second ship was nearing completion. Three ships were reported in commission by 2004.

NEW ENTRY

771
12/2004 * / 0581402

PATROL FORCES

6 HOUXIN (TYPE 037/1G) CLASS
(FAST ATTACK CRAFT—GUN) (PTG)

MAGA 471   DUWA 473   475-476
SAITTRA 472   ZEYDA 474

Displacement, tons: 478 full load
Dimensions, feet (metres): 206 × 23.6 × 7.9 *(62.8 × 7.2 × 2.4)*
Main machinery: 4 PR 230ZC diesels; 4,000 hp(m) *(2.94 MW)*; 4 shafts
Speed, knots: 28. **Range, n miles:** 1,300 at 15 kt
Complement: 71

Missiles: SSM: 4 YJ-1 (C-801) (2 twin); active radar homing to 40 km *(22 n miles)* at 0.9 Mach; warhead 165 kg; sea skimmer. C-802 may be fitted in due course.
Guns: 4—37 mm/63 Type 76A (2 twin); 180 rds/min to 8.5 km *(4.6 n miles)*; weight of shell 1.42 kg.
  4—14.5 mm Type 69 (2 twin).
Countermeasures: ESM/ECM: intercept and jammer.
Radars: Surface search: Square Tie; I-band.
Fire control: Rice Lamp; I-band.

Programmes: First pair arrived from China in December 1995, second pair in mid-1996 and last two in late 1997. The first four were wrongly reported as Hainan class.
Structure: Details given are for this class in Chinese service.
Operational: *475* damaged in a collision during sea trials in August 1996. All based at Rangoon.

VERIFIED

6 MYANMAR CLASS (COASTAL PATROL CRAFT) (PBO)

551-556

Displacement, tons: 213 full load
Dimensions, feet (metres): 147.3 × 23 × 8.2 *(45 × 7 × 2.5)*
Main machinery: 2 Mercedes-Benz diesels; 2 shafts
Speed, knots: 30+
Complement: 34 (7 officers)
Guns: 1—40 mm/60. 2—37 mm (1 twin). 4—23 mm (2 twin). 4-14.5 mm (2 twin).
Radars: Surface search: I-band.
Fire control: Rice Lamp; I-band.

Comment: First ship under construction at the Naval Engineering Depot, Rangoon in 1991. *551* launched on 2 January 1996 and *552* on 4 January 1996. Four further vessels reported in service by 2004. More may be built. Missiles (C-801) may be carried.

UPDATED

MYANMAR CLASS 556   *12/2004* * / 0581401

MYANMAR CLASS 553   *6/2001* / 0130746

ZEYDA   *6/2001* / 0130747

2 OSPREY CLASS (OFFSHORE PATROL VESSELS) (PBO)

Name	No	Builders	Commissioned
INDAW	FV 55	Frederikshavn Dockyard	30 May 1980
INYA	FV 57	Frederikshavn Dockyard	25 Mar 1982

Displacement, tons: 385 standard; 505 full load
Dimensions, feet (metres): 164 × 34.5 × 9 *(50 × 10.5 × 2.8)*
Main machinery: 2 Burmeister and Wain Alpha diesels; 4,640 hp(m) *(3.4 MW)*; 2 shafts; cp props
Speed, knots: 20. **Range, n miles:** 4,500 at 16 kt
Complement: 20 (5 officers)
Guns: 1 Bofors 40 mm/60. 2 Oerlikon 20 mm.

Comment: Operated by Burmese Navy for the People's Pearl and Fishery Department. Helicopter deck with hangar in *Indaw*. Carry David Still craft or RIBs capable of 25 kt. *Inya* reported to be in poor condition. Both based at Rangoon. A third of class, *Inma*, reported to have sunk in 1987. A similar ship is in service in Namibia.

UPDATED

INYA   *1980* / 0056642

10 HAINAN (TYPE 037) CLASS (COASTAL PATROL CRAFT) (PC)

Name	No	Name	No	Name	No
YAN SIT AUNG	441	YAN YE AUNG	445	YAN WIN AUNG	448
YAN MYAT AUNG	442	YAN MIN AUNG	446	YAN AYE AUNG	449
YAN NYEIN AUNG	443	YAN PAING AUNG	447	YAN ZWE AUNG	450
YAN KHWIN AUNG	444				

Displacement, tons: 375 standard; 392 full load
Dimensions, feet (metres): 192.8 × 23.6 × 7.2 *(58.8 × 7.2 × 2.2)*
Main machinery: 4 PCR/Kolomna Type 9-D-8 diesels; 4,000 hp(m) *(2.94 MW)* sustained; 4 shafts
Speed, knots: 30.5. **Range, n miles:** 1,300 at 15 kt
Complement: 69
Guns: 4 China 57 mm/70 (2 twin); 120 rds/min to 12 km *(6.5 n miles)*; weight of shell 6.31 kg. 4 USSR 25 mm/60 (2 twin); 270 rds/min to 3 km *(1.6 n miles)* anti-aircraft; weight of shell 0.34 kg.
A/S mortars: 4 RBU 1200 5-tubed fixed launchers; range 1,200 m; warhead 34 kg.
Depth charges: 2 BMB-2 projectors; 2 racks.
Mines: Rails fitted.
Countermeasures: ESM: Intercept.
Radars: Surface search: Pot Head; I-band.
Navigation: Raytheon Pathfinder; I-band.
IFF: High Pole.
Sonars: Stag Ear; hull-mounted; active search and attack; high frequency.

Comment: First six delivered from China in January 1991, four more in mid-1993. The first six originally had double figure pennant numbers which have been changed to three figures. These ships are the later variant of this class with tripod masts. Based at Rangoon.

VERIFIED

YAN WIN AUNG   *9/1993* / 0056641

YAN KHWIN AUNG   *12/1994, G Toremans*

3 PB 90 CLASS (COASTAL PATROL CRAFT) (PB)

424	425	426

Displacement, tons: 92 full load
Dimensions, feet (metres): 89.9 × 21.5 × 7.2 *(27.4 × 6.6 × 2.2)*
Main machinery: 3 diesels; 4,290 hp(m) *(3.15 MW)*; 3 shafts
Speed, knots: 32. **Range, n miles:** 400 at 25 kt
Complement: 17
Guns: 8—20 mm M75 (two quad). 2—128 mm launchers for illuminants.
Radars: Surface search: Decca 1226; I-band.

Comment: Built by Brodotechnika, Yugoslavia for an African country and completed in 1986-87. Laid up when the sale did not go through and shipped to Burma arriving in October 1990. All are active. Based at Rangoon.

VERIFIED

PB 90 (Yugoslav colours)   *1990, Yugoslav FDSP* / 0056643

6 BURMA PGM TYPE (COASTAL PATROL CRAFT) (PB)

PGM 412-415   THIHAYARZAR I and II

Displacement, tons: 168 full load
Dimensions, feet (metres): 110 × 22 × 6.5 *(33.5 × 6.7 × 2)*
Main machinery: 2 Deutz SBA16MB816 LLKR diesels; 2,720 hp(m) *(2 MW)*; 2 shafts
Speed, knots: 16. **Range, n miles:** 1,400 at 14 kt
Complement: 17
Guns: 2 Bofors 40 mm/60.

Comment: Built by Burma Naval Dockyard modelled on the US PGM 43 type. First two completed 1983. Two more craft with identical dimensions and named *Thihayarzar I* and *II* were delivered by Myanma Shipyard to the Customs on 27 June 1993. Both craft are armed and were taken over by the Navy.

VERIFIED

PGM 415   *4/1993* / 0056644

4 RIVER GUNBOATS (Ex-TRANSPORTS) (PBR)

SAGU	SEINDA	SHWETHIDA	SINMIN

Displacement, tons: 98 full load
Dimensions, feet (metres): 94.5 × 22 × 4.5 *(28.8 × 6.7 × 1.4)*
Main machinery: 1 Crossley ERL 6-cyl diesel; 160 hp *(119 kW)*; 1 shaft
Speed, knots: 12
Complement: 32
Guns: 1—40 mm/60 *(Sagu)*. 1—20 mm (3 in *Sagu*).

Comment: Built in mid-1950s. *Sinmin, Seinda* and *Shwethida* have a roofed-in upper deck with a 20 mm gun forward of the funnel. *Sagu* has an open upper deck aft of the funnel but with a 40 mm gun forward and mountings for 20 mm aft on the upper deck and midships either side on the lower deck. Based at Moulmein. Four other ships of the same type are unarmed and are listed under *Auxiliaries*.

VERIFIED

SEINDA   *8/1994* / 0056649

2 IMPROVED Y 301 CLASS (RIVER GUNBOATS) (PBR)

Y 311   Y 312

Displacement, tons: 250 full load
Dimensions, feet (metres): 121.4 × 24 × 3.9 *(37 × 7.3 × 1.2)*
Main machinery: 2 MTU MB diesels; 1,000 hp(m) *(735 kW)*; 2 shafts
Speed, knots: 12
Complement: 37
Guns: 2 Bofors 40 mm/60. 4 Oerlikon 20 mm.
Radars: Surface search: Raytheon; I-band.

Comment: Built at Simmilak in 1969 and based on similar Yugoslav craft which have been scrapped. Based at Sittwe. *VERIFIED*

Y 312   *8/1994 / 0056647*

6 CARPENTARIA CLASS (RIVER PATROL CRAFT) (PBR)

112-117

Displacement, tons: 26 full load
Dimensions, feet (metres): 51.5 × 15.7 × 4.3 *(15.7 × 4.8 × 1.3)*
Main machinery: 2 MTU 8V 331 TC92 diesels; 1,770 hp(m) *(1.3 MW)* sustained; 2 shafts
Speed, knots: 29. **Range, n miles:** 950 at 18 kt
Complement: 10
Guns: 1 Oerlikon 20 mm. 1—12.7 mm MG.

Comment: Built by De Havilland Marine, Sydney. First two delivered 1979, remainder in 1980. Similar to craft built for Indonesia. Based at Rangoon. *VERIFIED*

CARPENTARIA 113   *1991 / 0056651*

25 MICHAO CLASS (PBR)

001-025

Comment: Small craft, 52 ft *(15.8 m)* long, acquired from Yugoslavia in 1965. Also used to ferry troops and two are used as VIP launches. 1 to 7 based at Rangoon; 8 to 16 at Moulmein and 17 to 25 at Sittwe. *VERIFIED*

MICHAO CLASS   *5/1995 / 0056650*

2 CGC TYPE (RIVER GUNBOATS) (PBR)

MGB 102   MGB 110

Displacement, tons: 49 standard; 66 full load
Dimensions, feet (metres): 83 × 16 × 5.5 *(25.3 × 4.9 × 1.7)*
Main machinery: 4 GM diesels; 800 hp *(596 kW)*; 2 shafts
Speed, knots: 11
Complement: 16
Guns: 1 Bofors 40 mm/60. 1 Oerlikon 20 mm.

Comment: Ex-USCG type cutters with new hulls built in Burma. Completed in 1960. Based at Rangoon but have not been seen recently. *VERIFIED*

MGB 110

9 RIVER PATROL CRAFT (PBR)

RPC 11-19

Displacement, tons: 37 full load
Dimensions, feet (metres): 50 × 14 × 3.5 *(15.2 × 4.3 × 1.1)*
Main machinery: 2 Thornycroft RZ 6 diesels; 250 hp *(186 kW)*; 2 shafts
Speed, knots: 10. **Range, n miles:** 400 at 8 kt
Complement: 8
Guns: 1 Oerlikon 20 mm or 2—12.7 mm MGs (twin). 1—12.7 mm MG.

Comment: Built by the Naval Engineering Depot, Rangoon. First five in mid-1980s; second batch of a modified design in 1990-91. Sometimes used by the Naval Infantry and can carry up to 35 troops. Based at Rangoon. *VERIFIED*

10 Y 301 CLASS (RIVER GUNBOATS) (PBR)

Y 301-310

Displacement, tons: 120 full load
Dimensions, feet (metres): 104.8 × 24 × 3 *(32 × 7.3 × 0.9)*
Main machinery: 2 MTU MB diesels; 1,000 hp(m) *(735 kW)*; 2 shafts
Speed, knots: 13
Complement: 29
Guns: 2 Bofors 40 mm/60 or 1 Bofors 40 mm/60 and 1 Vickers 2-pdr.

Comment: All of these boats were completed in 1958 at the Uljanik Shipyard, Pula, Yugoslavia. Y 301, 303 and 307 based at Moulmein. The remainder at Rangoon. *VERIFIED*

Y 304   *1991 / 0056648*

3 SWIFT TYPE PGM (COASTAL PATROL CRAFT) (PB)

PGM 421-423

Displacement, tons: 128 full load
Dimensions, feet (metres): 103.3 × 23.8 × 6.9 *(31.5 × 7.2 × 3.1)*
Main machinery: 2 MTU 12V 331 TC81 diesels; 2,450 hp(m) *(1.8 MW)* sustained; 2 shafts
Speed, knots: 27. **Range, n miles:** 1,800 at 18 kt
Complement: 25
Guns: 2 Bofors 40 mm/60. 2 Oerlikon 20 mm. 2—12.7 mm MGs.
Radars: Surface search: Raytheon 1500; I-band.

Comment: Swiftships construction completed between March and September 1979. Acquired 1980 through Vosper, Singapore. *PGM 421* previously reported sunk in 1990s but reported to have been repaired. Based at Rangoon. *VERIFIED*

PGM   *6/1991 / 0056645*

6 PBR Mk II RIVER PATROL CRAFT (PBR)

PBR 211-216

Displacement, tons: 9 full load
Dimensions, feet (metres): 32 × 11 × 2.6 *(9.8 × 3.4 × 0.8)*
Main machinery: 2 GM 6V-53 diesels; 348 hp *(260 kW)* sustained; 2 water-jets
Speed, knots: 25. **Range, n miles:** 180 at 20 kt
Complement: 4 or 5
Guns: 2—12.7 mm (twin, fwd) MGs. 1—7.9 mm LMG (aft).
Radars: Surface search: Raytheon 1900; I-band.

Comment: Acquired in 1978. Built by Uniflite, Washington. GRP hulls. Based at Moulmein but not reported as active recently. *VERIFIED*

PBR 211   *1987*

6 PGM 43 TYPE (COASTAL PATROL CRAFT) (PB)

PGM 401-406

Displacement, tons: 141 full load
Dimensions, feet (metres): 101 × 21.1 × 7.5 *(30.8 × 6.4 × 2.3)*
Main machinery: 8 GM 6-71 diesels; 1,392 hp *(1.04 MW)* sustained; 2 shafts
Speed, knots: 17. **Range, n miles:** 1,000 at 15 kt
Complement: 17
Guns: 1 Bofors 40 mm/60. 2 Oerlikon 20 mm (twin). 2—12.7 mm MGs.
Radars: Surface search: Raytheon 1500 (PGM 405-406).
EDO 320 (PGM 401-404); I/J-band.

Comment: First four built by Marinette Marine in 1959; last pair by Peterson Shipbuilders in 1961. PGM 401-403 based at Moulmein and 404-405 at Rangoon. PGM 406 at Sittwe. *VERIFIED*

PGM 406   *3/1992* / 0056646

SURVEY SHIPS

1 SURVEY SHIP (AGS)

Name	No	Builders	Commissioned
— (ex-*Changi*)	802	Miho Shipyard, Shimizu	20 June 1973

Displacement, tons: 880 full load
Dimensions, feet (metres): 154.2 × 28.6 × 11.9 *(47 × 8.7 × 3.6)*
Main machinery: 1 Niigata diesel; 1 shaft
Speed, knots: 13
Complement: 45 (5 officers)
Guns: 2 Oerlikon 20 mm.
Radars: Navigation: I-band.

Comment: A fishery research ship of Singapore origin, arrested on 8 April 1974 and taken into service as a survey vessel in about 1981. Stern trawler type. Based at Rangoon. *VERIFIED*

802   *10/1993* / 0056657

1 SURVEY SHIP (AGS)

Name	No	Builders	Commissioned
—	801	Brodogradiliste Tito, Belgrade, Yugoslavia	1965

Displacement, tons: 1,059 standard
Dimensions, feet (metres): 204 × 36 × 11.8 *(62.2 × 11 × 3.6)*
Main machinery: 2 MTU 12V 493 TY7 diesels; 2,120 hp(m) *(1.62 MW)* sustained; 2 shafts
Speed, knots: 15
Complement: 99 (7 officers)
Guns: 2 Bofors 40 mm/60. 2 Oerlikon 20 mm (twin).
Radars: Navigation: Racal Decca; I-band.

Comment: Has two surveying motor boats. The after gun can be removed to provide a helicopter platform. This ship is sometimes referred to as Thu Tay Thi which means 'survey vessel'. Based at Rangoon. *VERIFIED*

801   *6/1993*

MINE WARFARE FORCES

Notes: Up to two Chinese-built minesweepers are expected to be acquired when funds are available.

AMPHIBIOUS FORCES

1 LCU

AIYAR LULIN 603

Displacement, tons: 360 full load
Dimensions, feet (metres): 119 × 34 × 6 *(36.3 × 10.4 × 1.8)*
Main machinery: 4 GM diesels; 600 hp *(448 kW)*; 2 shafts
Speed, knots: 10. **Range, n miles:** 1,200 at 8 kt
Complement: 14
Military lift: 168 tons
Guns: 1—12.7 mm MG.

Comment: Completed in Rangoon in 1966 to the US 1610 design. Based at Rangoon. *VERIFIED*

AIYAR LULIN   *1990* / 0056654

10 LCM 3 TYPE

LCM 701-710

Displacement, tons: 52 full load
Dimensions, feet (metres): 50 × 14 × 4 *(15.2 × 4.3 × 1.2)*
Main machinery: 2 Gray Marine 64 HN9 diesels; 330 hp *(246 kW)*; 2 shafts
Speed, knots: 9
Complement: 5

Comment: US-built LCM type landing craft. Used as local transports for stores and personnel. Cargo capacity, 30 tons. Guns have been removed. Based at Sittwe. *VERIFIED*

LCM 704   *5/1994* / 0056655

4 ABAMIN CLASS (LCU)

AIYAR MAI 604	**AIYAR MINTHAMEE** 606
AIYAR MAUNG 605	**AIYAR MINTHAR** 607

Displacement, tons: 250 full load
Dimensions, feet (metres): 125.6 × 29.8 × 4.6 *(38.3 × 9.1 × 1.4)*
Main machinery: 2 Kubota diesels; 600 hp(m) *(441 kW)*; 2 shafts
Speed, knots: 10
Complement: 10
Military lift: 100 tons
Guns: 1—12.7 mm MG.

Comment: All built by Yokohama Yacht in 1969. Based at Rangoon. *VERIFIED*

AIYAR MAUNG   *1991* / 0056653

3 LCU

001-003

Comment: Operated by the Army. Dimensions not known. *VERIFIED*

LANDING CRAFT 003   *7/1992* / 0056652

AUXILIARIES

Notes: As well as the ships listed below there is a small coastal oil tanker, a harbour tug and several harbour launches and personnel carriers.

1 TRANSPORT VESSEL (AK)

AYIDAWAYA

Displacement, tons: 805 full load
Dimensions, feet (metres): 163.4 × 27.6 × 12.1 *(49.8 × 8.4 × 3.7)*
Main machinery: 1 diesel; 600 hp(m) *(441 kW)*; 1 shaft
Speed, knots: 12
Complement: 30

Comment: Built in Norway in 1975. Acquired in 1991 and used as transport for stores and personnel.
VERIFIED

AYIDAWAYA *12/1991 /* 0056658

1 BUOY TENDER (ABU)

HSAD DAN

Displacement, tons: 706 full load
Dimensions, feet (metres): 130.6 × 37.1 × 8.9 *(39.8 × 11.3 × 2.7)*
Main machinery: 2 Deutz BA8M816 diesels; 1,341 hp(m) *(986 kW)*; 2 shafts
Speed, knots: 10
Complement: 23

Comment: Built by Italthai in 1986. Operated by the Rangoon Port Authority but manned by the Navy.
VERIFIED

HSAD DAN *5/1992 /* 0056662

8 MFVs

511	520-523	901	905	906

Comment: Armed vessels of approximately 200 tons *(901)*, 80 tons *(905, 906)* and 50 tons (remainder) with a 12.7 mm or 6.72 mm MG mounted above the bridge in some. All have navigational radars. Based at Rangoon.
VERIFIED

MFV *8/1990 /* 0104255

1 TRANSPORT VESSEL (AKL)

PYI DAW AYE

Displacement, tons: 850 full load
Dimensions, feet (metres): 163 × 27 × 11.5 *(49.7 × 8.3 × 3.5)*
Main machinery: 2 diesels; 600 hp *(447 kW)*; 2 shafts
Speed, knots: 11
Complement: 12

Comment: Completed in about 1975. Dimensions are approximate. Naval manned.
VERIFIED

PYI DAW AYE *1991 /* 0056663

1 DIVING SUPPORT VESSEL (YDT)

YAN LON AUNG 200

Displacement, tons: 536 full load
Dimensions, feet (metres): 179 × 30 × 8 *(54.6 × 9.1 × 2.4)*
Main machinery: 2 diesels; 2 shafts
Speed, knots: 12
Complement: 88
Guns: 1 Bofors 40 mm/60. 2—12.7 mm MGs.

Comment: Support diving ship acquired from Japan in 1967. Based at Rangoon.
VERIFIED

YAN LON AUNG *7/1993 /* 0056660

4 TRANSPORT VESSELS (AKL)

SABAN	SETHYA	SHWEPAZUN	SETYAHAT

Displacement, tons: 98 full load
Dimensions, feet (metres): 94.5 × 22 × 4.5 *(28.8 × 6.7 × 1.4)*
Main machinery: 1 Crossley ERL 6-cyl diesel; 160 hp *(119 kW)*; 1 shaft
Speed, knots: 12
Complement: 30

Comment: These are sister ships to the armed gunboats shown under *Patrol Forces*. It is possible that a 20 mm gun may be mounted on some occasions. Based at Rangoon.
VERIFIED

SHWEPAZUN *1991 /* 0056661

1 TANKER (AOT)

608 (ex-*Interbunker*)

Displacement, tons: 2,900 full load
Dimensions, feet (metres): 232 × 36.1 × 16.1 *(70.7 × 11 × 4.9)*
Main machinery: 2 Daihatsu diesels; 1,860 hp(m) *(1.37 MW)*; 1 shaft
Speed, knots: 11. Range, n miles: 5,000 at 11 kt
Complement: 15

Comment: Thai owned commercial tanker arrested in October 1991 and taken into the Navy.
VERIFIED

608 *12/1991 /* 0056659

SURVEY SHIPS

Notes: Thu Tay Thi means 'survey vessel'.

1 SURVEY CRAFT (AGSC)

Name	No	Builders	Commissioned
YAY BO	807	Damen, Netherlands	1958

Displacement, tons: 108 full load
Dimensions, feet (metres): 98.4 × 22.3 × 4.9 *(30 × 6.8 × 1.5)*
Main machinery: 2 diesels; 2 shafts
Speed, knots: 10
Complement: 34 (2 officers)
Guns: 1—12.7 mm MG.

Comment: Used for river surveys. Based at Rangoon.

VERIFIED

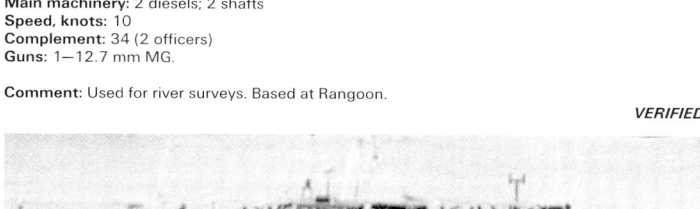

YAY BO   *1990* / 0056656

PRESIDENTIAL YACHT

1 TRANSPORT SHIP (YAC)

YADANABON

Comment: Built in Burma and used for VIP cruises on the Irrawaddy river and in coastal waters. Armed with 2—7.62 mm MGs and manned by the Navy.

VERIFIED

PRESIDENT'S YACHT   *1990* / 0056665

Namibia

Country Overview

Formerly South West Africa and governed by South Africa, Namibia gained independence in 1990 although South Africa continued to administer an enclave containing the principal seaport, Walvis Bay, until 1994. With an area of 318,252 square miles, it has borders to the north with Angola and to the south with South Africa. It has an 848 n mile coastline with the south Atlantic Ocean. The capital and largest city is Windhoek and there is another port at Lüderitz. Territorial seas (12 n miles) are claimed. It also claims a 200 n mile Exclusive Economic Zone (EEZ) but its limits have not been fully defined by boundary agreements.

Ships and aircraft belong to the Fishery Protection Service whose duties will be taken over by the Maritime Wing when it is formally established.

Headquarters Appointments

Head of Maritime Wing:
Captain Festus Sacharia

Bases

Walvis Bay, Luderitz

Personnel

2005: 120

Aviation

Five ex-US Air Force Cessna O-2A observation aircraft operate in a maritime surveillance role.

PATROL FORCES

Notes: Two 9 m Namacurra class inshore patrol vessels (ex-Y 1501 and Y 1510) were transferred from South Africa in November 2002.

1 IMPERIAL MARINHEIRO CLASS
(COASTAL PATROL SHIP) (PB)

Name	No	Builders	Commissioned
LIEUTENANT GENERAL	011	Smit, Kinderdijk, Netherlands	17 Apr 1955
DIMO HAMAAMBO (ex-*Purus*)	(ex-V 23)		

Displacement, tons: 911 standard; 1,025 full load
Dimensions, feet (metres): 184 × 30.5 × 11.7 *(56 × 9.3 × 3.6)*
Main machinery: 2 Sulzer 6TD36 diesels; 2,160 hp(m) *(1.59 MW)*; 2 shafts
Speed, knots: 16
Complement: 64 (6 officers)
Guns: 1—3 in *(76 mm)*/50 Mk 33; 50 rds/min to 12.8 km *(6.9 n miles)*; weight of shell 6 kg.
2 or 4 Oerlikon 20 mm.
Radars: Surface search: Racal Decca; I-band.

Comment: Built for Brazilian Navy as fleet tug but subsequently classified as a corvette. Withdrawn from Brazilian service in 2002 and recommissioned into the Namibian Navy on 27 August 2004.

NEW ENTRY

IMPERIAL MARINHEIRO CLASS (Brazilian colours)   *2/2000, van Ginderen Collection* / 0104229

1 PATROL SHIP (PBO)

Name	No	Builders	Commissioned
ORYX (ex-*S to S*)	—	Burmeister/Abeking & Rasmussen	May 1975

Displacement, tons: 406 full load
Dimensions, feet (metres): 149.9 × 28.9 × 7.9 *(45.7 × 8.8 × 2.4)*
Main machinery: 2 Deutz RSBA 16M diesels; 2,000 hp(m) *(1.47 MW)*; 1 shaft; cp prop; bow thruster
Speed, knots: 14. Range, n miles: 4,100 at 11 kt
Complement: 20 (6 officers)
Guns: 1—12.7 mm MG.
Radars: Surface search: Furuno ARPA FR 1525; I-band.
Navigation: Furuno FR 805D; I-band.

Comment: Built for the Nautical Investment Company, Panama and used as a yacht by the Managing Director of Fiat. Acquired in 1993 by Namibia. Replaced by *Nathanael Maxwilili* in fishery protection role and transferred to the navy as a patrol ship in 2002. Two are in service in the Myanmar navy.

UPDATED

ORYX   *6/1997* / 0081282

0 + 1 GRAJAÚ CLASS (LARGE PATROL CRAFT) (PBO)

Displacement, tons: 263 full load
Dimensions, feet (metres): 152.6 × 24.6 × 7.5 *(46.5 × 7.5 × 2.3)*
Main machinery: 2 MTU 16V 396 TB94 diesels; 5,800 hp(m) *(4.26 MW)* sustained; 2 shafts
Speed, knots: 26. Range, n miles: 2,200 at 12 kt
Complement: 29 (4 officers)
Guns: 1 Bofors 40 mm/70. 2 Oerlikon 20 mm.
Radars: Surface search: Racal Decca 1290A; I-band.

Comment: Reported to be delivered from Brazil in 2004 although it is unclear whether it is new build or second-hand. Details are for similar craft in Brazilian service. *NEW ENTRY*

0 + 4 TRACKER II CLASS (COASTAL PATROL CRAFT) (PB)

Displacement, tons: 31 standard; 45 full load
Dimensions, feet (metres): 68.6 × 17 × 4.8 *(20.9 × 5.2 × 1.5)*
Main machinery: 2 MTU 8V 396 TB83 diesels; 2,100 hp(m) *(1.54 MW)* sustained; 2 shafts
Speed, knots: 25. **Range, n miles:** 600 at 15 kt
Complement: 8 (2 officers)
Guns: 2—12.7 mm MGs.
Radars: Surface search: Racal Decca RM 1070A; I-band.

Comment: Reported to be delivered from Brazil in 2004 although it is unclear whether they are
new build or second-hand. Details are for similar craft in Brazilian service. ***NEW ENTRY***

GOVERNMENT MARITIME FORCES

Notes: There are also four research ships: *Benguela, Welwitschia, Nautilus II* and *Kuiseb.*

1 OSPREY FV 710 CLASS (PBOH)

Name	No	Builders	Commissioned
TOBIAS HAINYEKO (ex-*Havørnen*)	—	Frederikshavn Vaerft	July 1979

Displacement, feet: 505 full load
Dimensions, feet (metres): 164 × 34.5 × 9 *(50 × 10.5 × 2.8)*
Main machinery: 2 Burmeister & Wain Alpha 16V23L diesels; 4,640 hp(m) *(3.41 MW)*; 2 shafts; cp props
Speed, knots: 20. **Range, n miles:** 4,000 at 15 kt
Complement: 15 plus 20 spare
Radars: Surface search: Furuno ARPA FR 1525; I-band.
Navigation: Furuno FRM 64; I-band.

Comment: Donated by Denmark in late 1993, retaining some Danish crew. Recommissioned
15 December 1994. The helicopter deck can handle up to Lynx size aircraft and there is a
slipway on the stern for launching an RIB. Similar ships in service in Greece and Morocco.
VERIFIED

1 PATROL SHIP (PBOH)

Name	Builders	Commissioned
NATHANAEL MAXWILILI	Moen Slip AS, Kolvereid, Norway	14 May 2002

Measurement, tons: 380 dwt
Dimensions, feet (metres): 189.0 × 41.0 × 13.8 *(57.6 × 12.5 × 4.2)*
Main machinery: 2 Deutz SBV8M diesel; 4,063 hp *(3.03 MW)*; 2 shafts; Kamewa Ulstein bow thruster
Speed, knots: 17
Radars: Furuno FR-2125; I-band.
Helicopters: Platform only.

Comment: Ordered in 1999. Financed by NORAD (Norwegian Agency for Development Co-
Operation). Equipped with inspection craft for fishery protection role.
VERIFIED

1 PATROL SHIP (PBO)

Name	Builders	Commissioned
ANNA KAKURUKAZE MUNGUNDA	Freire Shipyards, Vigo	10 Feb 2004

Measurement, tons: 1,400 grt
Dimensions, feet (metres): 193.6 × 41.3 × 13.8 *(59.0 × 12.6 × 4.2)*
Main machinery: 2 Deutz SBV8M 628 diesels; 4,025 hp *(3.0 MW)*; 2 shafts; 1 bow thruster; 385 hp *(285 kW)*
Speed, knots: 17. **Range, n miles:** 8,200 at 16.8 kt
Radars: Navigation: I-band.
Helicopters: Platform for 1 medium.

Comment: Multipurpose fishery protection vessel financed by the Spanish government.
NEW ENTRY

ANNA KAKURUKAZE MUNGUNDA
2/2004*, Freire Shipyards / 0587786

TOBIAS HAINYEKO
12/1994, Harald Carstens / 0081283

NATO

Overview

The North Atlantic Treaty Organisation (NATO) was formed under Article 9 of the North Atlantic
Treaty signed on 4 April 1949. Now comprising 26 members, the original signatories were
Belgium, Canada, Denmark, France, Iceland, Italy, Luxembourg, Netherlands, Norway, Portugal, UK
and US. Greece and Turkey were admitted to the alliance in 1952, West Germany in 1955, and
Spain in 1982. In 1990 the newly unified Germany replaced West Germany. Three former
members of the Warsaw Pact, Czech Republic, Hungary and Poland were admitted in 1999. Seven
further countries: Bulgaria, Estonia, Latvia, Lithuania, Romania, Slovakia and Slovenia, became
members on 29 March 2004. A new NATO-Russia council was inaugurated on 28 May 2002. The
'Council of 20' replaced the 19 + 1 format of the Permanent Joint Council established in 1997.

RESEARCH SHIPS

1 RESEARCH SHIP (AGOR)

Name	No	Builders	Launched	Commissioned
ALLIANCE	A 1456	Fincantieri, Muggiano	9 July 1986	6 May 1988

Displacement, tons: 2,466 standard; 3,180 full load
Dimensions, feet (metres): 305.1 × 49.9 × 17.1 *(93 × 15.2 × 5.2)*
Main machinery: Diesel-electric; 2 Fincantieri GMT B 230.12 M diesels; 6,079 hp(m) *(4.47 MW)*
sustained; 2 AEG CC 3127 generators; 2 AEG motors; 4,039 hp(m) *(2.97 MW)* sustained;
2 shafts; bow thruster
Speed, knots: 16. **Range, n miles:** 7,200 at 11 kt
Complement: 24 (10 officers) plus 23 scientists
Radars: Navigation: 2 Kelvin Hughes ARPA; E/F- and I-bands.
Sonars: TVDS towed active VDS 200 Hz-4 kHz; medium and low frequency passive towed line
arrays.

Comment: Built at La Spezia. NATO's first wholly owned ship is a Public Service vessel of the
German Navy with a German, British and Italian crew. Designed for oceanography and acoustic
research. Based at La Spezia and operated by SACLANT Undersea Research Centre. Facilities
include extensive laboratories, position location systems, silent propulsion, and overside
deployment equipment. Can tow a 20 ton load at 12 kt. A Kongsberg gas turbine on 02 deck
provides silent propulsion power at 1,945 hp *(1.43 MW)* up to speeds of 12 kt. Atlas
hydrosweep side scan echo-sounder fitted in 1993. Qubit KH TRAC integrated navigational
system fitted in 1995. Carries two Watercraft R6 RIBs. Similar ships in Taiwan Navy and
building for Italy.
UPDATED

ALLIANCE
2/2004*, Martin Mokrus / 0583998

1 COASTAL RESEARCH VESSEL (AGOR(C))

Name	No	Builders	Commissioned
LEONARDO	A 5390	McTay Marine Ltd	6 Sep 2002

Displacement: tons: 393 full load
Dimensions, feet (metres): 93.8 × 29.5 × 8.2 *(28.6 × 9.0 × 2.5)*
Main machinery: Diesel-electric; 1,570 hp *(1,170 kW)*; 2 azimuth thrusters; 1—360° bow thruster
Speed, knots: 11. **Range, n miles:** 1,500 at 11 kt
Complement: 5 + 7 scientific staff
Radars: Navigation: 2 sets; I-band.
Sonars: Kongsberg Simrad multibeam echo-sounders.

Comment: The order for a coastal underwater research vessel was placed by SACLANT Undersea Research Centre in December 2000. Designed by Corlett and Partners, construction of the hull was undertaken by Remontowa in Poland while the superstructure and final assembly was undertaken by the prime contractor, McTay Marine Ltd. The ship is equipped with a moon pool, oceanographic winches, two cranes and Kongsberg navigation/research suite. A 20 ft container can be embarked to augment the main scientific laboratory. Based at La Spezia, the vessel is the first Italian Public Service vessel.

UPDATED

LEONARDO
6/2003*, SACLANTCEN / 0589004

Netherlands

Country Overview

The Kingdom of the Netherlands is situated in north-western Europe. With an area of 16,033 square miles, it is bordered to the east by Germany and to the south by Belgium. It has a 244 n mile coastline with the North Sea. The country also includes the self-governing Caribbean territories of Netherlands Antilles and Aruba. The seat of government is at The Hague while Amsterdam is the official capital, largest city and a major port. Rotterdam is one of the world's leading seaports. Both ports are linked both to the North Sea and to a comprehensive system of inland waterways whose total length is some 2,725 n miles. Territorial seas (12 n miles) are claimed. An EEZ and a Fishery Zone (200 n miles) have also been declared.

Headquarters Appointments

Commander in Chief:
    Vice Admiral R A A Klaver
Deputy Commander in Chief:
    Rear Admiral J W Kelder
Director, Material (Navy):
    Rear Admiral P S Bedet
Director, Personnel (Navy):
    Rear Admiral R T B Visser

Commands

Admiral Netherlands Fleet Command:
    Vice Admiral J van der Aa
Commander Belgium — Netherlands Task Group::
    Commodore H Ort
Commandant Royal Netherlands Marine Corps:
    Major General R L Zuiderwijk
Flag Officer Netherlands Forces Caribbean:
    Commodore F Sijtsma

Diplomatic Representation

Defence Attaché in Washington:
    Commodore M B Hijmans
Naval Attaché in London and Lisbon:
    Captain M C Wouters
Naval Attaché in Madrid and Rabat:
    Captain P P Metzelaar
Naval Attaché in Ankara:
    Captain R Bloemendaal
Naval Attaché in Washington:
    Captain V C Windt

Diplomatic Representation — *continued*

Naval Attaché in Oslo, Stockholm and Helsinki:
    Commander S C Barends
Naval Attaché in the Gulf:
    Commander R J C M van de Rijdt
Naval Attaché in Caracas, Brasilia, Georgetown and Paramaribo:
    Commander A Brokke
Naval Attaché in Baltic States and Finland:
    Commander B J Gerrits
Naval Attaché in Berlin:
    Commander M G M Hendriks Vettehen
Naval Attaché in Budapest and Sofia:
    Commander S J Hoekstra Bonnema
Naval Attaché in Kigali:
    Lieutenant Colonel E J van Broekhuizen

Personnel

(a)   2005: 7,800 naval and 3,100 Marines
(b)   3,900 civilians
(c)   Voluntary service

Bases

Naval HQ: The Hague
Main Base: Den Helder
Minor Bases: Flushing, Amsterdam, Rotterdam and Curaçao
NAS De Kooy (helicopters)
R Neth Marines: Rotterdam, Doorn and Texel, Aruba

Naval Air Arm

Squadron	Aircraft	Task
7	Lynx (SH-14)	Utility and Transport/SAR
320/321	P-3C Orion	LRMP
860	Lynx (SH-14D)	Embarked

Royal Netherlands Marine Corps

Four (one in reserve) Marine battalions; one combat support battalion, one logistic battalion and one amphibious support battalion. Based at Doorn and in the Netherlands Antilles and Aruba.

Prefix to Ships' Names

Hr Ms

Strength of the Fleet

Type	Active	Building (Projected)
Submarines	4	—
Frigates	12	—
Mine Hunters	10	—
Submarine Support Ship	1	—
Amphibious Transport Ship (LPD)	1	1
Landing Craft	11	—
Survey Ships	2	—
Combat Support Ships	2	(1)
Training Ships	2	—

Fleet Disposition

Operational Control of Belgium and Netherlands surface forces is under Admiral Benelux Command at Den Helder.

(1)   Two Task Groups, each with two air defence frigates, four Karel Doormans, one AOR, two SSK, 10 helicopters and five MPA.
(2)   One amphibious transport ship.
(3)   MCMV of 10 Alkmaar class.
(4)   Marine force of two battalions (arctic trained), one battalion for Antilles and Aruba, one battalion in reserve.
(5)   One hydrographic and one oceanographic vessel.

DELETIONS

Frigates

2002   *Philips van Almonde* (to Greece)
2003   *Bloys van Treslong* (to Greece)
2005   *Abraham van der Hulst, Jacob van Heemskerck* (both to Chile)

Mine Warfare Forces

2003   *Harlingen, Scheveningen*

Survey Ships

2003   *Buyskes*
2004   *Tydeman*

PENNANT LIST

Submarines

S 802	Walrus
S 803	Zeeleeuw
S 808	Dolfijn
S 810	Bruinvis

Frigates

F 802	De Zeven Provincien
F 803	Tromp
F 804	De Ruyter
F 805	Evertsen (bldg)
F 813	Witte de With
F 827	Karel Doorman
F 828	Van Speijk
F 829	Willem van der Zaan
F 830	Tjerk Hiddes
F 831	Van Amstel
F 833	Van Nes
F 834	Van Galen

Mine Warfare Vessels

M 853	Haarlem
M 856	Maassluis
M 857	Makkum
M 858	Middelburg
M 859	Hellevoetsluis
M 860	Schiedam
M 861	Urk
M 862	Zierikzee
M 863	Vlaardingen
M 864	Willemstad

Amphibious Forces

L 800	Rotterdam
L 801	Johan De Witt (bldg)

Auxiliaries

A 801	Pelikaan
A 802	Snellius
A 803	Luymes
A 832	Zuiderkruis
A 836	Amsterdam
A 851	Cerberus
A 852	Argus
A 853	Nautilus
A 854	Hydra
A 874	Linge

A 875	Regge
A 876	Hunze
A 877	Rotte
A 878	Gouwe
A 887	Thetis
A 900	Mercuur
A 902	Van Kinsbergen
Y 8005	Nieuwediep
Y 8018	Breezand
Y 8019	Balgzand
Y 8050	Urania
Y 8055	Schelde
Y 8056	Wierbalg
Y 8057	Malzwin
Y 8058	Zuidwal
Y 8059	Westwal
Y 8760	Patria

For details of the latest updates to *Jane's Fighting Ships* online and to discover the additional information available exclusively to online subscribers please visit

jfs.janes.com

SUBMARINES

Notes: The Moray class is a private design by Rotterdam Drydock with the government giving limited financial support on condition that the company collaborates with developers of air independent systems (AIP).

4 WALRUS CLASS (SSK)

Name	No	Builders	Laid down	Launched	Commissioned
WALRUS	S 802	Rotterdamse Droogdok Mij, Rotterdam	11 Oct 1979	26 Oct 1985 (13 Sep 1989)	25 Mar 1992
ZEELEEUW	S 803	Rotterdamse Droogdok Mij, Rotterdam	24 Sep 1981	20 June 1987	25 Apr 1990
DOLFIJN	S 808	Rotterdamse Droogdok Mij, Rotterdam	12 June 1986	25 Apr 1990	29 Jan 1993
BRUINVIS	S 810	Rotterdamse Droogdok Mij, Rotterdam	14 Apr 1988	25 Apr 1992	5 July 1994

Displacement, tons: 2,465 surfaced; 2,800 dived
Dimensions, feet (metres): 223.1 × 27.6 × 23
　(67.7 × 8.4 × 7)
Main machinery: Diesel-electric; 3 SEMT-Pielstick 12 PA4 200
　VG diesels; 6,300 hp(m) *(4.63 MW)*; 3 alternators; 2.88 MW;
　1 Holec motor; 6,910 hp(m) *(5.1 MW)*; 1 shaft
Speed, knots: 12 surfaced; 20 dived
Range, n miles: 10,000 at 9 kt snorting
Complement: 52 (7 officers)

Missiles: SSM: McDonnell Douglas Sub Harpoon; active radar
　homing to 130 km *(70 n miles)* at 0.9 Mach; warhead 227 kg.
Torpedoes: 4—21 in *(533 mm)* tubes. Honeywell Mk 48 Mod 4;
　wire-guided; active/passive homing to 38 km *(20.5 n miles)*
　active at 55 kt; 50 km *(27 n miles)* passive at 40 kt; warhead
　267 kg; 20 torpedoes or missiles carried. Mk 19 Turbine
　ejection pump. Mk 67 water-ram discharge.
Mines: 40 in lieu of torpedoes.
Countermeasures: ESM: ARGOS 700; radar warning.
Weapons control: Signaal SEWACO VIII action data automation.
　Signaal Gipsy data system. GTHW integrated Harpoon and
　Torpedo FCS.
Radars: Surface search: Signaal/Racal ZW07; I-band.
Sonars: Thomson Sintra TSM 2272 Eledone Octopus; hull-
　mounted; passive/active search and attack; medium
　frequency.
　GEC Avionics Type 2026; towed array; passive search; very
　low frequency.
　Thomson Sintra DUUX 5; passive ranging and intercept.

Programmes: Contract for the building of the first was signed
　16 June 1979, the second was on 17 December 1979. In
　1981 various changes to the design were made which
　resulted in a delay of one to two years. *Dolfijn* and *Bruinvis*
　ordered 16 August 1985; prefabrication started late 1985.
　Completion of *Walrus* delayed by serious fire 14 August
　1986; hull undamaged but cabling and computers destroyed.
　Walrus relaunched 13 September 1989.
Modernisation: A snort exhaust diffuser was fitted to *Zeeleeuw*
　in 1996. The rest of the class have been similarly modified.
Structure: These are improved Zwaardvis class with similar
　dimensions and silhouettes except for X stern. Use of H T steel

ZEELEEUW　　　　　　　　　　　　　　　　　　　*11/2001, B Sullivan* / 0534110

BRUINVIS　　　　　　　　　　　　　　　　　　　*7/2004*, H M Steele* / 1044146

increases the diving depth by some 50 per cent. Diving depth,
300 m *(984 ft)*. Pilkington Optronics CK 24 search and CH 74
attack periscopes.

Operational: Weapon systems evaluations completed 1990-93.
　Sub Harpoon is not carried.

UPDATED

WALRUS　　　　　　　　　　　　　　　　　　　*4/2004*, Derek Fox* / 1044164

BRUINVIS　　　　　　　　　　　　　　　*10/2003, Camil Busquets i Vilanova* / 0576320

FRIGATES

Notes: The size and shape of the future surface fleet is under consideration but construction of a new class of frigates is not expected to start before 2014.

4 DE ZEVEN PROVINCIEN CLASS (FFGHM)

Name	No	Builders	Laid down	Launched	Commissioned
DE ZEVEN PROVINCIEN	F 802	Royal Schelde, Vlissingen	1 Sep 1998	8 Apr 2000	26 Apr 2002
TROMP	F 803	Royal Schelde, Vlissingen	3 Sep 1999	7 Apr 2001	14 Mar 2003
DE RUYTER	F 804	Royal Schelde, Vlissingen	1 Sep 2000	13 Apr 2002	22 Apr 2004
EVERTSEN	F 805	Royal Schelde, Vlissingen	6 Sep 2001	19 Apr 2003	Mar 2005

Displacement, tons: 6,048 full load
Dimensions, feet (metres): 473.1 oa; 428.8 wl × 61.7 × 17.1 *(144.2; 130.7 × 18.8 × 5.2)*
Flight deck, feet (metres): 88.6 × 61.7 *(27 × 18.8)*
Main machinery: CODOG; 2 RR SM1C Spey; 52,300 hp *(39 MW)* sustained; 2 Stork-Wärtsilä 16V 26 ST diesels; 13,600 hp(m) *(10 MW)*; 2 shafts; LIPS; cp props
Speed, knots: 28. **Range, n miles:** 5,000 at 18 kt
Complement: 204 (32 officers) including staff

Missiles: SSM: 8 Harpoon ❶.
SAM: Mk 41 VLS (40 cells) ❷; 32 Standard SM2-MR (Block IIIA); command/inertial guidance; semi-active radar homing to 167 km *(90 n miles)* at 2 Mach.
32 Evolved Sea Sparrow (quad pack); semi-active radar homing to 18 km *(9.7 n miles)* at 3.6 Mach; warhead 39 kg.
Guns: 1 Otobreda 5 in *(127 mm)*/54 ❸; 45 rds/min to 23 km *(12.42 n miles)* anti-surface; weight of shell 32 kg. 2 Thales Goalkeeper 30 mm ❹; 4,200 rds/min to 1.5 km. 2 Browning 12.7 mm MGs ❺.
Torpedoes: 4—323 mm (2 twin) Mk 32 Mod 9 fixed launchers ❻. Mk 46 Mod 5 torpedoes.
Countermeasures: 4 SRBOC Mk 36 chaff launchers; Nixie torpedo decoy.
ESM/ECM: Racal Sabre ❼; intercept/jammer.
Combat data systems: CAMS Force Vision SEWACO XI; Link 11/16; SATCOMS ❽.
Weapons control: Thales Sirius IRST optronic director ❾. Thales Mirador Trainable Electro-Optical Observation System (TEOOS) ❿.

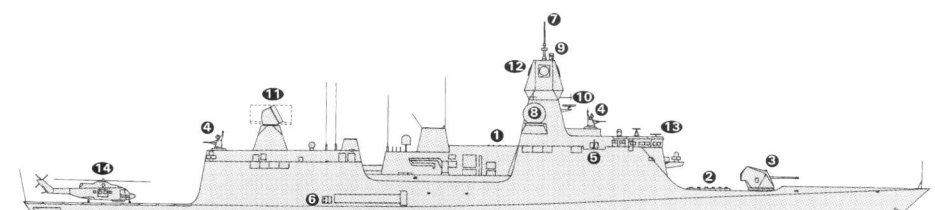

DE ZEVEN PROVINCIEN

(Scale 1 : 1,200), Ian Sturton / 0569256

Radars: Air search: Thales SMART L ⓫; 3D; D-band.
Air/surface search/fire control: Thales APAR ⓬; I/J-band.
Surface search: Thales Scout ⓭; I-band.
IFF: Mk XII.
Sonars: STN Atlas DSQS 24C; bow-mounted; active search and attack; medium frequency.

Helicopters: 1 NH90 NFH/Lynx ⓮.

Programmes: Project definition awarded to Royal Schelde on 15 December 1993 with a contract for first two ships and detailed design following on 30 June 1995. Second pair ordered 5 February 1997. Shipyards in Germany (ARGE for Type 124) collaborated to achieve some commonality of design and equipment.
Structure: As well as the listed equipment the ship is to have an electro-optic surveillance system and a navigation radar. The Scout radar is a Low Probability Intercept (LPI) set. High standards of stealth and NBC protection are part of the design. DCN Samahé helicopter handling system.
Operational: All ships fitted with command facilities. NFH 90 helicopter planned for 2007.

UPDATED

DE RUYTER

6/2004*, Maritime Photographic / 1044142

EVERTSEN

10/2004*, John Brodie / 1044161

7 KAREL DOORMAN CLASS (FFGHM)

Name	No	Builders	Laid down	Launched	Commissioned
KAREL DOORMAN	F 827	Koninklijke Maatschappij De Schelde, Flushing	26 Feb 1985	20 Apr 1988	31 May 1991
WILLEM VAN DER ZAAN	F 829	Koninklijke Maatschappij De Schelde, Flushing	6 Nov 1985	21 Jan 1989	28 Nov 1991
TJERK HIDDES	F 830	Koninklijke Maatschappij De Schelde, Flushing	28 Oct 1986	9 Dec 1989	3 Dec 1992
VAN AMSTEL	F 831	Koninklijke Maatschappij De Schelde, Flushing	3 May 1988	19 May 1990	27 May 1993
VAN NES	F 833	Koninklijke Maatschappij De Schelde, Flushing	10 Jan 1990	16 May 1992	2 June 1994
VAN GALEN	F 834	Koninklijke Maatschappij De Schelde, Flushing	7 June 1990	21 Nov 1992	1 Dec 1994
VAN SPEIJK	F 828	Koninklijke Maatschappij De Schelde, Flushing	1 Oct 1991	26 Mar 1994	7 Sep 1995

Displacement, tons: 3,320 full load
Dimensions, feet (metres): 401.2 oa; 374.7 wl × 47.2 × 14.1 *(122.3; 114.2 × 14.4 × 4.3)*
Flight deck, feet (metres): 72.2 × 47.2 *(22 × 14.4)*
Main machinery: CODOG; 2 RR Spey SM1C; 33,800 hp *(25.2 MW)* sustained (early ships of the class will initially only have SM1A gas generators and 30,800 hp *(23 MW)* sustained available); 2 Stork-Wärtsilä 12SW280 diesels; 9,790 hp(m) *(7.2 MW)* sustained; 2 shafts; LIPS cp props
Speed, knots: 30 (Speys); 21 (diesels)
Range, n miles: 5,000 at 18 kt
Complement: 156 (16 officers) (accommodation for 163)

Missiles: SSM: 8 McDonnell Douglas Harpoon Block 1C (2 quad) launchers **❶**; active radar homing to 130 km *(70 n miles)* at 0.9 Mach; warhead 227 kg.
SAM: Raytheon Sea Sparrow Mk 48 vertical launchers **❷**; semi-active radar homing to 14.6 km *(8 n miles)* at 2.5 Mach; warhead 39 kg; 16 missiles. Canisters mounted on port side of hangar.
Guns: 1—3 in *(76 mm)*/62 OTO Melara compact Mk 100 **❸**; 100 rds/min to 16 km *(8.6 n miles)* anti-surface; 12 km *(6.5 n miles)* anti-aircraft; weight of shell 6 kg. This is the version with an improved rate of fire.
1 Signaal SGE-30 Goalkeeper with General Electric 30 mm 7-barrelled **❹**; 4,200 rds/min combined to 2 km.
2 Oerlikon 20 mm; 800 rds/min to 2 km.
Torpedoes: 4—324 mm US Mk 32 Mod 9 (2 twin) tubes (mounted inside the after superstructure) **❺**. Honeywell Mk 46 Mod 5; anti-submarine; active/passive homing to 11 km *(5.9 n miles)* at 40 kt; warhead 44 kg.
Countermeasures: Decoys: 2 Loral Hycor SRBOC 6-tubed fixed Mk 36 quad launchers; IR flares and chaff to 4 km *(2.2 n miles)*.
SLQ-25 Nixie towed torpedo decoy.
ESM/ECM: Argo APECS II (includes AR 700 ESM) **❻**; intercept and jammers.
Combat data systems: Signaal SEWACO VIIB action data automation; Link 11. SATCOM **❼**. WSC-6 twin aerials.
Weapons control: Signaal IRSCAN infra-red detector (fitted in F 829 for trials and may be retrofitted in all in due course). Signaal VESTA helo transponder.
Radars: Air/surface search: Signaal SMART **❽**; 3D; F-band.
Air search: Signaal LW08 **❾**; D-band.
Surface search: Signaal Scout **❿**; I-band.
Navigation: Racal Decca 1226; I-band.
Fire control: 2 Signaal STIR **⓫**; I/J/K-band; range 140 km *(76 n miles)* for 1 m² target.
Sonars: Signaal PHS-36; hull-mounted; active search and attack; medium frequency.
Thomson Sintra Anaconda DSBV 61; towed array; passive low frequency. LFAS may be fitted in due course.

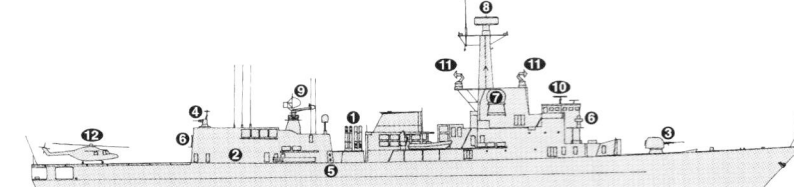

VAN SPEIJK

(Scale 1 : 1,200), Ian Sturton / 0012800

KAREL DOORMAN

5/2004 *, Per Körnefeldt / 1044162

Helicopters: 1 Westland SH-14 Lynx **⓬**.

Programmes: Declaration of intent signed on 29 February 1984 although the contract was not signed until 29 June 1985 by which time the design had been completed. A further four ordered 10 April 1986. Names were shuffled to make the new *Van Speijk* the last of the class but she retained her allocated pennant number.
Modernisation: SEWACO VII(A) operational from January 1992 and VII(B) from mid-1994. By 1994 all fitted with APECS II EW system and DSBV 61 towed array. IRSCAN infrared detector fitted on hangar roof in *Willem van der Zaan* for trials in 1993. SHF SATCOM based on the USN WSC-6, with twin aerials providing a 360° coverage even at high latitudes. Scout radar

fitted on bridge roof in 1997. Research into the fitting of low frequency active sonar is being conducted although a system is not expected to be fitted before 2010.
Structure: The VLS SAM is similar to Canadian Halifax and Greek MEKO classes. The ship is designed to reduce radar and IR signatures and has extensive NBCD arrangements. Full automation and roll stabilisation fitted. The APECS jammers are mounted starboard forward of the bridge and port aft corner of the hangar. The SAM launchers have been given added protection and better stealth features with a flat screen in some of the class.
Operational: F 832 sold to Chile and to be transferred in November 2005. F 830 to follow in April 2007.

UPDATED

VAN NES

6/2004 *, B Sullivan / 1121415

VAN SPEJK

4/2003*, M Declerck / 1044160

VAN NES

6/2004*, B Sullivan / 1044148

VAN AMSTEL

11/2004*, John Brodie / 1044163

TJERK HIDDES

6/2004*, H M Steele / 1044147

1 JACOB VAN HEEMSKERCK CLASS (FFGM)

Name	No	Builders	Laid down	Launched	Commissioned
WITTE DE WITH	F 813	Koninklijke Maatschappij De Schelde, Flushing	15 Dec 1981	25 Aug 1984	17 Sep 1986

Displacement, tons: 3,750 full load approx
Dimensions, feet (metres): 428 × 47.9 × 14.1 (20.3 screws) *(130.5 × 14.6 × 4.3 (6.2))*
Main machinery: COGOG; 2 RR Olympus TM3B gas turbines; 50,880 hp *(37.9 MW)* sustained; 2 RR Tyne RM1C gas turbines; 9,900 hp *(7.4 MW)* sustained; 2 shafts; LIPS cp props
Speed, knots: 30. **Range, n miles:** 4,700 at 16 kt on Tynes
Complement: 197 (23 officers)
Missiles: SSM: 8 McDonnell Douglas Harpoon (2 quad) launchers ❶; active radar homing to 130 km *(70 n miles)* at 0.9 Mach; warhead 227 kg.

SAM: 40 GDC Pomona Standard SM-1MR; Block IV; Mk 13 Mod 1 launcher ❷; command guidance; semi-active radar homing to 46 km *(25 n miles)* at 2 Mach.
Raytheon Sea Sparrow Mk 29 octuple launcher ❸; semi-active radar homing to 14.6 km *(8 n miles)* at 2.5 Mach; warhead 39 kg; 24 missiles.
Guns: 1 Signaal SGE-30 Goalkeeper ❹ with General Electric 30 mm 7-barrelled; 4,200 rds/min combined to 2 km. 2 Oerlikon 20 mm.
Torpedoes: 4—324 mm US Mk 32 (2 twin) tubes ❺. Honeywell Mk 46 Mod 5; anti-submarine; active/passive homing to 11 km *(5.9 n miles)* at 40 kt; warhead 44 kg.

Countermeasures: Decoys: 2 Loral Hycor Mk 36 SRBOC 6-tubed fixed quad launchers ❻; IR flares and chaff to 4 km *(2.2 n miles)*.
ESM/ECM: Ramses; intercept and jammer.
Combat data systems: Signaal SEWACO VI action data automation; Link 11. SHF SATCOM ❼ JMCIS.
Radars: Air search: Signaal LW08 ❽; D-band; range 264 km *(145 n miles)* for 2 m² target.
Air/surface search: Signaal Smart; 3D ❾; F-band.
Surface search: Signaal Scout ❿; I-band.
Fire control: 2 Signaal STIR 240 ⓫; I/J/K-band; range 140 km *(76 n miles)* for 1 m² target.
Signaal STIR 180 ⓬; I/J/K-band.

Sonars: Westinghouse SQS-509; hull-mounted; active search and attack; medium frequency.

Modernisation: Planned capability upkeep programme (CUP) cancelled. Twin SATCOM terminals fitted in 1994-95. SMART radar replaced DA05 in 1994-95. The SHF system is based on the USN WSC-6, with twin aerials providing a 360° coverage even at high latitudes.
Operational: Air defence frigate with command facilities for a task group commander and his staff. F 812 sold to Chile and to be transferred in November 2005. F 813 to follow in June 2006. **UPDATED**

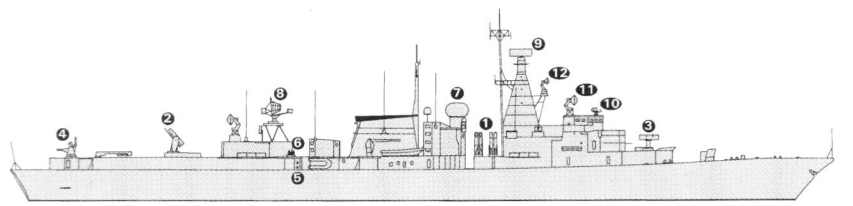

WITTE DE WITH   *(Scale 1 : 1,200), Ian Sturton* / 0114748

WITTE DE WITH   *4/2002, A Sharma* / 0534108

MINE WARFARE FORCES

10 ALKMAAR (TRIPARTITE) CLASS (MINEHUNTERS) (MHC)

Name	No	Laid down	Launched	Commissioned
HAARLEM	M 853	16 June 1981	6 May 1983	12 Jan 1984
MAASSLUIS	M 856	7 Nov 1982	5 May 1984	12 Dec 1984
MAKKUM	M 857	25 Feb 1983	27 Sep 1984	13 May 1985
MIDDELBURG	M 858	11 July 1983	23 Feb 1985	10 Dec 1986
HELLEVOETSLUIS	M 859	12 Dec 1983	18 July 1985	20 Feb 1987
SCHIEDAM	M 860	6 May 1984	20 Dec 1985	9 July 1986
URK	M 861	1 Oct 1984	2 May 1986	10 Dec 1986
ZIERIKZEE	M 862	25 Feb 1985	4 Oct 1986	7 May 1987
VLAARDINGEN	M 863	6 May 1986	4 Aug 1988	15 Mar 1989
WILLEMSTAD	M 864	3 Oct 1986	27 Jan 1989	20 Sep 1989

Displacement, tons: 562 standard; 595 full load
Dimensions, feet (metres): 168.9 × 29.2 × 8.5 *(51.5 × 8.9 × 2.6)*
Main machinery: 1 Stork Wärtsilä A-RUB 215X-12 diesel; 1,860 hp(m) *(1.35 MW)* sustained; 1 shaft; LIPS cp prop; 2 active rudders; 2 motors; 240 hp(m) *(179 kW)*; 2 bow thrusters
Speed, knots: 15 diesel; 7 electric. **Range, n miles:** 3,000 at 12 kt
Complement: 29-42 depending on task

Guns: 1 Giat 20 mm (an additional short-range missile system may be added for patrol duties).
Countermeasures: MCM: 2 PAP 104 remote-controlled submersibles. OD 3 mechanical minesweeping gear.
Combat data systems: Signaal Sewaco IX. SATCOM.
Radars: Navigation: Racal Decca TM 1229C or Consilium Selesmar MM 950; I-band.
Sonars: Thomson Sintra DUBM 21A; hull-mounted; minehunting; 100 kHz (± 10 kHz).

Programmes: The two Indonesian ships ordered in 1985 took the place of M 863 and 864 whose laying down was delayed as a result. This class is the Netherlands' part of a tripartite co-operative plan with Belgium and France for GRP hulled minehunters. The whole class built by van der Giessen-de Noord. Ships were launched virtually ready for trials.
Modernisation: An extensive modernisation programme is underway between mid-2003 and 2009 to extend service life to 2020. Upgrades include a MCM command and control system, an Integrated Mine Countermeasures System (comprising hull-mounted and self-propelled variable-depth sonar (installed in Double Eagle Mk III Mod 1 RoV)) and a Mine-Identification and Disposal System (MIDS) based on the Atlas Seafox. The equipment was first installed in M 859 2002-2004 after which a 12 month sea acceptance trials and operational evaluation is to be conducted. A number of platform improvements are also planned.
Structure: A 5 ton container can be shipped, stored for varying tasks-research; patrol; extended diving; drone control.
Operational: Endurance, 15 days. Automatic radar navigation system. Automatic data processing and display. EVEC 20. Decca Hi-fix positioning system. Alcatel dynamic positioning system. M 850-852 decommissioned in 2000. M 854 and M 855 decommissioned in 2003.
Sales: Two of a modified design to Indonesia, completed March 1988. **UPDATED**

WILLEMSTAD   *2/2004*, Martin Mokrus* / 1044157

ZIERIKZEE   *6/2004*, M Declerck* / 1044158

LAND-BASED MARITIME AIRCRAFT

Notes: The P-3C force was decommissioned on 31 December 2004. Eight modernised aircraft were sold to Germany and a letter of intent has been signed with Portugal for the remaining five aircraft.

SHIPBORNE AIRCRAFT

Notes: Up to 20 NFH 90 (14 NFH and six tactical transport) helicopters planned to replace Lynx from 2007.

Numbers/Type: 21 Westland Lynx Mks 25B/27A/81A.
Operational speed: 125 kt *(232 km/h)*.
Service ceiling: 12,500 ft *(3,810 m)*.
Range: 320 n miles *(590 km)*.
Role/Weapon systems: ASW, SAR and utility helicopter series all converted to SH-14D type. Mk 25B, Mk 27A and Mk 81A can all be embarked for ASW duties in escorts. Sensors: Ferranti Sea Spray radar, Alcatel DUAV-4 dipping sonar, FLIR Model 2000; Ferranti AWARE-3 ESM. Weapons: Two Mk 46 torpedoes or depth bombs.

UPDATED

L 9528   *7/2004*, Frank Findler /* 1044166

LYNX   *7/2004*, Frank Findler /* 1044159

6 LCVP Mk III (LCVP)

L 9536-9541

Displacement, tons: 30 full load
Dimensions, feet (metres): 55.4 × 15.7 × 3.6 *(16.9 × 4.8 × 1.1)*
Main machinery: 2 diesels; 750 hp(m) *(551 kW)*; 2 shafts
Speed, knots: 14 (full load); 16.5 (light). **Range, n miles:** 200 at 12 kt
Complement: 3
Military lift: 34 troops or 7 tons or 2 Land Rovers or 1 Snowcat
Guns: 1—7.62 mm MG.
Radars: Navigation: Racal Decca 110; I-band.

Comment: Ordered from van der Giessen-de Noord 10 December 1988. First one laid down 10 August 1989, commissioned 16 October 1990. Last one commissioned 19 October 1992.

UPDATED

AMPHIBIOUS FORCES

5 LCU Mk IX (LCU)

L 9525-9529

Displacement, tons: 200 (260 L 9526) full load
Dimensions, feet (metres): 89.6 (118.4 L 9526) × 22.4 × 4.3 *(27.3 (36.1) × 6.8 × 1.3)*
Main machinery: Diesel-electric; 2 Caterpillar 3412C diesel generators; 1,496 hp(m) *(1.1 MW)*; 2 Alconza D400 motors; 2 Schottel pumpjets; 2 pump jets
Speed, knots: 9. **Range, n miles:** 400 at 8 kt
Complement: 5 plus 2 spare
Military lift: 130 troops or 2 Warriors or 1 BARV or up to 3 trucks
Guns: 1—12.7 mm MG; 1—7.62 mm MG.
Radars: Navigation: I-band.

Comment: Ordered from Visser Dockyard, Den Helder on 19 July 1996. Steel vessels of which the first commissioned 7 April 1998. The others have been fabricated in Romania and fitted out by Visser in 1999/2000. Embarked in *Rotterdam*. L 9526 lengthened by 8.8 m at Visser dockyard in 2004.

UPDATED

L 9537   *9/2003*, A A de Kruijf /* 1044165

1 + 1 ROTTERDAM CLASS (LPD)

Name	No	Builders	Laid down	Launched	Commissioned
ROTTERDAM	L 800	Royal Schelde, Vlissingen	25 Jan 1996	22 Feb 1997	18 Apr 1998
JOHAN DE WITT	L 801	Royal Schelde, Vlissingen	18 June 2003	2005	2006

Displacement, tons: 12,750 full load (L 800); 16,680 full load (L 801)
Dimensions, feet (metres): 544.6 × 82 × 19.3 *(166 × 25 × 5.9)* (L 800)
577.5 × 95.8 × 19.3 *(176.0 × 29.2 × 5.9)*
Flight deck, feet (metres): 183.7 × 82 *(56 × 25)*
Main machinery: Diesel-electric; 4 Stork Wärtsilä 12SW28 diesel generators (L 800); 4 Wärtsilä 12V26A (L 801); 14.6 MW sustained; 2 Holec motors; 16,320 hp(m) *(12 MW)*; 2 shafts (2 Schottel podded propulsors L 801); bow thruster
Speed, knots: 19. **Range, n miles:** 6,000 at 12 kt
Complement: 113 (13 officers) + 611 (41 officers) Marines (L 800). 146 (17 officers) + 555 Marines or 402 CJTF (L 801)
Military lift: 611 troops; 170 APCs or 33 MBTs. 6 LCVP Mk 3 or 4 LCU Mk 9 or 4 LCM 8 (L 800); 4 LCVP and 2 LCU or 2 LCM (L 801)

Guns: 2 Signaal Goalkeeper 30 mm ❶. 8–12.7 mm MGs.
Countermeasures: Decoys: 4 SRBOC chaff launchers ❷; Nixie torpedo decoy system.
ESM/ECM: Intercept and jammer.
Combat data systems: SATCOM ❸; Link 11. MCCIS.
Weapons control: Signaal IRSCAN infra-red director.
Radars: Air/surface search: Signaal DA08 ❹; E/F-band.
Surface search: Signaal Scout/Kelvin Hughes ARPA ❺; I-band.
Navigation and CCA: 2 sets; I-band.

Helicopters: 6 NH90 ❻ or 4 Merlin/Sea King.

Programmes: Project definition for a joint design with Spain completed in December 1993. Contract for L 800 signed with Royal Schelde 25 April 1994. Contract for L 801 signed with Royal Schelde 3 May 2002. The hull is being constructed in Romania. To be fitted with command and control facilities for an afloat CJTF-HQ.

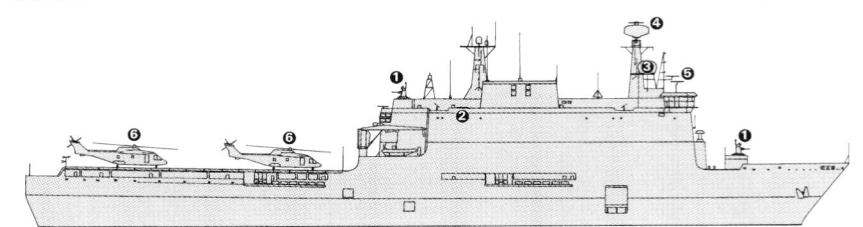

JOHAN DE WITT

(Scale 1 : 1,500), Ian Sturton / 0534121

ROTTERDAM

(Scale 1 : 1,500), Ian Sturton / 0534086

Structure: Facilities to transport a fully equipped Marine battalion with docking facilities for landing craft and a two spot helicopter flight deck with hangar space for six NH 90. 25 ton crane for disembarkation. Full hospital facilities. Built to commercial standards with military command and control and NBCD facilities. Can carry up to 30 torpedoes and 300 sonobuoys. L 801 is based on the L 800 design but is larger and wider. The flight deck is also stronger.

Operational: Alternative employment as an SAR ship for environmental and disaster relief tasks.

UPDATED

ROTTERDAM

7/2004*, Per Körnefeldt / 1044156

ROTTERDAM

6/2004*, H M Steele / 1044149

SURVEY SHIPS

2 SURVEY SHIPS (AGSH)

Name	No	Builders	Launched	Commissioned
SNELLIUS	A 802	Royal Schelde, Vlissingen	30 Apr 2003	11 Dec 2003
LUYMES	A 803	Royal Schelde, Vlissingen	22 Aug 2003	3 June 2004

Displacement, tons: 1,875 full load
Dimensions, feet (metres): 246.1 × 43.0 × 13.1 *(75 × 13.1 × 4)*
Main machinery: Diesel electric; 3 diesel generators; 2,652 hp(m) *(1.95 MW)*; 1 motor; 1,360 hp (m) *(1 MW)*; 1 shaft; cp prop
Speed, knots: 12. **Range, n miles:** 4,300 at 12 kt
Complement: 13 plus 5 scientists plus 24 spare
Radars: Navigation: E/F- and I-band.
Sonars: Multi and single beam; high frequency; active.

Comment: Designed for military and civil hydrographic surveys. Both laid down on 25 June 2002.
UPDATED

SNELLIUS   6/2004*, *Frank Findler* / 0583297

LUYMES   6/2004*, *Bram Plokker* / 1044167

AUXILIARIES

Notes: (1) In addition to the vessels listed there are large numbers of non self-propelled craft with Y pennant numbers, and six harbour launches Y 8200-8205.
(2) An Accommodation Ship *Thetis* (A 887) is based at Den Helder and provides harbour training for divers and underwater swimmers.
(3) Studies to define a multipurpose logistic support ship, capable of both fleet replenishment and strategic sealift, are to be completed in 2005. The ship is to replace *Zuiderkruis*.
(4) A programme to replace *Pelikaan* as a Marines support vessel in the Dutch Antilles is in progress. Following a contract in late 2004, delivery is to be made in mid-2006. The vessel is to be of the order of 64 m and 1,400 tons.

1 SUBMARINE SUPPORT SHIP and TORPEDO TENDER (ASL/YTT)

Name	No	Builders	Commissioned
MERCUUR	A 900	Koninklijke Maatschappij de Schelde	21 Aug 1987

Displacement, tons: 1,400 full load
Dimensions, feet (metres): 212.6 × 39.4 × 14.1 *(64.8 × 12 × 4.3)*
Main machinery: 2 Brons 61-20/27 diesels; 1,100 hp(m) *(808 kW)*; 2 shafts; bow thruster
Speed, knots: 14
Complement: 39 (6 officers)
Torpedoes: 3—324 mm (triple) tubes. 1—21 in *(533 mm)* underwater tube.
Mines: Can lay mines.
Radars: Navigation: Racal Decca 1229; I-band.
Sonars: SQR-01; hull-mounted; passive search.

Comment: Replacement for previous ship of same name. Ordered 13 June 1984. Laid down 6 November 1985. Floated out 25 October 1986. Can launch training and research torpedoes above and below the waterline. Services, maintains and recovers torpedoes.
UPDATED

MERCUUR   4/2003*, *Declerck and Steeghers* / 1044150

1 SUPPORT SHIP (AP)

Name	No	Builders	Commissioned
PELIKAAN (ex-*Kilindoni*)	A 801	Vinholmen, Arendal	1984

Displacement, tons: 505 full load
Dimensions, feet (metres): 151.6 × 34.8 × 9.2 *(46.2 × 10.6 × 2.8)*
Main machinery: 2 Caterpillar 3412T diesels; 1,080 hp *(806 kW)* sustained; 2 shafts
Speed, knots: 10
Complement: 15 plus 40 troops
Guns: 2—12.7 mm MGs.
Radars: Navigation: Racal Decca; I-band.

Comment: Ex-oil platform supply ship acquired 28 May 1990 after being refitted in Curaçao. Serves as tender and transport for marines in the Antilles. Capacity for 40 marines in five accommodation units. To be replaced in 2006.
UPDATED

PELIKAAN   5/2001 / 0114761

1 MODIFIED POOLSTER CLASS (FAST COMBAT SUPPORT SHIP) (AORH)

Name	No	Builders	Laid down	Launched	Commissioned
ZUIDERKRUIS	A 832	Verolme Shipyards, Alblasserdam	16 July 1973	15 Oct 1974	27 June 1975

Displacement, tons: 16,900 full load
Measurement, tons: 10,000 dwt
Dimensions, feet (metres): 556 × 66.6 × 27.6 *(169.6 × 20.3 × 8.4)*
Main machinery: 2 Stork-Werkspoor TM410 diesels; 21,000 hp (m) *(15.4 MW)*; 1 shaft; LIPS cp props
Speed, knots: 21
Complement: 266 (17 officers)
Cargo capacity: 10,300 tons including 8-9,000 tons oil fuel

Guns: 1 Signaal Goalkeeper 30 mm CIWS. 5 Oerlikon 20 mm.
Countermeasures: Decoys: 2 Loral Hycor SRBOC Mk 36 fixed 6-barrelled launchers; IR flares and chaff.
ESM: Ferranti AWARE-4; radar warning.
Weapons control: Signaal IRSCAN.
Radars: Air/surface search: Racal Decca 2459; F/I-band. Navigation: 2 Racal Decca TM 1226C; Signaal SCOUT; I-band.

Helicopters: 1 Westland UH-14A Lynx.

Structure: Helicopter deck aft. Funnel heightened by 4.5 m *(14.8 ft)*. 20 mm guns, containerised Goalkeeper CIWS and SATCOM, fitted for operational deployments.
Operational: Capacity for five helicopters with A/S weapons. Two fuelling stations each side for underway replenishment. Planned to remain in commission until replaced by a multipurpose logistic support ship in about 2010.
Sales: *Poolster* sold to Pakistan in June 1994.
UPDATED

ZUIDERKRUIS   8/2003*, *Harald Carstens* / 1044168

1 AMSTERDAM CLASS (FAST COMBAT SUPPORT SHIP) (AORH)

Name	No	Builders	Laid down	Launched	Commissioned
AMSTERDAM	A 836	Merwede, Hardinxveld, and Royal Schelde, Vlissingen	25 May 1992	11 Sep 1993	2 Sep 1995

Displacement, tons: 17,040 full load
Dimensions, feet (metres): 544.6 × 72.2 × 26.2 *(166 × 22 × 8)*
Main machinery: 2 Bazán/Burmeister & Wain 16V 40/45 diesels; 24,000 hp(m) *(17.6 MW)* sustained; 1 shaft; LIPS cp prop
Speed, knots: 20. **Range, n miles:** 13,440 at 20 kt
Complement: 160 (23 officers) including 24 aircrew plus 20 spare
Cargo capacity: 6,815 tons dieso; 1,660 tons aviation fuel; 290 tons solids

Guns: 2 Oerlikon 20 mm. 1 Signaal Goalkeeper 30 mm CIWS.
Countermeasures: Decoys: 4 SRBOC Mk 36 chaff launchers. Nixie towed torpedo decoy.
ESM: Ferranti AWARE-4; radar warning.
Weapons control: Signaal IRSCAN infrared director.
Radars: Surface search and helo control: 2 Kelvin Hughes; F-band.

Helicopters: 3 Lynx or 3 SH-3D or 3 NH90 or 2 EH 101.

Programmes: NP/SP AOR 90 replacement for *Poolster* ordered 14 October 1991. Hull built by Merwede, with fitting out by Royal Schelde from October 1993. A similar ship has been built for the Spanish Navy.
Structure: Close co-operation between Dutch Nevesbu and Spanish Bazán led to this design which has maintenance workshops as well as four abeam and one stern RAS/FAS station, and one Vertrep supply station. Built to merchant ship standards but with military NBC damage control.

VERIFIED

AMSTERDAM

6/2001, John Brodie / 0114762

1 TANKER (AOTL)

Name	No	Builders	Commissioned
PATRIA	Y 8760	De Hoop, Schiedam	9 June 1998

Displacement, tons: 681 full load
Dimensions, feet (metres): 145.3 × 22.4 × 8.9 *(44.4 × 6.9 × 2.8)*
Main machinery: 1 Volvo Penta TADM 122A; 381 hp(m) *(280 kW)*; 1 shaft
Speed, knots: 9.5
Complement: 2
Radars: Navigation: Furuno RHRS-2002R; I-band.

UPDATED

NAUTILUS

9/2002, A A de Kruijf / 0534091

PATRIA

6/2004*, RNLN / 1044143

1 SUPPORT CRAFT (YFL)

Name	No	Builders	Commissioned
NIEUWEDIEP	Y 8005	Akerboom, Leiden	Feb 1972

Displacement, tons: 27 full load
Dimensions, feet (metres): 58.4 × 14.1 × 4.9 *(17.8 × 4.3 × 1.5)*
Main machinery: 2 Volvo Penta diesels; 600 hp(m) *(441 kW)*; 2 shafts
Speed, knots: 10
Complement: 4

Comment: Acquired by the Navy in February 1992 as a passenger craft.

UPDATED

TRAINING SHIPS

Notes: Two Dokkum class minesweepers are used by Sea Cadets.

1 TRAINING SHIP (AXL)

Name	No	Builders	Commissioned
VAN KINSBERGEN	A 902	Damen Shipyards	2 Nov 1999

Displacement, tons: 630 full load
Dimensions, feet (metres): 136.2 × 30.2 × 10.8) *(41.5 × 9.2 × 3.3)*
Main machinery: 2 Caterpillar 3508 BI-TA; 1,572 hp(m) *(1.16 MW)* sustained; 2 shafts; bow thruster; 272 hp(m) *(200 kW)*
Speed, knots: 13
Complement: 5 plus 3 instructors and 16 students
Radars: Navigation: Consilium Selesmar; I-band.

Comment: Launched 30 August 1999. Has replaced *Zeefakkel* as the local training ship at Den Helder. Carries a 25 kt RIB.

UPDATED

NIEUWEDIEP

7/2004*, Frank Findler / 1044151

4 CERBERUS CLASS (DIVING TENDERS) (YDT)

Name	No	Builders	Commissioned
CERBERUS	A 851	Visser, Den Helder	28 Feb 1992
ARGUS	A 852	Visser, Den Helder	2 June 1992
NAUTILUS	A 853	Visser, Den Helder	18 Sep 1992
HYDRA	A 854	Visser, Den Helder	20 Nov 1992

Displacement, tons: 223 full load
Dimensions, feet (metres): 89.9 × 27.9 × 4.9 *(27.4 × 8.5 × 1.5)*
Main machinery: 2 Volvo Penta TAMD122A diesels; 760 hp(m) *(560 kW)*; 2 shafts
Speed, knots: 12. **Range, n miles:** 750 at 12 kt
Complement: 8 (2 officers)
Radars: Navigation: Racal Decca; I-band.

Comment: Ordered 29 November 1990. Capable of maintaining 10 kt in Sea State 3. Can handle a 2 ton load at 4 m from the ship's side. *Hydra* lengthened by 10.5 m to provide more accommodation and recommissioned on 13 March 1998.

VERIFIED

VAN KINSBERGEN

5/2004*, Maritime Photographic / 1044144

1 SAIL TRAINING SHIP (AXS)

Name	No	Builders	Commissioned
URANIA (ex-*Tromp*)	Y 8050	Haarlem	23 Apr 1938

Displacement, tons: 75 full load
Dimensions, feet (metres): 87.9 × 19.8 × 8.5 *(26.8 × 6.05 × 2.6)*
Main machinery: 1 Caterpillar diesel; 235 hp(m) *(186 kW)*; 1 shaft
Speed, knots: 10 diesel; 12 sail
Complement: 3 + 14 trainees

Comment: Schooner used for training in seamanship. Refit 2001-04 included a new hull and aluminium masts.

UPDATED

URANIA   6/2004*, RNLN / 1044145

TUGS

5 COASTAL TUGS (YTM)

Name	No	Builders	Commissioned
LINGE	A 874	Delta SY, Sliedrecht	20 Feb 1987
REGGE	A 875	Delta SY, Sliedrecht	6 May 1987
HUNZE	A 876	Delta SY, Sliedrecht	20 Oct 1987
ROTTE	A 877	Delta SY, Sliedrecht	20 Oct 1987
GOUWE	A 878	Delta SY, Sliedrecht	21 Feb 1997

Displacement, tons: 380 full load
Dimensions, feet (metres): 90.2 × 27.2 × 8.9 *(27.5 × 8.3 × 2.7)*
Main machinery: 2 Stork-Werkspoor or 2 Caterpillar (A 878) diesels; 1,600 hp(m) *(1.18 MW)*; 2 Kort nozzle props
Speed, knots: 11
Complement: 7
Radars: Racal Decca; I-band.

Comment: Order for first four placed in 1986. Based at Den Helder. A fifth of class was ordered in June 1996 to replace *Westgat*.

UPDATED

REGGE   7/2004*, Frank Findler / 1044152

7 HARBOUR TUGS (YTL)

BREEZAND Y 8018	WIERBALG Y 8056	ZUIDWAL Y 8058
BALGZAND Y 8019	MALZWIN Y 8057	WESTWAL Y 8059
SCHELDE Y 8055		

Comment: *Breezand* completed December 1989, *Balgzand* January 1990. The others are smaller pusher tugs and were completed December 1986 to February 1987. All built by Delta Shipyard.

UPDATED

WESTWAL   5/2004*, A A de Kruijf / 1044153

ARMY

Notes: Seven craft are operated by the Corps of Military Police: RV 165-166, RV 168-169, RV 175-177.

1 DIVING VESSEL (YDT)

RV 50

Dimensions, feet (metres): 137.3 × 31.2 × 4.9 *(41.8 × 9.5 × 1.5)*
Main machinery: 2 diesels; 476 hp(m) *(350 kW)*; 2 shafts; 1 bow thruster
Speed, knots: 8
Complement: 21
Radars: Navigation: JRC JMA 606; I-band.

Comment: Built by Vervako as a diving training ship and commissioned 3 November 1989. There is a moonpool aft with a 50 m diving bell, and a decompression chamber.

UPDATED

RV 50   10/2004*, Bram Plokker / 1047865

1 TANK TRANSPORT SHIP (AK)

JAN DE BOER RV 40

Dimensions, feet (metres): 176.8 × 31.2 × 7.2 *(53.8 × 9.5 × 2.2)*
Main machinery: 2 Mercedes OM424 diesels; 660 hp *(560 kW)*; 2 shafts
Speed, knots: 9.4
Complement: 4

Comment: Built by Grave BV and completed on 22 November 1979.

VERIFIED

COAST GUARD (KUSTWACHT)

Notes: (1) On 26 February 1987, many of the maritime services were merged to form a Coast Guard with its own distinctive colours. Included are assorted craft of the Ministries of Transport and Public Works, Finance, Defence, Justice and Agriculture, Nature Management and Food Quality. Also involved is the Ministry of Home Affairs. From 1 June 1995 the operational command of the Coast Guard has been exercised by the Navy.
(2) The following are the principal ships and craft:

Transport and Public Works: *Arca, Frans Naerebout, Nieuwe Diep, Rotterdam, Schuitengat, Terschelling, Vliestroom, Waddenzee, Waker*

Finance: *Visarend, Zeearend*

Defence: minehunters of the Alkmaar class

Justice: *P 41, P 42, P 44, P 48, P 49, P 96*

Agriculture, Nature Management and Food Quality: *Barend Biesheuvel*

(3) In addition, the Coast Guard can call upon 60 lifeboats in 39 stations from the Royal Netherlands Sea-Rescue Organisation.

WAKER (Transport)   4/2003, Per Körnefeldt / 0569212

P 49 (Justice)   6/2004*, Frank Findler / 1044154

BAREND BIESHEUVEL (Agriculture)   *6/2002, Imtech Marine and Offshore /* 0534130

VISAREND (Finance)   *7/2004*, Frank Findler /* 1044155

COAST GUARD (ANTILLES AND ARUBA)

Notes: Netherlands Antilles and Aruba Coast Guard (NAACG) formed 23 January 1996. Headquarters is co-located with the RNLN at Parera, Curaçao.

3 POLICE CRAFT (PB)

P 1   P 4   P 5

Displacement, tons: 35 full load
Dimensions, feet (metres): 57.4 × 15.7 × 5.6 *(17.5 × 4.8 × 1.7)*
Main machinery: 2 MTU 12V 183 TC91 diesels; 1,190 hp(m) *(875 kW)*; 2 shafts
Speed, knots: 18
Complement: 6

Comment: Built by Schottel in the 1970s. In addition there are six 40 kt RIBs which entered Police service in November 1997.

VERIFIED

P 1   *6/2003, Royal Netherlands Navy /* 0569211

3 STAN PATROL 4100 CUTTERS (PB)

Name	No	Builders	Commissioned
JAGUAR	P 810	Damen Shipyards	2 Nov 1998
PANTER	P 811	Damen Shipyards	18 Jan 1999
POEMA	P 812	Damen Shipyards	19 Mar 1999

Displacement, tons: 205 full load
Dimensions, feet (metres): 140.4 × 22.3 × 8.2 *(42.8 × 6.8 × 2.5)*
Main machinery: 2 Caterpillar 3516B diesels; 5,685 hp(m) *(4.18 MW)*; 2 shafts; LIPS cp props; bow thruster
Speed, knots: 26. **Range, n miles:** 2,000 at 12 kt
Complement: 11 plus 6 police
Guns: 1—12.7 mm MG.
Radars: Surface search: Signaal Scout; I-band.
Navigation: Kelvin Hughes; I-band.

Comment: Ordered from Damen shipyards in March 1997 for delivery in late 1998. Equipped with surveillance passive sensors. The cutters have a gas citadel. A 30 kt RIB is launched through a transom door. Based at Willemstad, Curaçao.

VERIFIED

POEMA   *3/1999, Damen/Flying Focus /* 0081315

New Zealand

Country Overview

New Zealand is an independent island country situated in the south Pacific Ocean with which it has a 8,170 n mile coastline. The British monarch, represented by a governor-general, is head of state. Situated about 865 n miles south-east of Australia, it comprises two main islands, North and South islands, which are separated by the Cook Strait. In addition there are numerous smaller islands including Stewart Island and the Auckland Islands. The overall area is 104,454 square miles. Overseas territories include Ross Dependency (Antarctica) and Tokelau (north of Samoa). In addition, the Cook Islands and Niue are self-governing territories in free association. The capital is Wellington and largest city is Auckland; both are ports located on North Island. Other principal ports are Tauranga, Lyttelton (near Christchurch), and Port Chalmers (Dunedin). Territorial seas (12 n miles) are claimed. An EEZ (200 n mile) is also claimed.

Headquarters Appointments

Chief of Navy:
   Rear Admiral D I Ledson, ONZM
Deputy Chief of Navy:
   Commodore D V Anson
Commander Joint Forces:
   Major General L J Gardiner
Maritime Component Commander:
   Commodore J R Steer, ONZM

Diplomatic Representation

Defence Adviser, Washington:
   Commodore P J Williams
Defence Adviser, London:
   Commodore A J Peck
Naval Adviser, London:
   Commander P D Mayer, MZNM
Naval Adviser, Canberra:
   Commander M H M Stumpel
Naval Adviser, Washington:
   Commander A H Keating
Defence Adviser, Singapore:
   Captain W J Tucker

Personnel

2005: 1,900 regulars and 400 reserves

Bases

Headquarters Joint Forces New Zealand (established 1 July 2001)
Naval Staff: HMNZS Wakefield (Wellington)
Fleet Support: HMNZS Philomel (Auckland)
Training: RNZN College Tamaki (Auckland)
Ship Repair: HMNZ Dockyard (Auckland)

RNZNVR Divisions

Auckland: HMNZS *Ngapona*
Wellington: HMNZS *Olphert*
Christchurch: HMNZS *Pegasus*
Dunedin: HMNZS *Toroa*

Prefix to Ships' Names

HMNZS

DELETIONS

Frigates

2005   *Canterbury*

For details of the latest updates to *Jane's Fighting Ships* online and to discover the additional information available exclusively to online subscribers please visit
jfs.janes.com

FRIGATES

2 ANZAC (MEKO 200) CLASS (FFHM)

Name	No	Builders	Laid down	Launched	Commissioned
TE KAHA	F 77	Transfield Defence Systems, Williamstown	19 Sep 1994	22 July 1995	22 July 1997
TE MANA	F 111	Tenix Defence Systems, Williamstown	28 June 1996	10 May 1997	10 Dec 1999

Displacement, tons: 3,600 full load
Dimensions, feet (metres): 387.1 oa; 357.6 wl × 48.6 × 14.3 *(118; 109 × 14.8 × 4.4)*
Main machinery: CODOG; 1 GE LM 2500 gas turbine; 30,172 hp *(22.5 MW)* sustained; 2 MTU 12V 1163 TB83 diesels; 8,840 hp(m) *(6.5 MW)* sustained; 2 shafts; cp props
Speed, knots: 27. **Range, n miles:** 6,000 at 18 kt
Complement: 163

Missiles: SAM: Raytheon Sea Sparrow RIM-7P; Lockheed Martin Marietta Mk 41 Mod 5 octuple cell vertical launcher ❶; semi-active radar homing to 14.6 km *(8 n miles)* at 0.9 Mach; warhead 39 kg. ESSM in due course.
Guns: 1 FMC 5 in *(127 mm)*/54 Mk 45 Mod 2 ❷; 20 rds/min to 23 km *(12.6 n miles)*; weight of shell 32 kg.
1 GE/GD 20 mm Vulcan Phalanx 6 barrelled Mk 15 Block 1 Baseline 2B ❸; 4,500 rds/min combined to 1.5 km.
Torpedoes: 6—324 mm US Mk 32 Mod 5 (2 triple) tubes ❹; Mk 46 Mod 2; anti-submarine; active/passive homing to 11 km *(5.9 n miles)* at 40 kt; warhead 44 kg.
Countermeasures: Decoys: 2 Loral Hycor Mk 36 Mod 1 chaff launchers ❺. SLQ-25A torpedo decoy system.
ESM: DASA Maigret; Racal Thorn Sceptre A; intercept (to be replaced by Racal Centaur in 2005).
Combat data systems: CelsiusTech 9LV 453 Mk 3. Link 11; GCCS-M.
Weapons control: CelsiusTech 9LV 453 optronic director ❻. Raytheon CWI Mk 73 Mod 1 (for SAM).

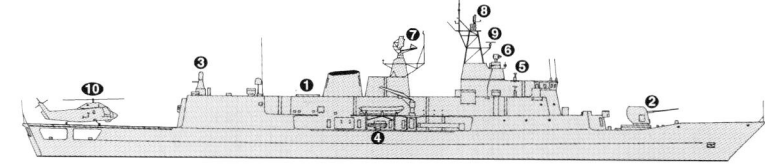

TE KAHA

(Scale 1 : 1,200), Ian Sturton / 0081317

Radars: Air search: Raytheon SPS-49(V)8 ❼; C/D-band.
Air/surface search: CelsiusTech 9LV 453 TIR (Ericsson Tx/Rx) ❽; G-band.
Navigation: Atlas Elektronik 9600 ARPA; I-band.
Fire control: CelsiusTech 9LV 453 ❾; G-band.
IFF: Cossor Mk XII.
Sonars: Thomson Sintra Spherion B Mod 5; hull-mounted; active search and attack; medium frequency.

Helicopters: 1 SH-2G (NZ) Super Seasprite ❿.

Programmes: Contract signed with Amecon consortium on 19 November 1989 to build eight Blohm + Voss designed MEKO 200 ANZ frigates for Australia and two for New Zealand. Options on a third of class were turned down in November 1998. Modules constructed at Newcastle,

Australia and Whangarei, New Zealand, and shipped to Melbourne for final assembly. The two New Zealand ships are the second and fourth of the class. First steel cut on *Te Kaha* on 11 February 1993. *Te Kaha* means Prowess. *Te Mana* means Power.
Modernisation: An upgrade programme is under consideration. Enhancements may include introduction of RIM-162 ESSM, radar and datalink improvements, a new torpedo and CIWS upgrade. The programme is likely to be phased between 2007 and 2012.
Structure: The ships include space and weight provision for considerable enhancement including canister-launched SSM, an additional fire-control channel and ECM. Signature suppression features are incorporated in the design. All-steel construction. Fin stabilisers. McTaggert Scott Trigon 3 helicopter traversing system. Two RHIBs are carried.

UPDATED

TE KAHA

10/2002, Mick Prendergast / 0567463

TE MANA

3/2002, Chris Sattler / 0567462

SHIPBORNE AIRCRAFT

Numbers/Type: 5 Kaman SH-2G (NZ) Super Seasprite.
Operational speed: 130 kt *(241 km/h).*
Service ceiling: 22,500 ft *(6,860 m).*
Range: 367 n miles *(679 km).*
Role/Weapon systems: Last of five delivered in February 2003. Sensors: Litton ASN 150 C2; Telephonics APS 143 radar; AAQ 32 Safire IRDS; ALR 100 ESM; ALE 47 ECM. Weapons: ASW; 2 Mk 46 torpedoes or Mk 11 depth bomb; ASV: 2 Hughes Maverick AGM 65D (NZ); 1—7.62 mm M60 MG.
VERIFIED

SUPER SEASPRITE　　　　　　　　　　5/2003, A Sharma / 0567466

LAND-BASED MARITIME AIRCRAFT

Numbers/Type: 6 Lockheed P-3K Orion.
Operational speed: 405 kt *(750 km/h).*
Service ceiling: 30,000 ft *(9,146 m).*
Range: 4,000 n miles *(7,410 km).*
Role/Weapon systems: Purchased in 1966. Long-range surveillance and reconnaissance patrol; updated 1984. Modernisation of airframes (Project Kestrel) undertaken 1995-2001 for 20 year extension. Upgrade project will modernise mission avionics, sensors and communication/ navigation systems. L-3 communications selected as preferred tenderer on 12 August 2004. Operated by RNZAF. Sensors: APS-134 radar, ASQ-10 MAD, acoustic processor, AYK 14 computers, IFF, ESM, SSQ 53/62 sonobuoys. Weapons: ASW; eight Mk 46 torpedoes, Mk 80 series depth bombs.
UPDATED

P-3K　　　　　　　　　　7/2004 *, Paul Jackson / 0589788

PATROL FORCES

0 + 4 INSHORE PATROL VESSELS (PBO)

Name	No	Builders	Laid down	Launched	Commissioned
—	—	Tenix Defence Systems, Whangerai	2005	2006	Feb 2007
—	—	Tenix Defence Systems, Whangerai	2005	2006	June 2007
—	—	Tenix Defence Systems, Whangerai	2005	2006	Oct 2007
—	—	Tenix Defence Systems, Whangerai	2006	2007	Jan 2008

Displacement, tons: 340
Dimensions, feet (metres): 180.4 × 29.5 × 9.5 *(55.0 × 9.0 × 2.9)*
Main machinery: To be announced
Speed knots: 25. **Range, n miles:** 3,000 at 15 kt
Complement: 20 plus 16 spare
Guns: 2—12.7 mm MGs.
Radars: Navigation: I-band.

Programmes: Following selection as 'Project Protector' prime contractor in April 2004, Tenix Defence awarded contract for final design and construction on 29 July 2004. The ships are to operate in support of civil agencies to meet patrol and surveillance requirements in New Zealand's inshore zone (out to 24 n miles), particularly around North Island, Marlborough Sounds and Tasman Bay. Manufacturing is to start at Tenix's Whangerai Shipyard in New Zealand in early 2005.
Structure: The Tenix design is based on the 56 m San Juan class built for the Philippines Coast Guard. Capable of operating in up to Sea State 5, they will be able to launch and recover rigid hull inflatable boats in up to Sea State 4.
NEW ENTRY

IPV (computer graphic)　　　　　　　　　　9/2004 *, RNZN / 0589790

---

0 + 2 OFFSHORE PATROL VESSELS (PBO)

Name	No	Builders	Laid down	Launched	Commissioned
—	—	Tenix Defence Systems, Williamstown	2005	2006	May 2007
—	—	Tenix Defence Systems, Williamstown	2005	2006	Nov 2007

Displacement, tons: 1,600
Dimensions, feet (metres): 278.9 × 45.9 × 11.8 *(85.0 × 14.0 × 3.6)*
Main machinery: To be announced
Speed, knots: 22. **Range, n miles:** 6,000 at 15 kt
Complement: 35 plus 44 spare
Guns: 1 Rafael Typhoon 25 mm. 2—12.7 mm MGs.
Radars: Navigation: I-band.
Helicopters: 1 SH-2G Super Seasprite.

Programmes: Following selection as 'Project Protector' prime contractor in April 2004, Tenix Defence awarded contract for final design and construction on 28 July 2004. The ships are to meet patrol and surveillance requirements in support of civil agencies in New Zealand's EEZ and the Southern Ocean and to assist South Pacific states to patrol their EEZs. Manufacturing of modules is to start at Tenix's Whangerai Shipyard in New Zealand in early 2005. Final assembly is to be undertaken at Williamstown, Victoria.
Structure: The design is a lengthened, helicopter-capable variant of a Kvaerner Masa Marine design in service in Ireland and Mauritius. They are to be ice-strengthened.
NEW ENTRY

OPV (computer graphic)　　　　　　　　　　9/2004 *, RNZN / 0589791

---

4 MOA CLASS (INSHORE PATROL CRAFT) (PB)

Name	No	Builders	Commissioned
MOA	P 3553	Whangarei Engineering and Construction Co Ltd	28 Nov 1983
KIWI	P 3554	Whangarei Engineering and Construction Co Ltd	2 Sep 1984
WAKAKURA	P 3555	Whangarei Engineering and Construction Co Ltd	26 Mar 1985
HINAU	P 3556	Whangarei Engineering and Construction Co Ltd	4 Oct 1985

Displacement, tons: 91.5 standard; 105 full load
Dimensions, feet (metres): 88 × 20 × 7.2 *(26.8 × 6.1 × 2.2)*
Main machinery: 2 Cummins KT-1105M diesels; 710 hp *(530 kW)*; 2 shafts
Speed, knots: 12. **Range, n miles:** 1,000 at 11 kt
Complement: 18 (5 officers (4 training))
Guns: 1 Browning 12.7 mm MG.
Radars: Surface search: Racal Decca Bridgemaster 2000; I-band.
Sonars: Klein 595 Tracpoint; side scan; active high frequency.

Comment: On 11 February 1982 the New Zealand Cabinet approved the construction of four inshore patrol craft. The four IPC are operated by the Reserve Divisions, *Moa* with *Toroa* (Dunedin), *Kiwi* with *Pegasus* (Lyttelton), *Wakakura* with *Olphert* (Wellington), *Hinau* with *Ngapona* (Auckland). Same design as Inshore Survey and Training craft *Kahu* but with a modified internal layout. MCM system fitted in 1993-94. Side scan sonar and MCAIS data system fitted to *Hinau* in 1993; the remainder in 1996. To be replaced by new vessels from 2007.
UPDATED

HINAU　　　　　　　　　　2002, RNZN / 0525918

AUXILIARIES

Note: In addition to vessels listed below there are three 12 m sail training craft used for seamanship training: *Paea II, Mako II, Manga II* (sail nos 6911-6913).

PAEA   *2002, RNZN* / 0525919

0 + 1 MULTIROLE VESSEL (AKRH/AX)

Name	No	Builders	Laid down	Launched	Commissioned
—	—	Merwede Shipyard, Netherlands	2005	2006	Dec 2006

Displacement, tons: 8,870
Dimensions, feet (metres): 429.8 × 76.8 × 18.4 *(131.0 × 23.4 × 5.6)*
Main machinery: To be announced
Speed, knots: 19. **Range, miles:** 6,000 at 15 kt
Complement: 53 + accommodation for 250 troops and 47 additional
Guns: 1 Rafael Typhoon 25 mm. 2—12.7 mm MGs.
Military lift: 1 infantry company including Light Armoured Vehicles and equipment
Radars: Navigation: 2 I-band.
Helicopters: 2 SH-2G Super Seasprites.

Programmes: Following selection as 'Project Protector' prime contractor in April 2004, Tenix Defence awarded contract for final design and construction on 29 July 2004. Manufacturing is to start in early 2005. The ship is to provide a limited tactical sealift capacity for disaster relief, humanitarian relief operations, peace support operations, military support activities and development assistance support. The ship will also be used as the principal sea training platform for the RNZN. After being built in the Netherlands, to be fitted out by Tenix in either Australia or New Zealand.
Structure: With a design based on a commercial roll-on/roll-off vessel, the ship is to be built to comply with Lloyds Register of Shipping rules. To be ice-strengthened for operations in the Southern Ocean and the Ross Sea. Staff facilities to be incorporated. ***NEW ENTRY***

MRV   *9/2004*, RNZN* / 0589789

1 REPLENISHMENT TANKER (AORH)

Name	No	Builders	Launched	Commissioned
ENDEAVOUR	A 11	Hyundai, South Korea	14 Aug 1987	6 Apr 1988

Displacement, tons: 12,390 full load
Dimensions, feet (metres): 453.1 × 60 × 23 *(138.1 × 18.4 × 7.3)*
Main machinery: 1 MAN-Burmeister & Wain 12V32/36 diesel; 5,780 hp(m) *(4.25 MW)* sustained; 1 shaft; LIPS cp prop
Speed, knots: 13.5. **Range, n miles:** 8,000 at 13.5 kt
Complement: 49 (10 officers)
Cargo capacity: 7,500 tons dieso; 100 tons Avcat; 20 containers
Radars: Navigation: Racal Decca 1290A/9; ARPA 1690S; I-band.
Helicopters: Platform only.

Comment: Ordered July 1986. Laid down 10 April 1987. Completion delayed by engine problems but arrived in New Zealand in May 1988. Two abeam RAS rigs (one QRC, one Probe) and one astern refuelling rig. Fitted with Inmarsat. Standard merchant design modified on building to provide a relatively inexpensive replenishment tanker. ***VERIFIED***

ENDEAVOUR   *9/2002, Bob Fildes* / 0567465

1 MOA CLASS (TRAINING SHIP) (AXL)

Name	No	Builders	Commissioned
KAHU (ex-*Manawanui*)	A 04 (ex-A 09)	Whangarei Engineering and Construction Co Ltd	28 May 1979

Displacement, tons: 91.5 standard; 105 full load
Dimensions, feet (metres): 88 × 20 × 7.2 *(26.8 × 6.1 × 2.2)*
Main machinery: 2 Cummins KT-1150M diesels; 710 hp *(530 kW)*; 2 shafts
Speed, knots: 12. **Range, n miles:** 1,000 at 11 kt
Complement: 16
Radars: Navigation: Racal Decca Bridgemaster 2000; I-band.

Comment: Same hull design as Inshore Survey Craft and Patrol Craft. Now used for navigation and seamanship training and as a standby diving tender. ***VERIFIED***

KAHU   *6/1999, RNZN* / 0081325

1 DIVING TENDER (YDT)

Name	No	Builders	Commissioned
MANAWANUI (ex-*Star Perseus*)	A 09	Cochrane, Selby	May 1979

Displacement, tons: 911 full load
Dimensions, feet (metres): 143 × 31.2 × 10.5 *(43.6 × 9.5 × 3.2)*
Main machinery: 2 Caterpillar D 379TA diesels; 1,130 hp *(843 kW)*; 2 shafts; cp props; bow thruster
Speed, knots: 10.7. **Range, n miles:** 5,000 at 10 kt
Complement: 24 (2 officers)
Radars: Surface search: Racal Decca Bridgemaster 2000; I-band.
Sonars: Klein 595 Tracpoint; side scan; active high frequency.

Comment: North Sea Oil Rig Diving support vessel commissioned into the RNZN on 5 April 1988. Completed conversion in December 1988 and has replaced the previous ship of the same name which proved to be too small for the role. Equipment includes two Phantom HDX remote-controlled submersibles, a decompression chamber (to 250 ft), wet diving bell and 13 ton crane. Fitted with Inmarsat. MCAIS data system, side scan sonar and GPS fitted in 1995. More modifications are planned to enable the ship to do some of the work previously undertaken by *Tui*. This includes a stern gantry and general purpose winches for research including MCM. Used to support RAN submarine trials in 1996/97. ***VERIFIED***

SURVEY AND RESEARCH SHIPS

1 STALWART CLASS (AGS)

Name	No	Builders	Commissioned
RESOLUTION (ex-*Tenacious*)	A 14 (ex-TAGOS 17)	Halter Marine, Moss Point	29 Sep 1989

Displacement, tons: 2,262 full load
Dimensions, feet (metres): 224 × 43 × 18.7 *(68.3 × 13.1 × 5.7)*
Main machinery: Diesel-electric; 4 Caterpillar D 398B diesel generators; 3,200 hp *(2.39 MW)*; 2 motors; 1,600 hp *(1.2 MW)*; 2 shafts; bow thruster; 550 hp *(410 kW)*
Speed, knots: 11. **Range, n miles:** 1,500 at 11 kt
Complement: 26 or 45 (when surveying)
Radars: Navigation: 2 Raytheon; I-band.

Comment: Laid up by USN in 1995 and acquired in September 1996. Reactivated in October 1996 and commissioned into RNZN 13 February 1997 for passage to New Zealand. Conversion commenced mid-1997 to suit the ship for hydrography with secondary role of acoustic research for about three months per year, replacing both *Tui* and *Monowai*. Second stage of conversion to fit Atlas Elektronik MD 2/30 multibeam echo-sounder, completed in January 1999. A fixed dome increased the ship's draught. A DGPS and a towed array fitted for acoustic research. A new survey boat with Atlas Elektronik MD20 multibeam echo sounder was embarked in 2001. The ship has been repainted grey. ***UPDATED***

RESOLUTION   *9/2004*, RNZN* / 0587566

Nicaragua
FUERZA NAVAL-EJERCITO DE NICARAGUA

Country Overview

The Republic of Nicaragua is the largest Central American republic. After many years of civil war, a 1989 peace plan introduced a more stable period of democratic government. With an area of 50,893 square miles, it is situated between Honduras to the north and Costa Rica to the south. It has a 381 n mile coastline with the Caribbean and a 225 n mile coastline with the Pacific Ocean. The capital and largest city is Managua while Corinto, on the Pacific coast, is the principal port. Nicaragua has not claimed an EEZ but is one of a few coastal states which claims a 200 n mile territorial sea.

Headquarters Appointments

Head of Navy:
   Captain Juan Santiago Estrada García

Personnel

2005: 750 officers and men

Bases

Pacific: Corito (HQ), San Juan del Sur, Puerto Sandino y Potosi
Atlantic: Bluefields (HQ), El Bluff, Puerto Cabezas, Corn Island, San Juan del Norte

PATROL FORCES

Notes: A programme to re-engine and return to service one Zhuk (Grif) class has been initiated. Work to renovate two North Korean built Sin Hung class patrol boats may also be funded.

3 DABUR CLASS (PB)

GC 201   GC 203   GC 205

Displacement, tons: 39 full load
Dimensions, feet (metres): 64.9 × 18 × 5.8 *(19.8 × 5.5 × 1.8)*
Main machinery: 2 GM 12V-71TS; 840 hp *(626 kW)* sustained; 2 shafts
Speed, knots: 15. **Range, n miles:** 450 at 13 kt
Complement: 8
Guns: 2—23 mm/80 (twin). 2—12.7 mm MGs.
Radars: Surface search: Decca; I-band.

Comment: Two delivered by Israel April 1978 and two more in May 1978. One lost by gunfire in 1985 and three severely damaged in 1988 by a hurricane. The current three were acquired from Israel in May 1996. *GC 201* and *GC 203* overhauled 2000-01 and *GC 205* in 2003-04. All are based on the Atlantic coast.

UPDATED

PBF   *6/1999, Nicaraguan Navy /* 0081330

GC 201   *5/2001, Julio Montes /* 0109945

EDUARDOÑO CLASS   *6/1999, Nicaraguan Navy /* 0081329

19 ASSAULT and RIVER CRAFT (PBF)

Comment: There are 19 Colombian-built Eduardoño class assault craft and 'Cigarette' craft of about 10 m length and powered by two Yamaha outboards; 400 hp *(298 kW)*.

VERIFIED

EDUARDOÑO class   *1/2000, Julio Montes /* 0109943

EDUARDOÑO class   *1/2000, Julio Montes /* 0109944

Nigeria

Country Overview

Formerly a British protectorate, the Federal Republic of Nigeria gained full independence in 1960. With an area of 356,669 square miles, it is situated in western Africa and is bordered to the north by Niger, to the east by Chad and Cameroon and to the west by Benin. It has a 459 n mile coastline with the Gulf of Guinea. Abuja is the capital while Lagos (the capital until 1991) is the largest city, commercial centre and one of its principal ports. There are other ports at Port Harcourt, Warri, Calabar, Bonny, and Burutu. Territorial Seas (12 n miles) are claimed. An EEZ (200 n miles) has been claimed but the limits have not been defined.

The Navy has suffered from chronic lack of investment over the last ten years but there are signs that a refit programme is attempting to restore a core seagoing capability for operations within the Nigerian EEZ. However, the operational status of weapon systems and sensors remains doubtful.

Headquarters Appointments

Chief of the Naval Staff:
   Vice Admiral Samuel Afolayan
Flag Officer Western Command:
   Commodore Musa Ajadi
Flag Officer Eastern Command:
   Commodore Ganiyu Adekeye

Personnel

(a)   2005: 5,600 (650 officers) including Coast Guard
(b)   Voluntary service

Naval Aviation

The official list includes two Lynx Mk 89, 12 MBB BO 105C, three Fokker F27 and 14 Dornier Do 128-6MPA. These aircraft are believed not to be operational. Four Agusta A 109 have been procured from Italy since 2003 for patrol duties.

Bases

Apapa-Lagos: Western Naval Command; Naval Base Lagos (NNS *Onaku*), Naval College (NNS *Onura*) and Naval Training (NNS *Quorra*).
Calabar: Eastern Naval Command; Naval Base Calabar (NNS *Anansu*), Naval Base Warri (NNS *Umalokun*) and Naval Base Port Harcourt (NNS *Okemini*). Plans for a further base at Bonny Island have not been confirmed.

Prefix to Ships' Names

NNS

Port Security Police

A separate force of 1,600 officers and men in Lagos.

FRIGATES

1 MEKO TYPE 360 (FFGHM)

Name	No	Builders	Laid down	Launched	Commissioned
ARADU (ex-*Republic*)	F 89	Blohm & Voss, Hamburg	1 Dec 1978	25 Jan 1980	20 Feb 1982

Displacement, tons: 3,360 full load
Dimensions, feet (metres): 412 × 49.2 × 19 (screws)
  (125.6 × 15 × 5.8)
Main machinery: CODOG; 2 RR Olympus TM3B gas turbines;
  50,880 hp *(37.9 MW)* sustained; 2 MTU 20V 956 TB92
  diesels; 10,420 hp(m) *(7.71 MW)* sustained; 2 shafts; 2
  Kamewa cp props
Speed, knots: 30.5. **Range, n miles:** 6,500 at 15 kt
Complement: 195 (26 officers)
Missiles: SSM: 8 OTO Melara/Matra Otomat Mk 1 **❶**; active
  radar homing to 80 km *(43.2 n miles)* at 0.9 Mach; warhead
  210 kg.

SAM: Selenia Elsag Albatros octuple launcher **❷**; 24 Aspide;
  semi-active radar homing to 13 km *(7 n miles)* at 2.5 Mach;
  warhead 30 kg.
Guns: 1 OTO Melara 5 in *(127 mm)*/54 **❸**; 45 rds/min to 23 km
  (12.4 n miles); weight of shell 32 kg.
  8 Breda Bofors 40 mm/70 (4 twin) **❹**; 300 rds/min to
  12.5 km *(6.8 n miles)* anti-surface; weight of shell 0.96 kg.
Torpedoes: 6—324 mm Plessey STWS-1B (2 triple) tubes **❺**. 18
  Whitehead A244S; anti-submarine; active/passive homing to
  7 km *(3.8 n miles)* at 33 kt; warhead 34 kg (shaped charge).
Depth charges: 1 rack.
Countermeasures: Decoys: 2 Breda 105 mm SCLAR 20-tubed
  trainable; chaff to 5 km *(2.7 n miles)*; illuminants to 12 km
  (6.6 n miles).

ESM: Decca RDL-2; intercept.
ECM: RCM-2; jammer.
Combat data systems: Sewaco-BV action data automation.
Weapons control: M20 series GFCS. Signaal Vesta ASW.
Radars: Air/surface search: Plessey AWS 5 **❻**; E/F-band.
Navigation: Racal Decca 1226; I-band.
Fire control: Signaal STIR **❼**; I/J/K-band. Signaal WM 25 **❽**; I/J-
  band.
Sonars: Atlas Elektronik EA80; hull-mounted; active search and
  attack; medium frequency.

Helicopters: 1 Lynx Mk 89 **❾**.

Modernisation: Refit started at Wilmot Point, Lagos with Blohm
  & Voss assistance in 1991 and completed in February 1994.
Operational: Had two groundings and a major collision in 1987
  and ran aground again during post refit trials in early 1994.
  Assessed as beyond economical repair in 1995 but managed
  to go to sea in early 1996, and again in 1997 when she broke
  down for several months in Monrovia. Back in Lagos on one
  engine in 1998 for further repairs. SSM system reported being
  refitted in 1999. Reported seagoing in 2003 but probably not
  fully operational.

VERIFIED

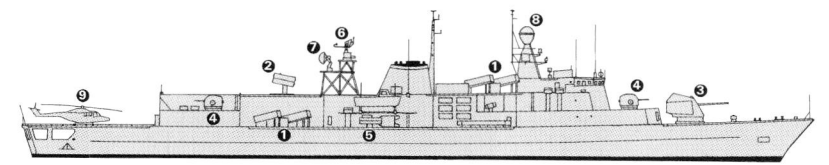

ARADU

(Scale 1 : 1,200), Ian Sturton / 0081331

ARADU

5/1999 / 0081332

CORVETTES

1 Mk 9 VOSPER THORNYCROFT TYPE (FSM)

Name	No	Builders	Commissioned
ENYMIRI	F 84	Vosper Thornycroft	2 May 1980

Displacement, tons: 680 standard; 780 full load
Dimensions, feet (metres): 226 × 31.5 × 9.8 *(69 × 9.6 × 3)*
Main machinery: 4 MTU 20V 956 TB92 diesels; 22,140 hp(m) *(16.27 MW)* sustained; 2 shafts;
  2 Kamewa cp props
Speed, knots: 27. **Range, n miles:** 2,200 at 14 kt
Complement: 90 (including Flag Officer)

Missiles: SAM: Short Brothers Seacat triple launcher.
Guns: 1 OTO Melara 3 in *(76 mm)*/62 Mod 6 compact; 85 rds/min to 16 km *(8.7 n miles)*; weight
  of shell 6 kg.
  1 Breda Bofors 40 mm/70 Type 350; 300 rds/min to 12.5 km *(6.8 n miles)*; weight of shell
  0.96 kg.
  2 Oerlikon 20 mm.
A/S mortars: 1 Bofors 375 mm twin launcher; range 1,600 or 3,600 m.
Countermeasures: ESM: Decca Cutlass; radar warning.
Weapons control: Signaal WM20 series.
Radars: Air/surface search: Plessey AWS 2; E/F-band.
Navigation: Racal Decca TM 1226; I-band.
Fire control: Signaal WM24; I/J-band; range 46 km *(25 n miles)*.
Sonars: Plessey PMS 26; lightweight; hull-mounted; active search and attack; 10 kHz.

Programmes: Ordered from Vosper Thornycroft 22 April 1975.
Operational: *Enymiri* reported operational in 1996 and again in 1998. Sister ship *Erinomi* assessed
  as beyond economical repair in 1996 and believed not to be operational.

VERIFIED

ENYMIRI

5/1999 / 0081333

PATROL FORCES

Notes: (1) All the Coastal Patrol Craft belong to the Coast Guard. Some 38 craft were acquired in
the mid-1980s from various shipbuilders including Simonneau, Damen, Swiftships, Intermarine,
Watercraft, Van Mill and Rotork. Few of these vessels have been reported at sea in recent years
although some are visible, laid up ashore, and are still serviceable.
(2) A Damen 2600 Mk II patrol craft was acquired from South Africa in 2001.
(3) Four 8 m Night Cat 27, capable of 70 kt, were delivered by Intercept Boats in 2003-04.

P 236 (Simonneau)

5/2002 / 0528302

P 240 (Rotork)

5/2002 / 0528301

3 EKPE (LÜRSSEN 57) CLASS (LARGE PATROL CRAFT) (PGF)

Name	No	Builders	Commissioned
EKPE	P 178	Lürssen, Vegesack	Aug 1980
DAMISA	P 179	Lürssen, Vegesack	Apr 1981
AGU	P 180	Lürssen, Vegesack	Apr 1981

Displacement, tons: 444 full load
Dimensions, feet (metres): 190.6 × 24.9 × 10.2 *(58.1 × 7.6 × 3.1)*
Main machinery: 4 MTU 16V 956 TB92 diesels; 17,700 hp(m) *(13 MW)* sustained; 2 shafts
Speed, knots: 42. **Range, n miles:** 2,000 at 10 kt
Complement: 40

Guns: 1 OTO Melara 3 in *(76 mm)*/62; 60 rds/min to 16 km *(8.7 n miles)*; weight of shell 6 kg.
2 Breda 40 mm/70 (twin); 4 Emerson Electric 30 mm (2 twin).
Radars: Surface search: Racal Decca TM 1226; I-band.
Fire control: Signaal WM28; I/J-band.

Programmes: Ordered in 1977. Major refit in 1984 at Vegesack.
Operational: P 178 refitted at Lagos in 1995 but broke down en route to Sierra Leone in 1997. P 179 believed to be operational but the operational status of the other two is doubtful.
VERIFIED

EKPE   3/1998 / 0052656

4 BALSAM CLASS (PBO)

Name	No	Builders	Commissioned
KYANWA	A 501	Marine Iron and Shipbuilding Corp, Duluth,	5 July 1944
(ex-*Sedge*)	(ex-WLB 402)	Minnesota	
OLOGBO	A 502	Marine Iron and Shipbuilding Corp, Duluth,	17 Oct 1942
(ex-*Cowslip*)	(ex-WLB 277)	Minnesota	
NWAMBA	A 503	Marine Iron and Shipbuilding Corp, Duluth	20 July 1944
(ex-*Firebush*)	(ex-WLB 393)	Minnesota	
OBULA	A 504	Marine Iron and Shipbuilding Corp, Duluth	23 May 1944
(ex-*Sassafras*)	(ex-WLB 401)	Minnesota	

Displacement, tons: 1,034 full load
Dimensions, feet (metres): 180 × 37 × 12 *(54.9 × 11.3 × 3.8)*
Main machinery: Diesel electric; 2 diesels; 1,402 hp *(1.06 MW)*; 1 motor; 1,200 hp *(895 kW)*; 1 shaft; bow thruster
Speed, knots: 13. **Range, n miles:** 8,000 at 12 kt
Complement: 53
Guns: 2—12.7 mm MGs.
Radars: Navigation: Raytheon SPS-64(V)1.

Comment: First ship transferred from the US Coast Guard on 30 September 2002, second on 30 December 2002, third on 30 June 2003 and fourth on 30 October 2003. *VERIFIED*

BALSAM CLASS (USCG colours)   7/1999, USCG / 0084209

1 COMBATTANTE IIIB CLASS
(FAST ATTACK CRAFT—MISSILE) (PGGF)

Name	No	Builders	Commissioned
AYAM	P 182	CMN, Cherbourg	11 June 1981

Displacement, tons: 385 standard; 430 full load
Dimensions, feet (metres): 184 × 24.9 × 7 *(56.2 × 7.6 × 2.1)*
Main machinery: 4 MTU 16V 956 TB92 diesels; 17,700 hp(m) *(13 MW)* sustained; 2 shafts
Speed, knots: 38. **Range, n miles:** 2,000 at 15 kt
Complement: 42

Missiles: SSM: 4 Aerospatiale MM 38 Exocet; inertial cruise; active radar homing to 42 km *(23 n miles)* at 0.9 Mach; warhead 165 kg; sea-skimmer.
Guns: 1 OTO Melara 3 in *(76 mm)*/62; 60 rds/min to 16 km *(8.7 n miles)*; weight of shell 6 kg.
2 Breda 40 mm/70 (twin); 300 rds/min to 12.5 km *(6.8 n miles)*; weight of shell 0.96 kg.
4 Emerson Electric 30 mm (2 twin); 1,200 rds/min combined to 6 km *(3.3 n miles)*; weight of shell 0.35 kg.
Countermeasures: ESM: Decca RDL; radar intercept.
Weapons control: Thomson-CSF Vega system. 2 CSEE Panda optical directors.
Radars: Air/surface search: Thomson-CSF Triton (TRS 3033); G-band.
Navigation: Racal Decca TM 1226; I-band.
Fire control: Thomson-CSF Castor II (TRS 3203); I/J-band.

Programmes: Ordered in late 1977. Finally handed over in February 1982 after delays caused by financial problems.
Modernisation: Major refit and repairs carried out at Cherbourg from March to December 1991 but the ships were delayed by financial problems.
Operational: *Ayam* believed to be operational but sister ships *Siri* and *Ekun* are almost certainly not.
VERIFIED

AYAM (outboard P 179)   5/2002 / 0528300

MINE WARFARE FORCES

2 LERICI CLASS (MINEHUNTERS/SWEEPERS) (MHSC)

Name	No	Builders	Commissioned
OHUE	M 371	Intermarine SY, Italy	28 May 1987
MARABA	M 372	Intermarine SY, Italy	25 Feb 1988

Displacement, tons: 540 full load
Dimensions, feet (metres): 167.3 × 32.5 × 9.2 *(51 × 9.9 × 2.8)*
Main machinery: 2 MTU 12V 396 TB83 diesels; 3,120 hp(m) *(2.3 MW)* sustained; 2 waterjets
Speed, knots: 15.5. **Range, n miles:** 2,500 at 12 kt
Complement: 50 (5 officers)
Guns: 2 Emerson Electric 30 mm (twin); 1,200 rds/min combined to 6 km *(3.3 n miles)*; weight of shell 0.35 kg.
2 Oerlikon 20 mm GAM-BO1.
Countermeasures: MCM: Fitted with 2 Pluto remote-controlled submersibles, Oropesa 'O' Mis 4 and Ibis V control system.
Radars: Navigation: Racal Decca 1226; I-band.
Sonars: Thomson Sintra TSM 2022; hull-mounted; mine detection; high frequency.

Comment: *Ohue* ordered in April 1983 and *Maraba* in January 1986. *Ohue* laid down 23 July 1984 and launched 22 November 1985, *Maraba* laid down 11 March 1985, launched 6 June 1986. GRP hulls but, unlike Italian and Malaysian versions they do not have separate hydraulic minehunting propulsion. Carry Galeazzi two-man decompression chambers. Both were refitted in 1999, after operations off Liberia.
VERIFIED

OHUE   7/1987, Marina Fraccaroli

AMPHIBIOUS FORCES

1 FDR TYPE RO-RO 1300 (LST)

Name	No	Builders	Commissioned
AMBE	LST 1312	Howaldtswerke, Hamburg	11 May 1979

Displacement, tons: 1,470 standard; 1,860 full load
Dimensions, feet (metres): 285.4 × 45.9 × 7.5 *(87 × 14 × 2.3)*
Main machinery: 2 MTU 16V 956 TB92 diesels; 8,850 hp(m) *(6.5 MW)* sustained; 2 shafts
Speed, knots: 17. **Range, n miles:** 5,000 at 10 kt
Complement: 56 (6 officers)
Military lift: 460 tons and 220 troops long haul; 540 troops or 1,000 troops seated short haul; can carry 5—40 ton tanks
Guns: 1 Breda 40 mm/70. 2 Oerlikon 20 mm.
Radars: Navigation: Racal Decca 1226; I-band.

Comment: Ordered September 1976. Built to a design prepared for the FGN. Has 19 m bow ramps and a 4 m stern ramp. Reported that bow ramps are welded shut. Second of class beyond repair but *Ambe* reported active in 2003.
VERIFIED

AMBE   7/1997 / 0012836

SURVEY SHIPS

1 SURVEY SHIP (AGS)

Name	No	Builders	Launched	Commissioned
LANA	A 498	Brooke Marine, Lowestoft	4 Mar 1976	18 July 1976

Displacement, tons: 1,088 full load
Dimensions, feet (metres): 189 × 37.5 × 12 *(57.8 × 11.4 × 3.7)*
Main machinery: 2 Lister Blackstone diesels; 2,640 hp *(1.97 MW)*; 2 shafts
Speed, knots: 16. **Range, n miles:** 4,500 at 12 kt
Complement: 52 (12 officers)
Radars: Navigation: Decca; I-band.

Comment: Similar to UK Bulldog class. Ordered in 1973. Not reported active since 1997, but back in reasonable condition in 1999. ***VERIFIED***

LANA   5/1999 / 0081334

TUGS

3 COASTAL TUGS (YTB/YTL)

COMMANDER APAYI JOE A 499   DOLPHIN MIRA   DOLPHIN RIMA

Comment: A 499 is of 310 tons and was built in 1983. The two Dolphin tugs are under repair. ***VERIFIED***

COMMANDER APAYI JOE   11/1983, Hartmut Ehlers

Norway

Country Overview

The Kingdom of Norway is a constitutional monarchy occupying the northwest part of the Scandinavian Peninsula. With an area of 125,016 square miles, it is bordered to the east by Sweden and to the northeast by Finland and Russia. The coastline of 11,842 n miles with the Atlantic Ocean (Norwegian Sea), Arctic Ocean (Barents Sea), North Sea and Skagerrak Strait contains numerous fjords and offshore islands. External territories in the Arctic Ocean include the Svalbard archipelago and Jan Mayen Island while the uninhabited Bouvet Island lies in the south Atlantic. Territorial claims in Antarctica include the territory known as Queen Maud Land and Peter I Island. The capital, largest city and principal port is Oslo. Other ports include Bergen, Trondheim and Stavanger. Territorial seas (12 n miles) and an EEZ (200 n miles) are claimed.

Headquarters Appointments

Chief of Naval Staff:
   Rear Admiral J E Finseth
Deputy Chief of Naval Staff:
   Commodore A I Skram
Commander Coast Guard:
   Commodore G A Osen
Commander Norwegian Fleet:
   Commodore T Grytting

Diplomatic Representation

Defence Attaché in Ankara:
   Colonel O Nordbø
Defence Attaché in Helsinki:
   Colonel I L Viddal
Defence Attaché in London:
   Colonel J P Ryste
Defence Attaché in Madrid:
   Captain S Hauger
Defence Attaché in Moscow:
   Brigadier General F S Hauen
Defence Attaché in Paris:
   Colonel J E Hynaas
Defence Attaché in Stockholm:
   Captain G Heløe
Defence Attaché in Washington:
   Major General J F Blom
Defence Attaché in Warsaw:
   Colonel L Lindalen

Diplomatic Representation — continued

Defence Attaché in Berlin:
   Colonel S O Åndal
Defence Attaché in the Hague:
   Captain H Smidt
Defence Attaché in the Baltic States:
   Colonel L W Strand-Torgersen
Defence Attaché in Rome:
   Captain J R Bakke

General

As a result of the Defence Analysis 2000 Plan, which was approved with some changes by the Norwegian parliament, the Royal Norwegian Navy is undergoing a thorough restructuring. This includes a shift of focus from a mainly anti-invasion oriented Navy towards a force capable of international operations as well as operations in national waters.

Personnel

(a)   2005: 4,950 officers and ratings
(b)   9 to 12 months' national service (up to 40 per cent of ships complement)

Coast Artillery

The fixed defence system of nine coastal forts and controlled minefields is in long-term storage. As a result, the Coastal Ranger Command was established in 2001 with a headquarters at Trondenes.

Coast Guard

Founded April 1977 with operational command held by Norwegian Defence Command. Main bases at Sortland (North) and Haakonsvern (South). Tasks include fishery protection, customs, police, SAR and environmental duties at sea.

Bases

Jåtta (Stavanger) Armed Forces HQ.
Reitan (Bodø) Regional Command North HQ.
Trondheim Regional Command South HQ.
Laksevag (Bergen)-Submarine Repair and Maintenance.
Ramsund-Supply/Repair/Maintenance.

Air Force Squadrons (see *Shipborne* and *Land-based Aircraft*)

Aircraft (Squadron)	Location	Duties
Sea King Mk 43 (330)	Bodø, Banak, Sola, Ørland	SAR
Orion P-3N/C (333)	Andøya	MPA
Lynx (337)	Coast Guard vessels/ Bardufoss	MP
Bell 412 (719, 339 & 720)	Bodø, Rygge, Bardufoss	Army Transport

Prefix to Ships' Names

KNM (Naval)
K/V (Coast Guard)

Strength of the Fleet

Type	Active	Building (Projected)
Submarines—Coastal	6	—
Frigates	3	5
Fast Attack Craft—Missile	15	5
Minelayers	1	—
Minesweepers/Hunters	6	—
Depot Ship	1	—
Auxiliaries	1	—
Naval District Auxiliaries	9	—
Coast Guard Vessels	19	—
Survey Vessels	6	—

DELETIONS

Mine Warfare Forces

2002   *Orkla*
2004   *Rauma, Glomma*
2005   *Vidar*

Amphibious Forces

2002   *Sørøysund, Maursund, Tjeldsun*

PENNANT LIST

Notes: Naval District Auxiliaries are listed on page 530.

Submarines					Patrol Forces					Auxiliaries		Coast Guard	

Submarines

S 300	Ula
S 301	Utsira
S 302	Utstein
S 303	Utvaer
S 304	Uthaug
S 305	Uredd

Frigates

F 301	Bergen
F 302	Trondheim
F 304	Narvik
F 310	Fridtjof Nansen (bldg)
F 311	Roald Amundsen (bldg)
F 312	Otto Sverdrup (bldg)
F 313	Helge Ingstad (bldg)
F 314	Thor Heyerdahl (bldg)

Minesweepers/Hunters

M 340	Oksøy
M 341	Karmøy
M 342	Måløy
M 343	Hinnøy
M 350	Alta
M 351	Otra

Minelayers

N 50	Tyr

Patrol Forces

P 358	Hessa
P 359	Vigra
P 960	Skjold
P 986	Hauk
P 987	Ørn
P 988	Terne
P 989	Tjeld
P 990	Skarv
P 991	Teist
P 992	Jo
P 993	Lom
P 994	Stegg
P 995	Falk
P 996	Ravn
P 997	Gribb
P 998	Geir
P 999	Erle

Auxiliaries

A 530	Horten
A 533	Norge
A 535	Valkyrien

Coast Guard

W 303	Svalbard
W 310	Eigun
W 312	Ålesund
W 313	Tromsö
W 315	Nordsjøbas
W 316	Malene Østervold
W 317	Lafjord
W 318	Harstad
W 319	Thorsteinson
W 320	Nordkapp
W 321	Senja
W 322	Andenes

SUBMARINES

Notes: Norway withdrew from the 'Viking' submarine project on 13 June 2003 at the end of the Project Definition Phase Step 1. It has retained observer status in the project which is being taken forward by Sweden. Following the reduction of submarine forces to the six Ula class, there is no longer a requirement for new submarines before 2020.

6 ULA CLASS (SSK)

Name	No	Builders	Laid down	Launched	Commissioned
ULA	S 300	Thyssen Nordseewerke, Emden	29 Jan 1987	28 July 1988	27 Apr 1989
UREDD	S 305	Thyssen Nordseewerke, Emden	23 June 1988	22 Sep 1989	3 May 1990
UTVAER	S 303	Thyssen Nordseewerke, Emden	8 Dec 1988	19 Apr 1990	8 Nov 1990
UTHAUG	S 304	Thyssen Nordseewerke, Emden	15 June 1989	18 Oct 1990	7 May 1991
UTSTEIN	S 302	Thyssen Nordseewerke, Emden	6 Dec 1989	25 Apr 1991	14 Nov 1991
UTSIRA	S 301	Thyssen Nordseewerke, Emden	15 June 1990	21 Nov 1991	30 Apr 1992

Displacement, tons: 1,040 surfaced; 1,150 dived
Dimensions, feet (metres): 193.6 × 17.7 × 15.1
   (59 × 5.4 × 4.6)
Main machinery: Diesel-electric; 2 MTU 16V 396 SB83 diesels; 2,700 hp(m) *(1.98 MW)* sustained; 1 Siemens motor; 6,000 hp(m) *(4.41 MW)*; 1 shaft
Speed, knots: 11 surfaced; 23 dived
Range, n miles: 5,000 at 8 kt
Complement: 21 (5 officers)

Torpedoes: 8—21 in *(533 mm)* bow tubes. 14 AEG DM 2A3 Sehecht; dual purpose; wire-guided; active/passive homing to 28 km *(15 n miles)* at 23 kt; 13 km *(7 n miles)* at 35 kt; warhead 260 kg; depth to 460 m.

Countermeasures: ESM: Racal Sealion; radar warning.
Weapons control: Kongsberg MSI-90(U) TFCS.
Radars: Surface search: Kelvin Hughes 1007; I-band.
Sonars: Atlas Elektronik CSU 83; active/passive intercept search and attack; medium frequency.
   Thomson Sintra; flank array; passive; low frequency.

Programmes: Contract signed on 30 September 1982. This was a joint West German/Norwegian effort known as Project 210 in Germany. Although final assembly was at Thyssen a number of pressure hull sections were provided by Norway.
Modernisation: MSI-90U being upgraded 2000-2005. A mid-life upgrade of all six boats is under consideration.

Structure: Diving depth, 250 m *(820 ft)*. The basic command and weapon control systems are Norwegian, the attack sonar is German but the flank array, based on piezoelectric polymer antenna technology, was developed in France and substantially reduces flow noise. Calzoni Trident modular system of non-penetrating masts has been installed. Zeiss periscopes.

UPDATED

ULA   *8/2004*, Derek Fox /* 1043505

UTVAER   *7/2004*, P Froud /* 1043512

UTSIRA   *8/2004*, E & M Laursen /* 1043506

FRIGATES

3 OSLO CLASS (FFGM)

Name	No	Builders	Laid down	Launched	Commissioned
BERGEN	F 301	Marinens Hovedverft, Horten	1964	23 Aug 1965	15 June 1967
TRONDHEIM	F 302	Marinens Hovedverft, Horten	1963	4 Sep 1964	2 June 1966
NARVIK	F 304	Marinens Hovedverft, Horten	1964	8 Jan 1965	30 Nov 1966

Displacement, tons: 1,650 standard; 1,950 full load
Dimensions, feet (metres): 317 × 36.8 × 18 (screws)
(96.6 × 11.2 × 5.5)
Main machinery: 2 Babcock & Wilcox boilers; 600 psi
(42.18 kg/cm²); 850°F *(454°C)*; 1 set De Laval Ljungstrom
PN20 geared turbines; 20,000 hp(m) *(14.7 MW)*; 1 shaft
Speed, knots: 25+. **Range, n miles:** 4,500 at 15 kt
Complement: 125 (11 officers)

Missiles: SSM: 4 Kongsberg Penguin Mk 1 ❶; IR homing to
20 km *(10.8 n miles)* at 0.7 Mach; warhead 120 kg.
SAM: Raytheon NATO RIM-7M Sea Sparrow Mk 29 octuple
launcher ❷; semi-active radar homing to 14.6 km *(8 n miles)* at
2.5 Mach; warhead 39 kg; 24 cell magazine.
Guns: 2 US 3 in *(76 mm)*/50 Mk 33 (twin) ❸; 50 rds/min to
12.8 km *(7 n miles)*; weight of shell 6 kg.
1 Bofors 40 mm/70 ❹; 300 rds/min to 12 km *(6.6 n miles)*;
weight of shell 0.96 kg
2 Rheinmetall 20 mm/20 (not in all); 1,000 rds/min to 2 km.
Torpedoes: 6—324 mm US Mk 32 (2 triple) tubes ❺. Marconi
Stingray; anti-submarine; active/passive homing to 11 km
(5.9 n miles) at 45 kt; warhead 32 kg (shaped charge); depth
to 750 m *(2,460 ft)*.
A/S mortars: Kongsberg Terne III 6-tubed trainable ❻; range
pattern from 400-5,000 m; warhead 70 kg. Automatic
reloading in 40 seconds.
Mines: Laying capability.
Countermeasures: Decoys: 2 chaff launchers.
ESM: Argo AR 700; intercept.
Combat data systems: NFT MSI-3100 (supplemented by
Siemens ODIN) action data automation; Link 11 and 14.
SATCOM ❼.

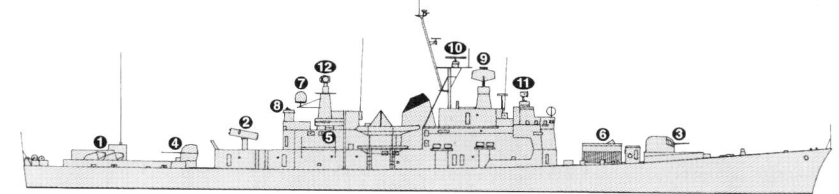

TRONDHEIM   *(Scale 1 : 900), Ian Sturton* / 0012838

Weapons control: Mk 91 MFCS. TVT 300 tracker ❽.
Radars: Air search: Siemens/Plessey AWS-9 ❾; 2D; E/F-band.
Surface search: Racal Decca TM 1226 ❿; I-band.
Fire control: NobelTech 9LV 218 Mk 2 ⓫; I-band (includes
search).
Raytheon Mk 95 ⓬, I/J-band (for Sea Sparrow).
Navigation: Decca; I-band.
Sonars: Thomson Sintra/Simrad TSM 2633; combined hull and
VDS; active search and attack; medium frequency.
Simrad Terne III; active attack; high frequency.

Programmes: Built under the five year naval construction
programme approved by the Norwegian Storting (Parliament)
late in 1960. Although all the ships of this class were
constructed in the Norwegian Naval Dockyard, half the cost
was borne by Norway and the other half by the US.
Modernisation: All ships modernised with improvements in
weapons control and habitability; new countermeasures

equipment includes two chaff launchers; Spherion TSM-2633
sonar (with VDS) (a joint Thomson Sintra/Simrad-Subsea
(Norway) project); the after 76 mm mounting replaced by a
Bofors 40 mm/70; and MSI 3100 action data automation.
Modernisation completion programme: F 302 30 November
1987, F 304 21 October 1988, F 301 4 April 1990. Plessey
AWS-9 radar fitted in 1996-98 together with Link 11 and GPS.
Structure: The hull and propulsion design of these ships is based
on that of the Dealey class destroyer escorts (now deleted) of
the US Navy, but considerably modified to suit Norwegian
requirements. The hulls became stressed by towing VDS in
heavy seas and were strengthened in 1995-96 increasing
displacement by over 200 tons.
Operational: The fifth of class *Oslo* sank under tow south of
Bergen in January 1994, after an engine failure had caused
her to run aground in heavy weather. *Stavanger* laid up in
1999 and sunk as target for torpedo firings in 2002.
UPDATED

TRONDHEIM   *6/2004*, J Ciślak* / 1043498

NARVIK   *7/2003*, Declerck/Steeghers* / 1043500

0 + 5 FRIDTJOF NANSEN CLASS (FFGHM)

Name	No	Builders	Laid down	Launched	Commissioned
FRIDTJOF NANSEN	F 310	Izar, Ferrol	9 Apr 2003	3 June 2004	Oct 2005
ROALD AMUNDSEN	F 311	Izar, Ferrol	3 June 2004	Apr 2005	Oct 2006
OTTO SVERDRUP	F 312	Izar, Ferrol	Apr 2005	2006	Oct 2007
HELGE INGSTAD	F 313	Izar, Ferrol	2006	2007	Oct 2008
THOR HEYERDAHL	F 314	Izar, Ferrol	2007	2008	Oct 2009

Displacement, tons: 5,290 full load
Dimensions, feet (metres): 433.1 × 55.1 × 16.1
 (132 × 16.8 × 4.9)
Main machinery: CODAG; 1 GE LM 2500 gas turbine; 26,112 hp
 (19.2 MW); 2 Bazán Bravo 12V diesels; 12,240 hp(m) *(9 MW)*;
 2 shafts; cp props; bow thruster; 1,360 hp(m) *(1 MW)*
Speed, knots: 26. **Range, n miles:** 4,500 at 16 kt
Complement: 120 plus 26 spare

Missiles: SSM: 8 Kongsberg NSM ❶.
 SAM: Mk 41 VLS (8 cells) ❷; 32 Evolved Sea Sparrow RIM 162B;
 semi-active radar homing to 18 km (9.7 n miles) at 3.6 Mach;
 warhead 39 kg
Guns: 1 Oto Melara 76 mm/62 Super Rapid ❸. 120 rds/min to
 15.75 km (8.5 n miles) anti-surface; 12 km (6.5 n miles) anti-
 aircraft; weight of shell 6 kg.
 4—12.7 mm MGs. Fitted for 1—40 mm/70.
Torpedoes: 4—324 mm (2 double) tubes ❹. Marconi Stingray;
 active/passive homing to 11 km (5.9 n miles) at 45 kt;
 warhead 35 kg shaped charge.
Countermeasures: Decoys: Terma SKWS chaff, IR and acoustic.
 ESM: Condor CS-3701; intercept ❺.
Combat data systems: AEGIS with ASW and ASuW segments
 from Kongsberg; Link 11 (fitted for Link 16/22).
Weapons control: Sagem VIGY 20 optronic director ❻.
Radars: Air search: Lockheed Martin SPY-1F ❼; E/F-band.
 Surface search: Litton; E/I-band ❽.
 Fire control: 2 Mk 82 (SPG-62); I/J-band ❾.
 Navigation: 2 Litton; I-band.
 IFF: Mk XII.
Sonars: Thomson Marconi Spherion MRS 2000 and Mk 2 ATAS;
 combined active/passive towed array.

Helicopters: 1 NH90 ❿.

Programmes: Design Definition for a new class of frigates
 started in March 1997. Izar and Lockheed Martin selected in
 March 2000 and contract signed 23 June 2000. Most of the
 construction is being undertaken by Izar. Two Norwegian
 shipyards, Bergen and Kvaerner Kleven, are collaborating to
 build six blocks for the aft section and three blocks for the bow
 section of each ship. These blocks are to be shipped to Ferrol
 where final assembly of the first three ships is to take place.
 The fourth and fifth ships may be built in Norway. Lockheed
 Martin is developing the combat system (in conjunction with
 Kongsberg Defence & Aerospace).
Structure: The design is based on the Alvaro de Bazan class.
Operational: Despite being fitted with Lockheed Martin SPY-1F
 radar and an AEGIS-derived combat system the primary roles
 of these ships will be anti-submarine and anti-surface warfare.
 UPDATED

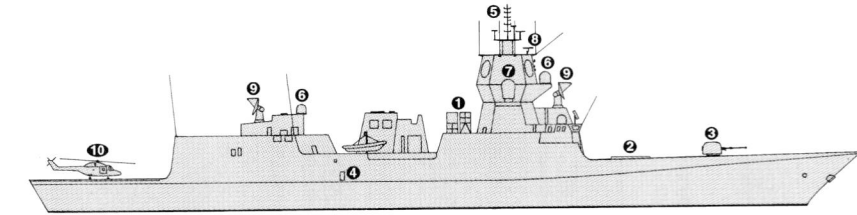

FRIDTJOF NANSEN

(Scale 1 : 1,200), Ian Sturton / 0105165

FRIDTJOF NANSEN

6/2004*, Camil Busquets i Vilanova / 1043493

FRIDTJOF NANSEN
6/2004*, Royal Norwegian Navy / 1043492

FRIDTJOF NANSEN

6/2004*, Royal Norwegian Navy / 1043494

SHIPBORNE AIRCRAFT

Numbers/Type: 6 NH Industries NH 90 NFH.
Operational speed: 157 kt *(291 km/h)*.
Service ceiling: 13,940 ft *(4,250 m)*.
Range: 621 n miles *(1,150 km)*.
Role/Weapon systems: Six ASW shipborne aircraft and eight with Coast Guard configuration to start entering service in 2005. Option for further ten SAR aircraft. ASW variant has crew of three. Sensors to be announced.

VERIFIED

NH 90 *2001, NH Industries / 0094462*

Numbers/Type: 6 Westland Lynx Mk 86.
Operational speed: 125 kt *(232 km/h)*.
Service ceiling: 12,500 ft *(3,810 m)*.
Range: 320 n miles *(590 km)*.
Role/Weapon systems: Operated by Air Force on behalf of the Coast Guard for fishery protection, offshore oil protection and SAR; embarked in CG vessels and shore-based. Sensors: Search radar, FLIR may be fitted, ESM. Weapons: Generally unarmed.

VERIFIED

LYNX *6/2002, Royal Norwegian Navy / 0572608*

LAND-BASED MARITIME AIRCRAFT

Notes: The Air Force has a total of 56 F-16 Falcons armed with Penguin 3 ASMs.

Numbers/Type: 4 Lockheed P-3C Orion.
Operational speed: 410 kt *(760 km/h)*.
Service ceiling: 28,300 ft *(8,625 m)*.
Range: 4,000 n miles *(7,410 km)*.
Role/Weapon systems: Long-range MR and oceanic surveillance duties in peacetime, with ASW added as a war role. Updated in 1998-99 with new radars and new tactical computers. P-3Ns used by Coast Guard paid off in 1999. Sensors: APS-137(V)5 radar, ASQ-81 MAD, AQS-212 processor and computer, IFF, AAR-36 IR detection; AAR-47 ESM; ALE 47 countermeasures; sonobuoys. Weapons: ASW; 8 MUSL Stingray torpedoes, depth bombs or mines. ASV; Penguin NFT Mk 3 ASM.

VERIFIED

P-3C *6/2001, A Sharma / 0130100*

Numbers/Type: 12 Westland Sea King Mk 43B.
Operational speed: 125 kt *(232 km/h)*.
Service ceiling: 10,500 ft *(3,200 m)*.
Range: 630 n miles *(1,165 km)*.
Role/Weapon systems: SAR, surface search and surveillance helicopter; supplemented by civil helicopters in wartime. Two 43B delivered in May 1996; remainder updated to 43B standard. Sensors: FLIR 2000 and dual Bendix radars RDR 1500 and RDR 1300. Weapons: Generally unarmed.

VERIFIED

SEA KING 43B *2001, GKN Westland / 0051448*

PATROL FORCES

14 HAUK CLASS (FAST ATTACK CRAFT—MISSILE) (PTGM)

Name	No	Builders (see *Programmes*)	Commissioned
HAUK	P 986	Bergens Mek Verksteder	17 Aug 1977
ØRN	P 987	Bergens Mek Verksteder	19 Jan 1979
TERNE	P 988	Bergens Mek Verksteder	13 Mar 1979
TJELD	P 989	Bergens Mek Verksteder	25 May 1979
SKARV	P 990	Bergens Mek Verksteder	17 July 1979
TEIST	P 991	Bergens Mek Verksteder	11 Sep 1979
JO	P 992	Bergens Mek Verksteder	1 Nov 1979
LOM	P 993	Bergens Mek Verksteder	15 Jan 1980
STEGG	P 994	Bergens Mek Verksteder	18 Mar 1980
FALK	P 995	Bergens Mek Verksteder	30 Apr 1980
RAVN	P 996	Westamarin A/S, Alta	20 May 1980
GRIBB	P 997	Westamarin A/S, Alta	10 July 1980
GEIR	P 998	Westamarin A/S, Alta	16 Sep 1980
ERLE	P 999	Westamarin A/S, Alta	10 Dec 1980

Displacement, tons: 120 standard; 160 full load
Dimensions, feet (metres): 120 × 20.3 × 5.9 *(36.5 × 6.2 × 1.8)*
Main machinery: 2 MTU 16V 538 TB92 diesels; 6,820 hp(m) *(5 MW)* sustained; 2 shafts
Speed, knots: 32. **Range, n miles:** 440 at 30 kt
Complement: 24 (6 officers)

Missiles: SSM: Up to 6 Kongsberg Penguin Mk 2 Mod 5; IR homing to 27 km *(14.6 n miles)* at 0.8 Mach; warhead 120 kg.
SAM: Twin Simbad launcher for Matra Sadral; IR homing to 4 km *(2.2 n miles)*; warhead 3 kg.
Guns: 1 Bofors 40 mm/70; 300 rds/min to 12 km *(6.6 n miles)*; weight of shell 0.96 kg.
Torpedoes: 2—21 in *(533 mm)* tubes. FFV Type 613; passive homing to 27 km *(14.5 n miles)* at 45 kt; warhead 240 kg.
Countermeasures: Decoys: Chaff launcher.
ESM: Argo intercept.
Combat data systems: DCN SENIT 2000 (from late 2001). Link 11.
Weapons control: Kongsberg MSI-80S or Sagem VIGY-20 optronic director.
Radars: Surface search/navigation: 2 Litton; I-band.

Programmes: Ordered 12 June 1975.
Modernisation: Simbad twin launchers for SAM fitted from 1994 to replace 20 mm gun. A further upgrade programme has extended the lives of these craft until replaced by the Skjold class or until 2015. 991, 992 and 994 were fitted with the SENIT 2000 combat data system in 2001 with the remainder of the class fitted by late 2003. Other aspects of the upgrade included a new navigation radar, an optronic director and new communications and bridge equipment. Trials on a wave-piercing bow have been conducted in 989 to improve hydrodynamic performance.
Operational: Penguin missiles are not normally embarked.

UPDATED

JO *4/2003, Frank Findler / 0572603*

SKARV *5/2004*, L-G Nilsson / 1043501*

1 + 5 SKJOLD CLASS (PTGMF)

Name	No	Builders	Launched	Commissioned
SKJOLD	P 960	Kvaerner Mandal	22 Sep 1998	17 Apr 1999
STORM	P 961	Umoe Mandal	2007	2008
SKUDD	P 962	Umoe Mandal	2007	2008
STEIL	P 963	Umoe Mandal	2007	2008
GLIMT	P 964	Umoe Mandal	2008	2009
GNIST	P 965	Umoe Mandal	2008	2009

Displacement, tons: 260 full load
Dimensions, feet (metres): 153.5 × 44.3 × 7.5; 2.6 on cushion *(46.8 × 13.5 × 2.3; 0.8)*
Main machinery: COGAG: 2 Pratt & Whitney ST 40 gas turbines; 10,730 hp *(8 MW)*; 2 Pratt & Whitney ST 18 gas turbines; 5,365 hp *(4 MW)*
2 MTU 12V 183 TE92 diesels (lift); 2,000 hp(m) *(1.47 MW)*
Speed, knots: 57; 44 in Sea State 3. **Range, n miles:** 800 at 40 kt
Complement: 15

Missiles: 8 SSM; 8 Kongsberg NSM.
SAM: Mistral; IR homing to 4 km *(2.2 n miles)* at 2.5 Mach; warhead 3 kg.
Guns: 1 Otobreda 76 mm/62. Super Rapid; 120 rds/min to 16 km *(8.7 n miles)*; weight of shell 6 kg.
Countermeasures: Buck Neue MASS decoys.
ESM: EDO CS 3701; intercept.
Combat data systems: DCN Senit 2000; Link 11/16.
Weapons control: Sagem VIGX-20 optronic director.
Radars: Air/surface search: Thales MRR; 3D-NG; G-band.
Navigation: I-band.
Fire control: CelsiusTech Ceros 2000; I/J-band.

Programmes: Project SMP 6081. A preproduction version (P 960) ordered 30 August 1996. This was tested by the Norwegian Navy from 1999-2001 and was under evaluation by the USN and USCG in 2001-02. The Norwegian parliament decided on 23 October 2003 that five additional vessels were to be built and that the preproduction vessel was to be rebuilt. Contract with Skjold Prime Consortium, comprising Umoe Mandal, Armaris and Kongsberg Defence & Aerospace, was signed 28 November 2003. Ships to be built at Umoe Mandal shipyard for delivery between 2008 and 2009.
Structure: SES hull with advanced stealth technology including anechoic coatings. Building on experience in US trials, a more raked bow has been adopted to improve performance into sea. The foredeck structure is also to be strengthened around the gun mounting. Two quadruple SSM launchers are to be recessed aft of the bridge. These will elevate to fire and then retract.

UPDATED

SKJOLD   *7/2003*, Declerck/Steeghers /* 1043502

SKJOLD   *7/2003*, Declerck/Steeghers /* 1043503

20 COMBATBOAT 90N (LCP)

TRONDENES KA 1	STANGENES KA 13	BRETTINGEN KA 21
HYSNES KA 2	KJØKØY KA 14	LØKHAUG KA 22
HELLEN KA 3	MØRVIKA KA 15	SØRVIKNES KA 23
TORAAS KA 4	KOPAAS KA 16	OSTERNES KA 31
MØVIK KA 5	TANGEN KA 17	FJELL KA 32
SKROLSVIK KA 11	ODDANE KA 18	LERØY KA 33
KRÅKENES KA 12	MALMØYA KA 19	

Displacement, tons: 19 full load
Dimensions, feet (metres): 52.2 × 12.5 × 2.6 *(15.9 × 3.8 × 0.8)*
Main machinery: 2 SAAB Scania DSI 14 diesels; 1,104 hp(m) *(812 kW)* or 1,251 hp(m) *(920 kW)* (KA 21-43) sustained; 2 FF 450 water-jets or 2 Kamewa FF 410 (KA 21-43)
Speed, knots: 35 or 40; 20 in Sea State 3
Range, n miles: 240 at 20 kt
Complement: 3
Military lift: 2.8 tons or 20 troops
Guns: 1—12.7 mm MG.
Radars: Navigation: I-band.

Comment: Ordered from Dockstavarvet, Sweden. Four Batch 1 units delivered for trials in July and October 1996. Three more of the class delivered in 1997, 13 in 1998. Used to carry mobile light missile units and prime method of transportation for new Coastal Ranger Commando. Similar in most details to the Swedish Coastal Artillery craft. Names are mostly taken from Coastal Fortresses.

VERIFIED

BRETTINGEN   *4/2002, P Froud /* 0529076

7 ALUSAFE 1290 CLASS (INSHORE PATROL CRAFT) (PB)

L 4540-4546

Displacement, tons: 7.6
Dimensions, feet (metres): 43.2 × 11.5 × 2.5 *(12.9 × 3.5 × 0.75)*
Main machinery: 2 Volvo Penta TAMD 74 EDC diesels; 900 hp *(670 kW)*; 2 Kamewa K28 waterjets
Speed, knots: 42
Complement: 2 (plus 13 troops)
Guns: 2—12.7 mm MGs.

Comment: Aluminium hull. Built by Maritime Partner, Ålesund and delivered in 2002. Designed for used by the Norwegian Naval Home Guard as multifunction assault and patrol vessels by the coastal rangers. The craft are also available to support police, customs, environmental and fishery authorities.

UPDATED

L 4541   *6/2004*, Royal Norwegian Navy /* 1043496

3 ALUSAFE 1300 CLASS (INSHORE PATROL CRAFT) (PB)

SHV 104-106

Displacement, tons: 10
Dimensions, feet (metres): 43.6 × 12.0 × 2.5 *(13.3 × 3.65 × 0.75)*
Main machinery: 2 Volvo Penta TAMD 74EDC diesels; 900 hp *(670 kW)*; 2 Kamewa K28 waterjets
Speed, knots: 40
Complement: 2 (plus 13 troops)
Guns: 2—12.7 mm MGs.

Comment: Aluminium hull. Built by Maritime Partner, Ålesund and delivered in 2003. Based at Stavanger, Bergen and Trondheim. Designed for use by the Norwegian Naval Home Guard as multifunction patrol vessels for the naval home guard. The craft are also available to support police, customs, environmental and fishery authorities.

UPDATED

SHV 104   *5/2004*, E & M Laursen /* 1043504

MINE WARFARE FORCES

6 OKSØY/ALTA CLASS
(MINEHUNTERS/SWEEPERS) (MHCM/MSCM)

Name	No	Builders	Commissioned
Hunters			
OKSØY	M 340	Kvaerner Mandal	24 Mar 1994
KARMØY	M 341	Kvaerner Mandal	24 Oct 1994
MÅLØY	M 342	Kvaerner Mandal	24 Mar 1995
HINNØY	M 343	Kvaerner Mandal	8 Sep 1995
Sweepers			
ALTA	M 350	Kvaerner Mandal	12 Jan 1996
OTRA	M 351	Kvaerner Mandal	8 Nov 1996

Displacement, tons: 375 full load
Dimensions, feet (metres): 181.1 × 44.6 × 8.2 (2.76 cushion) *(55.2 × 13.6 × 2.5 (0.84))*
Main machinery: 2 MTU 12V 396 TE84 diesels; 3,700 hp(m) *(2.72 MW)* sustained; 2 Kvaerner Eureka water-jets; 2 MTU 8V 396 TE54 diesels; 1,740 hp(m) *(1.28 MW/60 Hz)* sustained; lift engines
Speed, knots: 30. **Range, n miles:** 1,500 at 20 kt
Complement: 38 (12 officers) (minehunters); 32 (10 officers) (minesweepers)

Missiles: SAM: Matra Sadral twin launcher; Mistral; IR homing to 4 km *(2.2 n miles)*; warhead 3 kg.
Guns: 1 or 2 Rheinmetall 20 mm. 2—12.7 mm MGs.
Countermeasures: MCMV: 2 Pluto submersibles (minehunter); mechanical, AGATE (air gun and transducer equipment) acoustic and Elma magnetic sweep (minesweepers). Minesweeper mini torpedoes can be carried.
Radars: Navigation: 2 Racal Decca; I-band.
Sonars: Thomson Sintra/Simrad TSM 2023N; hull-mounted (minehunters); high frequency. Simrad Subsea SA 950; hull-mounted (minesweepers); high frequency.

Programmes: Order placed with Kvaerner on 9 November 1989. Four are minehunters, the remainder minesweepers.
Modernisation: Trials of the Kongsberg Simrad Hugin 1000 Autonomous Underwater Vehicle (AUV) conducted in M 341 2001-05. Designed to conduct route survey and forward mine reconnaissance. A full capability demonstration is planned for 2006. Hugin 1000 can dive to 600 m and has an endurance of 20 hours.
Structure: Design developed by the Navy in Bergen with the Defence Research Institute and Norsk Veritas and uses an air cushion created by the surface effect between two hulls. The hull is built of Fibre Reinforced Plastics (FRP) in sandwich configuration. The ROVs are carried in a large hangar and are launched by two hydraulic cranes. The minesweeper has an A frame aft for the sweep gear. SAM launcher mounted forward of the bridge.
Operational: Simrad Albatross tactical system including mapping; Cast/Del Norte mobile positioning system with GPS. The catamaran design is claimed to give higher transit speeds with lesser installed power than a traditional hull design. Other advantages are lower magnetic and acoustic signatures, clearer water for sonar operations and less susceptibility to shock. *Orkla* M 353 was lost after a catastrophic fire on 19 November 2002. M 352 and M 354 were decommissioned in 2004. ***UPDATED***

KARMØY   *11/2004*, Michael Nitz /* 1043497

HINNØY   *8/2004*, Martin Mokrus /* 1043499

ALTA   *5/2004*, L-G Nilsson /* 1043510

1 MINELAYER (ML)

Name	No	Builders	Commissioned
TYR (ex-*Standby Master*)	N 50	Alesund Mekaniske Verksted	1981

Displacement, tons: 495 full load
Dimensions, feet (metres): 138.8 × 33.1 × 11.5 *(42.3 × 10.1 × 3.5)*
Main machinery: 2 Deutz SBA12M816 diesels; 1,300 hp(m) *(956 kW)*; 1 shaft; cp prop; 1 MWM diesel; 150 hp(m) *(110 kW)*; bow and stern thrusters
Speed, knots: 12
Complement: 22 (7 officers)
Mines: 2 rails.
Radars: Navigation: Furuno 711 and Furuno 1011; I-band.

Comment: Former oil rig pollution control ship. Acquired in December 1993 and converted by Mjellum & Karlsen, Bergen. Recommissioned 7 March 1995 as a minelayer, and for the maintenance of controlled minefields. Carries a ROV. ***UPDATED***

TYR   *5/2004*, Per Körnefeldt /* 1043511

SURVEY AND RESEARCH SHIPS

1 RESEARCH SHIP (AGEH)

Name	Builders	Launched	Commissioned
MARJATA	Tangern Verft A/S	18 Dec 1992	July 1994

Displacement, tons: 7,560 full load
Dimensions, feet (metres): 267.4 × 130.9 × 19.7 *(81.5 × 39.9 × 6)*
Main machinery: Diesel-electric; 2 MTU Siemens 16V 396 TE diesels; 7,072 hp(m) *(5.2 MW)*; 2 Dresser Rand/Siemens gas-turbine generators; 9,792 hp(m) *(7.2 MW)*; 2 Siemens motors; 8,160 hp(m) *(6 MW)*; 2 Schottel 3030 thrusters. 1 Siemens motor; 2,720 hp(m); *(2 MW)*; 1 Schottel thruster (forward)
Speed, knots: 15
Complement: 14 plus 31 scientists

Helicopters: Platform for one medium

Comment: Ordered in February 1992 from Langsten Slip og Batbyggeri to replace the old ship of the same name. Called Project Minerva. Design developed by Ariel A/S, Horten. The three main superstructure-mounted cupolas contain ELINT and SIGINT equipment. Hull-reinforced to allow operations in fringe ice. Equipment includes Sperry radars, Elac sonars, Siemens TV surveillance, and a fully equipped helicopter flight deck. The unconventional hull which gives the ship an extraordinary length to beam ratio of 2:1 is said to give great stability and dynamic qualities. White hull and superstructure. ***VERIFIED***

MARJATA   *2000, Royal Norwegian Navy /* 0105173

5 SURVEY SHIPS (AGS)

Name	Displacement tons	Launched	Officers	Crew
OLJEVERN 01-04	200	1978	2	6
GEOFJORD	364	1958	2	6

Comment: Under control of Ministry of Environment based at Stavanger. *Oljevern 01* and *03* have red hulls and work for the Pollution Control Authority. ***UPDATED***

GEOFJORD   *5/2002, L-G Nilsson /* 0528972

TRAINING SHIPS

2 TRAINING SHIPS (AXL)

Name	No	Builders	Commissioned
HESSA (ex-*Hitra*, ex-*Marsteinen*)	P 358	Fjellstrand, Omastrand	Jan 1978
VIGRA (ex-*Kvarven*)	P 359	Fjellstrand, Omastrand	July 1978

Displacement, tons: 39 full load
Dimensions, feet (metres): 77 × 16.4 × 3.5 *(23.5 × 5 × 1.1)*
Main machinery: 2 GM 12V-71 diesels; 1,800 hp *(1.34 MW)*; 2 shafts
Speed, knots: 20
Complement: 5 plus 13 trainees
Guns: 1—12.7 mm Browning MG.
Radars: Navigation: Racal Decca; I-band.

Comment: The vessels are designed for training students at the Royal Norwegian Naval Academy in navigation, manoeuvring and seamanship. All-welded aluminium hulls. Also equipped with an open bridge and a blind pilotage position below deck. 18 berths.   *VERIFIED*

VIGRA   *4/2002, P Froud /* 0529133

AUXILIARIES

1 DEPOT SHIP (ASH/AGP)

Name	No	Builders	Commissioned
HORTEN	A 530	A/S Horten Verft	Apr 1978

Displacement, tons: 2,530 full load
Dimensions, feet (metres): 287 × 42.6 × 16.4 *(87.5 × 13 × 5)*
Main machinery: 2 Wichmann 7AX diesels; 4,200 hp(m) *(3.1 MW)*; 2 shafts; bow thruster
Speed, knots: 16.5
Complement: 86
Guns: 2 Bofors 40 mm/70.
Radars: Navigation: 2 Decca; I-band.
Helicopters: Platform only.

Comment: Contract signed 30 March 1976. Laid down 28 January 1977; launched 12 August 1977. Serves both submarines and fast attack craft. Quarters for 45 extra and can cater for 190 extra.   *VERIFIED*

HORTEN   *3/2001, A Sharma /* 0130096

1 SUPPLY AND RESCUE VESSEL (AKS/ATA)

Name	No	Builders	Commissioned
VALKYRIEN	A 535	Ulstein Hatlo	1981

Displacement, tons: 3,000 full load
Dimensions, feet (metres): 223.1 × 47.6 × 16.4 *(68 × 14.5 × 5)*
Main machinery: Diesel-electric; 4 diesels; 10,560 hp(m) *(7.76 MW)* sustained; 2 motors; 3.14 MW; 2 shafts; 2 bow thrusters; 1,600 hp(m) *(1.18 MW)*; 1 stern thruster; 800 hp(m) *(588 kW)*
Speed, knots: 16
Complement: 13
Radars: Navigation: 2 Furuno; H/I-band.

Comment: Tug/supply ship acquired in 1994 for supply and SAR duties. Bollard pull 128 tons. Can carry a 700 ton deck load. Oil recovery equipment is also carried.   *UPDATED*

VALKYRIEN   *5/2004*, L-G Nilsson /* 1043507

7 COASTAL VESSELS (YPT/YDT)

Notes: Due to re-organisation of the coastal vessels, the naval districts no longer operate many of the vessels previously assigned. The following remain in service and are prefaced by two letters as follows: HT (torpedo recovery), HM (multirole), HS (tugs), HD (diving), HP (personnel), HR (rescue). *Hitra* (HP 15) is also used for training cruises. All are less than 300 tons displacement.

Name	No	Speed, knots	Commissioned	Role
VIKEN	HD 2	12	1984	Cargo (4 tons)/Passengers (40) Diving vessel
TORPEN	HM 3	12	1977	Cargo (100 tons)/Passengers (15)
KJEØY	HM 7	10	1993	Training ship/Passengers (30)
HITRA	HP 15	—	—	Passengers (30)
KARLSØY	HT 3	10	1978	Torpedo fishing vessel
SLEIPNER	HS 4	11	2002	Tug/Cargo (10 tons)
MJØLNER	HS 5	11	2002	Tug/Cargo (10 tons)

UPDATED

HITRA   *8/1996, Erik Laursen /* 0081339

VIKEN   *7/2003*, Declerck/Steeghers /* 1043508

ROYAL YACHTS

1 ROYAL YACHT (YAC)

Name	No	Builders	Commissioned
NORGE (ex-*Philante*)	A 533	Camper & Nicholson's Ltd, Southampton	1937

Displacement, tons: 1,786 full load
Dimensions, feet (metres): 263 × 38 × 15.2 *(80.2 × 11.6 × 4.6)*
Main machinery: 2 Bergen KRMB-8 diesels; 4,850 hp(m) *(3.6 MW)* sustained; 2 shafts
Speed, knots: 17
Complement: 50 (18 officers)
Radars: Navigation: 2 Decca; I-band.

Comment: Built to the order of the late T O M Sopwith as an escort and store vessel for the yachts *Endeavour I* and *Endeavour II*. Launched on 17 February 1937. Served in the Royal Navy as an anti-submarine escort during the Second World War, after which she was purchased by the Norwegian people for King Haakon and reconditioned as a Royal Yacht at Southampton. Can accommodate about 50 people in addition to crew. Repaired after serious fire on 7 March 1985 when the ship was fitted with a bow-thruster.   *VERIFIED*

NORGE   *8/1999, Michael Nitz /* 0081352

COAST GUARD (KYSTVAKT)

1 ARCTIC CLASS (WPSOH)

Name	No	Builders	Commissioned
SVALBARD	W 303	Tangen Verft, Krager	5 Jan 2002

Displacement, tons: 6,300 full load
Dimensions, feet (metres): 340.3 × 62.7 × 21.3 *(103.7 × 19.1 × 6.5)*
Main machinery: Diesel electric; 4 diesel generators; 10 MW; 2 azimuth pods
Speed, knots: 17. **Range, n miles:** 10,000 at 13 kt
Complement: 50
Guns: 1 Bofors 57 mm/70.
Helicopters: 1 light.

Comment: Project definition completed in 1997 for an ice-reinforced vessel equipped with a helicopter. Built to Det Norske Veritas standards. Contract placed 15 December 1999 with Langsten Slip and Båtbyggeri A/S, Tomrefjord. Ship launched February 2001. Fitted for firefighting and counter-pollution work. There are two motor cutters and a sea-raider type dinghy. ***UPDATED***

SVALBARD   *7/2003, Freddie Philips /* 0572601

7 CHARTERED SHIPS (WPBO)

Name	No	Tonnage	Completion
EIGUN	W 310	342	1959
ÅLESUND	W 312	1,357	1996
TROMSÖ	W 313	1,970	1997
NORDSJØBAS	W 315	814	1978
MALENE ØSTERVOLD	W 316	1,678	1965
LAFJORD	W 317	814	1978
HARSTAD	W 318	3132	2004

Comment: *Lafjord* and *Nordsjøbas* chartered in 1980 and *Thorsteinson* in 1999. All armed with one 40 mm/60 gun. Two more ships have been built to Coast Guard requirements and leased for 10 years with an option to buy after five. *Tromsö* operates around north Norway while *Ålesund* is based in the south. Some ships are operated with two crews, changing over every three weeks. ***UPDATED***

EIGUN   *7/2003, Per Körnefeldt /* 0572602

3 NORDKAPP CLASS (WPSOH)

Name	No	Builders	Launched	Commissioned
NORDKAPP	W 320	Bergens Mek Verksteder	2 Apr 1980	25 Apr 1981
SENJA	W 321	Horten Verft	16 Mar 1980	6 Mar 1981
ANDENES	W 322	Haugesund Mek Verksted	21 Mar 1981	30 Jan 1982

Displacement, tons: 3,240 full load
Dimensions, feet (metres): 346 × 47.9 × 16.1 *(105.5 × 14.6 × 4.9)*
Main machinery: 4 Wichmann 9AXAG diesels; 16,163 hp(m) *(11.9 MW)*; 2 shafts
Speed, knots: 23. **Range, n miles:** 7,500 at 15 kt
Complement: 52 (6 aircrew)

Missiles: SSM: Fitted for 6 Kongsberg Penguin II but not embarked.
Guns: 1 Bofors 57 mm/70; 200 rds/min to 17 km *(9.3 n miles)*; weight of shell 2.4 kg.
    4 Rheinmetall 20 mm/20; 1,000 rds/min to 2 km.
Torpedoes: 6—324 mm US Mk 32 (2 triple) tubes. Honeywell Mk 46; anti-submarine; active/passive homing to 11 km *(5.9 n miles)* at 40 kt; warhead 44 kg. Mountings only in peacetime.
Depth charges: 1 rack.
Countermeasures: Decoys: 2 chaff launchers.
Combat data systems: Navkis or EDO (after modernisation). SATCOM can be carried.
Weapons control: THORN EMI MEOSS optronic director to be fitted.
Radars: Air/surface search: Plessey AWS 5; E/F-band or DRS Technologies SPS 67(V)3; G-band.
    Navigation: 2 Racal Decca 1226; I-band.
    Fire control: Philips 9LV 218 Mk 2; J-band.
Sonars: Simrad SS 105; hull-mounted; active search and attack; 14 kHz or SIMRAD SP 270 (after modernisation); hull mounted.

Helicopters: 1 Westland Lynx Mk 86.

Programmes: In November 1977 the Coast Guard budget was cut resulting in a reduction of the building programme from seven to three ships.
Modernisation: The ships are to be upgraded including an optronic director, new hull-mounted sonar, new air search radar and combat data system. W 321 was refitted in 2001 and both other ships were to have been completed by November 2002.
Structure: Ice strengthened. Fitted for firefighting, anti-pollution work, all with two motor cutters and a Gemini-type dinghy. SATCOM fitted for Gulf deployment.
Operational: Bunks for 109. War complement increases to 76. ***UPDATED***

ANDENES   *5/2004*, Per Körnefeldt /* 1043509

9 FISHERY PROTECTION SHIPS

Name	No	Tonnage	Completion
TITRAN	KV 1	184	1992
KONGSØY	KV 2	331	1958
AGDER	KV 5	140	1974
GARSØY	KV 6	195	1988
ÅHAV	KV 7	50	1981
BARENTSHAV	KV 22	318	1957
SJØVEIEN	KV 25	330	1964
NYSLEPPEN	KV 28	343	1967
THORSTEINSON	W 319	272	1960

Comment: An Inshore Patrol Force was established in January 1997. This comprises mostly chartered ships with KV pennant numbers. KV 1-7 are coastal cutters. Five new ships are to replace older ships from 2006. ***UPDATED***

ÅHAV   *8/1999, van Ginderen Collection /* 0081357

0 + 5 (5) OFFSHORE PATROL VESSELS (PBO)

Displacement, tons: 710 full load
Dimensions, feet (metres): 154.8 × 33.8 × 10.8 *(47.2 × 10.3 × 3.3)*
Main machinery: Diesel-electric; 2 azimuth thrusters
Speed, knots: 16
Complement: 20

Comment: Contract awarded in February 2005 to Remøy Management and Remøy Shipping for the construction of five new vessels with an option for a further five. The vessels are to be owned by the shipping companies and chartered to the Coast Guard for an initial period of 15 years. The design, developed by Skipsteknisk AS, is called ST-610. The ships are to be employed out to 24 n miles from the coast and are to be equipped to conduct towing, counter-pollution operations, fire-fighting and general patrol duties. Two fast rescue craft are to be carried. The ships are to enter service from 2006. ***NEW ENTRY***

ST-610 (artist's impression)   *2/2005*, Skipsteknisk AS /* 1043495

Oman

Country Overview

The Sultanate of Oman is an independent Middle-East state extending along the south-east coast of the Arabian Peninsula. It is bordered to the south-west by the Republic of Yemen, to the west by Saudi Arabia and to the north-west by the United Arab Emirates which separates a small exclave on the Musandam peninsula, on the south side of the Strait of Hormuz, from the rest of the country. Masirah island and the Khuriya Muriya Islands lie off the south-east coast. With an area of 82,030 square miles, it has a 1,129 n mile coastline with the Indian Ocean and Gulf of Oman. The capital, largest city and principal port is Muscat while there is a further port at Salalah. Territorial seas (12 n miles) are claimed. An EEZ (200 n miles) has also been claimed but its limits have only been partly defined by boundary agreements.

Headquarters Appointments

Commander Royal Navy of Oman:
Rear Admiral (Liwaa Rukn Bahry) Salim bin Abdullah bin Rashid al Alawi
Principal Staff Officer:
Commodore (Ameed) Abdullah Khamis Abdullah Al-Raisi
Director General Operations and Plans:
Commodore (Ameed) Abdullah Khamis Abdullah Al-Raisi
Commander Coast Guard:
Captain (Aqeed Bahry) Hamdan bin Marhoon Al Mamary
Commander Royal Yacht Squadron:
Commodore (Ameed) J M Knapp

Bases

Said bin Sultan, Widam A'Sahil (main base, dockyard and shiplift)
Ras Musandam
Muaskar al Murtafa'a (headquarters)

Personnel

(a) 2005: 4,500 officers and men
(b) Voluntary service

CORVETTES

2 QAHIR CLASS (FSGMH)

Name	No	Builders	Laid down	Launched	Commissioned
QAHIR AL AMWAJ	Q 31	Vosper Thornycroft, Woolston	21 May 1993	21 Sep 1994	3 Sep 1996
AL MUA'ZZAR	Q 32	Vosper Thornycroft, Woolston	4 Apr 1994	26 Sep 1995	13 Apr 1997

Displacement, tons: 1,450 full load
Dimensions, feet (metres): 274.6 oa; 249.3 wl × 37.7 × 11.8 *(83.7; 76 × 11.5 × 3.6)*
Main machinery: CODAD; 4 Crossley SEMT-Pielstick 16 PA6 V 280 STC; 28,160 hp(m) *(20.7 MW)* sustained; 2 shafts; Kamewa cp props
Speed, knots: 28. **Range, n miles:** 4,000 at 10 kt
Complement: 76 (14 officers) plus 3 spare

Missiles: SSM: 8 Aerospatiale MM 40 Block 2 Exocet ❶; inertial cruise; active radar homing to 70 km *(40 n miles)* at 0.9 Mach; warhead 165 kg; sea-skimmer.
SAM: Thomson-CSF Crotale NG octuple launcher ❷; 16 VT1; command line of sight guidance; radar/IR homing to 13 km *(7 n miles)* at 2.4 Mach; warhead 14 kg.
Guns: 1 OTO Melara 3 in *(76 mm)*/62 Super Rapid ❸; 120 rds/min to 16 km *(8.7 n miles)*; weight of shell 6 kg.
2 Oerlikon/Royal Ordnance 20 mm GAM-BO1 ❹.
2—7.62 mm MGs.
Torpedoes: 6—324 mm (2 triple) tubes may be fitted in due course.
Countermeasures: Decoys: 2 Barricade 12-barrelled chaff and IR launchers ❺.
ESM: Thomson-CSF DR 3000 ❻; intercept.
Combat data systems: Signaal/Thomson-CSF TACTICOS; Link Y; SATCOM.
Weapons control: Signaal STING optronic and radar tracker ❼; 2 Signaal optical directors.
Radars: Air/surface search: Signaal MW08 ❽; G-band.
Fire control: Signaal STING ❼; I/J-band.
Thomson-CSF DRBV 51C ❾; J-band (for Crotale).
Navigation: Kelvin Hughes 1007; I-band.
Sonars: Thomson Sintra/BAeSEMA ATAS; towed array; active search; 3 kHz (may be fitted).

Helicopters: Platform for 1 Super Lynx type ❿.

Programmes: Vosper Thornycroft signed the Muheet Project contract on 5 April 1992. First steel cut 23 September 1992. Q 31 accepted on 27 March 1996, and Q 32 on 26 November 1996. Commissioned after operational work up in the UK, and on return to Oman. Names mean Conqueror of the Waves, and The Supported.
Structure: The ship is based on the Vigilance class design with enhanced stealth features. It is possible lightweight torpedo tubes may be fitted. The towed array, if fitted, adds another 8 tons on the stern but does not affect the helicopter deck. RAM (Radar Absorbent Material) is widely used on the superstructure.
Operational: The helicopter platform can support a Super Puma sized aircraft. *UPDATED*

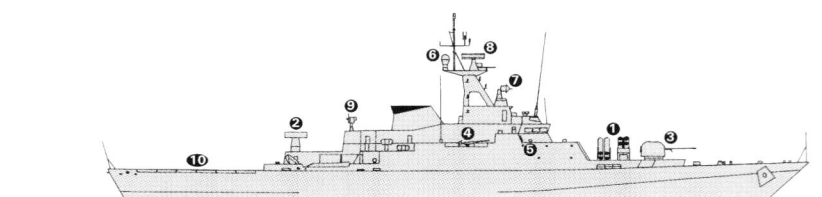

QAHIR AL AMWAJ   *(Scale 1 : 900), Ian Sturton*

QAHIR AL AMWAJ   *6/2003*, Royal Navy of Oman /* 0589792

QAHIR AL AMWAJ   *6/2001, Royal Navy of Oman /* 0114774

1 PATROL SHIP (FSH/AXL/AGS)

Name	No	Builders	Commissioned
AL MABRUKAH (ex-*Al Said*)	Q 30 (ex-A 1)	Brooke Marine, Lowestoft	1971

Displacement, tons: 900 full load
Dimensions, feet (metres): 203.4 × 35.1 × 9.8 *(62 × 10.7 × 3)*
Main machinery: 2 Paxman Valenta 12CM diesels; 5,000 hp *(3.73 MW)* sustained; 2 shafts
Speed, knots: 12
Complement: 39 (7 officers) plus 32 trainees
Guns: 1 Bofors 40 mm/70. 2 Oerlikon 20 mm A41A.
Countermeasures: Decoys: Wallop Barricade 18-barrelled chaff launcher.
ESM: Racal Cutlass; radar warning.
Radars: Surface search: Racal Decca TM 1226; I-band.
Helicopters: Platform only.

Comment: Built by Brooke Marine, Lowestoft. Launched 7 April 1970 as a yacht for the Sultan of Oman. Carried on board is one Rotork landing craft. Converted to training/patrol ship in 1983 with enlarged helicopter deck, additional accommodation and armament. Re-classified as a corvette and pennant number changed in 1997. Fitted with survey equipment in 2000, as an additional role.

UPDATED

AL MABRUKAH   6/2003*, Royal Navy of Oman / 0589799

SHIPBORNE AIRCRAFT

Numbers/Type: 16 GKN Westland Super Lynx.
Operational speed: 120 kt *(222 km/h).*
Service ceiling: 10,000 ft *(3,048 m).*
Range: 320 n miles *(593 km).*
Role/Weapon systems: Contract signed 19 January 2002. First three delivered 24 June 2004. Roles likely to include ASW and ASV. To be operated by Air Force.

UPDATED

SUPER LYNX
7/2004*, K Shaw/Jane's / 0566685

PATROL FORCES

Notes: 'Project Kareef' calls for the design, construction and support of three 80 m OPVs for delivery to the RNO from 2008. An initial statement of requirements was issued in February 2003. Following down-selection of a shortlist of bidders in late 2004, selection of a preferred contractor is expected in mid-2005.

3 AL BUSHRA CLASS (PBO)

Name	No	Builders	Laid down	Launched	Commissioned
AL BUSHRA	Z 1	CMN, Cherbourg/Wudam Dockyard	10 Nov 1993	3 May 1995	15 June 1995
AL MANSOOR	Z 2	CMN, Cherbourg/Wudam Dockyard	12 Apr 1994	3 May 1995	10 Aug 1995
AL NAJAH	Z 3	CMN, Cherbourg/Wudam Dockyard	27 June 1994	5 Mar 1996	15 Apr 1996

Displacement, tons: 475 full load
Dimensions, feet (metres): 178.6 × 26.2 × 8.9 *(54.5 × 8 × 2.7)*
Main machinery: 2 MTU 16V 538 TB93 diesels; 8,000 hp(m) *(5.88 MW)* sustained; 2 shafts
Speed, knots: 24. **Range, n miles:** 2,400 at 15 kt
Complement: 43 (8 officers)

Guns: 1 OTO Melara 76 mm/62 Compact; 85 rds/min to 16 km *(8.7 n miles)*; weight of shell 6 kg.
2 Oerlikon/Royal Ordnance 20 mm GAM-BO1. 2—12.7 mm MGs.

Countermeasures: Decoys: Plessey Barricade chaff launcher.
ESM: Thomson-CSF DR 3000; intercept.
Weapons control: CelsiusTech 9LV 207 Mk 3 command system and optronic director.
Radars: Surface search: Kelvin Hughes 1007 ARPA; I-band.

Programmes: Project Mawj order for three, with an option on five more, on 1 September 1993. The ships have had additional weapon systems fitted in Wudam dockyard.
Structure: Same hull design as the French P 400 class. 20 mm guns, and countermeasures were not fitted at Cherbourg and are planned to be installed in due course. 76 mm guns were fitted from 1998 from deleted Al Waafi class. The plan to fit torpedoes and sonars has been shelved.
Operational: First pair arrived in Oman on 28 September 1995, last one on 29 June 1996. Pennant numbers have been changed from B to Z.

UPDATED

AL MANSOOR   6/2003*, Royal Navy of Oman / 0589798

For details of the latest updates to *Jane's Fighting Ships* online and to discover the additional information available exclusively to online subscribers please visit
jfs.janes.com

4 DHOFAR (PROVINCE) CLASS
(FAST ATTACK CRAFT—MISSILE) (PGGF)

Name	No	Builders	Launched	Commissioned
DHOFAR	Z 10	Vosper Thornycroft	14 Oct 1981	7 Aug 1982
AL SHARQIYAH	Z 11	Vosper Thornycroft	2 Dec 1982	5 Dec 1983
AL BAT'NAH	Z 12	Vosper Thornycroft	4 Nov 1982	18 Jan 1984
MUSSANDAM	Z 14	Vosper Thornycroft	19 Mar 1988	31 Mar 1989

Displacement, tons: 311 light; 394 full load
Dimensions, feet (metres): 186 × 26.9 × 7.9 (56.7 × 8.2 × 2.4)
Main machinery: 4 Paxman Valenta 18CM diesels; 15,000 hp (11.2 MW) sustained; 4 shafts; auxiliary propulsion; 2 motors; 200 hp (149 kW)
Speed, knots: 38. **Range, n miles:** 2,000 at 18 kt
Complement: 45 (5 officers) plus 14 trainees

Missiles: SSM: 8 Aerospatiale MM 40 Exocet; inertial cruise; active radar homing to 70 km (40 n miles) at 0.9 Mach; warhead 165 kg; sea-skimmer.
Guns: 1 OTO Melara 3 in (76 mm)/62 compact; 85 rds/min to 16 km (8.7 n miles); weight of shell 6 kg.
2 Breda 40 mm/70 (twin); 300 rds/min to 12.5 km (6.8 n miles); weight of shell 0.96 kg.
2—20 mm.
Countermeasures: Decoys: 2 Wallop Barricade fixed triple barrels; for chaff and IR flares.
ESM: Racal Cutlass; radar warning.
ECM: Scorpion; jammer.
Weapons control: Sperry Sea Archer (B 10). Philips 9LV 307 (remainder).
Radars: Air/surface search: Plessey AWS 4 or AWS 6; E/F-band.
Fire control: Philips 9LV 307; I/J-band.
Navigation: KH 1007 ARPA; I-band.

Programmes: First ordered in 1980, two more in January 1981 and fourth in January 1986.
Structure: Similar to Kenyan Nyayo class. Mast structures are different dependent on radars fitted.
Operational: Pennant numbers have been changed from B to Z. **UPDATED**

AL BAT'NAH   6/2001, Royal Navy of Oman / 0114772

DHOFAR   6/2003, Royal Navy of Oman / 0567467

4 SEEB (VOSPER 25) CLASS (COASTAL PATROL CRAFT) (PB)

Name	No	Builders	Commissioned
SEEB	Z 20	Vosper Private, Singapore	15 Mar 1981
SHINAS	Z 21	Vosper Private, Singapore	15 Mar 1981
SADH	Z 22	Vosper Private, Singapore	15 Mar 1981
KHASSAB	Z 23	Vosper Private, Singapore	15 Mar 1981

Displacement, tons: 74 full load
Dimensions, feet (metres): 82.8 × 19 × 5.2 (25 × 5.8 × 1.6)
Main machinery: 2 MTU 12V 331 TC92 diesels; 2,660 hp(m) (1.96 MW) sustained; 2 shafts
1 Cummins N-855M diesel for slow cruising; 189 hp (141 kW) sustained; 1 shaft
Speed, knots: 25; 8 (Cummins diesel). **Range, n miles:** 750 at 14 kt
Complement: 13
Guns: 1 Oerlikon 20 mm GAM-BO1. 2—7.62 mm MGs.
Radars: Surface search: Racal Decca 1226; I-band.

Comment: Arrived in Oman on 19 May 1981 having been ordered one month earlier. The craft were built on speculation and completed in 1980. Pennant numbers have been changed from B to Z. **VERIFIED**

SHINAS   4/2001, Guy Toremans / 0114771

AMPHIBIOUS FORCES

Notes: (1) There are also some French-built Havas Mk 8 two-man SDVs in service.
(2) Following the issue of a restricted request for tender for a multipurpose high-speed sealift vessel, responses closed in March 2003. It is understood that the broad requirement is for a vessel, built to commercial standards, capable of transporting a company group (150 troops) and its equipment. Folding stern and bow ramps are thought to be required but the vessel is not to be fitted with a flight deck. Powered by two diesel engines, the vessel is to have a cruising speed of about 30 kt and a top speed of over 35 kt. With an unrefuelled range of about 500 n miles, the ship would operate in an intra-theatre sealift role. It is expected that the ship will be armed with self-defence weapons and have a complement of 10-12 people. A 40 m catamaran design is a potential contender.

1 LANDING SHIP—LOGISTIC (LSTH)

Name	No	Builders	Commissioned
NASR AL BAHR	A 2	Brooke Marine, Lowestoft	6 Feb 1985

Displacement, tons: 2,500 full load
Dimensions, feet (metres): 305 × 50.8 × 8.5 (93 × 15.5 × 2.6)
Main machinery: 2 Paxman Valenta 18 CM diesels; 7,500 hp (5.6 MW) sustained; 2 shafts; cp props
Speed, knots: 12. **Range, n miles:** 5,500 at 15 kt
Complement: 104 (13 officers)
Military lift: 7 MBT or 400 tons cargo; 190 troops; 2 LCVPs
Guns: 2 Breda 40 mm/70 (1 twin). 2 Oerlikon 20 mm GAM-BO1. 2—12.7 mm MGs.
Countermeasures: Decoys: Wallop Barricade double layer chaff launchers.
Weapons control: PEAB 9LV 107 GFCS and CSEE Lynx optical sight.
Radars: Surface search/navigation: 2 Racal Decca 1226; I-band.
Helicopters: Platform for Super Puma.

Comment: Ordered 18 May 1982. Launched 16 May 1984. Similar to Algerian LSLs. Carries one 16 ton crane. Bow and stern ramps. Full naval command facilities. The forward ramp is of two sections measuring length 59 ft (when extended) × 16.5 ft breadth (18 × 5 m), and the single section stern ramp measures 14 × 16.5 ft (4.3 × 5 m). Both hatches can support a 60 ton tank. The tank deck side bulkheads extend 7.5 ft (2.25 m) above the upper deck between the forecastle and the forward end of the superstructure, and provides two hatch openings to the tank deck below. Positioned between the hatches is a 2 ton crane with athwartship travel. New engine exhaust system and funnel fitted in 1997. Aft Oerlikon gun removed. Ship is also used as a ratings' training vessel. Pennant number has been changed from L to A. **UPDATED**

NASR AL BAHR   2/2002, A Sharma / 0533304

3 LCMs

Name	No	Builders	Commissioned
SABA AL BAHR	A 8 (ex-C 8)	Vosper Private, Singapore	17 Sep 1981
AL DOGHAS	A 9 (ex-C 9)	Vosper Private, Singapore	10 Jan 1983
AL TEMSAH	A 10 (ex-C 10)	Vosper Private, Singapore	12 Feb 1983

Displacement, tons: 230 full load
Dimensions, feet (metres): 108.2 (83.6, C 8) × 24.3 × 4.3 (33 (25.5) × 7.4 × 1.3)
Main machinery: 2 Caterpillar 3408TA diesels; 1,880 hp (1.4 MW) sustained; 2 shafts
Speed, knots: 8. **Range, n miles:** 1,400 at 8 kt
Complement: 11
Military lift: 100 tons
Radars: Navigation: Furuno 701; I-band.

Comment: First one launched 30 June 1981. Second pair of similar but not identical ships, launched 12 November and 15 December 1982. Pennant numbers have been changed from L to A. **VERIFIED**

AL TEMSAH   6/2003, Royal Navy of Oman / 0567468

1 LCU

Name	No	Builders	Commissioned
AL NEEMRAN	A 7 (ex-C 7)	Lewis Offshore, Stornoway	1979

Measurement, tons: 85 dwt
Dimensions, feet (metres): 84 × 24 × 6 *(25.5 × 7.4 × 1.8)*
Main machinery: 2 diesels; 300 hp *(220 kW)*; 2 shafts
Speed, knots: 7/8
Complement: 6
Radars: Navigation: Furuno; I-band.

Comment: Second of class deleted in 1993. Pennant number has been changed from L to A.
VERIFIED

AUXILIARIES

Notes: In addition to the listed vessels there are four 12 m Cheverton Work boats (W 41-44) and eight 8 m Work boats (W 4-11).

1 SUPPLY SHIP (AKS)

Name	No	Builders	Launched	Commissioned
AL SULTANA	T 1 (ex-A 2, ex-S 2)	Conoship, Groningen	18 May 1975	4 June 1975

Measurement, tons: 1,380 dwt
Dimensions, feet (metres): 215.6 × 35 × 13.5 *(65.7 × 10.7 × 4.2)*
Main machinery: 1 Mirrlees Blackstone diesel; 1,120 hp(m) *(835 kW)*; 1 shaft
Speed, knots: 11
Complement: 20
Radars: Navigation: Racal Decca TM 1226; I-band.

Comment: Major refit in 1992. Has a 1 ton crane. Pennant number changed in 1997 and again in 2002.
VERIFIED

AL SULTANA   *4/2002, Schaeffer/Marsan /* 0533305

1 SURVEY CRAFT (AGSC)

AL RAHMANNIYA M 1

Displacement, tons: 23.6 full load
Dimensions, feet (metres): 50.8 × 13.1 × 4.3 *(15.5 × 4 × 1.3)*
Main machinery: 2 Volvo TMD120A diesels; 604 hp(m) *(444 kW)* sustained; 2 shafts
Speed, knots: 13.5. **Range, n miles:** 500 at 12 kt
Complement: 10
Radars: Navigation: Furuno; I-band.

Comment: Built by Watercraft, Shoreham, UK and completed in April 1981. Pennant number changed from H to M.
VERIFIED

AL RAHMANNIYA   *10/1997, Royal Navy of Oman /* 0012860

TRAINING SHIPS

1 SAIL TRAINING SHIP (AXS)

Name	No	Builders	Recommissioned
SHABAB OMAN (ex-*Captain Scott*)	S 1	Herd and Mackenzie, Buckie, Scotland	1979

Displacement, tons: 386 full load
Dimensions, feet (metres): 144.3 × 27.9 × 15.1 *(44 × 8.5 × 4.6)*
Main machinery: 2 Gardner diesels; 460 hp *(343 kW)*; 2 shafts
Speed, knots: 10 (diesels)
Complement: 20 (5 officers) plus 3 officers and 24 trainees

Comment: Topsail schooner built in 1971 and taken over from Dulverton Trust in 1977 used for sail training. Name means Omani Youth.
VERIFIED

SHABAB OMAN   *4/2002, A Sharma /* 0533306

ROYAL YACHTS

Notes: The Royal Yacht Squadron of Oman is a distinct service that is not part of the Royal Navy of Oman. Based at Muscat, the squadron consists of three major units and a number of smaller craft.

1 ROYAL YACHT (YAC)

Name	No	Builders	Commissioned
AL SAID	—	Picchiotti SpA, Viareggio	July 1982

Displacement, tons: 3,800 full load
Dimensions, feet (metres): 340.5 × 62.4 × 15.4 *(103.8 × 19.0 × 4.7)*
Main machinery: 2 GMT A 420.6 H diesels; 8,400 hp(m) *(6.17 MW)* sustained; 2 shafts; cp props; bow thruster
Speed, knots: 18
Complement: 156 (16 officers)
Radars: Navigation: Decca TM 1226C; ACS 1230C; I-band.

Comment: Fitted with helicopter deck and fin stabilisers. Carries three Puma C service launches and one Rotork beach landing craft. A variety of small arms carried.
VERIFIED

AL SAID   *4/2002, A Sharma /* 0533307

1 SUPPORT SHIP (AKSH)

Name	No	Builders	Launched	Commissioned
FULK AL SALAMAH (ex-*Ghubat Al Salamah*)	—	Bremer-Vulkan	29 Aug 1986	3 Apr 1987

Measurement, tons: 10,797 grt; 3,239 net
Dimensions, feet (metres): 447.5 × 68.9 × 19.7 *(136.4 × 21 × 6)*
Main machinery: 4 Fincantieri GMT A 420.6 H diesels; 16,800 hp(m) *(12.35 MW)* sustained; 2 shafts; cp props
Speed, knots: 19.5
Military lift: 240 troops
Radars: Navigation: 2 Racal Decca; I-band.
Helicopters: Up to 2 AS 332C Super Pumas.

Comment: Primary role is to support the Royal Yacht on deployments. Secondary roles include government, environmental and training duties. Reported to be fitted with Javelin air-defence missile system.
VERIFIED

FULK AL SALAMAH   *10/2001, A Sharma /* 0126314

1 ROYAL DHOW (YAC)

Name	No	Builders	Commissioned
ZINAT AL BIHAAR	—	—	1988

Displacement, tons: 510 light
Dimensions, feet (metres): 200.2 × 32.2 × 12.8 *(61 × 9.8 × 3.9)*
Main machinery: 2 Siemens motors; 965 hp *(720 kW)*; 2 shafts
Speed, knots: 11.5

Comment: Three-masted wooden sailing vessel built in Oman on traditional lines.
UPDATED

ZINAT AL BIHAAR   *4/2004*, Derek Fox* / 0589797

POLICE

Notes: (1) In addition to the vessels listed below there are several harbour craft including a Cheverton 8 m work boat *Zahra 24*, *Zahra 16* and a fireboat pennant number *10*. There are also two Pilatus aircraft for SAR.
(2) 15 FPBs between 11 and 30 m may be ordered in due course. These could be for the Navy if it takes over Fishery Protection duties from the Police.

ZAHRA 16   *6/2003, Hartmut Ehlers* / 0567471

3 CG 29 TYPE (COASTAL PATROL CRAFT) (PB)

HARAS 7 H 7	HARAS 9 H 9	HARAS 10 H 10

Displacement, tons: 84 full load
Dimensions, feet (metres): 94.8 × 17.7 × 4.3 *(28.9 × 5.4 × 1.3)*
Main machinery: 2 MTU 12V 331 TC92 diesels; 2,660 hp(m) *(1.96 MW)* sustained; 2 shafts
Speed, knots: 25. **Range, n miles:** 600 at 15 kt
Complement: 13
Guns: 2 Oerlikon 20 mm GAM-BO1.
Radars: Navigation: Racal Decca 1226; I-band.

Comment: Built by Karlskrona Varvet. Commissioned in 1981-82. GRP Sandwich hulls.
UPDATED

HARAS 9   *12/2000* / 0114776

1 P 1903 TYPE (COASTAL PATROL CRAFT) (PB)

HARAS 8 H 8

Displacement, tons: 32 full load
Dimensions, feet (metres): 63 × 15.7 × 5.2 *(19.2 × 4.8 × 1.6)*
Main machinery: 2 MTU 8V 331 TC92 diesels; 1,770 hp(m) *(1.3 MW)*; 2 shafts
Speed, knots: 30. **Range, n miles:** 1,650 at 17 kt
Complement: 10
Guns: 2—12.7 mm MGs.
Radars: Navigation: Racal Decca 1226; I-band.

Comment: Built by Le Comte, Netherlands. Commissioned August 1981. Type 1903 Mk III.
UPDATED

HARAS 8   *10/1992, Hartmut Ehlers*

1 CG 27 TYPE (COASTAL PATROL CRAFT) (PB)

HARAS 6 H 6

Displacement, tons: 53 full load
Dimensions, feet (metres): 78.7 × 18 × 6.2 *(24 × 5.5 × 1.9)*
Main machinery: 2 MTU 12V 331 TC92 diesels; 2,660 hp(m) *(1.96 MW)* sustained; 2 shafts
Speed, knots: 25
Complement: 11
Guns: 1 Oerlikon 20 mm GAM-BO1.
Radars: Navigation: Furuno 701; I-band.

Comment: Completed in 1980 by Karlskrona Varvet. GRP hull.
VERIFIED

HARAS 6   *10/1992, Hartmut Ehlers*

14 RODMAN 58 CLASS (PB)

Displacement, tons: 19 full load
Dimensions, feet (metres): 59.0 × 16.0 × 3.9 *(18.0 × 4.9 × 1.2)*
Main machinery: 2 diesels; 2,000 hp *(1.49 MW)*; 2 waterjets
Speed, knots: 34. **Range, n miles:** 450 at 17 kt
Complement: 5
Radars: Navigation: I-band.

Comment: GRP hull. Built in 2002-03 by Rodman, Vigo.
UPDATED

RODMAN 58   *6/2002*, Rodman* / 0589793

1 P 2000 TYPE (COASTAL PATROL CRAFT) (PB)

DHEEB AL BAHAR 1   Z 1

Displacement, tons: 80 full load
Dimensions, feet (metres): 68.2 × 19 × 5 *(20.8 × 5.8 × 1.5)*
Main machinery: 2 MTU 12V 396 TB93 diesels; 3,260 hp(m) *(2.4 MW)* sustained; 2 shafts
Speed, knots: 40. **Range, n miles:** 423 at 36 kt; 700 at 18 kt
Guns: 1—12.7 mm MG.
Radars: Surface search: Furuno 701; I-band.

Comment: Delivered January 1985 by Watercraft Ltd, Shoreham, UK. GRP hull. Similar to UK Archer class. Carries SATNAV.
UPDATED

DHEEB AL BAHAR 1   6/2003*, *Hartmut Ehlers* / 0589794

2 D 59116 TYPE (COASTAL PATROL CRAFT) (PB)

DHEEB AL BAHAR 2 Z 2   **DHEEB AL BAHAR 3** Z 3

Displacement, tons: 65 full load
Dimensions, feet (metres): 75.5 × 17.1 × 3.9 *(23 × 5.2 × 1.2)*
Main machinery: 2 MTU 12V 396 TB93 diesels; 3,260 hp(m) *(2.4 MW)* sustained; 2 shafts
Speed, knots: 36. **Range, n miles:** 420 at 30 kt
Complement: 11
Guns: 1—12.7 mm MG.
Radars: Surface search: Furuno 711-2; Furuno 2400; I-band.

Comment: Built by Yokohama Yacht Co, Japan. Commissioned in 1988.
UPDATED

DHEEB AL BAHAR 3   6/2003, *Hartmut Ehlers* / 0567470

5 INSHORE PATROL CRAFT (PBI)

ZAHRA 14 Z 14   **ZAHRA 15** Z 15   **ZAHRA 17** Z 17   **ZAHRA 18** Z 18   **ZAHRA 21** Z 21

Displacement, tons: 16; 18 *(Zahra 18* and *21)* full load
Dimensions, feet (metres): 45.6 × 14.1 × 4.6 *(13.9 × 4.3 × 1.4)*
52.5 × 13.8 × 7.5 *(16 × 4.2 × 2.3)* *(Zahra 18* and *21)*
Main machinery: 2 Cummins VTA-903M diesels; 643 hp *(480 kW)*; 2 shafts
Speed, knots: 36. **Range, n miles:** 510 at 22 kt
Complement: 5-6
Guns: 1 or 2—7.62 mm MGs.
Radars: Navigation: Decca 101; I-band.

Comment: *Zahra 14, 15* and *17* built by Watercraft, Shoreham, UK and completed in 1981. *Zahra 21* completed by Emsworth SB in 1987 to a slightly different design. *Zahra 18* built by Lecomte in 1987.
VERIFIED

ZAHRA 17 (alongside Zahra 14)   6/2003, *Hartmut Ehlers* / 0567472

1 DIVING CRAFT (YDT)

ZAHRA 27 Z 27

Displacement, tons: 13 full load
Dimensions, feet (metres): 59 × 12.4 × 3.6 *(18 × 3.8 × 1.1)*
Main machinery: 2 Volvo Penta AQD70D diesels; 430 hp(m) *(316 kW)* sustained; 2 shafts
Speed, knots: 20
Complement: 4
Guns: 2—7.62 mm MGs.

Comment: Rotork Type, the last of several logistic support craft, delivered in 1981 and now used as a diving boat. Similar craft used by the Navy.
VERIFIED

ZAHRA 27   6/2003, *Hartmut Ehlers* / 0567469

5 VOSPER 75 ft TYPE (COASTAL PATROL CRAFT) (PB)

HARAS 1-5 H 1-5

Displacement, tons: 50 full load
Dimensions, feet (metres): 75 × 20 × 5.9 *(22.9 × 6.1 × 1.8)*
Main machinery: 2 Caterpillar D 348 diesels; 1,450 hp *(1.08 MW)* sustained; 2 shafts
Speed, knots: 24.5. **Range, n miles:** 1,000 at 11 kt
Complement: 11
Guns: 1 Oerlikon 20 mm GAM-BO1.
Radars: Navigation: Decca 101; I-band.

Comment: First four completed 22 December 1975 by Vosper Thornycroft. GRP hulls. *Haras 5* commissioned November 1978.
UPDATED

HARAS 3   3/2004*, *Bob Fildes* / 0589795

20 HALMATIC COUGAR ENFORCER 33
(FAST PATROL CRAFT) (PBF)

Displacement, tons: 5.4 full load
Dimensions, feet (metres): 35.7 × 9.3 × 2.5 *(10.88 × 2.84 × 0.75)*
Main machinery: 2 Yanmar diesels; 2 Hamilton waterjets
Speed, knots: 45. **Range, n miles:** 120 at 45 kt

Comment: Based on Cougar 33 deep Vee hull form, first batch of five craft supplied by Halmatic in March 2003 with further 15 delivered by late 2003. To be deployed in coastal patrol and interception role.
UPDATED

ENFORCER 33   3/2004*, *Bob Fildes* / 0589796

12 SEASPRAY ASSAULT BOATS (PB)

Displacement, tons: To be announced
Dimensions, feet (metres): 31.2 × 10.2 × 1.6 *(9.5 × 3.1 × 0.5)*
Main machinery: 2 outboards; 500 hp *(375 kW)*
Speed, knots: 50. **Range, n miles:** 450 at 17 kt
Complement: 5
Radars: Navigation: I-band.

Comment: Abu Dhabi Ship Building awarded contract in January 2004. Designed by SeaSpray Aluminium Boats. To be employed in policing, patrol and interception roles by the navy and police.
NEW ENTRY

Pakistan

Country Overview

The Islamic Republic of Pakistan gained independence in 1947. Situated in south Asia, it has an area of 307,293 square miles and is bordered to the west by Iran, to the north by Afghanistan and to the south by India. It has a 567 n mile coastline with the Arabian Sea. The former province of East Pakistan seceded in 1971 and assumed the name Bangladesh. The status of Jammu and Kashmir is disputed with India. The capital is Islamabad while Karachi is the largest city and principal port. There is a further port at Muhammad bin Qasim. Territorial waters (12 n miles) are claimed. A 200 n mile EEZ has been claimed but the limits have not been defined.

Headquarters Appointments

Chief of the Naval Staff:
   Admiral Shahid Karimullah, NI (M), SJ
Vice Chief of Naval Staff:
   Vice Admiral Farooq Rashid, HI (M), S.Bt
Deputy Chief of Naval Staff (Operations):
   Rear Admiral Iftikhar Ahmed, SI (M)

Senior Appointments

Commander Pakistan Fleet:
   Rear Admiral Vice Admiral Mohammed Haroon, HI (M), T.Bt
Commander Karachi:
   Rear Admiral Muhammad Asad Qureshi, HI (M)
Flag Officer Sea Training:
   Rear Admiral Saleem Khalid, HI (M)
Commander North:
   Commodore Raja Riaz Hussain
Director General Maritime Security Agency:
   Commodore Bakhtiar Mohsin, SI (M), T. Bt

Diplomatic Representation

Naval Adviser in London:
   Commodore Kamran Khan, TI (M)
Naval Attaché in Paris:
   Captain Adnan Nazir
Defence Attaché in Muscat (Oman):
   Captain Wasim Akram, PN
Naval Attaché in Tehran:
   Captain Khalid Saeed
Naval Attaché in Beijing:
   Captain Naveed Rizvi
Naval Attaché in Kuala Lumpur:
   Captain Muhammad Amjad Zaman
Naval Adviser in New Delhi:
   Captain Mateen-ur-Rehman

Personnel

(a)   2005: 25,100 (2,300 officers) including 1,200 Marines and 1,000 (75 officers) seconded to the MSA
(b)   Voluntary service
(c)   Reserves 5,000

Bases

PNS *Haider* (Naval HQ); PNS *Akram* (Gwadar Naval Base); PNS *Iqbal* (Commando Base); PNS *Mehran* (Karachi Naval Air Station); PNS *Qasim* (Marines HQ/Base), Jinnah Naval Base (Port Ormara)

The port of Gwadar has been upgraded and naval facilities may be added in the future.

Prefix to Ships' Names

PNS

Maritime Security Agency

Set up in 1986. Main purpose is to patrol the EEZ in co-operation with the Navy and the Army-manned Coast Guard.

Marines

A Marine Commando Unit was formed at PNS *Iqbal*, Karachi in 1991.

Strength of the Fleet

Type	Active	Building
Submarines—Patrol	8	1
Submarines—Midget	3	—
Destroyers/Frigates	7	—
Fast Attack Craft—Missile	3	—
Large Patrol Craft	2	—
Minehunters	3	—
Survey Ship	1	—
Tankers	5	—
Maritime Security Agency		
Destroyers	1	—
Large Patrol Craft	4	—
Fast Attack Craft—Gun	2	—

DELETIONS

Frigates

2003   *Shamsher*

Patrol Forces

2003   *Quwwatt*

PENNANT LIST

Submarines

S 131	Hangor
S 132	Shushuk
S 133	Mangro
S 134	Ghazi
S 135	Hashmat
S 136	Hurmat
S 137	Khalid
S 138	Saad
S 139	Hamza (bldg)

Destroyers/Frigates

D 181	Tariq
D 182	Babur

(Destroyers/Frigates continued)

D 183	Khaibar
D 184	Badr
D 185	Tippu Sultan
D 186	Shahjahan
F 262	Zulfiquar

Mine Warfare Forces

M 163	Muhafiz
M 164	Mujahid
M 166	Munsif

Patrol Forces

P 140	Rajshahi
P 157	Larkana

(Patrol Forces continued)

P 1026	Dehshat
P 1029	Jalalat
P 1030	Shujaat
P 1031	Haibat
P 1032	Jurat

Maritime Security Agency

D 156	Nazim
1060	Barkat
1061	Rehmat
1062	Nusrat
1063	Vehdat
1066	Subqat
1068	Rafaqat

Auxiliaries

A 20	Moawin
A 21	Kalmat
A 40	Attock
A 44	Bholu
A 45	Gama
A 47	Nasr
A 49	Gwadar
-	Janbaz
SV 48	Behr Paima

SUBMARINES

2 + 1 KHALID (AGOSTA 90B) CLASS (SSK)

Name	No	Builders	Laid down	Launched	Commissioned
KHALID	S 137	DCN, Cherbourg	15 July 1995	18 Dec 1998	6 Sep 1999
SAAD	S 138	DCN, Cherbourg/PN Dockyard, Karachi	2 Dec 1999	24 Aug 2002	12 Dec 2003
HAMZA (ex-*Ghazi*)	S 139	PN Dockyard, Karachi	2000	Dec 2005	2006

Displacement, tons: 1,510 surfaced; 1,760 dived (1,960 with MESMA)
Dimensions, feet (metres): 221.7; 252.7 (S 139) × 22.3 × 17.7 *(67.6; 77.0 (S 139) × 6.8 × 5.4)*
Main machinery: Diesel-electric; 2 SEMT-Pielstick 16 PA4 V 185 VG diesels; 3,600 hp(m) *(2.65 MW)*; 2 Jeumont Schneider alternators; 1.7 MW; 1 Jeumont motor; 2,992 hp(m) *(2.2 MW)*; 1 cruising motor; 32 hp(m) *(23 kW)*; 1 shaft
Speed, knots: 12 surfaced; 20 dived
Range, n miles: 8,500 at 9 kt snorting; 350 at 3.5 kt dived
Complement: 36 (7 officers)

Missiles: SSM: 4 Aerospatiale Exocet SM 39; inertial cruise; active radar homing to 50 km *(27 n miles)* at 0.9 Mach; warhead 165 kg.
Torpedoes: 4—21 in *(533 mm)* bow tubes. 16 ECAN F17P Mod 2; wire-guided; active/passive homing to 20 km *(10.8 n miles)* at 40 kt; warhead 250 kg. Total of 20 weapons.
Mines: Stonefish.
Countermeasures: ESM: Thomson-CSF DR-3000U; intercept.
Weapons control: Thomson Sintra SUBTICS Mk 2.
Radars: Surface search: KH 1007; I-band.
Sonars: Thomson Sintra TSM 2233 suite; bow cylindrical, passive ranging and intercept, and clip-on towed arrays.

Programmes: A provisional order for a second batch of three more Agostas was reported in September 1992 and this was confirmed on 21 September 1994. First one built in France. Parts for S 138 sent to Pakistan in April 1998 and for S 139 in September 1998.
Structure: The last of the class is to have a 200 kW MESMA liquid oxygen AIP system, thereby extending the hull by 9 m. The MESMA AIP system has a power output of 200 kW which will quadruple dived performance at 4 kt. Testing of the production system started in late 1999. The MESMA system

SAAD

9/2003, DCN / 0562934

is to be retrofitted in S 137 and S 138 during their next major refits. Hulls also have much improved acoustic quietening and a full integrated sonar suite including flank, intercept and towed arrays. SOPOLEM J 95 search and STS 95 attack periscopes. Sagem integrated navigation system. HLES 80 steel. Diving depth of 320 m *(1,050 ft)*.

Operational: *Khalid* completed 29 April 1999 and sailed for Pakistan in November 1999. Assigned to 5th Submarine Squadron. *Saad* completed its deep-dive test on 19 September 2003. The first submarine to be built in Pakistan, it was commissioned by the President.

UPDATED

2 HASHMAT (AGOSTA) CLASS (SSK)

Name	No	Builders	Laid down	Launched	Commissioned
HASHMAT (ex-*Astrant*)	S 135	Dubigeon Normandie, Nantes	15 Sep 1976	14 Dec 1977	17 Feb 1979
HURMAT (ex-*Adventurous*)	S 136	Dubigeon Normandie, Nantes	18 Sep 1977	1 Dec 1978	18 Feb 1980

Displacement, tons: 1,490 surfaced; 1,740 dived
Dimensions, feet (metres): 221.7 × 22.3 × 17.7
(67.6 × 6.8 × 5.4)
Main machinery: Diesel-electric; 2 SEMT-Pielstick 16 PA4 V 185 VG diesels; 3,600 hp(m) *(2.65 MW)*; 2 Jeumont Schneider alternators; 1.7 MW; 1 motor; 4,600 hp(m) *(3.4 MW)*; 1 cruising motor; 32 hp(m) *(23 kW)*; 1 shaft
Speed, knots: 12 surfaced; 20 dived
Range, n miles: 8,500 at 9 kt snorting; 350 at 3.5 kt dived
Complement: 59 (8 officers)

Missiles: SSM: McDonnell Douglas Sub Harpoon; active radar homing to 130 km *(70 n miles)* at 0.9 Mach; warhead 227 kg.

Torpedoes: 4—21 in *(533 mm)* bow tubes. ECAN F17P; wire-guided; active/passive homing to 20 km *(10.8 n miles)* at 40 kt; warhead 250 kg; water ram discharge gear. E14, E15 and L3 torpedoes are also available. Total of 20 torpedoes and missiles.
Mines: Stonefish.
Countermeasures: ESM: ARUD; intercept and warning.
Radars: Surface search: Thomson-CSF DRUA 33; I-band.
Sonars: Thomson Sintra TSM 2233D; passive search; medium frequency.
Thomson Sintra DUUA 2B; active/passive search and attack; 8 kHz active.
Thomson Sintra TSM 2933D towed array; passive; very low frequency.

Programmes: Purchased from France in mid-1978 after United Nations' ban on arms sales to South Africa. *Hashmat* arrived Karachi 31 October 1979, *Hurmat* arrived 11 August 1980.
Structure: Diving depth, 300 m *(985 ft)*. Both were modified to fire Harpoon in 1985 but may have had to acquire the missiles through a third party.
Operational: Assigned to 5th Submarine Squadron.

VERIFIED

HURMAT
6/1998 / 0052677

4 HANGOR (DAPHNE) CLASS (SSK)

Name	No	Builders	Laid down	Launched	Commissioned
HANGOR	S 131	Arsenal de Brest	1 Dec 1967	28 June 1969	12 Jan 1970
SHUSHUK	S 132	CN Ciotat, Le Trait	1 Dec 1967	30 July 1969	12 Jan 1970
MANGRO	S 133	CN Ciotat, Le Trait	8 July 1968	7 Feb 1970	8 Aug 1970
GHAZI (ex-*Cachalote*)	S 134	Dubigeon, Normandie, Nantes	12 May 1967	23 Sep 1968	1 Oct 1969

Displacement, tons: 700 standard; 869 surfaced; 1,043 dived
Dimensions, feet (metres): 189.6 × 22.3 × 15.1
(57.8 × 6.8 × 4.6)
Main machinery: Diesel-electric; 2 SEMT-Pielstick 12 PA4 V 185 diesels; 2,450 hp(m) *(1.8 MW)*; 2 Jeumont Schneider

alternators; 1.7 MW; 2 motors; 2,600 hp(m) *(1.9 MW)*; 2 shafts
Speed, knots: 13 surfaced; 15.5 dived
Range, n miles: 4,500 at 5 kt; 3,000 at 7 kt snorting
Complement: 53 (7 officers)

Missiles: SSM: McDonnell Douglas Sub Harpoon; active radar homing to 130 km *(70 n miles)* at 0.9 Mach; warhead 227 kg.
Torpedoes: 12—21.7 in *(550 mm)* (8 bow, 4 stern). E14, E15, L3 and Z16 torpedoes are available. Total of 12 weapons, no reloads.
Mines: Stonefish.
Countermeasures: ESM: ARUD; intercept and warning.
Radars: Surface search: Thomson-CSF DRUA 31; I-band.
Sonars: Thomson Sintra TSM 2233D; hull-mounted; passive search; medium frequency.
DUUA 1; active/passive search and attack.

Programmes: The first three were built in France. The Portuguese Daphne class *Cachalote* was bought by Pakistan in December 1975.
Structure: They are broadly similar to the submarines built in France for Portugal and South Africa and the submarines constructed to the Daphne design in Spain, but slightly modified internally. Diving depth 300 m *(985 ft)*. SSM capability added in late 1980s.
Operational: Assigned to 5th Submarine Squadron. Likely to be paid off as replaced by Khalid class.

VERIFIED

MANGRO
6/1998 / 0052678

3 MIDGET SUBMARINES (SSW)

X 01-03

Displacement, tons: 118 dived
Dimensions, feet (metres): 91.2 × 18.4 *(27.8 × 5.6)*
Speed, knots: 7 dived
Range, n miles: 2,200 surfaced; 60 dived
Complement: 8 + 8 swimmers
Torpedoes: 2—21 in *(533 mm)* tubes; 2 AEG SUT; wire-guided; active homing to 12 km *(6.5 n miles)* at 35 kt; passive homing to 28 km *(15 n miles)* at 23 kt; warhead 250 kg plus either

two short range active/passive homing torpedoes or two SDVs.
Mines: 12 Mk 414 Limpet type.
Sonars: Hull mounted; active/passive; high frequency.

Comment: MG 110 type built in Pakistan under supervision by Cosmos. These are enlarged SX 756 of Italian Cosmos design. Diving depth of 150 m and can carry eight swimmers with

2 tons of explosives as well as two CF2 FX 60 SDVs (swimmer delivery vehicles). Pilkington Optronics CK 39 periscopes. Reported as having a range of 1,000 n miles and an endurance of 20 days. All have been upgraded since 1995 with improved sensors and weapons. However, reports that *X 01* has been equipped with Harpoon are not considered likely. All are active. *VERIFIED*

X 03
5/2003 / 0569226

FRIGATES

Notes: (1) A contract to procure frigates (the F 22 programme) from China is expected in 2005. Up to four ships, probably of the Jiangwei II class, are required and some may be built in Pakistan. Delivery of the ships, initially to replace the Leander class, will probably start in 2009. The ships are likely to have a mix of Chinese and western systems.
(2) Procurement of up to four second-hand ships, to boost surface ship numbers, is also reported to be under consideration.

6 TARIQ (AMAZON) CLASS (TYPE 21) (FFHM/FFGH)

Name	No	Builders	Laid down	Launched	Commissioned	Recommissioned
TARIQ (ex-*Ambuscade*)	D 181 (ex-F 172)	Yarrow Shipbuilders, Glasgow	1 Sep 1971	18 Jan 1973	5 Sep 1975	28 July 1993
BABUR (ex-*Amazon*)	D 182 (ex-F 169)	Vosper Thornycroft, Woolston	6 Nov 1969	26 Apr 1971	11 May 1974	30 Sep 1993
KHAIBAR (ex-*Arrow*)	D 183 (ex-F 173)	Yarrow Shipbuilders, Glasgow	28 Sep 1972	5 Feb 1974	29 July 1976	1 Mar 1994
BADR (ex-*Alacrity*)	D 184 (ex-F 174)	Yarrow Shipbuilders, Glasgow	5 Mar 1973	18 Sep 1974	2 July 1977	1 Mar 1994
TIPPU SULTAN (ex-*Avenger*)	D 185 (ex-F 185)	Yarrow Shipbuilders, Glasgow	30 Oct 1974	20 Nov 1975	19 July 1978	23 Sep 1994
SHAHJAHAN (ex-*Active*)	D 186 (ex-F 171)	Vosper Thornycroft, Woolston	23 July 1971	23 Nov 1972	17 June 1977	23 Sep 1994

Displacement, tons: 3,100 standard; 3,700 full load
Dimensions, feet (metres): 384 oa; 360 wl × 41.7 × 19.5 (screws) *(117; 109.7 × 12.7 × 5.9)*
Main machinery: COGOG; 2 RR Olympus TM3B gas turbines; 50,000 hp *(37.3 MW)* sustained; 2 RR Tyne RM1C gas turbines (cruising); 9,900 hp *(7.4 MW)* sustained; 2 shafts; cp props
Speed, knots: 30; 18 on Tynes
Range, n miles: 4,000 at 17 kt; 1,200 at 30 kt
Complement: 175 (13 officers) (accommodation for 192)

Missiles: SSM: 4 McDonnell Douglas Harpoon 1C ❶ fitted in D 186, D 184 and D 182.
SAM: China LY 60N sextuple launchers ❷ semi-active radar homing to 13 km *(7 n miles)* at 2.5 Mach; warhead 33 kg (D 185, D 181 and D 183).
Guns: 1 Vickers 4.5 in *(114 mm)*/55 Mk 8 ❸; 25 rds/min to 22 km *(11.9 n miles)* anti-surface; 6 km *(3.3 n miles)* anti-aircraft; weight of shell 21 kg.
Hughes 20 mm Vulcan Phalanx Mk 15 ❹; 3,000 rds/min to 1.5 km (D 181, D 183, D 184 and D 186).
2 MSI DS 30B 30 mm/75 ❺ (D 182, D 185 and D 186).
Torpedoes: 6—324 mm Plessey STWS Mk 2 (2 triple) tubes ❼ (D 184 and D 186); others fitted with 2 Bofors Type 43X2 single launchers for Swedish Type 45 torpedoes.
Countermeasures: Decoys: Graseby Type 182; towed torpedo decoy.
2 Vickers Corvus 8-tubed trainable launchers ❽ Mk 36 SRBOC ❾ (D 185, D 181, D 183 and D 182).
ESM: Thomson-CSF DR 3000S; intercept.
Combat data systems: CAAIS combat data system with Ferranti FM 1600B computers (D 186 and D 184). CelsiusTech 9LV Mk 3 including Link Y (in remainder).
Weapons control: Ferranti WSA-4 digital fire-control system. CSEE Najir Mk 2 optronic director ❿ (D 182, D 185 and D 186).
Radars: Air/surface search: Marconi Type 992R ⓫; E/F-band (D 182, D 184 and D 186). Signaal DA08 ⓬; F-band (D 181, D 183 and D 185).
Surface search: Kelvin Hughes Type 1007 ⓭ or Type 1006 (D 184 and D 186); I-band.
Fire control: 1 Selenia Type 912 (RTN 10X) ⓮; I/J-band (D 182, D 184 and D 186).
1 China LL-1 ⓯ (for LY 60N); I/J-band (D 185, D 181 and D 183).
Sonars: Graseby Type 184P; hull-mounted; active search and attack; medium frequency.
Kelvin Hughes Type 162M; hull-mounted; bottom classification; 50 kHz.
Thomson Marconi ATAS; active; medium frequency (D 183 and D 185).

Helicopters: 1 Westland Lynx HAS 3 ⓰.

Programmes: Acquired from the UK in 1993-94. *Tariq* arrived in Karachi 1 November 1993 and the last pair in January 1995. These ships replaced the Garcia and Brooke classes and have been classified as destroyers.
Modernisation: Exocet, torpedo tubes and Lynx helicopter facilities were all added in RN service, but torpedo tubes were subsequently removed in all but *Badr* and *Shahjahan* and all retrofitted by Pakistan using Swedish equipment. Exocet was not transferred and the obsolete Seacat SAM system has been replaced by Phalanx taken from the Gearings. Chinese

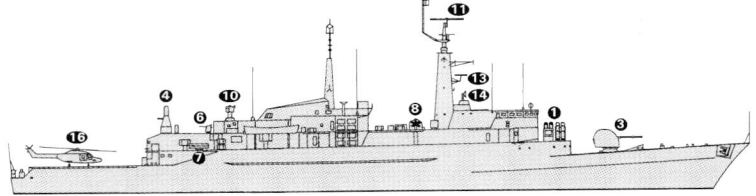

SHAHJAHAN

(Scale 1 : 1,200), Ian Sturton / 0114784

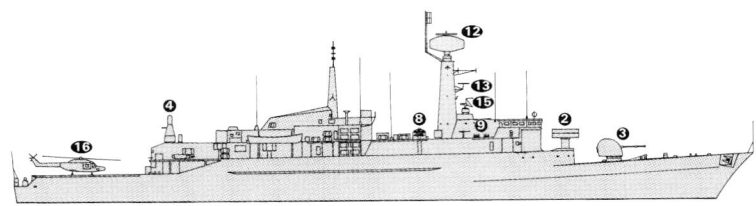

TARIQ

(Scale 1 : 1,200), Ian Sturton / 0572527

TIPPU SULTAN

6/2000, Pakistan Navy / 0105183

LY 60N, which is a copy of Aspide, has been fitted in three of the class, Harpoon in three others. New EW equipment has been installed and ATAS sonar has been acquired (two sets only). There are still plans to update the hull sonars. Other equipment upgrades include a DA08 search radar in three of the class, an optronic director, new 30 mm and 20 mm guns, SRBOC chaff launchers. An improved combat data system with a datalink to shore HQ is also fitted in four of the class.

Structure: Due to cracking in the upper deck structure large strengthening pieces have been fixed to the ships' side at the top of the steel hull as shown in the illustration. The addition of permanent ballast to improve stability has increased displacement by about 350 tons. Further hull modifications to reduce noise and vibration started in 1988 and completed in all of the class by 1992.
Operational: Form 25th Destroyer Squadron. **UPDATED**

BADR

6/2000, Pakistan Navy / 0114784

BABUR

6/2004 * / 1044174

TARIQ

6/2000, Pakistan Navy / 0105185

KHAIBAR

6/2000, Pakistan Navy / 0105186

1 LEANDER CLASS (FFH)

Name	No	Builders	Laid down	Launched	Commissioned
ZULFIQUAR (ex-*Apollo*)	F 262	Yarrows, Glasgow	1 May 1969	15 Oct 1970	28 May 1972

Displacement, tons: 2,500 standard; 2,962 full load
Dimensions, feet (metres): 360 wl; 372 oa × 43 × 14.8 (keel); 18 (screws) *(109.7; 113.4 × 13.1 × 4.5; 5.5)*
Main machinery: 2 Babcock & Wilcox boilers; 550 psi *(38.7 kg/cm²)*; 850°F *(454°C)*; 2 White/English Electric turbines; 30,000 hp *(22.4 MW)*; 2 shafts
Speed, knots: 28. **Range, n miles:** 4,000 at 15 kt
Complement: 235 (15 officers)

Guns: 2 Vickers 4.5 in *(114 mm)*/45 Mk 6 (twin) ❶; 20 rds/min to 19 km *(10.3 n miles)* anti-surface; 6 km *(3.3 n miles)* anti-aircraft; weight of shell 25 kg.
6—25 mm/60 (3 twin) ❷; 270 rds/min to 3 km *(1.6 n miles)*; weight of shell 0.34 kg.
A/S mortars: 3-barrelled UK MoD Mortar Mk 10 ❸; automatic loading; range 1 km; warhead 92 kg.
Countermeasures: Decoys: Graseby Type 182; towed torpedo decoy.
2 Vickers Corvus 8-barrelled trainable chaff launchers ❹.
ESM: UA-8/9/13; radar warning.
ECM: Type 668; jammer.
Weapons control: MRS 3 system for 114 mm guns.
Radars: Air search: Marconi Type 966 ❺; A-band; AKE-1.

Surface search: Plessey Type 994 ❻; E/F-band.
Navigation: 1 Kelvin Hughes Type 1006 and 1 Kelvin Hughes Type 1007; I-band.
Fire control: Plessey Type 904 (for 114 mm guns) ❼; I/J-band.
Sonars: Kelvin Hughes Type 162M; hull-mounted; bottom classification; 50 kHz.
Graseby Type 170B; hull-mounted; active search and attack; 15 kHz.
Graseby Type 184P; hull-mounted; active search and attack; 6-9 kHz.

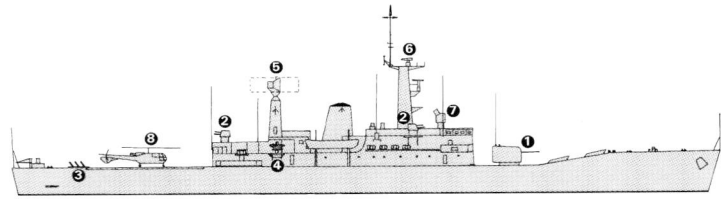

LEANDER CLASS (Scale 1 : 1,200), Ian Sturton

Helicopters: 1 SA 319B Alouette III ❽.

Programmes: Transferred from UK on 14 October 1988. From the Batch 3B broad-beamed group of this class.
Modernisation: Seacat and 20 mm guns replaced by twin 25 mm mountings. Additional navigation radar mounted on bridge roof.
Operational: Sailed for Pakistan in December 1988. Extensive refit carried out 1991-93. **VERIFIED**

LEANDER CLASS 3/1997 / 0012864

SHIPBORNE AIRCRAFT

Numbers/Type: 3 Westland Lynx HAS 3.
Operational speed: 120 kt *(222 km/h)*.
Service ceiling: 10,000 ft *(3,048 m)*.
Range: 320 n miles *(593 km)*.
Role/Weapon systems: Two delivered August 1994 and one in April 1995. Option on three more. Sensors: Ferranti Sea Spray radar, Orange Crop ESM. Weapons: ASW; two Mk 46 torpedoes. ASV; 2—12.7 mm MG pods; possibly Sea Skua ASM. **VERIFIED**

Numbers/Type: 6 Westland Sea King Mk 45.
Operational speed: 125 kt *(232 km/h)*.
Service ceiling: 10,500 ft *(3,200 m)*.
Range: 630 n miles *(1,165 km)*.
Role/Weapon systems: Sensors: MEL search radar, Marconi Type 2069 dipping sonar (two sets), AQS-928G acoustic processors. Weapons: ASW; two Mk 46 torpedoes; Mk 11 depth charges. ASV; one AM 39 Exocet missile. **VERIFIED**

LYNX HAS 3 6/2001, Pakistan Navy / 0114778

SEA KING 6/2003, Pakistan Navy / 0569229

Numbers/Type: 2/4 Aerospatiale SA 319B/SA 316 Alouette III.
Operational speed: 113 kt *(210 km/h).*
Service ceiling: 10,500 ft *(3,200 m).*
Range: 290 n miles *(540 km).*
Role/Weapon systems: Reconnaissance helicopter. Second four acquired in 1994. Sensors: Weather/search radar MAD (in two). Weapons: ASW; Mk 11 depth charges, one Mk 46 torpedo.
VERIFIED

ALOUETTE III   *6/2003, Pakistan Navy /* 0569234

LAND-BASED MARITIME AIRCRAFT

Notes: (1) Procurement of further maritime patrol aircraft is a high priority.
(2) The Maritime Security Agency operates three Britten-Norman Maritime Defenders. with Bendix RDR 1400C radars.

DEFENDER   *8/1996, MSA /* 0081375

Numbers/Type: 2 Lockheed P-3C Orion (Update II).
Operational speed: 410 kt *(760 km/h).*
Service ceiling: 28,300 ft *(8,625 m).*
Range: 4,000 n miles *(7,410 km).*
Role/Weapon systems: Order completed in 1991 but held up by the Pressler amendment, until delivery in December 1996. May be used for Elint. Sensors: APS-115 search radar; up to 100 sonobuoys; ASQ 81 MAD; ESM. Weapons: four Whitehead A 244 torpedoes or Mk 11 depth charges for ASW; Harpoon or Exocet AM 39 or C 802 for ASV.
VERIFIED

P-3C   *6/2001, Pakistan Navy /* 0114783

Numbers/Type: 5 Fokker F27-200.
Operational speed: 250 kt *(463 km/h).*
Service ceiling: 29,500 ft *(8,990 m).*
Range: 2,700 n miles *(5,000 km).*
Role/Weapon systems: Acquired in 1994-96 for maritime surveillance. Sensors: APS 504(V)2 radar, Thomson-CSF DR 3000A ESM.
VERIFIED

FOKKER F27-200   *6/2001, Pakistan Navy /* 0114780

Numbers/Type: 3 Breguet Atlantic 1.
Operational speed: 355 kt *(658 km/h).*
Service ceiling: 32,800 ft *(10,000 m).*
Range: 4,855 n miles *(8,995 km).*
Role/Weapon systems: Long-range MR/ASW cover for Arabian Sea; ex-French and Dutch stock. Upgraded in 1992-93. Three more acquired in 1994 for spares. Sensors: Thomson-CSF Ocean Master radar, Thomson-CSF DR 3000A ESM, MAD, sonobuoys, Sadang 1C sonobuoy signal processor. Weapons: ASW; nine Mk 46 torpedoes, Mk 11 depth bombs, mines. ASV: AM 39 Exocet missiles.
VERIFIED

ATLANTIC 1   *6/2001, Pakistan Navy /* 0114781

Numbers/Type: 12 AMD-BA Mirage III.
Operational speed: 750 kt *(1,390 km/h).*
Service ceiling: 59,055 ft *(18,000 m).*
Range: 740 n miles *(1,370 km).*
Role/Weapon systems: Operated by the Air Force, and all can be used for maritime strike. Sensors: Thomson-CSF radar. Weapons: ASV; two AM 39 Exocet or Harpoon; two 30 mm DEFA.
UPDATED

MIRAGE III   *6/2004*, Pakistan Navy /* 1044171

PATROL FORCES

Notes: Eight Mekat type catamarans ordered in late 1997. These may be for Customs.

2 + 2 JALALAT CLASS (FAST ATTACK CRAFT-MISSILE) (PTG)

Name	No	Builders	Launched	Commissioned
JALALAT	1029 (ex-1022)	PN Dockyard, Karachi	16 Nov 1996	14 Aug 1997
SHUJAAT	1030	PN Dockyard, Karachi	26 Mar 1999	30 Sep 1999
JURAT	—	Karachi Shipyards and Engineering Works	9 Sep 2004	2005
QUWWAT	—	Karachi Shipyards and Engineering Works	13 Sep 2004	2005

Displacement, tons: 185 full load
Dimensions, feet (metres): 128 × 22 × 5.7 *(39 × 6.7 × 1.8)*
Main machinery: 2 MTU diesels; 5,984 hp(m) *(4.4 MW)* sustained; 2 shafts
Speed, knots: 23. **Range, n miles:** 2,000 at 17 kt
Complement: 31 (3 officers)
Missiles: SSM: 4 China C 802 Saccade (2 twin); active radar homing to 120 km *(66 n miles)* at 0.9 Mach; warhead 165 kg; sea skimmer.
Guns: 2-37 mm/63 (twin); 180 rds/min to 8.5 km *(4.6 n miles)*; weight of shell 1.42 kg.
Countermeasures: Decoys: chaff launcher. ESM.
Radars: Surface search: Kelvin Hughes Type 756; I-band.
Fire control: Type 47G (for gun); Type SR-47 A/R (for SSM); I-band.

Comment: Designed with Chinese assistance to replace deleted Hegu class. Same hull as *Larkana.* Two further similar craft are under construction and have reportedly been developed in cooperation with the Thai company, Marsun.
UPDATED

JALALAT   *6/2000, Pakistan Navy /* 0105187

1 HUANGFEN CLASS (FAST ATTACK CRAFT—MISSILE) (PTFG)

DEHSHAT P 1026

Displacement, tons: 171 standard; 205 full load
Dimensions, feet (metres): 126.6 × 24.9 × 8.9 *(38.6 × 7.6 × 2.7)*
Main machinery: 3 Type 42-160 diesels; 12,000 hp(m) *(8.8 MW)* sustained; 3 shafts
Speed, knots: 35. **Range, n miles:** 800 at 30 kt
Complement: 28
Missiles: SSM: 4 HY-2; active radar or IR homing to 80 km *(43.2 n miles)* at 0.9 Mach; warhead 513 kg.
Guns: 4 Norinco 25 mm/80 (2 twin); 270 rds/min to 3 km *(1.6 n miles)*; weight of shell 0.34 kg.
Radars: Surface search/target indication: Square Tie (352); I-band.

Comment: Transferred from China April 1984. Chinese version of the Soviet Osa II class. One decommissioned in 2003. Assigned to 10th Patrol Squadron. *UPDATED*

DEHSHAT
6/2004*, Pakistan Navy / 1044172

1 TOWN CLASS (LARGE PATROL CRAFT) (PB)

Name	No	Builders	Commissioned
RAJSHAHI	P 140	Brooke Marine	1965

Displacement, tons: 115 standard; 143 full load
Dimensions, feet (metres): 107 × 20 × 6.9 *(32.6 × 6.1 × 2.1)*
Main machinery: 2 MTU 12V 538 diesels; 3,400 hp(m) *(2.5 MW)*; 2 shafts
Speed, knots: 24
Complement: 19
Guns: 2 Bofors 40 mm/60. 2-12.7 mm MGs.
Radars: Surface search: Pot Head; I-band.

Comment: The last survivor in Pakistan of a class of four built by Brooke Marine in 1965. Steel hull and aluminium superstructure. Assigned to 10th Patrol Squadron. *VERIFIED*

RAJSHAHI
6/2003, Pakistan Navy / 0569233

1 LARKANA CLASS (LARGE PATROL CRAFT) (PB)

Name	No	Builders	Commissioned
LARKANA	P 157	PN Dockyard, Karachi	6 June 1994

Displacement, tons: 180 full load
Dimensions, feet (metres): 128 × 22 × 5.4 *(39 × 6.7 × 1.7)*
Main machinery: 2 MTU diesels; 5,984 hp(m) *(4.4 MW)* sustained; 2 shafts
Speed, knots: 23. **Range, n miles:** 2,000 at 17 kt
Complement: 25 (3 officers)
Guns: 2 Type 76A 37 mm/63 (twin). 4—25 mm/60 (2 twin).
Depth charges: 2 Mk 64 launchers.
Radars: Surface search: Kelvin Hughes Type 756; I-band.

Comment: Ordered in 1991 and started building in October 1992. Has replaced the last of the Hainan class. The missile version on the same hull has taken priority but more may be built. Assigned to 10th Patrol Squadron. *VERIFIED*

LARKANA
6/2003, Pakistan Navy / 0569232

2 KAAN 15 (FAST INTERVENTION CRAFT) (PBF)

P 01   P 02

Displacement, tons: 19 full load
Dimensions, feet (metres): 54.2 × 13.2 × 3.15 *(15.4 × 4.04 × 0.96)*
Main machinery: 2 MTU 12V 183 TE93 diesels; 2,300 hp(m) *(1.69 MW)*; 2 Arneson ASD 12 B1L surface drives
Speed, knots: 54. **Range, n miles:** 350 at 35 kt
Complement: 4 plus 8 mission crew
Guns: 2—12.7 mm MGs.

Comment: Built by Yonca Shipyard, Turkey. The first delivered on 17 August 2004 and the second on 14 October 2004. To be operated by Special Services Group based at PNS Iqbar. Details based on those in Turkish Coast Guard service. *NEW ENTRY*

P 01
7/2004*, Selçuk Emre / 1044173

SURVEY SHIPS

Notes: Acquisition of a new oceanographic research vessel was reported in November 2002 to have received Presidential approval. It is not clear whether this is to be a specialist or a multipurpose vessel.

1 SURVEY SHIP (AGS/AGOR)

Name	No	Builders	Launched	Commissioned
BEHR PAIMA	SV 48	Ishikawajima, Japan	7 July 1982	17 Dec 1982

Measurement, tons: 1,183 gross
Dimensions, feet (metres): 200.1 × 38.7 × 12.1 *(61 × 11.8 × 3.7)*
Main machinery: 2 Daihatsu 6DSM-22 diesels; 2,000 hp(m) *(1.47 MW)*; 2 shafts; cp props; bow thruster
Speed, knots: 13.7. **Range, n miles:** 5,400 at 12 kt
Complement: 84 (16 officers)

Comment: Ordered in November 1981. Laid down 16 February 1982. Dynamic positioning system. Has seismic, magnetic and gravity survey equipment. DESO 20 deep echo sounder. There is a second survey ship *Jatli* under civilian control. *VERIFIED*

BEHR PAIMA
6/2003, Pakistan Navy / 0569231

MINE WARFARE FORCES

3 MUNSIF (ÉRIDAN) CLASS (MINEHUNTERS) (MHSC)

Name	No	Builders	Launched	Commissioned
MUNSIF (ex-*Sagittaire*)	M 166	Lorient Dockyard	9 Nov 1988	27 July 1989
MUHAFIZ	M 163	Lorient Dockyard	8 July 1995	15 May 1996
MUJAHID	M 164	Lorient/PN Dockyard, Karachi	28 Jan 1997	9 July 1998

Displacement, tons: 562 standard; 595 full load
Dimensions, feet (metres): 168.9 × 29.2 × 9.5 *(51.5 × 8.9 × 2.9)*
Main machinery: 1 Stork Wärtsilä A-RUB 215X-12 diesel; 1,860 hp(m) *(1.37 MW)* sustained; 1 shaft; LIPS cp prop; auxiliary propulsion; 2 motors; 240 hp(m) *(179 kW)*; 2 active rudders; 2 bow thrusters
Speed, knots: 15; 7 on auxiliary propulsion. **Range, n miles:** 3,000 at 12 kt
Complement: 46 (5 officers)
Guns: 1 GIAT 20F2 20 mm; 1-12.7 mm MG.
Countermeasures: MCM; 2 PAP 104 Mk 5 systems; mechanical sweep gear. Elesco MKR 400 acoustic sweep; MRK 960 magnetic sweep.
Combat data systems: Thomson-CSF TSM 2061 Mk 2 tactical system in the last pair.
Radars: Navigation: Racal Decca 1229 (M 166) or Kelvin Hughes 1007; I-band.
Sonars: Thomson Sintra DUBM 21B or 21D (163 and 164); hull-mounted; active; high frequency; 100 kHz (±10 kHz).
Thomson Sintra TSM 2054 MCM towed array may be included.

Comment: Contract signed with France 17 January 1992. The first recommissioned into the Pakistan Navy on 24 September 1992 after active service in the Gulf with the French Navy in 1991. Sailed for Pakistan in November 1992. The second was delivered in April 1996. The last one was transferred to Karachi by transporter ship in April 1995 with a final package following in November 1995. Form 21st Mine Countermeasures Squadron. *VERIFIED*

MUJAHID
6/2000, Pakistan Navy / 0105188

AUXILIARIES

1 FUQING CLASS (AORH)

Name	No	Builders	Commissioned
NASR (ex-*X-350*)	A 47	Dalian Shipyard	27 Aug 1987

Displacement, tons: 7,500 standard; 21,750 full load
Dimensions, feet (metres): 561 × 71.5 × 30.8 *(171 × 21.8 × 9.4)*
Main machinery: 1 Sulzer 8RLB66 diesel; 13,000 hp(m) *(9.56 MW)*; 1 shaft
Speed, knots: 18. **Range, n miles:** 18,000 at 14 kt
Complement: 130 (during visit to Australia in October 1988 carried 373 (23 officers) including 100 cadets)
Cargo capacity: 10,550 tons fuel; 1,000 tons dieso; 200 tons feed water; 200 tons drinking water
Guns: 1 GE/GD Vulcan Phalanx CIWS. 4—37 mm (2 twin). 2—12.7 mm MGs.
Countermeasures: Decoys: SRBOC Mk 36 chaff launcher.
Radars: Navigation: 1 Kelvin Hughes 1007; 1 SPS 66; I-band.
Helicopters: 1 SA 319B Alouette III.

Comment: Similar to Chinese ships of the same class. Two replenishment at sea positions on each side for liquids and one for solids. Phalanx fitted on the hangar roof in 1995. Assigned to 42nd Auxiliary Squadron. *VERIFIED*

NASR   5/2003 / 0569225

2 COASTAL TANKERS (AOTL)

Name	No	Builders	Commissioned
GWADAR	A 49	Karachi Shipyard	1984
KALMAT	A 21	Karachi Shipyard	29 Aug 1992

Measurement, tons: 831 grt
Dimensions, feet (metres): 206 × 37.1 × 9.8 *(62.8 × 11.3 × 3)*
Main machinery: 1 Sulzer diesel; 550 hp(m) *(404 kW)*; 1 shaft
Speed, knots: 10
Complement: 25
Cargo capacity: 340 m³ fuel or water

Comment: Assigned to 42nd Auxiliary Squadron. *VERIFIED*

GWADAR   6/2003, Pakistan Navy / 0569230

1 POOLSTER CLASS (AORH)

Name	No	Builders	Commissioned	Recommissioned
MOAWIN	A 20	Rotterdamse	10 Sep 1964	28 July 1994
(ex-*Poolster*)	(ex-A 835)	Droogdok Mij		

Displacement, tons: 16,800 full load
Measurement, tons: 10,000 dwt
Dimensions, feet (metres): 552.2 × 66.6 × 26.9 *(168.3 × 20.3 × 8.2)*
Main machinery: 2 boilers; 2 turbines; 22,000 hp(m) *(16.2 MW)*; 1 shaft
Speed, knots: 21
Complement: 200 (17 officers)
Cargo capacity: 10,300 tons including 8-9,000 tons oil fuel
Guns: 1 GE/GD 6-barrelled Vulcan Phalanx Mk 15 or 2 Oerlikon 20 mm.
Countermeasures: Decoys: SRBOC Mk 36 chaff launcher.
Radars: Air/surface search: Racal Decca 2459; F/I-band.
Navigation: Racal Decca TM 1229C; I-band.
Sonars: Signaal CWE 10; hull-mounted; active search; medium frequency.
Helicopters: 1 Sea King.

Comment: Acquired from the Netherlands Navy. Helicopter deck aft. Funnel heightened by 4.5 m *(14.8 ft)*. Capacity for five Lynx sized helicopters. Two fuelling stations each side for underway replenishment. Phalanx to be fitted in due course. Assigned to 42nd Auxiliary Squadron. *VERIFIED*

MOAWIN   3/2002, A Sharma / 0534060

1 TANKER (AOTL)

ATTOCK A 40

Displacement, tons: 1,200 full load
Dimensions, feet (metres): 177.2 × 32.3 × 15.1 *(54 × 9.8 × 4.6)*
Main machinery: 2 diesels; 800 hp(m) *(276 kW)*; 2 shafts
Speed, knots: 8
Complement: 18
Cargo capacity: 550 tons fuel
Guns: 2 Oerlikon 20 mm.

Comment: Built in Italy in 1957. Assigned to 42nd Auxiliary Squadron. *UPDATED*

ATTOCK   6/2004 *, Pakistan Navy / 1044169

TUGS

Notes: *Jandar* and *Jafakash* are two pusher tugs (10 ton bollard pull) built by Karachi Shipyard and commissioned in 2000.

JANDAR and JAFAKASH   6/2003 *, Pakistan Navy / 1044170

4 COASTAL TUGS (YTB)

Name	No	Builders	Commissioned
BHOLU	A 44	Giessendam Shipyard, Netherlands	Apr 1991
GAMA	A 45	Giessendam Shipyard, Netherlands	Apr 1991
JANBAZ	—	Karachi Shipyard	Sep 1990
JOSHILA	—	Karachi Shipyard	Sep 2000

Displacement, tons: 265 full load
Dimensions, feet (metres): 85.3 × 22.3 × 9.5 *(26 × 6.8 × 2.9)*
Main machinery: 2 Cummins KTA38-M diesels; 1,836 hp *(1.26 MW)* sustained; 2 shafts
Speed, knots: 12
Complement: 6

Comment: Details are for *Bholu* and *Gama*, built by Damen Shipyards and which entered service in 1991. *Janbaz* and *Joshila* were built by Karachi Shipyard and delivered in 1990 and 2000 respectively. *VERIFIED*

JOSHILA   5/2003 / 0569222

For details of the latest updates to *Jane's Fighting Ships* online and to discover the additional information available exclusively to online subscribers please visit

jfs.janes.com

MARITIME SECURITY AGENCY

Notes: (1) All ships are painted white with a distinctive diagonal blue and red band and MSA on each side.
(2) One Britten-Norman Maritime Defender acquired in 1993, a second in 1994 and a third in August 2004. Based near Karachi with 93 Squadron.
(3) Plans for new ships and aircraft are under consideration.

1 GEARING (FRAM 1) CLASS (DD)

Name	No	Builders	Commissioned
NAZIM (ex-*Tughril*)	D 156 (ex-D 167)	Todd Pacific	4 Aug 1945

Displacement, tons: 2,425 standard; 3,500 full load
Dimensions, feet (metres): 390.5 × 41.2 × 19 *(119 × 12.6 × 5.8)*
Main machinery: 4 Babcock & Wilcox boilers; 600 psi *(43.3 kg/cm²)*; 850°F *(454°C)*; 2 GE turbines; 60,000 hp *(45 MW)*; 2 shafts
Speed, knots: 32. **Range, n miles:** 4,500 at 16 kt
Complement: 180 (15 officers)
Guns: 1 US 5 in *(127 mm)*/38 Mk 38; 15 rds/min to 17 km *(9.3 n miles)* anti-surface; 11 km *(5.9 n miles)*; anti-aircraft; weight of shell 25 kg.
 4—25 mm (2 twin).
Torpedoes: 6—324 mm Mk 32 (2 triple) tubes.
Countermeasures: Decoys: 2 Plessey Shield 6-barrelled fixed launchers; chaff and IR flares in distraction, decoy or centroid modes.
Weapons control: Mk 37 for 5 in guns. OE 2 SATCOM.
Radars: Surface search: Raytheon/Sylvania; SPS-10; G-band.
 Navigation: KH 1007; I-band.
 Fire control: Western Electric Mk 25; I/J-band.

Comment: Transferred from the US on 30 September 1980 to the Navy. Passed on to the MSA in 1998 and renamed. This is the third Gearing to be renamed *Nazim*, the previous pair having been sunk as targets. All weapon systems removed except the torpedo tubes and main gun. Serves as the MSA Flagship.

UPDATED

NAZIM   *5/2003* / 0569224

2 SHANGHAI II CLASS (FAST ATTACK CRAFT—GUN) (PB)

SUBQAT P 1066   RAFAQAT P 1068

Displacement, tons: 131 full load
Dimensions, feet (metres): 127.3 × 17.7 × 5.6 *(38.8 × 5.4 × 1.7)*
Main machinery: 2 Type L12-180 diesels; 2,400 hp(m) *(1.76 MW)* (forward); 2 Type 12-D-6 diesels; 1,820 hp(m) *(1.34 MW)* (aft); 4 shafts
Speed, knots: 30. **Range, n miles:** 700 at 16.5 kt
Complement: 34
Guns: 2—37 mm/63 (twin). 2—25 mm/80 (twin).
Depth charges: 2 projectors; 8 weapons.
Mines: Fitted with mine rails for approx 10 mines.
Radars: Surface search: Anritsu ARC-32A; I-band.

Comment: Four of the class were transferred from the Navy in 1986 and two more in 1998. The last pair were then replaced by naval craft. All were originally acquired from China 1972-1976.

VERIFIED

SUBQAT   *5/2003* / 0569223

4 BARKAT CLASS (PBO)

Name	No	Builders	Commissioned
BARKAT	1060 (ex-P 60)	China Shipbuilding Corp	29 Dec 1989
REHMAT	1061 (ex-P 61)	China Shipbuilding Corp	29 Dec 1989
NUSRAT	1062 (ex-P 62)	China Shipbuilding Corp	13 June 1990
VEHDAT	1063 (ex-P 63)	China Shipbuilding Corp	13 June 1990

Displacement, tons: 435 full load
Dimensions, feet (metres): 190.3 × 24.9 × 7.5 *(58 × 7.6 × 2.3)*
Main machinery: 4 MTU 16V 396 TB93 diesels; 8,720 hp(m) *(6.4 MW)* sustained; 4 shafts
Speed, knots: 27. **Range, n miles:** 1,500 at 12 kt
Complement: 50 (5 officers)
Guns: 2—37 mm/63 (1 twin). 2—14.5 mm/60 (twin).
Radars: Surface search: 2 Anritsu ARC-32A; I-band.

Comment: Type P58A patrol craft built in China for the MSA. First two arrived in Karachi at the end of January 1990, second pair in August 1990. Some of this type of ship are in service with Chinese paramilitary forces.

VERIFIED

VEHDAT   *6/1994, Maritime Security Agency* / 0081380

COAST GUARD

Notes: (1) Unlike the Maritime Security Agency which comes under the Defence Ministry, the official Coast Guard was set up in 1985 and is manned by the Army and answerable to the Ministry of the Interior.
(2) The Customs Service is manned by naval personnel. It operates approximately 18 craft of which most are Crestifalia MV 55.

1 SWALLOW CLASS (PB)

SAIF

Displacement, tons: 52 full load
Dimensions, feet (metres): 65.6 × 15.4 × 4.3 *(20.0 × 4.7 × 1.3)*
Main machinery: 2 GM Detroit 12V71 T1 diesels; 2,120 hp *(1.58 MW)*; 2 shafts
Speed, knots: 25. **Range, n miles:** 500 at 20 kt
Complement: 8
Guns: 2—12.7 mm MGs.

Comment: Built by Swallowcraft/Kangnam and delivered in 1986.

VERIFIED

4 CRESTIFALIA MV 55 CLASS (PBF)

SADD P 551	SHABHAZ P 552	VAQAR P 553	BURQ P 554

Displacement, tons: 23 full load
Dimensions, feet (metres): 54.1 × 17.1 × 2.95 *(16.5 × 5.2 × 0.9)*
Main machinery: 2 MTU diesels; 2,200 hp *(1.64 MW)*; 2 shafts
Speed, knots: 35. **Range, n miles:** 425 at 25 kt
Complement: 5

Comment: Delivered in 1987.

VERIFIED

SHABHAZ   *5/2003* / 0569228

Palau

Country Overview

The Republic of Palau was a US-administered UN Trust territory from 1947 before becoming independent in 1994 when a Compact of Free Association, delegating to the US the responsibility for defence and foreign affairs, came into effect. Situated in the western Pacific Ocean, the country comprises about 200 of the Caroline Islands archipelago spread in a chain about 350 n miles long. These include Koror (the administrative centre), Babelthuap (the largest island), Arakabesan, Malakal and Peleliu. The capital is currently on Koror, but a new capital is being built in eastern Babelthuap. Territorial seas (3 n miles) are claimed. An extended fisheries zone (200 n miles) is also claimed but limits have not been fully defined.

Headquarters Appointments

Chief of Division of Marine Law Enforcement:
 Captain Ellender Ngirameketii
Maritime Surveillance Adviser:
 Commander Matthew Brown, RAN

PATROL FORCES

1 PACIFIC CLASS (LARGE PATROL CRAFT) (PB)

Name	No	Builders	Commissioned
PRESIDENT H I REMELIIK	001	Transfield Shipbuilding	May 1996

Displacement, tons: 162 full load
Dimensions, feet (metres): 103.3 × 26.6 × 6.9 *(31.5 × 8.1 × 2.1)*
Main machinery: 2 Caterpillar 3516TA diesels; 4,400 hp *(3.28 MW)* sustained; 2 shafts
Speed, knots: 20. **Range, n miles:** 2,500 at 12 kt
Complement: 17 (3 officers)
Guns: 2—7.62 mm MGs.
Radars: Surface search: Furuno 1011; I-band.

Comment: Ordered in 1995. This was the 21st hull in the Pacific class programme. Following the decision by the Australian government to extend the Pacific Patrol Boat project, the ship underwent a half-life refit at Gladstone in 2003. A life-extension refit will be required in 2012.

UPDATED

PRESIDENT H I REMELIIK
*6/2004 *, Division of Marine Law Enforcement, Palau / 1044175*

Panama

Country Overview

The Republic of Panama is an independent state situated on the isthmus linking South America with Central and North America. Bordered to the west by Costa Rica and to the east by Colombia, it has an area of 29,157 square miles and a 664 n mile coastline with the north Pacific Ocean and of 370 n miles with the Caribbean. The country is bisected by the Panama Canal. A new treaty in 1977 ended US operation, maintenance and defence of the canal in 1999. The capital is Panama City while the main ports are Balboa, Cristóbal, Coco Solo, Bahía Las Minas, Vacamonte, Almirante and Puerto Armuelles. Territorial seas (12 n miles) are claimed. An Exclusive Economic Zone (EEZ) (200 n miles) has been defined by boundary agreements. Reform of the security apparatus led to the creation of the Panamanian Public Forces, which includes the National Maritime Service, in 1994.

Headquarters Appointments

Director General National Maritime Service:
Captain Ricardo Traad Porras

Personnel

(a)   2005: 620
(b)   Voluntary service

Bases

Isla Flamenco (HQ) (Punta Brujas — HQ designate), Quebrada de Piedra, Largo Remo (under construction), Punta Cocos (air), Kuna Yala (air) (under construction)

PATROL FORCES

Notes: A further patrol craft *Cocle* P 814 has been reported.

2 VOSPER TYPE (COASTAL PATROL CRAFT) (PB)

Name	No	Builders	Commissioned
PANQUIACO	P 301 (ex-GC 10)	Vospers, Portsmouth	July 1971
LIGIA ELENA	P 302 (ex-GC 11)	Vospers, Portsmouth	July 1971

Displacement, tons: 96 standard; 145 full load
Dimensions, feet (metres): 103 × 18.9 × 5.8 *(31.4 × 5.8 × 1.8)*
Main machinery: 2 Detroit diesels; 5,000 hp *(3.73 MW)*; 2 shafts
Speed, knots: 18. **Range, n miles:** 1,500 at 14 kt
Complement: 17 (3 officers)
Guns: 2—7.62 mm MGs.
Radars: Surface search: Raytheon R-81; I-band.

Comment: *Panquiaco* launched on 22 July 1970, *Ligia Elena* on 25 August 1970. Hull of welded mild steel and upperworks of welded or buck-bolted aluminium alloy. Vosper fin stabiliser equipment. P 302 was sunk in December 1989, but subsequently recovered. Both vessels had major repairs in the Coco Solo shipyard from September 1992. This included new engines, a new radar and replacement guns. Pacific Flotilla. Similar craft in service in Malaysia.

UPDATED

LIGIA ELENA   *6/2003, Panama Maritime Service / 0568905*

1 COASTAL PATROL CRAFT (PB)

Name	No	Builders	Commissioned
NAOS (ex-*Erline*)	P 303 (ex-RV 821)	Equitable, NO	Dec 1964

Displacement, tons: 120 full load
Dimensions, feet (metres): 105 × 24.9 × 6.9 *(32 × 7.6 × 2.1)*
Main machinery: 2 Caterpillar diesels; 2 shafts
Speed, knots: 10. **Range, n miles:** 550 at 8 kt
Complement: 11 (2 officers)
Guns: 2—7.62 mm MGs.
Radars: Surface search: Raymarx 2600; I-band.

Comment: Served as a support/research craft at the US Underwater Systems establishment at Bermuda. Transferred from US in July 1992 and recommissioned in December 1992. Refitted in 1997 with new engines. Pacific Flotilla.

VERIFIED

NAOS   *6/2002, Panama Maritime Service / 0525006*

1 COASTAL PATROL CRAFT (PB)

ESCUDO DE VERAGUAS (ex-*Aun Sin Nombre*, ex-*Kathyuska Kelly*) P 305 (ex-P 206)

Displacement, tons: 158 full load
Dimensions, feet (metres): 90.5 × 24.1 × 6.1 *(27.6 × 7.3 × 1.9)*
Main machinery: 2 Detroit 12V-71 diesels; 840 hp *(627 kW)* sustained; 2 shafts
Speed, knots: 10
Complement: 10 (2 officers)
Guns: 1—12.7 mm MG.
Radars: Surface search: Raytheon; I-band.

Comment: Confiscated drug runner craft taken into service in 1996. Also used for transport duties. Caribbean Flotilla.

VERIFIED

ESCUDO DE VERAGUAS   *11/1998, Panama Maritime Service / 0052687*

1 COASTAL PATROL CRAFT (PB)

TABOGA P 306

Comment: Details not confirmed. Possibly a confiscated vessel.

VERIFIED

TABOGA   *6/2003, Panama Maritime Service* / 0568904

1 NEGRITA CLASS (COASTAL PATROL CRAFT) (PB)

CACIQUE NOME (ex-*Negrita*) P 203

Displacement, tons: 68 full load
Dimensions, feet (metres): 80 × 15 × 6 *(24.4 × 4.6 × 1.8)*
Main machinery: 2 Detroit 12V-71 diesels; 840 hp *(627 kW)*; 2 shafts
Speed, knots: 13. **Range, n miles:** 250 at 10 kt
Complement: 8 (2 officers)
Guns: 2—7.62 mm MGs.
Radars: Surface search: Raytheon 71; I-band.

Comment: Completely rebuilt in the Coco Solo shipyard and recommissioned 5 May 1993. Pacific Flotilla.

VERIFIED

CACIQUE NOME   *8/1998, Panama Maritime Service* / 0052688

5 POINT CLASS (COASTAL PATROL CRAFT) (PB)

Name	No	Builders	Commissioned
3 DE NOVIEMBRE (ex-*Point Barrow*)	P 204 (ex-82348)	CG Yard, MD	4 Oct 1964
10 DE NOVIEMBRE (ex-*Point Huron*)	P 206 (ex-82357)	CG Yard, MD	17 Feb 1967
28 DE NOVIEMBRE (ex-*Point Frances*)	P 207 (ex-82356)	CG Yard, MD	3 Feb 1967
4 DE NOVIEMBRE (ex-*Point Winslow*)	P 208 (ex-82360)	J M Martinac, Tacoma	3 Mar 1967
5 DE NOVIEMBRE (ex-*Point Hannon*)	P 209 (ex-82355)	J M Martinac, Tacoma	23 Jan 1967

Displacement, tons: 69 full load
Dimensions, feet (metres): 83 × 17.2 × 5.8 *(25.3 × 5.2 × 1.8)*
Main machinery: 2 Cummins V-12-900M diesels; 1,600 hp *(1.18 MW)*; 2 shafts
Speed, knots: 18. **Range, n miles:** 1,500 at 8 kt
Complement: 10 (2 officers)
Guns: 2—7.62 mm MGs.
Radars: Surface search: Raytheon Pathfinder; I-band.

Comment: First one transferred from US Coast Guard 7 June 1991 and recommissioned 10 July 1991. Further two transferred 22 April 1999. Fourth transferred 20 September 2000 and fifth on 11 January 2001. Carry a RIB with a 40 hp engine. Caribbean Flotilla.

VERIFIED

28 DE NOVIEMBRE   *6/2003, Panama Maritime Service* / 0568902

3 COASTAL PATROL CRAFT (PB)

CHIRIQUI P 841   **VERAGUAS** P 842   **BOCAS DEL TORO** P 843

Displacement, tons: 46 full load
Dimensions, feet (metres): 73.8 × 17.3 × 2.9 *(22.5 × 5.3 × 0.9)*
Main machinery: 3 Detroit 12V 71 diesels; 1,260 hp *(940 kW)* sustained; 3 shafts
Speed, knots: 20
Complement: 7 (1 officer)
Guns: 2—7.62 mm MGs.
Radars: Surface search: Furuno 1411; I-band.

Comment: Transferred as Grant-Aid from the US in March 1998. Used for drug prevention patrols in both Flotillas.

VERIFIED

BOCAS DEL TORO   *6/2003* / 0568903

2 HARBOUR PATROL CRAFT (PB)

PANAMA P 101   **CALAMAR** P 102 (ex-PC 3602)

Displacement, tons: 11 full load
Dimensions, feet (metres): 36 × 13 × 3 *(11 × 4 × 0.9)*
Main machinery: 1 Detroit 6-71T diesel; 300 hp *(224 kW)*; 1 shaft
Speed, knots: 15. **Range, n miles:** 160 at 12 kt
Complement: 5
Guns: 1—7.62 mm MG.

Comment: P 102 in service from December 1992, P 101 from February 1998. GRP construction. Pacific flotilla.

VERIFIED

CALAMAR   *8/1996, Panama Maritime Service*

6 FAST PATROL BOATS (PBF)

BPC 2201, 2203, 2206, 2207-2209

Dimensions, feet (metres): 22.3 × 7.5 × 2 *(6.8 × 2.3 × 0.6)*
Main machinery: 2 Johnson outboards; 280 hp *(209 kW)*
Speed, knots: 35
Complement: 4
Guns: 1—7.62 mm MG.

Comment: *BPC 2201-2205* are Boston Whaler Piraña class acquired between June 1991 and October 1992.

VERIFIED

BPC 2203   *11/1998, Panama Maritime Service* / 0052690

11 FAST PATROL BOATS (PBF)

BPC 3201, 3202, 3207-3209, 3214-3215, 3220, 3222-3223, 3225

Dimensions, feet (metres): 33.5 × 7.5 × 2 *(10.2 × 2.3 × 0.6)*
Main machinery: 2 Yamaha outboards; 400 hp(m) *(294 kW)*
Speed, knots: 35
Complement: 4
Guns: 1—7.62 mm MG.

Comment: Eduardoño class acquired between June 1995 and October 1998.
UPDATED

BPC 3202   *6/2003*, Panama Maritime Service /* 0587789

FLAMENCO (old number)   *12/1998, Panama Maritime Service /* 0052686

AUXILIARIES

Notes: (1) There are two auxiliary craft *Frailes del Norte* T 06 (ex-US LCM 8 class) and *Frailes del Sur* T 07.
(2) *General Esteban Huertas* (ex-YFU 81) has been reported with pennant number A 402 and may have replaced *Flamenco* in July 2004.

FRAILES DEL NORTE   *6/2003, Panama Maritime Service /* 0568901

1 BALSAM CLASS (PBO)

Name	No	Builders	Commissioned
INDEPENDENCIA	A 401	Marine Iron and Shipbuilding Corp, Duluth,	20 Nov 1943
(ex-*Sweetgum*)	(ex-WLB 309)	Minnesota	

Displacement, tons: 1,034 full load
Dimensions, feet (metres): 180 × 37 × 12 *(54.9 × 11.3 × 3.8)*
Main machinery: Diesel electric; 2 diesels; 1,402 hp *(1.06 MW)*; 1 motor; 1,200 hp *(895 kW)*; 1 shaft; bow thruster
Speed, knots: 13. **Range, n miles:** 8,000 at 12 kt
Complement: 53
Guns: 2—12.7 mm MGs.
Radars: Navigation: Raytheon SPS-64(V)1.

Comment: Transferred from US Coast Guard on 15 February 2002.
UPDATED

INDEPENDENCIA   *1/2004* /* 0587788

1 COASTAL PATROL CRAFT (YO)

FLAMENCO (ex-*Scheherazade*) A 402 (ex-P 304, ex-WB 831)

Displacement, tons: 220 full load
Dimensions, feet (metres): 105 × 25 × 6.9 *(32 × 7.6 × 2.1)*
Main machinery: 2 Caterpillar diesels; 2 shafts
Speed, knots: 10
Complement: 11 (2 officers)
Guns: 2—7.62 mm MGs.
Radars: Surface search: Furuno FCR 1411; I-band.

Comment: Built in 1963. Transferred from US 22 July 1992 and commissioned in December 1992. Former US wooden hulled COOP craft. Refitted in Panama in 1994. Now used as a refuelling auxiliary. May have been replaced by ex-YFU 81. *UPDATED*

1 MSB 5 CLASS (YAG)

NOMBRE DE DIOS (ex-*MSB 25*) L 16

Displacement, tons: 44 full load
Dimensions, feet (metres): 57.2 × 15.5 × 4 *(17.4 × 4.7 × 1.2)*
Main machinery: 2 Detroit diesels; 600 hp *(448 kW)*; 2 shafts
Speed, knots: 12
Complement: 6 (1 officer)
Guns: 1—7.62 mm MG.
Radars: Navigation: Raytheon Raystar; I-band.

Comment: Built between 1952 and 1956. Former US minesweeping boat. Served in the canal area until 1992 and transferred from US to Panama in December 1992 after refit. Wooden hull, new engine. Used as logistic craft. Pacific flotilla.
VERIFIED

NOMBRE DE DIOS   *6/2003 /* 0568899

6 SUPPORT CRAFT (YAG)

DORADO I BA 055   AGUACERO BA 057
DORADO II BA 056   PORTOBELO BA 058
DORADO III —   FANTASMA AZUL BA 059

Comment: *Dorado I* and *II* acquired in February 1998 and are used as 40 kt supply craft. *Aguacero* is a confiscated 50 kt power boat taken into service in November 1998.
VERIFIED

DORADO I   *12/1998, Panama Maritime Service /* 0052692

1 LOGISTIC CRAFT (YAG)

ISLA PARIDAS (ex-*Endeavour*) L 21

Displacement, tons: 120 full load
Dimensions, feet (metres): 75 × 14 × 7 *(22.9 × 4.3 × 2.1)*
Main machinery: 1 Caterpillar diesel; 365 hp *(270 kW)*; 1 shaft
Speed, knots: 12
Complement: 7 (1 officer)
Radars: Navigation: Furuno; I-band.

Comment: Acquired in September 1991. Pacific flotilla.

VERIFIED

LAND-BASED MARITIME AIRCRAFT

Numbers/Type: 1 Pilatus Britten-Norman Islander.
Operational speed: 150 kt *(280 km/h)*.
Service ceiling: 18,900 ft *(5,760 m)*.
Range: 1,500 n miles *(2,775 km)*.
Role/Weapon systems: Air Force operated coastal surveillance duties. Sensors: Search radar. Weapons: Unarmed.

VERIFIED

Numbers/Type: 3 CASA C-212 Aviocar.
Operational speed: 190 kt *(353 km/h)*.
Service ceiling: 24,000 ft *(7,315 m)*.
Range: 1,650 n miles *(3,055 km)*.
Role/Weapon systems: Air Force operated coastal patrol aircraft for EEZ protection and anti-smuggling duties. Sensors: APS-128 radar, limited ESM. Weapons: ASW; two Mk 44/46 torpedoes. ASV; two rocket or machine gun pods.

UPDATED

C-212

6/2003*, Adolfo Ortigueira Gil / 0587787

Papua New Guinea

Country Overview

Papua New Guinea lies north of Australia in the eastern half of New Guinea which it shares with the Indonesian province of Irian Jaya. An Australian-administered UN Trust territory from 1949, it became independent in 1975. Its head of state is the British sovereign, who is represented by a Governor-General. Its many island groups include the Bismarck and Louisiade Archipelagos, the Trobriand Islands, the D'Entrecasteaux Islands and Woodlark Island. Amongst other islands are Bougainville (a nine-year separatist conflict ended in 1997) and Buka. It has a 2,781 n mile coastline. The capital, principal city and port is Port Moresby. An archipelagic state, territorial seas (12 n miles) are claimed. A 200 n mile Exclusive Economic Zone (EEZ) has also been claimed but the limits have not been fully defined by boundary agreements.

Headquarters Appointments

Commander Defence Forces:
Commodore Peter Ilau, CBE
Director Naval Operations:
Commander Max S Aleale

Bases

Port Moresby (HQ PNGDF and PNGDF Landing Craft Base); Lombrum (Manus)

Prefix to Ships' Names

HMPNGS

PATROL FORCES

4 PACIFIC CLASS (LARGE PATROL CRAFT) (PB)

Name	No	Builders	Commissioned
TARANGAU	01	Australian Shipbuilding Industries	16 May 1987
DREGER	02	Australian Shipbuilding Industries	31 Oct 1987
SEEADLER	03	Australian Shipbuilding Industries	29 Oct 1988
BASILISK	04	Australian Shipbuilding Industries	1 July 1989

Displacement, tons: 162 full load
Dimensions, feet (metres): 103.3 × 26.6 × 6.9 *(31.5 × 8.1 × 2.1)*
Main machinery: 2 Caterpillar 3516TA diesels; 4,400 hp *(3.3 MW)* sustained; 2 shafts
Speed, knots: 20. **Range, n miles:** 2,500 at 12 kt
Complement: 17 (3 officers)
Guns: 1 Oerlikon GAM-BO1 20 mm. 2—7.62 mm MGs.
Radars: Surface search: Furuno 1011; I-band.

Comment: Contract awarded in 1985 to Australian Shipbuilding Industries (Hamilton Hill, West Australia) under Australian Defence co-operation. These are the first, third, sixth and seventh of the class and some of the few to be armed. All upgraded, during half-life refits in Australia with new radars and navigation support systems in 1997/98. Following the decision by the Australian government to extend the Pacific Patrol Boat project, *Tarangau* underwent a life-extension refit at Gladstone in 2003 and *Dreger* in 2004. Similar refits for *Seeadler* and *Basilisk* are due in 2005 and 2006 respectively.

UPDATED

TARANGAU
8/2003, John Mortimer / 0569236

AUXILIARIES

2 LANDING CRAFT (LSM)

Name	No	Builders	Commissioned
SALAMAUA	31	Walkers Ltd, Maryborough	19 Oct 1973
BUNA	32	Walkers Ltd, Maryborough	7 Dec 1973

Displacement, tons: 310 light; 503 full load
Dimensions, feet (metres): 146 × 33 × 6.5 *(44.5 × 10.1 × 1.9)*
Main machinery: 2 GM diesels; 2 shafts
Speed, knots: 10. **Range, n miles:** 3,000 at 10 kt
Complement: 15 (2 officers)
Military lift: 160 tons
Guns: 2—12.7 mm MGs.
Radars: Navigation: Racal Decca RM 916; I-band.

Comment: Transferred from Australia in 1975. Underwent extensive refits 1985-86. Both are still active.

VERIFIED

SALAMAUA
12/1990, James Goldrick / 0081510

Paraguay
ARMADA NACIONAL

Country Overview

The Republic of Paraguay is one of two landlocked countries in South America; Bolivia is the other. With an area of 157,048 square miles, it has borders to the north with Bolivia, to the east with Brazil and to the south with Argentina. There are some 1,500 n miles of internal waterways including the principal rivers, the Pilcomayo, Paraguay and Paraná. Navigable by large ships for much of their length, they link the capital, largest city and principal port, Asunción, with the Rio de la Plata estuary on the Atlantic Ocean. Other ports include Ciudad del Este, Encarnación and Concepción.

Headquarters Appointments

Commander-in-Chief of the Navy:
   Vice Admiral Julio Cesar Baez Acosta
Chief of Staff:
   To be confirmed
Fleet Commander:
   Rear Admiral Julian Paredes Morales

Personnel

2005: 3,600 including 300 Coast Guard, 800 marines and 100 naval air

Bases

Main Base: Puerto Sajonia, Asunción
Minor Bases: Base Naval de Bahia Negra (BNBN) (on upper Paraguay river)
Base Naval de Salto del Guaira (BNSG) (on upper Paraná river)
Base Naval de Ciudad del Este (BNCE) (on Paraná river)
Base Naval de Encarnacion (BNE) (on Paraná river)
Base Naval de Ita-Pirú (BNIP) (on Paraná river)

Training

Specialist training is done with Argentina (Exercise Sirena), Brazil (Exercise Ninfa) and US (Exercise Unitas).

Marine Corps

BIM 1:	Puerto Rosario
BIM 2:	Puerto Vallemi
BIM 3:	Asunción
BIM 5:	Bahia Negra
	Detachments at Pozo Hondo and Ita-Pirú
BIM 8:	Saltos del Guairá
	Detachments at Ciudad del Este and Encarnación

Naval Aviation

Fixed Wing	Asunción International Airport
Helicopters	Puerto Sajonia

Coast Guard

Prefectura General Naval

PATROL FORCES

1 RIVER DEFENCE VESSEL (PGR)

Name	No	Builders	Commissioned
PARAGUAY	C 1	Odero, Genoa	May 1931

Displacement, tons: 636 standard; 865 full load
Dimensions, feet (metres): 231 × 35 × 5.3 *(70 × 10.7 × 1.7)*
Main machinery: 2 boilers; 2 Parsons turbines; 3,800 hp *(2.83 MW)*; 2 shafts
Speed, knots: 17. **Range, n miles:** 1,700 at 16 kt
Complement: 86
Guns: 4—4.7 in *(120 mm)* (2 twin). 3—3 in *(76 mm).* 2—40 mm.
Mines: 6.
Radars: Navigation *(Paraguay)*; I-band.

Comment: Refitted in 1975. Has 0.5 in side armour plating and 0.3 in on deck. Still in restricted operational service with boiler problems. Plans to re-engine with diesels have not yet been implemented and the ship is probably non-operational. Based at Asunción. Gun tubs on either side of bridge can be fitted with single 20 mm guns.
UPDATED

ITAIPÚ   4/2003, *Hartmut Ehlers /* 0567473

PARAGUAY and TENIENTE FARINA   5/2003*, *Hartmut Ehlers /* 0587791

PARAGUAY   5/2000, *Hartmut Ehlers /* 0105192

1 ITAIPÚ CLASS (RIVER DEFENCE VESSEL) (PBR)

Name	No	Builders	Commissioned
ITAIPÚ	P 05 (ex-P 2)	Arsenal de Marinha, Rio de Janeiro	2 Apr 1985

Displacement, tons: 365 full load
Dimensions, feet (metres): 151.9 × 27.9 × 4.6 *(46.3 × 8.5 × 1.4)*
Main machinery: 2 MAN V6V16/18TL diesels; 1,920 hp(m) *(1.41 MW)*· 2 shafts
Speed, knots: 14. **Range, n miles:** 6,000 at 12 kt
Complement: 40 (9 officers) plus 30 marines
Guns: 1 Bofors 40 mm/60. 2—81 mm mortars. 4—12.7 mm MGs.
Radars: Navigation: I-band.
Helicopters: Platform for 1 HB 350B or equivalent.

Comment: Ordered late 1982. Launched 16 March 1984. Same as Brazilian Roraima class. Has some hospital facilities. Based at Asunción.
VERIFIED

2 BOUCHARD CLASS (PATROL SHIPS) (PBR)

Name	No	Builders	Commissioned
NANAWA (ex-*Bouchard* M 7)	P 02 (ex-P 01, ex-M 1)	Rio Santiago Naval Yard	27 Jan 1937
TENIENTE FARINA (ex-*Py* M 10)	P 04 (ex-P 03, ex-M 3)	Rio Santiago Naval Yard	1 July 1939

Displacement, tons: 450 standard; 620 normal; 650 full load
Dimensions, feet (metres): 197 × 24 × 8.5 *(60 × 7.3 × 2.6)*
Main machinery: 2 sets MAN 2-stroke diesels; 2,000 hp(m) *(1.47 MW)*; 2 shafts
Speed, knots: 16. **Range, n miles:** 6,000 at 12 kt
Complement: 70
Guns: 4 Bofors 40 mm/60 (2 twin). 2—12.7 mm MGs.
Mines: 1 rail.
Radars: Navigation: I-band.

Comment: Former Argentinian minesweepers of the Bouchard class. Launched on 20 March 1936 and 31 March 1938 respectively. Transferred from the Argentine Navy to the Paraguayan Navy; *Nanawa* recommissioned 14 March 1964; *Teniente Farina* 6 May 1968. Based at Asunción. A third ship, *Capitán Meza*, is used as a barracks ship.
UPDATED

NANAWA   6/1990, *Paraguay Navy /* 0081514

NANAWA   5/2000, *Hartmut Ehlers /* 0105194

1 RIVER PATROL CRAFT (PBR)

Name	No	Builders	Commissioned
CAPITÁN CABRAL	P 01 (ex-P 04, ex-A 1)	Werf-Conrad, Haarlem	1908
(ex-*Triunfo*)			

Displacement, tons: 180 standard; 206 full load
Dimensions, feet (metres): 107.2 × 23.5 × 6.7 *(32.7 × 7.2 × 2.0)*
Main machinery: 1 Caterpillar 3408 diesel; 360 hp *(269 kW)*; 1 shaft
Speed, knots: 9
Complement: 25
Guns: 1 Bofors 40 mm/60. 2 Oerlikon 20 mm. 2—12.7 mm MGs.
Radars: Navigation: I-band.

Comment: Former tug. Launched in 1907. Still in excellent condition. Vickers guns were replaced and a diesel engine fitted by Arsenal de Marina in 1984. Based at Asunción.
UPDATED

CAPITÁN CABRAL   4/2003, Hartmut Ehlers / 0567474

2 MODIFIED HAI OU CLASS (PBF)

CAPITÁN ORTIZ P 06   **TENIENTE ROBLES** P 07

Displacement, tons: 47 full load
Dimensions, feet (metres): 70.8 × 18 × 3.3 *(21.6 × 5.5 × 1)*
Main machinery: 2 MTU 12V 331 TC82 diesels; 2,605 hp(m) *(1.92 MW)* sustained; 2 shafts
Speed, knots: 36. **Range, n miles:** 700 at 32 kt
Complement: 10
Guns: 1—20 mm Type 75. 3—12.7 mm MGs.
Radars: Surface search: I-band.

Comment: Developed by Taiwan from Dvora class hulls and presented as a gift in 1996. It is possible that one of these craft was one of the two original Dvora hulls acquired by Taiwan.
UPDATED

CAPITÁN ORTIZ   4/2003, Hartmut Ehlers / 0567475

2 RIVER PATROL CRAFT (PBR)

YHAGUY P 08   **TEBICUARY** P 09

Displacement, tons: 25
Dimensions, feet (metres): 52.8 × 14.8 × 2.6 *(16.1 × 4.5 × 0.8)*
Main machinery: Caterpillar diesel; 800 hp *(596 kW)*
Speed, knots: 40
Radars: Surface search: Furuno; I-band.

Comment: Two former Taiwan coast guard patrol boats transferred 23 June 1999. Capable of 40 kt and armed with two 7.26 mm MGs. Two sister craft transferred to Gambia in 1999.
UPDATED

TEBICUARY   4/2003, Hartmut Ehlers / 0567476

5 RIVER PATROL CRAFT (PBR)

LP 7-11 (ex-P 7-11)

Displacement, tons: 18 full load
Dimensions, feet (metres): 48.2 × 10.2 × 2.6 *(14.7 × 3.1 × 0.8)*
Main machinery: 2 GM 6-71 diesels; 340 hp *(254 kW)*; 2 shafts
Speed, knots: 12. **Range, n miles:** 240 at 12 kt
Complement: 4
Guns: 2—12.7 mm MGs.

Comment: Built by Arsenal de Marina, Paraguay. *LP 7* launched March 1989, *LP 8-9* in February 1990 and *LP 10-11* in October 1991. The programme was then aborted.
UPDATED

LP 10   4/2003, Hartmut Ehlers / 0567477

5 TYPE 701 CLASS (PBR)

LP 101   LP 102   LP 103   LP 104   LP 106

Displacement, tons: 15 full load
Dimensions, feet (metres): 42.5 × 12.8 × 3 *(13 × 3.9 × 0.9)*
Main machinery: 2 diesels; 500 hp *(373 kW)*; 2 shafts
Speed, knots: 20
Complement: 7
Guns: 2—12.7 mm MGs.

Comment: Built by Sewart in 1970. Delivered 1967-71. *LP 105* is in reserve.
UPDATED

LP 104   4/2003, Hartmut Ehlers / 0567478

LAND-BASED MARITIME AIRCRAFT

Notes: The Naval Aviation inventory includes four fixed wing aircraft (one Cessna Centurion 210M, two Cessna 310 and one Cessna 410) in addition to the two Helibras Esquilo. Four further Robinson R44 helicopters have not been ordered, as previously reported.

Numbers/Type: 2 Helibras HB 350B Esquilo.
Operational speed: 125 kt *(232 km/h)*.
Service ceiling: 10,000 ft *(3,050 m)*.
Range: 390 n miles *(720 km)*.
Role/Weapon systems: Support helicopter for riverine patrol craft. Delivered in July 1985.
VERIFIED

ESQUILO   5/2000, Hartmut Ehlers / 0105198

AUXILIARIES

Notes: In addition to the craft listed, there are three LCVPs (EDVP 1-3), two service craft (Arsenal 1 and 2) one utility launch *(Teniente Cabrera)*, one suction dredger *(Teniente Oscar Carreras Saguier)*, one floating crane *(Grua Flotante)* and one floating dry dock *(Dique Flotante* (ex-AFDL 26)).

EDVP-03   4/2003*, Hartmut Ehlers / 0587790

1 TRAINING SHIP/TRANSPORT (AK/AX)

Name	No	Builders	Commissioned
GUARANI	—	Tomas Ruiz de Velasco, Bilbao	Feb 1968

Measurement, tons: 714 gross; 1,047 dwt
Dimensions, feet (metres): 240.3 × 36.3 × 11.9 *(73.6 × 11.1 × 3.7)*
Main machinery: 1 MWM diesel; 1,300 hp(m) *(956 kW)*; 1 shaft
Speed, knots: 13
Complement: 21
Cargo capacity: 1,000 tons

Comment: Refitted in 1975 after a serious fire in the previous year off the coast of France. Used to spend most of her time acting as a freighter on the Asunción-Europe run, commercially operated for the Paraguayan Navy. Since 1991 she has only been used for river service and for training cruises Asunción-Montevideo. Operational status is doubtful.

UPDATED

GUARANI   *4/2003, Hartmut Ehlers /* 0567480

1 RIVER TRANSPORT (AKL)

TENIENTE HERREROS (ex-*Presidente Stroessner*) T 1

Displacement, tons: 420 full load
Dimensions, feet (metres): 124 × 29.5 × 7.2 *(37.8 × 9 × 2.2)*
Main machinery: 2 MWM diesels; 330 hp(m) *(243 kW)*
Speed, knots: 10
Complement: 10
Cargo capacity: 120 tons

Comment: Built by Arsenal de Marina in 1964.

VERIFIED

TENIENTE HERREROS   *5/1991, Paraguay Navy /* 0081518

1 PRESIDENTIAL YACHT (MYAC)

3 de FEBRERO (ex-*26 de Febrero*)

Displacement, tons: 98.5 full load
Dimensions, feet (metres): 92.2 × 19.7 × 5.2 *(28.1 × 6.0 × 1.6)*
Main machinery: 2 Rolls Royce; 517 hp *(386 kW)*; 2 shafts
Speed, knots: 11. **Range, n miles:** 1,350 at 11 kt
Complement: 6 + 8 guests

Comment: Built by Naval Arsenal Asunción and launched in 1972. Entered service in 1982.

VERIFIED

3 de FEBRERO   *4/2003, Hartmut Ehlers /* 0567479

1 HYDROGRAPHIC LAUNCH (YGS)

SUBOFICIAL ROGELIO LESME LH 1

Displacement, tons: 16 full load
Dimensions, feet (metres): 65.5 × 10.2 × 2.6 *(14.7 × 3.1 × 0.8)*
Main machinery: 1 Mercedes-Benz diesel; 100 hp *(74 kW)*; 1 shaft
Speed, knots: 13
Complement: 5

Comment: Built in 1958.

VERIFIED

TUGS

3 TUGS (YTM/YTL)

TRIUNFO R 4 (ex-YTL 567)   **ESPERANZA** R 7
ANGOSTURA R 5 (ex-YTL 211)

Displacement, tons: 70 full load
Dimensions, feet (metres): 65 × 16.4 × 7.5 *(19.8 × 5 × 2.3)*
Main machinery: 1 Caterpillar 3408 diesel; 360 hp *(269 kW)*; 1 shaft
Speed, knots: 9
Complement: 5

Comment: Harbour tugs transferred under MAP in the 1960s and 1970s. Details given are for R 4 and R 5. R 7 is a 20 ton vessel.

UPDATED

TRIUNFO   *4/2003, Hartmut Ehlers /* 0567481

Peru
ARMADA PERUANA

Country Overview

The Republic of Peru is situated in western South America. With an area of 496,225 square miles it has borders to the north with Ecuador and Colombia, to the east with Brazil and Bolivia and to the south with Chile. It has a coastline of 1,303 n miles with the Pacific Ocean. Lima is capital and largest city and is served by the port of Callao. There are further ports at Salaverry, Paita, Chimbote, Matarani, Ilo and San Juan. Inland, Iquitos is linked to the Atlantic Ocean by the Amazon River. Lake Titicaca is also an important waterway. Peru has not claimed an EEZ but is one of a few coastal states which claims a 200 n mile territorial sea.

Headquarters Appointments

Commander of the Navy:
 Admiral José Luis Noriega Lores
Chief of the Naval Staff:
 Vice Admiral Frank Boyle Alvarado
Inspector General:
 Vice Admiral Jorge Ampuero Trabuco
Commander Pacific Operations (Callao):
 Vice Admiral Carlos Tubino Arias Schrieber
Commander Amazon Operations (Iquitos):
 Vice Admiral Juan Sierralta Fait
Commander Marines Force:
 Rear Admiral Manuel Perez Zumaeta

Commander, Naval Aviation:
 Rear Admiral José Quevedo de Lama
Commander, Submarines:
 Rear Admiral Oleg Kriljenko Arnillas
Commander Surface Forces:
 Rear Admiral Ernesto Schroth Mier y Proano

Personnel

(a)   2005: 25,000 (2,500 officers)
(b)   2 years' national service

Bases and Organisation

There are Five Naval Zones: 1st Piura, 2nd Callao, 3rd Arequipa, 4th Puerto Maldonado, 5th Iquitos
Pacific Naval Force (HQ Callao)
Amazon River Force (HQ Iquitos)
Lake Titicaca Patrol Force (HQ at Puno)
Callao-Main naval base; dockyard with shipbuilding capacity, 1 dry dock, 3 floating docks, 1 floating crane; training schools
Iquitos-River base for Amazon Flotilla; small building yard, repair facilities, floating dock
La Punta (naval academy), Chimbote, Paita, Talara, Puno (Lake Titicaca), Madre de Dios (river base), Piura, El Salto, Bayovar, Pimental, Pacasmayo, Salaverry, Mollendo, Matarani, Ilo, Puerto Maldonado, Inapari, Pucallpa and El Estrecho.

Marines

The Peruvian Marines comprise 3,500 men commanded by a Rear Admiral. Headquarters at Ancon. The force includes a Marine Brigade, the Amphibious Support Group and Marine Special Forces. The Marine Brigade has three battalions: First Battalion — Guarnición de Marina; Second Battalion — Guardia Chalaca; Third Battalion (including Fire Support Group armed with 122 mm howitzer and 120 mm mortar and Engineer Support company) — Vencedores de Punta Melpelo. The Amphibious Support Group is composed of the Vehicles and Motor Transport battalions. The Marines Special Forces include a Commando company and anti-terrorist unit. Additionally, the Peruvian Marines have jungle battalions at Iquitos and Pucallpa (BIS 1 and BIS 2).

Prefix to Ships' Names

BAP (Buque Armada Peruana).

Coast Guard

A separate service set up in 1975 with a number of light forces transferred from the Navy.

PENNANT LIST

Submarines			Frigates			CM 25	Larrea		Auxiliaries	
SS 31	Angamos		FM 51	Carvajal		CM 26	Sanchez Carrión		ABA 332	Barcaza Cisterna de Agua
SS 32	Antofagasta		FM 52	Villavisencio					ABH 302	Morona
SS 33	Pisagua		FM 53	Montero		**Amphibious Forces**			ABH 306	Puno
SS 34	Chipana		FM 54	Mariategui					ACA 111	Calayeras
SS 35	Islay		FM 55	Aguirre		DT 141	Paita		ACP 118	Noguera
SS 36	Arica		FM 56	Palacios		DT 142	Pisco		ACP 119	Gauden
						DT 143	Callao		ARB 120	Mejia
			Patrol Forces			DT 144	Eten		ARB 121	Huerta
Cruisers									ARB 123	Guardian Rios
			CF 11	Amazonas					ARB 126	Dueñas
CLM 81	Almirante Grau		CF 12	Loreto		**Survey Ships**			ARB 128	Olaya
			CF 13	Marañon					ARB 129	Selendon
			CF 14	Ucayali		AH 171	Carrasco		ATC 131	Mollendo
Destroyers			CM 21	Velarde		AH 172	Stiglich		ATP 152	Talara
			CM 22	Santillana		AEH 174	Macha		ATP 153	Lobitos
DM 74	Ferré		CM 23	De los Heros		AEH 175	Carrillo		ATP 157	Supe
			CM 24	Herrera		AEH 176	Melo		ART 322	San Lorenzo

SUBMARINES

Notes: Replacement of the current submarine flotilla is under consideration.

6 ANGAMOS/ISLAY CLASS (TYPES 209 and 1200) (SSK)

Name	No	Builders	Laid down	Launched	Commissioned
ANGAMOS (ex-*Casma*)	SS 31	Howaldtswerke, Kiel	15 July 1977	31 Aug 1979	19 Dec 1980
ANTOFAGASTA	SS 32	Howaldtswerke, Kiel	3 Oct 1977	19 Dec 1979	20 Feb 1981
PISAGUA	SS 33	Howaldtswerke, Kiel	15 Aug 1978	19 Oct 1980	12 July 1983
CHIPANA	SS 34	Howaldtswerke, Kiel	1 Nov 1978	19 May 1981	20 Sep 1982
ISLAY	SS 35	Howaldtswerke, Kiel	15 Mar 1971	11 Oct 1973	29 Aug 1974
ARICA	SS 36	Howaldtswerke, Kiel	1 Nov 1971	5 Apr 1974	21 Jan 1975

Displacement, tons: 1,185 surfaced; 1,290 dived
Dimensions, feet (metres): 183.7 × 20.3 × 17.9
(56 × 6.2 × 5.5)
Main machinery: Diesel-electric; 4 MTU 12V 493 AZ80 GA31L diesels; 2,400 hp(m) *(1.76 MW)* sustained; 4 Siemens alternators; 1.7 MW; 1 Siemens motor; 4,600 hp(m) *(3.38 MW)* sustained; 1 shaft
Speed, knots: 11 surfaced/snorting; 21.5 dived
Range, n miles: 240 at 8 kt
Complement: 35 (5 officers) *(Islay* and *Arica)*; 31 (others)

Torpedoes: 8—21 in *(533 mm)* tubes. 14 AEG SST4; wire-guided; active/passive homing to 12/28 km *(6.5/15 n miles)* at 35/23 kt; warhead 260 kg. Swim-out discharge.
Countermeasures: ESM: Radar warning.
Weapons control: Sepa Mk 3 or Signaal Sinbad M8/24 *(Angamos* and *Antofagasta).*
Radars: Surface search: Thomson-CSF Calypso; I-band.
Sonars: Atlas Elektronik CSU 3; active/passive search and attack; medium/high frequency.

Thomson Sintra DUUX 2C or Atlas Elektronik PRS 3; passive ranging.

Programmes: Two Type 209 (SS 35-36) ordered 1969. Two further Type 209 boats (SS 31-32) ordered 12 August 1976. Two Type 1200 (SS 33-34) ordered 21 March 1977.
Designed by Ingenieurkontor, Lübeck for construction by Howaldtswerke, Kiel and sale by Ferrostaal, Essen all acting as a consortium.
Modernisation: Sepa Mk 3 fire control fitted progressively from 1986. A programme to replace diesel generators has been reported but not confirmed. *Angamos* reported to have been modernised with new batteries, sonar and EW suite.
Structure: A single-hull design with two ballast tanks and forward and after trim tanks. Fitted with snort and remote machinery control. The single screw is slow revving, very high-capacity batteries with GRP lead-acid cells and battery cooling-by Wilh Hagen and VARTA. Fitted with two periscopes and Omega receiver. Foreplanes retract. Diving depth, 250 m *(820 ft)*.
Operational: Endurance, 50 days. Four are in service, two in refit or reserve at any one time. *Angamos* took part in multinational exercises in mid-2003 during which she achieved 126 days at sea. **UPDATED**

CHIPANA   *10/1997* / 0052694

ISLAY   *6/2004* *, Peruvian Navy* / 1121516

CRUISERS

1 DE RUYTER CLASS (CG/CLM)

Name	No	Builders	Laid down	Launched	Commissioned
ALMIRANTE GRAU (ex-De Ruyter)	CLM 81	Wilton-Fijenoord, Schiedam	5 Sep 1939	24 Dec 1944	18 Nov 1953

Displacement, tons: 12,165 full load
Dimensions, feet (metres): 624.5 × 56.7 × 22
 (190.3 × 17.3 × 6.7)
Main machinery: 4 Werkspoor-Yarrow boilers; 2 De Schelde-
 Parsons turbines; 85,000 hp (62.5 MW); 2 shafts
Speed, knots: 32. **Range, n miles:** 7,000 at 12 kt
Complement: 953 (49 officers)

Missiles: SSM: 8 OTO Melara/Matra Otomat Mk 2 (TG 1) ❶;
 active radar homing to 80 km (43.2 n miles) at 0.9
 Mach; warhead 210 kg; sea-skimmer for last 4 km (2.2 n
 miles).
Guns: 8 Bofors 6 in (152 mm)/53 (4 twin) ❷; 15 rds/min to
 26 km (14 n miles); weight of shell 46 kg.
 4 Otobreda 40 mm/70 (2 twin) ❸; 120 rds/min to 12.5 km
 (6.8 n miles); weight of shell 0.96 kg.
 4 Bofors 40 mm/70 ❹; 300 rds/min to 12 km (6.6 n miles);
 weight of shell 0.96 kg.
Countermeasures: Decoys: 2 Dagaie and 1 Sagaie chaff
 launchers.
Combat data systems: Signaal Sewaco PE SATCOM ❺.
Weapons control: 2 Lirod 8 optronic directors ❻.
Radars: Air search: Signaal LW08 ❼; D-band.
Surface search/target indication: Signaal DA08 ❽; E/F-band.
Navigation: Racal Decca 1226; I-band.
Fire control: Signaal WM25 ❾; I/J-band (for 6 in guns); range
 46 km (25 n miles).
 Signaal STIR ❿; I/J/K-band; range 140 km (76 n miles) for
 1 m² target.

Programmes: Transferred by purchase from Netherlands
 7 March 1973 and commissioned to Peruvian Navy 23 May
 1973.
Modernisation: Taken in hand for a two and a half year
 modernisation at Amsterdam Dry Dock Co in March 1985.
 This was to include reconditioning of mechanical
 and electrical engineering systems, fitting of SSM and
 SAM, replacement of electronics and fitting of one CSEE
 Sagaie and two Dagaie launchers. In 1986 financial
 constraints limited the work but much had been done
 to update sensors and fire-control equipment. Sailed for
 Peru 23 January 1988 without her secondary gun armament,
 which was completed at Sima Yard, Callao. The plan to
 retrofit Exocets from the Daring class to replace the
 Otomat launchers, appears to have been abandoned
 due to lack of funding. Sonar has been removed. SATCOM
 fitted aft.
Operational: Expected to be decommissioned in 2008.

UPDATED

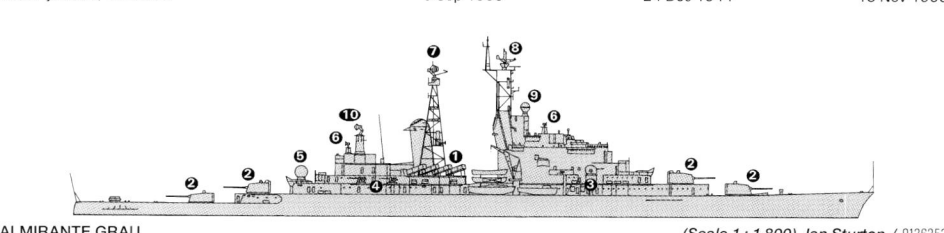

ALMIRANTE GRAU

(Scale 1 : 1,800), Ian Sturton / 0126352

ALMIRANTE GRAU
1/1999 / 0081521

DESTROYERS

1 DARING CLASS (DDGH)

Name	No	Builders	Laid down	Launched	Commissioned
FERRÉ (ex-Decoy)	DM 74	Yarrow, Glasgow	22 Sep 1946	29 Mar 1949	28 Apr 1953

Displacement, tons: 2,800 standard; 3,600 full load
Dimensions, feet (metres): 390 × 43 × 18
 (118.9 × 13.1 × 5.5)
Main machinery: 2 Foster-Wheeler boilers; 650 psi (45.7 kg/
 cm²); 850°F (454°C); 2 English Electric turbines; 54,000 hp
 (40 MW); 2 shafts
Speed, knots: 32. **Range, n miles:** 3,000 at 20 kt
Complement: 297

Missiles: SSM: 8 Aerospatiale MM 38 Exocet ❶; inertial cruise;
 active radar homing to 42 km (23 n miles) at 0.9 Mach;
 warhead 165 kg; sea-skimmer.
Guns: 6 (3 twin) Vickers 4.5 in (114 mm)/45 Mk 6 ❷; 20 rds/min
 to 19 km (10.4 n miles); weight of shell 25 kg.
 4 Breda 40 mm/70 (2 twin) ❸; 300 rds/min to 12.5 km (6.8 n
 miles); weight of shell 0.96 kg.
Radars: Air/surface search: Plessey AWS 1 ❹; E/F-band; range
 110 km (60 n miles).
Surface search: Racal Decca TM 1226; I-band.
Fire control: Selenia RTN 10X ❺; I/J-band.

Helicopters: Platform only.

Programmes: Purchased from UK in 1969 and refitted by
 Cammell Laird, Birkenhead.
Modernisation: A helicopter deck was fitted in the 1970s.
Operational: One of the class deleted in 1993. Expected to be
 decommissioned in 2006.

UPDATED

FERRÉ

(Scale 1 : 1,200), Ian Sturton / 0126349

FERRÉ
11/2004 *, Globke Collection / 1047864

FRIGATES

4 + 2 (2) CARVAJAL (LUPO) CLASS (FFGHM)

Name	No	Builders	Laid down	Launched	Commissioned
CARVAJAL	FM 51	Fincantieri, Riva Trigoso	8 Aug 1974	17 Nov 1976	5 Feb 1979
VILLAVISENCIO	FM 52	Fincantieri, Riva Trigoso	6 Oct 1976	7 Feb 1978	25 June 1979
MONTERO	FM 53	SIMA, Callao	Oct 1978	8 Oct 1982	25 July 1984
MARIATEGUI	FM 54	SIMA, Callao	1979	8 Oct 1984	10 Oct 1987
AGUIRRE (ex-Orsa)	FM 55 (ex-F 567)	Fincantieri, Muggiano	1 Aug 1977	1 Mar 1979	1 Mar 1980
PALACIOS (ex-Lupo)	FM 56 (ex-F 564)	Fincantieri, Riva Trigoso	11 Oct 1974	29 July 1976	12 Sep 1977

Displacement, tons: 2,208 standard; 2,500 full load
Dimensions, feet (metres): 371.3 × 37.1 × 12.1
(113.2 × 11.3 × 3.7)
Main machinery: CODOG; 2 GE/Fiat LM 2500 gas turbines;
50,000 hp *(37.3 MW)* sustained; 2 GMT A 230.20 M diesels;
8,000 hp(m) *(5.88 MW)* sustained; 2 shafts; LIPS cp props
Speed, knots: 35. **Range, n miles:** 3,450 at 20.5 kt
Complement: 185 (20 officers)

Missiles: SSM: 8 OTO Melara/Matra Otomat Mk 2 (TG 1) ❶;
active radar homing to 80 km *(43.2 n miles)* at 0.9 Mach;
warhead 210 kg; sea-skimmer for last 4 km *(2.2 n miles)*.
SAM: Selenia Elsag Albatros octuple launcher ❷; 8 Aspide; semi-
active radar homing to 13 km *(7 n miles)* at 2.5 Mach; height
envelope 15-5,000 m *(49.2-16,405 ft)*; warhead 30 kg.
An SA-N-10 launcher (MPG-86) may be fitted on the stern.
Raytheon NATO Sea Sparrow (FM 55-56) Mk 29 octuple
launcher; semi-active radar homing to 14.6 km *(8 n miles)* at
2.5 Mach; warhead 39 kg.
Guns: 1 OTO Melara 5 in *(127 mm)*/54 ❸; 45 rds/min to 16 km
(8.7 n miles); weight of shell 32 kg.

4 Breda 40 mm/70 (2 twin) ❹; 300 rds/min to 12.5 km *(6.8 n
miles)*; weight of shell 0.96 kg.
Torpedoes: 6—324 mm ILAS or Mk 32 (FM 55-56) (2 triple) tubes
❺. Whitehead A244; anti-submarine; active/passive homing
to 7 km *(3.8 n miles)* at 33 kt; warhead 34 kg (shaped
charge).
Countermeasures: Decoys: 2 Breda 105 mm SCLAR 20-
barrelled trainable launchers ❻; multipurpose; chaff to 5 km
(2.7 n miles); illuminants to 12 km *(6.6 n miles)*; HE
bombardment.
ESM: Elettronica Lambda; intercept.
Combat data systems: Selenia IPN-10 or IPN 20 (FM 55-56)
action data automation.
Weapons control: 2 Elsag Mk 10 Argo with NA-21 directors.
Dardo system for 40 mm.
Radars: Air search: Selenia RAN 10S (FM 52-56) ❼; E/F-band.
Surface search: Selenia RAN 11LX ❽; D/I-band. Signaal LW 08
(FM 51); D-band.
SMA SPQ-2F (FM 55-56); I-band.
Navigation: SMA 3 RM 20R; I-band.
SMA SPN-748 (FM 55-56); I-band.

Fire control: 2 RTN 10X ❾; I/J-band.
2 RTN 20X ❿; I/J-band (for Dardo).
Sonars: EDO 610E or Raytheon DE 1160B (FM 55-56); hull-
mounted; active search and attack; medium frequency.

Helicopters: 1 Agusta AB 212ASW ⓫. 1 Agusta ASH-3D Sea
King (deck only) (FM 51 and 54).

Programmes: *Montero* and *Mariategui* were the first major
warships to be built on the Pacific Coast of South America,
although some equipment was provided by Fincantieri.
Palacios and *Aguirre* formally transferred from the Italian Navy
on 3 November 2004, without ammunition, torpedoes, SSM
and helicopters, following eight-month refits at Fincantieri,
Muggiano. Two further decommissioned Lupo class (possibly
Sagittario and *Perseo*) may be similarly transferred at a later
date.
Modernisation: FM 51 and FM 54 have had flight deck
extensions in order to operate Sea Kings from the deck
although they cannot be stowed in the hangar. It is reported
that modernisation of the propulsion systems is also under
consideration SA-N-10 (MPG-86) may be fitted on the sterns
of two ships. LW 08 replaced RAN 10S in FM 51 in 2003.
Structure: FM 51-54 differ from those built for Italian service by
having a fixed hangar and higher 40 mm mounts. The SAM
system is also different. The ships were commissioned with a
step-down from the flight deck to the stern although this has
been modified in FM 51 and 54.
Operational: Helicopter provides an over-the-horizon targeting
capability for SSM. HIFR facilities fitted in 1989 allow
refuelling of Sea King helicopters.

UPDATED

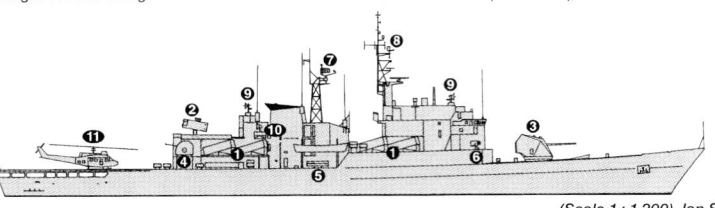

MARIATEGUI
(Scale 1 : 1,200), Ian Sturton / 0105275

MONTERO
7/2002, Chris Sattler / 0533237

MARIATEGUI
2000, Peruvian Navy / 0105206

SHIPBORNE AIRCRAFT

Notes: There are also five Bell 206B training helicopters.

Numbers/Type: 3 Agusta AB 212ASW.
Operational speed: 106 kt *(196 km/h).*
Service ceiling: 14,200 ft *(4,330 m).*
Range: 230 n miles *(425 km).*
Role/Weapon systems: ASW and surface search helicopter for smaller escorts. Sensors: Selenia search radar, Bendix ASQ-18 dipping sonar, ECM. Weapons: ASW; two Mk 46 or 244/S torpedoes or depth bombs.

VERIFIED

AB 212   *6/1994, Peruvian Navy /* 0081527

Numbers/Type: 3 Agusta-Sikorsky ASH-3D Sea King.
Operational speed: 120 kt *(222 km/h).*
Service ceiling: 12,200 ft *(3,720 m).*
Range: 630 n miles *(1,165 km).*
Role/Weapon systems: ASW helicopter; can be operated from two FFGs. Sensors: Selenia search radar, Bendix ASQ-18 dipping sonar, sonobuoys. Weapons: ASW; four Mk 46 or 244/S torpedoes or depth bombs or mines. ASV; two AM 39 Exocet missiles.

VERIFIED

SEA KING   *2000, Peruvian Navy /* 0105209

LAND-BASED MARITIME AIRCRAFT (FRONT LINE)

Notes: (1) There are also four Mi-8T transport helicopters.
(2) One Fokker F-27 is used by the Navy and two by the coast Guard for maritime surveillance.
(3) There are three Antonov AN-32 transport aircraft.
(4) Five Beech T-34C are used for training.

Mi-8T   *2000, Peruvian Navy /* 0105208

Numbers/Type: 3 Beech Super King Air B200T.
Operational speed: 282 kt *(523 km/h).*
Service ceiling: 35,000 ft *(10,670 m).*
Range: 2,030 n miles *(3,756 km).*
Role/Weapon systems: Coastal surveillance and EEZ patrol duties. Sensors: Search radar, cameras. Weapons: Unarmed.

UPDATED

PATROL FORCES

Notes: Procurement of ten hovercraft for river policing is under consideration.

6 VELARDE (PR-72P) CLASS
(FAST ATTACK CRAFT—MISSILE) (CM/PGGFM)

Name	No	Builders	Launched	Commissioned
VELARDE	CM 21	SFCN, France	16 Sep 1978	25 July 1980
SANTILLANA	CM 22	SFCN, France	11 Sep 1978	25 July 1980
DE LOS HEROS	CM 23	SFCN, France	20 May 1979	17 Nov 1980
HERRERA	CM 24	SFCN, France	16 Feb 1979	10 Feb 1981
LARREA	CM 25	SFCN, France	12 May 1979	16 June 1981
SANCHEZ CARRIÓN	CM 26	SFCN, France	28 June 1979	14 Sep 1981

Displacement, tons: 470 standard; 560 full load
Dimensions, feet (metres): 210 × 27.4 × 5.2 *(64 × 8.4 × 2.6)*
Main machinery: 4 SACM AGO 240 V16 M7 or 4 MTU 12V 595 diesels; 22,200 hp(m) *(16.32 MW)* sustained; 4 shafts
Speed, knots: 37. **Range, n miles:** 2,500 at 16 kt
Complement: 36 plus 10 spare

Missiles: SSM: 4 Aerospatiale MM 38 Exocet; inertial cruise; active radar homing to 42 km *(23 n miles)* at 0.9 Mach; warhead 165 kg; sea-skimmer.
SAM: An SA-N-10 launcher (MPG-86) may be fitted on the stern.
Guns: 1 OTO Melara 3 in *(76 mm)*/62; 85 rds/min to 16 km *(8.7 n miles)*; weight of shell 6 kg.
2 Breda 40 mm/70 (twin); 300 rds/min to 12.5 km *(6.8 n miles)*; weight of shell 0.96 kg.
Countermeasures: ESM: Thomson-CSF DR 2000; intercept.
Weapons control: CSEE Panda director. Vega system.
Radars: Surface search: Thomson-CSF Triton; G-band; range 33 km *(18 n miles)* for 2 m² target.
Navigation: Racal Decca 1226; I-band.
Fire control: Thomson-CSF/Castor II; I/J-band; range 15 km *(8 n miles)* for 1 m² target.

Programmes: Ordered late 1976. Hulls of *Velarde, De los Heros, Larrea* subcontracted to Lorient Naval Yard, the others being built at Villeneuve-la-Garenne. Classified as corvettes.
Modernisation: Three of the class re-engined in 2000. Remainder to follow.

UPDATED

VELARDE   *11/2004 *, Globke Collection /* 1047863

2 MARAÑON CLASS (RIVER GUNBOATS) (CF/PGR)

Name	No	Builders	Commissioned
MARAÑON	CF 13 (ex-CF 401)	John I Thornycroft & Co Ltd	July 1951
UCAYALI	CF 14 (ex-CF 402)	John I Thornycroft & Co Ltd	June 1951

Displacement, tons: 365 full load
Dimensions, feet (metres): 154.8 wl × 32 × 4 *(47.2 × 9.7 × 1.2)*
Main machinery: 2 British Polar M 441 diesels; 800 hp *(597 kW)*; 2 shafts
Speed, knots: 12. **Range, n miles:** 6,000 at 10 kt
Complement: 40 (4 officers)
Guns: 2—3 in *(76 mm)*/50. 2 Bofors 40 mm/60. 2 Oerlikon 20 mm.

Comment: Ordered early in 1950 and launched 7 March and 23 April 1951 respectively. Employed on police duties in Upper Amazon. Superstructure of aluminium alloy. Based at Iquitos.

VERIFIED

UCAYALI   *1993, Peruvian Navy /* 0081530

MARAÑON   *1995, Peruvian Navy /* 0081529

2 LORETO CLASS (RIVER GUNBOATS) (CF/PGR)

Name	No	Builders	Commissioned
AMAZONAS	CF 11 (ex-CF 403)	Electric Boat Co, Groton	1935
LORETO	CF 12 (ex-CF 404)	Electric Boat Co, Groton	1935

Displacement, tons: 250 standard
Dimensions, feet (metres): 145 × 22 × 4 *(44.2 × 6.7 × 1.2)*
Main machinery: 2 diesels; 750 hp(m) *(551 kW)*; 2 shafts
Speed, knots: 15. **Range, n miles:** 4,000 at 10 kt
Complement: 35 (5 officers)
Guns: 1—3 in *(76 mm)*. 3 Bofors 40 mm/60. 2 Oerlikon 20 mm.

Comment: Launched in 1934. In Upper Amazon Flotilla, based at Iquitos. The after 3 in gun has been replaced by a third 40 mm.

VERIFIED

LORETO *4/1997, Peruvian Navy /* 0012880

AMPHIBIOUS FORCES

Notes: There are plans for up to three 300 ft LSLs to be locally built when funds are available.

4 PAITA (TERREBONNE PARISH) CLASS (LSTH)

Name	No	Builders	Commissioned
PAITÁ (ex-*Walworth County* LST 1164)	DT 141	Ingalls SB	26 Oct 1953
PISCO (ex-*Waldo County* LST 1163)	DT 142	Ingalls SB	17 Sep 1953
CALLAO (ex-*Washoe County* LST 1165)	DT 143	Ingalls SB	30 Nov 1953
ETEN (ex-*Traverse County* LST 1160)	DT 144	Bath Iron Works	19 Dec 1953

Displacement, tons: 2,590 standard; 5,800 full load
Dimensions, feet (metres): 384 × 55 × 17 *(117.1 × 16.8 × 5.2)*
Main machinery: 4 GM 16-278A diesels; 6,000 hp *(4.48 MW)*; 2 shafts
Speed, knots: 15. **Range, n miles:** 15,000 at 9 kt
Complement: 116
Military lift: 2,000 tons; 395 troops
Guns: 5 Bofors 40 mm/60 (2 twin, 1 single).
Radars: Navigation: I-band.

Comment: Four transferred from USA on loan 7 August 1984, recommissioned 4 March 1985. Have small helicopter platform. Original 3 in guns replaced by 40 mm. Lease extended by grant aid in August 1989, again in August 1994, and again in April 1999. All are active.

UPDATED

PAITÁ *11/2004*, Globke Collection /* 1047862

SURVEY AND RESEARCH SHIPS

Notes: AH 177 is a 5 ton fast survey craft.

1 INSHORE SURVEY CRAFT (AGSC/EH)

Name	No	Builders	Commissioned
MACHA	AEH 174	SIMA, Chimbote	Apr 1982

Displacement, tons: 53 full load
Dimensions, feet (metres): 64.9 × 17.1 × 3 *(19.8 × 5.2 × 0.9)*
Main machinery: 2 diesels; 2 shafts
Speed, knots: 13
Complement: 8 (2 officers)

Comment: Side scan sonar for plotting bottom contours. EH (Embarcacion Hidrográfica).

VERIFIED

MACHA *2000, Peruvian Navy /* 0105213

1 DOKKUM CLASS (AGSC/EH)

Name	No	Builders	Commissioned
CARRASCO (ex-*Abcoude*)	AH 171 (ex-M 810)	Smulders, Schiedam	18 May 1956

Displacement, tons: 373 standard; 453 full load
Dimensions, feet (metres): 152.9 × 28.9 × 7.5 *(46.6 × 8.8 × 2.3)*
Main machinery: 2 Fijenoord MAN V64 diesels; 2,500 hp(m) *(1.84 MW)*; 2 shafts
Speed, knots: 16. **Range, n miles:** 2,500 at 10 kt
Complement: 27-36
Guns: 2 Oerlikon 20 mm/70 (1 twin).
Radars: Navigation: Racal Decca TM 1229C; I-band.

Comment: Service with the Netherlands Navy as a minesweeper included modernisation in the mid-1970s and a life prolonging refit in the late 1980s. *Carrasco* placed in reserve in 1993 and transferred to Peru 16 July 1994. The ship has been acquired for hydrographic duties. Two more were planned to follow in mid-1996 but the transfer was cancelled.

UPDATED

CARRASCO *2000, Peruvian Navy /* 0105211

2 VAN STRAELEN CLASS (AGSC/EH)

Name	No	Builders	Commissioned
CARRILLO (ex-*van Hamel*)	AEH 175	De Vries, Amsterdam	14 Oct 1960
MELO (ex-*van der Wel*)	AEH 176	De Vries, Amsterdam	6 Oct 1961

Displacement, tons: 169 full load
Dimensions, feet (metres): 108.6 × 18.2 × 5.2 *(33.1 × 5.6 × 1.6)*
Main machinery: 2 GM diesels; 1,100 hp(m) *(808 kW)* sustained; 2 shafts
Speed, knots: 13
Complement: 17 (2 officers)
Guns: 1—20 mm

Comment: Both built as inshore minesweepers. Acquired 23 February 1985 for conversion with new engines and survey equipment.

VERIFIED

MELO *2000, Peruvian Navy /* 0105212

2 RIVER VESSELS (AGSC/AH)

Name	No	Builders		Commissioned
MORONA	ABH 302	Sima, Iquitos		1976
STIGLICH	AH 172	Sima, Iquitos		1981

Displacement, tons: 230 full load
Dimensions, feet (metres): 112.2 × 25.9 × 5.6 *(34.2 × 7.9 × 1.7)*
Main machinery: 2 Detroit 12V-71TA diesels; 840 hp *(616 kW)* sustained; 2 shafts
Speed, knots: 15
Complement: 28 (2 officers)

Comment: *Stiglich* is based at Iquitos for survey work on the Upper Amazon. *Morona* is used as a hospital craft and has a red cross on her superstructure. *UPDATED*

STIGLICH *6/1999, Peruvian Navy /* 0081533

AUXILIARIES

Notes: (1) All auxiliaries may be used for commercial purposes if not required for naval use.
(2) There is also a 55 ton sail training ship *Marte*.
(3) There are three small river hospital craft: *Corrientes* (ABH 303), *Curaray* (ABH 304) and *Pastaza* (ABH 305).

1 MOLLENDO CLASS (TRANSPORT) (AOR)

Name	No	Builders	Commissioned
MOLLENDO (ex-*Ilo*)	ATC 131	SIMA, Callao	15 Dec 1971

Displacement, tons: 18,400 full load
Measurement, tons: 13,000 dwt
Dimensions, feet (metres): 507.7 × 67.3 × 27.2 *(154.8 × 20.5 × 8.3)*
Main machinery: 1 Burmeister & Wain 6K47 diesel; 11,600 hp(m) *(8.53 MW)*; 1 shaft
Speed, knots: 15.6
Complement: 60
Cargo capacity: 13,000 tons

Comment: Sister ship *Rimac* has been scrapped. *UPDATED*

MOLLENDO *6/2001, Maritime Photographic /* 0126392

1 TALARA CLASS (REPLENISHMENT TANKER) (AOT)

Name	No	Builders	Launched	Commissioned
TALARA	ATP 152	SIMA, Callao	9 July 1976	23 Jan 1978

Displacement, tons: 30,000 full load
Measurement, tons: 25,000 dwt
Dimensions, feet (metres): 561.5 × 82 × 31.2 *(171.2 × 25 × 9.5)*
Main machinery: 2 Burmeister & Wain 6K47EF diesels; 12,000 hp(m) *(8.82 MW)*; 1 shaft
Speed, knots: 15.5
Cargo capacity: 35,662 m³

Comment: Capable of underway replenishment at sea from the stern. *Bayovar*, of this class, launched 18 July 1977, having been originally ordered by Petroperu (State Oil Company) and transferred to the Navy while building. Sold back to Petroperu in 1979 and renamed *Pavayacu*. A third, *Trompeteros*, of this class has been built for Petroperu. *VERIFIED*

TALARA *2000, Peruvian Navy /* 0105214

1 SEALIFT CLASS (TANKER) (AOT)

Name	No	Builders	Commissioned
LOBITOS (ex-*Sealift Caribbean*)	ATP 153 (ex-ATP 159, ex-TAOT 174)	Bath Iron Works	10 Feb 1975

Displacement, tons: 34,100 full load
Measurement, tons: 27,736 dwt; 15,979 grt
Dimensions, feet (metres): 587 × 84 × 34.6 *(178.9 × 25.6 × 10.6)*
Main machinery: 2 diesels; 1 shaft; cp prop; bow thruster
Speed, knots: 12.5. **Range, n miles:** 7,500 at 15 kt
Complement: 57
Cargo capacity: 185,000 barrels oil

Comment: Originally built for the US Military Sealift Command. Returned to owners in 1995 and acquired by Peru in 1998. No RAS gear. *VERIFIED*

LOBITOS *6/2001, Maritime Photographic /* 0126391

1 SUPE CLASS TANKER (AOT)

Name	No	Builders	Commissioned
SUPE (ex-*Taxiarchos*)	ATP 157	Voldnes A/S Fosnaveg Noruega	1965

Measurement, tons: 1,138 grt
Dimensions, feet (metres): 219.2 × 31.2 × 12.9 *(66.8 × 9.52 × 3.95)*
Main machinery: Diesel propulsion; 1 shaft
Speed, knots: 11
Complement: 17
Cargo capacity: 7,500 barrels oil

Comment: Acquired from the USA in 1996. *VERIFIED*

SUPE *6/2001, Maritime Photographic /* 0126390

4 HARBOUR TANKERS (FUEL/WATER) (YW/YO)

CALAYERAS ACA 111 (ex-*YW 128*)   GAUDEN ACP 119 (ex-*YO 171*)
NOGUERA ACP 118 (ex-*YO 221*)   BARCAZA CISTERNA DE AGUA ABA 332 (ex-*113*)

Displacement, tons: 1,235 full load
Dimensions, feet (metres): 174 × 32 × 13.3 *(52.3 × 9.8 × 4.1)*
Main machinery: 1 GM diesel; 560 hp *(418 kW)*; 1 shaft
Speed, knots: 8
Complement: 23
Cargo capacity: 200,000 gallons
Radars: Navigation: Raytheon; I-band.

Comment: Details given are for first three. *YO 221* transferred from USA to Peru January 1975; *YO 171* 20 January 1981; *YW 128* 26 January 1985. *ABA 332* is a water tanker built in Peru in 1972, capacity 300 tons. *VERIFIED*

GAUDEN *1988, Peruvian Navy*

1 TORPEDO RECOVERY VESSEL (YPT)

Name	No	Builders	Commissioned
SAN LORENZO	ART 322	Lürssen/Burmeister	1 Dec 1981

Displacement, tons: 58 standard; 65 full load
Dimensions, feet (metres): 82.7 × 18.4 × 5.6 *(25.2 × 5.6 × 1.7)*
Main machinery: 2 MTU 8V 396 TC82 diesels; 1,740 hp(m) *(1.28 MW)* sustained; 2 shafts
Speed, knots: 19. **Range, n miles:** 500 at 15 kt
Complement: 9

Comment: Can carry four long or eight short torpedoes.
VERIFIED

SAN LORENZO　　　　　9/1981, *Lürssen Werft* / 0081534

1 LAKE HOSPITAL CRAFT (AH)

Name	No	Builders	Commissioned
PUNO (ex-*Yapura*)	ABH 306	J Watt Co, Thames Iron Works	18 May 1872

Comment: Stationed on Lake Titicaca. 500 grt and has a diesel engine. Sadly the second of the class was finally paid off in 1990.
VERIFIED

PUNO　　　　　8/1999, *A Campanera i Rovira* / 0081535

5 FLOATING DOCKS

ADF 104　　　　　106-109

Displacement, tons: 1,900 *(106)*; 5,200 *(107)*; 600 *(108)*; 18,000 *(109)*; 4,500 tons *(104)*

Comment: *106* (ex-US *AFDL 33*) transferred 1959; *107* (ex-US *ARD 8*) transferred 1961; *108* built in 1951; *109* built in 1979; *104* built in 1991.
VERIFIED

TUGS

Notes: (1) There are three river tugs *Tapuina* AER 180, *Gaudin* AER 186 and *Zambrano* AER 187.
(2) There are also five small harbour tugs *Mejia* ARB 120, *Huerta* ARB 121, *Dueñas* ARB 126, *Olaya* ARB 128 and *Selendon* ARB 129.

1 CHEROKEE CLASS (SALVAGE TUG) (ATS)

Name	No	Builders	Commissioned
GUARDIAN RIOS (ex-*Pinto* ATF 90)	ARB 123	Cramp, Philadelphia, PA	1 Apr 1943

Displacement, tons: 1,640 full load
Dimensions, feet (metres): 205 × 38.5 × 17 *(62.5 × 11.7 × 5.2)*
Main machinery: Diesel-electric; 4 GM 12-278 diesels; 4,400 hp *(3.28 MW)*; 4 generators; 1 motor; 3,000 hp *(2.24 MW)*; 1 shaft
Speed, knots: 16.5. **Range, n miles:** 6,500 at 16 kt
Complement: 99

Comment: Transferred from USA on loan in 1960, sold 17 May 1974. Fitted with powerful pumps and other salvage equipment.
VERIFIED

GUARDIAN RIOS　　　　　1993, *Peruvian Navy*

COAST GUARD

Notes: DCB 350-358 are 6 m Cougar 22 class inshore patrol craft. DCB 358 is a one ton harbour patrol launch.

5 RIO CANETE CLASS (LARGE PATROL CRAFT) (WPB)

Name	No	Builders	Commissioned
RIO NEPEÑA	PC 243	SIMA, Chimbote	1 Dec 1981
RIO TAMBO	PC 244	SIMA, Chimbote	1982
RIO OCOÑA	PC 245	SIMA, Chimbote	1983
RIO HUARMEY	PC 246	SIMA, Chimbote	1984
RIO ZAÑA	PC 247	SIMA, Chimbote	12 Feb 1985

Displacement, tons: 296 full load
Dimensions, feet (metres): 167 × 24.8 × 5.6 *(50.9 × 7.4 × 1.7)*
Main machinery: 4 Bazán MAN V8V diesels; 5,640 hp(m) *(4.15 MW)*; 2 shafts
Speed, knots: 23. **Range, n miles:** 3,050 at 17 kt
Complement: 39 (4 officers)
Guns: 1 Oerlikon 20 mm.
Radars: Surface search: Decca 1226; I-band.

Comment: Have aluminium alloy superstructures. The prototype craft was scrapped in 1990. *Rio Ocoña* completed refit in July 1996 and the rest of the class were refitted at one per year.
UPDATED

RIO OCOÑA　　　　　11/2004*, *Globke Collection* / 1047861

1 PGM 71 CLASS (LARGE PATROL CRAFT) (PB)

Name	No	Builders	Commissioned
RIO CHIRA	PC 223 (ex-*PGM 111*)	SIMA, Callao	June 1972

Displacement, tons: 147 full load
Dimensions, feet (metres): 118.2 × 21 × 6 *(36.0 × 6.4 × 1.8)*
Main machinery: 2 Detroit GN-71 diesels; 1,450 hp *(1.08 MW)*; 2 shafts
Speed, knots: 18. **Range, n miles:** 1,500 at 10 kt
Complement: 15
Guns: 1—12.7 mm MG.
Radars: Surface search: Raytheon; I-band.

Comment: Acquired from the Navy in 1975. Paid off in 1994 but back in service again in 1997, with refurbished engines.
VERIFIED

RIO CHIRA　　　　　2000, *Peruvian Coast Guard* / 0105218

1 RIVER PATROL CRAFT (PBR)

Name	No	Builders	Commissioned
RIO PIURA	PC 242 (ex-*P 252*)	Viareggio, Italy	5 Sep 1960

Displacement, tons: 55 full load
Dimensions, feet (metres): 65.7 × 17 × 3.2 *(20 × 5.2 × 1)*
Main machinery: 2 GM 8V-71 diesels; 460 hp *(344 kW)* sustained; 2 shafts
Speed, knots: 15. **Range, n miles:** 1,000 at 16 kt
Complement: 7 (1 officer)
Guns: 1—12.7 mm MG. 1 Oerlikon 20 mm.
Radars: Navigation: Raytheon; I-band.

Comment: Ordered in 1959. Armament changed in 1992. Refitted in 1996.
VERIFIED

RIO PIURA　　　　　2000, *Peruvian Coast Guard* / 0105219

16 LAKE and RIVER PATROL CRAFT (PBR)

RIO HULLAGA	PF 260	RIO ITAYA	PF 270
RIO SANTIAGO	PF 261	RIO PATAYACU	PF 271
RIO PUTUMAYO	PF 262	RIO ZAPOTE	PF 272
RIO NANAY	PF 263	RIO CHAMBIRA	PF 273
RIO NAPO	PF 264	RIO TAMBOPATA	PF 274
RIO YAVARI	PF 265	RIO RAMIS	PL 290
RIO MATADOR	PF 266	RIO ILAVE	PL 291
RIO PACHITEA	PF 267	RIO AZANGARO	PL 292

Displacement, tons: 5 full load
Dimensions, feet (metres): 32.8 × 11.2 × 2.6 *(10 × 3.4 × 0.8)*
Main machinery: 2 Perkins diesels; 480 hp *(358 kW)*; 2 shafts
Speed, knots: 15. **Range, n miles:** 450 at 28 kt
Complement: 3
Guns: 1—12.7 mm MG.
Radars: Surface search: Raytheon 2800; I-band.

Comment: Details given are for PL 290-292. Based at Puno on Lake Titicaca. GRP hulls built in 1982. The remainder are of various types. PF 260-263 are 4-5 ton craft, PF 264-267 are 2 tons and PF 270-274 are approximately 1 ton.

UPDATED

RIO ILAVE   *2000, Peruvian Coast Guard* / 0105220

RIO NAPO   *2000, Peruvian Coast Guard* / 0105221

10 PORT PATROL CRAFT (PBF/PB)

RIO SUPE PP 210		QUILCA PP 214		RIO MAJES PP 233
RIO VITOR PP 211		PUCUSANA PP 215		RIO VIRU PP 235
MANCORA PP 212		RIO SANTA PP 232		RIO LURIN PP 236
HUAURA PP 213				

Comment: Miscellaneous craft. PP 210-211 are Cougar 40 kt fast patrol craft of 2 tons, PP 212-214 are Cougar 25 (1 ton) and PP 215 of 4 tons. All six acquired in 1996. PP 232-233 and PP 235-236 are 14-15 ton craft. These were acquired in 1999.

UPDATED

RIO VIRU   *2000, Peruvian Coast Guard* / 0105222

RIO SUPE   *2000, Peruvian Coast Guard* / 0105223

6 DAUNTLESS CLASS (PBR)

CHICAMA PC 216	CHORRILLOS PC 218	CAMANA PC 220
HUANCHACO PC 217	CHANCAY PC 219	CHALA PC 221

Displacement, tons: 12 full load
Dimensions, feet (metres): 40 × 14 × 4.4 *(12.2 × 4.3 × 1.3)*
Main machinery: 2 Caterpillar 3208TA diesels; 870 hp *(650 kW)*; 2 shafts
Speed, knots: 27. **Range, n miles:** 600 at 18 kt
Complement: 5
Guns: 1—12.7 mm MG. 1—7.62 mm MG.
Radars: Surface search: Furuno 821; I-band.

Comment: Ordered in February 2000 under FMS funding. First pair delivered in August 2000 remainder in November 2000.

UPDATED

DAUNTLESS (US colours)   *8/2000, SeaArk Marine* / 0093586

5 + (7) ZORRITOS CLASS (RIVER PATROL CRAFT) (PBR)

ZORRITOS PC 222	PUNTA ARENAS PC 225	JULI PC 293
SANTA ROSA PC 224	PACASMAYO PC 226	

Displacement, tons: To be announced
Dimensions, feet (metres): 39.4 × 9.8 × ? *(12.0 × 3.0 × ?)*
Main machinery: 2 Caterpillar 275
Speed, knots: 25

Comment: Built by SIMA Chimbote and delivered 2003-04. A class of 12 is expected.

NEW ENTRY

Philippines

Country Overview

The Republic of the Philippines was formally proclaimed in 1946. Situated between Taiwan to the north and Indonesia and Malaysia to the south, the country comprises about 7,100 islands with a total coastline of 19,597 n miles with the South China, Philippine and Celebes Seas. Eleven islands, Bohol, Cebu, Leyte, Luzon, Masbate, Mindanao, Mindoro, Negros, Palawan, Panay, and Samar, contain the majority of the population. Most remaining islands are less than 1 square mile in area. The capital, principal city and port is Manila. Other important ports include Davao, Cebu and Zamboanga. An archipelagic state, territorial seas (12 n miles) are claimed. A 200 n mile EEZ has also been claimed but the limits have not been defined.

Headquarters Appointments

Flag Officer-in-Command:
    Vice Admiral Ernesto H De Leon
Commander Fleet:
    Rear Admiral Gilmer B Batestil
Commandant Coast Guard:
    Vice Admiral Arturo Gosingan
Commandant Marines:
    Major General Emmanuel Teodosio

Diplomatic Representation

Defence Attaché in London:
    Lieutenant Colonel C B Boquiren

Personnel

(a)   2005: 20,500 Navy; 8,700 Marines; 3,500 Coast Guard
(b)   Reserves: 17,000

Organisation

The Naval Headquarters is at Manila. The fleet is divided into functional units including the Ready Force, Patrol Force, Service Force, Assault Craft Force, Naval Air Group and Naval Special Warfare Group. There are six operational areas of responsibility: Southern Luzon; Northern Luzon; Central; West; Western Mindanao and Eastern Mindanao. The Coast Guard was transferred to the Department of Transport and Communication in 1998. There are eight Coast Guard Districts, 47 stations and 154 Coast Guard Detachment units.

Marine Corps

Marines comprise three tactical brigades composed of 10 tactical battalions, one support regiment, a service group, a guard battalion and a reconnaissance battalion. Headquarters at Ternate, Manila Bay. Deployed in Mindanao and Palawan.

Bases

Main: Cavite.
Operational: San Vicente, Mactan, Ternate.
Stations: Cebu, Davao, Legaspi, Bonifacio, Tacloban, San Miguel, Ulugan, Balabne, Puerto Princesa, Pagasa.

Prefix to Ships' Names

BRP: Barko Republika Pilipinas

Strength of the Fleet

Type	Active	Building
Frigates	(1)	—
Corvettes	13	(2)
Fast Attack Craft	6	—
Large Patrol Craft	5	1 (3)
Coastal Patrol Craft	37	2
LST/LSV Transports	8	—
LCM/LCU/RUC/LCVP	44	—
Repair Ship	1	—
Tankers	4	—
Coast Guard		
Tenders	4	—
Patrol Craft	58	1

PENNANT LIST

Frigates		PG 104	Bagong Silang	PG 388	Manuel Gomez	LT 516	Kalinga Apayao
		PG 110	Tomas Batilo	PG 389	Testimo Figuracion	LC 550	Bacolod City
PF 11	Rajah Humabon	PG 111	Bonny Serrano	PG 390	José Loor SR	LC 551	Dagupan City
		PG 112	Bienvenido Salting	PG 392	Juan Magluyan	AT 25	Ang Pangulo
Corvettes		PG 114	Salvador Abcede	PG 393	Florenca Nuno	AW 33	Lake Bulusan
		PG 115	Ramon Aguirre	PG 394	Alberto Navaret	AW 34	Lake Paoay
PS 19	Miguel Malvar	PG 116	Nicolas Mahusay	PG 395	Felix Apolinario	AF 72	Lake Taal
PS 20	Magat Salamat	PG 140	Emilo Aguinaldo	PG 396	Brigadier Abraham Campo	AF 78	Lake Buhi
PS 22	Sultan Kudarat	PG 141	Antonio Luna	PG 840	Conrado Yap	AC 90	Mactan
PS 23	Datu Marikudo	PG 370	José Andrada	PG 842	Tedorico Dominado Jr	AD 617	Yakal
PS 28	Cebu	PG 371	Enrique Jurado	PG 843	Cosme Acosta		
PS 29	Negros Occidental	PG 372	Alfredo Peckson	PG 844	José Artiaga Jr	**Coast Guard**	
PS 31	Pangasinan	PG 374	Simeon Castro	PG 846	Nicanor Jimenez		
PS 32	Iloilo	PG 375	Carlos Albert	PG 847	Leopoldo Regis	AE 46	Cape Bojeador
PS 35	Emilio Jacinto	PG 376	Heracleo Alano	PG 848	Leon Tadina	PG 61	Agusan
PS 36	Apolinario Mabini	PG 377	Liberato Picar	PG 849	Loreto Danipog	PG 62	Catanduanes
PS 37	Artemio Ricarte	PG 378	Hilario Ruiz	PG 851	Apollo Tiano	PG 63	Romblon
PS 38	General Mariano Alvares	PG 379	Rafael Pargas	PG 853	Sulpicio Hernandez	PG 64	Palawan
PS 70	Quezon	PG 380	Nestor Reinoso			AT 71	Mangyan
PS 74	Rizal	PG 381	Dioscoro Papa	**Auxiliaries**		AU 75	Bessang Pass
		PG 383	Ismael Lomibao			AE 79	Limasawa
Patrol Forces		PG 384	Leovigildo Gantioque	LT 86	Zamboanga Del Sur	AG 89	Kalinga
		PG 385	Federico Martir	LT 87	South Cotabato	AU 100	Tirad Pass
PG 101	Kagitingan	PG 386	Filipino Flojo	LT 501	Laguna	001	San Juan
PG 102	Bagong Lakas	PG 387	Anastacio Cacayorin	LT 504	Lanao Del Norte	002	Esda II

FRIGATES

Notes: *Rajah Lakandula*, paid off in 1988, is still afloat as an alongside HQ and depot ship.

1 CANNON CLASS (FF)

Name	No	Builders	Laid down	Launched	Commissioned
RAJAH HUMABON (ex-*Hatsuhi* DE 263, ex-*Atherton* DE 169)	PF 11 (ex-*PF 78*)	Norfolk Navy Yard, Portsmouth, VA	14 Jan 1943	27 May 1943	29 Aug 1943

Displacement, tons: 1,390 standard; 1,750 full load
Dimensions, feet (metres): 306 × 36.6 × 14
　(93.3 × 11.2 × 4.3)
Main machinery: Diesel-electric; 2 GM EMD 16V-645E7 diesels;
　5,800 hp *(4.32 MW)*; 4 generators; 2 motors; 2 shafts
Speed, knots: 18. **Range, n miles:** 6,000 at 14 kt
Complement: 165

Guns: 3 US 3 in *(76 mm)*/50 Mk 22; 20 rds/min to 12 km *(6.6 n
　miles)*; weight of shell 6 kg.
　6 US/Bofors 40 mm/56 (3 twin). 4 Oerlikon 20 mm/70;
　2—12.7 mm MGs.
Depth charges: 8 K-gun Mk 6 projectors; range 160 m; warhead
　150 kg; 1 rack.
Weapons control: Mk 52 GFCS with Mk 41 rangefinder for 3 in
　guns. 3 Mk 51 Mod 2 GFCS for 40 mm.
Radars: Surface search: Raytheon SPS-5; G/H-band.
Navigation: RCA/GE Mk 26; I-band.
Sonars: SQS-17B; hull-mounted; active search and attack;
　medium/high frequency.

Programmes: *Hatsuhi* originally transferred by the US to Japan
　14 June 1955 and paid off June 1975 reverting to US Navy.
　Transferred to Philippines 23 December 1978. Towed to
　South Korea 1979 for overhaul and modernisation.
　Recommissioned 27 February 1980. A sister ship *Datu
　Kalantiaw* lost during Typhoon Clara 20 September 1981.
Modernisation: Upgrade plans have been suspended.
Operational: Hedgehog A/S mortars have been reported.

UPDATED　RAJAH HUMABON

10/2001, Chris Sattler / 0126280

CORVETTES

3 JACINTO (PEACOCK) CLASS (FS)

Name	No	Builders	Launched	Commissioned	Recommissioned
EMILIO JACINTO (ex-*Peacock*)	PS 35 (ex-*P 239*)	Hall Russell, Aberdeen	1 Dec 1982	14 July 1984	4 Aug 1997
APOLINARIO MABINI (ex-*Plover*)	PS 36 (ex-*P 240*)	Hall Russell, Aberdeen	12 Apr 1983	20 July 1984	4 Aug 1997
ARTEMIO RICARTE (ex-*Starling*)	PS 37 (ex-*P 241*)	Hall Russell, Aberdeen	11 Sep 1983	10 Aug 1984	4 Aug 1997

Displacement, tons: 763 full load
Dimensions, feet (metres): 204.1 × 32.8 × 8.9
　(62.6 × 10 × 2.7)
Main machinery: 2 Crossley Pielstick 18 PA6 V 280 diesels;
　14,000 hp(m) *(10.6 MW)* sustained; 2 shafts; 1 retractable
　Schottel prop; 181 hp *(135 kW)*
Speed, knots: 25. **Range, n miles:** 2,500 at 17 kt
Complement: 31 (6 officers) plus 7 spare berths

Guns: 1—3 in *(76 mm)*/62 OTO Melara compact; 85 rds/min to
　16 km *(8.6 n miles)* anti-surface; 12 km *(6.5 n miles)* anti-
　aircraft; weight of shell 6 kg.
　4 FN 7.62 mm MGs.
Weapons control: BAe Sea Archer GSA-7 for 76 mm gun.
Radars: Navigation: Kelvin Hughes Type 1006; I-band.

Programmes: Letter of Intention to purchase from the UK signed
　in November 1996. Transferred 1 August 1997 after sailing
　from Hong Kong on 1 July 1997. Others of the class in service
　with the navy of the Irish Republic.
Modernisation: An upgrade programme was agreed in 2002.
　Phase one is being lead by QinetiQ in 2004-06 and is to
　include a new fire-control system, an upgrade for the 76 mm
　gun, the installation of a 25 mm gun aft and new navigation
　and communication systems. Phases two and three are to
　involve new propulsion and safety systems.
Structure: Fitted with telescopic cranes, loiter drive and
　replenishment at sea equipment. In UK service, two fast
　pursuit craft were carried.
Operational: These ships are the workhorses of the fleet. Based
　at Cavite.

UPDATED　EMILIO JACINTO

6/2002 / 0534067

2 AUK CLASS (FS)

Name	No	Builders	Laid down	Launched	Commissioned
RIZAL (ex-*Murrelet* MSF 372)	PS 74 (ex-PS 69)	Savannah Machine & Foundry Co, GA	24 Aug 1944	29 Dec 1944	21 Aug 1945
QUEZON (ex-*Vigilance* MSF 324)	PS 70	Associated Shipbuilders, Seattle, WA	28 Nov 1942	5 Apr 1943	28 Feb 1944

Displacement, tons: 1,090 standard; 1,250 full load
Dimensions, feet (metres): 221.2 × 32.2 × 10.8
  (67.4 × 9.8 × 3.3)
Main machinery: Diesel-electric; 2 GM EMD 16V-645E6 diesels;
  5,800 hp *(4.32 MW)*; 2 generators; 2 motors; 2 shafts
Speed, knots: 18. **Range, n miles:** 5,000 at 14 kt
Complement: 80 (5 officers)

Guns: 2 US 3 in *(76 mm)*/50 Mk 26; 20 rds/min to 12 km *(6.6 n
  miles)*; weight of shell 6 kg.
  4 US/Bofors 40 mm/56 (2 twin); 160 rds/min to 11 km
  (5.9 n miles); weight of shell 0.9 kg.
  2 Oerlikon 20 mm (twin). 2—12.7 mm MGs.
Depth charges: 2 Mk 9 racks.
Radars: Surface search: Raytheon SPS-5C; G/H-band.
Navigation: DAS 3; I-band.
Sonars: SQS-17B; hull-mounted; active search and attack; high
  frequency.

Programmes: *Rizal* transferred from the US to the Philippines on
  18 June 1965 and *Quezon* on 19 August 1967.
Modernisation: Upgrade plans have been suspended.
Structure: Upon transfer the minesweeping gear was removed
  and a second 3 in gun fitted aft.
Operational: Both ships were to have been deleted in 1994 but
  have been retained until new class of OPVs is built.

UPDATED   RIZAL

10/2001, *Chris Sattler* / 0534068

1 CYCLONE CLASS (COASTAL PATROL SHIP) (PB)

Name	No	Builders	Commissioned
GENERAL MARIANO ALVARES (ex-*Cyclone*)	PS 38 (ex-PC 1)	Bollinger, Lockport	7 Aug 1993

Displacement, tons: 386 full load
Dimensions, feet (metres): 179 × 25.9 × 7.9 *(54.6 × 7.9 × 2.4)*
Main machinery: 4 Paxman Valenta 16RP200CM diesels; 13,400 hp *(10 MW)* sustained; 4 shafts
Speed, knots: 35. **Range, n miles:** 2,500 at 12 kt
Complement: 28 (4 officers) plus 8
Guns: 1 Bushmaster 25 mm Mk 38. 1 Bushmaster 25 mm/87 Mk 96 (aft). 4—12.7 mm MGs.
  4—7.62 mm MGs. 2—40 mm Mk 19 grenade launchers (MGs and grenade launchers are
  interchangeable).
Countermeasures: Decoys: 2 Mk 52 sextuple and/or Wallop Super Barricade Mk 3 chaff
  launchers.
ESM: Privateer APR-39; radar warning.
Weapons control: Marconi VISTAR IM 405 IR system.
Radars: Surface search: 2 Sperry RASCAR; E/F/I/J-band.
Sonars: Wesmar; hull-mounted; active; high frequency.

Programmes: Transferred from the USN to the Philippines in February 2004 following refit at
  Bollinger. Recommissioned on 8 March 2004.
Structure: Design based on Vosper Thornycroft Ramadan class modified for USN requirements
  including 1 in armour on superstructure. The craft has a slow speed loiter capability and has
  been modified to incorporate a semi-dry well, boat ramp and stern gate to facilitate deployment
  and recovery of a fully loaded RIB while the ship is making way.

UPDATED

GENERAL MARIANO ALVARES   3/2004*, *US Embassy, Manila* / 0563762

CEBU   5/2000, *M Declerck* / 0105225

Radars: Surface search: SPS-50 (PS 23). SPS-21D (PS 19, 28). CRM-NIA-75 (PS 29, 31, 32).
  SPS-53A (PS 20).
Navigation: RCA SPN-18; I/J-band.

Programmes: Five transferred from the US to the Philippines in July 1948 (PS 28-32); PS 22 to
  South Vietnam from US Navy on 29 November 1961, PS 20 in April 1962, PS 19 on 11 July
  1966, and PS 23 in June 1970. PS 19, 20 and 22 to Philippines November 1975 and PS 23
  5 April 1976.
Modernisation: PS 19, 22, 31 and 32 refurbished in 1990-91, PS 23 and 28 in 1992 and the last
  pair in 1996/97.
Structure: First three were originally fitted as rescue ships (PCER). A/S equipment has been
  removed or is inoperable. PS 20 has some minor structural differences having been built as an
  Admirable class MSF.
Operational: PS 29 is probably not operational.

UPDATED

LAND-BASED MARITIME AIRCRAFT

Notes: There is a Cessna 177 Cardinal.

Numbers/Type: 7 PADC (Pilatus Britten-Norman) Islander F27MP.
Operational speed: 150 kt *(280 km/h)*.
Service ceiling: 18,900 ft *(5,760 m)*.
Range: 1,500 n miles *(2,775 km)*.
Role/Weapon systems: Short-range MR and SAR aircraft. First purchased in 1989. Three
  transferred from the Air Force. Upgrade of avionics and communications is planned. Sensors:
  Search radar, cameras. Weapons: Unarmed.

VERIFIED

8 PCE 827 CLASS (FS)

Name	No	Builders	Commissioned
MIGUEL MALVAR (ex-*Ngoc Hoi*, Ex-*Brattleboro* PCER 852)	PS 19	Pullman Standard Car Co, Chicago	26 May 1944
MAGAT SALAMAT (ex-*Chi Lang II*, ex-*Gayety* MSF 239)	PS 20	Winslow Marine Co, Seattle, WA	14 June 1944
SULTAN KUDARAT (ex-*Dong Da II*, ex-*Crestview* PCER 895)	PS 22	Willamette Iron & Steel Corporation, Portland, OR	30 Oct 1943
DATU MARIKUDO (ex-*Van Kiep II*, ex-*Amherst* PCER 853)	PS 23	Pullman Standard Car Co, Chicago	16 June 1944
CEBU (ex-*PCE 881*)	PS 28	Albina E and M Works, Portland, OR	31 July 1944
NEGROS OCCIDENTAL (ex-*PCE 884*)	PS 29	Albina E and M Works, Portland, OR	30 Mar 1944
PANGASINAN (ex-*PCE 891*)	PS 31	Willamette Iron & Steel Corp, Portland, OR	15 June 1944
ILOILO (ex-*PCE 897*)	PS 32	Willamette Iron & Steel Corp, Portland, OR	6 Jan 1945

Displacement, tons: 640 standard; 914 full load
Dimensions, feet (metres): 184.5 × 33.1 × 9.5 *(56.3 × 10.1 × 2.9)*
Main machinery: 2 GM 12-278A diesels; 2,200 hp *(1.64 MW)*; 2 shafts
Speed, knots: 15. **Range, n miles:** 6,600 at 11 kt
Complement: 85 (8 officers)

Guns: 1 US 3 in *(76 mm)*/50; 20 rds/min to 12 km *(6.6 n miles)*; weight of shell 6 kg.
  2 to 6 US/Bofors 40 mm/56 (single or 1—3 twin); 160 rds/min to 11 km *(5.9 n miles)*; weight
  of shell 0.9 kg.
  2 Oerlikon 20 mm/70; 800 rds/min to 2 km.

F-27MP   10/2001, *Adolfo Ortigueira Gil* / 0567482

Numbers/Type: 7 PADC (MBB) BO 105C.
Operational speed: 145 kt *(270 km/h)*.
Service ceiling: 17,000 ft *(5,180 m)*.
Range: 355 n miles *(657 km)*.
Role/Weapon systems: Sole shipborne helicopter; some shore-based for SAR; some commando
  support capability. Purchased at the rate of one per year from 1986 to 1992. Upgrade of
  avionics and communications is planned. Sensors: Some fitted with search radar. Weapons:
  Unarmed.

VERIFIED

PATROL FORCES

Notes: (1) The Navy operates one Swift Mk 1 (50 ft), three Mk 2 (51 ft) and four Mk 3 (65 ft) coastal patrol craft. Details as for identical craft operated by the Coast Guard.
(2) Plans to procure three offshore patrol craft have been suspended although they remain a long-term aspiration.

2 + 1 AGUINALDO CLASS (LARGE PATROL CRAFT) (PBO)

Name	No	Builders	Commissioned
EMILIO AGUINALDO	PG 140	Cavite, Sangley Point	21 Nov 1990
ANTONIO LUNA	PG 141	Cavite, Sangley Point	27 May 1999
—	PG 142	Cavite, Sangley Point	—

Displacement, tons: 236 full load
Dimensions, feet (metres): 144.4 × 24.3 × 5.2 *(44 × 7.4 × 1.6)*
Main machinery: 2 MTU 16V-396TB94 diesels; 3,480 hp *(2.59 MW)* sustained; 2 shafts
Speed, knots: 28. **Range, n miles:** 1,100 at 18 kt
Complement: 58 (6 officers)
Guns: 2 Bofors 40 mm/60. 2 Oerlikon 20 mm. 4—12.7 mm MGs.
Radars: Surface search: Raytheon; I-band.

Comment: Steel hulls of similar design to *Tirad Pass*. First of class launched 23 June 1984 but only completed in 1990. Second laid down 2 December 1990 and launched 23 June 1992. PG 142 laid down on 14 February 1994 and launched in April 2000. While the superstructure is 70 per cent completed, outfitting has been suspended due to budget constraints. For similar reasons, the outlook for completion of the six ship programme looks doubtful as do plans to upgrade existing ships with a SAM system and 76 mm gun.

UPDATED

EMILIO AGUINALDO   6/1993 / 0081540

2 POINT CLASS (PB)

Name	No	Builders	Commissioned
ALBERTO NAVARET (ex-*Point Evans*)	PG 394 (ex-82354)	CG Yard, Maryland	10 Jan 1967
BRIGADIER ABRAHAM CAMPO (ex-*Point Doran*)	PG 396 (ex-82375)	CG Yard, Maryland	1 June 1970

Displacement, tons: 67 full load
Dimensions, feet (metres): 83 × 17.2 × 5.8 *(25.3 × 5.2 × 1.8)*
Main machinery: 2 Caterpillar 3412 diesels; 1,600 hp *(1.19 MW)*; 2 shafts
Speed, knots: 23. **Range, n miles:** 1,500 at 8 kt
Complement: 10
Guns: 2—12.7 mm MGs.
Radars: Surface search: Furuno; I-band.

Comment: PG 394 transferred from US Coast Guard 16 November 1999. Second transferred 22 March 2001. This class is in service with many other navies.

VERIFIED

POINT CLASS (US colours)   4/1992, van Ginderen Collection / 0081549

3 KAGITINGAN CLASS (LARGE PATROL CRAFT) (PB)

Name	No	Builders	Commissioned
KAGITINGAN	P 101	Hamelin SY, Germany	9 Feb 1979
BAGONG LAKAS	PG 102 (ex-P 102)	Hamelin SY, Germany	9 Feb 1979
BAGONG SILANG	PG 104 (ex-P 104)	Hamelin SY, Germany	July 1979

Displacement, tons: 150 full load
Dimensions, feet (metres): 121.4 × 20.3 × 5.6 *(37 × 6.2 × 1.7)*
Main machinery: 2 MTU MB 16V-538TB91 diesels; 2,500 hp(m) *(1.86 MW)* sustained; 2 shafts
Speed, knots: 21
Complement: 30 (4 officers)
Guns: 2—30 mm (twin). 4—12.7 mm MGs. 2—7.62 mm MGs.
Radars: Surface search: I-band.

Comment: Based at Cavite. P 103 paid off and used for spares. All still in service.

UPDATED

BAGONG LAKAS   1993, Philippine Navy

6 TOMAS BATILO (SEA DOLPHIN) CLASS
(FAST ATTACK CRAFT) (PBF)

TOMAS BATILO PG 110	**BIENVENIDO SALTING** PG 112	**RAMON AGUIRRE** PG 115
BONNY SERRANO PG 111	**SALVADOR ABCEDE** PG 114	**NICOLAS MAHUSAY** PG 116

Displacement, tons: 150 full load
Dimensions, feet (metres): 121.4 × 22.6 × 5.6 *(37 × 6.9 × 1.7)*
Main machinery: 2 MTU 20V-538TB91 diesels; 9,000 hp(m) *(6.71 MW)* sustained; 2 shafts
Speed, knots: 38. **Range, n miles:** 600 at 20 kt
Complement: 31 (5 officers)
Guns: 2 Emerson Electric 30 mm (twin); 1,200 rds/min combined to 6 km *(3.2 n miles)*; weight of shell 0.35 kg.
  1 Bofors 40 mm/60. 2 Oerlikon 20 mm.
Weapons control: Optical director.
Radars: Surface search: Raytheon 1645; I-band.

Comment: Transferred from South Korea on 15 June 1995. Part of the PKM 200 series. Different armament to South Korean ships of the same class. Plans to modernise these craft during 2002-03 appear to have been suspended.

UPDATED

BIENVENIDO SALTING   6/1996, Philippine Navy

22 JOSÉ ANDRADA CLASS (COASTAL PATROL CRAFT) (PB)

JOSÉ ANDRADA PG 370	**RAFAEL PARGAS** PG 379	**ANASTACIO CACAYORIN** PG 387
ENRIQUE JURADO PG 371	**NESTOR REINOSO** PG 380	**MANUEL GOMEZ** PG 388
ALFREDO PECKSON PG 372	**DIOSCORO PAPA** PG 381	**TESTIMO FIGURACION** PG 389
SIMEON CASTRO PG 374	**ISMAEL LOMIBAO** PG 383	**JOSÉ LOOR SR** PG 390
CARLOS ALBERT PG 375	**LEOVIGILDO GANTIOQUE** PG 384	**JUAN MAGLUYAN** PG 392
HERACLEO ALANO PG 376	**FEDERICO MARTIR** PG 385	**FLORENCA NUNO** PG 393
LIBERATO PICAR PG 377	**FILIPINO FLOJO** PG 386	**FELIX APOLINARIO** PG 395
HILARIO RUIZ PG 378		

Displacement, tons: 56 full load
Dimensions, feet (metres): 78 × 20 × 5.8 *(23.8 × 6.1 × 1.8)*
Main machinery: 2 Detroit 16V-92TA diesels; 1,380 hp *(1.03 MW)* sustained; 2 shafts
Speed, knots: 28. **Range, n miles:** 1,200 at 12 kt
Complement: 8-12 (1 officer)
Guns: 1 Bushmaster 25 mm or Bofors 40 mm/60.
  4—12.7 mm Mk 26 MGs. 2—7.62 mm M60 MGs.
Radars: Surface search: Raytheon SPS-64(V)2; I-band.

Comment: There are four batches of this class. Batch I (PCF 370-378), Batch II (PCF 379-390), Batch III (PCF 392-393) and Batch IV (PCF 395). The main difference between batches include weapons, electronics and accommodation. First four ordered from Halter Marine in August 1989 under FMS and built at Equitable Shipyards, New Orleans, as were a further four ordered in 1990. Eight more ordered in March 1993 with co-production between Halter Marine and AG&P Shipyard, Batangas. An additional three were ordered in 1995. Built to US Coast Guard standards with an aluminium hull and superstructure. The main gun may be fitted in all after some minor modifications. PG 392 delivered in March 1998, PG 393 in July 1998 and PG 395 on 10 October 2000.

VERIFIED

FLORENCA NUNO   5/2000, van Ginderen Collection / 0105227

TESTIMO FIGURACION   5/2000, M Declerck / 0105226

10 CONRADO YAP (SEA HAWK/KILLER) CLASS
(COASTAL PATROL CRAFT) (PBF)

CONRADO YAP PG 840	**LEOPOLDO REGIS** PG 847
TEDORICO DOMINADO JR PG 842	**LEON TADINA** PG 848
COSME ACOSTA PG 843	**LORETO DANIPOG** PG 849
JOSÉ ARTIAGA JR PG 844	**APOLLO TIANO** PG 851
NICANOR JIMENEZ PG 846	**SULPICIO FERNANDEZ** PG 853

Displacement, tons: 74.5 full load
Dimensions, feet (metres): 83.7 × 17.7 × 6.2 *(25.5 × 5.4 × 1.9)*
Main machinery: 2 MTU 16V-538TB91 diesels; 5,000 hp(m) *(3.72 MW)*; 2 shafts
Speed, knots: 38. **Range, n miles:** 290 at 20 kt
Complement: 15 (3 officers)
Guns: 1 Bofors 40 mm/60. 2 Oerlikon 20 mm (twin) Mk 16.
Radars: Surface search: Raytheon 1645; I-band.

Comment: Type PK 181 built by Korea Tacoma and Hyundai 1975-78. Twelve craft transferred from South Korea 19 June 1993. Eight were commissioned 23 June 1993 and a further four on 23 June 1994. However PC 845 and PC 852 have not been reactivated and are probably used as spares.

VERIFIED

CONRADO YAP CLASS   *1993, Philippine Navy*

SURVEY AND RESEARCH SHIPS

Notes: (1) Survey ships are operated by Coast and Geodetic Survey of Ministry of National Defence and are not naval.
(2) Two research ships *Fort San Antonio* (AM 700) and *Fort Abad* (AM 701) were acquired in 1993.

AUXILIARIES

Notes: All LSTs, LSVs, LCMs and LCUs are classified as Transports.

2 BACOLOD CITY (FRANK S BESSON) CLASS (LSVH)

Name	No	Builders	Commissioned
BACOLOD CITY	LC 550	Moss Point Marine	1 Dec 1993
DAGUPAN CITY (ex-*Cagayan De Oro City*)	LC 551	Moss Point Marine	5 Apr 1994

Displacement, tons: 4,265 full load
Dimensions, feet (metres): 272.8 × 60 × 12 *(83.1 × 18.3 × 3.7)*
Main machinery: 2 GM EMD 16V-645E6 diesels; 5,800 hp *(4.32 MW)* sustained; 2 shafts; bow thruster; 250 hp *(187 kW)*
Speed, knots: 11.6. **Range, n miles:** 6,000 at 11 kt
Complement: 30 (6 officers)
Military lift: 2,280 tons (900 for amphibious operations) of vehicles, containers or cargo, plus 150 troops; 2 LCVPs on davits
Radars: Navigation: Raytheon SPS-64(V)2; I-band.
Helicopters: Platform for 1 BO 105C.

Comment: Contract announced by Trinity Marine 3 April 1992 for two ships with an option on a third which was not taken up. Ro-ro design with 10,500 sq ft of deck space for cargo. Capable of beaching with 4 ft over the ramp on a 1 : 30 offshore gradient with a 900 ton cargo. Similar to US Army vessels but with only a bow ramp. The stern ramp space is used for accommodation for 150 troops and a helicopter platform is fitted over the stern.

VERIFIED

DAGUPAN CITY   *12/1999, Sattler/Steele /* 0081543

DAGUPAN CITY   *12/1999, Sattler/Steele /* 0081544

5 LST 512-1152 CLASS (TRANSPORT SHIPS) (LST)

Name	No	Commissioned
ZAMBOANGA DEL SUR (ex-*Cam Ranh,* ex-*Marion County* LST 975)	LT 86	3 Feb 1945
SOUTH COTABATO (ex-*Cayuga County* LST 529)	LT 87	28 Feb 1944
LAGUNA (ex-*T-LST 230*)	LT 501	3 Nov 1943
LANAO DEL NORTE (ex-*T-LST 566*)	LT 504	29 May 1944
KALINGA APAYAO (ex-*Can Tho,* ex-*Garrett County* AGP 786, ex-LST 786)	LT 516 (ex-AE 516)	28 Aug 1944

Displacement, tons: 1,620 standard; 2,472 beaching; 4,080 full load
Dimensions, feet (metres): 328 × 50 × 14 *(100 × 15.2 × 4.3)*
Main machinery: 2 GM 12-567A diesels; 1,800 hp *(1.34 MW)*; 2 shafts
Speed, knots: 10
Complement: Varies-approx 60-110 (depending upon employment)
Military lift: 2,100 tons. 16 tanks or 10 tanks plus 200 troops

Guns: 6 US/Bofors 40 mm (2 twin, 2 single) or 4 Oerlikon 20 mm (in refitted ships).
Radars: Navigation: Raytheon SPS-64(V)2; I-band.

Programmes: Transferred from US Navy in 1976 with exception of LT 87 and LT 516 which were used as light craft repair ships in South Vietnam and have retained amphibious capability (transferred to Vietnam 1970 and to Philippines 1976, acquired by purchase 5 April 1976). LT 86 transferred (grant aid) 17 November 1975. LT 501 and 504 commissioned in Philippine Navy 8 August 1978 and LT 507 on 18 October 1978.
Modernisation: Several have had major refits including replacement of frames and plating as well as engines and electrics and provision for four 20 mm guns to replace the 40 mm guns.
Structure: Some of the later ships have tripod masts, others have pole masts.
Operational: All are used for general cargo work in Philippine service. Fourteen were deleted in 1989 and one sank in 1991. Two paid off in 1992 and one in 1993. *South Cotabato* was also paid off in 1993 but brought back in to service in 1994. *Benguet* broke down in the South China Sea in April 1995 and had to be taken in tow. *Benguet* again grounded in the Spratly Islands on 3 November 1999 and after a month on the rocks is probably beyond economical repair. One further ship, *Sierra Madre* is reported to be used as an observation post in the Spratly Islands. Replacements are needed but have not been given priority.

VERIFIED

LANAO DEL NORTE   *1993, Philippine Navy*

1 ACHELOUS CLASS (REPAIR SHIP) (ARL)

Name	No	Builders	Commissioned
YAKAL (ex-*Satyr* ARL 23, ex-*LST 852*)	AD 617 (ex-AR 517)	Chicago Bridge & Iron	20 Nov 1944

Displacement, tons: 4,342 full load
Dimensions, feet (metres): 328 × 50 × 14 *(100 × 15.2 × 4.3)*
Main machinery: 2 GM 12-567A diesels; 1,800 hp *(1.34 MW)*; 2 shafts
Speed, knots: 11.6
Complement: 220 approx
Guns: 4 US/Bofors 40 mm (quad). 10 Oerlikon 20 mm (5 twin).

Comment: Transferred from the US to the Philippines on 24 January 1977 by sale. (Originally to South Vietnam 30 September 1971.) Converted during construction. Extensive machine shop, spare parts stowage, and logistic support.

VERIFIED

YAKAL   *1994, Philippine Navy /* 0081545

42 LCM/LCU

Comment: Ex-US minor landing craft mostly transferred in the mid-1970s. 11 LCM 6, five LCM 8, eight LCU, 14 RUC and two LCVP. More LCVP are building at Cavite and two LCUs were reported delivered from South Korea in late 1995. More LCMs are planned. Used as transport vessels.

VERIFIED

LCU 286   5/1998, van Ginderen Collection / 0052706

1 ALAMOSA CLASS (SUPPLY SHIP) (AK)

Name	No	Builders	Commissioned
MACTAN (ex-*Kukui*, ex-*Colquith*)	AC 90 (ex-*TK 90*)	Froemming, Milwaukee	22 Sep 1944

Displacement, tons: 2,500 light; 7,570 full load
Dimensions, feet (metres): 338.5 × 50 × 18 *(103.2 × 15.2 × 5.5)*
Main machinery: 1 Nordberg diesel; 1,700 hp *(1.27 MW)*; 1 shaft
Speed, knots: 11
Complement: 85
Guns: 2—12.7 mm MGs.

Comment: Transferred from the US Coast Guard on 1 March 1972. Used to supply military posts and lighthouses in the Philippine archipelago. Was to have been paid off in 1994 but has been kept in service.

VERIFIED

MACTAN   4/1996, Philippine Navy

1 TRANSPORT VESSEL (AP)

Name	No	Builders	Commissioned
ANG PANGULO (ex-*The President*, ex-*Roxas*, ex-*Lapu-Lapu*)	AT 25 (ex-*TP 777*)	Ishikawajima, Japan	1959

Displacement, tons: 2,239 standard; 2,727 full load
Dimensions, feet (metres): 257.6 × 42.6 × 21 *(78.5 × 13 × 6.4)*
Main machinery: 2 Mitsui DE642/VBF diesels; 5,000 hp(m) *(3.68 MW)*; 2 shafts
Speed, knots: 18. **Range, n miles:** 6,900 at 15 kt
Complement: 81 (8 officers)
Guns: 3 Oerlikon 20 mm/70 Mk 4. 8—7.62 mm MGs.
Radars: Navigation: RCA CRMN-1A-75; I-band.

Comment: Built as war reparation; launched in 1958. Was used as presidential yacht and command ship with accommodation for 50 passengers. Originally named *Lapu-Lapu* after the chief who killed Magellan; renamed *Roxas* on 9 October 1962 after the late Manuel Roxas, the first President of the Philippines Republic. Renamed *The President* in 1967 and *Ang Pangulo* in 1975. In early 1987 was earmarked to transport President Marcos to Hong Kong and exile. The ship is now used as an attack transport, and still as a Presidential Yacht.

VERIFIED

ANG PANGULO   5/1998, John Mortimer / 0081546

2 YW TYPE (WATER TANKERS) (AWT)

Name	No	Builders	Commissioned
LAKE BULUSAN	AW 33 (ex-YW 111)	Marine Iron, Duluth	1 Aug 1945
LAKE PAOAY	AW 34 (ex-YW 130)	Leathem D Smith, Sturgeon Bay	28 Aug 1945

Displacement, tons: 1,237 full load
Dimensions, feet (metres): 174 × 32.7 × 13.2 *(53 × 10 × 4)*
Main machinery: 2 GM 8-278A diesels; 1,500 hp *(1.12 MW)*; 2 shafts
Speed, knots: 7.5
Complement: 29
Cargo capacity: 200,000 gallons
Guns: 1 Bofors 40/60. 1 Oerlikon 20 mm.

Comment: Basically similar to YOG type but adapted to carry fresh water. Transferred from the US to the Philippines on 16 July 1975.

VERIFIED

LAKE PAOAY   5/1998, van Ginderen Collection / 0052708

2 YOG TYPE (TANKERS) (YO)

Name	No	Builders	Commissioned
LAKE BUHI (ex-*YOG 73*)	AF 78 (ex-YO 78)	Puget Sound, Bremerton	28 Nov 1944
LAKE TAAL (ex-*YOG*)	AF 72 (ex-YO 72)	Puget Sound, Bremerton	14 Apr 1945

Displacement, tons: 447 standard; 1,400 full load
Dimensions, feet (metres): 174 × 32.7 × 13.2 *(53 × 10 × 4)*
Main machinery: 2 GM 8-278A diesels; 1,500 hp *(1.12 MW)*; 2 shafts
Speed, knots: 8
Complement: 28
Cargo capacity: 6,570 barrels dieso and gasoline
Guns: 2 Oerlikon 20 mm/70 Mk 4.

Comment: Former US Navy gasoline tankers. Transferred in July 1967 on loan and by purchase 5 March 1980.

VERIFIED

LAKE BUHI   1993, Philippine Navy

4 FLOATING DOCKS (YFD)

YD 200 (ex-*AFDL 24*)   YD 204 (ex-*AFDL 20*)   YD 205 (ex-*AFDL 44*)   — (ex-*AFDL 40*)

Comment: Floating steel dry docks built in the USA; all are former US Navy units with *YD 200* transferred in July 1948, *YD 204* in October 1961 (sale 1 August 1980), *YD 205* in September 1969 and *AFDL 40* in 1994.
Capacities: *YD 205*, 2,800 tons; *YD 200* and *YD 204*, 1,000 tons. In addition there are two floating cranes, *YU 206* and *YU 207*, built in US in 1944 and capable of lifting 30 tons.

VERIFIED

TUGS

Notes: A number of harbour tugs have been acquired from the US. The latest type is ex-Army of 390 tons, a speed of 12 kt and a bollard pull of 12 tons.

HARBOUR TUG   5/1998, John Mortimer / 0052709

COAST GUARD

Notes: (1) Some of the PCF craft listed are manned by the Navy as is the buoy tender *Mangyan*.
(2) The Coast Guard also operates one LCM 6 (BM 270), one LCVP (BV 182) and a River Utility Craft VU 463.
(3) A buoy tender *Corregidor* AG 891 of 1,100 tons was built and commissioned 2 March 1998 by Niigata Engineering.
(4) Ten Rodman 101 and four Rodman 38 have been ordered for delivery to the Police by 2005.

4 SAN JUAN CLASS (WPBO)

Name	No	Builders	Commissioned
SAN JUAN	001	Tenix Defence Systems	19 June 2000
EDSA II (ex-*Don Emilio*)	002 (ex-419)	Tenix Defence Systems	14 Dec 2000
PAMPANGA	003	Tenix Defence Systems	30 Jan 2003
BATANGAS	004	Tenix Defence Systems	8 Aug 2003

Displacement, tons: 500 full load
Dimensions, feet (metres): 183.7 × 34.5 × 9.8 *(56 × 10.5 × 3)*
Main machinery: 2 Caterpillar 3612 diesels; 4,800 hp(m) *(3.53 MW)* sustained; 2 shafts; cp props
Speed, knots: 24.5. **Range, n miles:** 3,000 at 15 kt
Complement: 38
Radars: Navigation: I-band.
Helicopters: Platform for one light.

Comment: First reported ordered in mid-1997. Construction of first of class started in February 1999. Steel hull and aluminium superstructure. Primarily used for SAR with facilities for 300 survivors. Fire-fighting and pollution control equipment included. A contract for a further two vessels was finalised in December 2001.

VERIFIED

SAN JUAN   *6/2000, Tenix Shipbuilding* / 0105228

4 + (10) ILOCOS NORTE CLASS (PATROL CRAFT) (PB)

Name	No	Builders	Commissioned
ILOCOS NORTE	3501	Tenix Defence Systems	9 May 2003
NUEVA VIZCAYA	3502	Tenix Defence Systems	8 Aug 2003
ROMBLON	3503	Tenix Defence Systems	20 Oct 2003
DAVAO DEL NORTE	3504	Tenix Defence Systems	16 Jan 2004

Displacement, tons: 115
Dimensions, feet (metres): 114.9 × 24.0 × 7.5 *(35.0 × 7.3 × 2.3)*
Main machinery: 2 diesels; 2 shafts. 1 loiter waterjet
Speed, knots: 23. **Range, n miles:** 2,000 at 12 kt
Complement: 11
Guns: 2—30 mm (1 twin). 2—12.7 mm MGs.
Radars: Navigation: I-band.

Comment: Contract on 9 December 2001 for the construction of four search and rescue vessels with an option for a further ten craft. Based on Bay class design with steel hull and aluminium superstructure.

VERIFIED

NUEVA VIZCAYA   *8/2003, Tenix* / 0569803

1 BALSAM CLASS (TENDER) (AKLH)

Name	No	Builders	Commissioned
KALINGA (ex-*Redbud, WAGL 398, ex-Redbud, T-AKL 398*)	AG 89	Marine Iron, Duluth	2 May 1944

Displacement, tons: 950 standard; 1,041 full load
Dimensions, feet (metres): 180 × 37 × 13 *(54.8 × 11.3 × 4)*
Main machinery: Diesel-electric; 2 diesels; 1,710 hp *(1.28 MW)*; 2 generators; 1 motor; 1,200 hp *(895 kW)*; 1 shaft
Speed, knots: 12. **Range, n miles:** 3,500 at 7 kt
Complement: 53
Guns: 2—12.7 mm MGs.
Radars: Navigation: Sperry SPS-53; I/J-band.
Helicopters: Platform for 1 light.

Comment: Originally US Coast Guard buoy tender (WAGL 398). Transferred to US Navy on 25 March 1949 as AG 398 and then to the Philippine Navy 1 March 1972. One 20 ton derrick. New engines fitted.

VERIFIED

KALINGA   *1994, Philippine Navy*

3 BUOY TENDERS (ABU)

CAPE BOJEADOR (ex-*FS 203*) AE 46 (ex-TK 46)
LIMASAWA (ex-*Nettle* WAK 129, ex-*FS 169*) AE 79 (ex-TK 79)
MANGYAN (ex-*Nasami*, ex-*FS 408*) AT 71 (ex-AE 71, ex-AS 71)

Displacement, tons: 470 standard; 950 full load
Dimensions, feet (metres): 180 × 32 × 10 *(54.9 × 9.8 × 3)*
Main machinery: 2 GM 6-278A diesels; 1,120 hp *(836 kW)*; 2 shafts
Speed, knots: 10. **Range, n miles:** 4,150 at 10 kt
Complement: 50
Cargo capacity: 400 tons
Guns: 1—12.7 mm MG can be carried.
Radars: Navigation: RCA CRMN 1A 75; I-band.

Comment: Former US Army FS 381 and FS 330 type freight and supply ships built in 1943-44. First two are employed as tenders for buoys and lighthouses. *Mangyan* transferred 24 September 1976 by sale. *Limasawa* acquired by sale 31 August 1978. One 5 ton derrick. *Cape Bojeador* paid off in 1988 but was back in service in 1991 after a major overhaul. *Mangyan* reclassified AT in 1993 and belongs to the Navy. Masts and superstructures have minor variations.

VERIFIED

CAPE BOJEADOR   *1993, Philippine Navy*

2 LARGE PATROL CRAFT (PB)

Name	No	Builders	Commissioned
TIRAD PASS	AU 100 (ex-SAR 100)	Sumidagawa, Japan	1974
BESSANG PASS	AU 75 (ex-SAR 99)	Sumidagawa, Japan	1974

Displacement, tons: 279 full load
Dimensions, feet (metres): 144.3 × 24.3 × 4.9 *(44 × 7.4 × 1.5)*
Main machinery: 2 MTU 12V 538 TB82 diesels; 4,050 hp(m) *(2.98 MW)*; 2 shafts
Speed, knots: 27.5. **Range, n miles:** 2,300 at 14 kt
Complement: 32
Guns: 4—12.7 mm (2 twin) MGs.

Comment: Paid for under Japanese war reparations. Similar type as *Emilio Aguinaldo*. *Bessang Pass* grounded in 1983 but was recovered.

VERIFIED

TIRAD PASS   *1992, Philippine Navy* / 0081548

4 PGM-39 CLASS (LARGE PATROL CRAFT) (PB)

Name	No	Builders	Commissioned
AGUSAN (ex-*PGM 39*)	PG 61	Tacoma, WA	Mar 1960
CATANDUANES (ex-*PGM 40*)	PG 62	Tacoma, WA	Mar 1960
ROMBLON (ex-*PGM 41*)	PG 63	Peterson Builders, WI	June 1960
PALAWAN (ex-*PGM 42*)	PG 64	Tacoma, WA	June 1960

Displacement, tons: 124 full load
Dimensions, feet (metres): 100.3 × 18.6 × 6.9 *(30.6 × 5.7 × 2.1)*
Main machinery: 2 MTU MB 12V 493 TY57 diesels; 2,200 hp(m) *(1.6 MW)* sustained; 2 shafts
Speed, knots: 17. **Range, n miles:** 1,400 at 11 kt
Complement: 26-30
Guns: 2 Oerlikon 20 mm. 2—12.7 mm MGs. 1—81 mm mortar.
Radars: Surface search: Alpelco DFR-12; I/J-band.

Comment: Steel-hulled craft built under US military assistance programmes. Assigned US PGM-series numbers while under construction. Transferred upon completion. These craft are lengthened versions of the US Coast Guard 95 ft Cape class patrol boat design. Operational status is doubtful.

VERIFIED

AGUSAN   *1994, Philippine Navy /* 0081550

10 PCF 46 CLASS (COASTAL PATROL CRAFT) (PB)

DB 411	DB 417	DB 422	DB 429	DB 432
DB 413	DB 419	DB 426	DB 431	DB 435

Displacement, tons: 21 full load
Dimensions, feet (metres): 45.9 × 14.5 × 3.3 *(14 × 4.4 × 1)*
Main machinery: 2 Cummins diesels; 740 hp *(552 kW)*; 2 shafts
Speed, knots: 25. **Range, n miles:** 1,000 at 15 kt
Complement: 8
Guns: 2—12.7 mm (twin) MGs. 1—7.62 mm M60 MG.
Radars: Surface search: Kelvin Hughes 17; I-band.

Comment: Survivors of a class built by de Havilland Marine, Sydney NSW between 20 November 1974 and 8 February 1975 (DF series). In August 1975 further craft of this design (DB series) were ordered from Marcelo Yard, Manila to be delivered 1976-78 at the rate of two per month. By the end of 1976, 25 more had been completed but a serious fire in the shipyard destroyed 14 new hulls and halted production. Many deleted.

VERIFIED

DB 435   *1993, Philippine Navy*

12 PCF 50 (SWIFT Mk 1 and Mk 2) CLASS
(COASTAL PATROL CRAFT) (PB)

DF 300-303	DF 305	DF 307-313

Displacement, tons: 22.5 full load
Dimensions, feet (metres): 50 × 13.6 × 4 *(15.2 × 4.1 × 1.2)* (Mk 1) 51.3 × 13.6 × 4 *(15.6 × 4.1 × 1.2)* (Mk 2)
Main machinery: 2 GM 12-71 diesels; 680 hp *(504 kW)* sustained; 2 shafts
Speed, knots: 28. **Range, n miles:** 685 at 16 kt
Complement: 6
Guns: 2—12.7 mm (twin) MGs. 2 M-79 40 mm grenade launchers.
Radars: Surface search: Decca 202; I-band.

Comment: Most built in the USA. Built for US military assistance programmes and transferred in the late 1960s. Some built in 1970 in the Philippines (ferro-concrete) with enlarged superstructure. *DF 300-303* are Swift Mk 1. *DF 305* and *DF 307-313* are Swift Mk 2. *DF 303, 309, 311* and *312* belong to the Navy.

VERIFIED

DF 308   *5/1998, van Ginderen Collection /* 0081552

14 PCF 65 (SWIFT Mk 3) CLASS (COASTAL PATROL CRAFT) (PB)

DF 325-332	DF 334	DF 347	DF 351-354

Displacement, tons: 29 standard; 37 full load
Dimensions, feet (metres): 65 × 16 × 3.4 *(19.8 × 4.9 × 1)*
Main machinery: 3 GM 12V-71TI diesels; 840 hp *(616 kW)* sustained; 3 shafts
Speed, knots: 25
Complement: 8
Guns: 2—12.7 mm MGs.
Radars: Surface search: Koden; I-band.

Comment: Improved Swift type inshore patrol boats built by Sewart for the Philippine Navy. Delivered 1972-76. *DF 351-354* belong to the Navy. Some that were laid up have been returned to service. New radars fitted.

VERIFIED

DF 354 (Navy)   *5/1998, John Mortimer /* 0081551

DF 347 (Coast Guard)   *5/1998, Sattler & Steele /* 0052711

3 DE HAVILLAND CLASS (PB)

DF 321-323

Displacement, tons: 25 full load
Dimensions, feet (metres): 54.8 × 16.4 × 4.3 *(16.7 × 5 × 1.3)*
Main machinery: 2 diesels; 740 hp *(552 kW)*; 2 shafts
Speed, knots: 25. **Range, n miles:** 450 at 14 kt
Complement: 8
Guns: 2—12.7 mm MGs.

Comment: Locally built in the mid-1980s. Others of this type have been paid off and numbers are uncertain.

VERIFIED

DF 321   *5/1998, van Ginderen Collection /* 0052713

11 CUTTERS (PBR)

CGC 103	CGC 110	CGC 115	CGC 128-130	CGC 132-136

Displacement, tons: 13 full load
Dimensions, feet (metres): 40 × 13.6 × 3 *(12.2 × 4.1 × 0.9)*
Main machinery: 2 Detroit diesels; 560 hp *(418 kW)*; 2 shafts
Speed, knots: 28
Complement: 5
Guns: 1—12.7 mm MG. 1—7.62 mm MG.

Comment: Built at Cavite Yard from 1984. One deleted in 1994. Used for harbour patrols. There are also some small unarmed Police craft.

VERIFIED

CGC 130
1994, Philippine Navy / 0081553

Poland
MARYNARKA WOJENNA

Country Overview

The modern democratic era of the Republic of Poland began in 1989 after forty-two years of communist rule. Situated in central Europe, the country has an area of 120,725 square miles and is bordered to the north by Russia (Kaliningrad), to the east by Lithuania, Belarus, and Ukraine, to the south by the Czech Republic and Slovakia and to the west by Germany. It has a 265 n mile coastline with the Baltic Sea. Warsaw is the capital and largest city while Gdansk, Szczecin and Gdynia are the principal ports. Territorial seas (12 n miles) are claimed but while it has claimed a 200 n mile EEZ, its limits have not been fully defined by boundary agreements.

Headquarters Appointments

Commander-in-Chief:
  Admiral Roman Krzvzelewski
Deputy Commander-in-Chief:
  Vice Admiral Jedrzej Czajkowski
Chief of Naval Staff:
  Rear Admiral Henryk Solkiewicz
Chief of Naval Logistics:
  Rear Admiral Tomasz Mathea
Chief of Naval Training:
  Rear Admiral Marek Bragoszewski

Diplomatic Representation

Defence and Naval Attaché in London:
  Colonel Tadeusz Jedrzejczak

Personnel

(a)   2005: 14,100
(b)   12 months' national service

Prefix to Ships' Names

ORP, standing for *Okret Rzeczypospolitej Polskiej*

Strength of the Fleet

Type	Active	Building
Submarines—Patrol	5	—
Frigates	2	—
Corvettes	8	2 (5)
Fast Attack Craft—Missile	2	—
Coastal Patrol Craft	11	—
Minesweepers—Coastal	2	—
Minehunters—Coastal	20	(14)
LSTs	6	—
LCUs	3	—
Survey and Research Ships	3	—
AGIs	2	—
Training Ships	4	—
Salvage Ships	4	—
Tankers	2	—
Logistic Support Ship	1	—

Sea Department of the Border Guard (MOSG)

A para-naval force, subordinate to the Minister of the Interior.

Bases

Gdynia (3rd Naval Flotilla), Hel (9th Coastal Defence Flotilla), Swinoujscie (8th Coastal Defence Flotilla), Kolobrzeg, Gdansk (Frontier Guard)

Naval Aviation

HQ at Gdynia
1st Wing at Gdynia (MiG-21, W-3, An-28)
2nd Wing at Darlowo (Mi-14, Mi-2, W-3)
3rd Wing at Siemirowice-Cewice (TS-11, B1-R, An-28)

Coast Defence

Two divisions with 24-57 mm guns.

DELETIONS

Submarines

2003   *Wilk, Dzik*

Destroyers

2003   *Warszawa*

Patrol Forces

2002   *Zwrotny, Nieueiety*
2003   *Puck, Darlowo*
2004   *Dziwnów, Grozny, Wytrwaly, Zreczny, Zwinny, Zawziety, Czujny*

Mine Warfare Forces

2002   *Rybitwa*

Survey and Research Ships

2003   *Zodiak*

Training Ships

2002   *Kadet, Elew*
2003   *Bryza*

Auxiliaries

2002   *Z3, Z9, Bolko, Semko*

PENNANT LIST

Submarines

291	Orzeł
294	Sókol
295	Sęp
296	Bielik
297	Kondor

Frigates

272	Generał Kazimierz Pułaski
273	Generał Tadeusz Kościuszko

Corvettes

240	Kaszub
421	Orkan
422	Piorun
423	Grom
434	Gornik
435	Hutnik
436	Metalowiec
437	Rolnik

Patrol Forces

431	Świnoujście
433	Władysławowo

Mine Warfare Forces

621	Flaming
623	Mewa
624	Czajka
625	TR 25
626	TR 26
630	Goplo
631	Gardno
632	Bukowo
633	Dabie
634	Jamno
635	Mielno
636	Wicko
637	Resko
638	Sarbsko
639	Necko
640	Naklo
641	Druzno
642	Hancza
643	Mamry
644	Wigry
645	Sniardwy
646	Wdzydze

Amphibious Forces

811	Grunwald
821	Lublin
822	Gniezno
823	Krakow
824	Poznan
825	Torun
851	KD 11
852	KD 12
853	KD 13

Survey Ships and AGIs

261	Kopernik
262	Nawigator
263	Hydrograf
265	Heweliusz
266	Arctowski

Auxiliaries

251	Wodnik
253	Iskra
281	Piast
282	Lech
511	Kontradmiral X Czernicki
711	Podchorazy
R 11	Gniewko
R 14	Zbyszko
R 15	Macko
SD 11	Wrona
SD 12	Rys
SD 13	—
Z 1	Baltyk
Z 8	Meduza

Maritime Frontier Guard

SG 311	Kaper I
SG 312	Kaper II
SG 323	Zefir
SG 325	Tecza

For details of the latest updates to *Jane's Fighting Ships* online and to discover the additional information available exclusively to online subscribers please visit
jfs.janes.com

SUBMARINES

4 SOKÓL (KOBBEN) CLASS (TYPE 207) (SSK)

Name	No	Builders	Laid down	Launched	Commissioned
SOKÓL (ex-*Stord*)	294 (ex-S 308)	Rheinstahl – Nordseewerke, Emden	1 Apr 1966	2 Sep 1966	14 Feb 1967
SĘP (ex-*Skolpen*)	295 (ex-S 306)	Rheinstahl – Nordseewerke, Emden	1 Nov 1965	24 Mar 1966	17 Aug 1966
BIELIK (ex-*Svenner*)	296 (ex-S 309)	Rheinstahl – Nordseewerke, Emden	8 Sep 1966	27 Jan 1967	12 Jun 1967
KONDOR (ex-*Kunna*)	297 (ex-S 319)	Rheinstahl – Nordseewerke, Emden	3 Mar 1964	16 Jul 1964	29 Oct 1964

Displacement, tons: 459 standard; 524 dived
Dimensions, feet (metres): 155.5 × 15 × 14
(47.4 × 4.6 × 4.3)
Main machinery: Diesel-electric; 2 MTU 12V 493 AZ80 GA31L
diesels; 1,200 hp(m) *(880 kW)* sustained; 1 motor; 1,800 hp
(m) *(1.32 MW)* sustained; 1 shaft
Speed, knots: 12 surfaced; 18 dived
Range, n miles: 5,000 at 8 kt (snorting)
Complement: 21 (5 officers)

Torpedoes: 8—21 in *(533 mm)* bow tubes.
Countermeasures: ESM: Argo radar warning.
Weapons control: Kongsberg MSI-70U TFCS.
Radars: Surface search: Kelvin Hughes 1007; I-band.
Sonars: Atlas Elektronik CSU 83; passive search and attack;
medium/high frequency.

Programmes: Commissioned into the Norwegian Navy from
1964, the building cost was shared between the Norwegian
and US governments. Decommissioned from the Norwegian
Navy in 2001. Following announcement on 18 January 2002
Sokól recommissioned on 25 May 2002 and *Sep* on 5 August
2002. *Bielik* recommissioned on 8 September 2003 and
Kondor on 20 October 2004. The ex-*Kobben* has been
transferred for spares and as a floating training base. The
contract also includes provision of in-service support. These
submarines are understood to be a stop-gap measure to
maintain a submarine capability until about 2012 when these
may be replaced.
Modernisation: All modernised at Urivale Shipyard, Bergen
between 1989-1992.
Structure: A development of the German Type 205 class, they
have a diving depth of 650 ft *(200 m)*. Pilkington optronics CK
30 search periscope.

UPDATED

BIELIK
7/2004*, B Sullivan / 1121522

BIELIK
7/2004*, B Sullivan / 1044468

1 KILO CLASS (PROJECT 877EM) (SSK)

Name	No	Builders	Commissioned
ORZEŁ	291	Sudomekh, Leningrad	21 June 1986

Displacement, tons: 2,457 surfaced; 3,076 dived
Dimensions, feet (metres): 243.8 × 32.8 × 21.7
(74.3 × 10 × 6.6)
Main machinery: Diesel-electric; 2 DL 42M diesels; 3,650 hp(m)
(2.68 MW); 2 generators; 6 MW; 1 PG 141 motor; 5,900 hp
(m) *(4.34 MW)*; 1 shaft; 2 auxiliary motors; 204 hp(m)
(150 kW); 1 economic speed motor; 130 hp *(95 kW)*
Speed, knots: 10 surfaced; 17 dived; 9 snorting
Range, n miles: 6,000 at 7 kt snorting; 400 at 3 kt dived
Complement: 60 (16 officers)

Missiles: SAM: 8 SA-N-5 (Strela 2M).
Torpedoes: 6—21 in *(533 mm)* tubes. Combination of 53-65;
anti-surface; passive/wake homing to 19 km *(10.3 n miles)*
at 45 kt; warhead 300 kg and TEST-71; anti-submarine;
active/passive homing to 15 km *(8.1 n miles)* at 40 kt;
warhead 205 kg. 53-56 WA and SET 53 M can also be
carried. Total of 18 torpedoes.
Mines: 24 in lieu of torpedoes.
Countermeasures: ESM: Brick Group (MRP-25); radar warning;
Quad Loop HF D/F.
Weapons control: Murena MWU 110 TFCS.
Radars: Surface search: Racal Decca Bridgemaster; I-band.

ORZEŁ
6/2004*, J Ciślak / 1044467

Sonars: Shark Teeth (MGK-400); hull-mounted; passive search
and attack (some active capability); low/medium frequency.
Mouse Roar (MG 519); active mine detection; high frequency.

Programmes: This was the second transfer of this class, the first
being to India and others have since gone to Romania, Algeria,
Iran and China. It was expected that more than one would be
acquired as part of an exchange deal with the USSR for Polish-
built amphibious ships, but this class is considered too large
for Baltic operations and subsequent transfers were of the
Foxtrot class.
Structure: Diving depth, 240 m *(787 ft)*. Has two torpedo tubes
modified for wire guided anti-submarine torpedoes.
Operational: Based at Gdynia.

UPDATED

FRIGATES

2 OLIVER HAZARD PERRY CLASS (FFGHM)

Name	No	Builders	Laid down	Launched	Commissioned
GENERAŁ KAZIMIERZ PUŁASKI (ex-*Clark*)	272 (ex-FFG 11)	Bath Iron Works	17 July 1978	24 Mar 1979	9 May 1980
GENERAŁ TADEUSZ KOŚCIUSZKO (ex-*Wadsworth*)	273 (ex-FFG 9)	Todd Shipyards, San Pedro	13 July 1977	29 July 1978	28 Feb 1980

Displacement, tons: 2,750 light; 3,638 full load
Dimensions, feet (metres): 445 × 45 × 14.8; 24.5 (sonar)
 (135.6 × 13.7 × 4.5; 7.5)
Main machinery: 2 GE LM 2500 gas turbines; 41,000 hp
 (30.59 MW) sustained; 1 shaft; cp prop
 2 auxiliary retractable props; 650 hp *(484 kW)*
Speed, knots: 29. **Range, n miles:** 4,500 at 20 kt
Complement: 200 (15 officers) including 19 aircrew

Missiles: SSM: 4 McDonnell Douglas Harpoon Block 1G; active
 radar homing to 130 km *(70 n miles)* at 0.9 Mach; warhead
 227 kg.
 SAM: 36 GDC Standard SM-1MR; command guidance; semi-
 active radar homing to 46 km *(25 n miles)* at 2 Mach.
 1 Mk 13 Mod 4 launcher for both SSM and SAM missiles ❶.
Guns: 1 OTO Melara 3 in *(76 mm)*/62 Mk 75 ❷; 85 rds/min to
 16 km *(8.7 n miles)* anti-surface; 12 km *(6.6 n miles)* anti-
 aircraft; weight of shell 6 kg.
 1 General Electric/General Dynamics 20 mm/76 6-barrelled
 Mk 15 Vulcan Phalanx ❸; 3,000 rds/min combined to 1.5 km.
 4—12.7 mm MGs.
Torpedoes: 6—324 mm Mk 32 (2 triple) tubes ❹. 24 Whitehead
 A244 Mod 3. To be replaced by Mu-90 Impact from 2002.
Countermeasures: Decoys: 2 Loral Hycor SRBOC 6-barrelled
 fixed Mk 36 ❺; IR flares and chaff to 4 km *(2.2 n miles)*.
 T-Mk 6 Fanfare/SLQ-25 Nixie; torpedo decoy.
ESM/ECM: SLQ-32(V)2 ❻; radar warning. Sidekick modification
 adds jammer and deception system.
Combat data systems: NTDS with Link 11 and 14. SATCOM
 SRR-1, WSC-3 (UHF).
Weapons control: SWG-1 Harpoon LCS. Mk 92 (Mod 2), WCS
 with CAS (Combined Antenna System). The Mk 92 is the US
 version of the Signaal WM28 system. Mk 13 weapon
 direction system. 2 Mk 24 optical directors.
Radars: Air search: Raytheon SPS-49(V)4 ❼; C/D-band.
 Surface search: ISC Cardion SPS-55 ❽; I-band.
 Fire control: Lockheed STIR (modified SPG-60) ❾; I/J-band.
 Sperry Mk 92 (Signaal WM28) ❿; I/J-band.
 Navigation: Furuno; I-band.
 Tacan: URN 25. IFF Mk XII AIMS UPX-29.
Sonars: SQQ 89(V)2 (Raytheon SQS 56 and Gould SQR 19); hull-
 mounted active search and attack; medium frequency and
 passive towed array; very low frequency.

Helicopters: 2 Kaman SH-2G Seasprite ⓫.

Programmes: *Puławski* approved for transfer from US by grant in
 1999. Recommissioned on 8 March 2000. *Kościuszko*
 recommissioned on 12 June 2002.
Structure: Details given are for the ship in service with the US
 Navy.
Operational: Based at Gdynia. Seasprite helicopters carried from
 2001.

UPDATED

PUŁASKI
(Scale 1 : 1,200), Ian Sturton / 0105229

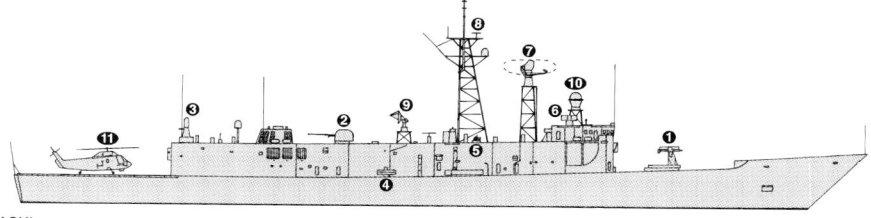

PUŁASKI
6/2004*, John Brodie / 1044470

PUŁASKI
6/2004*, B Prézelin / 1044469

KOŚCIUSZKO
9/2002, Harald Carstens / 0526828

0 + 2 (5) PROJECT 621 GAWRON II (MEKO A 100) CLASS (FSGHM)

Name	No	Builders	Laid down	Launched	Commissioned
—	—	Naval Shipyard, Gdynia	28 Oct 2001	2005	2008
—	—	Naval Shipyard, Gdynia	2003	2006	2009

Displacement, tons: 2,035 full load
Dimensions, feet (metres): 312.3 × 43.6 × 11.8
  (95.2 × 13.13 × 3.6)
Main machinery: CODAG; 1 gas turbine; 2 diesels; 2 shafts
Speed, knots: 30. **Range, n miles:** 4,000 at 15 kt
Complement: 74

Missiles: SSM: 8 RBS-15 Mk 3 ❶.
SAM: Evolved Sea Sparrow; VLS ❷.
Guns: 1—3 in *(76 mm)*/62 ❸. 2—35 mm. RAM ❹.
A/S mortars: 2 ASW 601 ❺.
Countermeasures: Decoys: 1—10 barrelled Jastrzab 122 mm;
  chaff and IR flares.
ESM: Radar warning.
TCM: C310 torpedo decoy system.
Combat data systems: Signaal TACTICOS or Saab Tech 9LV.
Radars: Air/surface search ❻; fire control ❼; navigation.
Sonars: Hull mounted; active; medium frequency.

Helicopters: Platform for 1 medium ❽.

Programmes: Design definition by German Corvette Consortium
  (Blohm + Voss, Lürssen, Thyssen and HDW) which is to act as
  subcontractor to the shipbuilder. There are options for a
  further five vessels. Details of the design and of the building
  programme have not been released but it is unlikely that the
  first of class will enter service before 2008.
Structure: The design is based on the MEKO A 100.
                                                    UPDATED

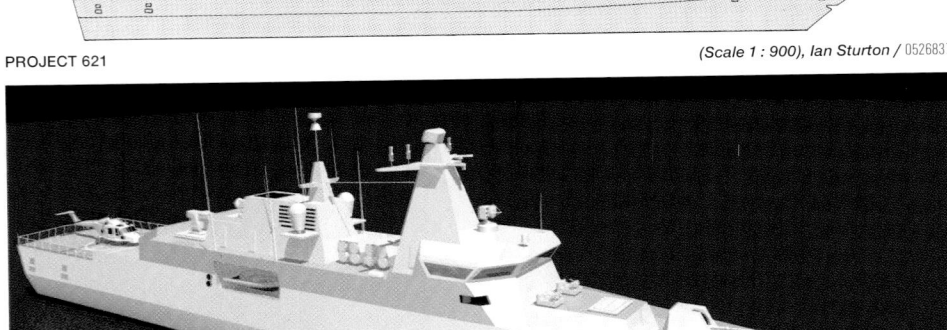

PROJECT 621   *(Scale 1 : 900), Ian Sturton /* 0526837

PROJECT 621
2001, Polish Navy / 0114788

CORVETTES

1 KASZUB CLASS (PROJECT 620) (FSM)

Name	No	Builders	Laid down	Launched	Commissioned
KASZUB	240	Northern Shipyard, Gdansk	9 June 1984	11 May 1986	3 Mar 1987

Displacement, tons: 1,051 standard; 1,183 full load
Dimensions, feet (metres): 270 × 32.8 × 10.2; 16.1 (sonar)
  (82.3 × 10 × 3.1; 4.9)
Main machinery: CODAD; 4 Cegielski-Sulzer AS 16V 25/30
  diesels; 16,900 hp(m) *(12.42 MW)*; 2 shafts; cp props
Speed, knots: 27. **Range, n miles:** 3,500 at 14 kt; 350 at 26 kt
Complement: 82 (10 officers)

Missiles: SAM: 2 SA-N-5 quad launchers ❶; IR homing to 10 km
  (5.5 n miles) at 1.5 Mach. VLS system to replace after 23 mm
  gun.
Guns: 1 USSR 3 in *(76 mm)*/66 AK 176 ❷; 120 rds/min to 12 km
  (6.4 n miles); weight of shell 7 kg.
  6 ZU-23-2M Wrobel 23 mm/87 (3 twin) ❸; 400 rds/min
  combined to 2 km.
Torpedoes: 4—21 in *(533 mm)* (2 twin) tubes ❹. SET-53M;
  passive homing to 15 km *(8.1 n miles)* at 29 kt; warhead
  100 kg.
A/S mortars: 2 RBU 6000 12-tubed trainable ❺; range 6,000 m;
  warhead 31 kg; 120 rockets.
Depth charges: 2 rails. 12 charges.
Countermeasures: Decoys: 1—10 barrelled 122 mm Jastrzab
  launcher ❻ for chaff.
ESM: Intercept.
Weapons control: Drakon TFCS.
Radars: Air/surface search: Strut Curve (MR 302) ❼; F-band.
Surface search: Racal Bridgemaster C-252 ❽; I-band.
Navigation: Racal Bridgemaster C-341; I-band.
IFF: RAWAR SA-10M2.
Sonars: MG 322T; hull-mounted; active search; medium
  frequency.
  MG 329M; stern-mounted dipping type mounted on the
  transom; active; high frequency.

Programmes: Second of class cancelled in 1989 and a class of
  up to ten more ships based on the Kaszub hull and specialised
  for anti-submarine warfare has been shelved.

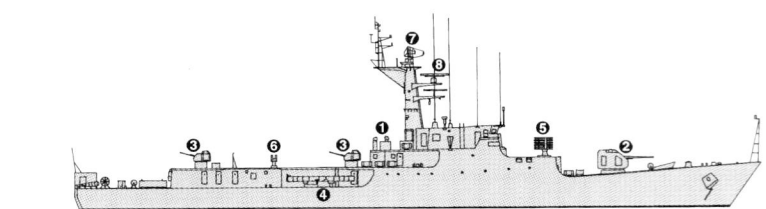

KASZUB   *(Scale 1 : 900), Ian Sturton /* 0081558

KASZUB   *6/2004*, Michael Winter /* 1044471

Structure: Design based on Grisha class but with many
  alterations. The 76 mm gun was fitted in late 1991. New
  decoy system fitted in 1999. There is space for a fire-control
  director on the bridge roof.

Operational: Finally achieved operational status in 1990. Based
  at Hel with the Border Guard in 1990 but returned to the Navy
  in 1991. The ship has to stop to use stern-mounted sonar.
                                                    UPDATED

KASZUB   *6/2004*, Harald Carstens /* 1044472

3 ORKAN (SASSNITZ) CLASS (PROJECT 660 (ex-151)) (FSGM)

Name	No	Builders	Launched	Commissioned
ORKAN	421	Peenewerft/Northern Shipyard, Gdansk	29 Sep 1990	18 Sep 1992
PIORUN	422	Peenewerft/Northern Shipyard, Gdansk	19 Oct 1990	11 Mar 1994
GROM (ex-*Huragan*)	423	Peenewerft/Northern Shipyard, Gdansk	11 Dec 1990	28 Mar 1995

Displacement, tons: 331 standard; 326 full load
Dimensions, feet (metres): 163.4 oa; 147.6 wl × 28.5 × 7.2
(49.8; 45 × 8.7 × 2.2)
Main machinery: 3 Type M 520T diesels; 16,000 hp(m)
(11.93 MW) sustained; 3 shafts
Speed, knots: 38. **Range, n miles:** 1,600 at 14 kt
Complement: 36 (4 officers)

Missiles: SSM: 8 (2 quad) launchers; RBS-15 Mk 3; active radar homing to 200 km *(108 n miles)* at 0.9 Mach; warhead 200 kg.
SAM: SA-N-5 Grail quad launcher; manual aiming; IR homing to 6 km *(3.2 n miles)* at 1.5 Mach; warhead 1.5 kg.
Guns: 1 USSR 3 in *(76 mm)*/66 AK 176; 120 rds/min to 12 km *(6.4 n miles)*; weight of shell 7 kg.
1—30 mm/65 AK 630; 6 barrels; 3,000 rds/min combined to 2 km.
Countermeasures: Decoys: 8—9 barrelled Jastrzab 81 mm and 1—10 barrelled Jastrzab 122 mm chaff and IR launchers.
ESM: PIT intercept.
Combat data systems: Signaal TACTICOS.
Weapons control: Thales STING optronic director.
Radars: Surface search: AMB Sea Giraffe; G-band.
Fire control: Bass Tilt MR-123; H/I-band.
Navigation: PIT; I-band.
IFF: Square Head; Salt Pot.

Programmes: Originally six of this former GDR Sassnitz class were to be built at Peenewerft for Poland. Three units were acquired and completed at Gdansk.
Modernisation: Contract with Thales Naval Nederland (TNNL) as prime contractor for upgrade of all three ships signed 29 June 2001. New equipment includes RBS-15 Mk 3 missiles, TACTICOS combat data system, STING optronic director, AMB Sea Giraffe surveillance radar, PIT navigational radar and ESM equipment, improved communications and Link 11. Refit

ORKAN   *2/2002, Frank Findler* / 0533232

of *Piorun* was completed by 2003 and the other two ships are to be completed by 2006.
Structure: The prototype vessel had two quadruple SSM launchers with an Exocet type (SS-N-25) of missile and the plan is to fit eight SSM in due course. Plank Shave radar has

been replaced by a Polish set. Unlike the German Coast Guard vessels of the same class, these ships have retained three engines.
Operational: Based at Gdynia.

UPDATED

GROM   *6/1999, L-G Nilsson* / 0081563

4 GORNIK (TARANTUL I) CLASS (PROJECT 1241RE) (FSGM)

Name	No	Builders	Commissioned
GORNIK	434	River Shipyard 341, Rybinsk	28 Dec 1983
HUTNIK	435	River Shipyard 341, Rybinsk	31 Mar 1984
METALOWIEC	436	River Shipyard 341, Rybinsk	13 Feb 1988
ROLNIK	437	River Shipyard 341, Rybinsk	4 Feb 1989

Displacement, tons: 385 standard; 455 full load
Dimensions, feet (metres): 184.1 × 37.7 × 8.2
(56.1 × 11.5 × 2.5)
Main machinery: COGAG; 2 Type M 70 gas turbines; 24,000 hp(m) *(17.65 MW)* sustained; 2 M 75 gas turbines with reversible gearbox; 8,000 hp(m) *(5.88 MW)* sustained; 2 shafts
Speed, knots: 42. **Range, n miles:** 1,650 at 14 kt
Complement: 45 (6 officers)

Missiles: SSM: 4 SS-N-2C Styx (2 twin) launchers; active radar or IR homing to 83 km *(45 n miles)* at 0.9 Mach; warhead 513 kg; sea-skimmer in terminal flight.
SAM: SA-N-5 Grail quad launcher; manual aiming; IR homing to 6 km *(3.2 n miles)* at 1.5 Mach; warhead 1.5 kg.
Guns: 1—3 in *(76 mm)*/60 AK 176; 120 rds/min to 12 km *(6.4 n miles)*; weight of shell 7 kg.
2—30 mm/65 AK 630 6-barrelled type; 3,000 rds/min combined to 2 km.
Countermeasures: Decoys: 2 PK 16 chaff launchers.
Weapons control: Korall-E WFCS; PMK 453 optronic director.
Radars: Air/surface search: Plank Shave (Garpun E); E-band.
Navigation: Racal Decca Bridgemaster; I-band.
Fire control: Bass Tilt (MR-123); H/I-band.
IFF: Square Head.
Sonars: Foal Tail; VDS; active; high frequency.

METALOWIEC   *10/2003*, J Ciślak* / 1044473

Programmes: Transferred from the USSR.
Modernisation: An upgrade of these ships is expected when the 'Orkan' modernisation programme is completed 2006.

Structure: Similar to others of the class exported to India, Yemen, Romania and Vietnam.
Operational: Based at Gdynia.

UPDATED

SHIPBORNE AIRCRAFT

Numbers/Type: 4 Kaman SH-2G (P) Seasprite.
Operational speed: 130 kt *(241 km/h).*
Service ceiling: 22,500 ft *(6,860 m).*
Range: 367 n miles *(697 km).*
Role/Weapon systems: First two delivered in 2002. Second pair in August 2003. Sensors: LN66/HP radar; ALR-66 ESM, ALE-39 ECM, AQS-81(V)2 MAD, AAQ-16 FLIR, ARR 57/84 sonobuoy receivers. Weapons: ASW: two A244S torpedoes (Mu 90 from 2002). ASV: one 7.62 mm MG. ***UPDATED***

SH-2G (P)   5/2004 *, Polish Navy / 0566178

LAND-BASED MARITIME AIRCRAFT (FRONT LINE)

Notes: In addition there are 6 TS training aircraft.

Numbers/Type: 8/2/3 PZL Mielec M-28 Bryza 1R/M-28E/M-28TD.
Operational speed: 181 kt *(335 km/h).*
Service ceiling: 13,770 ft *(4,200 m).*
Range: 736 n miles *(1,365 km).*
Role/Weapon systems: Based on the USSR Cash light transport and used for maritime patrol and SAR. First one delivered in January 1995. Sensors: Search radar ARS 400; ESM. Weapons: 2 SAB 100 bombs. ***VERIFIED***

M-28   6/2003, J Ciślak / 0567493

Numbers/Type: 10/3 Mil Mi-14PL Haze A/Mil Mi-14PS Haze C.
Operational speed: 120 kt *(222 km/h).*
Service ceiling: 16,000 ft *(4,670 m).*
Range: 500 n miles *(1,100 km).*
Role/Weapon systems: PL for ASW, PS for SAR. PL operates in co-operation with surface units. Adapted for landing and taking off from water. Sensors: I-2ME search radar; APM-60, MAD, sonobuoys, MGM 329M VDS. Weapons: ASW; Whitehead A 244 torpedoes, depth bombs and mines. Arming with Penguin ASM is also under consideration. ***VERIFIED***

Mi-14PL   6/2002, J Ciślak / 0567494

Numbers/Type: 2 Mi-17 Hip.
Operational speed: 124 kt *(230 km/h).*
Service ceiling: 16,400 ft *(5,000 m).*
Range: 324 n miles *(600 km).*
Role/Weapon systems: Transport aircraft. First one delivered in 2001. ***VERIFIED***

Mi-17   6/2003, J Ciślak / 0567496

Numbers/Type: 2/7 PZL Świdnik W-3 Sokol/W-3RM Anakonda.
Operational speed: 119 kt *(220 km/h).*
Service ceiling: 19,672 ft *(6,000 m).*
Range: 335 n miles *(620 km).*
Role/Weapon systems: W-3 for transport, W-3RM for SAR. Operates in co-operation with surface units. Adapted for landing and taking off from water. Sensors: RDS-82 VP Meteo, FLIR. ***UPDATED***

W-3RM   6/2004 *, J Ciślak / 1044474

Numbers/Type: 3/1 Mi-2RM Hoplite/Mi-2D.
Operational speed: 100 *(180 km/h).*
Service ceiling: 13,200 ft *(4,000 m).*
Range: 300 n miles *(550 km).*
Role/Weapon systems: Mi-2D is for transport aircraft and Mi-2RM is for SAR. ***VERIFIED***

Mi-2RM   7/2000, J Ciślak / 0105237

PATROL FORCES

2 PUCK (OSA I) CLASS (PROJECT 205)
(FAST ATTACK CRAFT—MISSILE) (PTFGM)

Name	No	Builders	Commissioned
ŚWINOUJŚCIE	431	Leningrad	13 Jan 1973
WŁADYSŁAWOWO	433	Leningrad	13 Nov 1975

Displacement, tons: 171 standard; 210 full load
Dimensions, feet (metres): 126.6 × 24.9 × 8.8 *(38.6 × 7.6 × 2.7)*
Main machinery: 3 Type M 503A diesels; 8,025 hp(m) *(5.9 MW)* sustained; 3 shafts
Speed, knots: 35. Range, n miles: 800 at 30 kt
Complement: 30

Missiles: SSM: 4 SS-N-2A Styx; active radar or IR homing to 46 km *(25 n miles)* at 0.9 Mach; warhead 513 kg.
SAM: SA-N-5 quad launcher.
Guns: 4—30 mm/65 (2 twin) AK 230 automatic; 1,000 rds/min to 6.5 km *(3.5 n miles)*; weight of shell 0.54 kg.
Radars: Surface search: Square Tie; I-band.
Fire control: Drum Tilt (MR-104); H/I-band.
Navigation: SRN 207M; I-band.

Modernisation: Retained as training vessels.
Operational: Based at Gdynia. 432 decommissioned in 2004. ***UPDATED***

WŁADYSŁAWOWO   6/2003, J Ciślak / 0567498

11 PILICA CLASS (PROJECT 918M)
(COASTAL PATROL CRAFT) (PB)

KP 166-176 166-176

Displacement, tons: 93 full load
Dimensions, feet (metres): 93.8 × 19 × 4.6 *(28.6 × 5.8 × 1.4)*
Main machinery: 3 M 50-F7 diesels; 3,604 hp(m) *(2.65 MW)*; 3 shafts
Speed, knots: 27. **Range, n miles:** 1,160 at 12 kt
Complement: 14 (1 officer)
Guns: 2 ZU-23-2M 23 mm/87 (twin); 400 rds/min to 2 km.
Torpedoes: 2—21 in *(533 mm)* tubes; SET 53M; active/passive homing to 14 km *(7.6 n miles)* at 29 kt; warhead 90 kg.
Radars: Surface search: SRN 301; I-band.
Sonars: MG 329M; dipping VDS.

Comment: Built at Naval Shipyard, Gdynia. First one commissioned 23 April 1977 and the last 24 February 1983. Based at Kolobrzeg. First batch of five, without torpedo tubes are part of the Maritime Frontier Guard. Guns replaced 1986-90.
UPDATED

KP 172   *5/2004*, J Cislak* / 1044475

MINE WARFARE FORCES

0 + (14) KORMORAN CLASS (PROJECT 257)
(MINEHUNTERS-COASTAL MHC)

Displacement, tons: 400 full load
Dimensions, feet (metres): 142.7 × 23.0 × 11.5 *(43.5 × 7.0 × 3.85)*
Main machinery: 2 diesels; 3,700 hp(m) *(2.76 MW)*; 2 shafts. 1 motor; 80 hp (minehunting)
Complement: 36
Missiles: twin 23 mm mount and SA-N-5 (Grail) SAM system.
Sonars: SHL-101/T; hull-mounted.

Comment: Details are speculative. Two batches of six and eight vessels are planned although the construction programme has not been announced. All to be built in a Polish shipyard in conjunction with a foreign partner. The minehunting sonar is under development by Centrum Techniki Morskiej (CTM) and is based on technology from the Thales Underwater Systems TSM 2022 Mk III.
UPDATED

3 KROGULEC CLASS (PROJECT 206FM) (MHCM)

Name	No	Builders	Commissioned
FLAMING	621	Gdynia Shipyard	3 Nov 1966
MEWA	623	Gdynia Shipyard	25 May 1967
CZAJKA	624	Gdynia Shipyard	17 June 1967

Displacement, tons: 550 full load
Dimensions, feet (metres): 190.9 × 25.3 × 6.9 *(58.2 × 7.7 × 2.1)*
Main machinery: 2 Sulzer/Cegielski 6AL 25/30 diesels; 2,203 hp(m) *(1.62 MW)*; 2 shafts; LIPS cp props
Speed, knots: 17. **Range, n miles:** 2,000 at 12 kt
Complement: 52 (6 officers)
Missiles: SAM: 2 Fasta-4M quad launchers. SA-N-5.
2 SA-N-10 (Grom) to be fitted in due course.
Guns: SAM/guns: 2 Wrobel ZU-23-2MR 23 mm (twin) with 2 SA-N-5 missiles.
Depth charges: 2 racks.
Mines: 6—12 depending on type.
Countermeasures: Decoys: 6—9 barrelled Jastrzab 2 launchers for chaff.
ECM: PIT Bren system being fitted.
MCM: 2 Bofors MT2W mechanical, 1 TEM-PE-2MA magnetic and 1 MTA-2 acoustic sweeps. CTM Ukwial ROV with sonar, TV and charges. 10 ZHH 230 sonobuoys.
Combat data systems: CTM Pstrokosz command support system.
Radars: Navigation: Racal Decca Bridgemaster; I-band.
IFF: RAWAR SC-10D2
Sonars: CTM SHL-100MA hull mounted; active minehunting; high frequency; Politechnica Gdansk SHL-200 VDS.

Comment: All taken out of service in 1997. New armament and minehunting equipment installed. Divers recompression chamber carried. *Mewa* returned to service in May 1999, *Czajka* in May 2000 and *Flaming* in 2001. Life extended by ten years.
UPDATED

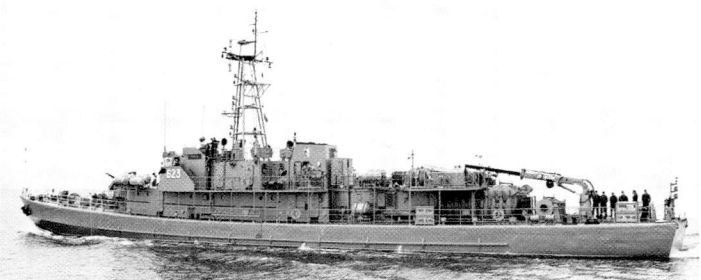

MEWA   *6/2004*, J Ciślak* / 1044478

13 GOPLO (NOTEC) CLASS (PROJECT 207P/207DM)
(MINESWEEPERS/HUNTERS—COASTAL) (MHC)

Name	No	Builders	Launched	Commissioned
GOPLO	630	Naval Shipyard, Gdynia	16 Apr 1981	13 Mar 1982
GARDNO	631	Naval Shipyard, Gdynia	23 June 1993	31 Mar 1984
BUKOWO	632	Naval Shipyard, Gdynia	19 Sep 1984	23 June 1985
DABIE	633	Naval Shipyard, Gdynia	12 June 1985	11 May 1986
JAMNO	634	Naval Shipyard, Gdynia	11 Feb 1986	11 Oct 1986
MIELNO	635	Naval Shipyard, Gdynia	27 June 1986	9 May 1987
WICKO	636	Naval Shipyard, Gdynia	20 Mar 1987	12 Oct 1987
RESKO	637	Naval Shipyard, Gdynia	1 Oct 1987	26 Mar 1988
SARBSKO	638	Naval Shipyard, Gdynia	10 May 1988	12 Oct 1988
NECKO	639	Naval Shipyard, Gdynia	21 Nov 1988	9 May 1989
NAKLO	640	Naval Shipyard, Gdynia	29 May 1989	2 Mar 1990
DRUZNO	641	Naval Shipyard, Gdynia	29 Nov 1989	17 Sep 1990
HANCZA	642	Naval Shipyard, Gdynia	9 July 1990	1 Mar 1991

Displacement, tons: 216 full load
Dimensions, feet (metres): 126.3 × 24.3 × 5.9 *(38.5 × 7.4 × 1.8)*
Main machinery: 2 M 401A1 diesels; 1,874 hp(m) *(1.38 MW)* sustained; 2 shafts
Speed, knots: 14. **Range, n miles:** 1,100 at 9 kt
Complement: 29 (6 officers)
Guns: 2 ZU-23-2MR 23 mm (twin); 400 rds/min combined to 2 km.
Depth charges: 24
Mines: 6-24
Countermeasures: MCM: MMTK1 mechanical; MTA 1 acoustic and TEM-PE 1 magnetic sweeps.
Radars: Navigation: Bridgemaster; I-band.
Sonars: MG 89 or MG 79; active minehunting; high frequency.

Comment: *Goplo* is an experimental prototype numbered 207D. The 23 mm guns have replaced the original 25 mm. GRP hulls. All are to be upgraded to 207DM for minehunting, and to carry divers. Named after lakes and based at Swinoujscie.
UPDATED

BUKOWO   *6/2004*, J Ciślak* / 1044476

4 MAMRY (NOTEC II) CLASS (PROJECT 207M)
(MINESWEEPERS/HUNTERS—COASTAL) (MHSCM)

Name	No	Builders	Launched	Commissioned
MAMRY	643	Naval Shipyard, Gdynia	30 Sep 1991	25 Sep 1992
WIGRY	644	Naval Shipyard, Gdynia	28 Nov 1992	14 May 1993
SNIARDWY	645	Naval Shipyard, Gdynia	18 June 1993	28 Jan 1994
WDZYDZE	646	Naval Shipyard, Gdynia	24 June 1994	2 Dec 1994

Displacement, tons: 216 full load
Dimensions, feet (metres): 126.3 × 24.3 × 5.9 *(38.5 × 7.4 × 1.8)*
Main machinery: 2 M 401A diesels; 1,874 hp(m) *(1.38 MW)*; 2 shafts
2 auxiliary motors; 816 hp(m) *(60 kW)*
Speed, knots: 14. **Range, n miles:** 865 at 14 kt
Complement: 27 (5 officers)
Missiles: SAM/Guns: 2 ZU-23-2MR 23 mm Wrobel II (twin); combination of 2 SA-N-5 missiles; IR homing to 6 km *(3.2 n miles)* at 1.5 Mach; warhead 1.5 kg and guns; 400 rds/min combined to 2 km.
Mines: 6-24 depending on type.
Countermeasures: MCM: MMTK 1m mechanical, MTA 2 acoustic and TEM-PE 1m magnetic sweeps.
Radars: Navigation: SRN 401XTA; I-band.
Sonars: SHL 100/200; hull mounted/VDS; active minehunting; high frequency.

Comment: Modified version of the 207P and equipped to carry divers. Identical hull to the 207P. All based at Hel. An enlarged design, the Type 207 MCMV with a length of 43.5 m, is a longer term project.
UPDATED

WIGRY   *6/2000, Curt Borgenstam* / 0105241

2 LENIWKA CLASS (PROJECT 410S)
(MINESWEEPERS—COASTAL) (MSC)

Name	No	Builders	Commissioned
TR 25	625	Ustka Shipyard	12 Oct 1983
TR 26	626	Ustka Shipyard	12 Oct 1983

Displacement, tons: 269 full load
Dimensions, feet (metres): 84.6 × 23.6 × 8.9 *(25.8 × 7.2 × 2.7)*
Main machinery: 1 Puck-Sulzer 6AL20/24 diesel; 570 hp(m) *(420 kW)*; 1 shaft; cp prop
Speed, knots: 11. **Range, n miles:** 3,200 at 10 kt
Complement: 16
Countermeasures: MCM: MT-3U mechanical, TEM-PE-3 (TR-25) magnetic and BAT 2 acoustic sweeps.
Radars: Navigation: SRN 311; I-band.

Comment: Project 410S modified stern trawlers. Sweeping is done by using strung-out charges. The ships can carry 40 tons of cargo or 40 people. Based at Swinoujscie.

VERIFIED

TR 26   9/2003, J Cislak / 0567500

AMPHIBIOUS FORCES

1 MODIFIED POLNOCHNY C CLASS (PROJECT 776) (LST)

Name	No	Builders	Commissioned
GRUNWALD	811	Northern Shipyard, Gdansk	28 Apr 1973

Displacement, tons: 1,253 full load
Dimensions, feet (metres): 266.7 × 31.8 × 7.9 *(81.3 × 9.7 × 2.4)*
Main machinery: 2 Type 40-D diesels; 4,400 hp(m) *(3.2 MW)* sustained; 2 shafts
Speed, knots: 18. **Range, n miles:** 1,000 at 18 kt; 2,600 at 12 kt
Complement: 45 plus 54 flag staff
Military lift: 2 light trucks
Missiles: SAM: 2 Fasta 4M quad launchers for SA-N-5.
Guns: 4—30 mm (2 twin). 2—140 mm rocket launchers.
Radars: Navigation: SRN 7453 Nogat; I-band.
Fire control: Drum Tilt (MR-104); H/I-band.

Comment: *Grunwald* converted to an amphibious command vessel. Command and electronic equipment fitted on the vehicle deck leaving a small area behind the bow doors for two light trucks or jeeps. Based at Swinoujscie.

VERIFIED

GRUNWALD   9/2003, J Cislak / 0567501

5 LUBLIN CLASS (PROJECT 767) (LST/MINELAYERS) (LST/ML)

Name	No	Builders	Launched	Commissioned
LUBLIN	821	Northern Shipyard, Gdansk	12 July 1988	12 Oct 1989
GNIEZNO	822	Northern Shipyard, Gdansk	7 Dec 1988	23 Feb 1990
KRAKOW	823	Northern Shipyard, Gdansk	7 Mar 1989	27 June 1990
POZNAN	824	Northern Shipyard, Gdansk	5 Jan 1990	8 Mar 1991
TORUN	825	Northern Shipyard, Gdansk	8 June 1990	24 May 1991

Displacement, tons: 1,350 standard; 1,745 full load
Dimensions, feet (metres): 313 × 35.4 × 6.6 *(95.4 × 10.8 × 2)*
Main machinery: 3 Cegielski 6ATL25D diesels; 5,390 hp(m) *(3.96 MW)* sustained; 3 shafts
Speed, knots: 16. **Range, n miles:** 1,400 at 16 kt
Complement: 50 (5 officers)
Military lift: 9 Type T-72 tanks or 9 APC or 17 medium or light trucks. 80 troops plus equipment (821-823); 125 troops plus equipment (824); 135 troops and equipment (825).
Missiles: SAM/Guns: 8 ZU-23-2MR 23 mm Wrobel II (4 twin); combination of 2 SA-N-5 missiles; IR homing to 6 km *(3.2 n miles)* at 1.5 Mach; warhead 1.5 kg and guns; 400 rds/min combined.
Depth charges: 9 throwers for counter-mining.
Mines: 50-134.
Countermeasures: Decoys: 2 12-barrelled 70 mm Derkacz chaff launchers (821 and 825). 2 12-barrelled Jastrzab chaff launchers (822-824).
Radars: Navigation: SRN 7453 and SRN 443XTA; I-band.

Comment: Designed with a through deck from bow to stern and can be used as minelayers as well as for amphibious landings. Folding bow and stern ramps and a stern anchor are fitted. The ship has a pressurised citadel for NBC defence and an upper deck washdown system. Mining capabilities upgraded in 1997/98. Based at Swinoujscie.

UPDATED

GNIEZNO   4/2004*, E & M Laursen / 1044477

3 DEBA CLASS (PROJECT 716) (LCU)

Name	No	Builders	Launched	Commissioned
KD 11	851	Naval Shipyard, Gdynia	13 Nov 1987	7 Aug 1988
KD 12	852	Naval Shipyard, Gdynia	2 July 1990	2 Jan 1991
KD 13	853	Naval Shipyard, Gdynia	26 Oct 1990	3 May 1991

Displacement, tons: 176 full load
Dimensions, feet (metres): 122 × 23.3 × 5.6 *(37.2 × 7.1 × 1.7)*
Main machinery: 3 Type M 401A diesels; 3,000 hp(m) *(2.2 MW)*; 3 shafts
Speed, knots: 20. **Range, n miles:** 430 at 16 kt
Complement: 12
Military lift: 1 tank or 2 vehicles up to 20 tons and 50 troops
Guns: 2 ZU-23-2M 23 mm (twin).
Radars: Surface search: SRN 207A; I-band.

Comment: The plan was to build 12 but the programme was suspended at three through lack of funds. A similar design has been assembled in Iran. Can carry up to six launchers for strung-out charges. Based at Swinoujscie.

UPDATED

KD 11   9/2003, J Cislak / 0567514

SURVEY AND RESEARCH SHIPS

2 MODIFIED FINIK 2 CLASS (PROJECT 874) (AGS)

Name	No	Builders	Launched	Commissioned
HEWELIUSZ	265	Northern Shipyard, Gdansk	11 Sep 1981	27 Nov 1982
ARCTOWSKI	266	Northern Shipyard, Gdansk	20 Nov 1981	27 Nov 1982

Displacement, tons: 1,135 standard; 1,218 full load
Dimensions, feet (metres): 202.1 × 36.7 × 10.8 *(61.6 × 11.2 × 3.3)*
Main machinery: 2 Cegielski-Sulzer 6AL25/30 diesels; 1,920 hp(m) *(1.4 MW)*; 2 auxiliary motors; 204 hp(m) *(150 kW)*; 2 shafts; cp props; bow thruster
Speed, knots: 13. **Range, n miles:** 5,900 at 11 kt
Complement: 49 (10 officers)
Radars: Navigation: SRN 7453 Nogat; SRN 743X; I-band.

Comment: Sister ships to Russian class which were built in Poland, except that *Heweliusz* and *Arctowski* have been modified and have no buoy handling equipment. Equipment includes Atlas Deso, Atlas Ralog and Atlas Dolog survey. Both ships are based at Gdynia. One sister ship, *Planeta*, is civilian operated and the other, *Zodiak*, was decommissioned in 2003.

UPDATED

ARCTOWSKI   7/2001, van Ginderen Collection / 0114792

HEWELIUSZ   6/2004*, J Cislak / 1044480

1 MOMA CLASS (PROJECT 861K) (AGS)

Name	No	Builders	Commissioned
KOPERNIK	261	Northern Shipyard, Gdansk	20 Feb 1971

Displacement, tons: 1,540 full load
Dimensions, feet (metres): 240.5 × 36.8 × 12.8 *(73.3 × 11.2 × 3.9)*
Main machinery: 2 Zgoda-Sulzer 6TD48 diesels; 3,300 hp(m) *(2.43 MW)* sustained; 2 shafts; cp props
Speed, knots: 17. **Range, n miles:** 9,000 at 12 kt
Complement: 20 (8 officers) plus 40 scientists
Radars: Navigation: SRN 7453 Nogat; SRN 743X; I-band.

Comment: Forward crane removed in 1983. Based at Gdynia.

UPDATED

KOPERNIK
10/2003*, J Ciślak / 1044479

2 SURVEY CRAFT (PROJECT 4234) (AGSC)

Name	Builders	Commissioned
K 10	Wisla, Gdansk	6 Feb 1989
K 4	Wisla, Gdansk	25 Sep 1989

Displacement, tons: 45 full load
Dimensions, feet (metres): 62 × 14.4 × 4.9 *(18.9 × 4.4 × 1.5)*
Main machinery: 1 Wola DM 150 diesel; 160 hp(m) *(117 kW)* sustained; 1 shaft
Speed, knots: 9
Complement: 10
Radars: Navigation: SRN 207A; I-band.

Comment: Coastal survey craft based at Gdynia. There are a number of survey launches and buoy tenders listed under *Auxiliaries*.

VERIFIED

K 10
5/2000, J Ciślak / 0105248

5 SURVEY CRAFT (PROJECT III/C) (AGSC)

M 35	M 37-40

Displacement, tons: 10 full load
Dimensions, feet (metres): 36.1 × 10.5 × 2.3 *(11 × 3.2 × 0.7)*
Main machinery: 1 Puck Rekin SW 400/MZ diesel; 95 hp(m) *(70 kW)*; 1 shaft
Speed, knots: 8. **Range, n miles:** 184 at 8 kt
Complement: 5
Radars: Navigation: SRN 207A; I-band.

Comment: Based at Gdynia and Swinoujscie (M 35).

VERIFIED

M 40
3/2003, J Ciślak / 0567515

INTELLIGENCE VESSELS

2 MODIFIED MOMA CLASS (PROJECT 863) (AGI)

Name	No	Builders	Commissioned
NAWIGATOR	262	Northern Shipyard, Gdansk	17 Feb 1975
HYDROGRAF	263	Northern Shipyard, Gdansk	8 May 1976

Displacement, tons: 1,677 full load
Dimensions, feet (metres): 240.5 × 35.4 × 12.8 *(73.3 × 10.8 × 3.9)*
Main machinery: 2 Zgoda-Sulzer 6TD48 diesels; 3,300 hp(m) *(2.43 MW)* sustained; 2 shafts
Speed, knots: 17. **Range, n miles:** 7,200 at 12 kt
Complement: 87 (10 officers)
Missiles: 2 Fasta-4M quad launchers. SA-N-5.
Guns: 4—25 mm (2 twin) (262).
Countermeasures: ESM/ECM: intercept and jammer.
Radars: Navigation: 2 SRN 7453 Nogat; I-band.

Comment: Much altered in the upperworks and unrecognisable as Momas. The forecastle in *Hydrograf* is longer than in *Nawigator* and one deck higher. *Hydrograf* fitted for but not with two twin 25 mm gun mountings. Forward radome replaced by a cylindrical type in *Nawigator* and after ones removed on both ships. Based at Gdynia.

VERIFIED

NAWIGATOR
7/2003, J Ciślak / 0567516

HYDROGRAF
10/2001, J Ciślak / 0126242

TRAINING SHIPS

Notes: The three masted sailing ship *Dar Mlodziezy* is civilian owned and operated but also takes naval personnel for training.

1 WODNIK CLASS (PROJECT 888) (AXTH)

Name	No	Builders	Launched	Commissioned
WODNIK	251	Northern Shipyard, Gdansk	19 Nov 1975	28 May 1976

Displacement, tons: 1,697 standard; 1,820 full load
Dimensions, feet (metres): 234.3 × 38.1 × 14.8 *(71.4 × 11.6 × 4.5)*
Main machinery: 2 Zgoda-Sulzer 6TD48 diesels; 2,650 hp(m) *(1.95 MW)* sustained; 2 shafts; cp props
Speed, knots: 16. **Range, n miles:** 7,200 at 11 kt
Complement: 56 (24 officers) plus 101 midshipmen
Guns: 4 ZU-23-2MR Wrobel 23 mm (2 twin). 2 *(Wodnik)* or 4 *(Gryf)* 30 mm AK 230 (1 or 2 twin).
Radars: Navigation: 2 SRN 7453 Nogat; I-band.
Helicopters: Platform for 1 light.

Comment: Sister to former GDR *Wilhelm Pieck* and two Russian ships. Converted to a hospital ship (150 beds) in 1990 for deployment to the Gulf. Armament removed as part of the conversion but partially restored in 1992. Based at Gdynia. Second of class in reserve from 1999.

UPDATED

WODNIK
10/2003*, J Ciślak / 1044481

1 PROJECT 722 CLASS (AXL)

Name	No	Builders	Launched	Commissioned
PODCHORAZY	711	Wisla Shipyard, Gdansk	6 Apr 1974	30 Nov 1974

Displacement, tons: 180 (167, *Bryza*) full load
Dimensions, feet (metres): 94.5 × 21.7 × 6.4 *(28.8 × 6.6 × 2)*
Main machinery: 2 Wola DM 300 diesels; 300 hp(m) *(220 kW)*; 2 shafts
Speed, knots: 11. **Range, n miles:** 1,410 at 11 kt
Complement: 11 plus 29 cadets
Radars: Navigation: 2 TRN 823; I-band.

Comment: Based at Hel.

VERIFIED

PROJECT 722 CLASS   9/2003, J Ciślak / 0567503

1 ISKRA CLASS (PROJECT B79) (SAIL TRAINING SHIP) (AXS)

Name	No	Builders	Launched	Commissioned
ISKRA	253	Gdansk Shipyard	6 Mar 1982	11 Aug 1982

Displacement, tons: 498 full load
Dimensions, feet (metres): 160.8 × 26.6 × 13.1 *(49 × 8.1 × 4.0)*
Main machinery: 1 Wola 75H12 diesel; 310 hp(m) *(228 kW)*; 1 auxiliary shaft; cp prop
Speed, knots: 9 (diesel)
Complement: 14 (6 officers) plus 50 cadets
Radars: Navigation: SRN 206; I-band.

Comment: Barquentine with 1,040 m² of sail. Used by the Naval Academy for training with a secondary survey role. Based at Gdynia.

VERIFIED

ISKRA   7/2003, J Ciślak / 0567504

AUXILIARIES

Note: Procurement of up to four Strategic Support Ships is reported to be under development. The broad requirement is for ships of approximately 10,000 tons with the capability of transporting about 500 troops plus some twenty vehicles and up to six helicopters. Funding is not thought to have been approved.

1 PROJECT 890 CLASS
(LOGISTICS SUPPORT VESSEL) (AKHM/APHM/AGI)

KONTRADMIRAL X CZERNICKI 511

Displacement, tons: 2,250 full load
Dimensions, feet (metres): 239.3 × 45.3 × 13.4 *(72.9 ×13.8 × 4.1)*
Main machinery: 2 Cegielski-Sulzer AL25D diesels; 2,934 hp(m) *(2.16 MW)* sustained; 2 shafts
Speed, knots: 14.1. **Range, n miles:** 7,000 at 12 kt
Complement: 38
Military lift: 140 troops with full individual armament or ten 20 ft containers or four 20 ft containers and six STAR 266 army trucks
Missiles: SAM/Guns: 1 ZU 23-2MR Wrobel I/II mounts: combination of 2 Strela 2M (Grail) missiles and 2—23 mm guns.
Countermeasures: Decoys: 4 WNP81/9 9 barrelled 81 mm Jastrzab chaff launchers.
ESM: PIT intercept.
Radars: Surface search: SRN; E/F-band.
Navigation: SRN; I-band.
Helicopters: Platform for 1 helicopter (up to ten ton).

Comment: Conversion from a Project 130 Degaussing Vessel to Logistic Support Ship in Northern Shipyard, Gdansk, has included new upper and forward hull sections, provision of a helicopter deck and NBC protection. The ship has a 16 ton hydraulic crane and after ramp. The multirole ship is capable of sealift, acting as a forward maintenance unit and maritime surveillance and reconnaissance (using containerised ESM sensors) and replenishment at sea. Commissioned on 1 September 2001.

VERIFIED

KONTRADMIRAL X CZERNICKI   4/2003, A Sharma / 0567505

1 BALTYK CLASS (PROJECT ZP 1200) (TANKER) (AORL)

Name	No	Builders	Commissioned
BALTYK	Z 1	Naval Shipyard, Gdynia	11 Mar 1991

Displacement, tons: 2,937 standard; 3,049 full load
Dimensions, feet (metres): 278.2 × 43 × 15.4 *(84.8 × 13.1 × 4.7)*
Main machinery: 2 Cegielski 8 ASL 25 diesels; 4,025 hp(m) *(2.96 MW)*; 2 shafts; cp props
Speed, knots: 15. **Range, n miles:** 4,250 at 12 kt
Complement: 34 (4 officers)
Cargo capacity: 1,184 tons fuel, 92.7 tons lub oil
Guns: 4 ZU-23-2M Wrobel 23 mm (2 twin).
Radars: Navigation: SRN 7453 and SRN 207A; I-band.

Comment: Beam replenishment stations, one each side. First of a projected class of four, of which the others were cancelled. Based at Gdynia.

UPDATED

BALTYK   6/2004*, J Ciślak / 1044482

1 MOSKIT CLASS (PROJECT B 199) (TANKER) (AOTL)

Name	No	Builders	Launched	Commissioned
MEDUZA	Z 8	Rzeczna, Wroclaw Shipyard	14 Sep 1969	21 July 1970

Displacement, tons: 1,225 full load
Dimensions, feet (metres): 190.3 × 30.5 × 10.8 *(58 × 9.3 × 3.3)*
Main machinery: 1 Magdeburg diesel; 965 hp(m) *(720 kW)*; 1 shaft
Speed, knots: 10. **Range, n miles:** 1,200 at 10 kt
Complement: 21 (3 officers)
Cargo capacity: 656.5 tons
Guns: 4 ZU-23-2M 23 mm (2 twin).
Radars: Navigation: TRN 823; I-band.

UPDATED

MEDUZA   7/2004*, J Ciślak / 1044484

2 KORMORAN CLASS (YPT)

Name	Builders	Launched	Commissioned
K 8	Naval Shipyard, Gdynia	26 Aug 1970	3 July 1971
K 11	Naval Shipyard, Gdynia	23 June 1971	11 Dec 1971

Displacement, tons: 150 full load
Dimensions, feet (metres): 114.8 × 19.7 × 5.2 *(35 × 6 × 1.6)*
Main machinery: 2 Type M 50F5 diesels; 2,200 hp(m) *(1.6 MW)*; 2 shafts
Speed, knots: 19. **Range, n miles:** 550 at 15 kt
Complement: 24
Guns: 2 ZU-23-2M Wrobel 23 mm (twin).
Radars: Navigation: SRN 206/301; I-band.

Comment: Armament updated in 1993. Both based at Gdynia.

VERIFIED

K 8   10/2001, J Ciślak / 0126239

3 MROWKA CLASS (PROJECT B 208)
(DEGAUSSING VESSELS) (YDG)

Name	No	Builders	Commissioned
WRONA	SD 11	Naval Shipyard, Gdynia	10 Oct 1971
RYS	SD 12	Naval Shipyard, Gdynia	25 June 1972
—	SD 13	Naval Shipyard, Gdynia	16 Dec 1972

Displacement, tons: 660 full load
Dimensions, feet (metres): 144.4 × 26.6 × 9.5 *(44 × 8.1 × 2.9)*
Main machinery: 1 6NV D36 diesel; 957 hp(m) *(704 kW)*; 1 shaft
Speed, knots: 9.5. **Range, n miles:** 2,230 at 9.5 kt
Complement: 37
Guns: 2—25 mm (twin) (SD 11 and 13); 2 ZU-23-2M Wrobel 23 mm (twin) (SD 12).
Radars: Navigation: SRN 206; I-band.

Comment: Names are unofficial. SD 12 based at Swinoujscie, remainder at Gdynia.

UPDATED

RYS　　　　　　　　　　　　　　*4/2004*, Hartmut Ehlers /* 1044483

2 PIAST CLASS (PROJECT 570) (SALVAGE SHIPS) (ARS)

Name	No	Builders	Commissioned
PIAST	281	Northern Shipyard, Gdansk	26 Jan 1974
LECH	282	Northern Shipyard, Gdansk	30 Nov 1974

Displacement, tons: 1,887 full load
Dimensions, feet (metres): 238.5 × 38.1 × 13.1 *(72.7 × 11.6 × 4)*
Main machinery: 2 Zgoda-Sulzer 6TD48 diesels; 3,300 hp(m) *(2.43 MW)* sustained; 2 shafts; cp props
Speed, knots: 15. **Range, n miles:** 3,000 at 12 kt
Complement: 56 (8 officers) plus 12 spare
Missiles: SAM: 2 Fasta 4M twin launchers for SA-N-5.
Guns: 4—25 mm (2 twin).
Radars: Navigation: 2 SRN 7453 Nogat; I-band.

Comment: Basically a Moma class hull with towing and firefighting capabilities. Ice-strengthened hulls. Wartime role as hospital ships. Carry three-man diving bells capable of 100 m depth and a decompression chamber. ROV added and other salvage improvements made in 1997/98. Based at Gdynia. Guns may not be carried.

UPDATED

LECH　　　　　　　　　　　　　　*5/2004*, J Cislak /* 1044485

2 ZBYSZKO CLASS (PROJECT B 823) (SALVAGE SHIPS) (ARS)

Name	No	Builders	Commissioned
ZBYSZKO	R 14	Ustka Shipyard	8 Nov 1991
MACKO	R 15	Ustka Shipyard	20 Mar 1992

Displacement, tons: 380 full load
Dimensions, feet (metres): 114.8 × 26.2 × 9.8 *(35 × 8 × 3)*
Main machinery: 1 Sulzer 6AL20/24D; 750 hp(m) *(551 kW)*; 1 shaft
Speed, knots: 11. **Range, n miles:** 3,000 at 10 kt
Complement: 15
Radars: Navigation: SRN 402X; I-band.

Comment: Type B-823 ordered 30 May 1988. Carries a decompression chamber and two divers. Mobile gantry crane on the stern. Based at Kolobrzeg.

UPDATED

MACKO　　　　　　　　　　　　*5/2004*, J Cislak /* 1044486

1 PLUSKWA CLASS (PROJECT R-30) (SALVAGE TUG) (ATS)

Name	No	Builders	Commissioned
GNIEWKO	R 11	Naval Shipyard, Gdynia	29 Sep 1981

Displacement, tons: 365 full load
Dimensions, feet (metres): 105 × 29.2 × 10.2 *(32 × 8.9 × 3.1)*
Main machinery: 1 Cegielski-Sulzer 6AL25/30 diesel; 1,470 hp(m) *(1.08 MW)*; 1 shaft
Speed, knots: 12. **Range, n miles:** 4,000 at 7 kt
Complement: 18
Radars: Navigation: SRN 443 XEA; I-band.

Comment: Based at Hel. Bollard pull 15 tons.

UPDATED

GNIEWKO　　　　　　　　　　　*6/2004*, J Cislak /* 1044489

1 DIVING TENDER AND TRANSPORT (YDT/YFB)

K 12

Displacement, tons: 46 full load
Dimensions, feet (metres): 61.7 × 14.4 × 4.9 *(18.8 × 4.4 × 1.5)*
Main machinery: 1 Wola diesel; 150 hp(m) *(112 kW)*; 1 shaft
Speed, knots: 9
Complement: 4-10
Radars: Navigation: SRN 2061; I-band.

Comment: Harbour craft used for various purposes. Built in 1979.

VERIFIED

HARBOUR CRAFT　　　　　　　*11/2000, J Cislak /* 0105257

3 TRANSPORT CRAFT (YFB)

M 1　　　　　　　　M 3　　　　　　　　M 32

Displacement, tons: 74 full load
Dimensions, feet (metres): 94.2 × 19 × 4.3 *(28.7 × 5.8 × 1.3)*
Main machinery: 3 M50F5 diesels; 3,600 hp(m) *(2.65 MW)*; 3 shafts
Speed, knots: 27
Complement: 7 plus 30
Radars: Navigation: SRN 207A; I-band.

Comment: Details given are for M 1 built at Gdynia. The other two are similar but slower and smaller. All can be used as emergency patrol craft. M 1 based at Gdynia as an Admirals' launch.

UPDATED

M 3　　　　　　　　　　　　　　*7/2004*, J Cislak /* 1044490

15 MISCELLANEOUS HARBOUR CRAFT

B 3, B 7, B 9, B 11-13, W 2, M 5, M 12, M 22, M 30, M 37-40

Comment: M numbers are patrol launches; B numbers are freighters and oil lighters; *W 2* is a floating workshop.

UPDATED

M 22   *3/2002, J Ciślak / 0567510*

B-12   *7/2004*, J Ciślak / 1044491*

TUGS

1 H 960 CLASS (ATA)

H 6

Displacement, tons: 340 full load
Dimensions, feet (metres): 91.2 × 26.2 × 12.1 *(27.8 × 8 × 3.7)*
Main machinery: 1 Sulzer GATL 25 D diesels; 1,306 hp(m) *(960 kW)*; 1 shaft
Speed, knots: 12. **Range, n miles:** 1,150 at 12 kt
Complement: 17 (1 officer)
Radars: Navigation: SRN 401 XTA; I-band.

Comment: Built at Nauta Ship Repair Yard, Gdynia and commissioned 25 September 1992. Based at Hel.

UPDATED

H 6   *6/2004*, J Ciślak / 1044488*

8 HARBOUR TUGS (PROJECTS H 900, H 800, H 820) (YTB/YTM)

H 1-2 (Type 800)   H 3-5, H 7 (Type 900)   H 9-10 (Type 820)

Displacement, tons: 218 full load
Dimensions, feet (metres): 84 × 22.3 × 11.5 *(25.6 × 6.8 × 3.5)*
Main machinery: 1 Cegielski-Sulzer 6AL20/24H diesel; 935 hp(m) *(687 kW)*; 1 shaft
Speed, knots: 11. **Range, n miles:** 1,500 at 10 kt
Complement: 17
Radars: Navigation: SRN 206; I-band.

Comment: Details given are for *H 3, 4, 5* and *7*. Completed 1979-81. Have firefighting capability except *H 9-10. H 1-2* completed 1970 and *H 9-10* in 1993.

UPDATED

H 7   *5/2004*, J Ciślak / 1044487*

SEA DETACHMENT OF THE BORDER GUARD (MOSG)

Headquarters Appointments

Commandant MOSG:
  Rear Admiral Konrad Wiśniowski
Deputy Commandant:
  Captain Marek Borkowski
Deputy Commandant:
  Commander Marec Ilnicki

Bases

Gdansk (HQ and Kaszubski Division)
Swinoujscie (Pomorski Division)

General

MOSG (Morski Oddzial Strazy Granicznej) formed on 1 August 1991. Vessels have blue hulls with red and yellow striped insignia. Superstructures are painted white. The use of ships' names was discontinued in 2004. MOSG also operates one Piper PA 34 Seneca II patrol aircraft.

PATROL FORCES

2 OBLUZE CLASS (PROJECT 912) (LARGE PATROL CRAFT) (WPB)

No	Builders	Commissioned
SG-323	Naval Shipyard, Gdynia	10 June 1967
SG-325	Naval Shipyard, Gdynia	31 Jan 1968

Displacement, tons: 236 full load
Dimensions, feet (metres): 135.5 × 21.3 × 7 *(41.3 × 6.5 × 2.1)*
Main machinery: 2 40DM diesels; 4,400 hp(m) *(3.24 MW)* sustained; 2 shafts
Speed, knots: 24. **Range, n miles:** 1,200 at 12 kt
Complement: 19
Guns: 4 AK 230 30 mm (2 twin) (SG 323). 2 AK 230 30 mm (twin) (SG 325).
Depth charges: 2 internal racks.
Radars: Surface search: SRN 207; I-band.
Sonars: Tamir II (MG 11); hull-mounted; active attack; high frequency.

Comment: First of class *Fala* is now a museum ship. Both based at Kolobrzeg.

UPDATED

SG-323   *5/2004*, J Ciślak / 1044492*

2 KAPER CLASS (PROJECT SKS-40)
(LARGE PATROL CRAFT) (WPB)

No	Builders	Commissioned
SG-311	Wisla Yard, Gdansk	21 Jan 1991
SG-312	Wisla Yard, Gdansk	3 Apr 1992

Displacement, tons: 470 full load
Dimensions, feet (metres): 139.4 × 27.6 × 9.2 *(42.5 × 8.4 × 2.8)*
Main machinery: 2 Sulzer 8ATL25/30 diesels; 4,720 hp(m) *(3.47 MW)*; 2 shafts; cp props
Speed, knots: 17. **Range, n miles:** 2,800 at 14 kt
Complement: 15
Guns: 2—7.62 mm MGs.
Radars: Surface search: SRN 207; I-band.
Navigation: Racal Decca; I-band.

Comment: *Kaper I* completed at Wisla Yard, Gdansk in January 1991, *Kaper II* on 1 October 1994. Have Simrad fish-finding sonars fitted. Used for Fishery Protection. 311 based at Gdansk and 312 at Kolobrzeg. *UPDATED*

SG-311   *5/2004*, J Ciślak /* 1044493

6 WISLOKA CLASS (PROJECT 90)
(COASTAL PATROL CRAFT) (WPB)

SG-142   SG-144-146   SG-150   SG-152

Displacement, tons: 45 full load
Dimensions, feet (metres): 69.6 × 14.8 × 5.2 *(21.2 × 4.5 × 1.6)*
Main machinery: 2 Wola 31 ANM28 H12A diesels; 1,000 hp(m) *(735 kW)*; 2 shafts
Speed, knots: 18. **Range, n miles:** 300 at 18 kt
Complement: 6
Guns: 2—12.7 mm MGs (twin) and 1 ZM rocket launcher.
Radars: Surface search: SRN 207; I-band.

Comment: Built at Wisla Shipyard, Gdansk and completed between October 1973 and August 1977. Three are based at Gdansk and three at Swinoujscie. *UPDATED*

SG-152   *4/2004*, Hartmut Ehlers /* 1044494

2 PILICA CLASS (PROJECT 918)
(COASTAL PATROL CRAFT) (WPB)

SG-161   SG-164

Displacement, tons: 93 full load
Dimensions, feet (metres): 94.8 × 19 × 4.6 *(28.9 × 5.8 × 1.4)*
Main machinery: 3 M 50-F6 diesels; 3,604 hp(m) *(2.65 MW)*; 3 shafts
Speed, knots: 27. **Range, n miles:** 1,160 at 12 kt
Complement: 12
Guns: 2 ZU-23-2M Wrobel 23 mm (twin).
Radars: Surface search: SRN 231; I-band.

Comment: Same as naval craft but without the torpedo tubes and sonar. Built by Naval Shipyard, Gdynia and completed between June 1973 and October 1974. One based at Swinoujscie and one at Gdansk. *VERIFIED*

SG-161   *3/2002, J Ciślak /* 0567519

1 PATROL LAUNCH (PROJECT M-35) (WYFL)

SG 036

Displacement, tons: 41 full load
Dimensions, feet (metres): 35.3 × 14.4 × 5.2 *(10.7 × 4.4 × 1.6)*
Main machinery: 1 Wola DM 150 diesel; 150 hp *(112 kW)*
Speed, knots: 8
Complement: 4

Comment: Built in 1985. Similar to those in Polish naval service. *UPDATED*

SG 036   *5/2003, J Ciślak /* 0567518

4 SPORTIS CLASS (PROJECT 7500)
FAST INTERCEPT CRAFT (WPBF)

SG-002-005

Displacement, tons: 2
Dimensions, feet (metres): 24.6 × 9.2 × 1.3 *(7.5 × 2.8 × 0.4)*
Main machinery: Volvo Penta 230 hp (170 kW)
Speed, knots: 42
Complement: 3

Comment: Built in Bojano in 1996. *VERIFIED*

SG-003   *5/2003, J Ciślak /* 0567484

SG-005   *5/2002, J Ciślak /* 0567485

1 PATROL CRAFT (PROJECT MI-6) (WPB)

SG-008

Displacement, tons: 16
Dimensions, feet (metres): 42.7 × 12.14 × 3.6 *(13.0 × 3.7 × 1.1)*
Main machinery: 1 Wola; 200 hp *(147 kW)*; 1 shaft
Speed, knots: 11
Complement: 4

Comment: Harbour craft built at Wisla Shipyard, Gdansk, 1989. *UPDATED*

SG-008   *5/2003, J Ciślak /* 0567483

2 STRAZNIK CLASS (PROJECT SAR-1500) (WPBF)

No	Builders	Commissioned
SG-211	Damen Yard, Gdynia	29 Apr 2000
SG-212	Damen Yard, Gdynia	7 July 2000

Displacement, tons: 26
Dimensions, feet (metres): 49.9 × 17.7 × 2.95 *(15.2 × 5.39 × 0.90)*
Main machinery: 2 MAN D2848 diesels; 1,360 hp *(1,000 kW)*; water jet system
Speed, knots: 35. **Range, n miles:** 200 at 30 kt
Complement: 4 (1 officer)
Guns: 1—7.62 mm MG.
Radars: Surface search: SIMRAD; I-band.

Comment: Contract between MOSG and Damen Shipyard signed 5 October 1999. Based on Dutch SAR 1500 lifeboat. Hull and superstructure of aluminium alloy.   *UPDATED*

SG-212   *4/2004*, Hartmut Ehlers /* 1044495

6 MODIFIED SPORTIS CLASS (PROJECT S-6100) (FAST INTERCEPT CRAFT) (WPBF)

SG 061-066

Displacement, tons: 1.9
Dimensions, feet (metres): 20.0 × 7.5 × 1.3 *(6.1 × 2.3 × 0.4)*
Main machinery: 2 Johnson outboard motors; 120 hp *(89.6 kW)*
Speed, knots: 35
Complement: 2

Comment: Built at Bojano in 2001. Located at Border units along the coast.   *UPDATED*

SG 063   *6/2003, MOSG /* 0567506

Portugal
MARINHA PORTUGUESA

Country Overview

The Republic of Portugal is situated in south-western Europe in the western portion of the Iberian Peninsula. It is bordered to the north and east by Spain and has a 967 n mile coastline with the Atlantic Ocean. The Azores and Madeira Islands in the Atlantic are integral parts of the republic, the total area of which is 35,553 square miles. Lisbon is the capital, largest city and principal port. There are further ports at Leixões (near Oporto), Setúbal, and Funchal (Madeira). Territorial seas (12 n miles) and an EEZ (200 n miles) are claimed.

Headquarters Appointments

Chief of Naval Staff:
    Admiral Francisco António Torres Vidal Abreu
Deputy Chief of Naval Staff:
    Vice Admiral João Manuel Lopes Pires Neves
Naval Commander:
    Vice Admiral Henrique Alexandre Machado da Silva da Fonseca
Azores Maritime Zone Commander:
    Rear Admiral António Alberto Rodrigues Cabral
Madeira Maritime Zone Commander:
    Captain Raúl Bernardo Mourato Ramos Gouveia
Marine Corps Commander:
    Rear Admiral Fernando Manuel Oliveira Vargas de Matos

Diplomatic Representation

Defence and Naval Attaché in London, Dublin and The Hague:
    Captain Augusto Mourão Ezequiel
Naval Attaché in Washington and Ottawa:
    Colonel Isidro de Morais Pereira
Defence Attaché in Luanda, Kinshasa, Brazzaville and Windhoek:
    Captain Luís Augusto Loureiro Nunes
Defence Attaché in Maputo, Lillongwe, Harare and Dar-Es-Salam
    Colonel Armandio Amador Pires Pinelo
Defence Attaché in Madrid, Cairo and Athens:
    Colonel Carlos Manuel Santos Gaudêncio

Diplomatic Representation — continued

Defence Attaché in S. Tomé and Libreville:
    Lieutenant Colonel Joaquim Epifânio Santana Santos
Defence Attaché in Bissau, Conakry and Dakar:
    Colonel João Pereira de Araújo
Defence Attaché in Brasilia:
    Colonel Duarte Veríssimo Pires Torrão
Defence Attaché in Berlin, Prague, Copenhagen, Stockholm and Oslo:
    Captain Jorge Alberto Araújo Cunha Serra
Defence Attaché in Warsaw, Budapest, Kiev, Bucharest and Bratislava:
    Colonel Pedro Miguel de Palhares Veloso da Silva
Defence Attaché in Canberra, Dili and Jakarta:
    Colonel Carlos Manuel Cristina de Aguiar
Defence Attaché in Paris, Luxembourg and Brussels:
    Colonel Vitor Manuel Nunes dos Santos
Defence Attaché in Rabat and Tunis:
    Colonel Alberto Jorge Crispim Gomes
Defence Attaché in Praia:
    Lieutenant Colonel José Maria Rebocho Pais Paula Santos
Defence Attaché in Pretoria:
    Commander José António Ruivo
Defence Attaché in Rome, Tel-Aviv and Ankara:
    Commander José Luis Santos Alcobia

Personnel

(a)   2005: 11,670 (1,500 officers) including 1,660 marines
(b)   4 months national service

Marine Corps

2 battalions, 1 special operations detachment, 1 naval police unit

Bases

Main Base: Lisbon—Alfeite
Dockyard: Arsenal do Alfeite
Fleet Support: Porto, Portimão, Funchal, Ponta Delgada, Tróia
Air Base: Montijo (Lisbon)

Naval Air

The helicopter squadron was formally activated on 23 September 1993 at Montijo air force base, Lisbon. Operational and logistic procedures are similar to the air force.

Prefix to Ships' Names

NRP (Navio da República Portuguesa)

Strength of the Fleet

Type	Active (Reserve)	Building (Projected)
Submarines (Patrol)	2	(3)
Frigates	6	(3)
Corvettes	7	—
Patrol Craft	4	(10)
Coastal/River Patrol Craft	12	(5)
LPD	—	(1)
LCTs/LST	1	—
Survey Ships and Craft	7	—
Sail Training Ships	4	—
Replenishment Tanker	1	(1)
Buoy Tenders	2	(2)

DELETIONS

Corvettes

2002   *Oliveira e Carmo, Honorio Barreto*
2004   *Augusto Castilho*

Patrol Forces

2002   *Geba, Cunene, Mandovi*
2003   *Zambeze, Limpopo, Andorinha*

Auxiliaries

2002   *Almeida Carvalho*

PENNANT LIST

Submarines

S 164	Barracuda
S 166	Delfim

Frigates

F 330	Vasco da Gama
F 331	Alvares Cabral
F 332	Corte Real
F 480	Comandante João Belo
F 481	Comandante Hermenegildo Capelo
F 483	Comandante Sacadura Cabral

Corvettes

F 471	Antonio Enes
F 475	João Coutinho
F 476	Jacinto Candido

F 477	Gen Pereira d'Eça
F 486	Baptista de Andrade
F 487	João Roby
F 488	Afonso Cerqueira

Patrol Forces

P 370	Rio Minho
P 1140	Cacine
P 1144	Quanza
P 1146	Zaire
P 1150	Argos
P 1151	Dragão
P 1152	Escorpião
P 1153	Cassiopeia
P 1154	Hidra
P 1155	Centauro
P 1156	Orion
P 1157	Pégaso
P 1158	Sagitario

P 1161	Save
P 1165	Aguia
P 1167	Cisne

Amphibious Forces

LDG 203	Bacamarte

Service Forces

A 520	Sagres
A 521	Schultz Xavier
A 522	D. Carlos I
A 523	Almirante Gago Coutinho
A 5201	Vega
A 5203	Andromeda
A 5204	Polar
A 5205	Auriga
A 5210	Bérrio
UAM 201	Creoula

SUBMARINES

0 + 2 TYPE 209PN CLASS (SSK)

Name	No	Builders	Laid down	Launched	Commissioned
—	—	—	2005	2007	2009
—	—	—	2006	2008	2010

Displacement, tons: 1,700 (surfaced); 1,800 (dived)
Dimensions, feet (metres): 222.8 × 20.7 × 21.6
(67.9 × 6.3 × 6.6)
Main machinery: 2 MTU 16V 396 diesels; 5,600 hp(m)
(4.17 MW); 1 Siemens Permasyn motor; 1 shaft; 2 HDW PEM
fuel cells; 240 kW
Speed, knots: 20 dived; 12 surfaced
Complement: 32 (5 officers)

Torpedoes: 8—21 in *(533 mm)* bow tubes. 16 weapons
including torpedoes and SSM.
Countermeasures: To be announced.
Weapons control: To be announced.
Radars: To be announced.
Sonars: Bow, flank and towed arrays.

Programmes: Contract signed on 21 April 2004 with German
Submarine Consortium (GSC) for construction and delivery of
two boats with option for a third. The consortium consists of
Howaldtswerke-Deutsche Werft, Kiel, Nordseewerke, Emden
(NSWE) and Ferrostaal, Essen.
Structure: Very similar to the Type 214 Air-Independent
Propulsion (AIP) submarines under construction for Greece.
Diving depth likely to be about 400 m *(1,300 ft)*.
NEW ENTRY

2 ALBACORA (DAPHNÉ) CLASS (SSK)

Name	No	Builders	Laid down	Launched	Commissioned
BARRACUDA	S 164	Dubigeon-Normandie, Nantes	19 Oct 1965	24 Apr 1967	4 May 1968
DELFIM	S 166	Dubigeon-Normandie, Nantes	14 May 1967	23 Sep 1968	1 Oct 1969

Displacement, tons: 869 surfaced; 1,043 dived
Dimensions, feet (metres): 189.6 × 22.3 × 17.1
(57.8 × 6.8 × 5.2)
Main machinery: Diesel-electric; 2 SEMT-Pielstick 12 PA4 V 185
diesels; 2,450 hp(m) *(1.8 MW)*; 2 Jeumont Schneider
alternators; 1.7 MW; 2 motors; 2,600 hp(m) *(1.9 MW)*; 2
shafts
Speed, knots: 13.5 surfaced; 16 dived
Range, n miles: 2,710 at 12.5 kt surfaced; 2,130 at 10 kt
snorting
Complement: 54 (7 officers)

Torpedoes: 12—21.7 in *(550 mm)* (8 bow, 4 stern) tubes. ECAN
E14; anti-surface; passive homing to 12 km *(6.6 n miles)* at
25 kt; warhead 300 kg or ECAN L3; anti-submarine; active
homing to 5.5 km *(3 n miles)* at 25 kt; warhead 200 kg. No
reloads.
Countermeasures: ESM: ARUR; radar warning.
Weapons control: DLT D3 torpedo control.
Radars: Surface search: Kelvin Hughes KH 1007; I-band.
Sonars: Thomson Sintra DSUV 2; passive search and attack;
medium frequency.
  DUUA 2; active search and attack; 8.4 kHz.

Modernisation: New radar fitted in 1993-94.
Structure: Diving depth, 300 m *(984 ft)*.
Operational: *Albacora* paid off mid-2000 and cannibalised for
spares. *Delfim* and *Barracuda* expected to remain in service
until 2006 and 2009 respectively.
UPDATED

DELFIM
6/2004*, B Sullivan / 1044177

DELFIM

6/2003, B Sullivan / 0569801

For details of the latest updates to *Jane's Fighting Ships* online and to discover the
additional information available exclusively to online subscribers please visit
jfs.janes.com

FRIGATES

Notes: It is planned to replace the João Belo class frigates with three frigates. This is likely to be fulfilled by the transfer of two decommissioned Oliver Hazard Perry class (possibly FFG 12 and FFG 14) in 2006 and a third ship in 2010. Further transfers are anticipated.

3 VASCO DA GAMA (MEKO 200) CLASS (FFGHM)

Name	No	Builders	Laid down	Launched	Commissioned
VASCO DA GAMA	F 330	Blohm + Voss, Hamburg	1 Feb 1989	26 June 1989	18 Jan 1991
ALVARES CABRAL	F 331	Howaldtswerke, Kiel	2 June 1989	6 June 1990	24 May 1991
CORTE REAL	F 332	Howaldtswerke, Kiel	24 Nov 1989	6 June 1990	22 Nov 1991

Displacement, tons: 2,700 standard; 3,300 full load
Dimensions, feet (metres): 380.3 oa; 357.6 pp × 48.7 × 20 (115.9; 109 × 14.8 × 6.1)
Main machinery: CODOG; 2 GE LM 2500 gas turbines; 53,000 hp (39.5 MW) sustained; 2 MTU 12V 1163 TB83 diesels; 8,840 hp(m) (6.5 MW); 2 shafts; cp props
Speed, knots: 32 gas; 20 diesel
Range, n miles: 4,900 at 18 kt; 9,600 at 12 kt
Complement: 182 (23 officers) (including aircrew of 16 (4 officers)) plus 16 Flag Staff

Missiles: SSM: 8 McDonnell Douglas Harpoon (2 quad) launchers ❶; active radar homing to 130 km (70 n miles) at 0.9 Mach; warhead 227 kg.
SAM: Raytheon Sea Sparrow Mk 29 Mod 1 octuple launcher ❷; RIM-7M; semi-active radar homing to 14.6 km (8 n miles) at 2.5 Mach; warhead 39 kg. Space left for VLS Sea Sparrow ❸.
Guns: 1 Creusot-Loire 3.9 in (100 mm)/55 Mod 68 CADAM ❹; 60 rds/min to 17 km (9 n miles) anti-surface; 8 km (4.4 n miles) anti-aircraft; weight of shell 13.5 kg.
1 General Electric/General Dynamics Vulcan Phalanx 20 mm Mk 15 Mod 11 ❺; 6 barrels per mounting; 3,000 rds/min combined to 1.5 km.
2 Oerlikon 20 mm (on VLS deck) ❻ can be carried.
Torpedoes: 6—324 mm US Mk 32 (2 triple) tubes ❼. Honeywell Mk 46 Mod 5; anti-submarine; active/passive homing to 11 km (5.9 n miles) at 40 kt; warhead 44 kg.
Countermeasures: Decoys: 2 Loral Hycor Mk 36 SRBOC 6-barrelled chaff launchers ❽. Sea Gnat.
SLQ-25 Nixie; towed torpedo decoy.
ESM/ECM: APECS II; intercept and jammer.
Combat data systems: Signaal SEWACO action data automation with STACOS tactical command; Link 11 and 14. Matra Marconi SCOT 3 SATCOM ❾ (1 set between 3 ships).
Weapons control: SWG 1A(V) for SSM. Vesta Helo transponder with datalink for OTHT.
Radars: Air search: Signaal MW08 (derived from Smart 3D) ❿; 3D; G-band.
Air/surface search: Signaal DA08 ⓫; F-band.
Navigation: Kelvin Hughes Type 1007; I-band.
Fire control: 2 Signaal STIR ⓬; I/J/K-band; range 140 km (76 n miles) for 1 m² target.
IFF Mk 12 Mod 4.
Sonars: Computing Devices (Canada) SQS-510(V); hull-mounted; active search and attack; medium frequency.

Helicopters: 2 Super Sea Lynx Mk 95 ⓭.

Programmes: The contract for all three was signed on 25 July 1986. These are Meko 200 type ordered from a consortium of builders. As well as Portugal, which provided 40 per cent of the cost, assistance was given by NATO with some missile, CIWS and torpedo systems being provided by the US.
Modernisation: Full mid-life refits are planned 2009-2014. Upgrades are likely to include CIWS improvements.
Structure: All-steel construction. Stabilisers fitted. Full RAS facilities. Space has been left for a sonar towed array and for VLS Sea Sparrow.

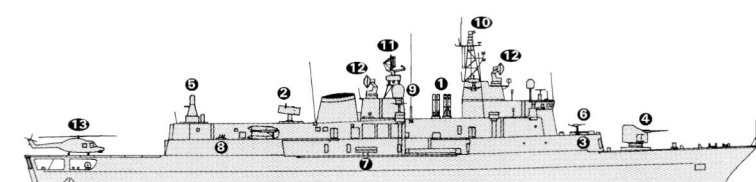

VASCO DA GAMA

(Scale 1 : 1,200), Ian Sturton / 0567520

CORTE REAL

6/2001, John Brodie / 0121386

ALVARES CABRAL

2/2003*, Freddy Philips / 1044179

Operational: Designed primarily as ASW ships. SCOT SATCOM rotated between the three ships. 20 mm guns can be mounted on the VLS deck. Three year running cycles include 18 months at full readiness, three months training and six months refit.

UPDATED

CORTE REAL

6/2004*, B Sullivan / 1044178

3 COMANDANTE JOÃO BELO CLASS (FF)

Name	No	Builders	Laid down	Launched	Commissioned
COMANDANTE JOÃO BELO	F 480	At et Ch de Nantes	6 Sep 1965	22 Mar 1966	1 July 1967
COMANDANTE HERMENEGILDO CAPELO	F 481	At et Ch de Nantes	13 May 1966	29 Nov 1966	26 Apr 1968
COMANDANTE SACADURA CABRAL	F 483	At et Ch de Nantes	18 Aug 1967	15 Mar 1968	25 July 1969

Displacement, tons: 1,750 standard; 2,250 full load
Dimensions, feet (metres): 336.9 × 38.4 × 14.4
 (102.7 × 11.7 × 4.4)
Main machinery: 4 SEMT-Pielstick 12 PC2.2 V 400 diesels;
 16,000 hp(m) *(11.8 MW)* sustained; 2 shafts
Speed, knots: 25. **Range, n miles:** 7,500 at 15 kt
Complement: 201 (15 officers)

Guns: 2 Creusot-Loire 3.9 in *(100 mm)*/55 Mod 1953 **❶**;
 60 rds/min to 17 km *(9 n miles)* anti-surface; 8 km *(4.4 n
 miles)* anti-aircraft; weight of shell 13.5 kg.
 2 Bofors 40 mm/60 **❷**; 300 rds/min to 12 km *(6.6 n miles)*;
 weight of shell 0.89 kg.
Torpedoes: 6—324 mm Mk 32 Mod 5 (2 triple) tubes **❸**;
 Honeywell Mk 46 Mod 5; active/passive homing to 11 km
 (5.9 n miles) at 40 kt; warhead 44 kg.
Countermeasures: Decoys: 2 Loral Hycor Mk 36 SRBOC
 6-barrelled chaff launchers.
 SLQ-25 Nixie; towed torpedo decoy.
 ESM: AR-700(V2); intercept.
Combat data systems: Link 11.
Weapons control: C T Analogique. Sagem DMA optical director.
Radars: Air search: Thomson-CSF DRBV 22A **❹**; D-band.
 Surface search: Thomson-CSF DRBV 50 **❺**; G-band.
 Navigation: Kelvin Hughes KH 1007; I-band.
 Fire control: Thomson-CSF DRBC 31D **❻**; I-band.

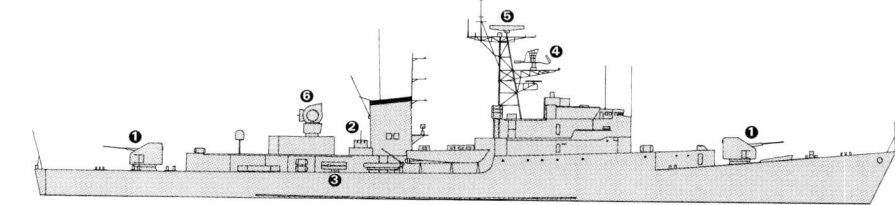

COMANDANTE HERMENEGILDO CAPELO

(Scale 1 : 900), Ian Sturton / 0121391

Sonars: CDC SQS-510; hull-mounted; active search and attack;
 medium frequency.
 Thomson Sintra DUBA 3A; hull-mounted; active search; high
 frequency.

Modernisation: Modernisation of external communications,
 sensors and electronics completed 1987-90. Chaff launchers
 installed in 1989. Further modernisation started in 1993. The
 hull sonar has been replaced, torpedo tubes updated, the A/S
 mortar removed, towed torpedo decoy installed and ESM
 equipment changed. F 481 completed in 1995, F 480 in 1996

and F 483 in 2000. A combat data system with Link 11 has
been added, compatible with the Vasco da Gama class. The
plan to have one or both after guns replaced either by flight
deck and hangar for helicopter or by SSM has been shelved,
but X turret is removed.
Operational: Designed for tropical service. A fourth of class has
 been cannibalised for spares. Expected to be replaced by ex-
 USN FFG 7 class from 2006.

UPDATED

COMANDANTE HERMENEGILDO CAPELO

2/2000, J Mortimer / 0105264

CORVETTES

3 BAPTISTA DE ANDRADE CLASS (FSH)

Name	No	Builders	Laid down	Launched	Commissioned
BAPTISTA DE ANDRADE	F 486	Empresa Nacional Bazán, Cartagena	1 Sep 1972	13 Mar 1973	19 Nov 1974
JOÃO ROBY	F 487	Empresa Nacional Bazán, Cartagena	1 Dec 1972	3 June 1973	18 Mar 1975
AFONSO CERQUEIRA	F 488	Empresa Nacional Bazán, Cartagena	10 Mar 1973	6 Oct 1973	26 June 1975

Displacement, tons: 1,203 standard; 1,380 full load
Dimensions, feet (metres): 277.5 × 33.8 × 10.2
 (84.6 × 10.3 × 3.1)
Main machinery: 2 OEW Pielstick 12 PC2.2 V 400 diesels;
 12,000 hp(m) *(8.82 MW)* sustained; 2 shafts
Speed, knots: 22. **Range, n miles:** 5,900 at 18 kt
Complement: 71 (7 officers)

Guns: 1 Creusot-Loire 3.9 in *(100 mm)*/55 Mod 1968; 80 rds/

min to 17 km *(9 n miles)* anti-surface; 8 km *(4.4 n miles)* anti-
 aircraft; weight of shell 13.5 kg.
 2 Bofors 40 mm/70; 300 rds/min to 12 km *(6.6 n miles)*;
 weight of shell 0.96 kg.
Radars: Navigation: 1 Racal Decca RM 316P and 1 KH 5000
 Nucleos 2; I-band.
Helicopters: Platform only.

Programmes: Reclassified as corvettes.

Modernisation: Communications equipment updated 1988-91.
 Previous modernisation programme was abandoned in 1998.
 Between 1999 and 2001 ASW and weapons control systems
 removed.
Operational: Class is used for Maritime Law Enforcement/SAR/
 Fishery Protection and for Humanitarian Operations. To be
 replaced by Viana do Castelo class. F 489 decommissioned in
 2002.

UPDATED

BAPTISTA DE ANDRADE CLASS (before adaptation to Fishery Protection role)

8/1998, Winter & Findler / 0052749

4 JOÃO COUTINHO CLASS (FSH)

Name	No	Builders	Laid down	Launched	Commissioned
ANTONIO ENES	F 471	Empresa Nacional Bazán, Cartagena	10 Apr 1968	16 Aug 1969	18 June 1971
JOÃO COUTINHO	F 475	Blohm + Voss, Hamburg	24 Dec 1968	2 May 1969	28 Feb 1970
JACINTO CANDIDO	F 476	Blohm + Voss, Hamburg	10 Feb 1969	16 June 1969	29 May 1970
GENERAL PEREIRA D'EÇA	F 477	Blohm + Voss, Hamburg	21 Apr 1969	26 July 1969	10 Oct 1970

Displacement, tons: 1,203 standard; 1,380 full load
Dimensions, feet (metres): 277.5 × 33.8 × 10.8
 (84.6 × 10.3 × 3.3)
Main machinery: 2 OEW Pielstick 12 PC2.2 V 400 diesels;
 12,000 hp(m) *(8.82 MW)* sustained; 2 shafts
Speed, knots: 22. **Range, n miles:** 5,900 at 18 kt
Complement: 70 (7 officers)

Guns: 2 US 3 in *(76 mm)*/50 (twin) Mk 33; 50 rds/min to
 12.8 km *(7 n miles)*; weight of shell 6 kg.
 2 Bofors 40 mm/60 (twin); 300 rds/min to 12 km *(6.6 n miles)*; weight of shell 0.89 kg.
Weapons control: Mk 51 GFCS for 40 mm.
Radars: Air/surface search: Kelvin Hughes 1007; I-band.
Navigation: Racal Decca RM 1226C; I-band.

Helicopters: Platform only.

Programmes: Reclassified as corvettes.
Modernisation: A programme for this class to include SSM and
 PDMS has been shelved. In 1989-91 the main radar was

updated and SATCOM (INMARSAT) installed. Also fitted with
SIFICAP which is a Fishery Protection data exchange system
by satellite to the main database ashore.

ANTONIO ENES   *7/1994, van Ginderen Collection* / 0052750

Operational: A/S equipment no longer operational and laid apart
 on shore. Crew reduced as a result. F 484 decommissioned in
 2004. To be replaced by Viana do Castelo class. **UPDATED**

SHIPBORNE AIRCRAFT

Notes: Procurement of three further Lynx helicopters is under consideration. Options include Mk
95 aircraft, Super Lynx 300 (including upgrade of current aircraft) or second-hand aircaft.

Numbers/Type: 5 Westland Super Navy Lynx Mk 95.
Operational speed: 125 kt *(231 km/h)*.
Service ceiling: 12,000 ft *(3,660 m)*.
Range: 320 n miles *(593 km)*.
Role/Weapon systems: Ordered 2 November 1990 for MEKO 200 frigates; two are updated HAS
 3 and three were new aircraft, all delivered in August and November 1993. Sensors: Bendix
 1500B radar; Bendix AQS-18V dipping sonar; Racal RNS 252 datalink. Weapons: Mk 46
 torpedoes. 1—12.7 mm MG.
                                                          UPDATED

SUPER LYNX   *9/2002, H M Steele* / 0534127

LAND-BASED MARITIME AIRCRAFT

Notes: (1) All Air Force manned.
(2) Twelve EH 101 utility helicopters for SAR and surveillance duties are to be procured by 2006. Of
these, four are to be marinised for amphibious lift duties.

Numbers/Type: 10 Aerospatiale SA 330C Puma.
Operational speed: 151 kt *(280 km/h)*.
Service ceiling: 15,090 ft *(4,600 m)*.
Range: 343 n miles *(635 km)*.
Role/Weapon systems: For SAR and surface search. Sensors: Omera search radar. Weapons:
 Unarmed except for pintle-mounted 12.7 mm machine guns.
                                                          VERIFIED

PUMA   *6/2002, Adolfo Ortigueira Gil* / 0567521

Numbers/Type: 5/2 CASA C-212-200 Aviocar/C-212-300 Aviocar.
Operational speed: 190 kt *(353 km/h)*.
Service ceiling: 24,000 ft *(7,315 m)*.
Range: 1,650 n miles *(3,055 km)*.
Role/Weapon systems: The first five are for short-range SAR support and transport operations.
 The last pair were ordered in February 1993 for maritime patrol and fisheries surveillance off the
 Azores and Madeira. Sensors: Search radar and MAD. FLIR and datalink (last pair). Weapons:
 Unarmed.
                                                          VERIFIED

CASA 212   *6/2001, Adolfo Ortigueira Gil* / 0529552

Numbers/Type: 6 Lockheed P-3P Orion.
Operational speed: 410 kt *(760 km/h)*.
Service ceiling: 28,300 ft *(8,625 m)*.
Range: 4,000 n miles *(7,410 km)*.
Role/Weapon systems: Long-range surveillance and ASW patrol aircraft; acquired with NATO
 funding from RAAF update programme and modernised by Lockheed to 3P standard starting in
 1987. A further five aircraft may be acquired from the Netherlands by 2006. Sensors: APS-134/
 137 radar, ASQ-81 MAD, AQS-901 sonobuoy processor, AQS-114 computer, IFF, ALR-66 ECM/
 ESM. Weapons: ASW; eight Mk 46 torpedoes, depth bombs or mines; ASV; 10 underwing
 stations for Harpoon. AGM 65 Maverick in due course.
                                                          UPDATED

P-3P   *3/2001, Adolfo Ortigueira Gil* / 0567522

PATROL FORCES

0 + 4 (8) VIANA DO CASTELO (NPO 2000) CLASS (PSOH)

Name	No	Builders	Commissioned
VIANA DO CASTELO	—	Viana do Castelo Shipyards	2005
PONTA DELGADA	—	Viana do Castelo Shipyards	2006

Displacement, tons: 1,500 full load
Dimensions, feet (metres): 272.6 × 42.5 × 12.1 *(83.1 × 12.95 × 3.69)*
Main machinery: 2 diesels; 10,460 hp *(7.8 MW)*; 2 shafts
Speed, knots: 20. **Range, n miles:** 5,000 at 15 kt
Complement: 35
Guns: 1—40 mm.
Weapons control: Mk 51 GFCS or Optronic director.
Helicopters: Platform for one medium.

Comment: Contract on 15 October 2002 with Viana do Castelo Shipyards for two Offshore Patrol
 vessels. Construction started in 2003. These ships are designed for EEZ patrol duties and a
 further eight are planned to be delivered by 2015 to replace the corvettes and Cacine class. Two
 further modified vessels, a Buoy Tender and a Pollution Control Ship, were ordered in May
 2004.
                                                          UPDATED

NPO 2000   *(Scale 1 : 900), Ian Sturton* / 0081605

4 CACINE CLASS (LARGE PATROL CRAFT) (PBO)

Name	No	Builders	Commissioned
CACINE	P 1140	Arsenal do Alfeite	May 1969
QUANZA	P 1144	Estaleiros Navais do Mondego	May 1969
ZAIRE	P 1146	Estaleiros Navais do Mondego	Nov 1970
SAVE	P 1161	Arsenal do Alfeite	May 1973

Displacement, tons: 292.5 standard; 310 full load
Dimensions, feet (metres): 144 × 25.2 × 7.1 *(44 × 7.7 × 2.2)*
Main machinery: 2 MTU 12V 538 TB80 diesels; 3,750 hp(m) *(2.76 MW)* sustained; 2 shafts
Speed, knots: 20. Range, n miles: 4,400 at 12 kt
Complement: 33 (3 officers)
Guns: 1 Bofors 40 mm/60. 1 Oerlikon 20 mm/65.
Radars: Surface search: Kelvin Hughes Type 1007; I/J-band.

Comment: Originally mounted a second Bofors aft but most have been removed as has the 37 mm rocket launcher. Have SIFICAP satellite data handling system for Fishery Protection duties. An RIB is carried. Two of the class are based at Madeira on a two month rotational basis. Re-engined in 1992-94. To be replaced by Viana do Castelo class from 2005.

UPDATED

CACINE   *12/2003*, Martin Mokrus /* 1044180

CACINE   *12/2003*, Martin Mokrus /* 1044181

2 ALBATROZ CLASS (RIVER PATROL CRAFT) (PBR)

Name	No	Builders	Commissioned
AGUIA	P 1165	Arsenal do Alfeite	28 Feb 1975
CISNE	P 1167	Arsenal do Alfeite	31 Mar 1976

Displacement, tons: 45 full load
Dimensions, feet (metres): 77.4 × 18.4 × 5.2 *(23.6 × 5.6 × 1.6)*
Main machinery: 2 Cummins diesels; 1,100 hp *(820 kW)*; 2 shafts
Speed, knots: 20. Range, n miles: 2,500 at 12 kt
Complement: 8 (1 officer)
Guns: 1 Oerlikon 20 mm/65. 2—12.7 mm MGs.
Radars: Surface search: Decca RM 316P; I-band.

Comment: One other is used for harbour patrol duties. Two transferred to East Timor in 2001. Replacement of these craft is under consideration.

VERIFIED

ALBATROZ CLASS   *1/2001, van Ginderen Collection /* 0121389

5 ARGOS CLASS (RIVER PATROL CRAFT) (PBR)

Name	No	Builders	Commissioned
ARGOS	P 1150	Arsenal do Alfeite	2 July 1991
DRAGÃO	P 1151	Arsenal do Alfeite	18 Oct 1991
ESCORPIÃO	P 1152	Arsenal do Alfeite	26 Nov 1991
CASSIOPEIA	P 1153	Conafi	11 Nov 1991
HIDRA	P 1154	Conafi	18 Dec 1991

Displacement, tons: 94 full load
Dimensions, feet (metres): 89.2 × 19.4 × 4.6 *(27.2 × 5.9 × 1.4)*
Main machinery: 2 MTU 12V 396 TE84 diesels; 3,700 hp(m) *(2.73 MW)* sustained; 2 shafts
Speed, knots: 26. Range, n miles: 1,350 at 15 kt
Complement: 12 (1 officer)
Guns: 2—12.7 mm MGs (1150-1154).
Radars: Navigation: Furuno 1505 DA or Furuno FR 1411; I-band.

Comment: First five ordered in 1989 and 50 per cent funded by the EC. Of GRP construction, capable of full speed operation up to Sea State 3. Carries a RIB with a 37 hp outboard engine. The boat is recoverable via a stern well at up to 10 kt.

UPDATED

ESCORPIÃO   *6/2002 /* 0567523

4 CENTAURO CLASS (RIVER PATROL CRAFT) (PBR)

Name	No	Builders	Commissioned
CENTAURO	P 1155	Arsenal do Alfeite	20 Mar 2000
ORION	P 1156	Arsenal do Alfeite	27 Mar 2001
PÉGASO	P 1157	Estaleiros Navals do Mondego	27 Mar 2001
SAGITARIO	P 1158	Estaleiros Navais do Mondego	27 Mar 2001

Displacement, tons: 89 full load
Dimensions, feet (metres): 93.2 × 19.5 × 4.6 *(28.4 × 5.95 ×1.4)*
Main machinery: 2 Cummins KTA-50-M2 diesels; 3,600 hp(m) *(2.64 MW)*; 2 shafts
Speed, knots: 26. Range, n miles: 640 at 20 kt
Complement: 8 (1 officer)
Guns: 1 Oerlikon 20 mm/65.
Radars: 1 Furuno FCR-1411 MK3.

Comment: Similar to Argos class but of aluminium hull. Capable of full speed operation up to Sea State 3. Carries a semi-rigid boat with a 50 hp outboard engine. The boat is recoverable via a stern well at up to 10 kt.

VERIFIED

ORION, PÉGASO and SAGITARIO   *3/2001, A Campanera i Rovira /* 0126192

1 RIO MINHO CLASS (RIVER PATROL CRAFT) (PBR)

Name	No	Builders	Commissioned
RIO MINHO	P 370	Arsenal do Alfeite	1 Aug 1991

Displacement, tons: 72 full load
Dimensions, feet (metres): 73.5 × 19.7 × 2.6 *(22.4 × 6 × 0.8)*
Main machinery: 2 KHD-Deutz diesels; 664 hp(m) *(488 kW)*; 2 Schottel pumpjets
Speed, knots: 9.5. Range, n miles: 420 at 7 kt
Complement: 8 (1 officer)
Guns: 1—7.62 mm MG.
Radars: Navigation: Furuno FR 1505DA; I-band.

VERIFIED

RIO MINHO   *8/1994, van Ginderen Collection /* 0081609

AMPHIBIOUS FORCES

Notes: (1) Studies into the procurement of a new amphibious ship are in progress. Such a vessel is likely to draw on the Royal Netherlands Navy's *Rotterdam* and the Spanish Navy's *Galicia* designs. With a projected length of 162 m and displacement of 10,000 tons the ship is expected to be capable of embarking a battalion-sized force (650 marines) and carrying 76 vehicles in the garage (including 40 light armoured vehicles), plus 22 more in the hangar. Four new LCMs would be accommodated in the dock and 53 light inflatable boats in the garage. In addition, it will have 3,000 cu m storage space in the hold. Vehicle loading will be aided through a door on the port side. The ship will probably be built at Viana do Castelo Shipyards for delivery in 2010.
(2) Subject to approval for funding of a new LPD, four new LCMs will be required to aid rapid offload of vehicles and their support.

1 BOMBARDA CLASS (LCU)

Name	No	Builders	Commissioned
BACAMARTE	LDG 203	Arsenal do Alfeite	Dec 1985

Displacement, tons: 652 full load
Dimensions, feet (metres): 184.3 × 38.7 × 6.2 *(56.2 × 11.8 × 1.9)*
Main machinery: 2 MTU MB diesels; 910 hp(m) *(669 kW)*; 2 shafts
Speed, knots: 9.5. **Range, n miles:** 2,600 at 9 kt
Complement: 21 (3 officers)
Military lift: 350 tons
Guns: 2 Oerlikon 20 mm.
Radars: Navigation: Decca RM 316P; I-band.

Comment: Similar to French EDIC.

VERIFIED

BACAMARTE   *2000, Portuguese Navy /* 0105267

ANDROMEDA   *8/1997, van Ginderen Collection /* 0012932

3 SURVEY CRAFT (YGS)

CORAL UAM 801   **ATLANTA** (ex-*Hidra*) UAM 802   **FISALIA** UAM 805

Comment: Craft are of 36 tons launched in 1980.

VERIFIED

FISALIA   *3/1992, van Ginderen Collection /* 0081611

SURVEY SHIPS

2 STALWART CLASS (AGS)

Name	No	Builders	Commissioned
D. CARLOS I (ex-*Audacious*, ex-*Dauntless*)	A 522 (ex-T-AGOS 11)	Tacoma Boat	18 June 1989
ALMIRANTE GAGO COUTINHO (ex-*Assurance*)	A 523 (ex-T-AGOS 5)	Tacoma Boat	1 May 1985

Displacement, tons: 2,285 full load
Dimensions, feet (metres): 224 × 43 × 15.9 *(68.3 × 13.1 × 4.6)*
Main machinery: Diesel-electric; 4 Caterpillar D 398B diesel generators; 3,200 hp *(2.39 MW)*; 2 GE motors; 1,600 hp *(1.2 MW)*; 2 shafts; bow thruster; 550 hp *(410 kW)*
Speed, knots: 11. **Range, n miles:** 4,000 at 11 kt; 6,450 at 3 kt
Complement: 31 (6 officers) plus 15 scientists
Radars: Navigation: 2 Raytheon; I-band.

Comment: Paid off from USN in November 1995. First one acquired 21 July 1996. Refitted to serve as a hydrographic ship, operating predominantly off the west coast of Africa. Recommissioned 9 December 1996. A second of class acquired by gift 30 September 1999, has been similarly refitted and recommissioned 26 January 2000.

UPDATED

D. CARLOS I   *6/2004*, Portuguese Navy /* 1044176

2 ANDROMEDA CLASS (AGSC)

Name	No	Builders	Commissioned
ANDROMEDA	A 5203	Arsenal do Alfeite	1 Feb 1987
AURIGA	A 5205	Arsenal do Alfeite	1 July 1987

Displacement, tons: 245 full load
Dimensions, feet (metres): 103.3 × 25.4 × 8.2 *(31.5 × 7.7 × 2.5)*
Main machinery: 1 MTU 12V 396 TC62 diesel; 1,200 hp(m) *(880 kW)* sustained; 1 shaft
Speed, knots: 12. **Range, n miles:** 1,980 at 10 kt
Complement: 17 (3 officers)
Radars: Navigation: Decca RM 914C; I-band.

Comment: Both ordered in January 1984. *Auriga* has a research submarine ROV Phantom S2 and a Klein side scan sonar. Mostly used for oceanography.

VERIFIED

TRAINING SHIPS

1 SAIL TRAINING SHIP (AXS)

Name	No	Builders	Commissioned
SAGRES (ex-*Guanabara*, ex-*Albert Leo Schlageter*)	A 520	Blohm + Voss, Hamburg	10 Feb 1938

Displacement, tons: 1,725 standard; 1,940 full load
Dimensions, feet (metres): 231 wl; 295.2 oa × 39.4 × 17 *(70.4; 90 × 12 × 5.2)*
Main machinery: 2 MTU 12V 183 TE92 auxiliary diesels; 1 shaft
Speed, knots: 10.5. **Range, n miles:** 5,450 at 7.5 kt on diesel
Complement: 162 (12 officers)
Radars: Navigation: 1 Racal Decca and 1 KH 1500 Nucleos 2; I-band.

Comment: Former German sail training ship launched 30 October 1937. Sister of US Coast Guard training ship *Eagle* (ex-German *Horst Wessel*) and Soviet *Tovarisch* (ex-German *Gorch Fock*). Taken by the USA as a reparation after the Second World War in 1945 and sold to Brazil in 1948. Purchased from Brazil and commissioned in the Portuguese Navy on 2 February 1962 at Rio de Janeiro and renamed *Sagres*. Sail area, 20,793 sq ft. Height of main mast, 142 ft. Phased refits 1987-88 and again in 1991-92 which included new engines, improved accommodation, hydraulic crane and updated navigation equipment.

VERIFIED

SAGRES   *7/2003, B Prézelin /* 0567524

1 SAIL TRAINING SHIP (AXS)

Name	No	Builders	Commissioned
CREOULA	UAM 201	Lisbon Shipyard	1937

Displacement, tons: 818 standard; 1,055 full load
Dimensions, feet (metres): 221.1 × 32.5 × 13.8 *(67.4 × 9.9 × 4.2)*
Main machinery: 1 MTU 8V 183 TE92 auxiliary diesel; 665 hp(m) *(490 kW)*; 1 shaft

Comment: Ex-deep sea sail fishing ship used off the coast of Newfoundland for 36 years. Bought by Fishing Department in 1976 to turn into a museum ship but because she was still seaworthy it was decided to convert her to a training ship. Recommissioned in the Navy in 1987. Refit completed in 1992 including a new engine and improved accommodation. A life-extension refit is under consideration.

VERIFIED

CREOULA
9/1991, van Ginderen Collection / 0081613

2 SAIL TRAINING YACHTS (AXL)

VEGA (ex-*Arreda*) A 5201   POLAR (ex-*Anne Linde*) A 5204

Displacement, tons: 70 (60, *Vega*)
Dimensions, feet (metres): 75 × 16 × 8.2 *(22.9 × 4.9 × 2.5) (Polar)*
    65 × 14.1 × 8.2 *(19.8 × 4.3 × 2.5) (Vega)*
Radars: Navigation: Raytheon; I-band.

Comment: Sail numbers are displayed. *Vega* is P-165 and *Polar* is P-551.

UPDATED

VEGA
8/2004*, Diego Quevedo / 1047860

AUXILIARIES

Notes: (1) Two craft are employed on Pollution Control tasks. *Vazante* (UAM 687) is 14 tons and *Enchente* (UAM 688) is 65 tons. *Barrocas* (UAM 854) is an accommodation barge. *Marateca* (UAM 304) and *Meuro* (UAM 305) are fuel lighters.
(2) Studies for the procurement fo a new AOR, to enter service after 2010, are in progress.

1 ROVER CLASS (REPLENISHMENT TANKER) (AORLH)

Name	No	Builders	Launched	Commissioned
BÉRRIO (ex-*Blue Rover*)	A 5210 (ex-A 270)	Swan Hunter	11 Nov 1969	15 July 1970

Displacement, tons: 4,700 light; 11,522 full load
Dimensions, feet (metres): 461 × 63 × 24 *(140.6 × 19.2 × 7.3)*
Main machinery: 2 SEMT-Pielstick 16 PA4 185 diesels; 15,360 hp(m) *(11.46 MW)*; 1 shaft; Kamewa cp prop; bow thruster
Speed, knots: 19. **Range, n miles:** 15,000 at 15 kt
Complement: 54 (7 officers)
Cargo capacity: 6,600 tons fuel
Guns: 2 Oerlikon 20 mm.
Countermeasures: Decoys: 2 Vickers Corvus launchers. 2 Plessey Shield launchers. 1 Graseby Type 182; towed torpedo decoy.
Radars: Navigation: Kelvin Hughes Type 1006; I-band.
Helicopters: Platform for 1 medium.

Comment: Transferred from UK and recommissioned 31 March 1993. Small fleet tanker designed to replenish oil and aviation fuel, fresh water, limited dry cargo and refrigerated stores under all conditions while under way. Full refit in 1990-91 gave a service life expectancy until about 2004 and a further refit is to be undertaken to prolong life to 2010. No hangar but helicopter landing platform is served by a stores lift, to enable stores to be transferred at sea by 'vertical lift'. Capable of HIFR. Can pump fuel at 600 m³/h. Others of the class in service in Indonesia and the UK.

UPDATED

BÉRRIO
4/2000, Maritime Photographic / 0105268

1 BUOY TENDER (ABU)

Name	No	Builders	Commissioned
SCHULTZ XAVIER	A 521	Alfeite Naval Yard	14 July 1972

Displacement, tons: 900 full load
Dimensions, feet (metres): 184 × 33 × 12.5 *(56 × 10 × 3.8)*
Main machinery: 2 diesels; 2,400 hp(m) *(1.76 MW)*; 2 shafts
Speed, knots: 14.5. **Range, n miles:** 3,000 at 12.5 kt
Complement: 54 (4 officers)

Comment: Used for servicing navigational aids and as an occasional tug. A replacement ship is projected.

VERIFIED

SCHULTZ XAVIER
5/1998, Diego Quevedo / 0052756

1 BUOY TENDER (ABU)

Name	No	Builders	Commissioned
GUIA	UAM 676	S Jacinto, Aveiro	30 Jan 1985

Displacement, tons: 70 full load
Dimensions, feet (metres): 72.2 × 25.9 × 7.2 *(22 × 7.9 × 2.2)*
Main machinery: 1 Deutz MWM SBA6M816 diesel; 465 hp(m) *(342 kW)* sustained; 1 Schottel Navigator prop
Speed, knots: 8.5 (3.5 on auxiliary engine)
Complement: 6

Comment: Belongs to the Lighthouse Service.

VERIFIED

GUIA
5/1993, van Ginderen Collection / 0081614

8 CALMARIA CLASS (HARBOUR PATROL CRAFT) (YP)

CALMARIA UAM 642	MONÇÃO UAM 645	PREIA-MAR UAM 648
CIRRO UAM 643	SUÃO UAM 646	BAIXA-MAR UAM 649
VENDAVAL UAM 644	MACAREU UAM 647	

Displacement, tons: 12 full load
Dimensions, feet (metres): 39 × 12.5 × 2.3 *(11.9 × 3.8 × 0.7)*
Main machinery: 2 Bazán MAN 2866 LXE diesels; 881 hp(m) *(648 kW)*; 2 water-jets
Speed, knots: 32. **Range, n miles:** 275 at 20 kt
Complement: 3
Guns: 1—7.62 mm MG.
Radars: Surface search: Furuno 1830; I-band.

Comment: Harbour patrol craft similar to Spanish Guardia Civil del Mar Saetta II craft. Ordered from Bazán, Cadiz on 8 January 1993. First pair completed 30 November 1993, third one on 18 January 1994. Remainder delivered between August and December 1994. GRP hulls.

UPDATED

BAIXA-MAR
3/2004* / 1044182

56 MISCELLANEOUS SERVICE CRAFT (YAG)

UAM 650-652	UAM 780
UAM 655-675	UAM 831
UAM 677-679	UAM 851-853
UAM 681-682	UAM 901
UAM 685	UAM 907-916
UAM 689-696	

Displacement, tons: 14 full load
Dimensions, feet (metres): 45.9 × 12.5 × 3.9 *(14.0 × 3.8 × 1.2)*
Main machinery: 2 diesels; 650 hp *(485 kW)*; 2 waterjets
Speed, knots: 23. **Range, n miles:** 300 at 15 kt

Comment: Details are for UAM 689-696, Rodman 46 SAR craft commissioned 1997-2000. The remaining craft are personnel and other service craft.

UPDATED

UAM 692   *9/1998, Schaeffer/Marsan* / 0081616

UAM 852   *3/2002, Diego Quevedo* / 0534049

GOVERNMENT MARITIME FORCES
(GUARDIA NACIONAL REPUBLICANA)

POLICE

12 CONAFI 55 CLASS (PBF)

Displacement, tons: 18 full load
Dimensions, feet (metres): 55.8 × 12.5 × 2.9 *(17.0 × 3.8 × 0.9)*
Main machinery: 2 MTU 12V 183TE93 diesels; 2,400 hp(m) *(1.8 MW)*; 2 waterjets
Speed, knots: 48. **Range, n miles:** 400 at 18 kt
Complement: 5

Comment: Built at Conafi Shipyards with collaboration with Rodman and delivered between 2000 and 2002.

NEW ENTRY

CONAFI 55   *6/2000*, Conafi* / 1044183

4 RODMAN 38 CLASS (PB)

Displacement, tons: 10 full load
Dimensions, feet (metres): 36.1 × 12.8 × 2.3 *(11.0 × 3.9 × 0.7)*
Main machinery: 2 diesels; 400 hp *(300 kW)*; 2 waterjets
Speed, knots: 28. **Range, n miles:** 300 at 15 kt
Complement: 4

Comment: GRP hull. Built by Rodman, Vigo in 1985-87.

NEW ENTRY

Qatar

Country Overview

Formerly a British protectorate from 1916, the State of Qatar gained its independence in 1971. Situated on the eastern side of the Arabian Peninsula, it occupies the Qatar Peninsula which has a 304 n mile coastline with the Gulf. With an area of 4,416 square miles, it is bordered to the south by Saudi Arabia and the United Arab Emirates. The sovereignty of the Hawar islands is disputed with Bahrain as is their joint maritime boundary. The capital, largest city and principal port is Doha. Territorial seas (12 n miles) are claimed. An EEZ (200 n miles) has been claimed but the limits are not defined.

Headquarters Appointments

Commander Naval Force:
  Brigadier Rabeea' Fadal Sa'eed al-Ka'bee

Personnel

(a)   2005: 1,800 officers and men (including Marine Police)
(b)   Voluntary service

Bases

Doha (main); Halul Island (secondary)

Coast Defence

Two truck-mounted batteries of Exocet MM 40 quad launchers.

Prefix to Ships' Names

QENS (Qatar Emiri Navy)

PATROL FORCES

Notes: A requirement for two 49 m patrol craft has been reported.

3 DAMSAH (COMBATTANTE III M) CLASS
(FAST ATTACK CRAFT—MISSILE) (PGGF)

Name	No	Builders	Launched	Commissioned
DAMSAH	Q 01	CMN, Cherbourg	17 June 1982	10 Nov 1982
AL GHARIYAH	Q 02	CMN, Cherbourg	23 Sep 1982	10 Feb 1983
RBIGAH	Q 03	CMN, Cherbourg	22 Dec 1982	11 May 1983

Displacement, tons: 345 standard; 395 full load
Dimensions, feet (metres): 183.7 × 26.9 × 7.2 *(56 × 8.2 × 2.2)*
Main machinery: 4 MTU 20V 538 TB93 diesels; 18,740 hp(m) *(13.8 MW)* sustained; 4 shafts
Speed, knots: 38.5. **Range, n miles:** 2,000 at 15 kt
Complement: 41 (6 officers)

Missiles: SSM: 8 Aerospatiale MM 40 Exocet; inertial cruise; active radar homing to 70 km *(40 n miles)* at 0.9 Mach; warhead 165 kg; sea-skimmer.
Guns: 1 OTO Melara 3 in *(76 mm)*/62; 60 rds/min to 16 km *(8.7 n miles)*; weight of shell 6 kg.
  2 Breda 40 mm/70 (twin); 300 rds/min to 12.5 km *(6.8 n miles)*; weight of shell 0.96 kg.
  4 Oerlikon 30 mm/75 (2 twin); 650 rds/min to 10 km *(5.5 n miles)*.
Countermeasures: Decoys: CSEE Dagaie trainable single launcher; 6 containers; IR flares and chaff; H/J-band.
ESM/ECM: Racal Cutlass/Cygnus.
Weapons control: Vega system. 2 CSEE Naja optical directors.
Radars: Surface search: Thomson-CSF Triton; G-band.
Navigation: Racal Decca 1226; I-band.
Fire control: Thomson-CSF Castor II; I/J-band; range 15 km *(8 n miles)* for 1 m² target.

Programmes: Ordered in 1980. All arrived at Doha July 1983. All refitted in 1996/98.

UPDATED

AL GHARIYAH   *10/2001* / 0121393   RBIGAH   *7/2001, Ships of the World* / 0121396

4 BARZAN (VITA) CLASS (PGGFM)

Name	No	Builders	Laid down	Launched	Commissioned
BARZAN	Q04	Vosper Thornycroft	Feb 1994	1 Apr 1995	9 May 1996
HUWAR	Q05	Vosper Thornycroft	Aug 1994	15 July 1995	10 June 1996
AL UDEID	Q06	Vosper Thornycroft	Mar 1995	21 Mar 1996	16 Dec 1996
AL DEEBEL	Q07	Vosper Thornycroft	Aug 1995	31 Aug 1996	3 July 1997

Displacement, tons: 376 full load
Dimensions, feet (metres): 185.7 × 29.5 × 8.2
(56.3 × 9 × 2.5)
Main machinery: 4 MTU 20V 538 TB93 diesels; 18,740 hp(m)
(13.8 MW) sustained; 4 shafts
Speed, knots: 35. **Range, n miles:** 1,800 at 12 kt
Complement: 35 (7 officers)

Missiles: SSM: 8 Aerospatiale MM 40 Exocet (Block II) ❶; inertial
cruise; active radar homing to 70 km *(40 n miles)* at 0.9 Mach;
warhead 165 kg; sea-skimmer.
SAM: Matra Sadral sextuple launcher for Mistral ❷; IR homing to
4 km *(2.2 n miles)*; warhead 3 kg.
Guns: 1 OTO Melara 76 mm/62 Super Rapid ❸; 120 rds/min to
16 km *(8.7 n miles)*; weight of shell 6 kg.
1 Signaal Goalkeeper 30 mm ❹; 7 barrels; 4,200 rds/min
combined to 2 km. 2—12.7 mm MGs.
Countermeasures: Decoys: CSEE Dagaie Mk 2 ❺ for chaff and IR
flares.
ESM: Thomson-CSF DR 3000S ❻; intercept.
ECM: Dassault Salamandre ARBB 33 ❼; jammer.
Combat data systems: Signaal SEWACO FD with Thomson-CSF
TACTICOS; Link Y.
Weapons control: Signaal STING optronic director. Signaal
IRSCAN electro-optical tracker ❽.
Radars: Air/surface search; Thomson-CSF MRR ❾; G-band.
Navigation: Kelvin Hughes 1007 ❿; I-band.
Fire control: Signaal STING ⓫; I/J-band.

Programmes: Order announced on 4 June 1992 by Vosper
Thornycroft. First steel cut 20 July 1993.
Structure: Vita design derivative based on the hull used for
Oman and Kenya in the 1980s. Steel hull and aluminium
superstructure. CSEE Sidewind EW management system is
installed and a Racal Thorn data distribution system is used.
Baffles have been added around the ECM aerials to prevent
mutual interference with other sensors. An advanced
machinery control and surveillance system allows one-man
operation of main propulsion, electrical generation and
auxiliary systems from the bridge. The bridge staff are also
able to monitor the state of all compartments for damage
control purposes.
Operational: First pair arrived in the Gulf in August 1997, second
pair in May 1998. All of the class carry 40 kt RIBs with twin
60 hp outboards.

UPDATED

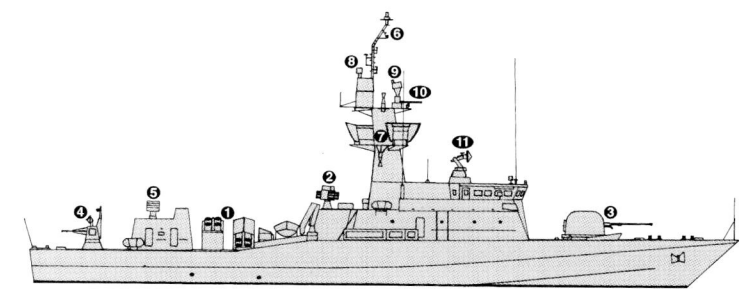

BARZAN

(Scale 1 : 600), Ian Sturton / 0012934

AL DEEBEL
10/2000 / 0121394

AL DEEBEL

7/2001, Ships of the World / 0121395

3 DAMEN POLYCAT 1450 CLASS
(COASTAL PATROL CRAFT) (PB)

Q 31-36 series

Displacement, tons: 18 full load
Dimensions, feet (metres): 47.6 × 15.4 × 4.9 *(14.5 × 4.7 × 2.1)*
Main machinery: 2 Detroit 12V-71TA diesels; 840 hp *(627 kW)* sustained; 2 shafts
Speed, knots: 26
Complement: 11
Guns: 1 Oerlikon 20 mm.
Radars: Navigation: Racal Decca; I-band.

Comment: Three remain of six delivered February-May 1980.

DV 15 class   *2/2003, CMN /* 0531701

3 HALMATIC M 160 CLASS (PB)

Displacement, tons: 20 full load
Dimensions, feet (metres): 52.5 × 15.4 × 4.6 *(16 × 4.7 × 1.4)*
Main machinery: 2 MTU diesels; 520 hp(m) *(388 kW)* sustained; 2 shafts
Speed, knots: 27. **Range, n miles:** 500 at 17 kt
Complement: 6
Guns: 1—7.62 mm MG.
Radars: Surface search: Racal Decca; I-band.

Comment: Order confirmed on 11 October 1995. Delivered to Police in November 1996. Similar to Police craft obtained by Caribbean countries.

VERIFIED

Q 33   *3/1980, Damen SY /* 0081617

AUXILIARIES

Notes: There are a number of amphibious craft including an LCT *Rabha* of 160 ft *(48.8 m)* with a capacity for three tanks and 110 troops, acquired in 1986-87. Also four Rotork craft and 30 Sea Jeeps in 1985. It is not clear how many of the smaller craft are for civilian use.

POLICE

Notes: (1) Requirements have been reported for patrol craft, two of 24 m, two of 22 m and 19 of 12 m. Also for two hovercraft.
(2) Two Halmatic 18 m pilot boats (based on Arun class lifeboat hull) delivered in 2000.

M 160   *11/1996, Halmatic /* 0081619

4 + (2) DV 15 FAST INTERCEPT CRAFT (HSIC)

Displacement, tons: 12 full load
Dimensions, feet (metres): 50.9 × 9.8 × 2.6 *(15.5 × 3.0 × 0.8)*
Main machinery: 2 diesels; 2 surface drives
Speed, knots: 55. **Range, n miles:** 400 at 30 kt
Complement: 4
Guns: 1—12.7 mm MG.
Radars: Surface search: I-band.

Comment: Built by CMN Cherbourg and one delivered every two months from October 2003 to March 2004 to replace P 1200 class. Composite hull construction similar to those in Yemeni service. Roles include coastal protection and security of offshore oil and gas installations.

UPDATED

4 CRESTITALIA MV-45 CLASS (PB)

RG 91-94

Displacement, tons: 17 full load
Dimensions, feet (metres): 47.6 × 12.5 × 2.6 *(14.5 × 3.8 × 0.8)*
Main machinery: 2 diesels; 1,270 hp(m) *(933 kW)*; 2 shafts
Speed, knots: 32. **Range, n miles:** 275 at 29 kt
Complement: 6
Guns: 1 Oerlikon 20 mm. 2—7.62 mm MGs.
Radars: Surface search: I-band.

Comment: Built by Crestitalia and delivered in mid-1989. GRP construction.

VERIFIED

Romania

Country Overview

Situated in south-eastern Europe, the Republic of Romania has an area of 91,700 square miles and is bordered to the north by Ukraine and Moldova, to the west by Hungary and Serbia, and to the south by Bulgaria. The River Danube forms much of the southern border. Romania has a coastline of 121 n miles with the Black Sea on which Constanta, linked to the Danube port of Cernavodà by canal, is the principal seaport. Prominent river ports include Galati and Bràila on the lower Danube, and Giurgiu, which has pipeline connections to the Ploiesti oil fields. The capital and largest city is Bucharest. Territorial waters (12 n miles) are claimed. An EEZ (299 n miles) is claimed but the limits have not been defined.

Headquarters Appointments

Commander-in-Chief of the Navy:
    Rear Admiral Gheorghe Marin

Personnel

(a)   2005: 8,135 Navy
(b)   Reserves: 8,000

Organisation

The Navy is composed of the Naval Staff (Bucharest), the Naval Operational Command, Naval Academy, Hydrographic Directorate, one River Flotilla, Maritime Logistic Base and one Naval Infantry Battalion.

Border Guard

Responsible for land and sea borders and has four brigades, two of which have sea forces based at Orsova and Constanta.

Strength of the Fleet

Type	Active (Reserve)
Frigates	3
Corvettes	7
River Monitors	8
Fast Attack Craft (Missile)	3
Fast Attack Craft (Torpedo)	9
Minelayer/MCM Support	1
Minesweepers (Coastal and River)	10
Training Ships	3
Survey Ships	2

Bases

Black Sea—Mangalia (Training); Constanta (Naval Operational Command and Logistic Base)
Danube—Bràila (HQ River Flotilla), Tulcea (River Logistic Base)

DELETIONS

Corvettes

2004   *Vice Admiral Vasile Scodrea, Vice Admiral Vasile Urseanu*

Patrol Forces

2002   *Grivita*
2003   *Virtejul, Trasnetul, Tornada, Soimul, Eretele, Albatrosul*
2004   18 VB 76 river monitors

Mine Warfare Forces

2004   *Vice Admiral Ioan Murgescu,* 6 VD 141 river minesweepers

Auxiliaries

2002   *Eugen Stihi, Ion Ehiculescu, Automatica, Energetica*

PENNANT LIST

Submarines		260	Admiral Petre Barbuneanu	179	Posada	**Survey Ships**	
		263	Vice Admiral Eugeniu Rosca	180	Rovine		
521	Delfinul	264	Contre Admiral Eustatiu Sebastian	202	Smeul	75	Grigore Antipa
		265	Admiral Horia Macelariu	204	Vijelia	115	Emil Racovita
Frigates				209	Vulcanul		
						Auxiliaries	
111	Marasesti	**Patrol Forces**					
221	Regele Ferdinand					281	Constanta
222	Regina Maria			**Mine Warfare Forces**		283	Midia
		45	Mikhail Kogalniceanu			288	Mircea
Corvettes		46	I C Bratianu	24	Lieutenant Remus Lepri	296	Electronica
		47	Lascar Catargiu	25	Lieutenant Lupu Dunescu	298	Magnetica
188	Zborul	176	Rahova	29	Lieutenant Dimitrie Nicolescu	500	Grozavu
189	Pescarusul	177	Opanez	30	Sub Lieutenant Alexandru Axente	501	Hercules
190	Lastunul	178	Smardan	274	Vice Admiral Constantin Balescu	532	Tulcea

SUBMARINES

1 KILO CLASS (PROJECT 877E) (SSK)

DELFINUL 521

Displacement, tons: 2,325 surfaced; 3,076 dived
Dimensions, feet (metres): 238.2 × 32.8 × 21.7
  (72.6 × 10 × 6.6)
Main machinery: Diesel-electric; 2 diesels; 3,650 hp(m)
  (2.68 MW); 2 generators; 1 motor; 5,900 hp(m) *(4.34 MW)*;
  1 shaft; 2 auxiliary MT-168 motors; 204 hp(m) *(150 kW)*;
  1 economic speed motor; 130 hp(m) *(95 kW)*
Speed, knots: 10 surfaced; 17 dived; 9 snorting
Range, n miles: 6,000 at 7 kt surfaced; 400 at 3 kt dived
Complement: 52 (12 officers)

Torpedoes: 6—21 in *(533 mm)* tubes. Combination of Russian
  53-65; anti-surface; passive wake homing to 19 km *(10.3 n
  miles)* at 45 kt; warhead 300 kg or TEST-71; anti-submarine;
  active/passive (optional wire-guided) homing to 15 km *(8.1 n
  miles)* warhead 205 kg. Total of 18 weapons.
Mines: 24 in lieu of torpedoes.
Countermeasures: ESM: Brick Group; radar warning. Quad Loop
  D/F.
Radars: Surface search: Snoop Tray; I-band.
Sonars: Shark Teeth/Shark Fin; hull-mounted; passive search
  and attack; medium frequency.
  Mouse Roar; active attack; high frequency.

Programmes: Transferred from USSR in December 1986.
Structure: Diving depth, 240 m *(785 ft)*. Two torpedo tubes
  probably capable of firing wire-guided torpedoes. There is a
  well between the snort and W/T masts for a containerised
  portable SAM launcher for SA-N-5/8.
Operational: Based at Constanta. Used as a training ship. There
  are no plans for a refit.

UPDATED DELFINUL

2001, Romanian Navy / 0114544

DELFINUL

6/2003 / 0052761

FRIGATES

1 MARASESTI CLASS (FFGH)

Name	No	Builders	Laid down	Launched	Commissioned
MARASESTI (ex-Muntenia)	111	Mangalia Shipyard	7 Aug 1979	4 June 1981	3 June 1985

Displacement, tons: 5,790 full load
Dimensions, feet (metres): 474.4 × 48.6 × 23
(144.6 × 14.8 × 7)
Main machinery: 4 diesels; 32,000 hp(m) *(23.5 MW)*; 4 shafts
Speed, knots: 27
Complement: 270 (25 officers)

Missiles: SSM: 8 SS-N-2C Styx ❶; active radar or IR homing to
83 km *(45 n miles)* at 0.9 Mach; warhead 513 kg.
Guns: 4 USSR 3 in *(76 mm)*/60 (2 twin) ❷; 90 rds/min to 15 km
(8 n miles); weight of shell 6.8 kg.
4—30 mm/65 ❸; 6 barrels per mounting; 3,000 rds/min to
2 km.
Torpedoes: 6—21 in *(533 mm)* (2 triple) tubes ❹. Russian 53-65;
passive/wake homing to 25 km *(13.5 n miles)* at 50 kt;
warhead 300 kg.
A/S mortars: 2 RBU 6000 ❺; 12-tubed trainable; range
6,000 m; warhead 31 kg.
Countermeasures: Decoys: 2 PK 16 chaff launchers.
ESM/ECM: 2 Watch Dog; intercept. Bell Clout and Bell Slam.
Radars: Air/surface search: Strut Curve ❻; F-band.
Surface search: Plank Shave ❼; E-band.
Fire control: Two Drum Tilt ❽; H/I-band.
Hawk Screech ❾; I-band.
Navigation: Nayada (MR 212); Racal Decca; I-band.
IFF: High Pole B.
Sonars: Hull-mounted; active search and attack; medium
frequency.

Helicopters: 2 IAR-316 Alouette III ❿.

Modernisation: Attempts have been made to modernise some of
the electronic equipment. Also topweight problems have
been addressed by reducing the height of the mast structures
and lowering the Styx missile launchers by one deck. Two
RBU 6000s have replaced the RBU 1200. A series of
upgrades are planned including communications, Link 11 and
improved weapons and sensors. Precise details of the

package are not known but implementation is expected by
2006.
Structure: A distinctive Romanian design. Originally thought to
be powered by gas turbines but a diesel configuration
including four shafts is now confirmed.
Operational: Deactivated in June 1988 due to manpower and
fuel shortages but modernisation work was done from 1990

to 1992 and sea trials started in mid-1992. Carried out a major
naval exercise in September 1993, which included firing the
Styx missile. Deployed to the Mediterranean in September
1994 for a short cruise, in 1995 on two occasions and again in
March 1998. Reclassified as frigate in 2001. Based at
Constanta.

UPDATED

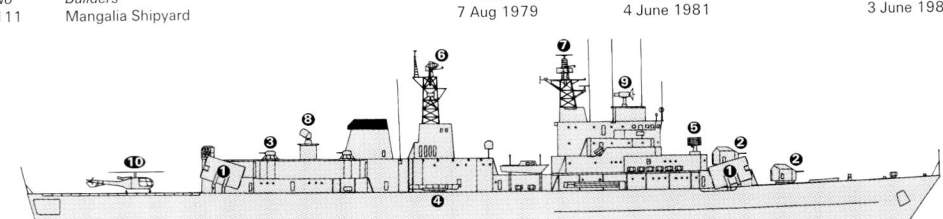

MARASESTI

(Scale 1 : 1,200), Ian Sturton / 1044186

MARASESTI

1/2001, van Ginderen Collection / 0106855

MARASESTI

6/2004*, C D Yaylali / 0589801

MARASESTI

7/1995, Diego Quevedo / 0052762

1 + 1 BROADSWORD CLASS (TYPE 22) (FFHM)

Name	No	Builders	Laid down	Launched	Commissioned	Recommissioned
REGINA MARIA (ex-London)	222 (ex-F 95)	Yarrow Shipbuilders, Glasgow	7 Feb 1983	27 Oct 1984	5 June 1987	Apr 2005
REGELE FERDINAND (ex-Coventry)	221 (ex-F 98)	Swan Hunter Shipbuilders, Wallsend-on-Tyne	29 Mar 1984	8 Apr 1986	14 Oct 1988	9 Sep 2004

Displacement, tons: 4,100 standard; 4,800 full load
Dimensions, feet (metres): 480.5 × 48.5 × 21
  (146.5 × 14.8 × 6.4)
Main machinery: CODOG: 2 RR Olympus TM3B gas turbines;
  50,000 hp (37.3 MW) sustained; 2 RR Tyne RM1C gas
  turbines; 9,900 hp (7.4 MW); 2 shafts; cp props
Speed, knots: 30; 18 on Tynes
Complement: 203

Guns: 1 OTO Melara 3 in (76 mm)/62 Super Rapid ❶; 120 rds/
  min to 16 km (8.7 n miles); weight of shell 6 kg.
Countermeasures: Decoys: 2 Terma 130 mm DL-12 12-barrelled
  chaff launchers ❷.
Combat data systems: Ferranti CACS 1.
Weapons control: Radamec 2500 optronic director ❸. Nautis 3
  fire-control system.
Radars: Air/Surface search: Marconi Type 967/968 ❹; D/E-
  band.
  Navigation: Kelvin-Hughes Type 1007 ❺; I-band.
Sonars: Ferranti/Thomson Sintra Type 2050; hull-mounted
  search and attack.

Helicopters: Platform for 1 medium.

Programmes: Originally successors to the UK Leander class,
  these ships entered RN service in 1987 but were withdrawn,
  half-way through their ships' lives, as a result of the 1998 UK
  Defence Review. Sale agreement signed on 14 January 2003
  included platform overhaul, installation of reconditioned
  engines and combat system modernisation. Training is also
  included in the package. Following trials and sea training,
  Regele Ferdinand arrived in Romania on 10 December 2004.
Modernisation: BAE Systems was prime contractor and FSL sub-
  contractor for reactivation and modernisation. CACS
  command system upgraded and 76 mm gun installed. A
  second-phase upgrade is to be undertaken in Romania
  2008-09. This is expected to include improved command and
  control, air-defence and anti-ship weapons and an improved
  EW suite.
Structure: Broadsword Batch 2 ships were stretched versions of
  Batch 1. The flight decks are capable of embarking medium
  helicopters.
Opinion: Acquisition of these ships gives Romania its first
  modern combatants. As well as providing an increase in
  capability, the ships will enable a higher degree of
  interoperability with NATO forces.

UPDATED

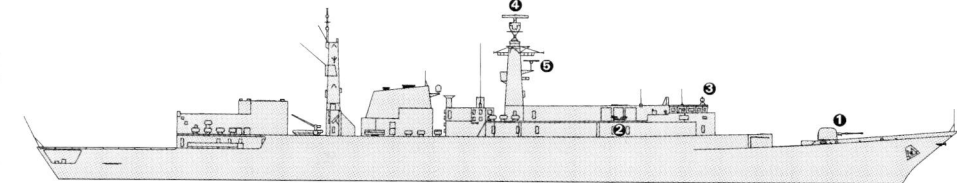

REGELE FERDINAND (Scale 1 : 1,200), Ian Sturton / 1044184

REGELE FERDINAND 7/2004*, BAE Systems / 0577658

REGELE FERDINAND 7/2004*, W Sartori / 0589803

CORVETTES

2 TETAL CLASS (FS)

Name	No	Builders	Launched	Commissioned
ADMIRAL PETRE BARBUNEANU	260	Mangalia Shipyard	23 May 1981	4 Feb 1983
VICE ADMIRAL EUGENIU ROSCA	263	Mangalia Shipyard	11 July 1985	23 Apr 1987

Displacement, tons: 1,440 full load
Dimensions, feet (metres): 303.1 × 38.4 × 9.8
  (92.4 × 11.7 × 3)
Main machinery: 4 diesels; 13,000 hp(m) (9.6 MW); 4 shafts
Speed, knots: 24
Complement: 98

Guns: 4 USSR 3 in (76 mm)/60 (2 twin) ❶; 90 rds/min to 15 km
  (8 n miles); weight of shell 6.8 kg.
  4 USSR 30 mm/65 (2 twin) ❷; 500 rds/min to 4 km (2.2 n
  miles); weight of shell 0.54 kg.
  2—14.5 mm MGs.
Torpedoes: 4—21 in (533 mm) (2 twin) tubes ❸. Russian 53-65;
  passive/wake homing to 25 km (13.5 n miles) at 50 kt;
  warhead 300 kg.
A/S mortars: 2 RBU 2500 16-tubed trainable ❹; range 2,500 m;
  warhead 21 kg.
Countermeasures: Decoys: 2 PK 16 chaff launchers.
  ESM: 2 Watch Dog; intercept.

Radars: Air/surface search: Strut Curve ❺; F-band.
  Fire control: Drum Tilt ❻; H/I-band. Hawk Screech ❼; I-band.
  Navigation: Nayada; I-band.
  IFF: High Pole.
Sonars: Hercules (MG 322); Hull-mounted; active search and
  attack; medium frequency.

Programmes: Building terminated in 1987 in favour of the
  improved design with a helicopter platform.
Structure: A modified Soviet Koni design.
Operational: Both based at Constanta. Two decommissioned in
  2004.

UPDATED

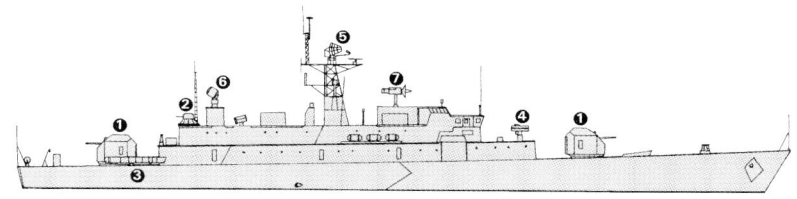

ADMIRAL PETRE BARBUNEANU (Scale 1 : 900), Ian Sturton

ADMIRAL PETRE BARBUNEANU 6/2001 / 0126335

2 IMPROVED TETAL CLASS (FSH)

Name	No	Builders	Launched	Commissioned
CONTRE ADMIRAL EUSTATIU SEBASTIAN	264	Mangalia Shipyard	12 Apr 1988	30 Dec 1989
ADMIRAL HORIA MACELARIU	265	Mangalia Shipyard	15 May 1994	29 Sep 1997

Displacement, tons: 1,500 full load
Dimensions, feet (metres): 303.1 × 38.4 × 10
(92.4 × 11.7 × 3.1)
Main machinery: 4 diesels; 13,000 hp(m) *(9.6 MW)*; 4 shafts
Speed, knots: 24
Complement: 95

Guns: 1 USSR 3 in *(76 mm/60)* ❶; 120 rds/min to 15 km *(8 n miles)*; weight of shell 6.8 kg.
2—30 mm/65 AK 630 ❷; 6 barrels per mounting; 3,000 rds/min to 2 km.
2—30 mm/65 AK 306 ❸; 6 barrels per mounting; 3,000 rds/min to 2 km.
Torpedoes: 4—21 in *(533 mm)* (2 twin) tubes ❹. Russian 53—65; passive/wake homing to 25 km *(13.5 n miles)* at 50 kt; warhead 300 kg.
A/S mortars: 2 RBU 6000 ❺; 12-tubed trainable; range 6,000 m; warhead 31 kg.
Countermeasures: Decoys: 2 PK 16 chaff launchers ❻.
ESM: 2 Watch Dog; intercept.

ADMIRAL HORIA MACELARIU   *(Scale 1 : 900), Ian Sturton /* 1044187

Radars: Air/surface search; Strut Curve ❼; F-band.
Fire control: Drum Tilt ❽; H/I-band.
Navigation: Nayada; I-band.
IFF: High Pole.
Sonars: Hull-mounted; active search and attack; medium frequency.

Helicopters: 1 IAR-316 Aloutte III ❾.

Programmes: Follow on to Tetal class. Second of class was delayed when work stopped for a time in 1993-94.
Structure: As well as improved armament and a helicopter deck, there are superstructure changes from the original Tetals, but the hull and propulsion machinery are the same.
Operational: Both based at Mangalia.

UPDATED

ADMIRAL HORIA MACELARIU   *9/2002, C D Yaylali /* 0533314   CONTRE ADMIRAL EUSTATIU SEBASTIAN   *9/2003*, C D Yaylali /* 0589802

3 ZBORUL (TARANTUL I) CLASS (PROJECT 1241 RE) (FSG)

Name	No	Builders	Commissioned
ZBORUL	188	Petrovsky Shipyard	Dec 1990
PESCARUSUL	189	Petrovsky Shipyard	Feb 1992
LASTUNUL	190	Petrovsky Shipyard	Feb 1992

Displacement, tons: 385 standard; 455 full load
Dimensions, feet (metres): 184.1 × 37.7 × 8.2 *(56.1 × 11.5 × 2.5)*
Main machinery: COGAG; 2 Type DR 77 gas turbines; 16,016 hp(m) *(11.77 MW)* sustained; 2 Nikolayev Type DR 76 gas turbines with reversible gearboxes; 4,993 hp(m) *(3.67 MW)* sustained; 2 shafts
Speed, knots: 36. **Range, n miles:** 2,000 at 20 kt; 400 at 36 kt
Complement: 41 (5 officers)
Missiles: 4 SS-N-2C Styx (2 twin); active radar or IR homing to 83 km *(45 n miles)* at 0.9 Mach; warhead 513 kg.
Guns: 1 USSR 3 in *(76 mm)*/60; 120 rds/min to 15 km *(8 n miles)*; weight of shell 7 kg.
2—30 mm/65 AK 630; 6 barrels per mounting; 3,000 rds/min to 2 km.
Countermeasures: 2 PK 16 chaff launchers.
ESM: 2 Watch Dog; intercept.
Weapons control: Hood Wink optronic director.
Radars: Air/surface search: Plank Shave; E-band.
Fire control: Bass Tilt; H/I-band.

LASTUNUL   *6/1998, Valentino Cluru /* 0052766

Navigation: Spin Trough; I-band.
IFF: Square Head. High Pole.

Comment: Built in 1985 and later transferred from the USSR. Export version similar to those built for Poland, India and Yemen. Based at Mangalia.   *UPDATED*

SHIPBORNE AIRCRAFT

Numbers/Type: 6 IAR-316B Alouette III.
Operational speed: 113 kt *(210 km/h)*.
Service ceiling: 10,500 ft *(3,200 m)*.
Range: 290 n miles *(540 km)*.
Role/Weapon systems: ASW helicopter. Sensors: Nose-mounted search radar. Weapons: ASW; two lightweight torpedoes.

VERIFIED

LAND-BASED MARITIME AIRCRAFT

Numbers/Type: 5 Mil Mi-14PL Haze A.
Operational speed: 124 kt *(230 km/h)*.
Service ceiling: 15,000 ft *(4,570 m)*.
Range: 432 n miles *(800 km)*.
Role/Weapon systems: Medium-range ASW helicopter. Sensors: Short Horn search radar, dipping sonar, MAD, sonobuoys. Weapons: ASW; internally stored torpedoes, depth mines and bombs.

VERIFIED

PATROL FORCES

Notes: There is a total of about 20 river patrol boats. These include three 27 ft Boston Whalers presented by the US in March 1993 for Customs/Police patrols on the Danube in support of UN sanctions operations. There is also a hovercraft built at Mangalia in 1998.

5 BRUTAR II CLASS (RIVER MONITORS) (PGR)

Name	No	Builders	Commissioned
RAHOVA	176	Mangalia Shipyard	14 Apr 1988
OPANEZ	177	Mangalia Shipyard	24 July 1990
SMARDAN	178	Mangalia Shipyard	24 July 1990
POSADA	179	Mangalia Shipyard	14 May 1992
ROVINE	180	Mangalia Shipyard	30 June 1993

Displacement, tons: 410 full load
Dimensions, feet (metres): 150 × 26.4 × 4.9 *(45.7 × 8 × 1.5)*
Main machinery: 2 diesels; 2,700 hp(m) *(2 MW)*; 2 shafts
Speed, knots: 16
Guns: 1—100 mm (tank turret). 2—30 mm (twin). 10—14.5 mm (2 quad, 2 single) MGs. 2—122 mm BM-21 rocket launchers; 40-tubed trainable.
Radars: Navigation; I-band.

Comment: Operational with the Danube Flotilla. The first is a Brutar I. The next pair are Brutar IIs based at Tulcea and the last two are Brutar IIs based at Mangalia.   *UPDATED*

RAHOVA   *10/2003, Freddy Philips /* 0567529

OPANEZ   *10/2003*, Freddy Philips /* 0589804

HAZE PL (Polish colours)   *6/2000 /* 0105235

3 KOGALNICEANU CLASS (RIVER MONITORS) (PGR)

Name	No	Builders	Recommissioned
MIKHAIL KOGALNICEANU	45	Drobeta Santierul, Turnu Severin	19 Dec 1993
I C BRATIANU	46	Drobeta Santierul, Turnu Severin	28 Dec 1994
LASCAR CATARGIU	47	Drobeta Santierul, Turnu Severin	22 Nov 1996

Displacement, tons: 575 full load
Dimensions, feet (metres): 170.6 × 29.5 × 5.6 *(52 × 9 × 1.7)*
Main machinery: 2 24-H-165 RINS diesels; 4,400 hp(m) *(3.3 MW)*; 2 shafts
Speed, knots: 18
Guns: 2—100 mm (tank turrets). 4—30 mm (2 twin). 4—14.5 mm (2 twin).
   2—122 mm BM-21 rocket launchers.
Radars: Navigation: I-band.

Comment: Based at Braila.

UPDATED

I C BRATIANU   *6/1999, Romanian Navy /* 0081622

6 HUCHUAN CLASS (FAST ATTACK CRAFT—TORPEDO) (PTK)

320, 321, 324, 325, 353, 354

Displacement, tons: 39 standard; 45 full load
Dimensions, feet (metres): 71.5 × 20.7 oa; 11.8 hull × 3.3 *(21.8 × 6.3; 3.6 × 1)*
Main machinery: 3 Type M 50 diesels; 3,300 hp(m) *(2.4 MW)* sustained; 3 shafts
Speed, knots: 50 foilborne. **Range, n miles:** 500 at 30 kt
Complement: 11
Guns: 4—14.5 mm/93 (2 twin) MGs.
Torpedoes: 2—21 in *(533 mm)* tubes; anti-surface.
Depth charges: 2 rails.
Radars: Surface search: Type 753; I-band.

Comment: Hydrofoils of the same class as the Chinese. *353-354* built at Mangalia Shipyard 1974-
   1983. *320-325* was a repeat order also built at Mangalia 1988-1990. Following 19 deletions,
   these vessels remain in service although further deletions are expected. Based at Mangalia.

VERIFIED

HUCHUAN CLASS   *6/1999, Romanian Navy /* 0081624

3 NALUCA CLASS (FAST ATTACK CRAFT—TORPEDO) (PTF)

Name	No	Builders	Commissioned
SMEUL	202	Mangalia Shipyard	25 Oct 1979
VIJELIA	204	Mangalia Shipyard	7 Feb 1980
VULCANUL	209	Mangalia Shipyard	26 Oct 1981

Displacement, tons: 215 full load
Dimensions, feet (metres): 120.7 × 24.9 × 5.9 *(36.8 × 7.6 × 1.8)*
Main machinery: 3 Type M 503A diesels; 8,025 hp *(5.9 MW)* sustained; 3 shafts
Speed, knots: 36. **Range, n miles:** 500 at 35 kt
Complement: 22 (4 officers)
Guns: 4—30 mm/65 (2 twin) AK 230.
Torpedoes: 4—21 in *(533 mm)* tubes; anti-surface.
Radars: Surface search: Pot Drum; H/I-band.
Fire control: Drum Tilt; H/I-band.
IFF: High Pole A.

Comment: Based on the Osa class hull with torpedo tubes in lieu of SSMs. Last remaining three
   craft. Based at Mangalia.

UPDATED

SMEUL   *6/1999, Romanian Navy /* 0081625

MINE WARFARE FORCES

1 CORSAR CLASS
(MINELAYER/MCM SUPPORT SHIP) (ML/MCS)

Name	No	Builders	Commissioned
VICE ADMIRAL CONSTANTIN BALESCU	274	Mangalia Shipyard	16 Nov 1981

Displacement, tons: 1,450 full load
Dimensions, feet (metres): 259.1 × 34.8 × 11.8 *(79 × 10.6 × 3.6)*
Main machinery: 2 diesels; 6,400 hp(m) *(4.7 MW)*; 2 shafts
Speed, knots: 19
Complement: 75
Guns: 1—57 mm/70. 4—30 mm/65 (2 twin) AK 230. 8—14.5 mm (2 quad) MGs.
A/S mortars: 2 RBU 1200 5-tubed fixed; range 1,200 m; warhead 34 kg.
Mines: 200.
Countermeasures: ESM: Watch Dog; intercept.
Radars: Air/surface search: Strut Curve; F-band.
Navigation: Don 2; I-band.
Fire control: Muff Cob; G/H-band. Drum Tilt; H/I-band.
Sonars: Tamir II; hull-mounted; active search; high frequency.

Comment: Has a large crane on the after deck. Similar to survey ship *Grigore Antipa*. Based at
   Constanta.

UPDATED

VICE ADMIRAL CONSTANTIN BALESCU   *6/1999, Romanian Navy /* 0081627

4 MUSCA CLASS (MINESWEEPERS—COASTAL) (MSC)

Name	No	Builders	Commissioned
LIEUTENANT REMUS LEPRI	24	Mangalia Shipyard	23 Apr 1987
LIEUTENANT LUPU DUNESCU	25	Mangalia Shipyard	6 Jan 1989
LIEUTENANT DIMITRIE NICOLESCU	29	Mangalia Shipyard	7 Dec 1989
SUB LIEUTENANT ALEXANDRU AXENTE	30	Mangalia Shipyard	7 Dec 1989

Displacement, tons: 790 full load
Dimensions, feet (metres): 194.2 × 31.1 × 9.2 *(59.2 × 9.5 × 2.8)*
Main machinery: 2 diesels; 4,800 hp(m) *(3.5 MW)*; 2 shafts
Speed, knots: 17
Complement: 60
Missiles: SAM: 2 quad SA-N-5 launchers.
Guns: 4—30 mm/65 (2 twin) AK 230.
A/S mortars: 2 RBU 1200 5-tubed fixed; range 1,200 m; warhead 34 kg.
Radars: Surface search: Krivach; I-band.
Fire control: Drum Tilt; H/I-band.
Navigation: Nayada; I-band.
Sonars: Hull-mounted; active search; high frequency.

Comment: Reported as having a secondary mining capability but this is not confirmed. Based at
   Mangalia.

UPDATED

SUB LIEUTENANT ALEXANDRU AXENTE   *3/2003, B Lemachko /* 0567527

6 VD 141 CLASS (RIVER MINESWEEPERS) (MSR)

141-165 series

Displacement, tons: 97 full load
Dimensions, feet (metres): 109 × 15.7 × 2.8 *(33.3 × 4.8 × 0.9)*
Main machinery: 2 diesels; 870 hp(m) *(640 kW)*; 2 shafts
Speed, knots: 13
Guns: 4—14.5 mm (2 twin) MGs.
Mines: 6.
Radars: Navigation: Nayada; I-band.

Comment: Built in Romania at Dobreta Severin Shipyard 1976-84. Belong to the Danube Flotilla.

UPDATED

VD 150   *6/1999, Romanian Navy /* 0081629

SURVEY AND RESEARCH SHIPS

1 CORSAR CLASS (AGOR/AGI)

Name	No	Builders	Commissioned
GRIGORE ANTIPA	75	Mangalia Shipyard	25 May 1980

Displacement, tons: 1,450 full load
Dimensions, feet (metres): 259.1 × 34.8 × 11.8 *(79 × 10.6 × 3.6)*
Main machinery: 2 diesels; 6,400 hp(m) *(4.7 MW)*; 2 shafts
Speed, knots: 19
Complement: 75
Radars: Navigation: Nayada; I-band.

Comment: Large davits aft for launching manned submersible. Same hull as Corsar class. Used mostly as an AGI. Based at Constanta.

VERIFIED

GRIGORE ANTIPA   *5/1998, Diego Quevedo /* 0052770

1 RESEARCH SHIP (AGOR/AGI)

Name	No	Builders	Commissioned
EMIL RACOVITA	115	Drobeta Severin Shipyard	30 Oct 1977

Displacement, tons: 1,900 full load
Dimensions, feet (metres): 229.9 × 32.8 × 12.7 *(70.1 × 10 × 3.9)*
Main machinery: 1 diesel; 3,285 hp(m) *(2.4 MW)*; 1 shaft
Speed, knots: 11
Complement: 80

Comment: Modernised in the mid-1980s. Similar design to *Grigore Antipa*. Used mostly as an AGI. Based at Constanta.

VERIFIED

EMIL RACOVITA   *2001, Romanian Navy /* 0114548

TRAINING SHIPS

Notes: *Neptun* belongs to the Merchant Navy.

1 SAIL TRAINING SHIP (AXS)

Name	No	Builders	Launched	Commissioned
MIRCEA	288	Blohm + Voss, Hamburg	29 Sep 1938	29 Mar 1939

Displacement, tons: 1,604 full load
Dimensions, feet (metres): 206; 266.4 (with bowsprit) × 39.3 × 16.5 *(62.8; 81.2 × 12 × 5.2)*
Main machinery: 1 MaK 6M 451 auxiliary diesel; 1,000 hp(m) *(735 kW)*; 1 shaft
Speed, knots: 8. **Range, n miles:** 5,000 at 8 kt
Complement: 83 (5 officers) plus 140 midshipmen
Radars: Navigation: Decca 202; I-band.

Comment: Refitted at Hamburg in 1966. Sail area, 5,739 m² *(18,830 sq ft)*. A smaller version of US Coast Guard cutter *Eagle*, German *Gorch Fock* and Portuguese *Sagres*. Based at Constanta.

UPDATED

MIRCEA   *9/2004*, B Prézelin /* 1044185

AUXILIARIES

2 CROITOR CLASS (LOGISTIC SUPPORT SHIPS) (AETLMH)

Name	No	Builders	Commissioned
CONSTANTA	281	Braila Shipyard	15 Sep 1980
MIDIA	283	Braila Shipyard	26 Feb 1982

Displacement, tons: 2,850 standard; 3,500 full load
Dimensions, feet (metres): 354.3 × 44.3 × 12.5 *(108 × 13.5 × 3.8)*
Main machinery: 2 diesels; 6,500 hp(m) *(4.8 MW)*; 2 shafts
Speed, knots: 16
Missiles: SAM: 2 SA-N-5 Grail quad launchers; manual aiming; IR homing to 6 km *(3.2 n miles)* at 1.5 Mach; warhead 1.5 kg.
Guns: 2—57 mm/70 (twin). 4—30 mm/65 (2 twin). 8—14.5 mm (2 quad) MGs.
A/S mortars: 2 RBU 1200 5-tubed fixed; range 1,200 m; warhead 34 kg.
Countermeasures: ESM: 2 Watch Dog; intercept.
Radars: Air/surface search: Strut Curve; F-band.
Navigation: Krivach; I-band.
Fire control: Muff Cob; G/H-band. Drum Tilt; H/I-band.
Sonars: Tamir II; hull-mounted; active attack; high frequency.
Helicopters: 1 IAR-316 Alouette III type.

Comment: These ships are a scaled down version of Soviet Don class. Forward crane for ammunition replenishment. Some ASW escort capability. Can carry Styx missiles and torpedoes. Based at Constanta.

VERIFIED

CONSTANTA   *6/2001, Schaeffer/Marsan /* 0533268

2 DEGAUSSING SHIPS (ADG/AGI)

Name	No	Builders	Commissioned
ELECTRONICA	296	Braila Shipyard	6 Aug 1973
MAGNETICA	298	Mangalia Shipyard	18 Dec 1989

Displacement, tons: 299 full load
Dimensions, feet (metres): 134 × 21.6 × 10.7 *(40.8 × 6.6 × 3.2)*
Main machinery: Diesel-electric; 1 diesel generator; 600 kW; 1 shaft
Speed, knots: 12.5
Complement: 18
Guns: 2—14.5 mm (twin) MGs. 2—12.7 mm MGs.

Comment: Built for degaussing ships up to 3,000 tons displacement. Electronica is used as an AGI. Based at Tulcea.

VERIFIED

MAGNETICA   *6/1999, Romanian Navy /* 0081632

1 TANKER (AOT)

TULCEA 532

Displacement, tons: 2,170 full load
Dimensions, feet (metres): 250.4 × 41 × 16.4 *(76.3 × 12.5 × 5)*
Main machinery: 2 diesels; 4,800 hp(m) *(3.5 MW)*; 2 shafts
Speed, knots: 16
Cargo capacity: 1,200 tons oil
Guns: 2—30 mm/65 (twin). 4—14.5 mm (2 twin) MGs.

Comment: First one built by Tulcea Shipyard and commissioned 24 December 1992. Second of class reported in 1997 but not confirmed. Based at Constanta. *VERIFIED*

TULCEA   *2001, Romanian Navy /* 0114542

2 COASTAL TANKERS (AOTL)

530   531

Displacement, tons: 1,042 full load
Dimensions, feet (metres): 181.2 × 30.9 × 13.4 *(55.2 × 9.4 × 4.1)*
Main machinery: 2 diesels; 1,800 hp(m) *(1.3 MW)*; 2 shafts
Speed, knots: 12.5
Cargo capacity: 500 tons oil
Guns: 1—37 mm. 2—12.7 mm MGs.

Comment: Built by Braila Shipyard and both commissioned 15 June 1971. Based at Constanta. *VERIFIED*

531   *6/1999, Romanian Navy /* 0081636

1 FLAG OFFICERS BARGE

RINDUNICA

Displacement, tons: 40 full load
Dimensions, feet (metres): 78.7 × 16.4 × 3.6 *(24 × 5 × 1.1)*
Main machinery: 2 diesels; 2,200 hp(m) *(1.6 MW)*; 2 shafts
Speed, knots: 28
Complement: 6

Comment: Used as a barge by the Commander-in-Chief. *VERIFIED*

TUGS

Notes: There are also a number of harbour and river tugs, some of which are armed. These include two Roslavl (101 and 116) at Mangalia.

HARBOUR TUG 570   *12/1994 /* 0081638

2 OCEAN TUGS (ATA)

GROZAVU 500   **HERCULES** 501

Displacement, tons: 3,600 full load
Dimensions, feet (metres): 212.6 × 47.9 × 18 *(64.8 × 14.6 × 5.5)*
Main machinery: 2 diesels; 5,000 hp(m) *(3.7 MW)*; 2 shafts
Speed, knots: 12
Guns: 2—30 mm (twin). 8—14.5 mm (2 quad) MGs.

Comment: First one built at Oltenitza Shipyard and commissioned 29 June 1993. Second of class completed in 1995. Based at Constanta. *VERIFIED*

GROZAVU   *2001, Romanian Navy /* 0114543

RINDUNICA   *1/1995 /* 0081637

Russian Federation
ROSIYSKIY VOENNOMORSKY FLOT

Country Overview

Formerly a constituent republic of the Soviet Union, the Russian Federation was established as an independent state in 1991. The largest country in the world with an area of 6,592,850 square miles, it is bordered to the south by North Korea, China, Mongolia, Kazakhstan, Azerbaijan and Georgia and to the west by Norway, Finland, Latvia, Estonia, Ukraine and Belarus, which with Lithuania separates the Kaliningrad oblast (formerly Königsberg) from the rest of Russia. It has a 20,331 n mile coastline with the Arctic and Pacific Oceans and the Caspian, Baltic and Black Seas. These three seas are inter-connected by an extensive inland waterway system whose main components are the Volga and Don rivers, the Volga-Don canal and the Volga-Baltic Waterway. A canal also links the system to the capital and largest city, Moscow. The Amur River is the most important navigable river in the far east region. Offshore, principal islands in the Arctic Ocean include the Franz Josef Land and Severnaya Zemlya archipelagos, Novaya Zemlya, Vaygach Island, the New Siberian Islands and Wrangel Island. In the Pacific lie the Kuril Islands, which extend from the Kamchatka Peninsula, and Sakhalin Island. Principal seaports include Novorossiysk (Black Sea), St Petersburg and Kaliningrad (Baltic), Nakhodka, Vostochnyy, Vladivostok, and Vanino (Pacific) and Murmansk and Archangel (Arctic). Major river ports include Rybinsk, Nizhniy Novgorod, Samara, Volgograd, Astrakhan and Rostov-on-Don. Territorial waters (12 n miles) are claimed. An EEZ (200 n miles), is also claimed and the limits have been partly defined by boundary agreements.

Headquarters Appointments

Commander-in-Chief:
　Admiral of the Fleet V Kuroyedov
First Deputy Commander-in-Chief:
　Admiral Victor Kravchenko
Deputy Commander-in-Chief:
　Admiral M G Zakharenko
Deputy Chief of Naval Staff:
　Vice Admiral V A Ilim
Chief of Operations:
　Vice Admiral V Patrushev
Commander Naval Units Border Guard:
　Vice Admiral I I Nalyotov

Northern Fleet
Commander:
　Vice Admiral Mikhail Abramov

Pacific Fleet
Commander:
　Admiral Victor Fyodorov

Black Sea Fleet
Commander:
　Vice Admiral Vladimir Masorin

Baltic Fleet
Commander:
　Admiral V Valuyev

Caspian Flotilla
Commander:
　Vice Admiral Yuriy Startsev

Personnel

(a)　2005: 80,000 not including naval aviation and naval infantry. The approximate division is 27,000 in the North, 22,000 in the Pacific, 15,000 in the Baltic, 11,000 in the Black Sea and 5,000 in the Caspian.
(b)　Approximately 30 per cent volunteers (officers and senior ratings) — remainder two years' national service (or three years if volunteered)

Associated Navies

The Soviet Union was dissolved in December 1991. In 1992 a Commonwealth of Independent States was formed from the Republics of the former Union, but without the Baltic States. In the Baltic the Russian flotilla had withdrawn from the former East German and Polish ports by 1993 and from the Baltic Republics by the end of 1994. The Caspian flotilla divided with some units going to Azerbaijan, Kazakhstan and Turkmenistan. In the Black Sea the division of the Fleet between Russia and Ukraine was finally implemented in 1997. Facilities are shared in some Crimean ports.

Main Bases

North: Severomorsk (HQ), Polyarny, Gremika, Zapandaya Litsa, Gadzhievo, Vidyayevo
Baltic: Kaliningrad (HQ), St Petersburg, Kronshtadt, Baltiysk
Black Sea: Sevastopol (HQ) (Crimea), Tuapse, Novorssiysk, Feodosiya
Caspian: Astrakhan (HQ), Makhachkala
Pacific: Vladivostok (HQ), Sovetskaya Gavan, Magadan, Petropavlovsk, Komsomolsk, Racovaya

Operational

From 1991 a shortage of funds to pay for dockyard repairs, spare parts and fuel meant that many major surface warships were rarely at sea, and few operated away from their local exercise areas. Activity levels temporarily rose from 1996 but many ships, although technically in commission, remained in harbour. Activity reached a low point in 2002 but, in recent years, improvements in the budgetary situation and the publication of a new naval doctrine have led to a higher operational tempo. A busier pattern of exercises and operations was initiated in 2003 and activity levels were maintained during 2004.

Coast Defence

The Command of Naval Infantry and Coastal Artillery and Missiles includes a Division of Coastal Artillery and three Mechanised Infantry (Coastal Defence Troops) Brigades, an Artillery Self-Propelled Brigade, plus the units of Naval Infantry (five Brigades and one Division) and a number of minor units. The force of Coastal Artillery includes 19 Missile Batallions (SSC-1 Sepal SS-C-3 Styx) and 11 Gun Batallions (130 mm and 152 mm). Many of these units are in reserve. The Naval Infantry were deployed in Chechnya in 2000.

Pennant Numbers

There have been no major changes to pennant numbers since 1993 except when ships transfer fleets. Some submarines have temporary numbers on the side of the fin.

Class and Weapon Systems Names

Most Russian ship class names differ from those allocated by NATO. In such cases the Russian name is placed in brackets after the NATO name. Type or Project numbers are also placed in brackets. Weapon systems retain their NATO names with the Russian name, when known, placed in brackets. Some equipment now has three names — NATO, Russian Navy and Russian export.

Civilian Support Ships

Previously, civilian manned research ships and some icebreakers were effectively under naval control and were therefore included in the former Soviet/Russian section. These ships have been removed as all are now employed solely for commercial purposes.

Strength of the Fleet

Type	Active	Building
Submarines (SSBN)	16	2 (1)
Submarines (SSGN/SSN)	22	3
Submarines (SSK)	17	2
Auxiliary Submarines (SSA(N))	9	—
Aircraft Carriers (CV)	1	—
Battle Cruisers (CGN)	1	—
Cruisers (CG)	4	—
Destroyers (DDG)	17	—
Frigates (FFG)	10	4
Frigates (FF and FFL)	34	—
Corvettes	41	1 (9)
Patrol Forces	6	1
Minesweepers—Ocean	12	1
Minesweepers—Coastal	31	—
LPDs	1	—
LSTs	22	—
Hovercraft (Amphib)	12	—
Replenishment Tankers	22	—
Hospital Ships	3	—

Notes: There are large numbers of most classes 'in reserve', and flying an ensign so that skeleton crews may still be paid. The list above reflects only those units assessed as having some realistic operational capability or some prospect of returning to service after refit.

Fleet Disposition (1 January 2005)

Type	Northern	Baltic	Black Sea	Pacific	Caspian
SSBN	11	—	—	5	—
SSGN/SSN	16	—	—	6	—
SSK	6	2	1	8	—
SSA(N)	9	—	—	—	—
CV	1	—	—	—	—
CGN	1	—	—	—	—
CG	1	—	2	1	—
DDG	6	3	1	7	—
FFG	3	3	2	1	1
FF and FFL	8	11	6	9	—
Corvettes	5	12	5	18	1
LPD	1	—	—	—	—
LST	4	6	7	5	—

Notes: MCMV are divided evenly between the four main Fleets plus a few in the Caspian Sea.

DELETIONS

Notes: Some of the ships listed are still theoretically 'laid up', and some still fly an ensign so that skeleton crews may be paid.

Submarines

2002　1 Delta IV (SSBN), 2 Victor III (SSN), 4 Tango, 1 Foxtrot
2003　1 Delta I
2004　1 Yankee Notch

Cruisers

2003　1 Kynda (*Admiral Golovko*)

Frigates

2002　1 Grisha IV, 1 Krivak (*Druzhny*)

Corvettes (Missile)

2002　1 Tarantul I

Patrol Forces

2002　1 Ivan Susanin, 1 Matka, 2 Turya, 7 Yaz, 1 Piyavka, 5 Shmel, 2 Vosh

Mine Warfare Vessels

2002　1 Natya I

AGIs

2002　SSV 590, SSV 591

Naval Survey and Research Ships

2002　1 Moma

PENNANT LIST

Submarines
Ballistic Missile Submarines

Typhoon class
TK 17	Arkhangelsk
TK 29	Severstal
TK 208	Dmitry Donskoy

Delta IV class
K 18	Karelia
K 51	Verchoture
K 84	Ekaterinburg
K 114	Tula
K 117	Briansk
K 407	Novomoskvosk

Delta III class
K 44	Borisoglebsk
K 180	Ryazan
K 211	Podolsk
K 433	Syvatoy Giorgiy Pobedonosets
K 490	Voskresensk
K 496	Zelenograd
K 506	Petrapavlovsk-Kamchatsky

Attack Submarines

Oscar II class
K 119	Voronezh
K 139	Belgorod
K 186	Omsk
K 266	Orel
K 410	Smolensk
K 442	Cheliabinsk
K 526	Tomsk

Yasen class
K 329	Severodvisnk (bldg)

Akula I and II classes
K 152	Nerpa (bldg)
K 154	Tigr
K 157	Vepr
K 267	Samara
K 328	Leopard
K 331	Magadan
K 335	Gepard
K 337	Cougar (bldg)
K 419	Kuzbass
K 461	Volk

Sierra I and II classes
K 276	Kostroma
K 336	Pskov

Victor III class
B 138	Obninsk
B 292	Perm
B 388	Petrozavodsk
B 414	Danil Moskovskiy
B 448	Tambov

Auxiliary Submarines
BS 129	-
AS 13	-
AS 14	-
AS 15	-
KS 403	Kazan
KS 411	Orienburg

Patrol Submarines
S 100	Saint Petersburg (bldg)
B 177	Lipetsk
B 187	—
B 190	—
B 227	-
B 248	-
B 260	Razboynik
B 345	-
B 401	Novosibirsk
B 402	Vologda
B 405	-
B 437	Magnetogorsk
B 439	-
B 445	-
B 471	Kaluga
B 494	Ust-Bolsheretsk
B 800	Magneto-Gorsk
B 806	Tur
B 808	Jaroslavl
B 871	Alrosa

Aircraft Carriers
063	Admiral Kuznetsov

Battle Cruisers
099	Pyotr Velikiy

Cruisers
011	Varyag
055	Marshal Ustinov
121	Moskva
713	Kerch

Destroyers
400	Vitse admiral Kulakov
434	Marshal Ushakov
543	Marshal Shaposhnikov
548	Admiral Panteleyev
564	Admiral Tributs
572	Admiral Vinogradov
605	Admiral Levchenko
610	Nastoychivy
619	Severomorsk
620	Bespokoiny
650	Admiral Chabanenko
678	Admiral Kharlamov

687	Marshal Vasilevsky
715	Bystry
754	Bezboyaznennyy
778	Burny
810	Smetlivy

Frigates
-	Steregushchiy (bldg)
052	Vorovsky (BG)
060	Anadyr (BG)
097	Dzerzhinksky (BG)
103	Kedrov (BG)
104	Pskov (BG)
113	Menzhinsky (BG)
156	Orel (BG)
661	Letuchy
691	Tatarstan
702	Pylky
712	Neustrashimy
731	Neukrotimy
801	Ladny
808	Pytlivy
930	Legky
937	Zharky
955	Zadorny

Corvettes
409	Moroz
423	Smerch
450	Razliv
505	Uragan
520	Rassvet
526	Nakat
533	Tusha
535	Aysberg
551	Liven
555	Geyzer
560	Zyb
570	Passat
590	Meteor
615	Bora
616	Samum
617	Mirazh
620	Shtyl

Mine Warfare Forces
610	Svyazist
718	MT 265
738	MT 264
762	V Gumavenko
770	Valentin Pikul
-	Vitse admiral Zacharin
806	Motorist
855	Kontradmiral Vlasov
901	A Zheleznyakov
912	Turbinist

913	Kovrovets
919	Snayper

Amphibious Forces
012	Olenegorskiy Gorniak
016	Georgiy Pobedonosets
020	Mitrofan Moskalenko
027	Kondopoga
031	Alexander Otrakovskiy
055	BDK-98
066	Mukhtar Avezov
070	Bobruysk
077	Nikolay Korsakov
081	Nikolay Vilkov
102	Kaliningrad
110	Alexander Shabalin
119	Minsk
125	BDK-105
127	Minsk
130	Korolev
148	Orsk
150	Saratov
151	Azov
152	Nikolay Filchenkov
156	Yamal
158	Tsesar Kunikov

Auxiliaries
154	Berezina
SFP 177	Akademik Isanin
SFP 183	Akademir Seminikhin
200	Perekop
208	Sevan
210	Smolny
212	Yamal
506	Dauriya
SB 921	Paradoks
SB 922	Shakhter

Intelligence Collection Ships
GS 39	Syzran
SSV 080	Pribaltika
SSV 169	Tavriya
SSV 175	Odograf
SSV 201	Priazove
SSV 208	Kurily
SSV 231	Vassily Tatischev
SSV 418	Ekvator
SSV 506	Nakhoda
SSV 512	Kildin
SSV 520	Meridian
SSV 535	Kareliya
SSV 571	Belomore
SSV 824	Liman

SUBMARINES

Notes: Temporary pennant numbers are shown on the fins of some submarines.

NOVOMOSKOVSK

6/2003*, B Lemachko / 1121441

Strategic Missile Submarines (SSBN)

Notes: A Borey (Project 955) class submarine *Yuri Dolgoruky* was laid down on 2 November 1996. The plan was to fit a new strategic missile SS-NX-28. When , following three unsuccessful tests, this missile programme was cancelled in 1998, construction of the boat slowed and may have stopped while development of a new missile, a navalised version of the SS-27 Topol-M (known as SS-NX-30 'Bulava') was undertaken. A test

launch of an unpowered mock-up of this missile was conducted on 23 September 2004 from the Typhoon class SSBN, *Dmitriy Donskoy* and full flight tests are expected to continue in 2005. It is also expected that *Yuri Dolgoruky*, the original design of which is likely to have been modified to accommodate 12 of the smaller 47 tonne 'Bulava', is to be launched in 2005. The second of class, *Alexander Nevsky*, was laid down at the Sevmash plant in

Severodvinsk on 19 March 2004 and if new units are to enter service at two year intervals, it is possible that three may have entered service by 2011. At least six of the class are expected. Overall, a force level of up to 15 SSBNs remains an aspiration although this may prove difficult to sustain.

3 TYPHOON (AKULA) CLASS (PROJECT 941/941U) (SSBN)

Name	No	Builders	Laid down	Launched	Commissioned
DMITRIY DONSKOY	TK 208	Severodvinsk Shipyard	30 June 1976	23 Sep 1979	12 Dec 1981
ARKHANGELSK	TK 17	Severodvinsk Shipyard	24 Feb 1985	Aug 1986	6 Nov 1987
SEVERSTAL	TK 20	Severodvinsk Shipyard	6 Jan 1986	July 1988	4 Sep 1989

Displacement, tons: 18,500 surfaced; 26,500 dived
Dimensions, feet (metres): 562.7 oa; 541.3 wl × 80.7 × 42.7 *(171.5; 165 × 24.6 × 13)*
Main machinery: Nuclear; 2 VM-5 PWR; 380 MW; 2 GT3A turbines; 81,600 hp(m) *(60 MW)*; 2 emergency motors; 517 hp(m) *(380 kW)*; 2 shafts; shrouded props; 2 thrusters (bow and stern); 2,860 hp(m) *(1.5 MW)*
Speed, knots: 25 dived; 12 surfaced
Complement: 175 (55 officers)

Missiles: SLBM: 20 Makeyev SS-N-20 (RSM 52/3M20) Sturgeon; three-stage solid fuel rocket; stellar inertial guidance to 8,300 km *(4,500 n miles)*; warhead nuclear 10 MIRV each of 200 kT; CEP 500 m. 2 missiles can be fired in 15 seconds.
SAM: SA-N-8 SAM capability when surfaced.
A/S: Novator SS-N-15 Starfish; inertial flight to 45 km *(24.3 n miles)*; warhead nuclear 200 kT or Type 40 torpedo.
Torpedoes: 6—21 in *(533 mm)* tubes. Combination of torpedoes (see table at front of section). The weapon load includes a total of 22 torpedoes and A/S missiles.
Mines: Could be carried in lieu of torpedoes.

Countermeasures: Decoys: MG 34/44 tube launched decoys.
ESM: Rim Hat (Nakat M); radar warning. Park Lamp D/F.
Weapons control: 3R65 data control system.
Radars: Surface search: Snoop Pair (Albatros); I/J-band.
Sonars: Shark Gill; hull-mounted; passive/active search and attack; low/medium frequency.
Shark Rib flank array; passive; low frequency.
Mouse Roar; hull-mounted; active attack; high frequency.
Pelamida towed array; passive search; very low frequency.

Modernisation: First of class TK 208 started refit at Severodvinsk in 1994, was relaunched on 26 June 2002 and started sea trials in August 2004. It is expected to conduct test-firings of the new 'Bulava' missile system and subsequently continue in service as an operational unit. TK 17 and TK 20 may be similarly upgraded to remain in service beyond 2010.
Structure: This is the largest type of submarine ever built. Two separate 7.2 m diameter hulls covered by a single outer free-flood hull with anechoic Cluster Guard tiles plus separate 6 m diameter pressure-tight compartments in the fin and fore-ends. There is a 1.2 m separation between the outer and inner

hulls along the sides. The unique features of Typhoon are her enormous size and the fact that the missile tubes are mounted forward of the fin. The positioning of the launch tubes mean a fully integrated weapons area in the bow section leaving space abaft the fin for the provision of two nuclear reactors, one in each hull. The fin configuration indicates a designed capability to break through ice cover up to 3 m thick; the retractable forward hydroplanes, the rounded hull and the shape of the fin are all related to under-ice operations. Diving depth, 1,000 ft *(300 m)*.
Operational: Strategic targets are within range from anywhere in the world. Two VLF/ELF communication buoys are fitted. VLF navigation system for under-ice operations. Pert Spring SATCOM mast, Cod Eye radiometric sextant and Kremmny 2 IFF. All based in the Northern Fleet at Litsa Guba. Of six boats completed, the second and third, TK 202 and TK 12 have been formally decommissioned while the fourth of class TK 13 is expected to follow. TK 17 was damaged by fire during a missile loading accident in 1991 but was subsequently repaired. Old hulls are being disposed of under the Co-operative Threat Reduction Programme.

UPDATED

SEVERSTAL

1/1997 / 0081639

6 DELTA IV (DELFIN) CLASS (PROJECT 667BDRM) (SSBN)

Name	No	Builders	Laid down	Launched	Commissioned
VERCHOTURE	K 51	Severodvinsk Shipyard	23 Feb 1981	Jan 1984	29 Dec 1984
EKATERINBURG	K 84	Severodvinsk Shipyard	Nov 1983	Dec 1984	Feb 1985
TULA	K 114	Severodvinsk Shipyard	Dec 1985	Sep 1986	Jan 1987
BRIANSK	K 117	Severodvinsk Shipyard	Sep 1986	Sep 1987	Mar 1988
KARELIA	K 18	Severodvinsk Shipyard	Sep 1987	Nov 1988	Sep 1989
NOVOMOSKOVSK	K 407	Severodvinsk Shipyard	Nov 1988	Oct 1989	1991

Displacement, tons: 10,800 surfaced; 13,500 dived
Dimensions, feet (metres): 544.6 oa; 518.4 wl × 39.4 × 28.5 *(166; 158 × 12 × 8.7)*
Main machinery: Nuclear; 2 VM-4 PWR; 180 MW; 2 GT3A-365 turbines; 37,400 hp(m) *(27.5 MW)*; 2 emergency motors; 612 hp(m) *(450 kW)*; 2 shafts
Speed, knots: 24 dived; 14 surfaced
Complement: 130 (40 officers)

Missiles: SLBM: 16 Makeyev SS-N-23 (RSM 54) Skiff (Shtil); 3-stage liquid fuel rocket; stellar inertial guidance to 8,300 km *(4,500 n miles)*; warhead nuclear 4-10 MIRV each of 100 kT; CEP 500 m. Same diameter as SS-N-18 but longer.
A/S: Novator SS-N-15 Starfish; inertial flight to 45 km *(24.3 n miles)*; warhead nuclear 200 kT or Type 40 torpedo.
Torpedoes: 4—21 in *(533 mm)* tubes. Combination of 53 cm torpedoes (see table at front of section). Total of 18 weapons.
Countermeasures: ESM: Brick Pulp/Group; radar warning. Park Lamp D/F.
Radars: Surface search: Snoop Tray; I-band.

Sonars: Shark Gill; hull-mounted; passive/active search and attack; low/medium frequency.
Shark Hide flank array; passive; low frequency.
Mouse Roar; hull-mounted; active attack; high frequency.
Pelamida towed array; passive search; very low frequency.

Programmes: Construction first ordered 10 December 1975. This programme completed in late 1990 and included seven boats.
Modernisation: Tests of the modified SS-N-23 missile, known as Sineva, were conducted from K 84 during 2004. It is expected to be fitted to the later boats of the class.
Structure: A slim fitting is sited on the after fin which is reminiscent of a similar tube in one of the November class in the early 1980s. This is a dispenser for a sonar thin line towed array. The other distinguishing feature, apart from the size being greater than Delta III, is the pressure-tight fitting on the after end of the missile tube housing, which may be a TV camera to monitor communications buoy and wire retrieval operations. This is not fitted in all of the class. Brick Spit

optronic mast. Diving depth, 1,300 ft *(400 m)*. The outer casing has a continuous acoustic coating and fewer free flood holes than the Delta III.
Operational: Two VLF/ELF communication buoys. Navigation systems include SATNAV, SINS, Cod Eye. Pert Spring SATCOM. A modified and more accurate version of SS-N-23 was tested at sea in 1988 bringing the CEP down from 900 to 500 m. These improvements are likely to be incorporated in the 'Sineva' missile, which is planned to be fitted throughout the class. Missile launch is conducted at keel depth 55 m and at a speed of 6 kt. All operational units based in the Northern Fleet at Saida Guba. Long refits have been completed as follows: K 51 (1999); K 84 (2002); K 114 (2004). K 117 started refit at Severodvinsk in 2002 and K 18 in 2004. It is likely that K 407 will follow in due course. K 64 has been paid off, although such a young hull has the potential to be re-roled for other tasks. K 407 launched a German commercial satellite on 7 July 1998 from the Barents Sea.

UPDATED

DELTA IV

6/2003*, B Lemachko / 1042306

KARELIA and VERCHOTURE

9/2000, B Lemachko / 0126226

7 DELTA III (KALMAR) CLASS (PROJECT 667BDR) (SSBN)

Name	No	Builders	Laid down	Launched	Commissioned
VOSKRESENSK	K 490	Severodvinsk Shipyard	Nov 1976	Mar 1977	30 Oct 1977
BORISOGLEBSK	K 44	Severodvinsk Shipyard	23 Feb 1977	Aug 1977	Apr 1978
ZELENOGRAD	K 496	Severodvinsk Shipyard	May 1978	Sep 1978	Aug 1979
PETROPAVLOVSK KAMCHATSKY	K 506	Severodvinsk Shipyard	Sep 1978	Mar 1979	Nov 1979
PODOLSK	K 211	Severodvinsk Shipyard	Apr 1979	Dec 1979	Aug 1980
SYVATOY GIORGIY POBEDONOSETS	K 433	Severodvinsk Shipyard	Apr 1980	Nov 1980	Aug 1981
RYAZAN	K 180	Severodvinsk Shipyard	Sep 1981	Mar 1982	Dec 1982

Displacement, tons: 10,550 surfaced; 13,250 dived
Dimensions, feet (metres): 524.9 oa; 498.7 wl × 39.4 × 28.5
(160; 152 × 12 × 8.7)
Main machinery: Nuclear; 2 VM-4 PWR; 180 MW; 2 GT3A-635 turbines; 37,400 hp(m) *(27.5 MW)*; 2 emergency motors; 612 hp(m) *(450 kW)*; 2 shafts
Speed, knots: 24 dived; 14 surfaced
Complement: 130 (20 officers)

Missiles: SLBM: 16 Makeyev SS-N-18 (RSM 50) Stingray (Volna); 2-stage liquid fuel rocket with post boost vehicle (PBV); stellar inertial guidance; 3 variants:
  Mod 1; range 6,500 km *(3,500 n miles)*; warhead nuclear 3 MIRV each of 200 kT; CEP 900 m.
  Mod 2; range 8,000 km *(4,320 n miles)*; warhead nuclear 450 kT; CEP 900 m.
  Mod 3; range 6,500 km *(3,500 n miles)*; warhead nuclear 7 MIRV 100 kT; CEP 900 m.
  Mods 1 and 3 were the first MIRV SLBMs in Soviet service.
Torpedoes: 4—21 in *(533 mm)* and 2—400 mm tubes. Combination of torpedoes (see table at front of section). Total of 16 weapons.
Countermeasures: ESM: Brick Pulp/Group; radar warning. Park Lamp D/F.
Radars: Surface search: Snoop Tray; I-band.
Sonars: Shark Teeth; hull-mounted; passive/active search and attack; low/medium frequency.
  Shark Hide flank array; passive; low frequency.
  Mouse Roar; hull-mounted; active attack; high frequency.
  Pelamida towed array; passive search; very low frequency.

Modernisation: The dispenser tube on the after fin has been fitted to most of the class. It was planned to retrofit SS-N-23 but this was shelved.
Structure: The missile casing is higher than in Delta I class to accommodate SS-N-18 missiles which are longer than the SS-N-8. The outer casing has a continuous 'acoustic' coating but is less streamlined and has more free flood holes than the Delta IV. Brick Spit optronic mast. Diving depth, 1,050 ft *(320 m)*.
Operational: ELF/VLF communications with floating aerial and buoy; UHF and SHF aerials. Navigation equipment includes Cod Eye radiometric sextant, SATNAV, SINS and Omega. Pert Spring SATCOM. Kremmny 2 IFF. Of the 14 hulls completed, the first of class (K 441) paid of in 1996, three more in 1997 and another two by 1999. Four of these (K 449, K 455, K 487 and K 223) are laid up in fleet bases. The operational state of the remaining seven has been variously reported but it must be assumed that they can still fire missiles. K 211 is reported to have test-fired an SS-N-18 on 2 September 2003. K 44 and K 180 are based at Saida Guba in the Northern Fleet and remainder at Tarya Bay in the Pacific. The last hull of the class K 129 converted to a DSRV carrier with missile tubes removed. It is expected that the earlier unit of the class will be paid off as the Borey class enter service.

UPDATED

DELTA III
6/1998, S Breyer / 0081641

DELTA III

2000, B Lemachko / 0126264

Attack Submarines (SSN/SSGN)

Notes: (1) Attack submarines are coated with Cluster Guard anechoic tiles. All submarines are capable of laying mines from their torpedo tubes. All SSNs are fitted with non-acoustic environmental sensors for measuring discontinuities caused by the passage of a submarine in deep water.
(2) The cost of recovering the wreck of *Kursk* is estimated to have been of the order of US$100 million.

0 + 1 YASEN CLASS (PROJECT 885) (SSN/SSGN)

Name	No	Builders	Laid down	Launched	Commissioned
SEVERODVINSK	K 329	Severodvinsk Shipyard	21 Dec 1993	2005	2007

Displacement, tons: 5,900 surfaced; 8,600 dived
Dimensions, feet (metres): 364.2 × 39.4 × 27.6
(111 × 12 × 8.4)
Main machinery: Nuclear; 1 PWR; 195 MW; 2 GT3A turbines; 43,000 hp(m) *(31.6 MW)*; 1 shaft; pump-jet propulsor; 2 spinners
Speed, knots: 28 dived; 17 surfaced

Complement: 80 (30 officers)

Missiles: SLCM/SSM: Novator Alfa SS-N-27.
  8 VLS launchers in after casing. Total of 24 missiles.
A/S: SS-N-15. Fired from torpedo tubes.
Torpedoes: 8—21 in *(533 mm)* tubes. Inclined outwards. Total of about 30 weapons.

Countermeasures: ESM: Radar warning.
Radars: Surface search: I-band.
Sonars: Irtysh Amfora system includes bow-mounted spherical array; passive/active search and attack; low frequency.
  Flank and towed arrays; passive; very low frequency.

Programmes: Malakhit design. Confirmed building in 1993. Reported plans were for seven of the class to replace the Victor III class. It was initially reported that these were to be multipurpose SSNs derived from the Akula II class, but the length of the programme suggests that there has been considerable scope for technical ugprade.
Structure: Some of the details given are speculative. VLS launchers for SSMs, canted torpedo tubes and spherical bow sonars are all new to Russian designs.

UPDATED

SEVERODVINSK
1996, US Navy / 0019006

2 SIERRA II (KONDOR) CLASS (PROJECT 945B) (SSN)

Name	No	Builders	Laid down	Launched	Commissioned
PSKOV (ex-Okun)	K 336	Nizhny Novgorod	May 1990	June 1992	12 Aug 1993
NIZHNY NOVGOROD (ex-Zubatka)	K 534	Nizhny Novgorod	June 1986	June 1988	28 Dec 1990

Displacement, tons: 7,600 surfaced; 9,100 dived
Dimensions, feet (metres): 364.2 × 46.6 × 28.9
(111 × 14.2 × 8.8)
Main machinery: Nuclear; 1 VM-5 PWR; 190 MW; 1 GT3A
turbine; 47,500 hp(m) (70 MW); 2 emergency motors;
2,004 hp(m) (1.5 MW); 1 shaft; 2 spinners; 1,006 hp(m)
(740 kW)
Speed, knots: 32 dived; 10 surfaced
Complement: 61 (31 officers)

Missiles: SLCM: Raduga SS-N-21 Sampson (Granat) fired from
21 in (533 mm) tubes; land-attack; inertial/terrain-following
to 3,000 km (1,620 n miles) at 0.7 Mach; warhead nuclear
200 kT. CEP 150 m. Flies at a height of about 200 m.
SAM: SA-N-5/8 Strela portable launcher; 12 missiles.

A/S: Novator SS-N-15 Starfish (Tsakra) fired from 53 cm tubes;
inertial flight to 45 km (24.3 n miles); warhead nuclear 200 kT
or Type 40 torpedo.
Novator SS-N-16 Stallion fired from 65 cm tubes; inertial flight
to 100 km (54 n miles); payload nuclear 200 kT (Vodopad) or
Type 40 torpedo (Veder).
Torpedoes: 4—25.6 in (650 mm) and 4—21 in (533 mm) tubes.
Combination of 65 and 53 cm torpedoes (see table at front of
section). Total of 40 weapons.
Mines: 42 in lieu of torpedoes.
Countermeasures: ESM: Rim Hat; intercept. Park Lamp D/F.
Radars: Surface search: Snoop Pair with back-to-back ESM
aerial.
Sonars: Shark Gill; hull-mounted; passive/active search and
attack; low/medium frequency.

Shark Rib flank array; passive; low frequency.
Mouse Roar; hull-mounted; active attack; high frequency.
Skat 3 towed array; passive; very low frequency.

Programmes: A third of class K 536 Mars, was scrapped before
completion in July 1992.
Structure: Titanium hull. The towed communications buoy has
been recessed. A 10 point environmental sensor is fitted at
the front end of the fin. The standoff distance between hulls is
considerable and has obvious advantages for radiated noise
reduction and damage resistance. Diving depth, 2,460 ft (750
m). Numbers and sizes of torpedo tubes are uncertain with
different figures given by Russian sources. There are seven
watertight compartments.
Operational: Based in the Northern Fleet, at Ara Guba. K 534
returned to service in 2003. *UPDATED*

SIERRA II 6/1997 / 0019009

SIERRA II 8/1998 / 0050009

PSKOV 6/2002, B Lemachko / 0570928

6 + 1 OSCAR II (ANTYEY) (PROJECT 949B) (SSGN)

Name	No	Builders	Laid down	Launched	Commissioned
VORONEZH	K 119	Severodvinsk Shipyard	1984	1986	1988
SMOLENSK	K 410	Severodvinsk Shipyard	1986	1988	1990
CHELIABINSK	K 442	Severodvinsk Shipyard	1987	1989	29 Dec 1990
OREL (ex-*Severodvinsk*)	K 266	Severodvinsk Shipyard	1989	22 May 1992	Dec 1992
OMSK	K 186	Severodvinsk Shipyard	1990	8 May 1993	15 Dec 1993
TOMSK	K 526	Severodvinsk Shipyard	1993	18 July 1996	28 Feb 1997
BELGOROD	K 139	Severodvinsk Shipyard	1994	2005	2007

Displacement, tons: 13,900 surfaced; 18,300 dived
Dimensions, feet (metres): 505.2 × 59.7 × 29.5
(154 × 18.2 × 9)
Main machinery: Nuclear; 2 VM-5 PWR; 380 MW; 2 GT3A
turbines; 98,000 hp(m) *(72 MW)*; 2 shafts; 2 spinners
Speed, knots: 28 dived; 15 surfaced
Complement: 107 (48 officers)

Missiles: SSM: 24 Chelomey SS-N-19 Shipwreck (Granit); inertial
with command update guidance; active radar homing to 20-
550 km *(10.8-300 n miles)* at 2.5 Mach; warhead 750 kg HE
or 500 kT nuclear. Novator Alfa SS-N-27 may be carried in due
course.
A/S: Novator SS-N-15 Starfish (Tsakra) fired from 53 cm tubes;
inertial flight to 45 km *(24.3 n miles)*; warhead nuclear 200 kT
or Type 40 torpedo.
Novator SS-N-16 Stallion fired from 65 cm tubes; inertial flight
to 100 km *(54 n miles)*; payload nuclear 200 kT (Vodopad) or
Type 40 torpedo (Veder).
Torpedoes: 4—21 in *(533 mm)* and 2—26 in *(650 mm)* tubes.
Combination of 65 and 53 cm torpedoes (see table at front of

section). Total of 28 weapons including tube-launched A/S
missiles.
Mines: 32 can be carried.
Countermeasures: ESM: Rim Hat; intercept.
Weapons control: Punch Bowl for third party targeting.
Radars: Surface search: Snoop Pair or Snoop Half; I-band.
Sonars: Shark Gill; hull-mounted; passive/active search and
attack; low/medium frequency.
Shark Rib flank array; passive; low frequency.
Mouse Roar; hull-mounted; active attack; high frequency.
Pelamida towed array; passive search; very low frequency.

Programmes: Building of a class of 14 began in 1978. Two
Oscar Is and 11 Oscar IIs were completed. Work on the 12th
Oscar II (K 139, *Belgorod*) was thought to have stopped but it
was announced by the Defence Minister on 16 July 2004 that
the boat would be completed although funding of the project
remains problematical.
Structure: SSM missile tubes are in banks of 12 either side and
external to the 8.5 m diameter pressure hull; they are inclined
at 40° with one hatch covering each pair, the whole resulting

in the very large beam. The position of the missile tubes
provides a large gap of some 4 m between the outer and inner
hulls. Diving depth, 1,000 ft *(300 m)* although 2,000 ft
(600 m) is claimed. There are 10 watertight compartments.
Operational: ELF/VLF communications buoy. All have a tube on
the rudder fin as in Delta IV which is used for dispensing a thin
line towed sonar array. Pert Spring SATCOM. K 119, K 410
and K 266 are based at Litsa South in the Northern Fleet and
K 442, K 186 and K 526 at Tarya Bay in the Pacific. In 1999
one Northern Fleet unit deployed for the first Russian SSGN
patrol in the Mediterranean for ten years. At the same time a
Pacific Fleet unit sailed to the western seaboard of the United
States. The two Oscar Is (K 206 and K 525) are laid up in the
Northern Fleet. K 148, K 173, K 132 and K 456 are laid up
awaiting disposal and K 141 *(Kursk)* sunk as the result of an
internal weapon explosion on 12 August 2000. The
submarine was raised in late 2001 and broken up ashore.
UPDATED

OSCAR II

11/2001, *Ships of the World* / 0528392

OSCAR II

3/2002, B Lemachko / 0570930

VORONEZH and DANIL MOSKOVSKIY (VICTOR III)

6/2002, B Lemachko / 0570929

OREL

9/2001, Ships of the World / 0126366

OSCAR II

8/2002, Ships of the World / 0528336

8 + 2 AKULA (BARS) CLASS (PROJECT 971/971U/09710) (SSN)

Name	No	Builders	Laid down	Launched	Commissioned
MAGADAN (ex-*Narwhal*)	K 331	Komsomolsk Shipyard	1984	1986	1990
VOLK	K 461	Severodvinsk Shipyard	1986	11 June 1991	30 Dec 1991
KUZBASS (ex-*Morzh*)	K 419	Komsomolsk Shipyard	1984	1989	1991
LEOPARD	K 328	Severodvinsk Shipyard	Oct 1988	28 July 1992	Dec 1992
TIGR	K 154	Severodvinsk Shipyard	1989	10 June 1993	Dec 1993
SAMARA (ex-*Drakon*)	K 267	Komsomolsk Shipyard	1985	15 July 1994	29 July 1995
NERPA	K 152	Komsomolsk Shipyard	1986	May 1994	2006
VEPR (II)	K 157	Severodvinsk Shipyard	1991	10 Dec 1994	Dec 1995
GEPARD (II)	K 335	Severodvinsk Shipyard	1991	18 Aug 1999	29 July 2001
COUGAR (II)	K 337	Severodvinsk Shipyard	1993	2005	2006

Displacement, tons: 7,500 surfaced; 9,100 (9,500 Akula II) dived

Dimensions, feet (metres): 360.1 oa; 337.9 wl × 45.9 × 34.1 *(110; 103 × 14 × 10.4)*

Main machinery: Nuclear; 1 VM-5 PWR; 190 MW; 2 GT3A turbines; 47,600 hp(m) *(35 MW)*; 2 emergency propulsion motors; 750 hp(m) *(552 kW)*; 1 shaft; 2 spinners; 1,006 hp (m) *(740 kW)*

Speed, knots: 28 dived; 10 surfaced

Complement: 62 (31 officers)

Missiles: SLCM/SSM: Reduga SS-N-21 Sampson (Granat) fired from 21 in *(533 mm)* tubes; land-attack; inertial/terrain-following to 3,000 km *(1,620 n miles)* at 0.7 Mach; warhead nuclear 200 kT. CEP 150 m. Flies at a height of about 200 m. Novator Alfa SS-N-27 subsonic flight with supersonic boost for terminal flight; 180 km *(97 mm)*; warhead 200 kg. May be fitted in due course.

SAM: SA-N-5/8 Strela portable launcher. 18 missiles.

A/S: Novator SS-N-15 Starfish (Tsakra) fired from 53 cm tubes; inertial flight to 45 km *(24.3 n miles)*; warhead nuclear 200 kT or Type 40 torpedo.

Novator SS-N-16 Stallion fired from 650 mm tubes; inertial flight to 100 km *(54 n miles)*; payload nuclear 200 kT (Vodopad) or Type 40 torpedo (Veder).

Torpedoes: 4—21 in *(533 mm)* and 4—25.6 in *(650 mm)* tubes. Combination of 53 and 65 cm torpedoes (see table at front of section). Tube liners can be used to reduce the larger diameter tubes to 533 mm. Total of 40 weapons. In addition the Improved Akulas and Akula IIs have six additional 533 mm external tubes in the upper bow area.

Countermeasures: ESM: Rim Hat; intercept.

Radars: Surface search: Snoop Pair or Snoop Half with back-to-back aerials on same mast as ESM.

Sonars: Shark Gill (Skat MGK 503); hull-mounted; passive/active search and attack; low/medium frequency.

Mouse Roar; hull-mounted; active attack; high frequency.

Skat 3 towed array; passive; very low frequency.

Programmes: Malakhit design. From K 461 onwards, the Akula were 'improved'. K 157 was the first Akula II to complete. Little is known of K 337; she may be the last of class before the Yasen class enters service and is likely to incorporate further developments. She may be leased to the Indian Navy. Akula I K 152 has been building for 15 years at Komsomolsk but there has been speculation that she may be completed and also leased to the Indian Navy.

Structure: The very long fin is particularly notable. Has the same broad hull as Sierra and has reduced radiated noise levels by comparison with Victor III of which she is the traditional follow-on design. A number of prominent non-acoustic sensors appear on the fin leading-edge and on the forward casing in the later Akulas. The engineering standards around the bridge and casing are noticeably to a higher quality than other classes. The design has been incrementally improved with reduced noise levels, boundary layer suppression and active noise cancellation reported in the later units. The Improved hulls have an additional six external torpedo tubes and the first two Akula IIs have been lengthened by 3.7 m to incorporate further noise reduction developments. There are six watertight compartments. Operational diving depth, 1,476 ft *(450 m)*.

Operational: Pert Spring SATCOM. Both Akula IIs and the three Severodvinsk built Akula Is serve in the Northern Fleet and are based at Saida Guba and Sevr. The three Komsomolsk built Akula Is serve in the Pacific Fleet and are based at Tarya Bay. These submarines are the core units of the Russian SSN force. *Vepr* visited Brest in September 2004, the first visit by a Russian nuclear submarine to a foreign port.

UPDATED

VEPR   9/2004*, B Prézelin / 1042292

VEPR (Akula II)   9/2004*, B Prézelin / 1042291

GEPARD (Akula II)   6/2002, S Breyer / 0528327

AKULA II   *6/2003*, B Lemachko /* 1042293

1 SIERRA I (BARRACUDA) CLASS (PROJECT 945) (SSN)

Name	No	Builders	Laid down	Launched	Commissioned
KOSTROMA (ex-*Krab*)	K 276	Nizhny Novgorod/Severodvinsk Shipyard	8 May 1982	29 June 1983	21 Sep 1984

Displacement, tons: 7,200 surfaced; 8,100 dived
Dimensions, feet (metres): 351 × 41 × 28.9
   (107 × 12.5 × 8.8)
Main machinery: Nuclear; 1 VM-5 PWR; 190 MW; 1 GT3A
   turbine; 47,500 hp(m) *(70 MW)*; 2 emergency motors;
   2,004 hp(m) *(1.5 MW)*; 1 shaft; 2 spinners; 1,006 hp(m)
   (740 kW)
Speed, knots: 34 dived; 10 surfaced
Complement: 61 (31 officers)

Missiles: SLCM: Raduga SS-N-21 Sampson (Granat) fired from
   21 in *(533 mm)* tubes; land-attack; inertial/terrain-following
   to 3,000 km *(1,620 n miles)* at 0.7 Mach; warhead nuclear
   200 kT. CEP 150 m. Probably flies at a height of about 200 m.
A/S: Novator SS-N-15 Starfish (Tsakra) fired from 53 cm tubes;
   inertial flight to 45 km *(24.3 n miles)*; warhead nuclear 200 kT
   or Type 40 torpedo.

Novator SS-N-16 Stallion fired from 65 cm tubes; inertial flight
   to 100 km *(54 n miles)*; payload nuclear 200 kT (Vodopad) or
   Type 40 torpedo (Veder).
Torpedoes: 4—25.6 in *(650 mm)* and 4—21 in *(533 mm)* tubes.
   Combination of 65 and 53 cm torpedoes (see table at front of
   section). Total of 40 weapons.
Mines: 42 in lieu of torpedoes.
Countermeasures: ESM: Rim Hat/Bald Head; intercept. Park
   Lamp D/F.
Radars: Surface search: Snoop Pair with back-to-back ESM
   aerial.
Sonars: Shark Gill; hull-mounted; passive/active search and
   attack; low/medium frequency.
   Shark Rib flank array; passive; low frequency.
   Mouse Roar; hull-mounted; active attack; high frequency.
   Skat 3 towed array; passive; very low frequency.

Programmes: Launched at Gorky (Nizhny Novgorod) and
   transferred by river/canal to be fitted out at Severodvinsk.
Structure: Based on design experience gained with deleted Alfa
   class, pressure hull constructed of titanium alloy, providing
   deep diving capability. Magnetic signature also reduced.
   Distance between hulls increases survivability and reduces
   radiated noise. There are six watertight compartments. The
   pod on the after fin is larger than that in 'Victor III'. Bulbous
   casing at the after end of the fin is for a towed
   communications buoy. Diving depth 2,460 ft *(750 m)*.
Operational: Pert Spring SATCOM. Based in the Northern Fleet
   at Ara Guba. It is believed that K 276 was in a collision with
   USS *Baton Rouge* on 11 February 1992. A second of class
   K 239 *Karp* is laid up.

                                                                    UPDATED

KOSTROMA   *6/2002, B Lemachko /* 0547070

5 VICTOR III (SCHUKA) CLASS (PROJECT 671 RTMK) (SSN)

Name	No	Builders	Laid down	Launched	Commissioned
PERM	B 292	Admiralty, Leningrad	15 Apr 1986	29 Apr 1987	27 Nov 1987
PETROZAVODSK (ex-*Snezhnogorsk*)	B 388	Admiralty, Leningrad	8 Sep 1987	3 June 1988	30 Nov 1988
OBNINSK	B 138	Admiralty, Leningrad	7 Dec 1988	5 Aug 1989	10 May 1990
DANIL MOSKOVSKIY	B 414	Admiralty, Leningrad	1 Dec 1988	31 Aug 1990	30 Dec 1990
TAMBOV	B 448	Admiralty, Leningrad	31 Jan 1991	17 Oct 1991	24 Sep 1992

Displacement, tons: 4,850 surfaced; 6,300 dived
Dimensions, feet (metres): 351.1 × 34.8 × 24.3
  (107 × 10.6 × 7.4)
Main machinery: Nuclear; 2 VM-4 PWR; 150 MW; 2 turbines;
  31,000 hp(m) *(22.7 MW)*; 1 shaft; 2 spinners; 1,020 hp(m)
  (750 kW)
Speed, knots: 30 dived; 10 surfaced
Complement: 98 (17 officers)

Missiles: SLCM: Raduga SS-N-21 Sampson (Granat) fired from
  21 in *(533 mm)* tubes; land-attack; inertial/terrain-following
  to 3,000 km *(1,620 n miles)* at 0.7 Mach. CEP 150 m or
  Novator Alfa SS-N-27 (B 244 only); to 180 km *(97 n miles)*;
  warhead 200 kg.
  A/S: Novator SS-N-15 Starfish (Tsakra) fired from 53 cm tubes;
  inertial flight to 45 km *(24.3 n miles)*; Type 40 torpedo.
  Novator SS-N-16 Stallion fired from 65 cm tubes; inertial flight
  to 100 km *(54 n miles)*; payload nuclear 200 kT (Vodopad) or
  Type 40 torpedo (Veder).
Torpedoes: 4—21 in *(533 mm)* and 2—25.6 in *(650 mm)* tubes.

Combination of 53 and 65 cm torpedoes (see table at front of
section). Can carry up to 24 weapons. Liners can be used to
reduce 650 mm tubes to 533 mm.
Mines: Can carry 36 in lieu of torpedoes.
Countermeasures: ESM: Brick Group (Brick Spit and Brick Pulp);
  intercept. Park Lamp D/F.
Radars: Surface search: Snoop Tray 2; I-band.
Sonars: Shark Gill; hull-mounted; passive/active search and
  attack; low/medium frequency.
  Shark Rib flank array; passive; low frequency.
  Mouse Roar; hull-mounted; active attack; high frequency.
  Scat 3 towed array; passive; very low frequency.

Programmes: The first of class was completed at Komsomolsk in
1978. With construction also being carried out at Admiralty
Yard, Leningrad, there was a very rapid building programme
up to the end of 1984. Construction then continued only at
Leningrad and at a rate of about one per year which
terminated in 1991. The last of the class of 26 boats
completed sea trials in October 1992. Of these, the first 21

hulls were designated Type 671RTM. The final five hulls were
designated Type 671RTMK to reflect modifications to fire
cruise missiles. The last four of these are in service and B 292,
having been named, is to be re-activated.
Structure: The streamlined pod on the stern fin is a towed sonar
array dispenser. Water environment sensors are mounted at
the front of the fin and on the forward casing as in the Akula
and Sierra classes. Diving depth, 1,300 ft *(400 m)*.
Operational: VLF communications buoy. VHF/UHF aerials.
Navigation equipment includes SINS and SATNAV. Pert
Spring SATCOM. Kremmny 2 IFF. Much improved acoustic
quietening puts the radiated noise levels at the upper limits of
the USN Los Angeles class. All remaining operational units are
based in the Northern Fleet at Litsa South or Ara Guba
although they rarely go to sea. Twenty two have paid off so far
although up to nine of these are in reserve and laid up at
anchorages in both Fleets.

UPDATED

DANIL MOSKOVSKY

7/2004* / 1042331

VICTOR III

2000, B Lemachko / 0126230

Patrol Submarines (SSK)

17 KILO CLASS (PROJECT 877K/877M/636) (SSK)

Name	No	Builders	Laid down	Launched	Commissioned
RAZBOYNIK	B 260	Komsomolsk Shipyard	Sep 1980	19 Aug 1981	Dec 1981
—	B 227	Komsomolsk Shipyard	Sep 1981	Sep 1982	Dec 1982
—	B 405	Komsomolsk Shipyard	1983	1984	Dec 1984
VOLOGDA	B 402	Nizhny Novgorod	Feb 1983	1984	27 Dec 1984
—	B 439	Komsomolsk Shipyard	1985	1986	1986
TUR	B 806	Nizhny Novgorod			1986
—	B 445	Komsomolsk Shipyard	1986	1987	Dec 1987
JAROSLAVL	B 808	Nizhny Novogorod			1988
MAGNETO-GORSK	B 800	Nizhny Novgorod			1989
NOVOSIBIRSK	B 401*	Nizhny Novgorod	June 1988	Aug 1989	4 Jan 1990
KALUGA	B 471	Nizhny Novgorod			1990
UST-BOLSHERETSK	B 494	Komsomolsk Shipyard	1989	1990	1990
ALROSA	B 871	Nizhny Novgorod	May 1998	Aug 1989	Dec 1990
LIPETSK	B 177	Nizhny Novgorod			1991
—	B 187*	Komsomolsk Shipyard	1990	1990	1991
—	B 190*	Komsomolsk Shipyard	8 May 1992	1993	1993
—	B 345	Komsomolsk Shipyard	22 Apr 1993	1993	22 Jan 1994

* indicates Project 636

Displacement, tons: 2,325 surfaced; 3,076 dived
Dimensions, feet (metres): 238.2; 242.1 (Project 636 × 32.5 × 21.7 *(72.6; 73.8 × 9.9 × 6.6)*
Main machinery: Diesel-electric; Type 4-2DL-42M 2 diesels (Type 4-2AA-42M in Project 636); 3,650 hp(m) *(2.68 MW)*; 2 generators; 1 motor; 5,900 hp(m) *(4.34 MW)*; 1 shaft; 2 auxiliary MT-168 motors; 204 hp(m) *(150 kW)*; 1 economic speed motor; 130 hp(m) *(95 kW)*
Speed, knots: 17 dived; 10 surfaced; 9 snorting
Range, n miles: 6,000 at 7 kt snorting; 400 at 3 kt dived
Complement: 52 (13 officers)

Missiles: SSM: Novator Alfa SS-N-27 may be fitted in due course.
SAM: 6-8 SA-N-5/8; IR homing from 600 to 6,000 m at 1.65 Mach; warhead 2 kg; portable launcher stowed in a well in the fin between snort and W/T masts.
Torpedoes: 6—21 in *(533 mm)* tubes. 18 combinations of 53 cm torpedoes (see table at front of section). USET-80 is wire-guided in the 4B version (from 2 tubes).
Mines: 24 in lieu of torpedoes.
Countermeasures: ESM: Squid Head or Brick Pulp; radar warning. Quad Loop D/F.
Weapons control: MVU-110EM or MVU-119EM Murena torpedo fire-control system.
Radars: Surface search: Snoop Tray (MRP-25); I-band.
Sonars: Shark Teeth/Shark Fin (MGK-400); hull-mounted; passive/active search and attack; medium frequency.
Mouse Roar; hull-mounted; active attack; high frequency.

Programmes: Also known as the Vashavyanka class, first launched in 1979 at Konsomolsk and commissioned 12 September 1980. Subsequent construction also at Nizhny Novgorod. A total of 24 were built for Russia of which six were of the improved Project 636 variant.
Structure: Had a better hull form than the now deleted Tango class but was nevertheless considered fairly basic by comparison with contemporary western designs. Diving depth 790 ft *(240 m)* normal. Battery has a 9,700 kW/h capacity. The basic 'Kilo' was the Project 877; 877K has an improved fire-control system and 877M includes wire-guided torpedoes from two tubes. Project 636 is an improved design with uprated diesels, a propulsion motor rotating at half the speed (250 rpm), higher standards of noise reduction and an automated combat information system capable of providing simultaneous fire-control data on five targets. Pressure hull length is 170 ft *(51.8 m)* or 174 ft *(53 m)* for Project 636. Foreplanes on the hull are just forward of the fin. Project 636 can be identified by a vertical cut off to the after casing. B 871 has been fitted with a pump jet propulsor.
Operational: With a reserve of buoyancy of 32 per cent and a heavily compartmented pressure hull, this class is capable of being holed and still surviving. B 401, B 402, B 808, B 471, B 800 and B 177 are based in the Northern Fleet, B 439, B 405, B 260, B 445, B 494, B 190, B 345 and B 187 are based in the Pacific, B 806 and B 227 in the Baltic and

ALROSA   9/2004*, *Hartmut Ehlers* / 1042295

TUR   8/2004*, *E & M Laursen* / 1042296

B 871 in the Black Sea. Russian made batteries have been a source of problems in warm water operations.
Sales: Exports of Project 877 have been to Poland (one), Romania (one), India (ten), Algeria (two), Iran (three) and China (two). The only exports of Project 636 have been to China (two). A further eight were ordered by China in 2002. Export versions have the letter E after the project number.

UPDATED

VOLOGDA

5/2003, *Jürg Kürsener* / 0570899

0 + 2 LADA CLASS (PROJECT 677) (SSK)

Name	No	Builders	Laid down	Launched	Commissioned
SAINT PETERSBURG	S 100	Admiralty, St Petersburg	26 Dec 1997	28 Oct 2004	2005
—		Admiralty, St Petersburg	26 Dec 1997	2005	2006

Displacement, tons: 1,765 surfaced; 2,650 dived
Dimensions, feet (metres): 219.2 × 23.6 × 14.4
  (66.8 × 7.2 × 4.4)
Main machinery: Diesel-electric; 2 diesel generators; 3,400 hp
  (m) *(2.5 MW)*; 1 motor; 5,576 hp(m) *(4.1 MW)*; 1 shaft
Speed, knots: 21 dived; 10 surfaced
Range, n miles: 6,000 at 7 kt snorting
Complement: 37

Torpedoes: 6—21 in *(533 mm)* tubes. 18 weapons.
Mines: In lieu of torpedoes.
Countermeasures: ESM: Intercept.
Radars: Surface search: I-band.
Sonars: Hull and flank arrays; active/passive; medium frequency.

Programmes: The national variant of this submarine is known as
  the Lada class. Work began on the first of class in 1996 and
  construction started in St Petersburg in 1987. Progress has
  been slow but further orders are likely. The export version of
  the submarine is known as the Amur class of which there are
  six designs based on different surface displacements (550,
  750, 950, 1450, 1650 and 1850). The 'Amur 1650' probably
  has the most export potential and it was possibly in
  anticipation of an order from India and China that work began
  on such a submarine at the same time as the similar Lada
  class. Although work is believed to have been temporarily
  suspended in 1998, it is expected that the boat will be
  completed and operated by the Russian Navy as a technology-
  demonstrator.
Structure: The first Russian single-hulled submarine, built to a
  Rubin design based on the 'Amur 1650'. A fuel cell plug (for
  AIP) of about 12 m can be inserted to allow installation of AIP
  although this is unlikely in the near future. Diving depth: 820 ft
  (250 m). A non-hull penetrating optronic periscope supplied
  by Elektropribor, is fitted.
Operational: Sea trials are expected to start in 2005.

UPDATED

SAINT PETERSBURG
10/2004 *, *Rubin Design Bureau* / 0590179

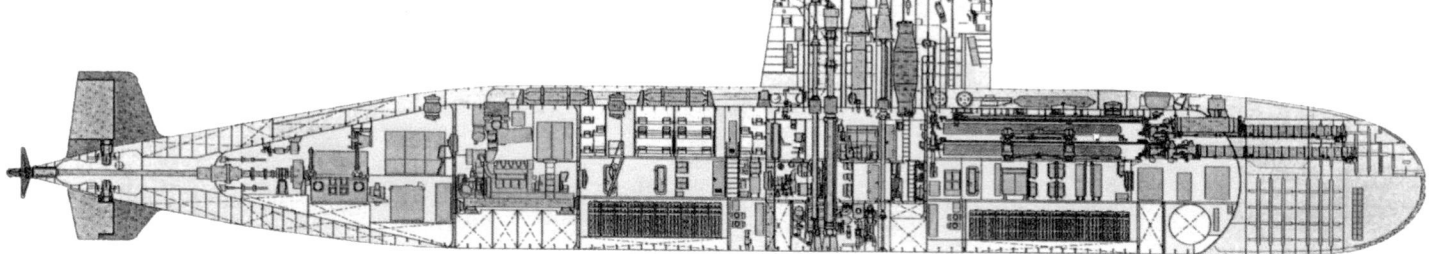

AMUR 1650
1996, *Rubin* / 0019012

Auxiliary Submarines (SSA(N))

Notes: (1) There are a number of Swimmer Delivery Vessels (SDV) in service including Siren (three-man) and Triton, Sever and Elbrus types.
(2) Rescue submersibles in service include AS 22 and AS 26 both of which were used during attempts to recover survivors from the sunken *Kursk* in the Barents Sea in August 2000.
(3) A new auxiliary submarine (SSAN) was launched at Severodvinsk Shipyard on 6 August 2003. Nicknamed 'Losharik', this is likely to be used for scientific research and there has been speculation that it is similar to the Uniform class. It is known as Project 210.

1 DELTA III STRETCH (PROJECT 667 BDR) (SSAN)

Name	No	Builders	Laid down	Launched	Commissioned
—	BS 129	Severodvinsk Shipyard	Feb 1979	Mar 1981	5 Nov 1981

Dimensions, feet (metres): 534.9 × 39.4 × 28.5
  (163 × 12 × 8.7)
Main machinery: Nuclear: 2 VM-4 PWR; 180 MW; 2 GT 3A-635
  turbines; 37,400 hp(m) *(27.5 MW)*; 2 emergency motors;
  612 hp(m) *(450 kW)*; 2 shafts
Speed, knots: 24 dived 14 surfaced
Complement: 130 (40 officers)
Torpedoes: 4—21 in *(533 mm)* and 2—400 mm tubes.

Countermeasures: ESM: Brick Pulp/Group; radar warning.
Radars: Surface search: Snoop Tray; I-band.
Sonars: Shark Teeth; hull mounted; active/passive search; low/
  medium frequency.
  Shark Hide; flank array; passive low frequency.
  Mouse Roar; hull mounted; active high frequency.

Comment: Originally launched in 1981, this former SSBN has
  been converted by replacing the central section with a 43 m
  plug, extending the overall hull length by 3 m. The submarine
  was reported to have returned to service in 2003 and is
  expected to replace the Yankee Stretch as the Paltus mother-
  ship. Both submarines can operate as a mother ship for the
  Paltus class. Based in the Northern Fleet.

UPDATED

DELTA III (before conversion)
8/1997, *JMSDF* / 0019003

2 YANKEE SSAN/STRETCH (PROJECT 09780/09744) CLASS (SSAN)

Name	No	Builders	Laid down	Launched	Commissioned
ORIENBURG	KS 411	Severodvinsk Shipyard	25 May 1968	16 Jan 1970	31 Aug 1970
KAZAN	KS 403	Severodvinsk Shipyard	18 Aug 1969	25 Mar 1971	12 Aug 1971

Displacement, tons: 9,800 dived (Yankee SSAN); 11,700 dived (Yankee Stretch)
Dimensions, feet (metres): 498.2 (SSAN); 528.2 (Stretch) × 38 × 26.6 *(151.8; 161 × 11.6 × 8.1)*
Main machinery: Nuclear; 2 VM-4 PWR; 180 MW; 2 turbines; 37,400 hp(m) *(27.5 MW)*; 2 shafts
Speed, knots: 26 dived; 20 surfaced
Complement: 120 (20 officers)
Torpedoes: 4—21 in *(533 mm)* bow tubes.

Countermeasures: ESM: Brick Pulp; radar warning.
Radars: Surface search: Snoop Tray; I-band.
Sonars: Shark Gill; hull-mounted; passive/active search and attack.

Comment: As well as the Yankee Notch SSN and inactive Yankee SSGN conversions, two other hulls have been converted for research and development roles. The Yankee SSAN *(Kazan)* formerly known as the Yankee Pod was used for sonar trials (Project Akson-1) from 1983. The prototype Pod towed array was fitted at the stern in 1984 but after refit in 1993-94 the submarine emerged in 1995 without the Pod but with a bulbous bow sonar (Project Akson-2) which is likely to be the prototype for the *Severodvinsk* (Project 885). The other conversion is the Yankee Stretch *(Orienburg)* completed conversion in June 1990, and has a lengthened central section extending the hull which is used for underwater research. This includes submarine operations using the Paltus class. The Yankee SSAN and the Yankee Stretch are based in the Northern Fleet at Yagri Island and Olenya Guba respectively. Both are expected to decommission soon.
UPDATED

ORIENBURG and PALTUS (artist's impression) 1997 / 0019014

KAZAN 6/2000*, Rubin Design Bureau / 1042322

ORIENBURG 6/2003*, B Lemachko / 0580530

ORIENBURG 2000, S Breyer / 0126225

3 PALTUS/X-RAY (PROJECT 1851) CLASS (SSAN/SSA)

AS 23   AS 35   AS 21 (X-Ray)

Displacement, tons: 730 dived
Dimensions, feet (metres): 173.9 × 12.5 × 13.8 *(53 × 3.8 × 4.2)*
Main machinery: Nuclear; 1 reactor; 10 MW; 1 shaft; ducted thrusters
Speed, knots: 6 dived
Complement: 14

Comment: Details given are for the two Paltus (Nelhma) class (A 21, AS 35). The first was launched at Sudomekh, St Petersburg in April 1991, a second of class in September 1994 and a third was started but not completed. This is a follow-on to the single 520 ton X-Ray (AS 21) class which was first seen in 1984 and after a long spell out of service was back in operation in 1999. Paltus probably owes much to the USN NR 1. Paltus is associated with the Yankee Stretch SSAN which acts as a mother ship for special operations. The Delta Stretch SSAN which reportedly entered service in 2003, probably serves the same role. Titanium hulled and very deep diving to 1,000 m *(3,280 ft)*. Paltus based in the Northern Fleet at Olenya Guba, X-Ray at Yagri Island.
UPDATED

PALTUS (artist's impression) 1994

3 UNIFORM (KACHALOT) CLASS (PROJECT 1910) (SSAN)

No	Builders	Laid down	Launched	Commissioned
AS 13	Sudomekh, Leningrad	20 Oct 1977	25 Nov 1982	31 Dec 1986
AS 14	Sudomekh, Leningrad	23 Feb 1983	29 Apr 1988	30 Dec 1991
AS 15	Sudomekh, St Petersburg	16 July 1990	26 Aug 1995	Feb 1998

Displacement, tons: 1,340 surfaced; 1,580 dived
Dimensions, feet (metres): 226.4 × 23.0 × 17.0
(69.0 × 7.0 × 5.2)
Main machinery: Nuclear; 1 PWR; 15 MW; 2 turbines;
10,000 hp(m) *(7.35 MW)*; 1 shaft; 2 thrusters
Speed, knots: 10 surfaced; 28 dived
Complement: 36

Radars: Navigation: Snoop Slab; I-band.

Comment: Research and development nuclear-powered submarines. Have single hulls and 'wheel' arches either side of the fin which house side thrusters. These are titanium hulled and very deep diving submarines (possibly down to 700 m *(2,300 ft)*), based in the Northern Fleet at Olenya Guba, and are used mainly for ocean bed operations. Plans to build more of the class were thought to have been shelved. It is not clear whether an auxiliary submarine launched on 6 August 2003 is a fourth 'Uniform' or a different design.
VERIFIED

UNIFORM   *8/1996 /* 0016660

AIRCRAFT CARRIERS

Note: Of the former aircraft carriers of the Kiev class, *Kiev* was sold to China for scrap in 2000; *Minsk* and *Novorossiysk* were sold to a South Korean Corporation in 1994. *Minsk* later became a tourist attraction in Shenzen, China, while *Novorossiysk* was scrapped in India. *Admiral Gorshkov* (ex-*Baku*) and is being refitted is to be sold to the Indian Navy.

ADMIRAL KUZNETSOV   *10/2004*, Ships of the World */* 1042330

1 KUZNETSOV (OREL) CLASS (PROJECT 1143.5/6) (CVGM)

Name	No	Builders	Laid down	Launched	Commissioned
ADMIRAL KUZNETSOV (ex-*Tbilisi*, ex-*Leonid Brezhnev*)	063	Nikolayev South, Ukraine	1 Apr 1982	16 Dec 1985	25 Dec 1990

Displacement, tons: 45,900 standard; 58,500 full load
Dimensions, feet (metres): 999 oa; 918.6 wl × 229.7 oa; 121.4 wl × 34.4 *(304.5; 280 × 70; 37 × 10.5)*
Flight deck, feet (metres): 999 × 229.7 *(304.5 × 70)*
Main machinery: 8 boilers; 4 turbines; 200,000 hp(m) *(147 MW)*; 4 shafts
Speed, knots: 30. **Range, n miles:** 3,850 at 29 kt; 8,500 at 18 kt
Complement: 1,960 (200 officers) plus 626 aircrew plus 40 Flag staff

Missiles: SSM: 12 Chelomey SS-N-19 Shipwreck (3M-45) launchers (flush mounted) ❶; inertial guidance with command update; active radar homing to 20-550 km *(10.8-300 n miles)* at 2.5 Mach; warhead 500 kT nuclear or 750 kg HE.
SAM: 4 Altair SA-N-9 Gauntlet (Klinok) sextuple vertical launchers (192 missiles) ❷; command guidance and active radar homing to 12 km *(6.5 n miles)* at 2 Mach; warhead 15 kg. 24 magazines; 192 missiles; 4 channels of fire.
SAM/Guns: 8 Altair CADS-N-1 (Kortik/Kashtan) ❸; each has a twin 30 mm Gatling combined with 8 SA-N-11 (Grisson) and Hot Flash/Hot Spot fire-control radar/optronic director. Laser beam-riding guidance for missiles to 8 km *(4.4 n miles)*; warhead 9 kg; 9,000 rds/min combined to 2 km (for guns).
Guns: 6—30 mm/65 ❹ AK 630; 6 barrels per mounting; 3,000 rds/min combined to 2 km. Probably controlled by Hot Flash/Hot Spot on CADS-N-1.
A/S mortars: 2 RBU 12,000 ❺; range 12,000 m; warhead 80 kg. UDAV-1M; torpedo countermeasure.
Countermeasures: Decoys: 10 PK 10 and 4 PK 2 chaff launchers.
ESM/ECM: 8 Foot Ball. 4 Wine Flask (intercept). 4 Flat Track. 10 Ball Shield A and B.
Weapons control: 3 Tin Man optronic trackers. 2 Punch Bowl SATCOM datalink ❻. 2 Low Ball SATNAV ❼. 2 Bell Crown and 2 Bell Push datalinks.
Radars: Air search: Sky Watch; four Planar phased arrays ❽; 3D.
Air/surface search: Top Plate B ❾; D/E-band.
Surface search: 2 Strut Pair ❿; F-band.
Navigation: 3 Palm Frond; I-band.
Fire control: 4 Cross Sword (for SAM) ⓫; K-band. 8 Hot Flash; J-band.

ADMIRAL KUZNETSOV   *6/2003*, B Lemachko /* 1042294

Aircraft control: 2 Fly Trap B; G/H-band.
Tacan: Cake Stand ⓬.
IFF: 4 Watch Guard.
Sonars: Bull Horn and Horse Jaw; hull-mounted; active search and attack; medium/low frequency.

Fixed-wing aircraft: 18 Su-27 Flanker D; 4 Su-25 UTG Frogfoot.
Helicopters: 15 Ka-27 Helix. 2 Ka-31 RLD Helix AEW.

Programmes: This was a logical continuation of the deleted Kiev class. The full name of *Kuznetsov* is *Admiral Flota Sovietskogo Sojuza Kuznetsov*. The second of class, *Varyag*, was between 70 and 80 per cent complete by early 1993 at Nikolayev in the Ukraine. Building was then terminated after an unsuccessful attempt by the Navy to fund completion. Subsequently the ship was bought by Chinese interests and, having been towed through the Bosporus on 2 November 2001, arrived at Dalian in March 2002.
Structure: The hangar is 183 × 29.4 × 7.5 m and can hold up to 18 Flanker aircraft. There are two starboard side lifts, a ski jump of 14° and an angled deck of 7°. There are four arrester wires. The SSM system is in the centre of the flight deck

forward with flush deck covers. The ship has some 16.5 m of freeboard. There is no Bass Tilt radar and the ADG guns are controlled by Kashtan fire-control system. The ship suffers from severe water distillation problems.
Operational: AEW, ASW and reconnaissance tasks undertaken by Helix helicopters. The aircraft complement listed is based on the number which might be embarked for normal operations but the Russians claim a top limit of 60. *Kuznetsov* conducted extensive flight operations throughout the second half of both 1993 and 1994, and was at sea again by September 1995 after a seven month refit. Deployed to the Mediterranean for 80 days in early 1996 before returning to the Northern Fleet. Based alongside at Rosta from mid-1996 to mid-1998. Sailed for a VIP demonstration in August 1998 and then continued trials and training in-area. There were limited local exercises in 2000 but no activity in 2001 and 2002. The ship left the jetty for Navy Days in 2003. The ship made a short deployment to the Norwegian Sea in October 2004 but operations were tentative and the effectiveness of the ship is questionable.

UPDATED

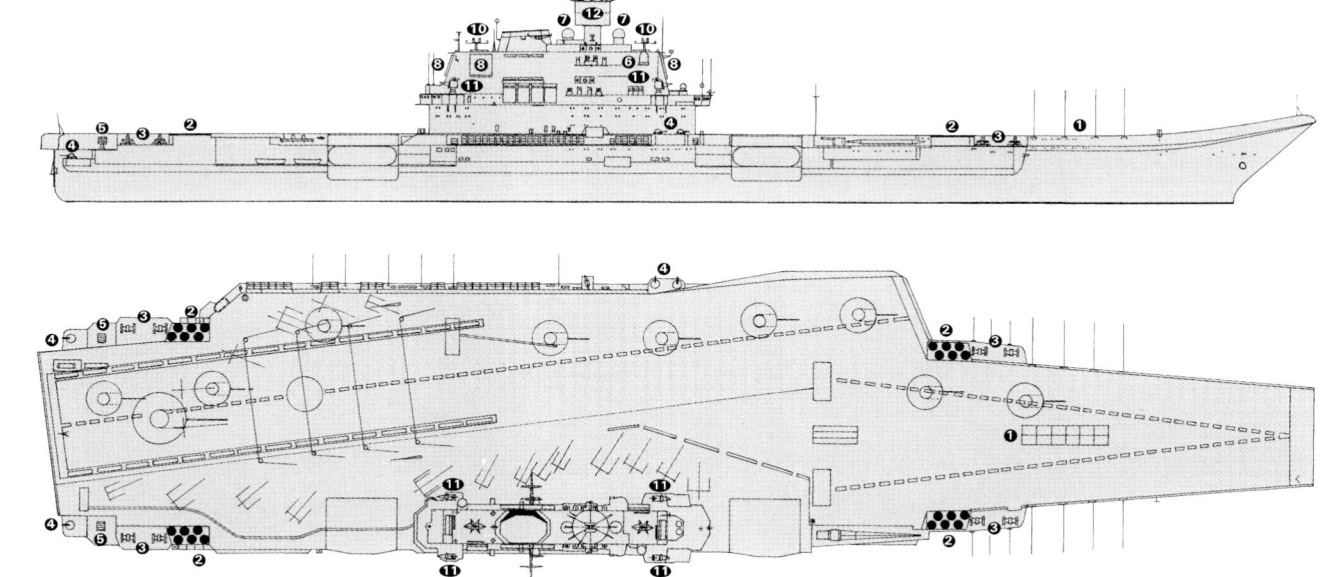

ADMIRAL KUZNETSOV   *6/2000*, S Breyer /* 1042308

ADMIRAL KUZNETSOV   *(Scale 1 : 1,800), Ian Sturton*

BATTLE CRUISERS

1 KIROV (ORLAN) CLASS (PROJECT 1144.1/1144.2) (CGHMN)

Name	No	Builders	Laid down	Launched	Commissioned
PYOTR VELIKIY (ex-*Yuri Andropov*)	099 (ex-183)	Baltic Yard 189, St Petersburg	11 Mar 1986	29 Apr 1989	9 Apr 1998

Displacement, tons: 19,000 standard; 24,300 full load

Dimensions, feet (metres): 826.8; 754.6 wl × 93.5 × 29.5 *(252; 230 × 28.5 × 9.1)*

Main machinery: CONAS; 2 KN-3 PWR; 300 MW; 2 oil-fired boilers; 2 GT3A-688 turbines; 140,000 hp(m) *(102.9 MW)*; 2 shafts

Speed, knots: 30. **Range, n miles:** 14,000 at 30 kt

Complement: 726 (82 officers) plus 18 aircrew

Missiles: SSM: 20 Chelomey SS-N-19 Shipwreck (3M 45) (P-700 Granit) (improved SS-N-12 with lower flight profile) ❶; inertial guidance with command update; active radar homing to 20-450 km *(10.8-243 n miles)* at 1.6 Mach; warhead 350 kT nuclear or 750 kg HE; no reloads.

SAM: 12 SA-N-6/SA-N-20 Grumble (Fort/Fort M) vertical launchers ❷; 8 rounds per launcher; command guidance; semi-active radar homing to 100 km *(54 n miles)*; warhead 90 kg (or nuclear?); 96 missiles

2 SA-N-4 Gecko twin launchers ❸; semi-active radar homing to 15 km *(8 n miles)* at 2.5 Mach; warhead 50 kg; altitude 9.1-3,048 m *(30-10,000 ft)*; 40 missiles.

2 SA-N-9 Gauntlet (Kinzhal) octuple vertical launchers ❹; command guidance; active radar homing to 12 km *(6.5 n miles)* at 2 Mach; warhead 15 kg; altitude 3.4-12,192 m *(10-40,000 ft)*; 128 missiles; 4 channels of fire.

SAM/Guns: 6 CADS-N-1 (Kortik/Kashtan) ❺; each has a twin 30 mm Gatling combined with 8 SA-N-11 (Grisson) and Hot Flash/Hot Spot fire-control radar/optronic director. Laser beam-riding guidance for missiles to 8 km *(4.4 n miles)*; warhead 9 kg; 9,000 rds/min combined to 2 km (for guns).

A/S: Novator SS-N-15 (Starfish); inertial flight to 45 km *(24.3 n miles)*; payload Type 40 torpedo or nuclear warhead; fired from fixed torpedo tubes behind shutters in the superstructure.

Guns: 2—130 mm/70 (twin) AK 130 ❻; 35/45 rds/min to 29 km *(16 n miles)*; weight of shell 33.4 kg.

Torpedoes: 10—21 in *(533 mm)* (2 quin) tubes. Combination of 53 cm torpedoes (see table at front of section). Mounted in the hull adjacent the RBU 1000s on both quarters. Fixed tubes behind shutters can fire either SS-N-15 (see *Missiles A/S*) or Type 40 torpedoes.

A/S mortars: 1 RBU 12,000 ❼; 10 tubes per launcher; range 12,000 m; warhead 80 kg.

2 RBU 1000 6-tubed aft ❽; range 1,000 m; warhead 55 kg. UDAV-1M; torpedo countermeasures.

Countermeasures: Decoys: 2 twin PK 2 150 mm chaff launchers. Towed torpedo decoy.

ESM/ECM: 8 Foot Ball. 4 Wine Flask (intercept). 8 Bell Bash. 4 Bell Nip. Half Cup (laser intercept).

Combat Data Systems: Lesorub-44.

Weapons control: 4 Tin Man optronic trackers ❾. 2 Punch Bowl C SATCOM ❿. 4 Low Ball SATNAV. 2 Bell Crown and 2 Bell Push datalinks.

Radars: Air search: Top Pair (Top Sail + Big Net) ⓫; 3D; C/D-band; range 366 km *(200 n miles)* for bomber, 183 km *(100 n miles)* for 2 m² target.

Air/surface search: Top Plate ⓬; 3D; D/E-band.

Navigation: 3 Palm Frond; I-band.

Fire control: Cross Sword ⓭; K-band (for SA-N-9). Top Dome for SA-N-6 ⓮; Tomb Stone E-band (for Fort M) ⓯. 2 Pop Group, F/H/I-band, (for SA-N-4) ⓰. Kite Screech ⓱; H/I/K-band (for main guns). 6 Hot Flash for CADS-N-1; I/J-band.

Aircraft control: Flyscreen B; I-band.

IFF: Salt Pot A and B.

Tacan: 2 Round House B ⓲.

Sonars: Horse Jaw (Polinom); hull-mounted; active search and attack; low/medium frequency.

Horse Tail; VDS; active search; medium frequency. Depth to 150-200 m *(492.1-656.2 ft)* depending on speed.

Helicopters: 3 Ka-27 Helix ⓳.

Programmes: Design work started in 1968. Type name is *atomny raketny kreyser* meaning nuclear-powered missile cruiser. A fifth of class was scrapped before being launched in 1989.

Structure: The Kirov class were the first Russian surface warships with nuclear propulsion. In addition to the nuclear plant a unique maritime combination with an auxiliary oil-fuelled system has been installed. This provides a superheat capability, boosting the normal steam output by some 50 per cent. The SS-N-19 tubes are set at an angle of about 45°. CADS-N-1 with a central fire-control radar on six mountings, each of which has two cannon and eight missile launchers. Two are mounted either side of the SS-N-19 forward and four on the after superstructure. Same A/S system as the frigate *Neustrashimy* with fixed torpedo tubes in ports behind shutters in the superstructure for firing SS-N-15 or Type 45 torpedoes. There are reported to be about 500 SAM of different types. *Velikiy*, the only operational ship, has a Tomb Stone fire-control radar instead of a forward Top Dome for SA-N-20 which is a maritime variant of SA-10C.

Operational: Based in the Northern Fleet. Over-the-horizon targeting for SS-N-19 provided by Punch Bowl SATCOM or helicopter. The first ship of the class of four, *Admiral Ushakov*, was formally decommissioned in 2004 and is to be scrapped. The second ship, *Admiral Lazarev* has been inactive in the Pacific Fleet for some years and is also likely to be scrapped. Plans to refit the third ship, *Admiral Nakhimov*, laid up since 1999, appear to have been abandoned.

UPDATED

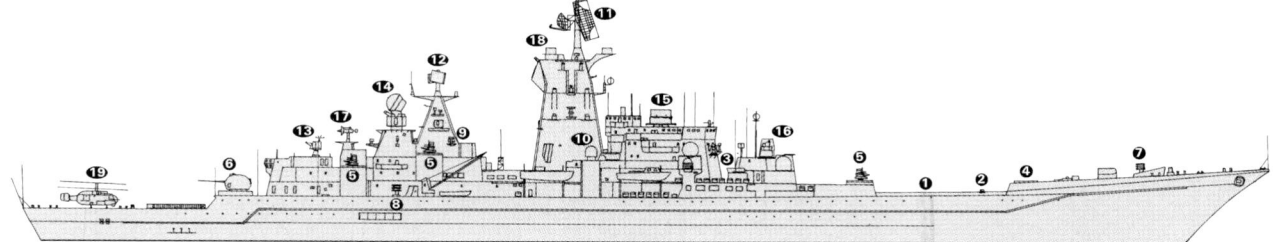

PYOTR VELIKIY

(Scale 1 : 1,500), Ian Sturton / 0528401

PYOTR VELIKIY

6/2001, S Breyer /* 1042309

PYOTR VELIKIY

8/2003 / 0570896

CRUISERS

3 SLAVA (ATLANT) CLASS (PROJECT 1164) (CGHM)

Name	No	Builders	Laid down	Launched	Commissioned
MOSKVA (ex-*Slava*)	121	Nikolayev North (61 Kommuna), Ukraine	5 Dec 1976	27 July 1979	30 Dec 1982
MARSHAL USTINOV	055	Nikolayev North (61 Kommuna), Ukraine	5 Oct 1978	25 Feb 1982	15 Sep 1986
VARYAG (ex-*Chervona Ukraina*)	011	Nikolayev North (61 Kommuna), Ukraine	31 July 1979	28 Aug 1983	25 Dec 1989

Displacement, tons: 9,380 standard; 11,490 full load
Dimensions, feet (metres): 611.5 × 68.2 × 27.6
(186.4 × 20.8 × 8.4)
Main machinery: COGAG; 4 gas-turbines; 88,000 hp(m)
(64.68 MW); 2 M-70 gas-turbines; 20,000 hp(m) *(14.7 MW)*;
2 shafts
Speed, knots: 32. **Range, n miles:** 2,200 at 30 kt; 7,500 at 15 kt
Complement: 476 (62 officers)

Missiles: SSM: 16 Chelomey SS-N-12 (8 twin) Sandbox (Bazalt)
launchers ❶; inertial guidance with command update; active
radar homing to 550 km *(300 n miles)* at 1.7 Mach; warhead
nuclear 350 kT or HE 1,000 kg.
SAM: 8 SA-N-6 Grumble (Fort) vertical launchers ❷; 8 rounds per
launcher; command guidance; semi-active radar homing to
100 km *(54 n miles)*; warhead 90 kg (or nuclear?); altitude
27,432 m *(90,000 ft)*. 64 missiles.
2 SA-N-4 Gecko twin retractable launchers ❸; semi-active
radar homing to 15 km *(8 n miles)* at 2.5 Mach; warhead
50 kg; altitude 9.1-3,048 m *(30-10,000 ft)*; 40 missiles.
Guns: 2—130 mm/70 (twin) AK 130 ❹; 35/45 rds/min to
29 km *(16 n miles)*; weight of shell 33.4 kg.
6—30 mm/65 AK 650; ❺ 6 barrels per mounting; 3,000 rds/
min to 2 km.
Torpedoes: 10—21 in *(533 mm)* (2 quin) tubes ❻. Combination
of 53 cm torpedoes (see table at front of section).
A/S mortars: 2 RBU 6000 12-tubed trainable ❼; range 6,000 m;
warhead 31 kg.
Countermeasures: Decoys: 2 PK 2 chaff launchers.
ESM/ECM: 8 Side Globe (jammers). 4 Rum Tub (intercept).
Weapons control: 2 Tee Plinth and 3 Tilt Pot optronic directors. 2
Punch Bowl satellite data receiving/targeting systems. 2 Bell
Crown and 2 Bell Push datalinks.
Radars: Air search: Top Pair (Top Sail + Big Net) ❽; 3D; C/D-
band; range 366 km *(200 n miles)* for bomber, 183 km *(100 n
miles)* for 2 m² target.
Air/surface search: Top Steer ❾ or Top Plate *(Varyag)*; 3D; D/E-
band.
Navigation: 3 Palm Frond; I-band.
Fire control: Front Door ❿; F-band (for SS-N-12). Top Dome ⓫;
J-band (for SA-N-6). 2 Pop Group ⓬; F/H/I-band (for SA-N-4).
3 Bass Tilt ⓭; H/I-band (for Gatlings). Kite Screech ⓮; H/I/
K-band (for 130 mm).
IFF: Salt Pot A and B. 2 Long Head.
Sonars: Bull Horn and Steer Hide (Platina); hull-mounted; active
search and attack; low/medium frequency.

Helicopters: 1 Ka-27 Helix ⓯.

MOSKVA *9/2004*, Selim San / 1042297*

MOSKVA *9/2003, Giorgio Ghiglione / 0570926*

Programmes: Built at the same yard as the Kara class. This is a
smaller edition of the dual-purpose surface warfare/ASW
Kirov, designed as a conventionally powered back-up for that
class. The fourth of class, originally being completed for
Ukraine, was transferred to Russia in July 1995 but returned
to Ukraine in February 1999 for completion. However, work
was not finished due to lack of funds. Re-sale back to Russia is
unlikely. A fifth of class was started but cancelled in October
1990.
Structure: The notable gap abaft the twin funnels (SA-N-6 area)
is traversed by a large crane which stows between the
funnels. The hangar is recessed below the flight deck with an
inclined ramp. The torpedo tubes are behind shutters in the
hull below the Top Dome radar director aft. Air conditioned
citadels for NBCD. There is a bridge periscope.

Operational: The SA-N-6 system effectiveness is diminished by
having only one radar director. Over-the-horizon targeting for
SS-N-12 provided by helicopter or Punch Bowl SATCOM.
Moskva is based in the Black Sea Fleet at Sevastopol and
conducted an Indian Ocean deployment in 2003. Her nine-
year refit was beset by payment problems. Some funds were
provided by the city of Moscow. *Marshal Ustinov* deployed to
the Northern Fleet in March 1987 and completed refit at St
Petersburg in May 1995 where she remained until January
1998, when she transferred back to the Northern Fleet and is
based at Severomorsk and is active. *Varyag* transferred to
Petropavlovsk in the Pacific in October 1990 and deployed to
Japan and the Pacific during 2002.

UPDATED

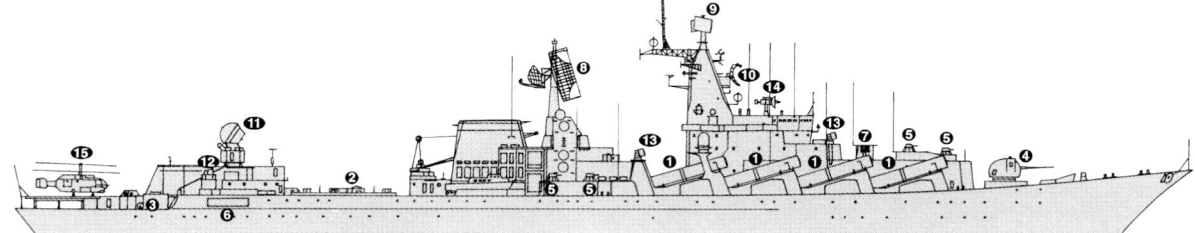

VARYAG *(Scale 1 : 1,200), Ian Sturton / 0050017*

VARYAG *12/2004, Ships of the World / 1042329*

1 KARA (BERKOT-B) CLASS (PROJECT 1134B) (CGHM)

Name	No	Builders	Laid down	Launched	Commissioned
Name	*No*	*Builders*	*Laid down*	*Launched*	*Commissioned*
KERCH	713 (ex-711)	Nikolayev North (61 Kommuna), Ukraine	30 Apr 1971	21 July 1972	25 Dec 1974

Displacement, tons: 7,650 standard; 9,900 full load
Dimensions, feet (metres): 568 × 61 × 22
(173.2 × 18.6 × 6.7)
Main machinery: COGAG; 4 gas turbines; 108,800 hp(m)
(80 MW); 2 gas turbines; 13,600 hp(m) *(10 MW)*; 2 shafts
Speed, knots: 32
Range, n miles: 9,000 at 15 kt cruising turbines; 3,000 at 32 kt
Complement: 390 (49 officers)

Missiles: SAM: 2 SA-N-3 Goblet twin launchers ❶; semi-active
radar homing to 55 km *(30 n miles)* at 2.5 Mach; warhead
80 kg; altitude 91.4-22,860 m *(300-75,000 ft)*; 72 missiles.
2 SA-N-4 Gecko twin launchers (twin either side of mast) ❷;
semi-active radar homing to 15 km *(8 n miles)* at 2.5 Mach;
warhead 50 kg; altitude 9.1-3,048 m *(30-10,000 ft)*; 40
missiles.
A/S: 2 Raduga SS-N-14 Silex (Rastrub) quad launchers ❸;
command guidance to 55 km *(30 n miles)* at 0.95 Mach;
payload nuclear 5 kT or Type 40 torpedo or E53-72 torpedo.
SSM version; range 35 km *(19 n miles)*; warhead 500 kg.
Guns: 4—3 in *(76 mm)*/60 (2 twin) ❹; 90 rds/min to 15 km *(8 n
miles)*; weight of shell 6.8 kg.
4—30 mm/65 ❺; 6 barrels per mounting; 3,000 rds/min
combined to 2 km.
Torpedoes: 10—21 in *(533 mm)* (2 quin) tubes ❻. Combination
of 53 cm torpedoes (see table at front of section).
A/S mortars: 2 RBU 6000 12-tubed trainable ❼; range 6,000 m;
warhead 31 kg.
2 RBU 1000 6-tubed (aft) ❽; range 1,000 m; warhead 55 kg;
torpedo countermeasures.
Countermeasures: Decoys: 2 PK 2 chaff launchers. 1 BAT-1
torpedo decoy.
ESM/ECM: 8 Side Globe (jammers). 2 Bell Slam. 2 Bell Clout.
4 Rum Tub (intercept) (fitted on mainmast).
Weapons control: 4 Tilt Pot optronic directors. Bell Crown, Bike
Pump and Hat Box datalinks.
Radars: Air search: Flat Screen ❾; E/F-band.
Air/surface search: Head Net C ❿; 3D; E-band; range 128 km
(70 n miles).
Navigation: 2 Don Kay; I-band. Don 2 or Palm Frond; I-band.
Fire control: 2 Head Light B/C ⓫; F/G/H-band (for SA-N-3 and
SS-N-14). 2 Pop Group ⓬; F/H/I-band (for SA-N-4). 2 Owl
Screech ⓭; G-band (for 76 mm). 2 Bass Tilt ⓮; H/I-band (for
30 mm).
Tacan: Fly Screen A or Fly Spike.
IFF: High Pole A. High Pole B.
Sonars: Bull Nose (Titan 2-MG 332); hull-mounted; active search
and attack; low/medium frequency.
Mare Tail; VDS (Vega-M 325) ⓯; active search; medium
frequency.

Helicopters: 1 Ka-27 Helix ⓰.

Programmes: Type name is *bolshoy protivolodochny korabl*,
meaning large anti-submarine ship.
Modernisation: The Flat Screen air search radar, replaced Top
Sail.
Structure: The helicopter is raised to flight deck level by a lift. In
addition to the 8 tubes for the SS-N-14 A/S system and the

KERCH   *(Scale 1 : 1,500), Ian Sturton /* 0081651

KERCH   *7/2000, Hartmut Ehlers /* 0105541

pair of twin launchers for SA-N-3 system with Goblet missiles,
Kara class mounts the SA-N-4 system in 2 silos, either side of
the mast. The SA-N-3 system has only 2 loading doors per
launcher and a larger launching arm.
Operational: Two of the class started refits in July 1987 and have
been scrapped by the Ukraine. One more was scrapped in the
Pacific in 1996. Two others *Ochakov* and *Petropavlovsk* are
laid up in the Black Sea and Pacific respectively and are

unlikely to go to sea again. There have been several reports of
work being done on *Ochakov*, most recently in 2004, but
these have not been confirmed. Formally, she remains in
service. *Azov* was cannibalised for spares in 1998. *Kerch* is
based in the Black Sea at Sevastopol and is believed to be
operational although there have been no recent reports of
activity.

UPDATED

DESTROYERS

1 KASHIN (PROJECT 61) CLASS (DDGM)

Name	No	Builders	Laid down	Launched	Commissioned
Name	*No*	*Builders*	*Laid down*	*Launched*	*Commissioned*
SMETLIVY	810	Nikolayev North, Ukraine	15 July 1966	26 Aug 1967	25 Sep 1969

Displacement, tons: 4,010 standard; 4,750 full load
Dimensions, feet (metres): 472.4 × 51.8 × 15.4
(144 × 15.8 × 4.7)
Main machinery: COGAG: 4 DE 59 gas turbines; 72,000 hp(m)
(52.9 MW); 2 shafts
Speed, knots: 32. **Range, n miles:** 4,000 at 18 kt; 1,520 at 32 kt
Complement: 280 (25 officers)

Missiles: SSM: 8 Zvezda SS-N-25 (KH 35 Uran) (2 quad) ❶.
SAM: 2 SA-N-1 Goa twin launchers ❷; command guidance to
31.5 km *(17 n miles)* at 2 Mach; warhead 72 kg; altitude 91.4-
22,860 m *(300-75,000 ft)*; 32 missiles.
Guns: 2—3 in *(76 mm)*/60 (1 or 2 twin) ❸; 90 rds/min to 15 km
(8 n miles); weight of shell 6.8 kg.
Torpedoes: 5—21 in *(533 mm)* (quin) tubes ❹. Combination of
53 cm torpedoes (see table at front of section).
A/S mortars: 2 RBU 6000 12-tubed trainable ❺; range 6,000 m;
warhead 31 kg; 120 rockets.
Countermeasures: Decoys: PK 16 chaff launchers (modified).
2 towed torpedo decoys.
ESM/ECM: 2 Bell Shroud. 2 Watch Dog.
Weapons control: 3 Tee Plinth and 4 Tilt Pot optronic directors.
Radars: Air/surface search: Head Net C ❻; 3D; E-band.
Big Net ❼; C-band.
Navigation: 2 Don 2/Don Kay/Palm Frond; I-band.
Fire control: 2 Peel Group ❽; H/I-band (for SA-N-1). 1 Owl Screech
❾; G-band (for guns).
IFF: High Pole B.
Sonars: Bull Nose (MGK 336) or Wolf Paw; hull-mounted; active
search and attack; medium frequency.
Vega; VDS; active search; medium frequency.

Programmes: The first class of warships in the world to rely
entirely on gas-turbine propulsion. Type name is *bolshoy
protivolodochny korabl*, meaning large anti-submarine ship.
Modernisation: Modernised with a VDS aft, vice the after gun,
and fitted for SS-N-25 in place of the RBU 1000 launchers.
Operational: Based in the Black Sea. Refitted from 1990 to 1996
but back in service in 1997. Deployed to the Indian Ocean in
2003 and remains active.

SMETLIVY   *(Scale 1 : 1,200), Ian Sturton /* 0126351

SMETLIVY   *6/2004 *, Selim San /* 1042300

Sales: Additional ships of a modified design built for India. First
transferred September 1980, the second in June 1982, the
third in 1983, the fourth in August 1986 and the fifth and last

in January 1988. *Smely* transferred to Poland 9 January 1988.

UPDATED

1 UDALOY II (FREGAT) CLASS (PROJECT 1155.1) (DDGHM)

Name	No	Builders	Laid down	Launched	Commissioned
ADMIRAL CHABANENKO	650 (ex-437)	Yantar, Kaliningrad 820	15 Sep 1988	14 Dec 1992	20 Feb 1999

Displacement, tons: 7,700 standard; 8,900 full load
Dimensions, feet (metres): 536.4 × 63.3 × 24.6
(163.5 × 19.3 × 7.5)
Main machinery: COGAG; 2 gas turbines; 48,600 hp(m)
(35.72 MW); 2 gas turbines; 24,200 hp(m) *(17.79 MW)*; 2
shafts
Speed, knots: 28. **Range, n miles:** 4,000 at 18 kt
Complement: 249 (29 officers)

Missiles: SSM: 8 Raduga SS-N-22 Sunburn (3M-82 Moskit)
(2 quad) ❶; active/passive radar homing to 160 km *(87 n
miles)* at 2.5 Mach (4.5 for attack); warhead nuclear or HE
300 kg; sea-skimmer.
SAM: 8 SA-N-9 Gauntlet (Klinok) vertical launchers ❷; command
guidance; active radar homing to 12 km *(6.5 n miles)* at
2 Mach; warhead 15 kg. 64 missiles; 4 channels of fire.
SAM/Guns: 2 CADS-N-1 (Kashtan) ❸; each with twin 30 mm
Gatling; combined with 8 SA-N-11 (Grisson) and Hot Flash/
Hot Spot fire-control radar/optronic director. Laser beam
guidance for missiles to 8 km *(4.4 n miles)*; warhead 9 kg;
9,000 rds/min combined to 1.5 km for guns.
A/S: Novator SS-N-15 (Starfish); inertial flight to 45 km *(24.3 n*

miles); payload Type 40 torpedo or nuclear, fired from torpedo
tubes.
Guns: 2—130 mm/70 (twin) AK 130 ❹; 35-45 rds/min to
29.5 km *(16 n miles)*; weight of shell 33.4 kg.
Torpedoes: 8—21 in *(533 mm)* (2 quad tubes) ❺. Combination of
53 cm torpedoes (see table at front of section). The tubes are
protected by flaps in the superstructure.
A/S mortars: 2 RBU 6000 ❻; 12-tubed trainable; range
6,000 m; warhead 31 kg.
Countermeasures: 8 PK 10 and 2 PK 2 chaff launchers ❼.
ESM/ECM: 2 Wine Glass (intercept). 2 Bell Shroud. 2 Bell Squat.
4 Half Cup laser warner. 2 Shot Dome.
Weapons control: M 145 radar and optronic system. 2 Bell
Crown datalink. Band Stand ❽ datalink for SS-N-22; 2 Light
Bulb, 2 Round House and 1 Bell Nest datalinks.
Radars: Air Search: Strut Pair II ❾; F-band.
Top Plate ❿; 3D; D/E-band.
Surface Search: 3 Palm Frond ⓫; I-band.
Fire control: 2 Cross Swords ⓬; K-band (for SA-N-9). Kite
Screech ⓭; H/I/K-band (for 100 mm gun).
CCA: Fly Screen B ⓮.
IFF: Salt Pot B and C.

Sonars: Horse Jaw (Polinom); hull-mounted; active search and
attack; medium/low frequency.
Horse Tail; VDS; active search; medium frequency.

Helicopters: 2 Ka-27 Helix A ⓯.

Programmes: A single ship follow-on class from the Udaloys.
NATO designator Balcom 12. At least two more were
projected with names *Admiral Basisty* and *Admiral Kucherov*;
Basisty was scrapped in March 1994, and *Kucherov* was
never started.
Structure: Similar size to the Udaloy and has the same
propulsion machinery. Improved combination of weapon
systems owing something to both the Sovremenny and the
Neustrashimy classes. The distribution of SA-N-9 launchers
may be the same as Udaloy class. The torpedo tubes are
protected by a hinged flap in the superstructure.
Operational: Sea trials started on 14 September 1995 from
Baltiysk. Deployed to the Northern Fleet in March 1999 when
the pennant number changed. Based at Severomorsk and is
very active.

UPDATED

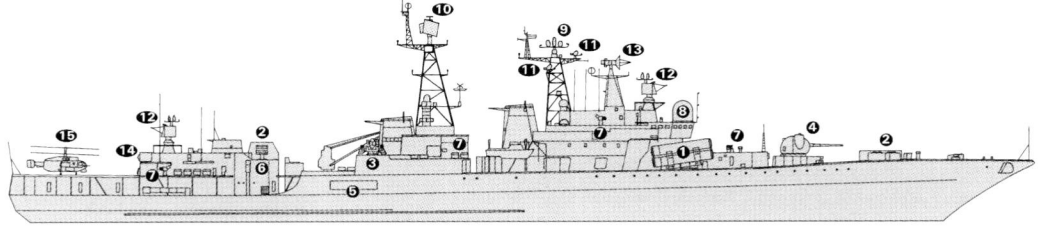

ADMIRAL CHABANENKO

(Scale 1 : 1,200), Ian Sturton / 0569929

ADMIRAL CHABANENKO

9/2004 *, B Prézelin / 1042301

ADMIRAL CHABANENKO

8/2002 / 0528328

For details of the latest updates to *Jane's Fighting Ships* online and to discover the
additional information available exclusively to online subscribers please visit
jfs.janes.com

9 UDALOY (FREGAT) CLASS (PROJECT 1155) (DDGHM)

Name	No	Builders	Laid down	Launched	Commissioned
VITSE ADMIRAL KULAKOV	400	Zhdanov Yard, Leningrad 190	7 Nov 1977	16 May 1980	29 Dec 1981
MARSHAL VASILEVSKY	687	Zhdanov Yard, Leningrad 190	22 Apr 1979	29 Dec 1981	8 Dec 1983
ADMIRAL TRIBUTS	564	Zhdanov Yard, Leningrad 190	19 Apr 1980	26 Mar 1983	30 Dec 1985
MARSHAL SHAPOSHNIKOV	543	Yantar, Kaliningrad 820	25 May 1983	27 Dec 1984	30 Dec 1985
SEVEROMORSK (ex-*Simferopol*, ex-*Marshal Budienny*)	619	Yantar, Kaliningrad 820	12 June 1984	24 Dec 1985	30 Dec 1987
ADMIRAL LEVCHENKO (ex-*Kharbarovsk*)	605	Zhdanov Yard, Leningrad 190	27 Jan 1982	21 Feb 1985	30 Sep 1988
ADMIRAL VINOGRADOV	572	Yantar, Kaliningrad 820	5 Feb 1986	4 June 1987	30 Dec 1988
ADMIRAL KHARLAMOV	678	Yantar, Kaliningrad 820	7 Aug 1986	29 June 1988	30 Dec 1989
ADMIRAL PANTELEYEV	548	Yantar, Kaliningrad 820	28 Jan 1988	7 Feb 1990	19 Dec 1991

Displacement, tons: 6,700 standard; 8,500 full load
Dimensions, feet (metres): 536.4 × 63.3 × 24.6
(163.5 × 19.3 × 7.5)
Flight deck, feet (metres): 65.6 × 59 *(20 × 18)*
Main machinery: COGAG; 2 gas turbines; 55,500 hp(m)
(40.8 MW); 2 gas turbines; 13,600 hp(m) *(10 MW)*; 2 shafts
Speed, knots: 29. **Range, n miles:** 2,600 at 30 kt; 7,700 at 18 kt
Complement: 249 (29 officers)

Missiles: SAM: 8 SA-N-9 Gauntlet (Klinok) vertical launchers ❶;
command guidance; active radar homing to 12 km *(6.5 n miles)* at 2 Mach; warhead 15 kg; altitude 3.4-12,192 m *(10-40,000 ft)*; 64 missiles; four channels of fire.
The launchers are set into the ships' structures with 6 ft diameter cover plates-4 on the forecastle, 2 between the torpedo tubes and 2 at the forward end of the after deckhouse between the RBUs.
A/S: 2 Raduga SS-N-14 Silex (Rastrub) quad launchers ❷;
command guidance to 55 km *(30 n miles)* at 0.95 Mach; payload nuclear 5 kT or Type 40 torpedo or Type E53-72 torpedo. SSM version; range 35 km *(19 n miles)*; warhead 500 kg.
Guns: 2—3.9 in *(100 mm)*/59 ❸; 60 rds/min to 15 km *(8.2 n miles)*; weight of shell 16 kg.
4—30 mm/65 AK 630 ❹; 6 barrels per mounting; 3,000 rds/min combined to 2 km.
Torpedoes: 8—21 in *(533 mm)* (2 quad) tubes ❺. Combination of 53 cm torpedoes (see table at front of section).
A/S mortars: 2 RBU 6000 12-tubed trainable ❻; range 6,000 m; warhead 31 kg.
Mines: Rails for 26 mines.
Countermeasures: Decoys: 2 PK-2 and 8 PK-10 chaff launchers. US Masker type noise reduction.
ESM/ECM: 2 Foot Ball B (*Levchenko* onwards); 2 Wine Glass (intercept). 6 Half Cup laser warner (*Levchenko* onwards); 2 Bell Squat (jammers).
Weapons control: MP 145 radar and optronic system. 2 Bell Crown and Round House C datalink.
Radars: Air search: Strut Pair ❼; F-band.
Top Plate ❽; 3D; D/E-band.
Surface search: 3 Palm Frond ❾; I-band.
Fire control: 2 Eye Bowl ❿; F-band (for SS-N-14). 2 Cross Sword ⓫; K-band (for SA-N-9). Kite Screech ⓬; H/I/K-band (for 100 mm guns). 2 Bass Tilt ⓭; H/I/K-band (for 30 mm guns).
IFF: Salt Pot A and B. Box Bar A and B.
Tacan: 2 Round House.
CCA: Fly Screen B (by starboard hangar) ⓮. 2 Fly Spike B.
Sonars: Horse Jaw (Polinom); hull-mounted; active search and attack; low/medium frequency.
Mouse Tail; VDS; active search; medium frequency.

Helicopters: 2 Ka-27 Helix A ⓯.

Programmes: Design approved in October 1972. Successor to Kresta II class but based on Krivak class. Type name is *bolshoy protivolodochny korabl* meaning large anti-submarine ship. Programme stopped at 12 in favour of Udaloy II class (Type 1155.1).

ADMIRAL KHARLAMOV   6/2003 *, B Lemachko / 1042299

ADMIRAL PANTELEYEV   7/2004 *, Ships of the World / 1042290

Structure: The two hangars are set side by side with inclined elevating ramps to the flight deck. Has pre-wetting NBCD equipment and replenishment at sea gear. Active stabilisers are fitted. The chaff launchers are on both sides of the foremast and inboard of the torpedo tubes. Cage Flask aerials are mounted on the mainmast spur and on the mast on top of the hangar. There are indications of a nuclear release mechanism, or interlock, on the lower tubes of the SS-N-14 launchers.
Operational: These general purpose ships have good sea-keeping and endurance and are the backbone of the fleet. Based as follows: Northern Fleet—*Severomorsk, Kharlamov, Vasilevsky* and *Levchenko*; Pacific Fleet—*Shaposhnikov,*

Panteleyev, Vinogradov and *Tributs*. Baltic Fleet – *Kulakov*. *Vinogradov* was in collision in April 2000 but was quickly repaired. *Severomorsk* deployed to St Petersburg for refit in June 1998 completing in late 2000, and *Levchenko* followed in November 1999 completing in 2001. The fourth of class, *Zakharov* was scrapped after a fire in March 1992. *Tributs* was in reserve in 1994 and had a machinery space fire in September 1995, was back in service in mid-1999. *Udaloy* and *Spiridonov* have been laid up or scrapped. *Kulakov* started refit in 1990 and returned to service in 2004. *Vasilevsky* was in refit during 2001 and completed in 2004.
Opinion: Obvious efforts are being made to keep this class in service at the expense of the Sovremennys. **UPDATED**

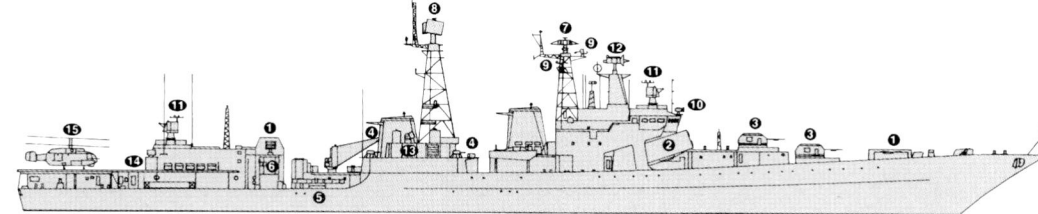

SEVEROMORSK   (Scale 1 : 1,200), Ian Sturton

MARSHAL SHAPOSHNIKOV   9/2004 *, Hachiro Nakai / 1042307

6 SOVREMENNY (SARYCH) CLASS (PROJECT 956/956A) (DDGHM)

Name	No	Builders	Laid down	Launched	Commissioned
BURNY	778	Zhdanov Yard, Leningrad (190)	4 Nov 1983	30 Dec 1986	30 Sep 1988
BYSTRY	715	Zhdanov Yard, Leningrad (190)	29 Oct 1985	28 Nov 1987	30 Sep 1989
BEZBOYAZNENNYY	754	Zhdanov Yard, Leningrad (190)	8 Jan 1987	18 Feb 1989	28 Nov 1990
BESPOKOINY	620	Zhdanov Yard, Leningrad (190)	18 Apr 1987	22 Feb 1992	29 Dec 1993
NASTOYCHIVY (ex-Moskowski Komsomolets)	610	Zhdanov Yard, Leningrad (190)	7 Apr 1988	15 Feb 1992	27 Mar 1993
MARSHAL USHAKOV (ex-Besstrashny)	434	Zhdanov Yard, Leningrad (190)	16 Apr 1988	31 Dec 1992	17 Apr 1994

Displacement, tons: 6,500 standard; 7,940 full load
Dimensions, feet (metres): 511.8 × 56.8 × 21.3
(156 × 17.3 × 6.5)
Main machinery: 4 KVN boilers; 2 GTZA-674 turbines;
99,500 hp(m) *(73.13 MW)* sustained; 2 shafts; bow thruster
Speed, knots: 32
Range, n miles: 2,400 at 32 kt; 6,500 at 20 kt; 4,000 at 14 kt
Complement: 296 (25 officers) plus 60 spare

Missiles: SSM: 8 Raduga SS-N-22 Sunburn (3M-80 Zubr)
(2 quad) launchers ❶; active/passive radar homing to 110 km
(60 n miles) at 2.5 (4.5 for attack) Mach; warhead nuclear
200 kT or HE 300 kg; sea-skimmer. From *Bespokoiny*
onwards the launchers are longer and fire a modified missile
(3M-82 Moskit) with a range of 160 km *(87 n miles)*.
SAM: 2 SA-N-7 Gadfly 3S 90 (Uragan) ❷; command/semi-active
radar and IR homing to 25 km *(13.5 n miles)* at 3 Mach;
warhead 70 kg; altitude 15-14,020 m *(50-46,000 ft)*; 44
missiles. Multiple channels of fire. From *Bespokoiny* onwards
the same launcher is used for the SA-N-17 Grizzly/SA-N-12
Yezh.
Guns: 4—130 mm/70 (2 twin) AK 130 ❸; 35-45 rds/min to
29.5 km *(16 n miles)*; weight of shell 33.4 kg.
4—30 mm/65 AK 630 ❹; 6 barrels per mounting; 3,000 rds/
min combined to 2 km.
Torpedoes: 4—21 in *(533 mm)* (2 twin) tubes ❺. Combination of
53 cm torpedoes (see table at front of section).

A/S mortars: 2 RBU 1000 (Smerch 3) 6-barrelled ❻; range
1,000 m; warhead 100 kg; 120 rockets carried. Torpedo
countermeasure.
Mines: Mine rails for up to 22.
Countermeasures: Decoys: 8 PK 10 and 2 PK 2 chaff launchers.
ESM/ECM: 4 Foot Ball (some variations including 2 Bell Shroud
and 2 Bell Squat). 6 Half Cup laser warner.
Combat data systems: Sapfir-U.
Weapons control: 1 Squeeze Box optronic director and laser
rangefinder ❼. Band Stand ❽ datalink for SS-N-22. Bell Nest,
2 Light Bulb and 2 Tee Pump datalinks.
Radars: Air search: Top Plate (MR-750 Fregat) ❾; 3D; D/E-band.
Surface search: 3 Palm Frond (MR 212/201) ❿; I-band.
Fire control: 6 Front Dome ⓫; F-band (for SA-N-7/17). Kite
Screech (MR-184) ⓬; H/I/K-band (for 130 mm guns). 2 Bass
Tilt ⓭; H/I-band (for 30 mm guns).
IFF: Salt Pot A and B. High Pole A and B. Long Head.
Tacan: 2 Light Bulb.
Sonars: Bull Horn (MGK-335 Platina) and Whale Tongue; hull-
mounted; active search and attack; medium frequency.

Helicopters: 1 Kamov Ka-27 Helix ⓮.

Programmes: Type name is *eskadrenny minonosets* meaning
destroyer. From *Bespokoiny* onwards the class is known as
956A. Total of 17 built for Russia, two (hulls 18 and 19) for
China, and one more *(Bulny)* which is unlikely to be completed
unless for export.
Structure: Telescopic hangar. The fully automatic 130 mm gun
was first seen in 1976. Chaff launchers are fitted on both sides
of the foremast and either side of the after SAM launcher. A
longer range version of SS-N-22 has been introduced in the
Type 956A. This has slightly longer launch tubes. Also the
SAM system has been improved to take the SA-N-17. There
are also some variations in the EW fit.
Operational: A specialist surface warfare ship complementing
the ASW-capable Udaloy class. Based as follows: Northern
Fleet—*Marshal Ushakov*. Pacific Fleet—*Burny*,
Bezboyaznennyy and *Bespokoiny*. Baltic Fleet-*Nastoychivy* and
Bespokoiny. So far 11 others have paid off or are non-
operational although *Bystry* completed refit in 2002 and
Bezboyaznennyy in 2004. 434 renamed *Marshal Ushakov* in
2004. *Rastoropny* was reportedly in refit at Severnaya Verf in
2004 and may return to service in 2005. Steam-plant
reliability has been a class problem.
Sales: Two others which were near completion in 1996, were
sold to China and sailed in December 1999 and December
2000 respectively, from the Baltic to the South China Sea. A
contract for the procurement of two new ships was signed by
the Chinese government on 3 January 2002.

UPDATED

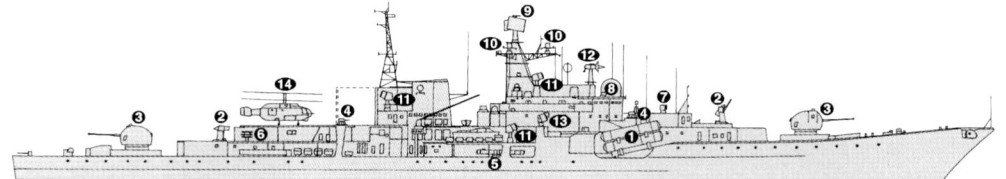

BURNY

(Scale 1 : 1,200), Ian Sturton / 0019027

BYSTRY

10/2004*, Ships of the World / 1042289

NASTOYCHIVY

6/2003, Martin Mokrus / 0570893

FRIGATES

0 + 1 GROM CLASS (PROJECT 1244.1) (FFG)

Name	No	Builders	Laid down	Launched	Commissioned
NOVIK	—	Yantar, Kaliningrad	26 July 1997	2001	2005

Displacement, tons: 3,600 full load
Dimensions, feet (metres): 400.3 × 49.2 × 31.2 (sonar)
(122.0 × 15.0 × 9.5)
Main machinery: CODAG; 2 gas turbines; 2 diesels; 2 shafts
Speed, knots: 30

Missiles: SSM: Space for eight or 16 Zvezda SS-N-25 (KH 35 Uran) ❶ (2 quad); active radar homing to 130 km *(70.2 n miles)* at 0.9 Mach; warhead 145 kg; sea skimmer.
SAM: Space for VLS system ❷.
Guns: 1—3 in *(76 mm)*/60 ❸.
2—30 mm AK 630 ❹.
Radars: Air/surface search: Top Plate (Fregate M) ❺; 3D; D/E-band.
Surface search: Cross Dome; E/F-band ❻.
Fire control: Bass Tilt; H/I-band ❼.
Sonars: Hull mounted and VDS.

Helicopters: 1 Ka-29 Helix.

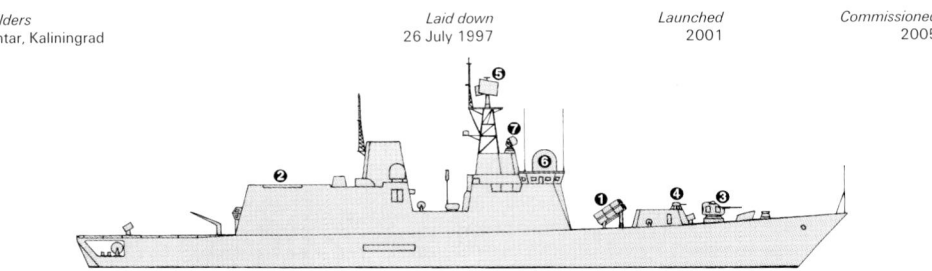

NOVIK
(Scale 1 : 1,200), Ian Sturton / 0019031

Programmes: Designed by Almaz. Considerable publicity when keel laid down in 1997 but the project stalled due to budget cuts. However, building was reported to have been restarted in 2003 and it is speculated that the ship was completed as an unarmed training ship in 2004. The ship may be commissioned in 2005.

Structure: Most details are speculative and are based on the original published export design.
Operational: Likely to be based in the Baltic Fleet.
NEW ENTRY

1 NEUSTRASHIMY (JASTREB) CLASS (PROJECT 1154) (FFHM)

Name	No	Builders	Laid down	Launched	Commissioned
NEUSTRASHIMY	712	Yantar, Kaliningrad	27 Mar 1987	25 May 1988	24 Jan 1993

Displacement, tons: 3,450 standard; 4,250 full load
Dimensions, feet (metres): 425.3 oa; 403.5 wl × 50.9 × 15.7
(129.6; 123 × 15.5 × 4.8)
Main machinery: COGAG; 2 gas turbines; 48,600 hp(m) *(35.72 MW)*; 2 gas turbines; 24,200 hp(m) *(17.79 MW)*; 2 shafts
Speed, knots: 30. **Range, n miles:** 4,500 at 16 kt
Complement: 210 (35 officers)

Missiles: SSM: Fitted for but not with 16 SS-N-25 (4 quad). SS-CX-5 Sapless (possibly a version of SS-N-22 (Moskit M)) may be carried (see *Torpedoes*).
SAM: 4 SA-N-9 Gauntlet (Klinok) octuple vertical launchers ❶; command guidance; active radar homing to 12 km *(6.5 n miles)* at 2 Mach; warhead 15 kg. 32 missiles.
SAM/Guns: 2 CADS-N-1 (Kortik/Kashtan) (3M87) ❷; each has a twin 30 mm Gatling combined with 8 SA-N-11 (Grisson) and Hot Flash/Hot Spot fire-control radar/optronic director. Laser beam guidance for missiles to 8 km *(4.4 n miles)*; warhead 9 kg; 9,000 rds/min (combined) to 1.5 km (for guns).
A/S: SS-N-15/16; inertial flight to 120 km *(65 n miles)*; payload Type 40 torpedo or nuclear warhead; fired from torpedo tubes.
Guns: 1—3.9 in *(100 mm)*/59 A 190E ❸; 60 rds/min to 15 km *(8.2 n miles)*; weight of shell 16 kg.
Torpedoes: 6—21 in *(533 mm)* tubes combined with A/S launcher ❹; can fire SS-N-15/16 missiles with Type 40 anti-submarine torpedoes or 53 cm torpedoes (see table at front of section).
A/S mortars: 1 RBU 12,000 ❺; 10-tubed trainable; range 12,000 m; warhead 80 kg.
Mines: 2 rails.
Countermeasures: Decoys: 8 PK 10 and 2 PK 16 chaff launchers. ESM/ECM: Intercept and jammers. 2 Foot Ball; 2 Half Hat; 4 Half Cup laser intercept.
Weapons control: 2 Bell Crown datalink.
Radars: Air search: Top Plate ❻; 3D; D/E-band.
Air/Surface search: Cross Dome ❼; E/F-band.
Navigation: 2 Palm Frond; I-band.
Fire control: Cross Sword ❽ (for SAM); K-band. Kite Screech B ❾ (for SSM and guns); I-band.
IFF: 2 Salt Pot; 4 Box Bar.
Sonars: Ox Yoke and Whale Tongue; hull-mounted; active search and attack; medium frequency.
Ox Tail; VDS ❿ or towed sonar array.

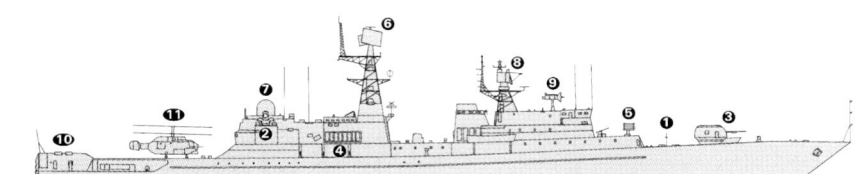

NEUSTRASHIMY
(Scale 1 : 1,200), Ian Sturton / 0569927

NEUSTRASHIMY
6/2004 *, Michael Winter / 1042302

Helicopters: 1 Ka-27 Helix ⓫.

Programmes: At least four of the class were planned. The first of the class started sea trials in the Baltic in December 1990. Second of class (*Yaroslav Mudryy*) was launched in May 1991, but in October 1988 the shipyard stated that the hull would be sold for scrap. However, after several years' inaction, it was reported in 2002 that work had recommenced and there has been continuing speculation that the ship is to be completed. The export version of the ship is known as 'Korsar'. The third ship (*Tuman*) was launched in July 1993 with only the hull completed and work stopped in December 1997 without any work being done. The future of the ship is uncertain.
Structure: Slightly larger than the Krivak and has a helicopter which is a standard part of the armament of modern Western

frigates. There are two horizontal launchers at main deck level on each side of the ship, angled at 18° from forward. These double up for A/S missiles of the SS-N-15/16 type using a 'plunge-fly-plunge' launch and flight and normal torpedoes. Similar launchers are behind shutters in the last three of the Kirov class. The helicopter deck extends across the full width of the ship. The after funnel is unusually flush decked but both funnels have been slightly extended after initial sea trials. Attempts have been made to incorporate stealth features. Main propulsion is the same as the Udaloy II class. Reported as having a basic computerised combat data system.
Operational: Based in the Baltic at Baltiysk. Active in 2003.
UPDATED

NEUSTRASHIMY
8/2004 * / 1042328

8 KRIVAK (PROJECT 1135/1135M/1135MP) CLASS (FFM)

Name	Type	No	Builders	Laid down	Launched	Commissioned
ZHARKY	I Mod	937	Zhdanov, Leningrad (190)	16 Apr 1974	3 Nov 1975	29 June 1976
LEGKY	I Mod	930	Zhdanov, Leningrad (190)	22 Apr 1976	1 Apr 1977	29 Sep 1977
NEUKROTIMY	II	731	Yantar, Kaliningrad	22 Jan 1976	27 June 1977	30 Dec 1977
LETUCHY	I	661	Zhdanov, Leningrad	9 Mar 1977	19 Mar 1978	10 Aug 1978
PYLKY	I Mod	702	Zhdanov, Leningrad	16 May 1977	20 Aug 1978	28 Dec 1978
ZADORNY	I	955	Zhdanov, Leningrad	10 Nov 1977	25 Mar 1979	15 Sep 1979
LADNY	I	801	Kamish-Burun, Kerch	25 May 1979	7 May 1980	29 Dec 1980
PYTLIVY	II	808	Yantar, Kaliningrad	27 June 1979	16 Apr 1981	30 Nov 1981

Displacement, tons: 3,100 standard; 3,650 full load
Dimensions, feet (metres): 405.2 × 46.9 × 24 (sonar)
(123.5 × 14.3 × 7.3)
Main machinery: COGAG; 2 M8K gas-turbines; 55,500 hp(m)
(40.8 MW); 2 M 62 gas-turbines; 13,600 hp(m) *(10 MW)*;
2 shafts
Speed, knots: 32. **Range, n miles:** 4,000 at 14 kt; 1,600 at 30 kt
Complement: 194 (18 officers)

Missiles: SSM: 8 Zvezda SS-N-25 (KH 35 Uran) (2 quad) ❶;
(Krivak I after modernisation); fitted for but not with.
SAM: 2 SA-N-4 Gecko (Zif 122) twin launchers ❷; Osa-M semi-
active radar homing to 15 km *(8 n miles)* at 2.5 Mach;
warhead 50 kg; altitude 9.1-3,048 m *(30-10,000 ft)*; 40
missiles (20 in Krivak III).
A/S: Raduga SS-N-14 Silex quad launcher ❸; command
guidance to 55 km *(30 n miles)* at 0.95 Mach; payload nuclear
5 kT or Type 40 torpedo or Type E53-72 torpedo. SSM
version; range 35 km *(19 n miles)*; warhead 500 kg.
Guns: 4—3 in *(76 mm)*/60 (2 twin) (Krivak I) ❹; 90 rds/min to
15 km *(8 n miles)*; weight of shell 6.8 kg.
2—3.9 in *(100 mm)*/59 (Krivak II) ❺; 60 rds/min to 15 km
(8.2 n miles); weight of shell 16 kg.
Torpedoes: 8—21 in *(533 mm)* (2 quad) tubes ❻. Combination of
53 cm torpedoes (see table at front of section).
A/S mortars: 2 RBU 6000 12-tubed trainable ❼; (not in
modernised Krivak I); range 6,000 m; warhead 31 kg.
Mines: Capacity for 16.
Countermeasures: Decoys: 4 PK 16 or 10 PK 10 chaff launchers.
Towed torpedo decoy.
ESM/ECM: 2 Bell Shroud. 2 Bell Squat. Half Cup laser warning (in
some).
Radars: Air search: Head Net C ❽; 3D; E-band; or Half Plate
(Krivak I mod) ❾.
Surface search: Don Kay or Palm Frond or Don 2 or Spin Trough
❿; I-band.

Fire control: 2 Eye Bowl ⓫; F-band (for SS-N-14). 2 Pop Group
(one in Krivak III) ⓬; F/H/I-band (for SA-N-4). Owl Screech
(Krivak I) ⓭; G-band. Kite Screech (Krivak II and III) ⓮; H/I/K-
band. Plank Shave (Harpun B) (for SS-N-25) not fitted.
IFF: High Pole B.
Sonars: Bull Nose (MGK-335S or MG-332); hull-mounted; active
search and attack; medium frequency.
Mare Tail (MGK-345) or Steer Hide (some Krivak Is after
modernisation); VDS (MG 325) ⓯; active search; medium
frequency.

Programmes: Type name was originally *bolshoy protivolodochny
korabl*, meaning large anti-submarine ship. Changed in
1977-78 to *storozhevoy korabl* meaning escort ship. The naval

Krivaks I and II are known as the Burevestnik class and the
border guard ships Krivak III (listed separately) as Nerey class.
Modernisation: Top Plate radar has replaced Head Net in some
and a more modern VDS is also fitted. Ships converted with
SS-N-25 launchers are *Legky* and *Pylky*. This programme has
stopped and missiles are not embarked. The launchers
replaced the RBU mountings.
Structure: The modified Krivak I class has a larger bow. Krivak II
class has Y-gun mounted higher than in Krivak I and the break
to the quarterdeck further aft apart from other variations
noted above.
Operational: Northern Fleet: *Zadorny, Legky, Zharky*. Black Sea:
Ladny, Pytlivy. Baltic: *Pylky, Neukrotimy*. Pacific: *Letuchy*.

UPDATED

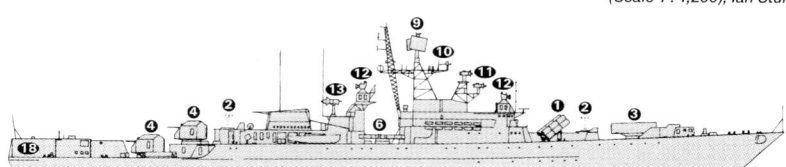

KRIVAK II

(Scale 1 : 1,200), Ian Sturton / 0506084

KRIVAK I (mod)

(Scale 1 : 1,200), Ian Sturton / 0506083

ZHARKY

6/2003 *, B Lemachko / 1042298

LADNY (I)

7/2000, Hartmut Ehlers / 0105545

PYTLIVY (II)

8/2003 / 0570894

1 + 1 GEPARD (PROJECT 11661) CLASS (FFGM)

Name	No	Builders	Laid down	Launched	Commissioned
TATARSTAN (ex-*Albatros*)	691	Zelenodolsk, Kazan, Tartarstan	15 Sep 1992	July 1993	12 July 2002
DAGESTAN (ex-*Burevestnik*)	—	Zelenodolsk, Kazan, Tartarstan			2005

Displacement, tons: 1,560 standard; 1,930 full load
Dimensions, feet (metres): 334.6 × 44.6 × 14.4
(102 × 13.6 × 4.4)
Main machinery: CODOG; 2 gas turbines; 30,850 hp(m)
(23.0 MW); 1 Type 61D diesel; 7,375 hp(m) *(5.5 MW)*; 2
shafts; cp props
Speed, knots: 26 (18 on diesels)
Range, n miles: 5,000 at 10 kt
Complement: 110 (accommodation for 131)

Missiles: SSM: 8 Zvezda SS-N-25 (KH 35 Uran) (2 quad) ❶; IR or
radar homing to 130 km *(70.2 n miles)* at 0.9 Mach; warhead
145 kg; sea-skimmer.
SAM: 1 SA-N-4 Gecko twin launcher ❷; semi-active radar homing
to 15 km *(8 n miles)* at 2.5 Mach; warhead 50 kg. 20
weapons.
Guns: 1—3 in *(76 mm)*/60 AK-176 ❸; 120 rds/min to 15 km *(8 n
miles)*; weight of shell 5.9 kg.
2—30 mm/65 AK-630 ❹; 6 barrels per mounting; 3,000 rds/
min combined to 2 km.
Torpedoes: 4—21 in *(533 mm)* (2 twin) tubes ❺ (probably not
fitted).
A/S mortars: 1 RBU 6000 12-tubed trainable ❻ (probably not
fitted).
Mines: 2 rails. 48 mines.
Countermeasures: Decoys: 4 PK 16 chaff launchers.
ESM/ECM: 2 Bell Shroud. 2 Bell Squat. Intercept and jammers.
Weapons control: 2 Light Bulb datalink. Hood Wink and Odd Box
optronic systems. Band Stand ❼ datalink.
Radars: Air/surface search: Cross Dome ❽; E/F-band.
Fire control: Bass Tilt ❾; H/I-band (for guns). Pop Group ❿; F/H/
I-band (for SAM). Garpun-B (for SSM); I/J-band.
Navigation: Nayada; I-band.
IFF: 2 Square Head. 1 Salt Pot B.
Sonars: Ox Yoke; hull-mounted; active search and attack;
medium frequency (probably not fitted).
Ox Tail (probably not fitted); VDS; active search and attack;
medium frequency.

Programmes: Intended as a successor to the Koni class, the
Gepard family of ships, of which there were some five
variants, was developed with export in mind. The first of class
Yastreb was laid down in 1988 but was later broken up in

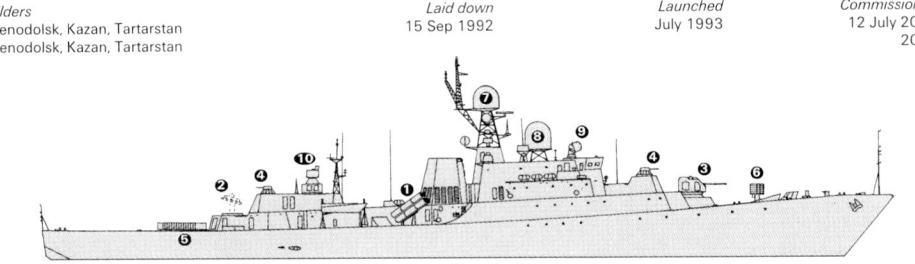

TATARSTAN   *(Scale 1 : 900), Ian Sturton /* 1042094

TATARSTAN   *7/2002, Military Parade /* 0528304

1992. The second and third of class were to have been
exported abroad but, following the completion of *Tatarstan*
for the Russian Navy, a second ship *Dagestan* (ex-*Burevestnik*)
is expected to follow in 2005.

Operational: Flagship of the Caspian Flotilla, the newly
commissioned *Tatarstan* took part in the large Caspian naval
exercise in August 2002.
UPDATED

11 PARCHIM II CLASS (PROJECT 1331) (FFLM)

MPK 67	301	**MPK 205**	311	**MPK 227**	243
MPK 99	308	**MPK 213**	222	**MPK 228**	244
MPK 105	245	**MPK 216**	258	**MPK 229**	232
MPK 192	304	**MPK 224**	218		

Displacement, tons: 769 standard; 960 full load
Dimensions, feet (metres): 246.7 × 32.2 × 14.4
(75.2 × 9.8 × 4.4)
Main machinery: 3 Type M 504A diesels; 10,812 hp(m)
(7.95 MW) sustained; 3 shafts
Speed, knots: 26. **Range, n miles:** 2,500 at 12 kt
Complement: 70 (8 officers)

Missiles: SAM: 2 SA-N-5 Grail quad launchers ❶; manual aiming;
IR homing to 6 km *(3.2 n miles)* at 1.5 Mach; altitude to
2,500 m *(8,000 ft)*; warhead 1.5 kg.
Guns: 1—3 in *(76 mm)*/66 AK 176 ❷; 120 rds/min to 12 km
(6.4 n miles); weight of shell 5.9 kg.
1—30 mm/65 AK 630 ❸; 6 barrels; 3,000 rds/min combined
to 2 km.
Torpedoes: 4—21 in *(533 mm)* (2 twin) tubes ❹. Combination of
53 cm torpedoes (see table at front of section).
A/S mortars: 2 RBU 6000 12-tubed trainable ❺; range 6,000 m;
warhead 31 kg. 96 weapons.
Depth charges: 2 racks.
Mines: Rails fitted.
Countermeasures: Decoys: 2 PK 16 chaff launchers.
ESM: 2 Watch Dog; intercept.
Weapons control: Hood Wink and Odd Box optronic systems.
Radars: Air/surface search: Cross Dome ❻; E/F-band.
Navigation: TSR 333 or Nayala or Kivach III; I-band.
Fire control: Bass Tilt ❼; H/I-band.
IFF: High Pole A.

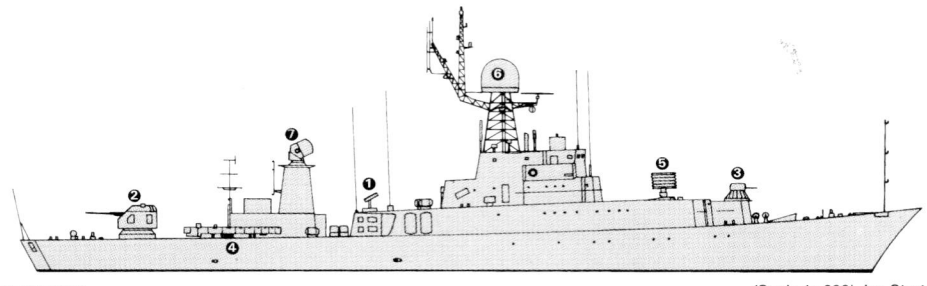

PARCHIM II   *(Scale 1 : 600), Ian Sturton*

Sonars: Bull Horn; hull-mounted; active search and attack;
medium frequency.
Lamb Tail; helicopter type VDS; high frequency.

Programmes: Built in the GDR at Peenewerft, Wolgast for the
USSR. First one commissioned 19 December 1986 and the
last on 6 April 1990. MPK 228 is named *Bashkortostan*, MPK
229 is named *Kalmykia*, MPK 224 is named *Yunga* and MPK
205 *Kazanets*.

Structure: Similar design to the ex-GDR Parchim I class now
serving with the Indonesian Navy but some armament
differences.
Operational: All operate in the Baltic and are based at Baltiysk or
Kronshtadt. All of the class refitted at Rostock in 1994-95.
MPK 228 damaged by fire in 1999 but has been repaired.
UPDATED

MPK 229 KALMYKIA   *6/2004 *, J Cislak /* 1042323

0 + 2 (2) STEREGUSHCHIY CLASS (PROJECT 20380) (FFGHM)

Name	No	Builders	Laid down	Launched	Commissioned
STEREGUSHCHIY	—	Severnaya, St Petersburg	21 Dec 2001	2005	2007
SOOBRAZITELNY	—	Severnaya, St Petersburg	20 May 2003	2007	2008

Displacement, tons: 1,900 full load
Dimensions, feet (metres): 366.2 × 45.9 × 12.1 (111.6 × 14.0 × 3.7)
Main machinery: CODOG: 2 gas turbines; 20,000 hp (14.9 MW); 2 diesels; 7,300 hp (5.44 MW); 2 shafts
Speed, knots: 26. **Range, n miles:** 4,000 at 14 kt

Missiles: SSM: 8 Zvezda SS-N-25 (KH 35 Uran) (2 quad) ❶; active radar homing to 130 km (70.2 n miles) at 0.9 Mach; warhead 145 kg; sea skimmer. VLS Silo.
SAM: 2 CADS-N-1 (Kashtan) ❷; each with twin 30 mm Gatling; combined with 8 SA-N-11 (Grisson) and Hot Flash/Hot Spot fire-control radar/optronic director. Laser beam guidance for

missiles to 8 km (4.4 n miles); warhead 9 kg; 9,000 rds/min combined to 1.5 km for guns.
A/S: Medvedka (SS-N-29); inertial flight to 25 km (13.5 n miles); payload Type 40 torpedo.
Guns: 1—100 mm A-190 ❸. 2—14.5 mm MGs.
Countermeasures: Decoys/ESM/ECM.
Radars: Air/surface search: Top Plate; 3D ❹; E/F-band.
Fire control: Plank Shave (Granit Harpun B (for SSM)); I-band.
Fire control: Kite Screech; H/I/K-band (for 100 mm gun)
Sonars: Bow fitted.

Helicopters: Platform for 1 Ka-27 Helix ❺.

Programmes: Multipurpose frigate designed to replace the Grisha class. Variants to be offered for export under Projects 20382 and 20383. First of class to commission in 2007 with follow on ships at annual intervals. The first batch consists of four ships, a reduction from the ten initially envisaged.
Structure: Not all details are known but measures to reduce radar cross section are reported.

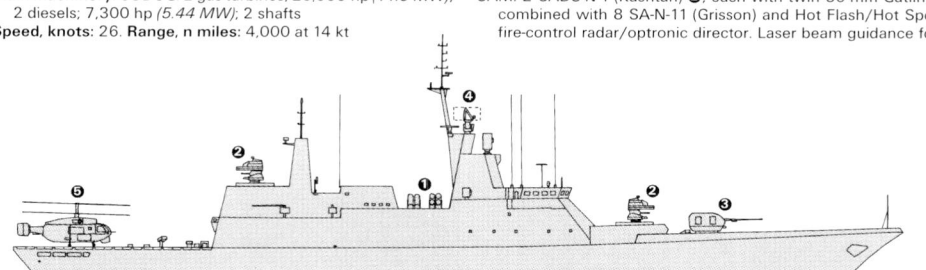

STEREGUSHCHIY   (Scale 1 : 900), Ian Sturton / 0569928   *UPDATED*

23 GRISHA (ALBATROS) (PROJECT 1124/1124M/1124K/1124EM) CLASS (FFLM)

North
MPK 7 164
MPK 14 190
MPK 56 155
MPK 59 196
MPK 113 171
MPK 130 138
MPK 139 129
MPK 194 199

Pacific
MPK 17 362
MPK 28 396
MPK 64 323
MPK 82 375
MPK 107 332
MPK 191 (III) 369
MPK 214 350
MPK 221 354
MPK 222 390

Black Sea
MPK 49 (III) 059
SUZDALETS MPK 118 071
MPK 127 (III) 078
MUROMETS MPK 134 064
KASIMOV MPK 199 055
EISK MPK 217 054

Displacement, tons: 950 standard; 1,200 full load
Dimensions, feet (metres): 233.6 × 32.2 × 12.1 (71.2 × 9.8 × 3.7)
Main machinery: CODAG; 1 gas-turbine; 15,000 hp(m) (11 MW); 2 diesels; 16,000 hp(m) (11.8 MW); 3 shafts
Speed, knots: 30. **Range, n miles:** 2,500 at 14 kt; 1,750 at 20 kt diesels; 950 at 27 kt
Complement: 70 (5 officers) (Grisha III); 60 (Grisha I)

Missiles: SAM: SA-N-4 Gecko twin launcher ❶; semi-active radar homing to 15 km (8 n miles) at 2.5 Mach; warhead 50 kg; altitude 9.1-3,048 m (30-10,000 ft); 20 missiles (see *Structure* for SA-N-9).
Guns: 2—57 mm/80 (twin) ❷; 120 rds/min to 6 km (3.3 n miles); weight of shell 2.8 kg.
1—3 in (76 mm)/60 (Grisha V) ❸; 120 rds/min to 15 km (8 n miles); weight of shell 7 kg.
1—30 mm/65 (Grisha III and V classes) ❹; 6 barrels; 3,000 rds/min combined to 2 km.
Torpedoes: 4—21 in (533 mm) (2 twin) tubes ❺. Combination of 53 cm torpedoes (see table at front of section).
A/S mortars: 2 RBU 6000 12-tubed trainable ❻; range 6,000 m; warhead 31 kg. (Only 1 in Grisha Vs.)
Depth charges: 2 racks (12).
Mines: Capacity for 18 in lieu of depth charges.
Countermeasures: Decoys: 4 PK 10 or 2 PK 16 chaff launchers. ESM: 2 Watch Dog.
Radars: Air/surface search: Strut Curve (Strut Pair in early Grisha Vs) ❼; F-band; range 110 km (60 n miles) for 2 m² target.
Half Plate Bravo (in later Grisha Vs); E/F-band.
Navigation: Don 2; I-band.
Fire control: Pop Group ❽; F/H/I-band (for SA-N-4). Bass Tilt (Grisha III and V) ❿; H/I-band (for 57/76 mm and 30 mm).
IFF: High Pole A or B. Square Head. Salt Pot.
Sonars: Bull Nose; hull-mounted; active search and attack; high/medium frequency.
Elk Tail; VDS ⓫; active search; high frequency. Similar to Hormone helicopter dipping sonar.

Programmes: Grisha III 1973-85 (two remaining); Grisha V 1982-1996 onwards (20 remaining). All were built at Kiev, Kharbarovsk and Zelenodolsk. Type name is *maly protivolodochny korabl* meaning small anti-submarine ship.
Structure: Grisha III class has Muff Cob radar removed, Bass Tilt and 30 mm ADG (fitted aft), and Rad-haz screen removed from abaft funnel as a result of removal of Muff Cob. Grisha V is similar to Grisha III with the after twin 57 mm mounting replaced by a single Tarantul type 76 mm gun.
Operational: Eight Grisha Vs are stationed in the Northern Fleet, one Grisha III and eight Vs in the Pacific and two IIIs and four Vs in the Black Sea. The modified Grisha III, known as Grisha IV, has been decommissioned.
Sales: Two Grisha III to Lithuania in November 1992. One Grisha V in 1994 and four Grisha II in 1996 to Ukraine.

UPDATED

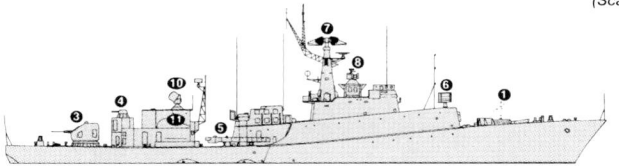

GRISHA III   (Scale 1 : 900), Ian Sturton

GRISHA V   (Scale 1 : 900), Ian Sturton

GRISHA III MPK 49   7/2000, Hartmut Ehlers / 0105546

GRISHA V MPK 17   10/2004 *, Ships of the World / 1042282

MPK 59   6/2003 *, B Lemachko / 1042304

SHIPBORNE AIRCRAFT

Notes: (1) A smaller variant of the Kamov Ka-60 is reported to have been offered to the Russian Navy. The Ka-40 anti-submarine helicopter has been under development since 1990. Loosely based on the 'Helix', it is likely to have a similar main rotor and tail-fin configuration.
(2) Haze B helicopters have all been placed in reserve as have all Ka-25 Hormones. The latter remain active and probably have a training role.

Numbers/Type: 17/2 Sukhoi Su-33 Flanker D/Su-33 UB.
Operational speed: 1,345 kt *(2,500 km/h).*
Service ceiling: 59,000 ft *(18,000 m).*
Range: 2,160 + n miles *(4,000 km).*
Role/Weapon systems: Fleet air defence fighter. 20 production aircraft delivered of which 3 have been lost. 10 are believed to be operational. All based in the Northern Fleet. Most training is done from a simulated flight deck ashore. Sensors: Track-while-scan pulse Doppler radar, IR scanner. Weapons: One 30 mm cannon, 10 AAMs (AA-12, AA-11, AA-8). *UPDATED*

FLANKER   *2/1996*

Numbers/Type: 4 Sukhoi Su-25UT Frogfoot UTG.
Operational speed: 526 kt *(975 km/h).*
Service ceiling: 22,965 ft *(7,000 m).*
Range: 675 n miles *(1,250 km).*
Role/Weapon systems: The UTG version is the two seater ground attack aircraft used for deck training in the carrier *Kuznetsov.* About 40 more of these aircraft are Air Force. Sensors: Laser rangefinder, ESM, ECM. Weapons: One 30 mm cannon, AAMs (AA-8), rockets, bombs. *VERIFIED*

FROGFOOT   *2/1996*

Numbers/Type: 2 Kamov Ka-31 Helix RLD.
Operational speed: 119 kt *(220 km/h).*
Service ceiling: 11,480 ft *(3,500 m).*
Range: 162 n miles *(300 km).*
Role/Weapon systems: AEW conversions with a solid-state radar under the fuselage. Four sold to India. Sensors: Oko E-801 Surveillance radar, datalinks. Weapons: Unarmed. *VERIFIED*

HELIX RLD   *9/1995*

Numbers/Type: 67/12/4 Kamov Ka-27PL Helix A/Ka-29 Helix B/Ka-32 Helix D.
Operational speed: 135 kt *(250 km/h).*
Service ceiling: 19,685 ft *(6,000 m).*
Range: 432 n miles *(800 km).*
Role/Weapon systems: ASW helicopter; three main versions—'A' for ASW, 'B' for assault and D for SAR; deployed to surface ships and some shore stations. Sensors: Osminog Splash Drop search radar, VGS-3 dipping sonar, sonobuoys, MAD, ESM. Weapons: ASW; three APR-2 torpedoes, nuclear or conventional S3V depth bombs or mines. Assault type: Two UV-57 rocket pods (2 × 32). *UPDATED*

HELIX A   *9/2004*, Hachiro Nakai /* 1042312

LAND-BASED MARITIME AIRCRAFT (FRONT LINE)

Notes: (1) The MiG-29 Fulcrum D has been abandoned by the Navy and the Ka-34 Hokum is not in production. Yak-41 Freestyle is not being developed but the prototype is for sale. Fitter C/D, Badgers and Bear D aircraft were out of service by 1995, Blinders and Bear G by 1997, and Mail and Haze A/C by 1999 (except for three still active in the Black Sea Fleet).
(2) Tu-204P has been proposed as an ASW/reconnaissance aircraft to replace the 'May'. It would be a development of the commercial transport aircraft.

Numbers/Type: 2 Ilyushin Il-20 Coot A.
Operational speed: 364 kt *(675 km/h).*
Service ceiling: 32,800 ft *(10,000 m).*
Range: 3,508 n miles *(6,500 km).*
Role/Weapon systems: Long-range Elint and MR for naval forces' intelligence gathering. Sensors: SLAR, weather radar, cameras, Elint equipment. Weapons: Unarmed. *UPDATED*

COOT A   *6/1995, Sergey Sergeyev /* 0574163

Numbers/Type: 26 Ilyushin Il-38 May.
Operational speed: 347 kt *(645 km/h).*
Service ceiling: 32,800 ft *(10,000 m).*
Range: 3,887 n miles *(7,200 km).*
Role/Weapon systems: Long-range MR and ASW. 12 in the North, 14 in the Pacific. Test flights of an upgraded version started in 2002 and continued in 2003. Sensors: Wet Eye search/weather radar, MAD, sonobuoys. Weapons: ASW; internal storage for 6 tons weapons. *UPDATED*

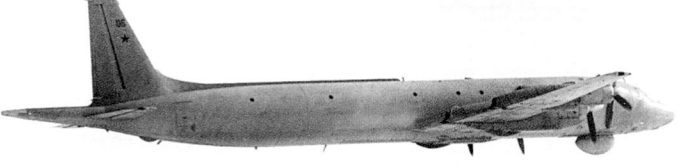

MAY   *6/1999 /* 0081667

Numbers/Type: 45 Tupolev Tu-22 M Backfire C.
Operational speed: 2.0 Mach.
Service ceiling: 60,000 ft *(18,300 m).*
Range: 2,500 n miles *(4,630 km).*
Role/Weapon systems: Medium-range nuclear/conventional strike and reconnaissance. About 20 are operational. Sensors: Down Beat search/Fan Tail attack radars, EW. Weapons: ASV; 12 tons of 'iron' bombs or standoff missiles AS-4 Kitchen (Kh 22N(A)) and AS-6 Kickback (Kh 15P). Self-defence; two 23 mm cannon. *VERIFIED*

BACKFIRE   *6/2003, Paul Jackson /* 0547316

Numbers/Type: 5 Antonov An-12 Cub ('Cub B/C/D') ('Cub C/D' ECM/ASW).
Operational speed: 419 kt *(777 km/h)*.
Service ceiling: 33,500 ft *(10,200 m)*.
Range: 3,075 n miles *(5,700 km)*.
Role/Weapon systems: Used either for intelligence gathering (B) or electronic warfare (C, D); is versatile with long range. Sensors: Search/weather radar, three EW blisters (B), tail-mounted EW/Elint equipment in addition (C/D). Weapons: Self-defence; two 23 mm cannon (B and D only). *VERIFIED*

Numbers/Type: 20/12 Tupolev Tu-142 Bear F/Bear J.
Operational speed: 500 kt *(925 km/h)*.
Service ceiling: 60,000 ft *(18,300 m)*.
Range: 6,775 n miles *(12,550 km)*.
Role/Weapon systems: Multimission long-range aircraft (ASW and communications variants). 36 in the North, remainder Pacific. Sensors: Wet Eye search radar, ESM; search radar, sonobuoys, EW, MAD (F), ELINT systems (J). The Bear J is reported to be equipped with VLF communications for SSBN connectivity. Weapons: ASW; various torpedoes, depth bombs and/or mines (F). Self-defence; some have two 23 mm or more cannon. *VERIFIED*

Numbers/Type: 40 Sukhoi Su-24 Fencer D/E.
Operational speed: 1.15 Mach.
Service ceiling: 57,400 ft *(17,500 m)*.
Range: 950 n miles *(1,755 km)*.
Role/Weapon systems: Fitted for maritime reconnaissance (19) and strike (27). Sensors: Radar and EW. Weapons: 30 mm Gatling gun; various ASM missiles and bombs; some have 23 mm cannon. *VERIFIED*

FENCER E   *6/1999, Jane's /* 0048910

BEAR 'J'   *6/1994, G Jacobs /* 0564718

CORVETTES

15 NANUCHKA III (VETER) (PROJECT 1234.1) (FSGM) and 1 NANUCHKA IV (NAKAT) (PROJECT 1234.7) CLASSES (FSG)

North	Baltic	Pacific	Black Sea
URAGAN 505	LIVEN 551	MOROZ 409	SHTYL 620
RASSVET 520	GEYZER 555	RAZLIV 450	MIRAZH 617
TUSHA 533	ZYB 560	SMERCH 423	
AYSBERG 535	PASSAT 570		
NAKAT (IV) 526	METEOR 590		

Displacement, tons: 660 full load
Dimensions, feet (metres): 194.5 × 38.7 × 8.5
(59.3 × 11.8 × 2.6)
Main machinery: 6 M 504 diesels; 26,112 hp(m) *(19.2 MW)*; 3 shafts
Speed, knots: 33. **Range, n miles:** 2,500 at 12 kt; 900 at 31 kt
Complement: 42 (7 officers)

Missiles: SSM: 6 Chelomey SS-N-9 Siren (Malakhit) (2 triple) launchers ❶; command guidance and IR and active radar homing to 110 km *(60 n miles)* at 0.9 Mach; warhead nuclear 250 kT or HE 500 kg. Nanuchka IV has 2 sextuple launchers for trials of SS-NX-26; radar homing to 300 km *(161.9 n miles)* at Mach 2-3.5.
SAM: SA-N-4 Gecko twin launcher ❷; semi-active radar homing to 15 km *(8 n miles)* at 2.5 Mach; warhead 50 kg; altitude 9.1-3,048 m *(30-10,000 ft)*; 20 missiles. Some anti-surface capability.
Guns: 1—3 in *(76 mm)*/60 ❸; 120 rds/min to 15 km *(8 n miles)*; weight of shell 7 kg.
1—30 mm/65 ❹; 6 barrels; 3,000 rds/min combined to 2 km.
Countermeasures: Decoys: 4 PK 10 chaff launchers ❺.
ESM: Foot Ball and Half Hat A and B. 4 Half Cup laser warners.
Weapons control: 2 Bell Nest or Light Bulb (datalinks). Band Stand ❻ datalink for SS-N-9.
Radars: Air/surface search: Peel Pair ❼; I-band or Plank Shave; E/F-band.
Fire control: Bass Tilt ❽; H/I-band. Pop Group ❾; F/H/I-band (for SA-N-4).
Navigation: Nayada; I-band.
IFF: High Pole. Square Head. Spar Stump. Salt Pot A and B.

Programmes: Built from 1969 onwards at Petrovsky, Leningrad and in the Pacific. Nanuchka III, first seen in 1978. Nanuchka IV completed in 1987 as a trials ship. Type name is *maly raketny korabl* meaning small missile ship.
Structure: The Nanuchka IV is similar in detail to Nanuchka III except that she is the trials vehicle for SS-NX-26.
Operational: Intended for deployment in coastal waters although formerly deployed in the Mediterranean (in groups of two or three), North Sea and Pacific.
UPDATED

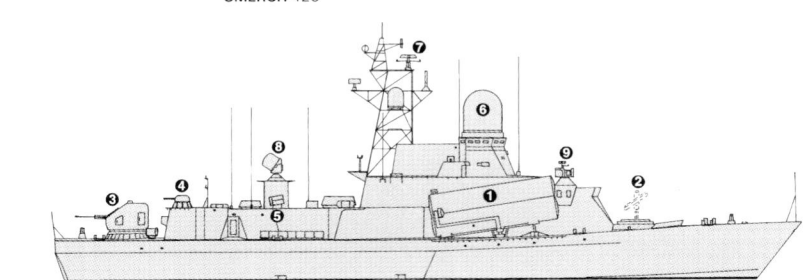

NANUCHKA III   *(Scale 1 : 600), Ian Sturton /* 0105552

NAKAT (NANUCHKA IV)   *8/2002 /* 0528321

AYSBERG
*6/2003 *, B Lemachko /* 1042303

2 DERGACH (SIVUCH) (PROJECT 1239) CLASS (PGGJM}

Name	No	Builders	Launched	Commissioned
BORA (MRK 27)	615	Zelenodolsk, Kazan	1987	20 May 1997
SAMUM (MRK 17)	616 (ex-575, ex-890)	Zelenodolsk, Kazan	1992	31 Dec 1995

Displacement, tons: 1,050 full load
Dimensions, feet (metres): 211.6 × 55.8 × 12.5
(64.5 × 17 × 3.8)
Main machinery: CODOG; 2 gas turbines; 55,216 hp(m)
(40.6 MW); 2 diesels; 10,064 hp(m) *(7.4 MW)*; 2 hydroprops;
2 auxiliary diesels; 2 props on retractable pods
Speed, knots: 53 foil; 12 hullborne
Range, n miles: 600 at 50 kt; 2,500 at 12 kt
Complement: 67 (8 officers)

Missiles: SSM: 8 SS-N-22 (2 quad) Sunburn (3M-82 Moskit)
launchers ❶; active radar homing to 160 km *(87 n miles)* at
2.5 Mach; warhead nuclear or 200 kT or HE 300 kg; sea-
skimmer.
SAM: SA-N-4 Gecko twin launcher ❷; semi-active radar homing
to 15 km *(8 n miles)* at 2.5 Mach; warhead 50 kg; 20 missiles.
Guns: 1—3 in *(76 mm)*/60 AK 176 ❸; 120 rds/min to 12 km
(6.4 n miles); weight of shell 7 kg.
2—30 mm/65 AK 630 ❹; 6 barrels per mounting; 3,000 rds/
min combined to 2 km.
Countermeasures: Decoys: 2 PK 16 and 2 PK 10 chaff launchers.
ESM/ECM: 2 Foot Ball A. 2 Half Hats.
Weapons control: 2 Light Bulb datalink ❺. Band Stand ❻
datalink for SS-N-22; Bell Nest.
Radars: Air/surface search: Cross Dome ❼; E/F-band.
Fire control: Bass Tilt ❽; H/I-band (for guns).
Pop Group ❾; F/H/I-band (for SAM).
Navigation: SRN-207; I-band.
IFF: Square Head. Salt Pot.

Programmes: Almaz design approved 24 December 1980.
Classified as a PGGA (Guided Missile Patrol Air Cushion
Vessels). Both did trials from 1989 (Bora) and 1993 *(Samum)*
before being accepted into service.
Structure: Twin-hulled surface effect design. The auxiliary
diesels are for slow speed operations.
Operational: The design was unreliable but efforts were made in
1996/97 to restore both to an operational state. Both based
at Sevastopol. SS-N-22 missiles were test-fired in April 2003.
Bora has a camouflaged hull and is based at Sevastopol but
operational status is doubtful.

UPDATED

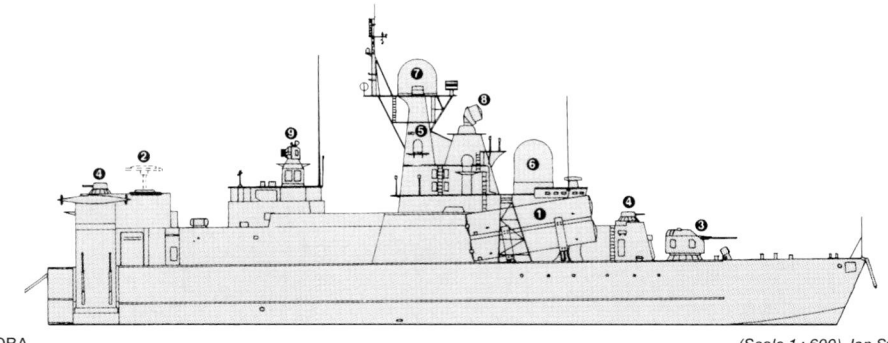

BORA   *(Scale 1 : 600), Ian Sturton*

BORA
6/2003 *, B Lemachko / 0580531

SAMUM   *6/2003 *, B Lemachko /* 0580532

24 TARANTUL (MOLNYA) (PROJECT 1241.1/ 1241.1M/1241.1MP/1242.1) CLASS (FSGM)

TARANTUL II		R 293	874	R 109	952
R 125	833	R 20	921	R 239	953
		R 29	916	R 334	954
TARANTUL III		R 14	924	R 60	955
R 42	819	R 104	927	R 298	971
R 291	825	R 18	937	R 19	978
R 129	852	R 11	940	—	991
R 187	855	R 24	946	—	994
R 2	870	R 297	951	R 79	995

Displacement, tons: 385 standard; 455 full load
Dimensions, feet (metres): 184.1 × 37.7 × 8.2 *(56.1 × 11.5 × 2.5)*
Main machinery: COGAG; 2 Nikolayev Type DR 77 gas turbines; 16,016 hp(m) *(11.77 MW)* sustained; 2 Nikolayev Type DR 76 gas turbines with reversible gearboxes; 4,993 hp(m) *(3.67 MW)* sustained; 2 shafts or CODOG with 2 CM 504 diesels; 8,000 hp(m) *(5.88 MW)*, replacing second pair of gas-turbines in Tarantul IIIs
Speed, knots: 36. **Range, n miles:** 400 at 36 kt; 1,650 at 14 kt
Complement: 34 (5 officers)

Missiles: SSM: 4 Raduga SS-N-2D Styx (2 twin) launchers (Tarantul II); active radar or IR homing to 83 km *(45 n miles)* at 0.9 Mach; warhead 513 kg; sea-skimmer at end of run.
4 Raduga SS-N-22 Sunburn (3M-82 Moskit) (2 twin) launchers (Tarantul III); active radar homing to 160 km *(87 n miles)* at 2.5 Mach; warhead nuclear 200 kT or HE 300 kg; sea-skimmer. Modified version in Type 1242.1.
SAM: SA-N-5 Grail quad launcher; manual aiming; IR homing to 6 km *(3.2 n miles)* at 1.5 Mach; altitude to 2,500 m *(8,000 ft)*; warhead 1.5 kg.
Guns: 1—3 in *(76 mm)*/60; 120 rds/min to 15 km *(8 n miles)*; weight of shell 7 kg.
2—30 mm/65; 6 barrels per mounting; 3,000 rds/min to 2 km.
Countermeasures: Decoys: 2 PK 16 or 4 PK 10 (Tarantul III) chaff launchers.
ESM: 2 Foot Ball, 2 Half Hat (in some).
Weapons control: Hood Wink optronic director. Light Bulb datalink. Band Stand; datalink for SSM; Bell Nest.
Radars: Air/surface search: Plank Shave or Positiv E (Tarantul 874); I-band.
Navigation: Kivach III; I-band.
Fire control: Bass Tilt; H/I-band.
IFF: Square Head. High Pole B.
Sonars: Foal Tail; VDS; active search; high frequency.

Programmes: Tarantul II were built at Kolpino, Petrovsky, Leningrad and in the Pacific in 1980-86. Production of Taruntul IIIs then continued until 1995 although a single experimental Tarantul III with four SS-N-22 had been completed at Petrovsky in 1981. One more was launched in September 1997 at Rybinsk, and a Tarantul III at Kolpino completed in December 1999 for the Baltic Fleet. Type name is *raketny kater* meaning missile cutter.
Modernisation: Tarantul III 874 served as a trials platform for a modified version of SS-N-22 with a longer range; the missile is distinguished by end caps on the launcher doors. Tarantul II 962 (now decommissioned) served as a trials platform for the CADS-N-1 point defence system in the Black Sea.
Structure: Basically same hull as Pauk class, without extension for sonar. The single Type 1242.1 has a Positiv E radar.
Operational: One Tarantul II (R 125) and six IIIs (R 129, R 42, R 291, R 187, R 2, R 293) are in the Baltic, one III (R 239) is in the Black Sea, one III in the Caspian, and 15 in the Pacific.
Sales: Tarantul I class-one to Poland 28 December 1983, second in April 1984, third in March 1988 and fourth in January 1989. One to India in April 1987, second in January 1988, third in December 1988, fourth in November 1989 and fifth in January 1990. Two to Yemen in November 1990 and January 1991. One to Romania in December 1990, two more in February 1992. One Tarantul II to Bulgaria in March 1990. Two Tarantul Is to Vietnam in 1996 and two more in 1999.

UPDATED

R 29   *9/2004*, Hachiro Nakai* / 1042313

R 18   *10/2002, Hachiro Nakai* / 0528322

MOD TARANTUL III   *6/1999* / 0081670

0 + 1 (9) SCORPION (PROJECT 12300) CLASS (FSGM)

Displacement, tons: 470 standard
Dimensions, feet (metres): 186.4 × 35.4 × 17.1 *(56.8 × 10.8 × 5.2)*
Main machinery: CODAG: 1 M-70FR gas turbine; 8,775 hp *(6.54 MW)*; 2 MTU 16V4000 M 90 diesels; 16,320 hp *(12.2 MW)*; 2 shafts; 1 hydrojet
Speed, knots: 38. **Range, n miles:** 1,500 at 12 kt
Complement: 37

Missiles: SSM: 8 NPOMash SS-N-26 (2 quad) 3M55 Oniks/Yakhont launchers ❶; inertial guidance and active/passive radar homing to 300 km *(160 n miles)* at 3.5 Mach; warhead 250 kg; sea skimmer (to be confirmed).
SAM/Guns: 1 CADS-N (Kortik/Kashtan) (3M87) ❷; twin 30 mm Gatling combined with 8 SA-N-11 (Grison) and Hot Flash/Hot Spot fire-control radar/optronic director. Laser beam guidance for missiles to 8 km *(4.4 n miles)*; warhead 9 kg; 9,000 rds/min to 1.5 km (for guns).
Guns: 1—3.9 in *(100 mm)* A 190 (L 59) ❸; 60 rds/min to 15 km *(8.2 n miles)*; weight of shell 16 kg.
Countermeasures: Decoys: 4 PK10 chaff launchers ❹. ESM.
Radars: Air/surface search: Positiv M1 ❺; I-band.
Fire control: Plank Shave (Garpun B) ❻; I/J-band.
Ratep 5P-10E Puma ❼; I-band (for 100 mm gun).

Programmes: Designed by ALMAZ, St Petersburg, as successor to the Tarantul class. First vessel laid down 5 June 2001 at Vympel shipyard, Rybinsk, with delivery planned for 2005. A class of 10 vessels was envisaged for the Russian Navy and further sales abroad but the status of the project is uncertain.
Structure: Designed with stealth features and equipped with stabilising automatic control system to reduce rolling.

UPDATED

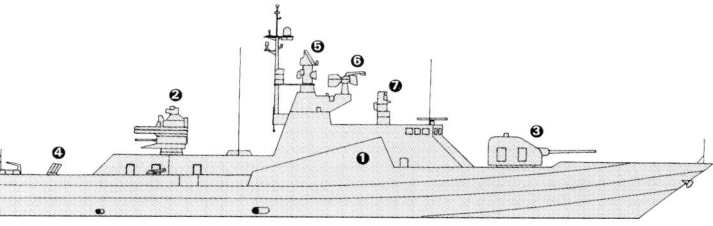

SCORPION   *(Scale 1 : 600), Ian Sturton* / 0569930

PATROL FORCES

Notes: The keel of a Project 21630 Buyan class gunboat of 500 tons was laid down at Almaz, St Petersburg, on 30 January 2004. It is expected to become operational in the Caspian Sea in late 2005. The first unit is to be named *Astrakhan* and there are to be at least four further units.

2 MATKA (VEKHR) CLASS (PROJECT 206MP) (FAST ATTACK CRAFT—MISSILE HYDROFOIL) (PGGK)

R 27 STUPINETS 701   R 44 KOSAR 966

Displacement, tons: 225 standard; 260 full load
Dimensions, feet (metres): 129.9 × 24.9 (41 over foils) × 6.9 (13.1 over foils) *(39.6 × 7.6 (12.5) × 2.1 (4))*
Main machinery: 3 Type M 504 diesels; 10,800 hp(m) *(7.94 MW)* sustained; 3 shafts
Speed, knots: 40. **Range, n miles:** 600 at 35 kt foilborne; 1,500 at 14 kt hullborne
Complement: 33

Missiles: SSM: 2 SS-N-2C/D Styx; active radar or IR homing to 83 km *(45 n miles)* at 0.9 Mach; warhead 513 kg; sea-skimmer at end of run.
8 SS-N-25 (in *966*); radar homing to 130 km *(70.2 n miles)* at 0.9 Mach; warhead 145 kg; sea-skimmer.
Guns: 1—3 in *(76 mm)*/60; 120 rds/min to 15 km *(8 n miles)*; weight of shell 7 kg.
1—30 mm/65 AK 630; 6 barrels per mounting; 3,000 rds/min to 2 km.
Countermeasures: Decoys: 2 PK 16 chaff launchers.
ESM: Clay Brick; intercept.
Weapons control: Hood Wink optronic directors.
Radars: Air/surface search: Plank Shave; E-band.
Navigation: SRN-207; I-band.
Fire control: Bass Tilt; H/I-band.
IFF: High Pole B or Salt Pot B and Square Head.

Programmes: In early 1978 the first of class was seen. Built at Kolpino Yard, Leningrad. Production stopped in 1983 being superseded by Tarantul class. Type name is *raketny kater* meaning missile cutter.
Structure: Similar hull to the deleted Osa class with similar single hydrofoil system to Turya class. The combination has produced a better sea-boat than the Osa class. *R 44* in the Black Sea was the trials craft for the SS-N-25.
Operational: *R 44* is based in the Black Sea and *R 27* is based in the Caspian. Five units transferred to Ukraine in 1996.

VERIFIED

KOSAR   *6/2003, B Lemachko* / 0570912

3 SVETLYAK (PROJECT 1041Z) CLASS (PGM)

Displacement, tons: 375 full load
Dimensions, feet (metres): 159.1 × 30.2 × 11.5 *(48.5 × 9.2 × 3.5)*
Main machinery: 3 diesels; 14,400 hp(m) *(10.58 MW)* sustained; 3 shafts
Speed, knots: 31. **Range, n miles:** 2,200 at 13 kt
Complement: 36 (4 officers)
Missiles: SAM: SA-N-5 Grail quad launcher; manual aiming; IR homing to 6 km *(3.2 n miles)* at 1.5
Mach; warhead 1.5 kg.
Guns: 1—3 in *(76 mm)*/60; 120 rds/min to 15 km *(8 n miles)*; weight of shell 7 kg.
1 or 2—30 mm/65 AK 630; 6 barrels; 3,000 rds/min combined to 2 km.
Torpedoes: 2—16 in *(406 mm)* tubes.
Depth charges: 2 racks; 12 charges.
Countermeasures: Decoys: 2 chaff launchers.
Weapons control: Hook Wink optronic director.
Radars: Air/surface search: Peel cone; E-band.
Fire control: Bass Tilt; H/I-band.
Navigation: Palm Frond B; I-band.
Sonars: Rat Tail; VDS; active search; high frequency.

Comment: Most of this class are in service with the Federal Border Guard but three were reported
to have been allocated to the navy.
VERIFIED

SVETLYAK 040   *10/2002, B Lemachko /* 0570904

1 MUKHA (SOKOL) (PROJECT 1145) CLASS
(FAST ATTACK CRAFT—PATROL HYDROFOIL) (PGK)

MPK 220 VLADIMIRETS 060

Displacement, tons: 400 full load
Dimensions, feet (metres): 164 × 27.9 (33.5 over foils) × 13.1 (19.4 foils)
(50 × 8.5 (10.2.) × 4 (5.9))
Main machinery: CODOG; 2 Type NK-12M gas turbines; 23,046 hp(m) *(16.95 MW)* sustained;
2 diesels; 2,400 hp(m) *(1.76 MW)*; 2 shafts
Speed, knots: 40; 12 hullborne
Complement: 45

Guns: 1—3 in *(76 mm)*/60; 120 rds/min to 15 km *(8 n miles)*; weight of shell 7 kg.
2—30 mm/65 AK 630; 6 barrels per mounting; 3,000 rds/min combined to 2 km.
Torpedoes: 8—16 in *(406 mm)* (2 quad) tubes. SAET-40; anti-submarine; active/passive homing to
10 km *(5.4 n miles)* at 30 kt; warhead 100 kg.
Countermeasures: Decoys: 2 PK 16 chaff launchers.
ESM: Radar warning.
Radars: Surface search: Peel Cone; E-band.
Navigation: SRN 206; I-band.
Fire control: Bass Tilt; H/I-band.
Sonars: Foal Tail; VDS; active search; high frequency.

Programmes: Built in 1986 at Feodosuja.
Structure: Features include a hydrofoil arrangement with a single fixed foil forward, large gas-
turbine exhausts aft, and trainable torpedo mountings.
Operational: Based in the Black Sea and probably soon to be paid off. The only ship of the class,
which was used as a trials platform for the Medveka ASW guided weapon, is based in the Black
Sea.
UPDATED

VLADIMIRETS   *9/2004*, Hartmut Ehlers /* 1042314

MINE WARFARE FORCES

Notes: (1) All remaining Yevgenya (Korond) class MHCs were laid up by 2001, except for two in the
Caspian Sea which may still be used as patrol craft.
(2) Some 40 to 50 craft of various dimensions, some with cable reels, some self-propelled and
unmanned, some towed and unmanned are reported including the 8 m Kater and Volga unmanned
mine clearance craft.

10 NATYA I (AKVAMAREN) (PROJECT 266M) CLASS
(MINESWEEPERS—OCEAN) (MSOM)

KOVROVETS 913	MT-265 718	TURBINIST 912
MOTORIST 806	KONTRADMIRAL VLASOV 855	SNAYPER 919
VALENTIN PIKUL 770	MT-264 738	
SVYAZIST 610	VITSEADMIRAL ZACHARIN	

Displacement, tons: 804 full load
Dimensions, feet (metres): 200.1 × 33.5 × 9.8 *(61 × 10.2 × 3)*
Main machinery: 2 Type M 504 diesels; 5,000 hp(m) *(3.67 MW)* sustained; 2 shafts; cp props
Speed, knots: 16. **Range, n miles:** 3,000 at 12 kt
Complement: 67 (8 officers)

Missiles: SAM: 2 SA-N-5/8 Grail quad launchers (in some); manual aiming; IR homing to 6 km
(3.2 n miles) at 1.5 Mach; altitude to 2,500 m *(8,000 ft)*; warhead 1.5 kg; 18 missiles.
Guns: 4—30 mm/65 (2 twin) AK 230; 500 rds/min to 6.5 km *(3.5 n miles)*; weight of shell 0.54 kg
or 2—30 mm/65 AK 306; 6 barrels per mounting; 3,000 rds/min combined to 2 km.
4—25 mm/80 (2 twin); 270 rds/min to 3 km *(1.6 n miles)*; weight of shell 0.34 kg.
A/S mortars: 2 RBU 1200 5-tubed fixed; range 1,200 m; warhead 34 kg.
Depth charges: 62.
Mines: 10.
Countermeasures: MCM: 1 or 2 GKT-2 contact sweeps; 1 AT-2 acoustic sweep; 1 TEM-3 magnetic
sweep.
Radars: Surface search: Don 2 or Long Trough; I-band.
Fire control: Drum Tilt; H/I-band (not in all).
IFF: 2 Square Head. High Pole B.
Sonars: MG 79/89; hull-mounted; active minehunting; high frequency.

Programmes: First reported in 1970. Built at Kolpino and Khabarovsk. Type name is *morskoy
tralshchik* meaning seagoing minesweeper. MT 264 and MT 265 were a new variant
commissioned in 1989 in which the AK 306 mounts replaced the twin AK 230 mounts. Two
further units, known as Natya III started construction in 1994. The first, *Valentin Pikul*, left St
Petersburg for the Black Sea in July 2002 while the second, *Vitseadmiral Zacharin*, was
launched at the Kolpino Yard in June 2002.
Structure: Some have hydraulic gantries aft. Have aluminium/steel alloy hulls. Some have Gatling
30 mm guns and a different radar configuration without Drum Tilt.
Operational: Usually operate in home waters but have deployed to the Mediterranean, Indian
Ocean and West Africa. Sweep speed is 14 kt. Of the remaining operational units, three are
based in the North, three in the Pacific, two in the Baltic and two in the Black Sea.
Sales: India (two in 1978, two in 1979, two in 1980, one in August 1986, two in 1987, three in
1988). Libya (two in 1981, two in February 1983, one in August 1983, one in January 1984, one
in January 1985, one in October 1986). Syria (one in 1985). Yemen (one in 1991). Ethiopia (one
in 1991). Some have been deleted.
UPDATED

VALENTIN PIKUL   *6/2003, B Lemachko /* 0570913

TURBINIST   *9/2004*, Hartmut Ehlers /* 1042305

For details of the latest updates to *Jane's Fighting Ships* online and to discover the
additional information available exclusively to online subscribers please visit
jfs.janes.com

2 GORYA (TYPE 12660) CLASS
(MINEHUNTERS—OCEAN) (MHOM)

Name	No	Builders	Laid down	Launched	Commissioned
A ZHELEZNYAKOV	901	Kolpino Yard, Leningrad	28 Feb 1985	17 July 1986	30 Dec 1988
V GUMANENKO	762	Kolpino Yard, Leningrad	15 Sep 1985	4 Mar 1991	9 Jan 1994

Displacement, tons: 1,130 full load
Dimensions, feet (metres): 216.5 × 36.1 × 10.8 *(66.0 × 11.0 × 3.3)*
Main machinery: 2 M 503 diesels; 5,000 hp(m) *(3.7 MW)*; 2 shafts
Speed, knots: 15. **Range, n miles:** 3,000 at 12 kt
Complement: 65 (7 officers)

Missiles: 2 SA-N-5 Grail quad launchers; IR homing to 6 km *(3.2 n miles)* at 1.5 Mach.
Guns: 1—3 in *(76 mm)*/60 AK 176; 120 rds/min to 12 km *(6.4 n miles)*; weight of shell 7 kg.
 1—30 mm/65 AK 630; 6 barrels; 3,000 rds/min to 2 km.
Countermeasures: Decoys: 2 PK 16 chaff launchers.
ESM: Cross Loop; Long Fold.
Radars: Surface search: Palm Frond; I-band.
Navigation: Nayada; I-band.
Fire control: Bass Tilt; H/I-band.
IFF: Salt Pot C. 2 Square Head.
Sonars: Hull-mounted; active search; high frequency.

Programmes: A third of class was started but has been scrapped.
Structure: Appears to carry mechanical, magnetic and acoustic sweep gear and may have accurate positional fixing equipment. A remote-controlled submersible is housed behind the sliding doors in the superstructure below the AK 630 mounting. Two 406 mm torpedo tubes are reported as used for mine countermeasures.
Operational: 901 is based in the Black Sea. 762 transferred from the Baltic to the Northern Fleet in 2000.

UPDATED

V GUMANENKO 8/2002 / 0528319

22 SONYA (YAKHONT) (PROJECT 12650/1265M) CLASS
(MINESWEEPERS—HUNTERS/COASTAL) (MHSC/MHSCM)

North		Pacific		Baltic		Caspian	
—	402	BT 245	533	BT 212	501	BT 88	107
BT 111	411	BT 96	554	—	505		
BT 22	418	BT 132	580	BT 230	510		
BT 31	425	BT 114	592	BT 213	522		
BT 21	433	BT 215	593	BT 115	561		
—	436			BT 44	563		
BT 152	443			BT 94	568		
—	454			BT 262	577		

Displacement, tons: 450 full load
Dimensions, feet (metres): 157.4 × 28.9 × 6.6 *(48 × 8.8 × 2)*
Main machinery: 2 Kolomna Type 9-D-8 diesels; 2,000 hp(m) *(1.47 MW)* sustained; 2 shafts
Speed, knots: 15. **Range, n miles:** 3,000 at 10 kt
Complement: 43 (5 officers)
Missiles: SAM: 2 quad SA-N-5 launchers (in some).
Guns: 2—30 mm/65 AK 630 or 2—30 mm/65 (twin) and 2—25 mm/80 (twin).
Mines: 8.
Radars: Surface search: Don 2 or Kivach or Nayada; I-band.
IFF: 2 Square Head. High Pole B.
Sonars: MG 69/79; hull-mounted; active minehunting; high frequency.

Comment: Wooden hull with GRP sheath. Built at about two a year at Petrozavodsk and at Ulis, Vladivostok (Pacific). First reported 1973 and the last one commissioned in January 1995. Type name is *bazovy tralshchik* meaning base minesweeper. Some have two twin 30 mm Gatling guns, others one 30 mm/65 (twin) plus one 25 mm (twin). In addition there are a further 50 in reserve.
 Transfers: Bulgaria, four in 1981-85. Cuba, four in 1980-85. Syria, one in 1986. Vietnam, one in February 1987, one in February 1988, one in July 1989 and one in February 1990. Ethiopia, one in February 1991. Ukraine, two in 1996.

VERIFIED

BT 115 6/2003, Guy Toremans / 0570933

9 LIDA (SAPFIR) (PROJECT 10750) CLASS
(MINEHUNTERS—COASTAL) (MHC)

RT 249	206	RT 252	239	RT 341	331
RT 273	210	RT 231	302	RT 248	348
RT 233	219	RT 57	316	RT 234	372

Displacement, tons: 135 full load
Dimensions, feet (metres): 103.3 × 21.3 × 5.2 *(31.5 × 6.5 × 1.6)*
Main machinery: 3 D12MM diesels; 900 hp(m) *(690 kW)*; 3 shafts
Speed, knots: 12. **Range, n miles:** 650 at 10 kt
Complement: 14 (1 officer)
Guns: 1—30 mm/65 AK 630; 6 barrels; 3,000 rds/min to 2 km.
Countermeasures: MCM: AT-6 acoustic, SEMT-1 magnetic and GKT-3M wire sweeps.
Radars: Surface search: Pechora; MR241; I-band.
Sonars: Kabarga I; minehunting; high frequency.

Comment: Type name *Reydnyy Tralshchik* meaning roadstead minesweeper. A follow-on to the Yevgenya class started construction in 1989 at Kolpino Yard, St Petersburg. Similar in appearance to Yevgenya. Building rate was about three a year to 1992 and then slowed to one a year until 1995. Some are painted a blue/grey colour. All are in the Baltic except *RT 233* which is in the Caspian.

VERIFIED

LIDA (SAPHIR) CLASS 6/1999, S Breyer / 0528317

AMPHIBIOUS FORCES

6 POLNOCHNY B CLASS (PROJECT 771) (LSM)

578

Displacement, tons: 760 standard; 834 full load
Dimensions, feet (metres): 246.1 × 31.5 × 7.5 *(75 × 9.6 × 2.3)*
Main machinery: 2 Kolomna Type 40-D diesels; 4,400 hp(m) *(3.2 MW)* sustained; 2 shafts
Speed, knots: 19. **Range, n miles:** 1,000 at 18 kt
Complement: 40-42
Military lift: 180 troops; 350 tons including 6 tanks
Missiles: SAM: 4 SA-N-5 Grail quad launchers.
Guns: 2 or 4—30 mm (1 or 2 twin).
 2—140 mm WM-18 rocket launchers; 18 barrels.
Weapons control: PED-1 system.
Radars: Surface search: Spin Trough; I-band.
Fire control: Drum Tilt; H/I-band (for 30 mm guns).
IFF: High Pole A. Square Head.

Comment: Built at Northern Shipyard, Gdansk, Poland in 1970. 578 serves in the Northern Fleet as a logistic support ship. Five have been reported active in the Caspian while others of the class are in reserve, derelict or scrapped. Tank deck 45.7 × 5.2 m *(150 × 17 ft)*. *UPDATED*

POLNOCHNY 7/1996, van Ginderen Collection / 0019047

2 ONDATRA (AKULA) (PROJECT 1176) CLASS (LCMs)

DKA 286 713 DKA 325 799

Displacement, tons: 145 full load
Dimensions, feet (metres): 78.7 × 16.4 × 4.9 *(24 × 5 × 1.5)*
Main machinery: 2 diesels; 300 hp(m) *(220 kW)*; 2 shafts
Speed, knots: 10. **Range, n miles:** 500 at 5 kt
Complement: 5
Military lift: 1 MBT

Comment: First completed in 1979 and associated with *Ivan Rogov*. 33 deleted so far. Tank deck of 45 × 13 ft. Two to Yemen in 1983. These last two are in the Baltic. *VERIFIED*

ONDATRA 7/1996, B Lemachko / 0570917

1 IVAN ROGOV (YEDNOROG) (PROJECT 1174) CLASS (LPDHM)

Name	*No*	*Builders*	*Launched*	*Commissioned*
MITROFAN MOSKALENKO	020	Yantar, Kaliningrad	July 1989	Mar 1991

Displacement, tons: 8,260 standard; 14,060 full load
Dimensions, feet (metres): 516.7 × 80.2 × 21.2 (27.8 flooded)
  (157.5 × 24.5 × 6.5 (8.5))
Main machinery: 2 M8K gas-turbines; 39,998 hp(m) *(29.4 MW)*;
  2 shafts
Speed, knots: 19. **Range, n miles:** 7,500 at 14 kt
Complement: 239 (37 officers)
Military lift: 522 troops (battalion); 20 tanks or equivalent weight
  of APCs and trucks; 3 ACVs or 6 LCM in docking bay

Missiles: SAM: SA-N-4 Gecko twin launcher ❶; semi-active radar
  homing to 15 km *(8 n miles)* at 2.5 Mach; warhead 50 kg;
  altitude 9.1-3,048 m *(30-10,000 ft)*; 20 missiles.
  2 SA-N-5 Grail quad launchers; manual aiming; IR homing to
  6 km *(3.2 n miles)* at 1.5 Mach; warhead 1.5 kg.
Guns: 2—3 in *(76 mm)*/60 (twin) ❷; 60 rds/min to 15 km *(8 n
  miles)*; weight of shell 6.8 kg.
  1—122 mm UMS-22 Grad-M; 2—40-barrelled rocket launcher;
  range 20 km *(10.8 n miles)*.
  4—30 mm/65 AK 630 ❸; 6 barrels per mounting; 3,000 rds/
  min combined to 2 km.
Countermeasures: Decoys: 16 PK 10 and 4 PK 16 chaff
  launchers.
  ESM: 3 Bell Shroud; intercept.
  ECM: 2 Bell Squat; jammers.
Weapons control: 2 Squeeze Box optronic directors ❹.
Radars: Air/surface search: Top Plate A ❺; 3D; E-band.
  Navigation: 2 Don Kay or 2 Palm Frond; I-band.
  Fire control: Owl Screech ❻; G-band (for 76 mm). 2 Bass Tilt ❼;
  H/I-band (for 30 mm). Pop Group ❽; F/H/I-band (for SA-N-4).
  CCA: Fly Screen ❾ and Fly Spike; I-band.
  IFF: Salt Pot B.
  Tacan: 2 Round House ❿.
Sonars: Mouse Tail VDS; active search; high frequency.

Helicopters: 4 Ka-29 Helix B ⓫.

Programmes: This was the third of class. Fourth of class was not
  completed. Type name is *bolshoy desantny korabl* meaning
  large landing ship.

MITROFAN MOSKALENKO   *10/1996 /* 0019045

Structure: Has bow ramp with beaching capability leading from
  a tank deck 200 ft long and 45 ft wide. Cargo capacity
  2,500 tons Stern doors open into a docking bay 250 ft long
  and 45 ft wide. A helicopter spot forward has a flying-control
  station and the after helicopter deck and hangar is similarly
  fitted. Helicopters can enter the hangar from both front and
  rear. Positions arranged on main superstructure for
replenishment of both fuel and solids. Has been reported
streaming a VDS from the stern door and has also conducted
unidentified missile trials.
Operational: Based in the Northern Fleet at Severomorsk. Two
Pacific Fleet units were paid off in 1996 and 1997. The refit
and sale of one of these ships to Indonesia was not concluded.
UPDATED

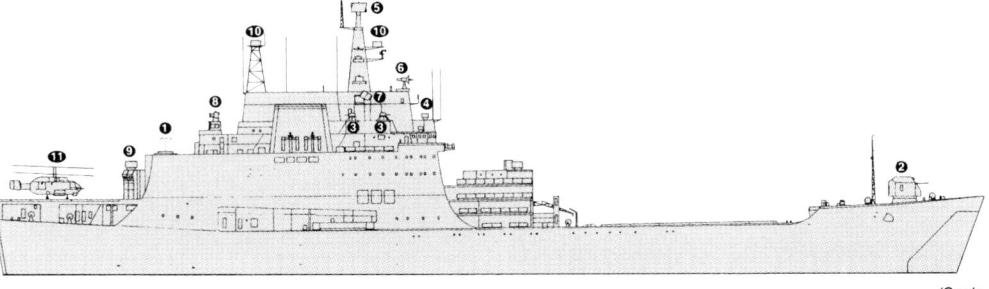

MITROFAN MOSKALENKO

(Scale 1 : 1,200), Ian Sturton

17 ROPUCHA (PROJECT 775/775M) CLASS (LSTM)

North:	OLENEGORSKIY GORNIAK	012	GEORGIY POBEDONOSETS	016	KONDOPOGA	027	ALEXANDER OTRAKOVSKIY	031		
Baltic:	KALININGRAD	102	ALEXANDER SHABALIN	110	BDK 105	125	MINSK	127	KOROLEV	130 (II)
Black:	AZOV	151 (II)	YAMAL	156	TSESAR KUNIKOV	158	NOVOCHERKASSK	142		
Pacific	BDK-98	055	MUKHTAR AVEZOV	066	BOBRUYSK	070	NIKOLAY KORSAKOV	077 (II)		

Displacement, tons: 4,400 full load
Dimensions, feet (metres): 369.1 × 49.2 × 12.1
  (112.5 × 15 × 3.7)
Main machinery: 2 Zgoda-Sulzer 16ZVB40/48 diesels;
  19,230 hp(m) *(14.14 MW)* sustained; 2 shafts
Speed, knots: 17.5. **Range, n miles:** 3,500 at 16 kt; 6,000 at
  12 kt
Complement: 95 (7 officers)
Military lift: 10 MBT plus 190 troops or 24 AFVs plus 170 troops
  or mines

Missiles: SAM: 4 SA-N-5 Grail quad launchers (in at least two
  ships); manual aiming; IR homing to 6 km *(3.2 n miles)* at
  1.5 Mach; altitude to 2,500 m *(8,000 ft)*; warhead 1.5 kg; 32
  missiles.

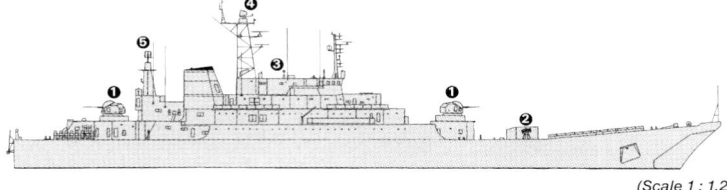

ROPUCHA I   *(Scale 1 : 1,200)*, Ian Sturton

Guns: 4—57 mm/80 (2 twin) ❶ (Ropucha I); 120 rds/min to
  6 km *(3.3 n miles)*; weight of shell 2.8 kg.
  1—76 mm/60 (Ropucha II); 60 rds/min to 15 km *(8 n miles)*;
  weight of shell 6.8 kg.
  2—30 mm/65 AK 630 (Ropucha II).
  2—122 mm UMS-73 Grad-M (in some) ❷. 2—40-barrelled
  rocket launchers; range 9 km *(5 n miles)*.
Mines: 92 contact type.
Weapons control: 2 Squeeze Box optronic directors ❸. Hood
  Wink and Odd Box.
Radars: Air/surface search: Strut Curve ❹ (Ropucha I) or Cross
  Dome (Ropucha II); F-band.
  Navigation: Don 2 or Nayada; I-band.

Fire control: Muff Cob ❺ (Ropucha I); G/H-band.
  Bass Tilt (Ropucha II); H/I-band.
IFF: 2 High Pole A or Salt Pot A.
Sonars: Mouse Tail VDS can be carried.

Programmes: Ropucha Is completed at Northern Shipyard,
  Gdansk, Poland in two spells from 1974-78 (12 ships) and
  1980-88. Ropucha IIs started building in 1987 with the first
  one commissioning in May 1990. The third and last of the
  class completed in January 1992. Type name is *bolshoy
  desantny korabl* (BDK) meaning large landing ship.
Structure: A Ro-Ro design with a tank deck running the whole
  length of the ship. All have very minor differences in
  appearance. These ships have a higher troop-to-vehicle ratio
  than the Alligator class. At least five of the class have rocket
  launchers at the after end of the forecastle. The second type
  have a 76 mm gun forward in place of one twin 57 mm and an
  ADG aft instead of the second. Radar and EW suites are also
  different. The after mast has been replaced by a solid
  extension to the superstructure.
Operational: Eight more have been deleted so far.
Sales: One to South Yemen in 1979, returned to Russia in late
  1991 for refit and was back in Aden in 1993. One to Ukraine in
  1996.
UPDATED

KALININGRAD (ROPUCHA I)   *8/2004 * /* 1042327

NIKOLAY KORSAKOV (ROPUCHA II)
9/2004*, Hachiro Nakai / 1042311

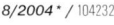

5 + 1 ALLIGATOR (TAPIR) (PROJECT 1171) CLASS (LSTM)

SARATOV (ex-*Voronezhsky Konsomolets*) 150   **MINSK** (ex-*Donetsky Shakhter*) 119   **NIKOLAY FILCHENKOV** 152   **NIKOLAY VILKOV** 081 (IV)   **ORSK** (ex-*Nicolay Obyekov*) 148

Displacement, tons: 3,400 standard; 4,700 full load
Dimensions, feet (metres): 370.7 × 50.8 × 14.7
(113 × 15.5 × 4.5)
Main machinery: 2 diesels; 9,000 hp(m) *(6.6 MW)*; 2 shafts
Speed, knots: 18. **Range, n miles:** 10,000 at 15 kt
Complement: 100
Military lift: 300 troops; 1,750 tons including about 20 tanks and various trucks; 40 AFVs

Missiles: SAM: 2 or 3 SA-N-5 Grail twin launchers; manual aiming; IR homing to 6 km *(3.2 n miles)* at 1.5 Mach; altitude to 2,500 m *(8,000 ft)*; warhead 1.5 kg; 16 missiles.
Guns: 2—57 mm/70 (twin); 120 rds/min to 8 km *(4.4 n miles)*; weight of shell 2.8 kg.
4—25 mm/80 (2 twin) (Type 4); 270 rds/min to 3 km *(1.6 n miles)*; weight of shell 0.34 kg.
1—122 mm UMS-72 Grad-M; 2—40-barrelled rocket launchers (in Types 3 and 4); range 9 km *(5 n miles)*.
Weapons control: 1 Squeeze Box optronic director (Types 3 and 4).
Radars: Surface search: 2 Don 2; I-band.

Programmes: First ship commissioned in 1966 at Kaliningrad. Last of class in service completed in 1976. Type name is *bolshoy desantny korabl* meaning large landing ship. One more Type 3 in service with Ukraine. A new ship, to be called *Ivan Gren* was laid down at Yantar, Kaliningrad, on 23 December 2004.
Structure: These ships have ramps on the bow and stern. In Type 3 the bridge structure has been raised and a forward deck house has been added to accommodate shore bombardment rocket launchers. Type 4 is similar to Type 3 with the addition of two twin 25 mm gun mountings on centreline abaft the bridge superstructure. As well as a tank deck 300 ft long stretching right across the hull there are two smaller deck areas and a hold.
Operational: In the 1980s the class operated regularly off West Africa, in the Mediterranean and in the Indian Ocean, usually with Naval Infantry units embarked. Half the class have been scrapped or laid up. Of the remainder, *Vilkov* is in the Pacific, *Minsk* in the Baltic, and the others in the Black Sea.
Sales: One to Ukraine in 1995.

UPDATED

SARATOV   9/2004 *, Hartmut Ehlers / 1042316

1 SERNA CLASS (LCU)

DKA-67 747

Displacement, tons: 105 full load
Dimensions, feet (metres): 86.3 × 19 × 5.2 *(26.3 × 5.8 × 1.6)*
Main machinery: 2 M 503A3 diesels; 5,522 hp(m) *(4.06 MW)*; 2 shafts
Speed, knots: 30. **Range, n miles:** 100 at 30 kt; 600 at 22 kt
Complement: 6
Military lift: 45 tons or 100 troops

Comment: High-speed utility landing craft capable of beaching and in service in May 1995. Has an 'air-lubricated' hull. Designed for both military and civilian use by the R Alexeyev Central Design Bureau and built at Nizhny Novgorod. Can be armed. Operational in the Baltic Fleet. Three others have been sold commercially.
UPDATED

DKA-67   6/2003 *, B Lemachko / 1042315

3 POMORNIK (ZUBR) (PROJECT 1232.2) CLASS (ACVM/LCUJM)

MDK-118 YEVGENIY KOCHESHKOV 770   **MDK-94 MORDOVIYA** 782   **MDK-108** 795

Displacement, tons: 550 full load
Dimensions, feet (metres): 189 × 84 *(57.6 × 25.6)*
Main machinery: 5 Type NK-12MV gas-turbines; 2 for lift, 23,672 hp(m) *(17.4 MW)* nominal; 3 for drive, 35,508 hp(m) *(26.1 MW)* nominal
Speed, knots: 63. **Range, n miles:** 300 at 55 kt
Complement: 31 (4 officers)
Military lift: 3 MBT or 10 APC plus 230 troops (total 130 tons)
Missiles: SAM: 2 SA-N-5 Grail quad launchers; manual aiming; IR homing to 6 km *(3.2 n miles)* at 1.5 Mach; altitude to 2,500 m *(8,000 ft)*; warhead 1.5 kg.
Guns: 2—30 mm/65 AK 630; 6 barrels per mounting; 3,000 rds/min combined to 2 km.
2—140 mm A-22 Ogon 22-barrelled rocket launchers.
Mines: 2 rails can be carried for 80.
Countermeasures: Decoys: MS227 chaff launcher.
ESM: Tool Box; intercept.
Weapons control: Quad Look (DWU-3) (modified Squeeze Box) optronic director.
Radars: Surface search: Curl Stone; I-band.
Fire control: Bass Tilt; H/I-band.
IFF: Salt Pot A/B. Square Head.

Comment: First of class delivered 1986, commissioned in 1988. Last of class launched December 1994. Produced at St Petersburg and at Feodosiya. Bow and stern ramps for ro-ro working. Last survivors are based at Baltiysk and two more are held by Ukraine. One (plus two from Ukraine) transferred and one new build for Greece in 2001. These are the first Former Soviet Union (FSU) naval platform sales to a NATO country.
UPDATED

6 AIST (DZHEYRAN) (PROJECT 1232.1) CLASS (ACV/LCUJ)

608   **609**   **610**   **700**   **MDK 113** 722   **MDK 89** 730

Displacement, tons: 298 full load
Dimensions, feet (metres): 155.2 × 58.4 *(47.3 × 17.8)*
Main machinery: 2 Type NK-12M gas turbines driving 4 axial lift fans and 4 propeller units for propulsion; 19,200 hp(m) *(14.1 MW)* nominal
Speed, knots: 70. **Range, n miles:** 120 at 50 kt
Complement: 15 (3 officers)
Military lift: 80 tons or 4 light tanks plus 50 troops or 2 medium tanks plus 200 troops or 3 APCs plus 100 troops
Guns: 4—30 mm/65 (2 twin) AK 630; 6 barrels per mounting; 3,000 rds/min combined to 2 km.
Countermeasures: Decoys: 2 PK 16 chaff launchers.
Radars: Surface search: Kivach; I-band.
Fire control: Drum Tilt; H/I-band.
IFF: High Pole B. Square Head.

Comment: First produced at Leningrad in 1970, subsequent production at rate of about six every four years. The first large hovercraft for naval use. Similar to UK SR. N4. Type name is *maly desantny korabl na vozdushnoy podushke* meaning small ACV. Modifications have been made to the original engines and some units have been reported as carrying two SA-N-5 quadruple SAM systems and chaff launchers. Three (700 series) are based in the Baltic, the other three are in the Caspian. *609* participated in the Caspian Sea exercise in 2002.
UPDATED

MDK 89   9/2000, J Ciślak / 0105561

3 LEBED (KALMAR) (PROJECT 1206) CLASS (ACV/LCUJ)

533   **639**   **640**

Displacement, tons: 87 full load
Dimensions, feet (metres): 80.1 × 36.7 *(24.4 × 11.2)*
Main machinery: 2 Ivchenko AI-20K gas turbines for lift and propulsion; 8,000 hp(m) *(5.88 MW)*
Speed, knots: 50. **Range, n miles:** 100 at 50 kt
Complement: 6 (2 officers)
Military lift: 2 light tanks or 40 tons cargo or 120 troops
Guns: 2—30 mm/65 AK 630; 6 barrels per mounting; 3,000 rds/min combined to 2 km.
Radars: Navigation: Kivach; I-band.

Comment: First entered service 1975. Can be carried in Ivan Rogov class. Has a bow ramp with gun on starboard side and the bridge to port. *533* is in the Northern Fleet and *639* and *640* took part in the Caspian Sea exercise in July 2002.
UPDATED

MDK-118   6/2003, Guy Toremans / 0570932

LEBED CLASS   4/1995, Eric Grove / 0081691

70 T-4 CLASS (PROJECT 1785) (LCM)

Displacement, tons: 35 light; 93 full load
Dimensions, feet (metres): 66.9 × 17.7 × 3.9 *(20.4 × 5.4 × 1.2)*
Main machinery: 2 diesels; 2 shafts
Speed, knots: 10
Complement: 2
Military lift: 50 tons cargo

Comment: Numbers are approximate and many may now be employed for civilian use.

VERIFIED

T-4 class   *7/2001, B Lemachko /* 0570916

3 GUS (SKAT) (PROJECT 1205) CLASS (ACV/LCMJ)

631   633   634

Displacement, tons: 17 light; 27 full load
Dimensions, feet (metres): 69.9 × 27.5 × 0.6 *(21.3 × 8.4 × 0.2)*
Main machinery: 3 TVD 10 gas turbines for lift and propulsion
Speed, knots: 49. **Range, miles:** 200 at 49 kt
Complement: 7 + 24 troops

Comment: Last survivors of an original class of 32 which entered service 1969-76.

NEW ENTRY

GUS   *6/1992*, B Lemachko /* 0583302

INTELLIGENCE VESSELS

Notes: (1) About half the AGIs are fitted with SA-N-5/8 SAM launchers.
(2) SSV in pennant numbers of some AGIs is a contraction of *sudno svyazy* meaning communications vessel.
(3) GS in pennant numbers of some AGIs is a contraction of *gidrograficheskoye sudno* meaning survey ship.
(4) Activity reported in all Fleet areas, as well as in the Mediterranean, in 2004.

2 BALZAM (ASIA) (PROJECT 1826) CLASS (AGIM)

Name	No	Builders	Commissioned
PRIBALTIKA	SSV 080	Yantar, Kaliningrad	July 1984
BELOMORE	SSV 571	Yantar, Kaliningrad	Dec 1987

Displacement, tons: 4,500 full load
Dimensions, feet (metres): 344.5 × 50.9 × 16.4 *(105 × 15.5 × 5)*
Main machinery: 2 diesels; 18,000 hp(m) *(13.2 MW)*; 2 shafts
Speed, knots: 20. **Range, n miles:** 7,000 at 16 kt
Complement: 200
Missiles: SAM: 2 SA-N-5 Grail quad launchers; manual aiming; IR homing to 6 km *(3.2 n miles)* at 1.5 Mach; altitude to 2,500 m *(8,000 ft)*; warhead 1.5 kg; 16 missiles.
Guns: 1—30 mm/65 AK 630; 6 barrels per mounting.
Radars: Surface search: Palm Frond and Don Kay; I-band.
Sonars: Lamb Tail/Mouse Tail VDS can be fitted.

Comment: Notable for twin radomes. Full EW and optronic fits. The first class of AGI to be armed. SSV 571 based in the Northern fleet and SSV 080, having been laid up in Vladivostok, has been re-activated and may have redeployed to the Northern Fleet. Capable of underway replenishment.

UPDATED

PRIBALTIKA   *7/2004* /* 1042324

7 VISHNYA (PROJECT 864) CLASS (AGIM)

Name	No	Builders	Commissioned
TAVRIYA	SSV 169	Northern Shipyard, Gdansk	Dec 1987
ODOGRAF	SSV 175	Northern Shipyard, Gdansk	July 1988
PRIAZOVE	SSV 201	Northern Shipyard, Gdansk	Jan 1987
KURILY	SSV 208	Northern Shipyard, Gdansk	Apr 1987
VASSILY TATISCHEV (ex-*Pelengator*)	SSV 231	Northern Shipyard, Gdansk	Apr 1989
MERIDIAN	SSV 520	Northern Shipyard, Gdansk	July 1986
KARELIYA	SSV 535	Northern Shipyard, Gdansk	July 1987

Displacement, tons: 3,470 full load
Dimensions, feet (metres): 309.7 × 47.9 × 14.8 *(94.4 × 14.6 × 4.5)*
Main machinery: 2 Zgoda 12AV25/30 diesels; 4,406 hp(m) *(3.24 MW)* sustained; 2 auxiliary electric motors; 286 hp(m) *(210 kW)*; 2 shafts; cp props
Speed, knots: 16. **Range, n miles:** 7,000 at 14 kt
Complement: 146
Missiles: SAM: 2 SA-N-5 Grail quad launchers; manual aiming; IR homing to 6 km *(3.2 n miles)* at 1.5 Mach; altitude to 2,500 m *(8,000 ft)*; warhead 1.5 kg.
Guns: 2—30 mm/65 AK 630; 6 barrels per mounting. 2—72 mm 4-tubed rocket launchers.
Radars: Surface search: 2 Nayada; I-band.
Sonars: Lamb Tail VDS can be carried.

Comment: SSV 231 and 520 based in the Baltic, SSV 201 in the Black Sea, SSV 169 and SSV 175 in the Northern Fleet and SSV 208 in the Pacific. All have a full EW fit plus optronic systems and datalinks. Punch Bowl is fitted in SSV 231 and possibly in others. Some superstructure differences in all of the class. NBC pressurised citadels. Ice-strengthened hulls. All are comparatively active.

UPDATED

KURILY   *10/2004*, Ships of the World /* 1042288

PRIAZOVE   *9/2000, B Lemachko /* 0126220

ODOGRAF   *6/2001 /* 0126311

4 MOMA (PROJECT 861M) CLASS (AGI/AGIM)

EKVATOR SSV 418　　**LIMAN** SSV 824　　**KILDIN** (mod) SSV 512　　**NAKHODA** SSV 506

Displacement, tons: 1,240 standard; 1,600 full load
Dimensions, feet (metres): 240.5 × 36.8 × 12.8 *(73.3 × 11.2 × 3.9)*
Main machinery: 2 Zgoda-Sulzer 6TD48 diesels; 3,300 hp(m) *(2.43 MW)* sustained; 2 shafts; cp props
Speed, knots: 17. **Range, n miles:** 9,000 at 11 kt
Complement: 66 plus 19 scientists
Missiles: SAM: 2 SA-N-5 Grail quad launchers in some.
Radars: Surface search: 2 Don 2; I-band.

Comment: Modernised ships have a foremast in the fore well-deck and a low superstructure before the bridge. Non-modernised ships retain their cranes in the forward well-deck. Similar class operates as survey ships. Built at Gdansk, Poland between 1968-72. Three based in the Black Sea and SSV 506 in the Northern Fleet. One to Ukraine in 1996.
UPDATED

KILDIN　　　　　　　　　　　　　　　6/2002, Globke Collection / 0528331

3 ALPINIST (PROJECT 503M/R) CLASS (AGIM)

GS 7　　　　**GS 19**　　　　**SYZRAN** GS 39

Displacement, tons: 1,260 full load
Dimensions, feet (metres): 177.1 × 34.4 × 13.1 *(54 × 10.5 × 4)*
Main machinery: 1 SKL 8 NVD 48 A2U diesel; 1,320 hp(m) *(970 kW)* sustained; 1 shaft; bow thruster
Speed, knots: 13. **Range, n miles:** 7,000 at 13 kt
Complement: 50
Missiles: SAM: 1 SA-N-5 Grail quad launcher *(GS 39)*.
Countermeasures: ESM: 2 Watch Dog; intercept.
Radars: Surface search: Nayada and Kivach; I-band.
Sonars: Paltus; active; high frequency.

Comment: Similar to Alpinist stern-trawlers which were built at about 10 a year at the Leninskaya Kuznitsa yard at Kiev and at the Volvograd shipyard. These AGIs were built at Kiev. In 1987 and 1988 forecastle was extended further aft and the electronics fit upgraded. *GS 7* in the Pacific, the other two in the Baltic. A fourth of class converted for ASW training was laid up in 1997.
VERIFIED

SYZRAN　　　　　　　　　　　　　6/2002, Frank Findler / 0528314

SURVEY AND RESEARCH SHIPS

Notes: Civilian research ships are now all used for commercial purposes only or are laid up, and are no longer naval associated, although some can still be leased for short operations. The former section has therefore been deleted since 1998.

2 SIBIRIYAKOV (PROJECT 865) CLASS (AGOR)

SIBIRIYAKOV　　　　　**ROMZUALD MUKLEVITCH**

Displacement, tons: 3,422 full load
Dimensions, feet (metres): 281.2 × 49.2 × 16.4 *(85.7 × 15 × 5)*
Main machinery: Diesel-electric; 2 Cegielski-Sulzer 12AS25 diesels; 6,480 hp(m) *(4.44 MW)* sustained; 2 motors; 2 shafts; cp props; bow and stern thrusters
Speed, knots: 14. **Range, n miles:** 11,000 at 14 kt
Complement: 58 plus 12 scientists
Guns: 1—30 mm AK 630 can be carried.
Radars: Navigation: 2 Nayada; I-band.

Comment: Built in Northern Shipyard, Gdansk 1990-92. Has a pressurised citadel for NBC defence, and a degaussing installation. Six separate laboratories for hydrographic and geophysical research. Two submersibles can be embarked. Both ships are very active, *Sibiriyakov* in the Baltic at Kronstadt, and *Muklevitch* in the North.
VERIFIED

SIBIRIYAKOV　　　　　　　　　　　5/2000 / 0105564

2 AKADEMIK KRYLOV (PROJECT 852/856) CLASS (AGORH)

LEONID DEMIN　　　　　　　　**ADMIRAL VLADIMIRSKIY**

Displacement, tons: 9,100 full load
Dimensions, feet (metres): 482.3 × 60.7 × 20.3 *(147 × 18.5 × 6.2)*
Main machinery: 4 Sulzer diesels; 14,500 hp(m) *(10.7 MW)*; 2 shafts; bow and stern thrusters
Speed, knots: 20. **Range, n miles:** 36,000 at 15 kt
Complement: 120
Radars: Navigation: Nayada, Palm Frond and Don 2; I-band.
Helicopters: 1 Hormone.

Comment: Built in Szczecin 1974-79. Carry two survey launches and have 26 laboratories. Based in the Baltic at Kronstadt. *Akademik Krylov* sold to a Greek company in 1993 and now flies the Cyprus flag. *Admiral Vladimirskiy*, previously inactive at Kronstadt, may have been reactivated.
VERIFIED

LEONID DEMIN　　　　　　　　　　5/1994 / 0081696

3 NIKOLAY ZUBOV (PROJECT 850) CLASS (AGOR)

ANDREY VILKITSKY　　　**BORIS DAVIDOV**　　　**SEMEN DEZHNEV**

Displacement, tons: 2,674 standard; 3,021 full load
Dimensions, feet (metres): 294.2 × 42.7 × 15 *(89.7 × 13 × 4.6)*
Main machinery: 2 Zgoda-Sulzer 8TD48 diesels; 4,400 hp(m) *(3.23 MW)* sustained; 2 shafts
Speed, knots: 16.5. **Range, n miles:** 11,000 at 14 kt
Complement: 50
Radars: Navigation: Palm Frond or Don 2; I-band.

Comment: Oceanographic research ships built at Szczecin Shipyard, Poland in 1964-68. Also employed on navigational, sonar and radar trials. Has nine laboratories and small deck aft for hydromet-balloon work. Carry two to four survey launches. Based in the Northern Fleet. *Bellingsgauzen* to Ukraine in 1995.
VERIFIED

BORIS DAVIDOV　　　　　　　　　　6/1994 / 0081698

10 MOMA (PROJECT 861) CLASS (AGS)

ANTARES	ASKOLD	SEVER (AGE)	ARTIKA	OKEAN
ANTARKTYDA	MARS	KRILON	ANDROMEDA	CHELEKEN

Displacement, tons: 1,550 full load
Dimensions, feet (metres): 240.5 × 36.8 × 12.8 *(73.3 × 11.2 × 3.9)*
Main machinery: 2 Zgoda-Sulzer 6TD48 diesels; 3,300 hp(m) *(2.43 MW)* sustained; 2 shafts; cp props
Speed, knots: 17. **Range, n miles:** 9,000 at 11 kt
Complement: 55
Radars: Navigation: Nayada and Don 2; I-band.
IFF: High Pole A.

Comment: Built at Northern Shipyard, Gdansk 1967-72. Some of the class are particularly active in ASW research associated operations. Four laboratories. One survey launch and a 7 ton crane. The AGE is fitted with bow probes. Two in the Northern Fleet, three in the Pacific, three in the Black and two in the Baltic. One transferred to Ukraine.

UPDATED

ANTARKTYDA   6/1998, B Lemachko / 0570919

22 FINIK (PROJECT 872) CLASS (AGS/AGE/AE)

North:	GS 87	GS 260	GS 278	GS 297	GS 392	GS 405	
Pacific:	GS 44	GS 47	GS 84	GS 272	GS 296	GS 397	GS 404
Black:	GS 86	GS 402	VTR 75				
Baltic:	GS 270	GS 301	GS 388	GS 399	GS 400	GS 403	

Displacement, tons: 1,200 full load
Dimensions, feet (metres): 201.1 × 35.4 × 10.8 *(61.3 × 10.8 × 3.3)*
Main machinery: 2 Cegielski-Sulzer 6AL25/30 diesels; 1,920 hp(m) *(1.4 MW)*; auxiliary propulsion; 2 motors; 204 hp(m) *(150 kW)*; 2 shafts; cp props; bow thruster
Speed, knots: 13. **Range, n miles:** 3,000 at 13 kt
Complement: 26 (5 officers) plus 9 scientists
Radars: Navigation: Kivach B; I-band.

Comment: Improved Biya class. Built at Northern Shipyard, Gdansk 1978-83. Fitted with 7 ton crane for buoy handling. Can carry two self-propelled pontoons and a boat on well-deck. Some have been used commercially. Ships of same class serve in the Polish Navy. Three transferred to Ukraine in 1997. Some may be laid up. VTR 75, originally built as a survey ship, was converted for use as an ammunition carrier in 2000.

UPDATED

GS 403   10/2002, B Lemachko / 0570911

VTR 75   5/2003*, B Lemachko / 0580533

14 YUG (PROJECT 862) CLASS (AGS/AGI/AGE)

North	Pacific	Baltic	Black
PLUTON	V ADM VORONTSOV	PERSEY	DONUZLAV
STRELETS	(ex-*Briz*)	NIKOLAY MATUSEVICH	STVOR
GORIZONT	PEGAS		
GIDROLOG	MARSHAL GELOVANI		
VIZIR			
SENEZH			
TEMRYUK (ex-*Mangyshlak*)			
SSV 700			

Displacement, tons: 2,500 full load
Dimensions, feet (metres): 270.6 × 44.3 × 13.1 *(82.5 × 13.5 × 4)*
Main machinery: 2 Zgoda-Sulzer 6TD48 diesels; 3,300 hp(m) *(2.43 MW)* sustained; 2 auxiliary motors; 272 hp(m) *(200 kW)*; 2 shafts; cp props; bow thruster; 300 hp *(220 kW)*
Speed, knots: 15. **Range, n miles:** 9,000 at 12 kt
Complement: 46 (8 officers) plus 20 scientists
Guns: 6—25 mm/80 (3 twin) (fitted for but not with).
Radars: Navigation: Palm Frond or Nayada; I-band.

Comment: Built at Northern Shipyard, Gdansk 1977-83. Have 4 ton davits at the stern and two survey craft. Others have minor variations around the stern area. *Pluton* and *Strelets* have been taken over by the Arctic Border Guard.. SSV 700 classified as AGE. The Pacific units are understood to be non-operational.

UPDATED

DONUZLAV   6/2003*, B Lemachko / 1042319

60 GPB-480 (PROJECT 1896) CLASS (INSHORE SURVEY CRAFT) (YGS)

BGK series

Displacement, tons: 116 full load
Dimensions, feet (metres): 93.8 × 17.1 × 5.6 *(28.6 × 5.2 × 1.7)*
Main machinery: 1 diesel; 300 hp(m) *(223 kW)*; 1 shaft
Speed, knots: 12

Comment: Entered service from 1955. Numbers approximate. Inshore survey craft equipped with two 1.5 ton derricks.

NEW ENTRY

BGK 785   7/2001*, B Lemachko / 0580537

7 BIYA (PROJECT 870/871) CLASS (AGS)

North	Pacific	Baltic
GS 193	GS 200	GS 204
	GS 210	GS 208
	GS 269	GS 214

Displacement, tons: 766 full load
Dimensions, feet (metres): 180.4 × 32.1 × 8.5 *(55 × 9.8 × 2.6)*
Main machinery: 2 diesels; 1,200 hp(m) *(882 kW)*; 2 shafts; cp props
Speed, knots: 13. **Range, n miles:** 4,700 at 11 kt
Complement: 25
Radars: Navigation: Don 2; I-band.

Comment: Built at Northern Shipyard, Gdansk 1972-76. With laboratory and one survey launch and a 5 ton crane. Two transferred to Ukraine in 1997.

VERIFIED

BIYA   4/1997, Riku Lehtinen / 0081700

6 KAMENKA (PROJECT 870/871) CLASS (AGS)

| GS 66 | GS 113 | GS 118 | GS 199 | GS 207 | GS 211 |

Displacement, tons: 760 full load
Dimensions, feet (metres): 175.5 × 29.8 × 8.5 *(53.5 × 9.1 × 2.6)*
Main machinery: 2 Sulzer diesels; 1,800 hp(m) *(1.32 MW)*; 2 shafts; cp props
Speed, knots: 14. **Range, n miles:** 4,000 at 10 kt
Complement: 25
Radars: Navigation: Don 2; I-band.
IFF: High Pole.

Comment: Built at Northern Shipyard, Gdansk 1968-69. A 5 ton crane forward. They do not carry a survey launch but have facilities for handling and stowing buoys. Two in the Baltic and four in the Pacific. One transferred to Vietnam in 1979, one to Estonia in 1996 and one to Ukraine in 1997. *VERIFIED*

GS 118　　　　　　　　　　　　　　　　　　6/2003, B Lemachko / 0570901

2 ONEGA (PROJECT 1806) CLASS (AGS)

| AKADEMIK SEMINIKHIN SFP 183 | AKADEMIK ISANIN SFP 177 |

Displacement, tons: 2,150 full load
Dimensions, feet (metres): 265.7 × 36 × 13.7 *(81 × 11 × 4.2)*
Main machinery: 2 gas turbines; 8,000 hp(m) *(5.88 MW)*; 1 shaft
Speed, knots: 20
Complement: 45
Radars: Navigation: Nayada; I-band.

Comment: Built at Zelenodolsk and first seen in September 1973. Helicopter platform but no hangar in earlier ships of the class but in later hulls the space is taken up with more laboratory accommodation. Used as hydroacoustic monitoring ships. *Akademik Seminikhin* was completed in October 1992. One to Ukraine in 1997. *Akademik Seminikhin* based in the Black Sea and *Akademik Isanin* in the Northern Fleet. Others are laid up in each of the four Fleets. *UPDATED*

AKADEMIK SEMINIKHIN　　　　　　　7/2000, Hartmut Ehlers / 0105566

2 VINOGRAD CLASS (AGOR)

| GS 525 | GS 526 |

Displacement, tons: 498 full load
Dimensions, feet (metres): 108.3 × 34.1 × 9.1 *(33 × 10.4 × 2.8)*
Main machinery: Diesel-electric; 2 diesels; 2 motors; 1,200 hp(m) *(882 kW)*; 2 trainable props
Speed, knots: 9. **Range, n miles:** 1,000 at 6 kt
Complement: 19

Comment: Built by Rauma-Repola, Finland, 1985-87 as hydrographic research ships. *GS 525* commissioned 12 November 1985 and *GS 526* on 17 December 1985. *525* is in the Baltic and *526* in the North. Both have side scan sonars. A similar ship has been reported operating with the Northern Fleet. *VERIFIED*

GS 525　　　　　　　　　　　　　　　　6/1998, Hartmut Ehlers / 0050054

1 MARSHAL NEDELIN (PROJECT 1914) CLASS
(MISSILE RANGE SHIP) (AGMH)

MARSHAL KRYLOV

Displacement, tons: 24,500 full load
Dimensions, feet (metres): 695.5 × 88.9 × 25.3 *(212 × 27.1 × 7.7)*
Main machinery: 2 gas turbines; 54,000 hp(m) *(40 MW)*; 2 shafts
Speed, knots: 20. **Range, n miles:** 22,000 at 16 kt
Complement: 450
Radars: Air search: Top Plate.
Navigation: 3 Palm Frond; I-band.
Helicopter control: Fly Screen B; I-band.
Space trackers: End Tray (balloons). Quad Leaf. 3 Quad Wedge. 4 smaller aerials.
Tacan: 2 Round House.
Helicopters: 2-4 Ka-32 Helix C.

Comment: Completed at Admiralty Yard, Leningrad 23 February 1990. Fitted with a variety of space and missile associated electronic systems. Fitted for but not with six twin 30 mm/65 ADG guns and three Bass Tilt fire-control radars. Naval subordinated, the task is monitoring missile tests with a wartime role of command ship. The Ship Globe radome is for SATCOM. Based in the Pacific and active. Second of class deleted. *VERIFIED*

MARSHAL KRYLOV　　　　　　　10/1995, van Ginderen Collection

TRAINING SHIPS

Notes: The Mir class sail training ships have no military connections.

2 SMOLNY (PROJECT 887) CLASS (AX)

| PEREKOP 200 | SMOLNY 210 |

Displacement, tons: 9,150 full load
Dimensions, feet (metres): 452.8 × 53.1 × 21.3 *(138 × 16.2 × 6.5)*
Main machinery: 2 Zgoda Sulzer 12ZV 40/48 diesels; 15,000 hp(m) *(11 MW)*; 2 shafts
Speed, knots: 20. **Range, n miles:** 9,000 at 15 kt
Complement: 137 (12 officers) plus 330 cadets
Guns: 4–3 in *(76 mm)*/60 (2 twin). 4–30 mm/65 (2 twin).
A/S mortars: 2 RBU 2500.
Countermeasures: ESM: 2 Watch Dog; radar warning.
Radars: Air/surface search: Head Net C; 3D; E-band; range 128 km *(70 n miles)*.
Navigation: 4 Don 2; I-band. Don Kay *(Perekop)*; I-band.
Fire control: Owl Screech; G-band. Drum Tilt; H/I-band.
IFF: 2 High Pole A. Square Head.
Sonars: Mouse Tail VDS; active; high frequency.

Comment: Built at Szczecin, Poland. *Smolny* completed in 1976, *Perekop* in 1977. Have considerable combatant potential. Both are active in the Baltic. *UPDATED*

SMOLNY　　　　　　　　　　　　　　　7/2002, M Declerck / 0528296

10 PETRUSHKA (UK-3) CLASS (AXL)

| MK 157 | MK 194 | MK 207 | MK 288 | MK 1277 |
| PSK 1304 | PSK 1519 | PSK 1556 | PSK 1562 | PSK 2017 |

Displacement, tons: 335 full load
Dimensions, feet (metres): 129.3 × 27.6 × 7.2 *(39.4 × 8.4 × 2.2)*
Main machinery: 2 Wola H12 diesels; 756 hp(m) *(556 kW)*; 2 shafts
Speed, knots: 11. **Range, n miles:** 1,000 at 11 kt
Complement: 13 plus 30 cadets

Comment: Training vessels built at Wisla Shipyard, Poland; first one commissioned in 1989. Very similar to the SK 620 class used as ambulance craft. Used for seamanship and navigation training and may be commercially owned. Three are active in the Pacific, and three in the Baltic with a further eight laid up. *VERIFIED*

PETRUSHKA CLASS　　　　　　　　　6/2003, E & M Laursen / 0570909

AUXILIARIES

Notes: (1) Lama class *Voronesh* has been renamed VTR-33 and is an alongside civilian manned support ship in the Black Sea.
(2) Two Belyanka class tankers *Amur* and *Pinega* are used for stowing low level radioactive waste.

2 AMGA (PROJECT 1791) CLASS
(MISSILE SUPPORT SHIPS) (AEM)

VETLUGA   DAUGAVA

Displacement, tons: 6,100 *(Vetluga)*, 6,350 *(Daugava)* full load
Dimensions, feet (metres): 354.3 × 59 × 14.8 *(108 × 18 × 4.5) (Vetluga)*
Main machinery: 2 diesels; 9,000 hp(m) *(6.6 MW)*; 2 shafts
Speed, knots: 16. **Range, n miles:** 4,500 at 14 kt
Complement: 210
Guns: 4—25 mm/80 (2 twin).
Radars: Surface search: Strut Curve; F-band.
Navigation: Don 2; I-band.
IFF: High Pole B.

Comment: Built at Gorkiy. Ships with similar duties to the Lama class. Fitted with a large 55 ton crane forward and thus capable of handling much larger missiles than their predecessors. Each ship has a different length and type of crane to handle later types of missiles. Designed for servicing submarines. *Vetluga* completed in 1976 and *Daugava* (5 m longer than *Vetluga*) in 1981. Both are in the Pacific Fleet. A third of class is laid up in the North.
VERIFIED

DAUGAVA   *3/2003, B Lemachko /* 0573518

11 AMUR (PROJECT 304/304M) CLASS (REPAIR SHIPS) (AR)

AMUR I
PM 40, PM 56, PM 64, PM 82, PM 138, PM 140, PM 156
AMUR II
PM 59, PM 69, PM 86, PM 97

Displacement, tons: 5,500 full load
Dimensions, feet (metres): 400.3 × 55.8 × 16.7 *(122 × 17 × 5.1)*
Main machinery: 1 Zgoda 8 TAD-48 diesel; 3,000 hp(m) *(2.2 MW)*; 1 shaft
Speed, knots: 12. **Range, n miles:** 13,000 at 8 kt
Complement: 145
Radars: Navigation: Kivach or Palm Frond or Nayada; I-band.

Comment: Amur I class general purpose depot and repair ships completed 1968-83 in Szczecin, Poland. Successors to the Oskol class. Carry two 5 ton cranes and have accommodation for 200 from ships alongside. Amur II class has extra deckhouse forward of the funnel. Built at Szczecin 1983-85. Three Amur IIs are based in the Pacific and one in the North. Three are laid up in the Baltic. PM 9 transferred to Ukraine. PM 56, based in the Black Sea, visited Tartous in 2002.
UPDATED

AMUR II PM 86   *9/2000, J Cislak /* 0105571

AMUR I PM 138   *9/2002, Globke Collection /* 0528330

1 MALINA (PROJECT 2020) CLASS
(NUCLEAR SUBMARINE SUPPORT SHIP) (AS)

PM 74

Displacement, tons: 10,500 full load
Dimensions, feet (metres): 449.5 × 68.9 × 18.4 *(137 × 21 × 5.6)*
Main machinery: 4 gas turbines; 60,000 hp(m) *(44 MW)*; 2 shafts
Speed, knots: 17
Complement: 260
Radars: Navigation: 2 Palm Frond or 2 Nayada; I-band.

Comment: Built at Nikolayev. First deployed to Pacific in 1986. PM is an abbreviation of Plavuchaya Masterskaya (Floating workshop). A fourth of class *(PM 16)* launched early in 1992, was not completed. Designed to support nuclear-powered submarines and surface ships. Carry two 15 ton cranes. Two others (PM 914 (ex-PM 12) and PM 921 (ex-PM 63)) in the North are inactive and used for storing low level radioactive waste.
VERIFIED

PM 74   *7/1996 /* 0081704

4 VYTEGRALES II (PROJECT 596P) CLASS
(SUPPLY SHIPS) (AKH/AGF)

APSHERON	DAURIYA	SEVAN	YAMAL
(ex-*Vagales*) —	(ex-*Vyborgles*) 506	(ex-*Siverles*) 208	(ex-*Tosnoles*) 212

Displacement, tons: 6,150 full load
Dimensions, feet (metres): 400.3 × 55.1 × 22.3 *(122.1 × 16.8 × 6.8)*
Main machinery: 1 Burmeister & Wain 950VTBF diesel; 5,200 hp(m) *(3.82 MW)*; 1 shaft
Speed, knots: 15
Complement: 46
Radars: Navigation: Nayada or Palm Frond or Spin Trough; I-band.
CCA: Fly Screen.
Helicopters: 1 Ka-25 Hormone C (not in *Yamal*).

Comment: Standard timber carriers of a class of 27. These ships were modified for naval use in 1966-68 with helicopter flight deck. Built at Zhdanov Yard, Leningrad between 1963 and 1966. *Sevan* fitted as squadron Flagship in support of Indian Ocean detachments. Variations exist between ships of this class; *Dauriya* has a deckhouse over the aft hold. All have two Vee Cone communications aerials. *Yamal* had her flight deck removed in 1995. The first of class, completed in 1962, was originally *Vytegrales*, but this was later changed to *Kosmonaut Pavel Belyayev* and, with three other ships of this class, converted to Space Support Ships. Four others (*Borovichi* and so on) received a different conversion for the same purpose. The civilian-manned ships together with these naval ships are often incorrectly called Vostok or Baskunchak class. *Apsheron* and *Dauriya* are in the Black Sea, *Sevan* and *Yamal* in the Baltic. Two others transferred to Ukraine in 1996.
VERIFIED

DAURIYA   *7/2000, Hartmut Ehlers /* 0105572

SEVAN   *6/1998 /* 0050059

1 BEREZINA (PROJECT 1833) CLASS (AORH)

Name	No	Builders	Laid down	Launched	Commissioned
BEREZINA	154	Nikolayev (61 Kommuna)	18 Aug 1972	20 Apr 1975	30 Dec 1977

Displacement, tons: 35,000 full load
Dimensions, feet (metres): 695.5 × 85.3 × 38.7
  (212.0 × 26.0 × 11.8)
Main machinery: 2 diesels; 47,500 hp(m) *(35.0 MW)*; 2 shafts
Speed, knots: 22. **Range, n miles:** 15,000 at 16 kt
Complement: 200 approx
Cargo capacity: 16,000 tons (approx) fuel; 2,000 tons
  provisions; 500 tons fresh water

Missiles: SAM: Originally equipped with SA-N-4 Gecko but
  probably non-operational.
Guns: Originally equipped with 4—57 mm/80 (2 twin) and
  4—30 mm/65.
A/S mortars: Originally equipped with 2 RBU 100 launchers.
Radars: Air/surface search: Strut Curve; E/F-band.
  Fire control: Muff Cob (for 57 mm); G/H-band.
  Navigation: 2 Don Kay; I-band.
Sonars: Hull mounted (probably non-operational).

Helicopters: 2 Ka-25 Hormone C.

Programmes: After being inactive since 1991, the ship was
  decommissioned in 1997 and subsequently stripped of
  armament. Returned to service in 2000 as a civilian-manned
  auxiliary and has been reported operational again in 2004. It is
  unlikely that she has been re-armed.
Structure: Fitted with two storing gantries, four 10 ton cranes,
  liquid-fuelling gantry (midships) and stern refuelling. Also
  capable of conducting vertrep.
Operational: Based in the Black Sea.

NEW ENTRY

BEREZINA (in former service)
8/1991*

5 BORIS CHILIKIN (PROJECT 1559V) CLASS (REPLENISHMENT SHIPS) (AOR)

BORIS BUTOMA   IVAN BUBNOV   SEGEY OSIPOV (ex-*Dnestr*)   VLADIMIR KOLECHITSKY   GENRICH GASANOV

Displacement, tons: 23,450 full load
Dimensions, feet (metres): 531.5 × 70.2 × 33.8
  (162.1 × 21.4 × 10.3)
Main machinery: 1 diesel; 9,600 hp(m) *(7 MW)*; 1 shaft
Speed, knots: 17. **Range, n miles:** 10,000 at 16 kt
Complement: 75 (without armament)
Cargo capacity: 13,000 tons oil fuel and dieso; 400 tons
  ammunition; 400 tons spares; 400 tons victualling stores;
  500 tons fresh water

Guns: 4—57 mm/80 (2 twin). Most are fitted for but not with the
  guns.
Radars: Air/surface search/fire control: Strut Curve (fitted for
  but not with).
  Muff Cob (fitted for but not with).
  Navigation: 2 Nayada or Palm Frond (plus Don 2 in *V
  Kolechitsky*); I-band.
IFF: High Pole B.

Programmes: Based on the Veliky Oktyabr merchant ship tanker
  design. Built at the Baltic Yard, Leningrad; *Vladimir Kolechitsky*
  completed in 1972, *Osipov* in 1973, *Ivan Bubnov* in 1975,
  Genrich Gasanov in 1977. Last of class *Boris Butoma*
  completed in 1978.
Structure: This is the only class of purpose-built underway fleet
  replenishment ships for the supply of both liquids and solids.

BORIS BUTOMA   8/2000, B Lemachko / 0126221

Although most operate in merchant navy paint schemes, all
wear naval ensigns.
Operational: Earlier ships can supply solids on both sides
  forward. Later ships supply solids to starboard, liquids to port
  forward. All can supply liquids either side aft and astern.

Osipov and *Gasanov* are based in the North, *Bubnov* in the
Black Sea, *Butoma* and *Kolechitsky* in the Pacific. Most are
used for commercial purposes. *Boris Chilikin* transferred to
Ukraine in 1997.   ***VERIFIED***

VLADIMIR KOLECHITSKY   3/2001, Ships of the World / 0126357

30 BOLVA (PROJECT 688/688A) CLASS
(BARRACKS SHIPS) (YPB)

Displacement, tons: 6,500
Dimensions, feet (metres): 560.9 × 45.9 × 9.8 *(110 × 14 × 3)*
Cargo capacity: 350-400 tons

Comment: A total of 59 built by Valmet Oy, Helsinki between 1960 and 1984. Of the remaining 30 ships, six are Bolva 1, 16 are Bolva 2 and eight are Bolva 3. Used for accommodation of ships' companies during refit and so on. The Bolva 2 and 3 have a helicopter pad. Have accommodation facilities for about 400 people. No means of propulsion but can be steered. In addition there are several other types of Barracks Ships including five ex-Atrek class depot ships as well as converted merchant ships and large barges. At least 18 have been scrapped.
VERIFIED

IMATRA (at Sevastopol)   3/2002, Hartmut Ehlers / 0529803

3 DUBNA CLASS (REPLENISHMENT TANKERS) (AOL/AOT)

DUBNA   PECHENGA   IRKUT

Displacement, tons: 11,500 full load
Dimensions, feet (metres): 426.4 × 65.6 × 23.6 *(130 × 20 × 7.2)*
Main machinery: 1 Russkiy 8DRPH23/230 diesel; 6,000 hp(m) *(4.4 MW)*; 1 shaft
Speed, knots: 16. **Range, n miles:** 7,000 at 16 kt
Complement: 70
Cargo capacity: 7,000 tons fuel; 300 tons fresh water; 1,500 tons stores
Radars: Navigation: 2 Nayada; I-band.

Programmes: Completed 1974 at Rauma-Repola, Finland.
Structure: *Dubna* has 1 ton replenishment stations forward. Normally painted in merchant navy colours.
Operational: *Dubna* can refuel on either beam and astern. *Pechenga* has had RAS gear removed. Based in North. One of the class transferred to Ukraine in 1997. *Irkut* was believed to have been sold commercially in 1999 but is reported operational in the Pacific.
UPDATED

DUBNA   7/1996, van Ginderen Collection / 0019061

PECHENGA   3/2003, B Lemachko / 0573517

6 MOD ALTAY CLASS (PROJECT 160)
(REPLENISHMENT TANKERS) (AOL)

PRUT   YELNYA   IZHORA   KOLA   ILIM   YEGORLIK

Displacement, tons: 7,250 full load
Dimensions, feet (metres): 348 × 51 × 22 *(106.2 × 15.5 × 6.7)*
Main machinery: 1 Burmeister & Wain BM550VTBN110 diesel; 3,200 hp(m) *(2.35 MW)*; 1 shaft
Speed, knots: 14. **Range, n miles:** 8,600 at 12 kt
Complement: 60
Cargo capacity: 4,400 tons oil fuel; 200 m³ solids
Radars: Navigation: 2 Don 2 or 2 Spin Trough; I-band.

Comment: Built from 1967-72 by Rauma-Repola, Finland. Modified for alongside replenishment. This class is part of 38 ships, being the third group of Rauma types built in Finland in 1967. *Ilim* and *Yegorlik* transferred to civilian companies in 1996/97 and operate in the Pacific with *Izhora*. *Prut* and *Kola* in the North and *Yelnya* in the Baltic.
UPDATED

KOLA   1/1997, van Ginderen Collection / 0019062

2 OLEKMA CLASS (PROJECT 92)
(REPLENISHMENT TANKERS) (AORL)

OLEKMA   IMAN

Displacement, tons: 7,300 full load
Dimensions, feet (metres): 344.5 × 47.9 × 22 *(105.1 × 14.6 × 6.7)*
Main machinery: 1 Burmeister & Wain diesel; 2,900 hp(m) *(2.13 MW)*; 1 shaft
Speed, knots: 14. **Range, n miles:** 8,000 at 14 kt
Complement: 40
Cargo capacity: 4,500 tons oil fuel; 180 m³ solids
Radars: Navigation: Don 2 or Nayada and Spin Trough; I-band.

Comment: Built by Rauma-Repola, Finland in 1966. Modified for replenishment with refuelling rig abaft the bridge as well as astern refuelling. *Olekma* based in the Baltic and *Iman* in the Black Sea.
VERIFIED

IMAN   3/2002, Hartmut Ehlers / 0529802

4 UDA CLASS (PROJECT 577D)
(REPLENISHMENT TANKERS) (AOL)

LENA   TEREK   VISHERA   KOYDA

Displacement, tons: 5,500 standard; 7,126 full load
Dimensions, feet (metres): 400.3 × 51.8 × 20.3 *(122.1 × 15.8 × 6.2)*
Main machinery: 2 diesels; 9,000 hp(m) *(6.6 MW)*; 2 shafts
Speed, knots: 17. **Range, n miles:** 4,000 at 15 kt
Complement: 85
Cargo capacity: 2,900 tons oil fuel; 100 m³ solids
Radars: Navigation: 2 Don 2 or Nayada/Palm Frond; I-band.
IFF: High Pole A.

Comment: Built between 1962 and 1967 at Vyborg Shipyard. All have a beam replenishment capability. Guns removed. *Vishera* in the Pacific, *Terek* in the Northern Fleet, *Koyda* in the Black Sea and *Lena* in the Baltic.
UPDATED

LENA   8/2004* / 1042325

1 MANYCH (PROJECT 1549) CLASS (WATER TANKER) (AWT)

MANYCH

Displacement, tons: 7,700 full load
Dimensions, feet (metres): 380.5 × 51.5 × 23.0 *(116.0 × 15.7 × 7.0)*
Main machinery: 2 diesels; 9,000 hp(m) *(6.6 MW)*; 2 shafts
Speed, knots: 18. **Range, n miles:** 7,500 at 16 kt
Complement: 90
Cargo capacity: 4,400 tons
Radars: Air/surface search: Strut Curve; E/F-band.
Navigation: Don Kay; I-band.

Comment: Distilled water carrier built at Vyborg and completed in 1972. Decommissioned and disarmed in 1996 but returned to service in 1998 after refit in Bulgaria. Based in the Black Sea.
NEW ENTRY

MANYCH   9/2004*, Hartmut Ehlers / 1042287

1 KALININGRADNEFT CLASS (SUPPORT TANKER) (AORL)

VYAZMA (ex-*Katun*)

Displacement, tons: 8,600 full load
Dimensions, feet (metres): 380.5 × 56 × 21 *(116 × 17 × 6.5)*
Main machinery: 1 Russkiy Burmeister & Wain 5DKRP50/110-2 diesel; 3,850 hp(m) *(2.83 MW)*; 1 shaft
Speed, knots: 14. Range, n miles: 5,000 at 14 kt
Complement: 32
Cargo capacity: 5,400 tons oil fuel and other liquids
Radars: Navigation: Okean; I-band.

Comment: Built by Rauma-Repola, Finland in 1982. Can refuel astern. At least an additional 20 of this class operate with the fishing fleets. Operational in the Northern Fleet.
VERIFIED

KALININGRADNEFT   *11/1991, G Jacobs*

35 TOPLIVO CLASS (PROJECT 1844/1844D) CLASS (YO)

VTN series

Displacement, tons: 1,180 full load
Dimensions, feet (metres): 178.1 × 24.3 × 10.5 *(54.3 × 7.4 × 3.2)*
Main machinery: 1 diesel; 600 hp *(450 kW)*; 1 shaft
Speed, knots: 10. Range, n miles: 1,500 at 10 kt
Complement: 20
Radars: Navigation: Don-2; I-band.

Comment: Details given are for the Toplivo-2 class, some of which were built in Egypt but the majority in the USSR. The Toplivo-3 class, built in the USSR, are slightly larger at 1,300 tons full load. Numbers remaining in service are approximate. All the original Toplivo-1 class are believed to have been decommissioned.
NEW ENTRY

VTN-28 (Toplivo-2)   *9/2001*, B Lemachko* / 1042318

7 KHOBI CLASS (PROJECT 437M) CLASS (YO)

CHEREMSHA   SHACHA   SISOLA   MOKSHA   SOSHA   ORSHA   INDIGA (ex-*Seyma*)

Displacement, tons: 1,520 full load
Dimensions, feet (metres): 221.1 × 33.1 × 11.8 *(67.4 × 10.1 × 3.6)*
Main machinery: 1 diesel; 1,600 hp *(1.2 MW)*; 2 shafts
Speed, knots: 13. Range, n miles: 2,000 at 10 kt
Complement: 30
Radars: Navigation: Don-2; I-band.

Comment: *Cheremsha, Shacha* and *Sisola* based in the North, *Moksha* in the Pacific, *Sosha* and *Orsha* in the Baltic and *Seyma* in the Black Sea. Used for the transport of all forms of liquids.
NEW ENTRY

INDIGA   *9/2004*, Hartmut Ehlers* / 1042284

3 OB (PROJECT 320) CLASS (HOSPITAL SHIPS) (AHH)

YENISEI   SVIR   IRTYSH

Displacement, tons: 11,570 full load
Dimensions, feet (metres): 499.7 × 63.6 × 20.5 *(152.3 × 19.4 × 6.3)*
Main machinery: 2 Zgoda-Sulzer 12ZV40/48; 15,600 hp(m) *(11.47 MW)* sustained; 2 shafts; cp props
Speed, knots: 19. Range, n miles: 10,000 at 18 kt
Complement: 124 plus 83 medical staff
Radars: Navigation: 3 Don 2 or 3 Nayada; I-band.
IFF: High Pole A.
Helicopters: 1 Ka-25 Hormone C.

Comment: Built at Szczecin, Poland. *Yenisei* completed 1981 and is based in the Black Sea. *Svir* completed in early 1989 and transferred to the Northern Fleet in September 1989. *Irtysh* completed in June 1990, was stationed in the Gulf in 1990-91 and is now based in the Pacific. A fourth of class is derelict and a fifth was cancelled. Have 100 beds and seven operating theatres. The first purpose-built hospital ships in the Navy, a programme which may have been prompted by the use of several merchant ships off Angola for Cuban casualties in the 'war of liberation.' NBC pressurised citadel. Ship stabilisation system. Decompression chamber. All are in use, mostly as alongside medical facilities.
UPDATED

SVIR   *6/2003*, B Lemachko* / 1042321

4 KLASMA (PROJECT 1274) CLASS (CABLE SHIPS) (ARC)

DONETS   INGURI   YANA   ZEYA

Displacement, tons: 6,000 standard; 6,900 full load
Measurement, tons: 3,400 dwt; 5,786 gross
Dimensions, feet (metres): 427.8 × 52.5 × 19 *(130.5 × 16 × 5.8)*
Main machinery: Diesel-electric; 5 Wärtsilä Sulzer 624TS diesel generators (4 in *Ingul* and *Yana*); 5,000 hp(m) *(3.68 MW)*; 2 motors; 2,150 hp(m) *(1.58 MW)*; 2 shafts
Speed, knots: 14. Range, n miles: 12,000 at 14 kt
Complement: 118
Radars: Navigation: Spin Trough and Nayada; I-band.

Comment: *Yana* built by Wärtsilä, Helsingforsvarvet, Finland in 1962; *Donets* at the Wärtsilä, Åbovarvet in 1968-69; *Zeya* in 1970. *Donets* and *Zeya* are of slightly modified design. *Inguri* completed in 1978. All are ice strengthened and can carry 1,650 miles of cable. *Yana* is distinguished by gantry right aft. *Donets* is in the Baltic, and the other three are in the North. All are active and can be leased for commercial use. One to Ukraine in 1997.
VERIFIED

KLASMA   *3/1992* / 0081709

4 EMBA (PROJECT 1172/1175) CLASS (CABLE SHIPS) (ARC)

SETUN (I)   NEPRYADAVA (I)   KEM (II)   BIRIUSA (II)

Displacement, tons: 2,050 full load (Group I); 2,400 (Group II)
Dimensions, feet (metres): 249 × 41.3 × 9.8 *(75.9 × 12.6 × 3)* (Group I)
   282.4 × 41.3 × 9.9 *(86.1 × 12.6 × 3)* (Group II)
Main machinery: Diesel-electric; 2 Wärtsilä Vasa 6R22 diesel alternators; 2,350 kVA 60 Hz; 2 motors; 1,360 hp(m) *(1 MW)*; 2 shafts (Group I)
   2 Wärtsilä Vasa 8R22 diesel alternators; 3,090 kVA 60 Hz; 2 motors; 2,180 hp(m) *(1.6 MW)*; 2 shafts (Group II)
   The 2 turnable propulsion units can be inclined to the ship's path giving, with a bow thruster, improved turning movement
Speed, knots: 11
Complement: 40
Radars: Navigation: Kivach and Don 2; I-band.

Comment: Both Emba Is built in 1981. Designed for shallow water cable-laying. Carry 380 tons of cable. Order placed with Wärtsilä in January 1985 for two larger (Group II) ships; *Kem* completed on 23 October 1986. Can lay about 600 tons of cable. Designed for use off Vladivostok but also capable of operations in inland waterways. *Setun* is based in the Black Sea, *Nepryadava* in the Baltic, and *Kem* and *Biriusa* are in the Pacific.
VERIFIED

SETUN   *6/2003, B Lemachko* / 0573515

4 MIKHAIL RUDNITSKY (PROJECT 05360/1) CLASS
(SALVAGE AND MOORING VESSELS) (ARS)

MIKHAIL RUDNITSKY	GEORGY KOZMIN	GEORGY TITOV	SAYANY

Displacement, tons: 10,700 full load
Dimensions, feet (metres): 427.4 × 56.7 × 23.9 *(130.3 × 17.3 × 7.3)*
Main machinery: 1 S5DKRN62/140-3 diesel; 6,100 hp(m) *(4.48 MW)*; 1 shaft
Speed, knots: 16. **Range, n miles:** 12,000 at 15.5 kt
Complement: 72 (10 officers)
Radars: Navigation: Palm Frond; Nayada; I-band.
Sonars: MG 89 *(Sayany)*.

Comment: Built at Vyborg, based on Moskva Pionier class merchant ship hull. First completed 1979, second in 1980, third in 1983 and fourth in 1984. Fly flag of Salvage and Rescue Service. Have two 40 ton and one 20 ton lift with cable fairleads forward and aft. This lift capability is adequate for handling small submersibles, one of which is carried in the centre hold. *Sayany* is also described as a research ship and has a high-frequency sonar. *Rudnitsky* and mini-submarine AS 26 took part in the *Kursk* rescue attempts in August 2000. *Rudnitsky* and *Titov* in the Northern Fleet and the other two in the Pacific.

VERIFIED

MIKHAIL RUDNITSKY   6/2003, B Lemachko / 0573514

8 KASHTAN (PROJECT 141) CLASS
(BUOY TENDERS) (ABU/AGL/ARS)

ALEXANDR PUSHKIN (ex-*KIL 926*)	KIL 143	KIL 164	KIL 498
KIL 927	KIL 158	SS 750 (ex-*KIL 140*)	KIL 168

Displacement, tons: 4,600 full load
Dimensions, feet (metres): 313.3 × 56.4 × 16.4 *(95.5 × 17.2 × 5)*
Main machinery: 4 Wärtsilä diesels; 29,000 hp(m) *(2.31 MW)*; 2 shafts
Speed, knots: 13.5
Complement: 51 plus 20 spare berths
Radars: Navigation: 2 Nayada; I-band.

Comment: Enlarged Sura class built at the Neptun Shipyard, Rostock. Ordered 29 August 1986; *Alexandr Pushkin* handed over in June 1988 and is classified as an AGL in the Baltic; *927* to the Pacific in July 1989; *143* to the North in July 1989; *158* to the Black Sea in November 1989; *164* to the North in January 1990; *498* to the Pacific in November 1990 and *168* to the Pacific in mid-1991. Lifting capacity: one 130 ton lifting frame, one 100 ton derrick, one 12.5 ton crane and one 10 ton derrick. All are civilian operated except *SS 750* in the Baltic which is used to support submersibles AS 22 and AS 26. *158* deployed to Tartous for several months in late 2002.

UPDATED

SS 750 (with mini-sub AS 22)   8/2004*, E & M Laursen / 1042320

4 SURA (PROJECT 145) CLASS (BUOY TENDERS) (ABU)

KIL 22	KIL 27	KIL 29	KIL 31

Displacement, tons: 2,370 standard; 3,150 full load
Dimensions, feet (metres): 285.4 × 48.6 × 16.4 *(87 × 14.8 × 5)*
Main machinery: Diesel-electric; 4 diesel generators; 2 motors; 2,240 hp(m) *(1.65 MW)*; 2 shafts
Speed, knots: 12. **Range, n miles:** 2,000 at 11 kt
Complement: 40
Cargo capacity: 900 tons cargo; 300 tons fuel for transfer
Radars: Navigation: 2 Don 2; I-band.

Comment: Heavy lift ships built as mooring and buoy tenders at Rostock in East Germany between 1965 and 1976. Lifting capacity: one 65 ton derrick and one 65 ton stern cage. Have been seen to carry 12 m DSRVs. *KIL 27* is in the Pacific, *KIL 29* in the Baltic and *KIL 22* and *KIL 31* in the North. Four others are laid up. One to Ukraine in 1997.

VERIFIED

KIL 31   4/1996, van Ginderen Collection

1 ELBRUS (OSIMOL) (PROJECT 537) CLASS
(SUBMARINE RESCUE SHIP) (ASRH)

ALAGEZ

Displacement, tons: 19,000 standard; 22,500 full load
Dimensions, feet (metres): 575.8 × 80.4 × 27.9 *(175.5 × 24.5 × 8.5)*
Main machinery: Diesel-electric; 4 diesel generators; 2 motors; 20,000 hp(m) *(14.7 MW)*; 2 shafts
Speed, knots: 17. **Range, n miles:** 14,500 at 15 kt
Complement: 420
Guns: 4—30 mm/65 (2 twin).
Radars: Navigation: 2 Nayada and 2 Palm Frond; I-band.
Helicopters: 1 Ka-25 Hormone C.

Comment: Very large submarine rescue and salvage ship with icebreaking capability, possibly in view of under-ice capability of some SSBNs. Built at Nikolayev, and completed in 1982. Can carry two submersibles in store abaft the funnel which are launched from telescopic gantries. Based in the Pacific. Repairs of the Nepa class ASR *Karpaty* based in the Baltic have been discontinued.

UPDATED

ALAGEZ   6/2004*, Ships of the World / 0583298

24 SHELON I/II (PROJECT 1388/1388M) CLASS (YPT/YAG)

TL and KRH series

Displacement, tons: 270 full load
Dimensions, feet (metres): 150.9 × 19.7 × 6.6 *(46 × 6 × 2)*
Main machinery: 2 diesels; 8,976 hp(m) *(6.6 MW)*; 2 shafts
Speed, knots: 26. **Range, n miles:** 1,500 at 10 kt
Complement: 14
Radars: Navigation: Spin Trough or Kivach; I-band.

Comment: Type I built 1978-84. Built-in weapon recovery ramp aft. Type II built 1985-87. Type IIs can be used as environmental monitoring ships. One is an Admirals' yacht in the Baltic, and others are used as personnel transports.

UPDATED

SHELON I   7/2004* / 1042326

SHELON II   6/2003, B Lemachko / 0573513

FLAMINGO (TANYA) (PROJECT 1415) CLASS (TENDERS) (YDT)

Displacement, tons: 42 full load
Dimensions, feet (metres): 72.8 × 12.8 × 4.6 *(22.2 × 3.9 × 1.4)*
Main machinery: 1 Type 3-D-12 diesel; 300 hp(m) *(220 kW)* sustained; 1 shaft
Speed, knots: 12
Complement: 8

Comment: Successor to Nyryat II. There are some 27 with RVK numbers (diving tenders). There are also about 20 (PSKA numbers) assigned to the Border Guard for harbour patrol duties. These are known as the Kulik class. Other craft have BSK, RK (workboats) PRDKA (counterswimmer cutter) and BGK (inshore survey) numbers.

VERIFIED

FLAMINGO (Diving Tender)   *6/2003, A Sheldon-Duplaix / 0573512*

POLUCHAT I, II and III CLASSES (PROJECT 368) (YPT)

TL series

Displacement, tons: 70 standard; 100 full load
Dimensions, feet (metres): 97.1 × 19 × 4.8 *(29.6 × 5.8 × 1.5)*
Main machinery: 2 M 50 diesels; 2,200 hp(m) *(1.6 MW)* sustained; 2 shafts
Speed, knots: 20. **Range, n miles:** 1,500 at 10 kt
Complement: 15
Guns: 2—14.5 mm (twin) MGs (in some).
Radars: Navigation: Spin Trough; I-band.

Comment: Employed as specialised or dual-purpose torpedo recovery vessels and/or patrol boats. They have a stern slipway. Several exported as patrol craft. Some used by the Border Guard. Transfers: Algeria, Angola, Congo (three), Ethiopia (one), Guinea-Bissau, India, Indonesia (three), Iraq (two), Mozambique, Somalia (six), Syria, Tanzania, Vietnam (five), North Yemen (two), South Yemen. About 13 remain in service.

VERIFIED

POLUCHAT I   *11/1991, MoD Bonn / 0081713*

PO 2 (PROJECT 501) and NYRYAT 2 (PROJECT 1896) CLASSES (TENDERS) (YDT)

Displacement, tons: 56 full load
Dimensions, feet (metres): 70.5 × 11.5 × 3.3 *(21.5 × 3.5 × 1)*
Main machinery: 1 Type 3-D-12 diesel; 300 hp(m) *(220 kW)* sustained; 1 shaft
Speed, knots: 12
Complement: 8
Guns: Some carry 1—12.7 mm MG on the forecastle.

Comment: This 1950s design of hull and machinery has been used for a wide and diverse number of adaptations. Steel hull. Nyryat 2 have the same characteristics but are used as diving tenders (RVK), inshore survey craft (MGK) and workboats (RK).
Transfers: Albania, Bulgaria, Cuba, Guinea, Iraq. Many deleted.

VERIFIED

NYRYAT 2   *7/2000, Hartmut Ehlers / 0105577*

NYRYAT I (PROJECT 522) CLASS (TENDERS) (YDT)

Displacement, tons: 120 full load
Dimensions, feet (metres): 93 × 18 × 5.5 *(28.4 × 5.5 × 1.7)*
Main machinery: 1 diesel; 450 hp(m) *(331 kW)*; 1 shaft
Speed, knots: 12.5. **Range, n miles:** 1,500 at 10 kt
Complement: 15
Guns: 1—12.7 mm MG (in some).

Comment: Built from 1955. Can operate as patrol craft or diving tenders with recompression chamber. Similar hull and propulsion used for inshore survey craft. Some have BGK, VM or GBP (survey craft) numbers.
Transfers: Albania, Algeria, Cuba, Egypt, Iraq, North Yemen. Many deleted.

VERIFIED

NYRYAT I   *7/2000, Hartmut Ehlers / 0105578*

16 SK 620 CLASS (DRAKON) (TENDERS) (YH/YFL)

MK 391	MK 1409	PSK 1411	SN 128
MK 1303	PSK 382	PSK 1518	SN 401
MK 1407	PSK 405	SN 109	SN 1318
MK 1408	PSK 673	SN 126	SN 1520

Displacement, tons: 236 full load
Dimensions, feet (metres): 108.3 × 24.3 × 6.9 *(33 × 7.4 × 2.1)*
Main machinery: 2 56ANM30-H12 diesels; 620 hp(m) *(456 kW)* sustained; 2 shafts
Speed, knots: 12. **Range, n miles:** 1,000 at 12 kt
Complement: 14 plus 3 spare

Comment: Built at Wisla Shipyard, Poland as a smaller version of the Petrushka class training ship. PSK series serve as harbour ferries. Mostly used as hospital tenders capable of carrying 15 patients.

VERIFIED

PSK 405   *7/2001, J Cislak / 0528310*

29 YELVA (KRAB) (PROJECT 535M) CLASS (DIVING TENDERS) (YDT)

VM 20	VM 263	VM 413-416	VM 807
VM 72	VM 266	VM 420	VM 809
VM 143	VM 268	VM 425	VM 907-910
VM 146	VM 270	VM 429	VM 915
VM 153-154	VM 277	VM 725	VM 919
VM 250	VM 409		

Displacement, tons: 295 full load
Dimensions, feet (metres): 134.2 × 26.2 × 6.6 *(40.9 × 8 × 2)*
Main machinery: 2 Type 3-D-12A diesels; 630 hp(m) *(463 kW)* sustained; 2 shafts
Speed, knots: 12.5. **Range, n miles:** 1,870 at 12 kt
Complement: 30
Radars: Navigation: Spin Trough; I-band.

Comment: Diving tenders built 1971-83. Carry a 1 ton crane and diving bell. Some have submersible recompression chamber. Ice strengthened. One to Cuba 1973, one to Libya 1977. Some have probably been decommissioned.

UPDATED

VM 154   *6/2003 *, B Lemachko / 0580535*

1 + 2 PROJECT 11980 (DIVING TENDER) (YDT)

VM 596

Displacement, tons: 330 full load
Dimensions, feet (metres): To be announced
Main machinery: To be announced
Speed, knots: To be announced

Comment: A new class of diving vessel designed by Almaz Central Design Bureau and built at Vympel Shipyard, Rybinsk. Construction started in the early 1990s but the building programme was suspended until new funds were assigned in 2002. The ship is designed to support diving and salvage operations down to a depth of 60 m and is equipped with the Falkon remote-controlled underwater equipment, which can work at depths up to 300 m. It also carries hydrological instruments and welding equipment for deep-sea work, a satellite television system and a barochamber. The lead vessel was commissioned in the Northern Fleet on 28 November 2004 and is based at Severomorsk. Two further units are planned.
NEW ENTRY

1 SALVAGE LIFTING SHIP (YS)

Name	Builders	Launched	Commissioned
KOMMUNA (ex-*Volkhov*)	De Schelde, Vlissingen	30 Nov 1913	27 July 1915

Displacement, tons: 2,450 full load
Dimensions, feet (metres): 315.0 × 66.9 × 15.4 *(96.0 × 20.4 × 4.7)*
Main machinery: 2 diesels; 2 shafts
Speed, knots: 10. **Range, n miles:** 1,700 at 6 kt
Complement: 250
Radars: Navigation: I-band.

Comment: Catamaran-hulled vessel fitted with four lifting rigs to enable sunken submarines to be lifted between the hulls. Laid down in 1912, the vessel was thought to have been decommissioned in 1978 but returned to service after a refit from 1980-84. Now based at Sevastopol to support the operation of submersibles.
VERIFIED

KOMMUNA   *3/2003, B Lemachko /* 0570910

27 POZHARNY I (PROJECT 364) CLASS
(FIREFIGHTING CRAFT) (YTR)

PZHK 3	PZHK 5	PZHK 17	PZHK 30-32	PZHK36-37
PZHK 41-47	PZHK 49	PZHK 53-55	PZHK 59	PZHK 64
PZHK 66	PZHK 68	PZHK 79	PZHK 82	PZHK84
PZHK86				

Displacement, tons: 180 full load
Dimensions, feet (metres): 114.5 × 20 × 6 *(34.9 × 6.1 × 1.8)*
Main machinery: 2 Type M 50 diesels; 2,200 hp(m) *(1.6 MW)* sustained; 2 shafts
Speed, knots: 12. **Range:** 250 at 12 kt
Complement: 26
Guns: 4—12.7 mm (2 twin) MGs (in some).

Comment: Total of 84 built from mid-1950s to mid-1960s. Harbour fire boats but can be used for patrol duties. One transferred to Iraq (now deleted) and two to Ukraine.
VERIFIED

POZHARNY I   *8/2000, B Lemachko /* 0126224

16 IVA (MORKOV) (PROJECT 1461.3) CLASS (YTR)

PZHK 415	PZHK 417	PZHK 638	PZHK 900	PZHK 1296	PZHK 1378
PZHK 1514-1515	PZHK 1544-1547	PZHK 1560	PZHK 1580	PZHK 1859	PZHK 2055

Displacement, tons: 320 full load
Dimensions, feet (metres): 119.8 × 25.6 × 7.2 *(36.5 × 7.8 × 2.2)*
Main machinery: 2 diesels; 1,040 hp(m) *(764 kW)*; 2 shafts
Speed, knots: 12.5. **Range:** 250 at 12 kt
Complement: 20

Comment: Carry four water monitors. Completed in 1984-86 at Rybinsk. Can be used for patrol/towage. Some are under civilian control.
VERIFIED

PZHK 2055   *6/2003*, B Lemachko /* 0580534

23 PELYM (PROJECT 1799) CLASS (DEGAUSSING SHIPS) (YDG)

SR 26	SR 180	SR 218	SR 241	SR 334	SR 407
SR 77	SR 188	SR 220	SR 276	SR 344	SR 409
SR 111	SR 203	SR 222	SR 280	SR 370	SR 455
SR 179	SR 215	SR 233	SR 281	SR 373	

Displacement, tons: 1,370 full load
Dimensions, feet (metres): 214.8 × 38 × 11.2 *(65.5 × 11.6 × 3.4)*
Main machinery: 1 diesel; 1,536 hp(m) *(1.13 MW)*; 1 shaft
Speed, knots: 14. **Range, n miles:** 1,000 at 13 kt
Complement: 70
Radars: Navigation: Don 2; I-band.

Comment: Built from 1970 to 1987 at Khabarovsk and Gorokhovets. Earlier ships have stump mast on funnel, later ships a tripod main mast and a platform deck extending to the stern. Type name is *sudno razmagnichivanya* meaning degaussing ship. One to Cuba 1982. Several in reserve.
VERIFIED

SR 111   *3/2004, B Lemachko /* 0576441

16 BEREZA (PROJECT 130) CLASS
(DEGAUSSING SHIPS) (YDG)

North	Baltic	Black	Caspian
SR 74	SR 28	SR 137	SR 933
SR 216	SR 120	SR 541	
SR 478	SR 245	SR 939	
SR 548	SR 479		
SR 569	SR 570		
SR 938	SR 936		

Displacement, tons: 1,850 standard; 2,051 full load
Dimensions, feet (metres): 228 × 45.3 × 13.1 *(69.5 × 13.8 × 4)*
Main machinery: 2 Zgoda-Sulzer 8AL25/30 diesels; 2,938 hp(m) *(2.16 MW)* sustained; 2 shafts; cp props
Speed, knots: 13. **Range, n miles:** 1,000 at 13 kt
Complement: 48
Radars: Navigation: Kivach; I-band.

Comment: First completed at Northern Shipyard, Gdansk 1984-1991. One transferred to Bulgaria in 1988. Have NBC citadels and three laboratories. Several in reserve. One to Ukraine in 1997. SR 938 converted to a logistic ship for service in the Polish Navy.
UPDATED

SR 541   *9/2004*, Hartmut Ehlers /* 1042283

HARBOUR CRAFT (YFL/YFU)

Comment: There are numerous types of officers' yachts, harbour launches, training cutters and trials vessels in all of the major Fleet bases. Class names include *Bryza* (Project 722), *Nazhimovets* (Project 286), *Admiralets* (Project 371), *Slavyanka* (Project 20150), *Albatros* (Project 183), Project 14670 and Project 360. ***VERIFIED***

BURUN (Project 14670) 8/2001*, B Lemachko / 0583299

ICEBREAKERS

Notes: Only military icebreakers are shown in this section. Other icebreakers come under civilian management and are now used predominantly for commercial purposes. Civilian ships include the nuclear powered *Taymyr, Vaygach, Arktika, Rossiya, S Soyuz, Yamal*, all of which are operated by the Murmansk Shipping Company. Diesel powered ships include: 20,000 tons: *Ermak, Admiral Makarov, Krasin*; 14,600 tons: *Kapitan Sorokin, Kapitan Dranitsyn, Kapitan Nikolayev, Kapitan Khlebnikov*; 7,700 tons: *Mudyug*; 6,200 tons: *Magadan, Dikson*; 2,240 tons: *Kapitan Bukayev, Kapitan Chadayev, Kapitan Chechkin, Kapitan Krutov, Kapitan Plakhin, Kapitan Zarubin*; 2,200 tons: *Kapitan Babichev, Kapitan Borodkin, Kapitan Chudinov, Kapitan Demidov, Kapitan Evdokimov, Kapitan Metsayk, Kapitan Moshkin, Kapitan Yevdokimov, Avraamiy Zavenyagin*; 2,100 tons: *Kapitan A Radzhabov, Kapitan Kosolabov, Kapitan M Izmaylov*. The growing demand for oil tanker shipments in the Arctic region means that there is a potential shortage of icebreakers. This may be met by completing *50 Let Pobeda*, a nuclear-powered vessel which has lain unfinished at Baltic Shipyard, St Petersburg since 1989. Construction may be completed by 2006 at which stage *Arktika, Rossiya* and *Taymyr* will require life-extension refits.

2 DOBRYNYA NIKITICH CLASS (AGB)

VLADIMIR KAVRAYSKY   BURAN

Displacement, tons: 2,995 full load
Measurement, tons: 2,254 gross; 1,118 dwt; 50 net
Dimensions, feet (metres): 222.1 × 59.4 × 20 *(67.7 × 18.1 × 6.1)*
Main machinery: Diesel-electric; 3 Type 13-D-100 or 3 Wärtsilä 6L 26 *(Kruzenshtern)* diesel generators; 3 motors; 5,400 hp(m) *(4 MW)*; 3 shafts (1 fwd, 2 aft)
Speed, knots: 14.5. **Range, n miles:** 5,500 at 12 kt
Complement: 45
Guns: 2—57 mm/70 (twin). 2—37 mm/63.
Radars: Navigation: 2 Don 2; I-band.

Comment: Built at Admiralty Yard, Leningrad between 1960 and 1971. *Kavraysky* is in the Northern Fleet and *Buran* in the Baltic. Of the 18 others originally built, some have been decommissioned and others transferred to civilian service. ***UPDATED***

BURAN 10/2002*, B Lemachko / 1042317

TUGS

Notes: (1) SB means *Spasatelny Buksir* or Salvage Tug. MB means *Morskoy Buksir* or Seagoing Tug.
(2) Two Baklazhan (Project 5757) class tugs are owned by a civilian salvage company.

1 PRUT (PROJECT 527M) CLASS (RESCUE TUG) (ATS)

EPRON

Displacement, tons: 2,120 standard; 2,800 full load
Dimensions, feet (metres): 295.9 × 46.9 × 18.0 *(90.2 × 14.3 × 5.5)*
Main machinery: Diesel-electric; 4 diesel generators; 2 motors; 10,000 hp(m) *(7.35 MW)*; 2 shafts
Speed, knots: 20. **Range, n miles:** 9,000 at 16 kt
Complement: 140
Radars: Navigation: Don-2; I-band.

Comment: Large rescue tug built at Nikolayev, Ukraine and completed in 1968. Carries two heavy-duty derricks, submersible recompression chambers, rescue chambers and bells. Last survivor of the class which is based in the Black Sea. ***NEW ENTRY***

EPRON 9/2004*, Hartmut Ehlers / 1042286

3 INGUL (PROJECT 1453) CLASS (SALVAGE TUGS) (ATS)

PAMIR   MASHUK   ALTAY (ex-*Karabakh*)

Displacement, tons: 4,050 full load
Dimensions, feet (metres): 304.4 × 50.5 × 19 *(92.8 × 15.4 × 5.8)*
Main machinery: 2 Type 58-D-4R diesels; 6,000 hp(m) *(4.4 MW)*; 2 shafts; cp props
Speed, knots: 19. **Range, n miles:** 9,000 at 19 kt
Complement: 71 plus salvage party of 18
Radars: Navigation: 2 Palm Frond; I-band.
IFF: High Pole. Square Head.

Comment: Built at Admiralty Yard, Leningrad in 1975-84. NATO class name the same as one of the Klasma class cable-ships. Naval-manned arctic salvage and rescue tugs. Two more, *Yaguar* (Murmansk) and *Bars* (Vladivostok), operate with the merchant fleet. Carry salvage pumps, diving and firefighting gear as well as a high-line for transfer of personnel. Fitted for guns but these are not carried. *Pamir* and *Altay* in the North, *Mashuk* in the Pacific. ***VERIFIED***

ALTAY 4/1997 / 0019072

3 SLIVA (PROJECT 712) CLASS (SALVAGE TUGS) (ATS)

SB 406   PARADOKS SB 921   SHAKHTER SB 922

Displacement, tons: 3,050 full load
Dimensions, feet (metres): 227 × 50.5 × 16.7 *(69.2 × 15.4 × 5.1)*
Main machinery: 2 Russkiy SEMT-Pielstick 6 PC2.5 L 400 diesels; 7,020 hp(m) *(5.2 MW)* sustained; 2 shafts; cp props; bow thruster
Speed, knots: 16. **Range:** 6,000 at 16 kt
Complement: 43 plus 10 salvage party
Radars: Navigation: 2 Nayada; I-band.

Comment: Built at Rauma-Repola, Finland. Based on Goryn design. *SB 406* completed 20 February 1984. Second pair ordered 1984 *SB 921* completed 5 July 1985 and *SB 922* on 20 December 1985. *SB 922* named *Shakhter* in 1989. A fourth of class, *Iva SB 408*, was sold illegally to a Greek company in March 1993 and now flies the flag of Cyprus but is operated as a 'joint venture' with the Russian Navy. Diving facilities to 60 m. Bollard pull 60 tons. *SB 406* based in the Northern Fleet, *SB 921* in the Baltic and *SB 922* in the Black Sea. ***UPDATED***

PARADOKS 6/2003*, B Lemachko / 0580536

8 KATUN I (PROJECT 1893) and 4 KATUN II (PROJECT 1993) CLASSES (SALVAGE TUGS) (ATS)

Katun I: PZHS 96, 98, 123, 209, 273, 279, 282, 551
Katun II: PZHS 64, 92, 95, 219

Displacement, tons: 1,005 (Katun I); 1,220 (Katun II) full load
Dimensions, feet (metres): 205.3 × 33.1 × 11.5 *(62 × 10.1 × 3.5)* (Katun I)
Main machinery: 2 diesels; 5,000 hp(m) *(3.68 MW)*; 2 shafts
Speed, knots: 17. **Range, n miles:** 2,000 at 17 kt
Complement: 32
Radars: Navigation: Spin Trough or Kivach (Katun II); I-band.
IFF: High Pole A.

Comment: Katun I built at Kolpino 1970-78. Equipped for firefighting and rescue. *PZHS 64, 92, 95 and 219*, all Katun II, were completed in 1982 and are 3 m *(9.8 ft)* longer than Katun I with an extra bridge level and lattice masts. *273* and *279* are based in the Caspian Sea, *92, 95, 209* and *219* in the Pacific, *64* and *98* in the North, *96, 551* and *282* in the Baltic and *123* in the Black Sea.   ***UPDATED***

282 (Katun I)   8/2004* / 1042310

10 GORYN (PROJECT 714) CLASS (ARS/ATA)

MB 15	MB 18	MB 32	MB 35
MB 36	MB 38	MB 119	MB 241
SB 522 (ex-*MB 62*)	MB 523 (ex-*MB 64*)		

Displacement, tons: 2,240 standard; 2,600 full load
Dimensions, feet (metres): 208.3 × 46.9 × 16.7 *(63.5 × 14.3 × 5.1)*
Main machinery: 2 Russkiy SEMT-Pielstick 6 PC2.5 L 400 diesels; 7,020 hp(m) *(5.2 MW)* sustained; 2 shafts; cp props; bow thruster
Speed, knots: 15
Complement: 43 plus 16 spare berths
Radars: Navigation: 2 Don 2 or Nayada or Kivach; I-band.

Comment: Built by Rauma-Repola 1977-83. Have sick-bay. First ships have goalpost mast with 10 and 5 ton derricks and bollard pull of 35 tons. Remainder have an A-frame mast with a 15 ton crane and bollard pull of 45 tons. SB number indicates a 'rescue' tug. Three in the North, four in the Pacific, two in the Baltic and one in the Black Sea. One transferred to Ukraine in 1997.   ***VERIFIED***

GORYN SB 522   2/2001, B Lemachko / 0126222

12 SORUM (PROJECT 745) CLASS (ATA)

MB 4	MB 37	MB 58	MB 76	MB 100	MB 148
MB 28	MB 56	MB 61	MB 99	MB 110	MB 236

Displacement, tons: 1,660 full load
Dimensions, feet (metres): 190.2 × 41.3 × 15.1 *(58 × 12.6 × 4.6)*
Main machinery: Diesel-electric; 2 Type 5—2-DW2 diesel generators; 2,900 hp(m) *(2.13 MW)*; 1 motor; 2,000 hp(m) *(1.47 MW)*; 1 shaft
Speed, knots: 14. **Range, n miles:** 3,500 at 13 kt
Complement: 35
Guns: 4—30 mm/65 (2 twin) (all fitted for, but only Border Guard ships carry them).
Radars: Navigation: 2 Don 2 or Nayada; I-band.
IFF: High Pole B.

Comment: A class of ocean tugs with firefighting and diving capability. Built in Yaroslavl and Oktyabskoye from 1973 to 1989, design used for Ministry of Fisheries rescue tugs.   ***VERIFIED***

MB 37   10/2002, Ships of the World / 0529808

14 OKHTENSKY (PROJECT 733/733S) CLASS (ARS/ATA)

AYANKA SB 3	SPUTNIK MB 52	MB 172
MOSHCHNY SB 6	MB 162	MB 174
MB 12	SERDITY MB 165	SATURN MB 178
MB 21	POCHETNYY MB 169	MB 241
MB 23	LOKSA MB 171	

Displacement, tons: 948 full load
Dimensions, feet (metres): 156.1 × 34 × 13.4 *(47.6 × 10.4 × 4.1)*
Main machinery: Diesel-electric; 2 BM diesel generators; 1 motor; 1,500 hp(m) *(1.1 MW)*; 1 shaft
Speed, knots: 13. **Range, n miles:** 8,000 at 7 kt; 6,000 at 13 kt
Complement: 40
Guns: 2—57 mm/70 (twin) or 2—25 mm/80 (twin) (Border Guard only).
Radars: Navigation: 1 or 2 Don 2 or Spin Trough; I-band.
IFF: High Pole B.

Comment: Ocean-going salvage (MB) and rescue tugs (SB). First of a total of 62 completed 1958. Fitted with powerful pumps and other apparatus for salvage. A number of named ships are operated by the Border Guard and are armed. Two to Ukraine in 1997. Many have been scrapped.   ***UPDATED***

MB 174   9/2004*, Hartmut Ehlers / 1042285

18 PROMETEY (PROJECT 498/04983/04985) CLASS (TUGS) (YTB)

RB 1	RB 98	RB 179	RB 217	RB 265	RB 327
RB 7	RB 158	RB 201	RB 239	RB 296	RB 360
RB 57	RB 173	RB 202	RB 262	RB 314	RB 362

Displacement, tons: 360 full load
Dimensions, feet (metres): 96.1 × 27.2 × 10.5 *(29.3 × 8.3 × 3.2)*
Main machinery: 2 diesels; 1,200 hp(m) *(895 kW)*; 2 shafts
Speed, knots: 11

Comment: Entered service 1973-83. Bollard pull 14 tons. Later versions have more powerful engines. Based in the Northern, Pacific, Baltic and Black Sea Fleets.   ***NEW ENTRY***

RB 296   6/2003*, B Lemachko / 0580540

35 SIDEHOLE I and II (PROJECT 737 K/M) CLASS (TUGS) (YTB)

BUK 600	RB 25	RB 49	RB 194	RB 233	RB 248
RB 2	RB 26	RB 51	RB 197	RB 237	RB 250
RB 5	RB 29	RB 52	RB 198	RB 240	RB 256
RB 17	RB 43	RB 168	RB 199	RB 244	RB 310
RB 20	RB 44	RB 192	RB 212	RB 246	RB 311
RB 23	RB 46	RB 193	RB 232	RB 247	

Displacement, tons: 206 full load
Dimensions, feet (metres): 79.4 × 23.0 × 11.1 *(24.2 × 7.0 × 3.4)*
Main machinery: 2 diesels; 900 hp(m) *(670 kW)*; 2 shafts
Speed, knots: 10

Comment: Entered service 1973-83. Bollard pull 10 tons. Based in all fleets.   ***NEW ENTRY***

11 STIVIDOR (PROJECT 192) CLASS (TUGS) (YTB)

RB 22	RB 108	RB 136	RB 244	RB 280	RB 326
RB 40	RB 109	RB 167	RB 247	RB 325	

Displacement, tons: 575 full load
Dimensions, feet (metres): 117.1 × 31.1 × 15.1 *(35.7 × 9.5 × 4.6)*
Main machinery: 2 diesels; 2,400 hp(m) *(1.78 MW)*; 2 shafts; bow-thruster
Speed, knots: 12

Comment: Entered service 1980-90. Bollard pull 35 tons. Equipped with three water cannons. Based in the Northern, Pacific and Black Sea Fleets.

NEW ENTRY

RB 325
3/2004*, B Lemachko / 0580541

RUSSIAN FEDERAL BORDER GUARD SERVICE (EX-MARITIME BORDER GUARD)

General

(1) The Border Guard would be integrated with naval operations in a crisis. Formerly run by the KGB, the force came under the Ministry of Defence in October 1991 and was then given to the Ministry of Interior in December 1993. It merged with the Federal Security Service on 11 March 2003.

(2) From 1993 the Border Guard started to fly its own ensign which is the St Andrews Cross with a white border on a green background. Diagonal stripes are painted on the hull in white/blue/red. So far these colours have been seen in all Fleets except the Black Sea.
(3) Roles include Law Enforcement, Port Security, Counter Intelligence, Counter Terrorism and Fishery Protection.

Personnel

2005: 10,000 approx

FRIGATES

Notes: In addition to the frigates listed, there are believed to be a 'Grisha V' *(Stelyak)*, three 'Grisha III' *(Bezuprechniy, Zorkiy* and *Smelyy)*, which operate in the Pacific region. However, their operational status is not known. Four 'Grisha II' *(Izumrud, Predaniy, Nadezhnyy* and *Dozorniy)* are reported to be based in the Northern region but their operational status is also unknown.

7 KRIVAK III (NEREY) (PROJECT 1135MP) CLASS (FFHM)

Name	No	Builders	Laid down	Launched	Commissioned
MENZHINSKY	113	Kamish-Burun, Kerch	14 Aug 1981	31 Dec 1982	29 Dec 1983
DZERZHINSKY	097	Kamish-Burun, Kerch	11 Jan 1984	2 Mar 1984	29 Dec 1984
OREL (ex-*Imeni XXVII Sezda KPSS*)	156	Kamish-Burun, Kerch	26 Sep 1983	2 Nov 1985	30 Sep 1986
PSKOV (ex-*Imeni LXX Letiya VCHK-KGB*)	104	Kamish-Burun, Kerch		1987	30 Dec 1987
ANADYR (ex-*Imeni LXX Letiya Pogranvoysk*)	060	Kamish-Burun, Kerch	22 Oct 1987	28 Mar 1988	16 Aug 1989
KEDROV	103	Kamish-Burun, Kerch	5 Nov 1988	30 Apr 1989	20 Nov 1989
VOROVSKY	052	Kamish-Burun, Kerch	20 Feb 1990	28 July 1990	29 Dec 1990

Displacement, tons: 3,100 standard; 3,650 full load
Dimensions, feet (metres): 405.2 × 46.9 × 24 (sonar) *(123.5 × 14.3 × 7.3)*
Main machinery: COGAG; 2 M8K gas-turbines; 55,500 hp(m) *(40.8 MW)*; 2 M 62 gas-turbines; 13,600 hp(m) *(10 MW)*; 2 shafts
Speed, knots: 32. **Range, n miles:** 4,000 at 14 kt; 1,600 at 30 kt
Complement: 194 (18 officers)

Missiles: SAM: 1 SA-N-4 Gecko (Zif 122) twin launchers ❶; Osa-M semi-active radar homing to 15 km *(8 n miles)* at 2.5 Mach; warhead 50 kg; altitude 9.1-3,048 m *(30-10,000 ft)*; 20 missiles.

Guns: 1—3.9 in *(100 mm)*/59 ❷; 60 rds/min to 15 km *(8.2 n miles)*; weight of shell 16 kg.
2—30 mm/65 ❸; 6 barrels per mounting; 3,000 rds/min combined to 2 km.
Torpedoes: 8—21 in *(533 mm)* (2 quad) tubes ❹. Combination of 53 cm torpedoes (see table at front of section).
A/S mortars: 2 RBU 6000 12-tubed trainable ❺; range 6,000 m; warhead 31 kg. MRG-7 55 mm grenade launcher.
Mines: Capacity for 16.
Countermeasures: Decoys: 4 PK 16 or 10 PK 10 chaff launchers. Towed torpedo decoy.
ESM/ECM: 2 Bell Shroud. 2 Bell Squat. Half Cup laser warning (in some).

Radars: Air search: Top Plate ❻; 3D; D/E-band.
Surface search: Peel Cone ❼; I-band.
Fire control: Pop Group ❽; F/H/I-band (for SA-N-4). Kite Screech ❾; H/I/K-band. Bass Tilt ❿; H/I-band.
IFF: High Pole B. Salt Pot.
Sonars: Bull Nose (MGK-335S or MG-332); hull-mounted; active search and attack; medium frequency.

Helicopters: 1 Ka-27 Helix ⓫.

Programmes: Type name was originally *bolshoy protivolodochny korabl*, meaning large anti-submarine ship. Changed in 1977-78 to *storozhevoy korabl* meaning escort ship. The naval Krivaks are known as the Burevestnik class.
Structure: Krivak III class built for the former KGB but now under Border Guard Control. The removal of SS-N-14 and one SA-N-4 mounting compensates for the addition of a hangar and flight deck.
Sales: The Talwar class is an improved version of the Krivak III being built for India. Three of the Krivak III class transferred to Ukraine in July 1997.

VERIFIED

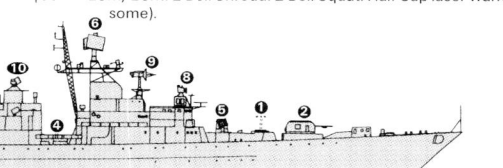

KRIVAK III

(Scale 1 : 1,200), Ian Sturton / 0506085

VOROVSKY (III)

10/1991

PATROL FORCES

Notes: In addition to the patrol forces listed, *Pluton* and *Strelets*, two Yug class former research vessels, are operated as patrol craft in Arctic waters.

8 ALPINIST (PROJECT 503) CLASS (PBO)

| ANTIAS | ARGAL | BARS | DIANA | PALIYA | PARELLA | +2 |

Displacement, tons: 1,150 full load
Dimensions, feet (metres): 176.2 × 34.4 × 13.4 *(53.7 × 10.5 × 4.1)*
Main machinery: 1 diesel; 1 shaft; cp prop
Speed, knots: 12. **Range, n miles:** 7,000 at 12 kt
Complement: 44

Comment: Trawler design adapted for use as fishery protection role. The named ships were built at Volgograd and at Khabarovsk between 1997 and 2000 while the latest two (unnamed) ships were built at Yarslavl and entered service in late 2001.
VERIFIED

4 KOMANDOR CLASS (PSO)

| KOMANDOR | SHKIPER GYEK | HERLUF BIDSTRUP | MANCHZHUR |

Displacement, tons: 2,435 full load
Dimensions, feet (metres): 289.7 × 44.6 × 15.4 *(88.3 × 13.6 × 4.7)*
Main machinery: 2 Russkiy SEMT-Pielstick 6 PC2.5 L400 diesels; 7,020 hp(m) *(5.2 MW)*; 1 shaft; cp prop; bow thruster
Speed, knots: 20. **Range, n miles:** 7,000 at 19 kt
Complement: 42
Radars: Navigation: Furuno; I-band.
Helicopters: 2 Ka-32 Helix D for SAR.

Comment: Specialist offshore patrol vessels ordered in December 1987 from Danyard, Frederikshaven, Denmark and delivered 1989-1990. The hangar is below the helicopter deck. Transferred from the Ministry of Fisheries to the Federal Border Guard and based in the Pacific.
UPDATED

MANCHZHUR   6/2003*, B Lemachko / 0580538

1 AKADEMIK FERSMAN CLASS (PSO)

SPASTEL PROCONCHIK (ex-*Akademik Fersman*)

Displacement, tons: 3,250 full load
Dimensions, feet (metres): 268.8 × 48.6 × 16.4 *(81.9 × 14.8 × 5.0)*
Main machinery: 1 Zgoda-Sulzer diesel; 1 shaft; cp prop
Speed, knots: 14. **Range, n miles:** 12,000 at 14 kt
Complement: 30 + 29 spare

Comment: Former seismic research ship which was built at Szczecin, Poland and originally entered service in 1986. Transferred to Russian government service in 1997.
VERIFIED

4 IVAN SUSANIN (PROJECT 97P) CLASS (PATROL SHIPS) (PGH)

| AISBERG 150 | MURMANSK (ex-*Dunay*) 018 | NEVA 170 | VOLGA 183 |

Displacement, tons: 3,567 full load
Dimensions, feet (metres): 229.7 × 59.4 × 21 *(70 × 18.1 × 6.4)*
Main machinery: Diesel-electric; 3 Type 13-D-150 diesel generators; 3 motors; 5,400 hp(m) *(4 MW)*; 3 shafts (1 fwd, 2 aft)
Speed, knots: 14.5. **Range, n miles:** 5,500 at 12.5 kt
Complement: 45
Guns: 2—3 in *(76 mm)*/60 (twin). 2—30 mm/65 AK 630 (not in all).
Radars: Surface search: Strut Curve; F-band.
Navigation: 2 Don Kay or Palm Frond; I-band.
Fire control: Hawk Screech; I-band.
Helicopters: Platform only.

Comment: Built at Admiralty Yard, Leningrad between 1974 and 1981. Generally similar to Dobrynya Nikitich class though larger with a tripod mast and different superstructure. Former icebreakers operated primarily as patrol ships. Two in the Pacific and two in the Northern Fleet. Two deleted so far.
VERIFIED

MURMANSK   8/2002 / 0528312

17 SORUM (PROJECT 745P) CLASS (PBO)

ALDAN (ex-*Bug*) 142	KARELIA 103	ZAPOLARYE 038
AMUR 043	SAKHALIN 185	PRIMORYE 172
BREST 106	URAL 016	LADOGA 058
CHUKOTKA 067	BAYKAL (ex-*Yan Berzin*) 105	TVER 022
DON 103	YENISEY 030	VYATKA (ex-*Victor Kingisepp*) 035
KAMCHATKA 198	ZABAYKALYE 073	

Displacement, tons: 1,660 full load
Dimensions, feet (metres): 190.2 × 41.3 × 15.1 *(58 × 12.6 × 4.6)*
Main machinery: Diesel-electric; 2 Type 5—2-DW2 diesel generators; 2,900 hp(m) *(2.13 MW)*; 1 motor; 2,000 hp(m) *(1.47 MW)*; 1 shaft
Speed, knots: 14. **Range, n miles:** 3,500 at 13 kt
Complement: 35
Guns: 4—30 mm/65 (2 twin) (all fitted for, but only Border Guard ships carry them).
Radars: Navigation: 2 Don 2 or Nayada; I-band.
IFF: High Pole B.

Comment: A class of ocean tugs armed for use as patrol vessels in the North, Pacific, Baltic and Caspian. Built in Yaroslavl and Oktyabskoye from 1973 to 1989, design used for Ministry of Fisheries rescue tugs.
VERIFIED

PRIMORYE   9/2002, Hachiro Nakai / 0528309

0 + 1 (9) SPRUT (PROJECT 6457S) CLASS (PSO)

Displacement, tons: 900 standard
Dimensions, feet (metres): 216.2 × 34.8 × 11.5 *(65.9 × 10.6 × 3.5)*
Main machinery: 1 MTU 16V 1163 diesel; 7,000 hp(m) *(5.2 MW)*; 1 shaft; fixed propeller
Speed, knots: 21.5. **Range, n miles:** 12,000 at 12 kt
Complement: 15 + 10 in temporary accommodation
Radars: Surface search: I-band.
Navigation: I-band.
Helicopters: Platform for 1 light.

Comment: Specialist Fishery Protection vessel based on German Coast Guard Bad Bramstedt design. First of class laid down at Yantar, Kaliningrad on 27 May 2002. Delivery is expected in 2005. Steel hull with aluminium superstructure. Equipped with a high speed RHIB for interception. A class of ten is planned.
UPDATED

SPRUT   4/2003, Military Parade / 0570892

2 PAUK II (PROJECT 1241 PE) CLASS (PCM)

| NOVOROSSIYSK 043 | | KUBAN 149 |

Displacement, tons: 495 full load
Dimensions, feet (metres): 191.9 × 33.5 × 11.2 *(58.5 × 10.2 × 3.4)*
Main machinery: 2 Type M 521 diesels; 16,184 hp(m) *(11.9 MW)* sustained; 2 shafts
Speed, knots: 32. **Range, n miles:** 2,400 at 14 kt
Complement: 32

Missiles: SAM: SA-N-5 quad launcher; manual aiming, IR homing to 10 km *(5.4 n miles)* at 1.5 Mach; warhead 1.1 kg.
Guns: 1 USSR 76 mm/60; 120 rds/min to 7 km *(3.8 n miles)*; weight of shell 16 kg.
1—30 mm/65 AK 630; 6 barrels; 3,000 rds/min combined to 2 km.
Torpedoes: 4—21 in *(533 mm)* (2 twin) fixed tubes.
A/S mortars: 2 RBU 1200 5-tubed fixed; range 1,200 m; warhead 34 kg.
Radars: Air/surface search: Cross Dome (Positiv E); E/F-band.
Navigation: Pechora; I-band.
Fire control: Bass Tilt; H/I-band.
Sonars: Rat Tail; VDS (on transom); attack; high frequency.

Comment: Built at Yaroslav Shipyard and entered service in 1997-98 when they were transferred to the Border Guard. Originally intended for Iraq, export Pauk II variant of the type sold to India and Cuba. Has a longer superstructure than the Pauk I with a radome similar to the Parchim II class. The torpedo tubes must be trained out to launch. Both operate in the Black Sea.
NEW ENTRY

18 PAUK I (MOLNYA) (PROJECT 1241-2) CLASS
(FAST ATTACK CRAFT—PATROL) (PCM)

PSKR-804	TOLYATTI	021
PSKR-818	NAKHODKA	023
PSKR-802	KALININGRAD	024
PSKR-810	YAROSLAVL	031
PSKR-816	YASTREB	037
PSKR-811	SARYCH	040
PSKR-808	GRIF	041
PSKR-814	ORLAN	042
PSKR-817	CHEBOKSARY	052
PSKR-812	SOKOL	063
PSKR-806	MINSK	065
PSKR-815	NIKOLAY KAPLUNOV	077
PSKR-803	KONDOR	130
PSKR-801	VORON	132
PSKR-809	KRECHET	134
PSKR-800	BERKUT	136
PSKR-807	KOBCHIK	152
PSKR-805	KORSHUN	182

Displacement, tons: 440 full load
Dimensions, feet (metres): 189 × 33.5 × 10.8 *(57.6 × 10.2 × 3.3)*
Main machinery: 2 Type M 521 diesels; 16,184 hp(m) *(11.9 MW)* sustained; 2 shafts
Speed, knots: 32. **Range, n miles:** 2,400 at 14 kt
Complement: 38

Missiles: SAM: SA-N-5 Grail quad launcher; manual aiming; IR homing to 6 km *(3.2 n miles)* at 1.5 Mach; altitude to 2,500 m *(8,000 ft)*; warhead 1.5 kg; 8 missiles.
Guns: 1—3 in *(76 mm)*/60; 120 rds/min to 15 km *(8 n miles)*; weight of shell 7 kg.
1—30 mm/65 AK 630; 6 barrels; 3,000 rds/min combined to 2 km.
Torpedoes: 4—16 in *(406 mm)* tubes. For torpedo details see table at front of section.
A/S mortars: 2 RBU 1200 5-tubed fixed; range 1,200 m; warhead 34 kg.
Depth charges: 2 racks (12).
Countermeasures: Decoys: 2 PK 16 or 4 PK 10 chaff launchers.
ESM: 3 Brick Plug and 2 Half Hat; radar warning.
Weapons control: Hood Wink optronic director.
Radars: Air/surface search: Peel Cone; E/F-band.
Surface search: Kivach or Pechora or SRN 207; I-band.
Fire control: Bass Tilt; H/I-band.
Sonars: Foal Tail; VDS (mounted on transom); active attack; high frequency.

Programmes: First laid down in 1977 and completed in 1979. In series production at Yaroslavl in the Black Sea and at Vladivostok until 1988 when the Svetlyak class took over. Type name is *maly protivolodochny korabl* meaning small anti-submarine ship. An improved version building at Kharbarovsk in 1995 was not completed.
Structure: An ASW version of the Tarantul class having the same hull form with a 1.8 m extension for dipping sonar. *Berkut*, *Voron* and *Kaliningrad* have a lower bridge than others. A modified version (Pauk II) with a longer superstructure, two twin 533 mm torpedo tubes and a radome similar to the Parchim class built for export.
Operational: Five in the Baltic, two in the Black Sea and the remainder in the Pacific. In addition five naval craft are laid up in the Baltic and one in the Black Sea.
Sales: One to Bulgaria in September 1989 and a second in December 1990. Two to Ukraine in 1996. A variant design built for Vietnam.

UPDATED

MINSK
7/2003, Freddy Philips / 0570908

3 TERRIER (PROJECT 14170) CLASS (PB)

001-003

Displacement, tons: 8.3 full load
Dimensions, feet (metres): 38.4 × 10.2 × 1.6 *(11.7 × 3.1 × 0.5)*
Main machinery: 2 diesels; 2 waterjets
Speed, knots: 32. **Range, n miles:** 120 at 30 kt
Complement: 6

Comment: Built at Zelenodolsk in 2000.

NEW ENTRY

TERRIER 002
*6/2003 *, B Lemachko /* 0580539

3 MURAVEY (ANTARES) (PROJECT 133) CLASS (PCK)

DELFIN   RYBA   TUAPSE

Displacement, tons: 212 full load
Dimensions, feet (metres): 126.6 × 24.9 × 6.2; 14.4 (foils) *(38.6 × 7.6 × 1.9; 4.4)*
Main machinery: 2 gas turbines; 22,600 hp(m) *(16.6 MW)*; 2 shafts
Speed, knots: 60. **Range, n miles:** 410 at 12 kt
Complement: 30 (5 officers)

Guns: 1—3 in *(76 mm)*/60; 120 rds/min to 15 km *(8 n miles)*; weight of shell 7 kg.
1—30 mm/65 AK 630; 6 barrels; 3,000 rds/min combined to 2 km.
Weapons control: Hood Wink optronic director.
Radars: Surface search: Peel Cone; E-band.
Fire control: Bass Tilt; H/I-band.
Sonars: Rat Tail; VDS; active attack; high frequency; dipping sonar.

Comment: Thirteen hydrofoil craft built at Feodosiya in the mid-1980s for the USSR Border Guard. Three transferred to Ukraine and the remainder decommissioned.

NEW ENTRY

MURAVEY CLASS
3/1998, Ukraine Coast Guard / 0050319

1 MUSTANG (PROJECT 18623) CLASS (PBF)

Displacement, tons: 35.5 full load
Dimensions, feet (metres): 65.6 × 14.8 × 3.6 *(20.0 × 4.5 × 1.1)*
Main machinery: 2 Zvezda M-470 diesels; 2,950 hp *(2.2 MW)*; 2 Kamewa waterjets
Speed, knots: 45. **Range, n miles:** 350 at 40 kt
Complement: 6

Comment: Designed by Redan Bureau, St Petersburg and built at Yaroslavl in 2000.

NEW ENTRY

MUSTANG (foreground)
*7/2003 *, R Scott/Jane's /* 0551879

11 SOBOL (PROJECT 12200) CLASS (PBF)

BSK 1-11

Displacement, tons: 57 full load
Dimensions, feet (metres): 91.7 × 16.0 × 4.4 *(27.9 × 4.9 × 1.4)*
Main machinery: 2 diesels; 3,600 hp *(2.6 MW)*; 2 shafts
Speed, knots: 47. **Range, n miles:** 700 at 40 kt
Complement: 6

Comment: Built at Almaz St Petersburg and at Soznovka Zavod, Rybinsk and delivered 2000-03.

NEW ENTRY

SOBOL CLASS (artist's impression)
*6/2004 *, S Breyer /* 1042412

20 SVETLYAK (PROJECT 1041Z) CLASS
(FAST ATTACK CRAFT—PATROL) (PGM)

PSKR-921	OTVAZHNIY	011
PSKR-914	KORSAKOV	013
PSKR-920	PODOLSK	014
PSKR-915	NEVELSK	023
PSKR-918	YUZHNO-SAKHALINSK	026
PSKR-906	SOCHI	028
PSKR-911	SIKTIVKAR	042
PSKR-908	BRIZ	065
PSKR-903	CHOLMSK	088
PSKR-902	STAVROPOL	100
PSKR-913	KIZLJAR	139
PSKR-912	DERBENT	137
PSKR-909	VYBORG	141
PSKR-913	ALMAZ	143
+6		

Displacement, tons: 375 full load
Dimensions, feet (metres): 159.1 × 30.2 × 11.5 *(48.5 × 9.2 × 3.5)*
Main machinery: 3 diesels; 14,400 hp(m) *(10.58 MW)*; 3 shafts
Speed, knots: 31. **Range, n miles:** 2,200 at 13 kt
Complement: 36 (4 officers)
Missiles: SAM; SA-N-5 Grail quad launcher; manual aiming; IR homing to 6 km *(3.2 n miles)* at 1.5 Mach; warhead 1.5 kg.
Guns: 1—3 in *(76 mm)*/60; 120 rds/min to 15 km *(8 n miles)*; weight of shell 7 kg.
  1 or 2—30 mm/65 AK 630; 6 barrels; 3,000 rds/min combined to 2 km; 12 missiles.
Torpedoes: 2—16 in *(406 mm)* tubes; SAET-40; anti-submarine; active/passive homing to 10 km *(5.4 n miles)* at 30 kt; warhead 100 kg.
Depth charges: 2 racks; 12 charges.
Countermeasures: Decoys: 2 PK 16 chaff launchers.
Weapons control: Hood Wink optronic director.
Radars: Air/surface search: Peel Cone; E-band.
  Fire control: Bass Tilt (MP 123); H/I-band.
  Navigation: Palm Frond B; I-band.
  IFF: High Pole B. Square Head.
Sonars: Rat Tail; VDS; active search; high frequency.

Comment: A class of attack craft for the Border Guard built at Vladivostok, St Petersburg and Yaroslavl. Series production after first of class trials in 1989. Although deliveries have been very slow in recent years, the class may still be building with the most recent launch in May 2000. One has a second AK 630 gun vice the 76 mm and no Bass Tilt radars. Six in the Northern Fleet, three in the Baltic, seven in the Pacific, two in the Caspian and two in the Black Sea are all known to be active. Two have been built for Vietnam. A further craft, *Yamalets* completed by Almaz for the Border Guard in late 2004. Three additional craft operated by the Navy. **UPDATED**

ALMAZ   *6/2003, A Sheldon-Duplaix /* 0570905

15 STENKA (TARANTUL) (PROJECT 205P) CLASS
(FAST ATTACK CRAFT—PATROL) (PTF)

PSKR-714 014	PSKR-717 078	PSKR-641 133
PSKR-660 044	PSKR-712 132	PSKR-725 134
PSKR-700 047	PSKR-665 113	PSKR-631 137
PSKR-715 048	PSKR-657 126	PSKR-659 139
PSKR-718 053	PSKR-690 129	PSKR-723 143

Displacement, tons: 211 standard; 253 full load
Dimensions, feet (metres): 129.3 × 25.9 × 8.2 *(39.4 × 7.9 × 2.5)*
Main machinery: 3 Type M 517 or M 583 diesels; 14,100 hp(m) *(10.36 MW)*; 3 shafts
Speed, knots: 37. **Range, n miles:** 800 at 24 kt; 500 at 35 kt; 2,300 at 14 kt
Complement: 25 (5 officers)
Guns: 4—30 mm/65 (2 twin) AK 230.
Torpedoes: 4—16 in *(406 mm)* tubes.
Depth charges: 2 racks.
Radars: Surface search: Pot Drum or Peel Cone; H/I- or E-band.
  Fire control: Drum Tilt; H/I-band.
  Navigation: Palm Frond; I-band.
  IFF: High Pole. 2 Square Head.
Sonars: Stag Ear or Foal Tail; VDS; high frequency; Hormone type dipping sonar.

Comment: Based on the hull design of the Osa class. Construction started in 1967 and continued at a rate of about five a year at Petrovsky, Leningrad and Vladivostok for the Border Guard. Programme terminated in 1989 at a total of 133 hulls. Type name is *pogranichny storozhevoy korabl* meaning border patrol ship. Four based in the Baltic, five in the Black Sea, one in the Pacific, and five in the Caspian Sea.
  Transfers include: Cuba, two in February 1985 and one in August 1985. Four to Cambodia in October 1985 and November 1987. Five transferred to Azerbaijan control in November 1992 and 10 more to Ukraine. **VERIFIED**

STENKA   *6/2000 /* 0126309

1 + 1 (18) MIRAZH (PROJECT 14310) CLASS (PBF)

PK 500

Displacement, tons: 126 full load
Dimensions, feet (metres): 111.5 × 21.7 × 8.9 *(34.0 × 6.6 × 2.7)*
Main machinery: 2 Zvezda M-521 diesels; 16,184 hp(m) *(11.9 MW)*; 2 shafts
Speed, knots: 48. **Range, n miles:** 1,500 at 8 kt
Complement: 12 (2 officers)
Guns: 1—30 mm AK 306.
  2—7.62 mm MGs.
Radars: Surface search: I-band.

Comment: Designed by Almaz and built by Vympel Shipbuilding, Rybinsk. Aluminium-magnesium alloy construction. Three vessels authorised for construction in 1993 but only first of class was completed in 1998. It entered service with the Border Guard in 2001 and has been based in the Caspian Sea. A second craft is reported to have been laid down in 2001. While a class of 20 craft is thought to be required, funding has not been confirmed. **VERIFIED**

MIRAZH   *6/2003, B Lemachko /* 0570906

MIRAZH   *6/2003, E & M Laursen /* 0570907

20 ZHUK (GRIF) (PROJECT 1400/1400M) CLASS
(COASTAL PATROL CRAFT) (PB)

PSKA series

Displacement, tons: 39 full load
Dimensions, feet (metres): 78.7 × 16.4 × 3.9 *(24 × 5 × 1.2)*
Main machinery: 2 Type M 401B diesels; 2,200 hp(m) *(1.6 MW)* sustained; 2 shafts
Speed, knots: 30. **Range, n miles:** 1,100 at 15 kt
Complement: 11 (3 officers)
Guns: 2—14.5 mm (twin, fwd) MGs. 1—12.7 mm (aft) MG.
Radars: Surface search: Spin Trough; I-band.

Comment: Under construction from 1976. Manned by the Border Guard. Export versions have twin (over/under) 14.5 mm aft. Some have twin guns forward and aft. Ukraine Border Guard and has received 12 from the Russians. 16 are in the Baltic and four in the Black Sea. These are the last operational units.
  Transfers: Algeria (one in 1981), Angola (one in 1977), Benin (four in 1978-80), Bulgaria (five in 1977), Cape Verde (one in 1980), Congo (three in 1982), Cuba (40 in 1971-88), Equatorial Guinea (three in 1974-75), Ethiopia (two in October 1982 and two in June 1990), Guinea (two in July 1987), Iraq (five in 1974-75), Cambodia (three in 1985-87), Mauritius (two in January 1990), Mozambique (five in 1978-80), Nicaragua (eight in 1982-86), Seychelles (one in 1981, one in October 1982), Somalia (one in 1974), Syria (six in 1981-84), Vietnam (nine in 1978-88 (at least one passed on to Cambodia), five in 1990 and two in 1995), North Yemen (five in 1978-87), South Yemen (two in 1975). Many have been deleted. **VERIFIED**

ZHUK   *7/1996, J Cíślak /* 0081678

1 + (3) SOKZHOI CLASS (PROJECT 14232) (PBF)

ALBATROS

Displacement, tons: 97.7 full load
Dimensions, feet (metres): 106.6 × 12.9 × 7.9 *(32.5 × 3.92 × 2.4)*
Main machinery: 2 Zvezda M535 diesels; 9,923 hp(m) *(7.4 MW)* sustained; 2 shafts
Speed, knots: 50. **Range, n miles:** 800
Complement: 16
Guns: 2—30 mm AK-306. 1—14.5 mm MG.

Comment: First of a new class of patrol craft launched at Volga Yard, Nizhny Novgorod on 23 June 2000. A feature of the design is that an air-cushion is generated below the hull to produce a planing effect to reduce drag. The machine-gun is mounted in a barbette in the forward part of the craft. Project 14232 is a family of high-speed air-cavern vessels based on a unified platform design developed by the Alekseyev Hydrofoil Design Bureau, Nizhny Novgorod. Other variants of the design have different superstructure configuration, armament and equipment. These include two unarmed vessels of the sister Mercury class *(Petr Matveyev (TS-100) and Pavel Vereshchagin (TS-101))* built for the customs service at Yaroslavl between 1996-2000. Two more of this type ship are to be completed in Yaroslavl and Khabarovsk.
VERIFIED

SOKZHOY CLASS (launch)   6/2000, S Breyer / 0106551

PAVEL VERESHCHAGIN (Customs)   6/2003, E & M Laursen / 0570902

1 + (9) MANGUST (PROJECT 12150) CLASS (PBF)

Displacement, tons: 27 standard
Dimensions, feet (metres): 64.0 × 15.1 × 3.6 *(19.5 × 4.6 × 1.1)*
Main machinery: 2 Zvezda M-470 diesels; 2 Arneson dive props
Speed, knots: 53. **Range, n miles:** 250 at 40 kt
Complement: 6
Guns: 2—14.5 mm MGs.
Radars: Navigation: I-band.

Comment: Prototype TS 300 built by Vympel, Rybinsk for the Customs service and completed in 1998. GRP construction. First Border Guard unit entered service in 2001. Further orders are expected.
VERIFIED

MANGUST   4/2003, Military Parade / 0570900

RIVER PATROL FORCES

Notes: Attached to Black Sea and Pacific Fleets for operations on the Danube, Amur and Usuri Rivers, and to the Caspian Flotilla.

2 YAZ (SLEPEN) (PROJECT 1208) CLASS (PGR)

BLAGOVESHCHENSK 066   SHKVAL 106

Displacement, tons: 440 full load
Dimensions, feet (metres): 180.4 × 29.5 × 4.9 *(55 × 9 × 1.5)*
Main machinery: 3 diesels; 11,400 hp(m) *(8.39 MW)*; 3 shafts
Speed, knots: 24. **Range, n miles:** 1,000 at 10 kt
Complement: 32 (4 officers)
Guns: 2—115 mm tank guns (TB 62) or 100 mm/56.
   2—30 mm/65 AK 630; 6 barrels per mounting.
   4—12.7 mm MGs (2 twin).
   2—40 mm mortars on after deckhouse.
Radars: Surface search: Spin Trough; I-band.
Fire control: Bass Tilt; H/I-band.
IFF: High Pole B. Square Head.

Comment: First entered service in Amur Flotilla 1978. Built at Khabarovsk until 1987. All but these last two have been placed in reserve.
VERIFIED

BLAGOVESHCHENSK   6/1995, B Lemachko / 0570903

7 PIYAVKA (PROJECT 1249) CLASS (PBR)

PSKR 52 117	**PSKR 56** 093
PSKR 53 065	**PSKR 57** 058
PSKR 54 146	**PSKR 58** 123
PSKR 55 013	

Displacement, tons: 229 full load
Dimensions, feet (metres): 136.5 × 20.7 × 2.9 *(41.6 × 6.3 × 0.9)*
Main machinery: 3 diesels; 3,300 hp(m) *(2.42 MW)*; 2 shafts
Speed, knots: 17
Complement: 30 (4 officers)
Guns: 1—30 mm/65 AK 630; 6 barrels. 2—14.5 mm (twin) MGs.
Radars: Surface search: Spin Trough; I-band.

Comment: Built at Khabarovsk 1979-84. Based in Amur Flotilla mostly for logistic support.
UPDATED

PSKR 58   6/2003*, B Lemachko / 0580529

3 OGONEK (PROJECT 12130) CLASS (PBR)

Displacement, tons: 98 full load
Dimensions, feet (metres): 109.6 × 13.8 × 2.6 *(33.4 × 4.2 × 0.8)*
Main machinery: 2 diesels; 2 shafts
Speed, knots: 25
Complement: 17 (2 officers)
Guns: 2—30 mm AK 630.

Comment: A smaller version of the Piyavka class built at Khabarovsk from 1999. Numbers in service are uncertain.
VERIFIED

OGONEK   6/2003, B Lemachko / 0576442

15 SHMEL (PROJECT 1204) CLASS (PGR)

PSKR 341-377 (not all in service)

Displacement, tons: 77 full load
Dimensions, feet (metres): 90.9 × 14.1 × 3.9 *(27.7 × 4.3 × 1.2)*
Main machinery: 2 Type M 50 diesels; 2,200 hp(m) *(1.6 MW)* sustained; 2 shafts
Speed, knots: 25. **Range, n miles:** 600 at 12 kt
Complement: 12 (4 officers)
Guns: 1—3 in *(76 mm)*/48 (tank turret). 1—25 mm/70 (later ships). 2—14.5 mm (twin) MGs (earlier ships). 5—7.62 mm MGs. 1 BP 6 rocket launcher; 18 barrels.
Mines: Can lay 9.
Radars: Surface search: Spin Trough; I-band.

Comment: Completed at Kerch and Nikolayev North (61 Kommuna) 1967-74. Some of the later ships also mount one or two multibarrelled rocket launchers amidships. The 7.62 mm guns fire through embrasures in the superstructure with one mounted on the 76 mm. Can be carried on land transport. Type name is *artillerisky kater* meaning artillery cutter. About 70 have been scrapped or laid up so far including the last naval units. These last survivors are based on the Amur River and belong to the Border Guard.
    Transfers: Four to Cambodia (1984-85) (since decommissioned). Some have been taken over by Belorussian forces, and others allocated to Ukraine.

UPDATED

SHMEL   *6/2000, B Lemachko /* 0106875

5 VOSH (MOSKIT) (PROJECT 1248) CLASS (PGR)

PSKR 300-310 series

Displacement, tons: 229 full load
Dimensions, feet (metres): 140.1 × 20.7 × 3.3 *(42 × 6.3 × 1)*
Main machinery: 3 diesels; 3,300 hp(m) *(2.42 MW)*; 3 shafts
Speed, knots: 17
Complement: 34 (3 officers)
Guns: 1—3 in *(76 mm)*/48 (tank turret). 1—30 mm/65 AK 630. 2—12.7 mm (twin) MGs.
Countermeasures: 1 twin barrel decoy launcher.
Radars: Surface search: Spin Trough; I-band.

Comment: Built at Sretensk on the Shilka river 1980-84. Based on Amur River. Same hull as Piyavka.

VERIFIED

VOSH CLASS   *6/2001, B Lemachko /* 0126218

AUXILIARIES

10 NEON ANTONOV (PROJECT 1595) CLASS
(TRANSPORTS) (AK)

VASILIY SUNTZOV 154	**VYACHESLAV DENISOV** 176	**IVAN YEVTEYEV** 105
IVAN LEDNEV 115	**MIKHAIL KONOVALOV** 184	**SERGEY SUDETSKY** 143
NIKOLAY SIPYAGIN 090	**NIKOLAY STARSHINOV** 119	**DVINA**
IRBIT		

Displacement, tons: 6,400 full load
Dimensions, feet (metres): 311.7 × 48.2 × 21.3 *(95 × 14.7 × 6.5)*
Main machinery: 2 diesels; 7,000 hp(m) *(5.15 MW)*; 2 shafts
Speed, knots: 17. **Range, n miles:** 8,500 at 13 kt
Complement: 45
Cargo capacity: 2,500 tons
Missiles: SAM: 2 SA-N-5 Grail twin launchers; manual aiming; IR homing to 6 km *(3.2 n miles)* at 1.5 Mach; altitude to 2,500 m *(8,000 ft)*; warhead 1.5 kg.
Guns: 2—30 mm/65 (twin). 4—14.5 mm (2 twin) MGs. 4—12.7 mm MGs.
Radars: Navigation: Don Kay; Spin Trough or Palm Frond; I-band.

Comment: Ten of the class built at Nikolayev from 1975 to early 1980s. All in the Pacific except *Dvina* and *Irbit* which are operated by the Russian Navy. Have two small landing craft aft. Armament is not normally mounted.

VERIFIED

MIKHAIL KONOVALOV   *4/1996 /* 0019068

6 KANIN CLASS (PROJECT 16900A) (AKL)

CHANTIJ-MANSISK	**JURGA**	**ARCHANGELSK**	**KANIN**	**URENGOY**	**OGRA**

Displacement, tons: 920 full load
Dimensions, feet (metres): 149.6 × 28.9 × 8.2 *(45.6 × 8.8 × 2.5)*
Main machinery: 2 diesels; 800 hp(m) *(558 kW)*; 2 shafts
Speed, knots: 9. **Range, n miles:** 3,500 at 9 kt
Complement: 22

Comment: Built in the Pacific since 1996 for the Border Guard. Others may be building for commercial service. Ice reinforced bows for Arctic service. *Chantij-Mansisk* based in the Black Sea.

VERIFIED

KANIN CLASS   *12/2002, S Breyer /* 0576443

AMPHIBIOUS FORCES

4 CZILIM (PROJECT 20910) CLASS (ACV/UCAC)

Displacement, tons: 8.6 full load
Dimensions, feet (metres): 39.4 × 19 *(12 × 5.8)*
Main machinery: 2 Deutz BF 6M 1013 diesels; 435 hp(m) *(320 kW)* sustained; for lift and propulsion
Speed, knots: 40. **Range, n miles:** 300 n miles at 30 kt
Complement: 2 + 6 Border Guard
Guns: 1—7.62 mm MG. 1—40 mm RPG.
Radars: Navigation: I-band.

Comment: Ordered from Jaroslawski Sudostroiteinyj Zawod to an Almaz design for Special Forces of the Border Guard. First one laid down 24 February 1998 and in service in early 2001. Further vessels are expected.

VERIFIED

CZILIM   *6/2001, S Breyer /* 0126219

8 TSAPLYA (MURENA) (PROJECT 12061) CLASS (ACV)

D-143 659	D-453 668	D-285 680	D-142 690
D-259 665	D-325 670	D-458 688	D-447 699

Displacement, tons: 149 full load
Dimensions, feet (metres): 103.7 × 47.6 × 5.2 *(31.6 × 14.5 × 1.6)*
Main machinery: 2 PR-77 gas turbines for lift and propulsion; 8,000 hp *(5.88 MW)*
Speed, knots: 50. **Range, n miles:** 500 at 50 kt
Complement: 11 (3 officers) + 100 troops
Guns: 2—300 mm AK 306M. 2—30 mm grenade launchers. 2—12.7 mm MGs.

Comment: Larger version of the Lebed class designed for river patrol. Built at Khabarovsk between 1987 and 1992. Operated on Amur river system.

UPDATED

TSAPLYA
6/2003 *, B Lemachko / 0580542

St Kitts and Nevis

The Federation of St Kitts and Nevis gained independence in 1983; the British monarch, represented by a governor-general, is the head of state. Located at the northern end of the Leeward Islands in the Lesser Antilles chain, the country comprises St Kitts (formerly Saint Christopher) (68 square miles) and, 2 n miles to the southeast, Nevis (36 square miles). The constitution allows for the secession of Nevis from the federation. The capital of St Kitts and of the federation is Basseterre; Charlestown is the capital and largest town on Nevis. Territorial seas (12 n miles) are claimed. A 200 n mile Exclusive Economic Zone (EEZ) has been claimed but the limits are not defined. The Coast Guard was part of the Police Force until 1997 when it transferred to the Regular Corps of the Defence Force.

Headquarters Appointments

Commanding Officer Coast Guard:
Major Patrick Wallace

Bases

Basseterre

Personnel

2005: 45

COAST GUARD

Notes: (1) There is a 40 kt RIB number *C 420*.
(2) A 920 Zodiac RHIB was donated by the US government in 2003.

1 SWIFTSHIPS 110 ft CLASS (PB)

STALWART C 253

Displacement, tons: 100 normal
Dimensions, feet (metres): 116.5 × 25 × 7 *(35.5 × 7.6 × 2.1)*
Main machinery: 4 Detroit 12V-71TA diesels; 1,680 hp *(1.25 MW)* sustained; 4 shafts
Speed, knots: 21. **Range, n miles:** 1,800 at 15 kt
Complement: 14
Guns: 2—12.7 mm MGs. 2—7.62 mm MGs.
Radars: Surface search: Raytheon; I-band.
Navigation: Furuno; I-band

Comment: Built by Swiftships, Morgan City, and delivered August 1985. Aluminium alloy hull and superstructure.

VERIFIED

STALWART
3/1998 / 0050088

1 DAUNTLESS CLASS (PB)

ARDENT C 421

Displacement, tons: 11 full load
Dimensions, feet (metres): 40 × 14 × 4.3 *(12.2 × 4.3 × 1.3)*
Main machinery: 2 Caterpillar 3208TA diesels; 870 hp *(650 kW)*; 2 shafts
Speed, knots: 27. **Range, n miles:** 600 at 18 kt
Complement: 4
Guns: 1—7.62 mm MG.
Radars: Surface search: Raytheon; I-band.

Comment: Built by SeaArk Marine under FMS funding and commissioned 8 August 1995.

VERIFIED

ARDENT
8/1996, St Kitts-Nevis Police / 0081725

1 FAIREY MARINE SPEAR CLASS (PB)

RANGER I

Displacement, tons: 4.3 full load
Dimensions, feet (metres): 29.8 × 9.5 × 2.8 *(9.1 × 2.8 × 0.9)*
Main machinery: 2 Ford Mermaid diesels; 360 hp *(268 kW)*; 2 shafts
Speed, knots: 20
Complement: 2
Guns: Mountings for 2—7.62 mm MGs.

Comment: Ordered for the police in June 1974 and delivered 10 September 1974. Refitted 1986. Considerably slower than when new but still in service.

VERIFIED

RANGER I
1992, St Kitts-Nevis Police / 0081726

2 BOSTON WHALERS (PBF)

ROVER I C 087   ROVER II C 088

Displacement, tons: 3 full load
Dimensions, feet (metres): 22 × 7.5 × 2 *(6.7 × 2.3 × 0.6)*
Main machinery: 1 Johnson outboard; 223 hp *(166 kW)*
Speed, knots: 35. **Range, n miles:** 70 at 35 kt
Complement: 2

Comment: Delivered in May 1988.

VERIFIED

ROVER I   *1990, St Kitts-Nevis Police /* 0081727

St Lucia

Country Overview

St Lucia gained independence in 1979; the British monarch, represented by a governor-general, is the head of state. The island (617 square miles) is one of the Windward Islands of the Lesser Antilles chain and is located between Martinique to the north and St Vincent to the south. The capital, main town and principal port is Castries, on the northwestern coast. Territorial seas (12 n miles) are claimed. Exclusive Economic Zone (EEZ) limits will not be fully defined until outstanding boundary disagreements have been resolved.

Headquarters Appointments

Coast Guard Commander:
    Assistant Superintendent Stephen Mitille

Bases

Castries, Vieux-Fort

Personnel

2005: 49

COAST GUARD

1 POINT CLASS (PB)

ALPHONSE REYNOLDS (ex-*Point Turner*)   P 01 (ex-WPB 82365)

Displacement, tons: 66 full load
Dimensions, feet (metres): 83 × 17.2 × 5.8 *(25.3 × 5.2 × 1.8)*
Main machinery: 2 Caterpillar 3412 diesels; 1,600 hp *(1.19 MW)*; 2 shafts
Speed, knots: 23. **Range, n miles:** 1,500 at 8 kt
Complement: 10
Guns: 2—12.7 mm MGs.
Radars: Surface search: Raytheon SPS-64(V)1; I-band.

Comment: Ex-US Coast Guard ship transferred on 3 April 1998. Originally built at Curtis Bay and first commissioned 14 April 1967.   *UPDATED*

ALPHONSE REYNOLDS   *12/2004*, Margaret Organ /* 1042343

1 SWIFT 65 ft CLASS (PB)

DEFENDER P 02

Displacement, tons: 42 full load
Dimensions, feet (metres): 64.9 × 18.4 × 6.6 *(19.8 × 5.6 × 2)*
Main machinery: 2 Detroit 12V-71 diesels; 680 hp *(507 kW)* sustained; 2 shafts
Speed, knots: 22. **Range, n miles:** 1,500 at 18 kt
Complement: 7
Radars: Surface search: Furuno; I-band.

Comment: Ordered from Swiftships, Morgan City in November 1983. Commissioned 3 May 1984. Similar to craft supplied to Antigua and Dominica. Painted grey instead of original blue and white.   *VERIFIED*

DEFENDER   *10/1999, St Lucia CG /* 0081729

1 DAUNTLESS CLASS (PB)

PROTECTOR P 04

Displacement, tons: 11 full load
Dimensions, feet (metres): 40 × 14 × 4.3 *(12.2 × 4.3 × 1.3)*
Main machinery: 2 Caterpillar 3208TA diesels; 870 hp *(650 kW)*; 2 shafts
Speed, knots: 27. **Range, n miles:** 600 at 18 kt
Complement: 4
Radars: Surface search: Raytheon; I-band.

Comment: Ordered October 1994. Built by SeaArk Marine under FMS funding and commissioned 9 October 1995.   *VERIFIED*

PROTECTOR (alongside Defender)   *7/1997, St Lucia CG /* 0019078

4 HARBOUR CRAFT

P 03   P 05   P 06   P 07

Comment: *P* 03 is a 9 m Zodiac 920 RHIB donated by the United States in 2004. *P 05* is a 35 kt Hurricane RIB acquired in June 1993 and *P 06* and *P 07* are 45 kt Mako craft acquired in November 1995.   *UPDATED*

P 07   *7/1997, St Lucia CG /* 0019080

St Vincent and the Grenadines

Country Overview

St Vincent and the Grenadines gained independence in 1979; the British monarch, represented by a governor-general, is the head of state. Lying between St Lucia to the north and Grenada to the south, they form part of the Windward Islands in the Lesser Antilles chain and comprise the island of St Vincent (133 square miles) and the 32 northernmost islands and cays of the Grenadines group including (north to south): Bequia, Mustique, Canouan, Mayreau, Union Island, Palm (formerly Prune) Island, and Petit St Vincent. The capital, largest town, and principal port is Kingstown, St Vincent. An archipelagic state, territorial seas (12 n miles) are claimed. A 200 n mile Exclusive Economic Zone (EEZ) has been claimed but the limits are not defined.

Headquarters Appointments

Coast Guard Commander:
   Lieutenant Commander Marcus Richards

Bases

Calliaqua, Bequia, Union Island

Personnel

2005: 56

COAST GUARD

Notes: A 920 Zodiac RHIB was donated by the US government in 2003.

1 SWIFTSHIPS 120 ft CLASS (PB)

CAPTAIN MULZAC SVG 01

Displacement, tons: 101
Dimensions, feet (metres): 120 × 25 × 7 *(36.6 × 7.6 × 2.1)*
Main machinery: 4 Detroit 12V-71TA diesels; 1,360 hp *(1.01 MW)* sustained; 4 shafts
Speed, knots: 21. **Range, n miles:** 1,800 at 15 kt
Complement: 14 (4 officers)
Guns: 2—12.7 mm MGs. 2—7.62 mm MGs.
Radars: Surface search: Furuno 1411 Mk II; I/J-band.

Comment: Ordered in August 1986. Built by Swiftships, Morgan City and delivered 13 June 1987. Carries a RIB with a 40 hp outboard.
VERIFIED

CAPTAIN MULZAC   *6/1994, St Vincent Coast Guard /* 0081730

1 VOSPER THORNYCROFT 75 ft CLASS (PB)

GEORGE McINTOSH SVG 05

Displacement, tons: 70 full load
Dimensions, feet (metres): 75 × 19.5 × 8 *(22.9 × 6 × 2.4)*
Main machinery: 2 Caterpillar D 348 diesels; 1,450 hp *(1.08 MW)* sustained; 2 shafts
Speed, knots: 24.5. **Range, n miles:** 1,000 at 11 kt; 600 at 20 kt
Complement: 11 (3 officers)
Guns: 1 Oerlikon 20 mm.
Radars: Surface search: Furuno 1411 Mk III; I/J-band.

Comment: Built by Vosper Thornycroft, Portchester. Handed over 23 March 1981. GRP hull.
VERIFIED

GEORGE McINTOSH   *6/1992, St Vincent Coast Guard /* 0081731

1 DAUNTLESS CLASS (PB)

HAIROUN SVG 04

Displacement, tons: 11 full load
Dimensions, feet (metres): 40 × 14 × 4.3 *(12.2 × 4.3 × 1.3)*
Main machinery: 2 Caterpillar 3208TA diesels; 870 hp *(650 kW)*; 2 shafts
Speed, knots: 27. **Range, n miles:** 600 at 18 kt
Complement: 4
Guns: 1—7.62 mm MG.
Radars: Surface search: Raytheon; I-band.

Comment: Ordered October 1984. Built by SeaArk Marine under FMS funding and commissioned 8 June 1995.
VERIFIED

HAIROUN   *7/1997 /* 0019082

2 HARBOUR CRAFT

SVG 03   **CHATHAM BAY** SVG 08

Comment: *SVG 03* is a 30 kt Zodiac RIB. *SVG 08* is a 30 kt Boston Whaler acquired in 1994. The Buhler craft *SVG 06* and *SVG 07* are no longer operational.
UPDATED

SVG 03   *2001, St Vincent Coast Guard /* 0109947

CHATHAM BAY   *2001, St Vincent Coast Guard /* 0109948

Samoa

Country Overview

Samoa was a New Zealand-administered UN Trust territory until it became independent in 1962. At the same time a Treaty of Friendship delegated responsibility to New Zealand for foreign affairs. An island nation, it lies in the south Pacific Ocean, approximately midway between Hawaii and New Zealand, in the western portion of the Samoan archipelago. There are two main islands, Savai'i and Upolu, and several smaller islands, of which only two, Apolima and Manono, are inhabited. The capital and chief port is Apia on Upolu. An archipelagic state, territorial seas (12 n miles) are claimed. An Exclusive Economic Zone (EEZ) (200 n miles) is also claimed but limits have not been fully defined by boundary agreements.

Headquarters Appointments

Head of Maritime Division:
   Commissioner Lorenese Neru
Marine Surveillance Adviser:
   Lieutenant Commander J Narbutas, RAN

Bases

Apia

PATROL FORCES

1 PACIFIC CLASS (LARGE PATROL CRAFT) (PB)

Name	No	Builders	Commissioned
NAFANUA	—	Australian Shipbuilding Industries	5 Mar 1988

Displacement, tons: 165 full load
Dimensions, feet (metres): 103.3 × 26.6 × 6.9 *(31.5 × 8.1 × 2.1)*
Main machinery: 2 Caterpillar 3516TA diesels; 4,400 hp *(3.28 MW)* sustained; 2 shafts
Speed, knots: 20. **Range, n miles:** 2,500 at 12 kt
Complement: 17 (3 officers)
Guns: Can carry 1 Oerlikon 20 mm and 2—7.62 mm MGs.
Radars: Surface search: Furuno 1011; I-band.

Comment: Under the Defence Co-operation Programme Australia has provided a number of these craft to Pacific islands. Training, operational and technical assistance is provided by the Royal Australian Navy. *Nafanua* ordered 3 October 1985. Refitted in 1996. Following the decision by the Australian government to extend the Pacific Patrol Boat programme, the ship will undertake a life extension refit in 2005.

UPDATED

NAFANUA

2000, RAN / 0105770

Saudi Arabia

Country Overview

The Kingdom of Saudi Arabia occupies most of the Arabian Peninsula and is bordered to the north by Jordan, Iraq, and Kuwait, to the south by Oman and the Republic of Yemen and to the east by Qatar and the United Arab Emirates. With an area of 864,869 square miles, it has coastlines with the Red Sea (972 n miles) and the Gulf (454 n miles). The capital and largest city is Riyadh while the principal ports are Jiddah and Yanbu al Bahr on the Red Sea, and the major oil-exporting ports of Al Jabayl, Ad Dammam, and Ras Tanura on the Gulf. Territorial seas (12 n miles) are claimed. An EEZ has not been claimed.

Headquarters Appointments

Chief of Naval Staff:
   Vice Admiral Talal Salem Al Mofadhi
Commander Eastern Fleet:
   Rear Admiral Mohammad Abdul Khalij Al Asseri
Commander Western Fleet:
   Rear Admiral Dakheel Allah Ahmed Al-Wakdani
Director Frontier Force (Coast Guard):
   Lieutenant General Mujib bin Muhammad Al-Qahtani

Personnel

(a)   2005: 13,500 officers and men (including 1,500 marines)
(b)   Voluntary service

Bases

Naval HQ: Riyadh
Main bases: Jiddah (HQ Western Fleet), Al Jubail (HQ Eastern Fleet), Aziziah (Coast Guard). Jizan (Red Sea) building from May 1996
Minor bases (Naval and Coast Guard): Ras Tanura, Al Dammam, Yanbo, Ras al-Mishab, Al Wajh, Al Qatif, Haqi, Al Sharmah, Qizan, Duba

General

Funding for the Navy has the lowest defence service priority. New programmes are slow to come forward, and the operational status of existing ships is variable.

Command and Control

The USA provided an update of command and control capabilities during the period 1991-95, including a commercial datalink to improve interoperability.

Coast Defence

Truck-mounted Otomat batteries.

Coast Guard

Part of the Frontier Force under the Minister for Defence and Aviation. 5,500 officers and men. It is not always clear which ships belong to the Navy and which to the Coast Guard.

Strength of the Fleet

Type	Active	Building
Frigates	7	—
Corvettes—Missile	4	—
Fast Attack Craft—Missile	9	—
Patrol Craft	56	—
Minehunters	3	—
Minesweepers—Coastal	4	—
Replenishment Tankers	2	—

SUBMARINES

Notes: (1) Orders for patrol submarines are a low priority although training has been done in France and Pakistan.
(2) Interest has been shown in the acquisition of Midget Submarines.

FRIGATES

AL RIYADH

9/2004, B Prézelin* / 1121520

3 AL RIYADH (MODIFIED LA FAYETTE) CLASS (TYPE F-3000S) (FFGHM)

Name	No	Builders	Laid down	Launched	Commissioned
AL RIYADH	812	DCN, Lorient	29 Sep 1999	1 Aug 2000	26 July 2002
MAKKAH	814	DCN, Lorient	25 Aug 2000	20 July 2001	3 Apr 2004
AL DAMMAM	816	DCN, Lorient	26 Aug 2001	7 Sep 2002	2005

Displacement, tons: 4,650 full load
Dimensions, feet (metres): 438.43 × 56.4 × 13.5 *(133.6 × 17.2 × 4.1)*
Main machinery: CODAD; 4 SEMT-Pielstick 16 PA6 STC diesels; 28,000 hp(m) *(20.58 MW)* sustained; 2 shafts; LIPS cp props; bow thruster
Speed, knots: 25. **Range, n miles:** 7,000 at 15 kt
Complement: 181 (25 officers); accommodation for 190

Missiles: SSM: 8 Aerospatiale MM 40 Block II Exocet ❶; inertial cruise; active radar homing to 70 km *(40 n miles)* at 0.9 Mach; warhead 165 kg; sea-skimmer.
SAM: Eurosam SAAM ❷; 2 octuple Sylver A43 VLS for Aster 15; command guidance active radar homing to 15 km *(8.1 n miles)* anti-missile, at 30 km *(16.2 n miles)* anti-aircraft. 16 missiles.
Guns: 1 OTO Melara 3 in *(76 mm)*/62 Super Rapid ❸; 120 rds/min to 16 km *(8.7 n miles)*; weight of shell 6 kg.
2 Giat 15B 20 mm ❹; 800 rds/min to 3 km; weight of shell 0.1 kg.
2—12.7 mm MGs.
Torpedoes: 4—21 in *(533 mm)* tubes; ECAN F17P; anti-submarine; wire-guided active/passive homing to 20 km *(10.8 n miles)* at 40 kt; warhead 250 kg.
Countermeasures: Decoys: 2 Matra Dagaie Mk 2 ❺; 10-barrelled trainable launchers; chaff and IR flares. SLAT anti-wake homing torpedoes system (when available).
ESM: Thomson-CSF (DR 3000-S2) ❻; intercept. Sagem Telegon 10.
CESM: Thales Altesse; intercept.
ECM: 2 Thales Salamandre; jammers.
Combat data systems: Thales.
Weapons control: Thomson-CSF CTM radar/IR system.

Radars: Air search: Thales DRBV 26C Jupiter II ❼; D-band.
Surveillance/Fire control: Thomson-CSF Arabel 3D ❽; I/J-band.
Fire control: Thomson-CSF Castor II UJ ❾; J-band; range 15 km *(8 n miles)* for 1 m² target.
Navigation: 2 Racal Decca 1226 ❿; I-band. A second set fitted for helicopter control.
Sonars: Thomson Marconi CAPTAS 20; active low frequency; towed array.

Helicopters: Hangar and platform for NH-90-sized helicopter ⓫.

Programmes: A provisional order was made on 11 June 1989, but this was not finally confirmed until 19 November 1994 when a contract for two ships was authorised under the Sawari II programme. Thomson-CSF was the prime contractor. On 25 May 1997 an order for a third ship was placed together with a substantial enhancement of the weapon systems in all three. First steel cut 13 December 1997. Following handover, 812 started an eight month training programme which concluded in March 2003. 814 started sea trials on 9 September 2002 and 816 in mid-2003. SAM successfully tested in 816 in April 2004.
Structure: The design is a development of the French La Fayette class. Some 10 m longer, space and weight included for two more octuple SAM launchers or A50 launcher for Aster 30. Provision is made for a larger NH 90 type helicopter in the future, DCN Samahé helo handling system. STAF stabilisers.
Operational: OTHT link for helicopters and Air Force F-15s. *Makkah* seriously damaged in a grounding incident off Jiddah in December 2004.

UPDATED

AL RIYADH *(Scale 1 : 1,200), Ian Sturton / 1044496*

AL DAMMAM *6/2004*, B Prézelin / 1044498*

MAKKAH *3/2004*, B Prézelin / 1044497*

For details of the latest updates to *Jane's Fighting Ships* online and to discover the additional information available exclusively to online subscribers please visit **jfs.janes.com**

4 MADINA (TYPE F 2000S) CLASS (FFGHM)

Name	No	Builders	Laid down	Launched	Commissioned
MADINA	702	Lorient (DTCN)	15 Oct 1981	23 Apr 1983	4 Jan 1985
HOFOUF	704	CNIM, Seyne-sur-Mer	14 June 1982	24 June 1983	31 Oct 1985
ABHA	706	CNIM, Seyne-sur-Mer	7 Dec 1982	23 Dec 1983	4 Apr 1986
TAIF	708	CNIM, Seyne-sur-Mer	1 Mar 1983	25 May 1984	29 Aug 1986

Displacement, tons: 2,000 standard; 2,870 full load
Dimensions, feet (metres): 377.3 × 41 × 16 (sonar)
  (115 × 12.5 × 4.9)
Main machinery: CODAD; 4 SEMT-Pielstick 16 PA6 280V BTC
  diesels; 38,400 hp(m) (28 MW) sustained; 2 shafts
Speed, knots: 30
Range, n miles: 8,000 at 15 kt; 6,500 at 18 kt
Complement: 179 (15 officers)

Missiles: SSM: 8 OTO Melara/Matra Otomat Mk 2 (2 quad) **①**;
  active radar homing to 160 km (86.4 n miles) at 0.9 Mach;
  warhead 210 kg; sea-skimmer for last 4 km (2.2 n miles).
  ERATO system allows mid-course guidance by ship's
  helicopter.
SAM: Thomson-CSF Crotale Naval octuple launcher **②**;
  command line of sight guidance; radar/IR homing to 13 km
  (7 n miles) at 2.4 Mach; warhead 14 kg; 26 missiles.
Guns: 1 Creusot-Loire 3.9 in (100 mm)/55 compact Mk 2 **③**; 20/
  45/90 rds/min to 17 km (9.3 n miles) weight of shell 13.5 kg.
  4 Breda 40 mm/70 (2 twin) **④**; 300 rds/min to 12.5 km (6.8 n
  miles); weight of shell 0.96 kg.
Torpedoes: 4—21 in (533 mm) tubes **⑤**. ECAN F17P; anti-
  submarine; wire-guided; active/passive homing to 20 km
  (10.8 n miles) at 40 kt; warhead 250 kg.
Countermeasures: Decoys: CSEE Dagaie double trainable
  mounting **⑥**; IR flares and chaff; H/J-band.
ESM: Thomson-CSF DR 4000; intercept; HF/DF.
ECM: Thomson-CSF Janet; jammer.

Combat data systems: Thomson-CSF TAVITAC action data
  automation; capability for Link W.
Weapons control: Vega system. 3 CSEE Naja optronic directors.
  Alcatel DLT for torpedoes.
Radars: Air/surface search/IFF: Thomson-CSF Sea Tiger (DRBV
  15) **⑦**; E/F-band; range 110 km (60 n miles) for 2 m² target.
  Navigation: 2 Racal Decca TM 1226; I-band.
  Fire control: Thomson-CSF Castor IIB/C **⑧**; I/J-band; range 15 km
  (8 n miles) for 1 m² target.
  Thomson-CSF DRBC 32 **⑨**; I/J-band (for SAM).
Sonars: Thomson Sintra Diodon TSM 2630; hull-mounted;
  active search and attack with integrated Sorel VDS **⑩**; 11, 12
  or 13 kHz.

Helicopters: 1 SA 365F Dauphin 2 **⑪**.

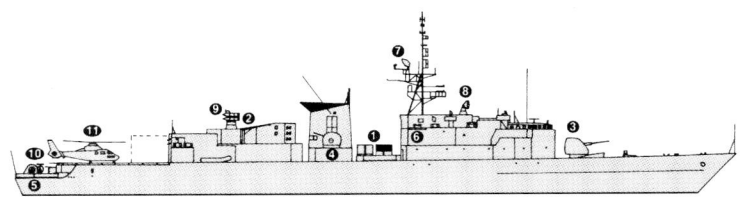

MADINA   (Scale 1 : 1,200), Ian Sturton

Programmes: Ordered in 1980, the major part of the Sawari I
  contract. Agreement for France to provide supplies and
  technical help.
Modernisation: The class have been upgraded by DCN Toulon,
  Madina completed in April 1997. Hofouf in mid-1998. Abha in
  late 1999, and Taif in March 2000. Improvements included
  updating TAVITAC, Otomat missiles, both sonars and fitting a
  Samahé 110 helo handling system.
Structure: Fitted with Snach/Saphir folding fin stabilisers.
Operational: Navigation: CSEE Sylosat. Helicopter can provide
  mid-course guidance for SSM. All based at Jiddah. Only a few
  weeks a year are spent at sea.

UPDATED

HOFOUF   9/2003, Hartmut Ehlers / 0567898

TAIF   6/2002 / 0526835

CORVETTES

4 BADR CLASS (FSG)

Name	No	Builders	Laid down	Launched	Commissioned
BADR	612	Tacoma Boatbuilding Co, Tacoma	6 Oct 1979	26 Jan 1980	30 Nov 1980
AL YARMOOK	614	Tacoma Boatbuilding Co, Tacoma	3 Jan 1980	13 May 1980	18 May 1981
HITTEEN	616	Tacoma Boatbuilding Co, Tacoma	19 May 1980	5 Sep 1980	3 Oct 1981
TABUK	618	Tacoma Boatbuilding Co, Tacoma	22 Sep 1980	18 June 1981	10 Jan 1983

Displacement, tons: 870 standard; 1,038 full load
Dimensions, feet (metres): 245 × 31.5 × 8.9
(74.7 × 9.6 × 2.7)
Main machinery: CODOG; 1 GE LM 2500 gas turbine; 23,000 hp
(17.2 MW) sustained; 2 MTU 12V 652 TB91 diesels; 3,470 hp
(m) *(2.55 MW)* sustained; 2 shafts; cp props
Speed, knots: 30 gas; 20 diesels. **Range, n miles:** 4,000 at 20 kt
Complement: 58 (7 officers)

Missiles: SSM: 8 McDonnell Douglas Harpoon (2 quad)
launchers ❶; active radar homing to 130 km *(70 n miles)* at
0.9 Mach; warhead 227 kg.
Guns: 1 FMC/OTO Melara 3 in *(76 mm)*/62 Mk 75 Mod 0 ❷;
85 rds/min to 16 km *(8.7 n miles)*; weight of shell 6 kg.
1 General Electric/General Dynamics 20 mm 6-barrelled
Vulcan Phalanx ❸; 3,000 rds/min combined to 2 km.
2 Oerlikon 20 mm/80 ❹.
1—81 mm mortar. 2—40 mm Mk 19 grenade launchers.
Torpedoes: 6—324 mm US Mk 32 (2 triple) tubes ❺. Honeywell
Mk 46; anti-submarine; active/passive homing to 11 km *(5.9 n
miles)* at 40 kt; warhead 44 kg.
Countermeasures: Decoys: 2 Loral Hycor SRBOC 6-barrelled
fixed Mk 36 ❻; IR flares and chaff to 4 km *(2.2 n miles)*.
ESM: SLQ-32(V)1 ❼; intercept.
Weapons control: Mk 24 optical director ❽. Mk 309 for
torpedoes. Mk 92 Mod 5 GFCS. FSI Safire FLIR.
Radars: Air search: Lockheed SPS-40B ❾; B-band; range 320 km
(175 n miles).
Surface search: ISC Cardion SPS-55 ❿; I/J-band.
Fire control: Sperry Mk 92 ⓫; I/J-band.
Sonars: Raytheon SQS-56 (DE 1164); hull-mounted; active
search and attack; medium frequency.

Modernisation: Refitting done in Saudi Arabia with US
assistance. FLIR being fitted from 1998.
Structure: Fitted with fin stabilisers.
Operational: All based at Al Jubail on the east coast and spend
little time at sea.

UPDATED

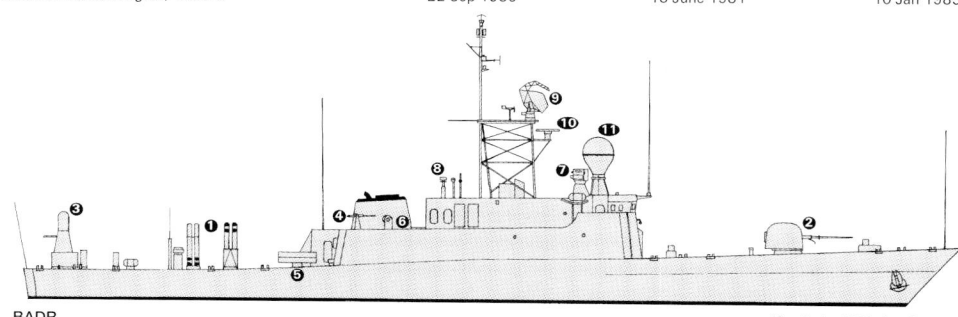

BADR *(Scale 1 : 600), Ian Sturton*

AL YARMOOK *2/1997, van Ginderen Collection /* 0019085

PATROL FORCES

9 AL SIDDIQ CLASS (PGGF)

Name	No	Builders	Launched	Commissioned
AL SIDDIQ	511	Peterson, WI	22 Sep 1979	15 Dec 1980
AL FAROUQ	513	Peterson, WI	17 May 1980	22 June 1981
ABDUL AZIZ	515	Peterson, WI	23 Aug 1980	3 Sep 1981
FAISAL	517	Peterson, WI	15 Nov 1980	23 Nov 1981
KHALID	519	Peterson, WI	23 Mar 1981	11 Jan 1982
AMYR	521	Peterson, WI	13 June 1981	21 June 1982
TARIQ	523	Peterson, WI	23 Sep 1981	11 Aug 1982
OQBAH	525	Peterson, WI	12 Dec 1981	18 Oct 1982
ABU OBAIDAH	527	Peterson, WI	3 Apr 1982	6 Dec 1982

Displacement, tons: 495 full load
Dimensions, feet (metres): 190.5 × 26.5 × 6.6
(58.1 × 8.1 × 2)
Main machinery: CODOG; 1 GE LM 2500 gas turbine; 23,000 hp
(17.2 MW) sustained; 2 MTU 12V 652 TB91 diesels; 3,470 hp
(m) *(2.55 MW)* sustained; 2 shafts; cp props
Speed, knots: 38 gas; 25 diesel. **Range, n miles:** 2,900 at 14 kt
Complement: 38 (5 officers)

Missiles: SSM: 4 McDonnell Douglas Harpoon (2 twin)
launchers; active radar homing to 130 km *(70 n miles)* at 0.9
Mach; warhead 227 kg.
Guns: 1 FMC/OTO Melara 3 in *(76 mm)*/62 Mk 75 Mod 0;
85 rds/min to 16 km *(8.7 n miles)*; weight of shell 6 kg.
1 General Electric/General Dynamics 20 mm 6-barrelled
Vulcan Phalanx; 3,000 rds/min combined to 2 km.
2 Oerlikon 20 mm/80; 800 rds/min to 2 km anti-aircraft.
2—81 mm mortars. 2—40 mm Mk 19 grenade launchers.

Countermeasures: Decoys: 2 Loral Hycor SRBOC 6-barrelled
fixed Mk 36; IR flares and chaff to 4 km *(2.2 n miles)*.
ESM: SLQ-32(V)1; intercept.
Weapons control: Mk 92 mod 5 GFCS. FSI Safire FLIR. Link W.
Radars: Surface search: ISC Cardion SPS-55; I/J-band.
Fire control: Sperry Mk 92; I/J-band.

Modernisation: Safire FLIR and Link W being fitted.
Operational: *Amyr* and *Tariq* operate from Jiddah, the remainder
are based at Al Jubail. *Faisal* damaged in the Gulf War in 1991
but was operational again in 1994. **UPDATED**

ABU OBAIDAH *6/2001, Ships of the World /* 0126360

17 HALTER TYPE (COASTAL PATROL CRAFT) (PB)

52-68

Displacement, tons: 56 full load
Dimensions, feet (metres): 78 × 20 × 5.8 *(23.8 × 6.1 × 1.8)*
Main machinery: 2 Detroit 16V-92TA diesels; 1,380 hp *(1.03 MW)* sustained; 2 shafts
Speed, knots: 28. **Range, n miles:** 1,200 at 12 kt
Complement: 8 (2 officers)
Guns: 2—25 mm Mk 38. 2—7.62 mm MGs.
Radars: Surface search: Raytheon SPS-64; I-band.

Comment: Ordered from Halter Marine 17th February 1991. Aluminium construction. Last delivered in January 1993. Same type for Philippines.

VERIFIED

HALTER TYPE   8/1990, Trinity Marine / 0080563

39 SIMONNEAU 51 TYPE (INSHORE PATROL CRAFT) (PBI)

Displacement, tons: 22 full load
Dimensions, feet (metres): 51.8 × 15.7 × 5.9 *(15.8 × 4.8 × 1.8)*
Main machinery: 4 outboards; 2,400 hp(m) *(1.76 MW)*
Speed, knots: 33. **Range, n miles:** 375 at 25 kt
Guns: 1—12.7 mm MG. 2—7.62 mm MGs.
Radars: Surface search: Furuno; I-band.

Comment: First 20 ordered from France in June 1988 and delivered in 1989-90. A second batch of 20 ordered in 1991. Aluminium construction. Used by naval commandos. These craft were also reported as Panhards. One deleted so far.

VERIFIED

SIMONNEAU TYPE   1989, Simonneau Marine

SHIPBORNE AIRCRAFT

Notes: Procurement of a new shipborne helicopter is under consideration. Up to ten are required for deployment to the Al Riyadh class frigates and for other tasks. The NH Industries NH 90 is reported to be a strong contender.

Numbers/Type: 15/6 Aerospatiale AS 365F Dauphin 2/Aerospatiale AS 365N Dauphin 2.
Operational speed: 140 kt *(260 km/h)*.
Service ceiling: 15,000 ft *(4,575 m)*.
Range: 410 n miles *(758 km)*.
Role/Weapon systems: AS 365F is the ASV/ASW helicopter; procured for embarked naval aviation force; surface search/attack is the primary role. Sensors: Thomson-CSF Agrion 15 radar; Crouzet MAD. Weapons: ASV; four AS/15TT missiles. ASW; 2 Mk 46 torpedoes. AS 365N is for SAR. Sensors: Omera DRB 32 search radar. Weapons: Unarmed.

VERIFIED

LAND-BASED MARITIME AIRCRAFT

Notes: (1) Six P-3C Orion or CASA CN-235 may be acquired in due course.
(2) Five Boeing E3-A AEW aircraft in service with Air Force.

Numbers/Type: 18 Aerospatiale AS 332SC Super Puma.
Operational speed: 150 kt *(280 km/h)*.
Service ceiling: 15,090 ft *(4,600 m)*.
Range: 335 n miles *(620 km)*.
Role/Weapon systems: First pair delivered in August 1989. Total of 18 by the end of 1996. Shared with the Coast Guard. Sensors: 12 have Omera search radar Safire AAQ-22 FLIR from 1998. Weapons: ASV; six have Giat 20 mm cannon; 12 have AM39 Exocet or Sea Eagle ASM.

VERIFIED

MINE WARFARE FORCES

4 ADDRIYAH (MSC 322) CLASS
(MINESWEEPERS/HUNTERS—COASTAL) (MHSC)

Name	No	Builders	Launched	Commissioned
ADDRIYAH	MSC 412	Peterson, WI	20 Dec 1976	6 July 1978
AL QUYSUMAH	MSC 414	Peterson, WI	26 May 1977	15 Aug 1978
AL WADEEAH	MSC 416	Peterson, WI	6 Sep 1977	7 Sep 1979
SAFWA	MSC 418	Peterson, WI	7 Dec 1977	2 Oct 1979

Displacement, tons: 320 standard; 407 full load
Dimensions, feet (metres): 153 × 26.9 × 8.2 *(46.6 × 8.2 × 2.5)*
Main machinery: 2 Waukesha L1616 diesels; 1,200 hp *(895 kW)*; 2 shafts
Speed, knots: 13
Complement: 39 (4 officers)
Guns: 1 Oerlikon 20 mm.
Radars: Surface search: ISC Cardion SPS-55; I/J-band.
Sonars: GE SQQ-14; VDS; active minehunting; high frequency.

Comment: Ordered on 30 September 1975 under the International Logistics Programme. Wooden structure. Fitted with fin stabilisers, wire and magnetic sweeps and also for minehunting. *Addriyah* based at Jiddah, the remainder at Al Jubail. Expected to be replaced by arrival of Sandowns but all are still in service mostly as patrol craft.

UPDATED

AL QUYSUMAH   6/1996, van Ginderen Collection / 0019090

3 AL JAWF (SANDOWN) CLASS
(MINEHUNTERS—COASTAL) (MHC)

Name	No	Builders	Launched	Commissioned
AL JAWF	420	Vosper Thornycroft	2 Aug 1989	12 Dec 1991
SHAQRA	422	Vosper Thornycroft	15 May 1991	7 Feb 1993
AL KHARJ	424	Vosper Thornycroft	8 Feb 1993	7 Aug 1997

Displacement, tons: 450 standard; 480 full load
Dimensions, feet (metres): 172.9 × 34.4 × 6.9 *(52.7 × 10.5 × 2.1)*
Main machinery: 2 Paxman 6RP200E diesels; 1,500 hp *(1.12 MW)* sustained; Voith-Schneider propulsion; 2 shafts; 2 Schöttel bow thrusters
Speed, knots: 13 diesels; 6 electric drive. **Range, n miles:** 3,000 at 12 kt
Complement: 34 (7 officers) plus 6 spare berths
Guns: 2 Electronics & Space Emerlec 30 mm (twin); 1,200 rds/min combined to 6 km *(3.3 n miles)*; weight of shell 0.35 kg.
Countermeasures: Decoys: 2 Loral Hycor SRBOC Mk 36 Mod 1 6-barrelled chaff launchers.
ESM: Thomson-CSF Shiploc; intercept.
MCM: ECA mine disposal system; 2 PAP 104 Mk 5.
Combat data systems: Plessey Nautis M action data automation.
Weapons control: Contraves TMEO optronic director (Seahawk Mk 2).
Radars: Navigation: Kelvin Hughes Type 1007; I-band.
Sonars: Plessey/MUSL Type 2093; VDS; high frequency.

Comment: Three ordered 2 November 1988 from Vosper Thornycroft. Option for three more unlikely to be taken up. GRP hulls. Combines vectored thrust units with bow thrusters and Remote Controlled Mine Disposal System (RCMDS). *Al Jawf* sailed for Saudi Arabia in November 1995, *Shaqra* in November 1996, and *Al Kharj* in August 1997. All based at Al Jubail.

UPDATED

AL KHARJ   8/1997, Vosper Thornycroft / 0050091

AUXILIARIES

2 MOD DURANCE CLASS (REPLENISHMENT SHIPS) (AORH)

Name	No	Builders	Launched	Commissioned
BORAIDA	902	La Ciotat, Marseilles	22 Jan 1983	29 Feb 1984
YUNBOU	904	La Ciotat, Marseilles	20 Oct 1984	29 Aug 1985

Displacement, tons: 11,200 full load
Dimensions, feet (metres): 442.9 × 61.3 × 22.9 *(135 × 18.7 × 7)*
Main machinery: 2 SEMT-Pielstick 14 PC2.5 V 400 diesels; 18,200 hp(m) *(13.4 MW)* sustained; 2 shafts; LIPS cp props
Speed, knots: 20.5. **Range, n miles:** 7,000 at 15 kt
Complement: 129 plus 11 trainees
Cargo capacity: 4,350 tons diesel; 350 tons AVCAT; 140 tons fresh water; 100 tons victuals; 100 tons ammunition; 70 tons spares
Guns: 4 Breda Bofors 40 mm/70 (2 twin); 300 rds/min to 12.5 km *(6.8 n miles)*; weight of shell 0.96 kg.
Weapons control: 2 CSEE Naja optronic directors. 2 CSEE Lynx optical sights.
Radars: Navigation: 2 Decca; I-band.
Helicopters: 2 SA 365F Dauphin or 1 AS 332SC Super Puma.

Comment: Contract signed October 1980 as part of Sawari I programme. Both upgraded by DCN at Toulon; *Boraida* in 1996/97, followed by *Yunbou*, in 1997/98. Refuelling positions: Two alongside, one astern. Also serve as training ships and as depot and maintenance ships. Helicopters can have ASM or ASW armament. Both based at Jiddah.

VERIFIED

YUNBOU   *1/1998 / 0016635*

BORAIDA   *9/2003, Hartmut Ehlers / 0567897*

4 LCU 1610 CLASS (TRANSPORTS) (YFU)

AL QIAQ (ex-*SA 310*) 212   AL ULA (ex-*SA 312*) 216
AL SULAYEL (ex-*SA 311*) 214   AFIF (ex-*SA 313*) 218

Displacement, tons: 375 full load
Dimensions, feet (metres): 134.9 × 29 × 6.1 *(41.1 × 8.8 × 1.9)*
Main machinery: 4 GM diesels; 1,000 hp *(746 kW)*; 2 Kort nozzles
Speed, knots: 11. **Range, n miles:** 1,200 at 8 kt
Complement: 14 (2 officers)
Military lift: 170 tons; 20 troops
Guns: 2—12.7 mm MGs.
Radars: Navigation: Marconi LN66; I-band.

Comment: Built by Newport Shipyard, Rhode Island. Transferred from US June/July 1976. Based at Al Jubail.

VERIFIED

LCU 1610 (US colours)   *9/1997, Hachiro Nakai / 0016483*

4 LCM 6 CLASS (TRANSPORTS) (YFU)

DHEBA 220   UMLUS 222   AL LEETH 224   AL QUONFETHA 226

Displacement, tons: 62 full load
Dimensions, feet (metres): 56.2 × 14 × 3.9 *(17.1 × 4.3 × 1.2)*
Main machinery: 2 GM diesels; 450 hp *(336 kW)*; 2 shafts
Speed, knots: 9. **Range, n miles:** 130 at 9 kt
Complement: 5
Military lift: 34 tons or 80 troops
Guns: 2—40 mm Mk 19 grenade launchers.

Comment: Four transferred July 1977 and four in July 1980. The first four have been cannibalised for spares. Based at Jiddah.

VERIFIED

TUGS

13 COASTAL TUGS (YTB/YTM)

RADHWA 1-6, 14-15   TUWAIG 113   DAREEN 111   RADHWA 12, 16, 17

Comment: *Radhwa 12, 16* and *17* are 43 m YTMs built in 1982-83. *Tuwaig* and *Dareen* are ex-US YTB transferred in October 1975. These two are used to tow targets for weapons firing exercises and are based at Al Jubail and Damman respectively. The remainder are all of about 35 m built in Singapore and the Netherlands between 1981 and 1983.

VERIFIED

YTB TYPE (US colours)   *9/1992, Jürg Kürsener / 0080564*

ROYAL YACHTS

1 ROYAL YACHT (YACH)

Name	No	Builders	Commissioned
AL YAMAMAH	—	Elsinore, Denmark	Feb 1981

Displacement, tons: 1,660 full load
Dimensions, feet (metres): 269 × 42.7 × 10.8 *(82 × 13 × 3.3)*
Main machinery: 2 MTU 12V 1163 TB82 diesels; 6,000 hp(m) *(4.41 MW)*; 2 shafts; cp props; bow thruster; 300 hp(m) *(221 kW)*
Speed, knots: 19
Complement: 42 plus 56 spare
Helicopters: Platform for 1 medium.

Comment: Ordered by Iraq but not delivered because of the war with Iran. Given to Saudi Arabia by Iraq in 1988. Based at Dammam.

VERIFIED

1 PEGASUS CLASS (HYDROFOIL) (YAGJ)

AL AZIZIAH

Displacement, tons: 115 full load
Dimensions, feet (metres): 89.9 × 29.9 × 6.2 *(27.4 × 9.1 × 1.9)*
Main machinery: 2 Allison 501-KF20A gas turbines; 8,660 hp *(6.46 MW)* sustained; 2 waterjets (foilborne); 2 Detroit 8V92 diesels; 606 hp *(452 kW)* sustained; 2 shafts (hullborne)
Speed, knots: 46. **Range, n miles:** 890 at 42 kt
Guns: 2 General Electric 20 mm Sea Vulcan.
Weapons control: Kollmorgen GFCS; Mk 35 optronic director.

Comment: Ordered in 1984 from Lockheed and subcontracted to Boeing, Seattle; delivered in August 1985. Mostly used as a tender to the Royal Yacht.

VERIFIED

1 ROYAL YACHT (YACH)

Name	No	Builders	Commissioned
ABDUL AZIZ	—	Halsingør Waerft, Denmark	12 June 1984

Displacement, tons: 5,200 full load
Measurement, tons: 1,450 dwt
Dimensions, feet (metres): 482.2 × 59.2 × 16.1 *(147 × 18 × 4.9)*
Main machinery: 2 Lindholmen-Pielstick 12 PC2.5 V diesels; 15,600 hp(m) *(11.47 MW)* sustained; 2 shafts
Speed, knots: 22
Complement: 65 plus 4 Royal berths and 60 spare
Helicopters: 1 Bell 206B JetRanger type.

Comment: Completed March 1983 for subsequent fitting out at Vosper's Ship Repairers, Southampton. Helicopter hangar set in hull forward of bridge-covers extend laterally to form pad. Swimming pool. Stern ramp leading to garage. Based at Jiddah. Operated by the Coast Guard.

VERIFIED

ABDUL AZIZ   *5/1994, van Ginderen Collection*

COAST GUARD

Notes: Three 32 m fireboats *Jubail I, Jubail 2* and *Jubail 3,* entered service in 1982.

2 SEA GUARD CLASS (WPBF)

AL RIYADH 304   **ZULURAB** 305

Displacement, tons: 56 full load
Dimensions, feet (metres): 73.8 × 18.4 × 5.6 *(22.5 × 5.6 × 1.7)*
Main machinery: 2 MTU 12V 331 TC92 diesels; 2,920 hp(m) *(21.46 MW)*; 2 shafts
Speed, knots: 35
Complement: 10
Guns: 2 Giat 20 mm (twin). 2—7.62 mm MGs.
Radars: Surface search: Racal Decca; I-band.
Fire control: Thomson-CSF Agrion; J-band.

Comment: Built by Simonneau Marine and delivered by SOFREMA in April 1992. Aluminium construction. Both based at Jiddah. The SSM launcher shown in the picture is not fitted.
VERIFIED

ZULURAB   *1992, Simonneau Marine* / 0080565

6 STAN PATROL 2606 CRAFT
(COASTAL PATROL CRAFT) (WPB)

ASSIR 317   **ALDHAHRAN** 318   **ALKAHRJ** 319   **ARAR** 320   + 2

Displacement, tons: 55 (approx) full load
Dimensions, feet (metres): 87.0 × 20.3 × 6.1 *(26.5 × 6.2 × 1.8)*
Main machinery: 2 MTU 12V 396 TE94 diesels; 4,429 hp *(3.3 MW)*; 2 shafts
Speed, knots: 28

Comment: Built by Damen Shipyards, Gorinchem and delivered 2002-03.
UPDATED

ARAR   *7/2002, A A de Kruijf* / 0533301

2 AL JUBATEL CLASS (WPB)

AL JUBATEL   **SALWA**

Displacement, tons: 95 full load
Dimensions, feet (metres): 86 × 19 × 6.9 *(26.2 × 5.8 × 2.1)*
Main machinery: 2 MTU 16V 396 TB94 diesels; 5,800 hp(m) *(4.26 MW)* sustained; 2 shafts
Speed, knots: 34. **Range, n miles:** 1,100 at 25 kt
Complement: 12 (4 officers)
Guns: 1 Oerlikon/GAM-BO1 20 mm. 2—12.7 mm MGs.
Radars: Surface search: Racal Decca AC 1290; I-band.

Comment: Built by Abeking & Rasmussen, completed in April 1987. Smaller version of Turkish SAR 33 Type. One based at Jizan and one at Al Wajh.
VERIFIED

AL JUBATEL   *1987, Abeking & Rasmussen* / 0080567

4 AL JOUF CLASS (WPBF)

AL JOUF 351   **TURAIF** 352   **HAIL** 353   **NAJRAN** 354

Displacement, tons: 210 full load
Dimensions, feet (metres): 126.6 × 26.2 × 6.2 *(38.6 × 8 × 1.9)*
Main machinery: 3 MTU 16 V 538 TB93 diesels; 11,265 hp(m) *(8.28 MW)* sustained; 3 shafts
Speed, knots: 38. **Range, n miles:** 1,700 at 15 kt
Complement: 20 (4 officers)
Guns: 2 Oerlikon GAM-BO1 20 mm. 2—12.7 mm MGs.
Radars: Surface search: Racal S 1690 ARPA; I-band.
Navigation: Racal Decca RM 1290A; I-band.

Comment: Ordered on 18 October 1987 from Blohm + Voss. First two completed 15 June 1989; second pair 20 August 1989. Steel hulls with aluminium superstructure. *Hail* and *Najran* based at Jiddah in the Red Sea and the others at Aziziah.
VERIFIED

AL JOUF   *6/1989, Blohm + Voss* / 0080566

3 SLINGSBY SAH 2200 HOVERCRAFT (UCAC)

Dimensions, feet (metres): 34.8 × 13.8 *(10.6 × 4.2)*
Main machinery: 1 Deutz BF6L913C diesel; 192 hp(m) *(141 kW)* sustained; lift and propulsion
Speed, knots: 40. **Range, n miles:** 500 at 40 kt
Complement: 2
Military lift: 2.2 tons or 16 troops
Guns: 1—7.62 mm MG.

Comment: Supplied by Slingsby Amphibious Hovercraft, York in December 1990. Have Kevlar armour. These craft have replaced the SRN type.
VERIFIED

SAH 2200   *1990, Slingsby* / 0080568

5 GRIFFON 8000 TD(M) CLASS (HOVERCRAFT) (LCAC)

Displacement, tons: 18.2; 24.6 full load
Dimensions, feet (metres): 69.5 × 36.1 × 1 *(21.15 × 11 × 0.32)*
Main machinery: 2 MTU 12V 183 TB32 V12 diesels; 800 hp *(597 kW)*
Speed, knots: 50. **Range, n miles:** 400 at 45 kt
Complement: 4 (2 officers) (accommodation for further 16)
Guns: 1—12.7 mm MG.
Radars: Raytheon R-80; I-band.

Comment: Five hovercraft ordered from Griffon in 2000 for delivery in 2001. Payload of about 8 tonnes. Similar to those supplied to Indian Coast Guard but with different superstructure. Three based on west coast and two on east coast.
UPDATED

GRIFFON 8000 (Indian colours)   *9/2000*, Indian Coast Guard* / 0104592

INSHORE PATROL CRAFT (PBI)

Type	Date	Speed
Rapier 15.2 m	1976	28
Enforcer, USA, 9.4 m	1980s	30
Simonneau SM 331, 9.3 m	1992	40
Boston Whalers, 8.3 m	1980s	30
Catamarans, 6.4 m	1977	30
Task Force Boats, 5.25 m	1976	20
Viper, 5.1 m	1980s	25
Cobra, 3.9 m	1984	40

Comment: About 500 mostly Task Force Boats. Many are based at Jiddah with the rest spread around the other bases. Most are armed with MGs and the larger craft have I-band radars.
VERIFIED

1 TRAINING SHIP (AXL)

TABBOUK

Displacement, tons: 585 full load
Dimensions, feet (metres): 196.8 × 32.8 × 5.8 *(60 × 10 × 1.8)*
Main machinery: 2 MTU MD 16V 538 TB80 diesels; 5,000 hp(m) *(3.68 MW)* sustained; 2 shafts
Speed, knots: 20. **Range, n miles:** 3,500 at 12 kt
Complement: 26 (6 officers) plus 70 trainees
Guns: 1 Oerlikon GAM-BO1 20 mm.
Radars: Surface search: Racal Decca TM 1226; I-band.
Navigation: Racal Decca 2690BT; I-band.

Comment: Built by Bayerische, Germany and commissioned 1 December 1977. Based at Jiddah.
VERIFIED

3 SMALL TANKERS (YO)

AL FORAT   **DAJLAH**   **AL NIL**

Displacement, tons: 233 full load
Dimensions, feet (metres): 94.2 × 21.3 × 6.9 *(28.7 × 6.5 × 2.1)*
Main machinery: 2 Caterpillar D343 diesels; 2 shafts
Speed, knots: 12. **Range, n miles:** 500 at 12 kt
Radars: Navigation: Decca 110; I-band.

Comment: *Al Nil* based at Aziziah, the others at Jiddah.
VERIFIED

SIMONNEAU SM 331
6/1992, Simonneau Marine / 0080569

Senegal
MARINE SÉNÉGALAISE

Country Overview

The Republic of Senegal was a French colony until 1960 when it gained independence. Situated in western Africa, it has an area of 75,750 square miles and is bordered to the north by Mauritania and to the south by Guinea and Guinea-Bissau. Its 286 n mile coastline with the Atlantic Ocean is divided in two by the coast of Gambia with which the country was united to form the confederation of Senegambia between 1981-89. The capital, largest city and principal port is Dakar. Territorial seas (12 n miles) are claimed. A 200 n mile Exclusive Economic Zone (EEZ) has been declared but its limits have only been partially defined by boundary agreements.

Headquarters Appointments

Head of Navy:
    Captain Ousmane Ibrahima Sall

Personnel

(a)   2005: 900 officers and men
(b)   2 years' conscript service

Bases

Dakar, Elinkine (Casamance)

PATROL FORCES

1 IMPROVED OSPREY 55 CLASS (LARGE PATROL CRAFT) (PBO)

Name	No	Builders	Commissioned
FOUTA	—	Danyard A/S, Fredrikshavn	1 June 1987

Displacement, tons: 470 full load
Dimensions, feet (metres): 180.5 × 33.8 × 8.5 *(55 × 10.3 × 2.6)*
Main machinery: 2 MAN Burmeister & Wain Alpha 12V23/30-DVO diesels; 4,400 hp(m) *(3.23 MW)* sustained; 2 shafts; cp props
Speed, knots: 20. **Range, n miles:** 4,000 at 16 kt
Complement: 38 (4 officers) plus 8 spare
Guns: 1 Hispano Suiza 30 mm. 1 Giat 20 mm.
Radars: Surface search: Furuno FR 1411; I-band.
Navigation: Furuno FR 1221; I-band.

Comment: Ordered in 1985. Intended for patrolling the EEZ rather than as a warship, hence the modest armament. A 25 kt rigid inflatable boat can be launched from a stern ramp which has a protective hinged door. Similar vessels built for Morocco and Greece. *VERIFIED*

FOUTA   *6/2003, Senegal Navy /* 0568323

1 PR 72M CLASS (PBO)

Name	No	Builders	Commissioned
NJAMBUUR	P 773	SFCN, Villeneuve-la-Garenne	Feb 1983

Displacement, tons: 451 full load
Dimensions, feet (metres): 191.0 × 26.9 × 7.2 *(58.2 × 8.2 × 2.2)*
Main machinery: 2 UD 33V16M6D diesels; 5,470 hp *(4.08 MW)* sustained; 2 shafts
Speed, knots: 16. **Range, n miles:** 2,160 at 15 kt
Complement: 46
Guns: 2 OTO Melara 3 in *(76 mm)*/62 compact; 85 rds/min to 16 km *(8.7 n miles)*; weight of shell 6 kg.
2—20 mm Oerlikon. 2—12.7 mm MGs.
Weapons control: 2 CSEE Naja optical directors.
Radars: Surface search: FR 7112 and FR 2105; I-band.

Comment: Ordered in 1979 and launched 23 December 1980. Completed September 1981 for shipping of armament at Lorient. Underwent overhaul at Lorient 2001-2002. *VERIFIED*

NJAMBUUR   *7/2003, B Prézelin /* 0561511

2 PR 48 CLASS (LARGE PATROL CRAFT) (PBO)

Name	No	Builders	Launched	Commissioned
POPENGUINE	—	SFCN, Villeneuve-la-Garenne	22 Mar 1974	10 Aug 1974
PODOR	—	SFCN, Villeneuve-la-Garenne	20 July 1976	13 July 1977

Displacement, tons: 250 full load
Dimensions, feet (metres): 156 × 23.3 × 8.1 *(47.5 × 7.1 × 2.5)*
Main machinery: 2 SACM AGO V12 CZSHR diesels; 4,340 hp(m) *(3.2 MW)*; 2 shafts
Speed, knots: 23. **Range, n miles:** 2,000 at 16 kt
Complement: 33 (3 officers)
Guns: 2 Bofors 40 mm/70. 2—7.62 mm MGs.
Radars: Surface search: Racal Decca 1226; I-band.

Comment: Ordered in 1973 and 1975. *Saint-Louis* decommissioned in 2003 and the operational status of the remaining two is doubtful. *UPDATED*

PR 48 CLASS   *6/2000, A Sharma /* 0105589

3 INTERCEPTOR CLASS (COASTAL PATROL CRAFT) (PB)

Name	No	Builders	Commissioned
SÉNÉGAL II	—	Les Bateaux Turbec Ltd, Sainte Catherine, Canada	Feb 1979
SINE-SALOUM II	—	Les Bateaux Turbec Ltd, Sainte Catherine, Canada	16 Nov 1979
CASAMANCE II	—	Les Bateaux Turbec Ltd, Sainte Catherine, Canada	Aug 1979

Displacement, tons: 62 full load
Dimensions, feet (metres): 86.9 × 19.3 × 5.2 *(26.5 × 5.8 × 1.6)*
Main machinery: 2 diesels; 2,700 hp *(2.01 MW)*; 2 shafts
Speed, knots: 32.5
Guns: 1—20 mm Giat.
Radars: Surface search: Furuno; I-band.

Comment: Used for EEZ patrol. All still in service with new radars.

UPDATED

SÉNÉGAL II   *4/2004 * / 0587793*

2 PETERSON Mk 4 CLASS (PB)

Name	No	Builders	Commissioned
MATELOT ALIOUNE SAMB	—	Peterson Builders Inc	28 Oct 1993
MATELOT OUMAR NDOYE	—	Peterson Builders Inc	4 Nov 1993

Displacement, tons: 22 full load
Dimensions, feet (metres): 51.3 × 14.8 × 4.3 *(15.6 × 4.5 × 1.3)*
Main machinery: 2 Detroit 6V-92TA diesels; 520 hp *(388 kW)*; 2 shafts
Speed, knots: 24. **Range, n miles:** 500 at 20 kt
Complement: 6
Guns: 2—12.7 mm (twin) MGs. 2—7.62 mm (twin) MGs.
Radars: Surface search: Furuno; I-band.

Comment: Ordered in September 1992. Same type delivered to Cape Verde, Gambia and Guinea-Bissau (since deleted) under FMS. Carries an RIB on the stern.

VERIFIED

MATELOT ALIOUNE SAMB   *6/1999 / 0080570*

LAND-BASED MARITIME AIRCRAFT

Numbers/Type: 1 de Havilland Canada DHC-6 Twin Otter.
Operational speed: 168 kt *(311 km/h)*.
Service ceiling: 23,200 ft *(7,070 m)*.
Range: 1,460 n miles *(2,705 km)*.
Role/Weapon systems: MR for coastal surveillance but effectiveness limited. Backed up by a French Navy Breguet Atlantique based at Dakar. Sensors: Search radar. Weapons: Unarmed.

VERIFIED

AMPHIBIOUS FORCES

1 CTM (LCM)

CTM (ex-*CTM 2*, ex-*CTM 5*)

Displacement, tons: 150 full load
Dimensions, feet (metres): 78 × 21 × 4.2 *(23.8 × 6.4 × 1.3)*
Main machinery: 2 Poyaud 520 V8 diesels; 225 hp(m) *(165 kW)*; 2 shafts
Speed, knots: 9.5. **Range, n miles:** 350 at 8 kt
Complement: 6
Military lift: 90 tons

Comment: Transferred from French Navy in September 1999. Has a bow ramp.

VERIFIED

CTM   *12/2001 / 0525011*

1 CDIC CLASS (LCT)

FALEMÉ 2 (ex-*Javeline* L 9070)

Displacement, tons: 710 full load
Dimensions, feet (metres): 194.9 × 39 × 5.9 *(59.4 × 11.9 × 1.8)*
Main machinery: 2 SACM Uni Diesel UD 30 VIZ M1 diesels; 1,200 hp(m) *(882 kW)* sustained; 2 shafts
Speed, knots: 10.5. **Range, n miles:** 1,000 at 10 kt
Complement: 12
Military lift: 336 tons
Guns: Fitted for 2 Giat 20F2 20 mm.
Radars: Navigation: Racal Decca 1229; I-band.

Comment: Acquired vice one Edic class on 7 February 1999.

UPDATED

FALEMÉ 2   *11/2003 * / 0587792*

1 EDIC 700 CLASS (LCT)

Name	No	Builders	Launched	Commissioned
KARABANE	841	SFCN, Villeneuve-la-Garenne	6 Mar 1986	30 Jan 1987

Displacement, tons: 736 full load
Dimensions, feet (metres): 193.5 × 39 × 5.6 *(59 × 11.9 × 1.7)*
Main machinery: 2 SACM MGO 175 V12 ASH diesels; 1,200 hp(m) *(882 kW)* sustained; 2 shafts
Speed, knots: 12; 8 *(Javeline)*. **Range, n miles:** 1,800 at 10 kt
Complement: 18 (33 spare billets)
Military lift: 12 trucks; 340 tons equipment
Guns: Fitted for 2 Giat 20 mm.
Radars: Navigation: Racal Decca 1226; I-band.

Comment: Ordered May 1985, delivered 23 June 1986 from France. Second of class from France in 1995 and returned again in 1996.

VERIFIED

KARABANE   *6/2000, A Sharma / 0093591*

AUXILIARIES

Notes: (1) A modified purse-seiner, *Itaf Deme* is used for fisheries research. Crewed by the Navy.
(2) There is also a harbour tug *Cheik Oumar Fall* (ex-Y 719) on loan from the French Navy since 1990.

Serbia and Montenegro
KORPUS RATNE MORNARIĆE

Country Overview

The Federal Republic of Serbia and Montenegro, known until February 2003 as the Federal Republic of Yugoslavia, was formed in 1992 following the secession of four of the six republics from the former Socialist Federal Republic of Yugoslavia (SFRY). With an area of 39,449 square miles, it is located in south-eastern Europe in the Balkan Peninsula and comprises Serbia (including the provinces of Kosovo and Vojvodina) and Montenegro. It is bordered to the north by Hungary, to the east by Romania and Bulgaria, to the south by Macedonia and Albania and to the west by Croatia and Bosnia and Herzegovina. It has a 107 n mile coastline with the Adriatic Sea on which Bar and Tivat are the principal ports. The capital and largest city is Belgrade. Territorial waters (12 n miles) are claimed but an EEZ has not been claimed.

Headquarters Appointments

Commander-in-Chief:
    Rear Admiral Jovan Grbavac
Chief of Staff:
    Captain Dragan Samardzić

Personnel

2005: 5,723 (2,543 conscripts)

Bases

Headquarters: Bar
Main base: Tivat
Minor base: Bar

Organisation

The Navy Corps forms part of the Armed Forces of Serbia and Montenegro and is subordinate to the Deputy Chief of General Staff for the Navy.

Flotilla: 88 Submarine Brigade, 18 Missile Boat Brigade, 38 Patrol Brigade and 376 Naval Logistics Command. The Riverine Flotilla (Novi Sad) is subordinate to the Maritime Department of the General Staff.

Coastal Defence Command: 81 Bde Coast Defence (Kumbor), 83 Bde Coast Defence (Bar), 110 Coastal Artillery Brigade and 108 Coastal Missile Brigade. The latter is equipped with 9-12 truck-mounted twin SS-C-3 Styx SSM launchers and a number of Coastal Artillery units with 130 mm, 122 mm and 88 mm guns.

DELETIONS

Submarines

2003	*Drava*

Frigates

2005	*Beograd*

Patrol Forces

2002	*Josip Mažar, Šoša, Nikola Martinović, Petar Drapšin Mirče, Acev, Fruška Gora*
2003	PC 303, 11 Type 15
2004	*Ramiz Sadiku, Jordan Nikolov Orče, Grimeč, Zelengora*

Mine Warfare Forces

2002	*Brseč, Mljet*

Amphibious Forces

2002	3 DSM 501
2003	11 Type 11
2004	1 Type 21

Auxiliaries

2002	*Vis*
2003	*Liganj*

SUBMARINES

1 SAVA CLASS (PATROL SUBMARINE) (SSC)

Name	No	Builders	Laid down	Launched	Commissioned
SAVA	831	S and DE Factory, Split	1975	1977	1978

Displacement, tons: 830 surfaced; 960 dived
Dimensions, feet (metres): 182.7 × 23.6 × 16.7 *(55.7 × 7.2 × 5.1)*
Main machinery: Diesel-electric; 2 Sulzer diesels; 1,600 hp(m) *(1.18 MW)*; 2 generators; 1 MW; 1 motor; 2,040 hp(m) *(1.5 MW)*; 1 shaft
Speed, knots: 10 surfaced; 16 dived
Complement: 35

Torpedoes: 6—21 in *(533 mm)* bow tubes. 10 TEST-71ME; wire-guided active/passive homing to 15 km *(8.1 n miles)* at 40 kt or 20 km *(10.8 n miles)* at 25 kt; warhead 220 kg.
Mines: 20 in lieu of torpedoes.
Countermeasures: ESM: Stop Light; radar warning.
Radars: Surface search: Snoop Group; I-band.
Sonars: Atlas Elektronik PRS-3; hull-mounted; active/passive search and attack; medium frequency.

Structure: An improved version of the Heroj class. Diving depth, 300 m *(980 ft)*, built with USSR electronic equipment and armament. Pressure hull is 42.7 × 5.05 m and the casing is made of GRP. Accommodation is poor, suggesting the design was intended for only short periods at sea.
Operational: Second of class *Drava* was decommissioned in 2003. Based at Tivat.

VERIFIED

SAVA   *2/1998, Yugoslav Navy /* 0050744

1 HEROJ CLASS (PATROL SUBMARINE) (SSC)

Name	No	Builders	Laid down	Launched	Commissioned
HEROJ	821	Uljanic Shipyard, Pula	1964	1967	1968

Displacement, tons: 615 surfaced; 705 dived
Dimensions, feet (metres): 165.4 × 15.4 × 14.8 *(50.4 × 4.7 × 4.5)*
Main machinery: Diesel-electric; 2 Rade Končar diesel generators; 1,200 hp(m) *(880 kW)*; 1 motor; 1,560 hp(m) *(1.15 MW)*; 1 shaft; auxiliary motor
Speed, knots: 10 surfaced; 15 dived
Range, n miles: 4,100 at 10 kt snorting
Complement: 28

Torpedoes: 4—21 in *(533 mm)* bow tubes. 6 SET-65E; active/passive homing to 15 km *(8.1 n miles)* at 40 kt; warhead 205 kg.
Mines: 12 in lieu of torpedoes.
Countermeasures: ESM: Stop Light; radar warning.
Radars: Surface search: Snoop Group; I-band.
Sonars: Thomson Sintra Eledone; hull-mounted; passive ranging; medium frequency.

Structure: Diving depth 150 m *(500 ft)*.
Operational: *Heroj* was assessed as having been placed in reserve in 2000 but was reported officially as being operational again in 2003. However, its effectiveness is likely to be mitigated by its age.

UPDATED

HEROJ   *1982*

3 UNA CLASS (MIDGET SUBMARINES) (SSW)

ZETA 913

KUPA 915   VARDAR 916

Displacement, tons: 76 surfaced; 88 dived
Dimensions, feet (metres): 61.7 × 9 × 8.2 *(18.8 × 2.7 × 2.5)*
Main machinery: 2 motors; 49 hp(m) *(36 kW)*; 1 shaft
Speed, knots: 6 surfaced; 7 dived
Range, n miles: 200 at 4 kt
Complement: 6 plus 4 swimmers
Sonars: Atlas Elektronik; passive/active search; high frequency.

Comment: Building yard, Split. First of class commissioned May
1985; last two in 1989. Exit/re-entry capability with mining
capacity. Can carry six combat swimmers, plus four Swimmer
Delivery Vehicles (SDV) and limpet mines. Diving depth:
120 m *(394 ft)*. Batteries can only be charged from shore or
from a depot ship. All based at Tivat and all reported
operational.

UPDATED

UNA CLASS
5/2004*, Sieche Collection / 1044499

2 R-2 MALA CLASS (TWO-MAN SWIMMER DELIVERY VEHICLES) (LDW)

Displacement, tons: 1.4
Dimensions, feet (metres): 16.1 × 4.6 × 4.3
(4.9 × 1.4 × 1.3)
Main machinery: 1 motor; 4.7 hp(m) *(3.5 kW)*; 1 shaft
Speed, knots: 4.4
Range, n miles: 18 at 4.4 kt; 23 at 3.7 kt
Complement: 2
Mines: 250 kg of limpet mines.

Comment: Free-flood craft with the main motor, battery,
navigation pod and electronic equipment housed in separate
watertight cylinders. Instrumentation includes aircraft type
gyrocompass, magnetic compass, depth gauge (with
0-100 m scale), echo-sounder, sonar and two searchlights.
Constructed of light aluminium and plexiglass, it is fitted with
fore and after-hydroplanes, the tail being a conventional
cruciform with a single rudder abaft the screw. Large perspex
windows give a good all-round view. Operating depth 60 m
(196.9 ft) maximum. Two operated by Croatia. Two reported
sold to Syria and one to Sweden.

Notes: There are also reported to be four R-1 craft. Transportable
in submarine torpedo tubes and crewed by one man, they are
3.7 m craft, powered by a 1 kW electric motor and 24V silver-
zinc batteries. Capable of 2.8 kt, they can dive to 60 m. They
have a range of 4 n miles. Of a total twelve reported to have
been manufactured. A further six units have probably been
deleted, one unit is in Croatia and one was exported to
Sweden.

VERIFIED R-2

6/2003, Serbia and Montenegro Navy / 0572433

FRIGATES

2 KOTOR CLASS (FFGM)

Name	No	Builders	Launched	Commissioned
KOTOR	33	Tito Shipyard, Kraljevica	21 May 1985	Jan 1987
NOVI SAD (ex-*Pula*)	34	Tito Shipyard, Kraljevica	18 Dec 1986	Nov 1988

Displacement, tons: 1,870 full load
Dimensions, feet (metres): 317.3 × 42 × 13.7
(96.7 × 12.8 × 4.2)
Main machinery: CODAG; 1 SGW Nikolayev gas turbine;
18,000 hp(m) *(13.2 MW)*; 2 SEMT-Pielstick 12 PA6 V 280
diesels; 9,600 hp(m) *(7.1 MW)* sustained; 3 shafts
Speed, knots: 27 gas; 22 diesel. **Range, n miles:** 1,800 at 14 kt
Complement: 110

Missiles: SSM: 4 SS-N-2C Styx ❶; active radar or IR homing to
83 km *(45 n miles)* at 0.9 Mach; warhead 513 kg; sea-
skimmer at end of run.
SAM: SA-N-4 Gecko twin launcher ❷; semi-active radar homing
to 15 km *(8 n miles)* at 2.5 Mach; height envelope 9-3,048 m
(29.5-10,000 ft); warhead 50 kg.
Guns: 4 USSR 3 in *(76 mm)*/60 (2 twin) (1 mounting only in 33
and 34) ❸; 90 rds/min to 15 km *(8 n miles)*; weight of shell
6.8 kg.
4 USSR 30 mm/65 (2 twin) ❹; 500 rds/min to 5 km *(2.7 n
miles)*; weight of shell 0.54 kg.
A/S mortars: 2 RBU 6000 12-barrelled trainable ❺; range
6,000 m; warhead 31 kg.
Mines: Can lay mines.
Countermeasures: Decoys: 2 Wallop Barricade double layer
chaff launchers.
Radars: Air/surface search: Strut Curve ❻; F-band.
Navigation: Palm Frond; I-band.
Fire control: PEAB 9LV200 ❽; I-band (for 76 mm and SSM).
Drum Tilt ❾; H/I-band (for 30 mm).
Pop Group ❿; F/H/I-band (for SAM).
IFF: High Pole; 2 Square Head.
Sonars: Bull Nose; hull-mounted; active search and attack;
medium frequency.

Programmes: Built under licence. Type name, VPB (Veliki
Patrolni Brod).
Modernisation: Combat data system fitted in *Novi Sad* in 2000
and may be retrofitted in the other. The intention is to upgrade
the missile system, possibly with Russian SS-N-25 or Chinese
C-802 SSMs.
Structure: The hulls are similar to the Russian Koni class but are
to a Yugoslavian design.
Operational: Based at Tivat.

UPDATED

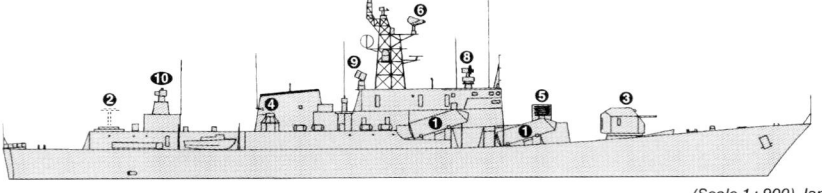

KOTOR   *(Scale 1 : 900), Ian Sturton*

KOTOR   *6/1998, Yugoslav Navy /* 0050746

LAND-BASED MARITIME AIRCRAFT

Notes: There are no operational land-based aircraft.

PATROL FORCES

3 KONČAR CLASS (TYPE 240) (PTFG)

Name	No	Builders	Launched	Commissioned
RADE KONČAR	401	Tito Shipyard, Kraljevica	16 Oct 1976	Apr 1977
HASAN ZAFIROVIĆ-LACA	404	Tito Shipyard, Kraljevica	9 Nov 1978	Dec 1978
ANTE BANINA	406	Tito Shipyard, Kraljevica	23 Nov 1979	Nov 1980

Displacement, tons: 271 full load
Dimensions, feet (metres): 147.6 × 27.6 × 8.5 *(45 × 8.4 × 2.6)*
Main machinery: CODAG; 2 RR Proteus gas turbines; 7,100 hp *(5.29 MW)* sustained; 2 MTU 16V 538 TB91 diesels; 6,000 hp(m) *(4.41 MW)* sustained; 4 shafts; cp props
Speed, knots: 38. **Range, n miles:** 490 at 38 kt; 870 at 23 kt (diesels)
Complement: 30 (5 officers)

Missiles: SSM: 2 SS-N-2B Styx; active radar or IR homing to 46 km *(25 n miles)* at 0.9 Mach; warhead 513 kg.
Guns: 2 Bofors 57 mm/70; 200 rds/min to 17 km *(9.3 n miles)*; weight of shell 2.4 kg.
2—128 mm rocket launchers for illuminants.
2—30 mm/65 (twin) or 1—30 mm/65 AK 630 may be fitted in place of the after 57 mm.
Countermeasures: 2 Wallop Barricade double layer chaff launchers.
Weapons control: PEAB 9LV 202 GFCS.
Radars: Surface search: Decca 1226; I-band.
Fire control: Philips TAB; I/J-band.

Programmes: Type name, Raketna Topovnjaca.
Structure: Aluminium superstructure. Designed by the Naval Shipping Institute in Zagreb based on Swedish Spica class with bridge amidships like Malaysian boats. The after 57 mm gun is replaced by a twin 30 mm mounting in at least one of the class.
Operational: 402 was taken by Croatia in 1991 and 401 was badly damaged but has been repaired. 401, 404 and 406 are reported active. 403 and 405 probably decommissioned. Three further craft cannibalised for spares. Based at Tivat. Names may no longer be in use.
UPDATED

HASAN ZAFIROVIĆ-LACA   *5/2004*, Sieche Collection /* 1044500

ANTE BANINA   *5/2004*, Sieche Collection /* 1044501

2 MIRNA CLASS (TYPE 140) (PCM)

UČKA 174   KOSMAJ 178

Displacement, tons: 142 full load
Dimensions, feet (metres): 104.9 × 22 × 7.5 *(32 × 6.7 × 2.3)*
Main machinery: 2 SEMT-Pielstick 12 PA4 200 VGDS diesels; 5,292 hp(m) *(3.89 MW)* sustained; 2 shafts
Speed, knots: 28. **Range, n miles:** 400 at 20 kt
Complement: 19 (3 officers)
Missiles: SAM: 1 SA-N-5 Grail quad mounting; manual aiming; IR homing to 6 km *(3.2 n miles)* at 1.5 Mach; altitude to 2,500 m *(8,000 ft)*; warhead 1.5 kg.
Guns: 1 Bofors 40 mm/70. 4—20 mm (quad). 2—128 mm illuminant launchers.
Depth charges: 8 DCs.
Radars: Surface search: Racal Decca 1216C; I-band.
Sonars: Simrad SQS-3D/3F; active; high frequency.

Comment: Builders, Kraljevica Yard. Launched between June 1981 and December 1983. An unusual feature of this design is the fitting of an electric outboard motor giving a speed of up to 6 kt. One sunk possibly by a limpet mine in November 1991. Four held by Croatia. Others have paid off and been cannibalised and the operational status of the remaining two is doubtful. Names probably discontinued. Gun armament aft changed in 1999/2000. *UPDATED*

UČKA   *5/2004*, Sieche Collection /* 1044502

6 TYPE 20 BISCAYA CLASS (RIVER PATROL CRAFT) (PBR)

PC 211-216

Displacement, tons: 55 standard
Dimensions, feet (metres): 71.5 × 17 × 3.9 *(21.8 × 5.3 × 1.2)*
Main machinery: 2 diesels; 1,156 hp(m) *(850 kW)*; 2 shafts
Speed, knots: 16. **Range, n miles:** 200 at 15 kt
Complement: 10
Guns: 2 Oerlikon 20 mm.
Radars: Surface search: Decca 110; I-band.

Comment: Completed in the late 1980s. Steel hull with GRP superstructure. All active with the Riverine Flotilla.
VERIFIED

PC 215   *1988, Yugoslav Navy /* 0084261

1 RIVER PATROL BOAT (PBR)

PC 111

Displacement, tons: 29 full load
Dimensions, feet (metres): 79.1 × 13.5 × 2.9 *(24.1 × 4.1 × 0.9)*
Main machinery: 2 diesels; 652 hp(m) *(486 kW)*; 2 shafts
Speed, knots: 17. **Range, n miles:** 720 at 17 kt
Complement: 6

Comment: Built for US Navy's Rhine River patrol and transferred in the 1950s.
VERIFIED

PC 111   *6/2001, Vojska /* 0528418

1 BOTICA CLASS (TYPE 16) (RIVER PATROL CRAFT) (PBR)

PC 302

Displacement, tons: 23 full load
Dimensions, feet (metres): 55.8 × 11.8 × 2.8 *(17.0 × 3.6 × 0.8)*
Main machinery: 2 diesels; 464 hp(m) *(340 kW)*; 2 shafts
Speed, knots: 15. **Range, n miles:** 340 at 14 kt
Complement: 7
Guns: 1 Oerlikon 20 mm (fitted for). 2—7.62 mm MGs.

Comment: Built in about 1970 and reactivated having been decommissioned in the 1990s. Used for riverine patrols. Can carry up to 30 troops.
NEW ENTRY

PC 302   *5/2004*, Sieche Collection /* 0583300

MINE WARFARE FORCES

2 SIRIUS CLASS (MINESWEEPERS/HUNTERS) (MSC/MHC)

Name	No	Builders	Commissioned
PODGORA (ex-Smeli)	M 152 (ex-D 26)	A Normand, France	Sep 1957
BLITVENICA (ex-Slobodni)	M 153 (ex-D 27)	A Normand, France	Sep 1957

Displacement, tons: 365 standard; 424 full load
Dimensions, feet (metres): 152 × 28 × 8.2 (46.4 × 8.6 × 2.5)
Main machinery: 2 SEMT-Pielstick PA1 175 diesels; 1,620 hp(m) (1.19 MW); 2 shafts
Speed, knots: 15. **Range, n miles:** 3,000 at 10 kt
Complement: 40
Guns: 2 Oerlikon 20 mm.
Countermeasures: MCMV: PAP 104 (minehunter); remote-controlled submersibles.
Radars: Navigation: Thomson-CSF DRBN 30; I-band.
Sonars: Thomson Sintra TSM 2022 (minehunter); hull-mounted; active minehunting; high frequency.

Comment: Built to a British design as US 'offshore' orders. *Blitvenica* converted to minehunter in 1980-81. Decca Hi-fix. These two are based at Tivat and are both operational.
VERIFIED

BLITVENICA *6/2003, Serbia and Montenegro Navy /* 0572435

7 NESTIN CLASS (RIVER MINESWEEPERS) (MSR)

Name	No	Builders	Commissioned
NESTIN	M 331	Brodotehnika, Belgrade	20 Dec 1975
MOTAJICA	M 332	Brodotehnika, Belgrade	18 Dec 1976
BELEGIŠ	M 333	Brodotehnika, Belgrade	1976
BOSUT	M 334	Brodotehnika, Belgrade	1979
VUČEDOL	M 335	Brodotehnika, Belgrade	1979
DJERDAP	M 336	Brodotehnika, Belgrade	1980
NOVI SAD	M 341	Brodotehnika, Belgrade	8 June 1996

Displacement, tons: 65 full load
Dimensions, feet (metres): 88.6 × 21.7 × 5.2 (27 × 6.3 × 1.6)
Main machinery: 2 diesels; 520 hp(m) (382 kW); 2 shafts
Speed, knots: 15. **Range, n miles:** 860 at 11 kt
Complement: 17
Guns: 6 Hispano 20 mm (quad fwd, 2 single aft). Some may still have a 40 mm gun forward. 8—20 mm (quad fwd and aft) (M 341).
Mines: 24 can be carried.
Countermeasures: MCMV: Magnetic, acoustic and explosive sweeping gear.
Radars: Surface search: Racal Decca 1226; I-band.

Comment: Some transferred to Hungary and Iraq. One more completed in 1996. The class is based at Novi Sad as part of the Riverine Flotilla. One deleted in 1997. One more building is unlikely to be completed. M 341, which replaced the previously deleted M 337, is to a modified design which includes different armament.
VERIFIED

DJERDAP *6/2003, Serbia and Montenegro Navy /* 0572436

AMPHIBIOUS FORCES

7 TYPE 22 (LCU)

DJC 627	DJC 412 (ex-DJC 625)	DJC 414 (ex-DJC 621)
DJC 628	DJC 413 (ex-DJC 630)	DJC 415 (ex-DJC 631)
DJC 411 (ex-DJC 632)		

Displacement, tons: 48 full load
Dimensions, feet (metres): 73.2 × 15.7 × 3.3 (22.3 × 4.8 × 1)
Main engines: 2 MTU diesels; 1,740 hp(m) (1.28 MW); 2 water-jets
Speed, knots: 30. **Range, n miles:** 320 at 22 kt
Complement: 8
Military lift: 40 troops or 15 tons cargo
Guns: 2—20 mm M71. 1—30 mm grenade launcher.
Radars: Navigation: Decca 101; I-band.

Comment: Built of polyester and glass fibre. Last one completed in 1987. Based in Danube Flotilla.
UPDATED

DJC 412 *6/2003, Serbia and Montenegro Navy /* 0572441

1 SILBA CLASS (LCT/ML)

Name	No	Builders	Commissioned
KRK (ex-Silba)	DBM 241	Brodosplit Shipyard, Split	1990

Displacement, tons: 880 full load
Dimensions, feet (metres): 163.1 oa; 144 wl × 33.5 × 8.5 (49.7; 43.9 × 10.2 × 2.6)
Main machinery: 2 Burmeister & Wain Alpha 10V23L-VO diesels; 3,100 hp(m) (2.28 MW) sustained; 2 shafts; cp props
Speed, knots: 12. **Range, n miles:** 1,200 at 12 kt
Complement: 33 (3 officers)
Military lift: 460 tons or 6 medium tanks or 7 APCs or 4—130 mm guns plus towing vehicles or 300 troops with equipment
Missiles: SAM: 1 SA-N-5 Grail quad mounting.
Guns: 4—30 mm/65 (2 twin) AK 230.
4—20 mm M75 (quad). 2—128 mm illuminant launchers.
Mines: 94 Type SAG-1.
Radars: Surface search: Racal Decca; I-band.

Comment: Ro-ro design with bow and stern ramps. Can be used for minelaying, transporting weapons or equipment and troops. Operational and based at Tivat. Two further craft, launched in 1992 and 1994, are in the Croatian Navy.
VERIFIED

KRK *6/1998, MoD Bonn /* 0050751

3 TYPE 21 (LCU)

DJC 614	DJC 616	DJC 618

Displacement, tons: 32 full load
Dimensions, feet (metres): 69.9 × 15.7 × 5.2 (21.3 × 4.8 × 1.6)
Main machinery: 1 diesel; 1,450 hp(m) (1.07 MW); 1 shaft
Speed, knots: 23. **Range, n miles:** 320 at 22 kt
Complement: 6
Military lift: 6 tons
Guns: 1—20 mm M71.

Comment: The survivors of a class of 20 built between 1976 and 1979. Four held by Croatia in 1991 of which three have paid off. Others sunk or scrapped. Some of these may be laid up.
UPDATED

DJC 616 *6/2003, Serbia and Montenegro Navy /* 0572440

AUXILIARIES

Notes: (1) Three 22 m inshore survey vessels BH 11, BH 12 and CH 1 are operated by the Naval Hydrological Institute.
(2) There are seven tenders BM 58, BM 65, BM 66, BM 67, BM 70 and BS 22.
(3) There are five diving tenders BRM 81, BRM 84, BRM 85, BRM 87 and BRM 88.
(4) PV 17 is a 44 m water tanker.

1 SAIL TRAINING SHIP (AXS)

JADRAN

Displacement, tons: 737 full load
Dimensions, feet (metres): 196.9 × 29.2 × 13.3 *(60.0 × 8.9 × 4.05)*
Main machinery: 1 Burmeister Alpha diesel; 353 hp *(263 kW)*
Speed, knots: 10.4
Radars: 1 FR 2120 and 1 FR 7061; I-band.

Comment: The contract for a barquentine sail training ship was signed on 4 September 1930 with the German shipbuilding company H C Silken Zon of Hamburg. She was launched on 25 June 1931 and arrived in Tivat on 16 July 1933. During the Second World War, she was used by the Italian Navy under the name of *Marco Polo* before being allowed to fall into disrepair. She returned to Yugoslavia in 1946 and was reconstructed in her original form at Tivat.
VERIFIED

JADRAN   *6/2001, Vojska* / 0528417

1 KOZARA CLASS (HEADQUARTERS SHIP) (PBR)

KOZARA (ex-*Oregon*, ex-*Kriemhild*) RPB 30

Displacement, tons: 695 full load
Dimensions, feet (metres): 219.8 × 31.2 × 4.6 *(67 × 9.5 × 1.4)*
Main machinery: 2 Deutz RV6M545 diesels; 800 hp(m) *(588 kW)*; 2 shafts
Speed, knots: 12
Guns: 9 Hispano Suiza 20 mm (3 triple).

Comment: Former Presidential Yacht on Danube. Built in Austria in 1940. Acts as Flagship of the Riverine Flotilla. A similar ship served in the Russian Black Sea Fleet before being transferred to Ukraine. Although previously believed to have been decommissioned, continues to be used to accommodate Riverine Flotilla Staff.
VERIFIED

KOZARA   *6/2003, Serbia and Montenegro Navy* / 0572439

1 LUBIN CLASS (TRANSPORT SHIP) (AKR)

LUBIN PO 91

Displacement, tons: 860 full load
Dimensions, feet (metres): 190.9 × 36.1 × 9.2 *(58.2 × 11.0 × 2.8)*
Main machinery: 2 diesels; 3,500 hp(m) *(2.57 MW)*; 2 shafts; cp props
Speed, knots: 16. **Range, n miles:** 1,500 at 16 kt
Complement: 43
Military lift: 150 troops; 6 tanks
Guns: 1 Bofors 40 mm/70. 4—20 mm M75. 128 mm rocket launcher for illuminants.

Comment: Fitted with bow doors and two upper-deck cranes. Ro-Ro cargo ship built in Split in the 1980s and used as an ammunition transport. Based at Tivat. This ship had been assessed decommissioned in the early 1990s but has been officially reported as being in good condition and operational.
UPDATED

LUBIN   *5/2004*, Sieche Collection* / 1044503

1 DRINA CLASS (AOTL)

SIPA PN 27

Displacement, tons: 430 full load
Dimensions, feet (metres): 151 × 23.6 × 10.2 *(46 × 7.2 × 3.1)*
Main machinery: 1 diesel; 300 hp(m) *(220 kW)*; 1 shaft
Speed, knots: 7
Complement: 12
Missiles: SAM: 1 SA-N-5.
Guns: 6 Hispano 20 mm (1 quad, 2 single).

Comment: Built at Kraljevic in mid-1950s. Based at Tivat.
UPDATED

SIPA   *6/2003*, Serbia and Montenegro Navy* / 1044504

1 SABAC CLASS (DEGAUSSING VESSEL) (YDG)

SABAC RSRB 36

Displacement, tons: 110 standard
Dimensions, feet (metres): 105.6 × 23.3 × 3.9 *(32.2 × 7.1 × 1.2)*
Main machinery: 1 diesel; 528 hp(m) *(388 kW)*; 1 shaft
Speed, knots: 10. **Range, n miles:** 660 at 10 kt
Complement: 20
Guns: 2—20 mm M71.
Radars: Navigation: Decca 101; I-band.

Comment: Built in 1985. Used to degauss River vessels up to a length of 50 m.
VERIFIED

SABAC   *6/2003, Serbia and Montenegro* / 0572437

TUGS

Notes: There are three coastal tugs PR 37, PR 38 and PR 41 and seven harbour tugs LR 23, LR 72, LR 74, LR 75, LR 77 and LR 80.

Seychelles

Country Overview

A former British colony, the Republic of the Seychelles became independent in 1976. Situated in the western Indian Ocean, northeast of Madagascar, the archipelago consists of some 90 islands, disposed over 13,000 square miles in two groups. The 40 islands of the northern group include the principal islands: Mahé (the largest), Praslin, Silhouette and La Digue. The 50 or so low-lying coral islands in the south are mostly uninhabited.

Victoria (Mahé) is the capital, largest town and principal port. Territorial seas (12 n miles) are claimed. A 200 n mile Exclusive Economic Zone (EEZ) has been declared but the limits have not been fully defined by boundary agreements.

Headquarters Appointments

Commander of the Coast Guard:
Lieutenant Colonel D Gertrude

Bases

Port Victoria, Mahé

Personnel

2005: 300 including 80 air wing and 100 marines

COAST GUARD

1 ZHUK (PROJECT 1400M) CLASS
(COASTAL PATROL CRAFT) (PB)

Name	No	Builders	Commissioned
FORTUNE	604	USSR	6 Nov 1982

Displacement, tons: 39 full load
Dimensions, feet (metres): 78.7 × 16.4 × 3.9 *(24 × 5 × 1.2)*
Main machinery: 2 Type M 401B diesels; 2,200 hp *(1.6 MW)* sustained; 2 shafts
Speed, knots: 30. **Range, n miles:** 1,100 at 15 kt
Complement: 12 (3 officers)
Guns: 4—14.5 mm (2 twin) MGs.
Radars: Surface search: Furuno; I-band.

Comment: Two transferred from USSR. Second of class paid off in 1996 and used for spares.
VERIFIED

FORTUNE 6/1998, Seychelles Coast Guard / 0050097

1 TYPE FPB 42 (LARGE PATROL CRAFT) (PB)

Name	No	Builders	Commissioned
ANDROMACHE	605	Picchiotti, Viareggio	10 Jan 1983

Displacement, tons: 268 full load
Dimensions, feet (metres): 137.8 × 26 × 8.2 *(41.8 × 8 × 2.5)*
Main machinery: 2 Paxman Valenta 16 CM diesels; 6,650 hp *(5 MW)* sustained; 2 shafts
Speed, knots: 26. **Range, n miles:** 3,000 at 16 kt
Complement: 22 (3 officers)
Guns: 1 Oerlikon 25 mm. 2—7.62 mm MGs.
Radars: Surface search: 2 Furuno; I-band.

Comment: Ordered from Inma, La Spezia in November 1981. A second of class reported ordered in 1991 but the order was not confirmed.
VERIFIED

ANDROMACHE 3/1997 / 0019096

1 COASTAL PATROL CRAFT (PB)

JUNON 602

Displacement, tons: 40 full load
Dimensions, feet (metres): 60.0 × 16.7 × 5.9 *(18.3 × 5.1 × 1.8)*
Speed, knots: 20
Complement: 5
Radars: Surface search: Furuno; I-band.

Comment: Former Port and Marine Services patrol boat reintegrated into the Coast Guard in 2003.
VERIFIED

JUNON 9/2003, Seychelles Coast Guard / 0568333

5 PATROL CRAFT (PB)

ARIES	VIRGO	LIBRA	TAURUS	PISCES

Displacement, tons: 17.7 full load
Dimensions, feet (metres): 44.0 × 12.5 × 3.9 *(13.4 × 3.8 × 1.2)*
Main machinery: 2 General Motors Detroit 6V53 diesels; 2 shafts
Speed, knots: 13. **Range, n miles:** 200 at 11 kt
Complement: 3
Radars: Surface search: Furuno; I-band.

Comment: Former US Coast Guard lifeboats (MLB) constructed in the 1960s. Three were transported to the Seychelles onboard USS *Anchorage* in October 2000 and a further two onboard USS *Tarawa* in December 2000.
VERIFIED

MLBs 9/2003, Seychelles Coast Guard / 0568334

LAND-BASED MARITIME AIRCRAFT

Numbers/Type: 1 Britten-Norman BN-2A-21 Maritime Defender.
Operational speed: 150 kt *(280 km/h)*.
Service ceiling: 18,900 ft *(5,760 m)*.
Range: 1,500 n miles *(2,775 km)*.
Role/Weapon systems: Coastal surveillance and surface search aircraft. Sensors: Search radar. Weapons: Provision for rockets or guns.
VERIFIED

BN2T-4S (Irish Police colours) 8/1997 / 0016662

Sierra Leone

Country Overview

A former British colony, Sierra Leone became independent in 1961. Located in west Africa, the country has an area of 27,699 square miles, a 217 n mile coastline with the Atlantic Ocean and is bordered to the north by Guinea and to the south by Liberia. The capital, largest city and principal port is Freetown. An EEZ has not been claimed and the country is one of a few that claims 200 n mile territorial seas.

Headquarters Appointments

Senior Officer, Navy:
   Captain A B Sessay

Personnel

(a)   2005: 230 (30 officers)
(b)   Voluntary service

Bases

Freetown (HQ), Yeliboya, Sulima and Mania (Turtle Is)

PATROL FORCES

Notes: Two RIBs and five other small craft have been acquired for inshore patrol. The acquisition of up to three larger craft is a possibility.

1 SHANGHAI III (TYPE 062/1) CLASS (COASTAL PATROL CRAFT) (PB)

ALIMAMY RASSIN PB 103

Displacement, tons: 170 full load
Dimensions, feet (metres): 134.5 × 17.4 × 5.9 *(41 × 5.3 × 1.8)*
Main machinery: 4 Chinese L12-180A diesels; 4,400 hp(m) *(3.22 MW)* sustained; 4 shafts
Speed, knots: 25. **Range, n miles:** 750 at 17 kt
Complement: 43
Guns: 4 China 37 mm/63 (2 twin); 180 rds/min to 8.5 km *(4.6 n miles)*; weight of shell 1.42 kg.
   4 China 14.5 mm (2 twin) Type 69 or 4 China 25 mm (2 twin).
Radars: Surface search: Pot Head or Anritsu 726; I-band.

Comment: Transferred from China in 1987. Two further craft laid up. Operational status doubtful.
UPDATED

ALIMAMY RASSIN   *4/2002* / 0528294

Singapore

Country Overview

Formerly under British rule, the Republic of Singapore became self-governing in 1959. It joined Malaysia in 1963, but separated from the Federation in 1965 to become a sovereign state. With an area of 247 square miles and a coastline of 104 n miles, the main island is separated from the southern tip of Malaysia by the narrow Johore Strait. There are 59 small adjacent islets. To the south the Singapore Strait, an important shipping channel linking the Indian Ocean with the South China Sea, separates the island from the Riau archipelago of Indonesia. Territorial seas (3 n miles) are claimed. An EEZ is not claimed.

Headquarters Appointments

Chief of the Navy:
   Rear Admiral Ronnie Tay
Chief of Staff:
   Rear Admiral Sim Gim Guan
Fleet Commander:
   Colonel Chew Men Leong
Commander Police Coast Guard:
   Deputy Assistant Commissioner Jerry See Buck Thye

Personnel

(a)   2005: 4,500 officers and men including 1,800 conscripts
(b)   National Service: two and a half years for Corporals and above; two years for the remainder
(c)   5,000 reservists (operationally trained)

Bases

Tuas (Jurong), Changi, Sembawang

Organisation

Five Commands: Fleet, Naval Diving Unit, Coastal, Naval Logistics and Training.
Fleet: First Flotilla (six Victory, six Sea Wolf, 12 Fearless).
   Third Flotilla (three LSTs, Fast Craft at Civil Squadron).
Coastal Command: (four Bedok, 12 PBs)
   Coastal Command operates five unmanned Giraffe 100 air/surface surveillance radar sites at Changi, Pedra Branca, St John's Island, Sultan Shoal Lighthouse and Raffles Lighthouse. Air and surface track data is passed to HQ RSN.

Prefix to Ships' Names

RSS

Special Forces

Singapore's special forces include the Naval Diving Unit, Singapore Army Special Operations Force and Singapore Police Special Tactics and Rescue unit.

Police Coast Guard

The Police Coast Guard is a unit of the Singapore Police Force and was first established in 1924. Its role is to maintain coastal security within Singapore territorial waters and to support the Singapore Armed Forces in emergencies. Its four regional commands are Kallang (SE sector), Gul (SW sector), Seletar (NE sector) and Lim Chu Kang (NW sector). The PCG HQ is currently co-located at Kallang but will move to a new site at Brani (near Sentosa) by 2006. The Coastal Patrol Squadron and Special Task Squadron operate under central control. All vessels have Police Coast Guard on the superstructure and a white-red-white diagonal stripe on the hull except for Interceptor craft which have dark blue hulls with grey superstructures. Personnel numbers are about 1,000.

Strength of the Fleet

Type	Active	Building (Projected)
Submarines	4	—
Frigates	—	6
Missile Corvettes	6	—
Offshore Patrol Vessels	11	—
Fast Attack Craft—Missile	6	—
Inshore Patrol Craft	12	—
Minehunters	4	—
LSL/LPD	4	—
LCMs	4	—

DELETIONS

Patrol Forces

2003   *Courageous*

Amphibious Forces

2002   *Perseverance* (re-roled as submarine rescue ship)

Auxiliaries

2002   *Jupiter* (to Indonesia)

SUBMARINES

CONQUEROR   *3/2000, Per Körnefeldt* / 0084434

4 CHALLENGER (SJÖORMEN) CLASS (SSK)

Name	Builders	Laid down	Launched	Commissioned
CHALLENGER (ex-*Sjöbjörnen*)	Karlskronavarvet	1967	6 Aug 1969	28 Feb 1969
CENTURION (ex-*Sjöörmen*)	Kockums	1965	25 Jan 1967	31 July 1968
CONQUEROR (ex-*Sjölejonet*)	Kockums	1966	29 June 1967	16 Dec 1968
CHIEFTAIN (ex-*Sjohunden*)	Kockums	1966	21 Mar 1968	25 June 1969

Displacement, tons: 1,130 surfaced; 1,210 dived
Dimensions, feet (metres): 167.3 × 20 × 19
(51 × 6.1 × 5.8)
Main machinery: Diesel-electric; 2 Hedemora-Pielstick V12A/
A2/15 diesels; 2,200 hp(m) *(1.62 MW)*; 1 ASEA motor;
1,500 hp(m) *(1.1 MW)*; 1 shaft
Speed, knots: 12 surfaced; 20 dived
Complement: 23 (7 officers)

Torpedoes: 4—21 in *(533 mm)* bow tubes. 10 FFV Type 613;
anti-surface; wire-guided; passive homing to 15 km *(8.2 n
miles)* at 45 kt; warhead 250 kg.
2—16 in *(400 mm)* tubes; 4 FFV Type 431; anti-submarine;
wire-guided; active/passive homing to 20 km *(10.8 n miles)* at
25 kt; warhead 45 kg shaped charge.
Mines: Minelaying capability.
Weapons control: UDS SUBTICS.
Radars: Navigation: Terma; I-band.
Sonars: Plessey Hydra; hull-mounted; passive search and attack;
medium frequency.

Programmes: It was announced on 23 September 1995 that a
submarine would be acquired from Sweden for training

CONQUEROR

9/2000, Sattler/Steele / 0105592

purposes only. Three more of the same class acquired in July
1997 for conversion plus one more for spares.
Structure: Albacore hull. Twin-decked. Diving depth, 150 m
(492 ft). Air conditioning added for tropical service, together
with battery cooling.
Operational: *Challenger* re-launched on 26 September 1997,
Conqueror and *Centurion* on 28 May 1999 and *Chieftain* on
22 May 2001. *Conqueror* was recommissioned in Singapore

on 24 July 2000 and *Chieftain* on 24 August 2002.
Challenger and *Centurion* remained in Sweden to support
training until January 2004 when they were transported to
Singapore. Ex-*Sjohasten* was also shipped as a source of
spares. *Centurion* was recommissioned on 26 June 2004.
The four submarines form 171 squadron. Based at Changi.
UPDATED

CORVETTES

6 VICTORY CLASS (FSGM)

Name	No	Builders	Launched	Commissioned
VICTORY	P 88	Lürssen Werft, Bremen	8 June 1988	18 Aug 1990
VALOUR	P 89	Singapore SB and Marine	10 Dec 1988	18 Aug 1990
VIGILANCE	P 90	Singapore SB and Marine	27 Apr 1989	18 Aug 1990
VALIANT	P 91	Singapore SB and Marine	22 July 1989	25 May 1991
VIGOUR	P 92	Singapore SB and Marine	1 Dec 1989	25 May 1991
VENGEANCE	P 93	Singapore SB and Marine	23 Feb 1990	25 May 1991

Displacement, tons: 595 full load
Dimensions, feet (metres): 204.7 oa; 190.3 wl × 27.9 × 10.2
(62.4; 58 × 8.5 × 3.1)
Main machinery: 4 MTU 16V 538 TB93 diesels; 15,020 hp(m)
(11 MW) sustained; 4 shafts
Speed, knots: 35. **Range, n miles:** 4,000 at 18 kt
Complement: 49 (8 officers)

Missiles: SSM: 8 McDonnell Douglas Harpoon ❶; active radar
homing to 130 km *(70 n miles)* at 0.9 Mach; warhead 227 kg.
SAM: 2 Octuple IAI/Rafael Barak I ❷ radar or optical guidance to
10 km *(5.5 m)* at 2 Mach; warhead 22 kg.
Guns: 1 OTO Melara 3 in *(76 mm)*/62 Super Rapid ❸; 120 rds/
min to 16 km *(8.7 n miles)*; weight of shell 6 kg.
4 CIS 50 12.7 mm MGs.
Torpedoes: 6—324 mm Whitehead B 515 (2 triple) tubes ❹.
Whitehead A 244S; anti-submarine; active/passive homing to
7 km *(3.8 n miles)* at 33 kt; warhead 34 kg (shaped charge).
Countermeasures: Decoys: 2 Plessey Shield 9-barrelled chaff
launchers ❺. 4 Rafael (2 twin) long-range chaff launchers to
be fitted below the bridge wings.
ESM: Elisra SEWS ❻; intercept.
ECM: Rafael RAN 1101; ❼ jammer.
Combat data systems: Elbit command system. SATCOM ❽.
Weapons control: Elbit MSIS optronic director ❾.
Radars: Surface search: Ericsson/Radamec Sea Giraffe 150HC
❿; G/H-band.
Navigation: Kelvin Hughes 1007; I-band.
Fire control: 2 Elta EL/M-2221(X) ⓫; I/J/K-band.
Sonars: Thomson Sintra TSM 2064; VDS ⓬; active search and
attack.

Programmes: Ordered in June 1986 to a Lürssen MGB 62 design
similar to Bahrain and UAE vessels.
Modernisation: Barak launchers fitted on either side of the VDS,
together with a second fire-control radar on the platform aft of
the mast and an optronic director on the bridge roof. Rudder
roll stabilisation retrofitted to improve sea-keeping qualities.
Unidentified EW antennae have been installed below RAN
1101.
Operational: Form 188 Squadron. Designated Missile Corvettes
(MCV). First live Barak firing in September 1997.
UPDATED

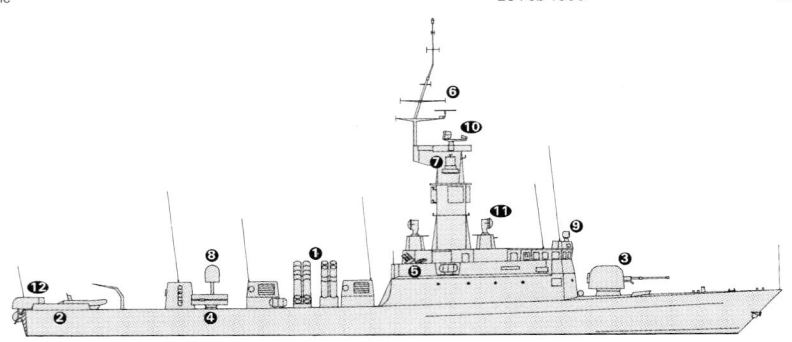

VICTORY

(Scale 1 : 600), Ian Sturton / 0114802

VENGEANCE

5/2004*, David Boey / 1044508

VIGILANCE
3/2004*, Bob Fildes / 1044507

FRIGATES

0 + 6 FORMIDABLE (PROJECT DELTA) CLASS (FFGHM)

Name	No	Builders	Laid down	Launched	Commissioned
FORMIDABLE	68	DCN, Lorient	14 Nov 2002	7 Jan 2004	May 2005
INTREPID	69	Singapore SB and Marine	8 Mar 2003	3 July 2004	2005
STEADFAST	70	Singapore SB and Marine	15 Nov 2003	28 Jan 2005	2006
TENACIOUS	71	Singapore SB and Marine	2004	2006	2007
STALWART	72	Singapore SB and Marine	2005	2007	2008
SUPREME	73	Singapore SB and Marine	2006	2008	2009

Displacement, tons: 3,200 full load
Dimensions, feet (metres): 374.0 × 52.5 × 16.4
(114 × 16.0 × 5.0)

Main machinery: CODAD; 4 MTU 20V 8000 M90 diesels; 48,276 hp *(36 kW)*; 2 shafts; cp props; bow thruster.
Speed, knots: 27

Range, n miles: 4,000 at 15 kt.
Complement: 71 + 15 aircrew

Missiles: SSM: 8 Boeing Harpoon ❶; active radar homing to 130 km *(70 n miles)* at 0.9 Mach; warhead 227 kg.
SAM: Eurosam SAAAM; 4 octuple Sylver A 43 VLS ❷ for MBDA Aster 15; command guidance active radar homing to 15 km *(8.1 n miles)* anti-missile and to 30 km *(16.2 n miles)* anti-aircraft. 32 missiles.
Guns: 1 OTO Melara 3 in *(76 mm)*/62 Super rapid ❸; 120 rds/min to 16 km *(8.7 n miles)*; weight of shell 6 kg.
2—20 mm. 2—12.7 mm MGs.
Torpedoes: 6—324 mm (2 triple (recessed)) ❹ tubes. Eurotorp A 244/S Mod 3; anti-submarine; active/passive homing to 7 km *(3.8 n miles)* at 33 kt; warhead 34 kg (shaped charge).
Countermeasures: Decoys: 3 EADS NGDS 12-barrelled chaff ❺, IR and anti-torpedo decoy launchers.
ESM: RAFAEL C-PEARL-M; intercept.
Combat Data Systems: DSTA/ST Electronics system.
Weapons Control: 2 EADS Nagir 2000 optronic directors ❻.
Radars: Air/search: Thales Herakles 3-D radar multifunction ❼; E/F-band
Surface search/Navigation: 2 Terma Scanter 2001 ❽; I-band.
Sonars: EDO 980 ALOFTS VDS; low frequency (2 kHz).

Helicopters: 1 S-70B Seahawk.

Programmes: Ordered from DCN International on 6 March 2000. First steel cut for hulls two and three on 2 October 2002. Prime Contractor is Singapore's Defence Science and Technology Agency (DSTA) who are also leading combat system integration in partnership with ST Electronics.
Structure: Derived from La Fayette class but there are notable differences to accommodate the weapon and sensor fit.
Operational: *Formidable* is planned to arrive in Singapore in May 2005 and to become operational in 2007. All ships to be operational by 2009.

UPDATED

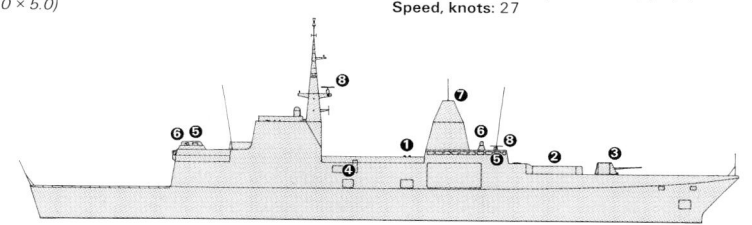

PROJECT DELTA

(Scale 1 : 1,200), Ian Sturton / 0533892

FORMIDABLE
10/2004*, B Prézelin / 1044521

FORMIDABLE
10/2004*, B Prézelin / 1044522

FORMIDABLE
10/2004*, B Prézelin / 1044523

PATROL FORCES

11 FEARLESS CLASS (PCM/PGM)

Name	No	Builders	Launched	Commissioned
FEARLESS	94	Singapore STEC	18 Feb 1995	5 Oct 1996
BRAVE	95	Singapore STEC	9 Sep 1995	5 Oct 1996
GALLANT	97	Singapore STEC	27 Apr 1996	3 May 1997
DARING	98	Singapore STEC	27 Apr 1996	3 May 1997
DAUNTLESS	99	Singapore STEC	23 Nov 1996	3 May 1997
RESILIENCE	82	Singapore STEC	23 Nov 1996	7 Feb 1998
UNITY	83	Singapore STEC	19 July 1997	7 Feb 1998
SOVEREIGNTY	84	Singapore STEC	19 July 1997	7 Feb 1998
JUSTICE	85	Singapore STEC	18 Oct 1997	7 Feb 1998
FREEDOM	86	Singapore STEC	18 Oct 1997	22 Aug 1998
INDEPENDENCE	87	Singapore STEC	18 Apr 1998	22 Aug 1998

Displacement, tons: 500 full load
Dimensions, feet (metres): 180.4 × 28.2 × 8.9 (55 × 8.6 × 2.7)
Main machinery: 2 MTU 12V 595 TE90 diesels; 8,554 hp(m) (6.29 MW) sustained; 2 Kamewa water-jets
Speed, knots: 20
Complement: 32 (5 officers)

Missiles: SAM: Matra Simbad twin launcher; Mistral; IR homing to 4 km (2.2 n miles); warhead 3 kg.
Guns: 1 OTO Melara 3 in (76 mm)/62 Super Rapid; 120 rds/min to 16 km (8.7 n miles); weight of shell 6 kg. 4 CIS 50 12.7 mm MGs.
1—25 mm Bushmaster (82).
Torpedoes: 6—324 mm Whitehead B515 (triple) tubes; (94-99) Whitehead A244S; active/passive homing to 7 km (3.8 m) at 33 kt; warhead 34 kg (shaped charge).
Countermeasures: Decoys: 2 GEC Marine Shield III 102 mm sextuple fixed chaff launchers.
ESM: Elisra NS-9010C; intercept.
Weapons control: ST 3100 WCS. Elbit MSIS optronic director.
Radars: Surface search and fire control: Elta EL/M-2228(X); I-band.
Navigation: Kelvin Hughes 1007; I-band.
Sonars: Thomson Sintra TSM 2362 Gudgeon; hull-mounted; active attack; medium frequency (94-99 only).
Towed array fitted in Brave.

Programmes: Contract awarded on 27 February 1993 for 12 patrol vessels to Singapore Shipbuilding and Engineering.
Structure: First six are ASW specialist ships. All have water-jet propulsion. Second batch were to have been fitted with Gabriel II SSMs but this plan has been shelved. MSIS director being fitted. Fearless modified with new EW radome on mainmast. Simbad SAM in Brave replaced by towed array and in Resilience by 25 mm Bushmaster. Sovereignty has deck crane to facilitate special forces operations.
Operational: First six serve with 189 Squadron; second six with 182 Squadron. Unity is to be used as a test bed for new technologies including an Indep 21 combat system. Courageous badly damaged in collision on 3 January 2003 and unlikely to be repaired. **UPDATED**

BRAVE (with VDS) 3/2004*, Bob Fildes / 1044509

INDEPENDENCE 4/2004*, John Mortimer / 1121521

SOVEREIGNTY 5/2004*, David Boey / 1044510

6 SEA WOLF CLASS (FAST ATTACK CRAFT—MISSILE) (PTGFM)

Name	No	Builders	Commissioned
SEA WOLF	P 76	Lürssen Werft, Vegesack	1972
SEA LION	P 77	Lürssen Werft, Vegesack	1972
SEA DRAGON	P 78	Singapore SBEC	1974
SEA TIGER	P 79	Singapore SBEC	1974
SEA HAWK	P 80	Singapore SBEC	1975
SEA SCORPION	P 81	Singapore SBEC	29 Feb 1976

Displacement, tons: 226 standard; 254 full load
Dimensions, feet (metres): 147.3 × 23 × 8.2 (44.9 × 7 × 2.5)
Main machinery: 4 MTU 16V 538 TB92 diesels; 13,640 hp(m) (10 MW) sustained; 4 shafts
Speed, knots: 35. **Range, n miles:** 950 at 30 kt; 1,800 at 15 kt
Complement: 41 (5 officers)

Missiles: SSM: 4 McDonnell Douglas Harpoon (2 twin); active radar homing to 130 km (70 n miles) at 0.9 Mach; warhead 227 kg.
4 IAI Gabriel I launchers; radar or optical guidance; semi-active radar homing to 20 km (10.8 n miles) at 0.7 Mach; warhead 75 kg.
SAM: 1 Matra Simbad twin launcher; Mistral; IR homing to 4 km (2.2 n miles); warhead 3 kg.
Guns: 1 Bofors 57 mm/70; 200 rds/min to 17 km (9.3 n miles); weight of shell 2.4 kg.
Countermeasures: Decoys: 2 Hycor Mk 137 sextuple RBOC chaff launchers. 4 Rafael (2 twin) long-range chaff launchers.
ESM/ECM: RQN 3B (INS-3) intercept and jammer. TDF-205 DF.
Weapons control: Elbit MSIS optronic director to be fitted.
Radars: Surface search: Racal Decca; I-band.
Fire control: Signaal WM28/5; I/J-band; range 46 km (25 n miles).

Programmes: Lürssen Werft FPB 45 type. Designated Missile Gunboats (MGB).
Modernisation: Sea Hawk was the first to complete refit in January 1988 with two sets of twin Harpoon launchers replacing the triple Gabriel launcher. The remainder were converted by December 1990 with the exception of Sea Wolf which finished refit in Spring 1991. ECM equipment has also been fitted on a taller mast. SATCOM and GPS installed. There have also been some superstructure changes. The Bofors 40 mm gun has been replaced with a Matra Simbad SAM launcher. Fitted with four launch pedestals for Gabriel and racks for eight Harpoons, though usual warload will consist of two Gabriel and four harpoons. Retrofitted with MSIS optronic director forward of the mainmast. Expected to pay off as new frigates enter service from 2005. **UPDATED**

SEA WOLF 5/2004*, David Boey / 1044511

SEA DRAGON 6/2004*, Singapore Navy / 1044505

12 INSHORE PATROL CRAFT (PB)

FB 31-42

Displacement, tons: 20 full load
Dimensions, feet (metres): 47.6 × 13.8 × 3.6 *(14.5 × 4.2 × 1.1)*
Main machinery: 2 MAN D2848 LE 401 diesels; 1,341 hp(m) *(1 MW)*; 2 Hamilton 362 water-jets
Speed, knots: 30
Complement: 5
Guns: 1—40 mm grenade launcher. 1—12.7 mm MG. 2—7.62 mm MGs.
Radars: Surface search: Racal Decca; I-band.

Comment: Built by Singapore SBEC and delivered in 1990-91. Based at Tuas. Designated Fast
Boats (FB). Some are kept in storage at Tuas. Similar to Police PT 1-19 class.

UPDATED

FB 31   *5/2004*, David Boey /* 1044512

SHIPBORNE AIRCRAFT

Numbers/Type: 6 Sikorsky S-70B Seahawk.
Operational speed: 135 kt *(250 km/h).*
Service ceiling: 10,000 ft *(3,050 m).*
Range: 600 n miles *(1,110 km).*
Role/Weapon systems: Contract placed 21 January 2005 for six new helicopters for operation
from Formidable class frigates. Delivery between 2008 and 2010. Roles ASW, ASV and
surveillance. Weapons and sensors to be announced.

NEW ENTRY

LAND-BASED MARITIME AIRCRAFT

Notes: (1) The Air Force also has 20 A4-SU Skyhawks, 40 F-5 S/T Tiger II and 60 F-16C/D.
(2) There are also 12 CH-47D used for maritime tasks.

Numbers/Type: 4 Grumman E-2C Hawkeye.
Operational speed: 323 kt *(598 km/h).*
Service ceiling: 30,800 ft *(9,390 m).*
Range: 1,000 n miles *(1,850 km).*
Role/Weapon systems: Delivered in 1987 for air control and surveillance of shipping in sea areas
around Singapore. Sensors: APS-138 radar; datalink for SSM targeting. Weapons: Unarmed.

VERIFIED

HAWKEYE   *9/2003, David Boey /* 0567531

Numbers/Type: 5 Fokker F50 Mk 2S Enforcer.
Operational speed: 220 kt *(463 km/h).*
Service ceiling: 29,500 ft *(8,990 m).*
Range: 2,700 n miles *(5,000 km).*
Role/Weapon systems: In service from September 1995. Part of Air Force 121 Squadron but with
mixed crews and under naval op con. Sensors: Texas Instruments APS-134(V)7 radar; GEC FLIR;
Elta ESM. Jammer fitted under wing-tip. Weapons: Harpoon ASM; mines; A-244S torpedoes.

VERIFIED

FOKKER F 50   *9/2003, David Boey /* 0567532

MINE WARFARE FORCES

4 BEDOK (LANDSORT) CLASS (MINEHUNTERS) (MHC)

Name	No	Builders	Launched	Commissioned
BEDOK	M 105	Kockums/Karlskrona	24 June 1993	7 Oct 1995
KALLANG	M 106	Singapore Shipbuilding	29 Jan 1994	7 Oct 1995
KATONG	M 107	Singapore Shipbuilding	8 Apr 1994	7 Oct 1995
PUNGGOL	M 108	Singapore Shipbuilding	16 July 1994	7 Oct 1995

Displacement, tons: 360 full load
Dimensions, feet (metres): 155.8 × 31.5 × 7.5 *(47.5 × 9.6 × 2.3)*
Main machinery: 4 Saab Scania diesels; 1,592 hp(m) *(1.17 MW)*; coupled in pairs to 2 Voith
Schneider props
Speed, knots: 15. **Range, n miles:** 2,000 at 10 kt
Complement: 31 (5 officers)

Guns: 1 Bofors 40 mm/70. 4—12.7 mm MGs.
Mines: 2 rails.
Weapons control: Thomson-CSF TSM 2061 Mk II minehunting and mine disposal system. Signaal
WM20 director.
Radars: Navigation: Norcontrol DB 2000; I-band.
Sonars: Thomson-CSF TSM 2022; hull-mounted; minehunting; high frequency.

Programmes: Kockums/Karlskrona design ordered in February 1991. *Bedok* started trials in
Sweden in December 1993, and was shipped to Singapore in early 1994 to complete.
Prefabrication work done for the other three in Sweden with assembly and fitting out in
Singapore at Benoi Basin.
Structure: GRP hulls. Two PAP 104 Mk V ROVs embarked. Racal Precision Navigation system. Two
sets of Swedish SAM minesweeping system. Magnavox GPS.
Operational: Based at Tuas.

UPDATED

BEDOK   *5/2004*, David Boey /* 1044519

BEDOK   *6/2004*, Chris Sattler /* 1044515

AMPHIBIOUS FORCES

Notes: (1) The Tiger 40 hovercraft acquired in 1997 is beyond repair but the design may be used
again for a repeat order.
(2) Trials of at least one hovercraft were reported in early 2005.

4 RPL TYPE (LCU)

RPL 60-63

Displacement, tons: 151 standard
Dimensions, feet (metres): 120.4 × 28 × 5.9 *(36.7 × 8.5 × 1.8)*
Main machinery: 2 MAN D2540MLE diesels; 860 hp(m) *(632 kW)*; 2 Schottel props
Speed, knots: 10.7
Complement: 6
Military lift: 2 tanks or 450 troops or 110 tons cargo (fuel or stores)

Comment: First pair built at North Shipyard Point, second pair by Singapore SBEC. First two
launched August 1985, next two in October 1985. Cargo deck 86.9 × 21.6 ft *(26.5 × 6.6 m)*.
Bow ramp suitable for beaching.

VERIFIED

RPL 60   *6/2001, John Mortimer /* 0126301

4 ENDURANCE CLASS (LPDM)

Name	No	Builders	Laid down	Launched	Commissioned
ENDURANCE	L 207	Singapore Technologies Marine, Banoi	26 Mar 1997	14 Mar 1998	18 Mar 2000
RESOLUTION	L 208	Singapore Technologies Marine, Banoi	22 Oct 1997	1 Aug 1998	18 Mar 2000
PERSISTENCE	L 209	Singapore Technologies Marine, Banoi	3 Apr 1998	13 Mar 1999	7 Apr 2001
ENDEAVOUR	L 210	Singapore Technologies Marine, Banoi	15 Oct 1998	12 Feb 2000	7 Apr 2001

Displacement, tons: 8,500 full load
Dimensions, feet (metres): 462.6 pp × 68.9 × 16.4
_(141 × 21 × 5)_
Main machinery: 2 Ruston 16RK 270 diesels; 12,000 hp(m)
_(8.82 MW)_; 2 shafts; Kamewa cp props; bow thruster
Speed, knots: 15. **Range, n miles:** 10,400 at 12 kt
Complement: 65 (8 officers)
Military lift: 350 troops; 18 tanks; 20 vehicles; 4 LCVP

Missiles: SAM: 2 Matra Simbad twin launchers for Mistral ❶; IR
homing to 4 km _(2.2 n miles)_; warhead 3 kg. 2 Barak octuple
launchers may be fitted in due course.
Guns: 1 Otobreda 76 mm/62 Super Rapid ❷; 120 rds/min to
16 km _(8.7 n miles)_; weight of shell 6 kg.
1—25 mm Bushmaster (can be fitted).
5—12.7 mm MGs.
Countermeasures: Decoys: 2 GEC Marine Shield III 102 mm
sextuple fixed chaff launcher.
ESM/ECM: Rafael RAN 1101; intercept and jammer.
Weapons control: CS Defense NAJIR 2000 optronic director ❸.
Radars: Air/surface search: Elta EL/M-2238 ❹; E/F-band.
Navigation: Kelvin Hughes Type 1007; I-band.

Helicopters: 2 Super Pumas.

Programmes: Ordered in September 1994 and confirmed in
mid-1996.
Structure: US drive through design with bow and stern ramps.
Single intermediate deck with three hydraulic ramps.
Helicopter platform aft. Indal ASIST helo handling system.
Dockwell for four LCUs and davits for four LCVPs. Two 25 ton
cranes. Four 36 m self-propelled pontoons can be secured to
winching points on the ships' sides.
Operational: _Endurance_ completed the RSN's first round-the-
world deployment in late 2000. _Resolution_ deployed in
November 2004 as part of coalition forces in northern Gulf.
Based at Changi. Form 191 Squadron.

UPDATED

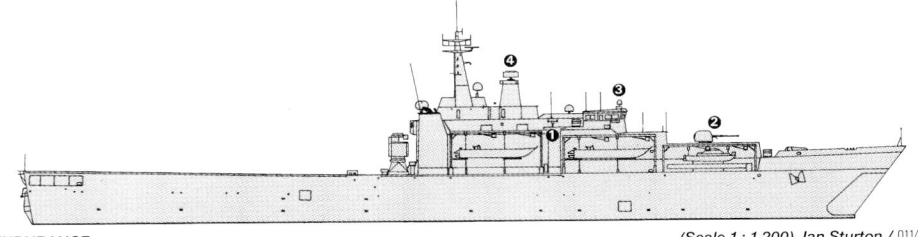

ENDURANCE   _(Scale 1 : 1,200), Ian Sturton_ / 0114801

ENDEAVOUR
5/2004 *, Chris Sattler / 1044514

RESOLUTION   _5/2004 *, David Boey_ / 1044513

30 LANDING CRAFT UTILITY (LCU)

Dimensions, feet (metres): 75.4 × 19.7 × 2.6 _(23 × 6 × 0.8)_
Main machinery: 2 MAN 2842 LZE diesels; 4,400 hp(m) _(3.23 MW)_; 2 Kamewa water-jets
Speed, knots: 20. **Range, n miles:** 180 at 15 kt
Complement: 4
Military lift: 18 tons
Guns: 2—12.7 mm MGs or 40 mm grenade launchers.

Comment: This is a larger and much faster version of the LCVPs. Being built from 1993. Designated
Fast Craft Utility (FCU). Pennant numbers in the 300 series.

UPDATED

100 LANDING CRAFT (LCVP/FCEP)

Displacement, tons: 4 full load
Dimensions, feet (metres): 44.6 × 12.1 × 2 _(13.6 × 3.7 × 0.6)_
Main machinery: 2 MAN D2866 LE diesels; 816 hp(m) _(600 kW)_; sustained; 2 Hamilton 362
water-jets
Speed, knots: 20. **Range, n miles:** 100 at 20 kt
Complement: 3
Military lift: 4 tons or 30 troops

Comment: Fast Craft, Equipment and Personnel (FCEP), built by Singapore SBEC from 1989 and
are used to transport troops around the Singapore archipelago. They have a single bow ramp
and can carry a rifle platoon. More than 25 are in service and the rest in storage. Have numbers
in the 500 and 800 series except for those carried in LSTs.

VERIFIED

LCU 371   _5/2004 *, John Mortimer_ / 1043180

LCVPs   _8/2003, David Boey_ / 0567534

10 DIVING SUPPORT CRAFT (YTB)

Comment: Boston Whalers used by the Naval Diving Unit. Armed with 7.62 mm MGs and 40 mm grenade launchers. *VERIFIED*

BOSTON WHALER · 8/2000, David Boey / 0105601

6 FAST INTERCEPT CRAFT (HSIC)

Displacement, tons: 12.5 full load
Dimensions, feet (metres): 47.6 × 9.4 × 4.4 *(14.5 × 2.85 × 1.35)*
Main machinery: Triple Seatek diesels coupled to Trimax drives
Speed, knots: 55+
Guns: 2 CIS 40 mm AGL.
2 CIS 50 12.7 mm MGs.
1—7.62 mm GPMG.
Radars: Raytheon SL 72.

Comment: Details are of craft used by Naval Diving Unit. The multistep planing hull design is similar to that in UK service. At least five other planing and wave-piercing craft are reported to be in service with special forces units. *VERIFIED*

FIC 145 · 9/2002, David Boey / 0554729

450 ASSAULT CRAFT (LCA)

Dimensions, feet (metres): 17.7 × 5.9 × 2.3 *(5.4 × 1.8 × 0.7)*
Main machinery: 1 outboard; 50 hp(m) *(37 kW)*
Speed, knots: 12
Military lift: 12 troops
Guns: 1—7.62 mm MG or 40 mm grenade launcher.

Comment: Built by Singapore SBEC. Man-portable craft which can carry a section of troops in the rivers and creeks surrounding Singapore island. Numbers are approximate. *VERIFIED*

ASSAULT CRAFT · 9/1995, David Boey / 0080588

AUXILIARIES

Notes: (1) An oil rig supply ship MV *Ocean Explorer* is chartered as an unarmed training vessel. Orange hull with white superstructure.
(2) There are two Floating Docks with a lift of 600 tons. *FD 1* (to be decommissioned in 2003) at Tuas and *FD 2* at Changi.
(3) A 2,700 ton oil rig supply ship MV *Kendrick*, built in Poland in 1985, is chartered to support submarine rescue operations. Additional roles include acting as a target ship for submarine torpedo firings and torpedo recovery.
(4) MV *Avatar*, a Ro-Ro vessel, is leased to support submarine rescue operations. It has an orange hull and white superstructure.

KENDRICK · 5/2004*, David Boey / 1044517

POLICE COAST GUARD

12 SHARK CLASS (WPB)

HAMMERHEAD SHARK (ex-*Swift Archer*)	PH 50 (ex-P 16)
MAKO SHARK (ex-*Swift Lancer*)	PH 51 (ex-P 12)
WHITE SHARK (ex-*Swift Swordsman*)	PH 52 (ex-P 14)
BLUE SHARK (ex-*Swift Combatant*)	PH 53 (ex-P 18)
TIGER SHARK (ex-*Swift Knight*)	PH 54 (ex-P 11)
BASKING SHARK (ex-*Swift Warrior*)	PH 55 (ex-P 15)
SANDBAR SHARK (ex-*Swift Chieftain*)	PH 56 (ex-P 23)
THRESHER SHARK (ex-*Swift Conqueror*)	PH 57 (ex-P 21)
WHITETIP SHARK (ex-*Swift Warlord*)	PH 58 (ex-P 17)
BLACKTIP SHARK (ex-*Swift Challenger*)	PH 59 (ex-P 19)
GOBLIN SHARK (ex-*Swift Cavalier*)	PH 60 (ex-P 20)
SCHOOL SHARK (ex-*Swift Centurion*)	PH 61 (ex-P 22)

Displacement, tons: 45.7 full load
Dimensions, feet (metres): 74.5 × 20.3 × 5.2 *(22.7 × 6.2 × 1.6)*
Main machinery: 2 Deutz BA16M816 diesels; 2,680 hp(m) *(1.96 MW)* sustained; 2 shafts
Speed, knots: 32. **Range, n miles:** 550 at 20 kt; 900 at 10 kt
Complement: 15
Guns: 1 Oerlikon 20 mm GAM-BO1. 2 CIS 90 12.7 mm MGs.
Radars: Surface search: Decca 1226; I-band.

Comment: Built by Singapore SBEC and all completed 20 October 1981 for the Navy. First four transferred from the Navy on 15 February 1993, second four on 8 April 1994, last four on 7 November 1996. All to be fitted with ARPA radar in due course. Employed on territorial waters patrol. *UPDATED*

HAMMERHEAD SHARK · 4/2004*, Bob Fildes / 1044518

2 COMMAND CRAFT (WPB)

MANTA RAY PT 20 · EAGLE RAY PT 30

Dimensions, feet (metres): 65.6 × 19.7 × 3.3 *(20.0 × 6.0 × 1.0)*
Main machinery: 2 MTU 16V 2000 M90 diesels; 2 Hamilton 521 water-jets
Speed, knots: 30
Complement: 5
Guns: 2—7.62 mm MGs.

Comment: Built by Asia-Pacific Geraldton, Singapore to a Geraldton, Australia design. These command craft are larger versions of the 18 m patrol craft. *VERIFIED*

MANTA RAY · 4/2002, David Boey / 0554731

25 PATROL CRAFT (WPB)

PT 21-29 · PT 31-39 · PT 61-67

Dimensions, feet (metres): 59.1 × 17.7 × 3 *(18 × 5.4 × 0.9)*
Main machinery: 2 MTU 16V 2000M 90 diesels; 2 Hamilton 521 waterjets
Speed, knots: 40
Complement: 5

Comment: 18 patrol craft built by Geraldton Boats, Australia in 1999. A further seven patrol craft delivered late 2000. *UPDATED*

PT 39 · 3/2004*, Bob Fildes / 1044516

19 PATROL CRAFT (WPB)

PT 1-19

Displacement, tons: 20 full load
Dimensions, feet (metres): 47.6 × 13.8 × 3.9 *(14.5 × 4.2 × 1.2)*
Main machinery: 2 MAN D2542MLE diesels; 1,076 hp(m) *(791 kW)*; or MTU 12V 183 TC91 diesels; 1,200 hp(m) *(882 kW)* maximum; 2 shafts
Speed, knots: 30. **Range, n miles:** 310 at 22 kt
Complement: 4 plus 8 spare berths
Guns: 1—7.62 mm MG.
Radars: Surface search: Furuno or Racal Decca Bridgemaster; I-band.

Comment: First 13 completed by Singapore SBEC between January and August 1984, two more completed February 1987 and eight more (including two Command Boats) in 1989. Of aluminium construction. Four are operated by Customs and Excise. There are differences in the deckhouses between earlier and later vessels. Employed on patrol duties in southern territorial waters.

UPDATED

PT 18   *4/2004*, Bob Fildes /* 1044520

11 INTERCEPTOR CRAFT (PBF)

SAILFISH PK 10	**STRIPED MARLIN** PK 23	**BILLFISH** PK 30
SPEARFISH PK 20	**BLACK MARLIN** PK 24	**SWORDFISH** PK 40
WHITE MARLIN PK 21	**BLUE MARLIN** PK 25	**SPIKEFISH** PK 50
SILVER MARLIN PK 22	**JUMPING MARLIN** PK 26	

Dimensions, feet (metres): 42 × 10.5 × 1.6 *(12.8 × 3.2 × 0.5)*
Main machinery: 3 Mercruiser 502 Magnum diesels; 3 shafts
Speed, knots: 50
Complement: 5
Guns: 1—7.62 mm MG.

Comment: First five built locally and delivered in 1995. Colours have been changed to dark blue hulls and grey superstructures, to make the craft less visible at sea. Six more ordered from Pro Marine/North Shipyard in 1999 to a slightly different design, with twin outboard motors.

VERIFIED

SWORDFISH   *4/2002, David Boey /* 0554733

WHITE MARLIN   *1/2000, David Boey /* 0105606

32 FAST RESPONSE CRAFT (PBF)

PC 201-232

Dimensions, feet (metres): 37.7 × 10.8 × 1.6 *(11.5 × 3.3 × 0.5)*
Main machinery: 3 Mercury outboard motors; 750 hp *(560 kW)*
Speed, knots: 40

Comment: Order placed August 2000 with Asia Pac Geraldton for 20 craft delivered in 2002. A further twelve craft were later added.

VERIFIED

PC 209   *4/2002, David Boey /* 0554734

HARBOUR CRAFT

Comment: There are large numbers of harbour craft, many of them armed, with PC numbers. These include four RHIBs with Yamaha 200 hp outboards capable of 43 kt, and with pennant numbers PJ 1-4.

UPDATED

RHIB   *9/2002*, David Boey /* 1044506

CUSTOMS

Notes: Customs Craft include CE 1-4 and CE 5-8, the latter being sisters to PT 1 Police Craft.

CE 8   *10/2002, Mick Prendergast /* 0533877

For details of the latest updates to *Jane's Fighting Ships* online and to discover the additional information available exclusively to online subscribers please visit

jfs.janes.com

Slovenia
SLOVENSKI MORNARJI

Country Overview

Formerly a constituent republic of Yugoslavia, the Republic of Slovenia proclaimed its independence in 1991. Situated in south-eastern Europe, it is bordered to the north by Austria and Hungary, to the south by Croatia and to the west by Italy. With an area of 7,820 square miles, it has a short 24 n mile coastline with the Adriatic Sea on which the port of Koper is located. The capital and largest city is Ljubljana. Territorial waters (12 n miles) are claimed.

Headquarters Appointments

Chief of General Staff:
  Major General Ladislav Lipič
Chief of Naval Section, Forces Command:
  Lieutenant Ludvik Kožar
Chief of Navy Detachment:
  Commander Ivan Žnidar

General

Navy formed in January 1993.

Personnel

2005: 61

Bases

Ankaran

PATROL FORCES

1 SUPER DVORA Mk II (PBF)

Name	No	Builders	Commissioned
ANKARAN	HPL 21	IAI Ramta	Aug 1996

Displacement, tons: 58 full load
Dimensions, feet (metres): 82 × 18.4 × 3.6 *(25 × 5.6 × 1.1)*
Main machinery: 2 MTU 12V 396 TE94 diesels; 4,570 hp(m) *(3.36 MW)*; 2 ASD 15 surface drives
Speed, knots: 45. **Range, n miles:** 700 at 30 kt
Complement: 10 (5 officers)
Guns: 2 Oerlikon 20 mm; 2—7.62 mm MGs.
Weapons control: Elop MSIS optronic director.
Radars: Surface search: Raytheon; I-band.

Comment: First one delivered in August 1996 at Isola base. Plans for a second craft have been cancelled. *UPDATED*

ANKARAN   *6/1999, Slovenian Navy /* 0080597

POLICE

Notes: In addition there is a 40 kt cabin cruiser *Sinji Galeb* (P 101) and two RIBs.

1 HARBOUR PATROL CRAFT (PBF)

Name	No	Builders	Commissioned
LADSE	P 111	Aviotechnica	21 June 1995

Displacement, tons: 44 full load
Dimensions, feet (metres): 65.3 × 16.4 × 3 *(19.9 × 5 × 0.9)*
Main machinery: 2 MTU 8V 396 TE84 diesels; 2,400 hp(m) *(1.76 MW)*; 2 shafts
Speed, knots: 40. **Range, n miles:** 270 at 38 kt
Complement: 10
Guns: 1—7.62 mm MG.
Radars: Surface search: I-band.

Comment: Acquired from Italy in 1995.

VERIFIED

LADSE   *10/1997 /* 0080598

Solomon Islands

Country Overview

Formerly a British protectorate, the Solomon Islands gained independence in 1978. Its head of state is the British sovereign, who is represented by a Governor-General. Situated in the southwest Pacific Ocean, east of New Guinea, the country comprises more than 35 islands and numerous atolls which extend some 650 n miles from east to west and includes most of the Solomon Islands group. The six main islands are: Guadalcanal, Malaita, New Georgia, San Cristobal (now Makira), Santa Isabel and Choiseul. Vella Lavella, Ontong Java, Rennell, Bellona and the Santa Cruz islands are also part of the group, together with the Florida, Russell, Reef and Duff island groups. Honiara, on Guadalcanal, is the capital and principal port. An archipelagic state, territorial seas (12 n miles) are claimed. An Exclusive Economic Zone (EEZ) (200 n miles) is also claimed but limits have not been fully defined by boundary agreements. Patrol boats are operated by the National Surveillance and Reconnaissance Force (NSRF).

Headquarters Appointments

Director of Maritime forces:
  Chief Superintendent Eddie Tokuru

Bases

Honiara (HQ NSRF)

Personnel

2005: 60 (14 officers)

Prefix to Ships' Names

RSIPV

DELETIONS

2000 *Tulagi*

POLICE

2 PACIFIC CLASS (LARGE PATROL CRAFT) (PB)

Name	No	Builders	Commissioned
LATA	03	Australian Shipbuilding Industries	3 Sep 1988
AUKI	04	Australian Shipbuilding Industries	2 Nov 1991

Displacement, tons: 162 full load
Dimensions, feet (metres): 103.3 × 26.6 × 7.5 *(31.5 × 8.1 × 2.3)*
Main machinery: 2 Caterpillar 3516TA diesels; 4,400 hp *(3.28 MW)* sustained; 2 shafts
Speed, knots: 20. **Range, n miles:** 2,230 at 12 kt
Complement: 14 (1 officer)
Guns: 3—12.7 mm MGs.
Radars: Surface search: Furuno 8100-D; I-band.

Comment: Built under the Australian Defence Co-operation Programme. Training, operational and technical assistance provided by the Royal Australian Navy. Aluminium construction. Nominal endurance of 10 days. The Australian government has extended the Pacific Patrol Boat programme but, following suspension of most of support of the Solomon Islands' craft in 2001, an overdue half-life refit was not completed for *Auki* until 2002. Life-extension refits for *Lata* and *Auki* are due in 2005 and 2009 respectively subject to Australian government decisions. *UPDATED*

LATA   *8/2004*, Chris Sattler /* 0589806

AUKI
8/2004*, Chris Sattler / 0589807

1 INSHORE PATROL CRAFT (PBR)

JACKPOT

Comment: Details are not known.

NEW ENTRY

JACKPOT
8/2004*, Chris Sattler / 0589805

South Africa

Country Overview

The Republic of South Africa is bordered to the north by Namibia, Botswana, Zimbabwe, Mozambique and Swaziland. With an area of 472,731 square miles, it has a 1,512 n mile coastline with the south Atlantic and Indian Oceans. South Africa also has sovereignty over the Prince Edward Islands which lie some 950 n miles south-east of Port Elizabeth. The independent country of Lesotho forms an enclave in the eastern part of the country. The administrative capital of South Africa is Pretoria and the judicial capital is Bloemfontein. Cape Town is the legislative capital, largest city and a prominent port. There are further ports at Mossel Bay, Port Elizabeth, East London, Durban, Saldanha, and Richards Bay. Territorial seas (12 n miles) are claimed. It also claims a 200 n mile EEZ but its limits have not been fully defined.

Headquarters Appointments

Chief of the Navy:
   Vice Admiral J F Retief
Chief of Naval Staff:
   Rear Admiral J Mudimu
Flag Officer Fleet:
   Rear Admiral E M Green

Diplomatic Representation

Naval Attaché in Washington:
   Captain D J Christian
Defence and Naval Attaché in Paris:
   Captain L O Reeders
Armed Forces Attaché in Buenos Aires:
   Captain L Van Dyk
Defence Attaché in Berlin:
   Captain R R Goveia

Personnel

2005:

(a)   4,728 naval
(b)   2,266 (Public Service Act Personnel)

Prefix to Ships' Names

SAS (South African Ship)

Bases

Simon's Town (main); Durban (naval station).
Saldanha Bay (ratings' training), Gordon's Bay (officer training).

DELETIONS

Submarines

2003   *Umkhonto, Assegaai*

Patrol Forces

2004   *Adam Kok, Job Maseko*

Auxiliaries

2003   *Fleur*, 6 Delta class
2004   *Outeniqua*

SUBMARINES

0 + 3 209 CLASS (TYPE 1400Mod (Sa)) (SSK)

Name	No	Builders	Laid down	Launched	Commissioned
—	S 101	Howaldswerke, Kiel	22 May 2001	15 June 2004	Sep 2005
—	S 102	Thyssen Nordseewerke, Emden	12 Nov 2003	4 May 2005	Sep 2006
—	—	Thyssen Nordseewerke, Emden	Nov 2004	2006	Sep 2007

Displacement, tons: 1,454 surfaced; 1,594 dived
Dimensions, feet (metres): 201.5 × 24.7 × 18.8
   (62 × 7.6 × 5.8)
Main machinery: Diesel electric: 4 MTU 12V 396 diesels; 3,800 hp(m) *(2.8 MW)*; 4 alternators; 1 Siemens motor; 5,032 hp(m) *(3.7 MW)*; 1 shaft
Speed, knots: 10 surfaced; 21.5 dived
Complement: 30

Torpedoes: 8—21 in *(533 mm)* bow tubes. 14 torpedoes.
Countermeasures: ESM: Grintek Avitronics; intercept.
Weapons control: STN Atlas ISUS 90 TFCS.
Radars: Surface search: I-band.
Sonars: STN Atlas CSU-90; hull mounted and flank arrays.

Programmes: Being acquired from the German Submarine Consortium. Final approval given on 15 September 1999. Contract signed on 7 July 2000. The first boat is being built by HDW for delivery in September 2005. The second and third boats are being built by TNSW and will follow at 12 month intervals.
Structure: Diving depth 250 m *(820 ft)*. Zeiss non-hull penetrating optronic mast.

UPDATED

S 101   *6/2004*, Michael Nitz /* 1044525

S 101   *6/2004*, Harald Carstens /* 1043179

FRIGATES

0 + 4 VALOUR CLASS (MEKO A-200 TYPE) (FFGHM)

Name	No	Builders	Laid down	Launched	Commissioned
AMATOLA	F 145	Blohm + Voss, Hamburg	2 Aug 2001	6 June 2002	25 Sep 2003
ISANDLWANA	F 146	Howaldswerke, Kiel	26 Oct 2001	5 Dec 2002	19 Dec 2003
SPIOENKOP	F 147	Blohm + Voss, Hamburg	28 Feb 2002	6 June 2003	15 Mar 2004
MENDI	F 148	Howaldswerke, Kiel	28 June 2002	15 June 2004	2005

Displacement, tons: 3,590 full load
Dimensions, feet (metres): 397 × 53.8 × 20.3
(121 × 16.4 × 6.2)
Main machinery: CODAG; 1 GE LM 2500 gas turbine 26,820 hp
(m) *(20 MW)*; 2 MTU 16V 1163 TB93 diesels 16,102 hp(m)
(11.84 MW); 2 shafts; LIPS cp props; 1 LIPS LJ210E waterjet
(centreline)
Speed, knots: 28. **Range, n miles:** 7,700 at 15 kt
Complement: 117 plus 25 spare

Missiles: SSM: 8 Exocet MM 40 Block 2 ❶.
SAM: Denel Umkhonto 32 cell VLS ❷ inertial guidance with mid-
course guidance and IR homing to 12 km *(6.5 n miles)* at 2.4
Mach; warhead 23 kg.
Guns: 1 Otobreda 76 mm/62 compact ❸.
2 LIW DPG 35 mm (twin) ❹. 2 Oerlikon 20 mm Mk 1.
Torpedoes: 4—324 mm tubes (2 twin) ❺.
Countermeasures: Decoys: 2 Super Barricade chaff launchers
❻.
ESM/ECM: intercept and jammer.
Combat data systems: ADS CMS.
Weapons control: 2 Reutech RTS 6400 optronic trackers.
Radars: Air/surface search: Thales MRR ❼ 3D; G-band.
Fire control: 2 Reutech RTS 6400 ❽; I/J-band.
Navigation/Helicopter control: 2 Racal Bridgemaster E ❾; I-band.
Sonars: Thomson Marconi 4132 Kingklip; hull mounted, active
search; medium frequency.

Helicopters: 1 Super Lynx ❿ in due course.

Programmes: Contract for four ships, with option for one further,
signed on 3 December 1999 with ESACC which includes
Blohm + Voss, HDW, TRT, African Defence Systems and
Thomson-CSF. Contract effective 28 April 2000. *Amatola*
arrived at Simon's Town on 4 November 2003 for weapon
systems integration by African Defence Systems. Sea
acceptance trials began in October 2004 and she is expected
to become fully operational in early 2006. The others are to
follow at six month intervals.
Operational: Westland Super Lynx helicopters selected, but not
ordered at the same time as the ships.

VALOUR CLASS   *(Scale 1 : 1,200), Ian Sturton /* 1044524

UPDATED SPIOENKOP   *3/2004*, Michael Nitz /* 1044526

SPIOENKOP   *5/2004*, Robert Pabst /* 1044528

ISANDLWANA   *9/2004*, Guy Toremans /* 1044527

PATROL FORCES

Notes: It is planned to replace the current patrol craft inventory with up to ten 60-75 m multipurpose craft.

3 WARRIOR (EX-MINISTER) CLASS
(FAST ATTACK CRAFT—MISSILE) (PGG)

Name	No	Builders	Commissioned
ISAAC DYOBHA (ex-*Frans Erasmus*)	P 1565	Sandock Austral, Durban	27 July 1979
GALESHEWE (ex-*Hendrik Mentz*)	P 1567	Sandock Austral, Durban	11 Feb 1983
MAKHANDA (ex-*Magnus Malan*)	P 1569	Sandock Austral, Durban	4 July 1986

Displacement, tons: 430 full load
Dimensions, feet (metres): 204 × 25 × 8 *(62.2 × 7.8 × 2.4)*
Main machinery: 4 Maybach MTU 16V 965 TB91 diesels; 15,000 hp(m) *(11 MW)* sustained; 4 shafts
Speed, knots: 32. **Range, n miles:** 1,500 at 30 kt; 3,600+ at economical speed
Complement: 52 (7 officers)

Missiles: SSM: 6 Skerpioen; active radar or optical guidance; semi-active radar homing to 36 km *(19.4 n miles)* at 0.7 Mach; warhead 75 kg. Another pair may be mounted. Skerpioen is an Israeli Gabriel II built under licence in South Africa.
Guns: 2 OTO Melara 3 in *(76 mm)*/62 compact; 85 rds/min to 16 km *(8.7 n miles)*; weight of shell 6 kg; 500 rounds per gun.
2 LIW Vektor 35 mm (twin); 550 rds/min; may replace one 76 mm gun in one craft for trials.
2 LIW Mk 1 20 mm. 2—12.7 mm MGs.
Countermeasures: Decoys: 4 ACDS launchers for chaff.
ESM: Delcon (ADS/Sysdel) EW system.
ECM: Elta Rattler; jammer.
Combat data systems: ADS Diamant (after upgrade). Mini action data automation with Link.
Radars: Air/surface search: Elta EL/M 2208; E/F-band.
Fire control: Selenia RTN 10X; I/J-band.

Programmes: Contract signed with Israel in late 1974 for this class, similar to Saar 4 class. Three built in Haifa and reached South Africa in July 1978. The ninth craft launched late March 1986. Three more improved vessels of this class were ordered but subsequently cancelled. The last of the class was finally christened in March 1992. Pennant numbers restored to the ships side and stern in 1994.
Modernisation: Ship life extension programme included a new communications refit, improvements to EW sensors, a third-generation target designation assembly, a computer-assisted action information system served by datalinks, improvements to fire control, a complete overhaul of the Skerpioen missiles, and a new engine room monitoring system. P 1565 completed upgrade in April 1999, P 1567 in March 2000 and P 1569 in mid-2000.
Operational: All are based at Simon's Town. Likely to be decommissioned by 2008.

UPDATED

MAKHANDA   9/2004 *, Guy Toremans / 1044529

3 T CRAFT CLASS (PB)

Name	No	Builders	Commissioned
TOBIE	P 1552	T Craft International, Cape Town	18 July 2003
TERN	P 1553	T Craft International, Cape Town	18 July 2003
TEKWANE	P 1554	T Craft International, Cape Town	22 July 2003

Displacement, tons: 36 full load
Dimensions, feet (metres): 72.2 × 23 × 3 *(22 × 7 × 0.9)*
Main machinery: 2 ADE 444 TI 12V diesels; 2,000 hp *(1.5 MW)*; 2 Hamilton waterjets
Speed, knots: 32. **Range, n miles:** 530 at 22 kt
Complement: 16 (1 officer)
Guns: 1—12.7 mm MG.
Weapons control: Hesis optical director.
Radars: Surface search: Racal Decca; I-band.

Comment: Twin hulled catamarans of GRP sandwich construction. Capable of carrying up to 15 people. Originally ordered in mid-1991 but not fully commissioned until 2003. Carries an RIB in the stern well. Three of this type built for Israel in 1997. Two based at Simon's Town and one at Durban.

VERIFIED

TOBIE   6/2001, South African Navy / 0114554

24 NAMACURRA CLASS (INSHORE PATROL CRAFT) (PB)

Y 1502-1505   Y 1508-1509   Y 1511-1519   Y 1521-1529

Displacement, tons: 5 full load
Dimensions, feet (metres): 29.5 × 9 × 2.8 *(9 × 2.7 × 0.8)*
Main machinery: 2 Yamaha outboards; 380 hp(m) *(279 kW)*
Speed, knots: 32. **Range, n miles:** 180 at 20 kt
Complement: 4
Guns: 1—12.7 mm MG. 2—7.62 mm MGs.
Depth charges: 1 rack.
Radars: Surface search: Furuno; I-band.

Comment: Built in South Africa in 1980-81. Can be transported by road. Y 1520 transferred to Malawi in October 1988 and Y 1506 has sunk at sea. Y 1501 and Y 1510 donated to Namibia on 29 November 2002 and Y 1507 and Y 1530 donated to Mozambique in 2004.

UPDATED

NAMACURRA   8/2001, van Ginderen Collection / 0132783

NAMACURRA   8/2001, van Ginderen Collection / 0132784

SHIPBORNE AIRCRAFT

Numbers/Type: 4 Agusta-Westland Super Lynx 300.
Operational speed: 120 kt *(222 km/h)*.
Service ceiling: 10,000 ft *(3.048 m)*.
Range: 320 n miles *(593 km)*.
Role/Weapon systems: Ordered on 14 August 2003 for delivery in 2007. Surveillance. Sensors: Telephonics APS-143 B(V)3 radar; ESM: Sea Raven 118; Cumulus Leo Mk II FLIR. Weapons: Unarmed (torpedoes and ASM may be fitted in future upgrades).

VERIFIED

SUPER LYNX   6/2003 / 0095707

Numbers/Type: 8 Aerospatiale SA 330E/H/J Oryx.
Operational speed: 139 kt *(258 km/h)*.
Service ceiling: 15,750 ft *(4,800 m)*.
Range: 297 n miles *(550 km)*.
Role/Weapon systems: Support helicopter; allocated by SAAF for naval duties and can be embarked in *Drakensberg, Outeniqua* and *Agulhas*. Sensors: Doppler navigation with search radar. Weapons: Unarmed but can mount Armscor 30 mm Rattler.

UPDATED

ORYX   12/2003 * / 1044530

LAND-BASED MARITIME AIRCRAFT

Notes: There are also some utility Alouette helicopters.

Numbers/Type: 5 Douglas Turbodaks.
Operational speed: 161 kt *(298 km/h).*
Service ceiling: 24,000 ft *(7,315 m).*
Range: 1,390 n miles *(2,575 km).*
Role/Weapon systems: A number of Dakotas has been converted for MR/SAR and other tasks. Additional fuel tanks extend the range to 2,620 n miles *(4,800 km).* Sensors: Elta M-2022 search radar and FLIR; Sysdel ESM; sonobuoy acoustic processor. Weapons: Unarmed.

VERIFIED

DOUGLAS DC-3   *6/2003, South African Navy /* 0568890

MINE WARFARE FORCES

Notes: Mine countermeasures capability is likely to be replaced by autonomous underwater vehicles rather than by specialist ships.

2 CITY (LINDAU) CLASS (TYPE 351, TROIKA)
(MINESWEEPERS—COASTAL) (MSCD)

Name	No	Builders	Commissioned
KAPA (ex-*Düren*)	M1223 (ex-M1079)	Burmester, Bremen	22 Apr 1959
TEKWINI (ex-*Wolfsburg*)	M1225 (ex-M1082)	Burmester, Bremen	8 Oct 1959

Displacement, tons: 465 full load
Dimensions, feet (metres): 154.5 × 27.2 × 9.2 *(47.1 × 8.3 × 2.8)*
Main machinery: 2 MTU MD 16V 538 TB90 diesels; 5,000 hp(m) *(3.68 MW)* sustained; 2 shafts
Speed, knots: 16.5
Complement: 44 (4 officers)

Guns: 1 Bofors 40 mm/70; 330 rds/min to 12 km *(6.5 n miles)*; weight of shell 0.96 kg.
  2—12.7 mm MGs.
  2—7.62 mm MGs (fitted for).
Radars: Navigation: Kelvin Hughes 14/9 or Atlas Electronik TRS N, or Raytheon SPS-64; I-band.
Sonars: Atlas Electronik DSQS-11; minehunting; high frequency.

Programmes: The contract for the acquisition of six German Type 351 minesweepers was finalised on 10 November 2000. After shipment to South Africa in 2001, *Kapa* (ex-*Düren*) and *Tekwini* (ex-*Wolfsburg*) were recommissioned on 5 September 2001 while *Tshwane* (ex-*Schleswig*) and *Mangaung* (ex-*Paderborn*) were placed in reserve. *Konstanz* (M1081) and *Ulm* (M1083) will be used for spares. The ships are based at Simon's Town and are an interim replacement for the four Ton class minesweepers.
Modernisation: Converted as guidance ships for Troika between 1981 and 1983. Each can guide three of these unmanned minesweeping vehicles as well as maintaining their moored-mine sweeping capabilities.
Structure: The hull is of wooden construction. Laminated with plastic glue. The engines are of non-magnetic materials.
Operational: Likely to be decommissioned in 2005.

UPDATED

KAPA   *11/2002, Robert Pabst /* 0568862

4 RIVER CLASS (COASTAL MINEHUNTERS) (MHC)

Name	No	Builders	Commissioned
UMKOMAAS (ex-*Navors I*)	M 1499	Abeking & Rasmussen/ Sandock Austral	13 Jan 1981
UMGENI (ex-*Navors II*)	M 1213	Abeking & Rasmussen/ Sandock Austral	1 Mar 1981
UMZIMKULU (ex-*Navors III*)	M 1142	Sandock Austral	30 Oct 1981
UMHLOTI (ex-*Navors IV*)	M 1212	Sandock Austral	15 Dec 1981

Displacement, tons: 380 full load
Dimensions, feet (metres): 157.5 × 27.9 × 8.2 *(48 × 8.5 × 2.5)*
Main machinery: 2 MTU 12V 652 TB81 diesels; 4,515 hp(m) *(3.32 MW)*; 2 Voith Schneider props
Speed, knots: 16. **Range, n miles:** 2,000 at 13 kt
Complement: 40 (7 officers)
Guns: 1 Oerlikon 20 mm GAM-BO1. 2—12.7 mm MGs. 2—7.62 mm MGs.
Countermeasures: MCM: 2 PAP 104 remote-controlled submersibles.
Radars: Navigation: Decca; I-band.
Sonars: Klein VDS; side scan; high frequency.

Comment: Ordered in 1978 as Research Vessels to be operated by the Navy for the Department of Transport. The lead ship *Navors I* was shipped to Durban from Germany in the heavy lift ship *Uhenfels* in June 1980 for fitting out, shortly followed by the second. The last pair were built in Durban. The vessels were painted blue with white upperworks and formed the First Research Squadron. Painted grey and renamed in 1982 but continued to fly the national flag and not the naval ensign. The prefix RV was only changed to SAS on 3 February 1988 when they were formally accepted as naval ships. Minehunting capability could be enhanced by substituting the diving container on the after deck with lightweight mechanical and acoustic sweeping gear. Carry an RIB and a decompression chamber. M1499 refitted in 2002 followed by M1213 in 2003. All to continue in service until about 2010.

UPDATED

UMZIMKULU   *9/2004*, Guy Toremans /* 1044532

SURVEY AND RESEARCH SHIPS

Notes: (1) As well as *S A Agulhas*, the Department of Environmental Affairs has three research vessels: *Africana* of 2,471 grt, *Algoa* of 760 grt and *Sardinops* of 255 grt. These ships are operated by Smit Pentow Marine. There are also three fishery protection vessels: *Patella, Pelargus* and *Jasus*. A contract was signed in 2003 with Damen Shipyards, Gorinchem for the construction of one offshore and three inshore Fishery and Environmental Protection vessels. *Sarah Baartman* is an 83 m offshore patrol vessel whose design is based on the Dutch fishery patrol vessel *Barend Biesheuve*. Built at Damen Shipyards, Okean (Ukraine) and outfitted at Royal Schelde Yard, Vlissingen, delivery was made in June 2004. *Lilian Ngoyi, Ruth First* and *Victoria Mxenge* are three 47 m inshore patrol vessels whose design is based on the Damen Stan Patrol 4207 in service with UK Customs. Built by Farocean Marine, Cape Town, deliveries were made in November 2004, February 2005 and May 2005 respectively.
(2) Two survey launches, *Malgas* and *Seemeeu* are carried by *Protea*.
(3) A 50 m trawler, *Eagle Star*, is used as a training ship for the Department of Environmental Affairs.

SARAH BAARTMAN   *1/2005*, Robert Pabst /* 1043178

1 HECLA CLASS (AGSH)

Name	No	Builders	Commissioned
PROTEA	A 324	Yarrow (Shipbuilders) Ltd	23 May 1972

Displacement, tons: 2,733 full load
Dimensions, feet (metres): 260.1 × 49.1 × 15.6 *(79.3 × 15 × 4.7)*
Main machinery: Diesel-electric; 3 MTU diesels; 3,840 hp *(2.68 MW)* sustained; 3 generators; 1 motor; 2,000 hp *(1.49 MW)*; 1 shaft; cp prop; bow thruster
Speed, knots: 14. **Range, n miles:** 12,000 at 11 kt
Complement: 124 (10 officers)
Radars: Navigation: Racal Decca; I-band.
Helicopters: 1 Alouette III.

Comment: Laid down 20 July 1970. Launched 14 July 1971. Equipped for hydrographic survey with limited facilities for the collection of oceanographical data and for this purpose fitted with special communications equipment, Polaris survey system, survey launches *Malgas* and *Seemeeu* and facilities for helicopter operations. Hull strengthened for navigation in ice and fitted with a passive roll stabilisation system. New engines and full overhaul in 1995-96. Carries EGNG sidescan sonar and two survey boats. Fitted for two 20 mm guns.

UPDATED

PROTEA   *5/2004*, Robert Pabst /* 1044531

1 ANTARCTIC SURVEY AND SUPPLY VESSEL (AGOBH)

Name	Builders	Launched	Commissioned
S A AGULHAS	Mitsubishi, Shimonoseki	30 Sep 1977	31 Jan 1978

Measurement, tons: 5,353 gross
Dimensions, feet (metres): 358.3 × 59 × 19 *(109.2 × 18 × 5.8)*
Main machinery: 2 Mirrlees-Blackstone K6 major diesels; 6,600 hp *(4.49 MW)*; 1 shaft; bow and
    stern thrusters
Speed, knots: 14. **Range, n miles:** 8,200 at 14 kt
Complement: 40 plus 92 spare berths
Radars: Navigation: Racal Decca; I-band.
Helicopters: 2 SA 330J Puma.

Comment: Red hull and white superstructure. A Department of Environmental Affairs vessel,
    civilian manned and operated by Smit Pentow Marine. Major refit March to October 1992;
    25 ton crane moved forward, transverse thrusters and roll damping fitted, improved navigation
    and communications equipment. A hinged hatch has been fitted at the stern to recover towed
    equipment.

VERIFIED

S A AGULHAS   *8/2001, Robert Pabst /* 0121399

AUXILIARIES

Notes: It is planned to replace *Drakensberg* in about 2017.

1 FLEET REPLENISHMENT SHIP (AORH)

Name	No	Builders	Launched	Commissioned
DRAKENSBERG	A 301	Sandock Austral, Durban	24 Apr 1986	11 Nov 1987

Displacement, tons: 6,000 light; 12,500 full load
Dimensions, feet (metres): 482.3 × 64 × 25.9 *(147 × 19.5 × 7.9)*
Main machinery: 2 diesels; 16,320 hp(m) *(12 MW)*; 1 shaft; cp prop; bow thruster
Speed, knots: 20+. **Range, n miles:** 8,000 at 15 kt
Complement: 96 (10 officers) plus 10 aircrew plus 22 spare
Cargo capacity: 5,500 tons fuel; 750 tons ammunition and dry stores; 2 Lima LCUs
Guns: 4 Oerlikon 20 mm GAM-BO1. 8—12.7 mm MGs.
Helicopters: 2 SA 330H/J Oryx.

Comment: The largest ship built in South Africa and the first naval vessel to be completely
    designed in that country. In addition to her replenishment role she is employed on SAR, patrol
    and surveillance with a considerable potential for disaster relief. As well as LCUs carries two
    diving support boats and two RIBs. Two abeam positions and astern fuelling, jackstay and
    vertrep. Two helicopter landing spots, one forward and one astern. Main secondary role is the
    transport of consumables, but can also be used to support small craft and transport a limited
    number of troops.

UPDATED

DRAKENSBERG   *6/2003, South African Navy /* 0568889

DRAKENSBERG   *6/2002, John Kinross /* 0533266

6 LIMA CLASS (LCU)

Displacement, tons: 7.3 full load
Dimensions, feet (metres): 29.8 × 11.6 × 2.3 *(9.1 × 3.55 × 0.7)*
Main machinery: 2 outboards; 400 hp *(298 kW)*
Speed, knots: 38. **Range, n miles:** 120 at 26 kt
Complement: 3

Comment: Built in 2003 by Stingray Marine, Cape Town. GRP construction. Capable of carrying 24
    troops or 2.5 tons of cargo. Two craft can be carried in *Drakensberg*.

UPDATED

L 27   *8/2003, Helmoed-Römer Heitman /* 0530510

TUGS

Notes: There is also a harbour tug *De Neys*.

1 COASTAL TUG (YTB)

DE MIST

Displacement, tons: 275 full load
Dimensions, feet (metres): 112.5 × 25.6 × 11.1 *(34.3 × 7.8 × 3.4)*
Main machinery: 2 Mirlees-Blackstone diesels; 2,440 hp *(1.82 MW)*; 2 Voith-Schneider props
Speed, knots: 12
Complement: 11

Comment: Completed by Dorbyl Long, Durban on 23 December 1978.

UPDATED

DE MIST   *9/2004*, Guy Toremans /* 1044534

1 COASTAL TUG (YTB)

UMALUSI (ex-*Golden Energy*)

Displacement, tons: 315 full load
Dimensions, feet (metres): 98.5 × 32.8 × 17.1 *(30 × 10 × 5.2)*
Main machinery: 2 Caterpillar V6 diesels
Speed, knots: 10
Complement: 10

Comment: Completed in 1995 by Jaya Holding Ltd. Acquired from Taikong Trading Company in
    January 1997.

UPDATED

UMALUSI   *5/2004*, Robert Pabst /* 1044533

Spain
ARMADA ESPAÑOLA

Country Overview

The Kingdom of Spain is a constitutional monarchy that occupies the greater part of the Iberian Peninsula in southwest Europe. It is bordered to the north by France and Andorra and to the west by Portugal. It has a 2,678 n mile coastline with the Atlantic Ocean and Mediterranean Sea. With a total area of 194,897 square miles, the country comprises the mainland, the Balearic Islands in the Mediterranean and the Canary Islands in the Atlantic Ocean. There are also two small exclaves in Morocco, Ceuta and Melilla and three island groups near the Moroccan coast, Peñón de Vélez de la Gomera, the Alhucemas and the Chafarinas. The British dependency of Gibraltar is situated at the southern extremity of Spain. Madrid is the capital and largest city while Barcelona, Algeciras, Valencia and Bilbao are the principal ports. Territorial seas (12 n miles) and an EEZ (200 n miles) are claimed.

Headquarters Appointments

Chief of the Naval Staff:
   Admiral Sebastián Zaragoza Soto
Second Chief of the Naval Staff:
   Fernando Armada Vadillo
Chief of Fleet Support:
   Admiral Juan José González-Irún Sánchez
Chief of Naval Personnel:
   Admiral Rafael Lapique Dobarro

Commands

Commander-in-Chief of the Fleet:
   Admiral José Antonio Balbás Otal
Commander-in-Chief, Maritime Action (Almart):
   Admiral Mario Rafael Sánchez-Barriga Fernández
Commander, Spanish Maritime Forces (SPMARFOR):
   Vice Admiral José Antonio Martinez Sainz-Rozas
Commander, Logistic Support (Cartagena):
   Vice Admiral Rafael Martín de la Escalera Mandillo
Commander, Logistic Support (Cadiz):
   Vice Admiral José Enrique de Benito Dorronzoro
Commander, Logistic Support (Ferrol):
   Vice Admiral Francisco Cañete Muñoz
Commander-in-Chief, Canary Islands Zone:
   Vice Admiral Manuel Calvo Freijomil
Marines General Commander:
   Major General Juan García Lizana
Commander, Fleet Task Group (COMGRUFLOT):
   Rear Admiral José María Treviño Ruiz
Commander, Northern Forces (ALFUNOR):
   Rear Admiral Juan Francisco Serón Martinez
Commander, Maritime Action (Cadiz):
   Rear Admiral Juan José Ollero Marín
Commander, Maritime Action (Ferrol):
   Rear Admiral Tomás Bolibar Piñeiro

Diplomatic Representation

Defence Attaché in Washington:
   Rear Admiral Teodoro de Leste Contreras
Naval Attaché in Berlin:
   Commander Javier Garzón Heydt
Naval Attaché in Brasilia:
   Captain Darío Lanza Carballo
Naval Attaché in Lisbon:
   Captain José Enrique Jarque Pérez
Naval Attaché in London and Dublin:
   Captain Alfredo Martinez Mahamud
Naval Attaché in Paris:
   Commander Ramón Barro Ordovás
Naval Attaché in Rabat:
   Commander Manuel Caridad Villaverde
Naval Attaché in Rome:
   Commander Antonio González Llanos López
Naval Attaché in Santiago, Lima and La Paz:
   Captain Antonio Carrasco Gómez
Naval Attaché in Washington:
   Captain Juan Carlos San Martin Naya
Naval Attaché in Oslo, Stockholm and Helsinki:
   Captain Alvaro Ollero Marín

Diplomatic Representation — *continued*

Naval Attaché in Bangkok, Manila and Singapore:
   Captain Ramón Díaz-Guevara Domínguez
Naval Attaché in Pretoria:
   Captain Mariano Mayo Consentino
Naval Attaché in Athens:
   Captain Angel Cabrera Juega

Personnel

2005: Navy: 20,612 (3,083 officers)
         Marines: 5,243 (473 officers)

Bases

Naval Zones are being re-organised into a single Area. Headquarters are to be in Cartagena with subordinate commands in Ferrol, Cádiz and Las Palmas.
Ferrol: Cantabrian Zone HQ-Ferrol arsenal, support centre at La Graña, naval school at Marín, Pontevedra
Cádiz: Straits Zone HQ-La Carraca arsenal, fleet command HQ and naval air base at Rota, amphibious base at Puntales. Marines Brigade (TEAR) HQ at San Fernando, Cádiz.
Cartagena: Maritime Action, HQ-Cartagena arsenal, underwater weapons and divers school at La Algameca; support base at Mahón, Minorca and at Soller and Porto Pi, Majorca, submarine weapons schools at La Algameca and Porto Pi base, Majorca. Naval Infantry school at Cartagena.
Las Palmas: Canaries Zone HQ-Las Palmas arsenal

Naval Air Service

The Naval Air Arm Flotilla is based at Rota.

Type	Escuadrilla
AB 212	3
Cessna Citation II	4
Sikorsky SH-3D/G Sea King	5
Sikorsky SH-3E Sea King (AEW)	
Hughes 500M (Training)	6
EAV-8B Harrier II/Harrier Plus	9
Sikorsky SH-60B Seahawk	10

Fleet Deployment

(1) Fleet (under Commander-in-Chief, Fleet)

   (a) Principe de Asturias (based at Rota)
   (b) Escuadrillas de Escoltas:
         31st Squadron; 5 Baleares class (based at Ferrol)
         41st Squadron; 6 Santa María class (based at Rota)
         51st Squadron; 4 Alvaro de Bazan class (based at Ferrol)
   (c) Amphibious Forces: (2 LST and 2 LPD at Rota, small units at Puntales, Cádiz).
         Naval infantry at San Fernando.
   (d) Fuerza de Medidas contra Minas: (based at Cartagena)
         6 MSCs, 1 MCCS
   (e) Flotilla de Submarinos: (based at Cartagena)
         All submarines
   (f) Flotilla de Aeronaves: (based at Rota)

(2) Maritime Action units (under Commander-in-Chief, Maritime Action)

   (a) Cantabrian Zone:
         1 Ocean Tug, 8 Tugs, 4 Patrol Ships, 5 Large Patrol Craft, 7 Coastal Patrol Craft, 1 Logistics Support Ship, 4 Sail Training Ships, 5 Training Craft
   (b) Straits Zone:
         1 oiler, 6 Oceanographic Ships, 1 Sail Training Ship, 4 Fast Attack Craft, 1 Transport, 1 Ocean Tug, 7 Tugs, 1 Water-boat
   (c) Mediterranean Zone:
         4 Fast Attack Craft, 3 Patrol Ships, 1 Water-boat, 7 Tugs, 1 Frogman Support Ship
   (d) Canaries Zone:
         4 Patrol Ships, 1 Ocean Tug, 2 Tugs, 1 LCT
   (e) Minor auxiliaries. Identified by 'Y' pennant numbers and form Tren Naval.

Guardia Civil del Mar

Started operations in 1992. For details, see end of section.

Prefix to Ships' Names

SPS (Spanish Ship)

Strength of the Fleet

Type	Active	Building (Planned)
Submarines—Patrol	6	4
Aircraft Carriers	1	—
Frigates	12	1 (2)
Corvettes	1	—
Offshore Patrol Vessels	12	(8)
Coastal Patrol Craft	21	—
Inshore Patrol Craft	3	—
LHD	—	1
LPDs	2	—
LSTs	2	—
Minehunters	5	1 (2)
MCM support ship	1	—
Survey and Research Ships	7	—
Replenishment Tankers	2	(1)
Tankers	9	—
Transport Ships	5	—
Training Ships	10	—
Ocean Tugs	2	—

DELETIONS

Submarines

2003	*Narval, Delfín*

Frigates

2004	*Cataluña*
2005	*Baleares*

Patrol Forces

2004	*Deva*

Mine Warfare Forces

2003	*Sil*
2004	*Genil, Odiel, Ebro*

Amphibious Warfare Forces

2004	2 LCT

Survey and Research Ships

2003	*Pollux*
2004	*Castor*

Auxiliaries

2002	*Proserpina*
2003	*Ferrol, Guardiamarina Barrutia*
2004	*Torpedista Hernández*

PENNANT LIST

Submarines

S 62	Tonina
S 63	Marsopa
S 71	Galerna
S 72	Siroco
S 73	Mistral
S 74	Tramontana

Aircraft Carriers

R 11	Príncipe de Asturias

Frigates

F 72	Andalucía
F 74	Asturias
F 75	Extremadura
F 81	Santa María

F 82	Victoria
F 83	Numancia
F 84	Reina Sofía
F 85	Navarra
F 86	Canarias
F 101	Alvaro de Bazán
F 102	Almirante Don Juan de Borbón
F 103	Blas de Lezo (bldg)
F 104	Mendez Nuñez (bldg)

Patrol Forces

F 33	Infanta Elena
P 11	Barceló
P 12	Laya
P 13	Javier Quiroga
P 14	Ordóñez
P 15	Acevedo
P 16	Cándido Pérez

P 21	Anaga
P 22	Tagomago
P 23	Marola
P 24	Mouro
P 25	Grosa
P 26	Medas
P 27	Izaro
P 28	Tabarca
P 30	Bergantín
P 31	Conejera
P 32	Dragonera
P 33	Espalmador
P 34	Alcanada
P 61	Chilreu
P 62	Alboran
P 63	Arnomendi
P 71	Serviola
P 72	Centinela
P 73	Vigía

P 74	Atalaya
P 75	Descubierta
P 77	Infanta Cristina
P 78	Cazadora
P 79	Vencedora
P 201	Cabo Fradera

Amphibious Forces

L 41	Hernán Cortés
L 42	Pizarro
L 51	Galicia
L 52	Castilla

Mine Warfare Forces

M 11	Diana
M 31	Segura
M 32	Sella

SUBMARINES

0 + 4 S 80 CLASS (SSK)

Name	No	Builders	Laid down	Launched	Commissioned
—	S 81	IZAR, Cartagena	20 Mar 2005	Feb 2010	Oct 2011
—	S 82	IZAR, Cartagena	Jan 2007	Aug 2011	Oct 2012
—	S 83	IZAR, Cartagena	Jan 2008	July 2012	Oct 2013
—	S 84	IZAR, Cartagena	Jan 2009	July 2013	Oct 2014

Displacement, tons: 2,190 surfaced; 2,345 dived
Dimensions, feet (metres): 233.0 × 27.6 × 23.9
   (71.0 × 8.4 × 7.3)
Main machinery: Diesel electric; 2 diesels; 3,600 hp *(2.65 MW)*;
   1 motor; 3,500 hp *(2.6 MW)*; 1 shaft; AIP (MESMA?) system
Speed, knots: 12 surfaced; 20 dived
Complement: 32

Missiles: To be announced.
Torpedoes: 6—21 in *(533 mm)* bow tubes.
Countermeasures: To be announced.
Weapons control: To be announced.
Radars: To be announced.
Sonars: To be announced.

Programmes: Approval for the procurement of four submarines
   was given by the Spanish Cabinet on 5 September 2003.
   Contract awarded on 25 March 2004 The national design is
   derived from the Scorpene class and includes Air Independent
   Propulsion (AIP). A land-attack capability, possibly Tomahawk,
   is to be fitted. A second batch of four boats may follow.

UPDATED

S 80
10/2004*, Spanish Navy / 1044543

2 DELFÍN (DAPHNE) CLASS (SSK)

Name	No	Builders	Laid down	Launched	Commissioned
TONINA	S 62	Bazán, Cartagena	2 Mar 1970	3 Oct 1972	10 July 1973
MARSOPA	S 63	Bazán, Cartagena	19 Mar 1971	15 Mar 1974	12 Apr 1975

Displacement, tons: 869 surfaced; 1,043 dived
Dimensions, feet (metres): 189.6 × 22.3 × 15.1
   (57.8 × 6.8 × 4.6)
Main machinery: Diesel-electric; 2 SEMT-Pielstick 12 PA4 V 185
   diesels; 2,450 hp(m) *(1.8 MW)*; 2 Jeumont Schneider
   alternators; 1.7 MW; 2 motors; 2,600 hp(m) *(1.9 MW)*; 2
   shafts
Speed, knots: 13.5 surfaced; 15.5 dived
Range, n miles: 4,300 snorting at 7.5 kt; 2,710 surfaced at
   12.5 kt
Complement: 47 (6 officers)

Torpedoes: 12—21.7 in *(550 mm)* (8 bow, 4 stern) tubes. 12
   combination of (a) ECAN L5 Mod 3/4; dual purpose; active/
   passive homing to 9.5 km *(5.1 n miles)* at 35 kt; warhead
   150 kg; depth to 550 m *(1,800 ft)*.
   (b) ECAN F17 Mod 2; wire-guided; active/passive homing to
   20 km *(10.8 n miles)* at 40 kt; warhead 250 kg; depth 600 m
   (1,970 ft).
Mines: 12 in lieu of torpedoes.
Countermeasures: ESM: THORN EMI Manta; radar warning.
Weapons control: DLT-3A TFCS.
Radars: Surface search: Thomson-CSF DRUA 31 or 33A; I-band.
Sonars: Thomson Sintra DSUV 22; passive search and attack;
   medium frequency.

TONINA
1/1999, Diego Quevedo / 0080606

Thomson Sintra DUUA 2A; active search and attack; 8 or
8.4 kHz active.

Programmes: First pair ordered 26 December 1966 and second
   pair in March 1970. Built with extensive French assistance.
Modernisation: The class was taken in hand from 1983 for
   modernisation of sonar (DUUA 2A for DUUA 1), updating of
   fire control and torpedo handling. Torpedo tubes were also

linered to 21 in *(533 mm)*. Last one completed at the end of
1988. ESM updated in 1990-91. Torpedo fire control updated
for the F17 wire-guided weapons in *Marsopa* in 1994-95.
Structure: Diving depth, 300 m *(984 ft)*.
Operational: Two of this class decommissioned in 2003, *Tonina*
expected to decommission in 2005 and *Marsopa* in 2006.
Based at Cartagena.

UPDATED

TONINA
9/2004*, Diego Quevedo / 1043177

4 GALERNA (AGOSTA) CLASS (SSK)

Name	No	Builders	Laid down	Launched	Commissioned
GALERNA	S 71	Bazán, Cartagena	5 Sep 1977	5 Dec 1981	22 Jan 1983
SIROCO	S 72	Bazán, Cartagena	27 Nov 1978	13 Nov 1982	5 Dec 1983
MISTRAL	S 73	Bazán, Cartagena	30 May 1980	14 Nov 1983	5 June 1985
TRAMONTANA	S 74	Bazán, Cartagena	10 Dec 1981	30 Nov 1984	27 Jan 1986

Displacement, tons: 1,490 surfaced; 1,740 dived
Dimensions, feet (metres): 221.7 × 22.3 × 17.7
　(67.6 × 6.8 × 5.4)
Main machinery: Diesel-electric; 2 SEMT-Pielstick 16 PA4 V 185
　VG diesels; 3,600 hp(m) *(2.7 MW)*; 2 Jeumont Schneider
　alternators; 1.7 MW; 1 motor; 4,600 hp(m) *(3.4 MW)*; 1
　cruising motor; 32 hp(m) *(23 kW)*; 1 shaft
Speed, knots: 12 surfaced; 20 dived; 17.5 sustained
Range, n miles: 8,500 snorting at 9 kt; 350 dived on cruising
　motor at 3.5 kt
Complement: 54 (6 officers)

Torpedoes: 4–21 in *(533 mm)* tubes. 20 combination of (a)
　ECAN L5 Mod 3/4; dual purpose; active/passive homing to
　9.5 km *(5.1 n miles)* at 35 kt; warhead 150 kg; depth to
　550 m *(1,800 ft)*.

(b) ECAN F17 Mod 2; wire-guided; active/passive homing to
　20 km *(10.8 n miles)* at 40 kt; warhead 250 kg; depth 600 m
　(1,970 ft).
Mines: 19 can be carried if torpedo load is reduced to 9.
Countermeasures: ESM: THORN EMI/Inisel Manta E; radar
　warning.
Weapons control: DLA-2A TFCS.
Radars: Surface search: Thomson-CSF DRUA 33C; I-band.
Sonars: Thomson Sintra DSUV 22; passive search and attack;
　medium frequency.
　Thomson Sintra DUUA 2A/2B; active search and attack; 8 or
　8.4 kHz active.
　DUUX 2A/5; passive; rangefinding. Eledone; intercept.
　SAES Solarsub towed passive array; low frequency.

Programmes: First two ordered 9 May 1975 and second pair
　29 June 1977. Built with some French advice. About 67 per
　cent of equipment and structure from Spanish sources.
Modernisation: Modernised with improved torpedo fire control,
　new ESM and IR enhanced periscopes. New main batteries
　installed with central control monitoring. *Galerna* started in
　April 1993 and completed in late 1994, *Siroco* in mid-1995,
　Tramontana in early 1997, and *Mistral* in 2000. The plan to fit
　SSM has been shelved. Solarsub towed arrays are being fitted
　to all of the class during overhauls. At least one submarine
　capable of being fitted with Dry Dock Shelter.
Structure: Diving depth, 300 m *(984 ft)*.
Operational: Endurance, 45 days. Based at Cartagena.
　　　　　　　　　　　　　　　　　　　　　　　UPDATED

GALERNA　　　　　　　　　　　　　　　　　　　　　　*8/2004*, Martin Mokrus /* 1044545

GALERNA　　　　　　　　　　　　　　　　　　　　　　*8/2004*, E & M Laursen /* 1044546

SIROCO　　　　　　　　　　　　　　　　　　　*12/2003*, Adolfo Ortigueira Gil /* 1044547

AIRCRAFT CARRIERS

1 PRÍNCIPE DE ASTURIAS CLASS (CV)

Name	No	Builders	Laid down	Launched	Commissioned
PRÍNCIPE DE ASTURIAS	R 11	Bazán, Ferrol	8 Oct 1979	22 May 1982	30 May 1988
(ex-*Almirante Carrero Blanco*)					

Displacement, tons: 17,188 full load
Dimensions, feet (metres): 642.7 oa; 615.2 pp × 79.7 × 30.8
(195.9; 187.5 × 24.3 × 9.4)
Flight deck, feet (metres): 575.1 × 95.1 *(175.3 × 29)*
Main machinery: 2 GE LM 2500 gas turbines; 46,400 hp
(34.61 MW) sustained; 1 shaft; LIPS cp prop; 2 motors;
1,600 hp(m) *(1.18 MW);* retractable prop
Speed, knots: 25 (4.5 on motors). **Range, n miles:** 6,500 at
20 kt
Complement: 555 (90 officers) plus 208 (Flag Staff (7 officers)
and Air Group)

Guns: 4 Bazán Meroka Mod 2A/2B 12-barrelled 20 mm/120 ❶;
3,600 rds/min combined to 2 km.
2 Rheinmetall 37 mm saluting guns.
Countermeasures: Decoys: 4 Loral Hycor SRBOC 6-barrelled
fixed Mk 36; IR flares and chaff to 4 km *(2.2 n miles).*
SLQ-25 Nixie; towed torpedo decoy.
US Prairie/Masker; hull noise/blade rate suppression.
ESM/ECM: Elettronica Nettunel; intercept and jammers.
Combat data systems: Tritan Digital Command and Control
System NTDS; Links 11 and 14. Marconi Matra SCOT 3
Secomsat ❷. SSR-1, WSC-3 (UHF).
Weapons control: 4 Selenia directors (for Meroka). Radamec
2000 series.
Radars: Air search: Hughes SPS-52C/D ❸; 3D; E/F-band; range
439 km *(240 n miles).*
Surface search: ISC Cardion SPS-55 ❹; I/J-band.
Aircraft control: ITT SPN-35A ❺; J-band.
Fire control: 4 Sperry/Lockheed VPS 2 ❻; I-band (for Meroka).
RTN 11L/X; I/J-band; missile warning.
Selenia RAN 12L (target designation); I/J-band.
Tacan: URN 25.

Fixed-wing aircraft: 6-12 AV-8B Harrier II/Harrier Plus.
Helicopters: 6-10 SH-3 Sea Kings; 2-4 AB 212EW.

PRÍNCIPE DE ASTURIAS

10/2004*, Carlos Pardo Gonzalez / 1044548

Programmes: Ordered on 29 June 1977. Associated US firms
were Gibbs and Cox, Dixencast, Bath Iron Works and Sperry
SM. Commissioning delays caused by changes to command
and control systems and the addition of a Flag Bridge.
Modernisation: After two years' service some modifications
were made to the port after side of the island, to improve
briefing rooms and provide sheltered parking space for FD
vehicles. Also improved accommodation has been added on
for six officers and 50 specialist ratings. Mid-life refit planned
for 2006-08.
Structure: Based on US Navy Sea Control Ship design. 12° ski-
jump of 46.5 m. Two flight deck lifts, one right aft. Two LCVPs
carried. Two pairs of fin stabilisers. The hangar is 24,748 sq ft

(2,300 m²). The Battle Group Commander occupies the lower
bridge. Two saluting guns have been mounted on the port
quarter.
Operational: Three Sea Kings have Searchwater AEW radar.
Aircraft complement could be increased to 37 (parking on
deck) in an emergency but maximum operational number is
29 (17 in hangar, 12 on deck). A typical air wing includes
eight/ten AV-8B/AV-8B Plus, five SH-3 (including two AEW)
and three/four AB-212. Based at Rota.
Sales: Modified design built for Thailand.

UPDATED

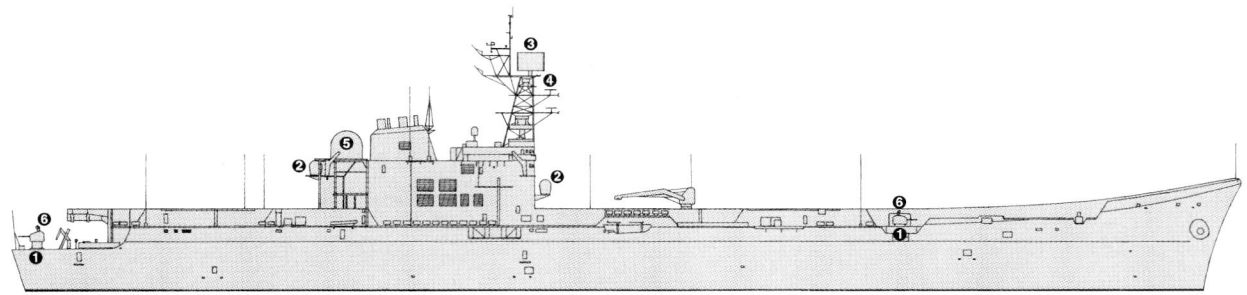

PRÍNCIPE DE ASTURIAS

(Scale 1 : 1,200), Ian Sturton

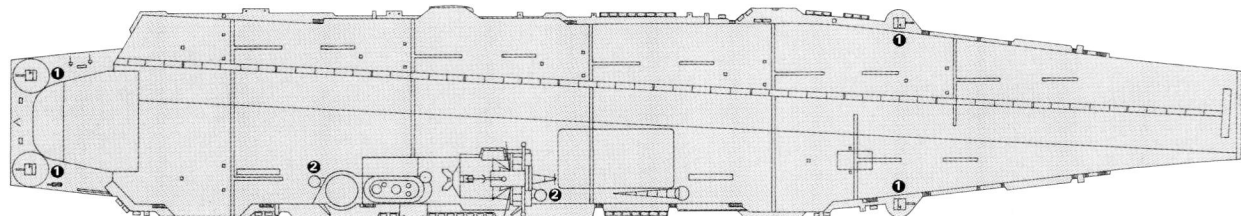

PRÍNCIPE DE ASTURIAS

(Scale 1 : 1,200), Ian Sturton / 0130391

PRÍNCIPE DE ASTURIAS

2/2004*, Guy Toremans / 1044535

PRÍNCIPE DE ASTURIAS

6/2003, Spanish Navy / 0570978

FRIGATES

Notes: Plans for a future frigate class have been suspended.

3 + 1 (2) ALVARO DE BAZÁN CLASS (FFGHM)

Name	No	Builders	Laid down	Launched	Commissioned
ALVARO DE BAZÁN	F 101	IZAR, Ferrol	14 June 1999	27 Oct 2000	19 Sep 2002
ALMIRANTE DON JUAN DE BORBÓN	F 102	IZAR, Ferrol	27 Oct 2000	28 Feb 2002	3 Dec 2003
BLAS DE LEZO	F 103	IZAR, Ferrol	28 Feb 2002	16 May 2003	16 Dec 2004
MENDEZ NUÑEZ	F 104	IZAR, Ferrol	16 May 2003	12 Nov 2004	Feb 2006

Displacement, tons: 5,853 full load
Dimensions, feet (metres): 481.3 oa; 437 pp × 61 × 16.1
(146.7; 133.2 × 18.6 × 4.9)
Flight deck, feet (metres): 86.6 × 56 *(26.4 × 17)*
Main machinery: CODOG; 2 GE LM 2500 gas turbines;
47,328 hp(m) *(34.8 MW)* sustained; 2 Bazán/Caterpillar
diesels; 12,240 hp(m) *(9 MW)* sustained; 2 shafts; LIPS cp
props
Speed, knots: 28. **Range, n miles:** 4,500 at 18 kt
Complement: 250 (35 officers)

Missiles: SSM: 8 Harpoon Block II **❶**; active radar homing to
130 km *(70 n miles)* at 0.9 Mach; warhead 227 kg.
SAM: Mk 41 VLS (48 cells) **❷** 32 GDC Standard SM-2MR (Block
IIIA); command/inertial guidance; semi-active radar homing
to 167 km *(90 n miles)* at 2 Mach. 64 Evolved Sea Sparrow
RIM 162B (in quadpacks); semi-active radar homing to 18 km
(9.7 n miles) at 3.6 Mach; warhead 39 kg.
Guns: 1 FMC 5 in *(127 mm)*/54 Mk 45 Mod 2 **❸** (ex-US); 20 rds/
min to 23 km *(12.6 n miles)*; weight of shell 32 kg.
1 Bazán 20 mm/120 Meroka 2B **❹**; 3,600 rds/min to 2 km
(fitted for but not with) 2 Oerlikon 20 mm.
Torpedoes: 4—323 mm (2 twin) Mk 32 Mod 9 fixed launchers
❺. Honeywell Mk 46 Mod 5; anti-submarine; active/passive
homing to 11 km *(5.9 n miles)* at 40 kt; warhead 44 kg.
A/S mortars: 2 ABCAS/SSTDS launchers.
Countermeasures: Decoys: 4 SRBOC Mk 36 Mod 2 chaff
launchers **❻**. SLQ-25A Nixie torpedo decoy.
ESM: Regulus Mk-9500; **❼** intercept.
ECM: Ceselsa Aldebaran **❽**; jammer.
Combat data systems: Lockheed Aegis Baseline 5 Phase III
(DANCS); Link 11/16. SCOT 3, SATURN 3S.
Weapons control: Sirius optronic director **❾**; FABA Dorna GFCS.
Sainsel DLT 309 TFCS. SQR-4 helo datalink.
Radars: Air/surface search: Aegis SPY-1D **❿**.
Surface search: DRS SPS-67 (RAN 12S) **⓫**.
Fire control: 2 Raytheon SPG-62 Mk 99 (for SAM) **⓬**.
Navigation: 1 Raytheon SPS-73(v); I-band.
Sonars: Raytheon DE 1160 LF; hull-mounted; active search and
attack; medium frequency. Possible ATAS active towed sonar.

Helicopters: 1 SH-60B Seahawk Lamps III **⓭**.

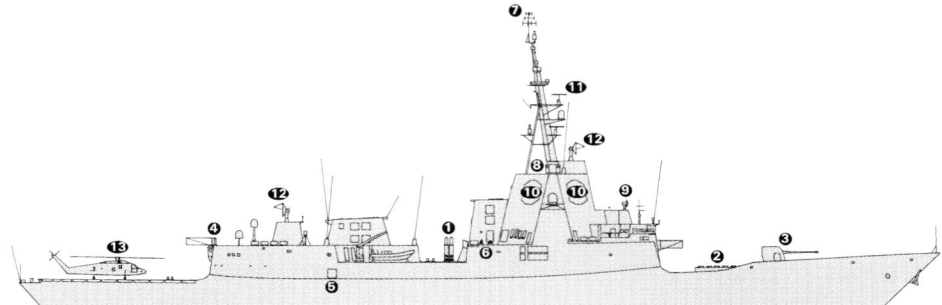

ALVARO DE BAZÁN   *(Scale 1 : 1,200), Ian Sturton /* 0529156

ALVARO DE BAZÁN   *6/2003, Spanish Navy /* 0570979

Programmes: Project definition from September 1992 to July
1995, and then extended to July 1996 to incorporate Aegis.
Design collaboration with German and Netherlands shipyards
started 27 January 1994. Spain withdrew from the APAR air
defence radar project in June 1995 and decided to
incorporate Aegis SPY-1D into the design. Production order
for four ships agreed on 21 October 1996 and building
approved 24 January 1997. FSC in November 1997. The
procurement of two further ships with enhanced capabilities
is under consideration.
Structure: The inclusion of SPY-1D radar increased the original
size of the ship and caused major changes to the shape of the
superstructure. Stealth technology incorporated. Indal RAST
helicopter system. Hangar for one helicopter. 127 mm gun for
gunfire support to land forces, taken from USN *Tarawa* class.
RAM may be fitted vice Meroka.
Operational: All based at Ferrol as the 51st Squadron. To
become the 31st Squadron when the Baleares class have
been decommissioned. F 101 completed the USN Combined
Combat Systems Ship Qualifications Trial (CSSQT) 17-22 July
2003. This included live SM-2 firings.   ***UPDATED***

ALVARO DE BAZÁN   *3/2003, Diego Quevedo /* 0570974

6 SANTA MARÍA CLASS (FFGHM)

Name	No	Builders	Laid down	Launched	Commissioned
SANTA MARÍA	F 81	Bazán, Ferrol	23 May 1982	24 Nov 1984	12 Oct 1986
VICTORIA	F 82	Bazán, Ferrol	16 Aug 1983	23 July 1986	11 Nov 1987
NUMANCIA	F 83	Bazán, Ferrol	8 Jan 1986	30 Jan 1987	8 Nov 1988
REINA SOFÍA (ex-*América*)	F 84	Bazán, Ferrol	12 Dec 1987	19 July 1989	18 Oct 1990
NAVARRA	F 85	Bazán, Ferrol	15 Apr 1991	23 Oct 1992	30 May 1994
CANARIAS	F 86	Bazán, Ferrol	15 Apr 1992	21 June 1993	14 Dec 1994

Displacement, tons: 3,610 standard; 3,969 full load
Dimensions, feet (metres): 451.2 × 46.9 × 24.6
  (137.7 × 14.3 × 7.5)
Main machinery: 2 GE LM 2500 gas turbines; 41,000 hp
  (30.59 MW) sustained; 1 shaft; cp prop
  2 auxiliary retractable props; 650 hp *(484 kW)*
Speed, knots: 29. **Range, n miles:** 4,500 at 20 kt
Complement: 223 (13 officers)

Missiles: SSM: 8 McDonnell Douglas Harpoon; active radar
  homing to 130 km *(70 n miles)* at 0.9 Mach; warhead 227 kg.
  SAM: 32 GDC Pomona Standard SM-1MR; Mk 13 Mod 4
  launcher ❶; command guidance; semi-active radar homing to
  46 km *(25 n miles)* at 2 Mach.
  Both missile systems share a common magazine.
Guns: 1 OTO Melara 3 in *(76 mm)*/62 ❷; 85 rds/min to 16 km
  (8.7 n miles); weight of shell 6 kg.
  1 Bazán 20 mm/120 12-barrelled Meroka Mod 2A or 2B ❸;
  3,600 rds/min combined to 2 km. 2–12.7 mm MGs.
Torpedoes: 6–324 mm US Mk 32 (2 triple) tubes ❹. Honeywell/
  Alliant Mk 46 Mod 5; anti-submarine; active/passive homing
  to 11 km *(5.9 n miles)* at 40 kt; warhead 44 kg.
Countermeasures: Decoys: 4 Loral Hycor SRBOC 6-barrelled
  fixed Mk 37 Mod 1/2 ❺; IR flares and chaff to 4 km *(2.2 n
  miles)*.
  Prairie/Masker: hull noise/blade rate suppression.
  SLQ-25 Nixie; torpedo decoy.
ESM/ECM: Elettronica Nettunel or Mk 3000 Neptune (F 84-86);
  intercept and jammer.
Combat data systems: IPN 10 action data automation; Link 11.
  SQQ 28 LAMPS III helo datalink. Saturn and SCOT 3
  Secomsat ❻ fitted.
Weapons control: Loral Mk 92 Mod 2 (Mod 6 with CORT in F 85
  and 86). Enosa optronic tracker for Meroka 2B.

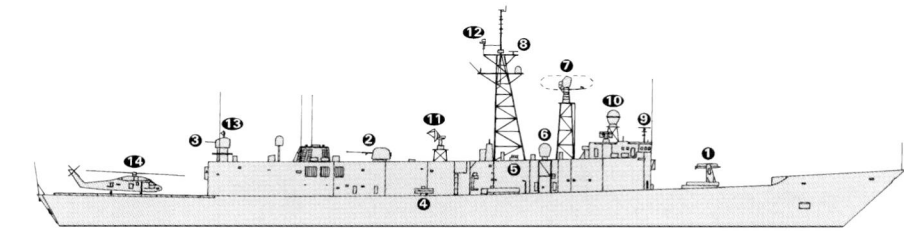

REINA SOFÍA   *(Scale 1 : 1,200), Ian Sturton* / 0130395

Radars: Air search: Raytheon SPS-49(V)4/6 ❼; C/D-band; range
  457 km *(250 n miles)*.
  Surface search: Raytheon SPS-55 ❽; I-band.
  Navigation: Raytheon 1650/9 ❾; I/J-band.
  Fire control: RCA Mk 92 Mod 4/6 ❿; I/J-band.
  Raytheon STIR ⓫; I/J-band.
  Selenia RAN 12L ⓬; D-band (for Meroka).
  Sperry/Lockheed VPS 2 ⓭; I-band (for Meroka).
  Tacan: URN 25.
Sonars: Raytheon SQS-56 (DE 1160); hull-mounted; active
  search and attack; medium frequency.
  Gould SQR-19(V)2; tactical towed array (TACTASS); passive;
  very low frequency.

Helicopters: 2 Sikorsky SH-60B ⓮ (only one normally
  embarked).

Programmes: Three ordered 29 June 1977. The execution of
  this programme was delayed due to the emphasis placed on

the carrier construction. The fourth ship was ordered on
19 June 1986, and numbers five and six on 26 December
1989.
Modernisation: F 85 and F 86 are fitted with the improved Mod
  2B Meroka CIWS which includes an Enosa optronic tracker.
  SCOT SATCOM fitted in F 85 and F 84. Others may be
  similarly fitted. All modernised with RAN 12L target
  designator for Meroka but VPS 2 fire control radar is yet to be
  replaced by RAN 30X radar. An update programme is to start
  in 2005. Details have not been confirmed but the aim is to
  adapt the ships for littoral warfare.
Structure: Based on the US FFG 7 Oliver Perry class although
  broader in the beam and therefore able to carry more
  topweight. Fin stabilisers fitted. RAST helicopter handling
  system. *Navarra* and *Canarias* have an indigenous combat
  data system thereby increasing national inputs to 75 per cent.
Operational: All based at Rota as the 41st Squadron.

           VERIFIED

NAVARRA   *2/2003, A Campanera i Rovira* / 0570940

REINE SOFIA   *4/2003, A Sharma* / 0570946

3 BALEARES (F 70) CLASS (FFGM)

Name	No	Builders	Laid down	Launched	Commissioned
ANDALUCÍA	F 72	Bazán, Ferrol	2 July 1969	30 Mar 1971	23 May 1974
ASTURIAS	F 74	Bazán, Ferrol	30 Mar 1971	13 May 1972	2 Dec 1975
EXTREMADURA	F 75	Bazán, Ferrol	3 Nov 1971	21 Nov 1972	10 Nov 1976

Displacement, tons: 3,350 standard; 4,177 full load
Dimensions, feet (metres): 438 × 46.9 × 15.4; 25.6 (sonar)
   (133.6 × 14.3 × 4.7; 7.8)
Main machinery: 2 Combustion Engineering V2M boilers; 1,200
   psi *(84.4 kg/cm²)*; 950°F *(510°C)*; 1 Westinghouse turbine;
   35,000 hp(m) *(25.7 MW)*; 1 shaft
Speed, knots: 28. **Range, n miles:** 4,500 at 20 kt
Complement: 256 (15 officers)

Missiles: SSM: 8 McDonnell Douglas Harpoon (4 normally
   carried) ❶; active radar homing to 130 km *(70 n miles)* at 0.9
   Mach; warhead 227 kg.
   SAM: 16 GDC Pomona Standard SM-1MR; Mk 22 Mod 0
   launcher ❷; command guidance; semi-active radar homing to
   46 km *(25 n miles)* at 2 Mach.
   A/S: Honeywell ASROC Mk 112 octuple launcher ❸ (except
   F 71); 8 reloads; inertial guidance to 1.6-10 km *(1-5.4 n miles)*;
   payload Mk 46 torpedo.
Guns: 1 FMC 5 in *(127 mm)*/54 Mk 42 Mod 9 ❹; dual purpose;
   20-40 rds/min to 24 km *(13 n miles)* anti-surface; 14 km
   (7.7 n miles) anti-aircraft; weight of shell 32 kg; 600 rounds in
   magazine.
   2 Bazán 20 mm/120 12-barrelled Meroka ❺; 3,600 rds/min
   combined to 2 km. 2—12.7 mm MGs.
Torpedoes: 4—324 mm US Mk 32 fixed tubes (fitted internally
   and angled at 45° ❻. Honeywell/Alliant Mk 46 Mod 5; anti-
   submarine; active/passive homing to 11 km *(5.9 n miles)* at
   40 kt; warhead 44 kg.
Countermeasures: Decoys: 4 Loral Hycor SRBOC Mk 36 Mod 2
   6-barrelled chaff launchers ❼.
   ESM: Ceselsa Deneb; or Mk 1600; intercept.
   ECM: Ceselsa Canopus; or Mk 1900; jammer.
Combat data systems: Tritan 1 action data automation; Link 11.
   Saturn SATCOM ❽.
Weapons control: Mk 68 GFCS. Mk 74 missile system with Mk
   73 director. Mk 114 torpedo control. Dorna GFCS on trial
   (F 71).
Radars: Air search: Hughes SPS-52B ❾; 3D; E/F-band; range
   439 km *(240 n miles)*.
   Surface search: Raytheon SPS-10 ❿; G-band.
   Navigation: Raytheon Marine Pathfinder ⓫; I/J-band.
   Fire control: Western Electric SPG-53B ⓬; I/J-band (for Mk 68).
   Raytheon SPG-51C ⓭; G/I-band (for Mk 73).
   Selenia RAN 12L ⓮; I-band (for Meroka). 2 Sperry VPS 2 ⓯
   (for Meroka).
   Tacan: SRN 15A ⓰.
Sonars: Raytheon SQS-56 (DE 1160); hull-mounted; active
   search and attack; medium frequency.
   EDO SQS-35V ⓱; VDS; active search and attack; medium
   frequency.

Programmes: This class resulted from a very close co-operation
   between Spain and the US. Programme was approved
   17 November 1964, technical support agreement with USA
   being signed 31 March 1966. US Navy supplied weapons and
   sensors. Major hull sections, turbines and gearboxes made at
   El Ferrol, superstructures at Alicante, boilers, distilling plants
   and propellers at Cádiz.
Modernisation: The mid-life update programme was done in two
   stages; all had completed the first stage by the end of 1987
   and *Asturias* was the first to be fully modernised in 1988;
   Extremadura completed in May 1989 and *Andalucía* in
   February 1991. Changes included the fitting of two Meroka
   20 mm CIWS, Link 11, Tritan data control, Deneb passive EW
   systems, four Mk 36 SRBOC chaff launchers and replacing
   SQS-23 with DE 1160 sonar. The stern torpedo tubes have
   been replaced by VDS.
Structure: Generally similar to US Navy's Knox class although
   they differ in the missile system and lack of helicopter
   facilities.
Operational: All based at Ferrol, forming the 31st Squadron. F 73
   decommissioned in 2004 and F 71 in March 2005. F 72 and
   F 75 to decommission in 2005 and F 74 in 2006.

UPDATED

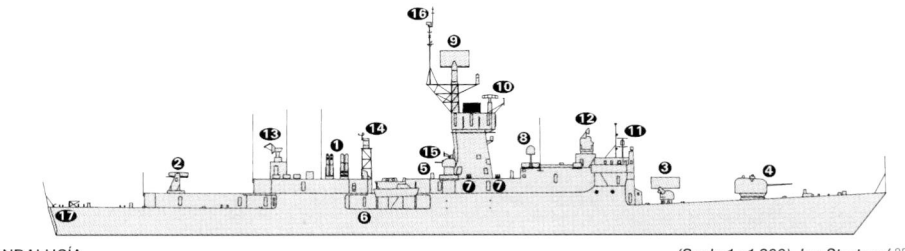

ANDALUCÍA   *(Scale 1 : 1,200), Ian Sturton /* 0529155

ANDALUCIA   *2/2004*, Guy Toremans /* 1044537

EXTREMADURA   *1/2004*, H M Steele /* 1044536

ASTURIAS   *3/2003, B Prézelin /* 0570969

SHIPBORNE AIRCRAFT

Numbers/Type: 4/12/1 BAe/McDonnell Douglas EAV-8B (Harrier II)/EAV-8B (Harrier Plus)/ TAV-8B.
Operational speed: 562 kt *(1,041 km/h)*.
Service ceiling: Not available.
Range: 480 n miles *(889 km)*.
Role/Weapon systems: First batch of nine delivered in 1987-88 and a further eight in 1996-97. All to be upgraded to AV-8B Harrier Plus standard with APG-65 radar plus FLIR by 2004. A further TAV-8B twin seat delivered in September 2000. Sensors: ECM; ALQ 164. Weapons: Strike; two 25 mm GAU-12/U cannon, two or four AIM-9L Sidewinders, two or four AGM-65E Mavericks; up to 16 GP bombs. AMRAAM AIM-120 in updated aircraft.

UPDATED

HARRIER PLUS   *2/2004 *, Guy Toremans /* 1044542

Numbers/Type: 8 Sikorsky SH-3D/G/H Sea King.
Operational speed: 118 kt *(219 km/h)*.
Service ceiling: 14,700 ft *(4,480 m)*.
Range: 542 n miles *(1,005 km)*.
Role/Weapon systems: Former ASW helicopters converted to tactical transport role. ASW equipment fitted for but not with. Can be replaced in 48 hours. Converted to 3H standard in 1996-97. APN-217 Doppler radar and IFF. Sensors: APS-124 search radar, Bendix AQS-18V dipping sonar, sonobuoys. Weapons: ASW; four Mk 46 torpedoes or depth bombs; 1—12.7 mm MG.

UPDATED

SEA KING   *2/2004 *, Guy Toremans /* 1044541

Numbers/Type: 3 Sikorsky SH-3D Sea King AEW.
Operational speed: 110 kt *(204 km/h)*.
Service ceiling: 14,700 ft *(4,480 m)*.
Range: 542 n miles *(1,005 km)*.
Role/Weapon systems: Three Sea King helicopters were taken in hand in 1986 for conversion to AEW role to provide organic cover; first entered service August 1987. Radar to be replaced by RACAL 2000. Sensors: THORN EMI Searchwater radar, ESM. Weapons: Unarmed.

VERIFIED

SEA KING AEW   *5/2001, Guy Toremans /* 0130130

Numbers/Type: 12 Sikorsky SH-60B Seahawk (LAMPS III).
Operational speed: 135 kt *(249 km/h)*.
Service ceiling: 10,000 ft *(3,050 m)*.
Range: 600 n miles *(1,110 km)*.
Role/Weapon systems: ASW helicopter; delivery in 1988-89 for FFG 7 frigates. Six more Block 1 acquired in 2002 for new frigates, and original six to be upgraded to Block 1 with first two aircraft to be completed by October 2003. Sensors: Search radar, FLIR (Block I), sonobuoys, ECM/ESM. Weapons: ASW; two Mk 46 torpedoes or depth bombs. ASV; AGM-119B Penguin and AGM-114B/K Hellfire.

UPDATED

SEA HAWK   *9/2004 *, Adolfo Ortigueira Gil /* 1044549

Numbers/Type: 9 Agusta AB 212.
Operational speed: 106 kt *(196 km/h)*.
Service ceiling: 14,200 ft *(4,330 m)*.
Range: 230 n miles *(426 km)*.
Role/Weapon systems: Surface search; four are equipped for Fleet ECM support and have a datalink and six for Assault operations with the Amphibious Brigade. All ASW equipment and ASM missiles removed. Sensors: Selenia search radar, Mk 2000 EW system (in four); Elmer ECM. Weapons: 1—12.7 mm MG and 70 mm rocket launchers.

UPDATED

AB 212   *2/2004 *, Guy Toremans /* 1044540

Numbers/Type: 10 Hughes 500MD.
Operational speed: 110 kt *(204 km/h)*.
Service ceiling: 10,000 ft *(3,050 m)*.
Range: 203 n miles *(376 km)*.
Role/Weapon systems: Used for training; secondary role is SAR and surface search. ASW role removed. Sensors: Some may have search radar, MAD.

VERIFIED

500 MD   *7/2002, Adolfo Ortigueira Gil /* 0528937

LAND-BASED MARITIME AIRCRAFT (FRONT LINE)

Notes: (1) The Air Force F/A-18 Hornet (C.15) and Eurofighter Typhoon (C.16) can be armed with Harpoon ASM. Air Force CN-235 are not used for maritime role.
(2) Three CASA C-212/400 are operated by the Fishery Department and are based at Torrejón (Madrid), Jerez and Alicante. Two Agusta A-109C and two Dauphin N3 helicopters are based at Alicante, Jerez and Santander and Canary Islands.

Numbers/Type: 3 Cessna Citation II (C-550).
Operational speed: 275 kt *(509 km/h)*.
Service ceiling: 27,750 ft *(8,458 m)*.
Range: 2,000 n miles *(3,704 km)*.
Role/Weapon systems: Used for transport, training and reconnaissance.

UPDATED

CESSNA CITATION   *6/2004 *, Adolfo Ortigueira Gil /* 1044550

Numbers/Type: 14 CASA C-212 Aviocar.
Operational speed: 190 kt *(353 km/h).*
Service ceiling: 24,000 ft *(7,315 m).*
Range: 1,650 n miles *(3,055 km).*
Role/Weapon systems: Operated by Air Force. Primary role SAR, secondary role surveillance. Based at Mallorca, Las Palmas and Madrid. Six are leased by Customs. Sensors: APS-128 radar, MAD, sonobuoys and ESM. Weapons (not SAR role): ASW; Mk 46 torpedoes or depth bombs. ASV; two rockets or machine gun pods.

VERIFIED

C-212   *5/2002, Adolfo Ortigueira Gil /* 0528933

Numbers/Type: 3 Fokker F27 Maritime.
Operational speed: 250 kt *(463 km/h).*
Service ceiling: 29,500 ft *(8,990 m).*
Range: 2,700 n miles *(5,000 km).*
Role/Weapon systems: Canaries and offshore patrol by Air Force. Are to be replaced by CN-235 in due course. Sensors: APS-504 search radar, cameras. Weapons: none.

UPDATED

F-27   *6/2004*, Adolfo Ortigueira Gil /* 1044552

Numbers/Type: 2/5 Lockheed P-3A/P-3B Plus Orion.
Operational speed: 410 kt *(760 km/h).*
Service ceiling: 28,300 ft *(8,625 m).*
Range: 4,000 n miles *(7,410 km).*
Role/Weapon systems: Air Force operation for long-range MR/ASW. Original P-3A aircraft supplemented in 1988 by P-3B Orions from Norway after Lockheed modernisation. P-3A aircraft upgraded to P-3A plus in 1995-97. The five P-3Bs are undergoing modernisation programme with improved acoustic signal processor, ALR-66 ESM, FLIR and new radar and communications. Sensors: APS-134 (Searchwater 2000 in due course); FLIR; search radar, AQS-81 MAD, ALR 66 V(3) ECM/ESM, 87 sonobuoys. Weapons: ASW; eight torpedoes or depth bombs internally; 10 underwing stations. ASV; four Harpoon or 127 mm rockets.

VERIFIED

P-3B   *7/2002, Adolfo Ortigueira Gil /* 0528935

Numbers/Type: 10 Eurocopter AS 332 Super Puma.
Operational speed: 130 kt *(240 km/h).*
Service ceiling: 15,090 ft *(4,600 m).*
Range: 672 n miles *(1,245 km).*
Role/Weapon systems: Air Force operated for SAR/CSAR. Based at Mallorca, Las Palmas and Madrid.

VERIFIED

AS 332   *7/2001, Adolfo Ortigueira Gil /* 0528932

PATROL FORCES

Notes: A programme for the procurement of a new class of up to eight multirole offshore patrol vessels known as Buques de Accion Maritima (BAM) was initiated in 2004. Outline plans are for a ship of up to 2,500 tons with a maximum speed of 22 kt and complemented by a crew of under 40. Variants of the BAM design would be capable of conducting surveillance, constabulary and environmental support tasks. Orders for the first batch of ships is expected in 2005 for entry into service in 2010.

4 SERVIOLA CLASS (OFFSHORE PATROL VESSELS) (PSOH)

Name	No	Builders	Laid down	Launched	Commissioned
SERVIOLA	P 71	Bazán, Ferrol	17 Oct 1989	10 May 1990	22 Mar 1991
CENTINELA	P 72	Bazán, Ferrol	12 Dec 1989	30 Mar 1990	24 Sep 1991
VIGÍA	P 73	Bazán, Ferrol	30 Oct 1990	12 Apr 1991	24 Mar 1992
ATALAYA	P 74	Bazán, Ferrol	14 Dec 1990	22 Nov 1991	29 June 1992

Displacement, tons: 1,147 full load
Dimensions, feet (metres): 225.4; 206.7 pp × 34 × 11 *(68.7; 63 × 10.4 × 3.4)*
Main machinery: 2 MTU-Bazán 16V 956 TB91 diesels; 7,500 hp (m) *(5.5 MW)* sustained; 2 shafts; LIPS cp props
Speed, knots: 19. **Range, n miles:** 8,000 at 12 kt
Complement: 42 (8 officers) plus 6 spare berths

Guns: 1 US 3 in *(76 mm)*/50 Mk 27; 20 rds/min to 12 km *(6.6 n miles)*; weight of shell 6 kg.
2—12.7 mm MGs.
Countermeasures: ESM: ULQ-13 (in P 71).

Weapons control: Bazán Alcor or MSP 4000 (P 73) optronic director. Hispano mini combat system. SATCOM.
Radars: Surface search: Racal Decca 2459; I-band.
Navigation: Racal Decca ARPA 2690 BT; I-band.

Helicopters: Platform for 1 AB 212.

Programmes: Project B215 ordered from Bazán, Ferrol in late 1988. The larger Milano design was rejected as being too expensive.
Modernisation: The guns are old stock refurbished but could be replaced by an OTO Melara 76 mm/62 or a Bofors 40 mm/70 Model 600 if funds can be found. Other equipment fits could include four Harpoon SSM, Meroka CIWS, Sea Sparrow SAM or a Bofors 375 mm ASW rocket launcher. No plans to carry out any of these improvements so far. EW equipment fitted in Serviola for training.
Structure: A modified Halcón class design similar to ships produced for Argentina and Mexico. Helicopter facilities enabling operation in up to Sea State 4 using non-retractable stabilisers. Three firefighting pumps.
Operational: For EEZ patrol. *Vigía* based at Cádiz, *Serviola* and *Atalaya* at Ferrol and *Centinela* at Las Palmas.

UPDATED

ATALAYA   *7/2003, Adolfo Ortigueira Gil /* 0570956

ATALAYA
6/2004*, Adolfo Ortigueira Gil / 1044551

6 DESCUBIERTA CLASS (PSOH/MCS/FSGM)

Name	No	Builders	Laid down	Launched	Commissioned
DESCUBIERTA	P 75 (ex-F 31)	Bazán, Cartagena	16 Nov 1974	8 July 1975	18 Nov 1978
DIANA	M 11 (ex-F 32)	Bazán, Cartagena	8 July 1975	26 Jan 1976	30 June 1979
INFANTA ELENA	F 33	Bazán, Cartagena	26 Jan 1976	14 Sep 1976	12 Apr 1980
INFANTA CRISTINA	P 77 (ex-F 34)	Bazán, Cartagena	11 Sep 1976	25 Apr 1977	24 Nov 1980
CAZADORA	P 78 (ex-F 35)	Bazán, Ferrol	14 Dec 1977	17 Oct 1978	20 July 1982
VENCEDORA	P 79 (ex-F 36)	Bazán, Ferrol	1 June 1978	27 Apr 1979	18 Mar 1983

Displacement, tons: 1,233 standard; 1,666 full load

Dimensions, feet (metres): 291.3 × 34 × 12.5
(88.8 × 10.4 × 3.8)

Main machinery: 4 MTU-Bazán 16V 956 TB91 diesels; 15,000 hp(m) *(11 MW)* sustained; 2 shafts; cp props

Speed, knots: 25. **Range, n miles:** 4,000 at 18 kt; 7,500 at 12 kt

Complement: 118 (10 officers) plus 30 marines

Missiles: SSM: 8 McDonnell Douglas Harpoon (2 quad) launchers ❶ (F 33) ; active radar homing to 130 km *(70 n miles)* at 0.9 Mach; warhead 227 kg. Normally only 2 pairs are embarked.
SAM: Selenia Albatros octuple launcher ❷ (F 33); 24 Raytheon Sea Sparrow/Selenia Aspide; semi-active radar homing to 14.6 km *(8 n miles)* at 2.5 Mach; height envelope 15-5,000 m *(49.2-16,405 ft)*; warhead 39 kg.

Guns: 1 OTO Melara 3 in *(76 mm)*/62 compact ❸; 85 rds/min to 16 km *(8.7 n miles)*; weight of shell 6 kg.
1 or 2 Bofors 40 mm/70 ❹ (F 33); 300 rds/min to 12.5 km *(6.8 n miles)*; weight of shell 0.96 kg.
2 Oerlikon 20 mm (P 75, P 77, P 78, P 79 and M 11).

Torpedoes: 6—324 mm US Mk 32 (2 triple) tubes ❺ (F 33). Honeywell/Alliant Mk 46 Mod 5; anti-submarine; active/passive homing to 11 km *(5.9 n miles)* at 40 kt; warhead 44 kg.

A/S mortars: 1 Bofors 375 mm twin-barrelled trainable ❻ (F 33-F 36); automatic loading; range 3,600 m.

Countermeasures: Decoys: 4 Loral Hycor SRBOC 6-barrelled Mk 36 for chaff and IR flares ❼ (F 33).
Acoustic decoys ❽.

ESM: Elsag Mk 1000 (part of Deneb system); or Mk 1600; intercept.

ECM: Ceselsa Canopus; or Mk 1900; jammer.

Combat data systems: Tritan IV; Link 11 (F 33). Saturn SATCOM ❾.

Weapons control: Signaal WM25; GM 101.

Radars: Air/surface search: Signaal DA05/2 ❿ (F 33); E/F-band; range 137 km *(75 n miles)* for 2 m² target.
Surface search: Signaal ZW06 ⓫ (F 33); I-band.
Navigation: 2 Furuno ⓬; I-band.
Fire control: Signaal WM22/41 or WM25 system (F 33) ⓭; I/J-band; range 46 km *(25 n miles)*.

Sonars: Raytheon 1160B (F 33); hull-mounted; active search and attack; medium frequency.

Programmes: Officially rated as Corvettes. *Diana* (tenth of the name) originates with the galley *Diana* of 1570. *Infanta Elena* and *Infanta Cristina* are named after the daughters of King Juan Carlos. Approval for second four ships given on 21 May 1976. First four ordered 7 December 1973 (83 per cent Spanish ship construction components) and two more from Bazán, Ferrol on 25 May 1976.

Structure: Original Portuguese 'João Coutinho' design by Comodoro de Oliveira PN developed by Blohm + Voss and considerably modified by Bazán including use of Y-shaped funnel. Noise reduction measures include Masker fitted to shafts, auxiliary gas-turbine generator fitted on upper deck, all main and auxiliary diesels sound-mounted. Fully stabilised. Automatic computerised engine and alternator control; two independent engine rooms; normal running on two diesels.

Operational: P 75 completed conversion to an OPV, with capability to act as helicopter platform, in 2000. Most major weapon systems removed. M 11 completed conversion to MCMV support role in 2000 and based at Cartagena. P 79 converted to OPV role in 2003 and P 77 and P 78 in 2004. F 33 to be converted in 2005. P 75 and P 77 based at Cartagena and P 78 and P 79 at Las Palmas.

Sales: F 37 and F 38 sold to Egypt prior to completion. One to Morocco in 1983.

UPDATED

INFANTA ELENA

(Scale 1 : 900), Ian Sturton / 0130393

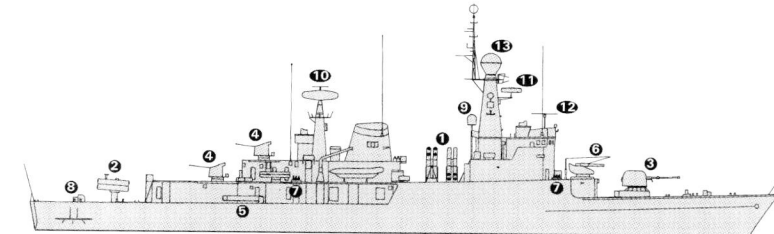

DIANA

2/2004*, Guy Toremans / 1044538

CAZADORA

7/2004*, Diego Quevedo / 1044553

INFANTA ELENA

2/2004*, Guy Toremans / 1044539

1 PESCALONSO CLASS (OFFSHORE PATROL CRAFT) (PSO)

Name	No	Builders	Commissioned
CHILREU (ex-Pescalonso 2)	P 61	Gijon, Asturias	30 Mar 1992

Displacement, tons: 2,101 full load
Dimensions, feet (metres): 222.4 × 36.1 × 15.4 (67.8 × 11 × 4.7)
Main machinery: 1 MaK 6M-453K diesel; 2,460 hp(m) (1.81 MW) sustained; 1 shaft; cp prop
Speed, knots: 12. **Range, n miles:** 1,500 at 12 kt
Complement: 35 (7 officers)
Guns: 1—12.7 mm MG.
Radars: Surface search: 2 Consilium Selesmar; E/F/I-band.

Comment: Launched 2 May 1988 and purchased by the Fisheries Department for the Navy to use as a Fishery Protection vessel based at Ferrol. Former stern ramp trawler. Inmarsat fitted.

UPDATED

CHILREU   2/2004*, Adolfo Ortigueira Gil / 1044554

3 ALBORAN CLASS (OFFSHORE PATROL CRAFT) (PSOH)

Name	No	Builders	Commissioned
ALBORAN	P 62	Freire, Vigo	8 Jan 1997
ARNOMENDI	P 63	Freire, Vigo	13 Dec 2000
TARIFA	P 64	Freire, Vigo	14 June 2004

Displacement, tons: 1,963 full load
Dimensions, feet (metres): 218.2 × 36.1 × 14.4 (66.5 × 11 × 4.4)
Main machinery: 1 Krupp MaK 6 M 453C diesel; 2,400 hp(m) (1.76 MW) sustained (P 62); 1 Krupp MaK 8M25 diesel; 3,250 hp(m) (2.39 MW) sustained (P 63); 1 diesel generator and motor for emergency propulsion; 462 hp(m) (340 kW); 1 shaft; bow thruster; 350 hp(m) (257 kW)
Speed, knots: 13 (P 62); 15.8 (P 63) (3.5 on emergency motor).
Range, n miles: 20,000 at 13 kt
Complement: 37 (7 officers) plus 9 spare
Guns: 2—12.7 mm MGs.
Radars: Surface search: Furuno FAR-2825; I-band.
Navigation: Furuno FR-2130S; I-band.
Helicopters: Platform for 1 light.

Comment: *Alboran* launched in 1991 and purchased by the Fisheries Department to use as a Fishery Protection vessel based at Cartagena. *Arnomendi*, with a slightly larger bridge and more powerful engine, based at Las Palmas, Canary Islands. *Tarifa* is fitted with anti-pollution equipment and is based at Cartagena.

UPDATED

TARIFA   4/2004*, SEGEPESCA / 1044555

6 BARCELÓ CLASS (LARGE PATROL CRAFT) (PB)

Name	No	Builders	Commissioned
BARCELÓ	P 11	Lürssen, Vegesack	20 Mar 1976
LAYA	P 12	Bazán, La Carraca	23 Dec 1976
JAVIER QUIROGA	P 13	Bazán, La Carraca	4 Apr 1977
ORDÓÑEZ	P 14	Bazán, La Carraca	7 June 1977
ACEVEDO	P 15	Bazán, La Carraca	14 July 1977
CÁNDIDO PÉREZ	P 16	Bazán, La Carraca	25 Nov 1977

Displacement, tons: 145 full load
Dimensions, feet (metres): 118.7 × 19 × 6.2 (36.2 × 5.8 × 1.9)
Main machinery: 2 MTU-Bazán MD 16V 538 TB90 diesels; 6,000 hp(m) (4.41 MW) sustained; 2 shafts
Speed, knots: 22. **Range, n miles:** 1,200 at 17 kt
Complement: 19 (3 officers)
Guns: 1 Breda 40 mm/70. 1 Oerlikon 20 mm/85. 2—12.7 mm MGs.
Torpedoes: Fitted for 2—21 in (533 mm) tubes.
Weapons control: CSEE optical director.
Radars: Surface search: Raytheon 1220/6XB; I/J-band.

Comment: Ordered 5 December 1973. All manned by the Navy although building cost was borne by the Ministry of Commerce. Of Lürssen TNC 36 design. Reported as able to take two or four surface-to-surface missiles instead of 20 mm gun and torpedo tubes. Plans to transfer to the Guardia Civil del Mar have been shelved. Speed much reduced from original 36 kt.

UPDATED

CÁNDIDO PÉREZ   7/2004*, Diego Quevedo / 1043176

9 ANAGA CLASS (LARGE PATROL CRAFT) (PB)

Name	No	Builders	Commissioned
ANAGA	P 21	Bazán, La Carraca	14 Oct 1980
TAGOMAGO	P 22	Bazán, La Carraca	30 Jan 1981
MAROLA	P 23	Bazán, La Carraca	4 June 1981
MOURO	P 24	Bazán, La Carraca	14 July 1981
GROSA	P 25	Bazán, La Carraca	15 Sep 1981
MEDAS	P 26	Bazán, La Carraca	16 Oct 1981
IZARO	P 27	Bazán, La Carraca	9 Dec 1981
TABARCA	P 28	Bazán, La Carraca	30 Dec 1981
BERGANTÍN	P 30	Bazán, La Carraca	28 July 1982

Displacement, tons: 319 full load
Dimensions, feet (metres): 145.6 × 21.6 × 8.2 (44.4 × 6.6 × 2.5)
Main machinery: 1 MTU-Bazán 16V 956 SB90 diesel; 4,000 hp(m) (2.94 MW) sustained; 1 shaft; cp prop
Speed, knots: 16. **Range, n miles:** 4,000 at 13 kt
Complement: 25 (3 officers)
Guns: 1 FMC 3 in (76 mm)/50 Mk 22. 1 Oerlikon 20 mm Mk 10. 2—7.62 mm MGs.
Radars: Surface search: 1 Racal Decca 1226; I-band.
Navigation: Consilium Selesmar SRL MM 950; F/I-band.

Comment: Ordered from Bazán, Cádiz on 22 July 1978. For fishery and EEZ patrol duties. Rescue and firefighting capability. Speed reduced from original 20 kt.

UPDATED

TABARCA   10/2004*, Adolfo Ortigueira Gil / 1044556

4 CONEJERA CLASS (COASTAL PATROL CRAFT) (PB)

Name	No	Builders	Commissioned
CONEJERA	P 31	Bazán, Ferrol	31 Dec 1981
DRAGONERA	P 32	Bazán, Ferrol	31 Dec 1981
ESPALMADOR	P 33	Bazán, Ferrol	10 May 1982
ALCANADA	P 34	Bazán, Ferrol	10 May 1982

Displacement, tons: 85 full load
Dimensions, feet (metres): 106.6 × 17.4 × 4.6 (32.2 × 5.3 × 1.4)
Main machinery: 2 MTU-Bazán MA 16V 362 SB80 diesels; 2,450 hp(m) (1.8 MW); 2 shafts
Speed, knots: 13. **Range, n miles:** 1,200 at 13 kt
Complement: 12
Guns: 1 Oerlikon 20 mm/120 Mk 10. 1—12.7 mm MG.
Radars: Surface search: Furuno; I-band.

Comment: Ordered in 1978, funded jointly by the Navy and the Ministry of Commerce. Naval manned. Speed reduced from original 25 kt.

UPDATED

ALCANADA   7/2004*, A Campanera i Rovira / 1044557

2 TORALLA CLASS (COASTAL PATROL CRAFT) (PB)

Name	No	Builders	Commissioned
TORALLA	P 81	Viudes, Barcelona	29 Apr 1987
FORMENTOR	P 82	Viudes, Barcelona	23 June 1988

Displacement, tons: 102 full load
Dimensions, feet (metres): 93.5 × 21.3 × 5.9 *(28.5 × 6.5 × 1.8)*
Main machinery: 2 MTU-Bazán 8V 396 TB93 diesels; 2,100 hp(m) *(1.54 MW)* sustained; 2 shafts
Speed, knots: 19. **Range, n miles:** 1,000 at 12 kt
Complement: 13
Guns: 1 Browning 12.7 mm MG.
Radars: Surface search: Racal Decca RM 1070; I-band.
Navigation: Racal Decca RM 270; I-band.

Comment: Wooden hull with GRP sheath. Very similar to Customs Alcaravan class. *Formentor* refitted in 1996-97. Based at Cartagena.

UPDATED

P 101 class   *6/2000, Adolfo Ortigueira Gil /* 0087858

1 INSHORE/RIVER PATROL LAUNCH (PBR)

Name	No	Builders	Commissioned
CABO FRADERA	P 201	Bazán, La Carraca	11 Jan 1963

Displacement, tons: 21 full load
Dimensions, feet (metres): 58.3 × 13.8 × 3 *(17.8 × 4.2 × 0.9)*
Main machinery: 2 diesels; 280 hp(m) *(206 kW)*; 2 shafts
Speed, knots: 11
Complement: 9
Guns: 1—7.62 mm MG.
Radars: Surface search: Furuno; I-band.

Comment: Based at Tuy on River Miño for border patrol with Portugal.

VERIFIED

TORALLA   *6/2004*, Diego Quevedo /* 1044558

2 P 101 CLASS (PBR)

P 114   P 111

Displacement, tons: 18.5 standard; 20.8 full load
Dimensions, feet (metres): 44.9 × 14.4 × 4.3 *(13.7 × 4.4 × 1.3)*
Main machinery: 2 Baudouin-Interdiesel DNP-350; 768 hp(m) *(564 kW)*; 2 shafts
Speed, knots: 23.3. **Range, n miles:** 430 at 18 kt
Complement: 6
Guns: 1—12.7 mm MG.
Radars: Surface search: Decca 110; I-band.

Comment: Ordered under the programme agreed 13 May 1977, funded jointly by the Navy and the Ministry of Commerce. Built to the Aresa LVC 160 design by Aresa, Arenys de Mar, Barcelona. GRP hull. Eight of the class conduct harbour auxiliary duties with Y numbers, the remainder paid off in 1993. P 111 transferred back again to patrol duties in 1996 and is based at Ayamonte (Huelva). P 114 is also used for patrol duties and is based at Ceuta.

VERIFIED

CABO FRADERA   *4/2003, Camil Busquets i Vilanova /* 0570981

AMPHIBIOUS FORCES

2 NEWPORT CLASS (LSTH)

Name	No	Builders	Laid down	Launched	Commissioned
HERNÁN CORTÉS (ex-*Barnstable County*)	L 41 (ex-L 1197)	National Steel, San Diego	19 Dec 1970	2 Oct 1971	27 May 1972
PIZARRO (ex-*Harlan County*)	L 42 (ex-L 1196)	National Steel, San Diego	7 Nov 1970	24 July 1971	8 Apr 1972

Displacement, tons: 4,975 light; 8,550 full load
Dimensions, feet (metres): 522.3 (hull) × 69.5 × 18.2 (aft) *(159.2 × 21.2 × 5.5)*
Main machinery: 6 Alco 16-251 diesels; 16,500 hp *(12.3 MW)* sustained; 2 shafts; cp props; bow thruster
Speed, knots: 20. **Range, n miles:** 14,250 at 14 kt
Complement: 255 (15 officers)
Military lift: 374 troops; (20 officers) 500 tons vehicles; 2 LCVPs and 2 LCPLs on davits

Guns: 1 General Electric/General Dynamics 20 mm Vulcan Phalanx Mk 15. 2 Oerlikon 20 mm/85. 4—12.7 mm MGs.
Countermeasures: ESM: Celesa Deneb.
Radars: Surface search: Raytheon SPS-10F/67; G-band.
Navigation: Marconi LN66; I-band.

Helicopters: Platform only for 3 AB 212.

Programmes: First one transferred from the USA and recommissioned 26 August 1994, the second one transferred 14 April 1995. Leases have been renewed.
Structure: The 3 in guns removed on transfer. The ramp is supported by twin derrick arms. A ramp just forward of the superstructure connects the lower tank deck with the main deck and a vehicle passage through the superstructure provides access to the parking area amidships. A stern gate to the tank deck permits unloading of amphibious tractors into the water, or unloading of other vehicles into an LCU or on to a pier. Vehicle stowage covers 19,000 sq ft. Length over derrick arms is 562 ft *(171.3 m)*; full load draught is 11.5 ft forward and 17.5 ft aft. Bow thruster fitted to hold position offshore

HERNÁN CORTÉS   *7/2003, B Sullivan /* 0570977

while unloading amphibious tractors. Meroka may replace Phalanx or be fitted on the former 3 in gun platform. SCOT 3 SATCOM fitted in 1995-96. Can carry four Mexeflotes, two of them powered.

Operational: Based at Rota. Both to be replaced in 2008 by the Strategic Projection Ship.

VERIFIED

2 GALICIA CLASS (LPD)

Name	No	Builders	Laid down	Launched	Commissioned
GALICIA	L 51	Bazán, Ferrol	31 May 1996	21 July 1997	30 Apr 1998
CASTILLA	L 52	Bazán, Ferrol	11 Dec 1997	14 June 1999	26 June 2000

Displacement, tons: 13,815 full load
Dimensions, feet (metres): 524.9 oa; 465.9 pp × 82 × 19.3
  (160; 142 × 25 × 5.9)
Flight deck, feet (metres): 196.9 × 82 *(60 × 25)*
Main machinery: 2 Bazán/Caterpillar 3612 diesels; 12,512 hp
  (m) *(9.2 MW)*; 2 shafts; LIPS cp props; bow thruster 680 hp
  (m) *(500 kW)*
Speed, knots: 20. **Range, n miles:** 6,000 at 12 kt
Complement: 115 plus 12 spare; 189 (L 52)
Military lift: 543 or 404 (L 52) fully equipped troops and 72 (staff
  and aircrew)
  6 LCVP or 4 LCM or 1 LCU and 1 LCVP. 130 APCs or 33 MBTs.

Guns: 1 Bazán 20 mm/120 12-barrelled Meroka (fitted for) ❶;
  3,600 rds/min combined to 2 km. 2 Oerlikon GAM-B01
  20 mm.
Countermeasures: Decoys: 4 SRBOC chaff launchers.
  ESM: Intercept.
Combat data systems: SICOA (L 52); SATCOM; Link 11.
Radars: Surface search: TRS 3D/16 (L 52) ❷; G-band.
  Surface search: Kelvin Hughes ARPA ❸; I-band.
  Navigation and helo control: I-band.

Helicopters: 6 AB 212 or 4 SH-3D Sea King ❹ or 4 Eurocopter
  Tiger.

Programmes: Originally started as a national project by the
  Netherlands. In 1990 the ATS was seen as a possible solution

CASTILLA   *(Scale 1 : 1,500), Ian Sturton /* 0106549

to fulfil the requirements for a new LPD. Joint project
definition study announced in July 1991 and completed in
December 1993 and the first ship was authorised on 29 July
1994. The second of class ordered 9 May 1997.
Modernisation: L 52 C² capabilities upgraded in 2002-03 to
support Flagship requirements. L 52 embarked the HQ of the
Spanish High Readiness Force (Maritime) in November 2003
as part of the NATO Response Force. Both ships are to be
fitted with RAM CIWS.
Structure: Able to transport a fully equipped battalion of marines
providing a built-in dock for landing craft and a helicopter
flight deck for debarkation in offshore conditions. Docking
well is 885 m²; vehicle area 1,010 m². Access hatch on the

starboard side. Hospital facilities. Built to commercial
standards with military command and control and NBCD
facilities. *Castilla* has improved command and control facilities
with two operations centres, one for amphibious and one for a
combat group.
Operational: Alternatively can also be used for a general logistic
support for both military and civil operations, including
environmental and disaster relief tasks. Based at Rota.
                                                                     VERIFIED

CASTILLA (without guns)   3/2001, Diego Quevedo / 0130138   GALICIA   1/2002, Adolfo Ortigueira Gil / 0528928

GALICIA (without main radar and guns)   5/2000, Camil Busquets i Vilanova / 0130137

0 + 1 STRATEGIC PROJECTION SHIP (LHD)

Name	No	Builders	Laid down	Launched	Commissioned
—	—	IZAR, Ferrol	2005	2007	2008

Displacement, tons: 26,800 full load
Dimensions, feet (metres): 759.2 × 105.0 × 23.0
(231.4 × 32.0 × 7.0)
Flight deck, feet (metres): 663.9 × 105.0 *(202.3 × 32)*
Main machinery: Diesel-electric; 4 diesels; 35,000 hp *(26 MW)*; 2 podded propulsors
Speed, knots: 21. **Range, n miles:** 9,000 at 15 kt
Complement: 243 (plus 1,220 including flag staff, air group and 900 landing force)
Helicopters: 6 landing spots for helicopter or AV-8 operations.

Programmes: Approval for the procurement of a Strategic Projection Ship was given by the Spanish Cabinet on 5 September 2003. Contract for design and construction was awarded in March 2004.
Structure: The hangar is 1,000 m². Below the hangar there is a 2,000 m² garage. Typical transport configurations include: 46 tanks and 42 Leopard; 70 containers of 20 tons; 32 NH-90 or 19 AV-8 or 12 CH-47 or 12 NH 90 and 11 AV-8. The landing

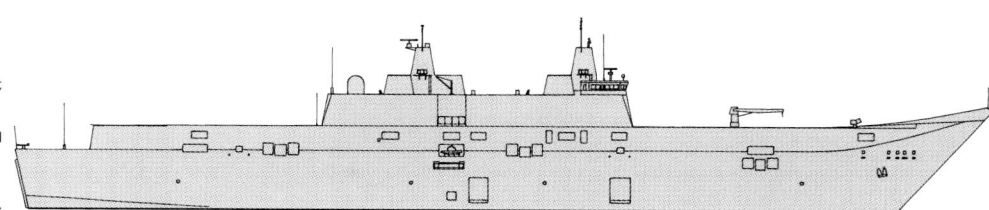

STRATEGIC PROJECTION SHIP   *(Scale 1 : 1,800), Ian Sturton /* 0569926

dock (69.3 × 16 m) will be capable of operating four LCM (1E) landing craft or at least one landing craft air cushion. Medical facilities will include operating rooms, intensive care unit and sick bay.

Operational: The principal roles are amphibious, strategic projection of land forces and disaster relief. The ship will also be capable of operating the fixed-wing aircraft of *Príncipe de Asturias*.

UPDATED

1 LCU

L 072 (ex-*LCU 12*, ex-*LCU 2*, ex-*LCU 1491*)

Displacement, tons: 360 full load
Dimensions, feet (metres): 119.7 × 31.5 × 5.2 *(36.5 × 9.6 × 1.6)*
Main machinery: 3 Gray Marine 64 YTL diesels; 675 hp *(504 kW)*; 3 shafts
Speed, knots: 8
Complement: 14
Military lift: 160 tons
Radars: Navigation: Furuno; I-band.

Comment: Transferred from US June 1972. Purchased August 1976. Refitted 1993-94. Based at Puntales (Cadiz). *VERIFIED*

LCU   *4/1999, Diego Quevedo /* 0080627

7 LCM 8

L 81-86 (ex-*LCM 81-86*, ex-*E 81-86*)   L 87

Displacement, tons: 115-120 full load
Dimensions, feet (metres): 74.5 × 21.7 × 5.9 *(22.7 × 6.6 × 1.8)*
Main machinery: 4 GM 6-71 diesels; 696 hp *(519 kW)* sustained; 2 shafts
Speed, knots: 11
Complement: 5

Comment: First six ordered from Oxnard, California in 1974. Assembled in Spain. Commissioned in June-September 1975. Two more built by Bazán, San Fernando, and completed in April 1989. *L 88* decommissioned 2003. Based at Puntales (Cadiz). *UPDATED*

LCM   *6/2004*, Camil Busquets i Vilanova /* 1043187

2 +12 LCM (1E)

L 601-602

Displacement, tons: 108 full load
Dimensions, feet (metres): 76.5 × 21 × 3.4 *(23.3 × 6.4 × 1.1)*
Main machinery: 2 MAN-D 2842-LE 402 diesels; 2,200 hp(m) *(1.62 MW)*; 2 MJP-650 DD waterjets
Speed, knots: 14. **Range, n miles:** 160 at 12 kt
Complement: 3
Military lift: 100 tons

Comment: L 601-602 built by IZAR, San Fernando, for LPDs and delivered in early 2001. Bow and stern ramps. Steel construction with wheelhouse of composites. Maximum speed in ballast is 22 kt. Based at Puntales. An order for a further 12 craft made in November 2004. To be constructed by IZAR, San Fernando, and delivered by 2009. *UPDATED*

L 601   *11/2002, A Campanera i Rovira /* 0570942

40 LANDING CRAFT

Comment: Apart from those used for divers there are 14 LCM 6 *(L 161-167, L 261-267)*, 16 LCVP and 8 LCPL. All of the LCM 6, eight of the LCVPs and most of the LCPs were built in Spanish Shipyards 1986-88. There are also two tug pontoons (mexeflotes) *(L 91-L 92)* completed in 1995. Most of these craft are laid up. *VERIFIED*

LCM L 162   *10/1993, Diego Quevedo*

MINE WARFARE FORCES

5 + 1 (2) SEGURA CLASS (MINEHUNTERS) (MHC)

Name	No	Builders	Launched	Commissioned
SEGURA	M 31	Bazán, Cartagena	25 July 1997	27 Apr 1999
SELLA	M 32	Bazán, Cartagena	6 July 1998	28 May 1999
TAMBRE	M 33	Bazán, Cartagena	5 Mar 1999	18 Feb 2000
TURIA	M 34	Bazán, Cartagena	22 Nov 1999	16 Oct 2000
DUERO	M 35	Izar, Cartagena	28 April 2003	5 July 2004
TAJO	M 36	Izar, Cartagena	10 June 2004	2005

Displacement, tons: 530 full load
Dimensions, feet (metres): 177.2 oa; 167.3 wl × 35.1 × 7.2 *(54; 51 × 10.7 × 2.2)*
Main machinery: 2 MTU-Bazán 6V 396 TB83 diesels; 1,523 hp(m) *(1.12 MW)*; 2 motors (for hunting); 200 kW; 2 Voith Schneider props; 2 side thrusters; 150 hp(m) *(110 kW)*
Speed, knots: 14; 7 (hunting). **Range, n miles:** 2,000 at 12 kt
Complement: 41 (7 officers)
Guns: 1 Bazán/Oerlikon 20 mm GAM-BO1.
Countermeasures: MCM: FABA/Inisel system. 2 Gayrobot Pluto Plus ROVs.
Combat data systems: FABA/SMYC Nautis.
Radars: Navigation: Kelvin-Hughes 1007; I-band.
Sonars: Raytheon/ENOSA SQQ-32 multifunction VDS mine detection; high frequency.

Comment: On 4 July 1989 a technology transfer contract was signed with Vosper Thornycroft to allow Bazán to design a new MCM vessel based on the Sandown class. The order for four of the class was authorised on 7 May 1993, and an agreement signed on 26 November 1993 between DCN and Bazán provided for training in GRP technology. The first of class laid down 30 May 1995. Two more ordered on 26 January 2001 with an option for a further two. Sonar includes side scanning, and a towed body tracking and positioning system. M 35 and M 36 are to be fitted with the Minesniper mine disposal system. M 31-34 are likely to be retrofitted in due course. Although a class of 12 was envisaged, higher priority may now be given to a minesweeper programme. Form 1st MCM Squadron based at Cartagena. *UPDATED*

SELLA   *5/2004*, A Campanera i Rovira /* 1044559

SURVEY AND RESEARCH SHIPS

1 DARSS CLASS (RESEARCH SHIP) (AGI/AGOR)

Name	No	Builder	Commissioned
ALERTA (ex-*Jasmund*)	A 111	Peenewerft, Wolgast	6 Dec 1992

Displacement, tons: 2,292 full load
Dimensions, feet (metres): 250.3 × 39.7 × 13.8 *(76.3 × 12.1 × 4.2)*
Main machinery: 1 Kolomna Type 40-DM diesel; 2,200 hp(m) *(1.6 MW)* sustained; 1 shaft; cp prop
Speed, knots: 11. **Range, n miles:** 1,000 at 11 kt
Complement: 60
Guns: Fitted for 3 twin 25 mm/70. 2—12.7 mm MGs.
Radars: Navigation: Racal Decca; I-band.

Comment: Former GDR depot ship launched on 27 February 1982 and converted to an AGI, with additional accommodation replacing much of the storage capacity. Was to have transferred to Ecuador in 1991 but the sale was cancelled. Commissioned in the Spanish Navy and sailed from Wilhelmshaven for a refit at Las Palmas prior to being based at Cartagena and used as an AGI and equipment trials ship. Saturn 35 SATCOM.

VERIFIED

ALERTA　　　　　　　　　　　　　　　　　　7/2002, Diego Quevedo / 0528951

1 RESEARCH SHIP (AGOBH)

Name	No	Builders	Commissioned
HESPÉRIDES (ex-*Mar Antártico*)	A 33	Bazán, Cartagena	16 May 1991

Displacement, tons: 2,738 full load
Dimensions, feet (metres): 270.7; 255.2 × 46.9 × 14.8 *(82.5; 77.8 × 14.3 × 4.5)*
Main machinery: Diesel-electric; 4 MAN-Bazán 14V20/27 diesels; 6,860 hp(m) *(5 MW)* sustained; 4 generators; 2 AEG motors; 3,800 hp(m) *(2.8 MW)*; 1 shaft; bow and stern thrusters; 350 hp (m) *(257 kW)* each
Speed, knots: 15. **Range, n miles:** 12,000 at 13 kt
Complement: 39 (9 officers) plus 30 scientists
Radars: Surface search: Racal/Hispano ARPA 2690; I-band.
Navigation: Racal 2690 ACS; F-band.
Helicopters: 1 AB 212.

Comment: Ordered in July 1988 from Bazán, Cartagena, by the Ministry of Education and Science. Laid down in 1989, launched 12 March 1990. Has 330 sq m of laboratories, Simbad ice sonar. Dome in keel houses several sensors. Ice-strengthened hull capable of breaking first year ice up to 45 cm at 5 kt. Based at Cartagena, the main task is to support the Spanish base at Livingston Island, Antarctica. Manned and operated by the Navy. Has a telescopic hangar. Modifications made to superstructure in 2004 to increase accommodation for scientific staff.

UPDATED

HESPÉRIDES　　　　　　　　　　　　　　　10/2004*, Diego Quevedo / 1043175

2 CASTOR CLASS (SURVEY SHIPS) (AGS)

Name	No	Builders	Commissioned
ANTARES	A 23	Bazán, La Carraca	21 Nov 1974
RIGEL	A 24	Bazán, La Carraca	21 Nov 1974

Displacement, tons: 363 full load
Dimensions, feet (metres): 125.9 × 24.9 × 10.2 *(38.4 × 7.6 × 3.1)*
Main machinery: 1 Sulzer 4TD36 diesel; 720 hp(m) *(530 kW)*; 1 shaft
Speed, knots: 11.5. **Range, n miles:** 3,620 at 8 kt
Complement: 36 (4 officers)
Radars: Navigation: Raytheon 1620; I/J-band.

Comment: Fitted with Raydist, Omega and digital presentation of data. Likely to be decommissioned in the near future. Based at Cadiz.

UPDATED

ANTARES　　　　　　　　　　　1/2003, Camil Busquets i Vilanova / 0570982

1 RESEARCH SHIP (AGOB)

Name	No	Builders	Commissioned
LAS PALMAS (ex-*Somiedo*)	A 52	Astilleros Atlántico, Santander	1978

Displacement, tons: 1,450 full load
Dimensions, feet (metres): 134.5 × 38.1 × 18 *(41 × 11.6 × 5.5)*
Main machinery: 2 AESA/Sulzer 16ASV25/30 diesels; 7,744 hp(m) *(5.69 MW)*; 2 shafts
Speed, knots: 13. **Range, n miles:** 27,000 at 12 kt
Complement: 33 (8 officers) plus 45 scientists
Guns: 2—12.7 mm MGs.
Radars: Navigation: 2 Racal Decca; I-band.

Comment: Built as a tug for Compania Hispano Americana de Offshore SA. Commissioned in the Navy 30 July 1981. Converted in 1988 for Polar Research Ship duties in Antarctica with an ice strengthened bow, an enlarged bridge and two containers aft for laboratories. Based at Cartagena.

NEW ENTRY

LAS PALMAS　　　　　　　　　　　　　　　5/2004*, Diego Quevedo / 1044561

2 MALASPINA CLASS (SURVEY SHIPS) (AGS)

Name	No	Builders	Commissioned
MALASPINA	A 31	Bazán, La Carraca	21 Feb 1975
TOFIÑO	A 32	Bazán, La Carraca	23 Apr 1975

Displacement, tons: 820 standard; 1,090 full load
Dimensions, feet (metres): 188.9 × 38.4 × 12.8 *(57.6 × 11.7 × 3.9)*
Main machinery: 2 San Carlos MWM TbRHS-345-61 diesels; 3,600 hp(m) *(2.64 MW)*; 2 shafts; LIPS cp props
Speed, knots: 15. **Range, n miles:** 4,000 at 12 kt; 3,140 at 14.5 kt
Complement: 63 (9 officers)
Guns: 2 Oerlikon 20 mm.
Radars: Navigation: Raytheon 1220/6XB; I/J-band.

Comment: Ordered mid-1972. Both named after their immediate predecessors. Developed from British Bulldog class. Fitted with two Atlas DESO-10 AN 1021 (280-1,400 m) echo-sounders, retractable Burnett 538-2 sonar for deep sounding, Egg Mark B side scan sonar, Raydist DR-S navigation system, Hewlett Packard 2100A computer inserted into Magnavox Transit satellite navigation system, active rudder with fixed pitch auxiliary propeller. *Malaspina* used for a NATO evaluation of a Ship's Laser Inertial Navigation System (SLINS) produced by British Aerospace. Based at Cadiz.

UPDATED

MALASPINA　　　　　　　　　　6/2004*, Camil Busquets i Vilanova / 1043186

2 LHT-130 CLASS (SURVEY MOTOR BOATS) (YGS)

Name	No	Builders	Commissioned
ASTROLABIO	A 91	Rodman, Vigo	30 Nov 2001
ESCANDALLO	A 92	Rodman, Vigo	27 Feb 2004

Displacement, tons: 8 full load
Dimensions, feet (metres): 41.3 × 13.8 × 1.6 *(12.6 × 4.2 × 0.5)*
Main machinery: 2 diesels; 700 hp *(522 kW)*; 2 shafts
Speed, knots: 30

Comment: Support craft of the Hydrographic Flotilla. Based at Puntales and transportable by road, rail, ship or aircraft.

UPDATED

ESCANDALLO   *10/2004*, Diego Quevedo /* 1043174

AUXILIARIES

0 + 1 FLEET REPLENISHMENT SHIP (AORH)

Name	No	Builders	Laid down	Launched	Commissioned
CANTABRIA	—	IZAR, Puerto Real	2005	2006	2007

Displacement, tons: 19,500 full load
Dimensions, feet (metres): 564.3 × 75.5 × 26.2 *(172.0 × 23.0 × 8.0)*
Main machinery: 2 diesels; 1 shaft
Speed, knots: 21. **Range, n miles:** 6,000 at 15 kt
Complement: To be announced
Cargo capacity: To be announced
Guns: To be announced.
Countermeasures: To be announced.
Combat data systems: To be announced.
Radars: To be announced.
Helicopters: 2 SH-3D Sea King or 3 AB 212.

Comment: Similar in design to Patiño class with improved capabilities including double-hull, container cargo capacity, enhanced sensors and a combat data system. Construction is to start in 2005.
NEW ENTRY

REPLENISHMENT SHIP   *10/2004*, Spanish Navy /* 1044544

1 PATIÑO CLASS (FLEET LOGISTIC TANKER) (AORH)

Name	No	Builders	Launched	Commissioned
PATIÑO	A 14	Bazán, Ferrol	22 June 1994	16 June 1995

Displacement, tons: 5,762 light; 17,045 full load
Dimensions, feet (metres): 544.6 × 72.2 × 26.2 *(166 × 22 × 8)*
Main machinery: 2 Bazán/Burmeister & Wain 16V40/45 diesels; 24,000 hp(m) *(17.6 MW)* sustained; 1 shaft; LIPS cp prop
Speed, knots: 20. **Range, n miles:** 13,440 at 20 kt
Complement: 146 plus 19 aircrew plus 20 spare
Cargo capacity: 6,815 tons dieso; 1,660 tons aviation fuel; 500 tons solids
Guns: 2 Bazán 20 mm/120 Meroka CIWS (fitted for). 2 Oerlikon 20 mm/90.
Countermeasures: Decoys: 4 SRBOC chaff launchers. Nixie torpedo decoy.
ESM/ECM: Aldebaran intercept and jammer.
Radars: 3 navigation/helo control; I-band.
Helicopters: 2 SH-3D Sea King or 3 AB 212.

Comment: The Bazán design AP 21 was rejected in favour of this joint Netherlands/Spain design. Ordered on 26 December 1991. Laid down 1 July 1993. Two supply stations each side for both liquids and solids. Stern refuelling. One Vertrep supply station, and workshops for aircraft maintenance. Medical facilities. Built to merchant ship standards with military NBC. Accommodation for up to 50 female crew members. SCOT 3 SATCOM to be fitted. Based at Ferrol.
VERIFIED

PATIÑO   *10/2003, A E Galarce /* 0570955

1 TRANSPORT SHIP (AKRH)

Name	No	Builders	Commissioned
MARTÍN POSADILLO	A 04 (ex-ET-02)	Duro Felguera, Gijon	1973
(ex-*Rivanervión*, ex-*Cala Portals*)			

Displacement, tons: 1,920 full load
Dimensions, feet (metres): 246.1 × 42.7 × 14.1 *(75 × 13 × 4.3)*
Main machinery: 1 BMW diesel; 2,400 hp(m) *(1.77 MW)*; 1 shaft
Speed, knots: 10
Complement: 18
Military lift: 42 trucks plus 25 jeeps
Helicopters: Platform for 1 Chinook.

Comment: Ro-Ro ship taken on by the Army in 1990 and transferred to the Navy on 14 February 2000. Based at Cartagena.
VERIFIED

MARTÍN POSADILLO   *5/2001, Camil Busquets i Vilanova /* 0130116

1 TRANSPORT SHIP (APH)

Name	No	Builders	Recommissioned
CONTRAMAESTRE CASADO	A 01	Eriksberg-Göteborg, Sweden	15 Dec 1982
(ex-*Thanasis-K*, ex-*Fortuna Reefer*, ex-*Bonzo*, ex-*Bajamar*, ex-*Leeward Islands*)			

Displacement, tons: 4,965 full load
Dimensions, feet (metres): 343.4 × 46.9 × 29.2 *(104.7 × 14.3 × 8.9)*
Main machinery: 1 Burmeister & Wain diesel; 3,600 hp(m) *(2.65 MW)*; 1 shaft
Speed, knots: 14. **Range, n miles:** 8,000 at 14 kt
Complement: 72
Guns: 2 Oerlikon 20 mm.
Radars: Navigation: Racal Decca 1226 and 626; I-band.

Comment: Built in 1953. Impounded as smuggler. Delivered after conversion 6 December 1983. Has a helicopter deck. Since 2001, based at La Carraca (Cadiz).
VERIFIED

CONTRAMAESTRE CASADO   *3/2001, Diego Quevedo /* 0132007

7 HARBOUR TANKERS (YO)

No	Displacement, tons	Dimensions metres	Cargo, tons fuel	Commissioned
Y 231	524	34 × 7 × 2.9	300	1981
Y 237	344	34.3 × 6.2 × 2.5	193	1965
Y 251	830	42.8 × 8.4 × 3.1	500	1981
Y 252	337	34.3 × 6.2 × 2.5	193	1965
Y 253	337	34.3 × 6.2 × 2.5	193	1965
Y 254	214.7	24.5 × 5.5 × 2.2	100	1981
Y 255	524	34 × 7 × 2.9	300	1981

Comment: All built by Bazán at Cádiz and Ferrol.
VERIFIED

Y 251   *11/2003, Diego Quevedo /* 0570953

1 TRANSPORT SHIP (AKR)

Name	No	Builders	Commissioned
EL CAMINO ESPAÑOL	A 05 (ex-ET 03)	Maua, Rio de Janeiro	Oct 1984
(ex-*Araguary*, ex-*Cyndia*)			

Displacement, tons: 5,804 full load
Dimensions, feet (metres): 313.6 × 59.8 × 15.2 *(95.5 × 18.3 × 4.6)*
Main machinery: 2 Sulzer diesels; 6,482 hp(m) *(4.76 MW)*; 2 shafts
Speed, knots: 12
Complement: 24 (3 officers) plus 40 Army
Military lift: 24 tanks plus 15 trucks and 102 jeeps
Radars: Navigation: I-band.

Comment: Acquired by the Army in early 1999 but commissioned into the Navy on 21 September 1999. Ro-Ro design converted for military use by Bazán in Cartagena. Used for logistic support of armed forces. Has two 25 ton cranes. Based at Cartagena.
UPDATED

EL CAMINO ESPAÑOL
7/2004*, Diego Quevedo / 1043184

1 FLEET TANKER (AORLH)

Name	No	Builders	Launched	Commissioned
MARQUÉS DE LA ENSENADA	A 11	Bazán, Ferrol	5 Oct 1990	3 June 1991
(ex-*Mar del Norte*)				

Displacement, tons: 13,592 full load
Dimensions, feet (metres): 403.9 oa; 377.3 wl × 64 × 25.9 *(123.1; 115 × 19.5 × 7.9)*
Main machinery: 1 MAN-Bazán 18V40/50A; 11,247 hp(m) *(8.27 MW)* sustained; 1 shaft
Speed, knots: 16. **Range, n miles:** 10,000 at 15 kt
Complement: 80 (11 officers)
Cargo capacity: 7,498 tons dieso; 1,746 tons JP-5; 120 tons deck cargo
Guns: 2—12.7 mm MGs.
Radars: Surface search: Racal Decca 2459; I/F-band.
Navigation: Racal Decca ARPA 2690/9; I-band.
Helicopters: 1 AB 212 or similar.

Comment: Ordered 30 December 1988; laid down 16 November 1989. The deletion of the *Teide* left a serious deficiency in the Fleet's at sea replenishment capability which has been restored by the *Patiño*. In addition, and as a stop gap, this tanker was built at one third of the cost of the larger support ship. Two Vertrep stations and a platform for a Sea King size helicopter. Replenishment stations on both sides and one astern. Provision for Meroka CIWS four chaff launchers as well as ESM. Has a small hospital. Based at Rota.
VERIFIED

MARQUÉS DE LA ENSENADA
5/2001, Giorgio Ghiglione / 0130115

1 WATER TANKER (AWT)

Name	No	Builders	Commissioned
MARINERO JARANO	A 65 (ex-AA 31)	Bazán, Cádiz	16 Mar 1981

Displacement, tons: 549 full load
Dimensions, feet (metres): 123 × 23 × 9.8 *(37.5 × 7 × 3)*
Main machinery: 1 diesel; 600 hp(m) *(441 kW)*; 1 shaft
Speed, knots: 10
Complement: 13
Cargo capacity: 300 tons

Comment: Similar to Y 231 and Y 255 (harbour tankers). Based at Cartagena.
VERIFIED

MARINERO JARANO
9/2003, Diego Quevedo / 0570951

2 LOGISTIC SUPPORT SHIPS (ATF/AGDS)

Name	No	Builders	Commissioned
MAR CARIBE (ex-*Amatista*)	A 101	Duro Felguera, Gijon	24 Mar 1975
NEPTUNO (ex-*Mar Rojo*, ex-*Amapola*)	A 20 (ex-A 102)	Duro Felguera, Gijon	24 Mar 1975

Displacement, tons: 1,860 full load
Dimensions, feet (metres): 176.4 × 38.8 × 14.8 *(53.8 × 11.8 × 4.5)*
Main machinery: 2 Echevarria-Burmeister & Wain 18V23HU diesels; 4,860 hp(m) *(3.57 MW)*; 2 shafts; bow thruster
Speed, knots: 12. **Range, n miles:** 6,000 at 10 kt
Complement: 44

Comment: Two offshore oil rig support tugs were acquired and commissioned into the Navy 14 December 1988. Bollard pull, 80 tons. *Neptuno* converted as a diver support vessel. She has a dynamic positioning system and carries a side scan mine detection high-frequency sonar as well as a semi-autonomous remote-controlled DSRV. The control cable restricts operations to within 75 m of an auxiliary diving unit. The DSRV is launched and recovered by a hydraulic arm. *Mar Caribe* works with Amphibious Forces and is based at Cadiz. *Neptuno* based at Cartagena.
UPDATED

MAR CARIBE
5/2004*, Diego Quevedo / 1044560

NEPTUNO
11/2002, A Campanera i Rovira / 0570952

1 WATER TANKER (AWT)

Name	No	Builders	Commissioned
CONDESTABLE ZARAGOZA	A 66 (ex-AA 41)	Bazán, Cádiz	16 Oct 1981

Displacement, tons: 895 full load
Dimensions, feet (metres): 152.2 × 27.6 × 11.2 *(46.4 × 8.4 × 3.4)*
Main machinery: 1 diesel; 700 hp(m) *(515 kW)*; 1 shaft
Speed, knots: 10
Complement: 16
Cargo capacity: 600 tons

Comment: Based at Puntales (Cadiz).
VERIFIED

CONDESTABLE ZARAGOZA
2/1995, Diego Quevedo / 0080637

47 BARGES (YO/YE)

Comment: Have Y numbers. 11 in 200 series carry fuel, 30 in 300 for ammunition and general stores, six in 400 for anti-pollution. Some floating pontoons have L numbers.
UPDATED

Y 365
7/2004*, Diego Quevedo / 1043183

36 HARBOUR LAUNCHES (YDT/YFL)

Y 502-512	Y 521-527	Y 540	Y 556
Y 515	Y 529-535	Y 545	Y 579-580
Y 519	Y 539	Y 548	Y 583-584

Comment: Some used as diving tenders, others as harbour ferries. Some are former patrol craft of the P 101 and P 202 class. *Y 540* is an Admirals' Yacht.

UPDATED

Y 507   *3/2004*, Diego Quevedo /* 1044562

Y 526   *1/2003, Diego Quevedo /* 0570967

TRAINING SHIPS

6 SAIL TRAINING SHIPS (AXS)

Name	No	Builders	Commissioned
JUAN SEBASTIÁN DE ELCANO	A 71	Echevarrieta, Cádiz	17 Aug 1928
AROSA	A 72	Inglaterra	1 Apr 1981
LA GRACIOSA (ex-*Dejá Vu*)	A 74	Inglaterra	30 June 1988
GIRALDA (ex-*Southern Cross*)	A 76	Morris & Mortimer, Argyll	26 Aug 1993
SISARGAS	A 75	Novo Glass, Polinya	18 May 1995
SÁLVORA	A 77	—	29 May 2001

Displacement, tons: 3,420 standard; 3,656 full load
Dimensions, feet (metres): 308.5 oa × 43.3 × 24.6 *(94.1 × 13.15 × 7.46)*
Main machinery: 1 Deutz MWM KHD 6M diesel; 1,950 hp(m) *(1.43 MW)*; 1 shaft
Speed, knots: 9. **Range, n miles:** 10,000 at 9 kt
Complement: 347 (students 120)
Guns: 2—37/80 mm Bazán saluting guns.
Radars: Navigation: 2 Racal Decca; I-band.

Comment: Details are for A 71 (based at La Carraca) which is a four masted top-sail schooner-near sister of Chilean *Esmeralda*. Named after the first circumnavigator of the world (1519-22) who succeeded to the command of the expedition led by Magellan after the latter's death. Laid down 24 November 1925. Launched on 5 March 1927. Carries 230 tons oil fuel. Engine replaced in 1992. Five further are based at the Naval School, Marín. A ketch (A 72) (52 tons and 22.84 m in length), a schooner (A 74) (16.8 m in length), a 90 tons ketch (A 75) launched in 1958 and formerly owned by the father of King Juan Carlos I and presented to the Naval School in 1993, an ex-yacht (A 76) and a yacht (A 77).

VERIFIED

JUAN SEBASTIÁN DE ELCANO   *3/2002, A Campanera i Rovira /* 0529022

4 TRAINING CRAFT (AXL)

Name	No	Builders	Commissioned
GUARDIAMARINA SALAS	A 82	Cartagena	10 May 1983
GUARDIAMARINA GODÍNEZ	A 83	Cartagena	4 July 1984
GUARDIAMARINA RULL	A 84	Cartagena	11 June 1984
GUARDIAMARINA CHEREGUINI	A 85	Cartagena	11 June 1984

Displacement, tons: 56 full load
Dimensions, feet (metres): 62 × 16.7 × 5.2 *(18.9 × 5.1 × 1.6)*
Speed, knots: 13
Complement: 15; 22 (A 81)
Radars: Navigation: Halcon 948; I-band.

Comment: Tenders to Naval School. A 81 has been decommissioned.

VERIFIED

GUARDIAMARINA CLASS   *2/2002, Diego Quevedo /* 0528943

TUGS

1 OCEAN TUG (ATA)

Name	No	Builders	Commissioned
MAHÓN (ex-*Circos*)	A 51	Astilleros Atlántico, Santander	1978

Displacement, tons: 1,450 full load
Dimensions, feet (metres): 134.5 × 38.1 × 18 *(41 × 11.6 × 5.5)*
Main machinery: 2 AESA/Sulzer 16ASV25/30 diesels; 7,744 hp(m) *(5.69 MW)*; 2 shafts
Speed, knots: 13. **Range, n miles:** 27,000 at 12 kt (A 52)
Complement: 33 (8 officers) plus 45 scientists
Guns: 2—12.7 mm MGs.
Radars: Navigation: 2 Racal Decca; I-band.

Comment: Built for Compania Hispano Americana de Offshore SA. Commissioned in the Navy 30 July 1981. Based at Ferrol.

UPDATED

MAHÓN   *7/2000, Adolfo Ortigueira Gil /* 0105651

1 OCEAN TUG (ATA)

Name	No	Builders	Commissioned
LA GRAÑA (ex-*Punta Amer*)	A 53 (ex-Y 119)	Astilleros Luzuriaga, San Sebastian	1982

Displacement, tons: 664 full load
Dimensions, feet (metres): 102.4 × 27.6 × 10.5 *(31.2 × 8.4 × 3.2)*
Main machinery: 1 diesel; 3,240 hp(m) *(2.38 MW)*; 1 Voith Schneider prop
Speed, knots: 13. **Range, n miles:** 1,750 at 12 kt
Complement: 28

Comment: Former civilian tug acquired by Navy on 20 October 1987. Now designated as ocean-going. Based at Cadiz.

VERIFIED

LA GRAÑA   *4/2003, Camil Busquets i Vilanova /* 0570983

28 COASTAL and HARBOUR TUGS (YTB/YTM/YTL)

No	Displacement tons (full load)	HP/speed	Commissioned
Y 116-117	422	1,500/12	1981
Y 118, Y 121-126	236	1,560/14	1989-91
Y 120 (ex-*Punta Roca*)	260	1,750/12	1973
Y 132, Y 140	70	200/8	1965-67
Y 141-142	229	800/11	1981
Y 143	133	600/10	1961
Y 144-145	195	2,030/11	1983
Y 147-148	87	400/10	1987/1999
Y 171-179	10	440/11	1982/1985

Comment: *Y 143* has a troop carrying capability. *Y 171-176* are pusher tugs for submarines. *Y 118, Y 121-126* have Voith Schneider propulsion.

UPDATED

Y 122   *5/2003, Camil Busquets i Vilanova /* 0570984

GOVERNMENT MARITIME FORCES

POLICE (GUARDIA CIVIL — MARITIME SERVICE)

Notes: Created by Royal decree on 22 February 1991 and owned by the Ministry of Interior. Bases at La Coruña, Barcelona, Valencia, Tarragona, Baleares, Cantabria, Almeira, Malaga, Murcia and, Algeciras. Personnel strength 1,000 (35 officers). The force has taken over the anti-terrorist role and some general patrol duties as a peacetime paramilitary organisation coming under the Ministry of Defence in war. In addition to the craft listed there are some 42 smaller craft (under 9 m). 18 BO 105, 8 BK-117 and 31 Eurocopter EC-135 helicopters are used for coastal patrols and are based at Tenerife, Seville, Valencia, Mallorca, Huesca, Logroño, Leon and La Coruña. All vessels are armed.

BK 117   *6/2002, Adolfo Ortigueira Gil /* 0528940

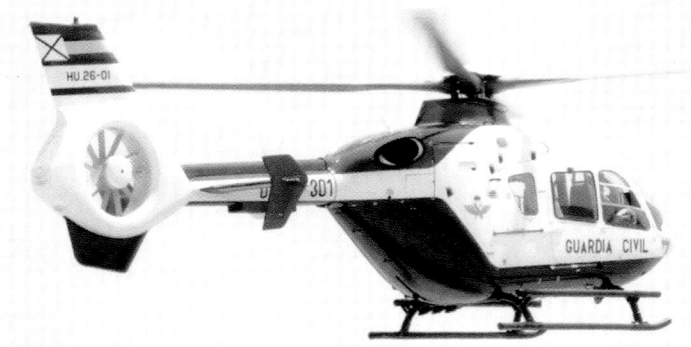

EC-135   *6/2004 *, Oris /* 1044563

1 IZAR IVP-22 CLASS (WPB)

SALEMA A 01

Displacement, tons: 52 full load
Dimensions, feet (metres): 80.4 × 19.6 × 5.9 *(24.5 × 5.96 × 1.8)*
Main machinery: 2 MAN diesels; 1,100 hp *(820 kW)*
Speed, knots: 20. **Range, n miles:** 400 at 12 kt
Complement: 8
Guns: 1—12.7 mm MG.
Radars: Navigation: I-band.

Comment: Built by Bazán, San Fernando. Steel hull. Commissioned on 24 June 1999 having been procured by Agriculture and Fisheries Ministry for operation by Guardia Civil. Hull lengthened in 2003 to facilitate operation of RIB. Based at Algeciras.

UPDATED

SALEMA   *9/2003, Adolfo Ortigueira Gil /* 0570966

3 RODMAN 82 CLASS (WPB)

RIO GUADIARO (ex-*Seriola*) A 02   **RIO PISUERGA** A 03   **RIO NALON** A 04

Displacement, tons: 93 full load
Dimensions, feet (metres): 85.3 × 19.4 × 4.3 *(26.0 × 5.9 × 1.3)*
Main machinery: 2 diesels; 1,400 hp *(1.04 MW)*; 2 waterjets
Speed, knots: 30. **Range, n miles:** 720 at 17 kt
Complement: 9
Guns: 1 LAG 40 mm grenade launcher.
Radars: Navigation: I-band.

Comment: Built in 2001 by Rodman, Vigo. A 02 based at Alicante, A 03 at Algeciras and A 04 at Asturias. A 02 purchased by Fisheries department.   *UPDATED*

RIO GUADIARO   *4/2002, Diego Quevedo /* 0528939

10 RODMAN 101 CLASS (WPB)

RIO PALMA A 05	**RIO GUADALAVIAR** A 10
RIO ANDARAX A 06	**RIO CABRIEL** A 11
RIO GUADALOPE A 07	**RIO CERVANTES** A 12
RIO ALMANZORA A 08	**RIO ARA** A 13
RIO NERVION A 09	**RIO ADAJA** A 14

Displacement, tons: 109 full load
Dimensions, feet (metres): 98.4 × 19.4 × 4.3 *(30.0 × 5.9 × 1.3)*
Main machinery: 2 Caterpillar 3412C diesels; 2,800 hp *(2.06 MW)*; 2 Hamilton waterjets
Speed, knots: 30. **Range, n miles:** 800 at 12 kt
Complement: 9
Guns: 1 LAG 40 mm grenade launcher.
Radars: Navigation: I-band.

Comment: GRP hull. Built by Rodman, Vigo and delivered in 2002 (A 05), 2003 (A 06-08), 2004 (A 09-13) and 2005 (A 14). A 05 A 06 and A 08 purchased by Agriculture and Fisheries Ministry. All operated by Guardia Civil.   *UPDATED*

RIO ALMANZORA   *5/2004 *, Diego Quevedo /* 1044567

13 RODMAN 55M CLASS (WPBF)

M 02-14

Displacement, tons: 15.7 full load
Dimensions, feet (metres): 54.1 × 12.5 × 2.3 *(16.5 × 3.8 × 0.7)*
Main machinery: 2 MAN D2848-LXE diesels; 1,360 hp(m) *(1 MW)* sustained; 2 Hamilton waterjets
Speed, knots: 35. **Range, n miles:** 500 at 25 kt
Complement: 7
Guns: 1—12.7 mm MG.
Radars: Surface search: Ericsson; I-band.

Comment: GRP hulls built by Rodman, Vigo. First five in service in 1992, three in 1993, six more in 1995-96. M 01 sunk in 2002. Known as Baltic class.   *VERIFIED*

M 14   *1/2003, Diego Quevedo /* 0528950

2 RODMAN 55 CANARIAS CLASS (WPBF)

TINEYCHEIDE M 15   ALMIRANTE DIAZ PIMIENTA M 16

Displacement, tons: 18.5
Dimensions, feet (metres): 57.1 × 12.5 × 2.6 (17.4 × 3.8 × 0.8)
Main machinery: 2 MAN D2848 LXE406 diesels; 2,300 hp (1.71 MW); 2 Hamilton waterjets
Speed, knots: 48. Range, n miles: 400 at 25 kt
Complement: 5
Guns: 1—12.7 mm MG.
Radars: Navigation: I-band.

Comment: GRP hull built by Rodman, Vigo. Purchased in 1999 by Canary Islands Agriculture and Fishery Department. Based at Lanzarote. Same class sold to Cyprus.
UPDATED

TINEYCHEIDE   8/1999, Rodman / 0570948

12 RODMAN 55HJ CLASS (PB)

RIO ARBA M 17	RIO LADRA M 23
RIO CAUDAL M 18	RIO CERVERA M 24
RIO BERNESGA M 19	RIO ARAGON M 25
RIO MARTIN M 20	RIO ALFAMBRA M 26
RIO GUADALOBON M 21	RIO ULLA M 27
RIO CEDENTA M 22	RIO JUCAR M 28

Displacement, tons: 20 full load
Dimensions, feet (metres): 55.8 × 12.5 × 2.9 (17.0 × 3.8 × 0.9)
Main machinery: 2 MAN D2848 LXE406 diesels; 2,300 hp (1.71 MW); 2 Hamilton waterjets
Speed, knots: 52. Range, n miles: 400 at 25 kt
Complement: 5
Radars: Navigation: I-band.

Comment: GRP hull built by Rodman, Vigo. Similar to Colimbo class of Spanish Customs. M 17-24 delivered in 2004 and M 25-28 in 2005.
NEW ENTRY

RIO ARBA   5/2004*, Diego Quevedo / 1044564

11 SAETA-12 CLASS (WPBF)

L 01-02   L 04-12

Displacement, tons: 14 full load
Dimensions, feet (metres): 39 × 12.5 × 2.3 (11.9 × 3.8 × 0.7)
Main machinery: 2 MAN D2848-LXE diesels; 1,360 hp(m) (1 MW) sustained; 2 Hamilton water-jets
Speed, knots: 38. Range, n miles: 300 at 25 kt
Complement: 4
Guns: 1—7.62 mm MG.
Radars: Surface search: Ericsson; I-band.

Comment: GRP hulls built by Bazán and delivered in 1993-97. Known as Aegean class. L 03 deleted in 2004 following an accident.
UPDATED

L 12   5/2003, Diego Quevedo / 0570964

RESEARCH SHIPS

Notes: Nine civilian research ships are owned by the Government Science and Technology Ministry and by the Agriculture Fishery and Food Ministry. Those operated by the Instituto Español de Oceanografía (IEO) are *Vizconde de Eza* (1,400 tons), *Cornide de Saavedra* (1,113 tons), *F P Navarro* (178 tons), *Odón de Buen* (64 tons), *Lura* (34 tons), *José Rioja* (32 tons) and *J M Navaz* (30 tons). Two further ships are expected in 2006. Those operated by CSIC are *Garcia del Cid* (539 tons) and *Mytilus* (170 tons). The ships operate in co-operation with the Spanish Navy ship *Hespérides* and the French research ship *Thalassa*. A new ship is to be delivered to CSIC in 2005.

VIZCONDE DE EZA   6/2003*, Adolfo Ortigueira Gil / 1044565

CORNIDE DE SAAVEDRA   6/2003*, Adolfo Ortigueira Gil / 1044566

CUSTOMS

Notes: Customs service is the responsibility of the Ministry of Treasure. All carry ADUANAS on ships' sides. Some of the larger vessels are armed with machine guns. Ships are based at 17 ports including Ceuta and Melilla in north Africa. There are also three MBB-105, one MBB-117 and one AS 365 Dauphin helicopters. Six CASA C-212 patrol aircraft were transferred to the Air Force in 1997 and are operated by the 37th Air Wing.

CASA C-212   3/2003, Adolfo Ortigueira Gil / 0570963

45 PATROL CRAFT (PB)

Name	Displacement tons (full load)	HP/speed	Commissioned
ÁGUILA	80	2,700/29	1974
ALBATROS II and III	85	2,700/29	1964-69
ALCA I and III	24	2,000/45	1987-88
ALCAUDON II/ALCOTÁN/FENIX	18.5	1,200/55	1997-99
ALCAVARÁN I-V	85	3,920/28	1984-87
COLIMBO II	17	2,400/50	1999-03
CORMORÁN/HJ 1/ COLIMBO III-IV	17	2,400/52	1986-03
GAVILÁN II-IV	65	3,200/26	1983-87
ARAO/GERIFALTE/ DÉCIMO ANIVERSARIO	46	2,366/35	2001/2003
HJA	12	2,200/55	1994
HJ III-X	20	2,300/50	1986-89
HALCÓN II-III	68	3,200/28	1980-83
IMP I-II	5	600/40	1989
IPP I and III	2	200/50	1989
MILANO II	15	2,000/50	1999
PETREL I	1,600	1,200/12	1994
VA II-V	23	1,400/27	1985

Comment: These craft are also listed as auxiliary ships of the Navy. Flagship is *Petrel I* for which replacement is under construction at Astilleros Gondan for delivery in 2006.
UPDATED

COLIMBO IV   7/2004*, Diego Quevedo / 1044568

VA II   *9/2004*, Adolfo Ortigueira Gil* / 1044569

PUNTA SALENAS   *8/2002, Adolfo Ortigueira Gil* / 0528948

ARAO   *7/2003, Rodman* / 0570976

SALVAMAR GADIR   *9/2004*, Adolfo Ortigueira Gil* / 1044570

PETREL I   *11/2003, Javier Somavilla* / 0570950

ESPERANZA DEL MAR   *9/2001, Adolfo Ortigueira Gil* / 0130140

MARITIME RESCUE, SAFETY AND LOGISTIC SUPPORT

Notes: These roles are discharged by two services: SASEMAR (Sociedad Estatal de Salvamento y Seguridad Marítima) is under the direction of the Merchant Marine but may come under a Coast Guard service in due course. It operates 11 salvage tugs (two more are being built), 38 small rescue craft, five harbour pollution craft and five Sikorsky S 61 helicopters. All ships are painted red with a white stripe on the hull. ISM (Instituto Social de la Marina) operates one specialised medical and logistic ship for support of fishing vessels. *Esperenza del Mar* (5,000 tons) is based at Las Palmas (Canary Islands) Naval Base. A second ship is to be based at Santander from 2005.

S 61   *8/2002, Adolfo Ortigueira Gil* / 0528949

Sri Lanka

Country Overview

Formerly known as Ceylon, the Democratic Socialist Republic of Sri Lanka gained independence in 1948. Situated off the southeast coast of India, from which it is separated by the Palk Strait and Gulf of Mannar, it has an area of 25,326 square miles and a coastline of 723 n miles with the Indian Ocean. The capital of Sri Lanka is Sri Jayavardhanapura (Kotte) while Colombo is the largest city and principal port. There are further ports at Trincomalee, Kankasanthurai and Galle. Territorial waters (12 n miles) are claimed. A 200 n mile EEZ has been claimed although the limits have only been partly defined by boundary agreements.

Headquarters Appointments

Commander of the Navy:
   Vice Admiral D W K Sandagiri, VSV, USP
Chief of Staff:
   Rear Admiral D S M Wijewickrama, RSP, USP
Director General, Operations:
   Rear Admiral S P Weerasekara, RWP, USP

Area Commanders

Commander Western Naval Area:
   Rear Admiral W K J Karannagoda, RSP, USP
Commander North Central Naval Area:
   Rear Admiral H S Rathnakeerthi, USP
Commander Northern Naval Area:
   Rear Admiral L D Dharmapriya, RSP, USP

Area Commanders — *continued*

Commander Eastern Naval Area:
   Commodore J H U Ranaweera, RWP, RSP, USP
Commander Southern Naval Area:
   Rear Admiral T M W K B Thennakoon, RSP, USP

Personnel

2005:

(a)   24,782 (1,387 officers) regulars
(b)   SLVNF: 3,283 (196 officers)
(c)   Reserve force (regular): 198 (7 officers)
(d)   Reserve force (volunteer): 31 (5 officers)

Bases

Navy HQ: Colombo.
Western Command HQ: Colombo port (other bases at Welisara and Kalpitiya).
Eastern Command HQ: Trincomalee port (other bases at Nilaweli and Thiriyaya, Naval Academy at Trincomalee).
Southern Command HQ: Galle port (other bases at Tangalle, Boossa training centre and Kirinda harbour).
Northern Command HQ: Kankasanthurai port (other bases at Madagal, Karainagar, Velerni Island, Mandathive Island, Nagadeepa Island and Pungudathive Island).
North Central Command HQ: Medawachchiya (other bases at Punewa training centre, Thalaimannar and Mannar Island).

Pennant Numbers

Pennant numbers were reviewed in 1996 and 2002.

Prefix to Ships' Names

SLNS.

DELETIONS

Amphibious Forces

2002   *Lihiniya*

PATROL FORCES

1 + 1 SUKANYA CLASS OFFSHORE PATROL VESSEL (PSOH)

Name	No	Builders	Launched	Commissioned
SAYURA (ex-*Saryu*)	P 620 (ex-54)	Hindustan SY, Vishakapatnam	16 Oct 1989	8 Oct 1991

Displacement, tons: 1,890 full load
Dimensions, feet (metres): 331.7 oa; 315 wl × 37.7 × 14.4
  (101.1; 96 × 11.5 × 4.4)
Main machinery: 2 SEMT-Pielstick 16 PA6 V 280 diesels;
  12,800 hp(m) *(9.41 MW)* sustained; 2 shafts
Speed, knots: 21. **Range, n miles:** 5,800 at 15 kt
Complement: 140 (15 officers)
Guns: 1 Bofors 40 mm/60. 2 China 14.5 mm (twin).
Radars: Surface search: Racal Decca 2459; I-band.
Navigation: Bharat1245; I-band.

Comment: Transferred from India and recommissioned on 9 December 2000. Plans to acquire a second refurbished ship have been reported.
UPDATED

SAYURA *10/2001, Chris Sattler /* 0130149

1 RELIANCE CLASS (PSOH)

Name	No	Builders	Commissioned
— (ex-*Courageous*)	P 621 (ex-WMEC 622)	Coast Guard Yard, Baltimore	8 Dec 1967

Displacement, tons: 1,129 full load
Dimensions, feet (metres): 210.5 × 34 × 10.5 *(64.2 × 10.4 × 3.2)*
Main machinery: 2 Alco 16V-251 diesels; 6,480 hp *(4.83 MW)* sustained; 2 shafts; LIPS cp props
Speed, knots: 18. **Range, n miles:** 6,100 at 14 kt; 2,700 at 18 kt
Complement: 75 (12 officers)
Guns: 1 Boeing 25 mm/87 Mk 38 Bushmaster; 200 rds/min to 6.8 km *(3.4 n miles).* 2—12.7 mm MGs.
Radars: Surface search: Hughes/Furuno SPS-73; I-band.
Helicopters: Platform for one medium.

Comment: Transferred to Sri Lanka on 24 June 2004. During 34 years in USCG service, underwent Major Maintenance Availability (MMA) in 1989. The exhausts for main engines, ship service generators and boilers were run in a vertical funnel which reduced flight deck size. Capable of towing ships up to 10,000 tons. *UPDATED*

RELIANCE CLASS *10/2002, M Mazumdar /* 0530032

1 JAYASAGARA CLASS (OFFSHORE PATROL VESSEL) (PB)

Name	No	Builders	Launched	Commissioned
JAYASAGARA	P 601	Colombo Dockyard	26 May 1983	9 Dec 1983

Displacement, tons: 330 full load
Dimensions, feet (metres): 130.5 × 23 × 7 *(39.8 × 7 × 2.1)*
Main machinery: 2 MAN 8L20/27 diesels; 2,180 hp(m) *(1.6 MW)* sustained; 2 shafts
Speed, knots: 15. **Range, n miles:** 3,000 at 11 kt
Complement: 52 (4 officers)
Guns: 2 China 25 mm/80 (twin). 2 China 14.5 mm (twin) MGs. 2—12.7 mm MGs. 2—40 mm AGL. 2—7.62 mm MGs.
Radars: Surface search: Anritsu RA 723; I-band.

Comment: Ordered from Colombo Dockyard on 31 December 1981. Second of class sunk by Tamil forces in September 1994. *UPDATED*

JAYASAGARA *6/2004 *, Sri Lanka Navy /* 1044193

5 SHANGHAI II (TYPE 062) CLASS
(FAST ATTACK CRAFT—GUN) (PB)

WEERAYA P 311 (ex-P 3141)	JAGATHA P 315 (ex-P 3146)	ABEETHA II P 316	EDITHARA II P 317	WICKRAMA II P 318

Displacement, tons: 139 full load
Dimensions, feet (metres): 127.3 × 17.7 × 5.2 *(38.8 × 5.4 × 1.6)*
Main machinery: 4 Type L12-180 diesels; 4,800 hp(m) *(3.53 MW);* 4 shafts
Speed, knots: 28. **Range, n miles:** 750 at 16 kt
Complement: 44
Guns: 4 (2 in P 311, P 315) Royal Ordnance GCM-A03 30 mm (2 (1 in P 311, P 315) twin).
  2—37 mm (twin) (P 311, P 315).
  4 China 14.5 mm (2 twin) MG.
  2—7.62 mm MGs.
  2—40 mm AGL (P 311, P 315).
Radars: Surface search: Koden MD 3220 Mk 2; I-band.
Navigation: Furuno 825 D; I-band.

Comment: Five transferred by China in 1971 of which four since decommissioned and *Weeraya* remains in service. Two further craft transferred in 1980 of which *Jagatha* remains in service. Three further craft (*Abeetha II*, *Edithara II* and *Wickrama II*) are modified craft with improved habitability but similar specifications. These were built at Qinxin Shipyard and commissioned on 11 June 2000.
UPDATED

WEERAYA *6/2001, Sri Lanka Navy /* 0130146

EDITHARA II *6/2003, Sri Lanka Navy /* 0570992

1 HAIQING (TYPE 037/1) CLASS
(FAST ATTACK CRAFT—GUN) (PC)

Name	No	Builders	Commissioned
PARAKRAMABAHU	P 351	Qingdao Shipyard	22 May 1996

Displacement, tons: 478 full load
Dimensions, feet (metres): 206 × 23.6 × 7.9 *(62.8 × 7.2 × 2.4)*
Main machinery: 4 PR 230ZC diesels; 4,000 hp(m) *(2.94 MW)*; 4 shafts
Speed, knots: 28. **Range, n miles:** 1,300 at 15 kt
Complement: 71
Guns: 4 China 37 mm/63 (2 twin) Type 76A; 4 China 14.5 mm (2 twin) MGs Type 69.
A/S mortars: 2 Type 87 6-tubed launchers.
Radars: Surface search: Anritsu RA 273; I-band.
Sonars: Stag Ear; hull mounted; active search and attack; high frequency.

Comment: Ordered in March 1994 from China and delivered on 13 December 1995. This is a Hainan hull with a similar armament to the Chinese vessels of that class. **VERIFIED**

PARAKRAMABAHU   6/2003, Sri Lanka Navy / 0570990

2 SAAR 4 CLASS (FAST ATTACK CRAFT—MISSILE) (PGG)

Name	No	Builders	Launched	Commissioned
NANDIMITHRA (ex-*Moledt*)	P 701	Israel Shipyard, Haifa	22 Mar 1979	May 1979
SURANIMALA (ex-*Komemiut*)	P 702	Israel Shipyard, Haifa	19 July 1978	Aug 1980

Displacement, tons: 415 standard; 450 full load
Dimensions, feet (metres): 190.6 × 25 × 8 *(58 × 7.8 × 2.4)*
Main machinery: 4 MTU/Bazán 16V 956 TB91 diesels; 15,000 hp(m) *(11.03 MW)* sustained; 4 shafts
Speed, knots: 32. **Range, n miles:** 1,650 at 30 kt; 4,000 at 17.5 kt
Complement: 75

Missiles: 3 Gabriel II; radar or TV optical guidance; semi-active radar plus anti-radiation homing to 36 km *(20 n miles)* at 0.7 Mach; warhead 75 kg.
Guns: 1 OTO Melara 3 in *(76 mm)*/62 compact; 85 rds/min to 16 km *(8.7 n miles)*; weight of shell 6 kg. Adapted for shore bombardment.
1—40 mm.
2 Rafael Typhoon 20 mm. 2—20 mm. 2—12.7 mm MGs. 2—40 mm AGL.
Radars: Air/surface search: Thomson-CSF TH-D 1040 Neptune; G-band; range 33 km *(18 n miles)* for 2 m² target.
Fire control: Selenia Orion RTN 10X; I-band.

Comment: Transferred from Israel and recommissioned on 9 December 2000. **VERIFIED**

NANDIMITHRA   10/2003, Hartmut Ehlers / 0570991

1 MOD SHANGHAI II CLASS
(FAST ATTACK CRAFT—GUN) (PB)

Name	No	Builders	Commissioned
RANARISI	P 322	Guijiang Shipyard	14 July 1992

Displacement, tons: 150 full load
Dimensions, feet (metres): 134.5 × 17.7 × 5.2 *(41 × 5.4 × 1.6)*
Main machinery: 4 diesels; 4,800 hp(m) *(3.53 MW)*; 4 shafts
Speed, knots: 29. **Range, n miles:** 750 at 16 kt
Complement: 44 (4 officers)
Guns: 2 China 37 mm/63 (1 twin) Type 76 or 2 Royal Ordnance GCM-AO3 30 mm (1 twin).
4 China 14.5 mm (twin) Type 69. 2—12.7 mm MGs. 2—40 mm AGL.
Radars: Surface search: Racal Decca; I-band.

Comment: Acquired from China in September 1991. Automatic guns and improved habitability. *Ranaviru* and *Ranasuvu* destroyed by Tamil guerrillas. **VERIFIED**

RANARISI   6/2003, Sri Lanka Navy / 0570988

3 HAIZHUI (TYPE 062/1G) CLASS (PB)

Name	No	Builders	Commissioned
RANAJAYA	P 330	Guijiang Shipyard	22 May 1996
RANADEERA	P 331	Guijiang Shipyard	22 May 1996
RANAWICKRAMA	P 332	Guijiang Shipyard	22 May 1996

Displacement, tons: 170 full load
Dimensions, feet (metres): 134.5 × 17.4 × 5.9 *(41 × 5.3 × 1.8)*
Main machinery: 4 Type L12-180A diesels; 4,400 hp(m) *(3.22 MW)* sustained; 4 shafts
Speed, knots: 21
Complement: 44
Guns: 4 China 37 mm/63 (2 twin). 4 China 25 mm/60 (2 twin). 2—12.7 mm MGs. 2—40 mm AGL.
Radars: Surface search: Anritsu 726UA; I-band.

Comment: Transferred from China by lift ship after delivery in 1995. **UPDATED**

RANAJAYA   9/1996, Sri Lanka Navy / 0080693

2 MOD HAIZHUI (LUSHUN) (TYPE 062/1G) CLASS
(FAST ATTACK CRAFT—GUN) (PB)

PRATHPA P 340   UDARA P 341

Displacement, tons: 212 full load
Dimensions, feet (metres): 149 × 21 × 5.6 *(45.5 × 6.4 × 1.7)*
Main machinery: 4 Type Z12V 190 BCJ diesels; 4,800 hp(m) *(3.53 MW)*; 4 shafts
Speed, knots: 28. **Range, n miles:** 750 at 16 kt
Complement: 44 (3 officers)
Guns: 2 China 37 mm/63 (1 twin) Type 76.
2 China 14.5 mm (1 twin) Type 82 MGs.
2—12.7 mm MGs.
2—40 mm AGL.
Radars: Surface search: Racal Decca RM 1070A; I-band.

Comment: Built at Lushun Dockyard, Darlin. Commissioned on 2 March 1998. Larger version of Haizhui class. **VERIFIED**

PRATHPA   3/1999, Sri Lanka Navy / 0080699

22 COLOMBO MK I/II/III CLASS
(FAST ATTACK CRAFT—GUN) (PBF)

P 410-419   P 420-424   P 450-451   P 490-492   P 494   P 497

Displacement, tons: 56 full load
Dimensions, feet (metres): 81.4 × 19.7 × 3.9 *(24.8 × 6 × 1.2)*
Main machinery: 2 MTU 12V 396 TE94 diesels; 4,570 hp(m) *(3.36 MW)*; ASD 16 surface drives
Speed, knots: 45. **Range, n miles:** 850 at 16 kt
Complement: 20
Guns: 1 Rafael Typhoon 23 mm. 1 Oerlikon 20 mm. 4—12.7 mm MGs. 8—7.62 mm MGs. 2—40 mm AGL.
Weapons control: Elop MSIS optronic director; Typhoon GFCS.
Radars: Surface search: Furuno FR 8250 or Corden Mk 2; I-band.

Comment: Built by Colombo Dockyard to the Israeli Shaldag design. First deliveries 1997. Mk I (P 450-451); Mk II (P 490-492, P 494, P 497); Mk III (P 410-424). P 493 and P 496 sunk in action in 2000. **VERIFIED**

COLOMBO MK II   11/1999 / 0080695

7 SHALDAG CLASS (FAST ATTACK CRAFT—GUN) (PBF)

P 470 (ex-P 491)   P 471 (ex-P 492)   P 472-476

Displacement, tons: 58 full load
Dimensions, feet (metres): 81.4 × 19.7 × 3.9 *(24.8 × 6 × 1.2)*
Main machinery: 2 Deutz 620 TB 16V or MTU 396 TE diesels; 5,000 hp(m) *(3.68 MW)*; 2 LIPS or MJP water-jets
Speed, knots: 50. **Range, n miles:** 700 at 32 kt
Complement: 20
Guns: 1 Rafael Typhoon 23 mm. 1—20 mm. 2—12.7 mm MGs. 6—7.62 mm MGs. 2—40 mm AGL.
Weapons control: ELOP compass optronic director. Typhoon GFCS.
Radars: Surface search: MD 3220 Mk II; I-band.

Comment: Originally launched in December 1989, first one acquired from the Israeli Shipyards, Haifa on 24 January 1996, second 20 July 1996 and third on 16 February 2000. Four more followed. Same hull used for the Colombo class. Also in service in Cyprus. *UPDATED*

SHALDAG CLASS   *6/2003, Sri Lanka Navy /* 0570989

5 SUPER DVORA MK II CLASS
(FAST ATTACK CRAFT—GUN) (PBF)

P 460 (ex-P 441)   P 461 (ex-P 496)   P 462 (ex-P 497)   P 464-465

Displacement, tons: 64 full load
Dimensions, feet (metres): 82 × 18.4 × 3.6 *(25 × 5.6 × 1.1)*
Main machinery: 2 MTU 12V 396 TE94 diesels; 4,570 hp(m) *(3.36 MW)*; ASD16 surface drives
Speed, knots: 50. **Range, n miles:** 700 at 30 kt
Complement: 20 (1 officer)
Guns: 1 Rafael Typhoon 20 mm or 2 Royal Ordnance GCM-AO3 30 mm (twin). 4—12.7 mm MGs. 6—7.62 mm MGs. 2—40 mm AGL.
Weapons control: Elop MSIS optronic director; Typhoon GFCS.
Radars: Surface search: Koden MD 3220; I-band.

Comment: First four ordered from Israel Aircraft Industries Ramta in early 1995. A slightly larger version of the Mk 1. First one delivered 5 November 1995, second 30 April 1996, third 22 June 1996 and fourth in December 1996. Two more were acquired on 9 June 1999 and 15 September 1999 respectively. The engines are an improved version of those fitted in the Israeli Navy craft. P 463 sunk in action in 2000. *VERIFIED*

SUPER DVORA Mk II   *11/1999 /* 0080697

4 SUPER DVORA MK I CLASS (FAST ATTACK CRAFT—GUN) (PBF)

P 440-443 (ex-P 465-468)

Displacement, tons: 54 full load
Dimensions, feet (metres): 73.5 × 18 × 5.8 *(22.4 × 5.5 × 1.8)*
Main machinery: 2 MTU 12V 396 TB93 diesels; 3,260 hp(m) *(2.4 MW)* sustained; 2 shafts
Speed, knots: 46. **Range, n miles:** 1,200 at 17 kt
Complement: 20 (1 officer)
Guns: 2 Oerlikon 20 mm. 2—12.7 mm MGs. 4—40 mm AGL.
Radars: Surface search: Decca 926; I-band.

Comment: Ordered from Israel Aircraft Industries in October 1986 and delivered in 1987-88. A more powerful version of the Dvora class. P 464 was destroyed by Tamil guerrillas on 29 August 1993 and P 463 on 29 August 1995. These craft have a deeper draft than the Mk II version with surface drives. *VERIFIED*

P 443 (ex-P 468)   *1995, Sri Lanka Navy /* 0130147

3 DVORA CLASS (FAST ATTACK CRAFT—GUN) (PBF)

P 401-403 (ex-P 420 (ex-P 453)-P 422 (ex-P 456))

Displacement, tons: 47 full load
Dimensions, feet (metres): 70.8 × 18 × 5.8 *(21.6 × 5.5 × 1.8)*
Main machinery: 2 MTU 12V 331 TC81 diesels; 2,605 hp(m) *(1.91 MW)* sustained; 2 shafts
Speed, knots: 36. **Range, n miles:** 1,200 at 17 kt
Complement: 18
Guns: 2 Oerlikon 20 mm. 2—12.7 mm MGs. 2—40 mm AGL.
Radars: Surface search: Anritsu 721UA; I-band.

Comment: 'Dvora' class, first pair of which transferred from Israel early 1984, next four in October 1986. Built by Israel Aircraft Industries. One sunk by Tamil forces on 29 August 1995 and second on 30 March 1996. One more deleted in late 1996. Not downgraded to patrol craft as previously reported but speed may have been reduced. *VERIFIED*

DVORA CLASS   *6/2003, A Sharma /* 0570995

3 SOUTH KOREAN KILLER CLASS
(FAST ATTACK CRAFT—GUN) (PBF)

P 404-406 (ex-P 430 (ex-P 473)-P 432 (ex-P 475))

Displacement, tons: 56 full load
Dimensions, feet (metres): 75.5 × 17.7 × 5.9 *(23 × 5.4 × 1.8)*
Main machinery: 2 MTU 396 TB93 diesels; 3,260 hp(m) *(2.4 MW)* sustained; 2 shafts
Speed, knots: 40
Complement: 18
Guns: 2 Oerlikon 20 mm. 4—12.7 mm MGs.
Radars: Surface search: Racal Decca 926; I-band.

Comment: 'South Korean Killer' class, built by Korea SB and Eng, Buson. All commissioned February 1988. Not downgraded to patrol craft as previously reported but speed may have been reduced. *UPDATED*

KILLER CLASS   *6/2003, A Sharma /* 0570994

5 TRINITY MARINE CLASS (FAST ATTACK CRAFT—GUN) (PBF)

P 480   P 481   P 483-485

Displacement, tons: 68 full load
Dimensions, feet (metres): 81.7 × 17.7 × 4.9 *(24.9 × 5.4 × 1.5)*
Main machinery: 2 MTU 12V 396 TE94 diesels; 4,570 hp(m) *(3.36 MW)* sustained; 2 water-jets
Speed, knots: 47. **Range, n miles:** 600 at 17 kt
Complement: 20
Guns: 2 Oerlikon 20 mm. 2—12.7 mm MGs. 2—7.62 mm MGs. 1 Grenade launcher.
Radars: Surface search: Raytheon R 1210; I-band.

Comment: All built at Equitable Shipyard, New Orleans. First three delivered in January 1997; second three in September 1997. All aluminium construction. P 482 sunk in action in 2000. *VERIFIED*

P 480   *1/1997, Sri Lanka Navy /* 0080701

5 COASTAL PATROL CRAFT (PB)

P 201	P 211	P 214	P 215	P 233

Displacement, tons: 21 full load
Dimensions, feet (metres): 46.6 × 12.8 × 3.3 *(14.2 × 3.9 × 1)*
Main machinery: 2 Detroit 8V-71TA diesels; 460 hp *(343 kW)*; 2 shafts
Speed, knots: 20. **Range, n miles:** 450 at 14 kt
Complement: 15 (1 officer)
Guns: 2—12.7 mm MGs.
Radars: Surface search: Furuno FR 2010; I-band.

Comment: Built by Colombo DY and commissioned in 1982 *(P 201)*, June 1986 *(P 211* and *P 214)*
and 1993 *(P 215)*. P 241 and P 243 decommissioned in 2001.

UPDATED

P 201   *6/2003, Sri Lanka Navy* / 0570987

4 CHEVERTON CLASS COASTAL PATROL CRAFT (PB)

P 221-224 (ex-*P 421-424*)

Displacement, tons: 22 full load
Dimensions, feet (metres): 55.9 × 14.8 × 3.9 *(17 × 4.5 × 1.2)*
Main machinery: 2 Detroit 8V-71TA diesels; 460 hp *(343 kW)*; 2 shafts
Speed, knots: 23. **Range, n miles:** 1,000 at 12 kt
Complement: 15
Guns: 1—12.7 mm MG.
Radars: Surface search: Racal Decca 110; I-band.

Comment: Used for general patrol duties. Built by Cheverton Workboats, UK and commissioned in
1977. One paid off in 1996.

UPDATED

P 222   *6/2004*, Sri Lanka Navy* / 1044192

3 SIMONNEAU CLASS (PBF)

P 250 (ex-*P 410,* ex-*P 483*)	P 252 (ex-*P 412,* ex-*P 485*)	P 253 (ex-*P 413,* ex-*P 486*)

Displacement, tons: 28 full load
Dimensions, feet (metres): 56.8 × 16.1 × 4.6 *(17.3 × 4.9 × 1.4)*
Main machinery: 2 MTU 12V 183 TE93 diesels; 2,300 hp(m) *(1.69 MW)*; 2 Hamilton water-jets
Speed, knots: 42. **Range, n miles:** 500 at 35 kt
Complement: 15
Guns: 1 DCN 20 mm. 2—12.7 mm MGs. 2—7.62 mm MGs.
Radars: Surface search: Racal Decca; I-band.

Comment: Simonneau Marine Type 508 craft. First pair completed in December 1993 and
shipped to Colombo in 1994. Second pair built in Colombo and completed in 1995. The plan to
build more was shelved. Downgraded to patrol craft on 1 August 2000. Speed likely to have
been reduced. *P 251* sunk in 2001.

UPDATED

SIMONNEAU CLASS   *6/2004*, Sri Lanka Navy* / 1044191

41 INSHORE PATROL CRAFT (PBR)

P 120-131	P 169	P 180-181
P 151-152	P 171-173	P 183-194
P 162-167	P 175	P 196-197

Displacement, tons: 10 full load
Dimensions, feet (metres): 44.3 × 9.8 × 1.6 *(13.5 × 3 × 0.5)*
Main machinery: 2 Cummins 6BTA5.9-M2; 584 hp *(436 kW)* sustained; 2 water-jets
Speed, knots: 33. **Range, n miles:** 330 at 25 kt
Complement: 5
Guns: 2—12.7 mm MGs. 2—7.62 mm MGs.
Radars: Surface search: Furuno 1941; I-band.

Comment: First pair (P 151, 152) built by TAOS Yacht Company, Colombo, and delivered in 1991.
Next 27 built by Blue Star Marine, Colombo and delivered between 1994 and 1998. There are
minor superstructure differences between the first pair and the rest. P 120-131 built by SLN,
IPCCP Welisara. More are being built. P 168, P 174 and P 182 sunk in action. P 101 and P 104
decommissioned.

UPDATED

INSHORE PATROL CRAFT   *6/2003, A Sharma* / 0570993

4 INSHORE PATROL CRAFT (TYPE BSM) (PBR)

P 145-147	P 149

Displacement, tons: 3.5 full load
Dimensions, feet (metres): 42 × 8 × 1.6 *(12.8 × 2.4 × 0.5)*
Main machinery: 2 outboard motors; 280 hp *(209 kW)*
Speed, knots: 30
Complement: 9
Guns: 1—12.7 mm MG.

Comment: Acquired in 1988 from Blue Star Marine. Similar to *P 111* but with outboard engines.
P 143 (ex-P 150) was mined and sunk in August 1991 and again sunk in 1995.

UPDATED

INSHORE PATROL CRAFT   *6/2004*, Sri Lanka Navy* / 1044190

4 INSHORE PATROL CRAFT (TYPE CME) (PBR)

P 110-113

Displacement, tons: 5 full load
Dimensions, feet (metres): 44 × 9.8 × 1.6 *(13.4 × 3 × 0.5)*
Main machinery: 2 Yamaha D 343 diesels; 730 hp(m) *(544 kW)* sustained; 2 shafts
Speed, knots: 26
Complement: 5
Guns: 1—12.7 mm MG. 1—7.62 mm MG.
Radars: Surface search: Furuno FR 1941; I-band.

Comment: Built by Consolidated Marine Engineers, Sri Lanka. First nine delivered in 1988; four more in 1992 and two more in 1994. Most of these craft have been destroyed.
UPDATED

P 111   *6/2003, Sri Lanka Navy /* 0570986

AMPHIBIOUS FORCES

1 YUHAI (WUHU-A) (TYPE 074) CLASS (LSM)

Name	No	Builders	Commissioned
SHAKTHI	L 880	China	22 May 1996

Displacement, tons: 799 full load
Dimensions, feet (metres): 191.6 × 34.1 × 8.9 *(58.4 × 10.4 × 2.7)*
Main machinery: 2 MAN 8 L 20/27 diesels; 4,900 hp(m) *(3.6 MW)*; 2 shafts
Speed, knots: 14. **Range, n miles:** 1,000 at 12 kt
Complement: 60
Military lift: 150 tons
Guns: 10—14.5 mm/93 (5 twin) MGs. 6—12.7 mm MGs.
Radars: Navigation: Racal Decca; I-band.

Comment: Transferred by lift ship from China arriving 13 December 1995. A planned second of class was built but not acquired.
VERIFIED

SHAKTHI   *5/1996, Sri Lanka Navy /* 0080710

2 LANDING CRAFT (LCM)

Name	No	Builders	Commissioned
RANAGAJA	L 839	Colombo Dockyard	15 Nov 1991
RANAVIJAYA	L 836	Colombo Dockyard	21 July 1994

Displacement, tons: 268 full load
Dimensions, feet (metres): 108.3 × 26 × 4.9 *(33 × 8 × 1.5)*
Main machinery: 2 Caterpillar diesels; 1,524 hp *(1.14 MW)*; 2 shafts
Speed, knots: 8. **Range, n miles:** 1,800 at 8 kt
Complement: 28 (2 officers)
Guns: 4 China 14.5 mm (2 twin) *(Ranagaja)* or 2 Oerlikon 20 mm. 2—12.7 mm MGs.
Radars: Navigation: Furuno FCR 1421; I-band.

Comment: Two built in 1983 and acquired in October 1985. Third of the class taken over by the Navy in September 1991 and a fourth in March 1992. *Kandula* sank in October 1992 and the hulk was salvaged in mid-December. *Pabbatha* sank in action in February 1998.
UPDATED

RANAVIJAYA   *6/2004*, Sri Lanka Navy /* 1044189

2 YUNNAN CLASS (TYPE 067) (LCU)

L 820   L 821

Displacement, tons: 135 full load
Dimensions, feet (metres): 93.8 × 17.7 × 4.9 *(28.6 × 5.4 × 1.5)*
Main machinery: 2 diesels; 600 hp(m) *(441 kW)*; 2 shafts
Speed, knots: 12. **Range, n miles:** 500 at 10 kt
Complement: 22 (2 officers)
Military lift: 46 tons
Guns: 4—14.5 mm (2 twin) MGs.
Radars: Surface search: Fuji; I-band.

Comment: First one acquired from China in May 1991, second in May 1995.
UPDATED

L 820   *6/2004*, Sri Lanka Navy /* 1044188

1 M 10 CLASS HOVERCRAFT (UCAC)

A 530

Displacement, tons: 18 full load
Dimensions, feet (metres): 67.6 × 28.9 *(20.6 × 8.8)*
Main machinery: 2 Deutz diesels; 1,050 hp(m) *(772 kW)*
Speed, knots: 40; 7 with cushion deflated. **Range, n miles:** 600 at 30 kt
Complement: 10
Military lift: 56 troops or 20 troops plus 2 vehicles
Guns: 1—12.7 mm MG.
Radars: Navigation: Furuno; I-band.

Comment: Acquired from ABS Hovercraft/Vosper Thornycroft in April 1998 and designated a Utility Craft Air Cushion (UCAC). Has a Kevlar superstructure. More may be ordered in due course.
VERIFIED

A 530   *4/1998, ABS Hovercraft /* 0033561

3 FAST PERSONNEL CARRIERS (LCP)

Name	No	Builders	Commissioned
HANSAYA (ex-*Offshore Pioneer*)	A 540	Sing Koon Seng, Singapore	20 Dec 1987
— (ex-*Lanka Rani*)	A 542	Kvaerner Fielistrand Ltd, Singapore	2000
— (ex-*Lanka Devi*)	A 543	Kvaerner Fielistrand Ltd, Singapore	2000

Displacement, tons: 154 full load
Dimensions, feet (metres): 98.4 × 36.8 × 7.7 *(30 × 11.2 × 2.3)*
Main machinery: 2 Paxman Vega 12 diesels; 1,800 hp(m) *(1.32 MW)*; 2 shafts
Speed, knots: 30
Complement: 15 (2 officers)
Military lift: 60 tons; 120 troops
Guns: 1 Oerlikon 20 mm. 2—12.7 mm MGs.
Radars: Navigation: Furuno FR 1012; I-band.

Comment: A 540 acquired in January 1986 from Aluminium Shipbuilders. Catamaran hull built as oil rig tender. Now used as fast transport. A 541 decommissioned in 2002.
UPDATED

HANSAYA   *2001, Sri Lanka Navy /* 0130145

Sudan

Country Overview

The Republic of Sudan is situated in north-eastern Africa. The largest country in Africa, it has an area of 967,500 square miles and is bordered to the north by Egypt, to the east by Eritrea and Ethiopia, to the south by Kenya, Uganda and the Democratic Republic of the Congo and to the west by the Central African Republic, Chad, and Libya. It has a 459 n mile coastline with the Red Sea. Khartoum is the capital and largest city and Port Sudan is the principal port. There are about 2,867 n miles of navigable waterways. Territorial waters (12 n miles) are claimed. An EEZ has not been claimed.

Headquarters Appointments

Commander, Naval Forces:
  Rear Admiral Abbas al-Sayyid Uthman

Personnel

(a)   2005: 1,750 officers and men
(b)   Voluntary service

Establishment

The Navy was established in 1962 to operate on the Red Sea coast and on the River Nile.

Bases

Port Sudan (HQ). Flamingo Bay (Red Sea), Khartoum (Nile), Kosti (Nile).

General

The Navy has low budgetary priority. There have been no reports of Iranian assistance since the mid-1990s. Patrols on the Nile have not been observed and activity levels are very low.

PATROL FORCES

7 ASHOORA I CLASS (INSHORE PATROL CRAFT) (PBR)

Displacement, tons: 3 full load
Dimensions, feet (metres): 26.6 × 8 × 1.6 *(8.1 × 2.4 × 0.5)*
Main machinery: 2 Yamaha outboards; 400 hp(m) *(294 kW)*
Speed, knots: 42
Complement: 2
Guns: 1—7.62 mm MG.

Comment: Acquired from Iran in 1992-94. Four based at Flamingo Bay and three at Khartoum but operational status is doubtful.   *UPDATED*

ASHOORA I   *1992, IRI Marine Industries /* 0080715

4 SEWART CLASS (INSHORE PATROL CRAFT) (PBR)

MAROUB 1161	FIJAB 1162	SALAK 1163	HALOTE 1164

Displacement, tons: 9.1 full load
Dimensions, feet (metres): 40 × 12.1 × 3.3 *(12.2 × 3.7 × 1)*
Main machinery: 2 GM diesels; 348 hp *(260 kW)*; 2 shafts
Speed, knots: 31
Complement: 6
Guns: 1—12.7 mm MG.

Comment: Transferred by Iranian Coast Guard in 1975. All are based at Flamingo Bay but operational status is doubtful.   *VERIFIED*

4 KURMUK (TYPE 15) CLASS (INSHORE PATROL CRAFT) (PBR)

KURMUK 502	QAYSAN 503	RUMBEK 504	MAYOM 505

Displacement, tons: 19.5 full load
Dimensions, feet (metres): 55.4 × 12.8 × 2.3 *(16.9 × 3.9 × 0.7)*
Main machinery: 2 diesels; 330 hp(m) *(243 kW)*; 2 shafts
Speed, knots: 16. **Range, n miles:** 160 at 12 kt
Complement: 6
Guns: 1 Oerlikon 20 mm; 2—7.62 mm MGs.

Comment: Delivered by Yugoslavia on 18 May 1989 for operations on the White Nile. All based at Flamingo Bay.   *UPDATED*

KURMUK   *1989, G Jacobs*

AUXILIARIES

Notes: (1) In addition there are two small miscellaneous support ships. *Baraka* 21 a water boat, and a Rotork 512 craft. Both restored with Iranian assistance.
(2) Five Type II LCVPs were delivered from Yugoslavia in 1991 and are based at Kosti.

2 SUPPLY SHIPS (AFL)

SOBAT 221	DINDER 222

Displacement, tons: 410 full load
Dimensions, feet (metres): 155.1 × 21 × 7.5 *(47.3 × 6.4 × 2.3)*
Main machinery: 3 Gray Marine diesels; 495 hp *(369 kW)*; 3 shafts
Speed, knots: 9
Complement: 15
Guns: 1 Oerlikon 20 mm. 2—12.7 mm MGs.

Comment: Two Yugoslav MFPD class LCTs transferred in 1969. Used for transporting ammunition, petrol and general supplies.   *VERIFIED*

Suriname

Country Overview

Formerly known as Dutch Guiana, the Republic of Suriname gained full independence in 1975. With an area of 63,037 square miles it has borders to the east with French Guiana, to the west with Guyana and to the south with Brazil; its 208 n mile coastline is on the Atlantic Ocean. The capital, largest city and chief port is Paramaribo. Territorial seas (12 n miles) and a fisheries zone (200 n miles) are claimed. There are further ports at Nieuw-Nickerie, Moengo,

Paranam and Smalkalden. Territorial waters (12 n miles) are claimed. A 200 n mile Exclusive Economic Zone (EEZ) has also been claimed but the limits are not defined.

Personnel

2005: 240 (25 officers)

Bases

Kruktu Tere

Aircraft

Two CASA C-212-400 Aviocar aircraft acquired for maritime patrol in 1998/99.

PATROL FORCES

3 RODMAN 101 CLASS (PB)

JARABAKKA P 01	SPARI P 02	GRAMORGU P 03

Displacement, tons: 72 full load
Dimensions, feet (metres): 98.4 × 19.4 × 4.3 *(30.0 × 5.9 × 1.3)*
Main machinery: 2 MTU 12V 2000 diesels; 2,900 hp *(2.16 MW)* sustained; 2 Hamilton 571 water-jets
Speed, knots: 26. **Range, n miles:** 800 at 12 kt
Complement: 9
Guns: 1—40 mm grenade launcher.
Radars: Surface search: 2 Furuno; I-band.

Comment: Ordered in December 1997, from Rodman, Vigo. First one delivered in February 1999, second and third on 3 July 1999. Carry a RIB with twin outboards. Operational status doubtful.

                                                    VERIFIED

SPARI   *9/1999, A Campanera i Rovira /* 0080716

5 RODMAN 55M CLASS (PBR)

P 04-08

Displacement, tons: 16 full load
Dimensions, feet (metres): 57.1 × 12.8 × 2.3 *(17.4 × 3.9 × 0.7)*
Main machinery: 2 MAN D2848-LXE diesels; 1,360 hp(m) *(1 MW)* sustained; 2 Hamilton water-jets
Speed, knots: 35. **Range, n miles:** 500 at 25 kt
Complement: 7
Guns: 1—12.7 mm MG.
Radars: Surface search: Furuno; I-band.

Comment: Ordered in December 1997 from Rodman, Vigo. First one delivered in October 1998, remainder in April 1999. Carry a RIB with a single outboard engine. Operational status doubtful.
VERIFIED

P 06
4/1999 / 0080717

Sweden
SVENSKA MARINEN

Country Overview

The Kingdom of Sweden is a constitutional monarchy occupying the eastern part of the Scandinavian Peninsula. With an area of 173,730 square miles, it is bordered to the north and west by Norway and to the north-east by Finland. It has a 1,740 n mile coastline with the Gulf of Bothnia, the Baltic Sea, the Öresund, Kattegatt, and Skagerrak. The country comprises the mainland and the islands of Gotland and Öland in the Baltic Sea. The capital and largest city is Stockholm which is also a leading port. Others include Göteborg, Malmö and Norrköping. Territorial seas (12 n miles) and an EEZ (200 n miles) are claimed.

Headquarters Appointments

Inspector General of the Navy:
    Rear Admiral Jörgen Ericsson

Diplomatic Representation

Defence Attaché in London and Dublin:
    Colonel Frank Fredriksson
Naval Attaché in Moscow:
    Commander Tage Andersson
Naval Attaché in Washington:
    Colonel Göran Boijsen
Defence Attaché in Paris:
    Lt Col Bertil Olson
Defence Attaché in Oslo:
    Colonel Tommy Johansson
Defence Attaché in Singapore:
    Captain Lennart Månsson
Navl Attaché in Berlin:
    Colonel Mats Hansson
Defence Attaché in Canberra:
    Commander Rolf Löthman
Defence Attaché in Sarajevo:
    Colonel Per-Ove Eriksson
Defence Attaché in Santiago:
    Colonel Thomas Karlsson
Defence Attaché in Helsinki:
    Colonel Ossi Koukkula
Defence Attaché in Athens and Budapest:
    Colonel Peter Backland

Diplomatic Representation — *continued*

Defence Attaché in Rome and Bern:
    Colonel Lars-Gunnar Nilsson
Defence Attaché in Riga:
    Colonel Ulf Persson
Defence Attaché in Warsaw:
    Colonel Krister Edvardson
Defence Attaché in Pretoria:
    Colonel Lars Ljung
Defence Attaché in Ankara:
    Colonel Håkan Swedin
Defence Attaché in Vienna:
    Lieutenant Colonel Lars-Gunnar Sjölin

Organisation

The Navy consists of the Fleet and the Amphibious Corps (ex-Coastal Artillery). The Navy has one submarine flotilla, two surface flotillas, one MCM flotilla, two amphibious regiments, two naval bases, one naval school and one amphibious school.

Coast Defence

The Amphibious Corps is divided into two amphibious regiments and is the basis for the Amphibious Brigade. The Amphibious Brigade consists of three amphibious battalions with 800 troops divided into two Coast Ranger Companies, one Mortar Company (81 mm) and one Amphibious Company with remote-controlled mines, underwater surveillance systems and short-range land-sea missiles (RBS 17 Hellfire). Each Amphibious Battalion has 35 combat boat 90H, 13 Combat Boat 90E, four small tenders, 26 group landing craft and 19 Klepper collapsible kayaks.

Personnel

(a)   2005: 8,140 including 2,500 officers, 540 civilians, 2,500 reserve officers and 2,600 national servicemen
(b)   10-17¼ months' national service

Bases

Muskö (Stockholm), Karlskrona.

Strength of the Fleet

Type	Active	Building (Planned)
Submarines—Patrol	7	—
Missile Corvettes	7	4
Fast Attack Craft—Missile	4	—
Inshore Patrol Craft	13	—
Minelayers/Command Ship	1	—
MCM Support Ship	1	—
Minelayers—Coastal	4	—
Minelayers—Small	2	—
Minesweepers/Hunters—Coastal	7	—
Minesweepers—Inshore	5	—
Sonobuoy Craft	4	—
LCMs	17	—
Survey Ships	2	—
Electronic Surveillance Ship	1	—
Transport Ships	1	—
Repair and Support Ships	8	—

DELETIONS

Patrol Forces

2002   *Lulgå, Halmstad, Jägaren, Dalarö, Sandhamm, Östhammar*
2003   *Väktaren, Snapphanen, Nynäshamn, Piteå, Ekeskär, Skifteskär, Gråskär, Altaskär, Hojskär, Bredskär, Hamnskär*
2004   *Kaparen, Styrbjörn, Starkodder, Tordön*

Mine Warfare Forces

2002   *Skramsösund*
2004   *Kalmarsund, Barösund*, M 508, M 513, M 515

Amphibious Forces

2004   *Bore, Räfsnäs, Heimdal,* LCU 208, LCU 225, LCU 228

Auxiliaries

2002   *Utö*
2003   26 LCU, *Urd, Hercules*

PENNANT LIST

Corvettes

K 11	Stockholm
K 12	Malmö
K 21	Göteborg
K 22	Gävle
K 23	Kalmar
K 24	Sundsvall
K 31	Visby
K 32	Helsingborg
K 33	Härnösand (bldg)
K 34	Nyköping (bldg)
K 35	Karlstad (bldg)

Patrol Forces

P 162	Spejaren
P 166	Tirfing
R 131	Norrköping
R 142	Ystad
77	Huvudskär
81	Tapper
82	Djärv

83	Dristig
84	Händig
85	Trygg
86	Modig
87	Hurtig
88	Rapp
89	Stolt
90	Ärlig
91	Munter
92	Orädd

Mine Warfare Forces

M 04	Carlskrona
M 11	Styrsö
M 12	Spårö
M 13	Skaftö
M 14	Sturkö
M 33	Viksten
M 71	Landsort
M 72	Arholma
M 73	Koster
M 74	Kullen

M 75	Vinga
M 76	Ven
M 77	Ulvön
MRF 01	Sökaren
MUL 12	Arkösund
MUL 13	Kalmarsund
MUL 15	Grundsund
MUL 18	Fårösund
MUL 19	Barösund
MUL 20	Furusund
B 01	Ejdern
B 02	Krickan
B 03	Svärtan
B 04	Viggen

Auxiliaries

A 201	Orion
A 212	Ägir
A 213	Nordanö
A 214	Belos III
A 241	Urd
A 247	Pelikanen

A 248	Pingvinen
A 251	Achilles
A 253	Hermes
A 262	Skredsvik
A 265	Visborg
A 270	Trossö
A 271	Gålö
A 322	Heros
A 324	Hera
A 343	Sleipner
A 344	Loke

Training ships

M 67	Nämndö
S 01	Gladan
S 02	Falken

SUBMARINES

Notes: (1) The Swedish requirement for two new submarines to be delivered from 2010 is being taken forward via the 'Viking' project. This had been a bilateral programme with Denmark, following the withdrawal of Norway at the end of Project Definition Phase (PDP) Step 1 on 13 June 2003, but in wake of approval of the 2005-09 Defence Plan by the Danish parliament on 10 June 2004, Denmark has also decided to end participation. PDP Step 2, the contract for which was signed on 6 October 2003, is to be completed in 2005 when decisions on whether to progress to the manufacturing stage will be made. It remains to be seen whether Sweden will continue with the project alone or whether programme costs might be shared with another partner. Nations holding observer status include Singapore and Poland.

(2) *Näcken* was decommissioned from the Swedish Navy in 2001. Subsequently, she was leased to the Danish Navy and renamed *Kronborg*. She was returned to Sweden in 2004 and is likely to be sold or scrapped.

2 VÄSTERGÖTLAND (A 17) CLASS (SSK)

Name	No	Builders	Laid down	Launched	Commissioned
VÄSTERGÖTLAND	—	Kockums, Malmö	10 Jan 1983	17 Sep 1986	27 Nov 1987
HÄLSINGLAND	—	Kockums, Malmö	1 Jan 1984	31 Aug 1987	20 Oct 1988

Displacement, tons: 1,070 surfaced; 1,143 dived
Dimensions, feet (metres): 159.1 × 20 × 18.4 *(48.5 × 6.1 × 5.6)*
Main machinery: Diesel-electric; 2 Hedemora V12A/15 diesels; 2,200 hp(m) *(1.62 MW)*; 1 Jeumont Schneider motor; 1,800 hp(m) *(1.32 MW)*; 1 shaft; LIPS prop
Speed, knots: 10 surfaced; 20 dived
Complement: 27 (5 officers)

Torpedoes: 6—21 in *(533 mm)* tubes. 12 FFV Type 613; anti-surface; wire-guided; passive homing to 20 km *(10.8 n miles)* at 45 kt; warhead 240 kg. Swim-out discharge.
3—15.75 in *(400 mm)* tubes. 6 FFV Type 431/451; anti-submarine; wire-guided; active/passive homing to 20 km *(10.8 n miles)* at 25 kt; warhead 45 kg shaped charge or a small charge anti-intruder version is available.
Mines: 12 Type 47 swim-out mines in lieu of torpedoes.
Countermeasures: ESM: Argo AR-700-S5; or Condor CS 3701; intercept.
Weapons control: Ericsson IPS-17 (Sesub 900A) TFCS.
Radars: Navigation: Terma; I-band.
Sonars: Atlas Elektronik CSU 83; hull-mounted; passive search and attack; medium frequency.
Flank array; passive search; low frequency.

Programmes: Design contract awarded to Kockums, Malmö on 17 April 1978. Contract for construction of these boats signed 8 December 1981. Kockums built midship section and carried out final assembly while Karlskrona built bow and stern sections. A second pair of boats have been upgraded with AIP propulsion.
Structure: Single hulled with an X type rudder/after hydroplane design. Reported that SSM were considered but rejected as non-cost-effective weapons in the context of submarine operations in the Baltic. Equipped with Pilkington Optronics CK 38 electro-optic search periscope, enhanced with night vision capability. Diving depth 300 m *(984 ft)*. Anechoic coating.
Operational: To be decommissioned on 1 January 2006.

UPDATED

VÄSTERGÖTLAND

6/2003, E & M Laursen / 0572635

VÄSTERGÖTLAND

6/2003, E & M Laursen / 0572634

VÄSTERGÖTLAND

3/2004*, John Brodie / 1043520

2 SÖDERMANLAND (A 17) CLASS (SSK)

Name	No	Builders	Laid down	Launched	Commissioned
SÖDERMANLAND	—	Kockums, Malmö	2 Feb 1985	12 Apr 1988	21 Apr 1989
ÖSTERGÖTLAND	—	Kockums, Malmö	15 Oct 1985	9 Dec 1988	10 Jan 1990

Displacement, tons: 1,500 surfaced; 1,600 dived
Dimensions, feet (metres): 198.5 × 20 × 18.4
  (60.5 × 6.1 × 5.6)
Main machinery: Diesel-Stirling-electric; 2 Hedemora V12A/15 diesels; 2,200 hp(m) *(1.62 MW)*; 2 Kockums Stirling Mk III AIP; 204 hp *(150 kW)*; 1 Jeumont Schneider motor; 1,800 hp (m) *(1.32 MW)*; 1 shaft; LIPS prop
Speed, knots: 10 surfaced; 20 dived
Complement: 27 (5 officers)

Torpedoes: 6—21 in *(533 mm)* tubes. 12 FFV Type 613; anti-surface; wire-guided; passive homing to 20 km *(10.8 n miles)* at 45 kt; warhead 240 kg. Swim-out discharge.
  3—15.75 in *(400 mm)* tubes. 6 FFV Type 431/451; anti-submarine; wire-guided; active/passive homing to 20 km *(10.8 n miles)* at 25 kt; warhead 45 kg shaped charge or a small charge anti-intruder version is available.
Mines: 12 Type 47 swim-out mines in lieu of torpedoes.
Countermeasures: ESM: Argo AR-700-S5; or Condor CS 3701; intercept.
Weapons control: Ericsson IPS-17 (Sesub 900A) TFCS.
Radars: Navigation: Terma; I-band.
Sonars: Atlas Elektronik CSU 83; hull-mounted; passive search and attack; medium frequency.
  Flank array; passive search; low frequency.

Programmes: Design contract awarded to Kockums, Malmö on 17 April 1978. Contract for construction of these boats signed 8 December 1981. Kockums built midship section and carried out final assembly while Karlskrona built bow and stern sections.
Modernisation: Modernised variants of the Västergötland class. Mid-life refit of *Södermanland* began at Kockums in late 2000 and included the installation of Air Independent Propulsion (Stirling Mk 3 AIP) by the insertion of a 12 m plug in the pressure hull. Other work included the installation of a pressurised diver's lock-out in the base of the sail to facilitate special forces operations. The refit also included a new climate control system. Thales Optronics CK 038 periscope has been upgraded with a thermal imaging camera and an improved image intensifier. Communications may be upgraded in the future to improve interoperability. A new active sonar suite, Subac, is to be installed by 2008.
Structure: Single hulled with an X type rudder/after hydroplane design. Diving depth 300 m *(984 ft)*. Anechoic coating.
Operational: *Södermanland* relaunched on 8 September 2003 and, after six-months sea trials, returned to service in mid-2004, *Östergötland* was relaunched on 3 September 2004 and is to return to service in 2005.

UPDATED

SÖDERMANLAND

6/2004*, Kockums / 1043515

SÖDERMANLAND

6/2004*, Kockums / 1043514

SÖDERMANLAND

6/2004*, Kockums / 1043513

3 GOTLAND (A 19) CLASS (SSK)

Name	No	Builders	Laid down	Launched	Commissioned
GOTLAND	—	Kockums, Malmö	20 Nov 1992	2 Feb 1995	2 Sep 1996
UPPLAND	—	Kockums, Malmö	14 Jan 1994	9 Feb 1996	1 May 1997
HALLAND	—	Kockums, Malmö	21 Oct 1994	27 Sep 1996	1 Oct 1997

Displacement, tons: 1,494 surfaced; 1,599 dived
Dimensions, feet (metres): 198.2 × 20.3 × 18.4
   (60.4 × 6.2 × 5.6)
Main machinery: Diesel-stirling-electric; 2 MTU diesels;
   2 Kockums V4-275R Stirling AIP; 204 hp(m) *(150 kW)*;
   1 Jeumont Schneider motor; 1 shaft; LIPS prop
Speed, knots: 10 surfaced; 20 dived
Complement: 27 (5 officers)

Torpedoes: 4—21 in *(533 mm)* bow tubes; 12 FFV Type
   613/62; anti-surface; wire-guided; passive homing to 20 km
   (10.8 n miles) at 45 kt; warhead 240 kg or Bofors Type 62
   (2000); wire-guided; active/passive homing to 50 km *(27 n
   miles)* at 20-50 kt; warhead 250 kg. Swim-out discharge.
   2—15.75 in *(400 mm)* bow tubes; 6 Swedish Ordnance Type
   432/451; anti-submarine; wire-guided; active/passive
   homing to 20 km *(10.8 n miles)* at 25 kt; warhead 45 kg.
   Shaped charge or a small charge anti-intruder version.
Mines: 12 Type 47 swim-out mines in lieu of torpedoes.
Countermeasures: ESM: Racal THORN Manta S; radar warning.
Weapons control: CelsiusTech IPS-19 (Sesub 940A); TFCS.
Radars: Navigation: Terma Scanter; I-band.
Sonars: STN/Atlas Elektronik CSU 90-2; hull-mounted; bow,
   flank and intercept arrays; passive search and attack.

Programmes: In October 1986 a research contract was
   awarded to Kockums for a design to replace the Sjöormen
   class. Ordered on 28 March 1990.
Modernisation: A new active sonar suite, Subac, is to be installed
   by 2008.
Structure: The design has been developed on the basis of the
   Type A 17 series but this class is the first to be built with Air
   Independent Propulsion as part of the design. This type of AIP
   runs on liquid oxygen and diesel in a helium environment.
   Space has been reserved to fit two more V4-275R engines in
   due course. Single electro-optic periscope. The periscope is
   the only hull penetrating mast. Anechoic coatings are being
   applied. The four 21 in torpedo tubes are mounted over the
   smaller 15.75 in tubes. The smaller tubes can be tandem-
   loaded with two torpedoes per tube.
Operational: Reported as being able to patrol at 5 kt for several
   weeks without snort charging. The Type 47 mine swims out
   to a predetermined position before laying itself on the bottom.
   Gotland is to participate in exercises with the USN on both the
   east and west coasts of the USA during 2005.

UPDATED

UPPLAND   *6/2003, L-G Nilsson /* 0572636

UPPLAND   *7/2004*, E & M Laursen /* 1043523

HALLAND   *8/2004*, L-G Nilsson /* 1043522

1 MIDGET SUBMARINE (SSW)

SPIGGEN II

Displacement, tons: 17 dived
Dimensions, feet (metres): 34.8 × 5.6 × 4.6 *(10.6 × 1.7 × 1.4)*
Main machinery: 1 Volvo Penta diesel; 1 shaft
Speed, knots: 5 dived; 6 surfaced
Complement: 4

Comment: Built by Försvarets Materielverk and commissioned
   on 19 June 1990. Has an endurance of 14 days and a diving
   depth of 100 m *(330 ft)*, and is used as a target for ASW
   training. Refitted by Kockums and back in service in
   December 1996.

VERIFIED

SPIGGEN II
8/1998, Per Körnefeldt / 0050201

CORVETTES

1 + 4 VISBY CLASS (FSGH)

Name	No	Builders	Laid down	Launched	Commissioned
VISBY	K 31	Karlskronavarvet	Dec 1996	8 June 2000	Jan 2005
HELSINGBORG	K 32	Karlskronavarvet	June 1997	27 June 2003	July 2005
HÄRNÖSAND	K 33	Karlskronavarvet	Dec 1997	16 Dec 2004	Jan 2006
NYKÖPING	K 34	Karlskronavarvet	June 1998	Sep 2004	July 2006
KARLSTAD	K 35	Karlskronavarvet	Dec 1999	Nov 2005	Jan 2007

Displacement, tons: 620 full load
Dimensions, feet (metres): 236.2 × 34.1 × 8.2
  (72 × 10.4 × 2.5)
Main machinery: CODOG; 4 AlliedSignal TF 50A gas turbines;
  21,760 hp(m) *(16 MW)*; 2 MTU 16V N90 diesels; 3,536 hp
  (m) *(2.6 MW)*; 2 Kamewa 125 water-jets; bow thruster
Speed, knots: 35; 15 (diesels)
Complement: 43 (6 officers)

Missiles: SSM: 8 RBS 15 Mk II (Batch 2) inertial guidance; active
  radar homing to 110 km *(54 n miles)* at 0.8 Mach; warhead
  150 kg.
Guns: 1 Bofors 57 mm/70 SAK Mk 3 ❶. 220 rds/min to 17 km
  (9.3 n miles); weight of shell 2.4 kg.
Torpedoes: 4 fixed 400 mm tubes ❷. Type 43/45 anti-
  submarine.
A/S mortars: Saab Alecto 601 127 mm rocket-powered
  grenades ❸; range 1,200 m.
Mines: Can be carried.
Countermeasures: Decoys: Chaff launcher ❹. Decoys can also
  be fired from the ASW mortar.
  MCMV: STN Atlas Seafox Combat (C) sonar/TV sensor; range
  500 m at 6 kt; shaped charge.
ESM/ECM: Condor Systems CS 701; intercept and jammer.
Combat data systems: CelsiusTech 9LV Mk 3E CETRIS with
  Link.
Weapons control: Optronic director.
Radars: Air/surface search: Ericsson Sea Giraffe AMB 3D ❺;
  G-band.
Surface search: CelciusTech Pilot; I-band.
Fire control: CEROS 200 Mk 3 ❻; I/J-band.
Sonars: Computing Devices Canada (CDC) Hydra; bow mounted
  active high frequency plus passive towed array and VDS
  active.

Helicopters: 1 Agusta A 109M ❼.

Programmes: Order for first two with an option for two more on
  17 October 1995. Second pair ordered 17 December 1996
  and third pair in mid-1999. However, due to cost overruns,
  order reduced on 9 October 2001 to five ships.
Structure: Stealth features developed from the trials vessel
  Smyge but without the twin hull design for which this ship was
  considered too large. The hull is of Fibre Reinforced Plastic used in a sandwich
  construction and the superstructure is covered with RAM. A

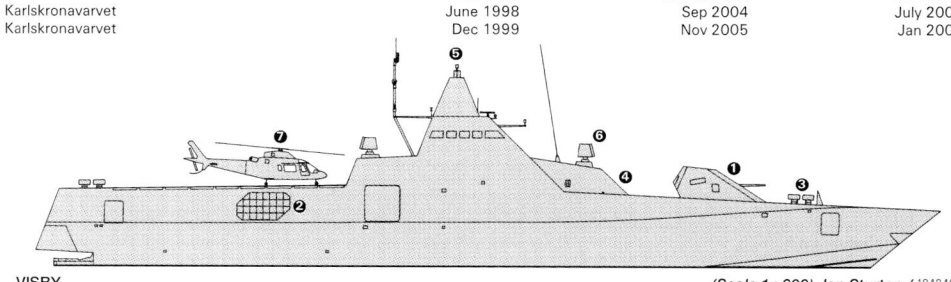

VISBY   *(Scale 1 : 600), Ian Sturton /* 1043484

HELSINGBORG   *6/2003, L-G Nilsson /* 0572616

Double Eagle ROV with active sonar is carried in the MCM role
as well as expendable mini torpedoes for mine
countermeasures. There is provision for a SAM system to be
installed.
Operational: The first of class, was launched at Karlskrona
  shipyard on 8 June 2000 but delays in outfitting schedule led
  to 10-month slippage of contractor's sea trials until

7 December 2001. Initially, ships fitted with 57 mm gun, A/S
mortars and torpedo tubes. The combat system installed from
late 2002 followed by trials from late 2003. *Visby* is planned
to enter operational service in early 2005 followed by the
other ships at six-month intervals. The first helicopter-fitted
ship is to be *Härnösand*.

UPDATED

VISBY   *6/2002, Michael Nitz /* 0525009

HELSINGBORG   *6/2003, L-G Nilsson /* 0576438

4 GÖTEBORG CLASS (FSG)

Name	No	Builders	Laid down	Launched	Commissioned
GÖTEBORG	K 21	Karlskronavarvet	10 Feb 1987	14 Apr 1989	15 Feb 1990
GÄVLE	K 22	Karlskronavarvet	21 Mar 1988	23 Mar 1990	1 Feb 1991
KALMAR	K 23	Karlskronavarvet	21 Nov 1988	1 Nov 1990	1 Sep 1991
SUNDSVALL	K 24	Karlskronavarvet	20 Nov 1989	29 Nov 1991	7 July 1993

Displacement, tons: 300 standard; 399 full load
Dimensions, feet (metres): 187 × 26.2 × 6.6
  (57 × 8 × 2)
Main machinery: 3 MTU 16V 396 TB94 diesels; 8,700 hp(m)
  (6.4 MW) sustained; Kamewa 80562-6 water-jets. Bow
  thruster in K 22
Speed, knots: 30
Complement: 36 (7 officers) plus 4 spare berths

Missiles: SSM: 8 Saab RBS 15 Mark II (4 twin) launchers **❶**;
  inertial guidance; active radar homing to 110 km *(59.4 n
  miles)* at 0.8 Mach; warhead 150 kg.
Guns: 1 Bofors 57 mm/70 Mk 2 **❷**; 220 rds/min to 17 km *(9.3 n
  miles)*; weight of shell 2.4 kg.
  1 Bofors 40 mm/70 (stealth dome in K 22) **❸**; 330 rds/min to
  12.5 km *(6.8 n miles)*; weight of shell 0.96 kg.
Torpedoes: 4—15.75 in *(400 mm)* tubes can be fitted **❹**.
  Swedish Ordnance Type 43/45; anti-submarine.
A/S mortars: 4 Saab 601 **❺** 9-tubed launchers; range 1,200 m;
  shaped charge.
Depth charges: On mine rails.
Mines: Minelaying capability.
Countermeasures: Decoys: 4 Philips Philax fixed launchers; IR
  flares and chaff grenades. A/S mortars have also been
  adapted to fire IR/chaff decoys.
ESM: Condor; intercept.
Combat data systems: CelsiusTech 9LV Mk 3 SESYM. Datalink.
Weapons control: 2 Bofors Electronics 9LV 200 Mk 3 Sea Viking
  or Signaal IRST (K 24) optronic directors. Bofors Electronics
  9LV 450 GFCS. RC1-400 MFCS. 9AU-300 ASW control
  system with AQS 928G/SM sonobuoy processor. Bofors
  9EW 400 EW control.
Radars: Air/surface search: Ericsson Sea Giraffe 150 HC **❻**;
  G-band.
  Navigation: Terma PN 612; I-band.
  Fire control: 2 Bofors Electronics 9GR 400 **❼**; I/J-band.
Sonars: Hydra multisonar system (K 22) **❽**; bow-mounted active
  high-frequency plus passive towed array and active VDS.
  Thomson Sintra TSM 2643 Salmon (K 21, K 23-24); VDS;
  active search; medium frequency.
  Simrad SA 950; hull-mounted; active attack.
  STN Atlas passive towed array; low frequency.

Programmes: Ordered 1 December 1985 as replacements for
  Spica I class.

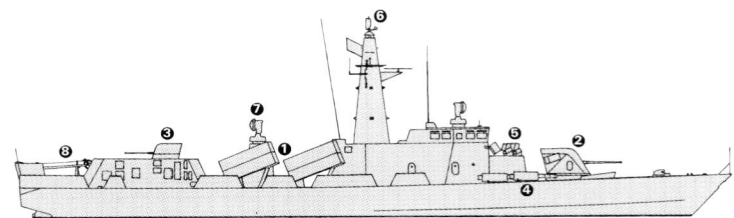

GÄVLE   *(Scale 1 : 600), Ian Sturton /* 0131370

SUNDSVALL   *8/2004*, B Prézelin /* 1043521

Modernisation: *Gävle* refitted to accommodate Hydra towed
array/VDS aft and 40 mm gun with stealth dome. Bridge
wings removed and topmast modified. Similar upgrade
expected for rest of class but timetable has not been
confirmed. An IRST director is fitted aft of the mast for trials in
K 24.

Structure: Efforts have been made to reduce radar and IR
signatures.

UPDATED

GÄVLE   *6/2004*, B Sullivan /* 1043546

2 STOCKHOLM CLASS (FSG)

Name	No	Builders	Laid down	Launched	Commissioned
STOCKHOLM	K 11	Karlskronavarvet	1 Aug 1982	24 Aug 1984	22 Feb 1985
MALMÖ	K 12	Karlskronavarvet	14 Mar 1983	22 Mar 1985	10 May 1985

Displacement, tons: 350 standard; 372 full load
Dimensions, feet (metres): 164 × 24.6 × 10.8
 (50 × 7.5 × 3.3)
Main machinery: CODAG; 1 Allied Signal TF50A gas turbine;
 5,440 hp(m) *(4.0 MW)* sustained; 2 MTU 16V 396 TB94
 diesels; 5,277 hp(m) *(3.9 MW)* sustained; 3 shafts; Kamewa
 props
Speed, knots: 32 gas; 20 diesel
Complement: 33 (7 officers)

Missiles: SSM: 8 Saab RBS 15 Mk II (4 twin) launchers ❶; inertial
 guidance; active radar homing to 110 km *(54 n miles)* at 0.8
 Mach; warhead 150 kg.
Guns: 1 Bofors 57 mm/70 Mk 2 ❷; 220 rds/min to 13.5 km
 (7.3 n miles); weight of shell 2.4 kg.
Torpedoes: 4—15.75 in *(400 mm)* tubes; Swedish Ordnance
 Type 43/45; anti-submarine can be fitted.
A/S mortars: 4 Saab 601 ❸ 9-tubed launchers; range 1,200 m;
 shaped charge.
Countermeasures: Decoys: 2 quadruple Bofors 57 mm rocket
 launchers for chaff/illumination ❹.
ESM: Condor CS-5460; intercept and warning.
ECM: to be fitted.
Combat data systems: SAAB Tech 9LV Mk 3E Cetris; datalink
Weapons control: Philips 9LV 300 GFCS including a 9LV 100
 optronic director and laser range-finder ❺.
Radars: Air/surface search: Ericsson Sea Giraffe 50HC ❻;
 G-band.
 Navigation: Terma Scanter; I-band.
 Fire control: Philips 9LV 200 Mk 3 ❼; J-band.
Sonars: Simrad SA 950; hull-mounted; active attack.
 Thomson Sintra TSM 2642 Salmon ❽; VDS; search; medium
 frequency.

Programmes: Orders placed in September 1981. Developed
 from Spica II class.
Modernisation: RBS 15 missile upgraded to Mk II from 1994.
 Improved A/S mortar fitted in 1998-99. Extensive mid-life
 upgrade carried out 1999-2002. Modernisation included
 removal of the 21 in torpedo tubes and the aft 40 mm
 mounting and modification of the superstructure to reduce
 radar and IR signatures. The bridge wings have been removed
 and a pylon mast has replaced a lattice structure. Upgrades
 include a new propulsion system, combat data system and
 EW systems. The decoys are situated on either side of the gun
 turret. Both ships are to be fitted with CDC Hydra sonar.
Operational: Both ships are expected to remain in service until
 2015.

UPDATED

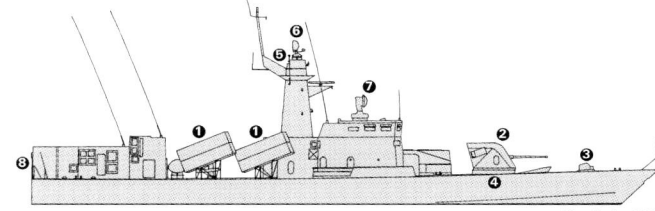

STOCKHOLM

(Scale 1 : 600), Ian Sturton / 0530053

MALMÖ

6/2004, Michael Nitz /* 1043519

STOCKHOLM

6/2004, Michael Nitz /* 1043518

SHIPBORNE AIRCRAFT

Numbers/Type: 8 Agusta A 109M.
Operational speed: 152 kt *(280 km/h).*
Service ceiling: 16,500 ft *(5,029 m).*
Range: 447 n miles *(827 km).*
Role/Weapon systems: Military version of A 109E with Arrius 2K2 engine. Swedish Armed Forces to receive 20 from September 2002 of which eight are to be 'navalised' for operation from Visby class and from shore bases. ASW and ASV roles.

VERIFIED

A 109E (USCG livery)   *2002, USCG /* 0102391

Numbers/Type: 5 Agusta-Bell 206A JetRanger (HKP-6B).
Operational speed: 115 kt *(213 km/h).*
Service ceiling: 13,500 ft *(4,115 m).*
Range: 368 n miles *(682 km).*
Role/Weapon systems: Operated from shore bases in ASW and surface search roles. To be replaced by A 109M. Weapons: ASW; four Type 11 or depth charges.

UPDATED

AB 206   *2000, Andreas Karlsson, Swedish Defence image /* 0106563

LAND-BASED MARITIME AIRCRAFT (FRONT LINE)

Notes: (1) In addition 11 AS 332 Super Puma helicopters are used for SAR.
(2) SH-37 Viggen aircraft are Air Force operated and can be used for maritime strike.

Numbers/Type: 7/7 Boeing 107-II-15/Kawasaki KV 107-II.
All to HKP-4C/D standard.
Operational speed: 137 kt *(254 km/h).*
Service ceiling: 8,500 ft *(2,590 m).*
Range: 300 n miles *(555 km).*
Role/Weapon systems: ASW and surface search helicopter; updated with avionics, TM2D engines, radar, sonar, datalink for SSM targeting, MARIL 920 combat information system. The Nordic Standard helicopter is expected to start replacing these aircraft from 2005. Sensors: BEAB Omera radar, Thomson Sintra DUAV-4 dipping sonar. Weapons: ASW; six Type 11/51 depth charges and/or two Type 45 torpedoes.

UPDATED

KV 107   *5/2004 *, Per Körnefeldt /* 1043525

Numbers/Type: 1 CASA C-212-200 Aviocar.
Operational speed: 190 kt *(353 km/h).*
Service ceiling: 24,000 ft *(7,315 m).*
Range: 1,650 n miles *(3,055 km).*
Role/Weapon systems: For ASW and surface surveillance. Two others belong to the Coast Guard. Sensors: Omera radar, Sonobuoys/Lofar, CDC sonobuoy processor, FLIR, datalink. Weapons: ASW; depth charges.

UPDATED

CASA C-212   *6/2002, Adolfo Ortigueira Gil /* 0572619

Numbers/Type: 10 NH Industries NH 90.
Operational speed: 157 kt *(291 km/h).*
Service ceiling: 13,940 ft *(4,250 m).*
Range: 621 n miles *(1,150 km).*
Role/Weapon systems: Eighteen aircraft to be procured for tactical troop transport and ASW role. Modular construction is to enable rapid re-roling. There are to be 10 sets of ASW sensors for delivery between 2007 and 2010. Sensors: Telephonics APS-143B(V) ocean eye radar, Galileo Avionica FLIR, Thales FLASH-S dipping sonar. Weapons: To be announced.   *NEW ENTRY*

NH 90   *5/2004 * /* 0094463

PATROL FORCES

2 NORRKÖPING CLASS (FAST ATTACK CRAFT—MISSILE) (PTFG)

Name	No	Builders	Commissioned
NORRKÖPING	R 131	Karlskronavarvet	11 May 1973
YSTAD	R 142	Karlskronavarvet	10 Jan 1976

Displacement, tons: 190 standard; 230 full load
Dimensions, feet (metres): 143 × 23.3 × 7.4 *(43.6 × 7.1 × 2.4)*
Main machinery: 3 RR Proteus gas turbines; 12,750 hp *(9.5 MW)* sustained; 3 shafts; Kamewa props
Speed, knots: 40.5. **Range, n miles:** 500 at 40 kt
Complement: 27 (7 officers)

Missiles: SSM: 8 Saab RBS 15; active radar homing to 70 km *(37.8 n miles)* at 0.8 Mach; warhead 150 kg.
Guns: 1 Bofors 57 mm/70 Mk 1; 200 rds/min to 17 km *(9.3 n miles)*; weight of shell 2.4 kg. 8 launchers for 57 mm illuminants on side of mounting.
Torpedoes: 6—21 in *(533 mm)* tubes (2-6 can be fitted at the expense of missile armament); Swedish Ordnance Type 613; anti-surface; wire-guided passive homing to 15 km *(8.2 n miles)* at 45 kt; warhead 240 kg.
Mines: Minelaying capability.
Countermeasures: Decoys: 2 Philips Philax fixed launchers; IR flares and chaff.
ESM: Argo Systems AR 700 or MEL Susie; radar intercept.
Combat data systems: MARIL 2000 datalink.
Radars: Air/surface search: Ericsson Sea Giraffe 50HC; G/H-band.
Fire control: Philips 9LV 200 Mk 1; J-band.

Modernisation: Programme included missile launchers, new fire-control equipment, modernised electronics and new 57 mm guns. A/S mortars removed. All completed by late 1984. These last two have had some weapon systems upgrading to keep them in service. All completed by August 1999. The plan to fit new engines has been shelved.
Structure: Similar to the original Spica class from which they were developed.
Operational: The six unmodernised craft have paid off. A further two decommissioned in 2002 and two in 2003. These last two are expected to be decommissioned in September 2005.

UPDATED

YSTAD   *4/2003, Per Körnefeldt /* 0572621

KAPAREN CLASS (FAST ATTACK CRAFT—MISSILE) (PCGF)

Name	No	Builders	Commissioned
SPEJAREN	P 162	Bergens MV, Norway	21 Mar 1980
TIRFING	P 166	Bergens MV, Norway	23 Jan 1982

Displacement, tons: 120 standard; 170 full load
Dimensions, feet (metres): 120 × 20.7 × 5.6 (36.6 × 6.3 × 1.7)
Main machinery: 2 MTU 16V 396 TB94; 5,800 hp(m) (4.26 MW) sustained; 2 shafts; 2 hydraulic motors for slow speed propulsion
Speed, knots: 36. **Range, n miles:** 550 at 36 kt
Complement: 22 (3 officers)

Missiles: SSM: 6 Kongsberg Penguin Mk 2; IR homing to 27 km (14.6 n miles) at 0.8 Mach; warhead 120 kg.
Guns: 1 Bofors 57 mm/70 Mk 1; 200 rds/min to 17 km (9.3 n miles); weight of shell 2.4 kg. 57 mm illuminant launchers on either side of mounting.
Torpedoes: 4—15.75 in (400 mm) Swedish Ordnance Type 43/45; anti-submarine.
A/S mortars: 4 Saab Elma 9-tubed launchers; range 400 m; warhead 4.2 kg shaped charge or Saab 601 9-tubed launchers; range 1,200 m; shaped charge.
Depth charges: 2 racks.
Mines: 24. Mine rails extend from after end of bridge superstructure with an extension over the stern. Use of these would mean the removal of missiles in the ratio of 8 mines for each pair of missiles.
ESM: Saab Scania EWS 905; radar intercept.
Radars: Surface search: Skanter 16 in Mk 009; I-band.
Fire control: Philips 9LV 200 Mk 2; J-band.
Sonars: Simrad SA 950; hull-mounted; active attack; high frequency.
    Simrad ST 570 VDS; active; high frequency.

Programmes: In the early 1970s it was decided to build fast attack craft similar to the Norwegian Hauk class. Prototype *Jägaren* underwent extensive trials and, on 15 May 1975, an order for a further 16 was placed.
Modernisation: Half-life modernisation between 1991 and 1994. The update included new engines with improved loiter capability and new sonars including the Simrad Toadfish VDS. Saab ASW-601 multibarrel grenade launcher being fitted.
Operational: The VDS can be operated down to 100 m as a dipping sonar, or towed at speeds up to 16 kt. These last two craft to remain in service until at least late 2005 after which they may be retained in a training role.   *UPDATED*

SPEJAREN   6/2004*, Frank Findler / 1043524

1 TYPE 72 INSHORE PATROL CRAFT (PBR)

HUVUDSKÄR 77

Displacement, tons: 30 full load
Dimensions, feet (metres): 69.2 × 15 × 4.3 (21.1 × 4.6 × 1.3)
Main machinery: 3 diesels; 3 shafts
Speed, knots: 18 (61-67); 22 (68-77)
Guns: 1—20 mm.
Depth charges: Carried in all of the class.
Radars: Surface search: Decca RM 914; I-band.
Sonars: Simrad; hull-mounted; active search; high frequency.

Comment: Last remaining vessel of a class built in 1966-67. Modernised in the 1980s with a tripod mast and radar mounted over the bridge.   *VERIFIED*

TYPE 72   6/1994, Curt Borgenstam / 0080731

12 TAPPER CLASS (PBR)

TAPPER 81	HÄNDIG 84	HURTIG 87	ÄRLIG 90
DJÄRV 82	TRYGG 85	RAPP 88	MUNTER 91
DRISTIG 83	MODIG 86	STOLT 89	ORÄDD 92

Displacement, tons: 57 full load
Dimensions, feet (metres): 71.9 × 17.7 × 4.9 (21.9 × 5.4 × 1.5)
Main machinery: 2 MWM TBD234V16 diesels; 1,812 hp(m) (1.33 MW) sustained; 2 shafts
Speed, knots: 25
Complement: 9
Guns: 2—12.7 mm MGs.
A/S mortars: 4 Elma/Saab grenade launchers; range 300 m; warhead 4.2 kg shaped charge.
Depth charges: 18.
Mines: 2 rails (in four of the class).
Radars: Surface search: 2 Racal Decca; I-band.
Sonars: Simrad; hull-mounted; active search; high frequency.

Comment: Seven Type 80 ordered from Djupviksvarvet in early 1992, and delivered between February 1993 and December 1995. Five more ordered in 1995 for delivery at six month intervals between December 1996 and January 1999. A Phantom HD-2 ROV is carried. This is equipped with a Tritech ST 525 imaging sonar.   *UPDATED*

ÄRLIG   4/2004*, E & M Laursen / 1043526

RAPP   4/2004*, Per Körnefeldt / 1043527

MINE WARFARE FORCES

Notes: A transportable COOP system was ordered in 1991. The unit can be shifted from one ship to another and comprises a container, processing module and tactical display, an underwater positioning system, sonar, Sea Eagle ROV and mine disposal charge. Optimised for shallow water surveillance and can be used in conjunction with other MCM systems. The primary role is route survey.

1 FURUSUND CLASS (COASTAL MINELAYER) (MLC)

Name	No	Builders	Launched	Commissioned
FURUSUND	MUL 20	ASI Verken	16 Dec 1982	10 Oct 1983

Displacement, tons: 225 full load
Dimensions, feet (metres): 106.9 × 26.9 × 7.5 (32.6 × 8.2 × 2.3)
Main machinery: Diesel-electric; 2 Scania GAS 1 diesel generators; 2 motors; 416 hp(m) (306 kW); 2 shafts
Speed, knots: 11.5
Complement: 24
Guns: 1—12.7 mm MG.
Mines: 22 tons.
Radars: Navigation: Racal Decca 1226; I-band.

Comment: Built for the former Coastal Artillery. Carries a Mantis type ROV.

                                    VERIFIED

FURUSUND   5/2001, Per Körnefeldt / 0131140

1 CARLSKRONA CLASS (MINELAYER) (AXH/MLH)

Name	No	Builders	Launched	Commissioned
CARLSKRONA	M 04	Karlskronavarvet	28 June 1980	11 Jan 1982

Displacement, tons: 3,600 full load
Dimensions, feet (metres): 346.7 × 49.9 × 13.1 (105.7 × 15.2 × 4)
Main machinery: 4 Nohab F212 D825 diesels; 10,560 hp(m) (7.76 MW); 2 shafts; cp props
Speed, knots: 20
Complement: 50 plus 136 trainees. Requires 118 as operational minelayer
Guns: 2 Bofors 57 mm/70. 2 Bofors 40 mm/70.
Mines: Can lay 105.
Countermeasures: 2 Philips Philax chaff/IR launchers.
ESM: Argo AR 700; intercept.
Radars: Air/surface search: Ericsson Sea Giraffe 50HC; G/H/I-band.
Surface search: Raytheon; E/F-band.
Fire control: 2 Philips 9LV 200 Mk 2; I/J-band.
Navigation: Terma Scanter 009; I-band.
Helicopters: Platform only.

Comment: Ordered 25 November 1977, laid down in sections late 1979 and launched at the same time as Karlskrona celebrated its tercentenary. Midshipmen's Training Ship as well as a minelayer; also a 'padded' target for exercise torpedoes. Sonar removed. Name is an older form of Karlskrona. Completed a 12-month refit in December 2002. **UPDATED**

CARLSKRONA 5/2003, Diego Quevedo / 0572622

3 ARKÖSUND CLASS (COASTAL MINELAYERS) (MLC)

ARKÖSUND MUL 12 GRUNDSUND MUL 15 FÅRÖSUND (ex-Öresund) MUL 18

Displacement, tons: 200 standard; 245 full load
Dimensions, feet (metres): 102.3 × 24.3 × 10.2 (31.2 × 7.4 × 3.1)
Main machinery: Diesel-electric; 2 Nohab/Scania diesel generators; 2 motors; 460 hp(m) (338 kW); 2 shafts
Speed, knots: 12
Complement: 24
Guns: 4—7.62 mm MGs.
Mines: 2 rails; 26 tons.
Radars: Navigation: Racal Decca 1226; I-band.

Comment: All completed by 1954-1957. Former Coastal Artillery craft for laying and maintaining minefields. One deleted in 1992, one in 1996, Skramsösund in 1998 and Kalmarsund and Barösund in 2004. 40 mm guns removed. **UPDATED**

GRUNDSUND 8/2004*, Per Körnefeldt / 1043528

2 SMALL MINELAYERS (MLI)

M 502 M 505

Displacement, tons: 15 full load
Dimensions, feet (metres): 47.9 × 13.8 × 2.9 (14.6 × 4.2 × 0.9)
Main machinery: 2 diesels; 2 shafts
Speed, knots: 14
Complement: 7
Mines: 2 rails; 12.
Radars: Navigation: Racal Decca; I-band.

Comment: Ordered in 1969. Mines are laid from rails or by crane on the stern. Former Coastal Artillery craft. **UPDATED**

SMALL MINELAYER 5/1995, Curt Borgenstam / 0080735

5 SAM CLASS (MCM DRONES) (MSD)

SAM 01-02, SAM 04, SAM 06-07

Displacement, tons: 20 full load
Dimensions, feet (metres): 59.1 × 20 × 5.2 (18 × 6.1 × 1.6)
Main machinery: 1 Volvo Penta TAMD70D diesel; 210 hp(m) (154 kW); 1 Schottel prop
Speed, knots: 8. **Range, n miles:** 330 at 8 kt

Comment: Built by Karlskronavarvet in 1983. SAM 03 and 05 sold to the USA for Gulf operation in March 1991 and replaced in 1992/93. Remote-controlled catamaran magnetic and acoustic sweepers operated by the Landsort and Styrsö classes. Six sold to Japan. **VERIFIED**

SAM 07 4/2003, Per Körnefeldt / 0572624

7 LANDSORT CLASS (MINEHUNTERS) (MHSCDM)

Name	No	Builders	Launched	Commissioned
LANDSORT	M 71	Karlskronavarvet	2 Nov 1982	19 Apr 1984
ARHOLMA	M 72	Karlskronavarvet	2 Aug 1984	23 Nov 1984
KOSTER	M 73	Karlskronavarvet	16 Jan 1986	30 May 1986
KULLEN	M 74	Karlskronavarvet	15 Aug 1986	28 Nov 1986
VINGA	M 75	Karlskronavarvet	14 Aug 1987	27 Nov 1987
VEN	M 76	Karlskronavarvet	10 Aug 1988	12 Dec 1988
ULVÖN	M 77	Karlskronavarvet	4 Mar 1992	9 Oct 1992

Displacement, tons: 270 standard; 360 full load
Dimensions, feet (metres): 155.8 × 31.5 × 7.3 (47.5 × 9.6 × 2.2)
Main machinery: 4 Saab-Scania DSI 14 diesels; 1,592 hp(m) (1.17 MW) sustained; coupled in pairs to 2 Voith Schneider props
Speed, knots: 15. **Range, n miles:** 2,000 at 12 kt
Complement: 29 (12 officers) plus 4 spare

Missiles: SAM: Saab Manpads.
Guns: 1 Bofors 40 mm/70 Mod 48; 240 rds/min to 12.5 km (6.8 n miles); weight of shell 0.96 kg. Bofors Sea Trinity CIWS trial carried out in Vinga (fitted in place of 40 mm/70). 2—7.62 mm MGs.
A/S mortars: 4 Saab Elma 9-tubed launchers; range 400 m; warhead 4.2 kg shaped charge or Saab 601 9-tubed launchers; range 1,200 m; shaped charge.
Countermeasures: Decoys: 2 Philips Philax fixed launchers can be carried with 4 magazines each holding 36 grenades; IR/chaff.
ESM: Matilda; intercept.
MCM: This class is fitted for mechanical sweeps for moored mines as well as magnetic and acoustic sweeps. In addition it is possible to operate two SAM drones (see separate entry). Fitted with 2 Sutec Sea Eagle or Double Eagle remote-controlled units with 600 m tether and capable of 350 m depth.
Weapons control: Philips 9LV 100 optronic director. Philips 9 MJ 400 minehunting system.
Radars: Navigation: Thomson-CSF Terma; I-band.
Sonars: Thomson-CSF TSM-2022; Racal Decca 'Mains' control system; hull-mounted; minehunting; high frequency.

Programmes: The first two of this class ordered in early 1981. Second four in 1984 and the seventh in 1989.
Modernisation: Following the mid-life upgrade of Kullen and Ven in 2003, refit of the remaining five of the class is to start in 2005 and continue until 2009.
Structure: The GRP mould for the hull has also been used for the Coast Guard former KBV 171 class.
Operational: The integrated navigation and action data automation system developed by Philips and Racal Decca.
Sales: Four built for Singapore. **UPDATED**

KULLEN 9/2004*, Per Körnefeldt / 1043529

KULLEN 5/2004*, E & M Laursen / 1043530

1 MSF MK 1 CLASS (MSD)

SÖKAREN MRF 01

Displacement, tons: 128 full load
Dimensions, feet (metres): 86.9 × 23 × 6.9 *(26.5 × 7 × 2.1)*
Main machinery: 2 Scania DSI 14 diesels; 1,000 hp(m) *(736 kW)*; 2 Schottel azimuth thrusters
Speed, knots: 12
Complement: 6
Radars: Navigation: Bridgewater E; I-band.
Sonars: STS 2054 side scan active; high frequency.

Comment: MCMV drone with GRP hull transferred from Denmark in 2001 for evaluation following cancellation of SAM II drone project. Further orders are not expected. ***UPDATED***

SÖKAREN 5/2004*, E & M Laursen / 1043531

4 STYRSÖ CLASS
(MINESWEEPERS/HUNTERS—INSHORE) (MHSDI)

Name	No	Builders	Launched	Commissioned
STYRSÖ	M 11	Karlskronavarvet	8 Mar 1996	20 Sep 1996
SPÅRÖ	M 12	Karlskronavarvet	30 Aug 1996	21 Feb 1997
SKAFTÖ	M 13	Karlskronavarvet	20 Jan 1997	13 June 1997
STURKÖ	M 14	Karlskronavarvet	27 June 1997	19 Dec 1997

Displacement, tons: 205 full load
Dimensions, feet (metres): 118.1 × 25.9 × 7.2 *(36 × 7.9 × 2.2)*
Main machinery: 2 Saab Scania DSI 14 diesels; 1,104 hp(m) *(812 kW)*; 2 shafts; bow thruster
Speed, knots: 13
Complement: 17 (9 officers)
Guns: 2—12.7 mm MGs.
Countermeasures: MCM: AK-90 acoustic, EL-90 magnetic, and mechanical sweeps.
  2 Sutec Sea Eagle/Double Eagle ROVs equipped with Tritech SE 500 sonar and mine disposal charges.
Combat data systems: Ericsson tactical data system with datalink.
Radars: Navigation: Racal Bridgemaster; I-band.
Sonars: Reson Sea Bat 8100; mine avoidance; active; high frequency.
  EG & G side scan; active for route survey; high frequency.

Comment: Contract awarded to KKV and Erisoft AB on 11 February 1994. Capable of operating two SAM drones. These ships are also used for inshore surveillance patrols. ***VERIFIED***

STURKÖ 5/2001, L-G Nilsson / 0131148

1 GÅSSTEN CLASS (MINESWEEPER—INSHORE) (MSI)

Name	No	Builders	Commissioned
VIKSTEN	M 33	Karlskronavarvet	15 July 1974

Displacement, tons: 120 standard; 135 full load
Dimensions, feet (metres): 78.7 × 21.3 × 11.5 *(24 × 6.5 × 3.5)*
Main machinery: 1 diesel; 460 hp(m) *(338 kW)*; 1 shaft
Speed, knots: 11
Complement: 9
Guns: 1 Bofors 20 mm.
Radars: Navigation: Terma; I-band.

Comment: Ordered 1972 built of GRP. ***UPDATED***

VIKSTEN 4/2004*, E & M Laursen / 1043532

4 EJDERN CLASS (SONOBUOY CRAFT) (MSI)

EJDERN B 01	KRICKAN B 02	SVÄRTAN B 03	VIGGEN B 04

Displacement, tons: 39 full load
Dimensions, feet (metres): 65.6 × 15.7 × 4.3 *(20 × 4.8 × 1.3)*
Main machinery: 2 Volvo Penta TAMD122 diesels; 366 hp(m) *(269 kW)* sustained; 2 shafts
Speed, knots: 15
Complement: 9 (3 officers)
Guns: 1—7.62 mm MG.
Radars: Navigation: Terma; I-band.

Comment: Built by Djupviksvarvet and completed in 1991. GRP hulls. Classified as mine warfare vessels. Used for laying and monitoring sonobuoys to detect intruders in Swedish territorial waters. Carry an AQS-928 acoustic processor. ***VERIFIED***

VIGGEN 8/2003, E & M Laursen / 0572625

AMPHIBIOUS FORCES

54 COMBATBOAT 90E (STRIDSBÅT) (YH)

101-154

Displacement, tons: 9 full load
Dimensions, feet (metres): 39 × 9.5 × 2.3 *(11.9 × 2.9 × 0.7)*
Main machinery: 1 Scania AB DSI 14 diesel; 398 hp(m) *(293 kW)* sustained; FFJet 410 water-jet
Speed, knots: 40; 37 (laden)
Complement: 2
Military lift: 2 tons or 6-10 troops
Radars: Navigation: Furuno 8050; I-band.

Comment: First batch ordered from Storebro Royal Cruiser AB in 1995 for delivery from August 1995-98. Second batch ordered in 1997 for delivery in 1998-99. These are ambulance boats but may also be used for stores. Two more ordered for Chinese Customs and three for Malaysian Customs in April 1997. ***VERIFIED***

COMBATBOAT 145 8/2002, E & M Laursen / 0529915

For details of the latest updates to *Jane's Fighting Ships* online and to discover the additional information available exclusively to online subscribers please visit
jfs.janes.com

145 COMBATBOAT 90H/90HS (STRIDSBÅT) (LCPFM)

803-946   BLÅTUNGA 947

Displacement, tons: 19 full load
Dimensions, feet (metres): 52.2 × 12.5 × 2.6 *(15.9 × 3.8 × 0.8)*
Main machinery: 2 Saab Scania DSI 14 diesels; 1,250 hp(m) *(935 kW)* (1,350 hp(m) in 90HS *(1,000 kW)*); 2 Kamewa water-jets
Speed, knots: 35-50; 20 (Sea State 3). **Range, n miles:** 240 at 30 kt
Complement: 3
Military lift: 20 troops plus equipment or 2.8 tons
Missiles: SSM: Rockwell RBS 17 Hellfire; semi-active laser guidance to 5 km *(3 n miles)* at 1.0 Mach; warhead 8 kg.
Guns: 3—12.7 mm MGs.
Mines: 4 (or 6 depth charges).
Radars: Navigation: Racal Decca; RD 360 or Furuno 8050; I-band.

Comment: The first two prototypes (801-802) ordered in January 1988 are no longer in service. Twelve more (803-814) built in 1991-92. There were 63 (815-877) ordered from Dockstavarvet and Gotlands Varv in mid-January 1992, with an option for 30 more (878-907) which was taken up in 1994. The building period for these completed in mid-1997. A further 40 (908-947) were ordered in August 1996 and delivery was completed in October 2003. Of these, the last 27 (90HS) units were all modified to undertake international peacekeeping operations by the inclusion of armoured protection, an NBC citadel and air conditioning. All have a 20° deadrise and all carry four six-man inflatable rafts. Some of these craft are on loan to the Swedish Police. 947 is equipped as VIP craft and 914 is used as a trials craft and is fitted with a twin 120 mm mortar. There were 22 (90N) of the class delivered to Norway 1999, 40 (90 HEX) to Mexico, 17 (90H) to Malaysia and three (90 HEX) to the Hellenic Coast Guard.

UPDATED

COMBATBOAT 884   *5/2004*, Michael Winter /* 1043533

COMBATBOAT 946   *5/2004*, E & M Laursen /* 1043534

17 LCMs (TROSSBÅT)

603-612   652-658

Displacement, tons: 55 full load
Dimensions, feet (metres): 68.9 × 19.7 × 4.9 *(21 × 6 × 1.5)*
Main machinery: 2 Saab DSI 11/40 M2 diesels; 340 hp(m) *(250 kW)*; 2 Schottel props
Speed, knots: 10
Military lift: 30 tons
Radars: Navigation: Racal Decca 914C; I-band.

Comment: Completed from 1980-88. Classified as Trossbåt (support boat). Built by Djupviksvarvet.

UPDATED

LCM 609   *5/2004*, E & M Laursen /* 1043535

23 LCUs (LCU)

208	234, 235, 237	261, 263, 264
215	241, 244	267-269
223	247, 248	281, 283
230-232	252, 258	

Displacement, tons: 31 full load
Dimensions, feet (metres): 70.2 × 13.8 × 4.2 *(21.4 × 4.2 × 1.3)*
Main machinery: 3 diesels; 600 hp(m) *(441 kW)*; 3 shafts
Speed, knots: 18
Military lift: 40 tons; 40 troops
Guns: 3 to 8—6.5 mm MGs.
Mines: Minelaying capability *(281-283* only).

Comment: Built between 1960 and 1987. 35 modernised 1989-93; 12 more 1997-98. Many have been decommissioned and some sold to private buyers.

UPDATED

LCU 281   *6/2002, Per Körnefeldt /* 0529912

2 TRANSPORTBÅT 2000 (AGF/YFLB)

451   452

Displacement, tons: 43 full load
Dimensions, feet (metres): 77.1 × 16.7 × 3.3 *(23.5 × 5.1 × 1)*
Main machinery: 3 Saab Scania DSI 14 diesels or 3 Volvo Penta 163 diesels; 1,194 hp(m) *(878 kW)* sustained; 3 FFJet 450 or 3 Kamewa K40 waterjets
Speed, knots: 25
Complement: 3
Military lift: 45 troops or 10 tons
Guns: 2—12.7 mm MGs.
Radars: Navigation: Terma; I-band.

Comment: Two similar prototypes ordered from Djupviks Shipyard in 1997. *452* is configured for troop carrying, and *451* as a command boat. Each has a different propulsion system which are in competition to be fitted in seven more troop carriers and three more command boats. Further orders are not expected.

VERIFIED

TRANSPORTBÅT 452   *6/2000, Michael Nitz /* 0106583

1 M 10 CLASS HOVERCRAFT (UCAC)

Displacement, tons: 26 full load
Dimensions, feet (metres): 61.8 × 29 *(18.8 × 8.8)*
Main machinery: 2 Scania diesels; 1,224 hp(m) *(900 kW)*
Speed, knots: 40; 7 with cushion deflated. **Range, n miles:** 600 at 30 kt
Complement: 3
Military lift: 10 tons or 50 troops
Radars: Navigation: I-band.

Comment: ABS design built at Karlskronavarvet under licence and started trials in June 1998 which included ice trials in early 1999. Has a Kevlar superstructure. Three more were planned but further orders are not now expected.

UPDATED

M 10   *5/2004*, E & M Laursen /* 1043536

102 RAIDING CRAFT (GRUPPBÅT) (LCP)

Displacement, tons: 3 full load
Dimensions, feet (metres): 26.2 × 6.9 × 1 *(8 × 2.1 × 0.3)*
Main machinery: 1 Volvo Penta TAMD 42WJ diesel; 230 hp(m) *(169 kW)*; 1 Kamewa 240 waterjet
Speed, knots: 30
Complement: 2
Military lift: 1 ton

Comment: Small raiding craft are used throughout the Archipelago. Some have been replaced.
UPDATED

GRUPPBÅT   *5/2004*, E & M Laursen /* 1043537

ICEBREAKERS

Notes: All Icebreakers have been transferred to and are manned by the National Maritime Administration.

1 ODEN CLASS (AGB/ML)

Name	Builders	Launched	Commissioned
ODEN	Gotaverken Arendal, Göteborg	25 Aug 1988	29 Jan 1989

Displacement, tons: 12,900 full load
Dimensions, feet (metres): 352.4 × 102 × 27.9 *(107.4 × 31.1 × 8.5)*
Main machinery: 4 Sulzer ZAL40S8L diesels; 23,940 hp(m) *(17.6 MW)* sustained; 2 shafts; cp props
Speed, knots: 17. **Range, n miles:** 30,000 at 13 kt
Complement: 32 plus 17 spare berths
Guns: 4 Bofors 40 mm/70 can be fitted.

Comment: Ordered in February 1987, laid down 19 October 1987. Can break 1.8 m thick ice at 3 kt. Towing winch aft with a pull of 150 tons. Helicopter platform 73.5 × 57.4 ft *(22.4 × 17.5 m)*. The main hull is only 25 m wide but the full width is at the bow. Fitted with water-jet and heeling pump system to assist with ice-breaking operations. Also equipped as a minelayer.
VERIFIED

ODEN (with water-jets)   *10/2000, Per Körnefeldt /* 0106584

3 ATLE CLASS (AGBH/ML)

Name	Builders	Launched	Commissioned
ATLE	Wärtsilä, Helsinki	27 Nov 1973	21 Oct 1974
FREJ	Wärtsilä, Helsinki	3 June 1974	30 Sep 1975
YMER	Wärtsilä, Helsinki	3 Sep 1976	25 Oct 1977

Displacement, tons: 7,900 standard; 9,500 full load
Dimensions, feet (metres): 343.1 × 78.1 × 23.9 *(104.6 × 23.8 × 7.3)*
Main machinery: 5 Wärtsilä-Pielstick diesels; 22,000 hp(m) *(16.2 MW)*; 4 Strömberg motors; 22,000 hp(m) *(16.2 MW)*; 4 shafts (2 fwd, 2 cp aft)
Speed, knots: 19
Complement: 50 (16 officers)
Guns: 4 Bofors 40 mm/70.
Helicopters: 2 light.

Comment: Similar to Finnish Urho class. Also equipped as minelayers.
VERIFIED

FREJ   *8/1994, E & M Laursen /* 0080744

1 ALE CLASS (AGB/AGS)

Name	Builders	Launched	Commissioned
ALE	Wärtsilä, Helsinki	1 June 1973	19 Dec 1973

Displacement, tons: 1,550
Dimensions, feet (metres): 154.2 × 42.6 × 16.4 *(47 × 13 × 5)*
Main machinery: 2 diesels; 4,750 hp(m) *(3.49 MW)*; 2 shafts
Speed, knots: 14
Complement: 32 (8 officers)
Guns: 1 Bofors 40 mm/70 (not embarked).

Comment: Built for operations on Lake Vänern. Is also used as a survey ship.
VERIFIED

ALE   *9/2002, E & M Laursen /* 0529911

INTELLIGENCE VESSELS

1 ELECTRONIC SURVEILLANCE SHIP (AGIH)

Name	No	Builders	Launched	Commissioned
ORION	A 201	Karlskronavarvet	30 Nov 1983	7 June 1984

Displacement, tons: 1,400 full load
Dimensions, feet (metres): 201.1 × 32.8 × 9.8 *(61.3 × 10 × 3)*
Main machinery: 2 Hedemora V8A diesels; 1,800 hp(m) *(1.32 MW)* sustained; 2 shafts; cp props
Speed, knots: 15
Complement: 35
Radars: Navigation: Terma Scanter 009; I-band.
Helicopters: Platform for 1 light.

Comment: Ordered 23 April 1982. Laid down 28 June 1982. The communications aerials are inside the elongated dome.
VERIFIED

ORION   *4/2003, Per Körnefeldt /* 0572628

SURVEY SHIPS

Notes: (1) Owned and manned (since 1 January 2002) by the National Maritime Administration.
(2) There is a research ship *Argos*. Civilian manned and owned by the National Board of Fisheries. A second civilian ship *Ocean Surveyor* belongs to the Geological Investigation but has been leased as a Support Ship on occasions.
(3) The Board of Navigation owns two buoy tenders *Scandica* and *Baltica* built in 1982 and two lighthouse tenders *Fyrbyggaren* and *Fyrbjörn*.

SCANDICA   *6/2000, Curt Borgenstam* / 0106585

1 SURVEY SHIP (AGS)

JACOB HÄGG

Displacement, tons: 192 standard
Dimensions, feet (metres): 119.8 × 24.6 × 5.6 *(36.5 × 7.5 × 1.7)*
Main machinery: 4 Saab Scania DSI 14 diesels; 1,592 hp(m) *(1.17 MW)* sustained; 2 shafts
Speed, knots: 16
Complement: 13 (5 officers)

Comment: Laid down April 1982 at Djupviks Shipyard. Launched 12 March 1983. Completed 16 May 1983. Aluminium hull.
VERIFIED

JACOB HÄGG   *5/1998, J Cíślak* / 0050199

1 SURVEY SHIP (AGS)

NILS STRÖMCRONA

Displacement, tons: 210 full load
Dimensions, feet (metres): 98.4 × 32.8 × 5.9 *(30 × 10 × 1.8)*
Main machinery: 4 Saab Scania DSI 14 diesels; 1,592 hp(m) *(1.17 MW)* sustained; 2 shafts; bow and stern thrusters
Speed, knots: 12
Complement: 14 (5 officers)

Comment: Completed 28 June 1985. Of catamaran construction-each hull of 3.9 m made of aluminium.
VERIFIED

NILS STRÖMCRONA   *9/2001, Per Körnefeldt* / 0131143

RESCUE VEHICLES

Notes: (1) Kockums is developing S-SRV, a replacement submarine rescue vehicle. A further development of URF, it will be capable of rescuing 35 people in a single mission. With a speed of 4.5 kt, it will be capable of operating down to depths of 700 m and mating with the hull of a submarine at angles up to 60°. Rescued personnel will be transferred to a recompression chamber on a surface vessel. Navigational aids will include sonars and underwater cameras. It will be road/air/ship transportable and be capable of operating from another submarine.
(2) Early in 1995 Sweden signed an agreement with Norway to provide submarine rescue.

S-SRV   *6/2004*, Kockums* / 1043516

1 RESCUE SUBMERSIBLE (DSRV)

URF

Displacement, tons: 52
Dimensions, feet (metres): 45.6 × 10.5 × 9.2 *(13.9 × 3.2 × 2.8)*
Main machinery: Electric/hydraulic: single shaft
Speed, knots: 3
Complement: 4

Comment: Rescue submersible URF (*Ubåts Räddnings Farkost*) was launched by Kockums on 17 April 1978 and commissioned in 1979. The double-hulled vehicle is capable of operating down to 460 m with an endurance of 85 hours. The URF can mate with the hull of a submarine at angles up to 45° and is equipped with a lockout chamber that can support two divers to 300 m. It has a rescue capacity of 35 per dive and submariners can be transferred directly from the pressurised hull of the submarine to a compression chamber on board the support ship *Belos*. The vehicle is normally based at the Naval Diving Centre at Berga but can be transported by road using a specially designed trailer to a site suitable for loading on to the support ship. URF has recently been refitted and will remain in service until 2005.
UPDATED

URF   *8/2004*, E & M Laursen* / 1043538

AUXILIARIES

1 TRANSPORT (AKR)

Name	No	Builders	Commissioned
SLEIPNER (ex-*Ardal*)	A 343	Bergen	1980

Displacement, tons: 1,049 full load
Dimensions, feet (metres): 163.1 × 36.1 × 11.5 *(49.7 × 11 × 3.5)*
Main machinery: 1 Normo diesel; 1,300 hp(m) *(956 kW)*; 1 shaft
Speed, knots: 12
Complement: 12
Cargo capacity: 260 tons

Comment: Former Ro-Ro vessel acquired in 1992 from a Norwegian Shipping Company. There is a stern ramp and side door.
VERIFIED

SLEIPNER   *7/2003, E & M Laursen* / 0572629

1 TROSSÖ CLASS (SUPPORT SHIP) (AGP)

Name	No	Builders	Commissioned
TROSSÖ (ex-*Arnold Viemer*, ex-*Livonia*)	A 264	Valmet, Finland	1 Jan 1984

Displacement, tons: 2,140 full load
Dimensions, feet (metres): 234.9 × 42 × 14.8 *(71.6 × 12.8 × 4.5)*
Main machinery: 2 Russkiy G74 36/45 diesels; 3,084 hp(m) *(2.27 MW)*; 2 shafts
Speed, knots: 14
Complement: 64

Comment: Built as a survey ship for the USSR and used in the Baltic as an AGOR. Taken on by the Estonian Marine Institute and then transferred to Sweden on 23 September 1996. Converted as a depot ship for corvettes and patrol craft and back in service in 1997. Can act as a Headquarters Ship. A second vessel, *Ornö*, was purchased in 2001 but rebuilding of the ship was abandoned in November 2001 due to its poor material state.

VERIFIED

TROSSÖ 5/2002, Curt Borgenstam / 0529913

1 ÄLVSBORG CLASS (SUPPORT SHIP) (AKH)

Name	No	Builders	Launched	Commissioned
VISBORG	A 265	Karlskronavarvet	25 Jan 1975	6 Feb 1976

Displacement, tons: 2,400 standard; 2,650 full load
Dimensions, feet (metres): 303.1 × 48.2 × 13.2 *(92.4 × 14.7 × 4)*
Main machinery: 2 Nohab-Polar diesels; 4,200 hp(m) *(3.1 MW)*; 1 shaft; cp prop; bow thruster; 350 hp(m) *(257 kW)*
Speed, knots: 16
Complement: 95
Guns: 3 Bofors 40 mm/70 SAK 48.
ESM: Argo 700; intercept.
Radars: Surface search: Raytheon; E/F-band.
Fire control: Philips 9LV 200 Mk 2; I/J-band.
Navigation: Terma Scanter 009; I-band.
Helicopters: Platform only.

Comment: Laid down on 16 October 1973. Formerly a minelayer, now supply ship for second surface flotilla. Sister ship transferred to Chile in 1996.

VERIFIED

VISBORG 8/2003, E & M Laursen / 0572630

1 MCM SUPPORT (MCS)

Name	No	Builders	Commissioned
GÅLÖ (ex-*Herjolfur*)	A 263	Bergen	1976

Displacement, tons: 1,400 full load
Dimensions, feet (metres): 196.9 × 39.4 × 14.4 *(60 × 12 × 4.4)*
Main machinery: 1 Wichman diesel; 2,400 hp(m) *(1.76 MW)*; 1 shaft; 2 bow thrusters
Speed, knots: 12
Complement: 24

Comment: Built as a ferry for Iceland. Acquired in 1993 as a repair and maintenance ship for patrol craft. Role changed to MCM support ship in 2000.

VERIFIED

GÅLÖ 7/2003, E & M Laursen / 0572631

1 DIVER SUPPORT SHIP (YDT/AGFH)

Name	No	Builders	Commissioned
SKREDSVIK (ex-*KBV 172*)	A 262 (ex-M 70)	Karlskronavarvet	Oct 1981

Displacement, tons: 375 full load
Dimensions, feet (metres): 164 × 27.9 × 7.9 *(50 × 8.5 × 2.4)*
Main machinery: 2 Hedemora V16A diesels; 4,500 hp(m) *(3.3 MW)*; 2 shafts
Speed, knots: 18. **Range, n miles:** 3,000 at 12 kt
Complement: 15
Radars: Navigation: 2 Racal Decca; I-band.
Helicopters: Platform for 1 light.

Comment: Transferred from the Coast Guard in 1991 and used as a Diver Support ship for mine clearance operations after a refit from October 1992 to February 1993. Employed for Command and Control in war. Has a Simrad high-frequency active sonar.

VERIFIED

SKREDSVIK 7/1996, Per Körnefeldt / 0080752

1 DIVER SUPPORT SHIP (YDT/AGF)

ÄGIR (ex-*Bloom Syrveyor*) A 212

Displacement, tons: 117 full load
Dimensions, feet (metres): 82 × 24.9 × 6.6 *(25 × 7.6 × 2)*
Main machinery: 2 GM diesels; 2 shafts
Speed, knots: 11
Complement: 15
Radars: Navigation: Terma; I-band.

Comment: Built in Norway in 1984. Acquired in 1989.

VERIFIED

ÄGIR 7/2003, E & M Laursen / 0572632

1 DIVER SUPPORT SHIP (YDT)

NORDANÖ (ex-*Sjöjungfrun*) A 213

Displacement, tons: 148 full load
Dimensions, feet (metres): 80.1 × 24.9 × 8.9 *(24.4 × 7.6 × 2.7)*
Main machinery: 2 Volvo Penta TAMD diesels; 767 hp(m) *(564 kW)*; 2 shafts
Speed, knots: 10
Complement: 15

Comment: Launched in 1983 and bought by the Navy in 1992.

VERIFIED

NORDANÖ 6/2002, Swedish Navy / 0530069

1 SALVAGE SHIP (ARSH)

Name	No	Builders	Recommissioned
BELOS III (ex-Energy Supporter)	A 214	De Hoop, Netherlands	Nov 1992

Measurement, tons: 5,096 grt
Dimensions, feet (metres): 344.2 × 59.1 × 16.7 (104.9 × 18 × 5.1)
Main machinery: 5 MAN 9ASL 25/30 diesel alternators; 8.15 MW; 2 motors; 5,110 hp(m) (3.76 MW); 2 azimuth thrusters and 3 bow thrusters
Speed, knots: 14
Complement: 50 (22 officers)

Comment: Bought from Midland and Scottish Resources in mid-1992 and arrived in Sweden in November 1992. Replaced the previous ship of the same name which paid off in April 1993. Ice-strengthened hull and fitted with a helicopter platform. Acts as the support ship for the rescue submersible URF. Equipped with Dynamic Positioning System MOSHIP. Life-extension refit to be completed in December 2005. *UPDATED*

BELOS III *8/2004*, E & M Laursen /* 1043539

1 TORPEDO AND MISSILE RECOVERY VESSEL (YPT)

Name	No	Builders	Commissioned
PELIKANEN	A 247	Djupviksvarvet	26 Sep 1963

Displacement, tons: 144 full load
Dimensions, feet (metres): 108.2 × 19 × 7.2 (33 × 5.8 × 2.2)
Main machinery: 2 MTU MB diesels; 1,040 hp(m) (764 kW); 2 shafts
Speed, knots: 14
Complement: 14
Radars: Navigation: Terma; I-band.

Comment: Torpedo recovery and rocket trials vessel. *VERIFIED*

PELIKANEN *9/1998, J Ciślak /* 0050204

1 TORPEDO AND MISSILE RECOVERY VESSEL (YPT)

Name	No	Builders	Commissioned
PINGVINEN	A 248	Lunde Varv-och Verstads	20 Mar 1975

Displacement, tons: 191 full load
Dimensions, feet (metres): 109.3 × 20 × 7.2 (33.3 × 6.1 × 2.2)
Main machinery: 2 MTU diesels; 1,140 hp(m) (838 kW); 2 shafts
Speed, knots: 13
Complement: 14
Radars: Navigation: Terma; I-band.

Comment: Recovery vessel for trials firings. *VERIFIED*

PINGVINEN *6/2002, Swedish Navy /* 0530068

1 SUPPORT SHIP (AKL)

Name	No	Builders	Commissioned
LOKE	A 344	Oskarsham Shipyard	Sep 1994

Displacement, tons: 455 full load
Dimensions, feet (metres): 117.8 × 29.5 × 8.6 (35.9 × 9 × 2.7)
Main machinery: 2 Scania diesels; 2 shafts
Speed, knots: 12
Complement: 8
Cargo capacity: 50 tons or 50 passengers
Radars: Navigation: Terma; I-band.

Comment: General support craft which can be used as a ferry. Landing craft bow. *UPDATED*

LOKE *4/2004*, E & M Laursen /* 1043540

16 SUPPORT VESSEL (TROSSBÅT) (YAG)

662-677

Displacement, tons: 60 full load
Dimensions, feet (metres): 80.1 × 17.7 × 4.6 (24.4 × 5.4 × 1.4)
Main machinery: 3 Saab Scania DSI 14 diesels; 1,194 hp(m) (878 kW) sustained; 3 FFJet 450 water-jets
Speed, knots: 25; 13 (laden)
Complement: 3
Military lift: 22 tons
Guns: 1—12.7 mm MG.
Radars: Navigation: Terma; I-band.

Comment: Prototype Trossbåt-built at Holms Shipyard in 1991 and capable of carrying 15 tons of deck cargo and 9 tons internal cargo or 17 troops plus mines. Aluminium hull with a bow ramp. Some ice capability. A second prototype delivered in late 1993, and the first production vessel in 1996. Eight vessels have been modified to undertake international peacekeeping operations by the inclusion of armoured protection, an NBC citadel and air conditioning. *UPDATED*

TROSSBÅT 668 *8/2002, E & M Laursen /* 0529906

TRAINING SHIPS

2 SAIL TRAINING SHIPS (AXS)

Name	No	Builders	Commissioned
GLADAN	S 01	Naval Dockyard, Stockholm	1947
FALKEN	S 02	Naval Dockyard, Stockholm	1947

Displacement, tons: 225 standard
Dimensions, feet (metres): 112.8 × 23.6 × 13.8 (34.4 × 7.2 × 4.2)
Main machinery: 1 diesel; 120 hp(m) (88 kW); 1 shaft

Comment: Sail training ships. Two masted schooners. Sail area, 512 sq m. Both had major overhauls in 1986-88 in which all technical systems were replaced. *VERIFIED*

FALKEN *5/1997, Marek Twardowski /* 0019218

1 ARKÖ CLASS TRAINING VESSEL (YXT)

Name	No	Builders	Commissioned
NÄMNDÖ	M 67	Karlskronavarvet, Karlskrona	1964

Displacement, tons: 300 full load
Dimensions, feet (metres): 145.7 × 24.6 × 8.2 *(44.4 × 7.5 × 2.5)*
Main machinery: 2 MTU 12V493 diesels; 2 shafts
Speed, knots: 14

Comment: Former minesweeper of wooden construction. Converted to navigational training ship during 1990s.

UPDATED

NÄMNDÖ   *4/2004*, E & M Laursen /* 1043541

5 M 15 CLASS (MINESWEEPERS—INSHORE) (MSI/AXL)

M 20   M 21   M 22   M 24   M 25

Displacement, tons: 70 full load
Dimensions, feet (metres): 90.9 × 16.5 × 6.6 *(27.7 × 5 × 2)*
Main machinery: 2 diesels; 320 hp(m) *(235 kW)* sustained; 2 shafts
Speed, knots: 12
Complement: 10
Radars: Navigation: Terma; I-band.

Comment: All launched in 1941 and now used as platforms for mine clearance divers. *M 20* of this class was re-rated as tender and renamed *Skuld*, but was converted back again in 1993. All five were modernised in 1992-93 and are also used for navigation training.

UPDATED

M 25   *4/2004*, Per Körnefeldt /* 1043542

TUGS

1 OCEAN TUG (ATA/AGB)

ACHILLES A 251

Displacement, tons: 450 full load
Dimensions, feet (metres): 108.2 × 28.9 × 15.1 *(33 × 8.8 × 4.6)*
Main machinery: 1 diesel; 1,650 hp(m) *(1.2 MW)*; 1 shaft
Speed, knots: 12
Complement: 12
Radars: Navigation: Racal Decca 1226C; I-band.

Comment: Launched in 1962. Icebreaking tug.

VERIFIED

ACHILLES   *6/2000, E & M Laursen /* 0106590

3 COASTAL TUGS (YTM)

HERMES A 253   HEROS A 322   HERA A 324

Displacement, tons: 185 standard; 215 full load
Dimensions, feet (metres): 80.5 × 22.6 × 13.1 *(24.5 × 6.9 × 4)*
Main machinery: 1 diesel; 600 hp(m) *(441 kW)*; 1 shaft
Speed, knots: 11
Complement: 8

Comment: Details given for the first pair launched 1953-57. Third is smaller at 127 tons and was launched in 1969-71. All are icebreaking tugs.

UPDATED

HERA   *6/2000, E & M Laursen /* 0106591

9 COASTAL TUGS (YTL)

A 702-705   A 751   A 753-756

Displacement, tons: 42 full load
Dimensions, feet (metres): 50.9 × 16.4 × 8.9 *(15.5 × 5 × 2.7)*
Main machinery: 1 diesel; 1 shaft
Speed, knots: 9.5
Complement: 6

Comment: Can carry 40 people. Icebreaking tugs. *702-703* used by Amphibious Corps. All can carry mines.

UPDATED

A 753   *6/2003, L-G Nilsson /* 0572637

COAST GUARD (KUSTBEVAKNING)

Establishment: Established in 1638, and for 350 years was a part of the Swedish Customs administration. From 1 July 1988 the Coast Guard became an independent civilian authority with a Board supervised by the Ministry of Defence. Organised in four regions with a central Headquarters.
Duties: Responsible for civilian surveillance of Swedish waters, fishery zone and continental shelf. Supervises and enforces fishing regulations, customs, dumping and pollution regulations, environmental protection and traffic regulations. Also concerned with prevention of drug running and forms part of the Swedish search and rescue organisation.

Headquarters Appointments

Director General:
  Marie Hafstrom

Personnel: 2005: 583

Aircraft: Four CASA 212.

Ships: Tv pennant numbers replaced by KBV in 1988 but the KBV is not displayed. Vessels are not normally armed.

1 KBV 181 CLASS (HIGH ENDURANCE CUTTER) (WHEC/PBO)

KBV 181

Displacement, tons: 991 full load
Dimensions, feet (metres): 183.7 oa; 167.3 wl × 33.5 × 15.1 *(56; 51 × 10.2 × 4.6)*
Main machinery: 2 Wärtsilä Vasa 8R22 diesels; 3,755 hp(m) *(2.76 MW)* sustained; 1 shaft;
     Kamewa cp prop; bow thruster
Speed, knots: 16. Range, n miles: 2,800 at 15 kt
Complement: 11
Guns: 1 Oerlikon 20 mm (if required).
Radars: Navigation: Furuno FAR 2830; I-band.
Sonars: Simrad Subsea; active search; high frequency.

Comment: Ordered from Rauma Shipyards in August 1989 and built at Uusikaupunki.
     Commissioned 30 November 1990. Unarmed in peacetime. Equipped as a Command vessel for
     SAR and anti-pollution operations. All-steel construction similar to Finnish *Tursas*.
                                                                                    VERIFIED

KBV 181   *6/2000, B Sullivan /* 0106592

2 KBV 101 CLASS (MEDIUM ENDURANCE CUTTERS) (WMEC/PB)

KBV 104-105

Displacement, tons: 65 full load
Dimensions, feet (metres): 87.6 × 16.4 × 7.2 *(26.7 × 5 × 2.2)*
Main machinery: 2 Cummins KTA38-M diesels; 2,120 hp *(1.56 MW)*; 2 shafts
Speed, knots: 21. Range, n miles: 1,000 at 15 kt
Complement: 5 plus 2 spare
Sonars: Hull-mounted; active search; high frequency.

Comment: Built 1969-73 at Djupviksvarvet. Class A cutters. All-welded aluminium hull and
     upperworks. Equipped for salvage divers. Modernisation with new diesels, a new bridge and
     new electronics completed in 1988. *KBV 101* transferred to Lithuania in 1996.
                                                                                    VERIFIED

KBV 101 class   *5/1994, E & M Laursen /* 0080758

2 KBV 201 CLASS (HIGH ENDURANCE CUTTERS) (WHEC/PBO)

KBV 201   KBV 202

Displacement, tons: 476 full load
Dimensions, feet (metres): 170.6 × 28.2 × 7.9 *(52 × 8.6 × 2.4)*
Main machinery: 2 MWM 610 diesels; 5,440 hp(m) *(4 MW)*; 2 MWM 616 diesels; 1,904 hp(m)
     (1.4 MW); 2 shafts; Kamewa cp props; 2 bow thruster 424 hp(m) *(312 kW)*
Speed, knots: 21. Range, n miles: 1,340 at 16 kt
Complement: 9
Radars: Navigation: E/F- and I-band.

Comment: Ordered from Kockums in January 1999 and built at Karlskrona. First one delivered in
     March 2001 and second in September 2001. Steel hulls. Multirole vessels for surveillance and
     environmental protection. Stern ramp for launching a RIB.
                                                                                    UPDATED

KBV 201   *5/2004 *, E & M Laursen /* 1043543

6 KBV 281 CLASS (MEDIUM ENDURANCE CUTTERS) (WMEC/PB)

KBV 281-283   KBV 285-287

Displacement, tons: 45 full load
Dimensions, feet (metres): 71.5 × 16.4 × 6.2 *(21.8 × 5 × 1.9)*
Main machinery: 2 Cummins KTA38-M or MWM diesels; 2,120 hp *(1.56 MW)*; 2 shafts
Speed, knots: 27
Complement: 4
Radars: Navigation: Furuno; I-band.

Comment: Built by Djupviksvarvet and delivered at one a year from 1979. Last one commissioned
     in 1990. Aluminium hulls. Some of the class have an upper bridge.   *UPDATED*

KBV 287   *5/2004 *, P Marsan /* 1043544

3 KBV 288 CLASS
(MEDIUM ENDURANCE CUTTERS) (WMEC/PBO)

KBV 288-290

Displacement, tons: 53 full load
Dimensions, feet (metres): 71.5 × 17.7 × 5.9 *(21.8 × 5.4 × 1.8)*
Main machinery: 2 Cummins KTA38-M or MWM diesels; 2,120 hp *(1.56 MW)*; 2 shafts
Speed, knots: 24
Complement: 5
Radars: Navigation: Furuno; I-band.

Comment: An improved design of the KBV 281 class which entered service 1990-93.
                                                                                    UPDATED

KBV 290   *8/2004 *, Harald Carstens /* 1043545

3 KBV 591 (GRIFFON 2000 TDX) CLASS (HOVERCRAFT) (UCAC)

KBV 591-593

Displacement, tons: 3.5 full load
Dimensions, feet (metres): 38.4 × 19.4 *(11.7 × 5.9)*
Main machinery: 1 Deutz BF8L diesel; 350 hp(m) *(235 kW)*
Speed, knots: 50. Range, n miles: 450 at 35 kt
Complement: 3
Radars: Navigation: Furuno 7010 D; I-band.

Comment: Built by Griffon Hovercraft, Southampton and delivered in 1992-93. Aluminium hulls.
     Based at Stockholm, Lutea and Umea.   *VERIFIED*

KBV 591   *6/2003, Swedish Coast Guard /* 0572610

16 KBV 301 CLASS
(MEDIUM ENDURANCE CUTTERS) (WMEC/PB)

KBV 301-316

Displacement, tons: 35 full load
Dimensions, feet (metres): 65.6 × 15.1 × 3.6 *(20 × 4.6 × 1.1)*
Main machinery: 2 MTU 183 TE92 diesels; 1,830 hp(m) *(1.35 MW)* sustained; 2 MTP 7500S or Kamewa water-jets
Speed, knots: 34. **Range, n miles:** 500 at 25 kt
Complement: 4
Radars: Navigation: 2 Kelvin Hughes 6000; I-band.

Comment: Built at Karlskronavarvet. First one delivered in May 1993 and the remainder ordered in December 1993. Three delivered in 1995, four in 1996 and the last three in 1997.   **VERIFIED**

KBV 301   *3/2003, Per Körnefeldt /* 0572638

60 COAST GUARD PATROL CRAFT (SMALL) (PB)

KBV 401-408   + 52

Displacement, tons: 2.2 full load
Dimensions, feet (metres): 29.7 × 8.5 × 2.9 *(9.05 × 2.6 × 0.9)*
Main machinery: 2 Yamaha outboard engines; 500 hp *(372 kW)*
Speed, knots: 55. **Range, n miles:** 100 at 35 kt
Complement: 3

Comment: Details are for *KBV 401-408* built in 1994-95. There is a total of some 60 speed boats with Raytheon radars.   **VERIFIED**

KBV 408   *6/2003, Swedish Coast Guard /* 0572612

POLLUTION CONTROL CRAFT (YPC)

Number	Displacement (tons)	Comment
KBV 004	450	Built by Lunde in 1978. Has helipad and carries salvage divers.
KBV 005	990	Ice Class 1A built in 1980 and acquired in 1993.
KBV 006	450	Built by Lunde in 1985.
KBV 010	400	Built by Lunde in 1985. Oil spill clean-up craft.
KBV 020	60	Catamaran design built by Djupviks in 1982.
KBV 044	100	Class B Sea Trucks built by Djupviks in 1976. Oil spill clean-up craft.
KBV 045-049	230	Pollution control craft built by Lunde 1980-83. Have bow ramp.
KBV 050-051	340	Enlarged version of KBV 045 class with bow ramp. Built by Lunde in 1983.

VERIFIED

KBV 047   *6/2003, Per Körnefeldt /* 0572639

KBV 005   *6/2003, Swedish Coast Guard /* 0572611

Switzerland

Country Overview

A landlocked western European country, the Swiss Confederation has an area of 15,940 square miles and is bordered by France, Germany, Austria, Liechtenstein and Italy. The largest city is Zurich and the capital is Bern. The principal lakes are Lake Geneva in the southwest and Lake Constance in the northeast. Others not wholly within Swiss borders are Lake Lugano and Lake Maggiore. The river Rhine, whose source is in the Swiss Alps, is navigable northwards and downstream from the port of Basel. The patrol boats are split between lakes Constance, Geneva and Maggiore; one company to each.

Diplomatic Representation

Defence Attaché in London:
Colonel B Stoll

ARMY

Notes: There are also large numbers of flat bottomed raiding craft powered by single 40 hp outboard engines.

11 AQUARIUS CLASS (Patrouillenboot 80) (PBR)

ANTARES	AQUARIUS	ORION	SATURN
URANUS	CASTOR	PERSEUS	SIRIUS
VENUS	MARS	POLLUX	

Displacement, tons: 7 full load
Dimensions, feet (metres): 35.1 × 10.8 × 3 *(10.7 × 3.3 × 0.9)*
Main machinery: 2 Volvo KAD 3 diesels; 460 hp(m) *(338 kW)*; 2 shafts
Speed, knots: 35
Complement: 7
Guns: 2—12.7 mm MGs.
Radars: Surface search: JFS Electronic 364; I-band.

Comment: Builders Müller AG, Spiez. GRP hulls, wooden superstructure. *Aquarius* commissioned in 1978, *Pollux* in 1984, the remainder in 1981. Re-engined with diesels which have replaced the former petrol engines.

VERIFIED

AQUARIUS
10/1997, Swiss Army / 0019223

Syria

Country Overview

The Syrian Arab Republic was proclaimed in 1961 following brief federation with Egypt as the United Arab Republic from 1958. Situated in the Middle East, the country has an area of 71,498 square miles and is bordered to the north by Turkey, to the east by Iraq, to the south by Jordan and Israel and to the west by Lebanon. It has a 104 n mile coastline with the Mediterranean Sea. The capital and largest city is Damascus while the principal ports are Latakia and Tartus. It is the only country to claim 35 n mile Territorial seas. An EEZ is not claimed.

Headquarters Appointments

Commander-in-Chief Navy:
Vice Admiral Tayyara
Director of Naval Operations:
Rear Admiral A Meidar
Chief of Staff:
Vice Admiral Kassim Muhammed Baidun

Organisation

Naval Forces come under the command of the Chief of General Staff, Commander of Land Forces.

Personnel

(a) 2005: 3,200 officers and men (2,500 reserves)
(b) 18 months' national service

Bases

Latakia, Tartous, Al-Mina-al-Bayda, Baniyas

Coast Defence

Coastal defence has been under naval control since 1984. A missile brigade is equipped with SS-C-1 Sepal and SS-C-3 Styx with sites at Tartous (2), Baniyas and Latakia. Two artillery battalions have a total of 36—130 mm guns and 12—100 mm guns. Coastal observation sites are manned by an Observation Battalion. There are two infantry brigades each of which is assigned to a coastal zone.

DELETIONS

Mine Warfare Forces

2004   Sonya 532

FRIGATES

2 PETYA III (PROJECT 159A) CLASS (FFL)

1-508 (ex-*12*)   **AL HIRASA**   2-508 (ex-*14*)

Displacement, tons: 950 standard; 1,180 full load
Dimensions, feet (metres): 268.3 × 29.9 × 9.5
(81.8 × 9.1 × 2.9)
Main machinery: CODAG; 2 gas turbines; 30,000 hp(m)
(22 MW); 1 Type 61V-3 diesel; 5,400 hp(m) *(3.97 MW)*
sustained (centre shaft); 3 shafts
Speed, knots: 32. **Range, n miles:** 4,870 at 10 kt; 450 at 29 kt
Complement: 98 (8 officers)

Guns: 4—3 in *(76 mm)*/60 (2 twin) ❶; 90 rds/min to 15 km *(8 n miles)*; weight of shell 6.8 kg.

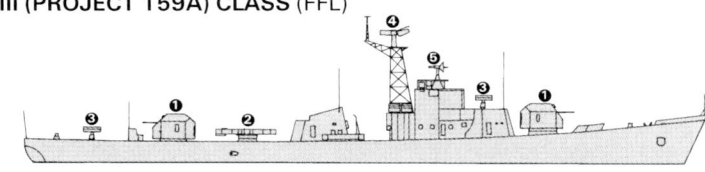
PETYA 1-508   *(Scale 1 : 900), Ian Sturton*

Torpedoes: 3—21 in *(533 mm)* (triple) tubes ❷. SAET-60; active/passive homing to 15 km *(8.1 n miles)* at 40 kt; warhead 100 kg.

A/S mortars: 4 RBU 2500 16-tubed trainable ❸; range 2,500 m; warhead 21 kg.
Depth charges: 2 racks.
Mines: Can carry 22.
Radars: Surface search: Slim Net ❹; E/F-band.
Navigation: Don 2; I-band.
Fire control: Hawk Screech ❺; I-band.
IFF: High Pole B. 2 Square Head.
Sonars: Herkules; hull-mounted; active search and attack; high frequency.

Programmes: Transferred by the USSR in July 1975 and March 1975.
Operational: Based at Tartous. *2-508* in dock in mid-1998 to 2000 and reported to be sea-going. The operational status of *1-508* is not known.   *VERIFIED*

AL HIRASA   *6/2001* / 0121400

PATROL FORCES

10 OSA (PROJECT 205) CLASS
(FAST ATTACK CRAFT—MISSILE) (PTFG)

31-32 (OSA I)   33-40 (OSA II)

Displacement, tons: 245 full load
Dimensions, feet (metres): 126.6 × 24.9 × 8.8 *(38.6 × 7.6 × 2.7)*
Main machinery: 3 Type M 504 (Osa II)/M 503 (Osa I) diesels; 8,025/10,800 hp(m) *(6.0/8.1 MW)* sustained; 3 shafts
Speed, knots: 35 (Osa I), 37 (Osa II). **Range, n miles:** 500 at 35 kt
Complement: 25 (3 officers)

Missiles: SSM: 4 SS-N-2C; active radar or IR homing to 83 km *(43 n miles)* at 0.9 Mach; warhead 513 kg; sea-skimmer at end of run.
Guns: 4—30 mm/65 (2 twin); 500 rds/min to 5 km *(2.7 n miles)*; weight of shell 0.54 kg.
Countermeasures: Decoys: PK 16 chaff launcher.
Radars: Surface search: Square Tie; I-band.
Fire control: Drum Tilt; H/I-band.
IFF: 2 Square Head. High Pole A or B.

Programmes: Delivered: October 1979 (two), November 1979 (two), August 1982 (one), September 1982 (one) and May 1984 (two). Others have already been deleted.
Structure: Two are modified (Nos 39 and 40).
Operational: OSA II based at Latakia. All are still fully operational and active. Two Osa Is (31 and 32) at Tartous also reported to be operational again. The Osa Is may be fitted with SSN 2A/B.   *VERIFIED*

OSA II 38   *6/1998* / 0050214

8 ZHUK (GRIF) (PROJECT 1400M) CLASS
(COASTAL PATROL CRAFT) (PB)

1-8   2-8   3-8   4-8   5-8   6-8   7-8   8-8

Displacement, tons: 39 full load
Dimensions, feet (metres): 78.7 × 16.4 × 3.9 *(24 × 5 × 1.2)*
Main machinery: 2 Type M 401B diesels; 2,200 hp(m) *(1.6 MW)* sustained; 2 shafts
Speed, knots: 30. **Range, n miles:** 1,100 at 15 kt
Complement: 11 (3 officers)
Guns: 4—14.5 mm (2 twin) MGs.
Radars: Surface search: Spin Trough; I-band.

Comment: Three transferred from USSR in August 1981, three on 25 December 1984 and two more in the late 1980s. All based at Tartous but most are non-operational.   *VERIFIED*

ZHUK 5-8   *6/1998* / 0050215

LAND-BASED MARITIME AIRCRAFT

Numbers/Type: 11/2 Mil Mi-14P Haze A/Mil Mi-14P Haze C.
Operational speed: 124 kt *(230 km/h)*.
Service ceiling: 15,000 ft *(4,570 m)*.
Range: 432 n miles *(800 km)*.
Role/Weapon systems: Medium-range ASW helicopter. Sensors: Short Horn search radar, dipping sonar, MAD, sonobuoys. Weapons: ASW; internally stored torpedoes, depth mines and bombs.   *VERIFIED*

Numbers/Type: 2 Kamov Ka-28 Helix.
Operational speed: 135 kt *(250 km/h)*.
Service ceiling: 19,685 ft *(6,000 m)*.
Range: 432 n miles *(800 km)*.
Role/Weapon systems: ASW helicopter. Delivered in February 1990. Sensors: Splash Drop search radar, dipping sonar, sonobuoys, MAD, ECM. Weapons: ASW; 3 torpedoes, depth bombs, mines.   *VERIFIED*

AMPHIBIOUS FORCES

3 POLNOCHNY B CLASS (PROJECT 771) (LSM)

1-114	2-114	3-114

Displacement, tons: 760 standard; 834 full load
Dimensions, feet (metres): 246.1 × 31.5 × 7.5 *(75 × 9.6 × 2.3)*
Main machinery: 2 Kolomna Type 40-D diesels; 4,400 hp(m) *(3.2 MW)* sustained; 2 shafts
Speed, knots: 19. **Range, n miles:** 1,500 at 15 kt
Complement: 40
Military lift: 180 troops; 350 tons cargo
Guns: 4—30 mm/65 (2 twin); 500 rds/min to 5 km *(2.7 n miles)*; weight of shell 0.54 kg.
    2—140 mm rocket launchers; 18 barrels per launcher; range 9 km *(5 n miles)*.
Radars: Surface search: Spin Trough; I-band.
Fire control: Drum Tilt; H/I-band.

Comment: Built at Northern Shipyard, Gdansk. First transferred from USSR January 1984, two in February 1985 from Black Sea. All based at Tartous and still active. *VERIFIED*

POLNOCHNY B (Russian colours)   *1988*

MINE WARFARE FORCES

1 NATYA (PROJECT 266M) CLASS (MSC/AGORM)

642

Displacement, tons: 804 full load
Dimensions, feet (metres): 200.1 × 33.5 × 10.8 *(61 × 10.2 × 3)*
Main machinery: 2 Type 504 diesels; 5,000 hp(m) *(3.67 MW)* sustained; 2 shafts
Speed, knots: 16. **Range, n miles:** 3,000 at 12 kt
Complement: 65
Missiles: SAM: 2 SA-N-5 Grail quad launchers; manual aiming; IR homing to 6 km *(3.2 n miles)* at 1.5 Mach; altitude to 2,500 m *(8,000 ft)*; warhead 1.5 kg; 16 missiles.
Guns: 4—30 mm/65 (2 twin) can be fitted.
Radars: Surface search: Don 2; I-band.
Fire control: Drum Tilt; H/I-band.

Comment: Arrived in Tartous from USSR in January 1985. Has had sweeping gear and guns removed and converted to serve as an AGOR. Painted white. Based at Latakia in reasonable condition. Reported active. *VERIFIED*

NATYA 642   *6/1996* / 0080764

3 YEVGENYA (PROJECT 1258) CLASS
(MINESWEEPERS—INSHORE) (MSI/PC)

6-507	7-507	8-507

Displacement, tons: 77 standard; 90 full load
Dimensions, feet (metres): 80.7 × 18 × 4.9 *(24.6 × 5.5 × 1.5)*
Main machinery: 2 Type 3-D-12 diesels; 600 hp(m) *(444 kW)*; 2 shafts
Speed, knots: 11. **Range, n miles:** 300 at 10 kt
Complement: 10
Guns: 2—14.5 mm (twin) MGs (first pair). 2—25 mm/80 (twin) (second pair).
Radars: Surface search: Spin Trough; I-band.
IFF: High Pole.
Sonars: MG-7; stern-mounted VDS; active; high frequency.

Comment: First transferred from USSR 1978, two in 1985 and two in 1986. Second pair by Ro-flow from Baltic in February 1985 being new construction with tripod mast. Based at Tartous, at least two are operational. Two others have been deleted. *VERIFIED*

YEVGENYA (Russian colours)   *1991*

TRAINING SHIPS

1 TRAINING SHIP (AX/AKR)

AL ASSAD

Displacement, tons: 3,500 full load
Dimensions, feet (metres): 344.5 × 56.4 × 13.1 *(105 × 17.2 × 4)*
Main machinery: 2 Zgoda-Sulzer 6ZL40/48 diesels; 8,700 hp(m) *(6.4 MW)*; 2 shafts; bow thruster
Speed, knots: 16. **Range, n miles:** 4,500 at 15 kt
Complement: 56 plus 140 cadets
Radars: Navigation: Decca Seamaster; E/F- and I-band.

Comment: Built in Polnochny Shipyard, Gdansk and launched 18 February 1987. Delivered in late 1988. Ro-ro design used as a naval training ship. Unarmed but has minelaying potential. Based at Latakia and occasionally deploys on cruises. *VERIFIED*

AL ASSAD   *7/2003, B Prézelin* / 0570996

AL ASSAD   *7/2003, B Prézelin* / 0570997

Taiwan
REPUBLIC OF CHINA

Country Overview

The Republic of China was established in 1949 when the Nationalist government of China withdrew to Taiwan (Formosa) and established its headquarters. Though in practice an autonomous state, Taiwan is still formally a province of China and, as such, is claimed by the People's Republic of China. The country comprises the island of Taiwan (area 13,900 square miles), the Pescadores, or P'eng-hu Islands, the Quemoy Islands off the mainland city of Amoy (Xiamen), and the Matsu group off Fuzhou (Foochow). It has a 783 n mile coastline with East China Sea, Pacific Ocean and South China Sea. The capital and largest city of Taiwan is Taipei while Chi-lung (Keelung), Hualien, Kaohsiung and T'ai-chung are the principal ports. Territorial seas (12 n miles) are claimed. A 200 n mile EEZ and Fishery Zone have also been claimed.

Headquarters Appointments

Commander-in-Chief:
Admiral Miao Yung-Ching
Deputy Commander-in-Chief:
Vice Admiral Shen Fang-Hsiang
Commandant of Marine Corps:
Lieutenant General Chen Bon-Chih
Director of Logistics:
Vice Admiral Jin Feng-Shiang

Senior Flag Officers

Fleet Commander:
Vice Admiral Hu Chai-Kwei
Director of Logistics:
Vice Admiral Gan Ke-Chiang

Personnel

(a)   2005: 31,500 in Navy, 15,000 in Marine Corps
(b)   2 years' conscript service

Bases

Tsoying: HQ First Naval District (Southern Taiwan, Pratas and Spratly). Main Base, HQ of Fleet Command, Naval Aviation Group and Marine Corps. Base of southern patrol and transport squadrons. Officers and ratings training, Naval Academy, Naval shipyard.

Kaohsiung; Naval shipyard.
Makung (Pescadores): HQ Second Naval District (Pescadores, Quemoy and Wu Ch'iu). Base for attack squadrons. Naval shipyard and training facilities.
Keelung: HQ Third Naval District (Northern Taiwan and Matsu group). Base of northern patrol and transport squadrons. Naval shipyard.
Hualien: Naval Aviation Command.
Suao: East Coast Command, submarine depot and shipyard.
Minor bases at Hualien, Tamshui, Hsinchu, Wuchi and Anping.
Building: Taitung.

Organisation

1. Fleet Command:
124th Attack squadron, based at Tsoying
131st Patrol squadron, based at Keelung
142nd Support squadron, based at Kaohsiung
146th Attack squadron, based at Pescadores
151st Amphibious squadron, based at Tsoying
168th Patrol squadron, based at Suao
192nd Mine Warfare squadron, based at Tsoying
256th Submarine Unit, based at Tsoying.

2. Naval Aviation Command: There are two Groups. The fixed-wing Group consists of two squadrons (133 and 134) and the helicopter Group of three squadrons (501, 701 and 702). Land bases include Tsoying, Hualien, Hsinchu and Pintung.

Coast Defence

The land-based SSM command has six squadrons equipped with Hsiung-Feng II SSM at Tonying Island of the Matsu Group, Siyu Island of the Pescadores, Shiao Liuchiu off Kaohsiung, north of Keelung harbour, Tsoying naval base and Hualien. The ROCMC deploy eight SAM Platoons, equipped with Chaparral SAM quad-launchers, to the offshore island of Wuchiu, and Pratas islets in the South China Sea. There are also a number of 127 mm guns.

Marine Corps

Reduced to two brigades by mid-2000 supported by one amphibious regiment and one logistics regiment. Equipped with M-116, M-733, LARC-5, LVTP5 (to be replaced by AAV-7A-IRAM/RS) personnel carriers and LVTH6 armour tractors. Based at Tsoying and in southern Taiwan. Spratly detachment provided by the Coast Guard from 1 January 2000 and Marine Corps detachment withdrawn from Pratas Islands at the same time.

Coast Guard

Formerly the Maritime Security Police but name changed on 1 January 2000. Comes under the Minister of the Interior but its numerous patrol boats are integrated with the Navy for operational purposes.

Strength of the Fleet

Type	Active (Reserve)	Building/ Transfer (Planned)
Submarines	4	(6)
Destroyers	7	4
Frigates	21	1
Corvettes	—	(10)
Fast Attack Craft (Missile)	48	(29)
Large Patrol Craft	19	—
Ocean Minesweepers	4	—
Coastal Minesweepers/Hunters	8	(2)
LSD	2	(1)
Landing Ships (LST and LSM)	17	4
LCUs	18	—
Survey Ships	1	—
Combat Support Ships	1	—
Transports	4	—
Salvage Ships	4	—
Support Tankers	2	—
Customs	10+	—

PENNANT LIST

Submarines

791	Hai Shih
792	Hai Bao
793	Hai Lung
794	Hai Hu

Destroyers

912	Chien Yang
921	Liao Yang
923	Shen Yang
925	Te Yang
927	Yun Yang
928	Chen Yang
929	Shao Yang
1801	Chi Teh
1802	Ming Teh
1803	Tong Teh
1805	Wu Teh

Frigates

932	Chin Yang
933	Fong Yang
934	Feng Yang
935	Lan Yang
936	Hae Yang
937	Hwai Yang
938	Ning Yang
939	Yi Yang
1101	Cheng Kung
1103	Cheng Ho
1105	Chi Kuang
1106	Yueh Fei
1107	Tzu-I
1108	Pan Chao
1109	Chang Chien
1110	Tien Tan (bldg)
1202	Kang Ding
1203	Si Ning
1205	Kun Ming
1206	Di Hua
1207	Wu Chang
1208	Chen Te

Patrol Forces

PCL 1	Ning Hai
PCL 2	An Hai
601	Lung Chiang
602	Sui Chang
603	Jin Chiang
605	Tan Chiang
606	Hsin Chiang
607	Feng Chiang
608	Tseng Chiang
609	Kao Chiang
610	Jing Chiang
611	Hsian Chiang
612	Tsi Chiang
614	Po Chiang
615	Chan Chiang
617	Chu Chiang

Amphibious Forces

191	Chung Cheng
193	Shiu Hai
201	Chung Hai
205	Chung Chien
208	Chung Shun
216	Chung Kuang
217	Chung Chao
218	Chung Chi
221	Chung Chuan
226	Chung Chih
227	Chung Ming
230	Chung Pang
231	Chung Yeh
232	Chung Ho
233	Chung Ping
341	Mei Chin
347	Mei Sung
353	Mei Ping
356	Mei Lo
401	Ho Chi
402	Ho Huei
403	Ho Yao
406	Ho Chao
481	Ho Shun
484	Ho Chung
488	Ho Shan

489	Ho Chuan
490	Ho Seng
491	Ho Meng
492	Ho Mou
493	Ho Shou
494	Ho Chun
495	Ho Yung
LCC1	Kao Hsiung
SB 1	Ho Chie
SB 2	Ho Ten

Mine Warfare Forces

158	Yung Chuan
162	Yung Fu
167	Yung Ren
168	Yung Sui
1301	Yung Feng
1302	Yung Chia
1303	Yung Ting
1305	Yung Shun
1306	Yung Yang
1307	Yung Tzu
1308	Yung Ku
1309	Yung Teh

Auxiliaries and Survey Ships

507	Chang Bai
515	Lung Chuan
517	Hsin Lung
524	Yuen Feng
525	Wu Kang
526	Hsin Kang
530	Wu Yi
552	Ta Hu
1601	Ta Kuan

Tugs

ATF 551	Ta Wan
ATF 553	Ta Han
ATF 554	Ta Kang
ATF 555	Ta Fung
ATF 563	Ta Tai

SUBMARINES

Notes: Project Kwang Hua 8: Following the announcement in 2001 by the US government that it will support the acquisition of eight diesel submarines, debate has centred on how these will be procured. Northrop Grumman has reportedly offered a modernised version of the Barbel class, which dates from the 1950s, but this would probably require a non-US partner. The licence of a design from a third country has proved to be problematic in view of the re-affirmation of earlier decisions by the governments of the Netherlands (1992) and Germany (1993) not to grant export licences for Taiwan. Efforts to sell the Agosta class were similarly discouraged by the French government while Australia has rejected expressions of interest in the Collins class. An indigenous build programme remains a possibility although this would present significant technical and financial challenges.

2 HAI LUNG CLASS (SSK)

Name	No	Builders	Laid down	Launched	Commissioned
HAI LUNG	793	Wilton Fijenoord, Netherlands	Dec 1982	6 Oct 1986	9 Oct 1987
HAI HU	794	Wilton Fijenoord, Netherlands	Dec 1982	20 Dec 1986	9 Apr 1988

Displacement, tons: 2,376 surfaced; 2,660 dived
Dimensions, feet (metres): 219.6 × 27.6 × 22 *(66.9 × 8.4 × 6.7)*
Main machinery: Diesel-electric; 3 Bronswerk D-RUB 215-12 diesels; 4,050 hp(m) *(3 MW)*; 3 alternators; 2.7 MW; 1 Holec motor; 5,100 hp(m) *(3.74 MW)*; 1 shaft
Speed, knots: 12 surfaced; 20 dived
Range, n miles: 10,000 at 9 kt surfaced
Complement: 67 (8 officers)

Torpedoes: 6—21 in *(533 mm)* bow tubes. 20 AEG SUT; dual purpose; wire-guided; active/passive homing to 12 km *(6.6 n miles)* at 35 kt; warhead 250 kg.

Countermeasures: ESM: Argo AR 700SF and Elbit Timnex 4CH (V)2; intercept.
Weapons control: Sinbads M TFCS.
Radars: Surface search: Signaal ZW06; I-band.
Sonars: Signaal SIASS-Z; hull-mounted; passive/active intercept search and attack; low/medium frequency.
Fitted for but not with towed passive array.

Programmes: Order signed with Wilton Fijenoord in September 1981 for these submarines with variations from the standard Netherlands Zwaardvis design. Construction was delayed by the financial difficulties of the builders but was resumed in 1983. Sea trials of *Hai Lung* in March 1987 and *Hai Hu* in January 1988 and both submarines were shipped out on board a heavy dock vessel. The names mean *Sea Dragon* and *Sea Tiger*.
Structure: The four horns on the forward casing are Signaal sonar intercept transducers. Torpedoes manufactured under licence in Indonesia.
Operational: Hsiung Feng II submerged launch SSMs are planned to be part of the weapons load and a torpedo tube launched version is being developed, although no recent progress has been reported. Belong to 256th Submarine Unit based at Tsoying.

UPDATED

HAI HU and HAI LUNG

11/2004*, Ships of the World / 1044575

2 GUPPY II CLASS (SS)

Name	No	Builders	Laid down	Launched	Commissioned
HAI SHIH (ex-*Cutlass* SS 478)	791 (ex-SS 91)	Portsmouth Navy Yard	22 July 1944	5 Nov 1944	17 Mar 1945
HAI BAO (ex-*Tusk* SS 426)	792 (ex-SS 92)	Federal SB & DD Co, Kearney, New Jersey	23 Aug 1943	8 July 1945	11 Apr 1946

Displacement, tons: 1,870 standard; 2,420 dived
Dimensions, feet (metres): 307.5 × 27.2 × 18 *(93.7 × 8.3 × 5.5)*
Main machinery: Diesel-electric; 3 Fairbanks-Morse diesels; 4,500 hp *(3.3 MW)*; 2 Elliott motors; 5,400 hp *(4 MW)*; 2 shafts
Speed, knots: 18 surfaced; 15 dived
Range, n miles: 8,000 at 12 kt surfaced
Complement: 75 (7 officers)

Torpedoes: 10—21 in *(533 mm)* (6 fwd, 4 aft) tubes. AEG SUT; active/passive homing to 12 km *(6.5 n miles)* at 35 kt; 28 km *(15 n miles)* at 23 kt; warhead 250 kg.
Countermeasures: ESM: WLR-1/3; radar warning.
Radars: Surface search: US SS 2; I-band.
Sonars: EDO BQR 2B; hull-mounted; passive search and attack; medium frequency.
Raytheon/EDO BQS 4C; adds active capability to BQR 2B.
Thomson Sintra DUUG 1B; passive ranging.

Programmes: Originally fleet-type submarines of the US Navy's Tench class; extensively modernised under the Guppy II programme. *Hai Shih* transferred in April 1973 and *Hai Bao* in October the same year.
Structure: After 56 years in service diving depth is very limited.
Operational: Kept in service because of difficulty in buying replacements, but operational status doubtful. Likely to have an alongside training role only. Belong to the 256th Submarine Unit based at Tsoying.

UPDATED

HAI BAO

11/2004*, Ships of the World / 1044574

DESTROYERS

Notes: (1) Fully operational numbers of destroyers, frigates, corvettes and PGGs are restricted to a total of 48. As a new ship commissions, one of the older ships is paid off.
(2) Acquisition of the Aegis Combat System remains a firm aspiration but, following the decision to procure the Kidd class DDGs as an interim measure, this is unlikely before 2012.

7 GEARING (WU CHIN III CONVERSION) (FRAM I) CLASS (DDGHM)

Name	No	Builders	Laid down	Launched	Commissioned
CHIEN YANG (ex-*James E Kyes*)	912 (ex-DD 787)	Todd Pacific SY, Seattle, WA	27 Dec 1944	4 Aug 1945	8 Feb 1946
LIAO YANG (ex-*Hanson*)	921 (ex-DD 832)	Bath Iron Works Corporation	7 Oct 1944	11 Mar 1945	11 May 1945
SHEN YANG (ex-*Power*)	923 (ex-DD 839)	Bath Iron Works Corporation	26 Feb 1945	30 June 1945	13 Sep 1945
TE YANG (ex-*Sarsfield*)	925 (ex-DD 837)	Bath Iron Works Corporation	15 Jan 1945	27 May 1945	31 July 1945
YUN YANG (ex-*Hamner*)	927 (ex-DD 718)	Federal SB and DD Co	5 Apr 1945	24 Nov 1945	11 July 1946
CHEN YANG (ex-*Johnston*)	928 (ex-DD 821)	Consolidated Steel Corporation	6 May 1945	19 Oct 1945	10 Oct 1946
SHAO YANG (ex-*Hollister*)	929 (ex-DD 788)	Todd Pacific SY, Seattle, WA	18 Jan 1945	9 Oct 1945	26 Mar 1946

Displacement, tons: 2,425 standard; 3,540 full load
Dimensions, feet (metres): 390.5 × 41.2 × 19
(119 × 12.6 × 5.8)
Main machinery: 4 Babcock & Wilcox boilers; 600 psi *(43.3 kg/ cm²)*; 850°F *(454°C)*; 2 GE turbines; 60,000 hp *(45 MW)*; 2 shafts
Speed, knots: 30. **Range, n miles:** 6,080 at 15 kt
Complement: 275 approx

Missiles: SSM: 4 Hsiung Feng II (quad) ❶; active radar/IR homing to 80 km *(43.2 n miles)* at 0.85 Mach; warhead 190 kg.
SAM: 10 General Dynamics Standard SM1-MR (2 triple ❷; 2 twin ❸); command guidance; semi-active radar homing to 46 km *(25 n miles)* at 2 Mach.
A/S: Honeywell ASROC Mk 112 octuple launcher ❹; inertial guidance to 1.6-10 km *(1-5.4 n miles)*; payload Mk 46 torpedo.
Guns: 1 OTO Melara 3 in *(76 mm)*/62 ❺; 85 rds/min to 16 km *(8.7 n miles)*; weight of shell 6 kg.
1 GE/GD 20 mm Vulcan Phalanx Block 1 6-barrelled Mk 15 ❻; 3,000 rds/min combined to 1.5 km.
2 Bofors 40 mm/70 ❼. 4 or 6—12.7 mm MGs.
Torpedoes: 6—324 mm US Mk 32 (2 triple) tubes ❽. Honeywell Mk 46; anti-submarine; active/passive homing to 11 km *(5.9 n miles)* at 40 kt; warhead 44 kg. Some Mk 44s are still in service.
Countermeasures: Decoys: 4 Kung Fen 6 16-tubed chaff launchers ❾.
Mk T-6 Fanfare torpedo decoy.
ESM/ECM: Chang Feng III (Hughes SLQ-17/SLQ-31) intercept and jammers.
Combat data systems: Ta Chen tactical datalink.
Weapons control: Honeywell H 930 MFCS Mk 114 system (for ASROC). Signaal LIOD Mk 2 optronic director. ❿
Radars: Air search: Signaal DA08 (with DA05 aerial) ⓫; E/F-band.
Surface search: Raytheon SPS-10 ⓬; G/I-band.
Fire control: Signaal STIR ⓭; I/J-band (for Standard and 76 mm).
Westinghouse W-160 ⓮; I-band (for Bofors).
Navigation: I-band.
Tacan: SRN 15.
Sonars: Raytheon SQS-23H; hull-mounted; active search and attack; medium frequency.

Helicopters: 1 McDonnell Douglas MD 500 ⓯.

Programmes: *Chien Yang* transferred 18 April 1973; *Liao Yang*, 18 April 1973; *Te Yang* and *Shen Yang*, 1 October 1977 by sale; *Yun Yang*, December 1980; *Chen Yang* by sale 27 February 1981; *Shao Yang* by sale 3 March 1983. There has been some confusion over the English translation of some names. These are now correct.
Modernisation: All ships converted to area air defence ships under the Wu Chin III programme. This upgrade involved the installation of the H 930 Modular Combat System (MCS) with a Signaal DA08 air search radar (employing a lightweight DA05 antenna) and a Signaal STIR missile control radar directing 10 box-launched Standard SM-1 surface-to-air missiles (two twin in 'B' position, two triple facing either beam

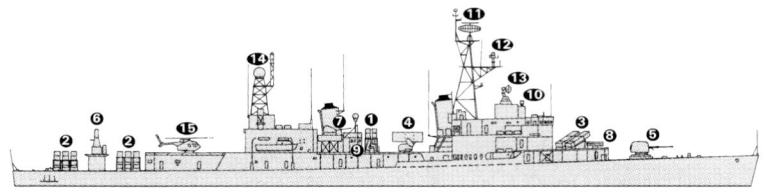

LIAO YANG

(Scale 1 : 1,200), Ian Sturton / 0126186

CHEN YANG

9/2001, Ships of the World / 0121407

aft). The system can track 24 targets simultaneously and attack four with an 8 second response time. An OTO Melara 76 mm is fitted to 'A' position, one Bofors 40 mm/70 is mounted forward of the seaboat on the starboard side, one abaft the ASROC magazine on the port side and a Mk 15 Block 1 Phalanx CIWS is aft between two banks of Standard launchers. A Westinghouse W-160 is mounted on a lattice mast on the hangar to control the Bofors. The amidships ASROC launcher is retained, its Mk 114 fire-control system is integrated with the H 930 MCS via a digital-analogue interface. The SQS-23 sonar has also been upgraded to the H standard using a Raytheon solid-state transmitter. The Chang Feng III EW system was developed jointly by Taiwan's Chung-

Shan Institute of Science and Technology (CSIST) with the assistance of Hughes. The Chang Feng III employs phased-array antennas which resemble those of the Hughes SLQ-17, it is capable of both deception and noise jamming. An eighth ship, *Lao Yang*, was to have been given the same conversion but missed the programme after a grounding incident in 1987 and has since been deleted. Ta Chen datalink fitted from 1994. Four Hsiung Feng II SSM launchers (quad) are fitted aft of the ASROC launcher.
Operational: Having paid off all the other ex-US Second World War destroyers, these ships are likely to be replaced by the Kidd class from 2005. Form 131 Squadron. Based at Keelung. *UPDATED*

CHIEN YANG

9/2001, Ships of the World / 0121404

0 + 4 KIDD CLASS (DDGHM)

Name	No	Builders	Laid down	Launched	Commissioned
CHI TEH (ex-*Scott*)	1801 (ex-DD 995)	Ingalls Shipbuilding	12 Feb 1979	1 Mar 1980	24 Oct 1981
MING TEH (ex-*Kidd*)	1802 (ex-DD 993)	Ingalls Shipbuilding	26 June 1978	11 Aug 1979	27 June 1981
TONG TEH (ex-*Chandler*)	1803 (ex-DD 996)	Ingalls Shipbuilding	7 May 1979	24 May 1980	13 Mar 1982
WU TEH (ex-*Callaghan*)	1805 (ex-DD 994)	Ingalls Shipbuilding	23 Oct 1978	1 Dec 1979	29 Aug 1981

Displacement, tons: 6,950 light; 9,574 full load
Dimensions, feet (metres): 563.3 × 55 × 20
  (171.7 × 16.8 × 6.2)
Main machinery: 4 GE LM 2500 gas turbines; 86,000 hp
  (64.16 MW) sustained; 2 shafts
Speed, knots: 33. **Range, n miles:** 6,000 at 20 kt
Complement: 363 (31 officers)

Missiles: SSM: 8 McDonnell Douglas Harpoon (2 quad)
  launchers; active radar homing to 130 km *(70 n miles)* at
  0.9 Mach; warhead 227 kg.
  SAM: 52 GDC Standard SM-2 Block IIIA; command/inertial
  guidance; semi-active radar homing to 167 km *(90 n miles)* at
  2 Mach. 2 twin Mk 26 launchers.
  A/S: 16 Honeywell ASROC; inertial guidance to 1.6-10 km *(1-
  5.4 n miles)*; payload Mk 46 Mod 5 Neartip. Fired from SAM
  launchers.
Guns: 2 FMC 5 in *(127 mm)*/54 Mk 45 Mod 0; 20 rds/min to
  23 km *(12.6 n miles)*; weight of shell 32 kg.
  2 General Electric/General Dynamics 20 mm Vulcan Phalanx
  6-barrelled Mk 15; 3,000 rds/min (4,500 in Block 1).
  4-12.7 mm MGs.
Torpedoes: 6-324 mm Mk 32 (2 triple) tubes. Honeywell Mk 46
  Mod 5; anti-submarine; active/passive homing to 11 km
  (5.9 n miles) at 40 kt; warhead 44 kg. Torpedoes fired from
  inside the hull under the hangar.
Countermeasures: Decoys: 4 Loral Hycor SRBOC 6-barrelled
  fixed Mk 36; IR flares and chaff to 4 km *(2.2 n miles)*. SLQ-25
  Nixie; torpedo decoy.
Combat data systems: ACDS Block 1 Level 1 with datalinks.
Weapons control: SWG-1A Harpoon LCS. 2 Mk 74 MFCS. Mk
  86 Mod 5 GFCS. Mk 116 FCS for ASW. Mk 14 WDS. SYS
  2(V)2 IADT. 4 SYR 3393 for SAM mid-course guidance.
Radars: Air search: ITT SPS-48E; 3D; E/F-band.
  Raytheon SPS-49(V)5; C/D-band.

KIDD class (US colours)   *4/1997, H M Steele* / 0016364

Air/Surface search: ISC Cardion SPS-55; I/J-band.
Navigation: Raytheon SPS-64; I/J-band.
Fire control: 2 Raytheon SPG-51D, 1 Lockheed SPG-60, 1
  Lockheed SPQ-9A.
Sonars: General Electric/Hughes SQS-53A; bow-mounted;
  search and attack; medium frequency.
  Gould SQR-19 (TACTAS); passive towed array (may be fitted).

Helicopters: 2 medium size.

Programmes: Originally ordered by the Iranian government in
  1974, the contracts were taken over by the US Navy on

25 July 1979. All paid off from USN service in 1998-99.
  Offered to the Taiwan government, intention to buy confirmed
  on 2 October 2001. Entry into service is expected to begin in
  2005.
Modernisation: All received major modernisation from 1988-90.
  Further package to be completed prior to transfer. This may
  include replacement of the Mk 26 launchers by VLS.
Structure: Optimised for general warfare, mainmast and radar
  aerials are in different configuration than Spruance class.
  Details given are for the ships in USN service.

UPDATED

FRIGATES

Notes: The Kuang Hua 7 programme has superseded the former Kuang Hua 5 programme for the procurement of a new class of frigates/corvettes to replace the Knox class frigates.
It is understood that there is a requirement for up to eight new ships of above 2,000 tons with a main armament of Hsiung Feng-II missiles.
It is not clear whether the ships are to be procured abroad (ex-US Spruance class are a possibility) or built locally.

8 KNOX CLASS (FFGH)

Name	No	Builders	Laid down	Launched	Commissioned	Recommissioned
CHIN YANG (ex-*Robert E Peary*)	932 (ex-FF 1073)	Lockheed Shipbuilding	20 Dec 1970	23 June 1971	23 Sep 1972	6 Oct 1993
FONG YANG (ex-*Brewton*)	933 (ex-FF 1086)	Avondale Shipyards	2 Oct 1970	24 July 1971	8 July 1972	6 Oct 1993
FENG YANG (ex-*Kirk*)	934 (ex-FF 1087)	Avondale Shipyards	4 Dec 1970	25 Sep 1971	9 Sep 1972	6 Oct 1993
LAN YANG (ex-*Joseph Hewes*)	935 (ex-FF 1078)	Avondale Shipyards	15 May 1969	7 Mar 1970	22 Apr 1971	4 Aug 1995
HAE YANG (ex-*Cook*)	936 (ex-FF 1083)	Avondale Shipyards	20 Mar 1970	23 Jan 1971	18 Dec 1971	4 Aug 1995
HWAI YANG (ex-*Barbey*)	937 (ex-FF 1088)	Avondale Shipyards	5 Feb 1971	4 Dec 1971	11 Nov 1972	4 Aug 1995
NING YANG (ex-*Aylwin*)	938 (ex-FF 1081)	Avondale Shipyards	13 Nov 1969	29 Aug 1970	18 Sep 1971	18 Oct 1999
YI YANG (ex-*Valdez*)	939 (ex-FF 1096)	Avondale Shipyards	30 June 1972	24 Mar 1973	27 July 1974	18 Oct 1999

Displacement, tons: 3,011 standard; 3,877 (932, 935), 4,260
  (933, 934) full load
Dimensions, feet (metres): 439.6 × 46.8 × 15; 24.8 (sonar)
  (134 × 14.3 × 4.6; 7.8)
Main machinery: 2 Combustion Engineering/Babcock & Wilcox
  boilers; 1,200 psi *(84.4 kg/cm²)*; 950°F *(510°C)*; 1 turbine;
  35,000 hp *(26 MW)*; 1 shaft
Speed, knots: 27. **Range, n miles:** 4,000 at 22 kt on 1 boiler
Complement: 288 (17 officers) including aircrew

Missiles: SSM: 8 McDonnell Douglas Harpoon; active radar
  homing to 130 km *(70 n miles)* at 0.9 Mach; warhead 227 kg.
  A/S: Honeywell ASROC Mk 16 octuple launcher with reload
  system (has 2 cells modified to fire Harpoon) ❶; inertial
  guidance from 1.6-10 km *(1-5.4 n miles)*; payload Mk 46 Mod
  5 Neartip.
Guns: 1 FMC 5 in *(127 mm)*/54 Mk 42 Mod 9 ❷; 20-40 rds/min
  to 24 km *(13 n miles)* anti-surface; 14 km *(7.7 n miles)* anti-
  aircraft; weight of shell 32 kg.
  1 General Electric/General Dynamics 20 mm/76 6-barrelled
  Mk 15 Vulcan Phalanx ❸; 3,000 rds/min combined to 1.5 km.
  4 Type 75 20 mm.
Torpedoes: 4—324 mm Mk 32 (2 twin) fixed tubes ❹. 22
  Honeywell/Alliant Mk 46 Mod 5; anti-submarine; active/
  passive homing to 11 km *(5.9 n miles)* at 40 kt; warhead
  44 kg. May be replaced by 2 triple tubes.
Countermeasures: Decoys: 2 Loral Hycor SRBOC 6-barrelled
  fixed Mk 36 ❺; IR flares and chaff to 4 km *(2.2 n miles)*. T Mk 6
  Fanfare/SLQ-25 Nixie; torpedo decoy. Prairie Masker hull and
  blade rate noise suppression.
ESM/ECM: SLQ-32(V)2 ❻; radar warning. Sidekick modification
  adds jammer and deception system.
Combat data systems: Link 14 receive only. Link W may be
  fitted. FFISTS (Frigate Integrated Shipboard Tactical System).
  RADDS (Radar Displays and Distribution System).
Weapons control: SWG-1A Harpoon LCS. Mk 68 GFCS. Mk 114
  ASW FCS. Mk 1 target designation system. SRQ-4 for
  LAMPS I.
Radars: Air search: Lockheed SPS-40B ❼; B-band.
  Surface search: Raytheon SPS-10 or Norden SPS-67 ❽; G-band.
  Navigation: Marconi LN66; I-band.
  Fire control: Western Electric SPG-53A/D/F ❾; I/J-band.
  Tacan: SRN 15. IFF: UPX-12.
Sonars: EDO/General Electric SQS-26CX; bow-mounted; active
  search and attack; medium frequency.
  EDO SQR-18A(V)1; passive towed array.

Helicopters: 1 MD 500 ❿.

Programmes: *Fong Yang* leased from the US on 23 July 1992.
  Chin Yang 7 August 1992 and *Feng Yang* 6 August 1993. *Hae*

CHIN YANG   *(Scale 1 : 1,200), Ian Sturton* / 0121406

YI YANG   *9/1999, van Ginderen Collection* / 0080777

Yang leased 31 May 1994; *Hwai Yang* 21 June 1994 and *Lan
  Yang* 30 June 1994. The second batch of three were
  overhauled and upgraded by Long Beach Shipyard, California.
  Ning Yang and *Yi Yang* transferred by sale on 29 April 1998,
  and refitted at Denton Shipyard, South Carolina. The transfer
  of a third (ex-*Pharris* 1094) was declined as were further offers
  of ex-*Whipple* (1062) and ex-*Downes* (1070) for use as spares.
Structure: ASROC-torpedo reloading capability (note slanting
  face of bridge structure immediately behind ASROC). Four Mk

32 torpedo tubes are fixed in the midships structure, two to a
  side, angled out at 45°. The arrangement provides improved
  loading capability over exposed triple Mk 32 torpedo tubes. A
  4,000 lb lightweight anchor is fitted on the port side and an
  8,000 lb anchor fits into the after section of the sonar.
Operational: Seasprite helicopters were planned to be embarked
  but this now seems unlikely. All of the class are assigned to
  168 Patrol Squadron at Suao. *Lan Yang* is the Flagship.

VERIFIED

8 CHENG KUNG CLASS (KWANG HUA 1 PROJECT) (FFGHM)

Name	No	Builders	Laid down	Launched	Commissioned
CHENG KUNG	1101	China SB Corporation, Kaohsiung	7 Jan 1990	5 Oct 1991	7 May 1993
CHENG HO	1103	China SB Corporation, Kaohsiung	21 Dec 1990	15 Oct 1992	28 Mar 1994
CHI KUANG	1105	China SB Corporation, Kaohsiung	4 Oct 1991	27 Sep 1993	4 Mar 1995
YUEH FEI	1106	China SB Corporation, Kaohsiung	5 Sep 1992	26 Aug 1994	7 Feb 1996
TZU-I	1107	China SB Corporation, Kaohsiung	7 Aug 1994	13 July 1995	9 Jan 1997
PAN CHAO	1108	China SB Corporation, Kaohsiung	25 July 1995	4 July 1996	16 Dec 1997
CHANG CHIEN	1109	China SB Corporation, Kaohsiung	4 Dec 1995	14 May 1997	1 Dec 1998
TIEN TAN	1110	China SB Corporation, Kaohsiung	21 Feb 2001	15 Oct 2002	11 Mar 2004

Displacement, tons: 2,750 light; 4,105 full load
Dimensions, feet (metres): 453 × 45 × 14.8; 24.5 (sonar)
  (138.1 × 13.7 × 4.5; 7.5)
Main machinery: 2 GE LM 2500 gas turbines; 41,000 hp
  (30.59 MW) sustained; 1 shaft; cp prop
  2 auxiliary retractable props; 650 hp *(484 kW)*
Speed, knots: 29. **Range, n miles:** 4,500 at 20 kt
Complement: 234 (15 officers) including 19 aircrew

Missiles: SSM: 8 Hsiung Feng II ❶ (2 quad); inertial guidance;
  active radar/IR homing to 80 km *(43.2 n miles)* at 0.85 Mach;
  warhead 190 kg or Harpoon Block II in due course.
  SAM: 40 GDC Standard SM1-MR; Mk 13 launcher ❷; command
  guidance; semi-active radar homing to 46 km *(25 n miles)* at
  2 Mach.
Guns: 1 OTO Melara 76 mm/62 Mk 75 ❸; 85 rds/min to 16 km
  (8.7 n miles); weight of shell 6 kg.
  2 Bofors 40 mm/70 ❹. 3—20 mm Type 75 (on hangar roof
  when fitted).
  1 GE/GD 20 mm/76 Vulcan Phalanx 6-barrelled Mk 15 ❺;
  3,000 rds/min combined to 1.5 km.
Torpedoes: 6—324 mm Mk 32 (2 triple) tubes ❻. Honeywell/
  Alliant Mk 46 Mod 5; anti-submarine; active/passive homing
  to 11 km *(5.9 n miles)* at 40 kt; warhead 44 kg.
Countermeasures: Decoys: 4 Kung Fen 6 chaff launchers or
  locally produced version of RBOC (114 mm). SLQ-25A Nixie;
  torpedo decoy.
  ESM/ECM: Chang Feng IV (locally produced version of
  SLQ-32(V)2 with Sidekick); combined radar warning and
  jammers.
Combat data systems: Norden SYS-2(V)2 action data
  automation with UYK 43 computer. Ta Chen link (from *Chi
  Kuang* onwards and being backfitted).

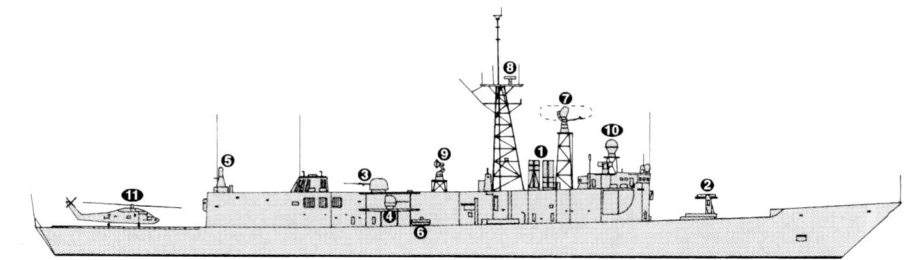

CHENG KUNG

(Scale 1 : 1,200), Ian Sturton / 0019226

Weapons control: Loral Mk 92 Mod 6. Mk 13 Mod 4 weapon
  direction system. Mk 114 ASW. 2 Mk 24 optical directors. Mk
  309 TFCS.
Radars: Air search: Raytheon SPS-49(V)5 or SPS-49A (1108-9)
  ❼; C/D-band.
  Surface search: ISC Cardion SPS-55 ❽ or Raytheon Chang Bai;
  I/J-band.
  Fire control: USN UD 417 STIR ❾; I/J-band.
  Unisys Mk 92 Mod 6 ❿; I/J-band.
Sonars: Raytheon SQS-56/DE 1160P; hull-mounted; active
  search and attack; medium frequency.
  SQR-18A(V)2; passive towed array or BAe/Thomson Sintra
  ATAS active towed array (from *Chi Kuang* onwards).

Helicopters: 2 Sikorsky S-70C(M) ⓫ (only 1 embarked).

Programmes: First two ordered 8 May 1989. Named after
  Chinese generals and warriors. An eighth of class was ordered
  in late July 1999. Originally this ship was planned to be the
  first of a Flight II design, which was scrapped.
Modernisation: It is reported that Harpoon is to replace Hsiung
  Feng in due course. A PDMS may be fitted vice the 40 mm
  guns. RAM is a possibility.
Structure: Similar to the USS *Ingraham*. RAST helicopter
  hauldown. The area between the masts had to be
  strengthened to take the Hsiung Feng II missiles. Prairie
  Masker hull acoustic suppression system fitted.
Operational: Form the 146th Squadron based at Makung
  (Pescadores).

UPDATED

CHENG HO

10/2001, Chris Sattler / 0534104

YUEH FEI

4/2004, Chris Sattler /* 1044571

6 KANG DING (LA FAYETTE) CLASS (KWANG HUA 2 PROJECT) (FFGHM)

Name	No	Builders	Laid down	Launched	Commissioned
KANG DING	1202	Lorient Dockyard/Kaohsiung Shipyard	26 Aug 1993	12 Mar 1994	24 May 1996
SI NING	1203	Lorient Dockyard/Kaohsiung Shipyard	27 Apr 1994	5 Nov 1994	15 Sep 1996
KUN MING	1205	Lorient Dockyard/Kaohsiung Shipyard	7 Nov 1994	13 May 1995	26 Feb 1997
DI HUA	1206	Lorient Dockyard/Kaohsiung Shipyard	1 July 1995	27 Nov 1995	14 Aug 1997
WU CHANG	1207	Lorient Dockyard/Kaohsiung Shipyard	1 July 1995	27 Nov 1995	16 Dec 1997
CHEN TE	1208	Lorient Dockyard/Kaohsiung Shipyard	27 Dec 1995	2 Aug 1996	16 Jan 1998

Displacement, tons: 3,800 full load
Dimensions, feet (metres): 407.5 × 50.5 × 18 (screws)
  (124.2 × 15.4 × 5.5)
Main machinery: CODAD; 4 SEMT-Pielstick 12 PA6 V 280 STC
  diesels; 23,228 hp(m) *(17.08 MW)*; 2 shafts; LIPS cp props
Speed, knots: 25. **Range, n miles:** 7,000 at 15 kt
Complement: 134 (15 officers) plus 25 spare

Missiles: SSM: 8 Hsiung Feng II (2 quad) ❶; inertial guidance;
  active radar/IR homing to 80 km *(43.2 n miles)* at 0.85 Mach;
  warhead 190 kg.
  SAM: 1 Sea Chaparral quad launcher ❷; IR homing to 3 km *(1.6 n
  miles)* supersonic; warhead 5 kg.
Guns: 1 OTO Melara 76 mm/62 Mk 75 ❸; 85 rds/min to 16 km
  (8.7 n miles); weight of shell 6 kg.
  1 Hughes 20 mm/76 Vulcan Phalanx Mk 15 Mod 2 ❹.
  2 Bofors 40 mm/70 ❺. 2 CS 20 mm Type 75.
Torpedoes: 6—324 mm Mk 32 (2 triple) tubes ❻; Alliant Mk 46
  Mod 5; active/passive homing to 11 km *(5.9 n miles)* at 40 kt;
  warhead 44 kg.
Countermeasures: Decoys: 2 CSEE Dagaie chaff launchers ❼.
  ESM/ECM: Thomson-CSF DR 3000S; intercept and jammer.
  Chang Feng IV (1206); intercept and jammer.
Combat data systems: Thomson-CSF TACTICOS. Link W (Ta
  Chen).
Weapons control: CSEE Najir Mk 2 optronic director ❽.
Radars: Air/surface search: Thomson-CSF DRBV-26D Jupiter II
  (with LW08 aerial) ❾; D-band.

Surface search: Thomson-CSF Triton G ❿; G-band.
Fire control: 2 Thomson-CSF Castor IIC ⓫; I/J-band.
Navigation and helo control: 2 Racal Decca 20V90; I-band.
Sonars: BAe/Thomson Sintra ATAS (V)2; active towed array.
  Thomson Sintra Spherion B; bow-mounted; active search;
  medium frequency.

Helicopters: 1 Sikorsky S-70C(M)1 ⓬ Thunderhawk.

Programmes: Sale of up to 16 of the class authorised by the
French government in August 1991. Contract for six signed
with Thomson-CSF in early 1992, manufactured in France
with some weapon assembly by China SB Corporation at
Kaohsiung in Taiwan. First one to Taiwan in March 1996 and

the last in January 1998. Names are those of Chinese cities.
Second batch of 10 to be built by China SB Corporation was
planned but this now seems unlikely.
Modernisation: There are plans to move Phalanx to the bridge
roof and fit two 10-round RAM launchers on the hangar.
Structure: There are considerable differences with the French
'La Fayette' design in both superstructure and weapon
systems. A comprehensive ASW fit has been added as well as
additional gun armament. There is also no stern hatch for
launching RIBs. Some of the weapons were fitted after arrival
in Taiwan. DCN Samahé helicopter landing gear installed.
Operational: Form 124 Squadron based at Tsoying.

UPDATED

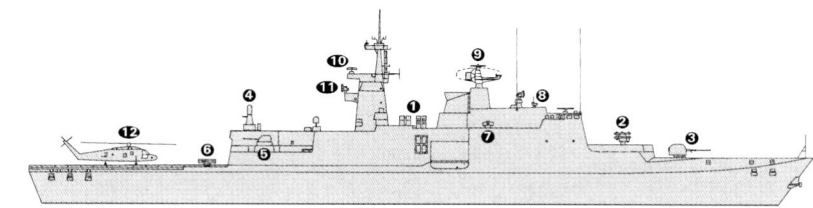

KANG DING

(Scale 1 : 1,200), Ian Sturton / 0121405

KUN MING

*3/2004 *, Chris Sattler /* 1044572

SI NING

6/2000, Sattler/Steele / 0106598

SHIPBORNE AIRCRAFT

Notes: Negotiations to acquire SH-2F Seasprite helicopters for the Knox class, conducted for several years, have not been satisfactorily concluded.

Numbers/Type: 21 Sikorsky S-70C(M)1 Thunderhawks.
Operational speed: 145 kt *(269 km/h).*
Service ceiling: 19,000 ft *(5,790 m).*
Range: 324 n miles *(600 km).*
Role/Weapon systems: First delivered in 1991. This is a variant of the SH-60B and became seaborne with the first Cheng Kung and Kang Ding class frigates. 701 and 702 Squadrons. Two modified for EW and Sigint role. Another 14 S-70B/C SAR and assault aircraft belong to the Air Force. Sensors: APS 128 search radar; Litton ALR 606(V)2 ESM; ARR 84 sonobuoy receiver with Litton ASN 150 datalink; Allied AQS 18(V)3 dipping sonar; ASQ 504 MAD. Ta Chen datalink to be fitted. Weapons: ASW; two Hughes Mk 46 Mod 5 torpedoes or two Mk 64 depth bombs. ASV; could carry ASM. *VERIFIED*

THUNDERHAWK
1/2000, C Chung / 0106599

Numbers/Type: 9 Hughes MD 500/ASW.
Operational speed: 110 kt *(204 km/h).*
Service ceiling: 16,000 ft *(4,880 m).*
Range: 203 n miles *(376 km).*
Role/Weapon systems: Short-range ASW helicopter with limited surface search capability. 501 ASW Squadron. Sensors: Search radar, Texas Instruments ASQ 81(V)2 MAD. Weapons: ASW; one Mk 46 Mod 5 torpedo or two depth bombs. ASV; could carry machine gun pods. *VERIFIED*

MD 500
1/1995, L J Lamb / 0080778

LAND-BASED MARITIME AIRCRAFT

Notes: Four Grumman E-2T Hawkeye AEW aircraft were acquired by the Air Force in February 1995.

Numbers/Type: 3/21 Grumman S-2E/S-2T (Turbo) Trackers.
Operational speed: 130 kt *(241 km/h).*
Service ceiling: 25,000 ft *(7,620 m).*
Range: 1,350 n miles *(2,500 km).*
Role/Weapon systems: Patrol and ASW tasks transferred to the Navy in July 1998; 21 aircraft updated with turboprop engines and new sensors. Based at Pintung. To be replaced by P-3C when they enter service. Sensors: APS 504 search radar, ESM, MAD, AAS 40 FLIR, SSQ-41B, SSQ-47B sonobuoys; AQS 902F sonobuoy processor; ASN 150 datalink. Weapons: ASW; four Mk 44 torpedoes, Mk 54 depth charges or Mk 64 depth bombs or mines. ASV; Hsiung Feng II ASM; six 127 mm rockets. *VERIFIED*

TRACKER
6/2002, Adolfo Ortigueira Gil / 0569245

Numbers/Type: 12 Lockheed Martin P-3C Orion.
Operational speed: 411 kt *(761 km/h).*
Service ceiling: 28,300 ft *(8,625 m).*
Range: 4,000 n miles *(7,410 km).*
Role/Weapon systems: Authorised by US government in 2001 for sale to Taiwan. Delivery date uncertain. Equipment based on aircraft in US service. Sensors: APS-115 search radar, ASQ-81 MAD, UYS-1 acoustic suite, up to 100 sonobuoys, AAR-36 FLIR, cameras, ALR-66 ESM. AAR 47 ESM and ALE 47 chaff/IR dispenser in Update II. Weapons: ASW; four Mk 46/50 torpedoes or depth bombs. ASUW; four AGM-84C Harpoon, AGM-65F Maverick, AGM-84E SLAM, six Mk 55/56 mines. *VERIFIED*

ORION
4/2000, Hachiro Nakai / 0106787

Numbers/Type: 12 Sikorsky MH-53E Sea Dragon.
Operational speed: 170 kt *(315 km/h).*
Service ceiling: 18,500 ft *(5,640 m).*
Range: 1,000 n miles *(1,850 km).*
Role/Weapon systems: Authorised in 2001 by US government for sale to Taiwan. Delivery date uncertain. Three-engined AMCM helicopter (HM) similar to Super Stallion; tows ALQ-166 MCM sweep equipment; self-deployed if necessary. Sensors: Northrop Grumman AQS-14A dipping sonar. Weapons: Two 12.7 mm guns for self-defence. *VERIFIED*

SEA DRAGON
5/1999, US Navy / 0106785

PATROL FORCES

Notes: All coastal patrol craft were transferred to the Maritime Police on 8 December 1992. The Maritime Police became the Coast Guard 1 January 2000.

12 + (12) JIN CHIANG CLASS (LARGE PATROL CRAFT) (PCG)

Name	No	Builders	Launched	Commissioned
JIN CHIANG	603	Lien-Ho, Kaohsiung	1 May 1994	1 Dec 1994
TAN CHIANG	605	China SB, Kaohsiung	18 June 1998	7 Sep 1999
HSIN CHIANG	606	China SB, Kaohsiung	14 Aug 1998	7 Sep 1999
FENG CHIANG	607	China SB, Kaohsiung	22 Oct 1998	29 Oct 1999
TSENG CHIANG	608	China SB, Kaohsiung	16 Nov 1998	29 Oct 1999
KAO CHIANG	609	China SB, Kaohsiung	15 Dec 1998	29 Oct 1999
JING CHIANG	610	China SB, Kaohsiung	13 May 1999	15 Feb 2000
HSIAN CHIANG	611	China SB, Kaohsiung	16 July 1999	15 Feb 2000
TSI CHIANG	612	China SB, Kaohsiung	22 Dec 1999	15 Feb 2000
PO CHIANG	614	China SB, Kaohsiung	22 Dec 1999	21 July 2000
CHAN CHIANG	615	China SB, Kaohsiung	21 Jan 2000	21 July 2000
CHU CHIANG	617	China SB, Kaohsiung	25 Feb 2000	21 July 2000

Displacement, tons: 680 full load
Dimensions, feet (metres): 201.4 × 31.2 × 9.5 *(61.4 × 9.5 × 2.9)*
Main machinery: 2 MTU 20V 1163 TB93 diesels; 20,128 hp(m) *(14.79 MW)*; 2 shafts
Speed, knots: 25. **Range, n miles:** 4,150 at 15 kt
Complement: 50 (7 officers)

Missiles: SSM: 4 Hsiung Feng I; radar or optical guidance to 36 km *(19.4 n miles)* at 0.7 Mach; warhead 75 kg.
Guns: 1 Bofors 40 mm/70. 1 CS 20 mm Type 75. 2—12.7 mm MGs.
Depth charges: 2 racks.
Mines: 2 rails for Mk 6.
Weapons control: Honeywell H 930 Mod 2 MFCS. Contraves WCS. Rafael Sea Eye FLIR; range out to 3 km.
Radars: Air/surface search: Marconi LN66; I-band.
Fire control: Hughes HR-76C5; I/J-band.
Navigation: Racal Decca Bridgemaster; I-band.
Sonars: Simrad; search and attack; high frequency.

Programmes: Kwang Hua Project 3 design by United Ship Design Centre. First one laid down 25 June 1993. Eleven more ordered 26 June 1997. A further 12 are expected by 2010. *VERIFIED*

HSIN CHIANG
6/2002, Ships of the World / 0569244

2 LUNG CHIANG CLASS (FAST ATTACK CRAFT—MISSILE) (PGGF)

Name	No	Builders	Commissioned
LUNG CHIANG	601 (ex-PGG 581)	Tacoma Boatbuilding, WA	15 May 1978
SUI CHIANG	602 (ex-PGG 582)	China SB Corporation, Kaohsiung	31 Dec 1981

Displacement, tons: 270 full load
Dimensions, feet (metres): 164.5 × 23.1 × 9.5 *(50.2 × 7.3 × 2.9)*
Main machinery: CODAG; 3 Avco Lycoming TF-40A gas turbines; 12,000 hp *(8.95 MW)* sustained; 3 Detroit 12V-149TI diesels; 2,736 hp *(2.04 MW)* sustained; 3 shafts; cp props
Speed, knots: 20 kt diesels; 38 kt gas. **Range, n miles:** 3,100 at 12 kt on 1 diesel; 800 at 36 kt
Complement: 38 (5 officers)

Missiles: SSM: 4 Hsiung Feng I; radar or optical guidance to 36 km *(19.4 n miles)* at 0.7 Mach; warhead 75 kg.
Guns: 1 OTO Melara 3 in *(76 mm)*/62; 60 rds/min to 16 km *(8.7 n miles)*; weight of shell 6 kg.
 1 Bofors 40 mm/70. 2—12.7 mm MGs.
Countermeasures: Decoys: 4 Israeli AV2 (601) or SMOC-4 (602) chaff launchers.
ESM: WD-2A; intercept.
Combat data systems: IPN 10 action data automation.
Weapons control: NA 10 Mod 0 GFCS. Honeywell H 930 Mod 2 MFCS (602).
Radars: Surface/air search: Selenia RAN 11 L/X; D/I-band.
Fire control: RCA HR 76; I/J-band (for SSM) (602).
 Selenia RAN IIL/X; I/J-band for SSM (601).
Navigation: SPS-58(A); I-band.

Programmes: Similar to the US Patrol Ship Multi-Mission Mk 5 (PSMM Mk 5). Second of class was built to an improved design. A much larger number of this class was intended, all to be armed with Harpoon. However at that time the US ban on export of Harpoon to Taiwan coupled with the high cost and doubts about seaworthiness caused the cancellation of this programme.
Structure: Fin stabilisers were fitted to help correct the poor sea-keeping qualities of the design. Both have had engine room fires caused by overheating in GT gearboxes.
Operational: *Lung Chiang* may be non-operational and could be sold.
VERIFIED

SUI CHIANG   *4/1997, Ships of the World /* 0019231

47 HAI OU CLASS (FAST ATTACK CRAFT—MISSILE) (PTG)

FABG 7-12 FABG 14-21 FABG 23-30 FABG 32-39 FABG 41-45
FABG 47-57 FABG 59

Displacement, tons: 47 full load
Dimensions, feet (metres): 70.8 × 18 × 3.3 *(21.6 × 5.5 × 1)*
Main machinery: 2 MTU 12V 331 TC82 diesels; 2,605 hp(m) *(1.92 MW)* sustained; 2 shafts
Speed, knots: 30. **Range, n miles:** 700 at 32 kt
Complement: 10 (2 officers)

Missiles: SSM: 2 Hsiung Feng I; radar or optical guidance to 36 km *(19.4 n miles)* at 0.7 Mach; warhead 75 kg.
Guns: 1 CS 20 mm Type 75. 2—12.7 mm MGs.
Countermeasures: Decoys: 4 Israeli AV2 chaff launchers.
ESM: WD-2A; intercept.
Weapons control: Kollmorgen Mk 35 optical director.
Radars: Surface search: Marconi LN66; I-band.
Fire control: RCA R76 C5; I-band.

Programmes: This design was developed by Sun Yat Sen Scientific Research Institute from the basic Israeli Dvora plans. Built by China SB Corporation (Tsoying SY), Kaohsiung except for the first pair (FABG 5-6) which were the original Dvora class hulls and were commissioned on 31 December 1977.
Structure: Aluminium alloy hulls. The first series had a solid mast and the missiles were nearer the stern. Second series changed to a lattice mast and moved the missiles further forward allowing room for two 12.7 mm MGs right aft. One 20 mm has been added on the stern.
Operational: The prototype reached 45 kt on trials but top speeds are now reported as being much reduced. These craft often carry shoulder-launched SAMs. One task is to provide exercise high-speed targets in shallow waters. From 1997 organised in five divisions based at Makung, Tamsui, Tsoying, Suao and Keelung. Not all are operational and the class is likely to be paid off in the short term.
Sales: Two similar craft to Paraguay in 1996.
VERIFIED

FABG 56   *6/2000, DTM /* 0126194

0 + 1 (28) KWANG HUA 6 CLASS (PTG)

Displacement, tons: 180 standard
Dimensions, feet (metres): 128.0 × 24.9 × 6.2 *(39.0 × 7.6 × 1.9)*
Main machinery: 2 diesels; 2 shafts
Speed, knots: 30. **Range, n miles:** 1,150 at 22 kt
Complement: 14
Missiles: SSM: 4 Hsiung Feng II.
Guns: 1 CS 20 mm Type 75.
Countermeasures: Decoys: Chaff launchers. ESM.
Weapons control: Optronic director.
Radars: Surface search. Fire control.

Comment: Funds allocated for the budget period July 1998 to June 2003 to build these craft in Taiwan to replace the Hai Ou class. First of class laid down in early 2001 and launched on 26 September 2002. Commissioned in October 2003.
UPDATED

KWANG HUA 6   *9/2002, DTM /* 0529547

8 NING HAI CLASS (LARGE PATROL CRAFT) (PCF)

NING HAI PCL 1   AN HAI PCL 2   PCL 3   PCL 5-9

Displacement, tons: 143 full load
Dimensions, feet (metres): 105 × 29.5 × 5.9 *(32 × 9 × 1.8)*
Main machinery: 3 MTU 12V 396 TB93 diesels; 4,890 hp(m) *(3.6 MW)* sustained; 3 shafts
Speed, knots: 40
Complement: 18 (2 officers)
Guns: 1 Bofors 40 mm/60. 1 CS 20 mm Type 75.
Depth charges: 2 racks.
Radars: Surface search: Decca; I-band.
Sonars: Hull-mounted; active search and attack; high frequency.

Comment: Built to Vosper QAF design by China SB Corporation, Kaohsiung in 1987-90. Previously reported numbers had been exaggerated. They are used mainly for harbour defence against midget submarines and frogmen and also for Fishery protection tasks.
VERIFIED

PCL 5   *6/2002, Ships of the World /* 0569243

AMPHIBIOUS FORCES

1 CABILDO CLASS (LSDM)

Name	No	Builders	Commissioned
CHUNG CHENG (ex-*Comstock*)	191 (ex-LSD 19)	Newport News, Virginia	2 July 1945

Displacement, tons: 4,790 standard; 9,375 full load
Dimensions, feet (metres): 475 × 76.2 × 18 *(144.8 × 23.2 × 5.5)*
Main machinery: 2 boilers; 435 psi *(30.6 kg/cm²)*; 740°F *(393°C)*; 2 turbines; 7,000 hp *(5.22 MW)*; 2 shafts
Speed, knots: 15.4. **Range, n miles:** 8,000 at 15 kt
Complement: 316
Military lift: 3 LCUs or 18 LCMs or 32 LVTs in docking well
Missiles: SAM: 1 Sea Chaparral quadruple launcher.
Guns: 12 Bofors 40 mm/56 (2 quad, 2 twin).
Weapons control: US Mk 26 Mod 4.
Radars: Surface search: Raytheon SPS-5; G/H-band.
Navigation: Marconi LN66; I-band.

Comment: Launched 28 April 1945 and transferred to Taiwan on 1 October 1985 having been bought from a ship breaker. SAM system fitted in 1992. Collision with merchant ship on 28 June 2001 resulted in five months repair work. Second of class scrapped in mid-1999.
VERIFIED

CHUNG CHENG   *3/2001, C Chung /* 0126197

1 ANCHORAGE CLASS (LSDH)

Name	No	Builders	Commissioned
SHIU HAI (ex-*Pensacola*)	LSD 193 (ex-LSD 38)	General Dynamics, Quincy	27 Mar 1971

Displacement, tons: 8,600 light; 13,700 full load
Dimensions, feet (metres): 553.3 × 84 × 20 *(168.6 × 25.6 × 6)*
Main machinery: 2 Foster-Wheeler boilers; 600 psi *(42.3 kg/cm²)*; 870°F *(467°C)*; 2 De Laval turbines; 24,000 hp *(18 MW)*; 2 shafts
Speed, knots: 22. **Range, n miles:** 14,800 at 12 kt
Complement: 374 (24 officers)
Military lift: 366 troops (18 officers); 2 LCU or 18 LCM 6 or 9 LCM 8 or 50 LVT; 1 LCM 6 on deck; 2 LCPL and 1 LCVP on davits. Aviation fuel, 90 tons
Guns: 2 General Electric/General Dynamics 20 mm/76 6-barrelled Vulcan Phalanx Mk 15; 3,000 rds/min combined to 1.5 km.
2—25 mm Mk 38. 6—12.7 mm MGs.
Countermeasures: Decoys: 4 Loral Hycor SRBOC 6-barrelled Mk 36; IR flares and chaff to 4 km *(2.2 n miles)*.
ESM: SLQ-32(V)1; intercept.
Radars: Air search: Lockheed SPS-40B; B-band.
Surface search: Raytheon SPS-10F; G-band.
Navigation: Marconi LN66; I-band.
Helicopters: Platform only.

Comment: First one acquired from US Navy 30 September 1999 and arrived in Taiwan on 2 June 2000. Transfer of ex-*Anchorage* (LSD 36) did not take place as expected in 2004 although procurement of a further amphibious ship remains a requirement. Has a docking well 131.1 × 15.2 m and two 50 ton cranes. Based at Tsoying. *UPDATED*

SHIU HAI   6/2000, DTM / 0126195

2 NEWPORT CLASS (LSTH)

Name	No	Builders	Commissioned
CHUNG HO (ex-*Manitowic*)	232 (ex-LST 1180)	Philadelphia Shipyard	24 Jan 1970
CHUNG PING (ex-*Sumter*)	233 (ex-LST 1181)	Philadelphia Shipyard	20 June 1970

Displacement, tons: 4,975 light; 8,450 full load
Dimensions, feet (metres): 522.3 × 69.5 × 17.5 *(159.2 × 21.2 × 5.3)*
Main machinery: 6 ALCO 16-251 diesels; 16,500 hp *(12.3 MW)* sustained; 2 shafts; cp props; bow thruster
Speed, knots: 20. **Range, n miles:** 14,250 at 14 kt
Complement: 257 (13 officers)
Military lift: 400 troops; 500 tons vehicles; 3 LCVPs and 1 LCPL on davits
Guns: 1 General Electric/General Dynamics 20 mm Vulcan Phalanx Mk 15.
4—40 mm/60 (2 twin).
Countermeasures: ESM: WD-2A (233); intercept.
ESM/ECM: Chang Feng III (232); intercept and jammer.
Radars: Surface search: Raytheon SPS-67; G-band.
Navigation: Marconi LN66; I-band.
Helicopters: Platform only.

Comment: First pair transferred from USA by lease confirmed for both ships on 1 July 1995. Refitted at Newport News and recommissioned 8 May 1997, sailing for Taiwan after a short operational work-up. Purchased outright on 29 September 2000. Transfer of further ships is unlikely. These ships unload by a 112 ft ramp over their bow. The ramp is supported by twin derrick arms. A ramp just forward of the superstructure connects the lower tank deck with the main deck and a vehicle passage through the superstructure provides access to the parking area amidships. A stern gate to the tank deck permits unloading of amphibious tractors into the water, or unloading of other vehicles into an LCU or on to a pier. Vehicle stowage covers 19,000 sq ft. Length over derrick arms is 562 ft *(171.3 m)*; full load draught is 11.5 ft forward and 17.5 ft aft. Bow thruster fitted to hold position offshore while unloading amphibious tractors. *VERIFIED*

CHUNG HO   6/2000, Sattler/Steele / 0106602

4 LSM 1 CLASS (LSM)

MEI CHIN (ex-*LSM 155*) 341 (ex-649)		MEI SUNG (ex-*LSM 457*) 347 (ex-694)	
MEI PING (ex-*LSM 471*) 353 (ex-659)		MEI LO (ex-*LSM 362*) 356 (ex-637)	

Displacement, tons: 1,095 full load
Dimensions, feet (metres): 203.5 × 34.2 × 8.3 *(62.1 × 10.4 × 2.5)*
Main machinery: 2 Fairbanks Morse 38D8-1/8-10 diesels; 3,540 hp *(2.64 MW)* sustained (356 and 353); 4 GM 16-278A diesels; 3,000 hp *(2.24 MW)* (341 and 347); 2 shafts
Speed, knots: 13. **Range, n miles:** 2,500 at 12 kt
Complement: 65-75
Guns: 2 Bofors 40 mm/56 (twin). 4 or 8 Oerlikon 20 mm (4 single or 4 twin).
Radars: Surface search: SO 8; I-band.

Comment: All built in 1945. *Mei Chin* and *Mei Sung* transferred from US 1946, *Mei Ping* in 1956 and *Mei Lo* in 1962. Rebuilt in Taiwan and bear little resemblance to 1970s photographs. *VERIFIED*

MEI SUNG   6/2000, C Chung / 0126198

11 LST 1-510 and 512-1152 CLASSES (LST)

CHUNG HAI (ex-*LST 755*) 201 (ex-697)	CHUNG CHIH (ex-*Sagadahoc County* LST 1091)
CHUNG CHIEN (ex-*LST 716*) 205 (ex-679)	226 (ex-655)
CHUNG SHUN (ex-*LST 732*) 208 (ex-624)	CHUNG MING (ex-*Sweetwater County*
CHUNG KUANG (ex-*LST 503*) 216 (ex-646)	LST 1152) 227 (ex-681)
CHUNG SUO (ex-*Bradley County* LST 400)	CHUNG PANG (ex-*LST 578*) 230 (ex-629)
217 (ex-667)	CHUNG YEH (ex-*Sublette County* LST 1144)
CHUNG CHI (ex-*LST 279*) 218	231 (ex-699)
CHUNG CHUAN (ex-*LST 1030*) 221 (ex-651)	

Displacement, tons: 1,653 standard; 4,080 (3,640, 1-510 class) full load
Dimensions, feet (metres): 328 × 50 × 14 *(100 × 15.2 × 4.3)*
Main machinery: 2 GM 12-567A diesels; 1,800 hp *(1.34 MW)*; 2 shafts
Speed, knots: 11.6. **Range, n miles:** 15,000 at 10 kt
Complement: Varies-100-125 in most ships
Guns: Varies-up to 10 Bofors 40 mm/56 (2 twin, 6 single) with some modernised ships rearmed with 2 USN 3 in *(76 mm)*/50 and 6—40 mm (3 twin).
Several Oerlikon 20 mm (twin or single).
Radars: Navigation: US SO 1, 2 or 8; I-band.

Comment: Constructed between 1943 and 1945. These ships have been rebuilt in Taiwan. Six transferred from US in 1946; two in 1947; one in 1948; eight in 1958; one in 1959; two in 1960; and one in 1961. Some have davits forward and aft. Pennant numbers have reverted to those used in the 1960s. One deleted in 1990, six more in 1993, one more in 1995 after going aground, and two more in 1997. The midships deck is occasionally used as a helicopter platform. These last 11 may be retained due to the cancellation of the programme for more locally built AKs. *VERIFIED*

CHUNG SUO   6/2000, DTM / 0126196

1 LST 512-1152 CLASS (FLAGSHIP) (AGF)

Name	No	Builders	Commissioned
KAO HSIUNG (ex-*Chung Hai*, ex-*Dukes County* LST 735)	LCC 1 (ex-219, ex-663)	Dravo Corporation, Neville Island, Penn	26 Apr 1944

Displacement, tons: 1,653 standard; 3,675 full load
Dimensions, feet (metres): 328 × 50 × 14 *(100 × 15.2 × 4.3)*
Main machinery: 2 GM 12-567A diesels; 1,800 hp *(1.34 MW)*; 2 shafts
Speed, knots: 11.6. **Range, n miles:** 11,200 at 10 kt
Complement: 195
Guns: 8 Bofors 40 mm/56 (3 twin, 2 single).
Radars: Air search: Raytheon SPS 58; D-band.
Surface search: Raytheon SPS-10; G-band.

Comment: Launched on 11 March 1944. Transferred from US in May 1957 for service as an LST. Converted to a flagship for amphibious operations and renamed and redesignated (AGC) in 1964. Purchased November 1974. Note lattice mast above bridge structure, modified bridge levels, and antenna mountings on main deck. Redesignated as Command and Control Ship LCC 1. *VERIFIED*

KAO HSIUNG   6/1999 / 0080783

10 LCU 501 CLASS (LCU)

HO CHI (ex-*LCU 1212*) 401   HO SHUN (ex-*LCU 1225*) 481   HO YUNG (ex-*LCU 1271*) 495
HO HUEI (ex-*LCU 1218*) 402   HO CHUNG (ex-*LCU 849*) 484   HO CHIE (ex-*LCU 700*) SB 1
HO YAO (ex-*LCU 1244*) 403   HO CHUN (ex-*LCU 892*) 494   HO TEN (ex-*LCU 1367*) SB 2
HO CHAO (ex-*LCU 1429*) 406

Displacement, tons: 158 light; 309 full load
Dimensions, feet (metres): 119 × 32.7 × 5 *(36.3 × 10 × 1.5)*
Main machinery: 3 GM 6-71 diesels; 522 hp *(390 kW)* sustained; 3 shafts
Speed, knots: 10
Complement: 10-25
Guns: 2 Oerlikon 20 mm. Some also may have 2—12.7 mm MGs.

Comment: Built in US in the 1940s and transferred in 1959. *SB 1* and *SB 2* are used as auxiliaries. *Ho Feng* 405 converted for ferry duties in 1998 and serves Matzu island.   ***VERIFIED***

HO SHUN   *6/2000, DTM /* 0569238

6 LCU 1466 CLASS (LCU)

HO SHAN (ex-*LCU 1596*) 488   HO SENG (ex-*LCU 1598*) 490   HO MOU (ex-*LCU 1600*) 492
HO CHUAN (ex-*LCU 1597*) 489   HO MENG (ex-*LCU 1599*) 491   HO SHOU (ex-*LCU 1601*) 493

Displacement, tons: 180 light; 360 full load
Dimensions, feet (metres): 119 × 34 × 6 *(36.3 × 10.4 × 1.8)*
Main machinery: 3 Gray Marine 64 YTL diesels; 675 hp *(504 kW)*; 3 shafts
Speed, knots: 10. **Range, n miles:** 800 at 11 kt
Complement: 15-25
Military lift: 167 tons or 300 troops
Guns: 3 Oerlikon 20 mm. Some may also have 2—12.7 mm MGs.

Comment: Built by Ishikawajima Heavy Industries Co, Tokyo, Japan, for transfer to Taiwan; completed in March 1955. All originally numbered in 200 series; subsequently changed to 400 series.   ***VERIFIED***

HO CHUAN   *1991*

2 TAIWAN TYPE LCU (LCU)

HO FONG LCU 497   HO HU LCU 498

Displacement, tons: 190 light; 439 full load
Dimensions, feet (metres): 135.5 × 29.9 × 6.9 *(41.3 × 9.1 × 2.1)*
Main machinery: 4 Detroit diesels; 1,200 hp *(895 kW)*; 2 Kort nozzle props
Speed, knots: 11. **Range, n miles:** 1,200 at 10 kt
Complement: 16
Military lift: 180 tons or 350 troops
Guns: 2—12.7 mm MGs.

Comment: Locally built versions of US types. Ramps at both ends.   ***VERIFIED***

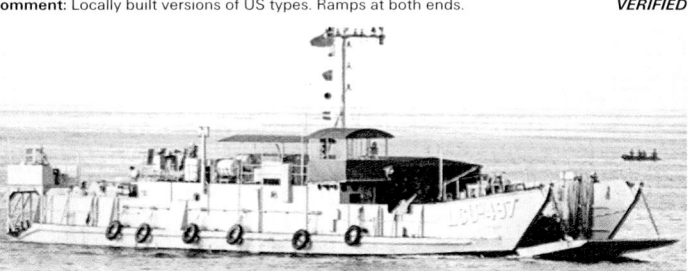

HO FONG   *6/2000, DTM /* 0569237

170 LCM 6 CLASS (LCM)

Displacement, tons: 57 full load
Dimensions, feet (metres): 56.4 × 13.8 × 3.9 *(17.2 × 4.2 × 1.2)*
Main machinery: 2 diesels; 450 hp *(336 kW)*; 2 shafts
Speed, knots: 9
Military lift: 34 tons
Guns: 1—12.7 mm MG.

Comment: Some built in the US, some in Taiwan. 20 were exchanged for torpedoes with Indonesia. Some 55 have been deleted in the last four years. Form part of 151 Squadron.   ***VERIFIED***

LCM 6   *7/2000, C Chung /* 0106603

100 LCVPs and ASSAULT CRAFT

Comment: Some ex-US, and some built in Taiwan. Most are armed with one or two 7.62 mm MGs. Two transferred to Indonesia in 1988. About 20 deleted in the last three years and 30 transferred to Honduras in 1996 for River operations. There are also a number of amphibious reconnaissance boats in the ARP 1000, 2000 and 3000 series. Form part of 151 Squadron.   ***VERIFIED***

TYPE 272   *1989, DTM (Raymond Cheung)*

MINE WARFARE FORCES

Notes: A new class of minehunters is planned. Tenders may be invited in when funds are available.

4 AGGRESSIVE CLASS (MINESWEEPERS) (MSO)

Name	No	Builders	Commissioned
YUNG YANG (ex-*Implicit*)	1306 (ex-455)	Wilmington Boat	10 Mar 1954
YUNG TZU (ex-*Conquest*)	1307 (ex-488)	Martenac, Tacoma	20 July 1955
YUNG KU (ex-*Gallant*)	1308 (ex-489)	Martenac, Tacoma	14 Sep 1955
YUNG TEH (ex-*Pledge*)	1309 (ex-492)	Martenac, Tacoma	20 Apr 1956

Displacement, tons: 720 standard; 780 full load
Dimensions, feet (metres): 172.5 × 35.1 × 14.1 *(52.6 × 10.7 × 4.3)*
Main machinery: 4 Packard ID-1700 or Waukesha diesels; 2,280 hp *(1.7 MW)*; 2 shafts; cp props
Speed, knots: 14. **Range, n miles:** 3,000 at 10 kt
Complement: 86 (7 officers)
Guns: 2—12.7 mm MGs.
Radars: Navigation: Sperry SPS-53L; I-band.
Sonars: General Electric SQQ-14; VDS; active minehunting; high frequency.

Comment: Transferred by sale to Taiwan from the USN 3 August and 30 September 1994. Delivery was delayed into 1995 while replanking work was carried out in the US. All recommissioned 1 March 1995. Second batch of three planned to transfer but were subsequently scrapped after cannibalisation for spares. All are fitted with SLQ-37 mechanical acoustic and magnetic sweeps and can carry an ROV. The plan is to update the class with a Unisys SYQ-12 minehunting system and Pluto ROVs.   ***VERIFIED***

YUNG KU   *6/2000, DTM /* 0569242

For details of the latest updates to *Jane's Fighting Ships* online and to discover the additional information available exclusively to online subscribers please visit
jfs.janes.com

4 ADJUTANT and MSC 268 CLASSES
(MINESWEEPERS—COASTAL) (MSC)

YUNG CHUAN (ex-*MSC 278*) 158   YUNG REN (ex-*St Nicholas*, ex-*MSC 64*)) 167
YUNG FU (ex-*Macaw*, ex-*MSC 77*) 162   YUNG SUI (ex-*Disksmude*, ex-*MSC 65*) 168

Displacement, tons: 375 full load
Dimensions, feet (metres): 144 × 27.9 × 8 *(43.9 × 8.5 × 2.4)*
Main machinery: 2 GM 8-268A diesels; 880 hp *(656 kW)*; 2 shafts
Speed, knots: 13. **Range, n miles:** 2,500 at 12 kt
Complement: 35
Guns: 2 Oerlikon 20 mm (twin).
Radars: Navigation: Decca 707; I-band.
Sonars: Simrad 950; hull-mounted; minehunting; high frequency.

Comment: Non-magnetic, wood-hulled minesweepers built in the US in the 1950s specifically for transfer to allied navies. All refitted 1984-86. All are in very poor condition. Several deleted so far. Two put back in service in 1996 and one in 1997 to replace three others paid off.
VERIFIED

YUNG CHUAN   *6/2000, DTM /* 0569241

4 YUNG FENG (MWV 50) CLASS
(MINEHUNTERS—COASTAL) (MHC)

YUNG FENG 1301   YUNG CHIA 1302   YUNG TING 1303   YUNG SHUN 1305

Displacement, tons: 500 full load
Dimensions, feet (metres): 163.1 × 28.5 × 10.2 *(49.7 × 8.7 × 3.1)*
Main machinery: 2 MTU 8V 396 TB93 diesels; 2,180 hp(m) *(1.6 MW)* sustained; 2 shafts
Speed, knots: 14. **Range, n miles:** 3,500 at 14 kt
Complement: 45 (5 officers)
Guns: 1—20 mm. 2—12.7 mm MGs.
Radars: Navigation: I-band.
Sonars: TSM-2022; hull-mounted; active minehunting; high frequency.

Comment: Built for the Chinese Petroleum Corporation by Abeking & Rasmussen at Lemwerder, Germany. First four delivered in 1991 as offshore oil rig support ships and then converted for minehunting in Taiwan. Thomson Sintra IBIS V minehunting system is fitted and two STN Pinguin B3 ROVs are carried. The civilian livery has been abandoned. These ships spend little time at sea.
VERIFIED

YUNG FENG   *6/2000, DTM /* 0569240

SURVEY AND RESEARCH SHIPS

1 ALLIANCE CLASS (AGOR)

Name	No	Builders	Launched	Commissioned
TA KUAN	1601	Fincantieri, Muggiano	17 Dec 1994	27 Sep 1995

Displacement, tons: 2,466 standard; 3,180 full load
Dimensions, feet (metres): 305.1 × 49.9 × 16.7 *(93 × 15.2 × 5.1)*
Main machinery: Diesel-electric; 3 MTU/AEG diesel generators; 5,712 hp(m) *(4.2 MW)*; 2 AEG motors; 5,100 hp(m) *(3.75 MW)*; 2 shafts; bow thruster; stern trainable and retractable thruster
Speed, knots: 15. **Range, n miles:** 12,000 at 12 kt
Complement: 82
Guns: 2—12.7 mm MGs.
Radars: Navigation: H/I-band.

Comment: Ordered in June 1993 and laid down 8 April 1994. Almost identical to the NATO vessel. Designed for oceanography and hydrographic research. Facilities include laboratories, position location systems, and overside deployment equipment. Equipment includes a Simrad side scan sonar EM 1200, deep and shallow echo-sounders, two radars, Navsat and Satcom, an ROV for remote inspection, and a dynamic positioning system with bow thruster and stern positioning propeller.
VERIFIED

TA KUAN   *8/1997, C Chung /* 0019239

AUXILIARIES

1 COMBAT SUPPORT SHIP (AOEHM)

Name	No	Builders	Launched	Commissioned
WU YI	530	China SB Corporation, Keelung	4 Mar 1989	23 June 1990

Displacement, tons: 7,700 light; 17,000 full load
Dimensions, feet (metres): 531.8 × 72.2 × 28 *(162.1 × 22 × 8.6)*
Main machinery: 2 MAN 14-cyl diesels; 25,000 hp(m) *(18.37 MW)*; 2 shafts
Speed, knots: 21. **Range, n miles:** 9,200 at 10 kt
Cargo capacity: 9,300 tons
Missiles: SAM: 1 Sea Chaparral quad launcher.
Guns: 2 Bofors 40 mm/70. 2 Oerlikon 20 mm GAM-BO1. 4—12.7 mm MGs.
Countermeasures: Decoys: 2 chaff launchers.
ESM: Radar warning.
Radars: 2 navigation; I-band.
Helicopters: Platform for CH-47 or S-70C(M)1.

Comment: Largest unit built so far for the Taiwanese Navy. Design assisted by the United Shipping Design Center in the US. Beam replenishment rigs on both sides. SAM system on forecastle, 40 mm guns aft of the funnels.
UPDATED

WU YI   *3/2004*, Chris Sattler / 1044573

3 WU KANG CLASS (ATTACK TRANSPORTS) (AKM)

Name	No	Builders	Commissioned
YUEN FENG	524	China SB Corporation, Keelung	10 Sep 1982
WU KANG	525	China SB Corporation, Keelung	9 Oct 1984
HSIN KANG	526	China SB Corporation, Keelung	30 Nov 1988

Displacement, tons: 2,804 standard; 4,845 full load
Dimensions, feet (metres): 334 × 59.1 × 16.4 *(101.8 × 18 × 5)*
Main machinery: 2 diesels; 2 shafts; bow thruster
Speed, knots: 20. **Range, n miles:** 6,500 at 12 kt
Complement: 61 (11 officers)
Military lift: 1,400 troops
Missiles: SAM: 1 Sea Chaparral quad launcher.
Guns: 2 Bofors 40 mm/60. 2 or 4—12.7 mm MGs.
Countermeasures: ESM: WD-2A (524 only); intercept.

Comment: First three were built and then the programme stopped. Restarted with the fourth of class laid down in July 1994. The plan was to build at about one a year to a final total of seven, but the programme has been cancelled without the fourth ship being completed. With a helicopter platform, stern docking facility and davits for four LCVP, the design resembles an LPD. Used mostly for supplying garrisons in offshore islands, and on the Spratley and Pratas islands in the South China Sea. SAM launcher is mounted aft of the foremast. Accommodation is air conditioned. *Hsin Kang* was badly damaged in harbour in collision with a merchant ship in March 1996, but was back in service by mid-1999.
VERIFIED

HSIN KANG   *6/2002, Ships of the World /* 0569239

2 PATAPSCO CLASS (SUPPORT TANKERS) (AOTL)

Name	No	Builders	Commissioned
CHANG BAI (ex-*Pecatonia* AOG 57)	507	Cargill, Inc, Savage, Minnesota, MN	12 Feb 1944
HSIN LUNG (ex-*Elkhorn* AOG 7)	517	Cargill, Inc, Savage, Minnesota, MN	28 Nov 1945

Displacement, tons: 1,850 light; 4,335 full load
Dimensions, feet (metres): 310.8 × 48.5 × 15.7 *(94.8 × 14.8 × 4.8)*
Main machinery: 2 GM 16-278A diesels; 3,000 hp *(2.24 MW)*; 2 shafts
Speed, knots: 14. **Range, n miles:** 7,000 at 12 kt
Complement: 124
Cargo capacity: 2,000 tons
Guns: 1 USN 3 in *(76 mm)*/50. 2 Bofors 40 mm/60.
Radars: Navigation: Raytheon SPS-21; G/H-band.

Comment: *Chang Bai* was launched on 6 December 1943 and was transferred to Taiwan in 1971. *Hsin Lang* was launched on 7 September 1942 and was transferred in 1972.

UPDATED

CHANG BAI   *5/1999* / 0080789

1 SALVAGE SHIP (ARS)

Name	No	Builders	Commissioned
TA HU (ex-*Grapple*)	552 (ex-ARS 7)	Basalt Rock, USA	16 Dec 1943

Displacement, tons: 1,557 standard; 1,745 full load
Dimensions, feet (metres): 213.5 × 39 × 15 *(65.1 × 11.9 × 4.6)*
Main machinery: Diesel-electric; 4 Cooper Bessemer GSB-8 diesels; 3,420 hp *(2.55 MW)*; 2 generators; 2 motors; 2 shafts
Speed, knots: 14. **Range, n miles:** 8,500 at 13 kt
Complement: 85
Guns: 2 Oerlikon 20 mm.
Radars: Navigation: SPS-53; I-band.

Comment: Fitted for salvage, towing and compressed-air diving. *Ta Hu* transferred from US 1 December 1977 by sale. The reported transfer from the US of ex-*Conserver* did not take place.

VERIFIED

TA HU   *1/1999, van Ginderen Collection* / 0050231

6 FLOATING DOCKS (YFD)

HAY TAN (ex-*AFDL 36*) AFDL 1	**FO WU 5** (ex-*ARD 9*) ARD 5
KIM MEN (ex-*AFDL 5*) AFDL 2	**FO WU 6** (ex-*Windsor* ARD 22) ARD 6
HAN JIH (ex-*AFDL 34*) AFDL 3	**FO WU 7** (ex-*AFDM 6*)

Comment: Former US Navy floating dry docks. *Hay Tan* transferred in March 1947, *Kim Men* in January 1948, *Han Jih* in July 1959, *Fo Wu 5* in June 1971, *Fo Wu 6* in June 1971. *Fo Wu 6* by sale 19 May 1976 and *Fo Wu 5* on 12 January 1977. *Fo Wu 7* transferred by sale in 1999.

VERIFIED

TUGS

11 HARBOUR TUGS (YTL)

YTL 16-17	YTL 27-30	YTL 32-36

Comment: Replacements for the old US Army type which were scrapped in 1990/91. Some are used for fire fighting.

VERIFIED

YTL 36   *5/1997, van Ginderen Collection* / 0019242

5 CHEROKEE CLASS (ATF/ARS)

TA WAN (ex-*Apache*) ATF 551	**TA FUNG** (ex-*Narragansett*) ATF 555
TA HAN (ex-*Tawakoni*) ATF 553	**TA TAI** (ex-*Shakori*) ATF 563
TA KANG (ex-*Achomawi*) ATF 554	

Displacement, tons: 1,235 standard; 1,731 full load
Dimensions, feet (metres): 205 × 38.5 × 17 *(62.5 × 11.7 × 5.2)*
Main machinery: Diesel-electric; 4 GM 12-278 diesels; 4,400 hp *(3.28 MW)*; 4 generators; 1 motor; 3,000 hp *(2.24 MW)*; 1 shaft
Speed, knots: 15. **Range, n miles:** 6,000 at 14 kt
Complement: 85
Guns: 1 Bofors 40 mm/60. Several 12.7 mm MGs.

Comment: All built between 1943 and 1945. *Ta Wan* transferred from US in June 1974; *Ta Han* in June 1978; and the last three in June 1991 together with two more which were cannibalised for spares.

VERIFIED

TA HAN   *4/1995* / 0080790

19 LARGE HARBOUR TUGS (YTB)

YTB 37-39	YTB 41-43	YTB 45-49	YTB 150-157

Comment: Various types of about 30 m length.

VERIFIED

COAST GUARD

Headquarters Appointments

Director General of the Coast Guard:
  Dr Wang Chun
Deputy Director General (Law Enforcement):
  You Chian-Tshiz
Deputy Administrator (Coastal Patrol):
  Lieutenant General Bi
Deputy Administrator (Oceanic Patrol):
  Vice Admiral Liu Chih-Yuan
Director, General Customs:
  Chung Huo-Cheng

Notes: The Coast Guard Agency was restructured on 28 January 2001 by merging the former agencies of Maritime Police, Customs and Coastal Defence Command. It is responsible for coastal and harbour security, maritime law enforcement, anti-smuggling, anti-terrorism, SAR, fishery protection, and pollution control. It consists of two major wings. The Oceanic Patrol wing has 2,500 personnel in 20 patrol detachments around the coast. The Coastal Patrol wing has 14,701 personnel in 24 battalions. In times of crisis, the Oceanic and Coastal patrol wings would be integrated into the Navy and Army respectively.

Bases

HQ: Tamsui

2 HO HSING CLASS (LARGE PATROL CRAFT) (WPSO)

HO HSING 509	**WEI HSING** 510

Displacement, tons: 1,823 full load
Dimensions, feet (metres): 270 × 38.1 × 13.5 *(82.3 × 11.6 × 4.1)*
Main machinery: 2 MTU 16V 1163 TB93 diesels; 13,310 hp(m) *(9.78 MW)* sustained; 2 shafts; cp props; bow thruster
Speed, knots: 22. **Range, n miles:** 7,000 at 16 kt
Complement: 80 (18 officers)
Guns: 2—12.7 mm MGs.
Radars: Surface search: Racal Decca; I-band.

Comment: Built by the China SB Corporation, Keelung, to a Tacoma design and both delivered 26 December 1991. Four high-speed interceptor boats are carried on individual davits. This is a variant of the US Coast Guard Bear class.

VERIFIED

WEI HSING   *6/2001, Ships of the World* / 0121408

5 + (4) OFFSHORE PATROL VESSELS (WPBO)

Name	Tonnage	Length	SP	Commissioned
SHUN HU 1	800	60 m	12	1992
SHUN HU 2	400	51 m	16	1992
SHUN HU 3	400	51 m	16	1992
SHUN HU 5	100	31 m	—	1992
SHUN HU 6	200	38 m	—	1993

Comment: Five former Ministry of Agriculture vessels conduct fishery protection duties. Ocean operations are conducted by: *Shun Hu 1* and *Shun Hu 2-3*; Coastal operations are conducted by: *Shun Hu 5* and *Shun Hu 6*. There are plans to replace the four larger vessels but a contract has not been confirmed. *UPDATED*

SHUN HU 1   *6/2001, Mitsuhiro Kadota* / 0121409

SHUN HU 2   *6/2001, Ships of the World* / 0121410

SHUN HU 5   *6/2001, C Chung* / 0126199

1 YUN HSIUNG CLASS (COASTAL PATROL CRAFT) (ABU)

YUN HSIUNG

Displacement, tons: 964 full load
Dimensions, feet (metres): 213.3 × 32.8 × 9.5 *(65 × 10 × 2.9)*
Main machinery: 2 MAN 12V 25/30 diesels; 7,183 hp(m) *(5.28 MW)*; 2 shafts
Speed, knots: 18
Complement: 67
Guns: 2—12.7 mm MGs.
Radars: Surface search: JRC; I-band.

Comment: Built by China SB Corporation and delivered 28 December 1987. Operated by Customs as a light-house tender. *UPDATED*

YUN HSING   *1/2000, C Chung* / 0106606

1 OFFSHORE PATROL VESSEL (WPSO)

TAIPEI 116

Displacement, tons: 680 full load
Dimensions, feet (metres): 201.4 × 31.2 × 9.2 *(61.4 × 9.5 × 2.8)*
Main machinery: 2 MTU 20V 1163 TB93 diesels; 20,128 hp(m) *(14.79 MW)*; 2 shafts
Speed, knots: 25. **Range, n miles:** 4,150 at 15 kt
Complement: 40
Guns: 2—20 mm T75. 3—12.7 mm MGs.
Radars: Navigation: I-band.

Comment: Based on the naval Jin Chiang class, ship built by Chung-Hsin Ship Building Corporation, launched in November 1999 and commissioned on 20 March 2000. *VERIFIED*

TAIPEI   *9/2002, C Chung* / 0534124

4 OFFSHORE PATROL CRAFT (WPSO)

TAICHUNG 117	KEELUNG 118	HUALIEN 119	PENGHU 120

Displacement, tons: 630 full load
Dimensions, feet (metres): 208.3 × 30.4 × 8.9 *(63.5 × 9.28 × 2.7)*
Main machinery: 2 MTU 1163 TB93 diesels; 15,608 hp(m) *(11.48 MW)* sustained; 2 shafts
Speed, knots: 30. **Range, n miles:** 2,400 at 18 kt
Complement: 46 (9 officers)
Guns: 1—20 mm T 75.
Radars: Surface search: JRC E/F- and I-bands.

Comment: Built by Ching-Fu SB Corporation in Kaohsiung, to Lürssen Asia design. Delivered 28 June 2001. Two high-speed interceptor boats are carried on individual davits. *VERIFIED*

PENGHU   *10/2001, C Chung* / 0126201

3 PAO HSING CLASS (COASTAL PATROL CRAFT) (WPBO)

PAO HSING	CHIN HSING	TEH HSING 109

Displacement, tons: 550 full load
Dimensions, feet (metres): 189.6 × 25.6 × 6.9 *(57.8 × 7.8 × 2.1)*
Main machinery: 2 MAN 12V25/30 diesels; 7,183 hp(m) *(5.28 MW)* sustained; 2 shafts
Speed, knots: 20
Complement: 40
Guns: 2—12.7 mm MGs.
Radars: Surface search: JRC; I-band.

Comment: First delivered 20 May 1980; second 23 May 1985. Built by Keelung yard of China SB Corporation. *VERIFIED*

PAO HSING (Customs colours)   *1994, Taiwan Customs* / 0080793

2 COASTAL PATROL CRAFT (WPSO)

MOU HSING	FU HSING 106

Displacement, tons: 917 full load
Dimensions, feet (metres): 214.6 × 31.5 × 10.5 *(65.4 × 9.6 × 3.2)*
Main machinery: 2 MTU 16V 1163 TB93 diesels; 13,310 hp(m) *(9.78 MW)* sustained; 2 shafts
Speed, knots: 28. **Range, n miles:** 4,500 at 12 kt
Complement: 54
Guns: 2—12.7 mm MGs.
Radars: Surface search: Racal Decca; I-band.

Comment: Ordered from Wilton Fijenoord in September 1986, and commissioned 14 June 1988.
VERIFIED

FU HSING   *9/2002, C Chung /* 0534122

1 HSUN HSING CLASS (COASTAL PATROL CRAFT) (WPBO)

HSUN HSING

Displacement, tons: 264 full load
Dimensions, feet (metres): 146 × 24.6 × 5.8 *(44.5 × 7.5 × 1.7)*
Main machinery: 3 MTU 16V 396 TB93 diesels; 6,540 hp(m) *(4.81 MW)* sustained; 3 shafts
Speed, knots: 28. **Range, n miles:** 1,500 at 20 kt
Complement: 41
Guns: 3 Oerlikon 20 mm.
Radars: Surface search: JRC; I-band.

Comment: Built by China SB Corporation and delivered 15 December 1985.
VERIFIED

HSUN HSING   *10/2001, C Chung /* 0126200

13 COASTAL PATROL CRAFT (WPBF)

PP 6001-6003	6005-6007	6009-6012	6014-6016

Displacement, tons: 91 full load
Dimensions, feet (metres): 91.9 × 20.3 × 7.9 *(28 × 6.2 × 2.4)*
Main machinery: 2 Paxman 12V P185 diesels; 6,645 hp(m) *(4.89 MW)* sustained; 2 shafts; cp props
Speed, knots: 40. **Range, n miles:** 600 at 25 kt
Complement: 12
Guns: 2—12.7 mm MGs.
Radars: Surface search: 2 Racal Decca; I-band.

Comment: Built by Lung Teh Shipyard, Taiwan from March 1996. First six delivered in 1997 and following seven in 2001. GRP hulls with some Kevlar protection. Can carry a 6.5 m RIB. May be given new pennant numbers.
VERIFIED

PP 6002   *10/2001, C Chung /* 0126203

11 LARGE PATROL CRAFT (WPB)

PP 10001-10002	10005-10010	10017-10019

Displacement, tons: 140 full load
Dimensions, feet (metres): 90 × 28.6 × 6 *(27.4 × 8.7 × 1.8)*
Main machinery: 2 MTU diesels; 6,000 hp(m) *(4.4 MW)*; 2 shafts
Speed, knots: 30
Guns: 2—12.7 mm MGs (aft).
Radars: Surface search: Decca; I-band.

Comment: The first pair were former naval craft transferred 8 December 1992. Two more completed in October 1994, three more by February 1995. The construction programme continues.
VERIFIED

PP 10017   *10/2001, C Chung /* 0126205

4 HAI YING CLASS (INSHORE PATROL CRAFT) (WPB)

HAI YING	HAI TUNG	HAI KO	HAI TA

Displacement, tons: 99.43 full load
Dimensions, feet (metres): 82.8 × 19.0 × 10.7 *(25.25 × 5.8 × 3.3)*
Main machinery: 2 Deutz MWM TBD 620 V12 diesels; 4,314 bhp *(2,646 kW)*; 2 shafts
Speed, knots: 32.6
Complement: 7
Guns: 2—9 mm T75.
Radars: Surface search: Furuno; I-band.

Comment: Transferred to Customs on 28 December 2000.
UPDATED

HAI YING CLASS   *12/2000, Taiwan Customs /* 0114555

4 HAI CHENG CLASS (INSHORE PATROL CRAFT) (WPB)

HAI CHENG	HAI EN	HAI LIANG	HAI CHING

Displacement, tons: 147 full load
Dimensions, feet (metres): 100 × 22.3 × 11.6 *(30.5 × 6.8 × 3.6)*
Main machinery: 2 MTU diesels; 6,000 hp(m) *(4.4 MW)*; 2 shafts
Speed, knots: 30
Complement: 8
Guns: 2—9 mm T75.
Radars: Surface search: Decca; I-band.

Comments: Transferred to Customs on 26 December 2000.
UPDATED

HAI CHENG CLASS   *12/2000, Taiwan Customs /* 0114556

71 COASTAL PATROL CRAFT (WPB)

No/Type	Displacement, tons	Speed, knots	Pennant Numbers series
10 PP 5500	55		PP 5501-5503, 5505-5511
15 PP 5000	50	32	PP 5001-5003, 5005-5008, 5010-5013, 5015-5016, 5022-5023, 5025-5026, 5028-5032
9 PP 3550	35		PP 3550, 3552-3553, 3555-3559, 3561
23 PP 3500	35	28	PP 3503, 3505, 3507, 3510-3513, 3516-3522, 3525, 3527, 3530-3531, 3535-3539
14 PP 3000	30	45	PP 3002-3003, 3005-3009, 3011-3012, 3015-3019

Comment: These are all armed with 12.7 mm MGs and capable of speeds up to 35 kt. The pennant numbers are not consecutive, as some have been deleted. *P 805* sunk in November 1997. May have been given new pennant numbers.   ***VERIFIED***

PP 3539   *10/2001, C Chung* / 0126208

PP 3005   *10/2001, C Chung* / 0126204

PP 5025   *10/2001, C Chung* / 0126207

PP 5511   *6/2002, C Chung* / 0534123

Tanzania

Country Overview

The United Republic of Tanzania was formed by the federation of the former British protectorates of Tanganyika and Zanzibar in 1964. It also includes Pemba, Mafia and other offshore islands. Situated in south-eastern Africa, it has a total area of 364,900 square miles and is bordered to the north by Uganda and Kenya, to the west by Rwanda, Burundi, Democratic Republic of Congo and Zambia and to the south by Mozambique and Malawi. It has a 767 n mile coastline with the Indian Ocean. The country also includes parts of Lake Tanganyika, Lake Victoria and Lake Malawi. Dodoma is the capital while the former capital, Dar es Salaam, is the largest city and principal port. Territorial seas (12 n miles) are claimed. A 200 n mile EEZ has also been claimed but the limits are not fully defined by boundary agreements.

Headquarters Appointments

Commander of the Navy:
   Major General L G Sande

General

There is a small Coastguard Service (KMKM), based on Zanzibar, which uses small boats for anti-smuggling patrols.

Personnel

(a)   2005: 1,050 (including Zanzibar)
(b)   Voluntary service

Coast Defence

85 mm mobile gun battery.

Bases

Dar Es Salaam, Zanzibar, Mtwara. Kigoma (Lake Tanganyika) and Mwanza (Lake Victoria).

PATROL FORCES

Notes: The Police have four Yulin class patrol boats which are probably non-operational.

2 SHANGHAI II CLASS (FAST ATTACK CRAFT—GUN) (PB)

MZIZI P 67   **MZIA** P 68

Displacement, tons: 134 full load
Dimensions, feet (metres): 127.3 × 17.7 × 5.6 *(38.8 × 5.4 × 1.7)*
Main machinery: 2 Type L12-180 diesels; 2,400 hp(m) *(1.76 MW)* (forward); 2 Type 12-D-6 diesels; 1,820 hp(m) *(1.34 MW)* (aft); 4 shafts
Speed, knots: 30. **Range, n miles:** 700 at 16.5 kt
Complement: 38
Guns: 4—37 mm/63 (2 twin). 4—25 mm/80 (2 twin).
Radars: Surface search: Skin Head; E/F-band.

Comment: Six transferred by the People's Republic of China in 1971-72, two more in June 1992.   ***VERIFIED***

MZIA
5/2002, Ships of the World / 0525923

2 VOSPER THORNYCROFT 75 ft TYPE
(COASTAL PATROL CRAFT) (PB)

Displacement, tons: 70 full load
Dimensions, feet (metres): 75 × 19.5 × 8 *(22.9 × 6 × 2.4)*
Main machinery: 2 Caterpillar D 348 diesels; 1,450 hp *(1.08 MW)* sustained; 2 shafts
Speed, knots: 24.5. **Range, n miles:** 800 at 20 kt
Complement: 11
Guns: 2 Oerlikon 20 mm GAM-BO1.
Radars: Surface search: Furuno; I-band.

Comment: First pair delivered 6 July 1973, second pair 1974. Used for anti-smuggling patrols off Zanzibar. Two still operational.   ***VERIFIED***

VOSPER 75 ft (Omani colours)   *1984, N Overington*

2 HUCHUAN CLASS (FAST ATTACK CRAFT—TORPEDO) (PTK)

P 43   P 44

Displacement, tons: 39 standard; 45.8 full load
Dimensions, feet (metres): 71.5 × 20.7 oa × 11.8 (hullborne) *(21.8 × 6.3 × 3.6)*
Main machinery: 3 Type M 50 diesels; 3,300 hp(m) *(2.4 MW)* sustained; 3 shafts
Speed, knots: 50. **Range, n miles:** 500 at 30 kt
Complement: 16
Guns: 4—14.5 mm (2 twin) MGs.
Torpedoes: 2—21 in *(533 mm)* tubes.
Radars: Surface search: Skin Head; E/F-band.

Comment: Four transferred from the People's Republic of China 1975. After a major effort in 1992, were all operational and reported to be in good condition but by 1998 two had been laid up. Present operational status is unclear but one at least appears to have had torpedo tubes removed.   ***UPDATED***

HUCHUAN   *6/2003* * / 0587794*

AUXILIARIES

2 YUCH'IN (TYPE 069) CLASS (LCU)

PONO L 08   KIBUA L 09

Displacement, tons: 85 full load
Dimensions, feet (metres): 81.2 × 17.1 × 4.3 *(24.8 × 5.2 × 1.3)*
Main machinery: 2 diesels; 600 hp(m) *(441 kW)*; 2 shafts
Speed, knots: 12. **Range, n miles:** 450 at 11.5 kt
Complement: 12
Military lift: 46 tons
Guns: 4—14.5 mm (2 twin) MGs.
Radars: Navigation: Fuji; I-band.

Comment: Transferred from China in 1995 probably to replace the Police Yuchai transport craft. Based at Dar-es-Salaam. *Pono* reported to be operational.   ***VERIFIED***

PONO   *1/2001 / 0109946*

Thailand

Country Overview

The Kingdom of Thailand (formerly Siam) is a constitutional monarchy in South East Asia. With an area of 198,114 square miles, it is bordered to the west by Burma, to the east by Laos and Cambodia and to the south by Malaysia. It has a 1,739 n mile coastline with the Gulf of Thailand and with the Andaman Sea. The capital, largest city and principal port (which also serves neighbouring Laos) is Bangkok. Territorial seas (12 n miles). An EEZ (200 n miles) is claimed and the limits have been partly defined by boundary agreements.

Headquarters Appointments

Commander-in-Chief of the Navy:
  Admiral Sampop Amrapala
Deputy Commander-in-Chief:
  Admiral Pirasak Watcharamul
Assistant Commander-in-Chief:
  Admiral Phan-Charoon Vichayapai Bonnag
Chief of Staff:
  Admiral Surin Reone-Arom
Deputy Chiefs of Staff:
  Vice Admiral Suchart Yanothai
  Vice Admiral Nopporn Ajavakom

Senior Appointments

Commander-in-Chief, Fleet:
  Admiral Vichai Yuwanangoon
Deputy Commanders-in-Chief, Fleet:
  Vice Admiral Yongyot Huangnikorn
  Vice Admiral Verawatch Wongdontri
Chief of Staff, Fleet:
  Vice Admiral Somded Tongpiam

Diplomatic Representation

Naval Attaché in London:
  Captain P Rujites
Naval Attaché in Washington:
  Captain Chonlathis Navanugraha
Naval Attaché in Paris:
  Captain Wittanarat Gaiaseni
Naval Attaché in Canberra:
  Captain Adoong Pan-lam

Diplomatic Representation — *continued*

Naval Attaché in Madrid:
  Captain Chatchai Srivorakan
Naval Attaché in New Delhi:
  Captain Graivut Vattanatham
Naval Attaché in Singapore:
  Captain Jiamsak Chantarasena
Naval Attaché in Kuala Lumpur:
  Captain Nuttapol Diewvanich
Naval Attaché in Beijing:
  Captain Ranat Debavalya
Naval Attaché in Rome:
  Captain Banjerd Sripraram

Personnel

(a)   2005: Navy, 74,000 (including 2,000 Naval Air Arm, 11,000 Marines and Coastal Defense Command)
(b)   2 years' national service (28,000 conscripts)

Organisation

First naval area command (Upper Thai Gulf)
Second naval command (Lower Thai Gulf)
Third naval command (Andaman Sea)

Bases

Bangkok, Sattahip, Songkhla, Phang-Nga (west coast)

Naval Aviation

First air wing (U-Tapao)
Second air wing (Songkhla)
101 Sqdn   MPA/ASW
102 Sqdn   MPA/ASuW
103 Sqdn   Utility
104 Sqdn   Maritime Strike
105 Sqdn   Matador
201 Sqdn   Central Patrol
202 Bell Helos
203 Sikorsky Helos

Pennant Numbers

Ships over 150 tons displacement which had single digit numbers, had new numbers allocated in 1994-95.

Prefix to Ships' Names

HTMS

Strength of the Fleet

Type	Active	Building (Projected)
Aircraft Carrier	1	—
Frigates	8	—
Corvettes	7	—
Fast Attack Craft (Missile)	6	—
Fast Attack Craft (Gun)	3	—
Offshore Patrol Craft	9	2
Coastal Patrol Craft	52	—
MCM Support Ship	1	—
Minehunters	4	—
Coastal Minesweepers	2	—
MSBs	12	—
LSTs	6	—
LSM	1	—
Hovercraft	3	—
Survey Vessels	5	—
Replenishment Ship	1	—
MCMV Depot Ship	1	—
Tankers/Transports	8	—
Training Ships	3	(1)

Coast Defence

Coastal Defence Command was rapidly expanded to the 1992 two Division level after the government charged the RTN with the responsibility of defending the entire Eastern Seaboard. Ships and aircraft are rotated monthly from the Navy. Equipment includes 10 batteries of truck-mounted Exocet MM 40, 155 and 130 mm guns for coastal defence, 76, 40, 37, 20 mm guns and PL-9B SAM for air defence.

Marine Police

Acts as a Coast Guard in inshore waters with some 60 armed patrol craft and another 65 equipped with small arms only.

PENNANT LIST

Aircraft Carriers		Patrol Forces		Amphibious Forces		Survey and Research Ships	
911	Chakri Naruebet	311	Prabparapak	712	Chang	811	Chanthara
		312	Hanhak Sattru	713	Pangan	812	Suk
		313	Suphairin	714	Lanta		
		321	Ratcharit	715	Prathong		
		322	Witthayakhom	721	Sichang		
		323	Udomdet	722	Surin	**Auxiliaries**	
Frigates		331	Chon Buri	731	Kut		
421	Naresuan	332	Songkhla	741	Prab	821	Suriya
422	Taksin	333	Phuket	742	Satakut	831	Chula
455	Chao Phraya	521	Sattahip	761	Mataphon	832	Samui
456	Bangpakong	522	Klongyai	762	Rawi	833	Prong
457	Kraburi	523	Takbai	763	Adang	834	Proet
458	Saiburi	524	Kantang	764	Phetra	835	Samed
461	Phuttha Yotfa Chulalok	525	Thepha	765	Kolam	841	Chuang
462	Phuttha Loetla Naphalai	526	Taimuang	766	Talibong	842	Chik
		541	Hua Hin	771	Thong Kaeo	851	Klueng Badaan
		542	Klaeng	772	Thong Lang	852	Marn Vichai
		543	Si Racha	773	Wang Nok	853	Rin
				774	Wang Nai	854	Rang
Corvettes		**Mine Warfare Forces**		781	Man Nok	855	Samaesan
				782	Man Klang	856	Raet
431	Tapi	612	Bangkeo	783	Man Nai	861	Kled Keo
432	Khirirat	613	Donchedi			871	Similan
441	Rattanakosin	621	Thalang	**Training Ships**			
442	Sukothai	631	Bang Rachan				
531	Khamronsin	632	Nongsarai	413	Pin Klao		
532	Thayanchon	633	Lat Ya	433	Makut Rajakumarn		
533	Longlom	634	Tha Din Daeng	611	Phosamton		

SUBMARINES

Notes: Acquisition of a submarine force remains a high priority but funding difficulties continue to frustrate plans.

AIRCRAFT CARRIERS

1 CHAKRI NARUEBET CLASS (CVM)

Name	No	Builders	Laid down	Launched	Commissioned
CHAKRI NARUEBET	911	Bazán, Ferrol	12 July 1994	20 Jan 1996	27 Mar 1997

Displacement, tons: 11,485 full load
Dimensions, feet (metres): 599.1 oa; 538.4 wl × 100.1 oa; 73.8 wl × 20.3 *(182.6; 164.1 × 30.5; 22.5 × 6.2)*
Flight deck, feet (metres): 572.8 × 90.2 *(174.6 × 27.5)*
Main machinery: CODOG; 2 GE LM 2500 gas turbines; 44,250 hp *(33 MW)* sustained; 2 MTU 16V 1163 TB83 diesels; 11,780 hp(m) *(8.67 MW)*; 2 shafts; LIPS cp props
Speed, knots: 26; 16 (diesels). **Range, n miles:** 10,000 at 12 kt
Complement: 455 (62 officers) plus 146 aircrew plus 4 (Royal family)

Missiles: SAM: 1 Mk 41 LCHR 8 cell VLS launcher (fitted for but not with) ❶.
3 Matra Sadral sextuple launchers for Mistral ❷; IR homing to 4 km *(2.2 n miles)*; warhead 3 kg.
Guns: 2—30 mm. To be fitted.
Combat data systems: Tritan derivative with Unisys UYK-3 and 20 computers.
Radars: Air search: Hughes SPS-52C ❸; E/F-band.

Surface search: SPS-64 ❹; I-band. To be fitted.
Fire control: to be fitted.
Navigation: Kelvin Hughes; I-band.
Aircraft control: Kelvin Hughes; E/F-band.
Tacan: URN 25.

Fixed-wing aircraft: 6 AV-8S Matador (Harrier).
Helicopters: 6 S-70B-7 Seahawk; Chinook capable.

Programmes: An initial contract for a 7,800 ton vessel with Bremer Vulcan was cancelled on 22 July 1991 and replaced on 27 March 1992 with a government to government contract for a larger ship to be built by Bazán. Fabrication started in October 1993. Sea trials conducted from November 1996 to January 1997 followed by an aviation work-up at Rota from April 1997. The ship arrived in Thailand on 10 August 1997.
Structure: Similar to Spanish *Príncipe de Asturias*. 12° ski jump and two 20 ton aircraft lifts. Provision made to fit a Mk 41 VLS launcher, a surface search radar, EW systems, a hull mounted sonar and CIWS. Matra Sadral fitted in 2001. Hangar can take up 10 Sea Harrier or Seahawk aircraft.
Operational: Main tasks are SAR co-ordination and EEZ surveillance. Secondary role is air support for all maritime operations. Due to funding shortages, the ship rarely goes to sea and fixed-wing flying has been conducted from shore bases.

UPDATED

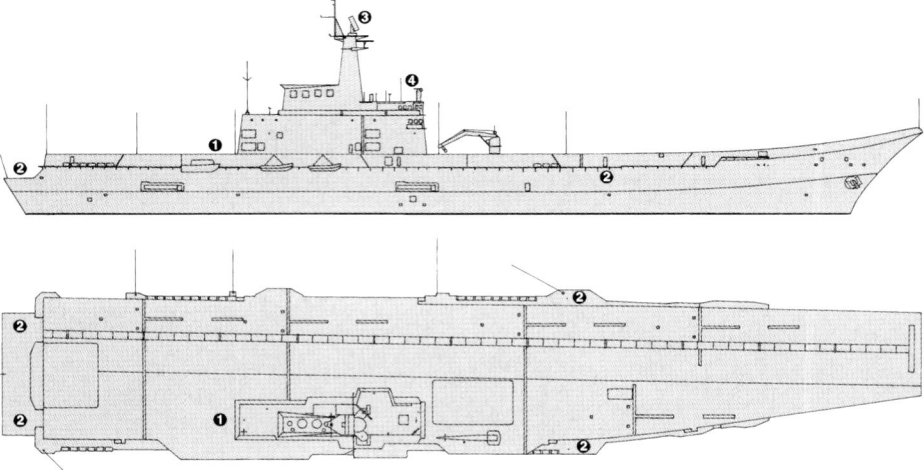

CHAKRI NARUEBET   (Scale 1 : 1,500), Ian Sturton / 0080799

CHAKRI NARUEBET   5/1997, S G Gaya / 0019250

CHAKRI NARUEBET   1/2004*, Thai Navy League / 0589817

CHAKRI NARUEBET   1/2004*, Thai Navy League / 0589816

FRIGATES

Notes: It was announced on 22 July 2003 that two frigates are to be procured from the UK. These are likely to be based on the design of those acquired by the Royal Malaysian Navy. Subject to successful negotiations with BAE, a contract may be let in 2005 if funding becomes available.

2 NARESUAN CLASS (TYPE 25T) (FFGHM)

Name	No	Builders	Laid down	Launched	Commissioned
NARESUAN	421 (ex-621)	Zhonghua SY, Shanghai	Feb 1992	24 July 1993	15 Dec 1994
TAKSIN	422 (ex-622)	Zhonghua SY, Shanghai	Nov 1992	14 May 1994	28 Sep 1995

Displacement, tons: 2,500 standard; 2,980 full load
Dimensions, feet (metres): 393.7 × 42.7 × 12.5
(120 × 13 × 3.8)
Main machinery: CODOG; 2 GE LM 2500 gas turbines; 44,250 hp *(33 MW)* sustained; 2 MTU 20 V 1163 TB83 diesels; 11,780 hp(m) *(8.67 MW)* sustained; 2 shafts; LIPS cp props
Speed, knots: 32. **Range, n miles:** 4,000 at 18 kt
Complement: 150

Missiles: SSM: 8 McDonnell Douglas Harpoon (2 quad) launchers ❶; active radar homing to 130 km *(70 n miles)* at 0.9 Mach; warhead 227 kg.
SAM: Mk 41 LCHR 8 cell VLS launcher ❷ Sea Sparrow; semi-active radar homing to 14.6 km *(8 n miles)* at 2.5 Mach; warhead 39 kg (fitted for but not with).
Guns: 1 FMC 5 in *(127 mm)*/54 Mk 45 Mod 2 ❸; 20 rds/min to 23 km *(12.6 n miles)*; weight of shell 32 kg.
4 China 37 mm/76 (2 twin) H/PJ 76 A ❹; 180 rds/min to 8.5 km *(4.6 n miles)* anti-aircraft; weight of shell 1.42 kg.
Torpedoes: 6—324 mm Mk 32 Mod 5 (2 triple) tubes ❺. Honeywell Mk 46; active/passive homing to 11 km *(5.9 n miles)* at 40 kt; warhead 44 kg.
Countermeasures: Decoys: 4 China Type 945 GPJ 26-barrelled launchers ❻; chaff and IR.
ESM/ECM: Elettronica Newton Beta EW System; intercept and jammer.
Weapons control: 1 JM-83H Optical Director ❼.
Radars: Air search: Signaal LW08 ❽; D-band.
Surface search: China Type 360 ❾; E/F-band.
Navigation: 2 Raytheon SPS-64(V)5; I-band.
Fire control: 2 Signaal STIR ❿; I/J/K-band (for SSM and 127 mm). After one to be fitted.
China 374 G ⓫ (for 37 mm).
Sonars: China SJD-7; hull-mounted; active search and attack; medium frequency.

Helicopters: 1 Super Lynx ⓬ in due course or 1 Sikorsky S-70B-7 Seahawk.

Programmes: Contract signed 21 September 1989 for construction of two ships by the China State SB Corporation

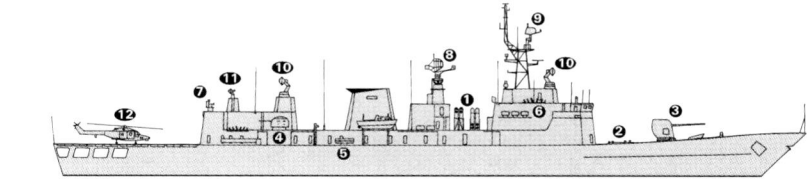

NARESUAN
(Scale 1 : 1,200), Ian Sturton / 0543398

NARESUAN
2/2004*, Bob Fildes / 0589808

(CSSC) with delivery in 1994. US and European weapon systems were fitted as funds became available. The first ship sailed for Bangkok without most weapon systems in January 1995 with the second following in October 1995.

Structure: Jointly designed by the Royal Thai Navy and China State Shipbuilding Corporation (CSSC). This is a design incorporating much Western machinery and equipment and provides enhanced capabilities by comparison with the four Type 053 class. The anti-aircraft guns are Breda 40 mm types with 37 mm ammunition and they are controlled by a Chinese RTN-20 Dardo tracker.

Operational: *Naresuan* acted as one of the escorts for the aircraft carrier during her aviation work-up in Spanish waters in 1997.
UPDATED

TAKSIN
7/1998, Hachiro Nakai / 0050234

NARESUAN
10/2002, Hachiro Nakai / 0530029

4 CHAO PHRAYA CLASS (TYPES 053 HT and 053 HT (H)) (FFG/FFGH)

Name	No	Builders	Laid down	Launched	Commissioned
CHAO PHRAYA	455	Hudong SY, Shanghai	1989	24 June 1990	5 Apr 1991
BANGPAKONG	456	Hudong SY, Shanghai	1989	25 July 1990	20 July 1991
KRABURI	457	Hudong SY, Shanghai	1990	28 Dec 1990	16 Jan 1992
SAIBURI	458	Hudong SY, Shanghai	1990	27 Aug 1991	4 Aug 1992

Displacement, tons: 1,676 standard; 1,924 full load
Dimensions, feet (metres): 338.5 × 37.1 × 10.2
 (103.2 × 11.3 × 3.1)
Main machinery: 4 MTU 20V 1163 TB83 diesels; 29,440 hp(m)
 (21.6 MW) sustained; 2 shafts; LIPS cp props
Speed, knots: 30. **Range, n miles:** 3,500 at 18 kt
Complement: 168 (22 officers)

Missiles: SSM: 8 Ying Ji (Eagle Strike) (C-801) ❶; active radar/IR homing to 85 km *(45.9 n miles)* at 0.9 Mach; warhead 165 kg; sea-skimmer. This is the extended range version.
 SAM: 1 HQ-61 launcher for PL-9 or Matra Sadral for Mistral to be fitted.
Guns: 2 (457 and 458) or 4 China 100 mm/56 (1 or 2 twin) ❷; 25 rds/min to 22 km *(12 n miles)*; weight of shell 15.9 kg.
 8 China 37 mm/76 (4 twin) H/PJ 76 A ❸; 180 rds/min to 8.5 km *(4.6 n miles)* anti-aircraft; weight of shell 1.42 kg.
A/S mortars: 2 RBU 1200 (China Type 86) 5-tubed fixed launchers ❹; range 1,200 m.
Depth charges: 2 BMB racks.
Countermeasures: Decoys: 2 China Type 945 GPJ 26-barrelled chaff launchers.
 ESM: China Type 923(1); intercept.
 ECM: China Type 981(3); jammer.
Combat data systems: China Type ZKJ-3 or STN Atlas mini COSYS action data automation being fitted.
Radars: Air/surface search: China Type 354 Eye Shield ❺; G-band.
 Surface search/fire control: China Type 352C Square Tie ❻; I-band (for SSM).
 Fire control: China Type 343 Sun Visor ❼; I-band (for 100 mm). China Type 341 Rice Lamp ❽; I-band (for 37 mm).
 Navigation: Racal Decca 1290 A/D ARPA and Anritsu RA 71CA ❾; I-band.
 IFF: Type 651.
Sonars: China Type SJD-5A; hull-mounted; active search and attack; medium frequency.

Helicopters: Platform for 1 Bell 212 (457 and 458) ❿.

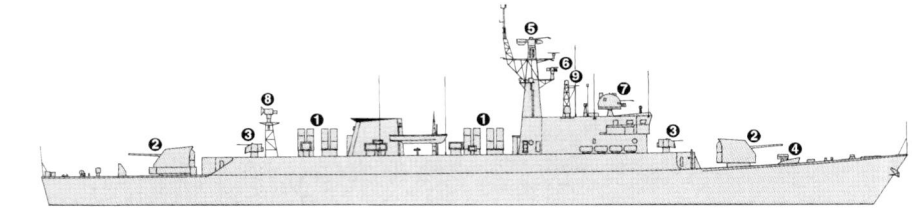

CHAO PHRAYA · *(Scale 1 : 900), Ian Sturton /* 0080802

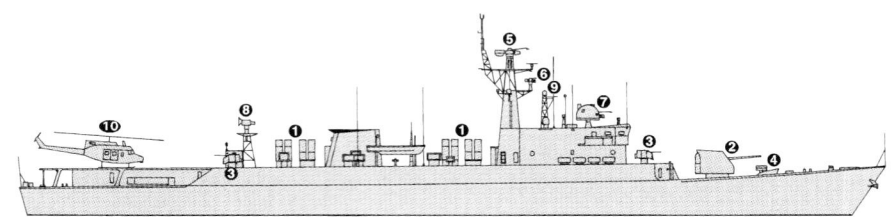

KRABURI · *(Scale 1 : 900), Ian Sturton /* 0080803

Programmes: Contract signed 18 July 1988 for four modified Jianghu class ships to be built by the China State SB Corporation (CSSC).
Modernisation: A mini COSYS system was acquired for two of the class in 1999.
Structure: Thailand would have preferred only the hulls but China insisted on full armament. Two of the ships are the Type III variant with 100 mm guns, fore and aft, and the other two are a variation with a helicopter platform replacing the after 100 mm gun. German communication equipment fitted. The EW fit is Italian designed.

Operational: On arrival in Thailand each ship was docked to make good poor shipbuilding standards and improve damage control capabilities. The ships are mostly used for rotating monthly to the Coast Guard, and for training, although *Kraburi* was part of the escort force for the aircraft carrier in Spanish waters in 1997. *Kraburi* badly damaged by the Tsunami on 26 December 2004.

UPDATED

KRABURI · *5/1998, Ships of the World /* 0050235

BANGPAKONG · *10/2002, John Mortimer /* 0529998

2 KNOX CLASS (FFGHM)

Name	No	Builders	Laid down	Launched	Commissioned
PHUTTHA YOTFA CHULALOK (ex-*Truett*)	461 (ex-FF 1095)	Avondale Shipyards	27 Apr 1972	3 Feb 1973	1 June 1974
PHUTTHA LOETLA NAPHALAI (ex-*Ouellet*)	462 (ex-FF 1077)	Avondale Shipyards	15 Jan 1969	17 Jan 1970	12 Dec 1970

Displacement, tons: 3,011 standard; 4,260 full load
Dimensions, feet (metres): 439.6 × 46.8 × 15; 24.8 (sonar)
(134 × 14.3 × 4.6; 7.8)
Main machinery: 2 Combustion Engineering/Babcock & Wilcox
boilers; 1,200 psi *(84.4 kg/cm²)*; 950°F *(510°C)*; 1 turbine;
35,000 hp *(26 MW)*; 1 shaft
Speed, knots: 27. **Range, n miles:** 4,000 at 22 kt on 1 boiler
Complement: 288 (17 officers)

Missiles: SSM: 8 McDonnell Douglas Harpoon; active radar
homing to 130 km *(70 n miles)* at 0.9 Mach; warhead 227 kg.
A/S: Honeywell ASROC Mk 16 octuple launcher with reload
system (has 2 starboard cells modified to fire Harpoon) **❶**;
inertial guidance to 1.6-10 km *(1-5.4 n miles)*; payload Mk 46.
Guns: 1 FMC 5 in *(127 mm)*/54 Mk 42 Mod 9 **❷**; 20-40 rds/min
to 24 km *(13 n miles)* anti-surface; 14 km *(7.7 n miles)* anti-
aircraft; weight of shell 32 kg.
1 General Electric/General Dynamics 20 mm/76 6-barrelled
Mk 15 Vulcan Phalanx **❸**; 3,000 rds/min combined to 1.5 km.
Torpedoes: 4—324 mm Mk 32 (2 twin) fixed tubes **❹**. 22
Honeywell Mk 46; anti-submarine; active/passive homing to
11 km *(5.9 n miles)* at 40 kt; warhead 44 kg.
Countermeasures: Decoys: 2 Loral Hycor SRBOC 6-barrelled
fixed Mk 36 **❺**; IR flares and chaff to 4 km *(2.2 n miles)*. T Mk-6
Fanfare/SLQ-25 Nixie; torpedo decoy. Prairie Masker hull and
blade rate noise suppression.
ESM/ECM: SLQ-32(V)2 **❻**; radar warning. Sidekick modification
adds jammer and deception system.
Combat data systems: Link 14 receive only.
Weapons control: SWG-1A Harpoon LCS. Mk 68 GFCS. Mk 114
ASW FCS. Mk 1 target designation system. MMS target
acquisition sight (for mines, small craft and low-flying aircraft).
Radars: Air search: Lockheed SPS-40B **❼**; B-band; range 320 km
(175 n miles).
Surface search: Raytheon SPS-10 or Norden SPS-67 **❽**; G-band.
Navigation: Marconi LN66; I-band.
Fire control: Western Electric SPG-53A/D/F **❾**; I/J-band.
Tacan: SRN 15. IFF: UPX-12.
Sonars: EDO/General Electric SQS-26CX; bow-mounted; active
search and attack; medium frequency.
EDO SQR-18(V) TACTASS; passive; low frequency.

Helicopters: 1 Bell 212 **❿**.

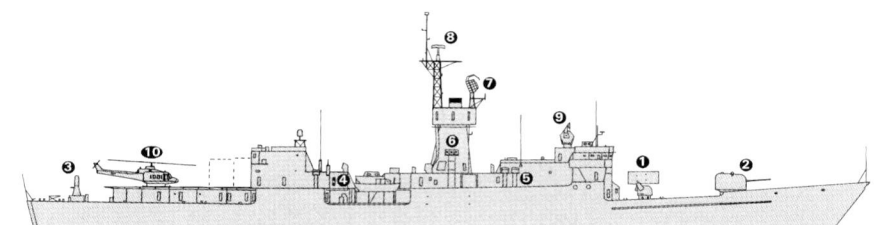

PHUTTHA YOTFA CHULALOK   *(Scale 1 : 1,200), Ian Sturton* / 0543397

PHUTTHA LOETLA NAPHALAI   *1/2001, Thai Navy League* / 0105841

Programmes: The first ship transferred on five year lease from
the USA on 30 July 1994. This was renewed by grant in 1999.
The second transferred on lease 27 November 1996 and
arrived in Thailand in November 1998.
Structure: Four Mk 32 torpedo tubes are fixed in the midships
structure, two to a side, angled out at 45°. The arrangement
provides improved loading capability over exposed triple Mk
32 torpedo tubes. A 4,000 lb lightweight anchor is fitted on
the port side and an 8,000 lb anchor fits into the after section
of the sonar dome.

VERIFIED

CORVETTES

2 RATTANAKOSIN CLASS (FSGM)

Name	No	Builders	Laid down	Launched	Commissioned
RATTANAKOSIN	441 (ex-1)	Tacoma Boatbuilders, WA	6 Feb 1984	11 Mar 1986	26 Sep 1986
SUKHOTHAI	442 (ex-2)	Tacoma Boatbuilders, WA	26 Mar 1984	20 July 1986	10 June 1987

Displacement, tons: 960 full load
Dimensions, feet (metres): 252 × 31.5 × 8
(76.8 × 9.6 × 2.4)
Main machinery: 2 MTU 20V 1163 TB83 diesels; 14,730 hp(m)
(10.83 MW) sustained; 2 shafts; Kamewa cp props
Speed, knots: 26. **Range, n miles:** 3,000 at 16 kt
Complement: 87 (15 officers) plus Flag Staff

Missiles: SSM: 8 McDonnell Douglas Harpoon (2 quad)
launchers **❶**; active radar homing to 130 km *(70 n miles)* at
0.9 Mach; warhead 227 kg (84A) or 258 kg (84B/C).
SAM: Selenia Elsag Albatros octuple launcher **❷**; 24 Aspide;
semi-active radar homing to 13 km *(7 n miles)* at 2.5 Mach;
height envelope 15-5,000 m *(49.2-16,405 ft)*; warhead
30 kg.
Guns: 1 OTO Melara 3 in *(76 mm)*/62 **❸**; 60 rds/min to 16 km
(8.7 n miles); weight of shell 6 kg.
2 Breda 40 mm/70 (twin) **❹**; 300 rds/min to 12.5 km *(6.8 n
miles)*; weight of shell 0.96 kg.
2 Rheinmetall 20 mm **❺**.
Torpedoes: 6—324 mm US Mk 32 (2 triple) tubes **❻**. MUSL
Stingray; active/passive homing to 11 km *(5.9 n miles)* at
45 kt; warhead 35 kg (shaped charge); depth to 750 m
(2,460 ft).
Countermeasures: Decoys: CSEE Dagaie 6- or 10-tubed
trainable; IR flares and chaff; H- to J-band.
ESM: Elettronica; intercept.
Weapons control: Signaal Sewaco TH action data automation.
Lirod 8 optronic director **❼**.
Radars: Air/surface search: Signaal DA05 **❽**; E/F-band; range
137 km *(75 n miles)* for 2 m² target.
Surface search: Signaal ZW06 **❾**; I-band.
Navigation: Decca 1226; I-band.
Fire control: Signaal WM25/41 **❿**; I/J-band; range 46 km *(25 n
miles)*.
Sonars: Atlas Elektronik DSQS-21C; hull-mounted; active search
and attack; medium frequency.

Programmes: Contract signed with Tacoma on 9 May 1983.
Intentions to build a third were overtaken by the Vosper
corvettes.
Structure: There are some similarities with the missile corvettes
built for Saudi Arabia five years earlier. Space for Phalanx aft
of the Harpoon launchers, but there are no plans to fit.

VERIFIED

RATTANAKOSIN   *(Scale 1 : 600), Ian Sturton*

SUKHOTHAI
6/2001, Royal Thai Navy / 0130170

3 KHAMRONSIN CLASS (FS)

Name	No	Builders	Laid down	Launched	Commissioned
KHAMRONSIN	531 (ex-1)	Ital Thai Marine, Bangkok	15 Mar 1988	15 Aug 1989	29 July 1992
THAYANCHON	532 (ex-2)	Ital Thai Marine, Bangkok	20 Apr 1988	7 Dec 1989	5 Sep 1992
LONGLOM	533 (ex-3)	Bangkok Naval Dockyard	15 Mar 1988	8 Aug 1989	2 Oct 1992

Displacement, tons: 630 full load
Dimensions, feet (metres): 203.4 oa; 186 wl × 26.9 × 8.2
  (62; 56.7 × 8.2 × 2.5)
Main machinery: 2 MTU 12V 1163 TB93; 9,980 hp(m)
  (7.34 MW) sustained; 2 Kamewa cp props
Speed, knots: 25. **Range, n miles:** 2,500 at 15 kt
Complement: 57 (6 officers)

Guns: 1 OTO Melara 76 mm/62 Mod 7 ❶; 60 rds/min to 16 km
  (8.7 n miles); weight of shell 6 kg.
  2 Breda 30 mm/70 (twin) ❷; 800 rds/min to 12.5 km (6.8 n
  miles); weight of shell 0.37 kg.
  2—12.7 mm MGs.
Torpedoes: 6 Plessey PMW 49A (2 triple) launchers ❸; MUSL
  Stingray; active/passive homing to 11 km (5.9 n miles) at
  45 kt; warhead 35 kg shaped charge.
Combat data systems: Plessey Nautis P action data automation.
Weapons control: British Aerospace Sea Archer 1A Mod 2
  optronic GFCS ❹.
Radars: Air/surface search: Plessey AWS 4 ❺; E/F-band.
  Navigation: Racal Decca 1226; I-band.

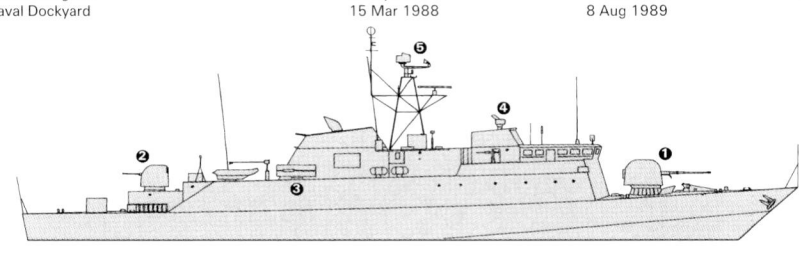

KHAMRONSIN   *(Scale 1 : 600), Ian Sturton /* 0572649

Sonars: Atlas Elektronik DSQS-21C; hull-mounted; active search
  and attack; medium frequency.

Programmes: Contract signed on 29 September 1987 with Ital
  Thai Marine of Bangkok for the construction of two ASW
  corvettes and for technical assistance with a third to be built in
  Bangkok Naval Dockyard. A fourth of the class with a different

superstructure and less armament was ordered by the Police
in September 1989.
Structure: The vessels are based on a Vosper Thornycroft
  Province class 56 m design stretched by increasing the frame
  spacing along the whole length of the hull. Depth charge racks
  and mine rails may be added.

VERIFIED

LONGLOM   *9/2003, Hartmut Ehlers /* 0572641

2 TAPI (PF 103) CLASS (FS)

Name	No	Builders	Laid down	Launched	Commissioned
TAPI	431 (ex-5)	American SB Co, Toledo, OH	1 July 1970	17 Oct 1970	19 Nov 1971
KHIRIRAT	432 (ex-6)	Norfolk SB & DD Co	18 Feb 1972	2 June 1973	10 Aug 1974

Displacement, tons: 885 standard; 1,172 full load
Dimensions, feet (metres): 275 × 33 × 10; 14.1 (sonar)
  (83.8 × 10 × 3; 4.3)
Main machinery: 2 Fairbanks-Morse 38TD8-1/8-9 diesels;
  5,250 hp (3.9 MW) sustained; 2 shafts
Speed, knots: 20. **Range, n miles:** 2,400 at 18 kt
Complement: 135 (15 officers)

Guns: 1 OTO Melara 3 in (76 mm)/62 compact ❶; 85 rds/min to
  16 km (8.7 n miles) anti-surface; 12 km (6.6 n miles) anti-
  aircraft; weight of shell 6 kg.
  1 Bofors 40 mm/70 ❷; 300 rds/min to 12.5 km (6.8 n miles);
  weight of shell 0.96 kg.
  2 Oerlikon 20 mm ❸. 2—12.7 mm MGs.
Torpedoes: 6—324 mm US Mk 32 (2 triple) tubes ❹. Honeywell
  Mk 46; anti-submarine; active/passive homing to 11 km (5.9 n
  miles) at 40 kt; warhead 44 kg.
Depth charges: 1 rack.
Combat data systems: Signaal Sewaco TH.
Radars: Air/surface search: Signaal LW04 ❺; D-band; range
  137 km (75 n miles) for 2 m² target.
  Surface search: Raytheon SPS-53E ❻; I-band.
  Fire control: Signaal WM22-61 ❼; I/J-band; range 46 km (25 n
  miles).
IFF: UPX-23.
Sonars: Atlas Elektronik DSQS-21C; hull-mounted; active search
  and attack; medium frequency.

Programmes: *Tapi* was ordered on 27 June 1969. *Khirirat* was
  ordered on 25 June 1971.
Modernisation: *Tapi* completed 1983 and *Khirirat* in 1987. This
  included new gunnery and radars and a slight heightening of
  the funnel. Further modernisation in 1988—89 mainly to
  external and internal communications.
Structure: Of similar design to the Iranian ships of the Bayandor
  class.
Operational: Used for EEZ patrols.

VERIFIED

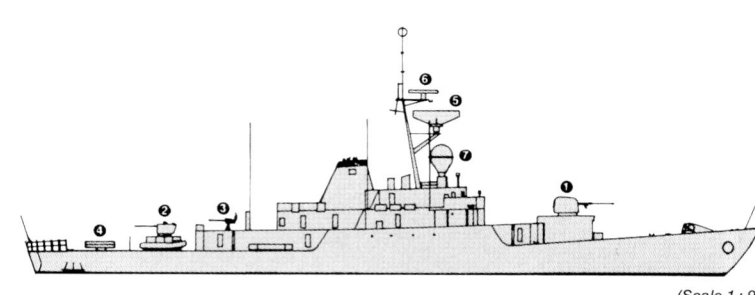

TAPI   *(Scale 1 : 900), Ian Sturton*

TAPI
6/2001, Royal Thai Navy / 0130171

0 + 2 (2) OFFSHORE PATROL VESSELS (PBOH)

Name	No	Builders	Laid down	Launched	Commissioned
—	—	Hudong Shipyard, Shanghai	2003	19 Sep 2004	2005
—	—	Hudong Shipyard, Shanghai	2004	Mar 2005	2006

Displacement, tons: 1,300; 1,440 full load
Dimensions, feet (metres): 313.3 × 38.0 × 10.2
(95.5 × 11.6 × 3.1)
Main machinery: 2 Ruston diesels; 15,660 hp *(11.7 MW)*; 2
shafts; cp props
Speed, knots: 25. **Range, n miles** 3.500 at 15 kt
Complement: 78 (18 officers)
Guns: 1 OTO Melara 3 in *(76 mm)*/62; 85 rds/min to 16 km
(8.6 n miles). 2—200 mm.
Combat data systems: COSYS.
Weapons control: Optronic director combined with TMX.
Radars: Surface search: Alenia Marconi RAN 30X; I-band.
Fire control: Oerlikon/Contraves TMX; I/J-band.
Navigation: I-band.

Helicopters: Platform for one medium.

Comment: The contract for two Offshore Patrol Vessels was
signed with China Shipbuilding Trading Company on 20
December 2002. Space and weight provision for the addition
of eight SSM, CIWS and ASW capabilities at a later date. A
further two vessels are projected.
NEW ENTRY

OPV (under construction)
10/2004 * / 1044194

SHIPBORNE AIRCRAFT

Numbers/Type: 4 Bell 214 ST.
Operational speed: 120 kt *(228 km/h)*.
Service ceiling: 13,200 ft *(4,025 m)*.
Range: 400 n miles *(740 km)*.
Role/Weapon systems: Procured in 1987 for maritime surveillance and utility roles.
NEW ENTRY

BELL 214
6/2004 *, Royal Thai Navy / 1044195

Numbers/Type: 2 AgustaWestland Super Lynx 300.
Operational speed: 125 kt *(231 km/h)*.
Service ceiling: 12,000 ft *(3,660 m)*.
Range: 340 n miles *(630 km)*.
Role/Weapon systems: Two helicopters ordered 7 August 2001 for ASW, ASV and surveillance
roles. Delivered expected in 2005.
UPDATED

SUPER LYNX
9/2004 *, AgustaWestland / 0566704

Numbers/Type: 7/2 BAe/McDonnell Douglas AV-8A/TAV-8A (Harrier).
Operational speed: 640 kt *(1,186 km/h)*.
Service ceiling: 51,200 ft *(15,600 m)*.
Range: 800 n miles *(1,480 km)*.
Role/Weapon systems: AV-8S supplied via USA to Spain and transferred in 1996. Sensors: None.
Weapons: Strike; two 30 mm Aden cannon, two AIM-9 Sidewinder or 20 mm/127 mm rockets
and 'iron' bombs.
VERIFIED

HARRIER
1/2001, Thai Navy League / 0130153

Numbers/Type: 6 Sikorsky S-70B7 Seahawk.
Operational speed: 135 kt *(250 km/h)*.
Service ceiling: 10,000 ft *(3,050 m)*.
Range: 600 n miles *(1,110 km)*.
Role/Weapon systems: Multimission helicopters delivered by June 1997. Plans to acquire ASW
equipment have been abandoned. Sensors: Telephonics APS-143(V)3 radar; ASN 150 databus;
provision for sonobuoys and dipping sonar; ALR 606(V)2 ESM. Weapons: Provision for ASM and
MUSL Stingray torpedoes.
VERIFIED

SEAHAWK
8/1997, Thai Navy League / 0019260

Numbers/Type: 4 Bell 212.
Operational speed: 100 kt *(185 km/h)*.
Service ceiling: 13,200 ft *(4,025 m)*.
Range: 200 n miles *(370 km)*.
Role/Weapon systems: Commando assault and general support. At least two transferred from
Army. May be sold to help pay for new shipborne helicopter. Mostly based ashore but operate
from Normed class and frigates. Weapons: Pintle-mounted M60 machine guns. *UPDATED*

BELL 212
2000, Thai Navy League / 0105842

LAND-BASED MARITIME AIRCRAFT (FRONT LINE)

Notes: There are also five Cessna Bird Dog light reconnaissance aircraft, 9 Cessna Skywagon and two UH-1H helicopters.

Numbers/Type: 4 Sikorsky S-76B
Operational speed: 145 kt *(269 km/h).*
Service ceiling: 6,500 ft *(1,980 m).*
Range: 357 n miles *(661 km).*
Role/Weapon systems: Six originally acquired in 1996 for maritime surveillance and utility purposes. Sensors: Weather radar. Weapons: Unarmed.

VERIFIED

S-76 *8/1996, Royal Thai Navy /* 0050241

Numbers/Type: 2/1 Lockheed P-3T/UP-3T Orion.
Operational speed: 411 kt *(761 km/h).*
Service ceiling: 28,300 ft *(8,625 m).*
Range: 4,000 n miles *(7,410 km).*
Role/Weapon systems: Delivered in 1996. Two for ASW and one utility. Two more are required. Sensors: APS-115 radar, ECM/ESM. Weapons: ASW; Mk 46 or Stingray torpedoes. ASV; four Harpoon.

VERIFIED

ORION *8/1997, Royal Thai Navy /* 0019261

Numbers/Type: 14/4 A-7E Corsair II/TA-7E Corsair II.
Operational speed: 600 kt *(1,112 km/h).*
Service ceiling: 50,000 ft *(15,240 m).*
Range: 2,000 n miles *(3,705 km).*
Role/Weapon systems: Delivered in 1996-97 from the US. Reconditioning programme in progress 2004. Weapons: AIM-9L Sidewinder; 1—20 mm cannon.

UPDATED

CORSAIR II *8/1996, Royal Thai Navy /* 0053451

Numbers/Type: 3/2 Fokker F27 Maritime 200ME/400M.
Operational speed: 250 kt *(463 km/h).*
Service ceiling: 25,000 ft *(7,620 m).*
Range: 2,700 n miles *(5,000 km).*
Role/Weapon systems: Increased coastal surveillance and response is provided, including ASW and ASV action by 200ME. 400M is for transport. Sensors: APS-504 search radar, Bendix weather radar, ESM and MAD equipment. Weapons: ASW; four Mk 46 or Stingray torpedoes or depth bombs or mines. ASV; two Harpoon ASM.

VERIFIED

FOKKER 400 *1994, Royal Thai Navy /* 0053452

Numbers/Type: 5 GAF N24A Searchmaster B (Nomad).
Operational speed: 168 kt *(311 km/h).*
Service ceiling: 21,000 ft *(6,400 m).*
Range: 730 n miles *(1,352 km).*
Role/Weapon systems: Short-range MR for EEZ protection and anti-smuggling operations. Sensors: Search radar, cameras. Weapons: Unarmed.

VERIFIED

NOMAD (US colours) *2/2004, ASTA /* 0010107

Numbers/Type: 6 Dornier 228.
Operational speed: 200 kt *(370 km/h).*
Service ceiling: 28,000 ft *(8,535 m).*
Range: 940 n miles *(1,740 km).*
Role/Weapon systems: Coastal surveillance and EEZ protection. Three acquired in 1991, three more in 1996. Sensors: APS-128/504 search radar.

VERIFIED

DORNIER 228 *6/1996, Royal Thai Navy /* 0019262

Numbers/Type: 2 Canadair CL-215.
Operational speed: 206 kt *(382 km/h).*
Service ceiling: 10,000 ft *(3,050 m).*
Range: 1,125 n miles *(2,085 km).*
Role/Weapon systems: Used for general purpose transport, SAR and fire-fighting.

VERIFIED

CL-215 *1993, Royal Thai Navy /* 0053453

Numbers/Type: 7/2 Summit T-337SP/T-337G.
Operational speed: 200 kt *(364 km/h).*
Service ceiling: 20,000 ft *(6,100 m).*
Range: 900 n miles *(1,650 km).*
Role/Weapon systems: Maritime surveillance and targeting. Weapons: LAU-32 and 59A rocket launchers, CBU-14 bomblets and 12.7 mm MG.

NEW ENTRY

PATROL FORCES

3 HUA HIN CLASS (PSO)

Name	No	Builders	Laid down	Launched	Commissioned
HUA HIN	541	Asimar, Samut Prakarn	Mar 1997	3 Mar 1999	25 Mar 2000
KLAENG	542	Asimar, Samut Prakarn	May 1997	19 Apr 1999	17 Jan 2001
SI RACHA	543	Bangkok Naval Dockyard	Dec 1997	6 Sep 1999	17 Jan 2001

Displacement, tons: 645 full load
Dimensions, feet (metres): 203.4 × 29.2 × 8.9
 (62 × 8.9 × 2.7)
Main machinery: 3 Paxman 12VP 185 diesels; 10,372 hp(m)
 (7.63 MW) sustained; 3 shafts; 1 LIPS cp prop (centreline)
Speed, knots: 25. **Range, n miles:** 2,500 at 15 kt
Complement: 45 (11 officers)

Guns: 1—3 in *(76 mm)*/50 Mk 22 ❶; 50 rds/min to 12 km *(6.5 n miles)*; weight of shell 6 kg.
 1 Bofors 40 mm/60 ❷; 2 Oerlikon 20 mm GAM-BO1 ❸.
 2—12.7 mm MGs.
Weapons control: Optronic director ❹.
Radars: Surface search: Sperry Rascar ❺; E/F-band.
Navigation: Sperry Apar; I-band.

Programmes: Ordered in September 1996 from Asian Marine. Delayed and reported cancelled by the Thai Navy in late 1997

HUA HIN
(Scale 1 : 600), Ian Sturton / 0587563

but, despite being beset by building delays, all three ships had entered service by 2001.
Structure: Derived from the Khamronsin design.

Operational: Due to budgetary constraints, older weapon systems have been installed as a temporary measure.

UPDATED

KLAENG
6/2001, Royal Thai Navy / 0130174

3 RATCHARIT CLASS (FAST ATTACK CRAFT—MISSILE) (PGGF)

Name	No	Builders	Commissioned
RATCHARIT	321 (ex-4)	CN Breda (Venezia)	10 Aug 1979
WITTHAYAKHOM	322 (ex-5)	CN Breda (Venezia)	12 Nov 1979
UDOMDET	323 (ex-6)	CN Breda (Venezia)	21 Feb 1980

Displacement, tons: 235 standard; 270 full load
Dimensions, feet (metres): 163.4 × 24.6 × 7.5 *(49.8 × 7.5 × 2.3)*
Main machinery: 3 MTU MD 20V 538 TB91 diesels; 11,520 hp(m) *(8.47 MW)* sustained; 3 shafts; Kamewa cp props
Speed, knots: 37. **Range, n miles:** 2,000 at 15 kt
Complement: 45 (7 officers)

Missiles: SSM: 4 Aerospatiale MM 38 Exocet; inertial cruise; active radar homing to 42 km *(23 n miles)* at 0.9 Mach; warhead 165 kg; sea-skimmer.
Guns: 1 OTO Melara 3 in *(76 mm)*/62 compact; 85 rds/min to 16 km *(8.7 n miles)* anti-surface; 12 km *(6.6 n miles)* anti-aircraft; weight of shell 6 kg.
 1 Bofors 40 mm/70; 300 rds/min to 12.5 km *(6.8 n miles)*; weight of shell 0.96 kg.
 2—12.7 mm MGs.
Countermeasures: ESM: Racal RDL-2; intercept.
Radars: Surface search: Decca; I-band.
Fire control: Signaal WM25; I/J-band; range 46 km *(25 n miles)*.

Programmes: Ordered June 1976. *Ratcharit* launched 30 July 1978, *Witthayakhom* 2 September 1978 and *Udomdet* 28 September 1978.
Structure: Standard Breda BMB 230 design.

UPDATED

3 PRABPARAPAK CLASS (FAST ATTACK CRAFT—MISSILE) (PTFG)

Name	No	Builders	Commissioned
PRABPARAPAK	311 (ex-1)	Singapore SBEC	28 July 1976
HANHAK SATTRU	312 (ex-2)	Singapore SBEC	6 Nov 1976
SUPHAIRIN	313 (ex-3)	Singapore SBEC	1 Feb 1977

Displacement, tons: 224 standard; 268 full load
Dimensions, feet (metres): 149 × 24.3 × 7.5 *(45.4 × 7.4 × 2.3)*
Main machinery: 4 MTU 16V 538 TB92 diesels; 13,640 hp(m) *(10 MW)* sustained; 4 shafts
Speed, knots: 40. **Range, n miles:** 2,000 at 15 kt; 750 at 37 kt
Complement: 41 (5 officers)

Missiles: SSM: 5 IAI Gabriel I (1 triple, 2 single) launchers; radar or optical guidance; semi-active radar homing to 20 km *(10.8 n miles)* at 0.7 Mach; warhead 75 kg.
Guns: 1 Bofors 57 mm/70; 200 rds/min to 17 km *(9.3 n miles)*; weight of shell 2.4 kg. 8 rocket illuminant launchers on either side of 57 mm gun.
 1 Bofors 40 mm/70; 300 rds/min to 12 km *(6.6 n miles)*; weight of shell 2.4 kg.
Countermeasures: ESM: Racal RDL-2; intercept.
Radars: Surface search: Kelvin Hughes Type 17; I-band.
Fire control: Signaal WM28/5 series; I/J-band.

Programmes: Ordered June 1973. Built under licence from Lürssen. Launch dates-*Prabparapak* 29 July 1975, *Hanhak Sattru* 28 October 1975, *Suphairin* 20 February 1976.
Modernisation: There are plans to replace Gabriel possibly by RBS 15.
Structure: Same design as Lürssen standard 45 m class built for Singapore. Normally only three Gabriel SSM are carried.

VERIFIED

UDOMDET
3/2004*, Bob Fildes / 0589809

PRABPARAPAK
6/2001, Royal Thai Navy / 0130172

3 CHON BURI CLASS (FAST ATTACK CRAFT—GUN) (PG)

Name	No	Builders	Commissioned
CHON BURI	331 (ex-1)	CN Breda (Venezia) Mestre	22 Feb 1983
SONGKHLA	332 (ex-2)	CN Breda (Venezia) Mestre	15 July 1983
PHUKET	333 (ex-3)	CN Breda (Venezia) Mestre	13 Jan 1984

Displacement, tons: 450 full load
Dimensions, feet (metres): 198 × 29 × 15 *(60.4 × 8.8 × 4.5)*
Main machinery: 3 MTU 20V 538 TB92 diesels; 12,795 hp(m) *(9.4 MW)* sustained; 3 shafts; cp props
Speed, knots: 30. **Range, n miles:** 2,500 at 18 kt; 900 at 30 kt
Complement: 41 (6 officers)
Guns: 2 OTO Melara 3 in *(76 mm)*/62; 85 rds/min to 16 km *(8.7 n miles)*; weight of shell 6 kg.
2 Breda 40 mm/70 (twin).
Countermeasures: Decoys: 4 Hycor Mk 135 chaff launchers.
ESM: Elettronica Newton; intercept.
Weapons control: Signaal Lirod 8 optronic director.
Radars: Surface search: Signaal ZW06; I-band.
Fire control: Signaal WM22/61; I/J-band; range 46 km *(25 n miles)*.

Comment: Ordered in 1979 (first pair) and 1981. Laid down—*Chon Buri* 15 August 1981 (launched 29 November 1982), *Songkhla* 15 September 1981 (launched 6 September 1982), *Phuket* 15 December 1981 (launched 3 February 1983). Steel hulls, alloy superstructure. Can be adapted to carry SSMs.

VERIFIED

PHUKET 10/2001, *Chris Sattler* / 0130157

CHON BURI 9/2003, *Hartmut Ehlers* / 0572642

6 SATTAHIP (PSMM Mk 5) CLASS (LARGE PATROL CRAFT) (PG)

Name	No	Builders	Commissioned
SATTAHIP	521 (ex-4)	Ital Thai (Samutprakarn) Ltd	16 Sep 1983
KLONGYAI	522 (ex-5)	Ital Thai (Samutprakarn) Ltd	7 May 1984
TAKBAI	523 (ex-6)	Ital Thai (Samutprakarn) Ltd	18 July 1984
KANTANG	524 (ex-7)	Ital Thai (Samutprakarn) Ltd	14 Oct 1985
THEPHA	525 (ex-8)	Ital Thai (Samutprakarn) Ltd	17 Apr 1986
TAIMUANG	526 (ex-9)	Ital Thai (Samutprakarn) Ltd	17 Apr 1986

Displacement, tons: 270 standard; 300 full load
Dimensions, feet (metres): 164.5 × 23.9 × 5.9 *(50.1 × 7.3 × 1.8)*
Main machinery: 2 MTU 16V 538 TB92 diesels; 6,820 hp(m) *(5 MW)* sustained; 2 shafts
Speed, knots: 22. **Range, n miles:** 2,500 at 15 kt
Complement: 56
Guns: 1 OTO Melara 3 in *(76 mm)*/62 (in 521-523). 1 USN 3 in *(76 mm)*/50 Mk 26 (in 524-526).
1 Bofors 40 mm/70 or 40 mm/60. 2 Oerlikon 20 mm GAM-BO1. 2—12.7 mm MGs.
Weapons control: NA 18 optronic director (in 521-523).
Radars: Surface search: Decca; I-band.

Comment: First four ordered 9 September 1981, *Thepha* on 27 December 1983 and *Taimuang* on 31 August 1984.

VERIFIED

TAKBAI 10/1999, *Royal Thai Navy* / 0080815

3 T 81 CLASS (COASTAL PATROL CRAFT) (PB)

T 81   T 82   T 83

Displacement, tons: 120 full load
Dimensions, feet (metres): 98.8 × 20.7 × 5.6 *(30.1 × 6.3 × 1.7)*
Main machinery: 2 MTU 16V 2000 TE90 diesels; 3,600 hp(m) *(2.56 MW)*; 2 shafts
Speed, knots: 25. **Range, n miles:** 1,300 at 15 kt
Complement: 28 (3 officers)
Guns: 1 Bofors 40 mm/70. 1 Oerlikon 20 mm. 2—12.7 mm MGs.
Radars: Surface search: Sperry SM 5000; I-band.

Comment: Ordered in October 1996 from ASC Silkline in Pranburi. First one commissioned 5 August 1999, second 9 December 1999 and the third in 2000. Plans for seven more have been shelved.

UPDATED

T 83 3/2004*, *Bob Fildes* / 0589810

10 PGM 71 CLASS (COASTAL PATROL CRAFT) (PB)

T 11-19   T 110

Displacement, tons: 130 standard; 147 full load
Dimensions, feet (metres): 101 × 21 × 6 *(30.8 × 6.4 × 1.9)*
Main machinery: 2 GM diesels; 1,800 hp *(1.34 MW)*; 2 shafts
Speed, knots: 18.5. **Range, n miles:** 1,500 at 10 kt
Complement: 30
Guns: 1 Bofors 40 mm/60. 1 Oerlikon 20 mm. 2—12.7 mm MGs.
In some craft the 20 mm gun has been replaced by an 81 mm mortar/12.7 mm combined mounting aft.
Radars: Surface search: Decca 303 *(T 11 and 12)* or Decca 202 (remainder); I-band.

Comment: Built by Peterson Inc between 1966 and 1970. Transferred from US. Some likely to be retained in service in view of curtailment of T 81 building programme.

UPDATED

T 16 10/1999, *Royal Thai Navy* / 0080816

9 T 91 CLASS (COASTAL PATROL CRAFT) (PB)

T 91-99

Displacement, tons: 87.5 *(T 91-92)*, 117 (remainder) standard
Dimensions, feet (metres): 103.4 × 17.6 × 4.9 *(31.5 × 5.4 × 1.5) (T 91-92)*
111.6 × 18.7 × 4.9 *(34.0 × 5.7 × 1.5)* (remainder)
Main machinery: 2 MTU 12V 538 TB81/82 diesels; 3,300 hp(m) *(2.43 MW)*/4,430 hp(m) *(3.26 MW)* sustained; 2 shafts
Speed, knots: 25. **Range, n miles:** 700 at 21 kt
Complement: 21 *(T 91-92)*; 25 (remainder)
Guns: 2 or 1 Bofors 40 mm/60 *(T 91 and T 99)*. 1 Oerlikon 20 mm GAM-BO1 *(T 91 and T 99)*. 2—12.7 mm MGs *(T 93-99)*.
Weapons control: Sea Archer 1A optronic director *(T 99 only)*.
Radars: Surface search: Raytheon SPS-35 (1500B); I-band.

Comment: Built by Royal Thai Naval Dockyard, Bangkok. *T 91* commissioned in 1965; *T 92-93* in 1973; *T 94-98* between 1981 and 1984; *T 99* in 1987. *T 91* has an extended upperworks and a 20 mm gun in place of the after 40 mm. *T 99* has a single Bofors 40/70, one Oerlikon 20 mm and two MGs. Major refits from 1983-86 for earlier vessels of the class.

VERIFIED

T 96 10/2001, *Chris Sattler* / 0130457

3 SEA SPECTRE MK III CLASS (PB)

T 210-212

Displacement, tons: 28; 37 full load
Dimensions, feet (metres): 65.0 × 18.0 × 5.9 *(19.8 × 5.5 × 1.8)*
Main machinery: 3 Detroit diesels; 1,800 hp *(1.34 MW)*; 3 shafts
Speed, knots: 30. **Range, n miles:** 450 at 20 kt
Complement: 9 (1 officer)
Guns: 2 Oerlikon 20 mm. 1—12.7 MG.
Radars: Surface search: Raytheon; I-band.

Comment: Aluminium hulled craft built by Peterson. Transferred from the US in 1975.
VERIFIED

9 SWIFT CLASS (COASTAL PATROL CRAFT) (PB)

T 21-29

Displacement, tons: 22 full load
Dimensions, feet (metres): 50 × 13 × 3.5 *(15.2 × 4 × 1.1)*
Main machinery: 2 Detroit diesels; 480 hp *(358 kW)*; 2 shafts
Speed, knots: 25. **Range, n miles:** 400 at 25 kt
Complement: 8 (1 officer)
Guns: 2—81 mm mortars. 2—12.7 mm MGs.
Radars: Surface search: Raytheon Pathfinder; I-band.

Comment: Transferred from US Navy from 1967-75.
VERIFIED

T 26   8/1998, P Marsan / 0080818

14 T 213 CLASS (COASTAL PATROL CRAFT) (PB)

T 213-226

Displacement, tons: 35 standard
Dimensions, feet (metres): 64 × 17.5 × 5 *(19.5 × 5.3 × 1.5)*
Main machinery: 2 MTU diesels; 715 hp(m) *(526 kW)*; 2 shafts
Speed, knots: 25
Complement: 8 (1 officer)
Guns: 1 Oerlikon 20 mm. 1—81 mm mortar with 12.7 mm MG.
Radars: Surface search: Racal Decca 110; I-band.

Comment: Built by Ital Thai Marine Ltd. Commissioned-*T 213-215*, 29 August 1980; *T 216-218*, 26 March 1981; *T 219-223*, 16 September 1981; *T 224*, 19 November 1982; *T 225* and *T 226*, 28 March 1984. Construction of *T 227-230* is not to have been completed. Of alloy construction. Used for fishery patrol and coastal control duties.
VERIFIED

T 219   9/2003, Hartmut Ehlers / 0572643

3 SEAL ASSAULT CRAFT (LCP)

Comment: Locally built for special forces operations. Details are not known but reported to be larger and faster than PBR Mk II craft. Equipped with stern ramp.
VERIFIED

T 242 (SEAL)   5/1997, A Sharma / 0050242

13 PBR Mk II (RIVER PATROL CRAFT) (PBR)

Displacement, tons: 8 full load
Dimensions, feet (metres): 32.1 × 11.5 × 2.3 *(9.8 × 3.5 × 0.7)*
Main machinery: 2 Detroit diesels; 430 hp *(321 kW)*; 2 Jacuzzi water-jets
Speed, knots: 25. **Range, n miles:** 150 at 23 kt
Complement: 4
Guns: 2—6.72 mm MGs. 1—60 mm mortar.
Radars: Surface search: Raytheon SPS-66; I-band.

Comment: Transferred from US from 1967-73. Employed on Mekong River. Reported to be getting old, numbers are reducing and maximum speed has been virtually halved. All belong to the Riverine and SEAL Squadron.
VERIFIED

PBR MK II   6/2002, Thai Navy League / 0543390

90 ASSAULT BOATS (LCP)

Displacement, tons: 0.4 full load
Dimensions, feet (metres): 16.4 × 6.2 × 1.3 *(5 × 1.9 × 0.4)*
Main machinery: 1 outboard; 150 hp *(110 kW)*
Speed, knots: 24
Complement: 2
Guns: 1—7.62 mm MG.

Comment: Part of the Riverine Squadron with the PBRs and two PCFs. Can carry six people. Numbers uncertain.
VERIFIED

ASSAULT BOAT   6/2002, Thai Navy League / 0530060

MINE WARFARE FORCES

1 MCM SUPPORT SHIP (MCS)

Name	No	Builders	Commissioned
THALANG	621 (ex-1)	Bangkok Dock Co Ltd	4 Aug 1980

Displacement, tons: 1,000 standard
Dimensions, feet (metres): 185.5 × 33 × 10 *(55.7 × 10 × 3.1)*
Main machinery: 2 MTU diesels; 1,310 hp(m) *(963 kW)*; 2 shafts
Speed, knots: 12
Complement: 77
Guns: 1 Bofors 40 mm/70. 2 Oerlikon 20 mm. 2—12.7 mm MGs.
Radars: Surface search: Racal Decca 1226; I-band.

Comment: Has minesweeping capability. Two 3 ton cranes provided for change of minesweeping gear in MSCs-four sets carried. Design by Ferrostaal, Essen. Has dormant minelaying capability.
VERIFIED

THALANG   11/2001, Maritime Photographic / 0130165

2 LAT YA (GAETA) CLASS
(MINEHUNTERS/SWEEPERS) (MHSC)

Name	No	Builders	Launched	Commissioned
LAT YA	633	Intermarine, Sarzana	30 Mar 1998	18 June 1999
THA DIN DAENG	634	Intermarine, Sarzana	31 Oct 1998	18 Dec 1999

Displacement, tons: 680 full load
Dimensions, feet (metres): 172.1 × 32.4 × 9.4 *(52.5 × 9.9 × 2.9)*
Main machinery: 2 MTU 8V 396 TE74K diesels; 1,600 hp(m) *(1.18 MW)* sustained; 2 Voith
  Schneider props; auxiliary propulsion; 2 hydraulic motors
Speed, knots: 14. **Range, n miles:** 2,000 at 12 kt
Complement: 50 (8 officers)
Guns: 1 MSI 30 mm.
Countermeasures: MCM: Atlas MWS 80-6 minehunting system. Magnetic, acoustic and
  mechanical sweeps; ADI Mini Dyad, Noise Maker, Bofors MS 106, 2 Pluto Plus ROVs.
Radars: Navigation: Atlas Elektronik 9600M (ARPA); I-band.
Sonars: Atlas Elektronik DSQS-11M; hull-mounted; active; high frequency.

Comment: Invitations to tender lodged by 3 April 1996. Ordered 19 September 1996.
  Specifications include hunting at up to 6 kt and sweeping at 10 kt. No further ships are planned.
  VERIFIED

THA DIN DAENG   *1/2000, Giorgio Ghiglione /* 0080821

2 BANG RACHAN CLASS
(MINEHUNTERS/SWEEPERS) (MHSC)

Name	No	Builders	Commissioned
BANG RACHAN	631 (ex-2)	Lürssen Vegesack	29 Apr 1987
NONGSARAI	632 (ex-3)	Lürssen Vegesack	17 Nov 1987

Displacement, tons: 444 full load
Dimensions, feet (metres): 161.1 × 30.5 × 8.2 *(49.1 × 9.3 × 2.5)*
Main machinery: 2 MTU 12V 396 TB83 diesels; 3,120 hp(m) *(2.3 MW)* sustained; 2 shafts;
  Kamewa cp props; auxiliary propulsion; 1 motor
Speed, knots: 17; 7 (electric motor). **Range, n miles:** 3,100 at 12 kt
Complement: 33 (7 officers)
Guns: 3 Oerlikon GAM-BO1 20 mm.
Countermeasures: MCM: MWS 80R minehunting system. Acoustic, magnetic and mechanical
  sweeps. 2 Gaymarine Pluto 15 remote-controlled submersibles.
Radars: Navigation: 2 Atlas Elektronik 8600 ARPA; I-band.
Sonars: Atlas Elektronik DSQS-11H; hull-mounted; minehunting; high frequency.

Comment: First ordered from Lürssen late 1984, arrived Bangkok 22 October 1987. Second
  ordered 5 August 1985 and arrived in Bangkok May 1988. Amagnetic steel frames and
  deckhouses, wooden hull. Motorola Miniranger MRS III precise navigation system. Draeger
  decompression chamber. Plans to upgrade the sonar have not been confirmed.
  UPDATED

NONGSARAI   *3/2004*, Chris Sattler /* 0589811

2 BLUEBIRD CLASS (MINESWEEPERS—COASTAL) (MSC)

Name	No	Builders	Commissioned
BANGKEO	612 (ex-6)	Dorchester SB Corporation,	9 July 1965
(ex-*MSC 303*)		Camden	
DONCHEDI	613 (ex-8)	Peterson Builders Inc,	17 Sep 1965
(ex-*MSC 313*)		Sturgeon Bay, WI	

Displacement, tons: 317 standard; 384 full load
Dimensions, feet (metres): 145.3 × 27 × 8.5 *(44.3 × 8.2 × 2.6)*
Main machinery: 2 GM 8-268 diesels; 880 hp *(656 kW)*; 2 shafts
Speed, knots: 13. **Range, n miles:** 2,750 at 12 kt
Complement: 43 (7 officers)
Guns: 2 Oerlikon 20 mm/80 (twin).
Countermeasures: MCM: US Mk 4 (V). Mk 6. US Type Q2 magnetic.
Radars: Navigation: Decca TM 707; I-band.
Sonars: UQS-1; hull-mounted; minehunting; high frequency.

Comment: Constructed for Thailand. One paid off in 1992, one in 1995 and the last two are in
  limited operational service.
  VERIFIED

DONCHEDI   *11/2001, Maritime Photographic /* 0130164

12 MSBs (MSR)

MLM 6-10   **MSB 11-17**

Displacement, tons: 25 full load
Dimensions, feet (metres): 50.2 × 13.1 × 3 *(15.3 × 4 × 0.9)*
Main machinery: 1 Gray Marine 64 HN9 diesel; 165 hp *(123 kW)*; 1 shaft
Speed, knots: 8
Complement: 10
Guns: 2—7.62 mm MGs.

Comment: Three transferred from USA in October 1963 and two in 1964. More were built locally
  from 1994. Wooden hulled, converted from small motor launches. Operated on Chao Phraya
  river.
  VERIFIED

MSB 11   *10/1995, Royal Thai Navy /* 0080822

AMPHIBIOUS FORCES

Note: There are approximately 24 landing craft of about 100 tons operated by the Army.

2 NORMED CLASS (LSTH)

Name	No	Builders	Launched	Commissioned
SICHANG	721 (ex-LST 6)	Ital Thai	14 Apr 1987	9 Oct 1987
SURIN	722 (ex-LST 7)	Bangkok Dock Co Ltd	12 Apr 1988	16 Dec 1988

Displacement, tons: 3,540 standard; 4,235 full load
Dimensions, feet (metres): 337.8; 357.6 (722) × 51.5 × 11.5 *(103; 109 × 15.7 × 3.5)*
Main machinery: 2 MTU 20V 1163 TB82 diesels; 11,000 hp(m) *(8.1 MW)* sustained; 2 shafts;
  cp props
Speed, knots: 16. **Range, n miles:** 7,000 at 12 kt
Complement: 53
Military lift: 348 troops; 14 tanks or 12 APCs or 850 tons cargo; 3 LCVP; 1 LCPL
Guns: 2 Bofors 40 mm/70. 2 Oerlikon GAM-CO1 20 mm. 2—12.7 mm MGs. 1—81 mm mortar.
Weapons control: 2 BAe Sea Archer Mk 1A optronic directors.
Radars: Navigation: Racal Decca 1226; I-band.
Helicopters: Platform for 2 Bell 212.

Comment: First ordered 31 August 1984 to a Chantier du Nord (Normed) design. Second ordered
  to a modified design and lengthened to accommodate a battalion. The largest naval ships yet
  built in Thailand. Have bow doors and a 17 m ramp.
  UPDATED

SURIN   *11/2001, Maritime Photographic /* 0130163

SICHANG   *2/2004*, Bob Fildes /* 0589812

4 LST 512-1152 CLASS (LST)

Name	No	Builders	Commissioned
CHANG	712 (ex-LST 2)	Dravo Corporation	29 Dec 1944
(ex-*Lincoln County* LST 898)			
PANGAN	713 (ex-LST 3)	Chicago Bridge and Iron	7 Apr 1945
(ex-*Stark County* LST 1134)		Co, ILL	
LANTA	714 (ex-LST 4)	Chicago Bridge and Iron	9 May 1945
(ex-*Stone County* LST 1141)		Co, ILL	
PRATHONG	715 (ex-LST 5)	Jefferson B & M Co, Ind	13 Sep 1944
(ex-*Dodge County* LST 722)			

Displacement, tons: 1,650 standard; 3,640/4,145 full load
Dimensions, feet (metres): 328 × 50 × 14 *(100 × 15.2 × 4.4)*
Main machinery: 2 GM 12-567A diesels; 1,800 hp *(1.34 MW)*; 2 shafts
Speed, knots: 11.5. **Range, n miles:** 9,500 at 9 kt
Complement: 80; 157 (war)
Military lift: 1,230 tons max; 815 tons beaching
Guns: 8 Bofors 40 mm/60 (2 twin, 4 single) (can be carried).
  2—12.7 mm MGs *(Chang)*. 2 Oerlikon 20 mm (others).
Weapons control: 2 Mk 51 GFCS. 2 optical systems.
Radars: Navigation: Racal Decca 1229; I/J-band.

Comment: *Chang* transferred from USA in August 1962. *Pangan* 16 May 1966, *Lanta* on 15 August 1973 (by sale 1 March 1979) and *Prathong* on 17 December 1975. *Chang* has a reinforced bow and waterline. *Lanta*, *Prathong* and *Chang* have mobile crane on the well-deck. All have tripod mast.

VERIFIED

LANTA

5/2002, Mick Prendergast / 0530001

1 LSM 1 CLASS (LSM)

Name	No	Builders	Commissioned
KUT (ex-*LSM 338*)	731 (ex-LSM 1)	Pullman Std Car Co, Chicago	10 Jan 1945

Displacement, tons: 743 standard; 1,107 full load
Dimensions, feet (metres): 203.5 × 34.5 × 9.9 *(62 × 10.5 × 3)*
Main machinery: 2 Fairbanks-Morse 38D8-1/8-10 diesels; 3,540 hp *(2.64 MW)* sustained; 2 shafts
Speed, knots: 12.5. **Range, n miles:** 4,500 at 12.5 kt
Complement: 91 (6 officers)
Military lift: 452 tons beaching; 50 troops with vehicles
Guns: 2 Bofors 40 mm/60 Mk 3 (twin). 4 Oerlikon 20 mm/70.
Weapons control: Mk 51 Mod 2 optical director *(Kram)*.
Radars: Surface search: Raytheon SPS-5 *(Kram)*; G/H-band.
Navigation: Raytheon 1500 B; I-band.

Comment: Former US landing ship of the LCM, later LSM (Medium Landing Ship) type. Transferred in 1946. *Kram* reported sunk as a target in February 2003.

VERIFIED

KUT

6/1998, Royal Thai Navy / 0050244

3 MAN NOK CLASS (LCU)

Name	No	Builders	Launched	Commissioned
MAN NOK	781	Sahai Sant, Pratum Thani	1 May 2001	6 Dec 2001
MAN KLANG	782	Sahai Sant, Pratum Thani	1 May 2001	14 Nov 2001
MAN NAI	783	Sahai Sant, Pratum Thani	1 May 2001	6 Dec 2001

Displacement, tons: 170 light; 550 full load
Dimensions, feet (metres): 172 × 36.7 × 5.9 *(52.4 × 11.2 × 1.8)*
Main machinery: 2 Caterpillar 3432 DITA diesels; 700 hp(m) *(515 kW)*; 2 shafts
Speed, knots: 12. **Range, n miles:** 1,500 at 10 kt
Complement: 30 (3 officers)
Military lift: 2 M60 tanks or 25 tons vehicles
Guns: 2 Oerlikon 20 mm.
Radars: Navigation: I-band.

Comment: Ordered from Silkline ASC in 1997. All three craft launched 1 May 2000.

UPDATED

MAN NAI

5/2002, Mick Prendergast / 0530000

2 LSIL 351 CLASS (LCM)

PRAB 741 (ex-LSIL 1)   **SATAKUT** 742 (ex-LSIL 2)

Displacement, tons: 230 standard; 399 full load
Dimensions, feet (metres): 157 × 23 × 6 *(47.9 × 7 × 1.8)*
Main machinery: 4 GM diesels; 2,320 bhp *(1.73 MW)*; 2 shafts
Speed, knots: 15. **Range, n miles:** 5,600 at 12.5 kt
Complement: 49 (7 officers)
Military lift: 101 tons or 76 troops
Guns: 1 US 3 in *(76 mm)*/50. 1 Bofors 40 mm/60. 2 Oerlikon 20 mm/70.
Radars: Surface search: Raytheon SPS-35 (1500B); I-band.

Comment: *Prab* transferred to Thailand in October 1946. *Satakut* was refitted in the mid-1990s.

VERIFIED

PRAB

2/1998, Thai Navy League / 0019275

4 THONG KAEO CLASS (LCU)

Name	No	Builders	Commissioned
THONG KAEO	771 (ex-7)	Bangkok Dock Co Ltd	23 Dec 1982
THONG LANG	772 (ex-8)	Bangkok Dock Co Ltd	19 Apr 1983
WANG NOK	773 (ex-9)	Bangkok Dock Co Ltd	16 Sep 1983
WANG NAI	774 (ex-10)	Bangkok Dock Co Ltd	11 Nov 1983

Displacement, tons: 193 standard; 396 full load
Dimensions, feet (metres): 134.5 × 29.5 × 6.9 *(41 × 9 × 2.1)*
Main machinery: 2 GM 16V-71 diesels; 1,400 hp *(1.04 MW)*; 2 shafts
Speed, knots: 10. **Range, n miles:** 1,200 at 10 kt
Complement: 31 (3 officers)
Military lift: 3 lorries; 150 tons equipment
Guns: 2 Oerlikon 20 mm. 2—7.62 mm MGs.

Comment: Ordered in 1980.

VERIFIED

WANG NAI

5/1997, Maritime Photographic / 0019276

6 MATAPHON CLASS (LCM/LCVP/LCP)

MATAPHON 761 (ex-LCU 1260)   **ADANG** 763 (ex-LCU 861)   **KOLAM** 765 (ex-LCU 904)
RAWI 762 (ex-LCU 800)   **PHETRA** 764 (ex-LCU 1089)   **TALIBONG** 766 (ex-LCU 753)

Displacement, tons: 145 standard; 330 full load
Dimensions, feet (metres): 120.4 × 32 × 4 *(36.7 × 9.8 × 1.2)*
Main machinery: 3 Gray Marine 65 diesels; 675 hp *(503 kW)*; 3 shafts
Speed, knots: 10. **Range, n miles:** 650 at 8 kt
Complement: 13
Military lift: 150 tons or 3-4 tanks or 250 troops
Guns: 4 Oerlikon 20 mm (2 twin).
Radars: Navigation: Raytheon Pathfinder; I-band.

Comment: Transferred from US 1946-47. Employed as transport ferries.

VERIFIED

TALIBONG

11/2001, Maritime Photographic / 0130160

40 LANDING CRAFT (LCM/LCVP/LCA)

Displacement, tons: 56 full load
Dimensions, feet (metres): 56.1 × 14.1 × 3.9 *(17.1 × 4.3 × 1.2)*
Main machinery: 2 Gray Marine 64 HN9 diesels; 330 hp *(264 kW)*; 2 shafts
Speed, knots: 9. **Range, n miles:** 135 at 9 kt
Complement: 5
Military lift: 34 tons

Comment: Details given are for the 24 ex-US LCMs delivered in 1965-69. The 12 ex-US LCVPs can lift 40 troops and are of 1960s vintage. The four LCAs can lift 35 troops and were built in 1984 in Bangkok.

VERIFIED

3 GRIFFON 1000 TD HOVERCRAFT (UCAC)

401-403

Dimensions, feet (metres): 27.6 × 12.5 *(8.4 × 3.8)*
Main machinery: 1 Deutz BF6L913C diesel; 190 hp(m) *(140 kW)*
Speed, knots: 33. **Range, n miles:** 200 at 27 kt
Complement: 2
Cargo capacity: 1,000 kg plus 9 troops
Radars: Navigation: Raytheon; I-band.

Comment: Acquired in mid-1990 from Griffon Hovercraft. Although having an obvious amphibious capability they are also used for rescue and flood control.

VERIFIED

LCM 208   *11/1998, Thai Navy League* / 0050247

GRIFFON 401   *6/1999, Royal Thai Navy* / 0084413

TRAINING SHIPS

Notes: A 3,500 ton ship Carlskrona type is wanted for training and as a support ship.

1 CANNON CLASS (FFT)

Name	No	Builders	Laid down	Launched	Commissioned
PIN KLAO (ex-*Hemminger* DE 746)	413 (ex-3, ex-1)	Western Pipe & Steel Co	1943	12 Sep 1943	30 May 1944

Displacement, tons: 1,240 standard; 1,930 full load
Dimensions, feet (metres): 306 × 36.7 × 14 *(93.3 × 11.2 × 4.3)*
Main machinery: Diesel-electric; 4 GM 16-278A diesels; 6,000 hp *(4.5 MW)*; 4 generators; 2 motors; 2 shafts
Speed, knots: 20. **Range, n miles:** 10,800 at 12 kt; 6,700 at 19 kt
Complement: 192 (14 officers)

Guns: 3 USN 3 in *(76 mm)*/50 Mk 22; 20 rds/min to 12 km *(6.6 n miles)*; weight of shell 6 kg.
6 Bofors 40 mm/60 (3 twin); 120 rds/min to 10 km *(5.5 n miles)*; weight of shell 0.89 kg.

Torpedoes: 6—324 mm US Mk 32 (2 triple) tubes; anti-submarine.
A/S mortars: 1 Hedgehog Mk 10 multibarrelled fixed; range 250 m; warhead 13.6 kg; 24 rockets.
Depth charges: 8 projectors; 2 racks.
Countermeasures: ESM: WLR-1; radar warning.
Weapons control: Mk 52 radar GFCS for 3 in guns. Mk 63 radar GFCS for aft gun only. 2 Mk 51 optical GFCS for 40 mm.
Radars: Air/surface search: Raytheon SPS-5; G/H-band.
Navigation: Raytheon SPS-21; G/H-band.
Fire control: Western Electric Mk 34; I/J-band.
RCA/General Electric Mk 26; I/J-band.
IFF: SLR 1.

Sonars: SQS-11; hull-mounted; active attack; high frequency.

Programmes: Transferred from US Navy at New York Navy Shipyard in July 1959 under MDAP and by sale 6 June 1975.
Modernisation: The three 21 in torpedo tubes were removed and the 20 mm guns were replaced by 40 mm. The six A/S torpedo tubes were fitted in 1966.
Operational: Used mostly as a training ship.

VERIFIED

PIN KLAO   *6/1997, Royal Thai Navy* / 0019254

1 YARROW TYPE (TRAINING SHIP) (FFH/AX)

Name	No	Builders	Laid down	Launched	Commissioned
MAKUT RAJAKUMARN	433 (ex-7)	Yarrow Shipbuilders	11 Jan 1970	18 Nov 1971	7 May 1973

Displacement, tons: 1,650 standard; 1,900 full load
Dimensions, feet (metres): 320 × 36 × 18.1
  (97.6 × 11 × 5.5)
Main machinery: CODOG; 1 RR Olympus TM3B gas turbine;
  22,500 hp *(16.8 MW)* sustained; 1 Crossley-SEMT-Pielstick
  12 PC2.2 V 400 diesel; 6,000 hp(m) *(4.4 MW)* sustained; 2
  shafts
Speed, knots: 26 gas; 18 diesel
Range, n miles: 5,000 at 18 kt; 1,200 at 26 kt
Complement: 140 (16 officers)

Guns: 2 Vickers 4.5 in *(114 mm)*/55 Mk 8 ❶; 25 rds/min to
  22 km *(12 n miles)* anti-surface; 6 km *(3.3 n miles)* anti-
  aircraft; weight of shell 21 kg.
  2 Breda 40 mm/70 (twin) ❷; 300 rds/min to 12.5 km *(6.8 n
  miles)*; weight of shell 0.96 kg.
  2 Oerlikon 20 mm.
Torpedoes: 6 Plessey PMW 49A tubes ❸ Mk 46; active/passive
  homing to 11 km *(5.9 n miles)* at 40 kt; warhead 44 kg.
A/S mortars: 1 Limbo 3-tubed Mk 10 ❹.
Depth charges: 1 rack.
Countermeasures: Decoys: 2 Loral Mk 135 chaff launchers.
  ESM/ECM: Elettronica Newton ❺; intercept and jammer. WLR-1;
  radar warning.
Combat data systems: Signaal Sewaco TH.
Radars: Air/surface search: Signaal DA05 ❻; E/F-band; range
  137 km *(75 n miles)* for 2 m² target.
Surface search: Signaal ZW06 ❼; I-band.

Fire control: Signaal WM22/61 ❽; I/J-band; range 46 km *(25 n
  miles)*.
Navigation: Racal Decca; I-band.
Sonars: Atlas Elektronik DSQS-21C; hull-mounted; active search
  and attack; medium frequency.

Helicopters: A small helicopter can land when the Mortar Mk 10
  well is closed.

Programmes: Ordered on 21 August 1969.
Modernisation: A severe fire in February 1984 resulted in
  extensive work including replacement of the Olympus gas

turbine, a new ER control room and central electric
switchboard. Further modifications included the removal of
Seacat SAM system and the installation of new EW
equipment in 1993. In 1997 two Bofors 40 mm were fitted on
the old Seacat mounting, and torpedo tubes replaced the old
Bofors abreast the funnel.
Operational: The ship is largely automated with a consequent
saving in complement, and has been most successful in
service. Has lost its Flagship role to one of the Chinese-built
frigates and become a training ship.

UPDATED

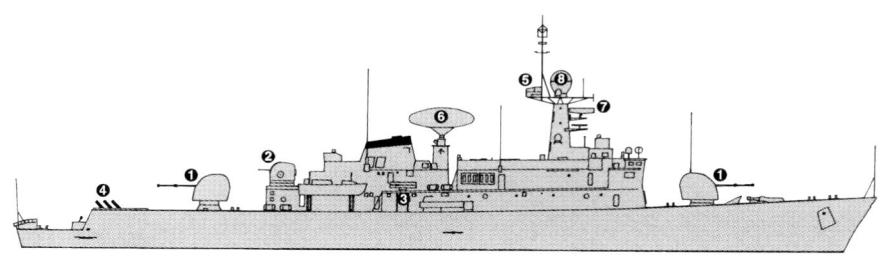

MAKUT RAJAKUMARN   *(Scale 1 : 900), Ian Sturton /* 0080806

MAKUT RAJAKUMARN   *2/2004*, Bob Fildes /* 0589813

1 ALGERINE CLASS (AXL)

Name	No	Builders	Commissioned
PHOSAMTON (ex-*Minstrel*)	611 (ex-415, ex-MSF 1)	Redfern Construction Co	9 June 1945

Displacement, tons: 1,040 standard; 1,335 full load
Dimensions, feet (metres): 225 × 35.5 × 11.5 *(68.6 × 10.8 × 3.5)*
Main machinery: 2 boilers; 2 reciprocating engines; 2,000 ihp *(1.49 MW)*; 2 shafts
Speed, knots: 16. **Range, n miles:** 4,000 at 10 kt
Complement: 103
Guns: 1 USN 3 in *(76 mm)*/50. 1 Bofors 40 mm/60. 4 Oerlikon 20 mm.
Radars: Navigation: Raytheon Pathfinder; I-band.

Comment: Transferred from UK in April 1947. Received engineering overhaul in 1984.
  Minesweeping gear replaced by a deckhouse to increase training space. Vickers 4 in gun
  replaced.

VERIFIED

SURVEY AND RESEARCH SHIPS

Notes: There is also a civilian research vessel *Chulab Horn* which completed in 1986.

1 OCEANOGRAPHIC SHIP (AGOR)

Name	No	Builders	Commissioned
SUK	812	Bangkok Dock Co Ltd	3 Mar 1982

Displacement, tons: 1,450 standard; 1,526 full load
Dimensions, feet (metres): 206.3 × 36.1 × 13.4 *(62.9 × 11 × 4.1)*
Main machinery: 2 MTU diesels; 2,400 hp(m) *(1.76 MW)*; 2 shafts
Speed, knots: 15
Complement: 86 (20 officers)
Guns: 2 Oerlikon 20 mm. 2—7.62 mm MGs.
Radars: Navigation: Racal Decca 1226; I-band.

Comment: Laid down 27 August 1979, launched 8 September 1981. Designed for oceanographic
  and survey duties.

VERIFIED

PHOSAMTON   *8/2002, John Mortimer /* 0529999

SUK   *5/1999, van Ginderen Collection /* 0080828

For details of the latest updates to *Jane's Fighting Ships* online and to discover the
additional information available exclusively to online subscribers please visit
jfs.janes.com

1 SURVEY SHIP (AGS)

Name	No	Builders	Commissioned
CHANTHARA	811 (ex-AGS 11)	Lürssen Werft	30 May 1961

Displacement, tons: 870 standard; 996 full load
Dimensions, feet (metres): 229.2 × 34.5 × 10 *(69.9 × 10.5 × 3)*
Main machinery: 2 KHD diesels; 1,090 hp(m) *(801 kW)*; 2 shafts
Speed, knots: 13.25. **Range, n miles:** 10,000 at 10 kt
Complement: 68 (8 officers)
Guns: 2 Bofors 40 mm/60.

Comment: Laid down on 27 September 1960. Launched on 17 December 1960. Has served as a Royal Yacht.

VERIFIED

CHANTHARA   *12/2001, Thai Navy League* / 0130158

AUXILIARIES

1 SIMILAN (HUDONG) CLASS (TYPE R22T)
(REPLENISHMENT SHIP) (AORH)

Name	No	Builders	Launched	Commissioned
SIMILAN	871	Hudong Shipyard, Shanghai	9 Nov 1995	12 Sep 1996

Displacement, tons: 23,000 full load
Dimensions, feet (metres): 562.3 × 80.7 × 29.5 *(171.4 × 24.6 × 9)*
Main machinery: 2 HD-SEMT-Pielstick 16 PC2 6V400; 24,000 hp(m) *(17.64 MW)*; 2 shafts; Kamewa cp props
Speed, knots: 19. **Range, n miles:** 10,000 at 15 kt
Complement: 157 (19 officers) plus 26
Cargo capacity: 9,000 tons fuel, water, ammunition and stores
Radars: Air/surface search: Eye Shield (Type 354); E/F-band.
Navigation: Racal Decca 1290 ARPA; I-band.
Helicopters: 1 Seahawk type.

Comment: Contract signed with China State Shipbuilding Corporation on 29 September 1993. Fabrication started in December 1994. Two replenishment at sea positions each side and facilities for Vertrep. This ship complements the carrier and the new frigates to give the Navy a full deployment capability. Four twin 37 mm guns (Type 354) and associated Rice Lamp FC radar were not fitted.

VERIFIED

SIMILAN   *10/1998, Thai Navy League* / 0050248

1 REPLENISHMENT TANKER (AORL)

Name	No	Builders	Launched
CHULA	831 (ex-2)	Singapore SEC	24 Sep 1980

Displacement, tons: 2,000 full load
Measurement, tons: 960 dwt
Dimensions, feet (metres): 219.8 × 31.2 × 14.4 *(67 × 9.5 × 4.4)*
Main machinery: 2 MTU 12V 396 TC62 diesels; 2,400 hp(m) *(1.76 MW)* sustained; 2 shafts
Speed, knots: 14
Complement: 39 (7 officers)
Cargo capacity: 800 tons oil fuel
Guns: 2 Oerlikon 20 mm.
Radars: Navigation: Racal Decca 1226; I-band.

Comment: Replenishment is done by a hose handling crane boom.

VERIFIED

CHULA   *6/1998, Royal Thai Navy* / 0050249

4 HARBOUR TANKERS (YO)

PRONG 833 (ex-YO 5)   **PROET** 834 (ex-YO 9)   **SAMED** 835 (ex-YO 10)   **CHIK** 842 (ex-YO 11)

Displacement, tons: 360 standard; 485 full load
Dimensions, feet (metres): 122.7 × 19.7 × 8.7 *(37.4 × 6 × 2.7)*
Main machinery: 1 GM 8-268A diesel; 500 hp(m) *(368 kW)*; 1 shaft
Speed, knots: 9
Cargo capacity: 210 tons

Comment: Details are for 834, 835 and 842. Built by Bangkok Naval Dockyard. 834 commissioned 27 January 1967, remainder the same year. Details of 833 not known but reported to be approximately 180 tons.

VERIFIED

SAMED   *5/1999* / 0080829

1 HARBOUR TANKER (YO)

SAMUI 832 (ex-YOG 60, ex-YO 4)

Displacement, tons: 1,420 full load
Dimensions, feet (metres): 174.5 × 32 × 15 *(53.2 × 9.7 × 4.6)*
Main machinery: 1 Union diesel; 600 hp *(448 kW)*; 1 shaft
Speed, knots: 8
Complement: 29
Cargo capacity: 985 tons fuel
Guns: 2 Oerlikon 20 mm can be carried.
Radars: Navigation: Raytheon Pathfinder; I-band.

VERIFIED

SAMUI   *12/1995*

1 WATER TANKER (YW)

Name	No	Builders	Commissioned
CHUANG	841 (ex-YW 5)	Royal Thai Naval Dockyard, Bangkok	1965

Displacement, tons: 305 standard; 485 full load
Dimensions, feet (metres): 136 × 24.6 × 10 *(42 × 7.5 × 3.1)*
Main machinery: 1 GM diesel; 500 hp *(373 kW)*; 1 shaft
Speed, knots: 11
Complement: 29
Guns: 1 Oerlikon 20 mm.

Comment: Launched on 14 January 1965.

VERIFIED

CHUANG (alongside Proet)   *5/1997, Maritime Photographic* / 0019284

1 TRANSPORT SHIP (AKS)

Name	No	Builders	Commissioned
KLED KEO	861 (ex-AF-7)	Norfjord, Norway	1948

Displacement, tons: 450 full load
Dimensions, feet (metres): 150.1 × 24.9 × 14 *(46 × 7.6 × 4.3)*
Main machinery: 1 CAT diesel; 900 hp(m) *(662 kW)*; 1 shaft
Speed, knots: 12
Complement: 54 (7 officers)
Guns: 3 Oerlikon 20 mm.

Comment: Former Norwegian transport acquired in 1956. Paid off in 1990 but back in service in 1997. Operates with the patrol boat squadron.
VERIFIED

KLED KEO   *6/1998, Royal Thai Navy* / 0050250

1 BUOY TENDER (ABU)

Name	No	Builders	Commissioned
SURIYA	821	Bangkok Dock Co Ltd	15 Mar 1979

Displacement, tons: 690 full load
Dimensions, feet (metres): 177.8 × 33.5 × 10.2 *(54.2 × 10.2 × 3.1)*
Main machinery: 2 MTU diesels; 1,310 hp(m) *(963 kW)*; 2 shafts; bow thruster; 135 hp(m) *(99 kW)*
Speed, knots: 12
Complement: 60 (12 officers)
Radars: Navigation: Racal Decca; I-band.

Comment: Can carry 20 mm guns.
VERIFIED

SURIYA   *11/2001, Maritime Photographic* / 0130167

TUGS

2 COASTAL TUGS (YTB)

RIN 853 (ex-ATA 5)   RANG 854 (ex-ATA 6)

Displacement, tons: 350 standard
Dimensions, feet (metres): 106 × 29.7 × 15.2 *(32.3 × 9 × 4.6)*
Main machinery: 1 MWM TBD441V/12K diesel; 2,100 hp(m) *(1.54 MW)*; 1 shaft
Speed, knots: 12. Range, n miles: 1,000 at 10 kt
Complement: 19

Comment: Launched 12 and 14 June 1980 at Singapore Marine Shipyard. Both commissioned 5 March 1981. Bollard pull 22 tons.
VERIFIED

RANG   *1992, Royal Thai Navy* / 0080830

2 SAMAESAN CLASS (COASTAL TUGS) (YTR)

SAMAESAN 855   RAET 856

Displacement, tons: 300 standard
Dimensions, feet (metres): 82 × 27.9 × 7.9 *(25 × 8.5 × 2.4)*
Main machinery: 2 Caterpillar 3512TA diesels; 2,350 hp(m) *(1.75 MW)* sustained; 2 Aquamaster US 901 props
Speed, knots: 10
Complement: 6

Comment: Contract signed 23 September 1992 for local construction at Thonburi Naval dockyard. Completed in December 1993. Equipped for firefighting.
VERIFIED

RAET   *5/1997, A Sharma* / 0050251

2 YTL 422 CLASS (YTL)

KLUENG BADAAN 851 (ex-YTL 2)   MARN VICHAI 852 (ex-YTL 3)

Displacement, tons: 63 standard
Dimensions, feet (metres): 64.7 × 16.5 × 6 *(19.7 × 5 × 1.8)*
Main machinery: 1 diesel; 240 hp *(179 kW)*; 1 shaft
Speed, knots: 8

Comment: Built by Central Bridge Co, Trenton and bought from Canada 1953.
VERIFIED

KLUENG BADAAN   *11/2001, Maritime Photographic* / 0130166

POLICE

Notes: (1) There is also a Customs service which operates unarmed patrol craft with CUSTOMS on the hull, and a Fishery Patrol Service also unarmed but vessels are painted blue with broad white and narrow gold diagonal stripes on the hull. Two Hydrofoil craft are on loan from the Police to the Customs service.
(2) There are large numbers of RIBs in service.

1 VOSPER THORNYCROFT TYPE (LARGE PATROL CRAFT) (PSO)

SRINAKARIN 1804

Displacement, tons: 630 full load
Dimensions, feet (metres): 203.4 × 26.9 × 8.2 *(62 × 8.2 × 2.5)*
Main machinery: 2 Deutz MWM BV16M628 diesels; 9,524 hp(m) *(7 MW)* sustained; 2 shafts; Kamewa cp props
Speed, knots: 25. Range, n miles: 2,500 at 15 kt
Complement: 45
Guns: 1 Breda 30 mm/82. 2 Oerlikon 20 mm. GAM-BO1.
Radars: Surface search: Racal Decca 1226; I-band.

Comment: Ordered in September 1989 from Ital Thai Marine. Same hull as the Khamronsin class corvettes for the Navy but much more lightly armed. Delivered in April 1992.
VERIFIED

SRINAKARIN   *6/2003, Royal Thai Navy* / 0572648

2 HAMELN TYPE (LARGE PATROL CRAFT) (PBO)

DAMRONG RACHANUPHAP 1802   LOPBURI RAMES 1803

Displacement, tons: 430 full load
Dimensions, feet (metres): 186 × 26.6 × 8 *(56.7 × 8.1 × 2.4)*
Main machinery: 4 MTU diesels; 4,400 hp(m) *(3.23 MW)*; 2 shafts
Speed, knots: 23
Complement: 45
Guns: 2 Oerlikon 30 mm/75 (twin). 2 Oerlikon 20 mm (twin).
Radars: Surface search: Racal Decca 1226; I-band.

Comment: Delivered by Schiffwerft Hameln, Germany, on 3 January 1969 and 10 December 1972 respectively.

VERIFIED

LOPBURI RAMES   *6/2003, Royal Thai Navy /* 0572647

2 SUMIDAGAWA TYPE (COASTAL PATROL CRAFT) (PB)

CHASANYABADEE 1101   PHROMYOTHEE 1103

Displacement, tons: 130 full load
Dimensions, feet (metres): 111.5 × 19 × 9.1 *(34 × 5.8 × 2.8)*
Main machinery: 3 Ikegai diesels; 4,050 hp(m) *(2.98 MW)*; 3 shafts
Speed, knots: 32
Complement: 23
Guns: 2—12.7 mm MGs.
Radars: Surface search: Racal Decca; I-band.

Comment: Commissioned in August 1972 and May 1973 respectively.

VERIFIED

PHROMYOTHEE   *1990, Marine Police /* 0080833

1 YOKOHAMA TYPE (COASTAL PATROL CRAFT) (PB)

CHAWENGSAK SONGKRAM 1102

Displacement, tons: 190 full load
Dimensions, feet (metres): 116.5 × 23 × 11.5 *(35.5 × 7 × 3.5)*
Main machinery: 4 Ikegai diesels; 5,400 hp(m) *(3.79 MW)*; 2 shafts
Speed, knots: 32
Complement: 18
Guns: 2 Oerlikon 20 mm.

Comment: Commissioned 13 April 1973. A second of class operates for the Customs with the number 1201.

VERIFIED

CHAWENGSAK SONGKRAM   *1990, Marine Police /* 0080834

1 ITAL THAI MARINE TYPE (COASTAL PATROL CRAFT) (PB)

SRIYANONT 901

Displacement, tons: 52 full load
Dimensions, feet (metres): 90 × 16 × 6.5 *(27.4 × 4.9 × 2)*
Main machinery: 2 Deutz BA16M816 diesels; 2,680 hp(m) *(1.97 MW)* sustained; 2 shafts
Speed, knots: 23
Complement: 14
Guns: 1 Oerlikon 20 mm. 2—7.62 mm MGs.
Radars: Surface search: Racal Decca; I-band.

Comment: Commissioned 12 June 1986.

VERIFIED

SRIYANONT   *12/2001, Thai Navy League /* 0130155

1 BURESPADOONGKIT CLASS (COASTAL PATROL CRAFT) (PB)

BURESPADOONGKIT 813

Displacement, tons: 65 full load
Dimensions, feet (metres): 80.5 × 19.4 × 6 *(24.5 × 5.9 × 1.8)*
Main machinery: 2 SACM UD 23 V12 M5D diesels; 2,534 hp(m) *(1.86 MW)* sustained; 2 shafts
Speed, knots: 28. **Range, n miles:** 650 at 20 kt
Complement: 14
Guns: 1 Oerlikon GAM-CO1 20 mm; 2—7.62 mm MGs.

Comment: Built by Matsun, Thailand and commissioned 9 August 1995. Badly damaged in the Tsunami of 26 December 2004.

UPDATED

BURESPADOONGKIT   *6/1999, Marine Police /* 0080835

3 CUTLASS CLASS (COASTAL PATROL CRAFT) (PB)

PHRAONGKAMROP 807   PICHARNPHOLAKIT 808   RAMINTHRA 809

Displacement, tons: 34 full load
Dimensions, feet (metres): 65 × 17 × 8.3 *(19.8 × 5.2 × 2.5)*
Main machinery: 3 Detroit 12V-71TA diesels; 1,020 hp(m) *(761 kW)* sustained; 3 shafts
Speed, knots: 25
Complement: 14
Guns: 1 Oerlikon 20 mm. 2—7.62 mm MGs.

Comment: Delivered by Halter Marine, New Orleans, and all commissioned on 9 March 1969. Aluminium hulls.

VERIFIED

PICHARNPHOLAKIT   *6/1999 /* 0080836

3 TECHNAUTIC TYPE (COASTAL PATROL CRAFT) (PB)

810-812

Displacement, tons: 50 full load
Dimensions, feet (metres): 88.6 × 19.4 × 6.2 *(27 × 5.9 × 1.9)*
Main machinery: 3 Isotta Fraschini diesels; 2,500 hp(m) *(1.84 MW)*; 3 Castoldi hydrojets
Speed, knots: 27
Complement: 14
Guns: 1 Oerlikon 20 mm GAM-BO1. 2—7.62 mm MGs.

Comment: Delivered by Technautic, Bangkok in 1984.

VERIFIED

812   *1990, Marine Police /* 0080837

5 ITAL THAI MARINE TYPE (COASTAL PATROL CRAFT) (PB)

625-629

Displacement, tons: 42 full load
Dimensions, feet (metres): 64 × 17.5 × 5 *(19.5 × 5.3 × 1.5)*
Main machinery: 2 MAN D2842LE diesels; 1,350 hp(m) *(992 kW)* sustained; 2 shafts
Speed, knots: 27
Complement: 14
Guns: 1—12.7 mm MG.

Comment: Built in Bangkok 1987-90. Aluminium hulls. More of the class operated by the Fishery Patrol Service.

UPDATED

ITAL THAI 626   *3/2004*, Bob Fildes /* 0589814

6 MARSUN TYPE (COASTAL PATROL CRAFT) (PB)

630-635

Displacement, tons: 38 full load
Dimensions, feet (metres): 65.6 × 18.2 × 5 *(20 × 5.6 × 1.5)*
Main machinery: 2 MAN D2840LXE diesels; 1,640 hp(m) *(1.2 MW)* sustained; 2 shafts
Speed, knots: 25
Complement: 11
Guns: 1—12.7 mm MG.

Comment: Built by Marsun, Thailand and commissioned from 2 August 1994. *UPDATED*

MARSUN 634   *3/2004*, Bob Fildes /* 0589815

17 TECHNAUTIC TYPE (COASTAL PATROL CRAFT) (PB)

608-624

Displacement, tons: 30 full load
Dimensions, feet (metres): 60 × 16 × 2.9 *(18.3 × 4.9 × 0.9)*
Main machinery: 2 Isotta Fraschini ID 36 SS 8V diesels; 1,760 hp(m) *(1.29 MW)* sustained; 2 Castoldi hydrojets
Speed, knots: 27
Complement: 11
Guns: 1—12.7 mm MG.

Comment: Built from 1983-87 in Bangkok. Operational status of some of these craft doubtful.
VERIFIED

TECHNAUTIC 609   *11/2001, Maritime Photographic /* 0130168

2 MARSUN TYPE (PB)

539-540

Displacement, tons: 30 full load
Dimensions, feet (metres): 57 × 16 × 3 *(17.4 × 4.9 × 0.9)*
Main machinery: 2 Detroit 12V-71TA diesels; 840 hp *(627 kW)* sustained; 2 shafts
Speed, knots: 25
Complement: 8
Guns: 1—12.7 mm MG.

Comment: Built in Thailand. Both commissioned 26 March 1986.
VERIFIED

MARSUN 539   *11/2001, Maritime Photographic /* 0130169

26 SUMIDAGAWA TYPE (RIVER PATROL CRAFT) (PBR)

513-538

Displacement, tons: 18 full load
Dimensions, feet (metres): 54.1 × 12.5 × 2.3 *(16.5 × 3.8 × 0.7)*
Main machinery: 2 Cummins diesels; 800 hp *(597 kW)*; 2 shafts
Speed, knots: 23
Complement: 6
Guns: 1—12.7 mm MG.

Comment: First 21 built by Sumidagawa, last five by Captain Co, Thailand 1978-79.
VERIFIED

SUMIDAGAWA 526   *6/2003, Royal Thai Navy /* 0572646

SUMIDAGAWA 529   *6/1999, Marine Police /* 0080841

20 CAMCRAFT TYPE (RIVER PATROL CRAFT) (PBR)

415-440 series

Displacement, tons: 13 full load
Dimensions, feet (metres): 40 × 12 × 3.2 *(12.2 × 3.7 × 1)*
Main machinery: 2 Detroit diesels; 540 hp *(403 kW)*; 2 shafts
Speed, knots: 25
Complement: 6

Comment: Delivered by Camcraft, Louisiana. Aluminium hulls. Numbers uncertain.
VERIFIED

CAMCRAFT 435   *6/1999, Marine Police /* 0080842

34 RIVER PATROL CRAFT (PBR)

Displacement, tons: 5 full load
Dimensions, feet (metres): 37 × 11 × 6 *(11.3 × 3.4 × 1.8)*
Main machinery: 2 diesels; 2 shafts
Speed, knots: 25
Complement: 4

Comment: Numbers in the 300 series. Details given are for *339* built in 1990. The remainder are similar.

VERIFIED

RIVER PATROL CRAFT 339 (alongside Technautic 609)
7/2000 / 0106613

Togo

Country Overview

Formerly French Togoland, the Togolese Republic gained full independence in 1960 having rejected proposals to be united with Ghana. Situated in west Africa, it has an area of 21,925 square miles and borders to the east with Benin and to the west with Ghana. Togo has a short coastline of 30 n miles with the Gulf of Guinea. Lomé is the capital, largest town and principal port.

Togo is the only coastal state to claim territorial seas of 30 n miles. A 200 n mile Exclusive Economic Zone (EEZ) is also claimed but this has not been defined by boundary agreements.

Headquarters Appointments

Commanding Officer, Navy:
    Commander Lucien Laval

Personnel

(a)   2005: 120
(b)   Conscription (2 years)

Bases

Lomé

PATROL FORCES

2 COASTAL PATROL CRAFT (PB)

Name	No	Builders	Launched
KARA	P 761	Chantiers Navals de l'Esterel, Cannes	18 May 1976
MONO	P 762	Chantiers Navals de l'Esterel, Cannes	16 June 1976

Displacement, tons: 80 full load
Dimensions, feet (metres): 105 × 19 × 5.3 *(32 × 5.8 × 1.6)*
Main machinery: 2 MTU MB 12V 493 TY60 diesels; 2,000 hp(m) *(1.47 MW)* sustained; 2 shafts
Speed, knots: 30. **Range, n miles:** 1,500 at 15 kt
Complement: 17 (1 officer)
Missiles: SSM: Aerospatiale SS 12M; wire-guided to 5 km *(3 n miles)* subsonic; warhead 30 kg.
Guns: 1 Bofors 40 mm/70. 1 Oerlikon 20 mm.
Radars: Surface search: Decca 916; I-band.

Comment: Both craft operational.

VERIFIED

MONO
6/1998 / 0050252

Tonga

Country Overview

A former British protectorate, the Kingdom of Tonga became a sovereign state in 1970. Situated in the southwestern Pacific Ocean some 1,080 n miles northeast of New Zealand, the country consists of more than 170 islands and islets running generally north-south. There are three main groups, Tongatapu, Ha'apai and Vava'u, and several outlying islands. Nuku'alofa, on Tongatapu Island, is the capital, largest town and principal port. Territorial seas (12 n miles) are claimed. An Exclusive Economic Zone (EEZ) (200 n miles) is claimed but limits have not been fully defined by boundary agreements.

Headquarters Appointments

Commanding Officer, Navy:
    Lieutenant Commander Sione Fifita

Personnel

2005: 115

Bases

Touliki Base, Nuku'alofa (HMNB *Masefield*)

Prefix to Ships' Names

VOEA (Vaka O Ene Afio)

PATROL FORCES

Notes: A Beech 18 aircraft was acquired in May 1995 for maritime surveillance.

3 PACIFIC CLASS (LARGE PATROL CRAFT) (PB)

Name	No	Builders	Commissioned
NEIAFU	P 201	Australian Shipbuilding Industries	28 Oct 1989
PANGAI	P 202	Australian Shipbuilding Industries	30 June 1990
SAVEA	P 203	Australian Shipbuilding Industries	23 Mar 1991

Displacement, tons: 162 full load
Dimensions, feet (metres): 103.3 × 26.6 × 6.9 *(31.5 × 8.1 × 2.1)*
Main machinery: 2 Caterpillar 3516TA diesels; 2,820 hp *(2.1 MW)* sustained; 2 shafts
Speed, knots: 20. **Range, n miles:** 2,500 at 12 kt
Complement: 17 (3 officers)
Guns: 2—12.7 mm MGs.
Radars: Surface search: Furuno 1101; I-band.

Comment: Part of the Pacific Forum Australia Defence co-operation. First laid down 30 January 1989, second 2 October 1989, third February 1990. *Savea* has a hydrographic survey capability. Following half-life refits 1998-99 and the decision of the Australian government to extend the Pacific Patrol Boat programme, *Neiafu*, *Pangai* and *Savea* will be due life-extension refits in 2007, 2008 and 2008 respectively.

UPDATED

PANGAI
2/2003, Chris Sattler / 0558665

AUXILIARIES

1 LCM

Name	No	Builders	Commissioned
LATE (ex-*1057*)	C 315	North Queensland, Cairns	1 Sep 1982

Displacement, tons: 116 full load
Dimensions, feet (metres): 73.5 × 21 × 3.3 *(22.4 × 6.4 × 1)*
Main machinery: 2 Detroit 12V-71 diesels; 680 hp *(507 kW)* sustained; 2 shafts
Speed, knots: 10. **Range, n miles:** 480 at 10 kt
Complement: 5
Cargo capacity: 54 tons
Guns: 1—7.62 mm MG can be carried.
Radars: Surface search: Koden MD 305; I-band.

Comment: Acquired from the Australian Army for inter-island transport.

VERIFIED   LATE   *6/1999, Tongan Navy* / 0084414

Trinidad and Tobago

Country Overview

Trinidad and Tobago gained independence in 1962 and became a republic in 1976. The country lies at the southern end of the Lesser Antilles chain and comprises the main islands of Trinidad (1,864 square miles), Tobago (116 square miles) and 21 minor islands and rocks. Trinidad is close to the northeastern coast of Venezuela and the mouth of the Orinoco River. The capital, largest town, and principal port is Port-of-Spain, Trinidad. An archipelagic state, territorial seas (12 n miles) are claimed. While a 200 n mile Exclusive Economic Zone (EEZ) has

been claimed, the limits have only been partly defined by boundary agreements.

Headquarters Appointments

Commanding Officer, Coast Guard:
    Captain Garnet Best

Aircraft

The Coast Guard operates three Cessna (Types 172, 402B and 310R) for surveillance and two C26B acquired in 1999. These aircraft can be

backed by Air Division Gazelle and Sikorsky S-76 helicopters when necessary.

Personnel

(a)   2005: 650 (55 officers)
(b)   Voluntary service

Bases

Staubles Bay (HQ)
Hart's Cut, Tobago, Point Fortin (all from 1989)
Piarco (Air station), Cedros (from 1992)
Galeota (from 1995)

Coast Defence

There are plans to install a coastal radar system.

Prefix to Ships' Names

TTS

COAST GUARD

Notes: The 2003 Budget announced plans to acquire two fast patrol vessels by 2005, each capable of launching fast interceptor boats and carrying a helicopter with attack capability.

1 ISLAND CLASS (PBO)

Name	No	Builders	Commissioned
NELSON (ex-*Orkney*)	CG 20 (ex-P 299)	Hall Russell	25 Feb 1977

Displacement, tons: 925 standard; 1,260 full load
Dimensions, feet (metres): 176 wl; 195.3 oa × 36 × 15 *(53.7; 59.5 × 11 × 4.5)*
Main machinery: 2 Ruston 12RKC diesels; 5,640 hp *(4.21 MW)* sustained; 1 shaft; cp prop
Speed, knots: 16.5. **Range, n miles:** 7,000 at 12 kt
Complement: 35 (5 officers)
Guns: 2—7.62 mm MGs can be carried.
Radars: Navigation: Kelvin Hughes Type 1006; I-band.

Comment: Transferred from the UK Navy on 18 December 2000 and recommissioned on 22 February 2001. Based at Port of Spain.

UPDATED

NELSON   *1/2001, H M Steele* / 0106616

2 TYPE CG 40 (LARGE PATROL CRAFT) (PB)

Name	No	Builders	Commissioned
BARRACUDA	CG 5	Karlskronavarvet	15 June 1980
CASCADURA	CG 6	Karlskronavarvet	15 June 1980

Displacement, tons: 210 full load
Dimensions, feet (metres): 133.2 × 21.9 × 5.2 *(40.6 × 6.7 × 1.6)*
Main machinery: 2 Paxman Valenta 16CM diesels; 6,700 hp *(5 MW)* sustained; 2 shafts
Speed, knots: 30. **Range, n miles:** 3,000 at 15 kt
Complement: 25
Guns: 1 Bofors 40 mm/70. 1 Oerlikon 20 mm.
Weapons control: Optronic GFCS.
Radars: Surface search: Racal Decca 1226; I-band.

Comment: Ordered in Sweden mid-1978. Laid down early 1979. Fitted with foam-cannon oil pollution equipment and for oceanographic and hydrographic work. Nine spare berths. The hull is similar to Swedish Spica class but with the bridge amidships. One refitted in 1988, the other in 1989. *Cascadura* refitted again in 1998/99.

VERIFIED

CASCADURA   *1/1994, Maritime Photographic*

4 POINT CLASS (COASTAL PATROL CRAFT) (PB)

Name	No	Builders	Commissioned
COROZAL POINT (ex-*Point Heyer*)	CG 7 (ex-82369)	J Martinac, Tacoma	3 Aug 1967
CROWN POINT (ex-*Point Bennett*)	CG 8 (ex-82351)	Coast Guard Yard, Curtis Bay	19 Dec 1966
GALERA POINT (ex-*Point Bonita*)	CG 9 (ex-82347)	J Martinac, Tacoma	12 Sep 1966
BARCOLET POINT (ex-*Point Highland*)	CG 10 (ex-82333)	Coast Guard Yard, Curtis Bay	27 June 1962

Displacement, tons: 66 full load
Dimensions, feet (metres): 83 × 17.2 × 5.8 *(25.3 × 5.2 × 1.8)*
Main machinery: 2 Caterpillar 3412 diesels; 1,600 hp *(1.19 MW)*; 2 shafts
Speed, knots: 23. **Range, n miles:** 1,500 at 8 kt
Complement: 10
Guns: 2—7.62 mm MGs.
Radars: Surface search: Raytheon SPS-64(V)I and Raytheon SPS 69AN; I-band.

Comment: CG 7 and CG 8 transferred from US Coast Guard 12 February 1999 and CG 9 on 14 November 2000. CG 10 transferred on 24 July 2001.

UPDATED

CROWN POINT   *6/1999, Trinidad and Tobago Coast Guard* / 0080846

4 SOUTER WASP 17 METRE CLASS
(COASTAL PATROL CRAFT) (PB)

Name	No	Builders	Commissioned
PLYMOUTH	CG 27	WA Souter, Cowes	27 Aug 1982
CARONI	CG 28	WA Souter, Cowes	27 Aug 1982
GALEOTA	CG 29	WA Souter, Cowes	27 Aug 1982
MORUGA	CG 30	WA Souter, Cowes	27 Aug 1982

Displacement, tons: 20 full load
Dimensions, feet (metres): 55.1 × 13.8 × 4.6 *(16.8 × 4.2 × 1.4)*
Main machinery: 2 MANN 8V diesels; 1,470 hp *(1.1 MW)*; 2 shafts
Speed, knots: 32. **Range, n miles:** 500 at 18 kt
Complement: 7 (2 officers)
Guns: 1—7.62 mm MG.
Radars: Surface search: Raytheon SPS 69AN; I-band.

Comment: GRP hulls. All refitted from September 1997 with new engines.
VERIFIED

MORUGA   *6/1996, Trinidad and Tobago Coast Guard*

2 WASP 20 METRE CLASS (COASTAL PATROL CRAFT) (PB)

Name	No	Builders	Commissioned
KAIRI (ex-*Sea Bird*)	CG 31	WA Souter, Cowes	Dec 1982
MORIAH (ex-*Sea Dog*)	CG 32	WA Souter, Cowes	Dec 1982

Displacement, tons: 32 full load
Dimensions, feet (metres): 65.8 × 16.5 × 5 *(20.1 × 5 × 1.5)*
Main machinery: 2 MANN 12V diesels; 2,400 hp *(1.79 MW)*; 2 shafts
Speed, knots: 30. **Range, n miles:** 450 at 30 kt
Complement: 6 (2 officers)
Guns: 2—7.62 mm MGs.
Radars: Surface search: Decca 150; I-band.

Comment: Ordered 30 September 1981. Aluminium alloy hull. Transferred from the Police in June 1989. New engines in 1999.
VERIFIED

KAIRI   *7/2001, Margaret Organ /* 0114370

1 SWORD CLASS (COASTAL PATROL CRAFT) (PB)

Name	No	Builders	Commissioned
MATELOT (ex-*Sea Skorpion*)	CG 33	SeaArk Marine	May 1979

Displacement, tons: 15.5 full load
Dimensions, feet (metres): 44.9 × 13.4 × 4.3 *(13.7 × 4.1 × 1.3)*
Main machinery: 2 GM diesels; 850 hp *(634 kW)*; 2 shafts
Speed, knots: 28. **Range, n miles:** 500 at 20 kt
Complement: 6
Guns: 1—7.62 mm MG.
Radars: Surface search: Decca 150; I-band.

Comment: Two transferred from the Police 30 June 1989, one scrapped in 1990. Refitted in 1998.
VERIFIED

MATELOT   *1/1994, Maritime Photographic*

3 DAUNTLESS CLASS (PB)

Name	No	Builders	Commissioned
SOLDADO	CG 38	SeaArk Marine, Monticello	14 June 1995
ROXBOROUGH	CG 39	SeaArk Marine, Monticello	14 June 1995
MAYARO	CG 40	SeaArk Marine, Monticello	14 June 1995

Displacement, tons: 11 full load
Dimensions, feet (metres): 40 × 14 × 4.3 *(12.2 × 4.3 × 1.3)*
Main machinery: 2 Caterpillar 3208TA diesels; 870 hp *(650 kW)* sustained; 2 shafts
Speed, knots: 27. **Range, n miles:** 600 at 18 kt
Complement: 4
Guns: 1—7.62 mm MG.
Radars: Surface search: Raytheon R40; I-band.

Comment: Several of these craft provided to Caribbean navies through FMS funding. Used as utility boats.
VERIFIED

SOLDADO   *6/1995, SeaArk Marine /* 0080848

2 BOWEN CLASS and 3 RHIB (FAST INTERCEPTOR CRAFT) (PBF)

CG 001   CG 002   CG 003-005

Comment: The first pair are 31 ft fast patrol boats acquired in May 1991. Capable of 40 kt. The last three are 25 ft RHIBs with twin Johnson outboards acquired from the US in 1993. Capable of 45 kt.
VERIFIED

CG 001   *1991, Trinidad and Tobago Coast Guard /* 0080849

2 AUXILIARY VESSELS

NAPARIMA (ex-*CG 26*) A 01   REFORM A 04

Comment: Used for Port Services and other support functions.
VERIFIED

CUSTOMS

Notes: Among other craft, the Customs service operate a High Speed Interception craft *Kenneth Mohammed*.

KENNETH MOHAMMED   *2/2001, van Ginderen Collection /* 0114369

Tunisia

Country Overview

Formerly a French protectorate, the Tunisian Republic gained independence in 1957. Situated in northern Africa, it has an area of 63,170 square miles and is bordered to the west by Algeria and to the south by Libya. It has a 619 n mile coastline with the Mediterranean Sea. The capital and largest city is the seaport of Tunis. There are further ports at Bizerta, Sousse, Sfax and Gabès while as-Sukhayrah, specialises in petroleum bunkering. Territorial seas (12 n miles) are claimed. An EEZ has not been claimed.

Headquarters Appointments

Naval Chief of Staff:
Commodore Tarek Faouzi El Arbi

Personnel

(a) 2005: 4,800 officers and men (including 800 conscripts)
(b) 1 year's national service

Bases

Bizerte, Sfax, La Goulette, Kelibia

PATROL FORCES

Notes: Two 30 m and two 12 m patrol craft are required as soon as funds are available.

3 COMBATTANTE III M CLASS
(FAST ATTACK CRAFT—MISSILE) (PGGF)

Name	No	Builders	Launched	Commissioned
LA GALITÉ	501	CMN, Cherbourg	16 June 1983	27 Feb 1985
TUNIS	502	CMN, Cherbourg	27 Oct 1983	27 Mar 1985
CARTHAGE	503	CMN, Cherbourg	24 Jan 1984	29 Apr 1985

Displacement, tons: 345 standard; 425 full load
Dimensions, feet (metres): 183.7 × 26.9 × 7.2 *(56 × 8.2 × 2.2)*
Main machinery: 4 MTU 20V 538 TB93 diesels; 18,740 hp(m) *(13.8 MW)* sustained; 4 shafts
Speed, knots: 38.5. **Range, n miles:** 700 at 33 kt; 2,800 at 10 kt
Complement: 35

Missiles: SSM: 8 Aerospatiale MM 40 Exocet (2 quad) launchers; inertial cruise; active radar homing to 70 km *(40 n miles)* at 0.9 Mach; warhead 165 kg; sea-skimmer.
Guns: 1 OTO Melara 3 in *(76 mm)*/62; 55-65 rds/min to 16 km *(8.7 n miles)*; weight of shell 6 kg.
2 Breda 40 mm/70 (twin); 300 rds/min to 12.5 km *(6.8 n miles)*; weight of shell 0.96 kg.
4 Oerlikon 30 mm/75 (2 twin); 650 rds/min to 10 km *(5.5 n miles)*; weight of shell 1 kg or 0.36 kg.
Countermeasures: Decoys: 1 CSEE Dagaie trainable launcher; IR flares and chaff.
ESM: Thomson-CSF; DR 2000; intercept.
Combat data systems: Tavitac action data automation.
Weapons control: 2 CSEE Naja optronic directors for 30 mm. Thomson-CSF Vega II for SSM, 76 mm and 40 mm.
Radars: Air/surface search: Thomson-CSF Triton S; G-band; range 33 km *(18 n miles)* for 2 m² target.
Fire control: Thomson-CSF Castor II; I/J-band; range 31 km *(17 n miles)* for 2 m² target.

Programmes: Ordered 27 June 1981.
Operational: CSEE Sylosat navigation system. All three ships operating but reported in need of refits.

UPDATED

CARTHAGE   *8/2004*, Schaeffer/Marsan* / 1044197

CARTHAGE   *8/2004*, B Prézelin* / 1044198

3 MODIFIED HAIZHUI CLASS (LARGE PATROL CRAFT) (PB)

UTIQUE P 207   JERBA P 208   KURIAT P 209

Displacement, tons: 120 full load
Dimensions, feet (metres): 114.8 × 17.7 × 5.9 *(35 × 5.4 × 1.8)*
Main machinery: 4 MWM TB 604 BV12 diesels; 4,400 hp(m) *(3.22 MW)* sustained; 4 shafts
Speed, knots: 28. **Range, n miles:** 750 at 17 kt
Complement: 39
Guns: 4 China 25 mm/80 (2 twin).
Radars: Surface search: Pot Head; I-band.

Comment: Delivered from China in March 1994. These craft resemble a smaller version of the Haizhui class in service with the Chinese Navy but with a different armament and some superstructure changes. Built to Tunisian specifications.

VERIFIED

KURIAT   *4/1995* / 0080852

3 BIZERTE CLASS (TYPE PR 48) (LARGE PATROL CRAFT) (PBOM)

Name	No	Builders	Commissioned
BIZERTE	P 301	SFCN, Villeneuve-la-Garenne	10 July 1970
HORRIA (ex-*Liberté*)	P 302	SFCN, Villeneuve-la-Garenne	Oct 1970
MONASTIR	P 304	SFCN, Villeneuve-la-Garenne	25 Mar 1975

Displacement, tons: 250 full load
Dimensions, feet (metres): 157.5 × 23.3 × 7.5 *(48 × 7.1 × 2.3)*
Main machinery: 2 MTU 16V 652 TB81 diesels; 4,600 hp(m) *(3.4 MW)* sustained; 2 shafts
Speed, knots: 20. **Range, n miles:** 2,000 at 16 kt
Complement: 34 (4 officers)
Missiles: SSM: 8 Aerospatiale SS 12M; wire-guided to 5.5 km *(3 n miles)* subsonic; warhead 30 kg.
Guns: 4—37 mm/63 (2 twin). 2—14.5 mm MGs.
Radars: Surface search: Thomson-CSF DRBN 31; I-band.

Comment: First pair ordered in 1968, third in August 1973. Guns changed in 1994. All are active.

UPDATED

HORRIA   *10/2001* / 0533311

BIZERTE   *3/2002, van Ginderen Collection* / 0141859

0 + 6 ALBATROS CLASS (TYPE 143B) (PG)

Name	No	Builders	Commissioned
— (ex-*Geier*)	— (ex-P 6113)	Lurssen, Vegesack	2 June 1976
— (ex-*Sperber*)	— (ex-P 6115)	Kroger, Rendsburg	27 Sep 1976
— (ex-*Greif*)	— (ex-P 6116)	Lurssen, Vegesack	25 Nov 1976
— (ex-*Seeadler*)	— (ex-P 6118)	Lurssen, Vegesack	28 Mar 1977
— (ex-*Habicht*)	— (ex-P 6119)	Kroger, Rendsburg	23 Dec 1977
— (ex-*Kormoran*)	— (ex-P 6120)	Lurssen, Vegesack	29 July 1977

Displacement, tons: 398 full load
Dimensions, feet (metres): 189 × 25.6 × 8.5 *(57.6 × 7.8 × 2.6)*
Main machinery: 4 MTU 16V 956 TB91 diesels; 17,700 hp(m) *(13 MW)* sustained; 4 shafts
Speed, knots: 40. **Range, n miles:** 1,300 at 30 kt
Complement: 40 (4 officers)

Missiles: SSM: 4 Aerospatiale MM 38 Exocet (2 twin) launchers; inertial cruise; active radar homing to 42 km *(23 n miles)* at 0.9 Mach; warhead 165 kg; sea-skimmer.
Guns: 2 OTO Melara 3 in *(76 mm)*/62 compact; 85 rds/min to 16 km *(8.6 n miles)* anti-surface; 12 km *(6.5 n miles)* anti-aircraft; weight of shell 6 kg.
2—12.7 mm MGs (may be fitted).
Torpedoes: 2—21 in *(533 mm)* aft tubes. AEG Seeal; wire-guided; active homing to 13 km *(7 n miles)* at 35 kt; passive homing to 28 km *(15 n miles)* at 23 kt; warhead 260 kg.
Countermeasures: Decoys: Buck-Wegmann Hot Dog/Silver Dog; IR/chaff dispenser.
ESM/ECM: Racal Octopus (Cutlass intercept, Scorpion jammer).
Combat data systems: AEG/Signaal command and fire-control system; Link 11.
Weapons control: ORG7/3 optronics GFCS. STN Atlas WBA optronic sensor to be fitted.
Radars: Surface search/fire control: Signaal WM27; I/J-band.
Navigation: SMA 3 RM 20; I-band.

Programmes: To be sold to Tunisia on being decommissioned from the German Navy in 2005.
Structure: Wooden hulled craft.

NEW ENTRY

ALBATROS CLASS   9/2004*, Per Körnefeldt / 1044199

4 COASTAL PATROL CRAFT (PB)

Name	No	Builders	Commissioned
ISTIKLAL (ex-*VC 11, P 761*)	P 201	Ch Navals de l'Esterel	Apr 1957
JOUMHOURIA	P 202	Ch Navals de l'Esterel	Jan 1961
AL JALA	P 203	Ch Navals de l'Esterel	Nov 1963
REMADA	P 204	Ch Navals de l'Esterel	July 1967

Displacement, tons: 60 standard; 80 full load
Dimensions, feet (metres): 104 × 19 × 5.3 *(31.5 × 5.8 × 1.6)*
Main machinery: 2 MTU MB 12V 493 TY70 diesels; 2,200 hp(m) *(1.62 MW)* sustained; 2 shafts
Speed, knots: 30. **Range, n miles:** 1,500 at 15 kt
Complement: 17 (3 officers)
Guns: 2 Oerlikon 20 mm.
Radars: Surface search: Racal Decca 1226; I-band.

Comment: *Istiklal* transferred from France March 1959. Wooden hulls. At least one may belong to the Coast Guard.

VERIFIED

REMADA   4/1997 / 0019295

6 COASTAL PATROL CRAFT (PB)

V 101-106

Displacement, tons: 38 full load
Dimensions, feet (metres): 83 × 15.6 × 4.2 *(25 × 4.8 × 1.3)*
Main machinery: 2 Detroit 12V-71TA diesels; 840 hp *(627 kW)* sustained; 2 shafts; LIPS cp props
Speed, knots: 23. **Range, n miles:** 900 at 15 kt
Complement: 11
Guns: 1 Oerlikon 20 mm.
Radars: Surface search: Racal Decca 1226; I-band.

Comment: Built by Chantiers Navals de l'Esterel and commissioned in 1961-63. Two further craft of the same design (*Sabaq el Bahr* T 2 and *Jaouel el Bahr* T 1) but unarmed were transferred to the Fisheries Administration in 1971-same builders. Refitted in 1997/98. *V 102* is Coast Guard.

VERIFIED

V 106   1/1995 / 0080853

TRAINING/SURVEY SHIPS

Notes: *Degga* A 707 and *El Jem* A 708 are converted fishing vessels used for divers' training.

EL JEM   7/2002, Schaeffer/Marsan / 0533312

1 WILKES CLASS (AGS)

Name	No	Builders	Launched	Commissioned
KHAIREDDINE	A 700	Defoe SB Co, Bay City, MI	31 July 1969	28 June 1971
(ex-*Wilkes*)	(ex-T-AGS 33)			

Displacement, tons: 2,843 full load
Dimensions, feet (metres): 285.3 × 48 × 15.1 *(87 × 14.6 × 4.6)*
Main machinery: Diesel-electric; 2 Alco diesel generators; 1 Westinghouse/GE motor; 3,600 hp *(2.69 MW)*; 1 shaft; bow thruster; 350 hp *(261 kW)*
Speed, knots: 15. **Range, n miles:** 8,000 at 13 kt
Complement: 37
Radars: Navigation: RM 1650/9X; I-band.

Comment: Decommissioned on 29 August 1995 and transferred from the USA by grant aid on 29 September 1995. Designed specifically for surveying operations. Bow propulsion unit for precise manoeuvrability and station keeping. Second of class planned for transfer but not confirmed.

VERIFIED

KHAIREDDINE   7/2001, Diego Quevedo / 0126191

1 ROBERT D CONRAD CLASS (AGOR/AX)

Name	No	Builders	Launched	Commissioned
N N O SALAMMBO	A 701	Northwest Iron Works,	13 June 1966	28 Feb 1969
(ex-*De Steiguer*)	(ex-T-AGOR 12)	Portland		

Displacement, tons: 1,370 full load
Dimensions, feet (metres): 208.9 × 40 × 15.3 *(63.7 × 12.2 × 4.7)*
Main machinery: Diesel-electric; 2 Cummins diesel generators; 1 motor; 1,000 hp *(746 kW)*; 1 shaft; bow thruster; 350 hp *(257 kW)*
Speed, knots: 13. **Range, n miles:** 12,000 at 12 kt
Complement: 40
Radars: Navigation: Raytheon 1650/6X; I-band.

Comment: Transferred from USA on 2 November 1992 and recommissioned on 11 February 1993. Built as an oceanographic research ship. Special features include a 10 ton boom, and a gas turbine for quiet propulsion up to 6 kt. Used primarily for training having replaced the frigate *Inkadh*, which is now an accommodation hulk.

VERIFIED

N N O SALAMMBO   7/1997, Camil Busquets i Vilanova / 0019296

AUXILIARIES

Notes: *Sidi Bou Said* A 802 is a 38 m buoy tender.

2 WHITE SUMAC CLASS (BUOY TENDERS) (ABU)

Name	No	Launched	Recommissioned
TABARKA (ex-*White Heath*)	A 804 (ex-WLM 545)	21 July 1943	31 Mar 1998
TAGUERMESS (ex-*White Lupine*)	A 805 (ex-WLM 546)	28 July 1943	31 Mar 1998

Displacement, tons: 485 full load
Dimensions, feet (metres): 133 × 31 × 9 *(40.5 × 9.5 × 2.7)*
Main machinery: 2 Caterpillar diesels; 600 hp *(448 kW)*; 2 shafts
Speed, knots: 9
Complement: 24
Radars: Navigation: Raytheon; I-band.

Comment: Former US Coast Guard vessels transferred by gift on 10 June 1998. Arrived in Tunisia one month later.

UPDATED

TAGUERMESS (US colours)   *9/1997, Harald Carstens* / 0012986

1 COASTAL TUG (YTB)

SIDI DAOUD (ex-*Porto D'Ischia*) — (ex-Y436)

Displacement, tons: 412 full load
Measurement, tons: 122 dwt
Dimensions, feet (metres): 106.3 × 27.9 × 12.8 *(32.4 × 8.5 × 3.9)*
Main machinery: 1 GMT B 230.8 M diesels; 1,600 hp(m) *(1.18 MW)* sustained; 1 shaft; cp prop
Speed, knots: 12.7. **Range, n miles:** 4,000 at 12 kt
Complement: 13
Radars: Navigation: GEM BX 132; I-band.

Comment: Built in 1970. Transferred from Italy in November 2002.

UPDATED

COASTAL TUG (Italian colours)   *5/2001, Giorgio Ghiglione* / 0130337

1 SIMETO CLASS (WATER TANKER) (AWT)

AIN ZAGHOUAN (ex-*Simeto*) — (ex-A 5375)

Displacement, tons: 1,858 full load
Dimensions, feet (metres): 229 × 33.1 × 14.4 *(69.8 × 10.1 × 4.1)*
Main machinery: 2 GMT B 230.6 BL diesels; 2,530 hp(m) *(1.86 MW)* sustained; 2 shafts; cp props; bow thruster; 300 hp(m) *(220 kW)*
Speed, knots: 13. **Range, n miles:** 1,800 at 12 kt
Complement: 36 (3 officers)
Cargo capacity: 1,130 tons
Guns: 1—20 mm/70. 2—7.62 mm MGs can be carried.
Radars: Navigation: 2 SPN-753B(V); I-band.

Comment: Built by Cinet, Molfetta and originally commissioned on 9 July 1988. Transferred from Italy on 30 June 2003.

VERIFIED

SIMETO CLASS (Italian colours)   *2/2000, van Ginderen Collection* / 0104888

NATIONAL GUARD

Notes: (1) *Tazarke* P 205 and *Menzel Bourguiba* P 206 may have transferred from the Navy to the National Guard but this is not confirmed.
(2) There are four further patrol craft; *GN 1704, GN 1705, GN 1105, GN 1401* and *GN 907*.

GN 1704   *3/2002, van Ginderen Collection* / 0141821

GN 1105   *3/2002, van Ginderen Collection* / 0141834

6 KONDOR I CLASS (PBO)

RAS EL BLAIS (ex-*Demmin*) 601	**RAS EL MANOURA** (ex-*Templin*) 604
RAS AJDIR (ex-*Malchin*) 602	**RAS ENGHELA** (ex-*Ahrenshoop*) 605
RAS EL EDRAK (ex-*Altentreptow*) 603	**RAS IFRIKIA** (ex-*Warnemunde*) 606

Displacement, tons: 377 full load
Dimensions, feet (metres): 170.3 × 23.3 × 7.2 *(51.9 × 7.1 × 2.2)*
Main machinery: 2 Russki/Kolomna 40-DM; 4,408 hp(m) *(3.24 MW)* sustained; 2 shafts; cp props
Speed, knots: 20. **Range, n miles:** 1,800 at 15 kt
Complement: 24
Guns: 2—25 mm (twin) can be carried.
Radars: Navigation: TSR 333 or Racal Decca 360; I-band.

Comment: Former GDR minesweepers built at Peenewerft, Wolgast in 1969. First four transferred in May 1992, one in August 1997 and the last one in May 2000. In German service they were fitted with a twin 25 mm gun and a hull-mounted sonar. *Ras Ifrikia* has more extensive superstructure. These ships may belong to the Navy. Ships of the same class acquired by Cape Verde, Malta and Guinea-Bissau. Reported operational.

VERIFIED

RAS ENGHELA   *8/1997, Diego Quevedo* / 0050258

RAS IFRIKIA   *5/2000, Kristian Lundgren* / 0567538

5 BREMSE CLASS (PB)

SBEITLA (ex-*G 32*)	UTIQUE (ex-*G 37*)	SELEUTA (ex-*G 39*)
BULLARIJIA (ex-*G 36*)	UERKOUANE (ex-*G 38*)	

Displacement, tons: 42 full load
Dimensions, feet (metres): 74.1 × 15.4 × 3.6 *(22.6 × 4.7 × 1.1)*
Main machinery: 2 SKL 6VD 18/5 AL-1 diesels; 944 hp(m) *(694 kW)* sustained; 2 shafts
Speed, knots: 14
Complement: 6
Guns: 2—14.5 mm (twin) MGs can be carried.
Radars: Navigation: TSR 333; I-band.

Comment: Built in 1971-72 for the ex-GDR GBK. Transferred from Germany in May 1992. Others of the class sold to Malta and Cyprus.

VERIFIED

BREMSE (Malta colours)   *10/1995, van Ginderen Collection /* 0080855

11 COASTAL PATROL CRAFT (PB)

ASSAD BIN FOURAT	MOHAMMED BRAHIM REJEB	GN 2004	GN 2005	+7

Displacement, tons: 32 full load
Dimensions, feet (metres): 67.3 × 15.4 × 4.3 *(20.5 × 4.7 × 1.3)*
Main machinery: 2 diesels; 1,000 hp(m) *(735 kW)*; 2 shafts
Speed, knots: 28. **Range, n miles:** 500 at 20 kt
Complement: 8
Guns: 1—12.7 mm MG.

Comment: Details given are for the first one built by Socomena Bizerte, with assistance from South Korea, and completed 2 March 1986. The remainder are probably ARCOR 38 craft of which five belong to the Coast Guard and five to the National Guard.

VERIFIED

COASTAL PATROL CRAFT   *3/2002, van Ginderen Collection /* 0141833

4 GABES CLASS (PB)

GABES	JERBA	KELIBIA	TABARK

Displacement, tons: 18 full load
Dimensions, feet (metres): 42.3 × 12.5 × 3 *(12.9 × 3.8 × 0.9)*
Main machinery: 2 diesels; 800 hp(m) *(588 kW)*; 2 shafts
Speed, knots: 38. **Range, n miles:** 250 at 15 kt
Complement: 6
Guns: 2—12.7 mm MGs.

Comment: Built by SBCN, Loctudy in 1988-89.

VERIFIED

GABES   *1/1995 /* 0080857

4 RODMAN 38 CLASS (PB)

Displacement, tons: 11.2 full load
Dimensions, feet (metres): 38.7 × 12.8 × 2.82 *(11.6 × 3.9 × 0.86)*
Main machinery: 2 diesels; 2 shafts
Speed, knots: 28. **Range, n miles:** 300 at 28 kt
Complement: 4
Radars: Navigation: I-band.

Comment: GRP hull. Two supplied in 2000 and two in 2002. Built by Rodman, Vigo for Customs Service.

VERIFIED

RODMAN 38   *9/2002, Martin Mokrus /* 0567539

Turkey
TÜRK DENIZ KUVVETLERI

Country Overview

The modern Republic of Turkey was founded in 1923. Situated in south-east Europe and south-west Asia, the country has an area of 300,948 square miles and is bordered to the north-west by Bulgaria and Greece, to the north-east by Georgia and Armenia, to the east by Iran and to the south with Iraq and Syria. It has a 739 n mile coastline with the Black Sea and 3,149 n mile coastline with the Aegean and Mediterranean Seas. The capital is Ankara, while the leading ports are Istanbul (largest city) and Izmir. In addition, Black Sea ports include Trabzon, Giresun, Samsun, and Zonguldak while Iskenderun and Mersin lie on the Mediterranean. Territorial waters for Black Sea and Mediterranean (12 n miles) are claimed and for Aegean (6 n miles). An EEZ (200 n miles) is claimed in the Black Sea only.

Headquarters Appointments

Commander-in-Chief, Turkish Naval Forces:
  Admiral Özer Örnek
Chief of Naval Staff:
  Vice Admiral Uğur Yiğit
Chief of Coast Guard:
  Rear Admiral Engin Heper

Flag Officers

Comturfleet (Gölcük):
  Admiral Yener Karahanoğlu
Comtursarnorth (Istanbul):
  Vice Admiral Metin Ataç

Flag Officers — *continued*

Comtursarsouth (Izmir):
  Vice Admiral Lütfü Sancar
Comturnavtrain (Istanbul):
  Vice Admiral Alev Gümüşoğlu
Comturampgroup (Foça):
  Rear Admiral M Tayfun Uraz
Comtursuracgroup (Gölcük):
  Rear Admiral Feyyaz Öğütcü
Comturfastgroup (Istanbul):
  Rear Admiral Halit Özkoç
Comturcoguard (Ankara):
  Rear Admiral Engin Heper
Comturminegroup (Erdek):
  Rear Admiral Özer Karabulut
Comtursubgroup (Gölcük):
  Rear Admiral A Can Erenoğlu
Comturespatgroup (Izmir):
  Rear Admiral Omer Akdağli
Comiststrait (Istanbul):
  Rear Admiral Kadir Sağdiç
Comcanstrait (Çanakkale):
  Rear Admiral Veysel Kösele
Comturageanzone (Izmir):
  Rear Admiral Hüseyin Çiftçi
Comturmedzone (Mersin):
  Rear Admiral Necati Kurt
Comtursouthtskgrp (Aksaz):
  Rear Admiral Baha Eren
Comturnavgolbase (Gölcük):
  Rear Admiral Atilla Kazez

Flag Officers — *continued*

Comturnaviskbase (Iskenderun):
  Rear Admiral Serdar Akinsel
Comturnavaksbase (Aksaz):
  Rear Admiral Engin Baykal
Comturmairbase (Topel):
  Rear Admiral Deniz Dağlilar

Diplomatic Representation

Defence and Naval Attaché in London:
  Commander A Kenanoglu

Personnel

(a) 2005: 55,000 (5,500 officers) including 31,000 conscripts, 3,000 Marines and 900 Air Arm (reserves 70,000)
(b) 15 months' national service

Organisation

Fleet HQ (Ankara), Fleet Command (Gölcük), Northern Area Command (Black Sea and Marmara), Southern Area Command (Aegean and Mediterranean), Naval Training Command (Istanbul).

Bases

Headquarters: Ankara
Black Sea: Ereğli, Bartin, Samsun, Trabzon
Marmara: Istanbul, Erdek, Çanakkale, Gölcük
Mediterranean: Izmir, Foça, Antalya, Mersin, Iskenderun, Aksaz
Dockyards: Gölcük, Pendik (Istanbul), Izmir

Strength of the Fleet (including Coast Guard)

Type	Active	Building (Planned)
Submarines—Patrol	11	3
Frigates	20	—
Corvettes	6	—
Fast Attack Craft—Missile	22	5
Large Patrol Craft	17	—
Minesweepers/Hunters—Coastal	14	5
Minesweepers—Inshore	4	—
LSTs/Minelayers	5	—
LCTs	25	—
Survey Vessels	3	—
Training Ships	11	—
Fleet Support Ships	2	—
Tankers	4	—
Transports—Large and small	22	—
Salvage Ships	3	—
Boom Defence Vessels	2	—

Prefix to Ships' Names

TCG (Turkish Republic Ship)
TCSG (Turkish Republic Coast Guard)

Pennant Numbers

From mid-1997 all pennant numbers have been repainted in non-reflective paint.

Marines

Total: 3,000
One brigade of HQ company, three infantry battalions, one artillery battalion, support units.

Coast Guard (Sahil Güvenlik)

Formed in July 1982 from the naval wing of the Jandarma. Prefix J replaced by SG and paint scheme is very light grey with a diagonal stripe forward. About 1,700 officers and men.

DELETIONS

Submarines

2004 *Hizirreis, Pirireis*

Frigates

2002 *Kocatepe*
2003 *Trakya*

Mine Warfare Forces

2004 *Sinop, Sürmene*

Patrol Forces

2002 *Kochisar, Sultanhisar, AB 25*
2003 *Bora*

Amphibious Forces

2003 *Bayraktar, Ç 113, Ç 117, Ç 122, Ç 124, Ç 302, Ç 303, Ç 309*

Auxiliaries

2002 *Inebolu*

PENNANT LIST

Submarines

S 347	Atilay
S 348	Saldiray
S 349	Batiray
S 350	Yildiray
S 351	Doğanay
S 352	Dolunay
S 353	Preveze
S 354	Sakarya
S 355	18 Mart
S 356	Anafartalar
S 357	Gür
S 358	Çanakkale (bldg)
S 359	Burakreis (bldg)
S 360	Birinci Inönü (bldg)

Frigates

F 240	Yavuz
F 241	Turgutreis
F 242	Fatih
F 243	Yildirim
F 244	Barbaros
F 245	Orucreis
F 246	Salihreis
F 247	Kemalreis
F 250	Muavenet
F 253	Zafer
F 255	Karadeniz
F 256	Ege
F 490	Gaziantep
F 491	Giresun
F 492	Gemlik
F 493	Gelibolu
F 494	Gökçeada
F 495	Gediz
F 496	Gökova
F 497	Göksu
F 500	Bozcaada
F 501	Bodrum
F 502	Bandirma
F 503	Beykoz
F 504	Bartin
F 505	Bafra

Mine Warfare Forces (Sweepers/Hunters)

M 260	Edincik
M 261	Edremit
M 262	Enez
M 263	Erdek
M 264	Erdemli
M 265	Alanya
M 266	Amasra (bldg)
M 267	Ayvalik (bldg)
M 268	Akçakoca
M 269	Anamur
M 270	Akçay
M 500	Foça
M 501	Fethiye
M 502	Fatsa
M 503	Finike
M 514	Silifke
M 515	Saros
M 516	Sigacik
M 517	Sapanca
M 518	Sariyer
M 520	Karamürsel
M 521	Kerempe
M 522	Kilimli
P 313-314	MTB 3-4
P 316-319	MTB 6-9

Amphibious Forces

L 401	Ertuğrul
L 402	Serdar
NL 123	Sarucabey
NL 124	Karamürselbey
NL 125	Osman Gazi

Patrol Forces

P 113	Yarhisar
P 114	Akhisar
P 121	AB 21
P 122	AB 22
P 123	AB 23
P 124	AB 24
P 127	AB 27
P 128	AB 28
P 129	AB 29
P 131	AB 31
P 133	AB 33
P 135	AB 35
P 136	AB 36
P 301	Kozlu
P 302	Kuşadasi
P 321	Denizkuşu
P 322	Atmaca
P 323	Şahin
P 324	Kartal
P 326	Pelikan
P 327	Albatros
P 328	Şimşek
P 329	Kasirga
P 330	Kiliç
P 331	Kalkan
P 332	Mizrak
P 333	Tufan
P 334	Meltem (bldg)
P 335	Imbat (bldg)
P 336	Zipkin (bldg)
P 337	Atak (bldg)
P 338	Bora (bldg)
P 340	Doğan
P 341	Marti
P 342	Tayfun
P 343	Volkan
P 344	Rüzgar
P 345	Poyraz
P 346	Gurbet
P 347	Firtina
P 348	Yildiz
P 349	Karayel
P 530	Trabzon
P 531	Terme

Auxiliaries

P 305	AG 5
P 306	AG 6
A 570	Taşkizak
A 571	Albay Hakki Burak
A 572	Yuzbasi Ihsan Tolunay
A 573	Binbaşi Sadettin Gürcan
A 576	Değirmendere
A 577	Sokullu Mehmet Paşa
A 578	Darica
A 579	Cezayirli Gazi Hasan Paşa
A 580	Akar
A 581	Çinar
A 582	Kemer
A 583	Aksaz
A 585	Akin
A 586	Akbas
A 587	Gazal
A 588	Çandarli
A 589	Isin
A 592	Karadeniz Ereğlisi
A 593	Eceabat
A 594	Çubuklu
A 595	Yarbay Kudret Güngör
A 596	Ulubat
A 597	Van
A 598	Söğüt
A 599	Çeşme
A 600	Kavak
A 1531	E 1
A 1532	E 2
A 1533	E 3
A 1534	E 4
A 1535	E 5
A 1536	E 6
A 1537	E 7
A 1538	E 8
A 1539	Samsun
A 1542	Söndüren 2
A 1543	Söndüren 3
A 1544	Söndüren 4
A 1600	Iskenderun
Y 50	Gölcük
Y 95	Torpido Tenderi
Y 98	Takip 1
Y 99	Takip 2
Y 107	Layter-7
Y 111-116	Pinar 1-6
Y 139	Yakit
Y 140-142	H 500-502

SUBMARINES

PREVEZE

10/2003, C D Yaylali / 0567543

5 + 3 PREVEZE (209) CLASS (TYPE 1400) (SSK)

Name	No	Builders	Laid down	Launched	Commissioned
PREVEZE	S 353	Gölcük, Kocaeli	12 Sep 1989	22 Oct 1993	28 July 1994
SAKARYA	S 354	Gölcük, Kocaeli	1 Feb 1990	28 July 1994	21 Dec 1995
18 MART	S 355	Gölcük, Kocaeli	28 July 1994	25 Aug 1997	28 June 1998
ANAFARTALAR	S 356	Gölcük, Kocaeli	1 Aug 1995	1 Sep 1998	22 July 1999
GÜR	S 357	Pendik Naval Shipyard, Istanbul	24 July 1998	May 2002	24 July 2003
ÇANAKKALE	S 358	Pendik Naval Shipyard, Istanbul	22 July 1999	Aug 2002	2005
BURAKREIS	S 359	Pendik Naval Shipyard, Istanbul	25 July 2001	2005	2006
BIRINCI INÖNÜ	S 360	Gölcük, Kocaeli	25 July 2002	2005	2007

Displacement, tons: 1,454 surfaced; 1,586 dived
Dimensions, feet (metres): 203.4 × 20.3 × 18
(62 × 6.2 × 5.5)
Main machinery: Diesel-electric; 4 MTU 12V 396 SB83 diesels; 3,800 hp(m) *(2.8 MW)* sustained; 4 alternators; 1 Siemens motor; 4,000 hp(m) *(3.38 MW)* sustained; 1 shaft
Speed, knots: 10 surfaced/snorting; 21.5 dived
Range, n miles: 8,200 at 8 kt surfaced; 400 at 4 kt dived
Complement: 30 (8 officers)

Missiles: SSM: McDonnell Douglas Sub Harpoon; active radar homing to 130 km *(70 n miles)* at 0.9 Mach; warhead 227 kg.
Torpedoes: 8—21 in *(533 mm)* bow tubes. GEC-Marconi Tigerfish Mk 24 Mod 2; wire-guided; active/passive homing to 13 km *(7 n miles)* at 35 kt active; 29 km *(15.7 n miles)* at

24 kt passive; warhead 134 kg or STN Atlas DM 2A4 (S 357 onwards). Total of 14 torpedoes and missiles.
Mines: In lieu of torpedoes.
Countermeasures: ESM: Racal Porpoise or Racal Sealion (UAP) (S 357 onwards); intercept.
Weapons control: Atlas Elektronik ISUS 83-2 TFCS. Link 11 receive only.
Radars: Surface search; I-band.
Sonars: Atlas Elektronik CSU 83; passive/active search and attack; medium/high frequency.
Atlas Elektronik TAS-3; towed array; passive low frequency.
STN Atlas flank array; passive low frequency.

Programmes: Order for first two signed in Ankara on 17 November 1987 with option on two more taken up in

1993. Four more ordered 22 July 1998. All built with HDW prefabrication and assembly at Gölcük. The last four are to complete at one a year from 2004 and are called the Gür class. Despite the earthquake damage to Gölcük in 1999, the building programme is reported to be keeping to schedule. The first four names commemorate Turkish victories. 18 March 1915 marks a victory in the Dardanelles and *Anafartalar* is a hill in Gallipoli.
Structure: Single hull design. Diving depth, 280 m *(820 ft)*. Kollmorgen masts. Four torpedo tubes can be used for SSM. STN Atlas flank arrays fitted in 1998/99 to the first four.
Operational: Endurance, 50 days.

UPDATED

18 MART

7/2000, Michael Nitz / 0106618

SAKARYA

12/2001, M Declerck / 0533241

ANAFARTALAR

4/2001, Selçuk Emre / 0132789

6 ATILAY (209) CLASS (TYPE 1200) (SSK)

Name	No	Builders	Laid down	Launched	Commissioned
ATILAY	S 347	Howaldtswerke, Kiel	1 Dec 1972	23 Oct 1974	12 Mar 1976
SALDIRAY	S 348	Howaldtswerke, Kiel	2 Jan 1973	14 Feb 1975	15 Jan 1977
BATIRAY	S 349	Howaldtswerke, Kiel	1 June 1975	24 Oct 1977	7 Nov 1978
YILDIRAY	S 350	Gölcük, Izmit	1 May 1976	20 July 1979	20 July 1981
DOĞANAY	S 351	Gölcük, Izmit	21 Mar 1980	16 Nov 1983	16 Nov 1984
DOLUNAY	S 352	Gölcük, Izmit	9 Mar 1981	22 July 1988	29 June 1990

Displacement, tons: 980 surfaced; 1,185 dived
Dimensions, feet (metres): 200.8 × 20.3 × 17.9
  (61.2 × 6.2 × 5.5)
Main machinery: Diesel-electric; 4 MTU 12V 493 TY60 diesels;
  2,400 hp(m) *(1.76 MW)* sustained; 4 alternators; 1.7 MW; 1
  Siemens motor; 4,600 hp(m) *(3.38 MW)* sustained; 1 shaft
Speed, knots: 11 surfaced; 22 dived
Range, n miles: 7,500 at 8 kt surfaced
Complement: 38 (9 officers)

Torpedoes: 8—21 in *(533 mm)* tubes. 14 AEG SST 4; wire-
  guided; active/passive homing to 28 km *(15.3 n miles)* at
  23 kt; 12 km *(6.6 n miles)* at 35 kt; warhead 260 kg. Swim-
  out discharge.
Countermeasures: ESM: Thomson-CSF DR 2000 or Racal
  Sealion (UAP) or Racal Porpoise; intercept
Weapons control: Signaal M8 (S 347-348). Signaal Sinbads
  (remainder). Link 11 receive.
Radars: Surface search: S 63B; I-band.
Sonars: Atlas Elektronik CSU 3; hull-mounted; passive/active
  search and attack; medium/high frequency.

Programmes: Designed by Ingenieurkontor, Lübeck for
  construction by Howaldtswerke, Kiel and sale by Ferrostaal,
  Essen, all acting as a consortium. Last three built in Turkey
  with assistance given by HDW.
Modernisation: Mid-life upgrades planned to include fire-control
  system starting with the first pair. Possibly by STN Elektronik
  to Preveze class standards. Programme drawn up in July
  1995 and started with first pair being fitted with a new ESM in
  1999. Further modernisation, including the installation of AIP
  propulsion, is under consideration.
Structure: A single-hull design with two ballast tanks and
  forward and after trim tanks. Fitted with snort and remote
  machinery control. The single screw is slow revving. Very
  high-capacity batteries with GRP lead-acid cells and battery
  cooling-by Wilh Hagen. Active and passive sonar, sonar
  detection equipment, sound ranging gear and underwater
  telephone. Fitted with two periscopes, radar and Omega
  receiver. Fore-planes retract. Diving depth, 250 m *(820 ft)*.
Operational: Endurance, 50 days. Some US Mk 37 torpedoes
  may also be carried.

UPDATED

BATIRAY
12/2001, M Declerck / 0533242

YILDIRAY
1/2002, M Declerck / 0132790

FRIGATES

Notes: (1) The Turkish Frigate 2000 (TF 2000) project has been
delayed due to economic difficulties. Three batches of two ships
are envisaged with a mix of in-country and foreign construction.
TF 2000 is to meet the requirement for an area air defence
capability.

(2) Project MILGEM is for up to 12 corvettes optimised for patrol
and anti-submarine warfare duties. The 90 m ships are expected
to displace about 1,500 tons, to be helicopter-capable and to
have a complement of about 80. This is to be a national building
programme with the first of class to be built in Istanbul for

delivery in about 2012. The construction of follow-on ships is to
be shared between several shipyards. A request for proposals
was made in August 2004.

6 BURAK (TYPE A 69) CLASS (FFGM)

Name	No	Builders	Laid down	Launched	Commissioned	Recommissioned
BOZCAADA (ex-*Commandant de Pimodan*)	F 500 (ex-F 787)	Lorient Naval Dockyard	15 July 1975	7 Aug 1976	20 May 1978	22 June 2001
BODRUM (ex-*Drogou*)	F 501 (ex-F 783)	Lorient Naval Dockyard	16 Oct 1973	30 Nov 1974	1 Oct 1976	18 Oct 2001
BANDIRMA (ex-*Quartier Maitre Anquetil*)	F 502 (ex-F 786)	Lorient Naval Dockyard	1 Aug 1975	7 Aug 1976	4 Feb 1978	15 Oct 2001
BEYKOZ (ex-*d'Estienne d'Orves*)	F 503 (ex-F 781)	Lorient Naval Dockyard	1 Sep 1972	1 June 1973	10 Sep 1976	18 Mar 2002
BARTIN (ex-*Amyot d'Inville*)	F 504 (ex-F 782)	Lorient Naval Dockyard	2 July 1973	30 Nov 1974	13 Oct 1976	3 June 2002
BAFRA (ex-*Second Maitre Le Bihan*)	F 505 (ex-F 788)	Lorient Naval Dockyard	1 Nov 1976	13 Aug 1977	7 July 1979	26 June 2002

Displacement, tons: 1,175 standard; 1,250 (1,330 later ships)
  full load
Dimensions, feet (metres): 264.1 × 33.8 × 18 (sonar)
  (80.5 × 10.3 × 5.5)
Main machinery: 2 SEMT-Pielstick 12 PC2 V 400 diesels;
  12,000 hp(m) *(8.82 MW)*; 2 shafts; LIPS cp props
Speed, knots: 23. **Range, n miles:** 4,500 at 15 kt
Complement: 104 (10 officers)

Missiles: SSM: 2 Aerospatiale MM 38 Exocet ❶; inertial cruise;
  active radar homing to 70 km *(40 n miles)* or 42 km *(23 n
  miles)* at 0.9 Mach; warhead 165 kg; sea-skimmer.
SAM: Matra Simbad twin launcher for Mistral ❷; IR homing to
  4 km *(2.2 n miles)*; warhead 3 kg. This may be replaced by
  Stinger.
Guns: 1 DCN/Creusot-Loire 3.9 in *(100 mm)*/55 Mod 68
  CADAM automatic ❸; 80 rds/min to 17 km *(9 n miles)* anti-
  surface; 8 km *(4.4 n miles)* anti-aircraft; weight of shell
  13.5 kg.
  2 Giat 20 mm ❹; 720 rds/min to 10 km *(5.5 n miles)*.
  4—12.7 mm MGs.
Torpedoes: 4 fixed tubes ❺. ECAN L5; dual purpose; active/
  passive homing to 9.5 km *(5.1 n miles)* at 35 kt; warhead
  150 kg; depth to 550 m *(1,800 ft)*.
A/S mortars: 1 Creusot-Loire 375 mm Mk 54 6-tubed trainable
  launcher ❻; range 1,600 m; warhead 107 kg.
Countermeasures: Decoys: 2 CSEE Dagaie 10-barrelled
  trainable launchers ❼; chaff and IR flares; H- to J-band.
  Nixie torpedo decoy.
ESM: ARBR 16; radar warning.
Weapons control: Thomson-CSF Vega system; CSEE Panda
  optical secondary director.
Radars: Air/surface search: Thomson-CSF DRBV 51A ❽; G-band.
  Navigation: Racal Decca 1226; I-band.
  Fire control: Thomson-CSF DRBC 32E ❾; I-band.

Sonars: Thomson Sintra DUBA 25; hull-mounted; search and
  attack; medium frequency.

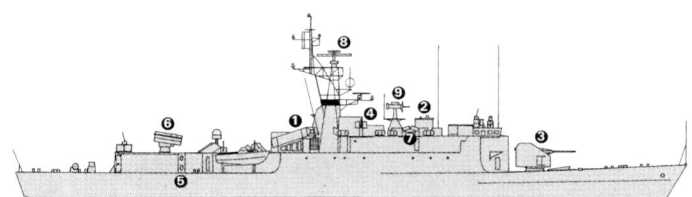
BOZCAADA
(Scale 1 : 900), Ian Sturton / 0114803

BAFRA
7/2003, Selçuk Emre / 0567549

Comment: Six Type A 69 class bought second-hand from France
in October 2000. All, except *Bafra*, refitted at Brest. Work
done on propulsion and weapons systems. Exocet MM 38

SSMs procured under separate contract. Operational use is
coastal patrol duties, for which they were designed, in order to
release more capable ships for front-line service. *UPDATED*

4 BARBAROS CLASS (MODIFIED MEKO 200 TYPE) (FFGHM)

Name	No	Builders	Laid down	Launched	Commissioned
BARBAROS	F 244	Blohm + Voss, Hamburg	18 Mar 1993	29 Sep 1993	16 Mar 1995
ORUCREIS	F 245	Gölcük, Kocaeli	15 Sep 1993	28 July 1994	10 May 1996
SALIHREIS	F 246	Blohm + Voss, Hamburg	24 July 1995	26 Sep 1997	17 Dec 1998
KEMALREIS	F 247	Gölcük, Kocaeli	4 Apr 1997	24 July 1998	8 June 2000

Displacement, tons: 3,380 full load
Dimensions, feet (metres): 387.1 × 48.6 × 14.1; 21 (sonar)
  (118 × 14.8 × 4.3; 6.4)
Main machinery: CODOG; 2 GE LM 2500 gas turbines;
  60,000 hp *(44.76 MW)* sustained; 2 MTU 16V 1163 TB83
  diesels; 11,780 hp(m) *(8.67 MW)* sustained; 2 shafts; Escher
  Wyss; cp props
Speed, knots: 32. **Range, n miles:** 4,100 at 18 kt
Complement: 187 (22 officers) plus 9 aircrew plus 8 spare

Missiles: SSM: 8 McDonnell Douglas Harpoon (2 quad)
  launchers ❶; active radar homing to 130 km *(70 n miles)* at
  0.9 Mach; warhead 227 kg.
  SAM: Raytheon Sea Sparrow RIM-7M Mk 29 Mod 1 octuple
  launcher ❷ (F 244 and F 245) and VLS Mk 41 Mod 8 ❸ (F 246
  and F 247); 24 Selenia Elsag Aspide; semi-active radar
  homing to 13 km *(7 n miles)* at 2.5 Mach; warhead 39 kg.
  Evolved Sea Sparrow (ESSM) in due course.
Guns: 1 FMC 5 in *(127 mm)*/54 Mk 45 Mod 1/2 ❹; 20 rds/min
  to 23 km *(12.6 n miles)* anti-surface; 15 km *(8.2 n miles)* anti-
  aircraft; weight of shell 32 kg.
  3 Oerlikon-Contraves 25 mm Sea Zenith ❺; 4 barrels per
  mounting; 3,400 rds/min combined to 2 km.
Torpedoes: 6—324 mm Mk 32 Mod 5 (2 triple) tubes ❻.
  Honeywell Mk 46 Mod 5; anti-submarine; active/passive
  homing to 11 km *(5.9 n miles)* at 40 kt; warhead 44 kg.
Countermeasures: Decoys: 2 Loral Hycor 6-tubed fixed Mk 36
  Mod 1 SRBOC ❼; IR flares and chaff to 4 km *(2.2 n miles)*.
  Nixie SLQ-25; towed torpedo decoy.
ESM/ECM: Racal Cutlass/Scorpion; intercept and jammer.
Combat data systems: Thomson-CSF/Signaal STACOS Mod 3;
  Link 11. WSC 3V(7) SATCOMs. Marisat.
Weapons control: 2 Siemens Albis optronic directors ❽.
  SWG-1A for Harpoon.
Radars: Air search: Siemens/Plessey AWS 9 (Type 996) ❾; 3D;
  E/F-band.
  Air/surface search: Plessey/BAe AWS 6 Dolphin ❿; G-band.
  Fire control: 1 or 2 (F 246-247) Signaal STIR ⓫; I/J/K-band (for
  SAM); range 140 km *(76 n miles)* for 1 m² target.
  Contraves TMKu (F 244-245) ⓬; I/J-band (for SSM and
  127 mm).
  2 Contraves Seaguard ⓭; I/J-band (for 25 mm).
  Navigation: Racal Decca 2690 BT ARPA; I-band.
  Tacan: URN 25. IFF Mk XII Mod 4.
Sonars: Raytheon SQS-56 (DE 1160); hull-mounted; active
  search and attack; medium frequency.

Helicopters: 1 AB 212ASW ⓮ or S-70B Seahawk.

Programmes: First pair ordered 19 January 1990, second pair
  authorised 14 December 1992. Programme started
  5 November 1991 with construction commencing in June
  1992 in Germany. Completion of the last one delayed by the
  Gölcük earthquake in 1999.
Structure: An improvement on the Yavuz class. Mk 29 Sea
  Sparrow launchers fitted in the first two, while the second pair

have Mk 41 VLS aft of the funnel, which will be retrofitted in
the first two in due course. The ships have CODOG propulsion
for a higher top speed. Other differences with *Yavuz* include a
full command system, improved radars and a citadel for NBCD

protection. A bow bulwark has been added in the second pair.
Operational: Helicopter has Sea Skua anti-ship missiles. The first
pair used as Flagships.

UPDATED

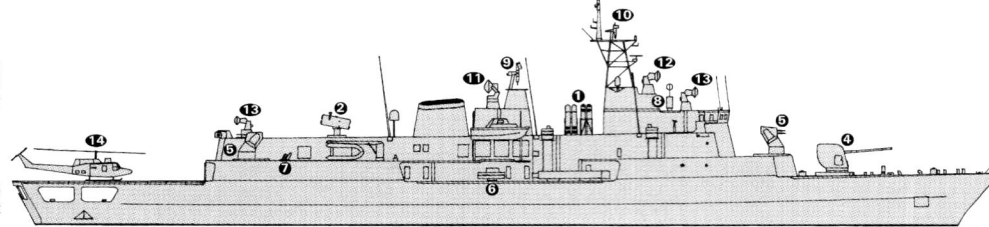

BARBAROS   (Scale 1 : 900), Ian Sturton / 0019315

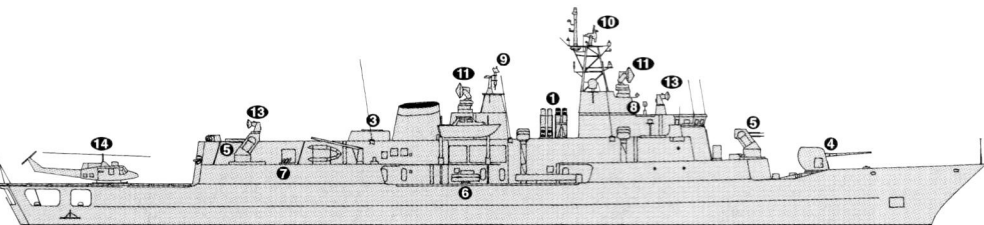

SALIHREIS   (Scale 1 : 900), Ian Sturton / 0050273

KEMALREIS   9/2004 *, Selim San / 0567560

SALIHREIS   7/2001, Michael Nitz / 0533247

KEMALREIS   9/2004 *, Selim San / 0587561

4 YAVUZ CLASS (MEKO 200 TYPE) (FFGHM)

Name	No	Builders	Laid down	Launched	Commissioned
YAVUZ	F 240	Blohm + Voss, Hamburg	30 May 1985	7 Nov 1985	17 July 1987
TURGUTREIS (ex-*Turgut*)	F 241	Howaldtswerke, Kiel	20 May 1985	30 May 1986	4 Feb 1988
FATIH	F 242	Gölcük, Izmit	1 Jan 1986	24 Apr 1987	28 Aug 1988
YILDIRIM	F 243	Gölcük, Izmit	24 Apr 1987	22 July 1988	17 Nov 1989

Displacement, tons: 2,414 standard; 2,919 full load
Dimensions, feet (metres): 378.9 × 46.6 × 13.5
  (115.5 × 14.2 × 4.1)
Main machinery: CODAD; 4 MTU 20V 1163 TB93 diesels;
  29,940 hp(m) *(22 MW)* sustained; 2 shafts; cp props
Speed, knots: 27. **Range, n miles:** 4,100 at 18 kt
Complement: 180 (24 officers)

Missiles: SSM: 8 McDonnell Douglas Harpoon (2 quad)
  launchers **❶**; active radar homing to 130 km *(70 n miles)* at
  0.9 Mach; warhead 227 kg.
  SAM: Raytheon Sea Sparrow RIM-7M Mk 29 Mod 1 octuple
  launcher **❷**; 24 Selenia Elsag Aspide; semi-active radar
  homing to 13 km *(7 n miles)* at 2.5 Mach; warhead 39 kg.
Guns: 1 FMC 5 in *(127 mm)*/54 Mk 45 Mod 1 **❸**; 20 rds/min to
  23 km *(12.6 n miles)* anti-surface; 15 km *(8.2 n miles)* anti-
  aircraft; weight of shell 32 kg.
  3 Oerlikon-Contraves 25 mm Sea Zenith **❹**; 4 barrels per
  mounting; 3,400 rds/min combined to 2 km.
Torpedoes: 6—324 mm Mk 32 (2 triple) tubes **❺**. Honeywell Mk
  46 Mod 5; anti-submarine; active/passive homing to 11 km
  (5.9 n miles) at 40 kt; warhead 44 kg.
Countermeasures: Decoys: 2 Loral Hycor 6-tubed fixed Mk 36
  Mod 1 SRBOC **❻**; IR flares and chaff to 4 km *(2.2 n miles)*.
  Nixie SLQ-25; towed torpedo decoy.
ESM/ECM: Signaal Rapids/Ramses; intercept and jammer.
Combat data systems: Signaal STACOS-TU; action data
  automation; Link 11. WSC 3V(7) SATCOMs. Marisat.
Weapons control: 2 Siemens Albis optronic directors (for Sea
  Zenith). SWG-1A for Harpoon.
Radars: Air search: Signaal DA08 **❼**; F-band.
  Air/surface search: Plessey AWS 6 Dolphin **❽**; G-band.
  Fire control: Signaal STIR **❾**; I/J/K-band (for SAM); range 140 km
  (76 n miles) for 1 m² target.
  Signaal WM25 **❿**; I/J-band (for SSM and 127 mm).
  2 Contraves Seaguard **⓫**; I/J-band (for 25 mm).
  Navigation: Racal Decca TM 1226; I-band.
  Tacan: URN 25. IFF Mk XII.
Sonars: Raytheon SQS-56 (DE 1160); hull-mounted; active
  search and attack; medium frequency.

Helicopters: 1 AB 212ASW **⓬**.

Programmes: Ordered 29 December 1982 with builders and
Thyssen Rheinstahl Technik of Dusseldorf. Meko 200 type
similar to Portuguese frigates. *Turgutreis* was renamed on
14 February 1988.

Operational: Helicopter has Sea Skua anti-ship missiles.
  UPDATED

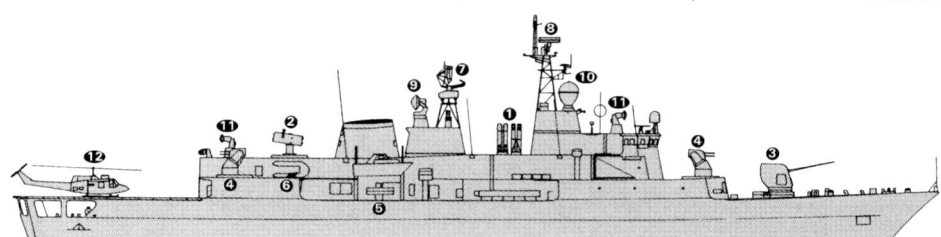
YAVUZ *(Scale 1 : 900), Ian Sturton*

FATIH *5/2003, Selim San / 0567546*

TURGUTREIS *10/1999, H M Steele / 0080864*

YILDIRIM *9/2003, C D Yaylali / 0567547*

8 GAZIANTEP (OLIVER HAZARD PERRY) CLASS (FFGHM)

Name	No	Builders	Laid down	Launched	Commissioned	Recommissioned
GAZIANTEP (ex-*Clifton Sprague*)	F 490 (ex-FFG 16)	Bath Iron Works	30 Sep 1979	16 Feb 1980	21 Mar 1981	24 July 1998
GIRESUN (ex-*Antrim*)	F 491 (ex-FFG 20)	Todd Shipyards, Seattle	21 June 1978	27 Mar 1979	26 Sep 1981	24 July 1998
GEMLIK (ex-*Flatley*)	F 492 (ex-FFG 21)	Bath Iron Works	13 Nov 1979	15 May 1980	20 June 1981	24 July 1998
GELIBOLU (ex-*Reid*)	F 493 (ex-FFG 30)	Todd Shipyards, San Pedro	8 Oct 1980	27 June 1981	19 Feb 1983	22 July 1999
GÖKÇEADA (ex-*Mahlon S Tisdale*)	F 494 (ex-FFG 27)	Todd Shipyards, San Pedro	19 Mar 1980	7 Feb 1981	27 Nov 1982	8 June 2000
GEDIZ (ex-*John A Moore*)	F 495 (ex-FFG 19)	Todd Shipyards, San Pedro	19 Dec 1978	20 Oct 1979	14 Nov 1981	25 July 2000
GOKOVA (ex-*Samuel Eliot Morison*)	F 496 (ex-FFG 13)	Bath Iron Works	4 Aug 1978	14 July 1979	11 Oct 1980	11 Apr 2002
GÖKSU (ex-*Estocin*)	F 497 (ex-FFG 15)	Bath Iron Works	2 Apr 1979	3 Nov 1979	10 Jan 1981	4 Apr 2003

Displacement, tons: 2,750 light, 3,638 full load
Dimensions, feet (metres): 453 × 45 × 14.8; 24.5 (sonar)
  (138.1 × 13.7 × 4.5; 7.5)
Main machinery: 2 GE LM 2500 gas turbines; 41,000 hp
  (30.59 MW) sustained; 1 shaft; cp prop
  2 auxiliary retractable props; 650 hp *(484 kW)*
Speed, knots: 29. **Range, n miles:** 4,500 at 20 kt
Complement: 206 (13 officers) including 19 aircrew

Missiles: SSM: 4 McDonnell Douglas Harpoon; active radar
  homing to 130 km *(70 n miles)* at 0.9 Mach; warhead 227 kg.
  SAM: 36 GDC Standard SM-1MR; command guidance; semi-
  active radar homing to 46 km *(25 n miles)* at 2 Mach.
  1 Mk 13 Mod 4 launcher for both SSM and SAM missiles **❶**.
Guns: 1 OTO Melara 3 in *(76 mm)*/62 Mk 75 **❷**; 85 rds/min to
  16 km *(8.7 n miles)* anti-surface; 12 km *(6.6 n miles)* anti-
  aircraft; weight of shell 6 kg.
  1 General Electric/General Dynamics 20 mm/76 6-barrelled
  Mk 15 Vulcan Phalanx **❸**; 3,000 rds/min combined to 1.5 km.
  4—12.7 mm MGs.
Torpedoes: 6—324 mm Mk 32 (2 triple) tubes **❹**. 24 Honeywell
  Mk 46 Mod 5; anti-submarine; active/passive homing to
  11 km *(5.9 n miles)* at 40 kt; warhead 44 kg.
Countermeasures: Decoys: 2 Loral Hycor SRBOC 6-barrelled
  fixed Mk 36 **❺**; IR flares and chaff to 4 km *(2.2 n miles)*.
  T-Mk-6 Fanfare/SLQ-25 Nixie; torpedo decoy.
  ESM/ECM: SLQ-32(V)2 **❻**; radar warning. Sidekick modification
  adds jammer and deception system.
Combat data systems: NTDS with Link 11 and 14. SATCOM.

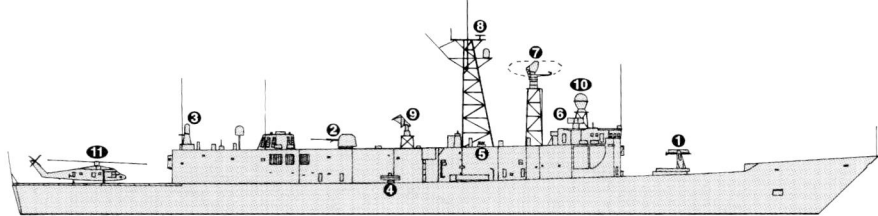

GÖKÇEADA　　　　　　　　　　　　　　　*(Scale 1 : 1,200), Ian Sturton /* 0587565

Weapons control: SWG-1 Harpoon LCS. Mk 92 Mod 4 WCS with
  CAS (Combined Antenna System). The Mk 92 is the US
  version of the Signaal WM28 system. Mk 13 weapon
  direction system. 2 Mk 24 optical directors.
Radars: Air search: Raytheon SPS-49(V)4 **❼**; C/D-band.
Surface search: ISC Cardion SPS-55 **❽**; I-band.
Fire control: Lockheed STIR (modified SPG-60) **❾**; I/J-band.
  Sperry Mk 92 (Signaal WM28) **❿**; I/J-band.
Navigation: Furuno; I-band.
Tacan: URN 25.
Sonars: Raytheon SQS-56; hull-mounted; active search and
  attack; medium frequency.

Helicopters: 1 S-70B Seahawk **⓫**.

Programmes: Three approved for transfer by grant aid. Transfer
  delayed by Greek objections, and Turkish sailors were sent
  home from the US in mid-1996. Congress authorised the go-
  ahead again on 27 August 1997. Two more approved for
  transfer by sale 30 September 1998, one in February 2000,
  one in April 2002 and one in April 2003. At least one other
  Duncan FFG 10 for spares.
Modernisation: Project 'Genesis' includes plans to upgrade the
  combat data system.
Structure: A flight deck extension programme, to enable S-70
  helicopters has been completed. The work involved angling
  the transom as in later USN ships of the class.
Operational: Sonar towed arrays were not transferred.

　　　　　　　　　　　　　　　　　　　　　　　　　UPDATED

GÖKÇEADA　　　　　　　　　　　　　　　*10/2003, C D Yaylali /* 0567548

GEDIZ　　　　　　　　　　　　　　　*6/2004*, John Brodie /* 0589818

4 TEPE (KNOX) CLASS (FFGH)

Name	No	Builders	Laid down	Launched	Commissioned	Recommissioned
MUAVENET (ex-*Capodanno*)	F 250 (ex-1093)	Avondale Shipyards	12 Oct 1971	21 Oct 1972	17 Nov 1973	12 Sep 1993
ZAFER (ex-*Thomas C Hart*)	F 253 (ex-1092)	Avondale Shipyards	8 Oct 1971	12 Aug 1972	28 July 1973	30 Aug 1993
KARADENIZ (ex-*Donald B Beary*)	F 255 (ex-1085)	Avondale Shipyards	24 July 1970	22 May 1971	22 July 1972	20 May 1994
EGE (ex-*Ainsworth*)	F 256 (ex-1090)	Avondale Shipyards	11 June 1971	15 Apr 1972	31 Mar 1973	27 May 1994

Displacement, tons: 3,011 standard; 4,260 full load
Dimensions, feet (metres): 439.6 × 46.8 × 15; 24.8 (sonar) *(134 × 14.3 × 4.6; 7.8)*
Main machinery: 2 Combustion Engineering/Babcock & Wilcox boilers; 1,200 psi *(84.4 kg/cm²)*; 950°F *(510°C)*; 1 Westinghouse turbine; 35,000 hp *(26 MW)*; 1 shaft
Speed, knots: 27. **Range, n miles:** 4,000 at 22 kt on 1 boiler
Complement: 288 (20 officers)

Missiles: SSM: 8 McDonnell Douglas Harpoon; active radar homing to 130 km *(70 n miles)* at 0.9 Mach; warhead 227 kg.
A/S: Honeywell ASROC Mk 16 octuple launcher with reload system (has 2 cells modified to fire Harpoon) ❶; inertial guidance to 1.6-10 km *(1-5.4 n miles)*; payload Mk 46 Mod 5 Neartip.
Guns: 1 FMC 5 in *(127 mm)*/54 Mk 42 Mod 9 ❷; 20-40 rds/min to 24 km *(13 n miles)* anti-surface; 14 km *(7.7 n miles)* anti-aircraft; weight of shell 32 kg.

1 General Electric/General Dynamics 20 mm/76 6-barrelled Mk 15 Vulcan Phalanx ❸; 3,000 rds/min combined to 1.5 km.
Torpedoes: 4—324 mm Mk 32 (2 twin) fixed tubes ❹. 22 Honeywell Mk 46 Mod 5; anti-submarine; active/passive homing to 11 km *(5.9 n miles)* at 40 kt; warhead 44 kg.
Countermeasures: Decoys: 2 Loral Hycor SRBOC 6-barrelled fixed Mk 36 ❺; IR flares and chaff to 4 km *(2.2 n miles)*. T Mk-6 Fanfare/SLQ-25 Nixie; torpedo decoy. Prairie Masker hull and blade rate noise suppression.
ESM: SLQ-32(V)2 ❻; intercept.
Combat data systems: Signaal Sigma K5 with Link 11.
Weapons control: SWG-1A Harpoon LCS. Mk 68 Mod 3 GFCS. Mk 114 Mod 6 ASW FCS. Mk 1 target designation system. MMS target acquisition sight (for mines, small craft and low-flying aircraft).
Radars: Air search: Lockheed SPS-40B ❼; B-band.
Surface search: Raytheon SPS-10 or Norden SPS-67 ❽; G-band.
Navigation: Marconi LN66; I-band.

Fire control: Western Electric SPG-53D/F ❾; I/J-band.
Tacan: SRN 15.
Sonars: EDO/General Electric SQS-26CX; bow-mounted; active search and attack; medium frequency.

Helicopters: 1 AB 212ASW ❿.

Programmes: In late 1992 the US offered Turkey four of the class. A proposal was put to Congress in June 1993 and four approved for transfer on a five year lease, plus one more, *Elmer Montgomery* for spares, on a grant basis under the Foreign Assistance Act. The latter replaced the former destroyer *Muavenet* which was scrapped after being hit by a Sea Sparrow missile. The second batch of four transferred in 1994. All eight purchased outright in 1999. F 251 decommissioned in 2000, F 257 in 2001, F 252 in 2002 and F 254 in 2003.
Modernisation: Hangar and flight deck enlarged. In 1979 a programme was initiated to fit 3.5 ft bow bulwarks and spray strakes adding 9.1 tons to a displacement. Sea Sparrow SAM replaced by Phalanx 1982-88. Link 11 fitted after transfer. Project 'Kalyon-5' integrated new multipurpose consoles into the combat data system.
Structure: Improved ASROC torpedo reloading capability (note slanting face of bridge structure immediately behind ASROC). Four Mk 32 torpedo tubes are fixed in the midships structure, two to a side, angled out at 45°. The arrangement provides improved loading capability over exposed triple Mk 32 torpedo tubes. A 4,000 lb lightweight anchor is fitted on the port side and an 8,000 lb anchor fits into the after section of the sonar dome.

UPDATED

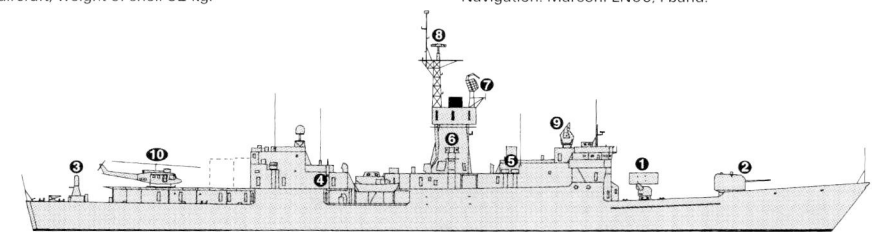

MUAVENET
(Scale 1 : 1,200), Ian Sturton

ZAFER
1/2002, M Declerck / 0533245

KARADENIZ
1/2002, M Declerck / 0533246

SHIPBORNE AIRCRAFT

Note: There are three AB 204 utility aircraft.

Numbers/Type: 9 Agusta AB 212ASW.
Operational speed: 106 kt *(196 km/h)*.
Service ceiling: 14,200 ft *(4,330 m)*.
Range: 230 n miles *(426 km)*.
Role/Weapon systems: ASV/ASW helicopter with updated systems. Sensors: BAe Ferranti Sea Spray Mk 3 radar, ECM/ESM, MAD, Bendix ASQ-18 dipping sonar. Weapons: ASW; two Mk 46 or 244/S torpedoes. AVS; two Sea Skua missiles.

UPDATED

AB 212   *1/2002, M Declerck* / 0533250

Numbers/Type: 7 Sikorsky S-70B Seahawk.
Operational speed: 135 kt *(250 km/h)*.
Service ceiling: 10,000 ft *(3,050 m)*.
Range: 600 n miles *(1,110 km)*.
Role/Weapon systems: Contracts placed 3 June 1998 for first four. Second contract for four further aircraft on 31 December 1998. First three delivered 26 April 2002 and second four on 24 July 2003. One aircraft lost in accident. Helras ASW weapon systems ordered. Sensors: APS-124 search radar; Helras dipping sonar. Weapons: ASW: 2 Mk 46 torpedoes; AGM-114B Hellfire II ASM.

VERIFIED

SEAHAWK S-70B   *6/2002, Selçuk Emre* / 0533251

LAND-BASED MARITIME AIRCRAFT

Notes: Three Agusta AB 212EW helicopters are reported to be based at Topel.

Numbers/Type: 6 CASA CN-235 D/K MPA.
Operational speed: 240 kt *(445 km/h)*.
Service ceiling: 26,600 ft *(8,110 m)*.
Range: 669 n miles *(1,240 km)*.
Role/Weapon systems: Initial batch of four delivered 25 July 2002 and a further two on 24 July 2003. A total of nine expected by 2006. Long-range maritime patrol for surface surveillance and ASW. Sensors: Ocean Master radar (SAR, ISAR, MTI and air-to-air modes); FLIR; AAR-60 missile warning; DR 3000 A ESM; MAD; Link 11. Weapons: 2 Mk-46 torpedoes.

VERIFIED

CN-235MPA   *1997, Paul Jackson* / 0056630

PATROL FORCES

Notes: *Kozlu* P 301 and *Kuşadası* P 302 are listed under Vegesack class in Mine Warfare section.

4 + 5 KILIÇ CLASS (FAST ATTACK CRAFT—MISSILE) (PGGF)

Name	No	Builders	Launched	Commissioned
KILIÇ	P 330	Lürssen, Vegesack	15 July 1997	17 Mar 1998
KALKAN	P 331	Taşkizak, Istanbul	22 Sep 1998	22 July 1999
MIZRAK	P 332	Taşkizak, Istanbul	5 Apr 1999	8 June 2000
TUFAN	P 333	Lürssen, Vegesack	25 July 2001	2004
MELTEM	P 334	Pendik Naval Shipyard, Istanbul	1 Sep 2004	2005
IMBAT	P 335	Pendik Naval Shipyard, Istanbul	2004	2005
ZIPKIN	P 336	Pendik Naval Shipyard, Istanbul	2005	2006
ATAK	P 337	Pendik Naval Shipyard, Istanbul	2006	2007
BORA	P 338	Pendik Naval Shipyard, Istanbul	2006	2007

Displacement, tons: 550 full load
Dimensions, feet (metres): 204.6 × 27.2 × 8.5 *(62.4 × 8.3 × 2.6)*
Main machinery: 4 MTU 16V 956 TB91 diesels; 15,120 hp(m) *(11.1 MW)* sustained; 4 shafts
Speed, knots: 38. **Range, n miles:** 1,050 at 30 kt
Complement: 46 (12 officers)

Missiles: SSM: 8 McDonnell Douglas Harpoon (2 quad) launchers; active radar homing to 130 km *(70 n miles)* at 0.9 Mach; warhead 227 kg.
Guns: 1 Otobreda 3 in *(76 mm)*/62 compact; 85 rds/min to 16 km *(8.7 n miles)* anti-surface; 12 km *(6.6 n miles)* anti-aircraft; weight of shell 6 kg.
2 Otobreda 40 mm/70 (twin); 300 rds/min to 12 km *(6.6 n miles)*; weight of shell 0.96 kg.
Countermeasures: Decoys: 2 Mk 36 SRBOC chaff launchers.
ESM: Racal Cutlass; intercept.
Combat data systems: Signaal/Thomson-CSF STACOS.
Weapons control: LIROD Mk 2 optronic director; Vesta helo datalink/transponder.
Radars: Surface search: Signaal MW08; G-band.
Fire control: Signaal STING; I/J-band.
Navigation: KH 1007; I-band.

Programmes: Contract for first three signed in May 1993 but there was a delay in confirming it. Further four ordered 19 June 2000 since when a further two craft have been ordered.
Structure: A development of the Yildiz class but with reduced radar cross-section mast and a redesigned bow to improve sea-keeping. The after gun and radars are also different to *Yildiz*.
Operational: First of class arrived in Turkey in April 1998.   *UPDATED*

KALKAN   *10/2003, C D Yaylali* / 0567550

TUFAN   *4/2003, Martin Mokrus* / 0567551

8 KARTAL CLASS (FAST ATTACK CRAFT-MISSILE) (PTGF)

Name	No	Builders	Commissioned
DENIZKUŞU	P 321 (ex-*P 336*)	Lürssen, Vegesack	9 Mar 1967
ATMACA	P 322 (ex-*P 335*)	Lürssen, Vegesack	9 Mar 1967
ŞAHIN	P 323 (ex-*P 334*)	Lürssen, Vegesack	3 Nov 1966
KARTAL	P 324 (ex-*P 333*)	Lürssen, Vegesack	3 Nov 1966
PELIKAN	P 326	Lürssen, Vegesack	11 Feb 1970
ALBATROS	P 327 (ex-*P 325*)	Lürssen, Vegesack	18 Mar 1970
ŞIMŞEK	P 328 (ex-*P 332*)	Lürssen, Vegesack	6 Nov 1969
KASIRGA	P 329 (ex-*P 338*)	Lürssen, Vegesack	25 Nov 1967

Displacement, tons: 160 standard; 190 full load
Dimensions, feet (metres): 139.4 × 23 × 7.9 *(42.5 × 7 × 2.4)*
Main machinery: 4 MTU MD 16V 538 TB90 diesels; 12,000 hp(m) *(8.82 MW)* sustained; 4 shafts
Speed, knots: 42. **Range, n miles:** 500 at 40 kt
Complement: 39 (4 officers)

Missiles: SSM: 2 or 4 Kongsberg Penguin Mk 2; IR homing to 27 km *(14.6 n miles)* at 0.8 Mach; warhead 120 kg.
Guns: 2 Bofors 40 mm/70; 300 rds/min to 12 km *(6.6 n miles)*; weight of shell 0.96 kg.
Torpedoes: 2-21 in *(533 mm)* tubes; anti-surface.
Mines: Can carry 4.
Radars: Surface search: Racal Decca 1226; I-band.

Operational: *Meltem* sunk in collision with Soviet naval training ship *Khasan* in Bosphorus in 1985. Subsequently salvaged but beyond repair. Although these craft are getting old, there are no plans to decommission them in the short term.   *UPDATED*

ŞIMŞEK   *11/2001, Selim San* / 0126283

8 DOĞAN CLASS (FAST ATTACK CRAFT-MISSILE) (PGGF)

Name	No	Builders	Commissioned
DOĞAN	P 340	Lürssen, Vegesack	23 Dec 1977
MARTI	P 341	Taşkizak Yard, Istanbul	1 Aug 1978
TAYFUN	P 342	Taşkizak Yard, Istanbul	9 Aug 1979
VOLKAN	P 343	Taşkizak Yard, Istanbul	25 July 1980
RÜZGAR	P 344	Taşkizak Yard, Istanbul	24 May 1985
POYRAZ	P 345	Taşkizak Yard, Istanbul	28 Aug 1986
GURBET	P 346	Taşkizak Yard, Istanbul	24 July 1988
FIRTINA	P 347	Taşkizak Yard, Istanbul	14 Oct 1988

Displacement, tons: 436 full load
Dimensions, feet (metres): 190.6 × 25 × 8.8 *(58.1 × 7.6 × 2.7)*
Main machinery: 4 MTU 16V 956 TB92 diesels; 17,700 hp(m) *(13 MW)* sustained; 4 shafts
Speed, knots: 38. **Range, n miles:** 1,050 at 30 kt
Complement: 40 (5 officers)

Missiles: SSM: 8 McDonnell Douglas Harpoon (2 quad) launchers; active radar homing to 130 km *(70 n miles)* at 0.9 Mach; warhead 227 kg.
Guns: 1 OTO Melara 3 in *(76 mm)*/62 compact; 85 rds/min to 16 km *(8.7 n miles)* anti-surface; 12 km *(6.6 n miles)* anti-aircraft; weight of shell 6 kg.
2 Oerlikon 35 mm/90 (twin); 550 rds/min to 6 km *(3.3 n miles)*; weight of shell 1.55 kg.
Countermeasures: Decoys: 2 Mk 36 SRBOC chaff launchers.
ESM: MEL Susie (344-347); intercept.
Combat data systems: Signaal mini TACTICOS (344-347).
Weapons control: LIOD Mk 2 optronic director.
Radars: Surface search: Racal Decca 1226; I-band.
Fire control: Signaal WM28/41; I/J-band.

Programmes: First ordered 3 August 1973 to a Lürssen FPB 57 design.
Modernisation: A mid-life programme includes upgrade to the combat data system, communications and ESM. Work on the first four was completed in 2002 while work on the second four is to start in 2004.
Structure: Aluminium superstructure; steel hulls. The last pair were built with optronic directors which are being retrofitted in all, together with an improved Signaal combat data system.
UPDATED

MARTI
10/2003, C D Yaylali / 0567553

DOĞAN
4/2000, Selim San / 0106630

2 YILDIZ CLASS (FAST ATTACK CRAFT—MISSILE) (PGGF)

Name	No	Builders	Commissioned
YILDIZ	P 348	Taşkizak Yard, Istanbul	3 June 1996
KARAYEL	P 349	Taşkizak Yard, Istanbul	19 Sep 1996

Displacement, tons: 433 full load
Dimensions, feet (metres): 189.6 oa; 178.5 wl × 25 × 8.8 *(57.8; 54.4 × 7.6 × 2.7)*
Main machinery: 4 MTU 16V 956 TB91 diesels; 15,120 hp(m) *(11.1 MW)* sustained; 4 shafts
Speed, knots: 38. **Range, n miles:** 1,050 at 30 kt
Complement: 45 (6 officers)

Missiles: SSM: 8 McDonnell Douglas Harpoon (2 quad) launchers; active radar homing to 130 km *(70 n miles)* at 0.9 Mach; warhead 227 kg.
Guns: 1 OTO Melara 3 in *(76 mm)*/62 compact; 85 rds/min to 16 km *(8.7 n miles)* anti-surface; 12 km *(6.6 n miles)* anti-aircraft; weight of shell 6 kg.
2 Oerlikon 35 mm/90 (twin); 550 rds/min to 6 km *(3.3 n miles)*; weight of shell 1.55 kg.
Countermeasures: Decoys: 2 Mk 36 SRBOC chaff launchers.
ESM: Racal Cutlass; intercept.
Combat data systems: Signaal/Thomson-CSF TACTICOS.
Weapons control: LIOD Mk 2 optronic director; Vesta helo datalink/transponder.
Radars: Surface search: Siemens/Plessey AW 6 Dolphin; G-band.
Fire control: Oerlikon/Contraves TMX; I/J-band.
Navigation: Racal Decca TM 1226; I-band.

Programmes: Ordered in June 1991. *Karayel* launched 20 June 1995.
Structure: Doğan class hull with much improved weapon systems. A second batch is sufficiently different to merit a separate entry.
UPDATED

YILDIZ
10/2003, C D Yaylali / 0567552

2 HISAR (PC 1638) CLASS (LARGE PATROL CRAFT) (PBO)

Name	No	Builders	Commissioned
YARHISAR (ex-*PC 1640*)	P 113	Gunderson, Portland	Sep 1964
AKHISAR (ex-*PC 1641*)	P 114	Gunderson, Portland	Dec 1964

Displacement, tons: 325 standard; 477 full load
Dimensions, feet (metres): 173.7 × 23 × 10.2 *(53 × 7 × 3.1)*
Main machinery: 2 Fairbanks-Morse diesels; 2,800 hp *(2.09 MW)*; 2 shafts
Speed, knots: 19. **Range, n miles:** 6,000 at 10 kt
Complement: 31 (3 officers)
Guns: 2 Bofors 40 mm/60.
Depth charges: 4 projectors; 1 rack (9).
Radars: Surface search: Decca 707; I-band.

Comment: Transferred from the US on build. ASW equipment removed. Three paid off in 2000. One in 2002 and the remainder are expected to follow as the Kiliç class enter service.
UPDATED

YARHISAR
6/2002, C D Yaylali / 0533253

2 TRABZON CLASS (LARGE PATROL CRAFT) (PBO/AGI)

TRABZON (ex-*Gaspe*) P 530 (ex-M 530)   TERME (ex-*Trinity*) P 531 (ex-M 531)

Displacement, tons: 370 standard; 470 full load
Dimensions, feet (metres): 164 × 30.2 × 9.2 *(50 × 9.2 × 2.8)*
Main machinery: 2 GM 12-278A diesels; 2,200 hp *(1.64 MW)*; 2 shafts
Speed, knots: 15. **Range, n miles:** 4,500 at 11 kt
Complement: 35 (4 officers)
Guns: 1 Bofors 40 mm/60. 2—12.7 mm MGs.
Radars: Surface search: Racal Decca 1226; I-band.

Comment: All transferred from Canada and recommissioned 31 March 1958. Built by Davie SB Co 1951-53. Of similar type to British Ton class. Pennant numbers changed in 1991 reflecting use as patrol ships with all minesweeping gear removed.
UPDATED

TRABZON CLASS
6/1999, Selim San / 0080879

7 TURK CLASS (LARGE PATROL CRAFT) (PC)

Name	No	Builders	Commissioned
AB 27	P 127 (ex-P 1227)	Haliç Shipyard	27 June 1969
AB 28	P 128 (ex-P 1228)	Haliç Shipyard	Apr 1969
AB 29	P 129 (ex-P 1229)	Haliç Shipyard	21 Feb 1969
AB 31	P 131 (ex-P 1231)	Haliç Shipyard	17 Nov 1971
AB 33	P 133 (ex-P 1233)	Camialti Shipyard	15 May 1970
AB 35	P 135 (ex-P 1235)	Taşkizak Shipyard	13 Apr 1976
AB 36	P 136 (ex-P 1236)	Taşkizak Shipyard	13 Apr 1976

Displacement, tons: 170 full load
Dimensions, feet (metres): 132 × 21 × 5.5 *(40.2 × 6.4 × 1.7)*
Main machinery: 4 SACM-AGO V16CSHR diesels; 9,600 hp(m) *(7.06 MW)*
2 cruise diesels; 300 hp(m) *(220 kW)*; 2 shafts
Speed, knots: 22
Complement: 31 (3 officers)
Guns: 1 or 2 Bofors 40 mm/70.
1 Oerlikon 20 mm (in those with 1—40 mm). 2—12.7 mm MGs.
A/S mortars: 1 Mk 20 Mousetrap 4 rocket launcher; range 200 m; warhead 50 kg.
Depth charges: 1 rack.
Radars: Surface search: Racal Decca; I-band.
Sonars: Plessey PMS 26; hull-mounted; active search and attack; high frequency.

Comment: Pennant numbers changed in 1991. Similar to *SG 21* Coast Guard class. One to Georgia (AB 30) in December 1998, one to Azerbaijan (AB 34) in July 2000 and one to Kazakhstan (AB 26) in July 2001.
UPDATED

AB 33
5/2004, *Martin Mokrus* / 0589827

4 PGM 71 CLASS (LARGE PATROL CRAFT) (PC)

AB 21-24 (ex-*PGM 104-108*) P 121-124 (ex-P 1221-1224)

Displacement, tons: 130 standard; 147 full load
Dimensions, feet (metres): 101 × 21 × 7 *(30.8 × 6.4 × 2.1)*
Main machinery: 8 GM diesels 2,040 hp *(1.52 MW)*; 2 shafts
Speed, knots: 18.5. **Range, n miles:** 1,500 at 10 kt
Complement: 31 (3 officers)
Guns: 1 Bofors 40 mm/60. 4 Oerlikon 20 mm (2 twin). 1—7.62 mm MG.
A/S mortars: 2 Mk 22 Mousetrap 8 rocket launchers; range 200 m; warhead 50 kg.
Depth charges: 2 racks (4).
Radars: Surface search: Raytheon 1500B; I-band.
Sonars: EDO SQS-17A; hull-mounted; active attack; high frequency.

Comment: Built by Peterson, Sturgeon Bay and commissioned 1967-68. Transferred from US almost immediately after completion. Pennant numbers changed in 1991.

UPDATED

AB 23   *3/2001, Selim San* / 0114810

MINE WARFARE FORCES

Notes: Minelayers: see *Sarucabey, Karamürselbey* and *Osmangazi* under Amphibious Forces.

1 + 5 AYDIN CLASS (TYPE MHV 54-014) (MHSC)

Name	No	Builders	Commissioned
ALANYA	M 265	Abeking & Rasmussen	2004
AMASRA	M 266	Pendik Naval Shipyard, Istanbul	2005
AYVALIK	M 267	Pendik Naval Shipyard, Istanbul	2005
AKÇAKOCA	M 268	Pendik Naval Shipyard, Istanbul	2006
ANAMUR	M 269	Pendik Naval Shipyard, Istanbul	2006
AKÇAY	M 270	Pendik Naval Shipyard, Istanbul	2007

Displacement, tons: 715 full load
Dimensions, feet (metres): 178.8 × 31.8 × 8.5 *(54.5 × 9.7 × 2.6)*
Main machinery: 2 MTU 8V 396 TB84 diesels; 2 Voith-Schneider props; 2 Schottel bow thrusters
Speed, knots: 14
Complement: 53 (6 officers)
Guns: 1 Otobreda 30 mm. 2—12.7 mm MGs.
Countermeasures: MCM: 2 ECA PAP 104 Mk 5. 1 Oropesa mechanical sweep.
Combat data systems: Alenia Marconi Nautis-M.
Radars: Navigation: KH 1007; I-band.
Sonars: Thomson Marconi Type 2093; VDS; active high frequency.

Comment: Ordered from Abeking & Rasmussen and Lürssen on 30 July 1999. First one built in Bremen, remainder in Turkey. The design is based on the German Type 332 but with different propulsion and mine countermeasures equipment. First of class laid down 6 November 2000, second on 25 July 2001, third on 25 July 2002 and fourth on 24 July 2003. All to complete by 2007.

UPDATED

ALANYA   *5/2004*, Frank Findler* / 0589819

ALANYA   *6/2003, Martin Mokrus* / 0567555

2 MINELAYER TENDERS (MLI)

ŞAMANDIRA MOTORU-1 Y 91   ŞAMANDIRA MOTORU-2 Y 92
  (ex-Y 132, ex-Y 1149)   (ex-Y 1150)

Displacement, tons: 72 full load
Dimensions, feet (metres): 64.3 × 18.7 × 5.9 *(19.6 × 5.7 × 1.8)*
Main machinery: 1 Gray Marine 64 HN9 diesel; 225 hp *(168 kW)*; 1 shaft
Speed, knots: 10
Complement: 8

Comment: Acquired in 1959. Used for laying and recovering mine distribution boxes.

UPDATED

ŞAMANDIRA MOTORU-1   *6/2004*, Turkish Navy* / 0589830

5 EDINCIK (CIRCÉ) CLASS (MINEHUNTERS) (MHC)

Name	No	Builders	Commissioned	Recommissioned
EDINCIK (ex-*Cybèle*)	M 260 (ex-M 712)	CMN, Cherbourg	28 Sep 1972	24 July 1998
EDREMIT (ex-*Calliope*)	M 261 (ex-M 713)	CMN, Cherbourg	28 Sep 1972	28 Aug 1998
ENEZ (ex-*Cérès*)	M 262 (ex-M 716)	CMN, Cherbourg	7 Mar 1973	30 Oct 1998
ERDEK (ex-*Circé*)	M 263 (ex-M 715)	CMN, Cherbourg	18 May 1972	4 Dec 1998
ERDEMLI (ex-*Clio*)	M 264 (ex-M 714)	CMN, Cherbourg	18 May 1972	15 Jan 1999

Displacement, tons: 460 standard; 495 normal; 510 full load
Dimensions, feet (metres): 167 × 29.2 × 11.2 *(50.9 × 8.9 × 3.4)*
Main machinery: 1 MTU diesel; 1,800 hp(m) *(1.32 MW)*; 2 active rudders; 1 shaft
Speed, knots: 15. **Range, n miles:** 3,000 at 12 kt
Complement: 48 (5 officers)
Guns: 1 Oerlikon 20 mm.
Countermeasures: MCM: DCN Mintac minehunting system with PAP Plus ROV.
Radars: Navigation: Racal Decca 1229; I-band.
Sonars: Thomson Sintra DUBM 20B; hull-mounted; active search; high frequency.

Comment: Acquired from France on 24 September 1997. Full refits included installation of Mintac system before being handed over.

UPDATED

ENEZ   *1/2004*, Giorgio Ghiglione* / 0589820

6 VEGESACK CLASS (MSC/AGS/PBO)

Name	No	Builders	Commissioned
KARAMÜRSEL (ex-*Worms* M 1253)	M 520	Amiot, Cherbourg	30 Apr 1960
KEREMPE (ex-*Detmold* M 1252)	M 521	Amiot, Cherbourg	20 Feb 1960
KILIMLI (ex-*Siegen* M 1254)	M 522	Amiot, Cherbourg	9 July 1960
KOZLU (ex-*Hameln* M 1251)	P 301 (ex-M 523)	Amiot, Cherbourg	15 Oct 1959
KUŞADASI (ex-*Vegesack* M 1250)	P 302 (ex-M 524)	Amiot, Cherbourg	19 Sep 1959
KEMER (ex-*Passau* M 1255)	A 582 (ex-M 525)	Amiot, Cherbourg	15 Oct 1960

Displacement, tons: 362 standard; 378 full load
Dimensions, feet (metres): 155.1 × 28.2 × 9.5 *(47.3 × 8.6 × 2.9)*
Main machinery: 2 MTU MB diesels; 1,500 hp(m) *(1.1 MW)*; 2 shafts; cp props
Speed, knots: 15
Complement: 33 (2 officers)
Guns: 2 Oerlikon 20 mm (twin).
Radars: Navigation: Decca 707; I-band.
Sonars: Simrad; active mine detection; high frequency.

Comment: Transferred by West Germany and recommissioned in the mid-1970s. Sonars were fitted from 1989. *Kemer* paid off in 1998 but returned as a survey ship in 1999. *Kozlu* and *Kuşadasi* similarly refitted as patrol ships in 1999.

UPDATED

KILIMLI   *4/1996, van Ginderen Collection* / 0080884

5 MSC 289 CLASS (MINESWEEPERS—COASTAL) (MSC)

SILIFKE (ex-*MSC 304*) M 514   **SAPANCA** (ex-*MSC 312*) M 517
SAROS (ex-*MSC 305*) M 515   **SARIYER** (ex-*MSC 315*) M 518
SIGACIK (ex-*MSC 311*) M 516

Displacement, tons: 320 standard; 370 full load
Dimensions, feet (metres): 141 × 26 × 8.3 *(43 × 8 × 2.6)*
Main machinery: 4 GM 6-71 diesels; 696 hp *(519 kW)* sustained; 2 shafts (M 510-M 513)
     2 Waukesha L 1616 diesels; 1,200 hp *(895 kW)*; 2 shafts (M 514-M 518)
Speed, knots: 14. **Range, n miles:** 2,500 at 10 kt
Complement: 35 (2 officers)
Guns: 2 Oerlikon 20 mm (twin).
Radars: Navigation: Racal Decca 1226; I-band.
Sonars: UQS-1D; hull-mounted mine search; high frequency.

Comment: Built 1965-67. Transferred from US. Commissioning dates in the Turkish Navy were
     respectively: 21 March 1966, 25 October 1966, 20 December 1965, 20 December 1965 and
     7 December 1967.
UPDATED

SARIYER   *10/2003, C D Yaylali /* 0567557

SAPANCA   *10/2003, C D Yaylali /* 0567556

4 COVE CLASS (MINESWEEPERS—INSHORE) (MSI)

Name	No	Builders	Commissioned
FOÇA (ex-*MSI 15*)	M 500	Peterson, WI	19 Apr 1968
FETHIYE (ex-*MSI 16*)	M 501	Peterson, WI	24 Apr 1968
FATSA (ex-*MSI 17*)	M 502	Peterson, WI	21 Mar 1968
FINIKE (ex-*MSI 18*)	M 503	Peterson, WI	26 Apr 1968

Displacement, tons: 180 standard; 235 full load
Dimensions, feet (metres): 111.9 × 23.5 × 7.9 *(34 × 7.1 × 2.4)*
Main machinery: 4 GM 6-71 diesels; 696 hp *(520 kW)* sustained; 2 shafts
Speed, knots: 13. **Range, n miles:** 900 at 11 kt
Complement: 25 (3 officers)
Guns: 1—12.7 mm MG.
Radars: Navigation: I-band.

Comment: Built in US and transferred under MAP at Boston, Massachusetts, August-December
     1967.
UPDATED

FINIKE   *5/2001, Selim San /* 0114808

6 MINEHUNTING TENDERS (YAG/YDT)

MTB 3 P 313   **MTB 6** P 316   **MTB 8** P 318
MTB 4 P 314   **MTB 7** P 317   **MTB 9** P 319

Displacement, tons: 70 standard
Dimensions, feet (metres): 71.5 × 13.8 × 8.5 *(21.8 × 4.2 × 2.6)*
Main machinery: 2 diesels; 2,000 hp(m) *(1.47 MW)*; 2 shafts
Speed, knots: 20
Guns: 1 Oerlikon 20 mm or 1—12.7 mm MG (aft) (in some).

Comment: All launched in 1942. Now employed as minehunting base ships.
UPDATED

MTB 6   *7/1995, van Ginderen Collection /* 0080886

AMPHIBIOUS FORCES

Notes: (1) The prefix 'Ç' for smaller amphibious vessels stands for 'Çikartma Gemisi' (landing
vessel) and indicates that the craft are earmarked for national rather than NATO control.
(2) Subject to funding, there are plans to procure a landing platform dock capable of both military
and humanitarian assistance operations. It is envisaged that the ship will be capable of carrying a
battalion and of operating helicopters.

1 OSMAN GAZI CLASS (LSTH/ML)

Name	No	Builders	Launched	Commissioned
OSMAN GAZI	NL 125	Taşkizak Yard, Istanbul	20 July 1990	27 July 1994

Displacement, tons: 3,773 full load
Dimensions, feet (metres): 344.5 × 52.8 × 15.7 *(105 × 16.1 × 4.8)*
Main machinery: 2 MTU 12V 1163 TB73 diesels; 8,800 hp(m) *(6.47 MW)*; 2 shafts
Speed, knots: 17. **Range, n miles:** 4,000 at 15 kt
Military lift: 900 troops; 15 tanks; 4 LCVPs
Guns: 2 Oerlikon 35 mm/90 (twin). 4 Bofors 40 mm/70 (2 twin). 2 Oerlikon 20 mm.
Radars: Navigation: Racal Decca; I-band.
Helicopters: Platform for 1 large.

Comment: Laid down 7 July 1989. Full NBCD protection. Equipped with a support weapons co-
     ordination centre to control amphibious operations. The ship has about a 50 per cent increase in
     military lift capacity compared with the Sarucabey class. Secondary role as minelayer. Second
     of class cancelled in 1991 and *Osman Gazi* took a long time to complete. Marisat fitted.
UPDATED

OSMAN GAZI   *5/1998, Diego Quevedo /* 0050291

OSMAN GAZI   *5/1998, Diego Quevedo /* 0050292

For details of the latest updates to *Jane's Fighting Ships* online and to discover the
additional information available exclusively to online subscribers please visit
jfs.janes.com

2 ERTUĞRUL (TERREBONNE PARISH) CLASS (LST)

Name	No	Builders	Commissioned
ERTUĞRUL (ex-*Windham County* LST 1170)	L 401	Christy Corporation	15 Dec 1954
SERDAR (ex-*Westchester County* LST 1167)	L 402	Christy Corporation	10 Mar 1954

Displacement, tons: 2,590 light; 5,800 full load
Dimensions, feet (metres): 384 × 55 × 17 *(117.1 × 16.8 × 5.2)*
Main machinery: 4 GM 16-278A diesels; 6,000 hp *(4.48 MW)*; 2 shafts; cp props
Speed, knots: 15
Complement: 163 (14 officers)
Military lift: 395 troops; 2,200 tons cargo; 4 LCVPs
Guns: 6 USN 3 in *(76 mm)*/50 (3 twin).
Weapons control: 2 Mk 63 GFCS.
Radars: Surface search: Racal Decca 1226; I-band.
Fire control: 2 Western Electric Mk 34; I/J-band.

Comment: Transferred by US and recommissioned 3 October 1973 and 24 February 1975 respectively. Purchased outright 6 August 1987. Marisat fitted.

UPDATED

ERTUĞRUL   *9/1998, Selim San /* 0050293

2 SARUCABEY CLASS (LSTH/ML)

Name	No	Builders	Launched	Commissioned
SARUCABEY	NL 123	Taşkizak Naval Yard	30 July 1981	17 July 1984
KARAMÜRSELBEY	NL 124	Taşkizak Naval Yard	26 July 1984	19 June 1987

Displacement, tons: 2,600 full load
Dimensions, feet (metres): 301.8 × 45.9 × 7.5 *(92 × 14 × 2.3)*
Main machinery: 3 diesels; 4,320 hp *(3.2 MW)*; 3 shafts
Speed, knots: 14
Military lift: 600 troops; 11 tanks; 12 jeeps; 2 LCVPs
Guns: 3 Bofors 40 mm/70. 4 Oerlikon 20 mm (2 twin).
Mines: 150 in lieu of amphibious lift.
Radars: Navigation: Racal Decca 1226; I-band.
Helicopters: Platform only.

Comment: *Sarucabey* is an enlarged Çakabey design more suitable for naval requirements. Dual-purpose minelayers. NL 124 has superstructure one deck lower.

UPDATED

SARUCABEY   *7/2000, Michael Nitz /* 0106640

1 EDIC TYPE (LCT)

120

Displacement, tons: 580 full load
Dimensions, feet (metres): 186.9 × 39.4 × 4.6 *(57 × 12 × 1.4)*
Main machinery: 3 GM 6-71 diesels; 522 hp *(390 kW)* sustained; 3 shafts
Speed, knots: 8.5. **Range, n miles:** 600 at 10 kt
Complement: 15
Military lift: 100 troops; up to 5 tanks
Guns: 2 Oerlikon 20 mm. 2—12.7 mm MGs.
Radars: Navigation: Racal Decca; I-band.

Comment: Vessel built at Gölcük Naval Shipyard in 1973. French EDIC type.

UPDATED

EDIC TYPE   *2/1996, C D Yaylali /* 0080888

24 LCT

Ç 123, Ç 125-129, Ç 132-135, Ç 137-150

Displacement, tons: 600 standard
Dimensions, feet (metres): 195.5 × 38 × 10.5 *(59.6 × 11.6 × 3.2)*
Main machinery: 3 GM 6-71 diesels; 522 hp *(390 kW)* sustained; 3 shafts *(119-138)* or 3 MTU diesels; 900 hp(m) *(662 kW)*; 3 shafts *(139-150)*
Speed, knots: 8.5. **Range, n miles:** 600 at 8 kt
Complement: 17 (1 officer)
Military lift: 100 troops; 5 tanks
Guns: 2 Oerlikon 20 mm. 2—12.7 mm MGs.
Radars: Navigation: Racal Decca; I-band.

Comment: Follow-on to the Ç 107 type started building in 1977. Ç 130 and Ç 131 transferred to Libya January 1980 and Ç 136 sunk in 1985. The delivery rate was about two per year from the Taşkizak and Gölcük yards until 1987. Then two launched in July 1987 and commissioned in 1991. Last three completed in 1992. Dimensions given are for Ç 139 onwards, earlier craft are 3.6 m shorter and have less freeboard.

VERIFIED

Ç 129   *8/1995, Selçuk Emre /* 0080889

17 LCM 8 TYPE

Ç 305, 308, 312-314, 316, 319, 321-327, 329-331

Displacement, tons: 58 light; 113 full load
Dimensions, feet (metres): 72 × 20.5 × 4.8 *(22 × 6.3 × 1.4)*
Main machinery: 4 GM 6-71 diesels; 696 hp *(520 kW)* sustained; 2 shafts
Speed, knots: 9.5
Complement: 9
Military lift: 60 tons or 140 troops
Guns: 2—12.7 mm MGs.

Comment: Up to Ç 319 built by Taşkizak and Haliç in 1965-66. Ç 321-331 built by Taşkizak and Naldöken in 1987-89.

UPDATED

Ç 308   *6/2003, Turkish Navy /* 0567540

2 FAST INTERVENTION CRAFT (HSIC)

Displacement, tons: 18 full load
Dimensions, feet (metres): 55.1 × 13.2 × 3.2 *(16.8 × 4.04 × 1.0)*
Main machinery: 2 MTU 12V 183 TE94 diesels; 2,588 hp(m) *(1.93 MW)*; 2 Arneson ASD 12 B1L surface drives
Speed, knots: 50+
Complement: 2 plus 10 mission crew

Comment: Capable of carrying a RIB and used by Special Forces. Details are speculative.

VERIFIED

FAST INTERVENTION CRAFT   *10/2000 /* 0106641

SURVEY SHIPS

Notes: *Kemer* A 582 (ex-M 525) is listed under Vegesack class in Mine Warfare Forces.

2 SILAS BENT CLASS (AGS)

Name	No	Builders	Commissioned
ÇEŞME (ex-*Silas Bent*)	A 599 (ex-TAGS 26)	American SB Co, Lorain	23 July 1965
ÇANDARLI (ex-*Kane*)	A 588 (ex-TAGS 27)	Christy Corp, Sturgeon Bay	19 May 1967

Displacement, tons: 2,843 full load
Dimensions, feet (metres): 285.3 × 48 × 15.1 *(87 × 14.6 × 4.6)*
Main machinery: Diesel-electric; 2 Alco diesel generators; 1 Westinghouse/GE motor; 3,600 hp *(2.69 MW)*; 1 shaft; cp prop; bow thruster 350 hp *(261 kW)*
Speed, knots: 15. **Range, n miles:** 12,000 at 14 kt
Complement: 31 plus 28 spare
Radars: Navigation: RM 1650/9X; I-band.

Comment: *Çeşme* transferred from US on 28 October 1999 and *Canadarli* on 14 March 2001.
UPDATED

ÇEŞME   *6/2003, Selçuk Emre /* 0567558

1 SURVEY SHIP (AGS)

Name	No	Builders	Launched	Commissioned
ÇUBUKLU (ex-*Y 1251*)	A 594	Gölcük	17 Nov 1983	24 June 1987

Displacement, tons: 680 full load
Dimensions, feet (metres): 132.8 × 31.5 × 10.5 *(40.5 × 9.6 × 3.2)*
Main machinery: 1 MWM diesel; 820 hp(m) *(603 kW)*; 1 shaft; cp prop
Speed, knots: 11
Complement: 37 (6 officers)
Guns: 2 Oerlikon 20 mm.
Radars: Navigation: Racal Decca; I-band.

Comment: Qubit advanced integrated navigation and data processing system fitted in 1991.
VERIFIED

ÇUBUKLU   *6/1997, Turkish Navy /* 0050296

2 SURVEY CRAFT (AGSC)

MESAHA 1 Y 35   **MESAHA 2** Y 36

Displacement, tons: 38 full load
Dimensions, feet (metres): 52.2 × 14.8 × 4.3 *(15.9 × 4.5 × 1.3)*
Main machinery: 2 GM 6-71 diesels; 348 hp *(260 kW)* sustained; 2 shafts
Speed, knots: 10. **Range, n miles:** 600 at 10 kt
Complement: 9

Comment: Completed in 1994 and took the names and pennant numbers of their deleted predecessors.
VERIFIED

MESAHA 2   *9/1994, C D Yaylali /* 0080890

TRAINING SHIPS

Notes: Adjutant class minesweeper *Samsun* A 1539 (ex-M 510) was converted to a training role in 2004.

2 RHEIN CLASS (AG/AX)

Name	No	Builders	Commissioned
CEZAYIRLI GAZI HASAN PAŞA (ex-*Elbe*)	A 579	Schliekerwerft, Hamburg	17 Apr 1962
SOKULLU MEHMET PAŞA (ex-*Donau*)	A 577	Schlichting, Travemünde	23 May 1964

Displacement, tons: 2,370 standard; 2,940 full load
Dimensions, feet (metres): 322.1 × 38.8 × 14.4 *(98.2 × 11.8 × 4.4)*
Main machinery: Diesel-electric; 6 MTU MD diesels; 14,400 hp(m) *(10.58 MW)*; 2 Siemens motors; 11,400 hp(m) *(8.38 MW)*; 2 shafts
Speed, knots: 20.5. **Range, n miles:** 1,625 at 15 kt
Complement: 188 (15 officers)
Guns: 2 Creusot-Loire 3.9 in *(100 mm)*/55. 4 Bofors 40 mm/60.
Radars: Surface search: Signaal DA02; E/F-band.
Fire control: 2 Signaal M 45; I/J-band.

Comment: *Elbe* transferred from Germany on 15 March 1993, taking over the same name and pennant number as the former *Ruhr*. *Donau* transferred 13 March 1995 taking the same name and pennant number as the deleted *Isar*.
VERIFIED

SOKOLLU MEHMET PAŞA   *3/2002, Schaeffer/Marsan /* 0533256

8 TRAINING CRAFT (AXL)

Name	No	Builders	Commissioned
E 1	A 1531	Bora-Duzgit	22 July 1999
E 2	A 1532	Bora-Duzgit	22 July 1999
E 3	A 1533	Bora-Duzgit	8 June 2000
E 4	A 1534	Bora-Duzgit	8 June 2000
E 5	A 1535	Bora-Duzgit	8 June 2000
E 6	A 1536	Bora-Duzgit	8 June 2000
E 7	A 1537	Bora-Duzgit	8 June 2000
E 8	A 1538	Bora-Duzgit	8 June 2000

Displacement, tons: 94 full load
Dimensions, feet (metres): 94.5 × 19.7 × 6.2 *(28.8 × 6 × 1.9)*
Main machinery: 1 MTU diesel; 1 shaft
Speed, knots: 12. **Range, n miles:** 240 at 12 kt
Complement: 15

Comment: Naval Academy training craft ordered in 1998.
VERIFIED

E 1   *7/1999, Selçuk Emre /* 0080892

AUXILIARIES

1 SUPPORT TANKER (AOTL)

Name	No	Builders	Launched	Commissioned
TAŞKIZAK	A 570	Taşkizak Naval DY, Istanbul	28 July 1983	14 Aug 1985

Displacement, tons: 1,440 full load
Dimensions, feet (metres): 211.9 × 30.8 × 11.5 *(64.6 × 9.4 × 3.5)*
Main machinery: 1 diesel; 1,400 hp(m) *(1.03 MW)*; 1 shaft
Speed, knots: 13
Complement: 57
Cargo capacity: 800 tons
Guns: 1 Bofors 40 mm/60. 2 Oerlikon 20 mm.
Radars: Navigation: Racal Decca 1226; I-band.

Comment: Laid down 20 July 1983.
VERIFIED

TAŞKIZAK (Doğan class in background)   *5/1990, A Sheldon Duplaix /* 0080893

1 TRANSPORT SHIP (AK)

Name	No	Builders	Commissioned	Recommissioned
ISKENDERUN	A 1600	Camialti Shipyard, Istanbul	1991	25 July 2002

Displacement, tons: 10,583 gross; 3,872 net
Dimensions, feet (metres): 418.4 × 64.0 × 17.7
(127.5 × 19.5 × 5.4)
Main machinery: 4 Skoda and Sulzer diesels; 16,800 hp
(12.52 MW); 2 shafts
Speed, knots: 15.5
Complement: 129 (11 officers)

Comment: Car ferry (214 cars and passengers) built to Polish
design. Commissioned in Turkish Navy on 25 July 2002.
UPDATED

ISKENDERUN
3/2004, Selim San / 0587559

2 FLEET SUPPORT SHIPS (AORH)

Name	No	Builders	Laid down	Launched	Commissioned
AKAR	A 580	Gölcük Naval Dockyard	5 Aug 1982	17 Nov 1983	9 Sep 1987
YARBAY KUDRET GÜNGÖR	A 595	Sedef Shipyard, Istanbul	5 Nov 1993	15 Nov 1994	24 Oct 1995

Displacement, tons: 19,350 full load
Dimensions, feet (metres): 475.9 × 74.8 × 27.6
(145.1 × 22.8 × 8.4)
Main machinery: 1 diesel; 6,500 hp(m) *(4.78 MW)*; 1 shaft
Speed, knots: 16. **Range, n miles:** 6,000 at 14 kt
Complement: 203 (14 officers)
Cargo capacity: 16,000 tons oil fuel (A 580); 9,980 tons oil fuel
(A 595); 2,700 tons water (A 595); 80 tons hub oil (A 595);
500 m³ stores (A 595)
Guns: 2—3 in *(76 mm)*/50 (twin) Mk 34 (A 580). 1—20 mm/76
Mk 15 Vulcan Phalanx (A 595). 2 Bofors 40 mm/70.
Weapons control: Mk 63 GFCS (A 580).
Radars: Fire control: SPG-34; I-band (A 580).
Navigation: Racal Decca 1226; I-band.
Helicopters: Platform for 1 medium.

Comment: Helicopter flight deck aft. *Akar* is primarily a tanker
whereas the second ship of the same type is classified as
logistic support vessel. *Güngör* was the first naval ship to be
built at a civilian yard in Turkey.
UPDATED

AKAR

6/2003, Selçuk Emre / 0589821

2 SUPPORT TANKERS (AOT)

Name	No	Builders	Commissioned
ALBAY HAKKI BURAK	A 571	RMK Tuzla, Istanbul	21 Nov 1999
YUZBASI IHSAN TOLUNAY	A 572	RMK Tuzla, Istanbul	8 June 2000

Displacement, tons: 3,300 full load
Dimensions, feet (metres): 267.1 × 40 × 16.4 *(81.4 × 12.2 × 5)*
Main machinery: 2 Caterpillar 3606TA diesels; 5,522 hp(m) *(4.06 MW)*; 2 shafts
Speed, knots: 15
Complement: 50
Cargo capacity: 2,355 m³ dieso

Comment: Ordered in 1998.

VERIFIED

ALBAY HAKKI BURAK
6/2002, Selçuk Emre / 0533258

1 SUPPORT TANKER (AORL)

Name	No	Builders	Commissioned
BINBAŞI SADETTIN GÜRCAN	A 573	Taşkizak Naval DY, Istanbul	4 Sep 1970

Displacement, tons: 1,505 standard; 4,460 full load
Dimensions, feet (metres): 294.2 × 38.7 × 17.7 *(89.7 × 11.8 × 5.4)*
Main machinery: Diesel-electric; 4 GM 16-567A diesels; 5,600 hp *(4.12 MW)*; 4 generators;
2 motors; 4,400 hp *(3.28 MW)*; 2 shafts
Speed, knots: 16
Complement: 63
Guns: 2 Oerlikon 20 mm.
Radars: Navigation: E/F-band.

Comment: Main armament removed. Can be used for replenishment at sea.
VERIFIED

BINBAŞI SADETTIN GÜRCAN
6/1995, Turkish Navy / 0080895

3 WATER TANKERS (AWT)

SÖGÜT (ex-*FW 2*) A 598 (ex-Y 1217)   KAVAK (ex-*FW 4*) A 600   ÇINAR (ex-*FW 1*) A 581

Displacement, tons: 626 full load
Dimensions, feet (metres): 144.4 × 25.6 × 8.2 *(44.1 × 7.8 × 2.5)*
Main machinery: 1 MWM diesel; 230 hp(m) *(169 kW)*; 1 shaft
Speed, knots: 9.5. **Range, n miles:** 2,150 at 9 kt
Complement: 12
Cargo capacity: 340 tons

Comment: *Sögüt* acquired from West Germany and commissioned 12 March 1976. Pennant
number changed in 1991. *Kavak* transferred from Germany 12 April 1991 and *Çinar* in early
1996.
VERIFIED

KAVAK
4/1997, C D Yaylali / 0019332

2 WATER TANKERS (AWT)

Name	No	Builders	Commissioned
VAN	A 597 (ex-Y 1208)	Camialti Shipyard	12 Aug 1968
ULUBAT	A 596 (ex-Y 1209)	Camialti Shipyard	3 July 1969

Displacement, tons: 1,250 full load
Dimensions, feet (metres): 174.2 × 29.5 × 9.8 *(53.1 × 9 × 3)*
Main machinery: 1 diesel; 650 hp(m) *(478 kW)*; 1 shaft
Speed, knots: 14
Complement: 39 (3 officers)
Cargo capacity: 700 tons
Guns: 1 Oerlikon 20 mm.
Radars: Racal Decca 707; I-band.

Comment: Pennant numbers changed in 1991.

VERIFIED

VAN
6/2003, Turkish Navy / 0567542

6 WATER TANKERS (YW)

PINAR 1-6 Y 111-116 (ex-Y 1211-1216)

Displacement, tons: 300 full load
Dimensions, feet (metres): 110.2 × 27.9 × 5.9 *(33.6 × 8.5 × 1.8)*
Main machinery: 1 GM diesel; 225 hp *(168 kW)*; 1 shaft
Speed, knots: 11
Complement: 12
Cargo capacity: 150 tons

Comment: Built by Taşkizak Naval Yard. Details given for last four, sisters to harbour tankers H 500-502. First pair differ from these particulars and are individually different. *Pinar 1* (launched 1938) of 490 tons displacement with one 240 hp *(179 kW)* diesel, and *Pinar 2* built in 1958 of 1,300 tons full load, 167.3 × 27.9 ft *(51 × 8.5 m)*.

UPDATED

PINAR 6 *12/2000, Selim San /* 0106643

3 HARBOUR TANKERS (YW)

H 500-502 Y 140-142 (ex-Y 1231-1233)

Displacement, tons: 300 full load
Dimensions, feet (metres): 110.2 × 27.9 × 5.9 *(33.6 × 8.5 × 1.8)*
Main machinery: 1 GM diesel; 225 hp(m) *(165 kW)*; 1 shaft
Speed, knots: 11
Complement: 12
Cargo capacity: 150 tons

Comment: Sisters of water tankers of Pinar series. Built at Taşkizak in early 1970s.

VERIFIED

H 500 *6/1994, Turkish Navy /* 0080901

1 HARBOUR TANKER (YO)

GÖLCÜK Y 50

Displacement, tons: 310 full load
Dimensions, feet (metres): 108.8 × 19.2 × 9.2 *(33.2 × 5.8 × 2.8)*
Main machinery: 1 diesel; 550 hp(m) *(404 kW)*; 1 shaft
Speed, knots: 12
Complement: 12

VERIFIED

GÖLCÜK *7/1992, Selçuk Emre /* 0080898

1 DIVER CLASS (SALVAGE SHIP) (ARS)

Name	No	Builders	Launched	Commissioned
IŞIN (ex-*Safeguard* ARS 25)	A 589	Basalt Rock, Napa	20 Nov 1943	31 Oct 1944

Displacement, tons: 1,530 standard; 1,970 full load
Dimensions, feet (metres): 213.5 × 41 × 13 *(65.1 × 12.5 × 4)*
Main machinery: Diesel-electric; 4 Cooper-Bessemer GSB-8 diesels; 3,420 hp *(2.55 MW)*; 4 generators; 2 motors; 2 shafts
Speed, knots: 14.8
Complement: 110
Guns: 2 Oerlikon 20 mm.

Comment: Transferred from US 28 September 1979 and purchased outright 6 August 1987.

UPDATED

IŞIN *6/2004*, C D Yaylali /* 0589822

1 CHANTICLEER CLASS (SUBMARINE RESCUE SHIP) (ASR)

Name	No	Builders	Launched	Commissioned
AKIN (ex-*Greenlet* ASR 10)	A 585	Moore SB & DD Co	12 July 1942	29 May 1943

Displacement, tons: 1,653 standard; 2,321 full load
Dimensions, feet (metres): 251.5 × 44 × 16 *(76.7 × 13.4 × 4.9)*
Main machinery: Diesel-electric; 4 Alco 539 diesels; 3,532 hp *(2.63 MW)*; 4 generators; 1 motor; 1 shaft
Speed, knots: 15
Complement: 111 (9 officers)
Guns: 1 Bofors 40 mm/60. 4 Oerlikon 20 mm (2 twin).
Radars: Navigation: Racal Decca 1226; I-band.

Comment: Transferred from US, recommissioned 23 December 1970 and purchased 15 February 1973. Carries a Diving Bell.

VERIFIED

AKIN *4/1996, van Ginderen Collection /* 0080902

2 TRANSPORTS (AKS/AWT)

Name	No	Builders	Commissioned
KARADENIZ EREĞLISI	A 592 (ex-Y 1157)	Erdem	30 Aug 1982
ECEABAT	A 593 (ex-Y 1165)	Taşkazik	4 Nov 1968

Displacement, tons: 820 full load
Dimensions, feet (metres): 166.3 × 26.2 × 9.2 *(50.7 × 8 × 2.8)*
Main machinery: 1 diesel; 1,440 hp *(1.06 MW)*; 1 shaft
Speed, knots: 10
Complement: 23 (3 officers)
Cargo capacity: 300 tons
Guns: 1 Oerlikon 20 mm.

Comment: Funnel-aft coaster type. Pennant numbers changed in 1991. *Karadeniz Ereğlisi* is a stores ship, *Eceabat* is a water carrier.

VERIFIED

ECEABAT *11/1999, Selim San /* 0084417

2 BARRACK SHIPS (YPB)

YÜZBAŞI NAŞIT ÖNGÖREN (ex-US *APL 47*) Y 38 (ex-Y 1204)
BINBAŞI METIN SÜLÜŞ (ex-US *APL 53*) Y 39 (ex-Y 1205)

Comment: Ex-US barrack ships transferred on lease: Y 1204 in October 1972 and Y 1205 on 6 December 1974. Y 1204 based at Ereğli and Y 1205 at Gölcük. Purchased outright June 1987. Pennant numbers changed in 1991.

UPDATED

15 SMALL TRANSPORTS (YFB/YE)

ŞALOPA 11-12, 15, 18, 22-24, 27, 30-35
LAYTER 7 Y 107
PONTON 7 Y 137
YAKIT Y 139

Comment: Of varying size and appearance. Pennant numbers changed in 1991.

UPDATED

1 BOOM DEFENCE VESSEL (ABU)

Name	No	Builders	Commissioned
AG 6 (ex-*AN 93*, ex-Netherlands *Cerberus* A 895)	P 306	Bethlehem Steel Corporation, Staten Island, NY	10 Nov 1952

Displacement, tons: 780 standard; 855 full load
Dimensions, feet (metres): 165 × 33 × 10 *(50.3 × 10.1 × 3)*
Main machinery: Diesel-electric; 2 GM 8-268A diesels; 880 hp *(656 kW)*; 2 generators; 1 motor; 1 shaft
Speed, knots: 12.8. **Range, n miles:** 5,200 at 12 kt
Complement: 32 (3 officers)
Guns: 1 USN 3 in *(76 mm)*/50. 4 Oerlikon 20 mm.
Radars: Navigation: Racal Decca 1226; I-band.

Comment: Netlayer. Transferred from US to Netherlands in December 1952. Used first as a boom defence vessel and latterly as salvage and diving tender since 1961 but retained her netlaying capacity. Handed back to US Navy on 17 September 1970 but immediately turned over to the Turkish Navy under grant aid.

VERIFIED

AG 6   *7/1995, Frank Behling /* 0080906

1 BOOM DEFENCE VESSEL (ABU)

Name	No	Builders	Commissioned
AG 5 (ex-*AN 104*)	P 305	Kröger, Rendsburg	25 Feb 1962

Displacement, tons: 960 full load
Dimensions, feet (metres): 173.8 × 35 × 13.5 *(53 × 10.7 × 4.1)*
Main machinery: Diesel-electric; 1 MAN G7V40/60 diesel generator; 1 motor; 1,470 hp(m) *(1.08 MW)*; 1 shaft
Speed, knots: 12. **Range, n miles:** 6,500 at 11 kt
Complement: 32 (3 officers)
Guns: 1 Bofors 40 mm/60. 3 Oerlikon 20 mm.
Radars: Navigation: Racal Decca 1226; I-band.

Comment: Netlayer P 305 built in US offshore programme for Turkey.

UPDATED

AG 5   *6/2003*, Selçuk Emre /* 0589823

3 TORPEDO RETRIEVERS (YPT)

TORPIDO TENDERI Y 95 (ex-Y 1051)   **TAKIP 1** Y 98 (ex-Y 1052)   **TAKIP 2** Y 99

Comment: Of different types.

UPDATED

Y 44   *9/1998, C D Yaylali /* 0050297

2 OFFICERS' YACHTS (YAC)

GÜL   **NEVCIHAN**

Comment: Pennant numbers not displayed.

VERIFIED

NEVCIHAN   *10/2003, C D Yaylali /* 0567559

GÜL   *6/2003, Turkish Navy /* 0567541

15 FLOATING DOCKS/CRANES

Name	Lift	Name	Lift
LEVENT Y 59 (ex-Y 1022)	—	**HAVUZ 5** Y 125 (ex-Y 1085)	400 tons
ALGARNA 1 Y 58		**HAVUZ 6** Y 126 (ex-Y 1086)	3,000 tons
ALGARNA 3 Y 60 (ex-Y 1021)	—	**HAVUZ 8** Y 128 (ex-Y-1088)	700 tons
		HAVUZ 9 Y 129 (ex-Y-1089)	4,500 tons
HAVUZ 1 Y 121 (ex-Y 1081)	16,000 tons	**HAVUZ 10** Y 130 (ex-Y-1090)	3,500 tons
HAVUZ 2 Y 122 (ex-Y 1082)	12,000 tons	**HAVUZ 11** Y 134	14,500 tons
HAVUZ 3 Y 123 (ex-Y 1083) (ex-US AFDL)	2,500 tons	**HAVUZ 12** Y 135	5,000 tons
HAVUZ 4 Y 124 (ex-Y 1084)	4,500 tons	**HAVUZ 13** Y 136	7,500 tons

Comment: *Algarna* and *Levent* are ex-US floating cranes.

UPDATED

TUGS

1 CHEROKEE CLASS (ATF)

GAZAL (ex-*Sioux* ATF 75) A 587

Displacement, tons: 1,235 standard; 1,675 full load
Dimensions, feet (metres): 205 × 38.5 × 17 *(62.5 × 11.7 × 5.2)*
Main machinery: Diesel-electric; 4 GM 12-278 diesels; 4,400 hp *(3.28 MW)*; 4 generators; 1 motor; 3,000 hp *(2.24 MW)*; 1 shaft
Speed, knots: 16. **Range, n miles:** 15,000 at 8 kt
Complement: 85
Guns: 1 USN 3 in *(76 mm)*/50. 2 Oerlikon 20 mm.
Radars: Navigation: Racal Decca; I-band.

Comment: Originally completed on 6 December 1942. Transferred from US and commissioned 9 March 1973. Purchased 15 August 1973. Can be used for salvage. 3 in gun removed in 1987 but has since been restored.

VERIFIED

GAZAL   *6/1995, Turkish Navy /* 0080908

1 TENACE CLASS (ATA)

Name	No	Builders	Commissioned
DEĞIRMENDERE (ex-Centaure)	A 576 (ex-A 674)	Chantiers de la Rochelle	14 May 1974

Displacement, tons: 1,454 full load
Dimensions, feet (metres): 167.3 × 37.8 × 18.6 *(51 × 11.5 × 5.7)*
Main machinery: 2 SACM AGO 240 V12 diesels; 4,600 hp(m) *(3.38 MW)*; 1 shaft; Kort nozzle
Speed, knots: 13. **Range, n miles:** 9,500 at 13 kt
Complement: 37 (3 officers)
Radars: Navigation: Racal Decca RM 1226 and Racal Decca 060; I-band.

Comment: Transferred from French Navy 16 March 1999. Recommissioned after refit 22 July 1999. Bollard pull 60 tons.

VERIFIED

DEĞIRMENDERE   *1/2002, M Declerck /* 0533259

16 COASTAL/HARBOUR TUGS (YTB/YTM/YTL)

Name	No	Displacement, tons/ Speed, knots	Commissioned
AKBAŞ	A 586	1660/14	1978
SÖNDÜREN 2-4	A 1542-1544	385/12	1999/2000
SÖNDÜREN 1	Y 51 (ex-Y 1117)	128/12	1954
KUVVET	Y 53 (ex-Y 1122)	390/10	1962
DOĞANARSLAN	Y 52 (ex-Y 1123)	500/12	1985
ATIL	Y 55 (ex-Y 1132)	300/10	1962
PENDIK	Y 56	238/10	2000
ERSEV BAYRAK	Y 64 (ex-Y 1134)	30/9	1946
AKSAZ (ex-Koos)	(ex-Y 1651, ex-A 08)	320/11	1962
ÖNDER	Y 160	230/12	1998
ÖNCÜ	Y 161	230/12	1998
ÖZGEN	Y 162	230/12	1999
ÖDEV	Y 163	230/12	1999
ÖZGÜR	Y 164	230/12	2000

Comment: In addition there are 47 Katir pusher berthing tugs. *Koos* was transferred from Germany on 7 October 1996.

UPDATED

ÖZGEN   *1/2002, M Declerck /* 0533260

SÖNDÜREN 4   *6/2004*, C D Yaylali /* 0589824

1 OCEAN TUG (ATR)

DARICA A 578 (ex-Y 1125)

Displacement, tons: 750 full load
Dimensions, feet (metres): 134.2 × 32.2 × 12.8 *(40.9 × 9.8 × 3.9)*
Main machinery: 2 ABC diesels; 4,000 hp *(2.94 MW)*; 2 shafts
Speed, knots: 14. **Range, n miles:** 2,500 at 14 kt

Comment: Built at Taşkizak Naval Yard and commissioned 13 June 1991. Equipped for firefighting and as a torpedo tender. Pennant number changed in 1991.

VERIFIED

DARICA   *11/1994, van Ginderen Collection /* 0080910

COAST GUARD (SAHIL GÜVENLIK)

Notes: (1) Patrol craft based in North Cyprus include KKTCSG 101 *(Raif Denktas)*, KKTCSG 02, 2 Kaan 15 class, KKTCSG 11 and 12 and a converted cabin cruiser KKTCSG 104.
(2) Four Vigilante class Boston Whalers were acquired by the Police in September 1999.

12 LARGE PATROL CRAFT (WPB)

SG 80-91

Displacement, tons: 195 full load
Dimensions, feet (metres): 133.5 × 23.3 × 7.2 *(40.7 × 7.1 × 2.2)*
Main machinery: 2 diesels; 5,700 hp(m) *(4.19 MW)*; 2 shafts
Speed, knots: 27
Complement: 25
Guns: 1 Breda 40 mm/70. 2—12.7 mm MGs.
Radars: Surface search: Racal Decca; I-band.

Comment: All built at Taşkizak Shipyard except SG 89 which was built at Istanbul Shipyard. SG 80-82 commissioned in 1996, 83-84 in 1997, 85 in 1998, 86-87 in 2000, 89-90 in 2001, 88 in 2002 and 91 in 2004. One, *(Erenköy)*, based in northern Cyprus with a pennant number KKTCSG 02.

UPDATED

KKTCSG 02   *6/2004*, Selçuk Emre /* 1044200

SG 90   *10/2003, C D Yaylali /* 0567560

14 LARGE PATROL CRAFT (WPB)

SG 121-134

Displacement, tons: 180 full load
Dimensions, feet (metres): 132 × 21 × 5.5 *(40.2 × 6.4 × 1.7)*
   131.2 × 21.3 × 4.9 *(40 × 6.5 × 1.5)* (SG 30-34)
Main machinery: 2 SACM AGO 195 V16 CSHR diesels; 4,800 hp(m) *(3.53 MW)* sustained
   2 cruise diesels; 300 hp(m) *(220 kW)*; 2 shafts
Speed, knots: 22
Complement: 25
Guns: 1 or 2 Bofors 40 mm/60. 2—12.7 mm MGs.
Radars: Surface search: Racal Decca 1226; I-band.

Comment: *SG 121* and *122* built by Gölcük Naval Yard, remainder by Taşkizak Naval Yard. *SG 134* commissioned in 1977, remainder 1968-71. *SG 130-134* have minor modifications-knuckle at bow, radar stirrup on bridge and MG on superstructure sponsons. These are similar craft to the Turk class listed under *Patrol Forces* for the Navy.

UPDATED

SG 134   *6/2004*, C D Yaylali /* 0589825

10 SAR 33 TYPE (LARGE PATROL CRAFT) (WPB)

SG 61-70

Displacement, tons: 180 full load
Dimensions, feet (metres): 113.5 × 28.3 × 9.7 *(34.6 × 8.6 × 3)*
Main machinery: 3 SACM AGO 195 V16 CSHR diesels; 7,200 hp(m) *(5.29 MW)* sustained; 3 shafts; cp props
Speed, knots: 33. **Range, n miles:** 450 at 24 kt; 550 at 18 kt
Complement: 24
Guns: 1 Bofors 40 mm/60. 2—12.7 mm MGs.
Radars: Surface search: Racal Decca; I-band.

Comment: Prototype Serter design ordered from Abeking & Rasmussen, Lemwerder in May 1976. The remainder were built at Taşkızak Naval Yard, Istanbul between 1979 and 1981. Fourteen of this class were to have been transferred to Libya but the order was cancelled. Two delivered to Saudi Arabia. The engines have been governed back and the top speed correspondingly reduced from the original 12,000 hp and 40 kt.

UPDATED

SG 67   5/2004*, Martin Mokrus / 0589828

4 SAR 35 TYPE (LARGE PATROL CRAFT) (WPB)

SG 71-74

Displacement, tons: 210 full load
Dimensions, feet (metres): 120 × 28.3 × 6.2 *(36.6 × 8.6 × 1.9)*
Main machinery: 3 SACM AGO 195 V16 CSHR diesels; 7,200 hp(m) *(5.29 MW)* sustained; 3 shafts; cp props
Speed, knots: 33. **Range, n miles:** 450 at 24 kt; 550 at 18 kt
Complement: 24
Guns: 1 Bofors 40 mm/60. 2—12.7 mm MGs.
Radars: Surface search: Racal Decca 1226; I-band.

Comment: A slightly enlarged version of the Serter designed SAR 33 Type built by Taşkızak Shipyard between 1985 and 1987. Restricted to less than designed speed.

VERIFIED

SG 71   8/2000, C D Yaylali / 0106645

9 KAAN 29 CLASS (LARGE PATROL CRAFT) (WPBF)

SG 101-109

Displacement, tons: 95 full load
Dimensions, feet (metres): 103.7 × 22.0 × 4.6 *(31.6 × 6.7 × 1.4)*
Main machinery: 2 MTU 16V 400 M90 diesels; 7,398 hp(m) *(5.44 MW)*; 2 MJP 753DD waterjets
Speed, knots: 45. **Range, n miles:** 750 at 20 kt
Complement: 13 (2 officers)
Guns: 1 Oerlikon 20 mm.
Radars: Surface search: I-band.

Comment: All built at Yonca Shipyard. Onuk MRTP 29 design. *TCSG 101-103* commissioned 25 July 2001, *TCSG 104-105* on 25 July 2002, *TCSG 106-108* in 2003 and *TCSG 109* in February 2004. A further five craft are expected.

UPDATED

SG 106   7/2004*, Selçuk Emre / 0589826

1 KAAN 33 CLASS (LARGE PATROL CRAFT) (WPBF)

SG 301

Displacement, tons: 124 full load
Dimensions, feet (metres): 116.8 × 22.0 × 4.7 *(35.6 × 6.7 × 1.4)*
Main machinery: 2 MTU 16V 4000 M90 diesels; 7,396 hp(m) *(5.44 MW)*; 2 MJP 753DD waterjets
Speed, knots: 43. **Range, n miles:** 650 at 20 kt
Complement: 18 (2 officers)
Guns: 1 Bofors 40 mm/70. 1 Oerlikon 20 mm.

Comment: Onuk MRTP 33 design built at Yonca shipyard. Commissioned in July 2004.

UPDATED

SG 301   7/2004*, Selçuk Emre / 0589829

7 KW 15 CLASS (LARGE PATROL CRAFT) (WPB)

SG 113-116, 118-120

Displacement, tons: 70 full load
Dimensions, feet (metres): 94.8 × 15.4 × 4.6 *(28.9 × 4.7 × 1.4)*
Main machinery: 2 MTU diesels; 2,700 hp(m) *(1.98 MW)*; 2 shafts
Speed, knots: 20. **Range, n miles:** 550 at 16 kt
Complement: 16
Guns: 1 Bofors 40 mm/60. 2 Oerlikon 20 mm.
Radars: Surface search: Racal Decca; I-band.

Comment: Built by Schweers, Bardenfleth. Commissioned 1961-62.

VERIFIED

SG 119   9/2002, Selim San / 0533264

18 KAAN 15 CLASS (FAST INTERVENTION CRAFT) (WPBF)

SG 1-18

Displacement, tons: 19 full load
Dimensions, feet (metres): 56.8 × 13.2 × 3.9 *(17.3 × 4.04 × 1.2)*
Main machinery: 2 MTU 12V 183 TE93 diesels; 2,300 hp(m) *(1.69 MW)*; 2 Arneson ASD 12 B1L surface drives
Speed, knots: 54. **Range, n miles:** 350 at 35 kt
Complement: 4 plus 8 mission crew
Guns: 2—12.7 mm MGs.
Radars: Surface search: Raytheon; I-band.

Comment: Contract for first six with Yonca Technical Investment signed in May 1997, second order for six more in February 1999, and third for 12 more in August 2000. All built at Tuzla-Istanbul shipyard. Three delivered in 1998, seven in 1999, two in April 2000, four in July 2001 and two in July 2002. Onuk MRTP 15 design. Fibreglass hulls. SG 11 and SG 12 are based in northern Cyprus.

UPDATED

KKTCSG 11 (North Cyprus)   6/2004*, Selçuk Emre / 1044202

SG 10   10/2003, C D Yaylali / 0567562

12 COASTAL PATROL CRAFT (WPB)

SG 50-59   SG 102-103

Displacement, tons: 29 full load
Dimensions, feet (metres): 47.9 × 13.7 × 3.6 *(14.6 × 4.2 × 1.1)*
Main machinery: 2 diesels; 700 hp(m) *(514 kW)*; 2 shafts
Speed, knots: 15
Complement: 7
Guns: 1—12.7 mm MG or 1 Oerlikon 20 mm *(SG 102-103)*.
Radars: Surface search: Raytheon; I-band.

Comment: SG 102 and 103 were built for North Cyprus and have been based there since August 1990 and July 1991 respectively. Both these craft were given a heavier gun in 1992. Second batch of three completed by Taşkizak in October 1992, three more in June 1993, four more in December 1993.

UPDATED

SG 55   9/2004*, Selim San / 0587558

2 COASTAL PATROL CRAFT (WPB)

SG 42-43

Displacement, tons: 45 full load
Dimensions, feet (metres): 55.1 × 13.9 × 3.6 *(16.8 × 4.2 × 1.1)*
Main machinery: 2 Gray Marine 64HN9 diesels; 450 hp *(335 kW)*; 2 shafts
Speed, knots: 12. **Range, n miles:** 200 at 12 kt
Complement: 7
Guns: 1—12.7 mm MG.

Comment: Former US Mk 5 craft built in Second World War. SG 45 and 46 transferred to north Cyprus in August 1993.

VERIFIED

COASTAL PATROL CRAFT   3/2001, Selim San / 0114805

1 INSHORE PATROL CRAFT (WPBI)

RAIF DENKTAŞ 101 (ex-74)

Displacement, tons: 10 full load
Dimensions, feet (metres): 38 × 11.5 × 2.4 *(11.6 × 3.5 × 0.7)*
Main machinery: 2 Volvo Aquamatic AQ200F petrol engines; 400 hp(m) *(294 kW)*; 2 shafts
Speed, knots: 28. **Range, n miles:** 250 at 25 kt
Complement: 6
Guns: 1—12.7 mm MG.
Radars: Surface search: Raytheon; I-band.

Comment: Built by Protekson, Istanbul. Transferred to North Cyprus 23 September 1988. Can be equipped with a rocket launcher.

UPDATED

RAIF DENKTAŞ   6/2004*, Selçuk Emre / 1044201

1 HARBOUR PATROL CRAFT (WPBI)

SG 41

Displacement, tons: 35 *(SG 1)*; 15 *(SG 2)* full load
Dimensions, feet (metres): 55.8 × 16.4 × 3.3 *(17 × 5 × 1) (SG 1)*
47.5 × 14.4 × 2.3 *(14.5 × 4.4 × 0.7) (SG 2)*
Main machinery: 2 diesels; 1,050 hp(m) *(771 kW) (SG 1)*
2 diesels; 700 hp(m) *(515 kW) (SG 2)*
Speed, knots: 20 *(SG 1)*; 18 *(SG 2)*
Complement: 7
Radars: Surface search: I-band.

Comment: High-speed patrol boat for anti-smuggling duties. Probably confiscated drug smuggling craft.

UPDATED

SG 41   6/2004*, Selim San / 0587557

LAND-BASED MARITIME AIRCRAFT

Numbers/Type: 9 Agusta AB 412 EP.
Operational speed: 122 kt *(226 km/h)*.
Service ceiling: 17,000 ft *(5,180 m)*.
Range: 374 n miles *(656 km)*.
Role/Weapon systems: Nine aircraft ordered 15 April 1999. Operated by Coast Guard/Frontier Force for patrol SAR. Sensors: Radar and FLIR. Weapons: Unarmed.

VERIFIED

AB 412   10/2003, C D Yaylali / 0567563

Numbers/Type: 3 CASA CN-235.
Operational speed: 240 kt *(445 km/h)*.
Service ceiling: 26,600 ft *(8,110 m)*.
Range: 669 n miles *(1,240 km)*.
Role/Weapon systems: Three delivered in July 2002. Long range maritime patrol for surveillance.

VERIFIED

CN-235   2/2003, CASA / 0531695

Turkmenistan

Country Overview

Formerly part of the USSR, the Republic of Turkmenistan declared its independence in 1991. Situated in Central Asia, it has an area of 188,460 square miles and is bordered to the north by Kazakhstan, to the east by Uzbekistan and Afghanistan and to the south by Iran. It has a 562 n mile coastline with the Caspian Sea. Türkmenbashi, the principal port, is linked by rail to Ashgabat, the capital and largest city. Maritime claims in the Caspian Sea are yet to be resolved. The Navy acts under the operational control of the Border Guard but is the weakest component of the Turkmen armed forces.

Headquarters Appointments

Minister of Defence:
  Mammet Geldiyev

Personnel

2005: 3,000

Base

Türkmenbashi (formerly Krasnovodsk)

PATROL FORCES

1 POINT CLASS (WPB)

Name	No	Builders	Commissioned
MERJEN (ex-*Point Jackson*)	PB-129 (ex-82378)	USCG Yard Curtis Bay	3 Aug 1970

Displacement, tons: 66; 69 full load
Dimensions, feet (metres): 83 × 17.2 × 5.8 *(25.3 × 5.2 × 1.8)*
Main machinery: 2 Caterpillar 3412 diesels; 1,600 hp *(1.19 MW)*; 2 shafts
Speed, knots: 23.5. **Range, n miles:** 1,200 at 8 kt
Complement: 10 (1 officer)
Guns: 2—12.7 mm MGs.
Radars: Surface search: Hughes/Furuno SPS-73; I-band.

Comment: Steel hulled craft with aluminium superstructure. Transferred from United States on 30 May 2000.

VERIFIED

KALKAN   *6/2003, Morye /* 0573698

10 ZHUK (GRIF) CLASS (PROJECT 1400M) (PB)

Displacement, tons: 39 full load
Dimensions, feet (metres): 78.7 × 16.4 × 3.9 *(24 × 5 × 1.2)*
Main machinery: 2 Type M 401B diesels; 2,200 hp(m) *(1.6 MW)* sustained; 2 shafts
Speed, knots: 30. **Range, n miles:** 1,100 at 15 kt
Complement: 13 (1 officer)
Guns: 2—14.5 mm (twin, fwd) MGs. 1—12.7 mm (aft) MG.
Radars: Surface search: Spin Trough; I-band.

Comment: Reported to have been acquired from Ukraine in 2004.

UPDATED

MERJEN (inboard ship)   *11/2000, Selim San /* 0104495

10 KALKAN (PROJECT 50030) M CLASS
(INSHORE PATROL CRAFT) (PBI)

Displacement, tons: 8.5 full load
Dimensions, feet (metres): 38.1 × 10.8 × 2.0 *(11.6 × 3.3 × 0.6)*
Main machinery: 1 Type 475K diesel; 496 hp *(370 kW)*; 1 waterjet
Speed, knots: 34
Complement: 2

Comment: Seven craft delivered during 2002 with a further three to follow. Built by Morye Feodosiya (Ukraine) and constructed with aluminium hulls and GRP superstructure. Can be armed with 7.62 mm or 12.7 mm MGs.

VERIFIED

ZHUK (Ukraine colours)   *8/2000, Hartmut Ehlers /* 0106674

Tuvalu

Country Overview

Tuvalu, formerly the Ellice Islands, is a south Pacific island group which gained independence in 1978; the other part of the former British colony, the Gilbert Islands, became independent as Kiribati the following year. Situated some 1,600 n miles east of Papua New Guinea, the country comprises nine atolls of which Funafuti is the location of the capital, Fongafale, and home to more than 30 per cent of the population. An archipelagic state, territorial seas (12 n miles) are claimed. An Exclusive Economic Zone (EEZ) (200 n miles) is also claimed but limits have not been fully defined by boundary agreements.

Headquarters Appointments

Commander Maritime Wing:
  Inspector Sele Fusi

Bases

Funafuti

PATROL FORCES

1 PACIFIC CLASS (LARGE PATROL CRAFT) (PB)

Name	No	Builders	Commissioned
TE MATAILI	801	Transfield Shipbuilding, WA	8 Oct 1994

Displacement, tons: 165 full load
Dimensions, feet (metres): 103.3 × 26.6 × 6.9 *(31.5 × 8.1 × 2.1)*
Main machinery: 2 Caterpillar 3516TA diesels; 4,400 hp *(3.28 MW)* sustained; 2 shafts
Speed, knots: 18. **Range, n miles:** 2,500 at 12 kt
Complement: 18 (3 officers)
Guns: Can carry 1—12.7 mm MG but is unarmed.
Radars: Navigation: Furuno 1011; I-band.

Comment: This is the 18th of the class to be built by the Australian Government for Exclusive Economic Zone (EEZ) patrols in the Pacific islands. The programme originally terminated at 15 but was re-opened on 19 February 1993 to include construction of five more craft for Fiji, Kiribati and Tuvalu. Training and support assistance is given by the Australian Navy. Half-life refit completed at Gladstone in 2001. Following the decision by the Australian government to extend the Pacific Patrol Boat programme, *Te Mataili* will require a life-extension refit in 2011 in order to achieve a 30-year ship life.

UPDATED

Ukraine

Country Overview

Formerly part of the USSR, Ukraine declared its independence in 1991. Situated in eastern Europe, it has an area of 233,090 square miles and is bordered to the north by Belarus, to the east by Russia, to the south-west by Romania and Moldova and to the west by Hungary, Slovakia and Poland. It has a 1,501 n mile coastline with the Black Sea and the Sea of Azov. Kiev is the capital and largest city while Sevastopol, Odessa, Kerch, and Mariupol are the principal ports. Territorial Seas (12 n miles) have been claimed. An EEZ (200 n miles) has been claimed but the limits have not been defined.

Division of the former Soviet Black Sea Fleet between Russia and Ukraine had been achieved by 1997.

Headquarters Appointments

Commander of the Navy:
Vice Admiral Igor Knyaz
First Deputy Commander of the Navy and Chief of Staff:
Rear Admiral Mykola Kostrov

Bases

Sevastopol (HQ), Donuzlav (Southern Region), Odessa (Western Region), Mikolaiv, Feodosiya, Izmail, Balaklava, Kerch

Personnel

2005: 12,500 navy

Border Guard

The Maritime Border Guard is an independent subdivision of the State Committee for Border Guards, and is not part of the Navy. It has three cutter brigades, based in Kerch, Odessa and Balaklava, to patrol the 827 mile coastline and two river brigades, which include a gunship squadron, a minesweeping squadron, an auxiliary ship group and a training division. Pennant numbers changed in July 1999.

DELETIONS

Notes: Some of these ships are laid up and may return to service, but this is unlikely.

Cruisers

2001   *Ukraina* (not completed)

Frigates

2001   *Mikolaiv*

Patrol Forces

2003   *Kremenchuk*

Mine Warfare Forces

2003   *Mariupol*

Amphibious forces

2001   *Kramatorsk*

Auxiliaries

2001   *Kerch, Novy Bug*

PENNANT LIST

Submarines

U 01	Zaporizya

Frigates

U 130	Hetman Sagaidachny
U 200	Lutsk
U 202	Ternopil (bldg)
U 205	Chernigiv
U 206	Vinnitsa

Patrol Forces

U 120	Skadovsk
U 153	Priluki
U 154	Kahovka
U 155	Nikopol
U 208	Khmelnitsky

Mine Warfare Forces

U 310	Zhovti Vody
U 311	Cherkasy
U 330	Melitopol
U 360	Genichesk

Amphibious Forces

U 400	Rivne
U 410	Kirovograd
U 402	Konstantin Olshansky
U 420	Donetsk
U 424	Artemivsk
U 862	Korosten
U 904	Bilyaivka

Auxiliaries

U 240	Feodisiya
U 510	Slavutich

U 540	Chigirin
U 541	Smila
U 542	Darnicha
U 635	Skvyra
U 700	Netisin
U 705	Kremenets
U 706	Izyaslav
U 707	Brodi
U 721	V Volnidsky
U 722	Borshev
U 728	Evpatoriya
U 731	Mirgood
U 733	Tokmak
U 753	Kriviy Rig
U 756	Sudak
U 757	Makivka
U 759	Bahmach
U 760	Fastiv
U 782	Sokal
U 783	Illichivsk
U 803	Krasnodon

U 811	Balta
U 830	Korets
U 831	Kovel
U 852	Shostka
U 860	Kamyankha
U 890	Malin
U 891	Kherson
U 947	Krasnoperekovsk
U 953	Dubno

Survey Ships

U 511	Simferopol
U 512	Pereyaslav
U 601	Alchevsk
U 754	Dzhankoi

SUBMARINES

1 FOXTROT CLASS (PROJECT 641) (SS)

Name	No	Builders	Laid down	Launched	Commissioned
ZAPORIZYA (ex-B 435)	U 01	Sudomekh, Leningrad	24 Mar 1970	29 May 1970	6 Nov 1970

Displacement, tons: 1,952 surfaced; 2,475 dived
Dimensions, feet (metres): 299.5 × 24.6 × 19.7 *(91.3 × 7.5 × 6)*
Main machinery: Diesel-electric; 3 Type 37-D diesels; 6,000 hp (m) *(4.4 MW);* 3 motors (1 × 2,700 and 2 × 1,350); 5,400 hp (m) *(3.97 MW);* 3 shafts; 1 auxiliary motor; 140 hp(m) *(103 kW)*
Speed, knots: 16 surfaced; 15 dived; 9 snorting
Range, n miles: 20,000 at 8 kt surfaced; 380 at 2 kt dived
Complement: 75

Torpedoes: 10—21 in *(533 mm)* (6 bow, 4 stern) tubes. Combination of 22—53 cm torpedoes.
Mines: 44 in lieu of torpedoes.
Countermeasures: ESM: Stop Light; radar warning.
Radars: Surface search: Snoop Tray; I-band.
Sonars: Pike Jaw; hull-mounted; passive/active search and attack; high frequency.

Programmes: Transferred from Russia in August 1997 with three others of the class.

Operational: Based at Balaklava and, following a three year refit which completed in 2000, further work is required to restore hydraulic systems and replace batteries. The boat was expected to return to operational service in 2004 but there have been no reports of activity and the boat may have been relegated to a training role.

UPDATED

ZAPORIZYA

8/2000, Hartmut Ehlers / 0106649

FRIGATES

1 KRIVAK III (NEREY) CLASS (PROJECT 1135.1) (FFHM)

Name	No	Builders	Laid down	Launched	Commissioned
HETMAN SAGAIDACHNY (ex-*Kirov*)	U 130 (ex-201)	Kamysh-Burun, Kerch	5 Oct 1990	29 Mar 1992	5 July 1993

Displacement, tons: 3,100 standard; 3,650 full load
Dimensions, feet (metres): 405.2 × 46.9 × 16.4
  (123.5 × 14.3 × 5)
Main machinery: COGAG; 2 gas turbines; 55,500 hp(m)
  (40.8 MW); 2 gas turbines; 13,600 hp(m) *(10 MW)*; 2 shafts
Speed, knots: 32
Range, n miles: 4,600 at 20 kt; 1,600 at 30 kt
Complement: 180 (18 officers)

Missiles: SAM: 1 SA-N-4 Gecko twin launcher ❶; semi-active
  radar homing to 15 km *(8.1 n miles)* at 2.5 Mach; warhead
  50 kg; altitude 9.1-3,048 m *(30-10,000 ft)*; 20 missiles. The
  launcher retracts into the mounting for stowage and
  protection, rising to fire and retracting to reload. The two
  mountings are forward of the bridge and abaft the funnel.
Guns: 1—3.9 in *(100 mm)*/59 ❷; 60 rds/min to 15 km *(8.1 n
  miles)*; weight of shell 16 kg.
  2—30 mm/65 ❸; 6 barrels per mounting; 3,000 rds/min
  combined to 2 km.
Torpedoes: 8—21 in *(533 mm)* (2 quad) tubes ❹. Combination of
  Russian 53 cm torpedoes.
A/S mortars: 2 RBU 6000 12-tubed trainable ❺; range 6,000 m;
  warhead 31 kg.

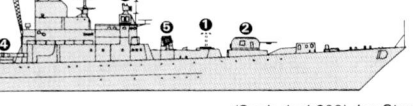

HETMAN SAGAIDACHNY   *(Scale 1 : 1,200), Ian Sturton* / 0506208

Countermeasures: Decoys: 4 PK 16 chaff launchers. Towed
  torpedo decoy.
ESM: 2 Bell Shroud; intercept.
ECM: 2 Bell Squat; jammers.
Radars: Air search: Top Plate ❻; 3D; D/E-band.
  Surface search: Spin Trough ❼; I-band. Peel Cone ❽; E-band.
  Fire control: Pop Group ❾; F/H/I-band (for SA-N-4). Kite Screech
    ❿; H/I/K-band. Bass Tilt (Krivak III) ⓫; H/I-band.
  Navigation: Kivach; I-band.
IFF: Salt Pot (Krivak III).
Sonars: Bull Nose (MGK 335MS); hull-mounted; active search
  and attack; medium frequency.

Helicopters: 1 Ka-27 Helix ⓬.

Programmes: This is the last of the 'Krivak IIIs' originally
  designed for the USSR Border Guard. The seven others are
  based in the Russian Pacific Fleet. A ninth of class was not
  completed.
Operational: *Sagaidachny* has so far not been sighted with a
  helicopter embarked. Deployed to the Mediterranean in 1994
  and late 1995, to the Indian Ocean in early 1995 and to the US
  in late 1996. 'Krivak II' *Sevastopol* (U 132) is probably being
  used for spares while 'Krivak I' *Mikolaiv* and *Dnipropetrovsk*
  are to be scrapped.

VERIFIED

HETMAN SAGAIDACHNY   *7/2000* / 0106650

HETMAN SAGAIDACHNY   *6/2003, Ships of the World* / 0572651

3 + 1 GRISHA CLASS (PROJECT 1124EM/P) (FFLM)

Name	No	Builders	Laid down	Launched	Commissioned
LUTSK	U 200	Leninskaya Kuznitsa, Kiev	—	12 May 1993	12 Feb 1994
TERNOPIL	U 202	Leninskaya Kuznitsa, Kiev	—	20 Mar 2002	2005
CHERNIGIV (ex-*Izmail*)	U 205	Zrelenodolsk	12 Sep 1978	22 June1980	28 Dec 1980
VINNITSA (ex-*Dnepr*)	U 206	Zrelenodolsk	23 Dec 1975	12 Sep 1976	31 Dec 1976

Displacement, tons: 950 standard; 1,150 full load
Dimensions, feet (metres): 233.6 × 32.2 × 12.1
(71.2 × 9.8 × 3.7)
Main machinery: CODAG; 1 gas turbine; 15,000 hp(m) *(11 MW)*;
2 diesels; 16,000 hp(m) *(11.8 MW)*; 3 shafts
Speed, knots: 30. **Range, n miles:** 2,500 at 14 kt; 1,750 at 20 kt
diesels; 950 at 27 kt
Complement: 70 (5 officers)

Missiles: SAM: SA-N-4 Gecko twin launcher ❶ *(Lutsk)*; semi-
active radar homing to 15 km *(8 n miles)* at 2.5 Mach;
warhead 50 kg; altitude 9.1-3,048 m *(30-10,000 ft)*; 20
missiles.
Guns: 1—3 in *(76 mm)*/60 ❷; *(Lutsk)*; 120 rds/min to 15 km *(8 n
miles)*; weight of shell 7 kg.
4—57 mm/80 (twin); 120 rds/min to 6 km *(3.3 n miles)*;
weight of shell 2.8 kg.
1—30 mm/65 ❸; *(Lutsk)*; 6 barrels; 3,000 rds/min combined
to 2 km.
Torpedoes: 4—21 in *(533 mm)* (2 twin) tubes ❹. SAET-60;
passive homing to 15 km *(8.1 n miles)* at 40 kt; warhead
400 kg.
A/S mortars: 1 or 2 RBU 6000 12-tubed trainable ❺; range
6,000 m; warhead 31 kg.

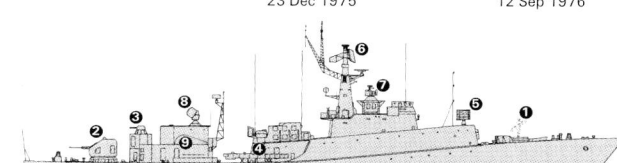

LUTSK

Depth charges: 2 racks (12).
Mines: Capacity for 18 in lieu of depth charges.
Countermeasures: ESM: 2 Watch Dog. 2 PK 16 chaff launchers.
Radars: Air/surface search: Half Plate B ❻; *(Lutsk)*; E/F-band.
Strut Curve (remainder); F-band.
Navigation: Don 2; I-band.
Fire control: Pop Group ❼; *(Lutsk)*; F/H/I-band (for SA-N-4). Bass
Tilt ❽; *(Lutsk)*; H/I-band (for 76 mm and 30 mm).
Muff Cobb (remainder); G/H-band.
IFF: High Pole A or B. Square Head. Salt Pot.
Sonars: Bull Nose (MGK 335MS); hull-mounted; active search
and attack; high/medium frequency.
Elk Tail VDS ❾; active search; high frequency.

(Scale 1 : 900), Ian Sturton

Programmes: *Lutsk* is a 'Grisha V' (Type 1124EM) launched
12 May 1993 and completed 27 November 1993. *Ternopil* is
under construction and is the first new ship to join the fleet
since 1992. It is expected to commission in 2005. *Chernigiv*
and *Vinnitsa* are both 'Grisha II' (Type 1124P) ex-Russian
Border Guard ships transferred in 1996. Two 'Grisha I' were
also transferred but have been deleted.
Operational: Both active. **UPDATED**

VINNITSA
7/2000, Hartmut Ehlers / 0106552

LUTSK
8/2003, Guy Toremans / 0572654

LAND-BASED MARITIME AIRCRAFT

Notes: There are two principal formations. OSAE (Otdelnaya Smeshannaya Aviotsionnaya Eskadrilya) comprises both fixed- and rotary-wing aircraft and is based at Sevastopol. OPVLP (Otdelnyi Protivolodochnyi VV Polk) comprises fixed-wing aircraft with a secondary anti-shipping role and is based at Ochakov. The majority of aircraft is thought to be unserviceable but the current operational inventory is assessed to include 16 Tu-22 Backfire, 18 Su-24 Fencer, 5 Mi-14 Haze, 3 Be-12 Mail, 18 Ka-25 Hormone and 16 Ka 27/29 Helix.

PATROL FORCES

1 PAUK I (MOLNYA) (PROJECT 1241P) CLASS (PCM)

KHMELNITSKY (ex-*MPK 116*) U 208

Displacement, tons: 440 full load
Dimensions, feet (metres): 189 × 33.5 × 10.8 *(57.6 × 10.2 × 3.3)*
Main machinery: 2 Type M 521 diesels; 16,184 hp(m) *(11.9 MW)* sustained; 2 shafts
Speed, knots: 32. **Range, n miles:** 2,400 at 14 kt
Complement: 32

Missiles: SAM; SA-N-5 Grail quad launcher; manual aiming; IR homing to 6 km *(3.2 n miles)* at 1.5 Mach; altitude to 2,500 m *(8,000 ft)*; warhead 1.5 kg; 8 missiles.
Guns: 1—3 in *(76 mm)*/60; 120 rds/min to 15 km *(8 n miles)*; weight of shell 7kg.
  1—30 mm/65 AK 630; 6 barrels; 3,000 rds/min combined to 2 km.
Torpedoes: 4—16 in *(406 mm)*.
A/S mortars: 2 RBU 1200 5-tubed fixed; range 1,200 m; warhead 34 kg.
Depth charges: 2 racks (12).
Countermeasures: Decoys: 2 PK 16 or 4 PK 10 chaff launchers.
ESM: 3 Brick Plug and 2 Half Hat; radar warning.
Weapons control: Hood Wink optronic director.
Radars: Air/surface search: Peel Cone; E/F-band.
Surface search: Kivach or Pechora; I-band.
Fire control: Bass Tilt; H/I-band.
Sonars: Foal Tail; VDS (mounted on transom); active attack; high frequency.

Programmes: Built at Yaroslavl in 1985. Transferred from Black Sea Fleet Border Guard in 1996. Others of this class are in the Ukraine Border Guard.
Structure: ASW version of the Russian Tarantul class.
Operational: A second of class (*Uzhgorod* U 207) is laid up.

VERIFIED

KHMELNITSKY   *7/2000, Hartmut Ehlers* / 0106653

1 TARANTUL II (MOLNYA) (PROJECT 1241.1/2) CLASS (FSGM)

PRIDNEPROVYE (ex-*Nikopol*, ex-*R-54*) U 155

Displacement, tons: 385 standard; 455 full load
Dimensions, feet (metres): 184.1 × 37.7 × 8.2 *(56.1 × 11.5 × 2.5)*
Main machinery: COGAG: 2 Nikolayev Type DR 77 gas turbines; 16,016 hp(m) *(11.77 MW)*; 2 Nikolayev DR 76 gas turbines with reversible gearboxes; 4,993 hp(m) *(3.67 MW)* sustained 2 shafts; cp props
Speed, knots: 36. **Range, n miles:** 1,650 at 14 kt
Complement: 34 (5 officers)

Missiles: SSM: 4 Raduga SS-N-2D Styx (2 twin); active radar or IR homing to 83 km *(45 n miles)* at 0.9 Mach; warhead 513 kg; sea skimmer at end of run
SAM: 1 SA-N-5 Grail quad launcher; manual aiming; IR homing to 6 km *(3.2 n miles)* at 1.5 Mach; warhead 1.5 kg.
Guns: 1—3 in *(76 mm)*/60; AK-176; 120 rds/min to 15 km *(8 n miles)*; weight of shell 7 kg.
  2—30 mm/65 AK-630; 6 barrels per mounting; 3,000 rds/min to 2 km.
Countermeasures: Decoys: 4 PK 16 chaff launchers.
Weapons control: Hood Wink optronic director. Light bulb datalink. Band Stand for SSM.
Radars: Air/surface search: Plank Shave; I-band.
Fire control: Bass Tilt; H/I-band (for guns).
Navigation: Kivach III; I-band.
IFF: High Pole B.

Programmes: Built at Kolpino and originally commissioned in 1983. Transferred in 1997 and recommissioned in 2002 following a refit.

UPDATED

PRIDNEPROVYE   *3/2002, Hartmut Ehlers* / 0529997

2 MATKA (VEKHR) CLASS (PROJECT 206MP) (FAST ATTACK CRAFT—MISSILE HYDROFOIL) (PGGK)

PRILUKI (ex-R-262) U 153   **KAHOVKA** (ex-R-265) U 154

Displacement, tons: 225 standard; 260 full load
Dimensions, feet (metres): 129.9 × 24.9 (41 over foils) × 6.9 (13.1 over foils) *(39.6 × 7.6 (12.5) × 2.1 (4))*
Main machinery: 3 Type M 504 diesels; 10,800 hp(m) *(7.94 MW)* sustained; 3 shafts
Speed, knots: 40. **Range, n miles:** 600 at 35 kt foilborne; 1,500 at 14 kt hullborne
Complement: 33
Missiles: SSM: 2 SS-N-2C/D Styx; active radar or IR homing to 83 km *(45 n miles)* at 0.9 Mach; warhead 513 kg; sea-skimmer at end of run.
Guns: 1—3 in *(76 mm)*/60; 120 rds/min to 15 km *(8 n miles)*; weight of shell 7 kg.
  1—30 mm/65 AK 630; 6 barrels per mounting; 3,000 rds/min to 2 km.
Countermeasures: Decoys: 2 PK 16 chaff launchers.
ESM: Clay Brick; intercept.
Weapons control: Hood Wink optronic directors.
Radars: Air/surface search: Plank Shave; E-band.
Navigation: SRN-207; I-band.
Fire control: Bass Tilt; H/I-band.
IFF: High Pole B or Salt Pot B and Square Head.

Comment: Five Russian Black Sea Fleet units transferred in 1996. Built between 1978 and 1983 with similar hulls to the Osa class. One was transferred to Georgia in 1999 and two others (*Uman* and *Tsurupinsk*) have been cannibalised for spares.

VERIFIED

PRILUKI   *7/2000* / 0106654

1 ZHUK (GRIF) CLASS (PROJECT 1400M) (PB)

SKADOVSK U 120

Displacement, tons: 39 full load
Dimensions, feet (metres): 78.7 × 16.4 × 3.9 *(24 × 5 × 1.2)*
Main machinery: 2 Type M 401B diesels; 2,200 hp(m) *(1.6 MW)* sustained; 2 shafts
Speed, knots: 30. **Range, n miles:** 1,100 at 15 kt
Complement: 13
Guns: 2—14.5 mm (twin). 1—12.7 mm MG.
Radars: Surface search: Spin Trough; I-band.

Comment: Transferred from Russia in 1997 and became operational in 2000. Others of the class are in service with the Border Guard.

VERIFIED

SKADOVSK   *7/2000, Hartmut Ehlers* / 0106655

AMPHIBIOUS FORCES

Notes: Two Vydra class LCUs, *Korosten* U 862 and *Bilyaivka* U 904, are used as trials and transport craft. There are also two Ondatra class LCM, *Svatove* U 430 and *Vil* U 537 and a T-4 LCM *Tarpan* U 538 which are laid up.

1 POLNOCHNY C (PROJECT 773 I) CLASS (LSM)

KIROVOGRAD (ex-SDK 137) U 401

Displacement, tons: 1,120 standard; 1,150 full load
Dimensions, feet (metres): 266.7 × 31.8 × 7.9 *(81.3 × 9.7 × 2.4)*
Main machinery: 2 Kolomna Type 40-D diesels; 4,400 hp(m) *(3.2 MW)* sustained; 2 shafts
Speed, knots: 18. **Range, n miles:** 2,000 at 12 kt
Complement: 40-42
Military lift: 350 tons including 6 tanks; 180 troops
Missiles: 4 SA-N-5 Grail quad launchers; manual aiming; IR homing to 6 km *(3.2 n miles)* at 1.5 Mach; warhead 1.5 kg; 32 missiles.
Guns: 4—30 mm/65 (2 twin). 2—140 mm 18-tubed rocket launchers.
Radars: Surface search: Spin Trough; I-band.
Fire control: Drum Tilt; H/I-band (for 30 mm guns).

Comment: Built in 1970s and transferred from Russian Fleet in 1994. Reported operational again in 2001 following refit.

VERIFIED

KIROVOGRAD   *6/2003, Ships of the World* / 0572650

1 ROPUCHA I (PROJECT 775) CLASS (LST)

KONSTANTIN OLSHANSKY (ex-BDK 56) U 402

Displacement, tons: 4,400 full load
Dimensions, feet (metres): 370.7 × 47.6 × 11.5 *(113 × 14.5 × 3.6)*
Main machinery: 2 Zgoda-Sulzer 16ZVB40/48 diesels; 19,230 hp(m) *(14.14 MW)* sustained;
   2 shafts
Speed, knots: 17.5. **Range, n miles:** 3,500 at 16 kt
Complement: 95 (7 officers)
Military lift: 10 MBT plus 190 troops or 24 AFVs plus 170 troops
Missiles: SAM: 4 SA-N-5 Grail quad launchers.
Guns: 4—57 mm/80 (2 twin); 120 rds/min to 6 km *(3.3 n miles)*; weight of shell 2.8 kg.
Weapons control: 2 Squeeze Box optronic directors.
Radars: Air/surface search: Strut Curve; F-band.
Navigation: Don 2; I-band.
Fire control: Muff Cob; G/H-band.
IFF: High Pole B.

Comment: Built at Gdansk, Poland in 1978 and transferred from Russia in 1996. Can be used to
   carry mines. Ro-Ro design with 540 m² of parking space between the stern gate and the bow
   doors.   ***VERIFIED***

KONSTANTIN OLSHANSKY   *8/2000, B Lemachko /* 0131165

1 ALLIGATOR (PROJECT 1171) CLASS (LST)

RIVNE (ex-*BDK 104*) U 400

Displacement, tons: 4,700 full load
Measurement, tons: 370.7 × 50.8 × 14.7 *(113 × 15.5 × 4.5)*
Main machinery: 2 diesels; 9,000 hp(m) *(6.6 MW)*; 2 shafts
Speed, knots: 18. **Range, n miles:** 10,000 at 15 kt
Complement: 100
Military lift: 300 troops; 1,700 tons including about 20 tanks and various trucks; 40 AFVs
Radars: Surface search: 2 Don 2; I-band.

Comment: Launched in 1971 and transferred from Russian Fleet in 1994. This is the Type 2 variant
   of this class. Disarmed in 1997, refitted in 1999 and reported undergoing repairs in dock in
   2004.   ***UPDATED***

RIVNE   *4/1997, Ukraine Navy /* 0019345

2 POMORNIK (ZUBR) (PROJECT 1232.2) CLASS (ACV/LCUJM)

DONETSK U 420   **ARTEMIVSK** (ex-MDK-93) U 424

Displacement, tons: 550 full load
Dimensions, feet (metres): 189 × 70.5 *(57.6 × 21.5)*
Main machinery: 5 Type NK-12MV gas turbines; 2 for lift, 23,672 hp(m) *(17.4 MW)* nominal; 3 for
   drive, 35,508 hp(m) *(26.1 MW)* nominal
Speed, knots: 60. **Range, n miles:** 300 at 55 kt
Complement: 31 (4 officers)
Military lift: 3 MBT or 10 APC plus 230 troops (total 170 tons)
Missiles: SAM: 2 SA-N-5 Grail quad launchers; manual aiming; IR homing to 6 km *(3.2 n miles)* at
   1.5 Mach; altitude to 2,500 m *(8,000 ft)*; warhead 1.5 kg.
Guns: 2—30 mm/65 AK 630; 6 barrels per mounting; 3,000 rds/min combined to 2 km.
   2 retractable 122 mm rocket launchers.
Mines: 80.
Countermeasures: Decoys: TSP 41 chaff.
ESM: Tool Box; intercept.
Weapons control: Quad Look (modified Squeeze Box) (DWU 3) optronic director.
Radars: Air/surface search: Cross Dome (Ekran); I-band.
Fire control: Bass Tilt MR 123; H/I-band.
IFF: Salt Pot A/B. Square Head.

Comment: *Donetsk* was completed by Morye, Feodosiya on 20 July 1993. Sister *U 421* was
   incomplete in 1999 when procured by Greece, delivery being made in 2001. Three further craft
   were transferred from Russia in 1996. Of these, *U 423* (ex-*MDK 123*) was also sold to Greece,
   U 422 (ex-*MDK 57*) is not operational and U 424 remains in service.   ***VERIFIED***

DONETSK   *8/2000, B Lemachko /* 0131164

MINE WARFARE FORCES

2 NATYA I CLASS (PROJECT 266M) (MSO)

ZHOVTI VODY (ex-*Zenitchik*) U 310   **CHERKASY** (ex-*Razvedchik*) U 311

Displacement, tons: 804 full load
Dimensions, feet (metres): 200.1 × 33.5 × 9.8 *(61 × 10.2 × 3)*
Main machinery: 2 Type M 504 diesels; 5,000 hp(m) *(3.67 MW)* sustained; 2 shafts; cp props
Speed, knots: 16. **Range, n miles:** 3,000 at 12 kt
Complement: 67 (8 officers)

Guns: 4—30 mm/65 (2 twin) AK 306 or 2—30 mm/65 AK 630; 4—25 mm/80 (2 twin).
A/S mortars: 2 RBU 1200 5-tubed fixed.
Depth charges: 62.
Mines: 10.
Countermeasures: MCM: 1 or 2 GKT-2 contact sweeps; 1 AT-2 acoustic sweep.
   1 TEM-3 magnetic sweep.
Radars: Surface search: Long Trough; E-band.
Fire control: Drum Tilt; H/I-band.
IFF: 2 Square Head. High Pole B.
Sonars: MG 79/89; hull-mounted; active minehunting; high frequency.

Comment: Built in the mid-1970s. Transferred from Russia in 1996. Both are operational.
                                                                                               VERIFIED

ZHOVTI VODY   *9/2002, C D Yaylali /* 0530030

1 SONYA (YAKHONT) (PROJECT 1265) CLASS (MHSC)

MELITOPOL (ex-BT 79) U 330

Displacement, tons: 460 full load
Dimensions, feet (metres): 157.4 × 28.9 × 6.6 *(48 × 8.8 × 2)*
Main machinery: 2 Kolomna diesels; 2,000 hp(m) *(1.47 MW)* sustained; 2 shafts
Speed, knots: 15. **Range, n miles:** 3,000 at 10 kt
Complement: 43
Guns: 2—30 mm/65 (twin). 2—25 mm/80 (twin).
Mines: 8.
Radars: Surface search: Don 2; I-band.
IFF: Two Square Head.
Sonars: MG 69/79; hull-mounted; active; high frequency.

Comment: Built in 1978. Transferred from Russia in 1996. Wooden hull.

                                                                                               UPDATED

MELITOPOL   *6/2003, B Lemachko /* 0572653

1 YEVGENYA (KOROND) (PROJECT 1258) CLASS (MHC)

GENICHESK (ex-RT 214) U 360

Displacement, tons: 77 standard; 90 full load
Dimensions, feet (metres): 80.7 × 18 × 4.9 *(24.6 × 5.5 × 1.5)*
Main machinery: 2 Type 3-D-12 diesels; 600 hp(m) *(440 kW)* sustained; 2 shafts
Speed, knots: 11. **Range, n miles:** 300 at 10 kt
Complement: 10
Guns: 2—14.5 mm (twin) MGs.
Mines: 8 racks.
Radars: Surface search: Spin Trough or Mius; I-band.
IFF: Salt Pot.
Sonars: A small MG-7 sonar is lifted over stern on crane; a TV system may also be used.

Comment: Transferred from Russia in 1996. Reported as being operational.

VERIFIED

SUDAK   *3/2002, Hartmut Ehlers /* 0529996

1 BEREZA CLASS (PROJECT 18061) (ADG)

BALTA U 811 (ex-*SR 568*)

Displacement, tons: 1,850 standard; 2,051 full load
Dimensions, feet (metres): 228 × 45.3 × 13.1 *(69.5 × 13.8 × 4)*
Main machinery: 2 Zgoda-Sulzer 8AL25/30 diesels; 2,938 hp(m) *(2.16 MW)* sustained; 2 shafts
Speed, knots: 14. **Range, n miles:** 1,000 at 14 kt
Complement: 88
Radars: Navigation: Kivach; I-band.

Comment: Built at Northern Shipyard, Gdansk in 1987. Transferred from Russia in 1997. Degaussing vessel with an NBC citadel and three laboratories. Not seen at sea since being transferred but may be in use.

VERIFIED

GENICHESK   *6/2003, Ships of the World /* 0572652

AUXILIARIES

Notes: Other ships transferred from Russia in 1997, and possibly still in limited service, are a Keyla II class tanker, *Kriviy Rig* U 753, two Toplivo class tankers, *Fastiv* U 760 and *Bahmach* U 759, two Pozharny class, ATR *Borshev* U 722 and *Evpatoriya* U 728 and a Shalanda class trials craft *Kamyankha* U 860.

KAMYANKHA   *6/2003*, B Lemachko /* 1043552

1 AMUR (PROJECT 304) CLASS SUPPORT SHIP (AGF/AR)

DONBAS (ex-*Krasnodon*) U 500 (ex-U 803)

Displacement, tons: 5,500 full load
Dimensions, feet (metres): 400.3 × 55.8 × 16.7 *(122 × 17 × 5.1)*
Main machinery: 1 Zgoda 8 TAD-48 diesel; 3,000 hp(m) *(2.2 MW)*; 1 shaft
Speed, knots: 12. **Range, n miles:** 13,000 at 8 kt
Complement: 145
Radars: Navigation: Don 2; I-band.

Comment: Transferred in 1977. Completed refit in 2001 to serve as command ship and support ships for surface ships and submarines based at Sevastopol. Has two 3-ton cranes and one 1.5-ton crane.

VERIFIED

BALTA   *9/1998, Hartmut Ehlers /* 0050313

1 BAMBUK (PROJECT 12884) CLASS (AGFHM)

Name	No	Builders	Launched	Commissioned
SLAVUTICH	U 510 (ex-800, ex-SSV 189)	Nikolayev	12 Oct 1990	28 July 1992

Displacement, tons: 5,403 full load
Dimensions, feet (metres): 350.1 × 52.5 × 19.7 *(106.7 × 16 × 6)*
Main machinery: 2 Skoda 6L2511 diesels; 6,100 hp(m) *(4.5 MW)*; 2 shafts
Speed, knots: 16. **Range, n miles:** 8,000 at 12 kt
Complement: 178
Missiles: SAM: 2 SA-N-5/8 Grail quad launchers; manual aiming; IR homing to 6 km *(3.2 n miles)* at 1.5 Mach; altitude to 2,500 m *(8,000 ft)*; warhead 1.5 kg.
Guns: 2—30 mm/65 AK 630; 6 barrels per mounting.
Countermeasures: Decoys: 2 PK 16 chaff launchers.
Radars: Navigation: 3 Palm Frond; I-band.
CCA: Fly Screen; I-band.
Tacan: 2 Round House.

Comment: Laid down on 20 March 1988. Second of a class built for acoustic research but taken over before completion and used as a command ship by the Ukrainian Navy. The ship is not capable of helicopter operations as previously reported.

UPDATED

DONBAS   *7/2003, B Lemachko /* 0576461

1 VODA (PROJECT 561) CLASS WATER TANKER (AWT)

SUDAK (ex-*Sura*) U 756

Displacement, tons: 982 standard; 2,250 full load
Dimensions, feet (metres): 266.8 × 37.4 × 11.3 *(81.3 × 11.4 × 3.44)*
Main machinery: 2 diesels; 2 shafts
Speed, knots: 12. **Range, n miles:** 2,900 at 10 kt
Complement: 22
Radars: Navigation: Don 2; I-band.

Comment: Transferred in 1977. Has a three ton derrick.

VERIFIED

SLAVUTICH   *9/2004*, Hartmut Ehlers /* 1043551

SLAVUTICH   *7/2000, Hachiro Nakai /* 0106664

1 SURA (PROJECT 145) CLASS (ABU)

SHOSTKA (ex-*Kil 33*) U 852

Displacement, tons: 2,370 standard; 3,150 full load
Dimensions, feet (metres): 285.4 × 48.6 × 16.4 *(87 × 14.8 ×5)*
Main machinery: Diesel-electric; 4 diesel generators; 2 motors; 2,240 hp(m) *(1.65 MW)*; 2 shafts
Speed, knots: 12. **Range, n miles:** 2,000 at 11 kt
Complement: 40
Cargo capacity: 900 tons cargo; 300 tons fuel for transfer
Radars: Navigation: 2 Don 2; I-band.

Comment: Transferred from Russia in 1997. Heavy lift ship built at Rostock in 1973. Lifting capacity includes one 65 ton derrick and one 65 ton stern cage. Can carry a 12 m DSRV, although this has not been seen in Ukrainian service. ***VERIFIED***

SHOSTKA   *8/2000, Hartmut Ehlers /* 0106666

1 YELVA (PROJECT 535M) CLASS (DIVING TENDER) (YDT)

NETISIN U 700

Displacement, tons: 295 full load
Dimensions, feet (metres): 134.2 × 26.2 × 6.6 *(40.9 × 8 × 2)*
Main machinery: 2 Type 3-D-12A diesels; 630 hp(m) *(463 kW)* sustained; 2 shafts
Speed, knots: 12.5. **Range, n miles:** 1,870 at 12 kt
Complement: 30
Radars: Navigation: Spin Trough; I-band.

Comment: Diving tender built in mid-1970s. Transferred from Russia in 1997. Carries a 1 ton crane and diving bell. Operational. ***VERIFIED***

YELVA CLASS (to left)   *6/1998, van Ginderen Collection /* 0050315

20 HARBOUR CRAFT (YDT/YFL/YPT)

FEODOSIYA U 240	**V VOLNIDSKY** U 721	**SHULYAVKA** U 853
U 241	**MIRGOOD** U 731	**MALIN** U 890
U 631-634	**U 732**	**KHERSON** (ex-*Monastirishze*) U 891
SKVYRA U 635	**TOKMAK** U 733	**U 926**
BRODI U 707	**ILLICHIVSK** U 783	**+3**

Displacement, tons: 42 full load
Dimensions, feet (metres): 72.8 × 12.8 × 4.6 *(22.2 × 3.9 × 1.4)*
Main machinery: 1 diesel; 300 hp(m) *(220 kW)* sustained; 1 shaft
Speed, knots: 12
Complement: 8

Comment: Details given are for the Flamingo class harbour patrol craft of which there are five (U 240, U 634, U 721, U 732, U 733). There are also five 'Nyryat 1' diving tenders and inshore survey craft (U 631, U 632, U 633, U 635, U 707) and six PO 2 class tenders (U 731, U 926 plus four). There are also two Shelon class YPTs (U 890, U 891), an ambulance craft U 783 and a flag officers' yacht U 853. ***UPDATED***

KHERSON   *9/2004 *, Hartmut Ehlers /* 1043553

ILLICHIVSK   *7/2003 *, B Lemachko /* 1043554

SURVEY SHIPS

Notes: (1) Also transferred in 1997 were two Muna class AGIs, *Pereyaslav* U 512 and *Dzhankoi* U 754. Both are used as transports, mostly for commercial goods.
(2) Ten former Russian civilian research ships were transferred in 1996/97. All are now in commercial service.

2 MOMA (PROJECT 861M) CLASS (AGS)

SIMFEROPOL (ex-*Jupiter*) U 511   — (ex-*Berezan*) U 602

Displacement, tons: 1,600 full load
Dimensions, feet (metres): 240.5 × 36.8 × 12.8 *(73.3 × 11.2 × 3.9)*
Main machinery: 2 Zgoda-Sulzer diesels; 3,300 hp(m) *(2.43 MW)* sustained; 2 shafts; cp props
Speed, knots: 17. **Range, n miles:** 9,000 at 11 kt
Complement: 56
Radars: Navigation: Don 2; I-band.

Comment: U 511 transferred from Russia in February 1996 and is active. A second of class *U 602* has also been reported as active. ***VERIFIED***

SIMFEROPOL   *7/2000, Hartmut Ehlers /* 0106669

2 BIYA (PROJECT 870) CLASS (AGS)

ALCHEVSK U 601 (ex-*GS 212*)   **U 603** (ex-GS 273)

Displacement, tons: 766 full load
Dimensions, feet (metres): 180.4 × 32.1 × 8.5 *(55 × 9.8 × 2.6)*
Main machinery: 2 diesels; 1,200 hp(m) *(882 kW)*; 2 shafts; cp props
Speed, knots: 13. **Range, n miles:** 4,700 at 11 kt
Complement: 25
Radars: Navigation: Don 2; I-band.

Comment: Built at Northern Shipyard, Gdansk 1972-76. Transferred from Russia in 1997. Laboratory and one survey launch, and a 5 ton crane. ***VERIFIED***

ALCHEVSK   *8/1997, W Globke /* 0019356

For details of the latest updates to *Jane's Fighting Ships* online and to discover the additional information available exclusively to online subscribers please visit
jfs.janes.com

TUGS

6 TUGS (ATA/YTM)

KREMENETS U 705	KOVEL U 831
IZYASLAV U 706	KRASNOPEREKOPSK U 947
KORETS U 830	DUBNO U 953

Comment: All transferred from Russia in 1997. *U 706* and *U 831* are Okhtensky class coastal tugs built in 1958. *U 705* is a Goryn class ocean going tug with a bollard pull of 45 tons, *U 830* is a Sorum class and *U 947* a Prometey class large tug. *U 953* is a Sidehole II class harbour tug.

UPDATED

KORETS

6/2003*, B Lemachko / 1043550

TRAINING SHIPS

Note: In addition there are two Bryza class training cutters *U 543* and *U 544*.

4 PETRUSHKA (DRAKON) CLASS (AXL)

CHIGIRIN U 540	SMILA U 541	DARNICHA U 542	SOKAL U 782

Displacement, tons: 335 full load
Dimensions, feet (metres): 129.3 × 27.6 × 7.2 *(39.4 × 8.4 × 2.2)*
Main machinery: 2 Wola H12 diesels; 756 hp(m) *(556 kW)*; 2 shafts
Speed, knots: 11. **Range, n miles:** 1,000 at 11 kt
Complement: 13 plus 30 cadets

Comment: Training vessels built at Wisla Shiyard, Poland in 1989. Transferred from Russia in 1997. Used for seamanship and navigation training. *U 782* acts as a tender.

VERIFIED

SMILA

4/2001, B Lemachko / 0131162

1 MIR CLASS (SAIL TRAINING SHIP) (AXS)

Name	No	Builders	Commissioned
KHERSONES	—	Northern Shipyard, Gdansk	10 June 1988

Measurement, tons: 2,996 grt
Dimensions, feet (metres): 346.1 × 45.9 × 19.7 *(105.5 × 14 × 6)*
Main machinery: 1 Sulzer 8AL20/24 diesels; 1,500 hp(m) *(1.1 MW)*; 1 shaft
Speed, knots: 17
Complement: 55 plus 144 trainees

Comment: One of five of a class ordered in July 1985. Civilian manned.

UPDATED

KHERSONES

6/2003*, Martin Mokrus / 1043549

BORDER GUARD (MORSKA OKHORONA)

Notes: (1) There are plans to build new patrol cutters of the 'Kordon' (47 m) and 'Afalina' (44 m) classes, and new patrol boats of the 'Scif' (26 m) class. These new designs are also on offer for export by the Feodosiya Shipbuilding Association Morye. A 67 m OPV design, by Nikolayev 61 Kommuna shipyard, is also in the export market.
(2) The river brigades also include four minesweeping boats, and 16 training craft. Not all of these are operational.
(3) Border Guard vessels are painted dark grey with a thick yellow and thin blue diagonal line on the hull. From July 1999, pennant numbers were changed and are preceded by the letters BG.

3 PAUK I (MOLNYA) CLASS (PROJECT 1241) (PC)

GRIGORY KUROPIATNIKOV BG 50 (ex-PSKR 817)	POLTAVA BG 51 (ex-PSKR 813)
GRIGORY GNATENKO BG 52 (ex-PSKR 815)	

Displacement, tons: 475 full load
Dimensions, feet (metres): 189 × 33.5 × 10.8 *(57.6 × 10.2 × 3.3)*
Main machinery: 2 Type M 521 diesels; 16,184 hp(m) *(11.9 MW)* sustained; 2 shafts
Speed, knots: 32. **Range, n miles:** 1,260 at 14 kt
Complement: 44 (7 officers)

Guns: 1—3 in *(76 mm)*/60; 120 rds/min to 15 km *(8 n miles)*; weight of shell 7 kg.
  1—30 mm/65 AK 630; 6 barrels; 3,000 rds/min combined to 2 km.
Torpedoes: 4—16 in *(406 mm)* tubes. SAET-40; anti-submarine; active/passive homing to 10 km *(5.4 n miles)* at 30 kt; warhead 100 kg.
A/S mortars: 2 RBU 1200 5-tubed fixed; range 1,200 m; warhead 34 kg.
Depth charges: 2 racks (12).
Countermeasures: Decoys: 2 PK 16 or 4 PK 10.
ESM: Brick Plug and Half Hat; radar warning.
Weapons control: Hood Wink optronic director.
Radars: Air/surface search: Peel Cone; E/F-band.
Surface search: Kivach or Pechora or SRN 207; I-band.
Fire control: Bass Tilt; H/I-band.
Sonars: Foal Tail; VDS (mounted on transom); active attack; high frequency.

Comment: Built at Yaroslavl in the early 1980s and transferred from Russian Black Sea Fleet. Pennant numbers changed from July 1999. All are based at Balaklava.

UPDATED

POLTAVA

9/2004*, Hartmut Ehlers / 1043548

GRIGORY GNATENKO

6/2003, B Lemachko / 0576460

1 SSV-10 CLASS (SUPPORT SHIP) (AGF)

DUNAI BG 80 (ex-500)

Displacement, tons: 340 full load
Dimensions, feet (metres): 129.3 × 23 × 3.9 *(39.4 × 7 × 1.2)*
Main machinery: 2 diesels; 2 shafts
Speed, knots: 12
Complement: 20 (4 officers)

Comment: Headquarters ship built in 1940. Taken over from the Russian Danube Flotilla and now acts as the command ship for the river brigades. Based at Odessa.

VERIFIED

DUNAI (old number)

4/1998, Ukraine Coast Guard / 0050322

13 STENKA (TARANTUL) CLASS (PROJECT 205P) (PCF)

PEREKOP BG 30	**ZAKARPATTIYA**	**BUKOVINA**
DONBAS BG 32	(ex-031, ex-PSKR 648)	(ex-034, ex-PSKR 702)
(ex-PSKR 705)	**ZAPORIZKAYA SEC**	**PODILLIYA** (ex-036, ex-PSKR 709)
MIKOLAIV BG 57	(ex-032, ex-PSKR 650)	**PAVEL DERZHAVIN**
(ex-PSKR 722)	**ODESSA**	(ex-037, ex-PSKR 720)
VOLIN (ex-020, ex-PSKR 637)	(ex-033, ex-PSKR 652)	+ 3

Displacement, tons: 253 full load
Dimensions, feet (metres): 129.3 × 25.9 × 8.2 *(39.4 × 7.9 × 2.5)*
Main machinery: 3 Type M 517 or M 583 diesels; 14,100 hp(m) *(10.36 MW)*; 3 shafts
Speed, knots: 37. **Range, n miles:** 500 at 35 kt; 1,540 at 14 kt
Complement: 30 (5 officers)
Guns: 4—30 mm/65 (2 twin) AK 230.
Torpedoes: 4—16 in *(406 mm)* tubes.
Depth charges: 2 racks (12).
Radars: Surface search: Pot Drum or Peel Cone; H/I- or E-band.
Fire control: Drum Tilt; H/I-band.
Navigation: Palm Frond; I-band.
IFF: High Pole. 2 Square Head.
Sonars: Stag Ear or Foal Tail; VDS; high frequency; Hormone type dipping sonar.

Comment: Similar hull to the Osa class. Built in the 1970s and 1980s. Transferred from Russia. Others have been cannibalised for spares. Based at Kerch (three), Odessa (eight) and Balaklava (two).

UPDATED

STENKA class
6/2001, B Lemachko / 0131160

3 MURAVEY (ANTARES) CLASS (PROJECT 133) (PCK)

BG 53 (ex-PSKR 105)	**BG 54** (ex-PSKR 108)	**GALICHINA** BG 55 (ex-PSKR 115)

Displacement, tons: 212 full load
Dimensions, feet (metres): 126.6 × 24.9 × 6.2; 14.4 (foils) *(38.6 × 7.6 × 1.9; 4.4)*
Main machinery: 2 gas turbines; 22,600 hp(m) *(16.6 MW)*; 2 shafts
Speed, knots: 60. **Range, n miles:** 410 at 12 kt
Complement: 30 (5 officers)

Guns: 1—3 in *(76 mm)*/60; 120 rds/min to 15 km *(8 n miles)*; weight of shell 7 kg.
1—30 mm/65 AK 630; 6 barrels; 3,000 rds/min combined to 2 km.
Torpedoes: 2—16 in *(406 mm)* tubes; SAET-40; anti-submarine; active/passive homing to 10 km *(5.4 n miles)* at 30 kt; warhead 100 kg.
Depth charges: 6.
Weapons control: Hood Wink optronic director.
Radars: Surface search: Peel Cone; E-band.
Fire control: Bass Tilt; H/I-band.
Sonars: Rat Tail; VDS; active attack; high frequency; dipping sonar.

Comment: Built at Feodosiya in the mid-1980s for the USSR Border Guard. High speed hydrofoil craft. All based at Balaklava. Pennant numbers changed from July 1999.

VERIFIED

BG 53 (old number)
3/1998, Ukraine Coast Guard / 0050319

27 ZHUK (GRIF) CLASS (PROJECT 1400M) (PB)

BG 100-126

Displacement, tons: 39 full load
Dimensions, feet (metres): 78.7 × 16.4 × 3.9 *(24 × 5 × 1.2)*
Main machinery: 2 Type M 401B diesels; 2,200 hp(m) *(1.6 MW)* sustained; 2 shafts
Speed, knots: 30. **Range, n miles:** 1,100 at 15 kt
Complement: 13 (1 officer)
Guns: 2—14.5 mm (twin, fwd) MGs. 1—12.7 mm (aft) MG.
Radars: Surface search: Spin Trough; I-band.

Comment: Russian Border Guard vessels built in the 1980s and transferred in 1996. Pennant numbers changed from the 600 series in mid-1999.

UPDATED

BG 119
9/2004*, Hartmut Ehlers / 1043547

4 SHMEL CLASS (PROJECT 1204) (PGR)

LUBNY BG 81 (ex-171)   **KANIV** BG 82 (ex-173)   **NIZYN** BG 83 (ex-172)   **IZMAYL** BG 84 (ex-174)

Displacement, tons: 77 full load
Dimensions, feet (metres): 90.9 × 14.1 × 3.9 *(27.7 × 4.3 × 1.2)*
Main machinery: 2 Type M 50 diesels; 2,200 hp(m) *(1.6 MW)* sustained; 2 shafts
Speed, knots: 25. **Range, n miles:** 600 at 12 kt
Complement: 12 (4 officers)
Guns: 1—3 in *(76 mm)*/48 (tank turret). 2—14.5 mm (twin) MGs. 5—7.62 mm MGs. 1 BP 6 rocket launcher; 18 barrels.
Mines: Can lay 9.
Radars: Surface search: Spint Trough; I-band.

Comment: Built at Kerch from 1967-74. Now part of the river brigade having been transferred from the Russian Danube flotilla. New pennant numbers are unconfirmed and there is doubt about the operational status of this class.

VERIFIED

LUBNY (old number)
4/1998, Ukraine Coast Guard / 0050323

17 KALKAN (PROJECT 50030) M CLASS
(INSHORE PATROL CRAFT) (PBR)

BG 303-304	BG 503-504	BG 807-808	+ 8
BG 604	BG 320	BG 333	

Displacement, tons: 8.5 full load
Dimensions, feet (metres): 38.1 × 10.8 × 2.0 *(11.6 × 3.3 × 0.6)*
Main machinery: 1 Type 475K diesel; 496 hp *(370 kW)*; 1 waterjet
Speed, knots: 34
Complement: 2

Comment: Built by Morye Feodosiya and entered service from 1996. Aluminium hulls and GRP superstructure. Can be armed with 7.62 mm or 12.7 mm MGs and 'Strela' shoulder launched missile.

VERIFIED

KALKAN class
6/2003, Morye / 0572655

2 PO2 CLASS (COASTAL PATROL CRAFT) (PB)

BG 501   BG 503

Displacement, tons: 56 full load
Dimensions, feet (metres): 70.5 × 11.5 × 3.3 *(21.5 × 3.5 × 1)*
Main machinery: 1 Type 3-D-12 diesel; 150 hp(m) *(110 kW)* sustained; 1 shaft
Speed, knots: 12
Complement: 8

Comment: Based at Balaclava.

VERIFIED

United Arab Emirates

Country Overview

The United Arab Emirates was formed on 2 December 1971 by the federation of seven states (formerly the Trucial States) lying along the east-coast of the Arabian Peninsula. With an area of 30,000 square miles, the country includes Abu Dhabi, Ajman, Dubai, al-Fujairah, Ras al Khaimah, Sharjah and Umm al-Qaiwain. It is bordered to the north by Qatar and to the south by Saudi Arabia. To the east lies Oman which is separated from its small exclave on the Musandam peninsula. There is a coastline of 713 n miles with the Gulf and with the Gulf of Oman. The city of Abu Dhabi is the capital and largest city while Dubai is the principal port and commercial centre. Territorial Seas (12 n miles) are claimed. An EEZ (200 n miles) has also been claimed but its limits have not been defined.

Following a decision of the UAE Supreme Defence Council on 6 May 1976 the armed forces of the member states were unified and the organisation of the UAE armed forces was furthered by decisions taken on 1 February 1978.

Headquarters Appointments

Commander, Naval Forces:
  Brigadier Suhail Shaheen Al-Murar
Deputy Commander, Naval Forces:
  Colonel Mohammad Mehmood Al-Madni
Director General, Coast Guard:
  Colonel Abdullah Rahman Al-Shelwah

Personnel

(a)   2005: 2,400 (200 officers) Navy. 1,200 (110 officers) Coast Guard.
(b)   Voluntary service

Bases

Abu Dhabi (main base).
Mina Rashid and Mina Jebel Ali (Dubai),
Mina Saqr (Ras al Khaimah), Mina Sultan (Sharjah),
Khor Fakkan (Sharjah-East Coast).

SUBMARINES

Notes: Submarine training has been conducted in the past but acquisition of submarines is understood to be a long-term aspiration.

10 SWIMMER DELIVERY VEHICLES (SDV)

Comment: Two classes of indigenously built Long Range Submersible Carriers (LRSC) have been developed by Emirates Marine Technologies. The 7.35 × 0.95 m Class 4 variant, of which approximately ten are believed to have been in service with UAE Special Forces since 1998, is capable of deploying a 200 kg payload. These are likely to be augmented by the larger 9.1 × 1.15 m Class 5 variant which can deliver 450 kg. Constructed of glass and carbon fibres, both variants are manned by two people, have a top speed of 7 kt, a range of 60 n miles at 6 kt and an operational depth of 30 m. They are equipped with depth-sounder, sonar and built-in breathing system.

VERIFIED

CLASS 5 LRSC
2001, Emirates Marine Technologies / 0095256

FRIGATES

2 KORTENAER CLASS (FFGHM)

Name	No	Builders	Laid down	Launched	Commissioned
ABU DHABI (ex-*Abraham Crijnssen*)	F 01 (ex-F 816)	Koninklijke Maatschappij De Schelde, Flushing	25 Oct 1978	16 May 1981	27 Jan 1983
AL EMIRAT (ex-*Piet Heyn*)	F 02 (ex-F 811)	Koninklijke Maatschappij De Schelde, Flushing	28 Apr 1977	3 June 1978	14 Apr 1981

Displacement, tons: 3,050 standard; 3,630 full load
Dimensions, feet (metres): 428 × 47.9 × 14.1; 20.3 (screws) *(130.5 × 14.6 × 4.3; 6.2)*
Main machinery: GOGOG; 2 RR Olympus TM3B gas turbines; 50,880 hp *(37.9 MW)* sustained
  2 RR Tyne RM1C gas turbines; 9,900 hp *(7.4 MW)* sustained; 2 shafts; cp props
Speed, knots: 30. **Range, n miles:** 4,700 at 16 kt on Tynes
Complement: 176 (18 officers) plus 24 spare berths

Missiles: SSM: 8 McDonnell Douglas Harpoon (2 quad) launchers ❶; active radar homing to 130 km *(70 n miles)* at 0.9 Mach; warhead 227 kg.
SAM: Raytheon Sea Sparrow Mk 29 octuple launcher ❷; semi-active radar homing to 14.6 km *(8 n miles)* at 2.5 Mach; warhead 39 kg; 24 missiles.
Guns: 1 OTO Melara 3 in *(76 mm)*/62 compact ❸; 85 rds/min to 16 km *(8.6 n miles)* anti-surface; 12 km *(6.5 n miles)* anti-aircraft; weight of shell 6 kg.
  1 Signaal SGE-30 Goalkeeper with General Electric 30 mm ❹; 7-barrelled; 4,200 rds/min combined to 2 km.
  2 Oerlikon 20 mm.
Torpedoes: 4—324 mm US Mk 32 (2 twin) tubes ❺. Honeywell Mk 46 or Whitehead A-244S Mod 1.
Countermeasures: Decoys: 2 Loral Hycor SRBOC Mk 36 6-tubed launchers ❻; chaff distraction or centroid modes.
ESM/ECM: Ramses ❼; intercept and jammer.
Combat data systems: Signaal SEWACO II action data automation; Link 11.
Radars: Air search: Signaal LW08 ❽; D-band; range 264 km *(145 n miles)* for 2 m² target.
Surface search: Signaal Scout ❾; I-band.
Fire control: Signaal STIR ❿; I/J-band; range 140 km *(76 n miles)* for 1 m² target.
  Signaal WM25 ⓫; I/J-band; range 46 km *(25 n miles)*.
Sonars: Westinghouse SQS-505; bow-mounted; active search and attack; medium frequency.

Helicopters: 2 Eurocopter AS 565 Panther ⓬.

Programmes: Contract signed on 2 April 1996 to transfer two Netherlands frigates, after refits by Royal Schelde. First one recommissioned in December 1997, second one in May 1998. Further transfers are unlikely.
Structure: Harpoon SSM and Goalkeeper CIWS has been purchased separately, as have the Scout radars. Additional air conditioning has been fitted.
Operational: Crew training takes place in the Netherlands. Both based at Jebel Ali.

UPDATED

ABU DHABI   *(Scale 1 : 1,200), Ian Sturton* / 0121417

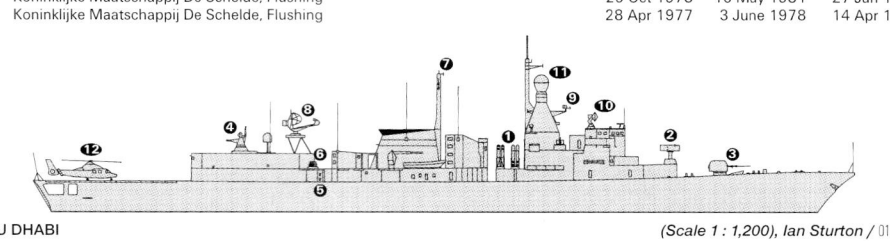

AL EMIRAT   *2/2002, A Sharma* / 0534078

ABU DHABI
1/2002, A Sharma / 0534062

CORVETTES

Notes: Project Fallah (LEWA 2) is for 90 m multirole corvettes.

2 MURAY JIB (MGB 62) CLASS (FSGHM)

Name	No	Builders	Launched	Commissioned
MURAY JIB	161 (ex-CM 01, ex-P 6501)	Lürssen, Bremen	Mar 1989	Nov 1990
DAS	162 (ex-CM 02, ex-P 6502)	Lürssen, Bremen	May 1989	Jan 1991

Displacement, tons: 630 full load
Dimensions, feet (metres): 206.7 × 30.5 × 8.2
(63 × 9.3 × 2.5)
Main machinery: 4 MTU 16V 538 TB92 diesels; 13,640 hp(m)
(10 MW) sustained; 4 shafts
Speed, knots: 32. **Range, n miles:** 4,000 at 16 kt
Complement: 43

Missiles: SSM: 8 Aerospatiale MM 40 Exocet (Block II) **❶**; inertial cruise; active radar homing to 70 km *(40 n miles)* at 0.9 Mach; warhead 165 kg; sea-skimmer.
SAM: Thomson-CSF modified Crotale Navale octuple launcher **❷**; radar guidance; IR homing to 13 km *(7 n miles)* at 2.4 Mach; warhead 14 kg.
Guns: 1 OTO Melara 3 in *(76 mm)*/62 Super Rapid **❸**; 120 rds/min to 16 km *(8.7 n miles)*; weight of shell 6 kg.

1 Signaal Goalkeeper with GE 30 mm 7-barrelled **❹**; 4,200 rds/min combined to 2 km.
2—12.7 mm MGs.
Countermeasures: Decoys: 2 Dagaie launchers **❺**; IR flares and chaff.
ESM/ECM: Racal Cutlass/Cygnus **❻**; intercept/jammer.
Weapons control: CSSE Najir optronic director **❼**.
Radars: Air/surface search: Bofors Ericsson Sea Giraffe 50HC **❽**; G-band.
Navigation: Racal Decca 1226; I-band.
Fire control: Bofors Electronic 9LV 223 **❾**; J-band (for gun and SSM).
Thomson-CSF DRBV 51C **❿**; J-band (for Crotale).

Helicopters: 1 Aerospatiale Alouette SA 316 **⓫**.

Programmes: Ordered in late 1986. Similar vessels to Bahrain craft. Delivery in October 1991.
Structure: Lürssen design adapted for the particular conditions of the Gulf. This class has good air defence and a considerable anti-ship capability. The helicopter hangar is reached by flight deck lift.
Operational: Pennant numbers changed in 2002.

VERIFIED

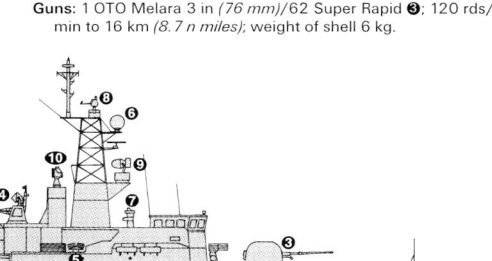

MURAY JIB

(Scale 1 : 600), Ian Sturton / 0080921

DAS (old number)

5/1998 / 0050325

MURAY JIB

5/2003, A Sharma / 0567564

0 + 4 (2) BAYNUNAH CLASS
(FAST ATTACK CRAFT—MISSILE) (PGGMH)

Name	No	Builders	Commissioned
—	—	CMN, Cherbourg	2008
—	—	Abu Dhabi Shipbuilding	2010
—	—	Abu Dhabi Shipbuilding	2011
—	—	Abu Dhabi Shipbuilding	2012

Displacement, tons: 630 full load
Dimensions, feet (metres): 229.6 × 36.1 × 9.2 *(70.0 × 11.0 × 2.8)*
Main machinery: 4 MTU 12V 595 TE 90 diesels; 22,500 hp *(16.8 MW)*; 3 (2—112 SII; 1—125 BII) Kamewa waterjets
Speed, knots: 32. **Range, n miles:** 2,400 at 15 kt
Complement: 37 (accommodation for 45)
Missiles: SSM: MBDA MM 40 Block III.
SAM: Raytheon Evolved Sea Sparrow RIM-162; Mk 56 VLS (8 missiles).
Guns: 1 OTO Melara 3 in *(76 mm)*/62 Super Rapid. 2—30 mm.
Countermeasures: Decoys: 2 MASS launchers.
RESM: Elettronica.
Combat data systems: Alenia Marconi Systems IPN-S. Link 11 and Link Y Mk 2.
Weapons control: VIGY-EOMS optronic director.
Radars: Air/surface search: Ericsson Sea Giraffe; G/H-band.
Surface search: Terma Scanter 2001; I-band.
Fire control: Alenia Marconi NA-25/XM; I-band.

Helicopters: 1 Eurocopter AS 565 Panther.

Comment: Project Baynunah succeeded Project LEWA 1 for the procurement of patrol boats and is a joint venture between Abu Dhabi Shipbuilding (ADSB) (Prime Contractor) and CMN of France. Contract signed 28 December 2003 for four ships with options for a further two. Based on a CMN BR67 design, it has a steel hull and aluminium superstructure. The first of class is to be built at Cherbourg for launch in 2006. CMN is to provide materials for follow-on vessels to be built by ADSB. *UPDATED*

BAYNUNAH PATROL CRAFT (artist's impression)   *1/2005*, CMN /* 0585982

PATROL FORCES

Notes: Three Fast Intercept Craft are reported to be in service.

2 MUBARRAZ CLASS (FAST ATTACK CRAFT—MISSILE) (PGGFM)

Name	No	Builders	Commissioned
MUBARRAZ	P 141 (ex-P 4401)	Lürssen, Bremen	Aug 1990
MAKASIB	P 142 (ex-P 4402)	Lürssen, Bremen	Aug 1990

Displacement, tons: 260 full load
Dimensions, feet (metres): 147.3 × 23 × 7.2 *(44.9 × 7 × 2.2)*
Main machinery: 2 MTU 20V 538 TB93 diesels; 9,370 hp(m) *(6.9 MW)* sustained; 2 shafts
Speed, knots: 40. **Range, n miles:** 500 at 38 kt
Complement: 40 (5 officers)

Missiles: SSM: 4 Aerospatiale MM 40 Exocet; inertial cruise; active radar homing to 70 km *(40 n miles)* at 0.9 Mach; warhead 165 kg; sea-skimmer.
SAM: 1 Matra Sadral sextuple launcher; Mistral; IR homing to 4 km *(2.2 n miles)*; warhead 3 kg.
Guns: 1 OTO Melara 3 in *(76 mm)*/62 Super Rapid; 120 rds/min to 16 km *(8.7 n miles)*; weight of shell 6 kg.
2 Rheinmetall 20 mm.
Countermeasures: Decoys: 2 Dagaie launchers; IR flares and chaff.
ESM/ECM: Racal Cutlass/Cygnus; intercept/jammer.
Weapons control: CSEE Najir optronic director (for SAM).
Radars: Air/surface search: Bofors Ericsson Sea Giraffe 50HC; G-band.
Navigation: Racal Decca 1226; I-band.
Fire control: Bofors Electronic 9LV 223; J-band (for gun and SSM).

Programmes: Ordered in late 1986 from Lürssen Werft at the same time as the two Type 62 vessels. Delivered in February 1991.
Modernisation: Mid-life refits for both vessels to be undertaken by Abu Dhabi Shipbuilding from 2004.
Structure: This is a modified TNC 38 design, with the first export version of Matra Sadral. The radome houses the jammer. The 20 mm guns are mounted on the bridge deck aft of the mast. *UPDATED*

MUBARRAZ   *9/2000 /* 0121415

MAKASIB   *2/2002, A Sharma /* 0534063

6 BAN YAS (TNC 45) CLASS
(FAST ATTACK CRAFT—MISSILE) (PGGF)

Name	No	Builders	Commissioned
BAN YAS	P 151 (ex-P 4501)	Lürssen Vegesack	Nov 1980
MARBAN	P 152 (ex-P 4502)	Lürssen Vegesack	Nov 1980
RODQM	P 153 (ex-P 4503)	Lürssen Vegesack	July 1981
SHAHEEN	P 154 (ex-P 4504)	Lürssen Vegesack	July 1981
SAGAR	P 155 (ex-P 4505)	Lürssen Vegesack	Sep 1981
TARIF	P 156 (ex-P 4506)	Lürssen Vegesack	Sep 1981

Displacement, tons: 260 full load
Dimensions, feet (metres): 147.3 × 23 × 8.2 *(44.9 × 7 × 2.5)*
Main machinery: 4 MTU 16V 538 TB92 diesels; 13,640 hp(m) *(10 MW)* sustained; 4 shafts
Speed, knots: 40. **Range, n miles:** 500 at 38 kt
Complement: 40 (5 officers)

Missiles: SSM: 4 Aerospatiale MM 40 Exocet; inertial cruise; active radar homing to 70 km *(40 n miles)* at 0.9 Mach; warhead 165 kg; sea-skimmer.
Guns: 1 OTO Melara 3 in *(76 mm)*/62; 60 rds/min to 16 km *(8.7 n miles)*; weight of shell 6 kg.
2 Breda 40 mm/70 (twin); 300 rds/min to 12.5 km *(6.8 n miles)*; weight of shell 0.96 kg.
2—7.62 mm MGs.
Countermeasures: Decoys: 1 CSEE trainable Dagaie; IR flares and chaff.
ESM: Racal Cutlass; intercept.
Combat data systems: CelsiusTech 9LV Mk 3E CETRIS
Weapons control: 1 CSEE Panda director for 40 mm. PEAB low-light USFA IR and TV tracker.
Radars: Surface search: Bofors Ericsson Sea Giraffe 50HC; G-band.
Navigation: Signaal Scout; I-band.
Fire control: Philips 9LV 200 Mk 2/3; J-band.

Programmes: Ordered in late 1977. First two shipped in September 1980 and four more in Summer 1981. This class was the first to be fitted with MM 40.
Modernisation: Upgrade contract for ship and propulsion systems given to Newport News. Work done by Abu Dhabi Shipbuilding Company. First pair completed in late 1998, second pair in mid-1999 and the third pair in mid-2000. Further modernisation is being undertaken under Project Tarif-45. This includes replacement of the combat data system, fire-control systems and improvement of Exocet to Block III configuration. Beginning in February 2004, two craft per year are to undergo the nine-month upgrades. *VERIFIED*

RODQM   *3/2003, A Sharma /* 0567565

6 ARDHANA CLASS (LARGE PATROL CRAFT) (PB)

Name	No	Builders	Commissioned
ARDHANA	P 3301 (ex-P 1101)	Vosper Thornycroft	24 June 1975
ZURARA	P 3302 (ex-P 1102)	Vosper Thornycroft	14 Aug 1975
MURBAN	P 3303 (ex-P 1103)	Vosper Thornycroft	16 Sep 1975
AL GHULLAN	P 3304 (ex-P 1104)	Vosper Thornycroft	16 Sep 1975
RADOOM	P 3305 (ex-P 1105)	Vosper Thornycroft	1 July 1976
GHANADHAH	P 3306 (ex-P 1106)	Vosper Thornycroft	1 July 1976

Displacement, tons: 110 standard; 175 full load
Dimensions, feet (metres): 110 × 21 × 6.6 *(33.5 × 6.4 × 2)*
Main machinery: 2 Paxman 12CM diesels; 5,000 hp *(3.73 MW)* sustained; 2 shafts
Speed, knots: 30. **Range, n miles:** 1,800 at 14 kt
Complement: 26
Guns: 2 Oerlikon/BMARC 30 mm/75 A32 (twin); 650 rds/min to 10 km *(5.5 n miles)*; weight of shell 1 kg or 0.36 kg.
1 Oerlikon/BMARC 20 mm/80 A41A; 800 rds/min to 2 km.
2—51 mm projectors for illuminants.
Radars: Surface search: Racal Decca TM 1626; I-band.

Comment: A class of round bilge steel hull craft. Originally operated by Abu Dhabi. New pennant numbers in 1996. To be replaced by the Project Baynunah craft from approximately 2004. *VERIFIED*

AL GHULLAN   *2/1997, A Sharma /* 0567566

8 ARCTIC 28 RIBs (PBF) and 12 AL-SHAALI TYPE

Displacement, tons: 4 full load
Dimensions, feet (metres): 27.9 × 9.7 × 2 *(8.5 × 3 × 0.6)*
Main machinery: 2 outboards; 450 hp *(336 kW)*
Speed, knots: 38
Complement: 1 plus 11 troops

Comment: RIBs ordered from Halmatic, Southampton in June 1992 and delivered in mid-1993. GRP hulls. Speed given is fully laden. Used by Special Forces. The Al-Shaali type were ordered in 1994 and built in Dubai.   **VERIFIED**

ARCTIC   *3/1995, H M Steele /* 0080926

SHIPBORNE AIRCRAFT

Numbers/Type: 4 Aerospatiale SA 316/319S Alouette.
Operational speed: 113 kt *(210 km/h)*.
Service ceiling: 10,500 ft *(3,200 m)*.
Range: 290 n miles *(540 km)*.
Role/Weapon systems: Reconnaissance and general purpose helicopters. Sensors: radar. Weapons: Unarmed.   **VERIFIED**

Numbers/Type: 7 Eurocopter AS 565 Panther.
Operational speed: 165 kt *(305 km/h)*.
Service ceiling: 16,700 ft *(5,100 m)*.
Range: 483 n miles *(895 km)*.
Role/Weapon systems: Ordered in March 1995 and delivered from 1998. Sensors: Thomson-CSF Agrion radar. Weapons: ASV; Aerospatiale AS 15TT ASM.   **VERIFIED**

LAND-BASED MARITIME AIRCRAFT

Notes: Procurement of five refurbished E-2C Hawkeye aircraft from the USA, under FMS funding arrangements, is under consideration.

Numbers/Type: 4 EADS CASA C-295M.
Operational speed: 256 kt *(474 km/h)*.
Service ceiling: 13,540 ft *(4,125 m)*.
Range: 2,250 n miles *(4,167 km)*.
Role/Weapon systems: Stretched version of CN-235M for maritime patrol and reconnaissance. Order announced in March 2001. Delivery to have started in 2004. Fitted with EADS Fully Integrated Tactical System (FITS), sensors and weapons to be selected.   **UPDATED**

C-295M   *2001, EADS CASA /* 0094291

Numbers/Type: 7 Aerospatiale AS 535 Cougar/Super Puma.
Operational speed: 150 kt *(280 km/h)*.
Service ceiling: 15,090 ft *(4,600 m)*.
Range: 335 n miles *(620 km)*.
Role/Weapon systems: Former transport helicopters. Five updated from 1995 with ASW equipment. Two used as VIP transports. Sensors: Omera ORB 30 radar; Thomson Marconi HS 312 dipping sonar. Weapons: ASV; one AM 39 Exocet ASM. ASW; A 244S torpedoes and mines.   **VERIFIED**

COUGAR   *6/1994 /* 0080927

Numbers/Type: 2 Pilatus Britten-Norman Maritime Defender.
Operational speed: 150 kt *(280 km/h)*.
Service ceiling: 18,900 ft *(5,760 m)*.
Range: 1,500 n miles *(2,775 km)*.
Role/Weapon systems: Coastal patrol and surveillance aircraft although seldom used in this role. Sensors: Nose-mounted search radar, underwing searchlight. Weapons: Underwing rocket and gun pods.   **VERIFIED**

AMPHIBIOUS FORCES

Notes: (1) There are also four civilian LCM ships, *El Nasirah 2, Baava 1, Makasib* and *Ghagha II*. Two Serna class LCUs are also civilian owned.
(2) A 42 m LCU was reportedly delivered by Abu Dhabi Shipbuilding in mid-2004 when a second unit was ordered.

4 LCT

L 64 (ex-6401)   L 65 (ex-6402)   L 66 (ex-6403)   L 67 (ex-6404)

Displacement, tons: 850 approx
Dimensions, feet (metres): 210 × 39.4 × 8.7 *(64.0 × 12.0 × 2.7)*
Main machinery: 2 Caterpillar diesels; 3,620 hp *(2.7 MW)*; 2 shafts
Speed, knots: 12.
Guns: 2—12.7 mm MGs.

Comment: Built at Abu Dhabi Naval Base and completed in 1996-99. Details are incomplete. Pennant numbers changed in 2001.   **UPDATED**

L 65 (old number)   *8/1996 /* 0080928

3 LANDING CRAFT (LCT)

Displacement, tons: 850 approx
Dimensions, feet (metres): 210 × 39.4 × 8.7 *(64.0 × 12.0 × 2.7)*
Main machinery: 2 Caterpillar 3508 diesels; 3,620 hp *(2.7 MW)*; 2 shafts
Speed, knots: 11.
Complement: 19 (plus 56 troops)
Military lift: military vehicles

Comment: Fully designed in the UAE, the vessels were ordered from ADSB in November 2001 and laid down in early 2002 for delivery by 2004. Details are speculative and based on those already in service. Weapons are expected to include medium calibre machine guns.   **UPDATED**

3 AL FEYI CLASS (LCU)

AL FEYI L 51 (ex-5401)   DAYYINAH L 52 (ex-5402)   JANANAH L 53 (ex-5403)

Displacement, tons: 650 full load
Dimensions, feet (metres): 164 × 36.1 × 9.2 *(50 × 11 × 2.8)*
Main machinery: 2 diesels; 1,248 hp *(931 kW)*; 2 shafts
Speed, knots: 11. **Range, n miles:** 1,800 at 11 kt
Complement: 10
Military lift: 4 vehicles
Guns: 2—12.7 mm MGs.

Comment: *Al Feyi* built by Siong Huat, Singapore; completed 4 August 1987. The other pair built by Argos Shipyard, Singapore to a similar design and completed in December 1988. Used mostly as transport ships. Pennant numbers changed in 2001.   **VERIFIED**

DAYYINAH (old number)   *6/1996 /* 0080929

12 + 4 TRANSPORTBÅT 2000 (LCP)

P 201-212

Displacement, tons: 43 full load
Dimensions, feet (metres): 77.1 × 16.7 × 3.3 *(23.5 × 5.1 × 1)*
Main machinery: 2 MTU 12V 2000 diesels; 2 Kamewa FF 550 waterjets
Speed, knots: 33.
Complement: 3
Military lift: 42 troops or 10 tons
Guns: 2—12.7 mm MGs.
Radars: Navigation: Terma; I-band.

Comment: Project 'Ghannatha' was for 12 amphibious transport craft based on the Transportbåt 2000 craft in service with the Royal Swedish Navy. Three craft were constructed at the Djupviks yard in Sweden while ADSB built the other nine. Details of the aluminium craft are based on those in Swedish service and have not been confirmed. Delivery was completed in 2004. A contract for a further four vessels was made with ADSB on 27 June 2004.
UPDATED

TRANSPORTBÅT (Swedish colours)   *6/2000, Michael Nitz /* 0106583

AUXILIARIES

1 DIVING TENDER (YDT)

AL GAFFA D 1051

Displacement, tons: 100 full load
Dimensions, feet (metres): 103 × 22.6 × 3.6 *(31.4 × 6.9 × 1.1)*
Main machinery: 2 MTU 12V 396 TB93 diesels; 3,260 hp(m) *(2.4 MW)* sustained; 2 water-jets
Speed, knots: 26. **Range, n miles:** 390 at 24 kt
Complement: 6

Comment: Ordered from Crestitalia in December 1985 for Abu Dhabi and delivered in July 1987. GRP hull. Used primarily for mine clearance but also for diving training, salvage and SAR. Fitted with a decompression chamber and diving bell. Lengthened version of Italian *Alcide Pedretti*.
VERIFIED

AL GAFFA   *3/1997 /* 0019363

1 COASTAL TUG (YTB)

ANNAD A 3501

Displacement, tons: 795 full load
Dimensions, feet (metres): 114.8 × 32.2 × 13.8 *(35 × 9.8 × 4.2)*
Main machinery: 2 Caterpillar 3606TA diesels; 4,180 hp *(3.12 MW)* sustained; 2 shafts; cp props; bow thruster; 362 hp *(266 kW)*
Speed, knots: 14. **Range, n miles:** 2,500 at 14 kt
Complement: 14 (3 officers)
Radars: Navigation: Racal Decca 2070; I-band.

Comment: Built by Dunston, Hessle, and completed in April 1989. Bollard pull, 55 tons. Equipped for SAR and is also used for logistic support.
VERIFIED

ANNAD   *6/1994 /* 0080930

2 HARBOUR TUGS (YTM)

TEMSAH A 51   UGAAB A 52

Displacement, tons: 90 full load
Dimensions, feet (metres): 54.1 × 16.4 × 5.9 *(16.5 × 5.0 × 1.8)*
Main machinery: 2 Volvo Penta TAMD-122A diesels; 760 hp *(560 kW)*; 2 shafts

Comment: Ordered from Damen shipyard, Gorinchem in 1996 and entered service in 1998. Main role to attend Kortenaer class frigates. Equipped with fire-fighting platform abaft the mainmast.
VERIFIED

UGAAB   *1/2002, A Sharma /* 0534119

COAST GUARD

Notes: Under control of Minister of Interior. In addition to the vessels listed below there is a number of Customs and Police launches including Barracuda craft, three Swedish Boghammar 13 m craft of the same type used by Iran and delivered in 1985, two Baglietto police launches acquired in 1988, about 10 elderly Dhafeer and Spear class of 12 and 9 m respectively, and two Halmatic Arun class Pilot craft delivered in 1990-91; some of these launches carry light machine guns.

POLICE BARRACUDA   *1/2002, A Sharma /* 0534111

2 PROTECTOR CLASS (WPB)

1101   1102

Displacement, tons: 180 full load
Dimensions, feet (metres): 108.3 × 22 × 6.9 *(33 × 6.7 × 2.1)*
Main machinery: 2 MTU 16V 396 TE94 diesels; 5,911 hp(m) *(4.35 MW)* sustained; 2 shafts; LIPS props
Speed, knots: 33
Complement: 14
Guns: 1 Mauser 20 mm. 2—12.7 mm MGs.
Weapons control: 1 SAGEM optronic director.
Radars: Surface search: I-band.

Comment: Ordered from FBM Marine, Cowes in 1998. Aluminium hulls. First one laid down 15 June 1998. Both delivered in late 1999. More may be built by Abu Dhabi Shipbuilders. Similar to Bahamas and Chilean naval craft.
VERIFIED

PROTECTOR 1101   *11/1999, UAE Coast Guard /* 0106675

5 CAMCRAFT 77 ft (COASTAL PATROL CRAFT) (WPB)

753-757

Displacement, tons: 70 full load
Dimensions, feet (metres): 76.8 × 18 × 4.9 *(23.4 × 5.5 × 1.5)*
Main machinery: 2 GM 12V-71TA diesels; 840 hp *(627 kW)* sustained; 2 shafts
Speed, knots: 25
Complement: 8
Guns: 2 Lawrence Scott 20 mm (not always embarked).
Radars: Surface search: Racal Decca; I-band.

Comment: Completed 1975 by Camcraft, New Orleans. Not always armed.

VERIFIED

CAMCRAFT 755   6/1997 / 0019364

16 CAMCRAFT 65 ft (COASTAL PATROL CRAFT) (WPB)

650-665

Displacement, tons: 50 full load
Dimensions, feet (metres): 65 × 18 × 5 *(19.8 × 5.5 × 1.5)*
Main machinery: 2 MTU 6V 396 TB93 diesels; 1,630 hp(m) *(1.2 MW)* sustained; 2 shafts (in 14)
  2 Detroit 8V-92TA diesels; 700 hp *(522 kW)* sustained; 2 shafts (in 2)
Speed, knots: 25
Complement: 8
Guns: 1 Oerlikon 20 mm GAM-BO1.
Radars: Surface search: Racal Decca; I-band.

Comment: Built by Camcraft, New Orleans and delivered by September 1978.

VERIFIED

CAMCRAFT 655   12/2001, A Sharma / 0534118

6 BAGLIETTO GC 23 TYPE (COASTAL PATROL CRAFT) (WPBF)

758-763

Displacement, tons: 50.7 full load
Dimensions, feet (metres): 78.7 × 18 × 3 *(24 × 5.5 × 0.9)*
Main machinery: 2 MTU 12V 396 TB93 diesels; 3,260 hp(m) *(2.4 MW)* sustained; 2 Kamewa
  water-jets
Speed, knots: 43. **Range, n miles:** 700 at 20 kt
Complement: 9
Guns: 1 Oerlikon 20 mm. 2—7.62 mm MGs.
Radars: Surface search: I-band.

Comment: Built by Baglietto, Varazze. First two completed in March and May 1986, second pair in
  July 1987 and two more in 1988. All were delivered to UAE Coast Guard in Dubai.

VERIFIED

BAGLIETTO 758   1987, UAE Coast Guard / 0080934

3 BAGLIETTO 59 ft (COASTAL PATROL CRAFT) (WPBF)

501-503

Displacement, tons: 22 full load
Dimensions, feet (metres): 59.4 × 13.9 × 2.3 *(18.1 × 4.3 × 0.7)*
Main machinery: 2 MTU 12V 183 TE92 diesels; 2 shafts
Speed, knots: 40
Complement: 6
Guns: 2—7.62 mm MGs.
Radars: Surface search: Racal Decca; I-band.

Comment: Ordered in 1992 and delivered in late 1993.

VERIFIED

BAGLIETTO 503   10/1993, UAE Coast Guard / 0080935

6 WATERCRAFT 45 ft (COASTAL PATROL CRAFT) (PB)

Displacement, tons: 25 full load
Dimensions, feet (metres): 45 × 14.1 × 4.6 *(13.7 × 4.3 × 1.4)*
Main machinery: 2 MAN D2542 diesels; 1,300 hp(m) *(956 kW)*; 2 shafts
Speed, knots: 26. **Range, n miles:** 380 at 18 kt
Complement: 5
Guns: Mounts for 2—7.62 mm MGs.
Radars: Surface search: Racal Decca; I-band.

Comment: Ordered from Watercraft, UK in February 1982. Delivery in early 1983. Four deleted.
  Two similar craft built by Halmatic were delivered to the Dubai Port Authority in October 1997.

VERIFIED

WATERCRAFT 45 ft   1984, UAE Coast Guard

35 HARBOUR PATROL CRAFT (PB/YDT)

Comment: The latest are 11 Shark 33 built by Shaali Marine, Dubai and delivered in 1993-94. The
  remainder are a mixture of Barracuda 30 ft and FPB 22 ft classes. All are powered by twin
  outboard engines and most carry a 7.62 mm MG and have a Norden radar. There are also two
  Rotork craft used as diving tenders. Customs boats are operated separately by each of the UAE
  states. Some have been built for Kuwait.

VERIFIED

BARRACUDA 271   12/2001, A Sharma / 0534117

24 + 30 SEASPRAY ASSAULT BOATS (PB)

Displacement, tons: To be announced
Dimensions, feet (metres): 31.2 × 10.2 × 1.6 *(9.5 × 3.1 × 0.5)*
Main machinery: 2 outboards; 500 hp *(375 kW)*
Speed, knots: 50. **Range, n miles:** 450 at 17 kt
Complement: 5
Radars: Navigation: I-band.

Comment: Initial batch of 24 craft delivered in September 2003. A further thirty were ordered in
  early 2004. Designed by Sea Spray Aluminium Boats.

NEW ENTRY

SEASPRAY   2/2004*, ADSB / 0563487

United Kingdom

Country Overview

The United Kingdom of Great Britain and Northern Ireland is situated in north-western Europe. It has a coastline of 6,700 n miles with the English Channel, the North Sea, the Irish Sea and the Atlantic Ocean. With an area of 93,341 square miles, it comprises the island of Great Britain (England, Scotland and Wales) and the six counties of Ulster that remained a constituent part of UK after Irish independence in 1922. It also includes the Isle of Wight, Anglesey, the Scilly, Orkney, Shetland, and Hebridean archipelagos and numerous smaller islands. The Isle of Man and the Channel Islands are direct dependencies but are not part of the UK. Other dependent territories are: Anguilla; Bermuda; British Antarctic Territory; British Indian Ocean Territory (BIOT); British Virgin Islands; Cayman Islands; Cyprus Sovereign Base Areas; Falkland Islands; Gibraltar; Montserrat; Ducie, Henderson and Oeno; St Helena and Dependencies (Ascension and Tristan da Cunha); South Georgia and South Sandwich Islands and the Turks and Caicos Islands. London is the capital, largest city and a major port. Major oil ports are at Forth, Sullom Voe and Milford Haven and non-oil ports at Tees and Hartlepool, Grimsby and Immingham, Southampton, Liverpool, Felixstowe, Medway, and Dover. Territorial seas of 12 n miles are claimed around the UK mainland and many dependencies. An EEZ (200 n miles) is claimed for Bermuda, South Georgia and South Sandwich Islands and Pitcairn. A Fishery Zone (200 n miles) is claimed for the mainland and some dependencies.

Headquarters Appointments

Chief of the Naval Staff and First Sea Lord:
   Admiral Sir Alan West, GCB, DSC, ADC
Commander-in-Chief, Fleet:
   Admiral Sir Jonathon Band, KCB
Chief of Naval Personnel and Commander-in-Chief, Naval Home Command:
   Vice Admiral Sir James Burnell-Nugent, KCB, CBE, ADC
Deputy Chief Executive, Warship Support Agency:
   Rear Admiral J Reeve, CB
Controller of the Navy:
   Rear Admiral R F Cheadle
Assistant Chief of the Naval Staff:
   Rear Admiral A J Johns, CBE

Flag Officers, Operational and National Commanders

Chief of Joint Operations:
   Air Marshal G Torphy, CBE
Deputy Commander-in-Chief, Fleet:
   Vice Admiral T P McClement, OBE
Chief of Staff (Warfare) (Rear Admiral Surface Ships):
   Rear Admiral D G Snelson, CB
Commander, Operations (Rear Admiral Submarines):
   Rear Admiral P Lambert
Rear Admiral, Fleet Air Arm:
   Rear Admiral A J Johns, CBE
Commander, UK Maritime Forces:
   Rear Admiral C R Style, CBE
Chief of Staff (Support):
   Rear Admiral T C Chittenden
Flag Officer, Sea Training:
   Rear Admiral R S Ainsley, ADC
Commander, UK Amphibious Forces:
   Major General J B Dutton, CBE
Flag Officer, Scotland, Northern England and Northern Ireland:
   Rear Admiral N H L Harris, MBE
Commander, British Forces Cyprus:
   Major General P T C Pearson, CBE
Commander United Kingdom Task Group:
   Commodore A J Rix
Commander, UK Maritime Component:
   Commodore P H Robinson
Commander Amphibious Task Group:
   Commodore C J Parry
Commander, Commando Brigade:
   Brigadier J G Rose
Commander, British Forces Gibraltar:
   Commodore A A S Adair
Commander, British Forces South Atlantic Islands:
   Air Commodore R Lacey

Flag Officers, Operational and National Commanders
— *continued*

Commodore Royal Fleet Auxiliaries:
   Commodore R C Thornton
Commodore Portsmouth Flotilla:
   Commodore R C Twitchen
Commodore Devonport Flotilla:
   Commodore J R Fanshawe
Captain Faslane Flotilla:
   Captain J A Boyd
Hydrographer to the Navy:
   Captain J Lye, OBE

Diplomatic Representation

Naval Attaché in Ankara:
   Wing Commander J Gillan, RAF
Naval Attaché in Athens:
   Captain J R Wills
Naval Attaché in Beijing:
   Group Captain K J Parkes
Naval Attaché in Berlin:
   Captain S R Gosden
Naval Attaché in Brasilia:
   Commander R A Harrison
Defence Adviser in Bridgetown:
   Captain S J Wilson
Defence Adviser in Brunei:
   Captain P H Watson
Naval Attaché in Cairo:
   Commander P W Holihead
Defence Adviser in Canberra:
   Commodore R T Love
Defence Attaché in Copenhagen:
   Commander C Wilson, OBE
Naval Attaché in The Hague:
   Captain A R Davies
Naval Adviser in Islamabad:
   Colonel M W Bibbey RM
Defence Attaché in Kiev:
   Captain S E Airey
Assistant Defence Adviser in Kuala Lumpur:
   Lieutenant Commander M P Davies
Defence Attaché in Lisbon:
   Commander R M Simmonds
Naval Attaché in Madrid:
   Captain N Dedman
Naval Attaché in Manama:
   Commander N P Smith
Defence Attaché in Manila:
   Group Captain Bailey
Naval Attaché in Moscow:
   Captain J Holloway
Naval Adviser in Ottawa:
   Group Captain T Brewer
Naval Attaché in Paris:
   Captain D Lombard
Naval Adviser in Pretoria:
   Commander P Lankester
Naval Attaché in Riyadh:
   Wing Commander M J Cole
Naval Attaché in Rome:
   Captain J H Hollidge
Defence Attaché in Santiago:
   Colonel I Campbell
Naval Attaché in Seoul:
   Group Captain C Greaves, OBE
Naval Adviser in Singapore:
   Commander B Boxall-Hunt
Defence Attaché in Stockholm:
   Wing Commander N Phillips
Naval Attaché in Tokyo:
   Captain S Chelton
Naval Attaché in Warsaw:
   Lieutenant Colonel S Croft
Naval Attaché in Washington:
   Commodore C J Gass

Royal Marines Operational Units

HQ 3 Commando Brigade RM; 40 Commando RM; 42 Commando RM; 45 Commando Group (RM/Army); Commando Logistic Regiment RM (RN/RM/Army); 3 Commando Brigade Command Support Group, EW Troop RM, Tactical Air Command Posts RM (3 regular, 1 reserve); 539 Assault Squadron RM (hovercraft, landing craft and raiding craft); Brigade Patrol Troop (reconnaissance); Special Boat Service RM; Fleet Royal Marines Protection Group (FRMPG); T Company RMR; 29 Commando Regiment RA (Army); 59 Independent Commando Squadron RE (Army); 20 Commando Battery RA; 131 Independent Squadron RE (Volunteers).

Bases

Northwood: C-in-C Fleet; CJO; Commander Operations
Portsmouth: C-in-C Navhome; DC-in-C Fleet; COS Warfare; COS Support; COMUKMARFOR; COMUKAMPHIBFOR; Com Portsmouth Flotilla; COMUKTG
Devonport: FOST; Com Devonport Flotilla; COMATG
Faslane: FOSNNI; Captain Faslane Flotilla

Prefix to Ships' Names

HMS (Her Majesty's Ship)

Personnel

2005:
(a)	Regulars:	RN 33,154 (6,976 officers)
		RM 6,585 (742 officers)
(b)	Volunteer Reserves:	RN 988 (198 officers)

Fleet Disposition

Portsmouth: 3 CV; 9 Type 42 DDG, 7 Type 23 FFG, 1st MCM Squadron; 2nd MCM Squadron; Fishery Protection Squadron; Antarctic Patrol Ship; 1st Patrol Boat Squadron; Fleet Diving Squadron
Devonport: 1 LPH; 2 LPD; 4 Type 22 FFG; 7 Type 23 FFG; 7 Trafalgar Class SSN; Surveying Squadron
Faslane: 4 SSBN; 4 Swiftsure Class SSN; 3rd MCM Squadron

Strength of the Fleet

Type	Active (Reserve)	Building (Projected)
SSBNs	4	—
Submarines-Attack	11	3 (3)
Aircraft Carriers	2 (1)	(2)
Destroyers	8	6 (2)
Frigates	18	—
Assault Ships (LPD)	2	—
Helicopter Carriers (LPH)	1	—
LSD/LSL (RFA)	3	4
Offshore Patrol Vessels	5	—
Patrol/Training Craft	18	—
Minehunters	19	—
Repair/Maintenance Ships (RFA)	1	—
Survey Ships	5	—
Antarctic Patrol Ships	1	—
Large Fleet Tankers (RFA)	2	—
Support Tankers (RFA)	4	—
Small Fleet Tankers (RFA)	3	—
Aviation Training Ships (RFA)	1	—
Fleet Replenishment Ships (RFA)	4	—
Transport Ro-Ro (RFR)	6	—

Fleet Air Arm Squadrons (see *Shipborne Aircraft* section) on 1 January 2005
FA = Fighter Attack. HMA = Helicopter Maritime Attack.

F/W	Aircraft	Role	Deployment	Squadron no	F/W	Aircraft	Role	Deployment	Squadron no
7	Sea Harrier	FA 2	Yeovilton, *Heron*	801	13	Jetstream	Aircrew Training	Culdrose, *Seahawk*	750
14	Sea Harrier	(FA. Mk 2/T8) Aircrew Training	RAF Wittering	OCU 20	5	Grob	Aircrew Training	Roborough	727

Helicopters		Role	Deployment	Squadron no	Helicopters		Role	Deployment	Squadron no
2	Merlin HM Mk 1	OEU	Culdrose, *Seahawk*	700M	7	Sea King HU Mk 5	SAR/Training	Culdrose, *Seahawk*	771
4	Merlin HM Mk 1	ASW/ASUW	Culdrose, *Seahawk*	814	4	Sea King HAS Mk 6	SAR/Utility	Culdrose, *Seahawk*	771
4	Merlin HM Mk 1	ASW/ASUW	Culdrose, *Seahawk*	820	10	Sea King HC 4	Commando Assault	Yeovilton, *Heron*	845
8	Merlin HM Mk 1	ASW/ASUW	Culdrose, *Seahawk*	824	10	Sea King HC 4	Commando Assault	Yeovilton, *Heron*	846
9	Merlin HM Mk 1	ASW/ASUW	Culdrose, *Seahawk*	829	10	Sea King HC 4	Commando Assault	Yeovilton, *Heron*	848
13 {	Sea King ASAC Mk 7	AEW	Culdrose, *Seahawk*	849 HQ	35	Lynx HMA. Mk 3/8	ASUW/ASW	Yeovilton, *Heron*	815
	Sea King ASAC Mk 7	AEW	Culdrose, *Seahawk*	849 A flight	12	Lynx HAS. Mk 3/8	Aircrew Training	Yeovilton, *Heron*	702
	Sea King ASAC Mk 7	AEW	Culdrose, *Seahawk*	849 B flight	6	Lynx Mk 7	Commando Support	Yeovilton, *Heron*	847 B
3	Sea King HU Mk 5	SAR	Prestwick, *Gannet*	Prestwick, SAR flight	6	Gazelle AH-1	Commando Recce	Yeovilton, *Heron*	847 A

Notes: (1) Joint Force Harrier (JFH) formed on 1 April 2000. Both RN Sea Harriers and RAF GR.7/9s form 3 Group whose Headquarters are at RAF High Wycombe. Following the withdrawal of the Sea Harrier from service in 2006, JFH will continue to be complemented by RN and RAF pilots. (2) 800 Squadron disbanded on 31 March 2004 and will rededicate in April 2006 with GR9A at RAF Cottesmore. (3) 801 Squadron is to disband on 31 March 2006 and rededicate in October 2006 with GR9A at RAF Cottesmore. (4) 899 Squadron disbanded on 31 March 2005 and joined Joint OCU 20 Squadron at RAF Wittering. (5) The Joint Helicopter Command (JHC) became operational on 1 April 2000 and brought all battlefield helicopters from all three services under one command at HQ Land, Wilton. Total helicopter assets number some 450. The command includes the Commando Helicopter Force (CHF), a group of four RN/RM squadrons, based at Yeovilton, which specialises in amphibious warfare. (6) Mirach 100/5 subsonic drones are operated by 792 Squadron at Culdrose.

DELETIONS

Submarines

2003   *Splendid*

Destroyers

2005   *Newcastle, Glasgow, Cardiff*

Frigates

2002   *Sheffield* (to Chile)
2003   *London* (to Romania)
2005   *Norfolk, Marlborough*

Amphibious Forces

2002   *Fearless*
2004   *Sir Geraint, Sir Percivale*

Mine Warfare Forces

2004   *Bridport*
2005   *Sandown, Inverness*

Patrol Forces

2002   *Alderney, Shetland* (both to Bangladesh)
2003   *Anglesey, Guernsey* (both to Bangladesh)
2004   *Lindisfarne* (to Bangladesh)

Survey Vessels

2002   *Beagle*

Auxiliaries

2002   *Bee, Mastiff, Collie, Milford*
2003   *Dalmatian, Joan, Norah, Seagull*
2004   *Cameron, Kinterbury*
2005   *Ladybird*

PENNANT LIST

Notes: Numbers are not displayed on Submarines.

Submarines

Ballistic Missile Submarines

S 28	Vanguard
S 29	Victorious
S 30	Vigilant
S 31	Vengeance

Attack Submarines

S 20	Astute (bldg)
S 21	Artful (bldg)
S 22	Ambush (bldg)
S 87	Turbulent
S 88	Tireless
S 90	Torbay
S 91	Trenchant
S 92	Talent
S 93	Triumph
S 104	Sceptre
S 105	Spartan
S 107	Trafalgar
S 108	Sovereign
S 109	Superb

Aircraft Carriers

R 05	Invincible
R 06	Illustrious
R 07	Ark Royal

Destroyers

D 32	Daring (bldg)
D 33	Dauntless (bldg)
D 34	Diamond (bldg)
D 35	Defender (bldg)
D 36	Dragon (bldg)
D 37	Duncan (bldg)
D 89	Exeter
D 90	Southampton
D 91	Nottingham
D 92	Liverpool
D 95	Manchester
D 96	Gloucester
D 97	Edinburgh
D 98	York

Frigates

F 78	Kent
F 79	Portland
F 80	Grafton
F 81	Sutherland
F 82	Somerset
F 83	St Albans
F 85	Cumberland
F 86	Campbeltown
F 87	Chatham
F 99	Cornwall
F 229	Lancaster
F 231	Argyll
F 234	Iron Duke
F 235	Monmouth
F 236	Montrose
F 237	Westminster
F 238	Northumberland
F 239	Richmond

Amphibious Warfare Forces

L 12	Ocean
L 14	Albion (bldg)
L 15	Bulwark (bldg)
L 105	Arromanches
L 107	Andalsnes
L 109	Akyab
L 110	Aachen
L 111	Arezzo
L 113	Audemer
L 3004	Sir Bedivere
L 3005	Sir Galahad
L 3006	Largs Bay (bldg)
L 3007	Lyme Bay (bldg)
L 3008	Mounts Bay (bldg)
L 3009	Cardigan Bay (bldg)
L 3505	Sir Tristram

Mine Warfare Forces

M 29	Brecon
M 30	Ledbury
M 31	Cattistock
M 32	Cottesmore
M 33	Brocklesby
M 34	Middleton
M 35	Dulverton
M 37	Chiddingfold
M 38	Atherstone
M 39	Hurworth
M 41	Quorn
M 104	Walney
M 106	Penzance
M 107	Pembroke
M 108	Grimsby
M 109	Bangor
M 110	Ramsey
M 111	Blyth
M 112	Shoreham

Patrol Forces

P 163	Express
P 164	Explorer
P 165	Example
P 167	Exploit
P 258	Leeds Castle
P 264	Archer
P 265	Dumbarton Castle
P 270	Biter
P 272	Smiter
P 273	Pursuer
P 274	Tracker
P 275	Raider
P 279	Blazer
P 280	Dasher
P 281	Tyne
P 282	Severn
P 283	Mersey
P 284	Clyde (bldg)
P 284	Scimitar
P 285	Sabre
P 291	Puncher
P 292	Charger
P 293	Ranger
P 294	Trumpeter

Survey Ships

H 86	Gleaner
H 87	Echo
H 88	Enterprise
H 130	Roebuck
H 131	Scott

Auxiliaries

A 81	Brambleleaf
A 83	Melton
A 84	Menai
A 87	Meon
A 96	Sea Crusader
A 98	Sea Centurion
A 109	Bayleaf
A 110	Orangeleaf
A 111	Oakleaf
A 132	Diligence
A 135	Argus
A 140	Tornado
A 142	Tormentor
A 146	Waterman
A 147	Frances
A 149	Florence
A 150	Genevieve
A 170	Kitty
A 171	Endurance
A 172	Lesley
A 178	Husky
A 182	Saluki
A 185	Salmoor
A 187	Salmaid
A 189	Setter
A 191	Bovisand
A 192	Cawsand
A 198	Helen
A 199	Myrtle
A 201	Spaniel
A 221	Forceful
A 222	Nimble
A 223	Powerful
A 224	Adept
A 225	Bustler
A 226	Capable
A 227	Careful
A 228	Faithful
A 229	Colonel Templer
A 231	Dexterous
A 232	Adamant
A 250	Sheepdog
A 269	Grey Rover
A 271	Gold Rover
A 273	Black Rover
A 280	Newhaven
A 281	Nutbourne
A 282	Netley
A 283	Oban
A 284	Oronsay
A 285	Omagh
A 286	Padstow
A 308	Ilchester
A 309	Instow
A 344	Impulse
A 345	Impetus
A 367	Newton
A 368	Warden
A 385	Fort Rosalie
A 386	Fort Austin
A 387	Fort Victoria
A 388	Fort George
A 389	Wave Knight
A 390	Wave Ruler
Y 21	Oilpress
Y 32	Moorhen
Y 33	Moorfowl

SUBMARINES

Notes: Three 6.7 m US-made Mk VIII Mod 1 Swimmer Delivery Vehicles were acquired in 1999. Battery-powered, they can transport six combat swimmers and have a radius of 67 km *(36 n miles).*

SDV Mk VIII   *1/2002, M Declerck /* 0132551

Strategic Missile Submarines (SSBN)

Notes: Decisions on whether to replace the Trident nuclear deterrent are expected in about 2008.

4 VANGUARD CLASS (SSBN)

Name	No	Builders	Laid down	Launched	Commissioned
VANGUARD	S 28	Vickers Shipbuilding & Engineering, Barrow-in-Furness	3 Sep 1986	4 Mar 1992	14 Aug 1993
VICTORIOUS	S 29	Vickers Shipbuilding & Engineering, Barrow-in-Furness	3 Dec 1987	29 Sep 1993	7 Jan 1995
VIGILANT	S 30	Vickers Shipbuilding & Engineering, Barrow-in-Furness	16 Feb 1991	14 Oct 1995	2 Nov 1996
VENGEANCE	S 31	Vickers Shipbuilding & Engineering, Barrow-in-Furness	1 Feb 1993	19 Sep 1998	27 Nov 1999

Displacement, tons: 15,900 dived
Dimensions, feet (metres): 491.8 × 42 × 39.4
(149.9 × 12.8 × 12)
Main machinery: Nuclear; 1 RR PWR 2; 2 GEC turbines; 27,500 hp *(20.5 MW)*; 1 shaft; pump jet propulsor; 2 auxiliary retractable propulsion motors; 2 WH Allen turbo generators; 6 MW; 2 Paxman diesel alternators; 2,700 hp *(2 MW)*
Speed, knots: 25 dived
Complement: 135 (14 officers)

Missiles: SLBM: 16 Lockheed Trident 2 (D5) 3-stage solid fuel rocket; stellar inertial guidance to 12,000 km *(6,500 n miles)*; thermonuclear warhead of up to 8 MIRV of 100-120 kT; cep 90 m. The D5 can carry up to 12 MIRV but under plans announced in November 1993 each submarine carried a maximum of 96 warheads (of UK manufacture). This reduced to 48 in 1999. Substrategic low yield nuclear warheads introduced in 1996.
Torpedoes: 4—21 in *(533 mm)* tubes. Marconi Spearfish; dual purpose; wire-guided; active/passive homing to 26 km *(14 n miles)* at 65 kt; or 31.5 km *(17 n miles)* at 50 kt; attack speed 55 kt; warhead 300 kg directed charge.
Countermeasures: Decoys: 2 SSE Mk 10 launchers for Submarine Countermeasures Acoustic Device (SCAD) Types 101, 102 and 200.
ESM: Racal UAP 3; intercept.

Combat data systems: Alenia Marconi Systems SMCS.
Weapons control: Ultra Electronics Outfit DCM 4.
Radars: Navigation: Kelvin Hughes Type 1007; I-band.
Sonars: TMSL Type 2054 composite multifrequency sonar suite includes Marconi/Ferranti Type 2046 towed array, Type 2043 hull-mounted active/passive search and Type 2082 passive intercept and ranging. Type 2081 Environmental Sensor System.

Programmes: On 15 July 1980 the decision was made to buy the US Trident I (C4) weapon system. On 11 March 1982 it was announced that the government had opted for the improved Trident II weapon system, with the D5 missile, to be deployed in a force of four submarines. *Vanguard* ordered 30 April 1986; *Victorious* 6 October 1987; *Vigilant* 13 November 1990 and *Vengeance* 7 July 1992.
Modernisation: *Vanguard* underwent LOP(R) at Devonport February 2002 to June 2004. *Victorious* to undergo LOP(R) from June 2004 to 2006. Invitations for tender for the upgrade of sonar Type 2054 were released in April 2005. A contract is expected in late 2005.
Structure: A new reactor core, Core H, has been fitted to *Vanguard* and is to be installed in the other three boats at their Long Overhaul Period (LOP(R)). No further reactor fuelling will be required during their service lives. The outer surface of the submarine is covered with conformal anechoic noise

reduction coatings. The limits placed on warhead numbers leaves spare capacity within the Trident system. This capacity is used for a non-strategic warhead variant which has been available since 1996. Fitted with Pilkington Optronics CK 51 search and CH 91 attack periscopes.
Operational: Three successful submerged launched firings of the D5 missile from USS *Tennessee* in December 1989 and the US missile was first deployed operationally in March 1990. *Vanguard* started sea trials in October 1992; the first UK missile firing was on 26 May 1994 and the first operational patrol in early 1995. At least one SSBN has been at immediate readiness to fire ballistic missiles since 1969, but as a result of the Strategic Defence Review in 1998, readiness to fire has been relaxed 'to days rather than minutes'. Plans to phase out the two crew system have been cancelled. Submarines on patrol can be given secondary tasks without compromising security. Based at Faslane.
Opinion: The intention not to use the full capacity of the Trident system, in the present international climate, has allowed the development of a number of sub-strategic options, while preserving the capability to increase the strategic weapon load at some time in the future. The UK Government has said it has no plans to deploy conventional warheads on Trident. Neither are there any known plans to modify some launch-tubes to accommodate cruise missiles.

UPDATED

VENGEANCE

6/2000 / 0106676

VENGEANCE

11/2004*, H M Steele / 1043570

VENGEANCE

6/2000 / 0106677

VIGILANT

6/2004*, H M Steele / 1043569

Attack Submarines (SSN)

Notes: Future submarine requirements are being taken forward in a twin-track approach. In the short-term, technology advances to an extended Astute class are under consideration. Conceptual studies are also investigating requirements for a 'Maritime Underwater Future Capability' (MUFC) post 2020. Decisions on the future of the Strategic Deterrent will also be a major factor.

7 TRAFALGAR CLASS (SSN)

Name	No	Builders	Laid down	Launched	Commissioned
TRAFALGAR	S 107	Vickers Shipbuilding & Engineering, Barrow-in-Furness	25 Apr 1979	1 July 1981	27 May 1983
TURBULENT	S 87	Vickers Shipbuilding & Engineering, Barrow-in-Furness	8 May 1980	1 Dec 1982	28 Apr 1984
TIRELESS	S 88	Vickers Shipbuilding & Engineering, Barrow-in-Furness	6 June 1981	17 Mar 1984	5 Oct 1985
TORBAY	S 90	Vickers Shipbuilding & Engineering, Barrow-in-Furness	3 Dec 1982	8 Mar 1985	7 Feb 1987
TRENCHANT	S 91	Vickers Shipbuilding & Engineering, Barrow-in-Furness	28 Oct 1985	3 Nov 1986	14 Jan 1989
TALENT	S 92	Vickers Shipbuilding & Engineering, Barrow-in-Furness	13 May 1986	15 Apr 1988	12 May 1990
TRIUMPH	S 93	Vickers Shipbuilding & Engineering, Barrow-in-Furness	2 Feb 1987	16 Feb 1991	12 Oct 1991

Displacement, tons: 4,740 surfaced; 5,208 dived
Dimensions, feet (metres): 280.1 × 32.1 × 31.2 *(85.4 × 9.8 × 9.5)*
Main machinery: Nuclear; 1 RR PWR 1; 2 GEC turbines; 15,000 hp *(11.2 MW)*; 1 shaft; pump jet propulsor; 2 WH Allen turbo generators; 3.2 MW; 2 Paxman diesel alternators; 2,800 hp *(2.09 MW)*; 1 motor for emergency drive; 1 auxiliary retractable prop
Speed, knots: 32 dived
Complement: 130 (18 officers)

Missiles: SLCM: Raytheon Tomahawk Block IV; Tercom and GPS aided inertial navigation system with DSMAC to 1,500 km *(810 n miles)* at 0.7 Mach; warhead 318 kg shaped charge or submunitions. Being fitted to all from 2005.
SSM: McDonnell Douglas UGM-84B Sub Harpoon Block 1C; active radar homing to 130 km *(70 n miles)* at 0.9 Mach; warhead 227 kg.
Torpedoes: 5—21 in *(533 mm)* bow tubes. Marconi Spearfish; wire-guided; active/passive homing to 26 km *(14 n miles)* at 65 kt; or 31.5 km *(17 n miles)* at 50 kt; attack speed 55 kt; warhead 300 kg directed charge; 20 reloads.
Mines: Can be carried in lieu of torpedoes.
Countermeasures: Decoys: SAWCS from 2002. 2 SSE Mk 8 launchers. Type 2066 and 2071 torpedo decoys.
RESM: Racal UAP; passive intercept.
CESM: Outfit CXA.
Combat data systems: Ferranti/Gresham/Dowty DCB/DCG or BAE Systems SMCS tactical data handling system.

Weapons control: BAE Systems SMCS.
Radars: Navigation: Kelvin Hughes Type 1007; I-band.
Sonars: Marconi 2072; hull-mounted; flank array; passive; low frequency.
Plessey Type 2020 or Marconi/Plessey 2074 or Thales Underwater Systems (TUS) 2076 (see *Modernisation*); hull-mounted; passive/active search and attack; low frequency.
Ferranti Type 2046 or TUS Type 2076; towed array; passive search; very low frequency.
Thomson Sintra Type 2019 PARIS or THORN EMI 2082; passive intercept and ranging.
Marconi Type 2077; short-range classification; active; high frequency.

Programmes: *Trafalgar* ordered 7 April 1977; *Turbulent* 28 July 1978; *Tireless* 5 July 1979; *Torbay* 26 June 1981; *Trenchant* 22 March 1983; *Talent* 10 September 1984; *Triumph* 3 January 1986.
Modernisation: *Trafalgar* completed refuel in December 1995 and was fitted with SMCS and Spearfish torpedoes. *Turbulent* refuelled by mid-1997 and was refitted with sonar 2074, SMCS and Spearfish. *Tireless* completed similar modernisation and refuelling in January 1999. Refuel periods for the last four boats are being undertaken in parallel with a major tactical modernisation programme, the main feature of which is installation of the sonar 2076 integrated sonar suite to replace the 2074 bow array and 2046 towed array. Other upgrades include enhancements to SMCS, a new command console and improved signature reduction measures. *Torbay*

and *Trenchant* were the first and second boats to complete a 2076 refit and refuel in 2003 and 2004 respectively. *Talent* is expected to complete her three-year refit in 2006 and *Triumph* is expected to start her refit in 2005. Meanwhile, an ongoing programme of software replacement will continue to realise capability improvements in the last four boats. As a parallel programme, Tomahawk cruise missiles are being fitted to the whole class. *Triumph* and *Trafalgar* were completed by mid-2001, *Turbulent* by the end of 2002 and *Trenchant* in 2004. *Tireless, Torbay* and *Talent* are to be fitted by 2006. Following successful US/UK development of an encapsulated torpedo-tube launch system for Block IV Tactical Tomahawk, Block III missiles are to be replaced. The Tactical Tomahawk Weapons Control (TTWC) and Tomahawk Strike Network (TSN) systems were installed in *Trafalgar* in 2004. *Turbulent* is to be the next to receive both systems. Replacement of the CESM system was initiated in 2002.
Structure: The pressure hull and outer surfaces are covered with conformal anechoic noise reduction coatings. Retractable forward hydroplanes and strengthened fins for under ice operations. Diving depth in excess of 300 m *(985 ft)*. Fitted with Pilkington Optronics CK 34 search and CH 84 or CM 010 attack optronic periscopes.
Operational: *Trafalgar* is to conduct test-firings of Tomahawk Block IV in May 2005. All of the class based at Devonport. The class is planned to pay off as follows: *Trafalgar* 2008; *Turbulent* 2009; *Tireless* 2011 and the remainder of the class by 2023.

UPDATED

TRENCHANT

6/2004 *, B Sullivan / 1043573

TIRELESS   8/2004*, B Sullivan / 1043571

TURBULENT   6/2004*, Tony Barclay-Jeffs, RAN / 1043572

TRAFALGAR   11/2004*, B Sullivan / 1043568

4 SWIFTSURE CLASS (SSN)

Name	No	Builders	Laid down	Launched	Commissioned
SOVEREIGN	S 108	Vickers Shipbuilding & Engineering, Barrow-in-Furness	18 Sep 1970	17 Feb 1973	11 July 1974
SUPERB	S 109	Vickers Shipbuilding & Engineering, Barrow-in-Furness	16 Mar 1972	30 Nov 1974	13 Nov 1976
SCEPTRE	S 104	Vickers Shipbuilding & Engineering, Barrow-in-Furness	19 Feb 1974	20 Nov 1976	14 Feb 1978
SPARTAN	S 105	Vickers Shipbuilding & Engineering, Barrow-in-Furness	26 Apr 1976	7 Apr 1978	22 Sep 1979

Displacement, tons: 4,000 light; 4,400 standard; 4,900 dived

Dimensions, feet (metres): 272 × 32.3 × 28 *(82.9 × 9.8 × 8.5)*

Main machinery: Nuclear; 1 RR PWR 1; 2 GEC turbines; 15,000 hp *(11.2 MW)*; 1 shaft; pump jet propulsor; 2 WH Allen turbo generators; 3.6 MW; 1 Paxman diesel alternator; 1,900 hp *(1.42 MW)*; 1 motor for emergency drive; 1 auxiliary retractable prop

Speed, knots: 30+ dived

Complement: 116 (13 officers)

Missiles: SLCM: Hughes Tomahawk Block III; Tercom aided inertial navigation system with GPS back-up to 1,700 km *(918 n miles)* at 0.7 Mach; warhead 318 kg shaped charge or submunitions. Fitted in S 105 only.

SSM: McDonnell Douglas UGM-84B Sub Harpoon; active radar homing to 130 km *(70 n miles)* at 0.9 Mach; warhead 227 kg.

Torpedoes: 5—21 in *(533 mm)* bow tubes. Marconi Spearfish; wire-guided; active/passive homing to 26 km *(14 n miles)* at 65 kt; or 31.5 km *(17 n miles)* at 50 kt; attack speed 55 kt; warhead 300 kg directed charge; 20 reloads.

Mines: Can be carried in lieu of torpedoes.

Countermeasures: Decoys: SAWCS from 2002. 2 SSE Mk 6 launchers. Type 2066 and 2071 torpedo decoys.

ESM: Racal UAP; passive intercept.

Combat data systems: Ferranti/Gresham/Dowty DCB/DCG or Dowty Sema SMCS tactical data handling system. Link 11 can be fitted.

Radars: Navigation: Kelvin Hughes Type 1007; I-band.

Sonars: Marconi/Plessey Type 2074; hull-mounted; active/passive search and attack; low frequency.

Marconi 2072; flank array; passive; low frequency.

Ferranti Type 2046; towed array; passive search; very low frequency.

SPARTAN (with DDH)   9/2004 *, H M Steele / 1043567

Thomson Sintra Type 2019 PARIS or THORN EMI 2082; passive intercept and ranging.

Marconi Type 2077; short-range classification; active; high frequency.

Programmes: *Sovereign* ordered 16 May 1969; *Superb*, 20 May 1970; *Sceptre*, 1 November 1971; *Spartan*, 7 February 1973; *Splendid*, 26 May 1976.

Modernisation: Each boat fitted with a PWR 1 Core Z during major refits to give a 8 to 10 year refit cycle (12 year life). *Sovereign* completed her first refit in 1984 and her second, with full tactical weapons system upgrade and Spearfish, in 1997. Similarly, *Superb* completed refits in 1986 and 1998 and *Sceptre* in 1987 and 2001. *Spartan* completed her first refit in 1989 and, having entered her second refit in 1999, completed early 2003. Other improvements to the class included acoustic elastomeric tiles, sonar processing equipment and improved decoys. Marconi Type 2077, short-

range classification sonar is also fitted and Marconi/Plessey 2074 bow sonar has replaced 2020. *Spartan* was equipped with Tomahawk Block III missiles in 2003 but the remainder of the class is not to be fitted. *Spartan* modified to operate with Dry Deck Hangar in 2003.

Structure: The pressure hull in the Swiftsure class maintains its diameter for much greater length than previous classes. Control gear by MacTaggart, Scott & Co Ltd for: attack and search periscopes, snort induction and exhaust, radar and ESM masts. The forward hydroplanes house within the casing. Fitted with Pilkington Optronics CK 33 search and CH 83 attack electro-optic periscopes.

Operational: All based at Faslane. As a result of budget cuts *Swiftsure* paid off in 1992 after less than 20 years' service and *Splendid* decommissioned in 2003. *Sovereign* and *Spartan* are to pay off in 2006, *Superb* in 2008 and *Sceptre* in 2010.

UPDATED

SOVEREIGN   5/2003, B Prézelin / 0572689

SPARTAN   3/2003, H M Steele / 0563246

0 + 3 (3) ASTUTE CLASS (SSN)

Name	No	Builders	Laid down	Launched	Commissioned
ASTUTE	S 20	BAE Systems, Barrow	31 Jan 2001	Aug 2007	2008
AMBUSH	S 21	BAE Systems, Barrow	22 Oct 2003	Feb 2009	2010
ARTFUL	S 22	BAE Systems, Barrow	11 Mar 2005	Aug 2009	2012

Displacement, tons: 6,500 surfaced; 7,800 dived
Dimensions, feet (metres): 318.2 × 37.0 × 32.8
(97 × 11.27 × 10)
Main machinery: Nuclear; 1 RR PWR 2; 2 Alsthom turbines;
   1 shaft; pump jet propulsor; 2 turbo generators; 2 diesel
   alternators; 2 motors for emergency drive; 1 auxiliary
   retractable prop
Speed, knots: 29 dived
Complement: 98 (12 officers) plus 12 spare

Missiles: SLCM: Tomahawk. Block IV (Tactical Tomahawk).
   SSM: Sub Harpoon.
Torpedoes: 6—21 in *(533 mm)* tubes for Tomahawk, Sub
   Harpoon and Spearfish torpedoes. Total of 38 weapons.
Mines: In lieu of torpedoes.
Countermeasures: Decoys. ESM: Racal UAP 4; intercept.
Combat data systems: BAE Systems ACMS tactical data
   handling system. Links 11/16.
Radars: Navigation: I-band.
Sonars: Thomson Marconi 2076 integrated suite with reelable
   towed array.

Programmes: Invitations to tender issued on 14 July 1994 to
   build three of the class with an option for two more. GEC-
   Marconi selected as prime contractor in December 1995.
   Contract to start building the first three placed on 17 March
   1997. First steel cut late 1999 but although formal keel-laying
   took place in 2001, design, engineering and programme
   management difficulties led to a three-year delay to the first of

ASTUTE (computer graphic)   *2001, BAE Systems /* 0131260

class. There are similar delays to the second and third of class.
Following the future force-structure announcements on
21 July 2004, a decision on the scope of Batch 2 remains
under consideration. Based on published decommissioning
plans, a fourth Astute class is required in about 2012 to
maintain an eight-strong SSN force. However, this date may
not be achievable. Thereafter, replacements for the last four
Trafalgar class will be required between about 2016 and
2022. It is possible, therefore, that there may be further
batches of three and two boats.

Structure: An evolution of the Trafalgar design with increased
   weapon load and reduced radiated noise but with overall
   performance similar to Trafalgar after full modernisation. The
   fin is slightly longer and there are two Pilkington Optronics
   CM010 non-hull-penetrating periscopes. The boats are to
   have a dry dock hangar capability.
Operational: Fitted with Core H, nuclear refuelling will not be
   necessary in the lifetime of the submarine. To be based at
   Faslane.

UPDATED

AIRCRAFT CARRIERS

0 + (2) QUEEN ELIZABETH CLASS (CV)

Name	No	Builders	Laid down	Launched	Commissioned
QUEEN ELIZABETH	—	—	2006	2009	2012
PRINCE OF WALES	—	—	2009	2012	2015

Displacement, tons: 65,000 full load
Dimensions, feet (metres): 931.7 × 127.9 × ?
(284 × 39.0 × ?)
Flight deck, feet (metres): to be announced
Main machinery: Integrated Full Electric Propulsion; 2 Rolls-
   Royce MT 30 gas turbine alternators; 96,500 hp *(72 MW)*; 4
   diesel generators; 53,640 hp *(40 MW)*; 2 induction motors;
   53,640 hp *(40 MW)*; 2 shafts
Speed, knots: 26+. **Range, n miles:** to be announced
Complement: 600 approx plus aircrew

Guns: to be announced.
Countermeasures: to be announced.
Combat data systems: to be announced.
Weapons control: to be announced.
Radars: Air search: to be announced.
   Surface search: to be announced.
   Navigation: to be announced.
   Fire Control: to be announced.
Tacan: to be announced.

Fixed-wing aircraft: Approximately 36: typically a mix of F35
   combat aircraft, Merlin anti-submarine aircraft and Maritime
   Airborne Surveillance & Control aircraft.

Programmes: Following competitive assessment work which
   included investigation of design options, an alliance approach
   involving BAE Systems, Thales UK and the MoD was
   announced on 30 January 2003 as the best method for
   delivering the ships to time and cost. This collaborative
   approach is to be enabled by formal contracting
   arrangements. The third stage of assessment (AP3) began
   formally on 5 September 2003 and was extended by 12
   months in July 2004 to accommodate further risk reduction
   work and to increase the maturity of the design. Following
   appointment on 7 February 2005 of Kellogg Brown & Root
   (UK) Ltd, a subsidiary of Halliburton, as its preferred 'physical
   integrator' to manage the construction programme of the two
   ships, it is planned to proceed to the Demonstration &
   Manufacture Phases later in 2005. The ship is likely to be built
   in three to five separate sections and four shipyards were
   identified in January 2003 as potential contenders to

undertake construction: BAE Naval Ships at Govan, Babcock
BES at Rosyth, Swan Hunter at Wallsend on Tyne and VT
Shipbuilding in Portsmouth. However, the involvement of
other yards has not been ruled out depending on the
shipbuilding strategy selected.

Structure: The initial 292 m 'Alpha' design solution was
   developed by Thales UK, for which BMT Defence Services
   was principal naval architect. This incorporated a 'ski-ramp' to
   enable operation of the STOVL variant of the Lockheed Martin
   F35 Joint Strike Fighter (selected on 30 September 2002)
   and a novel two-island arrangement, with flight control from
   the after island, and two deck-edge aircraft lifts. Allowance
   was also made for future conversion to facilitate CTOL
   operations although catapults, steam generation plant and
   arrestor gear will not be fitted in the initial STOVL
   configuration. Following consideration of smaller 265 m
   'Bravo' and 'Charlie' designs, it is likely that the final design will
   be based on a 280 m 'Delta' design which is reported to retain
   some of the original 'Alpha' features and have the capability to
   operate up to 36 aircraft.
Operational: To be based at Portsmouth.

UPDATED

CVF (initial design concept)   *6/2002, Thales /* 0523598

3 INVINCIBLE CLASS (CV)

Name	No	Builders	Laid down	Launched	Commissioned
INVINCIBLE	R 05	Vickers Shipbuilding & Engineering, Barrow-in-Furness	20 July 1973	3 May 1977	11 July 1980
ILLUSTRIOUS	R 06	Swan Hunter Shipbuilders, Wallsend	7 Oct 1976	1 Dec 1978	20 June 1982
ARK ROYAL	R 07	Swan Hunter Shipbuilders, Wallsend	14 Dec 1978	2 June 1981	1 Nov 1985

Displacement, tons: 20,600 full load
Dimensions, feet (metres): 685.8 oa; 632 wl × 118 oa; 90 wl × 26 (screws) *(209.1; 192.6 × 36; 27.5 × 8)*
Flight deck, feet (metres): 550 × 44.3 *(167.8 × 13.5)*
Main machinery: COGAG; 4 RR Olympus TM3B gas turbines; 97,200 hp *(72.5 MW)* sustained; 2 shafts
Speed, knots: 28. **Range, n miles:** 7,000 at 19 kt
Complement: 685 (60 officers) plus 366 (80 officers) aircrew plus up to 600 marines

Guns: 3 Signaal/General Electric 30 mm 7-barrelled Gatling Goalkeeper (R 05 and R 06); 4,200 rds/min to 1.5 km.
3 General Dynamics 20 mm Phalanx Mk 15 (R 07) ❶; 6 barrels/launcher; 4,500 rds/min to 1.5 km.
2 Oerlikon/BMARC 20 mm GAM-BO1 ❷.
Countermeasures: Decoys: Outfit DLH; 8 Sea Gnat 6-barrelled 130 mm/102 mm dispensers ❸. Prairie Masker noise suppression system.
ESM: Racal UAT ❹; intercept.
Combat data systems: ADAWS 20 action data automation; Link 11 and Siemens Plessey JTIDS Link 16. JMCIS. Matra Marconi SCOT 2D SATCOM ❺.
Weapons control: Rademec optronic director.
Radars: Air search: Marconi/Signaal Type 1022 ❻; D-band.
Air/Surface search: AMS Type 996 ❼; E/F-band.
Navigation: 2 Kelvin Hughes Type 1007 ❽; I-band.
1 Racal Decca 1008; E/F-band.

Fixed-wing aircraft: Tailored air group of up to 24 aircraft including: BAe Sea Harrier FA 2 ❾ and Harrier GR 7A/9A.
Helicopters: Westland Merlin HM.Mk 1 ❿; Westland Sea King ASAC Mk 7. Chinook HC2. Apache AH1.

Programmes: The first of class, the result of many compromises, was ordered from Vickers on 17 April 1973. The order for the second ship was placed on 14 May 1976, the third in December 1978.
Modernisation: In January 1989 R 05 completed modernisation which included a 12° ski ramp, space and support facilities for at least 21 aircraft, three Goalkeeper systems, Sea Gnat decoys, 996 radar, Flag and Command facilities and accommodation for an additional 120 aircrew and Flag Staff. In February 1994 R 06 completed a similar modernisation to bring her to the same standard, but with additional command and weapon system improvements. Ski ramp has been increased to 13°. Modifications to operate Harrier GR.7 have been completed in all three ships (R 06 – 1998, R 05 – 2000 and R 07 – 2001). This included the removal of Sea Dart, increasing the flight deck area by 7 per cent (23 × 18 m) and fitting GR.7 support facilities.
Structure: In 1976-77 an amendment was incorporated to allow for the transport and landing of an RM Commando. The forward end of the flight deck (ski ramp) allows STOVL aircraft of greater all-up weight to operate more efficiently. *Illustrious* fitted with a composite third mast at the after end of the island structure to provide mountings for additional communications. She has also had substantial internal changes to accommodate troops in the LPH role. *Ark Royal* is to receive a third advanced technology mast in 2005.
Operational: The primary role of this class is to operate STOVL aircraft and helicopters. Sea Harriers are to be phased out by 2006 by when the embarked fixed-wing air-group will migrate to an all Harrier GR. Mk 7A/9A ground-attack force. *Ark Royal* deployed the first front-line Merlin helicopter squadron. *Invincible*'s one year tailored refit completed in early 2003. *Illustrious* completed a two-year refit in late 2004 when she relieved *Ark Royal* (including the standby LPH role). Following a docking in 2006, *Ark Royal* will relieve *Invincible* in 2007. *Invincible* will then revert to extended readiness while *Illustrious* and *Ark Royal* continue in service until 2012 and 2013 respectively when the CVF enter service. All based at Portsmouth.

UPDATED

ILLUSTRIOUS   *12/2004*, Royal Navy /* 1043574

INVINCIBLE   *10/2003, Paul Smith, Royal Navy /* 0572721

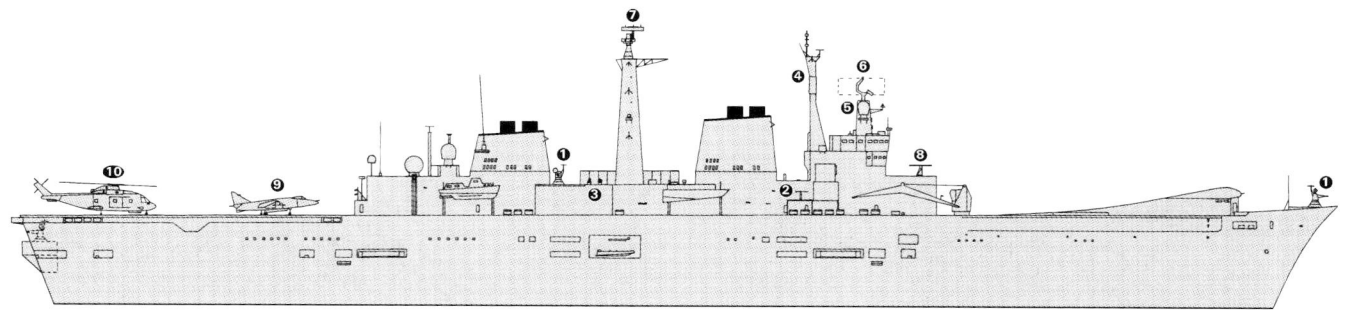

INVINCIBLE   *(Scale 1 : 1,200), Ian Sturton /* 1043487

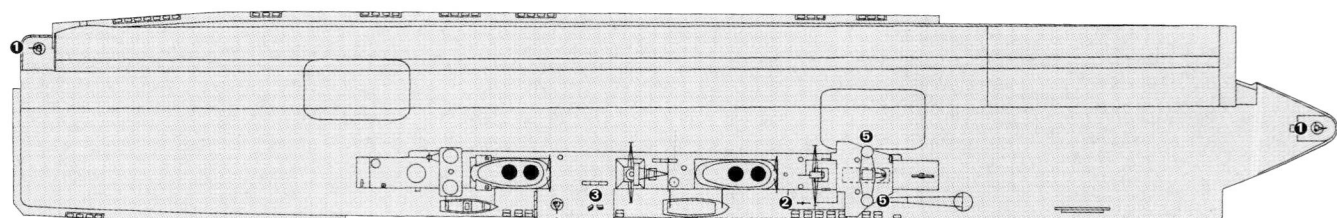

INVINCIBLE   *(Scale 1 : 1,200), Ian Sturton /* 1043486

INVINCIBLE

10/2003, Paul Smith, Royal Navy / 0572720

ILLUSTRIOUS

12/2004*, Royal Navy / 1043560

DESTROYERS

Notes: *Bristol* (D 23) is an immobile tender used for training in Portsmouth Harbour.

4 TYPE 42 CLASS (BATCH 2) (DDGH)

Name	No	Builders	Laid down	Launched	Commissioned
EXETER	D 89	Swan Hunter Shipbuilders, Wallsend-on-Tyne	22 July 1976	25 Apr 1978	19 Sep 1980
SOUTHAMPTON	D 90	Vosper Thornycroft, Woolston	21 Oct 1976	29 Jan 1979	31 Oct 1981
NOTTINGHAM	D 91	Vosper Thornycroft, Woolston	6 Feb 1978	18 Feb 1980	14 Apr 1983
LIVERPOOL	D 92	Cammell Laird, Birkenhead	5 July 1978	25 Sep 1980	1 July 1982

Displacement, tons: 4,500 standard; 4,800 full load
Dimensions, feet (metres): 412 oa; 392 wl × 47 × 19 (screws) *(125; 119.5 × 14.3 × 5.8)*
Main machinery: COGOG; 2 RR Olympus TM3B gas turbines; 50,000 hp *(37.3 MW)* sustained; 2 RR Tyne RM1C gas turbines (cruising); 9,900 hp *(7.4 MW)* sustained; 2 shafts; cp props
Speed, knots: 29. **Range, n miles:** 4,000 at 18 kt
Complement: 287 (24 officers) (accommodation for 312)

Missiles: SAM: BAE Systems Sea Dart twin launcher ❶; semi-active radar guidance to 40 km *(21.5 n miles)* at 2 Mach; height envelope 100-18,300 m *(328-60,042 ft)*; 22 missiles; limited anti-ship capability.
Guns: 1 Vickers 4.5 in *(114 mm)*/55 Mk 8 ❷; 25 rds/min to 27 km *(14.6 n miles)* anti-surface; weight of shell 21 kg.
2 BEMARC 20 mm GAM-BO1 ❸ or ❹; 1,000 rds/min to 2 km.
2 General Dynamics 20 mm Phalanx Mk 15 ❺; 6 barrels per launcher; 4,500 rds/min combined to 1.5 km.
Countermeasures: Decoys: Outfit DLH; 4 Sea Gnat 130 mm/102 mm 6-barrelled launchers ❻. Irvin DLF 3 offboard decoys.
Type 2070 (SLQ-25A); towed torpedo decoy.
ESM: Racal UAT 1; intercept.

SOUTHAMPTON   *(Scale 1 : 1,200), Ian Sturton /* 0572736

Combat data systems: ADAWS 20 Ed 2.4. 2 Matra Marconi SCOT 1C SATCOMs ❼; Link 11. JTIDS. Link 16 in due course.
Weapons control: GWS 30 Mod 2; GSA 1 secondary system. 2 Radamec 2100 series optronic surveillance directors.
Radars: Air search: Marconi/Signaal Type 1022 ❽; D-band.
Air/surface search: AMS Type 996 ❾; E/F-band.
Navigation: Kelvin Hughes Type 1007 ❿; I-band and Racal Decca Type 1008 ⓫; E/F-band.
Fire control: 2 Marconi Type 909 ⓬; I/J-band.
Sonars: Ferranti/Thomson Type 2050 or Plessey Type 2016; hull-mounted; active search and attack; medium frequency.

Helicopters: 1 Westland Lynx HAS 3/8 ⓭.

Programmes: Batch 1 ships decommissioned. All remaining ships Batch 2.
Modernisation: Phalanx replaced 30 mm guns in 1987-89. Batch 2 have had a command system update JTIDS (Link 16) improved ammunition.
Structure: Torpedo tubes removed.
Operational: First of class paid off in 1999. *Newcastle* and *Glasgow* in January 2005 and *Cardiff* is to be decommissioned in August 2005.   **UPDATED**

EXETER   *7/2004*, Hachiro Nakai /* 1043600

NOTTINGHAM   *10/2004*, W Sartori /* 1043599

SOUTHAMPTON   *6/2004*, John Brodie /* 1043598

4 TYPE 42 CLASS (BATCH 3) (DDGH)

Name	No	Builders	Laid down	Launched	Commissioned
MANCHESTER	D 95	Vickers Shipbuilding & Engineering, Barrow-in-Furness	19 May 1978	24 Nov 1980	16 Dec 1982
GLOUCESTER	D 96	Vosper Thornycroft, Woolston	29 Oct 1979	2 Nov 1982	11 Sep 1985
EDINBURGH	D 97	Cammell Laird, Birkenhead	8 Sep 1980	14 Apr 1983	17 Dec 1985
YORK	D 98	Swan Hunter Shipbuilders, Wallsend-on-Tyne	18 Jan 1980	21 June 1982	9 Aug 1985

Displacement, tons: 4,500 standard; 5,200 full load
Dimensions, feet (metres): 462.8 oa; 434 wl × 49.9 × 19
(screws) *(141.1; 132.3 × 15.2 × 5.8)*
Main machinery: COGOG; 2 RR Olympus TM3B gas turbines;
50,000 hp *(37.3 MW)* sustained; 2 RR Tyne RM1C gas
turbines (cruising); 10,680 hp *(8 MW)* sustained; 2 shafts; cp
props
Speed, knots: 30+. **Range, n miles:** 4,000 at 18 kt
Complement: 287 (26 officers)

Missiles: SAM: BAE Systems Sea Dart twin launcher ❶; semi-
active radar guidance to 40 km *(21 n miles)*; warhead HE; 22
missiles; limited anti-ship capability.
Guns: 1 Vickers 4.5 in *(114 mm)*/55 Mk 8 ❷; 25 rds/min to
27 km *(14.6 n miles)* anti-surface; weight of shell 21 kg. Mod
1 in D 97 and D 98 by 2005.
2 BEMARC 20 mm GAM-BO1 ❸; 1,000 rds/min to 2 km.
2 General Dynamics 20 mm Phalanx Mk 15 ❹; 6 barrels per
launcher; 4,500 rds/min combined to 1.5 km.
Countermeasures: Outfit DLH; 4 Sea Gnat 130 mm/102 mm
6-barrelled launchers ❺; DLF-3 offboard decoys.
Type 2070 (SLQ-25A); towed torpedo decoy.
ESM: Racal UAT; intercept.
Combat data systems: ADAWS 20 Ed 2.4 action data
automation. Matra Marconi SCOT 1C SATCOM ❻; Link 11.
JTIDS. Link 16.
Weapons control: GWS 30 Mod 2 (for SAM); GSA 1 secondary
system. 2 Radamec 2100 series optronic surveillance
directors.

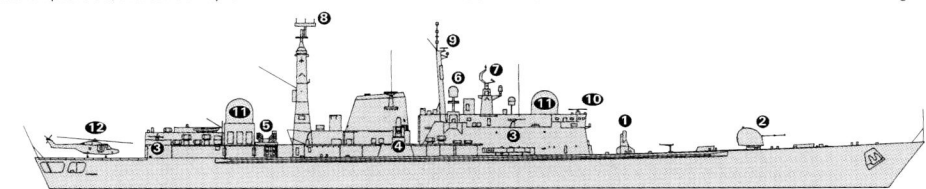

MANCHESTER   *(Scale 1 : 1,200)*, Ian Sturton / 0572735

Radars: Air search: Marconi/Signaal Type 1022 ❼; D-band.
Air/surface search: AMS Type 996 ❽; E/F-band.
Navigation: Kelvin Hughes Type 1007 ❾; I-band and Racal Decca
Type 1008 ❿; E/F-band.
Fire control: 2 Marconi Type 909 Mod 1 ⓫; I/J-band.
Sonars: Ferranti/Thomson Type 2050 or Plessey Type 2016;
hull-mounted; active search and attack.

Helicopters: 1 Westland Lynx HAS.Mk 3/8 ⓬.

Programmes: The completion of the last three ships was delayed
to allow for some modifications resulting from experience in
the Falklands' campaign (1982).
Modernisation: Vulcan Phalanx replaced 30 mm guns 1987-89.
D 97 had a partial conversion in 1990 with the Phalanx
moved forward and a protective visor fitted around the bow of
the ship but reverted to the standard armament in 1994. All

have had a command system update. Sea Gnat decoy
launchers can fire a variety of devices. JTIDS (Link 16) fitted in
D 97 in 1996, followed by the remainder by 1999. D 98
conducted trials of a Sea Ram SAM launcher vice port side
Phalanx in 2001. Mk 8 Mod 1 gun to be fitted in D 98 and
D 97 from 2004. Demonstration firings of Active Decoy
Round (ADR) in D 98 in 2003.
Structure: A strengthening beam has been fitted on each side
which increased displacement by 50 tons and beam by 2 ft.
Torpedo tubes removed.
Operational: The helicopter carries the Sea Skua air-to-surface
weapon for use against lightly defended surface ship targets.
Planned to pay off as Type 45 enter service. Based at
Portsmouth.

UPDATED

YORK   *11/2004**, Maritime Photographic / 1043603

EDINBURGH   *10/2004**, Marion Ferrette / 1043595

MANCHESTER   *6/2004**, John Brodie / 1043596

0 + 6 (2) D CLASS (TYPE 45) (DDGHM)

Name	No	Builders	Laid down	Launched	Commissioned
DARING	D 32	BAE Systems Marine/Vosper Thornycroft	Mar 2003	2006	May 2009
DAUNTLESS	D 33	BAE Systems Marine/Vosper Thornycroft	Aug 2004	2007	2010
DIAMOND	D 34	BAE Systems Marine/Vosper Thornycroft	2005	2007	2010
DEFENDER	D 35	BAE Systems Marine/Vosper Thornycroft	2005	2008	2011
DRAGON	D 36	BAE Systems Marine/Vosper Thornycroft	2006	2008	2012
DUNCAN	D 37	BAE Systems Marine/Vosper Thornycroft	2006	2009	2013

Displacement, tons: 5,800 standard; 7,350 full load

Dimensions, feet (metres): 500.1 oa; 462.9 wl × 69.6 × 17.4 *(152.4; 141.1 × 21.2 × 5.3)*

Main machinery: Integrated Electric Propulsion; 2 RR WR-21 gas turbine alternators; 42 MW; 2 diesel generators; 4 MW; 2 motors; 40 MW; 2 shafts; fixed props

Speed, knots: 29. **Range, n miles:** 7,000 at 18 kt

Complement: 187 plus 38 spare

Missiles: SSM: Space for 8 Harpoon (2 quad) ❶.
SAM: 6 DCN Sylver A 50 48 cell VLS ❷ PAAMS (principal anti-air missile system); 16 Aster 15 and 32 Aster 30 weapons or combination.

Guns: 1 Vickers 4.5 in *(114 mm)*/55 Mk 8 Mod 1 ❸.
2—20 mm Vulcan Phalanx (fitted for both not with) ❹.
2—30 mm ❺.

Countermeasures: Decoys: 4 DLH (chaff, IR); DLF offboard decoys; ❻. SSTD torpedo defence system.
ECM: to be decided.
RESM: Thales Type UAT (mod) ❼; intercept.
CESM: to be decided.

Combat data systems: SSCS-21 (based on DNA SSCS with additional AAW functions); Links 11, 16 STDL and 22. SATCOM ❽. CEC to be fitted.

Weapons control: 2 EOGCS (based on Radamec 2500).

Radars: Air/surface search: Signaal/Marconi S1850M ❾; D-band.
Surveillance/fire control: BAE Systems Sampson ❿; E/F-band; multifunction.
Surface search ⓫. E/F-band.
Navigation: E/F- and I-band.

Sonars: EDO/ULTRA MFS-7000; bow mounted; medium frequency.

Helicopters: Lynx Mk HMA 8 (first batch) or Merlin HM.Mk 1 ⓬.

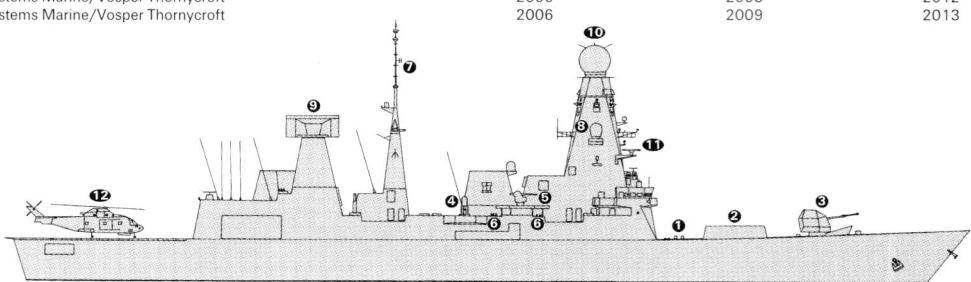

DARING   (Scale 1 : 1,200), Ian Sturton / 0131365

Programmes: This project has gone through many stages, the result of which has been slippage of the originally envisaged in-service date of 2000 to 2007 and the concomitant extension of the ship-lives of the ageing Type 42s. Starting life as NFR 90 in the 1980s, it was taken forward via the Anglo-French Future Frigate, the tri-nation Common New Generation Frigate (Horizon) and finally, when UK withdrew from the collaborative ship programme on 25 April 1999, a national Type 45 ship project. The contract for the design and build of the first three ships was placed with the prime contractor, BAE Systems, on 20 December 2000. Following extensive consultation this was amended in late 2001 to reflect a new procurement strategy in which commitment was made to the first six ships. Vosper Thornycroft is to build and outfit the forward section of each ship together with the masts and funnel. The remainder of the ship is to be built by BAE Marine. First steel of *Daring* cut at Govan on 28 March 2003 and of *Dauntless* on 26 August 2004. Final assembly started at

Scotstoun in February 2005. Assembly of follow-on ships is to be at Govan. Stage I trials are to be conducted by the final assembly shipyard and Stage II trials by BAE Systems. Procurement of the missile system is being pursued separately and a contract for full development and initial production of PAAMS was placed with the tri-national consortium, EUROPAAMS, in August 1999. Test firings are to be conducted from the trials barge *Longbow* from 2005. It was announced on 21 July 2004 that a class of eight ships is to be built.

Structure: Built to Lloyd's Naval Ship Rules. Provision for future installation of 155 mm gun or a 16-cell VLS silo, SSM, CIWS and magazine-launched torpedoes. An integrated technology mast is another potential modification. The ships are designed to support and deploy at least 30 troops. OTC facilities are to be included. A land-attack capability for D 35-37 is under consideration.

UPDATED

DARING (computer graphic)   9/2002, BAE Systems / 0523834

FRIGATES

Notes: Plans for a future class of warships have been placed in abeyance since the decision in 2004 to wind down the Future Surface Combatant (FSC) programme. These ships were to have entered service from about 2015. Options now under consideration include life-extension and capability upgrade programmes for the Type 22 Batch 3 and Type 23 frigates and the purchase of an off-the-shelf design, possibly derived from the Type 45 destroyer, to provide an interim capability post-2015. Studies are also likely to consider a high/low mix of ships to fulfil the FSC capability requirement. There are no plans for an all-new frigate class to enter service before at least 2020.

CHATHAM   10/2003, Maritime Photographic / 0576463

4 BROADSWORD CLASS (TYPE 22 BATCH 3) (FFGHM)

Name	No	Builders	Laid down	Launched	Commissioned
CORNWALL	F 99	Yarrow Shipbuilders, Glasgow	14 Dec 1983	14 Oct 1985	23 Apr 1988
CUMBERLAND	F 85	Yarrow Shipbuilders, Glasgow	12 Oct 1984	21 June 1986	10 June 1989
CAMPBELTOWN	F 86	Cammell Laird, Birkenhead	4 Dec 1985	7 Oct 1987	27 May 1989
CHATHAM	F 87	Swan Hunter Shipbuilders, Wallsend-on-Tyne	12 May 1986	20 Jan 1988	4 May 1990

Displacement, tons: 4,200 standard; 4,900 full load
Dimensions, feet (metres): 485.9 × 48.5 × 21
(148.1 × 14.8 × 6.4)
Main machinery: COGOG; 2 RR Spey SM1A gas turbines;
29,500 hp *(22 MW)* sustained; 2 RR Tyne RM3C gas turbines;
10,680 hp *(8 MW)* sustained; 2 shafts; LIPS; cp props
Speed, knots: 30; 18 on Tynes
Range, n miles: 4,500 at 18 kt on Tynes
Complement: 250 (31 officers) (accommodation for 301)

Missiles: SSM: 8 McDonnell Douglas Harpoon Block 1C (2 quad) launchers ❶; preprogrammed; active radar homing to 130 km *(70 n miles)* at 0.9 Mach; warhead 227 kg.
SAM: 2 British Aerospace Seawolf GWS 25 Mod 3 ❷; command line of sight (CLOS) with 2 channel radar tracking to 5 km *(2.7 n miles)* at 2+ Mach; warhead 14 kg.
Guns: 1 Vickers 4.5 in *(114 mm)*/55 Mk 8 ❸; 25 rds/min to 22 km *(11.9 n miles)*; 27.5 km *(14.8 n miles)* Mod 1 anti-surface; weight of shell 21 kg.
1 Signaal/General Electric 30 mm 7-barrelled Goalkeeper ❹; 4,200 rds/min combined to 1.5 km.
2 GAM-BO1-1 20 mm ❺; 700-900 rds/min to 1 km.
Countermeasures: Decoys: Outfit DLH; 4 Sea Gnat 6-barrelled 130 mm/102 mm fixed launchers ❻.
Graseby Type 182 or SLQ-25A; towed torpedo decoy.
RESM: Racal UAT; intercept.
CESM: CoBLU; intercept.
Combat data systems: CACS 5 action data automation; Link 11.
2 Matra Marconi SCOT 1C (being replaced by SCOT 5) SATCOMs ❼. ICS-3 integrated comms. INMARSAT.

CORNWALL   *(Scale 1 : 1,200), Ian Sturton /* 0572734

Weapons control: 2 BAe GSA 8B GPEOD Sea Archer optronic directors with TV and IR imaging and laser rangefinders ❽. GWS 25 Mod 3 (for SAM). GWS 60.
Radars: Air/surface search: Marconi Type 967/968 ❾; D/E-band.
Surface search: Racal Decca Type 2008; E/F-band; being fitted.
Navigation: Kelvin Hughes Type 1007; I-band.
Fire control: 2 Marconi Type 911 ❿; I/Ku-band (for Seawolf).
Sonars: Ferranti/Thomson Sintra Type 2050; hull-mounted; active search and attack.

Helicopters: 2 Westland Lynx HMA. Mk 3/8 ⓫.

Modernisation: Sonar 2016 replaced by Sonar 2050. EW fit updated and Sea Gnat is the standard launcher for a variety of distraction and seduction decoys. EDS Command Support System fitted by 2008. A major upgrade to the Seawolf system to be implemented from 2006; a rolling installation programme is planned to be completed by 2011. Enhancements include an improved I-band radar, an additional optronic tracker to improve low level performance and improved software. In a separate contract, the Mk 4 SWELL (Seawolf Enhanced Low Level) fuze is being incorporated into existing rounds and in Block 2 missiles from 2005. Mk 8 Mod 1 gun fitted in F 85 and is being progressively fitted to the rest of the class.
Structure: Batch 3 are stretched versions of original Batch 1. Flight decks enlarged to operate Sea King helicopters.
Operational: This class is primarily designed for ASW operations and is capable of acting as OTC. All have facilities for Flag and staff. One Lynx normally embarked. All are based at Devonport.
Sales: All Batch 1 to Brazil. Batch 2 ships *London* and *Coventry* to Romania and *Sheffield* to Chile.

UPDATED

CUMBERLAND   *10/2003, Maritime Photographic /* 0572726

CORNWALL   *3/2002, John Brodie /* 0530028

CAMPBELTOWN   *7/2003, B Sullivan /* 0572727

14 DUKE CLASS (TYPE 23) (FFGHM)

Name	No	Builders	Laid down	Launched	Commissioned
ARGYLL	F 231	Yarrow Shipbuilders, Glasgow	20 Mar 1987	8 Apr 1989	31 May 1991
LANCASTER	F 229 (ex-F 232)	Yarrow Shipbuilders, Glasgow	18 Dec 1987	24 May 1990	1 May 1992
IRON DUKE	F 234	Yarrow Shipbuilders, Glasgow	12 Dec 1988	2 Mar 1991	20 May 1993
MONMOUTH	F 235	Yarrow Shipbuilders, Glasgow	1 June 1989	23 Nov 1991	24 Sep 1993
MONTROSE	F 236	Yarrow Shipbuilders, Glasgow	1 Nov 1989	31 July 1992	2 June 1994
WESTMINSTER	F 237	Swan Hunter Shipbuilders, Wallsend-on-Tyne	18 Jan 1991	4 Feb 1992	13 May 1994
NORTHUMBERLAND	F 238	Swan Hunter Shipbuilders, Wallsend-on-Tyne	4 Apr 1991	4 Apr 1992	29 Nov 1994
RICHMOND	F 239	Swan Hunter Shipbuilders, Wallsend-on-Tyne	16 Feb 1992	6 Apr 1993	22 June 1995
SOMERSET	F 82	Yarrow Shipbuilders, Glasgow	12 Oct 1992	25 June 1994	20 Sep 1996
GRAFTON	F 80	Yarrow Shipbuilders, Glasgow	13 May 1993	5 Nov 1994	29 May 1997
SUTHERLAND	F 81	Yarrow Shipbuilders, Glasgow	14 Oct 1993	9 Mar 1996	4 July 1997
KENT	F 78	Yarrow Shipbuilders, Glasgow	16 Apr 1997	27 May 1998	8 June 2000
PORTLAND	F 79	Yarrow Shipbuilders, Glasgow	14 Jan 1998	15 May 1999	3 May 2001
ST ALBANS	F 83	Yarrow Shipbuilders, Glasgow	18 Apr 1999	6 May 2000	6 June 2002

Displacement, tons: 3,500 standard; 4,200 full load
Dimensions, feet (metres): 436.2 × 52.8 × 18 (screws);
  24 (sonar) *(133 × 16.1 × 5.5; 7.3)*
Main machinery: CODLAG; 2 RR Spey SM1A (F 229-F 236) or
  SM1C (F 237 onwards) gas turbines (see *Structure*);
  31,100 hp *(23.2 MW)*; 4 Paxman 12CM diesels;
  8,100 hp *(6 MW)*; 2 GEC motors; 4,000 hp *(3 MW)*; 2 shafts
Speed, knots: 28; 15 on diesel-electric
Range, n miles: 7,800 miles at 15 kt
Complement: 181 (13 officers)

Missiles: SSM: 8 McDonnell Douglas Harpoon (2 quad)
  launchers ❶; active radar homing to 130 km *(70 n miles)* at
  0.9 Mach; warhead 227 kg (84C). 4 normally carried.
SAM: British Aerospace Seawolf GWS 26 Mod 1 VLS ❷;
  command line of sight (CLOS) radar/TV tracking to 6 km
  (3.3 n miles) at 2.5 Mach; warhead 14 kg; 32 canisters.
Guns: 1 Vickers 4.5 in *(114 mm)*/55 Mk 8 ❸; 25 rds/min to
  22 km *(11.9 n miles)*; 27.5 km *(14.8 n miles)* Mod 1 anti-
  surface; weight of shell 21 kg. Mk 8 Mod 1 being progressively
  fitted.
  2 DES/MSI DS 30B 30 mm/75 ❹; 650 rds/min to 10 km
  (5.4 n miles) anti-surface; 3 km *(1.6 n miles)* anti-aircraft;
  weight of shell 0.36 kg.
Torpedoes: 4 Cray Marine 324 mm fixed (2 twin) tubes ❺.
  Marconi Stingray; active/passive homing to 11 km *(5.9 n
  miles)* at 45 kt; warhead 35 kg (shaped charge); depth to
  750 m *(2,460 ft)*. Reload in 9 minutes.
Countermeasures: Decoys: Outfit DLH; 4 Sea Gnat 6-barrelled
  130 mm/102 mm launchers ❻. DLF 2/3 offboard decoys.
  Type 2170 torpedo defence system.
  ESM: Racal UAT ❼; intercept.
Combat data systems: BAeSEMA Surface Ship Command
  System (DNA); Links 11 and JTIDS 16 in due course. 2 Matra
  Marconi SCOT 1D SATCOMs (being replaced by SCOT 5) ❽.
Weapons control: BAe GSA 8B/GPEOD optronic director ❾.
  GWS 60 (for SSM). GWS 26 (for SAM).
Radars: Air/surface search: Plessey Type 996(I) ❿; 3D; E/F-
  band.

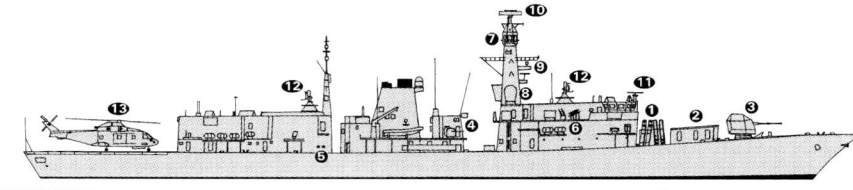

MARLBOROUGH   *(Scale 1 : 1,200)*, Ian Sturton / 0530055

Surface search: Racal Decca Type 1008 ⓫; E/F-band.
Navigation: Kelvin Hughes Type 1007; I-band.
Fire control: 2 Marconi Type 911 ⓬; I/Ku-band.
IFF: 1010/1011 or 1018/1019.
Sonars: Ferranti/Thomson Sintra Type 2050; bow-mounted;
  active search and attack.
  Dowty Type 2031Z (F 229, 231, 234-236, 238-239); towed
  array; passive search; very low frequency.
  Thales Type 2087 (F 237); active low-frequency (500 Hz)
  towed body with passive array (100 Hz).

Helicopters: 1 Westland Lynx HMA 3/8 or 1 Merlin HM 1 (F 229,
  F 235 and F 237) ⓭.

Programmes: The first of this class was ordered from Yarrows on
  29 October 1984. Further batches of three ordered in
  September 1986, July 1988, December 1989, January 1992
  and February 1996. Further orders are unlikely. F 229
  pennant number changed, because 232 was considered
  unlucky, as it is the RN report form number for collisions and
  groundings.
Modernisation: Major improvement programmes are planned.
  The Command System has been upgraded to Phase 5 with
  DNA(2) programmed to commence in 2008. The Mk 8 Mod 1
  gun is to be installed in the whole class. Modifications to
  improve the performance of Type 966 radar are being made.
  The Seawolf system is to be upgraded from 2006-2011;

enhancements include improved I-band radar, an additional
optronic tracker to improve low level performance and
improved software. In a separate contract the Mk 4 SWELL
(Seawolf Enhanced Low Level) fuze is being incorporated into
existing rounds and in Block 2 missiles from 2005. Surface
Ship Torpedo Defence, a development of Sonar 2070, is
being fitted. Co-operative Engagement Capability (CEC) is to
be fitted in ten ships from 2007. Low Frequency Active Sonar
(Type 2087) is to replace Type 2031 in eight ships with F 237
the first to have been fitted and F 238 to follow. Following
trials in F 229, Merlin HM1 is to be embarked in Type 2087
fitted ships.
Structure: Incorporates stealth technology to minimise acoustic,
magnetic, radar and IR signatures. The design includes a 7°
slope to all vertical surfaces, rounded edges, reduction of IR
emissions and a hull bubble system to reduce radiated noise.
The combined diesel electric and GT propulsion system
provides quiet motive power during towed sonar operations.
The SM1C engines although capable of 41 MW of power
combined are constrained by output into the gearbox.
MacTaggart Scott Helios helo landing system.
Operational: F 78, F 80, F 83, F 229, F 234, F 237 and F 239
are based at Portsmouth and the remainder, at Devonport.
Following announcements on 21 July 2004, *Norfolk* and
Marlborough decommissioned in 2005 and *Grafton* is to
follow in March 2006.

UPDATED

KENT   6/2004*, Marion Ferrette / 1043594

PORTLAND   7/2003, B Sullivan / 0572723

KENT

7/2004 *, H M Steele / 1043558

MONTROSE

11/2004 *, Maritime Photographic / 1043604

MONMOUTH

11/2004 *, H M Steele / 1043566

IRON DUKE

6/2004 *, John Brodie / 1043597

SHIPBORNE AIRCRAFT

Notes: (1) The STOVL (short take-off and landing) variant of the Lockheed Martin F-35 Joint Strike Fighter selected on 30 September 2002 to fulfil the Future Joint Combat Aircraft requirement. The aircraft is to replace the RAF/RN aircraft operated by the Joint Force Harrier for operation both from the future carriers and from land bases. There are 150 aircraft planned, first delivery of which is expected in 2010 to meet an in-service date of 2014.
(2) Sea Harrier to be phased out by 2006. Harrier force to migrate to an all GR force. 30 GR.7 aircraft upgraded to GR.7A standard (re-engined with Pegasus Mk.107) for carrier operations by 2005. In parallel, 60 GR.7 and GR.7A and 10 T.10 aircraft to receive an avionics and weapons upgrade to GR.9/GR.9A/T.12 standard by 2008.
(3) A programme to replace the current organic airborne early-warning capability in 2012 is in progress. Initial Gate for The Maritime Airborne Surveillance and Control (MASC) programme is planned in 2005 with Main Gate decisions to follow in December 2008. Potential solutions are likely to be based on rotary-wing platforms and UAVs.
(4) AgustaWestland Future Lynx selected on 27 March 2005 as preferred option to meet the requirement for the Maritime (Surface) Attack Helicopter.
(5) Dauphin 2 helicopters with Royal Naval markings are leased by FOST for staff transfers. Painted red.
(6) Proposals for a maritime version of the Chinook medium-lift helicopter were formalised in September 2004 with the launch of a three-year assessment phase.

DAUPHIN 2   7/2004*, B Sullivan / 1043559

Numbers/Type: 21/4 British Aerospace Sea Harrier FA 2/T 8.
Operational speed: 540 kt (1,000 km/h).
Service ceiling: 45,000 ft (13,716 m).
Range: 800 n miles (1,480 km).
Role/Weapon systems: Air defence, ground attacks, reconnaissance and maritime attack. Sensors: Blue Vixen radar, GEC-Marconi Sky Guardian ESM, ALE 40 chaff/IR flares, camera, IFF Mk 12. Weapons: ASV. Strike; 1,000 lb and 540 lb bombs; Paveway II laser-guided bombs. AD; four AIM-9M Sidewinder or Hughes AIM-120 AMRAAM (FA 2); two 30 mm Aden cannon.
UPDATED

SEA HARRIER FA 2   7/2004*, Paul Jackson / 1043605

Numbers/Type: 30 British Aerospace Harrier GR.7A/GR.9A.
Operational speed: 575 kt (1,065 km/h).
Service ceiling: 45,000 ft (13,716 m).
Range: 594 n miles (1,101 km).
Role/Weapon systems: RAF all weather single-seat close support, battlefield interdiction night attack and reconnaissance aircraft, first operated from HMS *Illustrious* in 1994. All are being upgraded to GR.9A standard by 2008. Sensors: FLIR, TIALD, VICON recce pod, ESM, ECM and chaff dispensers. Weapons: AAM; 2 or 4 AIM-9L Sidewinder. Ground attack: 'iron' bombs, Paveway II LGB, Paveway III LGB, Paveway IV Precision guided bomb, Maverick IR and TV, Brimstone advanced anti-armour weapon system.
UPDATED

HARRIER GR. 7A   6/2004*, Royal Navy / 1043610

Numbers/Type: 42 Westland/Agusta Merlin HM Mk 1.
Operational speed: 150 kt (277 km/h).
Service ceiling: 10,000 ft (3,048 m).
Range: 550 n miles (1,019 km).
Role/Weapon systems: Primary anti-submarine role with secondary anti-surface and troop-carrying capabilities. Contract for 44 signed 9 October 1991. In service with 700, 814, 820, 824 and 829 Squadrons. Sensors: GEC-Marconi Blue Kestrel 5000 radar, Thales Flash AQS 960 dipping sonar, GEC-Marconi sonobuoy acoustic processor AQS-903, Thales Orange Reaper ESM, ECM. Weapons: ASW; four Stingray torpedoes or Mk 11 Mod 3 depth bombs. ASV; OTHT for ship-launched SSM.
UPDATED

MERLIN HM Mk 1   10/2003, B Sullivan / 0572694

Numbers/Type: 22 Westland/Agusta Merlin HM Mk 3.
Operational speed: 150 kt (296 km/h).
Service ceiling: 13,125 ft (4,000 m).
Range: 550 n miles (1,019 km).
Role/Weapon systems: Operated by RAF. Roles include cargo and troop transport and combat SAR. Military lift is 24 troops and up to four tonnes underslung. Sensors: integrated defensive aids including Raytheon laser detection, BAE Systems Sky Guardian 2000 RWR, Doppler-based MAWS, Northrop Grumman AN/AAQ-24 Nemesis DIRCM and BAE Systems North America AN/ALE-47 chaff/flare dispensers. Weapons: machine guns.
UPDATED

MERLIN Mk 3   6/2003, Paul Jackson / 0572697

Numbers/Type: 21/27 Westland Lynx HAS 3/HMA 8.
Operational speed: 120 kt (222 km/h).
Service ceiling: 10,000 ft (3,048 m).
Range: 320 n miles (593 km).
Role/Weapon systems: Primarily anti-surface helicopter with short-range ASW capability; embarked in all modern RN escorts. Contract in 2001 for new Thales tactical situation displays to be fitted in all. 19 aircraft fitted with night vision goggles; all to be fitted in due course. All aircraft receiving a Saturn radio upgrade. Sensors: Ferranti Sea Spray Mk 1 radar, 'Orange Crop' ESM, chaff and flare dispenser. Weapons: ASW; two Stingray torpedoes or Mk 11 Mod 3 depth bombs. ASV; four Sea Skua missiles; two 12.7 mm MG pods.
UPDATED

LYNX HMA 8   3/2003, A Sharma / 0572680

LYNX HAS 3   7/2004*, Paul Jackson / 1043606

Numbers/Type: 10/7 Westland Sea King HU Mk 5/HAS Mk 6.
Operational speed: 112 kt *(207 km/h).*
Service ceiling: 10,000 ft *(3,050 m).*
Range: 400 n miles *(925 km).*
Role/Weapon systems: ASW capability removed from all Sea King HAS Mk 5 and 6 variants. Six converted to Commando role (similar performance to HC 4). Remaining aircraft employed in SAR role and two Mk 6 flights formed to provide support to the CVs. Programme underway to convert all Mk 5 aircraft to night vision goggles for SAR. Sensors: Sea Searcher radar, Orange Crop ESM (Mk 6 only). Weapons: Both the Mk 5 and 6 can be fitted with the 7.62 mm MGs. Mk 6 able to carry torpedoes and depth charges.

UPDATED

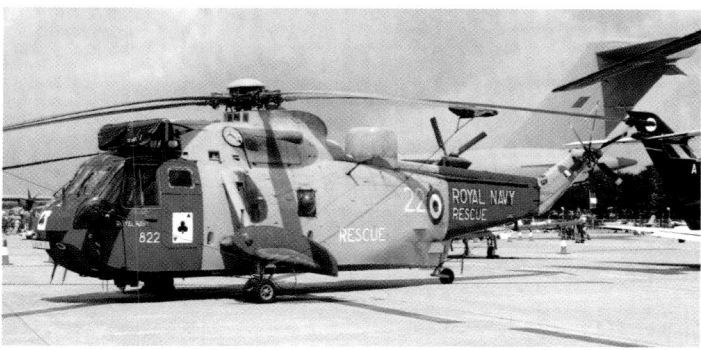

SEA KING HU Mk 5   *7/2003, Paul Jackson /* 0572729

Numbers/Type: 13 Westland Sea King ASAC Mk 7.
Operational speed: 112 kt *(207 km/h).*
Service ceiling: 10,000 ft *(3,050 m).*
Range: 400 n miles *(925 km).*
Role/Weapon systems: Primary role Airborne Surveillance and Control (ASaC), with capability for littoral and strike operations. Conversion contract awarded in October 1996 to upgrade AEW Mk 2 Fleet to ASAC Mk 7. Programme completed mid-2004. Two aircraft lost in Iraq conflict to be replaced by 2007. Sensors: Thales Searchwater 2000, Racal MIR-2 'Orange Crop' ESM, IFF Mk XII, Litton 100g navigation system and JTIDS/Link 16. Weapons: None.

UPDATED

SEA KING ASAC MK 7   *6/2004*, Royal Navy /* 1043607

Numbers/Type: 27 Westland Sea King HC Mk 4
Operational speed: 112 kt *(208 km/h).*
Service ceiling: 10,000 ft *(3,050 m).*
Range: 664 n miles *(1,230 km).*
Role/Weapon systems: 24 commando support and five training/special task helicopters; capable of carrying most Commando Force equipment underslung. Expected to remain in service to 2010. Engines of 10 aircraft upgraded in 2003 to improve hot weather performance. Six HAS 6 ASW aircraft undergoing conversion to Mk 6C commando assault configuration. Sensors: Prophet 2 plus AAR-47 ESM; jammer, chaff and IR flares ECM. Weapons: Can fit 7.62 mm GPMG or similar.

UPDATED

SEA KING HC MK 4   *8/2004*, B Sullivan /* 1043561

Numbers/Type: 35/6/8 Boeing Chinook HC Mk 2/HC Mk 2A/HC Mk 3.
Operational speed: 140 kt *(259 km/h).*
Service ceiling: 10,140 ft *(3,090 m).*
Range: 651 n miles *(1,207 km).*
Role/Weapon systems: All-weather medium-lift helicopter equivalent to CH-47D and operated by RAF. Operable from surface ships (CVS/LPH/LPD/LSL). Mk 2/2A capable of carrying 44 troops and up to ten tonnes cargo. Mk 3 is Special Forces version. Sensors: defensive aids suite including missile approach warning, IR jammers and chaff/flare dispensers. Weapons: machine guns.

UPDATED

CHINOOK   *6/2004*, Royal Navy /* 1043609

Numbers/Type: 40 Westland/Boeing WAH-64D AH Mk 1.
Operational speed: 150 kt *(278 km/h).*
Service ceiling: 21,000 ft *(6,400 m).*
Range: 260 n miles *(480 km).*
Role/Weapon systems: Westland selected on 13 July 1995 to build McDonnell Douglas (now Boeing) Apache. Total 67 ordered. All-weather attack helicopter with day and night capability. One squadron to be earmarked for amphibious operations. To be operable from surface ships (CVS/LPH/LPD/LSL). First embarkation in *Ocean* in 2004 with an embarkation of four aircraft for Exercise Argonaut 05 planned. Full operational capability expected early 2006. Sensors: Lockheed Martin/Northrop Grumman AN/APG-78 Longbow radar, Lockheed Martin target acquisition and designation sight (TV and direct view) and AN/AAQ-11 pilot's night vision FLIR sensor (TADS/PNVS), BAE Systems HIDAS helicopter integrated defensive aids system, including Sky Guardian 2000 RWR, Type 1223 Laser warning receiver, Thales (Vinten) Vicon 78 Srs 455 chaff/flare dispenser, BAE Systems AN/AAR-57(V) common missile warning system (CMWS) and Lockheed Martin AN/APR-48A radar frequency interferometer. Weapons: 16 Hellfire missiles or 76 CRV-7 70 mm rockets. 1—30 mm chain gun.

UPDATED

APACHE AH Mk 1   *3/2004*, Royal Navy /* 1043611

Numbers/Type: 9 Westland Gazelle AH Mk 1.
Operational speed: 130 kt *(242 km/h).*
Service ceiling: 9,350 ft *(2,850 m).*
Range: 361 n miles *(670 km).*
Role/Weapon systems: For observation with 847 Squadron. Sensors: Roof-mounted sight. Weapons: Normally unarmed.

VERIFIED

GAZELLE   *8/2001, A Sharma /* 0131189

For details of the latest updates to *Jane's Fighting Ships* online and to discover the additional information available exclusively to online subscribers please visit

jfs.janes.com

Numbers/Type: 6 Westland Lynx AH Mk 7.
Operational speed: 140 kt *(259 km/h)*.
Service ceiling: 10,600 ft *(3,230 m)*.
Range: 340 n miles *(630 km)*.
Role/Weapon systems: Military general purpose and anti-tank with 847 Squadron. Sensors: Day/ thermal imaging sight. Weapons: Up to eight Hughes TOW anti-tank missiles.

VERIFIED

LYNX AH MK 7   *3/2002, A Sharma /* 0530024

LAND-BASED MARITIME AIRCRAFT (FRONT LINE)

Notes: All Air Force manned. Training and Liaison aircraft not listed include Gazelle HT 2, Jetstream, Falcon 20 (under contract) and 16 Hawk (FRADU).

Numbers/Type: 19 Hawker Siddeley Nimrod MR 2/2P/4.
Operational speed: 500 kt *(926 km/h)*.
Service ceiling: 42,000 ft *(12,800 m)*.
Range: 5,000 n miles *(9,265 km)*.
Role/Weapon systems: ASW but with ASV, OTHT and control potential at long range from shore bases; three are modified for EW; duties include SAR, maritime surveillance. Contract in July 1996 to convert and upgrade 21 (reduced to 12 in 2004) aircraft to MRA.4 standard. First development aircraft flew in August 2004. To enter service in 2009. MRA 4 equipment includes Thales Searchwater 2000MR radar, Elta EL-8300 and ALR-56M ESM. Sensors: Thales Searchwater radar, ECM, Yellowgate ESM, cameras, CAE MAD, Ultra sonobuoys, Ultra/GDC AQS 971 acoustics suite, WESCAM EO system, cameras. Weapons: ASW; 6.1 tons of Stingray torpedoes or depth bombs or mines. ASV; four Harpoon missiles. Self-defence; four AIM-9L Sidewinder.

UPDATED

NIMROD MRA 4   *8/2004*, BAE Systems /* 0577851

NIMROD MR 2   *7/2003, Paul Jackson /* 0572731

Numbers/Type: 7 Boeing E-3D Sentry AEW Mk 1.
Operational speed: 460 kt *(853 km/h)*.
Service ceiling: 36,000 ft *(10,973 m)*.
Range: 870 n miles *(1,610 km)*.
Role/Weapon systems: Air defence early warning aircraft with secondary role to provide coastal AEW for the Fleet; 6 hours endurance at the range given above. Sensors: Westinghouse APY-2 surveillance radar, Bendix weather radar, Mk XII IFF, Yellow Gate, ESM, ECM. Weapons: Unarmed.

VERIFIED

E-3D   *10/2001, Ships of the World /* 0131206

MINE WARFARE FORCES

Notes: The application of Autonomous Underwater Vehicles (AUVs) to littoral mine-reconnaissance and mine-hunting is under investigation. Initial studies completed in 2004.

11 HUNT CLASS
(MINESWEEPERS/MINEHUNTERS—COASTAL) (MHSC/PP)

Name	No	Builders	Launched	Commissioned
BRECON	M 29	Vosper Thornycroft, Woolston	21 June 1978	21 Mar 1980
LEDBURY	M 30	Vosper Thornycroft, Woolston	5 Dec 1979	11 June 1981
CATTISTOCK	M 31	Vosper Thornycroft, Woolston	22 Jan 1981	16 June 1982
COTTESMORE	M 32	Yarrow Shipbuilders, Glasgow	9 Feb 1982	24 June 1983
BROCKLESBY	M 33	Vosper Thornycroft, Woolston	12 Jan 1982	3 Feb 1983
MIDDLETON	M 34	Yarrow Shipbuilders, Glasgow	27 Apr 1983	15 Aug 1984
DULVERTON	M 35	Vosper Thornycroft, Woolston	3 Nov 1982	4 Nov 1983
CHIDDINGFOLD	M 37	Vosper Thornycroft, Woolston	6 Oct 1983	10 Aug 1984
ATHERSTONE	M 38	Vosper Thornycroft, Woolston	1 Mar 1986	30 Jan 1987
HURWORTH	M 39	Vosper Thornycroft, Woolston	25 Sep 1984	2 July 1985
QUORN	M 41	Vosper Thornycroft, Woolston	23 Jan 1988	21 Apr 1989

Displacement, tons: 615 light; 750 full load
Dimensions, feet (metres): 187 wl; 197 oa × 32.8 × 9.5 (keel); 11.2 (screws) *(57; 60 × 10 × 2.9; 3.4)*
Main machinery: 2 Ruston-Paxman 9-59K Deltic diesels; 1,900 hp *(1.42 MW)*; 1 Deltic Type 9-55B diesel for pulse generator and auxiliary drive; 780 hp *(582 kW)*; 2 shafts; bow thruster
Speed, knots: 15 diesels; 8 hydraulic drive. **Range, n miles:** 1,500 at 12 kt
Complement: 45 (5 officers)

Guns: 1 DES/MSI DS 30B 30 mm/75; 650 rds/min to 10 km *(5.4 n miles)* anti-surface; 3 km *(1.6 n miles)* anti-aircraft; weight of shell 0.36 kg.
2 Oerlikon/BMARC 20 mm GAM-CO1 (enhancement); 900 rds/min to 2 km.
Dillon Aero M 134 7.62 mm Minigun; 6 barrels; 3,000 rds/min.
Countermeasures: MCM: 2 PAP 104 Mk 3/105 (RCMDS 1) remotely controlled submersibles, MS 14 magnetic loop, Sperry MSSA Mk 1 Towed Acoustic Generator and conventional Mk 8 Oropesa sweeps.
Combat data systems: AMS Nautis 3.
Radars: Navigation: Kelvin Hughes Type 1007; I-band.
Sonars: Thales 2193; hull-mounted; minehunting; 100/300 kHz.
Hull-mounted; active; high frequency.
Type 2059 to track PAP 104/105.

Programmes: A class of MCM Vessels combining both hunting and sweeping capabilities.
Modernisation: It is planned to replace RCMDS with an expendable mine disposal system. 30 mm gun has replaced the Bofors 40 mm. Drumgrange Precise Fixing System fitted 2003-04. A new minehunting sonar (Sonar 2193) and NAUTIS III command system have been fitted in eight ships 2004-05. M 134 Minigun CIWS is to be fitted in 2005. Plans for the influence minesweeping capability are to be replaced by a remote system, RIMS, have all been deferred.
Structure: GRP hull. Combines conventional propellers with bow thrusters. The eight MCM ships have been fitted with an improved two-man decompression chamber by 2004/05.
Operational: For operational deployments fitted with enhanced weapons systems. Ships of this class can be used for Fishery Protection duties as well as Northern Ireland patrols. M 33 and M 38 based at Faslane, the remainder at Portsmouth. Following announcements on 21 July 2004, the Northern Ireland patrol ships are to be decommissioned as follows: *Dulverton* (March 2006); *Brecon* (March 2007) and *Cottesmore* (March 2007).
Sales: *Bicester* and *Berkeley* to Greece in July 2000 and February 2001 respectively.

UPDATED

CHIDDINGFOLD   *11/2004*, R G Sharpe /* 1043592

BROCKLESBY   *7/2004*, W Sartori /* 1121518

BRECON   *11/2003*, Frank Findler /* 1043591

8 SANDOWN CLASS (MINEHUNTERS) (MHC/SRMH)

Name	No	Builders	Launched	Commissioned
WALNEY	M 104	Vosper Thornycroft, Woolston	25 Nov 1991	20 Feb 1993
PENZANCE	M 106	Vosper Thornycroft, Woolston	11 Mar 1997	14 May 1998
PEMBROKE	M 107	Vosper Thornycroft, Woolston	15 Dec 1997	6 Oct 1998
GRIMSBY	M 108	Vosper Thornycroft, Woolston	10 Aug 1998	25 Sep 1999
BANGOR	M 109	Vosper Thornycroft, Woolston	16 Apr 1999	26 July 2000
RAMSEY	M 110	Vosper Thornycroft, Woolston	25 Nov 1999	22 June 2001
BLYTH	M 111	Vosper Thornycroft, Woolston	4 July 2000	20 July 2001
SHOREHAM	M 112	Vosper Thornycroft, Woolston	9 Apr 2001	2 Sep 2002

Displacement, tons: 450 standard; 484 full load
Dimensions, feet (metres): 172.2 × 34.4 × 7.5 *(52.5 × 10.5 × 2.3)*
Main machinery: 2 Paxman Valenta 6RP200E/M diesels; 1,523 hp *(1.14 MW)* sustained; Voith-Schneider propulsion; 2 Schottel bow thrusters
Speed, knots: 13 diesels; 6.5 electric drive. **Range, n miles:** 2,500 at 12 kt
Complement: 34 (5 officers) plus 6 spare berths

Guns: 1 DES/MSI DS 30B 30 mm/75; 650 rds/min to 10 km *(5.4 n miles)* anti-surface; 3 km *(1.6 n miles)* anti-aircraft; weight of shell 0.36 kg.
Dillon Aero M 134 7.62 mm Minigun; 6 barrels; 3,000 rds/min.
Countermeasures: MCM: ECA mine disposal system, 2 PAP 104 Mk 5 (RCMDS 2). These craft can carry 2 mine wire cutters, a charge of 100 kg and a manipulator with TV/projector. Control cables are either 1,000 m (high capacity) or 2,000 m (low capacity) and the craft can dive to 300 m at 6 kt with an endurance of 5e 20 minute missions.
Combat data systems: Plessey Nautis M action data automation.
Radars: Navigation: Kelvin Hughes Type 1007; I-band.
Sonars: Marconi Type 2093; VDS; VLF-VHF multifunction with 5 arrays; mine search and classification.

Programmes: A class designed for hunting and destroying mines and for operating in deep and exposed waters. Single role minehunter (SRMH) complements the Hunt class. On 9 January 1984 the Vosper Thornycroft design for this class was approved. First one ordered August 1985, four more on 23 July 1987. A contract was to have been placed for a second batch in 1990 but this was deferred twice, until an order for seven more was placed in July 1994.
Modernisation: It is planned to replace RCMDS 2 with an expendable mine-disposal system. Drumgrange Precise Fixing System fitted in 2004. Nautis M combat system to be replaced by AMS Nautis 3 2005-07. M 134 Minigun CIWS to be fitted in 2005.
Structure: GRP hull. Combines vectored thrust units with bow thrusters and Remote-Control Mine Disposal System (RCMDS). The sonar is deployed from a well in the hull. Batch 2 have larger diameter (1.8 m) Voith-Schneider props and an improved two-man decompression chamber. The new ships have a larger Gemini crane for RCMDS vehicle launch and recovery. All of the class are to be retrofitted in due course.
Operational: M 106, M 111 based at Faslane, the remainder at Portsmouth. Following announcements on 21 July 2004, *Bridport* decommissioned in July 2004, *Sandown* in January 2005 and *Inverness* in April 2005.
Sales: Three to Saudi Arabia. *Cromer* decommissioned in October 2001 for conversion to a static training ship at Dartmouth.

UPDATED

WALNEY *7/2004*, Marian Ferrette / 1043589*

SHOREHAM *2/2004*, Derek Fox / 1043590*

AMPHIBIOUS FORCES

Notes: Further amphibious ships and craft covered in RFA and RLC (ex-RCT) sections. These include a Helicopter Support Ship, four LSD, five LSLs, and six LCLs.

10 LCU MK 10

L 1001-1010

Displacement, tons: 170 light; 240 full load
Dimensions, feet (metres): 97.8 × 24.3 × 5.6 *(29.8 × 7.4 × 1.7)*
Main machinery: 2 MAN diesels; 2 Schottel propulsors; bow thruster
Speed, knots: 10. **Range, n miles:** 600 at 12 kt
Complement: 7
Military lift: 1 MBT or 4 vehicles or 120 troops
Radars: Navigation: I-band.

Comment: Ordered in 1998 from Ailsa Troon Yard. First pair delivered in November 1999 and, following extensive trials, modifications made to ballast tanks to improve beach landing capabilities. This work carried out by BAE Systems Marine, Govan, from whom a further eight craft were ordered for delivery by mid-2003. Fitted with interlocking bow and stern ramps, they operate from the Albion class LPDs.

UPDATED

LCU MK 10 *6/2004*, Maritime Photographic / 1043612*

3 LCU Mk 9S

L 705	L 709	L 711

Displacement, tons: 115 light; 175 full load
Dimensions, feet (metres): 90.2 × 21.5 × 5 *(27.5 × 6.8 × 1.6)*
Main machinery: 2 Paxman or Dorman diesels; 474 hp *(354 kW)* sustained; Kort nozzles or Schottel propulsors
Speed, knots: 10. **Range, n miles:** 300 at 9 kt
Complement: 7
Military lift: 1 MBT or 60 tons of vehicles/stores or 90 troops
Radars: Navigation: Raytheon; I-band.

Comment: Last remaining craft of class of 14. Built in the mid-1960s and originally designated Mk 9M. Upgraded with Schottel propulsors in the 1990s and redesignated Mk 9S.

UPDATED

LCU Mk 9 *3/2003, A Sharma / 0572671*

4 GRIFFON 2000 TDX(M) (LCAC(L))

C 21-24

Displacement, tons: 6.8 full load
Dimensions, feet (metres): 36.1 × 15.1 *(11 × 4.6)*
Main machinery: 1 Deutz BF8L513 diesel; 320 hp *(239 kW)* sustained
Speed, knots: 33. **Range, n miles:** 300 at 25 kt
Complement: 2
Military lift: 16 troops plus equipment or 2 tons
Guns: 1—7.62 mm MG.
Radars: Navigation: Raytheon; I-band.

Comment: Ordered 26 April 1993. Design based on 2000 TDX(M) hovercraft. Aluminium hulls. Speed indicated is at Sea State 3 with a full load.

UPDATED

C 23 *8/2004*, B Sullivan / 1043564*

1 HELICOPTER CARRIER (LPH)

Name	No	Builders	Laid down	Launched	Commissioned
OCEAN	L 12	Vickers Shipbuilding/Kvaerner Govan	30 May 1994	11 Oct 1995	30 Sep 1998

Displacement, tons: 21,758 full load
Dimensions, feet (metres): 667.3 oa; 652.2 pp × 112.9 × 21.3 *(203.4; 198.8 × 34.4 × 6.6)*
Flight deck, feet (metres): 557.7 × 104 *(170 × 31.7)*
Main machinery: 2 Crossley Pielstick 12 PC2.6 V 400 diesels; 18,360 hp(m) *(13.5 MW)* sustained; 2 shafts; Kamewa fp props; bow thruster; 612 hp *(450 kW)*
Speed, knots: 19. **Range, n miles:** 8,000 at 15 kt
Complement: 285 plus 206 aircrew plus up to 830 Marines
Military lift: 4 LCVP Mk 5 (on davits); 2 Griffon hovercraft; 40 vehicles and equipment for most of a marine commando unit

Guns: 8 BEMARC 20 mm GAM-B03 (4 twin) ❶. 650 rds/min to 10 km *(5.4 n miles)* anti-surface; 3 km *(1.6 n miles)* anti-aircraft; weight of shell 0.36 kg.
3 General Dynamics 20 mm Phalanx Mk 15 ❷. 6 barrels per launcher; 4,500 rds/min combined to 1.5 km.
Countermeasures: Decoys: Outfit DLH; 8 Sea Gnat 130 mm/102 mm launchers ❸.
ESM: Racal UAT; intercept.
Combat data systems: Ferranti ADAWS 2000 Mod 1; Link 11, Link 16; Marconi Matra SCOT SATCOM 1D ❹.
Radars: Air/surface search: AMS Type 996 ❺; E/F-band.
Surface search: Racal Decca 1008 ❻; E/F-band.
Surface search/aircraft control: 2 Kelvin Hughes Type 1007 ❼; I-band.
IFF: Type 1016/1017.

Helicopters: 12 Sea King HC.Mk 4/Merlin plus 6 Lynx (or WAH-64 Apache by 2005).

Programmes: Initial invitations to tender were issued in 1987. Tenders submitted in July 1989 were allowed to lapse and it was not until 11 May 1993 that a contract was placed. The hull was built on the Clyde by Kvaerner Govan and sailed under its own power to Vickers at Barrow in November 1996 for the installation of military equipment.
Modernisation: Command and control facilities upgraded in 2002 to facilitate UKMCC role. Attack helicopter infrastructure fitted 2004-05.
Structure: The hull form is based on the Invincible class with a modified superstructure. The deck is strong enough to take Chinook helicopters. Six landing and six parking spots for the aircraft. Accommodation for 972 plus 303 bunk overload. A garage is situated at the after end of the hangar. This is accessible from the after aircraft lift and via ramps through the ship's stern. Hull 'blisters' were fitted at waterline level port and starboard during 2002 to improve deployment and recovery of LCVPs.

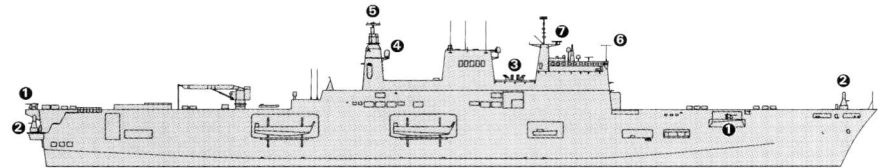

OCEAN

(Scale 1 : 1,800), Ian Sturton / 1043485

OCEAN

6/2004*, Royal Navy / 1043608

Operational: The LPH provides a helicopter lift and assault capability. The prime role of the vessel is embarking, supporting and operating a squadron of helicopters and carrying a Royal Marine Commando including vehicles, arms and ammunition. Up to 20 Sea Harriers can be carried but not supported. Operational sea trials started in June 1998 and completed in February 1999. Twin 20 mm guns are not always carried and may be replaced by single 20 mm. Based at Devonport.

UPDATED

OCEAN

5/2003, A Sharma / 0572732

OCEAN

4/2003, A Sharma / 0572673

2 ALBION CLASS (ASSAULT SHIPS) (LPD)

Name	No	Builders	Laid down	Launched	Commissioned
ALBION	L 14	BAE Systems, Barrow	22 May 1998	9 Mar 2001	19 June 2003
BULWARK	L 15	BAE Systems, Barrow	27 Jan 2000	15 Nov 2001	28 Apr 2005

Displacement, tons: 14,600 standard; 18,500 full load
Dimensions, feet (metres): 577.4 × 94.8 × 23.3
(176 × 28.9 × 7.1)
Main machinery: Diesel-electric; 2 Wärtsilä Vasa 16V 32E diesel
generators; 17,000 hp(m) *(12.5 MW)*; 2 Wärtsilä Vasa 4R
32LNE diesel generators; 4,216 hp(m) *(3.1 MW)*; 2 motors; 2
shafts; LIPS props; 1 bow thruster; 1,176 hp(m) *(865 kW)*
Speed, knots: 18. **Range, n miles:** 8,000 at 15 kt
Complement: 325
Military lift: 305 troops; 710 troops (including overload); 67
support vehicles; 4 LCU Mk 10 or 2 LCAC (dock); 4 LCVP Mk 5
(davits)

Guns: 2—20 mm ❶. 2 Signaal/General Dynamics 30 mm
7-barrelled Goalkeeper; 4,200 rds/min to 1.5 km ❷.
Countermeasures: Decoys: Outfit DLJ; 8 Sea Gnat launchers ❸
and DLH offboard decoys.
ESM/ECM: Racal Thorn UAT 1/4.
Combat data systems: ADAWS 2000. Thomson-CSF/Redifon/
BAeSEMA/CS comms system. Marconi Matra SCOT 2D
SATCOM ❹.
Radars: Air/surface search: Siemens Plessey Type 996 ❺; E/F-
band.
Surface search: Racal Decca 1008; E/F-band.
Navigation/aircraft control: 2 Racal Marine Type 1007 ❻; I-band.
IFF: Type 1016/1017.

Helicopters: Platform for 3 Sea King Mk 4 ❼. Chinook capable.

Programmes: A decision was taken in mid-1991 to replace both
existing LPDs. Project definition studies by YARD completed
in February 1994. Invitations to tender for design and build of
two ships were issued to VSEL and Yarrow on 18 August
1994 with an additional tender package to Vosper
Thornycroft in November 1994. In March 1995 it was
announced that only VSEL would bid, conforming to the rules
governing non-competitive tenders. The contract to build the
ships was awarded on 18 July 1996. First steel cut
17 November 1997.

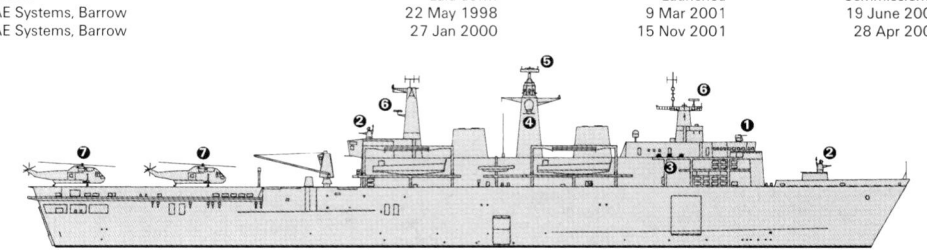

ALBION   *(Scale 1 : 1,500), Ian Sturton / 0572733*

ALBION   *5/2003, John Brodie / 0572691*

Structure: The design includes a floodable well dock, garage
(with capacity for six Challenger tanks), stern gate and side
ramp access. The Flight Deck has two helicopter landing
spots. A large joint operations room contains substantial
command and control facilities. The ships are built to military
damage control standards.
Operational: Based at Devonport. *Bulwark* started sea trials on
31 May 2004.   ***UPDATED***

ALBION   *9/2004*, B Sullivan / 1043565*

BULWARK   *10/2004*, R G Sharpe / 1043593*

11 LCVP Mk 4

8401	8407	8411	8413
8402	8409	8412	8619-8622

Displacement, tons: 10.5 light; 16 full load
Dimensions, feet (metres): 43.8 × 10.9 × 2.8 *(13.4 × 3.3 × 0.8)*
Main machinery: 2 Perkins T6.3544 diesels; 290 hp *(216 kW)*; 2 shafts
Speed, knots: 15. **Range, n miles:** 150 at 14 kt
Complement: 3
Military lift: 20 Arctic equipped troops or 5.5 tons

Comment: Built by Souters and McTays. Introduced into service in 1986. Fitted with removable arctic canopies across well-deck. Some Royal Marines' craft replaced by LCVP Mk 5. Six craft operated by Royal Logistics Corps. These serve in rotation between the Falklands and UK.
UPDATED

LCVP Mk 4   *6/2004*, Maritime Photographic /* 1043613

23 LCVP Mk 5

LCVP 9473	9673-9692	9707	9708

Displacement, tons: 25 full load
Dimensions, feet (metres): 50.9 × 13.8 × 3 *(15.5 × 4.2 × 0.9)*
Main machinery: 2 Volvo Penta TAMD 72 WJ diesels; 2 PP 170 water-jets
Speed, knots: 25. **Range, n miles:** 210 at 18 kt
Complement: 3
Military lift: 35 troops plus 2 tons equipment or 8 tons vehicles and stores
Radars: Navigation: Raytheon 40; I-band.

Comment: Contract placed with Vosper Thornycroft on 31 January 1995 for one craft which was handed over on 17 January 1996. Four more ordered on 23 October 1996 for *Ocean* were delivered 6 December 1997; and two more for RM Poole in October 1998. Sixteen more ordered from FBM Babcock Marine in August 2001. Can beach fully laden on a 1 : 120 gradient. Speed 18 kt at full load.
VERIFIED

LCVP Mk 5   *11/2002, W Sartori /* 0529940

FAST INTERCEPT CRAFT (HSIC)

Comment: Two main types in service. Both around 15 m in length and powered by two or three 660 hp diesels to give speeds up to 55 kt. One has a VSV wave piercing hull which can maintain fast speeds in high sea states. Both types can be transported by air.
VERIFIED

FIC 145   *5/2003, Maritime Photographic /* 0572700

VSV   *10/2003, Derek Fox /* 0572670

RRC and RIB

Comment: (1) 36 RRC Mk 3: 2.6 tons and 7.4 m *(24.2 ft)* powered by single Yamaha 220 hp *(162 kW)* diesel; 36 kt fully laden (40 light); carry 8 troops. Some used by the Army. In service 1996-98.
(2) RIBs: Halmatic Arctic 22/Pacific 22/Arctic 28/Pacific 28. Rolling contract for all four types. Capable of carrying 10 to 15 fully laden troops at speeds of 26 to 35 kt.
UPDATED

RIB   *4/2001, A Sharma /* 0131195

RRC   *8/2004*, B Sullivan /* 1043563

PATROL FORCES

Notes: Two Archer class, *Pursuer* and *Dasher* are based at Cyprus.

3 + 1 RIVER CLASS (OFFSHORE PATROL VESSELS) (PSO)

Name	No	Builders	Commissioned
TYNE	P 281	Vosper Thornycroft, Woolston	4 July 2003
SEVERN	P 282	Vosper Thornycroft, Woolston	31 July 2003
MERSEY	P 283	Vosper Thornycroft, Woolston	26 Mar 2004
CLYDE	P 284	VT Shipbuilding, Portsmouth	Sep 2006

Displacement, tons: 1,700 full load
Dimensions, feet (metres): 261.7 × 44.6 × 12.5 *(79.75 × 13.6 × 3.8)*
Main machinery: 2 MAN 12RK 270 diesels; 11,063 hp *(8.25 MW)*; 2 shafts; bow thruster; 375 hp *(280 kW)*
Speed, knots: 20. **Range, n miles:** 5,500 at 15 kt
Complement: 30 (plus 18 boarding party)
Guns: 1—20 mm Oerlikon/BMARC. 2—7.62 mm MGs.
Radars: Surface search: Kelvin Hughes Nucleus; E/F-band.
Navigation: Kelvin Hughes Nucleus; I-band.
Helicopters: Vertrep only.

Programmes: In the first agreement of its kind, Vosper Thornycroft contracted on 8 May 2001 for the construction, lease and support of three vessels over initial five-year period to replace five ships of Island class. VT Group selected on 13 December 2004 to build a fourth ship to a modified design to undertake Falkland Islands patrol duties. The ship will be chartered to the MoD for an initial period until 2012.
Structure: Based on Vosper Thornycroft EEZ Management Vessel concept design. The ships are capable of operating two RIBs. Fitted with a 3 tonne crane. The Batch 2 ship is to incorporate a flight deck, capable of accepting helicopters up to Merlin size, and an improved radar surveillance capability.
Operational: Part of Fishery Protection Squadron based at Portsmouth.
UPDATED

OPV Batch 2 (artist's impression)   *12/2004*, VT Shipbuilding /* 0585955

SEVERN   *10/2004*, B Sullivan /* 1043557

1 ANTARCTIC PATROL SHIP (AGOBH)

Name	No	Builders	Commissioned
ENDURANCE (ex-*Polar Circle*)	A 171 (ex-A 176)	Ulstein Hatlo, Norway	21 Nov 1991

Displacement, tons: 6,500 full load
Dimensions, feet (metres): 298.6 × 57.4 × 27.9 *(91 × 17.9 × 8.5)*
Main machinery: 2 Bergen BRM8 diesels, 8,160 hp(m) *(6 MW)* sustained; 1 shaft; cp prop; bow and stern thrusters
Speed, knots: 15. **Range, n miles:** 6,500 at 12 kt
Complement: 112 (15 officers) plus 14 Royal Marines
Radars: Surface search: Raytheon R 84 and M 34 ARG; E/F- and I-bands.
Navigation: Kelvin Hughes Type 1007; I-band.
IFF: Type 1011.
Helicopters: 2 Westland Lynx HAS.Mk 3.

Comment: Leased initially in late 1991 and then bought outright in early 1992 as support ship and guard vessel for the British Antarctic Survey. Hull is painted red. Inmarsat fitted. Main machinery is resiliently mounted. Ice-strengthened hull capable of breaking 1 m thick ice at 3 kt. Helicopter hangar is reached by lift from the flight deck. Accommodation is to high standards. Name and pennant number changed during refit in mid-1992. Sonars 2053, 2090 and 60-88. Based at Portsmouth.

UPDATED

ENDURANCE   *4/2004*, W Sartori /* 1043581

2 CASTLE CLASS (OFFSHORE PATROL VESSELS Mk 2) (PSOH)

Name	No	Builders	Launched	Commissioned
LEEDS CASTLE	P 258	Hall Russell, Aberdeen	29 Oct 1980	27 Oct 1981
DUMBARTON CASTLE	P 265	Hall Russell, Aberdeen	3 June 1981	26 Mar 1982

Displacement, tons: 1,427 full load
Dimensions, feet (metres): 265.7 × 37.7 × 11.8 *(81 × 11.5 × 3.6)*
Main machinery: 2 Ruston 12RKC diesels; 5,640 hp *(4.21 MW)* sustained; 2 shafts; cp props
Speed, knots: 19.5. **Range, n miles:** 10,000 at 12 kt
Complement: 45 (6 officers) plus austerity accommodation for 25 Royal Marines
Guns: 1 DES/MSI DS 30B 30 mm/75; 650 rds/min to 10 km *(5.4 n miles)*; weight of shell 0.36 kg.
Mines: Can lay mines.
Countermeasures: Decoys: Outfit DLE; 2 or 4 Plessey Shield 102 mm 6-tubed launchers.
ESM: 'Orange Crop'; intercept.
Combat data systems: Racal CANE DEA-3 action data automation; SATCOM.
Weapons control: Radamec 2000 series optronic director.
Radars: Surface search: Plessey Type 994; E/F-band.
Navigation: Kelvin Hughes Type 1006; I-band.
Helicopters: Platform for operating Sea King or Lynx.

Comment: Started as a private venture. Ordered 8 August 1980. Design includes an ability to lay mines. Inmarsat commercial terminals fitted. Two Avon Sea Rider high-speed craft are embarked. P 258 based at Portsmouth and employed on Fishery Protection tasks. P 265 based in the Falklands. The 76 mm (never fitted) gunbay has been converted to an operations room. Fitted with two cranes to facilitate launch/recovery of RIBs, during refit which completed in mid-2004.

UPDATED

DUMBARTON CASTLE   *5/2004*, Maritime Photographic /* 1043616

2 FAST PATROL CRAFT (PBF)

SCIMITAR (ex-*Grey Fox*) P 284   SABRE (ex-*Grey Wolf*) P 285

Displacement, tons: 26 full load
Dimensions, feet (metres): 52.5 × 14.43 × 3.9 *(16 × 4.4 × 1.2)*
Main machinery: 2 MAN V10 diesels; 740 hp *(603 kW)*; 2 shafts
Speed, knots: 32. **Range, n miles:** 260 at 19 kt
Complement: 5
Guns: 2—7.62 mm MGs.
Radars: Racal Decca Bridgemaster 360; I-band.

Comment: Halmatic M160 craft operated in Northern Ireland from 1988 but transferred to Gibraltar in September 2002 to augment the Gibraltar squadron. Both vessels renamed and commissioned on 31 January 2003. After mid-life refit and design modifications, the vessels replaced *Trumpeter* and *Ranger* as Gibraltar guard ships in 2004.

UPDATED

SABRE   *11/2003, W Sartori /* 0573697

TRAINING SHIPS

16 ARCHER CLASS (TRAINING and PATROL CRAFT) (PB/AXL)

EXPRESS P 163 (ex-A 163)	SMITER P 272	DASHER P 280
EXPLORER P 164 (ex-A 154)	PURSUER P 273	PUNCHER P 291
EXAMPLE P 165 (ex-A 153)	TRACKER P 274	CHARGER P 292
EXPLOIT P 167 (ex-A 167)	RAIDER P 275	RANGER P 293
ARCHER P 264	BLAZER P 279	TRUMPETER P 294
BITER P 270		

Displacement, tons: 54 full load
Dimensions, feet (metres): 68.2 × 19 × 5.9 *(20.8 × 5.8 × 1.8)*
Main machinery: 2 RR CV 12 M800T diesels; 1,590 hp *(1.19 MW)*; or 2 MTU diesels; 2,000 hp(m) *(1.47 MW)* (P 274-275); 2 shafts
Speed, knots: 22 or 25 (P 274-275). **Range, n miles:** 550 at 15 kt
Complement: 11 (3 officers)
Guns: 1 Oerlikon 20 mm (can be fitted). 2—7.62 mm MGs (P 293, P 294).
Radars: Navigation: Racal Decca 1216; I-band.

Comment: First 14 ordered from Watercraft Ltd, Shoreham. Commissioning dates: *Archer*, August 1985; *Example*, September 1985; *Explorer*, January 1986; *Biter* and *Smiter*, February 1986. The remaining nine were incomplete when Watercraft went into liquidation in 1986 and were towed to Portsmouth for completion in 1988 by Vosper Thornycroft. Initially allocated for RNR training but underused in that role and now employed as University Naval Units (URNU)- *Ranger* (Sussex), *Trumpeter* (Bristol), *Puncher* (London), *Blazer* (Southampton), *Smiter* (Glasgow), *Charger* (Liverpool), *Archer* (Aberdeen), *Biter* (Manchester and Salford), *Exploit* (Birmingham), *Express* (Wales), *Example* (Northumbria) and *Explorer* (Hull). The four ex-RNXS ships (P 163-P 167) are also part of the Inshore Training Squadron and are based at Gosport, Ipswich and Penarth. *Dasher* and *Pursuer* are based at Cyprus. Similar craft built for the Indian and Omani Coast Guards. Two more ordered from BMT in early 1997 to a modified design and built at Ailsa, Troon. *Tracker* and *Raider* commissioned January 1998 for Oxford and Cambridge University respectively.

UPDATED

EXPLOIT   *6/2004*, Maritime Photographic /* 1043615

RAIDER   *2/2004*, W Sartori /* 1043583

SURVEY SHIPS

Notes: In addition to the ships listed below, further survey work is undertaken by Naval Party 1016 embarked in a chartered vessel, *MV Confidante*. Naval Party 1008 was disbanded in December 2003.

1 SCOTT CLASS (AGSH)

Name	No	Builders	Launched	Commissioned
SCOTT	H 131	Appledore Shipbuilders, Bideford	13 Oct 1996	30 June 1997

Displacement, tons: 13,500 full load
Dimensions, feet (metres): 430.1 × 70.5 × 29.5 *(131.1 × 21.5 × 9)*
Main machinery: 2 Krupp MaK 9M32 9-cyl diesels; 10,800 hp(m) *(7.94 MW)*; 1 shaft; LIPS cp prop; retractable bow thruster
Speed, knots: 17.5
Complement: 62 (12 officers) (see *Comment*)
Radars: Navigation: Kelvin Hughes ARPA 1626; I-band.
Helicopters: Platform for 1 light.

Comment: Designed by BAeSEMA/YARD and ordered 20 January 1995 to replace *Hecla*. Ice-strengthened bow. Foredeck strengthened for helicopter operations. The centre of the OSV surveying operations consists of an integrated navigation suite, the Sonar Array Sounding System (SASS) and data processing equipment. Additional sensors include gravimeters, a towed proton magnetometer and the Sonar 2090 ocean environment sensor. The SASS IV multibeam depth-sounder is capable of gathering 121 individual depth samples concurrently over a 120° swathe, producing a three-dimensional image of the seabed. 8,000 tons of seawater ballast can be used to achieve a sonar trim. The ship is at sea for 300 days a year with a crew of 42 embarked, rotating with the other 20 ashore. Can be used as an MCMV support ship. *Scott* undertook a survey of the Indian Ocean tsunami epicentre in early 2005. Based at Devonport.

UPDATED

SCOTT   5/2004*, Robert Pabst / 1043582

2 ECHO CLASS (AGSH)

Name	No	Builders	Launched	Commissioned
ECHO	H 87	Appledore, Bideford	4 Mar 2002	7 Mar 2003
ENTERPRISE	H 88	Appledore, Bideford	2 May 2002	17 Oct 2003

Displacement, tons: 3,470 full load
Dimensions, feet (metres): 295.3 × 55.1 × 18 *(90 × 16.8 × 5.5)*
Main machinery: Diesel electric; 4.8 MW; 2 azimuth thrusters; 1 bow thruster
Speed, knots: 15. **Range, n miles:** 9,000 at 12 kt
Complement: 49 with accommodation for 81
Guns: 2—20 mm. 4—7.62 mm MGs.
Radars: Navigation: 2 sets; I-band.
Helicopters: Platform for 1 medium.

Comment: The order for two multirole Hydrographic and Oceanographic Survey Vessels was placed with the prime contractor, Vosper Thornycroft Ltd, on 19 June 2000. The ships were built by Appledore Shipbuilders in Devon. The contract covers the design, build and through-life support of the ships over their 25 year service. In addition to specialist surveying tasks, the ships' operational roles include Rapid Environmental Assessment, Amphibious Warfare surveys and Mine Countermeasures Tasking Support. The survey suite consists of hull mounted multibeam sonar, towed side scan sonar, towed undulating sensors, an adaptive survey planning system and a Survey Motor Launch. Both based at Devonport.

UPDATED

ENTERPRISE   6/2004*, John Brodie / 1043602

1 ROEBUCK CLASS (AGS)

Name	No	Builders	Launched	Commissioned
ROEBUCK	H 130	Brooke Marine, Lowestoft	14 Nov 1985	3 Oct 1986

Displacement, tons: 1,477 full load
Dimensions, feet (metres): 210 × 42.6 × 13 *(63.9 × 13 × 4)*
Main machinery: 4 Mirrlees Blackstone ESL8 Mk 1 diesels; 3,040 hp *(2.27 MW)*; 2 shafts; cp props
Speed, knots: 14. **Range, n miles:** 4,000 at 10 kt
Complement: 46 (6 officers)
Radars: Navigation: Kelvin Hughes Nucleus 2-6000; I-band.

Comment: Designed for hydrographic surveys to full modern standards on UK continental shelf. Air conditioned. Carries a 9 m surveying motor boat and one 4.5 m RIB. The decision to decommission in 2003 was cancelled and a Ship Life Extension Programme is planned to prolong service life to 2014. Roles include Rapid Environmental Assessment and Amphibious Warfare survey. The SIPS(S) integrated navigation and survey system is to be replaced by a survey suite consisting of a hull-mounted multibeam sonar, towed side-scan sonar and adaptive planning system.

UPDATED

ROEBUCK   11/2003, B Prézelin / 0572668

1 GLEANER CLASS (YGS)

Name	No	Builders	Launched	Commissioned
GLEANER	H 86	Emsworth Shipyard	18 Oct 1983	5 Dec 1983

Displacement, tons: 26 full load
Dimensions, feet (metres): 48.6 × 15.4 × 5.2 *(14.8 × 4.7 × 1.6)*
Main machinery: 2 Volvo Penta TMD 112; 524 hp(m) *(391 kW)*; 2 shafts
Speed, knots: 14 diesels; 7 centre shaft only
Complement: 5 plus 1 spare bunk
Radars: Navigation: Raytheon R40SX; I-band.

Comment: This craft is prefixed HMSML-HM Survey Motor Launch. Sonar 2094 and Kongsberg Simrad dual head multibeam echo sounder.

UPDATED

GLEANER   5/2004*, Derek Fox / 1043584

4 NESBITT CLASS (YGS)

NESBITT 9423	**PAT BARTON** 9424	**COOK** 9425	**OWEN** 9426

Displacement, tons: 11 full load
Dimensions, feet (metres): 34.8 × 9.4 × 3.3 *(10.6 × 2.9 × 1)*
Main machinery: 2 Perkins Sabre 185C diesels; 430 hp(m) *(316 kW)* sustained; 2 shafts
Speed, knots: 15. **Range, n miles:** 300 at 8 kt
Complement: 2 plus 10 spare

Comment: Delivered by Halmatic, Southampton by September 1996. Fitted with GPS, Ultra 3000 Side scan sonar and Qubit SIPS. In service at the Hydrographic School and can be carried in Echo class survey vessels.

UPDATED

NESBITT   11/1998, John Brodie / 0053244

RESCUE VEHICLES

Notes: Project definition of a NATO Submarine Rescue System (NSRS) was completed by W S Atkins Ltd in late 2001. The three partner nations are UK, Norway and France with the UK Defence Procurement Agency acting as contracting authority and host nation for project management. Following Invitations to Tender, award of the contract for the Design and Manufacture Phase was made in June 2004 to a team led by Rolls-Royce Naval Marine. The core of the service is to be a new free-swimming Submarine Rescue Vehicle (SRV) (Perry Slingsby LR7A) which will be capable of accommodating up to 15 rescued submariners. Operated by a crew of three and capable of reaching depths down to 600 m, the SRV will be launched and recovered from suitable commercial or military 'motherships' equipped with a portable launch-and-recovery installation. In addition, the contract provides for: an unmanned remotely operated vehicle (ROV) which will be used to locate the stricken submarine; decompression chambers; medical facilities and other support equipment. Achievement of an Initial Operating Capability is planned for December 2006 with a full operational capability following in the second quarter of 2007. The system is expected to remain in service for 25 years. It will be permanently maintained at HM Naval Base Clyde at 12 hours notice to meet a Time To First Rescue of no more than 72 hours worldwide. NSRS will complement other submarine rescue systems operated by Australia, Italy, Sweden and US.

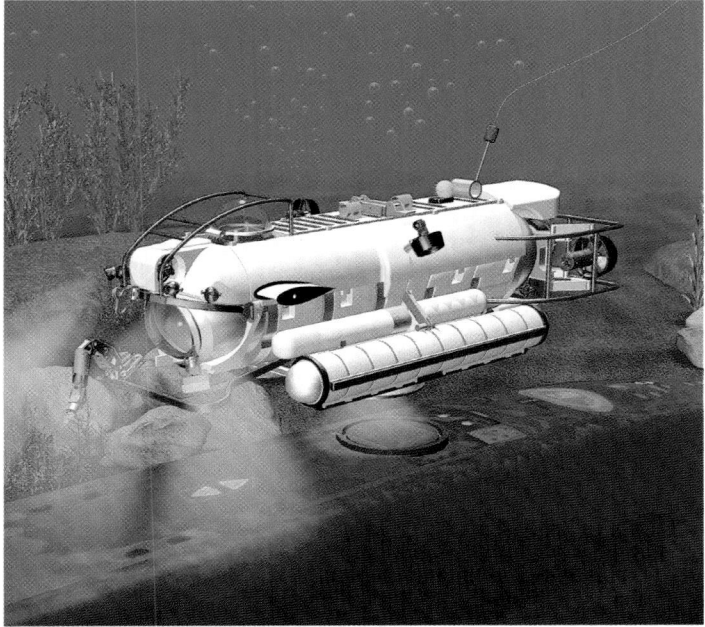

SUBMARINE RESCUE VEHICLE *6/2004*, Rolls-Royce /* 0566635

1 RESCUE SUBMERSIBLE (DSRV)

LR 5

Displacement, tons: 21.5
Dimensions, feet (metres): 30.2 × 9.8 × 11.5 *(9.2 × 3.0 × 3.5)*
Main machinery: 2 electric motors; 16 hp *(12 kW)*; 2 hydraulic transverse thrusters; 9 hp *(6.7 kW)*; 2 hydraulic tiltable side thrusters; 9 hp *(6.7 kW)*
Speed, knots: 2.5 dived
Complement: 2 pilots and 1 rescue chamber operator

Comment: LR 5 is the Rescue submersible that is part of the UK Submarine Rescue System (UKSRS). Manufactured in 1978 by Slingsby Ltd for North Sea commercial operations and subsequently purchased by the Royal Navy for whom it is managed, maintained (at 12 hours notice) and operated by James Fisher Rumic Ltd. Capable of operating down to 400 m depth, it can be deployed anywhere in the world and operated from the deck of any suitable mother ship. Its role is to rescue up to 16 survivors at a time from a disabled submarine on the seabed and bring them back to the surface. This can be done at normal atmospheric pressure and at increased pressure up to 5 bar. Mating with the disabled submarine can be achieved at up to 10° heel on LR 5 or at up to 60° bow up. LR 5 is complemented by an ROV, Scorpio 45, which is attached to a 1,000 m umbilical. This is used to locate the disabled submarine, clear obstructions from the escape hatches and replenish life support stores. LR 5 is due to be replaced by the NATO Submarine Rescue System (NSRS) in mid-2007.

UPDATED

LR 5 *6/2004*, Royal Navy /* 1043614

ROYAL FLEET AUXILIARY SERVICE

Personnel

1 January 2005: 2,274 (891 officers)

General

The Royal Fleet Auxiliary Service is a civilian-manned fleet under the command of the Commander in Chief Fleet from 1 April 1993. Its main task is to supply warships at sea with fuel, food, stores and ammunition. It also provides aviation platforms, amphibious support for the Navy and Marines and sea transport for Army units. All ships take part in operational sea training. An order in council on 30 November 1989 changed the status of the RFA service to government-owned vessels on non-commercial service.

Ships taken up from Trade

The following ships taken up from trade: *Confidante* operates in home waters under the Hydrographer. *Seabulk Condor* (supply), *St Brandan* (ferry), *Indomitable* (tug), all operate in the Falkland Islands.

New Construction

(1) The Joint Casualty Receiving Ship (JCRS), formerly known as the Primary Casualty Receiving Ship (PCRS) is due to enter service after 2010. The requirement for such a ship was identified in the 1998 Strategic Defence Review. The aviation support ship *Argus* was configured as a PCRS during the 1990-91 Gulf War and the 2003 Iraq War. The contract for the Assessment Phase was awarded to BMT Ltd in February 2002 since when the key drivers have been identified as a need for eight operating tables and a 150-200 bed hospital. A two-spot

flight deck and the ability to embark personnel by sea or land are also needed. Development of the Systems Requirement Document (SRD) by Atkins Aviation and Defence Systems has been completed and following the down-selection of two consortia in 2006, it is planned to select a single consortium in mid-2008 for demonstration and manufacture. Potential solutions range from a bespoke vessel to conversion/modification of an existing military or merchant hull. The ship will be manned by RFA personnel. The requirement for a second ship at 12 months notice is to be met by chartering a commercial hull. (2) Plans to provide the future afloat support capability are being taken forward through the Military Afloat Reach and Sustainability programme (MARS). The concept phase ended in 2004 and options to replace the Rover class and Leaf class are to be considered at Initial Gate. These take into account Joint Sea Based Logistics and the carrier programmes. An overall force of 8-12 ships is anticipated.

2 WAVE CLASS (LARGE FLEET TANKERS) (AORH)

Name	No	Builders	Laid down	Launched	Commissioned
WAVE KNIGHT	A 389	BAE Systems, Barrow	22 Oct 1998	29 Sep 2000	8 Apr 2003
WAVE RULER	A 390	BAE Systems, Govan	10 Feb 2000	9 Feb 2001	27 Apr 2003

Displacement, tons: 31,500 full load
Measurement, tons: 18,200 dwt

Dimensions, feet (metres): 643.2 × 91.2 × 32.8 *(196.0 × 27.8 × 10.0)*

Main machinery: Diesel-electric: 4 Wärtsilä 12V 32E/GECLM diesel generators; 25,514 hp(m) *(18.76 MW)*; 2 GECLM motors; 19,040 hp(m) *(14 MW)*; 1 shaft; Kamewa bow and stern thrusters
Speed, knots: 18. **Range, n miles:** 10,000 at 15 kt
Complement: 80 plus 22 aircrew
Cargo capacity: 16,000 m³ total liquids including 3,000 m³ aviation fuel; 8—20 ft refrigerated containers plus 500 m³ solids
Guns: 2 Vulcan Phalanx CIWS; fitted for but not with. 2—30 mm. 4—7.62 mm MGs.
Countermeasures: Decoys: Outfit DLJ.
Radars: Navigation: KH 1007; E/F/I-band.
IFF: Type 1017.
Helicopters: 1 Merlin HM.Mk 1.

Comment: Feasibility studies by BAeSEMA/YARD completed in early 1995. Draft invitation to tender issued 10 October 1995 followed by full tender on 26 June 1996. Contracts to build placed with VSEL (BAE Systems) on 12 March 1997. One spot flight deck with full hangar facilities for one Merlin. Enclosed bridge including bridge wings. Double hull construction. Inclined RAS gear with three rigs and two cranes.

VERIFIED

WAVE KNIGHT
11/2002, H M Steele / 0530076

1 OAKLEAF CLASS (SUPPORT TANKER) (AOR)

Name	No	Builders	Commissioned	Recommissioned
OAKLEAF (ex-*Oktania*)	A 111	Uddevalla, Sweden	1981	14 Aug 1986

Displacement, tons: 49,648 full load
Measurement, tons: 37,328 dwt
Dimensions, feet (metres): 570 × 105.6 × 36.7
(173.7 × 32.2 × 11.2)
Main machinery: 1 Burmeister & Wain 4L80MCE diesel;
10,800 hp(m) *(7.96 MW)* sustained; 1 shaft; cp prop; bow
and stern thrusters

Speed, knots: 14
Complement: 35 (14 officers)
Cargo capacity: 22,000 m³ fuel
Guns: 2—7.62 mm MGs.
Countermeasures: Decoys: 2 Plessey Shield chaff launchers can
be fitted.
Radars: Navigation: Racal Decca 1226 and 1229; I-band.

Comment: Acquired in July 1985 and converted by Falmouth
Ship Repairers to include full RAS rig and extra
accommodation. Handed over on completion and renamed.
Single hulled, ice strengthened. Marisat fitted. Major refit in
1994-95. To continue in service until at least 2010.
UPDATED

OAKLEAF   *8/2004*, B Prézelin /* 1043587

3 APPLELEAF CLASS (SUPPORT TANKERS) (AOR)

Name	No	Builders	Launched	Commissioned
BRAMBLELEAF (ex-*Hudson Deep*)	A 81	Cammell Laird, Birkenhead	22 Jan 1976	3 Mar 1980
BAYLEAF	A 109	Cammell Laird, Birkenhead	27 Oct 1981	26 Mar 1982
ORANGELEAF (ex-*Balder London*, ex-*Hudson Progress*)	A 110	Cammell Laird, Birkenhead	—	2 May 1984

Displacement, tons: 37,747 full load (A 109-110); 40,870 (A
81)
Measurement, tons: 20,761 gross; 11,573 net; 29,999 dwt
Dimensions, feet (metres): 560 × 85 × 36.1
(170.7 × 25.9 × 11)
Main machinery: 2 Pielstick 14 PC2.2 V 400 diesels; 14,000 hp
(m) *(10.29 MW)* sustained; 1 shaft
Speed, knots: 15.5; 16.3 (A 109)
Complement: 56 (19 officers)

Cargo capacity: 22,000 m³ dieso; 3,800 m³ Avcat
Guns: 2 Oerlikon 20 mm. 4—7.62 mm MGs.
Countermeasures: Decoys: 2 Corvus or 2 Plessey Shield
launchers.
Radars: Navigation: Racal Decca 1226 and 1229; I-band.

Comment: *Brambleleaf* chartered in 1979-80 and converted,
completing Autumn 1979. Part of a four-ship order cancelled
by Hudson Fuel and Shipping Co, but completed by the

shipbuilders, being the only mercantile order then in hand.
Bayleaf built under commercial contract to be chartered by
MoD. *Orangeleaf* major refit September 1985 to fit full RAS
capability and extra accommodation. Single-hull construction.
Appleleaf sold to Australia in September 1989.
Decommissioning dates: *Orangeleaf* 2008; *Brambleleaf* and
Bayleaf 2009.
UPDATED

BRAMBLELEAF   *4/2004*, Marian Ferrette /* 1043586

3 ROVER CLASS (SMALL FLEET TANKERS) (AORLH)

Name	No	Builders	Launched	Commissioned
GREY ROVER	A 269	Swan Hunter Shipbuilders, Wallsend-on-Tyne	17 Apr 1969	10 Apr 1970
GOLD ROVER	A 271	Swan Hunter Shipbuilders, Wallsend-on-Tyne	7 Mar 1973	22 Mar 1974
BLACK ROVER	A 273	Swan Hunter Shipbuilders, Wallsend-on-Tyne	30 Oct 1973	23 Aug 1974

Displacement, tons: 4,700 light; 11,522 full load
Measurement, tons: 6,692 (A 271, 273), 6,822 (A 269) dwt;
7,510 gross; 3,185 net
Dimensions, feet (metres): 461 × 63 × 24
(140.6 × 19.2 × 7.3)
Main machinery: 2 SEMT-Pielstick 16 PA4 185 diesels;
15,360 hp(m) *(11.46 MW)*; 1 shaft; Kamewa cp prop; bow
thruster
Speed, knots: 19. **Range, n miles:** 15,000 at 15 kt
Complement: 48 (17 officers) (A 269); 55 (18 officers) (A 271,
273)
Cargo capacity: 3,000 m³ fuel
Guns: 2 Oerlikon 20 mm. 2—7.62 mm MGs.
Countermeasures: Decoys: 2 Corvus and 2 Plessey Shield
launchers. 1 Graseby Type 182; towed torpedo decoy.
Radars: Navigation: Racal Decca 52690 ARPA; Racal Decca
1690; I-band.
Helicopters: Platform for Westland Sea King HAS. Mk 5 or
HC.Mk 4.

Comment: Single-hull construction. Small fleet tankers designed
to replenish HM ships at sea with fuel, fresh water, limited dry
cargo and refrigerated stores under all conditions while under
way. No hangar but helicopter landing platform is served by a

GREY ROVER   *11/2004*, Declerck/Cracco /* 1043585

stores lift, to enable stores to be transferred at sea by 'vertical
lift'. Capable of HIFR. Siting of SATCOM aerial varies. *Green
Rover* sold in September 1992 to Indonesia. *Blue Rover* to

Portugal in March 1993. *Grey Rover* to decommission in 2007
and the other two by 2010.
UPDATED

2 FORT VICTORIA CLASS (FLEET REPLENISHMENT SHIPS) (AORH)

Name	No	Builders	Laid down	Launched	Commissioned
FORT VICTORIA	A 387	Harland & Wolff/Cammell Laird	4 Apr 1988	12 June 1990	24 June 1994
FORT GEORGE	A 388	Swan Hunter Shipbuilders, Wallsend-on-Tyne	9 Mar 1989	1 Mar 1991	16 July 1993

Displacement, tons: 36,580 full load
Dimensions, feet (metres): 667.7 oa; 607 wl × 99.7 × 32 *(203.5; 185 × 30.4 × 9.8)*
Main machinery: 2 Crossley SEMT-Pielstick 16 PC2.6 V 400 diesels; 23,904 hp(m) *(17.57 MW)* sustained; 2 shafts
Speed, knots: 20
Complement: 134 (95 RFA plus 15 RN plus 24 civilian stores staff) plus 154 (28 officers) aircrew
Cargo capacity: 12,505 m³ liquids; 3,000 m³ solids

Guns: 2 DES/MSI DS 30B 30 mm/75.
2 Vulcan Phalanx 20 mm Mk 15.
Countermeasures: Decoys: 4 Plessey Shield or 4 Sea Gnat 6-barrelled 130 mm/102 mm launchers. Graseby Type 182; towed torpedo decoy.
ESM: Marconi Racal Thorn UAT; intercept.

Combat data systems: Marconi Matra SCOT 1D SATCOM.
Radars: Air search: Plessey Type 996; 3D; E/F-band (can be fitted).
Navigation: Kelvin Hughes Type 1007; I-band.
Aircraft control: Kelvin Hughes NUCLEUS; E/F-band.

Helicopters: 5 Westland Sea King/Merlin helicopters.

Programmes: The requirement for these ships is to provide fuel and stores support to the Fleet at sea. *Fort Victoria* ordered 23 April 1986 and *Fort George* on 18 December 1987. *Fort Victoria* delayed by damage during building and entered Cammell Laird Shipyard for post sea trials completion in July 1992. The original plan for six of this class was progressively eroded and no more of this type will be built.

Structure: Single-hull construction. Four dual-purpose abeam replenishment rigs for simultaneous transfer of liquids and solids. Stern refuelling. Repair facilities for Merlin helicopters. The plan to fit Seawolf GWS 26 VLS has been abandoned in favour of Phalanx CIWS fitted in 1998/99 to both ships.
Operational: Two helicopter spots. There is a requirement to provide an emergency landing facility for Harriers. To remain in service until 2015.

UPDATED

FORT VICTORIA
4/2003, A Sharma / 0572664

2 FORT GRANGE CLASS (FLEET REPLENISHMENT SHIPS) (AFSH)

Name	No	Builders	Laid down	Launched	Commissioned
FORT ROSALIE (ex-*Fort Grange*)	A 385	Scott-Lithgow, Greenock	9 Nov 1973	9 Dec 1976	6 Apr 1978
FORT AUSTIN	A 386	Scott-Lithgow, Greenock	9 Dec 1975	9 Mar 1978	11 May 1979

Displacement, tons: 23,384 full load
Measurement, tons: 8,300 dwt
Dimensions, feet (metres): 607.4 × 79 × 28.2 *(185.1 × 24.1 × 8.6)*
Main machinery: 1 Sulzer RND90 diesel; 23,200 hp(m) *(17.05 MW)*; 1 shaft; 2 bow thrusters
Speed, knots: 22. **Range, n miles:** 10,000 at 20 kt
Complement: 114 (31 officers) plus 36 RNSTS (civilian supply staff) plus 45 RN aircrew
Cargo capacity: 3,500 tons armament, naval and victualling stores in 4 holds of 12,800 m³
Guns: 2 Oerlikon 20 mm. 4—7.62 mm MGs.
Countermeasures: Decoys: 2 Corvus or 2 Plessey Shield launchers (upper bridge).
Radars: Navigation: Kelvin Hughes Type 1007; I-band.
Helicopters: 4 Westland Sea King.

Comment: Ordered in November 1971. Fitted for SCOT SATCOMs but carry Marisat. Normally only one helicopter is embarked. ASW stores for helicopters carried on board. Emergency flight deck on the hangar roof. There are six cranes, three of 10 tons lift and three of 5 tons.

UPDATED

FORT AUSTIN
10/2004*, B Sullivan / 1043556

1 STENA TYPE (FORWARD REPAIR SHIP) (ARH)

Name	No	Builders	Commissioned	Recommissioned
DILIGENCE (ex-*Stena Inspector*)	A 132	Oresundsvarvet AB, Landskrona, Sweden	1981	12 Mar 1984

Displacement, tons: 10,765 full load
Measurement, tons: 6,550 gross; 4,939 dwt
Dimensions, feet (metres): 367.5 × 67.3 × 22.3 *(112 × 20.5 × 6.8)*
Flight deck, feet (metres): 83 × 83 *(25.4 × 25.4)*
Main machinery: Diesel-electric; 5 V16 Nohab-Polar diesel generators; 2,650 kW; 4 NEBB motors; 6,000 hp(m) *(4.41 MW)*; 1 shaft; Kamewa cp prop; 2 Kamewa bow tunnel thrusters; 3,000 hp(m) *(2.2 MW)*; 2 azimuth thrusters (aft); 3,000 hp(m) *(2.2 MW)*
Speed, knots: 12. **Range, n miles:** 5,000 at 12 kt
Complement: 38 (15 officers) plus accommodation for 147 plus 55 temporary
Cargo capacity: Long-jib crane SWL 5 tons; maximum lift, 40 tons
Guns: 4 Oerlikon 20 mm. 3—7.62 mm MGs.
Countermeasures: Decoys: Outfit DLE; 4 Plessey Shield 102 mm 6-tubed launchers.

Helicopters: Facilities for up to Boeing Chinook HC. Mk 1 (medium lift) size.

Programmes: *Stena Inspector* was designed originally as a Multipurpose Support Vessel for North Sea oil operations, and completed in January 1981. Chartered on 25 May 1982 for use as a fleet repair ship during the Falklands War. Purchased from Stena (UK) Line in October 1983, and converted for use as Forward Repair Ship in the South Atlantic (Falkland Islands). Conversion by Clyde Dock Engineering Ltd, Govan from 12 November 1983 to 29 February 1984.
Modernisation: Following items added during conversion: large workshop for hull and machinery repairs (in well-deck); accommodation for naval Junior Rates (new accommodation block); accommodation for crew of conventional submarine (in place of Saturation Diving System); extensive craneage facilities; overside supply of electrical power, water, fuel, steam, air, to ships alongside; large naval store (in place of cement tanks); armament and magazines; Naval Communications System; decompression chamber.
Structure: Four 5 ton anchors for four-point mooring system. Strengthened for operations in ice (Ice Class 1A). Kongsberg Albatross Positioning System has been retained in full. Uses bow and stern thrusters and main propeller to maintain a selected position to within a few metres, up to Beaufort Force 9. Controlled by Kongsberg KS 500 computers.
Operational: Falkland Islands support team (Naval Party 2010) embarked when the ship is in the South Atlantic. Has also been used as MCMV support ship in the Gulf. Due to decommission by 2010.

UPDATED

DILIGENCE
3/2003, A Sharma / 0572704

1 AVIATION TRAINING SHIP (HSS/APCR)

Name	No	Builders	Commissioned	Recommissioned
ARGUS (ex-*Contender Bezant*)	A 135	CNR Breda, Venice	1981	1 June 1988

Displacement, tons: 18,280 standard; 26,421 full load
Measurement, tons: 9,965 dwt
Dimensions, feet (metres): 574.5 × 99.7 × 27
  (175.1 × 30.4 × 8.2)
Main machinery: 2 Lindholmen SEMT-Pielstick 18 PC2.5 V 400 diesels; 23,400 hp(m) *(17.2 MW)* sustained; 2 shafts
Speed, knots: 18. **Range, n miles:** 20,000 at 19 kt
Complement: 80 (22 officers) plus 35 permanent RN plus 137 RN aircrew
Military lift: 3,300 tons dieso; 1,100 tons aviation fuel; 138 4 ton vehicles in lieu of aircraft

Guns: 4 BMARC 30 mm Mk 1. 4—7.62 mm MGs.
Countermeasures: Decoys: Outfit DLB; 4 Sea Gnat 130 mm/102 mm launchers. Graseby Type 182; torpedo decoy.
ESM: THORN EMI Guardian; radar warning.
Combat data systems: Racal CANE DEB-1 data automation. Inmarsat SATCOM communications. Marisat.
Radars: Air search: Type 994 MTI; E/F-band.
Air/surface search: Kelvin Hughes Type 1006; I-band.
Navigation: Racal Decca Type 994; I-band.

Fixed-wing aircraft: Provision to transport 12 BAe Sea Harrier FA-2.
Helicopters: 6 Westland Sea King HAS.Mk 5/6 or similar.

ARGUS   *3/2003, A Sharma /* 0572661

Programmes: Ro-Ro container ship whose conversion for her new task was begun by Harland and Wolff in March 1984 and completed on 3 March 1988.
Structure: Uses former ro-ro deck as hangar with four sliding WT doors able to operate at a speed of 10 m/min. Can replenish other ships underway. One lift port midships, one abaft funnel. Domestic facilities are very limited if she is to be used in the Command support role. Flight deck is 372.4 ft *(113.5 m)* long and has a 5 ft thick concrete layer on its lower side. First RFA to be fitted with a command system. Ability to conduct subsidiary role as Primary Casualty Receiving Ship improved significantly following upgrade period completed late 2001. This included conversion of three decks into permanent 100-bed hospital with three operating theatres.
Operational: To decommission by 2010.

UPDATED

0 + 4 BAY CLASS LANDING SHIPS DOCK (AUXILIARY) (LSD)

Name	No	Builders	Laid down	Launched	Commissioned
LARGS BAY	L 3006	Swan Hunter (Tyneside) Ltd	1 Oct 2001	18 July 2003	2005
LYME BAY	L 3007	Swan Hunter (Tyneside) Ltd	2003	2005	2006
MOUNTS BAY	L 3008	BAE Systems Govan	24 Feb 2003	9 Apr 2004	2005
CARDIGAN BAY	L 3009	BAE Systems Govan	Apr 2004	8 Apr 2005	2006

Displacement, tons: 16,160 full load
Dimensions, feet (metres): 577.6 × 86.6 × 19
  (176 × 26.4 × 5.8)
Main machinery: Diesel-electric; 2 steerable propulsors
Speed, knots: 18. **Range, n miles:** 8,000 at 15 kt
Complement: 60 (plus 356 troops)
Military lift: 1,150 linear metres of space for vehicles equating to 36 Challenger MBTs or 150 light trucks plus 200 tons ammunition
Radars: Navigation: I-band.
Helicopters: Platform capable of operating Chinook.

Programmes: Two ships ordered from Swan Hunter on 18 December 2000 to enter Royal Fleet Auxiliary service to replace RFAs *Sir Percivale* and *Sir Geraint.* Contract for two further ships of the class, to replace RFAs *Sir Galahad* and *Sir Tristram* placed on 19 November 2001 with BAE Systems (Marine) at Govan.
Structure: Based on the Dutch LPD *Rotterdam*, the LSD(A)s are designed to transport troops, vehicles, ammunition and stores

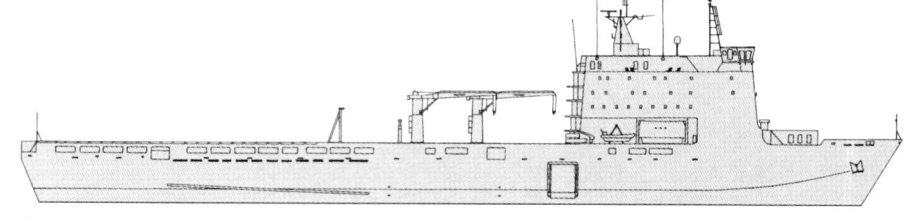

LARGS BAY   *(Scale 1 : 1,500), Ian Sturton /* 1043488

in support of amphibious operations. Offload is enabled by a flight deck capable of operating heavy helicopters, an amphibious dock capable of operating one LCU and mexeflotes which can be hung on the ships' sides. There is no beaching capability. Davit-launched infantry landing craft (LCVPs) are not fitted but two can be carried in the dock.

UPDATED

2 SIR BEDIVERE CLASS (LANDING SHIPS LOGISTIC) (LSLH)

Name	No	Builders	Laid down	Launched	Commissioned
SIR BEDIVERE	L 3004	Hawthorn Leslie, Hebburn-on-Tyne	Oct 1965	20 July 1966	18 May 1967
SIR TRISTRAM	L 3505	Hawthorn Leslie, Hebburn-on-Tyne	Feb 1966	12 Dec 1966	14 Sep 1967

Displacement, tons: 3,270 light; 5,674 full load
  6,700 full load (SLEP)
Dimensions, feet (metres): 412.1; 441.1 (SLEP) × 59.8 × 13
  (125.6; 134.4 × 18.2 × 4)
Main machinery: 2 Mirrlees 10-ALSSDM diesels; 9,400 hp *(7.01 MW)* or 2 Wärtsilä 280 V12 diesels; 9,928 hp(m) *(7.3 MW)* sustained (SLEP); 2 shafts; bow thruster; 980 hp(m) *(720 kW)* (SLEP)
Speed, knots: 17. **Range, n miles:** 8,000 at 15 kt
Complement: 51 (18 officers); 49 (15 officers) (SLEP)
Military lift: 340 troops (534 hard lying); 17 or 18 (SLEP) MBTs; 34 mixed vehicles; 120 tons POL; 30 tons ammunition; 1—25 ton crane; 2—4.5 ton cranes. Increased capacity for 20 helicopters (11 tank deck and 9 vehicle deck) after SLEP
Guns: 2 or 4 Oerlikon 20 mm. 4—7.62 mm MGs.
Countermeasures: Decoys: 2 Plessey Shield chaff launchers.
Radars: Navigation: Kelvin Hughes Type 1006 or Racal Decca 2690; I-band.
Aircraft control: Kelvin Hughes Type 1007; I-band (SLEP).
Helicopters: Platforms to operate Gazelle or Lynx or Chinook (SLEP) or Sea King.

Comment: Fitted for bow and stern loading with drive-through facilities and deck-to-deck ramps. Facilities provided for onboard maintenance of vehicles and for laying out pontoon

SIR TRISTRAM   *7/2004*, W Sartori /* 1043588

equipment. Mexeflote self-propelled floating platforms can be strapped one on each side. On 8 June 1982 *Sir Tristram* was severely damaged off the Falkland Islands. Tyne Shiprepairers were given a contract to repair and modify her. This included lengthening by 29 ft, an enlarged flight deck capable of taking Chinooks, a new bridge and twin masts. The aluminium superstructure was replaced by steel and the provision of new communications, an EMR, SATCOM, new navigation systems and helicopter control radar greatly increase her effectiveness. Completed 9 October 1985. *Sir Bedivere* had a similar SLEP in Rosyth from December 1994 to January 1998, including new main engines, a new bridge, and the helicopter platform lowered by one deck, which has reduced the size of the stern ramp. Planned decommissioning dates are: *Sir Tristram* in 2006; *Sir Bedivere* in 2011. Both based at Southampton.

UPDATED

SIR BEDIVERE   *7/2003, Maritime Photographic /* 0572705

6 TRANSPORT SHIPS (AKR)

Name	No	Builders	Commissioned
HURST POINT	—	Flensburger Schiffbau	16 Aug 2002
HARTLAND POINT	—	Harland & Wolff, Belfast	11 Dec 2002
EDDYSTONE	—	Flensburger Schiffbau	28 Nov 2002
ANVIL POINT	—	Harland & Wolff, Belfast	17 Jan 2003
LONGSTONE	—	Flensburger Schiffbau	24 Apr 2003
BEACHY HEAD	—	Flensburger Schiffbau	17 Apr 2003

Displacement, tons: 20,000 full load
Measurement, tons: 14,200 dwt
Dimensions, feet (metres): 633.4 × 85.3 × 24.3
(193.0 × 26.0 × 7.4)
Main machinery: 2 MaK 9M43 diesels; 21,700 hp *(16.2 MW)*; 2 cp props; bow thruster
Speed, knots: 21.5. **Range, n miles:** 9,200 at 21.5 kt
Complement: 18
Military lift: 2,650 linear metres of space for vehicles equating to 130 armoured vehicles plus 60 trucks and ammunition.
Radars: Navigation: I-band.

HURST POINT
4/2004*, John Brodie / 1043601

Comment: On 26 October 2000, it was announced that AWSR Ltd had been awarded the contract to provide a strategic sealift service in support of the Joint Rapid Reaction Force (JRRF) until late 2024. A key feature of the contract is that four Ro-Ro are in constant MoD use while the remaining ships are available for use by AWSR for the generation of commercial revenue. These can be called upon to support major operations and exercises.

UPDATED

1 SIR GALAHAD CLASS (LSLH)

Name	No	Builders	Laid down	Launched	Commissioned
SIR GALAHAD	L 3005	Swan Hunter Shipbuilders, Wallsend-on-Tyne	12 May 1985	13 Dec 1986	25 Nov 1987

Displacement, tons: 8,585 full load
Dimensions, feet (metres): 461 × 64 × 14.1
(140.5 × 19.5 × 4.3)
Main machinery: 2 Mirrlees-Blackstone diesels; 13,320 hp *(9.94 MW)*; 2 shafts; cp props
Speed, knots: 18. **Range, n miles:** 13,000 at 15 kt
Complement: 49 (15 officers)

Military lift: 343 troops (537 hard-lying); 18 MBT; 20 mixed vehicles; ammunition, fuel and stores
Guns: 2 Oerlikon 20 mm GAM-BO3. 2—12.7 mm MGs.
Countermeasures: Decoys: 4 Plessey Shield 102 mm 6-tubed launchers.
Combat data systems: Racal CANE data automation.
Radars: Navigation: Kelvin Hughes Type 1007; I-band.
Helicopters: 1 Westland Sea King HC.Mk 4.

Comment: Ordered on 6 September 1984 as a replacement for *Sir Galahad*, sunk as a war grave after air attack at Bluff Cove, Falkland Islands on 8 June 1982. Has bow and stern ramps with a visor bow gate. One 25 ton and three 8.6 ton cranes. Up to four (total) Mexeflote pontoons can be attached on both sides of the hull superstructure. To be replaced by Bay class LSD(A) in September 2006. Based at Southampton.

UPDATED

SIR GALAHAD
7/2003 / 0572706

ROYAL MARITIME AUXILIARY AND GOVERNMENT AGENCY SERVICES

Notes: (1) In early 2005, the following ships were operated by the RMAS or SERCo Denholm as part of the In Ports and Out of Ports contracts. These contracts are to be re-let as one Future Provisions of Marine Services (FPMS) contract that is due to start in October 2005.
(2) *Longbow* is a 12,000 ton trials barge whose conversion 2003-04 by FSL Portsmouth includes a mast, missile silo and firing system to facilitate PAAMS development trials.
(3) The research vessel *Triton*, used to prove the concept of trimaran design, was sold by QinetiQ to Gardline in January 2005. The ship is likely to be used for civilian hydrographic survey work.

LONGBOW
5/2004*, Derek Fox / 1043579

2 SAL CLASS (MOORING SHIPS) (ARSD)

Name	No	Builders	Commissioned
SALMOOR	A 185	Hall Russell, Aberdeen	12 Nov 1985
SALMAID	A 187	Hall Russell, Aberdeen	28 Oct 1986

Displacement, tons: 1,605 light; 2,225 full load
Dimensions, feet (metres): 253 × 48.9 × 12.5 *(77 × 14.9 × 3.8)*
Main machinery: 2 Ruston 8RKC diesels; 4,000 hp *(2.98 MW)* sustained; 1 shaft; cp prop
Speed, knots: 15
Complement: 17 (4 officers) plus 27 spare billets
Radars: Navigation: Racal Decca; I-band.

Comment: Ordered on 23 January 1984. *Salmoor* on the Clyde and *Salmaid* at Devonport. Lift, 400 tons; 200 tons on horns. Can carry submersibles including LR 5. Both operated by RMAS.

UPDATED

1 SUPPORT SHIP (AG)

Name	No	Builders	Commissioned
NEWTON	A 367	Scott-Lithgow, Greenock	17 June 1976

Displacement, tons: 3,140 light; 4,652 full load
Dimensions, feet (metres): 323.5 × 53 × 18.5 *(98.6 × 16 × 5.7)*
Main machinery: Diesel-electric; 3 Ruston 8 RK-215 diesels; 5,520 hp *(4.06 MW)*; 1 GEC motor; 2,040 hp *(1.52 MW)*; Kort nozzle; bow thruster
Speed, knots: 14. **Range, n miles:** 5,000 at 14 kt
Complement: 24
Radars: Navigation: Kelvin Hughes 1006; I-band.

Comment: Primarily used in support of RN training exercises. Limited support provided to trials. Refitted in 2001. Operated by the RMAS.

VERIFIED

SALMAID
11/2004*, Maritime Photographic / 1043620

NEWTON
7/2003, B Sullivan / 0572707

2 MOORHEN CLASS (MOORING SHIPS) (ARS)

Name	No	Builders	Commissioned
MOORHEN	Y 32	McTay, Bromborough	26 Apr 1989
MOORFOWL	Y 33	McTay, Bromborough	30 June 1989

Displacement, tons: 530 full load
Dimensions, feet (metres): 106 × 37.7 × 6.6 *(32.3 × 11.5 × 2)*
Main machinery: 2 Cummins KT19-M diesels; 730 hp *(545 kW)* sustained; 2 Aquamasters; bow thruster
Speed, knots: 8
Complement: 10 (2 officers) plus 2 divers

Comment: Classified as powered mooring lighters. The whole ship can be worked from a 'flying bridge' which is constructed over a through deck. Day mess for five divers. *Moorhen* at Portsmouth, *Moorfowl* at Devonport are RMAS vessels. *Cameron* sold to Briggs Marine in 2004.
UPDATED

MOORHEN   *2/2003, Maritime Photographic /* 0572708

1 OILPRESS CLASS (COASTAL TANKER) (AOTL)

Name	No	Builders	Launched
OILPRESS	Y 21	Appledore Ferguson	29 Aug 1968

Displacement, tons: 280 standard; 530 full load
Dimensions, feet (metres): 139.5 × 30 × 8.3 *(42.5 × 9 × 2.5)*
Main machinery: 1 Lister-Blackstone ES6 diesel; 405 hp *(302 kW)*; 1 shaft
Speed, knots: 9
Complement: 8 (3 officers)
Cargo capacity: 250 tons dieso

Comment: Ordered on 10 May 1967. GS ship on the Clyde.
VERIFIED

OILPRESS   *10/1998, M Declerck /* 0053258

1 RESEARCH SHIP (AGOR)

Name	No	Builders	Commissioned
COLONEL TEMPLER (ex-*Criscilla*)	A 229	Hall Russell, Aberdeen	1966

Displacement, tons: 1,300 full load
Dimensions, feet (metres): 185.4 × 36 × 18.4 *(56.5 × 11 × 5.6)*
Main machinery: Diesel-electric; 2 Cummins KTA-38G4M diesels; 2,557 hp(m) *(1.88 MW)* sustained; 2 Newage HC M734E1 generators; 1 Amsaldo DH 560S motor; 2,312 hp(m) *(1.7 MW)*; 1 Aquamaster azimuth thruster with contra rotating props
Speed, knots: 13.5
Complement: 14 plus 12 scientists
Radars: Navigation: Racal Decca 2690 ARPA; I-band.

Comment: Built as a stern trawler. Converted in 1980 for use at RAE Farnborough as an acoustic research ship. Major rebuild in 1992. Re-engined in early 1997 with a raft mounted diesel-electric plant to reduce noise and vibration. Carries a 9 m workboat *Quest* Q 26. Well equipped laboratories. Capable of deploying and recovering up to 5 tons of equipment from deck winches and a 5 ton hydraulic A frame. The ship is also used to support diving operations. GS vessel operated on the Clyde by SERCo Denholm.
UPDATED

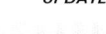

COLONEL TEMPLER   *7/2002, H M Steele /* 0530049

2 TORNADO CLASS (TORPEDO RECOVERY VESSELS) (YDT/YPT)

Name	No	Builders	Commissioned
TORNADO	A 140	Hall Russell, Aberdeen	15 Nov 1979
TORMENTOR	A 142	Hall Russell, Aberdeen	29 Apr 1980

Displacement, tons: 698 full load
Dimensions, feet (metres): 154.5 × 26.2 × 11.3 *(47.1 × 8.0 × 3.4)*
Main machinery: 2 Mirrlees-Blackstone ESL8 MCR diesels; 2,200 hp *(1.64 MW)*; 2 shafts
Speed, knots: 14. **Range, n miles:** 3,000 at 14 kt
Complement: 14
Radars: Navigation: Kelvin Hughes 1006; I-band.

Comment: Ordered on 1 July 1977. GS vessels on the Clyde. Both ships converted to support diving operations.
VERIFIED

TORNADO   *12/1999, W Sartori /* 0075841

1 WATERMAN CLASS (COASTAL TANKER) (AWT)

Name	No	Builders	Launched
WATERMAN	A 146	Dunston, Hessle	1978

Displacement, tons: 220 standard; 470 full load
Dimensions, feet (metres): 131.2 × 25.9 × 11.1 *(40.0 × 7.9 × 3.4)*
Main machinery: 1 Mirrlees Blackstone ERS8 diesel; 360 hp *(268 kW)*; 1 shaft
Speed, knots: 11. **Range, n miles:** 1,500 at 10 kt
Complement: 4
Cargo capacity: 250 tons fresh water

Comment: GS ship on the Clyde.
NEW ENTRY

WATERMAN   *6/2004*, Royal Navy /* 1043622

9 ADEPT CLASS (COASTAL TUGS) (YTB)

FORCEFUL A 221	ADEPT A 224	CAREFUL A 227
NIMBLE A 222	BUSTLER A 225	FAITHFUL A 228
POWERFUL A 223	CAPABLE A 226	DEXTEROUS A 231

Displacement, tons: 450
Dimensions, feet (metres): 127.3 × 30.8 × 11.2 *(38.8 × 9.4 × 3.4)*
Main machinery: 2 Ruston 6RKC diesels; 3,000 hp *(2.24 MW)* sustained; 2 Voith-Schneider props
Speed, knots: 12
Complement: 10

Comment: 'Twin unit tractor tugs' (TUTT). First four ordered from Richard Dunston (Hessle) on 22 February 1979 and next five on 8 February 1984. Primarily for harbour work with coastal towing capability. Nominal bollard pull, 27.5 tons. *Adept* accepted 28 October 1980, *Bustler* 15 April 1981, *Capable* 11 September 1981, *Careful* 12 March 1982, *Forceful* 18 March 1985, *Nimble* 25 June 1985, *Powerful* 30 October 1985, *Faithful* 21 December 1985, *Dexterous* 23 April 1986. *Powerful* and *Bustler* at Portsmouth, *Forceful, Faithful, Adept* and *Careful* at Devonport, *Nimble* and *Dexterous* on the Clyde. All are GS vessels except *Capable* at Gibraltar.
VERIFIED

POWERFUL   *10/2003, Maritime Photographic /* 0572711

5 DOG CLASS (YTM)

HUSKY A 178 **SETTER** A 189 **SHEEPDOG** A 250
SALUKI A 182 **SPANIEL** A 201

Displacement, tons: 248 full load
Dimensions, feet (metres): 94 × 24.5 × 12 *(28.7 × 7.5 × 3.7)*
Main machinery: 2 Lister-Blackstone ERS8 MCR diesels; 1,320 hp *(985 kW)*; 2 shafts
Speed, knots: 10. **Range, n miles:** 2,236 at 10 kt
Complement: 7

Comment: Harbour berthing tugs. Nominal bollard pull, 17.5 tons. Completed 1962-72. GS vessels serving at Portsmouth, Devonport and on the Clyde. Appearance varies considerably, some with mast, some with curved upper-bridge work, some with flat monkey-island.
UPDATED

SHEEPDOG *6/2004*, Maritime Photographic /* 1043619

3 TRITON CLASS (YTL)

KITTY A 170 **LESLEY** A 172 **MYRTLE** A 199

Displacement, tons: 107.5 standard
Dimensions, feet (metres): 57.7 × 18 × 7.9 *(17.6 × 5.5 × 2.4)*
Main machinery: 1 diesel; 330 hp *(264 kW)*; 1 shaft
Speed, knots: 7.5
Complement: 4

Comment: All completed by August 1974 by Dunstons. 'Water-tractors' with small wheelhouse and adjoining funnel. Later vessels have masts stepped abaft wheelhouse. Voith-Schneider vertical axis propellers. Nominal bollard pull, 3 tons. All are GS vessels divided between Devonport and Portsmouth.
VERIFIED

TRITON CLASS *6/2001, A Sharma /* 0131181

4 FELICITY CLASS (YTL)

FRANCES A 147 **FLORENCE** A 149 **GENEVIEVE** A 150 **HELEN** A 198

Displacement, tons: 144 full load
Dimensions, feet (metres): 70 × 21 × 9.8 *(21.5 × 6.4 × 3)*
Main machinery: 1 Mirrlees-Blackstone ESM8 diesel; 615 hp *(459 kW)*; 1 Voith-Schneider cp prop
Speed, knots: 10
Complement: 4
Radars: Navigation: Raytheon; I-band.

Comment: Four completed 1973 by Hancocks. A 147, 149 and 150 ordered early 1979 from Richard Dunston (Thorne) and completed by end 1980. Nominal bollard pull, 5.7 tons. All are GS vessels divided between Devonport, Portsmouth and the Clyde.
UPDATED

GENEVIEVE *4/2004*, Derek Fox /* 1043580

1 RANGE SUPPORT VESSEL (YFRT)

WARDEN A 368

Displacement, tons: 900 full load
Dimensions, feet (metres): 159.4 × 34.4 × 8.2 *(48.6 × 10.5 × 2.5)*
Main machinery: 2 Ruston 8RKC diesels; 4,000 hp *(2.98 MW)* sustained; 1 shaft; cp prop
Speed, knots: 15
Complement: 11 (4 officers)
Radars: Navigation: Racal Decca RM 1250; I-band.
Sonars: Dowty 2053; high frequency.

Comment: Built by Richards, Lowestoft and completed 20 November 1989. Reverted in 1998 to being an RMAS ship at Kyle of Lochalsh in support of BUTEC.
VERIFIED

WARDEN *5/1997, B Sullivan /* 0075845

2 RANGE SAFETY CRAFT (YFRT)

RSC 7713 (ex-*Samuel Morley VC*) **SIR WILLIAM ROE** 8127

Displacement, tons: 20.2 full load
Dimensions, feet (metres): 48.2 × 11.5 × 4.3 *(14.7 × 3.5 × 1.3)*
Main machinery: 2 RR C8M 410 or Volvo Penta TAMD-122A diesels; 820 hp *(612 kW)*; 2 shafts
Speed, knots: 22. **Range, n miles:** 300 at 20 kt
Complement: 3
Radars: Navigation: Furuno; I-band.

Comment: Range Safety Craft of the Honours and Sirs classes, built by Fairey Marine, A R P Whitstable and Halmatic. Completed 1982-86. Transferred from the RCT on 30 September 1988. *RSC 7713* operates at Kyle of Loch Alsh as RMAS Range trials vessel. *Sir William Roe* is based in Cyprus and has remained with the Royal Logistic Corps. New engines fitted from 1993.
UPDATED

RSC craft *10/2003, Maritime Photographic /* 0572713

2 SUBMARINE BERTHING TUGS (YTL)

Name	No	Builders	Commissioned
IMPULSE	A 344	Dunston, Hessle	11 Mar 1993
IMPETUS	A 345	Dunston, Hessle	28 May 1993

Displacement, tons: 530 full load
Dimensions, feet (metres): 106.7 × 34.2 × 11.5 *(32.5 × 10.4 × 3.5)*
Main machinery: 2 WH Allen 8S12 diesels; 3,400 hp *(2.54 MW)* sustained; 2 Aquamaster Azimuth thrusters; 1 Jastrom bow thruster
Speed, knots: 12
Complement: 6

Comment: Ordered 28 January 1992 for submarine berthing duties. There are two 10 ton hydraulic winches forward and aft with break capacities of 110 tons. Bollard pull 38.6 tons ahead, 36 tons astern. Fitted with firefighting and oil pollution equipment. Designed for one-man control from the bridge with all round vision and a comprehensive Navaids fit. *Impulse* launched 10 December 1992; *Impetus* 9 February 1993. GS vessels based on the Clyde.
UPDATED

IMPETUS *10/2004*, Maritime Photographic /* 1043618

For details of the latest updates to *Jane's Fighting Ships* online and to discover the additional information available exclusively to online subscribers please visit
jfs.janes.com

4 TOWED ARRAY TENDERS (YAG)

TARV 8611   **OHMS LAW** 8612   – 8613   **SAPPER** 8614

Dimensions, feet (metres): 65.9 × 19.7 × 7.9 *(20.1 × 6 × 2.4)*
Main machinery: 2 Perkins diesels; 400 hp *(298 kW)*; 2 Kort nozzles
Speed, knots: 12
Complement: 8
Radars: Navigation: Racal Decca; I-band.

Comment: First three built by McTay Marine, Bromborough in 1986. *Sapper* completed in February 1999. Used for transporting clip-on towed arrays from submarine bases, Faslane and Devonport. Also used as divers' support craft. Naval manned.

VERIFIED

OHMS LAW   6/2003, B Sullivan / 0572714

9 RANGE SAFETY CRAFT (YFRT)

SMIT STOUR   **SMIT ROTHER**   **SMIT ROMNEY**   **SMIT CERNE**   **SMIT WEY**
SMIT FROME   **SMIT MERRION**   **SMIT PENALLY**   **SMIT NEYLAND**

Displacement, tons: 6.1 full load
Dimensions, feet (metres): 37.1 × 11.2 × 3.9 *(11.3 × 3.4 × 1.2)*
Main machinery: 2 Volvo Penta KAD 42P diesels; 680 hp *(507 kW)*; 2 × Hamilton waterjets
Speed, knots: 35. **Range, n miles:** 160 at 21 kt
Complement: 2

Comment: MP-1111 class of vessels designed (based on a fast rescue boat) and built at Maritime Partners Ltd (Norway). Aluminium alloy hull and GRP superstructure. The order for the craft followed a contract awarded to Smit International (Scotland) Ltd for the provision of Range Clearance and Safety duties in and around the various sea danger areas of UK military ranges. Three based at Dover, Portland and Pembroke Dock.

NEW ENTRY

SMIT STOUR   6/2004*, Royal Navy / 1043617

8 AIRCREW TRAINING CRAFT (YXT)

SMIT DEE   **SMIT YARE**   **SMIT SPEY**   **SMIT TAMAR**
SMIT DON   **SMIT TOWY**   **SMIT DART**   **SMIT CYMRAN**

Displacement, tons: 55 full load
Dimensions, feet (metres): 90.5 × 21.6 × 4.9 *(27.6 × 6.6 × 1.5)*
Main machinery: 2 Cummins KTA 19M4 diesels; 1,400 hp *(1.04 MW)*; 2 shafts
   1 Ultrajet 305 centreline waterjet; 305 hp *(227 kW)*
Speed, knots: 21. **Range, n miles:** 650 at 21 kt
Complement: 6
Radars: Furuno FR-2115 EPA; I-band.

Comment: Vessels built at Babcock Engineering Services, Rosyth, and FBMA Babcock Marine, Cebu, Philippines *(Yare, Towy and Spey)*. All delivered on 11 July 2003. Of aluminium alloy construction, the design is an adaptation of FBM Babcock Marine's Protector class patrol vessel. The order for the craft followed a contract awarded to MoD and to Smit International for provision of marine support to aircrew training, high speed marine target towing and recovery of air-sea rescue apparatus. The craft have an after docking well for a daughter craft. Based at Buckie *(Dee)*, Blyth *(Don)*, Great Yarmouth *(Yare)*, Pembroke Dock *(Towy)* and Plymouth *(Spey* and *Dart)*. *Smit Dart* is employed as a passenger craft. *Tamar* (Plymouth) and *Cymran* (Holyhead) are similar second-hand craft used for passengers.

UPDATED

SMIT SPEY   7/2003, B Sullivan / 0572696

1 SUBMARINE TENDER (YFB)

Name	No	Builders	Commissioned
ADAMANT	A 232	FBM, Cowes	18 Jan 1993

Displacement, tons: 170 full load
Dimensions, feet (metres): 101 × 25.6 × 3.6 *(30.8 × 7.8 × 1.1)*
Main machinery: 2 Cummins KTA-19M2 diesels; 970 hp *(724 kW)* sustained; 2 water-jets
Speed, knots: 23. **Range, n miles:** 250 at 22 kt
Complement: 5 plus 36 passengers plus 1 ton stores

Comment: Twin-hulled support ship ordered in 1991 and launched 8 October 1992. A GS vessel used for personnel and stores transfers in the Firth of Clyde. In addition to the passengers, half a ton of cargo can be carried. Capable of top speed up to Sea State 3 and able to transit safely up to Sea State 6.

VERIFIED

ADAMANT   10/1998, M Verschaeve / 0053268

2 STORM CLASS (YFB)

Name	No	Builders	Commissioned
CAWSAND	A 192	FBM Marine, Cowes	July 1997
BOVISAND	A 191	FBM Marine, Cowes	Sep 1997

Displacement, tons: 225
Dimensions, feet (metres): 78.4 × 36.4 × 7.5 *(23.9 × 11.1 × 2.3)*
Main machinery: 2 diesels; 1,224 hp(m) *(900 kW)*; 2 shafts
Speed, knots: 14
Complement: 5 plus 75 passengers

Comment: GS vessels for the use of FOST staff at Devonport. Swath design with hydraulically operated telescopic gangways. Have replaced *Fionan of Skellig*.

VERIFIED

CAWSAND   5/2002, John Brodie / 0530050

3 OBAN CLASS (YFL)

OBAN A 283   **ORONSAY** A 284   **OMAGH** A 285

Displacement, tons: 230 full load
Dimensions, feet (metres): 90.9 × 24 × 12.3 *(27.7 × 7.3 × 3.8)*
Main machinery: 2 Cummins N14M diesels; 1,017 hp(m) *(748 kW)*; 2 Kort-Nozzles
Speed, knots: 10
Complement: 7

Comment: Built by McTay Marine and completed January to July 2000. Capable of carrying 60 passengers. *Oban* based at Devonport and the other two on the Clyde.

UPDATED

OMAGH   7/2004*, P Froud / 1043555

4 PADSTOW AND NEWHAVEN CLASSES (YFL)

PADSTOW A 286　　　**NEWHAVEN** A 280　　　**NUTBOURNE** A 281　　　**NETLEY** A 282

Displacement, tons: 125 full load
Dimensions, feet (metres): 60 × 22.3 × 6.2 *(18.3 × 6.8 × 1.9)*
Main machinery: 2 Cummins 6 CTA diesels; 710 hp(m) *(522 kW)*; 2 shafts
Speed, knots: 10
Complement: 4

Comment: Built by Aluminium Shipbuilders at Fishbourne, Isle of Wight and completed May to November 2000. Capable of carrying 60 passengers and based at Devonport (A 286) and Portsmouth. Catamaran hulls.

UPDATED

NUTBOURNE　　　　　　　　　　　　*8/2004*, Maritime Photographic* / 1043621

3 MANLY CLASS (YAG)

MELTON A 83　　　　　**MENAI** A 84　　　　　**MEON** A 87

Displacement, tons: 143 full load
Dimensions, feet (metres): 80 × 21 × 6.6 *(24.4 × 6.4 × 2)*
Main machinery: 1 Lister-Blackstone ESR4 MCR diesel; 320 hp *(239 kW)*; 1 shaft
Speed, knots: 10. **Range, n miles:** 600 at 10 kt
Complement: 6 (1 officer)

Comment: All built by Richard Dunston, Hessle. All completed by early 1983. *Melton* is an RMAS vessel at Kyle of Loch Alsh, the other two are GS ships at Falmouth.

UPDATED

MANLY CLASS　　　　　　　　　　　　　　*6/1999, A Sharma* / 0075851

1 FBM CATAMARAN CLASS (YFL)

8837

Displacement, tons: 21 full load
Dimensions, feet (metres): 51.8 × 18 × 4.9 *(15.8 × 5.5 × 1.5)*
Main machinery: 2 Mermaid Turbo 4 diesels; 280 hp *(209 kW)*; 2 shafts
Speed, knots: 13. **Range, n miles:** 400 at 10 kt
Complement: 2

Comment: Built by FBM Marine in 1989. Can carry 30 passengers or 2 tons stores. A catamaran type designed to replace some of the older harbour launches but no more have been ordered. GS vessel.

VERIFIED

8837　　　　　　　　　　　　　*4/2003, Maritime Photographic* / 0572716

ARMY (ROYAL LOGISTIC CORPS)

Notes: (1) Six Mk 4 LCVPs are listed in the RN section. One is based in the Falklands and one is in the Gulf.
(2) One Range Safety Craft is listed in RMAS section.
(3) 32 new Combat Support Boats delivered by 2002. These are 8.2 m craft, road transportable and with a top speed of 30 kt.
(4) The RLC is based at Marchwood, Southampton.

6 RAMPED CRAFT, LOGISTIC (RCL)

Name	No	Builders	Commissioned
ANDALSNES	L 107	James and Stone, Brightlingsea	22 May 1984
AKYAB	L 109	James and Stone, Brightlingsea	15 Dec 1984
AACHEN	L 110	James and Stone, Brightlingsea	12 Feb 1987
AREZZO	L 111	James and Stone, Brightlingsea	26 Mar 1987
ARROMANCHES	L 105	James and Stone, Brightlingsea	12 June 1987
(ex-*Agheila*)	(ex-L 112)		
AUDEMER	L 113	James and Stone, Brightlingsea	21 Aug 1987

Displacement, tons: 290 full load
Dimensions, feet (metres): 109.2 × 27.2 × 4.9 *(33.3 × 8.3 × 1.5)*
Main machinery: 2 Dorman 8JTCWM diesels; 504 hp *(376 kW)* sustained; 2 shafts
Speed, knots: 10. **Range, n miles:** 900 at 10 kt
Complement: 6 (2 NCOs)
Military lift: 96 tons
Radars: Navigation: Racal Decca; I-band.

Comment: *Andalsnes* and *Akyab* based in Cyprus, remainder at Southampton.

UPDATED

AUDEMER　　　　　　　　　　　　　*11/2004*, R G Sharpe* / 1043577

AREZZO　　　　　　　　　　　　　*12/2004*, R G Sharpe* / 1043576

4 WORK BOATS (YAG)

BREAM WB 03　　　**ROACH** WB 05　　　**PERCH** WB 06　　　**MILL REEF** WB 08

Displacement, tons: 27 full load
Dimensions, feet (metres): 48.6 × 14.1 × 4.2 *(14.8 × 4.3 × 1.3)*
Main machinery: 2 Gardner diesels; 640 hp *(460 kW)*; 2 shafts
Speed, knots: 10. **Range, n miles:** 500 at 10 kt
Complement: 4
Radars: Navigation: Raytheon; I-band.

Comment: First three built 1966-71; fourth one built in 1987 and last pair in late 1990s.

UPDATED

MILL REEF　　　　　　　　　　　　　*5/2004*, Derek Fox* / 1043575

SCOTTISH FISHERIES PROTECTION AGENCY

Notes: (1) The Agency has a complement of 275.
(2) There are two Cessna F-406 Caravan II aircraft with Bendix 1500 radars.
(3) A new 84 m patrol vessel, built by Ferguson Shipbuilders Port Glasgow, is to be delivered in December 2005.

3 SULISKER CLASS (PSO)

SULISKER	VIGILANT	NORNA

Displacement, tons: 1,566 (1,586 *Norna*) full load
Dimensions, feet (metres): 234.3 × 38 × 17.6 *(71.4 × 11.6 × 5.4)*
Main machinery: 2 Ruston 6AT350 diesels; 6,000 hp *(4.48 MW)* sustained *(Norna)*; 2 Ruston 12 RK 3 diesels; 5,600 hp *(4.18 MW)* sustained; 2 shafts; cp props; bow thruster; 450 hp *(336 kW)*
Speed, knots: 18. **Range, n miles:** 7,000 at 13 kt
Complement: 15 (6 officers) plus 6 spare bunks
Radars: Navigation: 2 Racal Decca Bridgemaster; I-band.

Comment: First pair built by Appledore Ferguson, Port Glasgow. *Sulisker* completed 1981, *Vigilant* completed June 1982. *Norna* built by Richards, Lowestoft and completed in June 1988. There are some structural differences in *Norna* including an A frame aft. It is planned to replace *Sulisker* in December 2005. **UPDATED**

VIGILANT   6/1998, SFPA / 0053280

NORNA   6/1998, SFPA / 0053279

1 MINNA CLASS (PBO)

MINNA

Displacement, tons: 855 full load
Dimensions, feet (metres): 156.5 × 32.8 × 14.8 *(47.7 × 10.0 × 4.5)*
Main machinery: 2 Wärtsilä Gensets; 2,896 hp *(2.16 MW)*; 2 Indar propulsion motors; 2,145 hp *(1.6 MW)*; 2 shafts; 1 Kamewa transverse thruster *(150 kW)*
Speed, knots: 14
Complement: 15 (6 officers)

Comment: Built by Ferguson Shipbuilders, Port Glasgow. Launched in February 2003 and accepted by SFPA on 31 July 2003 as replacement for *Westra*. **VERIFIED**

MINNA   6/2003, SFPA / 0561556

CUSTOMS

Notes: HM Customs and Excise Maritime Branch operates five offshore patrol vessels. The fleet comprises four Damen 42 m craft *(Seeker, Searcher, Vigilant, Valiant)*, one Vosper Thornycroft 36 m craft *(Sentinel)*.

VIGILANT   8/2003, Marian Ferrette / 0561512

SENTINEL   5/2001, A Sharma / 0131218

TRINITY HOUSE

Notes: The Corporation of Trinity House is the General Lighthouse Authority for England, Wales, The Channel Islands and Gibraltar. The Deep Sea Pilotage Authority for UK, it is a self-endowed charity which supports the education, welfare and training of mariners. The Trinity House Lighthouse Service (THLS) provides nearly 600 Aids to Navigation (AtoN) including lighthouses, buoys, beacons and a differential global positioning service. THLS is funded by dues levied on commercial shipping calling at UK ports. Currently administered from London, the majority of its operations are to be centralised at Harwich while a depot at Swansea is to serve the west coast. THLS operates a fleet of two Lighthouse tenders and three launches to service its offshore AtoN. On 22 January 2004, Burness Corlett and Partners were appointed to design and supervise construction of two multifunction tenders and a Rapid Response Vessel (to be based in the Strait of Dover). These will replace existing vessels.

PATRICIA

Displacement, tons: 3,139 full load
Dimensions, feet (metres): 284.0 × 46.0 × 14.0 *(86.3 × 13.8 × 4.3)*
Main machinery: 4 Ruston Oil diesels; 4,285 bhp *(3.2 MW)*; connected via 4 generators to 2 motors; 3,452 hp *(2.54 MW)*; 2 shafts
Speed, knots: 14. **Range, n miles:** 10,000 at 12 kt
Complement: 25 (8 officers)

Comment: Built by Henry Robb Ltd, Leith. Commissioned in May 1982. **VERIFIED**

PATRICIA   6/2003, Trinity House / 0567568

VECTIS

Displacement, tons: 76 full load
Dimensions, feet (metres): 65.0 × 18.0 × 6.0 *(19.8 × 5.4 × 1.8)*
Main machinery: 2 Volvo TMD 102 diesels
Speed, knots: 10. **Range, n miles:** 1,600 at 10 kt
Complement: 4 (2 officers)

Comment: Built by David Abels, Bristol. Commissioned in 1992. **VERIFIED**

VECTIS   6/2002, Trinity House / 0543402

MERMAID

Displacement, tons: 3,031 full load
Dimensions, feet (metres): 264.0 × 48.0 × 13.0 *(80.3 × 14.5 × 4.0)*
Main machinery: 4 Ruston Oil diesels; 5,329 bhp *(3.9 MW)*; connected via 4 generators to 2 motors; 2,338 hp *(1.72 MW)*; 2 shafts
Speed, knots: 13. **Range, n miles:** 10,000 at 12 kt
Complement: 25 (8 officers)

Comment: Built by Hyundai Heavy Industries, South Korea. Commissioned in March 1987.
VERIFIED

MERMAID   6/2002, *Trinity House* / 0543403

READY

Displacement, tons: 9.6 full load
Dimensions, feet (metres): 38.0 × 10.0 × 4.0 *(11.5 × 3.1 × 1.3)*
Main machinery: 2 Cummins 6CTA diesels.
Speed, knots: 32. **Range, n miles:** 1,600 at 10 kt
Complement: 2 (1 officer)

Comment: Built by Halmatic Ltd. Commissioned in 1995.
VERIFIED

READY   6/2003, *Trinity House* / 0567567

MARITIME & COASTGUARD AGENCY

Notes: The Maritime & Coastguard Agency is responsible for the development, promotion and enforcement of high standards of marine safety, response to maritime emergencies 24 hours a day, reduction of the risk of pollution of the marine environment from ships and, where pollution occurs, minimisation of its impact on the United Kingdom.

Response to maritime emergencies within the UK SAR region is undertaken by HM Coastguard and the MCA's Counter-pollution and Salvage Branch. SAR is co-ordinated through a network of 6 Maritime Rescue Co-ordination Centres (MRCCs) and 12 Maritime Rescue Sub Centres. Each MRCC provides continuous emergency telephone, radio and satellite communications distress watch plus safety information and radio medical advice services. The counter-pollution branch provides response to marine pollution and provides scientific and technical advice on shoreline clean up. The MCA has recently introduced a new fleet of patrol vessels for ship inspection, surveillance and accident prevention.

The MCA provides four civilian SAR helicopters (S61N) under contract from Bristow Helicopters. They are based at Sumburgh, Stornoway, Lee-on-Solent and Portland. Fixed-wing aircraft include a BN Islander which conducts surveillance patrols over the Dover Strait and forms part of the Channel Navigation Information Service while, for counter-pollution, a Cessna 404 and Cessna 406 are operated by Atlantic Reconnaissance of Coventry. Fitted with radar, IR and UV detection equipment. Additionally a Cessna 406 and two Lockheed Electra aircraft are available for dispersant spraying. Four emergency towing vessels for SAR, counter-pollution and salvage are under contract from Klyne Tugs Ltd: *Anglian Prince* (1,598 tons gwt), *Anglian Princess* and *Anglian Sovereign* (2,270 tons gwt) and *Anglian Monarch* (1,480 tons gwt).

HM Coastguard has its own corps of 3,100 volunteer Auxiliary Coastguards divided into 401 Coastguard Rescue Teams around the coast of UK. HM Coastguard also make significant use of Royal National Lifeboat Institution all-weather and inshore lifeboats and military SAR helicopters.

The MCA is also responsible for inspections and surveys of UK vessels, port state control inspections of non UK ships, the enforcement of merchant shipping legislation and the setting of ship and seafarer standards.

ANGLIAN PRINCESS   7/2004*, *B Sullivan* / 1043562

ANGLIAN PRINCESS   6/2002, *MCA* / 0529956

ANGLIAN PRINCE   10/2004*, *Maritime Photographic* / 1043623

United States

Country Overview

The United States of America is a federal republic which comprises 48 contiguous states (bounded to the north by Canada and to the south by Mexico) and the states of Alaska and Hawaii. External territories include Puerto Rico, American Samoa, Guam and the US Virgin Islands. With an area of 3,717,800 square miles, it occupies much of North America and has a coastline of 10,762 n miles with the Atlantic and Pacific Oceans and with the Gulf of Mexico. Washington, DC is the capital while New York, New York, is the largest city and a leading seaport. Other principal ports include New Orleans, Louisiana; Houston, Texas; Valdez, Alaska; Baton Rouge, Louisiana; Corpus Christi, Texas; Long Beach, California; Norfolk, Virginia; Tampa, Florida; Los Angeles, California; St Louis, Missouri; and Duluth, Wisconsin. There is an extensive inland waterway network, the three main components of which are the Mississippi river system (13,000 n miles long), the Great Lakes (ocean-going vessels can sail between the Great Lakes and the Atlantic Ocean via the St Lawrence Seaway (opened 1959)) and coastal waterways. Territorial seas (12 n miles) are claimed. A 200 n mile EEZ has been claimed but the limits have only been partly defined by boundary agreements.

Headquarters Appointments

Chief of Naval Operations:
   Admiral Michael G Mullen
Vice Chief of Naval Operations:
   Admiral Robert F Willard
Director, Naval Nuclear Propulsion:
   Admiral Kirland H Donald
Chief of Naval Personnel:
   Vice Admiral Gerald L Hoewing
Commander, Naval Sea Systems Command:
   Vice Admiral Phillip M Balisle
Commander, Naval Air Systems Command:
   Vice Admiral Walter B Massenburg
Commander, Space and Naval Warfare Systems Command:
   Rear Admiral Kenneth D Slaght

Unified Combatant Commanders

Commander, US Strategic Command:
   General James E Cartwright
Commander, US Pacific Command:
   Admiral William J Fallon
Commander, US Joint Forces Command:
   Admiral Edmund P Giambastiani Jr

Unified Combatant Commanders — *continued*

Commander, US European Command:
   General James L Jones
Commander, US Northern Command:
   Admiral Timothy J Keating
Commander, US Southern Command:
   General Bantz J Craddock
Commander, US Central Command:
   General John Abizaid

Fleet Commanders

Commander, US Fleet Forces Command:
   Admiral John B Nathman
Commander, US Pacific Fleet:
   Admiral Gary Roughead
Commander, US Joint Force Command, Naples and US Naval Forces Europe:
   Admiral Henry G Ulrich III
Commander, Military Sealift Command:
   Vice Admiral David L Brewer III

Flag Officers (Atlantic Area)

Commander, Second Fleet and Striking Fleet, Atlantic:
   Vice Admiral Mark P Fitzgerald
Commander, Surface Force, Atlantic Fleet:
   Rear Admiral Michael P Nowakowski
Commander, Sixth Fleet, Joint Command Lisbon and Striking Fleet NATO:
   Vice Admiral John D Stufflebeem
Commander, Submarine Force, Atlantic Fleet and Submarine Allied Command, Atlantic:
   Vice Admiral Charles L Munns
Commander, Naval Air Force, Atlantic Fleet:
   Rear Admiral Harold D Starling II
Commander, Fleet Air, Keflavik and Iceland Defense Force:
   Rear Admiral Noel G Preston
Commander, Mine Warfare Command:
   Rear Admiral Michael P Nowakowski

Flag Officers (Pacific Area)

Commander, Seventh Fleet:
   Vice Admiral Jonathan W Greenert
Commander, Naval Surface Force, Pacific Fleet:
   Vice Admiral Terrance T Etnyre

Flag Officers (Pacific Area) — *continued*

Commander, Third Fleet:
   Vice Admiral Barry M Costello
Commander, Naval Air Force, Pacific Fleet:
   Vice Admiral James M Zortman
Commander, US Naval Forces, Japan:
   Rear Admiral Frederick R Ruehe
Commander, Submarine Force, Pacific Fleet:
   Rear Admiral Jeffrey B Cassias
Commander, US Naval Forces, Korea:
   Rear Admiral Fred Byus
Commander, US Naval Forces, Marianas:
   Rear Admiral Arthur J Johnson
Commander, Fleet Anti-Submarine Warfare Command:
   Rear Admiral John J Waickwicz

Flag Officer (Central Area)

Commander, Fifth Fleet:
   Vice Admiral David C Nichols, Jr

Marine Corps

Commandant:
   General Michael W Hagee
Assistant Commandant:
   General William L Nyland
Commander, Fleet Marine Force, Atlantic:
   Lieutenant General Martin R Bernt
Commander, Fleet Marine Force, Pacific:
   Lieutenant General W C Gregson

Prefix to Ships' Names

USS (United States Ship) Warships
USNS (United States Naval Ship) Military Sealift Command

Personnel

	1 Jan 2003	1 Jan 2004	1 Jan 2005
Navy			
Officers	54,579	55,036	53,925
Enlisted	324,756	320,132	313,186
Marine Corps			
Officers	18,284	18,431	18,847
Enlisted	155,658	156,473	158,674

Strength of the Fleet (1 January 2005)

Type	Active (NRF) (Reserve)	Building (Projected) + Conversion/SLEP
SHIPS OF THE FLEET		
Strategic Missile Submarines		
SSBN (Ballistic Missile Submarines) (nuclear-powered)	14	—
Cruise Missile Submarines (SSGN) (nuclear-powered)	—	4
Attack Submarines		
SSN Submarines (nuclear-powered)	54	6 (3)
Aircraft Carriers		
CVN Multipurpose Aircraft Carriers (nuclear-powered)	10	1 (2)
CV Multipurpose Aircraft Carriers (conventionally powered)	2	—
Cruisers		
CG Guided Missile Cruisers	25	—
Destroyers		
DDG Guided Missile Destroyers	47	16 (1)
DD Destroyers	3	—
Frigates		
FFG Guided Missile Frigates	21 (9)	1 (3)
Patrol Forces		
PC Coastal Defense Ships	8	—
Command Ships		
LCC Command Ships	2	—
AGF Command Ships	1	—
Amphibious Warfare Forces		
LHA Amphibious Assault Ships (general purpose)	5	—
LHD Amphibious Assault Ships (multipurpose)	7	1
LPD Amphibious Transport Docks	12	8 (3)
LSD Dock Landing Ships	12	—
LSV Logistic Support Vessels	7	1
Mine Warfare Forces		
MCM Mine Countermeasures Ships	14	—
MHC Minehunters (Coastal)	12	—

Type	Active (NRF) (Reserve)	Building (Projected) + Conversion/SLEP
Auxiliaries		
AGSS Auxiliary Research Submarine	1	—
AOE Fast Combat Support Ships	1	—
ARS Salvage Ships	4	—
AS Submarine Tenders	2	—
MILITARY SEALIFT COMMAND INVENTORY		
Naval Fleet Auxiliary Force		
T-AOE Fast Combat Support	4	—
T-AKE Auxiliary Cargo and Ammunition	—	8 (4)
T-AE Ammunition	6	—
T-AFS Combat Stores	6	—
T-AH Hospital	2	—
T-AO Oilers	15	—
T-ATF Fleet Ocean Tugs	5	—
Special Mission Ships		
T-AG/T-AGM Miscellaneous	3	—
T-AGOS Surveillance/Patrol	4	—
T-AGS Surveying	7	—
T-ARC Cable Repair	1	—
Strategic Sealift Force		
T-AKR Fast Sealift	19	—
T-AOT Tankers	5	—
Prepositioning Programme		
HSV Theatre Support	2	—
T-AK Container Ro-Ro	4	—
T-AK Container	5	—
T-AK Vehicle Cargo	15	—
T-AKR Large, Medium-Speed, Ro-Ro	8	1
T-AVB Aviation Logistic	2	—
Ready Reserve Force		
T-AK Break Bulk	13	—
T-AKR Ro-ro	31	—
T-AOT/T-AOG Product Tankers	7	—
T-AK/T-AP Miscellaneous	7	—

Special Notes

To provide similar information to that included in other major navies' Deployment Tables the fleet assignment (abbreviated 'F/S') status of each ship in the US Navy has been included. The assignment appears in a column immediately to the right of the commissioning date. In the case of the Floating Dry Dock section this system is not used. The following abbreviations are used to indicate fleet assignments:

AA	active Atlantic Fleet
Active	active under charter with MSC
AR	in reserve Out of Commission, Atlantic Fleet
ASA	active In Service, Atlantic Fleet
ASR	in reserve Out of Service, Atlantic Fleet
Bldg	Building
CONV	ship undergoing conversion
LOAN	ship or craft loaned to another government, or non-government agency, but US Navy retains title and the ship or craft is on the NVR
MAR	in reserve Out of Commission, Atlantic Fleet and laid up in the temporary custody of the Maritime Administration
MPR	same as 'MAR', but applies to the Pacific Fleet
NRF	assigned to the Naval Reserve Force (ships so assigned are listed in a special table for major warships and amphibious ships)
Ord	the contract for the construction of the ship has been let, but actual construction has not yet begun
PA	active Pacific Fleet
PR	in reserve Out of Commission, Pacific Fleet
Proj	the ship is scheduled for construction at some time in the immediate future
PSA	active In Service, Pacific Fleet
PSR	in reserve Out of Service, Pacific Fleet
ROS	reduced Operating Status
TAA	active Military Sealift Command, Atlantic Fleet
TAR	in Ready Reserve, Military Sealift Command, Atlantic Fleet
TPA	active Military Sealift Command, Pacific Fleet
TPR	in Ready Reserve, Military Sealift Command, Pacific Fleet
TWWR	active Military Sealift Command, Worldwide Routes

Ship Status Definitions

In Commission: as a rule any ship, except a Service Craft, that is active, is in commission. The ship has a Commanding Officer and flies a commissioning pennant. 'Commissioning date' as used in this section means the date of being 'in commission' rather than 'completion' or 'acceptance into service' as used in some other navies.

In Service: all service craft (dry docks and with classifications that start with 'Y'), with the exception of *Constitution*, that are active, are 'in service'. The ship has an Officer-in-Charge and does not fly a commissioning pennant.

Ships in reserve, out of commission' or 'in reserve, out of service' are put in a state of preservation for future service. Depending on the size of the ship or craft, a ship in 'mothballs' usually takes from 30 days to nearly a year to restore to full operational service. The above status definitions do not apply to the Military Sealift Command.

Approved Fiscal Year 2005 Programme

	Appropriations (US dollars millions)
CVN 21 (R & D, advance procurement)	970
SSGN conversion (R & D, advance procurement)	537
Virginia class (SSN 779)	2,757
DD(X) Destroyer	304
3 Arleigh Burke (DDG 110-112)	3,559
1 San Antonio class (LPD 18)	1,227
Littoral Combat Ship (R & D)	453

Proposed Fiscal Year 2006 Programme

	Appropriations (US dollars millions)
CVN 21 (Advance procurement)	565
SSGN conversion (Advance procurement)	287
SSN Virginia class (SSN 780)	2,401
DD(X) Destroyer	716
San Antonio class (LPD 19)	1,345
Littoral Combat Ship	576

Naval Aviation

Naval Aviation had an active inventory of 3,836 aircraft as of 1 January 2005, with approximately 33 per cent of those being operated by the US Marine Corps. The principal aviation organisations are 10 active carrier air wings and one reserve, 12 active maritime patrol squadrons and seven reserve, and three Marine aircraft wings and one reserve.

Fighter Attack: 21 (3) Naval and 15 (3) Marine squadrons with F/A-18 Hornets. 12 Navy squadrons for F/A-18 Super Hornets and 7 Navy squadrons with F-14 Tomcats.
Attack: Eight Marine squadrons with AV-8B Harriers.
Airborne Early Warning: 11 (2) Navy squadrons with E-2C Hawkeyes.
Electronic Warfare: 14 (1) Navy and four Marine squadrons with EA-6B Prowlers.
Communication Relay: Two Navy squadrons of EA-6B Mercurys.
Anti-Submarine Warfare: 9 Navy squadrons with S-3B Vikings.
Maritime Patrol: 13 (7) Navy squadrons with P-3C Orions.
Electronic Reconnaissance: Two Navy squadrons with EP-3E Orions (Aries II).
Helicopter Anti-Submarine: 23 (4) Navy squadrons with SH/HH-60 Seahawks.
Helicopter Mine Countermeasures: 2 Navy squadrons with MH-53E Sea Dragons.

Helicopter Combat Support: Five Navy squadrons of MH-60S Knighthawks, one Navy squadron of MH-53E Sea Dragons and one (1) Navy squadron of UH-3H Sea Kings.
Helicopter Combat Support/Gunship: Seven (2) Marine squadrons with AH-1W Super Cobras.
Helicopter Transport: 15 (2) Marine squadrons of CH-46E Sea Knights, three with CH-53D Sea Stallions and six (two) of CH-53E Super Stallions.
Special or Composite: Two Navy special squadrons of HH-60H Seahawks.

Aircraft Procurement Plan FY2005-2007

	05	06	07
F/A-18E/F Super Hornet	42	38	30
MV-22 Osprey	8	9	14
T-45TS Goshawk	10	6	12
E-2C Hawkeye	2	2	2
EA-18G	–	4	12
AH-1Z/UH-1Y	7	10	18
MH-60S	15	26	26
MH-60R	6	12	25
KC-130J	4	12	–
C-35	2	–	1
C-40A	1		1
C-37	2		
V-XX	3	5	–
JPATS	2	–	24
F-5E	9	9	5

Naval Special Warfare

The Naval Special Warfare Command was commissioned 16 April 1987.

SEAL (Sea Air Land) teams are manned at a nominal 6 platoons per team, with 24 platoons on each coast based at Coronado, California (Group 1) and Little Creek, Virginia (Group 2). Platoons are allocated to theatre commanders during operational deployments. Total numbers approximately 5,700 of whom approximately 2,300 are SEALs and 600 are Special Warfare Combatant-craft Crewmen (SWCC) operators.

Bases

Naval Air Stations and Air Facilities

Naval Air Weapons Station (NAWS) China Lake, CA; Naval Air Facility (NAF) El Centro, CA; Naval Air Station (NAS) Lemoore, CA; NAF Washington, DC; NAS Jacksonville, FL; NAF Key West, FL; NAS Whiting Field (Milton), FL; NAS Pensacola, FL; NAS Atlanta (Marietta), GA; PMRF Barking Sands, HI; NAS Joint Reserve Base, New Orleans, LA; NAS Brunswick, ME; NAS Patuxent River, MD; NAS Meridian, MS; NAS Fallon, NV; Naval Air Engineering Station (NAES) Lakehurst, NJ; NAS Joint Reserve Base, Willow Grove, PA; NAS, Corpus Christi, TX; NAS Joint Reserve Base Fort Worth, TX; NAS Kingsville, TX; NAS Oceana, VA; NAS Whidbey Island (Oak Harbor), WA; NAS Keflavik, Iceland; NAS Sigonella, Italy; NAF Atsugi, Japan; NAF Misawa, Japan; NAF Mildenhall, UK.

Naval Stations and Naval Bases

Naval Station San Diego, CA; NB Coronado, CA; NB Ventura County, CA; NB Point Loma (San Diego), CA; NS Mayport, FL; Naval Station (NS) Pearl Harbor, HI; NS Great Lakes, ILL; NS Annapolis, MD; NS Pascagoula, MS; NS Ingleside, TX; NS Newport, RI; Naval Amphibious Base (NAB) (Amphibious) Little Creek, VA; NS Norfolk, VA; NB Kitsap, WA; NS Everett, WA; NS Guantanamo Bay, Cuba; Fleet Activities (FA) Okinawa, Japan; CFA Sasebo, Japan; CFA Chinhae, Korea; CFA Yokosuka, Japan; NS Rota, Spain; CBC Gulfport.

Naval Support Facilities

Naval Post Graduate School Monterey, CA; NSA Washington, DC; NSA New Orleans, LA; NSA Mechanicsburg, PA; NSA Mid-South (Millington), TN; NSA Norfolk, VA. NSF Diego Garcia, BIOT; Naval Forces Marianas Support Activity, Guam; NSA Souda Bay, Greece; NSA Gaeta, Italy; NSA La Maddalena, Italy; NSA Naples, Italy; Naval Activities (NA) United Kingdom (London), UK; NSA Bahrain; NAVREGCONTCTR Singapore; NCTAMS EASTPAC (Hawaii); NSGA Kunia; NUWC Keyport (WA); NAVMAG Indian Island (WA); NAVWPNSTA Seal Beach (CA); NSA Corona (CA); NSA Crane (IN); NAVWPNSTA Earle (NJ); NSU Saratoga Springs (NY); JMF St Mawgans (UK); NA Puerto Rico (PR); NSA Indian Head (MD); NSA Carderock (MD); NSF Thurmont (MD); NSA Dahlgren (VA); NAVWPNSTA Yorktown; NSWC Philadelphia; NSA Wallops Island; NSGA Sugar Grove; NAVWPNSTA Charleston; NSA Panama City; NAVSCSCOL Athens (GA); NSA Orlando; NUWC Bahamas (Andros Is).

Strategic Missile Submarine Bases

SUBASE Kings Bay, GA (East Coast); SUBASE New London, CT (East Coast).

Naval Shipyards

NSY/IMF Pearl Harbor, HI; Puget Sound NSY/IMF, Bremerton, WA; NSY Norfolk, VA; NSY Portsmouth, NH (located in Kittery, ME).

Marine Corps Air Stations and Helicopter Facilities

MCAS: Beaufort, SC; Yuma, AZ; Kaneohe Bay, Oahu, HI; Quantico, VA; Cherry Point, NC; Iwakuni, Honshu, Japan; New River (Jacksonville), NC. Futema, Okinawa, Miramar (San Diego), CA.

Marine Corps Bases

Camp Pendleton, CA; Twentynine Palms, CA; Camp H M Smith (Oahu), HI; Camp Lejeune, NC; Camp Smedley D Butler (Kawasaki), Okinawa, Japan.

Command and Control of US Naval Forces

Strategic and Operational Command

All US Military Forces operate under Title 10 of US Code and subsidiary Joint Force Doctrine publications. The President of the United States is the Commander-in-Chief of all US forces and exercises authority for the application of military force through the Secretary of Defense who is advised by the Chairman of the Joint Chiefs of Staff. The Unified Combatant Commanders are four-star officers who have broad geographic area of functional responsibilities. Exercising Combatant Command (COCOM), they have authority to employ forces as necessary to accomplish assigned military missions and are as follows:

Commander US European Command (Stuttgart-Vaihingen, Germany)
Commander US Northern Command (Peterson AFB, Colorado)
Commander US Pacific Command (Honolulu, Hawaii)
Commander US Southern Command (Miami, Florida)
Commander US Central Command (MacDill AFB, Florida)
Commander US Joint Forces Command (Norfolk, Virginia)
Commander US Special Operations Command (MacDill AFB, Florida)
Commander US Transportation Command (Scott AFB, Illinois)
Commander US Strategic Command (Offutt AFB, Nebraska)

The Unified Combatant Commanders may decide to exercise Operational (OPCON) command of naval forces directly. Alternatively, they may delegate such powers to another officer

THE WORLD 1:135,000,000 **THE WORLD WITH COMMANDERS' AREAS OF RESPONSIBILITY** EDITION 5-NIMA SERIES 1107 Based on Unified Command Plan February 2002

AREAS OF RESPONSIBILITY

4/2002, US DoD / 0127116

CHAIN OF COMMAND   0531964

who might be a subordinate Unified Commander (for example Commander, US Forces Korea), a service component commander (Army, Navy, Air Force, Marine Corps and so on), a functional component commander (air, maritime, land, special forces), a joint task force commander or a single service force commander.

Navy force commanders have a dual chain of command. They report to the Chief of Naval Operations for administrative matters such as training and equipping of forces and are also responsible to the combatant commanders for providing forces to accomplish missions. They include the following:

Commander US Fleet Forces Command
Commander Pacific Fleet
Commander US Naval Forces Europe
Commander US Naval Forces Central Command
Commander Naval Reserve Force
Commander Military Sealift Command

Once deployed in theatre, naval forces are operationally assigned to three-star numbered fleet commanders:

Commander Second Fleet (Atlantic)
Commander Third Fleet (Eastern Pacific)
Commander Fifth Fleet (Arabian Gulf and Indian Ocean)
Commander Sixth Fleet (Mediterranean)
Commander Seventh Fleet (Western Pacific)

These arrangements are intended to provide a framework that provides a clear chain of command while retaining the flexibility to be adapted to the operational circumstances. For example, it is feasible for a multimission naval task group, such as a carrier strike group (CSG) (baseline composition: 1 CVN/CV, 2 CG/DDG, 1 DD/FFG, 1 SSN and 1 logistic support ship) or an expeditionary strike group (ESG) (baseline composition: 3 amphibious ships (LHD/LHA, LPD and LSD), 2 CG/DDG, 1 DD/FFG and 1 SSN) to support service, component, and other superior commanders simultaneously.

Tactical Command and Composite Warfare Commander

US naval task groups and forces operate under Composite Warfare Commander (CWC) doctrine. The officer in tactical command (OTC) is responsible for accomplishing the missions of his assigned forces. The CWC directs the force and controls warfare functions. The OTC may designate a subordinate commander as CWC but, in general practice, the roles are combined. The OTC/CWC is supported by Principal Warfare Commanders (PWC), Functional Warfare Commanders (FWC) and Coordinators.

PWCs include the Air Defense Commander (ADC), Strike Warfare Commander (STWC), Information Warfare Commander (IWC), Anti-submarine Warfare Commander (ASWC), and Surface Warfare Commander (SUWC). ASW and SUW areas can be combined under a Sea Combat Commander (SCC). PWCs collect and distribute information pertinent to their warfare areas and can be delegated authority to respond to threats with assigned assets.

FWCs perform duties of a scope or duration more limited than that of a PWCs. Typical FWCs include Maritime Interception Operations Commander (MIOC), Mine Warfare Commander (MIWC), Operational Deception Group Commander, Screen Commander (SC) and Underway Replenishment Group (URG) Commander.

Coordinators are responsible to the OTC/CWC for managing assets and resources. Among assigned Coordinators are the Air Resource Element Coordinator (AREC), Air Control Authority (ACA), Cryptologic Resource Coordinator (CRC), Force Over-the-horizon Track Coordinator (FOTC), Force Track Coordinator (FTC), Helicopter Element Coordinator (HEC), Submarine Operations Coordinating Authority (SOCA), TLAM Launch Area Coordinator (LAC) and TLAM Strike Coordinator (TSC).

The OTC/CWC may activate any or all of these warfare commanders and coordinators as necessary. The guiding principle of CWC doctrine is flexibility to meet operational requirements.

Multinational Operations

US naval forces regularly participate in peacetime and wartime multinational operations. Although the President always retains command authority over US forces, he may place them under control of a foreign commander as required to achieve specific military objectives. Multinational operations may be conducted under the structure of a formal alliance (such as NATO) or of an ad hoc coalition (Operation Desert Shield/Desert Storm).

Complex Naval Task Forces

Complex Task Forces usually consist of multiple CSGs and/or ESGs and may also include naval assets of allied nations. Such forces may operate together under three generic command and control structures.

In Situation A, the forces integrate, the senior officer present becomes the overall OTC/CWC and a new single CWC organisation is established.

In Situation B, task groups do not integrate. The senior OTC/CWC coordinates the tactical operations of all naval forces and delegates responsibilities and TACON of specific forces to junior commanders as appropriate. The senior OTC/CWC may also designate junior commanders as sector OTC/CWCs.

In Situation C, each group retains its own OTC/CWC and its own set of warfare commanders and coordinators. The OTC/CWC of the supported force (or a common superior) draws on the assets of the entire force to achieve joint and combined force objectives.

Amphibious Operations

'Commander Amphibious Task Force' (CATF) and 'Commander Landing Force' (CLF) are historic naval command terms whose functional responsibilities are recognised by Joint Doctrine. The common superior establishes command relationships between CATF and CLF who are considered coequal in planning. CATF is responsible for operations at sea while CLF dictates landing force objectives and landing and drop zones.

US Marine Corps Organisation

Marine Corps Structure

Title 10 directs that the Marine Corps is to consist of three divisions and three air wings with their necessary logistics support and that there is to be a similar organisation in the reserves consisting of one division, one air wing, and their respective logistical support groups. MEFs I (Camp Pendleton, CA), II (Camp Lejeune, NC) and III (Okinawa, Japan) are the three standing Marine Expeditionary Forces (MEFs).

The MEF is the USMC's principle war-fighting organisation. Commanded by a lieutenant general, it consists of 50-60,000 personnel and includes, typically, a division, air wing, FSSG and headquarters group. MEFs can conduct a broad scope of missions in any environment for sixty days and are supported by amphibious shipping and/or Maritime Prepositioning Squadrons (MPS). Because of its size, the MEF is normally committed sequentially, building on a smaller operational unit such as a Marine Expeditionary Brigade (MEB) or Marine Expeditionary Unit (MEU).

The MEB is designed as the lead element for a MEF or for small-scale contingencies. Command by a major general or brigadier, it consists of 14-18,000 Marine and Navy personnel and has thirty days sustainability. The ground combat element consists of an infantry regiment reinforced by artillery, some armour, light armoured vehicles, assault amphibian vehicles, and combat engineers. These assets can be divided into four battalion-size manoeuvre elements, supported by three to six fixed- and rotary-wing aircraft squadrons.

MEUs routinely forward deploy on Amphibious Ready Groups (ARGs). Commanded by a colonel, they contain approximately 2,200 Marine and Navy personnel and can sustain operations for fifteen days. MEUs normally consist of a reinforced infantry Battalion Landing Team (BLT), a composite helicopter squadron

(with air command and control and six Harriers), and a MEU service support group (MSSG). Typically, such a force can act as the lead element for a larger force and/or provide shaping/ engagement activities, deterrence, and limited power projection. It has the capability to conduct company to battalion-sized raids to the range limits of assigned helicopters, roughly 70-100 miles from the ARG. An ARG typically consists of 1 LHD/LHA, 1 LPD and 1 LSD.

Marine Corps Operations

Operations are conducted by Marine Air Ground Task Forces (MAGTFs) whose size and composition will be dictated by operational circumstances. A MAGTF can be established by drawing ground, aviation, and combat service support assets from divisions, air wings, and their support groups. At the lower end of the scale, MEUs are available as immediately responsive, sea-based MAGTFs while, on a much greater scale, a full MEF might be required. This might be based on one of the standing MEFs or, as in Operation Desert Shield/Desert Storm, drawn from all three standing MEFs. A MAGTF always consists of a Command Element (CE), Ground Combat Element (GCE), Aviation Combat Element (ACE) and a Combat Service Support Element (CSSE).

Expeditionary Strike Group (ESG)

These are designed to give the Unified Combatant Commanders greater combat power and flexibility. A pilot deployment was conducted in early 2003. Under the ESG concept, an Amphibious Ready Group (LHA/LHD, LPD and LSD) and embarked MEU will be augmented by a cruiser, a destroyer, a frigate, and an attack submarine, together with a P-3 Orion ASW/ Recce aircraft. Ultimately, the Navy plans to maintain 12 ESGs and 12 CVBGs.

Embarked MEU

Marine Corps amphibious forces embarked on ARGs come under the OPCON of the naval or maritime component commander. They remain under the naval or maritime component commander throughout an amphibious operation if they will re-embark. If they transition to sustained operations ashore, they chop to either the Marine component commander or the land component commander. A Marine Corps component commander may be designated as the joint force maritime, land, or air component commander.

Communications and Data Systems

Advanced Combat Direction System (ACDS)

ACDS is a centralised, automated command and control system. An upgrade from the Naval Tactical Data System (NTDS) for aircraft carriers and large-deck amphibious ships, it provides the capability to identify and classify targets, prioritise and conduct engagements, and exchange targeting information and engagement orders within the battle group and among different service components in the joint theatre of operations. ACDS is a core Sea Shield component of non-Aegis/non-SSDS combat systems.

ACDS consists of two variants. The ACDS Block 0 system replaces obsolete NTDS computers and display consoles and incorporates new software. ACDS Block 0 is deployed on nine aircraft carriers, five Wasp (LHD-1) class amphibious assault ships, and all five Tarawa (LHA-1) class amphibious assault ships. ACDS Block 1 has been installed in five ships from 1996: Eisenhower (CVN 69), John F Kennedy (CV 67), Nimitz (CVN-68), Wasp and Iwo Jima (LHD-7). Following the OPEVAL failure of ACDS Block 1, it is to be replaced by the Ship Self Defense System (SSDS). Installation in Nimitz and Eisenhower is in progress and John F Kennedy and Iwo Jima are to be fitted in 2006 and 2007 respectively.

AEGIS Combat System

The AEGIS system is designed as a total weapon system, from detection to kill in the air, surface and sub-surface domains.
The SPY-1 radar system is the primary air and surface radar for the Aegis Combat System installed in the Ticonderoga (CG-47)

Composite Warfare Commander Structure

CWC STRUCTURE   0531963

and Arleigh Burke (DDG-51) class warships. It is a multifunction, phased-array radar capable of search, automatic detection, transition to track, tracking of air and surface targets, and missile engagement support. The third variant of this radar, SPY-1D(V), the Littoral Warfare Radar, improves the radar's capability against low-altitude, reduced radar cross-section targets in heavy clutter environments, and in the presence of intense electronic countermeasures. The SPY-1 Series radars also demonstrated the capability to detect and track theatre ballistic missiles. AEGIS equipped platforms include Spanish F-100 and Japanese DDG ship classes.

Automated Digital Network System (ADNS)

The Automated Digital Network System is responsible for the transport of all Wide Area Network (WAN) Internet Protocol (IP) services which connect afloat units to various global shore sites. It provides ship and shore IP connectivity and promotes efficient use of available satellite and line of sight communications bandwidth. ADNS converges all voice, video, and data communications between ship and shore to an IP medium and takes advantage of all shipborne RF to transmit data efficiently. Specifically, it automates routing and switching of tactical and strategic C4I data via Transmission Control Protocol/Internet Protocol (TCP/IP) networks linking deployed battle group units with each other and with the Defense Information Systems Network (DISN) ashore. ADNS uses Commercial Off-the-Shelf (COTS) and Non-Developmental Item (NDI) Joint Tactical Architecture (JTA) – compliant hardware (routers, processors and switches), and commercial-compliant software in a standardised, scalable, shock-qualified rack design.

Challenge Athena (WSC-8)

Challenge Athena is part of the Navy commercial wideband satellite program (CWSP). It is a full-duplex, high data-rate communications link that operates in the C-band spectrum up to 2.048 Mbps. The Challenge Athena terminal (AN/WSC-8(V)1,2) with modifications by the developer/manufacturer is also capable of operating in the Ku-band spectrum. Because of open ocean limitations, there are currently no plans to enhance Navy's commercial satellite terminal to include Ku coverage. CWSP provides access to voice, video, data and imagery circuit requirements. It supports fleet commander flagships (LCC/AGF), aircraft carriers (CV/CVN), amphibious ships (LHA/LHD/LPD) and other selected ships, including hospital ships (T-AH) and submarine tenders (AS). Terminals are also installed at training locations in San Diego, California, and Norfolk, Virginia. Examples of communications circuits that are provided include: Joint Service Imagery Processing System-Navy/Concentrator Architecture (JSIPS-N/JCA), Naval and Joint Fires Network (NFN), Video Tele-Conferencing (VTC), Video Information Exchange system (VIXS), Video Tele-Medicine (VTM), Video Tele-Training (VTT), Afloat Personal Telephone Service (APTS), Automated Digital Network System (ADNS), Integrated Digital Switching Network (IDSN) for voice/telephone, Secret/ Unclassified Internet Protocol Router Networks (SIPRNET/ NIPRNET), and Joint Worldwide Intelligence Communications System (JWICS). The CWSP terminal uses commercial satellite connectivity and COTS/NDI Equipment. In recent years, it has become an integral part of Navy's SATCOM architecture because of the overburdened military satellite communications systems.

Common Data Link – Navy (USQ-123 CDL-N)

This programme provides a common data terminal for the receipt of signal and imagery intelligence data from remote sensors and transmission of link and sensor control data to airborne platforms. CDL-N interfaces with shipboard processors of the Joint Services Imagery Processing System – Navy (JSIPS-N) and the Battle Group Passive Horizon Extension System Terminal (BGPHES), and links with BGPHES via CDL airborne terminal to airborne sensor systems. CDL-N is in production and the first ten systems are installed in Carriers. Future installations are planned for the remaining carriers, amphibious ships and command ships. The Shared Reconnaissance Pod (SHARP), which replaces the F-14 Tactical Airborne Reconnaissance Pod System (TARPS) and will be carried on the F/A-18F, supports strike warfare, amphibious warfare, and anti-surface warfare decision – making and utilises CDL-N for real time connectivity.

Co-operative Engagement Capability (CEC)

Co-operative Engagement Capability (CEC) improves battle force air-defense capabilities by integrating the sensor data of each co-operating ship and aircraft into a single, real-time, fire-control-quality composite track picture. CEC also interfaces the weapons capabilities of each CEC-equipped ship in the battle group to integrate engagement capability. By simultaneously distributing sensor data on airborne threats to each ship within a battle group, CEC extends the range at which a ship can engage hostile missiles to well beyond the radar horizon, thereby improving area, local, and self-defense capabilities. Operating under the direction of a designated commander, CEC enables a battle group or joint task force to act as a single, geographically dispersed combat system to confront the evolving threat of anti-ship cruise missiles and theatre ballistic missiles. As of 2005, CEC is installed on four aircraft carriers, *John F Kennedy*, *Nimitz*, *Eisenhower* and *Ronald Reagan*; six Aegis cruisers, *Princeton*, *Chosin*, *Hue City*, *Anzio*, *Vicksburg* and *Cape St George*; 12 new construction destroyers, including *McCampbell*, *Shoup*, *Mason*, *Mustin* and *Preble*; six amphibious ships including *Wasp* and *San Antonio*; and two E-2C Hawkeye 2000 air squadrons. It is planned to install CEC in all CV/CVN, CG 47, DDG 51, LHD, LPD 17, DD(X) and LCS class ships and in E-2C aircraft.

Global Broadcast Service (GBS)

The Global Broadcast Service augments and interfaces with other systems to provide virtual two-way Internet Protocol (IP) networked communications to deliver a continuous, high-speed, one-way flow of high-volume information broadcast to support: routine operations, training and military exercises, special activities, crisis, situational awareness, weapons targeting, intelligence, and the transition to and conduct of operations short of nuclear war. Homeland defensive operations are supported by a requirement for continental US coverage, which also provides exercise support, training and work-ups for deployment. GBS also supports military operations with US allies or coalition forces. GBS is an information technologies, mission-essential, national security system providing network-centric warfare communications, but does not incorporate nuclear survivability and hardening features. GBS provides a limited anti-jam capability and this may become a required capability in future. GBS will provide the capability to disseminate quickly large information products to various joint and small user platforms. With increased capacity, faster delivery of data, and near real-time receipt of imagery and data to the warfighter, it will reduced reliance on current MILSATCOM systems.

Global Command and Control System (GCCS)

GCCS is a comprehensive, worldwide network-centric system which provides the National Command Authority (NCA), Joint Chiefs of Staff, combatant and functional unified commands, Services, Defense Agencies, Joint Task Forces and their Service components, and others with information processing and dissemination capabilities necessary to conduct Command and Control (C^2) of forces. GCCS is a means to implement the Command, Control, Communications, Computers, and Intelligence for the Warrior (C^4IFTW) concept. GCCS provides the operational commanders with a near-realtime Common Operational Picture, intelligence information, collaborative joint operational planning and execution tools, and other information necessary for the execution of joint operations.

Global Command and Control System (Maritime) (GCCS-M) (ex-JMCIS)

GCCS-Maritime (GCCS-M) (formerly the Joint Maritime Command Information System (JMCIS)) is the designated command and control (C2) migration system for the Navy and is the naval implementation of the Global Command and Control System (GCCS). The evolutionary integration of previous C2 and intelligence systems, GCCS-M supports multiple warfighting and intelligence missions for commanders at every echelon, in all afloat, ashore, and tactical naval environments, and for joint, coalition, and allied forces. GCCS-M meets the joint and service requirements for a single, integrated, scalable Command and Control (C2) system that receives, displays, correlates, fuses, and maintains geo-locational track information on friendly, hostile, and neutral land, sea, and air forces and integrates it with available intelligence and environmental information.

- GCCS-M supports evolving concepts for Network-Centric Operations by receiving, displaying, correlating, fusing, and integrating all available track, intelligence and imagery information for the warfighter. In early 2004, more than 56 joint and Naval systems interfaced with GCCS-M to exchange data and support warfighter capabilities in 14 mission areas. Key capabilities include:Multisource information management
- Display and dissemination through extensive communications interfaces
- Multisource data fusion and analysis/decision making tools
- Force co-ordination.

GCCS-M is implemented afloat (formerly referred to as Navy Tactical Command System-Afloat (NTCS-A) and Joint Maritime Command Information System (JMCIS) Afloat), at ashore fixed command centers (formerly referred to as Operational Support System (OSS) and JMCIS Ashore), and as the command and control (C^2) portion of mobile command centers (formerly known as Tactical Support Center (TSC) and JMCIS Tactical-Mobile). GCCS-M now refers to all three GCCS-M implementations.

Integrated Broadcast Service/Joint Tactical Terminal (IBS/ JTT)

The Integrated Broadcast Service (IBS) is a system-of-systems that will migrate the Tactical Receive Equipment and Related Applications Data Dissemination System (TDDS), Tactical Information Broadcast Service (TIBS), Tactical Reconnaissance Intelligence Exchange System (TRIXS), and Near Real-Time Dissemination (NRTD) system into an integrated service with a common format. The IBS will send data via communications paths, such as UHF, SHF, EHF, GBS, and via networks. This program supports Indications Warning (I&W), surveillance, and targeting data requirements of tactical and operational commanders and targeting staffs across all warfare areas. It comprises broadcast-generation and transceiver equipment that provides intelligence data to tactical users. The Joint Tactical Terminal (JTT) will receive, decrypt, process, format, distribute, and transmit tactical data according to preset user-defined criteria across open-architecture equipment. JTT will be modular and will have the capability to receive all current tactical intelligence broadcasts (TDDS, TADIXS-B, TIBS, and TRIXS). JTT will also be interoperable with the follow-on IBS UHF broadcasts. However, the current JTT form factor does not meet space and weight constraints for a majority of the Navy and Air Force airborne platforms. Therefore, to ensure joint interoperability, the Navy and Air Force will continue to support the current Multimission Airborne Tactical Terminal (MATT) through a low cost Pre-Planned Product Improvement (P3I) program until the transition to an IBS capable JTRS airborne variant starting in FY 2007.

The US Army-procured Joint Tactical Terminal will receive, decrypt, process, format and distribute tactical data according to preset user-defined criteria across open-architecture equipment.

Integrated Radar Optical Surveillance and Sighting System (IROS3)

IROS3 is the Situational Awareness component of the Shipboard Protection System (SPS) Increment one. It employs COTS-based / Open Architecture products, and its key components include SPS-73 or equivalent surface search radar, electro-optical/infra-red devices, an integrated surveillance system, spotlights, long range acoustic devices, and remotely operated stabilised small arms mounts. SPS Increment I is designed to detect, classify and engage real-time asymmetric threats at close-range to ships in port, at anchor and while transiting choke points or operating in restricted waters. The system provides 360° Situational Awareness (SA) and employs COTS integration to support incremental modifications as needed to tailor the system to the mission. The system has undergone extensive testing in the laboratory and a prototype is being tested at sea in *Ramage*. The system is scheduled to be installed in most ship classes, including surface combatants, patrol boats, amphibious and auxiliary ships, and Coast Guard cutters from FY05.

Joint Service Imagery Processing System (JSIPS)

JSIPS-N is the component of the joint programme to receive and exploit infra-red and electro-optical imagery from tactical airborne reconnaissance systems. It is proposed to install the system in all aircraft carriers, amphibious assault ships and possibly in command ships. JSIPS will also provide the USMC with an afloat/ashore processing facility that can be deployed as an integral element of Marine Air-Ground Task Forces (MAGTFs). JSIPS upgrades under consideration include a common radar processor for both tactical and theatre-level radars, and an automated capability to insert and process mapping, charting and geodesy products.

Joint Surveillance Target Attack Radar System (JSTARS)

JSTARS is described as a 'bulletproof anti-jam datalink', utilising omnidirectional broadcast on UHF SATCOM. It receives and transmits real time MTI/FTI/SAR data via a secure uplink and downlink. It is used to demonstrate 'sensor to shooter' technology.

Joint Tactical Information Distribution System (JTIDS)

A joint program directed by the Office of the Secretary of Defense, JTIDS is a digital information-distribution system which provides rapid, crypto-secure, jam-resistant (frequency-hopping), and low-probability-of-exploitation tactical data and voice communication at a high data rate to Navy tactical aircraft and ships and Marine Corps units. JTIDS also provides capabilities for common-grid navigation and automatic communications relay. It has been integrated into numerous platforms and systems, including US Navy aircraft carriers, surface warships, amphibious assault ships, and E-2C Hawkeye aircraft; US Air Force Airborne Warning and Command System (AWACS) aircraft; and Marine Corps Tactical Air Operations Centers (TAOCs) and Tactical Air Command Centers (TACCs). Other service and foreign country participants include the US Army, UK and Canada. Additionally, JTIDS has been identified as the preferred communications link for Theatre Ballistic Missile Defense programs. JTIDS is the first implementation of the Link-16 Joint Message Standard (J – series) and provides the single, near real-time, joint datalink network for information exchange among joint and combined forces for command and control of tactical operations.

Land Attack Warfare System (LAWS)

This prototype system networks all shooters (tactical air, shore artillery and seaborne fire support) into a Battle Local Area Network (Battle LAN) known as the 'Ring of Fire'. This automatically assigns fire missions to the most capable unit in the Battle LAN. LAWS controls preplanned missions, including Tomahawk, as well as time critical calls for fire from land forces. Fleet Battle Experiment ALFA was the initial test of this system.

Link Eleven Improvement Program (LEIP)

This programme improves Link 11 connectivity and reliability and supports NILE (NATO Improved Link 11 and Link 22). Link 11 will be the common datalink for all US Navy and allied ships (seven NATO navies) not equipped with JTIDS/Link 16. NILE/ Link 22 is an element of the TADIL-J family. LEIP thus provides improvements in fleet support, training, commonality and interoperability. In addition to NILE, the programme comprises the Common Shipboard Data Terminal Set (CSDTS), Mobile Universal Link Translator System (MULTS), and Multiple Unit Link 11 Test and Operational Training System (MULTOTS). Some 190 CSDTS installations are planned.

Miniature Demand Assigned Multiple Access (Mini-DAMA)

Mini-DAMA is a communications system that supports the exchange of secure and non-secure Battle Group coordination data, tactical data and voice between base band processing equipment over UHF SATCOM, 25/5 kHz DAMA, 25/5 kHz Non-DAMA, and UHF LOS. In 2000, the Navy completed installations for submarines AV(2) and mine warfare ships V(2). Aircraft installations V(3) continue. These Mini-DAMA radio installations provide the channel utilisation efficiencies by employing Time Division Multiple Access (TDMA) methods that have been achieved for surface warfare ships and shore stations equipped with the larger version TD-1271 DAMA multiplexer.

Mission Data System (MDS)

This system allows planners to view Tomahawk Land Attack Missile information. MDS receives via TADIXS A or OTCIXS I digital Mission Data Updates (MDUs) from the Cruise Missile Support Activity (CMSA) and stores preplanned TLAM strike plans. Initial TLAM mission data fill is distributed via magnetic tape media provided by the CMSA.

Multifunctional Information Distribution – Low Volume Terminal (MIDS-LVT)

MIDS-LVT is a multinational co-operative development program to design, develop, and produce a tactical information distribution system equivalent in capability to Joint Tactical Information Distribution System (JTIDS), but in a low-volume, lightweight, compact terminal designed for fighter aircraft with applications in helicopters, ships, and ground sites. US Navy procurement, limited by available resources, is planned for F/A-18 Hornet aircraft as the lead aviation platform and surface craft. As a P3I of the JTIDS Class 2 Terminal, MIDS-LVT will employ the Link-16 (TADIL-J) message standard of US Navy/ NATO publications. MIDS-LVT is fully interoperable with JTIDS and was designed in response to current aircraft, surface ship, submarine, and ground host volume and weight constraints. The solution variants-MIDS-LVT (1), MIDS-LVT (2), and MIDS-LVT (3)-support US Navy, Marine Corps, and Air Force aircraft; US Navy ships; US Army Patriot, THAAD, MEADS and ground-based defense systems; USAF and USMC ground-based Command and Control platforms; and potentially other tactical aircraft and ground-based systems. MIDS-LVT is an international project involving the United States, Germany, Spain, Italy, and France.

Trusted Information Systems (TIS)

The Multi-Level Security (MLS) capabilities of the Navy's Ocean Surveillance Information System (OSIS) and Radiant Mercury are complementary systems which have been combined into a single TIS programme. The aim is to facilitate development and expansion of a Commander's capability automatically to exchange critical intelligence and operational information with all forces whether US, allied, or coalition.

The OSIS Evolutionary Development (OED) system is DoD's only PL-4 accredited C4I processing and dissemination system. It serves as the backbone automated information system supporting the Common Operational Picture (COP) at US and allied Joint Intelligence Centers (JICs). OED receives, processes, and disseminates timely all-source surveillance information on fixed and mobile targets of interest, both afloat and ashore, within an MLS environment. OED permits operators to collaborate in multiple domains, monitor, analyse, and support multiple views of the battle space corresponding to multiple security classification levels. Its robust correlation and communications subsystems ensure extremely rapid delivery of both record message traffic and intelligence broadcasts in support of the Unified Combatant Commanders, Joint Task Force commanders, individual units, and allies. The MLS capabilities in

OED are certified and accredited to support compartmented multilevel networks at the SCI level and are envisioned to serve as the core technology upon which future Navy networks and databases running at multiple classification levels can be effectively combined to allow appropriately cleared operators access to information from a single workstation.

Radiant Mercury (RM) provides the accredited capability to automatically sanitise, transliterate, and downgrade classified, formatted information to users at lower classification levels. RM helps ensure critical Indications and Warning intelligence is provided quickly to operational decision makers at various security and releasability levels. RM is currently fielded on Force Level ships bridging data transfer between SCI GCCS-M and GENSER GCCS-M. RM also serves as a sanitiser within OED. Radiant Mercury Imagery Guard (RMIG) combines a digital signature process with RM allowing the networked transfer of imagery between security domains.

Ship Self-Defense System (SSDS) Mk 1 and 2

SSDS provides the integrated combat system for aircraft carriers and amphibious ships, enabling them to keep pace with the anti-ship cruise missile (ASCM) threat. Moving toward an open-architecture distributed-processing system, SSDS integrates the detection and engagement elements of the combat system. With automated weapons control doctrine, Cooperative Engagement Capability (CEC), and enhanced battlespace awareness, SSDS provides these ships with a robust self-defense capability in support of Sea Shield.

SSDS Mk 1 provides doctrine-based, Quick Reaction Combat Capability (QRCC), plus automated detect through multithreat engagement capability. It enhances capabilities for Force Protection using own-ship and remote data in support of AAW capstone requirements.

SSDS Mk 2 integrates with Co-operative Engagement Capability (CEC) and provides the QRCC of SSDS Mk 1 and selected features of the Advanced Combat Direction System (ACDS) to support multiwarfare area capability, improve joint interoperability and provide an integrated, coherent real-time command and control system for CV/CVN, LPD and LHD class ships. SSDS Mk 1 has been installed in 12 LSDs; and SSDS Mk 2 in three CVNs (CVN 68, CVN 76, CVN 69) and one LPD (LPD 17).

Theatre Battle Management Core System (TBMCS)

TBMCS replaces the Contingency Theatre Automated Planning System (CTAPS) as the only command and control system authorised to produce the Air Tasking Order (ATO). TBMCS has the capability to plan and execute air operations in any theatre of operations and is considered the core system for the Air Force's Air Operation Center (AOC). All services use TBMCS and, within the USN, it is installed in carriers, command ships and large-deck amphibious ships (LHA/LHD).

Major Commercial Shipyards

Shipbuilders

General Dynamics Corporation, Bath Iron Works, Bath, Maine.
General Dynamics Corporation, Electric Boat Division, Groton, Connecticut.
General Dynamics Corporation, National Steel and Shipbuilding Company, San Diego, California.
Northrop Grumman, Ship Systems Sector, Avondale Operation, New Orleans, Louisiana.
Northrop Grumman, Ship Systems Sector, Ingalls Operation, Pascagoula, Mississippi.
Northrop Grumman, Newport News Shipbuilding, Newport News, Virginia.

Ship Repairers

Al Larson Boat Shop, Long Beach, California.
American Shipyard Co. L.L.C. Newport, Rhode Island.

Atlantic Drydock Corp., Jacksonville, Florida.
Atlantic Marine, Inc, Jacksonville, Florida.
Atlantic Marine, Inc, Mobile, Alabama.
Bay Ship & Yacht Co., San Francisco, California.
Bender Shipbuilding & Repair Co., Inc, Mobile, Alabama.
Cascade General Inc, Portland, Oregon.
Colonna's Shipyard, Inc, Norfolk, Virginia.
Continental Maritime of San Diego, San Diego, California.
Detyens Shipyards, Inc, Charleston, South Carolina.
Earl Industries, L.L.C., Norfolk, Virginia.
Halter Marine Inc, Gulfport, Mississippi.
Intermarine USA (Montedison Spa/Hercules, Inc.), Savannah, Georgia.
Lake Union Drydock Co., Seattle, Washington.
Marine Hydraulics International Inc, Norfolk, Virginia.
Metal Trades, Inc, Charleston, South Carolina.
Metro Machine Corp., Norfolk, Virginia.
Moon Engineering Co. Inc, Norfolk, Virginia.

Norfolk Shipbuilding & Drydock Corp., Norfolk, Virginia.
Norfolk Ship Repair & Drydock Co., Inc, Norfolk, Virginia.
North Florida Shipyards, Inc, Jacksonville, Florida.
Pacific Ship Repair & Fabrication, San Diego, California.
San Francisco Drydock, Inc, San Francisco, California.
Southwest Marine, Inc, San Diego, California.
Southwest Marine, Inc, (San Pedro Div) Long Beach, California.
Tampa Bay Shipbuilding & Repair Co., Tampa, Florida.
Tecnico Corporation, Norfolk, Virginia.
Todd Pacific Shipyards Corp., Seattle, Washington.

Notes: All the yards mentioned have been involved in naval shipbuilding, overhaul, or modernisation. General Dynamics/ Electric Boat yard is engaged only in submarine work and Newport News is the only US shipyard capable of building nuclear-powered aircraft carriers.

Major Warships Taken Out of Service 2002 to mid-2005

Submarines

2004 *Parche, Portsmouth*

Aircraft Carriers

2003 *Constellation*

Cruisers

2004 *Ticonderoga, Yorktown, Valley Forge*

Destroyers

2002 *David R Ray, John Young, Peterson, Nicholson*
2003 *Kinkaid, Fife, Paul F Foster, Arthur W Radford, Hayler, Oldendorf, Briscoe, Deyo, Elliot*
2004 *O'Brien, Stump, Thorn, Fletcher*

Frigates

2002 *Wadsworth* (to Poland), *Samuel Eliot Morison* (to Turkey)
2003 *Sides, George Philip, Estocin* (to Turkey)

Command Ships

2005 *La Salle*

Amphibious Forces

2002 *Frederick, Inchon*
2003 *Mount Vernon, Portland, Anchorage*

Auxiliaries

2002 *Butte*
2005 *Sacramento, Seattle, Detroit*

Special Mission Ships

2004 *Capable*

HULL NUMBERS

Notes: Ships in reserve not included.

SUBMARINES

Ballistic Missile Submarines

Ohio class
SSBN 730 Henry M Jackson
SSBN 731 Alabama
SSBN 732 Alaska
SSBN 733 Nevada
SSBN 734 Tennessee
SSBN 735 Pennsylvania
SSBN 736 West Virginia
SSBN 737 Kentucky
SSBN 738 Maryland
SSBN 739 Nebraska
SSBN 740 Rhode Island
SSBN 741 Maine
SSBN 742 Wyoming
SSBN 743 Louisiana

Cruise Missile Submarines

Ohio class
SSGN 726 Ohio (conversion)
SSGN 727 Michigan (conversion)
SSGN 728 Florida (conversion)
SSGN 729 Georgia (conversion)

Attack Submarines

Seawolf class
SSN 21 Seawolf

SSN 22 Connecticut
SSN 23 Jimmy Carter

Los Angeles class
SSN 688 Los Angeles
SSN 690 Philadelphia
SSN 691 Memphis
SSN 698 Bremerton
SSN 699 Jacksonville
SSN 700 Dallas
SSN 701 La Jolla
SSN 705 City of Corpus Christi
SSN 706 Albuquerque
SSN 708 Minneapolis-Saint Paul
SSN 709 Hyman G Rickover
SSN 710 Augusta
SSN 711 San Francisco
SSN 713 Houston
SSN 714 Norfolk
SSN 715 Buffalo
SSN 716 Salt Lake City
SSN 717 Olympia
SSN 718 Honolulu
SSN 719 Providence
SSN 720 Pittsburgh
SSN 721 Chicago
SSN 722 Key West
SSN 723 Oklahoma City
SSN 724 Louisville
SSN 725 Helena
SSN 750 Newport News
SSN 751 San Juan

SSN 752 Pasadena
SSN 753 Albany
SSN 754 Topeka
SSN 755 Miami
SSN 756 Scranton
SSN 757 Alexandria
SSN 758 Asheville
SSN 759 Jefferson City
SSN 760 Annapolis
SSN 761 Springfield
SSN 762 Columbus
SSN 763 Santa Fe
SSN 764 Boise
SSN 765 Montpelier
SSN 766 Charlotte
SSN 767 Hampton
SSN 768 Hartford
SSN 769 Toledo
SSN 770 Tucson
SSN 771 Columbia
SSN 772 Greeneville
SSN 773 Cheyenne

Virginia class
SSN 774 Virginia
SSN 775 Texas (bldg)
SSN 776 Hawaii (bldg)
SSN 777 North Carolina (bldg)
SSN 778 New Hampshire (bldg)
SSN 779 New Mexico (bldg)

SURFACE COMBATANTS

Aircraft Carriers

Kitty Hawk class
CV 63	Kitty Hawk

John F Kennedy class
CV 67	John F Kennedy

Enterprise class
CVN 65	Enterprise

Nimitz class
CVN 68	Nimitz
CVN 69	Dwight D Eisenhower
CVN 70	Carl Vinson
CVN 71	Theodore Roosevelt
CVN 72	Abraham Lincoln
CVN 73	George Washington
CVN 74	John C Stennis
CVN 75	Harry S Truman
CVN 76	Ronald Reagan
CVN 77	George H W Bush (bldg)

Cruisers

Ticonderoga class
CG 49	Vincennes
CG 51	Thomas S Gates
CG 52	Bunker Hill
CG 53	Mobile Bay
CG 54	Antietam
CG 55	Leyte Gulf
CG 56	San Jacinto
CG 57	Lake Champlain
CG 58	Philippine Sea
CG 59	Princeton
CG 60	Normandy
CG 61	Monterey
CG 62	Chancellorsville
CG 63	Cowpens
CG 64	Gettysburg
CG 65	Chosin
CG 66	Hue City
CG 67	Shiloh
CG 68	Anzio
CG 69	Vicksburg
CG 70	Lake Erie
CG 71	Cape St George
CG 72	Vella Gulf
CG 73	Port Royal

Destroyers

Spruance class
DD 963	Spruance
DD 985	Cushing
DD 987	O'Bannon

Arleigh Burke class
DDG 51	Arleigh Burke
DDG 52	Barry
DDG 53	John Paul Jones
DDG 54	Curtis Wilbur
DDG 55	Stout
DDG 56	John S McCain
DDG 57	Mitscher
DDG 58	Laboon
DDG 59	Russell
DDG 60	Paul Hamilton
DDG 61	Ramage
DDG 62	Fitzgerald
DDG 63	Stethem
DDG 64	Carney
DDG 65	Benfold
DDG 66	Gonzalez
DDG 67	Cole
DDG 68	The Sullivans
DDG 69	Milius
DDG 70	Hopper
DDG 71	Ross
DDG 72	Mahan
DDG 73	Decatur
DDG 74	McFaul
DDG 75	Donald Cook
DDG 76	Higgins
DDG 77	O'Kane
DDG 78	Porter
DDG 79	Oscar Austin
DDG 80	Roosevelt
DDG 81	Winston S Churchill
DDG 82	Lassen
DDG 83	Howard
DDG 84	Bulkeley
DDG 85	McCampbell
DDG 86	Shoup
DDG 87	Mason
DDG 88	Preble
DDG 89	Mustin
DDG 90	Chaffee
DDG 91	Pinckney
DDG 92	Momsen
DDG 93	Chung-Hoon
DDG 94	Nitze
DDG 95	James E Williams
DDG 96	Bainbridge
DDG 97	Halsey (bldg)
DDG 98	Forrest Sherman (bldg)
DDG 99	Farragut (bldg)
DDG 100	Kidd (bldg)
DDG 101	Gridley (bldg)
DDG 102	Sampson (bldg)
DDG 103	Truxtun (bldg)
DDG 104	Sterett (ord)
DDG 105	Dewey (ord)

Frigates

Oliver Hazard Perry class
FFG 8	McInerney
FFG 28	Boone (NRF)
FFG 29	Stephen W Groves (NRF)
FFG 32	John L Hall
FFG 33	Jarrett
FFG 36	Underwood
FFG 37	Crommelin (NRF)
FFG 38	Curts (NRF)
FFG 39	Doyle (NRF)
FFG 40	Halyburton
FFG 41	McClusky (NRF)
FFG 42	Klakring (NRF)
FFG 43	Thach
FFG 45	De Wert
FFG 46	Rentz
FFG 47	Nicholas
FFG 48	Vandegrift
FFG 49	Robert G Bradley
FFG 50	Taylor
FFG 51	Gary
FFG 52	Carr
FFG 53	Hawes
FFG 54	Ford
FFG 55	Elrod
FFG 56	Simpson (NRF)
FFG 57	Reuben James
FFG 58	Samuel B Roberts
FFG 59	Kauffman
FFG 60	Rodney M Davis (NRF)
FFG 61	Ingraham

Coastal Patrol Craft

Cyclone class
PC 2	Tempest
PC 3	Hurricane
PC 4	Monsoon
PC 5	Typhoon
PC 6	Sirocco
PC 7	Squall
PC 8	Zephyr
PC 9	Chinook
PC 10	Firebolt
PC 11	Whirlwind
PC 12	Thunderbolt
PC 13	Shamal
PC 14	Tornado

COMMAND SHIPS

Blue Ridge class
LCC 19	Blue Ridge
LCC 20	Mount Whitney

Raleigh and Austin class
AGF 11	Coronado

AMPHIBIOUS FORCES

Amphibious Assault Ships

Wasp class
LHD 1	Wasp
LHD 2	Essex
LHD 3	Kearsarge
LHD 4	Boxer
LHD 5	Bataan
LHD 6	Bonhomme Richard
LHD 7	Iwo Jima
LHD 8	Makin Island (bldg)

Tarawa class
LHA 1	Tarawa
LHA 2	Saipan
LHA 3	Belleau Wood
LHA 4	Nassau
LHA 5	Peleliu

Amphibious Transport Docks

Austin class
LPD 4	Austin
LPD 5	Ogden
LPD 6	Duluth
LPD 7	Cleveland
LPD 8	Dubuque
LPD 9	Denver
LPD 10	Juneau
LPD 12	Shreveport
LPD 13	Nashville
LPD 14	Trenton
LPD 15	Ponce

San Antonio class
LPD 17	San Antonio (bldg)
LPD 18	New Orleans (bldg)
LPD 19	Mesa Verde (bldg)
LPD 20	Green Bay (bldg)
LPD 21	New York (bldg)
LPD 22	San Diego (ord)
LPD 23	Anchorage (ord)
LPD 24	Arlington (ord)
LPD 25	Somerset (ord)

Amphibious Cargo Ships

Whidbey Island class
LSD 41	Whidbey Island
LSD 42	Germantown
LSD 43	Fort McHenry

Amphibious Cargo Ships — *continued*
LSD 44	Gunston Hall
LSD 45	Comstock
LSD 46	Tortuga
LSD 47	Rushmore
LSD 48	Ashland

Harpers Ferry class
LSD 49	Harpers Ferry
LSD 50	Carter Hall
LSD 51	Oak Hill
LSD 52	Pearl Harbor

MINE WARFARE FORCES

Mine Countermeasures Ships

Avenger class
MCM 1	Avenger (NRF)
MCM 2	Defender (NRF)
MCM 3	Sentry (NRF)
MCM 4	Champion (NRF)
MCM 5	Guardian
MCM 6	Devastator
MCM 7	Patriot
MCM 8	Scout
MCM 9	Pioneer
MCM 10	Warrior
MCM 11	Gladiator (NRF)
MCM 12	Ardent
MCM 13	Dextrous
MCM 14	Chief

Osprey class
MHC 51	Osprey (NRF)
MHC 52	Heron (NRF)
MHC 53	Pelican (NRF)
MHC 54	Robin (NRF)
MHC 55	Oriole (NRF)
MHC 56	Kingfisher (NRF)
MHC 57	Cormorant (NRF)
MHC 58	Black Hawk (NRF)
MHC 59	Falcon (NRF)
MHC 60	Cardinal
MHC 61	Raven
MHC 62	Shrike (NRF)

UNDERWAY REPLENISHMENT SHIPS

Fast Combat Support Ships

Sacramento class
AOE2	Camden

MATERIAL SUPPORT SHIPS

Submarine Tenders

Emory S Land class
AS 39	Emory S Land
AS 40	Frank Cable

Salvage Ships

Safeguard class
ARS 50	Safeguard
ARS 51	Grasp
ARS 52	Salvor
ARS 53	Grapple

MISCELLANEOUS

Auxiliary Research Submarine

Dolphin class
AGSS 555	Dolphin

Oceanographic Research Ships

AGOR 14	Melville
AGOR 15	Knorr
AGOR 23	Thomas G Thompson
AGOR 24	Roger Revelle
AGOR 25	Atlantis
AGOR 26	Kilo Moana
R 104	Ronald H Brown

MILITARY SEALIFT COMMAND

NAVAL FLEET AUXILIARY FORCE

Fast Combat Support Ships

T-AOE 6	Supply
T-AOE 7	Rainier
T-AOE 8	Arctic
T-AOE 10	Bridge

Ammunition Ships

T-AE 26	Kilauea
T-AE 28	Santa Barbara
T-AE 32	Flint
T-AE 33	Shasta
T-AE 34	Mount Baker
T-AE 35	Kiska

Cargo and Ammunition Ships

T-AKE 1	Lewis and Clark (bldg)
T-AKE 2	Sacagawea (bldg)

Combat Stores Ships

T-AFS 3	Niagara Falls
T-AFS 5	Concord
T-AFS 7	San Jose
T-AFS 8	Sirius
T-AFS 9	Spica
T-AFS 10	Saturn

Hospital Ships

T-AH 19	Mercy
T-AH 20	Comfort

Oilers

Henry J Kaiser class

T-AO 187	Henry J Kaiser (PREPO)
T-AO 188	Joshua Humphries
T-AO 189	John Lenthall
T-AO 193	Walter S Diehl
T-AO 194	John Ericsson
T-AO 195	Leroy Grumman
T-AO 196	Kanawha
T-AO 197	Pecos
T-AO 198	Big Horn
T-AO 199	Tippecanoe
T-AO 200	Guadalupe
T-AO 201	Patuxent
T-AO 202	Yukon
T-AO 203	Laramie
T-AO 204	Rappahannock

Fleet Ocean Tugs

Powhatan class

T-ATF 166	Powhatan
T-ATF 168	Catawba
T-ATF 169	Navajo
T-ATF 170	Mohawk
T-ATF 171	Sioux
T-ATF 172	Apache

SPECIAL MISSION SHIPS

Acoustic Survey Ship

T-AG 195	Hayes

Cable Repair Ship

T-ARC 7	Zeus

Missile Range Instrumentation Ships

T-AGM 23	Observation Island

Navigation Test/Launch Area Support Ships

T-AG 45	Waters

Ocean/Air Surveillance Ships

T-AGOS 19	Victorious
T-AGOS 21	Effective
T-AGOS 22	Loyal
T-AGOS 23	Impeccable

STRATEGIC SEALIFT FORCE

Fast Sealift Ships

T-AKR 287	Algol
T-AKR 288	Bellatrix
T-AKR 289	Denebola
T-AKR 290	Pollux
T-AKR 291	Altair
T-AKR 292	Regulus
T-AKR 293	Capella
T-AKR 294	Antares

Large, Medium-speed Ro-Ro

T-AKR 295	Shughart
T-AKR 296	Gordon
T-AKR 297	Yano
T-AKR 298	Gilliland
T-AKR 300	Bob Hope
T-AKR 301	Fisher
T-AKR 302	Seay
T-AKR 303	Mendonca
T-AKR 304	Pililaau
T-AKR 305	Brittin
T-AKR 306	Benavidez

Tankers

T-AOT 1121	Gus W Darnell
T-AOT 1122	Paul Buck
T-AOT 1123	Samuel L Cobb
T-AOT 1124	Richard G Matthiesen
T-AOT 1125	Lawrence H Gianella

Surveying Ships/Oceanographic Ships

T-AGS 51	John McDonnell
T-AGS 60	Pathfinder
T-AGS 61	Sumner
T-AGS 62	Bowditch
T-AGS 63	Henson

Surveying Ships/Oceanographic Ships — *continued*

T-AGS 64	Bruce C Heezen
T-AGS 65	Mary Sears

PREPOSITIONING PROGRAMME

Container Ships

T-AK 4296	Capt Steven L Bennett
T-AK 4396	Maj Bernard F Fisher
T-AK 4543	Lt Col John U D Page
T-AK 4544	SSGT Edward A Carter Jr
T-AK 4638	A1C William H Pitsenbarger
T-AK 323	Merlin

Large, Medium-Speed, Ro-Ro

T-AKR 310	Watson
T-AKR 311	Sisler
T-AKR 312	Dahl
T-AKR 313	Red Cloud
T-AKR 314	Charlton
T-AKR 315	Watkins
T-AKR 316	Pomeroy
T-AKR 317	Soderman (bldg)

Aviation Logistic Ships

T-AVB 3	Wright
T-AVB 4	Curtiss

Maritime Prepositioning Ships

T-AK 3000	CPL Louis J Hauge, Jr
T-AK 3001	PFC William B Baugh
T-AK 3002	PFC James Anderson, Jr
T-AK 3003	1st Lt Alex Bonnyman
T-AK 3004	PVT Franklin J Phillips
T-AK 3005	SGT Matej Kocak
T-AK 3006	PFC Eugene A Obregon
T-AK 3007	MAJ Stephen W Pless
T-AK 3008	2nd Lt John P Bobo
T-AK 3009	PFC Dewayne T Williams
T-AK 3010	1st Lt Baldomero Lopez
T-AK 3011	1st Lt Jack Lummus
T-AK 3012	SGT William R Button
T-AK 3015	1st Lt Harry L Martin
T-AK 3016	L/Cpl Roy M Wheat
T-AK 3017	GYSGT Fred W Stockham

READY RESERVE FORCE
(see pages 904–905)

SUBMARINES

Notes: (1) **Deep submergence vehicles:** The Deep Submergence Vehicles (DSV), including the nuclear-propelled *NR-1*, are listed following the 'Research Ships' section.

(2) **Seal Delivery Vehicles (SDVs):** There are 10 Mk VIII Mod 1 six-man mini wet submersibles in service for naval commando units. These SDVs can be carried by suitably modified SSNs. Range 35 n miles at up to 150 ft. All have been SLEPed from 1995 to improve performance. A new design ASDS (Advanced SEAL Delivery System) is a dry submersible (65 × 10 ft) with electrical propulsion and a crew of two. Designed to operate from a mother submarine, it can carry SEALs and combat gear clandestinely to and from hostile shores. Range is greater than 125 n miles. Optical and communications periscopes and a small sonar are fitted. A prototype was ordered from Northrop Grumman in September 1994 and contractor trials were completed in 2000. Final trials were successfully completed in

September 2002. These included launch and recovery of ASDS from a host submarine SSN 722 *Greeneville* over several days to validate its underwater capability. Operational evaluation followed in mid-2003. SSN 776 *Charlotte* is also configured to host ASDS and it is anticipated that two will be carried by the Ohio class SSGNs. The craft can be air-transported.

(3) **Unmanned Undersea Vehicles (UUVs):** Two prototype surface ship-launched vehicles were built under joint ARPA/Navy UUV programmes. Operational testing of the second vehicle, the Mine Search System (MSS), was completed in 1993. The MSS vehicle is 35 ft long and has a titanium hull with a diameter of 44 in. The payload is housed in an internal pressure hull. The propulsion motor is free flooding and develops 12 hp from two battery sections. MSS operational trials demonstrated the performance of mine detection sonars and the ability of a UUV to survey designated areas with precise navigation.

UUVs capable of remote control from either surface ships or submarines are being researched. Roles are limitless but remote sensing and acoustic deception are two obvious front runners. Operating difficulties should not be underestimated.

(4) **Autonomous Undersea Vehicles (AUVs):** Research work is being done with the aim of producing torpedo-launched remote-controlled vehicles for a range of tasks including surveillance, communications and mine warfare. The Long Term Mine Reconnaissance System (LMRS) is to be the USN's first autonomous UUV and is expected to enter service in 2005. Designed to be launched and recovered from a submarine's torpedo tubes, it will provide long range mine-reconnaissance capabilities. Development of Mission-Reconfigurable UUV (MRUUV) began in 2003 following analysis of alternatives for a large displacement vehicle. Initial procurement of two is planned for FY05. A total of ten is expected by FY09.

SDV Mk VIII

8/1999 / 0033436

ASDS

6/2003, US Navy / 0572767

Attack Submarines (SSN)

1 + 6 (3) VIRGINIA CLASS (SSN)

Name	No	Builders	Start date	Launched	Commissioned	F/S
VIRGINIA	SSN 774	General Dynamics (Electric Boat)	30 Sep 1998	16 Aug 2003	23 Oct 2004	AA
TEXAS	SSN 775	Northrop Grumman, Newport News Shipbuilding	7 Dec 1998	31 July 2004	2005	Bldg
HAWAII	SSN 776	General Dynamics (Electric Boat)	13 Nov 2000	2006	2007	Bldg
NORTH CAROLINA	SSN 777	Northrop Grumman, Newport News Shipbuilding	29 Jan 2002	2007	2008	Bldg
NEW HAMPSHIRE	SSN 778	General Dynamics (Electric Boat)	14 Aug 2003	2008	2009	Bldg
NEW MEXICO	SSN 779	Northrop Grumman, Newport News Shipbuilding	29 Jan 2004	2009	2010	Bldg
—	SSN 780	General Dynamics (Electric Boat)	2005	2010	2011	
—	SSN 781	Northrop Grumman, Newport News Shipbuilding	2006	2011	2012	
—	SSN 782	General Dynamics (Electric Boat)	2007	2012	2013	
—	SSN 783	Northrop Grumman, Newport News Shipbuilding	2008	2013	2014	

Displacement, tons: 7,800 dived
Dimensions, feet (metres): 377 × 34 × 30.5
(114.9 × 10.4 × 9.3)
Main machinery: Nuclear; 1 GE PWR S9G; 2 turbines; 40,000 hp
(29.84 MW); 1 shaft; pump jet propulsor; 1 secondary
propulsion submerged motor
Speed, knots: 34 dived
Complement: 134 (14 officers)

Missiles: SLCM Tomahawk. ATMS possibly.
SSM: 12 VLS tubes for cruise missiles.
Torpedoes: 4—21 in *(533 mm)* tubes. Mk 48 ADCAP Mod 6;
total of 38 weapons including SSM, torpedoes and UUVs.
Mines: Can lay Mk 67 Mobile and Mk 60 Captor mines (until new
mines are available).
Countermeasures: Decoys: 14 external and 1 internal
(reloadable). Anti-torpedo torpedo.
ESM: WLQ-4(V); BLQ-10; intercept.
Combat data systems: Lockheed Martin CCSM (Command and
Control System Module) compatible with JMCIS.
Radars: Navigation: BPS 16; I-band.
Sonars: Bow spherical active/passive array; wide aperture flank
passive arrays; high-frequency active keel and fin arrays; TB
16 and TB 29 towed arrays.

Programmes: In February 1997, a teaming agreement was
reached between General Dynamics (Electric Boat) and
Newport News to build the Virginia class co-operatively.
Electric Boat is the lead design yard, constructs engine room
and command and control modules and seven other sections
of each ship. It is also to perform final assembly, testing,
outfitting, and delivery of the even-numbered hulls. Newport
News constructs the sail, the habitability and auxiliary
machinery room modules, and six other sections. It is to
perform final assembly, testing, outfitting, and delivery of the
odd-numbered hulls. Reactor compartments built by the
delivery yard in each case. Advanced funding for first of class
in FY96 and this continued in to FY98. Second of class

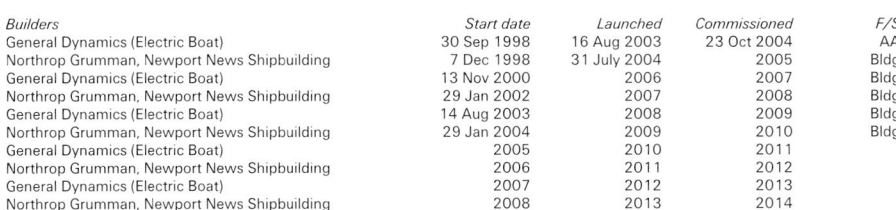

VIRGINIA
7/2004*, US Navy / 1043661

funding in FY99, third in FY01 and fourth in FY02. A follow-on
contract was awarded in August 2003 for construction of
SSNs 778-782 from 2003-07. In January 2004, the follow-on
contract was transitioned to a multi-year procurement for the
construction of one boat per year from 2004-08 (SSN 779-
783). A total of 30 hulls is planned and Batch 2 is expected to
start at hull ten of the programme. These are likely to be
substantially different in design and payload.
Structure: Seawolf level quietening. Acoustic hull cladding. The
reactor core will last the life of the ship. The integral nine-man
lock out chamber can be used with the Advanced SEAL

Delivery System mini submarine or a dry deck shelter, and the
torpedo magazine can be adapted to provide 2,400 ft³ of
space for up to 40 SEALs and equipment. Non-hull-
penetrating mast with eight antenna packages depending on
the role. Test depth 488 m *(1,600 ft)*. A scale version (0.294)
Cutthroat LSV-2 is listed under Deep Submergence Vehicles.
Operational: To be capable of both deep ocean ASW and
shallow water operations of all types. Advanced UUVs, wake
detection equipment and a deployable bistatic source are all
to be available.

UPDATED

VIRGINIA
7/2004*, US Navy / 1043660

50 LOS ANGELES CLASS (SSN)

Name	No	Builders	Laid down	Launched	Commissioned	F/S
LOS ANGELES	SSN 688	Newport News Shipbuilding	8 Jan 1972	6 Apr 1974	13 Nov 1976	PA
PHILADELPHIA	SSN 690	General Dynamics (Electric Boat Div)	12 Aug 1972	19 Oct 1974	25 June 1977	AA
MEMPHIS	SSN 691	Newport News Shipbuilding	23 June 1973	3 Apr 1976	17 Dec 1977	AA
BREMERTON	SSN 698	General Dynamics (Electric Boat Div)	8 May 1976	22 July 1978	28 Mar 1981	PA
JACKSONVILLE	SSN 699	Newport News Shipbuilding	21 Feb 1976	18 Nov 1978	16 May 1981	AA
DALLAS	SSN 700	General Dynamics (Electric Boat Div)	9 Oct 1976	28 Apr 1979	18 July 1981	AA
LA JOLLA	SSN 701	General Dynamics (Electric Boat Div)	16 Oct 1976	11 Aug 1979	24 Oct 1981	PA
CITY OF CORPUS CHRISTI	SSN 705	General Dynamics (Electric Boat Div)	4 Sep 1979	25 Apr 1981	8 Jan 1983	AA
ALBUQUERQUE	SSN 706	General Dynamics (Electric Boat Div)	27 Dec 1979	13 Mar 1982	21 May 1983	AA
MINNEAPOLIS-SAINT PAUL	SSN 708	General Dynamics (Electric Boat Div)	30 Jan 1981	19 Mar 1983	10 Mar 1984	AA
HYMAN G RICKOVER	SSN 709	General Dynamics (Electric Boat Div)	24 July 1981	27 Aug 1983	21 July 1984	AA
AUGUSTA	SSN 710	General Dynamics (Electric Boat Div)	1 Apr 1982	21 Jan 1984	19 Jan 1985	AA
SAN FRANCISCO	SSN 711	Newport News Shipbuilding	26 May 1977	27 Oct 1979	24 Apr 1981	PA
HOUSTON	SSN 713	Newport News Shipbuilding	29 Jan 1979	21 Mar 1981	25 Sep 1982	PA
NORFOLK	SSN 714	Newport News Shipbuilding	1 Aug 1979	31 Oct 1981	21 May 1983	AA
BUFFALO	SSN 715	Newport News Shipbuilding	25 Jan 1980	8 May 1982	5 Nov 1983	PA
SALT LAKE CITY	SSN 716	Newport News Shipbuilding	26 Aug 1980	16 Oct 1982	12 May 1984	PA
OLYMPIA	SSN 717	Newport News Shipbuilding	31 Mar 1981	30 Apr 1983	17 Nov 1983	PA
HONOLULU	SSN 718	Newport News Shipbuilding	10 Nov 1981	24 Sep 1983	6 July 1985	PA
PROVIDENCE	SSN 719	General Dynamics (Electric Boat Div)	14 Oct 1982	4 Aug 1984	27 July 1985	AA
PITTSBURGH	SSN 720	General Dynamics (Electric Boat Div)	15 Apr 1983	8 Dec 1984	23 Nov 1985	AA
CHICAGO	SSN 721	Newport News Shipbuilding	5 Jan 1983	13 Oct 1984	27 Sep 1986	PA
KEY WEST	SSN 722	Newport News Shipbuilding	6 July 1983	20 July 1985	12 Sep 1987	PA
OKLAHOMA CITY	SSN 723	Newport News Shipbuilding	4 Jan 1984	2 Nov 1985	9 July 1988	AA
LOUISVILLE	SSN 724	General Dynamics (Electric Boat Div)	16 Sep 1984	14 Dec 1985	8 Nov 1986	PA
HELENA	SSN 725	General Dynamics (Electric Boat Div)	28 Mar 1985	28 June 1986	11 July 1987	PA
NEWPORT NEWS	SSN 750	Newport News Shipbuilding	3 Mar 1984	15 Mar 1986	3 June 1989	AA
SAN JUAN	SSN 751	General Dynamics (Electric Boat Div)	16 Aug 1985	6 Dec 1986	6 Aug 1988	AA
PASADENA	SSN 752	General Dynamics (Electric Boat Div)	20 Dec 1985	12 Sep 1987	11 Feb 1989	PA
ALBANY	SSN 753	Newport News Shipbuilding	22 Apr 1985	13 June 1987	7 Apr 1990	AA
TOPEKA	SSN 754	General Dynamics (Electric Boat Div)	13 May 1986	23 Jan 1988	21 Oct 1989	PA
MIAMI	SSN 755	General Dynamics (Electric Boat Div)	24 Oct 1986	12 Nov 1988	30 June 1990	AA
SCRANTON	SSN 756	Newport News Shipbuilding	29 June 1986	3 July 1989	26 Jan 1991	AA
ALEXANDRIA	SSN 757	General Dynamics (Electric Boat Div)	19 June 1987	23 June 1990	29 June 1991	AA
ASHEVILLE	SSN 758	Newport News Shipbuilding	1 Jan 1987	28 Oct 1989	28 Sep 1991	PA
JEFFERSON CITY	SSN 759	Newport News Shipbuilding	21 Sep 1987	24 Mar 1990	29 Feb 1992	PA
ANNAPOLIS	SSN 760	General Dynamics (Electric Boat Div)	15 June 1988	18 May 1991	11 Apr 1992	AA
SPRINGFIELD	SSN 761	General Dynamics (Electric Boat Div)	29 Jan 1990	4 Jan 1992	9 Jan 1993	AA
COLUMBUS	SSN 762	General Dynamics (Electric Boat Div)	7 Jan 1991	1 Aug 1992	24 July 1993	PA
SANTA FE	SSN 763	General Dynamics (Electric Boat Div)	9 July 1991	12 Dec 1992	8 Jan 1994	PA
BOISE	SSN 764	Newport News Shipbuilding	25 Aug 1988	20 Oct 1990	7 Nov 1992	AA
MONTPELIER	SSN 765	Newport News Shipbuilding	19 May 1989	6 Apr 1991	13 Mar 1993	AA
CHARLOTTE	SSN 766	Newport News Shipbuilding	17 Aug 1990	3 Oct 1992	16 Sep 1994	PA
HAMPTON	SSN 767	Newport News Shipbuilding	2 Mar 1990	28 Sep 1991	6 Nov 1993	PA
HARTFORD	SSN 768	General Dynamics (Electric Boat Div)	27 Apr 1992	4 Dec 1993	10 Dec 1994	AA
TOLEDO	SSN 769	Newport News Shipbuilding	6 May 1991	28 Aug 1993	24 Feb 1995	AA
TUCSON	SSN 770	Newport News Shipbuilding	15 Aug 1991	19 Mar 1994	9 Sep 1995	PA
COLUMBIA	SSN 771	General Dynamics (Electric Boat Div)	24 Apr 1993	24 Sep 1994	9 Oct 1995	PA
GREENEVILLE	SSN 772	Newport News Shipbuilding	28 Feb 1992	17 Sep 1994	16 Feb 1996	PA
CHEYENNE	SSN 773	Newport News Shipbuilding	6 July 1992	4 Apr 1995	13 Sep 1996	PA

Displacement, tons: 6,082 standard; 6,927 dived
Dimensions, feet (metres): 362 × 33 × 32.3
(110.3 × 10.1 × 9.9)
Main machinery: Nuclear; 1 GE PWR S6G; 2 turbines; 35,000 hp
(26 MW); 1 shaft; 1 Magnetek auxiliary prop motor; 325 hp
(242 kW)
Speed, knots: 33 dived
Complement: 133 (13 officers)

Missiles: SLCM: GDC/Hughes Tomahawk (TLAM-N); land attack;
Tercom aided inertial navigation system (TAINS) to 2,500 km
(1,400 n miles) at 0.7 Mach; altitude 15-100 m; nuclear
warhead 200 kT; CEP 80 m. There are also 2 versions (TLAM-
C/D) with either a single 454 kg HE warhead or a single
warhead with submunitions; range 900 km *(485 n miles)*; CEP
10 m. Nuclear warheads are not normally carried. Block III
missiles, installed from 1994, increase TLAM-C range to
1,700 km *(918 n miles)*, add GPS back-up to TAINS and
substitute a 318 kg shaped charge warhead.
From SSN 719 onwards all are equipped with the Vertical
Launch System, which places 12 launch tubes external to the
pressure hull behind the BQQ 5 spherical array forward.
Torpedoes: 4—21 in *(533 mm)* tubes midships. Gould Mk 48;
ADCAP Mod 5/6; wire-guided (option); active/passive
homing to 50 km *(27 n miles)*/38 km *(21 n miles)* at 40/
55 kt; warhead 267 kg; depth to 900 m *(2,950 ft)*. Air Turbine
Pump discharge.
Total of 26 weapons can be tube-launched, for example-12
Tomahawk, 14 torpedoes.
Mines: Can lay Mk 67 Mobile and Mk 60 Captor mines.
Countermeasures: Decoys: Emerson Electric Mk 2; torpedo
decoy.
ESM: BRD-7; direction finding. WLR-1H (in 771-773).
WLR-8(V)2/6; intercept. WSQ-5 (periscope) and WLR-10;
radar warning. BLQ-10 being fitted in some.
Combat data systems: CCS Mk 2 (688-750) with UYK 7
computers; IBM BSY-1 (751-773) with UYK 43/UYK 44
computers. JOTS, BGIXS and TADIX-A can be fitted. USC-38
EHF (in some). Link 11; Link 16 being fitted.
Radars: Surface search/navigation/fire control: Sperry BPS 15
H/16; I/J-band.
Sonars: IBM BQQ 5D/E; passive/active search and attack; low
frequency. BSY-1 (SSN 751 onwards)
BQG 5D wide aperture flank array (SSN 710 and SSN 773).
TB 23/29 thin line array and TB 16; passive towed array.

JEFFERSON CITY

6/2002, Royal Australian Navy / 0530095

Ametek BQS 15; active close range including ice detection;
high frequency.
MIDAS (mine and ice detection avoidance system) (SSN 751
onwards); active high frequency.

Programmes: Various major improvement programmes and
updating design changes caused programme delays in the
late 1980s. From SSN 751 onwards the class is prefixed by an
'I' for 'improved'. Programme terminated at 62 hulls. Eleven
paid off by mid-1999. Plans to refuel the majority of the class
have delayed further de-activations but these will start again
with the decommissioning of SSN 716 in FY07.
Modernisation: Mk 117 TFCS backfitted in earlier submarines of
the class. EHF communications and Link 16 are being fitted.
Prototype HDR antenna fitted in SSN 719. Production
versions being installed from 2002. BQQ 10 and TB 29 fitted
in all. An ARCI (Acoustic Rapid COTS Insertion) AN/BQQ-10
programme from 1997 to 2002 to backfit BQQ 5 sonars with
open system architecture. Five of the class are fitted with DDS
(SSN 688, 690, 700, 701 and 715). Two others are fitted to
operate ASDS (SSN 772, 766). C-303 acoustic jammer in one
of the class for trials in 1998. 25 mm guns may be fitted in the
future.
Structure: Every effort has been made to improve sound quieting
and from SSN 751 onwards the class has acoustic tile
cladding to augment the 'mammalian' skin which up to then
had been the standard USN outer casing coating. Also from
SSN 751 the forward hydro planes are fitted forward instead
of on the fin. The planes are retractable mainly for surfacing
through ice. The S6G reactor is a modified version of the D2G
type. The towed sonar array is stowed in a blister on the side of
the casing. A single refuelling is required during the life of the

boat. Diving depth is 450 m *(1,475 ft)*. *Memphis* was
withdrawn from active service in late 1989 to become an
interim research platform for advanced submarine
technology. Early trials did not involve major changes to the
submarine but tests started in September 1990 for optronic
non-hull-penetrating masts and a major overhaul included
installation of a large diameter tube for testing UUVs and large
torpedoes. An after casing hangar is fitted for housing larger
UUVs and towed arrays. Many other ideas are being evaluated
with the main aim of allowing contractors easy access for
trials at sea of new equipment. *Augusta* was the trials platform
for the BQQ-5D wide aperture array passive sonar system.
Various staged design improvements have added some
220 tons to the class displacement between 688 and 773.
Operational: The Los Angeles class is the mainstay of the attack
submarine force. The land-attack mission has been a notable
feature of operations in Iraq, Kosovo and Afghanistan (SSN
719). Under-ice operations are still a priority and SSN 767
surfaced at the North Pole in 2004. Special forces and
intelligence gathering missions are also conducted. Normally
twelve Tomahawk missiles are carried internally (in addition
to the external tubes in 719 onwards) but this load can be
increased depending on the mission. Neither TASM nor
Harpoon are now deployed. Subroc phased out in 1990.
Nuclear weapons disembarked but still available. Predator
and Sea Ferret UAV trials done in two of the class in 1996-97
and may become available for launching from a sub-Harpoon
canister in due course. ASDS trials in SSN 772 during 2002.
SSN 713 seriously damaged in collision with an undersea
mountain south of Guam on 8 January 2005.

UPDATED

PHILADELPHIA (with DDS)

5/2001, H M Steele / 0131330

OLYMPIA

7/2004*, Michael Nitz / 1043662

LA JOLLA (with DSRV-1)

4/2002, Hachiro Nakai / 0530031

LOUISVILLE

6/2004*, US Navy / 1043663

MINNEAPOLIS-SAINT PAUL

10/2004*, B Sullivan / 1043701

3 SEAWOLF CLASS (SSN)

Name	No	Builders	Start date	Launched	Commissioned	F/S
SEAWOLF	SSN 21	General Dynamics (Electric Boat)	25 Oct 1989	24 June 1995	19 July 1997	AA
CONNECTICUT	SSN 22	General Dynamics (Electric Boat)	14 Sep 1992	1 Sep 1997	11 Dec 1998	AA
JIMMY CARTER	SSN 23	General Dynamics (Electric Boat)	12 Dec 1995	5 June 2004	19 Feb 2005	PA

Displacement, tons: 8,060 surfaced; 9,142; 12,139 (SSN 23) dived

Dimensions, feet (metres): 353; 453.2 (SSN 23) × 42.3 × 35.8 *(107.6; 138.1 × 12.9 × 10.9)* (see *Modernisation*)

Main machinery: Nuclear; 1 Westinghouse PWR S6W; 2 turbines; 45,000 hp *(33.57 MW)*; 1 shaft; pumpjet propulsor; 1 (4 in SSN 23) Westinghouse secondary propulsion submerged motor(s)

Speed, knots: 39 dived

Complement: 134 (14 officers)

Missiles: SLCM: Hughes Tomahawk (TLAM-N); land attack; Tercom Aided Inertial Navigation System (TAINS) to 2,500 km *(1,400 n miles)* at 0.7 Mach; altitude 15-100 m; nuclear warhead 200 kT; CEP 80 m. There are also 2 versions (TLAM-C/D) with either a single 454 kg HE warhead or a single warhead with submunitions; range 900 km *(485 n miles)*; CEP 10 m.
    Nuclear warheads are not normally carried. Block III missiles increase TLAM-C range to 1,700 km *(918 n miles)*, add GPS back-up to TAINS and substitute a 318 kg shaped charge warhead.

Torpedoes: 8—26 in *(660 mm)* tubes (external measurement is 30 in *(762 mm)*); Gould Mk 48 ADCAP; wire-guided (option); active/passive homing to 50 km *(27 n miles)*/38 km *(21 n miles)* at 40/55 kt; warhead 267 kg; depth to 900 m *(2,950 ft)*. Air turbine discharge. Total of 50 tube-launched missiles and torpedoes, and UUVs in due course.

Mines: 100 in lieu of torpedoes.

Countermeasures: Decoys: torpedo decoys. WLY-1 system in due course.
ESM: WLQ-4(V)1; BLD-1; intercept.

Combat data systems: General Electric BSY-2 system. USC-38 EHF. JMCIS.

Weapons control: Raytheon Mk 2 FCS

Radars: Navigation: BPS 16; I-band.

Sonars: BSY-2 suite with bow spherical active/passive array and wide aperture passive flank arrays; TB 16 and TB 29 surveillance and tactical towed arrays.

Programmes: First of class ordered on 9 January 1989; second of class on 3 May 1991 and third on 30 April 1996. Design changes to *Carter* contracted in late 1999 delayed the launch by four years.

Modernisation: The hull of SSN 23 is about 30 m longer to accommodate an hour-glass shaped Ocean Interface section with larger payload apertures to the sea. Modular architecture allows configuration for specific missions. Payloads could include standoff vehicles, distributed sensors and leave-behind weapons that would be activated after the submarine has left the area. It also supports Special Operations Forces including Dry Deck Shelter (DDS) and the Advanced SEAL Delivery System (ASDS). *Carter* retains all of the Seawolf class's original war-fighting capability.

Structure: The modular design has more weapons, a higher tactical speed, better sonars and an ASW mission effectiveness 'three times better than the improved Los Angeles class' according to the Navy. It is estimated that over a billion dollars has been allocated for research and development including the S6W reactor system. Full acoustic cladding fitted. Panels around wide aperture sonar array and torpedo tube doors were redesigned and refitted following sea-trials of SSN 21. Mk 21 Air turbine torpedo discharge pump. There are no external weapons. Emphasis has been put on sub-ice capabilities including retractable bow planes. SSN 23 equipped with a range of photonic masts and an advanced communications mast for covert operations. Test depth 1,950 ft *(594 m)*.

Operational: A quoted 'silent' speed of 20 kt. Other operational advantages include greater manoeuvrability and space for subsequent weapon systems development.

Opinion: This submarine was intended to restore the level of acoustic advantage (in the one to one nuclear submarine engagement against the Russians) which the USN had enjoyed for three decades. At the same time the larger capacity of the magazine enhances overall effectiveness in a number of other roles. The decision to discontinue building this very expensive design was the result of falling defence budgets at the end of the Cold War and technical problems.

UPDATED

JIMMY CARTER
6/2004*, *Ships of the World* / 0583667

SEAWOLF
4/2004*, *Ships of the World* / 1043702

Strategic Missile Submarines (SSBN)

Notes: The Trident missile fitted SSBN force provides the principal US strategic deterrent under the control of US Strategic Command at Offutt Air Force Base, Nebraska. The Strategic Arms Reduction Treaty (START), implemented in December 2001, limits the combined number of SLBM and ICBM re-entry bodies (RBs) to 4,900. Although there may be further bi-lateral

agreements with Russia to update verification regimes, the Bush administration has decided to pursue long-term strategic nuclear force reductions without further detailed arms control negotiations. The START II treaty has thus been overtaken. As part of the reduction, the first four Ohio class submarines are no longer required for strategic service. These boats are being

converted into conventionally-armed guided missile SSGNs, capable also of deploying Special Forces. Although the missile tubes on SSGNs will not contain SLBMs, they will continue to count against START treaty limits.

14 OHIO CLASS (SSBN)

Name	No	Builders	Launched	Commissioned	F/S
HENRY M JACKSON	SSBN 730	General Dynamics (Electric Boat Div)	15 Oct 1983	6 Oct 1984	PA
ALABAMA	SSBN 731	General Dynamics (Electric Boat Div)	19 May 1984	25 May 1985	PA
ALASKA	SSBN 732	General Dynamics (Electric Boat Div)	12 Jan 1985	25 Jan 1986	PA
NEVADA	SSBN 733	General Dynamics (Electric Boat Div)	14 Sep 1985	16 Aug 1986	PA
TENNESSEE	SSBN 734	General Dynamics (Electric Boat Div)	13 Dec 1986	17 Dec 1988	AA
PENNSYLVANIA	SSBN 735	General Dynamics (Electric Boat Div)	23 Apr 1988	9 Sep 1989	PA
WEST VIRGINIA	SSBN 736	General Dynamics (Electric Boat Div)	14 Oct 1989	20 Oct 1990	AA
KENTUCKY	SSBN 737	General Dynamics (Electric Boat Div)	11 Aug 1990	13 July 1991	PA
MARYLAND	SSBN 738	General Dynamics (Electric Boat Div)	10 Aug 1991	13 June 1992	AA
NEBRASKA	SSBN 739	General Dynamics (Electric Boat Div)	15 Aug 1992	10 July 1993	PA
RHODE ISLAND	SSBN 740	General Dynamics (Electric Boat Div)	17 July 1993	9 July 1994	AA
MAINE	SSBN 741	General Dynamics (Electric Boat Div)	16 July 1994	29 July 1995	AA/PA
WYOMING	SSBN 742	General Dynamics (Electric Boat Div)	15 July 1995	13 July 1996	AA
LOUISIANA	SSBN 743	General Dynamics (Electric Boat Div)	27 July 1996	6 Sep 1997	AA/PA

Displacement, tons: 16,600 surfaced; 18,750 dived
Dimensions, feet (metres): 560 × 42 × 36.4
(170.7 × 12.8 × 11.1)
Main machinery: Nuclear; 1 GE PWR S8G; 2 turbines; 60,000 hp *(44.8 MW)*; 1 shaft; 1 Magnetek auxiliary prop motor; 325 hp *(242 kW)*
Speed, knots: 24 dived
Complement: 155 (15 officers)

Missiles: SLBM: 24 Lockheed Trident C4 (730, 731); stellar inertial guidance to 7,400 km *(4,000 n miles)*; thermonuclear warhead of up to 8 MIRV Mk 4 with W76 warhead of 100 kT; CEP 450 m. Trident C4 is to be fully replaced by Trident D5 in 730-733 by 2005.
24 Lockheed Trident D5 (732-743); stellar inertial guidance to 12,000 km *(6,500 n miles)*; thermonuclear warheads of up to 12 MIRVs of either Mk 4 with W76 of 100 kT each, or Mk 5 with W88 of 300-475 kT each; CEP 90 m. A limit of 8 RVs was set in 1991 under the START counting rules.
Torpedoes: 4—21 in *(533 mm)* Mk 68 bow tubes. Gould/Westinghouse Mk 48; ADCAP; wire-guided (option); active/

passive homing to 50 km *(27 n miles)*/38 km *(21 n miles)* at 40/55 kt; warhead 267 kg; depth to 900 m *(2,950 ft)*.
Countermeasures: Decoys: 8 launchers for Emerson Electric Mk 2; torpedo decoy.
ESM: WLR-8(V)5; intercept. WLR-10; radar warning.
Combat data systems: DWS-118 and CCS Mk 2 Mod 3 with UYK 43/UYK 44 computers.
Weapons control: Mk 98 fire-control system.
Radars: Surface search/navigation/fire control: BPS 15A/H; I/J-band.
Sonars: IBM BQQ 6; passive search.
Raytheon BQS 13; spherical array for BQQ 6.
Ametek BQS 15; active/passive for close contacts; high frequency.
Western Electric BQR 15 (with BQQ 9 signal processor); passive towed array. TB 23 thin line array.
Raytheon BQR 19; active for navigation; high frequency.

Programmes: The size of the SSBN forces has been reduced to 14 hulls. *Ohio*, *Michigan*, *Florida* and *Georgia* are undergoing conversion to SSGN.

Modernisation: *Alaska* and *Nevada* have been upgraded to deploy Trident D5 missiles. *Jackson* is to complete conversion in 2007 and *Alabama* in 2008. Ohio class SSBNs are being upgraded with ARCI (Acoustic Rapid COTS insertion) sonar and CCS Mk 2 Block 1 fire-control systems. Installation in *Alaska*, *Nevada* and *Pennsylvania* is complete and is to be undertaken in *Kentucky* in 2005.
Structure: The size of the Trident submarine is dictated primarily by the 24 vertically launched Trident missiles and the larger reactor plant to drive the ship. The reactor has a nuclear core life of about 20 years between refuellings. Diving depth is 244 m *(800 ft)*. Kollmorgen Type 152 and Type 82 periscopes. Mk 19 Air Turbine Pump for torpedo discharge.
Operational: Pacific Fleet units are based at Bangor, Washington, while the Atlantic Fleet units are based at King's Bay, Georgia. SSBNs 741 and 743 are to transfer to Bangor on 1 October 2005. In the current state of worldwide tensions, a modified alert status has been implemented. Single crews were considered but rejected. Hull life of the class is extended to 42 years.

UPDATED

ALABAMA   *4/2004*, Ships of the World /* 1043703

ALABAMA   *4/2004*, Ships of the World /* 1043704

Cruise Missile Submarines (SSGN)

0 + 4 OHIO CLASS (SSGN)

Name	No
OHIO	SSGN 726 (ex-SSBN 726)
MICHIGAN	SSGN 727 (ex-SSBN 727)
FLORIDA	SSGN 728 (ex-SSBN 728)
GEORGIA	SSGN 729 (ex-SSBN 729)

Builders	Launched	Commissioned	F/S
General Dynamics (Electric Boat Div)	7 Apr 1979	11 Nov 1981	Conv/PA
General Dynamcis (Electric Boat Div)	26 Apr 1980	11 Sep 1982	Conv/PA
General Dynamics (Electric Boat Div)	14 Nov 1981	18 June 1983	Conv/AA
General Dynamics (Electric Boat Div)	6 Nov 1982	11 Feb 1984	Conv/AA

Displacement, tons: 16,764 surfaced; 18,750 dived
Dimensions, feet (metres): 560 × 42 × 36.4
(170.7 × 12.8 × 11.1)
Main machinery: Nuclear; 1 GE PWR S8G; 2 turbines; 60,000 hp
(44.8 MW); 1 shaft; 1 Magnetek auxiliary prop motor; 325 hp
(242 kW)
Speed, knots: 24 dived
Complement: 155 (15 officers)

Missiles: SLCM: up to 154 Tomahawk and Tactical Tomahawk.
Torpedoes: 4—21 in *(533 mm)* Mk 68 bow tubes. Gould/
Westinghouse Mk 48; ADCAP; wire-guided (option); active/
passive homing to 50 km *(27 n miles)*/38 km *(21 n miles)* at
40/55 kt; warhead 267 kg; depth to 900 m *(2,950 ft)*.
Countermeasures: Decoys: 8 launchers for Emerson Electric Mk
2; torpedo decoy.
ESM: WLR-8(V)5; intercept. WLR-10; radar warning.
Combat data systems: DWS-118 and CCS Mk 2 Mod 3 with
UYK 43/UYK 44 computers.
Weapons control: Mk 98 fire-control system.
Radars: Surface search/navigation/fire control: BPS 15A/H; I/J-
band.
Sonars: IBM BQQ 6; passive search.
Raytheon BQS 13; spherical array for BQQ 6.
Ametek BQS 15; active/passive for close contacts; high
frequency.
Western Electric BQR 15 (with BQQ 9 signal processor);
passive towed array. TB 23 thin line array.
Raytheon BQR 19; active for navigation; high frequency.

Programmes: Following the 1994 nuclear posture review which
recommended a 14-strong SSBN force, remaining four Ohio
class being converted to SSGN role. This includes land attack,
special forces insertion and support and ISR. Conversion
contract with General Dynamics Electric Boat in October
2002. *Ohio* started mid-life refuelling on 15 November 2002
and conversion work (at Puget Sound Naval Shipyard) on
19 November 2003. She is to complete conversion in
November 2005. *Florida* started mid-life refuelling in 2002
and conversion work (at Puget Sound) in April 2004.
Michigan started refuelling in February 2004 and conversion
work (at Norfolk Naval Shipyard) in October 2004. *Georgia*
started refuelling in March 2005 and is to start conversion (at
Norfolk) in October 2005.
Modernisation: Conversion to carry up to 154 Tomahawk or
Tactical Tomahawk missiles by enabling seven cruise missiles
to be fired from each of 22 of the current 24 Trident missile

Ohio Class - Reconfiguration

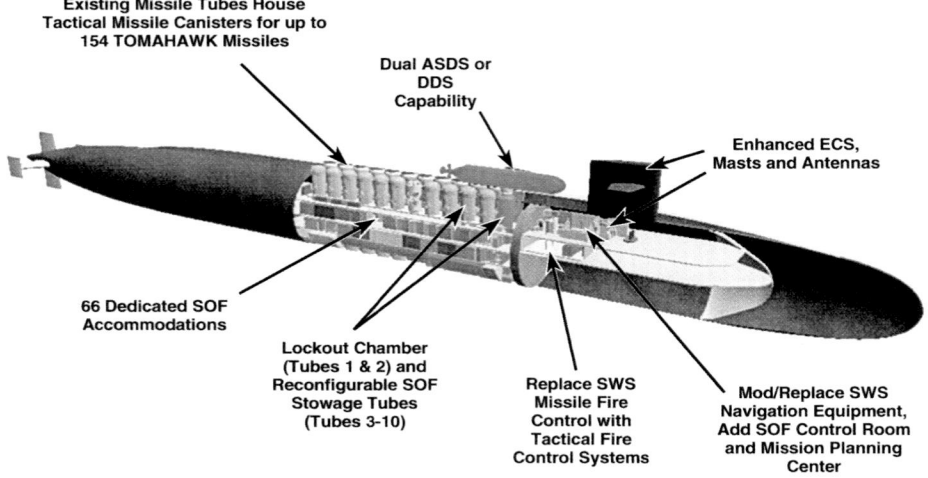

OHIO class reconfiguration

Existing Missile Tubes House Tactical Missile Canisters for up to 154 TOMAHAWK Missiles

Dual ASDS or DDS Capability

Enhanced ECS, Masts and Antennas

66 Dedicated SOF Accommodations

Lockout Chamber (Tubes 1 & 2) and Reconfigurable SOF Stowage Tubes (Tubes 3-10)

Replace SWS Missile Fire Control with Tactical Fire Control Systems

Mod/Replace SWS Navigation Equipment, Add SOF Control Room and Mission Planning Center

6/2001, US Navy / 0122978

tubes. Eight of these tubes are likely to be interchangeable
with Special Forces stowage canisters. The remaining two
tubes will be permanently configured for wet/dry launch of up
to 66 SEALs. The combat system is also to be upgraded and
future payloads are being developed to augment the baseline
configuration.
Structure: The size of the submarine was dictated primarily by
the 24 missile tubes and the larger reactor plant to drive the
ship. The reactor has a nuclear core life of about 20 years
between refuellings. Diving depth is 244 m *(800 ft)*.
Kollmorgen Type 152 and Type 82 periscopes. Mk 19 Air
Turbine Pump for torpedo discharge.

Operational: *Georgia* played the part of an SSGN during Exercise
'Silent Hammer' in mid-2004. This tested procedures for
strikes against time-critical targets and use of special
operations forces. A battle-centre onboard tested
communications and networking required to support them.
All boats will return to the fleet by 2007. *Ohio* and *Michigan*
are to be based at Bangor, WA, and perhaps later in Guam.
Florida and *Georgia* are to be based at King's Bay, GA.
UPDATED

AIRCRAFT CARRIERS

0 + 1 (1) FUTURE CARRIER (CVN 21) (CVN)

Name	No	Builders	Laid down	Launched	Commissioned
—	CVN 78	Northrop Grumman Newport News	2008	2012	2015
—	CVN 79	Northrop Grumman Newport News	2012	2016	2019

Displacement, tons: 100,000 approx
Dimensions, feet (metres): 836.6 × 265.1 × ?
(255.0 × 80.8 × ?)
Flight deck, feet (metres): to be announced
Main machinery: Nuclear; 2 reactors
Speed, knots: To be announced
Complement: To be announced

Missiles: Evolved Sea Sparrow.
Countermeasures: Decoys; ESM; ECM; torpedo defence.
Combat data systems: To be announced.
Weapons control: To be announced.
Radars: Air search: AN/SPY-3 MFR and S-VSR.
Navigation: To be announced.
Fire control: To be announced.
Tacan: To be announced.

Fixed-wing aircraft: Composition will depend on mission but will
include Joint Strike Fighter and Joint Unmanned Combat Air
Systems.

Programmes: Northrop Grumman Newport News awarded a
$1.39 billion contract in May 2004 to begin construction
preparation and continue design of the propulsion plant.
Long lead-time materials also are to be purchased. Advanced
construction is to begin for selected units. Building of lead ship
scheduled to begin in 2008 and of second of class ship in
2011. Further ships to follow at five year intervals.
Structure: The Nimitz class flight deck has been re-arranged to
increase sortie rates and improve weapons movement. To
achieve this, the island has been redesigned and relocated,
there are to be three rather than four aircraft lifts and there is
to be an advanced weapons elevator (AWE). Other features
include electromagnetic aircraft launching system (EMALS),
advanced arresting gear, new integrated warfare system
(IWS) and new, advanced life-of-the-ship nuclear power plant.
Significant habitability improvements are also likely. Electrical
generating capacity is to be at least 2½ times that of the
Nimitz class to support future war-fighting technologies,
including directed energy weapons and high bandwidth
communications.
Operational: CVN 21 is to require 500-900 fewer complement
than the Nimitz class. Increased sortie rates (by 20 per cent)
and reduced depot maintenance requirements to increase

operational availability by up to 25 percent. New command
centre to combine force networking with flexible, open
system architecture to support simultaneous multiple

CVN 21

6/2004*, Northrop Grumman Newport News / 1043671

missions, including integrated strike planning, joint/coalition
operations and special warfare missions.
NEW ENTRY

9 + 1 NIMITZ CLASS (CVNM)

Name	No	Builders	Laid down	Launched	Commissioned	F/S
NIMITZ	CVN 68	Newport News Shipbuilding	22 June 1968	13 May 1972	3 May 1975	PA
DWIGHT D EISENHOWER	CVN 69	Newport News Shipbuilding	15 Aug 1970	11 Oct 1975	18 Oct 1977	AA
CARL VINSON	CVN 70	Newport News Shipbuilding	11 Oct 1975	15 Mar 1980	13 Mar 1982	PA
THEODORE ROOSEVELT	CVN 71	Newport News Shipbuilding	13 Oct 1981	27 Oct 1984	25 Oct 1986	AA
ABRAHAM LINCOLN	CVN 72	Newport News Shipbuilding	3 Nov 1984	13 Feb 1988	11 Nov 1989	PA
GEORGE WASHINGTON	CVN 73	Newport News Shipbuilding	25 Aug 1986	21 July 1990	4 July 1992	AA
JOHN C STENNIS	CVN 74	Newport News Shipbuilding	13 Mar 1991	13 Nov 1993	9 Dec 1995	PA
HARRY S TRUMAN	CVN 75	Newport News Shipbuilding	29 Nov 1993	14 Sep 1996	25 July 1998	AA
RONALD REAGAN	CVN 76	Newport News Shipbuilding	12 Feb 1998	4 Mar 2001	12 July 2003	PA
GEORGE H W BUSH	CVN 77	Newport News Shipbuilding	19 May 2003	Mar 2006	Apr 2008	bldg

Displacement, tons: 72,916 (CVN 68-70), 73,973 (CVN 71) light; 91,487 (CVN 68-70), 96,386 (CVN 71), 102,000 (CVN 72-77) full load

Dimensions, feet (metres): 1,040 pp; 1,092 oa × 134 wl × 37 (CVN 68-70); 38.7 (CVN 71); 39 (CVN 72-76); 39.8 (CVN 77) *(317; 332.9 × 40.8 × 11.3; 11.8; 11.9; 12.1)*

Flight deck, feet (metres): 1,092; 779.8 (angled) × 252 *(332.9; 237.7 × 76.8)*

Main machinery: Nuclear; 2 Westinghouse/GE PWR A4W/A1G; 4 turbines; 280,000 hp *(209 MW)*; 4 emergency diesels; 10,720 hp *(8 MW)*; 4 shafts

Speed, knots: 30+

Complement: 3,200 (160 officers); 2,480 aircrew (320 officers); Flag 70 (25 officers)

Missiles: SAM: 2 (CVN 68, 69, 76, 77) or 3 Raytheon GMLS Mk 29 octuple launchers ❶; NATO Sea Sparrow RIM-7; semi-active radar homing to 14.6 km *(8 n miles)* at 2.5 Mach; warhead 39 kg. ESSM in due course.
2 GMLS Mk 49 RAM RIM-116 launchers ❷ (CVN 68, 69, 76 and 77); 21 rds/launcher; passive IR/anti-radiation homing to 9.6 km *(5.2 n miles)* at 2 Mach; warhead 9.1 kg.

Guns: 3 (CVN 71) or 4 (CVN 70, 72, 73, 74, 75) General Electric/General Dynamics 20 mm Vulcan Phalanx 6-barrelled Mk 15 (CVN 70-75); 3,000 rds/min (or 4,500 in Block 1) combined to 1.5 km.

Countermeasures: Decoys: SLQ 25 Torpedo Countermeasures Transmitting Set (Nixie).
ESM/ECM: SLQ-32(V)4 intercept and jammers. WLR-1H is being removed.

Combat data systems: ACDS Block 0 or 1 naval tactical and advanced combat direction systems; Links 4A, 11, 16 and Satellite Tadil J. GCCS (M) SATCOMS; SSR-1, WCS-3A (UHF DAMA), WSC-6 (SHF), WSC-8 (SHF), USC-38 (EHF), SSR-2A (GBS) (see Data Systems at front of section). SSDS Mk 2 (CVN 68, 69, 76. To be back fitted in all).

Weapons control: 3 Mk 91 Mod 1 MFCS directors (part of the NSSMS Mk 57 SAM system).

Radars: Air search: ITT SPS-48E ❸; 3D; E/F-band.
Raytheon SPS-49(V)5 (SPS-49A (V)1 (when SSDS installed)) ❹; C/D-band.
Hughes Mk 23 TAS ❺; D-band or SPQ-9B with SSDS in due course.
Surface search: Norden SPS-67(V)1; G-band.
CCA: SPN-41, SPN-43C, 2 SPN-46; J/F/J/K-band.
Navigation: Raytheon SPS-64(V)9 (being replaced by SPS-73); Furuno 900; I/J-band.
Fire control: 6 Mk 95; I/J-band (2 per GMLS Mk 29 launcher).
Tacan: URN 25.

Fixed-wing aircraft: Composition of air-wing depends on mission and typically includes: 12 F/A-18F Hornet; 36 F/A-18A/C/E Hornet; 4 EA-6B Prowler; 4 E-2C Hawkeye.

Helicopters: 4 SH-60F and 2 HH-60H Seahawk and up to 9 SH-60B Seahawk.

Programmes: *Nimitz* was authorised in FY67, *Dwight D Eisenhower* in FY70, *Carl Vinson* in FY74, *Theodore Roosevelt* in FY80 and *Abraham Lincoln* and *George Washington* in FY83. Construction contracts for *John C Stennis* and *Harry S Truman* were awarded in June 1988 and for *Ronald Reagan* in December 1994. Authorised in FY99, construction contract for *George H W Bush* awarded in January 2001.

Modernisation: CVN 68 completed a three-year Refuelling and Complex Overhaul (RCOH) in 2001. RCOH of CVN 69 started in 2001 and completed in January 2005. RCOH of CVN 70 is to be undertaken 2006-09. SSDS Mk 2 Mod 0 installed in CVN 68 (to be upgraded to Mk 2 Mod 1 in due course). This includes fitting two RAM systems and SPQ-9B radar vice Mk 23 TAS. SSDS Mk 2 Mod 1 fitted to CVN 69, CVN 76 and 77. RAM systems replace one Mk 29 and all Phalanx launchers on CVN 68 and 69. CVN 70 similarly fitted during RCOH but will retain upgraded CIWS (Phalanx) mounts as will CVN 71-75. CVN 71-75 to be upgraded to the SSDS-based combat system in due course.

Structure: Damage control measures include sides with system of full and empty compartments (full compartments can contain aviation fuel), approximately 2.5 in Kevlar plating over certain areas of side shell, box protection over magazine and machinery spaces. Aviation facilities include four lifts, two at the forward end of the flight deck, one to starboard abaft the island and one to port at the stern. There are four steam catapults (C13-1 (CVN 68-71), C13-2 (CVN 72-77)) and four (or three on CVN 76 and 77) Mk 7 Mod 3 arrester wires. Launch rate is one every 20 seconds. The hangar can hold less than half the full aircraft complement, deckhead is 25.6 ft. Aviation fuel, 8,500 tons. Tactical Flag Command Centre for Flagship role. During RCOH, CVN 68 and 69 fitted with reshaped island (the mainmast has three yardarms to support more antennas). Major structural differences in CVN 76 include: a stretched island which includes a bigger bridge and a three yardarm mainmast and which incorporates the after mast (separate in previous ships) and the internal bomb elevator. Other changes include a bulbous bow to reduce drag and a modified flight deck (angled deck increased by 0.1 degrees) to allow the use of two catapults while aircraft land.

Operational: Multimission role of 'strike/ASW'. From CVN 70 onwards ships have an A/S control centre and A/S facilities; CVN 68 and 69 are backfitted. Endurance of 16 days for aviation fuel (steady flying) with greater than 1 million miles between refuelling. Only one refuelling is required in the life of the ship. Ships' complements and air wings can be changed depending on the operational task.

UPDATED

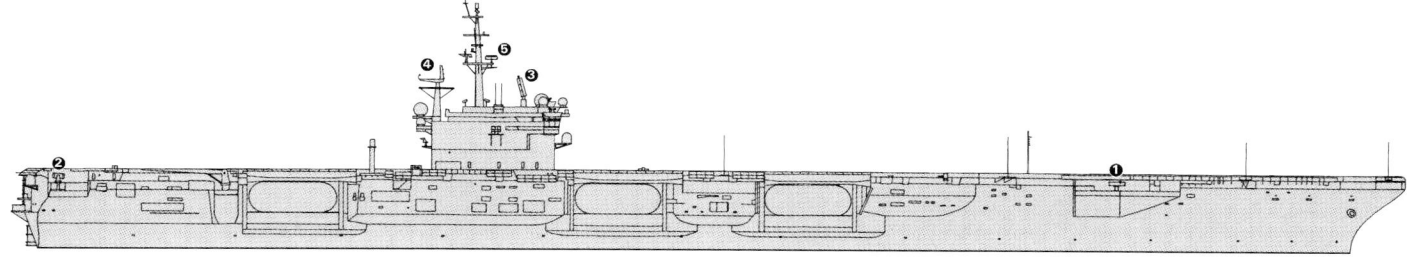

RONALD REAGAN
(Scale 1 : 1,800), Ian Sturton / 1043489

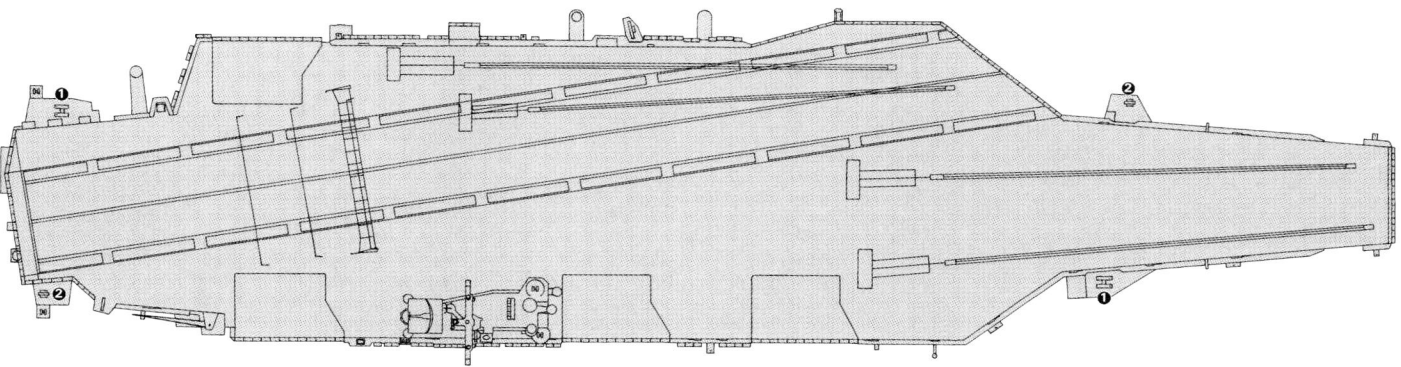

RONALD REAGAN
(Scale 1 : 1,800), Ian Sturton / 1043490

RONALD REAGAN
*7/2004 *, US Navy /* 1043664

NIMITZ

9/2003 *, *US Navy* / 1043665

GEORGE WASHINGTON

6/2004 *, *US Navy* / 1043666

HARRY S TRUMAN

6/2004*, Ships of the World / 1043700

GEORGE H W BUSH (computer graphic)

3/2003, Ships of the World / 0572799

JOHN C STENNIS

8/2004*, Hachiro Nakai / 1043694

2 KITTY HAWK and JOHN F KENNEDY CLASSES (CVM)

Name	No	Builders	Laid down	Launched	Commissioned	F/S
KITTY HAWK	CV 63	New York Shipbuilding	27 Dec 1956	21 May 1960	29 Apr 1961	PA
JOHN F KENNEDY	CV 67	Newport News Shipbuilding	22 Oct 1964	27 May 1967	7 Sep 1968	AA

Displacement, tons: 83,960 full load; 81,430 full load (CV 67)
Dimensions, feet (metres): 1,062.5 (CV 63); 1,072.5 (CV 64); 1,052 (CV 67) × 130 × 37.4
(323.6; 326.9; 320.6 × 39.6 × 11.4)
Flight deck, feet (metres): 1,046 × 252 *(318.8 × 76.8)*
Main machinery: 8 Foster-Wheeler boilers; 1,200 psi *(83.4 kg/cm²)*; 950°F *(510°C)*; 4 Westinghouse turbines; 280,000 hp *(209 MW)*; 4 shafts
Speed, knots: 32
Range, n miles: 4,000 at 30 kt; 12,000 at 20 kt
Complement: 2,930 (155 officers); aircrew 2,480 (320 officers); Flag 70 (25 officers)

Missiles: SAM: 2 Raytheon GMLS Mk 29 octuple launchers ❶; NATO Sea Sparrow RIM-7; semi-active radar homing to 14.6 km *(8 n miles)* at 2.5 Mach; warhead 39 kg. ESSM in due course.
2 GMLS Mk 49 RAM RIM-116 ❷; 21 rds/launcher; passive IR/anti-radiation homing to 9.6 km *(5.2 n miles)* at 2 Mach; warhead 9.1 kg.
Guns: 2 General Electric/General Dynamics 20 mm Vulcan Phalanx 6-barrelled Mk 15/14 Block I (CV 63), Mk 15/13 Block I (CV 67) ❸; 3,000 rds/min (or 4,500 in Block 1) combined to 1.5 km. Both mountings to be replaced by RAM in CV 63 and CV 67.
Countermeasures: Decoys: SLQ-25 Torpedo Countermeasures Transmitting Set (Nixie).
ESM/ECM: SLQ-32(V)4 intercept and jammers.
Combat data systems: ACDS Block 0 (Block 1 Level 1 in CV 67) naval tactical and advanced combat direction systems Links 4A, 11, 14, 16 and Satellite Tadil J. GCCS (M) SATCOMS; SSR-1, WSC-3A (UHF DAMA), WSC-6 (SHF), WSC-8 (SHF), USC-38 (EHF), SSR-2A (GBS) (see Data Systems at front of section). SSDS Mk 2 Mod 1 (CV 67 in 2005).
Weapons control: 2 Mk 91 MFCS directors (part of NSSMS Mk 57 SAM system).
Radars: Air search: ITT SPS-48E ❹; 3D; E/F-band.
Raytheon SPS-49(V)5, SPS-49A(V)1 (CV 67 when SSDS installed) ❺; C/D-band.
Hughes Mk 23/7 TAS ❻; D-band. SPQ-9B with SSDS in 2005 (CV 67).

Surface search: Norden SPS-67; G-band.
CCA: SPN-41, SPN-43A; 2 SPN-46; J/K/F-band.
Navigation: Furuno 900; I-band.
Fire control: 4 Mk 95; I/J-band (for SAM).
Tacan: URN 25.

Fixed-wing aircraft: Composition of air-wing depends on mission and typically includes: 12 F/A-18F Hornet; 36 F/A-18A/C/E Hornet; 4 EA-6B Prowler; 4 E-2C Hawkeye.
Helicopters: 4 SH-60F, 2 HH-60H Seahawk and up to 9 SH-60B Seahawk.

Programmes: *Kitty Hawk* was authorised in FY56 and *John F Kennedy* in FY63.
Modernisation: Service Life Extension Programme (SLEP): *Kitty Hawk* completed in February 1991. A 'complex overhaul' of *Kennedy* completed in September 1995 but a second 15-month complete overhaul scheduled to begin in 2005 has been cancelled. ACDS Block 1 trials done in *Kennedy* in 1998. SSDS Mk 2 Mod 1 to be fitted in CV 67 in due course including SPQ-9B radar. *Kennedy* battle group conducted successful CEC operational evaluation in 2001. *Kennedy* completed nine month maintenance period at Mayport, FL, in November 2003.
Structure: These ships were built to an improved Forrestal design. They have two deck-edge lifts forward of the superstructure, a third lift aft of the structure, and the port-side lift on the after quarter. Four C13 steam catapults (with one C13-1 in *Kennedy*) and four arrester wires. Aviation fuel of 5,882 tons carried.
Operational: *Kitty Hawk* based in Yokosuka from July 1998 and expected to remain in service until 2008 when it will be replaced in the force structure by CVN 77. *John F Kennedy* officially to pay off in 2018 when replaced by CVN 79 although, following cancellation of a complete overhaul, it is possible that she will be decommissioned in 2006. Air wings are changed depending on the operational role. CV 63 used as Special Forces base during operations off Afghanistan in 2001.

UPDATED   JOHN F KENNEDY   *7/2000, Hachiro Nakai* / 0106751

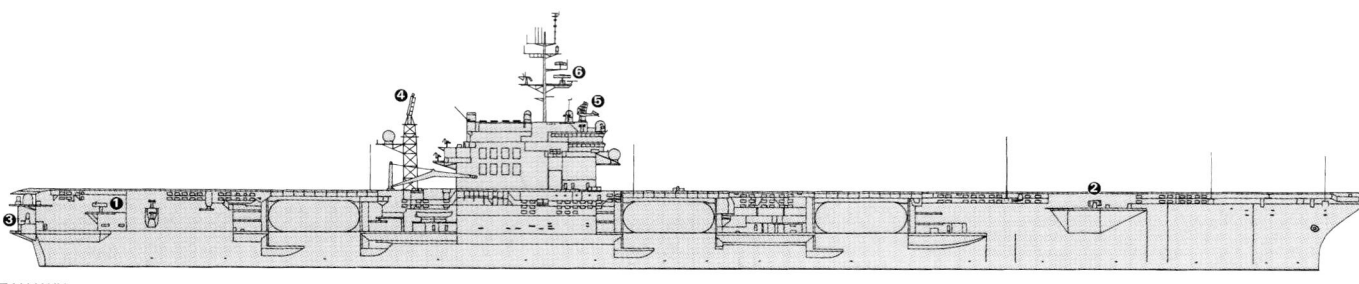

KITTY HAWK   *(Scale 1 : 1,800), Ian Sturton* / 0573701

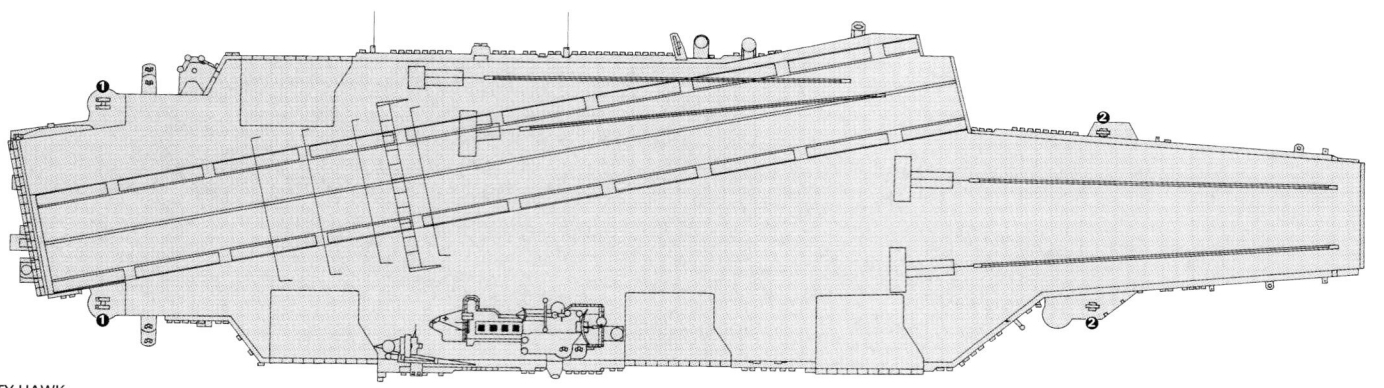

KITTY HAWK   *(Scale 1 : 1,800), Ian Sturton* / 0573700

KITTY HAWK   *8/2004 *, US Navy* / 1043670

KITTY HAWK
11/2002, Ships of the World / 0572770

JOHN F KENNEDY
1/2000, US Navy / 0131335

KITTY HAWK
5/2004*, US Navy / 1043669

1 ENTERPRISE CLASS (CVNM)

Name	No	Builders	Laid down	Launched	Commissioned	F/S
ENTERPRISE	CVN 65	Newport News Shipbuilding	4 Feb 1958	24 Sep 1960	25 Nov 1961	AA

Displacement, tons: 73,502 light; 75,700 standard; 89,600 full load

Dimensions, feet (metres): 1,123 × 133 × 39 *(342.3 × 40.5 × 11.9)*

Flight deck, feet (metres): 1,088 × 252 *(331.6 × 76.8)*

Main machinery: Nuclear; 8 Westinghouse PWR A2W; 4 Westinghouse turbines; 280,000 hp *(209 MW)*; 4 emergency diesels; 10,720 hp *(8 MW)*; 4 shafts

Speed, knots: 33

Complement: 3,350 (171 officers); 2,480 aircrew (358 officers); Flag 70 (25 officers)

Missiles: SAM: 2 Raytheon GMLS Mk 29 octuple launchers ❶; NATO Sea Sparrow RIM-7; semi-active radar homing to 14.6 km *(8 n miles)* at 2.5 Mach; warhead 39 kg. 2 GMLS Mk 49 RAM RIM-116 ❷; 21 rds/launcher; passive IR/anti-radiation homing to 9.6 km *(5.2 n miles)* at 2 Mach; warhead 9.1 kg.

Guns: 2 General Electric/General Dynamics 20 mm Vulcan Phalanx 6-barrelled Mk 15 ❸; 3,000 rds/min (or 4,500 in Block 1) combined to 1.5 km.

Countermeasures: Decoys: SLQ-25 Torpedo Countermeasures Transmitting Set (Nixie).

ESM/ECM: SLQ-32(V)4; intercept and jammers.

Combat data systems: ACDS Block 0 naval tactical and advanced combat direction systems; Links 4A, 11, 14, 16 and Satellite Tadil J. GCCS(M) SATCOMS; SSR-1, WSC-3 (UHF DAMA), WSC-6 (SHF), WSC-8 (SHF), USC-38 (EHF), SSR-2A (GBS) (see Data Systems at front of section).

Weapons control: 2 Mk 91 Mod 1 MFCS directors (part of NSSMS Mk 57 SAM system).

Radars: Air search: ITT SPS-48E ❹; 3D; E/F-band. Raytheon SPS-49(V)5 ❺; C/D-band. Hughes Mk 23 TAS ❻; D-band. SPQ-9B in due course.

Surface search: Norden SPS-67; G-band.

CCA: SPN-41, SPN-43C; 2 SPN-46; J/F/K-band.

Navigation: Raytheon SPS-64(V)9; Furuno 900; I/J-band.

Fire control: 4 Mk 95; I/J-band (for SAM).

Tacan: URN 25.

Fixed-wing aircraft: Composition of air-wing depends on mission and typically includes: 12 F/A-18F Hornet; 36 F/A-18A/C/E Hornet; 4 EA-6B Prowler; 4 E-2C Hawkeye.

Helicopters: 4 SH-60F, 2 HH-60H Seahawk and up to 9 SH-60B Seahawk.

Programmes: Authorised in FY58. Underwent a refit/overhaul at Puget Sound Naval SY, Bremerton, Washington from January 1979 to March 1982. Latest complex overhaul including refuelling started at Newport News in early 1991 and completed 27 September 1994. Minor refit in 1997 and again in 2002.

Modernisation: Mk 25 Sea Sparrow was installed in late 1967 and this has been replaced by two Mk 29 and supplemented with three 20 mm Mk 15 CIWS. A reshaping of the island took place in her 1979-82 refit. This included a replacement mast similar to the Nimitz class with SPS-48C and 49 radars. Improvements during latest overhaul included SPS-48E and Mk 23 TAS air search radars, SPN-46 precision approach and landing radar and C³ and EW systems. RAM to be fitted in 2004.

Structure: Built to a modified Forrestal class design. *Enterprise* was the world's second nuclear-powered surface warship (the cruiser *Long Beach* was completed a few months earlier). Aviation facilities include four deck edge lifts, two forward and one each side abaft the island. There are four 295 ft C 13 Mod 1 catapults. Hangars cover 216,000 sq ft with 25 ft deck head. Aviation fuel, 8,500 tons.

Operational: 12 days' aviation fuel for intensive flying. Air wing depends on the operational role. Expected to remain in service until 2014 when it will be replaced in the force structure by CVN 78.

UPDATED

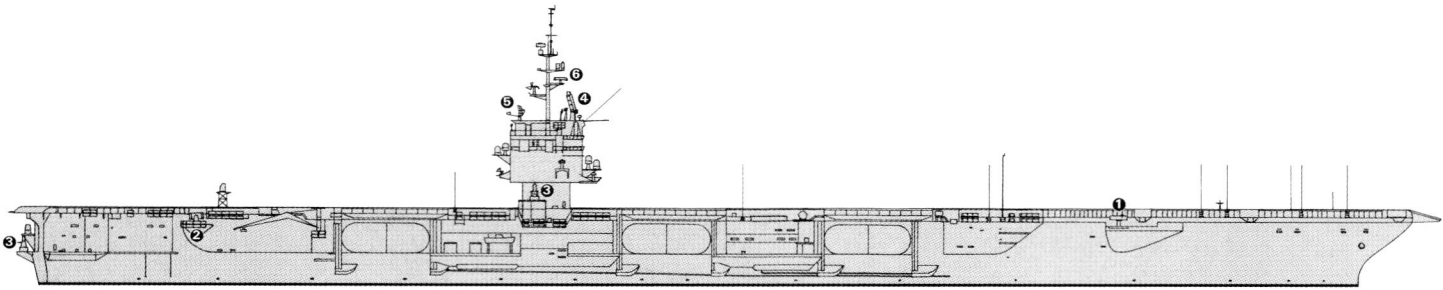

ENTERPRISE

(Scale 1 : 1,800), Ian Sturton / 0573702

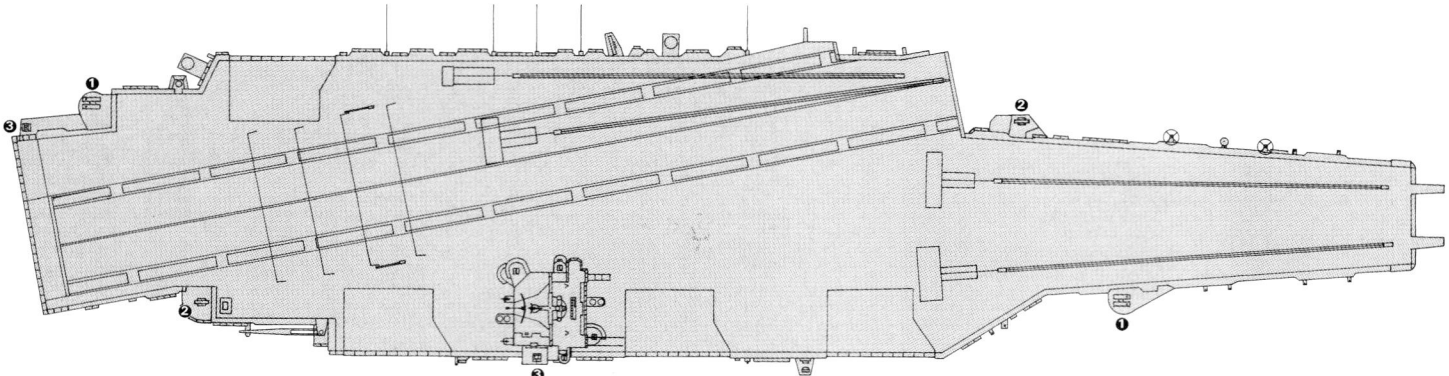

ENTERPRISE

(Scale 1 : 1,800), Ian Sturton / 0573703

ENTERPRISE

6/2004*, US Navy / 1043668

CRUISERS

Notes: (1) **Integrated Ship Controls.** Formerly known as Smart Ship, Integrated Ship Controls (ISC) began as Naval Research Advisory Committee recommendation in 1996 to reduce manning through technology. *Yorktown* (CG 48) was selected as first Smart Ship with implementation of 47 workload-reduction initiatives tested and evaluated during a five-month deployment completed in June 1997. Fourteen more initiatives were installed in July 1997. Core systems included: Integrated Bridge System (IBS), Integrated Condition Assessment System (ICAS), Machinery Control System (MCS), Damaged Control System (DCS), Fuel Control System (FCS), fibre optic Local Area Network (LAN) and Wireless Internal Communication System (WICS). *Yorktown's* experience validated these technologies, combined with changes in policies, procedures and new watch routines, to generate substantial reductions in workload. The first installation of ISC in *Ticonderoga* (CG 47) was very challenging and more costly than originally anticipated. All subsequent ISC installations have been completed successfully and within budget during 20-week installation periods: *Monterey* (CG 61) was fitted in 2000; *Valley Forge* (CG 50) and *Mobile Bay* (CG 53) in 2001; *Antietam* (CG 54) in 2002; *Hue City* (CG 66) in 2003; *Cape St George* (CG 71) in 2004. *San Jacinto* (CG 76) is scheduled for 2006. Remaining ships of class will receive ISC either as a stand-alone upgrade or during the Cruiser Modernisation Programme.
(2) CG(X) is the proposed replacement for the Ticonderoga (CG 47) class cruisers. It is expected to be a follow-on variant of the DD(X), incorporating an integrated power system and using a similar hull-form, but with enhanced missile-defence and air warfare capability. Entry into service is expected in about 2017.

25 TICONDEROGA CLASS: GUIDED MISSILE CRUISERS (AEGIS) (CGHM)

Name	*No*	*Builder/Programme*	*Laid down*	*Launched*	*Commissioned*	*F/S*
VINCENNES	CG 49	Ingalls Shipbuilding	20 Oct 1982	14 Jan 1984	6 July 1985	PA
VALLEY FORGE	CG 50	Ingalls Shipbuilding	14 Apr 1983	23 June 1984	18 Jan 1986	PA
THOMAS S GATES	CG 51	Bath Iron Works	31 Aug 1984	14 Dec 1985	22 Aug 1987	PA
BUNKER HILL	CG 52	Ingalls Shipbuilding	11 Jan 1984	11 Mar 1985	20 Sep 1986	PA
MOBILE BAY	CG 53	Ingalls Shipbuilding	6 June 1984	22 Aug 1985	21 Feb 1987	PA
ANTIETAM	CG 54	Ingalls Shipbuilding	15 Nov 1984	14 Feb 1986	6 June 1987	PA
LEYTE GULF	CG 55	Ingalls Shipbuilding	18 Mar 1985	20 June 1986	26 Sep 1987	AA
SAN JACINTO	CG 56	Ingalls Shipbuilding	24 July 1985	14 Nov 1986	23 Jan 1988	AA
LAKE CHAMPLAIN	CG 57	Ingalls Shipbuilding	3 Mar 1986	3 Apr 1987	12 Aug 1988	PA
PHILIPPINE SEA	CG 58	Bath Iron Works	8 May 1986	12 July 1987	18 Mar 1989	AA
PRINCETON	CG 59	Ingalls Shipbuilding	15 Oct 1986	2 Oct 1987	11 Feb 1989	PA
NORMANDY	CG 60	Bath Iron Works	7 Apr 1987	19 Mar 1988	9 Dec 1989	AA
MONTEREY	CG 61	Bath Iron Works	19 Aug 1987	23 Oct 1988	16 June 1990	AA
CHANCELLORSVILLE	CG 62	Ingalls Shipbuilding	24 June 1987	15 July 1988	4 Nov 1989	PA
COWPENS	CG 63	Bath Iron Works	23 Dec 1987	11 Mar 1989	9 Mar 1991	PA
GETTYSBURG	CG 64	Bath Iron Works	17 Aug 1988	22 July 1989	22 June 1991	AA
CHOSIN	CG 65	Ingalls Shipbuilding	22 July 1988	1 Sep 1989	12 Jan 1991	PA
HUE CITY	CG 66	Ingalls Shipbuilding	20 Feb 1989	1 June 1990	14 Sep 1991	AA
SHILOH	CG 67	Bath Iron Works	1 Aug 1989	8 Sep 1990	2 July 1992	PA
ANZIO	CG 68	Ingalls Shipbuilding	21 Aug 1989	2 Nov 1990	2 May 1992	AA
VICKSBURG	CG 69	Ingalls Shipbuilding	30 May 1990	2 Aug 1991	14 Nov 1992	AA
LAKE ERIE	CG 70	Bath Iron Works	6 Mar 1990	13 July 1991	24 July 1993	PA
CAPE ST GEORGE	CG 71	Ingalls Shipbuilding	19 Nov 1990	10 Jan 1992	12 June 1993	AA
VELLA GULF	CG 72	Ingalls Shipbuilding	22 Apr 1991	13 June 1992	18 Sep 1993	AA
PORT ROYAL	CG 73	Ingalls Shipbuilding	18 Oct 1991	20 Nov 1992	9 July 1994	PA

Displacement, tons: 9,590 (CG 47-48); 9,407 (CG 49-51); 9,957 (remainder) full load
Dimensions, feet (metres): 567 × 55 × 31 (sonar) *(172.8 × 16.8 × 9.5)*
Main machinery: 4 GE LM 2500 gas turbines; 86,000 hp *(64.16 MW)* sustained; 2 shafts; cp props
Speed, knots: 30+. **Range, n miles:** 6,000 at 20 kt
Complement: 358 (24 officers); accommodation for 405 total 312 (24 officers) (CG 48)

Missiles: SLCM: GDC Tomahawk (CG 52 onwards); Tercom aided guidance to 1,300 km *(700 n miles)* (TLAM-C and D) or 1,853 km *(1,000 n miles)* (TLAM-C Block III) at 0.7 Mach; warhead 454 kg (TLAM-C) or 347 kg shaped charge (TLAM-C Blocks II and III) or submunitions (TLAM-D). TLAM-C has GPS back-up to Tercom and a CEP of 10 m.
SSM: 8 McDonnell Douglas Harpoon (2 quad) (1); active radar homing to 130 km *(70 n miles)* at 0.9 Mach; warhead 227 kg. Extended range SLAM can be fired from modified Harpoon canisters.
SAM: 68 (CG 47-51); 122 (CG 52 onwards) GDC Standard SM-2MR; command/inertial guidance; semi-active radar homing to 167 km *(90 n miles)* at 2 Mach. SM-2 Block IIIB uses an IR seeker. SAM and ASROC missiles are fired from 2 twin Mk 26 Mod 5 launchers (2) (CG 47-51) and 2 Mk 41 Mod 0 vertical launchers (3) (61 missiles per launcher) (CG 52 onwards). Tomahawk is carried in CG 52 onwards.
A/S: Honeywell ASROC (CG 47-51) and Loral ASROC VLA (CG 52 onwards). VLA has a range of 16.6 km *(9 n miles)*; inertial guidance of 1.6-10 km *(1-5.4 n miles)*; payload Mk 46 Mod 5 Neartip or Mk 50.
Guns: 2 FMC 5 in *(127 mm)*/54 Mk 45 (Mod 0 (CG 47-50); Mod 1 (CG 51 onwards)) (4); 20 rds/min to 23 km *(12.6 n miles)* anti-surface; weight of shell 32 kg.
2 General Electric/General Dynamics 20 mm/76 Vulcan Phalanx 6-barrelled Mk 15 Mod 2 (5); 3,000 rds/min (4,500 in Block 1) combined to 1.5 km. To be fitted with high-definition thermal imagers (HDTI) for tracking small craft.
2 McDonnell Douglas 25 mm. 4—12.7 mm MGs.
Torpedoes: 6—324 mm Mk 32 (2 triple) Mod 14 tubes (fitted in the ship's side aft) (6). 36 Honeywell Mk 46 Mod 5; anti-submarine; active/passive homing to 11 km *(5.9 n miles)* at 40 kt; warhead 44 kg or Alliant/Westinghouse Mk 50; active/passive homing to 15 km *(8.1 n miles)* at 50 kt; warhead 45 kg shaped charge.
Countermeasures: Decoys: Up to 8 Loral Hycor SRBOC 6-barrelled fixed Mk 36 Mod 2 (7); IR flares and chaff. Nulka being acquired. SLQ-25 Nixie; towed torpedo decoy.
ESM/ECM: Raytheon SLQ-32V(3)/SLY-2 (8); intercept, jammers.
Combat data systems: CEC being fitted 1996-2007 starting with CG 66 and 69. NTDS with Links 4A, 11, 14. GCCS (M) and Link 16 being fitted. Link 22 in due course. SATCOM WRN-5, WSC-3 (UHF), USC-38 (EHF). UYK 7 and 20 computers (CG 49-58); UYK 43/44 (CG 59 onwards) COTS open-architecture based system to be installed as part of modernisation programme from 2008. SQQ-28 for LAMPS sonobuoy datalink (9) (see Data Systems at front of section).
Weapons control: SWG-3 Tomahawk WCS. SWG-1A Harpoon LCS. Aegis Mk 7 Mod 4 multitarget tracking with Mk 99 MFCS (includes 4 Mk 80 illuminator directors); has at least 12 channels of fire. Singer Librascope Mk 116 Mod 6 (53B) or Mod 7 (53C) FCS for ASW. Lockheed Mk 86 Mod 9 GFCS (to be replaced by Mk-160 Mod 11 from 2008).
Radars: Air search/fire control: RCA SPY-1A phased arrays (10); 3D; E/F-band (CG 47-58).
Raytheon SPY-1B phased arrays (11); 3D; E/F-band (CG 59 on).
Air search: Raytheon SPS-49(V)7 or 8 (12); C/D-band; range 457 km *(250 n miles)*.
Surface search: ISC Cardion SPS-55 (13); I/J-band.
Navigation: Raytheon SPS-64(V)9; I-band.
Fire control: Lockheed SPQ-9A/B (14); I/J-band.
Four Raytheon SPG-62 (15); I/J-band.
Tacan: URN 25. IFF Mk XII AIMS UPX-29.
Sonars: General Electric/Hughes SQS-53B (CG 47-51); bow-mounted; active search and attack; medium frequency.
Gould SQR-19 (CG 54-55); passive towed array (TACTAS).
Gould/Raytheon SQQ-89(V)3 (CG 52 onwards); combines hull-mounted active SQS-53B (CG 52-67) or SQS-53C (CG 68-73) and passive towed array SQR-19.

Helicopters: 2 SH-60B Seahawk LAMPS III (16); 2 SH-2F LAMPS I (CG 47-48) (17). UAV in due course.

Modernisation: Four major combat system Baselines define the broad capability categories of this class. Baseline 1 ships (CG 49-51) are fitted with SPY-1A radar, LAMPS I/III helicopter, Mk 26 missile launchers, Standard SM-2 MR Block I/II missiles, ASROC and the UYK-7 computing system. In the Baseline 2 ships (CG 52-58), Mk 26 launchers were replaced by Mk 41 Mod 0 Vertical Launch System (VLS) and Tomahawk was fitted. ASW improvements included SQR-19 in CG 54-55 and SQQ-89(V)3 integrated ASW suite with ASROC VLA in CG 56-58. ASROC VLA was later fitted in CG 52-55 in 1998. Baseline 3 ships (CG 59-64) incorporated the lighter SPY-1B radar, UYK-21 displays and UYK-43/44 computers in place of UYK-7s. Baseline 4 ships (CG 65-73) were fitted with SPY-1B (V) radar. While the basic hardware configurations have remained constant, the primary combat system computer has been progressively upgraded. Computer programmes currently in service are: Baseline 1 (1.4); Baseline 2 (2.10 which includes COTS Shipboard Advanced Radar Target Identification System (SARTIS)); Baselines 3 and 4 (5 phase 3 or 6 phase 1). Link 16 was also added to the Baseline 3 and 4 ships as part of the computer system upgrade. The Cruiser Modernisation Programme (formerly the Cruiser Conversion Programme) is an extensive upgrade that will be applied to CG 52-73 (Baselines 2, 3 and 4). The core components of the programme include modification of the Mk 41 VLS launchers to fire ESSM, Integrated Ship Controls (ISC), fitting of an additional SPQ-9B fire-control radar, installation of the SQQ-89A(V)15 sonar suite, upgrading all 127 mm/54 guns to Mod 4 version, upgrade of the Aegis computer to 7PH1C with open architecture, display upgrades, COTS SARTIS and further platform and habitability upgrades to prolong service life. The first modernisation period is to start in FY08. Plans to enable all Aegis ships to fire SM-3 ABM missiles are being pursued separately.
Structure: The Ticonderoga class design is a modification of the Spruance class. The same basic hull is used, with the same gas-turbine propulsion plant although the overall length is slightly increased. The design includes Kevlar armour to protect vital spaces. No stabilisers. *Vincennes* and later ships have a lighter tripod mainmast vice the square quadruped of the first two.
Operational: The continuing development of the sea-based element of the Ballistic Missile Defense Programme is known as Aegis BMD. Test firings of the experimental Standard SM-3 missile were made by CG 67 on 24 September 1999. Since then CG 70 has successfully fired an SM-3 missile with a Lightweight Exo-Atmospheric Projectile (LEAP) to intercept an 'Aries' missile target on five occasions: during FM-2 (Flight Mission-2) on 25 January 2002, FM-3 on 13 June 2002, FM-4 on 21 November 2002, FM-6 on 11 December 2003 and FTM-04-01 on 24 February 2005. Of the Baseline 1 ships, CG 47 and 48 decommissioned in 2004 and CG 49-51 are to pay off by 2006.

UPDATED

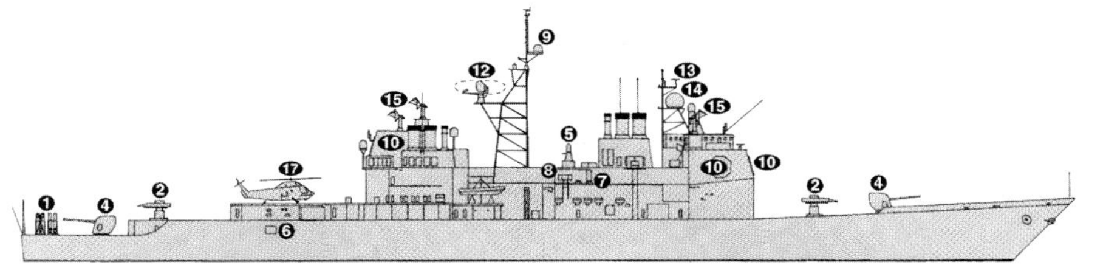

VINCENNES *(Scale 1 : 1,500), Ian Sturton /* 0016357

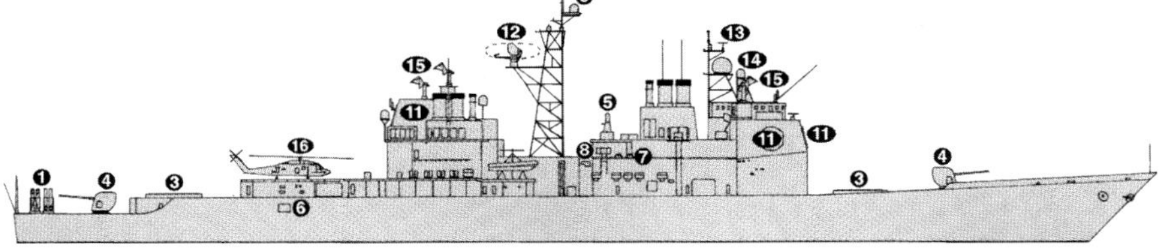

BUNKER HILL *(Scale 1 : 1,500), Ian Sturton /* 0016356

For details of the latest updates to *Jane's Fighting Ships* online and to discover the additional information available exclusively to online subscribers please visit jfs.janes.com

PORT ROYAL

11/2003*, US Navy / 1043641

SHILOH

12/2004*, US Navy / 1121523

SAN JACINTO

6/2004*, Frank Findler / 1121519

GETTYSBURG

7/2004*, US Navy / 1043640

LAKE CHAMPLAIN

7/2004*, Michael Nitz / 1043667

VINCENNES

10/2002, John Mortimer / 0529962

DESTROYERS

28 ARLEIGH BURKE CLASS (FLIGHTS I and II): GUIDED MISSILE DESTROYERS (AEGIS) (DDGHM)

Name	No	Builders	Laid down	Launched	Commissioned	F/S
ARLEIGH BURKE	DDG 51	Bath Iron Works	6 Dec 1988	16 Sep 1989	4 July 1991	AA
BARRY (ex-John Barry)	DDG 52	Ingalls Shipbuilding	26 Feb 1990	10 May 1991	12 Dec 1992	AA
JOHN PAUL JONES	DDG 53	Bath Iron Works	8 Aug 1990	26 Oct 1991	18 Dec 1993	PA
CURTIS WILBUR	DDG 54	Bath Iron Works	12 Mar 1992	16 May 1992	4 Apr 1994	PA
STOUT	DDG 55	Ingalls Shipbuilding	8 Aug 1991	16 Oct 1992	13 Aug 1994	AA
JOHN S McCAIN	DDG 56	Bath Iron Works	3 Sep 1991	26 Sep 1992	2 July 1994	PA
MITSCHER	DDG 57	Ingalls Shipbuilding	12 Feb 1992	7 May 1993	10 Dec 1994	AA
LABOON	DDG 58	Bath Iron Works	23 Mar 1992	20 Feb 1993	18 Mar 1995	AA
RUSSELL	DDG 59	Ingalls Shipbuilding	24 July 1992	20 Oct 1993	20 May 1995	PA
PAUL HAMILTON	DDG 60	Bath Iron Works	24 Aug 1992	24 July 1993	27 May 1995	PA
RAMAGE	DDG 61	Ingalls Shipbuilding	4 Jan 1993	11 Feb 1994	22 July 1995	AA
FITZGERALD	DDG 62	Bath Iron Works	9 Feb 1993	29 Jan 1994	14 Oct 1995	PA
STETHEM	DDG 63	Ingalls Shipbuilding	11 May 1993	17 June 1994	21 Oct 1995	AA
CARNEY	DDG 64	Bath Iron Works	3 Aug 1993	23 July 1994	13 Apr 1996	AA
BENFOLD	DDG 65	Ingalls Shipbuilding	27 Sep 1993	9 Nov 1994	30 Mar 1996	PA
GONZALEZ	DDG 66	Bath Iron Works	3 Feb 1994	18 Feb 1995	12 Oct 1996	AA
COLE	DDG 67	Ingalls Shipbuilding	28 Feb 1994	10 Feb 1995	8 June 1996	AA
THE SULLIVANS	DDG 68	Bath Iron Works	27 July 1994	12 Aug 1995	19 Apr 1997	AA
MILIUS	DDG 69	Ingalls Shipbuilding	8 Aug 1994	1 Aug 1995	23 Nov 1996	PA
HOPPER	DDG 70	Bath Iron Works	23 Feb 1995	6 Jan 1996	6 Sep 1997	PA
ROSS	DDG 71	Ingalls Shipbuilding	10 Apr 1995	23 Mar 1996	28 June 1997	AA
MAHAN	DDG 72	Bath Iron Works	17 Aug 1995	29 June 1996	14 Feb 1998	AA
DECATUR	DDG 73	Bath Iron Works	11 Jan 1996	10 Nov 1996	29 Aug 1998	PA
McFAUL	DDG 74	Ingalls Shipbuilding	26 Jan 1996	18 Jan 1997	25 Apr 1998	AA
DONALD COOK	DDG 75	Bath Iron Works	9 July 1996	3 May 1997	4 Dec 1998	AA
HIGGINS	DDG 76	Bath Iron Works	14 Nov 1996	4 Oct 1997	24 Apr 1999	PA
O'KANE	DDG 77	Bath Iron Works	5 May 1997	28 Mar 1998	23 Oct 1999	PA
PORTER	DDG 78	Ingalls Shipbuilding	2 Dec 1996	12 Nov 1997	20 Mar 1999	AA

Displacement, tons: 8,315 (DDG 51-71); 8,400 (DDG 72-78)
Dimensions, feet (metres): 504.5 oa; 466 wl × 66.9 × 20.7;
   32.7 (sonar) *(153.8; 142 × 20.4 × 6.3; 9.9)*
Main machinery: 4 GE LM 2500 gas turbines; 105,000 hp
   (78.33 MW) sustained; 2 shafts; cp props
Speed, knots: 32. **Range, n miles:** 4,400 at 20 kt
Complement: 346 (22 officers)

Missiles: SLCM: 56 GDC/Hughes Tomahawk; Tercom aided
   guidance to 1,300 km *(700 n miles)* (TLAM-C and D) or
   1,853 km *(1,000 n miles)* (TLAM-C Block III) at 0.7 Mach;
   warhead 454 kg (TLAM-C) or 347 kg shaped charge (TLAM-C
   Blocks II and III) or submunitions (TLAM-D). TLAM-C has GPS
   back-up to Tercom and a CEP of 10 m.
SSM: 8 McDonnell Douglas Harpoon (2 quad) ❶; active radar
   homing to 130 km *(70 n miles)*; warhead 227 kg.
SAM: GDC Standard SM-2MR Block IV; command/inertial
   guidance; semi-active radar homing to 167 km *(90 n miles)* at
   2 Mach. SM-2ER extended range to 137 km *(74 n miles)* from
   DDG 72 onwards. 2 Martin Marietta Mk 41 (Mod 0 forward,
   Mod 1 aft) Vertical Launch Systems (VLS) for Tomahawk,
   Standard and ASROC VLA ❷; 2 magazines; 29 missiles
   forward, 61 aft. Mod 2 from DDG 59 onwards.
A/S: Loral ASROC VLA; inertial guidance to 1.6-16.6 km *(1-9 n
   miles)*; payload Mk 46 Mod 5 Neartip.
Guns: 1 FMC/UDLP 5 in *(127 mm)*/54 Mk 45 Mod 1 or 2 ❸;
   20 rds/min to 23 km *(12.6 n miles)*; weight of shell 32 kg.
   Modified 5 in *(127 mm)*/62 barrel may be fitted from 2000 to
   take ERGM (extended-range guided munitions); GPS guidance
   to 116.7 km *(63 n miles)*; warhead 72 bomblets; cep 10 m.
   2 General Electric/General Dynamics 20 mm Vulcan Phalanx
   6-barrelled Mk 15 ❹; 3,000 rds/min (4,500 in Block 1)
   combined to 1.5 km. Being fitted with IR detectors for tracking
   small craft. To be replaced by RAM from 2004.
Torpedoes: 6—324 mm Mk 32 Mod 14 (2 triple) tubes ❺. Alliant
   Mk 46 Mod 5; anti-submarine; active/passive homing to
   11 km *(5.9 n miles)* at 40 kt; warhead 44 kg or Alliant/
   Westinghouse Mk 50; active/passive homing to 15 km *(8.1 n
   miles)* at 50 kt; warhead 45 kg shaped charge.
Countermeasures: Decoys: 2 Loral Hycor SRBOC 6-barrelled
   fixed Mk 36 Mod 12 ❻; IR flares and chaff to 4 km *(2.2 n
   miles)*. SLQ-25 Nixie; torpedo decoy. NATO Sea Gnat. SLQ-95
   AEB. SLQ-39 chaff buoy. Nulka being acquired.
ESM/ECM: Raytheon SLQ-32(V)2 ❼ or SLQ-32(V)3/SLY-2 (from
   DDG 72 and being backfitted); radar warning. Sidekick
   modification adds jammer and deception system to (V)2.
   SRS-1 DF (from DDG 72).

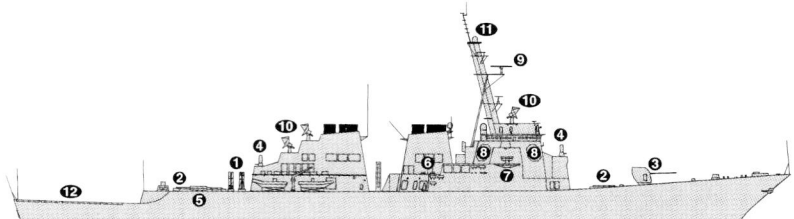

ARLEIGH BURKE   *(Scale 1 : 1,500), Ian Sturton /* 0053331

Combat data systems: CEC being fitted. NTDS Mod 5 with Links
   4A, 11, 14 and 16 (from DDG 72) and being back fitted.
   SATCOM SRR-1, WSC-3 (UHF), USC-38 (EHF). SQQ-28 for
   LAMPS processor datalink. TADIX B Tactical Information
   Exchange System (from DDG 72). Link 22 in due course (see
   Data Systems at front of section).
Weapons control: SWG-3 Mk 37 Tomahawk WCS. SWG-1A
   Harpoon LCS. Aegis multitarget tracking with Mk 99 Mod 3
   MFCS and three Mk 80 illuminators. GWS 34 Mod 0 GFCS
   (includes Mk 160 Mod 4 computing system and Kollmorgen
   Mk 46 Mod 0/1 optronic sight). Singer Librascope Mk 116
   Mod 7 FCS for ASW.
Radars: Air search/fire control: RCA SPY-1D phased arrays ❽;
   3D; E/F-band.
Surface search: Norden/DRS SPS-67(V)3 ❾; G-band.
Navigation: Raytheon SPS-64(V)9; I-band.
Fire control: Three Raytheon/RCA SPG-62 ❿; I/J-band.
Tacan: URN 25 ⓫. IFF Mk XII AIMS UPX-29.
Sonars: Gould/Raytheon/GE SQQ-89(V)6; combines SQS-53C;
   bow-mounted; active search and attack with SQR-19B passive
   towed array (TACTAS) low frequency.

Helicopters: Platform and facilities to fuel and rearm LAMPS III
   SH-60B/F helicopters ⓬. UAV in due course.

Programmes: First ship authorised in FY85, last pair in FY94. The
   first 21 are Flight I and the next seven are Flight II.
Modernisation: A half-life upgrade programme is planned to
   start in FY06.
Structure: The ship, except for the aluminium mast, is
   constructed of steel. 70 tons of armour provided to protect

vital spaces. This is the first class of US Navy warship
designed with a 'collective protection system for defense
against the fallout associated with NBC warfare'. The ship's
crew are protected by double air-locked hatches, fewer
accesses to the weatherdecks and positive pressurisation of
the interior of the ship to keep out contaminants. All incoming
air is filtered and more reliance placed on recirculating air
inside the ship. All accommodation compartments have
sprinkler systems. Stealth technology includes angled
surfaces and rounded edges to reduce radar signature and IR
signature suppression plus Prairie Masker hull/blade rate
suppression. The Ops room is below the waterline and
electronics are EMP hardened. The original upright mast
design has been changed to increase separation between
electronic systems and the forward funnel. Differences in
Flight II starting with DDG 72 include Link 16, SLQ-32(V)3 EW
suite, extended-range SAM missiles and improved tactical
information exchange systems. The topmast is vertical to take
the SRS-1. There is also an increase in displacement caused by
using more space to carry fuel.
Operational: Two of the class are based at Yokosuka in Japan.
   Repairs to *Cole*, damaged by a terrorist attack at Aden on
   12 October 2000, began in January 2001 at Ingalls and
   completed on 19 April 2002 when she returned to the fleet.
   Curtis Wilbur began missile-defence patrols in the Sea of
   Japan in October 2004.

UPDATED

PORTER   *10/2004 *, H M Steele /* 1043699

JOHN PAUL JONES

10/2004 *, Frank Findler / 1043682

ARLEIGH BURKE

11/2004 *, H M Steele / 1043698

MITSCHER

6/2004 *, US Navy / 1043643

19 + 15 ARLEIGH BURKE CLASS (FLIGHT IIA): GUIDED MISSILE DESTROYERS (AEGIS) (DDGHM)

Name	No	Builders	Laid down	Launched	Commissioned	F/S
OSCAR AUSTIN	DDG 79	Bath Iron Works	9 Oct 1997	7 Nov 1998	19 Aug 2000	AA
ROOSEVELT	DDG 80	Ingalls Shipbuilding	15 Dec 1997	10 Jan 1999	14 Oct 2000	AA
WINSTON S CHURCHILL	DDG 81	Bath Iron Works	7 May 1998	17 Apr 1999	10 Mar 2001	AA
LASSEN	DDG 82	Ingalls Shipbuilding	24 Aug 1998	16 Oct 1999	21 Apr 2001	PA
HOWARD	DDG 83	Bath Iron Works	9 Dec 1998	20 Nov 1999	20 Oct 2001	PA
BULKELEY	DDG 84	Ingalls Shipbuilding	10 May 1999	21 June 2000	8 Dec 2001	AA
McCAMPBELL	DDG 85	Bath Iron Works	15 July 1999	2 July 2000	17 Aug 2002	PA
SHOUP	DDG 86	Ingalls Shipbuilding	13 Dec 1999	22 Nov 2000	22 June 2002	PA
MASON	DDG 87	Bath Iron Works	20 Jan 2000	23 June 2001	12 Apr 2003	AA
PREBLE	DDG 88	Ingalls Shipbuilding	22 June 2000	1 June 2001	9 Nov 2002	PA
MUSTIN	DDG 89	Ingalls Shipbuilding	15 Jan 2001	12 Dec 2001	26 July 2003	PA
CHAFFEE	DDG 90	Bath Iron Works	12 Apr 2001	2 Nov 2002	20 Sep 2003	PA
PINCKNEY	DDG 91	Ingalls Shipbuilding	16 July 2001	29 June 2002	29 May 2004	PA
MOMSEN	DDG 92	Bath Iron Works	15 Oct 2001	9 Aug 2003	28 Aug 2004	PA
CHUNG-HOON	DDG 93	Ingalls, Shipbuilding	14 Jan 2002	11 Jan 2003	18 Sep 2004	PA
NITZE	DDG 94	Bath Iron Works	11 Aug 2002	17 Apr 2004	5 Mar 2005	AA
JAMES E WILLIAMS	DDG 95	Ingalls Shipbuilding	15 July 2002	28 June 2003	11 Dec 2004	AA
BAINBRIDGE	DDG 96	Bath Iron Works	16 Mar 2003	13 Nov 2004	May 2005	Bldg/PA
HALSEY	DDG 97	Ingalls Shipbuilding	13 Jan 2003	17 Jan 2004	Mar 2005	Bldg/PA
FORREST SHERMAN	DDG 98	Ingalls Shipbuilding	21 July 2003	2 Oct 2004	Aug 2005	Bldg/AA
FARRAGUT	DDG 99	Bath Iron Works	23 Nov 2003	23 July 2005	Jan 2006	Bldg
KIDD	DDG 100	Ingalls Shipbuilding	12 Jan 2004	22 Jan 2005	Feb 2006	Bldg
GRIDLEY	DDG 101	Bath Iron Works	30 July 2004	Dec 2005	Sep 2006	Bldg
SAMPSON	DDG 102	Bath Iron Works	Mar 2005	July 2006	2007	Bldg
TRUXTUN	DDG 103	Northrop Grumman Ship Systems	Apr 2005	Mar 2006	2007	Bldg
STERETT	DDG 104	Bath Iron Works	Dec 2005	Apr 2007	2008	Ord
DEWEY	DDG 105	Northrop Grumman Ship Systems	Mar 2006	Feb 2007	2008	Ord
—	DDG 106	Bath Iron Works	July 2006	Nov 2007	2009	Ord
—	DDG 107	Northrop Grumman Ship Systems	Mar 2007	Feb 2008	2009	Ord
—	DDG 108	Bath Iron Works	Feb 2007	Jun 2008	2009	Ord
—	DDG 109	Bath Iron Works	Oct 2007	Feb 2009	2010	Ord
—	DDG 110	Northrop Grumman Ship Systems	Mar 2007	Feb 2009	2010	Ord
—	DDG 111	Bath Iron Works	May 2008	Sep 2009	2011	Ord
—	DDG 112	Bath Iron Works	Dec 2008	May 2010	2011	Ord

Displacement, tons: 9,200 full load
Dimensions, feet (metres): 509.5 oa; 471 wl × 66.9 × 20.7; 32.7 (sonar) *(155.3; 143.6 × 20.4 × 6.3; 9.9)*
Main machinery: 4 GE LM 2500-30 gas turbines; 105,000 hp *(78.33 MW)* sustained; 2 shafts; cp props
Speed, knots: 31. **Range, n miles:** 4,400 at 20 kt
Complement: 344 plus 22 spare

Missiles: SLCM: GCD/Hughes Tomahawk; Tercom aided guidance to 1,300 km *(700 n miles)* (TLAM-C and D) or 1,853 km *(1,000 n miles)* (TLAM-C Block III) at 0.7 Mach; warhead 454 kg (TLAM-C) or 347 kg shaped charge (TLAM-C Blocks II and III) or submunitions (TLAM-D). TLAM-C has GPS back-up to Tercom and a CEP of 10 m. Tactical Tomahawk from 2004.
SAM: GDC Standard SM-2MR Block IV; command/inertial guidance; semi-active radar homing to 167 km *(90 n miles)*. 2 Lockheed Martin Mk 41 Vertical Launch Systems (VLS) for Tomahawk, Standard and ASROC VLS ❶; 2 magazines; 32 missile tubes forward, 64 aft. 32 Raytheon ESSM (4 quad forward, 4 quad aft) (DDG 85 onwards); semi-active radar homing to 18.5 km *(10 n miles)*. ESSM retro-fitted to DDG 82 (2004), DDGs 80, 83 and 84 (FY05) and DDGs 79 and 81 (FY06).
A/S: Loral ASROC VLA; inertial guidance to 1.6-16.6 km *(1-9 n miles)*; payload Mk 46 Mod 5 Neartip.
Guns: 1 United Defense 5 in *(127 mm)*/54 Mk 45 Mod 2 (DDG 79-80) ❷; 20 rds/min to 23 km *(12.6 n miles)*; weight of shell 32 kg.
United Defense 5 in *(127 mm)*/62 (DDG 81 onwards); 20 or 10 (ERGM) rds/min; GPS guidance to 116.7 km *(63 n miles)*; warhead 72 bomblets; cep 10 m.
2 Hughes 20 mm Vulcan Phalanx 6-barrelled Mk 15 ❸; 4,500 rds/min combined in 1.5 km. Fitted with IR detectors for tracking small craft. To be replaced by ESSM from DDG 85 forward during construction.
Torpedoes: 6—324 mm Mk 32 Mod 14 (2 triple) tubes ❹. Alliant Mk 46 Mod 5; anti-submarine; active/passive homing to 11 km *(5.9 n miles)* at 40 kt; warhead 44 kg or Alliant/ Westinghouse Mk 50; active/passive homing to 15 km *(8.1 n miles)* at 50 kt; warhead 45 kg shaped charge.
Countermeasures: Decoys: 2 Loral Hycor SRBOC 6-barrelled fixed Mk 36 Mod 12 ❺; IR flares and chaff to 4 km

(2.2 n miles). SLQ-25 Nixie; torpedo decoy. NATO Sea Gnat. SLQ-95 AEB. SLQ-39 chaff buoy.
ESM/ECM: Raytheon SLQ-32(V)3/SLY-2 ❻; intercept and jammer.
Combat data systems: TADIX-B and TADIL-J. CEC. Links 4, 11 and 16. (See Data Systems at front of section.) Link 22 in due course. CDS upgrade for DDG 91.
Weapons control: SWG-3 Mk 37 Tomahawk WCS. Aegis multitarget tracking with Mk 99 Mod 3 MFCS and three Mk 80 illuminators. GWS 34 GFCS (includes Mk 160 Mod 8 computing system and Kollmorgen Mk 46 optronic sight). Singer Librascope Mk 116 FCS for ASW.
Radars: Air search/fire control: RCA SPY-1D phased arrays ❼; 3D; E/F-band.
Surface search: DRS SPS-67(V)3 ❽; G-band.
Navigation: Raytheon SPS-64(V)9; I-band.
Fire control: Three Raytheon/SPG-62 ❾; I/J-band.
Tacan: URN 25 ❿. IFF Mk XII AIMS UPX-29.
Sonars: Lockheed Martin SQQ-89(V)10; underwater combat system with SQS-53C; bow-mounted; active search and attack. Kingfisher; mine detection; active; high frequency. Remote Minehunting System (DDG 92).

Helicopters: 2 LAMPS III SH-60B/F helicopters ⓫.

Programmes: DDG 79 was authorised in the FY94 budget. Funding for DDG 80-82 provided in FY95 and DDG 83-84 in FY96 plus partial funding for a third. Balance for DDG 85 plus

DDG 86-88 in FY97 and DDG 89-101 in FY98. On 6 March 1998, multi-year contract for six ships and one option (DDG 89) awarded to Ingalls Shipbuilding and contract for six ships awarded to Bath Iron Works. On 1 August 2002, contract awarded to Bath Iron Works for the construction of DDG 102 and on 13 September 2002 a fixed-price multi-year contract awarded to Bath Iron Works (DDGs 104, 106, 108, 109, 111 and 112) and Northrop Grumman Ship Systems (DDGs 103, 105, 107, 110) for the construction of ten ships. Further orders are not expected.
Modernisation: Future planned upgrades include Aegis Baseline 7.1, SPY 1D(V) radar, CEC, Nulka decoys, Remote Minehunting System (to be installed in DDG 92), Integrated Bridge System and Tactical Tomahawk.
Structure: The upgrade from Flight II includes two hangars for embarked helicopters and an extended transom to increase the size of a dual RAST fitted flight deck at the expense of SQR-19 TACTAS. Vertical launchers are increased at each end by three cells. Other changes include the Kingfisher minehunting sonar, a reconfiguration of the SPY-1D arrays and the inclusion of a Track Initiation Processor in the Aegis radar system. Use of fibre optic technology should reduce weight and improve reliability.
Operational: The helicopter carries Penguin and Hellfire missiles. ESSM fired from DDG 86 on 24 July 2002, the first to be fired from a USN ship.

UPDATED

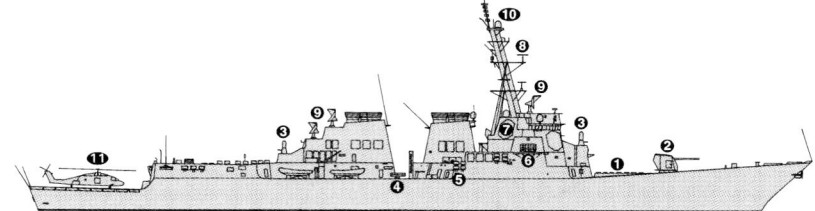

ROOSEVELT　　　　　　　　　　　　　　　　(Scale 1 : 1,500), Ian Sturton / 0106761

HOWARD　　　　　　　　　　　　　　　　　7/2004*, Michael Nitz / 1043642

BULKELEY

4/2004*, US Navy / 1043644

MOMSEN

10/2004*, Frank Findler / 1043683

OSCAR AUSTIN

6/2004*, Harald Carstens / 1043684

3 SPRUANCE CLASS: DESTROYERS (DDGHM)

Name	No	Builders	Laid down	Launched	Commissioned	F/S
SPRUANCE	DD 963	Ingalls Shipbuilding	17 Nov 1972	10 Nov 1973	20 Sep 1975	AA
CUSHING	DD 985	Ingalls Shipbuilding	27 Dec 1976	17 June 1978	20 Oct 1979	PA
O'BANNON	DD 987	Ingalls Shipbuilding	21 Feb 1977	25 Sep 1978	15 Dec 1979	AA

Displacement, tons: 5,770 light; 8,040 full load
Dimensions, feet (metres): 563.2 × 55.1 × 19; 29 (sonar)
(171.7 × 16.8 × 5.8; 8.8)
Main machinery: 4 GE LM 2500 gas turbines; 86,000 hp
(64.16 MW) sustained; 2 shafts; cp props
Speed, knots: 33. **Range, n miles:** 6,000 at 20 kt
Complement: 319-339 (20 officers)

Missiles: SLCM: GDC Tomahawk; Tercom aided guidance to
1,300 km *(700 n miles)* (TLAM-C and D) or 1,853 km *(1,000 n miles)* (TLAM-C Block III) at 0.7 Mach; warhead 454 kg
(TLAM-C) or 347 kg shaped charge (TLAM-C Blocks II and III)
or submunitions (TLAM-D). TLAM-C has GPS back-up to
Tercom and a CEP of 10 m.
Mk 41 Mod 0 VLS ❶ with one 61 missile magazine combining
45 Tomahawk and ASROC in some.
8 McDonnell Douglas Harpoon (2 quad) ❷; active radar
homing to 130 km *(70 n miles)* at 0.9 Mach; warhead 227 kg.
SAM: Raytheon GMLS Mk 29 octuple launcher ❸; 24 Sea
Sparrow; semi-active radar homing to 14.6 km *(8 n miles)* at
2.5 Mach; warhead 39 kg.
GDC RAM quadruple launcher ❹; passive IR/anti-radiation
homing to 9.6 km *(5.2 n miles)* at 2 Mach; warhead 9.1 kg.
Being fitted in 12 of the class.
A/S: Loral ASROC VLA can be carried; inertial guidance to
16.6 km *(9 n miles)*; payload Mk 46 Mod 5 Neartip.
Guns: 2 FMC 5 in *(127 mm)*/54 Mk 45 Mod 0/1 ❺; 20 rds/min
to 23 km *(12.6 n miles)* anti-surface; 15 km *(8.2 n miles)* anti-
aircraft; weight of shell 32 kg.
2 General Electric/General Dynamics 20 mm/76 6-barrelled
Mk 15 Vulcan Phalanx ❻; 3,000 rds/min (4,500 in Batch 1)
combined to 1.5 km.
4—12.7 mm MGs.
Torpedoes: 6—324 mm Mk 32 (2 triple) tubes ❼. 14 Honeywell
Mk 46; anti-submarine; active/passive homing to 11 km *(5.9 n miles)* at 40 kt; warhead 44 kg or Alliant/Westinghouse Mk
50; active/passive to 15 km *(8.1 n miles)* at 50 kt; warhead
45 kg shaped charge. The tubes are inside the superstructure
to facilitate maintenance and reloading. Torpedoes are fired
through side ports.
Countermeasures: Decoys: 4 Loral Hycor SRBOC 6-barrelled
fixed Mk 36 ❽; IR flares and chaff to 4 km *(2.2 n miles)*.
SLQ-39 chaff buoy.
SLQ-25 Nixie; torpedo decoy. Prairie/Masker hull/blade rate
noise suppression system. Remote Minehunting System
(RMS) can be carried; endurance 24 hours at 12 kt.
ESM/ECM: SLQ-32(V)2 ❾; radar warning. Sidekick modification
adds jammer and deception system. WLR-1 (in some).
Combat data systems: NTDS with Links 11 and 14. SATCOMS
❿ SRR-1, WSC-3 (UHF), USC-38 (EHF) (in some). SQQ-28 for
LAMPS datalink. SYQ-17 RAIDS (see *Modernisation*) being
fitted.
Weapons control: SWG-3 Tomahawk WCS. SWG-1A Harpoon
LCS. Mk 116 Mod 7 FCS ASW. Mk 86 Mod 3 GFCS. Mk 91
MFCS. SRQ-4 LAMPS III. SAR-8 IR director (DD 965).
Radars: Air search: Lockheed SPS-40B/C/D (not in DD 997) ⓫;
B-band; range 320 km *(175 n miles)*.
Raytheon SPS-49V (DD 997); C/D-band; range 457 km
(250 n miles).
Hughes Mk 23 TAS ⓬; D-band.
Surface search: ISC Cardion SPS-55 ⓭; I/J-band.
Navigation: Raytheon SPS-64(V)9; I-band.

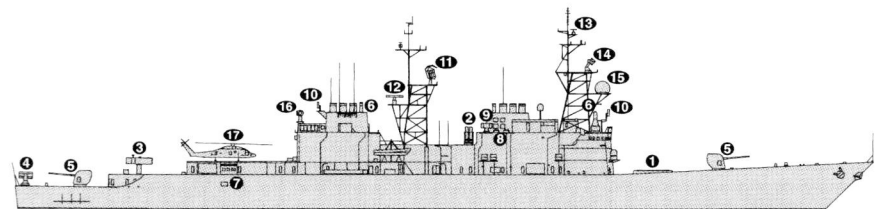

SPRUANCE CLASS
(Scale 1 : 1,500), Ian Sturton / 0016370

SPRUANCE
9/2002, H M Steele / 0530080

SPRUANCE
9/2003, H M Steele / 0572803

Fire control: Lockheed SPG-60 ⓮; I/J-band.
Lockheed SPQ-9A ⓯ or 9B (DD 972); I/J-band.
Raytheon Mk 95 ⓰; I/J-band (for SAM).
Tacan: URN 20 or URN 25 (D 997). IFF Mk XII AIMS UPX-25.
Sonars: SQQ-89(V)6 including GE/Hughes SQS-53B/C; bow-
mounted; active search and attack; medium frequency; and
Gould SQR-19 (TACTAS); passive towed array.

Helicopters: 2 SH-60B LAMPS III ⓱ or SH-2G Seasprite.

Programmes: Funds approved between FY70 and FY78.
Modernisation: Beginning with FY86 overhauls, major
improvements have been made. These included the
installation of VLS, upgrading of EW to SLQ-32V(2) plus
sidekick; LAMPS III and the recovery, assist, secure and
traverse system (RAST), the Halon 1301 firefighting system
and anti-missile and target acquisition systems. All are
capable of launching Standard SM-2MR for control by Aegis
fitted vessels. RAIDS (Rapid Anti-ship missile Integrated
Defence System) evaluated in DD 963, and installed in all of
the class in 1996-97.
Structure: Extensive use of the modular concept was used to
facilitate construction and block modernisation. There is a
high level of automation. These were the first large US
warships to employ gas-turbine propulsion and advanced self-
noise reduction features. Kevlar internal coating in all vital
spaces.
Operational: Last three ships of the class to decommission in
FY05. *UPDATED*

0 + (5) FUTURE DESTROYER (DDGH)

Name	No	Builders	Laid down	Launched	Commissioned
—	—	Northrop Grumman Ship Systems	Oct 2010	June 2012	July 2013

Displacement, tons: 14,264 full load
Dimensions, feet (metres): 600.0 × 79.0 × 28.0
(182.8 × 24.1 × 8.5)
Main machinery: Integrated electric propulsion; 2 large gas
turbines; 2 permanent magnet motors *(72 MW)*; 2 auxiliary
gas turbines *(8 MW)*
Speed, knots: 30+. **Range, n miles:** To be announced.
Complement: 114 + 36 aircrew

Missiles: 80 VLS cells.
SLCM: Tactical Tomahawk.
SAM: Standard SM-2 and Evolved Sea Sparrow.
A/S: Vertical launched ASROC.
Guns: 2—155 mm advanced gun systems capable of firing Long
Range Land Attack Projectiles (LRLAP) at ranges up to
100 miles. 2—57 mm.
Countermeasures: To be announced.
Combat data systems: To be announced.
Weapons control: To be announced.
Radars: Air/surface search: Dual band Multi-Function Radar and
Volume Search Radar.
Navigation: To be announced.
Sonars: Dual-frequency hull mounted and towed arrays.

Helicopters: 1 MH-60R and 3 VTUAVs, or 2 MH-60R.

Programmes: DDX programme instituted on November 2001.
Principal role of DDX is land-attack but the ship is multimission
for littoral combat. On 29 April 2002, Northrop Grumman
awarded the contract to lead the system design, engineering
prototype development and testing of DDX. Raytheon
Systems is combat systems integrator and other team
members include General Dynamics, Lockheed Martin and
United Defense. General Dynamics Bath Iron Works is also
performing design and test activities to provide competition in
ship construction. The contract for construction of the first of
class is expected in 2007 and fabrication is planned to start in
mid-2009.
Structure: Features of the ship include a wave-piercing
'tumblehome' hull designed to reduce radar signatures and

DDX
12/2003, US Navy / 0572812

provide stability in high sea states. Hull structure and
peripheral missile cells spread impacts outward to increase
survivability and reduce single-hit ship loss. Integrated
deckhouse and composite superstructure encloses masts,
sensors and antennas, bridge and exhaust silos. There are
shields for 155 mm guns. Engineering Development Models
(EDMs) for key systems include an Integrated Power System
(IPS), AGS, integrated undersea warfare suite and an open
architecture, total ship computing environment. IPS reduces
the number of prime movers from seven to four and is
designed to create sufficient reserve energy to power directed
energy weapons in the future.
Operational: The first of class is expected to become operational
in 2013 and a class of five is planned. *UPDATED*

FRIGATES

0 + 1 (1) LITTORAL COMBAT SHIP (LOCKHEED MARTIN) FLIGHT 0

Name	No	Builders	Laid down	Launched	Commissioned
FREEDOM	—	Marinette Marine, Wisconsin	2 June 2005	2007	2007
LCS 3	—		2006	2008	2008

Displacement, tons: 2,840 full load
Dimensions, feet (metres): 379.0 × 43.0 × 12.1
 (115.5 × 13.1 × 3.7)
Main machinery: CODAG: 2 Rolls Royce MT-30 gas turbines;
 2 Isotta Fraschini diesels; 4 Rolls Royce waterjets
Speed, knots: 45. **Range, n miles:** 3,500 at 18 kt
Complement: 50

Missiles: Raytheon RAM.
Guns: 1 United Defence 57 mm/70 Mk 2; 220 rds/min to 17 km
 (9 n miles); weight of shell 2.4 kg. 4—12.7 mm MGs.
Countermeasures: Decoy launching system: ESM/ECM.
Combat data systems: COMBATSS-21.
Weapons control: To be announced.
Radars: Air/surface search: EADS TRS-3D; C-band.
Navigation: To be announced.
Fire control: To be announced.
Sonars: To be announced.

Helicopters: 2 H-60 helicopters and 3 Firescout VTUAVs.

Programmes: Three winning LCS designs selected on 17 July 2003 to proceed to the Flight 0 preliminary design stage. Thereafter, General Dynamics and Lockheed Martin won competition on 27 May 2004 to build first four Flight 0 ships. Contract to build first ship awarded to Lockheed Martin on 15 December 2004 for delivery in 2007. First steel cut February 2005. This is to be followed by an order for a second ship for delivery in 2008. Following evaluation of both (General Dynamics and Lockheed Martin) Flight 0 variants, construction of three Flight I ships to begin in 2008 and of four further ships in 2009. Seven mission modules (three mine warfare; two ASW and two ASUW) are budgeted from FY05-07. Up to 60 ships of the class may ultimately be required.

LCS (Lockheed Martin)

4/2004*, Lockheed Martin / 0566824

Structure: Semi-planing steel monohull design. Steel hull and aluminium superstructure. The design incorporates a large reconfigurable space, a flight deck with integrated helicopter launch, recovery and handling system and the capability to launch and recover boats from both the stern and side.
Operational: Concept of operations for LCS includes deployment of two or three-ship team to operate near shore in support of surface strike groups. Role in homeland defense also likely. Principal capabilities to include shallow-water ASW, mine countermeasures and defence against attacking small boats.
NEW ENTRY

0 + (2) LITTORAL COMBAT SHIP (GENERAL DYNAMICS) FLIGHT 0

Name	No	Builders	Laid down	Launched	Commissioned
LCS 2	—		2005	2007	2008
LCS 4	—		2007	2009	2009

Displacement, tons: 2,650 full load
Dimensions, feet (metres): 419.3 × 93.2 × 14.8
 (127.8 × 28.4 × 4.5)
Main machinery: CODAG: 2 gas turbines, 2 diesels; 2 Azi thrusters
Speed, knots: 42. **Range, n miles:** 4,300 at 18 kt
Complement: 45

Missiles: Raytheon RAM.
Guns: 1 United Defence 57 mm/70 Mk 2; 220 rds/min to 17 km
 (9 n miles); weight of shell 2.4 kg. 4—12.7mm MGs.
Countermeasures: Decoys: ESM/ECM.
Combat data systems: Northrop Grumman Electronic Systems Integrated Combat Management System (ICMS).
Weapons control: To be announced.
Radars: Air/surface search: Ericson Sea Giraffe; G/H-band.

Navigation: To be announced.
Fire control: To be announced.
Sonars: To be announced.

Helicopters: 2 H-60 helicopters; multiple UAVs/VTUAVs; large flight deck area can accommodate CH-53.

Programmes: Three winning LCS designs selected on 17 July 2003 to proceed to the Flight 0 preliminary design stage. Thereafter, General Dynamics and Lockheed Martin won competition on 27 May 2004 to build first four Flight 0 ships. Contract to build first ship awarded to Lockheed Martin on 15 December 2004 for delivery in 2007. Contract for first General Dynamics ship expected in 2005 for delivery 2007. This is to be followed by a second ship for delivery in 2008. Following evaluation of both (General Dynamics and

Lockheed Martin) Flight 0 variants, construction of three Flight I ships to begin in 2008 and of four further ships in 2009. Seven mission modules (three mine warfare; two ASW and two ASUW) are budgeted from FY05-07. Up to 60 ships of the class may ultimately be required.
Structure: Trimaran hullform based on fast commercial ferry design for Fred Olsen Line. Aluminium construction. Large flight deck capable of operating Sea Stallion heavy-lift helicopter. Stern launch of boats. Side-ramp Ro-Ro capability.
Operational: Concept of operations for LCS includes deployment of two or three-ship team to operate near shore in support of surface strike groups. Role in homeland defense also likely. Principal capabilities to include shallow-water ASW, mine countermeasures and defence against attacking small boats.
NEW ENTRY

LSS (General Dynamics)

1/2005*, General Dynamics / 1043677

30 OLIVER HAZARD PERRY CLASS: GUIDED MISSILE FRIGATES (FFGHM)

Name	No	Builders	Laid down	Launched	Commissioned	F/S
McINERNEY	FFG 8	Bath Iron Works	7 Nov 1977	4 Nov 1978	15 Dec 1979	AA
BOONE	FFG 28	Todd Shipyards, Seattle	27 Mar 1979	16 Jan 1980	15 May 1982	NRF
STEPHEN W GROVES	FFG 29	Bath Iron Works	16 Sep 1980	4 Apr 1981	17 Apr 1982	NRF
JOHN L HALL	FFG 32	Bath Iron Works	5 Jan 1981	24 July 1981	26 June 1982	AA
JARRETT	FFG 33	Todd Shipyards, San Pedro	11 Feb 1981	17 Oct 1981	2 July 1983	PA
UNDERWOOD	FFG 36	Bath Iron Works	3 Aug 1981	6 Feb 1982	29 Jan 1983	AA
CROMMELIN	FFG 37	Todd Shipyards, Seattle	30 May 1980	1 July 1981	18 June 1983	NRF
CURTS	FFG 38	Todd Shipyards, San Pedro	1 July 1981	6 Mar 1982	8 Oct 1983	NRF
DOYLE	FFG 39	Bath Iron Works	16 Nov 1981	22 May 1982	21 May 1983	NRF
HALYBURTON	FFG 40	Todd Shipyards, Seattle	26 Sep 1980	15 Oct 1981	7 Jan 1984	AA
McCLUSKY	FFG 41	Todd Shipyards, San Pedro	21 Oct 1981	18 Sep 1982	10 Dec 1983	NRF
KLAKRING	FFG 42	Bath Iron Works	19 Feb 1982	18 Sep 1982	20 Aug 1983	NRF
THACH	FFG 43	Todd Shipyards, San Pedro	10 Mar 1982	18 Dec 1982	17 Mar 1984	PA
De WERT	FFG 45	Bath Iron Works	14 June 1982	18 Dec 1982	19 Nov 1983	AA
RENTZ	FFG 46	Todd Shipyards, San Pedro	18 Sep 1982	16 July 1983	30 June 1984	PA
NICHOLAS	FFG 47	Bath Iron Works	27 Sep 1982	23 Apr 1983	10 Mar 1984	AA
VANDEGRIFT	FFG 48	Todd Shipyards, Seattle	13 Oct 1981	15 Oct 1982	24 Nov 1984	PA
ROBERT G BRADLEY	FFG 49	Bath Iron Works	28 Dec 1982	13 Aug 1983	11 Aug 1984	AA
TAYLOR	FFG 50	Bath Iron Works	5 May 1983	5 Nov 1983	1 Dec 1984	AA
GARY	FFG 51	Todd Shipyards, San Pedro	18 Dec 1982	19 Nov 1983	17 Nov 1984	PA
CARR	FFG 52	Todd Shipyards, Seattle	26 Mar 1982	26 Feb 1983	27 July 1985	AA
HAWES	FFG 53	Bath Iron Works	22 Aug 1983	18 Feb 1984	9 Feb 1985	AA
FORD	FFG 54	Todd Shipyards, San Pedro	16 July 1983	23 June 1984	29 June 1985	PA
ELROD	FFG 55	Bath Iron Works	21 Nov 1983	12 May 1984	6 June 1985	AA
SIMPSON	FFG 56	Bath Iron Works	27 Feb 1984	21 Aug 1984	9 Nov 1985	NRF
REUBEN JAMES	FFG 57	Todd Shipyards, San Pedro	19 Nov 1983	8 Feb 1985	22 Mar 1986	PA
SAMUEL B ROBERTS	FFG 58	Bath Iron Works	21 May 1984	8 Dec 1984	12 Apr 1986	AA
KAUFFMAN	FFG 59	Bath Iron Works	8 Apr 1985	29 Mar 1986	21 Feb 1987	AA
RODNEY M DAVIS	FFG 60	Todd Shipyards, San Pedro	8 Feb 1985	11 Jan 1986	9 May 1987	NRF
INGRAHAM	FFG 61	Todd Shipyards, San Pedro	30 Mar 1987	25 June 1988	5 Aug 1989	PA

Displacement, tons: 2,750 light; 3,638 (FFG 33); 4,100 full load

Dimensions, feet (metres): 445 (FFG 33); 453 × 45 × 14.8; 24.5 (sonar)
(135.6; 138.1 × 13.7 × 4.5; 7.5)

Main machinery: 2 GE LM 2500 gas turbines; 41,000 hp *(30.59 MW)* sustained; 1 shaft; cp prop
2 auxiliary retractable props; 650 hp *(484 kW)*

Speed, knots: 29. **Range, n miles:** 4,500 at 20 kt

Complement: 200 (15 officers) including 19 aircrew

Guns: 1 OTO Melara 3 in *(76 mm)*/62 Mk 75 ❶; 85 rds/min to 16 km *(8.7 n miles)* anti-surface; 12 km *(6.6 n miles)* anti-aircraft; weight of shell 6 kg.
1 General Electric/General Dynamics 20 mm/76 6-barrelled Mk 15 Vulcan Phalanx ❷; 3,000 rds/min (4,500 in Block 1) combined to 1.5 km. Block 1B with anti-surface capability.
2 Boeing 25 mm Mk 38 guns can be fitted amidships. 4—12.7 mm MGs.

Torpedoes: 6—324 mm Mk 32 (2 triple) tubes ❸. 24 Honeywell Mk 46 Mod 5; anti-submarine; active/passive homing to 11 km *(5.9 n miles)* at 40 kt; warhead 44 kg or Alliant/Westinghouse Mk 50; active/passive homing to 15 km *(8.1 n miles)* at 50 kt; warhead 45 kg shaped charge.

Countermeasures: Decoys: 2 Loral Hycor SRBOC 6-barrelled fixed Mk 36 ❹; IR flares and chaff to 4 km *(2.2 n miles)*. Mk 34 (modified Mk 36) launcher fitted in FFGs 36, 37, 41, 43, 48, 50-55, 57-59; for Mk 234 Nulka decoys.
T-Mk 6 Fanfare/SLQ-25 Nixie; torpedo decoy.

ESM/ECM: SLQ-32(V)2 ❺; radar warning. Sidekick modification adds jammer and deception system.

Combat data systems: NTDS with Link 11 and 14. Link 14 only (NRF ships). SATCOM ❻ SRR-1, WSC-3 (UHF). SQQ-28 for LAMPS III datalink.

Weapons control: Mk 92 (Mod 4 or Mod 6 (FFG 61 and during modernisation in 11 others of the class)), WCS with CAS (Combined Antenna System). The Mk 92 is the US version of the Signaal WM28 system. SYS 2(V)2 IADT (FFG 61 and in 11 others of the class – see *Modernisation*). SRQ-4 for LAMPS III, SKR-4A for LAMPS I.

Radars: Air search: Raytheon SPS-49(V)4 or 5 (FFG 61 and during modernisation of others) ❼; C/D-band; range 457 km *(250 n miles)*.

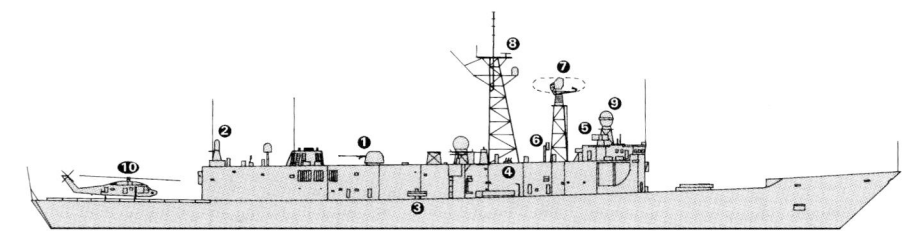

RENTZ   *(Scale 1 : 1,200), Ian Sturton* / 0572737

Surface search: ISC Cardion SPS-55 ❽; I-band.

Fire control: Sperry Mk 92 (Signaal WM28) ❾; I/J-band.

Navigation: Furuno; I-band.

Tacan: URN 25. IFF Mk XII AIMS UPX-29.

Sonars: SQQ 89(V)2 (Raytheon SQS 56 and Gould SQR 19); hull-mounted active search and attack; medium frequency and passive towed array; very low frequency.

Helicopters: 2 SH-60B LAMPS III ❿ in Flight III/IV and certified ships.

Programmes: The lead ship was authorised in FY73.

Modernisation: To accommodate the helicopter landing system (RAST), the overall length of the ship was increased by 8 ft *(2.4 m)* by increasing the angle of the ship's transom, between the waterline and the fantail, from virtually straight up to a 45° angle outwards. LAMPS III support facilities and RAST were fitted in all ships authorised from FFG 36 onwards, during construction and have been backfitted to all. The remainder can operate this aircraft without landing facilities. FFG 61 has much improved Combat Data and Fire-Control equipment which has been retrofitted in FFG 36, 47, 48, 50-55, 57 and 59. SQS-56 is modified for mine detection. Block 1B Phalanx fitted first in FFG 36 in October 1999. This has a capability against high speed surface craft at short range. Engineering and platform improvements (including new diesel generators, replacement of evaporators with reverse osmosis units and

new davit) completed in FFGs 47, 52, 53, 57, 59 and 61. Whole class to be similarly upgraded. Programme to install Nulka decoy system in progress.

Structure: The original single hangar has been changed to two adjacent hangars. Provided with 19 mm Kevlar armour protection over vital spaces. 25 mm guns can be fitted for some operational deployments.

Operational: Ships of this class were the first Navy experience in implementing a design-to-cost acquisition concept. Many of their limitations were manifest during the intense fires which resulted from *Stark* (FFG 31) being struck by two Exocet missiles in the Persian Gulf 17 May 1987. Since then there have been many improvements in firefighting and damage control doctrine and procedures and equipment to deal with residual missile propellant-induced fires. On 14 April 1988, *Samuel B Roberts* (FFG 58), was mined in the Gulf. Penguin SSM can be carried in some LAMPS III helicopters. Two of the class are based at Yokosuka. Nine ships are assigned to the Combatant Naval Reserve Force. SAM and SSM systems removed by the end of FY04.

Sales: Australia bought four of the class and has built two more. Spain has six and Taiwan eight. Transfers include six to Turkey plus one for spares, four to Egypt, one to Bahrain and one to Poland. In 2002, one sold to Turkey and one transferred to Poland. FFG 15 to Turkey in April 2003. Further transfers are likely.

UPDATED

DOYLE   *8/2003, Derek Fox* / 0572778

CROMMELIN
5/2004*, US Navy / 1043650

FORD
7/2004*, Michael Nitz / 1043651

KLAKRING
6/2004*, Camil Busquets i Vilanova / 1043649

SHIPBORNE AIRCRAFT (FRONT LINE)

Notes: (1) Numbers given are for 1 January 2005.
(2) **Joint Strike Fighter:** The Joint Strike Fighter programme is a family of aircraft to replace platforms in the US Navy, Air Force and Marine Corps. Now known as F-35, the conventional (angled deck) take off and landing (CTOL) variant (F-35B) will complement the US Navy's F/A-18E/F and replace the A6 Intruder. The Short Take Off and Vertical Landing (STOVL) variant (F-35C) will replace the Marine Corps' AV-8B and F/A-18A/C/D. The US government awarded the US$19 billion contract for the Systems Development and Demonstration phase on 26 October 2001 to the industry team of Lockheed Martin, Northrop Grumman and BAE Systems. First delivery of operational aircraft is expected in 2008. Pratt and Whitney was awarded a US$4 billion contract to develop an engine to compete in production with that developed by General Electric and Rolls Royce.
(3) **Electronic Warfare:** Navy EA-6B Prowlers are to be replaced by the EA-18G from 2009. A replacement for USMC EA-6Bs has not been decided but an EW variant of the F-35 is being studied.
(4) **Unmanned Aircraft:** The UAV strategy has been given significantly increased impetus since operations in Afghanistan in 2001. It is envisaged that a family of UAVs will be required to meet three principal capability requirements:

(a) Tactical Surveillance and Targeting: currently provided by 'Pioneer' which is to be transferred to the USMC while future requirements are developed.

(b) Long Dwell/Stand-off Intelligence Surveillance and Reconnaissance (ISR). Current development work on Unmanned Combat Air Vehicles (UCAV) is expected to be taken forward into a full acquisition programme with an ISR capability expected in about 2015.

(c) Penetrating Surveillance/Suppression of Enemy Air Defences (SEAD): Initial Operational Capability for a full weapons capability is not expected until 2020.
(5) **Tacair Integration:** Navy/Marine Corps tactical aviation (TACAIR) integration plan was approved in 2002 to optimize combat capability and efficiencies by relying on fewer but more capable aircraft. As part of the Navy's Fleet Response Plan, the Navy/Marine Corps in 2004 began integrating Marine Corps squadrons into carrier air wings and Navy squadrons into the Marine Corps' Unit Deployment Plan (UDP). TACAIR Integration is expected to reduce Navy and Marine Corps force structure by five squadrons from 64 to 59. As a result, new aircraft procurement is to fall by 497 from 1,637 to 1,140.
(6) **Advanced Hawkeye:** A contract was awarded Northrop Grumman Corporation in August 2003 for the Systems Development and Demonstration (SD&D) phase of the Advanced Hawkeye (AHE) programme. This began in 2003 and is to continue until 2013. The first developmental AHE aircraft is scheduled for delivery in 2007, followed by Initial Operating Capability (IOC) in FY11. AHE will use the Northrop Grumman-built E-2C Hawkeye 2000 configuration as a baseline but the existing AN/APS-145 radar system will be replaced along with other aircraft systems components that enable the radar upgrade.

Numbers/Type: 10/22/15 Grumman F-14A/B/D Tomcat.
Operational speed: 1,342 kt (2,485 km/h).
Service ceiling: 60,000 ft (18,283 m).
Range: 2,000 n miles (3,704 km).
Role/Weapon systems: Fleet strike fighter aircraft for long range air defense, precision strike and reconnaissance roles. Sensors: AWG-9 (A/B) or APG-71 (D). ALQ-126 jammer (A/B) or ALQ-165 (D). ASN-92 nav (A) embedded GPS-INS (B) or ASN-139/MAGR (D), LANTIRN (all models); ALR-67 (all models); IRST and JTIDS(D). Weapons: two AIM-9, one AIM-7, two GBU, 20 mm (strike). Two AIM-7, two AIM-9, 20 mm (OCA/DCA). Two AIM-7, two AIM-9, 20 mm (recce). *UPDATED*

F-14D   9/2004*, US Navy / 1043648

Numbers/Type: 138/18/396/139 McDonnell Douglas F/A-18A/F/A-18B/F/A-18C/F/A-18D Hornet.
Operational speed: 1,032 kt (1,910 km/h).
Service ceiling: 50,000 ft (15,240 m).
Range: 1,000 n miles (1,850 km).
Role/Weapon systems: Strike interdictor (VFA) for USN/USMC air groups. Some are used for EW support with ALQ-167 jammers. Sensors: ESM: Litton ALR 67(V)2, ALQ 165 ASPJ jammer (18C/D), ALQ-126B jammer, APG-65 or APG-73 radar, AAS-38 FLIR, AAR-50 Nav FLIR, ASQ-173 tracker. Weapons: ASV; four Harpoon or SLAM (18D) or AGM-88 HARM (18D) missiles. Strike; up to 7.7 tons of bombs (or LGM). AD; one 20 mm Vulcan cannon, nine AIM-7/AIM-9 missiles. Typical ASV load might include 20 mm gun, 7.7 ton bombs including AGM 154A JSOW, two AIM-9 missiles. Typical AAW load might include 20 mm gun, four AIM-7, two AIM-9 missiles. 18C/D includes AIM-120 AMRAAM and AGM-65 Maverick or AGM 62 Walleye capability. *UPDATED*

F/A-18C   9/2004*, US Navy / 1043647

Numbers/Type: 98/108 Boeing F/A-18E/F/A-18F Super Hornet.
Operational speed: 930 kt (1,721 km/h).
Service ceiling: 50,000 ft (15,240 m).
Range: 1,320 n miles (2,376 km).
Role/Weapon systems: Strike interdictor for USN. First one rolled out in September 1995. First 12 production aircraft ordered in FY97. First sea trials January 1997. Entered operational service November 1999. Initial deployment to CVN 72 in July 2002. 200th aircraft delivered in August 2004. A further 210 aircraft to be delivered by 2009. Sensors: APG-73 radar, ALR-67(V)3 RWR. ECM: ALQ-165 ASPJ, ALQ-214 RFCM, towed decoys. Weapons: 11 wing stations for 8,680 kg of weapons (same armament as C/D) plus 20 mm guns. *UPDATED*

F/A-18F   7/2004*, Paul Jackson / 1043646

Numbers/Type: 41/94/17 Boeing/British Aerospace AV-8B/AV-8B II Plus/TAV-8B Harrier II.
Operational speed: 585 kt (1,083 km/h).
Service ceiling: 50,000 ft (15,240 m).
Range: 800 n miles (1,480 km).
Role/Weapon systems: Attack and destroy surface and air targets in support of USMC. Operational since 1985, a total of 94 AV-8B Plus conversions completed in 2003. Sensors: Litening II targeting pod, Navigation FLIR, moving map, AN/AVS-9 night vision goggles, laser spot tracker and ECM; Litton ALR-67 ESM; APG-65 radar (AV-8B II Plus). Weapons: Strike; 500 and 1,000 lb general purpose bombs, Paveway II LGM, Joint Direct Attack Munition, AGM-65 Maverick, Cluster Bomb Units, 300-25 mm rounds, 2.75 in and 5.00 in rockets. Self-defence: one GAU-12/U 25 mm cannon and four AIM-9L Sidewinder. *UPDATED*

AV-8B   8/2004*, US Navy / 1043645

Numbers/Type: 101 Lockheed S-3B Viking/Shadow.
Operational speed: 450 kt (834 km/h).
Service ceiling: 35,000 ft (10,670 m).
Range: 2,000 n miles (3,706 km).
Role/Weapon systems: Originally configured as carrier-borne ASW aircraft. ASW capability removed since 1998 and primary roles are reconnaissance and in-flight refuelling. Sensors: APS-137(V)1 radar; APN-200 radar, FLIR, OR-263AA, ESM: ALR-76; ECM ALE 47; ALE 39 chaff. Weapons: ASV; two Harpoon Block 1C; mines. *UPDATED*

S-3B   10/2004*, Hachiro Nakai / 1043686

Numbers/Type: 119 Grumman EA-6B Prowler.
Operational speed: 566 kt *(1,048 km/h)*.
Service ceiling: 41,200 ft *(12,550 m)*.
Range: 955 n miles *(1,769 km)*.
Role/Weapon systems: EW and jamming aircraft (VAQ) to accompany strikes and armed reconnaissance. Block 89A avionics/computer upgrades first delivered 2001 and continuing. All aircraft to be modernised. Further update (ICAP III) planned with first in service in 2005. Sensors: APS-130 radar; ALQ-99, USQ-113 jammers. Weapons: AGM-88 HARM anti-radiation missile capable.

UPDATED

EA-6B
10/2004*, Hachiro Nakai / 1043685

Numbers/Type: 68 Grumman E-2C Hawkeye.
Operational speed: 323 kt *(598 km/h)*.
Service ceiling: 37,000 ft *(11,278 m)*.
Range: 1,540 n miles *(2,852 km)*.
Role/Weapon systems: Used for direction of AD and strike operations; (VAW); 16 Group II upgrades from 1995-2000 included APS 145 radar and Link 16. Sensors: ESM: ALR-73 PDS; Airborne tactical data system with Links 4A, 11 or 16; CEC from 2000. APS-125/138/145 radar; Mk XII IFF. Weapons: Unarmed.

UPDATED

E-2C
9/2004*, US Navy / 1043659

Numbers/Type: 12 Bell Boeing MV-22 Osprey.
Operational speed: 270 kt *(500 km/h)*.
Service ceiling: 25,000 ft *(7,620 m)*.
Range: 2,108 n miles *(2,110 km)*.
Role/Weapon systems: Replacement for legacy assault/support helicopters (CH-46, CH-53D) for Marines (MV), strike rescue for the Navy (Navy MV), and special ops for SOCOM (CV). Two CV-22s conducting developmental test at Edwards AFB; 7 MV-22s conducting developmental test at NAS Patuxent River (2 EMD, 4 LRIP, and 1 Block A); 12 Block A MV-22s conducting operational test with Marine Tiltrotor Test and Evaluation Squadron-22 (VMX-22) Marine Corps Air Station New River, NC. Flight operations were suspended in December 2000 after second of two fatal accidents but, following engineering changes, flight operations resumed in May 2002. Being reintroduced to fleet gradually through VMX-22. Nine aircraft procured in FY04. Final operational evaluation planned for Spring of 2005. IOC in late 2007. Subject to successful testing and industry posture, full rate production decision expected by the end of 2005. Full fleet of 360 MV, 50 CV and 48 Navy MV projected. Sensors: AAR-47 ESM; AN/ALQ-211 Suite of Integrated RF CounterMeasures (SIRFC), AN/AAQ-24(V) Nemesis Directional Infra-Red CounterMeasures (DIRCM); AN/AAQ-27 FLIR; APR 39A(V)2. Weapons: 12.7 mm MG or self-defence gun.

UPDATED

MV-22
12/2004*, US Navy / 1043658

Numbers/Type: 149 Sikorsky SH-60B Seahawk (LAMPS Mk III).
Operational speed: 145 kt *(268 km/h)*.
Service ceiling: 10,000 ft *(3,050 m)*.
Range: 450 n miles *(833 km)*.
Role/Weapon systems: LAMPS Mk III is airborne platform for ASW and ASUW: operated from cruisers, destroyers and frigates. First deployed in 1984. To be replaced by MH-60R. Sensors: APS-124 search radar, AAS-44 FLIR with laser designator, ASQ-81(V) MAD, 25 sonobuoys, ALQ-142 ESM, ALQ-144 IR suppressor and ALE-39 chaff and flare dispenser. UYS-1 Acoustic processor. Weapons: ASW; three Mk 46 or Mk 50 torpedoes. ASUW; one Penguin Mk 2 Mod 7 AGM-119B missile, one 7.62 mm MG M60 or 12.7 mm MG, AGM-114B/K Hellfire missile (87 from 1998).

UPDATED

SH-60B
1/2001, Darren Yates, RAN / 0131323

Numbers/Type: 7 Sikorsky MH-60R StrikeHawk.
Operational speed: 145 kt *(268 km/h)*.
Service ceiling: 10,000 ft *(3,050 m)*.
Range: 450 n miles *(833 km)*.
Role/Weapon systems: The plan is to replace the SH-60B/F fleet with the MH-60R which is to be the future tactical helicopter operated from carriers, cruisers, destroyers and frigates to enter front-line service in 2006. Sensors: Telephonics APS-147 long-range search radar with ISAR, ALQ-210 ESM, Raytheon AGS-22 ALFS sonar, acoustic processor, Raytheon AAS-44 2nd Gen FLIR with laser designation and Low Light Camera, Hawklink sensor datalink, Link 16, AAR-47 MAW with LWR, ALE-47 chaff/flare dispenser, and ALQ-144 IR Jammer. Weapons: ASW: three Mk 46/50 torpedoes. ASUW: four AGM-114B/K Hellfire missiles, one 7.62 mm MG M60 or 12.7 mm MG. Block One, which is scheduled to be fielded incrementally from 2006 to 2009, will upgrade the aircraft with the AAS-44A 3rd Gen FLIR, including Low Light Camera, CDL-N Ku-band sensor datalink, Link 16, and Mk 54 torpedoes.

UPDATED

MH-60R
12/2003, Lockheed Martin / 0111939

Numbers/Type: 73 Sikorsky SH-60F Seahawk (CV).
Operational speed: 145 kt *(268 km/h)*.
Service ceiling: 10,000 ft *(3,050 m)*.
Range: 600 n miles *(1,111 km)*.
Role/Weapon systems: Derivation of SH-60B that replaced SH-3H Sea King to provide close-in ASW protection to Carrier Battle Groups. First deployed in *Nimitz* 1991. To be replaced by MH-60R. Sonar: AQS-13F dipping sonar; ASQ-81 (V) MAD; UYS-2 acoustic processor; 14 sonobuoys. Weapons: ASW: three Mk 46/54 torpedoes. ASUW: One GAU 12.7 mm MG or one 7.62 mm MG.

UPDATED

SH-60F
6/2004*, US Navy / 1043657

Numbers/Type: 39 Sikorsky HH-60H Seahawk.
Operational speed: 147 kt *(272 km/h)*.
Service ceiling: 10,000 ft *(3,050 m)*.
Range: 500 n miles *(926 km)*.
Role/Weapon systems: Strike, special warfare support and SAR derivative (HCS) of the SH-60F. To be replaced by MH-60S. Sensors: AQS-44 FLIR with laser designator, APR-39A radar warning receiver, AVR-2 LWR and AA 47 MAW, ALE-47 chaff/flare dispenser, ALQ-144 ECM. Weapons: ASV; Hellfire AGM-114B/K; one GAU-16 12.7 mm MG or one M-240 7.62 mm MG. Can deploy eight SEAL to a range of 200 n miles.

UPDATED

HH-60H
10/2003*, US Navy / 1043656

Numbers/Type: 73 Sikorsky MH-60S Seahawk.
Operational speed: 147 kt *(272 km/h).*
Service ceiling: 10,000 ft *(3,050 m).*
Range: 400 n miles *(741 km).*
Role/Weapon systems: Derived from the UH-60 as a combat support helicopter to replace CH-46, HH-1, UH-3 and HH-60H in replenishment, armed helicopter, CSAR and Medevac roles. Organic airborne mine countermeasures capability from 2005. Approximately 271 are required. First deployment was in 2002. Sensors (Armed Helo role): AAS-44 FLIR with laser designator, APR-39A radar warning receiver, AAR-47 MAW with LWR, ALE-47 chaff/flare dispenser, ALQ-144 IR Jammer, Link 16. Weapons: 8 AGM-114B/K Hellfire missiles, two 7.62 mm MG M60 or 12.7 mm MGs.

UPDATED

MH-60S   *3/2005*, US Navy /* 1043655

Numbers/Type: 50 Sikorsky UH-3H Sea King.
Operational speed: 144 kt *(267 km/h).*
Service ceiling: 10,000 ft *(3,050 m).*
Range: 630 n miles *(1,166 km).*
Role/Weapon systems: Used for liaison and SAR tasks having been replaced in CV service by SH-60F. Sensors: ALR 606B ESM.

VERIFIED

UH-3H   *3/2002, Michael Nitz /* 0530016

Numbers/Type: 36 Sikorsky CH-53D Sea Stallion.
Operational speed: 160 kt *(294 km/h).*
Service ceiling: 10,000 ft *(3,048 m).*
Range: 578 n miles *(1,070 km).*
Role/Weapon systems: Assault, support and transport helicopters (MMH); can carry 37 Marines. Sensors: None. Weapons: Up to three 12.7 mm machine guns.

UPDATED

CH-53D   *7/2004*, US Navy /* 1043654

Numbers/Type: 6/225 Boeing HH-46D/CH-46E Sea Knight.
Operational speed: 137 kt *(254 km/h).*
Service ceiling: 8,500 ft *(2,590 m).*
Range: 180 n miles *(338 km).*
Role/Weapon systems: Support/assault (HMM) for 18 Marines and resupply (USN) helicopter. To be replaced by V-22 in due course. Can lift 1.3 or 4.5 tons in a cargo net or sling. Sensors: None. Weapons: Unarmed.

UPDATED

CH-46E   *9/2002*, Paul Jackson /* 0069856

Numbers/Type: 151 Sikorsky CH-53E Super Stallion.
Operational speed: 170 kt *(315 km/h).*
Service ceiling: 10,000 ft *(3,048 m).*
Range: 580 n miles *(1,074 km).*
Role/Weapon systems: Uprated, three-engined version of Sea Stallion (MMH) with heavy lift (USN) and logistics (USMC) roles. Carries 55 Marines. Sensors: None. Weapons: Up to three 12.7 mm machine guns.

UPDATED

CH-53E   *8/2000, US Navy /* 0106784

Numbers/Type: 37 Sikorsky MH-53E Sea Dragon.
Operational speed: 170 kt *(315 km/h).*
Service ceiling: 10,000 ft *(3,048 m).*
Range: 1,000 n miles *(1,850 km).*
Role/Weapon systems: Three-engined AMCM helicopter (HM) similar to Super Stallion; tows ALQ-166 Mod 4 MCM sweep equipment; self-deployed if necessary. Sensors: Northrop Grumman AQS-14A or AQS-20 (in due course) dipping sonar. Weapons: Two 12.7 mm guns for self-defence.

UPDATED

MH-53E   *5/1999, US Navy /* 0106785

Numbers/Type: 182/3 Bell AH-1W/AH-1Z Super Cobra.
Operational speed: 149 kt *(277 km/h).*
Service ceiling: 10,000 ft *(3,048 m).*
Range: 317 n miles *(587 km).*
Role/Weapon systems: Close support helicopter (HMLA) with own air-to-air capability. Four-bladed rotor upgrade to be fitted from 2004 to improve speed, range and lift. AH-1Z is remanufacture of AH-1W and features composite blades, new engines and gearboxes. To enter service in 2011. Sensors: NTS (laser and FLIR nightsight). Weapons: Strike/assault; one triple 20 mm cannon, eight TOW or Hellfire missiles and gun. AAW; two AIM-9L Sidewinder missiles. ***UPDATED***

AH-1W   6/2001, Jürg Kursener / 0131312

Numbers/Type: 23/115/2 Bell HH-1N/UH-1N/UH-1Y Iroquois. Twin Huey.
Operational speed: 110 kt *(204 km/h).*
Service ceiling: 10,000 ft *(3,048 m).*
Range: 230 n miles *(426 km).*
Role/Weapon systems: HH-1N is SAR, training, support and logistics helicopter for USN/USMC operations ashore. Can carry eight Marines. UH-1N is USMC Light Utility platform for all-weather assault, transport, airborne command and control, armed reconnaissance and SAR. Can carry eight marines. Four-bladed upgrade being fitted from 2004 to improve speed, range and lift. Can carry eight marines. UH-1Y is remanufacture of UH-1N featuring composite blades and new engines gearboxes. Service entry expected in 2009. Sensors: BRITE Star FLIR. Weapons: Can be armed with 12.7 mm or 7.62 mm machine guns and 2.75 in rockets. ***UPDATED***

UH-1N   5/1999, A Sharma / 0084120

Numbers/Type: 33 RQ-2 Pioneer UAV.
Operational speed: 65 kt *(120 km/h).*
Service ceiling: 15,000 ft *(4,570 m).*
Range: 100 n miles+ *(185 km+).*
Role/Weapon systems: Two systems in service with USMC UAV squadrons. One in service for training. Sensors: Optronic surveillance with Link terminal. ***UPDATED***

RQ-2   0106054

Numbers/Type: 6 Northrop Grumman RQ-8A Fire Scout UAV.
Operational speed: 115 kt *(213 km/h).*
Service ceiling: 20,000 ft *(6,094 m).*
Range: 110 n miles *(205 km).*
Role/Weapon systems: Six in service includes one prototype, two in EMD and three in LRIP. Selected in February 2000 as the next generation UAV but full production suspended until UAV strategy has been further developed. A candidate for the LCS. One system comprising three air vehicles with electro-optical/infra-red/laser designator range-finder payloads and two control stations in development. Endurance is over six hours. Will operate from any air capable ship as well as from confined land areas. 141 flights conducted by 30 September 2004. ***UPDATED***

RQ-8A   2000, Northrop Grumman / 0109059

LAND-BASED MARITIME AIRCRAFT (FRONT LINE)

Notes: There are also 32/12 Lockheed KC-130F/R Hercules tankers.

Numbers/Type: 11 Lockheed EP-3E Aries.
Operational speed: 411 kt *(761 km/h).*
Service ceiling: 28,300 ft *(8,625 m).*
Range: 2,380 n miles *(4,407 km).*
Role/Weapon systems: Electronic warfare and intelligence gathering aircraft (VQ). Sensors: EW equipment including AN/ALR-60, AN/ALQ-76, AN/ALQ-78, AN/ALQ-108 and AN/ASQ-114. Weapons: Unarmed.

UPDATED

EP-3E   10/2003, Paul Jackson / 0110197

Numbers/Type: 183/1 Lockheed P-3C/P-3B Orion.
Operational speed: 411 kt *(761 km/h).*
Service ceiling: 28,300 ft *(8,625 m).*
Range: 2,380 n miles *(4,407 km).*
Role/Weapon systems: 111 in active (49 in reserve) operational squadrons (VP) deployed worldwide; primary ASW/ASUW; ASUW update III conversions to 146 airframes by end 2000. Sensors: APS-115 search radar or APS-137(V)3 fitted in 18 aircraft, ASQ-81 MAD, UYS-1 acoustic suite, up to 100 sonobuoys, AAR-36 FLIR, cameras, ALR-66 ESM. AAR 47 ESM and ALE 47 chaff/IR dispenser in Update II. 18 Counter-Drug Upgrade (CDU) aircraft employ APG-66 air-to-air radar and AVX-1 Electro-Optics. Up to 68 ASUW Improvement Program (AIP) aircraft employ the APS-137B(V)5 ISAR/SAR radar, AVX-1 Electro-Optics, UYS-1 acoustics, OASIS III/OTCIXS communications suite, and SATCOM. Weapons: ASW; four Mk 46/50 torpedoes or depth bombs. ASUW; four AGM-84C Harpoon, AGM-65F Maverick, AGM-84E SLAM, six Mk 55/56 mines.

UPDATED

P-3C   7/2002, Hachiro Nakai / 0530043

Numbers/Type: Boeing Multimission Maritime Aircraft (MMA).
Operational speed: 490 kt *(907 km/h).*
Service ceiling: 41,000 ft *(12,500 m).*
Range: 1,380 n miles *(2,555 km).*
Role/Weapon systems: Contract for MMA System Development and Demonstration (SDD) awarded 14 June 2004. Two variants of MMA to replace P-3C and EP-C3 Aries aircraft. Design based on Boeing 737-800ERX. Crew of nine. To enter operational service in 2013. Sensors: To be equipped with modern ASW, ASUW and intelligence, surveillance and reconnaissance (ISR) sensors. Weapons: To be announced.

NEW ENTRY

BOEING 737 MMA   6/2004 *, US Navy / 1043653

Numbers/Type: 16 Boeing E-6B Mercury/TACAMO/ABNCP.
Operational speed: 455 kt *(842 km/h).*
Service ceiling: 42,000 ft *(12,800 m).*
Range: 6,350 n miles *(11,760 km/h).*
Role/Weapon systems: Strategic Command Airborne Command Post. First flown in E-6A configuration in 1990(VQ). E-6B first flown in TACAMO/ABNCP configuration in 1997 and assumed ABNCP role in 1998. Eleven aircraft converted from E-6A to E-6B with five remaining aircraft scheduled to be modified by 2003. Sensors: Radar Bendix APS-133; ALR-68(V)4 ESM; supports Trident Fleet radio communications with up to 28,000 ft of VLF trailing wire antenna. Weapons: Unarmed.

VERIFIED

TACAMO   8/2001, H M Steele / 0131325

COMMAND SHIPS

Notes: Options for replacement of the four in-service command ships remain under consideration.
They include new construction ships, service-life extensions of current ships and/or a mix of sea and land-based facilities.

2 BLUE RIDGE CLASS: COMMAND SHIPS (LCCH/AGFH)

Name	No	Builders	Laid down	Launched	Commissioned	F/S
BLUE RIDGE	LCC 19	Philadelphia Naval Shipyard	27 Feb 1967	4 Jan 1969	14 Nov 1970	PA
MOUNT WHITNEY	LCC 20	Newport News Shipbuilding	8 Jan 1969	8 Jan 1970	16 Jan 1971	AA

Displacement, tons: 13,077 light; 19,648 full load *(Blue Ridge)*
12,846 light; 19,760 full load *(Mount Whitney)*
Dimensions, feet (metres): 636.5 × 107.9 × 28.9
(194 × 32.9 × 8.8)
Main machinery: 2 Foster-Wheeler boilers; 600 psi *(42.3 kg/cm^2)*; 870°F *(467°C)*; 1 GE turbine; 22,000 hp *(16.4 MW)*;
1 shaft
Speed, knots: 23. **Range, n miles:** 13,000 at 16 kt
Complement: LCC 19: 786; 637 Flag staff. LCC 20: 819; 562
Flag staff
Military lift: 700 troops; 3 LCPs; 2 LCVPs; 2—7 m RHIBs.

Guns: 2 General Electric/General Dynamics 20 mm/76
6-barrelled Vulcan Phalanx Mk 15; 3,000 rds/min (4,500 in
Block 1) combined to 1.5 km.
2—25 mm Mk 38.
Countermeasures: Decoys: 4 Loral Hycor SRBOC 6-barrelled
fixed Mk 36; IR flares and chaff to 4 km *(2.2 n miles)*. SLQ-25
Nixie; torpedo decoy.
ESM/ECM: SLQ-32(V)3; combined radar intercept, jammer and
deception system.
Combat data systems: GCCS (M) Link 4A, Link 11, Link 14 and
JTIDS. Theatre Battle Management Core Systems (TBMCS).
Wide band commercial SATCOM, USC-38 SATCOM, WSC-3
EHF SATCOM, WSC-6(V)1 and 5, and WSC-6A(V)4 SHF
SATCOM. High Frequency Radio Group (HFRG). Mission
Display System (MDS). Demand Assigned Multiple Access
(DAMA QUAD). Area Air Defense Commander, Naval Fires
Network, Joint Service Imagery Processing System (JSIPS-N),
Common High Bandwidth Data Link, Shipboard Terminal
(CHBDL-ST), Ring Laser Gyro Network (RLGN), NITES 2000,
Joint Tactical Information Distribution System (JTIDS),
Navigational Sensor System Interface (NAVSSI). (See Data
Systems at front of section.)
Radars: Air search: Lockheed SPS-40E; B-band.
Surface search: Raytheon SPS-10B; G-band.
Navigation: Marconi LN66; Raytheon SPS-64(V)9; I-band.
Tacan: URN 25. IFF: Mk XII AIMS UPX-29.

Helicopters: 1 Sikorsky SH-3H Sea King.

Programmes: Authorised in FY65 and 1966. Originally
designated Amphibious Force Flagships (AGC); redesignated
Command Ships (LCC) on 1 January 1969.
Modernisation: Modernisation completed FY87. 3 in guns
removed in 1996/97 and Sea Sparrow missile launchers
have been disembarked. Mk 23 TAS and RAM are not now to
be fitted.
Structure: General hull design and machinery arrangement are
similar to the Iwo Jima class assault ships. Accommodation
for 250 officers and 1,300 enlisted men.
Operational: These are large force command ships of post-
Second World War design. They can provide integrated
command and control facilities for sea, air and land
commanders in all types of operations. *Blue Ridge* is the
Seventh Fleet flagship, based at Yokosuka, Japan. *Mount
Whitney* has served since January 1981 as flagship Second

BLUE RIDGE   *4/2002, Hachiro Nakai /* 0530045

MOUNT WHITNEY   *2/2002, Michael Nitz /* 0530046

Fleet, based at Norfolk, Virginia except during the period June
to November 1999 when she served as Sixth Fleet flagship.
Mount Whitney to undergo MSC conversion in 2005 prior to
resuming role as Sixth Fleet Flagship. The ship is to remain
under USN command but is to be largely civilian manned.
UPDATED

1 CONVERTED AUSTIN CLASS: COMMAND SHIP (AGFH)

Name	No	Builders	Laid down	Launched	Commissioned	F/S
CORONADO	AGF 11 (ex-LPD 11)	Lockheed SB & Construction Co	3 May 1965	30 July 1966	23 May 1970	PA

Displacement, tons: 11,482 light; 16,912 full load
Dimensions, feet (metres): 570 × 100 (84 hull) × 23
(173.8 × 30.5 (25.6) × 7)
Main machinery: 2 Foster-Wheeler boilers; 600 psi *(42.2 kg/cm^2)*; 870°F *(467°C)*; 2 De Laval turbines; 24,000 hp
(17.9 MW); 2 shafts
Speed, knots: 21. **Range, n miles:** 7,700 at 20 kt
Complement: 117 USN + 126 civilian + 491 Flag staff

Guns: 2 General Electric/General Dynamics 20 mm Vulcan
Phalanx Mk 15. 2—12.7 mm MGs.
Countermeasures: Decoys: 4 Loral Hycor SRBOC 6-barrelled Mk
36; IR flares and chaff.
ESM: SLQ-32V(2).
Combat data systems: GCCS (M) Link 11 and Link 16 (receive
only). Theatre Battle Management Core Systems (TBMCS).
Land Attack Warfare System (LAWS). Wide band commercial
SATCOM, USC-38 SATCOM, WSC-3 EHF SATCOM, WSC-6
SHF SATCOM. High Frequency Radio Group (HFRG). Single
Channel Ground Air Radio System, Joint Service Imagery
Processing System (JSIPS-N), Common High Bandwidth Data
Link, Shipboard Terminal (CHBDL-ST), Naval Fires Network
(NFN), Secure Voice System, TBMCS Host 1.1, COWAN,
TVDTS, SCI ADNS, EHF MDR, Tactical Switching System, Ring
Laser Gyro Network (RLGN), Navigational Sensor System
Interface (NAVSSI). (See Data Systems at front of section.)
Radars: Surface search: Norden SPS 67V(1); G-band.
Navigation: Raytheon SPS-73(V); I-band.
Tacan: URN 25.

Helicopters: 2 light.

Programmes: A former LPD of the Austin class. Authorised in
FY64.
Structure: The well deck was converted in 1997 to offices, with
a three deck command facility and additional accommodation
for up to four Flag Officers with a combined staff of 259. The
stern gate has been removed and sealed. 3 in guns removed.
Operational: Converted in late 1980 as a temporary
replacement for *La Salle* (AGF 3), as flagship, Middle East

CORONADO   *4/2004*, US Navy /* 1043652

Force and continued in flagship role thereafter. Served as
Sixth Fleet flagship for three years until relieved by *Belknap* in
July 1986. Then served as Third Fleet flagship in Hawaii and
subsequently San Diego. In 2000 *Coronado* was designated
as the US Navy's sea-based Battle-Lab to act as a testbed for
new IT systems. In 2002, acted as hub for major joint
exercises including 'Millenium Challenge', 'Rimpac 2002' and
'Fleet Battle Experiment 2002'. In 2003 underwent MSC
conversion and became largely civilian manned under USN
command. To be decommissioned in FY06.
UPDATED

AMPHIBIOUS FORCES

Notes: (1) The LHA Replacement Analysis of Alternatives was completed in September 2002. Options examined included: LHD 8 design, stretched LHD 8 design, new design, smaller designs and distributed solutions. On 6 April 2004, modified version of LHD 8 selected. The ship is to have improved aviation facilities (to house 20-25 F-35 STOVL aircraft) in lieu of a well-deck. The first of class is to enter service in 2012.

(2) Additional capacity is provided by the maritime pre-positioning ships (see listing under *Military Sealift Command* (MSC) section) which are either new construction or conversions of commercial ships. One squadron is maintained on station in the Atlantic, a second at Guam, and a third at Diego Garcia. Each squadron carries equipment to support a Marine Expeditionary Brigade.

(3) **Minesweeping:** Several of the larger amphibious ships have been used as operating bases for minesweeping helicopters.
(4) Five decommissioned LKAs and four LSTs are kept in an inactive reduced maintenance status (ROS). These are *Fresno* (LST 1182), *Tuscaloosa* (LST 1187), *Boulder* (LST 1190), *Racine* (LST 1191), *Charleston* (LKA 113), *Durham* (LKA 114), *Mobile* (LKA 115), *St Louis* (LKA 116) and *El Paso* (LKA 117).

1 + 8 (3) SAN ANTONIO CLASS: AMPHIBIOUS TRANSPORT DOCKS (LPDM)

Name	No	Builders	Laid down	Launched	Commissioned	F/S
SAN ANTONIO	LPD 17	Northrop Grumman Ship Systems (Avondale)	9 Dec 2000	19 July 2003	June 2005	Bldg/AA
NEW ORLEANS	LPD 18	Northrop Grumman Ship Systems (Avondale)	8 Nov 2002	20 Nov 2004	Dec 2005	Bldg/AA
MESA VERDE	LPD 19	Northrop Grumman Ship Systems (Ingalls)	25 Feb 2003	15 Jan 2005	Feb 2006	Bldg/AA
GREEN BAY	LPD 20	Northrop Grumman Ship Systems (Avondale)	26 Aug 2003	2005	Oct 2006	Bldg/PA
NEW YORK	LPD 21	Northrop Grumman Ship Systems (Avondale)	10 Sep 2004	2006	2007	Bldg/AA
SAN DIEGO	LPD 22	Northrop Grumman Ship Systems (Avondale)	2005	2007	2008	Ord/PA
ANCHORAGE	LPD 23	Northrop Grumman Ship Systems (Avondale)	2006	2007	2009	Ord/PA
ARLINGTON	LPD 24	Northrop Grumman Ship Systems (Avondale)	2007	2009	2010	Ord/PA
SOMERSET	LPD 25	Northrop Grumman Ship Systems (Avondale)	2008	2010	2011	Ord/AA

Displacement, tons: 25,300 full load
Dimensions, feet (metres): 683.7 × 104.7 × 23 *(208.4 × 31.9 × 7)*
Main machinery: 4 Colt Pielstick PC 2.5 diesels; 40,000 hp *(29.84 MW)*; 2 shafts; cp props
Speed, knots: 22
Complement: 360 (28 officers) plus 34 spare
Military lift: 720 troops; 2 LCACs, 14 AAAVs

Missiles: SAM: Mk 41 VLS for 2 octuple cell Evolved Sea Sparrow ❶. 64 missiles. May be fitted later. 2 Mk 31 Mod 1 RAM launchers ❷.
Guns: 2-30 mm Mk 46. 4-12.7 mm MGs.
Countermeasures: Decoys: 6 Mk 53 Mod 4 Nulka and chaff launcher ❸. SLQ-25A Nixie towed torpedo decoy.
ESM/ECM: SLQ-32A(V)2 ❹: intercept and jammer.
Combat data systems: SSDS Mk 2; GCCS (M), CEC, JTIDS (Link 16), AADS (see Data Systems at front of section).
Radars: Air search: ITT SPS-48E ❺; E/F-band.
Surface search/navigation: Raytheon SPS-73(V)13 ❻; I-band.
Fire control: Lockheed SPQ-9B ❼; I-band.

Helicopters: 1 CH-53E Sea Stallion or 2 CH-46E Sea Knight or 1 MV-22 Osprey.

Programmes: The LPD 17 (ex-LX) programme was first approved by the Defense Acquisition Board on 11 January 1993. It will replace four classes of amphibious ships: LPD 4s, LSTs, LKAs and LSD 36s. Contract for first ship, with an option on two more, awarded to Avondale on 17 December 1996. A protest

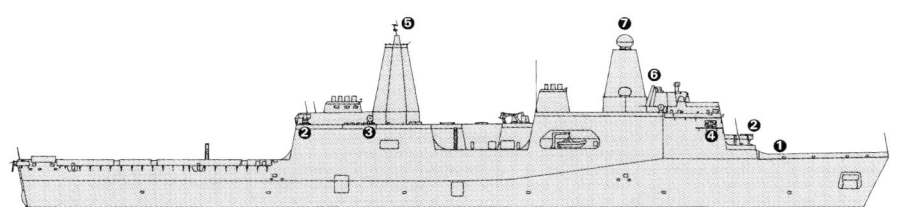

SAN ANTONIO
(Scale 1 : 1,800), Ian Sturton / 0573475

about the award delayed the effective contract date to April 1997. The lead ship contract options for FY99 and FY00 on LPD 18 and LPD 19 were exercised in December 1998 and February 2000 respectively. A negotiated modification added the second FY00 ship, LPD 20, to the lead ship contract in May 2000. Contract awarded for LPD 21 in November 2003 and for LPD 22 and 23 long-lead items in 2004. Under agreement reached in June 2002, NGSS is to build all 12 ships. Difficulties in design phase have led to two-year delay to original delivery date of lead ship. Delivery of final ship is planned for 2014.
Structure: Panama Canal-capable ships able to control and support landing forces disembarking either via surface craft such as LCACs or by VTOL aircraft, principally helicopters. The design supports a lift capability of 24,000 sq ft of deck space for vehicles, 34,000 cu ft of cargo below decks and 720

embarked Marines with surge lift capacity to 800 troops. The well-deck and stern gate arrangements are similar to those of the Wasp class; the well-deck can carry two LCACs or one LCU, or 14 Expeditionary Fighting Vehicles. The Flight deck can land/launch four CH-46s or two CH-53s or two MV-22s. The hangar will accommodate two CH-46s or one CH-53 or one MV-22. There is a 24-bed hospital. Although with similar capabilities as the classes they are to replace, the ships are not equipped with the flag facilities of some Austin class LPDs, the heavy over-the-side lift capability of LKAs and the ability of LSTs to beach. There is a crane for support of boat operations and an Advanced Enclosed Mast System, trialled in DD 968, is being fitted in all. On 9 September 2003, salvaged steel from the World Trade Centre was cast into the bow section of USS *New York*.

UPDATED

SAN ANTONIO
4/2005*, Northrop Grumman / 1121524

7 + 1 WASP CLASS: AMPHIBIOUS ASSAULT SHIPS (LHDM/MCSM)

Name	No	Builders	Laid down	Launched	Commissioned	F/S
WASP	LHD 1	Ingalls Shipbuilding	30 May 1985	4 Aug 1987	29 July 1989	AA
ESSEX	LHD 2	Ingalls Shipbuilding	16 Feb 1989	4 Jan 1991	17 Oct 1992	PA
KEARSARGE	LHD 3	Ingalls Shipbuilding	6 Feb 1990	26 Mar 1992	16 Oct 1993	AA
BOXER	LHD 4	Ingalls Shipbuilding	26 Mar 1991	13 Aug 1993	11 Feb 1995	PA
BATAAN	LHD 5	Ingalls Shipbuilding	16 Mar 1994	15 Mar 1996	20 Sep 1997	AA
BONHOMME RICHARD	LHD 6	Ingalls Shipbuilding	29 Mar 1995	14 Mar 1997	15 Aug 1998	PA
IWO JIMA	LHD 7	Ingalls Shipbuilding	12 Dec 1997	4 Feb 2000	30 June 2001	AA
MAKIN ISLAND	LHD 8	Northrop Grumman Ship Systems (Ingalls)	14 Feb 2004	Feb 2006	Aug 2007	Bldg

Displacement, tons: 40,650 (LHD 1-4); 40,358 (LHD 5-7); 41,772 (LHD 8) full load
Dimensions, feet (metres): 844 oa; 788 wl × 140.1 oa; 106 wl × 26.6 *(257.3; 240.2 × 42.7; 32.3 × 8.1)*
Flight deck, feet (metres): 819 × 106 *(249.6 × 32.3)*
Main machinery: 2 Combustion Engineering boilers; 600 psi *(42.3 kg/cm²)*; 900°F *(482°C)*; 2 Westinghouse turbines; 70,000 hp *(52.2 MW)*; 2 shafts
2 GE LM 2500+ gas turbines (LHD 8)
Speed, knots: 22. **Range, n miles:** 9,500 at 18 kt
Complement: 1,123 (65 officers)
Military lift: 2,000 troops; 12 LCM 6s or 3 LCACs; 1,232 tons aviation fuel; 4 LCPL

Missiles: SAM: 2 Raytheon GMLS Mk 29 octuple launchers ❶; 16 Sea Sparrow; semi-active radar homing to 14.6 km *(8 n miles)* at 2.5 Mach; warhead 39 kg. ESSM in due course.
2 GDC Mk 49 RAM; 21 rounds per launcher ❷; passive IR/anti-radiation homing to 9.6 km *(5.2 n miles)* at 2 Mach; warhead 9.1 kg.
Guns: 2 or 3 General Electric/General Dynamics 20 mm 6-barrelled Vulcan Phalanx Mk 15 ❸; 3,000 rds/min (4,500 in Batch 1) combined to 1.5 km.
3 Boeing Bushmaster 25 mm Mk 38. 4—12.7 mm MGs.
Countermeasures: Decoys: 4 or 6 Loral Hycor SRBOC 6-barrelled fixed Mk 36; IR flares and chaff to 4 km *(2.2 n miles)*.
SLQ-25 Nixie; acoustic torpedo decoy system. NATO Sea Gnat. SLQ-49 chaff buoys. AEB SSQ-95.
ESM/ECM: SLQ-32(V)3/SLY-2; intercept and jammers. Raytheon ULQ-20.

Combat data systems: ACDS Block 1 level 2 (LHD 1 and 7) and Block 0 (LHD 2-6). SSDS Mk 2 (LHD 8 on build and LHD 1 and 7 in 2007). Integrated Tactical Amphibious Warfare Data System (ITAWDS) and Marine Tactical Amphibious C² System (MTACCS). Links 4A, 11 (modified), 14 and 16. SATCOMS ❹ SSR-1, WSC-3 (UHF), USC-38 (EHF). SMQ-11 Metsat (see Data Systems at front of section). Advanced Field Artillery TDS (LHD 6 and 7).
Weapons control: 2 Mk 91 MFCS.
Radars: Air search: ITT SPS-48E ❺; 3D; E/F-band.
Raytheon SPS-49(V)9 ❻; C/D-band.
Hughes Mk 23 TAS ❼; D-band. SPQ-9B (LHD 8).
Surface search: Norden SPS-67 ❽; G-band.
Navigation: SPS-73; I-band.
CCA: SPN-35B and SPN-43C. SPN-46 in due course.
Fire control: 2 Mk 95; I/J-band. SPQ-9B to be fitted.
Tacan: URN 25. IFF: CIS Mk XV UPX-29.

Fixed-wing aircraft: 6-8 AV-8B Harriers or up to 20 in secondary role. V22 Osprey in due course.
Helicopters: Capacity for 42 CH-46E Sea Knight but has the capability to support: AH-1W Super Cobra, CH-53E Super Stallion, CH-53D Sea Stallion, UH-1N Twin Huey, AH-1T Sea Cobra, and SH-60B Seahawk helicopters. UAV in due course.

Programmes: LHD 8 being funded incrementally by Congress from FY99-FY06.
Modernisation: RAM launchers retrofitted in all vice the forward Phalanx (which may be moved to the bridge roof) and at the stern.

Structure: Two aircraft elevators, one to starboard and aft of the 'island' and one to port amidships; both fold for Panama canal transits. The well-deck is 267 × 50 ft and can accommodate up to three LCACs. The flight deck has nine helicopter landing spots. Cargo capacity is 125,000 cu ft total with an additional 20,000 sq ft to accommodate vehicles. Vehicle storage is available for five M1 tanks, 25 LAVs, eight M 198 guns, 68 trucks, 10 logistic vehicles and several service vehicles. The bridge is two decks lower than that of an LHA, command, control and communication spaces having been moved inside the hull to avoid 'cheap kill' damage. Fitted with a 64 bed capacity hospital and six operating rooms. HY-100 steel covers the flight deck. Three 32 ft monorail trains each carrying 6,000 lbs, deliver material to the well-deck at 6.8 mph. *Iwo Jima* is likely to be the last oil-fired steam turbine ship in the USN. LHD 8 is to be fitted with gas turbine propulsion.
Operational: A typical complement of aircraft is a mix of 30 helicopters and six to eight Harriers (AV-8B). In the secondary role as a sea control ship the most likely mix is 20 AV-8B Harriers and four to six SH-60B Seahawk helicopters. LHD 3 modified to provide interim Mine Countermeasures Command (MCS) capability following decommissioning of *Inchon* in June 2002. In 2003, LHA 5 deployed as the centrepiece of the first Expeditionary Strike Group (ESG). LHD 6 first amphibious ship to deploy with MH-60S helicopter. LHDs 1, 3, 5 and 7 based at Norfolk, Virginia, and LHDs 2, 4 and 6 at San Diego, California.

UPDATED

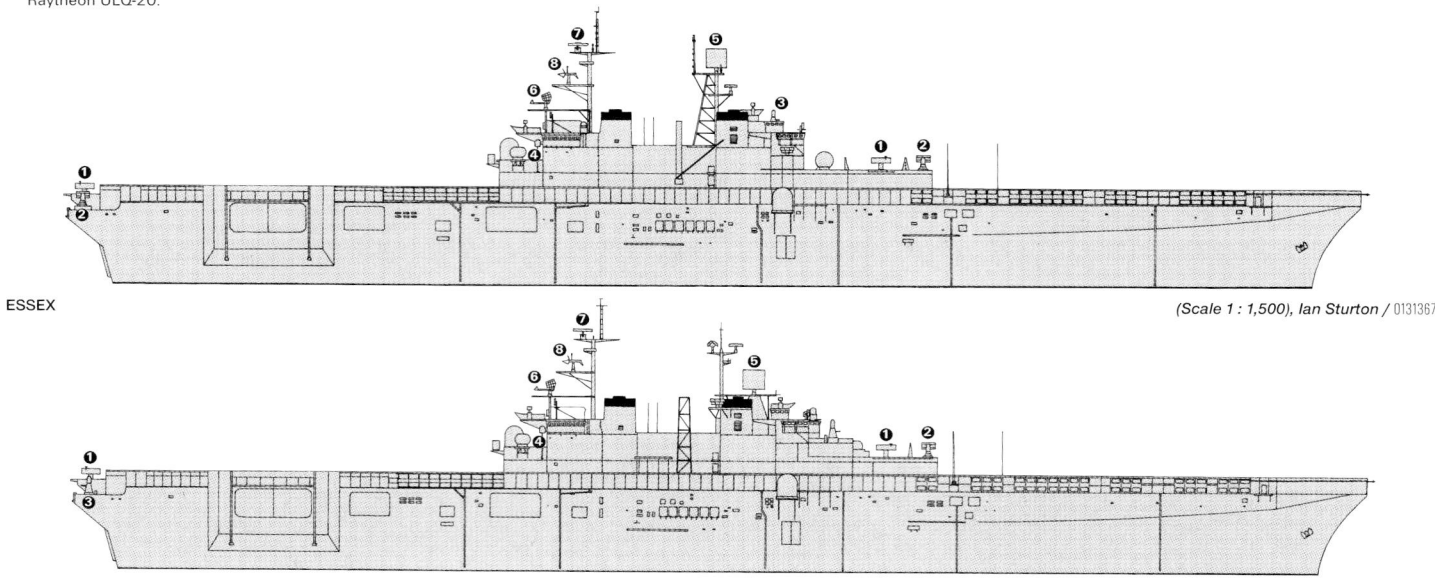

ESSEX

(Scale 1 : 1,500), Ian Sturton / 0131367

BONHOMME RICHARD

(Scale 1 : 1,500), Ian Sturton / 0131368

BATAAN

2/2002, Chris Sattler / 0529987

BOXER
6/2003, Mick Prendergast / 0572785

KEARSARGE
4/2003, A Sharma / 0572779

BONHOMME RICHARD
12/2004*, US Navy / 1043627

5 TARAWA CLASS: AMPHIBIOUS ASSAULT SHIPS (LHAM)

Name	No	Builders	Laid down	Launched	Commissioned	F/S
TARAWA	LHA 1	Ingalls Shipbuilding	15 Nov 1971	1 Dec 1973	29 May 1976	PA
SAIPAN	LHA 2	Ingalls Shipbuilding	21 July 1972	18 July 1974	15 Oct 1977	AA
BELLEAU WOOD	LHA 3	Ingalls Shipbuilding	5 Mar 1973	11 Apr 1977	23 Sep 1978	PA
NASSAU	LHA 4	Ingalls Shipbuilding	13 Aug 1973	21 Jan 1978	28 July 1979	AA
PELELIU (ex-Da Nang)	LHA 5	Ingalls Shipbuilding	12 Nov 1976	25 Nov 1978	3 May 1980	PA

Displacement, tons: 39,967 full load
Dimensions, feet (metres): 834 × 131.9 × 25.9
(254.2 × 40.2 × 7.9)
Flight deck, feet (metres): 820 × 118.1 *(250 × 36)*
Main machinery: 2 Combustion Engineering boilers; 600 psi
(42.3 kg/cm²); 900°F *(482°C)*; 2 Westinghouse turbines;
70,000 hp *(52.2 MW)*; 2 shafts; bow thruster; 900 hp
(670 kW)
Speed, knots: 24. **Range, n miles:** 10,000 at 20 kt
Complement: 930 (56 officers)
Military lift: 1,703 troops; 4 LCU 1610 type or 2 LCU and 2 LCM
8 or 17 LCM 6 or 45 Assault Amphibious Vehicles; 1,200 tons
aviation fuel. 1 LCAC may be embarked. 4 LCPL

Missiles: SAM: 2 GDC Mk 49 RAM ❶; 21 rounds per launcher;
passive IR/anti-radiation homing to 9.6 km *(5.2 n miles)* at 2
Mach; warhead 9.1 kg.
Guns: 2 General Electric/General Dynamics 20 mm/76
6-barrelled Vulcan Phalanx Mk 15 ❷; 3,000 rds/min (4,500 in
Block 1) combined to 1.5 km.
6 Mk 242 25 mm automatic cannons. 8—12.7 mm MGs.
Countermeasures: Decoys: 4 Loral Hycor SRBOC 6-barrelled
fixed Mk 36; IR flares and chaff to 4 km *(2.2 n miles)*.
SLQ-25 Nixie; acoustic torpedo decoy system. NATO Sea Gnat.
SLQ-49 chaff buoys. AEB SSQ-95.
ESM/ECM: SLQ-32V(3); intercept and jammers.
Combat data systems: ACDS Block 0. Advanced Combat
Direction System to provide computerised support in control
of helicopters and aircraft, shipboard weapons and sensors,
navigation, landing craft control and electronic warfare. Links
4A, 11 and 16. SATCOM SRR-1, WSC-3 (UHF), USC-38 (EHF).
SMQ-11 Metsat (see Data Systems at front of section).
Radars: Air search: ITT SPS-48E ❸; E/F-band.
Lockheed SPS-40E ❹; B-band.
Hughes Mk 23 TAS ❺; D-band.
Surface search: Raytheon SPS-67(V)3 ❻; G-band.
Navigation: Raytheon SPS-73; I-band.
CCA: SPN-35A; SPN-43B.
Tacan: URN 25. IFF: CIS Mk XV/UPX-36.

Fixed-wing aircraft: Harrier AV-8B VSTOL aircraft in place of
some helicopters as required. V22 Osprey in due course.
Helicopters: 19 CH-53D Sea Stallion or 26 CH-46D/E Sea Knight
UAV in due course.

TARAWA   *7/2004*, Michael Nitz /* 1043626

Programmes: Originally intended to be a class of nine ships. LHA
1 was authorised in FY69, LHA 2 and LHA 3 in FY70 and LHA
4 and LHA 5 in FY71.
Modernisation: Two Vulcan Phalanx CIWS replaced the GMLS
Mk 25 Sea Sparrow launchers. Programme completed in
early 1991. RAM launchers fitted to all of the class 1993-95.
One launcher is above the bridge offset to port, and the other
on the starboard side at the after end of the flight deck. Mk 23
TAS target acquisition radar fitted in LHA 3 and 5 in 1992,
LHA 4 in 1993 and the last pair in 1994. SPS-48E started
replacing SPS-52D in 1994 to improve low altitude detection
of missiles and aircraft. ACDS Block 0 in 1996. 5 in guns
removed in 1997/98. Plans to fit SSDS have been shelved.
Collective Protection Systems upgrade in progress. Fuel oil
compensation system has been installed to improve damaged
stability.
Structure: There are two lifts, one on the port side aft and one at
the stern. Beneath the after elevator is a floodable docking

well measuring 268 ft in length and 78 ft in width which is
capable of accommodating four LCU 1610 type landing craft.
Also included is a large garage for trucks and AFVs and troop
berthing for a reinforced battalion. 33,730 sq ft available for
vehicles and 116,900 cu ft for palletted stores. Extensive
medical facilities including operating rooms, X-ray room,
hospital ward, isolation ward, laboratories, pharmacy, dental
operating room and medical store rooms.
Operational: The flight deck can operate a maximum of nine
CH-53D Sea Stallion or 12 CH-46D/E Sea Knight helicopters
or a mix of these and other helicopters at any one time. With
some additional modifications, ships of this class can
effectively operate AV-8B aircraft. The normal mix of aircraft
allows for six AV-8Bs. The optimum aircraft configuration is
dependent upon assigned missions. Unmanned
Reconnaissance Vehicles (URVs) can be operated. LHA 3 to
be decommissioned in FY06 and LHA 2 in FY07.

UPDATED

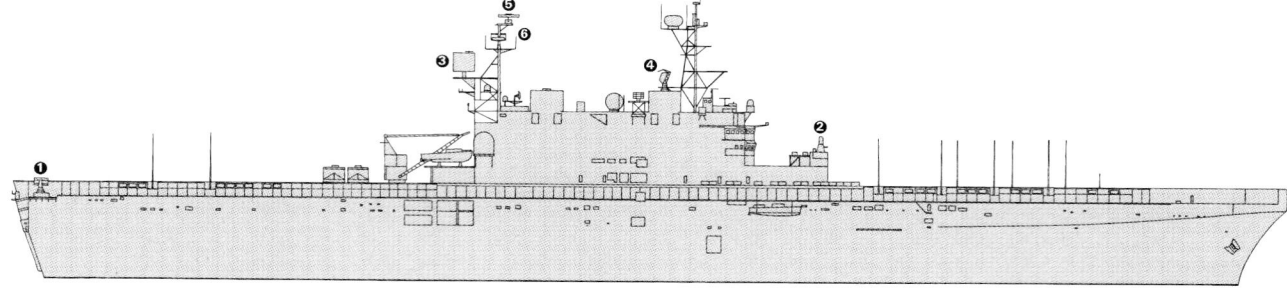

PELELIU   *(Scale 1 : 1,500), Ian Sturton /* 0131369

NASSAU   *7/2004*, US Navy /* 1043625

11 AUSTIN CLASS: AMPHIBIOUS TRANSPORT DOCKS (LPD)

Name	No	Builders	Laid down	Launched	Commissioned	F/S
AUSTIN	LPD 4	New York Naval Shipyard	4 Feb 1963	27 June 1964	6 Feb 1965	AA
OGDEN	LPD 5	New York Naval Shipyard	4 Feb 1963	27 June 1964	19 June 1965	PA
DULUTH	LPD 6	New York Naval Shipyard	18 Dec 1963	14 Aug 1965	18 Dec 1965	PA
CLEVELAND	LPD 7	Ingalls Shipbuilding	30 Nov 1964	7 May 1966	21 Apr 1967	PA
DUBUQUE	LPD 8	Ingalls Shipbuilding	25 Jan 1965	6 Aug 1966	1 Sep 1967	PA
DENVER	LPD 9	Lockheed SB & Construction Co	7 Feb 1964	23 Jan 1965	26 Oct 1968	PA
JUNEAU	LPD 10	Lockheed SB & Construction Co	23 Jan 1965	12 Feb 1966	12 July 1969	PA
SHREVEPORT	LPD 12	Lockheed SB & Construction Co	27 Dec 1965	25 Oct 1966	12 Dec 1970	AA
NASHVILLE	LPD 13	Lockheed SB & Construction Co	14 Mar 1966	7 Oct 1967	14 Feb 1970	AA
TRENTON	LPD 14	Lockheed SB & Construction Co	8 Aug 1966	3 Aug 1968	6 Mar 1971	AA
PONCE	LPD 15	Lockheed SB & Construction Co	31 Oct 1966	20 May 1970	10 July 1971 /	AA

Displacement, tons: 9,130 light; 16,500-17,244 full load
Dimensions, feet (metres): 570 × 100 (84 hull) × 23
 (173.8 × 30.5 (25.6) × 7)
Main machinery: 2 Foster-Wheeler boilers (Babcock & Wilcox in
 LPD 5, LPD 6 and LPD 12); 600 psi *(42.3 kg/cm²)*; 870°F
 (467°C); 2 De Laval (General Electric in LPD 9 and LPD 10)
 turbines; 24,000 hp *(18 MW)*; 2 shafts
Speed, knots: 21. **Range, n miles:** 7,700 at 20 kt.
Complement: 420 (24 officers); Flag 90 (in LPD 7-13)
Military lift: 930 troops (840 only in LPD 7-13); 9 LCM 6s or 4
 LCM 8s or 2 LCAC or 20 LVTs. 4 LCPL/LCVP

Guns: 2 General Electric/General Dynamics 20 mm/76
 6-barrelled Vulcan Phalanx Mk 15 ❶; 3,000 rds/min (4,500 in
 Block 1) combined to 1.5 km.
 2—25 mm Mk 38. 8—12.7 mm MGs.
Countermeasures: Decoys: 4 Loral Hycor SRBOC 6-barrelled Mk
 36; IR flares and chaff to 4 km *(2.2 n miles)*.
 ESM: SLQ-32(V)1; intercept. May be updated to (V)2.
Combat data systems: SATCOM ❷, WSC-3 (UHF), WSC-6 (SHF)
 (see Data Systems at front of section).
Radars: Air search: Lockheed SPS-40E ❸; B-band.
 Surface search: Norden SPS-67 ❹; G-band.
 Navigation: Raytheon SPS-73(V)12; I-band.
 Tacan: URN 25. IFF: Mk XII UPX-36.

Helicopters: Up to 6 CH-46D/E Sea Knight can be carried.
 Hangar for only 1 light (not in LPD 4).

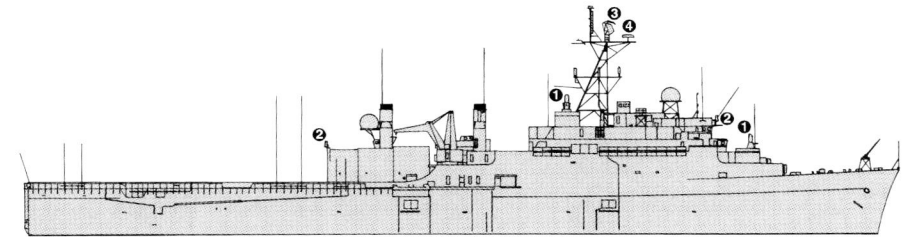

DENVER *(Scale 1 : 1,500), Ian Sturton /* 0016471

Programmes: LPD 4-6 were authorised in the FY62 new
construction programme, LPD 7-10 in FY63, LPD 12 and 13 in
FY64, LPD 14 and LPD 15 in FY65.
Modernisation: Modernisation carried out in normal
maintenance periods from FY87. This included fitting two
Phalanx, SPS-67 radar replacing SPS-10 and updating EW
capability. 3 in guns have been removed. Five ships to receive
machinery, electrical and habitability to extend service lives.
Structure: LPD 7-13 have an additional bridge and are fitted as
flagships. One small telescopic hangar. There are structural
variations in the positions of guns and electronic equipment in
different ships of the class. Flight deck is 168 ft *(51.2 m)* in

length. Well-deck 394 × 50 ft *(120.1 × 15.2 m)*.
Communications domes are not uniformly fitted.
Operational: A typical operational load might include one
Seahawk, two Sea Knight, two Twin Huey, four Sea Cobra
helicopters and one Cyclone patrol craft. LPD 10 based at
Sasebo, Japan. LPDs 5-9 at San Diego and LPDs 4 and 12-15
at Norfolk. LPD 6 to decommission in 2005, LPD 4 in 2006
and LPDs 5 and 14 in 2007.

UPDATED

DUBUQUE *6/2003, Mick Prendergast /* 0572790

CLEVELAND *10/2003, Frank Findler /* 0572791

8 WHIDBEY ISLAND CLASS and 4 HARPERS FERRY CLASS (DOCK LANDING SHIPS) (LSDHM and LSD-CV)

Name	No	Builders	Laid down	Launched	Commissioned	F/S
WHIDBEY ISLAND	LSD 41	Lockheed SB & Construction Co	4 Aug 1981	10 June 1983	9 Feb 1985	AA
GERMANTOWN	LSD 42	Lockheed SB & Construction Co	5 Aug 1982	29 June 1984	8 Feb 1986	PA
FORT McHENRY	LSD 43	Lockheed SB & Construction Co	10 June 1983	1 Feb 1986	8 Aug 1987	PA
GUNSTON HALL	LSD 44	Avondale Industries	26 May 1986	27 June 1987	22 Apr 1989	AA
COMSTOCK	LSD 45	Avondale Industries	27 Oct 1986	16 Jan 1988	3 Feb 1990	PA
TORTUGA	LSD 46	Avondale Industries	23 Mar 1987	15 Sep 1988	17 Nov 1990	AA
RUSHMORE	LSD 47	Avondale Industries	9 Nov 1987	6 May 1989	1 June 1991	PA
ASHLAND	LSD 48	Avondale Industries	4 Apr 1988	11 Nov 1989	9 May 1992	AA
HARPERS FERRY	LSD 49	Avondale Industries	15 Apr 1991	16 Jan 1993	7 Jan 1995	PA
CARTER HALL	LSD 50	Avondale Industries	11 Nov 1991	2 Oct 1993	30 Sep 1995	AA
OAK HILL	LSD 51	Avondale Industries	21 Sep 1992	11 June 1994	8 June 1996	AA
PEARL HARBOR	LSD 52	Avondale Industries	27 Jan 1995	24 Feb 1996	30 May 1998	PA

Displacement, tons: 11,125 light; 15,939 (LSD 41-48), 16,740 (LSD 49 onwards) full load

Dimensions, feet (metres): 609.5 × 84 × 20.5 *(185.8 × 25.6 × 6.3)*

Main machinery: 4 Colt SEMT-Pielstick 16 PC2.5 V 400 diesels; 33,000 hp(m) *(24.6 MW)* sustained; 2 shafts; cp props

Speed, knots: 22. **Range, n miles:** 8,000 at 18 kt

Complement: 413 (21 officers)

Military lift: 402 (+102 surge) troops; 2 (CV) or 4 LCACs, or 9 (CV) or 21 LCM 6, or 1 (CV) or 3 LCUs, or 64 LVTs. 2 LCPL

Cargo capacity: 5,000 cu ft for marine cargo, 12,500 sq ft for vehicles (including four preloaded LCACs in the well-deck). The cargo version' has 67,600 cu ft for marine cargo, 20,200 sq ft for vehicles but only two LCACs. Aviation fuel, 90 tons.

Missiles: 2 GDC/Hughes Mk 49 RAM ❶; passive IR/anti-radiation homing to 9.6 km *(5.2 n miles)* at 2 Mach; warhead 9.1 kg. Being fitted in all.

Guns: 2 General Electric/General Dynamics 20 mm/76 6-barrelled Vulcan Phalanx Mk 15 ❷; 3,000 rds/min (4,500 in Block 1) combined to 1.5 km.
2—25 mm Mk 38. 6—12.7 mm MGs.

Countermeasures: Decoys: 4 Loral Hycor SRBOC 6-barrelled Mk 36 and Mk 50; IR flares and chaff. SLQ-25 Nixie.
ESM: SLQ-32(V)1; intercept. SLQ-49.

Combat data systems: SATCOM SRR-1, WSC-3 (UHF) (see Data Systems at front of section). SSDS Mk 1.

Radars: Air search: Raytheon SPS-49(V)5 ❸; C-band.
Surface search: Norden SPS-67V ❹; G-band.
Navigation: Raytheon SPS-64(V)9 or SPS-73(V)12; I/J-band.
Tacan: URN 25. IFF: Mk XII UPX-29/UPX-36.

Helicopters: Platform only for 2 CH-53 Sea Stallion.

Programmes: Originally it was planned to construct six ships of this class as replacements for the Thomaston class LSDs. Eventually, the level of Whidbey Island class ships was established at eight, with four additional cargo-carrying variants of that class to provide increased cargo-carrying capability. The first cargo variant, LSD 49, was authorised and funded in the FY88 budget; LSD 50 in FY89 and LSD 51 in FY91. The fourth was authorised in FY92 but not ordered until 12 October 1993.

Modernisation: A Quick Reaction Combat Capability (QRCC)/Ship Self-Defense System (SSDS) was installed and successfully demonstrated in LSD 41 in 1993. During the QRCC demonstrations, the ship's SPS-49, SLQ-32, RAM and Phalanx were successfully integrated via SSDS. All ships of the class fitted with SSDS Mk 1.

Structure: Based on the earlier Anchorage class. One 60 and one 20 ton crane. Well-deck measures 440 × 50 ft *(134.1 × 15.2 m)* in the LSD but is shorter in the Cargo Variant (CV). The cargo version is a minimum modification to the LSD 41 design. Changes in that design include additional troop magazines, air conditioning, piping and hull structure; the forward Phalanx is forward of the bridge, RAM is on the bridge roof, and there is only one crane. There is approximately 90 per cent commonality between the two classes.

Operational: LSDs 43 and 49 are based at Sasebo.

UPDATED

ASHLAND   (Scale 1 : 1,500), Ian Sturton / 0053362

PEARL HARBOR   6/2003, Mick Prendergast / 0572788

TORTUGA   5/2003, A Sharma / 0572789

RUSHMORE   7/2004 *, US Navy / 1043624

74 LANDING CRAFT AIR CUSHION (LCAC)

Displacement, tons: 87.2 light; 170-182 full load
Dimensions, feet (metres): 88 oa (on-cushion) (81 between hard structures) × 47 beam (on-cushion) (43 beam hard structure) × 2.9 draught (off-cushion) *(26.8 (24.7) × 14.3 (13.1) × 0.9)*
Main machinery: 4 Allied-Signal TF40B marine gas turbines for propulsion and lift; 16,000 hp *(11.9 MW)* sustained; 2 shrouded reversible-pitch airscrews (propulsion); 4 double-entry fans, centrifugal or mixed-flow (lift). SLEP configuration, 4 Vericor Power Systems ETF40B marine gas turbines with Full Authority Digital Engine Control (FADEC) for propulsion and lift; 19,000 hp *(1.41 MW)* sustained; 2 shrouded reversible-pitch airscrews (propulsion); 4 double-entry fans, centrifugal or mixed-flow (lift)
Speed, knots: 40 (loaded). **Range, n miles:** 300 at 35 kt; 200 at 40 kt
Complement: 5
Military lift: 24 troops; 1 Main Battle Tank or 60-75 tons
Radars: Navigation: Marconi LN66 or Decca Bridgemaster E; I-band.

Programmes: Built by Textron Marine and Land Systems and Avondale Gulfport. A total of 90 craft delivered 1984-1997. The final craft LCAC 91 delivered in 2001 in SLEP configuration.
Modernisation: 72 in-service craft to receive Ship Life Extension Programme (SLEP) from 2002-2016. The programme includes the installation of more powerful engines to provide greater lift capacity, an improved deep skirt for better handling in heavier sea states and an integrated navigation suite for precise navigation, and advanced Multimode Integrated Communications System in either normal, secure modes. Four craft were upgraded in FY04, five in FY05 and six craft in subsequent metres.
Structure: Incorporates the best attributes of the JEFF(A) and JEFF(B) learned from over five years of testing the two prototypes. Bow ramp 28.8 ft, stern ramp 15 ft. Cargo space capacity is 1,809 sq ft. Noise and dust levels are high and if disabled the craft is not easy to tow. 30 mm Gatling guns can be fitted.
Operational: Ship classes capable of carrying the LCAC are Wasp (three), Tarawa (one), Austin (one), Whidbey Island (four), Harpers Ferry (two) and San Antonio (two). A portable transport module can be carried on the cargo deck to transport up to 180 troops. It is claimed that LCACs are capable of accessing 70 per cent of the world's coastlines compared to about 15 per cent for conventional landing craft. Some limitations in very rough seas. Shore bases on each coast at Little Creek, VA and Camp Pendleton, CA.
Sales: Six to Japan. One of a similar type built by South Korea.

UPDATED

LCAC 45   *3/2003, A Sharma* / 0572787

LCAC 53   *9/2002, Winter & Findler* / 0529980

7 + 1 (1) FRANK S BESSON CLASS: LOGISTIC SUPPORT VESSELS (LSV-ARMY)

Name	No	Builders	Commissioned
GENERAL FRANK S BESSON JR	LSV 1	Moss Point Marine, MS	18 Dec 1987
CW 3 HAROLD C CLINGER	LSV 2	Moss Point Marine, MS	20 Feb 1988
GENERAL BREHON B SOMERVELL	LSV 3	Moss Point Marine, MS	2 Apr 1988
LT GENERAL WILLIAM B BUNKER	LSV 4	Moss Point Marine, MS	18 May 1988
MAJOR GENERAL CHARLES P GROSS	LSV 5	Moss Point Marine, MS	30 Apr 1991
SPECIALIST 4 JAMES A LOUX	LSV 6	Moss Point Marine, MS	16 Dec 1994
SSGT ROBERT T KURODA	LSV 7	VT Halter Marine	2005
MAJOR GENERAL ROBERT SMAILS	LSV 8	VT Halter Marine	2006

Displacement, tons: 4,265 full load
Dimensions, feet (metres): 272.8 × 60 × 12 *(83.1 × 18.3 × 3.7)*
314 (LSV 7) × 60.0 × 12.0 *(95.7 × 18.3 × 36.6)*
Main machinery: 2 GM EMD 16-645E2 diesels; 3,900 hp *(2.9 MW)* sustained; 2 shafts; Schottel bow thruster; 250 hp *(187 kW)*
Speed, knots: 11.6. **Range, n miles:** 8,300 at 11 kt
Complement: 30 (6 officers)
Military lift: 2,280 tons of vehicles including 26 M-1 tanks, containers or general cargo
Radars: Navigation: Raytheon SPS-64(V)2; I-band.

Comment: First one approved in FY85, second in FY87, remainder from Army reserve funds. Army owned ro-ro design with 10,500 sq ft of deck space for cargo. Capable of beaching with 4 ft over the ramp on a 1:30 offshore gradient with a payload of 900 tons of cargo. Three are based at Fort Eustis, Virginia, two in Hawaii and LSV-3 is with the National Guard at Tacoma, Washington. Two modified ships of the class built for the Philippines Navy in 1993-94.

UPDATED

CW 3 HAROLD C CLINGER   *7/2002, Chris Sattler* / 0529979

13 MECHANISED LANDING CRAFT: LCM 6 TYPE

Displacement, tons: 64 full load
Dimensions, feet (metres): 56.2 × 14 × 3.9 *(17.1 × 4.3 × 1.2)*
Main machinery: 2 Detroit 6V-71 diesels; 348 hp *(260 kW)* sustained or 2 Detroit 8V-71 diesels; 460 hp *(344 kW)* sustained; 2 shafts
Speed, knots: 9. **Range, n miles:** 130 at 9 kt
Complement: 5
Military lift: 34 tons or 80 troops

Comment: Welded steel construction. Used for various utility tasks. Numbers are reducing.

UPDATED

LCM 6   *6/1997, J W Currie* / 0016482

35 LCU 2000 CLASS: UTILITY LANDING CRAFT (LCU-ARMY)

LCU 2001-2035

Displacement, tons: 1,102 full load
Dimensions, feet (metres): 173.8 × 42 × 8.5 *(53 × 12.8 × 2.6)*
Main machinery: 2 Cummins KTA50-M diesels; 2,500 hp *(1.87 MW)* sustained; 2 shafts; bow thruster
Speed, knots: 11.5. **Range, n miles:** 4,500 at 11.5 kt
Complement: 13 (2 officers)
Military lift: 350 tons
Radars: Navigation: 2 Raytheon SPS-64; I-band.

Comment: Order placed with Lockheed by US Army 11 June 1986. First one completed 21 February 1990 by Moss Point Marine. The 2000 series have names, some of which duplicate naval ships. These are the first LCUs to have been built to an Army specification. 14 are active, 12 in reserve, eight prepositioned and one used for training.

VERIFIED

LCU 2029   *6/2003, A Sharma* / 0572815

LCU 2024   *7/2003, A Sharma* / 0572813

89 MECHANISED LANDING CRAFT: LCM 8 TYPE

Displacement, tons: 65.6 light; 105 full load
Dimensions, feet (metres): 73.7 × 21 × 5.2 *(22.5 × 6.4 × 1.6)*
Main machinery: 2 Detroit 12V-71 diesels; 680 hp *(507 kW)* sustained; 2 shafts
Speed, knots: 12. **Range, n miles:** 190 at 9 kt full load
Complement: 5
Military lift: 1 M48 or 1 M60 tank or 200 troops

Comment: Naval craft are for use in amphibious ships. The last 12 were built in 1993-94. 22 similar craft are used by the Army.

UPDATED

LCM 8   *5/2003, A Sharma* / 0572786

37 LCU 1600 CLASS: UTILITY LANDING CRAFT
(LCU-ARMY (2) and NAVY (35))

Displacement, tons: 200 light; 375 full load
Dimensions, feet (metres): 134.9 × 29 × 6.1 *(41.1 × 8.8 × 1.9)*
Main machinery: 4 Detroit 6-71 diesels; 696 hp *(519 kW)* sustained; 2 shafts; Kort nozzles
2 Detroit 12V-71 diesels (LCU 1680-1681); 680 hp *(508 kW)* sustained; 2 shafts; Kort nozzles
Speed, knots: 11. **Range, n miles:** 1,200 at 8 kt
Complement: 14 (2 officers)
Military lift: 134 tons or 400 troops
Guns: 2—12.7 mm MGs.
Radars: Navigation: Furuno; I-band.

Comment: Steel hulled construction. Versatile craft used for a variety of tasks. Most were built between the mid-1960s and mid-1980s. There are no plans for more of this type and a replacement craft is under consideration. Three converted to Diver Support Craft (ASDV). LCU 1667 and 1675 operated by the US Army. Two USN craft are in reserve.
UPDATED

LCU 1651   *8/2004*, Hachiro Nakai /* 1043687

145 LANDING CRAFT PERSONNEL (LCPL)

Displacement, tons: 11 full load
Dimensions, feet (metres): 36 × 12.1 × 3.8 *(11 × 3.7 × 1.2)*
Main machinery: 1 GM 8V-71TI diesel; 425 hp *(317 kW)* sustained; 1 shaft
Speed, knots: 20. **Range, n miles:** 150 at 20 kt
Complement: 3
Military lift: 17 troops
Radars: Navigation: Marconi LN66; I-band.

Comment: There are 13 Mk 11, 105 Mk 12, 15 Mk 13 and 12 11 m LCPLs. Details given are for Mk 12 and 13. For use as control craft and carried aboard LHA, LPD and LSD classes. 33 are active.
VERIFIED

LCU 1600 class (Army)   *2/2001, M Declerck /* 0529975

LCPL Mk 13   *4/1991, Bollinger /* 0084143

MINE WARFARE FORCES

Notes: (1) There are no surface minelayers. Mining is done by carrier-based aircraft, land-based aircraft and submarines. The mine inventory includes Mk 56 moored influence mines, the Mk 67 submarine launched mobile mine (SLMM) and the Quickstrike series of bottom mines. Mk 56 is being phased out.
(2) NRF ships are manned by active and reserve crews.
(3) MH-53E Sea Stallion helicopters can be deployed in LHDs or transported by C-5 aircraft for mine countermeasures.
(4) The Long-term Mine Reconnaissance System (LMRS) will be an autonomous UUV, launched from a Los Angeles class submarine and thereafter from a Virginia class SSN, to provide covert mine reconnaissance capability. Boeing has been

contracted to design and develop the LMRS. An inventory of 12 is expected.
(5) Marine Mammal Systems (MMS) uses trained dolphins and sea lions for mine detection, swimmer protection, and recovery of exercise mines and torpedoes. The dolphins can be transported by C-5 aircraft or amphibious ships. The MMS is the only operational method of detecting and neutralising buried mines.
(6) A Remote Minehunting System (RMS) is being developed for Arleigh Burke Flight IIA destroyers (DDG 91-96). The system is an unmanned semi-submersible. There is an over-the-horizon radio link and an AQS-20A VDS. The RMS is 23 ft, weighs 14,000 lb

and has an endurance of 24 hours at 12 kt. RMS is in Engineering and Manufacturing Development Phase.
(7) Rapid Airborne Mine Clearance System (RAMICS) is under development. RAMICS is to be operated from a MH-60S helicopter and consists of an electro-optic detection and ranging system and a 30 mm gun system to destroy near-surface and floating moored mines.
(8) Organic Airborne and Surface Influence Sweep (OASIS) is under development with developmental testing to complete in 2005.

14 AVENGER CLASS: MINE COUNTERMEASURES VESSELS (MCM/MHSO)

Name	No	Builders	Laid down	Launched	Commissioned	F/S
AVENGER	MCM 1	Peterson Builders Inc	3 June 1983	15 June 1985	12 Sep 1987	NRF
DEFENDER	MCM 2	Marinette Marine Corp	1 Dec 1983	4 Apr 1987	30 Sep 1989	NRF
SENTRY	MCM 3	Peterson Builders Inc	8 Oct 1984	20 Sep 1986	2 Sep 1989	NRF
CHAMPION	MCM 4	Marinette Marine Corp	28 June 1984	15 Apr 1989	27 July 1991	NRF
GUARDIAN	MCM 5	Peterson Builders Inc	8 May 1985	20 June 1987	16 Dec 1989	PA
DEVASTATOR	MCM 6	Peterson Builders Inc	9 Feb 1987	11 June 1988	6 Oct 1990	AA
PATRIOT	MCM 7	Marinette Marine Corp	31 Mar 1987	15 May 1990	18 Oct 1991	PA
SCOUT	MCM 8	Peterson Builders Inc	8 June 1987	20 May 1989	15 Dec 1990	AA
PIONEER	MCM 9	Peterson Builders Inc	5 June 1989	25 Aug 1990	7 Dec 1992	AA
WARRIOR	MCM 10	Peterson Builders Inc	25 Sep 1989	8 Dec 1990	3 Apr 1993	AA
GLADIATOR	MCM 11	Peterson Builders Inc	7 July 1990	29 June 1991	18 Sep 1993	NRF
ARDENT	MCM 12	Peterson Builders Inc	22 Oct 1990	16 Nov 1991	18 Feb 1994	AE
DEXTROUS	MCM 13	Peterson Builders Inc	11 Mar 1991	20 June 1992	9 July 1994	AE
CHIEF	MCM 14	Peterson Builders Inc	19 Aug 1991	12 June 1993	5 Nov 1994	AA

Displacement, tons: 1,379 full load
Dimensions, feet (metres): 224.3 × 38.9 × 12.2
(68.4 × 11.9 × 3.7)
Main machinery: 4 Waukesha L-1616 diesels (MCM 1-2); 2,600 hp(m) *(1.91 MW)* or 4 Isotta Fraschini ID 36 SS 6V AM diesels (MCM 3 onwards); 2,280 hp(m) *(1.68 MW)* sustained; 2 Hansome Electric motors; 400 hp(m) *(294 kW)* for hovering; 2 shafts; cp props; 1 Omnithruster hydrojet; 350 hp *(257 kW)*
Speed, knots: 13.5. **Range, n miles:** 2,500 at 10 kt
Complement: 84 (8 officers)

Guns: 2—12.7 mm MGs.
Countermeasures: MCM: 2 SLQ-48; includes Honeywell/ Hughes ROV mine neutralisation system, capable of 6 kt (1,500 m cable with cutter (MP1), and countermining charge) (MP 2). SLQ-37(V)3; magnetic/acoustic influence sweep equipment. Oropesa SLQ-38 Type 0 Size 1; mechanical sweep.
Combat data systems: SATCOM SRR-1; WSC-3 (UHF). GEC/ Marconi Nautis M in last two ships includes SSN 2 PINS command system and control. USQ-119E(V), UHF Dama and OTCIXS provide JMCIS connectivity.
Radars: Surface search: ISC Cardion SPS-55; I/J-band.
Navigation: ARPA 2525 or LN66; I-band. Both to be replaced by SPS-73.
Sonars: Raytheon/Thomson Sintra SQQ-32(V)3; VDS; active minehunting; high frequency.

Programmes: The contract for the prototype MCM was awarded in June 1982. The last three were funded in FY90.
Modernisation: Integrated Combat Weapon System (ICWS) Block 1 upgrade. Integrated Ship Control System (ISCS) installed in all hulls.
Structure: The hull is constructed of oak, Douglas fir and Alaskan cedar, with a thin coating of fibreglass on the outside, to permit taking advantage of wood's low magnetic signature. A

AVENGER   *4/2004*, US Navy /* 1043631

problem of engine rotation on the Waukesha diesels in MCM 1-2 was resolved; however, those engines have been replaced in the rest of the class by low magnetic engines manufactured by Isotta-Fraschini of Milan, Italy. Fitted with SSN2(V) Precise Integrated Navigation System (PINS).
Operational: *Avenger* fitted with the SQQ-32 for Gulf operations in 1991 and all of the class have been retrofitted. Two

transferred to NRF in 1995, two more in 1996. *Ardent* and *Dextrous* permanently stationed in Bahrain from March 1996, and *Guardian* and *Patriot* are at Sasebo, Japan. The remainder are based at Ingleside, Texas.
UPDATED

12 OSPREY CLASS (MINEHUNTERS COASTAL) (MHC)

Name	No	Builders	Launched	Commissioned	F/S
OSPREY	MHC 51	Intermarine, Savannah	23 Mar 1991	20 Nov 1993	NRF
HERON	MHC 52	Intermarine, Savannah	21 Mar 1992	6 Aug 1994	NRF
PELICAN	MHC 53	Avondale Industries	27 Feb 1993	18 Nov 1995	NRF
ROBIN	MHC 54	Avondale Industries	11 Sep 1993	11 May 1996	NRF
ORIOLE	MHC 55	Intermarine, Savannah	22 May 1993	16 Sep 1995	NRF
KINGFISHER	MHC 56	Avondale Industries	18 June 1994	26 Oct 1996	NRF
CORMORANT	MHC 57	Avondale Industries	21 Oct 1995	12 Apr 1997	NRF
BLACK HAWK	MHC 58	Intermarine, Savannah	27 Aug 1994	11 May 1996	NRF
FALCON	MHC 59	Intermarine, Savannah	3 June 1995	26 Oct 1997	NRF
CARDINAL	MHC 60	Intermarine, Savannah	9 Mar 1996	18 Oct 1997	AE
RAVEN	MHC 61	Intermarine, Savannah	28 Sep 1996	5 Sep 1998	AE
SHRIKE	MHC 62	Intermarine, Savannah	24 May 1997	31 May 1999	NRF

Displacement, tons: 930 full load
Dimensions, feet (metres): 187.8 × 35.9 × 9.5
(57.2 × 11 × 2.9)
Main machinery: 2 Isotta Fraschini ID 36 SS 8V AM diesels;
1,600 hp(m) *(1.18 MW)* sustained; 2 Voith-Schneider props; 3
Isotta Fraschini ID 36 diesel generators; 984 kW
Speed, knots: 10. **Range, n miles:** 1,500 at 10 kt
Complement: 51 (5 officers)

Guns: 2—12.7 mm MGs.
Countermeasures: MCM: Alliant SLQ-48 mine neutralisation
system ROV (with 1,070 m cable). Degaussing DGM-4.

Combat data systems: Unisys SYQ 13 and SYQ 109; integrated
combat and machinery control system. USQ-119E(V), UHF
Dama, and OTCIXS provide GCCS connectivity.
Radars: Surface search: Raytheon SPS-64(V)9; I-band.
Navigation: R41XX; I-band.
Sonars: Raytheon/Thomson Sintra SQQ-32(V)3; VDS; active
minehunting; high frequency.

Programmes: A design contract for Lerici class mine hunters
was awarded in August 1986 followed by a construction
contract with Intermarine USA in May 1987 for eight of the 12
ships of the class. On 2 October 1989 Avondale, Gulfport was

named as the second construction source. Contracts were
awarded for four ships which were delivered between 1995
and 1997.
Structure: Construction is of monocoque GRP throughout hull,
with frames eliminated. Main machinery is mounted on GRP
cradles and provided with acoustic enclosures. SQQ-32 is
deployed from a central well forward. Fitted with Voith
cycloidal propellers which eliminate need for forward
thrusters during station keeping.
Operational: All but two are NRF based at Ingleside, Texas.
MHC-60 and MHC-61 are based at Bahrain.

UPDATED

RAVEN
4/2004*, US Navy / 1043630

PATROL FORCES

Notes: 'Spartan' is a technology demonstrator programme to prove utility of unmanned surface
craft. It is envisaged that such craft will be capable of conducting mine warfare, force protection
(including surveillance and reconnaissance) and anti-surface warfare. A prototype, a 7 m RHIB
installed with navigation, communications and remote control equipment, underwent sea trials in
2003 which included embarkation in USS *Gettysburg* as part of the *Enterprise* carrier strike group.

8 CYCLONE CLASS (PATROL COASTAL SHIPS) (PBFM)

Name	No	Builders	Commissioned	F/S
HURRICANE	PC 3	Bollinger, Lockport	15 Oct 1993	PA
TYPHOON	PC 5	Bollinger, Lockport	12 Feb 1994	AA
SIROCCO	PC 6	Bollinger, Lockport	11 June 1994	AA
SQUALL	PC 7	Bollinger, Lockport	4 July 1994	PA
CHINOOK	PC 9	Bollinger, Lockport	28 Jan 1995	AA
FIREBOLT	PC 10	Bollinger, Lockport	10 June 1995	AA
WHIRLWIND	PC 11	Bollinger, Lockport	1 July 1995	AA
THUNDERBOLT	PC 12	Bollinger, Lockport	7 Oct 1995	AA

Displacement, tons: 354 full load; 386 full load (PC 2, 8, 13, 14)
Dimensions, feet (metres): 170.3; 179 (PC 2, 8, 13, 14) × 25.9 × 7.9 *(51.9; 54.6 × 7.9 × 2.4)*

Main machinery: 4 Paxman Valenta 16RP200CM diesels; 13,400 hp *(10 MW)* sustained; 4 shafts
Speed, knots: 35. **Range, n miles:** 2,500 at 12 kt
Complement: 39 (4 officers) plus 9 SEALs
Missiles: SAM: 1 sextuple Stinger mounting.
Guns: 1 Bushmaster 25 mm Mk 38. 1 Bushmaster 25 mm/87 Mk 96 (aft). 4—12.7 mm MGs.
4—7.62 mm MGs. 2—40 mm Mk 19 grenade launchers (MGs and grenade launchers are
interchangeable).
Countermeasures: Decoys: 2 Mk 52 sextuple and/or Wallop Super Barricade Mk 3 chaff
launchers.
ESM: Privateer APR-39; radar warning. Sensytech Bobcat.
Weapons control: Marconi VISTAR IM 405 IR system.
Radars: Surface search: 2 Sperry RASCAR; E/F/I/J-band.
Sonars: Wesmar; hull-mounted; active; high frequency.

Programmes: Contract awarded for eight in August 1990 and five more in July 1991.
Structure: Design based on Vosper Thornycroft Ramadan class modified for USN requirements
including 1 in armour on superstructure. A stabilised weapon platform has been developed for
Stinger. The craft have a slow speed loiter capability. Swimmers can be launched from a
platform at the stern. Two SEAL raiding craft and one RIB are carried. PC 14 is fitted with
advanced ESM, a rocket launcher, better communications and a Mk 96 stabilised weapon
platform. It has also been modified, together with PC 2, PC 4, PC 8 and PC 13, to incorporate a
semi-dry well, boat ramp and stern gate to facilitate deployment and recovery of a fully loaded
RIB while the ship is making way.
Operational: Can be operated in pairs with a 12-man maintenance team in two vans ashore.
Tactical control of six ships (with USN crews) transferred to Coast Guard in November 2001 for
homeland security duties. The remaining seven similarly transferred in January 2002.
Operational control transferred from Special Operations Command to the Atlantic and Pacific
Fleets on 1 October 2002. A maintenance programme keeps the ships at full readiness. The
stern-gate fitted ships (PC 2, PC 4, PC 8, PC 13 and PC 14) are to have been transferred to the
USCG by 2005. Remaining ships are to be retained by the USN until at least 2008.
Sales: PC1 (*Cyclone*) transferred to the Philippines Navy for counter-terrorism duties.

UPDATED

CHINOOK
3/2003*, US Navy / 1043629

SQUALL
10/2002, M Mazumdar / 0529973

20 Mk V CLASS (HSIC)

Displacement, tons: 54 full load
Dimensions, feet (metres): 81.2 × 17.5 × 4.3 *(24.7 × 5.3 × 1.3)*
Main machinery: 2 MTU 12V 396 TE94 diesels; 4,506 hp *(3.36 MW)* sustained; 2 Kamewa water-jets
Speed, knots: 45. **Range, n miles:** 515 at 35 kt
Complement: 5
Military lift: 16 fully equipped troops
Guns: 5 Mk 46 Mod 4 mountings for twin 12.7 mm or 7.6 mm MGs, 1 Mk 19 40 mm grenade launcher.
Countermeasures: ESM: Sensytech Bobcat; radar intercept.
Radars: Navigation: Furuno; I-band.
IFF: APX-100(V).

Comment: This was the winning design of a competition held in 1994 to find a high-speed craft to insert and withdraw Navy SEAL teams and other special operations forces personnel. Fourteen delivered by mid-1998 and six more by mid-1999. All built at the Halter Marine Equitable Shipyard in New Orleans. The craft has an aluminium hull and is transportable by C-5 aircraft. Stinger missiles may be carried and gun armaments can be varied. A variant with three engines is in service with the Mexican Navy. *UPDATED*

MK V   *4/2003, A Sharma /* 0572743

20 SPECIAL OPERATIONS CRAFT RIVERINE (SOCR)

Displacement, tons: 9.1 full load
Dimensions, feet (metres): 33.0 × 9.0 × 2.0 *(10.1 × 2.7 × 0.6)*
Main machinery: 2 Yanmar 6LY2M-STE diesels; 440 hp *(328 kW)*; 2 Hamilton HJ292 waterjets
Speed, knots: 40+. **Range, n miles:** 195
Complement: 4
Military lift: 8 fully equipped troops
Guns: Combination of Mk 19 40 mm, 12.7 mm MG, 7.62 mm/M60, M240, GAU17 at 5 stations

Comment: Built by United States Marine, Inc. Aluminium hull. *NEW ENTRY*

SOCR   *2/2005*, US Navy /* 1043672

70 RIBs (RIGID INFLATABLE BOATS) (PBF)

Displacement, tons: 9 full load
Dimensions, feet (metres): 36.1 × 10.5 × 3 *(11 × 3.2 × 0.9)*
Main machinery: 2 Caterpillar 3126 diesels; 940 hp *(700 kW)*; 2 Kamewa FF 280 water-jets
Speed, knots: 35. **Range, n miles:** 200 at 33 kt
Complement: 4 plus 9 SEALs
Guns: 1—12.7 mm MG, 1—7.62 mm MG or Mk 19 Mod 3 grenade launcher.

Comment: Capable of carrying nine SEALS at 35 kt. Details given are for the latest type being built by USMI, New Orleans. Entered service between 1998 and 2002. *UPDATED*

RIB   *1/2002, M Declerck /* 0529972

RIB   *1/1998, US Navy /* 0016492

32 RIVERINE ASSAULT CRAFT (LCPF)

Displacement, tons: 7.5 full load
Dimensions, feet (metres): 35.1 × 9.2 × 2.3 *(10.7 × 2.8 × 0.7)*
Main machinery: 2 Cummins diesels; 600 hp *(447 kW)*; 2 waterjets
Speed, knots: 43
Complement: 4-5 plus 10-15 marines
Guns: 1—12.7 mm M2HB MG. 1—40 mm Mk 19 grenade launcher.

Comment: The first 14 were constructed by SeaArk Marine and delivered by 1992. A further 18 were constructed by Swiftships and delivered by 1994. Owned and operated by USMC and based at Camp Lejeune, NC. Can be transported in a C-130. *VERIFIED*

RAC   *2000, USMC /* 0106808

122 LIGHT PATROL BOATS (PBF)

Displacement, tons: 1.2 full load
Dimensions, feet (metres): 22.3 × 8.6 × 1.5 *(6.8 × 2.6 × 0.5)*
Main machinery: 2 OMC outboards; 300 hp *(224 kW)*
Speed, knots: 35
Complement: 3
Guns: 3—12.7 mm MGs. 1—7.62 mm MG.
Radars: Surface search: Furuno 1731; I-band.

Comment: Built by Boston Whaler in 1988 for US Special Operations Command. Air transportable. Glass fibre hulls. Replacement began in 2001. *VERIFIED*

PBL-CD   *1996, Boston Whaler /* 0084150

70 HARBOUR SECURITY CRAFT (YP)

Displacement, tons: 3.9 full load
Dimensions, feet (metres): 24 × 8 × 3.3 *(7.3 × 2.4 × 1)*
Main machinery: 1 Volvo Penta AQAD41A diesel; 330 hp(m) *(243 kW)* sustained; 1 Type 290 outdrive
Speed, knots: 22
Complement: 3
Guns: 1—7.62 mm MG.
Radars: Surface search: Furuno; I-band.

Comment: Built by Peterson, Wisconsin and delivered between 29 February 1988 and 12 May 1989 in batches of 50, 25 and 10. Used for protecting naval installations, ports, harbours and anchorages. Some have been deleted. In addition there are large numbers of other small craft used in similar roles. *VERIFIED*

HARBOUR SECURITY CRAFT   *1/2002, A Sharma /* 0131284

AUXILIARIES

Notes: (1) The current plan is to be able to provide support in two regional crises simultaneously. This plan is based on the availability of some storage depots on foreign territory, and the use of Military Sealift Ships to carry fuels, munitions and the stores from the USA or overseas sources for transfer to UNREP (Underway Replenishment) ships in overseas areas. Some 16 to 18 UNREP ships are normally forward deployed in the Mediterranean, Western Pacific and Indian Ocean areas in support of the 5th, 6th and 7th Fleets, respectively.

Fleet support ships provide primarily maintenance and related towing and salvage services at advanced bases and at ports in the USA. These ships normally do not provide fuel, munitions, or other supplies except when ships are alongside for maintenance. Most fleet support ships operate from bases in the USA.

(2) A few underway replenishment ships and fleet support ships are Navy manned and armed, but most are operated by the Military Sealift Command (MSC) with civilian crews and are unarmed. The latter ships have T-prefix before their designations and are listed in the MSC section.

1 SACRAMENTO CLASS: FAST COMBAT SUPPORT SHIP (AOEHM)

Name	No	Builders	Laid down	Launched	Commissioned	F/S
CAMDEN	AOE 2	New York Shipbuilding	17 Feb 1964	29 May 1965	1 Apr 1967	PA

Displacement, tons: 19,200 light; 51,400-53,600 full load
Dimensions, feet (metres): 793 × 107 × 39.3
(241.7 × 32.6 × 12)
Main machinery: 4 Combustion Engineering boilers; 600 psi *(42.2 kg/cm²)*; 900°F *(480°C)*; 2 GE turbines; 100,000 hp *(76.4 MW)*; 2 shafts
Speed, knots: 30. **Range, n miles:** 6,000 at 25 kt; 10,000 at 17 kt
Complement: 601 (24 officers)
Cargo capacity: 177,000 barrels of fuel; 2,150 tons munitions; 500 tons dry stores; 250 tons refrigerated stores
Missiles: SAM: Raytheon NATO Sea Sparrow Mk 29 octuple launcher.

Guns: 2 General Electric/General Dynamics 20 mm Vulcan Phalanx Mk 15. 4—12.7 mm MGs.
Countermeasures: Decoys: Loral Hycor SRBOC 6-barrelled Mk 36; IR flares and chaff to 4 km *(2.2 n miles)*. SLQ-25 Nixie towed torpedo decoy.
ESM/ECM: SLQ-32(V)3; combined intercept and jammer.
Weapons control: Mk 91 Mod 1 MFCS.
Radars: Air search: Lockheed SPS-40E, Westinghouse SPS-58A (AOE 1 and 2), SPS-58A only (AOE 4); E/F- and D-band (SPS-58). Hughes Mk 23 TAS (AOE 3); D-band.
Surface search: Raytheon SPS-10F; G-band.
Navigation: Raytheon SPS-64(V)9; I-band.

Fire control: 2 Raytheon Mk 95; I/J-band (for SAM).
Tacan: URN 25.
Helicopters: 2 UH-46E Sea Knight normally assigned.

Comment: Designed to provide rapid replenishment at sea of petroleum, munitions, provisions, and fleet freight. Fitted with large hangar for vertical replenishment operations (VERTREP). AOE 1, AOE 3 and AOE 4 decommissioned in 2005 and AOE 2 is to decommission in 2006. ***UPDATED***

SACRAMENTO CLASS
8/1999, John Mortimer / 0084154

2 EMORY S LAND CLASS: SUBMARINE TENDERS (ASH)

Name	No	Builders	Laid down	Launched	Commissioned	F/S
EMORY S LAND	AS 39	Lockheed SB & Construction Co, Seattle	2 Mar 1976	4 May 1977	7 July 1979	AA
FRANK CABLE	AS 40	Lockheed SB & Construction Co, Seattle	2 Mar 1976	14 Jan 1978	29 Oct 1979	PA

Displacement, tons: 13,911 standard; 22,978 full load
Dimensions, feet (metres): 643.8 × 85 × 28.5
(196.2 × 25.9 × 8.7)
Main machinery: 2 Combustion Engineering boilers; 620 psi *(43.6 kg/cm²)*; 860°F *(462°C)*; 1 De Laval turbine; 20,000 hp *(14.9 MW)*; 1 shaft
Speed, knots: 20. **Range, n miles:** 10,000 at 12 kt
Complement: 535 (52 officers) plus Flag Staff 69 (25 officers)
Guns: 4 Oerlikon 20 mm Mk 67.
Radars: Navigation: ISC Cardion SPS-55; I/J-band.
Helicopters: Platform only.

Comment: The first US submarine tenders designed specifically for servicing nuclear-propelled attack submarines. Each ship can simultaneously provide services to four submarines moored alongside. Carry one 30 ton crane and two 5 ton mobile cranes. Have a 23 bed sick bay. *Frank Cable* is based at Guam and *Emory S Land* in the Mediterranean.

UPDATED

EMORY S LAND
4/1999, Jürg Kürsener / 0084155

4 SAFEGUARD CLASS: SALVAGE SHIPS (ARS)

Name	No	Builders	Commissioned	F/S
SAFEGUARD	ARS 50	Peterson Builders	16 Aug 1985	PA
GRASP	ARS 51	Peterson Builders	14 Dec 1985	AA
SALVOR	ARS 52	Peterson Builders	14 June 1986	PA
GRAPPLE	ARS 53	Peterson Builders	15 Nov 1986	AA

Displacement, tons: 3,200 full load
Dimensions, feet (metres): 255 × 51 × 17 *(77.7 × 15.5 × 5.2)*
Main machinery: 4 Caterpillar diesels; 4,200 hp *(3.13 MW)*; 2 shafts; cp Kort nozzle props; bow thruster; 500 hp *(373 kW)*
Speed, knots: 14. **Range, n miles:** 8,000 at 12 kt
Complement: 99 (7 officers)
Guns: 2—25 mm Mk 38.
Radars: Navigation: SPS-64(V)9; I-band.

Comment: Prototype approved in FY81, two in FY82 and one in FY83. The procurement of the fifth ARS was dropped on instructions from Congress. The design follows conventional commercial and Navy criteria. Can support surface-supplied diving operations to a depth of 58 m. Equipped with recompression chamber. Bollard pull, 65.5 tons. Using beach extraction equipment the pull increases to 360 tons. 150 ton deadlift. ARS 52 took part in the salvage and recovery of *Ehime Maru* which sank off Hawaii after being struck by USS *Greeneville*.

UPDATED

SALVOR
6/2004*, Chris Sattler / 1043628

FLOATING DRY DOCKS

Notes: The US Navy operates a limited number of floating dry docks to supplement dry dock facilities at major naval activities. The larger floating dry docks are made sectional to facilitate movement and to render them self-docking. Some of the ARD-type docks have the forward end of their docking well closed by a structure resembling the bow of a ship to facilitate towing. Berthing facilities, repair shops and machinery are housed in sides of larger docks. None is self-propelled.

1 MEDIUM AUXILIARY FLOATING DRY DOCK (AFDM)

Name/No	Commissioned	Capacity (tons)	Construction	Status
AFDM 7	1945	13,500	Steel (3)	Commercial lease, Jacksonville, FL

Sales: AFDM 2 at Marad Reserve Fleet, Beaumont, Texas. To be donated or transferred to Federal or State government or non-profit making organisation. AFDM 3 sold to Bender Shipbuilding and Repair, Mobile, AL in 2002. AFDM 10 deactivated in May 2004.

VERIFIED

SMALL AUXILIARY FLOATING DRY DOCKS (AFDL)

Name/No	Completed	Capacity (tons)	Construction	Status
DYNAMIC (AFDL 6)	1944	950	Steel	Active, Norfolk, VA
ADEPT (AFDL 23)	1944	1,770	Steel	Commercial lease, Ingleside, TX
RELIANCE (AFDL 47)	1946	7,000	Steel	Commercial lease, Deytens, SC

Sales: AFDL 1 to Dominican Republic; 4, Brazil; 5, Taiwan; 11, Kampuchea; 20, Philippines; 22, Vietnam; 24, Philippines; 26, Paraguay; 28, Mexico; 33, Peru; 34 and 36, Taiwan; 39, Brazil; 40 and 44, Philippines.

VERIFIED

DYNAMIC   *6/1986, Giorgio Arra*

AUXILIARY REPAIR DRY DOCKS and MEDIUM AUXILIARY REPAIR DRY DOCKS (ARDM)

Name/No	Commissioned	Capacity (tons)	Construction	Status
SHIPPINGPORT (ARDM 4)	1979	7,800	Steel	Active, New London, CT
ARCO (ARDM 5)	1986	7,800	Steel	Active, San Diego, CA

Sales: ARD 2 to Mexico; 5, Chile; 6, Pakistan; 8, Peru; 9, Taiwan; 11, Mexico; 12, Turkey; 13, Venezuela; 14, Brazil; 15, Mexico; 17, Ecuador; 22 *(Windsor),* Taiwan; 23, Argentina; 24, Ecuador; 25, Chile; 28, Colombia; 29, Iran; 32, Chile. ARDM 1 (ex-ARD 19) awaiting disposal decision.

VERIFIED

ARCO   *8/2002, Hachiro Nakai /* 0530009

YARD FLOATING DRY DOCKS (YFD)

Name/No	Completed	Capacity (tons)	Construction	Status
YFD 54	1943	5,000	Wood	Commercial lease, Todd Pacific SY, Seattle, WA
YFD 69	1945	15,000	Steel (3)	Commercial lease, Port of Portland, OR
YFD 70	1945	15,000	Steel (3)	Commercial lease, Todd Pacific SY, Seattle, WA
YFD 83 (ex-AFDL 31)	1943	1,000	Steel	Loan, US Coast Guard

VERIFIED

UNCLASSIFIED MISCELLANEOUS (IX)

Notes: (1) In addition to the vessels listed below, one of the ex-Forrest Sherman class, *Decatur,* completed conversion on 21 October 1994 as a Self-Defence testing-ship, including high-energy laser trials. Tests with HFSWR (high-frequency surface wave radar) started mid-1997.
(2) *Mercer* APL 39 (ex-IX 502) and *Nueces* APL 40 (ex-IX 503) are barrack ships of mid-1940s vintage.
(3) IX 516 is a decommissioned SSBN used for propulsion plant training.
(4) IX 517 is a submarine sea trials escort vessel *(Gosport).*
(5) IX 519 is used as a boat platform for *La Salle.* IX 523 is used for security training, both at Norfolk, VA.
(6) IX 310 is an accommodation barge at Naval Undersea Warfare Center, Dresden, NJ.
(7) IX 521, IX 522 and IX 525 are individual drydock sections.
(8) IX 527 and IX 528 are submarine test platforms, IX 529 is a surface ship test platform and IX 531 a test platform for HM&E.
(9) IX 530 (ex-YFND 5) is a berthing barge.

IX 517   *7/2003 *, Declerck/Steeghers /* 1043688

1 CONSTITUTION CLASS (AXS)

Name	Builders	Launched	Under Way	F/S
CONSTITUTION	Hartt's Shipyard, Boston	21 Oct 1797	22 July 1798	AA

Displacement, tons: 2,200
Dimensions, feet (metres): 204 oa; 175 wl × 43.5 × 22.5 *(62.2; 53.3 × 13.2 × 6.8)*
Speed, knots: 13 under sail
Complement: 75 (4 officers)

Comment: The oldest ship remaining on the Navy List. One of six frigates authorised 27 March 1794. Best remembered for her service in the war of 1812, in which she earned the nickname 'Old Ironsides'. Following extensive restoration (1927-30), went on a three year goodwill tour around the United States (1931-34), travelling over 22,000 miles and receiving over 4 million visitors. Open to the public in her homeport of Boston, the ship receives over 400,000 visitors a year. The most recent overhaul was conducted at the Charlestown Navy Yard, Boston from 1992-96. Under fighting sails (jibs, topsails and spanker) *Constitution* sailed for the first time in 116 years on 21 July 1997 as part of her bicentennial celebration. Armament is 32 × 24 pounder guns, 20 × 32 pounder carronades and 2—24 pounder bow-chasers. Sail area 42,710 sq ft *(13,018 m²).*
VERIFIED

CONSTITUTION   *7/1997, Todd Stevens, US Navy /* 0016501

1 TRAINING SHIP (AXT)

Name	Builders	Commissioned
IX 514 (ex-*YFU 79*)	Pacific Coast Eng, Alameda	1968

Displacement, tons: 380 full load
Dimensions, feet (metres): 125 × 36 × 8.0 *(38.1 × 10.9 × 2.4)*
Main machinery: 4 GM 6-71 diesels; 696 hp *(519 kW)* sustained; 2 shafts
Speed, knots: 8
Radars: Navigation: Racal Decca; I-band.

Comment: Harbour utility craft converted in 1986 with a flight deck covering two thirds of the vessel and a new bridge and flight control position at the forward end. Used for basic helicopter flight training at Pensacola, Florida. Similar craft *IX 501* deleted.   *UPDATED*

IX 514   *12/1994, van Ginderen Collection*

1 RESEARCH SHIP (YAGK)

Name	Builders	Commissioned
IX 515 (SES-200) (ex-USCG *Dorado*)	Bell Halter, New Orleans	Feb 1979

Displacement, tons: 205 full load
Dimensions, feet (metres): 159.1 × 39.0 × 9; 3 on cushion *(48.5 × 11.9 × 2.7; 0.9)*
Main machinery: 2 MTU 16V 396 TB94 diesels (propulsion); 5,800 hp(m) *(4.26 MW)* sustained;
2 Kamewa 71 water-jets
2 MTU 6V 396 TB83 diesels (lift); 1,560 hp(m) *(1.15 MW)* sustained
Speed, knots: 45. **Range, n miles:** 2,950 at 30 kt
Complement: 22 (2 officers)

Comment: The *IX 515* is a waterborne, air-supported craft with catamaran-style rigid sidewalls. It uses a cushion of air trapped between the sidewalls and flexible bow and stern seals to lift a large part of the hull clear of the water to reduce drag. A portion of the sidewall remains in the water to aid in stability and manoeuvrability. Modifications to the advanced ride control system were made in 1989. Serves as a high-performance test platform for weapon system development programmes, and as operational demonstrator for an advanced naval vehicle hullform. Trials of ALISS (Advanced Lightweight Influence Sweep System) in 1996-97. Based at Office of Naval Research, Arlington, VA.

UPDATED

IX 515   *9/1998, Findler & Winter / 0053382*

1 RESEARCH SHIP (AGE)

SLICE

Displacement, tons: 180 full load
Dimensions, feet (metres): 105 × 55.5 × 14 *(32 × 16.9 × 4.3)*
Main machinery: 2 MTU 16V 396 TB94 diesels; 13,700 hp(m) *(10.07 MW)*; 2 shafts; LIPS cp props
Speed, knots: 30
Complement: 12

Comment: Technology demonstrator built by Pacific Marine and owned by Lockheed Martin. Participated as a littoral warfare combatant in Fleet Battle Experiment Juliet (FBE-J) (part of Millennium Challenge 2002) accompanied by *Joint Venture* (HSV-X1). Modular capability packages, carried to simulate Littoral Combat Ship (LCS), included Mine Countermeasures (MCM), Antisubmarine Warfare (ASW), Force Protection and Time Critical Targeting. Weapons tested during FBE-J included the Lockheed Martin/Oerlikon Contraves 35 mm Millennium Gun and the NetFires System and launcher.

VERIFIED

SLICE   *10/2002, US Navy / 0572739*

MINOR AUXILIARIES

Notes: As of January 2004, the US Navy had about 504 active and inactive service craft, primarily small craft, on the US Naval Vessel Register. A majority of these vessels provide services to the fleet in various harbours and ports. Others are ocean-going ships that provide services to the fleet for research purposes. Most of the service craft are rated as 'active, in service', while others are rated as 'in commission' and some are accommodation ships.

2 DIVING TENDERS (YDT)

NEPTUNE YDT 17   POSEIDON YDT 18

Displacement, tons: 275 full load
Dimensions, feet (metres): 132 × 27 × 6.0 *(40.2 × 8.2 × 1.8)*
Main machinery: 2 Caterpillar diesels; 2,600 hp *(1.91 MW)*; 2 Hamilton waterjets
Speed, knots: 20
Complement: 8 plus 7 divers

Comment: Tenders used to support shallow-water diving operations and are based at Panama City, FL. Ordered from Swiftships in July 1997 and delivered in April 1999.

VERIFIED

NEPTUNE   *8/1999, US Navy / 0084159*

1 HARBOUR UTILITY CRAFT LCU TYPE (YFU)

YFU 81

Comment: Former utility landing craft employed primarily as harbour and coastal cargo craft. Based at Roosevelt Roads.

VERIFIED

YFU   *5/1999, M Declerck / 0084160*

23 PATROL CRAFT (YP)

YP 663	YP 665	YP 680-692	YP 694-698	YP 700-702

Displacement, tons: 167 full load
Dimensions, feet (metres): 108 × 24 × 8 *(32.9 × 7.3 × 2.4)*
Main machinery: 2 Detroit 12V-71 diesels; 680 hp *(507 kW)* sustained; 2 shafts
Speed, knots: 13.3. **Range, n miles:** 1,500 at 12 kt
Complement: 6 (2 officers) plus 24 midshipmen
Radars: Navigation: I-band.

Comment: Built in the 1980s by Peterson Builders and Marinette Marine, both in Wisconsin. Nineteen are based at the Naval Academy, Annapolis; two at the Naval air station, Pensacola and two at Naval Underwater Warfare Centre, Keyport. Some earlier versions converted for mine countermeasures operations and assigned to the former COOP project which was discontinued in 1995.

UPDATED

YP 683   *7/2000, Hachiro Nakai / 0106812*

2 TORPEDO TRIALS CRAFT (YTT)

BATTLE POINT YTT 10   DISCOVERY BAY YTT 11

Displacement, tons: 1,168 full load
Dimensions, feet (metres): 186.5 × 40 × 10.5 *(56.9 × 12.2 × 3.2)*
Main machinery: 1 Cummins KTA50-M diesel; 1,250 hp *(932 kW)* sustained; 1 shaft; 1 bow thruster; 400 hp *(298 kW)*; 2 stern thrusters; 600 hp *(448 kW)*
Speed, knots: 11. **Range, n miles:** 1,000 at 10 kt
Complement: 31 plus 9 spare berths

Comment: Built by McDermott Shipyard, Morgan City, and delivered in 1991-92. Fitted with two 21 in Mk 59 and three (one triple) 12.75 in Mk 32 Mod 5 torpedo tubes. YTT 10 is used for torpedo trials and development at Keyport, Washington. A battery is fitted for limited duration operations during the diesel shutdown. YTT 11 assigned to National Oceanographic as a ROV carrier for research into US coastal waters until 2004. Underwater recovery vessels SORD 4, TROV and CURV. Both based at Naval Underwater Warfare Centre, Keyport, WA.

VERIFIED

YTT   *9/1999, van Ginderen Collection / 0084162*

24 TORPEDO RETRIEVERS (YPT)

Comment: Five different types spread around the Fleet bases and at AUTEC. There are 3 × 65 ft, 5 × 72 ft, 5 × 85 ft, 3 × 100 ft and 8 × 120 ft.

UPDATED

TR 6   7/2000, Sattler/Steele / 0106813

TUGS

28 LARGE HARBOUR TUGS (YTB)

MUSKEGON	YTB 763	ACCONAC	YTB 812
OKMULGEE	YTB 765	POUGHKEEPSIE	YTB 813
CHESANING	YTB 769	WAXAHACHIE	YTB 814
KEOKUK	YTB 771	NEODESHA	YTB 815
NIANTIC	YTB 781	MECOSTA	YTB 818
MANISTEE	YTB 782	WANAMASSA	YTB 820
REDWING	YTB 783	CANONCHET	YTB 823
KITTANNING	YTB 787	SANTAQUIN	YTB 824
MARINETTE	YTB 791	CATAHECASSA	YTB 828
TAMAQUA	YTB 797	DEKANAWIDA	YTB 831
OPELIKA	YTB 798	PETALESHARO	YTB 832
TUSKEGEE	YTB 806	NEWGAGON	YTB 834
MASSAPEQUA	YTB 807	SKENANDOA	YTB 835
WENATCHEE	YTB 808	POKAGON	YTB 836

Displacement, tons: 356 full load
Dimensions, feet (metres): 109 × 30 × 13.8 *(33.2 × 9.1 × 4.2)*
Main machinery: 1 Fairbanks-Morse 38D8-1/8 diesel; 2,000 hp *(1.49 MW)* sustained; 1 shaft
Speed, knots: 12. **Range, n miles:** 2,000 at 12 kt
Complement: 10-12
Radars: Navigation: Marconi LN66; I-band.

Comment: Built between 1959 and 1975. Two transferred to Saudi Arabia in 1975. Being withdrawn from service and tugs are being provided by MSC charter.

UPDATED

OPELIKA   7/2004*, US Navy / 1043634

RESEARCH SHIPS

Notes: There are many naval associated research vessels which are civilian manned and not carried on the US Naval Vessel Register. In addition civilian ships are leased for short periods to support a particular research project or trial. Some of those employed include *RSB-1* (missile booster recovery), *Acoustic Pioneer* and *Acoustic Explorer* (acoustic research).

ACOUSTIC PIONEER   3/1996, W H Clements / 0085302

1 HIGH SPEED VESSEL (HSVH)

JOINT VENTURE HSV-X1

Displacement, tons: 1,872
Dimensions, feet (metres): 318.9 × 87.2 × 13.4 *(97.2 × 26.6 × 4.1)*
Main machinery: 4 CAT 3618 diesels; 38,620 hp *(28.8 MW)*; 4 LIPS 150D waterjets
Speed, knots: 48 (light); 38 (full load). **Range, n miles:** 2,400 at 35 kt
Military lift: 815 tonnes cargo and 350 personnel

Comment: Built as *Incat 50* of aluminium construction and chartered by US Army Tank Automotive and Armament Command (TACOM) from Bollinger/Incat from October 2001 for trials. Modifications to the commercial design include deck strengthening, a 472 m² helicopter deck, suitable for SH-60 and CH-46, and a two-part hydraulically operated vehicle ramp to facilitate rapid loading/unloading from stern or alongside. Military communications have also been fitted. Manned by a joint Army/Navy crew, the ship is being used by both services in a series of exercises and trials. Typical roles being examined include those exercised in Millennium Challenge 2002: Mine Warfare Command and Control, Mine Countermeasures, Naval Special Warfare, Ship to Objective Manoeuvre and intra-theatre lift of an interim Brigade Combat Team. Further trials in 2003 included exploration of Littoral Combat Ship concepts. Based at Norfolk, Virginia. *UPDATED*

JOINT VENTURE   3/2003, A Sharma / 0572741

1 HIGH SPEED VESSEL (HSV)

SPEARHEAD HSV-1X

Displacement, tons: 1,872
Dimensions, feet (metres): 321.6 × 85.3 × 12.1 *(98.0 × 26.0 × 3.7)*
Main machinery: 4 MAN B&W 20 RK 270 diesels; 37,950 hp *(28.3 MW)*; 4 LIPS 150D waterjets
Speed, knots: 48 (light); 38 (full load). **Range, n miles:** 2,400 at 35 kt
Military lift: 815 tonnes cargo and 323 personnel

Comment: Built as *Incat 60* (Incat Evolution 10B) of aluminium construction and leased by US Army Tank Automotive and Armament Command (TACOM) from Bollinger/Incat of Lockport, Louisiana from November 2002. The ship is to be utilised for the movement of the Army's prepositioned stocks and for sustainment deliveries. Not fitted with helicopter facilities.
UPDATED

SPEARHEAD   4/2003, A Sharma / 0572740

1 EXPERIMENTAL CATAMARAN (X-CRAFT) (AGE)

SEA FIGHTER FSF-1

Displacement, tons: 1,150 full load
Dimensions, feet (metres): 262.0 × 72.2 × 11.5 *(79.9 × 22.0 × 3.5)*
Main machinery: CODOG; 2 GE LM 2500 gas turbines; 2 MTU 16V595 diesels; four Rolls-Royce Kamewa 125 SII waterjets
Speed, knots: 50
Complement: 26
Radars: Navigation: to be announced.
Helicopters: Platform for 2 SH-60R.

Comment: Titan Corporation of San Diego, California and Nigel Gee and Associates Ltd of Southampton, UK (later acquired by BMT) selected in September 2002 by the Office of Naval Research (ONR) to design an experimental vessel known as X-Craft. Contract for development and build awarded to Titan on 25 February 2003. Laid down 5 June 2003 and launched 5 February 2005 at Nichols Brothers Boat Builders, Whidbey Island, WA. For completion in 2005. The ship is to be based at San Diego and is to be used by ONR for hydrodynamic experimentation and operational concept development for high speed craft. Catamaran design of aluminium construction. A flush upper deck has a landing area for two helicopters. Access is via a large lift from the flight deck or over folding ramps at the stern. Propulsion is by waterjets driven by gas turbines for high speed operation and diesel engines for lower speed loitering and transit. *UPDATED*

SEA FIGHTER   2/2005*, US Navy / 0590764

1 RESEARCH SHIP (AGE)

SEA SHADOW

Displacement, tons: 560 full load
Dimensions, feet (metres): 164.0 × 68.0 × 14.5 *(50.0 × 20.7 × 4.4)*
Main machinery: Diesel-electric; 2 Detroit 12V-149TI diesels; 2 shafts
Speed, knots: 10. **Range, n miles:** 2,250 at 9 kt
Complement: 12

Comment: Built by Lockheed in 1983-84. Testing started in 1985 but the ship was then laid up until April 1993 when testing resumed off Santa Cruz island, southern California. This is a stealth ship prototype of SWATH design with sides angled at 45° to minimise the radar cross-section. Operated by Lockheed for testing all aspects of stealth technology and reactivated for a six year programme at San Diego in early 1999.

UPDATED

SEA SHADOW   *9/1999, van Ginderen Collection /* 0084164

3 ASHEVILLE CLASS (YFRT)

ATHENA (ex-*Chehalis*)   **ATHENA II** (ex-*Grand Rapids*)   **LAUREN** (ex-*Douglas*)

Displacement, tons: 245 full load
Dimensions, feet (metres): 164.5 × 23.8 × 9.5 *(50.1 × 7.3 × 2.9)*
Main machinery: CODOG; 1 GE LM 1500 gas-turbine; 12,500 hp *(9.3 MW)*; 2 Cummins VT12-875 diesels; 1,450 hp *(1.07 MW)*; 2 shafts; cp props
Speed, knots: 16. **Range, n miles:** 1,700 at 16 kt
Complement: 22

Comment: All built 1969-71. Work for the Naval Surface Warfare Center, at Panama City, Florida. Disarmed except *Lauren* which has maintained its military appearance.

UPDATED

ATHENA II   *6/1993, Giorgio Arra*

RESEARCH OCEANOGRAPHIC SHIPS

2 MELVILLE CLASS (AGOR)

Name	No	Builders	Commissioned	F/S
MELVILLE	AGOR 14	Defoe SB Co, Bay City, MI	27 Aug 1969	Loan
KNORR	AGOR 15	Defoe SB Co, Bay City, MI	14 Jan 1970	Loan

Displacement, tons: 2,944 full load
Dimensions, feet (metres): 278.9 × 46.3 × 16.5 *(85 × 14.1 × 5.0)*
Main machinery: Diesel-electric; 3 Caterpillar 3516 diesel generators; 1 Caterpillar 3508 diesel generator; 2 motors; 1,385 hp *(1 MW)*; 3 shafts (2 aft, 1 fwd)
Speed, knots: 14. **Range, n miles:** 10,060 at 11.7 kt
Complement: 23 (9 officers) plus 38 scientists

Comment: *Melville* operated by Scripps Institution of Oceanography and *Knorr* by Woods Hole Oceanography Institution for the Office of Naval Research, under technical control of the Oceanographer of the Navy. Fitted with internal wells for lowering equipment, underwater lights and observation ports. Problems with the propulsion system have led to major modifications including electric drive (vice the original mechanical) and the insertion of a 34 ft central section increasing the displacement from the original 1,915 tons and allowing better accommodation and improved laboratory spaces. The forward propeller is retractable. These ships are highly manoeuvrable for precise position keeping.

UPDATED

MELVILLE   *3/2003, Robert Pabst /* 0572738

1 AGOR-26 CLASS (AGOR)

Name	No	Builders	Commissioned
KILO MOANA	AGOR 26	Atlantic Marine, Jacksonville	3 Sep 2002

Displacement, tons: 2,542 full load
Dimensions, feet (metres): 186 × 88 × 25 *(56.7 × 26.8 × 7.6)*
Main machinery: Diesel-electric; 4 Caterpillar 3508B diesel generators; 2 Westinghouse motors; 4,025 hp *(3 MW)*; 1 bow thruster 1,100 hp *(820 kW)*
Speed, knots: 15. **Range, n miles:** 10,000 at 11 kt
Complement: 48 (31 scientists)

Comment: Replacement for R/V *Moana Wave*. Designed to commercial standards and constructed by Atlantic Marine, Jacksonville. Launched on 17 November 2001. The ship is a small waterplane area, twin hull (SWATH) oceanographic vessel capable of performing general purpose oceanographic research in coastal and deep ocean areas. The University of Hawaii School of Ocean and Earth Science and Technology operates the ship under a charter agreement with the Office of Naval Research (ONR). The survey suite consists of a Simrad EM 120 multibeam echosounder (12 kHz), a Simrad EM 1002 shallow water echo sounder (95 kHz), a Sontek current profiler and Simrad HPR-418 acoustic positioning system.

UPDATED

KILO MOANA   *6/2004*, University of Hawaii Marine Center /* 1043633

4 THOMAS G THOMPSON CLASS (AGOR)

Name	No	Builders	Launched	Commissioned	F/S
THOMAS G THOMPSON	AGOR-23	Halter Marine	27 July 1990	8 July 1991	Loan
ROGER REVELLE	AGOR-24	Halter Marine	20 Apr 1995	11 June 1996	Loan
ATLANTIS	AGOR-25	Halter Marine	1 Feb 1996	3 Mar 1997	Loan
RONALD H BROWN	R 104	Halter Marine	30 May 1996	25 Apr 1997	NOAA

Displacement, tons: 3,400 full load
Dimensions, feet (metres): 274 oa; 246.8 wl × 52.5 × 19 *(83.5; 75.2 × 16 × 5.6)*
Main machinery: Diesel-electric; 6 Caterpillar diesel generators; 6.65 MW (3—1.5 MW and 3—715 kW); 2 motors; 6,000 hp *(4.48 MW)*; 2 shafts; bow thruster; 1,140 hp *(850 kW)*
Speed, knots: 15. **Range, n miles:** 15,000 at 12 kt
Complement: 22 plus 37 scientists
Sonars: Simrad EM 120; Seabeam 2112.

Comment: *Thomas G Thompson* is the first of a class of oceanographic research vessels capable of operating worldwide in all seasons and suitable for use by navy laboratories, contractors and academic institutions. Dynamic positioning system enables station to be held within 300 ft of a point. 4,000 sq ft of laboratories. AGORs 23, 24 and 25 are operated by academic institutions for the Office of Naval Research through charter party agreements (AGOR 23-University of Washington; AGOR 24-Scripps Institution of Oceanography; AGOR 25-Woods Hole Oceanographic Institution). *Ronald H Brown* is operated by NOAA. Ships in this series are able to meet changing oceanographic requirements for general, year-round, worldwide research. This includes launching, towing and recovering a variety of equipment. The ships are also involved in hydrographic data collection. *Roger Revelle* was authorised in FY92, ordered 11 January 1993 and is operated by the Scripps Institute of Oceanography. The third and fourth were ordered from Trinity Marine on 15 February 1994. *Atlantis* can carry *Alvin* DSV 2.

UPDATED

THOMAS G THOMPSON   *6/2004*, Mitsuhiro Kadota /* 1043632

DEEP SUBMERGENCE VEHICLES

(Included in US Naval Vessel Register)

Notes: (1) Deep submergence vehicles and other craft and support ships are operated by Submarine Development Squadron Five at San Diego, California. The Squadron is a major operational command that includes advanced diving equipment; divers trained in 'saturation' techniques; DSRV-1; the submarine *Dolphin* (AGSS 555). DSRV-2 is inactive in lay-up condition. Two unmanned vessels CURV (Cable Controlled Underwater Remote Vehicle) Super Scorpios made test dives to 5,000 ft *(1,524 m)*.
(2) The Submarine Rescue, Diving and Recompression System (SRDRS) will replace the current DSRVs in 2006. A Remotely Operated Vehicle (ROV), launchable from a craft of opportunity in up to Sea State 4, will be based on the Australian 'Remora' system. Capable of rescuing up to 15 submariners at a time from depths up to 600 m, the ROV will be able to mate with decompression chambers. All components will fit into an ISO container to facilitate worldwide deployment.
(3) There are also four naval ROVs operated by the Supervisor of Salvage and Diving. These are air-transportable and can be operated from many warships and commercial vessels.

1 DOLPHIN CLASS (DSV)

Name	No	Builders	Launched	Commissioned	F/S
DOLPHIN	AGSS 555	Portsmouth Naval Shipyard, NH	8 June 1968	17 Aug 1968	PA

Displacement, tons: 860 standard; 948 full load
Dimensions, feet (metres): 165 × 19.3 × 18 *(50.3 × 5.9 × 5.5)*
Main machinery: Diesel-electric; 2 Detroit 12V42S diesels; 840 hp *(616 kW)* sustained; 2 generators; 1 motor; 1 shaft
Fitted with 246 cell VRLA battery
Speed, knots: 8 dived; 5 surfaced
Complement: 29 (3 officers) plus 4-7 scientists
Radars: Surface search: BPS-15; I/J-band.
Navigation: Furuno; I-band.
Sonars: Ametek BQS 15; active close-range detection; high frequency.
EDO BQR 2; passive search; low frequency; obstacle avoidance.

Comment: Authorised in FY61. Has a constant diameter cylindrical pressure hull approximately 15 ft in outer diameter closed at both ends with hemispherical heads. Pressure hull fabricated of HY-80 steel with aluminium and fibreglass used in secondary structures to reduce weight. No conventional hydroplanes are mounted, improved rudder design and other features provide manoeuvring control and hovering capability. Fitted for deep-ocean sonar and oceanographic research. There are several research stations for scientists and she is fitted to take water samples down to her operating depth. Assigned to Submarine Development Squadron Five at San Diego. Designed for deep diving operations and fitted with Helios seafloor imaging system, Draper Labs integrated navigation system and integrated science suite. Submerged endurance is approximately 24 hours with an at-sea endurance of 14 days. Refitted in 1993. Based at San Diego. Damaged by flooding in June 2002.

VERIFIED

DOLPHIN   5/2002, Ships of the World / 0530087

1 CUTTHROAT CLASS (DSV)

Name	No	Builders	Commissioned
CUTTHROAT	LSV-2	Newport News Shipbuilding and General Dynamics Electric Boat Division	Apr 2001

Displacement, tons: 205
Dimensions, feet (metres): 111 × 10 × 9 *(33.8 × 3.1 × 2.7)*
Main machinery: Permanent Magnet electric motor; 3,000 hp(m) *(2.23 MW)*
Speed, knots: 34 dived

Comment: The contract was placed with Newport News and Electric Boat in January 1999 to build *Cutthroat* LSV-2. The largest autonomous unmanned submarine in the world, it is a 1 : 3.4 scaled-down model of the Virginia class submarine and is to be used to test advanced submarine technologies, including hydroacoustics, hydrodynamics and manoeuvring. Its diving depth matches that of the Virginia class. The forward compartment contains 1,680 lead acid batteries and the after compartment contains the propulsion and auxiliary systems together with data recording and control systems. All appendages, including control surfaces and simulated sonar fairing, can be removed or relocated. LSV-2 is operated by the Acoustic Research Department at the instrumented range at Lake Pend Oreille in Bayview, Idaho. It is named after a species of trout indigenous to the lake.

VERIFIED

CUTTHROAT   2000, Newport News / 0105821

1 DEEP SUBMERGENCE RESCUE VEHICLE (DSRV)

Name	No	Builders	In service	F/S
MYSTIC	DSRV 1	Lockheed Missiles and Space Co,	7 Aug 1971	PSA

Displacement, tons: 30 surfaced; 38 dived
Dimensions, feet (metres): 49.2 × 8 *(15 × 2.4)*
Main machinery: Electric motors; silver/zinc batteries; 1 prop (movable control shroud); 4 ducted thrusters (2 fwd, 2 aft)
Speed, knots: 4. **Range, n miles:** 24 at 3 kt
Complement: 4 (pilot, co-pilot, 2 rescue sphere operators) plus 24 rescued men
Sonars: Search and navigational sonar, obstacle avoidance sonar and closed-circuit television (supplemented by optical devices) are installed in the DSRV to determine the exact location of a disabled submarine within a given area and for pinpointing the submarine's escape hatches. Side-looking sonar can be fitted for search missions.

Comment: The DSRV is intended to provide a quick-reaction worldwide, all-weather capability for the rescue of survivors in a disabled submarine. Transportable by road, aircraft (in C-141 and C-5 jet cargo aircraft), specially designed surface ship, and specially modified mother ships (MOSUBs).
The carrying submarine will launch and recover the DSRV while submerged and, if necessary, while under ice. A total of six DSRVs were planned, but only two were funded. DSRV 2 is now inactive.
The outer hull is constructed of formed fibreglass. Within this outer hull are three interconnected spheres which form the main pressure capsule. Each sphere is 7.5 ft in diameter and is constructed of HY-140 steel. The forward sphere contains the vehicle's control equipment and is manned by the pilot and co-pilot, the centre and after spheres accommodate 24 passengers and two additional life-support technicians. Under the DSRV's centre sphere is a hemispherical protrusion or 'skirt' which seals over the disabled submarine's hatch. During the mating operation the skirt is pumped dry to enable personnel to transfer. Operating depth, 1,525 m *(5,000 ft)*. Rescue depth limit 610 m *(2,000 ft)*. Name is not 'official'. Upgraded with modern electronics and navigation systems. In 2001, deployed from UK SSBN *Vanguard* to conduct simulated rescue from Swedish Submarine *Gotland* at 135 m in Raasay Sound, Scotland. To be replaced in 2006 by SRDRS.

VERIFIED

MYSTIC   4/2002, Hachiro Nakai / 0530003

1 DEEP SUBMERGENCE VEHICLE: ALVIN TYPE (DSV)

Name	No	Builders	F/S
ALVIN	DSV 2	General Mills Inc, Minneapolis	PSA

Displacement, tons: 18 full load
Dimensions, feet (metres): 26.5 × 8.5 *(8.1 × 2.6)*
Main machinery: 6 brushless DC motors; 6 thrusters; 2 vertical-motion thrusters (located near the centre of gravity); 2 horizontally (near stern) (1 directed athwartships, 1 directed longitudinally); 2 on rotatable shaft near stern for vertical or longitudinal motion
Speed, knots: 2. **Range, n miles:** 3 at 0.5 kt
Complement: 3 (1 pilot, 2 observers)

Comment: Built for operation by the Woods Hole Oceanographic Institution for the Office of Naval Research. Original configuration had an operating depth of 6,000 ft. Named for Allyn C Vine of Woods Hole Oceanographic Institution. *Alvin* accidentally sank in 5,051 ft of water on 16 October 1968; subsequently raised in August 1969; refurbished 1970-71 in original configuration. Placed in service on Navy List 1 June 1971. Subsequently refitted with titanium pressure sphere to provide increased depth capability and again operational in November 1973. She has two banks of lead acid batteries, 120 V DC system with 47 kW/h capacity. Operating depth, 4,000 m *(13,120 ft)*. Available for emergency use until 2005. Two other DSVs were placed out of service in 1997/98, one transferred to the Woods Hole Institute.

VERIFIED

ALVIN   10/2003, Rod Catanach, Woods Hole Oceanographic Institution / 0009310

1 NUCLEAR-POWERED OCEAN ENGINEERING AND RESEARCH VEHICLE (DSVN)

Name	Builders	In service	F/S
NR 1	General Dynamics (Electric Boat Div)	27 Oct 1969	ASA

Displacement, tons: 380 surfaced; 700 dived
Dimensions, feet (metres): 145.7 × 12.5 × 15.09 *(44.4 × 3.8 × 4.6)*
Main machinery: Nuclear; 1 PWR; 1 turbo-alternator; 2 motors (external to the hull); 2 props; 4 ducted thrusters (2 vertical, 2 horizontal)
Speed, knots: 3.5 dived
Complement: 13 (3 officers, 2 scientists)

Comment: NR 1 was built primarily to serve as a test platform for a small nuclear propulsion plant; however, the craft additionally provides an advanced deep submergence ocean engineering and research capability. She was the only Naval deep submergence vehicle to be used in the recovery of the wreckage of the space shuttle *Challenger* January to April 1986. Laid down on 10 June 1967; launched on 25 January 1969. Commanded by an officer-in-charge vice commanding officer. First nuclear-propelled service craft. Overhauled and refuelled in 1993. Refitted with a new bow which extended her length by 9.6 ft, and new sonars and cameras. In 1995 and again in 1997 she was used for an archaeological survey of the Carthage-Ostia trade route. In 2002, conducted deep water operations in Gulf of Mexico including study of *Mississippi Canyon* shipwreck site.
    The NR 1 has wheels beneath the hull to permit 'bottom crawling' and she is fitted with external lights, external television cameras, a remote-controlled manipulator, and various recovery devices. No periscopes, but fixed television mast. Diving depth, 3,000 ft *(914 m)*. A surface 'mother' ship is required to support her. Based at Groton, Connecticut. **VERIFIED**

NR 1   *7/2000, H M Steele /* 0106816

MILITARY SEALIFT COMMAND (MSC)

Notes: (1) The Military Sealift Command consists of the Naval Fleet Auxiliary Force, Special Missions Ships, Strategic Sealift Force and Prepositioning Programme.
(2) Headquarters are in the Washington Navy Yard, Washington DC. The organisation is commanded by a Vice Admiral, and its five principal area commands (Atlantic, Pacific, Europe, Far East, Central) by Captains.
(3) MSC ships are assigned standard hull designations with the added prefix 'T'. All are unarmed. Ships' funnels have black, grey, blue and gold horizontal bands.

NAVAL FLEET AUXILIARY FORCE

4 SUPPLY CLASS (FAST COMBAT SUPPORT SHIPS) (AOEH)

Name	No	Builder	Laid down	Launched	Commissioned	F/S
SUPPLY	T-AOE 6	National Steel & Shipbuilding Co	24 Feb 1989	6 Oct 1990	26 Feb 1994	AA
RAINIER	T-AOE 7	National Steel & Shipbuilding Co	31 May 1990	28 Sep 1991	21 Jan 1995	PA
ARCTIC	T-AOE 8	National Steel & Shipbuilding Co	2 Dec 1991	30 Oct 1993	16 Sep 1995	AA
BRIDGE	T-AOE 10	National Steel & Shipbuilding Co	16 Sep 1993	25 Aug 1996	5 Aug 1998	PA

Displacement, tons: 19,700 light; 49,000 full load
Dimensions, feet (metres): 753.7 × 107 × 38 *(229.7 × 32.6 × 11.6)*
Main machinery: 4 GE LM 2500 gas turbines; 105,000 hp *(78.33 MW)* sustained; 2 shafts
Speed, knots: 25. **Range, n miles:** 6,000 at 22 kt
Complement: 160 civilian; 28 military
Cargo capacity: 156,000 barrels of fuel; 1,800 tons ammunition; 400 tons refrigerated cargo; 250 tons general cargo; 20,000 gallons water
Helicopters: 2 MH-60.

Comment: Construction started in June 1988. *Supply* decommissioned and transferred to MSC in July 2001, *Arctic* in June 2002, *Rainier* in August 2003 and *Bridge* in June 2004. **UPDATED**

RAINIER   *8/2004 *, Hachiro Nakai /* 1043691

3 MARS CLASS: COMBAT STORE SHIPS (AFSH)

Name	No	Builders	Commissioned	F/S
NIAGARA FALLS	T-AFS 3	National Steel & Shipbuilding Co	29 Apr 1967	TPA
CONCORD	T-AFS 5	National Steel & Shipbuilding Co	27 Nov 1968	TPA
SAN JOSE	T-AFS 7	National Steel & Shipbuilding Co	23 Oct 1970	TPA

Displacement, tons: 9,200 light; 15,900-18,663 full load
Dimensions, feet (metres): 581 × 79 × 26 *(177.1 × 24.1 × 7.9)*
Main machinery: 3 Babcock & Wilcox boilers; 580 psi *(40.8 kg/cm²)*; 825°F *(440°C)*; 1 De Laval turbine (Westinghouse in AFS 6); 22,000 hp *(16.4 MW)*; 1 shaft
Speed, knots: 20. **Range, n miles:** 10,000 at 18 kt
Complement: 123-134 civilians plus 29-49 naval
Cargo capacity: 2,625 tons dry stores; 1,300 tons refrigerated stores (varies with specific loadings)
Radars: Navigation: 2 Raytheon; I-band.
Tacan: URN 25.
Helicopters: 2 MH-60.

Comment: *Concord* transferred to MSC on 15 October 1992 after disarming and conversion to a civilian crew. *San Jose* followed on 2 November 1993 and *Niagara Falls* on 23 September 1994. All have accommodation improvements and stores lifts installed. These ships carry comprehensive inventories of aviation spare parts as well as the cargo listed above. Two others of the class de-activated in 1997. **UPDATED**

SAN JOSE   *4/2001, Mick Prendergast /* 0131281

0 + 8 (4) LEWIS AND CLARK CLASS (DRY CARGO/AMMUNITION SHIPS) (AKEH)

Name	No	Builders	Commissioned
LEWIS AND CLARK	T-AKE 1	National Steel & Shipbuilding Co	Mar 2005
SACAGAWEA	T-AKE 2	National Steel & Shipbuilding Co	Sep 2005
—	T-AKE 3	National Steel & Shipbuilding Co	2006
—	T-AKE 4	National Steel & Shipbuilding Co	2007
—	T-AKE 5	National Steel & Shipbuilding Co	2008
—	T-AKE 6	National Steel & Shipbuilding Co	2008
—	T-AKE 7	National Steel & Shipbuilding Co	2008
—	T-AKE 8	National Steel & Shipbuilding Co	2008

Displacement, tons: 23,852 light; 40,298 full load
Dimensions, feet (metres): 689 × 106 × 29.5 *(210 × 32.3 × 9.0)*
Main machinery: Integrated electric propulsion; 4 FM/MAN B&W 9L and 8L 48/60 diesel generators (35.7 MW); 2 Alstom motors; 1 shaft; fixed pitch prop
Speed, knots: 20. **Range, n miles:** 14,000 at 20 kt
Complement: 172 (31 officers)
Cargo capacity: 5,463 tons dry; 2,390 tons fuel
Helicopters: 2 CH-46D Sea Knight or HH-60 Sea Hawk.

Comment: Design and construction contract placed on 18 October 2001 for delivery of first and second vessels in 2005. Contract for construction of the third of class in July 2002 and for the fourth in July 2003. A further two were ordered in January 2004 and two more on 11 January 2005. Four further ships are to be ordered by 2008. Capable of carrying 7,000 tons of dry cargo and 23,500 barrels of diesel and jet aviation fuel, the ships are to be built to commercial standards to replace existing AE and AFS. Three RAS stations are to be fitted each side. **UPDATED**

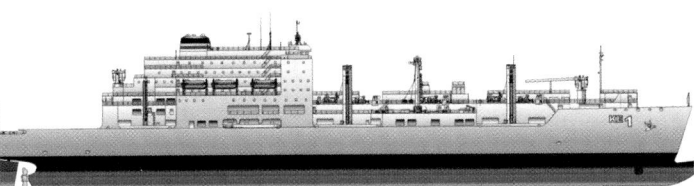

LEWIS AND CLARK (artist's impression)   *2001, NASSCO /* 0096365

6 KILAUEA CLASS: AMMUNITION SHIPS (AEH)

Name	No	Builders	Commissioned	F/S
KILAUEA	T-AE 26	General Dynamics, Quincy	10 Aug 1968	TAA/ROS
SANTA BARBARA	T-AE 28	Bethlehem Steel	11 July 1970	TAA/ROS
FLINT	T-AE 32	Ingalls Shipbuilding	20 Nov 1971	TPA
SHASTA	T-AE 33	Ingalls Shipbuilding	26 Feb 1972	TPA
MOUNT BAKER	T-AE 34	Ingalls Shipbuilding	22 July 1972	TAA
KISKA	T-AE 35	Ingalls Shipbuilding	16 Dec 1972	TPA

Displacement, tons: 9,340 light; 19,940 full load
Dimensions, feet (metres): 564 × 81 × 28 *(171.9 × 24.7 × 8.5)*
Main machinery: 3 Foster-Wheeler boilers; 600 psi *(42.3 kg/cm²)*; 870°F *(467°C)*; 1 GE turbine; 22,000 hp *(16.4 MW)*; 1 shaft
Speed, knots: 20. **Range, n miles:** 10,000 at 18 kt
Complement: 133 civilians
Radars: Navigation: 2 Raytheon; I-band.
Tacan: URN 25.
Helicopters: 2 CH-46E Sea Knight (cargo normally embarked).

Comment: *Kilauea* transferred to MSC 1 October 1980, *Flint* in August 1995, *Kiska* in August 1996, *Mount Baker* in December 1996, *Shasta* in October 1997 and *Santa Barbara* in September 1998. An eighth of class was to have transferred in 1999, but has been decommissioned. *Butte* decommissioned in 2002. With the exception of *Santa Barbara*, ships underwent a civilian modification overhaul during which accommodation was improved. Main armament taken out. Seven UNREP stations operational: four port, three starboard. *Kilauea* (Port Hueneme, CA), *Santa Barbara* (Charleston, SC) are kept in reduced operating status. **UPDATED**

SHASTA   *10/2003, Frank Findler /* 0572794

3 SIRIUS (LYNESS) CLASS: COMBAT STORES SHIP (AFSH)

Name	No	Builders	Commissioned	F/S
SIRIUS	T-AFS 8	Swan Hunter & Wigham Richardson	22 Dec 1966	TAA
(ex-*Lyness*)		Ltd, Wallsend-on-Tyne		
SPICA	T-AFS 9	Swan Hunter & Wigham Richardson	21 Mar 1967	TAA
(ex-*Tarbatness*)		Ltd, Wallsend-on-Tyne		
SATURN	T-AFS 10	Swan Hunter & Wigham Richardson	10 Aug 1967	TAA
(ex-*Stromness*)		Ltd, Wallsend-on-Tyne		

Displacement, tons: 9,010 light; 16,792 full load
Measurement, tons: 7,782 dwt; 12,359 gross; 4,744 net
Dimensions, feet (metres): 524 × 72 × 22 *(159.7 × 22 × 6.7)*
Main machinery: 1 Wallsend-Sulzer 8RD76 diesel; 11,520 hp *(8.59 MW)*; 1 shaft
Speed, knots: 18. **Range, n miles:** 12,000 at 16 kt
Complement: 108-119 civilians plus 29-49 naval
Cargo capacity: 8,313 m³ dry; 3,921 m³ frozen
Radars: Navigation: 2 Raytheon; I-band.
Tacan: URN 25.
Helicopters: 2 MH-60.

Comment: After a period of charter *Sirius* was purchased from the UK on 1 March 1982, *Spica* on 30 September 1982 and *Saturn* on 1 October 1983. All refitted from August 1992-96 to improve communications, RAS facilities and cargo handling equipment.

UPDATED

SIRIUS *7/2002, John Chaney /* 0529970

15 HENRY J KAISER CLASS: OILERS (AOH)

Name	No	Builders	Laid down	Commissioned	F/S
HENRY J KAISER	T-AO 187	Avondale	22 Aug 1984	19 Dec 1986	TPA/ROS
JOSHUA HUMPHRIES	T-AO 188	Avondale	17 Dec 1984	3 Apr 1987	TAA/ROS
JOHN LENTHALL	T-AO 189	Avondale	15 July 1985	2 June 1987	TAA
WALTER S DIEHL	T-AO 193	Avondale	8 July 1986	13 Sep 1988	TPA
JOHN ERICSSON	T-AO 194	Avondale	15 Mar 1989	18 Mar 1991	TPA
LEROY GRUMMAN	T-AO 195	Avondale	7 June 1987	2 Aug 1989	TAA
KANAWHA	T-AO 196	Avondale	13 July 1989	6 Dec 1991	TAA
PECOS	T-AO 197	Avondale	17 Feb 1988	6 July 1990	TPA
BIG HORN	T-AO 198	Avondale	9 Oct 1989	31 July 1992	TAA
TIPPECANOE	T-AO 199	Avondale	19 Nov 1990	26 Mar 1993	TPA
GUADALUPE	T-AO 200	Avondale	9 July 1990	26 Oct 1992	TPA
PATUXENT	T-AO 201	Avondale	16 Oct 1991	21 June 1995	TAA
YUKON	T-AO 202	Avondale	13 May 1991	11 Dec 1993	TPA
LARAMIE	T-AO 203	Avondale	1 Oct 1994	24 May 1996	TAA
RAPPAHANNOCK	T-AO 204	Avondale	29 June 1992	7 Nov 1995	TPA

Displacement, tons: 40,700; 42,000 (T-AO 201, 203-204) full load
Dimensions, feet (metres): 677.5 × 97.5 × 36 *(206.5 × 29.7 × 10.9)*
Main machinery: 2 Colt-Pielstick 10 PC4.2 V 570 diesels; 34,422 hp(m) *(24.3 MW)* sustained; 2 shafts; cp props
Speed, knots: 20. **Range, n miles:** 6,000 at 18 kt
Complement: 81 civilian (18 officers); 23 naval (1 officer) plus 22 spare
Cargo capacity: 180,000; 159,500 (T-AO 201, 203-204) barrels of fuel oil or aviation fuel
Countermeasures: Decoys: SLQ-25 Nixie; towed torpedo decoy.
Radars: Navigation: 2 Raytheon; I-band.
Helicopters: Platform only.

Comment: Construction was delayed initially by design difficulties, by excessive vibration at high speeds and other problems encountered in the first ship of the class. There are stations on both sides for underway replenishment of fuel and solids. Fitted with integrated electrical auxiliary propulsion. T-AOs 201, 203 and 204 were delayed by the decision to fit double hulls to meet the requirements of the Oil Pollution Act of 1990. This modification increased construction time from 32 to 42 months and reduced cargo capacity by 17 per cent although this can be restored in an emergency. Hull separation is 1.83 m at the sides and 1.98 m on the bottom. T-AOs 191 and 192 were transferred from Penn Ship (when the yard became bankrupt) to Tampa. Tampa's contract was also cancelled on 25 August 1993. Neither ship was completed. T-AO 187 is kept in reduced operating status on the west coast. T-AO 188 and 190 were laid up in mid-1996 and T-AO 189 in September 1997, but returned to service in January 1999. T-AO 188 returned to reduced operating status in April 2005.

UPDATED

PATUXENT *11/2004*, Frank Findler /* 1043690

JOHN LENTHALL *9/2003, Jürg Kürsener /* 0572793

2 MERCY CLASS: HOSPITAL SHIPS (AHH)

Name	No	Builders	Commissioned	F/S
MERCY	T-AH 19	National Steel & Shipbuilding Co	1976	ROS/TPA
(ex-SS *Worth*)				
COMFORT	T-AH 20	National Steel & Shipbuilding Co	1976	ROS/TAA
(ex-SS *Rose City*)				

Displacement, tons: 69,360 full load
Measurement, tons: 54,367 gross; 35,958 net
Dimensions, feet (metres): 894 × 105.6 × 32.8 *(272.6 × 32.2 × 10)*
Main machinery: 2 boilers; 2 GE turbines; 24,500 hp *(18.3 MW)*; 1 shaft
Speed, knots: 17. **Range, n miles:** 13,420 at 17 kt
Complement: 73 civilian crew; 820 naval medical staff; 387 naval support staff
Radars: Navigation: SPS-67; I-band.
Tacan: URN 25.
Helicopters: Platform only.

Comment: Converted San Clemente class tankers. *Mercy* was commissioned 19 December 1986; *Comfort* on 30 November 1987. Each ship has 1,000 beds and 12 operating theatres. Normally, the ships are kept in a reduced operating status in Baltimore, MD, and San Diego, CA, by a small crew of civilian mariners and active duty Navy medical and support personnel. Each ship can be fully activated and crewed within five days.

VERIFIED

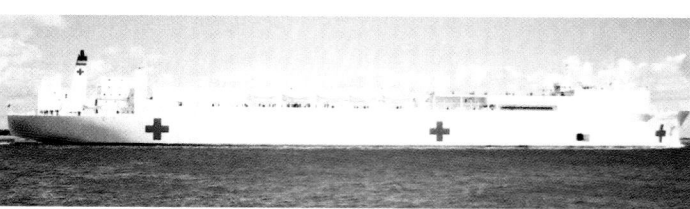

COMFORT *7/2002, Walter Sartori /* 0529969

5 POWHATAN CLASS: FLEET OCEAN TUGS (ATF)

Name	No	Laid down	Commissioned	F/S
CATAWBA	T-ATF 168	14 Dec 1977	28 May 1980	TPA
NAVAJO	T-ATF 169	14 Dec 1977	13 June 1980	TPA
MOHAWK	T-ATF 170	22 Mar 1979	16 Oct 1980	TAA
SIOUX	T-ATF 171	22 Mar 1979	1 May 1981	TPA
APACHE	T-ATF 172	22 Mar 1979	30 July 1981	TAA

Displacement, tons: 2,260 full load
Dimensions, feet (metres): 240.2 × 42 × 15 *(73.2 × 12.8 × 4.6)*
Main machinery: 2 GM EMD 20-645F7B diesels; 7,250 hp(m) *(5.41 MW)* sustained; 2 shafts; Kort nozzles; cp props; bow thruster; 300 hp *(224 kW)*
Speed, knots: 14.5. **Range, n miles:** 10,000 at 13 kt
Complement: 20 (16 civilians, 4 naval communications technicians)
Guns: Space provided to fit 2—20 mm and 2—12.7 mm MGs.
Radars: Navigation: Raytheon 1660/12 and SPS-64(V)9; I-band.

Comment: Built at Marinette Marine Corp, Wisconsin patterned after commercial offshore supply ship design. Originally intended as successors to the Cherokee and Abnaki class ATFs. All transferred to MSC upon completion. 10 ton capacity crane and a bollard pull of at least 54 tons. A 'deck grid' is fitted aft which contains 1 in bolt receptacles spaced 24 in apart. This allows for the bolting down of a wide variety of portable equipment. There are two GPH fire pumps supplying three fire monitors with up to 2,200 gallons of foam per minute. A deep module can be embarked to support naval salvage teams. Two others of the class deactivated for commercial lease in 1999.

UPDATED

APACHE *3/2004*, Diego Quevedo /* 1043689

SIOUX *7/2004*, Michael Nitz /* 1043636

SPECIAL MISSION SHIPS

Notes: (1) Civilian operated but some technical work, communications and research is done by naval personnel. Average of 25 days a month are spent at sea.

(2) There are also five chartered vessels: *Carolyn Chouest* (NR-1 support ship), *Kellie Chouest* and *Dolores Chouest* (both deep submergence submarine support ships), and *Cory Chouest* (support ship for IFA TAGOS trials). These ships are owned and operated by Edison Chouest. *C-Commando* supports Naval Special Warfare Command.

(3) The Command Ships *Coronado* AGF 11 and *Mount Whitney* LCC 20 became the first ships to be operated jointly by uniformed personnel and civil service mariners (CIVMARS) from MSC.

DOLORES CHOUEST   5/2002, L-G Nilsson / 0529993

1 HAYES CLASS: ACOUSTIC SURVEY SHIP (AGE)

Name	No	Builders	Commissioned	F/S
HAYES	T-AG 195 (ex-AGOR 16)	Todd Shipyards, Seattle, WA	21 July 1971	TAA

Displacement, tons: 4,037 full load
Dimensions, feet (metres): 256.5 × 75 (see *Comment*) × 22 *(78.2 × 22.9 × 6.7)*
Main machinery: Diesel-electric; 2 Caterpillar 3516TA diesels; 3,620 hp *(2.7 MW)* sustained; 2 generators; 2 Westinghouse motors; 2,400 hp *(1.79 MW)*; 2 auxiliary diesels (for creep speed); 330 hp *(246 kW)*; 2 shafts; LIPS cp props
Speed, knots: 10. **Range, n miles:** 2,000 at 10 kt
Complement: 20 civilian plus 30 scientists
Radars: Navigation: Raytheon TM 1650/6X and TM 1660/12S; I-band.

Comment: Catamaran hull. Laid down 12 November 1969; launched 2 July 1970. To Ready Reserve 10 June 1983 and transferred to James River (Maritime Administration) for lay-up in 1984 having been too costly to operate. Under FY86 programme was converted to Acoustic Research Ship (AG); reclassified T-AG 195 and completed in early 1992 after five years' work in two shipyards. Mission is to transport, deploy and retrieve acoustic arrays, to conduct acoustic surveys in support of the submarine noise reduction programme and to carry out acoustic testing. Catamaran hull design provides large deck working area, centre well for operating equipment at great depths, and removes laboratory areas from main propulsion machinery. Each hull is 246 ft long and 24 ft wide (maximum). There are three 36 in diameter instrument wells in addition to the main centre well. ***VERIFIED***

HAYES   2/1992

1 ZEUS CLASS: CABLE REPAIRING SHIP (ARC)

Name	No	Builders	Commissioned	F/S
ZEUS	T-ARC 7	National Steel & Shipbuilding Co	19 Mar 1984	TAA

Displacement, tons: 8,370 light; 14,934 full load
Dimensions, feet (metres): 513 × 73 × 25 *(156.4 × 22.3 × 7.6)*
Main machinery: Diesel-electric; 5 GM EMD 20-645F7B diesel generators; 14.32 MW sustained; 2 motors; 10,200 hp *(7.51 MW)*; 2 shafts; cp props; bow thrusters (forward and aft)
Speed, knots: 15.8. **Range, n miles:** 10,000 at 15 kt
Complement: 55 civilians plus 10 scientists

Comment: Ordered 7 August 1979. Remotely manned engineering room controlled from the bridge. Can lay up to 1,000 miles of cable in depths of 9,000 ft. As well as cable laying the ship is equipped for oceanographic survey. ***UPDATED***

ZEUS   3/1999, M Declerck / 0084175

1 CONVERTED COMPASS ISLAND CLASS: MISSILE RANGE INSTRUMENTATION SHIP (AGM)

Name	No	Builders	Commissioned	F/S
OBSERVATION ISLAND (ex-*Empire State Mariner*)	T-AGM 23 (ex-AG 154, ex-YAG 57)	New York Shipbuilding	5 Dec 1958	TPA

Displacement, tons: 13,060 light; 17,015 full load
Dimensions, feet (metres): 564 × 76 × 25 *(171.6 × 23.2 × 7.6)*
Main machinery: 2 Foster-Wheeler boilers; 600 psi *(42.3 kg/cm²)*; 875°F *(467°C)*; 1 GE turbine; 19,250 hp *(14.36 MW)*; 1 shaft
Speed, knots: 20. **Range, n miles:** 17,000 at 15 kt
Complement: 66 civilians plus 59 scientists
Missiles: SLBM: She fired the first ship-launched Polaris missile at sea on 27 August 1959. Refitted to fire the improved Poseidon missile in 1969 and launched the first Poseidon test missile fired afloat on 16 December 1969.
Radars: Navigation: Raytheon 1650/9X and 1660/12S; I-band.
Tacan: URN 25.

Comment: Built as a Mariner class merchant ship (C4-S-A1 type); launched on 15 August 1953; acquired by the Navy on 10 September 1956 for use as a Fleet Ballistic Missile (FBM) test ship. Converted at Norfolk Naval Shipyard. In reserve from September 1972. On 18 August 1977, *Observation Island* was reacquired by the US Navy from the Maritime Administration and transferred to the Military Sealift Command. Reclassified AGM 23 on 1 May 1979. Converted to Missile Range Instrumentation Ship from July 1979-April 1981 at Maryland SB and DD Co to carry an Air Force shipborne phased-array radar system for collection of data on foreign ballistic missile tests. Operated by the Navy in the North Pacific for the US Air Force Intelligence Command, Patrick Air Force Base, Florida. ***VERIFIED***

OBSERVATION ISLAND   8/2002, Hachiro Nakai / 0530006

1 WATERS CLASS: NAVIGATION TEST SUPPORT SHIP (AGM)

Name	No	Builders	Commissioned	F/S
WATERS	T-AG 45	Avondale Industries	26 May 1993	TAA

Displacement, tons: 12,208 full load
Dimensions, feet (metres): 455 × 68.9 × 21 *(138.7 × 21 × 6.4)*
Main machinery: Diesel-electric; 5 GM EMD diesels; 7,400 hp *(5.45 MW)*; 2 Westinghouse motors; 6,800 hp *(15.07 MW)*; 2 shafts; 4 thrusters
Speed, knots: 12. **Range, n miles:** 6,500 at 12 kt
Complement: 66 civilians plus 59 scientists
Radars: Navigation: 2 Raytheon; E/F- and I-bands.

Comment: Ordered 4 April 1990. Laid down 16 May 1991 and launched 6 June 1992. Carried out oceanographic and acoustic surveys in support of the Integrated Underwater Surveillance System. Converted in 1998 to support submarine navigation system testing and missile tracking. In 1999 has replaced *Vanguard* and *Range Sentinel*, both of which have been deactivated. ***VERIFIED***

WATERS   4/1994, Giorgio Arra

3 VICTORIOUS CLASS: OCEAN SURVEILLANCE SHIPS (AGOS)

Name	No	Builders	Commissioned	F/S
VICTORIOUS	T-AGOS 19	McDermott Marine	5 Sep 1991	TPA
EFFECTIVE	T-AGOS 21	McDermott Marine	27 Jan 1993	TPA
LOYAL	T-AGOS 22	McDermott Marine	1 July 1993	TAA

Displacement, tons: 3,396 full load
Dimensions, feet (metres): 234.5 × 93.6 × 24.8 *(71.5 × 28.5 × 7.6)*
Main machinery: Diesel-electric; 4 Caterpillar 3512TA diesels; 5,440 hp *(4 MW)* sustained; 2 GE motors; 3,200 hp *(2.39 MW)*; 2 shafts; 2 bow thrusters; 2,400 hp *(1.79 MW)*
Speed, knots: 16; 3 when towing
Complement: 22 civilian plus 5 Navy
Radars: Navigation: 2 Raytheon; I-band.
Sonars: SURTASS and LFA; towed array; passive/active surveillance.

Comment: All of SWATH design because of its greater stability at slow speeds in high latitudes under adverse weather conditions. A contract for the first SWATH ship, T-AGOS 19, was awarded in November 1986, and options for a further three were exercised in October 1988. T-AGOS 20 deactivated in July 2003. ***UPDATED***

EFFECTIVE   7/2004*, Michael Nitz / 1043635

1 IMPECCABLE CLASS: OCEAN SURVEILLANCE SHIP (AGOS)

Name	No	Builders	Commissioned	F/S
IMPECCABLE	T-AGOS 23	Tampa Shipyard/Halter Marine	Oct 2000	TAA

Displacement, tons: 5,370 full load
Dimensions, feet (metres): 281.5 × 95.8 × 26 *(85.8 × 29.2 × 7.9)*
Main machinery: Diesel-electric; 3 GM EMD 12-645F7B diesel generators; 5.48 MW *(60 Hz)* sustained; 2 Westinghouse motors; 5,000 hp *(3.73 MW)*; 2 shafts; 2 omni-thruster hydrojets; 1,800 hp *(1.34 MW)*
Speed, knots: 12; 3 when towing. **Range, n miles:** 3,000 at 12 kt
Complement: 20 civilians plus 5 scientists plus 20 Navy
Radars: Navigation: Raytheon; I-band.
Sonars: SURTASS; passive surveillance towed array.

Comment: Hull form based on that of *Victorious.* Acoustic systems include an active low frequency towed array (LFA), which has a series of modules each of which houses two high-powered active transducers. These can be used with either mono or bistatic receivers. The payload is lowered through a centre well. Laid down 2 February 1993. Ship was 60 per cent complete when shipyard encountered difficulties that led to termination in October 1993 of the contract for completion of two Kaiser class oilers. Work stopped on *Impeccable,* and the construction contract was also cancelled. The contract was assigned to Halter Marine on 20 April 1995 to complete the ship. Launched in April 1998. WSC-3(V)3 and WSC-6 communications fitted.
VERIFIED

IMPECCABLE
9/2002, Winter & Findler / 0529991

6 PATHFINDER CLASS: SURVEYING SHIPS (AGS)

Name	No	Builders	Launched	Commissioned	F/S
PATHFINDER	T-AGS 60	Halter Marine	7 Oct 1993	5 Dec 1994	TAA
SUMNER	T-AGS 61	Halter Marine	19 May 1994	30 May 1995	TPA
BOWDITCH	T-AGS 62	Halter Marine	15 Oct 1994	30 Dec 1995	TPA
HENSON	T-AGS 63	Halter Marine	21 Oct 1996	20 Feb 1998	TAA
BRUCE C HEEZEN	T-AGS 64	Halter Marine	17 Dec 1998	13 Jan 2000	TAA
MARY SEARS	T-AGS 65	Halter Marine	19 Oct 2000	17 Dec 2001	TAA

Displacement, tons: 4,762 full load
Dimensions, feet (metres): 328.5 × 58 × 19 *(100.1 × 17.7 × 5.8)*
Main machinery: Diesel-electric; 4 EMD/Baylor diesel generators; 11,425 hp *(8.52 MW)*; 2 GE CDF 1944 motors; 8,000 hp *(5.97 MW)* sustained; 6,000 hp *(4.48 MW)*; 2 LIPS Z drives; bow thruster; 1,500 hp *(1.19 MW)*
Speed, knots: 16. **Range, n miles:** 12,000 at 12 kt
Complement: 24 civilians plus 27 oceanographers

Comment: Contract awarded in January 1991 for two ships with an option for a third which was taken up on 29 May 1992. A fourth ship was ordered in October 1994 with an option for two more. Fifth ordered 15 January 1997 and sixth on 6 January 1999. There are three multipurpose cranes and five winches plus a variety of oceanographic equipment including multibeam echo-sounders, towed sonars and expendable sensors. ROVs may be carried.
UPDATED

BRUCE C HEEZEN
8/2004*, Hachiro Nakai / 1043693

HENSON
8/2002, John Brodie / 0530035

1 JOHN McDONNELL CLASS: SURVEYING SHIP (AGS)

Name	No	Builders	Commissioned	F/S
JOHN McDONNELL	T-AGS 51	Halter Marine	16 Dec 1991	TPA

Displacement, tons: 2,054 full load
Dimensions, feet (metres): 208 × 45 × 14 *(63.4 × 13.7 × 4.3)*
Main machinery: 1 GM EMD 12-645E6 diesel; 2,550 hp *(1.9 MW)* sustained; 1 auxiliary diesel; 230 hp *(172 kW)*; 1 shaft
Speed, knots: 12. **Range, n miles:** 13,800 at 12 kt
Complement: 22 civilians plus 11 scientists

Comment: Laid down on 3 August 1989 and launched on 15 August 1990. Carry 34 ft survey launches for data collection in coastal regions with depths between 10 and 600 m and in deep water to 4,000 m. A small diesel is used for propulsion at towing speeds of up to 6 kt. Simrad high-frequency active hull-mounted and side scan sonars are carried. *Littlehales* was deactivated in March 2003 and transferred to NOAA.
UPDATED

JOHN McDONNELL
8/2004*, Hachiro Nakai / 1043695

STRATEGIC SEALIFT FORCE

Notes: These ships provide ocean transportation for Defense and other government agencies. As well as those listed below, MSC also contracts additional tankers and dry cargo ships as needed. As a result these numbers vary with the operational requirement.

2 SHUGHART CLASS:
LARGE, MEDIUM-SPEED, RO-RO (LMSR) SHIPS (AKR)

Name	No	Commissioned	F/S
SHUGHART (ex-*Laura Maersk*)	T-AKR 295	7 May 1996	TWWR
YANO (ex-*Leise Maersk*)	T-AKR 297	8 Feb 1997	TWWR

Measurement, tons: 54,298 grt
Dimensions, feet (metres): 906.8 × 105.5 × 34.4 *(276.4 × 32.2 × 10.5)*
Main machinery: 1 Burmeister & Wain 12L90 GFCA diesel; 46,653 hp(m) *(34.29 MW)*; 1 shaft; bow and stern thrusters
Speed, knots: 24. **Range, n miles:** 12,000 at 24 kt
Complement: 21-44 civilian; up to 50 Navy
Cargo capacity: 255,064 sq ft plus 47,023 sq ft deck cargo
Radars: Navigation: 2 Sperry ARPA; I-band.

Comment: Both were container ships built in Denmark in 1981 and lengthened by Hyundai in 1987. Conversion contract awarded to National Steel and Shipbuilding in July 1993. Both fitted with a stern slewing ramp, side accesses and cranes for both roll-on/roll-off and lift-on/lift-off capabilities. Two twin 57 ton cranes. Conversion for *Shughart* started in June 1994; *Yano* in May 1995; *Soderman* underwent conversion to maritime prepositioning ship and renamed *Stockham.*
UPDATED

YANO
11/2004*, Frank Findler / 1043692

2 GORDON CLASS:
LARGE, MEDIUM-SPEED, RO-RO (LMSR) SHIPS (AKR)

Name	No	Commissioned	F/S
GORDON (ex-*Selandia*)	T-AKR 296	23 Aug 1996	TWWR
GILLILAND (ex-*Jutlandia*)	T-AKR 298	24 May 1997	TWWR

Measurement, tons: 55,422 grt
Dimensions, feet (metres): 956 × 105.8 × 36.3 *(291.4 × 32.2 × 11.9)*
Main machinery: 1 Burmeister & Wain 12K84EF diesel; 26,000 hp(m) *(19.11 MW)*; 2 Burmeister & Wain 9K84EF diesels; 39,000 hp(m) *(28.66 MW)*; 3 shafts (centre cp prop); bow thruster
Speed, knots: 24. **Range, n miles:** 12,000 at 24 kt
Complement: 21 civilian; 50 Navy
Cargo capacity: 276,109 sq ft plus 45,722 sq ft deck cargo
Radars: Navigation: 2 Sperry ARPA; I-band.

Comment: Built in Denmark in 1972 and lengthened by Hyundai in 1984. Conversion contract given to Newport News Shipbuilding on 30 July 1993. Both fitted with a stern slewing ramp, side accesses and improved craneage. Conversion started for both ships on 15 October 1993.
UPDATED

GILLILAND
2/2000, A Sharma / 0085304

8 ALGOL CLASS: FAST SEALIFT SHIPS (AKRH)

Name	No	Builders	Delivered
ALGOL	T-AKR 287	Rotterdamsche DD Mij NV, Rotterdam	7 May 1973
(ex-SS *Sea-Land Exchange*)			
BELLATRIX	T-AKR 288	Rheinstahl Nordseewerke, Emden, West Germany	6 Apr 1973
(ex-SS *Sea-Land Trade*)			
DENEBOLA	T-AKR 289	Rotterdamsche DD Mij NV, Rotterdam	4 Dec 1973
(ex-SS *Sea-Land Resource*)			
POLLUX	T-AKR 290	A G Weser, Bremen, West Germany	20 Sep 1973
(ex-SS *Sea-Land Market*)			
ALTAIR	T-AKR 291	Rheinstahl Nordseewerke, Emden, West Germany	17 Sep 1973
(ex-SS *Sea-Land Finance*)			
REGULUS	T-AKR 292	A G Weser, Bremen, West Germany	30 Mar 1973
(ex-SS *Sea-Land Commerce*)			
CAPELLA	T-AKR 293	Rotterdamsche DD Mij NV, Rotterdam	4 Oct 1972
(ex-SS *Sea-Land McLean*)			
ANTARES	T-AKR 294	A G Weser, Bremen, West Germany	27 Sep 1972
(ex-SS *Sea-Land Galloway*)			

Displacement, tons: 55,355 full load
Measurement, tons: 25,389 net; 27,051-28,095 dwt
Dimensions, feet (metres): 946.2 × 105.6 × 36.8 *(288.4 × 32.2 × 11.2)*
Main machinery: 2 Foster-Wheeler boilers; 875 psi *(61.6 kg/cm²)*; 950°F *(510°C)*; 2 GE MST-19 steam turbines; 120,000 hp *(89.5 MW)*; 2 shafts
Speed, knots: 33. **Range, n miles:** 12,200 at 27 kt
Complement: 43 (as merchant ship); 29 (minimum); 15 (ROS)
Helicopters: Platform only.

Comment: All were originally built as container ships for Sea-Land Services, Port Elizabeth, NJ, but used too much fuel to be cost-effective as merchant ships. Six ships of this class were approved for acquisition in FY81 and the remaining two in FY82. The purchase price included 4,000 containers and 800 container chassis for use in container ship configuration. All eight were converted to Fast Sealift Ships, which are vehicle cargo ships. Conversion included the addition of roll-on/roll-off features. The area between the forward and after superstructures allows for a helicopter flight deck. Capacities are as follows: (sq ft) 150,016 to 166,843 ro-ro; 43,407 lift-on/lift-off; and either 44 or 46 20 ft containers. In addition to one ro-ro ramp port and starboard, twin 35 ton pedestal cranes are installed between the deckhouses and twin 50 ton cranes are installed aft. Ninety-three per cent of a US Army mechanised division can be lifted using all eight ships. Seven of the class moved nearly 11 per cent of all the cargo transported between the US and Saudi Arabia during and after the Gulf War. Six were activated for the Somalian operation in December 1992 and all have been used in various operations and exercises since then. All based in Atlantic and Gulf of Mexico ports. *UPDATED*

REGULUS   5/2003, *Frank Findler* / 0572746

7 BOB HOPE CLASS: LARGE, MEDIUM-SPEED, RO-RO (LMSR) SHIPS (AKR)

Name	No	Builders	Launched	Commissioned	F/S
BOB HOPE	T-AKR 300	Avondale	27 Mar 1997	18 Nov 1998	TWWR
FISHER	T-AKR 301	Avondale	21 Oct 1997	4 Aug 1999	TWWR
SEAY	T-AKR 302	Avondale	25 June 1998	30 Mar 2000	TWWR
MENDONCA	T-AKR 303	Avondale	25 May 1999	30 Jan 2001	TWWR
PILILAAU	T-AKR 304	Avondale	18 Jan 2000	24 July 2001	TWWR
BRITTIN	T-AKR 305	Avondale	21 Oct 2000	11 July 2002	TWWR
BENAVIDEZ	T-AKR 306	Avondale	11 Aug 2001	10 Sep 2003	TWWR

Displacement, tons: 61,680 full load
Dimensions, feet (metres): 948.9 × 106 × 35 *(289.1 × 32.3 × 11)*
Main machinery: 4 Colt Pielstick 10 PC4.2 V diesels; 65,160 hp(m) *(47.89 MW)*; 2 shafts; cp props
Speed, knots: 24. **Range, n miles:** 12,000 at 24 kt
Complement: 26-45 civilian; up to 50 Navy
Cargo capacity: 317,510 sq ft plus 70,152 sq ft deck cargo

Comment: Contract awarded in 1993; options for additional ships exercised in 1994, 1995, 1996 and 1997. All fitted with a stern slewing ramp, side accesses and cranes for both roll-on/roll-off and lift-on/lift-off capabilities. Ramps extend to 130 ft *(40 m)*, and two twin 55 ton cranes are installed. *UPDATED*

SEAY   2/2003, *A Sharma* / 0572747

5 CHAMPION CLASS (TANKERS) (AOT/AOR)

Name	No	Builders	Commissioned
GUS W DARNELL	T-AOT 1121	American SB Co, Tampa, FL	11 Sep 1985
PAUL BUCK	T-AOT 1122	American SB Co, Tampa, FL	7 June 1985
SAMUEL L COBB	T-AOT 1123	American SB Co, Tampa, FL	15 Nov 1985
RICHARD G MATTHIESEN	T-AOT 1124	American SB Co, Tampa, FL	18 Feb 1986
LAWRENCE H GIANELLA	T-AOT 1125	American SB Co, Tampa, FL	22 Apr 1986

Displacement, tons: 39,624 full load
Dimensions, feet (metres): 615 × 90 × 36 *(187.5 × 27.4 × 10.8)*
Main machinery: 1 Sulzer 5RTA76 diesel; 18,400 hp(m) *(13.52 MW)* sustained; 1 shaft
Speed, knots: 16. **Range, n miles:** 12,000 at 16 kt
Complement: 23 (9 officers)
Cargo capacity: 238,400 barrels of oil fuel

Comment: Built for Ocean Carriers Inc, Houston, Texas specifically for long-term time charter to the Military Sealift Command (20 years) as Point-to-Point fuel tankers. AOT 1121 remains on long-term charter while the remainder were purchased in 2003 and are designated USNS. The last two are equipped with a modular fuel delivery system to allow them to rig underway replenishment gear. *UPDATED*

LAWRENCE H GIANELLA   1/2002, *A Sharma* / 0530036

PREPOSITIONING PROGRAMME

Notes: (1) Military Sealift Command's Afloat Prepositioning Force (APF) improves US capabilities to deploy forces rapidly to any area of conflict. The force includes long-term chartered commercial vessels, activated Ready Reserve Force ships and government ships and includes vehicle/cargo carriers, container ships, aviation logistics ships and Large, Medium-Speed, Roll-on/roll-off (LMSR) ships. Together these ships preposition equipment and supplies for the Marine Corps, Navy, Army, Air Force and the Defense Logistics Agency. The APF comprises: the Maritime Prepositioning Force (MPF), Logistics Prepositioning Ships (LPS) and Combat Prepositioning Ships (CPS).
(2) The MPF operate in forward-deployed squadrons: Squadron One is located in the Mediterranean, Squadron Two at Diego Garcia and Squadron Three in the Western Pacific. 16 ships are loaded with equipment and supplies for the US Marine Corps.
(3) Nine LPS are loaded with US Air Force and Navy ammunition and Defense Logistics Agency Petroleum products, while two are aviation logistic ships serving as USMC intermediate maintenance facilities.
(4) The 10 CPS preposition equipment and supplies for a US Army heavy brigade and combat support/combat service support elements.
(5) Maritime Prepositioning Force (Future): the US Navy's Seabasing initiative, in which manoeuvre forces are supported by logistics and combat fire support in staging bases in or near the theatre of operations, is under development. This is likely to require a variety of platforms that can assemble offshore and redeploy quickly. It is envisaged that a seabase would comprise an Expeditionary Strike Group (ESG), a Carrier Strike Group (CSG), and a Maritime Prepositioning Group (MPG) supported by a Combat Logistics Force. The seabase would itself require support from intermediate bases such as Guam and Diego Garcia. A key component of the seabasing concept is a new transport ship known as the Maritime Prepositioning Force (Future) (MPF(F)). Such vessels are to be capable of selectively offloading standard containerised loads and other equipment. Large deck areas suitable for flight operations are also necessary. MPF(F) could additionally assume some of the roles planned for the JCC(X) future command-and-control ship. Other potential applications of the MPF(F) include acting as intra-theatre shuttles, afloat medical care and mine-countermeasures support. Some 18 vessels are required and construction of the first ship could begin in FY08.
(6) Intra-theater Connectors: A new class of High Speed Connectors (HSC) required to move supplies and personnel to the seabase or austere ports from either advanced bases or intermediate staging bases further away. The vessels are likely to be high-speed, shallow-draft, self-deployable and to have a notional range of 2,500 n miles. A six-month analysis of alternatives (AoA) on which concepts to pursue is expected to be completed by late 2005. A winning design is expected to be chosen in FY08 followed by ship delivery in FY10. One contender is the Joint High Speed Vessel (JHSV) programme, the result of a merger between the army's Theater Support Vessel programme and the Navy/USMC High-Speed Connector. Part of the requirement could also be fulfilled by aircraft.
(7) Sea-Shore Connector: A second type of HSC which notionally is to have a range of 500 n miles or less and is intended to move supplies for the assaulting forces ashore from the seabase. Current examples include landing craft (LCUs) and LCACs. Replacement vessels are expected to be required to accommodate heavier loads.
(8) The US Army is developing plans for three ARFs in collaboration with the US Navy's seabasing concept. Each ARF would comprise five ships; two LMSR, a shallow draft Ro-Ro ship and two ammunition ships. The first ARF, to be based around Guam and Saipan in the Pacific Ocean, is already partly fielded. The second ARF will be based around Diego Garcia in the Indian Ocean, while the third is planned to be based in the Mediterranean.

1 THEATRE SUPPORT VESSEL (HSV)

WESTPAC EXPRESS

Measurement, tons: 750 dwt
Dimensions, feet (metres): 331.4 × 87.4 × 13.8 *(101.0 × 26.65 × 4.2)*
Main machinery: 4 Caterpillar 3618 diesels; 38,620 hp *(28.8 MW)*; 4 Kamewa waterjets
Speed, knots: 40. **Range, n miles:** 1,100 at 35 kt
Military lift: 550 tonnes of equipment and 970 personnel

Comment: Following trials which started in July 2001, chartered by Military Sealift Command from Austal Ships, West Australia. The current charter expires on 31 August 2005. Aluminium construction. Employed by US Marine Corps Third Expeditionary Force (III MEF) to transport equipment and troops from Okinawa for training exercises in Yokohama, Guam and other regional destinations. The benefits include reduced dependence on and cost of airlift. The vessel will retain commercial livery and markings. Based at Okinawa. *UPDATED*

WESTPAC EXPRESS   8/2001, *Mitsuhiro Kadota* / 0131282

1 CONTAINER SHIP (AK)

Name	No	Builders	Commissioned	F/S
A1C WILLIAM H PITSENBARGER	T-AK 4638	Chantiers de l'Atlantique, St Nazaire	28 Nov 2001	PREPO

Displacement, tons: 31,986 full load
Dimensions, feet (metres): 621.3 × 105.6 × 37.5 *(189.3 × 32.2 × 11.4)*
Main machinery: 1 Sulzer 7RLB 66 reversible diesel; one shaft
Speed, knots: 17.5. **Range, n miles:** 16,800 at 13 kt
Complement: 17
Cargo capacity: 1,670 TEU (808 under deck)

Comment: Completed in 1984. Chartered by MSC in October 2001 as LPS ship to carry Air Force munitions in Guam/Saipan. Reflagged and renamed. Owned by RR and VO LCC and operated by Red River Shipping Corporation.

UPDATED

A1C WILLIAM H PITSENBARGER   6/2003, US Navy / 0572807

1 RO-RO CONTAINER: CARGO SHIP (AK)

Name	No	Builders	Commissioned	F/S
MERLIN	AK 323	Chantiers, France	1978	PREPO

Displacement, tons: 26,378 full load
Dimensions, feet (metres): 669.8 × 86.9 × 34.5 *(204.1 × 26.5 × 10.5)*
Main machinery: Pielstick medium speed diesel; 1 shaft
Speed, knots: 16. **Range, n miles:** 23,200 at 13 kt
Complement: 19
Cargo capacity: 1,066 TEU

Comment: The ship is owned and operated by Sealift, Inc and is under charter to Military Sealift Command. Acquired in 2002, MV Merlin can carry more than 1,000 20 ft container equivalents of aviation munitions. Plans are to outfit the ship with a cocoon system in 2003. *Merlin* serves as an LPS ship.

VERIFIED

MERLIN   6/2003, US Navy / 0572808

1 HIGH SPEED LOGISTIC SUPPORT VESSEL (HSV/MCS)

SWIFT HSV-2

Displacement, tons: 1,800
Dimensions, feet (metres): 321.6 × 88.6 × 11.1 *(98.0 × 27.0 × 3.4)*
Main machinery: 4 CAT 3618 diesels; 38,620 hp *(28.8 MW)*; 4 LIPS 150D waterjets
Speed, knots: 48 (light); 38 (full load). **Range, n miles:** 2,400 at 35 kt
Complement: 42
Military lift: 615 tonnes cargo and 350 personnel
Guns: 1 Mk 96 25 mm.

Comment: Built as *Incat 61* (Incat Evolution 10B) of aluminium construction and leased for one year (with option to extend until 2008) by Military Sealift Command from Bollinger/Incat of Lockport, Louisiana from June 2003. The ship is to be utilised as a platform to conduct trials and exercises required by the Navy Warfare Development Command and also as an interim replacement for the Mine Countermeasures Support Ship *Inchon*. A stern ramp, loading directly astern or to the starboard quarter, is to be fitted. It will be capable of loading/unloading vehicles up to and including main battle tanks. The ramp will also be capable of launch and recovery of assault vehicles. The vessel is to be fitted with a helicopter deck capable of operating MH-60S, CH-46, UH-1 and AH-1 helicopters. A protected area for the storage and maintenance of two MH-60S helicopters will also be provided.

UPDATED

SWIFT   2/2004*, US Navy / 1043637

8 WATSON CLASS:
LARGE, MEDIUM-SPEED, RO-RO (LMSR) SHIPS (AKR)

Name	No	Builders	Launched	Commissioned	F/S
WATSON	T-AKR 310	NASSCO	26 July 1997	23 June 1998	PREPO
SISLER	T-AKR 311	NASSCO	28 Feb 1998	1 Dec 1998	PREPO
DAHL	T-AKR 312	NASSCO	2 Oct 1998	13 July 1999	PREPO
RED CLOUD	T-AKR 313	NASSCO	7 Aug 1999	18 Jan 2000	PREPO
CHARLTON	T-AKR 314	NASSCO	11 Dec 1999	23 May 2000	PREPO
WATKINS	T-AKR 315	NASSCO	28 July 2000	5 Dec 2000	PREPO
POMEROY	T-AKR 316	NASSCO	10 Mar 2001	14 Aug 2001	PREPO
SODERMAN	T-AKR 317	NASSCO	26 Apr 2002	25 Sep 2002	PREPO

Displacement, tons: 62,968 full load
Dimensions, feet (metres): 951.4 × 106 × 35 *(290 × 32.3 × 11)*
Main machinery: 2 GE Marine LM gas turbines; 64,000 hp *(47.7 MW)*; 2 shafts; cp props
Speed, knots: 24. **Range, n miles:** 12,000 at 22 kt
Complement: 26-45 civilian; up to 50 Navy
Cargo capacity: 394,673 sq ft. 13,000 tons

Comment: Contract awarded in 1993; options for additional ships exercised at one a year to 2002. All are fitted with a stern slewing ramp, side accesses and cranes for both roll-on/roll-off and lift-on/lift-off capabilities. Ramps extend to 130 ft *(40 m)*, and two twin 55 ton cranes are installed. Serve as Combat Prepositioning Ships.

UPDATED

SODERMAN   5/2003, A Sharma / 0572811

2 T-AVB 3 CLASS: AVIATION LOGISTIC SHIPS (AVB)

Name	No	Builders	Commissioned	F/S
WRIGHT (ex-SS *Young America*)	T-AVB 3	Ingalls Shipbuilding	1970	PREPO/ROS
CURTISS (ex-SS *Great Republic*)	T-AVB 4	Ingalls Shipbuilding	1969	PREPO/ROS

Displacement, tons: 23,872 full load
Measurement, tons: 11,757 gross; 6,850 net; 15,946 dwt
Dimensions, feet (metres): 602 × 90.2 × 29.8 *(183.5 × 27.5 × 9.1)*
Main machinery: 2 Combustion Engineering boilers; 2 GE turbines; 30,000 hp *(22.4 MW)*; 1 shaft
Speed, knots: 20. **Range, n miles:** 11,000 at 20 kt
Complement: 38 crew and 1 Aircraft Maintenance Detachment totalling 363 men

Comment: To reinforce the capabilities of the Maritime Prepositioning Ship programme, conversion of two ro-ro ships into maintenance aviation support ships was approved in FY85 and FY86. *Wright* was completed 14 May 1986, *Curtiss* 18 August 1987. Both conversions took place at Todd Shipyards, Galveston, Texas. Each ship has side ports and three decks aft of the bridge superstructure and has the capability to load the vans and equipment of a Marine Aviation Intermediate Maintenance Activity. The ships' mission is to service aircraft from an afloat platform. They can then revert to a standard sealift role if required. Maritime Administration hull design is C5-S-78a. These Logistics Prepositioning Ships are operated by American Overseas Marine and maintained in a reduced operating status.

UPDATED

WRIGHT   4/2003, A Sharma / 0572752

1 CONTAINER SHIP (AK)

Name	No	Builders	Commissioned	F/S
MAJ BERNARD F FISHER (ex-*Sea Fox*)	T-AK 4396	Odense	1985	PREPO

Displacement, tons: 48,012 full load
Dimensions, feet (metres): 652.2 × 105.6 × 36.1 *(198.1 × 32.2 × 11)*
Main machinery: 1 BMW diesel; 1 shaft
Speed, knots: 19
Complement: 24
Cargo capacity: 1,466 TEUs with 10,227 sq ft garage space for ro-ro cargo

Comment: An LPS ship owned and operated by Sealift, Inc. Acquired in 1998, it is used to preposition US Air Force war stocks at sea.

VERIFIED

CAPT STEVEN L BENNETT   *6/1999, US Navy* / 0016575

MAJ BERNARD F FISHER   *6/1999, US Navy* / 0084185

2 CONTAINER SHIPS (AK)

Name	No	Builders	Commissioned	F/S
LTC JOHN U D PAGE (ex-*Newark Bay*)	T-AK 4543	Daewoo Shipbuilding	1985	PREPO
SSGT EDWARD A CARTER (ex-*OOCL Innovation*)	T-AK 4544	Daewoo Shipbuilding	1984	PREPO

Displacement, tons: 81,284 full load
Dimensions, feet (metres): 950 × 106 × 38 *(289.5 × 32.3 × 11.6)*
Main machinery: 1 Sulzer RLB 90 diesel; 1 shaft
Speed, knots: 18
Complement: 20
Cargo capacity: 2,500 TEU

Comment: *LTC John U D Page* delivered to MSC in February 2001 and *SSGT Edward A Carter* in June 2001. Both operated by Maersk Sealand and are CPS ships for army prepositioning in the Indian Ocean.

VERIFIED

1 RO-RO CONTAINER: CARGO SHIP (AKR)

Name	No	Builders	Commissioned	F/S
1st LT HARRY L MARTIN (ex-*Tarago*)	T-AK 3015	Tarago Shipyard	20 Apr 2000	Sqn 3

Displacement, tons: 47,777 full load
Dimensions, feet (metres): 754.3 × 106 × 36.1 *(229.9 × 32.3 × 11)*
Main machinery: 1 MAN K7-SZ-90/160 diesel; 25,690 hp(m) *(18.88 MW)*; 1 shaft
Speed, knots: 18. **Range, n miles:** 17,000 at 17 kt
Complement: 23 plus 100 marines
Cargo capacity: 168,547 sq ft. 735 TEU

Comment: Acquired in February 1997 for conversion at Atlantic Drydock, Jacksonville. Carries USMC expeditionary airfield, fleet hospital package and construction equipment.

UPDATED

1st LT HARRY L MARTIN   *7/2003, Frank Findler* / 0572751

1 RO-RO CONTAINER: CARGO SHIP (AKR)

Name	No	Builders	Commissioned	F/S
L/CPL ROY M WHEAT (ex-*Bazaliya*)	T-AK 3016	Bender Shipbuilding	Oct 2001	Sqn 1

Displacement, tons: 50,101 full load
Dimensions, feet (metres): 863.8 × 98.4 × 34.8 *(263.3 × 30 × 10.6)*
Main machinery: 2 gas turbines; 47,020 hp(m) *(34.56 MW)*; 2 shafts
Speed, knots: 20. **Range, n miles:** 12,000 at 20 kt
Complement: 30 plus 100 marines
Cargo capacity: 109,170 sq ft. 846 TEU

Comment: Acquired in March 1997 for conversion for Maritime Prepositioning Force by Bender Shipbuilding, Mobile. The ship has been lengthened by 117 ft. Carries USMC expeditionary airfield, fleet hospital package and construction equipment.

UPDATED

SSGT EDWARD A CARTER   *2/2003, A Sharma* / 0572750

1 CONTAINER SHIP (AK)

Name	No	Builders	Commissioned	F/S
CAPT STEVEN L BENNETT (ex-*Sea Pride*)	T-AK 4296	Samsung Shipbuilding	1984	PREPO

Displacement, tons: 52,878 full load
Dimensions, feet (metres): 687 × 99.8 × 49.9 *(209.4 × 30.4 × 15.2)*
Main machinery: 1 diesel; 1 shaft
Speed, knots: 16.5
Complement: 21 civilian
Cargo capacity: 520 TEU (on deck within cocoons); 1,006 TEU (under deck); 422 TEU (reefer)

Comment: The ship is owned and operated by Sealift Inc, under charter to Military Sealift Command. When fully loaded, *Bennett* carries over 1,500 20 ft containers of various aviation munitions intended to resupply forward-deployed fighter and attack squadrons. The ship is fitted with an extensive cocoon system enabling it to maintain deck-loaded munitions in an environmentally controlled atmosphere. While lightweight and easy to maintain, the fabric cocoon system is able to withstand gusts of up to 90 kt. *Bennett* is an LPS ship.

UPDATED

L/CPL ROY M WHEAT   *9/2002, Winter & Findler* / 0529989

For details of the latest updates to *Jane's Fighting Ships* online and to discover the additional information available exclusively to online subscribers please visit
jfs.janes.com

1 RO-RO CONTAINER: CARGO SHIP (AKR)

Name	No	Builders	Commissioned	F/S
GYSGT FRED W STOCKHAM (ex-*Soderman*)	T-AK 3017	National Steel and Shipbuilding	July 2001	Sqn 2

Displacement, tons: 55,123 full load
Dimensions, feet (metres): 907 × 106 × 36 *(276.4 × 32.2 × 10.9)*
Main machinery: 1 Burmeister & Wain 12L90 GFCA diesel; 46,653 hp(m) *(34.29 MW)*; 1 shaft; bow and stern thrusters
Speed, knots: 24. **Range, n miles:** 12,000 at 24 kt
Complement: 28 crew, 12 cargo maintenance, 83 opp
Cargo capacity: 94,337 sq ft, 1,126 TEU

Comment: Ex-USNS *Soderman* joined the Maritime Prepositioning Force in July 2001 and is assigned to MPSRON Two. Carries USMC expeditionary airfield, fleet hospital package and construction equipment. *UPDATED*

GYSGT FRED W STOCKHAM   *8/2001, van Ginderen Collection* / 0131276

5 CPL LOUIS J HAUGE, JR CLASS: VEHICLE CARGO SHIPS (AKRH)

Name	No	Builders	Commissioned	F/S
CPL LOUIS J HAUGE, JR (ex-MV *Estelle Maersk*)	T-AK 3000	Odense Staalskibsvaerft A/S, Lindo	Oct 1979	Sqn 2
PFC WILLIAM B BAUGH (ex-MV *Eleo Maersk*)	T-AK 3001	Odense Staalskibsvaerft A/S, Lindo	Apr 1979	Sqn 2
PFC JAMES ANDERSON, JR (ex-MV *Emma Maersk*)	T-AK 3002	Odense Staalskibsvaerft A/S, Lindo	July 1979	Sqn 2
1st LT ALEX BONNYMAN (ex-MV *Emilie Maersk*)	T-AK 3003	Odense Staalskibsvaerft A/S, Lindo	Jan 1980	Sqn 2
PVT FRANKLIN J PHILLIPS (ex-Pvt *Harry Fisher*, ex-MV *Evelyn Maersk*)	T-AK 3004	Odense Staalskibsvaerft A/S, Lindo	Apr 1980	Sqn 2

Displacement, tons: 46,552 full load
Dimensions, feet (metres): 755 × 90 × 37.1 *(230 × 27.4 × 11.3)*
Main machinery: 1 Sulzer 7RND76M diesel; 16,800 hp(m) *(12.35 MW)*; 1 shaft; bow thruster
Speed, knots: 16.4. **Range, n miles:** 10,800 at 16 kt
Complement: 27 plus 10 technicians
Cargo capacity: Containers, 383; Ro-Ro, 121,595 sq ft; JP-5 bbls, 17,128; DF-2 bbls, 10,642; Mogas bbls, 3,865; stable water, 2,022; cranes, 3 twin 30 ton; 92,831 cu ft breakbulk
Helicopters: Platform only.

Comment: Converted from five Maersk Line ships by Bethlehem Steel, Sparrow Point, MD. Conversion work included the addition of 157 ft *(47.9 m)* amidships. All operated by Maersk Line Ltd. *UPDATED*

CPL LOUIS J HAUGE JR   *3/1998, A Sharma* / 0053410

3 SGT MATEJ KOCAK CLASS: VEHICLE CARGO SHIPS (AKRH)

Name	No	Builders	Commissioned	F/S
SGT MATEJ KOCAK (ex-SS *John B Waterman*)	T-AK 3005	Pennsylvania SB Co, Chester, PA	14 Mar 1981	Sqn 2
PFC EUGENE A OBREGON (ex-SS *Thomas Heywood*)	T-AK 3006	Pennsylvania SB Co, Chester, PA	1 Nov 1982	Sqn 1
MAJ STEPHEN W PLESS (ex-SS *Charles Carroll*)	T-AK 3007	General Dynamics Corp, Quincy, MA	14 Mar 1983	Sqn 3

Displacement, tons: 48,754 full load
Dimensions, feet (metres): 821 × 105.6 × 32.3 *(250.2 × 32.2 × 9.8)*
Main machinery: 2 boilers; 2 GE turbines; 30,000 hp *(22.4 MW)*; 1 shaft
Speed, knots: 20. **Range, n miles:** 13,000 at 20 kt
Complement: 29 plus 10 technicians
Cargo capacity: Containers, 562; Ro-Ro, 152,236 sq ft; JP-5 bbls, 20,290; DF-2 bbls, 12,355; Mogas bbls, 3,717; stable water, 2,189; cranes, 2 twin 50 ton and 1—30 ton gantry
Helicopters: Platform only.

Comment: Converted from three Waterman Line ships by National Steel and Shipbuilding, San Diego. Delivery dates T-AK 3005, 1 October 1984; T-AK 3006, 16 January 1985; T-AK 3007, 15 May 1985. Conversion work included the addition of 157 ft *(47.9 m)* amidships. All operated by Waterman SS Corp. *UPDATED*

SGT MATEJ KOJAK   *10/2002, H M Steele* / 0530097

5 2nd LT JOHN P BOBO CLASS: VEHICLE CARGO SHIPS (AKRH)

Name	No	Builders	Commissioned	F/S
2nd LT JOHN P BOBO	T-AK 3008	General Dynamics, Quincy	14 Feb 1985	Sqn 1
PFC DEWAYNE T WILLIAMS	T-AK 3009	General Dynamics, Quincy	6 June 1985	Sqn 1
1st LT BALDOMERO LOPEZ	T-AK 3010	General Dynamics, Quincy	20 Nov 1985	Sqn 2
1st LT JACK LUMMUS	T-AK 3011	General Dynamics, Quincy	6 Mar 1986	Sqn 3
SGT WILLIAM R BUTTON	T-AK 3012	General Dynamics, Quincy	27 May 1986	Sqn 2

Displacement, tons: 44,330 full load
Dimensions, feet (metres): 675.2 × 105.5 × 29.6 *(205.8 × 32.2 × 9)*
Main machinery: 2 Stork-Wärtsilä Werkspoor 16TM410 diesels; 27,000 hp(m) *(19.84 MW)* sustained; 1 shaft; bow thruster; 1,000 hp *(746 kW)*
Speed, knots: 17.7. **Range, n miles:** 12,840 at 18 kt
Complement: 30 plus 10 technicians
Cargo capacity: Containers, 578; Ro-Ro, 156,153 sq ft; JP-5 bbls, 20,776; DF-2 bbls, 13,334; Mogas bbls, 4,880; stable water, 2,357; cranes, 1 single and 2 twin 39 ton
Helicopters: Platform only.

Comment: Built for MPS operations. Owned and operated by American Overseas Marine. *UPDATED*

1st LT BALDOMERO LOPEZ   *6/2003, Bram Plokker* / 0572749

READY RESERVE FORCE (RRF)

Notes: (1) The Ready Reserve Force was created in 1976, due to a shrinking US-flagged commercial fleet, to support military sea transportation needs. Composed of merchant ships that are no longer commercially useful, the RRF is designed to be made available quickly for military sealift operations.
(2) On 1 January 2005 the RRF consisted of 57 ships, including Ro-Ro, breakbulk, auxiliary crane, heavy lift barge carriers and tankers. These are maintained in various stages of readiness and able to get underway in four, five, 10 or 20 days. They are located in various ports along the US East, West and Gulf Coasts and in Japan.
(3) The Department of Transportation's Maritime Administration (MARAD) is responsible for the maintenance and administration of the ships when they are not operating. Military Sealift Command assumes operational control once the ships are activated.
(4) RRF ships have red, white and blue funnel stripes.

10 KEYSTONE STATE CLASS: AUXILIARY CRANE SHIPS (AK)

Name	No	Builders	Conversion Commissioned
KEYSTONE STATE (ex-SS *President Harrison*)	T-ACS 1	Defoe SB Co, Bay City	1984
GEM STATE (ex-SS *President Monroe*)	T-ACS 2	Defoe SB Co, Bay City	1985
GRAND CANYON STATE (ex-SS *President Polk*)	T-ACS 3	Dillingham SR, Portland	1986
GOPHER STATE (ex-*Export Leader*)	T-ACS 4	Norshipco, Norfolk	Oct 1987
FLICKERTAIL STATE (ex-*Export Lightning*)	T-ACS 5	Norshipco, Norfolk	Dec 1987
CORNHUSKER STATE (ex-*Staghound*)	T-ACS 6	Norshipco, Norfolk	Mar 1988
DIAMOND STATE (ex-*President Truman*)	T-ACS 7	Tampa SY	Jan 1989
EQUALITY STATE (ex-*American Banker*)	T-ACS 8	Tampa SY	May 1989
GREEN MOUNTAIN STATE (ex-*American Altair*)	T-ACS 9	Norshipco, Norfolk	Sep 1990
BEAVER STATE (ex-*American Draco*)	T-ACS 10	Kreith Ship Repair, New Orleans	4 May 1997

Displacement, tons: 31,500 full load
Dimensions, feet (metres): 668.6 × 76.1 × 33.5 *(203.8 × 23.2 × 10.2)*
Main machinery: 2 boilers; 2 GE turbines; 19,250 hp *(14.4 MW)*; 1 shaft
Speed, knots: 20. **Range, n miles:** 13,000 at 20 kt
Complement: 89
Cargo capacity: 300+ standard containers

Comment: Auxiliary crane ships are container ships to which have been added up to three twin boom pedestal cranes which will lift containerised or other cargo from itself or adjacent vessels and deposit it on a pier or into lighterage. There are dimensional differences between some of the class. *UPDATED*

GOPHER STATE   *7/2003, W Sartori* / 0572766

7 PRODUCT TANKERS (AOT)

NODAWAY T-AOG 78	**MOUNT WASHINGTON** T-AOT 5076
ALATNA T-AOG 81	**CHESAPEAKE** T-AOT 5084
CHATTAHOOCHEE T-AOG 82	**PETERSBURG** T-AOT 9109
POTOMAC T-AOT 181	

Comment: *Chesapeake* and *Petersburg* are operational as LPS ships.

UPDATED

POTOMAC   *6/2000, Bob Fildes /* 0106835

3 BREAK BULK SHIPS (AK/AKR/AE)

CAPE NOME T-AK 1014   **CAPE GIRARDEAU** T-AK 2039   **CAPE GIBSON** T-AK 5051

UPDATED

BREAK BULK SHIP   *8/2002, Royal Australian Navy /* 0572796

31 RO-RO SHIPS (AKR)

COMET T-AKR 7	**CAPE HORN** T-AKR 5068
METEOR T-AKR 9	**CAPE EDMONT** T-AKR 5069
CAPE ISLAND (ex-*Mercury*) T-AKR 10	**CAPE INSCRIPTION** T-AKR 5076
CAPE INTREPID T-AKR 11	**CAPE LAMBERT** T-AKR 5077
CAPE TEXAS (ex-*Lyra*) T-AKR 112	**CAPE LOBOS** T-AKR 5078
CAPE TAYLOR (ex-*Cygnus*) T-AKR 113	**CAPE KNOX** T-AKR 5082
ADM WM H CALLAGHAN T-AKR 1001	**CAPE KENNEDY** T-AKR 5083
CAPE ORLANDO (ex-*American Eagle*) T-AKR 2044	**CAPE VINCENT** (ex-*Taabo Italia*) T-AKR 9666
CAPE DUCATO T-AKR 5051	**CAPE RISE** (ex-*Saudi Riyadh*) T-AKR 9678
CAPE DOUGLAS T-AKR 5052	**CAPE RAY** (ex-*Saudi Makkah*) T-AKR 9679
CAPE DOMINGO T-AKR 5053	**CAPE VICTORY** (ex-*Merzario Britania*) T-AKR 9701
CAPE DECISION T-AKR 5054	**CAPE TRINITY** (ex-*Santos*) T-AKR 9711
CAPE DIAMOND T-AKR 5055	**CAPE RACE** (ex-*G&C Admiral*) T-AKR 9960
CAPE ISABEL T-AKR 5062	**CAPE WASHINGTON** (ex-*Hual Transporter*)
CAPE HUDSON T-AKR 5066	T-AKR 9961
CAPE HENRY T-AKR 5067	**CAPE WRATH** (ex-*Hual Trader*) T-AKR 9962

UPDATED

CAPE DOUGLAS   *11/2004*, Frank Findler /* 1043678

6 MISCELLANEOUS HEAVY LIFT SHIPS
(AK/AP/AKR)

Lash ships	Heavy lift ships
CAPE FEAR T-AK 5061	**CAPE MAY** T-AKR 5063
CAPE FLATTERY T-AK 5070	**CAPE MOHICAN** T-AKR 5065
CAPE FLORIDA T-AK 5071	
CAPE FAREWELL T-AK 5073	

UPDATED

CAPE MOHICAN   *3/1999*, van Ginderen Collection /* 0084193

COAST GUARD

Headquarters Appointments

Commandant:
  Admiral Thomas Collins
Vice Commandant:
  Vice Admiral Terry M Cross
Commander, Atlantic Area:
  Vice Admiral Vivien S Crea
Commander, Pacific Area:
  Vice Admiral Harvey E Johnson Jr

Establishment

The United States Coast Guard was established by an Act of Congress approved 28 January 1915, which consolidated the Revenue Cutter Service (founded in 1790) and the Life Saving Service (founded in 1848). The act of establishment stated the Coast Guard "shall be a military service and a branch of the armed forces of the USA at all times. The Coast Guard shall be a service in the Treasury Department except when operating as a service in the Navy".

Congress further legislated that in time of national emergency or when the President so directs, the Coast Guard operates as a part of the Navy. The Coast Guard did operate as a part of the Navy during the First and Second World Wars.

The Lighthouse Service (founded in 1789) was transferred to the Coast Guard on 1 July 1939 and the Bureau of Navigation and Steamboat Inspection on 28 February 1942.

The Coast Guard was transferred from the Department of Transportation to the Department of Homeland Security on 1 March 2003.

Missions

The Coast Guard has five strategic aims:
Safety: Prevent deaths, injuries, and property damage associated with maritime transportation, fishing and recreational boating.
National Defense: Defend the nation as one of the five US Armed Services. Enhance regional stability in support of the National Security Strategy specifically maritime homeland security.
Maritime Security: Protect maritime borders from all intrusions by (a) halting the flow of illegal drugs, aliens, and contraband into the United States through maritime routes; (b) preventing illegal fishing; and (c) suppressing violations of federal law in the maritime arena.

Mobility: Facilitate maritime commerce and eliminate interruptions and impediments to the economical movement of goods and people, while maximizing recreational access and enjoyment of the water.
Protection of Natural Resources: Prevent environmental damage and natural resource degradation associated with maritime transportation, fishing, and recreational boating.

Personnel

2005: 6,051 officers, 1,493 warrant officers, 30,912 enlisted, 8,100 reserves

Integrated Deepwater System (IDS)

IDS is a 20-year programme to modernise and replace USCG ships and aircraft and to improve command and control and logistics systems. The first contract for the programme, was awarded in June 2002 to Integrated Coast Guard Systems (ICGS), a partnership of Lockheed Martin and Northrop Grumman. Northrop Grumman Ship Systems will conduct the design and build three classes of new cutters and associated small boats. Up to 91 vessels are planned. Lockheed Martin is responsible for the C4ISR and system integration aspects of the programme and for aircraft procurement.

Cutter Strength

All Coast Guard vessels over 65 ft in length and that have adequate crew accommodation are referred to as 'cutters'. All names are preceded by USCG. The first two digits of the hull number for all Coast Guard vessels under 100 ft in length indicates the approximate length overall.

Approximately 2,000 standard and non-standard boats are in service ranging in size from 11 ft skiffs to 55 ft aids-to-navigation craft.

Category/Classification		Active	Building (Projected)
Cutters			
WHEC	High Endurance Cutters	12	—
WMEC	Medium Endurance Cutters	30	—
Icebreakers			
WAGB	Icebreakers	4	1
WTGB	Icebreaking Tugs	9	—

Category/Classification		Active	Building (Projected)
Patrol Forces			
WPC	Patrol Coastal	4	—
WPB	Patrol Craft	105	—
Training Cutters			
WIX	Training Cutters	2	—
Buoy Tenders			
WLB	Buoy Tenders, Seagoing	17	—
WLM	Buoy Tenders, Coastal	14	—
WLI	Buoy Tenders, Inland	5	—
WLR	Buoy Tenders, River	18	—
Construction Tenders			
WLIC	Construction Tenders, Inland	13	—
Harbour Tugs			
WYTL	Harbour Tugs, Small	11	—

DELETIONS

Patrol Forces

2002 *Point Brower* (to Azerbaijan)

Tenders and Tugs

2002 *Sweetgum* (to Panama), *Madrona* (to El Salvador), *Cowslip* and *Sedge* (to Nigeria), *White Sumac*
2003 *Bramble*, *Firebush* and *Sassafras* (both to Nigeria)
2004 *Sundew*

HIGH ENDURANCE CUTTERS

0 + 2 (6) NATIONAL SECURITY CUTTERS (PSOH/WMSL)

Name	No	Builders	Laid down	Launched	Commissioned
—	WMSL 750	Northrop Grumman Ingalls Shipbuilding	29 Mar 2005	2006	2007
—	—	Northrop Grumman Ingalls Shipbuilding	2005	2006	2007

Displacement, tons: 3,086 standard; 4,112 full load
Dimensions, feet (metres): 421 × 54.2 × 20.1
   (128.3 × 16.5 × 6.1)

Main machinery: CODAG; 1 GE LM2500 gas turbine; 29,500 hp
   (22.0 MW); 2 MTU20V 1163 diesels; 19,310 hp *(14.4 MW)*;
   bow thruster; 2 shafts; cp props

Speed, knots: 29. **Range, n miles:** 12,000 at 9 kt
Complement: 126 (19 officers)
Guns: 1 Bofors 57 mm/70 Mk 3; 220 rds/min to 17 km *(9.3 n miles)*; weight of shell 2.4 kg.
   1 General Dynamics 20 mm Phalanx Mk 15 CIWS.
   4—12.7 mm MGs.
Countermeasures: Decoys: 6 Mk 53 Nulka.
ESM/ECM: SLQ 32.
Combat data systems: To be announced.
Radars: Surface search: TRS 3D/16; E/F-band.
Navigation: SPS 73; I-band.
Fire control: SPQ-9B; I/J-band.
Tacan: AN/URN 25.
Helicopters: 1 MCH and two Eagle Eye VUAV or 2 MCH.

Programmes: Contracts awarded to Northrop Grumman Ship
   Systems on 2 April 2003 for the design and long lead material
   procurement of the first of a class of eight Maritime (formerly
   National) Security Cutters to replace High Endurance Cutters.
   Lockheed Martin providing command/control/
   communications and intelligence integration and hardware.
   Contract for production and delivery of first ship on 28 June
   2004 and for second on 18 January 2005.
Structure: Can carry up to 11 m interceptor craft; stern ramps for
   rapid launch and recovery. Two helicopter hangars.
Operational: Although designed to deploy 230 days per year,
   deployments are likely to be of the order of 185 days away
   from homeport through an augmented crew concept.
                                                                    UPDATED

NATIONAL SECURITY CUTTER   *1/2005*, USCG /* 1043676

12 HAMILTON and HERO CLASSES (PSOH/WHEC)

Name	No	Builders	Laid down	Launched	Commissioned	F/S	Home Port
HAMILTON	WHEC 715	Avondale Shipyards	Jan 1965	18 Dec 1965	20 Feb 1967	PA	San Diego, CA
DALLAS	WHEC 716	Avondale Shipyards	7 Feb 1966	1 Oct 1966	1 Oct 1967	AA	Charleston, SC
MELLON	WHEC 717	Avondale Shipyards	25 July 1966	11 Feb 1967	22 Dec 1967	PA	Seattle, WA
CHASE	WHEC 718	Avondale Shipyards	27 Oct 1966	20 May 1967	1 Mar 1968	PA	San Diego, CA
BOUTWELL	WHEC 719	Avondale Shipyards	5 Dec 1966	17 June 1967	14 June 1968	PA	Alameda, CA
SHERMAN	WHEC 720	Avondale Shipyards	23 Jan 1967	23 Sep 1967	23 Aug 1968	PA	Alameda, CA
GALLATIN	WHEC 721	Avondale Shipyards	27 Feb 1967	18 Nov 1967	20 Dec 1968	AA	Charleston, SC
MORGENTHAU	WHEC 722	Avondale Shipyards	17 July 1967	10 Feb 1968	14 Feb 1969	PA	Alameda, CA
RUSH	WHEC 723	Avondale Shipyards	23 Oct 1967	16 Nov 1968	3 July 1969	PA	Honolulu, HI
MUNRO	WHEC 724	Avondale Shipyards	18 Feb 1970	5 Dec 1970	10 Sep 1971	PA	Alameda, CA
JARVIS	WHEC 725	Avondale Shipyards	9 Sep 1970	24 Apr 1971	30 Dec 1971	PA	Honolulu, HI
MIDGETT	WHEC 726	Avondale Shipyards	5 Apr 1971	4 Sep 1971	17 Mar 1972	PA	Seattle, WA

Displacement, tons: 3,300 full load
Dimensions, feet (metres): 378 × 42.8 × 20
   (115.2 × 13.1 × 6.1)
Flight deck, feet (metres): 88 × 40 *(26.8 × 12.2)*
Main machinery: CODOG; 2 Pratt & Whitney FT4A-6 gas
   turbines; 36,000 hp *(26.86 MW)*; 2 Fairbanks-Morse 38TD8-
   1/8-12 diesels; 7,000 hp *(5.22 MW)* sustained; 2 shafts; cp
   props; retractable bow propulsor; 350 hp *(261 kW)*
Speed, knots: 29. **Range, n miles:** 9,600 at 15 kt
Complement: 167 (19 officers)

Guns: 1 OTO Melara 3 in *(76 mm)*/62 Mk 75 Compact; 85 rds/
   min to 16 km *(8.7 n miles)* anti-surface; 12 km *(6.6 n miles)*
   anti-aircraft; weight of shell 6 kg.
   2 Boeing 25 mm/87 Mk 38 Bushmaster.
   1 GE/GD 20 mm Vulcan Phalanx 6-barrelled Mk 15;
   3,000 rds/min combined to 1.5 km. 4—12.7 mm MGs.
Countermeasures: Decoys: 2 Loral Hycor SRBOC 6-barrelled
   fixed Mk 36; IR flares and chaff.
ESM: WLR-1C, WLR-3; intercept.
Combat data systems: SCCS 378 includes OTCIXS satellite link.

Weapons control: Mk 92 Mod 1 GFCS.
Radars: Air search: Lockheed SPS-40B; B-band.
Surface search: Hughes/Furuno SPS-73; E/F- and I-bands.
Fire control: Sperry Mk 92; I/J-band.
Tacan: URN 25.

Helicopters: 1 HH-65A or 1 HH-60J or 1 MH-68A.

Programmes: In the Autumn of 1977 *Gallatin* and *Morgenthau*
   were the first of the Coast Guard ships to have women
   assigned as permanent members of the crew.
Modernisation: FRAM programme for all 12 ships in this class
   from October 1985 to October 1992. Work included
   standardising the engineering plants, improving the clutching
   systems, replacing SPS-29 air search radar with SPS-40 radar
   and replacing the Mk 56 fire-control system and 5 in/38 gun
   mount with the Mk 92 system and a single 76 mm OTO
   Melara Compact gun. In addition Harpoon and Phalanx CIWS
   fitted to five of the class by 1992 and CIWS to all by late 1993.
   The flight deck and other aircraft facilities upgraded to handle
   a Jay Hawk helicopter including a telescopic hangar. URN 25

Tacan added along with the SQR-4 and SQR-17 sonobuoy
receiving set and passive acoustic analysis systems. SRBOC
chaff launchers were also fitted but not improved ESM which
has been shelved. All missiles, torpedo tubes, sonar and ASW
equipment removed in 1993-94. 25 mm Mk 38 guns
replaced the 20 mm Mk 67. Shipboard Command and Control
System (SCCS) fitted to all of the class by 1996. Surface
search radar replaced 1997-99. First phase of C4ISR
upgrades, including access to SIPRNET and classified
networks, complete in 2004.
Structure: These ships have clipper bows, twin funnels
enclosing a helicopter hangar, helicopter platform aft. All are
fitted with elaborate communications equipment.
Superstructure is largely of aluminium construction. Bridge
control of manoeuvring is by aircraft-type joystick rather than
wheel.
Operational: Ten of the class are based in the Pacific, leaving
only two on the East Coast. The removal of SSMs and all ASW
equipment sensibly refocuses on Coast Guard roles.
                                                                    UPDATED

MUNRO   *10/2003, Frank Findler /* 0572763

MEDIUM ENDURANCE CUTTERS

13 FAMOUS CUTTER CLASS (PSOH/WMEC)

Name	No	Builders	Laid down	Launched	Commissioned	F/S	Home Port
BEAR	WMEC 901	Tacoma Boatbuilding Co	23 Aug 1979	25 Sep 1980	4 Feb 1983	AA	Portsmouth, VA
TAMPA	WMEC 902	Tacoma Boatbuilding Co	3 Apr 1980	19 Mar 1981	16 Mar 1984	AA	Portsmouth, VA
HARRIET LANE	WMEC 903	Tacoma Boatbuilding Co	15 Oct 1980	6 Feb 1982	20 Sep 1984	AA	Portsmouth, VA
NORTHLAND	WMEC 904	Tacoma Boatbuilding Co	9 Apr 1981	7 May 1982	17 Dec 1984	AA	Portsmouth, VA
SPENCER	WMEC 905	Robert E Derecktor Corp	26 June 1982	17 Apr 1984	28 June 1986	AA	Boston, MA
SENECA	WMEC 906	Robert E Derecktor Corp	16 Sep 1982	17 Apr 1984	4 May 1987	AA	Boston, MA
ESCANABA	WMEC 907	Robert E Derecktor Corp	1 Apr 1983	6 Feb 1985	27 Aug 1987	AA	Boston, MA
TAHOMA	WMEC 908	Robert E Derecktor Corp	28 June 1983	6 Feb 1985	6 Apr 1988	AA	New Bedford, MA
CAMPBELL	WMEC 909	Robert E Derecktor Corp	10 Aug 1984	29 Apr 1986	19 Aug 1988	AA	New Bedford, MA
THETIS	WMEC 910	Robert E Derecktor Corp	24 Aug 1984	29 Apr 1986	30 June 1989	AA	Key West, FL
FORWARD	WMEC 911	Robert E Derecktor Corp	11 July 1986	22 Aug 1987	4 Aug 1990	AA	Portsmouth, VA
LEGARE	WMEC 912	Robert E Derecktor Corp	11 July 1986	22 Aug 1987	4 Aug 1990	AA	Portsmouth, VA
MOHAWK	WMEC 913	Robert E Derecktor Corp	15 Mar 1987	5 May 1988	20 Mar 1991	AA	Key West, FL

Displacement, tons: 1,820 full load
Dimensions, feet (metres): 270 × 38 × 13.9
  (82.3 × 11.6 × 4.2)
Main machinery: 2 Alco 18V-251 diesels; 7,290 hp *(5.44 MW)*
  sustained; 2 shafts; cp props
Speed, knots: 19.5. **Range, n miles:** 12,700 at 15 kt
Complement: 100 (14 officers) plus 5 aircrew

Guns: 1 OTO Melara 3 in *(76 mm)*/62 Mk 75; 85 rds/min to
  16 km *(8.7 n miles)* anti-surface; 12 km *(6.6 n miles)* anti-
  aircraft; weight of shell 6 kg.
  2—12.7 mm MGs or 2—40 mm Mk 19 grenade launchers.
Countermeasures: Decoys: 2 Loral Hycor SRBOC 6-barrelled
  fixed Mk 36; IR flares and chaff.
ESM/ECM: SLQ-32(V)2; radar intercept.

Combat data systems: SCCS-270; OTCIXS satellite link.
Radars: Surface search: Hughes/Furuno SPS-73; I-band.
Fire control: Sperry Mk 92 Mod 1; I/J-band.
Tacan: URN 25.

Helicopters: 1 HH-65A or HH-60J or MH-68A or SH-60B.

Programmes: The contract for construction of WMEC 905-913
  was originally awarded to Tacoma Boatbuilding Co on
  29 August 1980. However, under lawsuit from the Robert E
  Derecktor Corp, Middletown, Rhode Island, the contract to
  Tacoma was determined by a US District Court to be invalid
  and was awarded to Robert E Derecktor Corp on 15 January
  1981.

Modernisation: OTCIXS satellite link fitted from 1992. C4ISR
  upgrades completed in 2004.
Structure: They are the only medium endurance cutters with a
  helicopter hangar (which is telescopic) and the first cutters
  with automated command and control centre. Fin stabilisers
  fitted. Plans to fit SSM and/or CIWS have been abandoned as
  has towed array sonar and sonobuoy datalinks. New radars
  fitted 1997-99.
Operational: Bases are at Portsmouth, New Bedford, Key West
  and Boston. Very lively in heavy seas because the length to
  beam ratio is unusually small for ships required to operate in
  Atlantic conditions.

UPDATED

TAHOMA
4/2003*, Declerck/Steeghers / 1043681

14 RELIANCE CLASS (PSOH/WMEC)

Name	No	Builders	Commissioned	MMA completion	F/S	Home Port
RELIANCE	WMEC 615	Todd Shipyards	20 June 1964	Jan 1989	AA	Kittery, ME
DILIGENCE	WMEC 616	Todd Shipyards	26 Aug 1964	Mar 1992	AA	Wilmington, NC
VIGILANT	WMEC 617	Todd Shipyards	3 Oct 1964	Aug 1990	AA	Cape Canaveral, FL
ACTIVE	WMEC 618	Christy Corp	17 Sep 1966	Feb 1987	PA	Port Angeles, WA
CONFIDENCE	WMEC 619	Coast Guard Yard, Baltimore	19 Feb 1966	June 1988	AA	Cape Canaveral, FL
RESOLUTE	WMEC 620	Coast Guard Yard, Baltimore	8 Dec 1966	Mar 1996	AA	St Petersburg, FL
VALIANT	WMEC 621	American Shipbuilding Co	28 Oct 1967	July 1993	AA	Miami, FL
STEADFAST	WMEC 623	American Shipbuilding Co	25 Sep 1968	Feb 1994	AA	Astoria, OR
DAUNTLESS	WMEC 624	American Shipbuilding Co	10 June 1968	Mar 1995	AA	Galveston, TX
VENTUROUS	WMEC 625	American Shipbuilding Co	16 Aug 1968	Sep 1995	AA	St Petersburg, FL
DEPENDABLE	WMEC 626	American Shipbuilding Co	22 Nov 1968	Aug 1997	AA	Cape May, NJ
VIGOROUS	WMEC 627	American Shipbuilding Co	2 May 1969	Jan 1993	AA	Cape May, NJ
DECISIVE	WMEC 629	Coast Guard Yard, Baltimore	23 Aug 1968	Sep 1998	AA	Pascagoula, MS
ALERT	WMEC 630	Coast Guard Yard, Baltimore	4 Aug 1969	Sep 1994	PA	Astoria, OR

Displacement, tons: 1,129 full load (WMEC 620-630)
  1,110 full load (WMEC 618, 619)
Dimensions, feet (metres): 210.5 × 34 × 10.5
  (64.2 × 10.4 × 3.2)
Main machinery: 2 Alco 16V-251 diesels; 6,480 hp *(4.83 MW)*
  sustained; 2 shafts; LIPS cp props
Speed, knots: 18. **Range, n miles:** 6,100 at 14 kt; 2,700 at 18 kt
Complement: 75 (12 officers)

Guns: 1 Boeing 25 mm/87 Mk 38 Bushmaster; 200 rds/min to
  6.8 km *(3.4 n miles)*. 2—12.7 mm MGs.
Combat data systems: SCCS-210.
Radars: Surface search: Hughes/Furuno SPS-73; I-band.
Helicopters: 1 HH-65A or MH-68A embarked as required.

Modernisation: All 14 cutters have undergone a Major
  Maintenance Availability (MMA). The exhausts for main
  engines, ship service generators and boilers have been run in
  a new vertical funnel which reduces flight deck size.
  Completion dates are listed above. 76 mm guns have been
  replaced by 25 mm Mk 38.
Structure: Designed for search and rescue duties. Design
  features include 360° visibility from bridge; helicopter flight
  deck (no hangar); and engine exhaust vent at stern which has
  been replaced by a funnel during MMA. Capable of towing
  ships up to 10,000 tons. Air conditioned throughout except
  engine room; high degree of habitability.
Operational: Normally operate within 500 miles of the coast.
  Primary roles are SAR, law enforcement homeland security
  and defence operations. *UPDATED*

ACTIVE
10/2003, Frank Findler / 0572762

0 + (25) OFFSHORE PATROL CUTTERS (PSOH/WMSM)

Displacement, tons: 2,921
Dimensions, feet (metres): 341 × 54 × 17.25 *(103.9 × 16.5 × 5.26)*
Main machinery: 2 diesels; 7,600 hp *(5.7 MW)*; bow thruster
Speed, knots: 22. **Range, n miles:** 9,000 at 9 kt
Complement: 94 (14 officers)
Radars: To be announced.
Helicopters: 1 HH-65A and two Eagle Eye VUAV or 2 HH-65A.

Programmes: Contract for accelerated design signed with Northrop Grumman Ship Systems on 10 June 2004. Lockheed Martin responsible for co-development of engineering design and system integration. Construction of the first of class is expected to start in 2009 for entry into service in 2012.
Structure: Similar capabilities and equipment as National Security Cutter (WMSL). Stern ramp for small boat launch/recover. Two helicopter hangars. Enhanced sea-keeping capability through roll stabilisation.
Operational: Although designed to deploy 230 days per year, deployments are likely to be of the order of 185 days away from homeport through an augmented crew concept.

NEW ENTRY

OFFSHORE PATROL CUTTER   *2/2005*, USCG /* 1043675

1 EDENTON CLASS (PSOH/WMEC)

Name	No	Builders	Commissioned	F/S	Home Port
ALEX HALEY	WMEC 39	Brooke Marine, Lowestoft	23 Jan 1971	PA	Kodiak, AK
(ex-*Edenton*)	(ex-*ATS 1*)				

Displacement, tons: 3,000 full load
Dimensions, feet (metres): 282.6 × 50 × 15.1 *(86.1 × 15.2 × 4.6)*
Main machinery: 4 Caterpillar 3516 DITAWJ diesels; 6,000 hp(m) *(4.41 MW)*; 2 shafts; cp props; bow thruster
Speed, knots: 18. **Range, n miles:** 10,000 at 13 kt
Complement: 99 (9 officers)
Guns: 2 McDonnell Douglas 25 mm/87 Mk 38; 200 rds/min to 6.8 km *(3.4 n miles)*. 2—12.7 mm MGs.
Radars: Surface search: Hughes/Furuno SPS-73; I-band.
Combat data systems: SCCS-282.
Helicopters: Platform for 1 HH-65A or 1 HH-60J.

Comment: Former Navy salvage ship paid off in 1996 and taken on by the Coast Guard in November 1997 for conversion. All diving and salvage gear removed, flight deck installed, and upgraded navigation and communications. Armed with 25 mm guns. Used in the Bering Sea, Gulf of Alaska and North Pacific as a multi-mission cutter from 16 December 1999.

UPDATED

ALEX HALEY   *12/1999, USCG /* 0084198

1 DIVER CLASS (PSO/WMEC)

Name	No	Builders	USN Comm	F/S	Home Port
ACUSHNET	WMEC 167	Basalt Rock Co,	5 Feb 1944	PA	Ketchikan, AK
(ex-*Shackle*)	(ex-WAGO 167,	Napa, CA			
	ex-WAT 167, ex-ARS 9)				

Displacement, tons: 1,557 standard; 1,745 full load
Dimensions, feet (metres): 213 × 41 × 15 *(64.9 × 12.5 × 4.6)*
Main machinery: 4 Fairbanks-Morse diesels; 3,000 hp *(2.24 MW)* sustained; 2 shafts
Speed, knots: 15.5. **Range, n miles:** 9,000 at 8 kt
Complement: 75 (9 officers)
Guns: 2—12.7 mm MGs.
Radars: Navigation: 2 Raytheon SPS-73; I-band.

Comment: Large, steel-hulled salvage ship transferred from the Navy to the Coast Guard and employed in tug and oceanographic duties. Modified for handling environmental data buoys and reclassified WAGO in 1968 and reclassified WMEC in 1980. Major renovation work completed in 1983 and now used for SAR homeland security and law enforcement operations.

UPDATED

ACUSHNET   *9/2000, Sattler/Steele /* 0105723

1 STORIS CLASS (PSO/WMEC)

Name	No	Builders	Commissioned	F/S	Home Port
STORIS	WMEC 38	Toledo Shipbuilding	30 Sep 1942	PA	Kodiak, AK
(ex-*Eskimo*)	(ex-WAGB 38)				

Displacement, tons: 1,715 standard; 1,925 full load
Dimensions, feet (metres): 230 × 43 × 15 *(70.1 × 13.1 × 4.6)*
Main machinery: Diesel-electric; 3 GM EMD diesel generators; 1 motor; 3,000 hp *(2.24 MW)*; 1 shaft
Speed, knots: 14. **Range, n miles:** 22,000 at 8 kt
Complement: 78 (10 officers)
Guns: 1 Boeing 25 mm/87 Mk 38 Bushmaster. 2—12.7 mm MGs.
Radars: Navigation: 2 Raytheon SPS-73; I-band.

Comment: Laid down on 14 July 1941; launched on 4 April 1942 as ice patrol tender. Strengthened for ice navigation but no longer employed as icebreaker. Employed in Alaskan service for search, rescue Homeland Security and law enforcement. Completed a major maintenance availability in June 1986, during which main engines were replaced with EMD diesels and living quarters expanded. Gun changed in 1994.

UPDATED

STORIS   *3/1998, van Ginderen Collection /* 0053416

SHIPBORNE AIRCRAFT

Notes: As part of the 'Deepwater' modernisation programme, it is planned to replace the HH-60J Jayhawks with an estimated 34 VTOL Recovery and Surveillance aircraft (VRS) from 2014. The Bell-Agusta AB 139 has been provisionally selected but the Eurocopter 155 and Bell-Textron 609 tilt-rotor craft are also under consideration.

Numbers/Type: 21/75 Eurocopter HH-65A/HH-65B Dolphin.
Operational speed: 165 kt *(300 km/h)*.
Service ceiling: 11,810 ft *(3,600 m)*.
Range: 300 n miles *(554 km)*.
Role/Weapon systems: Short-range rescue and recovery (SRR) helicopter. Sensors: Bendix RDR 1300 radar and Collins mission management system. HH-65B equipped with CDU-900G control displays and MFD-255 multifunctional displays. Following a series of in-flight engine failures which began in 2003, operational flight restrictions were imposed. A re-engineering programme with Turbomeca Arriel 2C2 engines initiated for entire fleet. First re-engined aircraft began tests 27 August 2004. It is planned to convert the HH-65 to a multimission cutter helicopter. Weapons: Unarmed.

UPDATED

HH-65A   *10/2003, Frank Findler /* 0572760

Numbers/Type: 42 Sikorsky HH-60J Jayhawk.
Operational speed: 146 kt *(270 km/h)*.
Service ceiling: 13,000 ft *(3,961 m)*.
Range: 300 n miles *(554 km)*.
Role/Weapon systems: Coast Guard version of Seahawk, first flew in 1988. 35 are operational. It has replaced HH-3F in MRR role. A life-extension programme and avionics upgrade is to begin in mid-2005 to extend service life until at least 2014. Sensors: Bendix RDR-1300C weather/search radar. AAQ-15 FLIR. Weapons: Unarmed. *VERIFIED*

HH-60J   *3/2003, Adolfo Ortigueira Gil / 0572761*

Numbers/Type: 8 Agusta A 109 MH-68A Stingray.
Operational speed: 168 kt *(310 km/h)*.
Service ceiling: 20,000 ft *(6,100 m)*.
Range: 200 n miles *(370 km)*.
Role/Weapon systems: All-weather, short-range interdiction helicopter selected as follow on to Enforcer to counter high-speed smuggling/drug-running craft. Last of eight aircraft procured in 2001. All based at Jacksonville, FL. *UPDATED*

MH-68A   *2000, USCG / 0105721*

LAND-BASED MARITIME AIRCRAFT

Notes: (1) The Bell Helicopter Textron HV-911 Eagle Eye tiltrotor selected by Integrated Coast Guard Systems as the Vertical Unmanned Aerial Vehicle (VUAV) to be part of the Integrated Deepwater System. Operable from land, it is also planned to deploy two VUAV in the new National Security and Offshore Patrol Cutters. Sensors are likely to include the FLIR Systems Safire and a radar. Capable of 210 kt, the aircraft can be recovered in up to Sea State 5. Having passed its Preliminary Design Review (PDR) phase in March 2004, up to 69 units are planned to enter service from 2006.
(2) A High Altitude Endurance Unmanned Air Vehicle (HAE-UAV) is planned to enter service from 2016. Equipped with high-resolution sensors (EO/FLIR, SAR, ISAR, GMT), the HAE-UAV is to provide long-range surveillance over large areas for extended periods of time. With a loiter altitude of up to 65,000 ft, they are to be capable of transmitting data and EO/IR imagery to shore-based command and control centres to contribute to the Common Operational Picture (COP). The programme is likely to be informed by the USN's BAMS (Broad Area Maritime Surveillance) programme, contenders for which include the Northrop Grumman RQ-4A Global Hawk.
(3) Four P-3B Orions are used for AEW by US Customs.

Numbers/Type: 2 Casa CN-235 200.
Operational speed: 210 kt *(384 km/h)*.
Service ceiling: 24,000 ft *(7,315 m)*.
Range: 2,000 n miles *(3,218 km)*.
Role/Weapon systems: First two of up to 34 maritime patrol aircraft ordered on 18 February 2004. Sensors: Synthetic aperture radar and electro-optic sensors. *NEW ENTRY*

CN-235   *6/2004*, CASA / 1043638*

Numbers/Type: 1 Gulfstream C-37A.
Operational speed: 459 kt *(850 km/h)*.
Service ceiling: 51,000 ft *(15,540 m)*.
Range: 5,600 n miles *(10,370 km)*.
Role/Weapon: Military version of Gulfstream V which replaced a C-20B Gulfstream III in May 2002. Based at Air Station Washington DC. Serves as a long-range command and control aircraft for Coast Guard command officials. *VERIFIED*

GULFSTREAM G 550   *6/2003, Paul Jackson / 0568402*

Numbers/Type: 4/7/6 AMD-BA HU-25 A/HU-25 C/HU-25 D Guardian Falcon.
Operational speed: 420 kt *(774 km/h)*.
Service ceiling: 42,000 ft *(12,800 m)*.
Range: 1,940 n miles *(3,594 km)*.
Role/Weapon systems: Medium-range maritime surveillance role. 17 are operational; 21 are in storage or support aircraft. Sensors: APS-127 weather/search radar. Weapons: unarmed. *VERIFIED*

HU-25 FALCON   *6/2001, Adolfo Ortigueira Gil / 0529903*

Numbers/Type: 23/8 Lockheed HC-130H/Lockheed HC-130J.
Operational speed: 325 kt *(602 km/h)*.
Service ceiling: 33,000 ft *(10,060 m)*.
Range: 4,250 n miles *(7,876 km)*; 5,020 n miles *(9,295 km)* (HC-130J).
Role/Weapon systems: Long-range maritime reconnaissance role. 22 are operational; five are in maintenance and modification. Delivery of first new HC-130J started in 2003. Sensors: APS-137 or APS-125 weather/search radar. FLIR. Weapons: unarmed. *UPDATED*

HC-130H   *10/2002, M Mazumdar / 0529904*

ICEBREAKERS

0 + 1 ICEBREAKER (WLBB)

Name	No	Builders	Commissioned	Home Port
MACKINAW	WLBB 30	Marinette Marine, Wisconsin	2006	Cheboygan, MI

Displacement, tons: 3,395 full load
Dimensions, feet (metres): 240 × 58 × 16 *(73.1 × 17.7 × 4.8)*
Main machinery: Diesel-electric; 3 diesel generators; 12,600 hp *(9.4 MW)*; 2 podded propulsors; 6,700 hp *(5 MW)*
Speed, knots: 15
Complement: 50 (8 officers)
Radars: Surface search: Kongsberg Data Bridge 10.
Navigation: Kongsberg Integrated Bridge System.

Comment: Contract to build new icebreaker/buoy tender awarded 15 October 2001. Keel laid 10 February 2004. Launch to be in April 2005 and delivery in late 2005. Icebreaker is to replace WAGB 83 in mid-2006 and assume same name. In addition to breaking ice (up to 32 in thick at 3 kt ahead, 2 kt astern) for the primary shipping lanes on the Great Lakes, the new ship will service aids to navigation, as well as performing search and rescue, pollution control, homeland security, and law enforcement duties from its homeport of Cheboygan, Michigan. Principal feature is 'podded' or protected propellers that can rotate 360° for greater manoeuvrability. Other features include fully integrated bridge system, robust communications suite and 3,200 sq ft of buoy deck space. A crane of 60 ft can recover buoys weighing up to 20 tons. *UPDATED*

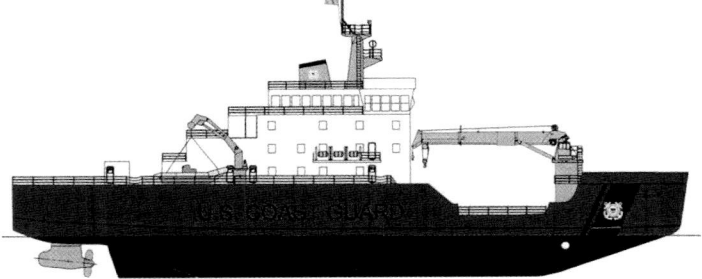

MACKINAW (artist's impression)   *1/2004, USCG / 0572806*

1 HEALY CLASS (WAGBH)

Name	No	Builders	Commissioned	F/S	Home Port
HEALY	WAGB 20	Avondale, New Orleans	29 Oct 1999	PA	Seattle, WA

Displacement, tons: 16,400 full load
Dimensions, feet (metres): 420 oa; 397.8 wl × 82 × 29 *(128; 121.2 × 25 × 8.9)*
Main machinery: Diesel-electric; 4 Westinghouse/Sulzer 12ZA 40S diesels; 42,400 hp *(31.16 MW)*; 4 Westinghouse alternators; 2 motors; 30,000 hp *(22.38 MW)*; 2 shafts; bow thruster; 2,200 hp *(1.64 MW)*
Speed, knots: 17. **Range, n miles:** 16,000 at 12.5 kt
Complement: 75 (12 officers) plus 45 scientists
Helicopters: 2 HH-65A or 1 HH-60J.

Comment: In response to the 1984 Interagency Polar Icebreaker Requirements Study and Congressional mandate, approval was given for the construction of a new icebreaker as a replacement for two Wind class which were then decommissioned in 1988. However, no action was taken to provide funds for the new ship until Congress included it in the Navy's FY91 ship construction budget and after further delays the ship was ordered 15 July 1993. Icebreaking capability of 4 ft at 3 kt. Launched 15 November 1997. Based at Seattle, WA. **UPDATED**

HEALY
1/2003, Bob Fildes / 0572759

2 POLAR CLASS (WAGBH)

Name	No	Builders	Launched	Commissioned	F/S	Home Port
POLAR STAR	WAGB 10	Lockheed SB	17 Nov 1973	19 Jan 1976	PA	Seattle, WA
POLAR SEA	WAGB 11	Lockheed SB	24 June 1975	23 Feb 1978	PA	Seattle, WA

Displacement, tons: 13,190 full load
Dimensions, feet (metres): 399 × 84 × 32 *(121.6 × 25.6 × 9.8)*
Main machinery: CODOG; diesel-electric (AC/DC); 6 Alco 16V-251F/Westinghouse AC diesel generators; 21,000 hp *(15.66 MW)* sustained; 3 Westinghouse DC motors; 18,000 hp *(13.42 MW)* sustained; 3 Pratt & Whitney FT4A-12 gas turbines; 60,000 hp *(44.76 MW)* sustained; 3 Philadelphia 75 VMGS gears; 60,000 hp *(44.76 MW)* sustained; 3 shafts; cp props
Speed, knots: 20. **Range, n miles:** 28,275 at 13 kt
Complement: 134 (15 officers) plus 33 scientists and 12 aircrew
Guns: 2—7.62 mm MGs.
Radars: Navigation: 2 Raytheon SPS-64; I-band.
Tacan: SRN 15.
Helicopters: 2 HH-65A or 1 HH-60J.

Comment: At a continuous speed of 3 kt, they can break ice 6 ft *(1.8 m)* thick, and by ramming can break 21 ft *(6.4 m)* pack. Conventional icebreaker hull form with 'White' cutaway bow configuration and well-rounded body sections to prevent being trapped in ice. The ice belt is 1.75 in *(44.45 mm)* thick supported by framing at 16 in *(0.4 m)* centres. Three heeling systems assist icebreaking and ship extraction. Two 15 ton capacity cranes fitted aft; one 3 ton capacity crane fitted forward. Two over-the-side oceanographic winches, one over-the-stern trawl/core winch. Deck fixtures for scientific research vans, and research laboratories provided for arctic and oceanographic research. Between 1986-92, science facilities were upgraded including habitability, lab spaces and winch capabilities. *Polar Sea* went to the North Pole in August 1994. **UPDATED**

POLAR SEA
6/2001, H M Steele / 0076792

1 MACKINAW CLASS (WAGB)

Name	No	Builders	Commissioned	F/S	Home Port
MACKINAW	WAGB 83	Toledo Shipbuilding	20 Dec 1944	GLA	Cheboygan, MI

Displacement, tons: 5,252 full load
Dimensions, feet (metres): 290 × 74 × 19 *(88.4 × 22.6 × 5.8)*
Main machinery: Diesel-electric; 6 Fairbanks-Morse 38D8-1/8-12 diesel generators; 8.7 MW sustained; Elliot electric drive; 10,000 hp *(7.46 MW)*; 3 shafts (1 fwd, 2 aft)
Speed, knots: 18.7. **Range, n miles:** 41,000 at 11.5 kt; 10,000 at 18.7 kt
Complement: 77 (13 officers)
Radars: Navigation: Hughes/Furuno SPS-73; I-band.

Comment: Specially designed and constructed for service as icebreaker on the Great Lakes. Equipped with two 5 ton capacity cranes. First scheduled to be paid off in FY88 but pressure from members of Congress from states bordering the Great Lakes resulted in her being placed in an 'In Commission, Special' status. Operational again in Spring 1988. Again scheduled for decommissioning in December 1994 after 50 years' service but extended to July 2006 when she will be replaced by a new ship of the same name. **UPDATED**

MACKINAW
2001, USCG / 0131334

9 BAY CLASS (TUGS—WTGB)

Name	No	Laid down	Commissioned	F/S	Home Port
KATMAI BAY	WTGB 101	7 Nov 1977	8 Jan 1979	GLA	Sault Sainte Marie, MI
BRISTOL BAY	WTGB 102	13 Feb 1978	5 Apr 1979	GLA	Detroit, MI
MOBILE BAY	WTGB 103	13 Feb 1978	2 Sep 1979	GLA	Sturgeon Bay, WI
BISCAYNE BAY	WTGB 104	29 Aug 1978	8 Dec 1979	GLA	St Ignace, MI
NEAH BAY	WTGB 105	6 Aug 1979	18 Aug 1980	GLA	Cleveland, OH
MORRO BAY	WTGB 106	6 Aug 1979	25 Jan 1981	AA	New London, CT
PENOBSCOT BAY	WTGB 107	24 July 1983	4 Sep 1984	AA	Bayonne, NJ
THUNDER BAY	WTGB 108	20 July 1984	29 Dec 1985	AA	Rockland, ME
STURGEON BAY	WTGB 109	9 July 1986	20 Aug 1988	AA	Bayonne, NJ

Displacement, tons: 662 full load
Dimensions, feet (metres): 140 × 37.6 × 12.5 *(42.7 × 11.4 × 3.8)*
Main machinery: Diesel-electric; 2 Fairbanks-Morse 38D8-1/8-10 diesel generators; 2.4 MW sustained; Westinghouse electric drive; 2,500 hp *(1.87 MW)*; 1 shaft
Speed, knots: 14.7. **Range, n miles:** 4,000 at 12 kt
Complement: 17 (3 officers)
Radars: Navigation: Raytheon SPS-64(V)1; I-band.

Comment: The size, manoeuvrability and other operational characteristics of these vessels are tailored for operations in harbours and other restricted waters and for fulfilling present and anticipated multimission requirements. All units are ice strengthened for operation on the Great Lakes, coastal waters and in rivers and can break 20 in of ice continuously and up to 8 ft by ramming. A self-contained portable bubbler van and system reduces hull friction. First six built at Tacoma Boatbuilding, Tacoma. WTGB 107-109 built in Tacoma by Bay City Marine, San Diego. *Bristol Bay* and *Mobile Bay* have had their bows reinforced to push the two aids-to-navigation barges on the Great Lakes. WTGB 106 was decommissioned in 1998 and re-activated on 4 February 2002. **UPDATED**

PENOBSCOT BAY
5/2002, van Ginderen Collection / 0144052

THUNDER BAY
7/2000, Hachiro Nakai / 0105725

PATROL FORCES

5 CYCLONE CLASS (PATROL COASTAL SHIPS) (WPC/PB)

Name	No	Builders	Commissioned	Home Port
TEMPEST	WPB 2 (ex-PC 2)	Bollinger, Lockport	21 Aug 1993	Pascagoula, MS
MONSOON	WPB 4 (ex-PC 4)	Bollinger, Lockport	22 Jan 1994	San Diego, CA
ZEPHYR	WPB 8 (ex-PC 8)	Bollinger, Lockport	15 Oct 1994	San Diego, CA
SHAMAL	WPC 13 (ex-PC 13)	Bollinger, Lockport	27 Jan 1996	Pascagoula, MS
TORNADO	WPC 14 (ex-PC 14)	Bollinger, Lockport	15 May 2000	Pascagoula, MS

Displacement, tons: 360 full load
Dimensions, feet (metres): 179 × 25.9 × 7.9 *(54.6 × 7.9 × 2.4)*
Main machinery: 4 Paxman Valenta 16RP 200M diesels; 13,400 hp *(10 MW)* sustained; 4 shafts
Speed, knots: 35. **Range, n miles:** 2,500 at 12 kt
Complement: 27 (2 officers)
Guns: 2 McDonnell Douglas 25 mm/87 Mk 38. 4—12.7 mm M60 MGs.
Combat data systems: SCCS-Lite.
Radars: Navigation: Hughes/Furuno SPS-73; I-band.

Comment: Contract awarded by USN for eight in August 1990 and five more in July 1991. Design based on Vosper Thornycroft Ramadan class modified for USN requirements including 1 in armour on superstructure. The craft have a slow speed loiter capability. Five vessels modified to incorporate a semi-dry well, boat ramp and stern gate to facilitate deployment and recovery of a fully loaded RIB while the ship is making way. All five transferred to the USCG 2004-05. Remaining vessels in USN service. ***NEW ENTRY***

CYCLONE CLASS   *9/2002, Winter & Findler* / 0529974

0 + (58) FAST RESPONSE CUTTERS (PBO/WPC)

Displacement, tons: 270 full load
Dimensions, feet (metres): 147 × 21.8 × 7.33 *(44.8 × 6.6 × 2.2)*
Main machinery: 2 diesels; 10,000 hp *(7.5 MW)*
Speed, knots: 30+. **Range, n miles:** 5,000 at 10 kt
Complement: 15
Guns: 1—30 mm. 2—12.7 mm MGs.

Comment: In July 2004, contract awarded to begin preliminary design phase of WPC cutters to enter service from 2012. The requirement is to be capable of independent deployment in support of law enforcement, port security, search and rescue, and defense operations missions. Typical missions to include offshore fishery protection, choke point interdiction, barrier patrols, and presence in high-risk areas. Design features are likely to include reduced signature through shaping, active fin stabilisation system, an integrated bridge with 360° visibility and a stern ramp to launch new Short Range Prosecutor (SRP) patrol craft. ***NEW ENTRY***

44 GUARDIAN CLASS (TPSB/YP)

Displacement, tons: 3 full load
Dimensions, feet (metres): 24.6 × 8.2 × 0.4 *(7.5 × 2.5 × 0.4)*
Main machinery: 2 Evinrude outboards; 350 hp *(261 kW)*
Speed, knots: 35
Complement: 4
Guns: 1—12.7 mm MG. 2—7.62 mm MGs.
Radars: Navigation: Raytheon; I-band.

Comment: Transportable Port Security Boats (TPSB) which serve with the six Port Security Units and a Training Detachment. Can be transported by aircraft. ***VERIFIED***

GUARDIAN   *7/2000, Hachiro Nakai* / 0105727

GUARDIAN   *4/2001, Guy Toremans* / 0131265

49 ISLAND CLASS (WPB)

Name	No	Commissioned	Home Port
FARALLON	WPB 1301	21 Feb 1986	Miami, FL
MANITOU	WPB 1302	28 Feb 1986	CONV
MATAGORDA	WPB 1303	24 Apr 1986	Key West, FL
MAUI	WPB 1304	9 May 1986	Miami, FL
MONHEGAN	WPB 1305	16 June 1986	CONV
NUNIVAK	WPB 1306	4 July 1986	CONV
OCRACOKE	WPB 1307	4 Aug 1986	San Juan, PR
VASHON	WPB 1308	15 Aug 1986	CONV
AQUIDNECK	WPB 1309	26 Sep 1986	Atlantic Beach, NC
MUSTANG	WPB 1310	3 Dec 1986	Seward, AK
NAUSHON	WPB 1311	5 Dec 1986	Ketchikan, AK
SANIBEL	WPB 1312	28 May 1987	Woods Hole, MA
EDISTO	WPB 1313	27 Mar 1987	San Diego, CA
SAPELO	WPB 1314	14 May 1987	Key West, FL
MATINICUS	WPB 1315	19 June 1987	San Juan, PR
NANTUCKET	WPB 1316	10 Aug 1987	Key West, FL
ATTU	WPB 1317	9 May 1988	Key West, FL
BARANOF	WPB 1318	25 May 1988	Miami, FL
CHANDELEUR	WPB 1319	8 June 1988	Miami, FL
CHINCOTEAGUE	WPB 1320	8 Aug 1988	Key West, FL
CUSHING	WPB 1321	8 Aug 1988	San Juan, PR
CUTTYHUNK	WPB 1322	5 Oct 1988	Port Angeles, WA
DRUMMOND	WPB 1323	19 Oct 1988	Key West, FL
KEY LARGO	WPB 1324	24 Dec 1988	Key West, FL
METOMPKIN	WPB 1325	12 Jan 1989	Key West, FL
MONOMOY	WPB 1326	19 May 1989	Woods Hole, MA
ORCAS	WPB 1327	14 Apr 1989	Coos Bay, OR
PADRE	WPB 1328	21 Apr 1989	Key West, FL
SITKINAK	WPB 1329	31 May 1989	Key West, FL
TYBEE	WPB 1330	4 Aug 1989	San Diego, CA
WASHINGTON	WPB 1331	6 Oct 1989	Honolulu, HI
WRANGELL	WPB 1332	15 Sep 1989	South Portland, ME
ADAK	WPB 1333	17 Nov 1989	Sandy Hook, NJ
LIBERTY	WPB 1334	22 Sep 1989	Auke Bay, AK
ANACAPA	WPB 1335	13 Jan 1990	Petersburg, AK
KISKA	WPB 1336	21 Apr 1990	Hilo, HI
ASSATEAGUE	WPB 1337	15 June 1990	Honolulu, HI
GRAND ISLE	WPB 1338	19 Apr 1991	Gloucester, MA
KEY BISCAYNE	WPB 1339	23 Apr 1991	St Petersburg, FL
JEFFERSON ISLAND	WPB 1340	16 Aug 1991	South Portland, ME
KODIAK ISLAND	WPB 1341	21 June 1991	St Petersburg, FL
LONG ISLAND	WPB 1342	27 Aug 1991	Valdez, AK
BAINBRIDGE ISLAND	WPB 1343	20 Sep 1991	Sandy Hook, NJ
BLOCK ISLAND	WPB 1344	22 Nov 1991	Atlantic Beach, NC
STATEN ISLAND	WPB 1345	22 Nov 1991	Atlantic Beach, NC
ROANOKE ISLAND	WPB 1346	8 Feb 1992	Homer, AK
PEA ISLAND	WPB 1347	29 Feb 1992	St Petersburg, FL
KNIGHT ISLAND	WPB 1348	22 Apr 1992	St Petersburg, FL
GALVESTON ISLAND	WPB 1349	5 June 1992	Apra Harbor, Guam

Displacement, tons: 168 (A series); 154 (B series); 134 (C series) full load
Dimensions, feet (metres): 110 (123 in Deepwater mod) × 21 × 7.3 *(33.5 (37.5) × 6.4 × 2.2)*
Main machinery: 2 Paxman Valenta 16RP 200M diesels (C series); 6,246 hp *(4.62 MW)*, sustained; 2 Caterpillar 3516 DITA diesels (A and B series); 5,596 hp *(4.17 MW)* sustained; 2 shafts
Speed, knots: 29. **Range, n miles:** 3,928 at 10 kt
Complement: 16 (2 officers)
Guns: 1 McDonnell Douglas 25 mm/87 Mk 38. 2—12.7 mm M60 MGs.
Combat data systems: SCCS-Lite.
Radars: Navigation: Hughes/Furuno SPS-73; I-band.

Comment: All built by the Bollinger Machine Shop and Shipyard at Lockport, Louisiana. The design is based upon the 110 ft patrol craft built by Vosper Thornycroft, UK, in service in Venezuela, UAE and UK Customs, but modified to meet Coast Guard needs. Vosper Thornycroft supplied design support, stabilisers, propellers, and steering gear. Batches: A 1301-1316, B 1317-1337, C 1338-1349. Radars replaced by 1999. As part of the Deepwater programme, it is planned progressively to upgrade some, possibly all 49, ships in batches. Modifications are being undertaken by Bollinger Shipyards of Lockport, LA. And VT Halter Marine of Gulfport, MS. The ships are being stretched to 123 ft by insertion of a 13 ft plug to enable installation of upgraded C4ISR systems, a stern launch and recovery system and various platform improvements. WPB 1303 was first to undergo conversion at Bollinger Shipyard in February 2004 and was followed by WPBs 1317, 1325 and 1328 in 2004. WPCs 1302, 1305 1306 and 1308 are to follow in 2005. VT Halter Marine Inc. is building a new superstructure and pilothouse with 360° visibility from the bridge. Habitability improvements include a new deckhouse, messdecks, galley and air-conditioning system. ***UPDATED***

MATAGORDA (after conversion)   *8/2004*, USCG* / 1043696

AQUIDNECK   *4/2003, A Sharma* / 0572758

65 MARINE PROTECTOR CLASS (WPB)

Name	No	Commissioned	Home Port
BARRACUDA	87301	24 Feb 1998	Eureka, CA
HAMMERHEAD	87302	17 May 1998	Woods Hole, MA
MAKO	87303	28 June 1998	Cape May, NJ
MARLIN	87304	2 Dec 1998	Fort Meyers, FL
STINGRAY	87305	13 Jan 1999	Mobile, AL
DORADO	87306	24 Feb 1999	Crescent City, CA
OSPREY	87307	7 Apr 1999	Port Townsend, WA
CHINOOK	87308	19 May 1999	New London, CT
ALBACORE	87309	30 June 1999	Little Creek, VA
TARPON	87310	11 Aug 1999	Tybee Island, GA
COBIA	87311	8 Sep 1999	Mobile, AL
HAWKSBILL	87312	6 Oct 1999	Monterey, CA
CORMORANT	87313	3 Nov 1999	Fort Pierce, FL
FINBACK	87314	1 Dec 1999	Cape May, NJ
AMBERJACK	87315	29 Dec 1999	Port Isabel, TX
KITTIWAKE	87316	26 Jan 2000	Nawiliwilli, HI
BLACKFIN	87317	23 Feb 2000	Santa Barbara, CA
BLUEFIN	87318	22 Mar 2000	Fort Pierce, FL
YELLOWFIN	87319	19 Apr 2000	Charleston, SC
MANTA	87320	17 May 2000	Freeport, TX
COHO	87321	14 June 2000	Pamana City, FL
KINGFISHER	87322	12 July 2000	Mayport, FL
SEAHAWK	87323	9 Aug 2000	Carrabelle, FL
STEELHEAD	87324	6 Sep 2000	Port Aransas, TX
BELUGA	87325	4 Oct 2000	Little Creek, VA
BLACKTIP	87326	1 Nov 2000	Oxnard, CA
PELICAN	87327	29 Nov 2000	Abbeville, LA
RIDLEY	87328	27 Dec 2000	Montauk, NY
COCHITO	87329	24 Jan 2001	Little Creek, VA
MANOWAR	87330	21 Feb 2001	Galveston, TX
MORAY	87331	21 Mar 2001	Jonesport, ME
RAZORBILL	87332	18 Apr 2001	Gulfport, MS
ADELIE	87333	16 May 2001	Port Angeles, WA
GANNET	87334	13 June 2001	Fort Lauderdale, FL
NARWHAL	87335	11 July 2001	Corona del Mar, CA
STURGEON	87336	8 Aug 2001	Grand Isle, LA
SOCKEYE	87337	5 Sep 2001	Bodega Bay, CA
IBIS	87338	3 Oct 2001	Cape May, NJ
POMPANO	87339	1 Nov 2001	Gulfport, MS
HALIBUT	87340	28 Nov 2001	Marina del Ray, CA
BONITO	87341	26 Dec 2001	Pensacola, FL
SHRIKE	87342	23 Jan 2002	Cape Canaveral, FL
TERN	87343	20 Feb 2002	San Francisco, CA
HERON	87344	20 Mar 2002	Sabine, TX
WAHOO	87345	17 Apr 2002	Port Angeles, WA
FLYINGFISH	87346	15 May 2002	Boston, MA
HADDOCK	87347	12 June 2002	San Diego, CA
BRANT	87348	10 July 2002	Corpus Christi, TX
SHEARWATER	87349	7 Aug 2002	Portsmouth, VA
PETREL	87350	4 Sep 2002	San Diego, CA
SEA LION	87352	19 Nov 2003	Bellingham, WA
SKIPJACK	87353	17 Dec 2003	Galveston, TX
DOLPHIN	87354	14 Jan 2004	Miami, FL
HAWK	87355	11 Feb 2004	St Petersburg, FL
SAILFISH	87356	10 Mar 2004	Sandy Hook, NJ
SAWFISH	87357	7 Apr 2004	Key West, FL
SWORDFISH	87358	9 Mar 2005	Port Angeles, WA
TIGER SHARK	87359	6 Apr 2005	Newport, RI
BLUE SHARK	87360	4 May 2005	Everett, WA
SEA HORSE	87361	1 June 2005	Portsmouth, VA
SEA OTTER	87362	29 June 2005	San Diego, CA
MANATEE	87363	27 July 2005	Ingleside, TX
DIAMONDBACK	87364	24 Aug 2005	Honolulu, HI
ALLIGATOR	87365	21 Sep 2005	San Francisco, CA
CROCODILE	87366	19 Oct 2005	Bellingham, WA

Displacement, tons: 91 full load
Dimensions, feet (metres): 86.9 × 19 × 5.2 *(26.5 × 5.8 × 1.6)*
Main machinery: 2 MTU 8V 396 TE94 diesels; 2,680 hp(m) *(1.97 MW)* sustained; 2 shafts
Speed, knots: 25. **Range, n miles:** 900 at 8 kt
Complement: 10 (1 officer)
Guns: 2—12.7 mm MGs.
Radars: Navigation: I-band.

Comment: Designed by David M Cannell based on the hull of the Damen Stan Patrol 2600 which is in service with the Hong Kong police. Steel hull built by Bollinger with GRP superstructure by Halmatic. A stern ramp is used for launching a 5.5 m RIB.   *UPDATED*

SHEARWATER   *7/2003, Jürg Kürsener /* 0572756

HAWKSBILL   *10/2003, Frank Findler /* 0572757

SEAGOING TENDERS

2 BALSAM CLASS (BUOY TENDERS—WLB/ABU)

Name	No	Launched	F/S
A Series			
GENTIAN	WIX 290	1942	AA
C Series			
ACACIA	WLB 406	1944	GLA

Displacement, tons: 757 standard; 1,034 full load
Dimensions, feet (metres): 180 × 37 × 12 *(54.9 × 11.3 × 3.8)*
Main machinery: Diesel-electric; 2 diesels; 1,402 hp *(1.06 MW)* or 1,800 hp *(1.34 MW)* (404); 1 motor; 1,200 hp *(895 kW)*; 1 shaft; bow thruster (except in 395 and 396)
Speed, knots: 13. **Range, n miles:** 8,000 at 12 kt
Complement: 48 (6 officers)
Guns: 2—12.7 mm MGs (except 392, 406 and 404).
Radars: Navigation: Raytheon SPS-64(V)1; I-band.

Comment: Seagoing buoy tenders. Built by Marine Iron & Shipbuilding Co, Duluth, Minnesota, or Zenith Dredge Co, Duluth, Minnesota. Completed 1943-45. Have 20 ton capacity booms.
Acacia underwent major renovation in the late 1970s. *Gentian* brought back into service on 27 July 1999 to support former USCG craft transferred to Caribbean countries. The complement includes people from eight different countries. *Acacia* is to decommission in 2006.   *UPDATED*

GENTIAN   *6/2004*, USCG /* 1043639

16 JUNIPER CLASS (BUOY TENDERS—WLB/ABU)

Name	No	Builders	Commissioned	Home Port
JUNIPER	WLB 201	Marinette Marine	12 Jan 1996	Newport, RI
WILLOW	WLB 202	Marinette Marine	27 Nov 1996	Newport, RI
KUKUI	WLB 203	Marinette Marine	9 Oct 1997	Honolulu, HI
ELM	WLB 204	Marinette Marine	29 June 1998	Atlantic Beach, NC
WALNUT	WLB 205	Marinette Marine	22 Feb 1999	Honolulu, HI
SPAR	WLB 206	Marinette Marine	9 Mar 2001	Kodiak, AK
MAPLE	WLB 207	Marinette Marine	21 June 2001	Sitka, AK
ASPEN	WLB 208	Marinette Marine	28 Sep 2001	San Francisco, CA
SYCAMORE	WLB 209	Marinette Marine	1 Mar 2002	Cordova, AK
CYPRESS	WLB 210	Marinette Marine	24 June 2002	Mobile, AL
OAK	WLB 211	Marinette Marine	17 Oct 2002	Charleston, SC
HICKORY	WLB 212	Marinette Marine	6 Mar 2003	Homer, AK
FIR	WLB 213	Marinette Marine	27 June 2003	Astoria, OR
HOLLYHOCK	WLB 214	Marinette Marine	15 Oct 2003	Port Huron, MI
SEQUOIA	WLB 215	Marinette Marine	21 Apr 2004	Apra Harbour, Guam
ALDER	WLB 216	Marinette Marine	2 Sep 2004	Duluth, MN

Displacement, tons: 2,064 full load
Dimensions, feet (metres): 225 × 46 × 13 *(68.6 × 14 × 4)*
Main machinery: 2 Caterpillar 3608 diesels; 6,200 hp *(4.6 MW)* sustained; 1 shaft; cp prop; bow; 460 hp *(343 kW)* and stern; 550 hp *(410 kW)* thrusters
Speed, knots: 15. **Range, n miles:** 6,000 at 12 kt
Complement: 40 (6 officers)
Guns: 2—12.7 mm MGs. 2—7.62 mm MGs.
Radars: Navigation: 2 Sperry/Litton BridgeMaster E340; I-band.

Comment: On 18 February 1993, the Coast Guard awarded Marinette Marine of Marinette, WI, a contract to construct the first of a new class of seagoing buoy tenders. Capable of breaking 14 in of ice at 3 kt or a minimum of 3 ft by ramming. Main hoist can lift 20 tons, secondary 5 tons. The class is named after the first *Juniper*, which was built in 1940 and decommissioned in 1975.   *UPDATED*

WALNUT   *4/2003, A Sharma /* 0572755

COASTAL TENDERS

14 KEEPER CLASS (BUOY TENDERS—WLM/ABU)

Name	No	Builders	Commissioned	Home Port
IDA LEWIS	WLM 551	Marinette Marine	1 Nov 1996	Newport, RI
KATHERINE WALKER	WLM 552	Marinette Marine	27 June 1997	Bayonne, NJ
ABIGAIL BURGESS	WLM 553	Marinette Marine	19 Sep 1997	Rockland, ME
MARCUS HANNA	WLM 554	Marinette Marine	26 Nov 1997	South Portland, ME
JAMES RANKIN	WLM 555	Marinette Marine	26 Aug 1998	Baltimore, MD
JOSHUA APPLEBY	WLM 556	Marinette Marine	20 Nov 1998	St Petersburg, FL
FRANK DREW	WLM 557	Marinette Marine	17 June 1999	Portsmouth, VA
ANTHONY PETIT	WLM 558	Marinette Marine	1 July 1999	Ketchikan, AK
BARBARA MABRITY	WLM 559	Marinette Marine	29 July 1999	Mobile, AL
WILLIAM TATE	WLM 560	Marinette Marine	16 Sep 1999	Philadelphia, PA
HARRY CLAIBORNE	WLM 561	Marinette Marine	28 Oct 1999	Galveston, TX
MARIA BRAY	WLM 562	Marinette Marine	6 Apr 2000	Mayport, FL
HENRY BLAKE	WLM 563	Marinette Marine	18 May 2000	Everett, WA
GEORGE COBB	WLM 564	Marinette Marine	22 June 2000	San Pedro, CA

Displacement, tons: 840 full load
Dimensions, feet (metres): 175 × 36 × 7.9 *(53.3 × 11 × 2.4)*
Main machinery: 2 Caterpillar 3508TA diesels; 1,920 hp *(1.43 MW)* sustained; 2 Ulstein Z-drives; bow thruster; 460 hp *(343 kW)*
Speed, knots: 12. **Range, n miles:** 2,000 at 10 kt
Complement: 18 (1 officer)
Radars: Navigation: Raytheon SPS-64; I-band.

Comment: Contract awarded 22 June 1993 for first of class with an option for 13 more. Capable of breaking 9 in of ice at 3 kt or 18 in by ramming. Named after Lighthouse Keepers for the Lighthouse Service, one of the predecessors of the modern Coast Guard. The ship is a scaled down model of the Juniper class for coastal service. Main hoist to lift 10 tons, secondary 3.75 tons. Able to skim and recover surface oil pollution using a vessel of opportunity skimming system.
VERIFIED

HARRY CLAIBORNE 2/2000, *van Ginderen Collection* / 0105729

BUOY TENDERS (INLAND—WLI)

2 BUOY TENDERS (WLI/ABU)

Name	No	Builders	Launched	Home Port
BLUEBELL	WLI 313	Birchfield Shipyard, Tacoma	28 Sep 1944	Portland, OR
BUCKTHORN	WLI 642	Mobile Ship Repair, Mobile	18 Aug 1963	Sault Sainte Marie, MI

Displacement, tons: 226 (174 *Bluebell*) full load
Dimensions, feet (metres): 100 × 24 × 5 *(30.5 × 7.3 × 1.5)* (*Buckthorn* draught 4 *(1.2)*)
Main machinery: 2 Caterpillar diesels; 600 hp *(448 kW)*; 2 shafts
Speed, knots: 11.9; 10.5 (*Bluebell*). **Range, n miles:** 2,700 at 10 kt
Complement: 15 (1 officer)

Comment: Different vintage but similar in design.
UPDATED

BUCKTHORN 3/2000, *US Coast Guard* / 0084213

3 BUOY TENDERS (WLI/ABU)

Name	No	Builders	Home Port
BLACKBERRY	WLI 65303	Dubuque Shipyard, Iowa	Long Beach, NC
BAYBERRY	WLI 65400	Reliable Shipyard, Olympia	Seattle, WA
ELDERBERRY	WLI 65401	Reliable Shipyard, Olympia	Petersburg, AK

Displacement, tons: 70 full load
Dimensions, feet (metres): 65 × 17 × 4 *(19.8 × 5.2 × 1.2)*
Main machinery: 1 (*Blackberry*) or 2 (*Bayberry* and *Elderberry*) GM diesels; 1 or 2 shafts
Speed, knots: 10
Complement: 8

Comment: First one completed in August 1946, two in June 1954.
UPDATED

BAYBERRY 5/1999, *Hartmut Ehlers* / 0084211

BUOY TENDERS (RIVER) (WLR)

Notes: (1) All are based on rivers of USA especially the Mississippi and the Missouri and its tributaries.
(2) Two ATON (aids to navigation) barges completed in 1991-92 by Marinette Marine. For use on the Great Lakes in conjunction with icebreaker tugs *Bristol Bay* and *Mobile Bay*.

6 RIVER TENDERS (WLR)

SANGAMON WLR 65506	**SCIOTO** WLR 65504	**OSAGE** WLR 65505
OUACHITA WLR 65501	**CIMARRON** WLR 65502	**OBION** WLR 65503

Displacement, tons: 146 full load
Dimensions, feet (metres): 65 × 21 × 4.5 *(19.8 × 6.4 × 0.4)*
Main machinery: 2 diesels; 750 hp *(560 kW)*; 2 shafts
Speed, knots: 10. **Range, n miles:** 3,500 at 8 kt
Complement: 13

Comment: WLR 65501 and 65502 built by Platzer Shipyard, Houston, TX; 65503-65506 by Gibbs Shipyard, Jacksonville, FL from 1960 to 1962. Deploy aids-to-navigation buoys on the inland river system.
UPDATED

SANGAMON 2/2005*, *USCG* / 1043674

12 RIVER TENDERS (WLR)

WEDGE WLR 75307	**KICKAPOO** WLR 75406
GASCONADE WLR 75401	**KANAWHA** WLR 75407
MUSKINGUM WLR 75402	**PATOKA** WLR 75408
WYACONDA WLR 75403	**CHENA** WLR 75409
CHIPPEWA WLR 75404	**KANKAKEE** WLR 75500
CHEYENNE WLR 75405	**GREENBRIAR** WLR 75501

Displacement, tons: 150 full load
Dimensions, feet (metres): 75 × 22 × 4 *(22.9 × 6.7 × 1.2)*
Main machinery: 2 Caterpillar diesels; 660 hp *(492 kW)*; 2 shafts
Speed, knots: 9. **Range, n miles:** 3,100 at 8 kt
Complement: 13

Comment: WLR 75401-75409 built 1964-70 by four different companies. WLR 75500 and 75501 were completed in early 1990. Details given are for the WLR 75401 series, but all are much the same size.
NEW ENTRY

GASCONADE 2/2005*, *USCG* / 1043673

TRAINING CUTTERS

Notes: *Gentian* WIX 290, a second training cutter, is listed under Balsam class.

1 EAGLE CLASS (WIX/AXS)

Name	No	Builders	Commissioned	Home Port
EAGLE (ex-*Horst Wessel*)	WIX 327	Blohm + Voss, Hamburg	15 May 1946	New London, CT

Displacement, tons: 1,816 full load
Dimensions, feet (metres): 231 wl; 293.6 oa × 39.4 × 16.1 *(70.4; 89.5 × 12 × 4.9)*
Main machinery: 1 Caterpillar D 399 auxiliary diesel; 1,125 hp *(839 kW)* sustained; 1 shaft
Speed, knots: 10.5; 18 sail. **Range, n miles:** 5,450 at 7.5 kt diesel only
Complement: 245 (19 officers, 180 cadets)
Radars: Navigation: SPS-73; I-band.

Comment: Former German training ship. Launched on 13 June 1936. Taken by the US as part of reparations after the Second World War for employment in US Coast Guard Practice Squadron. Taken over at Bremerhaven in January 1946; arrived at home port of New London, Connecticut, in July 1946. (Sister ship *Albert Leo Schlageter* was also taken by the USA in 1945 but was sold to Brazil in 1948 and re-sold to Portugal in 1962. Another ship of similar design, *Gorch Fock*, transferred to the USSR in 1946 and survives as *Tovarisch*.) *Eagle* was extensively overhauled 1981-82. When the Coast Guard added the orange-and-blue marking stripes to cutters in the 1960s *Eagle* was exempted because of their effect on her graceful lines; however, in early 1976 the stripes and words 'Coast Guard' were added in time for the July 1976 Operation Sail in New York harbour. During the Coast Guard's year long bicentennial celebration, which ended 4 August 1990, *Eagle* visited each of the 10 ports where the original revenue cutters were homeported: Baltimore, Maryland; New London, Connecticut; Washington, North Carolina; Savannah, Georgia; Philadelphia, Pennsylvania; Newburyport, Maryland; Portsmouth, New Hampshire; Charleston, South Carolina; New York, New York; and Hampton, Virginia. The cutter currently serves as a training ship for cadets and officer candidates.
　　Fore and main masts 150.3 ft *(45.8 m)*; mizzen 132 ft *(40.2 m)*; sail area, 25,351 sq ft.

VERIFIED

EAGLE　　　　　　　　　　　　　　　　　*7/2000, van Ginderen Collection /* 0105730

CONSTRUCTION TENDERS (INLAND) (WLIC)

Notes: All, although operating on inland waters, are administered by the Atlantic Area.

4 PAMLICO CLASS (WLIC)

PAMLICO WLIC 800	**HUDSON** WLIC 801	**KENNEBEC** WLIC 802	**SAGINAW** WLIC 803

Displacement, tons: 459 full load
Dimensions, feet (metres): 160.9 × 30 × 4 *(49 × 9.1 × 1.2)*
Main machinery: 2 Caterpillar diesels; 1,000 hp *(746 kW)*; 2 shafts
Speed, knots: 11
Complement: 14 (1 officer)
Radars: Navigation: Raytheon SPS-69; I-band.

Comment: Completed in 1976 at the Coast Guard Yard, Curtis Bay, Maryland. These ships maintain structures and buoys in bay areas along the Atlantic and Gulf coasts.

UPDATED

HUDSON　　　　　　　　　　　　　　　　　*12/1989, Giorgio Arra*

1 COSMOS CLASS (WLIC)

SMILAX WLIC 315

Displacement, tons: 218 full load
Dimensions, feet (metres): 100 × 24 × 5 *(30.5 × 7.3 × 1.5)*
Main machinery: 2 Caterpillar D 353 diesels; 660 hp *(492 kW)* sustained; 2 shafts
Speed, knots: 10.5
Complement: 14 (1 officer)
Radars: Navigation: Raytheon SPS-69; I-band.

Comment: Completed in 1944. Primary areas of operation are intercoastal waters from Virginia to Georgia. Based at Atlantic Beach, NC.

UPDATED

COSMOS　　　　　　　　　　　　　　　*7/1990, van Ginderen Collection*

8 ANVIL/CLAMP CLASSES (WLIC)

ANVIL WLIC 75301	**MALLET** WLIC 75304	**HATCHET** WLIC 75309
HAMMER WLIC 75302	**VISE** WLIC 75305	**AXE** WLIC 75310
SLEDGE WLIC 75303	**CLAMP** WLIC 75306	

Displacement, tons: 145 full load
Dimensions, feet (metres): 75 (76-WLIC 75306-75310) × 22 × 4 *(22.9 (23.2) × 6.7 × 1.2)*
Main machinery: 2 Caterpillar diesels; 750 hp *(559 kW)*; 2 shafts
Speed, knots: 10
Complement: 13 (1 officer in *Mallet, Sledge* and *Vise)*

Comment: Completed 1962-65. Primary areas of operation are intercoastal waters from Texas to New Jersey.

VERIFIED

CLAMP　　　　　　　　　　　　　　*10/1999, Marsan/Schaeffer /* 0084215

HARBOUR TUGS

11 65 ft CLASS (WYTL)

Name	No	Home Port
CAPSTAN	WYTL 65601	Philadelphia, PA
CHOCK	WYTL 65602	Portsmouth, VA
TACKLE	WYTL 65604	Rockland, ME
BRIDLE	WYTL 65607	Southwest Harbor, ME
PENDANT	WYTL 65608	Boston, MA
SHACKLE	WYTL 65609	South Portland, ME
HAWSER	WYTL 65610	Bayonne, NJ
LINE	WYTL 65611	Bayonne, NJ
WIRE	WYTL 65612	Saugerties, NY
BOLLARD	WYTL 65614	New Haven, CT
CLEAT	WYTL 65615	Philadelphia, PA

Displacement, tons: 72 full load
Dimensions, feet (metres): 65 × 19 × 7 *(19.8 × 5.8 × 2.1)*
Main machinery: 1 Caterpillar 3412TA diesel; 400 hp *(298 kW)* sustained; 1 shaft
Speed, knots: 10. **Range, n miles:** 2,700 at 10 kt
Complement: 6
Radars: Navigation: Raytheon SPS-69; I-band.

Comment: Built between 1961 and 1967. The tugs provide icebreaking services to several east coast areas. Re-engined 1993-96.

UPDATED

HAWSER　　　　　　　　　　　　　*7/2000, Hachiro Nakai /* 0105731

RESCUE AND UTILITY CRAFT

Notes: Craft of several different types. All carry five or six figure numbers of which the first two figures reflect the craft's length in feet.

140 UTILITY BOATS (YAG/UTB)

Displacement, tons: 13.4 full load
Dimensions, feet (metres): 41.3 × 14.1 × 4.1 *(12.6 × 4.3 × 1.3)*
Main machinery: 2 diesels; 680 hp *(507 kW)* sustained; 2 shafts
Speed, knots: 26. **Range, n miles:** 300 miles at 18 kt
Complement: 3

Comment: 205 built by Coast Guard Yard, Baltimore 1973-83. Aluminium hull with a towing capacity of 100 tons. Used for fast multimission response in weather conditions up to moderate.

UPDATED

41385   *4/2003*, Declerck/Steeghers /* 1043680

41422   *7/2000, Hachiro Nakai /* 0105732

117 MOTOR LIFEBOATS (MLB/SAR)

Displacement, tons: 20 full load
Dimensions, feet (metres): 47.9 × 14.5 × 4.5 *(14.6 × 4.4 × 1.4)*
Main machinery: 2 Detroit diesels; 850 hp *(634 kW)* sustained; 2 shafts
Speed, knots: 25. **Range, n miles:** 220 at 25 kt
Complement: 4

Comment: Built by Textron Marine, New Orleans. The prototype completed trials in mid-1991. Five production boats delivered in 1994. More ordered in September 1995. Replaced the fleet of 44 ft lifeboats. Aluminium hulls, self-righting with a 9,000 lb bollard pull and a towing capability of 150 tons. Primarily a lifeboat but it has a multimission capability.

UPDATED

MOTOR LIFEBOAT   *10/2003, Frank Findler /* 0572754

133 + 130 DEFENDER CLASS (RESPONSE BOATS) (PBF)

Displacement, tons: 2.7 full load
Dimensions, feet (metres): 25.0 × 8.5 × 8.8 *(7.6 × 2.6 × 2.7)*
Main machinery: 2 Honda outboard motors; 450 hp *(335 kW)*
Speed, knots: 46. **Range, n miles:** 175 at 35 kt
Complement: 4
Guns: 1—12.7 mm MG.
Radars: To be announced.

Comment: High-speed inshore patrol craft of aluminium construction and foam collar built by SAFE Boats International, Port Orchard, Washington. First delivery in July 2003 to replace nearly 300 non-standard shore based boats and provide a standardised platform for the USCG's new Maritime Safety and Security Teams (MSST), established as a result of the 11 September 2001 terrorist attacks. Up to 700 may be procured by 2010. Transportable in a C-130.

VERIFIED

DEFENDER CLASS   *10/2003, Frank Findler /* 0572753

18 + (64) SHORT RANGE PROSECUTOR CRAFT (SRP)

Displacement, tons: To be announced
Dimensions, feet (metres): 25.2 × ? × ? *(7.7 × ? × ?)*
Main machinery: 1 inboard diesel waterjet; 315 hp *(235 kW)*
Speed, knots: 33
Complement: 2 crew plus 8 passengers

Comment: Trials completed in April 2003 at Coast Guard Station Curtis Bay, MD. The first SRP was delivered to the newly converted 123 ft *Matagorda* on 1 March 2004. Launched and recovered via a stern launch and recovery system. Further to enter fleet as each modernised WPB delivered. A fleet of 82 is projected by 2021.

NEW ENTRY

SRP   *10/2004*, Jeff Murphy, USCG /* 1121007

0 + (42) LONG RANGE INTERCEPTOR CRAFT (LRI)

Displacement, tons: To be announced
Dimensions, feet (metres): 36 × ? × ? *(11.0 × ? × ?)*
Main machinery: Inboard diesel waterjet
Speed, knots: 45
Complement: 14 crew and passengers

Comment: First expected to enter service by 2007. A total of 42 is planned by 2021. Launched and recovered via a stern launch and recovery system.

NEW ENTRY

LRI   *10/2004*, Jeff Murphy, USCG /* 1121008

NATIONAL OCEANIC AND ATMOSPHERIC ADMINISTRATION (NOAA)

Headquarters Appointments

Under Secretary of Commerce for Oceans and Atmosphere:
  Vice Admiral Conrad Lautenbacher Jr USN (ret)
Director, NOAA Marine and Aviation Operations and NOAA Commissioned Officer Corps:
  Rear Admiral Samuel P de Bow Jr
Director, Marine and Aviation Operations Centers:
  Rear Admiral Richard R Behn

Establishment and Missions

NOAA is the largest bureau of the US Department of Commerce, with a diverse set of responsibilities in environmental sciences. NOAA components include NOAA Marine and Aviation Operations; the National Ocean Service; the National Weather Service; the National Marine Fisheries Service; the National Environmental Satellite, Data and Information Service; and the office of Oceanic and Atmospheric Research. NOAA's research vessels conduct operations in hydrography, bathymetry, oceanography, atmospheric research, fisheries assessments and research, and related programmes in marine resources. Larger research vessels operate in international waters, and smaller ones primarily in Atlantic and Pacific coastal waters, and the Gulfs of Mexico and Alaska. NOAA conduct diving operations. It also operates fixed-wing and rotary aircraft for hurricane research and reconnaissance; oceanographic and atmospheric research; marine mammal observations; hydrologic forecasts; and aerial mapping and charting.

NOAA's active fleet numbers 17 ships, and now includes seven former Navy ships. Two other ex-Navy T-AGOS ships *(Assertive* and *Capable)* have been acquired and will be converted to conduct research. A new oceanographic research ship, *Ronald H Brown* (ex-AGOR 26), was commissioned in 1997. Of the seven ex-naval ships, five are T-AGOS vessels: one has been converted for oceanographic research *(Ka'imimoana)*, two for fisheries research *(Gordon Gunter, Oscar Elton Sette)*, and two for coastal oceanographic research *(McArthur II* and *Hi'ialakai)*. *Sette* replaced *Townsend Cromwell* and *McArthur II* replaced *McArthur* in 2003. *Hi'ialakai* (formerly *Vindicator)*, homeported in Hawaii, was commissioned in 2004 and is an addition to the fleet. The former naval T-AGS hydrographic survey ship *Littlehales* was transferred to NOAA in 2003 and recommissioned *Thomas Jefferson*, replacing *Whiting*. A former naval YTT (Yard Torpedo Test) vessel was converted for coastal research and became operational in 2003 as *Nancy Foster*, replacing *Ferrel*. The hydrographic sruvey ship *Fairweather* was decommissioned in 1998, refurbished, and reactivated in 2004. A new class of fisheries survey vessels has been designed to NOAA specifications and international standards. *Oscar Dyson*, the first of four planned FSVs, was completed in late 2004 and the ship is to become operational in early 2005. Construction of the second FSV, *Henry B Bigelow*, started in October 2003 and she is expected to become operational in 2007. In 2004, NOAA exercised an option under an existing contract with VT Halter Marine Inc. to develop a contract design of a small-waterplane-area twin-hull ship that, if constructed, will conduct hydrographic surveys.

Ships

The following ships may be met at sea.
Oceanographic Research Ships: *Ronald H Brown, Ka'imimoana.*
Multipurpose Oceanographic/Coastal Research Ships: *McArthur II, Nancy Foster, Hi'ialakai*
Hydrographic Survey Ships: *Rainier, Rude, Thomas Jefferson, Fairweather.*
Fisheries Research Ships: *Miller Freeman, Oregon II, Albatross IV, Delaware II, David Starr Jordan, John N Cobb, Gordon Gunter, Oscar Elton Sette.*

Personnel

2005: 257 officers plus 12,000 civilians.

Bases

Major: Norfolk, VA and Seattle, WA.
Minor: Woods Hole, MA; Pascagoula, MS; Honolulu, HI; Charleston, SC; San Diego, CA; Ketchikan, AK.

OSCAR ELTON SETTE　　　　　　　　　　6/2002, *NOAA* / 0543400

ALBATROSS IV　　　　　　　　　　10/1997, *Harald Carstens* / 0053484

THOMAS JEFFERSON
6/2003, *NOAA* / 0572818

RONALD H BROWN　　　　　　　　　　　　　　　　12/2004 *, Globke Collection* / 1043679

Uruguay

Country Overview

The Oriental Republic of Uruguay is situated in south-eastern South America. With an area of 68,037 square miles it has borders to the north with Brazil and to the west with Argentina. It has a coastline of 356 n miles with the south Atlantic Ocean and River Plate. There are some 675 n miles of navigable internal waterways. The capital, largest city and principal port is Montevideo. Territorial Seas (12 n miles) and an EEZ (200 n miles) are claimed.

Headquarters Appointments

Commander-in-Chief of the Navy:
  Vice Admiral Tabare Daners Eyras
Fleet Commander:
  Rear Admiral Oscar Debali de Palleja
Commander Coast Guard:
  Rear Admiral Heber Fernández Maggio

Diplomatic Representation

Naval Attaché in London:
  Captain Ney Escandón

Personnel

(a) 2005: 5,700 (700 officers) (including 450 naval infantry, 300 naval air and 1,950 Coast Guard)
(b) Voluntary service

Prefectura Nacional Naval (PNN)

Established in 1934 primarily for harbour security and coastline guard duties. In 1991 it was integrated with the Navy, although patrol craft retain Prefectura markings. There are three regions: Atlantic, Rio de la Plata, and Rio Uruguay.

Bases

Montevideo: Main naval base with two dry docks (A new naval base is under construction at Punta Lobos and will replace the current harbour facilities.)
La Paloma: Naval station *(Ernesto Motto)*
Fray Bentos: River base
Laguna del Sauce, Maldonado: Naval air station *(Capitan Carlos A Curbelo)*

Marines

Cuerpo de Fusileros Navales consisting of 450 men in three rifleman companies and one combat support company plus a command company of 100.

Prefix to Ships' Names

ROU

DELETIONS

Survey and Research Ships

2002   *Comandante Pedro Campbell*

FRIGATES

Notes: Discussions about the acquisition of two Descubierta class corvettes from Spain took place in 2004 but are unlikely to lead to a contract.

3 COMMANDANT RIVIÈRE CLASS (FF)

Name	No	Builders	Laid down	Launched	Commissioned	Recommissioned
URUGUAY (ex-*Commandant Bourdais*)	1	Lorient Naval Dockyard	Apr 1959	15 Apr 1961	10 Mar 1962	20 Aug 1990
GENERAL ARTIGAS (ex-*Victor Schoelcher*)	2	Lorient Naval Dockyard	Oct 1957	11 Oct 1958	15 Oct 1962	9 Jan 1989
MONTEVIDEO (ex-*Amiral Charner*)	3	Lorient Naval Dockyard	Nov 1958	12 Mar 1960	14 Dec 1962	28 Jan 1991

Displacement, tons: 1,750 standard; 2,250 full load
Dimensions, feet (metres): 336.9 × 38.4 × 14.1 *(102.7 × 11.7 × 4.3)*
Main machinery: 4 SEMT-Pielstick 12 PC series diesels; 16,000 hp(m) *(11.8 MW)*; 2 shafts
Speed, knots: 25. **Range, n miles:** 7,500 at 15 kt
Complement: 159 (9 officers)

Guns: 2 DCN 3.9 in *(100 mm)*/55 Mod 1953 automatic ❶; dual purpose; 60 rds/min to 17 km *(9 n miles)* anti-surface; 8 km *(4.4 n miles)* anti-aircraft; weight of shell 13.5 kg.
  2 Hispano-Suiza 30 mm/70 ❷.
Torpedoes: 6—21.7 in *(550 mm)* (2 triple) tubes ❸ ECAN L3; anti-submarine; active homing to 5.5 km *(3 n miles)* at 25 kt; warhead 200 kg; depth to 300 m *(985 ft)*.
A/S mortars: 1 Mortier 305 mm 4-barrelled launcher ❹.
Countermeasures: ESM: NS 9010-UR; radar warning.
Weapons control: C T Analogique. Sagem DMAA optical director.
Radars: Air/surface search: Thomson-CSF DRBV 22A ❺; D-band.
  Navigation: Racal Decca 1226 ❻; I-band.
  Fire control: Thomson-CSF DRBC 32C ❼; I-band.
Sonars: EDO SQS-17; hull-mounted; active search; medium frequency.
  Thomson Sintra DUBA 3; active attack; high frequency.

Programmes: First one bought from France through SOFMA on 30 September 1988, second pair 14 March 1990. All refitted before transfer.
Structure: Exocet and Dagaie removed before transfer. SSM casings removed from *Montevideo* refit 1999/2000.
Operational: Can carry a Flag Officer and staff. In French service this class sometimes embarked up to 80 soldiers and two LCPs. The A/S mortar is non-operational. *Uruguay* was in refit 2000-2004. The ships form the Escort Division based at Montevideo.

UPDATED

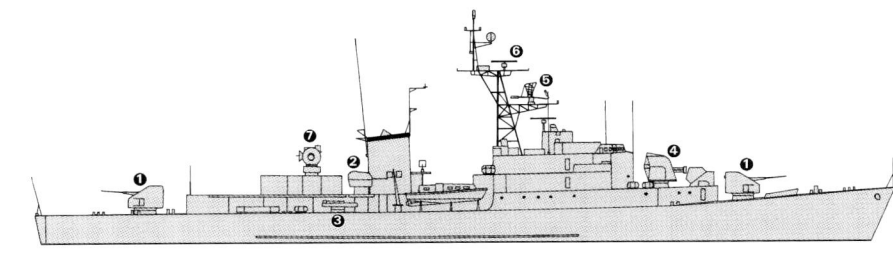
MONTEVIDEO   *(Scale 1 : 900), Ian Sturton /* 1044203

MONTEVIDEO   *3/2002, Robert Pabst /* 0534051

GENERAL ARTIGAS
2/2003, A E Galarce / 0569797

MONTEVIDEO
10/2002, Mario R V Carveiro / 0534050

LAND-BASED MARITIME AIRCRAFT

Notes: (1) There are plans to acquire a CN-235 Persuader maritime patrol aircraft which would complement the one Beech B 200T in service.
(2) In addition there are five helicopters (three Westland Wessex HC Mk 2, one Wessex W-60 and one Bell 47G). There are also three fixed-wing aircraft, two Beech T-34C and two Jetstream T2. The latter were bought from the UK in 1998.

WESSEX   2/2003, A E Galarce / 0569796

JET STREAM   2001, Uruguay Navy / 0121423

B-200T   2001, Uruguay Navy / 0121422

Numbers/Type: 1 Grumman S-2G Tracker.
Operational speed: 140 kt *(260 km/h)*.
Service ceiling: 25,000 ft *(7,620 m)*.
Range: 1,350 n miles *(2,500 km)*.
Role/Weapon systems: Ex-Israeli aircraft. ASW and surface search with improved systems. Sensors: Search radar, MAD, sonobuoys. Weapons: ASW; torpedoes, depth bombs or mines. ASV; rockets underwing.

UPDATED

GRUMMAN S-2G   2000, Uruguay Navy / 0105735

PATROL FORCES

Notes: There are plans to acquire three 35 m offshore patrol vessels.

3 VIGILANTE CLASS (LARGE PATROL CRAFT) (PBO)

Name	No	Builders	Commissioned
15 de NOVIEMBRE	5	CMN, Cherbourg	25 Mar 1981
25 de AGOSTO	6	CMN, Cherbourg	25 Mar 1981
COMODORO COÉ	7	CMN, Cherbourg	25 Mar 1981

Displacement, tons: 190 full load
Dimensions, feet (metres): 137 × 22.4 × 8.2 *(41.8 × 6.8 × 2.4)*
Main machinery: 2 MTU 12V 538 TB91 diesels; 4,600 hp(m) *(3.4 MW)* sustained; 2 shafts
Speed, knots: 28. **Range, n miles:** 2,400 at 15 kt
Complement: 28 (5 officers)
Guns: 1 Bofors 40 mm/70.
Weapons control: CSEE Naja optronic director.
Radars: Surface search: Racal Decca TM 1226C; I-band.

Comment: Ordered in 1979. Steel hull. First launched 16 October 1980, second 11 December 1980 and third 27 January 1981. Based at La Paloma. *VERIFIED*

15 DE NOVIEMBRE   7/2001, A E Galarce / 0534052

2 CAPE CLASS (LARGE PATROL CRAFT) (PB)

Name	No	Builders	Commissioned
COLONIA (ex-*Cape Higgon*)	10	Coast Guard Yard, Curtis Bay	14 Oct 1953
RIO NEGRO (ex-*Cape Horn*)	11	Coast Guard Yard, Curtis Bay	3 Sep 1958

Displacement, tons: 98 standard; 148 full load
Dimensions, feet (metres): 95 × 20.2 × 6.6 *(28.9 × 6.2 × 2)*
Main machinery: 2 GM 16V-149TI diesels; 2,322 hp *(1.73 MW)* sustained; 2 shafts
Speed, knots: 20. **Range, n miles:** 2,500 at 10 kt
Complement: 14 (1 officer)
Guns: 2—12.7 mm MGs.
Radars: Surface search: Raytheon SPS-64; I-band.

Comment: Designed for port security and search and rescue. Steel hulled. During modernisation in 1974 received new engines, electronics and deck equipment. Superstructure modified or replaced, and habitability improved. Transferred from the US Coast Guard 25 January 1990. Both based at Fray Bentos.

UPDATED

COLONIA   2/2004*, A E Galarce / 1044207

1 COASTAL PATROL CRAFT (PB)

Name	No	Builders	Commissioned
PAYSANDU	12 (ex-PR 12)	Sewart, USA	Nov 1968

Displacement, tons: 58 full load
Dimensions, feet (metres): 83 × 18 × 6 *(25.3 × 5.5 × 1.8)*
Main machinery: 2 GM 16V-71 diesels; 811 hp *(605 kW)* sustained; 2 shafts
Speed, knots: 22. **Range, n miles:** 800 at 20 kt
Complement: 8
Guns: 1—12.7 mm MGs. 2—7.62 mm MGs.
Radars: Surface search: Raytheon 1500B; I-band.

Comment: Formerly incorrectly listed under Coast Guard. Based at Montevideo.

UPDATED

PAYSANDU   2/2004*, A E Galarce / 1044206

3 COAST GUARD PATROL CRAFT (WPB)

70   71   72

Displacement, tons: 90 full load
Dimensions, feet (metres): 72.2 × 16.4 × 5.9 *(22 × 5 × 1.8)*
Main machinery: 2 GM diesels; 400 hp *(298 kW)*; 2 shafts
Speed, knots: 12
Complement: 8

Comment: Built in 1957 at Montevideo.

VERIFIED

PREFECTURA 70   2/2003, A E Galarce / 0569795

2 RIVER PATROL CRAFT (PBR)

URUGUAY 1   URUGUAY 2

Displacement, tons: 5 full load
Dimensions, feet (metres): 38.7 × 11.8 × 3.2 *(11.8 × 3.6 × 1)*
Main machinery: 3 Volvo AD41P 220MOP diesels
Speed, knots: 32. **Range, n miles:** 1,500 at 24 kt
Complement: 4
Guns: 2—12.7 mm MGs.

Comment: Built by Nuevos Ayres yacht builders. Deployed to Congo as part of UN force during 2001.

VERIFIED

URUGUAY 1   *2001, Uruguay Navy /* 0121420

3 RIVER PATROL CRAFT (PBR)

URUGUAY 3   URUGUAY 4

Displacement, tons: 6 full load
Dimensions, feet (metres): 37.1 × 10.7 × 2.6 *(11.3 × 3.25 × 0.8)*
Main machinery: 2 Volvo Penta diesels
Speed, knots: 32.
Complement: 6
Guns: 3—7.62 mm MGs.

Comment: Built by Astillero KLASE A, Buenos Aires and delivered to Uruguayan Navy in 2002. A third delivered in 2004. Deployed to Congo-Brazzaville as part of UN force.

UPDATED

9 TYPE 44 CLASS (WPB)

441-449

Displacement, tons: 18 full load
Dimensions, feet (metres): 44 × 12.8 × 3.6 *(13.5 × 3.9 × 1.1)*
Main machinery: 2 Detroit 6V-38 diesels; 185 hp *(136 kW)*; 2 shafts
Speed, knots: 14. **Range, n miles:** 215 at 10 kt
Complement: 3

Comment: Acquired from the US in 1999 and operated by the Coast Guard primarily as SAR craft.

VERIFIED

446   *5/2000, Hartmut Ehlers /* 0105801

442   *9/2003, A E Galarce /* 0569793

MINE WARFARE FORCES

3 KONDOR II CLASS (MINESWEEPERS—COASTAL) (MSC)

Name	No	Builders	Launched	Recommissioned
TEMERARIO (ex-*Riesa*)	31	Peenewerft, Wolgast	2 Oct 1972	11 Oct 1991
FORTUNA (ex-*Bernau*)	33	Peenewerft, Wolgast	3 Aug 1972	11 Oct 1991
AUDAZ (ex-*Eisleben*)	34	Peenewerft, Wolgast	2 Jan 1973	11 Oct 1991

Displacement, tons: 310 full load
Dimensions, feet (metres): 186 × 24.6 × 7.9 *(56.7 × 7.5 × 2.4)*
Main machinery: 2 Russki/Kolomna Type 40-DM diesels; 4,408 hp(m) *(3.24 MW)* sustained; 2 shafts; cp props
Speed, knots: 17. **Range, n miles:** 2,000 at 15 kt
Complement: 31 (6 officers)
Guns: 1 Hispano-Suiza 30 mm/70
Mines: 2 rails.
Radars: Surface search: TSR 333 or Raytheon 1900; I-band.

Comment: Belonged to the former GDR Navy. Transferred without armament. Minesweeping gear retained including MSG-3 variable depth sweep device. A fourth of class sunk after a collision with a merchant ship on 5 August 2000.

UPDATED

AUDAZ   *2/2004*, A E Galarce /* 1044205

SURVEY AND RESEARCH SHIPS

1 HELGOLAND (TYPE 720B) CLASS (AGS)

Name	No	Builders	Commissioned
OYARVIDE (ex-*Helgoland*)	22 (ex-A 1457)	Unterweser, Bremerhaven	8 Mar 1966

Displacement, tons: 1,310 standard; 1,643 full load
Dimensions, feet (metres): 223.1 × 41.7 × 14.4 *(68 × 12.7 × 4.4)*
Main machinery: Diesel-electric; 4 MWM 12-cyl diesel generators; 2 motors; 3,300 hp(m) *(2.43 MW)*; 2 shafts
Speed, knots: 17. **Range, n miles:** 6,400 at 16 kt
Complement: 34
Radars: Navigation: Raytheon; I-band.
Sonars: High definition, hull-mounted for wreck search.

Comment: Former German ocean-going tug launched on 25 November 1965. Paid off in 1997 and recommissioned on 21 September 1998 after being fitted out as a survey ship. Ice strengthened hull. Fitted for twin 40 mm guns. Sister ship *Fehmarn* may also be acquired from Germany.

VERIFIED

OYARVIDE   *6/2002, A E Galarce /* 0529549

1 INSHORE SURVEY CRAFT (AGSC)

TRIESTE

Displacement, tons: 12 full load
Dimensions, feet (metres): 39.7 × 11.8 × 3.3 *(12.1 × 3.6 × 1)*
Main machinery: 2 Kamewa waterjets
Speed, knots: 16. **Range, n miles:** 500 at 16 kt
Complement: 4
Sonars: Elac Compact Mk II; 180 kHz. Elac LAZ 4721; 200 kHz.

Comment: Formerly owned by the Academia Maritime Internacional de Trieste. Donated by Italian government in 2000.

VERIFIED

TRIESTE   *2001, Uruguay Navy /* 0121418

TRAINING SHIPS

Notes: Bonanza is a 26 ton sail training vessel built in UK in 1984 and commissioned in July 1997.

1 SAIL TRAINING SHIP (AXS)

Name	No	Builders	Commissioned
CAPITÁN MIRANDA	20 (ex-GS 10)	SECN Matagorda, Cádiz	1930

Displacement, tons: 839 full load
Dimensions, feet (metres): 209.9 × 26.3 × 12.4 *(64 × 8 × 3.8)*
Main machinery: 1 GM diesel; 750 hp *(552 kW)*; 1 shaft
Speed, knots: 10
Complement: 49
Radars: Navigation: Racal Decca TM 1226C; I-band.

Comment: Originally a diesel-driven survey ship with pronounced clipper bow. Converted for service as a three-masted schooner, commissioning as cadet training ship in 1978. Major refit by Bazán, Cadiz from June 1993 to March 1994, including a new diesel engine and a 5 m extension to the superstructure. Now has 853.4 m² of sail.

VERIFIED

CAPITÁN MIRANDA   *8/1999, A Campanera i Rovira /* 0084226

AUXILIARIES

Notes: ROU 73 *(Grito de Asencio)* is an 18 m harbour tug.

1 WANGEROOGE CLASS (AG)

Name	No	Builders	Commissioned
MALDONADO (ex-Norderney)	23 (ex-A1455)	Schichau, Bremerhaven	15 Oct 1970

Displacement, tons: 854 standard; 1,024 full load
Dimensions, feet (metres): 170.6 × 39.4 × 12.8 *(52 × 12.1 × 3.9)*
Main machinery: Diesel-electric; 4 MWM 16-cyl diesel generators; 2 motors; 2,400 hp(m) *(1.76 MW)*; 2 shafts
Speed, knots: 14. **Range, n miles:** 5,000 at 10 kt.
Complement: 24.
Guns: 1 Bofors 40 mm/70 (fitted for).

Comment: Built as a salvage tug with ice-strengthened hull. Transferred from the German Navy on 21 November 2002. Employed as a support ship.

UPDATED

MALDONADO (German colours)   *6/2001, Martin Mokrus /* 0130283

2 LCVPs

LD 45   LD 46

Displacement, tons: 15 full load
Dimensions, feet (metres): 46.5 × 11.6 × 2.7 *(14.1 × 3.5 × 0.8)*
Main machinery: 1 GM 4-71 diesel; 115 hp *(86 kW)* sustained; 1 shaft
Speed, knots: 9. **Range, n miles:** 580 at 9 kt
Military lift: 10 tons

Comment: Built at Naval Shipyard, Montevideo and completed 1980.

VERIFIED

LD 45 (PNN 631 alongside)   *12/1997, Hartmut Ehlers /* 0016610

1 PIAST CLASS (PROJECT 570) (SALVAGE SHIP) (ARS)

Name	No	Builders	Commissioned
VANGUARDIA	26 (ex-A 441)	Northern Shipyard, Gdansk	29 Dec 1976
(ex-*Otto Von Guericke*)			

Displacement, tons: 1,732 full load
Dimensions, feet (metres): 240 × 39.4 × 13.1 *(73.2 × 12 × 4)*
Main machinery: 2 Zgoda diesels; 3,800 hp(m) *(2.79 MW)*; 2 shafts; cp props
Speed, knots: 16. **Range, n miles:** 3,000 at 12 kt
Complement: 61
Radars: Navigation: 2 TSR 333; I-band.

Comment: Acquired from Germany in October 1991 and sailed from Rostock in January 1992 after a refit at Neptun-Warnow Werft. Carries extensive towing and firefighting equipment plus a diving bell forward of the bridge. Armed with four 25 mm twin guns when in service with the former GDR Navy.

UPDATED

VANGUARDIA   *8/2004*, A E Galarce /* 1044204

1 BUOY TENDER (ABU)

SIRIUS 21

Displacement, tons: 290 full load
Dimensions, feet (metres): 115.1 × 32.8 × 5.9 *(35.1 × 10 × 1.8)*
Main machinery: 2 Detroit 12V-71TA diesels; 840 hp *(626 kW)* sustained; 2 shafts
Speed, knots: 11
Complement: 15

Comment: Buoy tender built at Montevideo Naval Yard and completed on 5 February 1988. Endurance, five days.

VERIFIED

SIRIUS   *6/1999, Uruguay Navy /* 0084228

2 LCM CLASS

LD 41   LD 42

Displacement, tons: 24 light; 57 full load
Dimensions, feet (metres): 56.1 × 14.1 × 3.9 *(17.1 × 4.3 × 1.2)*
Main machinery: 2 Gray Marine 64 HN9 diesels; 330 hp *(264 kW)*; 2 shafts
Speed, knots: 9. **Range, n miles:** 130 at 9 kt
Complement: 5
Military lift: 30 tons

Comment: First one transferred on lease from USA October 1972. Lease extended in October 1986. Second built in Uruguay.

VERIFIED

LD 42   *2/2003, A E Galarce /* 0569791

TUGS

1 COASTAL TUG (YTB)

Name	No	Builders	Commissioned
BANCO ORTIZ	27 (ex-7, ex-Y 1655)	Peenewerft, Wolgast	10 Sep 1959
(ex-*Zingst*, ex-*Elbe*)			

Displacement, tons: 261 full load
Dimensions, feet (metres): 100 × 26.6 × 10.8 *(30.5 × 8.1 × 3.3)*
Main machinery: 1 R6 DV 148 diesel; 550 hp(m) *(404 kW)*; 1 shaft
Speed, knots: 10
Complement: 12
Guns: 1—12.7 mm MG

Comment: Ex-GDR Type 270 tug acquired in October 1991. 10 ton bollard pull.

VERIFIED   BANCO ORTIZ   *10/2000, A E Galarce* / 0105807

Vanuatu

Country Overview

The Republic of Vanuatu, formerly the New Hebrides, was jointly administered by Britain and France until it gained independence in 1980. Situated in the southwestern Pacific Ocean, some 1,100 n miles southeast of Papua New Guinea, the country comprises a group of about 80 islands, of which 67 are inhabited, which run generally north-south. The four main islands are Espíritu Santo (the largest), Malekula, Efate and Tanna. Others include Epi, Pentecost, Aoba, Maewo, Erromanga and Ambrym. The capital, largest town and principal port is Port-Vila on Efate. An archipelagic state, territorial seas (12 n miles) are claimed. An Exclusive Economic Zone (EEZ) (200 n miles) is also claimed but limits have not been fully defined by boundary agreements. Disputed sovereignty of Matthew and Hunter Islands, both uninhabited, is one complication.

Headquarters Appointments

Commander, Maritime Wing:
    Chief Inspector Tari Tamata

Bases

Port Vila, Efate Island

POLICE

1 PACIFIC CLASS (LARGE PATROL CRAFT) (PB)

Name	No	Builders	Commissioned
TUKORO	02	Australian Shipbuilding Industries	13 June 1987

Displacement, tons: 165 full load
Dimensions, feet (metres): 103.3 × 26.6 × 6.9 *(31.5 × 8.1 × 2.1)*
Main machinery: 2 Caterpillar 3516TA diesels; 4,400 hp *(3.28 MW)* sustained; 2 shafts
Speed, knots: 18. **Range, n miles:** 2,500 at 12 kt
Complement: 18 (3 officers)
Guns: 1—12.7 mm MG. 1—7.62 mm MG.
Radars: Navigation: Furuno 1011; I-band.

Comment: Under the Defence Co-operation Programme Australia has provided one Patrol Craft to the Vanuatu government. Training and operational and technical assistance is also given by the Royal Australian Navy. Ordered 13 September 1985 and launched 20 May 1987. A half-life refit was carried out in 1995 and, following extension of the Pacific Patrol Boat programme by the Australian government, a life-extension refit was carried out in 2004. The ship is employed on Exclusive Economic Zone (EEZ) fishery patrol and surveillance, including customs duties.

UPDATED   TUKORO   *8/2003, Chris Sattler* / 0568869

Venezuela

ARMADA DE VENEZUELA

Country Overview

The Republic of Venezuela is situated in northern South America. With an area of 352,144 square miles, it has borders to the east with Guyana, to the south with Brazil and to the west with Colombia. It has a 1,512 n mile coastline with the Caribbean Sea and Atlantic Ocean. Margarita is the principal offshore island, of which there are 70. The capital and largest city is Caracas which is served by the port of La Guaira. Other ports include Puerto Cabello, and Maracaibo. The chief port on the Orinoco River is Puerto Ordaz. Territorial Seas (12 n miles) are claimed. An EEZ (200 n miles) has also been claimed but the limits have not been fully defined by boundary agreements.

Headquarters Appointments

Commander General of the Navy (Chief of Naval Operations):
    Vice Admiral Armando Laguna Laguna

Diplomatic Representation

Naval Attaché in Washington:
    Rear Admiral Pedro Negrin Ruiz
Defence Attaché in London:
    Rear Admiral Luis Mérida Galindo
Naval Attaché in Rome:
    Captain Humberto Lazo Cividane
Naval Attaché in Bogota:
    Captain León González Maldonado

Personnel

(a)   2005: 15,800
(b)   2 years' national service

Fleet Organisation

The fleet is split into 'Type' squadrons — frigates (except GC 11 and 12), submarines, light and amphibious forces. Service Craft Squadron composed of RA 33, BO 11 and BE 11. The Fast Attack Squadron of the Constitución class is subordinate to the Fleet Command.

Marines

Following restructuring, the Marines are formed into a division, *General Simón Bolívar*, which consists of two amphibious brigades. The 1st Amphibious Brigade comprises four infantry battalions: *Rafael Urdaneta* (Puerto Cabello), *Francisco de Miranda* (Punto Fijo), *Ezequiel Zamora* (Maracaibo) and *Manuel Ponce Lugo* (Puerto Cabello). The 2nd Amphibious Brigade comprises three battalions: *General Simón Bolívar* (Maiquetía), *Mariscal Antonio José de Sucre* (Cumaná) and *General José Francisco Bermúdez* (Carúpano). Additionally, there is an Engineer Brigade with three construction battalions; a Fluvial Brigade with the Fluvial Frontier Command at Puerto Ayacucho (Amazonas State) and several posts on border rivers.

Coast Guard

Formed in August 1982. It is part of the Navy. Its Headquarters are at La Guaira (Vargas State). Its primary task is the surveillance of the 200 mile Exclusive Economic Zone and other jurisdictional areas of Venezuelan waters. Coast Guard Squadron includes the frigates GC 11 and GC 12 and several patrol craft.

Naval Aviation

Headquarters are at Puerto Cabello (Carabobo State). Under the command of a Rear Admiral, there are four Squadrons: Training, ASW, Patrol and Transport.

Naval Bases

Caracas: Navy Headquarters and *La Carlota* Naval Aviation Facility.
Vargas State: Division de Infantería HQ and Naval Academy at Mamo; OCHINA (Hydrography) and OCAMAR (Marines Support) HQ and the Coast Guard Command HQ at La Guaira. *Simón Bolívar* International Airport Naval Aviation Facility and Naval Police Training Centre at Maiquetía.

Puerto Cabello (Carabobo State): Fleet Command HQ. Two battalions of 1st Amphibious Brigade at *Contralmirante Agustín Armario* Naval Base and Naval Aviation Command at *General Salom* Airport, Naval Schools and Dockyard.
Punto Fijo (Falcón State): Western Naval Zone HQ, Patrol Ships Squadron and Infantry Battalion 'Francisco de Miranda' at *Mariscal Juán Crisóstomo Falcón* Naval Base.
Carúpano (Sucre State): Eastern Naval Zone HQ. Two battalions of the 2nd Amphibious Brigade and Marines Training Centre.
Ciudad Bolívar (Bolívar): Fluvial Brigade HQ at *Capitán de Fragata Tomás Machado* Naval Base, with several Naval Posts along the Orinoco River.
Turiamo (Aragua State): *Generalísimo Francisco de Miranda* Marines Special Operation Command at the *Capitán de Fragata Tomás Vega* Naval Station.
Puerto Ayacucho (Amazonas State): Fluvial Frontier Command *General de Brigada Franz Rízquez Iribarren* with several Naval Posts along Orinoco, Atabapo, Negro and Meta rivers.
El Amparo (Apure State): Fluvial Frontier Command HQ *Teniente de Navío Jacinto Muñoz*, with several Naval Posts along Arauca and Barinas rivers.
Puerto de Nutrias (Barinas State): River Post.
San José de Macuro (Delta Amacuro State): Atlantic naval post.
La Orchila (Caribbean Sea): Minor Naval Base and Naval Aviation Station.
Puerto Hierro (Sucre State): Minor Naval Base.
Maracaibo (Zulia State), Güiria (Sucre State), Guanta (Anzoátegui State) and Margarita Island (Nueva Esparta State): Main Coast Guard Stations.
Los Monjes (Gulf of Venezuela), Los Testigos, Aves de Sotavento, La Tortuga and La Blanquilla Island (Caribbean Sea): Secondary Coast Guard Stations.

Prefix to Ships' Names

ARV (Armada de la República de Venezuela)

SUBMARINES

2 SÁBALO (209) CLASS (TYPE 1300) (SSK)

Name	No	Builders	Laid down	Launched	Commissioned
SÁBALO	S 31 (ex-S 21)	Howaldtswerke, Kiel	2 May 1973	1 July 1975	6 Aug 1976
CARIBE	S 32 (ex-S 22)	Howaldtswerke, Kiel	1 Aug 1973	6 Nov 1975	11 Mar 1977

Displacement, tons: 1,285 surfaced; 1,600 dived
Dimensions, feet (metres): 200.1 × 20.3 × 18
(61.2 × 6.2 × 5.5)
Main machinery: Diesel-electric; 4 MTU 12V 493 AZ80 GA31L diesels; 2,400 hp(m) *(1.76 MW)* sustained; 4 alternators; 1.7 MW; 1 Siemens motor; 4,600 hp(m) *(3.38 MW)* sustained; 1 shaft
Speed, knots: 10 surfaced; 22 dived
Range, n miles: 7,500 at 10 kt surfaced
Complement: 33 (5 officers)

Torpedoes: 8—21 in *(533 mm)* bow tubes. AEG SST 4; anti-surface; wire-guided; active/passive homing to 12 km *(6.6 n miles)* at 35 kt or 28 km *(15.3 n miles)* at 23 kt; warhead 260 kg. 14 torpedoes carried. Swim-out discharge.
Countermeasures: ESM: Thomson-CSF DR 2000; intercept.
Weapons control: Atlas Elektronik ISUS TFCS.
Radars: Navigation: Terma Scanter Mil; I-band.
Sonars: Atlas Elektronik CSU 3-32; hull-mounted; passive/active search and attack; medium frequency.
Thomson Sintra DUUX 2; passive ranging.

Programmes: Type 209, IK81 designed by Ingenieurkontor Lübeck for construction by Howaldtswerke, Kiel and sale by Ferrostaal, Essen, all acting as a consortium. Both refitted at Kiel in 1981 and 1984 respectively. There are plans for two more of the class.

CARIBE   6/1999 / 0084231

Modernisation: Carried out by HDW at Kiel. *Sábalo* started in April 1990 and left in November 1992 without fully completing the refit. *Caribe* docked in Kiel throughout 1993 but was back in the water in mid-1994, and completed in 1995. The hull is slightly lengthened and new engines, fire control, sonar and attack periscopes fitted. Refit of *Sábalo* began at Dianca Shipyard in October 2003 and is to be followed by *Caribe*. The upgrade is to include new batteries and weapon control systems.

Structure: A single-hull design with two main ballast tanks and forward and after trim tanks. The additional length is due to the new sonar dome similar to German Type 206 system. Fitted with snort and remote machinery control. Slow revving single screw. Very high-capacity batteries with GRP lead-acid cells and battery-cooling. Diving depth 250 m *(820 ft)*.
Operational: Endurance, 50 days patrol.   *UPDATED*

FRIGATES

6 MODIFIED LUPO CLASS (FFGHM)

No	Builders	Laid down	Launched	Commissioned
F 21	Fincantieri, Riva Trigoso	19 Nov 1976	28 Sep 1978	10 May 1980
F 22	Fincantieri, Riva Trigoso	June 1977	22 Feb 1979	7 Mar 1981
F 23	Fincantieri, Riva Trigoso	23 Jan 1978	23 Mar 1979	8 Aug 1981
F 24	Fincantieri, Riva Trigoso	26 Aug 1978	4 Jan 1980	5 Dec 1981
F 25	Fincantieri, Riva Trigoso	7 Nov 1978	13 Jan 1980	3 Apr 1982
F 26	Fincantieri, Riva Trigoso	21 Aug 1979	4 Oct 1980	30 July 1982

Name
MARISCAL SUCRE
ALMIRANTE BRIÓN
GENERAL URDANETA
GENERAL SOUBLETTE
GENERAL SALOM
ALMIRANTE GARCIA (ex-*José Felix Ribas*)

Displacement, tons: 2,208 standard; 2,520 full load
Dimensions, feet (metres): 371.3 × 37.1 × 12.1
(113.2 × 11.3 × 3.7)
Main machinery: CODOG; 2 Fiat/GE LM 2500 gas turbines; 50,000 hp *(37.3 MW)* sustained; 2 GMT A230.20M or 2 MTU 20V 1163 (F 21 and F 22) diesels; 8,000 hp(m) *(5.97 MW)* sustained; 2 shafts; LIPS cp props
Speed, knots: 35; 21 on diesels. **Range, n miles:** 5,000 at 15 kt
Complement: 185

Missiles: SSM: 8 Otomat Teseo Mk 2 TG1 ❶; active radar homing to 80 km *(43.2 n miles)* at 0.9 Mach; warhead 210 kg; sea-skimmer for last 4 km *(2.2 n miles)*.
SAM: Selenia Elsag Albatros octuple launcher ❷; 8 Aspide; semi-active radar homing to 13 km *(7 n miles)* at 2.5 Mach; height envelope 15-5,000 m *(49.2-16,405 ft)*; warhead 30 kg.
Guns: 1 OTO Melara 5 in *(127 mm)*/54 ❸; 45 rds/min to 16 km *(8.7 n miles)*; weight of shell 32 kg.
4 Otobreda 40 mm/70 (2 twin) ❹; 300 rds/min to 12.5 km *(6.8 n miles)*; weight of shell 0.96 kg.
Torpedoes: 6—324 mm ILAS 3 (2 triple) tubes ❺. Whitehead A244S; anti-submarine; active/passive homing to 7 km *(3.8 n miles)* at 33 kt; warhead 34 kg (shaped charge).
Countermeasures: Decoys: 2 Breda 105 mm SCLAR 20-barrelled trainable ❻; chaff to 5 km *(2.7 n miles)*; illuminants to 12 km *(6.6 n miles)*. Can be used for HE bombardment.
ESM: Elisra NS 9003/9005; intercept.
Combat data systems: Selenia IPN 10. Elbit ENTCS 2000 (F 21 and F 22)
Weapons control: 2 Elsag NA 10 MFCS. 2 Dardo GFCS for 40 mm.
Radars: Air search: Selenia RAN 10S or Elta 2238 (F 21 and 22) ❼; E/F-band.
Air/surface search: Selenia RAN 11X; I-band.
Fire control: 2 Selenia Orion 10XP ❽; I/J-band.
2 Selenia RTN 20X ❾; I/J-band.
Navigation: SMA 3RM20; I-band.
Tacan: SRN 15A.
Sonars: EDO SQS-29 (Mod 610E) or Northrop Grumman 21 HS-7 (F 21 and F 22); hull-mounted; active search and attack; medium frequency.

Helicopters: 1 AB 212ASW ❿.

Programmes: All ordered on 24 October 1975. Similar to ships in the Italian and Peruvian navies.
Modernisation: F 21 and F 22 were scheduled to start a refit by Ingalls Shipyard in September 1992 but contractual problems delayed start until January 1998. Refits included upgrading the gas turbines, replacing the diesels, improving the combat data system, updating sonar and ESM, and overhauling all weapon systems. The ships were redelivered in mid-2002. Neither is believed to be fully operational. F 23 and F 24 have been upgraded by Dianca, Puerto Caballo, and returned to service in December and October 2003 respectively. Work included modernisation of the main machinery, air-conditioning and weapon systems. F 25 and F 26 may be similarly refitted.
Structure: Fixed hangar means no space for Aspide reloads. Fully stabilised.   *UPDATED*

ALMIRANTE BRIÓN
6/2001, Northrop Grumman Ingalls / 0096360

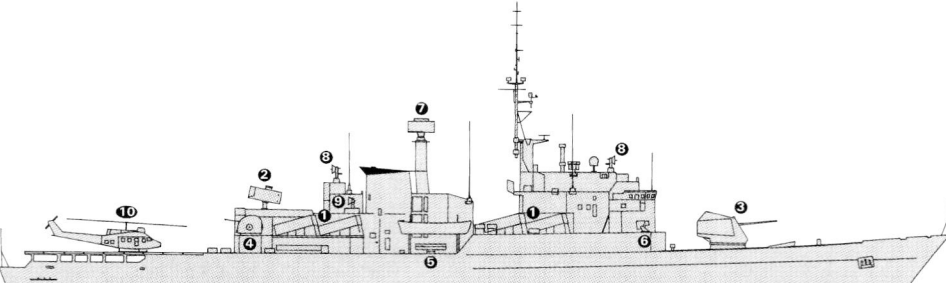
ALMIRANTE BRIÓN   *(Scale 1 : 900), Ian Sturton / 0529541*

GENERAL URDANETA   7/1996 / 0050730

SHIPBORNE AIRCRAFT

Notes: There are six operational Bell 412EP helicopters. Four acquired in 1999 and three more delivered in 2003 of which one has been lost. This was replaced in 2004. There are also three Bell 206B of which one is used for training.

Numbers/Type: 7 Agusta AB 212ASW.
Operational speed: 106 kt *(196 km/h).*
Service ceiling: 14,200 ft *(4,330 m).*
Range: 230 n miles *(426 km).*
Role/Weapon systems: ASW helicopter with secondary ASV role. Sensors: APS-705 search radar, Bendix AQS-18A dipping sonar. Weapons: ASW; two Mk 46 or A244/S torpedoes or depth bombs. ASV; mid-course guidance to Teseo Mk 2 missiles.

UPDATED

LAND-BASED MARITIME AIRCRAFT (FRONT LINE)

Notes: There are also two Beech King Air and three Cessnas used for training and transport.

Numbers/Type: 3/2/3 CASA C-212 S 43/S 200/S 400 Aviocar.
Operational speed: 190 kt *(353 km/h).*
Service ceiling: 24,000 ft *(7,315 m).*
Range: 1,650 n miles *(3,055 km).*
Role/Weapon systems: Medium-range MR and coastal protection aircraft; limited armed action. Acquired in 1981-82 and 1985-86. Three modernised and augmented in 1998 by S 400 type. Previous numbers have reduced. Sensors: APS-128 radar. Weapons: ASW; depth bombs. ASV; gun and rocket pods.

VERIFIED

AB 212ASW   *6/1999, Venezuelan Navy /* 0084233

C-212   *6/2002, CASA/EADS /* 0529548

PATROL FORCES

Notes: Replacement of the Constitución class began in February 1999 whan an Invitation to Tender was issued for six 30 kt craft. One was ordered from Bazán in September 1999 but funding withdrawn at last moment. A Memorandum of Understanding for the procurement of four patrol boats and four corvettes was signed with the Spanish government in April 2005.

6 CONSTITUCIÓN CLASS (FAST ATTACK CRAFT—MISSILE AND GUN) (PBG/PG)

Name	No	Builders	Laid down	Launched	Commissioned
CONSTITUCIÓN	PC 11	Vosper Thornycroft	Jan 1973	1 June 1973	16 Aug 1974
FEDERACIÓN	PC 12	Vosper Thornycroft	Aug 1973	26 Feb 1974	25 Mar 1975
INDEPENDENCIA	PC 13	Vosper Thornycroft	Feb 1973	24 July 1973	20 Sep 1974
LIBERTAD	PC 14	Vosper Thornycroft	Sep 1973	5 Mar 1974	12 June 1975
PATRIA	PC 15	Vosper Thornycroft	Mar 1973	27 Sep 1973	9 Jan 1975
VICTORIA	PC 16	Vosper Thornycroft	Mar 1974	3 Sep 1974	22 Sep 1975

Displacement, tons: 170 full load
Dimensions, feet (metres): 121 × 23.3 × 6 *(36.9 × 7.1 × 1.8)*
Main machinery: 2 MTU MD 16V 538 TB90 diesels; 6,000 hp (m) *(4.4 MW)* sustained; 2 shafts
Speed, knots: 31. **Range, n miles:** 1,350 at 16 kt
Complement: 20 (4 officers)

Missiles: SSM: 2 OTO Melara/Matra Teseo Mk 2 TG1 *(Federación, Libertad* and *Victoria);* active radar homing to 80 km *(43.2 n miles)* at 0.9 Mach; sea-skimmer for last 4 km *(2.2 n miles);* warhead 210 kg.

Guns: 1 OTO Melara 3 in *(76 mm)*/62 compact *(Constitución, Independencia* and *Patria);* 85 rds/min to 16 km *(8.7 n miles);* weight of shell 6 kg.
1 Breda 30 mm/70 *(Federación, Libertad* and *Victoria);* 800 rds/min; weight of shell 0.37 kg.
2—12.7 mm MGs.
Weapons control: Elsag NA 10 Mod 1 GFCS *(Constitución, Independencia* and *Patria).* Alenia Elsag Medusa optronic director *(Federación, Libertad* and *Victoria).*
Radars: Surface search: SMA SPQ-2D; I-band.
Fire control: Selenia RTN 10X (in 76 mm ships); I/J-band.
Navigation: Racal; I-band.

Programmes: Transferred from the Navy in 1983 to the Coast Guard but now back again with Fleet Command.
Modernisation: Single Breda 30 mm guns replaced the 40 mm guns in the missile craft in 1989. All were refitted at Puerto Cabello 1992-1995.
Operational: Plans to replace these ships are uncertain. Meanwhile it is reported that their propulsion systems have been refitted.

UPDATED

VICTORIA (missile craft)   *7/1999, Venezuelan Navy /* 0084235

AMPHIBIOUS FORCES

4 CAPANA (ALLIGATOR) CLASS (LSTH)

Name	No	Builders	Commissioned
CAPANA	T 61	Korea Tacoma Marine	24 July 1984
ESEQUIBO	T 62	Korea Tacoma Marine	24 July 1984
GOAJIRA	T 63	Korea Tacoma Marine	20 Nov 1984
LOS LLANOS	T 64	Korea Tacoma Marine	20 Nov 1984

Displacement, tons: 4,070 full load
Dimensions, feet (metres): 343.8 × 50.5 × 9.8 *(104.8 × 15.4 × 3)*
Main machinery: 2 SEMT-Pielstick 16 PA6 V 280 diesels; 12,800 hp(m) *(9.41 MW)*; 2 shafts
Speed, knots: 14. **Range, n miles:** 5,600 at 11 kt
Complement: 117 (13 officers)
Military lift: 202 troops; 1,600 tons cargo; 4 LCVPs
Guns: 2 Breda 40 mm/70 (twin). 2 Oerlikon 20 mm GAM-BO1.
Weapons control: Selenia NA 18/V; optronic director.
Helicopters: Platform only.

Comment: Ordered in August 1982. Version III of Korea Tacoma Alligator type. Each has a 50 ton tank turntable and a lift between decks. *Goajira* was out of service from June 1987 to May 1993 after a serious fire. T 62 and T 63 reported to have been refitted in 2003.

UPDATED

CAPANA   *6/1998, Venezuelan Navy* / 0084236

ESEQUIBO   *3/1999* / 0084237

2 AJEERA CLASS (LCU)

Name	No	Builders	Commissioned
MARGARITA	T 71	Swiftships Inc, Morgan City	20 Jan 1984
LA ORCHILA	T 72	Swiftships Inc, Morgan City	11 May 1984

Displacement, tons: 428 full load
Dimensions, feet (metres): 129.9 × 36.1 × 5.9 *(39.6 × 11 × 1.8)*
Main machinery: 2 Detroit 16V-149 diesels; 1,800 hp *(1.34 MW)* sustained; 2 shafts
Speed, knots: 13. **Range, n miles:** 1,500 at 10 kt
Complement: 26 (4 officers)
Military lift: 150 tons cargo; 100 tons fuel
Guns: 3—12.7 mm MGs.
Radars: Navigation: Raytheon 6410; I-band.

Comment: Both serve in Fluvial Command. Have a 15 ton crane.

VERIFIED

MARGARITA   *6/1999, Venezuelan Navy* / 0084238

AUXILIARIES

Notes: (1) There is one navigational aids tender *Macuro* BB-11.
(2) Construction of two new ocean tugs started at Dianca in 2003. The first is to be called *Almirante Manuel Ezequiel Bruzual* (RA 11). It is expected to replace RA 33 in 2005.

1 LOGISTIC SUPPORT SHIP (AORH)

Name	No	Builders	Commissioned
CIUDAD BOLÍVAR	T 81	Hyundai, Ulsan	2001

Displacement, tons: 9,750 full load
Dimensions, feet (metres): 451.8 × 59 × 21.7 *(137.7 × 18 × 6.6)*
Main machinery: 2 Caterpillar 3616 diesels; 2 shafts; LIPS cp props
Speed, knots: 18. **Range, n miles:** 4,500 at 15 kt
Complement: 104
Guns: 2 Bofors 40 mm/70. 2—12.7 mm MGs.

Comment: Ordered from Hyundai, South Korea, in February 1999. Delivered in October 2001. Capable of carrying 4,400 tons of fuel and 900 tons of cargo. Two replenishment stations on each beam. Hangar and deck for medium size helicopter. Replenishment operations reported conducted with both French and Netherlands units.

UPDATED

CIUDAD BOLÍVAR   *10/2002, Mario R V Carveiro* / 0534069

SURVEY SHIPS

1 CHEROKEE CLASS (AGS)

Name	No	Builders	Commissioned
MIGUEL RODRIGUEZ	RA 33	Charleston SB and DD Co	9 Nov 1945
(ex-*Salinan* ATF 161)	(ex-R 23)		

Displacement, tons: 1,235 standard; 1,675 full load
Dimensions, feet (metres): 205 × 38.5 × 17 *(62.5 × 11.7 × 5.2)*
Main machinery: Diesel-electric; 4 GM 16-278A diesels; 4,400 hp *(3.28 MW)*; 4 generators; 1 motor; 3,000 hp *(2.24 MW)*; 1 shaft
Speed, knots: 15. **Range, n miles:** 7,000 at 15 kt.
Complement: 85
Guns: 2—12.7 mm MGs.
Radars: Navigation: Sperry SPS-53; I/J-band.

Comment: Acquired from US on 1 September 1978. Last of a class of three. Assigned to OCHINA (Hydrographic department). Likely to be decommissioned in 2005.   *UPDATED*

MIGUEL RODRIGUEZ   *4/2001* / 0114819

1 SURVEY AND RESEARCH SHIP (AGOR)

Name	No	Builders	Launched	Commissioned
PUNTA BRAVA	BO 11	Bazán, La Carraca	9 Mar 1990	14 Mar 1991

Displacement, tons: 1,170 full load
Dimensions, feet (metres): 202.4 × 39 × 12.1 *(61.7 × 11.9 × 3.7)*
Main machinery: 2 Bazán-MAN 7L20/27 diesels; 2,500 hp(m) *(1.84 MW)*; 2 shafts; bow thruster
Speed, knots: 13. **Range, n miles:** 8,000 at 13 kt
Complement: 49 (6 officers) plus 6 scientists
Radars: Navigation: ARPA; I-band.

Comment: Ordered in September 1988. Developed from the Spanish Malaspina class. A multipurpose ship for oceanography, marine resource evaluation, geophysical and biological research. Equipped with Qubit hydrographic system. Carries two survey launches. EW equipment is fitted. Assigned to the OCHINA (Hydrographic department).

VERIFIED

PUNTA BRAVA   *4/2001* / 0114821

2 SURVEY CRAFT (AGSC)

Name	No	Builders	Commissioned
GABRIELA (ex-*Peninsula de Araya*)	LH 11	Abeking & Rasmussen	5 Feb 1974
LELY (ex-*Peninsula de Paraguana*)	LH 12	Abeking & Rasmussen	7 Feb 1974

Displacement, tons: 90 full load
Dimensions, feet (metres): 88.6 × 18.4 × 4.9 *(27 × 5.6 × 1.5)*
Main machinery: 2 MTU diesels; 2,300 hp(m) *(1.69 MW)*; 2 shafts
Speed, knots: 20
Complement: 9 (1 officer)

Comment: LH 12 laid down 28 May 1973, launched 12 December 1973 and LH 11 laid down 10 March 1973, launched 29 November 1973. Acquired in September 1986 from the Instituto de Canalizaciones. LH 11 is assigned to the Fluvial Command, LH 12 to the Coast Guard.
VERIFIED

GABRIELA (alongside *Alcatraz* PG 32)   *1/1994, Maritime Photographic*

TRAINING SHIPS

1 SAIL TRAINING SHIP (AXS)

Name	No	Builders	Launched	Commissioned
SIMÓN BOLÍVAR	BE 11	AT Celaya, Bilbao	21 Nov 1979	6 Aug 1980

Displacement, tons: 1,260 full load
Measurement, tons: 934 gross
Dimensions, feet (metres): 270.6 × 34.8 × 14.4 *(82.5 × 10.6 × 4.4)*
Main machinery: 1 Detroit 12V-149T diesel; 875 hp *(652 kW)* sustained; 1 shaft
Speed, knots: 10
Complement: 93 (17 officers) plus 102 trainees

Comment: Ordered in 1978. Three-masted barque; similar to *Guayas* (Ecuador), *Cuauhtemoc* (Mexico) and *Gloria* (Colombia). Sail area (23 sails), 1,650 m². Highest mast, 131.2 ft *(40 m)*. Has won several international sail competitions including Cutty Sark '96. A refit is to be completed in 2005.
UPDATED

SIMÓN BOLÍVAR   *6/2001, A Campanera i Rovira /* 0534070

COAST GUARD

Notes: Procurement of three new classes of patrol craft is in progress. Three 36 m and 15 23 m patrol craft are to be built in Venezuela while four patrol craft are under construction in the US.

7 RIVER PATROL CRAFT (PBR)

MANAURE PF 21	TAMANACO PF 24	SOROCAIMA PF 34
MARA PF 22	TEREPAIMA PF 31	
GUAICAIPURO PF 23	YARACUY PF 33	

Displacement, tons: 15 full load
Dimensions, feet (metres): 54.1 × 14.1 × 4.3 *(16.5 × 4.3 × 1.3)*
Main machinery: 2 diesels; 2 shafts
Speed, knots: 10
Complement: 8
Guns: 1—12.7 mm MG.
Radars: Surface search: Raytheon 6410; I-band.

Comment: River craft used by the Marines. Details given are for four Manaure class. There are also three Terepaima class which are 10 m long and capable of 45 kt. *VERIFIED*

MANAURE   *6/1998, Venezuelan Navy /* 0050737

4 PETREL (POINT) CLASS (WPB)

Name	No	Builders	Commissioned
PETREL (ex-*Point Knoll*)	PG 31	US Coast Guard Yard, Curtis Bay	26 June 1967
ALCATRAZ (ex-*Point Judith*)	PG 32	US Coast Guard Yard, Curtis Bay	26 July 1966
ALBATROS (ex-*Point Franklin*)	PG 33	US Coast Guard Yard, Curtis Bay	14 Nov 1966
PELÍCANO (ex-*Point Ledge*)	PG 34	US Coast Guard Yard, Curtis Bay	18 July 1962

Displacement, tons: 68 full load
Dimensions, feet (metres): 83 × 17.2 × 5.8 *(25.3 × 5.2 × 1.8)*
Main machinery: 2 Caterpillar diesels; 1,600 hp *(1.19 MW)*; 2 shafts
Speed, knots: 23.5. **Range, n miles:** 1,500 at 8 kt
Complement: 10 (1 officer)
Guns: 2—12.7 mm MGs.
Radars: Surface search: Raytheon SPS-64; I-band.

Comment: *Petrel* transferred from USCG on 18 November 1991 and *Alcatraz* on 15 January 1992, *Albatros* on 23 June 1998 and *Pelícano* on 3 August 1998. The transfer of four further craft is unlikely. Most of the class are believed to be operational.
UPDATED

ALCATRAZ   *4/1999 /* 0084242

12 GAVION CLASS (WPB)

GAVION	PG 401	CORMORAN	PG 405	NEGRON	PG 409
ALCA	PG 402	COLIMBO	PG 406	PIGARGO	PG 410
BERNACLA	PG 403	FARDELA	PG 407	PAGAZA	PG 411
CHAMAN	PG 404	FUMAREL	PG 408	SERRETA	PG 412

Displacement, tons: 45 full load
Dimensions, feet (metres): 80 × 17 × 4.8 *(24.4 × 5.2 × 1.5)*
Main machinery: 2 Detroit 12V-92TA diesels; 2,160 hp *(1.61 MW)* sustained; 2 shafts
Speed, knots: 25. **Range, n miles:** 1,000 at 12 kt
Complement: 10
Guns: 2—12.7 mm MGs. 2—7.62 mm MGs. 1—40 mm Mk 19 grenade launcher.
Radars: Surface search: Raytheon R1210; I-band.

Comment: Ordered from Halter Marine 24 April 1998 and delivered from late 1999 to early 2000. Aluminium construction. Four craft refitted in 2003 and all believed to be operational.
UPDATED

ALCA   *10/1999, Halter Marine /* 0084243

2 ALMIRANTE CLEMENTE CLASS (WFS)

Name	No	Builders	Laid down	Launched	Commissioned
ALMIRANTE CLEMENTE	GC 11	Ansaldo, Livorno	5 May 1954	12 Dec 1954	1956
GENERAL JOSÉ TRINIDAD MORAN	GC 12	Ansaldo, Livorno	5 May 1954	12 Dec 1954	1956

Displacement, tons: 1,300 standard; 1,500 full load
Dimensions, feet (metres): 325.1 × 35.5 × 12.2
(99.1 × 10.8 × 3.7)
Main machinery: 2 GMT 16-645E7C diesels; 6,080 hp(m)
(4.47 MW) sustained; 2 shafts
Speed, knots: 22. **Range, n miles:** 3,500 at 15 kt
Complement: 142 (12 officers)

Guns: 2 Otobreda 3 in *(76 mm)*/62 compact ❶; 85 rds/min to
16 km *(8.7 n miles)*; weight of shell 6 kg.
2 Breda 40 mm/70 (twin) ❷; 300 rds/min to 12.5 km *(6.8 n
miles)*; weight of shell 0.96 kg.
Torpedoes: 6—324 mm ILAS 3 (2 triple) tubes ❸. Whitehead A
244S; anti-submarine; active/passive homing to 7 km *(3.8 n
miles)* at 33 kt; warhead 34 kg (shaped charge).
Depth charges: 2 throwers.
Weapons control: Elsag NA 10 Mod 1 GFCS.
Radars: Air search: Plessey AWS 4 ❹; E/F-band.
Surface search: Racal Decca 1226 ❺; I-band.
Fire control: Selenia RTN 10X ❻; I/J-band.
Sonars: Plessey PMS 26; hull-mounted; active search and attack;
10 kHz.

Programmes: Survivors of a class of six ordered in 1953.
Modernisation: Both ships were refitted by Cammell Laird/
Plessey group in April 1968. 4 in guns replaced by 76 mm.
Both refitted again in Italy in 1984-85, prior to transfer to
Coast Guard duties in 1986.
Structure: Fitted with Denny-Brown fin stabilisers and air
conditioned throughout the living and command spaces.
Operational: Both reported operational.

UPDATED

ALMIRANTE CLEMENTE   *(Scale 1 : 900), Ian Sturton*

GENERAL JOSÉ TRINIDAD MORAN   *10/1998, E & M Laursen* / 0050734

ALMIRANTE CLEMENTE   *4/2001* / 0114820

2 UTILITY CRAFT (YAG)

LOS TAQUES LG 11   **LOS CAYOS** LG 12

Displacement, tons: 350 full load
Dimensions, feet (metres): 87.3 × 23.3 × 4.9 *(26.6 × 7.1 × 1.5)*
Main machinery: 1 diesel; 850 hp(m) *(625 kW)*; 1 shaft
Speed, knots: 8
Complement: 10
Guns: 1—12.7 mm MG.

Comment: Former trawlers. Commissioned 15 May 1981 and 17 July 1984 respectively. Used for
salvage and SAR tasks.

VERIFIED

18 INSHORE PATROL BOATS (PBR)

CONSTANCIA LRG 001	**HONESTIDAD** LRG 003	**INTEGRIDAD** LRG 005	+12
PERSEVERANCIA LRG 002	**TENACIDAD** LRG 004	**LEALTAD** LRG 006	

Displacement, tons: 11 full load
Dimensions, feet (metres): 39.4 × 9.2 × 5.6 *(12 × 2.8 × 1.7)*
Main machinery: 2 diesels; 640 hp(m) *(470 kW)*; 2 shafts
Speed, knots: 38
Complement: 4
Guns: 2—7.62 mm MGs.
Radars: Surface search: I-band.

Comment: First three speed boat type with GRP hulls delivered from a local shipyard in December
1991. Fourth completed in August 1993. Details given are for *Integridad* which is the first of two
built at Guatire, and delivered in 1997/98. GRP construction. The twelve un-named craft are
Boston Whaler Guardian class capable of 25 kt, mounting 2—12.7 mm and 2—6.72 mm MGs,
and with Raytheon radars. These were donated by the US. All of these craft are used by Marines.

VERIFIED

LOS CAYOS   *3/1999, Venezuelan Navy* / 0084244

GUARDIAN   *4/1999, Venezuelan Navy* / 0084245

8 PUNTA MACOLLA CLASS (PB)

PUNTA MACOLLA LSM 001	BAJO BRITO LSM 004	VELA DE COBO LSM 007
FARALLÓN CENTINELA LSM 002	BAJO ARAYA LSM 005	CAYO MACEREO LSM 008
CHARAGATO LSM 003	CARECARE LSM 006	

Displacement, tons: 5 full load
Dimensions, feet (metres): 41.7 × 9.2 × 6.6 *(12.7 × 2.8 × 2)*
Main machinery: 2 diesels; 2 shafts
Speed, knots: 30
Complement: 4
Guns: 1—12.7 mm MG.
Radars: Surface search: Raytheon; I-band.

Comment: Built in Venezuela by Intermarine. Used by OCHINA (Hydrographic department) and for SAR. First six delivered by 1997 and last two in 2000.
VERIFIED

BAJO ARAYA *3/1999, Venezuelan Navy /* 0084246

7 POLARIS CLASS (PBF)

POLARIS LG 21	ALDEBARAN LG 24	CANOPUS LG 26
SIRIUS LG 22	ANTARES LG 25	ALTAIR LG 27
RIGEL LG 23		

Displacement, tons: 5 full load
Dimensions, feet (metres): 26 wl × 8.5 × 2.6 *(7.9 × 2.6 × 0.8)*
Main machinery: 1 diesel outdrive; 400 hp(m) *(294 kW)*
Speed, knots: 50
Complement: 4
Guns: 1—12.7 mm MG.
Radars: Surface search: Raytheon; I-band.

Comment: Used by the Coast Guard for drug interdiction. Two more reported operational.
UPDATED

ALDEBARAN *4/1999, Venezuelan Navy /* 0084247

1 SUPPORT SHIP (AKSL)

Name	No	Builders	Commissioned
FERNANDO GOMEZ (ex-*José Felix Ribas*, ex-*Oswegatchie*)	RP 21 (ex-*R 13*)	Commercial Iron Works, Portland	14 Dec 1945

Displacement, tons: 245 full load
Dimensions, feet (metres): 100.1 × 25.9 × 9.5 *(30.5 × 7.9 × 2.9)*
Main machinery: 2 diesels; 1,270 hp *(947 kW)*; 1 shaft
Speed, knots: 10
Complement: 12
Guns: 2—12.7 mm MGs.
Radars: Navigation: Raytheon; I-band.

Comment: Former tug, originally acquired from the US in 1965. Out of service for some years but now employed as a logistic support ship and for occasional patrol and SAR.
VERIFIED

FERNANDO GOMEZ *6/1998, Venezuelan Navy /* 0050738

2 RIVER TRANSPORT CRAFT (LCM)

CURIAPO LC 21	YOPITO LC 01

Displacement, tons: 115 full load
Dimensions, feet (metres): 73.7 × 21 × 5.2 *(22.5 × 6.4 × 1.6)*
Main machinery: 2 Detroit diesels; 850 hp *(625 kW)*; 2 shafts
Speed, knots: 9
Complement: 5
Cargo capacity: 60 tons or 200 Marines
Guns: 2—12.7 mm MGs.

Comment: Details given are for *Curiapo* which is a former LCM. *Yopito* is a former LCU of 18 m. Both are used by the Marines. There are also 12 11 m LCVPs.
VERIFIED

YOPITO *7/1999, Venezuelan Navy /* 0084248

NATIONAL GUARD (GUARDIA NACIONAL)

Notes: (1) There are also a large number of US and Canadian built river craft of between 6 and 9 m length, which are armed with MGs.
(2) Four intercept launches were delivered in 2003; two in July 2003 and two in October 2003.
(3) Some 60 Pirana class river patrol craft have been ordered. The first 15 were delivered in August 2003.

10 RIO ORINOCO II CLASS (PBF)

B 9801 series

Displacement, tons: 30 full load
Dimensions, feet (metres): 54 × 14 × 4.6 *(16.5 × 4.3 × 1.4)*
Main machinery: 2 MTU 12V 183 TE93 diesels; 2,268 hp(m) *(1.67 MW)* sustained; 2 shafts
Speed, knots: 36. **Range, n miles:** 500 at 25 kt
Complement: 5
Guns: 2—12.7 mm MGs. 2—7.62 mm MGs.
Radars: Surface search: Raytheon R1210; I-band.

Comment: Ordered from Halter Marine 24 April 1998. All delivered by late 1999. Aluminium construction. Some of the similar sized Orinoco I craft built in the 1970s are still in limited use.
UPDATED

ORINOCO II *1/1999, Halter Marine /* 0050739

12 PUNTA CLASS (PB) and 12 PROTECTOR CLASS (PB)

Displacement, tons: 15 full load
Dimensions, feet (metres): 43 × 13.4 × 3.9 *(13.1 × 4.1 × 1.2)*
Main machinery: 2 MTU Series 183 diesels; 1,500 hp *(1.1 MW)*; 2 shafts
Speed, knots: 34. **Range, n miles:** 390 at 25 kt
Complement: 4
Guns: 2—12.7 mm MGs.
Radars: Navigation: Raytheon; I-band.

Comment: Details given for the Punta class. Ordered 24 January 1984. Built by Bertram Yacht, Miami, Florida. Names begin with *Punta*. Aluminium hulls. Completed from July-December 1984. Re-engined in 1996-97. The Protector class are the same size, but are slower at 28 kt. They were built by SeaArk Marine and completed in 1984. Some are probably non-operational. Names begin with *Rio*.
VERIFIED

PROTECTOR *2/1996, van Ginderen Collection /* 0084249

Vietnam

Country Overview

The Socialist Republic of Vietnam was established in 1976 when the Democratic Republic of Vietnam in the north and the Republic of Vietnam in the south became one nation. The country had been divided at the 17th parallel from the end of French colonial rule in 1954 and during the ensuing Vietnam War. Located on the east coast of the Indochina peninsula, it has an area of 127,844 square miles and is bordered to the north by China and to the west by Cambodia and Laos. It has a 1,858 n mile coastline with the South China Sea. Hanoi is the capital while Ho Chi Minh City (formerly Saigon) is the largest city and a major port. There are further ports at Haiphong and Da Nang. Territorial seas (12 n miles) are claimed. An EEZ (200 n miles) has also been claimed but the limits have not been defined.

Headquarters Appointments

Chief of Naval Forces:
  Vice Admiral Mai Xuan Binh
Deputy Chief of Naval Forces:
  Captain Tran Quang Khue

Personnel

(a) 2005: 9,000 regulars
(b) Additional conscripts on three to four year term (about 3,000)
(c) 27,000 naval infantry

Organisation and Bases

The fleet is organised into four regions based on, from north to south, Haiphong (HQ), Da Nang, Nha Trang and Cân Tho. There are other bases at Cam Ranh Bay, Hue and Ha Tou.

Coast Guard

A Coast Guard was formed on 1 September 1998. It is subordinate to the Navy and may take on Customs duties.

SUBMARINES

2 YUGO CLASS
(MIDGET SUBMARINES) (SSW)

Displacement, tons: 90 surfaced; 110 dived
Dimensions, feet (metres): 65.6 × 10.2 × 15.1 *(20 × 3.1 × 4.6)*
Main machinery: 2 diesels; 320 hp(m) *(236 kW)*; 1 shaft
Speed, knots: 12 surfaced; 8 dived
Range, n miles: 550 at 10 kt surfaced; 50 at 4 kt dived
Complement: 4 plus 6/7 divers

Comment: Transferred from North Korea in 1997. May be fitted with two short torpedo tubes and a snort mast, but used primarily for diver related operations. The conning tower acts as a wet/dry diver compartment. Operational status is doubtful.

VERIFIED

YUGO (North Korean colours)
6/1998, Ships of the World / 0052525

FRIGATES

Notes: Ex-Russian light frigates of the Parchim class may be acquired.

5 PETYA (PROJECT 159A) CLASS (FFL)

HQ 09,11 (Type III)   HQ 13, 15, 17 (Type II)

Displacement, tons: 950 standard; 1,180 full load
Dimensions, feet (metres): 268.3 × 29.9 × 9.5 *(81.8 × 9.1 × 2.9)*

Main machinery: CODAG; 2 gas turbines; 30,000 hp(m) *(22 MW)*; 1 Type 61V-3 diesel; 5,400 hp(m) *(3.97 MW)* sustained; centre shaft; 3 shafts
Speed, knots: 32. **Range, n miles:** 4,870 at 10 kt; 450 at 29 kt
Complement: 98 (8 officers)

Guns: 4 USSR 3 in *(76 mm)*/60 (2 twin); 90 rds/min to 15 km *(8 n miles)*; weight of shell 6.8 kg.
  4—37 mm (2 twin) (HQ 11). 4—23 mm (2 twin) (HQ 11).
Torpedoes: 3—21 in *(533 mm)* (triple) tubes (Petya III). SAET-60; passive homing to 15 km *(8.1 n miles)* at 40 kt; warhead 400 kg.
  5—16 in *(406 mm)* (1 quin) tubes (Petya II). SAET-40; active/passive homing to 10 km *(5.5 n miles)* at 30 kt; warhead 100 kg.
A/S mortars: 4 RBU 6000 12-tubed trainable (Petya II); range 6,000 m; warhead 31 kg.
  4 RBU 2500 16-tubed trainable (Petya III); range 2,500 m; warhead 21 kg.
Depth charges: 2 racks.
Mines: Can carry 22.
Countermeasures: ESM: 2 Watch Dog; radar warning.
Radars: Air/surface search: Strut Curve; F-band.
  Navigation: Don 2; I-band.
  Fire control: Hawk Screech; I-band.
  IFF: High Pole B. 2 Square Head.
Sonars: Vychada MG 311; hull-mounted; active attack; high frequency.

Programmes: Two Petya III (export version) transferred from USSR in December 1978 and three Petya IIs, two in December 1983 and one in December 1984. Petya II HQ 13 was reported decommissioned in 1996 but may have been refitted.
Modernisation: Refitted and updated 1994 to 1999. The RBUs replaced by 25 mm guns and the torpedo tubes by 37 mm guns in some of the class. *HQ 17* completed major overhaul at Ba Son Shipyard in 2001.
Structure: The Petya IIIs have the same hulls as the Petya IIs but different armament.
Operational: Reported active between the coast and the Spratly Islands.

UPDATED

HQ 09

9/1995, G Toremans

HQ 17 (PETYA II)
11/2001 / 0131341

1 BARNEGAT CLASS (FF)

Name	No	Builders	Commissioned
PHAM NGU LAO (ex-Absecon)	HQ 01 (ex-WHEC 374)	Lake Washington SY	28 Jan 1943

Displacement, tons: 1,766 standard; 2,800 full load
Dimensions, feet (metres): 310.8 × 41.1 × 13.5 *(94.7 × 12.5 × 4.1)*
Main machinery: 2 Fairbanks-Morse 38D8-1/8-10 diesels; 3,540 hp *(2.64 MW)* sustained; 2 shafts
Speed, knots: 18. **Range, n miles:** 20,000 at 12 kt
Complement: 200 approximately

Guns: 1 USN 5 in *(127 mm)*/38; 15 rds/min to 17 km *(9.3 n miles)*; weight of shell 25 kg.
3—37 mm/63. 4—25 mm (2 twin). 2—81 mm mortars.
Radars: Surface search: Raytheon SPS-28; A-band.
Fire control: RCA/GE Mk 26; I/J-band.

Programmes: Last of a group built as seaplane tenders for the US Navy. Transferred to US Coast Guard in 1948, initially on loan designated WAVP and then on permanent transfer, subsequently redesignated as high endurance cutter (WHEC). Transferred from US Coast Guard to South Vietnamese Navy in 1971.
Modernisation: Styx SSMs mounted aft and close-range armament fitted in the mid-1980s. The SSMs have probably been removed.
Operational: Still seaworthy but almost certainly non-operational.   *VERIFIED*

BARNEGAT (Italian colours)   *1991, van Ginderen Collection*

CORVETTES

Notes: (1) Ex-US Admirable class *HQ 07* is an alongside training hulk.
(2) The KBO 2000 project appears to have been abandoned.

2 BPS 500 (PROJECT 12418) CLASS (FSGM)

HQ 381   +1

Displacement, tons: 517 full load
Dimensions, feet (metres): 203.4 × 36.1 × 8.2 *(62 × 11 × 2.5)*
Main machinery: 2 MTU diesels; 19,600 hp(m) *(14.41 MW)*; 2 Kamewa waterjets
Speed, knots: 32. **Range, n miles:** 2,200 at 14 kt
Complement: 28

Missiles: SSM: 8 Zvezda SS-N-25 (KH-35 Uran) (2 quad) ❶; active radar homing to 130 km *(70.1 n miles)* at 0.9 Mach; warhead 145 kg.
SAM: SA-N-10. 24 missiles.
Guns: 1—3 in *(76 mm)*/60 ❷; 120 rds/min to 15 km *(8 n miles)*; weight of shell 7 kg.
1—30 mm/65 AK 630 ❸; 6 barrelled; 3,000 rds/min combined to 2 km.
2—12.7 mm MGs.
Mines: Rails fitted.
Countermeasures: Decoys: 2 chaff launchers ❹.
Weapons control: Optronic director ❺.
Radars: Air/surface search: Cross Dome ❻; E/F-band.
Navigation: I-band.
Fire control: Bass Tilt ❼; H/I-band.

Comment: Severnoye design (improved Pauk) ordered in 1996 and two ships subsequently delivered in kit form to Ba Son Shipyard, Ho Chi Minh City. First unit launched in June 1998 and became operational in late 2001. Missile systems yet to be fitted. The second unit may have been built but this has not been confirmed.   *VERIFIED*

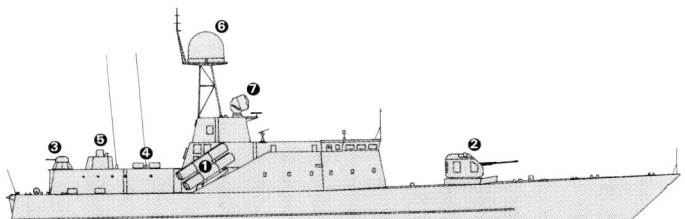

BPS 500   *(not to scale), Ian Sturton /* 0530054

HQ 381   *11/2001 /* 0131340

4 + 10 TARANTUL CLASS (PROJECT 1241) (FSGM)

HQ 371	HQ 372	HQ 373	HQ 374

Displacement, tons: 385 standard; 450 full load
Dimensions, feet (metres): 184.1 × 37.7 × 8.2 *(56.1 × 11.5 × 2.5)*
Main machinery: 2 Nikolayev Type DR 77 gas turbines; 16,016 hp(m) *(11.77 MW)* sustained; 2 Nikolayev Type DR 76 gas turbines with reversible gearboxes; 4,993 hp(m) *(3.67 MW)* sustained; 2 shafts
Speed, knots: 36. **Range, n miles:** 2,000 at 20 kt; 400 at 36 kt
Complement: 41 (5 officers)

Missiles: SSM: 4 SS-N-2D Styx; IR homing to 83 km *(45 n miles)* at 0.9 Mach; warhead 513 kg; sea-skimmer at end of run.
SAM: SA-N-5 Grail quad launcher; manual aiming; IR homing to 6 km *(3.2 n miles)* at 1.5 Mach; warhead 1.5 kg.
Guns: 1—3 in *(76 mm)*/60; 120 rds/min to 15 km *(8 n miles)*; weight of shell 7 kg.
2—30 mm/65 AK 630; 6 barrels per mounting; 3,000 rds/min combined to 2 km.
Countermeasures: Decoys: 2 PK 16 chaff launchers.
Weapons control: Hood Wink optronic director.
Radars: Air/surface search: Plank Shave; E-band.
Navigation: Pechora; I-band.
Fire control: Bass Tilt: H/I-band.
IFF: Salt Pot, Square Head A.
Sonars: Foal Tail; active; high frequency.

Programmes: First pair ordered in October 1994. These are new hulls exported at a favourable price and completed by 1996. Some delay in delivery because of late payments, but both were in service by April 1996. Two further vessels were reported to have been ordered in 1999 for delivery in 2000. Imagery of HQ 374 suggests that the contract may have been completed although this may be the result of a change in pennant numbers. Current numbers of vessels are thus uncertain. It was announced in November 2003 that up to ten further Tarantul IV, armed with SS-N-25 (Kh 35 Uran), had been ordered from Vympel Shipyard, Rybinsk.
Operational: Based at Da Nang.   *UPDATED*

HQ 374   *6/2004 *, M Mazumdar /* 1043705

LAND-BASED MARITIME AIRCRAFT

Notes: (1) Air Force Flankers and Filters can be used for maritime surveillance.
(2) Contract in October 2003 for the procurement of eight to 10 new Polskie Zaklady Lotnicze (PZL) M28 Skytruck short take-off and landing aircraft configured for maritime surveillance. The aircraft are variants of the PZL M28B-1R 'Bryza-1R' aircraft used by the Polish Navy. Vietnamese aircraft will be fitted with the latest version of the MSC-400 maritime surveillance system which utilises PIT's ARS-400 maritime surveillance radar and the CCS-400 command-and-control module with datalink. The initial contract involved two platforms to be delivered by the end of 2004.

Numbers/Type: 3 Beriev Be-12 Mail.
Operational speed: 328 kt *(608 km/h)*.
Service ceiling: 37,000 ft *(11,280 m)*.
Range: 4,050 n miles *(7,500 km)*.
Role/Weapon systems: Long-range ASW/MR amphibian. Sensors: Short Horn search radar, MAD, EW. Weapons: ASW; 5 tons of depth bombs, mines or torpedoes. ASV; limited missile and rocket armament.   *VERIFIED*

MAIL (Russian colours)   *7/1996, J Ciślak /* 0084250

PATROL FORCES

Notes: (1) At least one Shanghai II class PC may still be operational.
(2) Some of the craft listed may be transferred to the Coast Guard and Maritime Police.

14 ZHUK (PROJECT 1400M) CLASS (PB)

T 864	T 874	T 880	T 881	+10

Displacement, tons: 39 full load
Dimensions, feet (metres): 78.7 × 16.4 × 3.9 *(24 × 5 × 1.2)*
Main machinery: 2 Type M 401B diesels; 2,200 hp(m) *(1.6 MW)* sustained; 2 shafts
Speed, knots: 30. **Range, n miles:** 1,100 at 15 kt
Complement: 11 (3 officers)
Guns: 4—14.5 mm (2 twin) MGs.
Radars: Surface search: Spin Trough; I-band.

Comment: Transferred: three in 1978, three in November 1979, one in November 1981, one in May 1985, three in February 1986, two in December 1989, two in January 1990, three in January 1996, two in January 1998 and two in April 1998. So far seven have been deleted but operational numbers are uncertain. Some may be allocated to the Coast Guard.   *UPDATED*

2 + 2 (6) SVETLYAK (PROJECT 1041.2) CLASS (PGM)

HQ 261   HQ 262

Displacement, tons: 365 full load
Dimensions, feet (metres): 162.4 × 30.2 × 7.9 *(49.5 × 9.2 × 2.4)*
Main machinery: 3 diesels; 15,900 hp(m) *(11.85 MW)* sustained; 3 shafts; cp props
Speed, knots: 30. **Range, n miles:** 2,200 at 13 kt
Complement: 28 (4 officers)
Missiles: SAM: SA-N-10; shoulder launched and (manual aiming); IR homing to 5 km *(2.7 n miles)* at 1.7 Mach; warhead 1.5 kg.
Guns: 1–3 in *(76 mm)*/60 AK-176M; 120 rds/min to 15 km *(8 n miles)*; weight of shell 7 kg.
  1–30 mm/65 AK 630; 6 barrels; 3,000 rds/min combined to 2 km.
Countermeasures: Decoys: 2 chaff launchers.
Weapons control: Hood Wink optronic director.
Radars: Air/surface search: Peel Cone; E-band.
Fire control: Bass Tilt; H/I-band.
Navigation: Palm Frond B; I-band.

Comment: Contract for two craft signed with Almaz, St Petersburg in November 2001. First vessel launched on 17 July 2002 and second on 30 July 2002. Following acceptance on 17 October 2002, both vessels were shipped from St Petersburg on 14 December 2002. Two more are expected to be delivered in 2005 and there is reported to be an option for a further six vessels.
UPDATED

SVETLYAK 041   *9/2002, Almaz /* 0530061

8 OSA II CLASS (FAST ATTACK CRAFT—MISSILE) (PTFG)

HQ 354-361

Displacement, tons: 245 full load
Dimensions, feet (metres): 126.6 × 24.9 × 8.8 *(38.6 × 7.6 × 2.7)*
Main machinery: 3 Type M 504 diesels; 10,800 hp(m) *(7.94 MW)* sustained; 3 shafts
Speed, knots: 37. **Range, n miles:** 500 at 35 kt
Complement: 30
Missiles: SSM: 4 SS-N-2B Styx; active radar or IR homing to 46 km *(25 n miles)* at 0.9 Mach; warhead 513 kg.
Guns: 4 USSR 30 mm/65 (2 twin); 500 rds/min to 5 km *(2.7 n miles)*; weight of shell 0.54 kg.
Radars: Surface search: Square Tie; I-band.
Fire control: Drum Tilt; H/I-band.
IFF: High Pole. 2 Square Head.

Comment: Transferred from USSR: two in October 1979, two in September 1980, two in November 1980 and two in February 1981. All based at Da Nang. Operational status doubtful.
VERIFIED

OSA II 354   *5/2000, Bob Fildes /* 0105740

5 TURYA (PROJECT 206M) CLASS
(FAST ATTACK CRAFT—HYDROFOIL) (PCK)

HQ 321, HQ 331-332, HQ 334-335

Displacement, tons: 190 standard; 250 full load
Dimensions, feet (metres): 129.9 × 29.9 (41 over foils) × 5.9 (13.1 over foils)
  39.6 × 7.6 (12.5) × 1.8 (4))
Main machinery: 3 Type M 504 diesels; 10,800 hp(m) *(7.94 MW)* sustained; 3 shafts
Speed, knots: 40. **Range, n miles:** 600 at 35 kt foilborne; 1,450 at 14 kt hullborne
Complement: 30
Guns: 2 USSR 57 mm/70 (twin, aft); 120 rds/min to 8 km *(4.4 n miles)*; weight of shell 2.8 kg.
  2 USSR 25 mm/80 (twin, fwd); 270 rds/min to 3 km *(1.6 n miles)*; weight of shell 0.34 kg.
Torpedoes: 4—21 in *(533 mm)* tubes (not in all).
Depth charges: 2 racks.
Radars: Surface search: Pot Drum; H/I-band.
Fire control: Muff Cob; G/H-band.
IFF: High Pole B. Square Head.
Sonars: Foal Tail (not in all); VDS; high frequency.

Comment: Transferred from USSR: two in mid-1984, one in late 1984, two in January 1986. Two more acquired from Russia. Two of the five do not have torpedo tubes or sonar. Two scrapped so far, the remainder are probably non-operational.
VERIFIED

TURYA 331   *5/2000, Bob Fildes /* 0105741

3 SHERSHEN (PROJECT 206) CLASS
(FAST ATTACK CRAFT) (PTFM)

HQ 301 series

Displacement, tons: 145 standard; 170 full load
Dimensions, feet (metres): 113.8 × 22 × 4.9 *(34.7 × 6.7 × 1.5)*
Main machinery: 3 Type 503A diesels; 8,025 hp(m) *(5.9 MW)* sustained; 3 shafts
Speed, knots: 45. **Range, n miles:** 850 at 30 kt; 460 at 42 kt
Complement: 23
Missiles: SAM: 1 SA-N-5 Grail quad launcher; manual aiming; IR homing to 6 km *(3.2 n miles)* at 1.5 Mach; altitude to 2,500 m *(8,000 ft)*; warhead 1.5 kg.
Guns: 4 USSR 30 mm/65 (2 twin); 500 rds/min to 5 km *(2.7 n miles)*; weight of shell 0.54 kg.
Torpedoes: 4—21 in *(533 mm)* tubes (not in all).
Depth charges: 2 racks (12).
Mines: Can carry 6.
Radars: Surface search: Pot Drum; H/I-band.
Fire control: Drum Tilt; H/I-band.
IFF: High Pole A. Square Head.

Comment: A total of 16 transferred from USSR: two in 1973, two in April 1979 (without torpedo tubes), two in September 1979, two in August 1980, two in October 1980, two in January 1983 and four in June 1983. Most have been cannibalised for spares.
VERIFIED

SHERSHEN (refitting in Haiphong)   *8/2000, P Marsan /* 0105742

4 + (12) STOLKRAFT CLASS (PBR)

HQ 56-59

Displacement, tons: 44 full load
Dimensions, feet (metres): 73.5 × 24.6 × 3.9 *(22.4 × 7.5 × 1.2)*
Main machinery: 2 MTU 12V 183 TE93 diesels; 2,301 hp(m) *(1.69 MW)* sustained; 2 Doen waterjets
  1 Volvo Penta diesel; 360 hp(m) *(265 kW)*; 1 shaft
Speed, knots: 30
Complement: 7
Guns: 1 Oerlikon 20 mm.

Comment: First four built by Oceanfast Marine, Western Australia and delivered in early 1997. Trimaran construction forward, transforming into a catamaran at the stern. Shallow draft needed for inshore and river operations. The centreline single shaft is used for loitering. The craft show the colours of the Customs department. Up to 12 more may be built in Vietnam, when funds are available.
VERIFIED

RIVER PATROL CRAFT

Comment: There are large numbers of river patrol boats, mostly armed with MGs. A 14.5 m craft ordered from Singapore TSE in 1994. More are being built locally with Volvo Penta engines.
VERIFIED

RIVER PATROL BOAT   *8/2000, P Marsan /* 0105743

3 SO 1 (PROJECT 201M) CLASS (LARGE PATROL CRAFT) (PB)

HQ 3275   HQ 3277   HQ 3278

Displacement, tons: 170 standard; 215 full load
Dimensions, feet (metres): 137.8 × 19.7 × 5.9 *(42 × 6 × 1.8)*
Main machinery: 3 Kolomna Type 40-D diesels; 6,600 hp(m) *(4.8 MW)* sustained; 3 shafts
Speed, knots: 28. **Range, n miles:** 1,100 at 13 kt; 350 at 28 kt
Complement: 31
Guns: 4 USSR 25 mm/80 (2 twin); 270 rds/min to 3 km *(1.6 n miles)*; weight of shell 0.34 kg.
A/S mortars: 4 RBU 1200 5-tubed fixed; range 1,200 m; warhead 34 kg.
Depth charges: 2 racks (24).
Mines: 10.
Radars: Surface search: Pot Head; I-band.
IFF: High Pole A. Dead Duck.

Comment: Transferred from USSR: two in March 1980, two in September 1980, two in May 1981, and two in September 1983. Four more of this class were transferred in 1960-63 but have been deleted, as have four others. Based at Da Nang.
VERIFIED

SO 1   1984

4 MODIFIED ZHUK CLASS (PB)

HQ 37   HQ 55   +2

Displacement, tons: 38 full load
Dimensions, feet (metres): 95.1 ×? × ? *(29.0 × ? × ?)*
Main machinery: 2 Saab Scania diesels; 2,500 hp(m) *(18.64 MW)* sustained; 2 shafts
Speed, knots: 30
Complement: 11 (3 officers)
Guns: 2—12.7 mm MGs (2 twin).
Radars: Navigation: I-band.

Comment: Built in Vietnam to design based on Zhuk class.
VERIFIED

HQ 55 (under construction)   *8/2000, P Marsan* / 0105744

2 POLUCHAT (PROJECT 368) CLASS
(COASTAL PATROL CRAFT) (PB/YPT)

Displacement, tons: 100 full load
Dimensions, feet (metres): 97.1 × 19 × 4.8 *(29.6 × 5.8 × 1.5)*
Main machinery: 2 Type M 50 diesels; 2,200 hp(m) *(1.6 MW)* sustained; 2 shafts
Speed, knots: 20. **Range, n miles:** 1,500 at 10 kt
Complement: 15
Guns: 2—12.7 mm MGs.
Radars: Navigation: Spin Trough; I-band.

Comment: Both transferred from USSR in January 1990. Can be used as torpedo recovery vessels.
VERIFIED

POLUCHAT (Russian colours)   *7/1993, Hartmut Ehlers*

AMPHIBIOUS FORCES

3 POLNOCHNY (PROJECT 771) CLASS (LCM)

HQ 511   HQ 512   HQ 513

Displacement, tons: 760 standard; 834 full load
Dimensions, feet (metres): 246.1 × 31.5 × 7.5 *(75 × 9.6 × 2.3)*
Main machinery: 2 Kolomna Type 40-D diesels; 4,400 hp(m) *(3.2 MW)* sustained; 2 shafts
Speed, knots: 19
Complement: 40
Guns: 2 or 4 USSR 30 mm/65 (1 or 2 twin). 2—140 mm rocket launchers.
Radars: Surface search: Spin Trough; I-band.
Fire control: Drum Tilt; H/I-band.

Comment: Transfers from USSR: one in May 1979 (B), one in November 1979 (A) and one in February 1980 (B). Details are for Polnochny B class. All are reported to be in poor condition.
VERIFIED

HQ 512 and 513   *6/1995, Giorgio Arra* / 0084254

1 LST 1-510 CLASS (LST) and 2 LST 512-1152 CLASS (LST)

TRAN KHANH DU (ex-Da Nang, ex-*Maricopa County* LST 938) HQ 501
VUNG TAU (ex-*Cochino County* LST 603) HQ 503
QUI NONH (ex-*Bulloch County* LST 509) HQ 502

Displacement, tons: 2,366 beaching; 4,080 full load
Dimensions, feet (metres): 328 × 50 × 14 *(100 × 15.2 × 4.3)*
Main machinery: 2 GM 12-567A diesels; 1,800 hp *(1.34 MW)*; 2 shafts
Speed, knots: 11. **Range, n miles:** 6,000 at 10 kt
Complement: 110
Guns: 8 Bofors 40 mm/60 (2 twin, 4 single). 4 Oerlikon 20 mm.

Comment: Built in 1943-44. Transferred from US to South Vietnam in mid-1960s. Seldom seen at sea.
VERIFIED

TRAN KHANH DU   *8/2000* / 0105745

30 LANDING CRAFT (LCM and LCU)

Comment: About five LCUs, 12 LCM 8 and LCM 6, and three LCVPs remain of the 180 minor landing craft left behind by the USA in 1975. In addition there are about 10 T4 LCUs acquired from the USSR in 1979.
VERIFIED

LCU   *6/2001* / 0131363

MINE WARFARE FORCES

2 YURKA (RUBIN) (PROJECT 266) CLASS
(MINESWEEPER—OCEAN) (MSO)

HQ 851   HQ 885

Displacement, tons: 540 full load
Dimensions, feet (metres): 171.9 × 30.8 × 8.5 *(52.4 × 9.4 × 2.6)*
Main machinery: 2 Type M 503 diesels; 5,350 hp(m) *(3.91 MW)* sustained; 2 shafts
Speed, knots: 17. **Range, n miles:** 1,500 at 12 kt
Complement: 45
Guns: 4 USSR 30 mm/65 (2 twin); 500 rds/min to 5 km *(2.7 n miles)*; weight of shell 0.54 kg.
Mines: 10.
Radars: Surface search: Don 2; I-band.
Fire control: Drum Tilt; H/I-band.
Sonars: Stag Ear; hull-mounted; active minehunting; high frequency.

Comment: Transferred from USSR December 1979. Steel-hulled, built in early 1970s.
VERIFIED

YURKA (Egyptian colours)   *10/1998, F Sadek /* 0017818

4 SONYA (YAKHONT) (PROJECT 1265) CLASS
(MINESWEEPER/HUNTER—COASTAL) (MHSC)

HQ 861   HQ 862   HQ 863   HQ 864

Displacement, tons: 450 full load
Dimensions, feet (metres): 157.4 × 28.9 × 6.6 *(48 × 8.8 × 2)*
Main machinery: 2 Kolomna 9-D-8 diesels; 2,000 hp(m) *(1.47 MW)* sustained; 2 shafts
Speed, knots: 15. **Range, n miles:** 3,000 at 10 kt
Complement: 43
Guns: 2 USSR 30 mm/65 AK 630. 2—25 mm/80 (twin).
Mines: 8.
Radars: Surface search: Nayada; I-band.
Sonars: MG 69/79; active; high frequency.

Comment: First one transferred from USSR 16 February 1987, second in February 1988, third in July 1989, fourth in March 1990. Two based at Da Nang.   *VERIFIED*

SONYA 862   *5/2000, R Fildes /* 0105746

2 YEVGENYA (KOROND) (PROJECT 1258) CLASS
(MINEHUNTER—INSHORE) (MHI)

Displacement, tons: 90 full load
Dimensions, feet (metres): 80.7 × 18 × 4.9 *(24.6 × 5.5 × 1.5)*
Main machinery: 2 Type 3-D-12 diesels; 600 hp(m) *(440 kW)* sustained; 2 shafts
Speed, knots: 11. **Range, n miles:** 300 at 10 kt
Complement: 10
Guns: 2 USSR 25 mm/80 (twin).
Mines: 8.
Radars: Surface search: Spin Trough; I-band.
Sonars: MG 7; active; high frequency.

Comment: First transferred from USSR in October 1979; two in December 1986. One deleted in 1990.   *VERIFIED*

YEVGENYA (Russian colours)   *1993, B Lemachko*

5 K 8 (PROJECT 361T) CLASS (MINESWEEPING BOATS) (PBR)

Displacement, tons: 26 full load
Dimensions, feet (metres): 55.4 × 10.5 × 2.6 *(16.9 × 3.2 × 0.8)*
Main machinery: 2 Type 3-D-6 diesels; 300 hp(m) *(220 kW)* sustained; 2 shafts
Speed, knots: 18
Complement: 6
Guns: 2—14.5 mm (twin) MGs.

Comment: Transferred from USSR in October 1980. Probably used as river patrol craft.

VERIFIED

SURVEY SHIPS

Notes: A new survey ship of 2,500 tons is required.

1 KAMENKA (PROJECT 870) CLASS (AGS)

Displacement, tons: 760 full load
Dimensions, feet (metres): 175.5 × 29.8 × 8.5 *(53.5 × 9.1 × 2.6)*
Main machinery: 2 Sulzer diesels; 1,800 hp(m) *(1.32 MW)*; 2 shafts; cp props
Speed, knots: 14. **Range, n miles:** 4,000 at 10 kt
Complement: 25
Radars: Navigation: Don 2; I-band.

Comment: Transferred from USSR December 1979. Built at Northern Shipyard, Gdansk in the late 1960s. May be civilian manned.
VERIFIED

KAMENKA (Russian colours)   *1984*

AUXILIARIES

Notes: In addition to the vessels listed below there are two YOG 5 fuel lighters, two floating cranes, two ex-USSR unarmed Nyryat 2 diving tenders, two PO 2 harbour launches, a Sorum class ocean-going tug (BD 105) and approximately ten harbour tugs.

17 OFFSHORE SUPPLY VESSELS (AKL)

TRUONG HQ 966	HQ 618	HQ 669
BD 621-622	HQ 619	HQ 670
BD 630-632	HQ 643	HQ 671
HQ 601	HQ 651	HQ 673
HQ 614	HQ 661	

Measurement, tons: 1,000 dwt
Dimensions, feet (metres): 231.6 × 38.7 × 13.1 *(70.6 × 11.8 × 4)*
Main machinery: 1 diesel; 1 shaft
Speed, knots: 12
Complement: 30

Comment: Details are for HQ 966 launched at Halong Shipyard in June 1994. This is one of a group of 20 freighters reported as used by the Navy for coastal transport, and to service the Spratleys garrison. BD pennant numbers have been assigned to Spratly Islands service. The ships are of various sizes and include fishing vessels adapted for supply tasks. All are likely to be armed with machine guns.
VERIFIED

BD 621 (old number)   *3/1997 /* 0050742

2 FLOATING DOCKS

Comment: One has a lift capacity of 8,500 tons. Transferred from USSR August 1983. Second one *(Khersson)* has a lift capacity of 4,500 tons and was supplied in 1988.
VERIFIED

Virgin Islands (UK)

Country Overview

A British dependency, the British Virgin Islands are situated in the eastern Caribbean Sea at the northern end of the Leeward Islands chain. Puerto Rico lies some 52 n miles to the west. Comprising a group of 36 islands, 16 of them inhabited, and more than

20 islets and cays there are four main islands: Tortola (21 square miles); Anegada (15 square miles); Virgin Gorda (8 square miles); and Jost Van Dyke (3.5 square miles). Other inhabited islands include Peter Island, Cooper Island, Beef Island, Salt Island, and Norman Island. The capital, only town and principal port is Road Town, Tortola. Territorial seas (3 n miles) and a

Fishery Zone (200 n miles) are claimed. The remainder of the Virgin Islands form a separate external territory of the US.

Headquarters Appointments

Commissioner of British Virgin Islands Police:
  Barry W Webb

Bases

Road Town, Tortola

POLICE

Notes: (1) Following the withdrawal of *St Ursula* from service in 2003, procurement of a replacement craft is expected.
(2) There are also a 12 m Scarab, fitted with three 225 hp outboard motors, and a 10 m Mako with two 150 hp outboards.
(3) Two Dauntless class 12 m patrol boats are operated by the US Virgin Islands whose waters are also patrolled by USCG craft.

Yemen

Country Overview

The Republic of Yemen was formed in 1990 through the union of the People's Democratic Republic of Yemen and the Yemen Arab Republic. The country includes the islands of Socotra, Kamaran and Perim. With an area of 207,285 square miles, it is situated on the south-west coast of the Arabian Peninsula and is bordered to the north by Saudi Arabia and to the east by Oman. It has a 1,030 n mile coastline with the Red Sea and the Gulf of Aden, which are linked by a strategic strait, the Bab el Mandeb. The capital and largest city is Sanaa while the principal ports are Aden and Al Hudaydah. Territorial seas (12 n miles) are claimed. A 200 n mile EEZ has been claimed but the limits have only been partly defined by boundary agreements.

Headquarters Appointments

Chief of Navy:
  Admiral Abdel-Karim Yahya Moharrem

Personnel

2005: 1,700 naval plus 500 marines

Bases

Main: Aden, Hodeida
Secondary: Mukalla, Perim, Socotra, Al Katib
Coast Defence regions: Al Ghaydah, Aden and Cameron Island

Coast Defence

Two mobile SS-C-3 Styx batteries. Some 100 mm guns installed in tank turrets at Perim Island.

DELETIONS

Auxiliaries

2002   *Toplivos 135 and 140*

PATROL FORCES

Notes: (1) In addition there are two 'Osa IIs', 122 and 124. One is in a poor state of repair and may have been decommissioned. The other was sighted in a floating dock in mid-2002 and may be seaworthy, although the SSM system is probably not operational.
(2) Five ex-USCG life boats were reported to have been delivered in 2004.

2 ZHUK CLASS (PROJECT 1400M) (PB)

202   203

Displacement, tons: 39 full load
Dimensions, feet (metres): 78.7 × 16.4 × 3.9 *(24 × 5 × 1.2)*
Main machinery: 2 Type M 401B diesels; 2,200 hp(m) *(1.6 MW)* sustained; 2 shafts
Speed, knots: 30. **Range, n miles:** 1,100 at 15 kt
Complement: 11 (3 officers)
Guns: 4—14.5 mm (2 twin) MGs.
Radars: Surface search: Spin Trough; I-band.

Comment: Two delivered from USSR in December 1984 and three in January 1987. Three have been cannibalised for spares. One based at Aden, one at Al Katib. Used as utility craft including target towing.
VERIFIED

1 TARANTUL I CLASS (PROJECT 1241) (FSGM)

124 (ex-971)

Displacement, tons: 385 standard; 580 full load
Dimensions, feet (metres): 184.1 × 37.7 × 8.2 *(56.1 × 11.5 × 2.5)*
Main machinery: 2 Nikolayev Type DR 77 gas turbines; 16,016 hp(m) *(11.77 MW)* sustained; 2 Nikolayev Type DR 76 gas turbines with reversible gearboxes; 4,993 hp(m) *(3.67 MW)* sustained; 2 shafts
Speed, knots: 36. **Range, n miles:** 400 at 36 kt; 2,000 at 20 kt
Complement: 50

Missiles: SSM: 4 SS-N-2C Styx (2 twin) launchers; active radar or IR homing to 83 km *(45 n miles)* at 0.9 Mach; warhead 513 kg; sea-skimmer at end of run.
SAM: SA-N-5 Grail quad launcher; manual aiming; IR homing to 10 km *(5.4 n miles)* at 1.5 Mach; altitude to 2,500 m *(8,000 ft)*; warhead 1.1 kg.
Guns: 1—3 in *(76 mm)*/60; 120 rds/min to 7 km *(3.8 n miles)*; weight of shell 7 kg.
2—30 mm/65 AK 630; 6 barrels per mounting; 3,000 rds/min to 2 km.
Countermeasures: Decoys: 2 PK 16 chaff launchers.
Weapons control: Hood Wink optronic director.
Radars: Air/surface search: Plank Shave (also for missile control); E-band.
Navigation: Spin Trough; I-band.
Fire control: Bass Tilt; H/I-band.
IFF: Square Head. High Pole.

Programmes: First one delivered from USSR 7 December 1990, second on 15 January 1991. This is the standard export version.
Operational: Facilities for servicing missiles in Aden were destroyed in mid-1994 but one of the class is still in a reasonable state of repair although probably without missiles.
VERIFIED

ZHUK   *3/1990* / 0084258

TARANTUL 124   *10/1995* / 0016612

For details of the latest updates to *Jane's Fighting Ships* online and to discover the additional information available exclusively to online subscribers please visit
jfs.janes.com

3 HUANGFEN (TYPE 021) CLASS
(FAST ATTACK CRAFT—MISSILE) (PTFG)

126	127	128

Displacement, tons: 171 standard; 205 full load
Dimensions, feet (metres): 126.6 × 24.9 × 8.9 *(38.6 × 7.6 × 2.7)*
Main machinery: 3 Type 42-160 diesels; 12,000 hp(m) *(8.8 MW)* sustained; 3 shafts
Speed, knots: 35. **Range, n miles:** 800 at 30 kt
Complement: 28
Missiles: SSM: 4 YJ-1 (Eagle Strike) (C-801); inertial cruise; active radar homing to 40 km *(22 n miles)* at 0.9 Mach; warhead 165 kg; sea-skimmer.
Guns: 4 30 mm/(2 twin AK 230); 500 rds/min to 5 km *(2.7 n miles)*.
Radars: Surface search: Square Tie; I-band.
Fire control: Rice Lamp; H/I-band.
IFF: 2 Square Head. High Pole A.

Comment: Modified Huangfen type. Delivered on 6 June 1995 at Aden having been built by the China Shipbuilding Corporation and completed in 1993. Payment was delayed by the Yemeni civil war. Based at Al Katib. *128* ran aground in September 1997 but was salvaged and may be operational again. *126* is in a reasonable state of repair but is not armed with missiles.
VERIFIED

HUANGFEN 126   5/1995

6 BAKLAN CLASS (HSIC)

BAKLAN 1201	ZUHRAB 1203	HUNAISH 1205
SIYAN 1202	AKISSAN 1204	ZAKR 1206

Displacement, tons: 12 full load
Dimensions, feet (metres): 50.9 × 9.8 × 2.6 *(15.5 × 3 × 0.8)*
Main machinery: 2 diesels; 2 surface drives
Speed, knots: 55. **Range, n miles:** 400 at 30 kt
Complement: 4
Guns: 2—12.7 mm MGs.
Radars: Surface search: Furuno; I-band.

Comment: Ordered from CMN Cherbourg on 3 March 1996. First five were delivered 1 August 1996 and the last one in mid-1997. Top speed in Sea States up to 3. Composite hull construction.
VERIFIED

BAKLAN CLASS   8/1996, C M N Cherbourg / 0084259

10 AUSTAL PATROL SHIPS (PB)

P-1022	+9

Displacement, tons: To be announced
Dimensions, feet (metres): 123.0 × 23.6 × 7.2 *(37.5 × 7.2 × 2.2)*
Main machinery: 2 Caterpillar 3512 diesels; 3,500 hp *(2.61 MW)*; 2 shafts
Speed, knots: 29. **Range, n miles:** 1,000 at 25 kt
Complement: 19 (3 officers)
Guns: 1 Rafael Typhoon 25 mm; 2—12.7 mm MGs.

Comment: Contract with Austal Ships on 9 June 2003 for a total of 10 patrol craft, the first four of which were delivered in mid-2004 and the remainder by the end of 2004. Of aluminium construction, the design is based on the Bay class Australian Customs vessels. The contract includes engineering and practical training for 60 Yemeni crew.
UPDATED

P-1022   6/2004*, Kade Rogers, RAN / 0583301

AMPHIBIOUS FORCES

Notes: Ropucha 139 is an alongside hulk and is to be replaced by the NS-722 class.

1 NS-722 CLASS (LSMM)

BILQUIS

Displacement, tons: 1,383 full load
Dimensions, feet (metres): 295.4 × 31.8 × 7.9 *(90 × 9.7 × 2.4)*
Main machinery: 2 Caterpillar diesels; 5,670 hp *(4.2 MW)*; 2 shafts
Speed, knots: 18
Complement: 49
Military lift: 5 T-72 tanks and 111 marines
Missiles: SAM: SA-16 or ZM Mesko.
Guns: 4 ZSU-23-2MR Wrobel 23 mm (2 twin).

Comment: Ordered in late 1999 for delivery in 2002, development of the Polnochny class built by Naval Shipyard Gdynia, Poland. Shipped from Poland to Yemen on 24 May 2002. Roles include disaster relief and cadet training as well as amphibious warfare.
VERIFIED

BILQUIS   10/2001, J Ciślak / 0131343

3 DEBA CLASS (PROJECT NS-717) (LCU)

HIMYER (ex-*Dhaffar*)	SABA	ABDULKORI (ex-*Thamoud*)

Displacement, tons: 221 full load
Dimensions, feet (metres): 134.5 × 23.3 × 5.6 *(41 × 7.1 × 1.7)*
Main machinery: 2 Cummins diesels; 2 shafts
Speed, knots: 15. **Range, n miles:** 500 at 14.5 kt
Complement: 10
Military lift: 16 tons and 50 troops
Guns: 2 ZU-23-2MR Wrobel 23 mm/87 (1 twin).
2—12.7 mm MGs.
Radars: Navigation: I-band.

Comment: Ordered from Poland in October 1999 and delivered in mid-2001. AK-630 CIWS may also be fitted at a later date.
VERIFIED

ABDULKORI (on transport ship)   5/2001, J Ciślak / 0131342

2 ONDATRA (PROJECT 1176) CLASS (LCU)

13	14

Displacement, tons: 145 full load
Dimensions, feet (metres): 78.7 × 16.4 × 4.9 *(24 × 5 × 1.5)*
Main machinery: 1 diesel; 300 hp(m) *(221 kW)*; 1 shaft
Speed, knots: 10. **Range, n miles:** 500 at 5 kt
Complement: 4
Military lift: 1 MBT
Radars: Navigation: Spin Trough; I-band.

Comment: Transferred from USSR January 1983. Has a tank deck of 45 × 13 ft.
VERIFIED

ONDATRA 14   1990

1 T4 CLASS (LCM)

134

Displacement, tons: 93 full load
Dimensions, feet (metres): 65.3 × 18.4 × 4.6 *(19.9 × 5.6 × 1.4)*
Main machinery: 2 diesels; 316 hp(m) *(232 kW)*; 2 shafts
Speed, knots: 10
Complement: 4

Comment: Reported to have paid off in 1995 but still in service. *VERIFIED*

T4 134   *1/1990 / 0016614*

MINE WARFARE FORCES

1 NATYA CLASS (PROJECT 266ME)
(MINESWEEPER—OCEAN) (MSO)

201

Displacement, tons: 804 full load
Dimensions, feet (metres): 200.1 × 33.5 × 10.8 *(61 × 10.2 × 3)*
Main machinery: 2 Type M 504 diesels; 5,000 hp(m) *(3.67 MW)* sustained; 2 shafts; cp props
Speed, knots: 16. Range, n miles: 3,000 at 12 kt
Complement: 67
Guns: 4—30 mm/65 (2 twin); 500 rds/min to 5 km *(2.7 n miles)*; weight of shell 0.54 kg.
    4—25 mm/80 (2 twin); 270 rds/min to 3 km *(1.6 n miles)*; weight of shell 0.34 kg.
A/S mortars: 2 RBU 1200 five-tubed fixed launchers; range 1,200 m; warhead 34 kg.
Mines: 10
Countermeasures: MCM: Carries contact, acoustic and magnetic sweeps.
Radars: Surface search: Don 2; I-band.
Sonars: MG 69/79; hull-mounted; active minehunting; high frequency.

Comment: Transferred from USSR in February 1991. Operational status doubtful. A second of class was delivered to Ethiopia in October 1991 but sheltered in Aden for a time in 1992. *VERIFIED*

NATYA 201   *6/2002, Rahn/Globke / 0530089*

5 YEVGENYA (PROJECT 1258) CLASS (MINEHUNTERS) (MHC)

11   12   15   20   +1

Displacement, tons: 90 full load
Dimensions, feet (metres): 80.7 × 18 × 4.9 *(24.6 × 5.5 × 1.5)*
Main machinery: 2 Type 3-D-12 diesels; 600 hp(m) *(440 kW)* sustained; 2 shafts
Speed, knots: 11. Range, n miles: 300 at 10 kt
Complement: 10
Guns: 2—25 mm/80 (twin).
Radars: Navigation: Spin Trough; I-band.
Sonars: MG 7 small transducer lifted over stern on crane.

Comment: GRP hulls. Two transferred from USSR in May 1982, third in November 1987 and three more in March 1990. One deleted in 1994. Two based at Aden and three at Al Katib. At least two are fully operational. *VERIFIED*

YEVGENYA 20   *2/1997 / 0016615*

AUXILIARIES

Notes: (1) A 4,500 ton Floating Dock acquired from the USSR.
(2) A 14 m Hydrographic craft acquired from Cougar Marine in 1988.
(3) An oil-pollution control craft acquired in 1999.
(4) Two Toplivo class tankers, *135* and *140*, are reported to have been decommissioned.

Zimbabwe

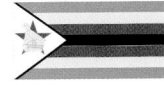

Country Overview

The Republic of Zimbabwe gained independence on 17 April 1980. Formerly the British colony of Southern Rhodesia and, between 1953 and 1963, part of the Federation of Rhodesia and Nyasaland (now Malawi), a unilateral declaration of independence on 11 November 1965 precipitated a turbulent period of guerilla war. This eventually led to a peace settlement in 1979 and elections in 1980. A landlocked country with an area of 150,873 square miles, it is situated in central southern Africa and is bordered to the north by Zambia, to the east by Mozambique, to the south by South Africa and to the west by Botswana and Namibia. It has a shoreline of approximately 350 n miles with Lake Kariba, artificially formed by the Kariba Dam, from which the country gets much of its electric power. The capital, largest city and commercial centre is Harare (formerly Salisbury). The railway system is linked to the port of Beira in Mozambique.

Bases

Kariba, Binga.

PATROL FORCES

2 RODMAN 46HJ CLASS (PB)

Displacement, tons: 12.5 full load
Dimensions, feet (metres): 45.9 × 12.5 × 2.0 *(14.0 × 3.8 × 0.6)*
Main machinery: 2 Caterpillar 3280 diesels; 850 hp *(633 kW)*
Speed, knots: 30. Range, n miles: 350 at 18 kt
Complement: 4

Comment: GRP hull. Built in 1999 by Rodman, Vigo. Operated by Zimbabwe Police. *VERIFIED*

RODMAN 46
6/1999, Rodman / 0570998

3 RODMAN 38 CLASS (PB)

Displacement, tons: 10 full load
Dimensions, feet (metres): 36.1 × 12.8 × 2.3 *(11.0 × 3.9 × 0.7)*
Main machinery: 2 diesels; 2 waterjets
Speed, knots: 28. **Range, n miles:** 300 at 15 kt
Complement: 4

Comment: GRP hull. Built in 1999 by Rodman, Vigo. Operated by Zimbabwe Police.

VERIFIED

RODMAN 38   *6/1999, Rodman /* 0571000

5 RODMAN 790 CLASS (PB)

Displacement, tons: 2.4 full load
Dimensions, feet (metres): 26.6 × 8.9 × 2.3 *(8.1 × 2.72 × 0.7)*
Main machinery: 2 Volvo Penta TAMD diesels
Speed, knots: 30
Complement: 2

Comment: GRP hull. Built in 1999 by Rodman, Vigo. Operated by Zimbabwe Police.

VERIFIED

RODMAN 790   *6/1999, Rodman /* 0570999

INDEXES

Indexes

Country abbreviations

Alb	Albania	DR	Dominican Republic	Lat	Latvia	Sen	Senegal
Alg	Algeria	Ecu	Ecuador	Lby	Libya	Ser	Serbia and Montenegro
Ana	Anguilla	Egy	Egypt	Leb	Lebanon	Sey	Seychelles
Ang	Angola	ElS	El Salvador	Lit	Lithuania	Sin	Singapore
Ant	Antigua and Barbuda	EqG	Equatorial Guinea	Mac	Macedonia, Former Yugoslav Republic of	SL	Sierra Leone
Arg	Argentina	Eri	Eritrea	Mad	Madagascar	Slo	Slovenia
Aus	Austria	Est	Estonia	Mex	Mexico	Sol	Solomon Islands
Aust	Australia	ETim	East Timor	MI	Marshall Islands	Spn	Spain
Az	Azerbaijan	Fae	Faroe Islands	Mic	Micronesia	Sri	Sri Lanka
Ban	Bangladesh	Fij	Fiji	Mld	Maldives	StK	St Kitts
Bar	Barbados	FI	Falkland Islands	Mlt	Malta	StL	St Lucia
Bel	Belgium	Fin	Finland	Mlw	Malawi	StV	St Vincent and the Grenadines
Ben	Benin	Fra	France	Mly	Malaysia	Sud	Sudan
Bhm	Bahamas	Gab	Gabon	Mor	Morocco	Sur	Suriname
Bhr	Bahrain	Gam	Gambia	Moz	Mozambique	Swe	Sweden
Blz	Belize	GB	Guinea-Bissau	Mrt	Mauritius	Swi	Switzerland
Bmd	Bermuda	Ger	Germany	Mtn	Mauritania	Syr	Syria
Bol	Bolivia	Geo	Georgia	Myn	Myanmar	Tan	Tanzania
Bru	Brunei	Gha	Ghana	Nam	Namibia	Tld	Thailand
Brz	Brazil	Gn	Guinea	NATO	NATO	Tkm	Turkmenistan
Bul	Bulgaria	Gra	Grenada	Nic	Nicaragua	Tog	Togo
Cam	Cameroon	Gre	Greece	Nig	Nigeria	Ton	Tonga
Can	Canada	Gua	Guatemala	Nld	Netherlands	TT	Trinidad and Tobago
Cay	Cayman Islands	Guy	Guyana	Nor	Norway	Tun	Tunisia
Chi	Chile	HK	Hong Kong	NZ	New Zealand	Tur	Turkey
CI	Cook Islands	Hon	Honduras	Omn	Oman	Tuv	Tuvalu
CtI	Côte d'Ivoire	Hun	Hungary	Pak	Pakistan	Twn	Taiwan
Cmb	Cambodia	Ice	Iceland	Pal	Palau	UAE	United Arab Emirates
Col	Colombia	Ind	India	Pan	Panama	UK	United Kingdom
Com	Comoros	Indo	Indonesia	Par	Paraguay	Ukr	Ukraine
ConD	Congo, Democratic Republic	Iran	Iran	Per	Peru	Uru	Uruguay
CPR	China, People's Republic	Iraq	Iraq	Plp	Philippines	USA	United States of America
CpV	Cape Verde	Ire	Irish Republic	PNG	Papua New Guinea	Van	Vanuatu
CR	Costa Rica	Isr	Israel	Pol	Poland	Ven	Venezuela
Cro	Croatia	Ita	Italy	Por	Portugal	VI	Virgin Islands
Cub	Cuba	Jam	Jamaica	Qat	Qatar	Vtn	Vietnam
Cypr	Cyprus (Republic)	Jor	Jordan	RoK	Korea, Republic of (South)	Yem	Yemen
Den	Denmark	Jpn	Japan	Rom	Romania	Zim	Zimbabwe
Dji	Djibouti	Kaz	Kazakhstan	Rus	Russian Federation		
Dom	Dominica	Ken	Kenya	SA	South Africa		
DPRK	Korea, Democratic People's Republic (North)	Kir	Kiribati	Sam	Samoa		
		Kwt	Kuwait	SAr	Saudi Arabia		

Named ships

* shows the vessel is listed in a **Note** at the beginning of the section. {|} shows the vessel is referred to in a **Comment** following the entry for another vessel. Vessel names commencing with initials are indexed under the surname. Ranks and forenames continue to be the primary entry.

1st Lt Alex Bonnyman (USA) ... 904
1st Lt Baldomero Lopez (USA) ... 904
1st Lt Harry L Martin (USA) ... 903
1st Lt Jack Lummus (USA) ... 904
2nd Lt John P Bobo (USA) ... 904
3 de Febrero (Par) ... 552
3 de Noviembre (Ecu) ... 193
3 de Noviembre (Pan) ... 548
4 de Noviembre (Pan) ... 548
5 de Agosto (Ecu) ... 192
5 de Noviembre (Pan) ... 548
6 of October (Egy) ... 196
9 de Octubre (Ecu) ... 192
10 de Agosto (Ecu) ... 193
10 de Noviembre (Pan) ... 548
15 de Enero (Gua) ... 299*
15 de Noviembre (Uru) ... 918
18 of June (Egy) ... 198
18 Mart (Tur) ... 776
21 of October (Egy) ... 198
23 of July (Egy) ... 198
24 de Mayo (Ecu) ... 193
25 de Agosto (Uru) ... 918
25 of April (Egy) ... 198
25 de Julio (Ecu) ... 193
27 de Febrero (Ecu) ... 192
27 de Octubre (Ecu) ... 192
28 de Noviembre (Pan) ... 548
50 Let Pobeda (Rus) ... 645*

A

A1C William H Pitsenbarger (USA) ... 902
A 33-A 35 (Lby) ... 458
A 41 (Ger) ... 277
A 72 (Ind) ... 329
A 110 (Alb) ... 3*
A 120 (Alb) ... 2
A 212 (Alb) ... 2
A 223 (Alb) ... 3

A 451 (Alb) ... 2†
A 530 (Sri) ... 711
A 641 (Alg) ... 7
A 701-2, 751, 753-6 (Swe) ... 729
Aachen (UK) ... 825
AB 21-24 (Tur) ... 784
AB 27-9, 31, 33, 35-6 (Tur) ... 783
AB 1050-1, 1053, 105-6, 1058-67 (Aust) ... 36
AB 1052, 1054, 1057 (Aust) ... 36†
AB 2000-2005 (Aust) ... 35
Abad (Iran) ... 357
Abadejo (Arg) ... 21
Abadia Medez (Col) ... 157*
Abalone (Can) ... 94
Abay (Kaz) ... 427
Abdul Aziz (PGGF) (SAr) ... 659
Abdul Aziz (YACH) (SAr) ... 661
Abdul Halim Perdanakusuma (Indo) ... 333
Abdul Rahman Al Fadel (Bhr) ... 42
Abeetha II (Sri) ... 707
Abeille Flandre (Fra) ... 255
Abeille Langedoc (Fra) ... 255
Aber-Wrach (Fra) ... 256
Abha (SAr) ... 658
Abhay (Ind) ... 320
Abigail Burgess (USA) ... 913
Abnegar (Iran) ... 356*
Abou Abdallah El Ayachi (Mor) ... 495
Aboubekr Ben Amer (Mtn) ... 477
Abraham Lincoln (USA) ... 859
Abraham van der Hulst (Nld) ... 107
Abrolhos (Brz) ... 72
Absalon (Den) ... 179
Abu Bakr (Ban) ... 47
Abu Dhabi (UAE) ... 804
Abu El Barakat Al Barbari (Mor) ... 496

Abu El Ghoson (Egy) ... 201
Abu Obaidah (SAr) ... 659
Abu Qir (Egy) ... 196
Abukuma (Abukuma class) (Jpn) ... 406
Abukuma (Bihoro class) (Jpn) ... 419
Aby (CtI) ... 162
Acacia (USA) ... 912
Acanthe (Fra) ... 252
Acchileus (Gre) ... 295
Acconac (USA) ... 894
Acevedo (Spn) ... 696
Acharné (Fra) ... 254
Achernar (Brz) ... 74
Achernar (Mex) ... 487
Achéron (Fra) ... 247
Achilles (Swe) ... 729
Achimota (Gha) ... 283
Aconcagua (Chi) ... 114
Aconit (Fra) ... 237
Acoustic Explorer (USA) ... 894*
Acoustic Pioneer (USA) ... 894*
Acrux (Mex) ... 487
Active (USA) ... 907
Acuario (Col) ... 157
Acuario (Mex) ... 488
Acushnet (USA) ... 908
Acute (Indo) ... 339†
Adak (USA) ... 911
Adamant (UK) ... 842
Adamastos (Gre) ... 295
Adang (Tld) ... 761
Addriyah (SAr) ... 660
Adelaide (Aust) ... 25
Adelie (USA) ... 912
Adept (UK) ... 840
Adept (USA) ... 892
ADF 104, 106-9 (Per) ... 560
Adhara II (Arg) ... 21
Adhara (Mex) ... 487
ADI 01-04 (Mex) ... 491
Aditya (Ind) ... 328
Adm Wm H Callaghan (USA) ... 905
Admiral Basisty (Rus) ... 619†

Admiral Branimir Ormanov (Bul) ... 84
Admiral Chabanenko (Rus) ... 619
Admiral Gorshkov (Rus) ... 310†, 614*
Admiral Horia Macellaru (Rom) ... 596
Admiral Kharlamov (Rus) ... 620
Admiral Kucherov (Rus) ... 619†
Admiral Kuznetsov (Rus) ... 615
Admiral Lazarev (Rus) ... 616†
Admiral Levchenko (Rus) ... 620
Admiral Makarov (Rus) ... 645*
Admiral Nakhimov (Rus) ... 616†
Admiral Panteleyev (Rus) ... 620
Admiral Petre Barbuneanu (Rom) ... 595
Admiral Pitka (Est) ... 210
Admiral Spiridonov (Rus) ... 620†
Admiral Tributs (Rus) ... 620
Admiral Ushakov (Rus) ... 616†
Admiral Vinogradov (Rus) ... 620
Admiral Vladimirskiy (Rus) ... 635
Admiral Zakharov (Rus) ... 620†
Adour (Fra) ... 256
ADRI XXXI (Indo) ... 347†
ADRI XXXII-ADRI LVIII (Indo) ... 347
Adrias (Gre) ... 287
Advent (Can) ... 100
Aegeon (Gre) ... 287, 293†
Aegeus (Gre) ... 295
Aegir (Ice) ... 306
AFDM 7 (USA) ... 892
AFDM 2 (Gre) ... 294†
AFDM 2 (USA) ... 892†
Afif (SAr) ... 661
Afonso Cerqueira (Por) ... 585
Africana (SA) ... 683*

AG 5 (Tur) ... 790
AG 6 (Tur) ... 790
Al Agami (Egy) ... 203
Agaral (Ind) ... 329
Agathos (Cypr) ... 170
Agder (Nor) ... 531
Agdlek (Den) ... 175
A G Efstratios (Gre) ... 296
AGI 201 (CPR) ... 146
Ägir (Swe) ... 727
Agnadeen (Iraq) ... 358*
Agon (Gre) ... 288
Agpa (Den) ... 175
Agradoot (Ban) ... 51
Agray (Ind) ... 320
Agu (Nig) ... 522
Aguacero (Pan) ... 549
Aguascalientes (Mex) ... 490
Aguia (Por) ... 587
Águila (Spn) ... 705
Aguila (Mex) ... 488
Aguirre (Per) ... 556
S A Agulhas (SA) ... 682†, 683*, 684
Agusan (Plp) ... 568
AH 177 (Per) ... 558*
Ahalya Bai (Ind) ... 331
Ähav (Nor) ... 531
Ahmad El Fateh (Bhr) ... 42
Ahmad Yani (Indo) ... 333
Al-Ahmadi (Kwt) ... 448
Ahmadi (Kwt) ... 449
Ahmed Es Sakali (Mor) ... 495
Ahti (Est) ... 210
Al Ahweirif (Lby) ... 458
Aias (Gre) ... 295
Aida IV (Egy) ... 202*
Aidon (Gre) ... 293
El Aigh (Mor) ... 496
Aigle (Fra) ... 247
Aiguière (Fra) ... 254
Ailette (Fra) ... 254*, 255
Ain Zaghouan (Tun) ... 773
Aina Vao Vao (Mad) ... 461
Aisberg (Rus) ... 648
Aishima (Jpn) ... 409

For details of the latest updates to *Jane's Fighting Ships* online and to discover the additional information available exclusively to online subscribers please visit

jfs.janes.com

Named classes

* shows the class is referred to in a **Note** at the beginning of the section. {||} shows the class is referred to in a **Comment** following the entry for another vessel or class. A Country abbreviation is only added if the nation is currently operating a vessel of the class.

Classes of Russian vessels previously found under Type and number are now shown as Project and number.

2nd Lt John P Bobo (USA) ... 904
10 De Agosto (Ecu) ... 192†, 193
31 metre Patrol Craft (Kris, Sabah) (Mly) ... 470
65 ft WYTL (USA) ... 914
300 Ton (Jpn) ... 410
430 Ton Class (PBO) (RoK) ... 446
570 ton MSC (Jpn) ... 410
1,800 ton class (Jpn) ... 417

A

A 17 (Södermanland) (Swe) ... 715
A 17 (Västergötland) (Swe) ... 714
A 19 (Swe) (Gotland) ... 716
A 69 (Arg, Fra, Tur) ... 14, 238, 777
Abamin (Myn) ... 501
Abeille Flandre (Fra) ... 255
Abhay (Pauk II) (Ind) ... 320
Abnaki (Mex) ... 491, 898†
Absalon (Den) ... 179
Abukuma (Jpn) ... 406
Achelous (Indo, Plp) ... 344, 565
Actif (Fra) ... 254
Acuario (Mex) ... 488
Addriyah (MSC 322) (SAr) ... 660
Adelaide (Oliver Hazard Perry) (Aust) ... 25
Adept (UK) ... 840
Aditya (Ind) ... 328
Adjutant (Gre, Tur, Twn) ... 292, 744, 787
Admirable (Bur, DR, Mex) ... 184, 491, 563†, 929*
Admiralets (Rus) ... 645†
Advanced SEAL Delivery System (USA) ... 852*, 853†, 854†, 856†
Aegean (Saeta-12) (Spn) ... 705†
Aegir (Ice) ... 306
Afalina (Ukr) ... 802*
AFDM (Gre, USA) ... 294†
Agdlek (Den) ... 175,175†
Aggressive (Fra, Twn) ... 247†, 743
AGOR-26 (USA) ... 895
Agosta (Fra, Mly, Pak, Spn) ... 228*, 463*, 463†, 539, 687, 735*
Agosta 90B (Pak) ... 538
Aguascalientes (Mex) ... 490
Aguinaldo (Plp) ... 564
Ahmad El Fateh (TNC 45) (Bhr) ... 42
Ahmad Yani (Van Speijk) (Indo) ... 333
Aigrette (Mad) ... 461*
Air Defence Ship (Ind) ... 310
Aist (Dzheyran) (Project 1398) (Geo, Rus) ... 259*, 261*, 633
Ajeera (Bhr, Ven) ... 42, 924
Akademic Fersman (Rus) ... 648
Akademik Krylov (Rus) ... 635
Akagi (Jpn) ... 419
Akizuki (Jpn) ... 421
Akshay (Ban) ... 49
Akula I (Bars) (Rus) ... 608, 610†
Akula II (Bars) (Rus) ... 308*, 604†, 608†
Akula (Ondatra) (Rus, Yem) ... 631, 934
Akula (Typhoon) (Rus) ... 602
Akvamaren (Natya) (Rus) ... 630
Al Bushra (Omn) ... 533
Al Feyi (UAE) ... 807
Al Hussein (Hawk) (Jor) ... 425
Al Jarim (FPB 20) (Bhr) ... 42
Al Jawf (Sandown) (SAr) ... 660
Al Jouf (SAr) ... 662
Al Jubatel (SAr) ... 662
Al Manama (MGB 62) (Bhr) ... 41
Al Riffa (FPB 38) (Bhr) ... 42
Al Riyadh (Modified La Fayette) (SAr) ... 657
Al Shaheed (Kwt) ... 449, 449†
Al Siddiq (SAr) ... 659
Al Tahaddy (Kwt) ... 450
Al Waafi (Omn) ... 533†
Al-Shaali (Kwt, UAE) ... 450, 807
Alamosa (Plp) ... 566
Albacora (Daphne) (Por) ... 583
Albatros (Grisha) (Lit, Rus, Ukr) ... 458, 625, 797
Albatros (Type 143B) (Ger, Tun) ... 273, 772
Albatros (YP) (Rus) ... 645†
Albatroz (ETim, Por) ... 186, 587
Albion (UK) ... 829†, 831
Alboran (Spn) ... 696
Alcaravan (Spn) ... 697†
Alcyon (Fra) ... 255
Ale (Swe) ... 725
Alfeite (GB) ... 300
Algerine (Tld) ... 763
Algol (USA) ... 901
Aliya (Saar 4.5) (Mex) ... 488
Alize (Fra) ... 250
Alkmaar (Tripartite) (Ger, Indo, Nld) ... 274†, 510, 510†, 515*
Alkyon (MSC 294) (Gre) ... 293
Allende (Knox) (Mex) ... 482
Alliance (NATO, Twn) ... 504, 744
Alligator (RoK, Ven) ... 443, 443†, 924
Alligator (Tapir) (Rus, Ukr) ... 632†, 633, 799
Almirante Brown (MEKO 360) (Arg) ... 12
Almirante Clemente (Ven) ... 926
Almirante Guilhem (Brz) ... 77
Almirante Padilla (Type FS 1500) (Col) ... 152
Alpinist (Project 503M) (Rus) ... 635
Alpino (AG) (Ita) ... 384
Alpino (MCS) (Ita) ... 385
Alta/Oksøy (Nor) ... 529
Altair (Col) ... 156†
Altay (modified) (Project 160) (Rus) ... 640
Alucat 850 (Arg) ... 21
Alucat 1050 (Arg) ... 21
Alusafe 1290 (Nor) ... 528
Alusafe 1300 (Nor) ... 528
Alvand (Vosper Mk 5) (Iran) ... 350
Alvaro De Bazán (F 100) (Spn) ... 526†, 690
Alvin (USA) ... 896
Älvsborg (Chi, Swe) ... 112, 727
Amami (Jpn) ... 419
Amazon (Type 21) (Pak) ... 540
Ambassador III (USA) ... 198*
Amga (Rus) ... 638
Amorim Do Valle (River) (Brz) ... 73
Amsterdam (Nld, Spn) ... 514
Amur I, II (Project 304) (Rus, Ukr) ... 638, 800
Amur (Lada) (Rus) ... 308*, 612†
Anaga (Spn) ... 696
Anchorage (Twn) ... 742, 886†
Andrea Doria (Horizon) (Ita) ... 375
Andrómeda (Col) ... 156
Andromeda (Por) ... 588
Angamos (Type 209) (Per) ... 554
Animoso (De La Penne) (Ita) ... 373
Anninos (La Combattante II) (Geo) ... 260
Antarès (BRS) (Fra) ... 247
Antares (Mlw) ... 461
Antares (Muravey) (Rus, Ukr) ... 649, 803
Antonio Zara (Ita) ... 389
Antyey (Oscar II) (Rus) ... 606
Anvil (USA) ... 914
Anzac (MEKO 200) (Aust, NZ) ... 26, 517
AP.1-88/100S (Can) ... 102
AP.1-88/200 (Can) ... 101
AP.1-88/400 (Can) ... 102
Apex RIB (CR) ... 161*
Appleleaf (UK) ... 836
Aquamaster/Drago (Ita) ... 391*
Aquarius (Swi) ... 731
Aquitane Explorer (Fra) ... 255
Aragosta (Ham) (Ita) ... 387*, 388
Aratú (Schütze) (Brz) ... 72
Arauca (Col) ... 154
Archer (Ind, Omn, UK) ... 331†, 479†, 537†, 832*, 833
Arcor 38 (Tun) ... 774†
Arcor 46 (Mor) ... 497
Arcor 53 (Mor) ... 497
Arctic 22 RIB (UK) ... 832†
Arctic 24 RIB (Bmd) ... 58†
Arctic 28 (UAE, UK) ... 807, 832†
Arctic (Nor) ... 531
ARD 12 (Ecu, Iran) ... 191, 356†
Ardhana (Vosper 110 ft) (UAE) ... 806
Aresa LVC 160 (P 101) (Spn) ... 697†
Argos (Por) ... 587, 587†
Arguin (Mtn) ... 478
Argus (Brz) ... 73
Argus (UK) ... 141†
Ariel (Fra) ... 251
Arkö (Swe) ... 729
Arkösund (Swe) ... 722
Arleigh Burke (Flight I and II) (USA) ... 868, 870†
Arleigh Burke (Flight IIA) (USA) ... 849*, 870, 888*
Arleigh Burke (improved) (Jpn) ... 398†
Armatolos (Gre) ... 290
Armidale (Aust) ... 31
Arrecife (Olmeca II) (Mex) ... 499
Arrow Post (Can) ... 99
Artigliere (Lupo) (Ita, Per, Ven) ... 376†, 377, 378†, 379†, 556, 922
Arun Type 300A (Can, Qat, UAE) ... 101†, 592*, 808*
Arvak (Den) ... 181
Asagiri (Jpn) ... 401†, 402, 414
ASDS (Advanced SEAL Delivery System) (USA) ... 852*, 853†, 854†, 856†
Ashdod (Eri, Isr) ... 209, 366
Asheville (Col, Gre, RoK, USA) ... 155, 290, 895
Ashoora I (Iran, Sud) ... 354†, 712
ASI 315 (HK) ... 304
Asia (Balzam) (Rus) ... 634
Asmar Protector (Chi, ElS) ... 114, 205†
Aso (Jpn) ... 417
Asogiri (Jpn) ... 421†, 422
Assad (Iraq, Mly) ... 358, 467
Astute (UK) ... 814*, 817
ASU 81 (Jpn) ... 412†
Asuka (Jpn) ... 413
Atilay (209) (Tur) ... 777
Atlant (Slava) (Rus) ... 617
Atle (Swe) ... 218†, 725
Atrek (Rus) ... 640†
Attack (Indo) ... 339
Attacker (Leb) ... 454
Audace (Ita) ... 373†, 374
Auk (Mex, Plp) ... 486, 563
Austal Patrol Ships (Yem) ... 934
Austin (converted) (USA) ... 880
Austin (USA) ... 880†, 881†, 885, 887†
AUV (Autonomous Undersea Vehicles) (USA) ... 852*
Avenger (USA) ... 409†, 888
Avon RIB (Dji, Ire) ... 182*, 359†
Avon Searider (Ger, UK) ... 281†, 833†
Aydin (MHV 54-014) (Tur) ... 784
Azteca (Mex) ... 487

B

Bacolod City (Frank S Besson) (Plp) ... 565
Bad Bramstedt (Ger) ... 280, 648†
Badr (SAr) ... 659
Baglietto 59 ft (UAE) ... 809
Baglietto GC 23 (UAE) ... 809
Baglietto Type 20 (Alg) ... 7
Bahamas (Vosper Europatrol) (Bhm) ... 39
Bakassi (P 48S) (Cam) ... 86
Baklan (Yem) ... 934
Baklazhan (Rus) ... 645*
Balcom 10 (Ger) ... 281†
Baldur (Ice) ... 306
Baleares (F 70) (Spn) ... 690†, 692
Balikpapan (Landing craft (Heavy)) (Aust) ... 29*, 30
Balsam (DR, ElS, Est, Gha, Nig, Pan, Plp, USA) ... 183, 207, 211, 283, 522, 549, 567, 912, 914*
Baltic (Rodman 55M) (Spn) ... 704†
Baltyk (Pol) ... 578
Balzam (Asia) (Rus) ... 634
Bambuk (Ukr) ... 800
Ban Yas (Lürssen TNC 45) (UAE) ... 806
Bang Rachan (Tld) ... 760
Bangabandhu (Ulsan modified) (Ban) .. 46
Bangalore (Project 15A) (Ind) ... 315†
Bango (Indo) ... 347*
Banna (Jpn) ... 420
Baptista de Andrade (Por) ... 585
Baradero (Dabur) (Arg) ... 16
Barbaros (Mod MEKO 200) (Tur) ... 778
Barbel (USA) ... 735*
Barceló (Spn) ... 696
Barkat (Pak) ... 546
Barnegat (Vtn) ... 929
Barracuda 30 ft (UAE) ... 808*, 809†
Barracuda (Fra) ... 228
Barracuda (Sierra I) (Rus) ... 609
Barroso (Brz) ... 66
Barroso Pereira (Brz) ... 76
Bars (Akula I, II) (Rus) ... 608
Baruna Jaya (Indo) ... 342, 346*
Barzan (Vita) (Qat) ... 591
Baskunchak (Rus) ... 638†
Batral (Chi, Fra, Gab, Mor) ... 112, 244, 258, 495
Battalion 17 (Dji, Eri) ... 182, 209
Bay (Aust, Plp) ... 37, 567†, 934†
Bay (Col) ... 156†
Bay (UK) ... 838, 839†
Bay (USA) ... 910
Bayandor (PF 103) (Iran) ... 349, 754†
Baynunah (UAE) ... 806, 806†
Bazan 39 (Arg) ... 21
Bear (USA) ... 745†
Beautemps Beaupré (Fra) ... 247†, 248, 250†
Bedok (Landsort) (Sin) ... 675
Bélier (Fra) ... 253
Bellatrix (DR) ... 184
Belyanka (Rus) ... 638*
Bendeharu (Bru) ... 80
Beograd (Koni) (Ser) ... 666†
Bereza (Bul, Rus, Ukr) ... 85, 644, 800
Berezina (Rus) ... 639
Bergamini (Ita) ... 382†
Berkot-B (Kara) (Rus) ... 618
Berlin (Ger) ... 275
J E Bernier (Can) ... 96
BH 2 (Fra) ... 249
BH 7 (Wellington) (Iran) ... 356
Bigliani (Ita) ... 389†, 390
Bihoro (Jpn) ... 419
Bima VIII (Indo) ... 345*
Bin Hai (CPR) ... 147
Biscaya (Type 20) (Ser) ... 667
Biya (Project 870/871) (Cub, Rus, Ukr) ... 169, 636, 636†, 801
Bizerte (PR 48) (Cam, Sen, Tun) ... 86, 663, 771
Black Swan (Egy) ... 203
Blue Ridge (USA) ... 880
Bluebird (Tld) ... 760
Bob Hope (USA) ... 901
Bodan (Ger) ... 280*, 280
Boghammar (Dji, Iran, UAE) ... 182*, 354, 808*
Bogomol (Gn) ... 300*
Bollinger (Egy) ... 204*
Bolva 1, 2, 3 (Rus) ... 640
Bombarda (Por) ... 588
Borey (Rus) ... 602*, 604†
Boris Chilikin (Rus) ... 639
Bormida (Ita) ... 385
Boston Whaler (Ana, Ant, Bar, Bhm, Bmd, Bol, Bul, Cay, Col, CR, Dom, Gra, Gre, Gua, Hon, HK, Iran, Jam, Kaz, Mex, Rom, StK, StV, SAr, Sin, Tur, USA, Ven) ... 8, 9*, 40, 53†, 58†, 59†, 82*, 103*, 157†, 161*, 182†, 296*, 298, 305†, 354†, 391*, 426*, 486†, 596*, 654, 655†, 663†, 677†, 791*, 890†, 911†, 926†
Botnica (Fin) ... 219
Botnica (Type 16) (Mac, Ser) ... 460, 667
Bouchard (Par) ... 551
Bougainville (Fra) ... 248, 248†
Bowen (TT) ... 770
BPS 500 (Project 12418) (Vtn) ... 929
Bracui (River) (Brz) ... 70
Brahmaputra (Project 16A) (Ind) ... 317
Brandenburg (Ger) ... 266
Braunschweig (K130) (Ger) ... 272
Bravo (Bronstein) (Mex) ... 483
Breda BMB 230 (Tld) ... 757†
Bredstedt (Ger) ... 281
Bremen (Ger) ... 268
Bremse (Ger, Mlt, Tun) ... 281, 476, 774
Briz (Sonya) (Bul) ... 83
Broadsword (Gua) ... 299
Broadsword (Type 22) (Brz, Chi, Rom, UK) ... 64, 107, 595, 823
Bronstein (Mex) ... 483
Brooke Marine 29 m (Mly) ... 473
Brooke Marine LSL (Omn) ... 533†
Brooke (Pak) ... 540†
BRS (Antarès) (Fra) ... 247

For details of the latest updates to *Jane's Fighting Ships* online and to discover the additional information available exclusively to online subscribers please visit

jfs.janes.com

Aircraft by countries

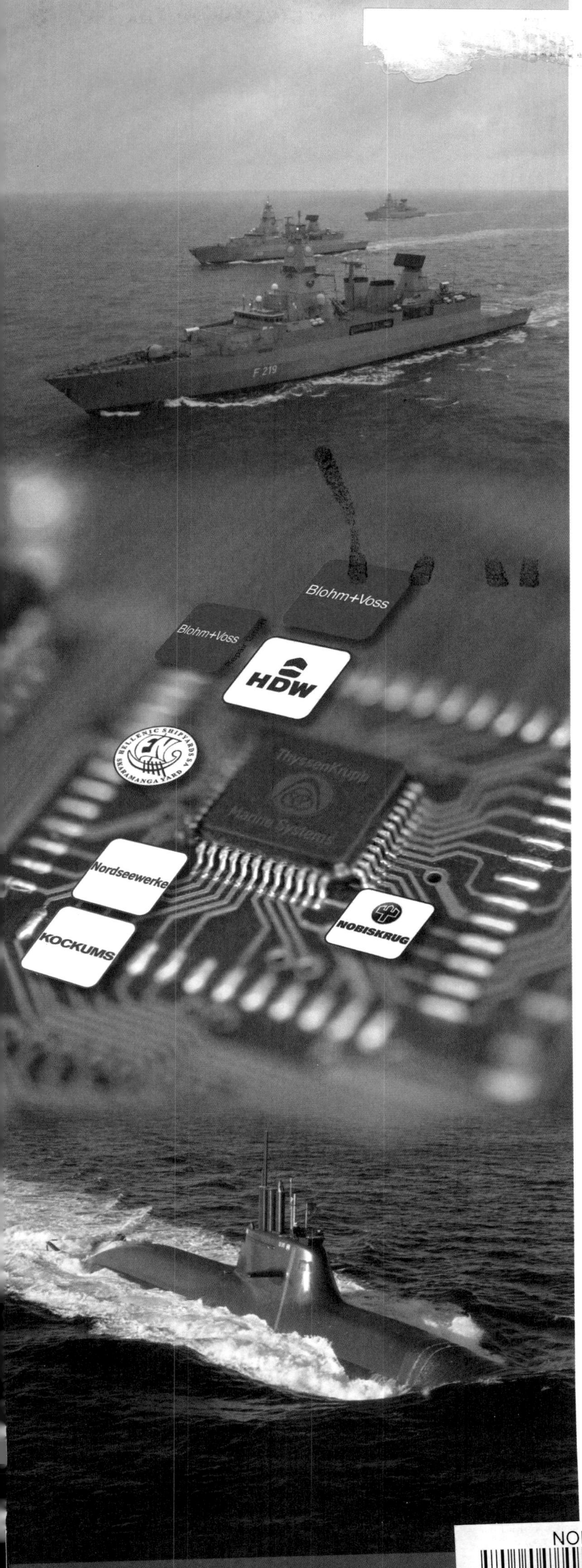